PHYSICAL CONSTANTS

Quantity	Symbol	Value	SI unit
Speed of light in vacuum	c	3.00×10^{8}	m/s
Permittivity of vacuum	ϵ_0	8.85×10^{-12}	F/m
Coulomb constant, $1/4\pi\epsilon_0$	k	9.0×10^{9}	$N \cdot m^2/C^2$
Permeability of vacuum	μ_0	1.26×10^{-6}	N/A^2
		($4\pi \times 10^{-7}$ exactly)	
Elementary charge	e	1.60×10^{-19}	C
Planck's constant	h	6.63×10^{-34}	J·s
	$\hbar = h/2\pi$	1.05×10^{-34}	J·s
Electron rest mass	m_e	9.11×10^{-31}	kg
		5.49×10^{-4}	u
Proton rest mass	m_p	1.67265×10^{-27}	kg
		1.007276	u
Neutron rest mass	m_n	1.67495×10^{-27}	kg
		1.008665	u
Avogadro constant	N_A	6.02×10^{23}	$(gmol)^{-1}$
Molar gas constant	R	8.31	J/mol·K
Boltzmann's constant	k	1.38×10^{-23}	J/K
Stefan-Boltzmann constant	σ	5.67×10^{-8}	$W/m^2 \cdot K^4$
Molar volume of ideal gas at STP	V	22.4	liters/mol
		2.24×10^{-2}	m^3/mol
Rydberg constant	R	1.10×10^{7}	m^{-1}
Bohr radius	a_0	5.29×10^{-11}	m
Electron Compton wavelength	$h/m_e c$	2.43×10^{-12}	m
Gravitational constant	G	6.67×10^{-11}	$m^3/kg \cdot s^2$
Standard gravity	g	9.80	m/s^2
Radius of earth (at equator)	R_e	6.38×10^{6}	m
Mass of earth	M_e	5.98×10^{24}	kg
Radius of moon	R_m	1.74×10^{6}	m
Mass of moon	M_m	7.36×10^{22}	kg

The values presented in this table are those used in computations in the text. Generally, the physical constants are known to much better precision.

College Physics

2nd Edition

COLLEGE

Second Edition

Saunders College Publishing
Complete Package for Teaching with

COLLEGE PHYSICS, Second Edition
by Raymond A. Serway and Jerry S. Faughn

INSTRUCTOR'S MANUAL with Solutions *J. Faughn and C. Teague*

STUDENT STUDY GUIDE with Computer Exercises *J. R. Gordon and R. A. Serway*

COURSEWARE DISK to Accompany Study Guide with Computer Exercises (Apple and IBM) *D. Oliver*

PRINTED TEST BANK *W. Junkin*

COMPUTERIZED TEST BANK (IBM) *W. Junkin*

OVERHEAD PROJECTION ACETATES

PHYSICS

Raymond A. Serway
James Madison University

Jerry S. Faughn
Eastern Kentucky University

SAUNDERS COLLEGE PUBLISHING
Philadelphia Fort Worth Chicago
San Francisco Montreal Toronto
London Sydney Tokyo

Text Typeface: Caledonia
Compositor: The Clarinda Company
Acquisitions Editor: John Vondeling
Developmental Editor: Ellen Newman
Project Editor: Sally Kusch
Copy Editor: Will Eaton
Managing Editor: Carol Field
Art Director: Carol Bleistine
Art Assistant: Doris Bruey
Text Designer: Edward A. Butler
Cover Designer: Lawrence R. Didona
Text Artwork: Tom Mallon, Rolin Graphics
Layout Artist: Pam Dooley
Production Manager: Harry Dean

Cover Credit: Soap Bubble Landscape by Peter Aprahamian/Science Photo Library © Photo Researchers, Inc.

Printed in the United States of America

COLLEGE PHYSICS, 2nd edition

ISBN 0-03-022952-9

Library of Congress Catalog Card Number: 88-043415

8901 061 98765432

Preface

This textbook, now in its second edition, is appropriate for a one-year course in introductory physics commonly taken by students majoring in biology and premedicine, as well as other disciplines such as health, environmental, earth, and social sciences. The mathematical techniques used in the book include algebra and trigonometry, but not calculus.

Changes in the Second Edition

A number of changes and improvements have been made in preparing the second edition of this text. Many of these changes are in response to comments and suggestions offered by users of the first edition and reviewers of the manuscript. The following represent the major changes in the second edition:

1. Portions of the text have been rewritten and revised to improve the clarity of presentation and to provide more discussions of recent developments in physics. Chapters that were substantially revised include Chapters 5, 7, 13, 19, and 20. In the optics portion of the text, several sections were rewritten, and new sections were added.
2. Two new chapters in modern physics have been added (Chapter 31, Molecules and Solids, and Chapter 34, Elementary Particles).
3. Guest Essays have been revised and expanded to include suggestions for additional reading, questions, and problems for added flexibility in covering these topics. Six of the essays are entirely new and cover black holes, superconductivity, the Big Bang theory, the scanning tunneling microscope, alternative energy resources, and tests of general relativity.
4. The problem sets have been extensively revised, and many new questions and problems have been added to the text. All problems have been checked for accuracy.
5. Full color photographs have been added that emphasize principles in mechanics, optics, and modern physics.
6. A new ancillary package includes an Instructor's Manual with solutions to *all* the problems in the text, Study Guide with Computer Exercises, Courseware Disk (Apple and IBM), Multiple Choice Test Bank (printed or computerized for IBM PC), Overhead Projection Acetates, and a free videotape on the latest findings in superconductivity.

Objectives

The main objectives of this introductory physics textbook are twofold: to provide the student with a clear and logical presentation of the basic concepts and principles of physics, and to strengthen an understanding of the concepts and principles through a broad range of interesting applications to the real world. In order to meet these objectives, emphasis is placed on sound physical arguments and discussions of everyday experiences. At the same time, we have attempted to motivate the student through practical examples that demonstrate the role of physics in other disciplines.

Coverage

The material covered in this book is concerned with standard topics in classical physics and modern physics. The book is divided into six parts: Part 1 (Chapters 1–10) deals with the fundamentals of Newtonian mechanics and the physics of fluids; Part 2 (Chapters 11–13) is concerned with heat and thermodynamics; Part 3 (Chapters 14–15) covers wave motion and sound; Part 4 (Chaptes 16–23) is concerned with electricity and magnetism; Part 5 (Chapters 24–27) treats the properties of light and the field of geometric and wave optics; and Part 6 (Chapters 28–34) represents an introduction to the theory of relativity and modern physics.

Features

Most instructors would agree that the textbook assigned in a course should be the student's major "guide" for understanding and learning the subject matter. With this in mind, we have included many pedagogic features in the textbook that enhance its usefulness to both the student and instructor. These are as follows:

Style We have attempted to write the book in a style that is clear, relaxed, and pleasing to the reader. New terms are carefully defined, and we have tried to avoid jargon.

Organization The book is divided into six parts: mechanics, thermodynamics, vibrations and wave motion, electricity and magnetism, light and optics, and modern physics, Each part includes an overview of the subject matter to be covered in that part and some historical perspectives.

Introductory Chapter An introductory chapter is included to "set the stage" for the text and to discuss the units of physical quantities, order-of-magnitude calculations, dimensional analysis, significant figures, mathematical notation, and the techniques of treating vector quantities.

Units The international system of units (SI) is used throughout the book. The British engineering system of units (conventional system) is used only to a limited extent in the early chapters on mechanics.

Previews Most chapters begin with a chapter preview, which includes a brief discussion of the chapter objectives and content.

Marginal Notes and Equations Important equations are enclosed in a color-ruled box, and marginal notes set in color are often used to describe their meaning. Marginal notes are also used to locate specific definitions and important statements.

Worked Examples A large number of worked examples (280) are presented as an aid in understanding concepts. In many cases, these examples serve as models for solving end-of-chapter problems. The examples are set off from the text by color rules for ease of location, and most examples are given titles to describe their content.

Worked Example Exercises Many of the worked examples (over 100) are followed immediately by exercises with answers. These examples are intended to make the textbook more interactive with the student, and to test the student's understanding of problem-solving techniques.

Special Topics Many chapters include special topic sections which are intended to expose the student to various practical and interesting applications of physical principles. Many of these are considered optional and are labeled with an asterisk(*).

Guest Essays In this second edition, we have included 14 essays written by guest authors. Six of the essays are new, while the remaining eight are revisions of essays that appeared in the first edition. The topics covered in the essays include arch structures, the circulatory system, the nervous system, applications of lasers in medicine, exponential growth, general relativity, superconductivity, atmospheric physics, vacuum tunneling spectroscopy, fiber optic communications, and alternate energy sources. Although the essays are intended as supplemental reading for the student, most include questions and problems for added flexibility in covering these topics. In this edition, the essays have been placed at the ends of the chapters to avoid interrupting the main textual material.

Important Statements Many important statements and definitions are set in color for added emphasis and ease of review.

Illustrations and Photographs The readability and effectiveness of the textual material and worked examples are enhanced by a large number of figures, diagrams, photographs, and tables. A second color is used to add clarity to the artwork. For example, vectors are color-coded, and curves in *xy* plots are drawn in color. Three-dimensional effects are produced with the use of airbrushed areas, where appropriate. Many new photographs have been added to this second edition, and captions added to all chapter opening photos to make them more useful as an instructional tool. Finally, a collection of 35 color photographs has been added to the text in four locations. Photographs were selected in which use of color is functional, and extensive captions add to the pedagogy. References to most of these color photos are provided in the text.

Summaries Each chapter contains a summary that reviews the important concepts and equations in that chapter.

Readings Each chapter contains a set of suggested additional readings, which have been selected on the basis of their level and the degree to which they supplement the text.

Thought Questions The list of questions requiring verbal answers given at the end of each chapter has been expanded by more than 25% for a total of 562. Some questions provide the student with a means of self-testing the concepts presented in the chapter. Others could serve as a basis for initiating classroom discussions. Answers to selected questions are included in the Student Study Guide With Computer Exercises accompanying the text.

Problems An extensive set of problems is included at the end of each chapter. In this second edition, many new problems have been added to the text, and most of the old problems have been revised or rewritten. The text contains 1771 problems, which is an increase of over 10% from the previous edition. Answers to odd-numbered problems are provided at the end of the book. For the convenience of both the student and instructor, most problems are keyed to specific sections of the chapter; the remaining problems, labeled "Additional Problems," are not keyed to specific sections. In general, there are three categories of problems in each chapter, corresponding to three levels of difficulty. Problems that are straightforward in nature are presented first and are unmarked. These are followed by problems of intermediate difficulty, marked with one dot (·). Finally, we include a small number of more challenging problems, marked with two dots (··). In our opinion, assignments should mainly consist of problems from the first two categories to help build self-confidence. As mentioned earlier, most of the guest essays now include several questions and problems for those instructors who choose to cover these topics.

Appendices Several appendices are provided at the end of the text. Most of the appendix material represents a review of mathematical techniques used in the text, including scientific notation, algebra, geometry, and trigonometry. Reference to these appendices is made throughout the text. Most mathematical review sections include worked examples and exercises with answers. The last few appendices supplement textural information. For easy reference, the front endpaper contains a table of conversion factors and a list of physical constants, and the back endpaper provides a periodic table of the elements.

Ancillaries

The ancillaries that are available with this text include an Instructor's Manual with Solutions to *all* problems in the text, a Computerized Test Bank (IBM), a Printed Test Bank containing 1600 multiple choice questions, a Student Study Guide With Computer Exercises (described below), a Courseware Disk (Apple and IBM) containing programs to be used with the computer exercises, a set of 100 Overhead Projection Acetates containing more than 200 figures that are useful for lecture presentations, and a free videotape on the latest findings in superconductivity.

The Student Study Guide With Computer Exercises is a unique student aid that combines the value of a problem-solving-oriented study guide with a group of integrated and interactive computer exercises. Each chapter of the study guide contains a list of objectives, a review and summary of important concepts, a few worked examples, answers to selected questions from the text, solutions to selected problems from the text, and further drill on problem-solving methods through the use of programmed exercises. The study guide also includes the option of using a number of computer programs (presented in special computer modules) that are interactive in nature. That is, the student's input will have direct and immediate effect on the output. This feature will enable students to work through many challenging numerical problems, and experience the power of the computer in scientific work. The computer exercises direct the student's use of the programs contained on the Courseware Disk. The disk is available for Apple and IBM computers upon adoption of the study guide.

Teaching Options

This book contains more than enough material for a one-year course in introductory physics. This serves two purposes. First, it gives the instructor more flexibility in choosing topics for a specific course. Second, the book becomes more useful as a resource for students. On the average, it should be possible to cover about one chapter each week. Many special topic sections containing interesting applications are considered optional and are therefore marked with an asterisk(*). Some of these could be covered to suit the needs of the students and instructors. For shorter courses, instructors may choose to omit several chaptes without loss of continuity. The chapters that we suggest could be omitted in such circumstances are Chapter 8 (Rotational Motion), Chapter 22 (Alternating Current Circuits), Chapter 23 (Electromagnetic Waves), Chapter 27 (Optical Instruments), Chapter 31 (Molecules and Solids), and Chapter 33 (Nuclear Physics Applications).

ACKNOWLEDGEMENTS

In preparing the second edition of this textbook, we have been guided by the expertise of many people who reviewed part or all of the manuscript. We wish to acknowledge the following scholars and express our sincere appreciation for their helpful suggestions, criticisms, and encouragement: John Anderson, University of Pittsburgh; Neil W. Ashcroft, Cornell University; Gordon Aubrecht, Ohio State University; Charles Bacon, Ferris State University; Louis Barrett, Western Washington University; Jeffrey Braun, University of Evansville; John Brennan, University of Central Florida; Joseph Catanzarite, Cypress College; Roger W. Clapp, University of South Florida; Giuseppe Colaccico, University of South Florida; Jorge Cossio, Miami-Dade Community College; Terry T. Crow, Mississippi State College; Albert Thomas Fromhold, Jr., Auburn University; Teymoor Gedayloo, California Polytechnic State University; Wlodzimierz Guryn, Brookhaven National Laboratory; James Harmon, Oklahoma State University; Fred Inman, Mankato State University; Drasko Jovanovic, Fermilab; Ronald E. Jodoin, Rochester Institute of Technology; Joan P. S. Kowalski, George Mason University; Sol Krasner, University of Chicago; James Linblad, Saddleback Community College; Michael Lieber, University of Arkansas; Jeffrey V. Mallow, Loyola University of Chicago; Lewis J. Oakland, University of Minnesota; Louis O'Kelly, Memphis State University; T. A. K. Pillai, University of Wisconsin—La Crosse; William D. Ploughe, Ohio State University; Joseph Priest, Miami University; Kurt Reibel, Ohio State University; Donald D. Snyder, Indiana University at South Bend; Carey E. Stronach, Virginia State University; Thomas W. Taylor, Cleveland State University; Larry Weaver, Kansas State University; and L. L. Van Zandt, Purdue University.

For their suggestions during the various stages of development of the first edition, we thank the following scholars: Albert Altman, University of

Lowell; Paul Bender, Washington State University; Roger W. Clapp, Jr., University of South Florida; John R. Gordon, James Madison University; Karl F. Kuhn, Eastern Kentucky University; Harvey S. Leff, California State Polytechnic University; Bill Lochslet, Pennsylvania State University; David Markowitz, The University of Connecticut; Joe McCauley, Jr., University of Houston; Bill F. Melton, The University of North Carolina at Charlotte; H. Kent Moore, James Madison University; Carl R. Nave, Georgia State University; Blaine Norum, University of Virginia; M. E. Oakes, The University of Texas at Austin; James Purcell, Georgia State University; William R. Savage, The University of Iowa; Howard G. Voss, Arizona State University; Donald H. White, Western Oregon State College; George A. Williams, The University of Utah; Robert M. Wood, University of Georgia; and Peter D. Zimmerman, Louisiana State University.

We wish to express our gratitude to Mary Lou Glick who typed various parts of the manuscript. We thank Sarah Evans, Henry Leap, and Jim Lehman for locating and providing many excellent photographs. We are especially indebted to John R. Gordon for writing the student study guide, and to Charles Teague for checking solutions to all the problems and for his assistance in preparing the Instructor's Manual.

We thank the following individuals for writing the interesting essays that appear throughout the text: Isaac D. Abella, University of Chicago; Albert A. Bartlett, University of Colorado at Boulder; Gordon Batson, Clarkson University; William G. Buckman, Western Kentucky University; Paul Davidovits, Boston College; Roger A. Freedman, University of California; Paul K. Hansma, University of California; Laurent Hodges, Iowa State University; Milton Kerker, Clarkson University; Edward A. Lacy, Satellite Beach, Florida; David Markowitz, University of Connecticut; John Meakin, University of Delaware; William Savage, University of Iowa; Clifford M. Will, Washington University; and Sidney Wolff, Kitt Peak National Observatory.

We wish to thank the professional staff at Saunders College Publishing for their expertise and perseverance in this project. In particular, we thank Ellen Newman, who worked on all aspects of the manuscript during its development, Sally Kusch, who was most instrumental in turning the manuscript into a most attractive book, and Will Eaton for his excellent copyediting. We thank John Vondeling for his expert guidance in this revision, and Carol Bleistine for her superb design work. Special thanks also go to Tom Mallon, Rolin Graphics, Linda Maugeri, and Larry Ward for an excellent set of illustrations.

Finally, we dedicate this book to our wives and children for their love, support, and long-term sacrifices.

RAYMOND A. SERWAY
James Madison University
Harrisonburg, Virginia 22807

JERRY S. FAUGHN
Eastern Kentucky University
Richmond, Kentucky 40475

To the Student

We feel it is appropriate to offer some words of advice that should be of benefit to you, the student. Before doing so, we shall assume that you have read the preface, which describes the various features of the text that will help you through the course.

How to Study

Very often we are asked "How should I study physics and prepare for examinations?" There is no simple answer to this question, but we would like to offer some suggestions based on our own experiences in learning and teaching over the years.

First and foremost, maintain a positive attitude towards the subject matter, keeping in mind that physics is the most fundamental of all natural sciences. Other science courses that follow will use the same physical principles, so it is important that you understand and be able to apply the various concepts and theories discussed in the text.

Concepts and Principles

It is essential that you understand the basic concepts and principles *before* attempting to solve assigned problems. This is best accomplished through a careful reading of the textbook before attending your lecture on that material. In the process, it is useful to jot down certain points that are not clear to you. Take careful notes in class, and then ask questions pertaining to those ideas that require clarification. Keep in mind that few people are able to absorb the full meaning of scientific material after one reading. Several readings of the text and notes may be necessary. Your lectures and laboratory work should supplement the text and clarify some of the more difficult material. You should reduce memorization of material to a minimum. Memorizing passages from a text, equations, and derivations does not necessarily mean you understand the material. Your understanding of the material will be enhanced through a combination of efficient study habits, discussions with other students and instructors, and your ability to solve the problems in the text. Ask questions whenever you feel it is necessary. If you are reluctant to ask questions in class, seek private consultation or initiate discussions with your classmates. Many individuals are able to speed up the learning process when the subject is discussed on a one-to-one basis.

Study Schedule

It is important to set up a regular study schedule, preferably on a daily basis. Make sure to read the syllabus for the course and adhere to the schedule set by your instructor. The lectures will be much more meaningful if you read the corresponding textual material *before* attending the lecture. As a general rule, you should devote about two hours of study time for every hour in class.

If you are having trouble with the course, seek the advice of the instructor or students who have already taken the course. You may find it necessary to seek further instruction from experienced students. Very often, instructors offer review sessions in addition to regular class periods. It is important that you avoid the practice of delaying study until a day or two before an exam. More often than not, this will lead to disastrous results. Rather than an all-night study session before an exam, it is better to briefly review the basic concepts and equations, followed by a good night's rest. If you feel you need additional help in understanding the concepts, preparing for exams, or in problem-solving, we suggest that you acquire a copy of the student study guide that accompanies the text, which should be available at your college bookstore.

Use the Features

You should make *full* use of the various features of the text discussed in the preface. For example, marginal notes are useful for locating and describing important equations, while important statements and definitions are highlighted in color. Many useful tables are contained in the appendices, but most are incorporated into the text where they are used most often. Appendix A is a convenient review of mathematical techniques. Answers to odd-numbered problems are given at the end of the text, and answers to most end-of-chapter questions are provided in the study guide. Exercises (with answers), which follow some worked examples, represent extensions of those examples, and in most cases you are expected to perform a simple calculation. Their purpose is to test your problem-solving skills as you read through the text. An overview of the entire text is given in the table of contents, while the index will enable you to locate specific material quickly. Footnotes are sometimes used to supplement the discussion or to cite other references on the subject. A list of suggested additional readings is given at the end of each chapter.

After reading a chapter, you should be able to define any new quantities introduced in that chapter, and discuss the principles and assumptions that were used to arrive at certain key relations. The chapter summaries and the review sections of the study guide should help you in this regard. In some cases, it will be necessary to refer to the index of the text to locate certain topics. You should be able to correctly associate with each physical quantity a symbol used to represent that quantity and the unit in which the quantity is specified. Furthermore, you should be able to express each important relation in a concise and accurate prose statement.

Problem Solving

R. P. Feynman, Nobel laureate in physics, once said, "You do not know anything until you have practiced." In keeping with this statement, we strongly advise that you develop the skills necessary to solve a wide range of problems. Your ability to solve problems will be one of the main tests of your knowledge of physics, and therefore you should try to solve as many problems as possible. It is essential that you understand basic concepts and principles before attempting to solve problems. It is good practice to try to find alternate solutions to the same problem. For example, problems in mechanics can be solved using Newton's laws, but very often an alternative method using energy considerations is more direct. You should not deceive yourself into thinking you understand the problem after seeing its solution in class. You must be able to solve the problem and similar problems on your own.

The method of solving problems should be carefully planned. A systematic plan is especially important when a problem involves several concepts. First, read the problem several times until you are confident you understand what is being asked. Look for any key words that will help you interpret the problem, and perhaps allow you to make certain assumptions. Your ability to interpret the question properly is an integral part of problem solving. You should acquire the habit of writing down the information given in a problem, and decide what quantities need to be found. You might want to construct a table listing quantities given, and quantities to be found. This procedure is sometimes used in the worked examples of the text. After you have decided on the method you feel is appropriate for the situation, proceed with your solution.

We often find that students fail to recognize the limitations of certain formulas or physical laws in a particular situation. It is very important that you understand and remember the assumptions underlying a particular theory or formalism. For example, we shall find that certain equations in kinematics apply only to a particle moving with constant acceleration. They are not valid for situations in which the acceleration is not constant, as in the cases of the motion of an object connected to a spring and the motion of an object through a fluid.

Experiments

Physics is a science based upon experimental observations. In view of this fact, we recommend that you make every effort to supplement the text through various types of "hands-on" experiments, either in the laboratory or at home. Many simple experiments can be used to test ideas and models discussed in class or in the text. For example, an object swinging on the end of a long string together with a wristwatch can be used to investigate pendulum motion; an object attached to the end of a vertical spring or a rubber band can be used to determine the nature of restoring forces and to investigate periodic motion; collisions between equal masses can be observed while playing billiards; an approximate value for the acceleration due to gravity can be obtained by dropping an object from a known height by simply measuring the time of its fall with a stopwatch; traveling waves can be investigated with the aid of a stretched rope or the common "Slinky" toy (a stretched spring); a pair of Polaroid sunglasses and some discarded lenses or a magnifying glass can be used to perform various experiments in optics. The list is endless. When physical models are not available, be imaginative and try to develop models of your own.

Closing Comments

Someone once said that there are only two professions in which people truly enjoy what they are doing: professional sports and physics. Although this statement is most likely an exaggeration, both professions are truly exciting and stretch your skills to the limit. It is our sincere hope that you too will find physics exciting and that you will benefit from this experience, regardless of your chosen profession.

Welcome to the exciting world of physics.

To see a World in a Grain of Sand
And a Heaven in a Wild Flower,
Hold infinity in the palm of your hand
And Eternity in an hour.

W. Blake, "Auguries of Innocence"

Contents Overview

Contents

*These sections are optional.

*These sections are optional.

*These sections are optional.

*These sections are optional.

*These sections are optional.

*These sections are optional.

*These sections are optional.

*These sections are optional.

*These sections are optional.

Part 1

MECHANICS

Facts which at first seem improbable will, even in scant explanation, drop the cloak which has hidden them and stand forth in naked and simple beauty.

GALILEO GALILEI

Nature and Nature's laws lay hid in night: God said, Let Newton be, and all was light.

ALEXANDER POPE

Physics, the most fundamental science, is concerned with the basic principles of the universe. It is one of the foundations upon which the other physical sciences—astronomy, chemistry, and geology—are based. The beauty of physics lies in the simplicity of the fundamental physical theories and in the way just a small number of fundamental concepts, equations, and assumptions can alter and expand our view of the world around us.

The myriad physical phenomena in our world are a part of one or more of the following five areas of physics:

1. Mechanics, which is concerned with the motion of material objects
2. Thermodynamics, which deals with heat, temperature, and the behavior of a large number of particles
3. Electromagnetism, which deals with charges, currents, and electromagnetic fields
4. Relativity, which is a theory describing particles moving at any speed
5. Quantum mechanics, a theory dealing with the behavior of particles at the submicroscopic level as well as the macroscopic world

The first part of this textbook deals with mechanics, sometimes referred to as classical mechanics or newtonian mechanics. This is an appropriate place to begin an introductory text since many of the basic principles used to understand mechanical systems can later be used to describe such natural phenomena as waves and heat transfer. Furthermore, the laws of conservation of energy and momentum introduced in mechanics retain their importance in the fundamental theories which follow, including the theories of modern physics.

The first serious attempts to develop a theory of motion were provided by the Greek astronomers and philosophers. Although they devised a complex model to describe the motions of heavenly bodies, their model lacked correlation between such motions and the motions of objects on earth. The study of mechanics was enhanced by a number of careful astronomical investigations by Copernicus, Brahe, and Kepler in the 16th century. In the 16th and 17th centuries, Galileo attempted to relate the motion of falling bodies and projectiles to the motion of planetary bodies, and Sevin and Hooke were studying forces and their relation to motion. A major development in the theory of mechanics was provided by Newton in 1687 when he published his *Principia*. Newton's elegant theory, which remained unchallenged for more than 200 years, was based on contributions made by Galileo and others, together with Newton's hypothesis of universal gravitation.

Today, mechanics is of vital importance to students from all disciplines. It is highly successful in describing the motions of material bodies, such as planets, rockets, and baseballs. In the first ten chapters of the text, we shall describe the laws of mechanics and examine a wide range of phenomena which can be understood with these fundamental ideas.

CHAPTER 1

Introduction

Weathervanes can be used to determine the direction of the wind velocity vector at any instant. (Photo Researchers, Inc. © Jules Bucher 1981)

The goal of physics is to provide a quantitative understanding of certain basic phenomena that occur in our universe. Physics is a science based on experimental observations and mathematical analyses. The main objective behind such experiments and analyses is to develop theories that explain the phenomenon being studied and to relate these theories to other established theories. Fortunately, it is possible to explain the behavior of various physical systems using relatively few fundamental laws. Analytical procedures require expressing these laws in the language of mathematics, the tool that provides a bridge between theory and experiment. In this chapter we shall discuss a few mathematical concepts and techniques that will be used throughout the text.

Since future chapters will be concerned with the laws of physics, it is necessary to provide a clear definition of the basic quantities involved in these laws. For example, such physical quantities as force, velocity, volume, and acceleration can be described in terms of more fundamental quantities. In the next several chapters we shall encounter three basic quantities: **length** (L), **time** (T), and **mass** (M). In later chapters we shall have to add two other standard units to our list, one for temperature (the kelvin), the other for electric current (the ampere). In our study of mechanics, however, we shall be concerned only with the units of mass, length, and time.

1.1 STANDARDS OF LENGTH, MASS, AND TIME

If we are to report the results of a measurement of a certain quantity to someone who wishes to reproduce this measurement, a unit for this quantity must be defined. For example, it would be meaningless for a visitor from another planet to talk to us about a length of 8 "gliches" if we do not know the meaning of the unit "glich." On the other hand, if someone familiar with our system of measurement and weights reports that a wall is 2.0 meters high and our unit of length is defined to be 1.0 meter, we then know that the height of the wall is twice our fundamental unit of length. Likewise, if we are told that a person has a mass of 75 kilograms and our unit of mass is defined as 1.0 kilogram, then that person has a mass 75 times greater than our fundamental unit of mass. In 1960, an international committee agreed on a set of definitions and standards to describe these fundamental physical quantities. The system that was established is called the **SI system** (Système International) of units. In this system, the units of mass, length, and time are the kilogram, meter, and second, respectively.

Length

In 1120 A.D. the king of England decreed that the standard of length in his country would be the yard and that the yard would be precisely equal to the distance from the tip of his nose to the end of his outstretched arm. Similarly, the original standard for the foot adopted by the French was the length of the royal foot of King Louis XIV. This standard prevailed until 1799, when the legal standard of length in France became the meter, defined as one ten-millionth of the distance from the equator to the north pole.

There have been many other systems developed in addition to those discussed above, but the advantages of the French system have caused it to become the prevailing system in most countries and in scientific circles everywhere. As recently as 1960, the length of the meter was preserved as the distance between two lines on a specific bar of platinum-iridium alloy stored under controlled conditions. This standard was abandoned for several reasons, a principal one being that the limited accuracy with which the separation between the lines can be determined does not meet the present requirements of science and technology. Until recently, the meter was defined as 1,650,763.73 wavelengths of orange-red light emitted from a krypton-86 lamp. However, in October 1983, the **meter** was redefined to be *the distance traveled by light in a vacuum during a time of 1/299,792,458 second*. In effect,

(Left) The National Standard Kilogram No. 20, an accurate copy of the International Standard Kilogram kept at Sèvres, France, is housed under a double bell jar in a vault at the National Bureau of Standards. (Right) The primary frequency standard (an atomic clock) at the National Bureau of Standards. This device keeps time with an accuracy of about 3 millionths of a second per year. (Photos courtesy of National Bureau of Standards, U.S. Dept. of Commerce)

this latest definition establishes that the speed of light in a vacuum is 299,792,458 meters per second.

Mass

The SI unit of mass, the **kilogram,** is defined as *the mass of a specific platinum-iridium alloy cylinder kept at the International Bureau of Weights and Measures at Sèvres, France.* At this point, we should add a word of caution. Most beginning students of physics tend to confuse the quantities called weight and mass. For the present, we shall not discuss the distinction between these two quantities, which will be clearly defined in later chapters. For now, however, you should note that these are two distinctly different physical quantities.

Time

Before 1960, the standard of time was defined in terms of the average length of a solar day. (A solar day is the time interval between successive appearances of the sun at the highest point it reaches in the sky each day.) The basic unit of time, the second, was defined to be (1/60)(1/60)(1/24) = 1/86,400 of the average solar day. In 1967, the second was redefined to take advantage of the high precision obtainable with a device known as an atomic clock, which uses the characteristic frequency of the cesium-133 atom as the "reference clock." The **second** is now defined as *9,192,631,770 times the period of one oscillation of the cesium atom.*

TABLE 1.1 Approximate Values of Various Masses

	Mass (kg)
Universe	10^{52}
Milky Way galaxy	7×10^{41}
Sun	2×10^{30}
Earth	6×10^{24}
Moon	7×10^{22}
Shark	1×10^{2}
Human	7×10^{1}
Frog	1×10^{-1}
Mosquito	1×10^{-5}
Bacterium	1×10^{-15}
Hydrogen atom	2×10^{-27}
Electron	9×10^{-31}

Approximate Values for Mass, Length, and Time

Approximate values of various masses, lengths, and time intervals are presented in Tables 1.1, 1.2, and 1.3, respectively. Note the wide range of values for these quantities. You should study these tables and get a feel for what is meant by a kilogram of mass (this book has a mass of about 2 kilograms) or a time interval of 10^{10} seconds (one year is about 3×10^{7} seconds). If you need to learn or review the powers of 10 notation, such as the designation of the number 50,000 in the form 5×10^{4}, you should study Appendix A.

Systems of units commonly used are the *SI system,* in which the units of mass, length, and time are the kilogram (kg), meter (m), and second (s), respectively; the *cgs or gaussian system,* in which the units of mass, length, and time are the gram (g), centimeter (cm), and second, respectively; and the *British engineering system* (sometimes called the conventional system), in which the units of mass, length, and time are the slug, foot (ft), and second, respectively. Throughout most of this text we shall use SI units since they are

TABLE 1.2 Approximate Values of Some Measured Lengths

	Length (m)
Distance from earth to most remote known quasar	1×10^{26}
Distance from earth to most remote known normal galaxies	4×10^{25}
Distance from earth to nearest large galaxy (M31 in Andromeda)	2×10^{22}
Distance from earth to nearest star (Proxima Centauri)	4×10^{16}
One lightyear	9×10^{15}
Mean orbit radius of the earth	2×10^{11}
Mean distance from earth to moon	4×10^{8}
Mean radius of the earth	6×10^{6}
Typical altitude of orbiting earth satellite	2×10^{5}
Length of a football field	9×10^{1}
Length of a housefly	5×10^{-3}
Size of smallest dust particles	1×10^{-4}
Size of cells of most living organisms	1×10^{-5}
Diameter of a hydrogen atom	1×10^{-10}
Diameter of an atomic nucleus	1×10^{-14}
Diameter of a proton	1×10^{-15}

TABLE 1.3 Approximate Values of Some Time Intervals

	Interval (s)
Age of the universe	5×10^{17}
Age of the earth	1×10^{17}
Average age of a college student	6×10^{8}
One year	3×10^{7}
One day (time for one revolution of earth about its axis)	9×10^{4}
Time between normal hearbeats	8×10^{-1}
Period[a] of audible sound waves	1×10^{-3}
Period of typical radio waves	1×10^{-6}
Period of vibration of an atom in a solid	1×10^{-13}
Period of visible light waves	2×10^{-15}
Duration of a nuclear collision	1×10^{-22}
Time for light to cross a proton	3×10^{-24}

[a]Period is defined as the time interval of one complete vibration.

almost universally accepted in science and industry. We shall make some limited use of British engineering units in the study of mechanics.

Some of the most frequently used prefixes defined within the SI system for the various powers of 10 and their abbreviations are listed in Table 1.4. For example, 10^{-3} m is equivalent to 1 millimeter (mm), and 10^3 m is 1 kilometer (km). Likewise, 1 kg is equal to 10^3 g, and 1 megavolt (MV) is equivalent to 10^6 volts (V).

TABLE 1.4 Some Prefixes for Powers of Ten Used in the SI System

Power	Prefix	Abbreviation
10^{-18}	atto	a
10^{-15}	femto	f
10^{-12}	pico	p
10^{-9}	nano	n
10^{-6}	micro	μ
10^{-3}	milli	m
10^{-2}	centi	c
10	deca	d
10^3	kilo	k
10^6	mega	M
10^9	giga	G
10^{12}	tera	T
10^{15}	peta	P
10^{18}	exa	E

1.2 DIMENSIONAL ANALYSIS

The word *dimension* has a special meaning in physics. It denotes the qualitative nature of a physical quantity. Whether the separation between two points is measured in units of feet or meters or furlongs, the measured quantity is a distance. We say that the dimension of distance is *length*.

The symbols that will be used to specify length, mass, and time are L, M, and T, respectively. We shall often use brackets [ ] to denote the dimensions of a physical quantity. For example, in this notation the dimensions of velocity, v, are written $v = [\mathrm{L}]/[\mathrm{T}]$, and the dimensions of area, A, are $A = [\mathrm{L}^2]$. The dimensions of area, volume, velocity, and acceleration are listed in Table 1.5, along with their units in the three common systems. The dimensions of other quantities, such as force and energy, will be described as they are introduced in the text.

In many situations, you may be faced with having to derive or check a specific formula. Although you may have forgotten the details of the derivation, there is a useful and powerful procedure called **dimensional analysis** that can be used to assist in the derivation or to check your final expression. Dimensional analysis makes use of the fact that *dimensions can be treated as algebraic quantities.* That is, quantities can be added or subtracted only if they have the same dimensions. Furthermore, the quantities on each side of an equation must have the same dimensions. By following these simple rules, you can use dimensional analysis to help determine whether or not you have the correct form of an expression, because the relationship can be correct only if the dimensions on each side of the equation are the same.

To illustrate this procedure, suppose that during an examination you wish to derive the formula for the distance x traveled by a car in a time t if the car starts from rest and moves with constant acceleration a. In Chapter 2 we shall find that the correct expression for this situation is $x = \frac{1}{2}at^2$. Suppose you suspect that this equation is in error and you wish to check its validity from a dimensional analysis approach.

We know that the quantity x on the left side has a dimension of length. In order for the equation to be dimensionally correct, the quantity on the

TABLE 1.5 Dimensions of Area, Volume, Velocity, and Acceleration

System	Area (L^2)	Volume (L^3)	Velocity (L/T)	Acceleration ($\mathrm{L/T}^2$)
SI	m^2	m^3	m/s	$\mathrm{m/s}^2$
cgs	cm^2	cm^3	cm/s	$\mathrm{cm/s}^2$
British engineering (conventional)	ft^2	ft^3	ft/s	$\mathrm{ft/s}^2$

right side must also have the dimension of length. To make a dimensional check, we substitute in the basic dimensions for acceleration, [L]/[T^2], and time, [T]. That is, the equation $x = \frac{1}{2}at^2$ can be written dimensionally as

$$[\mathrm{L}] = \frac{[\mathrm{L}]}{[\mathrm{T}^2]} \cdot [\mathrm{T}^2] = [\mathrm{L}]$$

The dimensions of time cancel as shown, leaving us with the dimensions of length. Note that the factor $\frac{1}{2}$ was ignored because it has no dimensions.

EXAMPLE 1.1 Check the Dimensions

Show that the expression $v = v_0 + at$ is dimensionally correct, where v and v_0 represent velocities, a is acceleration, and t is a time interval.

Solution Since the dimensions of velocity are [L]/[T] and those of acceleration are [L]/[T^2], the equation $v = v_0 + at$ can be written in dimensional terms as

$$\frac{[\mathrm{L}]}{[\mathrm{T}]} = \frac{[\mathrm{L}]}{[\mathrm{T}]} + \frac{[\mathrm{L}]}{[\mathrm{T}^2]} \cdot [\mathrm{T}]$$

$$= \frac{[\mathrm{L}]}{[\mathrm{T}]} + \frac{[\mathrm{L}]}{[\mathrm{T}]}$$

$$= \frac{[\mathrm{L}]}{[\mathrm{T}]}$$

In the second step, the two quantities to the right of the equal sign have the same dimensions; they can be added (or subtracted) to give a single quantity with those dimensions. Thus we see in the last step that [L]/[T] = [L]/[T], and the expression is dimensionally correct. On the other hand, if the expression were as $v = v_0 + at^2$, it would be dimensionally *incorrect*. Try it and see!

EXAMPLE 1.2 The Clever Student

In a desperate attempt to come up with an equation to use during an examination, a student tries the equation $v^2 = ax$. Use dimensional analysis to determine if the equation might be valid.

Solution The dimensions on the left side are [L^2]/[T^2], so the dimensions on the right side must be the same. Let us check to see if this is the case:

$$\frac{[\mathrm{L}^2]}{[\mathrm{T}^2]} = \frac{[\mathrm{L}]}{[\mathrm{T}^2]} \cdot [\mathrm{L}] = \frac{[\mathrm{L}^2]}{[\mathrm{T}^2]}$$

The dimensions do check, but the equation is incorrect nonetheless. The correct form is $v^2 = 2ax$. Dimensional analysis can tell us nothing about the numerical constants in an equation. You should bear in mind that even though dimensional analysis is a powerful tool, it does have its limitations.

1.3 SIGNIFICANT FIGURES

When one performs measurements on certain quantities, the measured values are only known to within the limits of the experimental uncertainty. The value of the uncertainty can depend on various factors such as the quality of the apparatus, the skill of the experimenter, and the number of measurements performed. Suppose that in a laboratory experiment we are asked to measure the area of a rectangular plate using a meter stick as a measuring instrument. Let us assume that the accuracy to which we can measure a particular dimension of the plate is ± 0.1 cm. If the length of the plate is measured to be 16.3 cm, we can only claim that its length lies somewhere between 16.2 cm and 16.4 cm. In this case, we say that the measured value has three significant figures. Likewise, if its width is measured to be 4.5 cm, the actual value lies between 4.4 cm and 4.6 cm. (This measured value has only two significant figures.) Note that the significant figures include the first estimated digit. Thus, we could write the measured values as 16.3 ± 0.1 cm and 4.5 ± 0.1 cm.

Suppose now that we would like to find the area of the plate by multiplying the two measured values together. If we were to claim that the area is (16.3 cm)(4.5 cm) = 73.35 cm^2, our answer would be unjustifiable since it contains four significant figures, which is greater than the number of significant figures in either of the measured lengths. A good "rule of thumb" to use as a guide in determining the number of significant figures that can be claimed is as follows:

When multiplying several quantities, the number of significant figures in the final answer is the same as the number of significant figures in the *least* accurate of the quantities being multiplied, where "least accurate" means "having the lowest number of significant figures." The same rule applies to division.

Applying this rule to the multiplication example above, we see that the answer for the area can have only two significant figures since the dimension of 4.5 cm has only two significant figures. Thus, we can only claim the area to be 73 cm^2, realizing that the value can range between (16.2 cm)(4.4 cm) = 71 cm^2 and (16.4 cm)(4.6 cm) = 75 cm^2.

The presence of zeros in an answer may be misinterpreted. For example, suppose the mass of an object is measured to be 1500 g. This value is ambiguous because it is not known whether the last two zeros are being used to locate the decimal point or whether they represent significant figures in the measurement. In order to remove this ambiguity, it is common to use scientific notation to indicate the number of significant figures. In this case, we would express the mass as 1.5×10^3 g if there are two significant figures in the measured value and 1.50×10^3 g if there are three significant figures. Likewise, a number such as 0.00015 should be expressed in scientific notation as 1.5×10^{-4} if it has two significant figures or as 1.50×10^{-4} if it has three significant figures. The three zeros between the decimal point and the digit 1 in the number 0.00015 are not counted as significant figures because they are present only to locate the decimal point. In general, a **significant figure** is a reliably known digit (other than a zero used to locate the decimal point).

For addition and subtraction, the number of decimal places must be considered.

When numbers are added (or subtracted), the number of decimal places in the result should equal the smallest number of decimal places of any term in the sum.

For example, if we wish to compute 123 + 5.35, the answer would be 128 and not 128.35.

Could this be the result of poor data analysis? Perhaps the improper handling of significant figures caused this disaster. What do you think? (Photo Mill Valley, CA, University Science Books, 1982)

Throughout this book, *we shall generally assume that the given data are precise enough to yield an answer having three significant figures.* Thus, if we state that a jogger runs a distance of 5 m, it is to be understood that the distance covered is 5.00 m. Likewise, if the speed of a car is given as 23 m/s, its value is understood to be 23.0 m/s.

EXAMPLE 1.3 Installing a Carpet

A carpet is to be installed in a room whose length is measured to be 12.71 m (four significant figures), and whose width is measured to be 3.46 m (three significant figures). Find the area of the room.

Solution If you multiply 12.71 m by 3.46 m on your calculator, you will get an answer of 43.9766 m^2. How many of these numbers should you claim? Our rule of thumb for multiplication tells us that you can only claim the number of significant figures in the least accurate of the quantities being measured. In this example, we have only three significant figures in our least accurate measurement, so we should express our final answer as 44.0 m^2. Note that in the answer given, we used a general rule for rounding off numbers which states that the last digit retained is to be increased by 1 if the first digit dropped was equal to 5 or greater.

EXAMPLE 1.4 Time to Go Fishing
A lucky fisherman catches two sturgeon. The smaller fish has a measured length of 93.46 cm (two decimal places, four significant figures), while the larger fish has a measured length of 135.3 cm (one decimal place, four significant figures). What is the total length of fish caught for the day?

Solution Adding the two lengths together we find an answer of 228.76 cm. However, the length known to the smallest number of decimal places is the 135.3 cm length, which has one decimal place. Thus, according to our rule of thumb for adding numbers, you can only claim your answer for the total length to be 228.8 cm.

1.4 CONVERSION OF UNITS

Sometimes it is necessary to convert units from one system to another. Conversion factors between the SI and conventional systems for units of length are as follows:

1 mile = 1609 m = 1.609 km   1 ft = 0.3048 m = 30.48 cm
1 m = 39.37 in. = 3.281 ft   1 in. = 0.0254 m = 2.54 cm

A more complete list of conversion factors can be found on the inside back cover of this book. Units can be treated as algebraic quantities that can cancel each other. For example, suppose we wish to convert 15.0 in. to cen-

Can you perform the conversion shown on this sign? (Photo Ohio Department of Transportation)

timeters. From the list of conversion factors above, we find that 1 in. = 2.54 cm.[1] We can now convert 15.0 in. to centimeters by multiplying 15.0 in. by 2.54 cm/1 in. The units cancel as follows:

$$15.0 \text{ in.} = 15.0 \not{\text{in.}} \times 2.54 \frac{\text{cm}}{\not{\text{in.}}} = 38.1 \text{ cm}$$

EXAMPLE 1.5 The Volume of a Cube

A cube is 5 in. on a side. What is its volume in SI units?

Solution: The conversion factor to use is 1 in. = 0.0254 m. Thus, the length of a side in meters is

$$L = (5 \not{\text{in.}}) \left(\frac{0.0254 \text{ m}}{1 \not{\text{in.}}} \right) = 0.127 \text{ m}$$

and the volume is

$$V = L^3 = (0.127 \text{ m})^3 = 2.05 \times 10^{-3} \text{ m}^3$$

1.5 ORDER-OF-MAGNITUDE CALCULATIONS

It is often useful to estimate an approximate answer to a problem in which information is available. Such results can then be used to determine whether or not a more precise calculation is necessary. These approximations are usually based on certain assumptions, which must be modified if more precision is needed. Thus, we shall sometimes refer to the **order of magnitude** of a certain quantity as the power of ten that is closest to the quantity. For example, the mass of a person might be 75 kg $\approx 10^2$ kg. Thus, we would say that the person's mass is of the order of 10^2 kg. Usually, when an order-of-magnitude calculation is made, the results are reliable to within a factor of 10.

EXAMPLE 1.6 How Much Gasoline Do We Use?

Estimate the number of gallons of gasoline used annually by all cars in the United States.

Solution Since there are about 200 million people in the United States, an estimate of the number of cars in the country is 40 million (assuming one car and five people per family). We shall also estimate that the average distance traveled per year is 10,000 miles. If we assume a gasoline consumption of 20 miles/gal, each car uses about 500 gal/year. Multiplying this by the total number of cars in the United States gives an estimated total consumption of 2×10^{10} gal. This corresponds to a yearly consumer expenditure of over 20 billion dollars! This is probably a low estimate since we haven't accounted for commercial consumption and for such factors as two-car families.

---

[1]The conversion 1 in. = 2.54 cm is *exact* and is the legal conversion in the United States.

1.6 MATHEMATICAL NOTATION

Many mathematical symbols will be used throughout this book, some of which you are aware of, such as the symbol = to denote the equality of two quantities.

The symbol $\propto$ denotes a proportionality. For example, $y \propto x^2$ means that y is proportional to the square of x.

The symbol $<$ means *less than,* and $>$ means *greater than.* For example, $x > y$ means x is greater than y.

The symbol $\ll$ means *much less than,* and $\gg$ means *much greater than.*

The symbol $\approx$ indicates that two quantities are *approximately equal* to each other.

The symbol $\equiv$ means *is defined as.* This is a stronger statement than a simple $=$.

It is convenient to use a symbol to indicate the *change in a quantity.* For example, Δx (read as "delta x") means "the change in the quantity x." (It does not mean "the product of Δ and x.") For example, suppose that a person out for a morning stroll starts measuring his distance away from home when he is 10 m from his doorway. He then walks along a straight-line path and stops strolling when he is 50 m from the door. His change in position during his walk is $\Delta x = 50 \text{ m} - 10 \text{ m} = 40 \text{ m}$, or in symbolic form

$$\Delta x = x_f - x_i$$

In this equation x_f is the *final position* and x_i is the *initial position.*

We shall often have occasion to add several quantities. A useful abbreviation for representing such a sum is the Greek Letter Σ (capital sigma). Suppose we wish to add a set of five numbers represented by x_1, x_2, x_3, x_4, and x_5. In the abbreviated notation, we would write the sum as

$$x_1 + x_2 + x_3 + x_4 + x_5 = \sum_{i=1}^{5} x_i$$

where the subscript i on x represents any one of the numbers in the set. For example, if there are five masses in a system, m_1, m_2, m_3, m_4, and m_5, the total mass of the system $M = m_1 + m_2 + m_3 + m_4 + m_5$ could be expressed

$$M = \sum_{i=1}^{5} m_i$$

Finally, the magnitude of a quantity x, written $|x|$, is simply the absolute value of that quantity. The sign of $|x|$ is always positive, regardless of the sign of x. For example, if $x = -5$, $|x| = 5$; if $x = 8$, $|x| = 8$.

1.7 COORDINATE SYSTEMS

In many of the interesting applications and problems in physics, it is necessary to specify the locations of objects. For example, the description of the motion of an airplane requires a method for describing the position of the

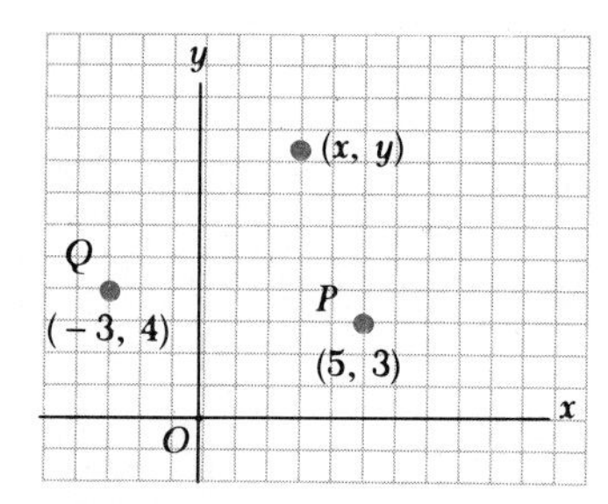

FIGURE 1.1 Designation of points in a cartesian coordinate system. Any point is labeled with coordinates (x, y).

plane. Locating an object in space is accomplished by means of a system of coordinates.

A **coordinate system** used to specify locations in space consists of:

1. A fixed reference point, called the origin.
2. A set of specified axes or directions.
3. Instructions that tell us how to label a point in space relative to the origin and axes.

One convenient coordinate system that we shall use frequently is the *cartesian coordinate system*, sometimes called the rectangular coordinate system. Such a system is illustrated in Figure 1.1. The point P shown in the figure has coordinates (5, 3). This notation means that if we start at the origin (O), we can reach P by moving 5 units along the positive x axis and then 3 units along the positive y axis. (Positive x is usually selected to the right of the origin and positive y upward from the origin. Negative x is usually to the left of the origin and negative y downward from the origin. This convention is not an absolute necessity, however. There may be instances in which you would like to take positive x to the left of the origin and negative x to the right. Feel free to do so.) The point Q has coordinates $(-3, 4)$, corresponding to going 3 units in the negative x direction and 4 units in the positive y direction.

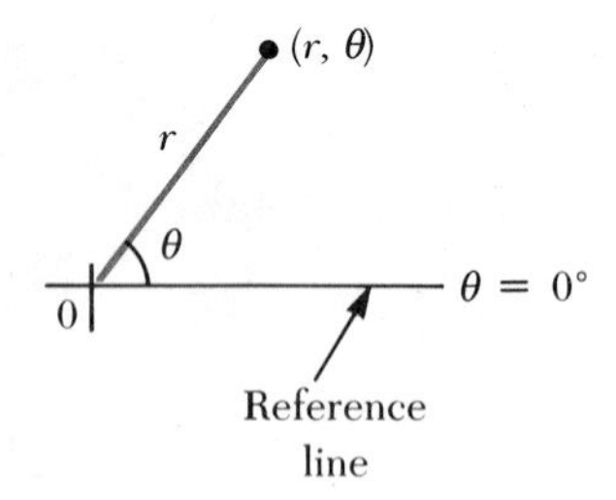

FIGURE 1.2 A polar coordinate system.

The cartesian coordinate system will be the one used far more often than any other in this text. However, sometimes it is more convenient to locate a point in space by its plane polar coordinates, (r, θ), as shown in Figure 1.2. When using this system, an origin and a reference line are selected as shown. A point is then specified by the distance r from the origin to the point and by the angle θ between r and the reference line. (Frequently the reference line is selected to be the positive x axis of a cartesian coordinate system, although this choice is not necessary.) For example, if a point is specified by the polar coordinates 3 m and 60°, one locates this point by moving out 3 m from the origin at an angle of 60° with respect to the reference line.

1.8 VECTORS AND SCALARS

The physical quantities that we shall encounter in this text can be placed in one of two categories: they are either scalar or vector quantities. A scalar is a quantity that is completely specified by a number with appropriate units. That is,

a **scalar** has only magnitude and no direction. On the other hand, a **vector** is a physical quantity that requires the specification of both direction and magnitude.

Temperature is an example of a scalar quantity. If someone tells you that the temperature of an object is −5°C, that information completely specifies the temperature of the object; no direction is required. Other examples of scalars are mass, time intervals, and the number of pages in this textbook. The rules of ordinary arithmetic are used to manipulate scalar quantities. For example, if you have 2 liters of water in a container (where 1 liter is defined to be 1000 cm^3) and you add 3 more liters, by ordinary arithmetic the amount of water you have in the container is 5 liters.

FIGURE 1.3 When a football player moves from O to P along any path, his displacement vector is the arrow drawn from O to P.

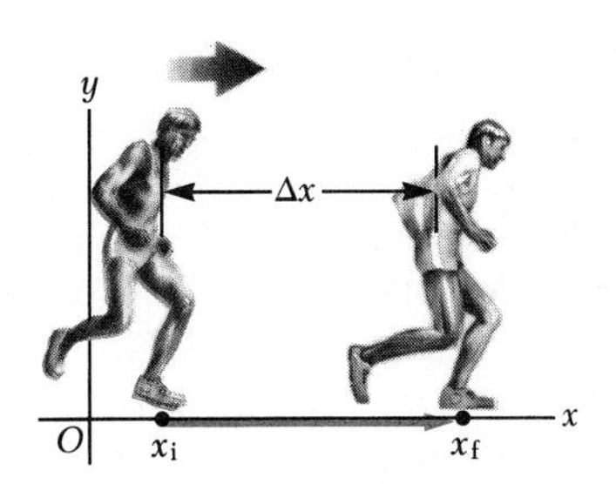

FIGURE 1.4 A sprinter moving along the x axis from x_i to x_f undergoes a displacement $\Delta x = x_f - x_i$.

An example of a vector quantity is force. If you are told that someone is going to exert a force of 10 lb on an object, that is not enough information to let you know what will happen to the object. The effect of a force of 10 lb exerted horizontally is different from the effect of a force of 10 lb exerted vertically. In other words, you need to know the direction of the force as well as its magnitude.

Another simple example of a vector quantity is the **displacement** of an object, defined as *the change in the position of the object.* Suppose a football player runs from position O to point P along a straight-line path, as in Figure 1.3. We pictorially represent the displacement by drawing an arrow from O to P, where the tip of the arrow indicates the direction of the displacement and the length of the arrow indicates the magnitude of the displacement. If the football player has to evade tacklers by running along some other path, such as the broken line in Figure 1.3, his displacement is still the same. In other words, the vector that represents the displacement of the player always starts at O and ends at P.

If a sprinter moves along the x axis from position x_i to position x_f, as in Figure 1.4, his displacement is given by $x_f - x_i$. As mentioned earlier, we use the Greek letter delta (Δ) to denote the change in a quantity. Therefore, we write the change in position (the displacement) as

Displacement

$$\Delta x = x_f - x_i \tag{1.1}$$

From this definition, we see that Δx is positive if x_f is greater than x_i and negative if x_f is less than x_i. If the sprinter moves from an initial position of $x_i = -3$ m to a final position of $x_f = 5$ m, his displacement is 8 m. (That is, $\Delta x = 5 - (-3) = 5 + 3 = 8$ m.)

1.9 SOME PROPERTIES OF VECTORS

Adding Vectors When two or more vectors are added together, all of them must have the same units. For example, it would be meaningless to add a velocity vector to a displacement vector because they are different physical quantities. Scalars also obey this same rule. For example, it would be meaningless to add temperatures and areas. (Remember dimensional analysis?)

Jennifer pointing in the right direction. (Photo by Raymond A. Serway)

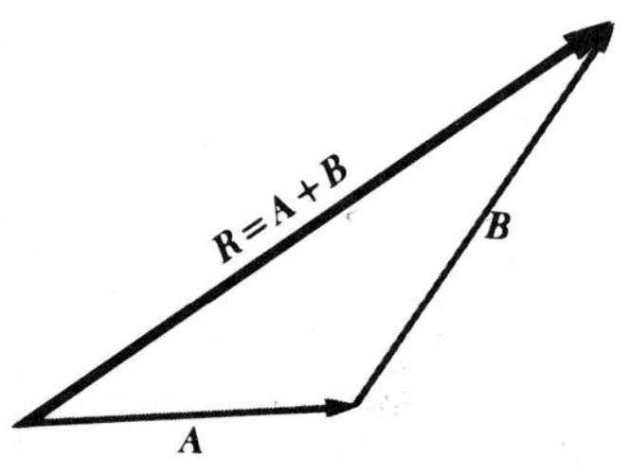

FIGURE 1.5 When vector **B** is added to vector **A**, the vector sum **R** is the vector that runs from the tail of **A** to the tip of **B**.

The rules for adding vectors are conveniently described by geometric methods. You should be aware, however, that adding vectors geometrically is done here to introduce you to the concepts involved. Later we shall develop a more precise technique for adding vectors that is much more convenient and will be the method used throughout the remainder of this text.

To add vector **B** to vector **A**, first draw **A** on a piece of graph paper to some scale, such as 1 cm equals 1 m. (Note that symbols for vector quantities in this book are represented by boldface print.) Also, the vector **A** must be drawn such that its direction is specified relative to a coordinate system. Then draw vector **B** using this same scale and with the tail of **B** starting from the tip of **A**, as in Figure 1.5. Vector **B** must be drawn along the direction which makes the proper angle relative to vector **A**. The resultant vector **R**, given by **R** = **A** + **B**, is the vector drawn from the tail of **A** to the tip of **B**. This is known as the *triangle method of addition*.

An alternative graphical procedure for adding two vectors, known as the *parallelogram rule of addition*, is shown in Figure 1.6. In this construction, the tails of the two vectors **A** and **B** are joined together and the resultant vector **R** is the diagonal of the parallelogram formed with **A** and **B** as its sides.

This same general approach can also be used to add more than two vectors. This is shown in Figure 1.7 for four vectors. The resultant vector sum **R** = **A** + **B** + **C** + **D** is the vector drawn from the tail of the first vector to the tip of the last vector. The order in which you add vectors is unimportant.

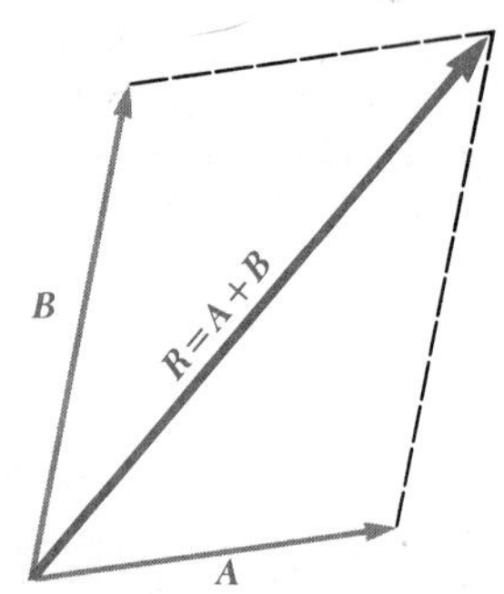

FIGURE 1.6 This construction shows that **A** + **B** = **B** + **A**. Note that the resultant vector **R** is the diagonal of a parallelogram with sides *A* and *B*.

Negative of a Vector The negative of the vector **A** is defined as the vector which when added to **A** gives a zero vector. This means that **A** and − **A** must have the same magnitude but point in opposite directions.

Subtraction of Vectors The operation of vector subtraction makes use of the definition of the negative of a vector. We define the operation **A** − **B** as vector − **B** added to vector **A**:

$$\mathbf{A} - \mathbf{B} = \mathbf{A} + (-\mathbf{B}) \qquad (1.2)$$

The geometric construction for subtracting two vectors is shown in Figure 1.8.

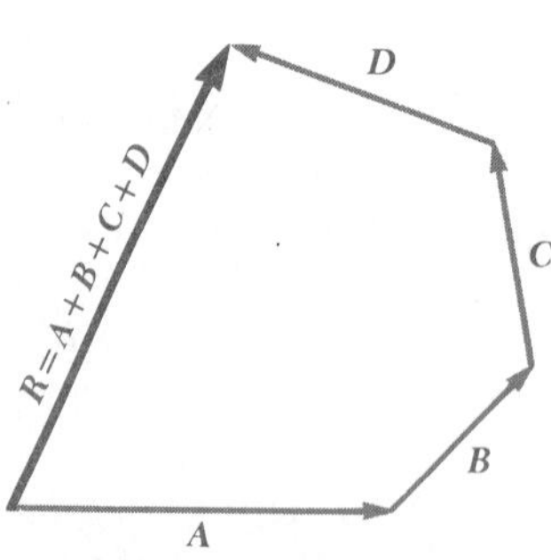

FIGURE 1.7 Geometric construction for summing four vectors. The resultant vector **R** is the vector that completes the polygon.

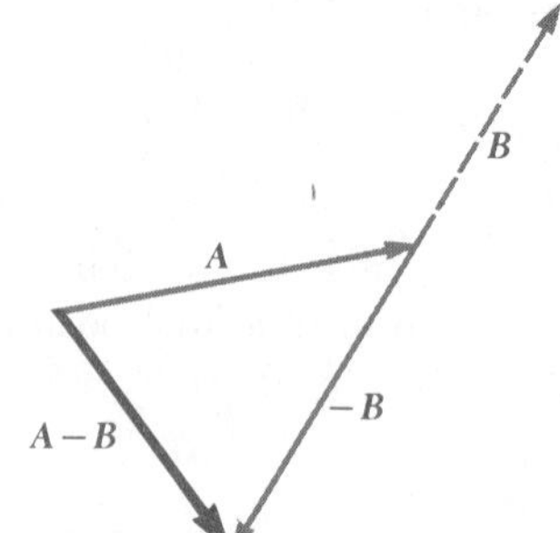

FIGURE 1.8 This construction shows how to subtract vector **B** from vector **A**. The vector −**B** has the same magnitude as the vector **B** but points in the opposite direction.

Multiplication and Division of Vectors by Scalars The multiplication or division of a vector by a scalar gives a vector. For example, if a vector ***A*** is multiplied by the scalar number 3, written as 3***A***, the result is a vector whose magnitude is three times that of the original vector ***A*** and pointing in the same direction as ***A***. On the other hand, if we multiply a vector ***A*** by the scalar -3, the result is a vector whose magnitude is three times that of ***A*** but pointing in the direction opposite ***A*** (because of the negative sign).

EXAMPLE 1.7 Taking a Trip

A car travels 20.0 km due north and then 35.0 km in a direction 60° west of north, as in Figure 1.9. Find the magnitude and direction of the car's resultant displacement.

Solution The problem can be solved geometrically using graph paper and a protractor, as shown in Figure 1.9. The resultant displacement ***R*** is the sum of the two individual displacements ***A*** and ***B***.

The length of ***R***, using the same scale factor used to draw ***A*** and ***B***, indicates that the displacement of the car is 48.2 km, and a measurement of the angle β shows that the displacement is approximately 39° west of north.

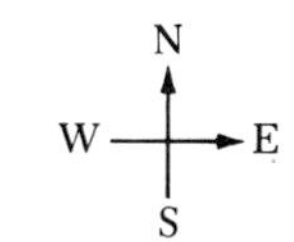

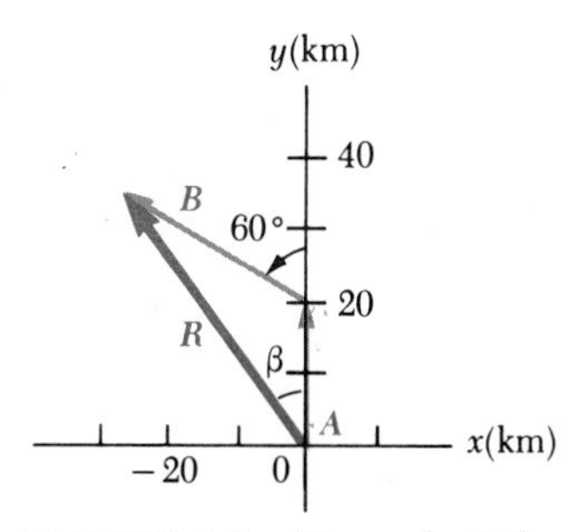

FIGURE 1.9 (Example 1.7) Graphical method for finding the resultant displacement vector ***R*** = ***A*** + ***B***.

1.10 COMPONENTS OF A VECTOR

As noted earlier, the geometric method of adding vectors is not the recommended procedure in situations where high precision is required. In this section, we describe a method of adding vectors that makes use of the projections of a vector along the axes of a rectangular coordinate system. These projections are called the **components** of the vector. Any vector can be completely described by its components.

Consider a vector ***A*** in a rectangular coordinate system, as shown in Figure 1.10. Note that the vector ***A*** can be expressed as the sum of two vectors, $\mathbf{A}_x$ parallel to the x axis, and $\mathbf{A}_y$ parallel to the y axis. That is,

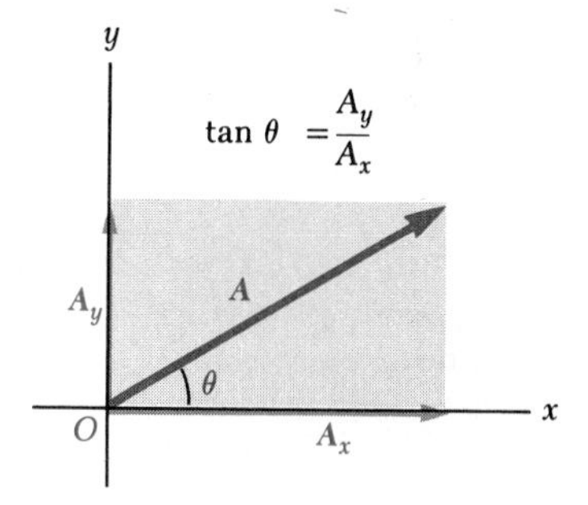

FIGURE 1.10 Any vector ***A*** lying in the xy plane can be represented by its rectangular components, A_x and A_y.

$$\mathbf{A} = \mathbf{A}_x + \mathbf{A}_y$$

where $\mathbf{A}_x$ and $\mathbf{A}_y$ are called the component vectors of **A**. The projection of **A** along the x axis, A_x, is called the x component of **A**, and the projection of **A** along the y axis, A_y, is called the y component of **A**. *These scalar components can be either positive or negative numbers with units.* From Figure 1.10 and the definitions of sine and cosine of an angle, we see that $\cos\theta = A_x/A$ and $\sin\theta = A_y/A$. (The sine, cosine, and tangent of an angle are defined in Figure 1.11 for your convenience. If you need to brush up on or be introduced to trigonometric functions such as sine, cosine, and tangent, please refer to Appendix A.4. At the same time, you should read Appendix A.2, which provides a review of basic algebra.) Hence, the **scalar components** of **A** are given by

$$\begin{aligned} A_x &= A\cos\theta \\ A_y &= A\sin\theta \end{aligned} \tag{1.3}$$

$\sin\theta = \frac{a}{c}$

$\cos\theta = \frac{b}{c}$

$\tan\theta = \frac{a}{b}$

c a θ b

FIGURE 1.11 The right triangle used to define the trigonometric functions sin θ, cos θ, and tan θ.

These components form two sides of a right triangle, the hypotenuse of which has a magnitude A. Thus, it follows that the magnitude of **A** and its direction are related to its scalar components through the Pythagorean theorem and the definition of the tangent

$$A = \sqrt{A_x^2 + A_y^2} \tag{1.4}$$

and

$$\tan\theta = \frac{A_y}{A_x} \tag{1.5}$$

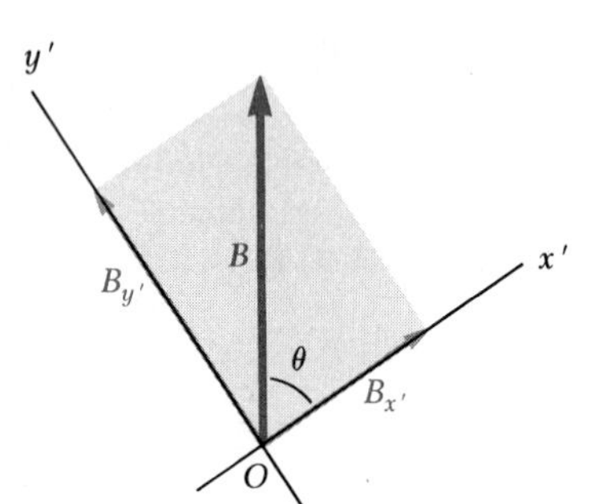

FIGURE 1.12 The components of vector **B** in a tilted coordinate system.

To solve for the angle θ, we can invert Equation 1.5 and write the expression as $\theta = \tan^{-1}(A_y/A_x)$, which is read "θ is the angle whose tangent equals the ratio A_y/A_x." (Again we recommend Appendix A.4 for a discussion of the Pythagorean theorem.)

If you choose a coordinate system other than the one shown in Figure 1.10, the components of the vector must be modified accordingly. In many applications it is more convenient to express the components of a vector in a coordinate system having axes that are not horizontal and vertical but still perpendicular to each other. Suppose a vector ***B*** makes an angle θ with the x' axis defined in Figure 1.12. The rectangular components of ***B*** along the axes of Figure 1.12 are given by $B_{x'} = B\cos\theta$ and $B_{y'} = B\sin\theta$, as in Equation 1.3. The magnitude and direction of ***B*** are obtained from expressions equivalent to Equations 1.4 and 1.5. Thus, we can express the components of a vector in any coordinate system that is convenient for a particular situation.

If you wish to add or subtract vectors analytically rather than graphically, first find the components of the vectors. Note that the sum of the x components of the vectors equals the x component of the resultant vector. Likewise, the sum of the y components of the vectors equals the y component of the resultant vector.

EXAMPLE 1.8 Help Is on the Way!

Find the horizontal and vertical components of the 100-m displacement of a superhero who flies from the top of a tall building following the path shown in Figure 1.13a.

Solution The triangle formed by the displacement and its components is shown in Figure 1.13b. Since $A = 100$ m and $\theta = -30°$ (θ is negative because it is measured clockwise from the x axis), we have

$$A_y = A\sin\theta = (100\text{ m})\sin(-30°) = -50.0\text{ m}$$

Note that $\sin(-\theta) = -\sin\theta$. The negative sign for A_y corresponds to the fact that displacement in the y direction is *downward* from the origin.

The x component of displacement is

$$A_x = A\cos\theta = (100\text{ m})\cos(-30°) = 86.6\text{ m}$$

Note that $\cos(-\theta) = \cos\theta$. Also, from an inspection of the figure, you should be able to see that A_x is positive in this case.

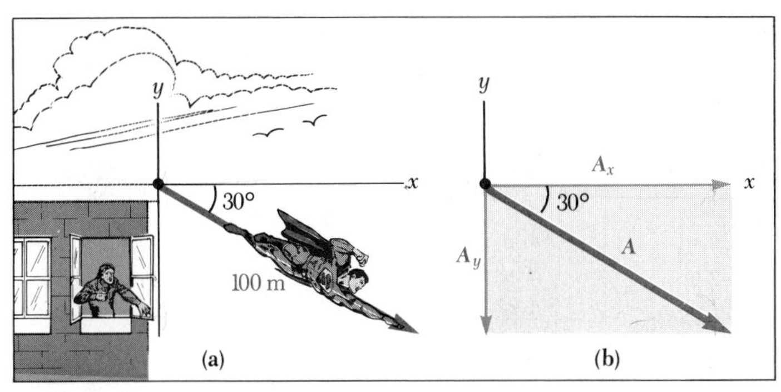

FIGURE 1.13 Example 1.8

EXAMPLE 1.9 Go and Take a Hike

On an extended hike, a camper follows the path shown in Figure 1.14a. The total trip consists of four straight-line paths. At the end of the hike, what is the hiker's resultant displacement from the tent site?

Solution Let us summarize the steps for approaching problems of this nature and then apply them to our particular situation. The step-by-step format is as follows:

1. Select a coordinate system.
2. Draw a sketch of the vectors to be added (or subtracted), with a label on each vector.
3. Find the x and y components of all vectors.
4. Find the resultant component (the sum of the components) in both the x and y directions.
5. Use the Pythagorean theorem to find the resultant vector.
6. Use a suitable trigonometric function to find the angle with respect to the x axis.

The coordinate system we shall use is shown in Figure 1.14a. The 100-m movement has only an x component, and the 300-m movement has only a y component

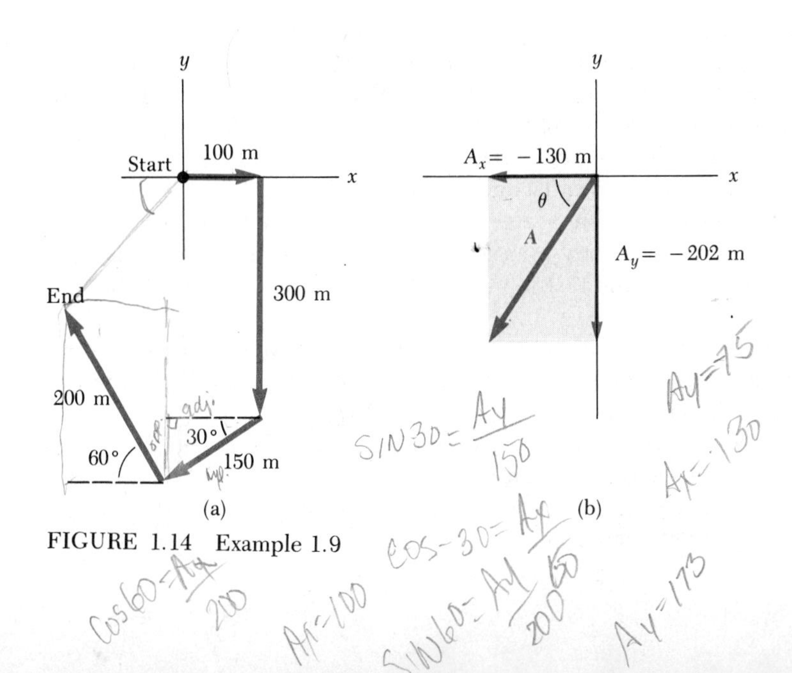

FIGURE 1.14 Example 1.9

(which is negative). You should use the procedures for finding components of vectors to show that the table below has the proper components for all the displacements.

Displacement	x Component	y Component
100 m	100 m	0
300 m	0	−300 m
150 m	−130 m	−75 m
200 m	−100 m	173 m
Total	−130 m	−202 m

The table also shows that the resultant x component is −130 m and the resultant y component is −202 m, as shown in Figure 1.14b.

Applying the Pythagorean theorem to Figure 1.14b gives the magnitude of the resultant displacement:

$$A = \sqrt{A_x^2 + A_y^2} = \sqrt{(-130 \text{ m})^2 + (-202 \text{ m})^2} = 240 \text{ m}$$

The direction is found by noting that

$$\tan \theta = \frac{-202 \text{ m}}{-130 \text{ m}} = 1.55$$

and that

$$\theta = \tan^{-1} 1.55 = 57°$$

The direction of the resultant is frequently specified as an angle measured relative to the positive x direction. If we follow this procedure for the above, the angle relative to the positive x direction is $180° + 57° = 237°$. You should be able to show that the total distance traveled is 750 m.

SUMMARY

The physical quantities we shall encounter in our study of mechanics can be expressed in terms of three fundamental quantities, mass, length, and time, which have the units kilograms (kg), meters (m), and seconds (s), respectively, in the SI system. It is often helpful to use dimensional analysis to check equations and to assist in deriving equations.

When inserting numerical factors into equations, you must be sure that the dimensions of these factors are consistent throughout the equation. In many cases it will be necessary to use the table of conversion factors given on the inside back cover of the text in order to convert from one system of units to another.

It is often useful to estimate an approximate answer to a problem in which little or no information is given. Such estimates are called **order-of-magnitude calculations.**

Vectors are quantities that have both magnitude and direction. **Scalars** are quantities that have only magnitude.

Two vectors ***A*** and ***B*** can be added geometrically by either the triangle method or the parallelogram rule. In the triangle method, the two vectors are drawn to scale, on graph paper, such that the tail of the second vector starts at the tip of the first. The resultant vector is the vector drawn from the tail of the first to the tip of the second. In the parallelogram method, the vectors are drawn to scale on graph paper with the tails of the two vectors joined together. The resultant vector is the diagonal of the parallelogram formed with the two vectors as its sides.

The negative of a vector $\mathbf{A}$ is a vector having the same magnitude as $\mathbf{A}$ but pointing in the direction opposite the direction of $\mathbf{A}$.

The x component of a vector is equivalent to its projection along the x axis of a coordinate system. Likewise, the y component is the projection of the vector along the y axis of this coordinate system. The resultant of two or more vectors can be found mathematically by resolving all vectors into their x and y components, finding the resultant x and y components, and then using the Pythagorean theorem to find the resultant vector. The angle of the resultant vector with respect to the x axis can be found by use of a suitable trigonometric function.

ADDITIONAL READING

A. V. Astin, "Standards of Measurements," *Sci. American*, June 1968, p. 50.

H. Butterfield, "The Scientific Revolution," *Sci. American*, September 1960, p. 173.

J. Friberg, "Numbers and Measures in the Earliest Written Records," *Sci. American*, February 1984, p. 110.

G. Goth, "Dimensional Analysis by Computer," *The Physics Teacher*, February 1986, p. 75.

L. M. Lederman, "The Value of Fundamental Science," *Sci. American*, November 1984, p. 40.

Philip and Phylis Morrison and the office of Charles and Ray Eames, *The Powers of Ten*, The Sci. American Library, New York, W. H. Freeman, 1982.

Lord Ritchie-Calder, "Conversion to the Metric System," *Sci. American*, July 1970, p. 17.

QUESTIONS

1. A book is moved once around the perimeter of a table of dimensions 1 m × 2 m. If the book ends up at its initial position, what is its displacement?
2. If $\mathbf{B}$ is added to $\mathbf{A}$, under what conditions does the resultant vector have a magnitude equal to $A + B$? Under what conditions is the resultant vector equal to zero?
3. Can a force directed vertically on an object ever cancel a force directed horizontally?
4. A student accurately uses the method for combining vectors. The two vectors which he combines have magnitudes of 50 and 25. The answer he gets is either 80, 20, or 55. Pick the correct answer and tell why it is the only one of the three that can be correct.
5. A vector $\mathbf{A}$ lies in the xy plane. For what orientations will both of its rectangular components be negative? For what orientations will its components have opposite signs?
6. Can a vector have a component equal to zero and still have a nonzero magnitude? Explain.
7. Can a vector have a component greater than its magnitude?
8. If an equation is dimensionally correct, does this mean that the equation must be true?
9. If the component of vector $\mathbf{A}$ along the direction of vector $\mathbf{B}$ is zero, what can you conclude about these two vectors?
10. Which of the following are vectors and which are not: force, temperature, the amount of water in a can, the weight of a book, the height of a building, the velocity of a sports car, the age of the universe?
11. Can the magnitude of a vector have a negative value?
12. If $\mathbf{A} + \mathbf{B} = 0$, what can you say about the components of the two vectors?
13. Under what circumstances would a vector have components which are equal in magnitude?
14. Is it possible to add a vector quantity to a scalar quantity? Explain.
15. Two vectors have unequal magnitudes. Can their sum be zero? Explain.

16. How many of the length or time intervals given in Table 1.2 and 1.3 could you verify using only equipment that might be found in a typical dormitory room?
17. An ancient unit of length called the cubit was equal to six palms, where a palm was the width of the four fingers of an open hand. Noah's ark was 300 cubits long, 50 cubits wide, and 30 cubits high. Estimate the volume of the arc in cubic feet. Also, estimate the volume of a typical home and compare it to the volume of the ark.
18. A space shuttle takes off from Florida and circles the globe several times, finally landing in California. While the shuttle is in flight, the wife of an astronaut flies from Florida to California so she can greet her husband when he steps off the shuttle. Who undergoes the greater displacement, the wife or the astronaut?

PROBLEMS

Section 1.2 Dimensional Analysis

1. Show that the equation $v^2 = v_0^2 + 2ax$ is dimensionally correct, where v and v_0 represent velocities, a is acceleration, and x is a distance.

2. (a) Suppose that the displacement of an object is related to the time according to the expression $x = ct^2$. What are the dimensions of c? (b) A displacement is related to the time as $x = A \sin(2\pi ft)$, where A and f are constants. Find the dimensions of A. *Hint:* A trigonometric function appearing in an equation must be dimensionless.

3. (a) One of the fundamental laws of motion states that the acceleration of an object is directly proportional to the resultant force on it and inversely proportional to its mass. From this statement, determine the dimensions of force. (b) The newton is the SI unit of force. According to the results for (a), how can you express a force having the units of newtons using the fundamental units of mass, length, and time?

4. Which of the equations below are dimensionally correct?
(a) $v = v_0 + ax$
(b) $y = (2 \text{ m}) \cos(kx)$, where $k = 2 \text{ m}^{-1}$.

Section 1.3 Significant Figures

5. The value of the speed of light is now known to be 2.997924574×10^8 m/s. Express the speed of light to (a) three significant figures, (b) five significant figures, and (c) seven significant figures.

6. How many significant figures are there in (a) 78.9 ± 0.2, (b) 3.788×10^9, (c) 2.46×10^{-6}, (d) 0.0032?

7. Carry out the following arithmetic operations: (a) the sum of the numbers 756, 37.2, 0.83, and 2.5; (b) the product 3.2×3.563; (c) the product $5.67 \times \pi$.

8. Carry out the following arithmetic operations: (a) 756 minus 37.2, (b) 3.2 divided by 1.4577.

9. Calculate (a) the circumference of a circle of radius 3.5 cm and (b) the area of a circle of radius 4.65 cm.

10. A farmer measures the distance around a rectangular field. The length of the long sides of the rectangle are found to be 38.44 m, and the length of the short sides are found to be 19.5 m. What is the total distance around the field?

Section 1.4 Conversion of Units

11. Estimate the age of the earth in years using the data in Table 1.3 and the appropriate conversion factors.

12. Estimate the distance to the nearest star in feet using the data in Table 1.2 and the appropriate conversion factors.

13. A painter is to cover the walls in a room 8 ft high and 12 ft along a side. What surface area in square meters must he cover?

14. (a) How many seconds are there in a year? (b) If one micrometeorite (a sphere with a diameter of 10^{-6} m) strikes each square meter of the moon each second, how many years would it take to cover the moon to a depth of 1 m? (*Hint:* Consider a cubic box on the moon 1 m on a side, and find how long it will take to fill the box.)

15. (a) Find a conversion factor to convert from mi/h to km/h. (b) Until recently, federal law mandated that highway speeds would be 55 mi/h. Use the conversion factor of part (a) to find the speed in km/h. (c) The maximum highway speed has been raised to 65 mi/h in some places. In km/h, how much increase is this over the 55 mi/h limit?

16. The speed of light is 3.00×10^8 m/s. Convert this to miles per hour.

17. A house is 50 ft long, 26 ft wide, and has 8-ft high ceilings. What is the volume of the house in cubic meters and in cubic centimeters?

18. The base of a pyramid covers an area of 13 acres (1 acre = 43,560 ft^2) and has a height of 481 ft. If the volume of a pyramid is given by the expression V = (1/3)Bh, where B is the area of the base and h is the height, find the volume of this pyramid in cubic meters.

19. The pyramid described in Problem 18 contains approximately two million stone blocks that average 2.5 tons each. Find the weight of this pyramid in pounds and in newtons.

20. Use the fact that the speed of light in free space is about 3.00×10^8 m/s to determine how many miles a pulse from a laser beam will travel in 1 hour.

Section 1.5 Order-of-magnitude Calculations

Note: In arriving at answers to the problems in this section, you should state your important assumptions, including the numerical values assigned to parameters used in the solution. Since only order-of-magnitude results are expected, do not be surprised if your results differ from those given in the answer section of the text.

21. A particular hamburger chain advertises that it has sold more than 50 billion hamburgers. Estimate how many pounds of hamburger meat must have been used by the restaurant chain and how many head of cattle were required to furnish the meat.

22. Imagine yourself to be the equipment manager of a professional baseball team. One of your jobs is to keep baseballs on hand for games. One of the ways by which balls are lost is by players hitting them into the stands either as home runs or foul balls. Estimate how many baseballs you have to buy per season in order to account for such losses. Assume your team plays an 81-game home schedule in a season.

23. Assume that you watch every pitch of every game of a 162-game major league baseball season. Approximately how many pitches would you see thrown?

• **24.** Estimate the number of piano tuners living in New York City. This question was raised by the physicist Enrico Fermi, who was well known for making order-of-magnitude calculations.

Section 1.7 Coordinate Systems

25. A certain corner of a room is selected as the origin of a rectangular coordinate system. If a fly is crawling on an adjacent wall at a point having coordinates (2.0, 1.0), where the units are in meters, what is the distance of the fly from the corner of the room?

26. Express the location of the fly in Problem 25 in polar coördinates.

27. Two points in a rectangular coordinate system have the coordinates (5, 3) and (−3, 4), where the units are in centimeters. Determine the distance between these points.

28. A point is located in a polar coordinate system by the coordinates $r = 2.5$ m and $\theta = 35°$. Find the x and y coordinates of this point, assuming the two coordinate systems have the same origin.

Section 1.9 Some Properties of Vectors

29. (a) What is the resultant displacement of a walk of 80 m followed by a walk of 125 m when both displacements are in the eastward direction? (b) What is the resultant displacement in a situation in which the 125-m walk is in the direction opposite the 80-m walk?

30. While traveling along a straight interstate highway you notice that the mile marker reads 260. You travel until you reach the 150-mile marker and note that you are low on gas. You then retrace your path to the 175-mile marker. What is the magnitude of your resultant displacement from the 260-mile marker?

31. A dog searching for a bone walks 3.5 m south, then 8.2 m at an angle 30° north of east, and finally 15 m west. Find the dog's resultant displacement vector using graphical techniques.

32. A roller coaster moves 200 ft horizontally, then rises 135 ft at an angle of 30° above the horizontal. It then travels 135 ft at an angle of 40° downward. What is its displacement from its starting point at the end of this movement? Use graphical techniques.

• **33.** A man lost in a maze makes three consecutive displacements such that at the end of the walks he is right back where he started. The first displacement is 8 m westward, and the second is 13 m northward. Find the magnitude and direction of the third displacement using the graphical method.

• **34.** A jogger runs 100 m due west, then changes direction for the second leg of the run. At the end of the run, she is 175 m away from the starting point at an angle of 15° north of west. What was the direction and length of her second displacement? Use graphical techniques.

Section 1.10 Components of a Vector

35. A submarine dives at an angle of 30° with the horizontal and follows a straight-line path for a total distance of 50 m. How far is the submarine below the surface of the water?

36. A roller coaster travels 135 ft at an angle of 40° above the horizontal. How far does it move horizontally and vertically?

37. An airplane flies from city A to city B in a direction due east for 800 miles. In the next part of the trip the airplane flies from city B to city C in a direction 40° north of east for 600 miles. What is the resultant displacement of the airplane between city A and city C?

38. A shopper pushing a cart through a store moves a distance 40 m down one aisle then makes a 90° turn and moves 15 m. He then makes another 90° turn and moves 20 m. How far is the shopper away from his original position in magnitude and direction? Note that you are not given the direction moved in any of the 90° turns. As a result, there could be more than one answer?

• **39.** A man pushing a mop across a floor causes it to undergo two displacements. The first has a magnitude of 150 cm and makes an angle of 120° with the positive x axis. The resultant displacement has a magnitude of 140 cm and is directed at an angle of 35° to the positive x axis. Find the magnitude and direction of the second displacement.

ADDITIONAL PROBLEMS

40. Use the standard prefixes in Table 1.4 and the well-established prefixes for the numbers 1 to 10 to determine appropriate words for the following conversions: (a) 10^{12} microphones, (b) 10^{21} picolos, (c) 10 rations, (d) 10^{6} bicycles, (e) 10^{12} pins, (f) $3\frac{1}{3}$ tridents, (g) 2000 mockingbirds, (h) 10^{-12} boo, (i) 10^{-9} goat. (Some of these amusing conversions are from a table published by Solomon W. Golomb of UCLA.)

41. A vector $\boldsymbol{R}$ has a magnitude of 100 units. Find its components in magnitude, including sign, when it makes the following angles with the $+x$ axis. (a) 60°, (b) 130°, (c) 200°, (d) 290°.

42. You close out your $1,000,000 checking account at a bank, and take your money all in one-dollar bills. Estimate the height of your stack of money.

43. The radius of the planet Saturn is 5.85×10^{7} m and its mass is 5.68×10^{26} kg. (a) Find the density of Saturn (its mass divided by its volume) in g/cm^3. (The volume of a sphere is given by $(4/3)\pi r^3$.) (b) Find the surface area of Saturn in square feet. (The surface area of a sphere is given by $4\pi r^2$.)

44. A taxi driver travels due south for 10 km and then moves 6 km in a direction 30° north of east. Find the magnitude and direction of the car's resultant displacement.

45. Soft drinks are commonly sold in aluminum containers. Estimate the number of such containers thrown away each year by U.S. consumers. Approximately how many tons of aluminum does this represent?

46. A person walks halfway around a circular path of radius 5 m. (a) Find the magnitude of the displacement vector. (b) How far does the person walk? (c) What will the magnitude of the displacement be if the circle is completed?

47. Two points in a rectangular coordinate system have coordinates (2.0, −4.0) and (−3.0, 3.0), where the units are in meters. Determine the distance between these points.

• **48.** While exploring a cave, a spelunker starts at the entrance and moves the following distances. She goes 75 m north, 250 m east, 125 m at an angle 30° north of east, and 150 m south. Find the resultant displacement from the cave entrance.

• **49.** The displacement of an object moving under uniform acceleration is some function of time and the acceleration. Suppose we write this displacement as $s = k\, a^m t^n$, where k is a dimensionless constant. Show by dimensional analysis that this expression is satisfied if $m = 1$ and $n = 2$. Can this analysis give the value of k?

CHAPTER 2

Motion in One and Two Dimensions

Fred MacMurray flies through the air in his magical car in the film "The Absent-Minded Professor." (Museum of Modern Art Film Stills Archives)

In some fashion, people have always been concerned with motion. For example, in today's world you consider motion when you describe to someone how fast your new car will go or how much pick-up it has. Likewise, Neanderthal people must have, in their own way, pondered about motion as they devised methods for capturing a rapidly moving antelope. *The branch of physics concerned with the study of the motion of an object and the relationship of this motion to such physical concepts as force and mass is called* **mechanics.** *That part of mechanics which describes motion without regard to the causes of the motion is called* **kinematics.** In this chapter we shall focus on kinematics and for the most part concentrate on one-dimensional motion, that is, motion along a straight line. Starting with the concept of displacement defined in Chapter 1, we shall define velocity and acceleration. These concepts allow us to study the motion of objects undergoing constant acceleration. At the end of the chapter we shall take a brief look at one kind of two-dimensional problem: the motion of a projectile.

Before we begin our study, it would be well to consider the long history of mechanics. The first recorded evidence of the study of mechanics can be traced to ancient Babylonian and Egyptian civilizations. The primary concern in those days was to understand the motion of heavenly bodies. The most systematic and detailed early studies of the heavens, however, were conducted by the Greeks during the period from about 300 B.C. to 300 A.D. For hundreds of years, scientists and lay people regarded the earth as the center of the universe. This geocentric model was accepted by such notables as Aristotle (384 to 322 B.C.) and Claudius Ptolemy (about 140 A.D.) Largely because of the authority of Aristotle, the geocentric model became the officially accepted theory of the universe and maintained this position until the seventeenth century. The Polish astronomer Nicolaus Copernicus (1473–1543) suggested that the earth and the other planets revolved in circular orbits around the sun (the heliocentric model). This early knowledge formed the foundation for the work of Galileo Galilei (1564–1642). Galileo stands out as perhaps the dominant figure in leading the world of physics into the modern era. In 1609 he became one of the first to make astronomical observations with a telescope. He observed mountains on the moon, the larger satellites of Jupiter, the rings of Saturn, and spots on the sun. His observations convinced him of the correctness of the Copernican theory. Galileo's work with motion is particularly well known, and because of his leadership, experimentation has become an important part of our search for knowledge.

2.1 AVERAGE VELOCITY

Consider a car moving along a highway (the x axis), as in Figure 2.1. Let the car's position be x_i at some time t_i, and let its position be x_f at time t_f. (The indices i and f refer to the initial and final locations, respectively.) In the time interval $\Delta t = t_f - t_i$, the displacement of the car is $\Delta x = x_f - x_i$.

The **average velocity,** $\bar{v}$, is defined as the displacement, Δx, divided by the time interval during which the displacement occurred:

Average velocity

$$\bar{v} = \frac{\Delta x}{\Delta t} = \frac{x_f - x_i}{t_f - t_i} \tag{2.1}$$

It is important to recognize that velocity is a vector quantity. As you will recall from Chapter 1, vector quantities are usually denoted by boldfaced symbols. In the case of one-dimensional motion, we often will represent vectors by using algebraic signs rather than boldfaced symbols. The average velocity of an object can be either positive or negative, depending on the sign of the displacement. (The time interval, Δt, is always positive.) For example, if we select a coordinate system such that a car moves from a position 100 m from the origin to a point 50 m from that origin in a time interval of 2 s, the average velocity is -25 m/s. If an object moves in the negative x direction during some time interval, its velocity in the x direction during that interval is necessarily negative. Similarly, an object moving toward negative y values has a negative y component of velocity.

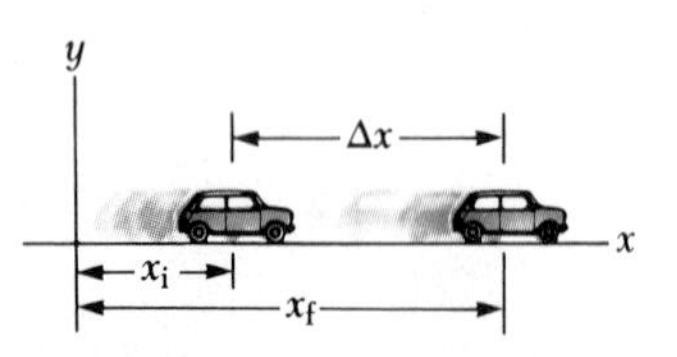

FIGURE 2.1 A car moving along the x axis. Its position at time t_i is x_i and its position at time t_f is x_f. Its displacement is Δx.

In order to understand the vector nature of velocity, consider the following situation. Suppose a friend tells you that she will be taking a trip in her car and will travel at a constant rate of 55 mi/h in a straight line for 1 h. If she starts from her home, where will she be at the end of the trip? Obviously, you cannot answer this question because the direction in which she will travel is not specified. All you can say is that she will be located 55 mi from her starting point. However, if she tells you that she will be driving at the rate of 55 mi/h directly northward, then her final location will be known exactly. Therefore, the velocity of an object is known only if its direction and its magnitude (speed) are specified. In general, *any physical quantity which is a vector must be characterized by both a magnitude and a direction.*

FIGURE 2.2 The view of a drag race from a blimp. One car follows the straight-line path from P to Q, and a second car follows the curved path.

As an example of the use of Equation 2.1, suppose the car in Figure 2.1 moves 100 m in 5 s. The substitution of these values into Equation 2.1 gives the average velocity in this time interval as

$$\bar{v} = \frac{\Delta x}{\Delta t} = \frac{100 \text{ m}}{5 \text{ s}} = 20 \text{ m/s}$$

Note that the units of average velocity are units of length divided by units of time, which is meters per second (m/s) in SI units. Other units for velocity might be feet per second (ft/s) in the conventional system or centimeters per second (cm/s) if we decide to measure distances in centimeters. Even furlongs per fortnight would be acceptable (but not very common); the only requirement is that velocity must have units of length divided by units of time.

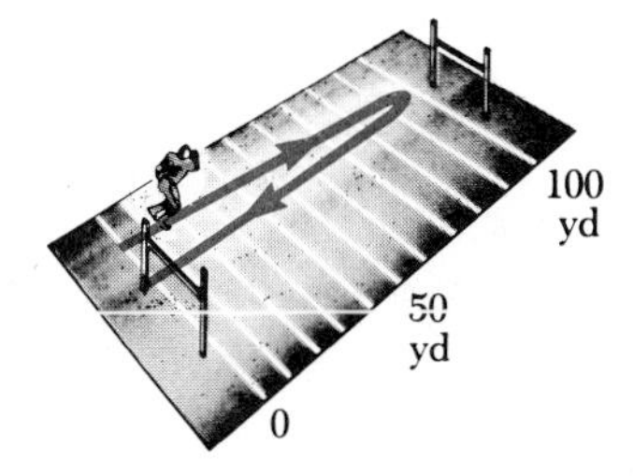

FIGURE 2.3 The path followed by a confused football player.

Let us assume we are watching a drag race from the Goodyear blimp. In one race we see a car follow the straight-line path from P to Q shown in Figure 2.2 during a time interval Δt, while in a second run a car follows the curved path shown in Figure 2.2 for the same time interval. If you examine our definition of average velocity carefully, you will see that both cars have the same average velocity. This is because both cars have the same displacement ($x_f - x_i$) during the same time interval (Δt), as indicated in Figure 2.2.

Figure 2.3 shows the unusual path of a confused football player. He receives a kickoff at his own goal, runs downfield to within inches of a touchdown, and then reverses his direction to race backward until he is tackled at the exact location where he first caught the ball. What is his average velocity during this run? From the definition of average velocity (Eq. 2.1), we see that his average velocity is zero because his displacement is zero. In other words, since x_i and x_f have the same value, $\bar{v} = \Delta x/\Delta t = 0$. From this we see that displacement should not be confused with distance traveled. The football player clearly traveled a long distance, almost 200 yards, yet his displacement is zero.

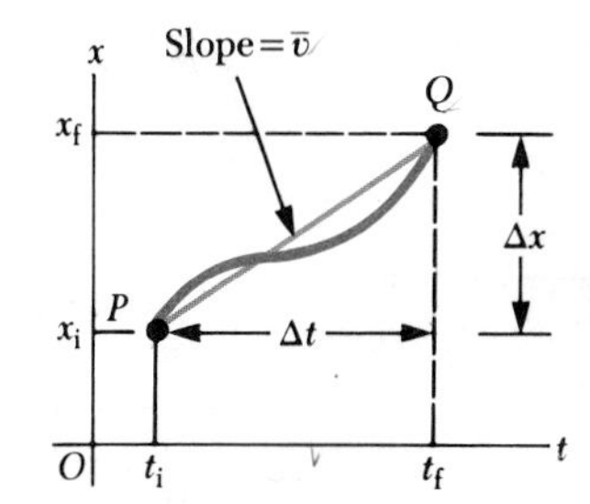

FIGURE 2.4 Position-time graph for an object moving along the x axis. The average velocity $\bar{v}$ in the time interval $\Delta t = t_f - t_i$ is the slope of the straight line connecting the points P and Q.

*Graphical Interpretation of Velocity

Figure 2.4 is a graph drawn for the motion of an object moving along a straight line path from the position x_i at time t_i to the position x_f at time t_f. (Note that the motion is along a straight line, yet the position-time graph is *not* a straight line. Why is that?) The straight line connecting points P and Q provides us with a geometric interpretation of average velocity. As indicated on the graph, the slope of the line is Δx $(= x_f - x_i)$ divided by the time interval for the motion, Δt $(= t_f - t_i)$. Therefore,

the average velocity of an object during the time interval t_i to t_f is equal to the slope of the straight line joining the initial and final points on a graph of the position of the object plotted versus time.

2.2 INSTANTANEOUS VELOCITY

Let's imagine that you take a trip in your car along a perfectly straight highway. At the end of your journey, it is a relatively simple task to calculate your average velocity. The car's odometer would give you the distance traveled, and a watch could supply the time interval. However, such a calculation would omit a great deal of information about what actually occurred on your trip. For example, if your calculation indicates that your average velocity was 55 mi/h, this does not necessarily mean that your velocity at every instant was 55 mi/h. Instead, you might have traveled at 55 mi/h for a short distance, stopped for lunch, and then made up some time by traveling at 70 mi/h, paused again as a police officer wrote out a ticket, and then raced forward again when the coast was clear.

More precisely, the instantaneous velocity is defined as the limit of the average velocity as the time interval Δt becomes infinitesimally short. In mathematical language this is written as

Instantaneous velocity

$$v \equiv \lim_{\Delta t \to 0} \frac{\Delta x}{\Delta t} \tag{2.2}$$

The notation lim $\Delta t \to 0$ means that the ratio $\Delta x/\Delta t$ is to be evaluated as the time interval Δt approaches zero. Note that we denote instantaneous velocities by the symbol v without a bar across the top.

To better understand the meaning of instantaneous velocity as expressed by Equation 2.2, consider the data shown in Table 2.1. Assume you have been observing a runner racing along a track. At a time 1 s after starting into motion, the runner has moved to a position 1 m from the starting point, at $t = 1.5$ s, the runner is 2.25 m from the starting point, and so on. After collecting these data, suppose you wish to determine the velocity of the runner at the time $t = 1$ s. Table 2.2 represents some of the calculations you might perform to determine the velocity in question. Let us use $t = 1$ s as

TABLE 2.1 Position of a Runner at Specific Instants of Time

t (s)	x (m)
1.00	1.00
1.01	1.02
1.10	1.21
1.20	1.44
1.50	2.25
2.00	4.00
3.00	9.00

TABLE 2.2 Calculated Values of the Time Intervals, Displacements, and Average Velocities for the Runner Using the Data in Table 2.1

Time Interval(s)	Δt (s)	Δx (m)	v (m/s)
1 to 1.01	0.01	0.02	2.0
1 to 1.10	0.10	0.21	2.1
1 to 1.20	0.20	0.44	2.2
1 to 1.50	0.50	1.25	2.5
1 to 2.00	1.00	3.00	3.0
1 to 3.00	2.00	8.00	4.0

our initial time, and first consider the bottom row of the table. For the observed portion of the run (from 1 s to 3 s), the time interval $\Delta t = 2$ s, and the displacement $\Delta x = 8$ m. Thus, the average velocity in this interval is $\Delta x/\Delta t = 4$ m/s. This gives only a rough approximation to the instantaneous velocity at $t = 1$ s. According to Equation 2.2, we can find a more reliable answer by letting the time interval become smaller and smaller. Therefore, consider the next to the last entry in Table 2.2, from 1 s to 2 s, corresponding to a time interval of $\Delta t = 1$ s. In this interval, $\Delta x = 3$ m, so the average velocity in this interval is 3 m/s. This is closer to the correct answer because the time interval is smaller than the first chosen interval. Now, let us consider a very short time interval indicated by the first entry in Table 2.2. In this case, $\Delta t = 0.01$ s and the displacement $\Delta x = 0.02$ m. Thus, the average velocity in this interval is 2 m/s. We could improve the reliability of our calculation by allowing the time interval to become even smaller, but we can state with some degree of confidence that the instantaneous velocity of the runner was 2 m/s at the time $t = 1$ s.

Before we consider some example problems to help clarify the concept of instantaneous velocity, there is one more definition we must discuss. In day-to-day usage, the terms *speed* and *velocity* are used interchangeably. In physics, however, there is a clear distinction between these two quantities.

The **instantaneous speed** of an object, which is a scalar quantity, is defined as the magnitude of the instantaneous velocity. Hence, by definition, speed can never be negative.

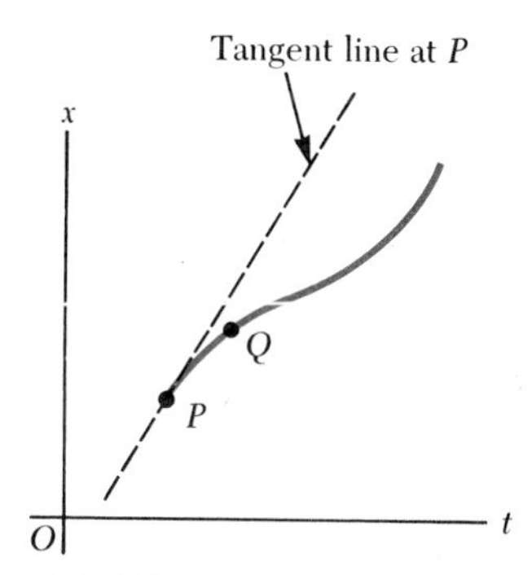

FIGURE 2.5 Geometric construction for obtaining the instantaneous velocity from the x versus t curve. The instantaneous velocity at P is defined as the slope of the line tangent to the curve at P.

*Graphical Interpretation of Instantaneous Velocity

Figure 2.5 is a repeat of the graph drawn earlier (Fig. 2.4) for the position of an object versus time. To find the instantaneous velocity of the object at point P, we must find the average velocity during an infinitesimally short time interval. This means that point Q on the curve must be brought closer and closer to point P until the two points are nearly overlapping each other. From this construction we see that the line joining P and Q is a line that is approaching the line tangent to the curve at point P.

The slope of the line tangent to the position-time curve at P is defined to be the instantaneous velocity at that time.

EXAMPLE 2.1 A Toy Train

A toy train moves slowly along a straight portion of track according to the graph of position versus time shown in Figure 2.6. Find (a) the average velocity for the total trip, (b) the average velocity during the first 4 s of motion, (c) the average velocity during the next 4 s of motion, (d) the instantaneous velocity at $t = 2$ s, and (e) the instantaneous velocity at $t = 5$ s.

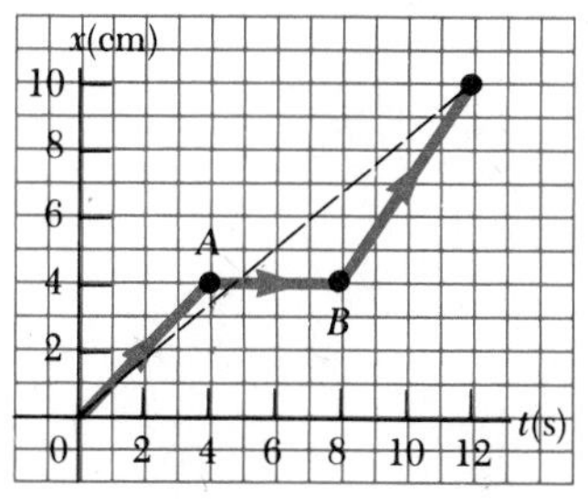

FIGURE 2.6 Example 2.1

Solution (a) The slope of the line joining the starting point and end point on the graph (the dashed line) gives us the average velocity for the total trip. A measurement of its slope gives

$$\bar{v} = \frac{\Delta x}{\Delta t} = \frac{10 \text{ cm}}{12 \text{ s}} = 0.83 \text{ cm/s}$$

(b) The slope of the line joining the starting point to the point on the curve at $t = 4$ s gives us the average velocity during the first 4 s:

$$v = \frac{\Delta x}{\Delta t} = \frac{4.0 \text{ cm}}{4.0 \text{ s}} = 1.0 \text{ cm/s}$$

(c) Following the same procedure for the next 4-s interval, we see that the slope of the line between points A and B is zero. During this time interval, the train has remained at the same location, 4 cm from the starting point.

(d) A line drawn tangent to the curve at the point corresponding to $t = 2$ s has the same slope as that found for the line in part (b). Thus, the instantaneous velocity at this time is 1.0 cm/s. This has to be true because the graph indicates that during the first 4 s of motion, the train is covering equal distances in equal intervals of time. In other words, the train is moving at a constant velocity during the first 4 s. Under these conditions the average velocity and the instantaneous velocity at all times are identical.

(e) At $t = 5$ s, the slope of the position-time curve is zero. Therefore, the instantaneous velocity is zero at this instant. In fact, the train is at rest in the entire time interval between 4 s and 8 s.

2.3 ACCELERATION

As you travel from place to place in your car, you normally do not travel long distances at a constant velocity. The velocity of the car will increase when you step on the gas, and it will slow down when you apply the brakes. Furthermore, the velocity of the car changes when you round a curve, corresponding to a change in direction of motion. The rate of change of velocity is referred to as acceleration. However, we need a more precise definition of acceleration than that.

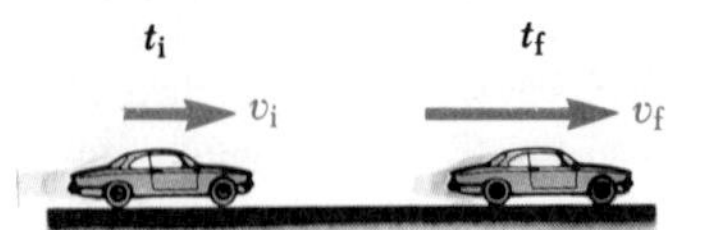

FIGURE 2.7 A car moving to the right accelerates from a velocity v_i to a velocity v_f in the time interval $\Delta t = t_f - t_i$.

Suppose a car moves along a straight highway as in Figure 2.7. At time t_i it has a velocity v_i, and at time t_f its velocity is v_f.

The **average acceleration** during this time interval is defined as the change in velocity divided by the time interval during which this change occurs:

Average acceleration

$$\bar{a} \equiv \frac{\Delta v}{\Delta t} = \frac{v_f - v_i}{t_f - t_i} \tag{2.3}$$

As an example of the computation of acceleration, suppose the car shown in Figure 2.7 accelerates from an initial velocity of $v_i = 10$ m/s to a final velocity of $v_f = 30$ m/s in a time interval of 2 s. (Note that both velocities are toward the right, the direction selected as the positive direction.) These values can be inserted into Equation 2.3 to give the average acceleration:

$$\bar{a} = \frac{30 \text{ m/s} - 10 \text{ m/s}}{2.0 \text{ s}} = 10 \text{ m/s}^2$$

Note that the dimensions of acceleration are those of length divided by $(\text{time})^2$. Some units of acceleration are meters per second per second (m/s^2), feet per second per second (ft/s^2), and miles per hour per second (mi/h · s). The acceleration we calculated in the above example was 10 m/s^2. This nota-

A racing car decelerating at the end of a run. (Courtesy of Greg Crisp, Crisp Images, Long Pond, PA)

tion means that, on the average, the car was moving such that its velocity increased at a rate of 10 m/s every second.

As a second example of the computation of acceleration, consider the car pictured in Figure 2.8. In this case, the velocity of the car has changed from an initial value of 30 m/s to a final value of 10 m/s in a time interval of 2 s. The average acceleration during this time interval is

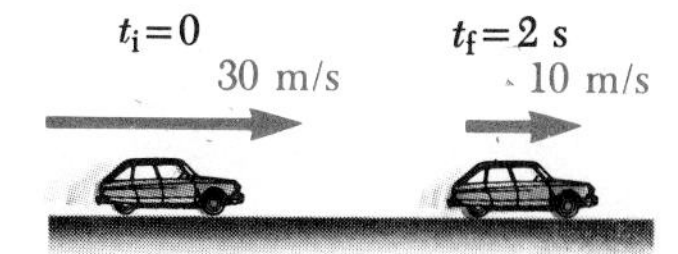

FIGURE 2.8 The velocity of the car decreases from 30 m/s to 10 m/s in a time interval of 2 s.

$$\bar{a} = \frac{10\text{ m/s} - 30\text{ m/s}}{2.0\text{ s}} = -10\text{ m/s}^2$$

Because acceleration is a vector quantity, the negative sign in the above example indicates that the acceleration vector is in the negative x direction (to the left). For the case of motion in a straight line, the direction of the velocity of an object and the direction of its acceleration are related as follows. *When the object's velocity and acceleration are in the same direction, the object is speeding up in that direction.* (The first example demonstrates this situation.) *When the object's velocity and acceleration are in opposite directions, the object is slowing down.* (Decreases in speed are sometimes called decelerations.)

To clarify this point, consider the following situation. Suppose the velocity of a car changes from -10 m/s to -30 m/s in a time interval of 2 s. The negative signs here indicate that the motion of the car is in the negative x direction. The average acceleration of the car in this time interval is

$$\bar{a} = \frac{-30\text{ m/s} - (-10\text{ m/s})}{2.0\text{ s}} = -10\text{ m/s}^2$$

The negative sign indicates that the acceleration is also in the negative x direction. Since the velocity and acceleration are in the same direction, the car must be speeding up as it moves to the left. Note that a negative value for the acceleration does not always indicate a deceleration.

Instantaneous Acceleration

In some situations, the value of the average acceleration may be different over different time intervals. It is therefore useful to define **instantaneous acceleration** as the *acceleration of an object at some particular instant of time.* This concept is analogous to the definition of instantaneous velocity discussed in Section 2.2. In mathematical terms, the instantaneous acceleration, a, is defined as the limit of the average acceleration as the time interval Δt becomes infinitesimally short. That is,

Instantaneous acceleration

$$a = \lim_{\Delta t \to 0} \frac{\Delta v}{\Delta t} \tag{2.4}$$

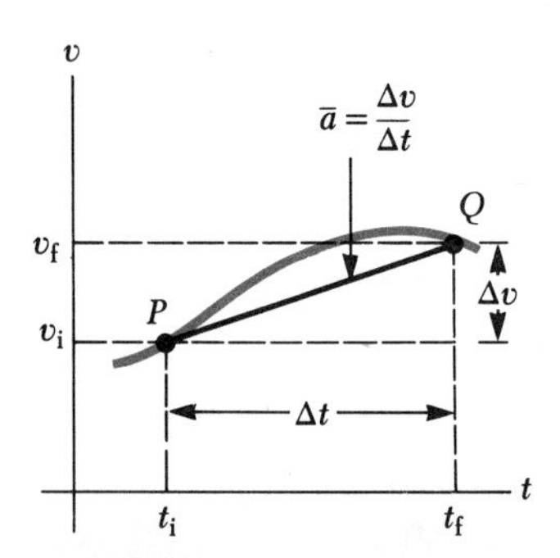

FIGURE 2.9 Velocity-time graph for an object moving in a straight line. The slope of the line connecting points P and Q is defined as the average acceleration in the time interval $\Delta t = t_f - t_i$.

Here again, the notation lim $\Delta t \to 0$ means that the ratio $\Delta v/\Delta t$ is to be evaluated as the time interval Δt approaches zero.

Figure 2.9 is a useful diagram for understanding the concept of instantaneous acceleration. Plotted here is the velocity of an object versus time. This could represent, for example, the motion of a car along a busy street. The average acceleration of the car between times t_i and t_f can be found by finding the slope of the line joining points P and Q. If we imagine that point Q is brought closer and closer to point P, the value that we find for the average acceleration between these points approaches the value of the acceleration of the car at point P. You should be able to see that the instantaneous acceleration at point P, for example, is the slope of the graph at the point. That is,

the instantaneous acceleration of an object at a certain time equals the slope of the velocity-time graph at that instant of time.

From now on we shall use the term *acceleration* to mean "instantaneous acceleration."

EXAMPLE 2.2 A Fly Ball Is Caught

A baseball player moves in a straight-line path in order to catch a fly ball hit to the outfield. His velocity as a function of time is shown in Figure 2.10. Find his instantaneous acceleration at points A, B, and C on the curve.

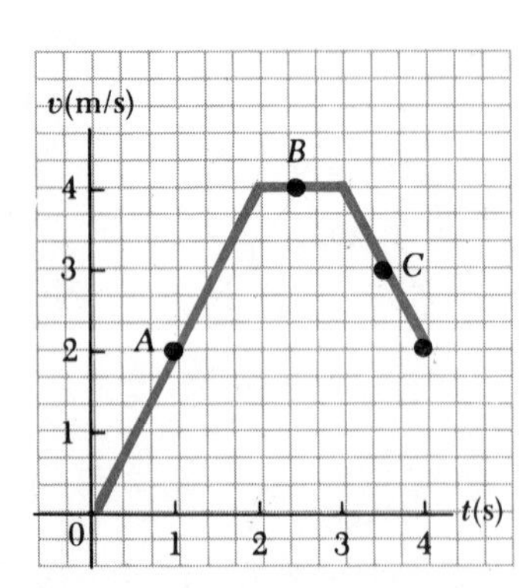

FIGURE 2.10 Example 2.2

Solution The instantaneous acceleration at any time is the slope of the velocity-time curve at that instant. The slope of the curve at point A is

$$a = \frac{\Delta v}{\Delta t} = \frac{4 \text{ m/s}}{2 \text{ s}} = 2 \text{ m/s}^2$$

At point B, the slope of the line is zero and hence the instantaneous acceleration at this time is also zero. You should note that even though the instantaneous acceleration has dropped to zero at this instant, the velocity of the player is not zero. (In general, if $a = 0$, the velocity is constant in time but *not necessarily* zero.) Instead, he is running at a constant velocity of 4 m/s. A value of zero for an instantaneous acceleration means that at the particular instant the velocity of the object is not changing.

Finally, at point C we calculate the slope of the velocity-time curve and find that the instantaneous acceleration is

$$a = \frac{\Delta v}{\Delta t} = \frac{-2\ \text{m/s}}{1\ \text{s}} = -2\ \text{m/s}^2$$

The negative sign here indicates that the player is decelerating as he approaches the location where he will attempt to catch the baseball.

2.4 ONE-DIMENSIONAL MOTION WITH CONSTANT ACCELERATION

Most of the applications in this text will be concerned with objects moving with *constant acceleration*. This type of motion is important because many objects in nature move with constant acceleration. For example, an object in free fall near the earth's surface moves in the vertical direction with constant acceleration, assuming that air resistance can be neglected. Furthermore, in many situations where the acceleration is not constant, one can obtain an approximate solution by assuming the acceleration to be constant and equal to its average value over some time interval. In almost all situations, one can approximate the actual motion by a series of motions all having a constant acceleration. When an object moves with constant acceleration, *the average acceleration equals the instantaneous acceleration*. Consequently, the velocity increases or decreases at the same rate throughout the motion.

Since the average acceleration equals the instantaneous acceleration when a = constant, we can eliminate the bar used to denote average values from our defining equation for acceleration. That is, since $\bar{a} = a$, we can write Equation 2.3 as

$$a = \frac{v_f - v_i}{t_f - t_i}$$

For convenience, let $t_i = 0$ and t_f be any arbitary time t. Also, we shall let $v_i = v_0$ (the initial velocity at $t = 0$) and $v_f = v$ (the velocity at any arbitrary time t). With this notation, we can express the acceleration as

$$a = \frac{v - v_0}{t}$$

or

$$\boxed{v = v_0 + at} \qquad (2.5)$$

One of the features of one-dimensional motion with constant acceleration is the manner in which the initial, final, and average velocities are related. Because the velocity is increasing or decreasing *uniformly* with time, we can express the **average velocity** in any time interval as the arithmetic average of the initial velocity, v_0, and the final velocity, v:

$$\boxed{\bar{v} = \frac{v_0 + v}{2}} \qquad \text{(for constant } a\text{)} \qquad (2.6)$$

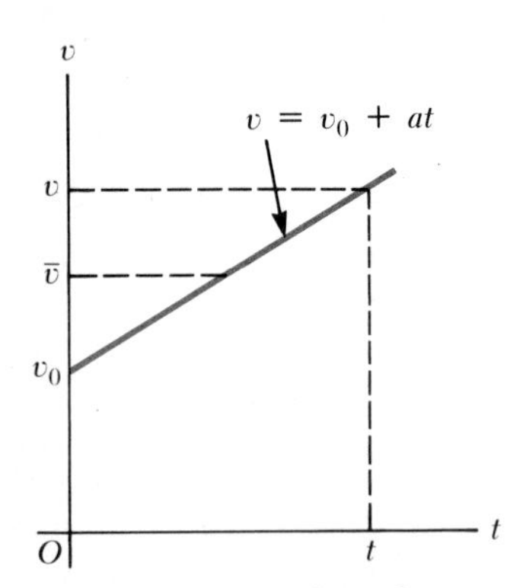

FIGURE 2.11 The velocity varies linearly with time for a particle moving with constant acceleration. The average velocity is just the mean value of the initial and final velocities.

Note that *this expression is valid only when the acceleration is constant, that is, when the velocity changes uniformly with time.* The graphical interpretation of $\bar{v}$ is shown in Figure 2.11. As you can see, the velocity varies linearly with time according to Equation 2.5.

We can now use this result along with the defining equation for average velocity, Equation 2.1, to obtain an expression for the displacement of an object as a function of time. Again we choose $t_i = 0$, and for convenience we shall also assume that we have selected our coordinate system such that the initial position of the object under consideration is at the origin, that is, $x_i = 0$. This gives

$$x = \bar{v}t = \left(\frac{v_0 + v}{2}\right)t$$

or

$$\boxed{x = \frac{1}{2}(v + v_0)t} \qquad (2.7)$$

We can obtain another useful expression for displacement by substituting the equation for v (Eq. 2.5) into Equation 2.7:

$$x = \frac{1}{2}(v_0 + at + v_0)t$$

or

$$\boxed{x = v_0 t + \frac{1}{2}at^2} \qquad (2.8)$$

Finally, we can obtain an expression that does not contain time by substituting the value of t from Equation 2.5 into Equation 2.7. This gives

$$x = \frac{1}{2}(v + v_0)\left(\frac{v - v_0}{a}\right) = \frac{v^2 - v_0^2}{2a}$$

or

$$\boxed{v^2 = v_0^2 + 2ax} \qquad (2.9)$$

Equations 2.5 through 2.9 may be used to solve any problem in one-dimensional motion with constant acceleration. Keep in mind that *these relationships were derived from the definitions of velocity and acceleration,* together with some algebraic manipulations and the requirement that the acceleration be constant. The four equations that are used most often are listed in Table 2.3 for convenience.

TABLE 2.3 Equations for Motion in a Straight Line Under Constant Acceleration

Equation	Information Given by Equation
$v = v_0 + at$	Velocity as a function of time
$x = \frac{1}{2}(v + v_0)t$	Displacement as a function of velocity and time
$x = v_0 t + \frac{1}{2}at^2$	Displacement as a function of time
$v^2 = v_0^2 + 2ax$	Velocity as a function of displacement

Note: Motion is along the x axis. At $t = 0$, the particle is at the origin ($x_0 = 0$) and its velocity is v_0.

The best way to gain confidence in the use of these equations is through working a number of problems. Many times you will discover that there is more than one method for obtaining a solution to a given problem.

EXAMPLE 2.3 The Indianapolis 500

A racing car starting from rest accelerates at a rate of 5 m/s^2. What is the velocity of the car after it has traveled 100 ft?

Solution This simple example has been selected in order to point out a few of the procedures you should follow in solving all the problems in this section. First, you must be sure that the units in the problem are consistent. For example, if you choose to measure distances in feet and time in seconds, then your units of velocity must be in ft/s and your units of acceleration in ft/s^2. As stated, the units in the example problem are not consistent. If we choose to leave the distance as 100 ft, we must change the length dimension of the units of acceleration from m/s^2 to ft/s^2. As an alternative, we can leave the units of acceleration above as m/s^2 and convert the distance traveled to meters. Let's do the latter. The table of conversion factors on the endsheet of this text gives 1 ft = 0.305 m; thus 100 ft = 30.5 m.

Second, you must choose a coordinate system. In Figure 2.12, a convenient one is shown. The origin of the coordinate system is at the initial location of the car, and the positive direction is to the right. Using this convention, we require that velocities, accelerations, and displacements to the right are positive and vice versa.

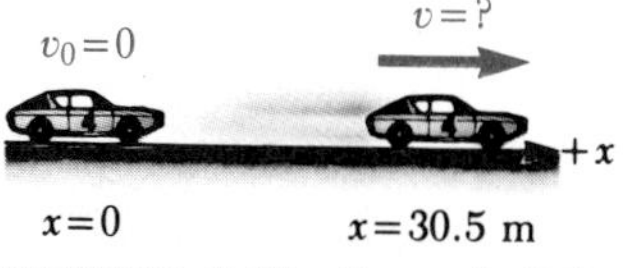

FIGURE 2.12 Example 2.3

Next, you will find it convenient to make a list of the quantities given in the problem and a separate list of those to be determined:

Given	To be Determined
$v_0 = 0$	v
$a = 5$ m/s^2	
$x = 30.5$ m	

The final step is to select from kinematic equations (Table 2.3) those that will allow you to determine the unknowns. In our present case, the equation

$$v^2 + v_0^2 + 2ax$$

is our best choice since it will give us a value for v directly:

$$v^2 = (0)^2 + 2(5 \text{ m/s}^2)(30.5 \text{ m}) = 305 \text{ m}^2/\text{s}^2$$

from which,

$$v = \sqrt{305 \text{ m}^2/\text{s}^2} = \pm 17.5 \text{ m/s}$$

Since the car is moving to the right, we choose +17.5 m/s as the correct solution for v. Alternatively, the problem can be solved by using $x = v_0t + \frac{1}{2}at^2$ to find t and then using the expression $v = v_0 + at$ to find v. Try it!

EXAMPLE 2.4 The Supercharged Sportscar

A certain automobile manufacturer claims that its super-deluxe sportscar will accelerate uniformly from rest to a speed of 87.0 mi/h in 8 s. (a) Determine the acceleration of the car.

Solution First note that $v_0 = 0$ and the velocity after 8 s is 87.0 mi/h = 38.9 m/s, where the conversion 1 mi/h = 0.447 m/s has been used. Using $v = v_0 + at$, we can find the acceleration:

$$a = \frac{v - v_0}{t} = \frac{38.9 \text{ m/s}}{8 \text{ s}} = 4.86 \text{ m/s}^2$$

(b) Find the distance the car travels in the first 8 s.

Solution The distance traveled by the car can be found from Equation 2.7:

$$x = \frac{1}{2}(v_0 + v)t = \frac{1}{2}(38.9 \text{ m/s})(8 \text{ s}) = 156 \text{ m}$$

Many examples in this text will be followed by an exercise. The purpose of these exercises is to test your understanding of the example by asking you a question related to it. Answers to these exercises will be provided at the end of the exercise, when appropriate. Here is your first exercise, which relates to Example 2.4.

EXERCISE What is the velocity of the car 10 s after it begins its motion, assuming it continues to accelerate at the rate of 4.86 m/s^2?

Answer 48.6 m/s, or 109 mi/h.

2.5 FREELY FALLING BODIES

It is well known that, in the absence of air resistance, all objects when dropped fall toward the earth with the same constant acceleration. However, it was not until about 1600 that this conclusion came to be accepted. Prior to that time, the teachings of the great philosopher Aristotle (384–322 B.C.) had held that heavier objects fell faster than lighter ones. It was Galileo, often called the father of modern science, who was responsible for our present ideas

Skydivers reach a terminal speed of about 160 km/h in the "spread-eagle" position. They can control their speed by adjusting the area they present to the atmosphere. (Courtesy of U.S. Air Force Academy)

Galileo Galilei (1564–1642), an Italian physicist and astronomer, investigated the motion of objects in free fall (including projectiles), and the motion of an object on an inclined plane, established the concept of relative motion, and noted that a swinging pendulum could be used to measure time intervals. Following the invention of the telescope, he said "I now have visual proof of what I already knew through my intellect." Galileo made several major discoveries in astronomy; he discovered four moons of Jupiter and many new stars, investigated the nature of the Moon's surface, observed sun spots and the phases of Venus, and proved that the Milky Way consists of an enormous number of stars. (Courtesy of AIP Niels Bohr Library)

concerning falling bodies. There is a legendary story that he discovered the law of falling bodies by observing that two different weights dropped simultaneously from the Leaning Tower of Pisa hit the ground at approximately the same time. Although there is some doubt that this particular experiment was carried out, it is well established that Galileo did perform many systematic experiments on objects moving on inclined planes. In his experiments he rolled balls down a slight incline and measured the distance they moved in successive time intervals. The purpose of the incline was to produce a smaller acceleration and enable him to make accurate measurements of the time intervals. By gradually increasing the slope of the incline, he was finally able to draw conclusions about freely falling objects because a falling ball is equivalent to a ball falling down a vertical incline. Galileo's achievements in the science of mechanics paved the way for Newton in his development of the laws of motion, which we shall study in Chapter 3.

You might want to try the following experiment. Drop a coin and a crumpled-up piece of paper simultaneously from the same height. If the effects of air friction are negligible, both will have the same motion and hit the floor at the same time. In the idealized case, where air resistance is absent, such motion is referred to as free fall. If this same experiment could be conducted in a vacuum, where air friction is truly negligible, the paper and coin would fall with the same acceleration, regardless of the shape of the paper. On August 2, 1971, such a demonstration was conducted on the moon by astronaut David Scott. He simultaneously released a hammer and a feather, and they fell with the same acceleration to the lunar surface. This demonstration would have surely pleased Galileo!

We shall denote the acceleration due to gravity by the symbol **g**. The magnitude of g decreases with increasing altitude. Furthermore, there are slight variations in the magnitude of g with latitude. However, at the surface of the earth the magnitude of g is approximately 9.8 m/s^2, or 980 cm/s^2, or 32 ft/s^2. Unless stated otherwise, we shall use the value 9.80 m/s^2 when doing calculations. Furthermore, we shall assume that the vector **g** is directed downward toward the center of the earth.

When we use the expression *freely falling body*, we do not necessarily refer to an object dropped from rest.

A freely falling body is an object moving freely under the influence of gravity only, regardless of its initial motion. Objects thrown upward or downward and those released from rest are all falling freely once they are released!

It is important to emphasize that any freely falling object experiences an acceleration directed downward. This is true regardless of the initial motion of the object. An object thrown upward (or downward) will experience the same acceleration as an object released from rest.

Once they are in free fall, all objects have an acceleration downward, equal to the acceleration due to gravity.

If we neglect air resistance and assume that the gravitational acceleration does not vary with altitude, then the motion of a freely falling body is equivalent to motion in one dimension under constant acceleration. Therefore, the equations developed in Section 2.4 for objects moving with constant acceleration can be applied. The only modification that we need to make in these equations for freely falling bodies is to note that the motion is in the vertical direction (the y direction) rather than along the horizontal (x direction), and that the acceleration is downwards and has a magnitude of 9.80 m/s^2. Thus, we always take $a = -g = -9.80$ m/s^2 to account for the fact that the acceleration of a freely falling body is downwards. We shall study how to deal with variations in g with altitude in Chapter 7, and the effects of air resistance on motion will be discussed in Chapter 10.

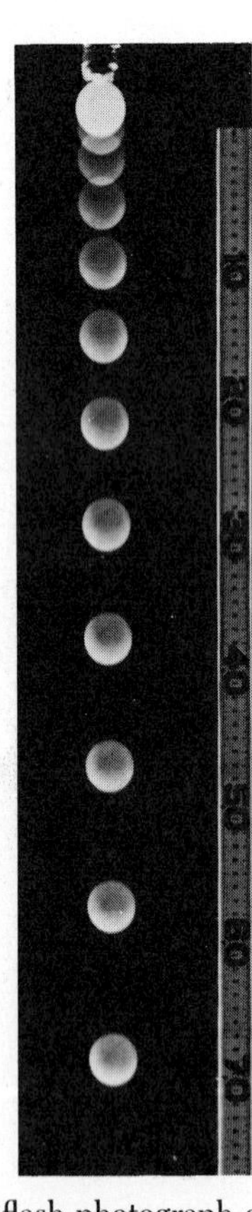

Multiflash photograph of a freely falling ball. The time interval between flashes is (1/30) s and the scale is in cm. Why not determine g from these data?

EXAMPLE 2.5 Look Out Below!

A golf ball is released from rest from the top of a very tall building. Neglecting air resistance, calculate the position and velocity of the ball after 1, 2, and 3 s.

Solution We choose our coordinates such that the starting point of the ball is at the origin ($y_0 = 0$ at $t = 0$) and remember that we have defined y to be positive upward. Since $v_0 = 0$, and $a = -g = -9.80$ m/s^2, Equations 2.5 and 2.8 become

$$v = at = -(9.80 \text{ m/s}^2)t$$

$$y = \frac{1}{2}at^2 = -\frac{1}{2}(9.80 \text{ m/s}^2)t^2$$

where t is in s, v is in m/s, and y is in m. These expressions give the velocity and displacement at any time t after the ball is released. Therefore, at $t = 1$ s,

$$v = -(9.80 \text{ m/s}^2)(1 \text{ s}) = -9.80 \text{ m/s}$$

$$y = -\frac{1}{2}(9.80 \text{ m/s}^2)(1 \text{ s})^2 = -4.90 \text{ m}$$

Likewise, at $t = 2$ s, we find that $v = -19.6$ m/s and $y = -19.6$ m. Finally, at $t = 3$ s, $v = -29.4$ m/s and $y = -44.1$ m. The minus signs for v indicate that the velocity vector is directed downward, and the minus signs for y indicate displacement in the negative y direction.

EXERCISE Calculate the position and velocity of the ball after 4 s.

Answer -78.4 m, -38.4 m/s.

EXAMPLE 2.6 Not a Bad Throw for a Rookie!

A stone is thrown from the top of a building with an initial velocity of 20 m/s straight upward. The building is 50 m high, and the stone just misses the edge of the roof on its way down, as in Figure 2.13. Determine (a) the time needed for the stone to reach its maximum height, (b) the maximum height, (c) the time needed for the stone to return to the level of the thrower, (d) the velocity of the stone at this instant, and (e) the velocity and position of the stone at $t = 5$ s.

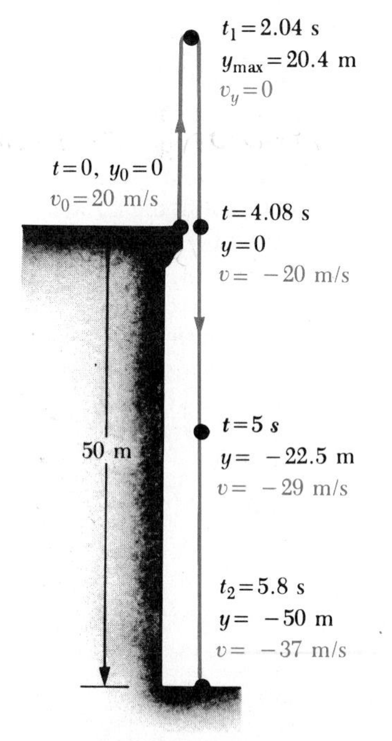

FIGURE 2.13 Example 2.6: Position and velocity versus time for a freely falling object thrown initially upward with a velocity $v_0 = 20$ m/s.

Solution (a) To find the time necessary to reach the maximum height, use Equation 2.5, $v = v_0 + at$, noting that $v = 0$ at maximum height:

$$20 \text{ m/s} - (9.80 \text{ m/s}^2)t_1 = 0$$

$$t_1 = \frac{20 \text{ m/s}}{9.80 \text{ m/s}^2} = 2.04 \text{ s}$$

(b) This value of time can be substituted into Equation 2.8, $y = v_0t + \frac{1}{2}at^2$, to give the maximum height as measured from the position of the thrower:

$$y_{max} = (20 \text{ m/s})(2.04 \text{ s}) - \frac{1}{2}(9.80 \text{ m/s}^2)(2.04 \text{ s})^2 = 20.4 \text{ m}$$

(c) When the stone is back at the height of the thrower, the y coordinate is zero. From the expression $y = v_0t + \frac{1}{2}at^2$ (Eq. 2.8), with $y = 0$, we obtain the expression

$$20t - 4.9t^2 = 0$$

This is a quadratic equation and has two solutions for t. The equation can be factored to give

$$t(20 - 4.9t) = 0$$

One solution is $t = 0$, corresponding to the time the stone starts its motion. The other solution is $t = 4.08$ s, which is the solution we are after.

(d) The value for t found in (c) can be inserted into $v = v_0 + at$ (Eq. 2.5) to give

$$v = 20 \text{ m/s} - (9.80 \text{ m/s}^2)(4.08 \text{ s}) = -20.0 \text{ m/s}$$

Note that the velocity of the stone when it arrives back at its original height is equal in magnitude to its initial velocity but opposite in direction. This indicates that the motion is symmetric.

(e) From $v = v_0 + at$ (Eq. 2.5), the velocity after 5 s is

$$v = 20 \text{ m/s} - (9.80 \text{ m/s}^2)(5 \text{ s}) = -29.0 \text{ m/s}$$

We can use $y = v_0t + \frac{1}{2}at^2$ (Eq. 2.8) to find the position of the particle at $t = 5$ s:

$$y = (20 \text{ m/s})(5 \text{ s}) - \frac{1}{2}(9.8 \text{ m/s}^2)(5 \text{ s})^2 = -22.5 \text{ m}$$

EXERCISE Find the velocity of the stone just before it hits the ground.

Answer -37 m/s.

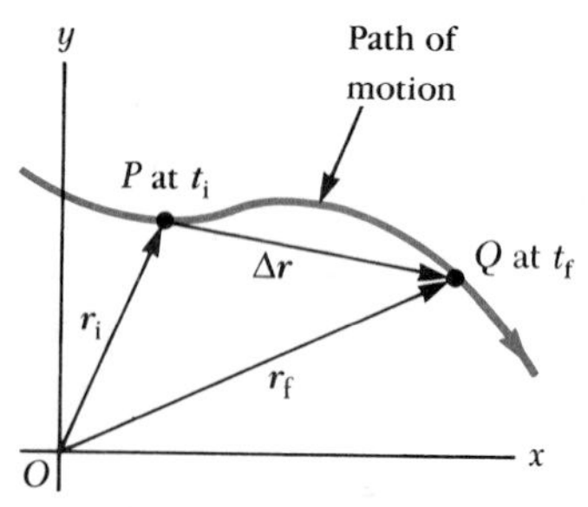

FIGURE 2.14 An object moving along some curved path between points P and Q. The displacement vector $\Delta \boldsymbol{r}$ is the difference in the position vectors. That is, $\Delta \boldsymbol{r} = \boldsymbol{r}_f - \boldsymbol{r}_i$.

2.6 VELOCITY AND ACCELERATION IN TWO DIMENSIONS

In the case of one-dimensional motion discussed in previous sections, the vector nature of displacement, velocity, and acceleration was taken into account by using positive and negative signs. In order to completely describe the motion of an object in two or three dimensions, we must make use of vectors.

Consider an object moving along some curve in space as shown in Figure 2.14. When the object is at some point P at time t_i, its position is described by the position vector $\boldsymbol{r}_i$ drawn from the origin to P. Likewise, when the particle has moved to some other point Q at time t_f, its position vector is $\boldsymbol{r}_f$. As you can see from the vector diagram in Figure 2.14, the final position vector is the sum of the initial position vector and $\Delta \boldsymbol{r}$. Since $\boldsymbol{r}_f = \boldsymbol{r}_i + \Delta \boldsymbol{r}$, the displacement of the object is defined as the change in the position vector

Displacement vector

$$\Delta \boldsymbol{r} = \boldsymbol{r}_f - \boldsymbol{r}_i \tag{2.10}$$

Note that the magnitude of the displacement vector is *not* equal to the distance traveled measured along the curved path. In fact, the displacement is less than this distance when the path is curved.

In analogy with our one-dimensional case, the average velocity of the object in the time interval $\Delta t = t_f - t_i$ is given by

Average velocity vector

$$\overline{\boldsymbol{v}} = \frac{\Delta \boldsymbol{r}}{\Delta t} \tag{2.11}$$

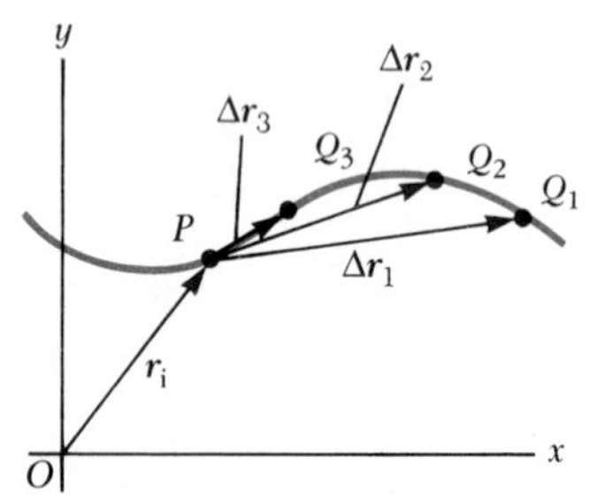

FIGURE 2.15 As the time interval between the final and initial positions approaches zero, the direction of the displacement vector approaches that of the line tangent to the path at P.

We now define the instantaneous velocity vector as the limiting value of the average velocity as the time interval approaches zero. With reference to Figure 2.15, we see that as the time interval gets smaller and smaller, and point Q approaches P, the direction of $\Delta \boldsymbol{r}$ approaches that of the line tangent to the curve at P. Thus, the direction of the instantaneous velocity vector at any point is along the line tangent to the path of the motion at that point.

The average acceleration of an object whose velocity changes by $\Delta \boldsymbol{v}$ in the time interval Δt is a vector defined as the ratio $\Delta \boldsymbol{v}/\Delta t$.

$$\overline{\boldsymbol{a}} = \frac{\Delta \boldsymbol{v}}{\Delta t} \tag{2.12}$$

The instantaneous acceleration vector is defined as the limit of the average acceleration vector as the time interval approaches zero.

You should note that an object is accelerating whenever the velocity vector changes in magnitude, direction, or both. The most familiar example of acceleration is motion in one dimension when the magnitude of the object's velocity (speed) changes. On the other hand,

an object moving with constant speed is accelerating if the direction of the velocity vector is changing.

One example of this is an object moving in a circular path with constant speed. In this case, the acceleration is directed towards the center of the circle. We shall return to this type of motion in Chapter 7.

2.7 PROJECTILE MOTION

The situations we considered in sections 2.1 through 2.6 were those in which an object moves along a straight-line path, such as the x axis. We now look at some cases in which the object moves in a plane. By this we mean that the object may move in both the x and the y direction simultaneously, corresponding to motion in two dimensions. The particular type of motion we shall concentrate on is called **projectile motion.** Anyone who has observed a baseball in motion (or for that matter, any object thrown into the air) has observed projectile motion. This very common form of two-dimensional motion is surprisingly simple to analyze if the following three assumptions are made:

1. The acceleration due to gravity, $g = 9.80 \text{ m/s}^2$, is constant over the range of motion and is directed downward.
2. The effect of air resistance is negligible.
3. The rotation of the earth does not affect the motion.

With these assumptions, we shall find that the path of a projectile is curved as shown in Figure 2.16. (Such a curve is called a parabola.)

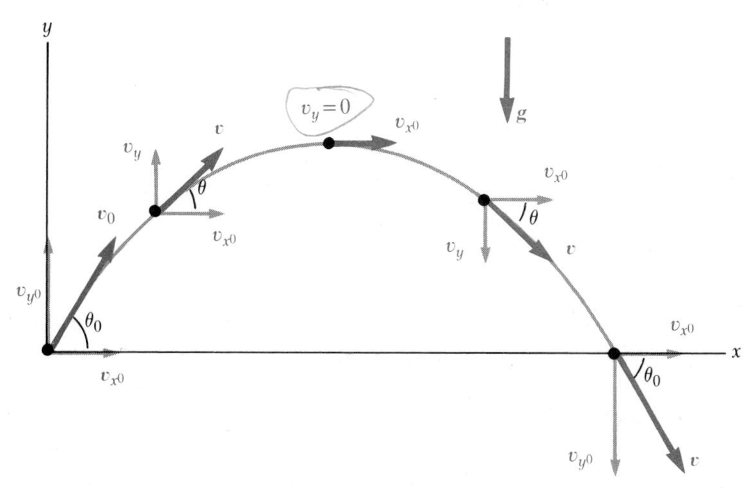

FIGURE 2.16 The parabolic trajectory of a particle that leaves the origin with a velocity $\mathbf{v}_0$. Note that the velocity, $\mathbf{v}$, changes with time. However, the x component of the velocity, v_x, remains constant in time. Also, $v_y = 0$ at the peak, but the acceleration is always equal to the acceleration of gravity, and acts vertically downwards.

A night-time photograph of the illuminated Milles water fountain in St. Louis, Missouri, illustrates the parabolic path followed by a stream of water. (Courtesy of Missouri Division of Tourism)

Let us choose our coordinate system such that the y direction is vertical and positive upward. In this case

the acceleration in the y direction is $-g$, just as in free fall, and the acceleration in the x direction is 0 (because air friction is neglected).

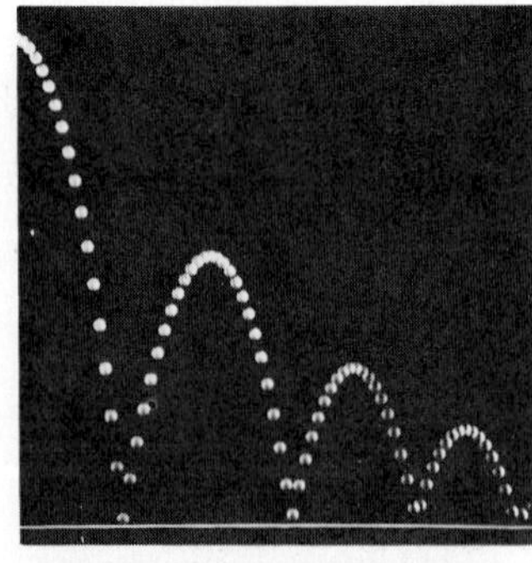

A ball undergoing several bounces off a hard surface. Note the parabolic path of the ball following each bounce. (Photo courtesy of Education Development Center, Newton, MA.)

Furthermore, let us assume that at $t = 0$, the projectile leaves the origin with a velocity $\mathbf{v}_0$, as shown in Figure 2.16. If the velocity vector makes an angle θ_0 with the horizontal, as shown, then the initial x and y components of velocity are given by

$$v_{x0} = v_0 \cos \theta_0 \quad \text{and} \quad v_{y0} = v_0 \sin \theta_0$$

In order to analyze projectile motion, we shall separate the motion into two parts, the x (or horizontal) motion, and the y (or vertical) motion and solve each part separately. We shall look first at the x motion. As noted before, motion along the x direction occurs with $a_x = 0$. This means that *the velocity component along the x direction never changes*. Thus, if the initial value of the velocity component in the x direction is $v_{x0} = v_0 \cos \theta_0$, this is also the value of the velocity at any later time. That is,

$$\boxed{v_x = v_{x0} = v_0 \cos \theta_0 = \text{constant}} \tag{2.13}$$

Equation 2.13 can be substituted into the defining equation for velocity (Eq. 2.1) to give us an expression for the horizontal position of the projectile as a function of time:

$$x = v_{x0}t = (v_0 \cos \theta_0)t \tag{2.14}$$

These equations tell us all we need to know about the motion in the x direction. Let us now consider the y motion. Since the motion in the y direction is motion with constant acceleration, the equations developed in Section 2.4 can be used. In these equations, we shall denote v_{y0} as the initial velocity in the y direction and $-g$ as the acceleration due to gravity. The negative sign for g assumes that the positive direction for the vertical motion is upward. With this choice of signs we have

$$v_y = v_{y0} - gt = v_0 \sin \theta_0 - gt \tag{2.15}$$

$$y = v_{y0}t - \frac{1}{2}gt^2 = (v_0 \sin \theta_0)t - \frac{1}{2}gt^2 \tag{2.16}$$

$$v_y^2 = v_{y0}^2 - 2gy = (v_0 \sin \theta_0)^2 - 2gy \tag{2.17}$$

Note that it is not necessary to assume that the upward direction is positive. If you choose the positive direction as downward, simply replace $-g$ with g in the equations above. The speed v of the projectile at any instant can be calculated from the components of velocity at that instant using the Pythagorean theorem:

$$v = \sqrt{v_x^2 + v_y^2}$$

Before we look at some numerical examples dealing with projectile motion, let us pause to summarize what we have learned so far about this kind of motion:

1. Provided air resistance is negligible, the horizontal component of velocity, v_x, remains constant since there is no horizontal component of acceleration.
2. The vertical component of acceleration is equal to the acceleration due to gravity, g.
3. The vertical component of velocity, v_y, and the displacement in the y direction are identical to those of a freely falling body.
4. Projectile motion can be described as a superposition of the two motions in the x and y directions.

EXAMPLE 2.7 The Stranded Explorers

An Alaskan rescue plane drops a package of emergency rations to a stranded party of explorers, as shown in Figure 2.17. If the plane is traveling horizontally at 40 m/s at a height of 100 m above the ground, (a) where does the package strike the ground relative to the point at which it was released?

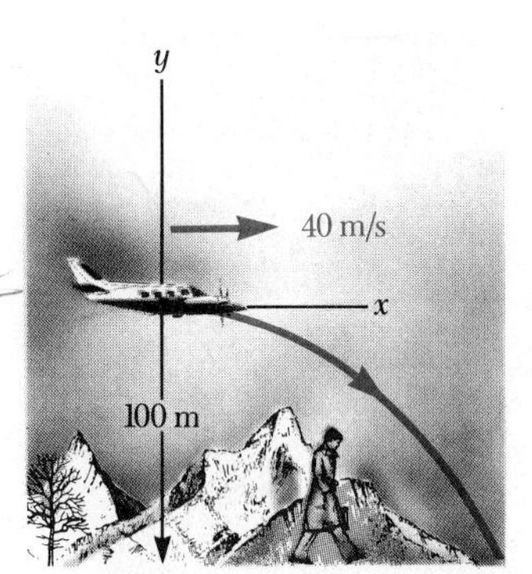

FIGURE 2.17 Example 2.7: To an observer on the ground, a package released from the rescue plane travels along the path shown. (This figure is not drawn to scale.)

Solution The coordinate system for this problem is selected as shown in Figure 2.17, with the positive x direction to the right and the positive y direction upward.

Consider first the horizontal motion of the package. The only equation available to us is

$$x = v_{x0}t$$

The initial x component of the package velocity is the same as the velocity of the plane when the package was released, 40 m/s. Thus, we have

$$x = (40 \text{ m/s})t$$

If we know t, the length of time the package is in the air, we can determine x, the distance traveled by the package along the horizontal. To find t, we move to the equations for the vertical motion of the package. We know that at the instant the package hits the ground its y coordinate is -100 m. We also know that the initial velocity of the package in the vertical direction, v_{y0}, is zero because the package was released with only a horizontal component of velocity. From Equation 2.16, we have

$$y = -\tfrac{1}{2}gt^2$$

$$-100 \text{ m} = -\tfrac{1}{2}(9.80 \text{ m/s}^2)t^2$$

$$t^2 = 20.4 \text{ s}^2$$

$$t = 4.51 \text{ s}$$

This value for the time of flight substituted into the equation for the x coordinate gives

$$x = (40 \text{ m/s})(4.51 \text{ s}) = 180 \text{ m}$$

(b) What are the horizontal and vertical components of the velocity of the package just before it hits the ground?

Solution We already know the horizontal component of the velocity of the package just before it hits, because the velocity in the horizontal direction remains constant at 40 m/s throughout the flight.

The vertical component of the velocity just before the package hits the ground may be found by using $v_y = v_0 \sin \theta_0 - gt$ (Eq. 2.15):

$$v_y = -(9.80 \text{ m/s}^2)(4.51 \text{ s}) = -44.1 \text{ m/s}$$

EXAMPLE 2.8 The Long-jump

A long-jumper leaves the ground at an angle of 20° to the horizontal and at a speed of 11 m/s. (a) How far does he jump? (Assume that the motion of the long-jumper is equivalent to that of a particle.)

Solution His horizontal motion is described by using $x = (v_0 \cos \theta_0)t$ (Eq. 2.14), or

$$x = (11 \text{ m/s})(\cos 20°)t$$

The value of x can be found if t, the total time of the jump, is known. We are able to find t using $v_y = v_0 \sin \theta_0 - gt$ (Eq. 2.15) by noting that at the top of the jump the vertical component of velocity goes to zero:

$$v_y = v_0 \sin \theta_0 - gt$$

$$0 = (11 \text{ m/s}) \sin 20° - (9.80 \text{ m/s}^2)t_1$$

$$t_1 = 0.384 \text{ s}$$

Note that t_1 is the time interval to reach the *top* of the jump. Because of the symmetry of the vertical motion, an identical time interval passes before the jumper returns to the ground. Therefore, the *total time* in the air is $t = 2t_1 = 0.768$ s and the distance jumped is

$$x = (11 \text{ m/s})(\cos 20°)(0.768 \text{ s}) = 7.94 \text{ m}$$

(b) What is the maximum height reached?

Solution The maximum height reached is found using $y = (v_0 \sin \theta_0)t - \frac{1}{2}gt^2$ (Eq. 2.16) with $t = t_1 = 0.384$ s.

$$y_{max} = (11 \text{ m/s})(\sin 20°)(0.384 \text{ s}) - \frac{1}{2}(9.80 \text{ m/s}^2)(0.384 \text{ s})^2$$
$$= 0.722 \text{ m}$$

The assumption that the motion of the long-jumper is that of a projectile is an oversimplification of the situation. Nevertheless, the values obtained are reasonable.

EXAMPLE 2.9 That's Quite an Arm

A stone is thrown upward from the top of a building at an angle of 30° to the horizontal and with an initial speed of 20 m/s, as in Figure 2.18. If the height of the building is 45 m, (a) how long is the stone "in flight"?

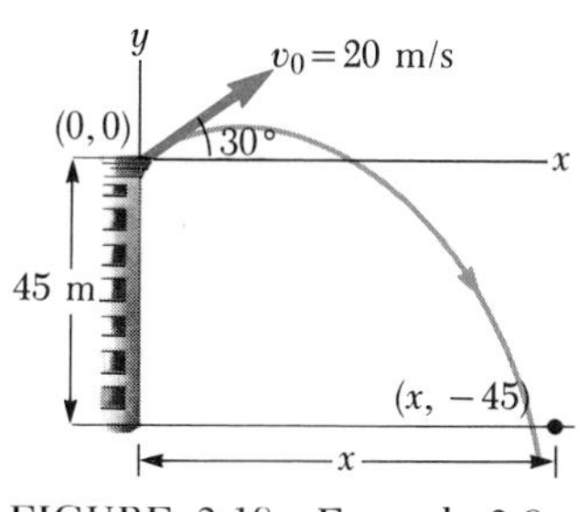

FIGURE 2.18 Example 2.9

Solution The initial x and y components of the velocity are

$$v_{x0} = v_0 \cos \theta_0 = (20 \text{ m/s})(\cos 30°) = 17.3 \text{ m/s}$$
$$v_{y0} = v_0 \sin \theta_0 = (20 \text{ m/s})(\sin 30°) = 10 \text{ m/s}$$

To find t, we can use $y = v_{y0}t - \frac{1}{2}gt^2$ (Eq. 2.16) with $y = -45$ m and $v_{y0} = 10$ m/s (we have chosen the top of the building as the origin, as shown in Fig. 2.18):

$$-45 \text{ m} = (10 \text{ m/s})t - \frac{1}{2}(9.80 \text{ m/s}^2)t^2$$

Solving the quadratic equation (see Appendix B) for t gives, for the positive root, the value $t = 4.22$ s. Does the negative root have any physical meaning? (Why not think of another way of finding t from the information given?)
(b) What is the speed of the stone just before it strikes the ground?

Solution The y component of the velocity just before the stone strikes the ground can be obtained using the equation $v_y = v_{y0} - gt$ (Eq. 2.15) with $t = 4.22$ s:

$$v_y = 10 \text{ m/s} - (9.80 \text{ m/s}^2)(4.22 \text{ s}) = 31.4 \text{ m/s}$$

Since $v_x = v_{x0} = 17.3$ m/s, the required speed is given by

$$v = \sqrt{v_x^2 + v_y^2} = \sqrt{(17.3)^2 + (-31.4)^2} \text{ m/s} = 35.9 \text{ m/s}$$

EXERCISE Where does the stone strike the ground?

Answer 73.0 m from the base of the building.

SUMMARY

The **average velocity** of an object moving along the x axis during some time interval is equal to the displacement of the object, Δx, divided by the time interval, Δt, during which the displacement occurred:

$$\bar{v} \equiv \frac{\Delta x}{\Delta t} = \frac{x_f - x_i}{t_f - t_i} \quad (2.1)$$

Average velocity

The average velocity is equal to the slope of the straight line joining the initial and final points on a graph of the displacement of the object versus time.

The velocity of an object at any instant of time is called the **instantaneous velocity.** The slope of the line tangent to the position-time curve at some point is defined to be the instantaneous velocity at that time. The **instantaneous speed** of an object is defined as the magnitude of the instantaneous velocity.

The **average acceleration** of an object during some time interval is defined as the change in velocity, Δv, divided by the time interval, Δt, during which the change occurred:

Average acceleration

$$\bar{a} \equiv \frac{\Delta v}{\Delta t} = \frac{v_f - v_i}{t_f - t_i} \tag{2.3}$$

The **instantaneous acceleration** of an object is its acceleration at some particular instant of time. The instantaneous acceleration of an object at a certain time equals the slope of a velocity time graph at that instant of time.

The equations that describe the motion of an object moving with constant acceleration along the x axis are

Equations of kinematics

$$v = v_0 + at \tag{2.5}$$

$$x = \frac{1}{2}(v + v_0)t \tag{2.7}$$

$$x = v_0 t + \frac{1}{2}at^2 \tag{2.8}$$

$$v^2 = v_0^2 + 2ax \tag{2.9}$$

An object falling in the presence of the earth's gravity experiences a gravitational acceleration directed toward the center of the earth. If air friction is neglected and if the altitude of the falling object is small compared with the earth's radius, then one can assume that the *acceleration of gravity*, **g**, is constant over the range of motion, where g is equal to 9.80 m/s^2, or 32 ft/s^2. Assuming the positive direction for y is chosen to be upward, the acceleration is $-g$ (downward) and the equations describing the motion of the falling object are the same as the equations above, with the substitutions $x \rightarrow y$ and $a \rightarrow -g$.

An object moving above the surface of the earth such that it moves in both the x and the y direction simultaneously is said to be undergoing **projectile motion.** Projectile motion is equivalent to two different motions occurring at the same time: (1) the object moves along the horizontal (x) direction such that its velocity in this direction, v_x, is a constant and (2) the object moves in the vertical (y) direction with a constant downward acceleration of magnitude $g = 9.80$ m/s^2. The equations describing the motion of a projectile are

Projectile motion

$$x = (v_0 \cos \theta_0)t \tag{2.14}$$

$$v_y = v_0 \sin \theta_0 - gt \tag{2.15}$$

$$y = (v_0 \sin \theta_0)t - \frac{1}{2}gt^2 \tag{2.16}$$

$$v_y^2 = (v_0 \sin \theta_0)^2 - 2gy \tag{2.17}$$

ADDITIONAL READING

P. Brancazio, "The Trajectory of a Fly Ball," *The Physics Teacher*, January 1985, p. 20.

I. B. Cohen, "Galileo," *Sci. American*, August 1949, p. 40.

S. Drake, "Galileo's Discovery of the Law of Free Fall," *Sci. American*, May 1973, p. 84.

S. Drake, and J. MacLachlan, "Galileo's Discovery of the Parabolic Trajectory," *Sci. American*, March 1975, p. 102.

Galileo Galilei, "Dialogues Concerning Two New Sciences," translated by H. Crew and A. de Salvio, Evanston, Ill., Northwestern University Press, 1939.

O. Gingerich, "The Galileo Affair," *Sci. American*, August 1982, p. 132.

W. F. Magie, *Source Book in Physics*, Cambridge, Mass., Harvard University Press, 1963. Contains many excerpts on projectile motion from Galileo.

QUESTIONS

1. Average velocity and instantaneous velocity are generally different quantities. Can they ever be equal for a specific type of motion?
2. A baseball is thrown vertically upwards into the air and returns to the starting point. Sketch graphs of (a) its velocity versus time and (b) its acceleration versus time.
3. Can the instantaneous velocity of an object ever be greater in magnitude than the average velocity? Can it ever be less?
4. If the velocity of a particle is not zero, can its acceleration ever be zero?
5. A ball is thrown vertically upward. What are its velocity and acceleration when it reaches its maximum altitude? What is its acceleration just before it hits the ground?
6. At the end of its arc, the velocity of a pendulum is zero. Is its acceleration also zero at this point?
7. If a rock is dropped from the top of a sailboat's mast, will it hit the deck at the same point whether or not the boat is at rest or in motion at constant velocity?
8. A stone is thrown upward from the top of a building. Does the stone's displacement depend on the location of the origin of the coordinate system? Does the stone's velocity depend on the origin?
9. A child throws a marble into the air with an initial velocity v_0. Another child drops a ball at the same instant. Compare the accelerations of the two objects while they are in flight.
10. Two children stand at the edge of a tall building. One drops a rock over the edge, while simultaneously the second throws a rock downward such that it has an initial speed of 10 m/s. Compare the accelerations of the two objects while in flight.
11. Can the sign of the acceleration ever be positive for an object which is slowing down? Explain.
12. If the average velocity of an object is zero in some time interval, what can you say about the displacement of the object for that interval?
13. Does a ball dropped out of the window of a moving car take longer to reach the ground than one dropped from a car at rest?
14. If a car is traveling eastward, can its acceleration be westward? Explain.
15. If an object is stationary, is its acceleration necessarily zero?
16. A ball is thrown vertically upward with an initial speed of 8 m/s. What will its speed be when it returns to its starting point?
17. A student at the top of a building of height h throws one ball upward with a speed v_0 and then throws a second ball downward with the same initial speed v_0. How do the final velocities of the balls compare when they reach the ground?
18. Suppose you are guilty of a speeding violation and are presented with a ticket. Will your fine be based on your average speed or your instantaneous speed?
19. A rock is dropped at the same instant a ball, at the same initial elevation, is thrown horizontally. Which will have the greater velocity when they reach ground level?

PROBLEMS

Section 2.1 Average Velocity and Section 2.2 Instantaneous Velocity

1. A football player makes a touchdown run of 100 yards in a time of 15 seconds. What was his average velocity in m/s during his run?

2. If the average speed of an orbiting space shuttle is 19,800 miles per hour, find the time required for it to circle the earth. Use the distance around the earth in your calculation and neglect the fact that the shuttle is about 200 miles above the earth's surface.

3. An athlete swims the length of a 50-m pool in 20 s and makes the return trip to the starting position in 22 s. Determine her average velocity in (a) the first half of the swim, and (b) the second half of the swim, and (c) the round trip.

4. A jogger runs eastward in a straight line with an average speed of 2 m/s for 5 min and then continues with an average speed of 1.5 m/s for 2 min. (a) What is her total displacement? (b) What is her average velocity during this time?

5. A runner moves such that his position at certain time intervals is given by the data in the table below. Use this data to construct a table like that given in the text as Table 2.2. From this table, find (a) the average velocity during the complete interval and (b) the instantaneous velocity of $t = 2$ s.

t(s)	x(m)
2	5.66
2.01	5.674
2.2	5.93
2.5	6.32
3	6.92
4	8.00

6. A tennis player moves in a straight-line path as shown in Figure 2.19. Find her average velocity in the time interval (a) 0 to 1 s, (b) 0 to 4 s, (c) 1 s to 5 s, (d) 0 to 5 s.

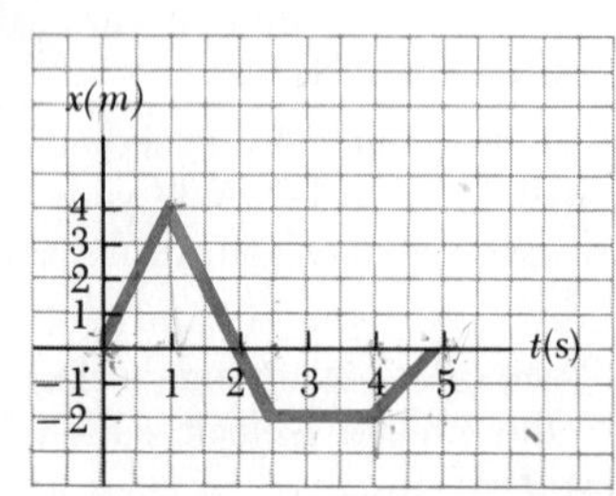

FIGURE 2.19 Problems 6 and 7

7. Find the instantaneous velocity of the tennis player described in Figure 2.19 at (a) 0.5 s, (b) 2 s, (c) 3 s, (d) 4.5 s.

8. The position-time graph for a bug crawling along the x axis is shown in Figure 2.20. Determine whether the velocity is positive, negative, or zero for the times (a) t_1, (b) t_2, (c) t_3, (d) t_4.

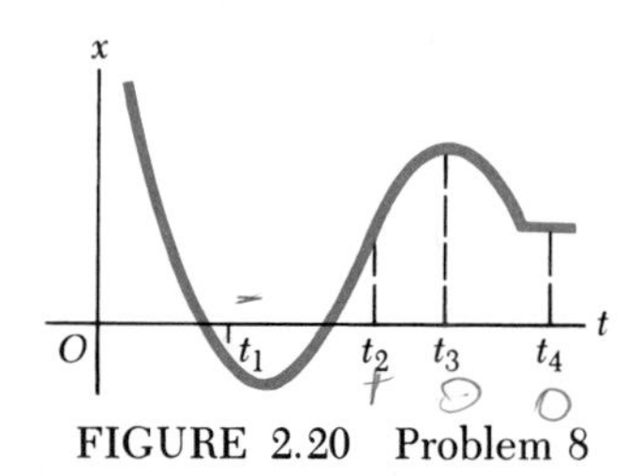

FIGURE 2.20 Problem 8

9. A shopper in a supermarket is in a great hurry. Plot a position-time graph for him as he moves along an aisle in a straight-line path. Use the following data and assume the origin of coordinates is at the initial position of the shopper. He moves from position 0 to -3 m at constant velocity in 1 s. He then moves from this position to $+3$ m at a constant velocity in 2 s. Finally, he pauses to catch his breath for 1 s. Use the graph you have plotted to find the average velocity during the total time interval, and the instantaneous velocity at 0.5 s, 2 s, and 3.5 s.

10. It is found that the position of a model airplane as a function of time is given for a portion of time by $x = 2t^2$. Plot a graph of this equation between $t = 0$ and $t = 3$ s. From your graph find (a) the average velocity during this 3 s interval, and (b) the instantaneous velocity at 2 s.

Section 2.3 Acceleration

11. A car traveling in a straight line has a speed of 5 m/s at some instant. After 4 s, its speed is 8 m/s. What is its average acceleration in this time interval?

12. A car traveling in a straight-line path has a speed of 10 m/s at some instant. After 3 s, its speed is 6 m/s. What is the average acceleration in this time interval?

13. A certain car is capable of accelerating at a rate of $0.6\ \text{m/s}^2$. How long a time is taken for this car to go from a speed of 55 mi/h to a speed of 60 mi/h?

14. A car traveling initially at 7 m/s accelerates at the rate of 0.8 m/s^2 for a time of 2 s. What is its velocity at the end of the acceleration?

• **15.** The velocity versus time graph for an object moving along a straight path is shown in Figure 2.21. (a) Find the average acceleration of this object during the time intervals 0 to 5 s, 5 s to 15 s, and 0 to 20 s. (b) Find the instantaneous acceleration at 2 s, 10 s, and 18 s.

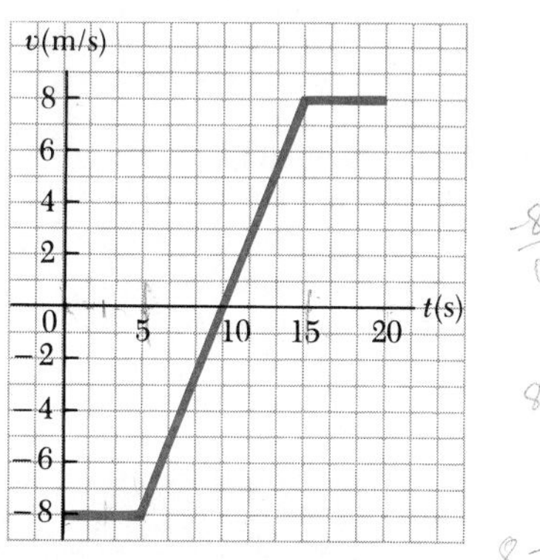

FIGURE 2.21 Problem 15

• **16.** The engine of a model rocket accelerates the rocket vertically upward for 2 s such that its speed is given by the following data. At $t = 0$, its speed is zero; at $t = 1$ s, its speed is 5 m/s; at $t = 2$ s, its speed is 16 m/s. Plot a velocity-time graph for this motion and from it determine (a) the average acceleration during the 2-s interval and (b) the instantaneous acceleration at $t = 1.5$ s.

Section 2.4 One-Dimensional Motion with Constant Acceleration

17. A speedboat increases its speed from 20 m/s to 30 m/s in a distance of 200 m. Find (a) the magnitude of its acceleration and (b) the time it takes the boat to travel this distance.

18. A car starts from rest and accelerates at 0.3 m/s^2. What is the speed of the car after it has traveled 25 m?

19. A racing car reaches a speed of 40 m/s. At this instant, it begins a uniform deceleration using a parachute and a braking system and comes to rest 5 s later. (a) Determine the deceleration of the car. (b) How far does the car travel after deceleration starts?

20. A car accelerates uniformly from rest to a speed of 40 mi/h in 12 s. (a) Find the distance the car travels during this time, and (b) the constant acceleration of the car.

21. An electron moving in a straight line has an initial speed of 3.0×10^5 m/s. If it undergoes an acceleration of 8.0×10^{14} m/s^2, (a) how long will it take to reach a speed of 5.4×10^5 m/s and (b) how far will it have traveled in this time?

22. A car travels at 20 m/s accelerates at a uniform rate of 4 m/s^2 over a distance of 50 m. How much time is required to cover this distance?

• **23.** In order to pass a physical education class at a university, a student must run one mile in twelve minutes. After running for 10 min, she still has 500 yd to go. If her maximum acceleration is 0.15 m/s^2 can she make it? If the answer is no, determine what acceleration she would have to have to be successful.

•• **24.** A hockey player is standing on his skates on a frozen pond when an opposing player moving with a uniform speed of 12 m/s skates by with the puck. After 3 s, the first player makes up his mind to chase his opponent. If he accelerates uniformly at 4 m/s^2, (a) how long does it take him to catch his opponent and (b) how far has he traveled in this time? (Assume the player with the puck remains in motion at constant speed.)

Section 2.5 Freely Falling Bodies

25. A ball is thrown vertically upward with a speed of 25 m/s. (a) How high does it rise? (b) How long does it take to reach its highest point? (c) How long does it take to hit the ground from its highest point? (d) What is its speed when it returns to the level from which it started?

26. A pebble is dropped into a deep water well, and the sound of the splash as the pebble reaches the bottom of the well is heard 3 s later. From this information, estimate the depth of the well. What assumptions did you have to make in your calculations?

27. A juggler performs in a room with a ceiling 2 m above hand level. (a) What is the maximum upward speed she can give a ball without letting the ball hit the ceiling? (b) How long is the ball in the air before it is caught?

28. A foul ball is hit into the stands at a baseball game. The ball rises to a height of 40 m and is caught by a fan at a height of 30 m as it drops back toward the field. What is its velocity in the vertical direction just before it is caught?

• **29.** A parachutist descending at a speed of 10 m/s drops a camera from an altitude of 50 m. (a) How long does it take the camera to reach the ground? (b) What is the velocity of the camera just before it hits the ground?

• **30.** A rocket moves upward starting from rest with an acceleration of 29.4 m/s^2 for 4 s. It runs out of fuel at the end of this 4 s, and continues to move upward, acted upon only by gravity. How high does it rise?

• **31.** The tallest volcano in the solar system is the 24-km tall Martian volcano, Olympus Mons. Assume an astronaut drops a ball off the rim of the crater and that the acceleration of gravity remains constant throughout its 24 km fall at a value of 0.38 the acceleration due to gravity on earth. (We assume the crater is as deep as the volcano is tall, which is not the case for real volcanoes.) Find (a) the time for the ball to reach the crater floor and (b) the velocity with which it hits. (In light of your answer for the velocity, does it seem reasonable that air resistance, even in Mars' thin atmosphere, can really be neglected in this problem?)

•• **32.** A "superball" is dropped from a height of 2 m above the ground. On the first bounce the ball reaches a height of 1.85 m, where it is caught. Find the velocity of the ball (a) just as it makes contact with the ground and (b) just as it leaves the ground on the bounce. (c) Neglecting the time the ball is in contact with the ground, find the total time required for the ball to move from the dropping point to the point where it is caught.

Section 2.6 Velocity and Acceleration in Two Dimensions and Section 2.7 Projectile Motion

33. A punter kicks a football at an angle of 30° with the horizontal at an initial speed of 20 m/s. Where should a punt returner position himself to catch the ball just before it strikes the ground?

34. Tom the cat is chasing Jerry the mouse across a table surface 1.5 m high. Jerry steps out of the way at the last second, and Tom slides off the edge of the table at a speed of 5 m/s. Where will Tom strike the floor and what velocity components will he have just before he hits?

35. In an ideal punt, the football has a "hangtime" (total time in the air) of 5 s. If a punter kicks the ball at an angle of 45° with the horizontal, what must be the initial velocity of the ball to achieve this?

• **36.** A brick is thrown upward from the top of a building at an angle of 25° to the horizontal and with an initial speed of 15 m/s. If the brick is in flight for 3 s, how tall is the building?

• **37.** A ball is thrown straight upward and returns to the thrower's hand after 3 s in the air. A second ball is thrown at an angle of 30° with the horizontal. What speed must the second ball be thrown so that it reaches the same height as the one thrown vertically?

•• **38.** A car is parked on a cliff overlooking the ocean on an incline that makes an angle of 37° below the horizontal. The negligent driver leaves the car in neutral, and the emergency brakes are defective. The car rolls from rest down the incline with a constant acceleration of 4 m/s^2 and travels 50 m to the edge of the cliff. The cliff is 30 m above the ocean. Find (a) the position of the car relative to the base of the cliff when it lands in the ocean, and (b) the time the car is in the air.

ADDITIONAL PROBLEMS

39. A baseball is thrown by an outfielder at a speed of 20 m/s at an angle of 35° above the horizontal. Find (a) the component of velocity along the ground at the instant the ball is released and (b) the upward component of velocity at this instant.

40. The multiflash photograph of Figure 2.22 shows two golf balls released simultaneously. The time interval between flashes is (1/30) s, and the white parallel lines were placed 15 cm apart. (a) Find the speed at which the right ball was projected, and (b) show that both balls should be expected to reach the floor simultaneously.

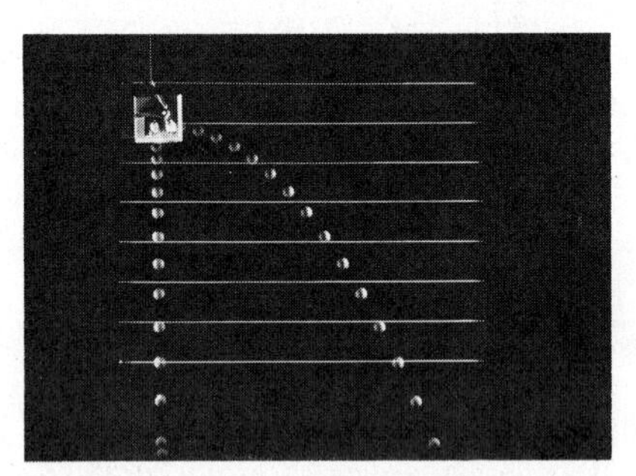

FIGURE 2.22 Problem 40

41. A bullet is fired through a board 10 cm thick in such a way that the bullet's line of motion is perpendicular to the face of the board. If the initial speed of the bullet is 400 m/s and it emerges from the other side of the board with a speed of 300 m/s, find (a) the deceleration of the bullet as it passes through the board and (b) the total time the bullet is in contact with the board.

42. A ball is thrown downward from the top of a tall cliff with an initial speed of 10 m/s. Determine the velocity and acceleration of the ball at $t = 2$ s.

43. If a person can jump a horizontal distance of 3 m on the earth, how far could the person jump on the moon where the acceleration due to gravity is $g/6$, where $g = 9.8$ m/s^2. Repeat for Mars where the acceleration due to gravity is $0.38g$.

• **44.** The velocity of an object is shown as a function of time in Figure 2.23. Find the average acceleration of the object in the intervals (a) 0 to 1 s, (b) 1 s to 3 s, and (c) 3 s to 4 s. (d) Find the instantaneous acceleration at 0.5 s, 2 s, and 3.5 s.

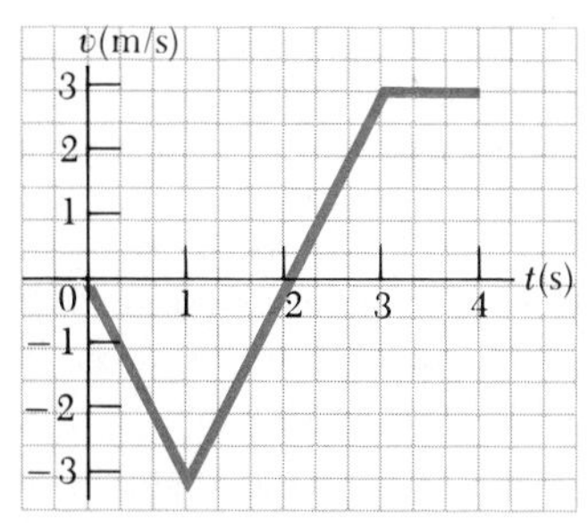

FIGURE 2.23 Problem 44

• **45.** One swimmer in a relay race has a 0.5 s lead and is swimming at a constant speed of 4 m/s. He has 50 m to swim before reaching the end of the pool. A second swimmer moves in the same direction as the leader. What constant speed must the second swimmer have in order to catch up to the leader at the end of the pool?

• **46.** A woman is reported to have fallen 44 m from the 17th floor of a building and to have landed on a metal ventilator box, which she crushed to a depth of 46 cm. She suffered only minor injuries. Neglecting air resistance, calculate (a) the speed of the woman just before she collided with the box, and (b) her deceleration while in contact with the box, and (c) the time it took to crush the box.

• **47.** A ball is thrown upward from the ground with an initial speed of 25 m/s at the same instant that a ball is dropped from a building 15 m high. At what time will the balls be at the same height?

• **48.** A baseball is hit in such a way that the ball just clears a wall 3 m tall 130 m from home plate. The ball leaves the bat at an angle of 35° to the horizontal and 1 m above the ground. Find its initial velocity.

•• **49.** A daredevil decides to jump a canyon of width 10 m. To do so, he drives a motorcycle up an incline sloped at an angle of 15°. What minimum speed must he have in order to clear the canyon?

• **50.** A ranger in a National Park is driving at 35 mph when a deer jumps into the road 200 ft ahead of her position. After a reaction time of t s, she applies the brakes and decelerates at $a = -9.0$ ft/s^2. What is the maximum reaction time allowed if she is to avoid hitting the deer?

• **51.** A mountain climber stands at the top of a 50-m cliff that overhangs a calm pool of water. He throws two stones vertically downward 1 s apart and observes that they cause a single splash. The first stone has an initial velocity of 2 m/s. (a) At what time after release of the first stone will the two stones hit the water? (b) What initial velocity must the second stone have if they are to hit simultaneously? (c) What will the velocity of each stone be at the instant they hit the water?

• **52.** A person sees a lightning bolt passing close to an airplane flying off in the distance. The person hears thunder 5 s after seeing the bolt and sees the airplane overhead 10 s after hearing the thunder. If the speed of sound in air is 1100 ft/s, (a) find the distance the airplane is from the person at the instant of the bolt. (Neglect the time it takes the light to travel from the bolt to the eye.) (b) Assuming the plane travels with a constant speed toward the person, find the velocity of the airplane. (c) Look up the speed of light in air and defend the approximation used in (a).

• **53.** A daredevil is shot out of a cannon at 45° to the horizontal with an initial speed of 25 m/s. A net is located at a horizontal distance of 50 m from the cannon. At what height above the cannon should the net be placed in order to catch the daredevil?

• **54.** A dart gun is fired while being held horizontally at a height of 1 m above ground level. With the gun at rest relative to the ground, the dart from the gun travels a horizontal distance of 5 m. A child holds the same gun in a horizontal position while sliding down a 45° incline at a constant speed of 2 m/s. How far will the dart travel if the gun is fired when it is 1 m above the ground?

•• **55.** A rocket is launched at an angle of 53° above the horizontal with an initial speed of 100 m/s. It moves along its initial line of motion with an acceleration of 30 m/s^2 for 3 s. At this time its engines fail and the rocket proceeds to move as a free body. Find (a) the maximum altitude reached by the rocket, (b) its total time of flight, and (c) its horizontal range.

•• **56.** A sportscar enthusiast buys a super deluxe machine that can accelerate at the rate of 16 ft/s^2. She decides to test her car in a drag race with another speedster in a souped-up stock car. Both start from rest, but the experienced stock car driver leaves 1 s before the sportscar. If the stock car moves with a constant acceleration of 12 ft/s^2, find (a) the time it takes the sportscar to overtake it, (b) the distance the two travel before they are side by side, and (c) the velocities of both cars at the instant they are side by side.

•• **57.** A projectile is fired with an initial speed of v_0 at an angle, θ_0, to the horizontal, as in Figure 2.16. When it reaches its peak, it has (x, y) coordinates given by $(R/2, h)$, and when it strikes the ground, its coordinates are $(R, 0)$, where R is called the horizontal range. (a) Show that it reaches a maximum height, h, given by

$$h = \frac{v_0^2 \sin^2 \theta_0}{2g}$$

(b) Show that its horizontal range is given by

$$R = \frac{v_0^2 \sin 2\theta_0}{g}$$

•• **58.** The determined coyote is out once more to try to capture the elusive roadrunner. The coyote wears a new pair of Acme power roller skates, which provide a constant horizontal acceleration of 15 m/s^2, as shown in Figure 2.24. The coyote starts off at rest 70 m from the edge of a cliff at the instant the roadrunner zips by in the direction of the cliff. (a) If the roadrunner moves with constant speed, determine the minimum speed he must have in order to reach the cliff before the coyote. (b) If the cliff is 100 m above the base of a canyon, determine where the coyote lands in the canyon. (Assume that his skates are still in operation when he is in "flight" and that his horizontal component of acceleration remains constant at 15 m/s^2.

FIGURE 2.24 Problem 58

•• **59.** An ice sled powered by a rocket engine starts from rest on a large frozen lake and accelerates at 40 ft/s^2. After some time t_1 the rocket engine is shut down and the sled moves with constant velocity v for a time t_2. If the total distance traveled by the sled is 17,500 ft and the total time is 90 s, find (a) the times t_1 and t_2 and (b) the velocity v. If, at the 17,500-ft mark, the sled begins to decelerate at 20 ft/s^2, (c) what is the final position of the sled when it comes to rest and (d) how long does it take to come to rest?

•• **60.** A certain cable car in San Francisco can stop in 10 s when traveling at maximum speed. On one occasion, the driver sees a dog a distance d m in front of the car and slams on the brakes instantly. The car reaches the dog 8 s later, and the dog jumps off the track just in time. If the car travels 4 m beyond the position of the dog before coming to a stop, how far was the car from the dog? (*Hint:* You will need three equations.)

•• **61.** An enemy ship is on the east side of a mountainous island as shown in Figure 2.25 (see below). The enemy ship can maneuver to within 2500 m of the 1800-m high mountain peak and can shoot projectiles with an initial speed of 250 m/s. If the western shoreline is horizontally 300 m from the peak, what are the distances from the western shore at which a ship can be safe from the bombardment of the enemy ship?

•• **62.** In a very popular lecture demonstration, a projectile is fired at a falling target as in Figure 2.26. The projectile leaves the gun at the same instant that the target dropped from rest. Assuming that the gun is initially aimed at the target, show that the projectile will hit the target. (One restriction of this experiment is that the projectile must reach the target before the target strikes the floor.)

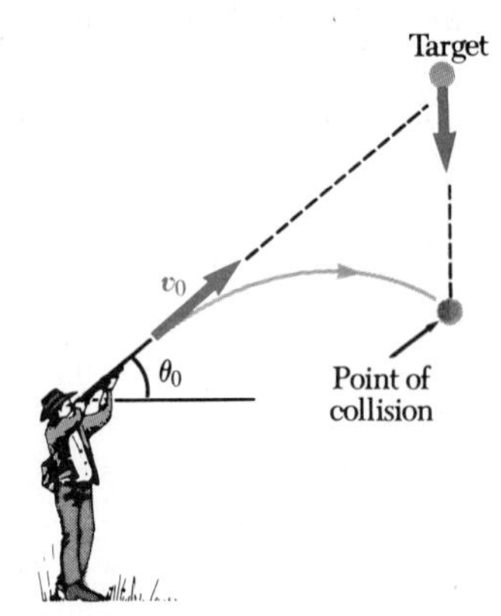

FIGURE 2.26 Problem 62

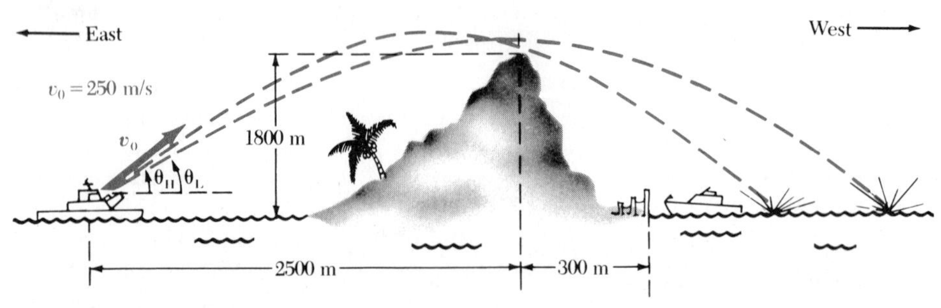

FIGURE 2.25 Problem 61

CHAPTER 3

The Laws of Motion

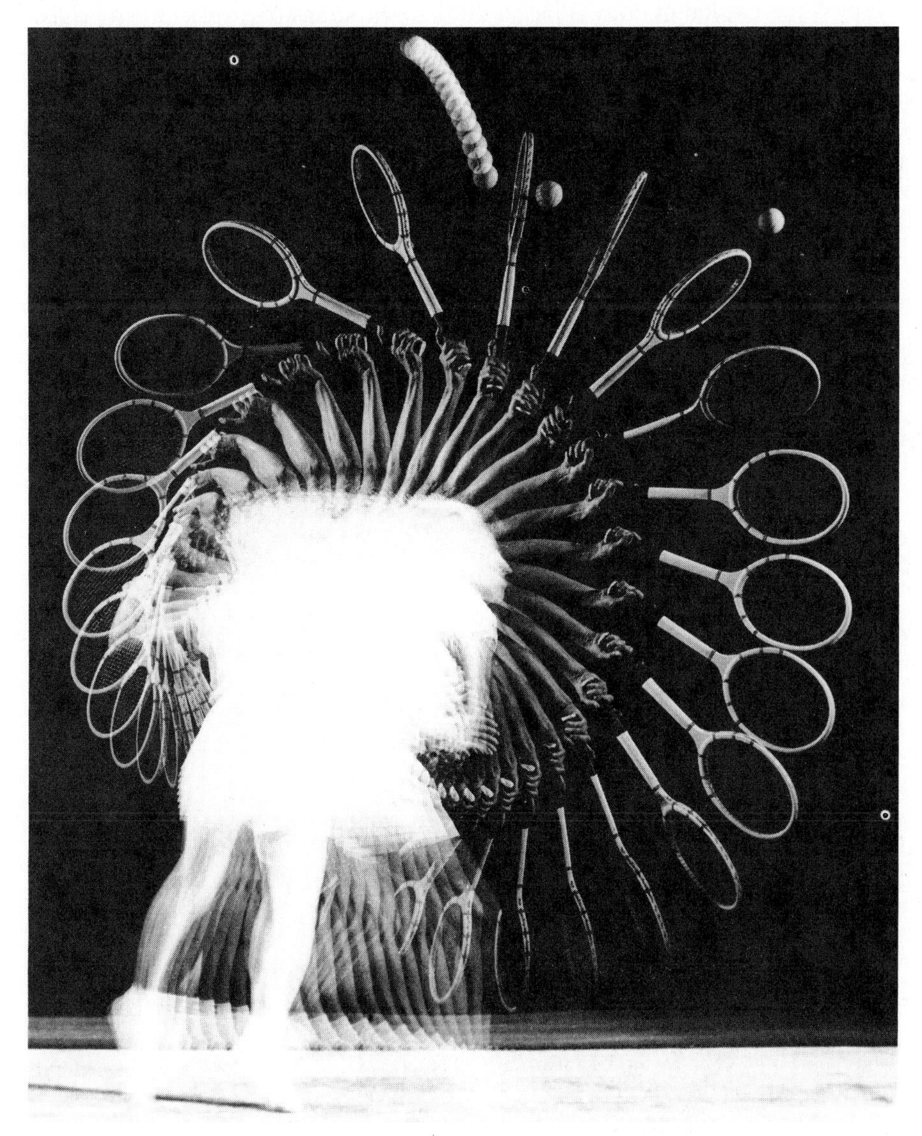

Multiflash exposure of a tennis player serving the ball. Such photographs can be used to study the quality of sports equipment and the performance of an athlete. (Photo courtesy of Dr. Harold T. Edgerton, M.I.T., Cambridge, MA)

3.1 INTRODUCTION TO CLASSICAL MECHANICS

Isaac Newton is considered by many to be the most brilliant scientist who ever lived. His genius came into full bloom in his early twenties when, during one 18-month period, he formulated the law of gravitation, invented calculus, and proposed theories of light and color. On his tomb in Westminster Abbey is the following epitaph: "Mortals, congratulate yourselves that so great a man lived for the honor of the human race." Newton's contributions to several branches of physics are truly remarkable. In this chapter we shall investigate only a few of the contributions he made to the field of mechanics. It is in this area that Newton is most widely known today, primarily because of his three laws of motion and his theory of universal gravitation.

As background to this chapter, recall that in our previous work we described the motion of objects based on the definitions of displacement, velocity, and acceleration. However, we have not yet been able to answer specific questions related to the nature of motion, questions such as "What causes motion?" and "Why do some objects accelerate at a greater rate than others?" In this chapter, we shall use the concepts of force and mass to describe the change in motion of objects. We shall see that it is possible to describe the acceleration of an object in terms of its mass and the external force acting on it. The mass of an object is a measure of the object's inertia, that is, the tendency of the object to resist an acceleration when a force acts on it.

We shall also discuss *force laws*, which describe how to calculate such quantities as the force on an object or its acceleration. We shall see that, although the force laws are rather simple in form, they successfully explain a wide variety of phenomena and experimental observations. These force laws, together with the laws of motion, are the foundations of classical mechanics.

The purpose of classical mechanics is to provide a connection between the acceleration of a body and the external forces acting on it. Keep in mind that classical mechanics deals with objects that (a) are large compared with the dimensions of atoms ($\approx 10^{-10}$ m) and (b) move at speeds that are much less than the speed of light (3×10^8 m/s). If either criterion is violated, the equations and results of this chapter do not apply. Our study of relativity and quantum mechanics in later chapters will enable us to cover the complete range of speeds and sizes.

3.2 THE CONCEPT OF FORCE

Everyone has a basic understanding of the concept of force from everyday experiences. When you push or pull an object, you exert a force on it. For example, you exert a force on a ball when you throw it or kick it. In each of these instances the word *force* is associated with the result of muscular activity and some form of motion. Although forces can cause motion, it does not necessarily follow that forces acting on an object will always cause it to move. For example, as you sit reading this book, the force of gravity acts on your body and yet you remain stationary. Also, you can push on a wall and not move it.

What force (if any) causes a distant star to move freely through space? Newton answered such questions by stating that the acceleration of an object is caused by forces. Therefore, if an object moves with uniform motion (constant velocity), no force is required to maintain the motion.

Now consider a situation in which several forces act simultaneously on an object. In this situation, the object will accelerate only if the *net* force acting on it is not equal to zero. We shall often refer to the net force as the *resultant* force or the unbalanced force. *If the net force on an object is zero, the acceleration is zero and the velocity of the object remains constant.* That is, if the net force acting on the object is zero, either the object will be at rest or it will move with constant velocity.

Kinds of Forces

If you pull on a coiled spring, as in Figure 3.1a, the spring stretches. If the spring is calibrated, the distance that it stretches can be used to measure the strength of the force. If you pull hard enough on a cart, as in Figure 3.1b, the cart will move. Finally, when a football is kicked, as in Figure 3.1c, it is deformed and set into motion. These are all examples of a class of forces called **contact forces** because they arise as the result of physical contact between two objects. Other examples of contact forces are the force that a gas exerts on the walls of a container (the result of gas molecules colliding with the walls) and the force of our feet on the floor.

Another class of forces are known as **action-at-a-distance forces.** These forces do not involve physical contact between an object and its surroundings, but act through empty space. The force of gravitational attraction between

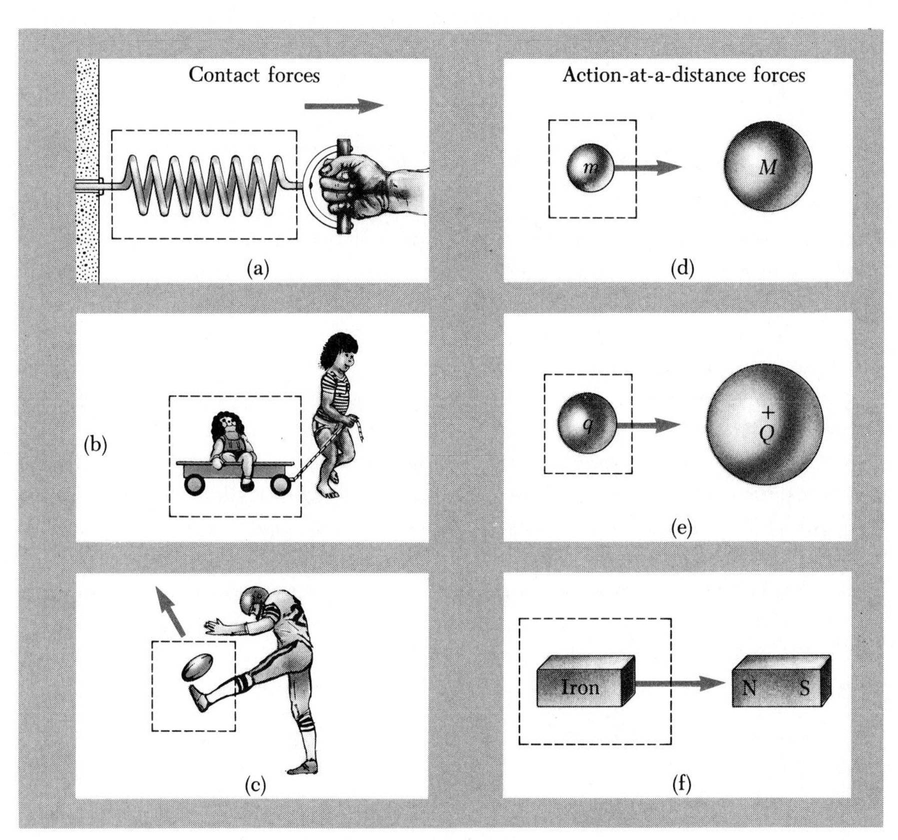

FIGURE 3.1 Some examples of forces applied to various objects. In each case a force is exerted on the particle or object within the dashed boxed area. Some agent or object external to the boxed area provides the force on the object.

two objects is an example of this class of force. Early scientists, including Newton himself, were uneasy with the concept of a force acting at a distance. To overcome this conceptual problem, Michael Faraday (1791–1867) introduced the concept of a *field*. According to this approach, when a mass m_1 is placed at some point P near a mass m_2, one can say that m_1 interacts with the gravitational field that exists at P produced by m_2. In Chapter 16, we shall see that the field concept is also useful in describing electric interactions between charged particles.

We should mention that the distinction between contact forces and action-at-a-distance forces is not as sharp as you may have been led to believe by the above discussion. At the atomic level, the so-called contact forces are actually due to repulsive electric forces between charges, which themselves are action-at-a-distance forces.

Now let us consider the force of attraction between any two objects. This force of attraction is called the gravitational force and is pictured in Figure 3.1d. This force keeps objects bound to the earth and gives rise to what we commonly call the weight of an object. The planets of our solar system move in their elliptical orbits under the action of gravitational forces exerted on them by the sun. Another common example of an action-at-a-distance force is the electric force that one electric charge exerts on another electric charge, as in Figure 3.1e. These charges might be an electron and proton pair forming the hydrogen atom. A third example of an action-at-a-distance force is the force that a bar magnet exerts on a piece of iron, as shown in Figure 3.1f.

3.3 NEWTON'S FIRST LAW

Before we begin our discussion, try the following simple experiment: slide a book across a table and observe what happens to it. You know what happens before you do the experiment: the book stops sliding after only a very short distance. Before about 1600, scientists chose to concentrate on this aspect of matter. They felt that the natural state of matter was the state of rest, a reasonable assumption in light of our experiment. Now, however, imagine shoving this same book across a smooth, highly polished floor. The book will again come to rest, but not as quickly as before. Finally, let your mind leap to the possibility of a floor so highly polished that friction is completely absent. Under these conditions, could it be that the book would slide forever? It was Galileo who developed such thought experiments as these and concluded that scientists were considering motion from an unproductive viewpoint. His conclusion was that it is not the nature of matter to stop once set in motion; instead, it is its nature to resist deceleration and acceleration.

This new approach to motion was later formalized by Newton in a form that has come to be known as **Newton's first law of motion:**

An object at rest will remain at rest and an object in motion will continue in motion with a constant velocity (that is, constant speed in a straight line) unless it experiences a net external force (called the resultant force).

In simpler terms, you can say that when the resultant force on a body is zero, its acceleration is zero. That is, when $\Sigma \boldsymbol{F} = 0$, then $\boldsymbol{a} = 0$. From the first law, we conclude that an isolated body (a body that does not interact with its

Isaac Newton (1642–1727), an English physicist and mathematician, was one of the most brilliant scientists in history. Before the age of 30, he formulated the basic concepts and laws of mechanics, discovered the law of universal gravitation, and invented the mathematical methods of calculus. As a consequence of his theories, Newton was able to explain the motion of the planets, the ebb and flow of the tides, and many special features of the motion of the moon and earth. He also interpreted many fundamental observations concerning the nature of light. His contributions to physical theories dominated scientific thought for two centuries and remain important today. (Courtesy AIP Niels Bohr Library)

environment) is either at rest or moving with constant velocity. Actually, Newton was not the first to state this law. Several decades earlier, in his publication *The New Sciences*, Galileo wrote, "Any velocity once imparted to a moving body will be rigidly maintained as long as the external causes of retardation are removed."

A common example of uniform motion on a nearly frictionless surface is the motion of a light disk on a column of air, as in Figure 3.2. If the disk is given an initial velocity, it will coast a great distance before coming to rest. This idea is used in the popular game of air hockey, in which the disk makes many collisions with the sides of the table before coming to rest.

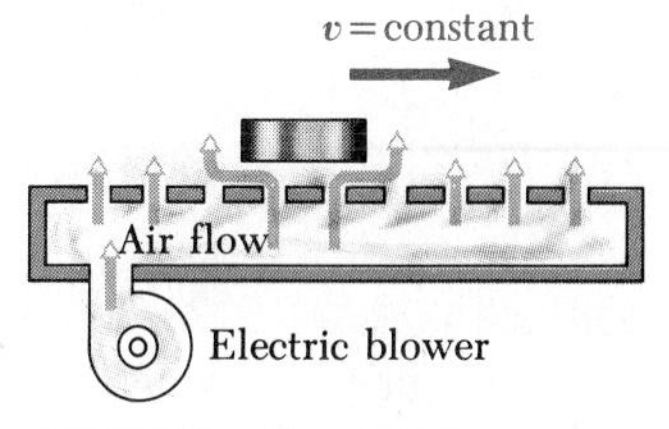

FIGURE 3.2 A disk moving on a column of air is an example of uniform motion, that is, motion in which the acceleration is zero and the velocity remains constant.

Finally, consider a moving spaceship far from any planets or other matter. The spaceship requires some propulsion system to change its velocity. However, if the propulsion system is turned off when the spaceship reaches a velocity $\boldsymbol{v}$, the spaceship will "coast" in space with the same velocity and the astronauts get a "free ride," that is, no propulsion system is required to keep them moving at the velocity $\boldsymbol{v}$.

Mass

If you attempt to change the state of motion of any object, the object will resist the change.

The resistance of an object to a change in its state of motion is called **inertia.**

For instance, imagine striking two objects with a golf club: a golf ball and a bowling ball. Both are initially at rest and, according to the first law, they will remain at rest unless a net force is applied. The golf club supplies this net force, but obviously the net force will have a much greater effect on the golf ball than on the bowling ball. The golf ball, if hit solidly, will fly many meters, whereas the bowling ball will barely move if the same force is applied. We conclude that the two balls have different amounts of inertia.

Mass is a term used to measure inertia, and the SI unit of mass is the kilogram. The greater the mass of a body, the less it will accelerate (change its state of motion) under the action of an applied force. For example, if a given force acting on a 3-kg mass produces an acceleration of 4 m/s^2, the same force applied to a 6-kg mass will produce an acceleration of only 2 m/s^2. This idea will be used shortly to obtain a quantitative description of the concept of mass.

It is important to point out that mass should not be confused with *weight*. Mass and weight are two different quantities.

The **weight** of a body is equal to the force of gravity acting on the body and varies with location.

Astronaut Edgar D. Mitchell walking on the moon following the Apollo 14 lunar landing. The weight of this astronaut on the moon is less than on earth, but his mass remains the same. (Courtesy of NASA)

For example, a person who weighs 180 lb on earth weighs only about 30 lb on the moon. On the other hand, the mass of a body is the same everywhere, regardless of location. An object having a mass of 2 kg on earth also has a mass of 2 kg on the moon.

Mass is an inherent property of a body and is independent of the body's surroundings and of the method used to measure the mass. It is an experimental fact that mass is a scalar quantity, since its value is independent of the direction of motion and independent of the coordinate system used in describing the motion. Finally, mass is a quantity that obeys the rules of ordinary arithmetic. That is, several masses can be combined in a simple numerical fashion. For example, if you combine a 3-kg mass with a 5-kg mass, the total mass is 8 kg. This can be verified experimentally by comparing the acceleration of each object produced by a known force with the acceleration of the combined system using the same force.

In the introduction to this chapter, we pointed out that much of the present discussion must be modified when we treat objects moving at speeds approaching the speed of light. One of these modifications is concerned with the nature of mass, which we assume to be constant in Newtonian mechanics. In our study of special relativity (Chapter 28), we shall find that the mass of an object increases with increasing speed. This effect, however, is perceptible only for speeds approaching 3×10^8 m/s, the speed of light.

3.4 NEWTON'S SECOND LAW

Newton's first law explains what happens to an object when the resultant force acting on it is zero. In such instances we find that the object either stays at rest or moves in a straight line with constant velocity. Newton's second law answers the question of what happens to an object that has a nonzero resultant force acting on it.

In order to gain some insight into what happens when a net force is applied to an object, consider a situation in which you are pushing a block of ice across a polished surface, smooth enough to make frictional forces negligible. When you exert a certain force on the block of ice, you find that it moves with an acceleration of, say, 2 m/s^2. If you push twice as hard, you find that the acceleration also doubles. Pushing three times as hard triples the acceleration, and so on. From observations such as this, we can conclude that *the acceleration of an object is directly proportional to the resultant force acting on it.* Common experience with pushing objects should indicate to you that the mass of the object also affects its acceleration. This can be seen as follows. Imagine that you stack identical blocks of ice on top of each other while pushing the stack with the same force. You will find that if the force produces an acceleration of 2 m/s^2 when you push one block, the acceleration will drop to half this value when two blocks are pushed, to one third this value for three blocks, and so on. From this we see that *the acceleration of an object is inversely proportional to its mass.* **Newton's second law** summarizes these observations:

The acceleration of an object is directly proportional to the resultant force acting on it and inversely proportional to its mass. The direction of the acceleration is in the direction of the resultant force.

In equation form, we can state Newton's second law as

$$\boxed{\sum \mathbf{F} = m\mathbf{a}} \qquad (3.1)$$

Newton's second law

where $\mathbf{a}$ is the acceleration of the object, m is its mass, and $\Sigma \mathbf{F}$ represents the *vector sum of all external forces acting on the object.* You should note that, because this is a vector equation, it is equivalent to the following three component equations:

$$\sum F_x = ma_x \qquad \sum F_y = ma_y \qquad \sum F_z = ma_z \qquad (3.2)$$

Note that if the resultant force is zero, then $\mathbf{a} = 0$, which corresponds to the equilibrium situation where $\mathbf{v}$ is either constant or zero. Hence, *the first law of motion is a special case of the second law.*

Units of Force and Mass

The SI unit of force is the **newton**, which is defined as the force that, when acting on a 1-kg mass, produces an acceleration of 1 m/s^2.

From this definition and Newton's second law, we see that the newton can be expressed in terms of the fundamental units of mass, length, and time:

$$\boxed{1 \text{ N} \equiv 1 \text{ kg} \cdot \text{m/s}^2} \qquad (3.3)$$

Definition of newton

The unit of force in the cgs system is called the **dyne** and is defined as the force that, when acting on a 1-g mass, produces an acceleration equal to 1 cm/s^2:

$$\boxed{1 \text{ dyne} \equiv 1 \text{ g} \cdot \text{cm/s}^2} \qquad (3.4)$$

Definition of dyne

In the conventional (British engineering) system, the unit of force is the **pound,** defined as the force that, when acting on a 1-slug mass, produces an acceleration of 1 ft/s^2:

Definition of pound

$$1 \text{ lb} \equiv 1 \text{ slug} \cdot \text{ft/s}^2 \tag{3.5}$$

The **slug** is the unit of mass in the British engineering system and is that system's counterpart of the SI kilogram. When we speak of going on a diet to lose a few pounds, we really mean that we want to lose a few slugs; that is, we want to reduce our mass. When we lose those few slugs, the force of gravity (pounds) on our reduced mass decreases and that is how we "lose a few pounds." Since most of the calculations we shall carry out in our study of classical mechanics will be in SI units, the slug will seldom be used in this text.

Since 1 kg $= 10^3$ g and 1 m $= 10^2$ cm, it follows that 1 N $= 10^5$ dynes. It is left as an exercise to show that 1 N $=$ 0.225 lb. The units of mass, acceleration, and force are summarized in Table 3.1.

Weight

We are well aware of the fact that bodies are attracted to the earth. The force exerted by the earth on a body is called the **weight** of the body, $\boldsymbol{w}$. This force is directed toward the center of the earth. Thus, weight is a vector while mass is a scalar.

We have seen that a freely falling body experiences an acceleration $\mathbf{g}$ acting toward the center of the earth. Applying Newton's second law to the freely falling body, with $\boldsymbol{a} = \mathbf{g}$ and $\boldsymbol{F} = \boldsymbol{w}$, gives

Weight

$$\boxed{\boldsymbol{w} = m\mathbf{g}} \tag{3.6}$$

Since it depends on $\mathbf{g}$, weight varies with geographic location. Bodies weigh less at higher altitudes than at sea level. This is because g decreases with increasing distance from the center of the earth. Hence, weight, unlike mass, is not an inherent property of a body. For example, if a body has a mass of 70 kg, then the magnitude of its weight in a location where $g = 9.80$ m/s^2 is $mg = 686$ N (about 154 lb). At the top of a mountain, where g might be 9.76 m/s^2, say, this weight would be 683 N. This corresponds to a decrease in weight of about 0.4 lb. Therefore, if you want to lose weight without going on a diet, move to the top of a mountain or weigh yourself at an altitude of 30,000 ft during a flight on a jet airplane.

TABLE 3.1 Units of Mass, Acceleration, and Force

System	Mass	Acceleration	Force
SI	kg	m/s^2	N = kg · m/s^2
cgs	g	cm/s^2	dyne = g · cm/s^2
British engineering (conventional)	slug	ft/s^2	lb = slug · ft/s^2

Note: 1 N $= 10^5$ dyne $=$ 0.225 lb.

An astronaut "floating" in a gravity-free environment of a space-shuttle cannot tell whether he is floating in free space or is in free-fall. The fact that one cannot tell the difference between gravity and acceleration is the *equivalence principle.* (From the film "2001 A Space Odyssey," Museum of Modern Art Film Stills Archives)

3.5 NEWTON'S THIRD LAW

In Section 3.2 we found that a force may be exerted on an object when it comes into contact with some other object. For example, consider the task of driving a nail into a block of wood as illustrated in Figure 3.3. In order to accelerate the nail and drive it into the block, a net force must be supplied to the nail by the hammer. Newton, however, recognized that a single isolated force (such as the force on the nail by the hammer) cannot exist. Instead, *forces in nature always exist in pairs*. According to Newton's point of view, the hammer does exert a force on the nail, but the nail also exerts a force on the hammer. Furthermore, there is clearly a net force on the hammer because it rapidly slows down (decelerates) after coming into contact with the nail. Newton stated this type of event in terms of his third law of motion which may be stated as follows:

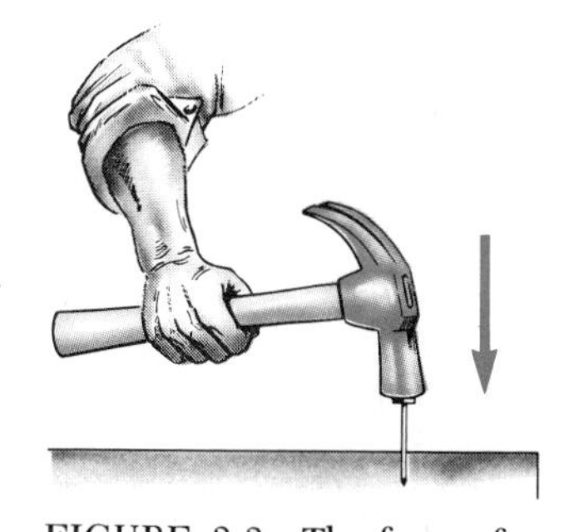

FIGURE 3.3 The force of the hammer on the nail is equal in magnitude and opposite in direction to the force of the nail on the hammer.

Newton's third law

If two bodies interact, the force exerted on body 1 by body 2 is equal to and opposite the force exerted on body 2 by body 1.

An alternative statement of this law is that *for every action there is an equal and opposite reaction*. It is important to note that action-reaction pairs are always exerted on *different* objects. In our hammer-nail example above, we could call the force the hammer exerts on the nail the action and the force that the nail exerts on the hammer the reaction. However, there is nothing special about this assignment of action-reaction pairs. If you choose to call the force the hammer exerts on the nail the reaction and the force that the nail exerts on the hammer the action, it is perfectly satisfactory to do so. Let us discuss some other examples of action-reaction pairs.

The force acting on a freely falling object is its weight $\mathbf{w}$. Let us call this the action. What is the reaction? Since the weight is the force exerted on the falling object by the earth, the reaction is the force exerted on the earth by

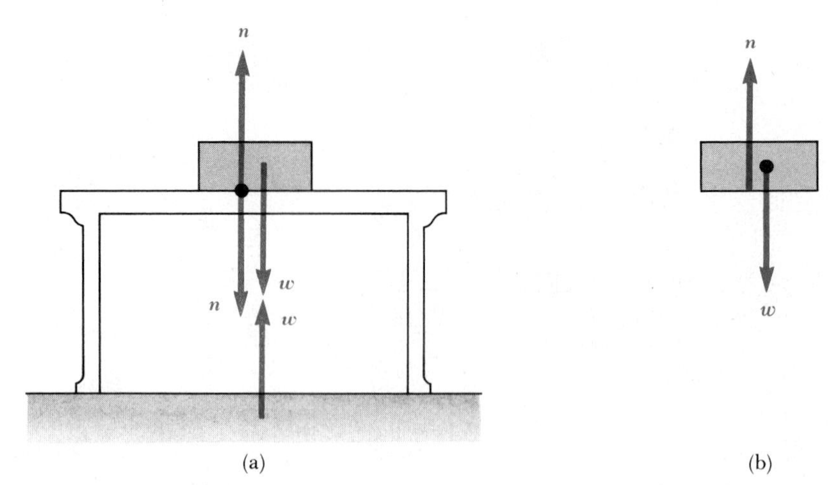

FIGURE 3.4 When a block is lying on a table, the forces acting on the block are the normal force, $\boldsymbol{n}$, and the force of gravity, $\boldsymbol{w}$, as illustrated in (b). The reaction to $\boldsymbol{n}$ is the force of the block on the table, $\boldsymbol{n}'$. The reaction to $\boldsymbol{w}$ is the force of the block on the earth, $\boldsymbol{w}'$.

the falling object. Thus, as the falling object accelerates toward the earth, the earth accelerates toward the object. However, since the earth has such a large mass compared to the object's mass, the acceleration of the earth due to this reaction force is negligibly small.

Figure 3.4a shows a block at rest on a table. There are two forces acting on the block indicated by $\boldsymbol{n}$ and $\boldsymbol{w}$ in the figure. The force $\boldsymbol{w}$ is the weight of the block and the force $\boldsymbol{n}$ is the upward force exerted on the block by the table. The force $\boldsymbol{n}$ is called the **normal** force. The word normal is used because the direction of $\boldsymbol{n}$ is perpendicular to the table surface and normal is the mathematical word for perpendicular. Because the block is not accelerating, we know from the first law that $\boldsymbol{w}$ and $\boldsymbol{n}$ must be equal in magnitude and opposite in direction. However, you should note that these two forces do not constitute an action-reaction pair because such forces always act on different objects. In our case, both $\boldsymbol{w}$ and $\boldsymbol{n}$ are exerted on the same object, the block. If we choose to call $\boldsymbol{w}$ the action, the reaction to this force is the force denoted as $\boldsymbol{w}'$ in Figure 3.4a, which is the force the block exerts on the earth. The reaction to $\boldsymbol{n}$, denoted by $\boldsymbol{n}'$ in the figure, is the force exerted on the table by the block. Note that the forces acting on the block are $\boldsymbol{w}$ and $\boldsymbol{n}$ as indicated in Figure 3.4b. When examining the motion of an object, we shall be interested only in such external forces.

You experience Newton's third law directly when you slam your fist against a wall or kick a football. You should try to identify the action and reaction forces in these cases.

3.6 SOME APPLICATIONS OF NEWTON'S LAWS

In this section we present some simple applications of Newton's laws to bodies moving under the action of constant external forces. As our model, we

Clint Eastwood experiences Newton's third law in a fight scene from "Hang 'em High." (Museum of Modern Art Film Stills Archives)

shall assume that the bodies behave as particles so that we need not worry about rotational motion. We shall also neglect the effects of friction for those problems involving motion. This is equivalent to stating that the surfaces are smooth. Finally, we shall neglect the mass of any ropes involved in a particular problem. In this approximation, the magnitude of the force exerted at any point along a rope is the same at all points along the rope.

When we apply Newton's laws to a body, we shall be interested only in those external forces that act on the body. For example, in Figure 3.4b the only external forces acting on the block are $\boldsymbol{n}$ and $\boldsymbol{w}$. The reactions to these forces, $\boldsymbol{n}'$ and $\boldsymbol{w}'$, act on the table and on the earth, respectively, and do not appear in Newton's second law as applied to the block.

Consider a block being pulled to the right on the smooth horizontal surface of a table, as in Figure 3.5a. Suppose you are asked to find the acceleration of the block and the force of the table on it. First, note that the horizontal force being applied to the block acts through the string. The force that the string exerts on the block is denoted by the symbol $\boldsymbol{T}$. The magnitude of $\boldsymbol{T}$ is called the **tension** in the string. A dashed circle is drawn around the block in Figure 3.5a to remind you to isolate the block from its surroundings. Since we are interested only in the motion of the block, we must be able to identify all external forces acting on it. These are illustrated in Figure 3.5b. In addition to the force $\boldsymbol{T}$, the force diagram for the block includes the weight, $\boldsymbol{w}$, and the normal force, $\boldsymbol{n}$. As before, $\boldsymbol{w}$ corresponds to the force of gravity pulling down on the block and $\boldsymbol{n}$ to the upward force of the table on the block. Such a force diagram is referred to as a **free-body diagram.** The construction of such a diagram is an important step in applying Newton's laws. The reactions to the forces we have listed, namely, the force of the string on the hand doing the pulling, the force of the block on the earth, and the force of the block on the table, are not included in the free-body diagram because they act on other bodies and not on the block.

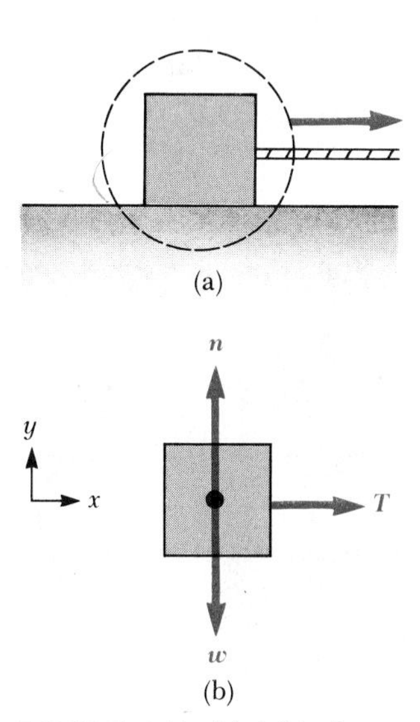

FIGURE 3.5 (a) A block being pulled to the right on a smooth surface. (b) The free-body diagram that represents the external forces on the block.

We are now in a position to apply Newton's second law to the block. First, we must choose an appropriate coordinate system. In this case it is convenient to use the coordinate system shown in Figure 3.5b, with the x axis as the horizontal and the y axis vertical. We are now ready to apply Newton's second law in the x direction, y direction, or both, depending on what we are asked to find in the problem. In addition, you may be able to make use of the equations of motion for constant acceleration found in Chapter 2. However, you should note that these equations can be used only when the acceleration is a constant. For example, if the force $\boldsymbol{T}$ in Figure 3.5 is constant, then it follows that the acceleration in the x direction is also a constant because $\boldsymbol{a}_x = \boldsymbol{T}/m$. Hence, if you need to find the displacement or the velocity of the object at some instant of time, the equations of motion with constant acceleration can be used. We shall illustrate the specifics of this approach through the example problems which follow this section.

Problem-solving Suggestion Because the basic steps discussed above for applying Newton's second law to specific problems are of such importance, let us repeat them here. We strongly suggest that you follow these steps when solving problems on your own.

1. Draw a simple, neat diagram of the system under consideration.
2. Isolate the object of interest whose motion is being analyzed. Draw a free-body diagram for this object, that is, a diagram showing all external forces acting on the object. For systems containing more than one object, draw a separate diagram for each object. Do not include forces that the object exerts on its surroundings.
3. Establish convenient coordinate axes for each body and find the components of the forces along these axes. Now, apply Newton's second law, $\Sigma \boldsymbol{F} = m\boldsymbol{a}$. Check your dimensions to make sure that all terms have units of force in one common system.
4. Solve the component equations for the unknowns. Remember that you must have as many independent equations as you have unknowns in order to obtain a complete solution.

EXAMPLE 3.1 A Boy and His Wagon

A boy pulls with a force of 20 N on the 300-N wagon of Figure 3.6. What acceleration does the wagon receive, and how far will it move in 2 s? Assume that the wagon starts from rest and assume there are no frictional forces opposing the motion of the wagon.

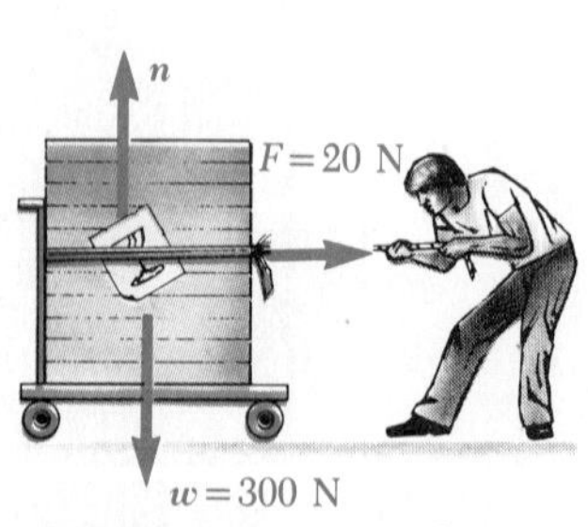

FIGURE 3.6 Example 3.1

Solution In order to apply Newton's second law to the wagon, we must first know its mass:

$$m = \frac{w}{g} = \frac{300\ \text{N}}{9.80\ \text{m/s}^2} = 30.6\ \text{kg}$$

The acceleration can now be found from the second law:

$$a_x = \frac{F_x}{m} = \frac{20\ \text{N}}{30.6\ \text{kg}} = 0.654\ \text{m/s}^2$$

Since the acceleration of the wagon is constant, we can find the distance it moves in 2 s using the relation $x = v_0t + \frac{1}{2}at^2$ with $v_0 = 0$:

$$x = \frac{1}{2}(0.654\ \text{m/s}^2)(2\ \text{s})^2 = 1.31\ \text{m}$$

EXAMPLE 3.2 Getting a Free Ride

An airplane of mass 1.5×10^4 kg tows a glider of mass 0.5×10^4 kg, as in Figure 3.7a. If the propellers produce a forward thrust of 7.5×10^4 N, what is the tension in the connecting cable (neglecting air resistance)?

Solution Problems involving connected bodies are usually most easily handled by considering the objects separately. In the present example, if we first focus our attention on the airplane, we find that two horizontal forces act on it, as shown in Figure 3.7b: ***F***, the forward thrust of the propellers, and ***T***, the tension in the connecting cable. Let us apply Newton's second law to the airplane, using m_p to denote its mass:

$$F - T = m_p a_x$$

$$(1) \qquad (7.5 \times 10^4 \text{ N}) - T = (1.5 \times 10^4 \text{ kg})a_x$$

With two unknowns and only a single equation, we must seek additional information in the form of another independent equation. This can be obtained by applying Newton's second law to the glider, as shown in Figure 3.7c. Since the tension, ***T***, in the cable is the only horizontal force acting on the glider, we get

$$(2) \qquad T = m_g a_x = (0.5 \times 10^4 \text{ kg})a_x$$

The equations for the plane and glider form a set of simultaneous equations that can be solved for T and a_x. Let us substitute the expression given in (2) for T into (1) to get

$$(7.5 \times 10^4 \text{ N}) - (0.5 \times 10^4 \text{ kg})a_x = (1.5 \times 10^4 \text{ kg})a_x$$

When the terms in a_x are combined, we get

$$(1.5 \times 10^4 \text{ kg} + 0.5 \times 10^4 \text{ kg})a_x = 7.5 \times 10^4 \text{ N}$$

$$a_x = \frac{7.5 \times 10^4 \text{ N}}{2.0 \times 10^4 \text{ kg}} = 3.75 \text{ m/s}^2$$

Substituting this value into (2) gives

$$T = (0.5 \times 10^4 \text{ kg})(3.75 \text{ m/s}^2) = 1.88 \times 10^4 \text{ N}$$

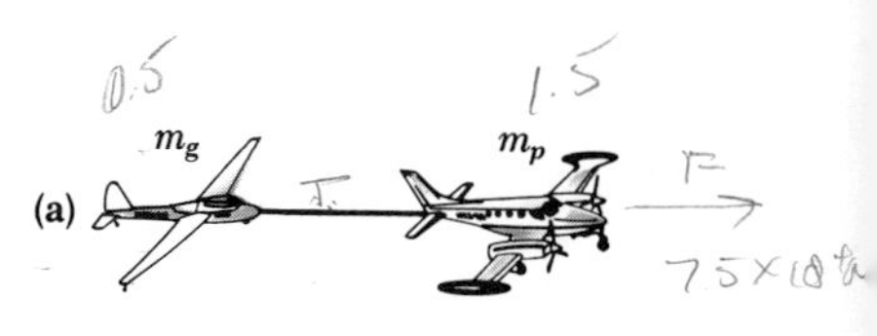

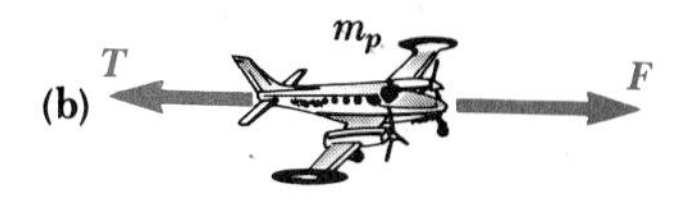

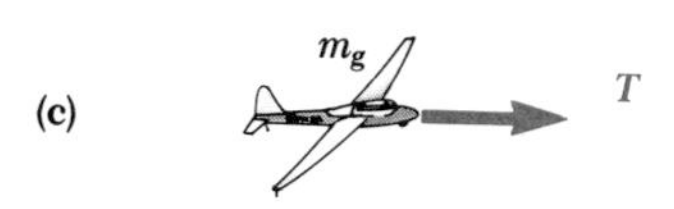

FIGURE 3.7 (Example 3.2) (a) A glider of mass m_g being towed by an airplane of mass m_p. (b) Horizontal forces on the airplane. (c) Horizontal forces on the glider. The vertical forces (not shown) on both airplane and glider are the weight (downward) and the net lift (upward). These vertical forces balance each other.

EXAMPLE 3.3 Two Connected Objects

Two objects of masses 10 kg and 5 kg are connected by a light string that passes over a light frictionless pulley, as in Figure 3.8a. The 5-kg object lies on a smooth incline of angle $\theta = 45°$. Find the acceleration of the two objects and the tension in the string.

Solution Since the two objects are connected by a string (which we assume doesn't stretch), they must have the same acceleration. The free-body diagrams for the two are shown in Figures 3.8b and 3.8c. Newton's second law applied to the 10-kg object in component form gives

$$(1) \qquad \sum F_x = 0$$

$$(2) \qquad \sum F_y = T - 10g = 10a$$

Note that a is positive (upward) only if $T > 10g$.

In describing the motion of the 5-kg object, we shall choose the positive x axis along the incline, as in Figure 3.8c. Newton's second law applied to this object in component form gives

$$(3) \qquad \sum F_x = 5g \sin 45° - T = 5a$$

$$(4) \qquad \sum F_y = n - 5g \cos 45° = 0$$

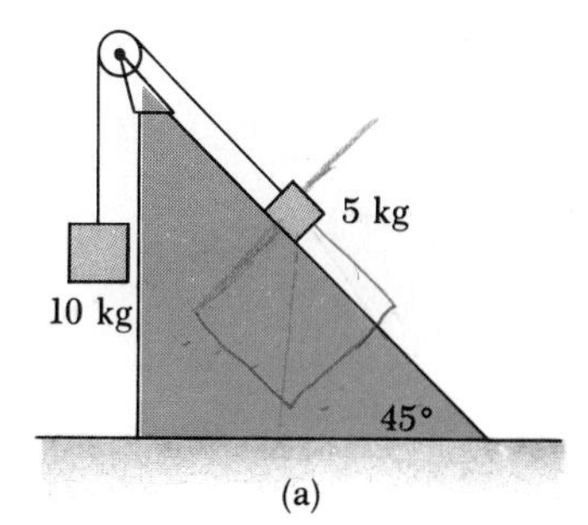

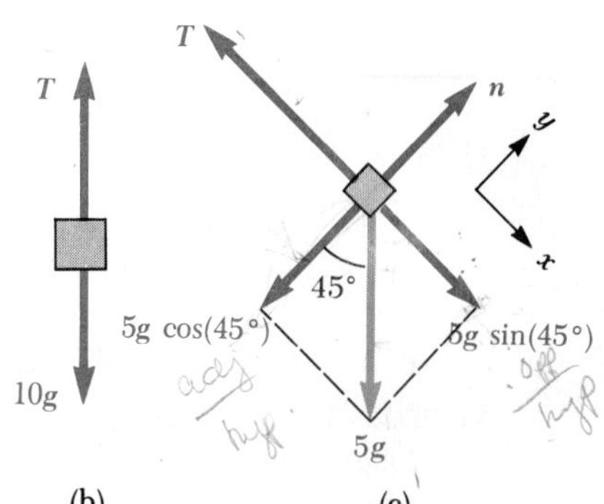

FIGURE 3.8 (Example 3.3) (a) Two objects connected by a light string passing over a frictionless pulley. (b) Free-body diagram for the 10-kg object. (c) Free-body diagram for the 5-kg object.

Equations (1) and (4) give no information on the acceleration. However, we can solve (2) and (3) simultaneously for a and T. By adding (2) and (3), we get

$$5g \sin 45° - 10g = 15a$$

$$a = \frac{(5 \text{ kg})(9.80 \text{ m/s}^2)(0.707) - (10 \text{ kg})(9.80 \text{ m/s}^2)}{15 \text{ kg}}$$

$$= -4.22 \text{ m/s}^2$$

The negative sign for a simply means that the acceleration is in the negative x direction. That is, the 10-kg object accelerates *downward*, whereas the 5-kg one accelerates *up* the incline.

Now that we know the acceleration, we can use Equation (2) to determine that $T = 55.8$ N.

EXERCISE Find the acceleration of the two objects if their positions are interchanged.

Answer 1.3 m/s² down the incline.

EXAMPLE 3.4 Block on a Smooth Incline

A block of mass m is placed on a smooth inclined plane of angle θ, as in Figure 3.9a. (a) Determine the acceleration of the block after it is released.

Solution The free-body diagram for the block is shown in Figure 3.9b. The only forces on the block are the normal force, $\mathbf{n}$, acting perpendicular to the plane, and the weight of the block, $\mathbf{w}$, acting vertically downward. It is convenient to choose the coordinate axes with x along the incline and y perpendicular to it. Then, we replace the weight vector by a component of magnitude $mg \sin \theta$ along the positive x axis and another of magnitude $mg \cos \theta$ in the negative y direction. Newton's second law applied in component form, with $a_y = 0$, gives

$$(1) \qquad \sum F_x = mg \sin \theta = ma_x$$

$$(2) \qquad \sum F_y = n - mg \cos \theta = 0$$

From (1) we see that the acceleration along the incline is provided by the component of weight down the incline:

$$(3) \qquad a_x = g \sin \theta$$

From (2) we conclude that the component of weight perpendicular to the incline is balanced by the normal force: $n = mg \cos \theta$. The acceleration given by (3) is independent of the mass of the block! It depends only on the angle of inclination and on g!

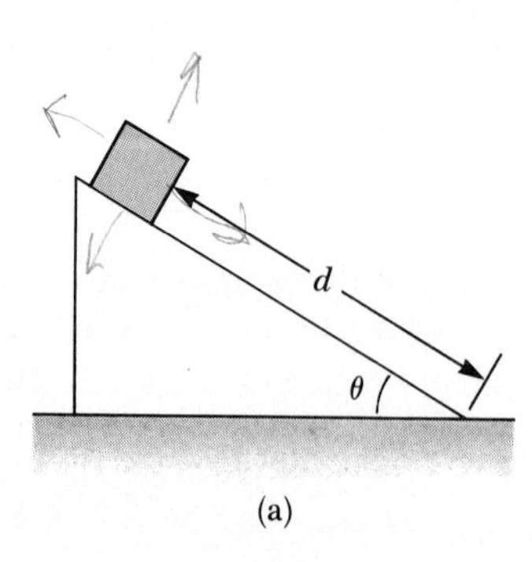

(a)

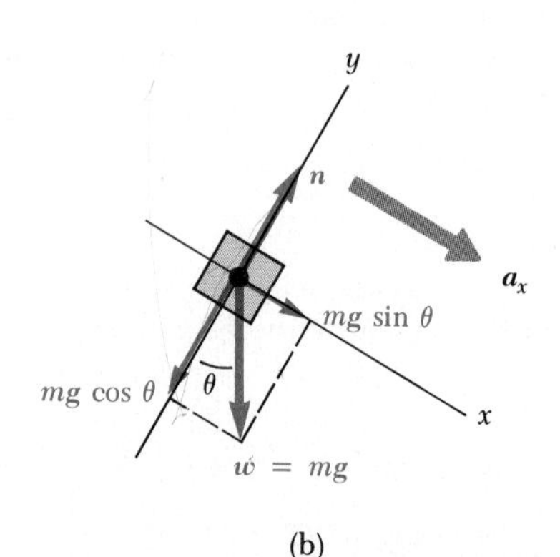

(b)

FIGURE 3.9 (Example 3.4) (a) A block sliding down a frictionless incline. (b) The free-body diagram for the block. Note that its acceleration along the incline is $g \sin \theta$. Also, the normal force, $\mathbf{n}$, is less than the weight of the block in this case. Why is this?

Special cases: We see that when $\theta = 90°$, $a = g$ and $n = 0$. This corresponds to the block in free fall. Also, when $\theta = 0$, $a_x = 0$ and $n = mg$ (its maximum value).

(b) Suppose the block is released from rest at the top and the distance from the block to the bottom is d. How long does it take the block to reach the bottom, and what is its speed just as it gets there?

Solution Since a_x = constant, we can apply the equation $x = v_{x0}t + \frac{1}{2}a_xt^2$ to the block. Since $x = d$ and $v_{x0} = 0$, we get

$$d = \frac{1}{2}a_xt^2$$

$$(4) \qquad t = \sqrt{\frac{2d}{a_x}} = \sqrt{\frac{2d}{g \sin \theta}}$$

Also, since $v_x^2 = v_{x0}^2 + 2a_xx$ and $v_{x0} = 0$, we find that

$$v_x^2 = 2a_xd$$

$$(5) \qquad v_x = \sqrt{2a_xd} = \sqrt{2gd \sin \theta}$$

Again t and v_x are independent of the mass of the block. This suggests a simple method of measuring g using an inclined air track or some other smooth incline. Simply measure the angle of inclination, the distance traveled by the block, and the time it takes to reach the bottom. The value of g can then be calculated from (4) and (5).

3.7 FORCES OF FRICTION

When a body is in motion on a rough surface or through a viscous medium such as air or water, there is resistance to the motion because of the interaction between the body and its surroundings. We call such resistance a **force of friction.** Forces of friction are very important in our everyday lives. They allow us to walk or run and are necessary for the motion of wheeled vehicles.

Consider a block on a table, as in Figure 3.10a. If we apply an external horizontal force $\boldsymbol{F}$ to the block, acting to the right, the block will remain stationary if $\boldsymbol{F}$ is not too large. The force that keeps the block from moving acts to the left and is called the **frictional force, f.** As long as the block is not moving, $f = F$. Since the block is stationary, we call this frictional force the **force of static friction, f_s.** Thus, if F is increased, the frictional force f also increases. Likewise, if F decreases, f also decreases. Experiments show that this force arises from the nature of the two surfaces; because of the roughness of the surface, contact is made only at a few points, as shown in the "magnified" view of the surfaces in Figure 3.10a. Actually, the frictional force is much more complicated than presented here since it ultimately involves forces between atoms or molecules where the surfaces are in contact.

If we increase the magnitude of $\boldsymbol{F}$ enough, as in Figure 3.10b, the block will eventually slip. When the block is on the verge of slipping, f_s is a maximum. When F exceeds $f_{s,max}$, the block moves and accelerates to the right. When the block is in motion, the retarding frictional force becomes less than $f_{s,max}$ (Fig. 3.10c). We call the retarding force for an object in motion the **force of kinetic friction, f_k.** The unbalanced force in the x direction, $\boldsymbol{F} - f_k$, produces an acceleration to the right. If $F = f_k$, the block moves to the right

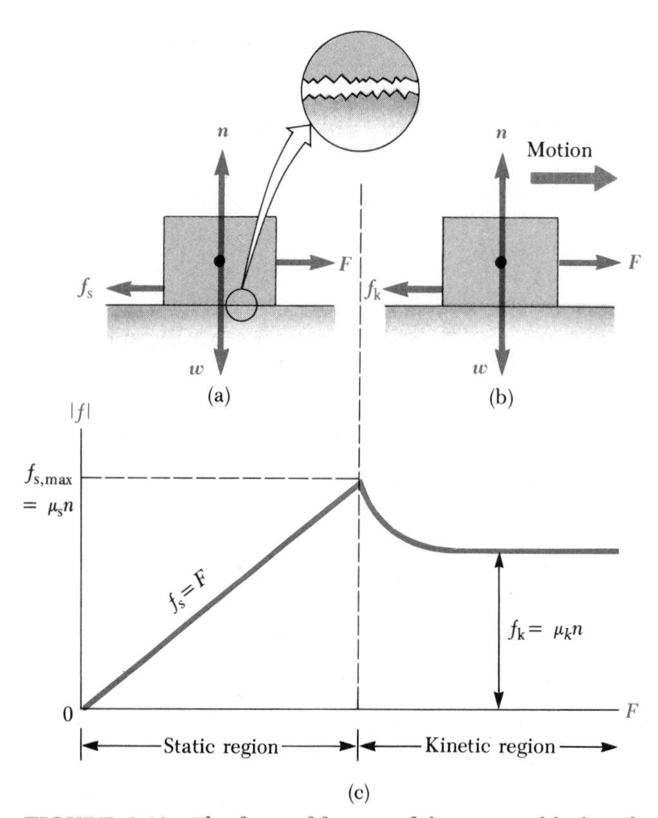

FIGURE 3.10 The force of friction, $\boldsymbol{f}$, between a block and a rough surface is opposite the applied force, $\boldsymbol{F}$. (a) The force of static friction equals the applied force. (b) When the magnitude of the applied force exceeds the maximum static friction force, $f_{s,max}$, the block accelerates to the right. (c) A graph of the applied force versus the magnitude of the frictional force. Note that $f_{s,max} > f_k$.

with constant speed. If the applied force is removed, then the frictional force acting to the left decelerates the block and eventually brings it to rest.

The kinetic friction is less than $f_{s,max}$ for the following reason. When the object is stationary, the contact points between the object and the surface are said to be cold-welded. While the object is in motion, these small welds can no longer form and the frictional force decreases.

Experimentally, one finds that both f_s and f_k are proportional to the normal force acting on the block and depend on the nature of the two surfaces in contact. The experimental observations can be summarized as follows:

1. The force of static friction between any two surfaces in contact is opposite the applied force and can have the values

$$f_s \leq \mu_s n \tag{3.7}$$

where the dimensionless constant μ_s is called the **coefficient of static friction** and n is the normal force. The equality in Equation 3.7 holds when the block is on the verge of slipping, that is, when $f_s = f_{s,max} \equiv \mu_s n$. The inequality holds when the applied force is less than this value. In general,

TABLE 3.2 Coefficients of Friction[a]

	μ_s	μ_k
Steel on steel	0.74	0.57
Aluminum on steel	0.61	0.47
Copper on steel	0.53	0.36
Rubber on concrete	1.0	0.8
Wood on wood	0.25–0.5	0.2
Glass on glass	0.94	0.4
Waxed wood on wet snow	0.14	0.1
Waxed wood on dry snow	—	0.04
Metal on metal (lubricated)	0.15	0.06
Ice on ice	0.1	0.03
Teflon on Teflon	0.04	0.04
Synovial joints in humans	0.01	0.003

[a]All values are approximate.

when an object is at rest relative to a surface, the frictional force always acts in such a way as to maintain a velocity of zero relative to the surface.

2. The force of kinetic friction is opposite the direction of motion and is given by

$$f_k = \mu_k n \tag{3.18}$$

where μ_k is the **coefficient of kinetic friction.**

3. The values of μ_k and μ_s depend on the nature of the surfaces, but μ_k is generally less than μ_s. Typical values of μ range from around 0.01 for smooth surfaces to 1.5 for rough surfaces. Table 3.2 lists some reported values.

Finally, the coefficients of friction are nearly independent of the area of contact between the surfaces. Although the coefficient of kinetic friction varies with speed, we shall neglect any such variations.

EXAMPLE 3.5 Moving Into the Dormitory

It is the beginning of a new school term, and you need to move a box of books into your dormitory room. To do so, you attach a rope to the box and pull on it with a force of 90 N at an angle of 30°, as shown in Figure 3.11. The box of books has a mass of 20 kg, and the coefficient of kinetic friction between the bottom of the box and the hallway surface is 0.5. Find the acceleration of the box.

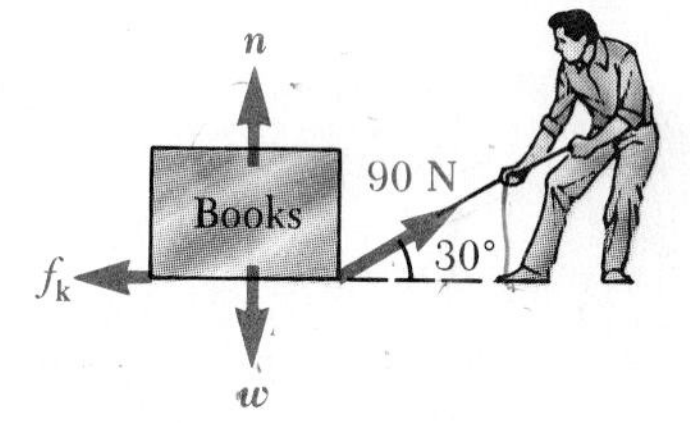

FIGURE 3.11 (Example 3.5) A box of books being pulled to the right at an angle of 30° with a rough horizontal surface.

Solution The box is not accelerating in the vertical direction, and so we find the normal force from $\Sigma F_y = 0$. The forces in the y direction are the weight, w, the normal force, n, and the vertical component of your 90 N pull. This vertical component has a magnitude equal to (90 N)(sin 30°). We find that

$$\sum F_y = n + (90\text{ N})(\sin 30°) - (20\text{ kg})(9.80\text{ m/s}^2) = 0$$

$$n = 151\text{ N}$$

Note that in this case the normal force is *not* equal to the weight of the box of books because the vertical component of your pull is helping to support some of the weight of the box.

Since the normal force is known, we can find the force of friction:

$$f_k = \mu_k n = (0.5)(151\text{ N}) = 75.5\text{ N}$$

Finally, we determine the horizontal acceleration using Newton's second law:

$$\sum F_x = (90\text{ N})(\cos 30°) - 75.5\text{ N} = (20\text{ kg})(a_x)$$

$$a_x = 0.122\text{ m/s}^2$$

EXERCISE If the initial speed of the box is zero, what is its speed after it has traveled 2 m? How long does it take you to pull it a distance of 2 m?

Answer 0.699 m/s, 5.73 s

EXAMPLE 3.6 The Sliding Hockey Puck

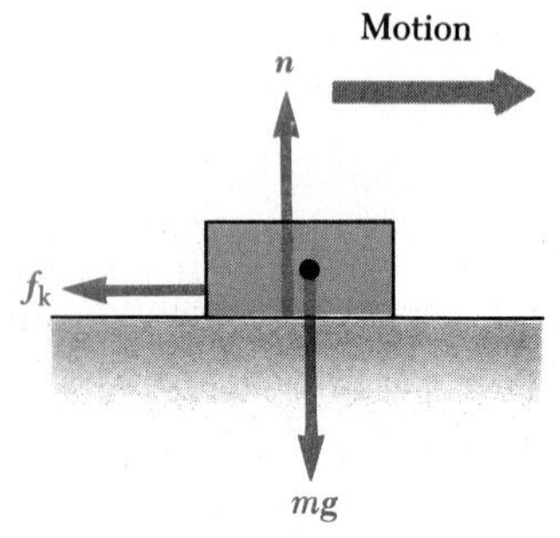

FIGURE 3.12 (Example 3.6) *After* the puck is given an initial velocity to the right, the external forces acting on it are its weight, **mg**, the normal force, **n**, and the force of kinetic friction, $\mathbf{f}_k$.

A hockey puck is given an initial speed of 20 m/s on a frozen pond as in Figure 3.12. The puck remains on the ice and slides 120 m before coming to rest. Determine the coefficient of friction between the puck and the ice.

Solution You can find the acceleration of the puck from $v^2 = v_0^2 + 2ax$, with the final speed, v, equal to zero, the initial speed $v_0 = 20$ m/s, and the distance traveled $x = 120$ m.

$$v^2 = v_0^2 + 2ax$$

$$0 = (20\text{ m/s})^2 + 2a\,(120\text{ m})$$

$$a = -1.67\text{ m/s}^2$$

The negative sign means that the acceleration is to the left, opposite the direction of the velocity.

The force of friction is found from $f_k = \mu_k n$, where n is found from $\Sigma F_y = 0$ as follows:

$$\sum F_y = n - w = 0$$

$$n = w = mg$$

Thus,

$$f_k = \mu_k n = \mu_k mg$$

Now we apply Newton's second law along the horizontal direction. We shall take the positive direction toward the right.

$$\sum F_x = -f_k = ma$$

$$-\mu_k mg = m(-1.67\text{ m/s}^2)$$

$$\mu_k = \frac{1.67\text{ m/s}^2}{9.80\text{ m/s}^2} = 0.170$$

EXAMPLE 3.7 Connected Objects

Two objects are connected by a light string that passes over a frictionless pulley, as in Figure 3.13a. The coefficient of sliding friction between the 4-kg object and the surface is 0.30. Find the acceleration of the two objects and the tension in the string.

Solution First, let us isolate each object in Figure 3.13a and determine the external forces on each. Newton's second law applied to the 4-kg object in component form with the positive x direction to the right gives

$$\sum F_x = T - f = (4\text{ kg})(a)$$

$$\sum F_y = n - (4\text{ kg})(g) = 0$$

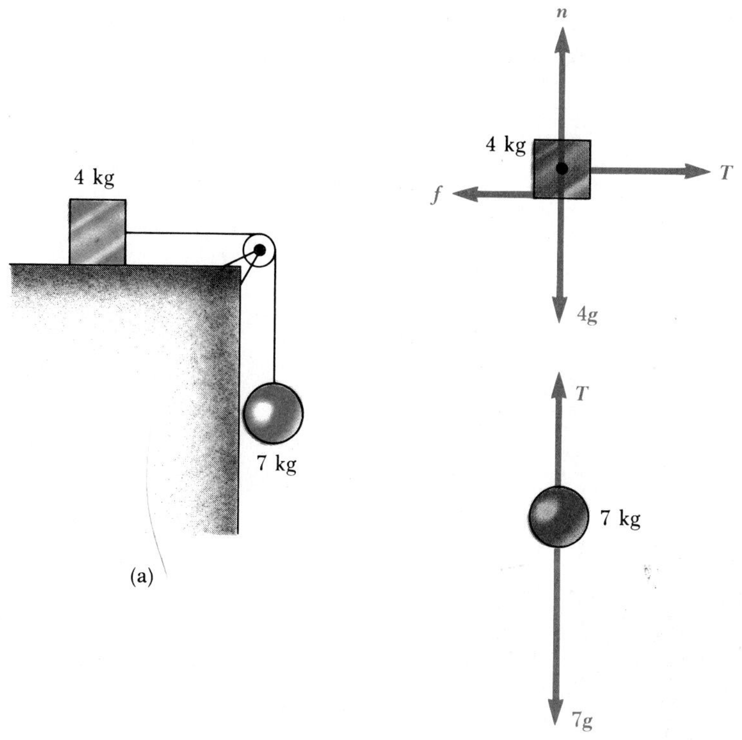

FIGURE 3.13 (Example 3.7) (a) Two objects connected by a light string that passes over a frictionless pulley. The surface is rough. (b) Free-body diagrams for the objects.

Since $f = \mu_k n$ and $n = 4g = (4\text{ kg})(9.80\text{ m/s}^2) = 39.2\text{ N}$, we have $f = (0.30)(39.2\text{ N}) = 11.8\text{ N}$. Therefore,

$$(1) \qquad T = f + (4\text{ kg})(a) = 11.8\text{ N} + (4\text{ kg})(a)$$

Now we apply Newton's second law to the 7-kg object moving in the vertical direction, where the downward direction is selected as the positive direction:

$$\sum F_y = (7\text{ kg})(g) - T = (7\text{ kg})(a)$$

or

$$(2) \qquad T = 68.6\text{ N} - (7\text{ kg})(a)$$

Subtracting (2) from (1) eliminates T:

$$56.8\text{ N} - (11\text{ kg})(a) = 0$$
$$a = 5.16\text{ m/s}^2$$

When this value for the acceleration is substituted into (1), we get

$$T = 32.5\text{ N}$$

EXAMPLE 3.8 Experimental Determination of μ_s and μ_k

In this example we describe a simple method of measuring the coefficients of friction between an object and a rough surface. Suppose the object is a small block placed on

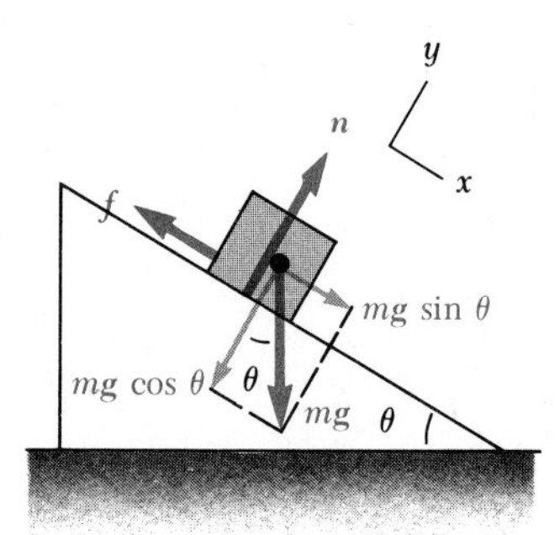

FIGURE 3.14 (Example 3.8) The external forces acting on a block on a rough incline are the weight of the block, **mg**, the normal force, **n**, and the force of friction, **f**. Note that the weight vector is resolved into a component along the incline, $mg \sin \theta$, and a component perpendicular to the incline, $mg \cos \theta$.

a surface inclined with respect to the horizontal, as in Figure 3.14. The angle of the inclined plane is increased until the block slips. By measuring the critical angle, θ_c, at which this slipping occurs, we obtain μ_s directly.

We note that the only forces acting on the block are its weight, **mg**, the normal force, **n**, and the force of static friction, $\mathbf{f}_s$. With x parallel to the plane and y perpendicular to the plane, Newton's second law applied to the block gives

$$(1) \qquad \sum F_x = mg \sin \theta - f_s = 0$$

$$(2) \qquad \sum F_y = n - mg \cos \theta = 0$$

We can eliminate mg by solving (2) for mg to get $mg = n \cos \theta$ and substituting this value into (1) to get

$$(3) \qquad f_s = mg \sin \theta = \left(\frac{n}{\cos \theta}\right)(\sin \theta) = n \tan \theta$$

When the inclined plane is at the critical angle, $f_s = f_{s,\max} = \mu_s n$, and so at this angle, (3) becomes

$$\mu_s n = n \tan \theta_c$$

$$\mu_s = \tan \theta_c$$

For example, if we find that the block just slips at $\theta_c = 20°$, then $\mu_s = \tan 20° = 0.364$. Once the block starts to move at $\theta \geq \theta_c$, it will accelerate down the incline and the force of friction is $f_k = \mu_k n$. However, once the block is in motion, if θ is reduced to a value less than θ_c, an angle θ_c' can be found such that the block moves down the incline with constant speed ($a_x = 0$). In this case, (1) and (2) with f_s replaced by f_k give

Block in motion

$$\mu_k = \tan \theta_c'$$

where $\theta_c' < \theta_c$.

You should try this simple experiment using a coin as the block and a book as the inclined plane. Also, you can try taping two coins together to prove that you still get the same critical angles as with one coin.

SUMMARY

Newton's first law states that an object at rest will remain at rest and an object in motion will continue in motion with a constant velocity unless it experiences a net force.

The resistance of an object to a change in its state of motion is called **inertia.** Mass is a physical quantity which is a measure of inertia.

Newton's second law states that the resultant force acting on an object is equal to the product of the mass of the object and its acceleration.

$$\sum \mathbf{F} = m\mathbf{a} \qquad (3.1)$$

The **weight** of an object is equal to the product of its mass and the acceleration due to gravity:

$$w = mg \qquad (3.6)$$

Newton's third law states that, if two bodies exert force on each other, the force of body 1 on body 2 is equal in magnitude and opposite in direction to the force of body 2 on body 1. Thus, an isolated force can never occur in nature.

The maximum force of static friction, $f_{s,max}$, between an object and a rough surface is proportional to the normal force acting on the object. This maximum force occurs when the object is on the verge of slipping. In general,

$$\boxed{f_s \leq \mu_s n} \tag{3.7}$$

where μ_s is the **coefficient of static friction.** When a body slides over a rough surface, the force of kinetic friction, f_k, is opposite the motion and is also proportional to the normal force. The magnitude of this force is

$$\boxed{f_k = \mu_k n} \tag{3.8}$$

Here, μ_k is the **coefficient of kinetic friction.** In general, $\mu_k < \mu_s$.

ADDITIONAL READING

P. Brancazio, "The Physics of Kicking a Football," *The Physics Teacher*, October 1985, p. 403.

B. I. Cohen, "Isaac Newton," *Sci. American*, December 1955, p. 73.

S. Drake, "Galileo's Discovery of the Law of Free Fall," *Sci. American*, May 1973, p. 17.

G. Gamow, "Gravity," *Sci. American*, March 1961, p. 94.

G. Gamow, *Gravity*, Science Study Series, Garden City, N.Y., Doubleday, 1962.

W. F. Magie, *Source Book in Physics*, Cambridge, Mass., Harvard University Press, 1963. Excerpts from Newton on the laws of motion.

M. McCloskey, "Intuitive Physics," *Sci. American*, April 1983, p. 122.

M. McCloskey, A. Caramaza, and B. Gross, "Curvilinear Motion in the Absence of External Forces: Naive Beliefs About the Motion of Objects," *Science*, December 1980, p. 1139.

I. Newton, *Mathematical Principles of Natural Philosophy (Principia)*, translated by A. Motte, revised by F. Cajori, Berkeley, University of California Press, 1947.

F. Palmer, "Friction," *Sci. American*, February 1951, p. 54.

R. Zimmerer, "The Measurement of Mass," *The Physics Teacher*, September 1983, p. 354.

QUESTIONS

1. If an object is at rest, can we conclude that there are no external forces acting on it?
2. If gold were sold by weight, would you rather buy it in Denver or in Death Valley? If sold by mass, in which of the two locations would you prefer to buy it? Why?
3. A passenger sitting in the rear of a bus claims that he was injured when the driver slammed on the brakes, causing a suitcase to come flying toward him from the front of the bus. If you were the judge in this case, what disposition would you make? Why?
4. A space explorer is moving through space far from any planet or star. He notices a large rock, taken as a specimen from an alien planet, floating around the cabin of his ship. Should he push it gently toward a storage compartment or kick it toward the compartment? Why?
5. How much does an astronaut weigh out in space, far from any planets?
6. Although the frictional force between two surfaces may decrease as the surfaces are smoothed, the force will again increase if the surfaces are made extremely smooth and flat. How do you explain this?
7. Why is it that the frictional force involved in the rolling of a body over another is less than for sliding motion?
8. A massive metal object on a rough metal surface may actually undergo contact welding to that surface. Discuss how this affects the frictional forces that arise between the object and the surface.
9. Analyze the motion of a rock dropped in water in terms of its speed and acceleration as it falls. Assume that there is a resistive force acting on the rock that increases as the velocity increases.

10. Identify the action-reaction pairs in the following situations: a man takes a step; a snowball hits a girl in the back; a baseball player catches a ball; a gust of wind strikes a window.
11. While a football is in flight, what forces act on it? What are the action-reaction pairs while the football is being kicked, and while it is in flight?
12. A ball is held in a person's hand. (a) Identify all the external forces acting on the ball and the reaction to each of these forces. (b) If the ball is dropped, what force is exerted on it while it is falling? Identify the reaction force in this case. (Neglect air resistance.)
13. Identify all the action-reaction pairs which exist for a horse pulling on a cart. Include the earth in your examination.
14. If a car is traveling with a constant speed of 20 m/s in the westward direction, what is the resultant force acting on it?
15. A large crate is placed on the bed of a truck without being tied to the truck. (a) As the truck accelerates forward, the crate remains at rest relative to the truck. What force causes the crate to accelerate forward? (b) If the truck driver slams on his brakes, what could happen to the crate?
16. A child pulls a wagon with some force, causing it to accelerate. Newton's third law says that the wagon exerts an equal and opposite reaction force on the child. How can the wagon accelerate?
17. A rubber ball is dropped onto a floor. What force causes the ball to bounce back into the air?
18. What is wrong with the following statement. "Since the car is at rest, there are no forces acting on it." How would you restate this sentence to make it correct?
19. Suppose you are driving a car along a highway at a high speed. Why should you avoid "slamming on" your brakes if you want to stop in the shortest distance?
20. A child in a car holds a helium-filled balloon attached to a string. What happens to the balloon when the car accelerates forward?
21. Is it possible to have motion in the absence of a force? Explain.
22. What causes a rotary lawn sprinkler to turn?
23. A janitor finds it quite easy to sweep a floor using a long-handled broom when the angle between the handle and floor is small. However, if the angle is large, it becomes very difficult to push the broom. Explain why this occurs.
24. The force of gravity is twice as great on a 20 N rock as it is on a 10 N rock. Why doesn't the 20 N rock have a greater free-fall acceleration?

PROBLEMS

Section 3.2 through Section 3.5

1. What is the weight in newtons of a 2-kg cannonball? What is its weight in pounds?

2. A bag of sugar weighs 5 lb on earth. What should it weigh in newtons on the moon where the acceleration due to gravity is 1/6 that on earth. Repeat for Jupiter where g is 2.64 times earth gravity. Find the mass in kilograms at all three locations.

3. A person weighs 120 lb. Determine (a) her weight in newtons and (b) her mass in kilograms.

4. (a) Draw a free-body diagram indicating all the forces acting on a freely falling baseball. For each force drawn, indicate the reaction force to it. (b) Repeat part (a) for the baseball while moving as a projectile toward an outfielder.

5. A 6-kg object undergoes an acceleration of 2 m/s^2. (a) What is the magnitude of the resultant force acting on it? (b) If this same force is applied to a 4-kg object, what acceleration will be produced?

6. A box weighing 20 N is to be pushed across a smooth floor with a horizontal force of 5 N. What acceleration will be produced?

7. A football punter accelerates a football from rest to a speed of 10 m/s during the time in which his toe is in contact with the ball (about 0.2 s). If the football has a mass of 0.5 kg, what average force does the punter exert on the ball?

8. The engine on a 0.2 kg model airplane exerts a forward force on the plane of 10 N. If the plane accelerates at 2 m/s^2, what is the magnitude of the resistive force due to the wind acting on the airplane?

• **9.** Two forces are applied to a car in an effort to move it, as shown in Figure 3.15. (a) What is the resultant of these two forces? (b) If the car has a mass of 3000 kg, what acceleration does it have?

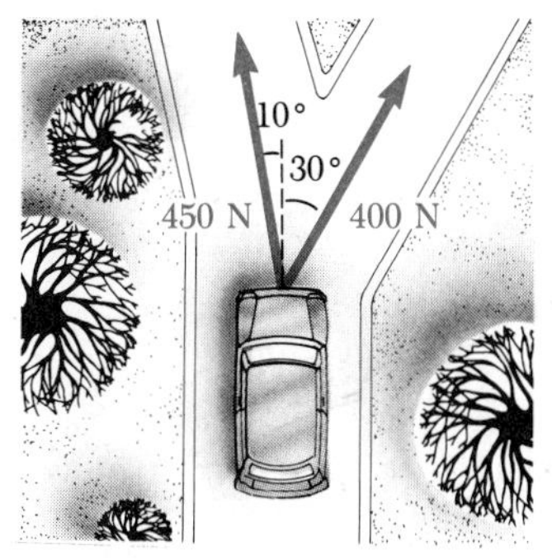

FIGURE 3.15 Problem 9

• **10.** A boat moves through the water with two forces acting on it. One is a 2000 N forward push by the motor, while the second is an 1800 N resistive force due to the water. (a) What is the acceleration of the 1000-kg boat? (b) If it starts from rest, how far will it move in 10 s? (c) What will be its velocity at the end of this time?

Section 3.6 Some Applications of Newton's Laws

11. A shopper in a supermarket pushes a loaded car with a horizontal force of 10 N. If the cart has a mass of 30 kg, how far will it move in 3 s, starting from rest? (Ignore friction.)

12. How far will the cart in Problem 11 move in the 3 s interval if the shopper places her 30-N child in the cart before she begins to push it?

13. A 0.15-kg baseball moving at 20 m/s strikes the glove of a catcher. The glove recoils a distance of 8 cm. What is the average force applied on the ball by the glove? What is the average force applied on the glove by the ball?

14. A 2-kg mass starts from rest and slides down an inclined plane 80 cm long in 0.5 s. What is the *net force* along the incline acting on the mass?

15. A roller coaster starts its descent with an initial speed of 4 m/s. It moves through a distance of 135 ft along an incline which makes an angle of 40° with the horizontal. Neglect friction and find its speed at the bottom of the incline.

16. A shopper pushes a 40-kg cart up a 20° smooth ramp. What force directed along the incline must the shopper exert to give the cart an acceleration of 0.5 m/s^2?

17. The parachute on a race car of weight 8820 N opens at the end of a quarter-mile run when the car is traveling at 35 m/s. What total retarding force must be supplied by the parachute to stop the car in a distance of 1000 m?

18. A 5-kg bucket of water is raised from a well by a rope attached to the bucket. If the upward acceleration of the bucket is 3 m/s^2, find the force exerted on the bucket by the rope.

• **19.** A 8000-kg tow truck pulls a 3000-kg car horizontally into a garage. If the motor of the tow truck exerts a forward thrust on the truck of 800 N, what is the tension in the cable connecting the truck to the car? (Neglect any forces of friction.)

• **20.** A 5-kg mass resting on a smooth horizontal table is connected to a cable which passes over a pulley and then is fastened to a hanging 10-kg mass as in Figure 3.16. Find the acceleration of the two objects and the tension in the string.

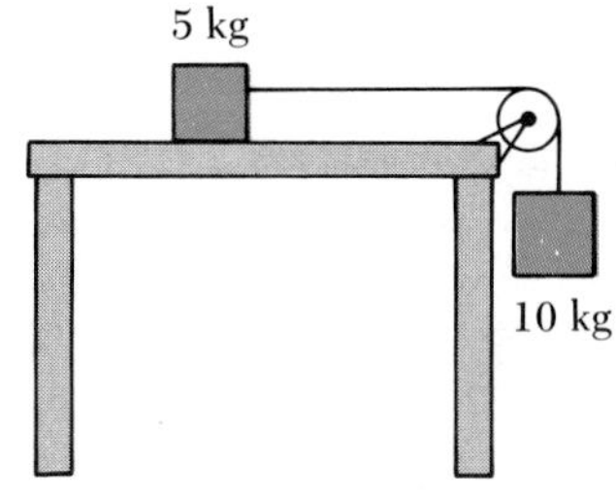

FIGURE 3.16 Problems 20 and 32

• **21.** Two masses of 3 kg and 5 kg are connected by a light string that passes over a frictionless pulley, as in Figure 3.17. Determine (a) the tension in the string, (b) the acceleration of each mass, and (c) the distance each mass will move in the first second of motion if they start from rest.

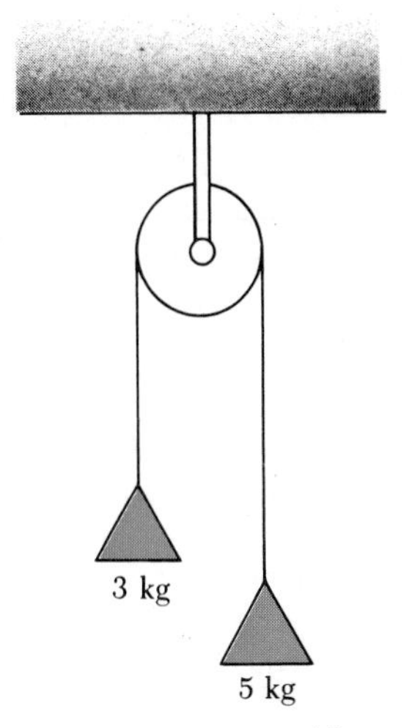

FIGURE 3.17 Problem 21

• **22.** Two blocks are fastened to the ceiling of an elevator as in Figure 3.18. The elevator accelerates upward at the rate of 2 m/s^2. Find the tension in each rope.

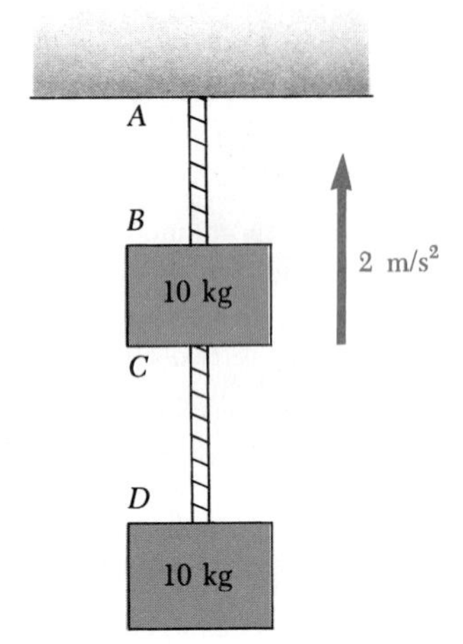

FIGURE 3.18 Problem 22

Section 3.7 Forces of Friction

23. A dockworker loading crates on a ship finds that a 20-kg crate, initially at rest on a rough horizontal surface, requires a 75 N horizontal force to set it into motion. However, after it is in motion, a horizontal force of 60 N is required to keep the crate moving with a constant speed. Find the coefficients of static and kinetic friction between the crate and the floor.

24. (a) A 10-kg box of books rests on a horizontal surface. If the coefficient of static friction between box and floor is 0.4, find the maximum horizontal force that can be applied to the box before it slips. (b) The coefficient of kinetic friction is 20% less than that of static friction. If the force found in part (a) is exerted on the box once it is in motion, find the acceleration of the box.

25. A 1000-N bobsled leaves a horizontal track with a speed of 30 m/s. It then moves onto a horizontal track lightly sprinkled with sawdust and comes to a stop in a distance of 150 m. Find the coefficient of kinetic friction between the runners of the sled and the sawdust covered surface.

26. (a) A student decides to move a box of books into her dormitory room by pulling on a rope attached to the box with a force of 80 N at an angle of 25° with the horizontal. The box of books has a mass of 25 kg, and the coefficient of kinetic friction between box and floor is 0.3. Find the acceleration of the box. (b) Along the way, the student must move the box up a ramp inclined at 10° with the horizontal. If the box starts from rest at the bottom of the incline and is pulled at an angle of 25° with respect to the incline and with the same force, can it be moved up the incline? If so, what is the acceleration up the ramp?

• **27.** A box slides down a 30° ramp with an acceleration 1.2 m/s^2. Determine the coefficient of kinetic friction between the box and the ramp.

• **28.** A car is traveling at 50 km/h on a flat highway. (a) If the coefficient of friction between the road and tires on a rainy day is 0.1, what is the minimum distance in which the car will stop? (b) What is the stopping distance when the surface is dry and the coefficient of friction is 0.6?

• **29.** A box weighing 5400 N is being pulled up an inclined plane that rises 1.3 m for every 7.5 m of length measured along the incline. If the coefficient of friction is 0.6, determine the force applied parallel to the incline necessary to move it up the incline at constant speed.

• **30.** A furniture crate of mass 60 kg is at rest on a loading ramp that makes an angle of 25° with the horizontal. The coefficient of kinetic friction between the crate and the ramp is 0.2. What force, applied parallel to the incline, is required to push the crate up the incline at constant speed?

• **31.** In order to determine the coefficients of friction between rubber and various surfaces, a student uses a rubber eraser and an inclined board. In one experiment, the eraser slips down the incline when the angle of inclination is 36°, and then moves down the incline with constant speed when the angle is reduced to 30°. From these data, determine the coefficients of static and kinetic friction for this experiment.

• **32.** Find the acceleration experienced by the two masses shown in Figure 3.16, if the coefficient of friction between the 5 kg mass and the table is 0.25.

• **33.** Find the acceleration experienced by the two masses shown in Figure 3.19, if the coefficient of friction between the 5 kg mass and the plane is 0.25.

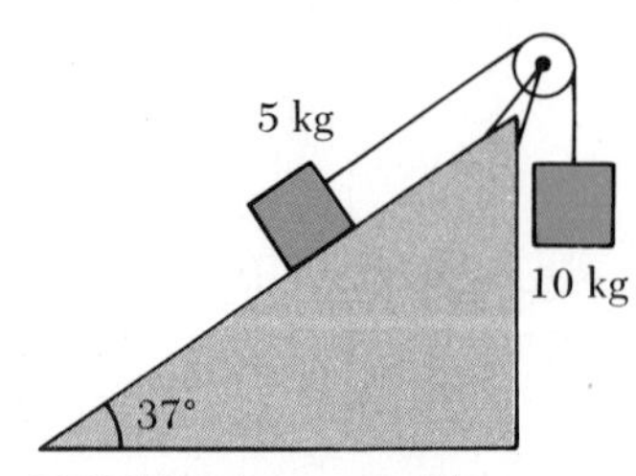

FIGURE 3.19 Problem 33

•• **34.** A sled weighing 60 N is pulled horizontally across snow such that the coefficient of kinetic friction between sled and snow is 0.1. A penguin, that looks amazingly like Opus, rides on the sled. (See Fig. 3.20.) If his weight is 70 N, and the coefficient of static friction between penguin and sled is 0.7, find the maximum horizontal force that can be exerted on the sled before the penguin begins to slide off.

FIGURE 3.20 Problem 34

ADDITIONAL PROBLEMS

35. A man attempting to train his dog is shown in Figure 3.21 pulling on him with a force of 70 N at an angle of 30° to the horizontal. Find the x and y components of this force.

FIGURE 3.21 Problem 35

36. Two people pulling on a mule are seen from the helicopter view shown in Figure 3.22. Find the magnitude and direction of the resultant force on the mule.

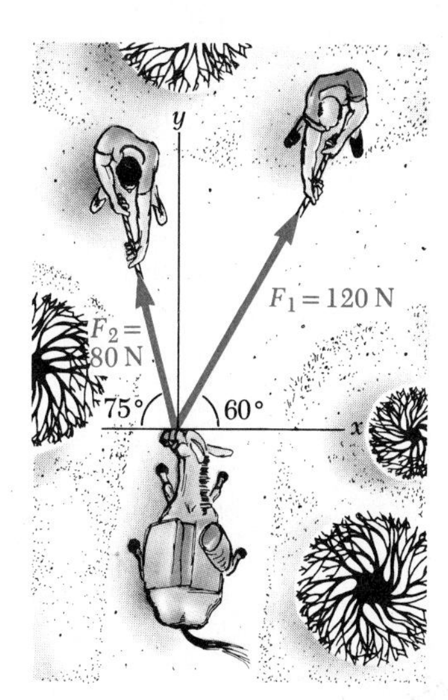

FIGURE 3.22 Problem 36

37. (a) What is the resultant force exerted by the two cables supporting the traffic light shown in Figure 3.23? (b) What is the weight of the light?

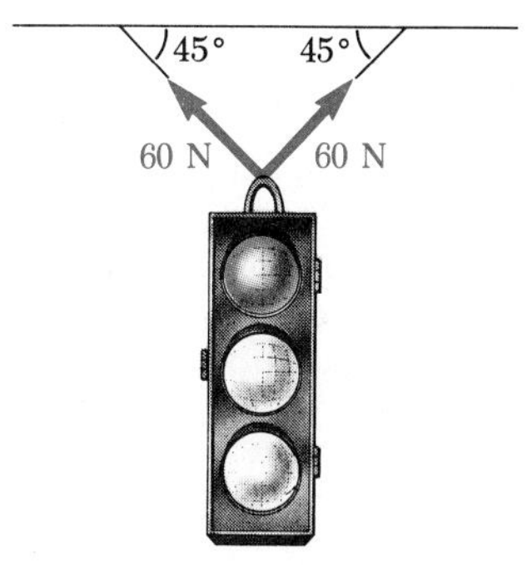

FIGURE 3.23 Problem 37

38. Four forces act on a boat shown from a top view in Figure 3.24. Find the magnitude and direction of the resultant force on the boat.

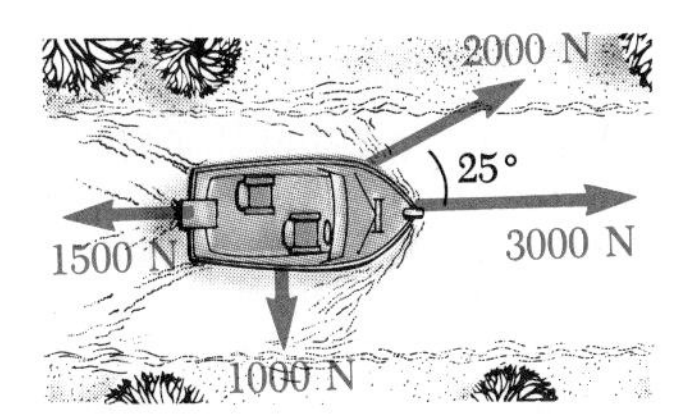

FIGURE 3.24 Problem 38

39. Some baseball pitchers are capable of throwing fastball at 100 mi/h. The pitcher achieves this speed by moving his arm through a distance of about 1.5 m. What average force must he exert on the 0.15 kg ball during this time?

40. A baseball pitcher throws a ball straight up into the air at a speed of 20 m/s as a protest against the umpire's calls. In the process, he moves his hand through a distance of 1.5 m. If the ball has a mass of 0.15 kg, find the force he exerts on the ball to give it this upward speed. Note that the force of gravity is acting against the motion of the ball.

41. A car is at rest at the top of a driveway that has a slope of 20°. If the brake of the car is released, find (a) the acceleration of the car down the drive and (b) the time it takes for the car to reach the street 10 m away.

42. A 2000-kg car is slowed down uniformly from a speed of 20 m/s to 5 m/s in a time of 4 s. What average force acted on the car during this time and how far did the car travel during the deceleration?

43. A girl coasts down a hill on a sled, reaching the bottom of the hill with a speed of 7 m/s. If the coefficient of friction between the steel runners and the snow is 0.05 and the girl and sled together weigh 600 N, how far does the sled travel on the level surface before coming to rest?

44. An elevator accelerates upward at 1.5 m/s^2. If the elevator has a mass of 200 kg, find the tension in the supporting cable.

45. Two barrels, of masses 4 kg and 7 kg, are pulled across the frictionless surface of a frozen pond by an ice fisherman. If he exerts a force of 30 N on the first barrel as shown in Figure 3.25, determine the acceleration of the system and the tension in the cord connecting the barrels.

FIGURE 3.25 Problem 45

• **46.** A box rests on the back of a truck. The coefficient of friction between the box and the surface is 0.3. (a) When the truck accelerates forward, what force accelerates the box? (b) Find the maximum acceleration the truck can have before the box slides.

• **47.** A 10,000-N force is exerted on a car by a cable pulling the car up a loading ramp which is inclined at 30° with the horizontal. If the car has a mass of 1500 kg, determine the acceleration of the system. (Ignore any effects of friction.)

•• **48.** The car of Problem 47 is pulled up the same ramp, but this time the pull is supplied by a 10,000-N counterweight passing over a pulley as shown in Figure 3.26. Find (a) the tension in the cable, and (b) the acceleration of the system. (c) What mass should the counterweight have in order for the car to move down the incline at an acceleration of 2 m/s^2? (Ignore any effects of friction.)

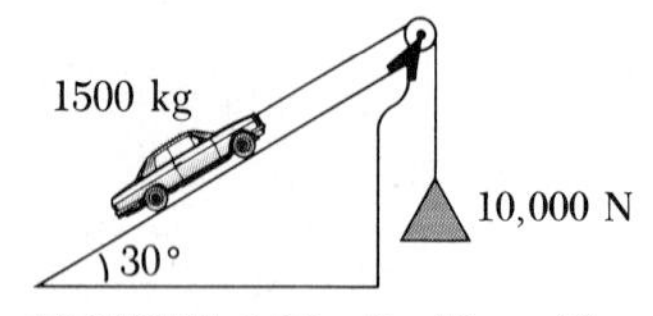

FIGURE 3.26 Problem 48

•• **49.** A 3-kg mass hangs at the end of a rope which is attached to a support on a railroad car. When the car accelerates the cord makes an angle of 4° with the vertical as shown in Figure 3.27. Find the acceleration of the car.

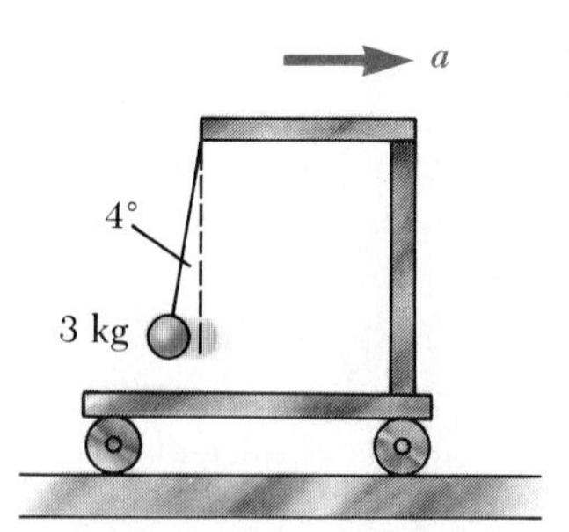

FIGURE 3.27 Problem 49

•• **50.** A bowling ball attached to a spring scale is suspended from the ceiling of an elevator. The scale reads 70 N when the elevator is at rest. (a) What will the scale read when the elevator accelerates upward at the rate of 3 m/s^2? (b) What will the scale read when the elevator accelerates downward at the rate of 3 m/s^2?

•• **51.** Two blocks on a frictionless horizontal surface are connected by a light string, as in Figure 3.28, where $m_1 = 10$ kg and $m_2 = 20$ kg. A force of 50 N is applied to the 20-kg block. (a) Determine the acceleration of each block and the tension in the connecting cable, assuming the surface is frictionless. (b) Repeat the problem assuming the coefficient of kinetic friction between each block and the surface is 0.1.

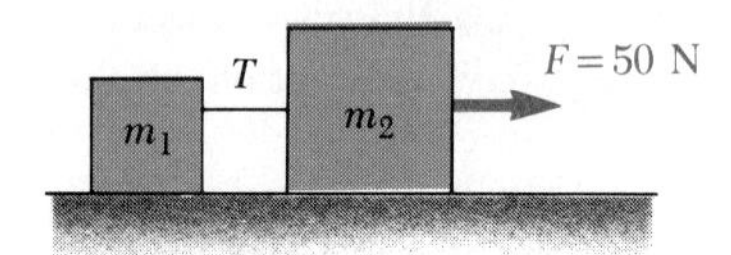

FIGURE 3.28 Problem 51

•• **52.** A 5-kg penguin sits on top of a 10-kg sled, as in Figure 3.29. A horizontal force of 45 N is applied to the sled, but the penguin attempts to impede the motion by holding onto a cord attached to a wall. The coefficient of kinetic friction between the moving surfaces is 0.2. (a) Draw force diagrams for the penguin and the sled, and identify the reaction force for each force you include. (b) Determine the tension in the string held by the penguin and the acceleration of the sled.

FIGURE 3.29 Problem 52

53. A book of mass 0.9 kg is projected up a 30° ramp with an initial speed of 3 m/s. The coefficient of a kinetic friction between the ramp and the book is 0.2. Determine (a) how far up the incline the book will go before coming to rest, (b) its acceleration as it slides back down the incline, and (c) its speed when it reaches the bottom of the incline.

•• **54.** In Figure 3.30, the coefficient of kinetic friction between the 2-kg and 3-kg blocks is 0.3. The horizontal surface and the pulleys are frictionless. (a) Draw free body diagrams for each block. (b) Determine the acceleration of each block. (c) Find the tension in the strings.

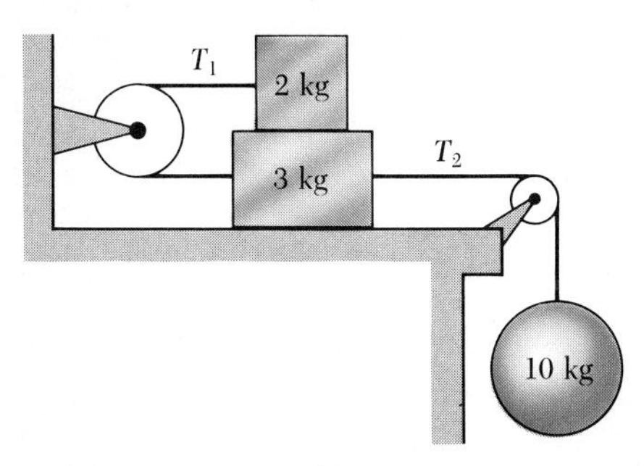

FIGURE 3.30 Problem 54

•• **55.** A horizontal force $\boldsymbol{F}$ is applied to a frictionless pulley of mass m_2 as in Figure 3.31. The horizontal surface is smooth. (a) Show that the acceleration of the block of mass m_1 is *twice* the acceleration of the pulley. Find (b) the acceleration of the pulley and the block and (c) the tension in the string. A constant supporting force is applied to the axle of the pulley equal to its weight.

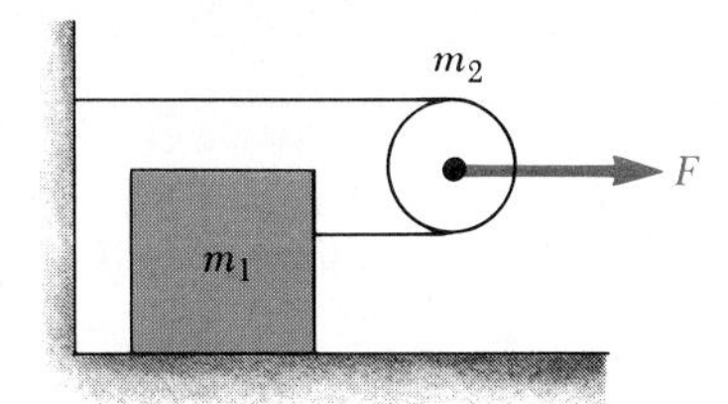

FIGURE 3.31 Problem 55

•• **56.** Three blocks are in contact with each other on a frictionless, horizontal surface as in Figure 3.32. A horizontal force F is applied to m_1. If m_1 = 2 kg, m_2 = 3 kg, m_3 = 4 kg, and F = 18 N, find (a) the acceleration of the blocks, (b) the resultant force on each block, and (c) the magnitude of the contact forces between the blocks.

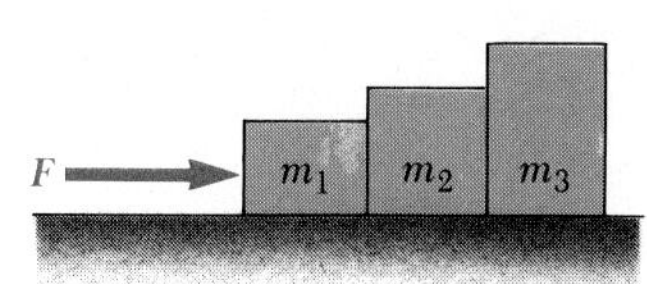

FIGURE 3.32 Problem 56

•• **57.** Repeat Problem 56 given that the coefficient of kinetic friction between the blocks and the surface is 0.1. Use the data given in Problem 56.

•• **58.** Two blocks are fastened to the top of an elevator as in Figure 3.18. The mass of each rope is 1 kg. The elevator accelerates upward at 4 m/s^2. Find the tensions in the ropes at (a) the ceiling of the elevator, (b) the point of contact of the top rope with the top block, (c) the point at which the lower rope attaches to the top block, and (d) the point of contact of the lower rope to the lower block.

•• **59.** The three blocks in Figure 3.33 are connected by light strings that pass over frictionless pulleys. The acceleration of the system is 2 m/s^2 to the left and the surfaces are rough. Find (a) the tension in each string and (b) the coefficient of kinetic friction between blocks and surface. (Assume the same μ for both blocks.)

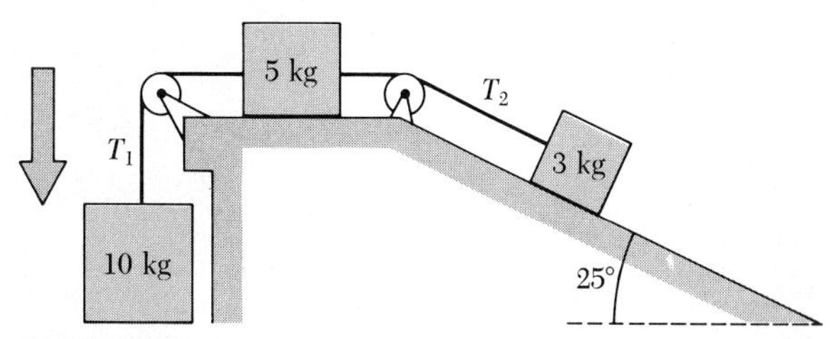

FIGURE 3.33 Problem 59

•• **60.** Two blocks on a rough incline are connected by a light string that passes over a frictionless pulley as in Figure 3.34. Assuming $m_1 > m_2$ and taking the coefficients of kinetic friction for each block to be μ, determine expressions for (a) the acceleration of the blocks and (b) the tension in the string. (Assume the system is in motion.)

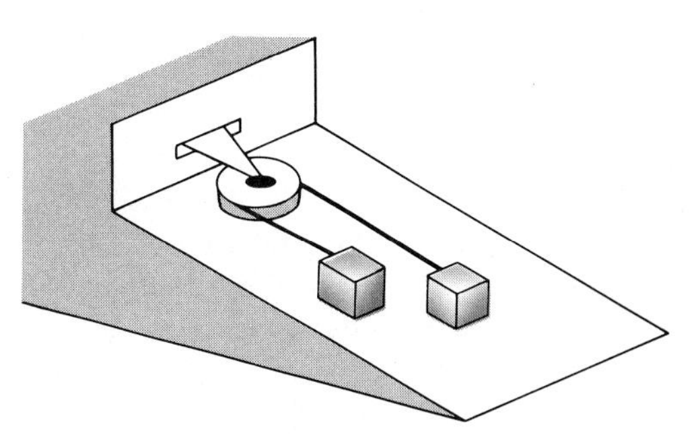

FIGURE 3.34 Problem 60

CHAPTER 4

Objects in Equilibrium

A large balanced rock photographed at the Garden of the Gods in Colorado Springs, Colorado. This is an example of unstable equilibrium. (Photo by David Serway)

In this chapter we shall study objects that are either at rest or moving with a constant velocity. Such objects are said to be in equilibrium, and an understanding of the conditions that prevail when an object is in equilibrium is important in a variety of fields. For example, students of architecture or industrial technology will benefit from an understanding of the forces that act on buildings or on large machines, and biology students should understand the forces at work in muscles and bones in the human body.

Newton's first law as discussed in Chapter 3 will form the basis for much of our work in this chapter. However, in order to fully understand objects in equilibrium, we must also consider the concepts of torque and center of gravity.

4.1 THE FIRST CONDITION FOR EQUILIBRIUM

Objects which are at rest or which are moving with constant velocity are said to be in *equilibrium*. **Newton's first law** is a statement of one condition that must be true for an object to be in equilibrium.

An object will remain at rest or move with constant velocity when no resultant force acts on it.

Newton's first law

In equation form, this can be expressed as

$$\Sigma \mathbf{F} = 0 \tag{4.1}$$

This statement signifies that the *vector* sum of all the forces (the net force) acting on the object is identically zero. Usually, the problems we encounter in our study of equilibrium will be more easily solved if we work with Equation 4.1 in terms of the components of the external forces acting on an object. By this we mean that, in a two-dimensional problem, the sum of all the external forces in the x and y directions must separately equal zero, that is,

$$\begin{aligned} \Sigma F_x &= 0 \\ \Sigma F_y &= 0 \end{aligned} \tag{4.2}$$

First condition for equilibrium

This set of equations is often referred to as the **first condition for equilibrium.** We shall not consider three-dimensional problems in this text, but the extension of Equation 4.2 to a three-dimensional situation can be made by adding a third equation, $\Sigma F_z = 0$.

4.2 PROBLEM-SOLVING TECHNIQUES

The following procedure is recommended when analyzing an object in equilibrium under the action of several external forces.

1. Make a sketch of the object showing all external forces acting on the object. As you recall from the last chapter, this sketch is called a free-body diagram.
2. In some problems the direction of a particular force may not be obvious. In this situation, try to guess the correct direction for this force. If your guess for the direction is incorrect, your answer for the force will be a negative number. The negative sign in your answer simply means that you guessed the incorrect direction for the force. However, the magnitude of the force will be correct.
3. Choose a convenient coordinate system and resolve all forces into their components along the axes of this coordinate system.
4. Apply the first condition of equilibrium along the x direction and along the y direction. Remember to keep track of the signs of the various forces. For example, forces in the positive x direction will have a positive sign while those in the negative x direction will have a negative sign.
5. The first condition of equilibrium will provide you with a set of simultaneous equations that can be solved for the unknown quantities.

The key to success with equilibrium problems is practice, and to this end we have provided a large number of worked examples to help you get started.

EXAMPLE 4.1 A Traffic Light at Rest

A traffic light weighing 100 N hangs from a cable tied to two other cables fastened to a support, as in Figure 4.1a. The upper cables make angles of 37° and 53° with the horizontal. Find the tension in the three cables.

Solution First we construct a free-body diagram for the traffic light, as in Figure 4.1b. The tension in the vertical cable, $\boldsymbol{T}_3$, supports the light, and so we see that $T_3 = w = 100$ N. Now we construct a free-body diagram for the knot that holds the three cables together, as in Figure 4.1c. This is a convenient point to choose because all forces in question act at this point. We choose the coordinate axes as shown in Figure 4.1c and resolve the forces into their x and y components:

Force	x Component	y Component
$\boldsymbol{T}_1$	$-T_1 \cos 37°$	$T_1 \sin 37°$
$\boldsymbol{T}_2$	$T_2 \cos 53°$	$T_2 \sin 53°$
$\boldsymbol{T}_3$	0	-100 N

The first condition for equilibrium gives us the equations

(1) $\sum F_x = T_2 \cos 53° - T_1 \cos 37° = 0$

(2) $\sum F_y = T_1 \sin 37° + T_2 \sin 53° - 100 \text{ N} = 0$

From (1) we see that the horizontal components of $\boldsymbol{T}_1$ and $\boldsymbol{T}_2$ must be equal in magnitude, and from (2) we see that the sum of the vertical components of $\boldsymbol{T}_1$ and $\boldsymbol{T}_2$ must balance the weight of the light. We can solve (1) for T_2 in terms of T_1 to give

$$T_2 = T_1\left(\frac{\cos 37°}{\cos 53°}\right) = T_1\left(\frac{0.800}{0.602}\right) = 1.33T_1$$

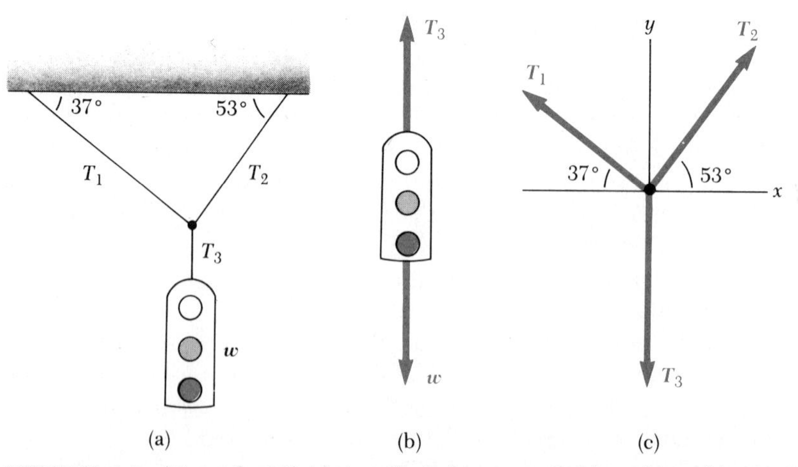

FIGURE 4.1 (Example 4.1) (a) A traffic light suspended by cables. (b) Free-body diagram for the traffic light. (c) Free-body diagram for the knot in the cable.

This value for T_2 can be substituted into (2) to give

$$T_1 \sin 37° + (1.33T_1)(\sin 53°) - 100\ \text{N} = 0$$

$$T_1[(0.602) + 1.33(0.800)] = 100\ \text{N}$$

$$T_1 = 60.0\ \text{N}$$

$$T_2 = 1.33T_1 = 1.33\ (60\ \text{N}) = 79.8\ \text{N}$$

EXAMPLE 4.2 Tug-a-Boat

Two fisherman are pulling a boat through the water as in Figure 4.2. Each exerts a force of 600 N directed at a 30° angle relative to the forward motion of the boat. If the boat moves with constant velocity, find the resistive force $\boldsymbol{F}$ of the water on the boat.

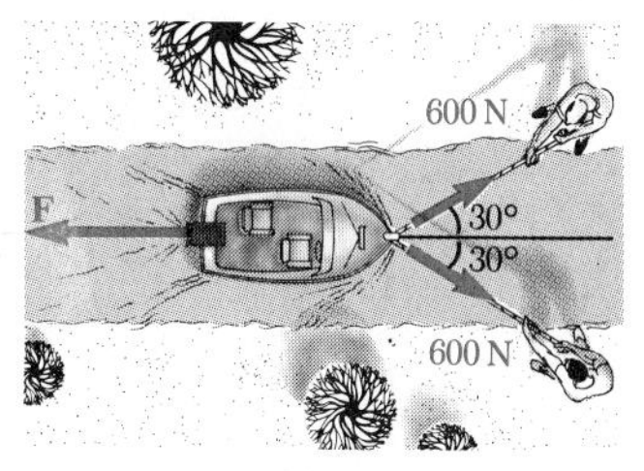

FIGURE 4.2 (Example 4.2) Two fishermen pulling a boat.

Solution If the boat moves with constant velocity, the first condition for equilibrium tells us that the resultant force on the boat in the x direction must be zero. The x component of each man's force is (600 N)(cos 30°), and the force of the water on the boat is $\boldsymbol{F}$. Hence, the net force on the boat in the x direction is

$$\sum F_x = (600\ \text{N})(\cos 30°) + (600\ \text{N})(\cos 30°) - F = 0$$

$$F = 2(600\ \text{N})(\cos 30°) = 1.04 \times 10^3\ \text{N}$$

EXAMPLE 4.3 Sled on a Slick Hill

A child holds a sled at rest on a frictionless snow-covered hill, as shown in Figure 4.3a. If the sled weighs 100 N, find the force the child must exert on the rope and the force the hill exerts on the sled.

Solution Figure 4.3b shows the forces acting on the sled and a convenient coordinate system to use for this type of problem. Note that $\boldsymbol{n}$, the force the ground exerts on the sled, is perpendicular to the hill. The hill can exert a component of force along the incline only if there is friction between the sled and the hill.

Applying the first condition for equilibrium to the sled, we find

$$\sum F_x = T - (100\ \text{N})(\sin 40°) = 0$$

$$T = 64.3\ \text{N}$$

$$\sum F_y = n - (100\ \text{N})(\cos 40°) = 0$$

$$n = 76.6\ \text{N}$$

Note that n is *less* than the weight of the sled in this case because the sled is on an incline and $\boldsymbol{n}$ is equal to and opposite the component of weight perpendicular to the incline.

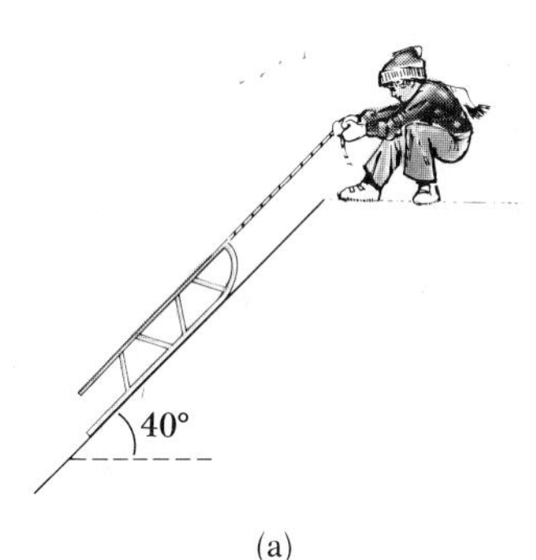

(a)

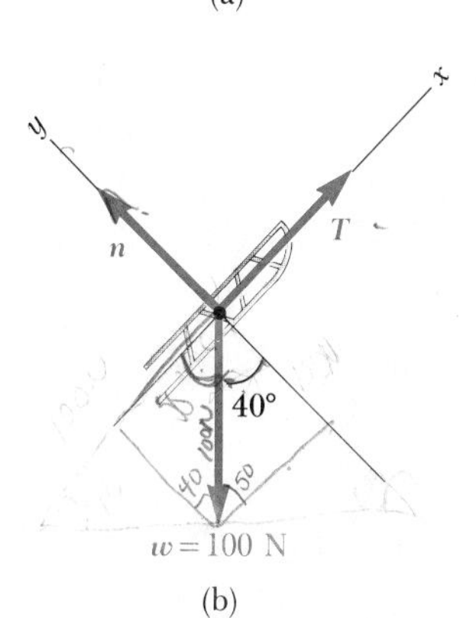

(b)

FIGURE 4.3 (Example 4.3) (a) A child holding a sled on a frictionless hill. (b) Free-body diagram for the sled.

EXERCISE What happens to the normal force as the angle of incline increases?

Answer It decreases.

EXERCISE Under what conditions would the normal force equal the weight of the sled?

Answer If the sled were on a horizontal surface and the applied force were either zero or along the horizontal.

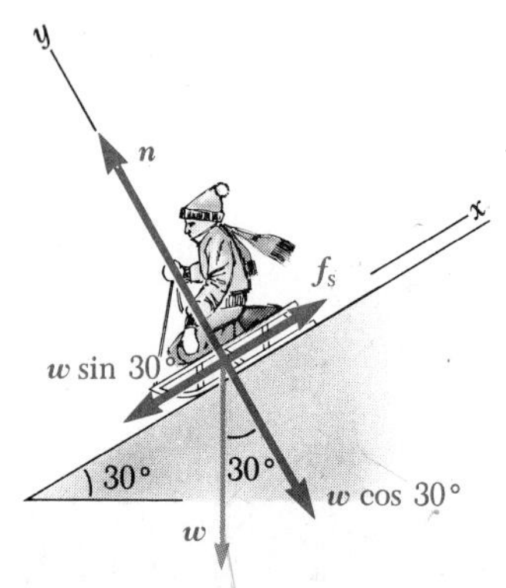

FIGURE 4.4 (Example 4.4) The sled is on the verge of slipping down the hill.

EXAMPLE 4.4 Sled on a Sticky Hill

The sled in Figure 4.4 is just on the verge of motion. If the combined weight of the sled and the child is w, find the coefficient of static friction between the sled and the hill.

Solution This example is similar to Example 4.3. Here, however, the force of static friction replaces the tension in the rope. The first condition for equilibrium in the x direction gives us the magnitude of the force of static friction, f_s:

$$\sum F_x = f_s - w \sin \theta = 0$$

$$f_s = w \sin \theta$$

The magnitude of the normal force is found from the first condition for equilibrium in the y direction:

$$\sum F_y = n - w \cos \theta = 0$$

$$n = w \cos \theta$$

When the sled is about to slip, the force of static friction is given by $f_s = \mu_s n$. Thus, we have

$$\mu_s = \frac{f_s}{n} = \frac{w \sin \theta}{w \cos \theta} = \tan \theta$$

Hence, if θ is known, one can calculate μ_s. For example, if θ is 30°, the coefficient of friction is 0.577. You should note that the expression for μ_s is independent of the mass or weight of the object.

EXERCISE If the child gives the sled a slight push to start it into motion, the sled will accelerate down the hill. Why is the velocity not constant in this situation?

Answer In this case, the component of weight down the incline is *greater* than the force of kinetic friction up the incline, so there is a net force acting down the incline which causes the sled to accelerate.

4.3 TORQUE

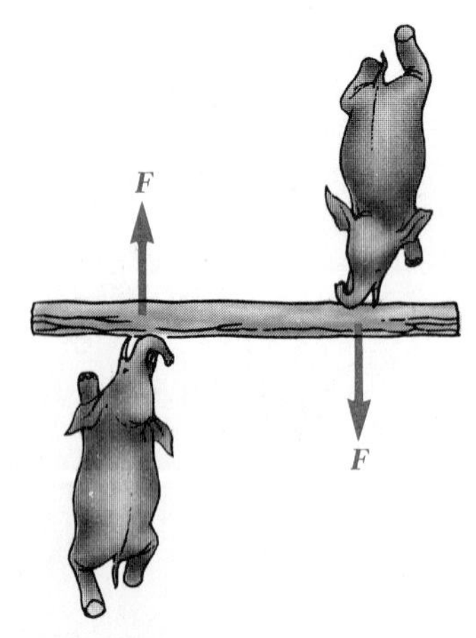

FIGURE 4.5 Top view of two elephants pushing a log with forces of equal magnitude but opposite in direction.

The first condition for equilibrium is not sufficient to ensure that an object is in complete equilibrium. This can be understood by considering the situation illustrated in Figure 4.5. Here we see a bird's-eye view of two elephants pushing on a large log with forces which are equal in magnitude but opposite in direction. The first condition for equilibrium is satisfied because the two equal-magnitude and oppositely directed external forces balance each other, and yet the log can still move. It will rotate clockwise because the forces do not act through a common point. This example should point out that, if one is to understand fully the effect of a force or group of forces on an object, not only must one know the magnitude and direction of the force(s), but the point of application of the force(s) must be considered as well.

The ability of a force to rotate a body about some axis is measured by a quantity called the **torque**, τ. The torque due to a force **F** has a magnitude given by the equation

$$\boxed{\tau = Fd} \qquad (4.3)$$

In this equation, τ (the Greek letter tau) is the torque and the distance d is called the **lever arm** (or moment arm) of the force $\boldsymbol{F}$.

The lever arm is the perpendicular distance from the axis of rotation to a line drawn along the direction of the force and through the point at which the force is applied.

For example, consider Figure 4.6, which represents a view looking down onto a door hinged at point O. A line perpendicular to the page and passing through O is the axis about which the door rotates. When the force $\boldsymbol{F}$ is applied as shown, the torque—and hence the rotational effect of the force—may be quite large. On the other hand, the same force applied at a point nearer the hinges produces a smaller torque (because d is smaller) and hence results in a smaller rotational effect on the door.

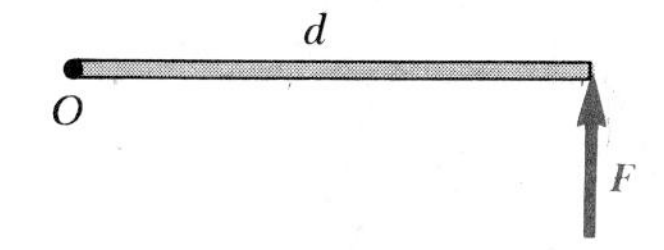

FIGURE 4.6 A bird's-eye view of a door hinged at O, with a force applied perpendicular to the door.

Next, consider the wrench pivoted about the axis O in Figure 4.7. In this case, the applied force $\boldsymbol{F}$ acts at an angle ϕ to the horizontal. If you examine the definition of lever arm given above, you will see that in this case the lever arm is the distance d shown in the figure and not L, the length of the wrench. That is, d is the perpendicular distance from the axis of rotation to the line along which the applied force acts. The distance d is related to L in this case by the expression $d = L \sin \phi$.

If there are two or more forces acting on an object, as in Figure 4.8, then each has a tendency to produce a rotation about the pivot O. For example, $\boldsymbol{F}_2$ has a tendency to rotate the object clockwise and $\boldsymbol{F}_1$ has a tendency to rotate the object counterclockwise. We shall use the convention that the sign of the torque resulting from a force is positive if its turning tendency is counterclockwise and negative if its turning tendency is clockwise. In Figure 4.8, then, the torque resulting from $\boldsymbol{F}_1$, which has a moment arm d_1, is positive and equal to $F_1 d_1$; the torque associated with $\boldsymbol{F}_2$ is negative and equal to $-F_2 d_2$. The *net torque* acting on the object about O is found by summing the torques:

$$\sum \tau = \tau_1 + \tau_2 = F_1 d_1 - F_2 d_2$$

Notice that the units of torque are units of force times length, such as newton-meter (N · m) or pound-foot (lb · ft).

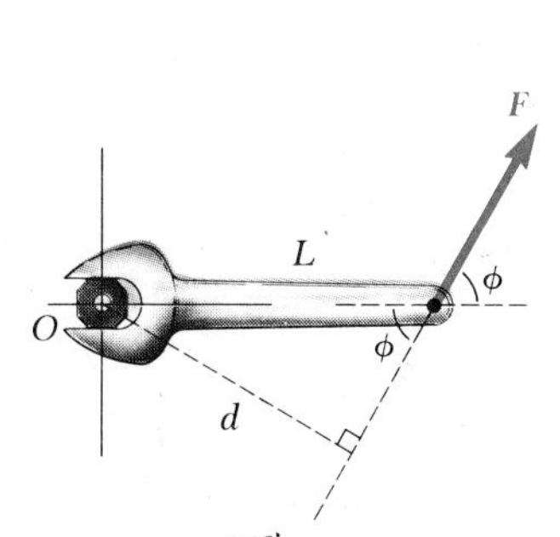

FIGURE 4.7 A force $\boldsymbol{F}$ acting at an angle ϕ to the horizontal produces a torque of magnitude $Fd = FL \sin \phi$ about the pivot O.

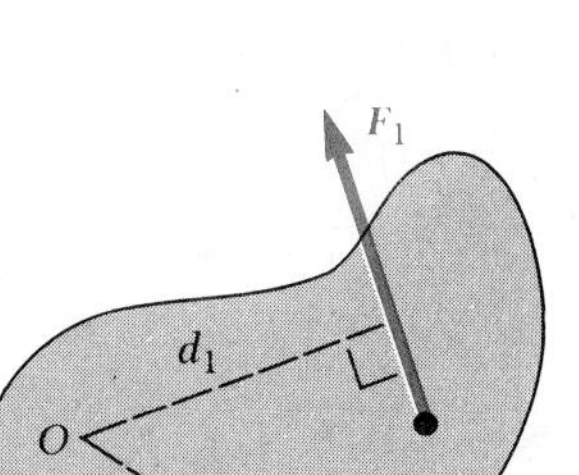

FIGURE 4.8 The force $\boldsymbol{F}_1$ tends to rotate the body counterclockwise about O, while $\boldsymbol{F}_2$ tends to rotate the body clockwise.

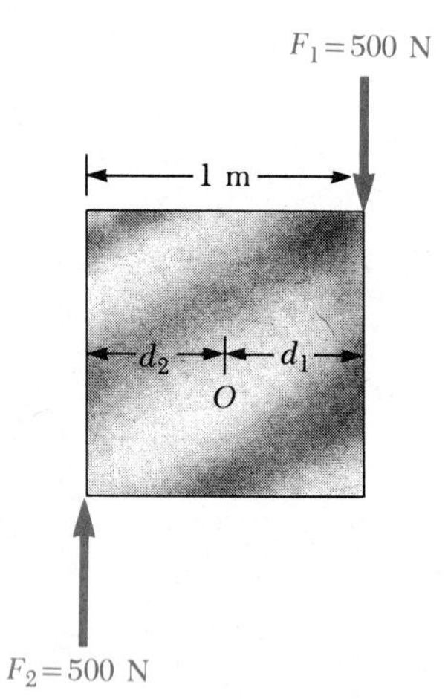

FIGURE 4.9 (Example 4.5) Top view of a safe being pushed with equal but opposite forces.

EXAMPLE 4.5 The Spinning Safe

Figure 4.9 shows a top view of a safe being pushed by two equal but opposite forces acting as shown. Find the net torque exerted on the safe if its width is 1 m. Assume an axis of rotation through the center of the safe.

Solution The torque produced by $\boldsymbol{F}_1$ is

$$\tau_1 = F_1 d_1 = -(500 \text{ N})(0.50 \text{ m}) = -250 \text{ N}\cdot\text{m}$$

and the torque produced by $\boldsymbol{F}_2$ is

$$\tau_2 = F_2 d_2 = -(500 \text{ N})(0.50 \text{ m}) = -250 \text{ N}\cdot\text{m}$$

Each force produces clockwise rotation, and hence the torques are both negative. Thus, the net torque is -500 N·m.

EXAMPLE 4.6 The Swinging Door

Find the torque produced by the 300-N force applied at an angle of 60° to the door of Figure 4.10a.

Solution The problem can be solved by proceeding along the lines indicated in Figure 4.10b. The lever arm d can be found from the dashed triangle shown in the figure. Since $\sin 60° = d/2$, we see that

$$d = (2 \text{ m})(\sin 60°) = 1.73 \text{ m}$$

Hence, the torque is

$$\tau = (300 \text{ N})(1.73 \text{ m}) = 520 \text{ N}\cdot\text{m}$$

Can you explain why the torque is positive in this case?

It is often simpler to solve a problem such as this by first resolving the applied force into components, as shown in Figure 4.10c. The horizontal component of the 300-N force is 150 N, and the vertical component is 260 N.

In this case, the 150-N force produces zero torque about the axis of rotation because the line along which the force acts passes through the axis of rotation and hence the lever arm is zero. The 260-N force has a lever arm of 2 m and hence produces a torque of

$$\tau = (260 \text{ N})(2 \text{ m}) = 520 \text{ N}\cdot\text{m}$$

This agrees with our previous calculation, as it must.

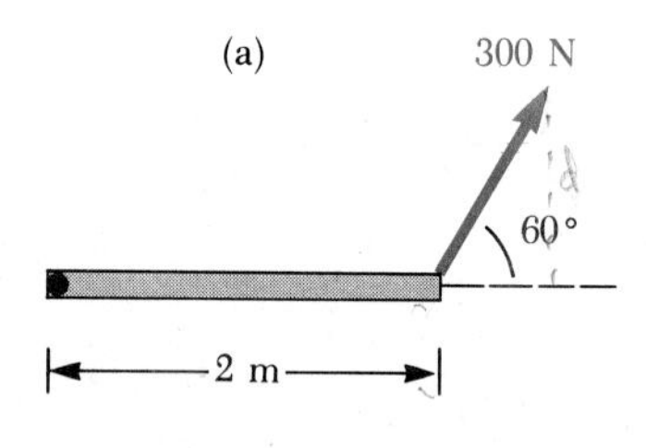

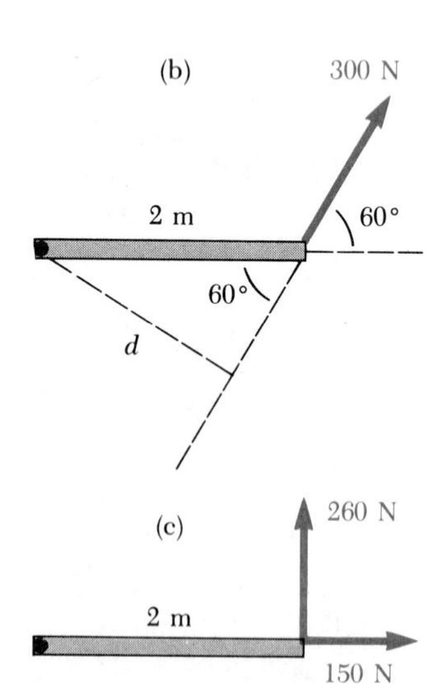

FIGURE 4.10 (Example 4.6) (a) Top view of a door being pulled by a 300-N force. (b) The moment arm of the 300-N force is d. (c) The components of the 300-N force.

4.4 THE SECOND CONDITION FOR EQUILIBRIUM

As we saw in Section 4.2, if an object meets the first condition for equilibrium, that object must be either at rest or moving with constant velocity. In other words, if the net force on the object is zero, its linear acceleration must be zero. Although this is a necessary condition for equilibrium, it is not a sufficient condition. Since real objects have a definite size, shape, and mass distribution, they tend to rotate under the action of external forces. For example, although the net force on the safe in Figure 4.9 is zero, the net torque is not zero and the safe rotates. We have already seen that torque is a measure of the rotational tendency of forces.

The **second condition for equilibrium** says that if an object is in rotational equilibrium, the net torque acting on it about any axis must be zero. That is,

$$\sum \tau = 0 \tag{4.4}$$

Thus we see that a body in static equilibrium must satisfy two conditions:

1. The resultant external force must be zero.

$$\sum \boldsymbol{F} = 0$$

2. The resultant external torque must be zero.

$$\sum \tau = 0$$

The first condition is a statement of translational equilibrium, and the second is a statement of rotational equilibrium. We shall provide a more detailed discussion of torque and its relation to rotational motion in Chapter 8.

Position of the Axis of Rotation

In the examples we have been using, an axis of rotation for calculating torques was selected without explanation. Often the nature of a problem will suggest a convenient location for this axis, but just as often no single location will stand out as being preferable. You might ask, "What axis should I choose in calculating the net torque?" The answer is that

it does not matter where you pick the axis of rotation for calculating the net torque if the object is in equilibrium; since the object is not rotating, the location of the axis is completely arbitrary.

To see that the location of the axis of rotation is completely arbitrary for an object in equilibrium, consider the object shown in Figure 4.11. Since the object is in equilibrium, the magnitudes of $\boldsymbol{F}_1$ and $\boldsymbol{F}_2$ must be equal. Furthermore, both forces have the same lever arm d relative to the arbitrary axis through O. Therefore, the sum of the counterclockwise torque F_1d and the clockwise torque $-F_2d$ is equal to zero.

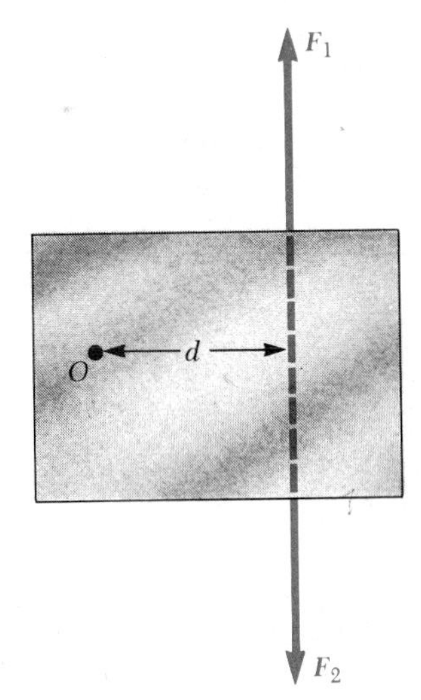

FIGURE 4.11 In order for the object to be in equilibrium, the magnitudes of $\boldsymbol{F}_1$ and $\boldsymbol{F}_2$ must be equal. The axis for calculating the torque is completely arbitrary.

4.5 THE CENTER OF GRAVITY

In all the problems we shall work, one of the forces that must be considered is the weight of the object being analyzed, that is, the force of gravity acting on the object. In order to compute the torque due to this weight, all of the weight can be considered as being concentrated at a single point called the **center of gravity.**

For uniform, symmetrical objects, the location of the center of gravity presents no difficulty. For example, a spherical object, such as a basketball, has its center of gravity at its geometric center. The same statement applies to uniform boxes, cylinders, and so on. Sometimes, though, the center of gravity lies outside the object, as in the case of a coat hanger or a doughnut.

When a system consists of several particles located at different points, the x coordinate of the center of gravity of the system can be found from the equation

$$x_{cg} = \frac{\sum m_i x_i}{\sum m_i} \tag{4.5}$$

where m_i is the mass of each particle and x_i is its x coordinate. Similarly, the y coordinate of the center of gravity of the system of particles can be found from

$$y_{cg} = \frac{\sum m_i y_i}{\sum m_i} \tag{4.6}$$

We shall withhold the derivation of these two equations until the problem set at the end of the chapter. There, we will give you several hints to guide you toward proving the validity of these equations.

EXAMPLE 4.7 Where is the Center of Gravity?

Three particles are located in a coordinate system as shown in Figure 4.12. Find the location of the center of gravity.

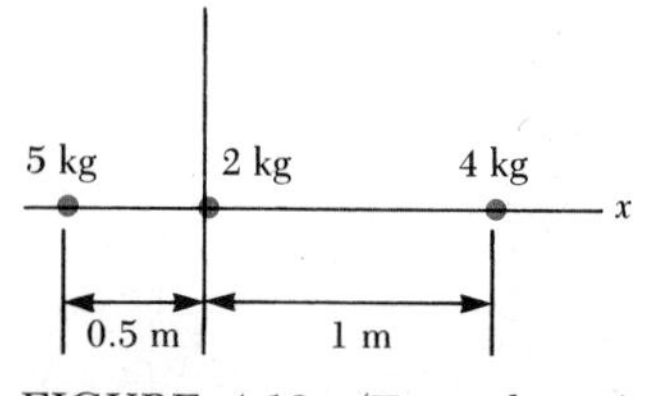

FIGURE 4.12 (Example 4.7) Locating the center of gravity for a system of three particles.

Solution The y coordinate of the center of gravity is zero because all particles are located on the x axis. To find the x coordinate of the center of gravity, we use Equation 4.5.

$$x_{cg} = \frac{\sum m_i x_i}{\sum m_i}$$

For the numerator, we find

$$\begin{aligned}\sum m_i x_i &= m_1 x_1 + m_2 x_2 + m_3 x_3 \\ &= (5\ \text{kg})(-0.5\ \text{m}) + (2\ \text{kg})(0\ \text{m}) + (4\ \text{kg})(1\ \text{m}) \\ &= 1.5\ \text{kg}\cdot\text{m}\end{aligned}$$

The denominator is $\sum m_i = 11$ kg, therefore

$$x_{cg} = \frac{1.5\ \text{kg}\cdot\text{m}}{11\ \text{kg}} = 0.136\ \text{m}$$

EXERCISE If a fourth particle of mass 2 kg is placed at the position $x = 0$ and $y = 0.25$ m, find the x and y coordinates of the center of gravity for this system of four particles.

Answer $x_{cg} = 0.115$ m, $y_{cg} = 0.038$ m.

4.6 EXAMPLES OF OBJECTS IN EQUILIBRIUM

In Section 4.2 we discussed some techniques that should be helpful in solving all problems dealing with objects in equilibrium. Now that we have added the second condition for equilibrium as an integral part of our approach to such problems, we need to add the following to our five-step problem-solving procedure of Section 4.2.

6. Choose a convenient origin for calculating the net torque on the object. Remember that the choice of the origin for the torque equation is arbitrary; therefore choose an origin that will simplify your calculation as much as possible. Note that a force that acts along a line passing through the point chosen as the axis of rotation produces zero contribution to the torque.
7. The first and second conditions for equilibrium will give a set of simultaneous equations with several unknowns. All that is left in your solution is to solve for the unknowns in terms of the known quantities.

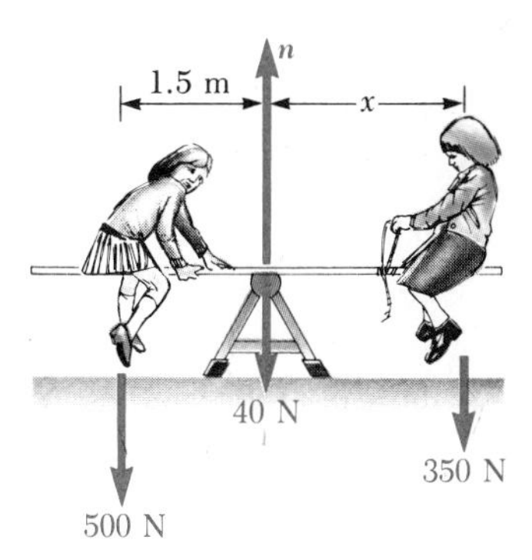

FIGURE 4.13 (Example 4.8) Two children balanced on a seesaw.

EXAMPLE 4.8 The Seesaw

A uniform board of weight 40 N supports two children weighing 500 N and 350 N, as shown in Figure 4.13. If the support (often called the *fulcrum*) is under the center of gravity of the board and if the 500 N child is 1.5 m from the center, (a) determine the upward force ***n*** exerted on the board by the support.

Solution First note that, in addition to ***n***, the external forces acting on the board are the weights of the children and the weight of the board, all of which act downward. We can assume that the center of gravity of the board is at its geometric center because we were told that the board is uniform. Since the system is in equilibrium, the upward force ***n*** must balance all the downward forces. From $\Sigma F_y = 0$, we have

$$n - 500 \text{ N} - 350 \text{ N} - 40 \text{ N} = 0 \qquad \text{or} \qquad n = 890 \text{ N}$$

It should be pointed out here that the equation $\Sigma F_x = 0$ also applies to this situation, but it is unnecessary to consider this equation because we have no forces acting horizontally on the board.

(b) Determine where the 350-N child should sit to balance the system.

Solution To find this position, we must invoke the second condition for equilibrium. Taking the center of gravity of the board as the axis for our torque equation, we see from $\Sigma\tau = 0$ that

$$(500 \text{ N})(1.5 \text{ m}) - (350 \text{ N})(x) = 0$$
$$x = 2.14 \text{ m}$$

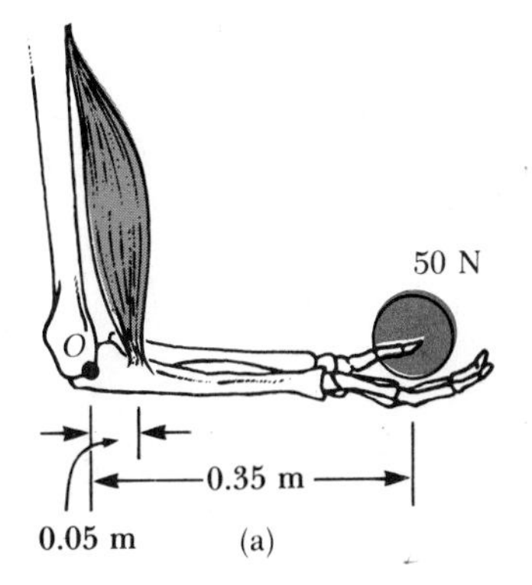

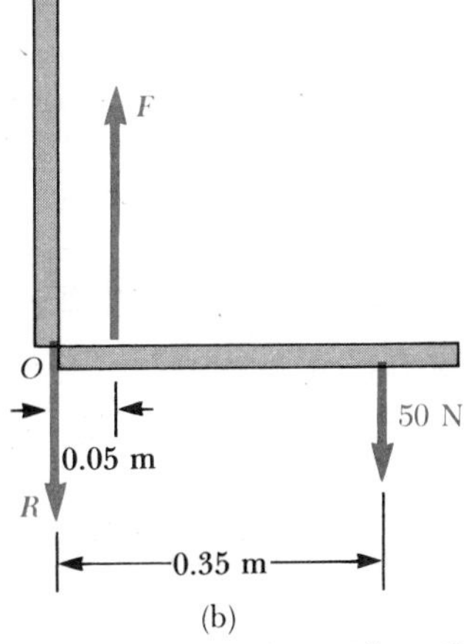

FIGURE 4.14 (Example 4.9) (a) A weight held with the forearm in the horizontal position. (b) The mechanical model for the system.

EXAMPLE 4.9 A Weighted Forearm

A 50-N weight is held in the hand with the forearm in the horizontal position, as in Figure 4.14a. The biceps muscle is attached 0.05 m from the joint, and the weight is 0.35 m from the joint. Find the upward force that the biceps exerts on the forearm (the ulna) and the downward force on the upper arm (the humerus) acting at the joint. Neglect the weight of the forearm.

Solution The forces acting on the forearm are equivalent to those acting on a bar of length 0.35 m, as shown in Figure 4.14b, where ***F*** is the upward force of the biceps and ***R*** is the downward force at the joint. From the first condition for equilibrium, we have

$$(1) \qquad \sum F_y = F - R - 50 \text{ N} = 0$$

From the second condition for equilibrium, we know that the sum of the torques about any point must be zero. With the joint O as the axis, we have

$$F(0.05 \text{ m}) - (50 \text{ N})(0.35 \text{ m}) = 0$$
$$F = 350 \text{ N}$$

This value for F can be substituted into (1) to give $R = 300$ N. These values correspond to $F = 79$ lb and $R = 68$ lb. Hence, the forces at joints and in muscles can be extremely large.

EXERCISE In reality, the biceps makes an angle of 15° with the vertical, so that $\boldsymbol{F}$ has both a vertical and a horizontal component. Find the value of F and the components of $\boldsymbol{R}$ including this fact in your analysis.

Answer $F = 362$ N, $R_x = 94.0$ N, $R_y = 300$ N.

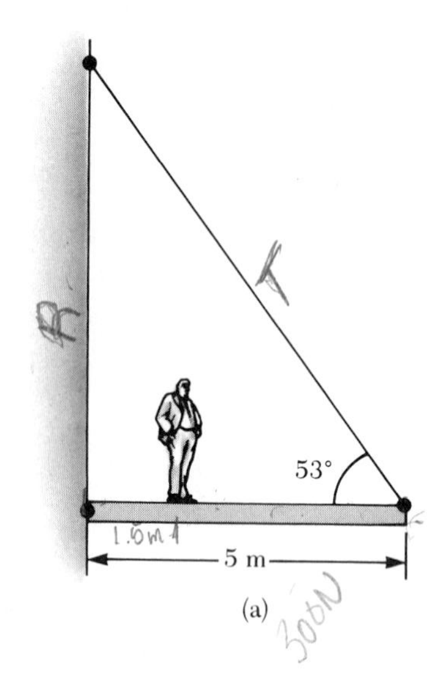

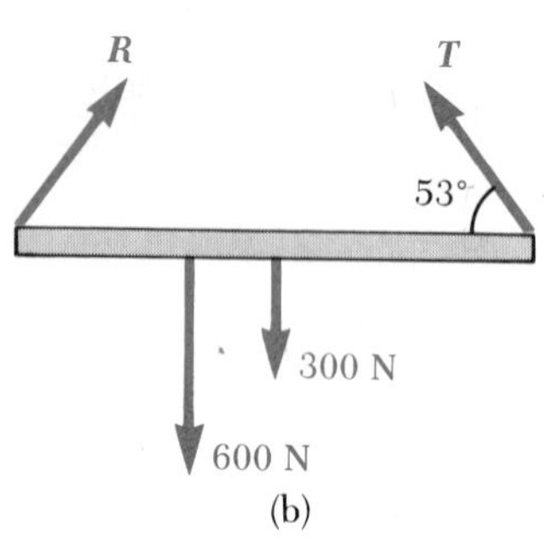

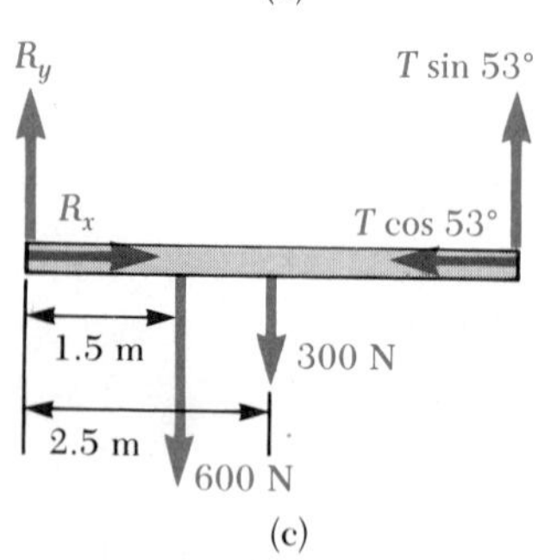

FIGURE 4.15 (Example 4.10) (a) A uniform beam attached to a wall and supported by a cable. (b) Free-body diagram for the beam. (c) Component form of the free-body diagram.

EXAMPLE 4.10 Walking a Horizontal Beam

A uniform horizontal beam of length 5 m and weight 300 N is attached to a wall by a pin connection that allows the beam to rotate. Its far end is supported by a cable that makes an angle of 53° with the horizontal (Fig. 4.15a). If a 600-N person stands 1.5 m from the wall, find the tension in the cable and the force exerted on the beam by the wall.

Solution First, we must identify all the external forces acting on the beam and sketch them on a free-body diagram. This is shown in Figure 4.15b. The forces on the beam consist of the weight of the beam, 300 N, acting downward, the downward force on the beam by the man which is equal in magnitude to his weight, 600 N, the tension force $\boldsymbol{T}$ in the cable, and the force of the wall on the beam, $\boldsymbol{R}$. We now resolve the forces $\boldsymbol{T}$ and $\boldsymbol{R}$ into their horizontal and vertical components as shown in Figure 4.15c. Note that the x component of the tension force ($T \cos 53°$) is to the left, while the y component ($T \sin 53°$) is upward. The horizontal and vertical components of $\boldsymbol{R}$ are denoted by R_x and R_y, respectively. Applying the first condition of equilibrium to the beam gives

$$(1) \qquad \sum F_x = R_x - T\cos 53° = 0$$

$$(2) \qquad \sum F_y = R_y + T\sin 53° - 600\text{ N} - 200\text{ N} = 0$$

Our unknowns are R_x, R_y, and T. Because there are three unknowns and only two equations, we are unable to find a solution from just the first condition of equilibrium.

Now let us use the second condition of equilibrium. A convenient axis to choose for our torque equation is the one that passes through the pivot at O. The feature that makes this point so convenient is that the forces R_x, R_y, and $T \cos 53°$ all have a lever arm of zero, and hence have zero torque about this pivot. Recalling our sign convention for the torque about an axis and noting that the lever arms of the 600 N, 300 N, and $T \sin 53°$ forces are 1.5 m, 2.5 m, and 5 m, respectively, we get

$$(3) \qquad \sum \tau_O = (T\sin 53°)(5\text{ m}) - (200\text{ N})(2.5\text{ m}) - (600\text{ N})(1.5\text{ m}) = 0$$

$$T = 350\text{ N}$$

Thus, the torque equation using this axis gives us one of the unknowns immediately! This value for T is then substituted into (1) and (2) to give

$$R_x = 211\text{ N}$$

$$R_y = 520\text{ N}$$

If we had selected some other axis for the torque equation, the solution would have been the same. For example, if the axis were to pass through the center of gravity of the beam, the torque equation would involve both T and R_y. However, this equation, together with (1) and (2), could still be solved for the unknowns. Try it!

EXERCISE Repeat this problem, but with the direction of R_x opposite to that shown in Figure 4.15c. What answers do you get for T, R_x, and R_y?

Answer $T = 350$ N, $R_x = -211$ N, and $R_y = 520$ N. The negative sign for R_x means its direction was chosen incorrectly. The direction of R_x must be to the right as shown in Figure 4.15c.

EXAMPLE 4.11 Don't Climb the Ladder

A uniform ladder of length 10 m and weight 50 N rests against a smooth vertical wall as in Figure 4.16a. If the ladder is just on the verge of slipping when the angle the ladder makes with the ground is 50°, find the coefficient of static friction between the ladder and ground.

Solution The free-body diagram for the ladder showing all external forces acting on it is illustrated in Figure 4.16b. At the base of the ladder, the earth exerts an upward normal force, $\boldsymbol{n}$, and a force of static friction, $\boldsymbol{f}$, acting to the right. The wall exerts a force $\boldsymbol{P}$ to the left. Note that the force $\boldsymbol{P}$ is horizontal because the wall is smooth. (If the wall were rough, there would be an upward frictional force exerted on the ladder.) From the first condition of equilibrium applied to the ladder, we have

$$(1) \qquad \Sigma F_x = f - P = 0$$

$$(2) \qquad \Sigma F_y = n - 50 = 0$$

From (2) we see that $n = 50$ N. Furthermore, when the ladder is on the verge of slipping, the force of static friction must be a maximum given by $f_{s,\max} = \mu_s n = \mu_s$ (50 N). Thus, (1) reduces to

$$(3) \qquad \mu_s(50 \text{ N}) = P$$

Let us now apply the second condition of equilibrium and take the torques about the axis O located at the bottom of the ladder as in Figure 4.16c. The force $\boldsymbol{P}$ and the weight of the ladder are the only forces that contribute to the torque about this axis, and their lever arms are shown in Figure 4.16c. Since the length of the ladder is 10 m, you should note that the lever arm for $\boldsymbol{P}$ is $d_1 = 10 \sin 50°$. Likewise, the lever arm for the 50 N weight force is $d_2 = 5 \cos 50°$, where the weight vector acts through the center because the ladder is uniform. Thus, we find

$$\sum \tau_O = P(10 \sin 50°) - (50 \text{ N})(5 \cos 50°) = 0$$

$$P = 21 \text{ N}$$

Now that P is known, we can substitute its value into (3) to find μ_s, giving

$$\mu_s = \frac{21 \text{ N}}{50 \text{ N}} = 0.42$$

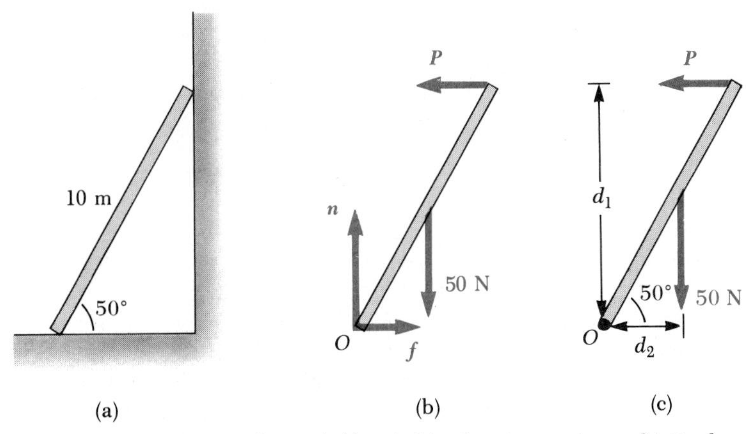

FIGURE 4.16 (Example 4.11) (a) A ladder leaning against a frictionless wall. (b) Free-body diagram for the ladder. (c) Lever arms for the forces $\boldsymbol{w}$ and $\boldsymbol{P}$.

SUMMARY

An object is in **equilibrium** when the following conditions are satisfied: (1) the resultant external force must be zero and (2) the resultant external torque must be zero about any origin. That is,

$$\sum \mathbf{F} = 0 \tag{4.1}$$

$$\sum \tau = 0 \tag{4.4}$$

The first equation is called the **first condition for equilibrium.** When this equation holds, an object will either be at rest or moving with a constant velocity. The second equation is called the **second condition for equilibrium.** When it holds, an object is said to be in rotational equilibrium.

The ability of a force to rotate an object about some axis is measured by a quantity called the **torque,** τ. The magnitude of the torque is given by

$$\tau = Fd \tag{4.3}$$

In this equation, d is the **lever arm,** which is the perpendicular distance from the axis of rotation to a line drawn along the direction of the force. The sign of the torque is negative if the turning tendency of the corresponding force is clockwise and positive if its turning tendency is counterclockwise.

The **center of gravity** of an object is the point at which the weight of the object can be considered to be concentrated. The x and y locations of the center of gravity are

$$x_{cg} = \frac{\Sigma m_i x_i}{\Sigma m_i} \qquad y_{cg} = \frac{\Sigma m_i y_i}{\Sigma m_i} \qquad \text{(4.5) and (4.6)}$$

ADDITIONAL READING

P. Davidovits, *Physics in Biology and Medicine*, Englewood Cliffs, N.J., Prentice-Hall 1975, chaps. 1 and 2.

J. G. Kemeny, "Man Viewed as a Machine," *Sci. American*, April 1955, p. 58.

R. Mark, "The Structural Analysis of Gothic Cathedrals," *Sci. American*, November 1972, p. 90.

M. Salvadori, *Why Buildings Stand Up: The Strength of Architecture*, New York, McGraw-Hill, 1982.

D. B. Steinman, "Bridges," *Sci. American*, November 1954, p. 60.

K. F. Wells, and J. Wessel, *Kinesiology*, Philadelphia, W.B. Saunders, 1971. The study of balance and body levers in humans.

QUESTIONS

1. Can an object be in equilibrium if only one force acts on it?
2. It is often said that Newton's first law is really a special case of his second law. Explain that statement.
3. An object thrown into the air comes to rest at the highest point in its path. Is it in equilibrium at this point? Explain this carefully.
4. Is the moon in equilibrium as it circles about the earth?
5. Under what conditions would a body falling through a viscous fluid be in equilibrium?
6. Give an example in which the net torque acting on an object is zero and yet the net force is nonzero.
7. Give an example in which the net force acting on an object is zero and yet the net torque is nonzero.
8. Can an object be in equilibrium if the only torques acting on it tend to produce clockwise rotation?

9. Can a body be in equilibrium if it is in motion? If so, give an example.
10. Locate the center of gravity for the following uniform objects: (a) a sphere, (b) a cube, (c) a cylinder, (d) a doughnut.
11. A male and a female student are asked to do the following task. Face a wall, step three foot lengths away from the wall, and then lean over and touch the wall with your nose, keeping your hands behind your back. The male usually fails, while the female succeeds. How would you explain this?
12. When lifting a heavy object, why is it recommended to straighten your back, as much as possible, rather than bend over and lift mainly with the arms?
13. A tall crate and a short crate of equal mass are placed side-by-side on an incline (without touching each other). As the incline angle is increased, which crate will tip first? Explain.
14. Why do you tend to lean backward when carrying a box of books in your arms but forward when the same books are carried in a backpack?
15. The mayor of a city decides to fire some city employees because they will not remove the sag from the cables which support the city traffic lights. If you were a lawyer, what defense would you give on behalf of the employees? Who do you think will win the case in court?
16. Why are sports cars usually built low to the ground and wide-tracked?

PROBLEMS

Section 4.1 The First Condition For Equilibrium

1. It is amateur night at the tightrope walkers' convention, and a 600-N performer finds himself in the awkward position shown in Figure 4.17. If the angle between the rope and the horizontal is 8°, find the tension in the rope on either side of the performer.

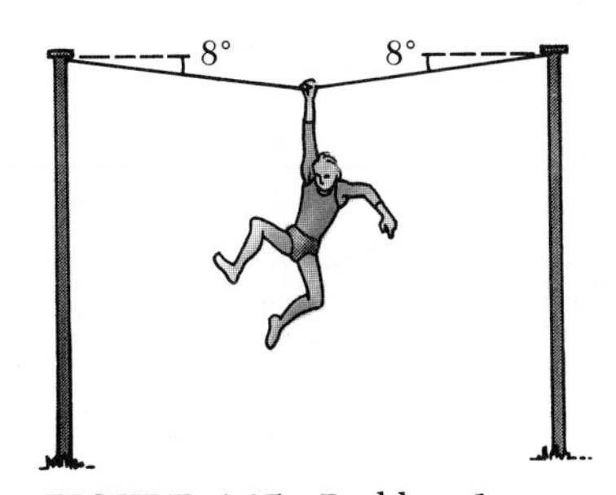

FIGURE 4.17 Problem 1

2. Two horses are pulling a sled along an icy road; assume that friction between the sled and the road is negligible. A farmer keeps the sled along the proper path by pulling on it with a rope attached at the point where the cables from the horses are attached. A noise startles the horses and they move toward opposite sides of the road as shown in Figure 4.18. What is the magnitude and direction of the force that the farmer will have to exert on the sled to keep it moving along the road at constant velocity?

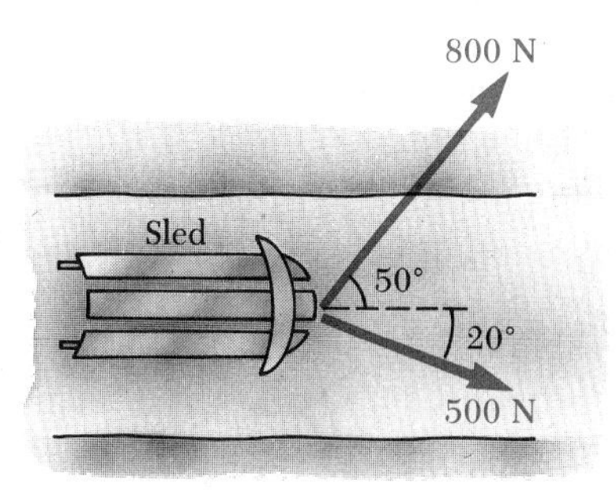

FIGURE 4.18 Problem 2

3. Find the tension in the two wires that support the 100-N light fixture shown in Figure 4.19.

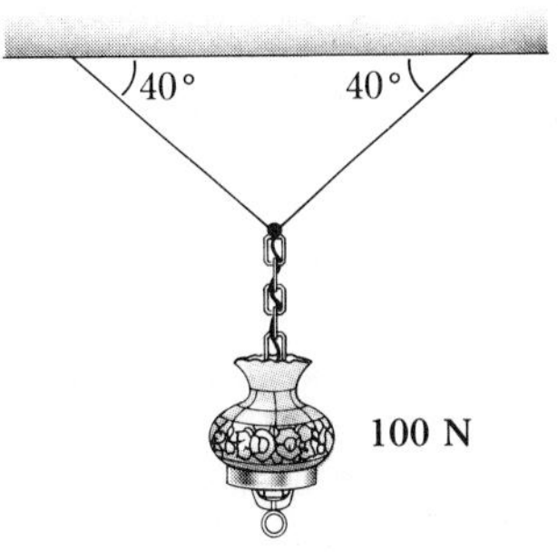

FIGURE 4.19 Problem 3

4. A 150-N bird feeder is supported by three cables as shown in Figure 4.20. Find the tension in each cable.

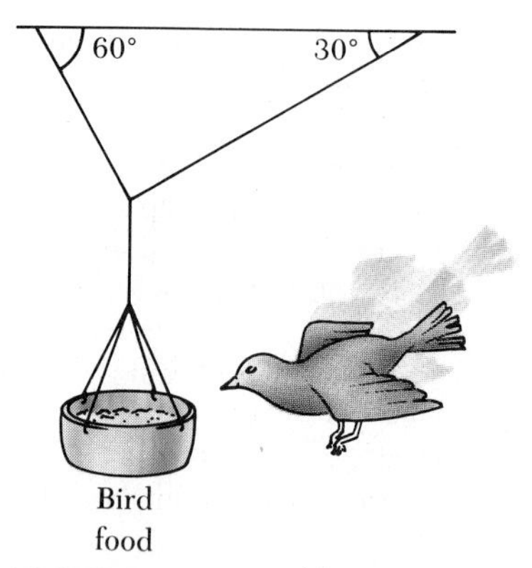

FIGURE 4.20 Problem 4

5. A 500-N figure skater moves across the floor of an ice rink, where the coefficient of friction between skates and ice is 0.1. What forward force must be exerted on her to keep her moving at a constant speed?

6. A man doing his spring cleaning pulls a 150-N vacuum cleaner across the floor at constant velocity by exerting a force on it at an angle of 30° with the horizontal. If he must pull with a force of 35 N to move the vacuum cleaner, what is the coefficient of friction between the vacuum cleaner and the floor?

• **7.** A boy drags a 100-N sled up a 20° slope at constant velocity. If the coefficient of friction between sled and hill is 0.2, what force must he exert at an angle of 35° with respect to the hill?

• **8.** A crate is held at rest on a frictionless 53° ramp by a rope parallel to the incline. If the tension in the rope is 7000 N, what is the mass of the crate?

Section 4.3 Torque

9. If the torque required to loosen a nut holding a flat tire in place on a car has a magnitude of 40 N·m, what *minimum* force must be exerted by a garage worker at the end of a 30-cm lug wrench to accomplish the task?

10. Find the torque exerted on the fishing pole by a fish pulling with a force of 100 N as shown in Figure 4.21. Choose the axis of rotation of the pole to be at the hand and perpendicular to the page.

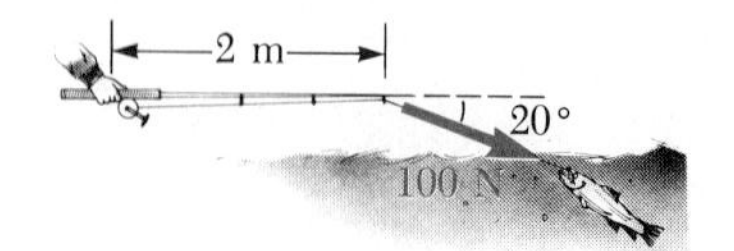

FIGURE 4.21 Problem 10

11. If the fishing pole of Problem 10 is inclined to the horizontal at an angle of 20°, as shown in Figure 4.22, what is the torque exerted by the fish about an axis perpendicular to the page and passing through the hand of the person holding the pole?

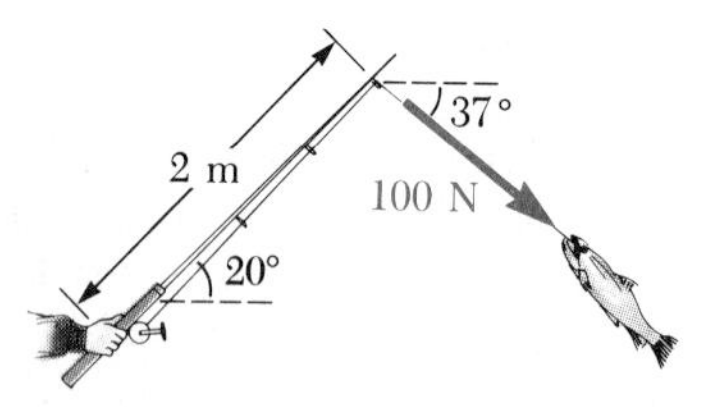

FIGURE 4.22 Problem 11

12. A 48-kg diver stands at the end of a 3-m long diving board. What torque does the weight of the diver produce about an axis perpendicular to and in the plane of the diving board through its midpoint?

Section 4.5 The Center of Gravity

13. Four objects are located along the y axis as follows. A 2-kg object is at +3 m, a 3-kg object is at +2.5 m, a 2.5-kg object is at the origin, and a 4-kg object is at −0.5 m. Where is the center of gravity of these objects located?

14. Where would a fifth object of mass 5 kg have to be placed along the y axis so that the center of gravity of the objects in problem 13 is at the origin?

15. Three masses are located in a rectangular coordinate system as follows: a 2-kg mass is at (3, −2), a 3-kg mass is at (−2, 4), and a 1-kg mass is at (2, 2), where all distances are in meters. Find the coordinates of the center of gravity for the three masses.

16. A uniform T square used by a draftsman has dimensions as shown in Figure 4.23. Locate the center of gravity with respect to point O. (*Hint:* Note that the mass of each rectangular part is proportional to its area.)

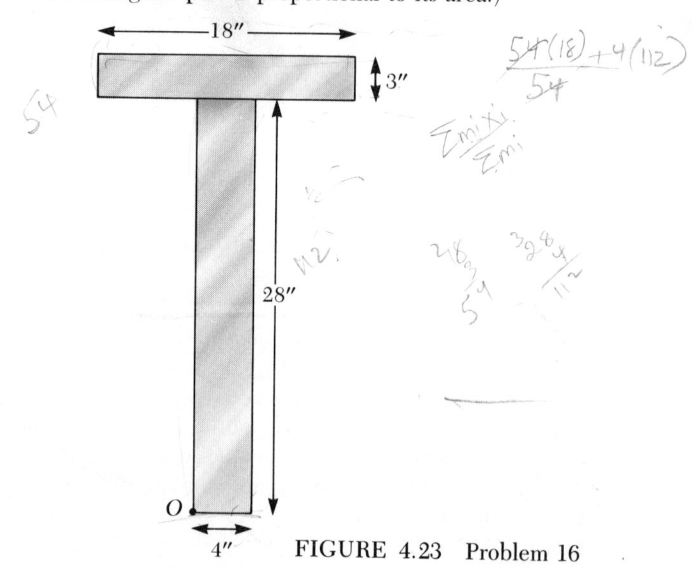

FIGURE 4.23 Problem 16

17. A uniform carpenter's square has the shape of an L, as in Figure 4.24. Locate the center of gravity relative to an origin at the lower left corner. (Note the hint for Problem 16.)

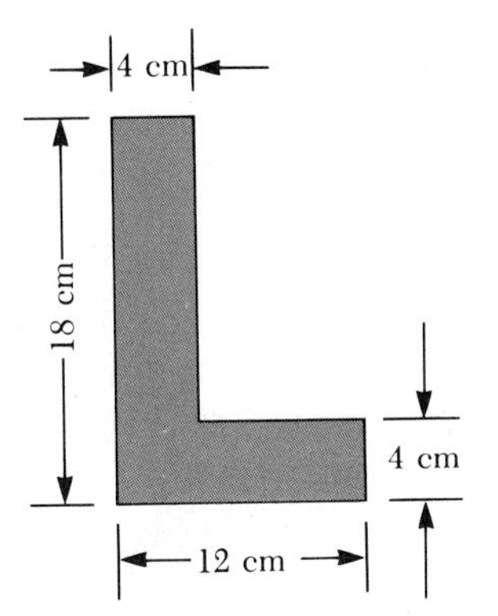

FIGURE 4.24 Problem 17

• **18.** The force of gravitational attraction on an object is exerted not on the object as a whole but on each molecule of the object. The resultant of all these forces is the weight of the object directed downward. In Figure 4.25 we show an object divided into many very small particles, such as the molecules mentioned above. The masses of these tiny parts are labeled m_1, m_2, m_3, . . . and have coordinates (x_1, y_1), (x_2, y_2), (x_3, y_3) . . . Also pictured are the vectors indicating the weight of each particle. (a) Find the torque on each of these particles about the origin and add these torques together. (b) The torque due to the total weight, $w = Mg$, of the object is located at the point (x_{cg}, y_{cg}). Find the torque of the weight about the origin. (c) The net torque due to the individual particles and the torque due to the weight considered to be acting at the center of gravity must be the same. Equate the expressions found in part (a) and part (b) to show that $x_{cg} = \Sigma m_i x_i / \Sigma m_i$.

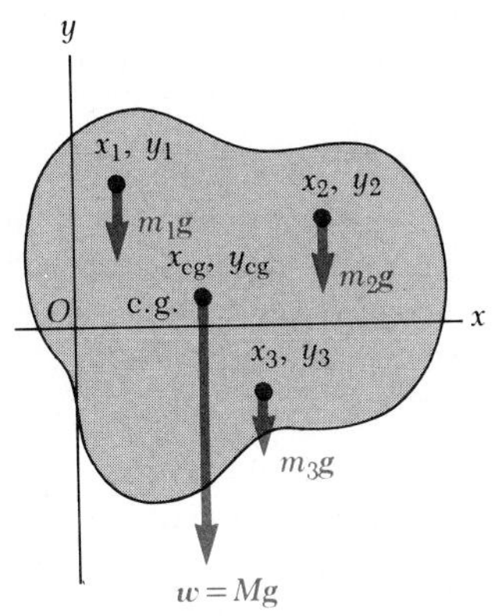

FIGURE 4.25 Problem 18

Section 4.6 Examples of Objects in Equilibrium

19. Two children sit on a seesaw such that a 400-N child is 2 m from the support (the fulcrum). Where should a second child of weight 475 N sit in order to balance the system if the support is at the center of the plank?

20. The 400-N child of Problem 19 decides that she would like to seesaw alone. To do so, she moves the board such that its weight is no longer directly over the fulcrum. She finds that she will be balanced when she is 1.5 m to the left of the fulcrum and the center of the plank is 0.5 m to the right of the fulcrum. What is the weight of the plank?

21. A window washer is standing on a scaffold supported by vertical ropes at each end. The scaffold weighs 200 N and is 3 m long. What is the tension in each rope when the 700-N worker stands 1 m from one end?

22. A college student decides to enter a contest on a popular television show called "Stupid People Tricks." His trick is to balance three objects on a metric ruler of negligible weight as shown in Figure 4.26. If the ruler is 20 cm long and uniform, where should the fulcrum be placed so that balance is achieved? (Neglect the dimensions of the objects.)

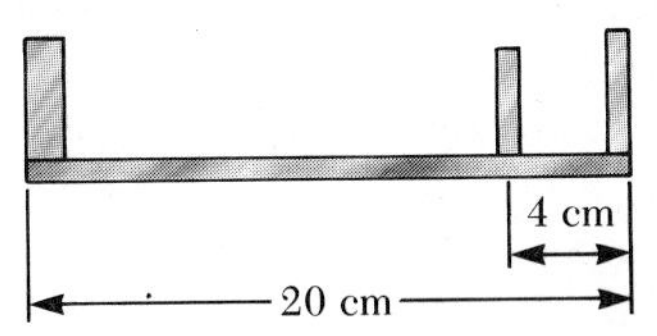

FIGURE 4.26 Problem 22

23. A student slides a 2000-N crate across a floor by pulling with a force of 750 N at an angle of 30° with the horizontal. If the crate moves at a constant speed, find the coefficient of sliding friction between the crate and the floor.

24. A 200-N floodlight in a park is supported at the end of a beam of negligible mass hinged to a pole as shown in Figure 4.27. A cable at an angle of 30° with the beam helps to support the light. Find (a) the tension in the cable, and (b) the horizontal and vertical forces exerted on the beam by the pole.

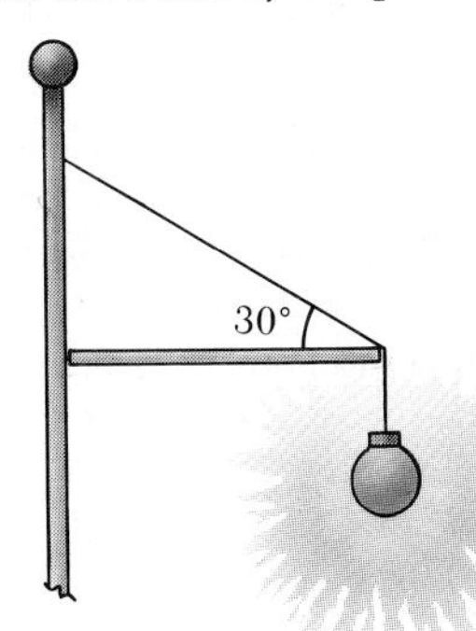

FIGURE 4.27 Problem 24

25. A uniform plank of length 2 m and mass 30 kg is supported by three ropes as in Figure 4.28. Find the tension in each rope when a 700-N person is 0.5 m from the left end.

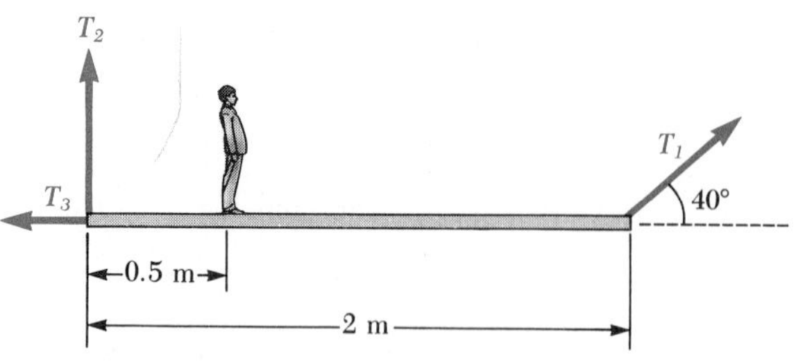

FIGURE 4.28 Problem 25

26. The muscle used for chewing, the masseter, is one of the strongest in the human body. It is attached to the mandible, or lower jawbone, as shown in Figure 4.29a. The jawbone is pivoted about a socket just in front of the auditory canal. The forces acting on the jawbone are equivalent to those acting on a curved bar as shown in Figure 4.29b. The force $\boldsymbol{C}$ is the force exerted against the jawbone by the food being chewed, $\boldsymbol{T}$ is the tension in the masseter, and $\boldsymbol{R}$ is the force exerted on the mandible by the socket. If you bite down on a piece of steak with a force of 50 N, find the forces $\boldsymbol{T}$ and $\boldsymbol{R}$.

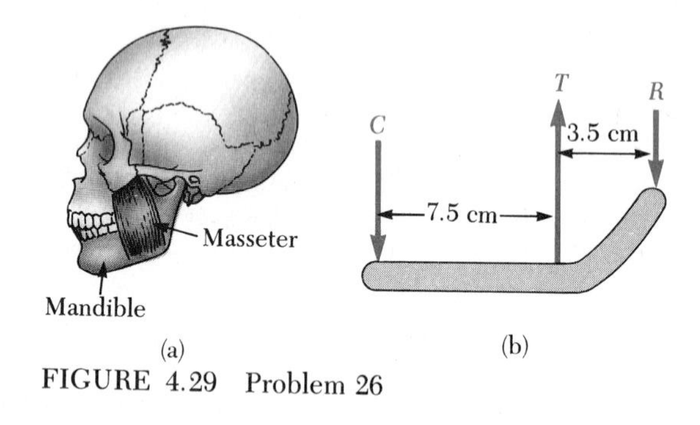

FIGURE 4.29 Problem 26

27. A racing car has a mass of 1600 kg. The distance between the front and rear axles is 3 m. If the center of gravity of the car is 2 m from the rear axle, what is the normal force on each tire?

28. The large quadriceps muscle in the upper leg terminates at its lower end in a tendon attached to the upper end of the tibia as shown in Figure 4.30a. The forces on the lower leg when the leg is extended are modeled as in Figure 4.30b, where $\boldsymbol{T}$ is the tension in the tendon, $\boldsymbol{C}$ is the weight of the lower leg, and $\boldsymbol{F}$ is the weight of the foot. Find the tension $\boldsymbol{T}$ when the tendon is at an angle of 25° with the tibia, assuming $C = 30$ N, $F = 12.5$ N, and the leg is extended at an angle of 40° with respect to the vertical ($\theta = 40°$). Assume the center of gravity of the lower leg is at its center, and that the tendon attaches to the lower leg at a point one fifth of the way down the leg.

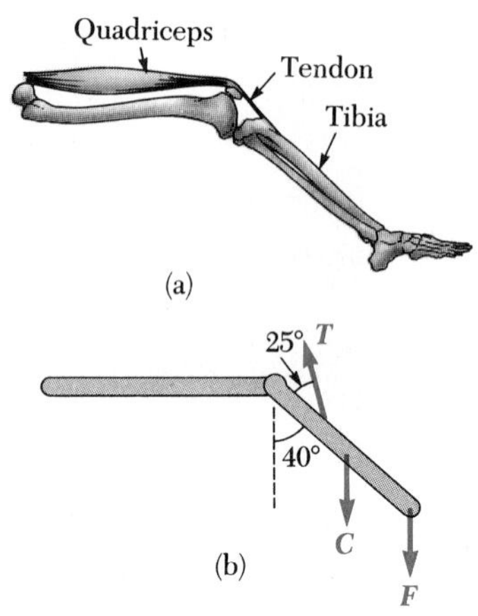

FIGURE 4.30 Problem 28

• 29. An 8-m uniform ladder weighing 200 N rests against a smooth wall. The coefficient of static friction between the ladder and the ground is 0.6, and the ladder makes a 50° angle with the ground. How far up the ladder can an 800-N person climb before the ladder begins to slip?

• 30. A 15-m uniform ladder weighing 500 N rests against a frictionless wall. The ladder makes an angle of 60° with the horizontal. (a) Find the horizontal and vertical forces that the earth exerts on the base of the ladder when an 800-N firefighter is 4 m from the bottom. (b) If the ladder is just on the verge of slipping when the firefighter is 9 m up, what is the coefficient of static friction between ladder and ground?

• 31. A 1200-N go-cart is being pulled up a 25° incline by a rope that makes an angle of 35° with the horizontal. Neglecting all frictional effects, determine the tension in the rope necessary to pull the cart up the incline at constant speed.

• 32. A dormitory door 2.5 m high and 1 m wide weighs 250 N, and its center of gravity is at its geometric center. The door is supported by hinges 0.25 m from top and bottom, each hinge carrying half the weight. Determine the horizontal component of the forces exerted by each hinge on the door.

•• 33. An iron trapdoor 1.25 m wide and 2 m long weighs 360 N and is hinged along the short dimension. Its center of gravity is at its geometric center. What force applied at right angles to the door is required to lift it (a) when it is horizontal and (b) when it has been opened so that it makes an angle of 30° with the horizontal? (Assume that the force is applied at the edge of the door opposite the hinges.)

ADDITIONAL PROBLEMS

34. A horizontal plank of negligible weight is supported by two scales. A woman 5′4″ tall lies down on the plank and the scale located directly under her head reads 60 lb while the scale directly under her feet reads 55 lb. Locate the center of gravity of the woman with respect to her feet.

35. A tightrope walker weighing 600 N stands at the center of a rope. (a) If the ends of the rope make an angle of 10° with the horizontal, what is the tension in the rope? (b) If the tightrope can withstand a tension of 2000 N without breaking, what is the minimum angle the rope can make with the horizontal without breaking?

36. Two men carry a 100-N bag of fertilizer out to their garden by placing it on a 1.5-m-long plank of negligible weight. If the bag is placed 0.25 m from one of the men, how much weight does each support?

37. A bicycle rider exerts a force of 90 N vertically downward on the pedal at a distance of 0.15 m from the center of the drive sprocket. What is the resulting torque about the axis of the sprocket when the pedal is (a) in the horizontal position and (b) in the vertical position?

38. Four masses are located in a rectangular coordinate system as follows: a 2-kg mass is at (0, 0), a 3-kg mass is at (0, 2), a 4-kg mass is at (2, 2), and a 5-kg mass is at (2, 0), where all distances are in meters. Find the location of the center of gravity of the four masses.

39. An 800-N person stands on a 9-m-long scaffold supported by two ropes, as shown in Figure 4.31. The weight of the scaffold is 240 N. If the person is 2 m from the left end, find the tensions in T_1 and T_2.

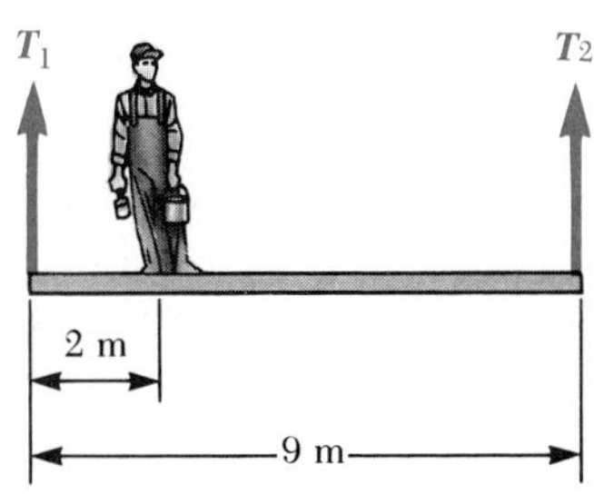

FIGURE 4.31 Problem 39

40. Find the tension in each cable supporting the 600-N cat burglar in Figure 4.32.

FIGURE 4.32 Problem 40

41. A 1.8-m-long plank is nonuniform such that its center of gravity is 0.4 m from the left end as shown in Figure 4.33. Find the tension in each cable if the weight of the plank is 500 N.

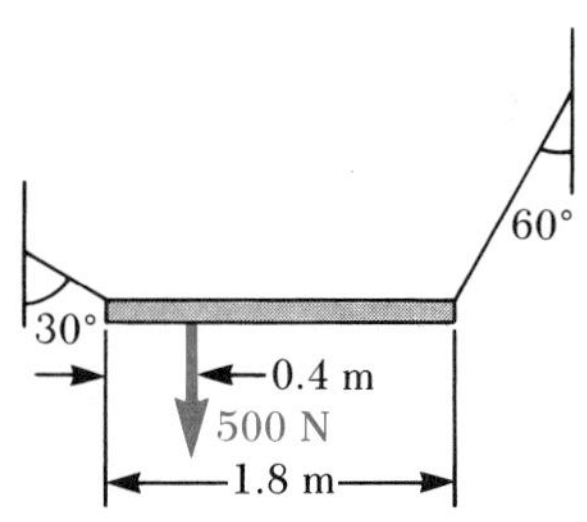

FIGURE 4.33 Problem 41

42. A uniform picture frame weighing 10 N is supported as shown in Figure 4.34. Find the tension in the cords and the magnitude of the horizontal force at P required to hold the frame in the position shown.

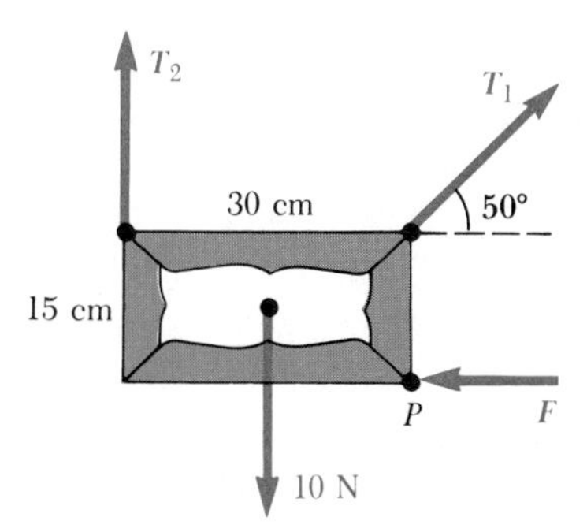

FIGURE 4.34 Problem 42

43. A hinged uniform flagpole weighing 500 N is supported in the horizontal position by a cable, as shown in Figure 4.35. Find the tension in the cable and the horizontal and vertical forces at the hinge.

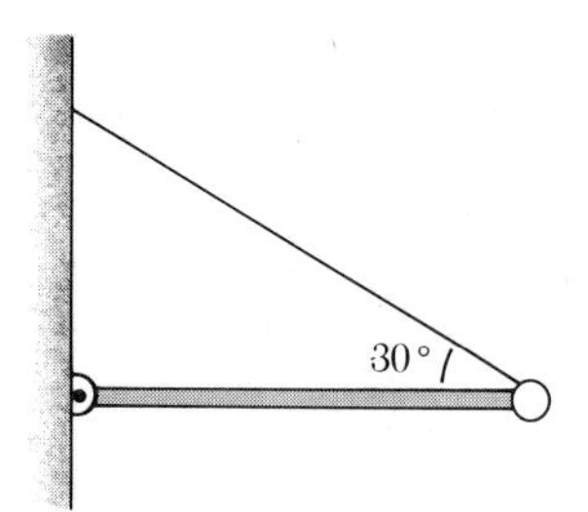

FIGURE 4.35 Problems 43 and 44

• **44.** If the maximum tension that the cable in Figure 4.35 can support without breaking is 2000 N, what is the maximum weight flagpole that can be supported?

• **45.** (a) What is the minimum force of friction required to hold the system of Figure 4.36 in equilibrium? (b) What coefficient of static friction between the 100-N block and the table will ensure equilibrium? (c) If the coefficient of kinetic friction between the 100-N block and the table is 0.25, what hanging weight should replace the 50-N weight to allow the system to move at a constant speed when set in motion?

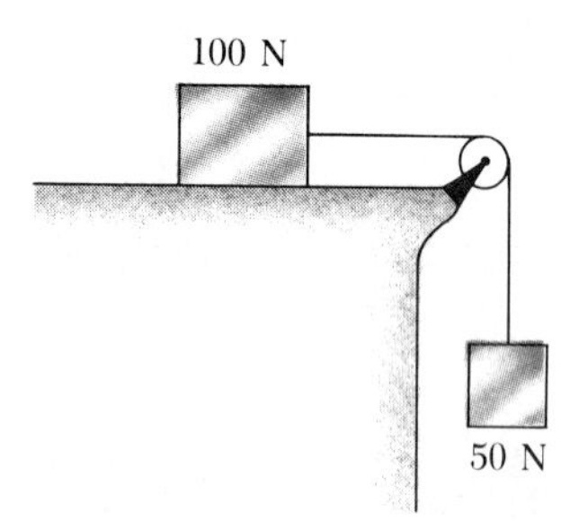

FIGURE 4.36 Problem 45

• **46.** A 150-kg mass rests on a 50-kg beam as in Figure 4.37. The weight is also connected to one end of the beam through a rope and pulley. Assuming the system is in equilibrium, (a) draw free-body diagrams for the weight and beam and (b) find the tension in the rope and the components of the reaction force at the pivot O.

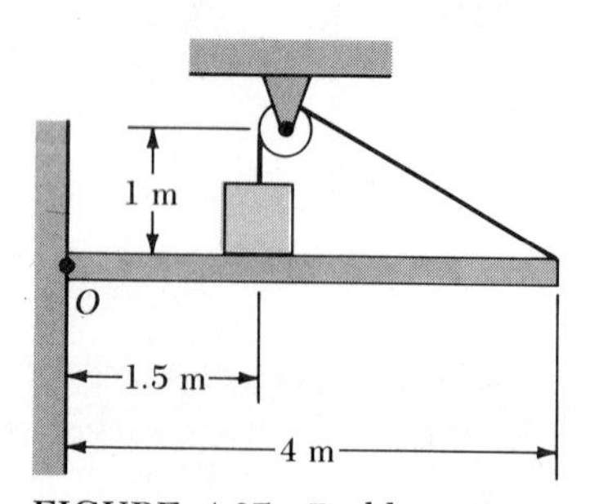

FIGURE 4.37 Problem 46

• **47.** A disk of mass m and of radius r rests on an inclined surface and is supported by a rope that is tangent to the disk and parallel to the inclined surface. The inclined surface makes an angle θ with the horizontal as shown in Figure 4.38. Find (a) the minimum value of the coefficient of static friction, in terms of θ, that will prevent the disk from slipping down the inclined surface and (b) the tension in the rope in terms of m, g, and θ.

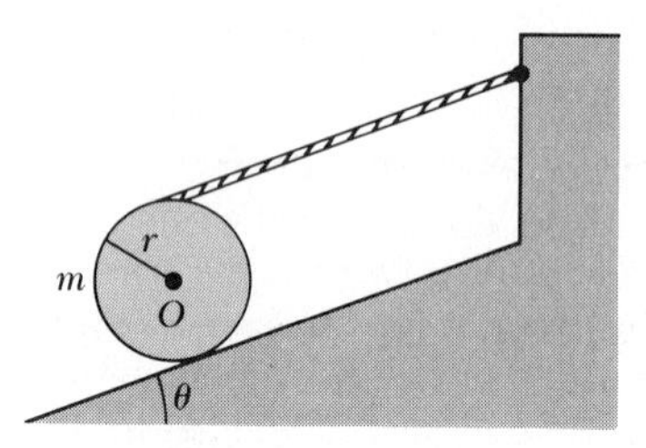

FIGURE 4.38 Problem 47

•• **48.** A 100-N monkey walks up a 125-N-uniform ladder of length 4 m, as in Figure 4.39. The upper and lower ends of the ladder rest on frictionless surfaces. The lower end of the ladder is fastened to the wall by a horizontal rope that can support a maximum tension of 110 N. Can the monkey make it halfway of the ladder?

FIGURE 4.39 Problem 48

•• **49.** A uniform beam of length 4 m and mass 10 kg, connected to the side of a building by a pivot hinge, supports a 20-kg light fixture, as shown in Figure 4.40. Determine the tension in the wire connecting the beam to the building and the horizontal and vertical forces exerted on the beam by the pivot hinge.

• **50.** A stepladder of negligible weight is constructed as shown in Figure 4.41. A painter of mass 70 kg stands on the ladder 3 m from the bottom. Assuming the floor is frictionless, find (a) the tension in the horizontal bar connecting the two halves of the ladder, (b) the normal forces at A and B, and (c) the components of the reaction force at the hinge C that the left half of the ladder exerts on the right half. (*Hint:* Treat each half of the ladder separately.)

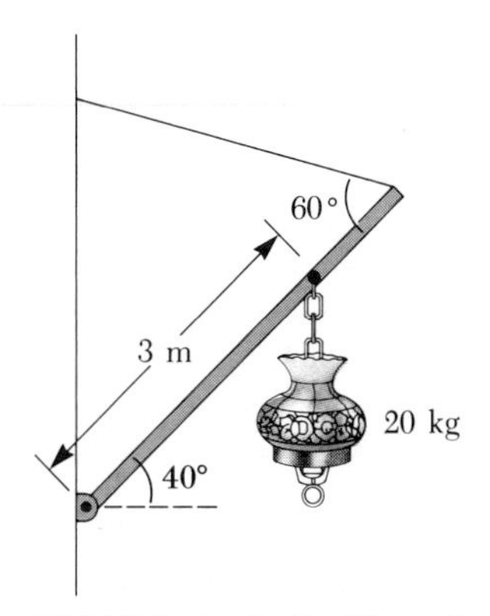

FIGURE 4.40 Problem 49

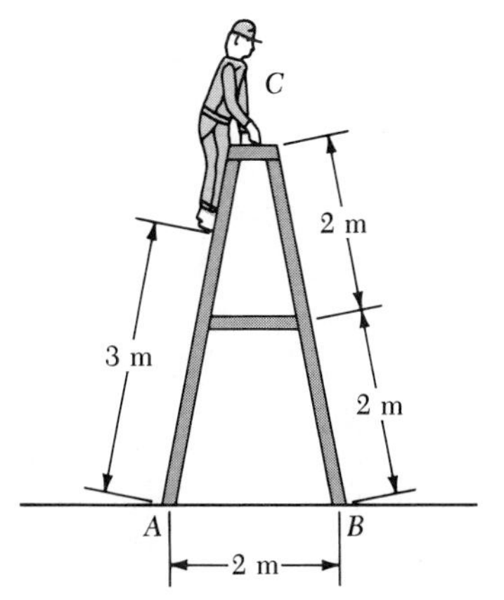

FIGURE 4.41 Problem 50

•• **51.** Two masses m and M are attached with strings as shown in Figure 4.42. If the system is in equilibrium, show that $\tan\theta = 1 + \dfrac{2M}{m}$.

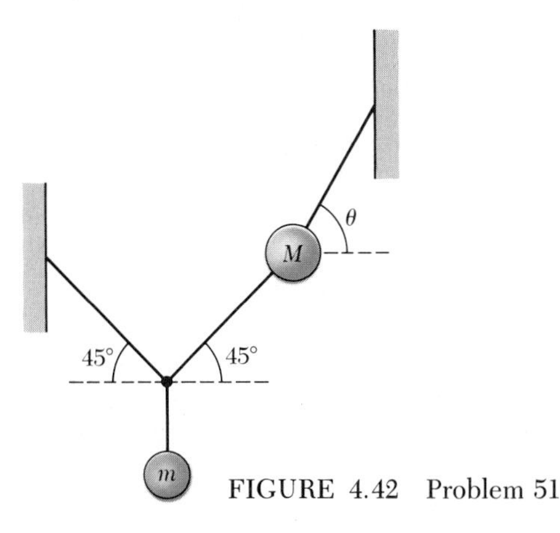

FIGURE 4.42 Problem 51

•• **52.** A uniform beam of weight w is inclined at an angle θ to the horizontal with its upper end supported by a horizontal rope tied to a wall and its lower end resting on a rough floor (Fig. 4.43). (a) If the coefficient of static friction between the beam and floor is μ_s, determine an expression for the maximum weight W that can be suspended from the top before the beam slips. (b) Determine the magnitude of the reaction force at the floor and the magnitude of the force of the beam on the rope at P in terms of w, W, and μ_s.

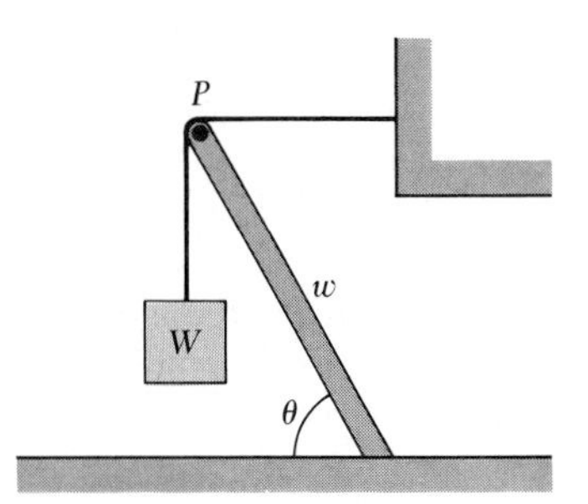

FIGURE 4.43 Problem 52

•• **53.** The cylinder shown in Figure 4.44 is held in position by a rope that supplies a force $\boldsymbol{F}$ and by static friction. What is the minimum value of μ_s such that the cylinder will remain in equilibrium when $\boldsymbol{F}$ is at the angle θ with the horizontal?

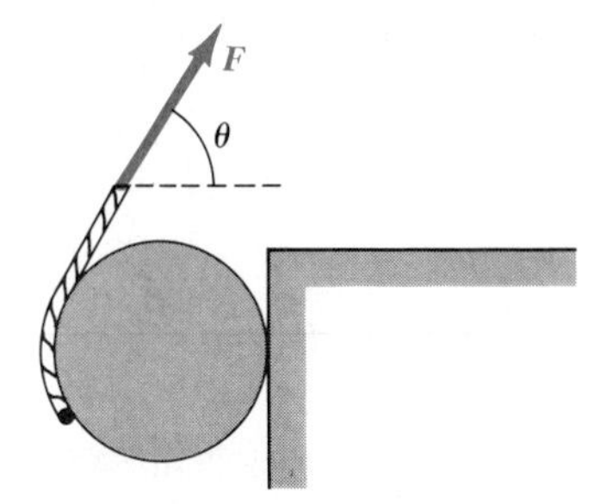

FIGURE 4.44 Problem 53

•• **54.** A disk of mass m and radius R rests on a surface which makes an angle θ with the horizontal and is supported by a horizontal rope as shown in Figure 4.45. (a) Find the minimum value of the coefficient of static friction, in terms of θ, that will prevent the disk from slipping down the inclined surface. (b) Find the tension in the rope in terms of m, g, and θ.

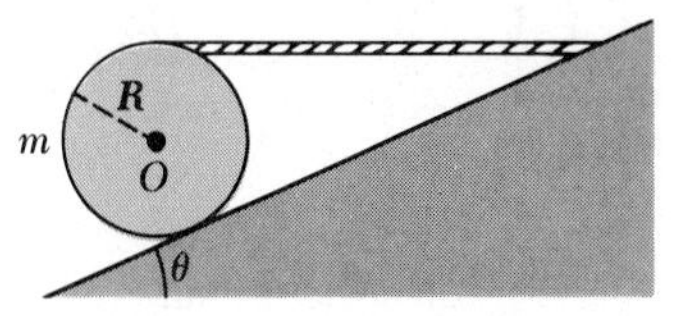

FIGURE 4.45 Problem 54

Essay

Arch Structures

Gordon Batson *Clarkson University*

Of all structures built for various utilitarian purposes, a bridge and its structural components are the most visible. The load-carrying tasks of the principal structural components can be comprehended easily; the supporting cables of a suspension bridge are under tension induced by the weight and loads on the bridge.

The arch is another type of structure whose shape indicates that the loads are carried by compression. The arch can be visualized as an upside-down suspension cable.

The stone arch is one of the oldest existing structures found in buildings, walls, and bridges. Other materials, such as timber, may have been used prior to stone, but nothing of these remains today most likely because of fires, warfare, and the decay processes of nature. Although stone arches were constructed prior to the Roman Empire, the Romans constructed some of the largest and most enduring stone arches.

Before the development of the arch, the principal method of spanning a space was the simple post-and-beam construction (Fig. 1a), in which a horizontal beam is supported by two columns. This type of construction was used to build the great Greek temples. The columns of these temples are closely spaced because of the limited length of available stones. Much larger spans can now be achieved using steel beams, but the spans are limited because the beams tend to sag under heavy loads.

The corbeled arch (or false arch) shown in Figure lb is another primitive structure; it is only a slight improvement over post-and-beam construction. The stability of this false arch depends upon the horizontal projection of one stone over another and the downward weight of stones from above.

The semicircular arch (Fig. 2a) developed by the Romans was a great technological achievement in architectural design. The stability of this true (or voussoir) arch depends on the compression between its wedge-shaped stones. (That is, the stones are forced to squeeze against each other.) This results in horizontal outward forces at

Post-and-beam
(a)

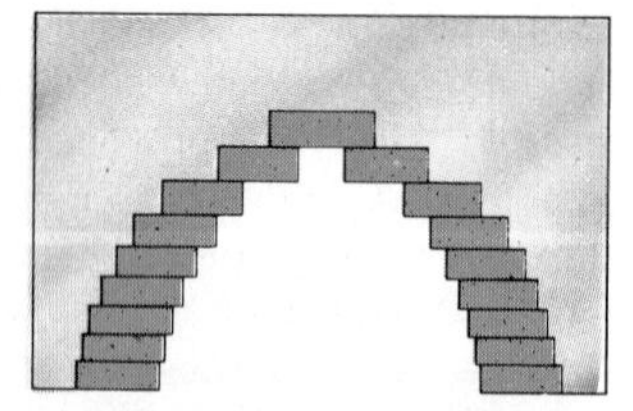

Corbeled (false) arch
(b)

FIGURE 1 Some methods of spanning a space: (a) simple post-and-beam structure and (b) corbeled, or false, arch.

Semicircular arch (Roman)
(a)

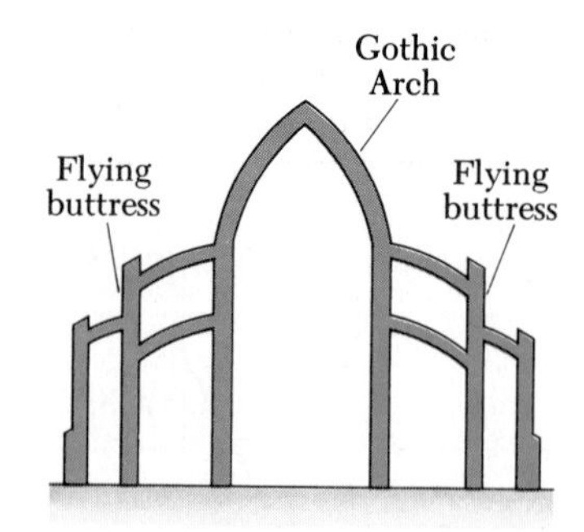

Pointed arch (Gothic)
(b)

FIGURE 2 (a) The semicircular arch developed by the Romans. (b) Gothic arch with flying buttresses to provide lateral support. (Typical cross-section of a church or cathedral.) The buttresses transfer the spreading forces of the arch by vertical loads to the foundation of the structure.

The St. Louis Gateway Arch seen from the Mississippi River. This beautiful structure has the shape of an inverted freely hanging cable, so that all of its members are under compression.

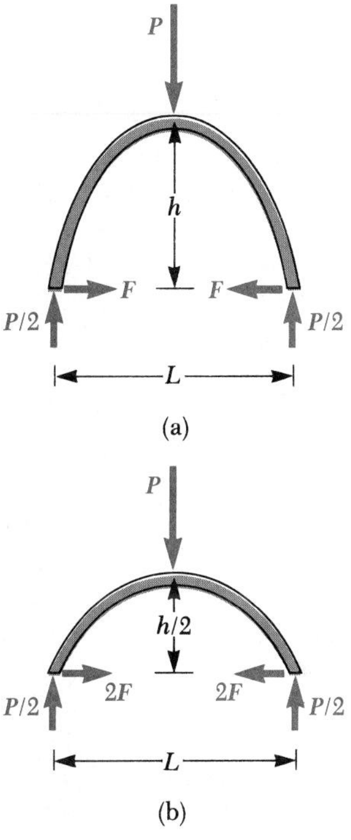

FIGURE 3 When the height of an arch is reduced by a factor of 2, and the load force **P** remains the same, the horizontal force at the base is doubled.

the springing of the arch (where it starts curving), which must be supported by the foundation (abutments) on the stone wall shown on the sides of the arch (Fig. 2a). It is common to use very heavy walls (buttresses) on either side of the arch to provide the horizontal stability. If the foundation of the arch should move, the compressive forces between the wedge-shaped stones may decrease to the extent that the arch collapses. The surfaces of the stones used in the semicircular arches constructed by the Romans were cut, or "dressed," to make a very tight joint; it is interesting to note that mortar was usually not used in these joints. The resistance to slipping between stones was provided by the compression force and the friction between the stone faces.

Another important architectural innovation was the pointed Gothic arch shown in Figure 2b. This type of structure was first used in Europe beginning in the 12th century, followed by the construction of several magnificent Gothic cathedrals in France in the 13th century. One of the most striking features of these cathedrals is their extreme height. For example, the cathedral at Chartres rises to 118 ft and the one at Reims has a height of 137 ft. It is interesting to note that such magnificent Gothic structures evolved over a very short period of time, without the benefit of any mathematical theory of structures. However, Gothic arches required flying buttresses to prevent the spreading of the arch supported by the tall, narrow columns. The fact that they have been stable for more than 700 years attests to the technical skill of their builders and architects, which was probably acquired through experience and intuition.

Figure 3 shows how the horizontal force at the base of an arch varies with arch height for an arch hinged at the peak. For a given load P, the horizontal force at the base is doubled when the height is reduced by a factor of 2. This explains why the horizontal force required to support a high pointed arch is less than that required for a circular arch. For a given span L, the horizontal force at the base is proportional to the total load P and inversely proportional to the height h. Therefore, in order to minimize the horizontal force at the base, the arch must be made as light and high as possible.

With the advent of more advanced methods of structural analysis, it has become possible to determine the optimum shape of an arch under given load conditions.

One of the most impressive modern arches, the St. Louis Gateway Arch, designed by Eero Saarinen, has a span of 192 m and a height of 192 m. The largest steel-truss arch bridge, the New River Gorge Bridge in Charleston, West Virginia, has a span of 520 m. Beautiful concrete arch bridges were designed and built in the 1920s and 1930s by Robert Maillart in Switzerland. The Sando Bridge in Sweden, a single arch of reinforced concrete, spans 264 m. Today, the arch is still the most common structure used to span large distances.

CHAPTER 5

Work and Energy

This multiflash photo of a pole-vaulter illustrates various forms of energy, including energy of motion (kinetic energy) and energy associated with position in space (potential energy). (Photo courtesy of Dr. Harold Edgerton, M.I.T., Cambridge, MA)

Several energy transformations occur during a pole vault event. Can you identify some of them?

5.1 INTRODUCTION

The concept of energy is one of the most important in the world of science. In everyday usage, we think of energy in terms of the cost of fuel for transportation and heating, electricity for lights and appliances, and the foods we consume. However, these ideas do not really define energy. They tell us only that fuels are needed to do a job and that those fuels provide us with something we call energy.

Energy is present in various forms, including mechanical energy, chemical energy, electromagnetic energy, heat energy, and nuclear energy. The various forms are related to each other such that, when energy is transformed from one form to another, the total amount of energy remains the same. This is the point that makes the energy concept so useful. If an isolated system loses energy in some form, then, by the law of conservation of energy, the system must gain an equal amount of energy in other forms. For example, when an electric motor is connected to a battery, chemical energy is converted to electrical energy, which in turn is converted to mechanical energy. The transformation of energy from one form to another is an essential part of the study of physics, chemistry, biology, geology, and astronomy.

In this chapter, we shall be concerned only with the mechanical forms of energy. We shall introduce the concept of *kinetic energy,* the energy associated with the motion of an object, and the concept of *potential energy,* the energy associated with the position of an object. We shall see that the ideas of work and energy can be used in place of Newton's laws to solve certain problems. In a complex situation, the "energy approach" can often provide a much simpler analysis than the direct application of Newton's second law.

We begin by defining *work,* a concept that provides a link between force and energy. With this as a foundation, we shall then be able to discuss the law of conservation of energy and apply it to various problems.

5.2 WORK

Almost all of the terms we have used thus far in our study have conveyed the same meaning in physics as they do in everyday life. Terms such as *velocity, acceleration,* and *force* have been used in virtually the same context here as you would use them if talking to a friend. We now encounter a term whose meaning in physics is distinctly different from its meaning in our day-to-day affairs. This new term is **work,** which can be defined with the help of Figure 5.1. Here we see an object that undergoes a displacement $\mathbf{s}$ along a straight line while acted on by a constant force $\mathbf{F}$, which makes an angle θ with $\mathbf{s}$.

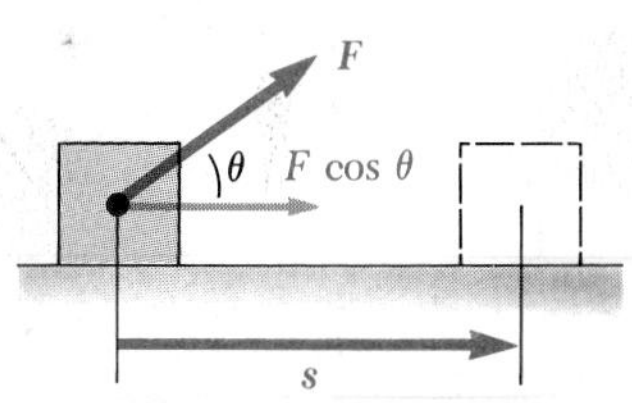

FIGURE 5.1 If an object undergoes a displacement $\mathbf{s}$, the work done by the force $\mathbf{F}$ is $(F \cos \theta)\mathbf{s}$.

The work done, W, by the constant force is defined as the product of the component of the force along the direction of displacement and the magnitude of the displacement:

$$W = (F \cos \theta)s \tag{5.1}$$

As an example of the distinction between this definition of *work* and what we mean in our everyday usage of the word, consider holding a heavy chair at arm's length for 10 min. At the end of this time interval, your tired arms might lead you to express the opinion that you have done a considerable amount of work. However, according to our definition, no work has been done on the chair.[1] You have exerted a force in order to support the chair,

[1]Actually, you do burn calories while holding a chair at arm's length. This is explained by the fact that the muscles are continually contracting and relaxing while the chair is being supported. Thus, work is being done on our body, but not on the chair.

FIGURE 5.2 No work is done when a bucket of water is moved horizontally because the applied force F is perpendicular to the displacement.

but you have not moved it. A force does no work on an object if the object does not move. This can be seen by noting that if $s = 0$, Equation 5.1 gives $W = 0$. Also note from Equation 5.1 that the work done by a force is zero when the force is perpendicular to the displacement. That is, if $\theta = 90°$, then $W = 0$ because $\cos 90° = 0$. For example, no work is done when a bucket of water is carried horizontally at constant velocity because the force exerted to support the bucket is perpendicular to the displacement of the bucket, as shown in Figure 5.2. Likewise, the work done by the force of gravity is also zero for the same reason.

The sign of the work depends on the direction of $\boldsymbol{F}$ relative to $\mathbf{s}$. The work done by an applied force is positive when the component $F \cos \theta$ is in the same direction as the displacement. For example, when you lift a box as in Figure 5.3a, the work done by the force you exert on the box is positive because the lifting force is upward, that is, in the same direction as the displacement. Work is negative when the component of the applied force is in the direction opposite the displacement. A common example of a situation in which work is always negative is the work done by a frictional force when a body slides over a rough surface. In Figure 5.3b, the work done by the frictional force on the sliding base runner is negative because the direction of the force is opposite the displacement. The negative sign comes from the fact that $\theta = 180°$ and $\cos 180° = -1$ which, from Equation 5.1, gives a negative value for W.

Finally, if an applied force acts along the direction of the displacement, then $\theta = 0°$. Since $\cos 0° = 1$, in this case Equation 5.1 becomes

$$W = Fs \tag{5.2}$$

Work is a scalar quantity, and its units are force multiplied by length. Therefore, the SI unit of work is the **newton-meter** (N · m). Another name for the newton-meter is the **joule** (J). The unit of work in the cgs system is the **dyne-centimeter** (dyne · cm), which is also called the **erg,** and the unit in the conventional (British engineering) system is the **foot-pound** (ft · lb). These are summarized in Table 5.1. Note that $1\ \text{J} = 10^7$ ergs.

Does the weight-lifter do any work as he holds the weight over his head? Does he do any work as he raises the weight?

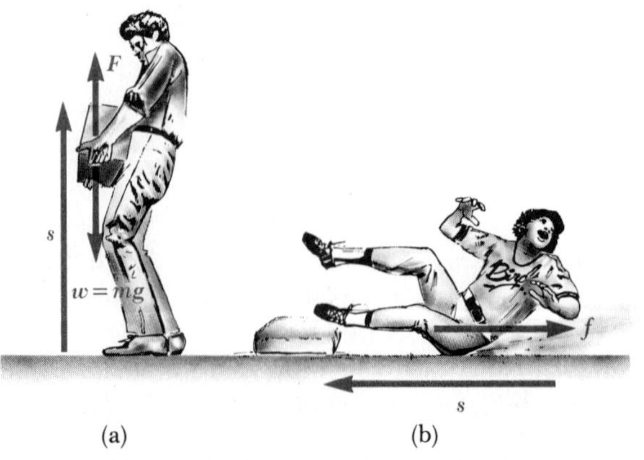

FIGURE 5.3 (a) Positive work is done by the person when the box is lifted because the applied force $\boldsymbol{F}$ is in the same direction as the displacement. (b) Negative work is done by the frictional force, $\boldsymbol{f}$, which is opposite the displacement.

TABLE 5.1 Units of Work in the Three Common Systems of Measurement

System	Unit of Work	Name of Combined Unit
SI	Newton-meter (N · m)	Joule (J)
cgs	Dyne-centimeter (dyne · cm)	Erg
British engineering (conventional)	Foot-pound (ft · lb)	Foot-pound

EXAMPLE 5.1 Mr. Clean

A man cleaning his apartment pulls a vacuum cleaner with a force of 50 N at an angle of 30°, as shown in Figure 5.4. A frictional force of 40 N retards the motion, and the vacuum is pulled a distance of 3 m. Calculate (a) the work done by the 50-N pull, (b) the work done by the frictional force, and (c) the net work done on the vacuum by all forces acting on it.

Solution (a) We can use $W = (F \cos \theta)s$ with $F = 50$ N, $\theta = 30°$, and $s = 3$ m to get

$$W_F = (50 \text{ N})(\cos 30°)(3 \text{ m}) = 130 \text{ J}$$

(b) In this case, we take $f = 40$ N and $\theta = 180°$ to get

$$W_f = (40 \text{ N})(\cos 180°)(3 \text{ m}) = -120 \text{ J}$$

(c) The normal force, $\boldsymbol{n}$, the weight, $m\boldsymbol{g}$, and the upward component of the applied force, 50 sin 30°, do *no* work because they are perpendicular to the displacement. Thus, the net work done is

$$W_{\text{net}} = W_F + W_f = 10 \text{ J}$$

EXERCISE Find the net work done on the vacuum cleaner if the man pulls it horizontally with a force of 50 N, assuming the frictional force is 40 N.

Answer 30 J.

FIGURE 5.4 (Example 5.1) A vacuum cleaner being pulled at an angle of 30° with the horizontal.

5.3 WORK AND KINETIC ENERGY

Figure 5.5 shows an object of mass m moving to the right under the action of a constant net force $\boldsymbol{F}$. Because the force is constant, we know from Newton's second law that the object will move with a constant acceleration $\boldsymbol{a}$. If the object is displaced a distance s, the work done by the net force $\boldsymbol{F}$ is

$$W_{\text{net}} = Fs = mas \tag{5.3}$$

However, in Chapter 2 we found that the following relationship holds when an object undergoes constant acceleration:

$$v^2 = v_0^2 + 2as \quad \text{or} \quad as = \frac{v^2 - v_0^2}{2}$$

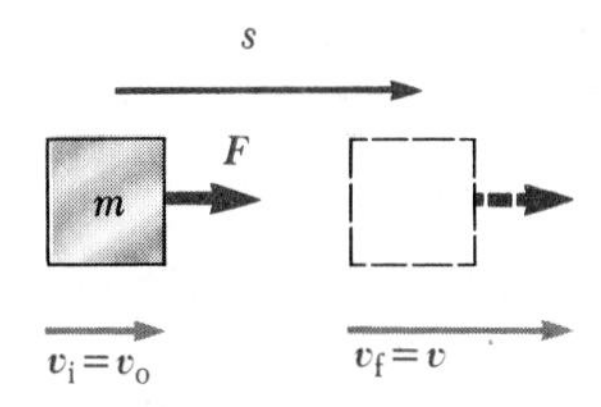

FIGURE 5.5 An object undergoing a displacement and change in velocity under the action of a constant net force $\boldsymbol{F}$.

This expression is now substituted into Equation 5.3 to give

$$W_{net} = m\left(\frac{v^2 - v_0^2}{2}\right)$$

$$W_{net} = \frac{1}{2}mv^2 - \frac{1}{2}mv_0^2 \qquad (5.4)$$

The **kinetic energy,** *KE*, of an object of mass *m* and speed *v* is defined as

Kinetic energy

$$KE \equiv \frac{1}{2}mv^2 \qquad (5.5)$$

Kinetic energy is a scalar quantity and has the same units as work. For example, a 1-kg mass moving with a speed of 4.0 m/s has a kinetic energy of 8.0 J. We can think of kinetic energy as the energy associated with the motion of an object. It is often convenient to write Equation 5.4 as

Work-energy theorem

$$W_{net} = KE_f - KE_i \qquad (5.6)$$

This equation says that, when work is done on an object, the effect of the work done by the *net* force is to change the kinetic energy from some initial value KE_i to some final value KE_f. Equation 5.6 is an important result known as the **work-energy theorem.** Thus, we conclude that

the net work done on an object by the force or forces acting on it is equal to the change in the kinetic energy of the object.

The work-energy theorem also says that the speed of the object will increase if the net work done on it is positive because the final kinetic energy will be greater than the initial kinetic energy. Its speed will decrease if the net work is negative because the final kinetic energy will be less than the initial kinetic energy. Notice that the speed and kinetic energy of an object will change only if work is done on the object by some external force or forces.

Consider the relationship between the work done on an object and the change in its kinetic energy as expressed by Equation 5.4. Because of this connection, we can also think of kinetic energy as the work the object can do in coming to rest. For example, suppose a hammer is on the verge of striking a nail as in Figure 5.6. The moving hammer has kinetic energy and is able to do work on the nail once it has struck it. The work done on the nail appears as the product *Fs*, where *F* is the average force exerted on the nail by the hammer and *s* is the distance it is driven into the wall.

FIGURE 5.6 The moving hammer has kinetic energy and thus is able to do work on the nail, driving it into the wall.

For convenience, Equation 5.6 was derived under the assumption that the net force acting on the object was constant. A more general derivation would show that this equation is valid under all circumstances, including that of a variable force.

EXAMPLE 5.2 Towing a Car

A 1400-kg car has a net forward force of 4500 N applied to it. The car starts from rest and travels down a horizontal highway. What are its kinetic energy and speed after it has traveled 100 m? (Ignore losses in energy due to friction, such as that caused by air resistance.)

Solution With the initial velocity given as zero, Equation 5.4 reduces to

$$(1) \qquad W_{net} = \frac{1}{2}mv^2$$

The work done by the net force on the car is

$$W_{net} = Fs = (4500\ \text{N})(100\ \text{m}) = 4.50 \times 10^5\ \text{J}$$

This work has all gone into changing the kinetic energy of the car; thus the final value of the kinetic energy, from (1), is also 4.50×10^5 J.

The speed of the car can be found from (1) as follows:

$$\frac{1}{2}mv^2 = 4.50 \times 10^5\ \text{J}$$

$$v^2 = \frac{2(4.50 \times 10^5\ \text{J})}{1400\ \text{kg}} = 643\ \text{m}^2/\text{s}^2$$

$$v = 25.4\ \text{m/s}$$

5.4 GRAVITATIONAL POTENTIAL ENERGY

In the previous section, we saw that an object with kinetic energy is able to do work on another object as illustrated by the moving hammer driving a nail into the wall. In this section, we shall see that an object can also do work on another object because of the energy it has due to its *position* in space.

In order to understand how an object can do work by virtue of its position in space, consider Figure 5.7 which shows a brick of mass m at a height h above a nail in a horizontal board lying on the ground. When the brick is released, it falls towards the ground, does work on the nail, and drives it into the plank.

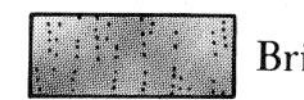

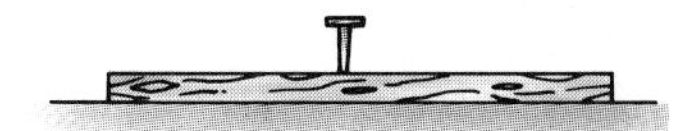

FIGURE 5.7 The brick at the top can do work on the nail because of its position in space.

The energy that an object has due to its position in space is called gravitational potential energy.

Let us now derive an expression for the gravitational potential energy of an object at a given location in space. To do this consider a block of mass m at an initial height y_i above the ground as in Figure 5.8. Neglecting air resistance, as the block falls the only force that does work on it is the gravitational force, mg. The work done by the gravitational force as the block undergoes a downward displacement $s = y_i - y_f$ is given by the product of the downward force mg times the displacement, or

$$W_g = mgs = mgy_i - mgy_f \tag{5.7}$$

We now define the gravitational potential energy, PE, as

$$\boxed{PE \equiv mgy} \tag{5.8}$$

If we substitute this expression for PE into Equation 5.7, we have

$$\boxed{W_g = PE_i - PE_f} \tag{5.9}$$

From this result we see that

The work done by the force of gravity is equal to the initial value of the potential energy minus the final value of the potential energy.

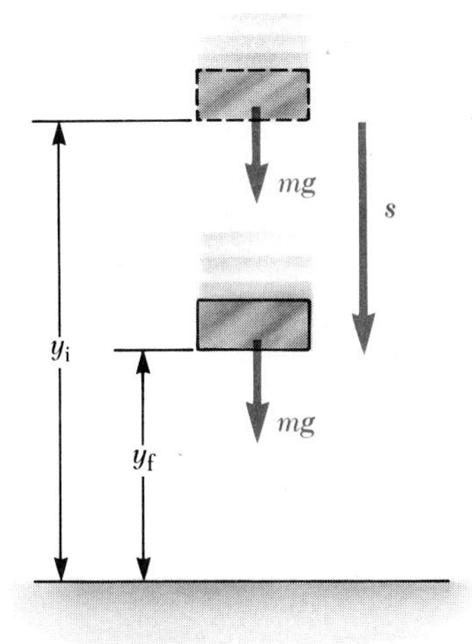

FIGURE 5.8 The work done by the gravitational force as the block falls from y_i to y_f is equal to $mgy_i - mgy_f$.

Note from Equation 5.7 that the units of gravitational potential energy must be the same as those of work. That is, potential energy has units of joules, ergs, or foot-pounds. Potential energy, like work and kinetic energy, is a scalar quantity.

Also, note that the gravitational potential energy of an object depends only on the vertical height of an object above the surface of the earth. From this result, we see that an object will do the same amount of work if it falls vertically to the earth as it will if it starts at the same point and slides down a frictionless incline to the earth. Finally, you should note that Equation 5.8 is valid only for objects near the surface of the earth where $\mathbf{g}$ is approximately constant. We shall derive a more general expression for potential energy in Chapter 7.

Reference Levels for Potential Energy

In working problems involving the concept of gravitational potential energy, it is always necessary to choose a location at which we set the gravitational potential energy equal to zero. We choose to call this the zero level for the gravitational potential energy. The choice of location of this zero level is completely arbitrary. This is true because the quantity of importance is the *difference* in potential energy, and the difference in potential energy between two points is independent of the choice of zero level.

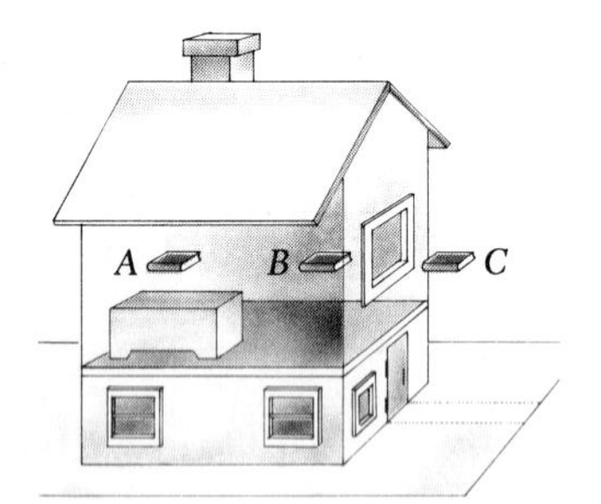

FIGURE 5.9 Any reference level could be used for measuring the gravitational potential energy of the book.

It is often convenient to choose the surface of the earth as the reference position for zero potential energy, but again this is not essential. Often, the statement of a problem will suggest a convenient level to use. As an example, consider a book at several possible locations as in Figure 5.9. When the book is at A, above the surface of a desk, a logical zero level for potential energy could be taken at the desk surface. When the book is at B, however, the floor surface might be a more appropriate zero reference level. Finally, a location such as C, where the book is flying out a window, would suggest choosing the ground level as the zero level of potential energy. The choice, however, makes no difference. Any of the reference levels mentioned could be used as the zero level, regardless of whether the book is at A, B, or C.

EXAMPLE 5.3 Let's Go Skiing

A novice skier of mass 60 kg is at the top of a slope as shown in Figure 5.10. At the initial point A, the skier is 10 m vertically above point B. (a) Choose the zero level for gravitational potential energy to be at B and find the gravitational potential energy of the skier at A and B, and the difference in potential energy between these two points.

FIGURE 5.10 (Example 5.3)

Solution The gravitational potential energy at the final point, B, is zero by choice. Hence, the potential energy at the initial point is

$$PE_i = mgy_i = (60\text{ kg})(9.80\text{ m/s}^2)(10\text{ m}) = 5880\text{ J}$$

Since $PE_f = 0$, the difference in potential energy is

$$\Delta PE = PE_i - PE_f = 5880\text{ J} - 0\text{ J} = 5880\text{ J}$$

(b) Repeat this problem with the zero level at point A.

Solution In this case, the initial potential energy is zero because of the choice of the reference level. The final potential energy is

$$PE_f = mgy_f = (60\text{ kg})(9.80\text{ m/s}^2)(-10\text{ m}) = -5880\text{ J}$$

Note that the distance y_f is -10 m because the final point is 10 m *below* the zero reference level. The difference in potential energy is

$$\Delta PE = PE_i - PE_f = 0\text{ J} - (-5880\text{ J}) = 5880\text{ J}$$

These calculations show that the potential energy of the skier at the top of the slope is greater than the potential energy at the bottom by 5880 J, *regardless of the zero level selected.*

EXERCISE If the zero level for gravitational potential energy is selected to be midway down the slope, at a height of 5 m, find the initial potential energy, the final potential energy, and the difference in potential energy between points A and B.

Answer 2940 J, -2940 J, 5880 J

Estimate the gravitational potential energy of this pole-vaulter at the top of his flight. How many forms of energy can you identify in this picture? (Photo courtesy of Dr. Harold Edgerton, M.I.T., Cambridge, MA)

5.5 CONSERVATIVE AND NONCONSERVATIVE FORCES

Conservative Forces

In Section 5.4 we found that the work done by the gravitational force acting on an object is given by

$$W_g = mgy_i - mgy_f.$$

In other words, *the work done by gravity depends only on the initial and final coordinates. The amount of work done does not depend on the path selected to move between these points.* When a force exhibits this property, it is called a **conservative force.** In addition to the gravitational force, other examples of conservative forces, which we shall encounter later, are the electrostatic force and the forces exerted by springs. In general,

Definition of a conservative force

a force is conservative if the work it does on an object moving between two points is independent of the path the object takes between the points. The work done on an object by a conservative force depends only on the initial and final positions of the object.

Conservative forces are important because we can always associate a potential energy function with them. For the gravitational force, we found the potential energy to be

$$PE = mgy$$

We are able to define potential energy functions only for conservative forces. In general, one always finds that the work done on an object by a conservative force, W_c, is given by the initial potential energy of the object minus its final potential energy. That is,

$$W_c = PE_i - PE_f \quad (5.10)$$

Nonconservative Forces

Definition of a nonconservative force

A force is nonconservative if the work it does on an object moving between two points depends on the path taken.

The force of sliding friction is a good example of a nonconservative force. If an object is moved over a rough horizontal surface between two points along various paths, the work done by the frictional force certainly depends on the path. The negative work done by the frictional force along any particular path between two points equals the force of friction multiplied by the length of the path. Different paths involve different amounts of work. The least amount of work will be done by the frictional force along a straight-line path between the two points. For example, suppose you were to displace a book between two points on a rough horizontal surface, such as a table. If the book is displaced in a straight line between points A and B in Figure 5.11, the work done by friction is simply $-fd$, where d is the distance between the two points. However, if the book is moved along any other path between the two points, the work done by friction is greater (in absolute magnitude) than $-fd$. For example, the work done by friction along the semicircular path in Figure 5.11 is equal to $-f(\pi d/2)$, where d is the diameter of the circle. Potential energy functions cannot be defined for nonconservative forces.

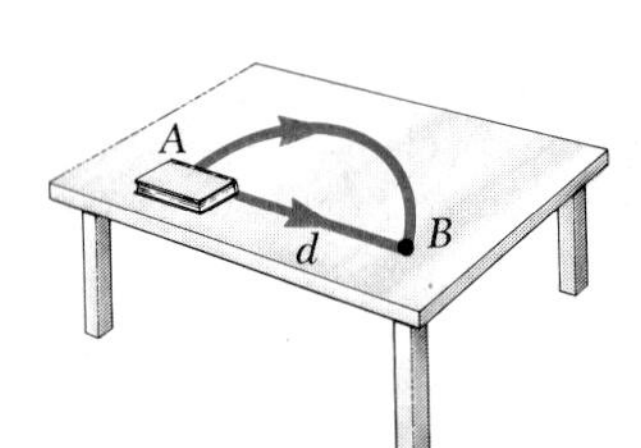

FIGURE 5.11 The work done by the force of friction depends on the path taken as the book is moved from A to B.

5.6 CONSERVATION OF MECHANICAL ENERGY

Conservation laws play a very important role in physics, and we shall encounter a number of these laws as we proceed through the remainder of this course. The first of these that we meet, the law of conservation of energy, is one of the most important. Before we describe the mathematical details of the law of conservation of energy, we must pause briefly to examine what is meant when we say something is conserved. When we say that a physical quantity is *conserved*, we simply mean that the value of this quantity does not change with time. Although the form of the quantity may change in some manner, its final value will be the same as its initial value. We shall find, for example, that the energy in a system may change from gravitational potential energy to kinetic energy or to one of a variety of other forms that we shall encounter soon, but energy is never lost from the system.

In order to develop the principle of conservation of energy, let us return to the work-energy relation, Equation 5.6, which says that the net work done on a system equals the change in the system's kinetic energy.

$$W_{net} = KE_f - KE_i$$

Now, let us assume that the only force doing work on the system is a conservative force. In this case the net work on the system is equal to W_c, and from Equation 5.10 we have

$$W_{net} = W_c = PE_i - PE_f$$

We now substitute this expression into the work energy equation for W_{net} to find

$$PE_i - PE_f = KE_f - KE_i$$

or

$$\boxed{KE_i + PE_i = KE_f + PE_f} \qquad (5.11)$$

Conservation of mechanical energy

The sum of the kinetic energy plus the potential energy is called the **total mechanical energy.** Equation 5.11 is a statement of **the law of conservation of mechanical energy.** We can describe this law as follows:

The law of conservation of mechanical energy states that the mechanical energy of a system remains constant if the only forces that do work on the system are conservative forces.

Therefore if a conservative system (one subject only to conservative forces) has some initial amount of kinetic and potential energy, the energy may later transform from one kind into the other, but the final total mechanical energy will remain the same. This is equivalent to saying that, if the kinetic energy of a conservative system increases (or decreases) by some amount, the potential energy of the system must decrease (or increase) by the same amount.

So far, we have encountered only one conservative force, and as a result we have only one potential energy function, that of gravitational potential energy. Thus, *if the force of gravity is the only force doing work on an object, then the total mechanical energy of the object is conserved,* and the law of conservation of mechanical energy takes the form

$$\boxed{\frac{1}{2}mv_i^2 + mgy_i = \frac{1}{2}mv_f^2 + mgy_f} \qquad (5.12)$$

As we go further into this course, we shall discover that there are other kinds of energy besides kinetic energy and gravitational potential energy. As a result, you should be aware of the fact that we shall have to modify Equation 5.12 to include these new forms of energy as they occur.

EXAMPLE 5.4 The Daring Diver

A diver weighing 750 N (mass $= w/g =$ 77 kg) drops from a board 10 m above the surface of a pool of water as in Figure 5.12. (a) Use the conservation of mechanical energy to find his speed at a point 5 m above the water surface.

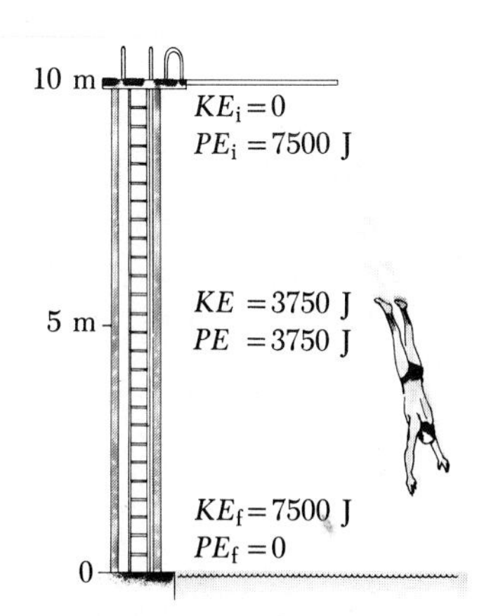

FIGURE 5.12 (Example 5.4) The kinetic energy and potential energy of a diver at various heights. The zero of potential energy is taken to be at the surface of the pool.

Solution As the diver falls toward the water, only one force acts on him: the force of gravity. (This assumes that air resistance can be neglected.) Therefore, we can be assured that no forces other than the force of gravity do any work on the diver and that mechanical energy is conserved. In order to find the diver's speed at the 5-m mark, let us choose the zero level for potential energy to be at the surface of the water. Also, note that the diver drops from the board, that is, he leaves with zero velocity and zero kinetic energy. Equation 5.12 gives

$$\frac{1}{2}mv_i^2 + mgy_i = \frac{1}{2}mv_f^2 + mgy_f$$

$$0 + (750\text{ N})(10\text{ m}) = \frac{1}{2}(77\text{ kg})v_f^2 + (750\text{ N})(5\text{ m})$$

$$v_f = 9.86\text{ m/s}$$

(b) Find the speed of the diver just before he strikes the water.

Solution With the final position of the diver at the surface of the pool, Equation 5.12 gives

$$0 + (750 \text{ N})(10 \text{ m}) = \frac{1}{2}(77 \text{ kg})v_f^2 + 0$$

$$v_f = 13.9 \text{ m/s}$$

EXERCISE If the diver pushes off such that he leaves the board with an initial speed of 2 m/s, find his speed when he strikes the water. Use conservation of energy.

Answer 14.1 m/s

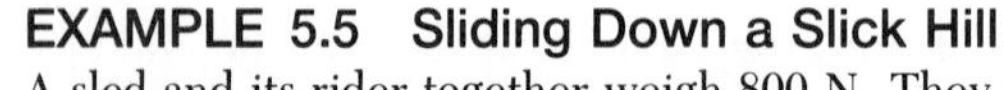

EXAMPLE 5.5 Sliding Down a Slick Hill

A sled and its rider together weigh 800 N. They move down a frictionless hill through a vertical distance of 10 m, as shown in Figure 5.13a. Use conservation of mechanical energy to find the speed of the sled-rider system at the bottom of the hill, assuming that the rider pushes off with an initial speed of 5 m/s.

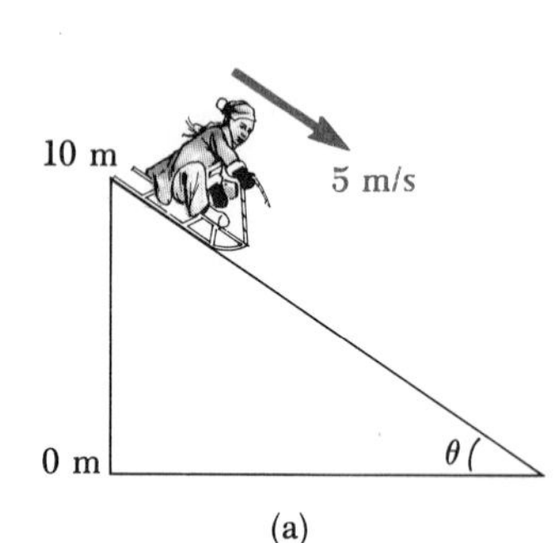

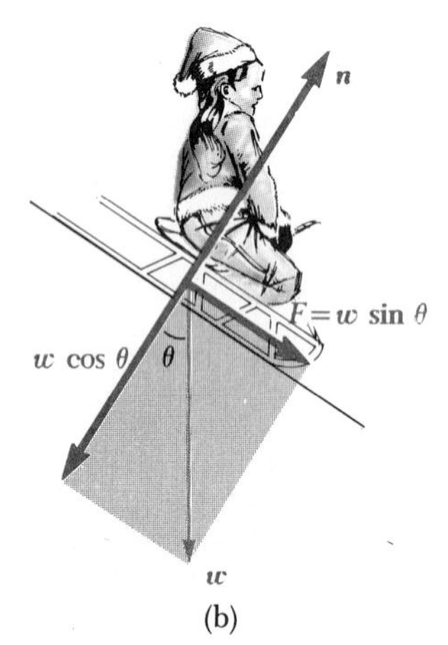

FIGURE 5.13 (Example 5.5) (a) A sled and rider start from the top of a frictionless hill with a speed of 5 m/s. (b) Free-body diagram for the system (sled plus rider).

Solution The forces acting on the sled and rider as they move down the hill are shown in Figure 5.13b. In the absence of a frictional force, the only forces acting are the normal force, $\boldsymbol{n}$, and the gravitational force. At all points along the path, $\boldsymbol{n}$ is perpendicular to the direction of travel and hence does no work. Likewise, the component of weight perpendicular to the incline ($w \cos \theta$) does no work. The only force that does any work is the component of the gravitational force $\boldsymbol{F}$ along the slope of the hill. As a result, the only force doing any work is the gravitational force and we are justified in using Equation 5.12. Note that in this case the initial energy includes kinetic energy because of the initial speed:

$$\frac{1}{2}mv_i^2 + mgy_i = \frac{1}{2}mv_f^2 + mgy_f$$

or, after canceling m throughout the equation,

$$\frac{1}{2}v_i^2 + gy_i = \frac{1}{2}v_f^2 + gy_f$$

If we select the origin at the bottom of the incline, we see that the initial and final y-coordinates are $y_i = 10$ m and $y_f = 0$. Thus we get

$$\frac{1}{2}(5 \text{ m/s})^2 + (9.80 \text{ m/s}^2)(10 \text{ m}) = \frac{1}{2}v_f^2 + (9.80 \text{ m/s}^2)(0)$$

$$v_f = 14.9 \text{ m/s}$$

EXERCISE If the sled and rider start at the bottom of the incline and are given an initial velocity of 5 m/s up the incline, how high will they rise vertically and what will their speed be when they return to the bottom of the hill?

Answer 1.28 m; 5 m/s.

5.7 NONCONSERVATIVE FORCES AND THE WORK-ENERGY THEOREM

In realistic situations, nonconservative forces, such as friction, are usually present. In such situations, the total mechanical energy of the system is not constant, and one cannot apply Equation 5.11. This is due to the fact that the law of conservation of mechanical energy holds only when the forces that do work on the system are conservative forces. In order to be able to account for nonconservative forces, let us return to the work-energy theorem.

$$W_{net} = \frac{1}{2}mv_f^2 - \frac{1}{2}mv_i^2$$

Let us separate the net work, W_{net}, into two parts, that due to the nonconservative forces, W_{nc}, and the work done by conservative forces, W_c. Thus, the work-energy relation becomes

$$W_{nc} + W_c = \frac{1}{2}mv_f^2 - \frac{1}{2}mv_i^2$$

Let us consider the special case in which the conservative force is the gravitational force. In this case,

$$W_c = mgy_i - mgy_f = PE_i - PE_f$$

and we have

$$W_{nc} = \left(\frac{1}{2}mv_f^2 - \frac{1}{2}mv_i^2\right) + (mgy_f - mgy_i) \qquad (5.13)$$

or

$$W_{nc} = (KE_f - KE_i) + (PE_f - PE_i) \qquad (5.14)$$

That is, the work done by all nonconservative forces equals the change in kinetic energy plus the change in potential energy.

According to Equation 5.13, the work done by nonconservative forces can be positive, negative, or zero. An example of a situation in which W_{nc} is positive is that of an object being pushed along the floor by an applied force which is greater than the force of friction. In this case the applied force and the displacement are in the same direction, and the work done by the applied force is positive. In such situations, the nonconservative force is adding mechanical energy to the system. The work done by a nonconservative force is negative when the direction of the force is opposite that of the displacement. Forces of sliding friction are nonconservative in nature. When nonconservative forces act on a system, they remove mechanical energy from the system. In reality, energy is not lost in the process; it is simply being transformed into some form of energy other than kinetic or potential energy. As we shall see in a later chapter, this other form of energy is usually heat energy. Because frictional forces produce a decrease in the total mechanical energy of a system, they are called *dissipative* forces.

Note that when there are no nonconservative forces present, $W_{nc} = 0$, and Equation 5.13 reduces to Equation 5.11. This is to be expected on the basis of the discussion in Section 5.6.

Finally, you should note that when Equation 5.13 is used, the term W_{nc} must involve only nonconservative forces. *This does not mean that conservative forces such as gravity are being ignored; they are already taken into account by the presence of the potential energy terms in Equation 5.13.* For example, the work done by the force of gravity appears in the terms mgy_i and mgy_f.

EXAMPLE 5.6 Crate on a Rough Ramp

A 3-kg crate slides down a rough ramp at a loading dock. The ramp is 1 m in length and is inclined at an angle of 30°, as shown in Figure 5.14. The crate starts from rest at the top and experiences a constant frictional force of magnitude 5 N. Use energy methods to determine the speed of the crate when it reaches the bottom of the ramp.

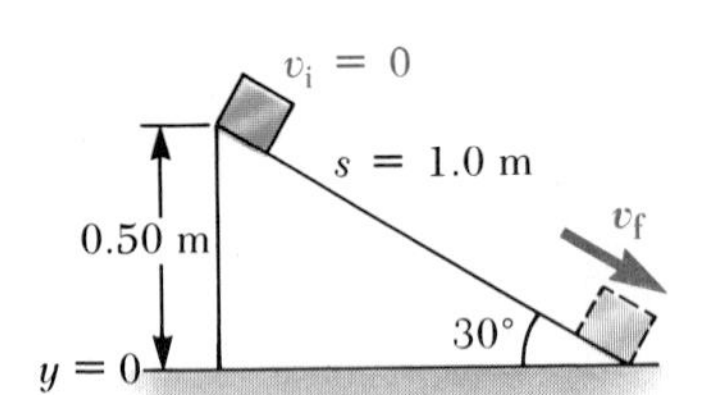

FIGURE 5.14 (Example 5.6) A crate slides down a rough incline under the influence of gravity. Its potential energy decreases while its kinetic energy increases.

Solution Because $v_i = 0$, the initial kinetic energy is zero. If the y coordinate is measured from the bottom of the incline, then $y_i = 0.50$ m. Therefore, the total mechanical energy of the crate at the top is all potential energy, given by

$$PE_i = (3 \text{ kg})(9.80 \text{ m/s}^2)(0.50 \text{ m}) = 14.7 \text{ J}$$

When the crate reaches the bottom, its potential energy is zero because its elevation is $y_f = 0$. Therefore, the total mechanical energy at the bottom is all kinetic energy:

$$KE_f = \frac{1}{2}mv_f^2$$

However, we cannot say that $PE_i = KE_f$ in this case because there is a nonconservative force that does work on the crate: the force of friction. In this case, $W_{nc} = -fs$, where s is the displacement along the ramp. (Recall that the forces normal to the ramp do no work on the crate because they are perpendicular to the displacement.) With $f = 5$ N and $s = 1$ m, we have

$$W_{nc} = -fs = (-5 \text{ N})(1 \text{ m}) = -5 \text{ J}$$

This says that some mechanical energy is lost because of the presence of the retarding frictional force. Applying the work-energy theorem in the form of Equation 5.13 gives

$$-fs = \frac{1}{2}mv_f^2 - mgy_i$$

$$\frac{1}{2}mv_f^2 = 14.7 \text{ J} - 5 \text{ J} = 9.7 \text{ J}$$

$$v_f^2 = \frac{19.4 \text{ J}}{3 \text{ kg}} = 6.47 \text{ m}^2/\text{s}^2$$

$$v_f = 2.54 \text{ m/s}$$

EXERCISE Find the final speed at the bottom of the ramp if the plane is assumed to be frictionless.

Answer $v_f = 3.13$ m/s.

EXAMPLE 5.7 Fun on the Slide

A child of mass 20 kg takes a ride on an irregularly curved slide of height 6 m, as in Figure 5.15. The child starts from rest at the top. Determine the speed of the child at the bottom, assuming there is no friction present.

Solution Note that the normal force, $\mathbf{n}$, does no work on the child because this force is always perpendicular to the displacement. Furthermore, because there is no friction, $W_{nc} = 0$ and we can apply the law of conservation of mechanical energy. If we measure the y coordinate from the bottom of the slide, then $y_i = 6$ m, $y_f = 0$, and we get

$$KE_i + PE_i = KE_f + PE_f$$

$$0 + (20 \text{ kg})(9.80 \text{ m/s}^2)(6 \text{ m}) = \frac{1}{2}(20 \text{ kg})(v_f^2) + 0$$

$$v_f = 10.8 \text{ m/s}$$

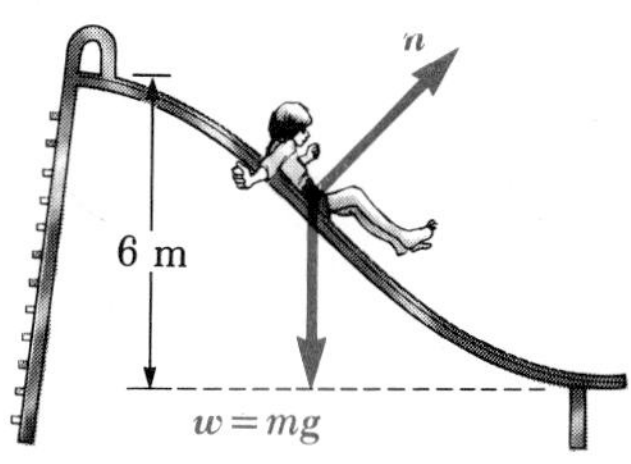

FIGURE 5.15 (Example 5.7) If the slide is very smooth, the speed of the child at the bottom of the slide depends essentially on the height of the slide, and is independent of its shape.

Note that this speed is the same as if the child had fallen vertically through a distance of 6 m! From the free-fall equation, $v_f^2 = v_i^2 - 2gy$, we get the same result:

$$v_f = \sqrt{2(9.80 \text{ m/s}^2)(6 \text{ m})} = 10.8 \text{ m/s}$$

The effect of the slide is to direct the motion at the end into a horizontal motion instead of the vertical motion of free fall.

EXERCISE If the speed of the child at the bottom of the slide is 8 m/s rather than 10.8 m/s found above, how much work would have been done by the force of friction?

Answer −536 J

EXAMPLE 5.8 Fasten Your Seat Belts

An airplane of mass 1.5×10^4 kg is moving at a speed of 60 m/s. The pilot then revs up the engine such that the forward thrust of the propeller becomes 7.5×10^4 N. If the force of air resistance on the airplane is 4×10^4 N, find the speed of the airplane after it has traveled 500 m. Assume the airplane is in level flight throughout this motion.

Solution There are two nonconservative forces doing work on the airplane. One of these is the forward thrust of the propeller on the airplane. The work done by this force, W_{nc1}, is

$$W_{nc1} = (7.5 \times 10^4 \text{ N})(500 \text{ m}) = 3.75 \times 10^7 \text{ J}$$

The second nonconservative force acting on the airplane is the force of air resistance. The work done by this force, W_{nc2}, is negative because this force and the displacement are in opposite directions. Hence, we have

$$W_{nc2} = -(4 \times 10^4 \text{ N})(500 \text{ m}) = -2 \times 10^7 \text{ J}$$

We choose the zero level of potential energy to be at the level of the airplane. Because the airplane remains in level flight during the motion, we see that $mgy_i = mgy_f = 0$. Therefore, Equation 5.13, together with the above values, gives

$$W_{nc1} + W_{nc2} = \frac{1}{2}mv_f^2 - \frac{1}{2}mv_i^2$$

$$3.75 \times 10^7 \text{ J} - 2 \times 10^7 \text{ J} = \frac{1}{2}(1.5 \times 10^4 \text{ kg})v_f^2 - \frac{1}{2}(1.5 \times 10^4 \text{ kg})(60 \text{ m/s})^2$$

$$v_f = 77 \text{ m/s}$$

5.8 POWER

Suppose you wish to climb several flights of stairs to the top of a tall building. You may be able to perform this task with little difficulty if you make the climb slowly. However, if you decide to sprint up the stairs, you will probably find yourself taking a rest after climbing a few flights. Although the total work done in both situations is the same, the *rate* at which work is being done is quite different. In fact, the rate at which work is being done is the key factor in such situations. **Power** is defined as the rate at which work is done, or as the rate at which energy is changed from one form to another:

Average power

$$\overline{P} = \frac{\Delta W}{\Delta t} \tag{5.15}$$

$\overline{P}$ is the average power, ΔW is the work done, and Δt is the time interval during which work is done. It is sometimes useful to rewrite Equation 5.15 by substituting $\Delta W = F\Delta s$ and noting that $\Delta s/\Delta t$ is the average velocity of the object during the time Δt:

$$\overline{P} = \frac{\Delta W}{\Delta t} = \frac{F\Delta s}{\Delta t} = F\overline{v} \tag{5.16}$$

This result says that the average power either delivered to an object or delivered by the object is equal to the product of the force acting on the object during some time interval and its average velocity during this time interval. In Equation 5.16, F is the component of force in the direction of the average velocity.

A result of the same form as Equation 5.16 is obtained for instantaneous values. That is, the instantaneous power delivered to an object is equal to the product of the force on the object at that instant and the instantaneous velocity, or $P = Fv$.

The unit of power in the SI system is joules per second, which is also called the **watt** (W), after James Watt:

$$1 \text{ W} = 1 \text{ J/s} = 1 \text{ kg}\cdot\text{m}^2/\text{s}^3 \tag{5.17}$$

The unit of power in the British engineering system is the horsepower (hp), where

$$1 \text{ hp} = 550 \frac{\text{ft}\cdot\text{lb}}{\text{s}} = 746 \text{ W} \tag{5.18}$$

The watt is commonly used in electrical applications, but it can be used in other scientific areas. For example, an automobile engine can be rated in watts as well as in horsepower. Likewise, the power consumption of an electric appliance can be expressed in horsepower. Use your skills to calculate the horsepower of a 100-W lightbulb and the wattage rating of a car engine.

When one refers to a car engine of 100 hp, one means that the engine is capable of changing 55,000 ft · lb of chemical or electrical energy to mechanical energy each second. Likewise, in 1 s, a 100-W lightbulb transforms 100 J of electrical energy into other forms, primarily heat and light. Returning to our stair-climbing example at the opening of this section, the reason that one may be unable to race up a flight of stairs is because the body may be unable to convert chemical energy to mechanical energy rapidly enough.

EXAMPLE 5.9 Drag Racing

A 2000-kg car starts from rest and accelerates to a final velocity of 20 m/s in a time of 15 s. Assume that the force of air resistance remains constant at a value of 500 N during this time. (a) Find the average power developed by the engine of the car.

Solution The equations of linear motion with constant acceleration allow us to find the acceleration of the car as follows:

$$v = v_0 + at$$

Since $v_0 = 0$, we find

$$20 \text{ m/s} = a(15 \text{ s})$$

$$a = 1.33 \text{ m/s}^2$$

The forces acting on the car along the direction of motion are the forward thrust of the engine, F, and the retarding 500-N force of air resistance. Let us now use Newton's second law as follows:

$$\sum F = ma$$

$$F - 500 \text{ N} = (2000 \text{ kg})(1.33 \text{ m/s}^2)$$

$$F = 3160 \text{ N}$$

The average velocity of the car during this time interval is

$$\bar{v} = \frac{v_f + v_i}{2} = \frac{20 \text{ m/s} + 0}{2} = 10 \text{ m/s}$$

Therefore, the average power can be found using Equation 5.15:

$$\overline{P} = F\bar{v} = (3160 \text{ N})(10 \text{ m/s}) = 3.16 \times 10^4 \text{ W} = 42.4 \text{ hp}$$

(b) Find the instantaneous power being developed by the engine when the car reaches its final velocity.

Solution The forward thrust of the car engine remains constant at the value calculated in part (a), $F = 3160$ N throughout the motion. Since the final velocity is 20 m/s, and the instantaneous power is given by $P = Fv$, we find

$$P = Fv = (3160 \text{ N})(20 \text{ m/s}) = 6.32 \times 10^4 \text{ W} = 84.8 \text{ hp}$$

EXERCISE This example is somewhat unrealistic because all of the horsepower developed by an engine does not actually appear as a forward thrust on the car. In reality, one must account for frictional and mechanical losses within the engine. Assuming that 75% of the power developed by the engine reaches the wheels, find the average horsepower the engine would have to develop to achieve the acceleration calculated in part (a).

Answer 56.5 hp

EXAMPLE 5.10 Power Delivered by an Elevator Motor

An elevator has a mass of 1000 kg and carries a maximum load of 800 kg. A constant frictional force of 4000 N retards its motion upward, as in Figure 5.16. (a) What minimum horsepower must the motor deliver to lift the fully loaded elevator at a constant speed of 3 m/s?

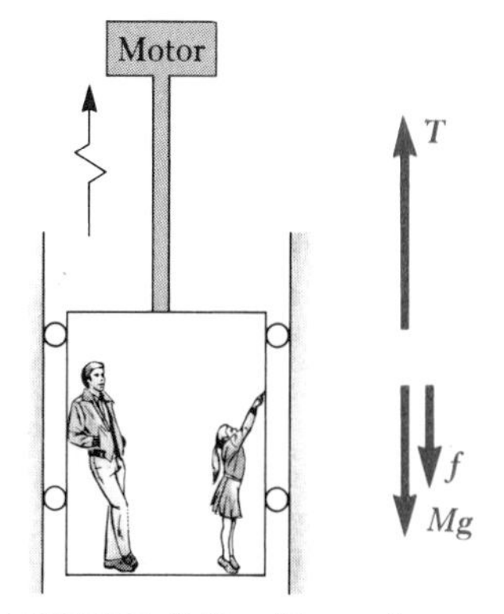

FIGURE 5.16 (Example 5.10) The motor provides a force $\boldsymbol{T}$ upward on the elevator. A frictional force $\boldsymbol{f}$ and the total weight Mg act downward.

Solution The motor must supply the force $\boldsymbol{T}$ that pulls the elevator upward. From Newton's second law and from the fact that $\boldsymbol{a} = 0$ because $\boldsymbol{v}$ is constant, we get

$$T - f - Mg = 0$$

where M is the total mass (elevator plus load), equal to 1800 kg. Therefore,

$$\begin{aligned} T &= f + Mg \\ &= 4 \times 10^3 \text{ N} + (1.8 \times 10^3 \text{ kg})(9.80 \text{ m/s}^2) \\ &= 2.16 \times 10^4 \text{ N} \end{aligned}$$

From $P = Fv$ and the fact that $\mathbf{T}$ is in the same direction as $\mathbf{v}$, we have

$$\begin{aligned} P &= Tv \\ &= (2.16 \times 10^4 \text{ N})(3 \text{ m/s}) = 6.48 \times 10^4 \text{ W} \\ &= 64.8 \text{ kW} = 86.9 \text{ hp} \end{aligned}$$

5.9 CONSERVATION OF ENERGY IN GENERAL

We have seen that the total mechanical energy of a system is conserved when only conservative forces act on the system. Furthermore, we have been able to associate a potential energy with each conservative force. In other words, mechanical energy is lost when nonconservative forces, such as friction, are present.

We can generalize the energy conservation principle to include all forces acting on the system, both conservative and nonconservative. In our study of heat, we shall find that mechanical energy can be transformed to thermal energy. For example, when a block slides over a rough surface, the mechanical energy lost is transformed to thermal energy stored in the block, as evidenced by a measurable increase in its temperature. On a submicroscopic scale, we shall see that this internal thermal energy is associated with the vibration of atoms about their equilibrium positions. Since this internal atomic motion has kinetic and potential energy, one can say that frictional forces arise fundamentally from conservative atomic forces. Therefore, if we include this increase in the internal energy of the system in our work-energy theorem, the total energy is conserved.

This is just one example of how you can analyze an isolated system and always find that the total energy does not change as long as you account for all forms of energy. That is, *energy can never be created or destroyed.* Energy may be transformed from one form to another, but *the total energy of an isolated system is always constant.* From a universal point of view, we believe that the total energy of the universe is constant. Therefore, if one part of the universe gains energy in some form, another part must lose an equal amount of energy. No violation of this principle has been found.

Other examples of energy transformations are the energy carried by sound waves resulting from the collision of two objects, the energy radiated by an accelerating charge in the form of electromagnetic waves (a radio antenna), and the elaborate sequence of energy conversions in a thermonuclear reaction.

In subsequent chapters, we shall see that the various parts of the energy concept, especially transformations of energy between various forms, join together the various branches of physics. In other words, one cannot completely separate the subjects of mechanics, thermodynamics, and electromagnetism. Finally, from a practical viewpoint, all mechanical and electronic devices rely on some form of energy transformation.

5.10 WORK DONE BY A VARYING FORCE

Consider an object being displaced along the x axis under the action of a varying force, as shown in Figure 5.17a. Such a situation would occur if a person attempted to pull an object at a constant speed along a surface that varied in roughness from point to point. The magnitude of the required force would change as the roughness increased or decreased. In such a situation, we cannot use $W = Fs \cos \theta$ to calculate the work done by the force because *this relationship applies only when the force is constant in magnitude and direction.* However, we can calculate the work done as the object undergoes a very small displacement, Δx, as shown in Figure 5.17a. During this small displacement the force is approximately constant, and so we can express the work done by the force as

$$\Delta W = F_x \Delta x$$

Note that this is just the area of the shaded rectangle in Figure 5.17a. Now, if we imagine that the force-displacement curve is divided into a large number of such intervals, as in Figure 5.17a, then the total work done for the displacement from x_i to x_f is approximately equal to the sum of a large number of such terms:

$$\boxed{W \approx \sum F_x \,\Delta x} \qquad (5.19)$$

The right side of this equation is the sum of the area of all the rectangles pictured, but if the rectangles are made very narrow by decreasing Δx, the right side of the equation turns out to be just the total area under the curve, which is shown in Figure 5.17. (Note that the word "area" here refers to the region under a curve of F vs. x, not a geometric area.) Thus, we can say that

the total work done by a varying force is equal to the area under the force-displacement curve.

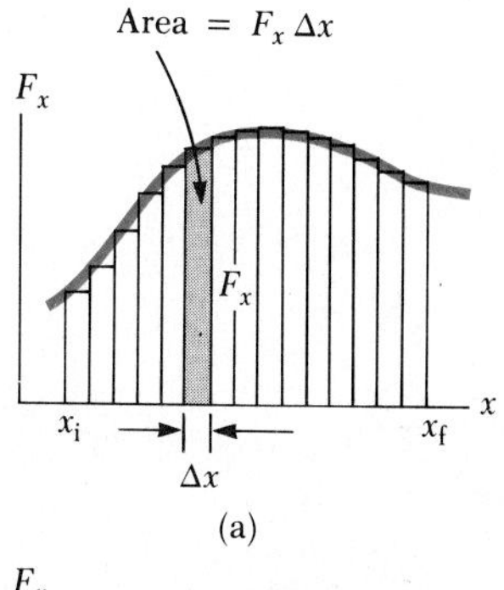

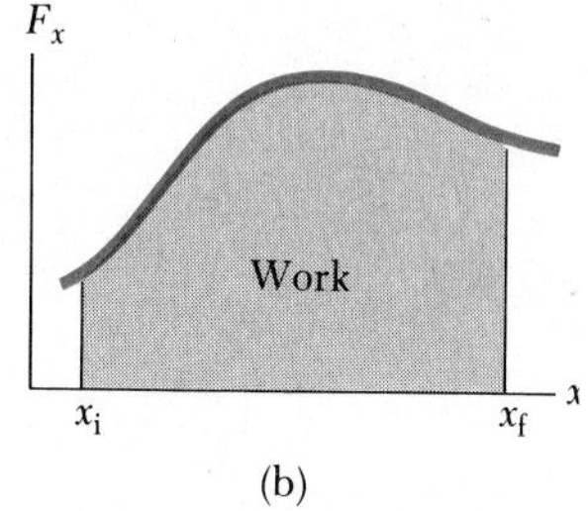

FIGURE 5.17 (a) The work done by a force F_x for the small displacement Δx is $F_x \Delta x$, which equals the area of the shaded rectangle. The total work done for the displacement x_i to x_f is approximately equal to the sum of the areas of all the rectangles. (b) The work done by a variable force F_x as the particle moves from x_i to x_f is *exactly* equal to the area under this curve.

EXAMPLE 5.11

A force acting on an object varies with x as shown in Figure 5.18. Calculate the work done by the force as the object moves from $x = 0$ to $x = 6$ m.

Solution The work done by the force is equal to the total area under the curve from $x = 0$ to $x = 6$ m. This area is equal to the area of the rectangular section between $x = 0$ to $x = 4$ m plus the area of the triangular section between $x = 4$ m to $x = 6$ m. The area of the rectangle is (4×5) N · m = 20 J, and the area of the triangle is $\frac{1}{2}(2 \times 5)$N · m = 5 J. Therefore, the total work done is 25 J.

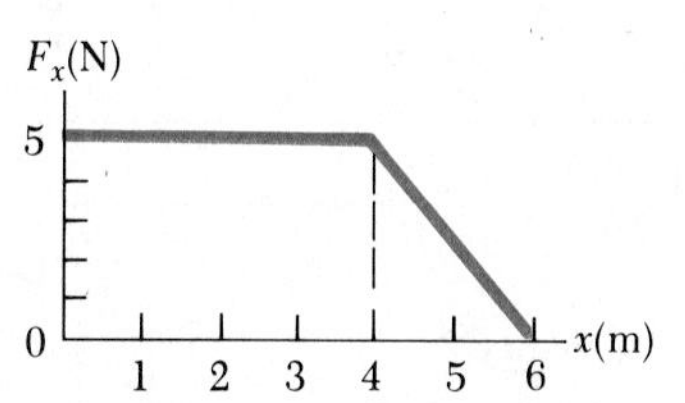

FIGURE 5.18 (Example 5.11) The force is constant for the first 4 m of motion and then decreases linearly with x from $x = 4$ m to $x = 6$ m. The net work done by this force is the area under this curve.

SUMMARY

The **work** done by a *constant* force $\boldsymbol{F}$ acting on an object is defined as the product of the component of the force in the direction of the displacement and the magnitude of the displacement. If the force makes an angle θ with the displacement $\boldsymbol{s}$, the work done by $\boldsymbol{F}$ is

$$\boxed{W \equiv (F \cos \theta) s} \qquad (5.1)$$

The **kinetic energy** of an object of mass m moving with a speed v is defined as

$$KE \equiv \frac{1}{2} mv^2 \tag{5.5}$$

The **work-energy theorem** states that the net work done on an object by external forces equals the change in kinetic energy of the object:

$$W_{net} = KE_f - KE_i \tag{5.6}$$

The **gravitational potential energy** of an object of mass m that is elevated a distance y above the earth's surface is given by

$$PE \equiv mgy \tag{5.8}$$

A force is **conservative** if the work it does depends only on the initial and final positions of an object and not on the path selected to move between these points. A force that does not follow this criterion is said to be **nonconservative.**

The **law of conservation of mechanical energy** states that if the only force acting on a system is conservative, total mechanical energy is conserved:

$$KE_i + PE_i = KE_f + PE_f \tag{5.11}$$

In the important case in which the only force doing any work on a system is the gravitational force, the law of conservation of mechanical energy reduces to

$$\frac{1}{2} mv_i^2 + mgy_i = \frac{1}{2} mv_f^2 + mgy_f \tag{5.12}$$

The **work-energy theorem** states that the work done by all nonconservative forces acting on a system equals the change in the total mechanical energy of the system:

$$W_{nc} = (KE_f - KE_i) + (PE_f - PE_i) \tag{5.14}$$

Average power is defined as the ratio of work done to the time interval during which the work is done:

$$\overline{P} = \frac{\Delta W}{\Delta t} \tag{5.15}$$

If a force $\boldsymbol{F}$ acts on an object moving with an average velocity $\overline{v}$, the average power delivered to the object is given by

$$\overline{P} = F\overline{v} \tag{5.16}$$

The total work done by a varying force is equal to the area under the force-displacement curve.

ADDITIONAL READING

Freeman J. Dyson, "Energy in the Universe," *Sci. American*, September 1971, p. 50.

A. Einstein and L. Infeld, *Evolution of Physics*, New York, Simon and Shuster, 1938.

"Energy and Power," *Sci. American*, September 1971. This entire issue is devoted to energy-related topics.

R. H. Romer, *Energy: An Introduction to Physics*, San Francisco, Freeman, 1976. This text contains many interesting energy applications.

G. H. Schurr, "Energy," *Sci. American*, September 1963, p. 110.

G. Waring, "Energy and the Automobile," *The Physics Teacher*, *18*, 1980, p. 494. An informative article on fuel consumption by automobiles.

R. Wilson and W. J. Jones, *Energy, Ecology, and the Environment*, New York, Academic Press, 1974.

QUESTIONS

1. Explain why the work done by the force of sliding friction is negative when an object undergoes a displacement on a rough surface.
2. A snowball is thrown against a wall with a velocity great enough that the ball melts on impact. Examine the energy changes in this collision.
3. Can the kinetic energy of an object have a negative value?
4. An investigation at the scene of a collision between two cars shows that one car left skid marks twice the length of those left by the other car. A bystander who viewed the collision claims that the cars applied their brakes at the same time. What conclusions can you draw?
5. A pile driver is a device used to drive objects into the earth by dropping a heavy weight on them. By how much does the energy of a pile driver increase when the weight it drops is doubled? (Assume the weight is dropped from the same height each time.)
6. What can be said about the speed of an object if the net work done on that object is zero?
7. A child riding in the back seat of a car exerts a constant force on the front seat in an effort to make the car go faster. The child exerts a force on the car and the car is in motion. Does the child do any work?
8. When a punter kicks a football, is he doing any work on the ball while his toe is in contact with the ball? Is he doing any work on the ball after it loses contact with his toe? Are there any forces doing work on the ball while it is in flight?
9. A ball is thrown straight up into the air. At what position is its kinetic energy a maximum? At what position is its gravitational potential energy a maximum?
10. Can the gravitational potential energy of an object ever have a negative value? Explain.
11. How can the work-energy theorem explain why the force of sliding friction always has the effect of reducing the kinetic energy of a particle?
12. A bowling ball is suspended from the ceiling of a lecture hall by a strong cord. The bowling ball is drawn from its equilibrium position and released from rest at the tip of the demonstrator's nose. If the demonstrator remains stationary, explain why she will not be struck by the ball on its return swing. Would this demonstrator be safe if the ball were given a slight push from this position?
13. A ball is dropped by a person from the top of a building, while another person at the bottom observes its motion. Will these two people always agree on the value of the ball's potential energy? On the change in potential energy of the ball? On the kinetic energy of the ball?
14. Two identical objects move with speeds of 5 m/s and 20 m/s. What is the ratio of their kinetic energies?
15. Discuss the work done by a pitcher when he throws a baseball. What is the approximate distance through which the force acts as the ball is thrown?
16. Discuss the energy transformations that occur during a pole vault event.
17. Estimate the time it takes you to climb a flight of stairs. Then approximate the power required to perform this task. Express your value in horsepower.
18. Does kinetic energy depend on the frame of reference, or coordinate system, in which it is measured?
19. Advertisements for "superballs" once stated that the ball would rebound to a height greater than the height from which it was dropped. Is this possible?
20. Many mountain roads are built so that they spiral around the mountain rather than go straight up toward the peak. Discuss this design from the viewpoint of energy and power.
21. Does the gravitational force exerted on the moon by the earth do any work on the moon? Explain.
22. As a pendulum swings back and forth, does the tension in the string do any work on the pendulum bob? Does the force of gravity do any work?

PROBLEMS

Section 5.2 Work

1. A tugboat exerts a constant force of 5000 N on a ship moving at constant speed through a harbor. How much work does the tugboat do on the ship if it moves a distance of 3 km?

2. A weight lifter lifts a 350-N set of weights from ground level to a position over his head, a vertical distance of 2 m. How much work does the weightlifter do, assuming he moves the weights at constant speed?

3. A shopper in a supermarket pushes a cart with a force of 35 N directed at an angle of 25° downward from the horizontal. Find the work done by the shopper as she moves down a 50-m length of aisle.

• **4.** A 70-kg baserunner begins his slide into second base when moving at a speed of 4 m/s. The coefficient of friction between his clothes and earth is 0.7. He slides such that his speed has been reduced to zero just as he reaches the base. (a) How much work was done by friction acting on the runner? (b) How far did he slide?

• **5.** A horizontal force of 150 N is used to push a 40-kg packing crate on a rough horizontal surface through a distance of 6 m. If the crate moves at constant speed, find (a) the work done by the 150-N force, (b) the work done by friction, and (c) the coefficient of kinetic friction.

• **6.** A stewardess pulls her 70-N flight bag a distance of 200 m along an airport floor at constant speed. The force she exerts is 40 N at an angle of 50° above the horizontal. Find (a) the work she does, (b) the work done by the force of friction, and (c) the coefficient of kinetic friction between her flight bag and the floor.

Section 5.3 Work and Kinetic Energy

7. What is the kinetic energy of a 3000-kg car moving at 55 mi/h? How much heat, in joules, will be lost through friction in the brake linings when the car is brought to a stop?

8. A very good major league pitcher can throw a baseball at 100 mi/h, while an average pitcher throws the ball at a speed of about 80 mi/h. If a baseball has a mass of 0.15 kg, find the kinetic energy of the ball in each case.

9. A gas molecule in the room you are sitting in may have a speed of about 500 m/s. If its mass is equal to 5×10^{-26} kg, what is its kinetic energy?

• **10.** A 2-g bullet leaves the barrel of a gun at a speed of 300 m/s. (a) Find its kinetic energy. (b) If the length of the barrel is 50 cm, find the average force exerted on the bullet by the expanding gases as it moves the length of the barrel.

• **11.** A 1200-kg vehicle is moving along a horizontal surface with a speed of 20 m/s. What work must be done by the brakes to bring the vehicle to rest in 20 s?

• **12.** A 2000-kg car moves down a level highway under the action of two forces. One is a forward thrust on the car of 1000 N from the motor; the second is a 950-N resistive force. Use the work-energy theorem to find the speed of the car after it has moved a distance of 20 m, assuming it starts from rest.

Section 5.4 Gravitational Potential Energy

Section 5.5 Conservative and Nonconservative Forces

13. What is the gravitational potential energy relative to the ground of a 0.15-kg baseball at the top of a 100-m tall building?

14. A 1000-kg roller coaster is initially at the top of a rise, point *A*. It then moves 135 ft at an angle of 40° below the horizontal to a lower point *B*. (a) Choose point *B* to be the zero level for gravitational potential energy, and find the potential energy of the roller coaster at points *A* and *B*, and find the difference in potential energy between these points. (b) Repeat part (a), choosing the zero reference level to be at point *A*.

15. Find the net force along the slope acting on the roller coaster of problem 14 and use this to calculate the work done as it moves from *A* to *B*. (Ignore any losses due to friction.) Compare your answer to the differences in potential energy found in Problem 14.

• **16.** A 40-N child is in a swing attached to ropes 2 m long. Find the gravitational potential energy of the child relative to the child's lowest position when (a) the ropes are horizontal, (b) the ropes make a 30° angle with the vertical, and (c) at the bottom of the circular arc.

Section 5.6 Conservation of Mechanical Energy

17. A baseball player decides to show his skill by catching a 0.15-kg baseball dropped from the top of a 100-m tall building. What is the speed of the baseball just before it strikes his glove?

18. A child and sled with a combined mass of 50 kg slide down a frictionless hill. If the sled starts from rest and has acquired a speed of 3 m/s by the time it reaches the bottom of the hill, what is the height of the hill?

19. A 70-kg high-jumper leaves the ground with a speed of 6 m/s. How high can he leap?

20. A baseball player throws a baseball straight up into the air with an initial speed of 20 m/s. Find (a) the maximum height to which the ball rises, and (b) the speed of the ball when it is halfway up to its maximum height.

• **21.** Tarzan swings on a vine 30 m long initially inclined at an angle of 37° to the vertical. (a) What is his speed at the bottom of the swing if he starts from rest? (b) If he pushes off with a speed of 4 m/s, what is his speed at the bottom of the swing?

•• **22.** The masses of the javelin, discus, and shot are 0.8 kg, 2.0 kg, and 7.2 kg, respectively, and record throws in the track events using these objects are about 89 m, 69 m, and 21 m, respectively. Neglecting air resistance, (a) calculate the minimum initial kinetic energies that would produce these throws, and (b) estimate the average force exerted on each object during the throw, assuming the force acts over a distance of 2 m. (c) Do your results suggest that air resistance is an important factor?

Section 5.7 Nonconservative Forces and the Work-Energy Theorem

• **23.** A child starts from rest at the top of a slide of height 4 m. (a) What is her speed at the bottom if the incline is frictionless? (b) If she reaches the bottom with a speed of 6 m/s, what percentage of her total energy at the top of the slide is lost as a result of friction?

• **24.** A 2000-kg car starts from rest at the top of a driveway 5 m long which is sloped at an angle of 20° with the horizontal. If an average friction force of 4000 N impedes the motion of the car, find the speed of the car at the bottom of the driveway.

• **25.** A 25-kg child on a swing 2 m long is released from rest when the swing supports make an angle of 30° with the vertical. (a) Neglecting friction, find the child's speed at the lowest position. (b) If the speed of the child at the lowest position is 2 m/s, what is the energy lost due to friction?

• **26.** An airplane of mass 1.5×10^4 kg is moving at a speed of 60 m/s. The pilot then increases the speed of the engine such that the forward thrust of the propeller becomes 7.5×10^4 N. The force of air resistance on the airplane is 4×10^4 N. The pilot then goes into a dive at an angle of 30° with the horizontal. Find the speed of the plane after it has traveled 500 m.

•• **27.** In a circus performance, a monkey on a sled is given an initial speed of 4 m/s up a 20° incline. The combined mass of the monkey and sled is 20 kg, and the coefficient of friction between sled and incline is 0.2. How far will the sled move up the incline?

•• **28.** A 2-kg block is projected up a 30° incline with an initial speed of 3 m/s at the bottom. The coefficient of kinetic friction between the block and the incline is 0.7. Find (a) the distance the block will travel up the incline before coming to rest, (b) the total work done by friction while the block is in motion, and (c) the change in potential energy and change in kinetic energy when the block has traveled 0.3 m up the incline.

Section 5.9 Power

29. A 1500-kg car accelerates uniformly from rest to a speed of 10 m/s in 7 s. Find (a) the work done on the car in this time and (b) the average power delivered by the engine in the first 7 s.

30. A 1500-kg car starts from rest and accelerates to a final speed of 18 m/s in a time of 12 s. Assume that the force of air resistance remains constant at 400 N during this time. Find (a) the average power developed by the engine of the car, and (b) the instantaneous power output of the engine at $t = 12$ s.

31. What average power is developed by an 800-N professor while running at constant speed up a flight of stairs rising 6 m if he takes 8 s to complete the climb?

• **32.** A motor is used to pull an 80 kg skier along a horizontal surface at a constant speed of 2 m/s to enable him to learn to keep his balance. If the coefficient of friction between the skis and the surface is 0.15, what horsepower motor is required?

• **33.** A machine lifts a 300-kg crate at constant speed through a height of 5 m in 8 s. Calculate the power output of the machine.

• **34.** A skier of mass 70 kg is pulled up a slope by a motor-driven cable. (a) How much work is required to pull him a distance of 60 m up a 30° slope (assumed frictionless) at a constant speed of 2 m/s? (b) What horsepower motor is required to perform this task?

• **35.** A 50-kg student climbs a rope 5 m in length and stops at the top. (a) What must her average speed have been in order to match the power output of a 200-W lightbulb? (b) How much work did she do?

• **36.** You have just found the "Lost Dutchman" goldmine, 80 ft below ground level. However, the mine is flooded, and you discover that in order to keep it dry enough to mine, you must pump water out at the rate of 60 lb every second. What is the minimum horsepower motor that can be used to perform this task?

Section 5.10 Work Done by a Varying Force

• **37.** A 1-kg object starts from rest, but then moves through a distance of 3 m when a nonconstant force begins to act on it. The force varies with position as shown in Figure 5.19. (a) How much work is done on the object by this force, and (b) what is its speed at $x = 3$ m?

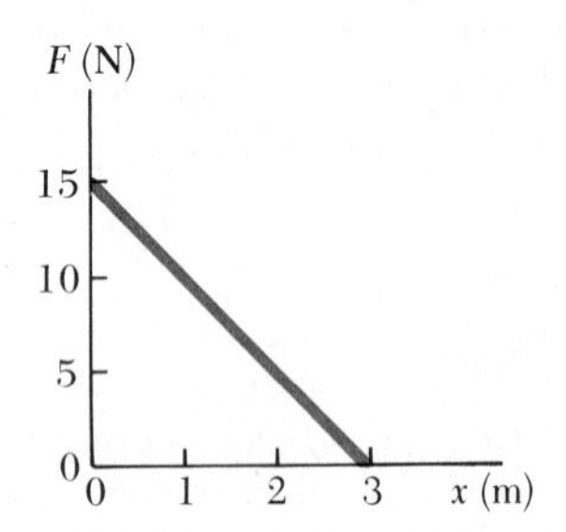

FIGURE 5.19 Problem 37

38. A body is subjected to a force that varies with position as in Figure 5.20. Find the work done by the force on the object as it moves (a) from $x = 0$ to $x = 5$ m, and (b) from $x = 5$ m to $x = 10$ m, and (c) from $x = 10$ m to $x = 15$ m. (d) What is the total work done by the force over the distance $x = 0$ to $x = 15$ m?

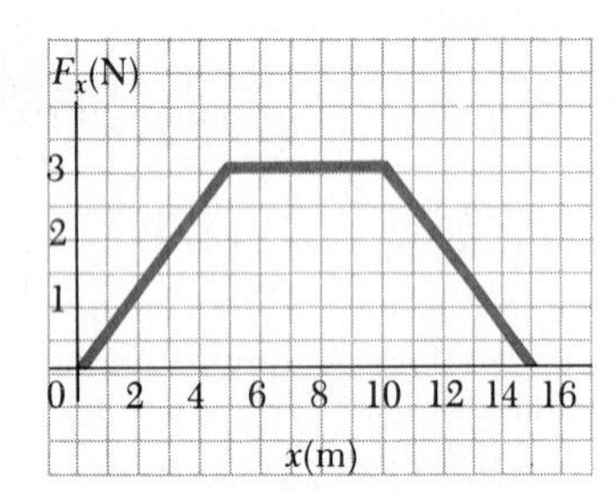

FIGURE 5.20 Problem 38

39. If the object of problem 38 had a mass of 3 kg and a speed of 0.5 m/s at $x = 0$ find its speed at (a) $x = 5$ m, (b) $x = 10$ m, and (c) $x = 15$ m.

ADDITIONAL PROBLEMS

40. Find the height from which you would have to drop this textbook so that it would have a speed of 10 m/s just before it strikes your professor's head.

41. Figure 5.21 shows the top surface of a rough table. A textbook is moved from position A, the origin of a coordinate system, to position B, which has coordinates (2, 0) where the units are meters. If the magnitude of the frictional force is 4 N, find the work done against friction as the book is moved from A to B along (a) the circular path, (b) the triangular path, and (c) the straight-line path.

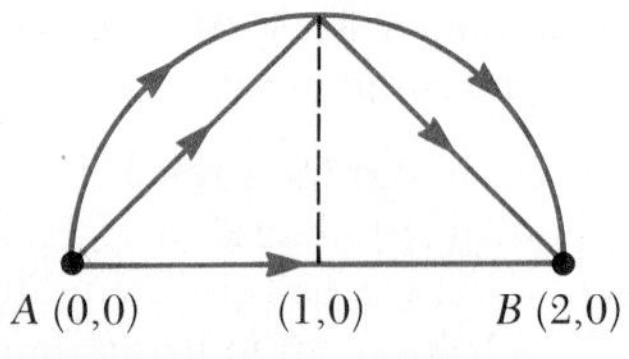

FIGURE 5.21 Problem 41

42. The plane shown in Figure 5.22 is designed for vertical takeoff and landing. If the plane has a mass of 8000 kg when fueled, find the net work done on the plane as it accelerates vertically at 1 m/s^2 for a distance of 30 m starting from rest.

FIGURE 5.22 Problem 42 (Courtesy of NASA)

43. A 0.4-kg bead slides on a curved wire, starting from rest at point A in Figure 5.23. If the wire is frictionless, find the speed of the bead (a) at B and (b) at C.

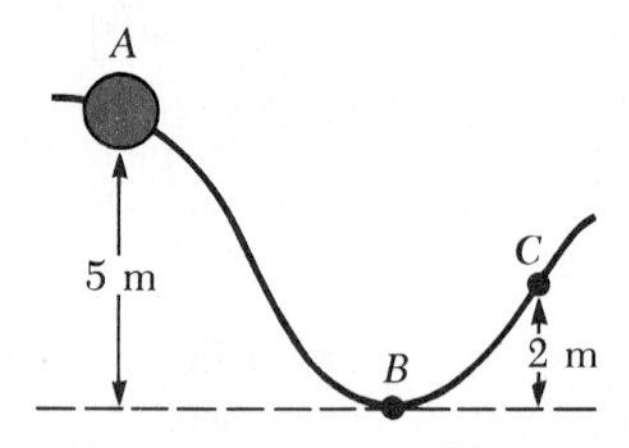

FIGURE 5.23 Problems 43 and 47

44. The engine of a jet aircraft develops a forward thrust of 4000 lbs. What horsepower is it developing when flying at its cruising speed of 500 mi/h?

45. A 200-kg crate is pulled parallel to a level surface by an engine. The coefficient of friction between crate and surface is 0.4. (a) How much power must the engine deliver to move the crate with a constant speed of 5 m/s? (b) How much work is done by the engine in 3 min?

46. A 90-horsepower engine is moving a car at a speed of 60 mi/h. What forward thrust is the engine exerting on the car?

47. If the wire in Problem 43 (Fig 5.23) is frictionless between points A and B and rough between points B and C, and the bead starts from rest at point A, (a) find the speed of the bead at B. (b) If the 0.4 kg bead comes to rest at C, find the total work done by friction as the bead goes from B to C. (c) What is the net work done by non-conservative forces as the bead moves from A to C?

• **48.** A catcher "gives" with the ball when he catches a 0.15-kg baseball moving at 25 m/s. (a) If he moves his catching glove through a distance of 2 cm, what is the average force acting on his hand? (b) Repeat for the case in which his hand moves 10 cm.

• **49.** A grocery cart weighing 98 N is pushed 12 m across the floor by a shopper who exerts a constant horizontal force of 40 N. If all frictional forces are neglected, what is the final speed of the cart if it starts from rest?

• **50.** If the speed of the cart of Problem 49 is half that found based on the prediction of no friction, find the coefficient of friction between cart and floor surface.

• **51.** (a) A 75-kg man jumps from a window 1 m above a sidewalk. What is his speed just before his feet strike the pavement? (b) If the man jumps with his knees and ankles locked, the only cushion for his fall is an approximate 0.5-cm give in the pads of his feet. Calculate the average force exerted on him by the ground in this situation. This average force is sufficient to cause cartilage damage in the joints or to break bones.

• **52.** Tarzan and Jane, whose total mass is 130 kg, start their swing on a vine 5 m long such that the vine is initially at an angle of 30° with the horizontal. At the bottom of the arc, Jane, whose mass is 50 kg, steps off. What is the maximum height of the tree branch that Tarzan can land on as his swing continues?

• **53.** A 6-kg object is lifted vertically through a distance of 5 m by a light string under a tension of 80 N. Find (a) the work done by the force of the tension, (b) the work done by gravity, and (c) the final speed of the object if it starts from rest.

• **54.** A 0.4-mg flea can jump to a maximum height of about 2.5 cm. Find the speed with which he takes off.

• **55.** A 65-kg athlete runs 600 m at constant speed up a mountain inclined at 20° to the horizontal. He performs this feat in 150 s. Assuming that air resistance is negligible, (a) how much work does he perform and (b) what is his power output during the run?

• **56.** A 3-kg mass is moving such that it has a component of velocity along the x direction of 5 m/s and a component of velocity along the y direction of -3 m/s. (a) What is the kinetic energy at this time? (b) Find the change in kinetic energy if the velocity changes such that the new x component of velocity is 8 m/s and its new y component of velocity is 4 m/s.

• **57.** A projectile of mass m is shot horizontally with initial velocity v_0 from a height h above a flat desert floor. At the instant before the projectile hits the desert floor, find (a) the work done on the projectile by gravity, (b) the change in kinetic energy since the projectile was fired, and (c) the final kinetic energy of the projectile.

• **58.** A 2-kg mass is suspended by means of a light string that passes over a light, frictionless pulley as shown in Figure 5.24. The other end of the string is connected to a 1-kg mass that rests upon a horizontal frictionless surface. The system starts in motion with the string connected to the 1-kg mass making an angle of 30° with the horizontal. When the string makes an angle of 45° with the horizontal, how much work has been done on the 1-kg mass? (The pulley is 2 m above the surface and the surface is frictionless.)

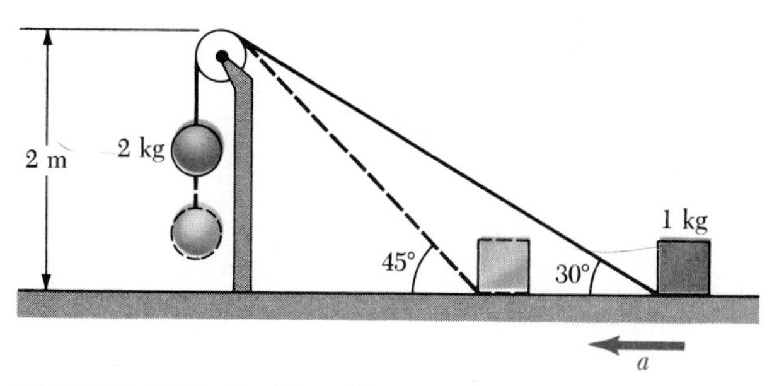

FIGURE 5.24 Problem 58

• **59.** A passenger car of mass 1500 kg accelerates from rest to 97 km/h in 10 s. (a) Find the acceleration of the car. (b) Show that the coefficient of friction between the rear tires and road must be at least 0.55. (c) Determine the average power delivered by the engine. (Assume that the normal force on each tire is ¼ mg.)

•• **60.** A 3-kg block is moved up a 37° incline under the action of a constant horizontal force of 40 N. The coefficient of kinetic friction is 0.1, and the block is pulled 2 m up the incline. Calculate (a) the work done by the 40 N force, (b) the work done by gravity, (c) the work done by friction, and (d) the change in kinetic energy of the block. (*Note:* the applied force is not parallel to the incline.)

•• **61.** A 0.4-kg mass slides on a horizontal circular track of radius 1.5 m. The mass is given an initial speed of 8 m/s. After one revolution, its speed drops to 6 m/s because of friction. Find (a) the work done by the force of friction in one revolution and (b) the coefficient of kinetic friction. (c) How many revolutions does the mass make before coming to rest?

•• **62.** A 200-g particle is released from rest at point A along the diameter of the inside of a smooth hemispherical bowl of radius $R = 30$ cm (Fig 5.25). Calculate (a) its gravitational potential energy at point A relative to point B, (b) its kinetic energy at point B, (c) its speed at point B, and (d) its kinetic energy and potential energy at point C.

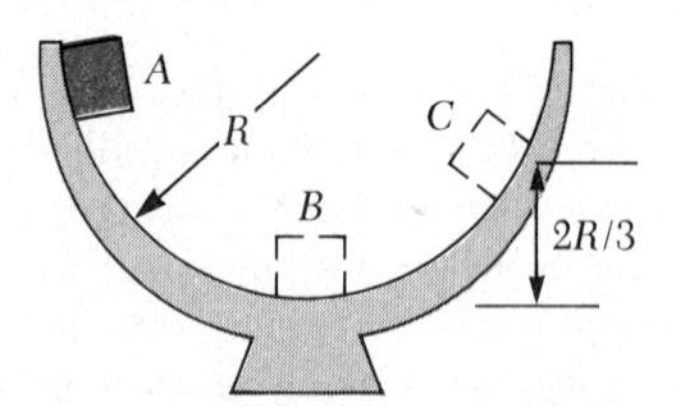

FIGURE 5.25 Problem 62

•• **63.** The particle described in Problem 62 (Fig. 5.25) is released from point A at rest. The speed of the particle at point B is 1.5 m/s. (a) What is its kinetic energy at B? (b) How much energy is lost as a result of friction as the particle goes from A to B? (c) Is it possible to determine μ from these results in any simple manner? Explain.

CHAPTER 6

Momentum and Collisions

As a result of this collision, the pins gain momentum while the bowling ball loses momentum. (Courtesy of Fair Lanes, Inc.)

Consider what happens when a golf ball is struck by a club. The ball is given a very large initial velocity as a result of the collision and consequently is able to travel several hundred feet through the air. The ball experiences a large change in velocity and a correspondingly large acceleration. Furthermore, because the ball experiences this acceleration over a very short time interval, the average force on the ball during collision with the club is very large. By Newton's third law, the club experiences a reaction force that is equal to and opposite the force on the ball. This reaction force produces a change in the velocity of the club. Since the club is much more massive than the ball, however, its change in velocity is much less than the change in velocity of the ball.

One of the main objectives of this chapter is to understand and analyze such events. As a first step, we shall introduce the concept of momentum, a term one often uses in describing objects in motion. For example, a very massive football player is often said to have a great deal of momentum as he runs down the field. A much less massive player, such as a halfback, can have equal or greater momentum if he moves with a higher velocity. This follows from the fact that momentum is defined as the product of mass and velocity.

Additionally, we shall find that the concept of momentum will lead us to a second conservation law, that of conservation of momentum. This law is especially useful for treating problems that involve collisions between objects.

6.1 MOMENTUM AND IMPULSE

The momentum of an object of mass m moving with a velocity $\boldsymbol{v}$ is defined as the product of the mass and the velocity:

Momentum

$$\boxed{\boldsymbol{p} \equiv m\boldsymbol{v}} \tag{6.1}$$

As its definition shows, momentum is a vector quantity, with its direction the same as that of the velocity. Momentum has dimensions of [M][L]/[T], and its SI units are kilogram-meters per second (kg · m/s).

Often we shall find it advantageous to work with the components of momentum. For two-dimensional motion, these are given by

$$p_x = mv_x \qquad p_y = mv_y \tag{6.2}$$

where p_x represents the momentum of an object in the x direction and p_y its momentum in the y direction.

The definition of momentum in Equation 6.1 coincides with our customary usage of the word in our daily affairs. When we think of a massive object moving with a high velocity, we often say that this object has a large momentum, in accordance with Equation 6.1. Likewise, a small object moving slowly is said to have a small momentum. On the other hand, a small object moving with a high velocity can have a large momentum.

When Newton first expressed the second law in mathematical form, he did not write it as $\boldsymbol{F} = m\boldsymbol{a}$; instead he wrote it as

$$\boxed{\boldsymbol{F} = \frac{\text{change in momentum}}{\text{time interval}} = \frac{\Delta \boldsymbol{p}}{\Delta t}} \tag{6.3a}$$

where Δt is the time interval during which the momentum changes by $\Delta \boldsymbol{p}$. This equation says that the rate of change of momentum of an object is equal to the resultant force acting on the object. To see that this is equivalent to $\boldsymbol{F} = m\boldsymbol{a}$ for an object of constant mass, consider a constant force $\boldsymbol{F}$ acting on an object and producing a constant acceleration. We can write Equation 6.3a as

$$\boldsymbol{F} = \frac{\Delta \boldsymbol{p}}{\Delta t} = \frac{m\boldsymbol{v}_f - m\boldsymbol{v}_i}{\Delta t} = \frac{m(\boldsymbol{v}_f - \boldsymbol{v}_i)}{\Delta t} \tag{6.3b}$$

Now recall that the velocity of an object moving with constant acceleration varies with time as

$$\boldsymbol{v}_f = \boldsymbol{v}_i + \boldsymbol{a}t$$

If we take $\Delta t = t$ and substitute for $\boldsymbol{v}_f$ in Equation 6.3b, we see that $\boldsymbol{F}$ reduces to the familiar equation

$$\boldsymbol{F} = m\boldsymbol{a}$$

Note from Equation 6.3a that

if the resultant force $\boldsymbol{F}$ is zero, the momentum of the object does not change. In other words, the linear momentum and velocity of an object are conserved when $\boldsymbol{F} = 0$.

This feature, or property, of momentum will be important to us in analyzing collisions, a subject we shall take up in a later section.

Equation 6.3a can be written as

$$\boldsymbol{F}\Delta t = \Delta \boldsymbol{p}$$

or

$$\boxed{\boldsymbol{F}\Delta t = \Delta \boldsymbol{p} = m\boldsymbol{v}_{\mathrm{f}} - m\boldsymbol{v}_{\mathrm{i}}} \qquad (6.4)$$

This result is often called the **impulse-momentum theorem.** The term on the left side of the equation, $\boldsymbol{F}\Delta t$, is called the **impulse** of the force $\boldsymbol{F}$ for the time interval Δt. According to this result, *the impulse of the force acting on an object equals the change in momentum of that object.*

This equation tells us that if we exert a force on an object for a time interval Δt, the effect of this force is to change the momentum of the object from some initial value $m\boldsymbol{v}_{\mathrm{i}}$ to some final value $m\boldsymbol{v}_{\mathrm{f}}$. For example, suppose a pitcher throws a baseball with a velocity $\boldsymbol{v}_{\mathrm{i}}$ and a batter hits the ball head-on so as to reverse the direction of its velocity. The force $\boldsymbol{F}$ that the bat exerts on the ball can change both the direction and the magnitude of the initial velocity to a higher value $\boldsymbol{v}_{\mathrm{f}}$.

It is necessary to add a word of caution here. If you were to try to solve a problem such as this using Newton's second law, you would encounter some difficulty in choosing a value for $\boldsymbol{F}$ because the force exerted on the ball is not constant. Instead, it might be represented by a curve like that of Figure 6.1a. The force starts out small as the bat comes in contact with the ball, rises to a maximum value when they are firmly in contact, and then drops off as the ball leaves the bat. In such instances, it is necessary to define an **average force** $\overline{\boldsymbol{F}}$, shown as the dashed line in Figure 6.1b. This average force can be thought of as the constant force that gives the same impulse to the object in the time interval Δt as the actual time-varying force gives over this same interval. The short collision between a bullet and an apple is illustrated in Color Plate 1.

The impulse imparted by a force during a time interval Δt is equal to the area under the force-time graph from the beginning to the end of the time interval.

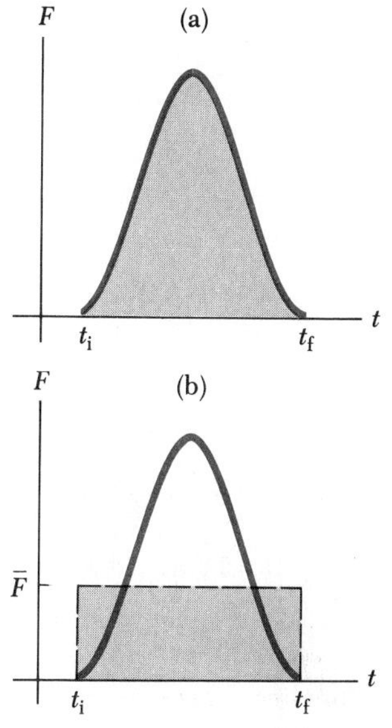

FIGURE 6.1 (a) A force acting on an object may vary in time. The impulse is the area under the force-time curve. (b) The average force (horizontal dashed line) gives the same impulse to the object in the time interval Δt as the real time-varying force described in (a).

EXAMPLE 6.1 Teeing Off

A golf ball of mass 50 g is struck with a club, as in Figure 6.2. The force on the ball varies from zero when contact is made up to some maximum value (where the ball is deformed) and then back to zero when the ball leaves the club. Thus the force-time graph is somewhat like that shown in Figure 6.1. Assume that the ball leaves the club face with a velocity of 44 m/s, and (a) estimate the impulse due to the collision.

Solution Before the club hit the ball, the ball was at rest on the tee, and as a result its initial momentum was zero. The momentum immediately after the collision is

$$p_{\mathrm{f}} = mv_{\mathrm{f}} = (50 \times 10^{-3}\ \mathrm{kg})(44\ \mathrm{m/s}) = 2.20\ \mathrm{kg \cdot m/s}$$

and thus the impulse imparted to the ball is

$$\Delta p = mv_{\mathrm{f}} - mv_{\mathrm{i}} = 2.20\ \mathrm{kg \cdot m/s}$$

(b) Estimate the length of time of the collision and the average force on the ball.

FIGURE 6.2 (Example 6.1) A golf ball being struck by a club. (Courtesy of Dr. Harold E. Edgerton, MIT.)

Solution From Figure 6.2, it appears that a reasonable estimate of the distance the ball travels while in contact with the club is the radius of the ball, about 2.0 cm. The time it takes the club to move this distance (the contact time) is then

$$\Delta t = \frac{\Delta x}{v_i} = \frac{2 \times 10^{-2} \text{ m}}{44 \text{ m/s}} = 4.5 \times 10^{-4} \text{ s}$$

Finally, the magnitude of the average force is estimated to be

$$\overline{F} = \frac{\Delta p}{\Delta t} = \frac{2.2 \text{ kg} \cdot \text{m/s}}{4.5 \times 10^{-4} \text{ s}} = 4.9 \times 10^3 \text{ N}$$

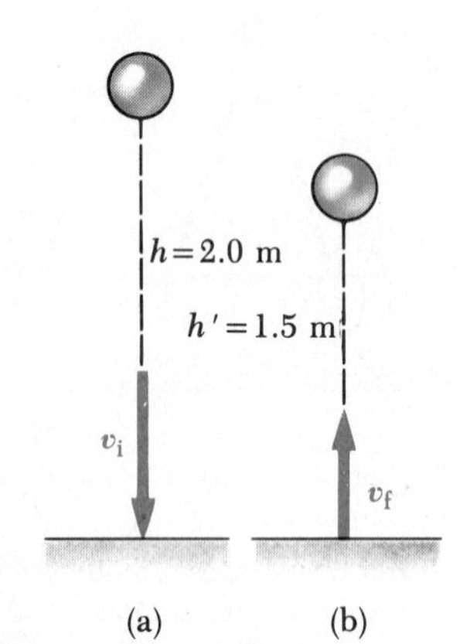

FIGURE 6.3 (Example 6.2) (a) A ball is dropped from a height h and reaches the floor with a velocity v_i. (b) The ball rebounds from the floor with a velocity v_f and reaches a height h'.

EXAMPLE 6.2 Follow the Bouncing Ball

A ball of mass 100 g is dropped from a height $h = 2$ m above the floor (Fig. 6.3). It rebounds vertically to a height $h' = 1.5$ m after colliding with the floor. (a) Find the momentum of the ball immediately before it collides with the floor, and immediately after it rebounds.

Solution We can find the velocity of the ball just before it strikes the floor by using the principle of conservation of mechanical energy. Equating the initial potential energy to the final kinetic energy, with the floor as the reference level, gives

$$mgh = \frac{1}{2}mv_i^2$$

$$v_i = \sqrt{2gh} = \sqrt{2(9.80 \text{ m/s}^2)(2 \text{ m})} = 6.26 \text{ m/s}$$

Likewise, v_f, the ball's velocity after colliding with the floor, is obtained from the energy expression

$$\frac{1}{2}mv_f^2 = mgh'$$

$$v_f = \sqrt{2gh'} = \sqrt{2(9.80 \text{ m/s}^2)(1.5 \text{ m})} = 5.42 \text{ m/s}$$

Because $m = 0.1$ kg, the initial and final momenta are

$$p_i = mv_i = -(0.1 \text{ kg})(6.26 \text{ m/s}) = -0.626 \text{ kg} \cdot \text{m/s}$$

and

$$p_f = mv_f = (0.1 \text{ kg})(5.42 \text{ m/s}) = 0.542 \text{ kg} \cdot \text{m/s}$$

The initial momentum is negative because the velocity, and hence the momentum, are directed downward, in the negative direction.

(b) Determine the average force exerted by the floor on the ball. Assume the time interval of the collision is 10^{-2}s (a typical value).

Solution From the impulse-momentum theorem, we find

$$\overline{F}\Delta t = mv_f - mv_i$$

$$\overline{F} = \frac{[0.542 - (-0.626)] \text{ kg} \cdot \text{m/s}}{10^{-2} \text{ s}}$$

$$= 1.17 \times 10^2 \text{ N}$$

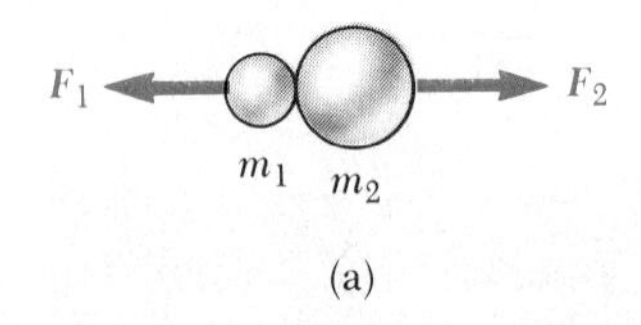

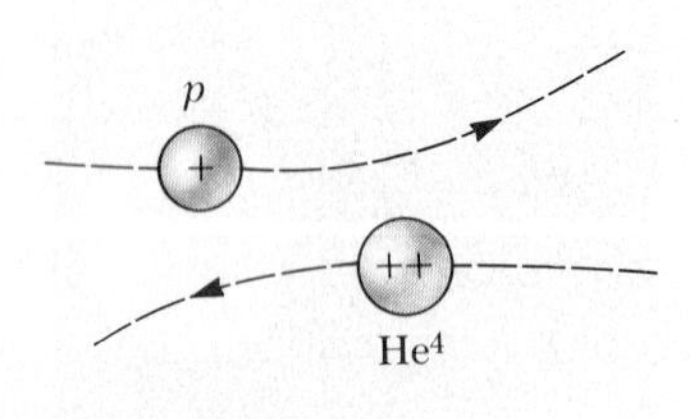

FIGURE 6.4 (a) The collision between two objects as the result of direct contact. (b) The collision between two charged objects.

6.2 CONSERVATION OF MOMENTUM

In this section we shall describe what happens when two objects collide with one another. The collision may be the result of physical contact between two objects, as described in Figure 6.4a. This is a common observation for large-scale objects, such as two billiard balls or a baseball and bat. However, the notion of what we mean by a collision must be generalized because "contact" in the atomic world is different from what we think of as contact in our everyday experiences. For example, Figure 6.4b shows the "collision" between a proton and an alpha particle (the nucleus of a helium atom). Because the two atomic particles are positively charged, they repel each other because of the electrostatic forces between them. Although the two particles never "touch" each other, the results of our analysis here will still apply to this situation.

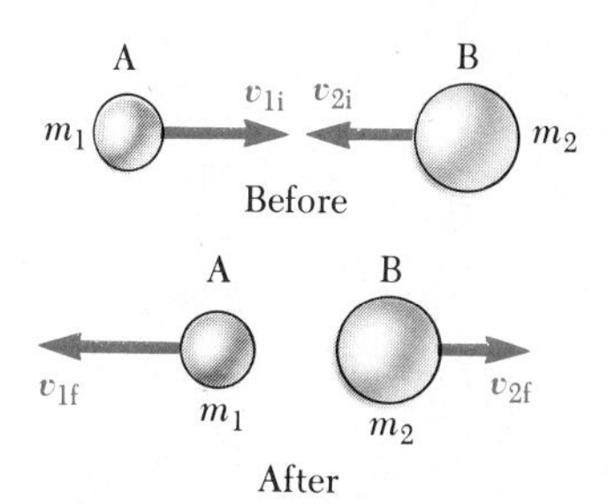

FIGURE 6.5 Before and after the head-on collision between two objects. The momentum of each object changes as a result of the collision, but the total momentum of the system remains constant.

Figure 6.5 shows a before-and-after picture of two objects colliding. Before the collision, the velocity of object A is $\boldsymbol{v}_{1i}$ and that of B is $\boldsymbol{v}_{2i}$; after the collision, the velocities are v_{1f} and $\boldsymbol{v}_{2f}$. The impulse-momentum theorem applied to object A becomes

$$\overline{F}_1\Delta t = m_1v_{1f} - m_1v_{1i}$$

Likewise, for object B we have

$$\overline{F}_2\Delta t = m_2v_{2f} - m_2v_{2i}$$

where $\overline{F}_1$ is the magnitude of the force exerted on A by B during the collision and $\overline{F}_2$ is the magnitude of the force exerted on B by A during the collision (Fig. 6.6).

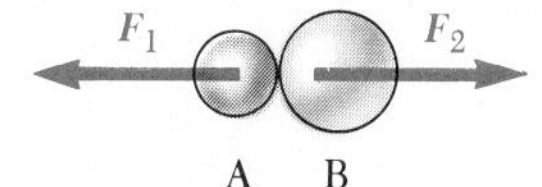

FIGURE 6.6 When two objects collide, the force $\boldsymbol{F}_1$ exerted on object A is equal to magnitude and opposite in direction to the force $\boldsymbol{F}_2$ exerted on object B.

We are using average values for $\overline{F}_1$ and $\overline{F}_2$ even though the actual forces may vary in time in a complicated way, such as described in Figure 6.7. However, regardless of how complicated the forces between the objects may be, Newton's third law states that at all times these two forces are equal in magnitude and opposite in direction ($\boldsymbol{F}_1 = -\boldsymbol{F}_2$), as shown in Figure 6.7. Additionally, both forces act for the same time interval. Thus,

$$\overline{F}_1\Delta t = -\overline{F}_2\Delta t$$

or

$$m_1v_{1f} - m_1v_{1i} = -(m_2v_{2f} - m_2v_{2i})$$

from which we find

$$\boxed{m_1v_{1i} + m_2v_{2i} = m_1v_{1f} + m_2v_{2f}} \quad (6.5)$$

Conservation of momentum

This result, known as the **law of conservation of momentum,** indicates that, when no external forces act on a system consisting of two objects, the total momentum of the system before the collision is equal to the total momentum of the system after the collision.

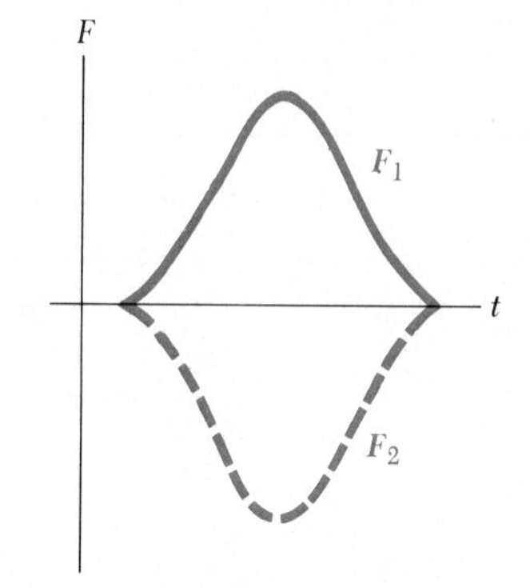

FIGURE 6.7 The force as a function of time for the two colliding particles described in Fig. 6.6. Note that $F_1 = -F_2$.

In the statement of conservation of momentum, note that momentum is said to be conserved for a *system* of objects. In the example used to derive Equation 6.5, the system was taken to be two colliding objects. More generally, a system includes all the objects that are interacting with one another. Additionally, we have assumed that the only forces acting during the collision were internal forces, meaning those forces which arise between the interacting objects of the system. For example, in our collision of two objects shown in Figure 6.6, the internal forces are the forces $\boldsymbol{F}_1$ and $\boldsymbol{F}_2$. If a third object outside our system (consisting of m_1 and m_2) were to exert a force on either

m_1 or m_2 (or both objects) during the collision, momentum would not be conserved for the system. In all the example problems that we shall consider, the system will be assumed to be *isolated.* An isolated system is a system which has no external forces exerted on it, or one in which the external forces are small enough to be considered negligible.

Our derivation has assumed only two objects interacting, but the result remains valid regardless of the number involved. In its most general form, we can state the law of conservation of momentum as follows:

The total momentum of an isolated system of objects is conserved regardless of the nature of the forces between the objects.

To understand what this statement means, consider the following circumstance. Imagine that you are initially standing at rest and then jump upward, leaving the ground with a speed v. Obviously, your momentum is not conserved because your momentum before the jump was zero and became mv as you began to rise. However, the total momentum of the system is conserved if the system selected includes all objects that exert forces on one another. In the present case, you must include the earth as part of the system because you exert a downward force on the earth when you jump. The earth in turn exerts on you an upward force of the same magnitude, as required by Newton's third law. Momentum is conserved for the system consisting of you and the earth. Thus, as you move upward with some momentum mv, the earth moves downward with equal but opposite momentum. The recoil velocity of the earth due to this event will be imperceptibly small, of course, because the earth is so massive, but its momentum does exist.

EXAMPLE 6.3 The Recoiling Pitching Machine

A baseball player attempts to use a pitching machine to help him improve his batting average. To do so, he places the 50 kg machine on a frozen pond as in Figure 6.8. The machine fires a 0.4 kg baseball with a speed of 35 m/s in the horizontal direction. What is the recoil velocity of the machine?

Solution In this example, the system consists of the baseball and the pitching machine. The system is not really isolated because of the force of gravity. However, this external force acts in the vertical direction while the motion of the system is in the horizontal direction. Therefore, the momentum of the system in the x direction is conserved because there are no external forces in this direction (assuming the surface is frictionless).

The total momentum of the system before firing the baseball is zero. Therefore, the total momentum after firing must also be zero, or

$$m_1v_1 + m_2v_2 = 0$$

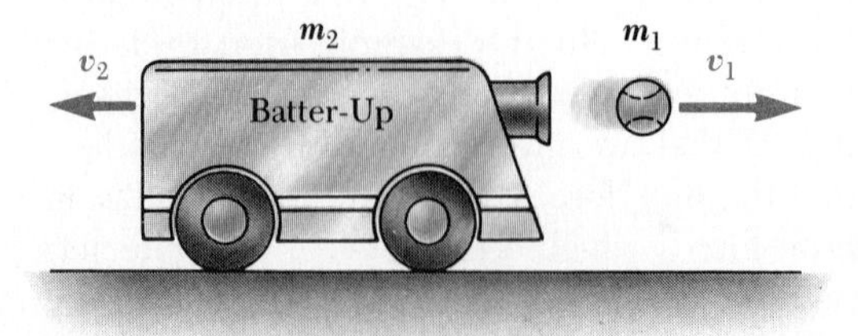

FIGURE 6.8 (Example 6.3) When the baseball is fired to the right, the pitching machine recoils to the left.

With $m_1 = 0.4$ kg, $v_1 = 35$ m/s, and $m_2 = 50$ kg, we find the recoil velocity of the pitching machine to be

$$v_2 = -\frac{m_1}{m_2} v_1 = -\left(\frac{0.4 \text{ kg}}{50 \text{ kg}}\right)(35 \text{ m/s}) = -0.28 \text{ m/s}$$

The negative sign for v_2 indicates that the pitching machine is moving to the left after firing, in the direction *opposite* the motion of the baseball.

6.3 COLLISIONS

We have seen that, for any type of collision, the total momentum of the system just before collision equals the total momentum just after collision. We can say that the total momentum is always conserved for any type of collision. However, the total kinetic energy is generally not conserved when a collision occurs because some of the kinetic energy is converted to heat or some other form of energy when the objects are deformed during the collision.

Inelastic collision

We define an **inelastic collision** as a collision in which momentum is conserved but kinetic energy is not.

For a general inelastic collision, we can apply the law of conservation of momentum in the form given by Equation 6.5. The collision of a rubber ball with a hard surface is inelastic because some of the kinetic energy is lost when the ball is deformed while it is in contact with the surface. When two objects collide and stick together, the collision is called **perfectly inelastic.** This is an extreme case of an inelastic collision. For example, if two pieces of putty collide, they stick together and move with some common velocity after the collision. If a meteorite collides head-on with the earth, it becomes buried in the earth and the collision is considered perfectly inelastic. However, you should note that not all of the initial kinetic energy is necessarily lost in a perfectly inelastic collision.

Elastic collision

An **elastic collision** is defined as a collision in which both momentum and kinetic energy are conserved.

Billiard ball collisions and the collisions of air molecules with the walls of a container at ordinary temperatures are almost elastic. In reality, collisions in the macroscopic world, such as those between billiard balls, can be only approximately elastic because in such collisions there is always some deformation of the objects; hence some kinetic energy is lost in the forms of heat and sound. On the other hand, truly elastic collisions do occur between atomic and subatomic particles.

We summarize the various types of collisions as follows:

1. An elastic collision is one in which both momentum and kinetic energy are conserved.
2. An inelastic collision is one in which momentum is conserved but kinetic energy is not.
3. A perfectly inelastic collision between two objects is an inelastic collision in which the two objects stick together after the collision, so that their final velocities are the same.

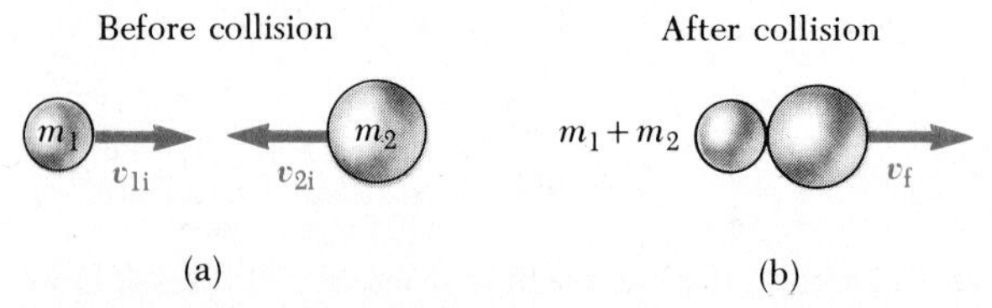

FIGURE 6.9 A perfectly inelastic head-on collision between two objects: (a) before the collision and (b) after the collision.

In the remainder of this section, we shall treat collisions in one dimension and consider two extreme types: perfectly inelastic and elastic.

Perfectly Inelastic Collisions

Consider two objects of masses m_1 and m_2 moving with initial velocities v_{1i} and v_{2i} along a straight line, as in Figure 6.9. We shall assume that the objects collide head-on so that they will be moving along the same line of motion after the collision. If the two objects stick together and move with some common velocity v_f after the collision, then only the momentum of the system is conserved. Therefore, we can say that the total momentum before the collision equals the total momentum after the collision:

$$m_1 v_{1i} + m_2 v_{2i} = (m_1 + m_2) v_f \tag{6.6}$$

It is important to note that v_{1i}, v_{2i}, etc. represent the x components of the vectors $\mathbf{v}_{1i}$, $\mathbf{v}_{2i}$, etc., so one must be careful with signs. For example, in Figure 6.9, v_{1i} would have a positive value (m_1 moving to the right), while v_{2i} would have a negative value (m_2 moving to the left).

In a typical inelastic collision problem, only one quantity in this equation will be unknown. As a result, conservation of momentum is sufficient to tell us what we need to know.

EXAMPLE 6.4 The Cadillac Versus the "Beetle"

A large luxury car with a mass of 1800 kg stopped at a traffic light is struck from the rear by a compact car with a mass of 900 kg. The two cars become entangled as a result of the collision. If the compact car was moving at 20 m/s before the collision, what is the velocity of the entangled mass after the collision?

Solution The momentum before the collision is that of the compact car alone because the large car was initially at rest. Thus, we have for the momentum before the collision

$$p_i = m_1 v_i = (900 \text{ kg})(20 \text{ m/s}) = 1.80 \times 10^4 \text{ kg} \cdot \text{m/s}$$

After the collision, the mass that moves is the sum of the masses of the large car plus that of the compact car, and the momentum of the combination is

$$p_f = (m_1 + m_2) v_f = (2700 \text{ kg})(v_f)$$

Equating the momentum before to the momentum after and solving for v_f, the velocity of the wreckage, we have

$$v_f = \frac{p_i}{m_1 + m_2} = \frac{1.80 \times 10^4 \text{ kg} \cdot \text{m/s}}{2700 \text{ kg}} = 6.67 \text{ m/s}$$

EXAMPLE 6.5 Here's Mud in Your Eye

Two balls of mud collide head-on in a perfectly inelastic collision, as in Figure 6.9. Suppose $m_1 = 0.5$ kg, $m_2 = 0.25$ kg, $v_{1i} = 4$ m/s, and $v_{2i} = -3$ m/s. (a) Find the velocity of the composite ball of mud after the collision. (b) How much kinetic energy is lost in the collision?

Solution (a) Writing Equation 6.6 for conservation of momentum with the positive direction for velocity to the right, we are able to find the velocity of the combined mass after the collision:

$$(0.5\text{ kg})(4\text{ m/s}) - (0.25\text{ kg})(3\text{ m/s}) = (0.75\text{ kg})(v_f)$$

$$v_f = 1.67\text{ m/s}$$

(b) The kinetic energy before the collision is

$$KE_i = KE_1 + KE_2 = \frac{1}{2}m_1v_{1i}^2 + \frac{1}{2}m_2v_{2i}^2$$

$$= \frac{1}{2}(0.5\text{ kg})(4\text{ m/s})^2 + \frac{1}{2}(0.25\text{ kg})(-3\text{ m/s})^2 = 5.13\text{ J}$$

The kinetic energy after the collision is

$$KE_f = \frac{1}{2}(m_1 + m_2)v_f^2 = \frac{1}{2}(0.75\text{ kg})(1.67\text{ m/s})^2 = 1.05\text{ J}$$

Hence, the loss in kinetic energy is

$$KE_i - KE_f = 4.08\text{ J}$$

Most of this lost energy is converted to heat as the objects collide and distort, and a small fraction is converted to energy in the form of sound waves.

EXAMPLE 6.6 The Ballistic Pendulum

The ballistic pendulum (Fig. 6.10) is a device used to measure the velocity of a fast-moving projectile, such as a bullet. The bullet is fired into a large block of wood suspended from some light wires. The bullet is stopped by the block, and the entire system swings through a height h. It is possible to obtain the initial velocity of the bullet by measuring h and the two masses. As an example of the technique, assume that the mass of the bullet, m_1, is 5 g, the mass of the pendulum, m_2, is 1 kg, and h is 5 cm. Find the initial velocity of the bullet, v_{1i}.

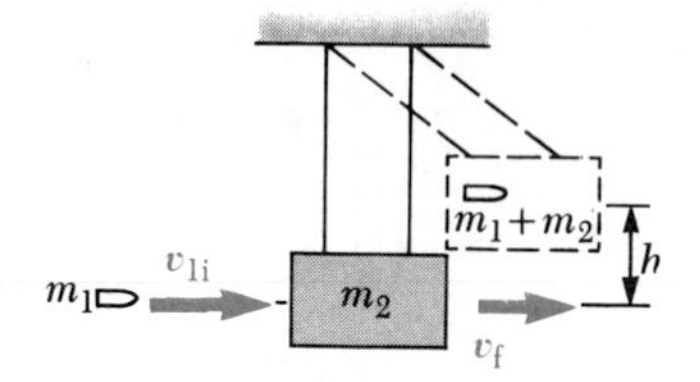

FIGURE 6.10 (Example 6.6) Diagram of a ballistic pendulum. Note that v_f is the velocity of the block and bullet right after the perfectly inelastic collision.

Solution The collision between the bullet and the block is perfectly inelastic. Writing the conservation of momentum for the collision in the form of Equation 6.6, we have

$$(1) \quad (5 \times 10^{-3}\text{ kg})(v_{1i}) = (1.005\text{ kg})(v_f)$$

There are two unknowns in this equation, v_{1i} and v_f, where the latter is the velocity of the block plus embedded bullet *immediately after the collision*. We must look for additional information if we are to complete the problem. Kinetic energy is not conserved during an inelastic collision. However, mechanical energy is conserved after the collision, and so the kinetic energy of the system at the bottom is transformed into the potential energy of the bullet plus block at the height h:

$$\frac{1}{2}(m_1 + m_2)v_f^2 = (m_1 + m_2)gh$$

$$\frac{1}{2}(1.005\text{ kg})(v_f^2) = (1.005\text{ kg})(9.80\text{ m/s}^2)(5 \times 10^{-2}\text{ m})$$

giving

$$v_f = 0.990 \text{ m/s}$$

With v_f now known, (1) yields v_{1i}:

$$v_{1i} = \frac{(1.005 \text{ kg})(0.990 \text{ m/s})}{5 \times 10^{-3} \text{ kg}} = 199 \text{ m/s}$$

EXERCISE Explain why it would be incorrect to equate the initial kinetic energy of the incoming bullet to the final gravitational potential energy of the bullet-block combination.

Elastic Collisions

Now consider two objects that undergo an elastic head-on collision (Fig. 6.11). In this situation, *both momentum and kinetic energy are conserved;* therefore we can write these conditions as

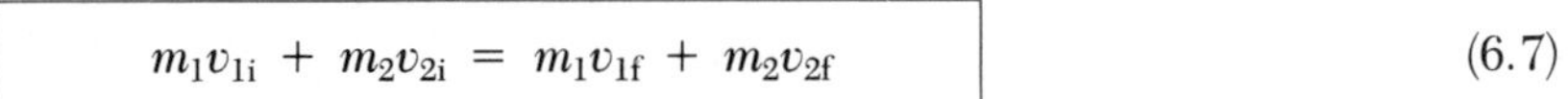

$$m_1v_{1i} + m_2v_{2i} = m_1v_{1f} + m_2v_{2f} \tag{6.7}$$

$$\frac{1}{2}m_1v_{1i}^2 + \frac{1}{2}m_2v_{2i}^2 = \frac{1}{2}m_1v_{1f}^2 + \frac{1}{2}m_2v_{2f}^2 \tag{6.8}$$

Before collision

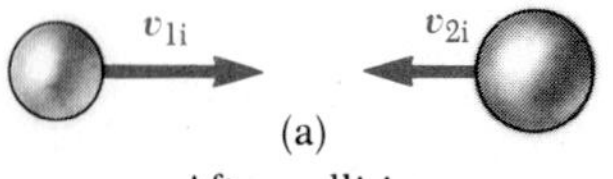

(a)

After collision

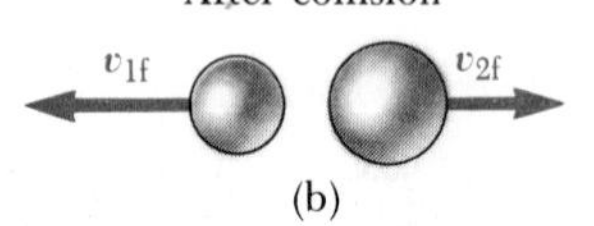

(b)

FIGURE 6.11 An elastic head-on collision between two hard spheres: (a) before the collision and (b) after the collision.

In a typical problem involving elastic collisions, there will be two unknown quantities and Equations 6.7 and 6.8 can be solved simultaneously to find these. However, an alternative approach, one that involves a little mathematical manipulation of Equation 6.8, often simplifies this process for head-on collisions. To see this, let's cancel the factor of $\frac{1}{2}$ in Equation 6.8 and rewrite it as

$$m_1(v_{1i}^2 - v_{1f}^2) = m_2(v_{2f}^2 - v_{2i}^2)$$

Here we have moved the terms containing m_1 to one side of the equation and those containing m_2 to the other. Next, let us factor both sides of the equation:

$$m_1(v_{1i} - v_{1f})(v_{1i} + v_{1f}) = m_2(v_{2f} - v_{2i})(v_{2f} + v_{2i}) \tag{6.9}$$

We now separate the terms containing m_1 and m_2 in the equation for the conservation of momentum (Eq. 6.7) to get

$$m_1(v_{1i} - v_{1f}) = m_2(v_{2f} - v_{2i}) \tag{6.10}$$

Our final result is obtained by dividing Equation 6.9 by Equation 6.10 to get

$$v_{1i} + v_{1f} = v_{2f} + v_{2i}$$

or

$$v_{1i} - v_{2i} = -(v_{1f} - v_{2f}) \tag{6.11}$$

This equation, in combination with the equation for conservation of momentum will be used to solve problems dealing with perfectly elastic, head-on collisions. Note that Equation 6.11 says that the relative velocity of the two objects before the collision, $v_{1i} - v_{2i}$, equals the negative of the relative velocity of the two objects after the collision, $-(v_{1f} - v_{2f})$.

EXAMPLE 6.7 Let's Play Pool

Two billiard balls move toward one another as shown in Figure 6.11. The balls have identical masses, and the collision between them is perfectly elastic. If the initial velocities of the balls are 30 cm/s and −20 cm/s, what is the velocity of each ball after the collision?

Solution We turn first to Equation 6.7. The masses cancel on each side, and after substituting the appropriate values for the initial velocities, we have

$$30 \text{ cm/s} + (-20 \text{ cm/s}) = v_{1f} + v_{2f}$$

$$(1) \quad 10 \text{ cm/s} = v_{1f} + v_{2f}$$

Since kinetic energy is also conserved, we can apply Equation 6.11, which gives

$$30 \text{ cm/s} - (-20 \text{ cm/s}) = v_{2f} - v_{1f}$$

$$(2) \quad 50 \text{ cm/s} = v_{2f} - v_{1f}$$

Solving (1) and (2) simultaneously, we find

$$v_{1f} = -20 \text{ cm/s} \qquad v_{2f} = 30 \text{ cm/s}$$

That is, the balls *exchange velocities!* This is always the case when two objects of equal mass collide head-on in an elastic collision.

EXERCISE Find the final velocity of the two balls if the ball with initial velocity −20 cm/s has a mass equal to half that of the ball with initial velocity 30 cm/s.

Answer $v_{1f} = -3$ cm/s; $v_{2f} = 47$ cm/s.

6.4 GLANCING COLLISIONS

The collision problems we have considered up until now have been head-on collisions. That is, the incident mass strikes a second mass head-on and both rebound along a straight-line path that coincides with the line of motion of the incident mass. Anyone who has ever played billiards knows that such collisions are the exception rather than the rule. A more common type of collision is a *glancing collision*, in which the colliding masses rebound at some angle relative to the line of motion of the incident mass. Figure 6.12 shows a white ball, traveling with an initial speed v_{iA}, striking a black ball obliquely (off-center). After the collision, the white ball caroms off at an angle A relative to its incident line of motion and the black ball rebounds at an angle B.

As we emphasized earlier, *momentum is conserved in all collisions*, and glancing collisions are no exception. Since momentum is a vector quantity, the conservation of momentum principle must be written as $\mathbf{p}_i = \mathbf{p}_f$. That is, the total initial momentum of the system (the two balls) must equal the total final momentum of the system. For a collision in two dimensions, as in Figure 6.12, this implies that the total momentum is conserved along the x direction *and* along the y direction. We can state this in equation form as

$$\sum p_{ix} = \sum p_{fx} \tag{6.12}$$

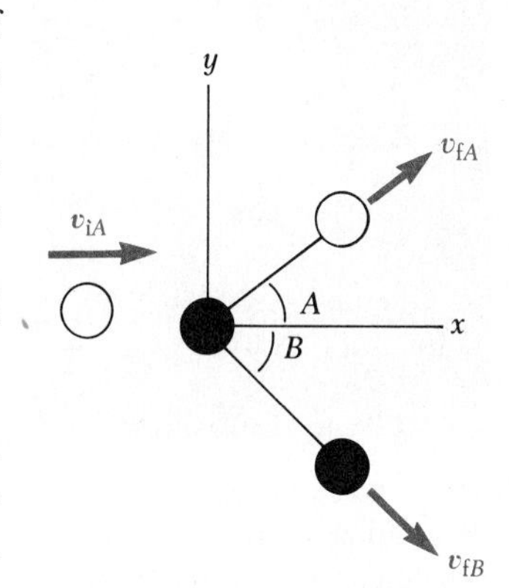

FIGURE 6.12 A glancing collision between two balls.

and

$$\sum p_{iy} = \sum p_{fy} \tag{6.13}$$

The following example illustrates how to use the principle of conservation of momentum to treat glancing collisions.

EXAMPLE 6.8 Collision at an Intersection

A 1500-kg car traveling east with a speed of 25 m/s collides at an intersection with a 2500-kg van traveling north at a speed of 20 m/s, as shown in Figure 6.13. Find the direction and magnitude of the velocity of the wreckage after the collision, assuming that the vehicles undergo a perfectly inelastic collision (that is, they stick together).

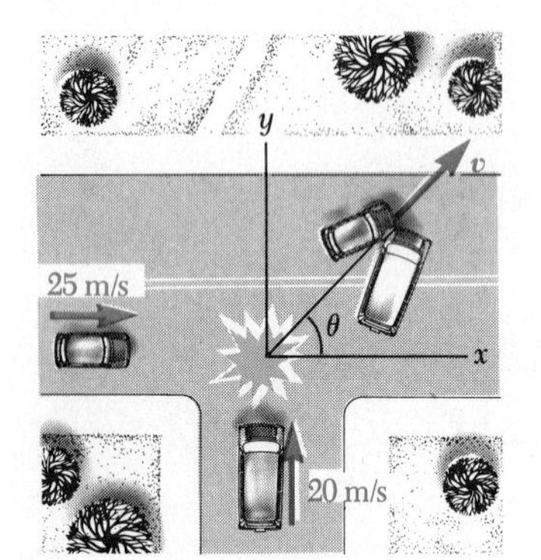

FIGURE 6.13 (Example 6.8) Top view of a perfectly inelastic collision between a car and a van.

Solution Let us choose east to be along the positive x direction and north to be along the positive y direction, as in Figure 6.13. Before the collision, the only object having momentum in the x direction is the car. Thus, the total initial momentum of the system (car plus van) in the x direction is

$$\sum p_{ix} = (1500 \text{ kg})(25 \text{ m/s}) = 37{,}500 \text{ kg} \cdot \text{m/s}$$

Now let us assume that the wreckage moves at an angle θ and speed v after the collision, as in Figure 6.13. The total momentum in the x direction after the collision is

$$\sum p_{fx} = (4000 \text{ kg})(v \cos \theta)$$

Because momentum is conserved in the x direction, we have

$$\sum p_{ix} = \sum p_{fx}$$

$$\text{(1)} \quad 37{,}500 \text{ kg} \cdot \text{m/s} = (4000 \text{ kg})(v \cos \theta)$$

Similarly, the total initial momentum of the system in the y direction is that of the van, which equals (2500 kg)(20 m/s). Applying conservation of momentum to the y direction, we have

$$\sum p_{iy} = \sum p_{fy}$$

$$(2500 \text{ kg})(20 \text{ m/s}) = (4000 \text{ kg})(v \sin \theta)$$

$$\text{(2)} \quad 50{,}000 \text{ kg} \cdot \text{m/s} = (4000 \text{ kg})(v \sin \theta)$$

If we divide (2) by (1), we get

$$\tan \theta = \frac{50{,}000}{37{,}500} = 1.33$$

$$\theta = 53°$$

When this angle is substituted into (2)—or alternatively into (1)—the value of v is

$$v = \frac{50{,}000 \text{ kg} \cdot \text{m/s}}{(4000 \text{ kg})(\sin 53°)} = 15.6 \text{ m/s}$$

The center of mass of this diver follows a parabolic path. (Photo courtesy of Dr. Harold Edgerton, M.I.T., Cambridge, MA)

6.5 CENTER OF MASS

In Chapter 4, we found that there is a special point for any object called the center of gravity and that we can consider the weight of the object as acting at this center. A related point is the **center of mass** of a system. There is a theoretical distinction between the two, but as long as the acceleration due to gravity is constant over the mass distribution, then the center of gravity coincides with the center of mass. Because those circumstances will always prevail in the situations we shall examine, we shall always assume that the center of mass coincides with the center of gravity. That is, in order to locate the center of mass of a system, we shall use the equations

$$x_c = \frac{\Sigma m_i x_i}{\Sigma m_i} \qquad y_c = \frac{\Sigma m_i y_i}{\Sigma m_i} \tag{6.14}$$

The physical significance and utility of the center of mass arise because *the net external force acting on a system equals the total mass of the system multiplied by the acceleration of the center of mass:*

$$\boxed{\mathbf{F}_{\text{ext}} = M\mathbf{a}_c} \tag{6.15}$$

where M is the total mass of the system, given by $M = \Sigma m_i$. This result shows that the center of mass moves like a particle of mass M under the influence of the resultant external force on the system.

From Equation 6.15, we see that, *if the resultant external force on a system of objects is zero, then* $\mathbf{a}_c$ *is also zero.* If the acceleration at the center of mass is zero, the velocity of the center of mass does not change and hence the momentum of the center of mass remains constant.

Suppose an isolated system consisting of two or more objects is at rest. *The center of mass of such a system will remain at rest unless acted upon by an external force.* For example, consider a system made up of a swimmer and a raft, with the system initially at rest. When the swimmer dives off the raft, the center of mass of the system will remain at rest (if we neglect friction between raft and water). Furthermore, the momentum of the diver will be equal in magnitude to the momentum of the raft, but opposite in direction. Next, imagine a pair of figure skaters initially at rest on ice. When the man pushes his female partner away from him, he will recoil with a momentum opposite that of the woman and the center of mass will remain at rest.

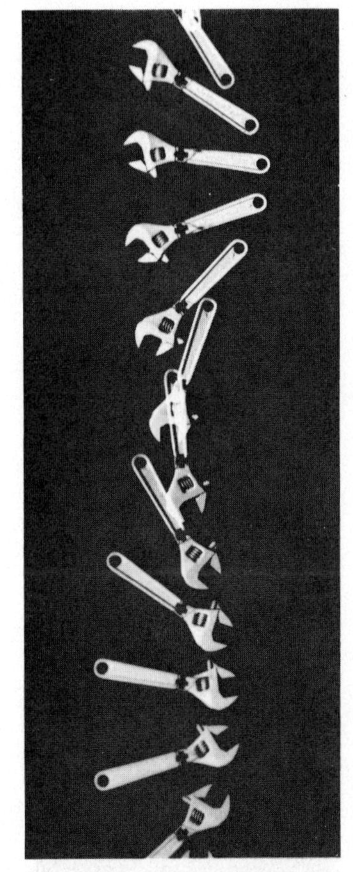

The center of mass of the wrench moves in a straight line as the wrench rotates about this point. (Courtesy Education Development Center, Newton, MA)

EXAMPLE 6.9 An Exploding Projectile

A projectile is fired into the air and suddenly explodes into several fragments (Fig. 6.14). What can be said about the motion of the fragments after the explosion? (Neglect air resistance).

Solution The only external force on the projectile is the force of gravity. Thus, the projectile follows a path like that shown in Figure 6.14. If the projectile did not explode, it would continue to move along the path indicated by the broken line. Because the forces due to the explosion are internal, they do not affect the motion of the center of mass. Thus, after the explosion, the fragments will fly away such that the motion of their center of mass follows the same trajectory the projectile would have followed if there had been no explosion. (Neglecting air resistance, the path is parabolic.)

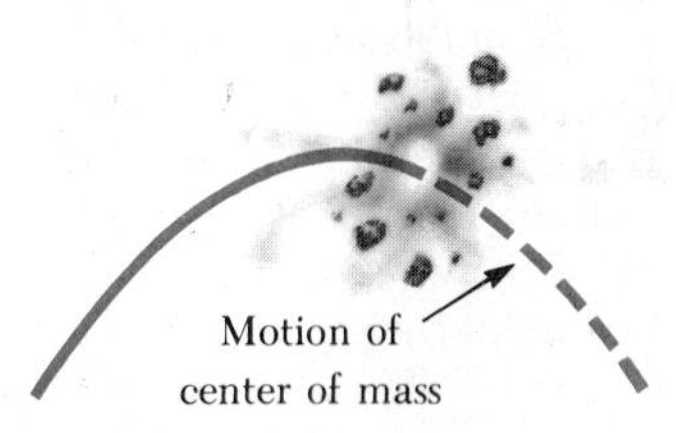

FIGURE 6.14 (Example 6.9) When a projectile explodes into several fragments, the center of mass of the fragments follows the same parabolic path as the projectile would have taken had it not exploded.

*6.6 ROCKET PROPULSION

When ordinary vehicles, such as automobiles, boats, and locomotives, are propelled, the driving force for the motion is one of friction. In the case of the automobile, the driving force is the force of friction between road and car.

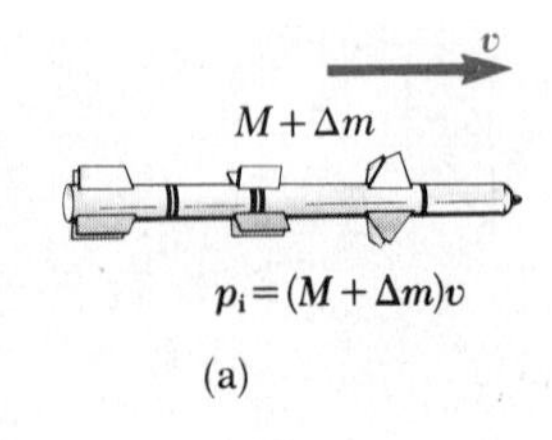

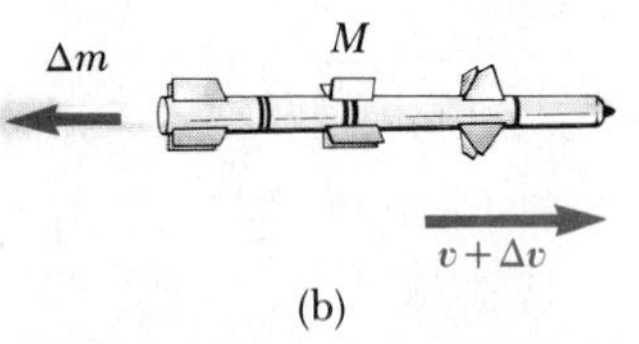

FIGURE 6.15 Rocket propulsion. (a) At some time t, the mass of the rocket is $M + \Delta m$, and its speed is v. (b) At a time $t + \Delta t$, the rocket's mass has reduced to M and an amount of fuel Δm has been ejected. The rocket's speed increases by an amount Δv, and its momentum increases by approximately $M\Delta v$.

Lift-off of the space shuttle Columbia. Enormous amounts of thrust are generated by the shuttle's liquid-fuel engines, aided by two solid-fuel boosters. (NASA)

A locomotive "pushes" against the tracks; hence, the driving force is the force of friction between the tracks and the wheels of the locomotive. A boat is propelled by the force of the water against a paddle or a rotating propeller. However, a rocket moving in space has no air, tracks, or water to push against. Therefore, the source of the propulsion of a rocket must be different. *The operation of a rocket depends upon the law of conservation of momentum as applied to the system comprised of the rocket and its ejected fuel.*

The propulsion of a rocket can be understood by first considering a machine gun mounted on a cart on wheels. As the machine gun is fired, each bullet receives a momentum mv in some direction. Each time a bullet is fired, the cart (along with the attached gun) must receive a momentum of equal magnitude but in the opposite direction, as illustrated in Example 6.3. If the gun fires N bullets in a time Δt, the total change in momentum of all the bullets is Nmv. Because the total momentum of the system (bullets plus gun and cart) is conserved, the change in momentum of the cart and gun must also have a magnitude Nmv. Thus, from Equation 6.3, the average force exerted on the cart and gun during the time Δt is

$$\overline{\boldsymbol{F}} = \frac{\Delta p}{\Delta t} = \frac{Nmv}{\Delta t}$$

This reaction force on the cart and gun produces an acceleration in the direction opposite to that of the motion of the bullets. One can view the motion of the cart and gun as a series of "kicks" it receives each time a bullet is fired.

In a similar manner, as a rocket moves through space, the momentum of the rocket changes when some of its mass is released in the form of ejected gases, as shown in Figure 6.15. You can easily demonstrate this phenomenon by releasing an inflated balloon with the stem left opened. As the air escapes, the balloon is propelled through the air in a somewhat erratic motion until it is completely deflated. Likewise, as the ejected gases of a rocket acquire some momentum to the left in Figure 6.15b, the rocket receives a momentum of equal magnitude to the right. The reaction force that accelerates the rocket to the right arises because of this change in momentum with time.

On the basis of the discussion in Section 6.5, you should note that in free space the center of mass of the entire system (rocket plus ejected gases) moves with constant velocity because the net external force on the system is zero.

SUMMARY

The **linear momentum** of an object of mass m moving with a velocity $\boldsymbol{v}$ is defined to be

$$\boxed{\boldsymbol{p} \equiv m\boldsymbol{v}} \tag{6.1}$$

The **impulse** of a force $\boldsymbol{F}$ acting on an object is equal to the product of the force and the time interval during which the force acts:

$$\boxed{\text{Impulse} = \boldsymbol{F}\Delta t}$$

The **impulse-momentum theorem** states that the impulse of a force on an object is equal to the change in momentum of the object:

$$\boxed{\mathbf{F}\Delta t = \Delta \mathbf{p} = m\mathbf{v}_f - m\mathbf{v}_i} \tag{6.4}$$

The **law of conservation of momentum** of two interacting objects states that, if the two objects form an isolated system, their total momentum is conserved regardless of the nature of the forces between them:

$$\boxed{m_1 v_{1i} + m_2 v_{2i} = m_1 v_{1f} + m_2 v_{2f}} \tag{6.5}$$

An **inelastic collision** is one in which momentum is conserved but kinetic energy is not. A **perfectly inelastic collision** is one in which the colliding objects stick together after the collision. An **elastic collision** is one in which both momentum and kinetic energy are conserved.

In glancing collisions, conservation of momentum can be applied along two perpendicular directions, that is, along an x and a y axis.

The x and y coordinates of the center of mass of an object are

$$x_c = \frac{\Sigma m_i x_i}{\Sigma m_i} \qquad y_c = \frac{\Sigma m_i y_i}{\Sigma m_i} \tag{6.14}$$

Newton's second law applied to a system of particles is given by

$$\boxed{\mathbf{F}_{ext} = M\mathbf{a}_c} \tag{6.15}$$

where $\mathbf{a}_c$ is the acceleration of the center of mass, $M = \Sigma m_i$ and $\mathbf{F}_{ext}$ is the net external force on the system. This equation says that the center of mass moves like an imaginary particle of mass M under the influence of the resultant external force on the system.

ADDITIONAL READING

A. Einstein and L. Infeld, *Evolution of Physics*, New York, Simon and Schuster, 1938.

H. W. Lewis, "Ballistocardiography," *Sci. American*, February, 1958, p. 89.

F. Ordway, "Principles of Rocket Engines," *Sky and Telescope*, *14*, 1954, p. 48. An introduction to the principles of rocket propulsion.

QUESTIONS

1. If the kinetic energy of a particle is zero, what is its momentum? If the total energy of a particle is zero, is its momentum necessarily zero? Explain.
2. If the forward momentum of a bullet is the same as the backward momentum of the gun, why isn't it as dangerous to be hit by the gun as by the bullet?
3. If the velocity of a particle is doubled, by what factor is its momentum changed?
4. A box slides across the frictionless icy surface of a frozen lake. What happens to the speed of the box as water collects in it from a rainshower? Explain.
5. Does a large force always produce a larger impulse on a body than a smaller force? Explain.
6. A piece of clay is thrown against a brick wall and sticks to the wall. What happens to the momentum of the clay? Is momentum conserved? Explain.
7. If two objects collide and one is initially at rest, is it possible for both to be at rest after the collision? Is it possible for one to be at rest after the collision? Explain.
8. Early in this century Dr. Robert Goddard proposed sending a rocket to the moon. Critics took the posi-

tion that in a vacuum, such as exists between earth and moon, the gases emitted by the rocket would have nothing to push against to propel the rocket. According to *Scientific American* (January 1975), Goddard placed a gun in a vacuum and fired a blank cartridge in it. (A blank cartridge fires only the hot gases of the burning gunpowder.) What happened when the gun was fired?

9. Is it possible to have a collision in which all of the kinetic energy is lost? If so, cite an example.
10. In a perfectly elastic collision between two objects, do both objects have the same kinetic energy after the collision? Explain.
11. An astronaut walking in space accidentally severs the safety cord attaching him to the spacecraft. If he happens to have with him a can of aerosol spray deodorant, how could he use this to return safely to his ship?
12. A pole-vaulter falls from a height of 15 ft onto a foam rubber pad. Could you calculate his velocity just before he reaches the pad? Would you be able to calculate the force exerted on the pole-vaulter due to the collision? Explain.
13. As a ball falls towards the earth, its momentum increases. How would you reconcile this fact with the law of conservation of momentum?
14. A man is at rest in a boat in the middle of a lake. If he walks to the opposite end of the boat towards the east, why does the boat move towards the west? What can you say about the center of mass of the system consisting of the boat and man?
15. Explain how you would use a balloon to demonstrate the mechanism responsible for rocket propulsion.
16. Gymnasts always perform with padded mats. Use the impulse-momentum theorem to discuss how these mats protect the athletes.
17. A boxer wisely moves his head backward just before receiving a punch to his nose. How does that help protect him from the punch?
18. Suppose you are out sailing when the wind calms. If you have a large electric fan on board, can you move the boat by directing the wind from the fan toward the sail? If you remove the sail, will the running fan cause the boat to move? Explain.
19. How do car bumpers which collapse on impact help to protect a driver?
20. A toy gun shoots rubber bullets at a target. Compare the impulse delivered to the target when the bullets embed in the target to the case in which they strike the target and bounce off.
21. An open bed sheet is loosely held at its sides by two students to form a "catching net." The instructor asks a third student to throw a raw egg into the sheet as hard as possible, being careful not to miss the sheet. Why is it that the egg does not break?
22. A magician places several dishes, cups, and silverware on a table containing a tablecloth. The magician proceeds to rapidly remove the tablecloth without disturbing the dishes, etc. Explain how this trick is possible on the basis of what you have learned in this chapter.

PROBLEMS

Section 6.1 Linear Momentum and Impulse

1. Calculate the magnitude of the linear momentum for the following cases: (a) a proton of mass 1.67×10^{-27} kg moving with a speed of 5×10^6 m/s, (b) a 15-g bullet moving with a speed of 300 m/s, (c) a 75-kg sprinter running with a speed of 10 m/s, and (d) the earth (mass = 5.98×10^{24} kg) moving with an orbital speed equal to 2.98×10^4 m/s.

2. A 0.15-kg baseball is thrown with a speed of 10 m/s. (a) How much more momentum does it have if its speed is doubled? (b) How much more kinetic energy will it have?

3. The momentum of a 1500-kg car is equal to the momentum of a 5000-kg truck traveling with a speed of 15 m/s. What is the speed of the car?

4. A 1500-kg car moving with a speed of 15 m/s collides with a utility pole and is brought to rest in 0.3 s. Find the average force exerted on the car during the collision.

5. An 18,000-kg van is moving with a speed of 15 m/s. If Superman is to stop the van in 0.5 s, what average force must he exert on the van?

6. A 0.10-kg ball is thrown straight up into the air with an initial speed of 15 m/s. Find the momentum of the ball (a) at its maximum height, and (b) halfway up to its maximum height.

• **7.** A 0.10-kg ball is thrown with a speed of 20 m/s at an angle of 30° with the horizontal. Find the momentum of the ball (a) at its maximum height, and (b) just before it strikes the ground.

• **8.** A 0.15-kg baseball is thrown with a speed of 20 m/s. It is hit straight back at the pitcher with a final speed of 22 m/s. (a) What is the impulse delivered to the ball? (b) Find the average force exerted by the bat on the ball if the two are in contact for 2×10^{-3} s.

• **9.** An 0.5-kg object is at rest at the origin of a coordinate system. A 3-N force in the $+x$ direction acts on the object for 1.5 s. (a) What is the velocity at the end of this time? (b) At the end of this time, a constant force of 4 N is applied in the $-x$ direction for 3 s. What is the velocity at the end of this time?

•• **10.** A prankster drops a plastic bag filled with 2 kg of water from the roof of a dormitory 25 m above street level. Find the force exerted on the head of someone walking along the street by assuming that the bag is spherical, falls freely, and stops in a distance equal to its diameter, which is 15 cm.

Section 6.2 Conservation of Momentum

11. A 40-kg child standing on a frozen pond throws a 0.5-kg stone to the east with a speed of 5 m/s. Neglecting friction between child and ice, find the recoil velocity of the child.

12. A rifle with a weight of 30 N fires a 5-g bullet with a speed of 300 m/s. (a) Find the recoil speed of the rifle. (b) If a 700-N man holds the rifle firmly against his shoulder, find the recoil speed of man and rifle.

13. A 65-kg boy and a 40-kg girl, both wearing skates, face each other at rest on a skating rink. The boy pushes the girl, sending her eastward with a speed of 4 m/s. Describe the subsequent motion of the boy. (Neglect friction.)

• **14.** A 730-N man stands in the middle of a frozen pond of radius 5 m. He is unable to get to the other side because of a lack of friction between his shoes and the ice. To overcome his difficulty, he throws his 1.2-kg physics textbook horizontally at a speed of 5 m/s toward the north shore. How long does it take him to reach the south shore?

Section 6.3 Collisions
Section 6.4 Glancing Collisions

15. A car moving at a speed of 5 m/s crashes into an identical car stopped at a light. What is the velocity of the wreckage immediately after the collision, assuming the cars stick together?

16. A bumper car at an amusement park ride traveling at 0.8 m/s collides with an identical car at rest. This second car moves away with a speed of 0.5 m/s. What is the velocity of the first car after the collision?

17. A 1.2-kg skateboard is coasting along the pavement at a speed of 5 m/s, when a 0.8-kg cat drops from a tree vertically down on the skateboard. What is the speed of the skateboard-cat combination?

18. A railroad car of mass 2×10^4 kg moving with a speed of 3 m/s collides and couples with two other coupled railroad cars each of the same mass as the single car and moving in the same direction at a speed of 1.2 m/s. (a) What is the speed of the three coupled cars after the collision? (b) How much energy is lost in the collision?

19. A 3-kg sphere makes a perfectly inelastic collision with a second sphere initially at rest. The composite system moves with a speed equal to one third the original speed of the 3-kg sphere. What is the mass of the second sphere?

• **20.** A 5-g object moving to the right with a speed of 20 cm/s makes an elastic head-on collision with a 10-g object initially at rest. Find (a) the velocity of each object after the collision and (b) the fraction of the energy transferred to the 10-g object.

• **21.** A 10-g object moving to the right with a speed of 20 cm/s makes an elastic head-on collision with a 15-g object moving in the opposite direction with a speed of 30 cm/s. Find the velocity of each object after the collision.

• **22.** A 25-g object moving to the right with a speed of 20 cm/s overtakes and collides elastically with a 10-g object moving in the same direction with a speed of 15 cm/s. Find the velocity of each object after the collision.

• **23.** An alpha particle of mass 4 u moving to the right with a speed of 10^6 m/s collides with a proton of mass 1 u that is at rest before the collision. Find (a) the speed of each particle after the collision, assuming a perfectly elastic collision, and (b) the kinetic energy of each particle before and after the collision (1 u = 1 atomic mass unit = 1.67×10^{-27} kg).

• **24.** An 8-g bullet is fired into a 2.5-kg ballistic pendulum and becomes embedded in it. If the pendulum rises a vertical distance of 6 cm, calculate the initial speed of the bullet.

• **25.** A 50-kg baseball pitching machine is placed on a frozen pond. It fires a 0.15-kg baseball at an angle of 30° with the vertical at a speed of 20 m/s. What is the recoil velocity of the machine?

• **26.** A 90-kg fullback moving east with a speed of 5 m/s is tackled by a 95-kg opponent running north with a speed of 3 m/s. If the collision is perfectly inelastic, calculate (a) the velocity of the players just after the tackle and (b) the energy lost as a result of the collision. Can you account for the missing energy?

• **27.** A 1500-kg car traveling east with a speed of 20 m/s collides at an intersection with a 2500-kg van traveling south at a speed of 15 m/s. The vehicles undergo a perfectly inelastic collision, and the wreckage slides 6 m before coming to rest. Find the magnitude and direction of the constant force that has decelerated them.

•• **28.** A neutron in a reactor makes an elastic head-on collision with a carbon atom initially at rest. (The mass of the carbon nucleus is about 12 times the mass of the neutron.) (a) What fraction of the neutron's kinetic energy is transferred to the carbon nucleus? (b) If the initial kinetic energy of the neutron is 1.6×10^{-13} J, find its final kinetic energy and the kinetic energy of the carbon nucleus after the collision.

•• **29.** Consider the ballistic pendulum described in Example 6.6 and shown in Figure 6.10. If the mass of the bullet is 8 g and the mass of the pendulum is 2 kg, find the ratio of the kinetic energy after the collision to the kinetic energy before collision. What accounts for this missing energy?

•• **30.** A cue ball traveling at a speed of 4 m/s makes a glancing, elastic collision with a target ball of equal mass initially at rest. It is found that the cue ball is deflected such that it makes an angle of 30° with respect to its original direction of travel. Find (a) the angle the second ball will move and (b) the speed of each ball after the collision.

Section 6.5 Center of Mass

31. The mass of the earth is 5.98×10^{24} kg and the mass of the moon is 7.36×10^{22} kg. The distance of separation measured from their centers is about 3.84×10^{8} m. Determine the location of the center of mass of the earth-moon system as measured from the center of the earth.

32. The nucleus of a hydrogen atom is a single proton of mass 1.67×10^{-27} kg. The mass of the electron which circles the proton at a radius of about 0.51×10^{-10} m is 9.1×10^{-31} kg. Find the center of mass of the proton-electron system as measured from the proton.

33. Three masses located in the xy plane have the following coordinates: a 2-kg mass has coordinates (3, −2) m, a 3-kg mass has coordinates (−2, 4) m, and a 1 kg mass has coordinates (2, 2) m. Find the coordinates of the center of mass of the system.

34. A 3-kg mass is at (2, 0) m, and a 4-kg mass is at (3, 0) m. Where would a 5-kg mass have to be placed so that the center of mass of the system is at (0, 0) m?

ADDITIONAL PROBLEMS

35. A soccer ball of mass 0.4 kg is approaching a player with a speed of 15 m/s in horizontal flight. The player illegally strikes the ball with her hand and causes it to move in the opposite direction with a speed of 22 m/s. What impulse was delivered to the ball by the player?

36. A 4-g marble has a head-on collision with a 2-g marble at rest on a playing surface. The speed of the 4-g marble is reduced from 0.8 m/s to 0.5 m/s in the collision. What is the speed of the 2-g marble after the collision?

37. A swimmer of mass 80 kg stands on the stern of a boat of mass 240 kg at rest on a lake. The swimmer dives off the boat with a velocity that has a horizontal component of 0.8 m/s in the westerly direction. Determine the magnitude and direction of the resulting velocity of the boat. (Neglect any frictional effects of the water.)

38. A 2-g particle moving at a speed of 8 m/s makes a perfectly elastic head-on collision with a resting object of mass 1 g. (a) Find the speed of each after the collision. (b) If the rest particle has a mass of 10 g, find the speed of each after the collision. (c) Find the final kinetic energy of the incident 2-g particle for the situations described in (a) and (b). In which case does the incident particle lose the most kinetic energy?

39. A diver of mass 70 kg drops from a 3 m diving board. (a) What is his momentum just before he hits the water? (b) What is his momentum halfway down?

• **40.** An 80-kg man drops from a 3-m diving board. In 2.0 s after reaching the water, the man comes to rest. What average force did the water exert on him?

• **41.** Two pieces of "Silly Putty" are thrown toward one another and stick together after the collision. If one piece has a mass of 2 g and is traveling with a speed of 4 m/s, and the second has a mass of 3 g and a speed of 3 m/s, (a) what is the speed of the combination after the collision? (b) What speed could the 3-g mass have such that the combination is brought to rest in the collision?

• **42.** A pool ball rolling across a table at 1.5 m/s makes a head-on collision with an identical ball. Assume the collision is elastic and find the speed of each ball after the collision (a) when the second ball is initially at rest, (b) when the second ball is in motion toward the first at a speed of 1 m/s, and (c) when the second ball is in motion away from the first at a speed of 1 m/s.

• **43.** A 0.03-kg bullet is fired vertically at a speed of 200 m/s into a 0.15-kg baseball initially at rest. How high will the combination rise after the collision, assuming the bullet embeds in the ball?

• **44.** A 2000-kg car moving eastward at a speed of 10 m/s collides with a 3000-kg car moving northward. The cars stick together and move as a unit after the collision at an angle of 40° north of east at a speed of 5.22 m/s. Find the velocity of the 3000-kg car before the collision.

• **45.** A 0.4-kg bead slides on a curved frictionless wire, starting from rest at point A in Figure 6.16. At point B, the bead collides elastically with a 0.6-kg ball at rest. Find the height that the ball moves up the wire.

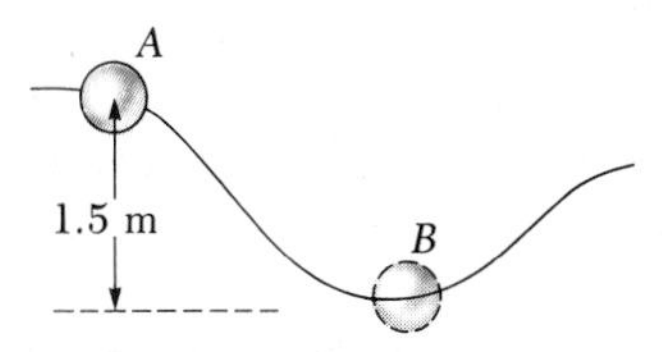

FIGURE 6.16 Problem 45

• **46.** Tarzan, whose mass is 80 kg, swings from a 3-m vine which is horizontal when he starts. At the bottom of his arc, he picks up 60-kg Jane, in an inelastic collision. What maximum height tree limb can they reach on their upward swing?

47. A truck which weighs 2.5×10^4 N is coasting at 7 m/s along a level roadway. A pallet of construction materials falls onto the truck from an overhead bridge and causes the speed of the truck to decrease by 10%. What is the weight of the pallet of materials?

• **48.** The force shown in the force-time diagram in Figure 6.17 acts on a 1.5-kg mass. Find (a) the impulse of the force, (b) the final velocity of the mass if it is initially at rest, and (c) the final velocity of the mass if it is initially moving along the x axis with a velocity of -2 m/s.

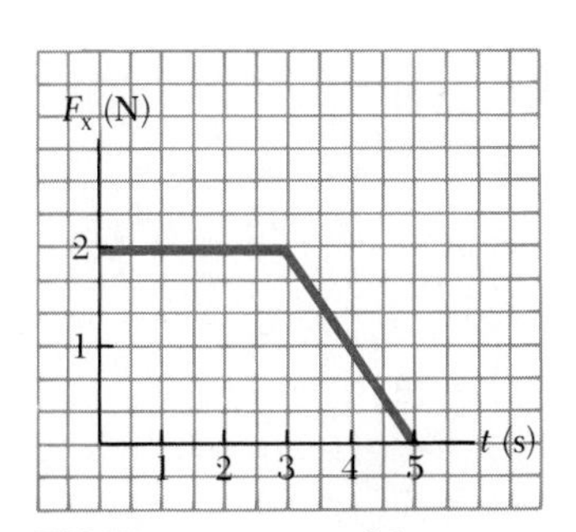

FIGURE 6.17 Problem 48

• **49.** A 1-kg mass moving with an initial speed of 5 m/s collides with and sticks to a 6-kg mass initially at rest. The combined mass then proceeds to collide with and stick to a 2-kg mass also at rest initially. If the collisions are all head-on, find (a) the final speed of the system and (b) the amount of kinetic energy lost.

• **50.** A 0.3-kg puck, initially at rest on a horizontal, frictionless surface, is struck by a 0.2-kg puck moving initially along the x axis with a velocity of 2 m/s. After the collision, the 0.2-kg puck has a speed of 1 m/s at an angle of $\theta = 53°$ to the positive x axis. (a) Determine the velocity of the 0.3-kg puck after the collision. (b) Find the fraction of kinetic energy lost in the collision.

•• **51.** The forces shown in the force-time diagram in Figure 6.18 act on a 1.5-kg mass. Find (a) the impulse for the interval $t = 0$ to $t = 3$ s, and (b) the impulse for the interval $t = 0$ to $t = 5$ s. (c) If the forces act on a particle of mass 1.5 kg which was initially at rest, find the speeds at $t = 3$ s and at $t = 5$ s.

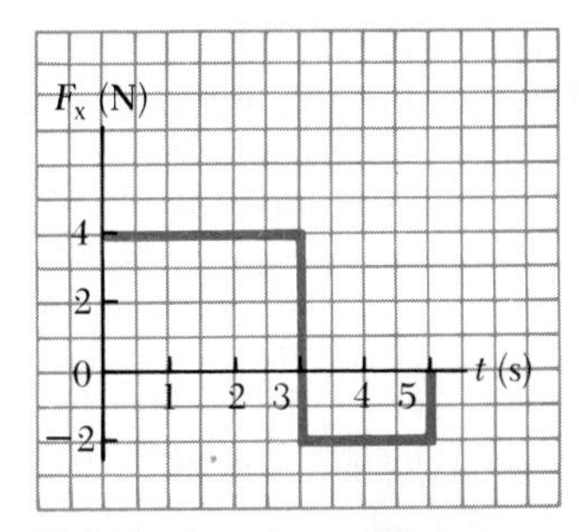

FIGURE 6.18 Problem 51

•• **52.** A 6-g bullet is fired into a 200 g block initially at rest at the edge of a table of height 1 m (Figure 6.19). The bullet remains in the block and after the impact the block lands 2 m from the bottom of the table. Determine the initial speed of the bullet.

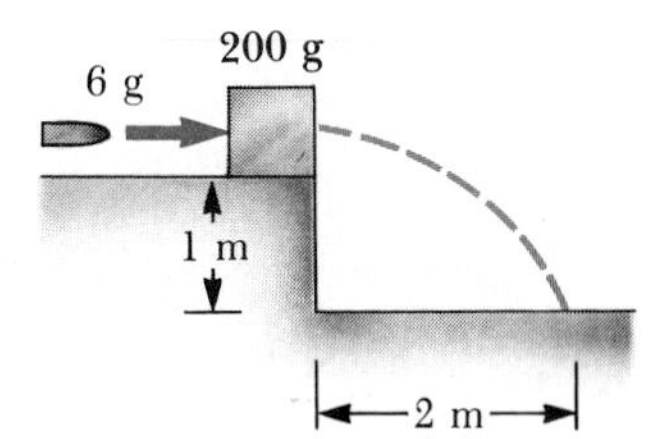

FIGURE 6.19 Problem 52

•• **53.** A candy store clerk drops 1-g pieces of candy from a height of 20 cm onto a scale at a rate of 4 per second. What is the scale reading after 10 s?

•• **54.** A 7-g bullet is fired into a 1.5-kg ballistic pendulum. The bullet emerges from the block with a speed of 200 m/s, and the block rises to a maximum height of 12 cm. Find the initial speed of the bullet.

•• **55.** An unstable nucleus of mass 17×10^{-27} kg initially at rest at the origin of a coordinate system disintegrates into three particles. One particle, having a mass of $m_1 = 5.0 \times 10^{-27}$ kg, moves along the y axis with a velocity $v_1 = 6 \times 10^6$ m/s. Another particle, of mass $m_2 = 8.4 \times 10^{-27}$ kg, moves along the x axis with a velocity of $v_2 = 4 \times 10^6$ m/s. Find the magnitude and direction of the velocity of the third particle.

•• **56.** Consider a machine gun mounted on a cart on wheels. The gun fires N bullets of mass m each second. (a) Show that the average force acting on the gun is Nmv. (b) If the gun-cart combination is free to move, find the average acceleration during an interval of 5 s in which the gun fires 50 bullets. Assume the mass of the cart is 70 kg, the mass of each bullet is 5 g, and the muzzle velocity of each bullet is 300 m/s.

•• **57.** A projectile of mass m collides with a flat surface. The angle between the incident path and the surface equals the angle θ between the reflected path and the surface. The speed of the projectile is unchanged by the collision. The time the projectile is in contact with the surface is t. Find (a) the change in momentum of the projectile in the x direction, (b) the change in momentum in the y direction, and (c) the average force exerted on the surface by the projectile during the collision.

•• **58.** A 40-kg child stands at one end of a 70-kg boat that is 4 m in length. (Fig. 6.20). The boat is initially 3 m from the pier. The child notices a turtle on a rock at the far end of the boat and proceeds to walk to that end to catch the turtle. Neglecting friction between the boat and water, (a) describe the subsequent motion of the system (child + boat). (b) Where will the child be relative to the pier when he reaches the far end of the boat? (c) Will he catch the turtle? (Assume he can reach out 1 m from the end of the boat.)

•• **59.** A block of mass M is given an initial velocity v_0 on a rough, horizontal surface. After traveling a distance d, it makes a head-on elastic collision with a block of mass $2M$. How far will the second block move before coming to rest? (Assume the coefficient of friction is the same for each block.)

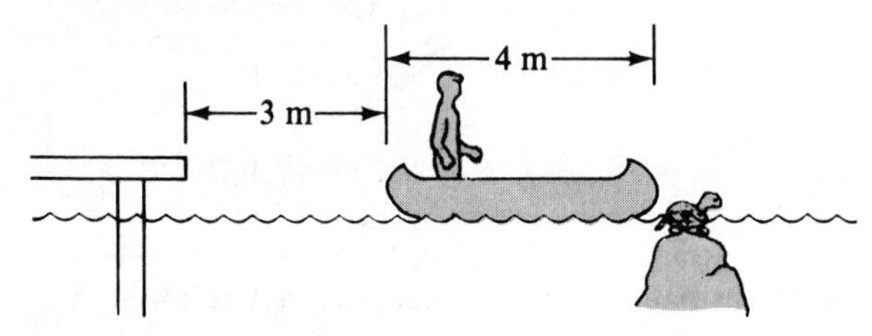

FIGURE 6.20 Problem 58

CHAPTER 7

Circular Motion and the Law of Gravity

King Cobra, the first roller coaster of its kind in America, designed to carry riders in a standing position, opened at Kings Island in April, 1984. The $3 million steel coaster features a 360-degree vertical loop and a horizontal loop. (Courtesy Kings Island)

In this chapter we shall look at circular motion, a specific type of two-dimensional motion. We shall encounter such terms as *angular velocity, angular acceleration, centripetal acceleration,* and *centripetal force.* The results we derive here will enable us to understand the motion of a diverse range of objects in our environment, from a car moving around a circular racetrack to clusters of galaxies orbiting a common center.

We shall also introduce Newton's universal law of gravitation, one of the fundamental laws in nature, and show how this law, together with Newton's laws of motion, enables us to understand a variety of familiar phenomena. Finally, we shall discuss Kepler's laws of planetary motion.

7.1 ANGULAR VELOCITY

We began our study of linear motion by defining the terms *displacement, velocity,* and *acceleration.* This same basic approach will be taken now as we move to a study of rotational motion. To begin, consider Figure 7.1a, which shows a top view of a phonograph record rotating on a turntable. The axis of rotation is at the center of the record at O. A point P on the record is at a distance r from the origin and rotates about O in a circle of radius r. In fact, every point on the record undergoes circular motion about O. To analyze such motion, it is convenient to set up a *fixed* reference line, as shown in Figure 7.1a. Let us assume that at time $t = 0$, the point P is on the reference line as in Figure 7.1a and a line is drawn on the record from the origin out to P. After a time Δt has elapsed, P has advanced to a new position, shown in Figure 7.1b. In this time interval, the line OP has rotated through an angle θ with respect to the reference line. Likewise, P has moved through a distance s measured along the circumference of the circle; the distance s is called an *arc length.*

In situations we have encountered thus far, angles have been measured in degrees. However, in scientific work we often measure angles in *radians* rather than degrees because it makes certain equations take on a simpler form. In fact, almost all of the equations we shall derive in this chapter and the next require that we measure angles in radians. With reference to Figure 7.1b, when the arc length s is equal to the radius r, the angle θ swept out by r is equal to one radian. In general, any angle θ, measured in radians, is defined by the relation

$$\theta \equiv s/r \tag{7.1}$$

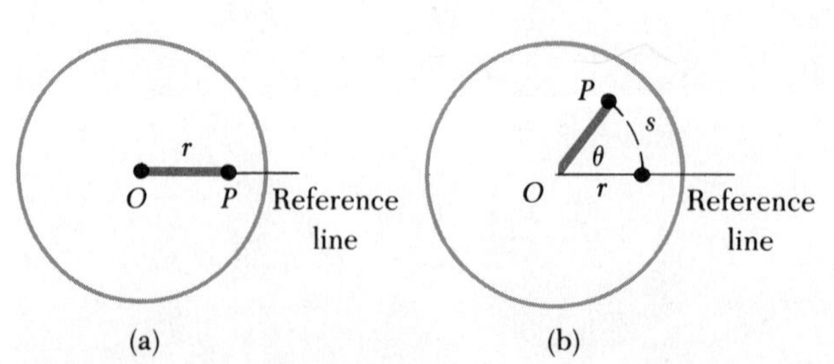

FIGURE 7.1 (a) The point P on a rotating record at $t = 0$. (b) As the record rotates, the point P rotates through an arc length s.

It is important to note that the radian is a pure number with no dimensions. This can be seen from Equation 7.1, since θ is the ratio of an arc length (a distance) and the radius of the circle (also a distance).

To find a conversion factor between degrees and radians, note that when point P in Figure 7.1 moves through an angle of 360° (one revolution), the arc length s is equal to the circumference of the circle, $2\pi r$. From Equation 7.1, we see that the corresponding angle in radians is $2\pi r/r = 2\pi$ rad. Hence,

$$1 \text{ rad} \equiv 360°/2\pi \approx 57.3°$$

From this definition, it follows that any angle in degrees can be converted to an angle in radians with the expression

$$\theta(\text{rad}) = \frac{\pi}{180°}\,\theta(\text{deg}) \qquad (7.2)$$

For example, 60° equals $\pi/3$ rad and 45° equals $\pi/4$ rad.

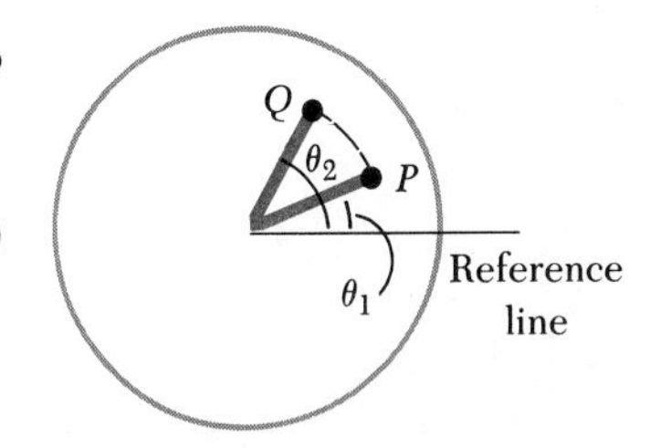

FIGURE 7.2 As a point on the rotating record moves from P to Q, the record rotates through the angle $\Delta\theta = \theta_2 - \theta_1$.

Returning to our phonograph record, we see from Figure 7.2 that, as the record rotates and our point moves from P to Q in a time Δt, the angle through which the record rotates is $\Delta\theta = \theta_2 - \theta_1$. We define $\Delta\theta$ as the **angular displacement.** The *average angular velocity,* $\overline{\omega}$ (ω is the Greek letter omega), of a rotating rigid object is the ratio of the angular displacement, $\Delta\theta$, to the time interval Δt it takes the object to rotate through the angle $\Delta\theta$:

Average angular velocity

$$\boxed{\overline{\omega} \equiv \frac{\theta_2 - \theta_1}{t_2 - t_1} = \frac{\Delta\theta}{\Delta t}} \qquad (7.3)$$

Instantaneous angular velocity

In analogy to linear velocity, the **instantaneous angular velocity,** ω, is defined as *the value of the angular velocity at any instant of time.* It is important to recognize that points at different radial distances on a rotating rigid body have *different* linear speeds along their circular paths. However, *every point on a rotating rigid body has the same value of angular velocity.* This fact demonstrates the convenience of defining and using angular velocity.

Angular velocity has dimensions of radians per second. You should note that the units for angular velocity are unusual in that the radian is a dimensionless unit. This means that we shall not always be able to cancel the radian unit in our equations as we do with such units as those of length, mass, and time. The radian can be viewed as a bookkeeping convenience in our equations. Evidence of this difference will appear soon.

EXAMPLE 7.1 The 45 rpm Record

Express the angular velocity of a 45 rpm (revolutions per minute) record in units of radians per second. (In this text, we shall sometimes use the abbreviation rpm, but in most cases we shall use the abbreviation rev/min.)

Solution First we use the fact that 1 rev corresponds to 2π rad and calculate the angular displacement:

$$\Delta\theta = (45 \text{ rev})\left(2\pi \frac{\text{rad}}{\text{rev}}\right) = 90\pi \text{ rad}$$

The time required for this rotation is 60 s, and therefore the angular velocity is

$$\bar{\omega} = \frac{\Delta\theta}{\Delta t} = \frac{90\pi \text{ rad}}{60 \text{ s}} = 1.5\pi \text{ rad/s}$$

You could also solve this problem by using the following conversion:

$$1\,\frac{\text{rev}}{\text{min}} = \left(1\,\frac{\text{rev}}{\text{min}}\right)\left(2\pi\,\frac{\text{rad}}{\text{rev}}\right)\left(\frac{1\text{ min}}{60\text{ s}}\right) = \frac{\pi}{30}\text{ rad/s}$$

Therefore

$$45\,\frac{\text{rev}}{\text{min}} = 45\left(\frac{\pi}{30}\right)\frac{\text{rad}}{\text{s}} = 1.5\pi \text{ rad/s}$$

7.2 ANGULAR ACCELERATION

Figure 7.3 shows a bicycle turned upside down so that a repairperson can work on the rear wheel. The bicycle peddles are turned such that at time t_1 the wheel has an angular velocity ω_1 (Fig. 7.3a), and at a later time t_2 it has an angular velocity ω_2 (Fig. 7.3b).

We define the **average angular acceleration** $\bar{\alpha}$ (α is the Greek letter alpha) of such an object as the ratio of the change in the angular velocity to the time interval Δt it takes the object to undergo the change:

$$\bar{\alpha} \equiv \frac{\text{change in angular velocity}}{\text{time interval}} = \frac{\omega_2 - \omega_1}{t_2 - t_1} = \frac{\Delta\omega}{\Delta t} \tag{7.4}$$

Instantaneous angular acceleration

The **instantaneous angular acceleration** is the value of the angular acceleration at a specific instant of time. Angular acceleration has units of radians per second per second (rad/s^2). Note that

when a rigid object rotates about a fixed axis, as does the bicycle wheel, every portion of the object has the same angular velocity and the same angular acceleration.

This, in fact, is precisely what makes these variables so useful in describing rotational motion.

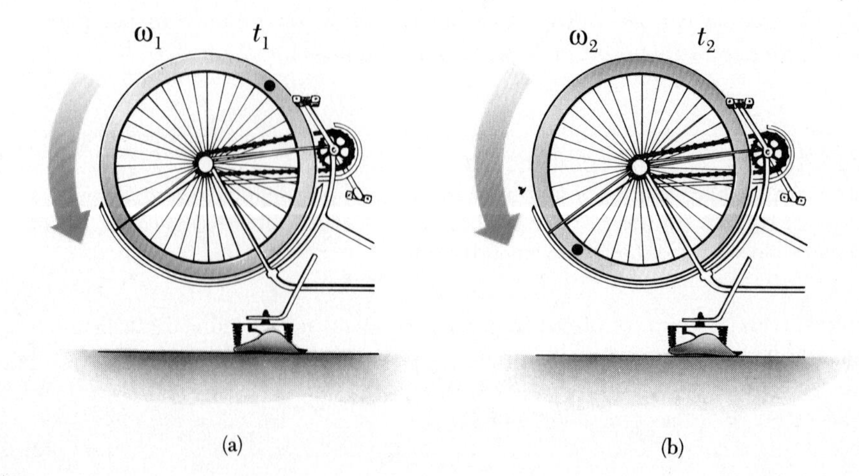

FIGURE 7.3 An accelerating bicycle wheel rotates with (a) an angular velocity ω_1 at time t_1 and (b) an angular velocity ω_2 at time t_2.

The following argument should convince you that ω and α are the same for every point on the wheel. If a point on the rim of the wheel had a larger angular velocity than a point nearer the center, the shape of the wheel would be changing. The wheel remains circular (symmetrically distributed about the axle) only if all points have the same angular velocity and the same angular acceleration.

7.3 ROTATIONAL MOTION UNDER CONSTANT ANGULAR ACCELERATION

Let us pause for a moment to consider some similarities between the equation we have found thus far for rotational motion and those we found in earlier chapters for linear motion. For example, compare the defining equation for average angular velocity,

$$\overline{\omega} \equiv \frac{\theta_f - \theta_i}{t_f - t_i} = \frac{\Delta\theta}{\Delta t}$$

with the defining equation for linear velocity,

$$\overline{v} \equiv \frac{x_f - x_i}{t_f - t_i} = \frac{\Delta x}{\Delta t}$$

The equations are similar in the sense that θ replaces x and ω replaces v. You should take more than a casual note of such similarities as you proceed through the study of rotational motion because virtually every linear quantity that we have seen thus far has a corresponding "twin" in rotational motion. Once you are adept at recognizing such analogies, you will find it unnecessary to memorize many of the equations in this chapter. Additionally, the techniques for solving rotational motion problems are quite similar to those you have already learned for linear motion. For example, problems concerned with objects that rotate with a constant angular acceleration can be solved in much the same manner as those dealing with linear motion under constant acceleration. Therefore, if you have a good understanding of problems involving objects moving with a constant linear acceleration, these new exercises should be little more than a review for you.

An additional analogy between linear motion and rotational motion arises when we compare the defining equation for average angular acceleration,

$$\overline{\alpha} \equiv \frac{\omega_f - \omega_i}{t_f - t_i} = \frac{\Delta\omega}{\Delta t}$$

with the defining equation for average linear acceleration,

$$\overline{a} \equiv \frac{v_f - v_i}{t_f - t_i} = \frac{\Delta v}{\Delta t}$$

Now that we have pointed out the analogies between variables in linear motion and those in rotational motion, it should not surprise you that the equations of rotational motion involve the variables θ, ω, and α. In Section 2.4, we developed a set of kinematic equations for linear motion under constant acceleration. The same procedure can be used to derive a similar set of equations for rotational motion under constant angular acceleration. The resulting equations of rotational kinematics, along with the corresponding equations for linear motion under constant acceleration, are

Rotational Motion About a Fixed Axis With α Constant (Variables: θ and ω)	Linear Motion With a Constant (Variables: x and v)	
$\omega = \omega_0 + \alpha t$	$v = v_0 + at$	(7.5)
$\theta = \omega_0 t + \frac{1}{2}\alpha t^2$	$x = v_0 t + \frac{1}{2}at^2$	(7.6)
$\omega^2 = \omega_0^2 + 2\alpha\theta$	$v^2 = v_0^2 + 2ax$	(7.7)

Again, note the one-to-one correspondence between the rotational equations involving the angular variables θ, and ω and the equations of linear motion involving the variables x and v.

EXAMPLE 7.2 The Rotating Wheel

The bicycle wheel of Figure 7.3 rotates with a constant angular acceleration of 3.5 rad/s^2. If the initial angular velocity of the wheel is 2.0 rad/s at $t_0 = 0$, (a) through what angle does the wheel rotate in 2 s?

Solution Since we are given that $\omega_0 = 2.0$ rad/s and $\alpha = 3.5$ rad/s^2, we use

$$\theta = \omega_0 t + \frac{1}{2}\alpha t^2$$

$$\theta = (2.0 \text{ rad/s})(2 \text{ s}) + \frac{1}{2}(3.5 \text{ rad/s}^2)(2 \text{ s})^2$$

$$= 11 \text{ rad} = 630°$$

(b) What is the angular velocity at $t = 2$ s?

Solution If we make use of Equation 7.5, we find

$$\omega = \omega_0 + \alpha t = 2.0 \text{ rad/s} + (3.5 \text{ rad/s}^2)(2 \text{ s}) = 9.0 \text{ rad/s}$$

EXERCISE Find the angular velocity of the wheel at $t = 2$ s by making use of Equation 7.7 and the results to part (a).

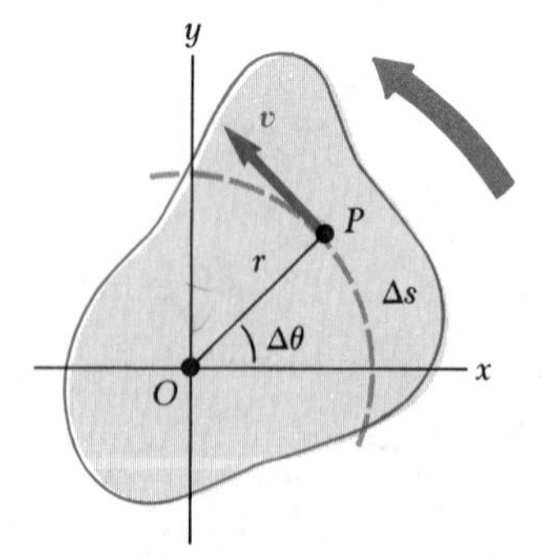

FIGURE 7.4 Rotation of an object about an axis through O which is perpendicular to the plane of the figure (the z axis). Note that a point P on the object rotates in a circle of radius r centered at O.

7.4 RELATIONS BETWEEN ANGULAR AND LINEAR QUANTITIES

In this section we shall derive some useful relations between the angular velocity and acceleration of a rotating object and the linear velocity and acceleration of an arbitrary point in the object. In order to do so, you should be aware of the fact that, when a rigid object rotates about a fixed axis, every point in the object moves in a circle whose center is along the axis of rotation.

Consider the arbitrarily shaped object shown in Figure 7.4 rotating about the z axis through the point O. Assume the object rotates through the angle $\Delta\theta$, and hence through the arc length Δs, in a time Δt. We know from the defining equation for radian measure that

$$\Delta\theta = \frac{\Delta s}{r}$$

Let us now divide both sides of this equation by Δt, the time during which the rotation occurred:

$$\frac{\Delta\theta}{\Delta t} = \frac{1}{r}\frac{\Delta s}{\Delta t}$$

If the time interval Δt is very small, then the angle $\Delta\theta$ through which the object rotates is small and the ratio $\Delta\theta/\Delta t$ is the instantaneous angular velocity, ω. Also, Δs is very small when Δt is very small, and the ratio $\Delta s/\Delta t$ equals the instantaneous linear speed, v. Hence, this equation is equivalent to

$$\omega = \frac{v}{r}$$

Figure 7.4 allows us to interpret this equation. The distance Δs is traversed along an arc of the circular path followed by the point P as it rotates during the time Δt. Thus, v must be the linear velocity of a point lying along this arc, a velocity that is *tangent to the circular path.* For this reason, we often refer to this linear velocity as the **tangential velocity** of a particle moving in a circular path and write

$$\boxed{v_t = r\omega} \qquad (7.8)$$

Tangential velocity

That is, the tangential velocity of a point on a rotating object equals the distance of that point from the axis of rotation multiplied by the angular velocity.

Note that although every point on the rotating object has the same angular velocity, not every point has the same linear, or tangential, velocity. In fact, Equation 7.8 shows that the linear velocity of a point on the rotating object increases as one moves outward from the center of rotation toward the rim, as you would intuitively expect.

A bit of caution should be exercised when using Equation 7.8. It has been derived using the defining equation for radian measure; hence, the equation is valid only when ω is measured in radians per unit time. Other measures of angular speed, such as degrees per second or revolutions per second, are not to be used in Equation 7.8.

To find a second equation relating linear and angular quantities, imagine that an object rotating about a fixed axis (Fig. 7.4) changes its angular velocity by $\Delta\omega$ in a time Δt. At the end of this time, the velocity of a point on the object, such as P, has changed by an amount Δv_t. From Equation 7.8, we have

$$\Delta v_t = r\,\Delta\omega$$

Dividing by Δt gives

$$\frac{\Delta v_t}{\Delta t} = r\frac{\Delta\omega}{\Delta t}$$

If the time interval Δt is very small, then the ratio $\Delta v_t/\Delta t$ is the tangential acceleration of that point and $\Delta\omega/\Delta t$ is the angular acceleration. Therefore, we see that

$$\boxed{a_t = r\alpha} \qquad (7.9)$$

Tangential acceleration

That is, the tangential acceleration of a point on a rotating object equals the distance of that point from the axis of rotation multiplied by the angular acceleration.

Again, radian measure must be used for the angular acceleration term in this equation.

There is one more equation that relates linear quantities to angular quantities, but we shall withhold the derivation of this relationship until the next section.

EXAMPLE 7.3 Going for a Spin

The turntable of a record player rotates initially at a rate of 33 rev/min and takes 20 s to come to rest. (a) What is the angular acceleration of the turntable, assuming the acceleration is uniform?

Solution Recalling that 1 rev $= 2\pi$ rad, we see that the initial angular velocity is

$$\omega_0 = \left(33\,\frac{\text{rev}}{\text{min}}\right)\left(2\pi\,\frac{\text{rad}}{\text{rev}}\right)\left(\frac{1\text{ min}}{60\text{ s}}\right) = 3.46\text{ rad/s}$$

If we use $\omega = \omega_0 + \alpha t$ and the fact that $\omega = 0$ at $t = 20$ s, we get

$$\alpha = -\frac{\omega_0}{t} = -\frac{3.46\text{ rad/s}}{20\text{ s}} = -0.173\text{ rad/s}^2$$

where the negative sign indicates an angular deceleration (ω is decreasing).

(b) How many rotations does the turntable make before coming to rest?

Solution Equation 7.6 enables us to find the angular displacement in 20 s:

$$\theta = \omega_0 t + \frac{1}{2}\alpha t^2$$

$$= (3.46\text{ rad/s})(20\text{ s}) + \frac{1}{2}(-0.173\text{ rad/s}^2)(20\text{ s})^2 = 34.6\text{ rad}$$

This corresponds to $34.6/2\pi$ rev $= 5.51$ rev.

(c) If the radius of the turntable is 0.14 m, what is the initial linear speed of a bug riding on the rim?

Solution The relation $v_t = r\omega$ and the value $\omega_0 = 3.46$ rad/s give

$$v_t = (0.14\text{ m})(3.46\text{ rad/s}) = 0.484\text{ m/s}$$

(d) What is the magnitude of the tangential acceleration of the bug at time $t = 0$?

Solution We can use $a_t = r\alpha$, which gives

$$a_t = (0.14\text{ m})(0.173\text{ rad/s}^2) = 0.0242\text{ m/s}^2$$

7.5 CENTRIPETAL ACCELERATION

Figure 7.5a shows a car moving in a circular path with *constant linear speed* v. It is often surprising to students to find that *even though the car moves at a constant speed, it still has an acceleration.* To see why this occurs, consider the defining equation for acceleration as

$$a = \frac{v_f - v_i}{t_f - t_i} \tag{7.10}$$

Note that the acceleration depends on *the change in the velocity vector*. Because velocity is a vector, there are two ways in which an acceleration can be produced: by a change in the *magnitude* of the velocity and by a change in the *direction* of the velocity. It is the latter situation that is occurring for the car moving in a circular path with constant speed (see Fig. 7.5b). We shall show that the acceleration vector in this case is perpendicular to the path and always points toward the center of the circle. An acceleration of this nature is called a **centripetal acceleration** (center-seeking) and is given by

$$a_c = \frac{v^2}{r} \tag{7.11}$$

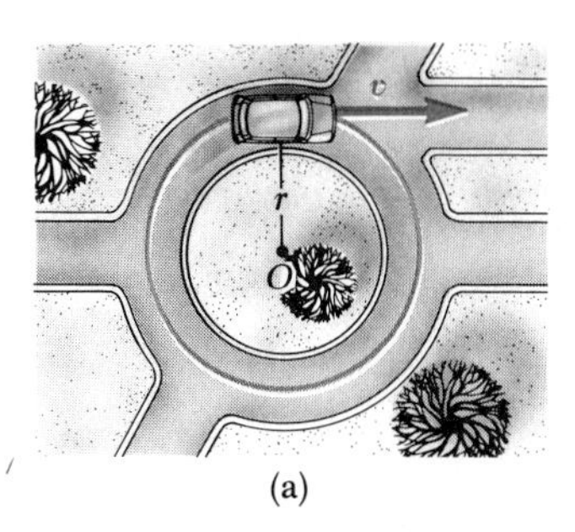

(a)

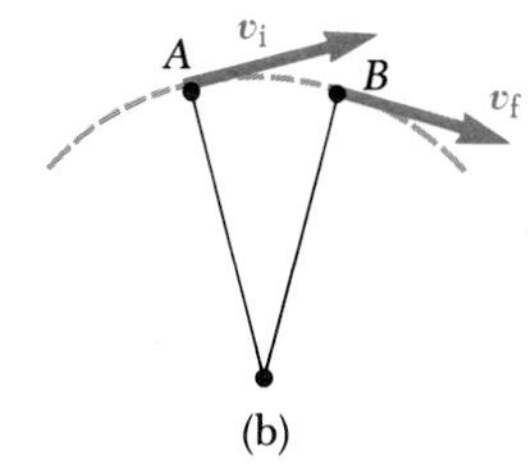

(b)

FIGURE 7.5 (a) Circular motion of a car moving with a constant speed. (b) As the car moves along the circular path from A to B, the direction of its velocity vector changes so the car undergoes a centripetal acceleration.

To derive Equation 7.11, consider Figure 7.6a. Here an object is seen first at point A with velocity $\mathbf{v}_i$ at time t_i and then at point B with velocity $\mathbf{v}_f$ at a later time t_f. Let us also assume here that $\mathbf{v}_i$ and $\mathbf{v}_f$ differ only in direction; their magnitudes are the same (that is, $v_i = v_f = v$). In order to calculate the acceleration, let us begin with Equation 7.10, which indicates that we must vectorially subtract $\mathbf{v}_i$ from $\mathbf{v}_f$:

$$\mathbf{a} = \frac{\mathbf{v}_f - \mathbf{v}_i}{t_f - t_i} = \frac{\Delta \mathbf{v}}{\Delta t} \tag{7.12}$$

where $\Delta\mathbf{v} = \mathbf{v}_f - \mathbf{v}_i$ is the change in the velocity. That is, $\Delta\mathbf{v}$ is obtained by adding to $\mathbf{v}_f$ the vector $-\mathbf{v}_i$. This can be accomplished graphically as shown by the vector triangle in Figure 7.6b. Note that when Δt is very small, Δs and $\Delta\theta$ are also very small. In this case, $\mathbf{v}_f$ will be almost parallel to $\mathbf{v}_i$ and the vector $\Delta\mathbf{v}$ will be approximately perpendicular to them, pointing toward the center of the circle. When one considers the limiting case where Δt becomes vanishingly small, $\Delta\mathbf{v}$ will point exactly toward the center of the circle. Furthermore, in this limiting case, the acceleration is also directed toward the center of the circle because it is in the direction of $\Delta\mathbf{v}$.

Now consider the triangle in Figure 7.6a, which has sides Δs and r. This triangle and the one with sides Δv and v in Figure 7.6b are similar. (Two triangles are similar if the angle between any two sides is the same for both triangles and if the ratio of lengths of these sides is the same.) This enables us to write a relationship between the lengths of the sides:

$$\frac{\Delta v}{v} = \frac{\Delta s}{r}$$

This equation can be solved for Δv and the expression so obtained can be substituted into Equation 7.12 to give $a\Delta t = v\Delta s/r$, or

$$a = \frac{v}{r}\frac{\Delta s}{\Delta t} \tag{7.13}$$

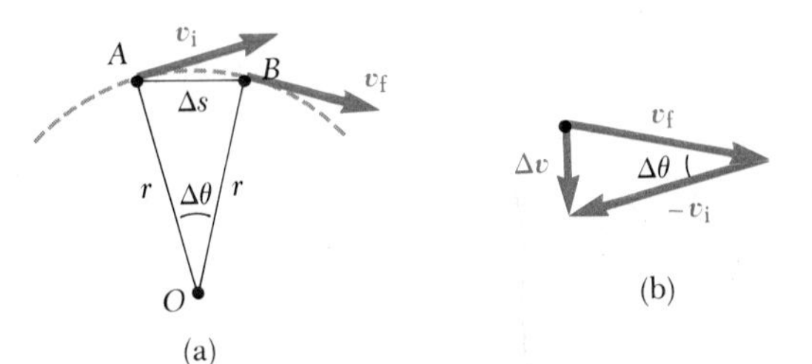

FIGURE 7.6 (a) As the particle moves from A to B, the direction of its velocity vector changes from $\mathbf{v}_i$ to $\mathbf{v}_f$. (b) The construction for determining the direction of the change in velocity, $\Delta\mathbf{v}$, which is toward the center of the circle.

In this situation, Δs is a small distance measured along the arc of the circle (a tangential distance), so $v = \Delta s/\Delta t$, where v is the tangential velocity. Therefore, Equation 7.13 reduces to

Centripetal acceleration

$$a_c = \frac{v^2}{r} \quad (7.11)$$

Since the tangential velocity is related to the angular velocity through the relation $v_t = r\omega$ (Eq. 7.8), an alternative form of Equation 7.11 is

$$a_c = \frac{r^2\omega^2}{r} = r\omega^2 \quad (7.14)$$

Thus we conclude that in circular motion the centripetal acceleration is directed inward toward the center of the circle and has a magnitude given either by v^2/r or $r\omega^2$.

You should show that the dimensions of a_c are $[L]/[T^2]$, as required.

In order to clear up any misconceptions that might exist concerning centripetal and tangential acceleration, let us consider a car moving around a circular racetrack. If the car is moving in a circular path, it always has a centripetal component of acceleration because the direction of travel of the car, and hence the direction of its velocity, are continuously changing. If the speed of the car is increasing or decreasing, it also has a tangential component of acceleration. To summarize, the tangential component of acceleration arises when the speed of the car is altered; the centripetal component of acceleration arises when the direction of travel is changed.

When both components of acceleration exist simultaneously, the tangential acceleration is tangent to the circular path and the centripetal acceleration points toward the center of the circular path. Because these components of acceleration are perpendicular to each other, we can find the **total acceleration** using the Pythagorean theorem:

Total acceleration

$$a = \sqrt{a_t^2 + a_c^2} \quad (7.15)$$

EXAMPLE 7.4 Let's Go For a Spin

A test car moves at a constant speed of 10 m/s around a circular road of radius 50 m. Find (a) the centripetal acceleration of the car, and (b) the angular velocity of the car.

Solution (a) The centripetal acceleration of the car is found from Equation 7.11 to be

$$a_c = \frac{v^2}{r} = \frac{(10 \text{ m/s})^2}{50 \text{ m}} = 2 \text{ m/s}^2$$

(b) The angular velocity of the car can be found using the expression $v_t = r\omega$, which gives

$$\omega = \frac{v_t}{r} = \frac{10 \text{ m/s}}{50 \text{ m}} = 0.2 \text{ rad/s}$$

Note that we can also find the centripetal acceleration by using the value of ω found above. This is an alternative method to that used in part (a). From Equation 7.12, we have

$$a_c = r\omega^2 = (50\ \text{m})(0.2\ \text{rad/s})^2 = 2\ \text{m/s}^2$$

As expected, both methods for finding the centripetal acceleration give the same answer.

EXERCISE Find the tangential acceleration and total acceleration of the car.

Answer The tangential acceleration of the car is zero because the speed of the car remains constant. Since $a_t = 0$, the total acceleration equals the centripetal acceleration, $2\ \text{m/s}^2$, found in part (a).

7.6 CENTRIPETAL FORCE

Consider a ball of mass m tied to a string of length r and being whirled in a horizontal circular path, as in Figure 7.7. Let us assume that the ball moves with a constant speed. Because the velocity vector, $\mathbf{v}$, changes its direction continuously during the motion, the ball experiences a centripetal acceleration directed toward the center of motion, as described in Section 7.5. This centripetal acceleration has a magnitude

$$a_c = \frac{v^2}{r}$$

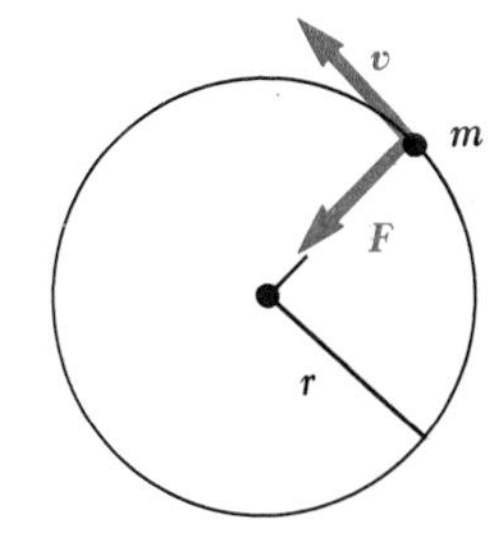

FIGURE 7.7 A ball attached to a string of length r, rotating in a circular path at constant speed.

The inertia of the ball tends to maintain motion in a straight-line path; however, the string prevents this by exerting a force on the ball to make it follow its circular path. This force (equal to the force of tension) is directed along the length of the string toward the center of the circle, as shown in Figure 7.7, and is an example of a class of forces called **centripetal forces.** Thus, the equation for Newton's second law along the radial direction is

Centripetal force

$$F_c = ma_c = m\frac{v^2}{r} \tag{7.16}$$

All centripetal forces act toward the center of the circular path along which the object moves.

Because they act toward the center of rotation, centripetal forces cause a change in the direction of the velocity. Beyond this, they are no different from any of the other forces that we have studied. For example, friction between the tires and the road provides the centripetal force that enables a racecar to travel in a circular path on a flat road, and the gravitational force exerted on the moon by the earth provides the centripetal force necessary to keep the moon in its orbit. We shall discuss the gravitational force in more detail in Section 7.8.

Regardless of the example used, if the centripetal force acting on an object moving initially in a circle should vanish, the object would no longer continue to move in its circular path; instead, it would move along a straight-line path tangent to the circle. To illustrate this point, consider a ball attached

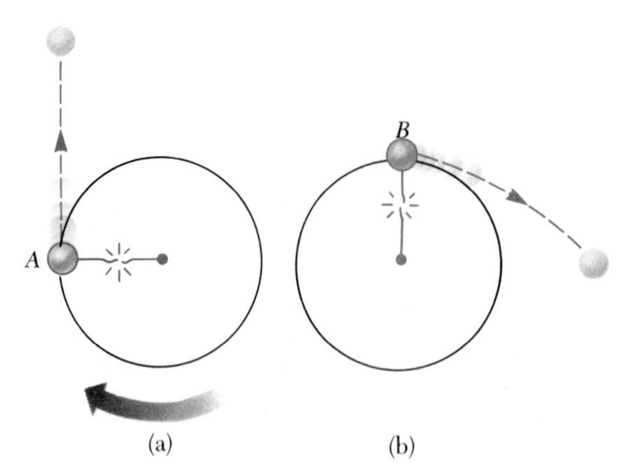

FIGURE 7.8 (a) When the string breaks at position A, the ball moves vertically upwards in free fall. (b) When the string breaks at position B, the ball moves along a parabolic path.

to a string being whirled in a vertical circle as in Figure 7.8a. If the string happens to break when the ball is at position A, the centripetal force (the tension in the string) would vanish and the ball would move vertically upward. Its subsequent motion would be that of a freely falling body. If the string happens to break when the ball is at the top of its circular path, shown as point B in Figure 7.8b, the ball would initially fly off horizontally in a direction tangent to the path, and would then move in a parabolic path according to the equations of a projectile.

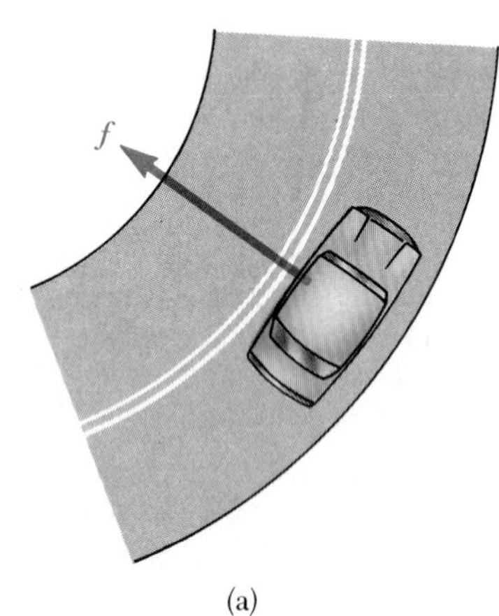

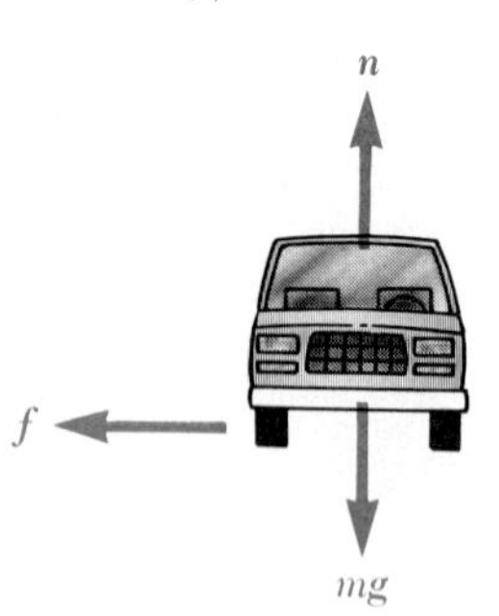

FIGURE 7.9 (Example 7.5) (a) Top view of a car on a curved path. (b) Free-body diagram for the car showing an end view.

EXAMPLE 7.5 Buckle Up For Safety

A car travels at a constant speed of 30 mi/h (13.4 m/s) on a level circular turn of radius 50 m as shown in the birds-eye view in Figure 7.9a. What is the minimum coefficient of static friction between the tires and roadway in order that the car make the circular turn without sliding?

Solution The centripetal force acting on the car is the force of static friction directed toward the center of the circular path as shown in Figure 7.9a. Thus, Equation 7.16 becomes

$$(1) \quad f = m\frac{v^2}{r}$$

All the forces acting on the car are shown in Figure 7.9b. Because we have equilibrium in the vertical direction, the normal force upward is balanced by the force of gravity downward, so that

$$n = mg$$

From this expression, we can find the minimum force of friction as follows:

$$(2) \quad f = \mu_s n = \mu_s mg$$

By setting the right-hand sides of (1) and (2) equal to each other, we find

$$\mu_s mg = m\frac{v^2}{r}$$

or

$$(3) \quad \mu_s = \frac{v^2}{rg}$$

Therefore, the minimum coefficient of static friction required for the car to make the turn without sliding outwards is

$$\mu_s = \frac{v^2}{rg} = \frac{(13.4 \text{ m/s})^2}{(50 \text{ m})(9.80 \text{ m/s}^2)} = 0.366$$

The value of μ_s for rubber on dry concrete is very close to 1. Thus, the car will be able to negotiate the curve with ease. However, if the road is wet or icy, the value for μ_s can be 0.2 or lower. Under this condition, the centripetal force provided by static friction would not be great enough to enable the car to follow the circular path, and it would slide off the roadway.

EXAMPLE 7.6 Having Fun with a Yo-Yo

A child swings a yo-yo of weight mg in a horizontal circle such that the cord makes an angle of 30° with the vertical, as in Figure 7.10a. Find the centripetal acceleration of the yo-yo.

Solution Two forces act on the yo-yo: its weight and the tension in the cord, $\boldsymbol{T}$, as shown in Figure 7.10b. Because the yo-yo is in equilibrium in the y direction, we can apply $\Sigma F_y = 0$ in order to find T:

$$\sum F_y = T \cos 30° - mg = 0$$

$$T = \frac{mg}{\cos 30°}$$

The net force acting toward the center of the circular path is the centripetal force, which for this case is the *horizontal component* of $\boldsymbol{T}$:

$$F_c = T \sin 30° = \frac{mg \sin 30°}{\cos 30°} = mg \tan 30°$$

The centripetal acceleration can now be found by applying Newton's second law along the horizontal direction:

$$F_c = ma_c$$

$$a_c = \frac{F_c}{m} = \frac{mg \tan 30°}{m} = g \tan 30° = (9.80 \text{ m/s}^2)(\tan 30°) = 5.66 \text{ m/s}^2$$

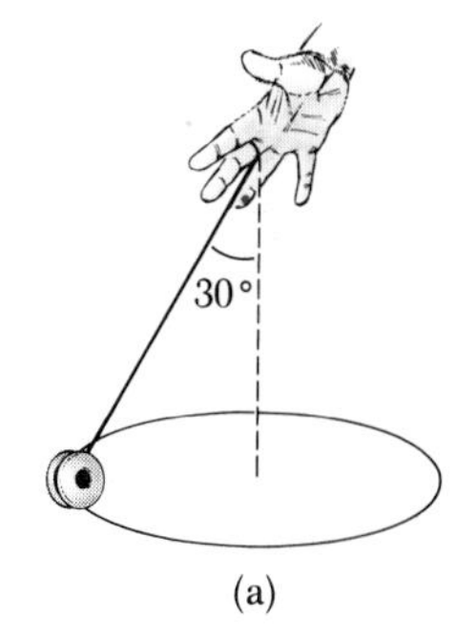

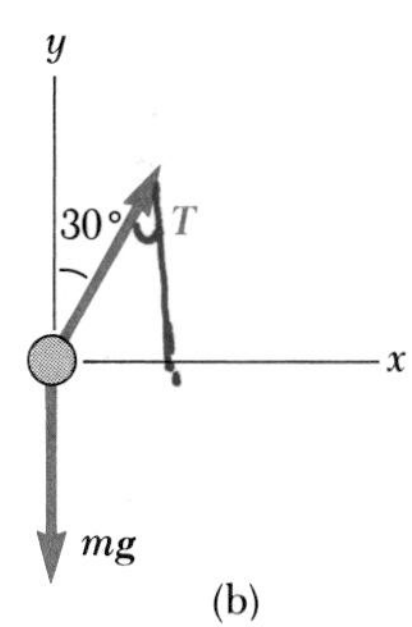

FIGURE 7.10 (Example 7.6) (a) A yo-yo swinging in a horizontal circle such that the cord makes a constant angle of 30° with the vertical. (b) Free-body diagram for the yo-yo.

7.7 DESCRIBING MOTION OF A ROTATING SYSTEM

We have seen that an object moving in a circle of radius r with constant speed v has a centripetal acceleration whose magnitude is v^2/r and whose direction is toward the center of rotation. The force necessary to maintain this centripetal acceleration, called a centripetal force, must also act toward the center of rotation. In the case of a ball rotating at the end of a string, the force exerted on the ball by the string (equal to the tension in the string) is the centripetal force. For a satellite in a circular orbit around the earth, the force of gravity is the centripetal force. The centripetal force acting on a car rounding a curve on a level road is the force of friction between the tires and the pavement, and so forth. Centripetal forces are no different from any other forces we have encountered. The term *centripetal* is used simply to indicate that *the force is directed toward the center of a circle.*

In order to better understand the motion of a rotating system, consider a car traveling along a highway at a high speed and approaching a curved exit

FIGURE 7.11 A car approaching a curved exit ramp.

ramp, as in Figure 7.11. As the car takes the sharp left turn onto the ramp, a person sitting in the passenger seat slides to the right across the seat and hits the door. At that point, the force of the door keeps him from being ejected from the car. What causes the passenger to move toward the door? A popular, but *improper*, explanation is that some mysterious force pushes him outward. (This is sometimes called the "centrifugal" force, but we shall not use this term since it always creates confusion.)

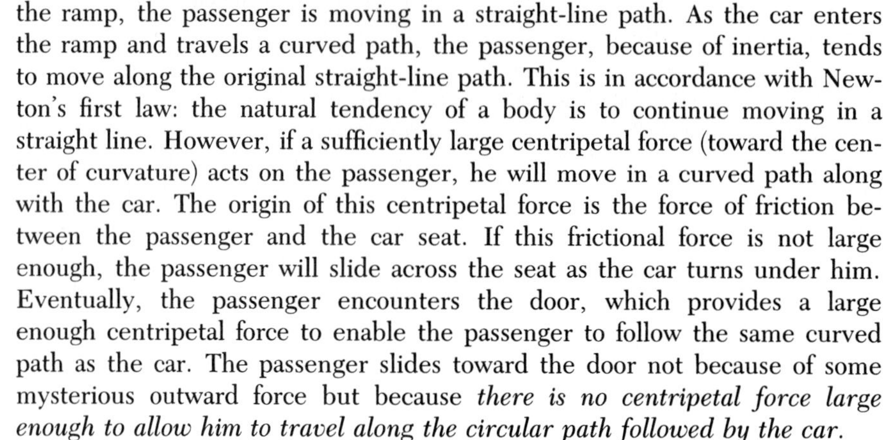

The phenomenon is correctly explained as follows. Before the car enters the ramp, the passenger is moving in a straight-line path. As the car enters the ramp and travels a curved path, the passenger, because of inertia, tends to move along the original straight-line path. This is in accordance with Newton's first law: the natural tendency of a body is to continue moving in a straight line. However, if a sufficiently large centripetal force (toward the center of curvature) acts on the passenger, he will move in a curved path along with the car. The origin of this centripetal force is the force of friction between the passenger and the car seat. If this frictional force is not large enough, the passenger will slide across the seat as the car turns under him. Eventually, the passenger encounters the door, which provides a large enough centripetal force to enable the passenger to follow the same curved path as the car. The passenger slides toward the door not because of some mysterious outward force but because *there is no centripetal force large enough to allow him to travel along the circular path followed by the car.*

As a second example, consider what happens when you run clothes through the rinse cycle of a washing machine. In the last phase of this cycle, the drum spins rapidly to remove water from the clothes. Why is the water thrown off? An *improper* explanation is that the rotating system creates some mysterious outward force on each drop of water and this force causes the water to be hurled to the outer drum of the machine. The correct explanation goes as follows. When the clothes are at rest in the machine, water is held to them by molecular forces between the water and the fabric. During the spin cycle, the clothes rotate and the molecular forces are not large enough to provide the necessary *centripetal* force to keep the water molecules moving in a circular path along with the clothes. Hence, the water drops, because of their inertia, move in a straight-line path until they encounter the sides of the spinning drum.

James Garner in a road race, from the film "Grand Prix." (Courtesy of Museum of Modern Art/Film Stills Archives)

In summary, one must be very careful to distinguish real forces from fictitious ones in describing motion in an accelerating frame. An observer in a car rounding a curve is in an accelerating frame and invents a fictitious outward force to explain why he or she is thrown outward. A stationary observer outside the car, however, considers only real forces on the passenger. To this observer, the mysterious outward force *does not exist!* The only real external force on the passenger is the centripetal (inward) force due to friction or the normal force of the door.

7.8 NEWTON'S LAW OF GRAVITATION

Prior to 1686, a great mass of data had been collected on the motion of the moon and the planets but a clear understanding of the forces that caused these celestial bodies to move the way they did was not available. In that year, however, Isaac Newton provided the key that unlocked the secrets of the heavens. He knew, from the first law, that a net force had to be acting on the moon. If not, it would move in a straight-line path rather than in its almost circular orbit. Newton reasoned that this force arose as a result of a gravitational attraction that the earth exerted on the moon. He also concluded that there could be nothing special about the earth-moon system or the sun and its planets that would cause gravitational forces to act on them alone. In other words, he saw that the same force of attraction that causes the moon to follow its path also causes an apple to fall to earth from a tree. He wrote, "I deduced that the forces which keep the planets in their orbs must be reciprocally as the squares of their distances from the centers about which they revolve; and thereby compared the force requisite to keep the Moon in her orb with force of gravity at the surface of the Earth; and found them answer pretty nearly."

In 1687 Newton published his work on the **universal law of gravitation.** This law states that

every particle in the universe attracts every other particle with a force that is directly proportional to the product of their masses and inversely proportional to the square of the distance between them.

If the particles have masses m_1 and m_2 and are separated by a distance r, the magnitude of this gravitational force is

$$\boxed{F = G\frac{m_1 m_2}{r^2}} \tag{7.17}$$

Universal law of gravitation

where G is a universal constant called the **gravitational constant,** which has been measured experimentally. Its value in SI units is

$$G = 6.673 \times 10^{-11}\,\frac{\text{N} \cdot \text{m}^2}{\text{kg}^2} \tag{7.18}$$

This force law is an example of an **inverse-square law,** in that it varies as the inverse square of the separation. The force acts such that the objects are always attracted to one another. From Newton's third law, we also know that the force on m_2 due to m_1, designated $\mathbf{F}_{21}$ in Figure 7.12, is equal in magnitude to the force on m_1 due to m_2, $\mathbf{F}_{12}$, and in the opposite direction. That is, these forces form an action-reaction pair.

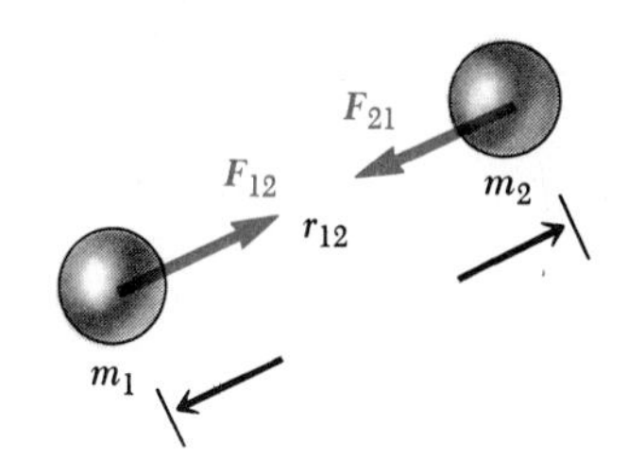

FIGURE 7.12 The gravitational force between two particles is attractive. Note that according to Newton's third law, $\mathbf{F}_{12} = -\mathbf{F}_{2f}$.

There are several features of the universal law of gravity that deserve some attention:

1. The gravitational force is an action-at-a-distance force that always exists between two particles regardless of the medium that separates them.
2. The force varies as the inverse square of the distance between the particles and therefore decreases rapidly with increasing separation.
3. The force is proportional to the product of their masses.

Newton's law of gravitation as given in Equation 7.17 applies to only two cases. It can be used to find the gravitational force between tiny objects (particles) or to find the gravitational force between a sphere and a particle located external to the sphere. In the latter case, the gravitational effect is as though the mass of the sphere were concentrated at its center. For example, the force on a particle of mass m at the earth's surface has the magnitude

$$F = G\frac{M_e m}{R_e^2}$$

where M_e is the earth's mass and R_e is its radius. This force is directed toward the center of the earth.

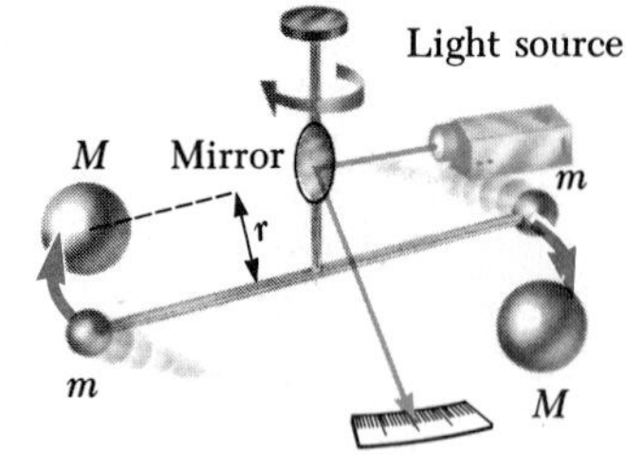

FIGURE 7.13 Schematic diagram of the Cavendish apparatus for measuring the constant G. The smaller spheres of mass m are attracted to the large spheres of mass M, and the bar rotates through a small angle. A light beam reflected from a mirror attached to the vertical suspension measures the angle of rotation.

Measurement of the Gravitational Constant

The gravitational constant was first measured in an important experiment by Henry Cavendish in 1798. The apparatus he used consisted of two small spheres, each of mass m, fixed to the ends of a light horizontal rod suspended by a thin metal wire, as in Figure 7.13. Two large spheres, each of mass M, were placed near the smaller spheres. The attractive force between the smaller and larger spheres caused the rod to rotate and the wire to twist. The angle through which the suspended rod rotated was measured by the deflection of a light beam reflected from a mirror attached to the vertical suspension. (Such a moving spot of light is an effective technique for amplifying the motion.) The experiment was carefully repeated with different masses at various separations. In addition to providing a value for G, the results showed that the force is attractive, proportional to the product mM, and inversely proportional to the square of the distance r.

EXAMPLE 7.7 Billiards, Anyone?

Three billiard balls each of mass 0.3 kg are placed on a table at the corners of a right triangle as shown in the overhead view of Figure 7.14. Find the net gravitational force on the ball denoted as m_1 due to the forces exerted on it by the other two balls.

Solution To find the net gravitational force on m_1, we first calculate the force exerted on m_1 due to m_2. Then, we find the force on m_1 due to m_3. Finally, we must add these two forces *vectorially* to obtain the net force on m_1.

The force exerted on m_1 due to m_2, denoted by $\mathbf{F}_1$ in Figure 7.14, is upwards. The magnitude of this force is calculated using Equation 7.17:

$$F_1 = G\frac{m_1 m_2}{r^2} = (6.67 \times 10^{-11}\ \text{N}\cdot\text{m}^2/\text{kg}^2)\frac{(0.3\ \text{kg})(0.3\ \text{kg})}{(0.4\ \text{m})^2}$$

$$= 3.75 \times 10^{-11}\ \text{N}$$

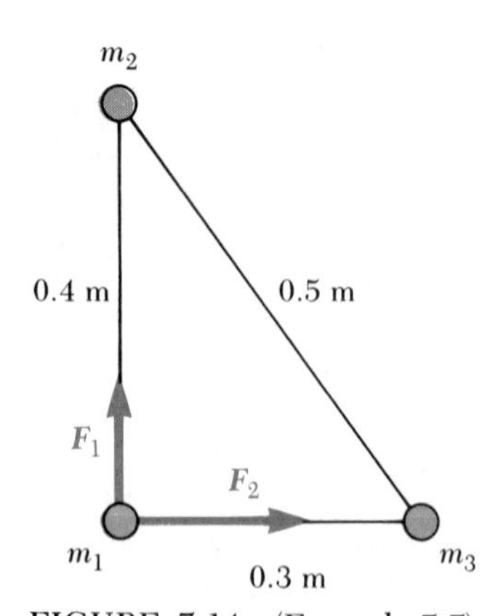

FIGURE 7.14 (Example 7.7)

This result shows that gravitational forces between common objects that surround us have extremely small magnitudes.

Now, let us calculate the gravitational force exerted on m_1 due to m_3. This force, denoted by $\mathbf{F}_2$ in Figure 7.14, is directed to the right, and its magnitude is

$$F_2 = G\frac{m_1 m_3}{r^2} = (6.67 \times 10^{-11}\ \text{N}\cdot\text{m}^2/\text{kg}^2)\frac{(0.3\ \text{kg})(0.3\ \text{kg})}{(0.3\ \text{m})^2}$$

$$= 6.67 \times 10^{-11}\ \text{N}$$

The net gravitational force exerted on m_1 is found by adding $\mathbf{F}_1$ and $\mathbf{F}_2$ as *vectors*. The magnitude of this net force is given by

$$F = \sqrt{F_1^2 + F_2^2} = \sqrt{(3.75)^2 + (6.67)^2} \times 10^{-11}\ \text{N} = 7.65 \times 10^{-11}\ \text{N}$$

EXERCISE Find the direction of the resultant force on m_1.

Answer The vector $\boldsymbol{F}$ makes an angle of 29.3° with respect to the line joining m_1 and m_3.

EXAMPLE 7.8 The Mass of the Earth

Use the gravitational force law to find an approximate value for the mass of the earth.

Solution Figure 7.15, obviously not to scale, shows a baseball falling toward the earth at a location where the acceleration due to gravity is g. We know that the gravitational force exerted on the baseball by the earth is the same as the weight of the ball and that this is given by $w = m_b g$. Since the force in the gravitational law is the weight of the ball, we find

$$m_b g = G\frac{M_e m_b}{R_e^{\,2}}$$

We can divide each side of this equation by m_b and solve for M_e:

$$M_e = \frac{gR_e^{\,2}}{G}$$

The falling baseball is close enough to the earth so that the distance of separation between the center of the ball and the center of the earth can be taken as the radius of the earth, 6.38×10^6 m. Thus, the mass of the earth is

$$M_e = \frac{(9.80\ \mathrm{m/s^2})(6.38 \times 10^6 \mathrm{m})^2}{6.67 \times 10^{-11}\ \mathrm{N \cdot m^2/kg^2}} = 5.98 \times 10^{24}\ \mathrm{kg}$$

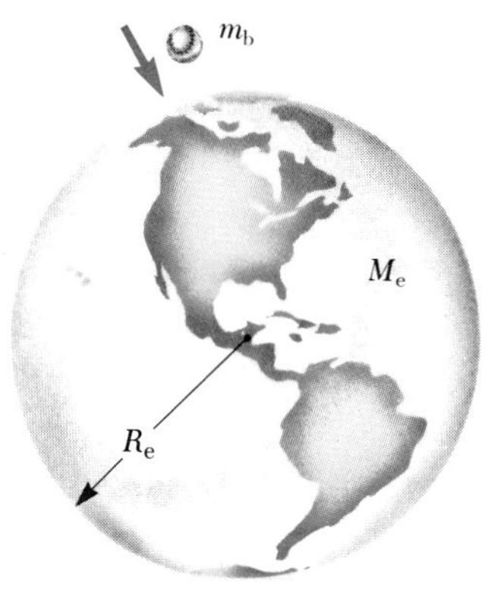

FIGURE 7.15 (Example 7.8) A baseball falling toward the earth (not drawn to scale).

EXAMPLE 7.9 Gravity and Altitude

Derive an expression that shows how the acceleration due to gravity varies with distance from the center of the earth at an exterior point.

Solution The falling baseball of Example 7.8 can be used here also. Now, however, assume that the ball is located at some arbitrary distance r from the earth's center. The first equation in Example 7.8, with r replacing R_e and m_b removed from both sides, becomes

$$g = G\frac{M_e}{r^2}$$

This indicates that the acceleration due to gravity at an exterior point decreases as the inverse square of the distance from the center of the earth. Our assumption of Chapter 6 that objects fall with a constant acceleration is obviously incorrect in light of our present example. For small distances of fall, however, this change in g is so small that neglecting the variation does not introduce a significant error in our results.

Because the true weight of an object is mg, we see that a change in the value of g produces a change in the weight of an object. For example, if you weigh 800 N at the surface of the earth, you will weigh only 200 N at a height above the earth equal to the radius of the earth. Also, we see that if the distance of an object from the earth becomes infinitely large, the true weight approaches zero. Values of g at various altitudes are given in Table 7.1.

EXERCISE If an object weighs 270 N at the earth's surface, what will it weigh at an altitude equal to twice the radius of the earth?

Answer 30 N.

TABLE 7.1 Acceleration Due to Gravity, g, at Various Altitudes

Altitude (km)[a]	g(m/s^2)
1000	7.33
2000	5.68
3000	4.53
4000	3.70
5000	3.08
6000	2.60
7000	2.23
8000	1.93
9000	1.69
10,000	1.49
50,000	0.13

[a] All values are distances above the earth's surface.

7.9 GRAVITATIONAL POTENTIAL ENERGY REVISITED

In Chapter 5 we introduced the concept of gravitational potential energy and found that the potential energy of an object could be calculated using the equation $PE = mgh$, where h is the height of the object above or below some reference level. This equation is actually valid only when the object is near the earth's surface. For objects high above the earth's surface, such as a satellite, an alternative expression must be used to compute the gravitational potential energy. The general expression for the gravitational potential energy for an object of mass m at a distance r *from the center of the earth* can be shown (using integral calculus) to be

Gravitational potential energy

$$PE = -G\frac{M_e m}{r} \tag{7.19}$$

where M_e is the mass of the earth.

This equation assumes that the zero level for potential energy is chosen to be at an infinite distance from the center of the earth. This point is a convenient one because the gravitational force goes to zero when r is set equal to infinity.

EXAMPLE 7.10 Does the Potential Energy Reduce to *mgh?*

(a) Find expressions for the gravitational potential energy of an object at the surface of the earth and for the same object when at a height h above the surface of the earth (see Fig. 7.16). (b) From the answers to part (a), show that the difference in potential energy between these two points reduces to the familiar expression $PE = mgh$ when h is small compared to the earth's radius.

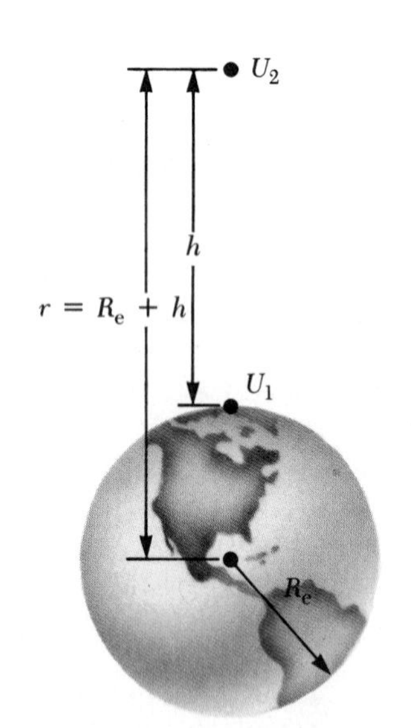

FIGURE 7.16 (Example 7.10)

Solution Equation 7.19 gives the potential energy at the surface of the earth as

$$PE_1 = -G\frac{M_e m}{R_e}$$

We can also use Equation 7.19 to find the potential energy of the object at the height h above the surface of the earth. Taking $r = R_e + h$, we find

$$PE_2 = -G\frac{M_e m}{(R_e + h)}$$

(b) The difference in potential energy between these two points is found as follows:

$$PE_2 - PE_1 = -G\frac{M_e m}{(R_e + h)} - \left(-G\frac{M_e m}{R_e}\right)$$

$$= -GM_e m\left[\frac{1}{(R_e + h)} - \frac{1}{R_e}\right]$$

After finding a common denominator and applying some algebra to the equation above, we find

$$PE_2 - PE_1 = \frac{GM_e mh}{R_e(R_e + h)}$$

When the height h is very small compared to R_e, the denominator in the expression above is approximately equal to R_e^2. Thus, we have

$$PE_2 - PE_1 \approx \frac{GM_e}{R_e^2} mh$$

Now note that the acceleration due to gravity at the surface of the earth is given by $g = GM_e/R_e^2$ (see Example 7.9). Thus,

$$PE_2 - PE_1 \approx mgh$$

*7.10 ESCAPE VELOCITY

Anyone who has ever watched a football game knows what happens to a football after the kickoff. It first goes up and then comes down, just as all common objects will do when projected in the air. The total time the ball remains in the air increases as its initial velocity increases. With this idea, one can imagine a situation in which the ball is kicked with such a large initial velocity that it would simply not come down. This particular velocity is called the escape velocity of an object from the earth. At this speed, any object will escape from the earth and soar forever in space until it is captured by the gravitational attraction of some other planet or celestial body. We shall now proceed to derive an expression which will enable us to calculate the escape velocity.

Suppose an object of mass m is projected vertically upward from the earth's surface with an initial speed v_i as in Figure 7.17. The initial mechanical energy (kinetic plus potential energy) is given by

$$KE_i + PE_i = \frac{1}{2}mv_i^2 - G\frac{M_e m}{R_e}$$

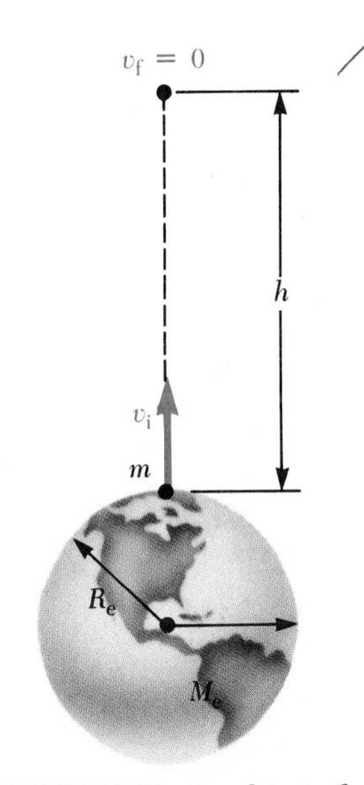

FIGURE 7.17 An object of mass m projected upward from the earth's surface with an initial speed v_i reaches a maximum altitude h (where $M_e >> m$).

We neglect air resistance and assume that the initial velocity is just large enough to allow the object to reach infinity with a speed of zero. We call this value of v_i the escape velocity v_e. When the object is at an infinite distance from the earth, its kinetic energy is zero, because $v_f = 0$, and the gravitational potential energy is also zero because our zero level of potential energy was selected at $r = \infty$. Hence, the total mechanical energy is zero, and the law of conservation of energy gives

$$\frac{1}{2}mv_e^2 - G\frac{M_e m}{R_e} = 0$$

and

$$\boxed{v_e = \sqrt{\frac{2GM_e}{R_e}}} \qquad (7.20)$$

Escape velocity

From this equation, one finds that the escape velocity for earth is about 11.2 km/s, which corresponds to about 25,000 mi/h (see Example 7.11). Note that this expression for v_e is independent of the mass of the object projected from the earth. For example, a spacecraft has the same escape velocity as a molecule. A list of escape velocities for the planets and the moon is given in Table 7.2. The data presented in this table helps in understanding why some planets have atmospheres while others do not. For example, on the very hot planet Mercury, gas molecules have an average speed which is greater than the escape velocity of 4.3 km/s. Consequently, any gases which might have been present on the surface of the planet at its formation have long since wandered off into space. Likewise, in the atmosphere of our own earth, hy-

TABLE 7.2 Escape Velocities for the Planets and the Moon

Planet	v_{esc} (km/s)
Mercury	4.3
Venus	10.3
Earth	11.2
Moon	2.3
Mars	5.0
Jupiter	60
Saturn	36
Uranus	22
Neptune	24

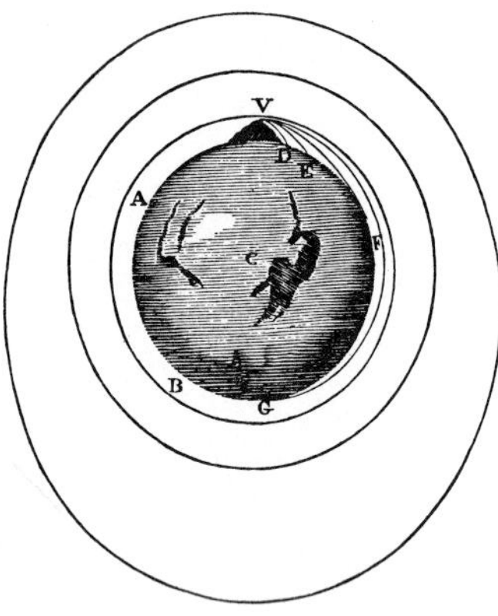

"The greater the velocity . . . with which [a stone] is projected, the farther it goes before it falls to the earth. We may therefore suppose the velocity to be so increased, that it would describe an arc of 1, 2, 5, 10, 100, 1000 miles before it arrived at the earth, till at last, exceeding the limits of the earth, it should pass into space without touching."—Newton, *System of the World.*

drogen and helium have average speeds greater than 11.2 km/s, the escape velocity on earth. Therefore, these gases are not retained in the earth's atmosphere. On the other hand, heavier gases such as oxygen and nitrogen in the earth's atmosphere have average speeds less than 11.2 km/s, and do not escape.

EXAMPLE 7.11 Escape Velocity of a Rocket
Calculate the escape velocity from the earth for a 5000-kg spacecraft.

Solution Using Equation 7.20 with $M_e = 5.98 \times 10^{24}$ kg and $R_e = 6.37 \times 10^6$ m gives

$$v_e = \sqrt{\frac{2GM_e}{R_e}} = \sqrt{\frac{2(6.67 \times 10^{-11})(5.98 \times 10^{24})}{6.37 \times 10^6}}$$
$$= 1.12 \times 10^4 \text{ m/s}$$

This corresponds to about 25,000 mi/h or about 7 mi/s. Note that the mass of the spacecraft was not required for this calculation.

*7.11 GRAVITY IN THE EXTREME, BLACK HOLES

The maternity wards of space are large clouds of gas and dust called nebulae. The mutual gravitational attraction between particles in the cloud pulls the particles together until eventually a huge collapsing spherical ball is formed. As the separation between the particles decreases, the gravitational force on each increases; the particles become part of a collapsing sphere, and fall together towards the center of the sphere at an increasing speed. Finally, their speed becomes great enough near the center of the collapsing sphere that nuclear reactions begin to occur. (We shall discuss these reactions more completely in Chapter 32.) One manifestation of these nuclear reactions is the continuous release of tremendous amounts of energy which prevents further gravitational collapse of the star. The end result is a star living out its active life in equilibrium between the inward pull of gravitational forces tending to collapse the star, and the outward radiation and thermal forces tending to expand the star.

During the first stages of nuclear burning at the center of a star, hydrogen nuclei are fused together to produce helium nuclei. When the supply of hydrogen is exhausted, the fusion process ceases. At this point, the balance of forces is destroyed since no energy is being released from fusion processes to prevent the collapse by gravitational forces. As the star collapses during its death struggle, the temperature inside the star increases further, and other nuclear reactions can occur involving elements heavier than hydrogen. The most spectacular deaths occur for stars which are originally very massive in nature. At some point in its flirtation with death, such a massive star may undergo a catastrophic explosion, hurling debris into space and leaving behind a central core. Such an exploding star is called a supernova. Supernovae are very rare events and the only ones that have been observed since the invention of the telescope in 1609 have been in distant galaxies. One exception to this is the one found in 1987 in the large Magellanic Cloud, an irregular galaxy close to our Milky Way galaxy. In 1054 A.D., Chinese observers recorded a "guest star" in the constellation Taurus, in our galaxy, which was so bright that it was visible in daylight. In this same region of space, we now find a nebula, called the Crab nebula, which is believed to be the remnants thrown off by the 1054 supernova.

As we mentioned earlier, when a star explodes as a supernova, most of the star's material is ejected, but a central core is left behind which continues to collapse. What will be the ultimate fate of the collapsing core? The answer to this intriguing question depends on the mass of the core. If the core has a mass less than about 1.4 times the mass of our sun, it will gradually cool down and end its life as a black dwarf star. However, if the core's mass is greater than about 1.4 solar masses, the star may collapse due to gravity such that the electrons and protons combine to form neutrons. Such a star, in which all the remaining matter is in the form of neutrons, is called a neutron star. The contraction of a neutron star eventually ceases when the neutron gas is compressed to a radius of about 10 km (about 6 miles). A teaspoonful of this material on earth would weigh about 5 billion tons!

An even more unusual death of a star may occur when the core has a mass greater than about 3 solar masses. No known forces in nature are strong enough to prevent the collapse of such a star. In fact, the collapse of the star may continue until the star becomes a mere point in space called a singularity. Such a singularity is commonly referred to as a **black hole.** *In effect, black holes are remains of stars that have collapsed under their own weight.*

Objects such as spaceships, atoms, and molecules can enter a black hole, but can never escape once in its grips. As soon as they enter a black hole, they are pulled deeper and deeper toward its center, and are trapped forever. In fact, even light cannot escape the tremendous gravitational forces set up inside a black hole; hence, the origin of the terminology "black hole."

Using the concept of escape velocity developed in Section 7.10, we can obtain some understanding of how particles become trapped by black holes. (A proper derivation of the results below requires some ideas and equations from relativity. These will be developed in Chapter 28.) Recall that the escape velocity from a spherical body of mass M and radius R is given by

$$v_e = \sqrt{\frac{2GM}{R}}$$

If the escape velocity exceeds the speed of light, $c = 3 \times 10^8$ m/s, light within the body will not be able to escape and the body will appear to be black. The critical radius for which this occurs is called the **Schwartzschild radius** after Karl Schwartzschild who predicted the existence of black holes in 1916 (see Fig. 7.18). Taking $v_e = c$ in the expression above, and solving for R, we find

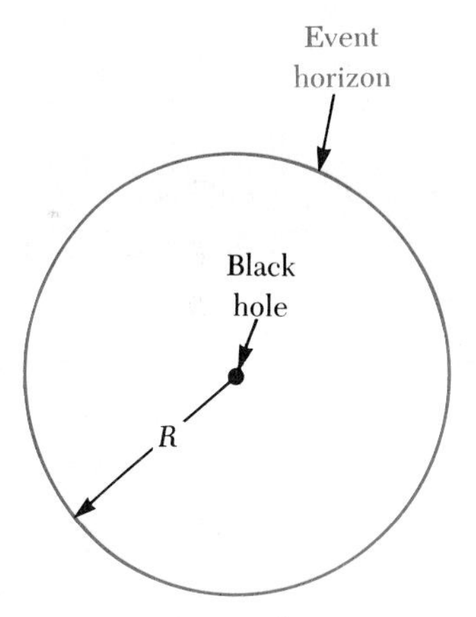

FIGURE 7.18 The event horizon surrounding a black hole. The distance R equals the Schwarzschild radius.

Schwartzschild radius

$$\boxed{R = \frac{2GM}{c^2}} \tag{7.21}$$

For example, the value of R for a black hole whose mass is equal to that of the sun is calculated to be 3.0 km (about 2 mi); a black hole whose mass equals that of the earth would have a radius of about 9 mm (about the size of a one-cent piece). The boundary of the region of radius R surrounding a black hole that permits entrance but prevents escape is called the **event horizon.** Any event occurring inside this horizon is invisible to an outside observer.

It is interesting to talk about exotic objects such as black holes, but do they really exist, and how does one go about detecting them? One way to find a black hole is to look for evidence of matter falling into it. As matter is pulled into a black hole, gravitational energy is converted into thermal energy, and the matter becomes hot enough to emit radiation in the form of x-rays. If these x-rays are emitted before the in-falling matter crosses the event horizon, they can be detected by astronomers. Although there are several objects in the sky that are suspected to contain black holes, the best known of these is Cygnus X-1, which is a strong source of x-rays in the constellation Cygnus. Astrophysicists believe that these objects consist of a black hole and a normal star orbiting about each other. X-rays are constantly emitted from the black hole as it "feeds on" the normal star. Although there is a great deal of observational evidence that black holes may exist in our Galaxy and other galaxies, the proof of their existence remains inconclusive.

Kepler's laws helped to explain the motion of all heavenly bodies. (Courtesy of NASA and the Jet Propulsion Laboratory)

*7.12 KEPLER'S LAWS

In this section we shall use Newton's law of universal gravity along with some of the ideas developed in this chapter to examine three important laws dealing with planetary motion and developed by the German astronomer Johannes Kepler (1571–1630). In order to place Kepler's three laws in their proper perspective, let us take a brief look at the history behind them.

The movements of the planets, stars, and other celestial bodies have been observed by people for thousands of years. In early history, scientists regarded the earth as the center of the universe. This *geocentric model* was developed extensively by the Greek astronomer Claudius Ptolemy in the second century A.D. and was accepted for the next 1400 years. In 1543, the Polish astronomer Nicolaus Copernicus (1473–1543) showed that the earth and the other planets revolved in circular orbits about the sun (the *heliocentric hypothesis*).

The Danish astronomer Tycho Brahe (1546–1601, pronounced Brah or BRAH-uh) made accurate astronomical measurements over a period of 20 years and provided the data on which our currently accepted model of the solar system was first established. It is interesting to note that these precise observations, made on the planets and 777 stars visible to the naked eye, were carried out with a large sextant and compass because the telescope had not yet been invented.

Kepler, who was Brahe's student, acquired Brahe's astronomical data and spent about 16 years trying to deduce a mathematical model for the motion of the planets. After many laborious calculations, he found that Brahe's precise data on the revolution of Mars about the sun provided the answer. Such data are difficult to sort out because the earth is also in motion about the sun. Kepler's analysis first showed that the concept of circular orbits about the sun had to be abandoned. He eventually discovered that the orbit of Mars could be accurately described by an ellipse with the sun at one focus. He then generalized this analysis to include the motion of all planets. The complete analysis is summarized in three statements known as **Kepler's laws.** These laws applied to the solar system are:

1. All planets move in elliptical orbits with the sun at one of the focal points.
2. A line drawn from the sun to any planet sweeps out equal areas in equal time intervals.
3. The square of the orbital period of any planet is proportional to the cube of the average distance from the planet to the sun.

About 100 years after Kepler's analysis, Newton demonstrated that these laws were the consequence of the gravitational force existing between any two masses. (See color plates 3–6 for good photographs of Earth, Saturn, and Jupiter.)

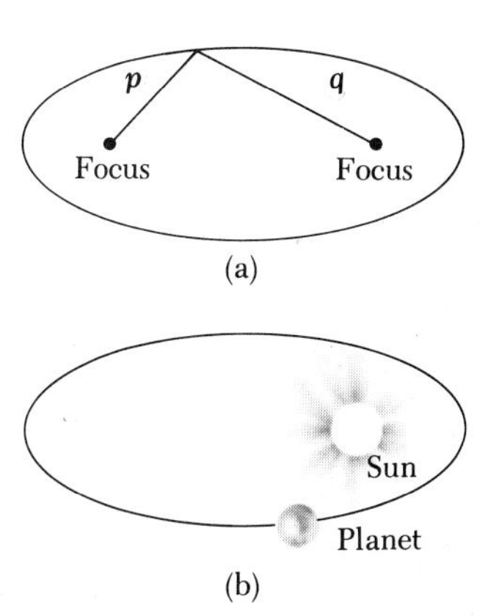

FIGURE 7.19 (a) The sum $p + q$ is the same for every point on the ellipse. (b) In the solar system, the sun is located at one focus of the elliptical orbit of each planet and the other focus is empty.

Kepler's First Law

We shall not attempt to derive Kepler's first law here. Instead, we simply state that it can be shown that the first law arises as a natural consequence of the inverse-square nature of Newton's law of gravitation. That is, any object bound to another by a force that varies as $1/r^2$ will move in an elliptical orbit. As shown in Figure 7.19a, an ellipse is a curve drawn such that the sum of the distances from any point on the curve to two internal points called focal points or foci (singular, focus) is always the same. For the sun-planet configuration, shown in Figure 7.19b, the sun is at one focus and the other focus is empty. Because the orbit is an ellipse, the distance from the sun to the planet continuously changes.

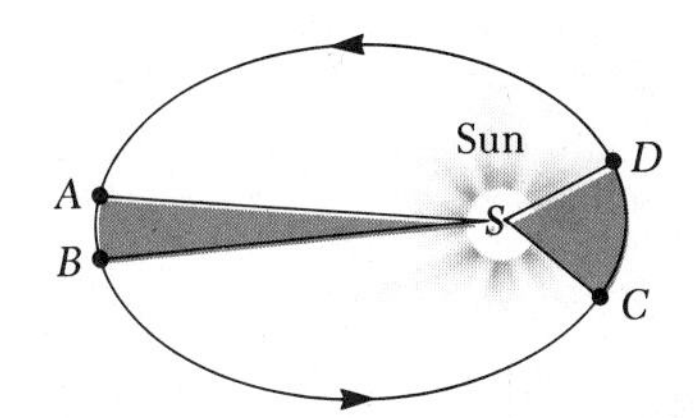

FIGURE 7.20 The two areas swept out by the planet in its elliptical orbit about the sun are equal if the time interval between points A and B is equal to the time interval between points C and D.

Kepler's Second Law

Kepler's second law states that a line drawn from the sun to any planet sweeps out equal areas in equal time intervals. Consider a planet in an elliptical orbit about the sun, as in Figure 7.20. Imagine that at some particular instant of time, we draw a line from the sun to a planet, tracing out the line AS where S represents the position of the sun. Exactly 30 days later, we repeat the process and draw the line BS. The area swept out is shown in color in the diagram. When the planet is at point C, we begin the process anew, drawing the line CS from the planet to the sun. We now wait 30 days, exactly the same time interval we waited before, and draw the line DS. The area swept out in this interval is also shown in color in the figure. These two areas are equal.

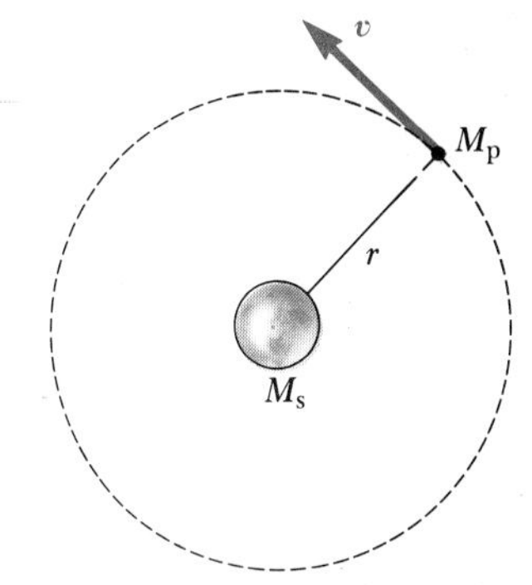

FIGURE 7.21 A planet of mass M_p moving in a circular orbit about the sun. The orbits of all planets except Mercury and Pluto are nearly circular.

Kepler's Third Law

Because of the simplicity of the derivation, we shall derive Kepler's third law here. Consider a planet of mass M_p moving about the sun of mass M_s in a circular orbit, as in Figure 7.21. (The assumption of a circular orbit rather than an elliptical orbit will not introduce serious error into our approach because the orbits of all planets except Mercury and Pluto are very close to being circular.) Because the gravitational force on the planet is equal to the centripetal force needed to keep it moving in a circle,

$$\frac{GM_sM_p}{r^2} = \frac{M_pv^2}{r}$$

TABLE 7.3 Useful Planetary Data

Body	Mass (kg)	Mean Radius (m)	Period (s)	Mean Distance From Sun (m)	$\frac{T^2}{r^3}\left[10^{-19}\left(\frac{s^2}{m^3}\right)\right]$
Mercury	3.18×10^{23}	2.43×10^{6}	7.60×10^{6}	5.79×10^{10}	2.97
Venus	4.88×10^{24}	6.06×10^{6}	1.94×10^{7}	1.08×10^{11}	2.99
Earth	5.98×10^{24}	6.37×10^{6}	3.156×10^{7}	1.496×10^{11}	2.97
Mars	6.42×10^{23}	3.37×10^{6}	5.94×10^{7}	2.28×10^{11}	2.98
Jupiter	1.90×10^{27}	6.99×10^{7}	3.74×10^{8}	7.78×10^{11}	2.97
Saturn	5.68×10^{26}	5.85×10^{7}	9.35×10^{8}	1.43×10^{12}	2.99
Uranus	8.68×10^{25}	2.33×10^{7}	2.64×10^{9}	2.87×10^{12}	2.95
Neptune	1.03×10^{26}	2.21×10^{7}	5.22×10^{9}	4.50×10^{12}	2.99
Pluto	$\approx 1 \times 10^{23}$	$\approx 3 \times 10^{6}$	7.82×10^{9}	5.91×10^{12}	2.96
Moon	7.36×10^{22}	1.74×10^{6}	—	—	—
Sun	1.991×10^{30}	6.96×10^{8}	—	—	—

For a more complete set of data, see, for example, *Handbook of Chemistry and Physics*, Cleveland, Chemical Rubber Publishing Co.

The speed v of the planet in its orbit is equal to the circumference of the orbit divided by the time for one revolution, T, called the **period** of the planet. That is, $v = 2\pi r/T$ and the above expression becomes

$$\frac{GM_s}{r^2} = \frac{(2\pi r/T)^2}{r}$$

Kepler's third law

$$T^2 = \left(\frac{4\pi^2}{GM_s}\right) r^3 = K_s r^3 \tag{7.22}$$

where K_s is a constant given by

$$K_s = \frac{4\pi^2}{GM_s} = 2.97 \times 10^{-19}\ \text{s}^2/\text{m}^3$$

This is Kepler's third law. It is valid for elliptical orbits if we replace r by a length equal to half the long axis of the ellipse. Note that K_s is independent of the mass of the planet. Therefore Equation 7.22 is valid for any planet. If we consider the orbit of a satellite about the earth, such as the moon, then the constant has a different value, with the mass of the sun replaced by the mass of the earth. In this case, K_e would equal $4\pi^2/GM_e$.

A collection of useful planetary data is given in Table 7.3. The last column of this table verifies that T^2/r^3 is a constant given by $K_s = 4\pi^2/GM_s = 2.97 \times 10^{-19}\ \text{s}^2/\text{m}^3$.

EXAMPLE 7.12 An Earth Satellite

A satellite of mass m moves in a circular orbit about the earth with a constant speed v and a height $h = 1000$ km above the earth's surface, as in Figure 7.22. (For clarity, this figure is not drawn to scale.) Find the orbital speed of the satellite. The radius of the earth is 6.38×10^6 m, and its mass is 5.98×10^{24} kg.

Solution The only external force on the satellite is that of the gravitational attraction exerted by the earth. This force is directed toward the center of the satellite's circular path and is the centripetal force acting on the satellite. Since the force of gravity is GM_em/r^2, we find

$$F_c = G\frac{M_e m}{r^2} = m\frac{v^2}{r}$$

$$v^2 = \frac{GM_e}{r}$$

The r term is the earth's radius plus the height of the satellite, that is, $r = R_e + h = 7.38 \times 10^6$ m, so that

$$v^2 = \frac{(6.67 \times 10^{-11}\ \text{N} \cdot \text{m}^2/\text{kg}^2)(5.98 \times 10^{24}\ \text{kg})}{7.38 \times 10^6\ \text{m}} = 5.40 \times 10^7\ \text{m}^2/\text{s}^2$$

Therefore,

$$v = 7.35 \times 10^3\ \text{m/s} \approx 16{,}400\ \text{mi/h}$$

Note that v *is independent of the mass of the satellite!*

EXERCISE Calculate the period of revolution, T, of the satellite (where T is the time for one revolution about the earth).

Answer 105 min.

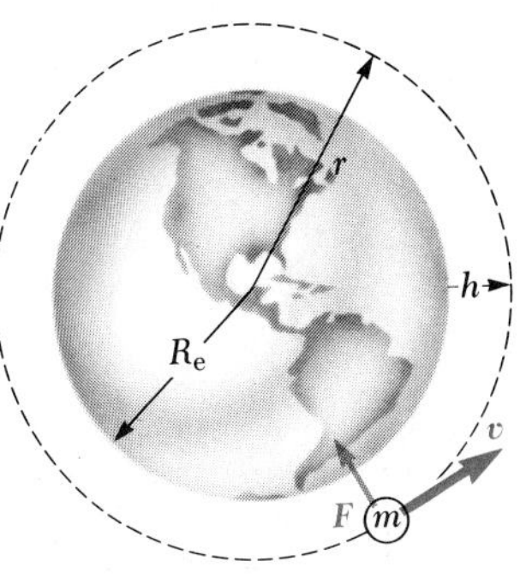

FIGURE 7.22 (Example 7.13) A satellite of mass m moving in a circular orbit of radius r and with constant speed v around the earth. The centripetal force is provided by the gravitational force acting on the satellite (not drawn to scale).

SUMMARY

The **average angular velocity,** $\bar{\omega}$, of a rigid object is defined as the ratio of the angular displacement $\Delta\theta$, to the time interval Δt:

$$\bar{\omega} \equiv \frac{\theta_2 - \theta_1}{t_2 - t_1} = \frac{\Delta\theta}{\Delta t} \qquad (7.3)$$

where $\bar{\omega}$ is in radians per second (rad/s).

The **average angular acceleration,** $\bar{\alpha}$, of a rotating object is defined as the ratio of the change in angular velocity, $\Delta\omega$, to the time interval Δt:

$$\bar{\alpha} \equiv \frac{\omega_2 - \omega_1}{t_2 - t_1} = \frac{\Delta\omega}{\Delta t} \qquad (7.4)$$

where $\bar{\alpha}$ is in radians per second per second (rad/s^2).

If an object undergoes rotational motion about a fixed axis under constant angular acceleration α, one can describe its motion by the following set of equations:

$$\omega = \omega_0 + \alpha t \qquad (7.5)$$

$$\theta = \omega_0 t + \frac{1}{2}\alpha t^2 \qquad (7.6)$$

$$\omega^2 = \omega_0^2 + 2\alpha\theta \qquad (7.7)$$

When an object rotates about a fixed axis, the angular velocity and angular acceleration are related to the tangential velocity and tangential acceleration through the relationships

$$v_t = r\omega \tag{7.8}$$

$$a_t = r\alpha \tag{7.9}$$

Any object moving in a circular path has an acceleration directed toward the center of the circular path. This acceleration is called a **centripetal acceleration** and its magnitude is given by

$$a_c = \frac{v^2}{r} = r\omega^2 \qquad \text{(7.11) and (7.14)}$$

Any object moving in a circular path must have a net force exerted on it that is directed toward the center of the circular path. This net force is called a **centripetal force.** From Newton's second law, the centripetal force and centripetal acceleration are related as

$$F_c = m\frac{v^2}{r} \tag{7.16}$$

Some examples of centripetal forces are the force of gravity (as in the motion of a satellite), and the force of tension in a string.

Newton's law of gravity states that every particle in the universe attracts every other particle with a force that is directly proportional to the product of their masses and inversely proportional to the square of the distance r between them:

$$F = G\frac{m_1 m_2}{r^2} \tag{7.17}$$

where $G = 6.673 \times 10^{-11}\ \text{N} \cdot \text{m}^2/\text{kg}^2$ is the **gravitational constant.**

The general expression for the gravitational potential energy for an object of mass m at a distance r from the center of the earth of mass M_e is given by

$$PE = -G\frac{M_e m}{r} \tag{7.19}$$

where the potential energy is chosen to be zero at $r = \infty$.

The escape velocity for any object at the earth's surface is independent of the mass of the object and is given by

$$v_e = \sqrt{\frac{2GM_e}{R_e}} \tag{7.20}$$

where R_e is the radius of the earth. This expression can also be used to calculate the escape velocity for any planet of mass M and radius R.

Kepler's laws of planetary motion state that

1. All planets move in elliptical orbits with the sun at one of the focal points.

2. A line drawn from the sun to any planet sweeps out equal areas in equal time intervals.
3. The square of the orbital period of a planet is proportional to the cube of the mean distance from the planet to the sun:

$$T^2 = \left(\frac{4\pi^2}{GM_s}\right)r^3 \tag{7.22}$$

ADDITIONAL READING

J. Beams, "Ultra-High-Speed Rotation," *Sci. American,* April 1961, p. 134.

I. B. Cohen, "Newton's Discovery of Gravity," *Sci. American,* March 1981, p. 166.

R. Feynman, *The Character of Physical Law,* Cambridge, Mass., MIT Press, 1965, chaps. 1, 2, and 3.

G. Gamow, *Gravity,* Science Study Series, Garden City, N.Y., Doubleday, 1962.

D. Layzer, *Constructing the Universe,* Scientific American Library, New York, N.Y., W. H. Freeman and Co., 1984.

F. Pipkin, "Gravity Up in the Air," *The Sciences,* July/August 1984, p. 24.

R. Smoluchowski, *The Solar System,* Scientific American Library, New York, N.Y., W. H. Freeman and Co., 1983.

G. Spetz, "Detection of Gravity Waves," *The Physics Teacher,* May 1984, p. 282.

C. Wilson, "How Did Kepler Discover His First Two Laws?" *Sci. American,* March 1972, p. 92.

QUESTIONS

1. When a wheel of radius R rotates about a fixed axis, do all points on the wheel have the same angular velocity? Do they all have the same linear velocity? If the angular velocity is constant and equal to ω_0, describe the linear velocities and linear accelerations of the points at $r = 0$, $r = R/2$, and $r = R$.
2. Why is it that an astronaut in a space capsule orbiting the earth experiences a feeling of weightlessness?
3. If a car's wheels are replaced with wheels of larger diameter, will the reading of the speedometer change? Explain.
4. A pail of water can be whirled in a vertical path such that none is spilled. Why does the water remain in the pail, even when the pail is above your head?
5. Imagine that you attach a heavy object to one end of a spring and then whirl the spring and object in a horizontal circle (by holding the other end of the spring). Does the spring stretch? If so, why? Discuss in terms of centripetal force.
6. It has been suggested that rotating cylinders about ten miles in length and five miles in diameter be placed in space for colonies. The purpose of the rotation is to simulate gravity for the inhabitants. Explain this concept for producing an effective gravity.
7. Why does a pilot tend to black out when he pulls out of a steep dive?
8. Cite an example of a situation in which an automobile driver can have a centripetal acceleration but no tangential acceleration.
9. Is it possible for a car to move in a circular path in such a way that it has a tangential acceleration but no centripetal acceleration?
10. Cite an example in which you could use Equation 7.6 to describe the motion of a record accelerating on a turntable and a situation in which this equation would not be valid.
11. Centrifuges are often used in dairies to separate the cream from the milk. Which remains on the bottom?
12. Use Kepler's second law to convince yourself that the earth must move faster in its orbit during the winter when it is closest to the sun than it does during the summer when it is at its greatest distance from the sun.
13. We often think of the brakes and the gas pedal on a car as the instruments which accelerate and decelerate the car. Could a steering wheel also fall into this category? Explain.
14. Because of the earth's rotation about its axis, you will weigh slightly less at the equator than at the poles. Why is this?

PROBLEMS

Section 7.1 Angular Velocity

Section 7.2 Angular Acceleration

1. Convert the following angles in degrees to radians: 30°, 45°, 60°, 90°, 180°, 270°, and 360°.

2. Convert the following angles in radians to degrees: $\pi/3$, 1.2π, 3π.

3. Find the angular velocity of the earth about the sun in units of rad/s and degrees/day.

4. Find the angular velocity of the earth about its axis in rad/s.

5. A record has an angular speed of 33 rev/min. (a) What is its angular speed in radians per second? (b) Through what angle in radians does it rotate in 1.5 s?

6. A potter's wheel moves from rest to an angular speed of 0.2 rev/s in a time of 30 s. Find its angular acceleration in rad/s^2.

Section 7.3 Rotational Motion Under Constant Angular Acceleration

7. The turntable of a record player rotates at $33\frac{1}{3}$ rev/min and takes 20 s to reach this speed from rest. Calculate (a) its angular acceleration and (b) the number of revolutions it makes before reaching its final speed.

8. A tire placed on a machine to balance it in a service station starts from rest and turns through 4.7 revolutions in 1.2 s before reaching its final angular speed. Calculate its angular acceleration.

9. A centrifuge in a medical laboratory is rotating at an angular speed of 3600 rev/min. When switched off, it rotates 50 times before coming to rest. Find the constant angular deceleration of the centrifuge.

10. A machine part is rotating at an angular velocity of 0.6 rad/s, and this speed is to be increased to 2.2 rad/s at an angular acceleration of 0.7 rad/s^2. Find the angle through which the part turns before reaching this final speed.

Section 7.4 Relations Between Angular and Linear Quantities

Section 7.5 Centripetal Acceleration

11. A racecar travels in a circular track of radius 200 m. If the car moves with a constant speed of 80 m/s, find (a) its angular velocity, and (b) its tangential acceleration.

12. The racecar of Problem 11 increases its speed at a constant linear acceleration from 80 m/s to 95 m/s in 10 s. (a) Find the constant angular acceleration, and (b) the angle the car moves through in this time.

13. Find the centripetal acceleration of (a) a point on the equator of the earth and (b) a point at the north pole of the earth.

14. It has been suggested that rotating cylinders about 10 mi in length and 5 mi in diameter be placed in space and used as colonies. What angular velocity must such a cylinder have so that the centripetal acceleration at the surface of the cylinder would equal earth's gravity?

• **15.** (a) What is the tangential acceleration of a bug on the rim of a 78 rpm record 10 in. in diameter if the record moves from rest to its final angular velocity in 3 s? (b) When the record is at its final speed, what is the tangential velocity of the bug? (c) One second after starting from rest, what is the bug's tangential acceleration, the bug's radial acceleration, and the bug's total acceleration?

• **16.** A wheel 60 cm in diameter rotates with a constant angular acceleration of 4 rad/s^2. The wheel starts at rest at $t = 0$, and a chalk line drawn to a point P on the rim of the wheel makes an angle of 57.3° with the horizontal at this time. At $t = 2$ s, find (a) the angular speed of the wheel, (b) the linear velocity and tangential acceleration of P, and (c) the position of P.

Section 7.6 Centripetal Force

17. A car of mass 2000 kg rounds a circular turn of radius 20 m. If the road is flat and the coefficient of friction is 0.7 between tires and road, how fast can the car go without skidding?

18. A rotating wheel at an amusement park shown in Figure 7.23 is 90 ft in diameter and makes one revolution in 6 s. Each of the 300 kg cars are held in place by two bolts. Find the force each bolt exerts on the car when the wheel is rotating in a horizontal plane.

FIGURE 7.23 Problem 18 (Courtesy Kings Island)

19. An airplane is flying in a horizontal circle at a speed of 100 m/s. The 80 kg pilot does not want his radial acceleration to exceed 7*g*. (a) What is the minimum radius of the circular path? (b) At this radius, what is the *net* centripetal force exerted on the pilot by the seat belts, friction between him and the seat, and so forth?

• **20.** A child stands at the rim of a rotating merry-go-round of radius 2 m. The child has a mass of 50 kg, and the carousel turns with an angular velocity of 3 rad/s. (a) What is the child's centripetal acceleration? (b) What is the minimum force required between her feet and the floor of the carousel to keep her in the circular path? (c) What minimum coefficient of static friction is required? Is the answer found reasonable? In other words, is she likely to be able to stay on the merry-go-round?

• **21.** In a swinging mishap, Tarzan, whose mass is 80 kg, finds himself circling around a tree in a horizontal circle at the end of a vine 3 m long which makes an angle of 5° with the vertical. Find (a) the centripetal force exerted on him by the vine, and (b) his centripetal acceleration.

•• **22.** An engineer wishes to design a curved exit ramp for a tollroad in such a way that a car will not have to rely on friction to round the curve without skidding. He does so by banking the road in such a way that the centripetal force needed will be supplied by the component of the normal force toward the center of the circular path. (a) Show that for a given speed v and radius r, the curve must be banked at an angle given by $\tan\theta = v^2/rg$. (b) Find the angle at which the curve should be banked if a typical car rounds at a 50 m radius and at a speed of 13.4 m/s.

•• **23.** A pail of water is rotated in a vertical circle of radius 1 m (the approximate length of a person's arm). What must be the minimum speed of the pail at the top of the circle if no water is to spill out?

•• **24.** A stuntman swings from the end of a 4 m long rope along the arc of a vertical circle. If his mass is 70 kg, find the tension in the rope required to make him follow his circular path, assuming he starts from rest when the rope is horizontal, at (a) the beginning of his motion, (b) at a height of 1.5 m above the bottom of the circular arc, and (c) at the bottom of his arc.

Section 7.8 Newton's Law of Gravitation

Section 7.12 Kepler's Laws

25. Use the data in Table 7.3 to find the gravitational force that the sun exerts on the earth.

26. A 30,000 kg spaceship is halfway between the earth and the moon. Use the data in Table 7.3 to find the net gravitational force exerted on the ship by the earth and the moon.

27. Two students sitting in adjacent seats in a lecture room have weights of 600 N and 700 N. Assume that Newton's law of gravitation can be applied to these students and find the gravitational force that one student exerts on the other when they are separated by 0.5 m.

28. A coordinate system is constructed on the surface of a pool table and three masses are placed on the coordinate system as follows: a 2-kg mass is at the origin, a 3-kg mass is at (0, 2), and a 4-kg mass is at (4, 0) where the coordinates are in meters. Find the resultant gravitational force exerted on the mass at the origin by the other two masses.

29. Use the data in Table 7.3 and the approach of Example 7.9 to find the acceleration due to gravity at the surface of Venus.

30. Communication satellites are placed at an altitude of about 25,000 mi above the surface of the earth. Find the acceleration due to gravity at this altitude.

31. Use the data of Table 7.3 to verify Kepler's third law for Venus, Mars, and Neptune.

• **32.** A satellite of mass 600 kg is in a circular orbit about the earth at a height above the earth equal to the earth's mean radius. Find (a) the satellite's orbital speed, (b) the period of its revolution, and (c) the gravitational force acting on it.

• **33.** Use the data of Table 7.3 to find the point between the earth and the sun that an object could be placed such that the net gravitational force exerted on it by these two objects is zero.

Section 7.9 Gravitational Potential Energy Revisited

Section 7.10 Escape Velocity

Section 7.11 Gravity in the Extreme, Black Holes

34. Sketch a graph of the gravitational potential energy, *PE*, versus *r* for a particle above the earth's surface. Your graph should extend from the earth's surface to infinity.

35. A satellite of the earth has a mass of 100 kg and is at an altitude of 2×10^6 m. (a) What is the potential energy of the satellite at this location? (b) What is the magnitude of the gravitational force on the satellite?

36. Use the data of Table 7.3 to find the escape velocity from (a) the moon, (b) Mercury, and (c) Jupiter.

37. (a) Use the approach of Example 7.9 and the data in Table 7.3 to find the acceleration due to gravity at the surface of a neutron star of mass 1.5 solar masses having a radius of 10 km. (b) Find the weight of a 0.12-kg baseball on this star. (c) Assume the equation $PE = mgh$ applies and calculate the energy that a 70-kg person would expend climbing a 1-cm tall mountain on this star.

38. Find the Schwartzschild radius (a) for a star having the mass of our sun; (b) for a star having a mass of 4 solar masses; (c) for a star whose mass is that of the earth.

ADDITIONAL PROBLEMS

39. An athlete swings a ball of mass 5 kg horizontally on the end of a rope. The ball moves in a circle of radius 0.8 m at an angular speed of 0.5 rev/s. What is (a) the tangential velocity of the ball and (b) its centripetal acceleration? (c) If the maximum tension the rope can withstand before breaking is 100 N, what is the maximum tangential velocity the ball can have?

40. A satellite of Mars has a period of 459 min. The mass of Mars is 6.42×10^{23} kg. From this information determine the radius of the satellite's orbit.

41. The Solar Maximum Mission Satellite shown in Figure 7.24 was placed at an orbit of about 150 mi above the earth. Determine (a) the orbital speed of the satellite, and (b) the time for one complete revolution.

FIGURE 7.24 Problem 41 (Courtesy NASA)

42. The angular velocity of a rotating bicycle wheel is 3 rad/s at some instant of time. It then is given an angular acceleration of 1.5 rad/s^2. A chalk line drawn on the wheel is horizontal at $t = 0$. (a) What angle does this line make with its original direction at $t = 2$ s? (b) What is the angular velocity of the wheel at $t = 2$ s?

43. A floppy disk drive on a computer has a period of 200.2 ms. (a) What is its angular speed? (b) If the floppy disk has a diameter of 5.25 in., what is the linear speed of a point on its rim?

• **44.** A high-speed sander has a disk 6 cm in radius which rotates at a constant rate of 1200 rev/min about its axis. Determine (a) the angular speed of the disk in rad/s, (b) the linear speed of a point 2 cm from its center, (c) the centripetal acceleration of a point on the rim, and (d) the total distance a point on the rim moves in 2 s.

• **45.** Three masses are aligned along the x axis of a rectangular coordinate system such that a 2-kg mass is at the origin, a 3-kg mass is at (2, 0), and a 4-kg mass is at (4, 0) m. Find (a) the gravitational force exerted on the 4-kg mass by the other two masses, and (b) the magnitude and direction of the gravitational force exerted on the 3-kg mass by the other two.

• **46.** The downward motion of an elevator is controlled by a cable that unwinds from a cylinder of radius 0.2 m. What is the angular velocity of the cylinder when the downward speed of the elevator is 1.2 m/s?

• **47.** A highway curve has a radius of 150 m and is designed for a traffic speed of 40 mi/h (17.9 m/s). If the curve is not banked, determine the minimum coefficient of friction between car and road needed to keep cars from skidding.

• **48.** A 4-kg mass is attached to a horizontal rod by two strings as in Figure 7.25. The strings are under tension when the rod rotates about its axis. If the speed of the mass is 4 m/s when observed at the following positions, find the tension in the string when the mass is (a) at its lowest point, (b) in the horizontal position.

• **49.** Suppose the rod in the system shown in Figure 7.25 is made vertical and rotates about this axis. If the mass rotates at a constant speed of 6 m/s in a horizontal plane, determine the tensions in the upper and lower strings.

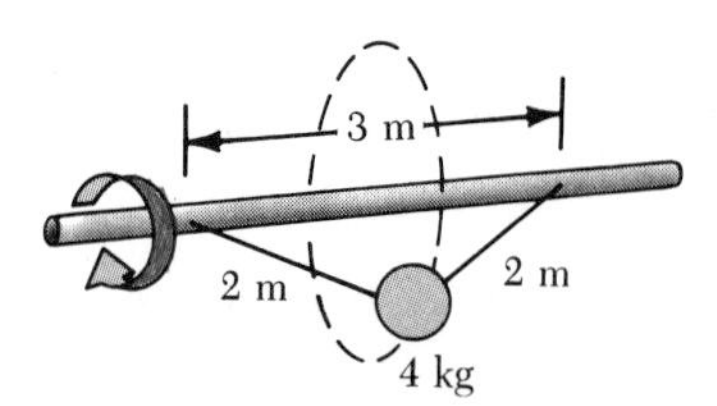

FIGURE 7.25 Problem 48

• **50.** Because of the earth's rotation about its axis, a point on the equator experiences a centripetal acceleration of 0.034 m/s^2, while a point at the poles experiences no centripetal acceleration. (a) Show that at the equator the gravitational force on an object (the true weight) must exceed the object's apparent weight. (b) What is the apparent weight at the equator and at the poles of a person having a mass of 75 kg? (Assume the earth is a uniform sphere and take $g = 9.800$ m/s^2.)

• **51.** A satellite of mass 600 kg is in a circular orbit about the earth at a height above the earth equal to the earth's mean radius. Find (a) the satellite's orbital speed, (b) the period of its revolution, and (c) the gravitational force acting on it.

•• **52.** At what angle should the curve of Problem 47 be banked if friction is not to be a factor in causing the car to follow its circular path?

•• **53.** A car moves across a bridge made in the shape of a circular arc of radius 30 m at a speed of 10 m/s. (a) Find the normal force acting on the car when it is at the top of the arc. (b) At what speed will the normal force go to zero? (Hint: The normal force becomes zero when the car loses contact with the road.)

•• **54.** A roller-coaster vehicle has a mass of 500 kg when fully loaded with passengers (Fig 7.26). (a) If the vehicle has a speed of 20 m/s at point A, what is the force of the track on the vehicle at this point? (b) What is the maximum speed the vehicle can have at B in order to remain on the track?

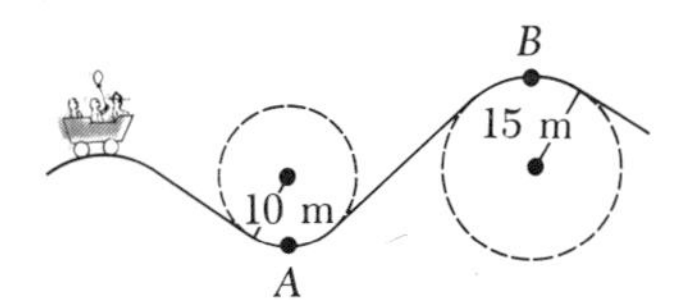

FIGURE 7.26 Problem 54

•• **55.** A skier starts at rest from the top of a large hill that is shaped like a hemisphere (Fig 7.27). Neglecting friction, show that the skier will leave the hill and become "air-borne" at a distance $h = R/3$ below the top of the hill. (*Hint:* At this point, the normal force goes to zero.)

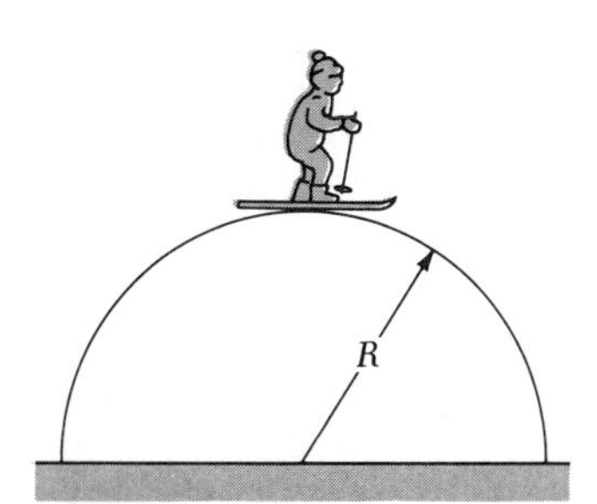

FIGURE 7.27 Problem 55

•• **56.** A frictionless roller coaster is given an initial velocity v_0 at a height h, as in Figure 7.28. The radius of curvature of the track at point A is R. (a) Find the maximum value of v_0 necessary in order that the roller coaster not leave the track at A. (b) Using the value of v_0 calculated in (a), determine the value of h' necessary if the roller coaster is to just make it to point B.

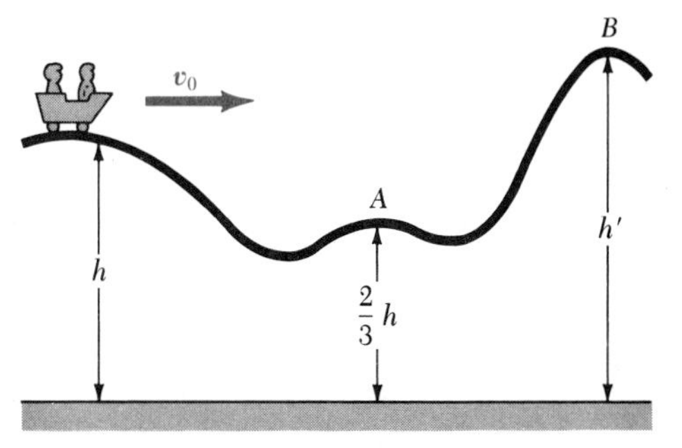

FIGURE 7.28 Problem 56

•• **57.** A car rounds a banked curve. The radius of curvature of the road is R, the banking angle is θ, and the coefficient of static friction is μ. (a) Determine the range of speeds the car can have without slipping up or down the road. (b) What is the range of speeds possible if $R = 100$ m, $\theta = 10°$, and $\mu = 0.1$ (slippery conditions).

•• **58.** In a popular amusement park ride, a rotating cylinder of radius 3 m is set into rotation at an angular velocity of 5 rad/s, as in Figure 7.29. The floor then drops away, leaving the riders suspended against the wall in a vertical position. What minimum coefficient of friction between a rider's clothing and the wall is needed to keep the rider from slipping? (*Hint:* Recall that the maximum force of static friction is equal to μn, where n is the normal force, which in this case is the centripetal force.)

FIGURE 7.29 Problem 58

Essay

The Big Bang

Sidney C. Wolff *National Optical Astronomy Observatory*

Newton's universal law of gravitation—today it is taught in every introductory physics course and is so familiar that it is easy to forget how profoundly this simple equation changed the nature of science. Newton not only offered a precise mathematical description of the motions of material objects, he also showed that this description applied far beyond the boundaries of the Earth. The same force that causes apples to fall to the ground also keeps the Moon revolving about the Earth and holds the planets in their orbits around the Sun. The heavens were transformed; no longer mysterious and unknowable, the Moon, the planets, and potentially even the stars, all subject to the same physical laws that are valid on Earth, became legitimate objects of scientific inquiry.

But is Newton's law of gravitation truly universal? Does it apply to the stars? Newton assumed that it did, but measurements available during his lifetime were not precise enough to offer proof. The influence of gravity is detected by analyzing the motions of objects, and apart from their rising and setting as a consequence of the Earth's rotation, stars did not appear to the astronomers of Newton's time to have motions of their own. The fact that there are clusters or groupings of stars, like the Pleiades and Hyades, suggested that attractive forces do influence stars, but it was meticulous measurements by Sir William Herschel, best known for his discovery of the planet Uranus, that offered the first concrete evidence. Figure 1 is a photograph of the spiral galaxy M74, which contains billions of individual stars.

FIGURE 1 The spiral galaxy M74, which is about 30 million lightyears distant from the Earth and has a diameter of about 80,000 lightyears. A galaxy like M74 contains billions of individual stars. (Kitt Peak National Observatory/National Optical Astronomy Observatories)

In January 1782, Herschel published a catalog of 269 pairs of stars that appear to be very close together in the sky. At that time, he did not know whether these double stars form physical systems or simply happen to appear in very nearly the same direction but at very different distances from the Earth. About 20 years later, Herschel repeated the observations of the double stars that he had found, and discovered that some pairs had changed position relative to each other in a way that could only be explained if they were revolving around each other. Several more decades of measurements by other astronomers proved that these stars do indeed follow orbits that can be described by the inverse square force of Newton's law of gravitation.

The conclusion that the motions of stars are controlled by gravity raises a problem that was recognized by Newton. Up until the beginning of the 20th century it was assumed that the universe is static, that is that it is neither expanding nor contracting. If the theory of gravitation applies to all of the objects in a static universe, and if the universe is finite in size, then it is difficult to see why the universe has not already collapsed to a single mass as all of the objects are attracted to one another. Newton thought that an infinite universe could perhaps be stable, since gravitational forces of all the stars spread uniformly throughout space might possibly cancel their mutual gravitational attractions.

The next major advance in gravitational theory is embodied in Einstein's general theory of relativity. In 1917, Einstein applied his new theory to the universe as a whole and found that there were no solutions to his equations that gave a homogeneous static universe, even if the universe was assumed to be infinite. Einstein consequently introduced a new term, called the cosmological constant, which represents a repulsion that can balance gravitational attraction over large distances and permit a static universe. There is, however, no evidence for such a repulsion in nature; the cosmological constant was introduced in order to make general relativity conform to the preconceived notion that the universe was static.

Subsequent theoretical work by the Russian mathematician Aleksandr Friedmann showed that general relativity is compatible with an infinite, homogeneous expanding universe for all values of the cosmological constant, including zero. At about the same time, observations proved that the universe is not static but is expanding. American physicist George Gamow reports that Einstein, upon learning that his relativity theory is compatible with an expanding universe and that the expansion had been directly observed, said that the introduction of the cosmological constant was "the biggest blunder of my life."

The universe is expanding. This fundamental observation underlies all of modern cosmological thought. In a classic paper published in 1931, Edwin Hubble and Milton Humason compared distances and velocities of remote galaxies, which are giant, gravitationally bound systems of billions of individual stars. They established that these galaxies are moving away from us with velocities that are proportional to their distances from us. This result, which has been substantiated and extended to still more distant galaxies by subsequent work, demonstrates that the universe is expanding.

This concept can be easily understood through use of an analogy. Suppose a cook is making raisin bread and suppose that after yeast has been mixed into the dough and the bread is set aside to rise, it doubles in size during the next hour, as in Figure 2. All of the raisins move further apart, and in fact the distance between each pair of raisins also doubles. Since each distance doubles during the hour, if one raisin is selected as the origin of a coordinate system, every other raisin must move away from the one selected at a speed proportional to its distance. The same is true, of course, no matter which raisin you select.

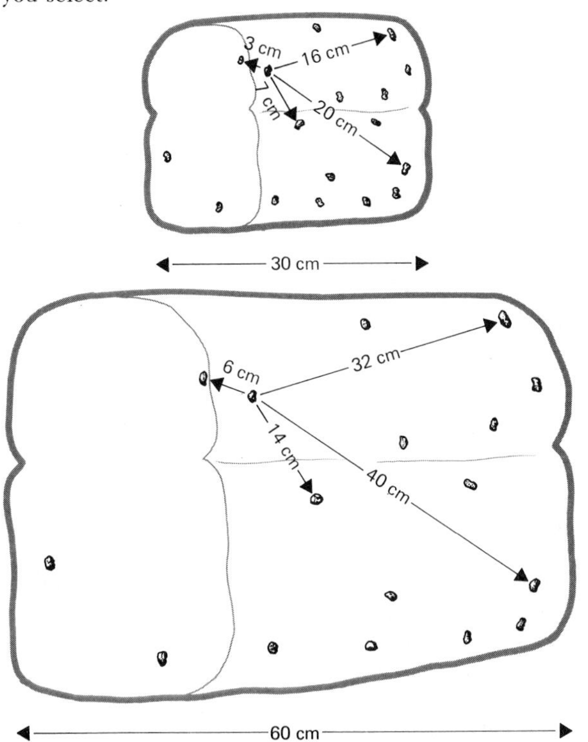

FIGURE 2 Expanding raisin bread. (From Abell, George O., David Morrison, and Sydney C. Wolff, *Exploration of the Universe*, 5th ed., Philadelphia, Saunders College Publishing, 1987; p. 659.)

From this analogy, it should be clear that if the universe is uniformly expanding, all observers, regardless of their location, must see all other galaxies moving away from them at speeds that are greater in proportion to their distances. As Hubble and Humason showed, that is precisely what we observe on earth.

As the universe expands, galaxies separate from each other. We can also, however, imagine what must have happened in the past. If we extrapolate backward in time, we find all of the galaxies coming together, until some time in the distant past when all matter was crowded to an extreme density. This is a condition that marks a unique beginning of the universe, or at least of that universe we can know about. At that beginning, the universe suddenly began its expansion with a phenomenon called the "big bang."

With the data now available, we can even estimate when our universe began. The total amount of matter and energy of the universe creates gravitation, whereby all objects pull on all other objects. This mutual attraction must slow the expansion, which means that in the past the expansion must have proceeded at a greater rate than it is today. How much greater depends on the importance of gravitation in decelerating the expansion. At the extreme, if the total mass-energy density is low enough that gravitation is ineffective (an essentially "empty" universe), the deceleration would be zero, and in that case the universe would always have been expanding at the present rate.

Clearly that extreme of an empty universe corresponds to the greatest age of the universe (since the big bang), because if the expansion had been faster in the past, galaxies would have reached their present separation in a smaller time. Consequently, we can obtain an estimate of the upper limit to the age of the universe by asking how long it would take for distant galaxies, always moving away from us at their present rates, to have reached those distances. Call that maximum possible age T_0. Now the relationship between the velocity of a galaxy v and its distance r is given by the relationship

$$v = Hr,$$

where H is the constant of proportionality between velocity and distance (the so-called Hubble constant). But since velocity is just distance divided by time, we can express v as $v = r/T_0$. Hence,

$$\frac{r}{T_0} = Hr,$$

or

$$T_0 = \frac{1}{H}.$$

We see, then, that the maximum age of the universe is just the reciprocal of the Hubble constant. The value of the Hubble constant is uncertain, primarily because of the uncertainties in deriving the distances to remote galaxies. The best estimate is that the big bang occurred 10 to 20 billion years ago. A plot of the velocity versus time relation adapted from a 1931 publication by Hubble and Humason is shown in Figure 3.

If our universe had a beginning, will it also have an end? Will the expansion go on forever? Or will gravity be strong enough to slow the expansion to a halt and force the galaxies to fall together again in a "big crunch?" A major focus of astronomical research at the present time is the effort to answer these questions. One approach relies on the application of the law of gravitation plus the cosmological principle. The cosmological principle asserts that on a large scale, that is, apart from local irregularities, the universe at any given time is the same everywhere. If this principle is valid, then we can reach conclusions about the universe as a whole from measurements of a small portion of it that is close enough to study in detail.

Let us consider a spherical region of the universe that contains a large number of galaxies. The rest of the universe can be regarded as a hollow spherical shell, since according to the cosmological principle, the matter in this shell is distributed in the same way in all directions. Newton showed that the gravitational forces exerted by a spherical shell on an object in its hollow interior all cancel exactly, so that the net force is zero. We can then consider the sphere as if it were isolated and uninfluenced by any gravitational forces from the rest of the universe. The cosmological principle also assures us that any conclusion that we reach about this sphere must apply to the rest of the universe. If the sphere chosen is

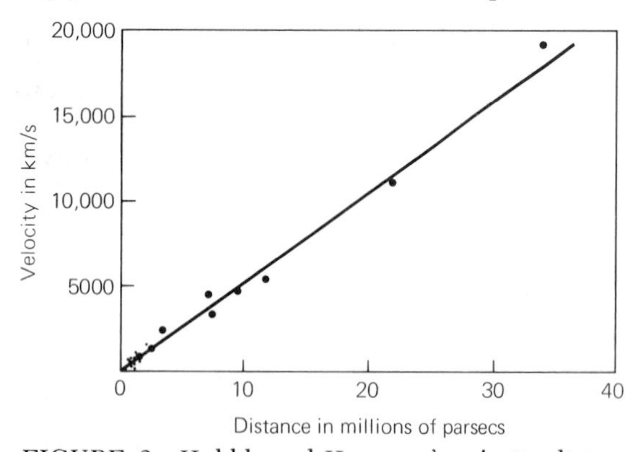

FIGURE 3 Hubble and Humason's velocity-distance relation, adapted from their 1931 paper in the *Astrophysical Journal.* Distance is measured in parsecs; one parsec is the distance that light travels in 3.26 years. (From Abell, Morrison, and Wolff, *Exploration of the Universe*, 5th ed., p. 658.)

a small one so that the velocities of the galaxies within it are small relative to the speed of light, then Newton's law of gravitation can be applied, although it has been shown that general relativity yields the same answer.

Consider then a galaxy located at the surface of the spherical region. It is subject to the gravitational attraction of all of the galaxies within the sphere and is decelerated. The situation is exactly analogous to an object launched upward from the surface of the earth. If the velocity of the object exceeds the escape velocity, the object will never return to the earth's surface. If the velocity is less than the escape velocity, the object will fall back to the ground. Similarly, if the galaxy has a velocity that is high enough to overcome the attractive force of all the other galaxies within the sphere, it will continue to move away from them, and the expansion of the universe will continue. If the velocity is too low, however, gravity will decelerate the galaxy until its velocity becomes zero, and the galaxy will then reverse direction and begin to move closer to the other galaxies within the sphere.

For an object launched from the earth, the escape velocity is determined by the mass of the earth. The escape velocity of a galaxy at the surface of the spherical region of galaxies is determined by the total mass within the sphere or by the average density of the universe (since by the cosmological principle, the density of the universe is everywhere the same). If the density of the universe is larger than some critical value, gravity will be strong enough to halt the expansion. If the density is smaller than this critical value, the expansion will go on forever.

The density of the universe can be estimated by counting up the mass associated with the galaxies within our imaginary sphere. (See, for example, Fig. 4.) Most galaxies occur in groups or clusters and can be "weighed" by measuring the gravitational influence that they exert on other nearby galaxies. For example, if two galaxies are in orbit around each other, and if their separations and velocities are known, their combined mass can be estimated from Kepler's third law. There are mathematical procedures that are not much more complex for estimating the total mass of a cluster containing not just two but many galaxies.

The best available observational estimates of the average density of the universe suggest that the amount of matter that it contains is no more than 10 to 20 percent of the value required to halt the expansion. There are some theoretical arguments for thinking that the density may be even higher and may just exactly equal the critical density required to slow the expansion to a halt, but the deceleration is so gradual in this case that an infinite amount of time is required to reduce the expansion velocity to zero. Therefore, both observation and theory suggest that galaxies will never reverse direction and move closer together.

FIGURE 4 The average density of the universe can be estimated by studying the velocities of galaxies in clusters or small groups. This group of five galaxies is known as Stephan's Quintet. (Kitt Peak National Observatory/National Optical Astronomy Observatories)

If this picture is right, there can be no rejuvenation of the universe. At the time of the big bang, the universe was very dense and hot. As it expanded, it cooled, and hydrogen and helium nuclei and then atoms formed. Somehow—theory is not yet very good at explaining this point—hydrogen and helium atoms came together to form concentrations of matter that would become galaxies. Within these protogalaxies stars formed, and in many galaxies star formation has continued to the present. Each star, some in a million years and others only after billions of years, exhausts its sources of nuclear energy. Some stars die gently, others explosively, hurling a portion of their mass into space to be incorporated into a new generation of stars. As they die, most stars also leave behind a portion of their mass in the form of a dense core that becomes a black dwarf, a neutron star, or a black hole. This material is forever lost to the universe; it radiates no energy and cannot, under any conditions that we now believe likely to occur, be broken apart into gas and dust particles that would provide the raw material for new stars. Ultimately, after the gas and dust are used up, star formation must cease, a final generation of stars will die, and the galaxies will fade into blackness.

Astronomy and physics as we understand them today offer no alternative to this outcome. Today's knowledge, however, may not be—almost certainly is not—complete. At the end of the 19th century, physics was thought to be totally understood. Theories of gravity, heat, electricity, and magnetism all seemed adequate to explain experimental data. Physicists expected to make no major new breakthroughs but simply to refine and improve the existing mathematical descriptions of physical phenomena. The revolution came in 1900 when Max Planck proposed the quantum principle.

With time, the picture presented here of the origin, evolution, and ultimate fate of the material universe may undergo a comparably radical transformation. Indeed, astronomers are attracted to this area of research in substantial numbers precisely because the subject is not fully understood and seems likely to yield its secrets to a systematic scientific attack. To learn something new, something never known before, is the ultimate challenge and satisfaction of scientific research.

SUGGESTED READING

Ferris, T., *The Red Limit*, 2nd ed., New York, Morrow, Quill, 1983.

Harrison, E. R., *Cosmology*, Cambridge, England, Cambridge University Press, 1981.

Wagoner, R., and Goldsmith, D., *Cosmic Horizons: Understanding the Universe*, New York, W. H. Freeman, 1983.

ESSAY QUESTIONS

1. When Hubble and Humason first reported their results for the relation between distance and velocity for galaxies, they thought that the constant of proportionality H was equal to about 200 km/s per million lightyears. How did their estimate for the age of the universe differ from the modern value based on H = 20 km/s per million lightyears? The oldest stars in the universe are about 15 billion years old. Is this consistent with Hubble and Humason's value for the Hubble constant H?
2. In the text, it was explained that it was possible to estimate the masses of two galaxies orbiting about each other by measuring their separations and velocities and applying Kepler's third law. Could this be done by measuring the orbital period directly? Why or why not? (Remember that galaxies contain billions of stars and have diameters of about 100,000 lightyears.)
3. Explain why the observation that all galaxies are moving away from us does not mean that we are at the center of the universe.

ESSAY PROBLEMS

1. Plot the "velocity-distance relation" for the raisins-in-the-bread analogy from the data given in Figure 3. Repeat but use some other raisin than A for a reference. Is your new plot the same as the last one?
2. Current estimates for the Hubble constant H range from 15 to 30 km/s per million lightyears, where one lightyear is the distance that light travels in one year. What is the corresponding range in the estimated age of the universe?

3. Derive an expression for escape velocity in terms of mean density. Now apply this result to two galaxies, separated by a distance R and moving away from each other at just the escape velocity and derive an expression for the critical density at which the escape and expansion velocities are equal. Assume that $H = 20$ km/s per million lightyears. How does your values for the critical density compare with the observed density, which is currently estimated to be less than about 10^{-30} g/cm^3. What does your calculation indicate about the ultimate fate of the universe?

Color Plate 1 What would William Tell think? An apple being pierced by a 30-caliber bullet traveling at a supersonic speed of 900 m/s. This collision was photographed with a microflash stroboscope using an exposure time of 0.33 μs. Shortly after the photograph was taken, the apple disintegrated completely. Note that the points of both entry and exit of the bullet are visually explosive. (Shooting the Apple, 1964. © Harold Edgerton. Courtesy of Palm Press, Inc.)

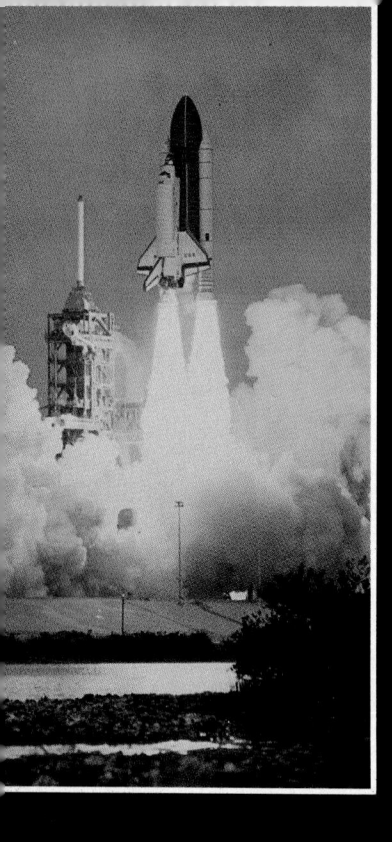

Color Plate 2 The space shuttle Columbia blasts off from Cape Kennedy. Many physical principles are involved in this complex operation in the areas of mechanics (the laws of motion, force, and momentum), thermodynamics (heat transfer and heat engines), and electricity and magnetism (electronic control and communication circuits). (NASA)

Color Plate 3 The Earth as viewed from space. (NASA)

Color Plate 4 The planet Jupiter photographed by Voyager 1 at a distance of 3.3×10^7 km. Jupiter is 318 times as massive as the Earth, and has the highest escape velocity of any planet, 60 km/s (see Section 7.10). Thus, it retains all kinds of gases in its atmosphere. Note the visible clouds consisting of alternating light and dark bands stretching around the planet parallel to the equator. The east-west wind patterns do not appear to change at all, even over several decades. (NASA/JPL)

Color Plate 5 Jupiter and two of its moons, Io and Europa. Io on the left (density 3.5 g/cm^3) is covered with sulfur and sulfur compounds, while Europa on the right (density 3.0 g/cm^3) appears to have a smooth icy surface. Two other satellites, Ganymede and Callisto, contain large amounts of ice, as evidenced by their much lower densities of about 1.9 g/cm^3.
(NASA/JPL)

Color Plate 6 Saturn and three of its satellites as photographed by Voyager 1 80 days before approaching the planet, when Voyager 1 was 106,250,000 km (66 million miles) away. By the time Voyager 1 had passed Saturn, thousands of rings became apparent. The particles making up the rings vary in size from about 1 μm to about 1 m, and are believed to be made up largely of ice. Other views of the rings reveal a radial pattern of "spokes" consisting of small particles that appear to be elevated about the plane of the rings.
(NASA/JPL)

Color Plate 7 The Crab Nebula in the constellation Taurus is the debris from a supernova explosion seen on earth in the year 1054 A.D. The red filaments represent hydrogen emission lines, while the blue regions correspond to sulfur emission.
(National Optical Astronomy Observatories)

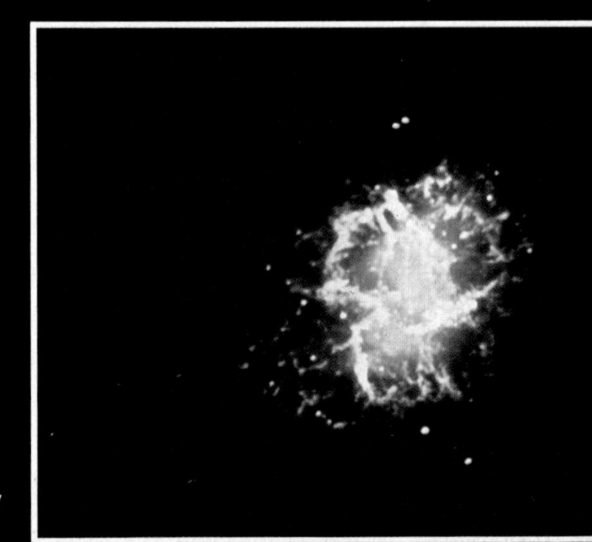

Color Plate 8 The Andromeda Galaxy, a spiral galaxy nearest to the Milky Way, is the most distant object that can be seen with the naked eye. By measuring the Doppler shifts of spectral lines, scientists are able to determine the rotational motion of such galaxies. Their masses can then be computed with the help of Kepler's third law. (Palomar Observatory, California Institute of Technology)

Color Plate 9 The Foucault pendulum at the Smithsonian institution in Washington, D.C. This type of pendulum was first used by the French physicist Jean Foucault to verify the earth's rotation experimentally. During its swinging motion, the pendulum's plane of oscillation appears to rotate, as evidenced by the fact that the bob successively knocks over the red indicators arranged in a horizontal circle. In reality, the pendulum's plane of motion is fixed in space, while the earth rotates beneath the swinging pendulum. (Courtesy of the Smithsonian Institution)

CHAPTER 8

Rotational Dynamics

The circular movement of Hurricane Elena photographed from space. (Courtesy NASA)

This chapter will complete our study of rotational motion. We shall build on the definitions of angular velocity and angular acceleration encountered in Chapter 7 by examining the relationship between these concepts and the forces that produce rotational motion. Specifically, we shall find the rotational analog of Newton's second law and define a new term that needs to be added into our equation for conservation of mechanical energy: rotational kinetic energy. One of the central points of this chapter will be to develop the concept of angular momentum, a quantity that plays a key role in rotational motion. Finally, just as we found that linear momentum is conserved, we shall also find that the angular momentum of any isolated system is always conserved.

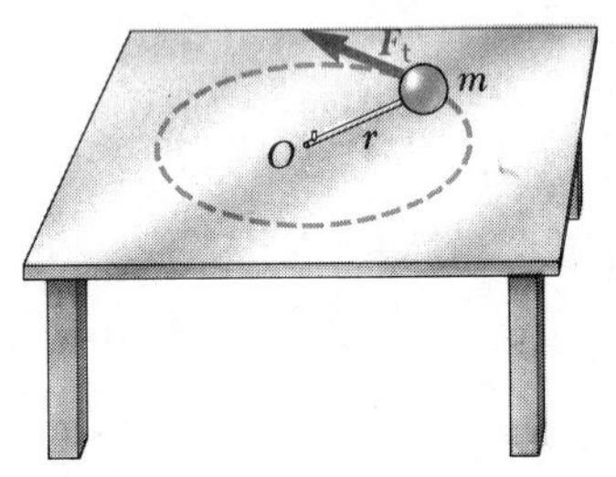

FIGURE 8.1 A mass m attached to a light rod of length r moves in a circular path on a horizontal frictionless surface while a tangential force F_t acts on it.

8.1 RELATIONSHIP BETWEEN TORQUE AND ANGULAR ACCELERATION

In Chapter 4 we considered the situation in which both the net force and the net torque acting on an object were equal to zero. Such objects were said to be in static equilibrium. We shall now describe the behavior of an object when the net torque acting on it is not zero. As we shall see, when there is a net torque acting on a rigid object, it will undergo an angular acceleration. Furthermore, we will find that the net torque is directly proportional to the angular acceleration. The end result of our investigation will be an expression that is analogous to $\mathbf{F} = m\mathbf{a}$ in linear motion.

Let us begin by considering the system shown in Figure 8.1, which consists of a mass m connected to a very light rod of length r. The rod is pivoted at the point O and its movement is confined to rotation on a *horizontal* frictionless table. Now let us assume there is a force $\mathbf{F}_t$ acting on m perpendicular to the rod and hence tangent to the circular orbit. Since there is no force to oppose this tangential force, the mass undergoes a tangential acceleration according to Newton's second law:

$$F_t = ma_t$$

Multiplying the left and right sides of this equation by r gives

$$F_t r = mra_t$$

In Chapter 7 we found that the tangential acceleration and angular acceleration for a particle rotating in a circular path are related by the expression

$$a_t = r\alpha$$

so now we find that

$$F_t r = mr^2\alpha \quad (8.1)$$

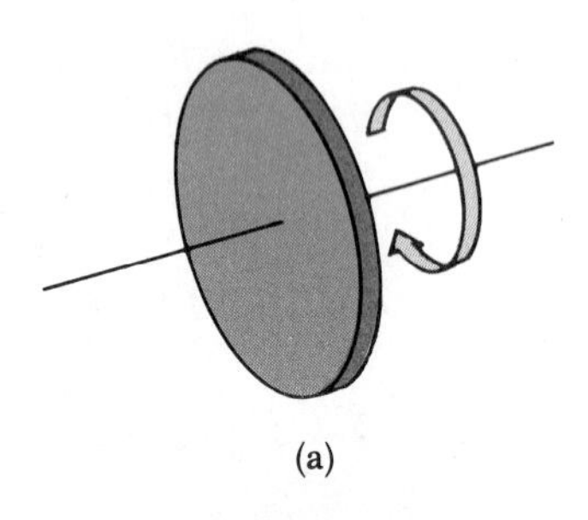

(a)

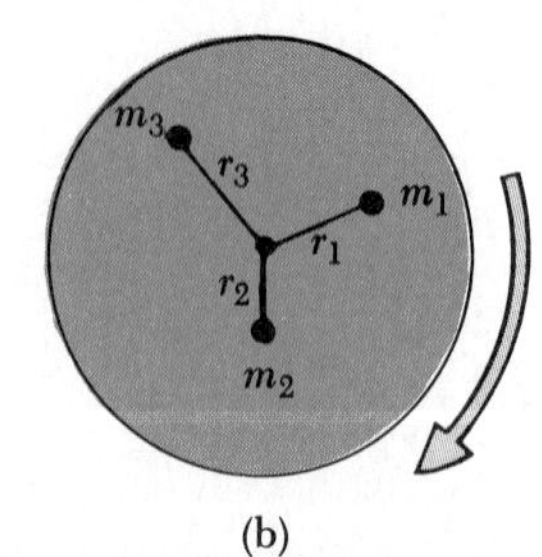

(b)

FIGURE 8.2 (a) A solid disk rotating about its axis. (b) The disk consists of many particles, and all the particles have the same angular acceleration.

The left side of Equation 8.1, which should be familiar to you, is the torque acting on the mass about its axis of rotation. That is, the torque is equal in magnitude to the force on m multiplied by the perpendicular distance from the pivot to the line of action of the force, or $\tau = F_t r$. Hence, we can write Equation 8.1 as

$$\boxed{\tau = mr^2\alpha} \quad (8.2)$$

Equation 8.2 shows that the torque on the system is proportional to angular acceleration, where the constant of proportionality mr^2 is called the **moment of inertia** of the mass m. (Because the rod is very light, its moment of inertia can be neglected.)

Torque on a Rotating Object

Now consider a solid disk rotating about its axis, as in Figure 8.2a. The disk consists of many particles located at various distances from the axis of rotation, as in Figure 8.2b. The torque on one of these particles is given by Equation

8.2. The *total* torque on the disk is given by the sum of the individual torques on all the particles:

$$\sum \tau = \left(\sum mr^2\right)\alpha \tag{8.3}$$

Note that, because the disk is rigid, all particles have the *same* angular acceleration so α is not involved in the sum. If the masses and distances of the particles are labeled with subscripts as in Figure 8.2b, then

$$\sum mr^2 = m_1 r_1^2 + m_2 r_2^2 + m_3 r_3^2 + \cdots$$

This quantity is called the **moment of inertia** of the whole body and is given the symbol I:

$$\boxed{I = \sum mr^2} \tag{8.4}$$

Moment of inertia

The moment of inertia has units of $\text{kg}\cdot\text{m}^2$ in SI units. Using this result in Equation 8.3, we see that the total torque on a rigid body rotating about a fixed axis is given by

$$\boxed{\sum \tau = I\alpha} \tag{8.5}$$

Net torque

The angular acceleration of an object is proportional to the net torque acting on it. The proportionality constant, I, between the net torque and angular acceleration is the moment of inertia.

It is important to note that the equation $\Sigma\tau = I\alpha$ (Eq. 8.5) is the rotational counterpart to Newton's second law, $F = ma$. Thus, the correspondence between rotational motion and linear motion continues. Recall that in Chapter 7 we found that the linear variables x, v, and a are replaced in rotational motion by the variables θ, ω, and α. Likewise, we now see that *the force and mass in linear motion correspond to torque and moment of inertia in rotational motion.* In this chapter, we shall develop other equations for the rotational kinetic energy and the angular momentum of a body rotating about a fixed axis. Based on the analogies already presented, you should be able to predict the form of these equations.

More on the Moment of Inertia

The moment of inertia of a body as defined by $I = \Sigma mr^2$ (Eq. 8.4), will be used throughout the remainder of this chapter. Thus it will be useful to examine this quantity in more detail before discussing other aspects of rotational motion.

As we saw earlier, a small object (or particle) orbiting about some axis has a moment of inertia equal to mr^2. Now consider a somewhat more complicated system, the baton being twirled by a majorette in Figure 8.3. Let us assume that the baton can be modeled as a very light rod of length 2ℓ and with a heavy mass at each end. (The rod of a real baton has significant mass relative to its ends. Our model neglects the mass of the rod.) Since we are neglecting the mass of the rod, the moment of inertia of the baton about an

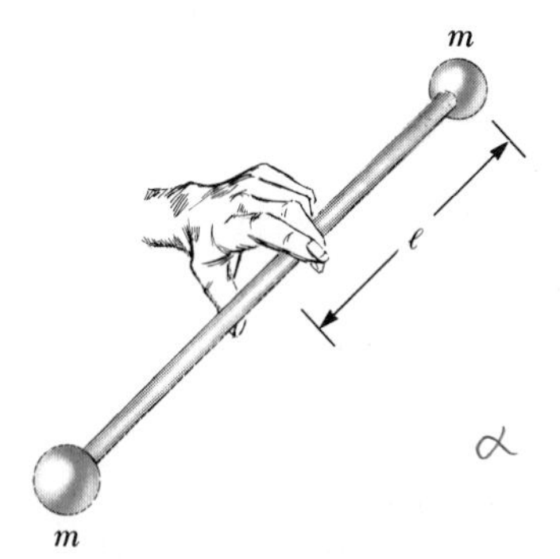

FIGURE 8.3 A baton of length 2ℓ and mass $2m$ (the mass of the connecting rod is neglected). The moment of inertia about the axis through the baton's center and perpendicular to its length is $2m\ell^2$.

axis through its center and perpendicular to its length is given by Equation 8.4:

$$I = \sum mr^2$$

Since in this system there are two equal masses equidistant from the axis of rotation, we see that $r = \ell$ for each mass and the sum is

$$I = \sum mr^2 = m\ell^2 + m\ell^2 = 2m\ell^2$$

We pointed out earlier that I is the rotational counterpart of m. However, there are some important distinctions between the two. For example, mass is an intrinsic property of an object that does not change. On the other hand, *the moment of inertia of a system depends upon the axis of rotation and upon the manner in which the mass is distributed.* Examples 8.1 and 8.2 illustrate this point.

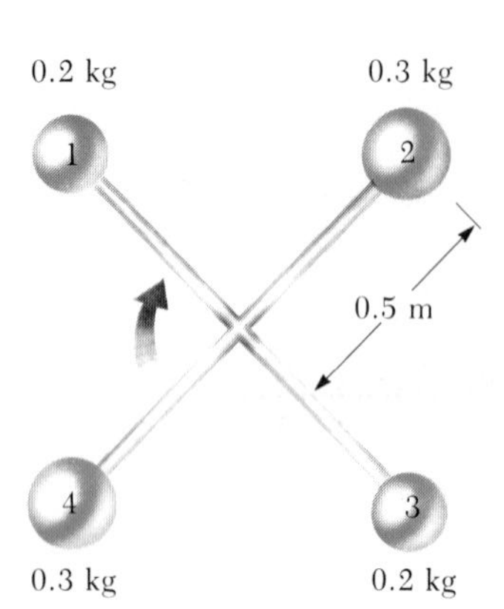

FIGURE 8.4 (Example 8.1) Four masses connected to light rods rotating in the plane of the paper.

EXAMPLE 8.1 The Baton Twirler

In an effort to be the star of the half-time show, a majorette twirls a highly unusual baton made up of four masses fastened to the ends of light rods as shown in Figure 8.4. Each rod is 1 m long. Find the moment of inertia of the system about an axis perpendicular to the page and through the point where the rods cross.

Solution Applying Equation 8.4, we get

$$\begin{aligned} I &= \sum mr^2 = m_1r_1^{\,2} + m_2r_2^{\,2} + m_3r_3^{\,2} + m_4r_4^{\,2} \\ &= (0.2 \text{ kg})(0.5 \text{ m})^2 + (0.3 \text{ kg})(0.5 \text{ m})^2 + (0.2 \text{ kg})(0.5 \text{ m})^2 + (0.3 \text{ kg})(0.5 \text{ m})^2 \\ &= 0.25 \text{ kg} \cdot \text{m}^2 \end{aligned}$$

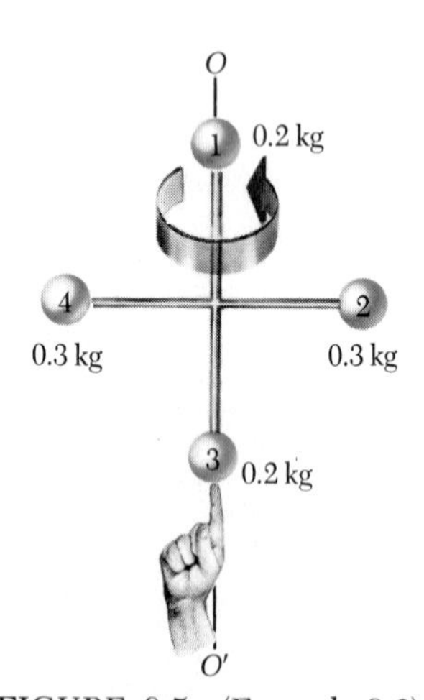

FIGURE 8.5 (Example 8.2) The double baton rotating about the axis OO'.

EXAMPLE 8.2 The Baton Twirler—Second Act

Not satisfied with the crowd reaction from the baton twirling of Example 8.1, the majorette tries spinning her strange baton about the axis OO', as shown in Figure 8.5. Calculate the moment of inertia about this axis.

Solution Again applying $I = \Sigma mr^2$, we have

$$\begin{aligned} I &= (0.2 \text{ kg})(0)^2 + (0.3 \text{ kg})(0.5 \text{ m})^2 + (0.2 \text{ kg})(0)^2 + (0.3 \text{ kg})(0.5 \text{ m})^2 \\ &= 0.15 \text{ kg} \cdot \text{m}^2 \end{aligned}$$

Calculation of Moments of Inertia for Extended Objects

The method used for calculating moments of inertia in Examples 8.1 and 8.2 is simple enough when you have only a few small masses rotating about an axis. However, the situation becomes much more complex when the object is an extended mass, such as a sphere, a cylinder, or a cone. One extended object that is amenable to a simple solution is a hoop rotating about an axis perpendicular to the plane of the hoop and passing through its center, as shown in Figure 8.6. For example, the object might be a bicycle tire.

To evaluate the moment of inertia for the hoop, we can still use the equation $I = \Sigma mr^2$ (Eq. 8.4) and imagine that the hoop is divided into a number of small segments having masses m_1, m_2, m_3, . . . , as in Figure 8.6. This approach is just an extension of the baton problem in our previous examples, except that now we have a large number of small masses in rotation instead of only four.

We can express the sum for I as

$$I = \sum mr^2 = m_1r_1^{\ 2} + m_2r_2^{\ 2} + m_3r_3^{\ 2} + \cdots$$

All of the segments around the hoop are at the *same distance* R from the axis of rotation; thus, we can drop the subscripts on the distances and factor out the common term R:

$$I = (m_1 + m_2 + m_3 + \cdots)R^2$$

We know, however, that the sum of the masses of all the segments must equal the total mass of the hoop, M:

$$M = m_1 + m_2 + m_3 + \cdots$$

and so we can express I as

$$I = Mr^2 \tag{8.6}$$

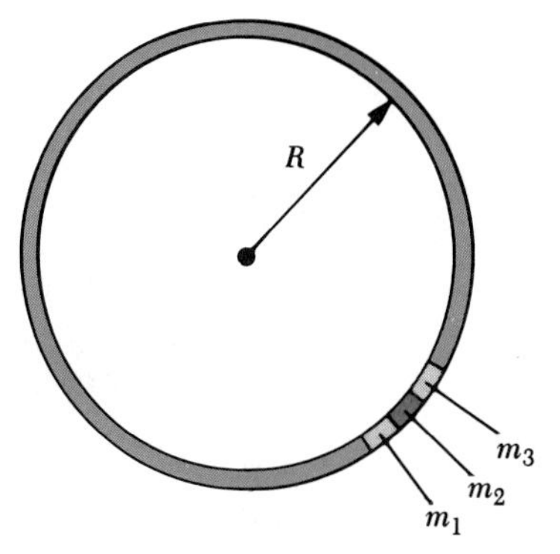

FIGURE 8.6 The uniform hoop can be divided into a large number of small segments that are equidistant from the center of the hoop.

This expression can be used for the moment of inertia of any ring-shaped object in any situation in which the ring rotates about an axis through its center and perpendicular to its plane. Note that the result is strictly valid only if the thickness of the ring is small relative to its inner radius.

The hoop we selected as an example is unique in that we were able to find an expression for its moment of inertia by using only simple algebra. Unfortunately, most extended objects are more difficult to work with, and the methods of integral calculus are required. Because calculus techniques are beyond the scope of this text, the moments of inertia for some common shapes are given without proof in Table 8.1. When the need arises, you should refer to this table for an appropriate expression for the moment of inertia.

TABLE 8.1 Moments of Inertia for Various Rigid Bodies of Uniform Composition

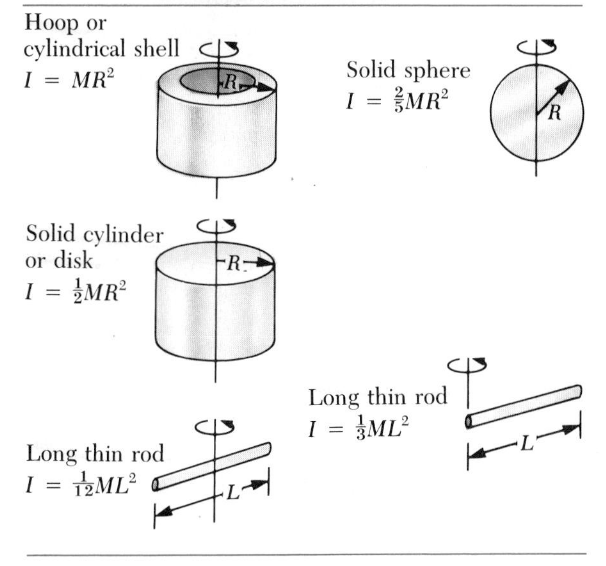

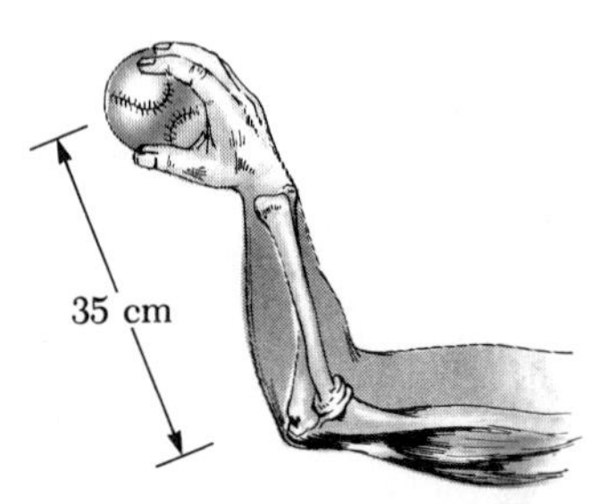

FIGURE 8.7 (Example 8.3) A ball being tossed by a pitcher. The forearm is being used to accelerate the ball.

EXAMPLE 8.3 Warming Up

A baseball pitcher loosening up his arm before a game tosses a 0.3-kg baseball using only the rotation of his forearm to accelerate the ball, as in Figure 8.7. The ball starts at rest and is released with a speed of 15 m/s in 0.3 s. (a) Find the constant angular acceleration of the arm and ball.

Solution During its acceleration, the ball moves through an arc of a circle having a radius of 0.35 m. We can determine the angular acceleration using $\omega = \omega_0 + \alpha t$. Since $\omega_0 = 0$, however, $\omega = \alpha t$, or

$$\alpha = \frac{\omega}{t}$$

We also know that $v = r\omega$, so that we get

$$\alpha = \frac{\omega}{t} = \frac{v}{rt} = \frac{15 \text{ m/s}}{(0.35 \text{ m})(0.3 \text{ s})} = 143 \text{ rad/s}^2$$

(b) Find the torque exerted on the ball to give it this angular acceleration.

Solution The moment of inertia of the ball about an axis through the elbow and perpendicular to the arm is

$$I = mr^2 = (0.3 \text{ kg})(0.35 \text{ m})^2 = 3.68 \times 10^{-2} \text{ kg} \cdot \text{m}^2$$

Thus, the torque required is

$$\tau = I\alpha = (3.68 \times 10^{-2} \text{ kg} \cdot \text{m}^2)(143 \text{ rad/s}^2) = 5.26 \text{ N} \cdot \text{m}$$

EXAMPLE 8.4 The Falling Bucket

A frictionless pulley in the shape of a solid cylinder of mass $M = 3$ kg and radius $R = 0.4$ m is used to draw water from a well (Fig. 8.8a). A bucket of mass $m = 2$ kg is attached to a cord wrapped around the cylinder. If the bucket starts from rest at the top of the well and falls for 3 s before hitting the water, how far does it fall?

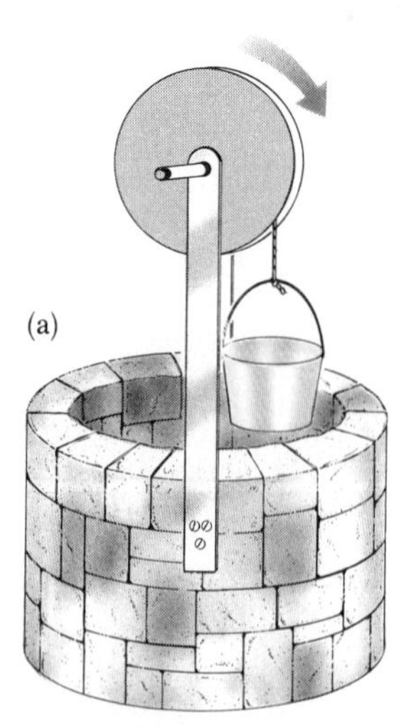

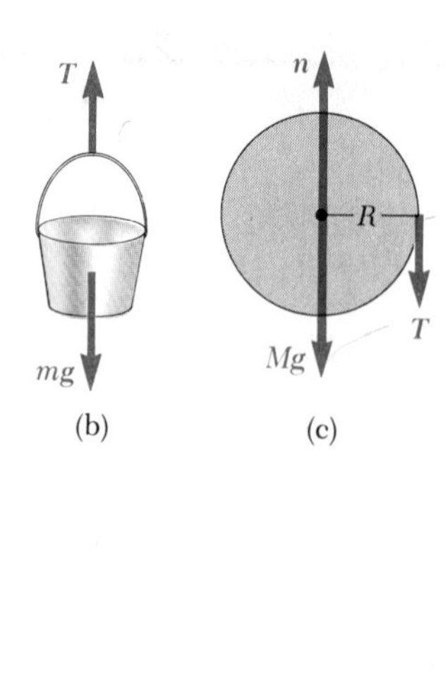

FIGURE 8.8 (Example 8.4) (a) A water bucket attached to a rope passing over a frictionless pulley. (b) Free-body diagram for the bucket. (b) The tension produces a torque on the cylinder about its axis of rotation.

Solution Figure 8.8b shows the two forces on the bucket as it falls: T is the tension in the cord, and mg is the weight of the bucket. We shall choose the positive direction downward and write Newton's second law for the bucket as

$$mg - T = ma$$

When the given quantities are substituted into this equation, we have

$$(1) \qquad (2\ \text{kg})(9.8\ \text{m/s}^2) - T = (2\ \text{kg})(a)$$

With one equation and two unknowns, we must develop an additional equation in order to complete the problem. To obtain this second equation, let us consider the cylinder's rotational motion. Equation 8.5 applied to this system gives the necessary expression:

$$\tau = I\alpha = \frac{1}{2}MR^2\alpha$$

Figure 8.8c shows that the only force producing a torque on the cylinder as it rotates about an axis through its center is $\boldsymbol{T}$, the tension in the cord. There are actually two other forces acting on the cylinder, its weight and an upward force that the axle exerts on the cylinder, but we do not have to consider these here because their lever arm about the axis of rotation is zero. Thus, we have

$$T(0.4\ \text{m}) = \frac{1}{2}(3\ \text{kg})(0.4\ \text{m})^2(\alpha)$$

$$(2) \qquad T = (0.6\ \text{kg} \cdot \text{m})(\alpha)$$

At this point, it is important to recognize that the downward acceleration of the bucket is equal to the tangential acceleration of a point on the rim of the cylinder. Therefore, the angular acceleration of the cylinder and the linear acceleration of the bucket are related by $a_t = r\alpha$. When this relation is used in (2), we get

$$(3) \qquad T = (1.5\ \text{kg})(a_t)$$

Equations (1) and (3) can now be solved simultaneously to find a_t and T. This procedure gives

$$a_t = 5.60\ \text{m/s}^2 \qquad T = 8.40\ \text{N}$$

Finally, we turn to the equations for motion with constant linear acceleration to find the distance d that the bucket falls:

$$d = v_0 t + \frac{1}{2}at^2$$

$$= \frac{1}{2}(5.60\ \text{m/s}^2)(3\ \text{s})^2 = 25.2\ \text{m}$$

8.2 ROTATIONAL KINETIC ENERGY

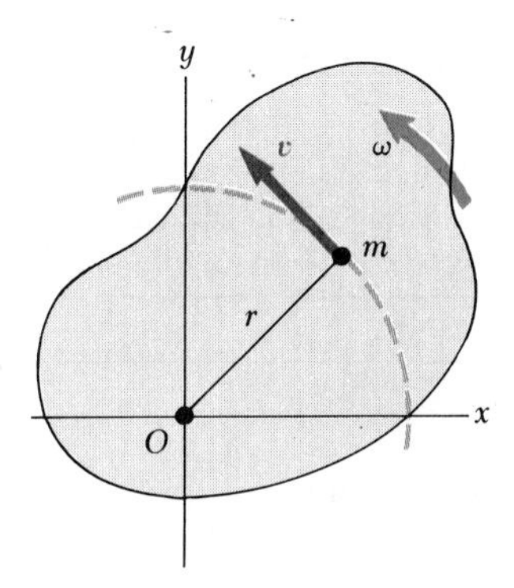

FIGURE 8.9 A rigid plane body rotating about the z axis with angular velocity ω. The kinetic energy of a particle of mass m is $\frac{1}{2}mv^2$. The total kinetic energy of the body is $\frac{1}{2}I\omega^2$.

In Chapter 5 we defined the kinetic energy of a particle moving through space with a speed v as the quantity $\frac{1}{2}mv^2$. In analogy with this,

a body rotating about some axis with an angular velocity ω is said to have rotational kinetic energy given by $\frac{1}{2}I\omega^2$.

To prove that this is true, consider a planar body rotating about some axis perpendicular to its plane surface, as in Figure 8.9. The body consists of many small particles, each of mass m. All these particles rotate in a circular path about the axis. If r is the distance of one of the particles from the axis of rotation, the velocity of this particle is $v = r\omega$. Since the *total* kinetic energy of the body is the sum of all the kinetic energies associated with the particles, we have

$$KE_r = \sum\left(\frac{1}{2}mv^2\right) = \sum\left(\frac{1}{2}mr^2\omega^2\right) = \frac{1}{2}\left(\sum mr^2\right)\omega^2$$

Rotational kinetic energy

$$\boxed{KE_r = \frac{1}{2}I\omega^2} \tag{8.7}$$

where $I = \Sigma mr^2$ is the moment of inertia of the body. Note that we were able to factor out the ω^2 term since it is the same for every particle.

In linear motion, we found the energy concept to be extremely useful in describing the motion of a system. The energy concept can be equally useful in simplifying the analysis of rotational motion. We now have developed expressions for three types of energy: *gravitational potential energy, PE_g, translational kinetic energy, KE_t,* and *rotational kinetic energy, KE_r.* We must include all these forms of energy in our equation for conservation of mechanical energy:

$$(KE_t + KE_r + PE_g)_i = (KE_t + KE_r + PE_g)_f \tag{8.8}$$

where "i" and "f" refer to initial and final values, respectively.

EXAMPLE 8.5 A Ball Rolling down an Incline

A ball of mass M and radius R starts from rest at a height of 2 m and rolls down a 30° slope, as shown in Figure 8.10. What is the linear velocity of the ball when it leaves the incline? Assume the ball rolls without slipping.

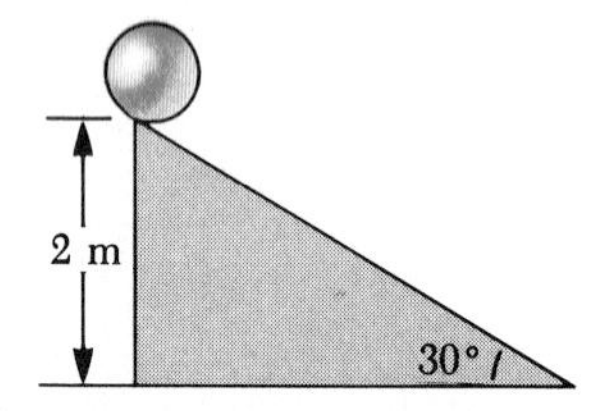

FIGURE 8.10 (Example 8.5) A ball starts from rest at the top of an incline and rolls to the bottom without slipping.

Solution The initial energy of the ball is gravitational potential energy, and when it reaches the bottom of the ramp, this potential energy has been converted to translational and rotational kinetic energy. The conservation of mechanical energy equation becomes

$$(PE_g)_i = (KE_t + KE_r)_f$$

or

$$Mgh = \frac{1}{2}Mv^2 + \frac{1}{2}\left(\frac{2}{5}MR^2\right)\omega^2$$

where we have used $I = \frac{2}{5}MR^2$ from Table 8.1 as the moment of inertia of the ball. If the ball rolls without slipping, a point on its surface must have the same instantaneous speed as its center of gravity.[1] Thus, we can relate the linear velocity of the ball to its rotational speed:

$$v = R\omega$$

[1]Note that a point on the surface of the ball travels a distance $2\pi R$ during one rotation. However, if no slipping occurs, this is also the distance that the center of gravity travels in the same time interval. Thus, the center of gravity has a speed equal to the speed of a point on the surface of the ball.

This expression can be used to eliminate ω from the equation for conservation of mechanical energy:

$$Mgh = \frac{1}{2}Mv^2 + \frac{1}{5}Mv^2$$

or, solving for v,

$$v = \sqrt{\frac{10gh}{7}} = \sqrt{\frac{10(9.80 \text{ m/s}^2)(2 \text{ m})}{7}} = 5.29 \text{ m/s}$$

EXERCISE Repeat this example for a solid cylinder of the same mass and radius as the ball and released at the same height. In a race between the two objects on the incline, which one would win?

Answer $v = \sqrt{4gh/3} = 5.11$ m/s. The ball would win.

8.3 ANGULAR MOMENTUM

In Figure 8.11, an object of mass m is located a distance r away from a center of rotation. Under the action of a constant torque on the object, its angular velocity will increase from a value ω_0 to a value ω in a time Δt. Thus, we can write

$$\tau = I\alpha = I\left(\frac{\omega - \omega_0}{\Delta t}\right) = \frac{I\omega - I\omega_0}{\Delta t}$$

If we define the product

$$L = I\omega \tag{8.9}$$

Angular momentum

as the angular momentum of the object, then we can write

$$\tau = \frac{\text{change in angular momentum}}{\text{time interval}} = \frac{\Delta L}{\Delta t} \tag{8.10}$$

Equation 8.10 is the rotational analog of Newton's second law, $F = \Delta p/\Delta t$, and states that

the torque acting on an object is equal to the time rate of change of its angular momentum.

When the net external torque acting on a system is zero, we see from Equation 8.10 that $\Delta L/\Delta t = 0$. In this case, the rate of change of the angular momentum of the system is zero. Therefore, *the product Iω remains constant in time.* That is, $L_i = L_f$

or
$$I_i\omega_i = I_f\omega_f \quad \text{if } \sum\tau = 0 \tag{8.11}$$

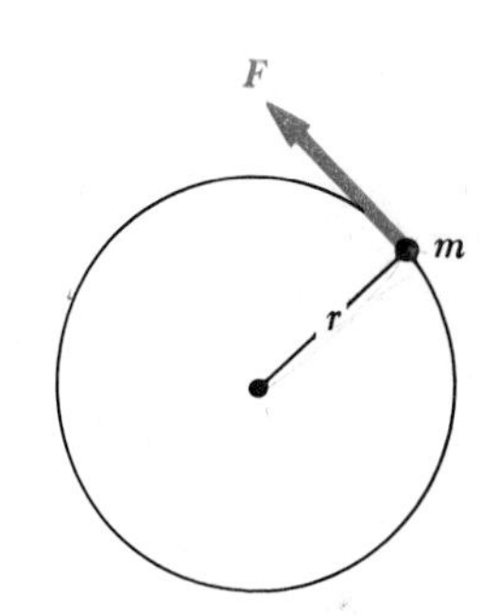

FIGURE 8.11 An object of mass m rotating in a circular path under the action of a constant torque.

The angular momentum of a system is conserved when the net external torque acting on the system is zero. That is, when $\Sigma\tau = 0$, the initial angular momentum equals the final angular momentum.

FIGURE 8.12 The angular velocity of the skater increases when she pulls her arms in close to her body demonstrating that angular momentum is conserved. (Photo, David Leonardi).

Conservation of angular momentum

In Equation 8.11, we have a third conservation law to add to our list: **the conservation of angular momentum.** We now can state that

the energy, linear momentum, and angular momentum of an isolated system all remain constant.

There are many examples that demonstrate conservation of angular momentum, some of which should be familiar to you. You may have observed a skater spinning in the finale of an act. The angular velocity of the skater increases upon pulling her hands and feet close to the trunk of her body, as in Figure 8.12. Neglecting friction between skater and ice, we see that there are no external torques on the skater. The moment of inertia of her body decreases as her hands and feet are brought in. The resulting change in angular velocity is accounted for as follows. Since angular momentum must be conserved, the product $I\omega$ remains constant and a decrease of the moment of inertia of the skater causes a corresponding increase in the angular velocity.

Similarly, when divers or acrobats wish to make several somersaults, they pull their hands and feet close to the trunk of their bodies in order to rotate at a greater angular velocity. In this case, the external force due to gravity acts through the center of gravity and hence exerts no torque about the axis of rotation. Therefore the angular momentum about the center of gravity is conserved. For example, when divers wish to double their angular velocity, they must reduce their moment of inertia to half its initial value.

EXAMPLE 8.6 The Spinning Stool

A student sits on a pivoted stool while holding a pair of weights, as in Figure 8.13. The stool is free to rotate about a vertical axis with negligible friction. The moment of inertia of student, weights, and stool is 2.25 kg · m^2. The student is set in rotation with an initial angular velocity of 5 rad/s with weights outstretched. As he rotates, he pulls the weights inward so that the new moment of inertia of the system (student, weight, and stool) becomes 1.80 kg · m^2. What is the new angular velocity of the system?

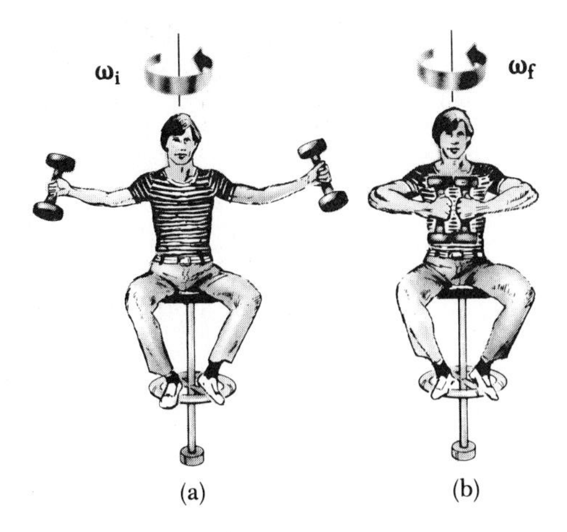

FIGURE 8.13 (Example 8.6) (a) The student is given an initial angular velocity while holding two weights as shown. (b) When the weights are pulled in close to the body, the angular velocity of the system increases. Why?

Solution The initial angular momentum of the system is

$$L_i = I_i\omega_i = (2.25\ \text{kg} \cdot \text{m}^2)(5\ \text{rad/s}) = 11.3\ \text{kg} \cdot \text{m}^2/\text{s}$$

When the weights are pulled in, they are closer to the axis of rotation, and as a result the moment of inertia of the system is reduced. The new angular momentum is

$$L_f = I_f\omega_f = (1.80\ \text{kg} \cdot \text{m}^2)(\omega_f)$$

Because the net external torque on the system is zero, angular momentum is conserved. Thus, we find

$$(11.3\ \text{kg} \cdot \text{m}^2/\text{s}) = (1.80\ \text{kg} \cdot \text{m}^2)(\omega_f)$$

$$\omega_f = 6.28\ \text{rad/s}$$

EXAMPLE 8.7 The Merry-go-round

A student is standing at the edge of a circular platform that rotates in a horizontal plane about a frictionless vertical axle (Fig. 8.14). The platform has a mass $M = 100$ kg and a radius $R = 2$ m. The student, whose mass is $m = 60$ kg, walks slowly from the rim of the disk toward the center. If the angular velocity of the system is 2 rad/s when the student is at the rim, calculate the angular velocity when the student has reached a point 0.5 m from the center.

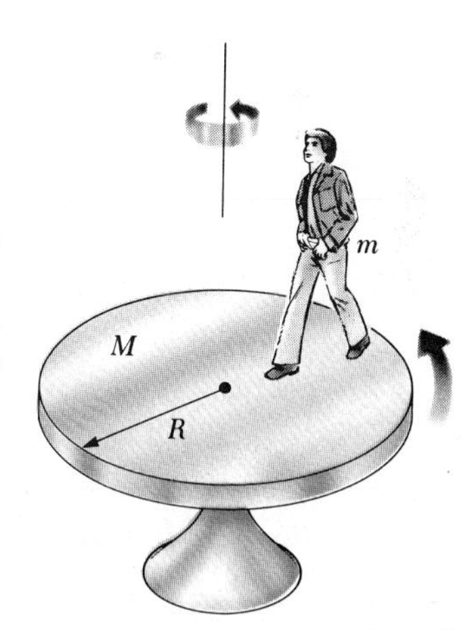

FIGURE 8.14 (Example 8.7) As the student walks toward the center of the rotating platform, the angular velocity of the system increases because the angular momentum of the system (student + platform) must remain constant.

Solution The moment of inertia of the platform, I_p, is

$$I_p = \frac{1}{2}MR^2 = \frac{1}{2}(100\ \text{kg})(2\ \text{m})^2 = 200\ \text{kg} \cdot \text{m}^2$$

Treating the student as a point mass, his initial moment of inertia is

$$I_s = mR^2 = (60\ \text{kg})(2\ \text{m})^2 = 240\ \text{kg} \cdot \text{m}^2$$

Thus, the initial angular momentum of the platform plus student is

$$L_i = (I_p + I_s)(\omega_i) = (200\ \text{kg} \cdot \text{m}^2 + 240\ \text{kg} \cdot \text{m}^2)(2\ \text{rad/s}) = 880\ \text{kg} \cdot \text{m}^2/\text{s}$$

When the student has walked to the position 0.5 m from the center, his moment of inertia is

$$I'_s = mr_f^2 = (60\ \text{kg})(0.5\ \text{m})^2 = 15\ \text{kg} \cdot \text{m}^2$$

There is no change in the moment of inertia of the platform. Because there are no external torques on the *system* (student plus platform) about the axis of rotation, we can apply the law of conservation of angular momentum:

$$L_i = L_f$$

$$880 \text{ kg} \cdot \text{m}^2/\text{s} = 200\omega_f + 15\omega_f = 215\omega_f$$

$$\omega_f = 4.09 \text{ rad/s}$$

EXERCISE Calculate the change in kinetic energy of the system (student plus platform) for this situation. What accounts for this change in energy?

Answer $KE_f - KE_i = 918$ J. The student must perform positive work in order to walk toward the center of the platform.

*8.4 ANGULAR MOMENTUM AS A FUNDAMENTAL QUANTITY

We have seen that the concept of angular momentum is very useful for describing the motion of macroscopic objects. However, the concept is also valid on a submicroscopic scale and has been used extensively in the development of modern theories of atomic, molecular, and nuclear physics. In these developments, it was found that the angular momentum of a system is a *fundamental* quantity. The word *fundamental* in this context implies that angular momentum is an inherent property of atoms, molecules, and their constituents.

In order to explain the results of a variety of experiments on atomic and molecular systems, it is necessary to assign discrete values to the angular momentum. These discrete values are some multiple of a fundamental unit of angular momentum, which equals $\hbar = h/2\pi$, where h is Planck's constant.

Fundamental unit of angular momentum

$$\hbar = 1.054 \times 10^{-34} \frac{\text{kg} \cdot \text{m}^2}{\text{s}^2} \tag{8.12}$$

Let us accept this postulate for the time being and show how it can be used to estimate the rotational frequency of a diatomic molecule. Consider the O_2 molecule as a rigid rotor, that is, two atoms separated by a fixed distance d and rotating about the center of mass (Fig. 8.15). Equating the rotational angular momentum to the fundamental unit $\hbar$, we can estimate the lowest rotational frequency:

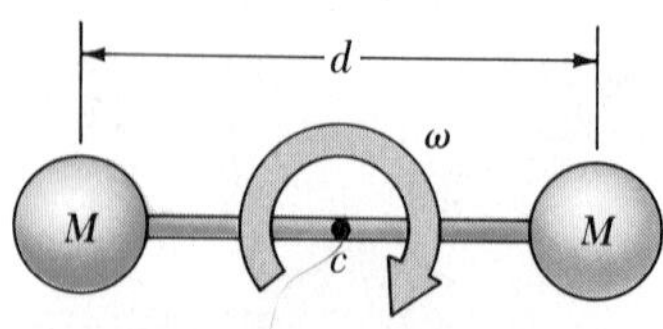

FIGURE 8.15 The rigid-rotor model of the diatomic molecule. The rotation occurs about the center of mass in the plane of the diagram.

$$I_c\omega \approx \hbar \quad \text{or} \quad \omega \approx \frac{\hbar}{I_c}$$

The moment of inertia of the O_2 molecule about this axis of rotation is 2.03×10^{-46} kg · m². Therefore,

$$\omega \approx \frac{\hbar}{I_c} = \frac{1.054 \times 10^{-34} \text{ kg} \cdot \text{m}^2/\text{s}}{2.03 \times 10^{-46} \text{ kg} \cdot \text{m}^2} = 5.19 \times 10^{11} \text{ rad/s}$$

This result is in good agreement with measured rotational frequencies. Furthermore, the rotational frequencies are much lower than the vibrational frequencies of the molecule, which are typically of the order of 10^{13} Hz.

This simple example shows that certain classical concepts and mechanical models might be useful in describing some features of atomic and molecular systems. However, a wide variety of phenomena on the submicroscopic scale can be explained only if one assumes discrete values of the angular momentum associated with a particular type of motion.

Historically, the Danish physicist Niels Bohr (1885–1962) was the first to suggest this radical idea in his theory of the hydrogen atom. Strictly classical models were unsuccessful in describing many properties of the hydrogen atom, such as the fact that the atom absorbs and emits radiation at discrete frequencies. Bohr postulated that the electron could only occupy circular orbits about the proton for which the orbital angular momentum was equal to $n\hbar$, where n is an integer. From this rather simple model, one can estimate the rotational frequencies of the electron in the various orbits.

Although Bohr's theory provided some insight concerning the behavior of matter at the atomic level, it is basically incorrect. Subsequent developments in quantum mechanics from 1924 to 1930 provided models and interpretations that are still accepted.

Later developments in atomic physics indicated that the electron also possesses another kind of angular momentum, called *spin*, which is also an inherent property of the electron. The spin angular momentum is also restricted to discrete values. We shall return to this important property in Chapter 30 and discuss its great impact on modern physical science.

SUMMARY

The **moment of inertia** of a group of particles is

$$I = \sum mr^2 \tag{8.4}$$

If a rigid body free to rotate about a fixed axis has a net external torque acting on it, the body will undergo an angular acceleration α, where

$$\sum \tau = I\alpha \tag{8.5}$$

If a rigid object rotates about a fixed axis with angular velocity ω, its **rotational kinetic energy** is

$$KE_r = \frac{1}{2}I\omega^2 \tag{8.7}$$

where I is the moment of inertia about the axis of rotation.

The **angular momentum** of a rotating object is

$$L = I\omega \tag{8.9}$$

If the net external torque acting on a system is zero, the total angular momentum of the system is constant. Applying the law of conservation of angular momentum to an object whose moment of inertia changes with time gives

$$I_i\omega_i = I_f\omega_f \tag{8.11}$$

ADDITIONAL READING

H. Brody, "The Moment of Inertia of a Tennis Racket," *The Physics Teacher*, April 1985, p. 213.

C. Frohlich, "Somersaulting and Twisting," *Sci. American*, March 1980, p. 154.

D. F. Griffing, *The Dynamics of Sports: Why That's the Way the Ball Bounces*, Mohican Publishing Co., 1982.

J. G. Kreifeldt and M. Chuang, "Moment of Inertia: Psychophysical Study of an Overlooked Sensation," *Science*, 1979, p. 588.

K. Laws, "The Physics of Dance," *Physics Today*, February 1935, p. 25.

R. F. Post and S. F. Post, "Flywheels," *Sci. American*, December 1973, p. 17.

QUESTIONS

1. Explain why changing the axis of rotation of an object should change its moment of inertia.
2. Is it possible to change the translational kinetic energy of an object without changing its rotational kinetic energy?
3. Under what conditions would the tension in the string attached to the bucket in Example 8.4 be equal to the weight of the bucket?
4. Stars originate as large bodies of slowly rotating gas. Because of gravity, these clumps of gas slowly decrease in size. What happens to the angular velocity of a star as it shrinks? Explain.
5. A cat usually lands on its feet regardless of the position from which it is dropped. A slow motion film of a cat falling shows that the upper half of its body twists in one direction while the lower half twists in the opposite direction. Why does this type of rotation occur?
6. Often when a high diver wants to turn a flip in midair, she will draw her legs up against her chest. Why does this make her rotate faster? What should she do when she wants to come out of her flip?
7. As a tetherball winds around a pole, what happens to its angular velocity? Explain.
8. Space colonies have been proposed which consist of large cylinders placed in space. Gravity would be simulated in these cylinders by setting them into rotation. Discuss the difficulties that would be encountered in attempting to set the cylinder into rotation. (*Hint:* consider problems presented by the conservation of angular momentum.)
9. Two cylinders having the same dimensions are set into rotation about their axes with the same angular velocity. One is hollow, while the other is filled with water. Which cylinder would be easier to stop rotating?
10. A mouse is initially at rest on a horizontal turntable mounted on a frictionless vertical axle. If the mouse begins to walk around the perimeter, what happens to the turntable? Explain.
11. A student sits on a stool that is free to rotate about its vertical axis. The student and stool are set into rotation while the student, with outstretched arms, holds a pair of weights. If she suddenly drops the weights to the floor, what happens to her angular velocity? Explain.
12. It is more difficult to do a situp with your hands behind your head than it is when you stretch your arms out in front of you. Why?
13. In order for a helicopter to be stable as it flies, it must have two propellers. Why?
14. Suppose you remove two eggs from the refrigerator, one which had been hard-boiled and the other uncooked. You wish to determine which is the hard-boiled egg without breaking the eggs. This is possible by spinning the two eggs on the floor and comparing the rotational motions. Which egg spins the fastest? Which will rotate more uniformly? Explain. (This is a neat trick for Mom.)
15. If angular momentum is conserved for a propeller-driven airplane, it seems that if the propeller is turning clockwise, the airplane should be turning counterclockwise. Why doesn't this occur?

PROBLEMS

Section 8.1 Relationship Between Torque and Angular Acceleration

1. A 4.5-kg ball on the end of a chain is whirled in a horizontal circle by an athlete. If the distance of the ball from the axis of rotation is 2.5 m, find the moment of inertia of the ball, assuming it can be considered as a point object.

2. What torque must the track star exert on the ball of Problem 1 to give it an angular acceleration of 2 rad/s^2?

3. Four masses are held in position at the corners of a rectangle by light rods as shown in Figure 8.16. Find the moment of inertia of the system about (a) the x axis, (b) the y axis, and (c) an axis through O and perpendicular to the page.

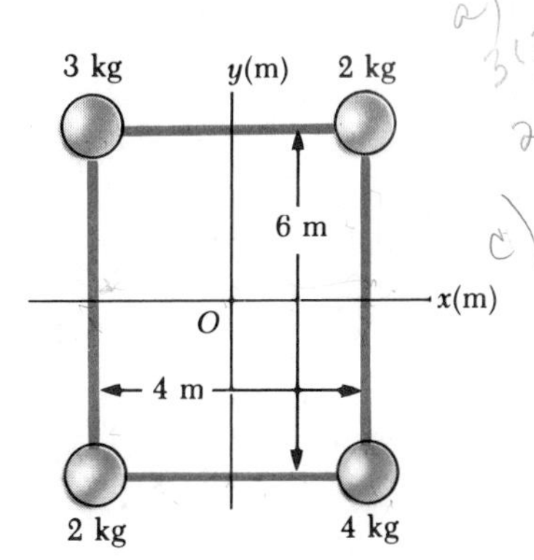

FIGURE 8.16 Problems 3, 4, 5 and 14

4. If the system shown in Figure 8.16 is set into rotation about each of the axes mentioned in Problem 3, find the torque which will produce an angular acceleration of 1.5 rad/s^2 in each case.

5. If a constant torque of 20 N · m acts on the system shown in Figure 8.16 causing it to rotate about the y axis, find its angular velocity after 3 s, assuming the system starts from rest.

6. (a) Find the moment of inertia of a solid cylinder of mass 1.5 kg and radius 30 cm about an axis through its center. (b) Repeat for a solid sphere of the same mass and radius about an axis through its center.

7. The cylinder of Problem 6 is rotating at an angular velocity of 2 rev/s. What torque is required to stop it in 15 s?

• **8.** A merry-go-round in the shape of a horizontal disk of mass 150 kg and radius 1.5 m is set into motion by wrapping a rope about the rim of the disk and pulling on the rope. What constant force would have to be exerted on the rope to bring it from rest to an angular speed of 0.5 rev/s in 2 s?

• **9.** A pulley in the shape of a cylinder of mass 5 kg and radius 0.6 m is used to lower a bucket of mass 3 kg into a well, as shown in Figure 8.17. The bucket starts from rest and falls for 4 s. (a) What is the linear acceleration of the falling bucket? (b) How far does it drop? (c) What is the angular acceleration of the cylinder?

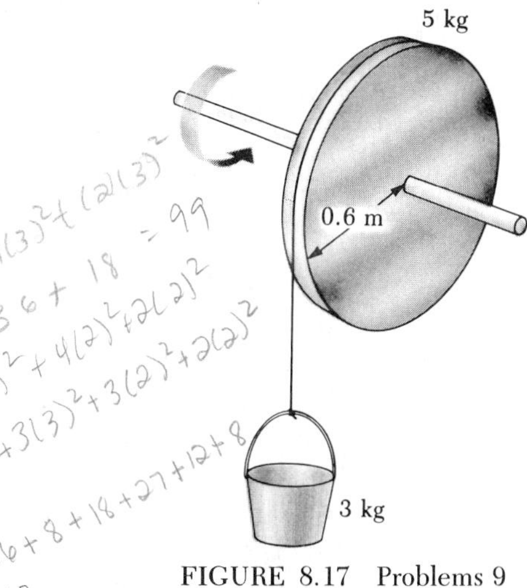

FIGURE 8.17 Problems 9 and 20

•• **10.** A string is wrapped around a uniform 2-kg cylinder of radius 15 cm as shown in Figure 8.18. One end of the string is attached to the ceiling and the cylinder is allowed to fall from rest. (a) Write down Newton's second law for the cylinder. (b) Find the net torque about the center of the cylinder and equate this to $I\alpha$. (c) Use the equations found in (a) and (b) along with $a = r\alpha$ to show that the linear acceleration of the cylinder is given by $a = (2/3)g$.

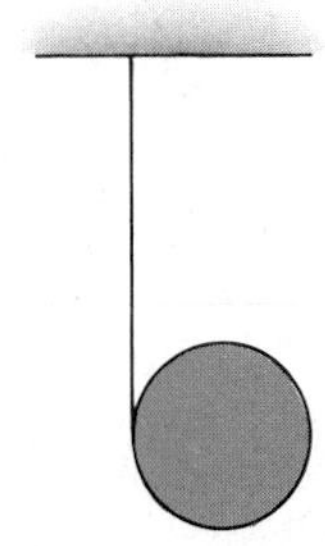

FIGURE 8.18 Problem 10

•• **11.** A grindstone used to sharpen tools is rotating at an angular speed of 8 rad/s when an ax is pressed against it. Because of the torque produced on the grindstone by the force of friction between it and the ax, it comes to rest in four rotations. (a) Find the frictional force produced, assuming the grindstone is a cylinder of mass 5 kg and radius 0.4 m. (b) If the coefficient of friction between ax and grindstone is 0.4, find the normal force with which the ax was pressed against the grindstone.

Section 8.2 Rotational Kinetic Energy

12. An automobile engine part is in the shape of a thin rod of mass 100 g and length 5 cm. When the rod is rotating at an angular velocity of 3 rad/s, find its kinetic energy when (a) rotating about an axis through a point 2.5 cm from each end. (b) Repeat when it is rotating about an axis through one end.

13. An automobile tire, considered as a solid disk, has a radius of 35 cm and a mass of 6 kg. Find its rotational kinetic energy when rotating about an axis through its center at an angular velocity of 2 rev/s.

14. If the system of masses shown in Figure 8.16 is set into rotation about the x axis with an angular velocity of 2.5 rad/s, (a) find the kinetic energy of the system. (b) Repeat the calculation for the system in rotation at the same speed about the y axis.

15. A horizontal 800-N merry-go-round of radius 1.5 m is started from rest by a constant horizontal force of 50 N applied tangentially to the cylinder. Find the kinetic energy of the solid cylinder after 3 s.

16. What angular velocity must a cylindrical wheel of radius 0.4 m and mass 3 kg have in order to have the same kinetic energy as a sphere of mass 3 kg and radius 0.4 m that is rotating at an angular velocity of 9 rad/s?

• **17.** A 2-kg solid ball of radius 0.5 m starts at a height of 3 m above the surface of the earth and *rolls* down a 20° slope. A solid disk and a ring start at the same time and at the same height. Both ring and disk have the same mass and radius as the ball. Which of the three wins the race to the bottom if they all roll without slipping?

•• **18.** A car is designed that gets its energy from a rotating flywheel. To use these cars, the flywheel is attached to an electric motor before leaving on a trip, which brings its rotational speed up to 5000 rev/min. The cylinder has a radius of 2 m and a mass of 500 kg. (a) Find the kinetic energy stored in the flywheel. (b) If the flywheel is to supply energy to the car as would a 10-hp motor, find the time the car could run before the flywheel had to be brought back up to speed.

•• **19.** In a circus performance, a large hoop of mass 5 kg and radius 3 m rolls without slipping. If the hoop is given an angular velocity of 3 rad/s while rolling on the horizontal and allowed to roll up a ramp inclined at 20° with the horizontal, how far (measured along the incline) will the hoop roll?

•• **20.** A bucket is attached to a pulley as shown in Figure 8.17. The cylinder has a mass of 5 kg, and a radius equal to 0.6 m, while the bucket has a mass of 3 kg. Starting from rest, the bucket falls for 3 m. At this point the rope comes free of the pulley. What is the kinetic energy of the pulley at this instant?

Section 8.3 Angular Momentum

21. (a) Calculate the angular momentum of the earth arising from its spinning motion on its axis and (b) the angular momentum of the earth arising from its orbital motion about the sun.

22. The movement of the moon about the earth is such that it revolves once in about every 28 days and it rotates on its axis once in this same time period. The net result of this is that the moon always keeps the same face pointed toward the earth. (a) Find the angular momentum of the moon in its orbit, and (b) its angular momentum due to rotation on its axis.

23. If the size of the earth should increase so that its radius is doubled but the mass is held constant, what would happen to the length of a day?

24. A figure skater rotates about a vertical axis through her center with both arms and one leg extended. Her rotation rate is 0.5 rev/s. She then pulls her arms and leg in closer to her body and begins to spin at 1.7 rev/s. What is the ratio of her new moment of inertia to her old?

• **25.** A student sits on a rotating stool holding two weights, each of mass 3 kg. When his arms are extended horizontally, the weights are 1 m from the axis of rotation and he rotates with an angular speed of 0.75 rad/s. The moment of inertia of the student plus stool is 3 kg · m^2 and is assumed to be constant. If the student pulls the weights horizontally to 0.3 m from the rotation axis, (a) find the new angular speed of the student. (b) Find the kinetic energy of the student before and after the weights are pulled in.

• **26.** A 45-rpm record in the shape of a solid disk 25 cm in diameter and mass 0.1 kg rotates freely about a vertical axis through its center. A 15 g spider drops onto the record at a point 2 cm from the center of the disk. What is the new angular velocity of the record?

•• **27.** A merry-go-round rotates at the rate of 0.2 rev/s with an 80-kg man standing at a point 2 m from the axis of rotation. (a) What is the new angular velocity when the man walks to a point 1 m from the center? Assume the merry-go-round is a solid cylinder of mass 25 kg and radius 2 m. (b) Calculate the change in kinetic energy due to this movement. How do you account for this change in kinetic energy?

•• **28.** The merry-go-round of Problem 27 rotates at the rate of 0.2 rev/s. (a) What is its new angular velocity if a 75-kg person sits on the outer edge? (b) Calculate the change in kinetic energy due to the man sitting down.

Section 8.4 Angular Momentum as a Fundamental Quantity

29. The diatomic oxygen molecule consists of two oxygen atoms separated by a distance of 1.21×10^{-10} m. The mass of each oxygen atom is 2.77×10^{-26} kg. Assume the molecule is in rotation in the xy plane about a z axis which passes through the center of the molecule and is perpendicular to its length. (a) Find the moment of inertia of this molecule about this axis. (b) If its angular momentum is $2\hbar$, find the rotational frequency. (c) How many fundamental units of angular momentum must the oxygen molecule have in order for its rotational frequency to be equal to the typical vibration frequency of this molecule, which is of the order of 10^{13} vibrations per second?

30. In the Bohr model of the hydrogen atom, the electron moves in a circular orbit of radius 0.529×10^{-10} m around the proton. Assuming the orbital angular momentum of the electron is equal to $\hbar$, calculate (a) the orbital speed of the electron, (b) the kinetic energy of the electron, and (c) the angular velocity of the electron.

ADDITIONAL PROBLEMS

31. In a race, a solid sphere, a solid disk, and a cylindrical shell are each set into rotation by a constant torque of 100 N · m. If all objects have a mass of 2 kg and a radius of 0.3 m, which reaches an angular speed of 2 rad/s first? Assume all objects start from rest.

32. A solid sphere of mass 2 kg and radius 0.5 m rotates about an axis through its center. (a) What is its moment of inertia? (b) At what distance, k, from this same axis could a point mass equal to the mass of the sphere be placed so that its moment of inertia is the same as that of the sphere? This distance k is called the *radius of gyration*.

33. A diver springs from a board with arms and legs extended such that her moment of inertia is 1.5 kg · m^2. She rotates at a rate of 0.1 rev/s. She then pulls her legs up against her chest, reducing her moment of inertia to 0.9 kg · m^2. What is her new angular velocity?

34. A seesaw is designed such that it can rotate in a horizontal plane. Two men each of mass 80 kg face one another on the seesaw, which has a moment of inertia of 3 kg · m^2. The men are 1.5 m from the axis of rotation and they are rotating at a speed of 2 rad/s. They then begin to move toward one another. How far apart will they be when the speed has increased to 2.5 rad/s?

• **35.** A sphere 0.2 m in radius and weighing 240 N rolls 6 m down a ramp inclined 37° to the horizontal. What is the angular velocity of the sphere at the bottom of the hill if it starts from rest?

• **36.** Two cylinders start from rest at the top of a plane 6 m long inclined at an angle of 30° with the horizontal. Both cylinders have a mass of 0.5 kg and a radius of 10 cm, but one of the cylinders is solid and the other is a cylindrical shell. When the first one reaches the bottom of the incline, how far down the incline has the other one moved?

• **37.** The system of point masses shown in Figure 8.19 is rotating at an angular velocity of 2 rev/s. The masses are connected by light flexible spokes that can be lengthened or shortened. What is the new angular velocity if the spokes are shortened to 0.5 m? (An effect similar to that illustrated in this problem occurred in the early stages of the formation of our galaxy. As the massive cloud of dust and gas that was the source of the stars and planets contracted, an initially small rotation increased with time.)

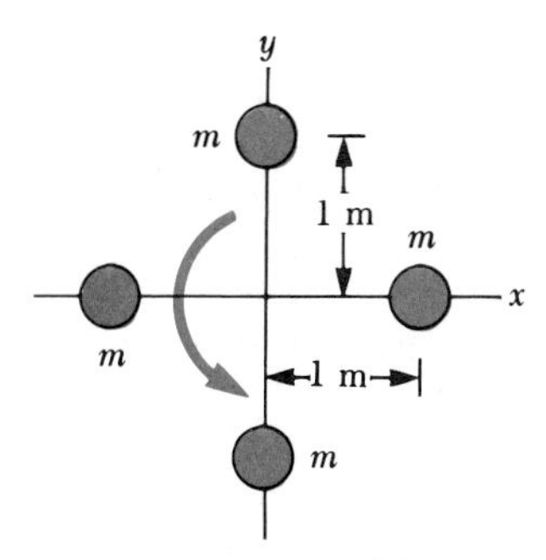

FIGURE 8.19 Problem 37

• **38.** As a planet moves about the sun in an elliptical orbit its angular momentum is a constant. Use this fact to show that the angular speed of the planet is greatest when it is closest to the sun.

• **39.** A spaceship of mass 5000 kg is fired vertically upward from the surface of the earth. Use conservation of angular momentum to show that the spaceship will not be directly overhead when it has risen to its orbit height of 200 km.

• **40.** A 12-kg mass is attached to a cord that is wrapped around a wheel of radius $r = 10$ cm (Fig. 8.20). The acceleration of the mass down the frictionless incline is measured to be 2.0 m/s^2. Assuming the axle of the wheel to be frictionless, determine (a) the tension in the rope, (b) the moment of inertia of the wheel, and (c) the angular speed of the wheel 2 s after it begins rotating, starting from rest.

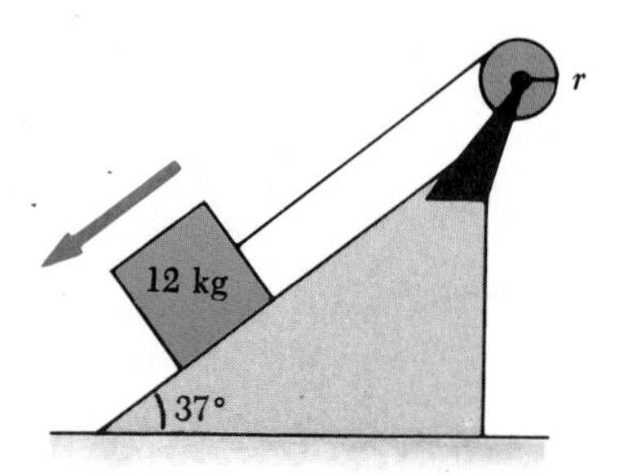

FIGURE 8.20 Problem 40

• **41.** A 3-m length of light nylon cord is wound around a uniform cylindrical spool of radius 0.6 m and 1-kg mass. The spool is mounted on a frictionless axle and is initially at rest. The cord is pulled from the spool with a constant acceleration of 2.5 m/s^2. (a) How much work has been done on the spool, when it reaches an angular speed, $\omega = 6$ rad/s? (b) Assuming there is enough cord on the spool, how long will it take the spool to reach an angular speed of 6 rad/s? (c) Is there enough cord on the spool to enable the spool to reach the angular speed of 6 rad/s?

• **42.** A uniform solid sphere of radius r is placed on the inside surface of a hemispherical bowl of radius R. The sphere is released from rest at an angle θ to the vertical and rolls without slipping (Fig. 8.21). Determine the angular speed of the sphere when it reaches the bottom of the bowl.

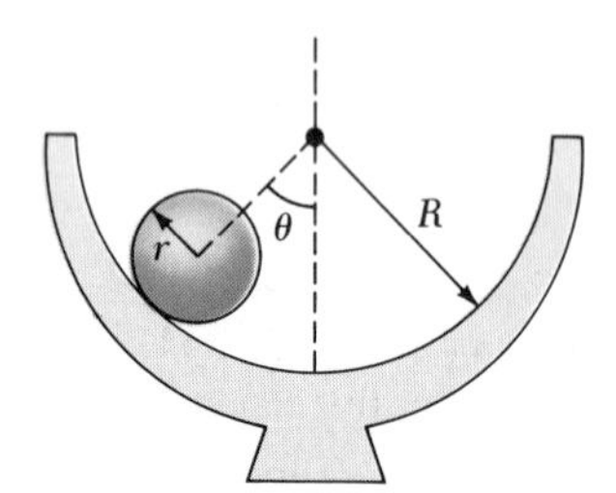

FIGURE 8.21 Problem 42

• **43.** A 50-g mass is attached to a cord passing through a small hole in a frictionless, horizontal surface (Fig. 8.22). The mass is initially orbiting in a circle of radius 0.3 m with velocity 1.5 m/s. The cord is then slowly pulled from below, decreasing the radius of the circle to 0.1 m. (a) What is the velocity of the mass when the radius is 0.1 m? (b) Find the tension in the cord when $r = 0.1$ m.

•• **44.** A string is wrapped around a cylinder of mass 0.5 kg and radius 0.1 m. The string is pulled upward, as shown in Figure 8.23 so that the center of mass of the cylinder does not move. (a) What force must be exerted on the string to cause this? (b) Assuming the cylinder is initially at rest, what angular velocity will it have after 0.2 s?

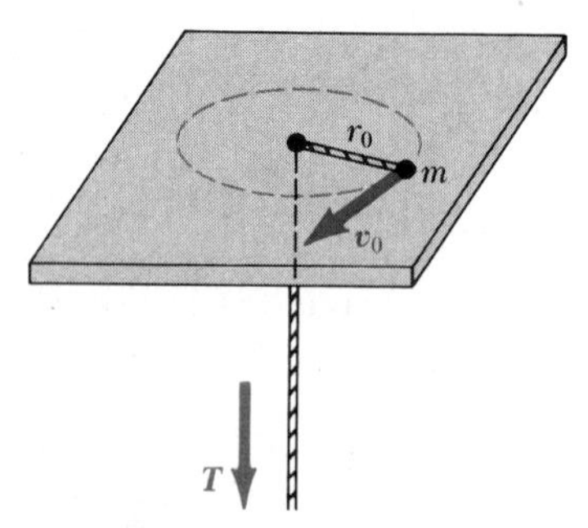

FIGURE 8.22 Problem 43

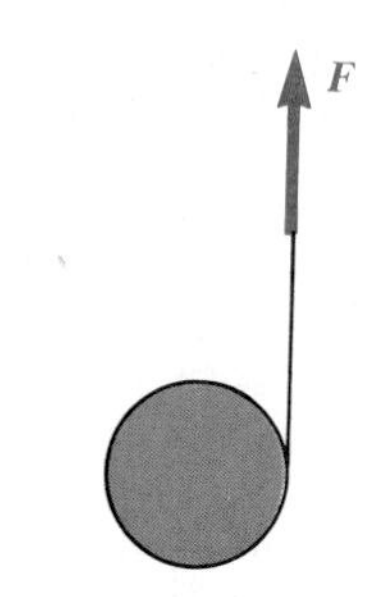

FIGURE 8.23 Problem 44

•• **45.** A large printing-press roller in the shape of a cylinder 0.7 m in radius and weighing 1800 N is brought uniformly from rest to an angular speed of 1400 rev/min in 150 s. What average power must be provided by the drive belt during this time interval?

•• **46.** A 50-kg child stands on the rim of a merry-go-round of radius 2 m and a mass of 150 kg, shaped as a solid cylinder. The merry-go-round is at rest when the child begins to run at a speed of 2 m/s relative to the earth clockwise around the rim of the disk. In what direction and with what angular speed does the turntable begin to rotate?

•• **47.** A 4-kg mass is connected by a light cord to a 3 kg mass on a smooth surface (Fig. 8.24). The pulley rotates about a frictionless axle and has a moment of inertia of 0.5 kg · m^2 and radius 0.3 m. Assuming the cord does not slip on the pulley, find (a) the acceleration of the two masses, and (b) the tensions T_1 and T_2.

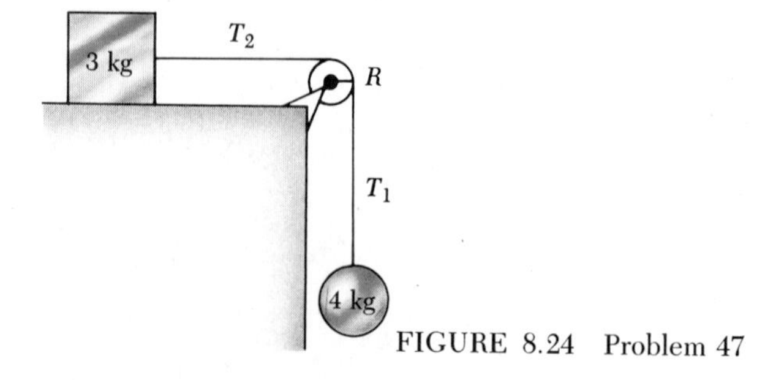

FIGURE 8.24 Problem 47

•• **48.** The pulley in Figure 8.25 has a moment of inertia of 5 kg · m^2 and radius 0.5 m. If the cord supporting the masses m_1 and m_2 does not slip and the axle is frictionless, find (a) the acceleration of the masses when $m_1 = 2$ kg and $m_2 = 5$ kg. Also, find (b) the tension in the cable supporting m_1 and the tension in the cable supporting m_2 (note that they are different).

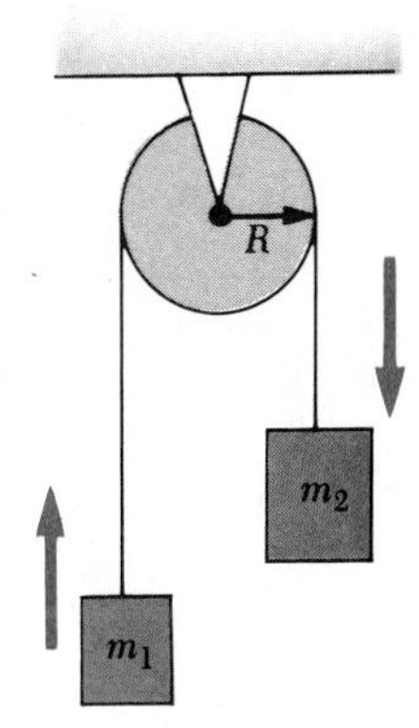

FIGURE 8.25 Problem 48

•• **49.** A small solid sphere of mass m and radius r rolls without slipping along the track shown in Figure 8.26. If it starts from rest at the top of the track, what is the minimum value of h (in terms of the radius of the loop R) such that the sphere completes the loop?

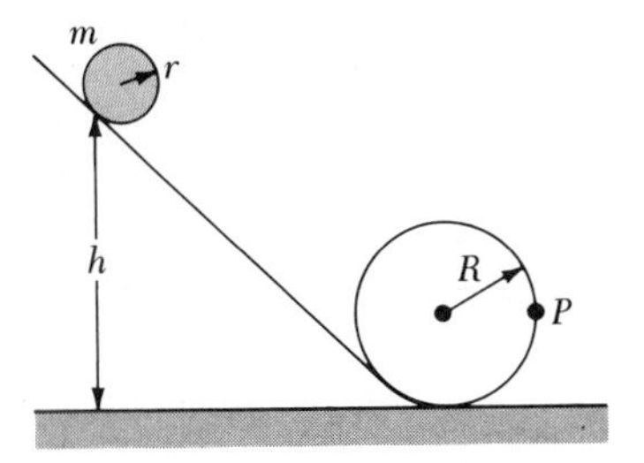

FIGURE 8.26 Problem 49

•• **50.** A uniform solid cylinder of mass M and radius R rotates on a horizontal, frictionless axle (Fig. 8.27). Two equal masses hang from light cords wrapped around the cylinder. If the system is released from rest, find (a) the tension in each cord and (b) the acceleration of each mass after the masses have descended a distance h.

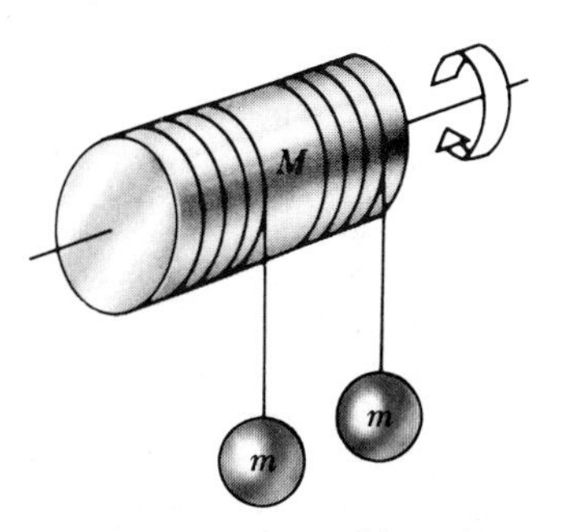

FIGURE 8.27 Problem 50

•• **51.** Suppose the pulley of Problem 50 has a moment of inertia $I = 5$ kg · m^2 and radius $R = 0.5$ m. If the cord supporting the masses $m_1 = 2$ kg and $m_2 = 5$ kg does not slip and the axle is frictionless, find (a) the acceleration of the masses, and (b) the tension in each cable.

CHAPTER 9

Solids and Fluids

A "crown" produced by a drop of milk falling on a plate covered with a thin layer of milk. Surface tension affects the delicate shapes that change too quickly for any eye to discern. However, the high-speed camera is able to capture this event. This classic photograph was taken by Professor Harold E. Edgerton, MIT.

In this chapter we shall consider some properties of solids and fluids (liquids and gases). We shall spend some time looking at properties that are peculiar to solids, but much of our emphasis here will be on the properties of fluids. We have taken this approach because an understanding of the behavior of fluids is of fundamental importance to students in the life sciences. In fact, the study of fluids is so important that we shall devote two chapters to this subject. The study of fluids in this chapter will be confined to a study of fluids at rest, while Chapter 10 will be concerned with the physics of fluids in motion.

9.1 STATES OF MATTER

Matter is normally classified as being in one of three states: solid, liquid, or gaseous. Often, this classification is extended to include a fourth state referred to as a plasma. This fourth state can occur when matter is heated to high temperatures. Under these conditions, many electrons surrounding each atom are freed from the nucleus. This resulting substance is a collection of free electrically charged particles: the negatively charged electrons and the positively charged ions. Such a highly ionized gas with equal amounts of positive and negative charges is called a **plasma.** The plasma state exists inside stars, for example. If we were to take a grand tour of our universe, we would find that there is far more matter in the plasma state than in the more familiar forms of solid, liquid, and gas because there are far more stars around than any other form of celestial matter. However, in this chapter we shall ignore this plasma state and concentrate instead on the more familiar solid, liquid, and gaseous forms that make up the environment on our planet.

Everyday experience tells us that a solid has a definite volume and shape. A brick maintains its familiar shape and size day in and day out. We also know that a liquid has a definite volume but no definite shape. For example, when you fill the tank on a lawn mower, the gasoline changes its shape from that of the original container to that of the tank on the mower, but if you have a gallon of gasoline before you pour, you will have a gallon after. Finally, a gas has neither definite volume nor definite shape. These definitions help us to picture the states of matter, but they are somewhat qualitative. For example, asphalt and plastics are normally considered solids, but over long periods of time they tend to flow like liquids. Likewise, water can be a solid, liquid, or gas (or combinations of these) depending on the temperature and pressure.

All matter consists of some distribution of atoms and molecules. The atoms in a solid are held at specific positions with respect to one another by forces that are mainly electrical in origin. The atoms of a solid vibrate about these equilibrium positions because of thermal agitation. However, at low temperatures, this vibrating motion is slight and the atoms can be considered to be almost fixed. As thermal energy (heat) is added to the material, the amplitude of these vibrations increases. One can view the vibrating motion of the atom as that which would occur if the atom were bound in its equilibrium position by springs attached to neighboring atoms. Such a vibrating collection of atoms and imaginary springs is shown in Figure 9.1. If a solid is compressed by external forces, we can picture these external forces as compressing these tiny internal springs. When the external forces are removed, the solid tends to return to its original shape and size. For this reason, a solid is said to have elasticity.

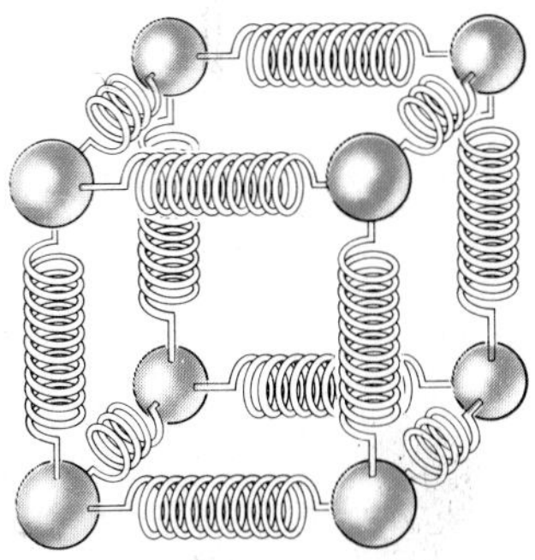

FIGURE 9.1 A model of a solid. The atoms (spheres) are imagined as being attached to each other by springs, which represent the elastic nature of the interatomic forces.

Solids can be classified as being either crystalline or amorphous. A **crystalline solid** is one in which the atoms have an ordered, periodic structure. For example, in the sodium chloride crystal (common table salt), sodium and chlorine atoms occupy alternate corners of a cube, as in Figure 9.2a. In an **amorphous solid,** such as glass, the atoms are arranged in a disordered, or random, fashion, as in Figure 9.2b.

In any given substance, the liquid state exists at a higher temperature than the solid state. Thermal agitation is greater in the liquid state than in the solid state. As a result, the molecular forces in a liquid are not strong enough to keep the molecules in fixed positions, and they wander through the liquid in a random fashion (Fig. 9.2c). Solids and liquids have the following property in common. When one tries to compress a liquid or a solid, strong repulsive atomic forces act internally to resist the deformation.

In the gaseous state, the molecules are in constant random motion and exert only weak forces on each other. The average separation distances between the molecules of a gas are quite large compared with the size of the molecules. Occasionally, the molecules collide with each other; however,

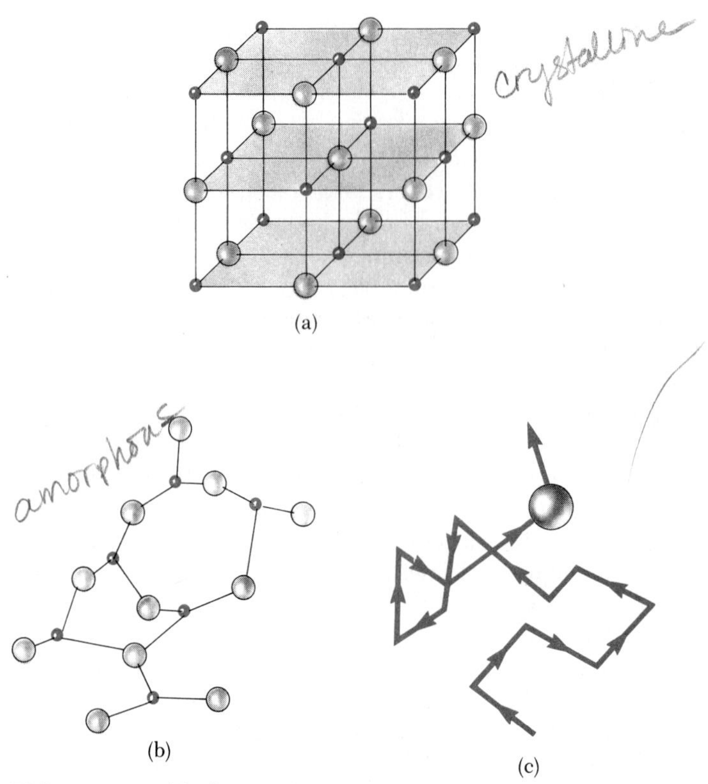

FIGURE 9.2 (a) The NaCl structure, with the Na^+ and Cl^- ions at alternate corners of a cube. The smaller spheres represent Na^+ ions, and the larger spheres represent Cl^- ions. (b) In an amorphous solid, the atoms are arranged in a random fashion. (c) Erratic motion of a molecule in a liquid.

most of the time they move as nearly free, noninteracting particles. We shall have more to say about the properties of gases in subsequent chapters.

9.2 BONDING IN SOLIDS

In this section we shall briefly describe how atoms or molecules combine to form solids. As more is learned about electrical forces and quantum mechanics in future chapters, we will fill in many details which must be omitted from our present discussion. For the present we shall focus on three important mechanisms by which atoms are bound together to form solids. These bonding mechanisms, in turn, will enable you to better understand certain characteristics of solids such as their melting point, and their electrical properties.

Ionic Bonds

Ionic bonds are fundamentally due to an attractive electrical force between oppositely charged ions. An ion is any atom that has excess electrons or is deficient in electrons. Common table salt, NaCl, is an excellent example of a solid held together by ionic bonds. When mixed together to form the compound NaCl, a sodium atom gives up its single outermost electron to a chlorine atom. As a result, the sodium ion is left with a single positive charge, a state designated as Na^+, and the chlorine ion is left with a single negative charge, represented as Cl^-. It is essentially the electrical force of attraction between these unlike charges that creates a stable solid structure. In the solid state, each Na^+ ion is surrounded by six Cl^- ions, and each Cl^- ion is surrounded by six Na^+ ions, as shown in Figure 9.3. Because of the orderly arrangement of atoms in a crystalline solid, some crystals can be split or cleaved with relative ease along certain planes in the solid. For example, NaCl can be cleaved (with a sharp razor struck by a hammer) along any plane parallel to the sides of its basic cubic structure shown in Figure 9.2a. On the other hand, if you try to cleave NaCl along a diagonal plane of the cube, it would shatter or crumble. The cleavage planes that form the sides of a cube for NaCl can be easily observed by viewing a table salt crystal under a microscope. Other crystals such as calcite ($CaCO_3$) and diamond have more complex cleavage planes because of their complex crystalline structure.

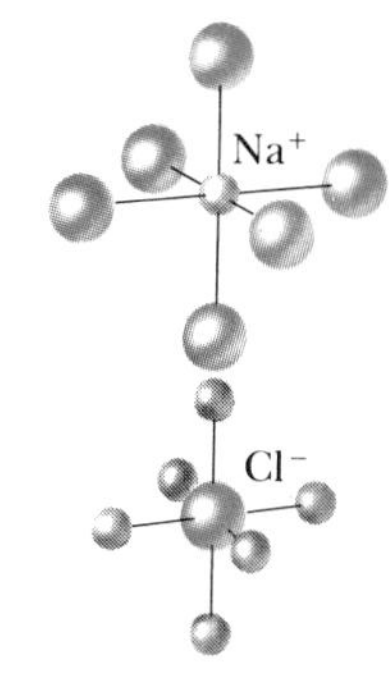

FIGURE 9.3 In the NaCl structure, each positive sodium ion is surrounded by six negative chlorine ions, while each chlorine ion is surrounded by six sodium ions.

Ionic

Certain properties of crystals can easily be understood solely on the basis of the strength of their bonds. For example, ionically bonded crystals are relatively stable and hard because ionic bonds are relatively strong. This explains why high temperatures are required to break the bonds apart and melt these crystals.

As we shall learn in our study of electricity, in order for a material to be a good conductor of electricity, it must have a large number of electrons that are free to move about its structure. An ionic solid has no free electrons because all of its electrons are either involved in the bonding process or are contained in the stable states of the normal atoms. For this reason, ionic crystals are poor conductors of electricity.

(2) Covalent Bonds

A **covalent bond** between two atoms can be visualized as the sharing of electrons supplied by one or both atoms that are bonded together. In many gases, atoms are bonded to other atoms by covalent bonds to form molecules. For example, the hydrogen molecule, H_2, is formed as shown in Figure 9.4. The constituent atoms which form this molecule are two neutral hydrogen atoms, each containing one electron. When the H_2 molecule is formed, the two electrons circle both atomic cores. (This is in contrast to the free atom in which the electron circles only its own nucleus.) Thus, in essence the bond system is the result of electron sharing. The actual mechanism that explains the attraction of two hydrogen atoms to form an H_2 molecule involves a quantum-mechanical effect.

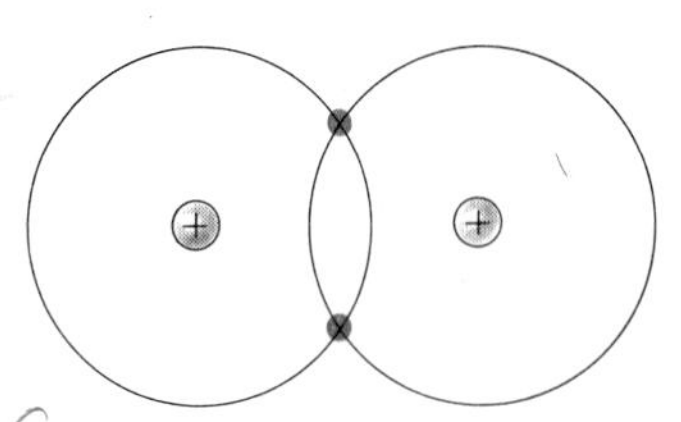

FIGURE 9.4 The covalent bond formed by the two electrons of the H_2 molecule.

Covalent

The covalent bond is also very strong, comparable to that of the ionic bond. Solid carbon, in the form of diamond, is an example of a crystal whose atoms are covalently bonded. Other examples of covalently bonded solids are silicon and germanium. In general, covalently bonded solids are very hard, have high melting points, and are poor electrical conductors.

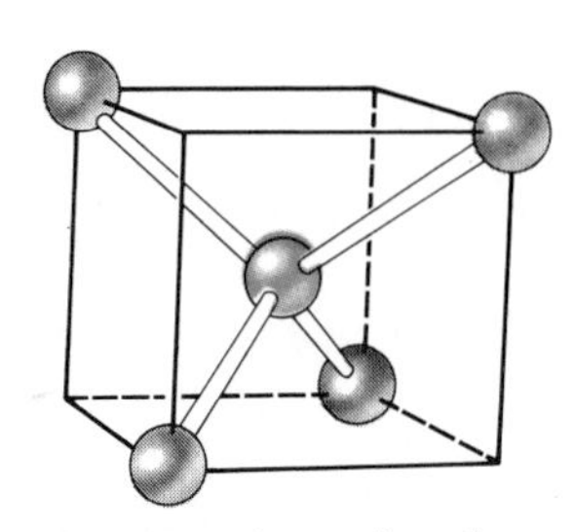

FIGURE 9.5 Each carbon atom in the diamond structure is covalently bonded to four other carbon atoms.

In the diamond structure, each carbon atom is covalently bonded to four other carbon atoms located at four corners of a cube (to form a tetrahedron) as in Figure 9.5. The crystalline structure of diamond is shown in Figure 9.6a.

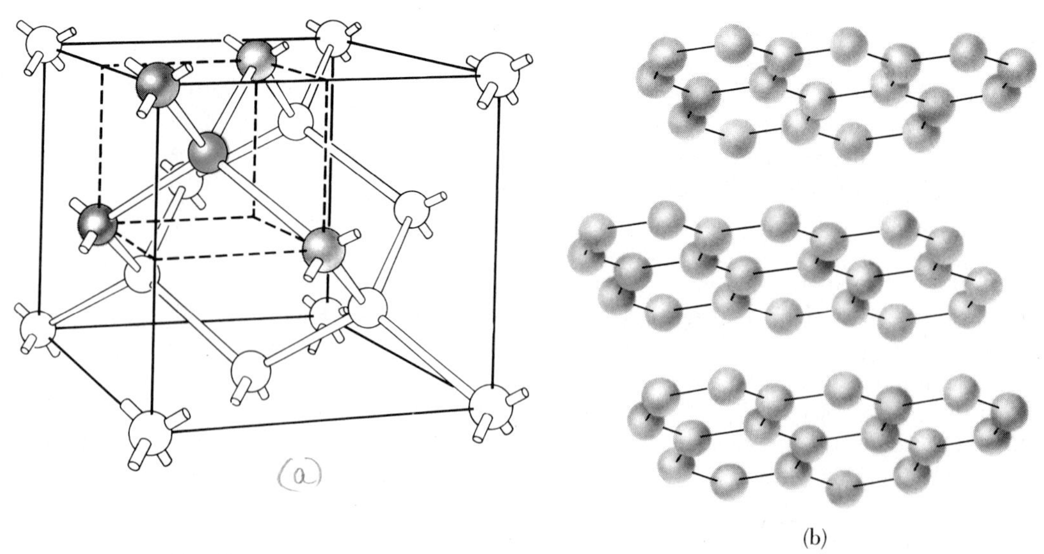

FIGURE 9.6 (a) The crystal structure of diamond. (b) The layered structure of graphite.

Each carbon atom has four valence electrons that are used in creating the covalent bonds. Each bond consists of a pair of electrons between an atom and each of its nearest neighbors. Likewise, each nearest neighbor of a given atom contributes an electron to each bond. Since all four valence electrons of a given atom are used in establishing the four covalent bonds, each atom is strongly bound to its nearest neighbors.

The cleavage planes of diamond are very complex as you might expect in view of its crystal structure shown in Figure 9.6a. For this reason, diamond cutters must be very careful and skilled when attempting to cut a typical diamond. One mistake could result in many small diamonds whose fate might be the grit on an industrial grinding tool. Graphite, another form of pure carbon, has a crystal structure which is considerably different than that of diamond as shown in Figure 9.6b. Note that the carbon atoms in the graphite structure are arranged in layers. Covalent bonding occurs between carbon atoms in a given plane; however, the bonding between atoms in adjacent layers is very weak, so the layers can readily slide past each other. This explains why graphite is very soft and slippery and can be used as a lubricant.

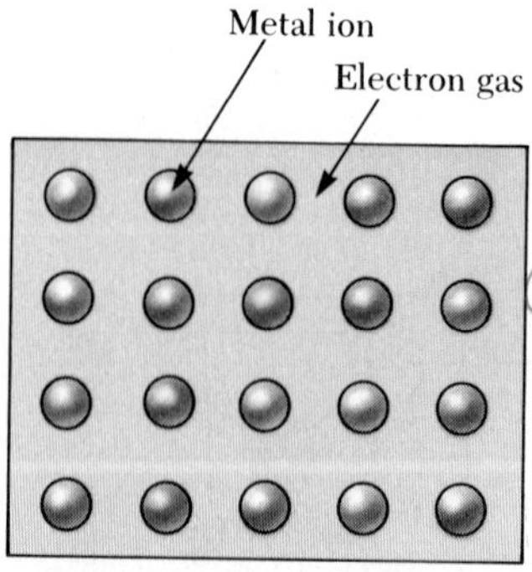

Metal ion
Electron gas

FIGURE 9.7 Schematic diagram of a metal. The colored area represents the electron gas, while the spheres represent the positive metal ion cores.

Metallic Bonding

The atoms in a metal are bonded in a manner quite different from the bonding of atoms in a molecule or a covalently bonded solid, where electrons are either exchanged or shared. Instead, in a metal, some of the outer electrons are freed from individual atoms and drift through the entire solid much like a gas. In effect, the so-called "free electrons" are shared by many atoms and the bonding is distributed throughout the metal. The positive ions left behind are held in position like buoys in a sea of electrons (see Fig. 9.7). The binding mechanism is the attractive force between the positive ions and the electron gas. These bonds are generally weaker than either ionic or covalent bonds.

Because metals have a large number of free electrons (of the order of 10^{28} electrons/m^3), they are good conductors of heat and electricity.

9.3 ELASTIC PROPERTIES OF SOLIDS

In our study of mechanics, we assumed that objects remain undeformed when external forces act on them. In reality, all objects are deformable. That is, it is possible to change the shape or size of a body (or both) through the application of external forces. Although these changes are observed as large-scale deformations, the internal forces that resist the deformation are due to short-range forces between atoms.

We shall discuss the elastic properties of solids in terms of the concepts of stress and strain. **Stress** is a quantity that is related to the force causing a deformation; **strain** is a measure of the degree of deformation. It is found that, for sufficiently small stresses, the stress is proportional to the strain and the constant of proportionality depends on the material being deformed and on the nature of the deformation. We call this proportionality constant the **elastic modulus.** The elastic modulus is therefore the ratio of stress to strain:

$$\text{Elastic modulus} \equiv \frac{\text{stress}}{\text{strain}} \tag{9.1}$$

We shall consider three types of deformation and define an elastic modulus for each:

1. **Young's modulus,** which measures the resistance of a solid to a change in its length
2. **Shear modulus,** which measures the resistance to displacement of the planes of a solid sliding past each other
3. **Bulk modulus,** which measures the resistance that solids or liquids offer to changes in their volume

Young's Modulus: Elasticity in Length

Consider a long bar of cross-sectional area A and length L_0 that is clamped at one end (Fig. 9.8). When an external force $\boldsymbol{F}$ is applied along the bar and perpendicular to the cross section, internal forces in the bar resist distortion ("stretching"), but the bar attains an equilibrium in which its length is greater and in which the external force is exactly balanced by internal forces. In such a situation, the bar is said to be stressed. We define the **tensile stress** as the ratio of the magnitude of the external force F to the cross-sectional area A. The SI unit of stress is the newton per square meter. This unit is also given a special name, the **pascal** (Pa).

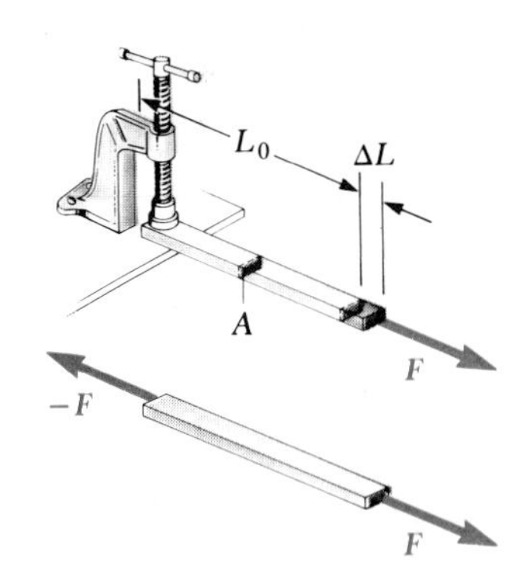

FIGURE 9.8 A long bar clamped at one end is stretched by an amount ΔL under the action of a force $\boldsymbol{F}$.

The pascal

$$1 \text{ Pa} \equiv \text{N/m}^2 \tag{9.2}$$

The **tensile strain** in this case is defined as the ratio of the change in length, ΔL, to the original length, L_0, and is therefore a dimensionless quantity.

TABLE 9.1 Typical Values for Elastic Modulus

Substance	Young's Modulus (Pa)	Shear Modulus (Pa)	Bulk Modulus (Pa)
Aluminum	7.0×10^{10}	2.5×10^{10}	7.0×10^{10}
Brass	9.1×10^{10}	3.5×10^{10}	6.1×10^{10}
Copper	11×10^{10}	4.2×10^{10}	14×10^{10}
Steel	20×10^{10}	8.4×10^{10}	16×10^{10}
Tungsten	35×10^{10}	14×10^{10}	20×10^{10}
Glass	$6.5–7.8 \times 10^{10}$	$2.6–3.2 \times 10^{10}$	$5.0–5.5 \times 10^{10}$
Quartz	5.6×10^{10}	2.6×10^{10}	2.7×10^{10}
Water	—	—	0.21×10^{10}
Mercury	—	—	2.8×10^{10}

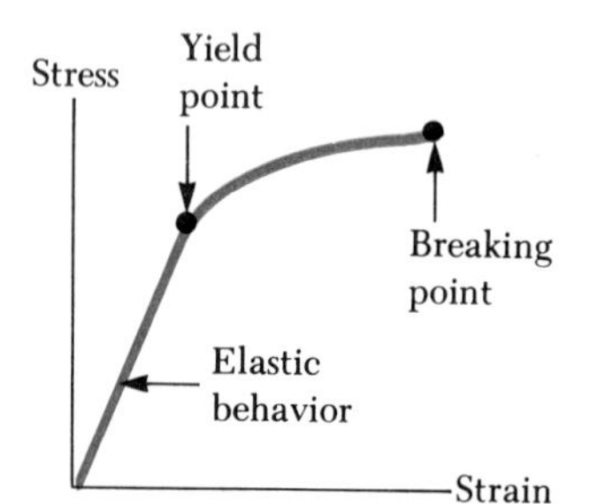

FIGURE 9.9 Stress-strain curve for an elastic solid.

Thus, we can use Equation 9.1 to define **Young's modulus,** Y:

$$Y \equiv \frac{\text{tensile stress}}{\text{tensile strain}} = \frac{F/A}{\Delta L/L_0} = \frac{FL_0}{A\Delta L} \tag{9.3}$$

This quantity is typically used to characterize a rod or wire stressed under *either tension or compression*. Note that because the strain is a dimensionless quantity, Y has units of Pa. Typical values are given in Table 9.1. Experiments show that (a) the change in length for a fixed applied force is proportional to the original length and (b) the force necessary to produce a given strain is proportional to the cross-sectional area. Both of these observations are in accord with Equation 9.3.

It is possible to exceed the *elastic limit* of a substance by applying a sufficiently large stress (Fig. 9.9). At the *yield point*, the stress-strain curve departs from a straight line. A material subjected to a stress beyond the yield point will ordinarily not return to its original length when the external force is removed. As the stress is increased even further, the material will ultimately break.

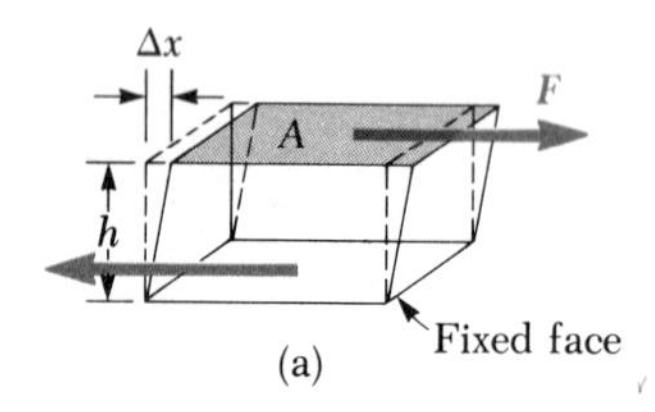

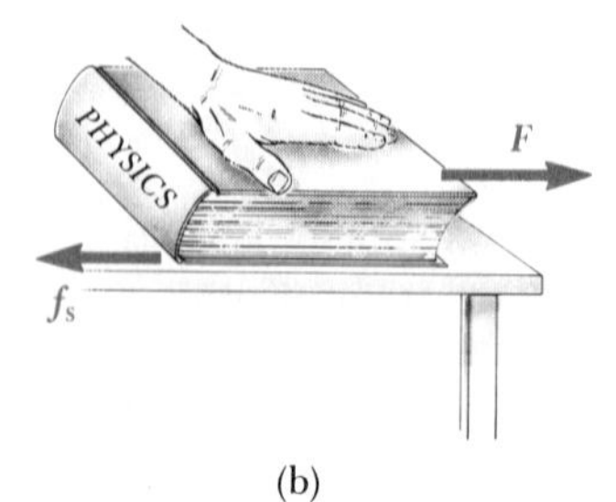

FIGURE 9.10 (a) A shear deformation in which a rectangular block is distorted by a force applied tangent to one of its faces. (b) A book under shear stress.

Shear Modulus: Elasticity of Shape

Another type of deformation occurs when a body is subjected to a force **F** tangential to one of its faces while the opposite face is held in a fixed position (Fig. 9.10a). If the object is originally a rectangular block, a shear stress results in a shape whose cross-section is a parallelogram. For this situation, the stress is called a shear stress. A book pushed sideways as in Figure 9.10b is an example of an object under a shear stress. There is no change in volume under this deformation. We define the **shear stress** as F/A, the ratio of the tangential force to the area, A, of the face being sheared. The **shear strain** is defined as the ratio $\Delta x/h$, where Δx is the horizontal distance the sheared face moves and h is the height of the object. In terms of these quantities, the **shear modulus,** S, is

$$S \equiv \frac{\text{shear stress}}{\text{shear strain}} = \frac{F/A}{\Delta x/h} \tag{9.4}$$

Values of the shear modulus for some representative materials are given in Table 9.1. Note that the units of shear modulus are force per unit area (Pa).

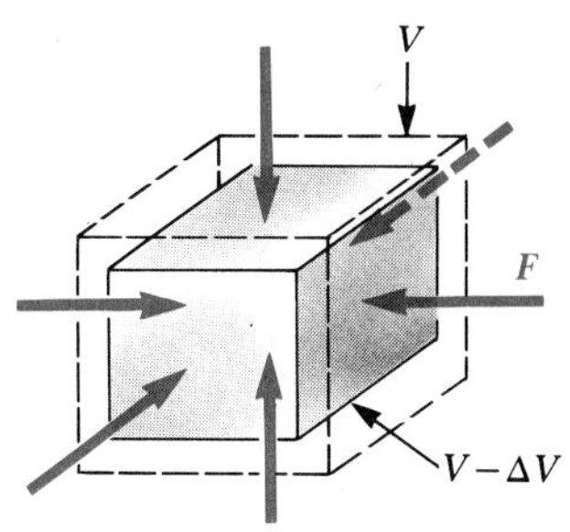

FIGURE 9.11 When a solid is under uniform pressure, it undergoes a change in volume but no change in shape. This cube is compressed on all sides by forces normal to its six faces.

Bulk Modulus: Volume Elasticity

Finally, we define the bulk modulus of a substance, which characterizes the response of the substance to uniform squeezing. Suppose that the external forces acting on an object are at right angles to all of its faces (Fig. 9.11) and distributed uniformly over all the faces. As we shall see later, this occurs when an object is immersed in a fluid. A body subject to this type of deformation undergoes a change in volume but no change in shape. The **volume stress, ΔP**, is defined as the ratio of the magnitude of the normal force, F, to the area, A. When dealing with fluids, we shall refer to this quantity $\Delta P = F/A$ as the **pressure.** The **volume strain** is equal to the change in volume, ΔV, divided by the original volume, V. Thus, from Equation 9.1 we can characterize a volume compression in terms of the **bulk modulus,** B, defined as

Bulk modulus

$$B \equiv \frac{\text{volume stress}}{\text{volume strain}} = -\frac{F/A}{\Delta V/V} = -\frac{\Delta P}{\Delta V/V} \tag{9.5}$$

Note that a negative sign is inserted in this defining equation so that B will always be a positive number. This is because an increase in pressure (positive ΔP) causes a decrease in volume (negative ΔV) and vice versa.

Table 9.1 lists bulk modulus values for some materials. If you look up such values in a different source, you will often find that the reciprocal of the bulk modulus is listed. The reciprocal of the bulk modulus is called the **compressibility** of the material. You should note from Table 9.1 that both solids and liquids have a bulk modulus. However, there is no shear modulus and no Young's modulus for liquids because a liquid will not sustain a shearing stress or a tensile stress (it will flow instead).

EXAMPLE 9.1 Built to Last

A vertical steel beam in a building supports a load of 7×10^4 N. If the length of the beam is 4 m and its cross-sectional area is 8×10^{-3} m^2, find the distance that the beam is compressed along its length.

Solution Since the beam is under compression, we can use Equation 9.3. Taking $Y = 20 \times 10^{10}$ Pa for steel, we have

B =7×10⁴N = F
L=4M
CS=8×10⁻³m

$$Y = \frac{F/A}{\Delta L/L_0}$$

or

$$\Delta L = \frac{FL_0}{YA} = \frac{(7 \times 10^4 \text{ N})(4 \text{ m})}{(20 \times 10^{10} \text{ Pa})(8 \times 10^{-3} \text{ m}^2)} = 1.75 \times 10^{-4} \text{ m}$$

EXAMPLE 9.2 Squeezing a Lead Sphere

A solid lead sphere of volume 0.5 m^3 is dropped in the ocean to a depth where the water pressure is 2×10^7 Pa. Lead has a bulk modulus of 7.7×10^9. What is the change in volume of the sphere?

Solution From the definition of bulk modulus, we have

$$B = -\frac{\Delta P}{\Delta V/V}$$

or

$$\Delta V = -\frac{V\Delta P}{B}$$

In this case, the change in pressure, ΔP, has the value 2×10^7 Pa. (This is large relative to atmospheric pressure, 1.01×10^5 Pa.) Taking $V = 0.5\ \text{m}^3$ and the bulk modulus $B = 7.7 \times 10^9$ Pa, we get

$$\Delta V = -\frac{(0.5\ \text{m}^3)(2 \times 10^7\ \text{Pa})}{7.7 \times 10^9\ \text{Pa}} = -1.3 \times 10^{-3}\ \text{m}^3$$

The negative sign indicates a *decrease* in volume.

9.4 Density and Pressure

The density of a substance of uniform composition is defined as *its mass per unit volume.*

In symbolic form, a substance of mass M and volume V has a density, ρ (Greek rho), given by

Density

$$\rho \equiv \frac{M}{V} \tag{9.6}$$

The units of density are kilograms per cubic meter in the SI system and grams per cubic centimeter in the cgs system. Table 9.2 lists the densities of various substances. These values vary slightly with temperature because the volume of a substance depends on the temperature (as we shall see in a later chapter). Note that under normal conditions the densities of solids and liquids are about 1000 times greater than the densities of gases. This implies that the average spacing between molecules in a gas under these conditions is about 10 times greater than in a solid or liquid.

TABLE 9.2 Density of Some Common Substances

Substance	ρ (kg/m^3)[a]	Substance	ρ (kg/m^3)[a]
Ice	0.917×10^3	Water	1.00×10^3
Aluminum	2.70×10^3	Glycerin	1.26×10^3
Iron	7.86×10^3	Ethyl alcohol	0.806×10^3
Copper	8.92×10^3	Benzene	0.879×10^3
Silver	10.5×10^3	Mercury	13.6×10^3
Lead	11.3×10^3	Air	1.29
Gold	19.3×10^3	Oxygen	1.43
Platinum	21.4×10^3	Hydrogen	8.99×10^{-2}
		Helium	1.79×10^{-1}

[a] All values are at standard atmospheric pressure and temperature (STP). To convert to grams per cubic centimeter, multiply by 10^{-3}.

The **specific gravity** of a substance is defined as the ratio of its density to the density of water at 4°C, which is 1.0×10^3 kg/m^3. By definition, specific gravity is a dimensionless quantity. For example, if the specific gravity of a substance is 3, its density is $3(1.0 \times 10^3 \text{ kg/m}^3) = 3.0 \times 10^3$ kg/m^3.

We have seen that fluids do not sustain shearing stresses, and thus the only stress that can exist on an object submerged in a fluid is one that tends to compress the object. The force exerted by the fluid on the object is always perpendicular to the surfaces of the object, as shown in Figure 9.12.

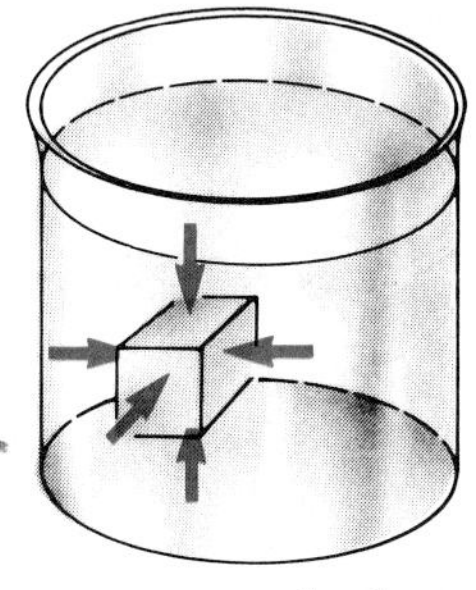

FIGURE 9.12 The force of the fluid on a submerged object at any point is perpendicular to the surface of the object. The force of the fluid on the walls of the container is perpendicular to the walls at all points.

The pressure at a specific point in a fluid can be measured with the device pictured in Figure 9.13. The device consists of an evacuated cylinder enclosing a light piston connected to a spring. As the device is submerged in a fluid, the fluid presses down on the top of the piston and compresses the spring until the inward force of the fluid is balanced by the outward force of the spring. The fluid pressure can be measured directly if the spring is calibrated in advance. This is accomplished by applying a known force to the spring to compress it a given distance.

If F is the magnitude of the force on the piston and A is the area of the piston, then the **average pressure**, P, of the fluid at the level to which the device has been submerged is defined as the ratio of force to area:

Pressure

$$P \equiv \frac{F}{A} \tag{9.7}$$

The technique just described represents a method for measuring average pressure. It is necessary to call this *average* pressure because the pressure in a fluid varies with depth. That is, the pressure at one point on the piston will be different from that at another point on the piston if the two points are at different depths in the fluid. However, if the measuring device is extremely small, its surface is also small and the variation in pressure over such small distances can be neglected. Thus, in this limit of an extremely small piston, one can assume that the device gives the value of the pressure at a specific point.

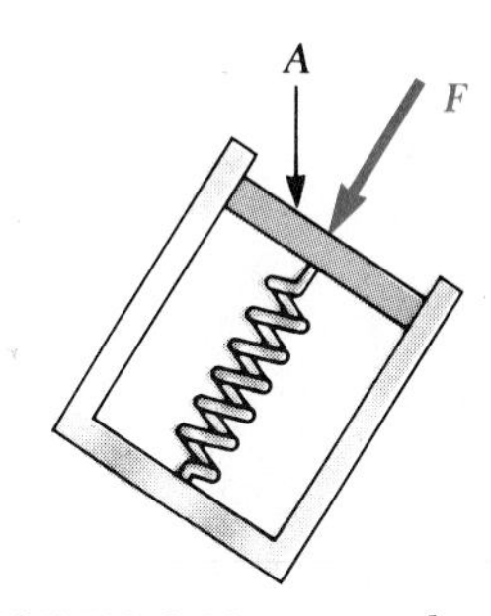

FIGURE 9.13 A simple device for measuring pressure in a fluid.

Since pressure is defined as force per unit area, it has units of Pa (N/m^2).

EXAMPLE 9.3 The Water Bed

A water bed is 2 m on a side and 30 cm deep. (a) Find its weight.

Solution Since the density of water is 1000 kg/m^3, and the bed's volume is equal to $(2 \times 2 \times 0.03)$ m^2 = 1.20 m^3, the mass of the bed is

$$M = \rho V = (1000 \text{ kg/m}^3)(1.20 \text{ m}^3) = 1.20 \times 10^3 \text{ kg}$$

and its weight is

$$w = Mg = (1.20 \times 10^3 \text{ kg})(9.80 \text{ m/s}^2) = 1.18 \times 10^4 \text{ N}$$

This is equivalent to approximately 2640 lb. In order to support such a heavy load, you would be well advised to keep your water bed in the basement or on a sturdy, well-supported floor.

(b) Find the pressure that the water bed exerts on the floor when the bed rests in its normal position. Assume that the entire lower surface of the bed makes contact with the floor.

Solution The weight of the water bed is 1.18×10^4 N. The cross-sectional area is 4 m² when the bed is in its normal position. This gives a pressure exerted on the floor of

$$P = \frac{1.18 \times 10^4 \text{ N}}{4 \text{ m}^2} = 2.95 \times 10^3 \text{ Pa}$$

EXERCISE Calculate the pressure exerted by the bed on the floor if the bed rests on its side.

Answer Since the area of its side is 0.6 m², the pressure is 1.96×10^4 Pa.

9.5 VARIATION OF PRESSURE WITH DEPTH

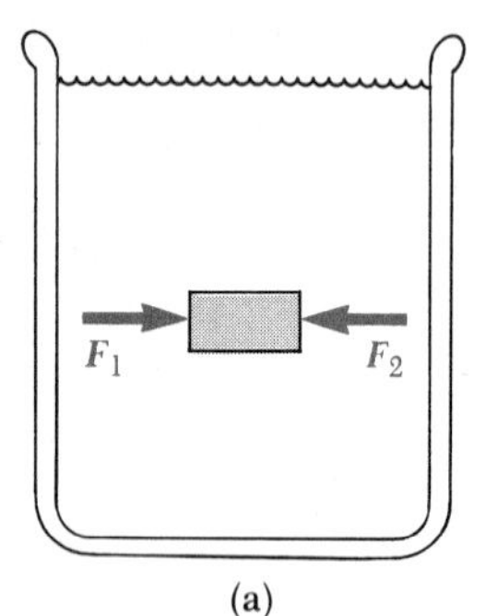

(a)

P_aA h Mg PA

(b)

FIGURE 9.14 (a) If the block of fluid is to be in equilibrium, the force F_1 must balance the force F_2. (b) The net force on the volume of water within the shaded region must be zero.

If a fluid is at rest in a container, *all* portions of the fluid must be in static equilibrium. Furthermore, *all points at the same depth must be at the same pressure*. If this were not the case, a given portion of the fluid would not be in equilibrium. For example, consider the small block of fluid shown in Figure 9.14a. If the pressure were greater on the left side of the block of fluid than on the right, F_1 would be greater than F_2 and the block would accelerate and move and therefore would not be in equilibrium.

Now let us examine the portion of the fluid contained within the volume shown in color in Figure 9.14b. This region has a cross-sectional area A and depth h below the surface of the water. Three external forces act on this volume of fluid: the weight Mg of the fluid, the upward force PA exerted by the fluid below the colored volume, and a downward force P_aA exerted by the atmosphere, where P_a is atmospheric pressure. Since this block of fluid is in equilibrium, these forces must add to zero and so we get

$$PA - Mg - P_aA = 0 \tag{9.8}$$

From the relation $M = \rho V = \rho A h$, the weight of the fluid in the volume is found to be

$$w = Mg = \rho g A h \tag{9.9}$$

When Equation 9.9 is substituted into Equation 9.8, we get

$$\boxed{P = P_a + \rho g h} \tag{9.10}$$

where normal atmospheric pressure is given by $P_a = 1.01 \times 10^5$ Pa (equivalent to 14.7 lb/in.² in the English system of units). According to Equation 9.10,

the absolute pressure, P, at a depth h below the surface of a liquid open to the atmosphere is greater than atmospheric pressure by an amount $\rho g h$. Moreover, the pressure is not affected by the shape of the vessel.

In view of the fact that the pressure in a given fluid depends only upon depth, any increase in pressure at the surface must be transmitted to every point in the fluid. This was first recognized by the French scientist Blaise Pascal (1623–1662) and is called **Pascal's principle:**

Pascal's principle

Pressure applied to an enclosed fluid is transmitted undiminished to every point of the fluid and to the walls of the containing vessel.

An important application of Pascal's principle is the hydraulic press, illustrated by Figure 9.15. A force $\boldsymbol{F}_1$ is applied to a small piston of area A_1. The pressure is transmitted through a fluid to a larger piston of area A_2. Because the pressure is the same on both sides, we see that $P = F_1/A_1 = F_2/A_2$. Therefore, the force $\boldsymbol{F}_2$ is larger than $\boldsymbol{F}_1$ by the factor A_2/A_1, and so a large load, such as a car, can be supported on the large piston by a much smaller force on the smaller piston. Hydraulic brakes, car lifts, hydraulic jacks, fork lifts, and so on make use of this principle.

FIGURE 9.15 Diagram of a hydraulic press. Since the pressure is the same at the left and right sides, a small force $\boldsymbol{F}_1$ at the left produces a much larger force $\boldsymbol{F}_2$ at the right.

EXAMPLE 9.4 The Car Lift

In a car lift used in a garage, compressed air exerts a force on a small piston having a radius of 5 cm. This pressure is transmitted to a second piston of radius 15 cm. What force must the compressed air exert in order to lift a car weighing 13,300 N? What air pressure will produce this force?

Solution Because the pressure exerted by the compressed air is transmitted undiminished throughout the fluid, we have

$$F_1 = \left(\frac{A_1}{A_2}\right)F_2 = \frac{\pi(5 \times 10^{-2}\text{ m})^2}{\pi(15 \times 10^{-2}\text{ m})^2}(13{,}300\text{ N}) = 1.48 \times 10^3\text{ N}$$

The air pressure that will produce this force is given by

$$P = \frac{F_1}{A_1} = \frac{1.48 \times 10^3\text{ N}}{\pi\,(5 \times 10^{-2}\text{ m})^2} = 1.88 \times 10^5\text{ Pa}$$

This pressure is approximately twice atmospheric pressure.

EXAMPLE 9.5 Pressure in the Ocean

Calculate the pressure at an ocean depth of 1000 m. Assume the density of water is 1.0×10^3 kg/m^3 and $P_a = 1.01 \times 10^5$ Pa.

Solution

$$\begin{aligned} P &= P_a + \rho g h \\ &= 1.01 \times 10^5\text{ Pa} + (1.0 \times 10^3\text{ kg/m}^3)(9.8\text{ m/s}^2)(10^3\text{ m}) \\ P &\approx 9.9 \times 10^6\text{ Pa} \end{aligned}$$

This is approximately 100 times greater than atmospheric pressure! Obviously, the design and construction of vessels that will withstand such enormous pressures are not trivial matters.

EXERCISE Calculate the total force exerted on the outside of a circular submarine window of diameter 30 cm at this depth.

Answer 7.0×10^5 N.

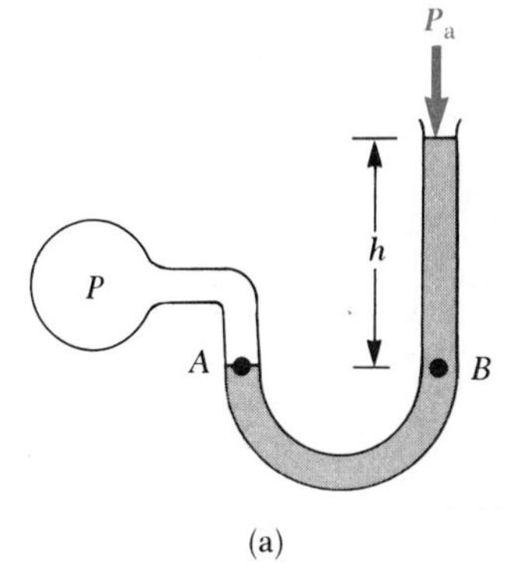

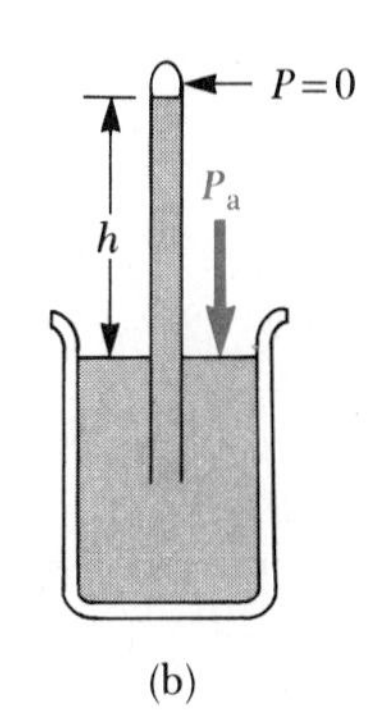

FIGURE 9.16 Two devices for measuring pressure: (a) the open-tube manometer; (b) the mercury barometer.

9.6 PRESSURE MEASUREMENTS

One simple device for measuring pressure is the open-tube manometer illustrated in Figure 9.16a. One end of a U-shaped tube containing a liquid is open to the atmosphere, and the other end is connected to a system of unknown pressure P. The pressure at point B equals $P_a + \rho g h$, where ρ is the

A diver can reach great depths in this modern diving suit. (Photo Courtesy of Oceaneering International, Inc.)

gauge pressure = P − Pa

density of the fluid. The pressure of B, however, equals the pressure at A, which is also the unknown pressure P. Therefore, we conclude that

$$P = P_a + \rho g h$$

absolute pressure

The pressure P is called the *absolute pressure*, and $P - P_a$ is called the *gauge pressure*. Thus, if the pressure P in the system is greater than atmospheric pressure, h is positive. If P is less than atmospheric pressure (a partial vacuum), h is negative.

Another instrument used to measure pressure is the *barometer* (Fig. 9.16b) invented by Evangelista Torricelli (1608–1647). A long tube closed at one end is filled with mercury and then inverted into a dish of mercury. The closed end of the tube is nearly a vacuum, and so its pressure can be taken as zero. Therefore, it follows that $P_a = \rho g h$, where ρ is the density of the mercury and h is the height of the mercury column. One atmosphere of pressure is defined to be the pressure equivalent of a column of mercury that is exactly 0.76 m in height at 0°C, with $g = 9.80665 \text{ m/s}^2$. At this temperature, mercury has a density of $13.595 \times 10^3 \text{ kg/m}^3$; therefore

$$P_a = \rho g h = (13.595 \times 10^3 \text{ kg/m}^3)(9.80665 \text{ m/s}^2)(0.7600 \text{ m})$$
$$= 1.013 \times 10^5 \text{ Pa}$$

It is interesting to note that the force of the atmosphere on our bodies is extremely large, on the order of 30,000 lb! (Assuming a body area of 2000 in.^{2}) A natural question to raise is, how can we exist under such large forces attempting to collapse our bodies? The answer is that our body cavities and tissues are permeated with fluids and gases pushing outward with this same atmospheric pressure. Consequently, our bodies are in equilibrium under the force of the atmosphere pushing in and an equal internal force pushing out.

9.7 BUOYANT FORCES AND ARCHIMEDES' PRINCIPLE

The Greek scientist and mathematician Archimedes invented a number of practical devices, including the pulley, the catapult, and a device for raising water called Archimedes' screw. According to legend, Archimedes was asked by King Hieron to determine whether the king's crown was made of pure gold or had been alloyed with some other metal. The task was to be performed without damaging the crown. Archimedes presumably arrived at a solution while taking a bath, noting a partial loss of weight after submerging his arms and legs in the water. As the story goes, he was so excited about his great discovery that he ran through the streets of Syracuse naked shouting, "Eureka!" which is Greek for "I have found it." **Archimedes' principle** can be stated as follows:

Archimedes' principle

Any body completely or partially submerged in a fluid is buoyed up by a force equal to the weight of the fluid displaced by the body.

Everyone has experienced Archimedes' principle. As an example of a common experience, recall that it is relatively easy to lift someone if you both are standing in a swimming pool, whereas lifting that same individual on dry land may be a very difficult task. Evidently, water provides partial support to any object placed in it. We say that an object placed in a fluid is buoyed up

Archimedes (287–212 B.C.), a Greek mathematician, physicist, and engineer, was perhaps the greatest scientist of antiquity. He was the first to accurately compute the ratio of a circle's circumference to its diameter and also showed how to calculate the volume and surface area of spheres, cylinders, and other geometric shapes. He is well known for discovering the nature of the buoyant force acting on floating objects and was also a gifted inventor. One of his practical inventions, still in use today, is the Archimedes screw, a rotating coiled tube used originally to lift water from the holds of ships. He also invented the catapult and devised systems of levers, pulleys, and weights for raising heavy loads. Such inventions were successfully used by the soldiers of his native city, Syracuse, during a two-year siege by the Romans.

by the fluid, and we call this upward force the **buoyant force**. According to Archimedes' principle, *the magnitude of the buoyant force always equals the weight of the fluid displaced by the object.* The buoyant force acts vertically upward through what was the center of gravity of the fluid before the fluid was displaced.

Archimedes' principle can be verified in the following manner. Suppose we focus our attention on the indicated cube of water in the container of Figure 9.17. This cube of water is in equilibrium under the action of the forces on it. One of these forces is the weight of the cube of water. What cancels this downward force? Apparently, the rest of the water inside the container is buoying up the cube and holding it in equilibrium. Thus, the buoyant force, B, on the cube of water is exactly equal in magnitude to the weight of the water inside the cube:

$$B = w$$

Now, imagine that the cube of water is replaced by a cube of steel of the same dimensions. What is the buoyant force on the steel? The water surrounding a cube will behave in the same way whether a cube of water or a cube of steel is being buoyed up; therefore, *the buoyant force acting on the steel is the same as the buoyant force acting on a cube of water of the same dimensions.* This result applies for a submerged object of any shape, size, or density.

A blimp floats because its density is less than that of the surrounding air. (Courtesy of Goodyear)

Let us show explicitly that the buoyant force is equal in magnitude to the weight of the displaced fluid. The pressure at the bottom of the cube in Fig-

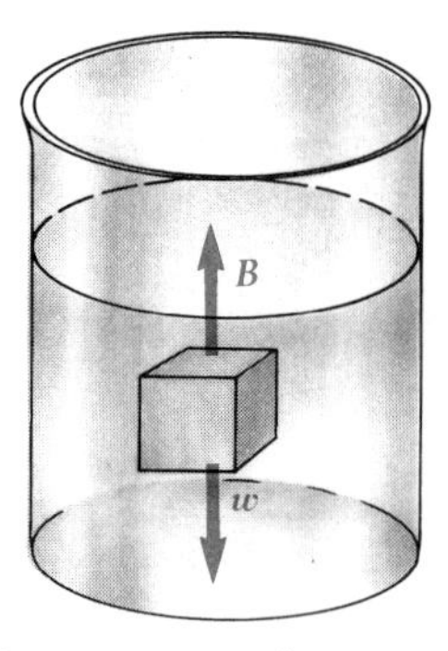

FIGURE 9.17 The external forces on the cube of water are its weight w and the buoyancy force B. Under equilibrium conditions, $B = W$.

ure 9.17 is greater than the pressure at the top by an amount $\rho_f gh$, where ρ_f is the density of the fluid and h is the height of the cube. Since the pressure difference, ΔP, is equal to the buoyant force per unit area, or, $\Delta P = B/A$, we see that $B = (\Delta P)(A) = (\rho_f gh)(A) = \rho_f gV$, where V is the volume of the cube. Since the mass of the water in the cube is $M = \rho_f V$, we see that

$$B = w = \rho_f Vg = Mg \tag{9.11}$$

where w is the weight of the displaced fluid.

Note that the weight of the submerged object is $\rho_o Vg$, where ρ_o is the density of the object. Therefore, if the density of the object is greater than the density of the fluid, the unsupported object will sink. If the density of the object is less than that of the fluid, the unsupported submerged object will accelerate upward and will ultimately float. When a *floating* object is in equilibrium, part of it is submerged. In this case, *the buoyant force equals the weight of the object*.

Under normal conditions, the average density of a fish is slightly greater than the density of water. This being the case, a fish would sink if it did not have some mechanism for adjusting its density. This mechanism is supplied by an internal gas bag. If a fish desires to move higher in the water, it causes this gas bag to expand. Likewise, in order to move lower in the water, the fish contracts the bag by expelling gas.

The human brain is immersed in a fluid of density 1007 kg/m^3, which is slightly less than the average density of the brain, 1040 kg/m^3. Consequently, most of the weight of the brain is supported by the buoyant force of the fluid surrounding the brain. In some clinical procedures it is necessary to remove a portion of this fluid for diagnostic purposes. During such procedures, the nerves and blood vessels in the brain are placed under great strain, which in turn can cause extreme discomfort and pain. One must exercise great care with such patients until the brain fluid has been restored by the body.

Mercury has a density about 13.6 times that of water. As a result, this steel ball is able to float in a pool of mercury.

EXAMPLE 9.6 A Red Tag Special on Crowns

A bargain hunter purchases a "gold" crown at a flea market. After she gets home, the bargain hunter (with some knowledge of Archimedes' principle) hangs the crown from a scale and finds its weight to be 7.84 N as shown in Figure 9.18a. She then weighs the crown while it is immersed in water of density 1000 kg/m^3 as in Figure 9.18b, and finds that the scale reads 6.86 N. Is the crown made of gold?

Solution The buoyant force on the crown is the difference between its weight in air and its weight in water. That is,

$$B = 7.84 \text{ N} - 6.86 \text{ N} = 0.98 \text{ N}$$

Because the buoyant force is equal in magnitude to the weight of the displaced fluid, w_f, we have

$$w_f = \rho_f g V_f = 0.98 \text{ N}$$

where V_f is the volume of the displaced fluid, and ρ_f is the density of the displaced fluid (1000 kg/m^3). Since the volume of the crown, V_c, is equal to the volume of the displaced fluid (for a completely submerged object), we find

$$V_c = V_f = \frac{0.98 \text{ N}}{g\rho_f} = \frac{0.98 \text{ N}}{(9.8 \text{ m/s}^2)(1000 \text{ kg/m}^3)} = 10^{-4} \text{ m}^3$$

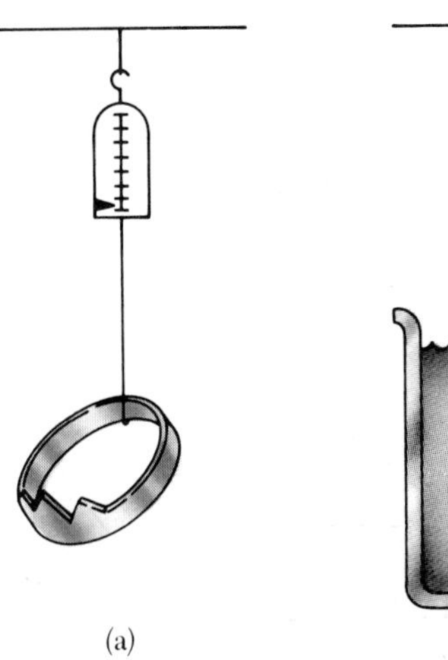

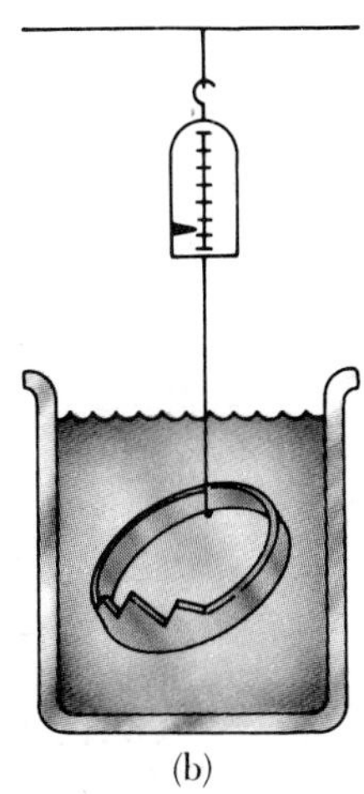

FIGURE 9.18 (Example 9.6) (a) The crown weighs 7.84 N in air, and (b) 6.86 N in water.

Finally, the density of the crown is given by

$$\rho_c = \frac{m_c}{V_c} = \frac{w_c}{gV_c} = \frac{7.84\ \text{N}}{(9.8\ \text{m/s}^2)(10^{-4}\ \text{m}^3)} = 8 \times 10^3\ \text{kg/m}^3$$

From Table 9.2 we see that the density of gold is 19.3×10^3 kg/m^3. Thus, the crown is either hollow or is made of an alloy. This was not a good day for the bargain hunter!

EXAMPLE 9.7 Floating Down the River

A raft is constructed of wood having a density of 600 kg/m^3. Its surface area is 5.9 m^2, and its volume is 0.6 m^3. When the raft is placed in fresh water of density 1000 kg/m^3 as in Figure 9.19, how much of the raft is below water level?

Solution The weight of the raft is calculated to be

$$w_r = \rho_r g V_r = (600\ \text{kg/m}^3)(9.80\ \text{m/s}^2)(0.6\ \text{m}^3) = 3530\ \text{N}$$

According to Archimedes' principle, the upward buoyant force acting on the raft equals (in magnitude) the weight of the displaced water, which, in turn, must equal the weight of the raft. Thus, we have

$$B = \rho_w g V_w = \rho_w g A h = 3530\ \text{N}$$

Since the area A and density, ρ_w, are known, we can use this result to find the depth, h, the raft sinks into the water:

$$h = \frac{3530\ \text{N}}{\rho_w g A} = \frac{3530\ \text{N}}{(1000\ \text{kg/m}^3)(9.8\ \text{m/s}^2)(5.9\ \text{m}^2)} = 0.06\ \text{m}$$

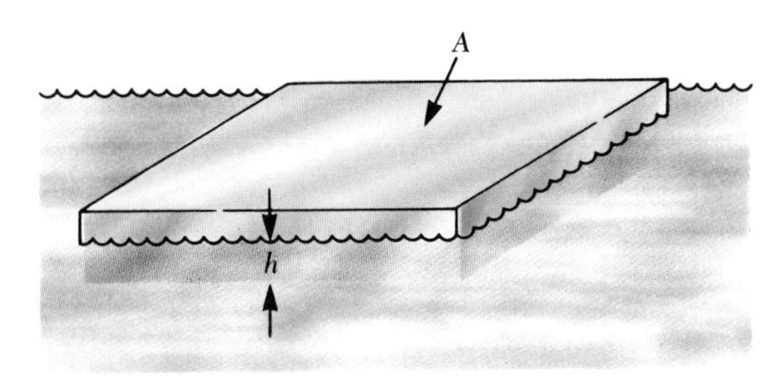

FIGURE 9.19 (Example 9.7) A raft is partially submerged in water.

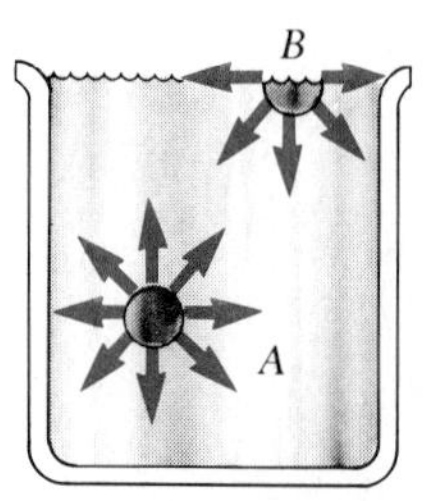

FIGURE 9.20 The net force on a molecule at A is zero because such a molecule is completely surrounded by other molecules. The net force on a surface molecule at B is downward because it is not completely surrounded by other molecules.

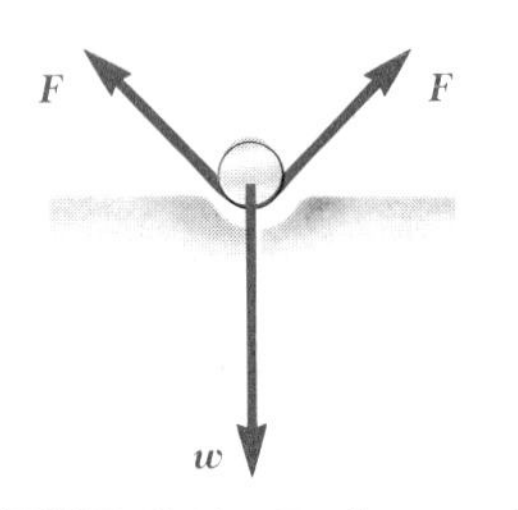

FIGURE 9.21 End view of a needle resting on the surface of water. The components of surface tension balance the weight force.

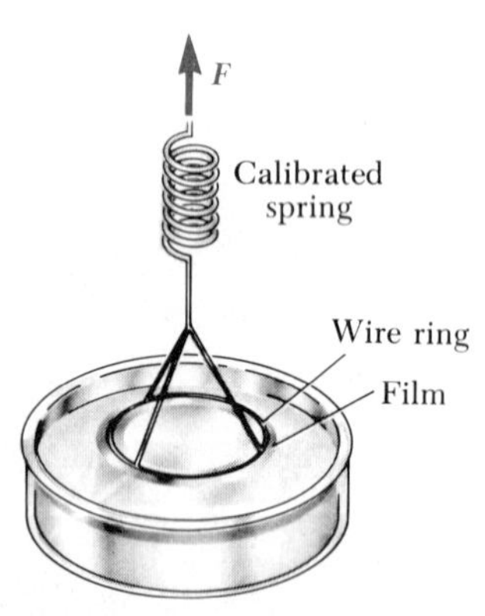

FIGURE 9.22 An apparatus for measuring the surface tension of liquids. The force on the wire ring is measured just before it breaks free of the liquid.

*9.8 SURFACE TENSION

If you look closely at a dewdrop sparkling in the morning sunlight, you will find that the drop is spherical. The drop takes this shape because of a property of liquid surfaces called **surface tension.** In order to understand the origin of surface tension, consider a molecule at point A in a container of water, as in Figure 9.20. Although nearby molecules exert forces on this molecule, the net force on it is zero because it is completely surrounded by other molecules and hence is attracted equally in all directions. The molecule at B, however, is not attracted equally in all directions. Since there are no molecules above it to exert upward forces, the molecule is pulled toward the interior of the liquid. The contraction at the surface of the liquid ceases when the inward pull exerted on the surface molecules is balanced by the outward repulsive forces that arise from collisions with molecules in the interior of the liquid. *The net effect of this pull on all the surface molecules is to make the surface of the liquid contract and consequently to make the surface area of the liquid as small as possible.* Drops of water take on a spherical shape because a sphere has the smallest surface area for a given volume.

If you place a sewing needle very carefully on the surface of a bowl of water, you will find that the needle floats even though the density of steel is about eight times that of water. This also can be explained by surface tension. A close examination of the needle shows that it actually rests in a depression in the liquid surface, as shown in Figure 9.21. The water surface acts like an elastic membrane under tension. The weight of the needle produces a depression, thus increasing the surface area of the film. Molecular forces now act at all points along the depression in an attempt to restore the surface to its original horizontal position. The vertical components of these forces act to balance w, the weight of the needle.

The **surface tension,** γ, in a film of liquid is defined as the ratio of the surface force, F, to the length along which the force acts:

$$\gamma \equiv \frac{F}{L} \tag{9.12}$$

The SI units of surface tension are newtons per meter, and values for a few representative materials are given in Table 9.3.

An apparatus used to measure the surface tension of liquids is shown in Figure 9.22. A circular wire with a circumference l is lifted from a body of liquid. The surface film clings to the inside and outside edges of the wire. This tends to hold back the wire, causing the spring to stretch. If the spring

TABLE 9.3 Surface Tensions for Various Liquids

Liquid	T (°C)	Surface Tension (N/m)
Ethyl alcohol	20	0.022
Mercury	20	0.465
Soapy water	20	0.025
Water	20	0.073
Water	100	0.059

is calibrated, one can measure the force required to overcome the surface tension of the liquid. In this case, the surface tension is given by

$$\gamma = \frac{F}{2l}$$

We must use $2l$ for the length because the surface film exerts forces on the inside and outside of the ring.

A razor blade floats on water because of surface tension.

The surface tension of liquids decreases with increasing temperature. This occurs because the faster moving molecules of a hot liquid are not bound together as strongly as are those in a cooler liquid. Furthermore, certain ingredients added to liquids decrease surface tension. For example, soap or detergent decreases the surface tension of water. This reduction in surface tension makes it easier for soapy water to penetrate the cracks and crevices of your clothes to clean them better than plain water. An effect similar to this occurs in the lungs. The surface tissue of the air sacs in the lungs contains a fluid that has a surface tension of about 0.050 N/m. A liquid with a surface tension this high would make it very difficult for the lungs to expand as one inhales. However, as the area of the lungs increases with inhalation, the body secretes into the tissue a substance that gradually reduces the surface tension of the liquid. At full expansion, the surface tension of the lung fluid can drop to as low as 0.005 N/m.

EXAMPLE 9.8 Walking on Water

In this example, we shall illustrate how an insect is supported on the surface of water by surface tension. Let us assume that the insect's "foot" is spherical. When the insect steps onto the water with all six legs, a depression is formed in the water around each foot, as shown in Figure 9.23. The surface tension of the water produces upward forces on the water which tend to restore the water surface to its normally flat shape. If the insect has a mass of 2×10^{-5} kg and if the radius of each foot is 1.5×10^{-4} m, find the angle θ.

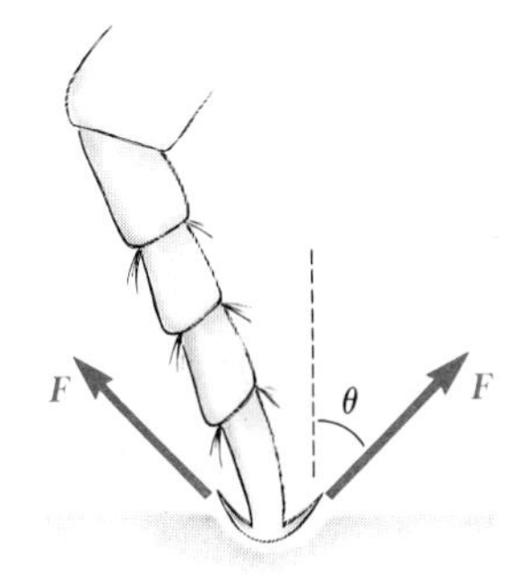

FIGURE 9.23 (Example 9.8) One foot of an insect resting on the surface of water.

Solution From the definition of surface tension, we can find the net force F directed tangential to the depressed part of the water surface:

$$F = \gamma L$$

The length L along which this force acts is equal to the distance around the insect's foot, $2\pi r$. (It is assumed that the insect depresses the water surface such that the radius of the depression is equal to the radius of the foot.) Thus,

$$F = \gamma\, 2\pi r$$

and the net vertical force is

$$F_v = \gamma\, 2\pi r \cos\theta$$

Since the insect has six legs, this upward force must equal one sixth the weight of the insect, assuming its weight is equally distributed on all six feet. Thus,

$$\gamma\, 2\pi r \cos\theta = \frac{1}{6} w$$

$$\text{(1)} \qquad \cos\theta = \frac{w}{12\pi r\gamma} = \frac{(2 \times 10^{-5}\ \text{kg})(9.8\ \text{m/s}^2)}{12\pi\ (1.5 \times 10^{-4}\ \text{m})(0.073\ \text{N/m})}$$

$$\theta = 62°$$

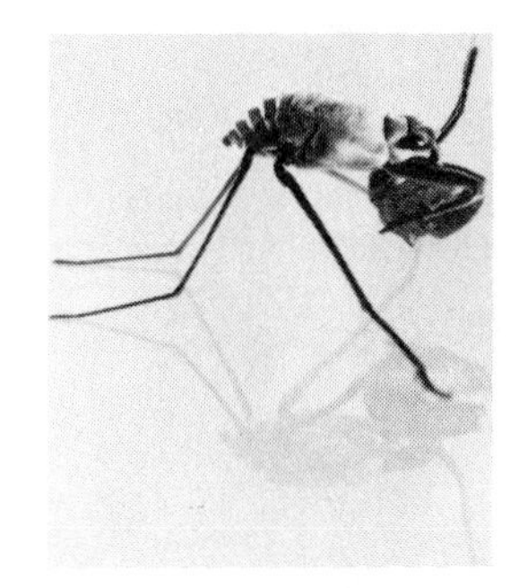

The water strider *Halobates* standing on the water. (Courtesy of Scripps Institution of Oceanography, University of California, San Diego.)

Note that if the weight of the insect were great enough to make the right side of (1) greater than unity, a solution for θ would be impossible because the cosine can never be greater than unity. Under these conditions, the insect would sink.

*9.9 THE SURFACE OF LIQUIDS

If you have ever closely examined the surface of water in a glass container, you may have noticed that the surface of the liquid near the walls of the glass curves upward, as shown in Figure 9.24a. However, if mercury is placed in a glass container, the mercury surface curves downward, as in Figure 9.24b. These surface effects can be explained by considering the forces between molecules. In particular, we must consider the forces that the molecules of the liquid exert on one another and the forces that the molecules of the glass surface exert on those of the liquid. In general terms, forces between like molecules, such as the forces between water molecules, are called *cohesive forces* and forces between unlike molecules, such as those of glass on water, are called *adhesive forces*.

Water tends to cling to the walls of the glass because the adhesive forces between the liquid molecules and the glass molecules are *greater* than the cohesive forces between the liquid molecules. In effect, the liquid molecules cling to the surface of the glass rather than fall back into the bulk of the liquid. When this condition prevails, the liquid is said to wet the glass surface. The surface of the mercury curves downward near the walls of the container because the cohesive forces between the mercury atoms are greater than the adhesive forces between mercury and glass. That is, a mercury atom near the surface is pulled more strongly toward other mercury atoms than toward the glass surface; hence mercury does not wet the glass surface.

The angle ϕ between the solid surface and a line drawn tangent to the liquid at the surface is called *the angle of contact* (Fig. 9.24a and 9.24b). Note that ϕ is less than 90° for any substance in which adhesive forces are stronger than cohesive forces and greater than 90° if cohesive forces predominate. For example, if a drop of water is placed on paraffin, the contact angle is approx-

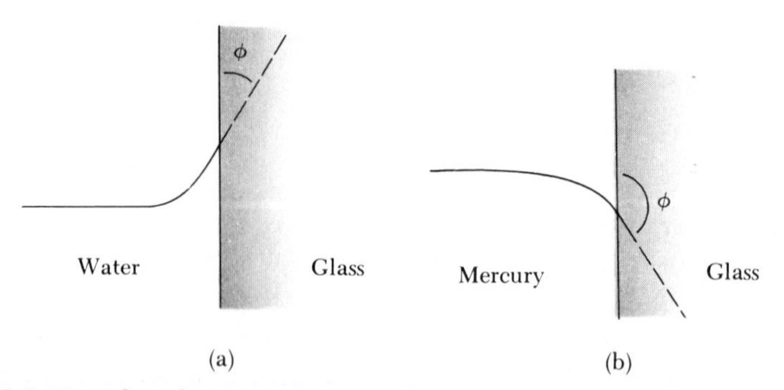

FIGURE 9.24 A liquid in contact with a solid surface. (a) For water, the adhesive force is greater than the cohesive force. (b) For mercury, the adhesive force is less than the cohesive force.

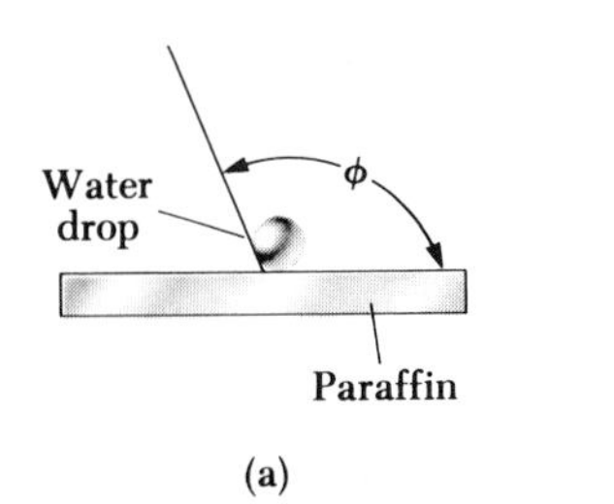

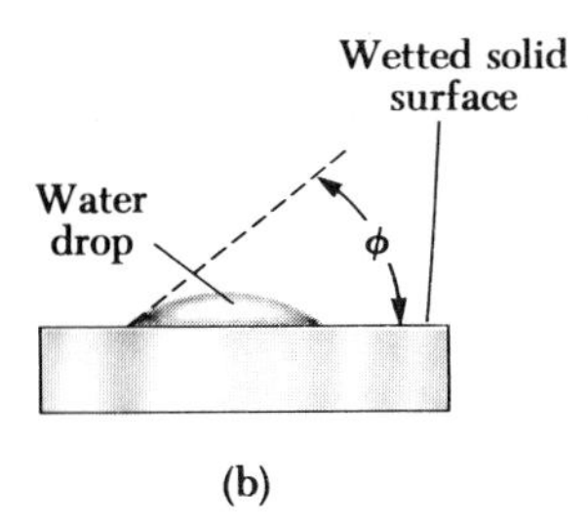

FIGURE 9.25 (a) The contact angle between water and paraffin is about 107°. In this case, the cohesive force is greater than the adhesive force. (b) When a chemical called a wetting agent is added to the water, it wets the paraffin surface, and $\phi < 90°$. In this case, the adhesive force is greater than the cohesive force.

imately 107° (Fig. 9.25a). If certain chemicals, called wetting agents or detergents, are added to the water, the contact angle becomes less than 90°, as shown in Figure 9.25b. The addition of such substances to water is of value when one wants to ensure that water makes intimate contact with a surface and penetrates it. For this reason, detergents are added to water to wash clothes or dishes. On the other hand, it is often necessary to keep water from making intimate contact with a surface, as in waterproofing clothing, where a situation somewhat the reverse of that shown in Figure 9.25 is called for. The clothing is sprayed with a waterproofing agent, which changes ϕ from less than 90° to greater than 90°. Thus, the water beads up on the surface and does not easily penetrate the clothing.

*9.10 CAPILLARITY

Capillary tubes are tubes in which the diameter of the opening is very small. In fact, the word *capillary* means "hair-like." If such a tube is inserted into a fluid for which adhesive forces dominate over cohesive forces, the liquid will rise into the tube, as shown in Figure 9.26. The rising of the liquid in the tube can be explained in terms of the shape of the surface of the liquid and in terms of the surface tension effects in the liquid. At the point of contact between liquid and solid, the upward force of surface tension is directed as shown in Figure 9.26. From Equation 9.12, the magnitude of this force is

$$F = \gamma L = \gamma\,(2\pi r)$$

We use $L = 2\pi r$ here because the liquid is in contact with the surface of the tube at all points around its circumference. The vertical component of this force due to surface tension is

$$F_v = \gamma(2\pi r)(\cos\phi) \tag{9.13}$$

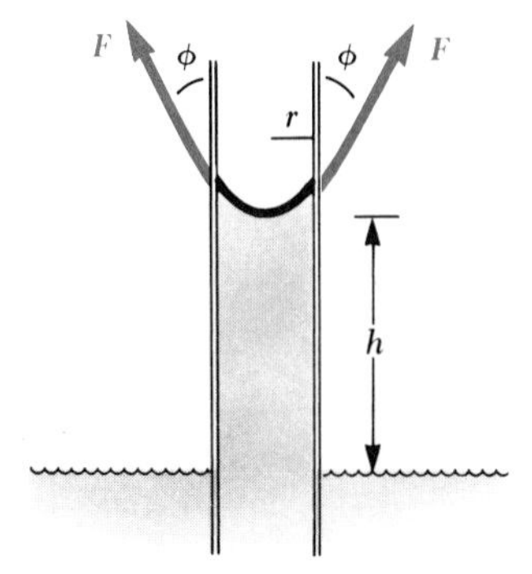

FIGURE 9.26 A liquid rises in a narrow tube because of capillary action, a result of surface tension and adhesive forces.

In order for the liquid in the capillary tube to be in equilibrium, this upward force must be equal to the weight of the cylinder of water of height h inside the capillary tube. The weight of this water is

$$w = Mg = \rho V g = \rho g \pi r^2 h \tag{9.14}$$

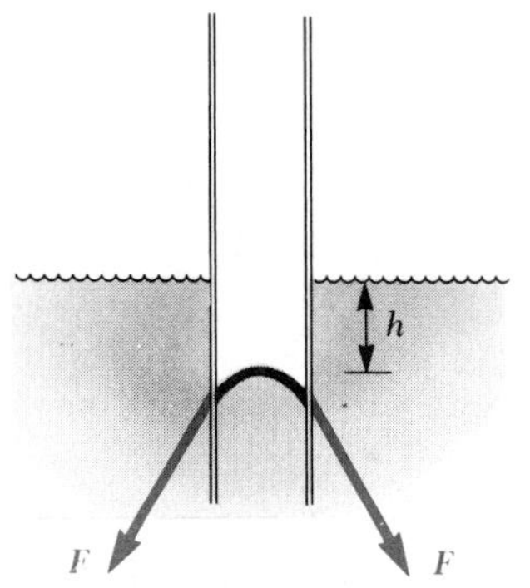

FIGURE 9.27 When cohesive forces between molecules of the liquid exceed adhesive forces, the liquid level in the capillary tube is below the surface of the surrounding fluid.

Equating F_v in Equation 9.13 to w in Equation 9.14, we have

$$\gamma\,(2\pi r)(\cos\phi) = \rho g \pi r^2 h$$

Thus, the height to which water is drawn into the tube is

$$h = \frac{2\gamma}{\rho g r}\cos\phi \qquad (9.15)$$

If a capillary tube is inserted into a liquid in which cohesive forces dominate over adhesive forces, the level of the liquid in the capillary tube will be below the surface of the surrounding fluid, as shown in Figure 9.27. An analysis similar to that done above would show that the distance h the surface is depressed is given by Equation 9.15.

Capillary tubes are often used to draw small samples of blood from a needle prick in the skin. Capillary action must also be considered in the construction of concrete-block buildings because water seepage through capillary pores in the blocks or the mortar may cause damage to the inside of the building. To prevent this, the blocks are usually coated with a waterproofing agent either outside or inside the building. Water seepage through a wall is an undesirable effect of capillary action, but paper towels use capillary action in a useful manner to absorb spilled fluids.

EXAMPLE 9.9 How High Does the Water Rise?

Find the height to which water would rise in a capillary tube with a radius equal to 5×10^{-5} m. Assume that the angle of contact between the water and the material of the tube is small enough to be considered zero.

Solution The surface tension of water is 0.073 N/m. For a contact angle of 0°, we have $\cos\phi = \cos 0° = 1$, so that Equation 9.15 gives

$$h = \frac{2\gamma}{\rho g r} = \frac{2\,(0.073\ \text{N/m})}{(10^3\ \text{kg/m}^3)(9.8\ \text{m/s}^2)(5 \times 10^{-5}\ \text{m})} = 0.29\ \text{m}$$

SUMMARY

Matter is normally classified as being in one of three states: solid, liquid, or gaseous.

Three primary bonding mechanisms that hold constituent atoms together in a solid are ionic bonds, covalent bonds, and metallic bonds.

The elastic properties of a solid can be described using the concepts of stress and strain. **Stress** is a quantity related to the force producing a deformation; **strain** is a measure of the degree of deformation. Stress is proportional to strain, and the constant of proportionality is the **elastic modulus:**

$$\text{Elastic modulus} \equiv \frac{\text{stress}}{\text{strain}} \qquad (9.1)$$

Three common types of deformation are: (1) the resistance of a solid to elongation or compression under a load, characterized by **Young's modulus,** Y; (2) the resistance of a solid to the planes in the solid sliding past each other, characterized by the **shear modulus,** S; (3) the resistance of a solid (or a liquid) to a volume change, characterized by the **bulk modulus,** B.

The **density,** ρ, of a substance of uniform composition is defined as its mass per unit volume and has units of kilograms per cubic meter (kg/m^3) in the SI system:

$$\rho \equiv \frac{M}{V} \tag{9.6}$$

The **pressure,** P, in a fluid is the force per unit area that the fluid exerts on an object immersed in the fluid:

$$P \equiv \frac{F}{A} \tag{9.7}$$

In the SI system, pressure has units of newtons per square meter, and $1\ N/m^2 = 1$ pascal (Pa).

The pressure in a fluid varies with depth h according to the expression

$$P = P_a + \rho g h \tag{9.10}$$

where P_a is atmospheric pressure (1.01×10^5 Pa) and ρ is the density of the fluid.

Pascal's principle states that, when pressure is applied to an enclosed fluid, the pressure is transmitted undiminished to every point of the fluid and to the walls of the containing vessel.

When an object is partially or fully submerged in a fluid, the fluid exerts an upward force on the object called the **buoyant force.** According to **Archimedes' principle,** the buoyant force is equal to the weight of the fluid displaced by the object.

The **surface tension,** γ, in a film of liquid is defined as the ratio of the surface force to the length along which the force acts:

$$\gamma \equiv \frac{F}{L} \tag{9.12}$$

The height to which a fluid will rise (or fall) in a capillary tube is given by

$$h = \frac{2\gamma}{\rho g r} \cos \phi \tag{9.15}$$

where ϕ is the **contact angle,** ρ is the density of the fluid, and r is the radius of the capillary tube.

ADDITIONAL READING

F. J. Almgren and J. E. Taylor, "The Geometry of Soap Films and Soap Bubbles," *Sci. American,* July 1976, p. 82.

E. Denton, "The Buoyancy of Marine Animals," *Sci. American,* July 1960, p. 118.

J. J. Gilman, "Fracture in Solids," *Sci. American,* February 1960, p. 94.

V. A. Greenlach, "The Rise of Water in Plants," *Sci. American,* October 1952, p. 78.

R. Hagen and L. Finger, "Crystals at High Pressures," *Sci. American,* May 1985, p. 110.

A. Holden and P. Singer, *Crystals and Crystal Growing,* Doubleday Anchor, 1960.

E. Rogers, *Physics for the Inquiring Mind,* Princeton, N.J., Princeton University Press, 1960. Chapter 6 treats surface tension in detail.

A. Stanger, "Reflections on the Electron Theory of the Chemical Bond: 1900–1925," *The Physics Teacher,* December 1981.

M. H. Zimmerman, "How Sap Moves in Trees," *Sci. American,* March 1963, p. 132.

QUESTIONS

1. What kind of deformation does a cube of Jello exhibit when it jiggles?
2. Two glass tumblers with different shapes and cross-sectional areas are filled to the same level with water. According to the expression $P = P_a + \rho gh$, the pressure is the same at the bottom of both tumblers. In view of this, why does one tumbler weigh more than the other?
3. How much force does the atmosphere exert on 1 mi^2 of land?
4. When you drink a liquid through a straw, you reduce the pressure in your mouth and let the atmosphere move the liquid. Explain how this works. Could you use a straw to sip a drink on the moon?
5. Indian fakirs do stretch out for a nap on a bed of nails. How is this possible?
6. It is dangerous to swim in a pool that is being drained. Why?
7. Pascal used a barometer with water as the working fluid. Why is it impractical to use water for a typical barometer?
8. A person sitting in a boat floating in a small pond throws a heavy anchor overboard. Does the level of the pond rise, fall, or remain the same?
9. Steel is much more dense than water. How, then, do boats made of steel float?
10. A helium-filled balloon will rise until its density becomes the same as that of the air. If a sealed submarine begins to sink, will it go all the way to the bottom of the ocean or will it stop when its density becomes the same as that of the surrounding water?
11. A fish rests on the bottom of a bucket of water while the bucket is being weighed. When it begins to swim around, does the weight change?
12. Will a ship ride higher in the water of an inland lake or in the ocean? Why?
13. If 1,000,000 N were placed on the deck of the World War II battleship North Carolina, it would sink only 2.5 cm lower in the water. What is the cross-sectional area of the ship at water level?
14. Lead has a greater density than iron, and both are denser than water. Is the buoyant force on a lead object greater than, less than, or equal to the buoyant force on an iron object of the same dimensions?

Question 18 (Photographed by Jim Lehman, James Madison University.)

15. In a popular TV advertisement, it is claimed that a certain paper towel absorbs a spilled liquid more readily than its competitors. What is the mechanism responsible for the absorption?
16. An ice cube is placed in a glass of water. What happens to the level of the water as the ice cube melts?
17. A woman wearing high-heeled shoes is invited into a home in which the kitchen has a newly installed vinyl floor covering. Why should the homeowner be concerned?
18. A typical silo on a farm has many bands wrapped around its perimeter, as shown in the photograph. Why is the spacing between successive bands smaller at the lower regions of the silo?
19. If an inflated beachball is placed beneath the surface of a pool of water and released, it will shoot upward out of the water. Use Archimedes' principle to explain why this occurs.
20. Will an ice cube float higher in water or in an alcoholic beverage?
21. A simple device that is used to test the strength of antifreeze in cars consists of a small tube which has

several small colored spheres inside it. When antifreeze is admitted into the tube, some of the spheres will float giving an indication of the strength of the antifreeze. How does this device work?

22. Why do your ears pop as you drive up a mountain rapidly?

23. A small amount of water is placed at the bottom of a flexible container and brought to a boil. The container is then capped and placed under cool water. The can collapses inward. Why?

PROBLEMS

Section 9.3 Elastic Properties of Solids

1. A mass of 2 kg is supported by a copper wire of length 4 m and diameter 4 mm. Determine (a) the stress in the wire and (b) the elongation of the wire.

2. A 6-m steel beam 1.5 cm in radius is one element used to support an overhead walkway. The beam is designed to not stretch more than 5×10^{-5} m. What is the maximum load in newtons that it can withstand?

3. A child slides across a floor in a pair of rubber-soled shoes. The friction force acting on each foot is 20 N, the cross-sectional area of one foot is about 14 cm^2, and the height of the soles is about 5 mm. Find the horizontal distance the sheared face of the sole moves. The shear modulus of the rubber is about 3×10^6 Pa.

4. The elastic limit of a material is defined as the maximum stress that can be applied to the material before it becomes permanently deformed. If the elastic limit of steel is 5×10^8 Pa, determine the minimum diameter a steel wire can have if it is to support a 70 kg circus performer without its elastic limit being exceeded.

5. A uniform pressure of 5×10^4 Pa is exerted on a copper block having a volume of 10^{-3} m^3. What is the change in volume of the block?

• **6.** What increase of pressure would be required to change the volume of a sample of water by 1%?

• **7.** During a tensile strength experiment, a small diameter fiber is elongated by a force of 0.025 N. The fiber has an initial length of 0.2 m and stretches 4×10^{-4} m. Young's modulus for the material being tested is equal to 7.6×10^{10} Pa. Determine the diameter of the fiber.

• **8.** If the shear stress in steel exceeds about 4.0×10^8 Pa, the steel ruptures. Determine the shear force necessary to punch a 1 cm diameter hole in a steel plate that is 0.5 cm thick.

Section 9.4 Density and Pressure

9. A solid cube 5.0 cm on an edge has a mass of 1.31 kg. What is the cube made of, assuming it consists of only one element? (*Hint:* Consult Table 9.2.)

10. A king orders a gold crown having a mass of 0.5 kg. When it arrives from the metalsmith, the volume of the crown is found to be 185 cm^3. Is the crown made of gold?

11. A 70-kg man in a 5-kg chair tilts back such that all the weight is balanced on two legs of the chair. Assume each leg of the chair makes contact with the floor over a circular area with radius 1 cm and find the pressure exerted by each leg on the floor.

12. A pipe contains water at a pressure 5×10^5 Pa, above atmospheric. If the only material you have available to patch a 4-mm diameter hole in the pipe is a piece of bubble gum, how much force must the gum be able to withstand?

Section 9.5 Variation of Pressure with Depth

Section 9.6 Pressure Measurements

13. Determine the absolute pressure at the bottom of a lake that is 30 m deep.

14. Water is to be pumped to the top of the Empire State building, which is 1200 ft high. What gauge pressure is needed in the water line at the base of the building to reach this height?

15. Engineers have developed a bathyscaph that can reach ocean depths of 7 mi. (a) What is the absolute pressure at this depth? (b) If the inside of the vessel is maintained at atmospheric pressure, what is the net force on a porthole of diameter 15 cm?

16. If the column of mercury in a barometer stands at a height of 74 cm, what is the atmospheric pressure in Pa?

17. To what height in meters would the water in a water barometer stand on a day when a mercury barometer stands at 77 cm?

18. The pressure in a sealed water pipe is 2.5×10^5 Pa above atmospheric on the first floor of an apartment building. In comparison to this value, the pressure on the top floor is only 1.5×10^5 Pa. How tall is the building?

19. A popular television show has a feature in which various objects are compressed by a 90-ton hydraulic press, which has a cross-sectional area of about 400 cm^2. This force is generated by oil pushing against a smaller piston of area 2 cm^2. What force must the oil exert on this smaller piston?

20. The U-shaped tube in Figure 9.16 contains mercury. (a) What is the absolute pressure on the left if the column height $h = 20$ cm? (b) What is the gauge pressure?

•• **21.** A container is filled to a depth of 20 cm with water. On top of the water floats a 30 cm thickness of oil, with specific gravity 0.7. What is the absolute pressure at the bottom of the container?

Section 9.7 Buoyant Forces and Archimedes' Principle

22. (a) Calculate the buoyant force on a solid object made of copper and having a volume of 0.2 m^3 if it is submerged in water. (b) What is the result if the object is made of steel?

23. A solid object has a weight in air of 5.0 N. When it is suspended from a spring scale and submerged in water, the scale reads 3.5 N. What is the density of the object?

24. A small boat weighing 1000 N has a surface area at water level of 3 m^2. It floats with only 5 cm above water level when in a fresh-water lake. How high out of the water will it ride in a salt-water lake? Assume the surface area of the boat does not change as it rises. (Salt water has a density of about 1.03×10^3 kg/m^3.)

25. Careful analyses of photographs of a "sea monster" show that its volume is 52 m^3. Assume that the "sea monster" is weightless when immersed in seawater (density 1030 kg/m^3), and calculate its mass in metric tons (1 metric ton equals 1000 kg).

• **26.** A ferry boat is 4 m wide and 6 m long. When a large loaded truck pulls onto the boat it sinks 4 cm lower in the water. What is the weight of the truck?

• **27.** An object is found to have a weight in air of 300 N. It is then tied to a string connected to a balance and immersed in water. The scale now reads 265 N. Finally, the object is immersed in oil and is found to have a weight of 275 N. (a) Find the density of the object and (b) the density of the oil.

•• **28.** A bar of soap 2 cm thick is floating on the surface of a container of water such that 1.5 cm of the soap is under water. Bath oil of specific gravity 0.6 is poured into the water and floats on top of the water. What is the depth of the oil layer when the bar of soap is just level with the liquid surface?

•• **29.** An object of volume 450 cm^3 hangs from the end of a copper wire of cross-sectional area 1.5×10^{-5} m^2. When the object is immersed in water the length of the wire decreases by 0.035 mm. What is the length of the wire?

Section 9.8 Surface Tension

30. A vertical force of 1.61×10^{-2} N is required to lift a wire ring of radius 1.75 cm from the surface of a container of blood plasma. Calculate the surface tension of blood plasma from this information.

31. Each of the six legs of an insect on a water surface makes a depression 0.25 cm in radius with a contact angle of 45°. Calculate the mass of the insect.

32. The surface tension of ethanol is 0.0227 N/m, and the surface tension of tissue fluid is 0.050 N/m. A force of 7.13×10^{-3} N is required to lift a 5 cm diameter wire ring vertically from the surface of ethanol. What should the diameter of a ring be so that this same force would lift it from the tissue fluid?

• **33.** A square metal sheet 5 cm on a side and of negligible thickness is attached to a balance and inserted into a container of fluid. The contact angle is found to be zero, as shown in Figure 9.28a and the balance to which the metal sheet is attached reads 0.40 N. A thin veneer of oil is then spread over the metal sheet and the contact angle becomes 180°, as shown in Figure 9.28b. The balance now reads 0.39 N. What is the surface tension of the fluid?

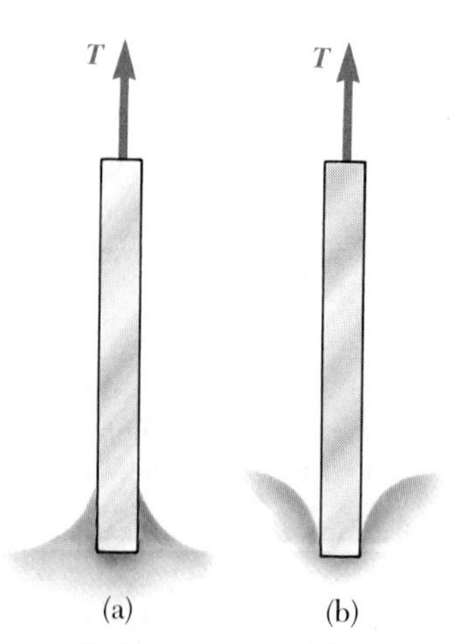

FIGURE 9.28 Problem 33

Section 9.10 Capillarity

34. A certain fluid has a density of 1080 kg/m^3 and is observed to rise to a height of 2.1 cm in a 1-mm diameter tube. The contact angle between the wall and the fluid is zero. Calculate the surface tension of the fluid.

35. Whole blood has a surface tension of 0.058 N/m and a density of 1050 kg/m^3. To what height can whole blood rise in a capillary blood vessel that has a radius of 2×10^{-6} m if the contact angle is zero?

36. Use density and surface tension values from Tables 9.2 and 9.3 to calculate the height to which water will rise in a capillary of diameter 10^{-4} m. Assume a contact angle of zero and a temperature of 20°C.

37. A staining solution used in a microbiology laboratory has a surface tension of 0.088 N/m and a density 1.035 times the density of water. What must the diameter of a capillary tube be so that this solution will rise to a height of 5 cm? (Assume a zero contact angle.)

• **38.** A capillary tube 1 mm in radius is immersed in a beaker of mercury. The mercury level inside the tube is found to be 0.536 cm *below* the level of the reservoir. Use the surface tension for mercury from Table 9.3 and the density from Table 9.2 to determine the contact angle between mercury and glass.

ADDITIONAL PROBLEMS

39. Bone has a Young's modulus of about 14.5×10^9 Pa. Under compression, a bone can withstand a stress of about 160×10^6 Pa before breaking. Estimate the length of your femur (thigh bone) and calculate the amount this bone can be compressed before breaking.

40. The tip of a phonograph needle has a diameter of about 2.5×10^{-4} m. The arm of the phonograph is balanced so that a force of about 0.1 N is exerted downward on the needle. Find the pressure exerted on the record.

41. A particular braking system is designed such that a piston is forced into a tube of diameter 1 cm with a force of about 80 N. This pressure is communicated to a piston 5 cm in diameter at each of the four wheels of a small car. What is the net braking force on the car?

42. Blood of density 1050 kg/m^3 is to be administered to a patient. To do so the blood is raised to a height of about 1 m higher than the level of the patient's arm. How much greater is the pressure of the blood at this level than if the container were at the same level as the arm?

43. The density of ice is 920 kg/m^3 and that of seawater is 1030 kg/m^3. What fraction of the total volume of an iceberg is exposed?

• **44.** A sample of an unknown material has a weight equal to 300 N in air and 200 N when immersed in alcohol of specific gravity 0.7. (a) What is the volume of the material, and (b) what is its density?

• **45.** The distortion of the earth's crustal plates is an example of shear on a large scale. A particular kind of crustal rock is determined to have a shear modulus equal to 1.5×10^{10} Pa. What shear stress is involved when a 10-km layer of rock is sheared through a distance of 5 m?

• **46.** A particular metal has a weight in air of 50 N. When placed in water it has a weight of 36 N and when placed in oil it has a weight of 41 N. (a) Find the density of the metal, and (b) the density of the oil.

• **47.** A U-tube open at both ends contains water to a level of 20 cm above the bottom of the U. A 20-cm column of ethyl alcohol (density 789.9 kg/m^3) is added to the right arm of the tube. What is the water level in each arm after equilibrium has been established?

• **48.** A weather balloon has a mass of 600 kg and is designed to lift an experiment package whose mass is equal to 4000 kg. What should the volume of the balloon be after being inflated with helium (density 0.178 kg/m^3) in order that the total load can be lifted? Use 1.29 kg/m^3 as the density of air.

• **49.** One method of measuring the density of a liquid is illustrated in Fig 9.29. One side of the U-shaped tube is in the liquid being tested; the other side is in water of density ρ_w. When the air is partially removed at the upper part of the tube, show that the density of the liquid on the left is given by $\rho = (h_w/h)\rho_w$.

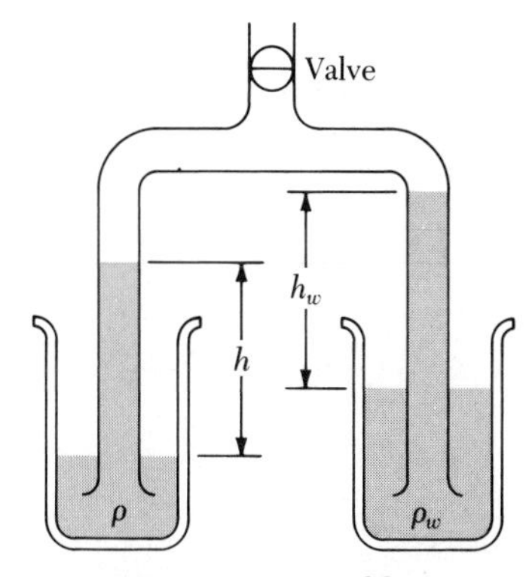

FIGURE 9.29 Problem 49

• **50.** A circular swimming pool has a flat bottom and a 6-m diameter. It is filled with water to a depth of 1.50 m. There is 1 atm of pressure on the top surface. What is the absolute pressure at the bottom of the pool? (b) Two people with a combined mass of 150 kg get into the pool and float there quietly. What is the resulting increase in the absolute pressure at the bottom of the pool?

•• **51.** A small ball of specific gravity 0.1 and volume 2 cm^3 is pulled below the surface of a pool of water. Find the acceleration of the ball at the instant it is released.

•• **52.** A block of wood has a weight of 50 N when weighed in air. A sinker is attached to the block and the weight of the wood-sinker combination is 200 N when the sinker alone is immersed in water. Finally, the wood-sinker combination is completely immersed and the weight is 140 N. Find the density of the block.

•• **53.** A high-speed lifting mechanism supports an 800 kg mass by a steel cable 25 m in length and 4 cm^2 in cross-sectional area. (a) Determine the elongation of the cable. (b) By what additional amount does the cable increase in length if the mass is accelerated upward at a rate of 3 m/s^2? (c) What is the greatest mass that could be accelerated upward at the rate found in part (b) if the stress in the cable is not to exceed the elastic limit of the cable, 2.2×10^8 Pa?

•• **54.** A small sphere of density 0.6 times the density of water is dropped from a height of 10 m above the surface of a smooth lake. Determine the maximum depth to which the sphere will sink. Neglect any energy transferred to the water during impact and sinking.

•• **55.** A balloon filled with helium at atmospheric pressure is designed to support a mass M (payload + empty balloon). Show that the volume of the balloon must be at least $V = M/(\rho_a - \rho_{He})$, where ρ_a is the density of air and ρ_{He} is the density of helium. (Ignore the volume of the payload.)

•• **56.** Consider a composite "raft" consisting of two square slabs each of side s, attached face to face. One slab has density ρ_1 and thickness h_1, while the other has thickness h_2 and density ρ_2 greater than ρ_1. (a) Find the average density of the raft. (b) Assume that the average density is less than that of water, so that the raft floats in water. The raft is placed in water with the denser slab on the bottom. Find d, the depth of the bottom surface of the raft. (c) If the raft is placed in water with the denser slab at the top, find d, the depth of the bottom surface of the raft. Comment on your answer.

•• **57.** In 1654 Otto von Guericke evacuated a sphere made of two brass hemispheres. Two teams of eight horses each could not pull the hemispheres apart (Fig. 9.30). (a) Show that the force required to pull the evacuated hemispheres apart is $\pi R^2(P_a - P)$, where R is the radius of the hemispheres and P is the pressure inside the hemispheres, which is much less than P_a. (b) Determine the force if $P = 0.1P_a$ and $R = 0.3$ m.

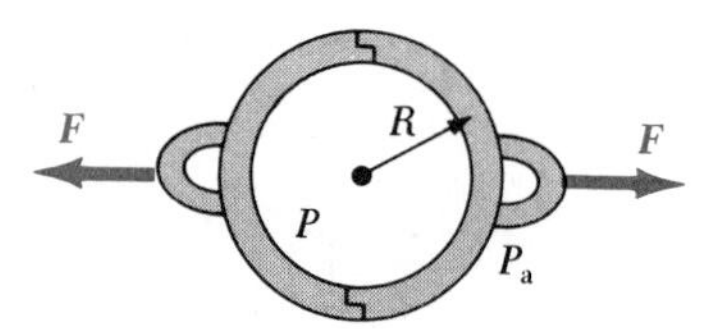

FIGURE 9.30 Problem 57

CHAPTER 10

Fluids in Motion

A surfer "rides the pipe" on an ocean wave. (Courtesy of California Dept. of Tourism)

Why does a baseball curve? How does one describe the motion of an object in a turbulent river? In order to answer such questions, we must examine the properties of fluids in motion, a subject that is quite complex and not fully understood. A fluid in motion can be described by a model in which certain simplifying assumptions are made. We shall use this model to analyze some situations of practical importance. For example, we shall treat the subject of fluid transport mechanisms in the human body and the physics of the circulatory system. An underlying principle known as the Bernoulli effect will enable us to determine relations between the pressure, density, and velocity at every point in a fluid.

10.1 LAMINAR AND TURBULENT FLOW

When a fluid is in motion, its flow can be characterized as being one of two types. The flow is said to be **streamline,** or **laminar,** if every particle that passes a particular point moves along the exact path followed by particles that passed that point earlier. In this case, each particle of the fluid moves along a smooth path called a *streamline,* as shown in Figure 10.1a. The various streamlines cannot cross each other under this steady-flow condition, and the streamline at any point coincides with the direction of fluid velocity at that

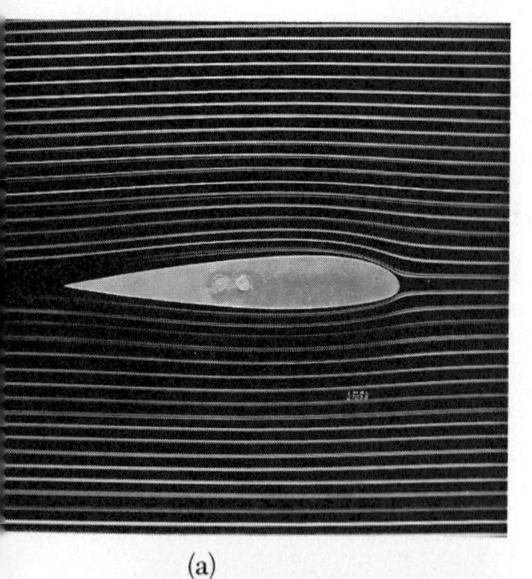

(a)

(b)

FIGURE 10.1 Streamline flow around an airfoil made visible by smoke particles moving along the streamlines. The flow is from right to left, simulating motion of the airfoil from left to right. (Courtesy of NASA.) (b) Hot gases from a cigarette made visible by smoke particles. Note that the smoke first moves in streamline flow at the bottom and then in turbulent flow above.

point. In contrast, the flow of a fluid becomes irregular, or **turbulent,** above a certain velocity or under conditions near its boundaries that can cause abrupt changes in velocity. Irregular motion of the fluid, called *eddy currents,* is characteristic in turbulent flow, as shown in Figure 10.1b.

Also, in the discussion of fluid flow, the term *viscosity* is used to characterize the degree of internal friction in the fluid. This internal friction is associated with the resistance of two adjacent layers of the fluid to move relative to each other. A fluid such as kerosene has a lower viscosity than crude oil or molasses.

10.2 FLUID DYNAMICS

Many features of fluid motion can be understood by considering the behavior of an **ideal fluid,** which satisfies the following conditions:

1. *The fluid is nonviscous,* that is, there is no internal friction force between adjacent fluid layers.
2. *The fluid is incompressible,* which means that its density is constant.
3. *The fluid motion is steady,* meaning that the velocity, density, and pressure at each point in the fluid do not change in time.
4. *The fluid moves without turbulence.* This implies that each element of the fluid has zero angular velocity about its center, that is, there can be no eddy currents present in the moving fluid.

Equation of Continuity

Figure 10.2 represents a fluid flowing through a pipe of nonuniform size. The particles in the fluid move along the streamlines in steady-state flow. In a small time interval Δt, the fluid entering the bottom end of the pipe moves a distance $\Delta x_1 = v_1 \Delta t$. If A_1 is the cross-sectional area in this region, then the mass contained in the bottom colored region is $\Delta M_1 = \rho_1 A_1 \Delta x_1 = \rho_1 A_1 v_1 \Delta t$, where ρ is the density of the fluid. Similarly, the fluid that moves out of the upper end of the pipe in the same time Δt has a mass $\Delta M_2 = \rho_2 A_2 v_2 \Delta t$. However, because mass is conserved and because the flow is

An experimental vehicle shown in a laboratory wind tunnel has an aerodynamic drag coefficient of .14 and requires only 2.1 hp to overcome wind resistance at 50 mi/h. (Courtesy of General Motors Corporation)

steady, the mass that flows into the bottom of the pipe through A_1 in a time Δt must equal the mass that flows out through A_2 in the same time. Therefore, $\Delta M_1 = \Delta M_2$, or

$$\rho_1 A_1 v_1 = \rho_2 A_2 v_2 \tag{10.1}$$

This expression is called the **equation of continuity.**

Since ρ is constant for an incompressible fluid, Equation 10.1 reduces to

$$\boxed{A_1 v_1 = A_2 v_2} \tag{10.2}$$

That is, the product of the cross-sectional area of the pipe and the fluid speed at any point along the pipe is a constant.

Therefore, as one would expect, the speed is high where the tube is constricted and low where the tube has a larger diameter. The product Av, which has the dimensions of volume per unit time, is called the *flow rate*.

The condition Av = constant is equivalent to the fact that the amount of fluid that enters one end of the tube in a given time interval equals the amount of fluid leaving the tube in the same time interval, assuming no leaks.

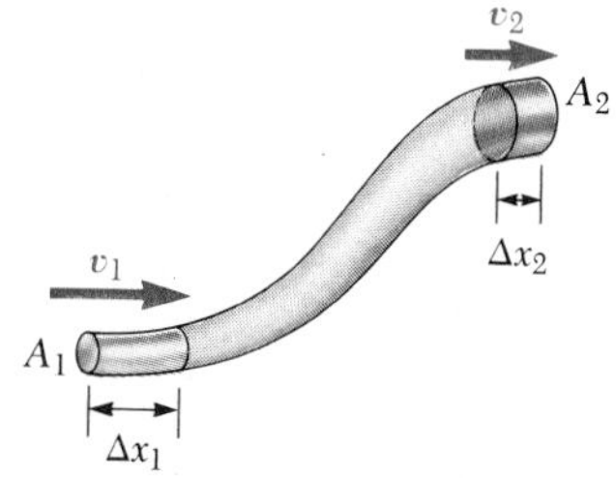

FIGURE 10.2 A fluid moving with streamline flow through a pipe of varying cross-sectional area. The volume of fluid flowing through A_1 in a time interval Δt must equal the volume flowing through A_2 in the same time interval. Therefore, $A_1V_1 = A_2V_2$.

EXAMPLE 10.1 Filling a Water Bucket

A water hose 2 cm in diameter is used to fill a 20-liter bucket. If it takes 1 min to fill the bucket, what is the speed v at which the water leaves the hose? (You should note that 1 liter = 10^3 cm^3.)

Solution The cross-sectional area of the hose is

$$A = \frac{\pi d^2}{4} = \frac{\pi(2\text{ cm})^2}{4} = \pi\text{ cm}^2$$

The flow rate of 20 liters/min is equal to the product Av. Thus,

$$Av = 20\,\frac{\text{liters}}{\text{min}} = \frac{20 \times 10^3\text{ cm}^3}{60\text{ s}}$$

$$v = \frac{20 \times 10^3\text{ cm}^3}{(\pi\text{ cm}^2)(60\text{ s})} = 106\text{ cm/s}$$

EXERCISE If the diameter of the hose is reduced to 1 cm, what will the speed of the water be as it leaves the hose, assuming the same flow rate?

Answer 424 cm/s.

Bernoulli's Equation

As a fluid moves through a pipe of varying cross section and elevation, the pressure will change along the pipe. In 1738 the Swiss physicist Daniel Bernoulli (1700–1782) derived a fundamental expression that relates pressure to fluid speed and elevation. Bernoulli's equation is not a free-standing law of physics. It is, instead, a consequence of energy conservation as applied to our ideal fluid.

In deriving Bernoulli's equation, we shall again assume that the fluid is incompressible and nonviscous and flows in a nonturbulent, steady-state man-

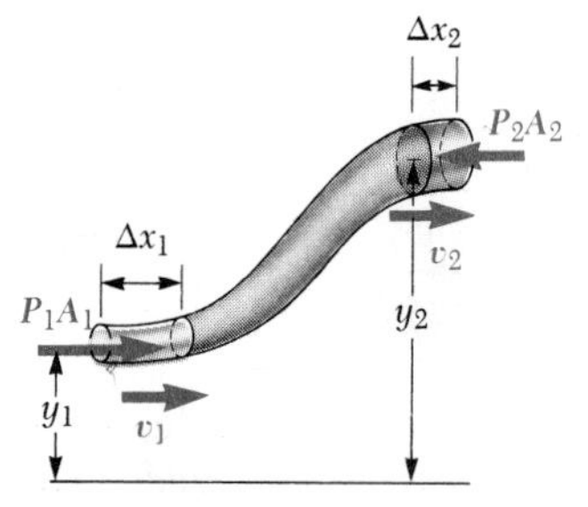

FIGURE 10.3 A fluid flowing through a constricted pipe with streamline flow. The fluid in the section of length Δx_1 moves to the section of length Δx_2. The volumes of fluid in the two sections are equal.

ner. Consider the flow through a nonuniform pipe in a time Δt, as illustrated in Figure 10.3. The force on the lower end of the fluid is P_1A_1, where P_1 is the pressure at the lower end. The work done on the lower end of the fluid by the fluid behind it is

$$W_1 = F_1\,\Delta x_1 = P_1A_1\,\Delta x_1 = P_1V$$

where V is the volume of the lower colored region in Figure 10.3. In a similar manner, the work done on the fluid on the upper portion in the time Δt is

$$W_2 = -P_2A_2\,\Delta x_2 = -P_2V$$

(Note that the volume of fluid that passes through A_1 in a time Δt equals the volume that passes through A_2 in the same time interval.) The work W_2 is negative because the force on the fluid at the top is opposite its displacement. Thus the net work done by these forces in the time Δt is

$$W = P_1V - P_2V$$

Part of this work goes into changing the kinetic energy of the fluid, and part goes into changing its gravitational potential energy. If m is the mass passing through the pipe in the time interval Δt, then the change in kinetic energy of the volume of fluid is

$$\Delta KE = \frac{1}{2}mv_2^2 - \frac{1}{2}mv_1^2$$

The change in its potential energy is

$$\Delta PE = mgy_2 - mgy_1$$

We can apply the work-energy theorem in the form $W = \Delta KE + \Delta PE$ (Chapter 5) to this volume of fluid to give

$$P_1V - P_2V = \frac{1}{2}mv_2^2 - \frac{1}{2}mv_1^2 + mgy_2 - mgy_1$$

If we divide each term by V and recall that $\rho = m/V$, this expression reduces to

$$P_1 - P_2 = \frac{1}{2}\rho v_2^2 - \frac{1}{2}\rho v_1^2 + \rho g y_2 - \rho g y_1$$

Let us move those terms that refer to point 1 to one side of the equation and those that refer to point 2 to the other side:

$$P_1 + \frac{1}{2}\rho v_1^2 + \rho g y_1 = P_2 + \frac{1}{2}\rho v_2^2 + \rho g y_2 \qquad (10.3)$$

This is **Bernoulli's equation.** It is often expressed as

Bernoulli's equation

$$P + \frac{1}{2}\rho v^2 + \rho g y = \text{constant} \qquad (10.4)$$

Bernoulli's equation says that the sum of the pressure (P), the kinetic energy per unit volume ($\frac{1}{2}\rho v^2$), and potential energy per unit volume ($\rho g y$), has the same value at all points along a streamline.

An important consequence of Bernoulli's law can be demonstrated by considering Figure 10.4. This figure shows water flowing through a horizontal constricted pipe from a region of large cross-sectional area into a region of smaller cross-sectional area. Let us compare the pressure at point 1 to the pressure at point 2. Because the pipe is horizontal, $y_1 = y_2$ and Equation 10.3 applied to points 1 and 2 gives

$$P_1 + \frac{1}{2}\rho v_1^2 = P_2 + \frac{1}{2}\rho v_2^2$$

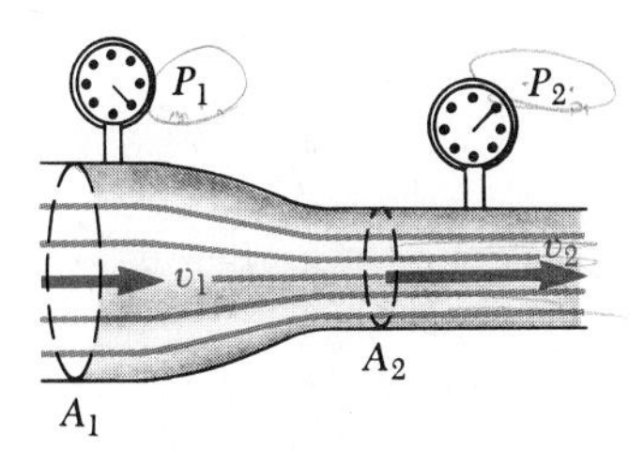

FIGURE 10.4 The pressure P_1 is greater than the pressure P_2 because v_1 is less than v_2.

The velocity of the liquid in the constriction, v_2, must be greater than the velocity v_1 if the water is not to back up in the pipe. This means that P_2 must be less than P_1. This result is often expressed by the statement that

swiftly moving fluids exert less pressure than do slowly moving fluids.

As we shall see in the next section, this important result will enable us to understand a wide range of everyday phenomena.

EXAMPLE 10.2 The Leaky Tank

A tank containing a liquid of density ρ has a small hole in its side at a distance y_1 from the bottom (Fig. 10.5). Assume the top of the tank is open to the atmosphere and determine the speed at which the fluid leaves the hole when the liquid level is a distance h above the hole.

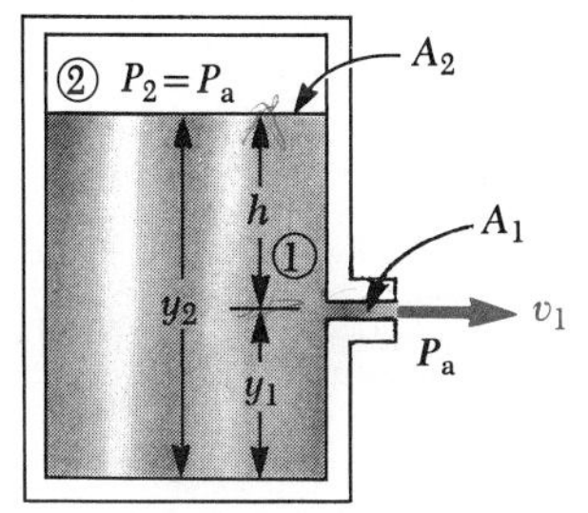

FIGURE 10.5 (Example 10.2) The water speed v_1, from the hole in the side of the container, is given by $v_1 = \sqrt{2gh}$.

Solution If we assume that the cross-sectional area of the tank is large relative to that of the hole ($A_2 \gg A_1$), then the fluid level will drop very slowly and we can assume that $v_2 \approx 0$. Let us apply Bernoulli's equation to points 1 and 2. If we note that $P_1 = P_a$ at the hole, we get

$$P_a + \frac{1}{2}\rho v_1^2 + \rho g y = P_a + \rho g y_2$$

$$v_1 = \sqrt{2g(y_2 - y_1)} = \sqrt{2gh}$$

This says that the speed of the water emerging from the hole is equal to the speed acquired by a body falling freely through a vertical distance h. This is known as **Torricelli's law.** If the height h is 50 cm, for example, the velocity of the stream is

$$v = \sqrt{2(9.80\text{ m/s}^2)(0.5\text{ m})} = 3.13\text{ m/s}$$

10.3 OTHER APPLICATIONS OF BERNOULLI'S EQUATION

In this section we shall give a qualitative description of some common phenomena that can be explained, at least in part, by Bernoulli's equation.

The curve of a spinning baseball is one example in which Bernoulli's equation arises. The ball in Figure 10.6 is moving toward the right and is rotating counterclockwise. From the point of view of the baseball, the air is streaming by it toward the left. However, because the ball is spinning, some air is "dragged" along with the ball because its surface is rough. The air below the ball is held back while that above the ball is helped along because of the direction in which the ball spins. Because the speed of the air is less at A than at B, it follows from Bernoulli's principle that the pressure at B is less

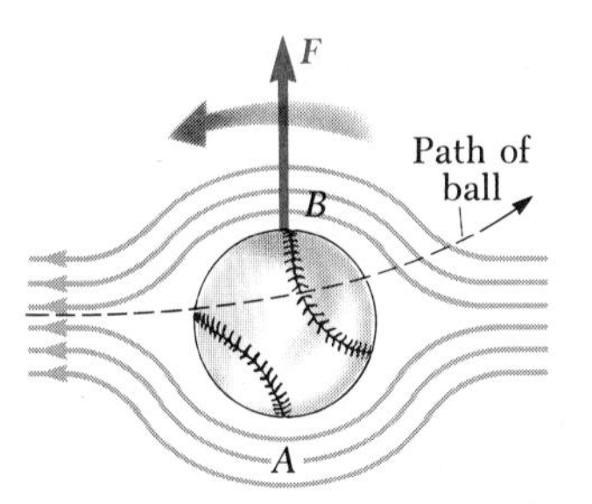

FIGURE 10.6 Streamline flow around a spinning ball. The ball will curve as shown because of a deflecting force $\boldsymbol{F}$ predicted by Bernoulli's equation.

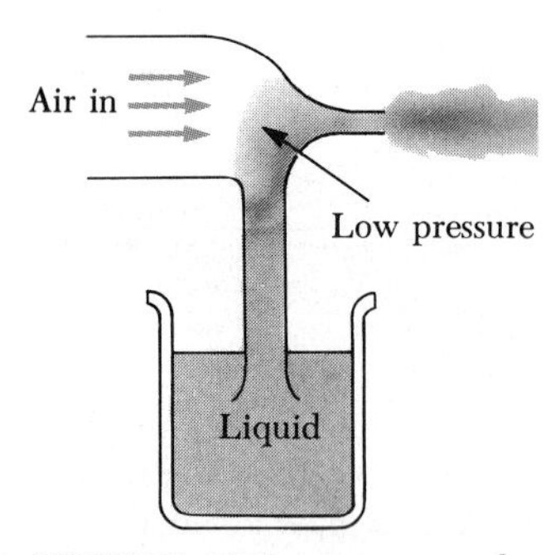

FIGURE 10.7 A stream of air passing over a tube dipped in a liquid will cause the liquid to rise in the tube as shown. This effect is used in perfume bottles and paint sprayers.

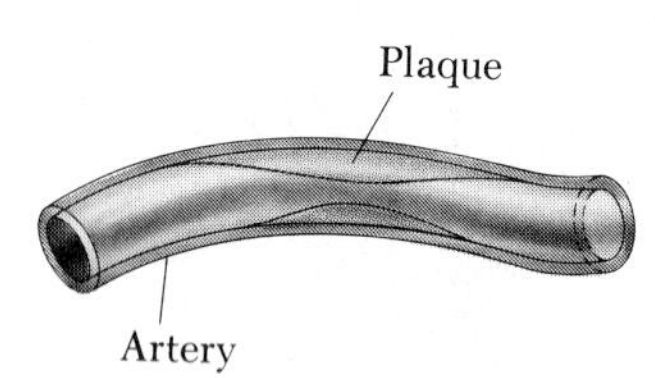

FIGURE 10.8 The flow of blood through a constricted artery.

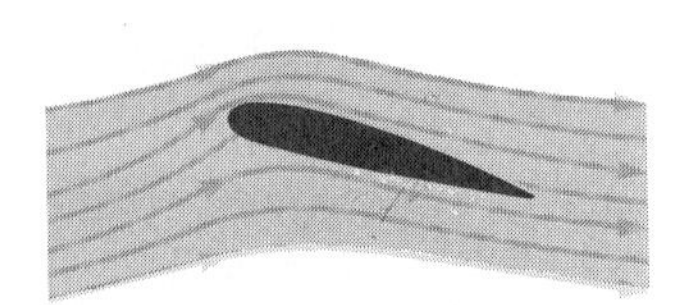
FIGURE 10.9 Airflow across an aircraft wing.

than at A. This pressure difference produces a net force on the ball as shown in Figure 10.6, causing it to follow the dashed path.

A number of devices operate in the manner described in Figure 10.7. A stream of air passing over an open tube reduces the pressure above the tube. This reduction in pressure causes the liquid to rise into the air stream. The liquid is then dispersed into a fine spray of droplets. You might recognize that this so-called atomizer is used in perfume bottles and paint sprayers. The same principle is used in the carburetor of a gasoline engine. In this case, the low-pressure region in the carburetor is produced by air drawn in by the piston through the air filter. The gasoline vaporizes, mixes with the air, and enters the cylinder of the engine for combustion.

If a person has advanced arteriosclerosis, the Bernoulli effect produces a symptom called vascular flutter. In this situation, the artery is constricted as a result of an accumulation of plaque on its inner walls, as in Figure 10.8. In order to maintain a constant flow rate through such a constricted artery, the driving pressure must increase. Such an increase in pressure requires a greater demand on the heart muscle. If the blood velocity is sufficiently high in the constricted region, the artery may collapse under external pressure, causing a momentary interruption in blood flow. At this point, there is no Bernoulli effect and the vessel reopens under arterial pressure. As the blood rushes through the constricted artery, the internal pressure drops and again the artery closes. Such variations in blood flow can be heard with a stethoscope. If the plaque becomes dislodged and ends up in a smaller vessel that delivers blood to the heart, the person can suffer a heart attack.

The lift on an aircraft wing can also be explained, in part, by the Bernoulli effect. Airplane wings are designed such that the air speed above the wing is greater than the air speed below the wing. As a result, the air pressure above the wing is less than the air pressure below the wing and there is a net upward force on the wing, called the "lift." Another factor influencing the lift on a wing is shown in Figure 10.9. The wing has a slight upward tilt which causes air molecules striking the bottom to be deflected downward. The air molecules bouncing off the wing at the bottom produces an upward force on the wing and a significant lift on the aircraft. Finally, turbulence also produces an effect. If the wing is tilted too much, the flow of air across the upper surface of the wing becomes turbulent, and the pressure difference across the wing is not as great as that predicted by Bernoulli's equation. In an extreme case, this turbulence may cause the aircraft to stall.

10.4 VISCOSITY

It is considerably easier to pour water out of a container than to pour syrup. This is because syrup has a higher viscosity than water. In a general sense, *viscosity refers to the internal friction of a fluid.* It is very difficult for layers of a viscous fluid to slide past one another. Likewise, it is very difficult for one solid surface to slide past another if there is a highly viscous fluid, such as soft tar, between them.

To better understand the concept of viscosity, consider a liquid layer placed between two solid surfaces, as in Figure 10.10. The lower surface is fixed in position, and the top surface moves to the right with a velocity v

under the action of an external force $\boldsymbol{F}$. Because of this motion, a portion of the liquid is distorted from its original shape, $ABCD$, at one instant to the shape $AEFD$ a moment later. With reference to Section 9.3, you will recognize that the liquid has undergone a constantly increasing shear strain. Previous sections in this chapter have assumed ideal fluids that have no internal frictional forces between adjacent layers. Such fluids could not have their shape distorted. However, in viscous fluids, there are cohesive forces between molecules in the various layers that can lead to strains which change with time. By definition, the shear stress on the liquid is

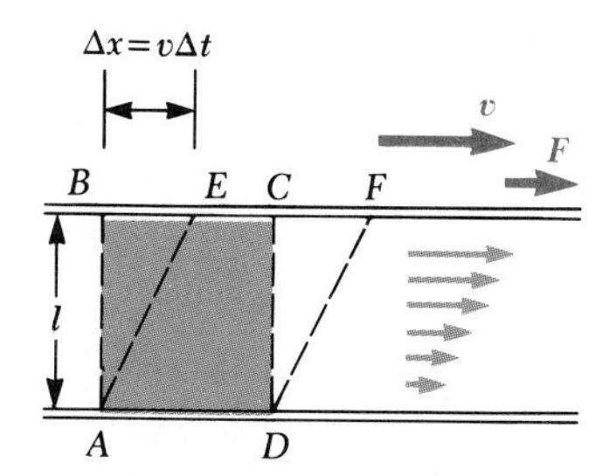

FIGURE 10.10 A layer of liquid between two solid surfaces in which the lower surface is fixed and the upper surface moves to the right with a velocity $\boldsymbol{v}$.

$$\text{Shear stress} \equiv \frac{F}{A}$$

where A is the area of the top plate. Furthermore, the shear strain is defined as

$$\text{Shear strain} \equiv \frac{\Delta x}{l}$$

The velocity of the fluid changes from zero at the lower plate to $\boldsymbol{v}$ at the upper. Thus, in a time Δt, the fluid at the upper plate moves a distance $\Delta x = v\Delta t$. Therefore,

$$\frac{\text{Shear strain}}{\Delta t} = \frac{\Delta x/l}{\Delta t} = \frac{v}{l}$$

This equation states that the rate of change of the shearing strain is v/l.

The **coefficient of viscosity,** η, for the fluid is defined as the ratio of the shearing stress to the rate of change of the shear strain:

$$\boxed{\eta \equiv \frac{Fl}{Av}} \qquad (10.5)$$

Coefficient of viscosity

The SI units of viscosity are $\text{N}\cdot\text{s/m}^2$. You should note that the units of viscosity in many reference sources are often expressed in $\text{dyne}\cdot\text{s/cm}^2$, called 1 **poise** in honor of the French scientist Poiseuille. The relationship between the SI unit of viscosity and the poise is

$$1 \text{ poise} = 10^{-1}\ \text{N}\cdot\text{s/m}^2 \qquad (10.6)$$

Small viscosities are often expressed in centipoise (cp), where 1 cp $= 10^{-2}$ poise. The coefficients of viscosity for some common substances are listed in Table 10.1.

TABLE 10.1 The Viscosities of Various Fluids

Fluid	**T(°C)**	**Viscosity η(N · s/m²)**
Water	20	1.0×10^{-3}
Water	100	0.3×10^{-3}
Whole blood	37	2.7×10^{-3}
Glycerine	20	1500×10^{-3}
10 wt motor oil	30	250×10^{-3}

*10.5 POISEUILLE'S LAW

Figure 10.11 shows a section of a tube containing a fluid under a pressure P_1 at the left end and a pressure P_2 at the right. Because of this pressure difference, the fluid will flow through the tube. The rate of flow (volume per unit time) depends on the pressure difference $(P_1 - P_2)$, the dimensions of the tube, and the viscosity of the fluid. The relationship between these quantities was derived by a French scientist, J. L. Poiseuille (1799–1869), who assumed that the flow was streamline. His result, known as **Poiseuille's law,** is

Poiseuille's law

$$\text{Rate of flow} = \frac{\Delta V}{\Delta t} = \frac{(P_1 - P_2)(\pi R^4)}{8L\eta} \tag{10.7}$$

where R is the radius of the tube, L is its length, and η is the coefficient of viscosity. We shall not attempt to derive this equation here because the methods of integral calculus are required. However, you should note that the equation does agree with common sense. That is, it is reasonable that the rate of flow should increase if the pressure difference across the tube or the tube radius increases. Likewise, the flow rate should decrease if the viscosity of the fluid or the length of the tube increases. Thus, the presence of R and the pressure difference in the numerator of Equation 10.7 and of L and η in the denominator makes sense.

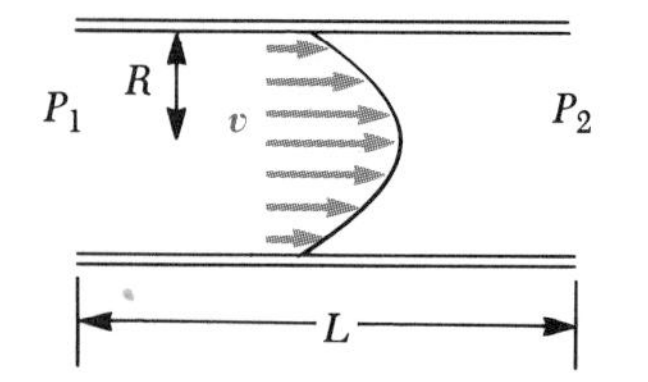

FIGURE 10.11 Velocity profile of a fluid flowing through a uniform pipe of circular cross-section. The rate of flow is given by Poiseuille's law. Note that the fluid velocity is greatest at the middle of the pipe.

From Poiseuille's law, we see that in order to maintain a constant flow rate, the pressure difference across the tube has to increase if the viscosity of the fluid increases. This is important when one considers the flow of blood through the circulatory system. The viscosity of blood increases as the number of red blood cells rises. Thus, blood with a high concentration of red blood cells requires greater pumping pressure from the heart to keep it circulating than does blood of lower red blood cell concentration.

Note that the flow rate varies as the radius of the tube raised to the fourth power. Consequently, if a constriction occurs in a vein or artery, the heart will have to work considerably harder in order to produce a higher pressure drop and hence to maintain the required flow rate.

EXAMPLE 10.3 A Blood Transfusion

A patient receives a blood transfusion through a needle of radius 0.2 mm and length 2 cm. The density of blood is 1050 kg/m^3. The bottle supplying the blood is 0.5 m above the patient's arm. What is the rate of flow through the needle?

Solution The pressure differential between the level of the blood and the patient's arm is

$$P_1 - P_2 = \rho g h = (1050\ \text{kg/m}^3)(9.80\ \text{m/s}^2)(0.5\ \text{m}) = 5.15 \times 10^3\ \text{Pa}$$

Thus, the rate of flow, from Poiseuille's law, is

$$\frac{\Delta V}{\Delta t} = \frac{(P_1 - P_2)(\pi R^4)}{8L\eta} = \frac{(5.15 \times 10^3\ \text{Pa})(\pi)(2 \times 10^{-4}\ \text{m})^4}{8(2 \times 10^{-2}\ \text{m})(2.7 \times 10^{-3}\ \text{N} \cdot \text{s/m}^2)}$$
$$= 5.98 \times 10^{-8}\ \text{m}^3/\text{s}$$

EXERCISE How long will it take to inject 1 pint (500 cm^3) of blood into the patient?

Answer 139 min.

---

*Reynolds Number

As we mentioned earlier, at sufficiently high velocities, fluid flow changes from simple streamline flow to turbulent flow, that is, flow characterized by a highly irregular motion of the fluid. Experimentally it is found that the onset of turbulence in a tube is determined by a dimensionless factor called the **Reynolds number**, given by

$$RN = \frac{\rho v d}{\eta} \tag{10.8}$$

Reynolds number

where ρ is the density of the fluid, v is the average velocity of the fluid along the direction of flow, d is the diameter of the tube, and η is the viscosity of the fluid. If RN is below about 2000, the flow of fluid through a tube is streamline; turbulence occurs if RN is above 3000. In the region between 2000 and 3000, the flow is unstable, meaning that the fluid can move in streamline flow but any small disturbance will cause its motion to change to turbulent flow.

EXAMPLE 10.4 Turbulent Flow of Blood

Determine the velocity at which blood flowing through an artery of diameter 0.2 cm would become turbulent. Assume that the density of blood is 1.05×10^3 kg/m^3 and that its viscosity is 2.7×10^{-3} N·s/m^2.

Solution At the onset of turbulence, the Reynolds number is 3000. Thus, the velocity of the blood would have to be

$$v = \frac{\eta(RN)}{\rho d} = \frac{(2.7 \times 10^{-3}\ \text{N}\cdot\text{s/m}^2)(3000)}{(1.05 \times 10^3\ \text{kg/m}^3)(0.2 \times 10^{-2}\ \text{m})} = 3.86\ \text{m/s}$$

---

*10.6 TRANSPORT PHENOMENA

When a fluid flows through a tube, the basic mechanism that produces the flow is a difference in pressure across the ends of the tube. This pressure difference is responsible for the transport of a mass of fluid from one location to another. The fluid may also move from place to place because of a second mechanism, one that depends on a concentration difference between two points in the fluid, as opposed to a pressure difference. When the concentration (the number of molecules per unit volume) is higher at one location than at another, molecules will flow from the point where the concentration is high to the point where it is lower. The two fundamental processes involved in fluid transport resulting from concentration differences are called *diffusion* and *osmosis*. This section examines the nature and importance of these processes.

*Diffusion

You can imagine what happens when someone wearing a strong shaving lotion or perfume strolls into a crowded room. All eyes turn to seek out the source of the delightful smell. The aroma spreads through the room by a process called diffusion.

In a **diffusion** process, molecules move from a region where their concentration is high to a region where their concentration is lower.

That is, the molecules of the lotion or perfume move from the source (near the person's face), where there are many molecules per unit volume, throughout the room, to regions where the concentration of these molecules is lower. Although the example used here is one of diffusion in air, the process also occurs in liquids and, to a lesser extent, in solids. For example, if a drop of food coloring is placed in a glass of water, the coloring soon spreads throughout the liquid by diffusion. In either case, diffusion ceases when there is a uniform concentration at all locations in the fluid.

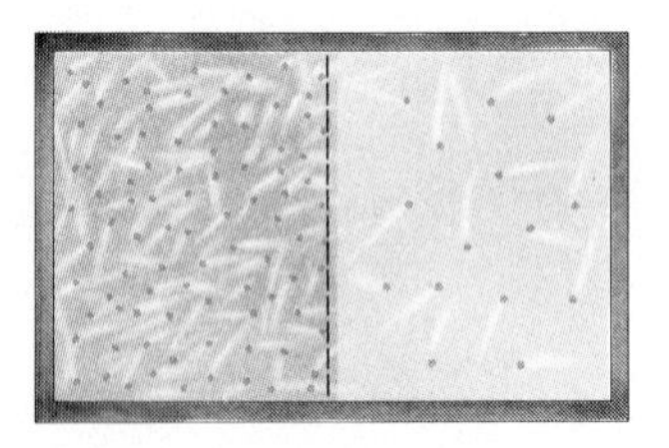

FIGURE 10.12 When the concentration of gas molecules on the left side of the container exceeds the concentration on the right side, there will be a net motion (diffusion) of molecules from left to right.

To understand why diffusion occurs, consider Figure 10.12, which represents a container in which a high concentration of molecules has been introduced into the left side. For example, this could be accomplished by releasing a few drops of perfume into the left side of the container. The dashed line in Figure 10.12 represents an imaginary barrier separating the region of high concentration from the region of lower concentration. Because the molecules are moving with high speeds in random directions, many of them will cross the imaginary barrier moving from left to right. Very few molecules of perfume will pass through his area moving from right to left simply because there are very few of them on the right side of the container at any instant. Thus, there will always be a *net* movement from the region where there are many molecules to the region where there are fewer molecules. For this reason, the concentration on the left side of the container will decrease in time and that on the right side will increase. There will be no *net* movement across the cross-sectional area once a concentration equilibrium has been reached. That is, when the concentration is the same on both sides, the number of molecules diffusing from right to left in a given time interval will equal the number moving from left to right in the same time interval.

The basic equation for diffusion is **Fick's law**, which in equation form is

Fick's law

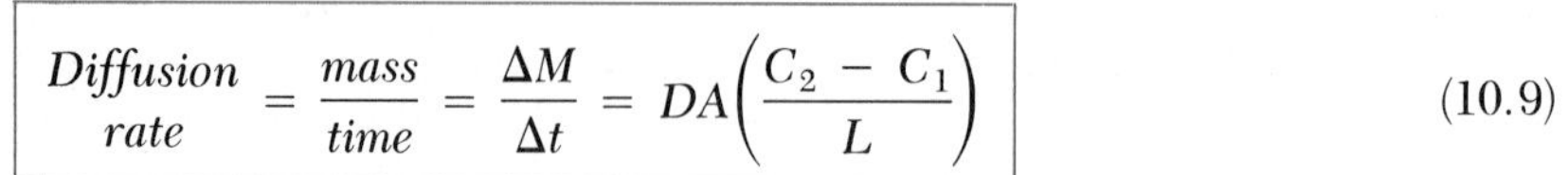

$$\boxed{\frac{\textit{Diffusion}}{\textit{rate}} = \frac{\textit{mass}}{\textit{time}} = \frac{\Delta M}{\Delta t} = DA\left(\frac{C_2 - C_1}{L}\right)} \tag{10.9}$$

where D is a constant of proportionality. The left side of this equation is called the diffusion rate and is a measure of the mass being transported per unit time. This equation says that

the rate of diffusion is proportional to the cross-sectional area A and to the change in concentration per unit distance, $(C_2 - C_1)/L$, which is called the concentration gradient.

The concentrations C_1 and C_2 are measured in kilograms per cubic meter. The proportionality constant D is called the **diffusion coefficient** and has units of square meters per second. Table 10.2 lists diffusion coefficients for a few substances.

*Diffusion and the Size of Cells

Diffusion through cell membranes is extremely vital in carrying oxygen to the cells of the body and in removing carbon dioxide and other waste products

TABLE 10.2 Diffusion Coefficients for Various Substances at 20°C

Substance	D (m^2/s)
Oxygen through air	6.4×10^{-5}
Oxygen through tissue	1×10^{-11}
Oxygen through water	1×10^{-9}
Sucrose through water	5×10^{-10}
Hemoglobin through water	7×10^{-11}

from them. Oxygen is required by the cells for those metabolic processes in which substances are either synthesized or broken down. In such metabolic processes, the cell uses up oxygen and produces carbon dioxide as a by-product. A fresh supply of oxygen diffuses from the blood, where its concentration is high, into the cell, where its concentration is low. Likewise, carbon dioxide diffuses from the cell into the blood, where it is in lower concentration. Water, ions, and other nutrients also pass into and out of cells by diffusion.

A common characteristic of cells in all plants and animals is their extremely small size. The adult human body contains literally trillions of cells. In order to understand why cells are so small, we must consider the relationship between the surface area of an object and its volume.

Let us consider a cube 2 cm on a side. The area of one of its faces is $2\text{ cm} \times 2\text{ cm} = 4\text{ cm}^2$, and because a cube has six sides, the total surface area is 24 cm^2. Its volume is $2\text{ cm} \times 2\text{ cm} \times 2\text{ cm} = 8\text{ cm}^3$. Hence, the ratio of surface area to volume is $24/8 = 3$. Now consider a larger cube, one measuring 3 cm on a side. Repeating the calculations gives us a surface area of 54 cm^2 and a volume of 27 cm^3. In this case, the ratio of surface area to volume is $54/27 = 2$. Thus, we see that as the size of an object decreases, the ratio of its surface area to its volume increases. This, of course, says that a small cell has a larger surface-area-to-volume ratio than a large cell. But how does this pertain to the operation of a cell?

A cell can function properly only if it can (a) rapidly receive vital substances such as oxygen and (b) rapidly eliminate waste products. If such substances are to readily move into and out of cells, the cells should have a large surface area. However, if the volume of the cell is too large, it could take a considerable period of time for the nutrients to diffuse into the interior of the cell where they are needed. Under optimum conditions, the surface area of the cell should be large enough so that the exposed membrane area can exchange materials effectively while at the same time the volume should be small enough so that materials can reach or leave particular locations rapidly. To reach these optimum conditions, a small cell with its high surface-area-to-volume ratio is necessary.

*Osmosis

As we have seen, the movement of material through cell membranes is necessary for the efficient functioning of cells. The diffusion of material through a membrane is partially determined by the size of the pores (holes) in the membrane wall. That is, small molecules, such as water, may pass through the pores easily while larger molecules, such as sugar, may pass through only with difficulty or not at all. A membrane that allows passage of some molecules but not others is called a selectively permeable membrane.

Osmosis is defined as the movement of water from a region where its concentration is high, across a selectively permeable membrane, into a region where its concentration is lower.

As in the case of diffusion, osmosis continues until the concentrations on the two sides of the membrane are equal. Osmosis is often described simply as the diffusion of water across a membrane.

To understand the effect of osmosis on living cells, let us consider a particular cell in the body that contains a sugar concentration of 1%. (That is, 1 g of sugar is dissolved in enough water to make 100 ml of solution.) Now assume that this cell is immersed in a 5% sugar solution (5 g of sugar dissolved in enough water to make 100 ml). In such a situation, water would diffuse from inside the cell, where its concentration is higher, across the cell wall membrane, to the outside solution, where the concentration of water is lower. This loss of water from the cell would cause it to shrink and perhaps become damaged through dehydration. If the concentrations were reversed, water would diffuse into the cell, causing it to swell and perhaps burst. It should be obvious from this description that normal osmotic relationships must be maintained in the body. If solutions are introduced into the body intravenously, care must be taken to ensure that these solutions do not disturb the osmotic balance of the body because such a disturbance could lead to cell damage. For example, if 9% saline solution surrounds a red blood cell, the cell will shrink. On the other hand, if the saline solution is about 1%, the cell will eventually burst.

Under normal circumstances, the cells of our bodies are in an environment such that there is no net movement of water into or out of them. However, certain one-celled organisms, such as protozoa, do not enjoy this osmotic equilibrium. These organisms usually live in fresh water, which obviously has a higher concentration of water than the solution inside the cell. To prevent an inflow of water to the point of bursting, these organisms possess an organelle (a tiny organ) that acts as a pump and continually forces water out of the cell.

Most plant cells are contained within a rigid wall, as shown in Figure 10.13a. If water accumulates in the cell, it expands (Fig. 10.13b) and exerts a pressure, called **turgor pressure,** against the rigid wall. The rigidity of the wall prevents the cell from bursting. If water within the cell is depleted, the rigid wall collapses inward slightly, as in Figure 10.13c. This causes the plant to wilt.

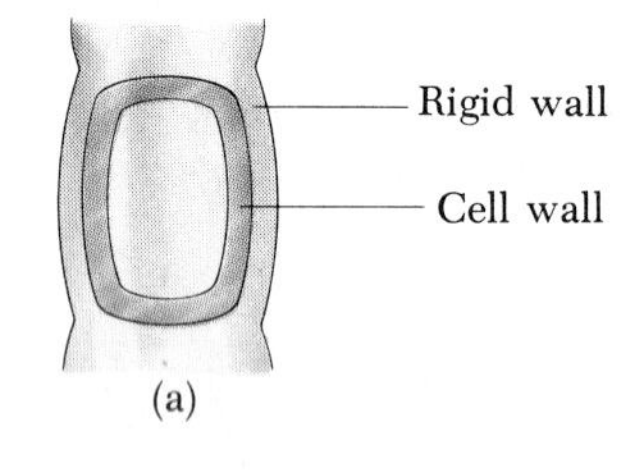

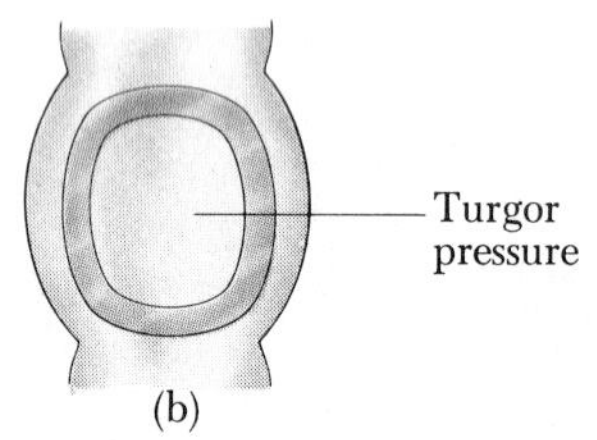

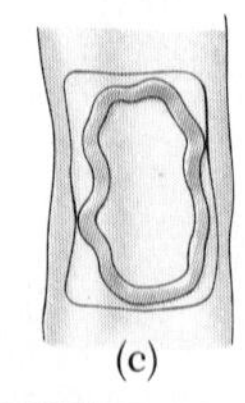

FIGURE 10.13 (a) The structure of a plant cell. (b) As water accumulates in its interior, the cell expands under turgor pressure.
(c) When water in the cell's interior is depleted, the walls of the cell collapse.

*10.7 MOTION THROUGH A VISCOUS MEDIUM

When an objects falls through air, its motion is impeded by the force of air resistance. In general, this force is dependent on the shape of the falling object and on its velocity. This viscous drag acts on all falling objects, but the exact details of the motion can be calculated only for a few cases in which the object has a simple shape, such as a sphere. In this section, we shall examine the motion of a tiny spherical object falling slowly through a viscous medium.

In 1845 a scientist named George Stokes found that the resistive force on a very small spherical object of radius r falling slowly through a fluid of viscosity η is given by

$$F_r = 6\pi\eta rv \tag{10.10}$$

This equation, called **Stokes's law**, has many important applications. For example, it describes the sedimentation of particulate matter in blood samples. It was used by Robert Millikan (1886–1953) to calculate the radius of charged oil droplets falling through air. From this, Millikan was ultimately able to determine the smallest known unit of electric charge. Millikan was awarded the Nobel prize in 1923 for this pioneering work on elemental charge.

As a sphere falls through a viscous medium, three forces act on it, as shown in Figure 10.14: $\mathbf{F}_r$ is the force of frictional resistance, $\mathbf{B}$ is the buoyant force of the fluid, and $\mathbf{w}$ is the weight of the sphere, whose magnitude is given by

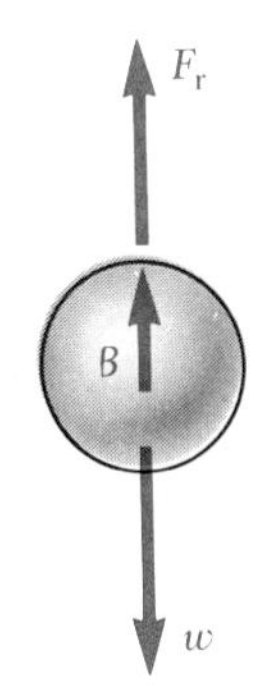

FIGURE 10.14 A sphere falling through a viscous medium. The forces acting on the sphere are the resistive frictional force, $\mathbf{F}_r$, the buoyant force, $\mathbf{B}$, and the weight of the sphere, $\mathbf{W}$.

$$w = \rho g V = \rho g\left(\frac{4}{3}\pi r^3\right)$$

where ρ is the density of the sphere and $\frac{4}{3}\pi r^3$ is its volume. According to Archimedes' principle, the buoyant force is equal to the weight of the fluid displaced by the sphere:

$$B = \rho_f g V = \rho_f g\left(\frac{4}{3}\pi r^3\right)$$

where ρ_f is the density of the fluid.

At the instant the sphere begins to fall, the force of frictional resistance is zero because the velocity of the sphere is zero. As it accelerates, the velocity increases and so does F_r. Finally, at a velocity called the **terminal velocity** v_t, *the resultant force goes to zero.* This occurs when the net upward force balances the downward weight force. Hence, the sphere reaches terminal velocity when

$$F_r + B = w$$

or

$$6\pi\eta r v_t + \rho_f g\left(\frac{4}{3}\pi r^3\right) = \rho g\left(\frac{4}{3}\pi r^3\right)$$

When this is solved for v_t, we get

$$v_t = \frac{2r^2 g}{9\eta}(\rho - \rho_f) \tag{10.11}$$

By spreading their arms and legs out and keeping the plane of the body parallel to the ground, skydivers experience maximum air drag to result in a terminal speed of about 60 m/s. (U.S. Air Force photo)

EXAMPLE 10.5 A Falling Pearl

A pearl of density 2×10^3 kg/m^3 and radius 2 mm falls through a liquid shampoo of density 1.4×10^3 kg/m^3 and viscosity 500×10^{-3} N · s/m^2. Find the terminal speed of the pearl.

Solution Substituting the values given into Equation 10.11, we have

$$\begin{aligned} v_t &= \frac{2r^2g}{9\eta}(\rho - \rho_f) \\ &= \frac{2(2 \times 10^{-3}\ \text{m})^2(9.80\ \text{m/s}^2)}{9(500 \times 10^{-3}\ \text{N}\cdot\text{s/m}^2)}(2 \times 10^3\ \text{kg/m}^3 - 1.4 \times 10^3\ \text{kg/m}^3) \\ &= 1.05 \times 10^{-2}\ \text{m/s} \end{aligned}$$

*Sedimentation and Centrifugation

If an object is not spherical, we can still use the basic approach just described to determine its terminal velocity. The only difference will be that we shall not be able to use Stokes's law for the resistive force. Instead, let us assume that the resistive force has a magnitude given by $F_r = kv$, where k is a coefficient of frictional resistance that must be determined experimentally. As we discussed above, the object reaches its terminal velocity when the weight downward is balanced by the net upward force, or

$$w = B + F_r \tag{10.12}$$

where B is the buoyant force, given by $B = \rho_f gV$.

We can use the fact that the volume, V, of the displaced fluid is related to the density of the falling object, ρ, by $V = m/\rho$. Hence, we can express the buoyant force as

$$B = \frac{\rho_f}{\rho}mg$$

Let us substitute this expression for B and $F_r = kv_t$ into Equation 10.12 (terminal velocity condition):

$$mg = \frac{\rho_f}{\rho}mg + kv_t$$

or

$$v_t = \frac{mg}{k}\left(1 - \frac{\rho_f}{\rho}\right) \tag{10.13}$$

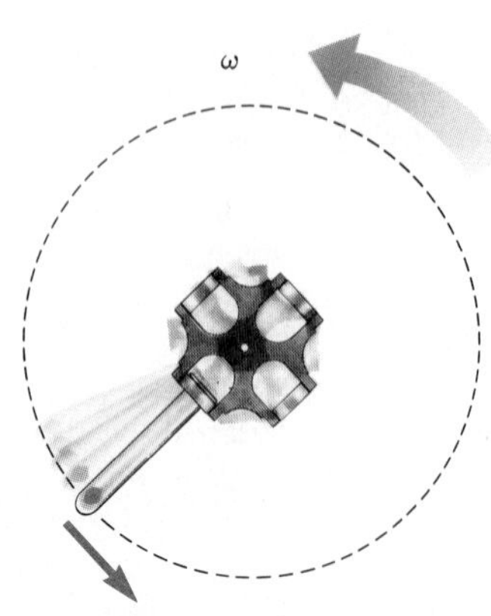

FIGURE 10.15 Simplified diagram of a centrifuge (top view).

The terminal velocity for particles in biological samples is usually quite small. For example, the terminal velocity for blood cells falling through plasma is about 5 cm/h in the gravitational field of the earth. The terminal velocities for the molecules that make up a cell are many orders of magnitude smaller than this because of their much smaller mass. The velocity at which materials fall through a fluid is called the *sedimentation rate*. This number is often important in clinical analysis.

It is often desired to increase the sedimentation rate in a fluid. A common method used to accomplish this is to increase the effective acceleration g which appears in Equation 10.13. A fluid containing various biological molecules is placed in a centrifuge and whirled at very high angular velocities (Fig. 10.15). Under these conditions, the particles experience a large radial acceleration, $a_r = v^2/r = \omega^2 r$, which is much greater than the acceleration due to gravity, and so we can replace g in Equation 10.13 by $\omega^2 r$:

$$v_t = \frac{m\omega^2 r}{k}\left(1 - \frac{\rho_f}{\rho}\right) \qquad (10.14)$$

This equation indicates that those particles having the greatest mass will have the largest terminal velocity. Therefore, the most massive particles will settle out on the bottom of a test tube first.

EXAMPLE 10.6 The Spinning Test Tube

A centrifuge rotates at 50,000 rev/min, which corresponds to an angular frequency of 5240 rad/s (a typical speed). A test tube placed in this device has its top 5 cm from the axis of rotation and its bottom 13 cm from this axis. Find the effective value of g at the midpoint of the test tube, which corresponds to a distance 9 cm from the axis of rotation.

Solution The acceleration experienced by the particles of the tube at a distance $r = 9$ cm from the axis of rotation is given by

$$a_r = \omega^2 r = \left(5240 \cdot \frac{\text{rad}}{\text{s}}\right)^2 (9 \times 10^{-2}\ \text{m}) = 2.47 \times 10^6\ \text{m/s}^2$$

EXERCISE If the mass of the contents of the test tube is 15 g, find the centripetal force that the bottom of the tube must exert on the contents of the tube. Assume a centripetal acceleration equal to that found at the midpoint of the tube.

Answer 3.71×10^4 N, or about 8000 lb! (Because of such large forces, the base of the tube in a centrifuge must be rigidly supported to keep the glass from shattering.)

SUMMARY

Various aspects of a fluid in motion can be understood by assuming that the fluid is nonviscous and incompressible and that its motion is in a steady state with no turbulence.

Under these assumptions, one obtains two important results regarding fluid flow through a pipe of nonuniform size:

1. The flow rate through the pipe is a constant, which is equivalent to stating that the product of the cross-sectional area, A, and the speed, v, at any point is constant:

$$\boxed{A_1 v_1 = A_2 v_2} \qquad (10.2)$$

   This relation is referred to as the **equation of continuity.**

2. The sum of the pressure, the kinetic energy per unit volume, and the potential energy per unit volume has the same value at all points along a streamline:

$$\boxed{P + \frac{1}{2}\rho v^2 + \rho g y = \text{constant}} \qquad (10.4)$$

   This is known as **Bernoulli's equation.**

Viscosity is a term used to describe the internal friction of a fluid. The **coefficient of viscosity,** η, for a fluid is defined as the ratio of the shearing stress to the rate of change of the shearing strain:

$$\eta = \frac{Fl}{Av} \tag{10.5}$$

The SI units of viscosity are $N \cdot s/m^2$.

The rate of flow of a fluid through a tube is given by **Poiseuille's law:**

$$\text{Rate of flow} = \frac{\Delta V}{\Delta t} = \frac{(P_1 - P_2)(\pi R^4)}{8L\eta} \tag{10.7}$$

The onset of turbulence in a fluid is determined by a factor called the **Reynolds number,** given by

$$RN = \frac{\rho v d}{\eta} \tag{10.8}$$

The movement of atoms or molecules from a region where their concentration is high to a region where their concentration is lower is called **diffusion.** The diffusion rate, which is a measure of the mass being transported per unit time, is given by

$$\text{Diffusion rate} = \frac{\text{mass}}{\text{time}} = DA\left(\frac{C_2 - C_1}{L}\right) \tag{10.9}$$

Osmosis is the movement of water from a region where its concentration is high, across a selectively permeable membrane, into a region where its concentration is lower.

ADDITIONAL READING

M. Greenwood, C. Hanna, and Rev. J. Melton, "Air Resistance Acting on a Sphere," *The Physics Teacher*, March 1986, p. 153.

D. C. Hazen and R. F. Lehnert, "Low-Speed Flight," *Sci. American*, April 1956, p. 46.

F. Hess, "The Aerodynamics of Boomerangs," *Sci. American*, November 1968, p. 124.

J. A. Lock, "The Physics of Air Resistance," *The Physics Teacher*, March 1982, p. 158.

N. Smith, "Bernoulli and Newton in Fluid Mechanics," *Physics Teacher, 10,* 1972, p. 451.

QUESTIONS

1. The water supply for a city is often provided from reservoirs built on high ground. Water flows from the reservoir, through pipes, and into your home when you turn the tap on your faucet. Why is the water flow more rapid out of a faucet on the first floor of a building than in an apartment on a higher floor?
2. Smoke rises in a chimney faster when a breeze is blowing. Use Bernoulli's principle to explain this phenomenon.
3. Why do many trailer trucks use wind deflectors on the top of their cabs? (See photograph.) How do such devices reduce fuel consumption?

The high cost of fuel has prompted many truck owners to install wind deflectors on their cabs to reduce air drag. (Photo by Lloyd Black)

4. Consider the cross-section of the wing on an airplane. The wing is designed such that the air travels faster over the top than under the bottom. Explain why there is a net upward force (lift) on the wing due to the Bernoulli effect.
5. When a fast-moving train passes a train at rest, the two tend to be drawn together. How does the Bernoulli effect explain this phenomenon?
6. A baseball moves past an observer from left to right spinning counterclockwise. In which direction will the ball tend to deflect?
7. A tornado or hurricane will often lift the roof of a house. Use the Bernoulli effect to explain why this occurs. Why should you keep your windows open during these conditions?
8. If you suddenly turn on your shower water at full speed, why is the shower curtain pushed inward?
9. If you hold a sheet of paper and blow across the top surface, the paper rises. Explain.
10. If air from a hair dryer is blown over the top of a ping-pong ball, the ball can be suspended in air. Explain how the ball can remain in equilibrium.
11. Two ships passing near each other in a harbor tend to be drawn together and run the risk of a sideways collision. How does the Bernoulli effect explain this?
12. When ski-jumpers are airborne, why do they bend their body forward and keep their hands at their sides?
13. How should a pitcher throw a baseball so that it rises as it reaches the batter?
14. Should an airplane take off into the wind or with the wind? Explain.
15. SAE 10 motor oil has less viscosity at a given temperature than does SAE 40 oil. Which should you use in your car during the winter months? (Actually, a multi-grade oil is best.)

PROBLEMS

Section 10.2 Fluid Dynamics

Section 10.3 Other Applications of Bernoulli's Equation

1. Each second, 0.25 liter of liquid flows through a tube 1 cm in diameter. (a) What is the flow velocity along the tube? (b) If the tube in part (a) is connected to a tube 0.75 cm in diameter, what will the flow velocity be in the smaller tube?

2. A cowboy at a dude ranch fills a horse trough which is 1.5 m long, 60 cm wide, and 40 cm tall. He uses a 2-cm diameter hose from which water emerges at 1.5 m/s. How long will it take him to fill the trough?

3. The velocity of flow of a liquid through a water pipe is 0.636 m/s. Determine the flow rate if the radius of the pipe is 1.5 cm.

4. A water tank open to the atmosphere contains water at a depth of 4 m. If a puncture occurs at the base, with what velocity does the stream of water leave the opening?

5. A liquid ($\rho = 1.65$ g/cm^3) flows through two horizontal sections of tubing joined end to end. In the first section the cross-sectional area is 10 cm^2, the flow velocity is 275 cm/s, and the pressure is 1.2×10^5 Pa. The liquid then enters the second section, where the cross-sectional area is 2.5 cm^2. Calculate (a) the flow velocity and (b) the pressure in the section of smaller diameter.

6. (a) If wind blows at a speed of 30 m/s over the roof of your house, what is the pressure difference on your roof between the air inside and the air outside? (b) What net force does this produce on a roof having an area of 175 m^2? (Use 1.3 kg/m^3 as the density of air.)

7. What is the net upward force on an airplane wing of area 20 m^2 if the velocity of flow across the top of the wing is 300 m/s and the velocity across the bottom of the wing is 280 m/s.

8. A liquid ($\rho = 1.05 \times 10^3$ kg/m^3) flows at constant speed through a plastic tube that forms a vertical loop 1.8 m in diameter. The pressure in the liquid at a point 0.5 m below the top of the loop is 1.3×10^4 Pa. Calculate the pressure (a) at the top of the loop and (b) at the bottom of the loop.

• **9.** Water flows through a pipe of radius 30 cm at the rate of 0.2 m^3/s. The pressure in the pipe is atmospheric. The pipe slants downhill and feeds into a second pipe of radius 15 cm which is located 60 cm lower. What is the gauge pressure in this lower pipe?

10. (a) Calculate the flow rate (in grams per second) of blood (ρ = 1 g/cm^3) in an aorta with a cross-sectional area of 2 cm^2 if the flow velocity is 40 cm/s. (b) Assume that the aorta branches to form a large number of capillaries with a combined cross-sectional area of 3×10^3 cm^2. What is the velocity of flow in the capillaries?

• **11.** You are visiting "fish world" while on vacation when a break occurs near the bottom of one of the tanks. If the break is 1.5 m below the surface of the water and 1 m above the floor, how far from the base of the tank will the water strike the floor?

• **12.** The water supply of a building is fed through a main entrance pipe 6 cm in diameter. A 2-cm diameter faucet tap located 2 m above the main pipe is observed to fill a 25 liter container in 30 s. (a) What is the speed at which the water leaves the faucet? (b) What is the gauge pressure in the main pipe? (Assume the faucet is the only outlet in the system.)

• **13.** What is the gauge pressure in a nozzle if the water leaving it sprays the water vertically upward to a height of 4 m?

• **14.** A large storage tank open to the atmosphere at the top and filled with water develops a small hole in the side at a point 16 m below the water level. If the rate of flow from the leak is 2.5×10^{-3} m^3/min, determine (a) the speed at which the water leaves the hole and (b) the diameter of the hole.

•• **15.** A water tank open to the atmosphere at the top has two holes punched in its side one above the other. The holes are 5 cm and 12 cm above the floor. How high does water stand in the tank if the two streams of water hit the floor at the same place?

•• **16.** A horizontal tube of radius 1 cm is joined to a second horizontal tube of radius 0.5 cm. There is a pressure difference of 6660 Pa between the two tubes. (a) Which tube has the higher water pressure? (b) What volume of water flows through the tubes per second?

Section 10.4 Viscosity

Section 10.5 Poiseuille's Law

17. A hypodermic needle is 3 cm in length and 0.3 mm in diameter. What excess pressure is required along the needle so that the flow rate of water through it will be 1 g/s? (Use 1×10^{-3} $N \cdot s/m^2$ as the viscosity of water.)

18. A metal block is pulled over a horizontal surface that has been coated with a layer of lubricant 1.0 mm thick. The face of the block in contact with the surface has dimensions 0.40 m by 0.12 m. A force of 1.9 N is required to move the block at a constant speed of 0.5 m/s. Calculate the coefficient of viscosity of the lubricant.

19. A thin 1.5 mm coating of glycerine has been placed between two microscope slides of width 1 cm and length 4 cm. Find the force required to pull one of the microscope slides at a speed of 0.3 m/s relative to the other.

20. Determine the velocity at which the flow of water through a 0.5 cm diameter pipe will become turbulent ($RN \geq 3000$).

21. Assume a value of 980 for the Reynolds number for blood in an artery and a viscosity of 4×10^{-3} $N \cdot s/m^2$ for whole blood. If the density of whole blood is equal to 1.05×10^3 kg/m^3, at what velocity does blood flow through an artery 0.45 cm in diameter?

22. A needle of radius 0.3 mm and length 3 cm is used to give a patient a blood transfusion. Assume the pressure differential across the needle is achieved by elevating the blood 1 m above the patient's arm. (a) What is the rate of flow of blood through the needle? (b) At this rate of flow, how long will it take to inject 1 pint (approximately 500 cm^3) of blood into the patient? The density of blood is about 1050 kg/m^3, and its coefficient of viscosity is 4×10^{-3} $N \cdot s/m^2$.

• **23.** The aorta in humans has a diameter of about 2 cm and, at certain times, the blood velocity through it is about 55 cm/s. Is the blood flow turbulent? (Values for several blood properties that you may need are given in Problem 22.)

24. What diameter needle should be used to inject a volume of 500 cm^3 of a solution into a patient in 30 min? Assume that the needle length is 2.5 cm and that the solution is elevated 1 m. Furthermore, assume the viscosity and density of the solution are those of pure water.

Section 10.6 Transport Phenomena

25. Sucrose is allowed to diffuse along a 10-cm length of tubing filled with water. The tube is 6 cm^2 in cross-sectional area. The diffusion coefficient is 5×10^{-10} m^2/s, and 8×10^{-14} kg is transported along the tube in 15 s. What is the difference in the concentration levels of sucrose at the two ends of the tube?

26. In a diffusion experiment, it is found that, in 60 s, 3×10^{-13} kg of sucrose will diffuse along a horizontal pipe of cross-sectional area 1 cm^2. If the diffusion coefficient is 5×10^{-10} m^2/s, what is the concentration gradient (that is, the change in concentration per unit length along the path)?

27. Glycerine in water diffuses along a horizontal column that has a cross-sectional area of 2 cm^2. The concentration gradient is 3×10^{-2} kg/m^4, and the diffusion rate is found to be 5.7×10^{-15} kg/s. Determine the diffusion coefficient.

28. Use the data for sucrose given in Problem 25 to calculate how much sucrose will diffuse down a horizontal pipe of cross-sectional area 4 cm^2 in 10 s if the concentration gradient is 0.2 kg/m^4.

Section 10.7 Motion Through a Viscous Medium

29. Small spheres of diameter 1 mm fall through water with a terminal velocity of 1.1 cm/s. Calculate the density of the spheres.

30. A test tube 10 cm tall is filled with water which contains spherical particles of density 1.8 g/cm^3 in suspension. If the radius of the particle is 2×10^{-4} cm, find the time it takes a particle near the top of the tube to reach the bottom.

31. Calculate the viscous force on a spherical oil droplet that is 2×10^{-5} cm in diameter and falling with a speed of 0.4 mm/s in air. (Use $\eta = 1.8 \times 10^{-5}$ N·s/m^2 as the coefficient of viscosity of air.)

32. The viscous force on an oil drop is measured to be 3×10^{-13} N when the drop is falling through air with a speed of 4.5×10^{-4} m/s. If the radius of the drop is equal to 2.5×10^{-6} m, what is the viscosity of air?

33. An oil drop (ρ = 800 kg/m^3) is falling with a terminal velocity of 0.04 mm/s in air of density 1.29 kg/m^3. Calculate the radius of the oil drop. (Use $\eta = 1.8 \times 10^{-5}$ N·s/m^2 as the coefficient of viscosity of air.)

• **34.** Spherical particles of a protein of density 1.8 g/cm^3 are shaken up in a solution of water. The solution is allowed to stand for 1 h. If the depth of water in the tube is 5 cm, find the radius of the largest particles still in solution at the end of the hour.

ADDITIONAL PROBLEMS

35. Water flows through a 2-cm diameter hose with a speed of 0.5 m/s. A person using the hose pinches the end such that the diameter becomes 0.25 cm. At what speed does the water now leave the hose?

36. Water at a pressure of 3×10^5 Pa flows through a horizontal pipe at a speed of 1 m/s. If the pipe narrows to one-fourth its original diameter, (a) find the speed of flow in the narrow section, and (b) the pressure in the narrow section.

37. Use dimensional analysis to convince yourself that the power required to drive a fluid through a pipe is given by the volume rate of flow, Av, times the pressure difference across the pipe, $P_1 - P_2$.

38. The water jet from an amusement park fountain reaches a height of 10 m. If the fountain nozzle has an opening 10 cm in diameter, what power is required to operate the fountain? (Assume that the length of the nozzle is negligible and that the velocity of the fluid in the supply line to the nozzle is also negligible.)

• **39.** One step in an industrial process requires the separation of very small spherical aluminum particles from a suspension of glycerine. If this separation is accomplished by natural sedimentation, calculate the terminal velocity for a particle diameter of 1×10^{-5} m. For glycerine use $\eta = 1500 \times 10^{-3}$ N·s/m^2 and ρ = 900 kg/m^3.

• **40.** A glass sphere of diameter 1.75 cm and density 2.5 g/cm^3 falls through water with a terminal velocity of 0.3 m/s. A wooden sphere of the same diameter and density 0.85 g/cm^3 is released from the bottom of a tank of water. Determine the terminal velocity with which the wooden sphere rises.

41. Ethanol (ρ = 791 kg/m^3) flows through a ground level horizontal pipe, 30 cm in diameter, at a gauge pressure of 2.5×10^4 Pa and a flow rate of 0.21 m^3/s. In order to cross a highway, the pipe has a U-shaped section that reaches 2 m below the ground-level section. What is the maximum diameter of the below-ground section such that the gauge pressure at the lowest point will not be greater than 3×10^4 Pa?

42. As part of an industrial process, a liquid for which $\rho = 7.5 \times 10^2$ kg/m^3 and η = 0.45 N·s/m^2 is pumped under pressure through a horizontal pipe 500 m in length. A flow rate of 0.15 m^3/s is required, and the Reynolds number is not to exceed 1000. Determine (a) the minimum pipe diameter and (b) the pressure difference across the ends of the pipe.

•• **43.** A siphon is a device that allows a fluid to seemingly defy gravity (Fig. 10.16). The flow must be initiated by a partial vacuum in the tube, as in a drinking straw. (a) Show that the speed at which the water emerges from the siphon is given by $v = \sqrt{2gh}$. (b) For what values of y will the siphon work?

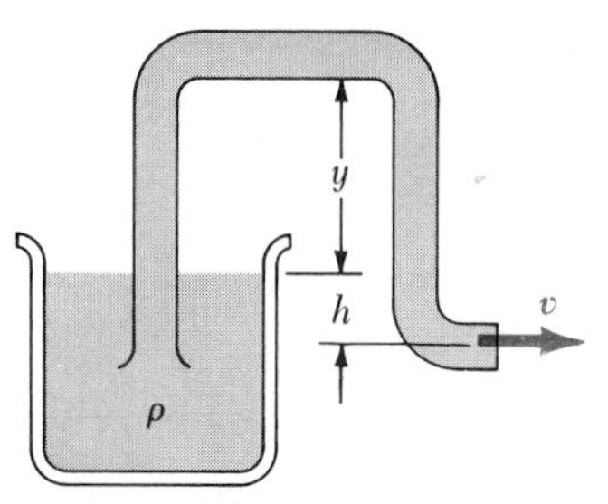

FIGURE 10.16 Problem 43

Essay

Physics of the Human Circulatory System

William G. Buckman, *Western Kentucky University*

The human circulatory system is an extremely complex and vital part of the human body. The blood supplies food and oxygen to the tissues of the body, carries away the waste products from the cells, distributes the heat generated by the cells to equalize the temperature of the body, carries hormones that stimulate and coordinate the activity of organs, distributes antibodies to fight infection, and performs numerous other functions.

William Harvey (1579–1657), an English physician and physiologist, studied blood flow and the action of the heart. He established the essential mechanics of the heart and found that the blood flows from the arterial system through capillary beds and into the veins to be returned to the heart.

The Physical Properties of Blood

Blood is a liquid tissue consisting of two principal parts: the plasma, which is the intercellular fluid, and the cells, which are suspended in the plasma. Plasma is about 90 percent water, 9 percent proteins, and 0.9 percent salts,

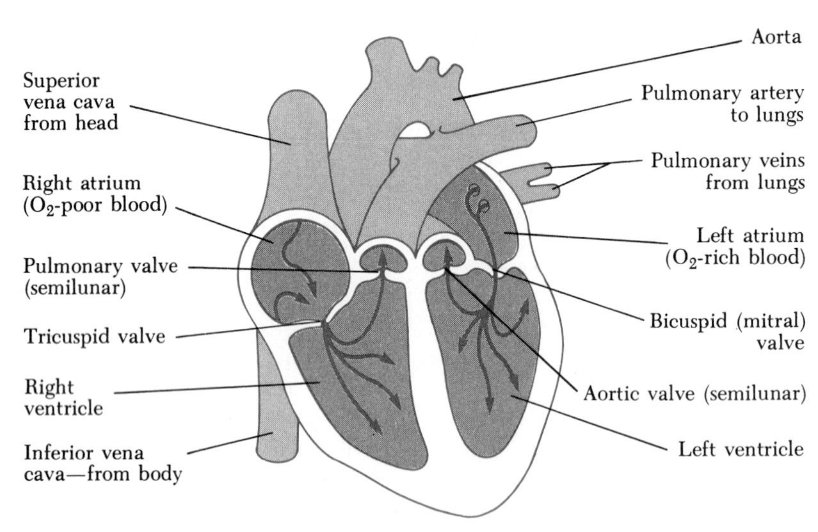

(a) Anatomy of the heart

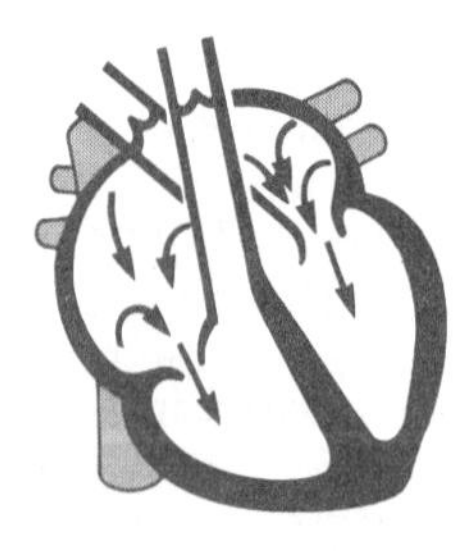

1. Blood fills both atria, some blood flows into ventricles—diastole phase of atria.

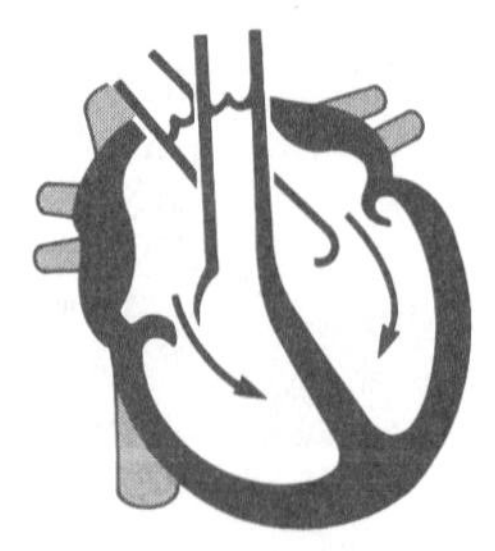

2. Atria contract, squeezing blood into ventricles—ventricular diastole.

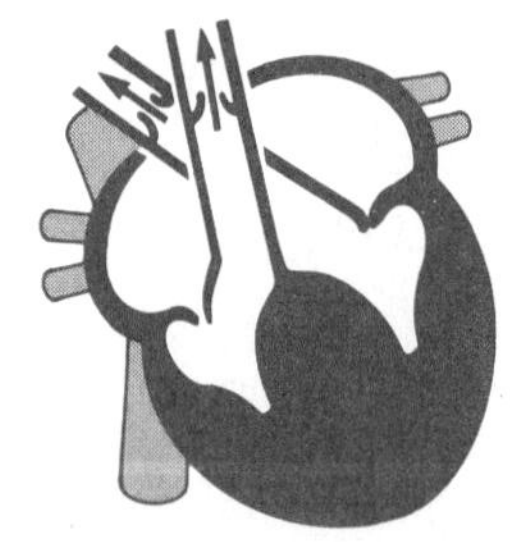

3. Ventricles contract, squeezing blood into aorta and pulmonary arteries—ventricular systole phase.

(b) Pumping cycle of the heart

FIGURE 1 (a) Anatomy of the heart. (b) Pumping cycle of the heart.

sugar, and traces of other materials. Blood contains white blood cells and red blood cells. The individual red blood cells are biconcave and have an average diameter of 7.5 μm. There are about 5×10^6 red blood cells per cubic millimeter of blood. The five types of white blood cells found in the blood have an average concentration of 8000 per cubic millimeter, with the concentration normally varying between 4500 and 11,000 per cubic millimeter. The density of blood is about 1.05×10^3 kg/m^3, and its viscosity varies from 2.5 to 4 times that of water.

The Heart as a Pump

The heart can be considered as a double pump, with each side consisting of an atrium and a ventricle (Fig. 10E.1a). Blood enters the right atrium, flows into the right ventricle, is pumped by the right ventricle to the lungs, and returns through the left atrium to the left ventricle. The left ventricle then pumps the oxygenated blood out through the aorta to the rest of the body. The heart has a system of one-way valves to assure that the blood flows in the proper direction. The heart's pumping cycle has the two ventricles pumping at the same time, as shown in Figure 10E.1b.

The pressure generated by the right ventricle is quite low (about 25 mm Hg), and the lungs offer a low resistance to blood flow. The left ventricle generates a larger pressure, typically greater than 120 mm Hg, at the peak (systole) of the pressure. During the resting stage (diastole) of the heartbeat, the pressure is typically about 80 mm Hg.

We shall now calculate the mechanical work done by the heart. Consider the fluid in the vessel shown in Figure 10E.2. The net force on the fluid is equal to the product of the pressure drop across the fluid, ΔP, and the cross-sectional area, A. The power expended is equal to the net force times the average velocity: $(\Delta PA)(\bar{v})$. Because $A\bar{v} = AL/t =$ volume/time, which is the flow rate, we may now write for the power expended by the heart

$$\text{Power} = (\text{flow rate})(\Delta P)$$

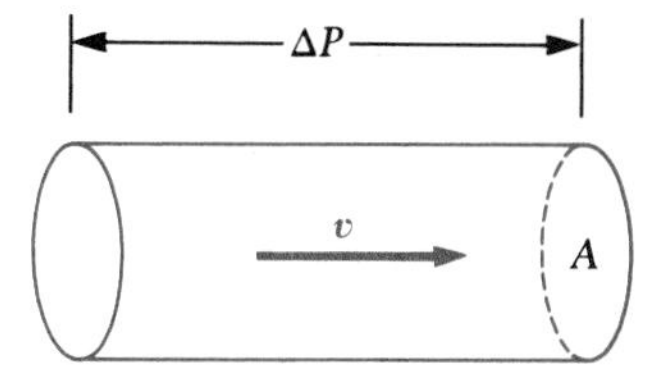

FIGURE 2 The power required to maintain blood flow against viscous forces.

If a normal heart pumps blood at the rate of 97 cm^3/s and the pressure drop from the arterial system to the venous system is 1.17×10^4 Pa, we then have

$$\begin{aligned}\text{Power} &= (97\ \text{cm}^3/\text{s})\ (10^6\ \text{m}^3/\text{cm}^3)\ (1.17 \times 10^4\ \text{Pa}) \\ &= 1.1\ \text{W}\end{aligned}$$

By measuring oxygen consumption, it is found that the heart of a 70 kg man at rest consumes about 10 W. In the calculation above, it was determined that 1.1 W is required to do the mechanical work of pumping blood; hence, the heart is typically about 10 percent efficient. During strenuous exercise, the blood pressure may increase by 50 percent and the blood volume pumped may increase by a factor of 5 to yield an increase of 7.5 times in the power generated by the left ventricle. Because the right ventricle has a systolic pressure about one fifth that of the left ventricle, its power requirement is about one fifth that of the left ventricle.

When we listen to a heart with a stethoscope, we hear two sharp sounds. The first corresponds with the closing of the tricuspid and mitral valves, and the second corresponds with the closing of the aortic and pulmonary valves. Other sounds that are heard are those associated with the flow and turbulence of the blood.

The Cardiovascular System

The cardiovascular system includes the heart to pump the blood; arteries to carry the blood to the organs, muscle, and skin; and veins to return the blood to the heart (Fig. 10E.3). The blood is pumped into the aorta by the left ventricle. The aorta branches to form smaller arteries, which in turn branch down to even smaller arteries, until finally the blood reaches the very small capillaries of the vascular bed. These capillaries are so small that the red blood cells must pass single file through them. After passing through the capillaries, where materials being carried by the blood are exchanged with the surrounding tissues, the blood flows to the veins and is returned to the heart.

The flow rate of the blood changes as it goes through this system. The cross-sectional area of the vascular bed, which is the product of the cross-sectional area and the number of capillaries, is much greater than the cross-sectional area of the aorta. Because the volume of the blood passing through a cross-sectional area per unit of time is Av, where v is the magnitude of the velocity of the blood, we may express the volume flow rate of the blood as

$$\text{Flow rate} = A_{\text{aorta}}v_{\text{aorta}}$$

Furthermore, because the total average flow rate through the aorta and the capillaries must be the same, we have

$$\text{Flow rate} = A_{\text{aorta}}v_{\text{aorta}} = A_{\text{cap}}v_{\text{cap}}$$

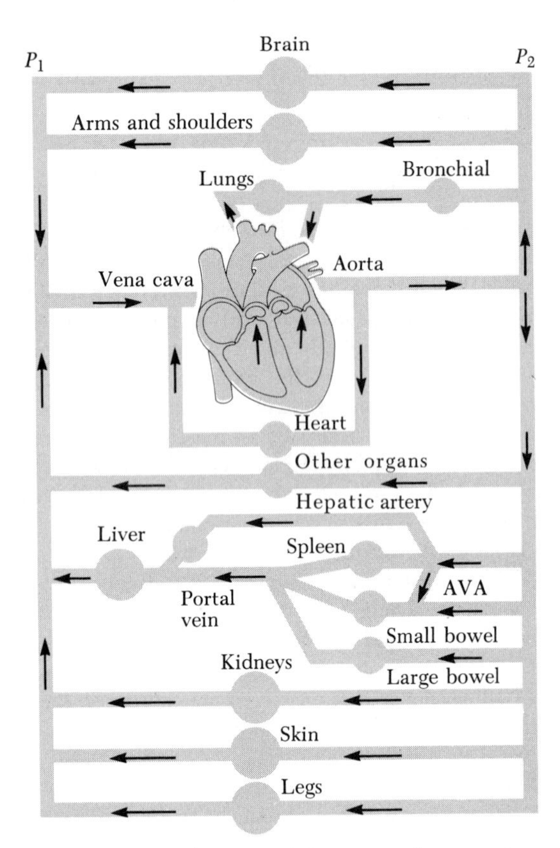

FIGURE 3 A diagram of the mammalian circulatory system. Pressure P_2 is that in the arterial system, and P_1 is that in the venous system. Arrows indicate the direction of blood flow.

EXAMPLE 10E.1 Flow of Blood in the Aorta and Capillaries

The velocity of blood in the aorta is 50 cm/s, and it has a radius of 1 cm. (a) What is the rate of flow of blood through this aorta? (b) If the capillaries have a total cross-sectional area of 3000 cm^2, what is the speed of the blood in the capillaries?

Solution (a) The area of the aorta is

$$A = \pi r^2 = \pi(1 \text{ cm})^2 = \pi \text{ cm}^2$$

$$\text{Flow rate} = Av = (\pi \text{ cm}^2)(50 \text{ cm/s}) = 50\pi \text{ cm}^3/\text{s}$$

(b) The flow rate in the capillaries $= 50\pi \text{ cm}^3/\text{s} = A_c v_c$,

$$v_c = \frac{\text{Flow rate}}{A_c} = \frac{50\pi \text{ cm}^3/\text{s}}{3000 \text{ cm}^2} = 0.05 \text{ cm/s}$$

This low blood velocity in the capillaries is necessary to enable the blood to exchange oxygen, carbon dioxide, and other nutrients with the surrounding tissues.

ESSAY QUESTIONS

1. Explain why some individuals tend to black out when they stand up rapidly.
2. If a person is standing and is at rest, what will be the relation between the blood pressure in the left arm and the left leg? What is the relation if the person is in a horizontal position?
3. At what upward acceleration would you expect the blood pressure in the brain to be zero? (Assume that no body mechanisms are operating to compensate for this condition.)
4. When using a sphygmomanometer to measure blood pressure, will the blood pressure readings depend upon the atmospheric pressure? If the atmospheric pressure decreases rapidly, how will this effect the blood pressure readings?
5. Why is it impractical to measure the pulse rate using a vein?
6. Assuming that an artery is clogged such that the effective radius is one-half its normal radius compared to a normal artery, by what factor must the pressure differential be increased to obtain the normal flow rate through the clogged artery?

ESSAY PROBLEMS

1. Determine the average velocity of the blood in the aorta if it has a radius of 1.2 cm, and the flow rate is 20 liters/min.
2. If the mean blood pressure in the aorta is 100 mm Hg, (a) determine the blood pressure in the artery located 2 ft above the heart. (b) One can not apply, without significant error, Bernoulli's principle in the smaller arteries and the capillaries. Why not?
3. When the flow rate is 5 liters/min, the velocity in the capillaries is 0.33 mm/s. Assuming the average diameter of a capillary to be 0.008 mm, calculate the number of capillaries in the circulatory system.
4. An artery has a length of 20 cm, and a radius of 0.5 cm, and blood is flowing at a rate of 6 liters/min. What is the difference in the pressure between the ends of the artery?
5. Assuming the internal pressure in the left ventricle is 100 mm Hg and the left ventricle of the heart has an effective radius of 3 cm, calculate the tension in the wall of the left ventricle. (Assume that the external pressure is zero gauge pressure.)
6. If the blood pressure is 10 mm Hg in a capillary which has a radius of 0.005 mm, determine the tension in the wall of the capillary.

Part 2

THERMODYNAMICS

When dining, I had often observed that some particular dishes retained their Heat much longer than others; and that apple pies, and apples and almonds mixed (a dish in great repute in England) remained hot a surprising length of time. Much struck with this extraordinary quality of retaining Heat, which apples appeared to possess, it frequently occurred to my recollection; and I never burnt my mouth with them, or saw others meet with the same misfortune, without endeavouring, but in vain, to find out some way of accounting, in a satisfactory manner, for this surprising phenomenon.

BENJAMIN THOMPSON (COUNT RUMFORD)

As we saw in the first part of this textbook, Newtonian mechanics explains a wide range of phenomena on a macroscopic scale, such as the motion of baseballs, rockets, and the planets of our solar system. We now turn to the study of thermodynamics, which is concerned with the concepts of heat and temperature. As we shall see, thermodynamics is very successful in explaining the bulk properties of matter and the correlation between these properties and the mechanics of atoms and molecules.

Historically, the development of thermodynamics paralleled the development of the atomic theory of matter. By the middle of the 19th century, chemical experiments provided solid evidence for the existence of atoms. At that time, scientists recognized that there must be a connection between the theory of heat and temperature, and the structure of matter. In 1827, the botanist Robert Brown reported that grains of pollen suspended in a liquid move erratically from one place to another, as if under constant agitation. In 1905, Albert Einstein developed a theory in which he used thermodynamics to explain the cause of this erratic motion, today called Brownian motion. Einstein explained this phenomenon by assuming that the grains of pollen are under constant bombardment by "invisible" molecules in the liquid, which themselves undergo an erratic motion. This important observation and Einstein's insight gave scientists a means of discovering vital information concerning molecular motion.

Have you ever wondered how a refrigerator is able to cool its contents or what types of transformations occur in a power plant or in the engine of your automobile or what happens to the kinetic energy of an object when it falls to the ground and comes to rest? The laws of thermodynamics and the concepts of heat and temperature will enable us to answer such practical questions.

Many things can happen to an object when it is heated. Its size will change slightly, but it may also melt, boil, ignite, or even explode. The outcome depends upon the composition of the object and the degree to which it is heated. In general, thermodynamics must concern itself with the physical and chemical transformations of matter in all of its forms: solid, liquid, and gas.

CHAPTER 11

Thermal Physics

The ceramic material being heated by a flame is a poor heat conductor but is able to withstand a large temperature gradient. Note that its left side is buried in ice (0° C) while the right side is very hot. (Courtesy Corning Glass Works)

Our study thus far has focused exclusively on the subject of mechanics. Such concepts as mass, force, and kinetic energy have been carefully defined in order to make the subject quantitative. We now move to a new branch of physics, thermal physics. Here we shall find that the quantitative description of thermal phenomena requires a careful definition of the concepts of temperature, heat, and internal energy. As the name **thermal physics** implies, we shall be concerned with such concepts as heat and temperature. We will be interested in finding answers to such questions as: What happens to an object when heat is added to it or removed from it? What physical changes occur in an object when its temperature increases or decreases? One familiar outcome of a temperature increase or decrease for an object is that its size may change. We shall examine the details of this process in our discussion of linear expansion.

An appreciable portion of this chapter will be devoted to a study of ideal gases. We shall approach this study on two levels. The first will examine ideal gases from a large-scale viewpoint. Here we shall be concerned with the relationships between such quantities as pressure, volume, and temperature. On the second level, we shall examine the behavior of gases from a point of view that pictures the components of a gas as small particles. This latter approach, called the kinetic theory of gases, will help us to understand what is happening on the atomic level to produce such large-scale effects as pressure and temperature.

11.1 TEMPERATURE AND THE ZEROTH LAW OF THERMODYNAMICS

When we speak of the **temperature** of an object, we often associate this concept with how hot or cold the object feels when we touch it. Thus, our senses provide us with a qualitative indication of temperature. However, our senses are unreliable and often misleading. For example, if we remove a metal ice tray and a package of frozen vegetables from the freezer, the ice tray feels colder to the hand than the vegetables even though both are at the same temperature. This is because metal is a better heat conductor than cardboard and so the ice tray conducts heat from our hand more efficiently than does the cardboard package. Hence, the ice tray feels colder. What we need is a reliable and reproducible method for establishing the relative "hotness" or "coldness" of objects. Scientists have developed various types of thermometers for making such quantitative measurements. Some typical thermometers will be described in Section 11.2.

We are all familiar with the fact that two objects at different initial temperatures will eventually reach some intermediate temperature when placed in contact with each other. For example, if you place two soft drinks, one hot and the other cold, in contact in an insulated container, the two will eventually reach an equilibrium with the cold one warming up and the hot one cooling off. Likewise, if a cup of hot coffee is cooled with an ice cube, the ice will eventually melt and the coffee's temperature will decrease.

In order to understand the concept of temperature, it is useful to first define two often-used phrases, *thermal contact* and *thermal equilibrium.* To understand the meaning of thermal contact, imagine two objects placed in an insulated container such that they interact with each other, but do not interact with the rest of the world. Furthermore, let us assume that neither object does work on the other. For example, they may exert forces on one another, but if neither undergoes a displacement, no work is done. If the objects are at different temperatures, energy can still be exchanged between them. The type of energy that is exchanged between the objects because of a temperature difference is called **heat energy.** We shall examine the concept of heat energy in more detail in the next chapter. For purposes of the current discussion, we shall assume that two objects are in **thermal contact** with each other if heat energy can be exchanged between them. **Thermal equilibrium** is a situation in which two objects in thermal contact with each other cease to have any heat energy exchange.

Now consider two objects, A and B, which are not in thermal contact, and a third object, C, which will be our thermometer. We wish to determine whether or not A and B would be in thermal equilibrium with each other if there was thermal contact between them. The thermometer (object C) is first placed in thermal contact with A until thermal equilibrium is reached. At that point, the thermometer's reading will remain constant. The thermometer is then placed in thermal contact with B, and its reading is recorded after thermal equilibrium is reached. If the two readings are the same, then A and B would be in thermal equilibrium with each other. We can summarize these results in a statement known as the **zeroth law of thermodynamics** (the law of equilibrium):

If bodies A and B are separately in thermal equilibrium with a third body, C, then A and B will be in thermal equilibrium with each other if placed in thermal contact.

This statement, insignificant and obvious as it may seem, is easily proved experimentally and is very important because it can be used to define temperature. We can think of temperature as the property that determines whether or not an object is in thermal equilibrium with other objects. *Two objects in thermal equilibrium with each other are at the same temperature.* Conversely, if two objects have different temperatures, they cannot be in thermal equilibrium with each other.

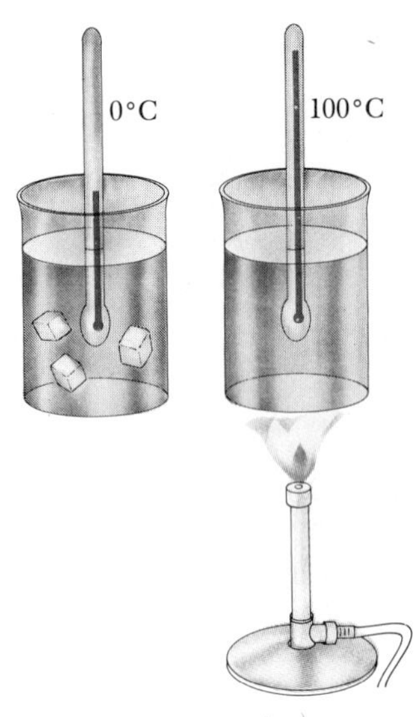

FIGURE 11.1 Schematic diagram of a mercury thermometer. Because of thermal expansion, the level of the mercury rises as the mercury is heated from 0°C (the ice point) to 100°C (the steam point).

11.2 THERMOMETERS AND TEMPERATURE SCALES

Thermometers are devices used to define and to measure the temperature of a system. All thermometers make use of a change in some physical property with temperature. Some of these physical properties are (1) the change in volume of a liquid, (2) the change in length of a solid, (3) the change in pressure of a gas held at constant volume, (4) the change in volume of a gas held at constant pressure, (5) the change in electric resistance of a conductor, and (6) the change in color of a very hot object. A temperature scale can be established for a given substance using any one of these physical quantities.

The most common thermometer in everyday use consists of a mass of mercury that expands into a glass capillary tube when heated (Fig. 11.1). Thus, the physical property in this case is the change in volume of a liquid. One can define any temperature change to be proportional to the change in length of the mercury column. The thermometer can be calibrated by placing it in thermal contact with some natural systems that remain at constant temperature. One such system is a mixture of water and ice in thermal equilibrium at atmospheric pressure, which is defined to have a temperature of zero degrees Celsius, written 0°C. (This temperature is called the ice point of water.) Another convenient system for calibrating a thermometer is a mixture of water and steam in thermal equilibrium at atmospheric pressure. The temperature of this system is 100°C (the steam point of water). Once the mercury levels have been established at these two points, the column is divided into 100 equal segments, each denoting a change in temperature of one Celsius

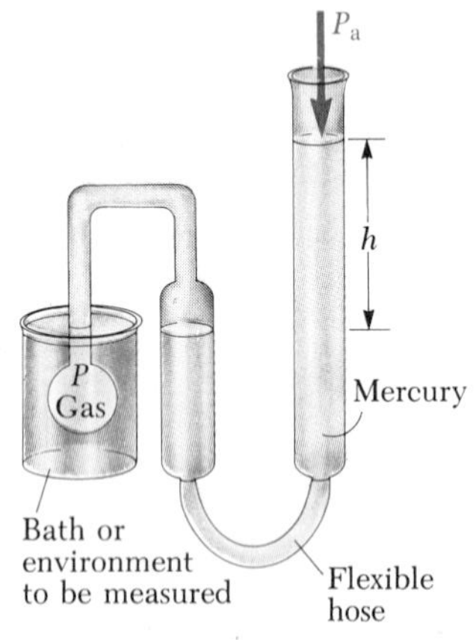

FIGURE 11.2 A constant-volume gas thermometer measures the pressure of the gas contained in the flask on the left. The volume of gas in the flask is kept constant by raising or lowering the column of mercury such that the mercury level remains constant.

degree. (The Celsius temperature scale was at one time called the centrigrade temperature scale.)

Thermometers calibrated in this way do present problems, however, when extremely accurate readings are needed. For instance, an alcohol thermometer calibrated at the ice and steam points of water might agree with a mercury thermometer only at the calibration points. Because mercury and alcohol have different thermal expansion properties, when one reads a temperature of 50°C, say, the other may indicate a slightly different value. The discrepancies between thermometers are especially large when the temperatures to be measured are far from the calibration points. An additional practical problem of any thermometer is its limited temperature range. A mercury thermometer, for example, cannot be used below the freezing point of mercury, which is −39°C. What we need is a universal thermometer whose readings are independent of the substance used. The gas thermometer meets this requirement.

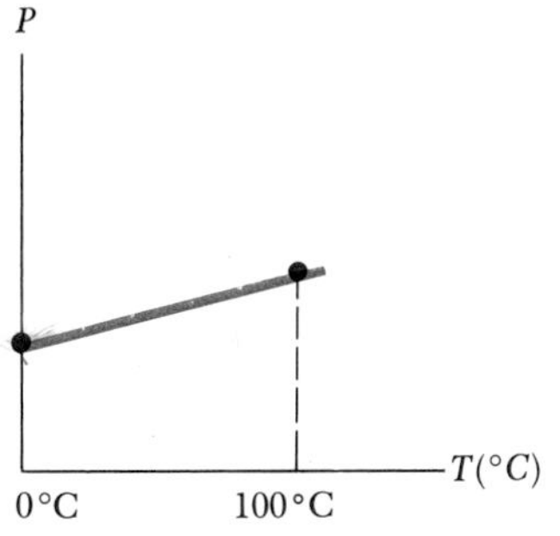

FIGURE 11.3 A typical graph of pressure versus temperature taken with a constant-volume gas thermometer. The dots represent known reference temperatures (the ice point and the steam point).

The Constant-Volume Gas Thermometer and the Kelvin Scale

In a **gas thermometer,** the temperature readings are nearly independent of the substance used in the thermometer. One type of gas thermometer is the constant-volume unit shown in Figure 11.2. The physical property used in this device is the pressure variation with temperature of a fixed volume of gas. When this thermometer was first developed, it was calibrated using the ice and steam points of water as follows. (A different calibration procedure, to be discussed shortly, is now used.) The gas flask is inserted into an ice bath, and the column of mercury raised or lowered until the volume of the gas is at some particular value. (Hence, we have a constant-volume device.) The height h of the mercury column indicates the pressure in the flask at 0°C. Then the flask is inserted into water at the steam point and the mercury level is readjusted until the gas volume is the same as it was in the ice bath. This gives a value for the pressure at 100°C. These values of pressure and temperatures are then plotted on a graph, as in Figure 11.3. The line connecting these points serves as a calibration curve for measuring unknown temperatures. If we wanted to measure the temperature of a substance, then, we would place the gas flask in thermal contact with the substance and adjust the column of mercury until the gas takes on its specified volume. The height of the mercury column tells us the pressure of the gas, and, since we know the pressure, we can find the temperature of the substance from the graph.

Now suppose that temperatures are measured with various gas thermometers containing different gases. Experiments show that the thermometer readings are nearly independent of the type of gas used, so long as the gas pressure is low and the temperature is well above the point at which the gas liquifies (Fig. 11.4). The agreement among thermometers using various gases improves as the pressure is reduced.

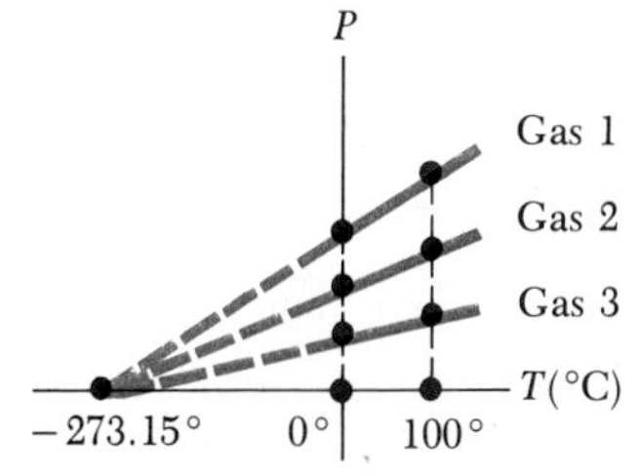

FIGURE 11.4 Pressure versus temperature for dilute gases. Note that, for all gases, the pressure extrapolates to zero at the unique temperature of −273.15°C.

An additional feature of the pressure-temperature curves may also be noted in Figure 11.4. If one extends these curves back toward negative temperatures, one finds that the pressure is zero when the temperature is −273.15°C. This significant temperature is used as the basis for the Kelvin temperature scale. This scale sets $-273.15°C$ as its zero point, 0 K. The size of a degree on the Kelvin scale is identical to the size of a degree on the Celsius scale. Thus, the relationship that enables one to convert between these temperatures is

$$T_c = T - 273.15 \tag{11.1}$$

where T_c is the **Celsius temperature** and T is the **Kelvin temperature** (which is sometimes called the **absolute temperature**).

Early gas thermometers made use of the ice point and the steam point as standard temperatures following the procedure just described. However, these points are experimentally difficult to duplicate because they are sensitive to dissolved impurities in the water. For this reason, a new procedure based on a single fixed point was adopted in 1954 by the International Committee on Weights and Measures. The **triple point of water,** which corresponds to *the single temperature and pressure at which water, water vapor, and ice can coexist in equilibrium,* was chosen as a convenient and reproducible reference temperature for the Kelvin scale. The triple point of water occurs at a temperature of 0.01°C and a pressure of 4.58 mm of mercury. The temperature at the triple point of water on the Kelvin scale was arbitrarily set at 273.16 kelvins, abbreviated 273.16 K. The SI unit of temperature, the **kelvin,** is defined as *1/273.16 of the temperature of the triple point of water.* Figure 11.5 shows the Kelvin temperatures for various physical processes and structures.

The kelvin

The Celsius, Fahrenheit, and Kelvin Temperature Scales

Equation 11.1 shows that the Celsius temperature, T, is shifted from the absolute (or Kelvin) temperature, T, by 273.15. From this we see that the size of a degree on the Kelvin scale is the same as on the Celsius scale. In other words, a temperature difference of 5 Celsius degrees, written 5 C°, is equal to a temperature difference of 5 kelvins. The two scales differ only in the choice of the zero point. Furthermore, the ice point (273.15 K) corresponds to 0.00°C, and the steam point (373.15 K) is equivalent to 100.00°C.

The most common temperature scale in everyday use in the United States is the **Fahrenheit scale.** This scale sets the temperature of the ice point at 32°F and the temperature of the steam point at 212°F. The relationship between the Celsius and Fahrenheit temperature scales is

$$T_F = \frac{9}{5}T_c + 32 \tag{11.2}$$

Equation 11.2 can easily be used to find a relationship between changes in temperature on the Celsius and Fahrenheit scales. It will be left as a problem for you to show that if the Celsius temperature changes by ΔT_c, the Fahrenheit temperature changes by an amount ΔT_F given by

$$\Delta T_F = \frac{9}{5}\Delta T_c \tag{11.3}$$

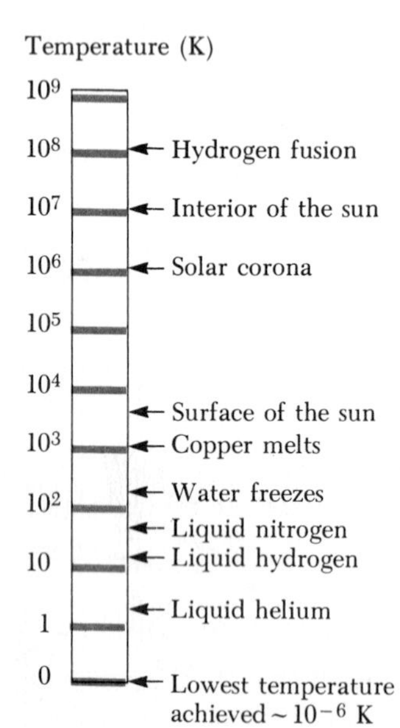

FIGURE 11.5 Absolute temperature at which various selected physical processes take place.

EXAMPLE 11.1 Converting Temperatures

On a day when the temperature reaches 50°F, what is the temperature in degrees Celsius and in kelvins?

Solution Let us solve Equation 11.2 for T_c and substitute $T_F = 50°F$:

$$T_c = \frac{5}{9}(T_F - 32) = \frac{5}{9}(50 - 32) = 10°C$$

From Equation 11.1, we find that

$$T = T_c + 273.15 = 283.15 \text{ K}$$

EXAMPLE 11.2 Heating a Pan of Water

A pan of water is heated from 25°C to 80°C. What is the change in its temperature on the Kelvin scale and on the Fahrenheit scale?

Solution From Equation 11.1, we see that the change in temperature on the Celsius scale equals the change on the Kelvin scale. Therefore,

$$\Delta T = \Delta T_c = 80 - 25 = 55 \text{ C°} = 55 \text{ K}$$

From Equation 11.2, we find that the change in temperature on the Fahrenheit scale is greater than the change on the Celsius scale by the factor $\frac{9}{5}$. That is,

$$\Delta T_F = \frac{9}{5}\Delta T_c = \frac{9}{5}(80 - 25) = 99 \text{ F°}$$

In other words, 55 C° = 99 F°, where the notations C° and F° refer to temperature *differences*, and are not to be confused with actual temperatures, which are written °C and °F.

11.3 THERMAL EXPANSION OF SOLIDS AND LIQUIDS

Our discussion of the mercury thermometer made use of one of the most well-known changes that occurs in a substance: as its temperature increases, its volume increases. (As we shall see shortly, in some materials the volume decreases when the temperature increases.) This phenomenon, known as **thermal expansion,** plays an important role in numerous applications. For example, thermal expansion joints must be included in buildings, concrete highways, and bridges to compensate for changes in dimensions with temperature variations.

Expansion joints allow bridges to expand and contract safely.

The overall thermal expansion of an object is a consequence of the change in the average separation between its constituent atoms or molecules. To understand this, consider a crystalline solid, which consists of a regular array of atoms held together by electric forces. At ordinary temperatures, the atoms vibrate about their equilibrium positions with an amplitude of about 10^{-11} m and a frequency of about 10^{13} cycles/s. The average spacing between the atoms is of the order of 10^{-10} m. As the temperature of the solid increases, the atoms vibrate with larger amplitudes and the average separation between them increases. Consequently, the solid as a whole expands with increasing temperature. If the expansion of an object is sufficiently small compared with its initial dimensions, then the change in any dimension (length,

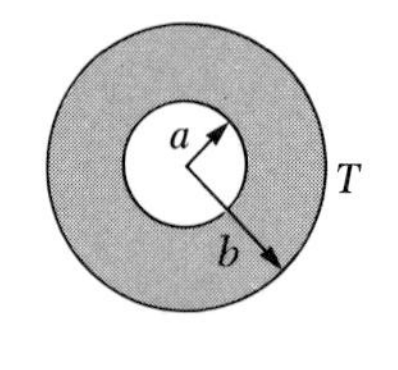

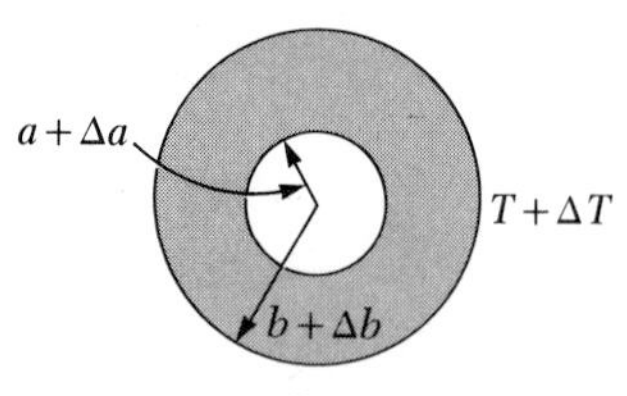

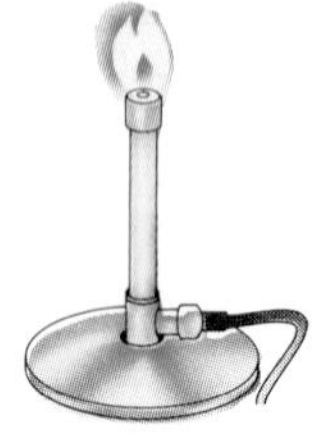

FIGURE 11.6 Thermal expansion of a homogeneous metal washer. As the washer is heated, all dimensions increase. (Note that the expansion is exaggerated in this figure.)

width, or thickness) is, to a good approximation, dependent on the first power of the temperature change.

Suppose an object has an initial length L_0 along some direction at some temperature. The length increases by an amount ΔL for a change in temperature ΔT. Experiments show that, when ΔT is small enough, the change in length is proportional to the temperature change and to the original length. Thus, the basic equation for the expansion of a solid is

$$\boxed{\Delta L = \alpha L_0 \Delta T} \tag{11.4a}$$

or

$$\boxed{L - L_0 = \alpha L_0(T - T_0)} \tag{11.4b}$$

where the constant α is called the **average coefficient of linear expansion** for a given material and has units of $(C°)^{-1}$.

It may be helpful to think of a thermal expansion as a magnification or as a photographic enlargement. For example, as a metal washer is heated (Fig. 11.6), all dimensions, including the radius of the hole, increase according to Equation 11.4. Table 11.1 lists the average coefficient of linear expansion for various materials. Note that for these materials α is positive, indicating an increase in length with increasing temperature. This is not always the case. For example, some substances, such as calcite ($CaCO_3$), expand along one dimension (positive α) and contract along another (negative α) with increasing temperature.

Because the linear dimensions of an object change with temperature, it follows that the area and volume also change with temperature. Consider a square having an initial length L_0 on a side and therefore an initial area L_0^2. As the temperature of the square is increased, the length of each side increases to

$$L = L_0 + \alpha L_0 \Delta T$$

The new area $A = L^2$ is

$$L^2 = (L_0 + \alpha L_0 \Delta T)(L_0 + \alpha L_0 \Delta T) = L_0^2 + 2\alpha L_0^2 \Delta T + \alpha^2 L_0^2 \Delta T^2$$

TABLE 11.1 Average Coefficients of Linear Expansion for Some Materials Near Room Temperature

Material	Average Coefficient of Linear Expansion $[(C°)^{-1}]$	Material	Average Coefficient of Volume Expansion $[(C°)^{-1}]$
Aluminum	24×10^{-6}	Ethyl alcohol	1.12×10^{-4}
Brass and bronze	19×10^{-6}	Benzene	1.24×10^{-4}
Copper	17×10^{-6}	Acetone	1.5×10^{-4}
Glass (ordinary)	9×10^{-6}	Glycerin	4.85×10^{-4}
Glass (Pyrex)	3.2×10^{-6}	Mercury	1.82×10^{-4}
Lead	29×10^{-6}	Turpentine	9.0×10^{-4}
Steel	11×10^{-6}	Gasoline	9.6×10^{-4}
Invar (Ni-Fe alloy)	0.9×10^{-6}	Air	3.67×10^{-3}
Concrete	12×10^{-6}	Helium	3.665×10^{-3}

The last term in this expression contains the quantity $\alpha\Delta T$ raised to the second power. Because $\alpha\Delta T$ is much less than unity, squaring it makes it even smaller. Therefore, we can neglect this term to get a simpler expression:

$$A = L^2 = L_0^2 + 2\alpha L_0^2\Delta T$$

$$A = A_0 + 2\alpha A_0\Delta T \tag{11.5a}$$

or

$$\boxed{\Delta A = A - A_0 = \gamma A_0\Delta T} \tag{11.5b}$$

where $\gamma = 2\alpha$. The quantity γ (Greek letter gamma) is called the **average coefficient of area expansion.**

By a similar procedure, we can show that the *increase in volume* of an object accompanying a change in temperature is given by

$$\boxed{\Delta V = \beta V_0\Delta T} \tag{11.6}$$

where β, the **average coefficient of volume expansion,** is given by $\beta = 3\alpha$.

EXAMPLE 11.3 Expansion of a Railroad Track

A steel railroad track has a length of 30 m when the temperature is 0°C. (a) What is its length on a hot day when the temperature is 40°C?

Solution If we make use of Table 11.1 and note that the change in temperature is 40 C°, we find that the *increase* in length is

$$\Delta L = \alpha L_0\Delta T = [11 \times 10^{-6}(\text{C°})^{-1}](30\text{ m})(40\text{ C°}) = 0.013\text{ m}$$

Therefore, its length at 40°C is 30.013 m.

(b) Suppose the ends of the rail are rigidly clamped at 0°C so as to prevent expansion. Calculate the thermal stress set up in the rail if its temperature is raised to 40°C. Assume the rail has a cross-sectional area of 30 cm^2 (0.003 m^2).

Solution From the definition of Young's modulus for a solid (Chapter 9), we have

$$\text{Tensile stress} = \frac{F}{A} = Y\frac{\Delta L}{L_0}$$

Since Y for steel is 20×10^{10} Pa, we have

$$\frac{F}{A} = (20 \times 10^{10}\text{ Pa})\left(\frac{0.013\text{ m}}{30\text{ m}}\right) = 8.7 \times 10^7\text{ Pa}$$

Therefore the force of compression in the rail is

$$F = (8.7 \times 10^7\text{ Pa})(0.003\text{ m}^2) = 2.6 \times 10^5\text{ N} \approx 59{,}000\text{ lb}$$

Obviously, it would be difficult to keep the rail clamped in place under such large forces. For this reason, expansion joints are used to allow rails to expand without bowing.

Spaces are left between rails to allow for thermal expansion. Failure to do so will often result in buckling of the tracks during very warm weather. (Courtesy of Gary Kleine)

As Table 11.1 indicates, each substance has its own characteristic coefficient of expansion. For example, a brass rod will expand more than a steel rod of equal length if their temperatures are raised by the same amount from some common initial value because brass has a larger coefficient of expansion

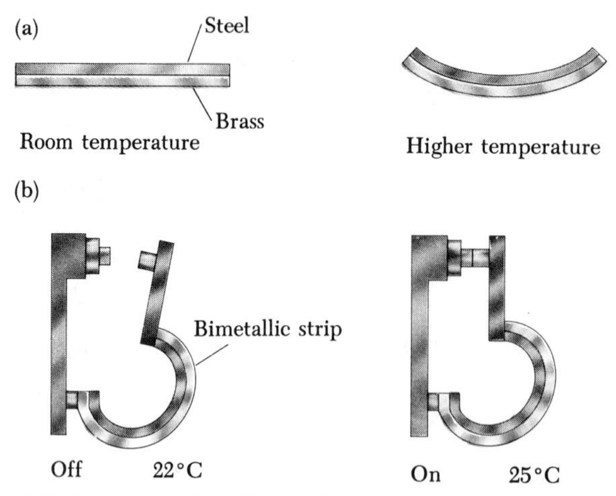

FIGURE 11.7 (a) A bimetallic strip bends as the temperature changes because the two metals have different expansion coefficients. (b) A bimetallic strip used in a thermostat to break or make electrical contact.

than steel. A simple device called a bimetallic strip utilizes this principle for practical applications. Such a strip is made by securely bonding two different metals together, as shown in Figure 11.7a. As the temperature of the strip increases, the two metals expand by different amounts and the strip bends. Figure 11.7b illustrates how a bimetallic strip is used in a thermostat.

EXAMPLE 11.4

A hole of cross-sectional area 100 cm^2 is cut in a piece of steel at 20°C. What is the area of the hole if the steel is heated from 20°C to 100°C?

Solution A hole in a substance expands in exactly the same way as would a piece of the substance having the same shape as the hole. The change in the area of the hole can be found by using Equation 11.5b:

$$\Delta A = \gamma A_0 \Delta T = [22 \times 10^{-6} (°C)^{-1}](100 \text{ cm}^2)(80 \text{ C°}) = 0.18 \text{ cm}^2$$

Therefore, the area of the hole at 100°C is

$$A = A_0 + \Delta A = 100.18 \text{ cm}^2$$

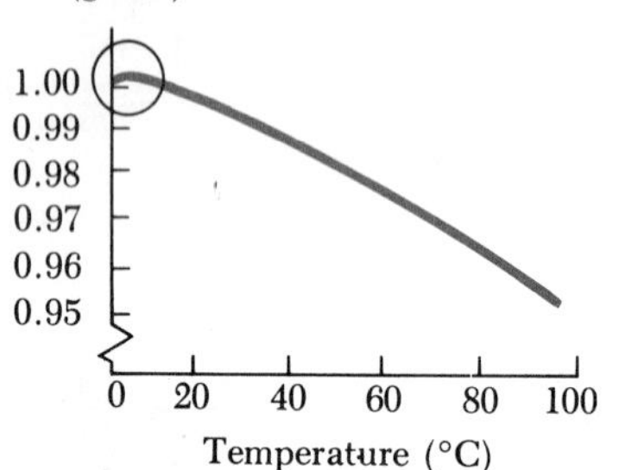

FIGURE 11.8 (a) The density of water as a function of temperature. (b) The maximum density of water occurs at 4°C.

The Unusual Behavior of Water

Liquids generally increase in volume with increasing temperature and have volume expansion coefficients about ten times greater than those of solids. Water is an exception to this rule, as we can see from its density versus temperature curve, in Figure 11.8. As the temperature increases from 0°C to 4°C, water contracts and thus its density increases. Above 4°C, water expands with increasing temperature. The density of water reaches a maximum value of 1000 kg/m^3 at 4°C.

We can explain why a pond or lake freezes at the surface from this unusual thermal expansion behavior of water. When the atmospheric temperature drops from, say, 7°C to 6°C, the water at the surface of the pool also cools and decreases in volume. This means that the surface water is denser than the water below it, which has not cooled and decreased in volume. As a result, the surface water sinks and warmer water from below is forced to the surface to be cooled. Between 4°C and 0°C, however, the surface water expands as it cools, becoming less dense than the water below it. The mixing process stops, and eventually the surface water freezes. As the water freezes, the ice remains on the surface because ice is less dense than water. The ice continues to build up on the surface, while water near the bottom of the pool remains at 4°C. If this did not happen, fish and other forms of marine life would not survive. In fact, if it were not for this peculiarity of water, life as we now know it would not exist.

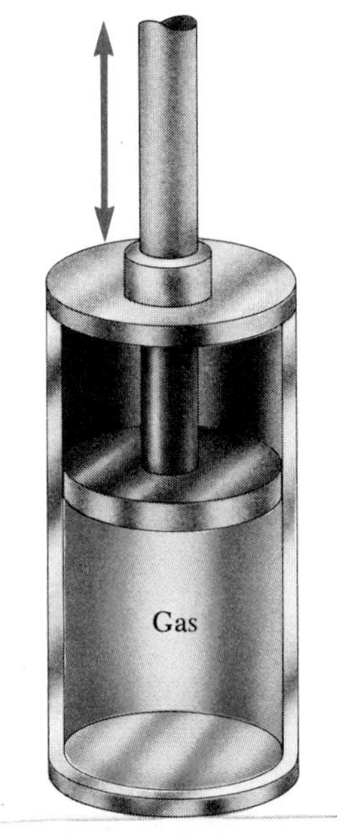

FIGURE 11.9 A gas confined to a cylinder whose volume can be varied with a movable piston.

11.4 IDEAL GASES

In this section we shall be concerned with the properties of a gas of mass m confined to a container of volume V at a pressure P and the temperature T. It would be useful to know how these quantities are related. If the gas is maintained at a very low pressure (or low density) and if the conditions of pressure and temperature are such that the gas does not condense into a liquid, the relationship is quite simple. Such a gas is commonly referred to as an *ideal gas*. (A more precise definition of an ideal gas will be given in a later section.) Simple substances that are gases at room temperature and atmospheric pressure behave as if they were ideal gases.

To determine how P, V, and T are related for an ideal gas, consider a gas confined to a cylindrical container whose volume can be varied by means of a movable piston, as in Figure 11.9. We shall assume that the cylinder does not leak and hence that the mass of the gas remains constant. For such a system, experiments provide the following information. First, when the gas is kept at constant temperature, its pressure is inversely proportional to its volume. This experimental fact is called **Boyle's law** and can be expressed as

$$P \propto \frac{1}{V} \tag{11.7}$$

Boyle's law

Second, when the pressure of the gas is kept constant, its volume is directly proportional to its temperature. This is called the **Charles–Gay-Lussac law** and can be expressed as

$$V \propto T \tag{11.8}$$

Charles–Gay-Lussac law

The relationships given in Equations 11.7 and 11.8 can be expressed in a single proportionality as

$$PV \propto T$$

Finally, it should be noted that the mass m of the gas must be considered in our analysis. Anyone who has ever pumped up a tire knows that the volume of the tire increases as more air is forced into the tire. Therefore, the volume of the gas is also directly proportional to the mass of the gas. Thus, we have

$$PV \propto mT \tag{11.9}$$

Proportionality relationships like the above can be expressed in equation form by experimentally determining a constant of proportionality which we shall denote as A. Thus, 11.9 can be written as

$$PV = AmT \tag{11.10}$$

The constant of proportionality A is different for different gases. However, if the amount of gas is measured in moles (defined below) rather than in terms of its mass, one finds that the constant A is the *same* for all gases. This constant of proportionality, R, is called the **universal gas constant.** Thus, we obtain the following important relationship for an ideal gas:

Ideal gas law

$$PV = nRT \tag{11.11}$$

In this equation, n is the number of moles of the gas. The number of moles of a substance is related to its mass m through the expression

$$n = \frac{m}{M} \tag{11.12}$$

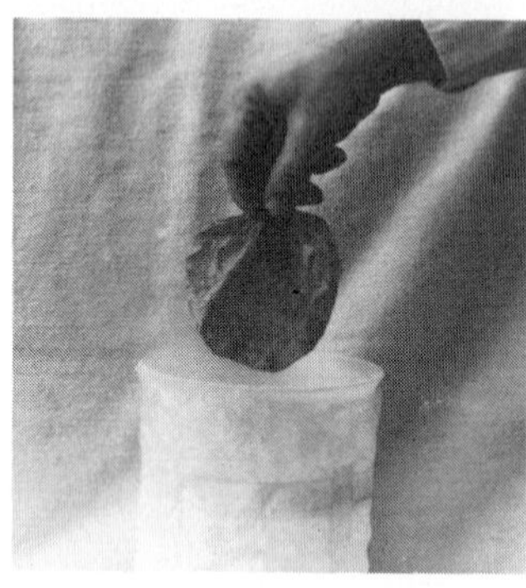

The balloon is filled with air. Why does it shrink when cooled with liquid nitrogen, which is at 77 K?

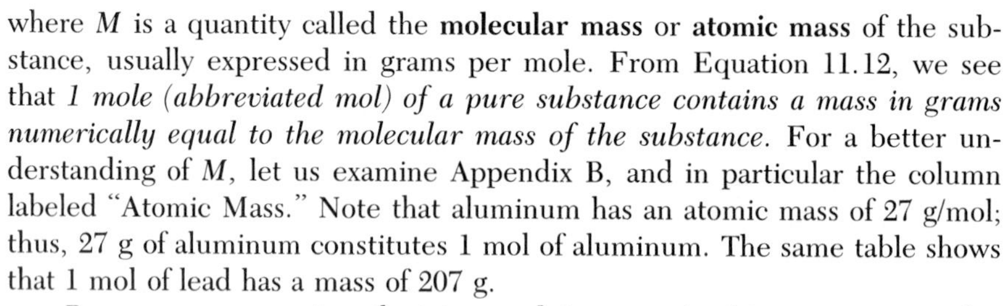

where M is a quantity called the **molecular mass** or **atomic mass** of the substance, usually expressed in grams per mole. From Equation 11.12, we see that *1 mole (abbreviated mol) of a pure substance contains a mass in grams numerically equal to the molecular mass of the substance.* For a better understanding of M, let us examine Appendix B, and in particular the column labeled "Atomic Mass." Note that aluminum has an atomic mass of 27 g/mol; thus, 27 g of aluminum constitutes 1 mol of aluminum. The same table shows that 1 mol of lead has a mass of 207 g.

Because we are primarily interested in gases in this section, consider oxygen, which has an atomic mass of approximately 16. In the gaseous state, oxygen consists of two oxygen atoms bound together to form a molecule, O_2. Therefore, the molecular mass of oxygen is 32 g/mol, and the mass of 1 mol of oxygen is 32 g.

In SI units, where pressure is expressed in Pa (N/m^2) and volume in cubic meters (m^3), the product PV has units of newton-meters, or joules, and R has the value

$$R = 8.31 \text{ J/mol} \cdot \text{K}$$

If the pressure is expressed in atmospheres and the volume in liters (1 liter $= 10^3 \text{ cm}^3 = 10^{-3} \text{ m}^3$), then R has the value

$$R = 0.0821 \text{ liter} \cdot \text{atm/mol} \cdot \text{K}$$

It is important to note that when using Equation 11.11, the temperature must always be expressed in kelvins.

Scientists have defined an ideal gas as one that obeys the ideal gas law $PV = nRT$ under all conditions. In reality, no gas is ideal. However, the concept is very useful in view of the fact that real gases behave as if they were ideal gases at and below atmospheric pressure.

EXAMPLE 11.5 How Many Moles in the Container?
An ideal gas occupies a volume of 100 cm^3 at 20°C and 10^{-3} atm. Determine the number of moles of gas in the container.

Solution The quantities given are volume, pressure, and temperature: that is, V = 100 cm^3 = 0.1 liter, $P = 10^{-3}$ atm, and T = 20°C = 293 K. Using Equation 11.11, we get

$$n = \frac{PV}{RT} = \frac{(10^{-3}\text{ atm})(0.1\text{ liter})}{(0.0821\text{ liter}\cdot\text{atm/mole}\cdot\text{K})(293\text{ K})} = 4.16 \times 10^{-6}\text{ mole}$$

Again, note that you must express T as an absolute temperature when using the ideal gas law.

EXAMPLE 11.6 Squeezing a Tank of Gas
Pure helium gas is admitted into a cylinder containing a movable piston. The initial volume, pressure, and temperature of the gas are 15 liters, 2 atm, and 300 K. If the volume is decreased to 12 liters and the pressure increased to 3.5 atm, find the final temperature of the gas. Assume it behaves like an ideal gas.

Solution If no gas escapes from the cylinder, the number of moles remains constant; therefore using $PV = nRT$ at the initial and final points gives

$$\frac{P_iV_i}{T_i} = \frac{P_fV_f}{T_f}$$

where "i" and "f" refer to the initial and final values. Solving for T_f, we get

$$T_f = \left(\frac{P_fV_f}{P_iV_i}\right)(T_i) = \frac{(3.5\text{ atm})(12\text{ liters})}{(2\text{ atm})(15\text{ liters})}(300\text{ K}) = 420\text{ K}$$

EXAMPLE 11.7 Heating a Bottle of Air
A sealed glass bottle at 27°C contains air at atmospheric pressure and has a volume of 30 cm^3. It is then tossed into an open fire. When the temperature of the air in the bottle reaches 200°C, what is the pressure inside the bottle? Assume any volume changes of the bottle are small enough to be negligible.

Solution This example is approached in the same fashion as that used in Example 11.6. We start with the expression

$$\frac{P_iV_i}{T_i} = \frac{P_fV_f}{T_f}$$

Since the initial and final volumes of the gas are assumed equal, this expression reduces to

$$\frac{P_i}{T_i} = \frac{P_f}{T_f}$$

This gives

$$P_f = \left(\frac{T_f}{T_i}\right)(P_i) = \left(\frac{473\text{ K}}{300\text{ K}}\right)(1\text{ atm}) = 1.58\text{ atm}$$

Obviously, the higher the temperature, the higher the pressure exerted by the trapped air. Of course, if the pressure rises high enough, the bottle will shatter.

EXERCISE In this example, we neglected the change in volume of the bottle. If the coefficient of volume expansion for glass is $27 \times 10^{-6}\ (\text{C}°)^{-1}$, find the magnitude of this volume change.

ANSWER $0.14\ \text{cm}^3$.

EXAMPLE 11.8 The Volume of 1 Mol of Gas

Verify that one mol of oxygen occupies a volume of 22.4 liters at 1 atm and 0°C.

Solution Let us solve the ideal gas equation for V:

$$V = \frac{nRT}{P}$$

In our problem, the mass of the gas, m, is assumed to be one mol, M. Thus

$$n = \frac{m}{M} = 1\ \text{mol}$$

Let us convert the temperature to the Kelvin scale and substitute into the ideal gas equation:

$$V = \frac{nRT}{P} = \frac{(1\ \text{mol})(0.0821\ \text{liter} \cdot \text{atm/mol} \cdot \text{K})(273\ \text{K})}{1\ \text{atm}} = 22.4\ \text{liters}$$

You should note that the answer above has general validity. *One mole of any gas at standard temperature and pressure (STP) occupies a volume of 22.4 liters.*

EXERCISE Repeat the calculations in this example problem for hydrogen gas to show that it also occupies a volume of 22.4 liters at STP.

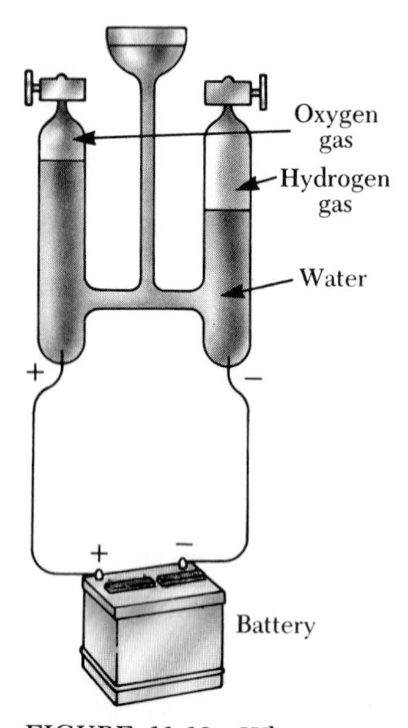

FIGURE 11.10 When water is decomposed by an electric current, the volume of hydrogen gas collected is twice that of oxygen.

11.5 AVOGADRO'S NUMBER AND THE IDEAL GAS LAW

In the early 1800s an important field of experimental investigation was performed in order to determine the relative masses of molecules. The equipment used in one such investigation is shown in Figure 11.10. The lower section of the glass vessel in this figure contains two electrodes which are connected to a battery so that an electric current can be passed through lightly salted water. (The purpose of the salt is to improve the electrical conductivity of the water.) One finds that bubbles of gas are produced at each electrode and become trapped in the column above it. An analytical examination of these gases reveals that the column above the positive electrode contains oxygen gas while the column above the negative electrode contains hydrogen gas. Obviously, the current decomposes the water into its constituent parts. Experimentally, it is found that the volume of hydrogen gas collected is always exactly twice the volume of oxygen gas collected, as you would expect in light of the chemical composition of water, H_2O. Additionally, it is found that if 9 g of water are decomposed, 8 g of oxygen and 1 g of hydrogen are collected. From this information, it is possible to determine the relative masses of oxygen and hydrogen molecules. Following such an experimental investigation, Amedeo Avogadro in 1811 made the following assumption:

Equal volumes of gas at the same temperature and pressure contain the same number of molecules.

With this assumption, and the fact that the ratio of hydrogen to oxygen molecules collected in the experiment is 2:1, one concludes that the mass of an oxygen molecule must be sixteen times that of a hydrogen molecule.

In Example 11.8, we noted that one mole of any gas occupies the same volume at 1 atm and at 0°C. Thus, a corollary to Avogadro's hypothesis is

one mole of any gas at standard temperature and pressure contains the same number of molecules.

Specifically, the number of molecules contained in one mole of any gas is 6.02×10^{23} molecules/mol. This number is known as **Avogadro's number,** and is given the symbol N_A

$$N_A = 6.02 \times 10^{23} \text{ molecules/mol} \tag{11.13}$$

It is of historical interest to note that Avogadro never knew the precise value for this number. An alternative method for calculating the number of moles of any gas in a container is to divide the total number of molecules present, N, by Avogadro's number, or

$$n = N/N_A \tag{11.14}$$

With this expression, we are able to rewrite the ideal gas law in the following alternative form:

$$PV = nRT = \frac{N}{N_A} RT$$

or

$$\boxed{PV = NkT} \tag{11.15}$$

Ideal gas law

where k is a constant called **Boltzmann's constant** and has the value

$$k = \frac{R}{N_A} = 1.38 \times 10^{-23} \text{ J/K} \tag{11.16}$$

Boltzmann's constant

EXAMPLE 11.9 What is Avogadro's Number?

One mole of hydrogen has a mass of 1.0078 g. In the early twentieth century, it was found that the mass of a hydrogen atom was approximately 1.673×10^{-24} g. Use these values to find Avogadro's number.

n = 1 mol
m = 1.0078 g
1.673 × 10⁻²⁴

Solution The number of molecules in 1 mol of hydrogen can be found by dividing the mass of 1 mol by the mass per atom. Hence, we find

$$N_A = \frac{1.0078 \text{ g/mol}}{1.673 \times 10^{-24} \text{ g/atom}} = 6.02 \times 10^{23} \text{ atoms/mol}$$

11.6 THE KINETIC THEORY OF GASES

In Section 11.4 we discussed the properties of an ideal gas using such quantities as pressure, volume, number of moles, and temperature. In this section we shall show that large-scale properties of a gas such as pressure and temperature can be understood on the basis of what is happening on the atomic scale. In addition, we shall re-examine the ideal gas law in terms of the behavior of individual molecules which compose the gas.

Because the interactions between the molecules of a gas are much weaker than in solids and liquids, our present discussion will be restricted to the molecular behavior of gases. In this view of gas behavior, called the kinetic theory, it is assumed that gas molecules move about in a random fashion, colliding with the walls of the container and with each other. Each molecule will be considered to be a hard sphere and each collision is assumed to be perfectly elastic. In practice, the assumptions of this section work well only for monatomic gases (those in which each particle is a single atom) because the energy of the atoms in a monatomic gas is solely translational kinetic energy. One must modify the theory for molecules consisting of two or more atoms bound together, such as O_2 and CO_2, to include the energy associated with any rotation or vibration of the molecules.

Molecular Model for the Pressure of an Ideal Gas

Assumptions of kinetic theory for an ideal gas

We shall first use the kinetic theory of gases to show that the pressure that a gas exerts on the walls of its container is a consequence of the collisions of the gas molecules with the walls. The following assumptions will be made:

1. *The number of molecules is large, and the average separation between them is large compared with their dimensions.* This means that the molecules occupy a negligible volume compared with the volume of the container.
2. *The molecules obey Newton's laws of motion, but as a whole they move in a random fashion.* By random fashion, we mean that any one molecule can move in any direction with any speed. We also assume that the distribution of velocities does not change in time, despite the collisions between molecules. That is, a certain percentage of molecules move at high speeds, a certain percentage move at low speeds, and so on.
3. *The molecules undergo elastic collisions with each other and with the walls of the container.* Thus, in the collisions *both kinetic energy and momentum are conserved.*
4. *The forces between molecules are negligible except during a collision.* The forces between molecules are short-range, so that the molecules interact with each other only during collisions.
5. *The gas under consideration is a pure substance,* that is, *all molecules are identical.*

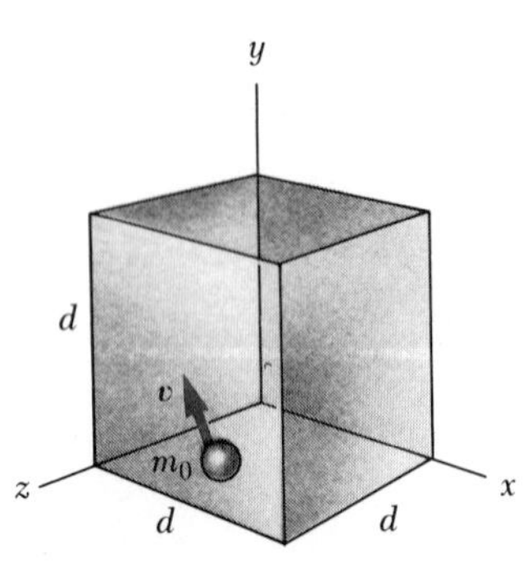

FIGURE 11.11 A cubical box of sides d containing an ideal gas. The molecule shown moves with velocity $\boldsymbol{v}$.

Now let us derive an expression for the pressure of N molecules of an ideal gas in a container of volume V. The container is assumed to be in the shape of a cube with edges of length d (Fig. 11.11). We shall focus our attention on one of these molecules, of mass m_0 and assumed to be moving such that its component of velocity in the x direction is v_x (see Fig. 11.12). As the molecule collides elastically with any wall, its velocity is reversed. Since the momentum, p, of the molecule is m_0v_x before the collision and $-m_0v_x$ after the collision, the *change in momentum of the molecule* is

$$\Delta p_x = mv_f - mv_i = -m_0v_x - m_0v_x = -2m_0v_x$$

Because the momentum of the system consisting of the wall and the molecule must be conserved, we see that, if the change in momentum of the molecule is $-2m_0v_x$, then the change in momentum of the wall must be $2m_0v_x$. Applying the impulse-momentum theorem to the wall gives

$$\overline{F}\Delta t = \Delta p = 2m_0v_x$$

In order for the molecule to make two collisions with the same wall, it must travel a distance $2d$ along the x direction. Therefore, the time between two collisions with the same wall is

$$\Delta t = \frac{2d}{v_x}$$

The substitution of this result into the impulse-momentum equation enables us to express the average force of a molecule on the wall:

$$\overline{F} = \frac{2m_0v_x}{\Delta t} = \frac{2m_0v_x^2}{2d} = \frac{m_0v_x^2}{d}$$

The net average force exerted on the wall by all the molecules is found by adding the forces exerted by the individual molecules:

$$\overline{F}_{\text{net}} = \frac{m_0}{d}(v_{x1}^2 + v_{x2}^2 + \cdots + v_{xN}^2)$$

In this equation, v_{x1} is the x component of velocity of molecule 1, v_{x2} is the x component of velocity of molecule 2, and so on. The summation terminates when we reach molecule N because there are N molecules in the container.

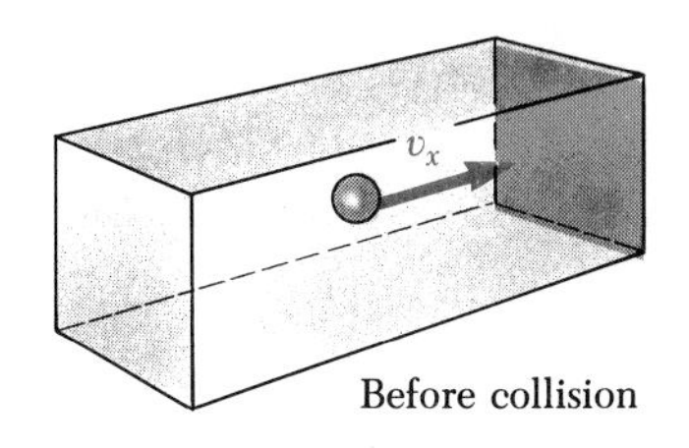

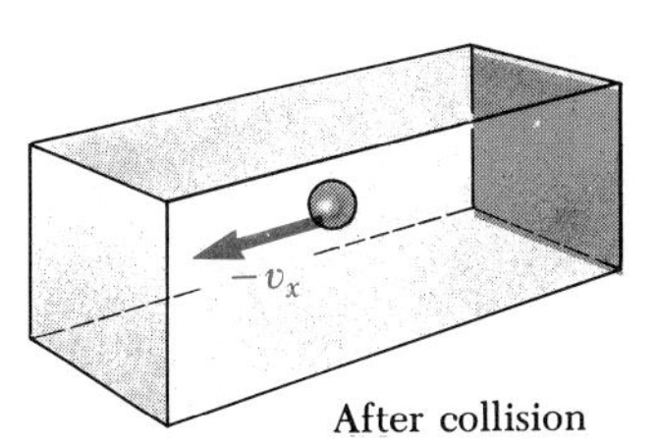

FIGURE 11.12. A molecule makes an elastic collision with the wall of the container. Its x component of momentum is reversed, and momentum is imparted to the wall.

To proceed further, note that the average value of the square of the velocity in the x direction for the N molecules is

$$\overline{v_x^2} = \frac{v_{x1}^2 + v_{x2}^2 + \cdots + v_{xN}^2}{N}$$

Thus, the net average force can be written as

$$\overline{F}_{\text{net}} = \frac{m_0}{d}N\overline{v_x^2}$$

Now let us focus our attention on one individual molecule in the container and say that this molecule has velocity components v_x, v_y, and v_z. The Pythagorean theorem relates the square of the velocity to the square of these components:

$$v^2 = v_x^2 + v_y^2 + v_z^2$$

Hence, the average value of v^2 for all the molecules in the container is related to the average values of v_x^2, v_y^2, and v_z^2 according to the expression

$$\overline{v^2} = \overline{v_x^2} + \overline{v_y^2} + \overline{v_z^2}$$

However, the average velocity is the same in any direction because the motion is completely random. Therefore,

$$\overline{v_x^2} = \overline{v_y^2} = \overline{v_z^2}$$

and we have

$$\overline{v^2} = 3\overline{v_x^2}$$

Thus, the net average force on the wall is given by

$$\overline{F}_{\text{net}} = \frac{N}{3}\left(\frac{m_0\overline{v^2}}{d}\right)$$

This expression allows us to find the pressure exerted on the wall:

$$P = \frac{\overline{F}_{\text{net}}}{A} = \frac{\overline{F}_{\text{net}}}{d^2} = \frac{1}{3}\left(\frac{N}{d^3} m_0\overline{v^2}\right) = \frac{1}{3}\left(\frac{N}{V}\right)(m_0\overline{v^2})$$

Pressure of an ideal gas

$$\boxed{P = \frac{2}{3}\left(\frac{N}{V}\right)\left(\frac{1}{2} m_0\overline{v^2}\right)} \qquad (11.17)$$

which tells us that *the pressure is proportional to the number of molecules per unit volume and to the average translational kinetic energy of the molecules.* With this simplified model of an ideal gas, we have arrived at an important result that relates the large-scale quantity of pressure to an atomic quantity, the average molecular speed. Thus, we have a key link between the atomic world and the large-scale world.

You should note that Equation 11.17 verifies some features of pressure that are probably familiar to you. One way to increase the pressure inside a container is to increase the number of molecules per unit volume in the container. For example, you do this when you add air to a tire. The pressure in the tire can also be increased by increasing the average translational kinetic energy of the molecules in the tire. As we shall see shortly, this can be accomplished by increasing the temperature of the gas inside the tire. This is why the pressure inside a tire increases as the tire heats up during long trips. The continuous flexing of the tires as they move along the road surface generates heat that is transferred to the air inside the tires. This added heat increases the temperature of the air, which in turn produces an increase in pressure in the tires.

Molecular Interpretation of Temperature

We can obtain some insight into the meaning of temperature by first writing Equation 11.17 in the more familiar form

$$PV = \frac{2}{3}N\left(\frac{1}{2} m_0\overline{v^2}\right)$$

Now let us compare this with the ideal gas equation (Eq. 11.15),

$$PV = NkT$$

Equating the right-hand sides of these two expressions and rearranging terms, we have

$$\frac{2}{3}N\left(\frac{1}{2} m_0\overline{v^2}\right) = NkT$$

TABLE 11.2 Some rms Speeds

Gas	Molecular Mass (g/mole)	v_{rms} at 20°C (m/s)[a]
H_2	2.02	1902
He	4.0	1352
H_2O	18	637
Ne	20.1	603
N_2 and CO	28	511
NO	30	494
CO_2	44	408
SO_2	48	390

[a]All values were calculated using Eq. 11.19.

$$\frac{1}{2} m_0 \overline{v^2} = \frac{3}{2} kT \tag{11.18}$$

Average kinetic energy

This important result says that the average kinetic energy of gas molecules is directly proportional to the absolute temperature of the gas. As the temperature increases, the molecules move with high average speeds. Conversely, temperature is a direct measure of the average molecular kinetic energy of an ideal gas.

The square root of $\overline{v^2}$ is called the **root mean square** (rms) speed of the molecules. From Equation 11.18, we find this to be

$$v_{rms} = \sqrt{\overline{v^2}} = \sqrt{\frac{3kT}{m_0}} = \sqrt{\frac{3RT}{M}} \tag{11.19}$$

Root mean square speed

This expression for the rms speed shows that at a given temperature lighter molecules move faster, on the average, than heavier molecules. For example, hydrogen, with a molecular mass of 2 g/mole, moves four times as fast as oxygen, whose molecular mass is 32 g/mole if they are at the same temperature. Note that the rms speed is not the speed at which a gas molecule moves across a room because a moving molecule undergoes several billion collisions per second with other molecules under standard conditions. Table 11.2 lists the rms speeds for various molecules at 20°C.

EXAMPLE 11.10 A Tank of Helium

A tank of volume 0.3 m³ contains 2 moles of helium gas at 20°C. Since the helium behaves like an ideal gas, (a) find the average kinetic energy per atom.

Solution From Equation 11.18, we see that the average kinetic energy per atom is

$$\frac{1}{2} m_0 \overline{v^2} = \frac{3}{2} kT = \frac{3}{2}(1.38 \times 10^{-23}\ \text{J/K})(293\ \text{K}) = 607 \times 10^{-21}\ \text{J}$$

(b) Determine the rms speed of the atoms.

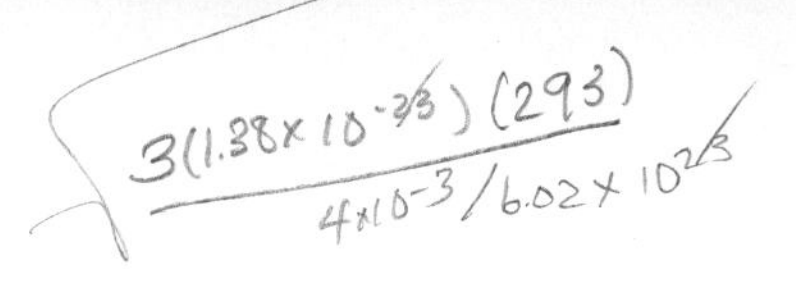

Solution Because the molecular mass of helium is 4 g/mol = 4×10^{-3} kg/mol, the rms speed is

$$v_{\text{rms}} = \sqrt{\frac{3RT}{M}} = \sqrt{\frac{3(8.317 \text{ J/mole} \cdot \text{K})(293 \text{ K})}{4 \times 10^{-3} \text{ kg/mole}}} = 1.35 \times 10^{3} \text{ m/s}$$

The same result is obtained using $v_{\text{rms}} = \sqrt{3kT/m_0}$. Remember that m_0, the mass of one molecule, is obtained by dividing the molecular mass by the number of particles in 1 mole of the gas (Avogadro's number). Try it!

*11.7 EVAPORATION, HUMIDITY, AND BOILING

If a container of water is left open, the water slowly evaporates. That is, it spontaneously goes from the liquid state to the gaseous state. The kinetic theory of gases enables us to explain why this phenomenon occurs.

The molecules of a liquid are in continuous random motion. Occasionally, the upward component of velocity of a molecule near the surface of the liquid will be sufficiently large to enable the molecule to escape from the liquid and become a gaseous molecule.

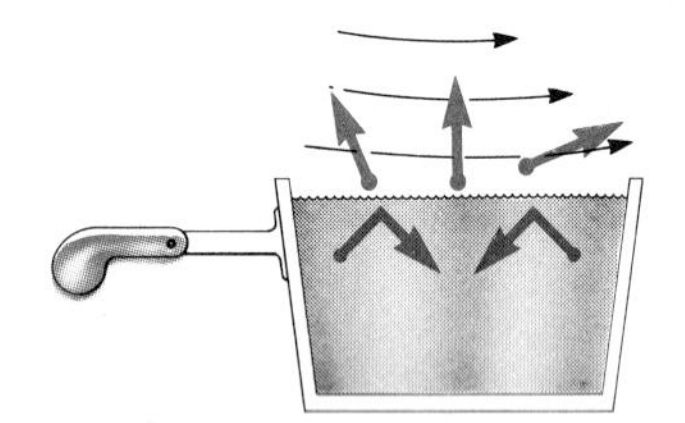

FIGURE 11.13 The evaporation rate increases as air moves over the surface of the liquid.

Equation 11.18 shows that the temperature of an ideal gas is a measure of the average kinetic energy of its molecules. This concept can be extended to liquids as well. The molecules that evaporate from the liquid are those that have sufficient energy (greater than the average kinetic energy) to break free of their neighbors on the surface. Since the evaporating molecules are very energetic, the molecules remaining in the liquid must have lower-than-average kinetic energy. For this reason, the liquid becomes cooler as a result of evaporation. You may have noticed this cooling due to evaporation after a swim or shower. Your body feels cooler when you are wet than when you are dry because the water is evaporating. As the rate of evaporation increases, so does the rate of cooling. In another familiar example, when a doctor swabs your arm with alcohol (which evaporates faster than water) before giving you an injection, the alcohol serves two purposes. It sterilizes the area to be injected, and it slightly numbs the area as a result of its rapid evaporation and subsequent cooling of the skin.

*Humidity

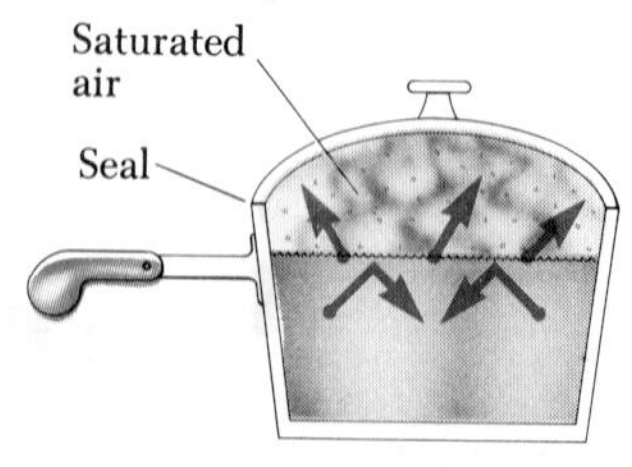

FIGURE 11.14 A liquid placed in a sealed container. The air above the liquid becomes saturated when the number of molecules leaving the liquid equals the number returning to the liquid in some time interval.

The rate of evaporation is enhanced if air is moved rapidly over the surface of the liquid. This occurs because the liquid molecules that vaporize are removed by the moving air before they have time to undergo collisions with air molecules near the surface that could deflect them back into the liquid (Fig. 11.13). For the same reason, fanning yourself on a hot day helps you cool down.

Let us now look at the exact opposite of fanning. Let us assume that water is placed in a sealed container such that the gaseous molecules cannot escape, as in Figure 11.14. In this case, the air above the liquid will absorb water molecules until it can hold no more. At this point, the air is said to be **saturated** with water vapor. Saturation occurs when the number of water molecules returning to the liquid in some time interval is equal to the number leaving the liquid in the same time interval. The amount of water vapor that air can hold increases as the temperature of the air increases.

The mass of water vapor in the air per unit volume is called the **absolute humidity.** Thus, absolute humidity is a measure of the dampness of the air. If there is a large amount of water vapor in the air, the air feels damp, or humid.

When weather forecasters refer to humidity, they are talking not about absolute humidity but about relative humidity. In order to understand the concept of *relative humidity,* we must first discuss partial pressures and vapor pressure.

We have seen that atmospheric pressure is about 760 mm Hg. The total atmospheric pressure at any location is the sum of the pressures exerted by all the gases that make up the atmosphere. That is, the oxygen molecules in the atmosphere exert a pressure on an object, as do the nitrogen molecules in the atmosphere, and so on for all the molecules contained in our atmosphere. The pressure exerted by each constituent gas is called the **partial pressure** of that gas. According to the **law of partial pressures**

the pressure exerted by each component of a mixture of gases is the same as the pressure that component would exert if it were alone in the volume occupied by the mixture. This means that each component acts virtually independently of the rest of the mixture.

Vapor pressure is *the pressure exerted by a gas when it is in equilibrium with its liquid form.* That is, in the case of the closed container of water discussed earlier, the partial pressure exerted by the gaseous water when saturation exists is equal to its vapor pressure. It should be evident that the partial pressure of a gas at a given temperature may equal but never exceed its vapor pressure. The vapor pressure of water at various temperatures is given in Table 11.3.

The **relative humidity,** RH, at a given temperature is defined as

$$RH \equiv \frac{\text{partial pressure of water vapor}}{\text{vapor pressure of water}} \times 100$$

If air saturated with water vapor is cooled, some of the water vapor must condense (liquify) in order to reduce the partial pressure of the water vapor. For example, from Table 11.3, we see that if the temperature of saturated air is reduced from 20°C to 10°C, enough water vapor must condense out to lower the partial pressure from 17.5 mm Hg to 8.9 mm Hg. This is the reason for the formation of fog, clouds, rain, dew, and frost.

Relative humidity can be measured by determining what is called the **dew point,** which is *the temperature at which the air becomes saturated with water vapor.* This temperature can be found by slowly adding ice to a metal

TABLE 11.3 Vapor Pressure of Water at Various Temperatures

Temperature (°C)	Vapor Pressure (mm Hg)
0	4.6
10	8.9
20	17.5
100	760.0

container filled with water. (The outside surface of the container should be clean, and the liquid should be stirred as the ice melts.) The temperatue at which moisture starts condensing on the outside surface of the container is the dew point.

*Boiling

Boiling can be thought of as evaporation taking place throughout the volume of a liquid. *Boiling begins when the vapor pressure is equal to atmospheric pressure.* From Table 11.3, we see that this occurs for water at 100°C. In order to understand the mechanism of boiling, consider what happens when a bubble of gas forms in the interior of a volume of liquid. If the vapor pressure inside the bubble is less than the pressure exerted on it by the liquid, the bubble collapses. If the pressure inside the bubble is equal to atmospheric pressure, however, the bubble is sustained and rises to the surface and the vapor escapes.

It should be obvious from this discussion that the boiling point of a liquid is influenced by the external pressure exerted on it. In general, as the pressure on the liquid is increased, the boiling point is also increased. That is, in order to reach the boiling point, the temperature of the liquid must be increased until the vapor pressure of the gas equals the external pressure.

Just as an increase in pressure increases the boiling point, a reduction of pressure reduces it. For example, at an altitude of 12 mi above sea level, atmospheric pressure is so low that blood boils at body temperature. Since you would not want to verify this experimentally, let us consider another way the effect of pressure on boiling can be observed. Imagine that water is placed in a container from which the air can be gradually pumped. Furthermore, assume that the water is at room temperature when placed in the container and hence is far below its boiling point. As the air is pumped out of the container, the pressure on the water is reduced, as is its boiling point. Eventually, when the pressure is reduced to about 17.5 mm Hg (Table 11.3), the boiling point is reduced to room temperature and the water boils. Because boiling is merely rapid evaporation (which is a cooling process), the water in the container is being cooled as it boils. If you reduce the pressure even further, the water will boil more rapidly. Finally, the temperature of the liquid will reach the freezing point even though it is boiling. What would the pressure inside the container have to be for this to occur? Refer to Table 11.3 for your answer.

SUMMARY

The **zeroth law of thermodynamics** states that if two objects A and B are separately in thermal equilibrium with a third object, then A and B are in thermal equilibrium with each other.

The relationship between T_c, the *Celsius temperature*, and T, the *Kelvin temperature*, is

$$\boxed{T_c = T - 273.15} \qquad (11.1)$$

The relationship between the *Fahrenheit* and *Celsius* temperatures is

$$T_F = \frac{9}{5}T_c + 32 \qquad (11.2)$$

When a substance is heated, it generally expands. If an object has an initial length L_0 at some temperature, and undergoes a change in temperature ΔT, its length changes by an amount ΔL which is proportional to its initial length and the temperature change:

$$\Delta L = \alpha L_0 \Delta T \qquad (11.4)$$

The parameter α is called the **average coefficient of linear expansion.**

Likewise, the change in volume of most substances is proportional to the initial volume, V_0, and the temperature change ΔT:

$$\Delta V = \beta V_0 \Delta T \qquad (11.6)$$

where β is the **average coefficient of volume expansion,** and is equal to 3α.

An **ideal gas** is one that obeys the equation

$$PV = nRT \qquad (11.11)$$

where P is the pressure of the gas, V is its volume, n is the number of moles of gas, R is the universal gas constant (8.31 J/mole·K), and T is the absolute temperature in kelvins. A real gas behaves approximately as an ideal gas at very low pressures.

The **pressure** of N molecules of an ideal gas contained in a volume V is given by

$$P = \frac{2}{3}\left(\frac{N}{V}\right)\left(\frac{1}{2}m_0\overline{v^2}\right) \qquad (11.17)$$

where $\frac{1}{2}m_0\overline{v^2}$ is the **average kinetic energy per molecule.**

The average kinetic energy of the molecules of a gas is directly proportional to the absolute temperature of the gas:

$$\frac{1}{2}m_0\overline{v^2} = \frac{3}{2}kT \qquad (11.18)$$

The **root mean square** (rms) speed of the molecules of gas is

$$v_{rms} = \sqrt{\frac{3kT}{m_0}} = \sqrt{\frac{3RT}{M}} \qquad (11.19)$$

The pressure exerted by each component in a mixture of gases is the pressure that component would exert if it were alone in the volume occupied by the mixture. This pressure exerted by a particular component is called the **partial pressure.**

Vapor pressure is the pressure exerted by a gas when it is in equilibrium with its liquid form.

The relative humidity at a given temperature is defined as

$$RH \equiv \frac{\text{partial pressure of water vapor}}{\text{vapor pressure of water}} \times 100$$

The **dew point** is the temperature at which the air becomes saturated with water vapor.

Boiling begins when the vapor pressure is equal to atmospheric pressure.

ADDITIONAL READING

T. B. Greenslade, "The Maximum Density of Water," *The Physics Teacher,* November 1985, p. 474.

M. B. Hall, "Robert Boyle," *Sci. American,* August 1967, p. 84.

E. Jones and R. Childers, "Observational Evidence for Atoms," *The Physics Teacher,* October 1984, p. 354.

F. Jones, "Fahrenheit and Celsius, A History," *Physics Today,* 18, 1980, p. 594.

R. H. Romer, "Temperature Scales: Celsius, Fahrenheit, Kelvin, Reaumur, and Romer," *The Physics Teacher,* October 1982, p. 450.

A. F. Scott, "The Invention of the Balloon and the Birth of Chemistry," *Sci. American,* January 1984, p. 126.

R. E. Wilson, "Standards of Temperature," *Physics Today,* January 1953, p. 10.

QUESTIONS

1. Is it possible for two objects to be in thermal equilibrium if they are not in thermal contact with each other? Explain.
2. A piece of copper is dropped into a beaker of water. If the water's temperature rises, what happens to the temperature of the copper? When will the water and copper be in thermal equilibrium?
3. Explain why a column of mercury in a thermometer first descends slightly and then rises when the thermometer is placed in hot water.
4. A steel wheel bearing is 1 mm smaller in diameter than an axle. How can the bearing be fit onto the axle without removing any material?
5. Markings to indicate length are placed on a steel tape in a room that has a temperature of 22°C. Are measurements made with the tape on a day when the temperature is 27°C too long, too short, or accurate? Defend your answer.
6. What would happen if the glass of a thermometer expanded more upon heating than did the liquid inside?
7. Determine the number of grams in 1 mole of (a) hydrogen, (b) helium, and (c) carbon monoxide.
8. Two identical cylinders at the same temperature each contain the same kind of gas. If cylinder A contains three times more gas than cylinder B, what can you say about the relative pressures in the cylinders?
9. Why do vapor bubbles in a pot of boiling water get larger as they approach the surface?
10. Although the average speed of gas molecules in thermal equilibrium at some temperature is greater than zero, the average velocity is zero. Explain.
11. One container is filled with helium gas and another with argon gas. If both containers are at the same temperature, which molecules have the higher rms speed?
12. The two metal hemispheres shown in the photograph are placed together, and the air in the cavity is removed with a vacuum pump. Why is it then almost impossible to pull the hemispheres apart? Estimate the force due to the atmosphere on one hemisphere.

Question 12

13. On a hot afternoon you often hear creaking noises in the attic of a house. These noises are also frequently heard at night. Why?
14. When drinking glasses become stuck together, an old trick is to fill the inner glass with water and to run water of a different temperature over the sides of the outer glass. Which water should be hot and which cold. Why?
15. A microwave oven is used to heat food in a sealed pouch. Why should the pouch be pricked with a fork before heating?
16. What does the ideal gas law predict about the volume of a gas at absolute zero? Why is this result incorrect?
17. Why is a power line more likely to break in winter rather than the summer even if it is loaded with the same weight?
18. If your water pipes freeze during the winter, it is almost always the hot water pipes that freeze first. Why? It is often said that if you keep a slow trickle of water running from your pipes they will not freeze. Is this true?
19. Chimneys are never used as a weight-bearing part of the structure of a building. Why is that the case?
20. As you go higher in the atmosphere, should you expect to find the percentage of nitrogen molecules to oxygen molecules increasing or decreasing? Explain.
21. After cooking food in a pressure cooker, why is it very important to cool the container with cold water before attempting to remove the lid?

PROBLEMS

Section 11.2 Thermometers and Temperature Scales

1. Convert the following temperatures to Celsius and kelvins: The normal human body temperature of 98.6°F, (b) the temperature of a cold day of −5°F.

2. Convert the following temperatures to Fahrenheit and kelvins: (a) the boiling point of liquid hydrogen −252.87°C, (b) the temperature of a room at 20°C.

3. The temperature in the interior of some stars is approximately 2.00×10^7 K. Express this temperature in (a) degrees Fahrenheit and (b) degrees Celsius.

4. Recently the temperature on a certain day changed by 54 F°. Express this change in temperature on (a) the Celsius scale and (b) the Kelvin scale.

5. Show that the temperature −40° is unique in that it has the same numerical value on the Celsius and Fahrenheit scales.

6. Show that if the temperature on the Celsius scale changes by ΔT_c, the Fahrenheit temperature changes by an amount ΔT_F given by $\Delta T_F = (9/5)\ \Delta T_c$.

Section 11.3 Thermal Expansion of Solids and Liquids

7. A structural steel beam is 20 m long when installed at 20°C. How much will its length change over the temperature extremes of −6°C to 30°C.

8. A copper steam pipe is 2 m long and is installed in a basement when the temperature is 20°C. What is the length of the pipe when it carries steam at 120°C?

9. The concrete sections of a certain highway are designed to have a length of 30 m. The sections are poured and cured at 10°C. What minimum spacing should the engineer leave between the sections to eliminate buckling if the concrete is to reach a temperature of 45°C?

10. A grandfather clock is controlled by a swinging brass pendulum of length 1.3 m at a temperature of 20°C. (a) What is the length of the pendulum rod when the temperature drops to 0°C? (b) If the period of a pendulum is given by $T = 2\pi\sqrt{L/g}$, where L is its length, does the change in length of the rod cause the clock to run fast or slow?

11. A cylindrical brass sleeve is to be shrink-fitted over a brass shaft whose diameter is 3.212 cm at 0°C. If the diameter of the sleeve is 3.196 cm at 0°C, (a) to what temperature must the sleeve be heated before it will slip over the shaft? (b) Alternatively, to what temperature must the shaft be cooled before it will slip into the sleeve?

12. Aluminum has a density of 2.70 g/cm^3 at 20°C. What is its density at 100°C?

13. A copper pipe has a length of 2 m when the temperature is 20°C. If it is rigidly clamped in place at 20°C, calculate the thermal stress set up in the pipe when the temperature is raised to 100°C.

14. What is the volume of 1 g of water at (a) 0°C and (b) 4°C? (*Hint:* See Figure 11.8b.)

• **15.** A certain steel measuring tape was designed to read correctly at 20°C. A parent then uses the tape to measure the height of a 1.2 m tall child. If the tape is used on a day when the temperature is 27°C, does the tape read the child to be taller or shorter than he really is and by how much?

• **16.** A construction worker uses a steel tape to measure the length of an aluminum support column. If the length is 18.7 m when the temperature is 21.2°C, what is the length when the temperature rises to 29.4°C? (*Note:* Do not neglect the expansion of the steel tape.)

•• **17.** Start with the relation $L - L_0 = \alpha L_0(T - T_0)$ to show that $\Delta V = \beta V_0 \Delta T$, where $\beta = 3\alpha$.

•• **18.** An underground fuel tank at a gasoline station at 54°F can hold 1000 gallons of gasoline. If a tanker truck fills the fuel tank on a day when the temperature is 90°F, how many gallons, according to his measure on the truck, can he pour into the tank? Assume the temperature of the gasoline cools to 54°F upon entering the tank.

Section 11.4 Ideal Gases

19. An ideal gas is held in a container at constant volume. Initially, its temperature is 20°C and its pressure is 3 atm. Find the pressure when its temperature is increased to 50°C.

20. (a) An ideal gas occupies a volume of 1 cm^3 at 20°C and atmospheric pressure. Determine the number of moles of gas in the container. (b) If the pressure of the 1 cm^3 volume is reduced to 10^{-11} Pa (an extremely good vacuum) while the temperature remains constant, how many moles of gas remain in the container?

21. The pressure of a gas in a container is tripled while its volume is halved. What is the ratio of the final to the original temperature of the gas?

22. (a) A 10-liter tank is to be filled with oxygen at 20°C to a pressure of 50 atm. Assume the ideal gas equation applies at this high pressure and find the mass of oxygen required. (b) Repeat assuming the gas is helium.

23. A cylinder with a movable piston contains gas at a temperature of 27°C, a volume of 1.5 m^3, and an absolute pressure of 0.2×10^5 Pa. What will be its final temperature if the gas is compressed to 0.7 m^3 and the absolute pressure increases to 0.8×10^5 Pa?

• **24.** Use the ideal gas equation and the relationship between moles and the mass of a gas to find an expression for the density of a gas.

• **25.** (a) Use the ideal gas equation to find the density of oxygen at standard temperature and pressure. (b) Repeat for helium.

• **26.** A bicycle tire is filled with air to a gauge pressure of 50 lb/in.^{2} at 20°C. What is the gauge pressure in the tire on a day when the temperature rises to 35°C? Assume the volume does not change, and recall that gauge pressure is absolute pressure in the tire minus atmospheric pressure. Furthermore assume that the atmospheric pressure remains constant at 14.7 lb/in.^{2}

• **27.** An air bubble has a volume of 1.5 cm^3 when released by a submarine 100 m below the surface of a lake. What is the volume of the bubble when it reaches the surface? Assume the temperature of the air in the bubble remains constant during ascent.

• **28.** Estimate the total mass of air inside a typical-size house on a day when the temperature is 0°F. (Assume a molecular weight of 28.8 g/mol for air.) How much mass must enter or leave the house if the temperature increases to 100° F?

• **29.** A weather balloon is designed to expand to a maximum radius of 20 m when in flight at its working altitude, where the air pressure is 0.03 atm and the temperature is 200 K. If the balloon is filled at atmospheric pressure and 300 K, what is the radius of the balloon at lift-off?

Section 11.5 Avogadro's Number and the Ideal Gas Law

Section 11.6 Kinetic Theory of Gases

30. (a) Find the number of molecules in 1 m^3 of air at atmospheric pressure and 27°C. (b) Repeat part (a) under the condition at which the pressure has been reduced to 10^{-11} Pa at 27°C.

31. Find the average distance of separation between molecules for the two situations described in Problem 30.

32. Use the definition of Avogadro's number to find the mass of a helium atom.

• **33.** A sealed cubical container 20 cm on a side contains three times Avogadro's number of molecules at a temperature of 20°C. Find the force exerted by the gas on one of the walls of the container.

34. In a 1-min interval, a machine gun fires 150 bullets, each of mass 8 g and speed 400 m/s. The bullets strike a stationary target that has an area of 5 m^2. Find the average force and pressure exerted on the target if the bullets are made of rubber and undergo an elastic, head-on collision with the stationary target.

35. In a period of 1 s, 5×10^{23} nitrogen molecules strike a wall of area 8 cm^2. If the molecules move with a speed of 300 m/s and strike the wall head-on in a perfectly elastic collision, find the pressure exerted on the wall. (The mass of one N_2 molecule is 4.68×10^{-26} kg.)

36. Find the average kinetic energy of a gas molecule at 27°C.

37. (a) Calculate the rms speed of an H_2 molecule when the temperature is 100°C. (b) Repeat the same calculation for an N_2 molecule.

38. Three moles of nitrogen gas, N_2, at 27°C are contained in a cylinder of volume 22.4 liters. Find the pressure the gas exerts on the cylinder walls.

39. At what temperature would the rms speed of helium atoms (mass $= 6.66 \times 10^{-27}$ kg) equal (a) the escape velocity from earth, 1.12×10^4 m/s, and (b) the escape velocity from the moon, 2.37×10^3 m/s?

40. Repeat Problem 34 for the case in which the bullets become embedded in the target in an inelastic collision.

41. Repeat Problem 35 in the case where the nitrogen molecules strike the wall at an angle of 30° to the normal with the wall.

ADDITIONAL PROBLEMS

42. How many kilograms of nitrogen are contained in a tank whose volume is 0.75 m^3 when the gauge pressure is 100 atm and the temperature is 27°C?

43. The active element of a certain laser is an ordinary glass rod 20 cm long and 1 cm in diameter. If the temperature of the rod increases by 75°C, find the increase in (a) its length, (b) its diameter, and (c) its volume.

44. An ideal gas is confined to a container at a pressure of 1.07×10^5 Pa and a temperature of 27°C. What is the temperature of the gas if the pressure is decreased to 8×10^4 Pa at constant volume?

45. If 2 moles of a gas are confined to a 5-liter vessel at a pressure of 8 atm, what is the average kinetic energy of a gas molecule?

• **46.** Calculate the density of helium stored at 300 K under a pressure of 1 atm.

• **47.** Before beginning a long trip on a hot day, a driver inflates an automobile tire to a gauge pressure of 1.8 atm at 300 K. At the end of the trip the pressure in the tire has increased to 2.2 atm. (a) Assuming the volume has remained constant, what is the temperature of the air inside the tire? (b) What volume of air measured at atmospheric pressure should be released from the tire so that the pressure returns to the initial value? Assume that the air is released during a short time interval during which the temperature remains at the value found in part (a).

• **48.** Calculate the number of molecules per cubic centimeter in an ideal gas at 500 K and pressure at 3 atm.

• **49.** A liquid with a coefficient of volume expansion β just fills a spherical shell of volume V at a temperature T (Fig. 11.15). The shell is made of a material that has a coefficient of linear expansion of α. The liquid is free to expand into a capillary of cross-sectional area A at the top. (a) If the temperature increases by ΔT, show that the liquid rises in the capillary by an amount Δh given by $\Delta h = \frac{V}{A}(\beta - 3\alpha)\,\Delta T$. (b) For a typical system, such as a mercury thermometer, why is it a good approximation to neglect the expansion of the shell?

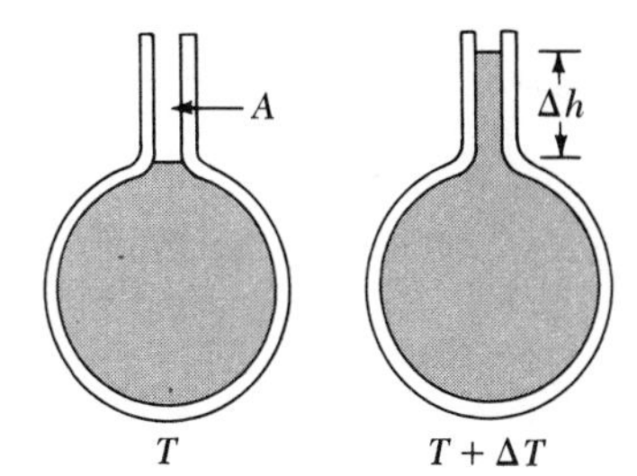

FIGURE 11.15 Problems 49 and 50

• **50.** A mercury thermometer is constructed as in Figure 11.15. The capillary tube has a diameter of 0.005 cm, and the bulb has a diameter of 0.30 cm. Neglecting the expansion of the glass, find the change in height of the mercury column for a temperature change of 25 C°.

•• **51.** A particular gas sample at 1 atm has a density of 3.75×10^{-3} g/cm^3. Determine the rms speed of the gas molecules.

•• **52.** A glass tube 1.5 m in length and closed at one end is weighted and lowered to the bottom of a freshwater lake. When the tube is recovered, a marking indicator shows that water rose to within 0.4 m of the closed end. Determine the depth of the lake. Assume constant temperature.

•• **53.** A hollow aluminum cylinder is to be fitted over a steel piston. At 20°C the inside diameter of the cylinder is 99% of the outside diameter of the piston. To what common temperature should the two pieces be heated in order that the cylinder just fits over the piston?

•• **54.** Two small containers of equal volume 100 cm^3 contain helium gas at 0°C and 1 atm pressure. The two containers are joined by a small open tube of negligible volume. What is the pressure of the gas in each container if the temperature of one of the containers is raised to 100°C while the other is kept at 0°C?

•• **55.** At $T = 0°C$, each one of three metal bars (two of aluminum and one of invar) is drilled with two holes a distance d apart. Pins are put through the holes to create an equilateral triangle. If the bars are then heated to 100°C, what will be the angle between the two aluminum bars?

•• **56.** When an ideal gas is held at constant pressure, a small change in volume, ΔV, associated with a small change in temperature, ΔT, are related as $P(\Delta V) = nR(\Delta T)$. (a) Show that the volume coefficient of thermal expansion for an ideal gas at constant pressure is given by $\beta = 1/T$, where T is the kelvin temperature. (b) What value does this equation predict for β at 0°C?

•• **57.** Show that the total pressure P in a container filled with a mixture of several different ideal gases is given by $P = P_1 + P_2 + P_3 + \ldots$, where P_1, P_2, etc., are the pressures that each gas would exert if it alone filled the container (or the partial pressures of the respective gases). This is known as Dalton's law of partial pressures.

•• **58.** An air bubble originating from a deep-sea diver has a radius of 2 mm at some depth h. When the bubble reaches the surface of the water, it has a radius of 3 mm. Assuming the temperature of the air in the bubble remains constant, determine (a) the depth h of the diver, and (b) the absolute pressure at this depth.

•• **59.** A vertical cylinder of cross-sectional area 0.05 m^2 is fitted with a tight-fitting, frictionless piston of mass 5 kg (Fig 11.16). If there are 3 moles of an ideal gas in the cylinder at 500 K, determine the height h at which the piston will be in equilibrium under its own weight.

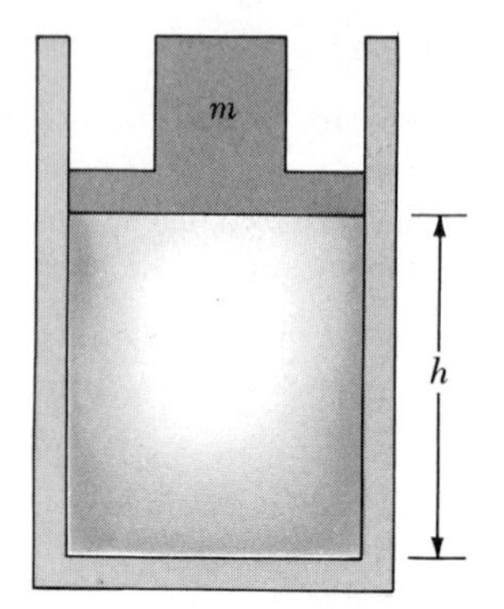

FIGURE 11.16 Problem 59

CHAPTER 12

Heat

A bullet passing through air being heated by a flame. This photo shows the nonuniform density of the air and the shock waves that accompany the supersonic bullet. (Photo, courtesy of H. E. Edgerton, MIT, Cambridge Mass.)

It is an experimental fact that when two objects at different temperatures are placed in thermal contact with each other, the temperature of the warmer object decreases and the temperature of the cooler object increases. If the two are left in contact for some time, they eventually reach a common equi-

Benjamin Thompson (1753–1814). "Being engaged, lately, in superintending the boring of cannon, in the workshops of the military arsenal at Munich, I was struck with the very considerable degree of Heat which a brass gun acquires, in a short time, in being bored; and with the still more intense Heat (much greater than that of boiling water, as I found by experiment) of the metallic chips separated from it by the borer."

librium temperature intermediate between the two initial temperatures. When such processes occur, we say that heat is transferred from the object at the higher temperature to the one at the lower temperature. But what is the nature of this heat transfer? Early investigators believed that heat was an invisible material substance called **caloric,** which was transferred from one object to another. According to this theory, caloric could be neither created nor destroyed. Although the caloric theory was successful in describing heat transfer, it eventually was abandoned when various experiments showed that caloric was in fact not conserved.

The first experimental observation suggesting that caloric was not conserved was made at the end of the 18th century by Benjamin Thompson. Thompson, an American-born scientist, emigrated to Europe during the Revolutionary War because of his Tory sympathies. Following his appointment as director of the Bavarian arsenal, he was given the title Count Rumford. While supervising the boring of an artillery cannon in Munich, Thompson noticed the great amount of heat generated by the boring tool, indicated by the fact that the water used to cool the tool had to be replaced continually as it boiled away. On the basis of the caloric theory, he reasoned that the ability of the metal filings produced by the boring tool to retain caloric should decrease as their size decreased. These heated filings, in turn, presumably transfer caloric to the cooling water, causing it to boil. To his surprise, Thompson discovered that the amount of water boiled away by a blunt boring tool was comparable to the quantity boiled away by a sharper tool. He then reasoned that if the tool were turned long enough, an almost infinite amount of heat could be produced from only a finite amount of metal filings. For this reason, Thompson rejected the caloric theory and suggested that heat is not a substance but rather some form of motion that is transferred from the boring tool to the cooling water. In another experiment, he showed that the heat generated by friction was proportional to the mechanical work done by the boring tool.

There are many other experiments that are at odds with the caloric theory. For example, if you rub two blocks of ice together on a day when the temperature is below 0°C, the blocks will melt. This experiment was first conducted by Humphry Davy (1778–1829). To properly account for this "creation of caloric," we note that mechanical work is done on the ice. Thus, we see that the equivalent effect is an increase in temperature. That is, heat and work are both forms of energy.

Although Thompson's observations provided evidence that heat energy is not conserved, it was not until the middle of the 19th century that the modern mechanical model of heat was developed. Before this time, heat and mechanics were considered to be two distinct branches of science and the law of conservation of energy seemed to be a rather specialized result used to describe certain kinds of mechanical systems. After the two disciplines were shown to be intimately related, the law of conservation of energy emerged as a universal law of nature. In this new view, heat is treated as just another form of energy, one that can be transformed into mechanical energy. Experiments performed by James Joule and others in this period showed that whenever heat is gained or lost by a system during some process, the gain or loss can be accounted for by an equivalent amount of work done on or by the system. Thus, by broadening the concept of energy to include heat, the law of energy conservation was extended.

12.1 HEAT AS A FORM OF ENERGY

The word *heat* should be used when describing energy transferred from one place to another. That is, *heat flow is an energy transfer that takes place as a consequence of temperature differences only.* For example, when a hot object is placed in thermal contact with a cooler object, heat is transferred from the hot object to the cold one, resulting in the hotter one cooling and the cooler one warming until the two arrive at some intermediate equilibrium temperature.

In this textbook, heat will usually be given in units of joules. It should be noted that before scientists arrived at a correct understanding of heat, the units in which heat was measured had already been developed. These unusual units were used because of the early misunderstandings of heat as exemplified by the caloric theory discussed in the introduction to this chapter. Because these units are still widely used in many applications, we shall discuss them briefly here. One of the most widely used of these units is the **calorie** (cal) which is defined as *the amount of heat necessary to raise the temperature of 1 g of water from 14.5°C to 15.5°C.* A related unit is the **kilocalorie** (kcal) defined as *the heat necessary to raise the temperature of 1 kg of water from 14.5°C to 15.5°C* (1 kcal = 10^3 cal). Note that the "Calorie" with a capital C, used in describing the energy equivalent of foods, is equal to 1 kcal.

Definition of the calorie

The unit of heat in the British engineering system is the **British thermal unit** (Btu), defined as *the heat required to raise the temperature of 1 lb of water from 63°F to 64°F.*

When the concept of mechanical energy was introduced in Chapter 5, we found that whenever friction is present in a mechanical system, some mechanical energy is lost. Experiments of various sorts show that this lost mechanical energy does not simply disappear; instead, it is transformed into thermal energy. Although this connection between mechanical and thermal energy was first suggested by Thompson's cannon boring experiment, it was James Prescott Joule (1818–1889) who first established the equivalence of the two forms of energy.

James Prescott Joule (1818–1889). "First: That the quantity of heat produced by the friction of bodies, whether solid or liquid, is always proportional to the quantity of energy expended. And second: That the quantity of heat capable of increasing the temperature of a pound of water . . . by 1°Fahr. requires for its evolution the expenditure of a mechanical energy represented by the fall of 772 lb through the distance of one foot."

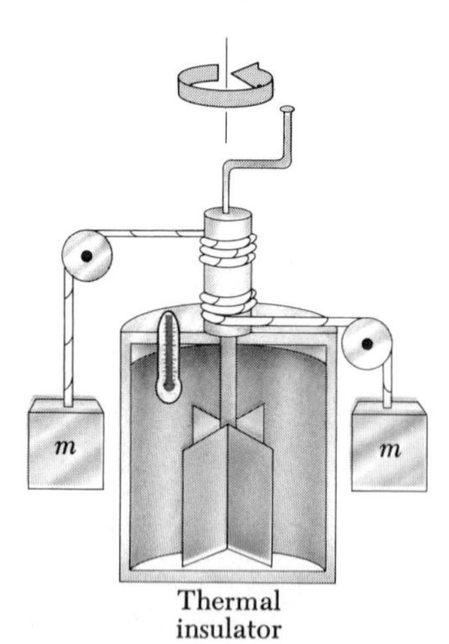

FIGURE 12.1 An illustration of Joule's experiment for measuring the mechanical equivalent of heat. The falling weights rotate the paddles, causing the temperature of the water to rise.

A schematic diagram of Joule's most famous experiment is shown in Figure 12.1. The system of interest is the water in a thermally insulated container. Work is done on the water by a rotating paddle wheel, which is driven by weights falling at a constant speed. The water, stirred by the paddles, warms up because of the friction between it and the paddles. If the energy lost in the bearings and through the walls is neglected, then the loss in potential energy of the weights equals the work done by the paddles on the water. If the two weights fall through a distance h, the loss in potential energy is $2mgh$, and it is this energy that is used to heat the water. By varying the conditions of the experiment, Joule found that the loss in mechanical energy, $2mgh$, is proportional to the increase in temperature of the water, ΔT, and to the mass of water used. The proportionality constant (the specific heat of water) was found to be 4.18 J/g·C°. Hence, 4.18 J of mechanical energy will raise the temperature of 1 g of water from 14.5°C to 15.5°C. One calorie is now defined to be *exactly* 4.186 J:

$$\boxed{1 \text{ cal} \equiv 4.186 \text{ J}} \tag{12.1}$$

This relation is called the **mechanical equivalent of heat.**

EXAMPLE 12.1 Losing Weight the Hard Way

A student eats a dinner rated at 2000 (food) Calories. He wishes to do an equivalent amount of work in the gymnasium by lifting a 50-kg mass. How many times must he raise the weight to expend this much energy? Assume he raises the weight a distance of 2 m each time and that no work is done when the weight is dropped to the floor.

Solution Since 1 Calorie = 10^3 cal, the work required is 2×10^6 cal. Converting this to Joules, we have for the total work required

$$W = (2 \times 10^6 \text{ cal})(4.186 \text{ J/cal}) = 8.37 \times 10^6 \text{ J}$$

The work done in lifting the weight once through a distance h is equal to mgh, and the work done in lifting the weight n times is $nmgh$. If we set $nmgh$ equal to the total work required, we have

$$W = nmgh = 8.37 \times 10^6 \text{ J}$$

Since $m = 50$ kg and $h = 2$ m, we get

$$n = \frac{8.37 \times 10^6 \text{ J}}{(50 \text{ kg})(9.80 \text{ m/s}^2)(2 \text{ m})} = 8.54 \times 10^3 \text{ times}$$

If the student is in good shape and lifts the weight, say, once every 5 s, it will take him about 12 h to perform this feat. Clearly, it is much easier to lose weight by dieting.

12.2 SPECIFIC HEAT

The quantity of heat energy required to raise the temperature of a given mass of a substance by some amount varies from one substance to another. For example, the heat required to raise the temperature of 1 kg of water by 1 C° is 4186 J, but the heat required to raise the temperature of 1 kg of copper by 1 C° is only 387 J. Every substance has a unique value for the amount of heat required to change the temperature of 1 kg of it by 1 C°, and this number is referred to as the specific heat of the substance.

Suppose that Q units of heat are added to m kg of a substance, thereby changing its temperature by ΔT. The **specific heat,** c, of the substance is defined as

Specific heat

$$c \equiv \frac{Q}{m\Delta T} \tag{12.2}$$

Table 12.1 lists specific heats for a few substances.

From the definition of specific heat, we can express the heat energy, Q, transferred between a system of mass m and its surroundings for a temperature change ΔT as

$$Q = mc\Delta T \tag{12.3}$$

For example, the heat energy required to raise the temperature of 0.5 kg of water by 3 C° is equal to (0.5 kg)(4186 J/kg · C°)(3 C°) = 6280 J. Note that when the temperature increases, ΔT and Q are taken to be *positive*, corresponding to heat flowing *into* the system. Likewise, when the temperature decreases, ΔT and Q are *negative* and heat flows *out* of the system.

Note from Table 12.1 that water has the highest specific heat of any substance we are likely to come in contact with on a routine basis. The high specific heat of water is responsible for the moderate temperatures found in regions near large bodies of water. As the temperature of a body of water decreases during the winter, the water gives off heat to the air, which carries the heat landward when prevailing winds are favorable. For example, the prevailing winds off the western coast of the United States are toward the land, and the heat liberated by the Pacific Ocean as it cools keeps coastal areas much warmer than they would be otherwise. This explains why the western coastal states generally have more favorable winter weather than the eastern coastal states, where the winds do not carry the heat toward land.

Table 12.1 Specific Heat of Some Materials at 25°C and Atmospheric Pressure

Material	J/kg · °C
Aluminum	900
Beryllium	1820
Cadmium	230
Copper	387
Germanium	322
Glass	837
Gold	129
Ice	2090
Iron	448
Lead	128
Mercury	138
Silicon	703
Silver	234
Steam	2010
Water	4186

FIGURE 12.2 Circulation of air near a large body of water. (a) On a hot day, the air above the warm land warms faster than the air above the cooler water. The cooler air over the water moves toward the beach to replace the rising warmer air. (b) At night, the land cools more rapidly than the water and hence the air currents reverse their direction.

That the specific heat of water is higher than that of land is responsible for the pattern of air flow at a beach. During the day, the sun adds roughly equal amounts of heat to beach and water, but the lower specific heat of sand causes the beach to reach a higher temperature than the water. Because of this, the air above the land reaches a higher temperature than that over the water. Consequently, cooler air from above the water is drawn in to displace this rising hot air, resulting in a breeze from ocean to land during the day. The hot air gradually cools as it rises and thus sinks, setting up the circulating pattern shown in Figure 12.2a. During the night, the land cools more quickly than the water and the circulating pattern reverses itself because the hotter air is now over the water (Fig. 12.2b). You should not expect to observe this pattern every time you go to the beach because prevailing winds caused by other factors often obscure it.

12.3 CONSERVATION OF ENERGY: CALORIMETRY

The principle of conservation of mechanical energy must now be expanded to include energy in the form of heat. Situations in which mechanical energy, such as gravitational energy or kinetic energy, is converted to heat occur frequently. We shall look at some of these in the examples following this section and in the problems at the end of the chapter, but most of our attention here will be directed toward a particular kind of conservation-of-energy situation. Problems using the procedure we shall describe are called *calorimetry* problems, and in them we consider only the transfer of heat energy inside an isolated system.

If a substance in a closed insulated container loses heat, something else in the container must gain this heat. We are ignoring the possibility of any heat transferred between the container and its surroundings. Devices in which this heat transfer occurs are called **calorimeters.** Calorimetry provides a technique for measuring the specific heat of solids and liquids. A known mass of a substance is heated to some known temperature and then placed in a vessel containing water (or some other liquid) of known mass, specific heat, and temperature. The temperature of the system is measured after equilibrium is reached. For example, suppose that m_x is the mass of a substance whose specific heat we wish to determine, c_x its specific heat, and T_x its initial temperature. Likewise, let m_w, c_w, and T_w represent the corresponding values for the water. If T is the final equilibrium temperature after everything is mixed, then from Equation 12.2, we find that the amount of heat gained by the water is $m_w c_w(T - T_w)$ and the amount of heat lost by the substance of unknown specific heat is $m_x c_x(T_x - T)$. Since energy is conserved, we can equate the heat lost to the heat gained:

$$m_w c_w(T - T_w) = m_x c_x(T_x - T)$$

When we solve this equation for c_x, we find

$$c_x = \frac{m_w c_w(T - T_w)}{m_x(T_x - T)} \tag{12.4}$$

You should not attempt to memorize this equation. Instead, you should always start from first principles in solving calorimetry problems. That is, determine which substances lose heat and which substances gain heat, and then equate heat loss to heat gain.

A word about sign conventions: In a later chapter we shall find it necessary to use a sign convention in which a positive sign for Q indicates heat gained by a substance and a negative sign indicates heat lost. However, for calorimetry problems it is less confusing if you ignore the positive and negative signs and instead equate heat loss to heat gain. That is, you should always write ΔT as a positive quantity. For example, if T_f is greater than T_i, let $\Delta T = T_f - T_i$. If T_f is less than T_i, let $\Delta T = T_i - T_f$.

EXAMPLE 12.2 Cooling a Hot Ingot

A 0.05-kg ingot of metal is heated to 200°C and then dropped into a beaker containing 0.4 kg of water initially at 20°C. If the final equilibrium temperature of the mixed system is 22.4°C, find the specific heat of the metal.

Solution Because the heat lost by the ingot equals the heat gained by the water, we can write

$$m_x c_x(T_i - T_f) = m_w c_w(T_f - T_i)$$

$$(0.05 \text{ kg})(c_x)(200°\text{C} - 22.4°\text{C}) = (0.4 \text{ kg})(4186 \text{ J/kg}\cdot\text{C}° (22.4°\text{C} - 20°\text{C})$$

from which we find that

$$c_x = 453 \text{ J/kg}\cdot\text{C}°$$

The ingot is most likely iron, as can be seen by comparing this result with the data in Table 12.1.

EXERCISE

Answer

EXAMPLE 12.3 Fun Time for a Cowboy

A cowboy fires a silver bullet of mass 2 g with a muzzle velocity of 200 m/s into the pine wall of a saloon. Assume that all the heat energy generated by the impact remains with the bullet. What is the temperature change of the bullet?

Solution The kinetic energy of the bullet is

$$\frac{1}{2}mv^2 = \frac{1}{2}(2 \times 10^{-3} \text{ kg})(200 \text{ m/s})^2 = 40 \text{ J}$$

All of this kinetic energy is transformed into heat, Q, as the bullet stops in the wall. Thus,

$$Q = mc\Delta T$$

Since the specific heat of silver is 234 J/kg·C° (Table 12.1), we get

$$\Delta T = \frac{Q}{mc} = \frac{40 \text{ J}}{(2 \times 10^{-3} \text{ kg})(234 \text{ J/kg·C°})} = 85.5 \text{ C°}$$

EXERCISE Suppose the cowboy runs out of silver bullets and fires a lead bullet of the same mass and velocity into the wall. What is the temperature change of the bullet?

Answer 157 C°.

12.4 LATENT HEAT

A substance usually undergoes a change in temperature when heat is added to it or extracted from it. There are situations, however, where the flow of heat does not result in a change in temperature. This occurs whenever a substance changes from one physical state to another. Such a change is referred to as a **phase change.** Two common phase changes are solid to liquid (melting) and liquid to gas (boiling).

Consider, for example, the heat required to convert a 1-g block of ice at −30°C to steam (water vapor) at 120°C. Figure 12.3 indicates the experimental results obtained when heat is gradually added to the ice. Let us examine each portion of the curve separately.

Part A During this portion of the curve, we are changing the temperature of the ice from −30°C to 0°C. Since the specific heat of ice is 2090 J/kg · C°, we can calculate the amount of heat added as follows:

$$Q = m_i c_i \Delta T = (10^{-3} \text{ kg})(2090 \text{ J/kg·C°})(30 \text{ C°}) = 62.7 \text{ J}$$

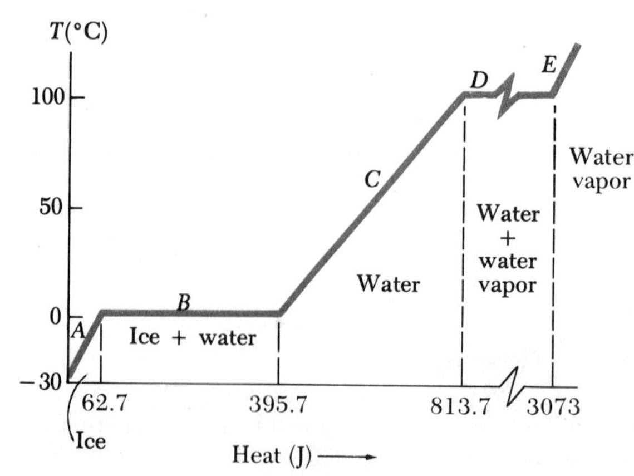

FIGURE 12.3 A plot of temperature versus heat added when 1 g of ice initially at −30°C is converted to steam.

TABLE 12.2 Latent Heats of Fusion and Vaporization

Substance	Melting Point (°C)	Latent Heat of Fusion (J/kg)	Boiling Point (°C)	Latent Heat of Vaporization (J/kg)
Helium	−269.65	5.23×10^3	−268.93	2.09×10^4
Nitrogen	−209.97	2.55×10^4	−195.81	2.01×10^5
Oxygen	−218.79	1.38×10^4	−182.97	2.13×10^5
Ethyl alcohol	−114	1.04×10^5	78	8.54×10^5
Water	0.00	3.33×10^5	100.00	2.26×10^6
Sulfur	119	3.81×10^4	444.60	3.26×10^5
Lead	327.3	2.45×10^4	1750	8.70×10^5
Aluminum	660	9.00×10^4	2450	1.14×10^7
Silver	960.80	8.82×10^4	2193	2.33×10^6
Gold	1063.00	6.44×10^4	2660	1.58×10^6
Copper	1083	1.34×10^5	1187	5.06×10^6

When the ice reaches 0°C, it remains at this temperature—even though heat is being added—until all the ice melts. The heat required to change the phase of a given mass m of a pure substance is

$$Q = mL \tag{12.5}$$

where L is the **latent heat** (hidden heat) of the substance. The value of L depends on the nature of the phase change as well as on the properties of the substance. The **latent heat of fusion,** L_f, is used when the phase change is from a solid to a liquid. From Table 12.2, which lists the latent heat of fusion for a few substances, we see that the latent heat of fusion for water at atmospheric pressure is 3.33×10^5 J/kg. Thus, the heat required to melt 1 g of ice at 0°C is

$$Q = mL_f = (10^{-3}\text{ kg})(3.33 \times 10^5\text{ J/kg}) = 333\text{ J}$$

Part C Between 0°C and 100°C, nothing surprising happens. No phase change occurs in this region. The heat added to the water is being used to increase its temperature. The amount of heat necessary to increase the temperature from 0°C to 100°C is

$$Q = m_w c_w \Delta T = (10^{-3}\text{ kg})(4.18 \times 10^3\text{ J/kg·C°})(100\text{ C°}) = 4.18 \times 10^2\text{ J}$$

Part D At 100°C, another phase change occurs as the water changes from water at 100°C to steam at 100°C. We can find the amount of heat required to produce this phase change by using Equation 12.5. In this case, we must set $L = L_v$, the **latent heat of vaporization.** The latent heat of vaporization for some substances is given in Table 12.2. Since the latent heat of vaporization for water is 2.26×10^6 J/kg, the amount of heat we must add to convert 1 g of water to steam at 100°C is

$$Q = mL_v = (10^{-3}\text{ kg})(2.26 \times 10^6\text{ J/kg}) = 2.26 \times 10^3\text{ J}$$

Part E On this portion of the curve, heat is being added to the steam with no phase change occurring. Using 2.01×10^3 J/kg·C° for the specific heat of steam (Table 12.1), we find that the heat we must add to raise the temperature of the steam to 120°C is

$$Q = m_s c_s \Delta T = (10^{-3}\text{ kg})(2.01 \times 10^3\text{ J/kg·C°})(20\text{ C°}) = 40.2\text{ J}$$

The *total amount of heat* that must be added to change one gram of ice at −30°C to steam at 120°C is about 3.11×10^3 J. It should be noted that this process is *reversible.* That is, if we cool steam at 120°C down to the point at which we have ice at −30°C, we must remove 3.11×10^3 J of heat.

Phase changes can be described in terms of a rearrangement of molecules when heat is added or removed from a substance. Consider first the liquid-to-gas phase change. The molecules in a liquid are close together, and the forces between them are stronger than in a gas, where the molecules are far apart. Therefore, work must be done on the liquid against these attractive molecular forces in order to separate the molecules. The latent heat of vaporization is the amount of energy that must be added to the liquid to accomplish this.

Similarly, at the melting point of a solid, we imagine that the amplitude of vibration of the atoms about their equilibrium position becomes large enough to overcome the attractive forces binding the atoms into their fixed positions. The heat energy required to totally melt a given mass of solid is equal to the work required to break these bonds and to transform the mass from the ordered solid phase to the disordered liquid phase.

Because the average distance between atoms in the gas phase is much larger than in either the liquid or the solid phase, we could expect that more work is required to vaporize a given mass of a substance than to melt it. Therefore, it is not surprising that the latent heat of vaporization is much larger than the latent heat of fusion for a given substance (Table 12.2).

EXAMPLE 12.4 Cooling the Steam

What mass of steam initially at 130°C is needed to warm 200 g of water in a 100-g glass container from 20° to 50°C?

Solution This is a heat transfer problem in which we must equate the heat lost by the steam to the heat gained by the water and glass container. There are three stages as the steam loses heat. In the first stage, the steam is cooled to 100°C. The heat liberated in the process is

$$Q_1 = m_x c_s \Delta T = m_x(2.01 \times 10^3\text{ J/kg · C°})(30\text{ C°}) = m_x(6.03 \times 10^4\text{ J/kg})$$

In the second stage, the steam is converted to water. In this case, to find the heat removed, we use the heat of vaporization and $Q = mL_v$:

$$Q_2 = m_x(2.26 \times 10^6\text{ J/kg})$$

In the last stage, the temperature of the water is reduced to 50°C. This liberates an amount of heat

$$Q_3 = m_x c_w \Delta T = m_x(4.18 \times 10^3\text{ J/kg · C°})(50\text{ C°}) = m_x(2.09 \times 10^5\text{ J/kg})$$

If we equate the heat lost by the steam to the heat gained by the water and glass and use the given information, we find

$$m_x(6.03 \times 10^4\text{ J/kg}) + m_x(2.26 \times 10^6\text{ J/kg}) + m_x(2.09 \times 10^5\text{ J/kg}) =$$
$$(0.2\text{ kg})(4.18 \times 10^3\text{ J/kg · C°})(30\text{ C°}) + (0.1\text{ kg})(837\text{ J/kg · C°})(30\text{ C°})$$
$$m_x = 1.09 \times 10^{-2}\text{ kg} = 10.9\text{ g}$$

EXAMPLE 12.5 Boiling Liquid Helium

Liquid helium has a very low boiling point, 4.2 K, and a very low heat of vaporization, 2.09×10^4 J/kg (Table 12.2). A constant power of 10 W (1 W = 1 J/s) is transferred to a container of liquid helium from an immersed electric heater. At this rate, how long does it take to boil away 1 kg of liquid helium? Liquid helium has a density of 0.125 g/cm^3, so that 1 kg corresponds to 8×10^3 cm^3 = 8 liters of liquid.

Solution Since $L_v = 2.09 \times 10^4$ J/kg for liquid helium, we must supply an energy of 2.09×10^4 J to boil away 1 kg. The power supplied to the helium is 10 W = 10 J/s. That is, in 1 s, 10 J of energy is transferred to the helium. Therefore, the time it takes to transfer 2.09×10^4 J is

$$t = \frac{2.09 \times 10^4 \text{ J}}{10 \text{ J/s}} = 2.09 \times 10^3 \text{ s} \approx 35 \text{ min}$$

He
b.p. 4.2K
hv = 2.09x10^{4}J/kg

Since 1 kg of helium corresponds to 8 liters, this corresponds to a boil-off rate of about 0.23 liter/min. In contrast, 1 kg of liquid nitrogen would boil away in about 3.4 h at the rate of 10 J/s!

EXERCISE If 10 W of power is supplied to 1 kg of water at 100°C, how long will it take for the water to boil away completely?

Answer 62.4 h.

EXAMPLE 12.6 What's Wrong With This Approach?

A copper container of mass 150 g contains 200 g of water and a 10-g ice cube all at 0°C. A 50-g block of lead at a temperature of 200°C is dropped into the container. Find the final temperature of the mixture if no heat is lost to the surroundings.

Solution At first glance, this seems to be a straightforward situation in which one simply equates the heat lost to the heat gained to find the final temperature. We shall first use this approach and discover that it is incorrect. This will be followed by the correct solution.

Let us assume that the final temperature of the mixture is T_f. Thus, the heat lost by the lead is

$$Q_{Pb} = (50 \times 10^{-3} \text{ kg})(128 \text{ J/kg} \cdot \text{C°})(200 - T_f)$$

The heat gained by the copper container is

$$Q_{Cu} = (0.150 \text{ kg})(387 \text{ J/kg})(T_f - 0\text{°C})$$

and the heat gained by the water is

$$Q_w = (0.2 \text{ kg})(4.18 \times 10^3 \text{ J/kg} \cdot \text{C°})(T_f - 0\text{°C})$$

Now consider the ice cube. First, the ice cube must melt, and the heat required to do this is

$$Q_{ice} = mL_f = (10 \times 10^{-3} \text{ kg})(3.35 \times 10^5 \text{ J/kg}) = 3.35 \times 10^3 \text{ J}$$

Finally, the heat required to increase the temperature of the melted ice cube to T_f is

$$Q_{ice\text{-}w} = (10 \times 10^{-3} \text{ kg})(4.18 \times 10^3 \text{ J/kg} \cdot \text{C°})(T_f - 0\text{°C})$$

If we now equate the heat lost to the heat gained, we find

$$(50 \times 10^{-3} \text{ kg})(128 \text{ J/kg} \cdot \text{C°})(200 - T_f) = (0.15 \text{ kg})(387 \text{ J/kg} \cdot \text{C°})T_f$$
$$+ 3.35 \times 10^3 \text{ J} + (10 \times 10^{-3} \text{ kg})(4.18 \times 10^3 \text{ J/kg} \cdot \text{C°})T_f$$
$$T_f = -2.2\text{°C}$$

This result makes no sense at all! When we add a hot piece of lead to a mixture at 0°C, the final temperature must be *greater* than 0°C, yet our calculations predict a final temperature which is *less* than the initial temperature by about 2°C.

The difficulty with this type of heat transfer problem that has not been faced as yet is that it involves a change of phase (solid to liquid) as the ice cube melts. To reconcile this difficulty, let us first calculate the maximum amount of heat that can be extracted from the block of lead assuming its final temperature is 0°C. This is

$$Q_{max} = mc\Delta T = (50 \times 10^{-3}\text{ kg})(128\text{ J/kg}\cdot\text{C°})(200\text{°C}) = 1.28 \times 10^{3}\text{ J}$$

However, the amount of heat required just to melt the entire ice cube would be

$$Q = mL_f = (10 \times 10^{-3}\text{ kg})(3.35 \times 10^{5}\text{ J/kg}) = 3.35 \times 10^{3}\text{ J}$$

Obviously, not enough heat energy is available from the lead to completely melt the ice cube. Therefore, if the ice cannot melt completely, the final temperature of the mixture must remain at 0°C. We can determine the mass of ice melted as follows. The heat added to the ice cube is $Q = 1.28 \times 10^3$ J. Thus, we have

$$Q = mL_f$$

$$m = Q/L_f = \frac{1.28 \times 10^{3}\text{ J}}{3.35 \times 10^{5}\text{ J/kg}} = 3.82 \times 10^{-3}\text{ kg} = 3.82\text{ g}$$

Thus, the end result is a copper container with a mixture of water, lead, and ice all at a temperature of 0°C.

12.5 HEAT TRANSFER BY CONDUCTION

There are three ways heat energy can be transferred from one location to another: conduction, convection, and radiation. Regardless of the process, however, there will be no net heat transfer between a system and its surroundings when the two are at the same temperature. In this section, we shall discuss heat transfer by conduction. Convection and radiation will be discussed in Sections 12.6 and 12.7.

Each of the methods of heat transfer can be examined by considering the various ways in which you can warm your hands over an open fire. If you insert a copper rod into the flame, as in Figure 12.4, you will find that the temperature of the metal in your hand increases rapidly. The heat reaches your hand through conduction. The manner in which heat is transferred from the flame, through the copper rod, and to your hand can be understood by examining what is happening to the atoms of the metal. Initially, before the rod is inserted into the flame, the copper atoms are vibrating about their equilibrium positions. As the flame heats the rod, those copper atoms near the flame begin to vibrate with larger and larger amplitudes. These wildly vibrating atoms collide with their neighbors and transfer some of their energy in the collisions. Slowly, copper atoms farther down the rod increase their amplitude of vibration, until those at the end being held are reached. The effect of this increased vibration results in an increase in temperature of the metal, and possibly a burned hand.

FIGURE 12.4 Heat reaches the hand by conduction through the copper rod.

Although the transfer of heat through a metal can be partially explained by atomic vibrations, the rate of heat conduction also depends on the properties of the substance being heated. For example, it is possible to hold a piece of asbestos in a flame indefinitely. This implies that very little heat is being conducted through the asbestos. In general, metals are good conductors of heat and materials such as asbestos, cork, paper, and fiber glass are poor conductors. Gases also are poor heat conductors because of their dilute nature. Metals are good conductors of heat because they contain large numbers of electrons that are relatively free to move through the metal and transport energy from one region to another. Thus, in a good conductor, such as copper, heat conduction takes place via the vibration of atoms and via the motion of free electrons.

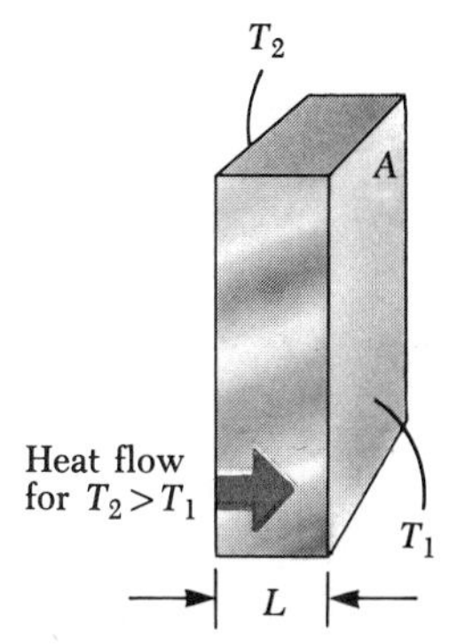

FIGURE 12.5 Heat transfer through a conducting slab of cross-sectional area A and thickness L. The opposite faces are at different temperatures T_1 and T_2.

In this section, we shall consider the rate at which heat is transferred from one location to another. If ΔQ is the amount of heat transferred from one location on an object to another in the time Δt, the **heat transfer rate,** H (sometimes called the heat current), is defined as

$$H \equiv \frac{\Delta Q}{\Delta t} \qquad (12.6)$$

Note that H has units of watts when ΔQ is in joules and Δt is in seconds (1 W = 1 J/s).

The conduction of heat occurs only if there is a difference in temperature between two parts of the conducting medium. Consider a slab of material of thickness L and cross-sectional area A, as in Figure 12.5. Suppose that one face is maintained at a temperature T_2 and the other face is held at a lower temperature T_1. Experimentally, one finds that the rate of heat flow—that is, the heat flow ΔQ per unit time, Δt—is proportional to the temperature difference $T_2 - T_1$ and the area A, and inversely proportional to the thickness of the slab. Specifically, the rate of flow of heat (or heat current) is given by

$$H = \frac{\Delta Q}{\Delta t} = kA\left(\frac{T_2 - T_1}{L}\right) \qquad (12.7)$$

where k is a constant called the **thermal conductivity** of the material. This constant is a property of the material. Table 12.3 lists some values of k for metals, gases, and nonmetals. Note that k is large for metals, which are *good heat conductors*, and small for gases and nonmetals, which are poor heat conductors (good *insulators*).

Melted snow pattern on a parking lot indicates the presence of underground steam pipes used to aid snow removal. Heat from the steam is conducted to the pavement from the pipes, causing the snow to melt. (Courtesy of Dr. Albert A. Bartlett, University of Colorado, Boulder)

The fact that different materials have different k values should help you understand the following phenomenon, first mentioned in Chapter 11. If you remove a metal ice tray and a package of frozen food from the freezer, which feels colder? Experience tells you that the metal tray feels colder even though it is at the same initial temperature as the cardboard package. This is explained by noting that, since metal has a much higher thermal conductivity than cardboard, it conducts heat more rapidly and hence removes heat from your hand at a higher rate. Hence, the metal tray *feels* colder than the carton even though it isn't. By use of a similar argument, you should be able to explain why a tile floor feels colder to a bare foot than a carpeted floor.

TABLE 12.3 Thermal Conductivities

Substance	Thermal Conductivity (J/s·m·C°)
Metals (at 25°C)	
Aluminum	238
Copper	397
Gold	314
Iron	79.5
Lead	34.7
Silver	427
Gases (at 20°C)	
Air	0.0234
Helium	0.138
Hydrogen	0.172
Nitrogen	0.0234
Oxygen	0.0238
Nonmetals (approximate values)	
Asbestos	0.08
Concrete	0.8
Glass	0.8
Ice	2
Rubber	0.2
Water	0.6
Wood	0.08

EXAMPLE 12.7 Heat Transfer Through a Concrete Wall

Find the amount of heat transferred by conduction in 1 h through a concrete wall 2 m high, 3.65 m long, and 0.2 m thick if one side of the wall is held at 20°C and the other side is at 5°C.

Solution Equation 12.7 gives the rate of heat transfer in joules per second. To find the amount of heat transferred in 1 h, we rewrite the equation as

$$\Delta Q = kA\Delta t\left(\frac{T_2 - T_1}{L}\right)$$

If we substitute the values given and consult Table 12.3, we find

$$\begin{aligned}\Delta Q &= (0.8 \text{ J/s·m·C°})(7.3 \text{ m}^2)(3600 \text{ s})\left(\frac{15 \text{ C°}}{0.2 \text{ m}}\right)\\ &= 1.6 \times 10^6 \text{ J}\end{aligned}$$

Early houses were insulated by constructing the walls of thick masonry blocks. This restricted the heat lost by conduction because k is relatively low for masonry. Furthermore, because the thickness, L, is large, the heat loss is also decreased, as shown by Equation 12.7.

*Home Insulation

If you would like to do some calculating to determine whether or not to add insulation to a ceiling or to some other portion of a building, what you have just learned about conduction needs to be modified slightly, for two reasons. (1) The insulating properties of materials used in buildings are usually expressed in engineering rather than SI units. For example, measurements stamped on a package of fiber glass insulating board will be in units such as British Thermal Units, feet, and degrees Fahrenheit. (2) In dealing with the insulation of a building, we must consider heat conduction through a compound slab, with each portion of the slab having a different thickness and a different thermal conductivity. For example, a typical wall in a house will consist of an array of materials, such as wood paneling, dry wall, insulation, sheathing, and wood siding.

It is found that the rate of heat transfer through a compound slab is given by

$$\frac{\Delta Q}{\Delta t} = \frac{A(T_2 - T_1)}{\sum_i L_i/k_i} \tag{12.8}$$

where T_1 and T_2 are the temperatures of the *outer extremities* of the slab and the summation is over all portions of the slab. For example, if the slab consists of three different materials, the denominator will consist of the sum of three terms. In engineering practice, the term L/k for a particular substance is referred to as the ***R* value** of the material. Thus, Equation 12.8 reduces to

$$\frac{\Delta Q}{\Delta t} = \frac{A(T_2 - T_1)}{\sum_i R_i} \tag{12.9}$$

The R values for a few common building materials are given in Table 12.4 (note the units).

Also, it should be noted that near any vertical surface there is a very thin, stagnant layer of air that must be considered when finding the total R value for a wall. The thickness of this stagnant layer on an outside wall depends on the velocity of the wind. As a result, heat loss from a house on a day when the wind is blowing hard is greater than heat loss on a day when the wind velocity is zero. A representative R value for this stagnant layer of air is given in Table 12.4.

TABLE 12.4 *R* Values for Some Common Building Materials

Material	***R* value ($ft^2 \cdot F° \cdot h/BTU$)**
Hardwood siding (1 in. thick)	0.91
Wood shingles (lapped)	0.87
Brick (4 in. thick)	4.00
Concrete block (filled cores)	1.93
Fiber glass batting (3.5 in. thick)	10.90
Fiber glass batting (6 in. thick)	18.80
Fiber glass board (1 in. thick)	4.35
Cellulose fiber (1 in. thick)	3.70
Flat glass (0.125 in. thick)	0.89
Insulating glass (0.25-in. space)	1.54
Vertical air space (3.5 in. thick)	1.01
Air film	0.17
Dry wall (0.5 in. thick)	0.45
Sheathing (0.5 in. thick)	1.32

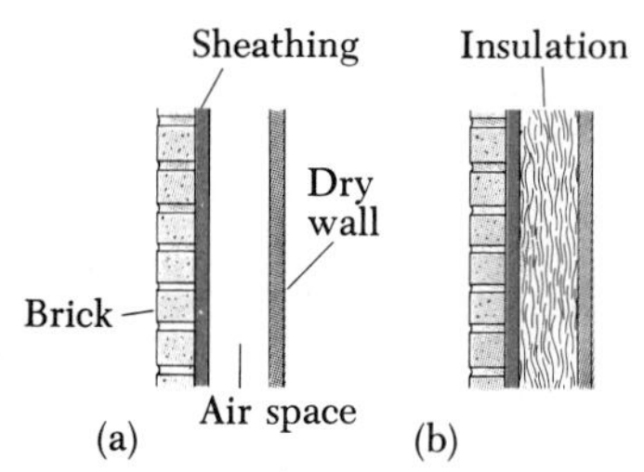

FIGURE 12.6 (Example 12.8) Cross-sectional view of an exterior wall containing (a) an air space and (b) insulation.

EXAMPLE 12.8 The *R* Value of a Typical Wall

Calculate the total R value for a wall constructed as shown in Figure 12.6a. Starting outside the house (to the left in Fig. 12.6a) and moving inward, the wall consists of brick, 0.5 in. of sheathing, a vertical air space 3.5 in. thick, and 0.5 in. of dry wall. Do not forget the dead-air layers inside and outside the house.

Solution Referring to Table 12.4, we find the total R value for the wall as follows:

R_1 (outside air film) $= 0.17\ \text{ft}^2\cdot\text{F}°\cdot\text{h/BTU}$
R_2 (brick) $= 4.00$
R_3 (sheathing) $= 1.32$
R_4 (air space) $= 1.01$
R_5 (dry wall) $= 0.45$
R_6 (inside air film) $= 0.17$

R_{total} $= 7.12\ \text{ft}^2\cdot\text{F}°\cdot\text{h/BTU}$

EXERCISE If a layer of fiberglass insulation 3.5 in. thick is placed inside the wall to replace the air space as in Figure 12.6b, what is the total R value of the wall? By what factor is the heat loss reduced?

Answer $R = 17\ \text{ft}^2\cdot\text{F}°\cdot\text{h/BTU}$; a factor of 2.5.

12.6 CONVECTION

At one time or another you probably have warmed your hands by holding them over an open flame, as illustrated in Figure 12.7. In this situation, the air directly above the flame is heated and expands. As a result, the density of the air decreases and the air rises. This warmed mass of air heats your hands as it flows by. *Heat transferred by the movement of a heated substance is said to have been transferred by* ***convection.*** When the movement results from differences in density, as in the example of air around a fire, it is referred to as *natural convection.* When the heated substance is forced to move by a fan or pump, as in some hot-air and hot-water heating systems, the process is called *forced convection.*

The circulating pattern of air flow at a beach pictured in Figure 12.2 is an example of convection. Likewise, the mixing that occurs as water is cooled and eventually freezes at its surface (Chapter 11) is an example of convection in nature. Recall that the mixing by convection currents ceases when the water temperature reaches 4°C. Since the water in the pool cannot be cooled by convection below 4°C, and because water is a relatively poor conductor of heat (Table 12.3), the water near the bottom remains near 4°C for a long time. As a result, fish have a comfortable temperature in which to live even in periods of prolonged cold weather.

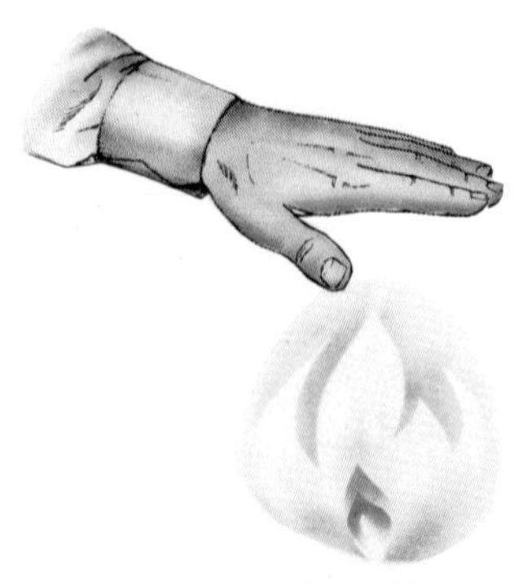

FIGURE 12.7 Heating your hand by convection.

If it were not for convection currents, it would be very difficult to boil water. As water is heated in a teakettle, the lower layers are warmed first. These heated regions expand and rise to the top because their density is lowered. At the same time, the denser cool water replaces the warm water at the bottom of the kettle so that it can be heated.

The same process occurs when a room is heated by a radiator. The hot radiator warms the air in the lower regions of the room. The warm air expands and rises to the ceiling because of its lower density. The denser regions of cooler air from above replace the warm air, setting up the continuous air current pattern shown in Figure 12.8.

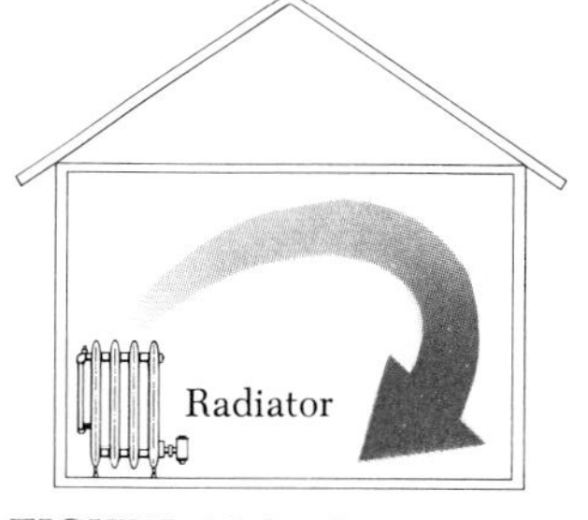

FIGURE 12.8 Convection currents are set up in a room heated by a radiator.

The rate at which heat is transferred by a fluid to a surface of area A is given by the equation

$$\frac{\Delta Q}{\Delta t} = hA\Delta T \tag{12.10}$$

where ΔT is the difference in temperature between the fluid and the surface, and h is a constant called the **convection coefficient.** It is found that h depends on several factors such as the shape of the surface and whether the surface is horizontal or vertical. For heated air in contact with a vertical surface, the convection coefficient is given by

$$h = 1.77(\Delta T)^{1/4} \text{ J/s·m}^2\text{·C°} \tag{12.11}$$

EXAMPLE 12.9 Heat Loss Through Windows

The temperature of the air inside a room is held at 20°C on a day when the outside air temperature is −10°C. Find the rate at which heat is transferred to a single-pane window by convection. Assume the window has a cross-sectional area of 0.5 m^2 and a thickness of 0.25 cm.

Solution On a cold day, the temperature of a glass window pane is much lower than the temperature of the air inside the room. In fact, the temperature of the glass is approximately equal to the average of the inside and outside air temperatures. In this example, this average temperature is 5°C. Thus, the temperature difference between the surface and outside air is 15°C, so the convection coefficient can be found from Equation 12.11 as

$$h = 1.77(\Delta T)^{1/4} = 1.77(15°\text{C})^{1/4} = (1.77)(1.968) = 3.48 \text{ J/s·m}^2\text{·C°}$$

We now can use this value of the convection coefficient in Equation 12.10 to find the rate at which heat is transferred to the inside surface of the glass by the air. This gives

$$\frac{\Delta Q}{\Delta t} = hA\Delta T = (3.48 \text{ J/s·m}^2\text{·C°})(0.5 \text{ m}^2)(15°\text{C}) = 26.1 \text{ J/s}$$

EXAMPLE 12.10 Conduction of Heat Through the Window

In the previous example, it was assumed that the temperature of the window is approximately 5°C. Obviously, the temperature of the window surface inside the room must be slightly greater than the temperature of the outside surface. Find the actual temperature of the inside and outside glass surfaces.

Solution In order to solve this problem, we must use the fact that the rate of heat transferred to the inside surface by convection must equal the rate of heat transferred through the glass by conduction. In Example 12.9 we found that the rate of heat transferred to the inside surface by convection is 26.1 J/s. This must also equal the rate of heat transferred by conduction through the glass. Recall that the conduction process requires a temperature difference $T_2 - T_1$ between the inside and outside surfaces of the glass. Using Equation 12.7, we find $T_2 - T_1$ as follows:

$$T_2 - T_1 = \left(\frac{\Delta Q}{\Delta t}\right)\frac{L}{kA} = (26.1 \text{ J/s})\frac{0.25 \times 10^{-2} \text{ m}}{(0.8 \text{ J/s·m·C°})(0.5 \text{ m}^2)} = 0.16 \text{ C°}$$

Thus, the temperature of the inside surface of the glass must be 5.08°C, while the temperature of the outside surface must be 4.92°C.

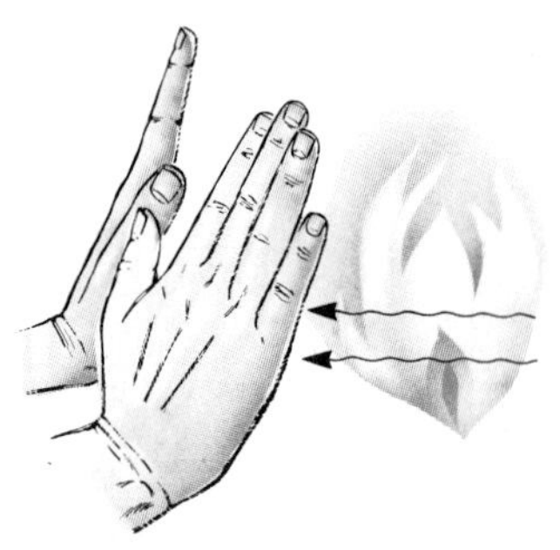

FIGURE 12.9 Warming your hands by radiation.

12.7 RADIATION

The third way of transferring heat is through **radiation.** Figure 12.9 shows how you can warm your hands at an open flame by means of radiant heat. In this case, the hands are placed to the side of the flame. You have most likely experienced radiant heat when sitting in front of a fireplace. The hands are not in physical contact with the flames either directly or indirectly. Therefore, conduction cannot account for the heat transfer. Furthermore, convection is not important in this case since the hands are not above the flame in the path of convection currents. The radiation of heat energy is the important process in this case.

All objects radiate energy continuously in the form of electromagnetic waves, which we shall discuss in Chapter 23. The type of radiation associated with the transfer of heat energy from one location to another is referred to as infrared radiation.

Through electromagnetic radiation, approximately 1340 J of heat energy strikes 1 m^2 of the top of the earth's atmosphere every second. Some of this energy is reflected back into space and some is absorbed by the atmosphere, but enough arrives at the surface of the earth each day to supply all of our energy needs on this planet hundreds of times over—if it could be captured and used efficiently. The growth in the number of solar houses in this country is one example of an attempt to make use of this free energy.

Radiant energy from the sun affects our day-to-day existence in a number of ways. For example, consider what happens to the atmospheric temperature at night. If there is a cloud cover above the earth, the water vapor in the clouds reflects back a part of the infrared radiation emitted by the earth and consequently the temperature remains at moderate levels. In the absence of this cloud cover, however, there is nothing to prevent this radiation from escaping into space, and thus the temperature drops more on a clear night than when it is cloudy.

The rate at which an object emits radiant energy is proportional to the fourth power of its absolute temperature. This is known as **Stefan's law** and is expressed in equation form as

Stefan's law

$$P = \sigma A e T^4 \tag{12.12}$$

where P is the power radiated by the body in watts (or joules per second), σ is a constant equal to 5.6696×10^{-8} W/m^2·K^4, A is the surface area of the object in square meters, e is a constant called the **emissivity,** and T is temperature in kelvins. The value of e can vary between zero and unity, depending on the properties of the surface.

An object radiates energy at a rate given by Equation 12.12. At the same time, the object also absorbs electromagnetic radiation. If the latter process did not occur, an object would eventually radiate all of its energy and its temperature would reach absolute zero. The energy that a body absorbs comes from its surroundings, which consists of other objects which radiate energy. If an object is at a temperature T and its surroundings are at a temperature T_0, the net energy gained or lost each second by the object as a result of radiation is given by

$$P_{net} = \sigma A e(T^4 - T_0^4) \tag{12.13}$$

When an object is in *equilibrium* with its surroundings, *it radiates and absorbs energy at the same rate, and so its temperature remains constant.* When an object is hotter than its surroundings, it radiates more energy than it absorbs and so it cools. An *ideal absorber* is defined as an object that absorbs all of the energy incident on it. The emissivity of an ideal absorber is equal to unity. Such an object is often referred to as a **black body.** An ideal absorber is also an ideal radiator of energy. In contrast, an object with an emissivity equal to zero absorbs none of the energy incident on it. Such an object reflects all the incident energy and so is a perfect reflector.

The amount of radiant energy emitted by a body can be measured with heat-sensitive recording equipment, using a technique called **thermography.** The level of radiation emitted by a body is greatest in the warmest regions, and the detected pattern, called a **thermogram,** is brightest in the warmest areas. Color plate 11 shows a thermogram of a house. The center portion of the door and windows and parts of the roof are red, signifying a temperature higher than that of surrounding areas. A higher temperature usually means that heat is escaping at these points. Thermograms can be useful in aiding energy conservation practices. For example, the owners of this house could save on their heating costs by adding insulation to the attic area and by installing thermal draperies over the windows.

EXAMPLE 12.11 Who Turned Down the Thermostat?

A student trying to decide what to wear is staying in a room that is at 20°C. If the skin temperature of the unclothed student is 37°C, how much heat is lost from his body in 10 min? Assume that the emissivity of skin is 0.90 that the surface area of the student is 1.5 m^2.

Solution Using Equation 12.13, the rate of heat loss from the skin is

$$P_{net} = \sigma A e(T^4 - T_0^4)$$
$$= (5.67 \times 10^{-8}\ \mathrm{W/m^2\cdot K^4})(1.5\ \mathrm{m^2})(0.90)[(310\ \mathrm{K})^4 - (293\ \mathrm{K})^4] = 143\ \mathrm{J/s}$$

Note that it was necessary to change the temperature to kelvins. At this rate of heat loss, the total heat lost by the skin in 10 min is

$$Q = P_{net} \times \text{time} = (143\ \mathrm{J/s})(600\ \mathrm{s}) = 8.58 \times 10^4\ \mathrm{J}$$

---

*12.8 HINDERING HEAT TRANSFER

The Thermos bottle, called a *Dewar flask* in scientific applications (after its inventor), is designed to minimize heat transfer by conduction, convection, and radiation. Such a container is used to store either cold or hot liquids for long periods of time. The standard construction (Fig. 12.10) consists of a double-walled Pyrex vessel with silvered inner walls. The space between the walls is evacuated to minimize heat transfer by conduction and convection. The silvered surfaces minimize heat transfer by radiation by reflecting most of the radiant heat. Very little heat is lost through the neck of the flask because Pyrex glass is a poor conductor. A further reduction in heat loss is obtained by reducing the size of the neck. A common usage of Dewar flasks in scientific applications is to store liquid nitrogen (boiling point 77 K) and liquid oxygen (boiling point 90 K). For substances that have a very low spe-

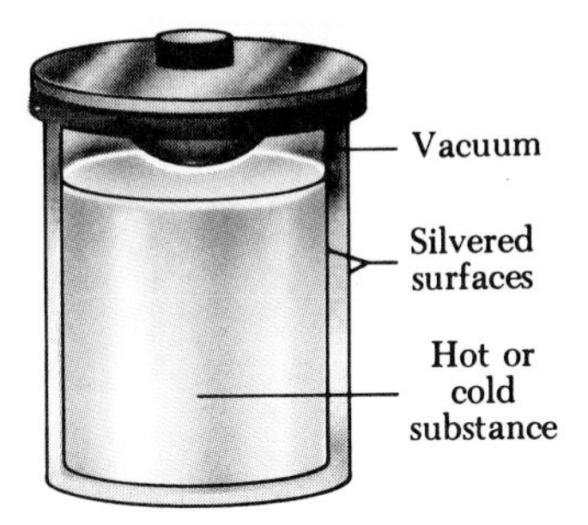

FIGURE 12.10 Cross-sectional view of a Dewar vessel used to store hot or cold liquids.

cific heat, such as liquid helium (boiling point 4.2 K), it is often necessary to use a double Dewar system in which the Dewar flask containing the liquid is surrounded by a second Dewar flask. The space between the two flasks is filled with liquid nitrogen.

Another application of the principles of minimizing heat transfer is the insulating of buildings, as discussed in Section 12.5. To achieve this, insulating materials are placed in the walls and above the ceilings. These materials, such as fiber glass, are poor conductors of heat. Furthermore, since they are loosely packed, there is air trapped in them in small cavities. Air is a poor heat conductor and cannot carry much heat by convection when it is trapped.

Wool sweaters and down jackets keep us warm by trapping the warmer air in regions close to our bodies and hence reducing heat loss by convection and conduction. In other words, what keeps us warm is the trapped air in the clothing, not the clothing itself.

*The Greenhouse Effect

Many of the principles of heat transfer, and the prevention of it, can be understood by considering a florist's greenhouse. One of the basic ideas underlying the operation of these buildings is the fact that glass will allow visible light to pass through but not infrared radiation. During the day, sunlight passes into the greenhouse and is absorbed by the walls, earth, plants, and so on. This absorbed visible light is subsequently re-radiated as infrared radiation, which cannot escape the enclosure. The increasing amount of trapped infrared radiation causes the temperature of the interior to rise.

Convection currents are also inhibited in a greenhouse. This prevents heated air from rapidly passing over those surfaces of the greenhouse exposed to the outside air, which would cause a heat loss through those surfaces. Many experts consider this to be an even more important effect in the operation of a greenhouse than the effect of trapped infrared radiation. In fact, experiments have shown that when the glass over a greenhouse is replaced by a special glass that transmits infrared light, the greenhouse temperature is only slightly lowered. Based on this evidence, one would have to conclude that the absorption of infrared radiation is not the primary mechanism that heats a greenhouse. It is the inhibition of air flow that occurs with any roof (such as in an attic) that is the main factor in warming a greenhouse.

A phenomenon known as the greenhouse effect can also play a major role in determining the earth's temperature. This can be understood by first noting that the earth's atmosphere is a good transmitter (hence a poor absorber) of visible radiation and a good absorber of infrared radiation. Carbon dioxide (CO_2) in the earth's atmosphere acts somewhat like the glass in a greenhouse in that it allows incoming visible radiation from the sun to pass through easily but it does not allow infrared radiation to pass through as readily. The visible light which reaches the earth's surface is absorbed and re-radiated as infrared light, which in turn is absorbed (trapped) by the earth's atmosphere. An extreme case is the warmest planet, Venus, which has a carbon-dioxide-rich atmosphere and temperatures approaching 850°F.

As fossil fuels (coal, oil, and natural gas) are burned, large amounts of carbon dioxide are released into the atmosphere, causing it to retain more heat. This is of great concern to scientists and governments throughout the world. Many scientists are convinced that the 10% increase in the amount of atmospheric carbon dioxide in the last 30 years could lead to drastic changes in world climate. According to one estimate, doubling the carbon dioxide content in the atmosphere will cause temperatures to increase by 2 C°! In temperate regions, such as Europe and the United States, billions of dollars per year would be saved in fuel costs. Unfortunately, the temperature increase would melt polar ice caps, which could cause flooding and destroy many coastal areas, and increase the frequency of droughts and consequently decrease already low crop yields in tropical and subtropical countries.

While this scenario is rather alarming, other scientists theorize that an atmosphere rich in carbon dioxide could bring on another Ice Age! According to this model, a heavy blanket of carbon dioxide would screen out a large percentage of the sun's incident rays, causing the earth's temperature to decrease in time. Another factor adding to the problem is the effect of large quantities of small particles released into the atmosphere by industries. Such particles produce a hazy atmosphere (smog) that reflects some of the incoming radiation back into space, producing a cooling effect.

Obviously, the problem is very complex, and the various models that have been offered are open to question and further study. Nonetheless, it is an important problem which all nations must address.

*12.9 SOLAR HOUSES

Because of the increasing cost of coal, oil, and natural gas (the fossil fuels), more and more people are considering alternative sources of energy. One attractive technique being used to circumvent high fuel costs is to rely on energy from the sun to heat houses. There are two distinct types of solar houses: *passive* and *active*.

*Passive Solar Houses

Passive solar houses are constructed in such a way that they are able to absorb a large fraction of the incident solar energy and store it in massive concrete walls, containers of water, or other storage media. The stored energy is then released in the house when the temperature drops. This heat can be dispersed throughout the house without the help of complicated ducting or pumps.

Figure 12.11 shows a unique wall design, called a **Trombe wall,** used in some of these houses. During the day, sunlight enters through the glass panel and is absorbed by the concrete wall behind it. The layer of air between the window and the concrete wall is heated, expands, and rises to pass through the vent at the top of the wall and into the room. At the same time, cool air near the floor in the room is drawn into a duct at the bottom of the wall. The convection currents set up are sufficient to heat the house during the day. At night, the window is shuttered and the drapes are closed to reduce the outward flow of heat through the glass. The warm concrete wall radiates energy into the room to provide warmth.

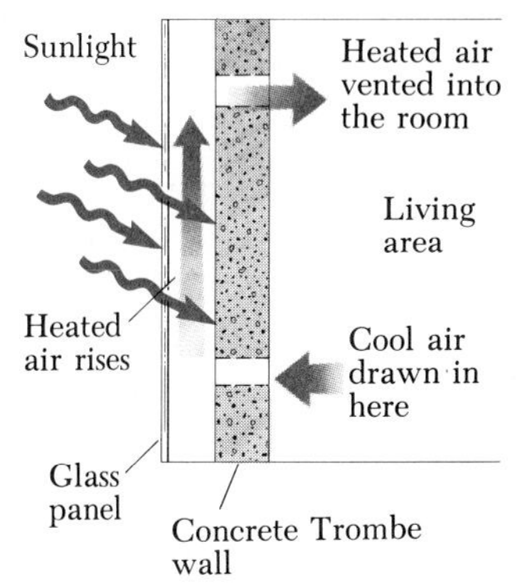

FIGURE 12.11 Cross-sectional view of a Trombe wall, one scheme for a passive solar heating system.

Active solar home. (From Jerry D. Wilson, *Practical Physics*, Saunders College Publishing, 1986.)

Heat locks are also used in many solar houses. A heat lock system works as follows. In order to enter the house, a person must pass through two doors arranged in such a way that the two are never open at the same time. Such a system reduces the infiltration of cold outside air into the house. It is particularly effective if the outside door includes a large window to allow sunlight to enter. This ensures that the air in the heat lock is always air that has been

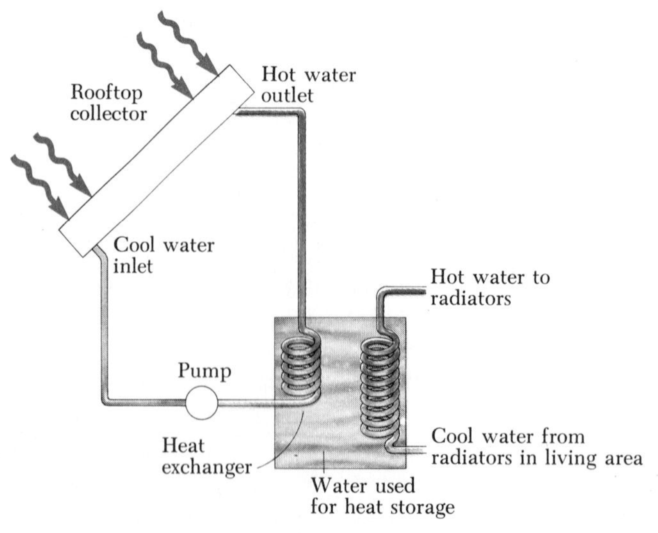

FIGURE 12.12 Diagram of an active solar heating system.

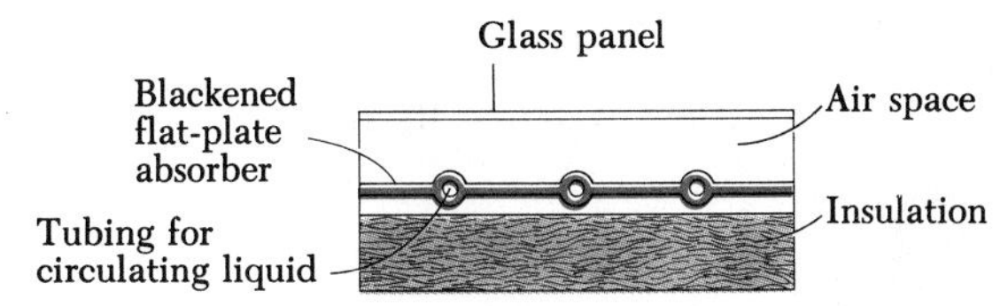

FIGURE 12.13 Cross-sectional view of a liquid-type flat-plate collector.

warmed by the greenhouse effect and by the fact that convection currents are inhibited in this region.

Proper shading is also used in most solar houses, whether active or passive. Awnings, trellises, or large overhangs are constructed such that sunlight is blocked from entering the house in summer, when the sun is high in the sky, but is allowed to enter during the winter, when the sun is low in the sky. Also, trees and shrubs that do not lose their leaves in winter are planted on the north side of the house to block off the cold northern windows.

*Active Solar Houses

Active solar houses use pumps to force a fluid, either liquid or air, through a rooftop collector, as shown in Figure 12.12. The heated fluid is then passed through a heat exchanger, where it gives up its heat to some storage medium. This heat is usually stored in water if the circulating fluid is a liquid or in a bed of rocks if air is the transfer agent. The heat is then drawn from storage as needed through a second heat exchanger.

A cutaway view of a liquid-type flat-plate collector is shown in Figure 12.13. Sunlight entering the glass cover is absorbed by the blackened flat plate. Heat is transferred from the absorber plate to the circulating liquid as it flows through tubing attached to the plate. The expected lowest outside temperature determines the type of fluid used. A layer of insulation is placed between the flat plate and the roof of the house to prevent heat loss by conduction along this pathway.

Very often, a single-panel solar collector (typically 4 ft by 8 ft) installed on the roof of a house is used to heat water for household use. In most installations, the warm water from the collector is fed to the hot water tank, and so the panel serves as part of the water-heating system. Swimming pool water can be heated by circulating it through such a panel. One inexpensive technique for heating swimming pool water is to circulate it through a coil of black garden hose placed on the roof of a garage or other structure.

EXAMPLE 12.12 Another Scheme for Passive Solar Heating

A room 6 m long by 4 m wide has a floor of poured concrete 20 cm thick. Assume that this slab is exposed to sunlight through a south-facing window on a winter day and that its temperature is 24°C at sunset. At sunrise the next morning the temperature of the slab has fallen to 18°C. How much heat has the slab released into the room? The density of concrete is approximately 2.3×10^3 kg/m^3, and its specific heat is about 920 J/kg · C°.

Solution The mass of the concrete is

$$m = \rho V = (2.3 \times 10^3\ \text{kg/m}^3)(6 \times 4 \times 0.2\ \text{m}^3) = 1.1 \times 10^4\ \text{kg}$$

Thus, the heat given up by the concrete is

$$Q = mc\Delta T = (1.1 \times 10^4\ \text{kg})(920\ \text{J/kg} \cdot \text{C}°)(6\ \text{C}°) = 6.1 \times 10^7\ \text{J}$$

EXERCISE If this much heat is released over a period of 12 h, how much power is radiated by the floor, on the average?

Answer 150 W.

SUMMARY

Heat flow is an energy transfer that takes place as a consequence of a temperature difference only.

The **calorie** is the amount of heat necessary to raise the temperature of 1 g of water from 14.5°C to 15.5°C.

The **mechanical equivalent of heat** is defined to be

$$\boxed{1\ \text{cal} \equiv 4.186\ \text{J}} \tag{12.1}$$

The **specific heat** of a substance is defined as

$$\boxed{c \equiv \frac{Q}{m\Delta T}} \tag{12.2}$$

The heat required to change the phase of a given mass m of a pure substance is given by

$$\boxed{Q = mL} \tag{12.5}$$

where L is the **latent heat** of the substance and depends on the nature of the phase change and the properties of the substance. The **latent heat of fusion,** L_f, is used when the phase change is from a solid to a liquid. The **latent heat of vaporization,** L_v, is used when the phase change is from a liquid to a gas.

Heat may be transferred by three fundamentally distinct processes: **conduction, convection,** and **radiation. Conduction** can be viewed as an exchange of kinetic energy between colliding molecules or by the motion of electrons. The rate at which heat flows by conduction through a slab of area A and thickness L is

$$\boxed{H = \frac{\Delta Q}{\Delta t} = kA\left(\frac{T_2 - T_1}{L}\right)} \tag{12.7}$$

where k is the **thermal conductivity** of the material. Heat transferred by the movement of a heated substance is said to have been transferred by **convection.**

All objects radiate and absorb energy in the form of electromagnetic waves. An object that is hotter than its surroundings radiates more energy than it absorbs, whereas an object that is cooler than its surroundings absorbs more energy than it radiates. The rate at which an object emits radiant energy is given by **Stefan's law:**

$$P = \sigma A e T^4 \quad (12.12)$$

where σ is a constant equal to 5.6696×10^{-8} W/m$^2 \cdot$ K^4 and e is a constant called the **emissivity.**

ADDITIONAL READINGS

P. B. Allen, "Conduction of Heat," *The Physics Teacher*, December 1983, p. 582.

E. Barr, "James Prescott Joule and the Quiet Revolution," *Physics Teacher*, April 1969, p. 199.

B. Chalmers, "How Water Freezes," *Sci. American*, February 1959, p. 144.

J. Dyson, "What Is Heat?" *Sci. American*, September 1954, p. 58.

J. Kelley, "Heat, Cold and Clothing," *Sci. American*, February 1956, p. 194.

R. Revelle, "Carbon Dioxide and World Climate," *Sci. American*, January 1982, p. 35.

M. G. Velarde, "Convection," *Sci. American*, January 1980, p. 92.

M. Wilson, "Count Rumford," *Sci. American*, October 1960, p. 158.

QUESTIONS

1. Ethyl alcohol has about one half the specific heat of water. If equal masses of alcohol and water in separate beakers are supplied with the same amount of heat, compare the temperature increases of the two liquids.
2. A small crucible is taken from a 200°C oven and immersed in a tub of water at room temperature (often referred to as quenching). What is the approximate final equilibrium temperature?
3. In a daring lecture demonstration, an instructor dips her wetted fingers into molten lead (327°C) and withdraws them quickly, without getting burned. How is this possible?
4. Why can you get a more severe burn from steam at 100°C than from water at 100°C?
5. Concrete has a higher specific heat than does soil. Use this fact to explain (partially) why cities have a higher average temperature than the surrounding countryside. If a city is hotter than the surrounding countryside, would you expect breezes to blow from city to country or from country to city? Explain.
6. Pioneers stored fruits and vegetables in underground cellars. Discuss as fully as possible this choice for a storage site.
7. In winter the pioneers mentioned in Question 6 would store an open barrel of water alongside their produce. Why?
8. During a cold spell, Florida orange growers often spray a mist of water over their trees during the night. Why does this help?
9. Why is it possible to hold a lighted match even when it is burned to within a few millimeters of your fingertips?
10. A piece of paper is wrapped around a rod made half of wood and half of copper. When held over a flame, the paper in contact with the wood burns but the half in contact with the metal does not. Explain.
11. If water is a poor conductor of heat, why can it be heated quickly when placed over a flame?
12. Why does a piece of metal feel colder than a piece of wood when they are at the same temperature?
13. Updrafts of air are familiar to all pilots. What causes these currents?
14. A tile floor in a bathroom may feel uncomfortably cold to your bare feet, but a carpeted floor in an adjoining room at the same temperature will feel warm. Why?
15. Why can potatoes be baked more quickly when a piece of metal has been inserted through them?
16. The U.S. penny is now made of copper-coated zinc. Can a calorimetric experiment be devised to test for the metal content in a collection of pennies? If so, describe the procedure you would use.
17. If you hold water in a paper cup over a flame, you can bring the water to a boil without burning the cup. How is this possible?
18. Some signs along a highway warn that bridges freeze faster than the roadway. Why is this so?
19. Thermopane windows are usually made with two layers of glass separated by an air space. One manufac-

turer claims that his panes are farther apart than those of his competitors, which gives them a better insulating factor. Give an argument which would falsify his claim.

20. On a very hot day it is possible to cook an egg on the hood of a car. Would you select a black car or a white car to cook your egg? Why?
21. Most of the heat from the water cooling the engine of your car is transferred to the atmosphere after passing the water through the radiator. Does the radiator transfer the heat by radiation as its name implies?
22. In the winter you might notice that some roofs are uniformly covered with snow while others have areas where the snow has melted, as shown in the photograph. Which houses would you say are better insulated? Why are there regular spacings between the melted areas in the photograph?

Question 22

PROBLEMS

Section 12.2 Specific Heat

1. A 50-g sample of copper is at 25°C. If 1200 J of heat energy is added to the copper, what is its final temperature?

2. How many joules of energy are required to raise the temperature of 100 g of gold from 20°C to 100°C?

3. As a part of your exercise routine you climb a 10 m rope. How many food calories do you expend in a single climb up the rope?

4. A 5-g lead bullet traveling with a speed of 300 m/s is stopped by a large tree. If half the kinetic energy of the bullet is transformed into heat energy and remains with the bullet while the other half is transmitted to the tree, what is the increase in temperature of the bullet?

5. A 75-kg weight-watcher wishes to climb a mountain to work off the equivalent of a large piece of chocolate cake rated at 500 (food) Calories. How high must the person climb?

6. Show that 1 BTU = 1054 J.

7. Find the specific heat of water, ice, and steam in cal/g · C°.

8. A 3-kg rock is initially at rest at the top of a cliff. If the energy equivalent of the rock is sufficient to raise the temperature of 1 kg of water by 0.10°C, how high is the cliff?

• 9. A 1.5-kg copper block is given an initial speed of 3 m/s on a rough horizontal surface. Because of friction, the block finally comes to rest. (a) If 85% of its initial kinetic energy is absorbed by the block in the form of heat, calculate the increase in temperature of the block. (b) What happens to the remaining energy?

• 10. An aluminum cup of mass 200 g contains 800 g of water in thermal equilibrium at 80°C. The combination of cup and water is cooled uniformly so that the temperature decreases by 1.5 C° per minute. At what rate is heat energy being removed? Express your answer in watts.

Section 12.3 Conservation of Energy: Calorimetry

11. A 0.4-kg iron horseshoe initially at 500°C is dropped into a bucket containing 20 kg of water at 22°C. What is the final equilibrium temperature? Neglect any heat transfer to or from the surroundings.

12. An unknown liquid of mass 400 g at a temperature of 80°C is poured into 400 g of water at 40°C. The final equilibrium temperature of the mixture is 49°C. What is the specific heat of the unknown liquid?

13. What mass of water at 25°C must be allowed to come to thermal equilibrium with a 3-kg gold bar at 100°C in order to lower the temperature of the bar to 50°C?

14. A block of copper of mass 200 g at a temperature of 90°C is dropped into 400 g of water at 27°C. The water is contained in a 300-g glass container. What is the final temperature of the mixture?

• 15. An aluminum cup contains 225 g of water at 27°C. A 400-g sample of silver at an initial temperature of 87°C is placed in the water. A 40-g copper stirrer is used to stir the mixture until it reaches its final equilibrium temperature of 32°C. Calculate the mass of the aluminum cup.

•• **16.** Three liquids are at temperatures of 10, 20, and 30°C, respectively. Equal masses of the first two liquids are mixed and the equilibrium temperature is 17°C. Equal masses of the second and third are then mixed and the equilibrium temperature is 28°C. Find the equilibrium temperature when equal masses of the first and third are mixed.

Section 12.4 Latent Heat

17. How much heat is required to change a 40-g ice cube from ice at −10°C to steam at 110°C?

18. A large block of ice at 0°C has a hole chipped into it and 400 g of aluminum pellets at a temperature of 30°C are poured into the hole. How much of the ice melts?

19. A 50-g ice cube at 0°C is heated until 45 g has become water at 100°C and 5 g has become steam at 100°C. How much heat was added to accomplish this?

20. What mass of steam initially at 120°C is needed to warm 350 g of water in a 300-g aluminum container from 20°C to 50°C?

• **21.** A 100-g ice cube at 0°C is placed in 650 g of water at 25°C. What is the final temperature of the mixture?

• **22.** A 300-g ice cube at 0°C is placed in 200 g of water at 10°C. The water is in a 100-g aluminum container. How much ice melts?

• **23.** Steam at 100°C is added to ice at 0°C. (a) Find the amount of ice melted and the final temperature when the mass of steam is 10 g and the mass of ice is 50 g. (b) Repeat when the mass of steam is 1 g and the mass of ice is 50 g.

• **24.** A 75-kg cross-country skier moves across snow such that the coefficient of friction between skis and snow is 0.2. Assume all the snow beneath his skis is at 0°C and that all the heat generated by friction is added to snow which sticks to his skis until melted. How far would he have to ski to melt 1 kg of snow?

Section 12.5 Heat Transfer by Conduction

Section 12.6 Convection

Section 12.7 Radiation

25. (a) Find the rate of heat flow through a copper block of cross-sectional area 15 cm^2 and length 8 cm when a temperature difference of 30°C is established across the block. Repeat the calculation assuming the material is (b) a block of air having the same dimensions and (c) a block of wood of these dimensions.

26. A window has a glass surface of 1600 cm^2 and a thickness of 3 mm. (a) Find the heat transferred by conduction through this pane when the temperature of the inside surface of the glass is 70°F and the outside temperature is 90°F. (b) Repeat for the same inside temperature and an outside temperature of 0°F.

27. Determine the R value for a wall constructed as follows: The outside of the house is covered by lapped wood shingles placed over 0.5 in. thick sheathing, followed by 3 in. of cellulose fiber, and finally 0.5 in. of dry wall.

28. (a) Find out the construction details of a typical outside wall in your home and estimate its R value. (b) Repeat your estimate for a typical dormitory room on your campus.

29. The temperature of the air inside a room is held at 23°C on a day when the outside temperature is 5°C. Find the rate at which heat is transferred to the window by convection. Assume the window has a cross-sectional area of 0.6 m^2 and a thickness of 0.125 cm.

30. In Problem 29, find the actual inside and outside temperatures of the glass.

31. Two identical objects are in the same surroundings at 0°C. One is at a temperature of 1200 K, and the other is at 1100 K. Find the ratio of the power emitted by the hotter object to the power emitted by the cooler object.

32. A sphere that is to be considered as a perfect black body radiator has a radius of 0.06 m and is at 200°C in a room where the temperature is 22°C. Calculate the rate at which the sphere radiates energy.

• **33.** A Styrofoam container is in the shape of a box of surface area 0.8 m^2 a thickness of 2 cm. The inside temperature is 5°C, and that outside is 25°C. If it takes 8 h for 5 kg of ice to melt in the container, determine the thermal conductivity of the styrofoam.

• **34.** Calculate the temperature at which a tungsten filament that has an emissivity of 0.25 and a surface area of 2.5×10^{-5} m^2 will radiate energy at the rate of 25 W in a room where the temperature is 22°C.

•• **35.** A copper rod and an aluminum rod of equal diameter are joined end-to-end in good thermal contact. The temperature of the free end of the copper rod is held constant at 100°C, and that of the far end of the aluminum rod is held at 0°C. If the copper rod is 0.15 m in length, what must be the length of the aluminum rod so that the temperature at the junction is 50°C?

ADDITIONAL PROBLEMS

36. In a showdown on the streets of Laredo, the good guy drops a 5-g silver bullet, at a temperature of 20°C, into a 100-cm^3 cup of water at 90°C. Simultaneously, the bad guy drops a 5-g copper bullet, at the same initial temperature, into an identical cup of water. Which one ends the showdown with the coolest cup of water in the west? Neglect any heat transfer into or away from the container.

37. A solar collector 1 m^2 in area collects radiation from the sun and focuses it on 250 g of water initially at 23°C. The average thermal energy arriving from the sun at the surface of the earth at this location is 550 W/m^2, and we assume that this is collected with 100% efficiency. Find the time required for the collector to raise the temperature of the water to 100°C.

38. The bottom of a copper kettle has a 10-cm radius and is 2 mm thick. The temperature of the outside surface is 102°C, and the water inside the kettle is boiling at 1 atm of pressure. Find the rate at which heat is being transferred through the bottom of the kettle.

39. Lead pellets, each of mass 1 g, are heated to 200°C. How many pellets must be added to 500 g of water initially at 20°C to make the equilibrium temperature 25°C? Neglect any heat transfer to or from the container.

40. A 40-g ice cube floats in 200 g of water in a 100-g copper cup all at a temperature of 0°C. A piece of lead at a temperature of 98°C is dropped into the cup, and the final equilibrium temperature is 12°C. What is the mass of the lead that was transferred?

• **41.** An ice cube of mass 10 g is in a copper cup of mass 200 g. Initially the ice and cup are at a temperature of −20°C. (a) What happens to the mixture when 500 cal of heat are added? (b) Repeat when 5000 cal are added.

• **42.** Calculate the temperature increase in 2 kg of water when it is heated with an 800-W immersion heater for a period of 5 min.

• **43.** A class of 10 students taking an exam has a power output per student of about 200 W. Assume the initial temperature of the room is 20°C and that its dimensions are 6 m by 15 m by 3 m. What will be the temperature of the room at the end of 1 h if all the heat remains in the air in the room and none is added by an outside source? The specific heat of air is 837 J/kg · C°, and its density is about 1.3×10^{-3} g/cm^3.

• **44.** In a foundry operation, molten lead of mass 90 g at 327.3°C is poured into an iron casting block that has a mass of 300 g and is initially at 20°C. What is the equilibrium temperature of the system? Assume no heat losses to the surroundings.

• **45.** An iron plate is held against an iron wheel so that there is a sliding frictional force of 50 N acting between the two pieces of metal. The relative speed at which the two surfaces slide over each other is 40 m/s. (a) Calculate the rate at which mechanical energy is converted to heat. (b) The plate and the wheel have a mass of 5 kg each, and each receives 50% of the frictional heat. If the system is run as described for 10 s and each object is then allowed to reach a uniform internal temperature, what is the resultant temperature increase?

• **46.** An automobile has a mass of 1500 kg, and its aluminum brakes have an overall mass of 60 kg. (a) Assuming that all of the frictional heat produced when the car stops is deposited in the brakes, and neglecting heat transfer, how many times could the car be braked to rest starting from 25 m/s (56 mph) before the brakes would begin to melt? (Assume an initial temperature of 20°C.) (b) Identify some effects that are neglected in part (a) but are likely to be important in a more realistic assessment of the heating of brakes.

• **47.** An aluminum rod, 1 m in length and of cross-sectional area 2 cm^2, is inserted vertically into a thermally insulated vessel containing liquid helium at 4.2 K. The rod is initially at 300 K. If one half of the rod is inserted into the helium, how many liters of helium boil off by the time the inserted half cools to 4.2 K?

• **48.** A "solar cooker" consists of a curved reflecting mirror that focuses sunlight onto the object to be heated (Fig. 12.14). The solar power per unit area reaching the earth at some location is 600 W/m^2, and a small solar cooker has a diameter of 0.5 m. Assuming that 50% of the incident energy is converted into heat energy, how long would it take to evaporate 1 liter of water initially at 20°C? (Neglect the specific heat of the container.)

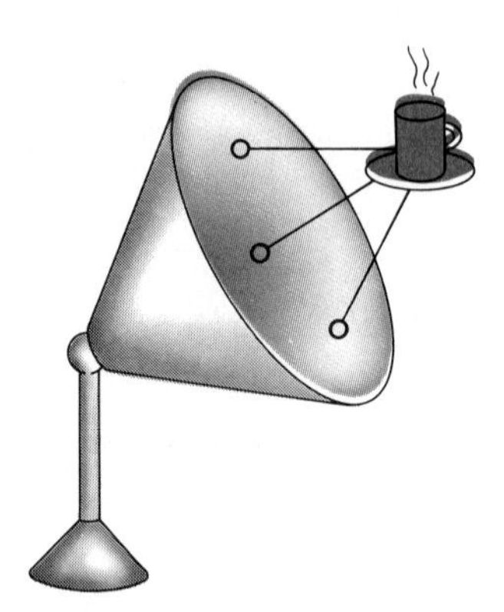

FIGURE 12.14 Problem 48

• **49.** An aluminum rod is 20 cm long at 20°C and has a mass of 350 g. If 10,000 J of heat energy are added to the rod, what is its new length?

•• **50.** An aluminum rod and an iron rod are joined end-to-end in a good thermal contact. The two rods have equal lengths and radii. The free end of the aluminum rod is maintained at a temperature of 100°C, and the free end of the iron rod is maintained at 0°C. (a) Determine the temperature of the interface between the two rods. (b) If each rod is 15 cm long and each has a cross-sectional area of 5 cm^2, what quantity of heat energy is conducted across the combination in 30 min?

•• **51.** A *flow calorimeter* is an apparatus used to measure the specific heat of a liquid. The technique is to measure the temperature difference between the input and output points of a flowing stream of the liquid while adding heat at a known rate. (a) Start with the equation $Q = mc\Delta T$ and $m = \rho V$, and convince yourself that the rate at which heat is added to the liquid is given by $\Delta Q/\Delta t = \rho c\Delta T(\Delta V/\Delta t)$. (b) In one particular experiment, a liquid of density 0.72 g/cm^3 flows through the calorimeter at the rate of 3.5 cm^3/s. At steady state, a temperature difference of 5.8 C° is established between the input and output points when heat is supplied at the rate of 40 J/s. What is the specific heat of the liquid?

•• **52.** Water is being boiled in an open kettle that has a 0.5-cm-thick aluminum bottom in the shape of a circle having a radius of 12 cm. If the water boils away at a rate of 0.5 kg/min, what is the temperature of the lower surface of the bottom of the kettle? Assume the top surface of the bottom of the kettle to be at 100°C.

CHAPTER 13

The Laws of Thermodynamics

Solar cells in an experimental irrigation project. The cells convert the radiant energy of the sun directly into electricity. (Photo, Department of Energy)

We now have a sufficient understanding of heat and its effect on matter to enable us to develop the laws of thermodynamics. The first law of thermodynamics is essentially the law of conservation of energy generalized to include heat as a form of energy. This law tells us that an increase in one form of energy must be accompanied by a decrease in some other form of energy. The first law, which considers both heat and work, places no restrictions on the types of energy conversions that occur. According to the first law, the internal energy of an object, a new concept to be discussed shortly, can be increased either by adding heat to it or by doing work on it.

The second law of thermodynamics, which can be stated in many equivalent ways, establishes which processes in nature can occur and which cannot. For example, the second law tells us that heat never flows spontaneously from a cold body to a hot body. One important application of the second law is in the study of heat engines, such as the internal combustion engine, and the principles that limit their efficiency.

13.1 INTRODUCTION

In discussing thermodynamic processes, we must be careful to first define the system under consideration. A system may be as simple as a flask of gas or as complicated as an internal combustion engine. In general, a *system* is whatever we decide to focus our attention on and the *environment*, or *surroundings*, is whatever interacts with the system. In the large-scale approach to thermodynamics, we describe the state of a system in terms of such quantities as pressure, volume, temperature, and internal energy. The number of such quantities needed to tell us what we need to know about the system depends on the nature of the system. For a simple system, such as a gas containing only one type of molecule, usually only two quantities are needed, such as pressure and volume. In this case, we assume that the gas is in thermal equilibrium internally. That is, we assume that every part of the container is at the same pressure and temperature.

Before we discuss the laws of thermodynamics, we must be sure that we understand three fundamental terms we shall encounter as we proceed: internal energy, work, and heat.

13.2 INTERNAL ENERGY

The **internal energy** of a substance is the total energy of all the atoms or molecules which make up the substance. For the case of an ideal gas, the internal energy is equal to the random translational kinetic energy of the constituent gas atoms. However, in a more complicated gas, the rotational and vibrational motions of the molecules also contribute to the internal energy of the gas (see Fig. 13.1). Atoms or molecules in a solid or a liquid have forces exerted on them by other nearby atoms that are similar to those which would be exerted by small "springs" between the particles. The stretching or compressing of these springs can be characterized by a potential energy which contributes to the internal energy of the system.

As noted above and discussed in Chapter 11, an ideal gas can be considered as a collection of very small particles that do not interact with each other. As a result, the internal energy of an ideal gas arises only from the translational kinetic energy of its constituent particles. Because temperature is related to the average kinetic energy of the particles of an ideal gas, we see that the greater the temperature of a gas, the greater is its internal energy.

It is important to note that energy can be added to a system even when no heat transfer takes place. For example, when two objects are rubbed to-

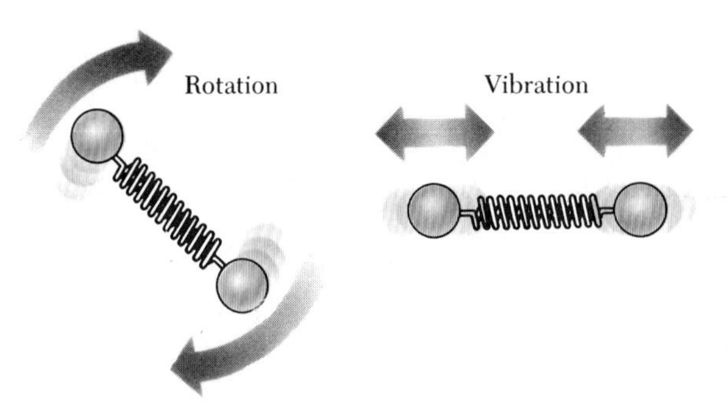

FIGURE 13.1 The rotational and vibrational energy of a diatomic gas contributes to the internal energy of the gas.

gether, their internal energy increases because mechanical work is done on them. The increase in temperature of the objects is an indicator of this increase in internal energy. Likewise, when an object slides across a surface and comes to rest because of frictional forces, the mechanical work done on the system is partially converted to internal energy, as indicated by the fact that the object and the surface both heat up.

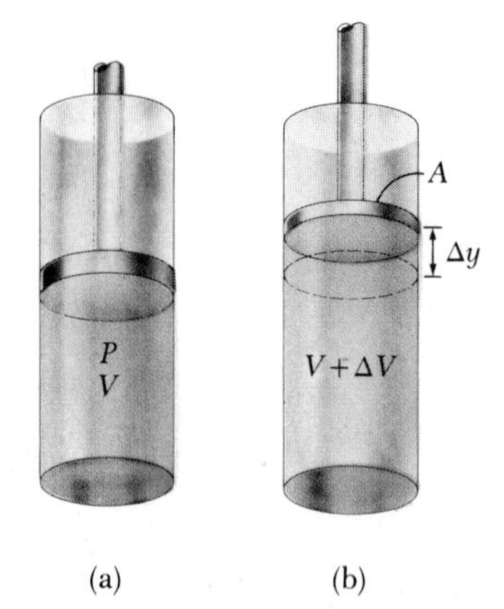

FIGURE 13.2 (a) A gas in a cylinder occupying a volume V and at a pressure P. (b) As the gas expands at constant pressure and the volume increases by an amount ΔV, the work done by the gas is $P\Delta V$.

13.3 WORK

Consider a system made up of a gas contained in a cylinder fitted with a movable piston (Fig. 13.2a). In equilibrium, the gas occupies a volume V and exerts a pressure P on the cylinder walls and piston. If the piston has a cross-sectional area A, the force exerted by the gas on the piston is $F = PA$. Now let us assume that the gas expands very slowly so that it always remains in equilibrium. As the piston moves up a distance Δy, the work done by the gas is

$$W = F\Delta y = PA\Delta y$$

Because $A\Delta y$ is the increase in volume of the gas (that is, $\Delta V = A\Delta y$), we can express the work done as

Work

$$\boxed{W = P\Delta V} \tag{13.1}$$

If the gas expands, as in Figure 13.2b, ΔV is positive and the work done by the gas is positive. If the gas is compressed, ΔV is negative and the work done by the gas is negative. In the latter case, negative work can be interpreted as work being done *on* the system. Clearly, the work done by (or on) the system is zero when the volume remains constant.

You should be aware that Equation 13.1 can be used to calculate the work done on or by the system only when the pressure of the gas remains constant during the expansion or compression. If the pressure changes, the methods of calculus are required to calculate the work done. We shall not attempt such calculations here.

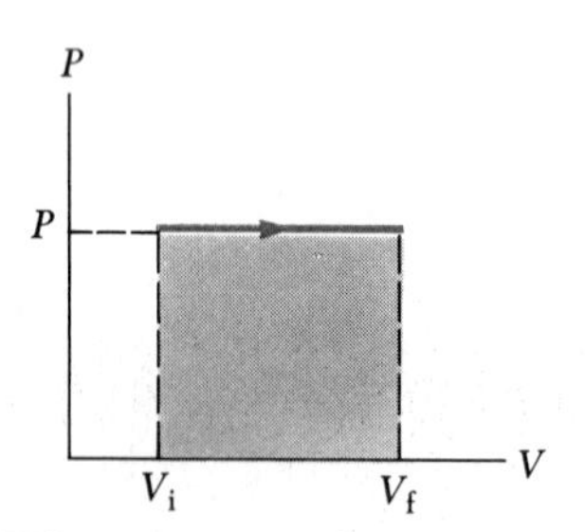

FIGURE 13.3 The PV diagram for a gas expanding at constant pressure. The work done by the gas equals the shaded area.

Consider the process represented by the pressure-volume diagram in Figure 13.3. We see here that the gas has expanded from an initial volume V_i to a final volume V_f at a constant pressure P. From Equation 13.1 we see that the work done by the gas is $P(V_f - V_i)$. Note that this is just the area under the pressure-volume curve.

In general, the work done in an expansion from some initial state to a final state is the area under the curve on a PV diagram.

This statement is true whether or not the pressure remains constant during the process.

As can be seen from Figure 13.4, the amount of work done in the expansion of a gas from an initial state I to a final state F depends on how the gas is allowed to expand. In the process described in Figure 13.4a, the pressure on the gas is first reduced from P_i to P_f by cooling at constant volume V_i and the gas then expands from V_i to V_f at constant pressure P_f. The work done along this path is $P_f(V_f - V_i)$. In Figure 13.4b, the gas first expands from V_i to V_f at constant pressure P_i and then its pressure is reduced to P_f at constant

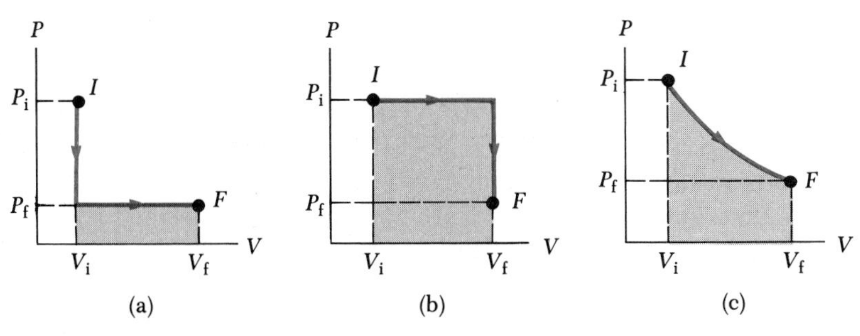

FIGURE 13.4 The work done by a gas as it is taken from an initial state to a final state depends on the path.

volume V_f. The work done along this path is $P_i(V_f - V_i)$, which is greater than that for the process described in Figure 13.4a. Finally, for the process described in Figure 13.4c, where both P and V change continuously, the work done has some value intermediate between the values obtained in the first two processes. Therefore, we see that

the amount of work done by a system depends on how the system goes from the initial to the final state. In other words, the work depends on the initial, final, and intermediate states of the system.

EXAMPLE 13.1 Work Done by an Expanding Gas

In the system shown in Figure 13.2, the gas in the cylinder is at a pressure of 8000 Pa and the piston has an area of 0.1 m². As heat is slowly added to the gas, the piston is pushed up a distance of 4 cm. Calculate the work done on the surroundings by the expanding gas. Assume the pressure remains constant.

Solution The change in volume of the gas is

$$\Delta V = A\Delta y = (0.1 \text{ m}^2)(4 \times 10^{-2} \text{ m}) = 4 \times 10^{-3} \text{ m}^3$$

and from Equation 13.1, the work done is

$$W = P\Delta V = (8000 \text{ Pa})(4 \times 10^{-3} \text{ m}^3) = 32 \text{ J}$$

EXAMPLE 13.2 Work Done Equals the Area Under the *PV* Curve

Find the work done during the thermodynamic process shown on the *PV* diagram of Figure 13.5.

Solution The work done during this process is equal to the area under the *PV* curve. This can be found by breaking the total area into two parts, one the rectangle outlined by the gray shaded area and the second the triangle indicated by the color-shaded regions.

The area of the rectangle (gray area) is

$$A_r = (1 \times 10^4 \text{ N/m}^2)[(10 \times 10^{-3} \text{ m}^3) - (5 \times 10^{-3} \text{ m}^3)] = 50.0 \text{ J}$$

and the area of the triangle (colored area) is

$$A_t = \frac{1}{2}(\text{base} \times \text{height})$$

$$= \frac{1}{2}[(10 \times 10^{-3} \text{ m}^3) - (7.5 \times 10^{-3} \text{ m}^3)][(1.5 \times 10^4 \text{ Pa}) - (1 \times 10^4 \text{ Pa})]$$

$$= 6.3 \text{ J}$$

Thus the total work done, which is the area under the curve, is

$$W = 50.0 \text{ J} + 6.3 \text{ J} = 56.3 \text{ J}$$

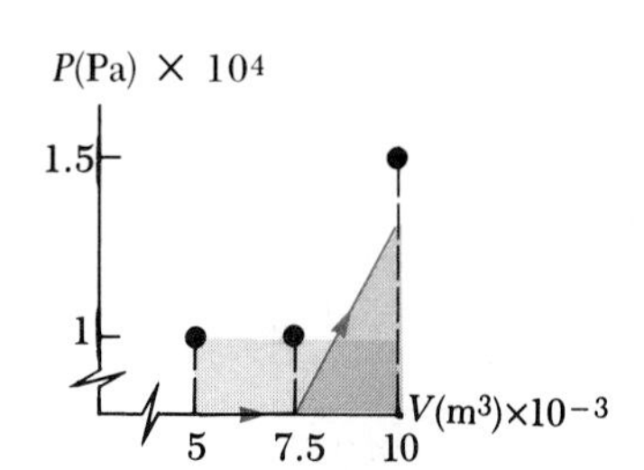

FIGURE 13.5 (Example 13.2) The work done equals the area under the *PV* curve.

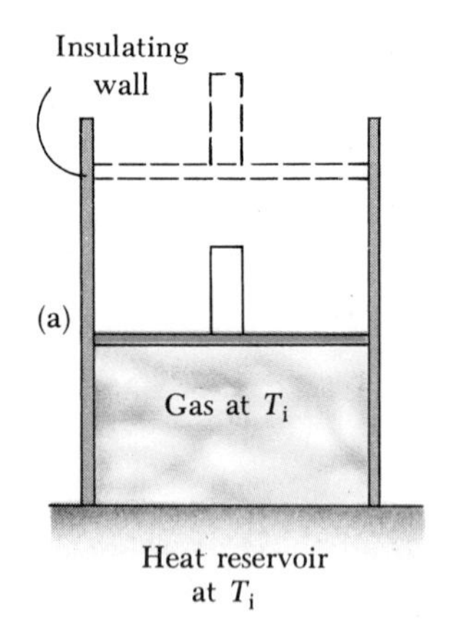

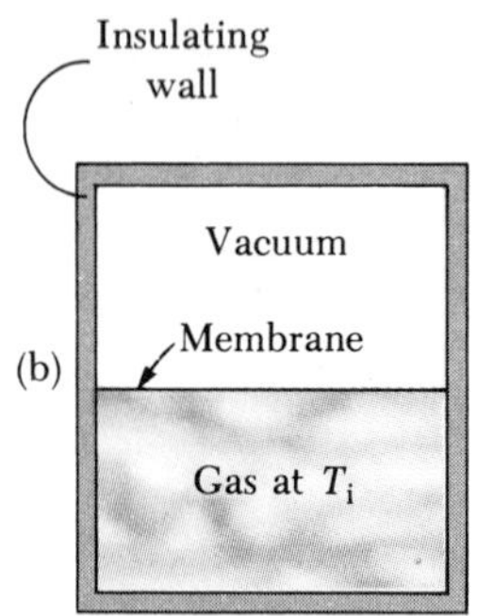

FIGURE 13.6 (a) A gas at temperature T_i expands slowly by absorbing heat from a reservoir at the same temperature. (b) A gas expands rapidly into an evacuated region after breaking a membrane separating the two regions.

13.4 HEAT

We found in Section 13.3 that the amount of work done on or by a system depends on the sequence used to take the system from the initial to the final state. In a similar manner, the heat transferred into or out of a system is also dependent on the sequence used. This can be demonstrated by considering the situations described in Figure 13.6. In each case, the gas has the same initial volume, temperature, and pressure and is assumed to be ideal. In Figure 13.6a, the gas is in thermal contact with a **heat reservoir,** which is *a source that is assumed to be capable of supplying large amounts of energy to a system without undergoing a temperature change*. For example, if one drops an ice cube into the ocean, the ocean can be considered a heat reservoir. The change in temperature of the ocean as it melts the cube is zero for all practical purposes since it has such an enormous capacity.

If the pressure of the gas in Figure 13.6a is slightly greater than atmospheric pressure, the gas will expand and cause the piston to rise. During this expansion to some final volume V_f, sufficient heat to maintain a constant temperature T_i will be transferred from the reservoir to the gas.

Now consider the thermally insulated system shown in Figure 13.6b. When the membrane is broken, the gas expands rapidly into the vacuum until it occupies a volume V_f. In this case, the gas does no work because there is no movable piston. Furthermore, no heat is transferred through the thermally insulated wall, which we call an *adiabatic* wall. (Any process in which no heat enters or leaves the system is called an **adiabatic process.**) The process illustrated in Figure 13.6b is referred to as an adiabatic free expansion, or simply a **free expansion.**

The important point to note here is that the initial values of the pressure, volume, and temperature are the same in both cases; likewise, the final values of the pressure, volume, and temperature are the same in both cases. However, in the first case heat was added to the system, while in the second case there was no heat transferred. Therefore, we conclude that

the amount of heat added or lost by a system depends on the initial, final, and intermediate states of the system.

13.5 THE FIRST LAW OF THERMODYNAMICS

When the law of conservation of energy was first introduced in Chapter 5, it was stated that the mechanical energy of a system is conserved in the absence of nonconservative forces, such as friction. Changes in the internal energy of the system were not included in our model at that time.

The first law of thermodynamics is a generalization of the law of conservation of mechanical energy that includes possible changes in internal energy. It is a universally valid law that can be applied to all kinds of processes.

We have seen that energy can be transferred between a system and its surroundings in two ways. One is *work done* by (or on) the system; the other is *heat transfer*. When energy has been transferred by either of these methods, we say that the system has undergone a **change in internal energy.**

Before we attempt to put this idea on a mathematical basis, let us explain the sign conventions for heat and work. We use the convention that Q is considered to be *positive* if heat is *added* to a system and *negative* if the system *loses* heat. For work, the work done is *positive* if the system does work *on its surroundings* and *negative* if work is done *on the system.*

Consider a gas that undergoes a pressure and volume change from P_i, V_i to P_f, V_f. *Regardless of the process* by which the system changes from the initial to the final state, it is found that the quantity $Q - W$ is *always the same*. Thus, we conclude that the quantity $Q - W$ is determined completely by the initial and final states of the system.

The quantity $Q - W$ is defined as **the change in the internal energy of the system.**

This definition has been substantiated experimentally. It is very important to note that, although both Q and W depend on the sequence used to change from the initial state to the final state, $Q - W$, the change in internal energy, does not. If we represent the internal energy by the letter U, then the change in internal energy, $\Delta U = U_f - U_i$, can be expressed as

$$\boxed{\Delta U = U_f - U_i = Q - W} \tag{13.2}$$

The first law of thermodynamics

where all the quantities must have the same units of energy. Equation 13.2 is known as the **first law of thermodynamics.** Again, note that this law is a generalization of the law of conservation of energy.

On the atomic level, the internal energy of a system includes the kinetic and potential energies of the molecules making up the system. In thermodynamics, we do not concern ourselves with the specific form of the internal energy. We simply use Equation 13.2 as a definition of the change in internal energy. One can make an analogy here between the potential energy associated with an object moving under the influence of gravity without friction. The potential energy is independent of the path, and it is only its change that is of concern. Likewise, the change in internal energy of a thermodynamic system is what matters because only differences in internal energy are important.

The following examples illustrate two important thermodynamic processes: isobaric (constant pressure) and isovolumetric (constant volume).

EXAMPLE 13.3 An Isobaric Process

A gas is enclosed in a container fitted with a piston of cross-sectional area 0.1 m^2. When the pressure of the gas is maintained at 8000 Pa while heat is slowly added, the piston is pushed up a distance of 4 cm. If 42 J of heat is added to the system during the expansion, what is the internal energy change of the system?

Solution Any process in which the pressure remains constant is called an **isobaric process.** The work done by the gas is

$$W = P\Delta V = (8000 \text{ Pa})(0.1 \text{ m}^2)(4 \times 10^{-2} \text{ m}) = 32 \text{ N} \cdot \text{m} = 32 \text{ J}$$

The change in internal energy is found from the first law:

$$\Delta U = Q - W = 42 \text{ J} - 32 \text{ J} = 10 \text{ J}$$

Note that *in an isobaric process the work done and the heat transferred are both nonzero.*

EXERCISE If 42 J of heat is added to the system with the piston clamped in a *fixed* position, what is the work done by the gas? What is the change in its internal energy?

Answer No work is done; $\Delta U = 42$ J.

EXAMPLE 13.4 An Isovolumetric Process

Water of mass 2 kg is held at constant volume in a container while 10,000 J of heat is slowly added by a flame. The container is not well insulated, and as a result 2000 J of heat leaks out to the surroundings. What is the temperature increase of the water?

Solution A process that takes place at constant volume is called an **isovolumetric process.** In such a process the work is clearly equal to zero. Thus, the first law reduces to

$$\Delta U = Q$$

This indicates that the net heat added to the water goes into increasing the internal energy of the water. The net heat added to the water is

$$Q = (10{,}000 \text{ J} - 2000 \text{ J}) = 8000 \text{ J}$$

The temperature increase of the water is

$$Q = mc\Delta T$$

$$\Delta T = \frac{Q}{mc} = \frac{8000 \text{ J}}{(2 \text{ kg})(4.18 \times 10^3 \text{ J/kg}\cdot\text{C}^\circ)} = 0.96 \text{ C}^\circ$$

EXAMPLE 13.5 Boiling Water

One gram of water occupies a volume of 1 cm^3 at atmospheric pressure (1.013×10^5 Pa). When this water is boiled, it becomes 1671 cm^3 of steam. Calculate the change in internal energy for this process.

Solution Since the heat of vaporization of water is 2.26×10^6 J/kg at atmospheric pressure, the heat required to boil 1 g is

$$Q = mL_v = (10^{-3} \text{ kg})(2.26 \times 10^6 \text{ J/kg}) = 2260 \text{ J}$$

The work done by the system is positive and equal to

$$W = P(V_{\text{steam}} - V_{\text{water}})$$
$$= (1.013 \times 10^5 \text{ Pa})[(1671 - 1) \times 10^{-6} \text{ m}^3] = 169 \text{ J}$$

Hence, the change in internal energy is

$$\Delta U = Q - W = 2260 \text{ J} - 169 \text{ J} = 2090 \text{ J}$$

The positive ΔU tells us that the internal energy of the system has increased. We see that most of the heat (93%) transferred to the liquid goes into increasing the internal energy. Only a small fraction (7%) goes into external work.

EXAMPLE 13.6 Heat Transferred to a Solid

The internal energy of a solid also increases when heat is transferred to it from its surroundings. To illustrate this point, suppose a 1-kg bar of copper is heated at atmo-

spheric pressure. If its temperature increases from 20°C to 50°C, (a) find the work done by the copper.

Solution The change in volume of the copper can be calculated using Equation 11.6 and the volume expansion coefficient for copper taken from Table 11.1 (remembering that $\beta = 3\alpha$):

$$\Delta V = \beta V \Delta T = [5.1 \times 10^{-5}\,(\text{C}°)^{-1}](50°\text{C} - 20°\text{C})V = (1.5 \times 10^{-3})V$$

The volume is equal to m/ρ, and the density of copper is 8.92×10^3 kg/m^3. Hence,

$$\Delta V = (1.5 \times 10^{-3})\left(\frac{m}{\rho}\right) = (1.5 \times 10^{-3})\left(\frac{1\text{ kg}}{8.92 \times 10^3\text{ kg/m}^3}\right)$$
$$= 1.7 \times 10^{-7}\text{ m}^3$$

Since the expansion takes place at constant pressure, the work done is

$$W = P\Delta V = (1.013 \times 10^5\text{ Pa})(1.7 \times 10^{-7}\text{ m}^3) = 1.9 \times 10^{-2}\text{ J}$$

(b) What quantity of heat is transferred to the copper?

Solution The specific heat of copper is given in Table 12.1, and from Equation 12.3 we find that the heat transferred is

$$Q = mc\Delta T = (1\text{ kg})(387\text{ J/kg}\cdot\text{C}°)(30\text{ C}°)$$
$$= 1.16 \times 10^4\text{ J}$$

EXERCISE What is the increase in internal energy of the copper?

Answer $\Delta U = 1.16 \times 10^4$ J.

13.6 ISOTHERMAL EXPANSION OF AN IDEAL GAS

A process in which an ideal gas expands (or is compressed) at a constant temperature is of such importance that we shall consider it separately from the processes described earlier. An expansion or compression of a substance at constant temperature is referred to as an **isothermal process.** The isothermal expansion of a gas can be achieved by placing the gas in thermal contact with a heat reservoir at the same temperature, as in Figure 13.6a.

The plot of the pressure of the gas versus its volume for an isothermal process is illustrated by the *PV* diagram shown in Figure 13.7. The work done during this isothermal expansion cannot easily be calculated because the pressure does not remain constant during the process. In fact, one must use the methods of integral calculus to calculate the work done in this situation. The result, which we state without proof, is given by the expression

$$W = nRT \ln\left(\frac{V_f}{V_i}\right) \tag{13.3}$$

where n is the number of moles of gas, R is the ideal gas constant, T is the kelvin temperature, and ln is the abbreviation for the natural logarithm function. Numerically, this work equals the shaded area under the *PV* curve in Figure 13.7. If the gas *expands* isothermally, then $V_f > V_i$ and the work calculated using Equation 13.3 is positive, as we would expect based on our previous sign convention. If the gas *compressed* isothermally, then $V_f < V_i$ and the work is found to be negative.

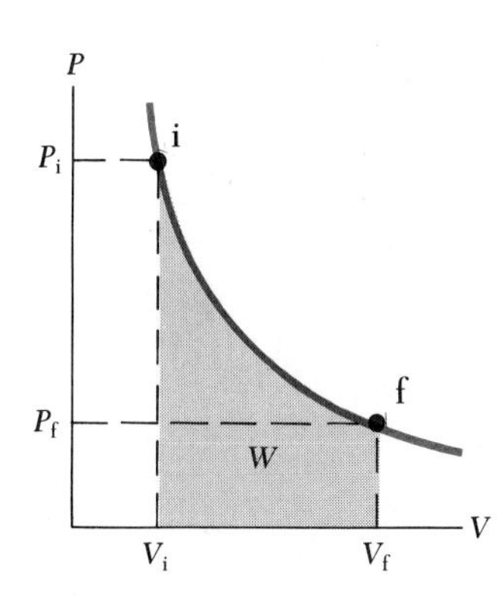

FIGURE 13.7 The *PV* diagram for an isothermal expansion of an ideal gas from an initial state to a final state.

As noted earlier, the internal energy of an ideal gas depends only on temperature. Hence, for an isothermal process $\Delta U = 0$. Thus, from the first law we have

$$\Delta U = 0 = Q - W$$

or

$$Q = W \quad \text{(Isothermal process)}$$

According to this result, we conclude that the heat that must be added to the gas to allow it to expand at constant temperature is equal to the work done by the gas.

EXAMPLE 13.7 Work Done During an Isothermal Expansion

Calculate the work done by 1 mol of an ideal gas that is kept at 0°C in an expansion from 3 liters to 10 liters.

Solution Substituting these values into Equation 13.3 gives

$$W = nRT \ln\left(\frac{V_f}{V_i}\right)$$

$$= (1 \text{ mol})(8.31 \text{ J/mol} \cdot \text{K})(273 \text{ K}) \ln\left(\frac{10}{3}\right)$$

$$= 2.73 \times 10^3 \text{ J}$$

The heat that must be supplied to the gas from the reservoir to keep T constant is also 2.73×10^3 J.

EXERCISE Calculate the work done if the gas above had been kept at 0°C and compressed from a volume of 10 liters to 3 liters.

Answer -2.73×10^3 J

13.7 HEAT ENGINES AND THE SECOND LAW OF THERMODYNAMICS

One of the oldest applications of thermodynamics, and one of importance today, is the study of heat engines. A **heat engine** is a device that converts thermal energy to other useful forms of energy, such as mechanical and electrical energy. More specifically, a heat engine can be represented as a device that carries a substance through a cycle during which (1) heat is absorbed from a source at a high temperature, (2) work is done by the engine, and (3) heat is expelled by the engine to a source at a lower temperature. In a typical process for producing electricity in a power plant, coal or some other fuel is burned and the heat produced is used to convert water to steam. This steam is then directed at the blades of a turbine, setting it into rotation. Finally, the mechanical energy associated with this rotation is used to drive an electric generator. The internal combustion engine in your automobile extracts heat from a burning fuel and converts a fraction of this energy to mechanical energy.

As was mentioned above, a heat engine carries some working substance through a cyclic process, defined as one in which the substance eventually returns to its initial state. As an example of a cyclic process, consider the operation of a steam engine in which the working substance is water. The water is carried through a cycle in which it first evaporates into steam in a boiler and then expands against a piston. After the steam is condensed with cooling water, it is returned to the boiler and the process is repeated.

It is useful to represent a heat engine schematically as in Figure 13.8. The engine, represented by the circle at the center, absorbs a quantity of heat Q_h from the heat reservoir at temperature T_h. It does work W and gives up heat Q_c to another heat reservoir at temperature T_c, where $T_c < T_h$. Because the working substance goes through a cycle in which it ends up in exactly the same state as it was when it entered the cycle, it follows that $\Delta U = 0$. Hence, from the first law of thermodynamics we see that *the net work done by the engine equals the net heat flowing into it*. As we can see from Figure 13.8, $Q_{net} = Q_h - Q_c$ and therefore

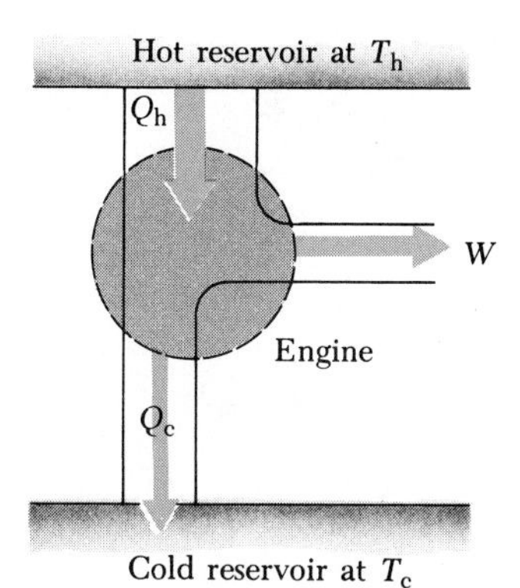

FIGURE 13.8 Schematic representation of a heat engine. The engine (in the circular area) absorbs heat Q_h from the hot reservoir, expels heat Q_c to the cold reservoir, and does work W.

$$W = Q_h - Q_c \tag{13.4}$$

where Q_h and Q_c are taken to be positive quantities. If the working substance is a gas, *the net work done for a cyclic process is the area enclosed by the curve representing the process on a PV diagram.* This is shown for an arbitrary cyclic process in Figure 13.9.

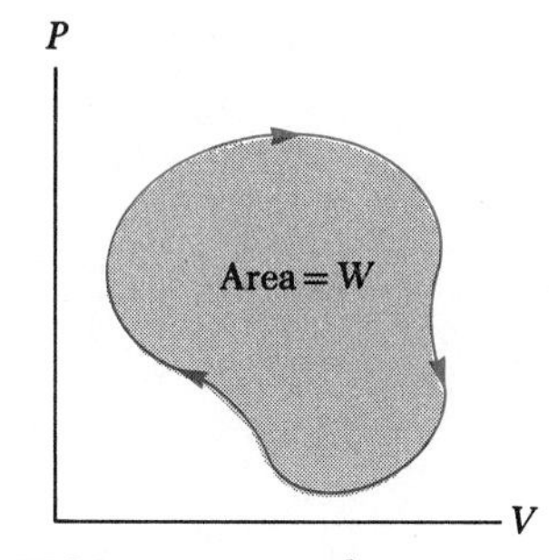

FIGURE 13.9 The *PV* diagram for an arbitrary cyclic process. The net work done in each cycle equals the area enclosed by the curve.

$$Eff \equiv \frac{W}{Q_h} = \frac{Q_h - Q_c}{Q_h} = 1 - \frac{Q_c}{Q_h} \tag{13.5}$$

We can think of the efficiency as the ratio of what you get (mechanical work) to what you pay for (energy). Equation 13.5 shows that a heat engine has 100% efficiency ($Eff = 1$) only if $Q_c = 0$, that is, if no heat is expelled to the cold reservoir. In other words, a heat engine with perfect efficiency would have to convert all of the absorbed heat energy to mechanical work. One of the consequences of the second law of thermodynamics is that this is impossible. The **second law of thermodynamics** can be stated as follows:

Second law of thermodynamics

It is impossible to construct a heat engine that, operating in a cycle, produces no other effect than the absorption of heat from a reservoir and the performance of an equal amount of work.

This form of the second law is useful in understanding the operation of heat engines. With reference to Equation 13.5, the second law says that, during the operation of a heat engine, W can never be equal to Q_h or, alternatively, that some heat, Q_c, must be rejected to the environment. As a result, it is theoretically impossible to construct an engine that will work with 100% efficiency. The net result of this assessment is that the first law says *we cannot get more work out of a process than the amount of energy we put in* and the second law says that *we cannot break even!*

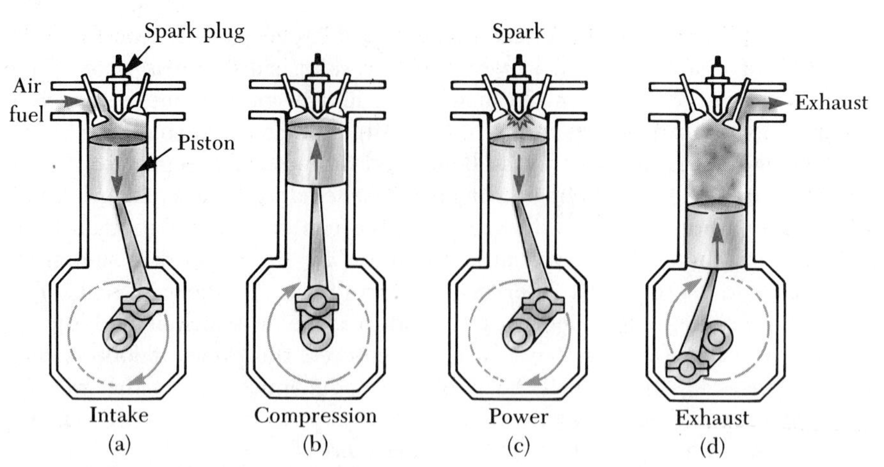

FIGURE 13.10 The four-stroke cycle of a conventional gasoline engine. (a) In the intake stroke, air is mixed with fuel. (b) The intake valve is then closed, and the air-fuel mixture is compressed by the piston. (c) The mixture is ignited by the spark plug in the power stroke. (d) Finally, the residual gases are expelled.

13.8 THE GASOLINE ENGINE

In this section we shall discuss the four-stroke gasoline engine. Four successive processes occur in each cycle, as illustrated in Figure 13.10. During the intake stroke of the piston shown in Figure 13.10a, the valve on the left opens and air that has been mixed with gasoline vapor in the carburetor is drawn into the cylinder. During the compression stroke (Fig. 13.10b) both valves are closed and the air-fuel mixture is compressed approximately adiabatically. At this point a spark from the sparkplug ignites the air-fuel mixture, causing a rapid increase in pressure and temperature at nearly constant volume. The burning gases expand approximately adiabatically and force the piston back, which produces the power stroke (Fig. 13.10c). Finally, during the exhaust stroke (Fig. 13.10d), the exhaust valve on the right is opened and the rising piston forces most of the remaining gas out of the cylinder. The cycle is repeated after the exhaust valve is closed and the intake valve is opened.

These four processes can be approximated by the *Otto cycle,* a *PV* diagram of which is illustrated in Figure 13.11:

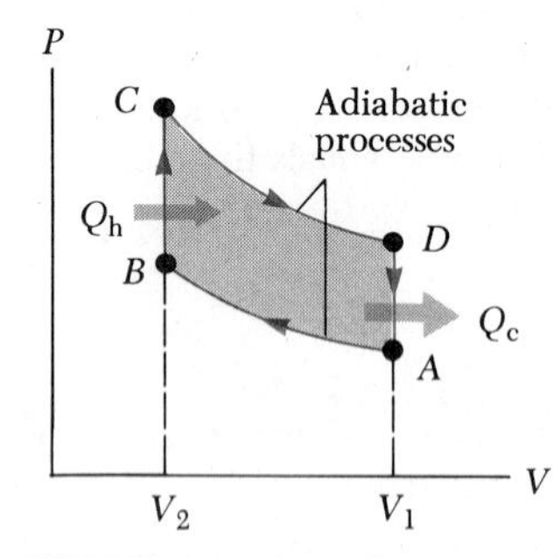

FIGURE 13.11 The *PV* diagram for the Otto cycle, which approximately represents the processes in the internal combustion engine. No heat is transferred during the adiabatic processes $A \rightarrow B$ and $C \rightarrow D$.

1. In the process $A \rightarrow B$ (compression stroke), the air-fuel mixture is compressed adiabatically from volume V_1 to V_2 and the temperature increases from T_A to T_B. The work done on the gas is the area under the curve *AB*.
2. In the process $B \rightarrow C$, combustion occurs and heat Q_h is added to the gas. During this time the pressure and temperature rise rapidly but the volume remains approximately constant. No work is done on the gas.
3. In the process $C \rightarrow D$ (power stroke, corresponding to Figure 13.10c), the gas expands adiabatically from V_2 to V_1, causing the temperature to drop from T_C to T_D. The work done by the gas equals the area under the curve *CD*.
4. In the final process, $D \rightarrow A$ (exhaust stroke, Figure 13.10d), heat Q_c is extracted from the gas as its pressure decreases at constant volume. (Hot

gas is replaced by cool gas, which is equivalent to taking the original substance through a cycle.) No work is done during this process.

EXAMPLE 13.8 The Efficiency of an Engine

Find the efficiency of an engine that introduces 2000 J of heat during the combustion phase and loses 1500 J at exhaust and through friction.

Solution The efficiency of the engine is given by Equation 13.5 as

$$Eff = 1 - \frac{Q_c}{Q_h} = 1 - \frac{1500\ \text{J}}{2000\ \text{J}} = 0.25, \text{ or } 25\%$$

EXERCISE If an engine has an efficiency of 20% and loses 3000 J through friction, how much work is done by the engine?

Answer 750 J.

EXAMPLE 13.9 The Efficiency of a Gasoline Engine

A gasoline engine is represented by the Otto cycle of Figure 13.12. If the input heat energy is 3000 J, estimate the efficiency of the engine.

Solution If we knew the work output of the engine, we could calculate its efficiency from $Eff = W/Q_h$. We can estimate the work output by recalling that the work done on or by a system is equal to the area under a *PV* curve. For the process pictured in Figure 13.12, work is done by the air-fuel mixture between points *C* and *D* and work is done on the mixture between points *A* and *B*. Thus, the net work is the area between the two curves, shown in color in the diagram. This is an irregular curve, and no simple equations enable us to calculate the area. We can estimate it, however, by finding the area of the rectangle shown in the figure:

$$\begin{aligned} A_r &= [(30 \times 10^5) - (1 \times 10^5)][(800 \times 10^{-6}) - (100 \times 10^{-6})]\ \text{N} \cdot \text{m} \\ &= 2030\ \text{J} \end{aligned}$$

In this example, one's judgment in estimation comes into play. Let us estimate that one half of the rectangular area is occupied by the Otto cycle. This estimate gives us the work done during the Otto cycle as

$$W = \frac{1}{2} A_r = \frac{1}{2}(2030\ \text{J}) = 1015\ \text{J}$$

and the efficiency is

$$Eff = \frac{W}{Q_h} = \frac{1015\ \text{J}}{3000\ \text{J}} = 0.34, \text{ or } 34\%$$

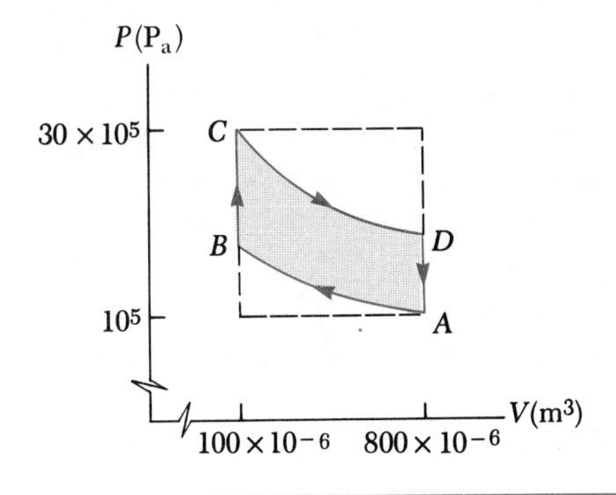

FIGURE 13.12 (Example 13.9) The *PV* diagram for a gasoline engine.

13.9 REVERSIBLE AND IRREVERSIBLE PROCESSES

In the next section we shall discuss a theoretical engine that is the most efficient engine possible. In order to understand the nature of that engine, we must first examine the meaning of reversible and irreversible processes.

A **reversible** process is one that can be performed such that, at its conclusion, both the system and its surroundings have been returned to their exact initial conditions. A process that does not satisfy these requirements is **irreversible.**

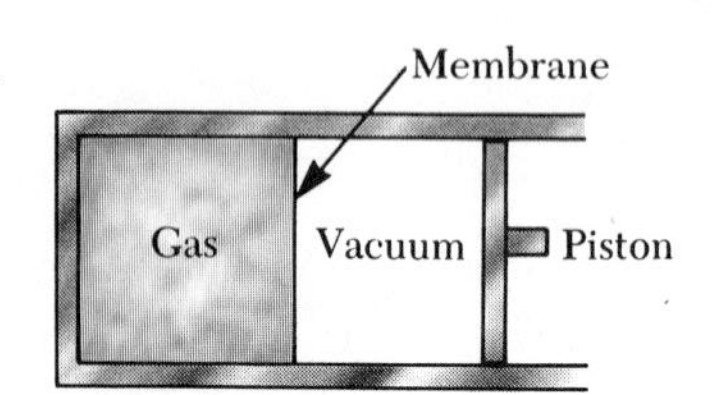

FIGURE 13.13 Free expansion of a gas.

All natural processes are known to be irreversible. From the endless number of examples that could be selected, let us examine the free expansion of a gas, discussed earlier in Section 13.4, and show that it cannot be reversible. A gas is contained in an insulated container, as shown in Figure 13.13, with a membrane separating the gas from a vacuum. If the membrane is punctured, the gas expands freely into the vacuum. Because the gas does not exert a force on anything, no work is done by the gas during the free expansion. In addition, no heat is transferred to or from the gas since the container is insulated from its surroundings. Thus, in this process, the system has changed, but the surroundings have not. Now imagine that we try to reverse the process by first compressing the gas to its original volume. To do this, let us imagine an engine being used to force the piston inward. Note, however, that in performing this task we are changing both the system and surroundings. The surroundings are changing because work is being done by an outside agent on the system, and the system is changing because the compression is increasing the temperature of the gas. We can lower the temperature of the gas by allowing it to come into contact with an external reservoir. Although this second procedure would return the gas to its original state, the surroundings would again be affected because heat would be added to the surroundings. If this heat could somehow be used to drive the engine and compress the gas, the system and its surroundings could be returned to their initial state. However, our statement of the second law says that this extracted heat cannot be completely converted into work. We must conclude that a reversible process has not occurred.

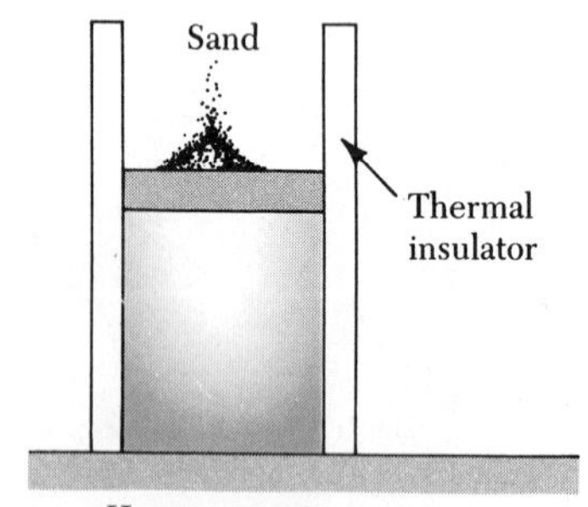

FIGURE 13.14 A gas in thermal contact with a heat reservoir is compressed slowly by dropping grains of sand onto a piston. The compression is isothermal and reversible.

Although real processes are always irreversible, there are some processes that are almost reversible. If a real process occurs very slowly so that the system is virtually always in equilibrium, the process can be considered reversible. For example, imagine compressing a gas very slowly by dropping some grains of sand onto a frictionless piston as in Figure 13.14. The compression process is made isothermal by placing the gas in thermal contact with a heat reservoir. The pressure, volume, and temperature of the gas are well defined during this isothermal compression. Each added grain of sand represents a change to a new equilibrium state. The process can be reversed by slowly removing grains of sand from the piston.

A general characteristic of a reversible process is that there can be no dissipative effects present such as turbulence or friction that produce heat. In reality, such effects are impossible to eliminate completely, hence it is not surprising that real processes in nature are irreversible.

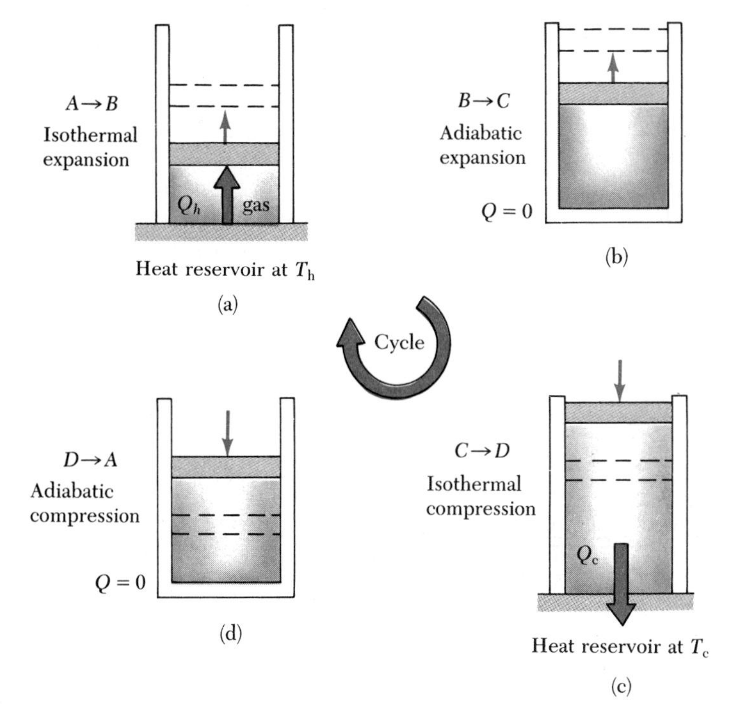

FIGURE 13.15 The Carnot cycle. In process $A \rightarrow B$, the gas expands isothermally while in contact with a reservoir at T_h. In process $B \rightarrow C$, the gas expands adiabatically ($Q = 0$). In process $C \rightarrow D$, the gas is compressed isothermally while in contact with a reservoir at $T_c < T_h$. In process $D \rightarrow A$, the gas is compressed adiabatically. The upward arrows on the piston indicate sand being removed during the expansions, and the downward arrows indicate the addition of sand during the compressions.

Read Over

13.10 THE CARNOT ENGINE

In 1824 a French engineer named Sadi Carnot (1796–1832) described a theoretical engine, now called a *Carnot engine,* that is of great importance from both a practical and a theoretical viewpoint. He showed that a heat engine operating in an ideal, reversible cycle, called a Carnot cycle, between two heat reservoirs is the most efficient engine possible. Such an ideal engine establishes an upper limit on the efficiencies of all engines. *That is, the net work done by a working substance taken through the Carnot cycle is the largest amount of work possible for a given amount of heat supplied to the substance.*

To describe the Carnot cycle, we shall assume that the substance working between temperatures T_c and T_h is an ideal gas contained in a cylinder with a movable piston at one end. The cylinder walls and the piston are thermally nonconducting. Four stages of the Carnot cycle are shown in Figure 13.15, and the *PV* diagram for the cycle is shown in Figure 13.16. The Carnot cycle consists of two adiabatic and two isothermal processes, all reversible.

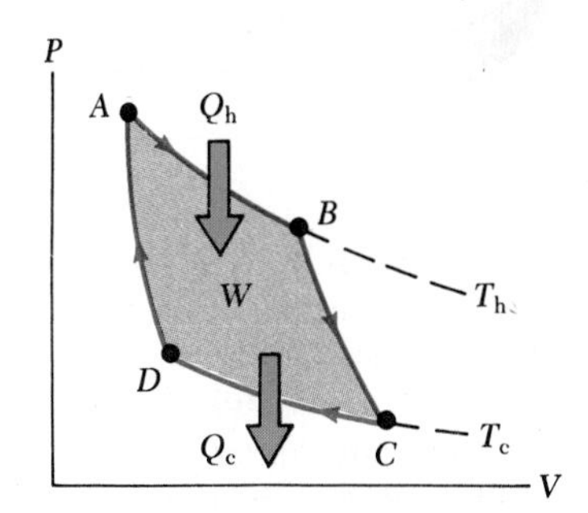

FIGURE 13.16 The *PV* diagram for the Carnot cycle.

1. The process $A \rightarrow B$ is an isothermal expansion at temperature T_h, in which the gas is placed in thermal contact with a heat reservoir at temperature T_h (Fig. 13.15a). During the process, the gas absorbs heat Q_h from the reservoir and does work W_{AB} in raising the piston.

2. In the process $B \rightarrow C$, the base of the cylinder is replaced by a thermally nonconducting wall and the gas expands adiabatically, that is, no heat enters or leaves the system (Fig. 13.15b). During the process, the temperature falls from T_h to T_c and the gas does work W_{BC} in raising the piston.
3. In the process $C \rightarrow D$, the gas is placed in thermal contact with a heat reservoir at temperature T_c (Fig. 13.15c) and is compressed isothermally at temperature T_c. During this time, the gas expels heat Q_c to the reservoir and the work done on the gas is W_{CD}.
4. In the final stage, $D \rightarrow A$, the base of the cylinder is replaced by a nonconducting wall (Fig. 13.15d) and the gas is compressed adiabatically. The temperature of the gas increases to T_h and the work done on the gas is W_{DA}.

Carnot showed that the thermal efficiency of a Carnot engine is given by

Carnot efficiency

$$Eff_C = \frac{T_h - T_c}{T_h} = 1 - \frac{T_c}{T_h} \tag{13.6}$$

It is important to note that in this equation, T must be in kelvins. From this result, we see that
all Carnot engines operating between the same two temperatures have the same efficiency. Furthermore, the efficiency of a Carnot engine operating between two temperatures is greater than the efficiency of any real engine operating between the same two temperatures.

Equation 13.6 can be applied to any working substance operating in a Carnot cycle between two heat reservoirs. According to this result, the efficiency is zero if $T_c = T_h$, as one would expect. The efficiency increases as T_c is lowered and as T_h is increased. However, the efficiency can be unity (100%) only if $T_c = 0$ K. Such reservoirs are not available, and so the maximum efficiency is always less than unity. In most practical cases, the cold reservoir is near room temperature, about 300 K. Therefore, one usually strives to increase the efficiency by raising the temperature of the hot reservoir. *All real engines are less efficient than the Carnot engine because they are all subject to such practical difficulties as friction and heat losses by conduction.*

Sadi Carnot (1796–1832). "The steam engine works our mines, impels our ships, excavates our ports and our rivers, forges iron. . . . Notwithstanding the work of all kinds done by steam engines, notwithstanding the satisfactory condition to which they have brought today, their theory is very little understood."

EXAMPLE 13.10 The Steam Engine

A steam engine has a boiler that operates at 500 K. The heat changes water to steam, which drives the piston. The exhaust temperature is that of the outside air, which is about 300 K. What is the maximum thermal efficiency of this steam engine?

Solution From the expression for the efficiency of a Carnot engine, we find the maximum thermal efficiency for any engine operating between these temperatures:

$$Eff_C = \frac{T_h - T_c}{T_h} = \frac{500 \text{ K} - 300 \text{ K}}{500 \text{ K}} = 0.4, \text{ or } 40\%$$

You should note that this is the highest theoretical efficiency of the engine. In practice, the efficiency will be considerably lower.

EXAMPLE 13.11 The Carnot Efficiency

The highest theoretical efficiency of a gasoline engine, based on the Carnot cycle, is 30%. If this engine expels its gases into the atmosphere, which has a temperature of 300 K, what is the temperature in the cylinder immediately after combustion?

Solution The Carnot efficiency is used to find T_h:

$$Eff_C = \frac{T_h - T_c}{T_h} = 1 - \frac{T_c}{T_h}$$

$$T_h = \frac{T_c}{1 - Eff_C} = \frac{300\text{ K}}{1 - 0.3} = 429\text{ K}$$

EXERCISE If the heat engine absorbs 837 J of heat from the hot reservoir during each cycle, how much work can it perform in each cycle?

Answer 251 J.

A **heat pump** is a mechanical device that is finding increased acceptance for use in heating and cooling homes and buildings. In the heating mode, a circulating fluid, such as Freon, absorbs heat from the outside and releases it to the interior of the structure. The Freon is usually in the form of a low pressure vapor when in the coils of a unit exterior to the structure, where it absorbs heat either from the air or the ground. This gas is then compressed and enters the structure as a hot, high-pressure vapor. In an interior unit, the gas condenses to a liquid and releases its stored heat. When the heat pump is used as an air conditioner, the cycle above is reversed.

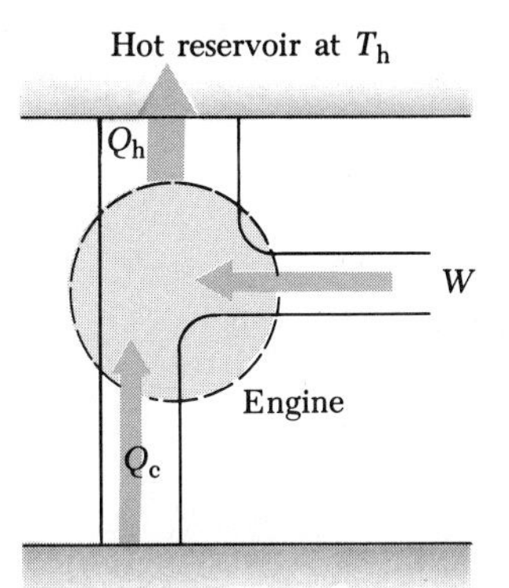

FIGURE 13.17 Schematic diagram of a heat pump, which absorbs heat Q_c from the cold reservoir and expels heat Q_h to the hot reservoir. The work done on the heat pump is W.

Figure 13.17 is a schematic representation of a heat pump used to heat a structure. The outside temperature is T_c, the inside temperature is T_h, and the heat absorbed by the circulating fluid is Q_c. The compressor does work W on the fluid, and the heat transferred from the pump into the structure is Q_h.

The effectiveness of a heat pump is described in terms of a number called the **coefficient of performance,** COP. This is defined as the ratio of the heat transferred into the hot reservoir and the work required to transfer that heat:

$$\text{COP (heat pump)} \equiv \frac{\text{heat transferred}}{\text{work done by pump}} = \frac{Q_h}{W} \qquad (13.7)$$

If the outside temperature is 25°F or higher, the COP for a heat pump is about 4. That is, the heat transferred into the house is about four times greater than the work done by the compressor in the heat pump. However, as the outside temperature decreases, it becomes more difficult for the heat pump to extract sufficient heat from the air and the COP drops. In fact, the COP can fall below unity for temperatures below the midteens.

Although heat pumps used in buildings are relatively new products in the heating and air conditioning field, the refrigerator has been a standard appliance in homes for years. The refrigerator works much like a heat pump, except that it cools its interior by pumping heat from the food storage compartments into the warmer air outside. During its operation, a refrigerator removes a quantity of heat Q_c from the interior of the refrigerator, and in the process its motor does work W. The coefficient of performance of a refrigerator or for a heat pump used in its cooling cycle is given by

$$\text{COP(refrigerator)} = \frac{Q_c}{W}$$

13.12 AN ALTERNATIVE STATEMENT OF THE SECOND LAW

The first law of thermodynamics says that energy is conserved in every process. There is more to the story than that, however, and the second law adds the finishing touches. To illustrate what we mean, suppose you wish to cool off a hot piece of pizza by placing it on a block of ice. You will certainly be successful because in every situation like this we have encountered, heat transfer always takes place from a hot object to a cooler one. If you wait a short while, 500 J of heat will be removed from the pizza and added to the ice. Energy is conserved and the first law is satisfied. Yet, there is nothing inherent in the first law that says that this heat transfer could not proceed in the opposite direction. Imagine your astonishment if some day you place a piece of hot pizza on ice and 500 J of heat moves from the cold ice to the pizza. Although this event will most likely never occur, there is nothing in the first law that says it could not. The second law of thermodynamics determines the direction of such natural phenomena.

In effect, the second law governs the order of the sequence of events that occur during any process. One can make an analogy with the impossible sequence of events seen in a movie film running backwards. Such events might be a person diving out of a swimming pool, an apple rising from the ground and latching onto the branch of a tree, or a pot of hot water becoming colder as it rests over an open flame. Such events running backwards in time are impossible because they violate the second law of thermodynamics. In other words, *real processes have a preferred direction of time,* often called the *arrow of time.*

There are many different statements of the second law, but all can be shown to be equivalent. The particular form of the second law you use depends on the application you have in mind. For example, if you were concerned about the heat transfer between pizza and ice, you might choose to concentrate on the second law in this form:

The second law of thermodynamics

Heat will not flow spontaneously from a cold object to a hot object.

At first glance, this statement of the second law seems to be radically different from the statement given in Section 13.7. They are, in fact, equivalent in all respects. Although we shall not prove it here, one can show that if either statement is false, so is the other.

13.13 ENTROPY

Thus far we have presented two different statements of the second law of thermodynamics. In this section we shall present yet a third statement of this law in terms of an abstract quantity called **entropy**, denoted by the symbol S.

The **change in entropy** of a system, ΔS, is equal to the heat, ΔQ, flowing into the system as the system changes from one state to another divided by the absolute temperature:

$$\Delta S = \frac{\Delta Q}{T} \tag{13.8}$$

This definition assumes that the process occurring is reversible and isothermal.

The concept of entropy was introduced into the study of thermodynamics by Rudolph Clausius in 1865. One reason it became useful and gained wide acceptance is because it provides another variable to describe the state of a system to go along with pressure, volume, and temperature. The concept of entropy reaches a position of even more significance when it was found that

the entropy of the universe increases in all natural processes.

This statement is yet another way of stating the second law of thermodynamics.

The statement of the second law given above must be interpreted with care. Although the statement says that the entropy of the universe always increases in all natural processes, there are processes for which the entropy decreases. That is, there are situations in which the entropy of some system may decrease but that decrease takes place at the expense of a net increase in entropy of some other system. For example, we may find a process for which the entropy of system A is decreasing. However, in all cases we would find that the process produces an increase in entropy of some other system (system B) such that the change in entropy of system B is greater than the change in entropy of system A.

The concept of entropy is satisfying because it enables us to present the second law of thermodynamics in the form of a mathematical statement. In the next section, we will find that entropy can also be interpreted in terms of probabilities, a relationship which has profound implications for our world.

Rudolph Clausius (1822–1888). "I propose . . . to call S the entropy of a body, after the Greek word 'transformation.' I have designedly coined the word 'entropy' to be similar to energy, for these two quantities are so analogous in their physical significance, that an analogy of denominations seems to be helpful." (AIP Niels Bohr Library, Lande Collection)

EXAMPLE 13.12 Melting a Piece of Lead

Calculate the change in entropy when 300 g of lead melts at 327°C (600 K). Lead has a latent heat of fusion of 2.45×10^4 J/kg.

Solution The amount of heat added to the lead to melt it is

$$Q = mL_f = (0.3 \text{ kg})(2.45 \times 10^4 \text{ J/kg}) = 7.35 \times 10^3 \text{ J}$$

The entropy change of the lead, from Equation 13.8, is

$$\Delta S = \frac{Q}{T} = \frac{7.35 \times 10^3 \text{ J}}{600 \text{ K}} = 12.3 \text{ J/K}$$

EXAMPLE 13.13 Which Way Does the Heat Flow?

A large cold object is at 273 K, and a large hot object is at 373 K. Show that it is impossible for a small amount of heat energy, say 8 J, to be transferred from the cold object to the hot object without decreasing the entropy of the universe and hence violating the second law.

Solution We assume here that during the heat transfer the two systems do not undergo a temperature change. This is not a necessary assumption; it is used to avoid using the techniques of integral calculus. The entropy change of the hot object is

$$\Delta S_h = \frac{Q}{T_h} = \frac{8 \text{ J}}{373 \text{ K}} = 0.0214 \text{ J/K}$$

The cold reservoir loses heat, and its entropy change is

$$\Delta S_c = \frac{Q}{T_c} = \frac{-8\text{ J}}{273\text{ K}} = -0.0293\text{ J/K}$$

The net entropy change of the universe is

$$\Delta S_u = \Delta S_c + \Delta S_h = -0.0079\text{ J/K}$$

This is in violation of the concept that the entropy of the universe always increases in natural processes. That is, *the spontaneous transfer of heat from a cold to a hot object cannot occur.*

EXERCISE In the above example, suppose that 8 J of heat is transferred from the hot to the cold object. What would be the net entropy change of the universe?

Answer + 0.0079 J/K.

13.14 ENTROPY AND DISORDER

As you look around at the beauties of nature, it is easy to recognize that the events of natural processes have in them a large element of chance. For example, the spacing between trees in a natural forest is quite random. On the other hand, if you were to discover a forest where all the trees are equally spaced, you would conclude that this was a man-made forest. Likewise, leaves fall to the ground with random arrangements. It would be highly unlikely to find the leaves laid out in perfectly straight rows. We can express the results of such observations by saying that **a disorderly arrangement is much more probable than an orderly one if the laws of nature are allowed to act without interference.**

Entropy originally found its place in thermodynamics, but its importance grew tremendously as the field of statistical mechanics developed. This analytical approach provides an alternative interpretation of entropy. In statistical mechanics, the behavior of a substance is described in terms of the statistical behavior of the atoms and molecules contained in the substance. One of the main results of this treatment is that

isolated systems tend toward disorder and entropy is a measure of that disorder.

In light of this new view of entropy, Boltzmann found that an alternative method for calculating entropy is through use of the important relation

$$S = k \ln W \qquad (13.9)$$

where k is Boltzmann's constant ($k = 1.38 \times 10^{-23}$ J/K) and W is a number proportional to the probability of the occurrence of a particular event.

Let us explore the meaning of this equation by presenting a specific example. Imagine that you have a bag of 100 marbles, where 50 are red and 50 are green. You are allowed to draw four marbles from the bag according to the following rules. Draw one marble, record its color, return it to the bag and draw again. Continue this process until four marbles have been drawn. Note that because each marble is returned to the bag before the next one is

TABLE 13.1 Possible Results of Drawing Four Marbles from a Bag

End Result	Possible Draws	Total Number of Same Results
All R	RRRR	1
1G, 3R	RRRG, RRGR, RGRR, GRRR	4
2G, 2R	RRGG, RGRG, GRRG, RGGR, GRGR, GGRR	6
3G, 1R	GGGR, GGRG, GRGG, RGGG	4
All G	GGGG	1

drawn, the probability of drawing a red marble is always the same as drawing a green one. The results of all the possible drawing sequences that could occur are shown in Table 13.1. For example, the result RRGR means that you drew a red marble on the first draw, a red one on the second, a green one on the third, and a red one on the fourth. This table indicates that there is only one possible way to draw four red marbles. However, there are four possible sequences that could give one green and three red marbles, six sequences that could give two green and two red, four sequences that produce three green and one red, and one sequence that gives all green. Thus, the most likely occurrence is that you will draw two red and two green marbles, which corresponds to the disordered state. There is a much lower probability that you will draw four red or four green marbles, these being the most ordered states. From Equation 13.9, we see that the state with the greatest disorder has the highest entropy because it is the most probable state. On the other hand, the most ordered states (all red or all green) are least likely to occur, and are states of lowest entropy.

In summary of this investigation, we see that the outcome of the draw can range from a highly ordered state (say all red marbles) which has the lowest entropy, to a highly disordered state (two green and two red marbles) which has the highest entropy. Thus, one can regard entropy as an index of how far the system has progressed from an ordered to a disordered state.

It should be noted here that *the second law of thermodynamics is really a statement of what is most probable rather than of what must be.* Let us return to the example of placing an ice cube in contact with a hot piece of pizza. There is nothing in nature that absolutely forbids the transfer of heat from the ice to the much warmer pizza. Statistically, it is possible for a slow-moving molecule in the ice to collide with a faster moving molecule in the pizza such that the slow one transfers some of its energy to the faster one. However, when one considers the large number of molecules present in the ice and pizza, the odds are overwhelmingly in favor of the transfer of energy from the faster-moving molecules to the slower-moving molecules. Furthermore, this example demonstrates that the system naturally tends to move from a state of order to a state of disorder. The initial state in which all the pizza molecules have a high kinetic energy and all the ice molecules have a lower kinetic energy is much more ordered than the final state after heat transfer has taken place and the ice has melted.

As another example, suppose you were able to measure the velocity of all the air molecules in a room at some instant. It is very unlikely that you would find all molecules moving in the same direction with the same speed. This would, indeed, be a highly ordered state. The most probable situation you would find is a system of molecules moving haphazardly in all directions with a distribution of speeds. This is a highly disordered state and also the most likely. Let us compare this example to that of drawing marbles from a bag. Consider a container of gas consisting of 10^{23} molecules. If all of them were found moving in the same direction with the same speed at some instant, the outcome would be similar to drawing marbles from the bag 10^{23} times and finding a red marble on every draw. This is clearly an unlikely set of events.

The tendency of nature to move toward a state of disorder affects the ability of a system to do work. Consider a ball thrown toward a wall. The ball has kinetic energy, and its state is an ordered one. That is, all of the atoms and molecules of the ball move in unison at the same speed and in the same direction (apart from their random thermal motions). When the ball hits the wall, however, part of this ordered energy is transformed to disordered energy. The temperature of the ball and the wall both increase slightly as part of the ball's kinetic energy is transformed into the random, disordered, thermal motion of the molecules in the ball and the wall. Before the collision, the ball is capable of doing work. It could drive a nail into the wall, for example. When part of the ordered energy is transformed to disordered thermal energy, this capability of doing work is reduced. That is, the ball rebounds with less kinetic energy than it had originally because the collision is inelastic.

An illustration from Flammarion's novel *La Fin du Monde*, depicting the heat death of the universe.

Various forms of energy can be converted to thermal energy, as in the collision between the ball and the wall, but the reverse transformation is never complete. In general, if two kinds of energy, A and B, can be completely interconverted, we say that they are the *same grade*. However, if form A can be completely converted to form B and the reverse is never complete, then form A is a *higher grade* of energy than form B. In the case of a ball hitting a wall, the kinetic energy of the ball is of a higher grade than the thermal energy contained in the ball and the wall after the collision. Therefore, when high-grade energy is converted to thermal energy, it can never be fully recovered as high-grade energy.

This conversion of high-grade energy to thermal energy is referred to as the *degradation of energy. The energy is said to be degraded because it takes on a form that is less useful for doing work.* In other words, *in all real processes where heat transfer occurs, the energy available for doing work decreases.*

Finally, it should be noted once again that the statement that entropy must increase in all natural processes is true only when one considers an isolated system. There are instances in which the entropy of some system may decrease but that decrease takes place at the expense of a net increase in entropy for some other system. When all systems are taken together as the universe, *the entropy of the universe always increases.*

Ultimately, the entropy of the universe should reach a maximum. At this point, the universe will be in a state of uniform temperature and density. All physical, chemical, and biological processes will have ceased because a state of perfect disorder implies no energy available for doing work. This gloomy state of affairs is sometimes referred to as an ultimate "heat death" of the universe.

*13.15 ENERGY CONVERSION AND THERMAL POLLUTION

In recent years, many thoughtful people have been concerned with the release of thermal energy into our environment. This **thermal pollution** is important to the welfare of our planet, and it is important that we understand its sources and develop methods to control it. There are many sources of thermal pollution, one of the primary ones being the waste heat from electric power plants. In this section, we shall focus on the reasons for the intrinsically low efficiency of power plants and on acceptable methods currently being used to dispose of thermal energy from these plants.

In the United States, about 85% of the electric power is produced by steam engines, which burn either fossil fuel (coal, oil, or natural gas) or use nuclear fuel (uranium-235). The remaining 15% is generated by water in hydroelectric plants. The overall thermal efficiency of a modern fossil-fuel plant is about 40%. The actual efficiency of any power plant must be lower than the theoretical efficiency derived from the second law of thermodynamics. Of course, one always seeks the highest efficiency possible for two reasons. First, higher efficiency results in lower fuel costs. Second, thermal pollution of the environment is reduced because there is less waste energy in a highly efficient power plant.

The burning of fossil fuels in a electric power plant involves three energy-conversion processes: (1) chemical to thermal energy, (2) thermal to mechanical energy, and (3) mechanical to electrical energy. These are indicated schematically in Figure 13.18.

During the first step, heat energy is transferred from the burning fuel to water, which is converted to steam. In this process about 12% of the available energy is lost up the chimney. In the second step, thermal energy in the form of steam at high pressure and temperature passes through a turbine and is converted to mechanical energy. A well-designed turbine has an efficiency of

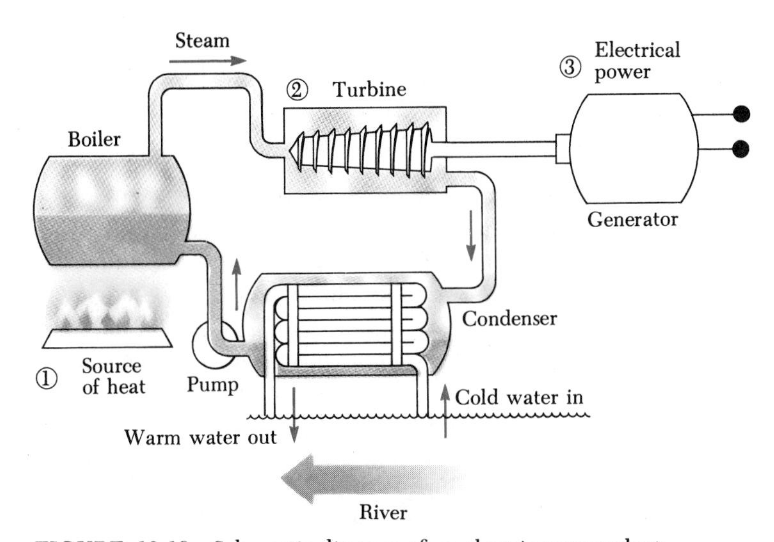

FIGURE 13.18 Schematic diagram of an electric power plant.

FIGURE 13.19 A cooling tower at a reactor site in Southwestern Washington. (From Jonathan Turk and Amos Turk, *Physical Science*, 2nd edition, Philadelphia, Saunders College Publishing, 1981.)

about 47%. The steam, which leaves the turbine at a lower pressure, is then condensed into water and gives up heat in the process. Finally, in the third step, the turbine drives an electric generator of very high efficiency, typically 99%. Hence, the overall efficiency is the product of the efficiencies of each step, which for the figures given is $(0.88)(0.47)(0.99) = 0.41$ or 41%. Therefore, if one accounts for the 12% of energy lost to the atmosphere, we conclude that the thermal energy transferred to the cooling water amounts to about 47% of the energy theoretically available from the fuel.

In nuclear power plants, the steam generated by the nuclear reactor is at a lower temperature than in a fossil-fuel plant. This is due primarily to material limitations in the reactor. Typical nuclear power plants have an overall efficiency of about 34%.

Cooling towers (Fig. 13.19) are commonly used to dispose of waste heat. These towers usually use the heat to evaporate water, which is then released to the atmosphere. Cooling towers also present environmental problems since evaporated water can cause increased precipitation, fog, and ice. Another type of tower is the dry cooling tower (nonevaporative), which transfers heat to the atmosphere by conduction. However, this type is more expensive and cannot cool to as low a temperature as an evaporative tower.

SUMMARY

The **work done** as a gas expands or contracts under a constant pressure is

$$\boxed{W = P\Delta V} \tag{13.1}$$

The work done is negative if the gas is compressed and positive if the gas expands. In general, the work done in an expansion from some initial state to a final state is the area under the curve on a *PV* diagram.

From the first law of thermodynamics, we see that when a system undergoes a change from one state to another, the **change in its internal energy,** ΔU, is

$$\boxed{\Delta U = U_f - U_i = Q - W} \tag{13.2}$$

where Q is the heat transferred into (or out of) the system and W is the work done by (or on) the system. The sign convention used for Q is that it is positive when heat enters the system and negative when the system loses heat.

The **first law of thermodynamics** is a generalization of the law of conservation of energy that includes heat transfer.

In a **cyclic process** (one in which the system ends up in the same state as it started), $\Delta U = 0$ and therefore $Q = W$. That is, the heat transferred into the system equals the work done during the cycle.

An **adiabatic process** is one in which no heat is transferred between the system and its surroundings ($Q = 0$). In this case, the first law gives $\Delta U = -W$. That is, the internal energy changes as a consequence of work being done by (or on) the system.

An **isobaric process** is one that occurs at constant pressure. The work done in such a process is $P\Delta V$.

An **isothermal process** is one that takes place at constant temperature. The work done during an isothermal process is given by

$$W = nRT \ln\left(\frac{V_f}{V_i}\right) \qquad (13.3)$$

where n is the number of moles of the gas, R is the ideal gas constant, T is the kelvin temperature, V_f is the final volume, and V_i is the initial volume.

A **heat engine** is a device that converts thermal energy to other forms of energy, such as mechanical and electrical energy. The work done by a heat engine in carrying a substance through a cyclic process ($\Delta U = 0$) is

$$W = Q_h - Q_c \qquad (13.4)$$

where Q_h is the heat absorbed from a hot reservoir and Q_c is the heat expelled to a cold reservoir.

The **thermal efficiency** of a heat engine is defined as the ratio of the net work done to the heat absorbed per cycle:

$$Eff \equiv \frac{W}{Q_h} = 1 - \frac{Q_c}{Q_h} \qquad (13.5)$$

No real heat engine operating between the temperatures T_h and T_c can be more efficient than an engine operating in a Carnot cycle between the same two temperatures. The efficiency of a heat engine operating in the **Carnot cycle** is

$$Eff_C = 1 - \frac{T_c}{T_h} \qquad (13.6)$$

Real processes proceed in an order governed by the **second law of thermodynamics.** Two ways of stating the **second law of thermodynamics** are

1. Heat will not flow spontaneously from a cold object to a hot object.
2. No heat engine operating in a cycle can absorb thermal energy from a reservoir and perform an equal amount of work.

The second law can also be stated in terms of a quantity called **entropy.** The *change in entropy* of a system is equal to the heat flowing into the system as the system changes from a state A to a state B divided by the absolute temperature:

$$\Delta S \equiv \frac{\Delta Q}{T} \qquad (13.8)$$

One of the primary results of statistical mechanics is that systems tend toward disorder and that entropy is a measure of this disorder. The entropy of the universe increases in all natural processes. This statement is an alternative statement of the second law.

ADDITIONAL READING

S. Angrist, "Perpetual Motion Machines," *Sci. American*, January 1968, p. 114.

P. W. Atkins, *The Second Law*, Scientific American Library, W. H. Freeman and Co., New York, 1984.

L. Bryant, "Rudolf Diesel and His Rational Engine," *Sci. American*, August 1969, p. 108.

D. A. Dicus, J. Letaw, D. Teplitz, and V. Teplitz, "The Future of the Universe," *Sci. American*, March 1983, p. 90.

W. Ehrenburg, "Maxwell's Demon," *Sci. American*, November 1967, p. 107.

K. Ford, "Probability and Entropy in Thermodynamics," *Physics Teacher*, February 1967, p. 77.

R. Giedel, "Real Otto and Diesel Engine Cycles," *The Physics Teacher*, January 1983, p. 29.

U. Haber-Schaim, "The Role of the Second Law of Thermodynamics in Energy Education," *The Physics Teacher*, January 1983, p. 17.

W. Scaife, "The Parsons Steam Turbine," *Sci. American*, April 1985, p. 132.

G. Walker, "The Stirling Engine," *Sci. American*, August 1973, p. 80.

S. Wilson, "Sadi Carnot," *Sci. American*, August 1981, p. 134.

QUESTIONS

1. Distinguish clearly between temperature, heat, and internal energy.
2. When a sealed Thermos bottle full of hot coffee is shaken, what are the changes, if any, in (a) the temperature of the coffee and (b) its internal energy?
3. Use the first law of thermodynamics to explain why the total energy of an isolated system is always conserved.
4. Is it possible to convert internal energy to mechanical energy?
5. What are some factors that affect the efficiency of automobile engines?
6. The statement was made in this chapter that the first law says we cannot get more out of a process than we put in but the second law says that we cannot break even. Explain.
7. Is it possible to cool a room by leaving the door of a refrigerator open? What happens to the temperature of a room in which an air conditioner is left running on a table in the middle of the room?
8. In practical heat engines, which do we have more control of, T_c or T_h? Explain.
9. A steam-driven turbine is one major component of an electric power plant. Why is it advantageous to increase the temperature of the steam as much as possible?
10. Is it possible to construct a heat engine that creates no thermal pollution?
11. Electrical energy can be converted to heat energy with an efficiency of 100%. Why is this number misleading with regard to heating a home? That is, what other factors must be considered in comparing the cost of electric heating with the cost of hot air or hot water heating?
12. Discuss three common examples of natural processes that involve an increase in entropy. Be sure to account for all parts of each system under consideration.
13. Discuss the change in entropy of a gas that expands (a) at constant temperature and (b) adiabatically.
14. Suppose the waste heat at a power plant is exhausted to a pond of water. Could the efficiency of the plant be increased by refrigerating the water?
15. A designer of an electric heating unit describes his product as being 100 percent efficient. Is his claim correct?
16. Energy is conserved; therefore, we should be able to recycle it. Defend or rebut this statement.
17. Does water have "more entropy" when frozen or when liquid?
18. It is often said that the conception and birth of a human child is an example of a situation in which entropy decreases. If so, how can this process be consistent with the second law?

PROBLEMS

Section 13.2 Internal Energy

Section 13.3 Work

Section 13.4 Heat

1. Steam moves into the cylinder of a steam engine at a constant pressure of 2×10^5 Pa. The diameter of the piston is 16 cm, and the piston travels 20 cm in one stroke. How much work is done during one stroke?

2. A container of volume 0.4 m^3 contains 3 moles of argon gas at 30°C. Assume that argon behaves like an ideal gas and find the total internal energy of the system. (*Hint:* You will have to review the discussion of ideal gases in Chapter 11 to find an expression for the energy of an ideal gas.)

3. Sketch a *PV* diagram of the following processes. (a) A gas expands at constant pressure P_1 from a volume V_1 to a volume V_2. It is then kept at constant volume while the pressure is reduced to P_2. (b) A gas is reduced in pressure from P_1 to P_2 while its volume is held constant at V_1. It is then expanded at constant pressure P_2 to a final volume V_2. (c) In which of the processes is more work done? Why?

4. A gas expands from *I* to *F* along the three paths indicated in Figure 13.20. Calculate the work done by the gas along paths (a) *IAF*, (b) *IF*, and (c) *IBF*.

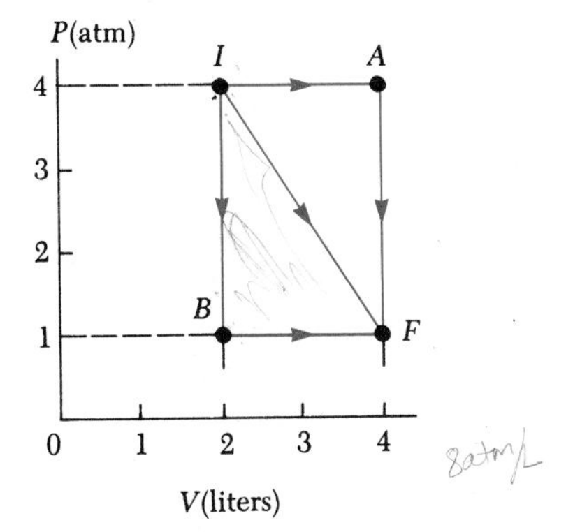

FIGURE 13.20 Problems 4 and 8

5. Sketch a *PV* diagram and find the work done by the gas during the following stages. (a) A gas is expanded from a volume of 1 liter to 3 liters at a constant pressure of 3 atm. (b) The gas is then cooled at constant volume until the pressure is reduced to 2 atm. (c) The gas is then compressed at a constant pressure of 2 atm from a volume of 3 liters to a volume of 1 liter. *Note:* Be careful of signs. (d) The gas is heated until its pressure increases from 2 atm to 3 atm at a constant volume. (e) Find the net work done during the complete cycle.

6. A sample of gas is compressed to one-half its initial volume at a constant pressure of 1.25×10^5 Pa. During the compression, 100 J of work is done on the gas. Determine the final volume of the gas.

Section 13.5 The First Law of Thermodynamics

Section 13.6 Isothermal Expansion of an Ideal Gas

7. A gas is compressed at a constant pressure of 0.3 atm from a volume of 8 liters to a volume of 3 liters. In the process, 400 J of heat energy flows out of the gas. (a) What is the work done by the gas? (b) What is the change in its internal energy?

8. A gas expands from *I* to *F* in Figure 13.20. The heat added to the gas is 418 J when the gas goes from *I* to *F* along the diagonal path. (a) What is the change in internal energy of the gas? (b) How much heat must be added to the gas for the indirect path *IAF* to give the same change in internal energy?

9. A gas is enclosed in a container fitted with a piston of cross-sectional area 0.15 m^2. The pressure of the gas is maintained at 6000 Pa as the piston moves inward 20 cm. (a) Calculate the amount of work done by the gas. (b) If the internal energy of the gas decreases by 8 J, find the amount of heat removed from the system during the compression.

10. Two cubic centimeters of water are boiled at atmospheric pressure to become 3342 cubic centimeters of steam also at atmospheric pressure. (a) Calculate the work done by the gas during this process. (b) Find the amount of heat added to the water to accomplish this process. (c) From (a) and (b), find the change in internal energy.

11. Two moles of an ideal gas expands isothermally at 27°C to three times its initial volume. Find (a) the work done by the gas and (b) the heat flow into the system.

12. Two moles of an ideal gas are compressed from a volume of 10 liters to a volume of 3 liters at a constant temperature of 27°C. Find (a) the work done by the gas during this process, (b) the change in internal energy of the gas, and (c) the amount of heat that must be added to or removed from the gas.

13. A 100-kg steel support rod in a building has a length of 2 m at a temperature of 20°C. The rod supports a load of 6000 kg. Find (a) the work done by the rod as the temperature increases to 40°C, (b) the heat added to the rod (assume the specific heat of steel is the same as that for iron), and (c) the change in internal energy of the rod.

14. A 1 kg block of aluminum is heated at atmospheric pressure such that its temperature increases from 22°C to 40°C. Find (a) the work done by the aluminum, (b) the heat added to the aluminum, and (c) the change in internal energy of the aluminum.

• **15.** One mole of helium gas, considered to be an ideal gas, is initially at a temperature of 300 K and pressure 0.2 atm. It is compressed isothermally to a pressure of 0.8 atm. (a) Use the ideal gas law to find the final volume of the gas. (b) Find the work done by the gas, and (c) the heat transferred to the gas.

• **16.** One mole of an ideal gas does 800 J of work during an isothermal expansion at a temperature of 300 K. If its initial volume is 2 liters, find the final volume.

• **17.** A container is placed in a water bath and held at constant volume as a mixture of fuel and oxygen is burned inside it. The temperature of the water is observed to rise during the burning and it also is held at constant volume. (a) Consider the burning mixture to be the system. What are the signs of Q, ΔU, and W? (b) What are the signs of these quantities if the water bath is considered to be the system?

• **18.** One mole of gas is initially at a pressure of 2 atm and a volume of 0.3 liter, and has an internal energy equal to 91 J. In its final state, the pressure is 1.5 atm, the volume is 0.8 liter, and the internal energy equals 182 J. For the paths *IAF*, *IBF*, and *IF* in Figure 13.21 (a) calculate the work done by the gas and (b) the net heat transferred to the gas in the process.

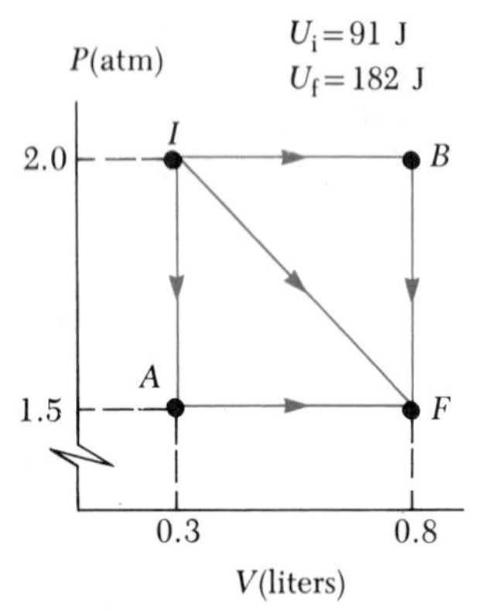

FIGURE 13.21 Problem 18

Section 13.7 Heat Engines and the Second Law of Thermodynamics

Section 13.8 The Gasoline Engine

19. In each cycle, a heat engine absorbs 375 J of heat and performs 25 J of work. Find (a) the efficiency of the engine and (b) the heat expelled in each cycle.

20. An engine absorbs 1700 J from a hot reservoir and expels 1200 J to a cold reservoir in each cycle. (a) What is the efficiency of the engine? (b) How much work is done in each cycle? (c) What is the power output of the engine if each cycle lasts for 0.3 s?

• **21.** A particular engine has a power output of 5 kW and an efficiency of 25%. If the engine expels 8000 J of heat in each cycle, find (a) the heat absorbed in each cycle and (b) the time for each cycle.

• **22.** The heat absorbed by an engine is three times greater than the work it performs. (a) What is its thermal efficiency? (b) What fraction of the heat absorbed is expelled to the cold reservoir?

Section 13.10 The Carnot Engine

23. A heat engine operates between two reservoirs at temperatures of 20°C and 300°C. What is the maximum efficiency possible for this engine?

24. A steam engine has a boiler that operates at 300°F and the temperature of the exhaust is 150°F. Find the maximum efficiency of this engine.

25. The exhaust temperature of a Carnot heat engine is 300°C. What is the intake temperature if the efficiency of the engine is 30%?

• **26.** In one cycle, a heat engine absorbs 500 J from the high-temperature reservoir and expels 300 J to a low temperature reservoir. If the efficiency of this engine is 60% of the efficiency of a Carnot engine, what is the ratio of the low temperature to the high temperature in the Carnot engine?

• **27.** A heat engine operates in a Carnot cycle between 80°C and 350°C. It absorbs 21,000 J of heat per cycle from the hot reservoir. The duration of each cycle is 1 s. (a) What is the maximum power output of this engine? (b) How much heat does it expel in each cycle?

• **28.** Water is used to cool the output of the engine of Problem 27. If the temperature of the water is not allowed to increase more than 2°C, find how much water must pass through the engine in one hour.

• **29.** A power plant that uses the temperature gradient in the ocean has been proposed. The system is to operate between 20°C (surface water temperature) and 5°C (water temperature at a depth of about 1 km). (a) What is the maximum efficiency of such a system? (b) If the power output of the plant is 75 MW, how much thermal energy is absorbed per hour? (c) In view of your results to (a) do you think such a system is worthwhile?

Section 13.13 Entropy

Section 13.14 Entropy and Disorder

30. What is the change in entropy of 1 kg of water at 100°C as it changes to steam at 100°C?

31. A 70-kg log falls from a height of 25 m into a lake. If the log, the lake, and the air are all at 300 K, find the change in entropy of the universe for this process.

32. Two 2000-kg cars, both traveling at 20 m/s, undergo a head-on collision and stick together. Find the entropy change of the universe during the collision if the temperature is 23°C.

33. An ice tray contains 500 g of water at 0°C. Calculate the entropy change of the water as it freezes completely and slowly at 0°C.

34. Prepare a table like Table 13.1 for the following occurrence. You toss four coins into the air simultaneously. Record the results of your tosses in terms of the numbers of heads and tails that result. For example, HHTH and HTHH are two possible ways in which three heads and one tail can be achieved. (a) On the basis of your table, what is the most probable result of a toss? (b) In terms of entropy what is the most ordered state, and (c) what is the most disordered?

35. Repeat the procedure used to construct Table 13.1 (a) for the case in which you draw three marbles from your bag rather than four and (b) for the case in which you draw five rather than four.

36. In tossing two dice what is the total number of ways in which you can obtain (a) a 12, (b) a 7?

ADDITIONAL PROBLEMS

37. A cylinder of gas is compressed by a piston from an initial volume of 125 liters to a final volume of 90 liters. The compression occurs at constant pressure, and the work done on the gas by the piston is 10^4 J. What is the gas pressure during the compression?

38. One mole of an ideal gas is compressed isothermally at 300 K until its volume is halved. Find the work done on the gas.

39. An engineer claims to have developed an engine that takes in 70,000 J of heat at 500 K and expels 20,000 J at 300 K, with 10,000 J of work being done. Would this be worth investing in? Why or why not?

40. A gas follows the path 123 on the *PV* diagram in Figure 13.22 and 418 J of heat flows into the system. Also, 167 J of work is done. (a) What is the internal energy change of the system? (b) How much heat flows into the system if the process follows the path 143? The work done by the gas along this path is 63 J. What net work would be done on or by the system if the system followed (c) the path 12341 and (d) the path 14321? (e) What is the change in internal energy of the system in the processes described in parts (c) and (d)?

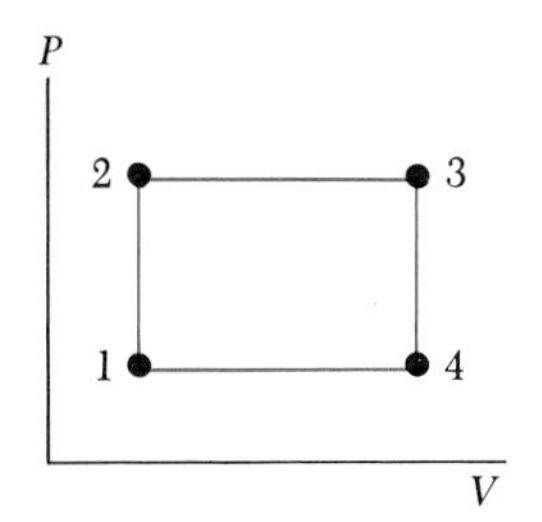

FIGURE 13.22 Problem 40

41. One mole of hydrogen gas is heated at constant pressure from 300 K to 420 K. Calculate (a) the heat energy transferred to the gas, (b) the change in the internal energy of the gas, and (c) the work done by the gas. Note that hydrogen has a specific heat c = 28.7 J/mol·K.

42. A 5-kg block of aluminum is heated at atmospheric pressure such that its temperature increases from 20°C to 90°C. (a) Find the work done by the aluminum, (b) the amount of heat transferred to it, and (c) the increase in its internal energy.

• 43. One mole of an ideal gas at 1 atm pressure and at 300°C expands isothermally until its volume is tripled. It then is compressed to its original volume at constant pressure. Sketch these two processes on a *PV* diagram.

• 44. Refer to Problem 43. (a) Calculate the work done by the gas during the isothermal expansion. (b) Find the work done on the gas during the second change.

• 45. A refrigerator has a coefficient of performance of 3. The ice tray compartment is at −20°C, and the room temperature is 22°C. The refrigerator can convert 30 g of water at 22°C to 30 g of ice at −20°C each minute. What input power is required? Ignore any cooling by the refrigerator and give your answer in watts.

• 46. During each cycle, a particular refrigerator absorbs 25 cal from the cold reservoir and expels 32 cal to the high-temperature reservoir. (a) If the refrigerator completes 60 cycles per second, what power is required? (b) What is the COP for this unit?

• 47. A 1500 kW heat engine operates at 25% efficiency. The heat energy expelled at the low temperature is absorbed by a stream of water that enters the cooling coils at 20°C. If 60 liters per second flows across the coils, determine the increase in temperature of the water.

• 48. One mole of an ideal gas is taken through the cycle shown in Figure 13.23. The process *AB* is an isothermal expansion. Calculate the net work done by the gas.

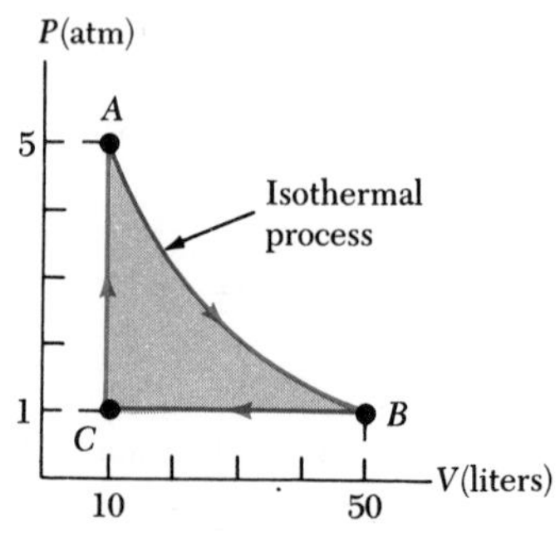

FIGURE 13.23 Problem 48

• **49.** An ideal refrigerator (or heat pump) is equivalent to a Carnot engine running in reverse. That is, heat Q_c is absorbed from a cold reservoir and heat Q_h is rejected to a hot reservoir. Show that the work that must be supplied to run the refrigerator is given by

$$W = \frac{T_h - T_c}{T_c} Q_c$$

•• **50.** One mole of an ideal gas is taken through the reversible cycle shown in Figure 13.24. At point A, the pressure, volume, and temperature are P_0, V_0, and T_0. In terms of R and T_0, find (a) the total heat entering the system per cycle, (b) the total heat leaving the system per cycle, (c) the efficiency of an engine operating in this reversible cycle, and (d) the efficiency of an engine operating in a Carnot cycle between the temperature extremes for this process. (*Hint:* Recall that work equals the area under a *PV* curve.)

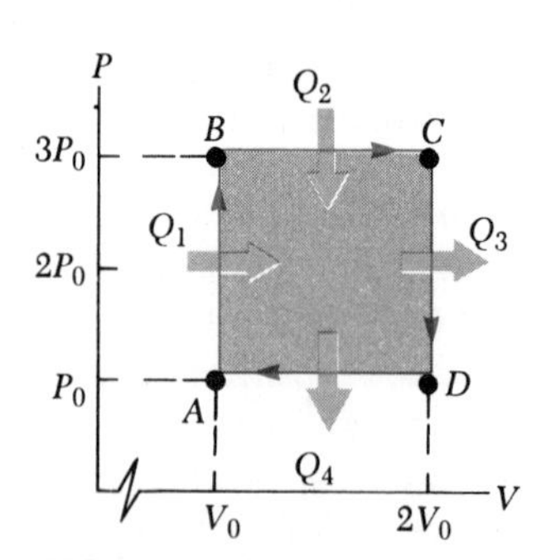

FIGURE 13.24 Problem 50

•• **51.** An ideal gas initially at pressure P_0, volume V_0, and temperature T_0 is taken through the cycle described in Figure 13.25. (a) Find the net work done by the gas per cycle in terms of P_0 and V_0. (b) What is the net heat added to the system per cycle? (c) Obtain a numerical value for the net work done per cycle for 1 mole of gas initially at 0°C. (See the hint for Problem 50.)

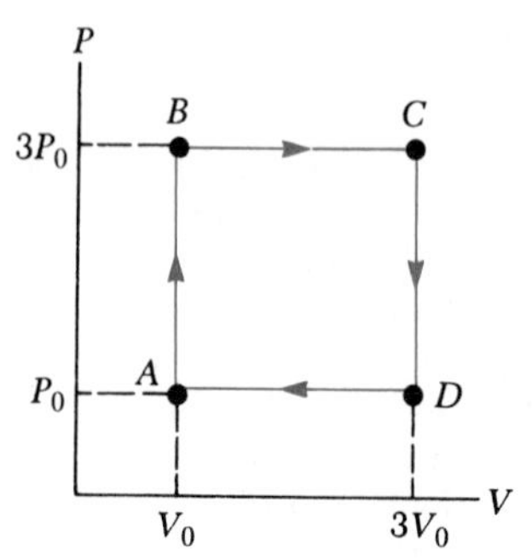

FIGURE 13.25 Problem 51

•• **52.** One mole of an ideal gas at an initial pressure of 4 atm and temperature 300 K is carried through the following reversible cycle: (a) It expands isothermally until its volume is doubled; (2) it is compressed to its original volume at constant pressure; (3) it is compressed isothermally to a pressure of 4 atm; and (4) it expands at constant pressure to its original volume. (a) Make a plot of the cyclic process on a *PV* diagram. (b) Calculate the work done by the gas per cycle.

•• **53.** An electrical power plant has an overall efficiency of 15%. The plant is to deliver 150 MW of power to a city, and its turbines use coal as the fuel. The burning coal produces steam at 190°C, which drives the turbines. This steam is then condensed into water at 25°C by passing it through coils in contact with river water. (a) How many metric tons of coal does the plant consume each day (metric ton = 10^3 kg)? (b) What is the total cost of the fuel per year if the delivered price is \$8/metric ton? (c) If the river water is delivered at 20°C, at what minimum rate must it flow over the cooling coils in order that its temperature not exceed 25°C? (Note: The heat of combustion of coal is 7.8×10^3 cal/g.)

Essay

Alternative Sources of Energy

Laurent Hodges *Iowa State University*

Today most of the world's energy is obtained from fossil fuels (coal, petroleum, and natural gas) and uranium. Despite their many advantages, these fuels have some significant disadvantages. All have become much costlier in recent years; new uranium power plants for example, have priced themselves out of the market. Petroleum and natural gas are becoming too scarce to maintain their position as energy sources for more than a few more decades; the production of these fuels in the United States actually peaked in the early 1970s. All of these energy sources have detrimental environmental effects, particularly air pollution. Consequently, society has begun to investigate alternative energy sources.

The term "alternative sources" has sometimes been applied to new fuels derived from old sources, such as hydrocarbon gases and liquids derived from coal or fissionable plutonium derived from uranium-238. However, the term is usually reserved for primary energy sources which have not been in widespread use in recent years. These include geothermal energy, fusion energy, and solar energy.

Geothermal Energy

One alternative energy source is underground heat, or geothermal energy. From a thermodynamic point of view, this energy source exploits the difference between the temperature of underground rocks or water and that of the earth. The higher underground temperatures are a source of thermal heat and can also be used to generate electricity.

Geothermal energy exists everywhere in the world, but it is being used only in places where it is available near the surface of the earth (within a few kilometers). These are generally, but not always, places with natural resources of geothermal steam or hot water.

Geothermal heat is used in Iceland for home heating (accounting for about one-third of all the energy used there) as well as in several other parts of the world, including Boise, Idaho and Klamath Falls, Oregon.

Geothermal energy can also be used to generate electricity. Geothermal electricity has been generated for many years in Larderello, Italy, The Geysers, California, (see Fig. 1) and several other places.

Geothermal energy is not limited to regions with steam or hot water. Research has been carried out in New Mexico with the aim of extracting geothermal heat from hot dry rocks deep underground. Two wells are dug, one to bring cold water down from the surface and the other to bring hot water back to the surface. The underground rocks are fractured with pressurized water, and the water that passes through the fractures is heated. This form of geothermal energy would be useful in many parts of the world.

FIGURE 1 An aerial view of The Geysers, a large geothermal electric plant located in Sonoma and Lake Counties in California. The plant produces over 900 MW of power and services about one million residents. (Photo, Department of Energy)

Fusion Energy

Nuclear power plants around the world today derive their energy from the fissioning of heavy nuclei such as uranium and plutonium. An alternative is the production of energy from the fusion of light nuclei into heavier nuclei of lower total mass (see Chapter 33). The decrease in mass in the fusion process appears as energy from Einstein's relationship $E = mc^2$.

An example is the fusion of hydrogen nuclei into a helium nucleus, a process which is the source of the radiant energy in sunlight and most starlight. Fusion is also the energy source in thermonuclear or "hydrogen" bombs.

Because nuclei are positively charged and thus electrically repel each other, fusion reactions require very high temperatures to give the nuclei sufficient kinetic energy to fuse. It is for this reason that the fusion reactions are referred to as *thermonuclear* reactions.

FIGURE 2 A view of the toroidal field coils on an experimental Tokomak fusion reactor (TFTR) at Princeton University. (Photo, Department of Energy)

There is an abundance on earth of naturally occurring isotopes suitable for use in fusion energy reactions, notably deuterium (H^2) and lithium. The difficulty is in creating conditions favorable for fusion. The temperatures must be very high, the plasma of ionized nuclei must be sufficiently dense, and the length of time during which nuclear reactions occur must be long enough for fusion to take place. Two approaches that have been extensively studied involve confinement to the plasma using magnetic fields or by bombardment with laser radiation.

Controlled fusion has not yet been demonstrated in the laboratory, much less been proved economical.

Forms of Solar Energy

The major alternative energy source is the sun. Solar energy comes to the earth in the form of electromagnetic radiation, mainly in the visible and near infrared parts of the spectrum.

Solar energy, broadly defined, includes a great variety of forms which can be classified as shown in Table 1. In addition to direct solar radiation, which can be used to heat or cool buildings, to heat water for domestic use, or to produce electricity, there are indirect forms of solar energy such as water, wind, and photosynthetic power. These are properly classified as solar energy because it is solar radiation that drives the hydrological cycle from which water power is derived and the circulation in the atmosphere from which wind power is derived, and that provides the energy for photosynthesis.

Solar energy is often regarded as a possible energy source of the future. It is often forgotten that solar energy was, in fact, the major energy source for humanity throughout almost all of recorded and unrecorded history. The great explorers all used solar energy (in the form of wind energy) on their sea voyages throughout the world. In the United States, the industrial revolution was powered by solar energy—in the form of mechanical water and wind power and wood fuel. Fossil fuels, originally in the form of coal and later in the form of petroleum and natural gas, did not surpass solar energy in importance until the 1880s.

TABLE 1. Forms and Uses of Solar Energy

1. DIRECT SOLAR RADIATION
   - a. Active solar systems
     - i. Wintertime space heating
     - ii. Water heating
     - iii. Space cooling (air conditioning)
     - iv. Process heat
   - b. Passive solar space heating
   - c. Electric generation systems
     - i. Photovoltaic systems
     - ii. Solar power towers
2. INDIRECT SOLAR ENERGY
   - a. Water power
     - i. Hydroelectricity
     - ii. Mechanical power
   - b. Wind power
     - i. Ocean transportation
     - ii. Mechanical power (windmills)
     - iii. Wind-generated electricity
   - c. Photosynthetic power
     - i. Solid fuels (wood)
     - ii. Liquid fuels (methanol, ethanol, etc.)
     - iii. Gaseous fuels (such as methane)
   - d. Ocean thermal energy conversion

Solar energy is still a significant energy source in the United States, ranking behind each of the three fossil fuels but ahead of nuclear energy or other energy sources. Currently, solar energy accounts for approximately 10 percent of the energy used in the United States. The major uses of solar power are hydroelectricity, passive solar energy, and photosynthetic materials (particularly wood).

Hydroelectricity

Hydroelectricity accounts for about 15 percent of the electricity generated in the United States, about the same as nuclear electricity. Hydroelectricity is an interesting example of several steps in energy conversion:

1. Electromagnetic radiation from the sun is absorbed by the oceans and thereby converted into sensible heat.

2. The sensible heat is converted into the latent heat of evaporated water, which is transported throughout the atmosphere and rains down in mountainous regions.
3. The gravitational potential energy of the water is converted into kinetic energy as the water flows down to sea level.
4. The kinetic energy of the water is converted into kinetic energy of the great hydroelectricity turbines at the generating plant.
5. The kinetic energy of the turbines is converted into electric energy at the turbine-generator.
6. The electric energy is transported to the end users, who may convert it, depending on their needs, into heat or light (electromagnetic radiation) or mechanical energy.

Where available, hydroelectric power is the cheapest form of electric power. It is a major source of electric systems in the Pacific Northwest, New England, New York, and the region served by the Tennessee Valley Authority.

The use of hydroelectric power has a major environmental impact, since it requires the damming of a river and often the creation of a large reservoir. Although many possible hydroelectric sites remain, most are unlikely to be developed because vast ecosystems would be adversely affected.

Solar Heating of Buildings

The most cost-effective use of solar energy is for space heating, or the winter heating of houses and other buildings. Space heating is the number one use of energy in American homes (see Fig. 3). The space heating requirements can be drastically reduced by energy conservation methods such as good wall and roof insulation, multipane windows, and air infiltration control by good caulking and weatherstripping. Solar energy can be used to meet a substantial portion of the remaining heating requirements.

A solar heating system requires three major components:

1. *Collection.* The function of the collector is to collect the solar radiation and convert it into heat.
2. *Storage.* Solar energy comes only during the daylight hours, often at a rate greater than it can be immediately used. The function of the storage system is to store excess heat for later use, such as at night or on a cloudy day.
3. *Distribution.* The function of the distribution system is to move the solar heat from where it is to where it is wanted. On a sunny day, the solar heat is distributed from the collector to the living space, and if there is an excess, to storage. During the night or on a cloudy day, the solar heat must be moved from storage to the living space.

Active Solar Heating

Active solar heating systems use mechanical means to distribute the solar energy. In active air systems, fans and ducts are used to move solar heat in the form of heated air; in active liquid systems, pumps and pipes are used to move solar heat in the form of heated water or antifreeze.

Active space heating systems were used in many houses during the 1970s. The collector normally consisted of a flat-plate collector of one or two panes of glass, behind which was a black absorber. In air systems, the storage was usually provided by a rockbed (a mass of uniformly sized rocks contained in a small room), and in liquid systems, by a large water tank.

Active space heating systems were not successful. They tended to be large, complex, and expensive, often costing $10,000 or more for a house of typical size. They required expert design, and had many parts requiring constant maintenance. They also tended to be conspicuous and made homes appear unconventional.

Passive Solar Heating

Passive solar systems have proved much more successful, particularly in new homes where they add little or nothing to the cost of the home. In passive solar systems the energy flows occur by natural means: natural conduction, natural convection, and natural radiation.

Passive solar heating systems typically consist of a large expanse of south-facing windows (which act as the collector) together with interior masonry to serve as ther-

FIGURE 3 Solar power provides hot water and heat for this 1200-square foot home in Shenandoah, Georgia. (Photo, Department of Energy)

mal storage by providing high heat capacity. Concrete, stone, brick, and adobe have been used for interior masonry in the form of walls, floors, or interior partitions. When masonry walls act as thermal storage, they must be insulated on the exterior, rather than the interior. Occasionally water-filled containers have been used instead.

Two common passive solar space heating systems are the direct gain and the thermal storage wall systems. In a direct gain system the sun shines directly into the living space through the windows. As long as there is sufficient heat capacity to the thermal storage, the house will not overheat during the day and will cool down very slowly at night. Direct-gain houses are well-lit by natural light during the day. As long as there is good air flow throughout the house, natural distribution of energy is satisfactory and no active system is required.

A thermal storage wall system has a concrete wall, typically about 30 cm (12 inches) thick, located just behind the south windows. The sunlight shines directly on the wall and is absorbed by its black surface. The wall acts as storage and also distribution, since the heat absorbed is eventually radiated back into the living space behind the wall.

There are several other types of passive heating systems. One popular system is the sunspace system, in which the main collection and storage of solar heat occurs in a separate south-facing room. Sunspaces can be used as sunporches, greenhouses, or swimming pool rooms.

Passive heating systems work very well in most parts of the United States, particularly in areas that have cold but fairly sunny winters. They have been effective in Arizona, New Mexico, Colorado, Wyoming, Iowa, Minnesota, and elsewhere. It is not uncommon for well-insulated passive solar homes in cold northern climates to be heated for an entire winter for only \$100 to \$200 worth of natural gas heat.

Photosynthetic Energy

Photosynthetic energy is a major source of energy in many homes—those that use wood for space heating. Wood and agricultural wastes are sometimes used by small industries to generate electricity or industrial process heat. It is worth remembering that wood was the main energy source in the United States during its first century. Not until the 1880s did coal supplant wood as the leading energy source, and wood remained a significant source through the rest of the 19th century.

Photovoltaic Cells

Photovoltaic or "solar" cells are made of semiconductor materials that develop a voltage when sunlight strikes them. Originally developed for the space program to provide orbiting satellites with a source of power, they are now widely available in small electronic devices such as "solar" calculators, but are also sometimes used to power remote installations or electric fences.

A house with photovoltaic cells covering a south-facing roof could collect enough electric energy to meet daily needs but would require a battery system in order to be able to store the energy for use at night. Such a system would currently cost perhaps \$20,000 or more, and would not be economical except in remote areas where the cost of electric transmission to the home would be excessive.

Research is under way to produce better photovoltaic cells, mainly cells that are higher in efficiency or cheaper to manufacture. It is hoped that they will be become economical for many purposes before the end of the century.

Ocean Thermal Energy Conversion

Ocean thermal energy conversion (OTEC) is the production of electricity using the small temperature difference between warm ocean surface water and cold deep ocean water. An OTEC plant would need to use a working substance other than steam; ammonia is one possibility. Although the temperature differences are only a few degrees and the plant efficiency would be very low, very large quantities of water and heat are available, and such a plant might be cost-effective.

OTEC plants have been tested intermittently for many decades. The Atlantic Ocean off the coast of Florida and the Pacific Ocean near Hawaii look like good locations for such a plant, and research has been carried out in both places. However, OTEC plants are currently far from being technologically and economically feasible and may never actually prove commercially successful.

Power Towers

The solar power tower or central receiver is an electric power plant located at the top of a large tower; its heat source is solar radiation reflected to the tower by a large field of heliostats (mirrors) surrounding it. The heliostats must be computer-controlled so they can reflect the sunlight to the proper place accurately and yet be defocused quickly in case of an emergency.

This concept has been thoroughly researched, and in 1982 an experimental version known as "Solar One" shown in Figure 4 went into operation at Barstow, California. Solar One is a small (10 MW) solar electric plant with a steam boiler situated in a field of 1,818 heliostats. The total land area covered is 130 acres.

If solar power towers prove to be economical, the most likely area in the United States for their widespread use would be the South.

FIGURE 4 Aerial view of a solar power plant at Barstow, California. The central receiver tower is surrounded by more than 1800 heliostats (mirrors) which reflect the heat to the tower to produce superheated steam and, ultimately, electricity. (Photo, Department of Energy)

Wind Power

Wind power, which has served humanity in many ways in the past, is currently under investigation as a source of electricity. Many wind generators were installed on midwestern farms in the 1920s and 1930s, and a surprising number are still in operation. Several small wind-generating systems are commercially available today; they are most economical in sizes that would provide electricity to several homes or a large farm.

In a good location, one with higher-than-average wind speeds, a modern wind generator will produce electricity at a cost higher than the current price of coal-generated electricity but less than the cost of some nuclear-generated electricity. A wind system is unlikely to be cost-effective in the short run, but could prove economical over a lifetime of 20 years or more.

A few large wind generators were built by utilities in the past, and since the 1970s, a considerable number of new designs for large machines have been developed and tested, as in Figure 5. Several "wind farms" have been built in the western states in regions with higher-than-average wind speeds, often by knowledgeable investors who expect to make a good return on their investment over the lifetime of the farm.

ESSAY QUESTIONS

1. What enables glass to act as a solar collector? Would transparent plastic materials such as Plexiglas work as well?
2. Passive solar homes normally use double- or triple-pane windows. Why not single-pane windows?
3. Does a solar collector have to face exactly south?

ESSAY PROBLEMS

1. The solar radiation reaching the earth, above the atmosphere, averages about 1350 W/m^2. Using the facts that the sun is about 150 million kilometers away and the earth has a diameter of 12.7 thousand kilometers, find: (a) The total power (in watts) of the solar radiation reaching the earth, (b) the total power (in watts) of the solar radiation emitted by the sun, assuming it is emitted equally in all directions, (c) the rate (in kilograms per second) at which the sun is losing mass, making use of Einstein's relation $E = mc^2$, and (d) the fraction of the sun's total mass of 2.0×10^{30} kg being lost each year.
2. At a midlatitude in the United States, the average amount of solar radiation falling on a tilted south-facing roof is about 200 W/m^2, averaging over the whole year, day and night, sunny weather and cloudy. Assuming that a homeowner uses solar cells which cover 80% of the roof and have an efficiency of 10% at converting solar radiation into electricity, how large an area (in square meters) would be needed to produce a total of 6000 kWh of electric energy? Is this a reasonable area to find on a real roof?
3. A passive solar home in Iowa has 40 m^2 of south-facing double-pane windows and 100 metric tons (100,000 kg) of concrete heat storage. It has a backup furnace whose heat output is 40 MJ/h. Concrete has a heat capacity of about 800 J/C°·kg (about one-fifth that of water). Suppose that on a sunny winter day each square meter of windows admits 16 MJ of solar energy into the home. (a) How many hours would the furnace have to run to provide the home with the same amount of heat as was admitted by the windows? (b) How much would the thermal storage warm up if 40% of the solar heat were lost through the windows, walls, and roof of the house and the other 60% went to heat the storage?

FIGURE 5 A vertical axis wind turbine being checked by a technician at Sandia National Laboratory in Albuquerque, New Mexico. This turbine has 55-foot diameter blades and is capable of producing 60 kW of power in a 30-mph wind. (Photo, Department of Energy)

Part 3

VIBRATIONS AND WAVE MOTION

The impetus is much quicker than the water, for it often happens that the wave flees the place of its creation, while the water does not; like the waves made in a field of grain by the wind, where we see the waves running across the field while the grain remains in place.

LEONARDO DA VINCI

As we look around us, we find many examples of objects that vibrate or oscillate: a pendulum, the strings of a guitar, an object suspended on a spring, the piston of an engine, the head of a drum, the reed of a saxophone. Most elastic objects will vibrate when an impulse is applied to them. That is, once they are distorted, their shape tends to be restored to some equilibrium configuration. Even at the atomic level, the atoms in a solid vibrate about some position as if they were connected to their neighbors by some imaginary springs.

Wave motion is closely related to the phenomenon of vibration. Sound waves, earthquake waves, waves on stretched strings, and water waves are all produced by some source of vibration. As a sound wave travels through some medium, such as air, the molecules of the medium vibrate back and forth; as a water wave travels across a pond, the water molecules vibrate up and down. As such waves travel through a medium, the particles of the medium move in repetitive cycles. Therefore, the motion of the particles bears a strong resemblance to the periodic motion of a vibrating pendulum or a mass attached to a spring.

There are many other phenomena in nature whose explanation requires us to first understand the concepts of vibrations and waves. Although many large structures, such as skyscrapers and bridges, appear to be rigid, they actually vibrate, a fact that must be taken into account by the architects and engineers who design and build them. To understand how radio and television work, we must understand the origin and nature of electromagnetic waves and how they propagate through space. Finally, much of what scientists have learned about atomic structure has come from information carried by waves. Therefore, we must first study waves and vibrations in order to understand the concepts and theories of atomic physics.

CHAPTER 14

Vibrations and Waves

Water waves formed by raindrops falling in a pond. (Courtesy of Jay Freedman)

This chapter marks a brief return to the subject of mechanics as we examine various forms of periodic motion. We shall concentrate on a very special type of motion, that which occurs when the force on an object is proportional to the displacement of the object from its equilibrium position. When this force acts only toward the equilibrium position, the result is a back-and-forth motion called simple harmonic motion. An object moving in this manner oscillates, or vibrates, between two extreme positions for an indefinite period of time with no loss of energy. The terms *harmonic motion* and *periodic motion* will be used interchangeably throughout this chapter. Both refer to some form of back-and-forth motion. You are most likely familiar with several types of periodic motion, such as the oscillations of a mass on a spring, the motion of a pendulum, and the vibrations of a stringed musical instrument.

Since vibrations can move through a medium, we shall also study wave motion in this chapter. There are many kinds of waves in nature, such as sound waves, seismic waves, and electromagnetic waves. We shall end this chapter with a brief discussion of some terms and concepts that are common to all types of waves, and in later chapters we shall focus our attention on specific categories of waves.

14.1 HOOKE'S LAW

One of the simplest types of vibrational motion is that of a mass attached to a spring, as in Figure 14.1. Let us assume that the mass moves on a horizontal smooth surface. If the spring is stretched or compressed a small distance x from its unstretched, or equilibrium, position and then released, it will exert a force on the mass. This spring force is found from experiment to obey the equation

$$F_s = -kx \tag{14.1}$$

Hooke's law

where x is the displacement of the mass from its unstretched ($x = 0$) position and k is a positive constant called the **spring constant.** This force law for springs was discovered by Robert Hooke in 1678 and is known as **Hooke's law.** The value of k is a measure of the stiffness of the spring. Stiff springs have large k values, and soft springs have small k values.

The negative sign in Equation 14.1 signifies that the force exerted by the spring is always directed *opposite* the displacement of the mass. When the mass is to the right of the equilibrium position, as in Figure 14.1a, x is positive and F_s is negative. This means that the force is in the negative direction, to the left. When the mass is to the left of the equilibrium position, as in Figure 14.1c, x is negative and F_s is positive, indicating that the direction of the force is to the right. Of course, when $x = 0$, as in Figure 14.1b, the spring is unstretched and $F_s = 0$. Because the spring force always acts toward the equilibrium position, it is sometimes called a restoring force. *The direction of the restoring force is such that the mass is being either pulled or pushed toward the equilibrium position.*

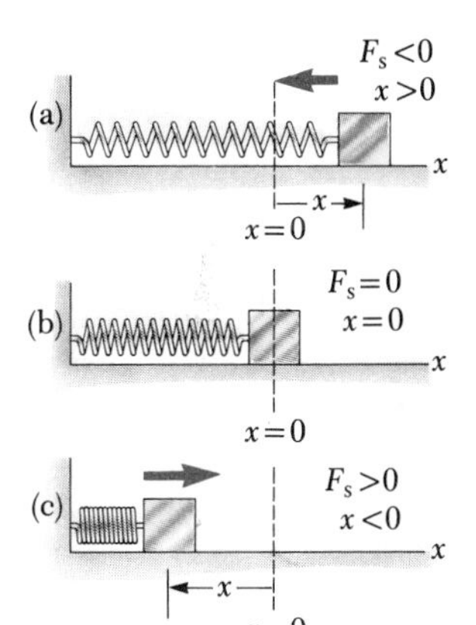

FIGURE 14.1 The force of a spring on a mass varies with the displacement of the mass from the equilibrium position $x = 0$. (a) When x is positive (stretched spring), the spring force is to the left. (b) When x is zero (unstretched spring), the spring force is zero. (c) When x is negative (compressed spring), the spring force is to the right.

Let us examine the motion of the mass if it is initially pulled a distance A to the right and released from rest. The force exerted on the mass by the spring pulls the mass back toward the equilibrium position. As the object moves toward $x = 0$, the magnitude of the force decreases (because x decreases) and reaches zero at $x = 0$. However, the speed of the mass increases as it moves toward the equilibrium position. In fact, the speed reaches its maximum value when $x = 0$. The momentum achieved by the mass causes it to overshoot the equilibrium position and to compress the spring. As the mass moves to the left of the equilibrium position (negative x values), the force acting on it begins to increase to the right and the speed of the mass begins to decrease. The mass finally comes to a stop at $x = -A$. The process is then repeated, and the mass continues to oscillate back and forth over the same path. This type of motion is referred to as **simple harmonic motion**. You should note that

simple harmonic motion occurs when the net force along the direction of motion is a Hooke's law type of force, that is, when the net force is proportional to the displacement and in the opposite direction.

Simple harmonic motion

It should be emphasized here that not all repetitive motion over the same path can be classified as simple harmonic motion. For example, a car could start at one end of a city block, accelerate to the other end of the block, stop, back up to its original position, and then repeat the process. The car is moving back and forth over the same path, but the motion is not simple harmonic. Unless the force acting on an object along the direction of motion is of the form of Equation 14.1, the object is not in simple harmonic motion.

The motion of a mass suspended from a vertical spring is also simple harmonic. In this case, the force of gravity acting on the attached mass stretches the spring until equilibrium is reached (where the mass is suspended and at rest). This position of equilibrium establishes the position of the mass for which $x = 0$. When the mass is stretched a distance x beyond this equilibrium position and then released, a net force acts (in the form of Hooke's law) toward the equilibrium position. Because the net force is proportional to x, the motion is simple harmonic.

Before we can discuss simple harmonic motion in more detail, it is necessary to define a few terms relative to such motion.

1. The **amplitude,** A, is the *maximum distance that an object moves away from its equilibrium position.* In the absence of friction, an object will continue in simple harmonic motion and reach a maximum displacement equal to the amplitude on each side of the equilibrium position during each cycle.
2. The **period,** T, is *the time it takes the object to execute one complete cycle of the motion.*
3. The **frequency,** f, is *the number of cycles or vibrations per unit of time.*

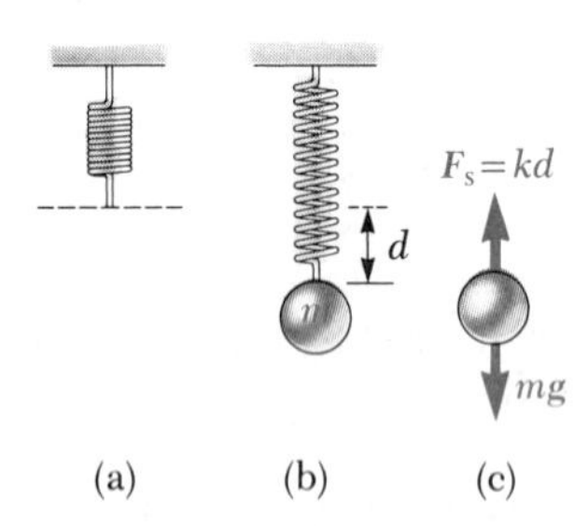

FIGURE 14.2 (Example 14.1) Determining the spring constant. The elongation, d, of the spring is due to the suspended weight, mg. Since the spring force upward balances the weight, it follows that $k = mg/d$.

EXAMPLE 14.1 Measuring the Spring Constant

A common technique used to evaluate the spring constant is illustrated in Figure 14.2. The spring is hung vertically as in Figure 14.2a, and a body of mass m is attached to the lower end of the spring as in Figure 14.2b. The spring stretches a distance d from its initial position under the action of the "load" mg. Since the spring force is upward, it must balance the weight mg downward when the system is at rest. In this case, we can apply Hooke's law to give

$$F_s = kd = mg$$

$$k = \frac{mg}{d}$$

For example, if a spring is stretched 2.0 cm by a mass of 0.55 kg, the force constant is

$$k = \frac{mg}{d} = \frac{(0.55\text{ kg})(9.8\text{ m/s}^2)}{2.0 \times 10^{-2}\text{ m}} = 2.7 \times 10^2\text{ N/m}$$

EXAMPLE 14.2 Simple Harmonic Motion on a Slick Surface

A 0.35-kg mass attached to a spring of spring constant 130 N/m is free to move on a horizontal frictionless surface, as in Figure 14.1. If the mass is released from rest at $x = 0.1$ m, find the force on it and its acceleration at (a) $x = 0.1$ m, (b) $x = 0.05$ m, (c) $x = 0$ m, and (d) $x = -0.05$ m.

Solution: (a) The point at which the mass is released defines the amplitude of the motion. In this case, $A = 0.1$ m. The mass moves continuously between the limits of 0.1 m and -0.1 m. When x is a maximum ($x = A$), the force on the mass is a maximum and is calculated as follows:

$$F = -kx$$

$$F_{max} = -kA = -(130\text{ N/m})(0.1\text{ m}) = -13.0\text{ N}$$

The negative sign indicates that the force acts to the left, in the negative x direction. We can use Newton's second law to calculate the acceleration at this position:

$$F_{max} = ma_{max}$$

$$a_{max} = \frac{F_{max}}{m} = -\frac{13.0 \text{ N}}{0.35 \text{ kg}} = -37.1 \text{ m/s}^2$$

Again, the negative sign indicates that the acceleration is to the left.

(b) We can use the same approach to find the force and acceleration at other positions. At $x = 0.05$ m, we have

$$F = -kx = -(130 \text{ N/m})(0.05 \text{ m}) = -6.50 \text{ N}$$

$$a = \frac{F}{m} = -\frac{6.50 \text{ N}}{0.35 \text{ kg}} = -18.5 \text{ m/s}^2$$

Note that the acceleration of an object moving with simple harmonic motion *is not constant* since F is not constant.

(c) At $x = 0$, the spring force is zero (since $F = -kx = 0$) and the acceleration is zero. In other words, when the spring is unstretched, it exerts no force on the mass attached to it.

(d) At $x = -0.05$ m, we have

$$F = -kx = -(130 \text{ N/m})(-0.05 \text{ m}) = 6.50 \text{ N}$$

$$a = \frac{F}{m} = \frac{6.50 \text{ N}}{0.35 \text{ kg}} = 18.5 \text{ m/s}^2$$

This result shows that the force and acceleration are positive when the mass is on the negative side of the equilibrium position. This indicates that the force of the spring on the mass is acting to the right as the spring is being compressed. At the same time, the mass is slowing down as it moves from $x = 0$ to $x = -0.05$ m.

EXERCISE Find the force and acceleration when $x = -0.1$ m.

Answer: 13.0 N; 37.1 m/s^2.

As indicated in Example 14.2, the acceleration of an object moving with simple harmonic motion can be found by using Hooke's law as the force in the equation for Newton's second law, $F = ma$. This gives

$$-kx = ma$$

$$\boxed{a = -\frac{k}{m}x} \quad (14.2)$$

Since the maximum value of x is defined to be the amplitude, A, we see that the acceleration ranges over the values $-kA/m$ to $+kA/m$. Equation 14.2 enables us to find the acceleration of the object as a function of its position. In subsequent sections, we shall find equations for the velocity as a function of position and the position as a function of time.

Earlier we stated that an object will move with simple harmonic motion when the net force acting on it is proportional to its displacement from equilibrium. Equation 14.2 provides an alternative definition of simple harmonic motion. An object moves with simple harmonic motion if its acceleration is proportional to its displacement.

14.2 ELASTIC POTENTIAL ENERGY

So far we have worked with three types of mechanical energy: gravitational potential energy, translational kinetic energy, and rotational kinetic energy. We now consider a fourth type of mechanical energy: **elastic potential energy.**

An object has potential energy by virtue of its shape or position. As we learned in Chapter 5, an object of mass m at a distance h above the ground has gravitational potential energy equal to mgh. This means that the object can do work after it is released. Likewise, a compressed spring has potential energy by virtue of its shape. In this case, the compressed spring, when allowed to expand, can move an object and thus do work on the object. As an example, Figure 14.3 shows a ball being projected from a spring-loaded toy gun, where the spring is compressed by a distance, x. As the gun is fired, the compressed spring does work on the ball and imparts kinetic energy to it.

The energy stored in a stretched or compressed spring or other elastic material is called **elastic potential energy.**

To find an expression for elastic potential energy, let us find the work required to compress a spring from its equilibrium position to some final arbitrary position x. The force that must be applied to compress the spring varies from $F = 0$ to $F = kx$ at maximum compression. Because the force increases linearly with position (that is, $F \propto x$), the average force that must be applied is

$$\overline{F} = \frac{F_0 + F_x}{2} = \frac{0 + kx}{2} = \frac{1}{2}kx$$

Therefore, *the work done by the applied force* is

$$W = \overline{F}x = \frac{1}{2}kx^2$$

This work is stored in the spring as elastic potential energy. Thus, we define the **elastic potential energy,** PE_s, as

Elastic potential energy

$$\boxed{PE_s \equiv \frac{1}{2}kx^2} \qquad (14.3)$$

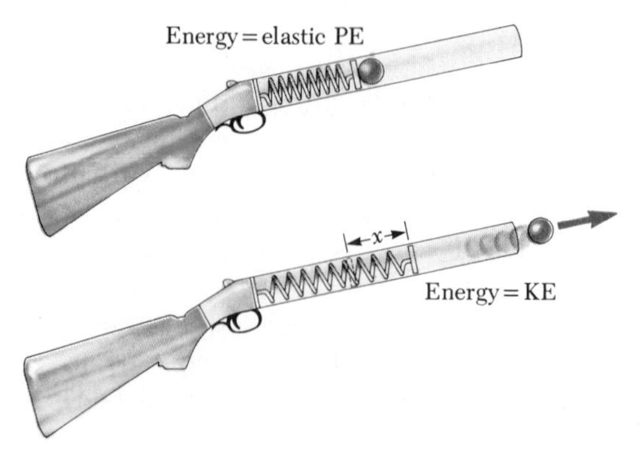

FIGURE 14.3 A ball projected from a spring-loaded gun. The elastic potential energy stored in the spring is transferred to the kinetic energy of the ball.

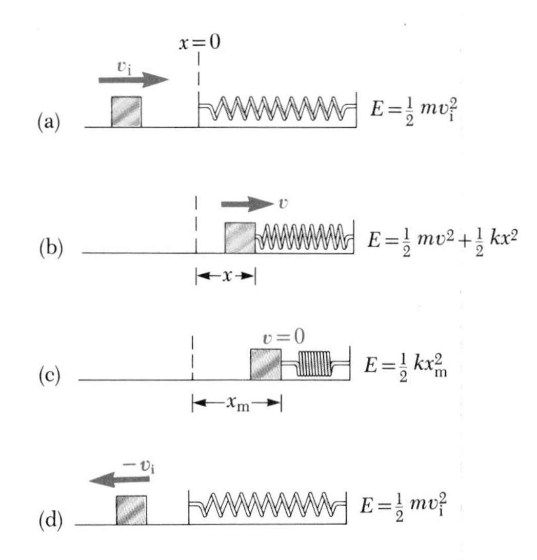

FIGURE 14.4 A block sliding on a frictionless horizontal surface collides with a light spring. (a) Initially, the mechanical energy is entirely the kinetic energy of the block. (b) The mechanical energy at some arbitrary position is the sum of the kinetic energy of the block and the elastic potential energy stored in the spring. (c) When the block comes to rest, the mechanical energy is entirely elastic potential energy stored in the compressed spring. (d) When the block leaves the spring, the mechanical energy is equal to its kinetic energy. Note that the total energy remains constant.

Note that the elastic potential energy stored in the spring is zero when the spring is unstretched or uncompressed ($x = 0$). *Energy is stored in the spring only when it is either stretched or compressed.* Furthermore, *the elastic potential energy is a maximum when the spring has reached its maximum compression or extension.* Finally, the potential energy in the spring is always positive because it is proportional to x^2. We now include this new form of energy in our equation for conservation of mechanical energy:

$$(KE + PE_g + PE_s)_i = (KE + PE_g + PE_s)_f \tag{14.4}$$

If there are nonconservative forces, such as friction, present, then the final mechanical energy will not equal the initial mechanical energy. In this case, the difference in the two energies must equal the work done by the nonconservative force, W_{nc}. According to the work-energy theorem, we have

$$W_{nc} = (KE + PE_g + PE_s)_f - (KE + PE_g + PE_s)_i \tag{14.5}$$

As an example of the energy conversions that take place when a spring is included in the system, consider Figure 14.4. A block of mass m slides on a frictionless horizontal surface with constant velocity v_i and collides with a coiled spring. The description that follows is greatly simplified by assuming that the spring is very light and therefore has negligible kinetic energy. As the spring is compressed, it exerts a force to the left on the block. At maximum compression, the block momentarily comes to rest (Fig. 14.4c). The initial total energy in the system before the collision (block plus spring) is the kinetic energy of the block. After the block collides with the spring and the spring is partially compressed, as in Figure 14.4b, the block has kinetic energy $\frac{1}{2}mv^2$ (where $v < v_i$) and the spring has potential energy $\frac{1}{2}kx^2$. When the block comes momentarily to rest after colliding with the spring, the kinetic energy is zero. Since the spring force is conservative and since there are no external forces that can do work on the system, *the total mechanical energy of the system consisting of the block and spring remains constant.* Thus, there is a transfer of energy from kinetic energy of the block to potential energy stored in the spring. As the spring expands, the block moves in the opposite direction and regains all of its initial kinetic energy, as in Figure 14.4d.

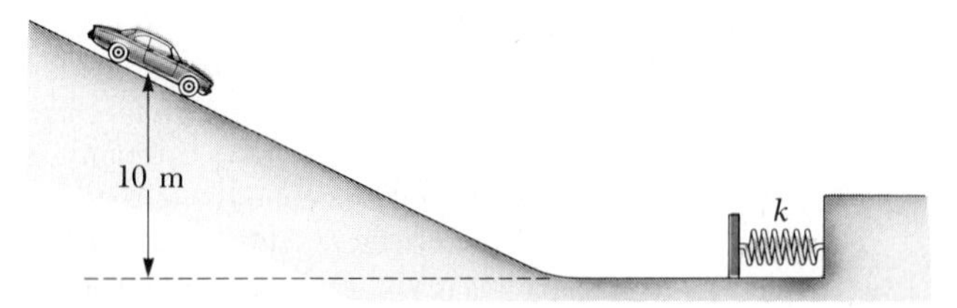

FIGURE 14.5 (Example 14.3) A car starts from rest on a hill at the position shown. When it reaches the bottom of the hill, the car collides with a spring-loaded guard rail.

EXAMPLE 14.3 Stop That Car!

A 13,000-N car starts at rest and rolls down a hill from a height of 10 m (Fig. 14.5). It then moves across a level surface and collides with a light spring-loaded guard rail. Neglecting any losses due to friction, find the maximum distance the spring is compressed. Assume the spring constant is 10^6 N/m.

Solution: The initial potential energy of the car is completely converted to elastic potential energy in the spring at the end of the trip. (This assumes we are neglecting any energy losses due to friction during the collision.) Thus, conservation of energy gives

$$mgh = \frac{1}{2}kx^2$$

Solving for x gives

$$x = \sqrt{\frac{2mgh}{k}} = \sqrt{\frac{2(13{,}000\ \text{N})(10\ \text{m})}{10^6\ \text{N/m}}} = 0.5\ \text{m}$$

Note that it was not necessary to calculate the velocity of the car at any point to obtain a solution. This demonstrates the power of the principle of conservation of energy. One has to work with the initial and final energy values only, without having to consider all the details in between.

EXERCISE What is the speed of the car just before it collides with the guard rail?

Answer: 14 m/s.

EXAMPLE 14.4 Motion With and Without Friction

A block of mass 1.6 kg is attached to a spring with a spring constant of 10^3 N/m, as in Figure 14.1. The spring is compressed a distance of 2.0 cm, and the block is released from rest. (a) Calculate the velocity of the block as it passes through the equilibrium position, $x = 0$, if the surface is frictionless.

Solution: By using Equation 14.3, we can find the initial elastic potential energy of the spring when $x = -2.0\ \text{cm} = -2 \times 10^{-2}$ m:

$$PE_s = \frac{1}{2}kx_i^2 = \frac{1}{2}(10^3\ \text{N/m})(-2 \times 10^{-2}\ \text{m})^2 = 0.20\ \text{J}$$

Since the block is always at the same height above the earth's surface, its gravitational potential energy remains constant. Hence, the initial potential energy stored in the spring is converted to kinetic energy at $x = 0$. That is,

$$\frac{1}{2}kx_1^2 = \frac{1}{2}mv_f^2$$

$$0.2\text{ J} = \frac{1}{2}(1.6\text{ kg})(v_f^2)$$

$$v_f = 0.5\text{ m/s}$$

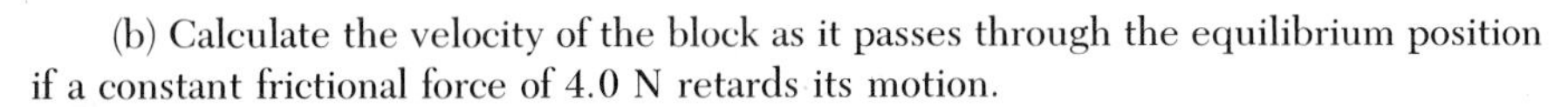

(b) Calculate the velocity of the block as it passes through the equilibrium position if a constant frictional force of 4.0 N retards its motion.

Solution: Since sliding friction is present in this situation, we know that the final mechanical energy will be less than the initial mechanical energy. The work done by the frictional force for a displacement of 2×10^{-2} m is

$$W_{nc} = W_f = -fs = -(4.0\text{ N})(2 \times 10^{-2}\text{ M}) = -0.08\text{ J}$$

Applying Equation 14.5 to this situation gives

$$-0.08\text{ J} = \frac{1}{2}(1.6\text{ kg})(v_f^2) - 0.20\text{ J}$$

$$v_f = 0.39\text{ m/s}$$

Note that this value for v_f is less than that obtained in the frictionless case. Does the result make sense?

EXERCISE How far does the block travel before coming to rest? Assume a constant friction force of 4.0 N and the same initial conditions as stated above.

Answer: 5 cm.

14.3 VELOCITY AS A FUNCTION OF POSITION

Conservation of energy provides us with a simple method for deriving an expression that gives the velocity of a mass undergoing periodic motion as a function of position. The mass in question is initially located at its maximum extension A, as in Figure 14.6a, and is then released from rest. In this situation, the initial energy of the system is entirely elastic potential energy stored in the spring, $\frac{1}{2}kA^2$. As the mass moves toward the origin to some new position x (Fig. 14.6b), part of this energy is transformed to kinetic energy and the potential energy stored in the spring is reduced to $\frac{1}{2}kx^2$. Since the total energy of the system is equal to $\frac{1}{2}kA^2$ (the initial energy stored in the spring), we can equate this to the sum of the kinetic and potential energies at the final position:

$$\frac{1}{2}kA^2 = \frac{1}{2}mv^2 + \frac{1}{2}kx^2$$

When this is solved for v, we get

$$v = \pm\sqrt{\frac{k}{m}(A^2 - x^2)} \tag{14.6}$$

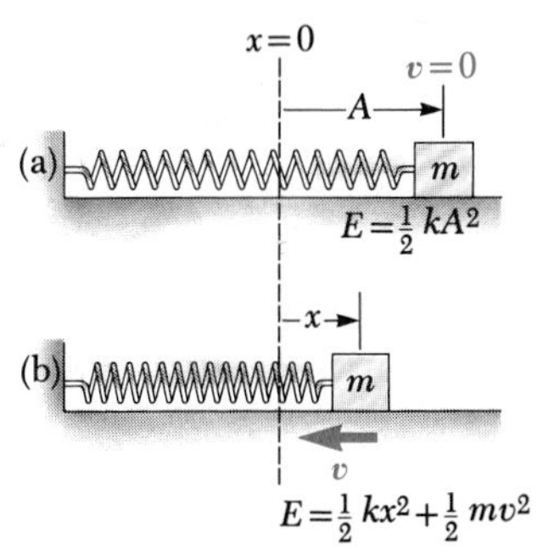

FIGURE 14.6 (a) A mass attached to a spring on a frictionless surface is released from rest with the spring extended a distance A. Just before release, the total energy is elastic potential energy, $kA^2/2$. (b) When the mass reaches position x, it has kinetic energy $mv^2/2$ and the elastic potential energy has decreased to $kx^2/2$.

This expression shows us that the speed is a maximum at $x = 0$ and zero at the extreme positions, $x = \pm A$.

The right side of Equation 14.6 is preceded by the $\pm$ sign because the square root of a number can be either positive or negative. The sign of v that should be selected depends on the circumstances of the motion. If the mass in Figure 14.6 is moving to the right, v must be positive; if the mass is moving to the left, v must be negative.

EXAMPLE 14.5 The Mass-Spring System Revisited

A mass of 0.5 kg connected to a light spring of spring constant 20 N/m oscillates on a horizontal frictionless surface. (a) Calculate the total energy of the system and the maximum speed of the mass if the amplitude of the motion is 3 cm.

Solution: From Equation 14.3, we have

$$E = PE_s = \frac{1}{2}kA^2 = \frac{1}{2}(20 \text{ N/m})(3 \times 10^{-2} \text{ m})^2 = 9.0 \times 10^{-3} \text{ J}$$

When the mass is at $x = 0$, $PE_s = 0$ and $E = \frac{1}{2}mv_{max}^2$; therefore

$$\frac{1}{2}mv_{max}^2 = 9 \times 10^{-3} \text{ J}$$

$$v_{max} = \sqrt{\frac{18 \times 10^{-3} \text{ J}}{0.5 \text{ kg}}} = 0.19 \text{ m/s}$$

(b) What is the velocity of the mass when the displacement is 2 cm?

Solution: We can apply Equation 14.6 directly:

$$v = \pm\sqrt{\frac{k}{m}(A^2 - x^2)} = \pm\sqrt{\frac{20}{0.5}(3^2 - 2^2) \times 10^{-4}} = \pm 0.14 \text{ m/s}$$

The $\pm$ sign indicates that the mass could be moving to the right or to the left at this instant.

(c) Compute the kinetic and potential energies of the system when the displacement equals 2 cm.

Solution: The results to (b) can be used to give

$$KE = \frac{1}{2}mv^2 = \frac{1}{2}(0.5 \text{ kg})(0.14 \text{ m/s})^2 = 5.0 \times 10^{-3} \text{ J}$$

$$PE_s = \frac{1}{2}kx^2 = \frac{1}{2}(20 \text{ N/m})(2 \times 10^{-2} \text{ m})^2 = 4.0 \times 10^{-3} \text{ J}$$

Note that the sum $KE + PE_s$ equals the total energy, E, found in part (a).

EXERCISE For what values of x does the speed of the mass equal 0.10 m/s?

Answer: ± 2.55 m.

14.4 COMPARING SIMPLE HARMONIC MOTION WITH UNIFORM CIRCULAR MOTION

We can better understand and visualize many aspects of simple harmonic motion along a straight line by looking at their relationships to uniform circular motion. Figure 14.7 shows an experimental arrangement useful for developing this concept. This figure represents a top view of a ball attached to the rim of a phonograph turntable of radius A, illuminated from the side by a lamp. Rather than concentrating on the ball, let us focus our attention on the shadow that the ball casts on the screen. We find that *as the turntable rotates with constant angular velocity, the shadow of the ball moves back and forth with simple harmonic motion.*

In order to understand why the shadow moves with simple harmonic motion, let us examine Equation 14.6 more closely. This equation says that the velocity of an object moving with simple harmonic motion is related to the displacement as

$$v = C\sqrt{A^2 - x^2}$$

where C is a constant. To see that the shadow also obeys this relation, consider Figure 14.8, which shows the ball moving with a constant speed, v_0, in a direction tangent to the circular path. At this instant, the velocity of the ball in the x direction, v, is given by $v = v_0 \sin\theta$, or

$$\sin\theta = \frac{v}{v_0}$$

Likewise, the larger triangle in Figure 14.8 which contains the angle θ enables us to write

$$\sin\theta = \frac{\sqrt{A^2 - x^2}}{A}$$

Equating the right-hand sides of the last two expressions, we see that v is related to the displacement x as

$$\frac{v}{v_0} = \frac{\sqrt{A^2 - x^2}}{A}$$

or

$$v = \frac{v_0}{A}\sqrt{A^2 - x^2} = C\sqrt{A^2 - x^2}$$

Thus, the velocity of the ball along the x direction is related to the displacement x in exactly the same manner as that of an object undergoing simple harmonic motion. Hence, the shadow moves with simple harmonic motion.

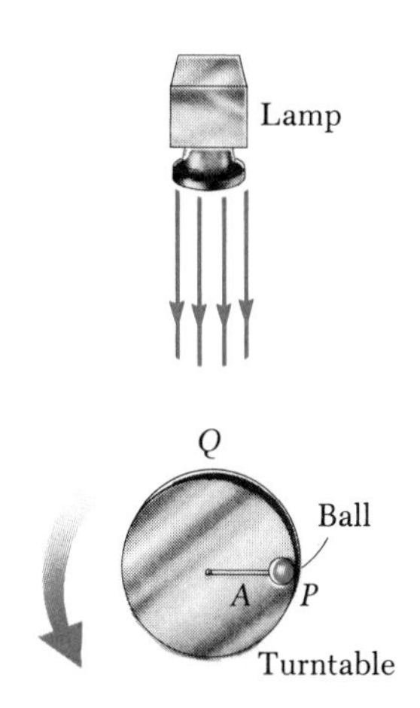

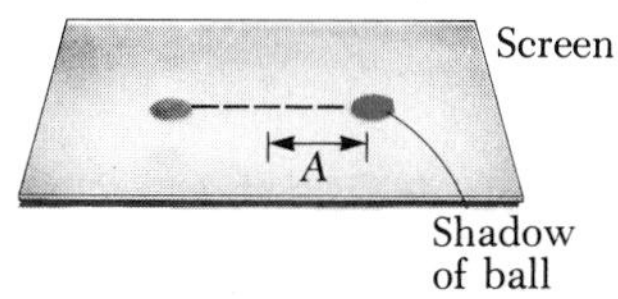

FIGURE 14.7 Experimental setup for demonstrating the connection between simple harmonic motion and uniform circular motion. As the ball rotates on the turntable with constant angular velocity, its shadow on the screen moves back and forth with simple harmonic motion.

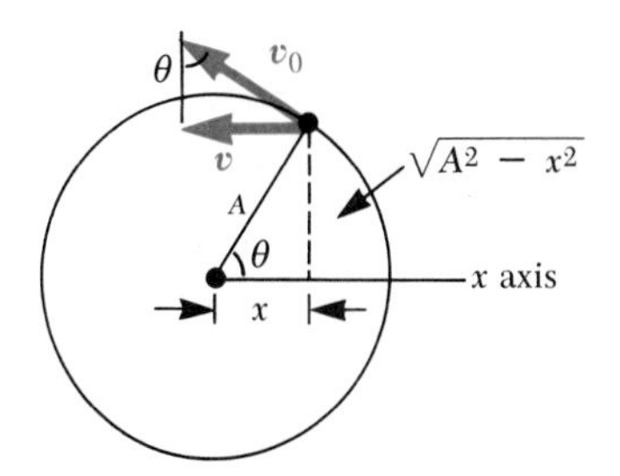

FIGURE 14.8 The ball rotates with constant speed v_0. The x component of the velocity of the ball equals the projection of v_0 on the x axis.

Period and Frequency

Note that the period, T, of the shadow, which equals the time for one complete trip back and forth, is also equal to the time it takes the ball to make one complete circular trip on the turntable. Since the ball moves through a distance $2\pi A$ (the circumference of the circle) in the time T, the speed v_0 of the ball around the circular path is

$$v_0 = \frac{2\pi A}{T}$$

$$T = \frac{2\pi A}{v_0}$$

However, for our purposes, let us consider only a fraction of the complete trip. Imagine that the ball moves from P to Q in Figure 14.7, which represents one quarter of a revolution. This requires a time equal to one fourth of the period, and the distance traveled by the ball is $2\pi A/4$. Therefore, we see that

$$\frac{T}{4} = \frac{2\pi A}{4v_0} \tag{14.7}$$

Now imagine that the motion of the shadow is equivalent to the motion of a mass on the end of a spring moving in the horizontal direction. During this quarter of a cycle, the shadow moves from a point where its energy is solely elastic potential energy to a point where its energy is solely kinetic energy. That is,

$$\frac{1}{2}kA^2 = \frac{1}{2}mv_0^{\,2}$$

$$\frac{A}{v_0} = \sqrt{\frac{m}{k}}$$

Substituting for A/v_0 in Equation 14.7, we find that the **period** is

Period

$$\boxed{T = 2\pi\sqrt{\frac{m}{k}}} \tag{14.8}$$

This expression gives the time required for an object to make a complete cycle of its motion. Now recall that the definition of frequency, f, is the number of cycles per unit of time. The symmetry in the units of period and frequency should lead you to see that the two must be related inversely as

$$\boxed{f = \frac{1}{T}} \tag{14.9}$$

Therefore, the **frequency** of the periodic motion is

Frequency

$$\boxed{f = \frac{1}{2\pi}\sqrt{\frac{k}{m}}} \tag{14.10}$$

The units of frequency are s^{-1}, or hertz (Hz).

EXAMPLE 14.6 That Car Needs a New Set of Shocks!

A car of mass 1300 kg is constructed using a frame supported by four springs. Each spring has a spring constant of 20,000 N/m. If two people riding in the car have a combined mass of 160 kg, find the frequency of vibration of the car when it is driven over a pot hole in the road.

Solution: We shall assume that the weight is evenly distributed. Thus, each spring supports one fourth of the load. The total mass supported by the springs is 1460 kg, and therefore each spring supports 365 kg. Hence, the frequency of vibration is

$$f = \frac{1}{2\pi}\sqrt{\frac{k}{m}} = \frac{1}{2\pi}\sqrt{\frac{20,000 \text{ N/m}}{365 \text{ kg}}} = 1.18 \text{ Hz}$$

EXERCISE How long does it take the car to execute three complete vibrations?

Answer: 2.54 s.

14.5 POSITION AS A FUNCTION OF TIME

We can obtain an expression for the position of an object moving with simple harmonic motion as a function of time by returning to the relationship between simple harmonic motion and uniform circular motion. Again, consider a ball on the rim of a rotating turntable of radius A, as in Figure 14.9. Let us assume that the turntable rotates at a constant angular velocity ω. We shall refer to this circle as the *reference circle* for the motion. As the ball rotates on the reference circle, the angle θ that the line OP makes with the x axis changes with time. Furthermore, as the ball rotates, the projection of P on the x axis, labeled point Q, moves back and forth along the axis with simple harmonic motion.

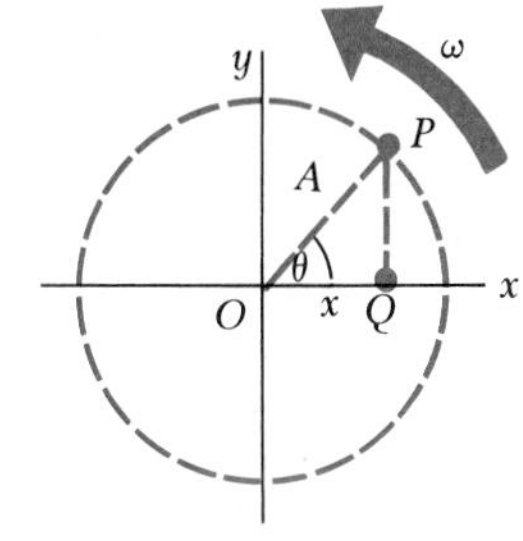

FIGURE 14.9 The reference circle. As the ball at P rotates in a circle with uniform angular velocity, its projection Q along the x axis moves with simple harmonic motion.

From the right triangle, OPQ, we see that $\cos\theta = x/A$. Therefore, the x coordinate of the ball is

$$x = A\cos\theta$$

Since the ball rotates with constant angular velocity, it follows that $\theta = \omega t$ (see Chapter 7). Therefore,

$$\boxed{x = A\cos(\omega t)} \qquad (14.11)$$

In one complete revolution, the ball rotates through an angle of 2π rad in a time equal to a period T. In other words, the motion repeats itself every T seconds. Therefore,

$$\omega = \frac{\Delta\theta}{\Delta t} = \frac{2\pi}{T} = 2\pi f \qquad (14.12)$$

where f is the frequency of the motion. Consequently, Equation 14.11 can be written

$$\boxed{x = A\cos(2\pi f t)} \qquad (14.13)$$

This equation represents the position of an object moving with simple harmonic motion as a function of time. A graph of this equation is shown in Figure 14.10a. The curve should be familiar to you from trigonometry. Note that x varies between A and $-A$ since the cosine function varies between 1 and -1.

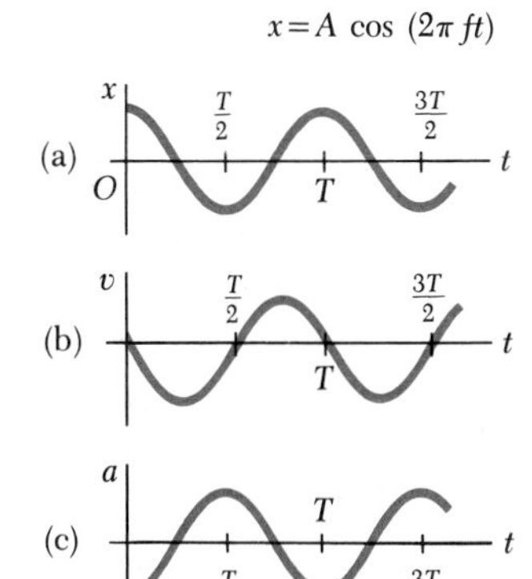

FIGURE 14.10 (a) Displacement, (b) velocity, and (c) acceleration versus time for an object moving with simple harmonic motion under the initial conditions that $x_0 = A$ and $v_0 = 0$ at $t = 0$.

Figure 14.10b and 14.10c represent curves for the velocity and acceleration as a function of time. Although we shall not prove it here, one can show that the velocity and acceleration are also sinusoidal functions of time. Note that when x is a maximum or minimum, the velocity is zero, and when x is zero, the magnitude of the velocity is a maximum. Furthermore, when x has its maximum positive value, the acceleration is a maximum but in the negative x direction, and when x is at its maximum negative position, the acceleration has its maximum value in the positive direction. These curves are consistent with what we discussed earlier concerning the points at which v and a reach their maximum, minimum, and zero values.

One experimental arrangement that demonstrates simple harmonic motion is illustrated in Figure 14.11. A mass connected to a spring has a marking pen attached to it. While the mass vibrates vertically, a sheet of paper is moved horizontally with constant speed, and the pen traces out a sinusoidal pattern.

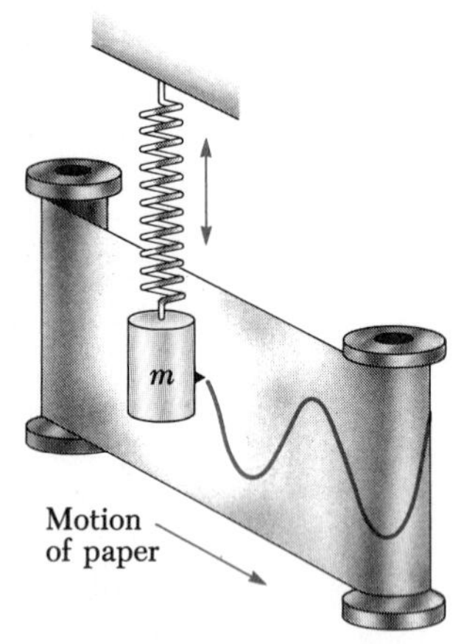

FIGURE 14.11 An experimental apparatus for demonstrating simple harmonic motion. A pen attached to the oscillating mass traces out a sinusoidal wave on the moving chart paper.

EXAMPLE 14.7 The Vibrating Mass-Spring System

Find the amplitude, frequency, and period of motion for an object vibrating at the end of a spring if the equation for its position as a function of time is

$$x = (0.25 \text{ m}) \cos\left(\frac{\pi}{8}t\right)$$

Solution: We can find two of our unknowns by comparing this equation with the general equation for such motion:

$$x = A \cos(2\pi f t)$$

By comparison, we see that

$$A = 0.25 \text{ m}$$

$$2\pi f = \frac{\pi}{8} \text{ s}^{-1}$$

$$f = \frac{1}{16} \text{ Hz}$$

Since the period $T = 1/f$, it follows that $T = 1/f = 16$ s

EXERCISE What is the position of the object after 2 s has elapsed?

Answer: 0.18 m.

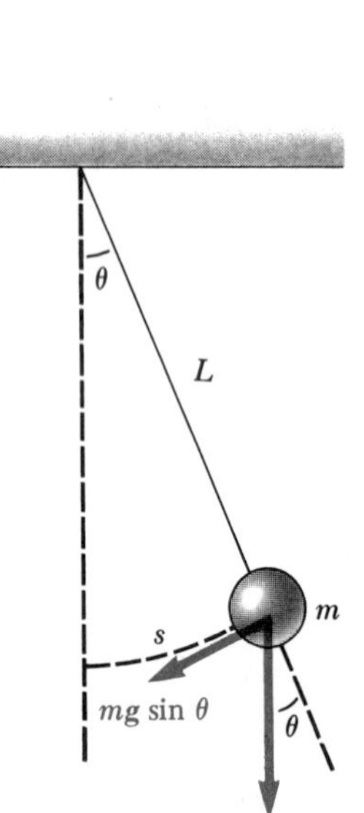

FIGURE 14.12 A simple pendulum consists of a mass m suspended by a light string of length L. The restoring force that causes the pendulum to undergo simple harmonic motion is the component of weight tangent to the path of motion, $mg \sin \theta$.

14.6 THE PENDULUM

A simple pendulum consists of a small mass m suspended by a light string of length L fixed at its upper end, as in Figure 14.12. (By a light string, we mean that its mass is assumed to be very small compared to the mass of the bob and hence can be ignored.) When released, the mass swings to and fro over the same path, but is its motion simple harmonic? In order to answer this question, we shall have to examine the force that acts as the restoring force on the pendulum. If this force is proportional to the displacement, s, then the force is of the Hooke's law form, $F = -ks$, and hence the motion is simple harmonic. Furthermore, since $s = L\theta$ in this case, we see that the motion is simple harmonic if F is proportional to the angle θ.

The component of weight tangential to the circular path is the force that acts to restore the pendulum to its equilibrium position. Thus, the restoring force is

$$F_t = -mg \sin \theta$$

From this equation, we see that the restoring force is proportional to sin θ rather than to θ. Thus, in general, the motion of a pendulum is not simple harmonic. However, for small angles, less than about 15 degrees, the angle θ measured in radians and the sine of the angle are approximately equal. Therefore, if we restrict the motion to small angles, the restoring force can be written as

$$F_t = -mg\theta$$

Because $s = L\theta$, we have

$$F_t = -\left(\frac{mg}{L}\right)s$$

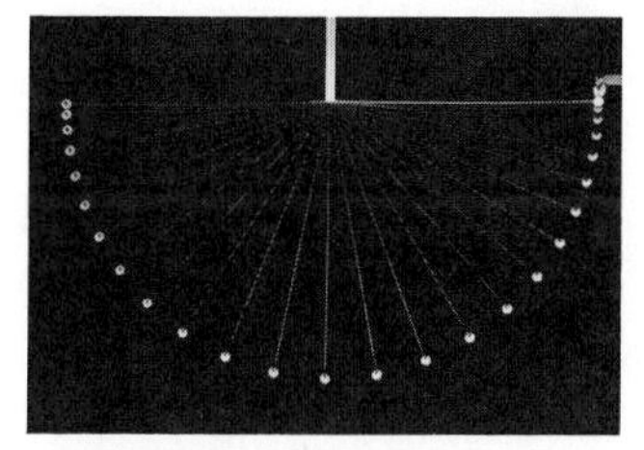

The motion of a simple pendulum captured with multiflash photography. Is the motion simple harmonic in this case? (Photograph © Bernice Abbott, 1963)

This equation is similar to the general form of the Hooke's law force, given by $F = -ks$, with $k = mg/L$. Thus, we are justified in saying that a pendulum undergoes simple harmonic motion only when it swings back and forth at very small amplitudes (or in this case, small values of θ, so that $\sin \theta \cong \theta$).

We can find the period of a pendulum by first recalling that the period of a mass-spring system is (Eq. 14.8)

$$T = 2\pi\sqrt{\frac{m}{k}}$$

If we replace k by its equivalent, mg/L, we see that the period of a simple pendulum is

$$T = 2\pi\sqrt{\frac{m}{mg/L}}$$

$$\boxed{T = 2\pi\sqrt{\frac{L}{g}}} \qquad (14.14)$$

This equation reveals the somewhat surprising result that the period of a simple pendulum does not depend on the mass. The period of a simple pendulum depends only on its length and on the acceleration due to gravity. Furthermore, the amplitude of the motion is not a factor as long as we restrict the motion to small amplitudes. The analogy between the motion of a simple pendulum and the mass-spring system is illustrated in Figure 14.13.

It is of historical interest to point out that it was Galileo who first noted that the motion of a pendulum was independent of its amplitude. He supposedly observed this while attending church services at a cathedral in Pisa. The pendulum he studied was a swinging chandelier that was set into motion when someone bumped it while lighting the candles. Galileo was able to measure its frequency, and hence its period, by timing the swings with his pulse.

Geologists often make use of the simple pendulum, and Equation 14.14, when prospecting for oil or minerals. Deposits beneath the earth's surface can

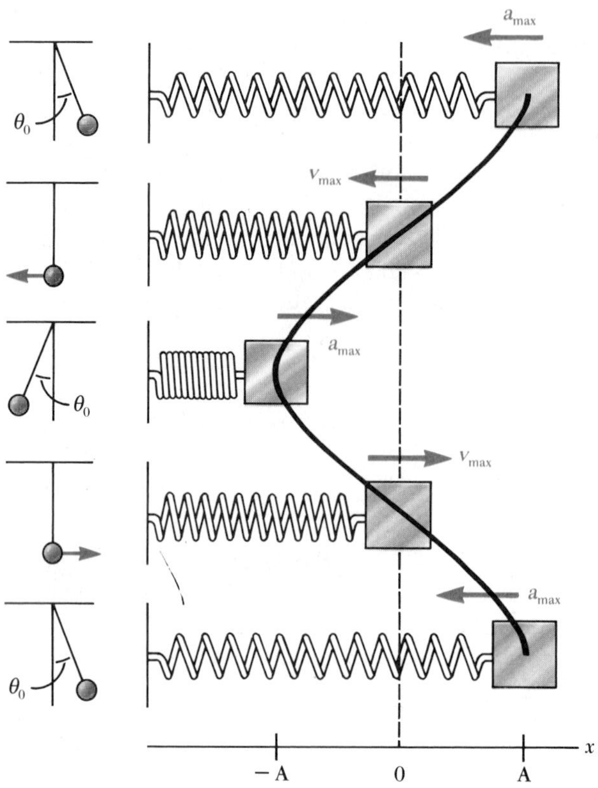

FIGURE 14.13 Simple harmonic motion for a mass-spring system and its analogy to the motion of a simple pendulum.

produce irregularities in the acceleration due to gravity over the region under investigation. In such studies, a specially designed pendulum of known length is used to measure the period, which in turn is used to calculate g. Although such a measurement in itself is inconclusive, it is an important tool for geological surveys.

EXAMPLE 14.8 What Is the Height of That Tower?
A man enters a tall tower. He needs to know the height of the tower, but darkness obscures the ceiling. He does note, however, that a long pendulum extends from the ceiling almost to the floor and that its period is 12 s. How tall is the tower?

Solution: If we use $T = 2\pi\sqrt{L/g}$ and solve for L, we get

$$L = \frac{gT^2}{4\pi^2} = \frac{(9.80 \text{ m/s}^2)(12 \text{ s})^2}{4\pi^2} = 35.8 \text{ m}$$

EXERCISE If the length of the pendulum is halved, what would its period of vibration be?

Answer: 8.49 s.

---

*14.7 DAMPED OSCILLATIONS

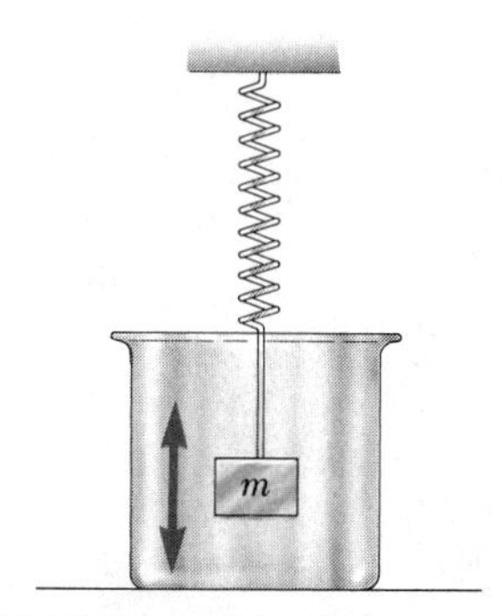

FIGURE 14.14 One example of a damped oscillator is a mass submerged in a liquid.

The vibrating motions we have discussed so far have dealt with ideal systems, that is, systems that *oscillate indefinitely* under the action of a linear restoring force. In real systems, forces of friction retard the motion and consequently the systems do not oscillate indefinitely. The presence of friction reduces the mechanical energy of the system as time progresses, and the motion is said to be **damped.** One system in which the motion is damped is a mass connected to a spring and moving through a liquid, as in Figure 14.14. If a small mass moves with low velocity through the liquid, a frictional force proportional to the velocity acts on the mass. That is, $f = -bv$, where b is called the coefficient of resistance.

There are various types of damped motion, depending on the magnitude of b. When b is small, the vibrating motion is preserved but the amplitude of vibration decreases in time and the motion ultimately ceases. This is known as an *underdamped* oscillator. The position-time curve for this case is shown in Figure 14.15. Note that the amplitude decreases with time, as you would expect. As b increases, the oscillations dampen more quickly. At some value of b, the mass returns rapidly to equilibrium when released and does not oscillate. At this value of b, the system is said to be *critically damped* (Fig. 14.16a); the mass reaches equilibrium in the shortest time possible without ever once passing through the equilibrium point. When b is larger than this critical value, the system is said to be *overdamped.* In this case, the mass, when released, again returns to equilibrium without ever passing through the equilibrium point, but the time required to reach equilibrium is greater than at critical damping, as shown by Figure 14.16b.

In any event, whenever friction is present, the mechanical energy of the system will eventually become zero. The lost mechanical energy is dissipated as thermal energy.

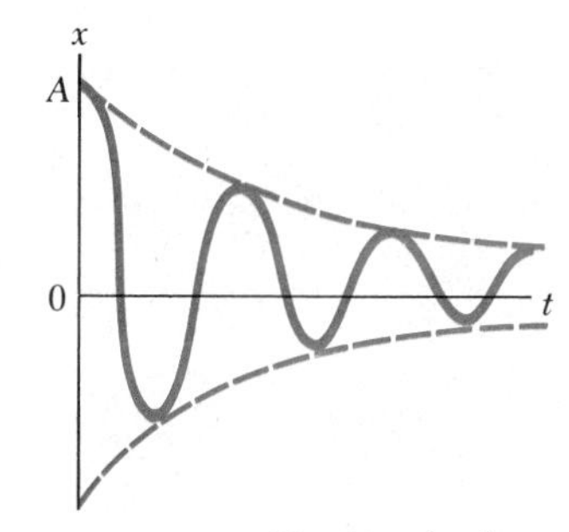

FIGURE 14.15 Graph of the displacement versus time for an underdamped oscillator. Note the decrease in amplitude with time.

14.8 WAVE MOTION

Most of us first saw waves when, as children, we dropped a pebble into a pool of water. The disturbance created by the pebble excites water waves, which move outward until they finally reach the edge of the pool. There are a wide variety of physical phenomena that have wave-like characteristics. The world is full of waves: sound waves, waves on a string, earthquake waves, and electromagnetic waves, such as visible light, radio waves, television signals, and x-rays. All of these waves have as their source a vibrating object. Thus, we shall use the terminology and concepts of simple harmonic motion as we move into the study of wave motion.

In the case of sound waves, the vibrations that produce waves arise from such sources as a person's vocal cords or a plucked guitar string. The vibrations of electrons in an antenna produce radio or television waves, and the simple up-and-down motion of a hand can produce a wave on a string. Regardless of the type of wave under consideration, there are certain concepts common to all varieties. In the remainder of this chapter, we shall focus our attention on a general study of wave motion. In later chapters we shall study specific types of waves, such as sound and electromagnetic waves.

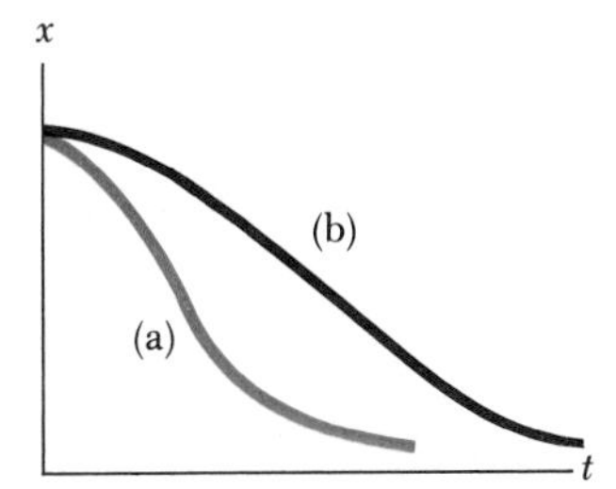

FIGURE 14.16 Plots of displacement versus time for (a) a critically damped oscillator and (b) an overdamped oscillator.

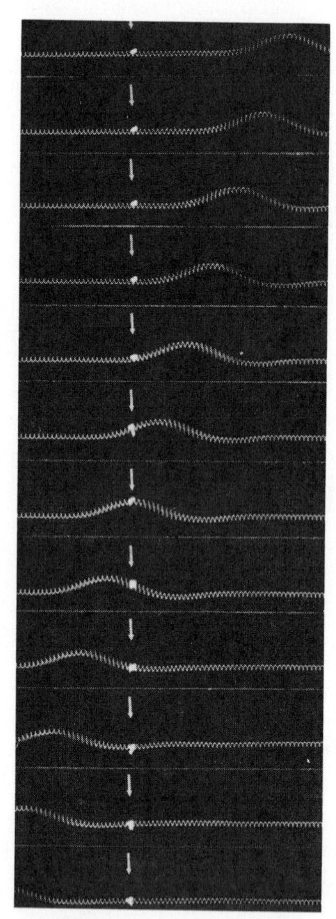

A disturbance traveling from right to left on a stretched spring.

What Is a Wave?

As we said before, when you drop a pebble into a pool of water, the disturbance produced by the pebble excites water waves, which move away from the point at which the pebble entered the water. If you were to examine carefully the motion of a leaf floating near the disturbance, you would see that it moves up and down and back and forth about its original position but does not undergo any net displacement attributable to the disturbance. That is, the water wave (or disturbance) moves from one place to another *but the water is not carried with it.*

Einstein and Infeld[1] make these remarks about wave phenomena:

> A bit of gossip starting in Washington reaches New York very quickly, even though not a single individual who takes part in spreading it travels between these two cities. There are two quite different motions involved, that of the rumor, Washington to New York, and that of the persons who spread the rumor. The wind, passing over a field of grain, sets up a wave which spreads out across the whole field. Here again we must distinguish between the motion of the wave and the motion of the separate plants, which undergo only small oscillations. . . . The particles constituting the medium perform only small vibrations, but the whole motion is that of a progressive wave. The essentially new thing here is that for the first time we consider the motion of something which is not matter, but energy propagated through matter.

When we observe what we call a water wave, what we see is a rearrangement of the water's surface. Without the water there would be no wave. A wave traveling on a string would not exist without the string. Sound waves travel through air as a result of pressure variations from point to point. (We shall discuss sound waves in Chapter 15.) The wave motion we shall consider in this chapter corresponds to the disturbance of a body or medium. Therefore, we can consider a wave *to be the motion of a disturbance.* In a later chapter, we shall discuss electromagnetic waves, which do not require a medium.

The mechanical waves discussed in this chapter require (1) some source of disturbance, (2) a medium that can be disturbed, and (3) some physical connection or mechanism through which adjacent portions of the medium can influence each other. All waves carry energy and momentum. The amount of energy transmitted through a medium and the mechanism responsible for the transport of energy will differ from case to case. For instance, the energy carried by ocean waves during a storm is much greater than that carried by a sound wave generated by a single human voice.

14.9 TYPES OF WAVES

One of the simplest ways to demonstrate wave motion is to flip one end of a long rope that is under tension and has its opposite end fixed, as in Figure 14.17. The bump (called a pulse) travels to the right with a definite speed. A disturbance of this type is called a **traveling wave.** Figure 14.7 shows three consecutive "snapshots" of the traveling wave.

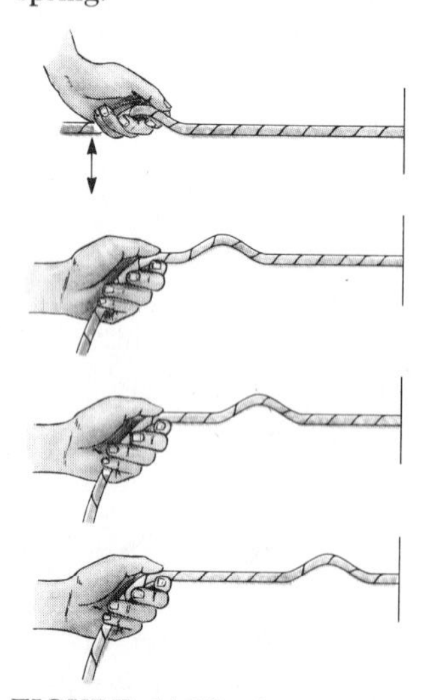

FIGURE 14.17 A wave pulse traveling down a stretched rope.

[1]Albert Einstein and Leopold Infeld, *The Evolution of Physics*, New York, Simon and Schuster, 1961.

An important point about a traveling wave is that, as the wave pulse travels along the rope, *each segment of the rope that is disturbed moves in a direction perpendicular to the wave motion.* Figure 14.18 illustrates this point for one particular segment, labeled P. Never does the rope move in the direction of the wave.

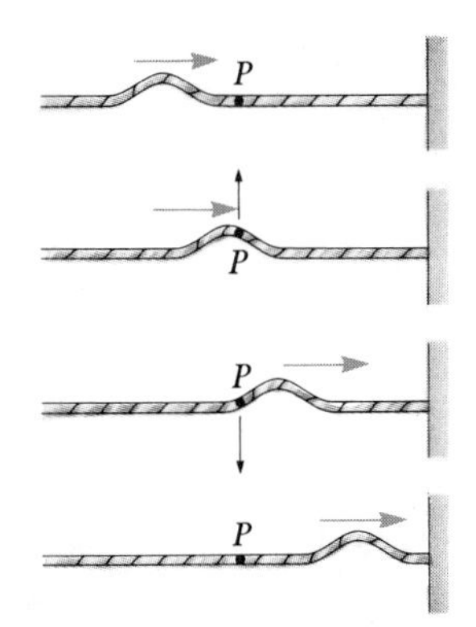

FIGURE 14.18 A pulse traveling on a stretched rope is a transverse wave. Any element P on the rope moves in a direction *perpendicular* to the motion of the pulse.

A traveling wave such as this, in which the particles of the disturbed medium move perpendicular to the wave velocity, is called a **transverse wave.**

Other examples of transverse waves are electromagnetic waves, such as light, radio, and television waves.

In another class of waves, called **longitudinal waves,** the particles of the medium undergo displacement in a direction parallel to the direction of travel of the wave.

Sound waves are longitudinal waves. The disturbance in a sound wave corresponds to a series of high- and low-pressure regions that travel through a material. A longitudinal pulse can be easily produced in a stretched spring, as in Figure 14.19. The left end of the spring is given a sudden jerk (consisting of a brief push to the right and an equally brief pull to the left) along the length of the spring; this creates a sudden compression of the coils. The compressed region C travels along the spring, and so we see that the disturbance is parallel to the wave motion. Region C is followed by region R, where the coils are spread farther apart than normal.

Picture of a Wave

The solid curve in Figure 14.20 shows the shape of a vibrating string. This particular pattern should be familiar to you from our study of simple harmonic motion; it is a sinusoidal curve. The solid curve can be thought of as a snapshot of a traveling wave taken at some instant of time; the dashed curve represents a snapshot of the same traveling wave at a later time. It is not difficult to imagine that the picture of the wave on a string in Figure 14.20 can as easily be used to represent a wave on water. In such a case, point A would correspond to the *crest* of the wave and point B would correspond to the low point, or *trough*, of the wave.

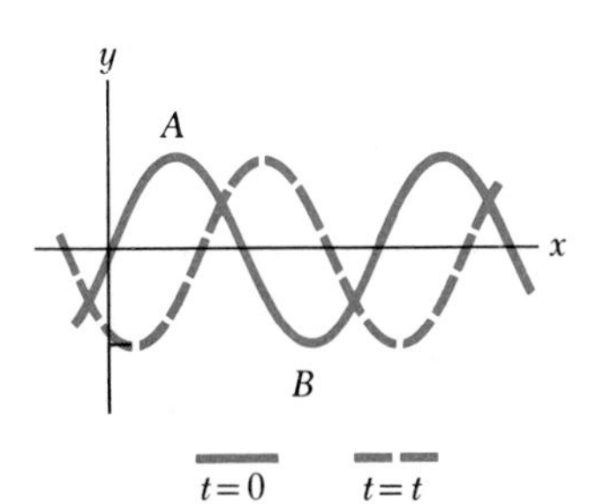

FIGURE 14.20 A harmonic wave traveling to the right. The solid curve represents a snapshot of the wave at $t = 0$, and the dashed curve is a snapshot of the wave at some later time t.

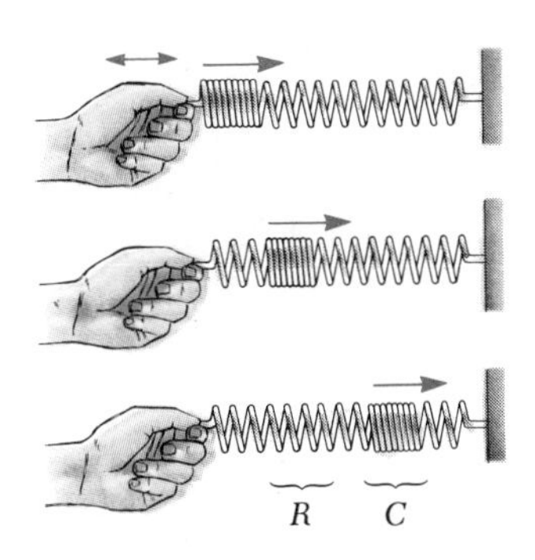

FIGURE 14.19 A longitudinal pulse moving along a stretched spring. The displacement of the coils is in the direction of the wave motion. For the starting motion described in the text, the compressed region C is followed by an extended region R.

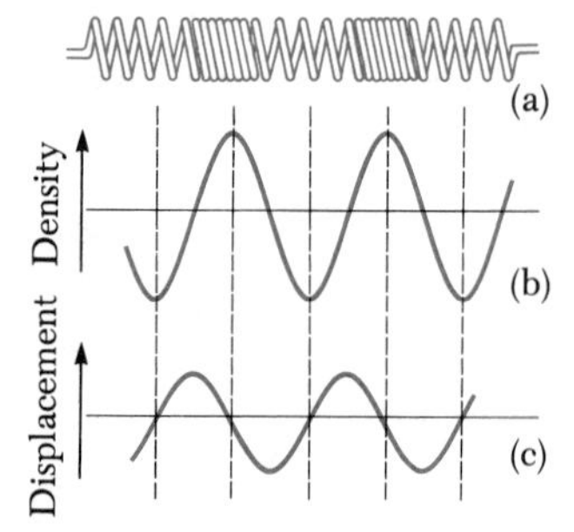

FIGURE 14.21 (a) A longitudinal wave on a spring. (b) The crests of the sinusoidal wave correspond to positions on the spring where the coils are close together, and the troughs of the wave correspond to positions of loosely spaced coils. (c) The displacement wave.

You may be surprised to find out that this same graphical representation can be used to describe a longitudinal wave. To see how this is done, consider a longitudinal wave traveling on a spring. Figure 14.21a represents a snapshot of the longitudinal wave at some instant. Figure 14.21b shows the sinusoidal curve that represents the wave. Points where the coils of the spring are packed tightly together correspond to the crests of the sinusoidal curve, and loosely spaced coils correspond to troughs.

The wave represented by the curve in Figure 14.21b is often referred to as a density or pressure wave. The reason for this terminology is that the crests of the curve indicate locations where the spring coils are close together (high density) and the troughs represent locations where the coils are far apart (low density).

An alternative method for representing wave motion along a spring is through the concept of a displacement wave, shown in Figure 14.21c. In this representation, coils that are displaced the greatest distance from equilibrium in one direction are indicated by crests and coils that are displaced the greatest distance from equilibrium in the opposite direction are represented by troughs. Both of these wave representations will be used in future sections.

14.10 FREQUENCY, AMPLITUDE, AND WAVELENGTH

One method of producing a wave on a very long string is shown in Figure 14.22. One end of the string is connected to a blade that is set into vibration. As the blade oscillates vertically with simple harmonic motion, a traveling wave moving to the right is set up on the string. Figure 14.22 represents snapshots of the wave at intervals of one quarter of a period. Note that *each particle of the string, such as P, oscillates vertically in the y direction with simple harmonic motion.* This must be the case because each particle follows the simple harmonic motion of the blade. Therefore, every segment of the string can be treated as a simple harmonic oscillator vibrating with a frequency equal to the frequency of vibration of the blade that drives the string.

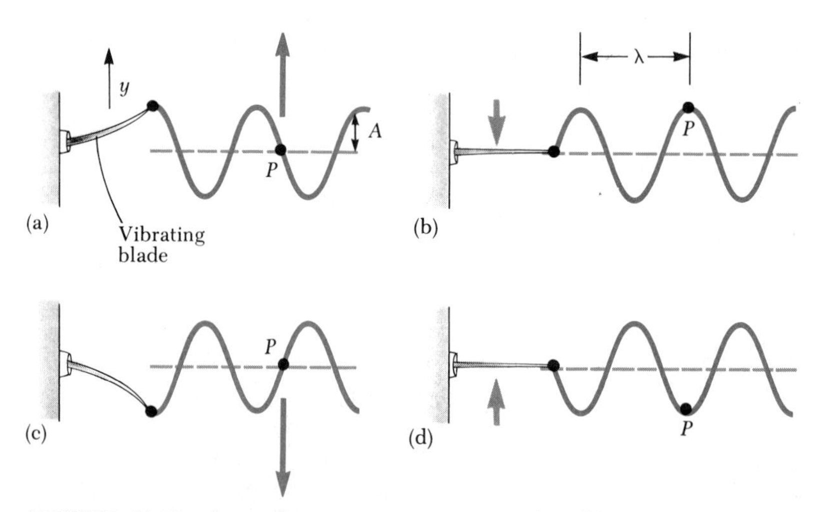

FIGURE 14.22 A traveling wave on a string produced by connecting one end of the string to a vibrating blade. The vertical bold arrows represent the direction of motion at the point P at some instant.

The frequencies of the waves we shall study will range from rather low values for waves on a string and waves on water to values between 20 and 20,000 Hz (recall that 1 Hz = 1 s^{-1}) for sound waves and to much higher frequencies for electromagnetic waves.

The horizontal dashed line in Figure 14.22 represents the shape of the string if no wave was present. The maximum distance that the string is raised above this equilibrium value is called the **amplitude,** A, of the wave. The amplitude can also be designated as the maximum distance that the string falls below the equilibrium value. Note that, for the waves we shall be working with, the amplitudes at the crest and the trough are identical.

Another characteristic of a wave is shown in Figure 14.22b. The horizontal arrows show the distance between two successive points that behave identically. This distance is called the **wavelength,** λ (Greek letter lambda).

We can use these definitions to derive an expression for the velocity of a wave. We start with the defining equation for velocity:

$$v = \frac{\Delta x}{\Delta t}$$

A little reflection should convince you that a wave will advance a distance equal to one wavelength in a time equal to one period of the vibration. Substitution in the velocity equation gives

$$v = \frac{\lambda}{T}$$

Since the frequency is equal to the reciprocal of the period, we have

$$\boxed{v = \lambda f} \tag{14.15}$$

Wave velocity

We shall apply this equation to many types of waves. For example, we shall use it often in our study of sound and electromagnetic waves.

EXAMPLE 14.9 A Traveling Wave

A wave traveling in the positive x direction is pictured in Figure 14.23. Find the amplitude, wavelength, period, and speed of the wave if it has a frequency of 8 Hz.

Solution: The amplitude and wavelength can be read directly off the figure:

$$A = 15 \text{ cm} \qquad \lambda = 40 \text{ cm} = 0.40 \text{ m}$$

The period of the wave is

$$T = \frac{1}{f} = \frac{1}{8} \text{ s} = 0.125 \text{ s}$$

and the speed is

$$v = \lambda f = (0.40 \text{ m})(8 \text{ s}^{-1}) = 3.2 \text{ m/s}$$

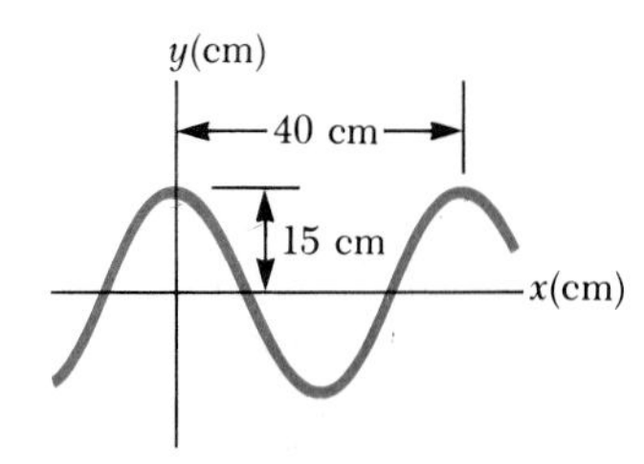

FIGURE 14.23 (Example 14.9) A harmonic wave of wavelength λ = 40 cm and amplitude A = 15 cm.

EXAMPLE 14.10 Give Me a "C" Note

The note middle C on a piano has a frequency of approximately 264 Hz and a wavelength in air of 1.31 m. Find the speed of sound in air.

Solution: By direct substitution into Equation 14.15, we find

$$v = \lambda f = (1.31 \text{ m})(264 \text{ s}^{-1}) = 346 \text{ m/s}$$

EXAMPLE 14.11 The Speed of Radio Waves

An FM station broadcasts at a frequency of 100 MHz (M = mega = 10^6) with a radio wave having a wavelength of 3 m. Find the speed of the radio wave.

Solution: As in the last example, we use Equation 14.15:

$$v = \lambda f = (3 \text{ m})(100 \times 10^6 \text{ s}^{-1}) = 3 \times 10^8 \text{ m/s}$$

This, in fact, is the speed of *all* electromagnetic waves traveling through empty space.

EXERCISE Find the wavelength of an electromagnetic wave whose frequency is 9 GHz = 9×10^9 Hz, which is the microwave range.

Answer: 0.033 m.

14.11 THE VELOCITY OF WAVES ON STRINGS

In this section, we shall focus our attention on the speed of a wave on a stretched string. Rather than deriving the equation, we shall use dimensional analysis to verify that the expression can be valid.

It is easy to understand why the wave velocity depends on the tension in the string. If a string under tension is pulled sideways and released, the tension in the string is responsible for accelerating a particular segment back toward its equilibrium position. The acceleration and wave velocity increase with increasing tension in the string. Likewise, the wave velocity is inversely proportional to the mass per unit length of the string. This is because it is more difficult to accelerate (and impart a large wave velocity) to a massive string compared to a light string. Thus, the wave velocity is directly proportional to the tension and is inversely proportional to the mass per unit length of the string. The exact relationship between the wave velocity v, the tension F, and the mass per length μ, is given by

$$\boxed{v = \sqrt{\frac{F}{\mu}}} \qquad (14.16)$$

From this we see that the velocity of mechanical waves, such as a wave on a string, depends only on the properties of the medium through which the disturbance travels.

Now, let us verify that this expression is dimensionally correct. The dimensions of F are [M][L]/[T^2], and the dimensions of μ are [M]/[L]. Therefore, the dimensions of F/μ are [L^2]/[T^2] and those of $\sqrt{F/\mu}$ are [L]/[T], which are indeed the dimensions of velocity. No other combination of F and μ is dimensionally correct, assuming these are the only variables relevant to the situation.

Equation 14.16 indicates that we can increase the velocity of a wave on a stretched string by increasing the tension in the string. It also shows that if we wrap a string with a metallic winding, as on the bass strings of a piano or a guitar, we decrease the velocity of a transmitted wave because the mass per unit length is increased.

EXAMPLE 14.12 A Pulse Traveling on a String

A uniform string has a mass, M, of 0.3 kg and a length, L, of 6 m. Tension is maintained in the string by suspending a 2-kg block from one end (Fig. 14.24). Find the speed of a pulse on this string.

Solution: The tension, F, in the string is equal to the mass, m, of the block multiplied by the gravitational acceleration:

$$F = mg = (2\text{ kg})(9.80\text{ m/s}^2) = 19.6\text{ N}$$

For the string, the mass per unit length, μ, is

$$\mu = \frac{M}{L} = \frac{0.3\text{ kg}}{6\text{ m}} = 0.05\text{ kg/m}$$

Therefore, the wave speed is

$$v = \sqrt{\frac{F}{\mu}} = \sqrt{\frac{19.6\text{ N}}{0.05\text{ kg/m}}} = 19.8\text{ m/s}$$

EXERCISE Find the time it takes the pulse to travel from the wall to the pulley.

Answer: 0.253 s.

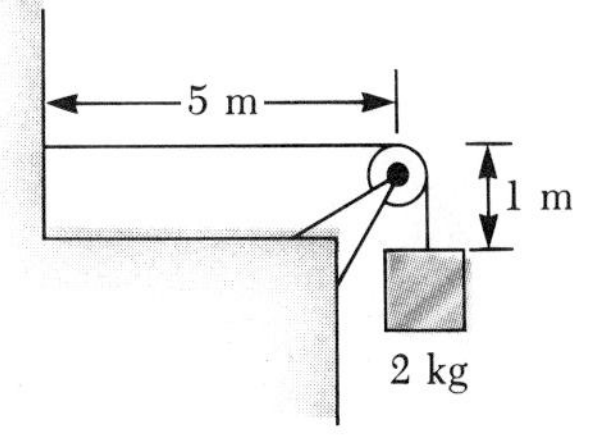

FIGURE 14.24 (Example 14.12) The tension, F, in the string is maintained by the suspended block. The wave speed is given by the expression $v = \sqrt{F/\mu}$.

14.12 SUPERPOSITION AND INTERFERENCE OF WAVES

Many interesting wave phenomena in nature cannot be described by a single moving wave. Instead, one must analyze what happens when two or more waves attempt to pass through the same region of space. To analyze such wave combinations, one can make use of the **superposition principle:**

If two or more traveling waves are moving through a medium, the resultant wave is found by adding the displacements of the individual waves together point by point.

Experiments show that the superposition principle is valid only when the individual waves have small amplitudes of displacement. This small amplitude approximation will be assumed in all our examples.

One consequence of the superposition principle is that *two traveling waves can pass through each other without being destroyed or even altered.* For instance, when two pebbles are thrown into a pond, the expanding circular waves do not destroy each other. In fact, the ripples pass through each other. Likewise, when sound waves from two sources move through air, they also pass through each other. The sound one hears at a given location is the result of both disturbances.

In Figures 14.25a and 14.25b, two waves of the same amplitude and frequency are shown. If at some instant of time, these two waves attempt to travel through the same region of space, the resultant wave at that instant would look like the wave shown in Figure 14.25c. For example, suppose these are water waves of amplitude 1 m. At the instant they overlap such that crest

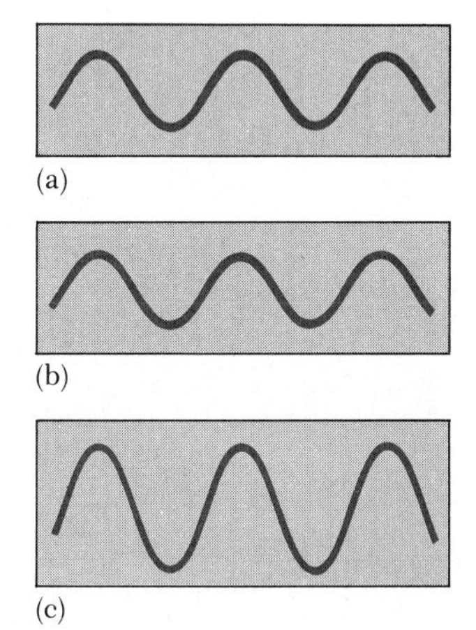

FIGURE 14.25 Constructive interference. If two waves having the same frequency and amplitude are in phase as in (a) and (b), the resultant wave (c) when they combine has the same frequency as the individual waves but twice their amplitude.

meets crest and trough meets trough, the resultant wave has an amplitude of 2 m. Waves coming together like this are said to be *in phase* and undergo **constructive interference.**

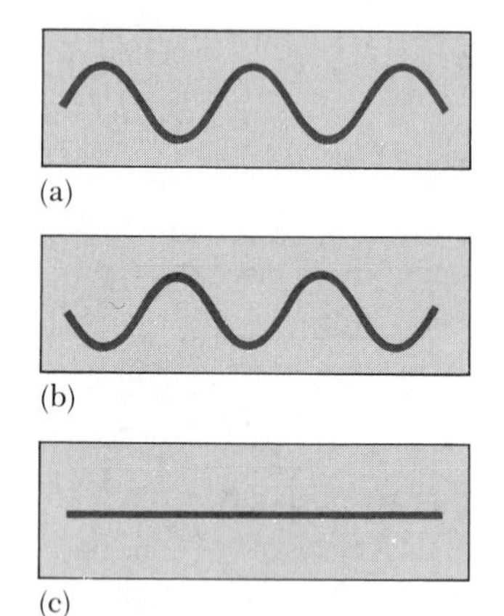

FIGURE 14.26 Destructive interference. When two waves having the same frequency and amplitude are 180° out of phase as in (a) and (b), the result when they combine is complete cancellation, as in (c).

In Figures 14.26a and 14.26b, two similar waves are shown. In this case, however, the crest of one coincides with the trough of the other, that is, one wave is *inverted* relative to the other. The resultant wave, shown in Figure 14.26c, is seen to be a state of complete cancellation. If these were water waves coming together, one of the waves would be trying to pull an individual drop of water upward at the same instant that the second wave was trying to pull the same drop downward. The result is no motion of the water at all. In this situation, the two waves are said to be 180° out of phase and undergo **destructive interference.**

Figure 14.27 shows constructive interference in two pulses moving toward each other along a stretched spring. Figure 14.28 shows destructive interference in two pulses. You should note from these figures that, when the two pulses separate, their shapes are unchanged, as if they had never met!

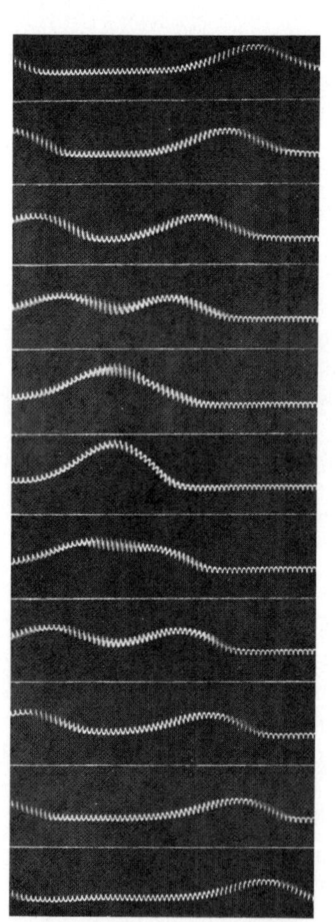

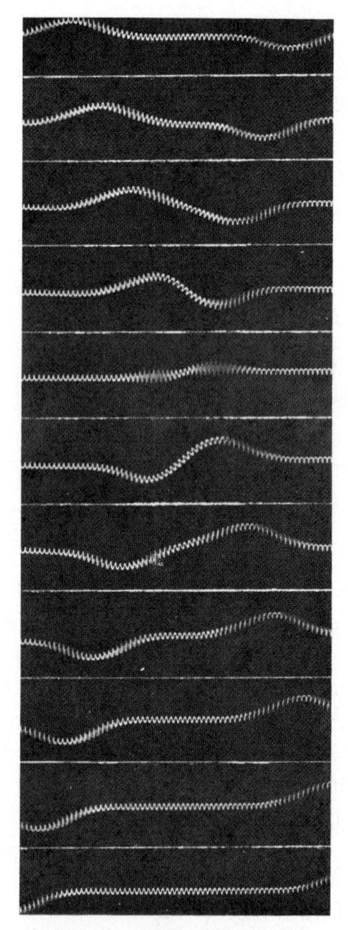

FIGURE 14.27 (left) Constructive interference. Superposition of two equal and symmetric pulses traveling in opposite directions on a stretched spring. FIGURE 14.28 (right) Destructive interference. Superposition of two symmetric pulses traveling in opposite directions, where one is inverted relative to the other. (Courtesy of Education Development Center, Newton, Mass.)

14.13 REFLECTION OF WAVES

In our discussion so far, we have assumed that the waves we are analyzing can travel indefinitely without striking anything. Obviously, such conditions are not realized in practice. Whenever a traveling wave reaches a boundary, part or all of the wave is reflected. For example, consider a pulse traveling on a string fixed at one end (Fig. 14.29). When the pulse reaches the wall, it will be reflected.

Note that *the reflected pulse is inverted.* This can be explained as follows. When the pulse meets the wall, the string exerts an upward force on the wall. By Newton's third law, the wall must exert an equal and opposite (downward) reaction force on the string. This downward force causes the pulse to invert upon reflection.

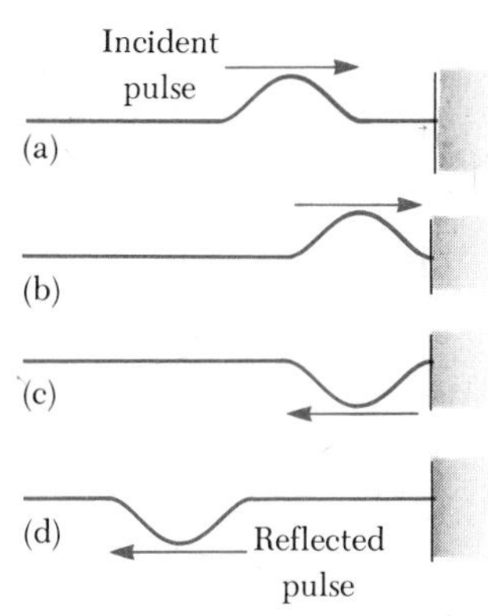

FIGURE 14.29 The reflection of a traveling wave at the fixed end of a stretched string. Note that the reflected pulse is inverted, but its shape remains the same.

Now consider another case, one where the pulse arrives at the end of the string shown in Figure 14.30, where the end of the string is attached to a ring of negligible mass that is free to slide along the post without friction. Again, the pulse will be reflected, but this time is not inverted. Upon reaching the post, the pulse exerts a force on the ring, causing it to accelerate upward. The ring is then returned to its original position by the downward component of the tension.

An alternative method of showing that a pulse is reflected without inversion when it strikes a free end of a string is to send a pulse down a string hanging vertically. When the pulse hits the free end, it is reflected without inversion similar to the situation shown in Figure 14.30.

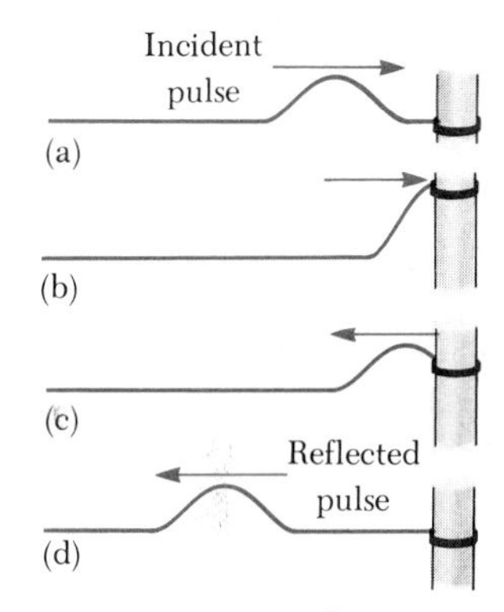

FIGURE 14.30 The reflection of a traveling wave at the free end of a stretched string. In this case, the reflected pulse is not inverted.

SUMMARY

Simple harmonic motion occurs when the net force along the direction of motion is a **Hooke's law** type of force, that is, when the net force is proportional to the displacement and in the opposite direction:

$$\boxed{F = -kx} \tag{14.1}$$

The time for one complete vibration is called the **period** of the motion. The inverse of the period is the **frequency** of the motion, which equals the number of oscillations per second.

When an object is moving with simple harmonic motion, its **acceleration** as a function of location is

$$\boxed{a = -\frac{k}{m}x} \tag{14.2}$$

The energy stored in a stretched or compressed spring or other elastic material is called **elastic potential energy:**

$$\boxed{PE_s \equiv \frac{1}{2}kx^2} \tag{14.3}$$

The **velocity** of an object as a function of position when the object is moving with simple harmonic motion is

$$v = \pm \sqrt{\frac{k}{m}(A^2 - x^2)} \tag{14.6}$$

The **period** of an object of mass m moving with simple harmonic motion while attached to a spring of spring constant k is

$$T = 2\pi \sqrt{\frac{m}{k}} \tag{14.8}$$

Because $f = 1/T$, the **frequency** of a mass-spring system is

$$f = \frac{1}{2\pi} \sqrt{\frac{k}{m}} \tag{14.10}$$

The **position** of an object as a function of time when the object is moving with simple harmonic motion is

$$x = A \cos(2\pi f t) \tag{14.13}$$

A **simple pendulum** of length L moves with simple harmonic motion for small angular displacements from the vertical, with a period given by

$$T = 2\pi \sqrt{\frac{L}{g}} \tag{14.14}$$

The period is independent of the suspended mass.

A **transverse wave** is a wave in which the particles of the medium move in a direction perpendicular to the direction of the wave velocity. An example is a wave on a stretched string.

A **longitudinal wave** is a wave in which the particles of the medium move in a direction parallel to the direction of the wave velocity. An example is a sound wave.

The relationship between the velocity, wavelength, and frequency of a wave is

$$v = \lambda f \tag{14.15}$$

The speed of a wave traveling on a stretched string of a mass per unit length μ and under tension F is

$$v = \sqrt{\frac{F}{\mu}} \tag{14.16}$$

The **superposition principle** states that, if two or more traveling waves are moving through a medium, the resultant wave is found by adding the individual waves together point by point. When waves meet crest to crest and trough to trough, they undergo **constructive interference.** If crest meets trough, the waves undergo complete **destructive interference.**

When a wave pulse reflects from a rigid boundary, the pulse is inverted. If the boundary is free, the reflected pulse is not inverted.

ADDITIONAL READING

W. Bascom, "Ocean Waves," *Sci. American*, August 1959, p. 74.

W. Bascom, *Waves and Beaches: The Dynamics of the Ocean Surface*, New York, Doubleday Anchor Books, 1980.

A. Einstein and L. Infeld, *The Evolution of Physics*, New York, Simon and Schuster, 1961.

B. Gilbert and P. Glanz, "Springs: Distorted and Combined," *The Physics Teacher*, October 1983, p. 430.

S. D. Kelby and R. P. Middleton, "The Vibrations of Hand Bells," *Physics Education*, Vol. 15, 1980, p. 320.

J. Oliver, "Long Earthquake Waves," *Sci. American*, March 1959, p. 14.

T. D. Rossing, "The Physics of Kettledrums," *Sci. American*, November 1982, p. 172.

R. A. Waldron, *Waves and Oscillations*, Momentum Series, Princeton, N.J., Van Nostrand, 1964.

QUESTIONS

1. What is the total distance traveled by an object moving with simple harmonic motion in a time equal to its period if its amplitude is A?
2. If a mass-spring system is hung vertically and set into oscillation, why does the motion eventually stop?
3. Explain why the kinetic and potential energies of a mass-spring system can never be negative.
4. What happens to the period of a simple pendulum if its length is doubled? What happens if the suspended mass is doubled?
5. How would you create a longitudinal wave in a stretched spring? Would it be possible to create a transverse wave in a spring?
6. By what factor would you have to increase the tension in a stretched string in order to double the wave speed?
7. When all the strings on a guitar are stretched to the same tension, will the velocity of a wave along the more massive bass strings be faster or slower than the velocity of a wave on the lighter strings?
8. Suppose two pulses are moving toward one another on a string. How could you tell whether such pulses reflect off or pass through one another?
9. If you were to periodically shake the end of a stretched rope up and down three times each second, what would be the period of the waves set up in the rope?
10. What happens to the wavelength of a wave on a string when the frequency is doubled? Assume the tension in the string remains the same.
11. What happens to the velocity of a wave on a string when the frequency is doubled? Assume the tension in the string remains the same.
12. Does the acceleration of a simple harmonic oscillator remain constant during its motion? Is the acceleration ever zero? Explain.
13. How do transverse waves differ from longitudinal waves?
14. Consider a wave traveling on a stretched rope. What is the difference, if any, between the speed of the wave and the speed of a small section of the rope?
15. If a long rope is hung from a ceiling and waves are sent up the rope from its lower end, the waves do not ascend with constant speed. Explain.
16. A pendulum bob is made with a ball filled with water. What happens to the frequency of vibration of this pendulum if there is a hole in the ball that allows water to slowly leak out?
17. If a pendulum clock keeps perfect time at the base of a mountain, will it also keep perfect time when moved to the top of the mountain? Explain.

PROBLEMS

Section 14.1 Hooke's Law

1. (a) A mass of 400 g is suspended from a spring hanging vertically, and the spring is found to stretch 8 cm. Find the spring constant. (b) How much will the spring stretch if the suspended mass is 575 g?

2. A 0.4-kg mass is attached to a spring of spring constant 160 N/m such that the mass is allowed to move on a horizontal frictionless surface. The mass is released from rest when the spring is compressed 0.15 m. Find

(a) the force on the mass and its acceleration at this instant, and (b) the maximum values of the force and acceleration.

3. A load of 50 N attached to a spring hanging vertically will stretch the spring 5 cm. The spring is now placed horizontally on a table and stretched 11 cm. (a) What force is required to stretch the spring by this amount? (b) Plot a graph of force (on the y axis) versus spring displacement from the equilibrium position along the x axis.

4. A 3-kg mass is attached to a spring and pulled out horizontally to a maximum displacement from equilibrium of 0.5 m. What spring constant must the spring have if the mass is to achieve an acceleration equal to that of gravity?

Section 14.2 Elastic Potential Energy

5. In an arcade game a 0.1 kg disk is shot across a horizontal frictionless surface by compressing it against a spring and releasing it. If the spring has a spring constant of 200 N/m and is compressed from its equilibrium position by 6 cm, find the speed with which the disk slides across the surface.

6. The disk of Problem 5 is projected up a 10° frictionless incline by the spring. Find (a) the distance the disk moves up the incline before coming to rest, and (b) its speed when it is halfway to the point where it comes to rest.

7. A spring having a spring constant of 19.6 N/m is stretched 0.4 m. Find the ratio of the energy stored in the spring at this extension to the energy stored when the extension is 0.2 m.

• **8.** A block of mass 1.5 kg is attached to a spring with spring constant of 2000 N/m. The spring is then stretched a distance of 0.3 cm and the block is released from rest. (a) Calculate the speed of the block as it passes through the equilibrium position if no friction is present. (b) Calculate the speed of the block as it passes through the equilibrium position if a constant frictional force of 2.0 N retards its motion. (c) What would be the strength of the frictional force if the block reaches the equilibrium position the first time with zero velocity?

• **9.** A bullet of mass 10 g is fired into and embeds in a 2-kg block attached to a spring of spring constant 19.6 N/m. How far will the spring be compressed if the speed of the bullet just before striking the block is 300 m/s and the block slides on a frictionless surface? (*Hint:* You must use conservation of momentum in this problem. Why?)

• **10.** The spring constant of the spring in Figure 14.31 is 19.6 N/m and the mass of the object is 1.5 kg. The spring is unstretched and the surface is frictionless. A constant 20-N force is applied to the object horizontally as shown. Find the speed of the object after it has moved a distance of 0.3 m.

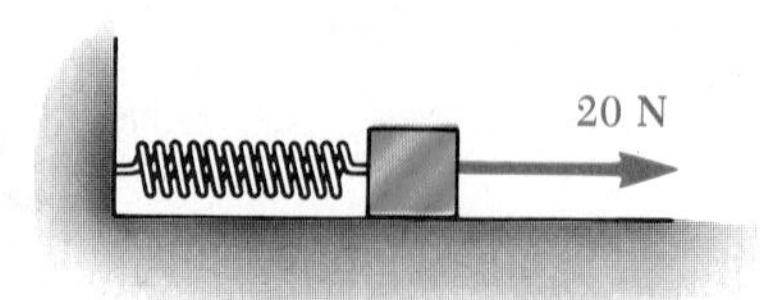

FIGURE 14.31 Problem 10

Section 14.3 Velocity as a Function of Position

11. A 50-g mass is attached to a horizontal spring of spring constant 10 N/m and released from rest with an amplitude of 25 cm. What is the velocity of the mass when it is halfway to the equilibrium position if the surface is frictionless?

12. A mass of 0.4 kg connected to a light spring of spring constant 19.6 N/m oscillates on a horizontal frictionless surface. If the spring is compressed 4 cm and released from rest, determine (a) the maximum speed of the mass, (b) the speed of the mass when the spring is compressed 1.5 cm, and (c) the speed of the mass when the spring is stretched 1.5 cm? (d) For what value of x does the speed equal one-half the maximum speed?

13. A spring of spring constant 19.6 N/m is compressed 5 cm. A mass of 0.3 kg is attached to the spring and released from rest. Find (a) the maximum elastic potential energy stored in the spring, and (b) the maximum speed of the mass.

• **14.** At an outdoor market, a bunch of bananas is set into oscillatory motion with an amplitude of 20 cm on a spring with a spring constant of 16 N/m. It is observed that the maximum speed of the bunch of bananas is 40 cm/s. What is the weight of the bananas in newtons?

Section 14.4 Comparing Simple Harmonic Motion With Uniform Circular Motion

15. A ball moves with constant speed of 5 m/s in a circular path of radius 0.4 m as shown in Figure 14.8. Find the x component of the velocity of the ball when θ equals (a) 0°, (b) 60°, (c) 90°, (d) 180°, (e) 270°.

16. (a) Find the frequency and (b) the period of the ball described in Problem 15.

17. The frequency of vibration of a mass-spring system is 5 Hz when a 4-g mass is attached to the spring. What is the spring constant of the spring?

18. A ball rotates in a circle of radius 20 cm making one complete revolution every 2 s. What is (a) the speed of the ball, (b) the frequency of motion in Hz, and (c) the angular velocity of the ball?

19. If the frequency of oscillation of the wave emitted by an FM radio station is 88 MHz, what is the period of vibration of the wave?

20. When four people having a combined mass of 320 kg sit down in a car they find that the car moves 0.8 cm lower on its springs. They then get out of the car and bounce it up and down. What is the frequency of its vibration if the mass of the car when empty is 2000 kg?

Section 14.5 Position as a Function of Time

21. An object attached to the end of a spring vibrates with an amplitude of 20 cm according to Equation 14.13. Find the position of the object at these times: 0, $T/8$, $T/4$, $3T/8$, $T/2$, $5T/8$, $3T/4$, $7T/8$, and T, where T is the period of vibration. Plot your results (position along the vertical axis and time along the horizontal axis).

22. The motion of an object is described by the equation

$$x = (0.3\text{ m})\cos(\pi t/3).$$

Find (a) the position of the object at $t = 0$ and $t = 0.2$ s, (b) the amplitude of the motion, (c) the frequency of the motion, and (d) the period of the motion.

23. A 0.5-kg object is attached to a spring of spring constant 30 N/m and released from rest at $t = 0$ from a position 0.25 m from equilibrium. Find the equation which describes its x location as a function of time.

• **24.** An object on a spring vibrates with a period of 3 s. The motion is initiated by releasing the object from its point of maximum displacement $x = A$. After what time will the object first be at (a) $x = A/2$, (b) $x = -A/2$, and (c) $x = 0$?

Section 14.6 The Pendulum

25. The period of a 50-cm-long pendulum is found to be 1.4 s. What is the acceleration due to gravity at this location?

26. A pendulum is to be used in a clock. What length should the pendulum be so that its period of vibration is 1 s?

27. (a) Find the ratio of the period of a pendulum on earth to the period of an identical pendulum on the moon, where the acceleration due to gravity is one-sixth that on earth. (b) If the period of the pendulum is 2.5 s on earth, what will be its period on the moon?

• **28.** A "seconds" pendulum is one that moves through its equilibrium position once each second. (The period of the pendulum is 2 s.) The length of a seconds pendulum is 0.9927 m at Tokyo and 0.9942 m at Cambridge, England. What is the ratio of the acceleration due to gravity at these two locations?

Section 14.10 Frequency, Amplitude, and Wavelength

29. A piano emits frequencies that range from a low of about 28 Hz to a high of about 4200 Hz. Find the range of wavelengths spanned by this instrument. The speed of sound in air is approximately 343 m/s.

30. A sound wave, traveling at 343 m/s, is emitted by the foghorn of a tugboat. An echo is heard 2.6 s later. How far away is the reflecting object?

31. (a) How long does it take light to reach us from the sun, 93,000,000 mi away? (The speed of light is 3×10^8 m/s.) (b) An astronaut communicates with earth from the moon (3.84×10^8 m away). How long does it take his signal to reach us?

32. How far in meters does light travel in one year? This distance is called a light-year.

33. A wave traveling in the positive x direction is pictured in Figure 14.32. Find (a) the amplitude, (b) wavelength, (c) period, and (d) speed of the wave if it has a frequency of 25 Hz.

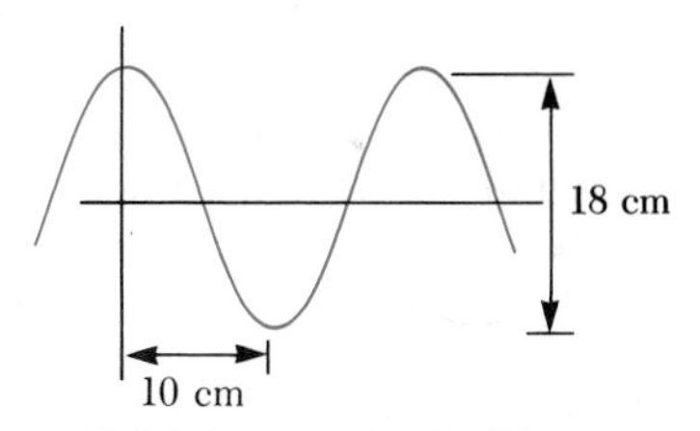

FIGURE 14.32 Problem 33

Section 14.11 The Velocity of Waves on Strings

34. A circus performer stretches a tightrope between two towers. He strikes one end of the rope and sends a wave along it toward the other tower. He notes that it takes the wave 0.8 s to reach the opposite tower 20 m away. If one meter of the wire has a mass of 0.35 kg, find the tension in the tightrope.

35. Transverse waves with a speed of 50 m/s are to be produced on a stretched string. A 5-m length of string with a total mass of 0.06 kg is used. (a) What is the required tension in the string? (b) Calculate the wave speed in the string if the tension is 8 N.

• **36.** One end of a string 3 m long is attached to a wall while the other end hangs over a pulley and is attached to a hanging 2.0 kg mass. The speed of a pulse on the string is observed to be 15 m/s. What is the mass of the string?

• **37.** Transverse waves travel with a speed of 20 m/s on a string under a tension of 6 N. What tension is required for a wave speed of 30 m/s in the same string?

• **38.** When a steel wire stretched between two clamps is plucked, waves travel along the wire with a speed of 80 m/s. A second wire of the same material but of twice the length and twice the radius of the first, is stretched between two points under the same tension as the first. At what speed will transverse waves travel along the second wire?

Section 14.12 Superposition and Interference of Waves

Section 14.13 Reflection of Waves

39. A series of pulses of amplitude 0.15 m are sent down a string attached to a post at one end. The pulses are reflected at the post and travel back along the string without loss of amplitude. What is the amplitude at a point on the string where two pulses are crossing (a) if the string is rigidly attached to the post and (b) if the end at which reflection occurs is free to slide up and down?

40. A wave of amplitude 0.3 m interferes with a second wave of amplitude 0.2 m. What is (a) the largest and (b) the smallest value of the absolute value of the resultant displacement that can occur and what will be the conditions in which these maxima and minima occur?

ADDITIONAL PROBLEMS

41. A string under a tension of 100 N has a wave on it traveling at 90 m/s. What is the speed of a wave on this string if the tension is doubled?

42. The motion of a 0.3 kg object attached to a spring vibrates such that its motion is described by

$$x = (0.25\ \text{m})\cos(0.4\pi t)$$

Find (a) the amplitude of the motion, (b) the spring constant, (c) the position at $t = 0.3$ s, and (d) its velocity at $t = 0.3$ s.

43. A mass of 1.5 kg is attached to a spring of spring constant 1000 N/m. Find the force on the mass and its acceleration at the positions (a) 0.5 m, (b) 0.1 m, (c) 0 m, (d) -0.20 m, and (3) -0.5 m. (The positions are measured from equilibrium, $x = 0$.)

44. A string is 50 cm long and has a mass of 3 g. A wave travels with a speed of 5 m/s along this string. A second string has the same length but a mass half that of the first. If both strings are under the same tension, what is the speed of a wave along the second string?

45. Tension is maintained in a string as shown in Figure 14.33. The observed wave speed is 24 m/s when the suspended mass is 3 kg. (a) What is the mass per unit length of the string? (b) What will be the wave speed when the suspended mass is 2 kg?

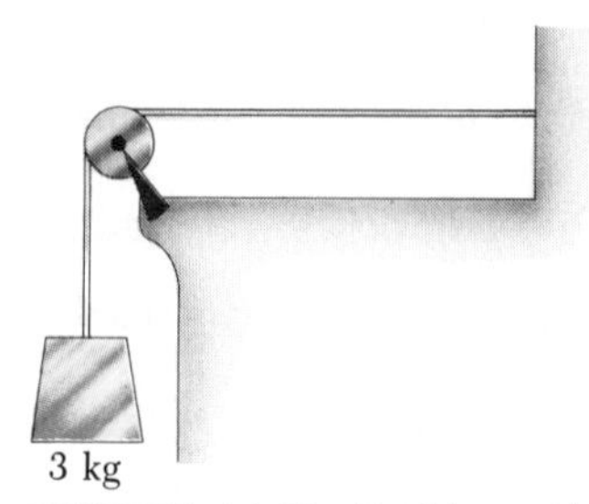

FIGURE 14.33 Problem 45

46. A spring having a spring constant of 30 N/m is stretched to a distance of 0.2 m from its equilibrium position. How much work must be done to stretch it an additional 0.1 m?

• **47.** A block of mass 500 g is released from rest and slides down a frictionless track of height 2 m above the horizontal, as shown in Figure 14.34. At the bottom of the track, where the surface is horizontal, the block strikes and sticks to a light spring of spring constant 20 N/m. Find the maximum distance the spring is compressed.

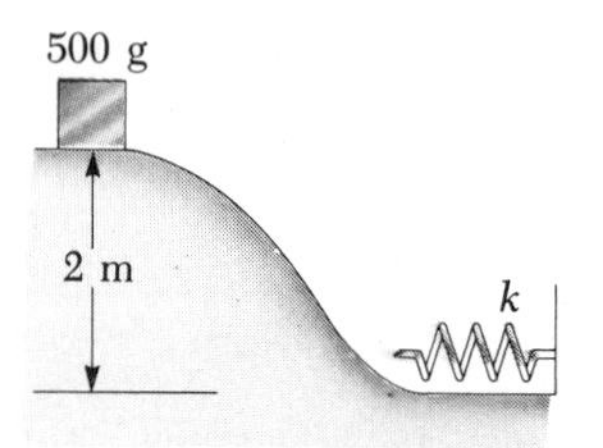

FIGURE 14.34 Problem 47

• **48.** One end of a hacksaw blade is clamped in a vise and the other end is set into vibration. The free end makes 5 complete vibrations each second with an amplitude of 0.4 cm. Find the speed of the end of the hacksaw blade when the displacement from the equilibrium position is 0.1 cm.

• **49.** A spring in a toy gun has a spring constant of 9.8 N/m and can be compressed 20 cm beyond the equilibrium position. A small pellet having a mass of 1 g resting against the spring is propelled forward when the spring is released. (a) Find the muzzle velocity of the pellet.

(b) If the pellet is fired horizontally from a height of 1 m above the floor, what is its range?

• **50.** When an object attached to the end of a metal rod or wire is suspended and then released, the object is found to vibrate with simple harmonic motion. Show that the effective spring constant for this wire is given by $k = YA/L$ where Y is Young's modulus, A is the cross-sectional area of the wire and L is its unstretched length.

• **51.** A 5-g bullet moving with an initial speed of 400 m/s is fired into and passes through a 1-kg block, as in Figure 14.35. The block, initially at rest on a frictionless, horizontal surface, is connected to a spring of spring constant 900 N/m. If the block moves a distance of 5 cm to the right after impact, find (a) the speed at which the bullet emerges from the block and (b) the energy lost in the collision.

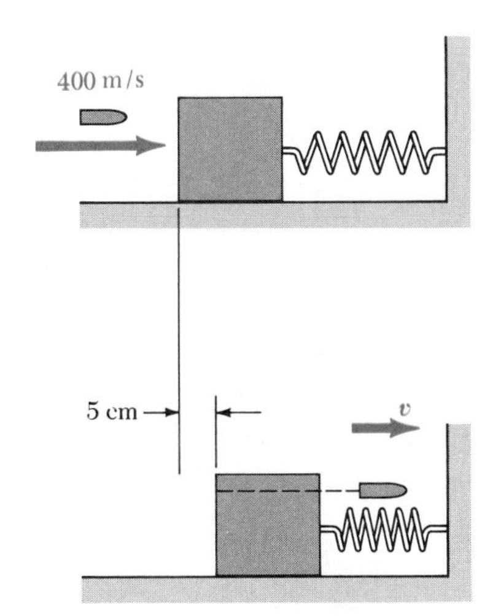

FIGURE 14.35 Problem 51

• **52.** A particle of mass m slides inside a hemispherical bowl of radius R. Show that for small displacements from equilibrium, the particle exhibits simple harmonic motion like that of a simple pendulum.

•• **53.** An 8-kg block travels on a rough horizontal surface and collides with a spring. The speed of the block just before the collision is 4 m/s. As it rebounds to the left with the spring uncompressed, the speed of the block is 3 m/s. If the coefficient of kinetic friction between the block and surface is 0.4, determine (a) the work done by friction while the block is in contact with the spring and (b) the maximum distance the spring is compressed.

•• **54.** Assume that a hole is drilled through the center of the earth. It can be shown that an object of mass m at a distance r from the center of the earth is pulled toward the center of the earth only by the mass in the shaded portion of Figure 14.36. Write down Newton's law of gravitation for an object at a distance r from the center of the earth and show that the force on it is of Hooke's law form, $F = kr$, where the effective force constant is $k = (4/3)\pi\rho Gm$, where ρ is the density of the earth, and G is the gravitational constant.

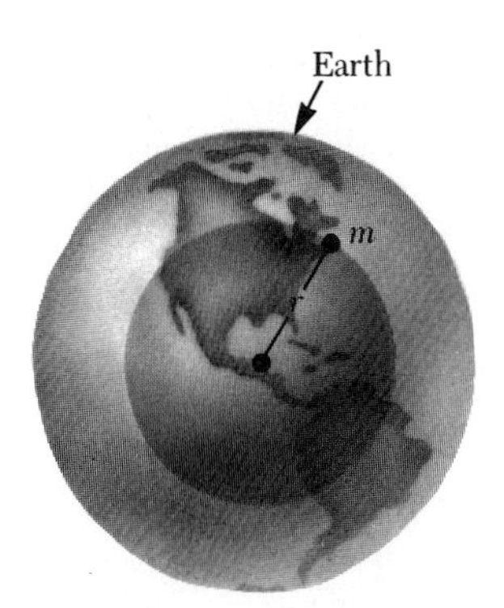

FIGURE 14.36 Problem 54

•• **55.** A 3-kg mass is fastened to a light spring that passes over a pulley, as in Figure 14.37. The pulley is frictionless and its inertia may be neglected. The mass is released from rest when the spring is unstretched. If the mass drops 10 cm before coming to rest, find (a) the spring constant of the spring and (b) the speed of the mass when it is 5 cm below its starting point.

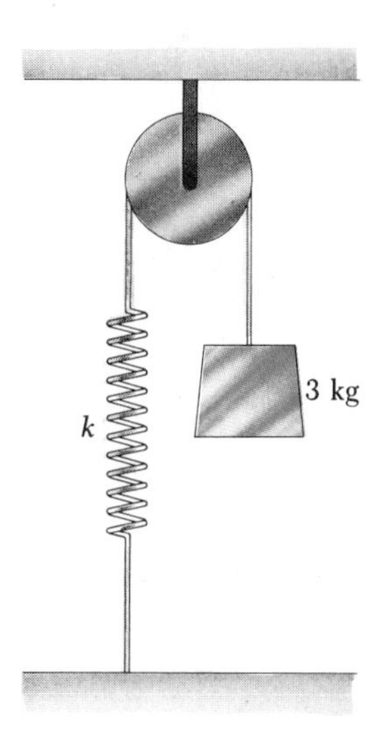

FIGURE 14.37 Problem 55

•• **56.** A 60-kg firefighter slides down a pole while a constant frictional force of 300 N retards his motion. A horizontal 20-kg platform is supported by a spring at the bottom of the pole to cushion the fall. The firefighter starts from rest 5 m above the platform, and the spring constant is 2500 N/m. Find (a) the firefighter's speed just before he collides with the platform and (b) the maximum distance the spring will be compressed. (Assume the frictional force acts during the entire motion.)

•• **57.** A 2-kg block situated on a rough incline is connected to a light spring having a spring constant of 100 N/m (Fig. 14.38). The block is released from rest when the spring is unstretched, and the pulley is frictionless. The block moves 20 cm down the incline before coming to rest. Find the coefficient of kinetic friction between the block and the incline.

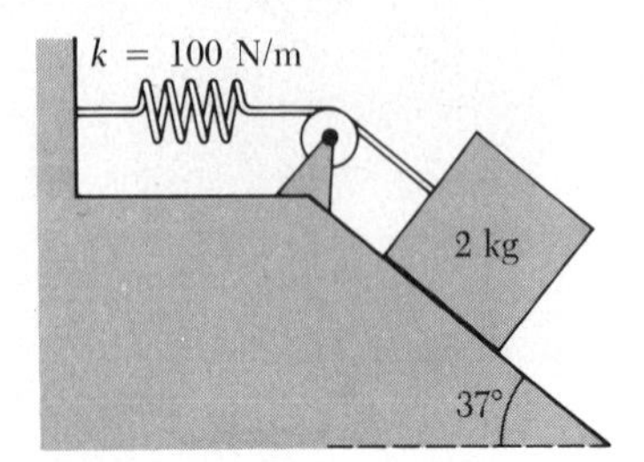

FIGURE 14.38 Problem 57

•• **58.** A 25-kg block is connected to a 30-kg block by a light string that passes over a frictionless pulley. The 30-kg block is connected to a light spring of force constant 200 N/m, as in Figure 14.39. The spring is unstretched when the system is as shown in the figure, and the incline is smooth. The 25-kg block is pulled a distance of 20 cm down the incline (so that the 30-kg block is 40 cm above the floor) and is released from rest. Find the speed of each block when the 30-kg block is 20 cm above the floor (that is, when the spring is unstretched).

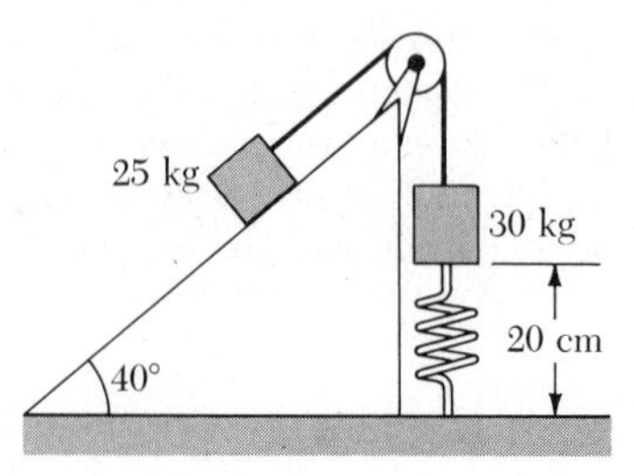

FIGURE 14.39 Problem 58

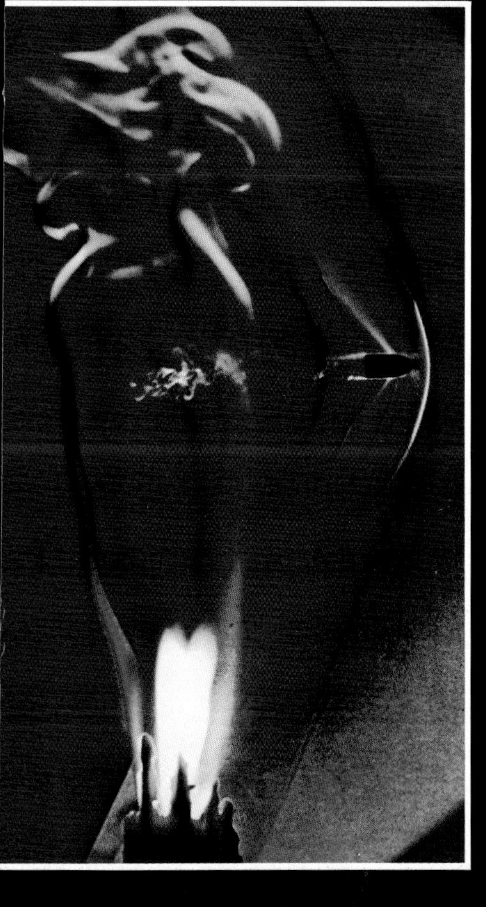

Color Plate 10 Supersonic candlelight. A stroboscopic shadow photograph showing a bullet traveling at supersonic speed passing through the hot air rising above a candle. This type of picture, called a *schlieren,* enables one to observe regions of nonuniform air density. Note the shock wave in the vicinity of the bullet. (Bullet through Flame, 1973. © Harold Edgerton. Courtesy of Palm Press, Inc.)

Color Plate 12 Fan blade vortex. The tip of a rotating fan blade (the dark region) forms a vortex in air that is being heated by an alcohol lamp (the wick is at the bottom). Note the air turbulence on both sides of the rotating blade. This picture was taken with a multicolor source of light using a stroboscope and an exposure time of 0.333 μs.(Fan and Flame Vortices, 1973. © Harold Edgerton. Courtesy Palm Press,Inc.)

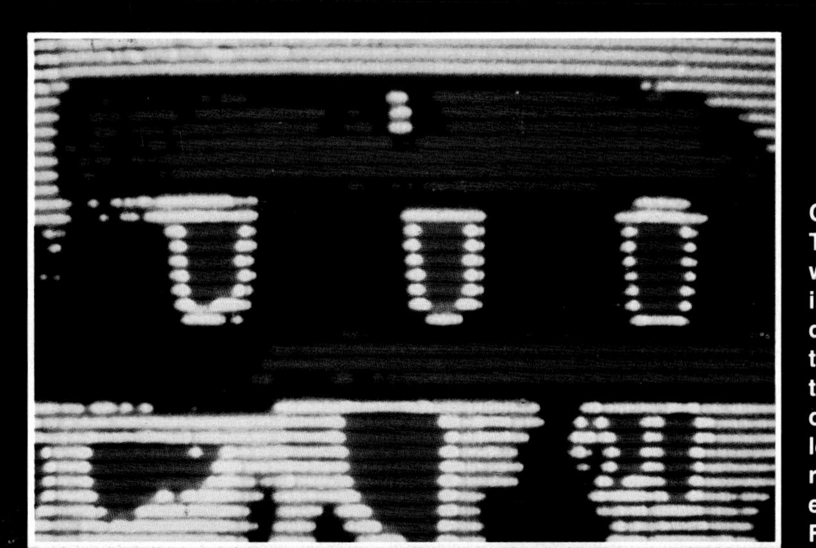

Color Plate 11. Thermogram of a house. This image of a house during cold weather was formed by measuring the infrared radiation emitted by the different surfaces and assigning a color to different intensities. In this picture, the red areas indicate the highest rate of radiation emitted in the infrared (high-loss regions) while the blue areas represent regions with low rates of emission. (Courtesy of Owens/Corning Fiberglas)

Color Plate 13 Jupiter's Great Red Spot. The Great Red Spot, a large anticyclonic storm in the southern hemisphere of Jupiter, rotates with a period of about six days and is many times larger than the earth. Such large vortices are disturbances produced by Jupiter's rapid rotation. (NASA)

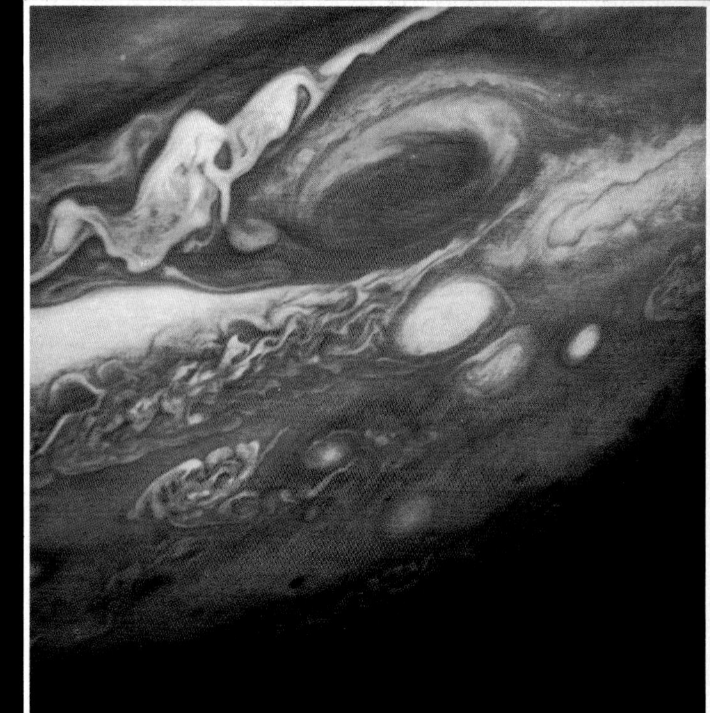

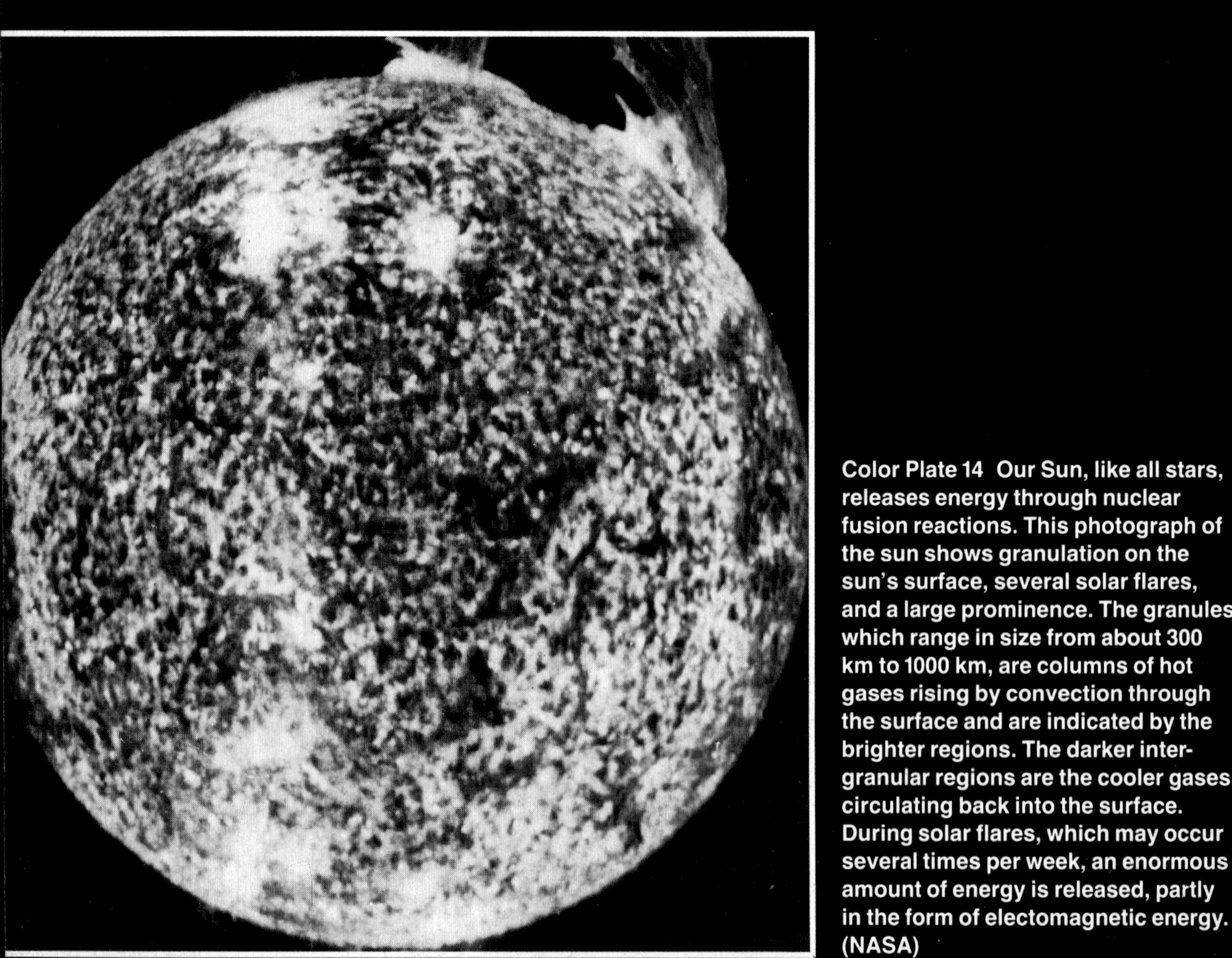

Color Plate 14 Our Sun, like all stars, releases energy through nuclear fusion reactions. This photograph of the sun shows granulation on the sun's surface, several solar flares, and a large prominence. The granules, which range in size from about 300 km to 1000 km, are columns of hot gases rising by convection through the surface and are indicated by the brighter regions. The darker intergranular regions are the cooler gases circulating back into the surface. During solar flares, which may occur several times per week, an enormous amount of energy is released, partly in the form of electomagnetic energy. (NASA)

Color Plate 15 This dramatic one-minute time exposure captures multiple lightning bolts illuminating Kitt Peak National Observatory in Arizona, illustrating electrical breakdown in the atmosphere. (© Gary Ladd, 1972)

Color Plate 16 Hot air balloons. These colorful balloons rise as the air inside them is heated by a large gas burner. Because warm air is less dense than cooler air, the buoyant force upward can exceed the total downward force, causing the balloons to rise (see Section 9.7). The vertical motion can also be controlled by releasing weights from the balloon. (Photo Researchers, © Margaret Durrance 1984)

Color Plate 17 Droplets of mercury lying on a glass surface. Mercury is the only metal that is a liquid at room temperature. Note that the small droplets are almost spherical, while the large droplets are flattened. This shows that the effect of surface tension has more influence on the shape of the small (lighter) droplets. (© Charles Steele)

Color Plate 18 Detecting radio waves from outer space. Several elements of the Very Large Array (VLA) radio telescope near Socorro, New Mexico. (National Radio Astronomy Observatory)

CHAPTER 15

Sound

String quartet. (Photograph by Jules Schick Photography, Inc., courtesy of the Curtis Institute of Music, Philadelphia)

Sound waves are the most important example of longitudinal waves. In this chapter, we shall describe how sound waves are produced, what they are, and how they travel through matter. After a discussion of the characteristics of sound waves, we shall investigate what happens when sound waves interfere with each other. The insights developed in this chapter will enable us to better understand the nature of sound waves and why we hear what we hear.

15.1 PRODUCING A SOUND WAVE

Whether it is the shrill whine of a jet engine or the soft melodies of a pop singer, the source of all sound waves is a vibrating object. Musical instruments produce sounds in a variety of ways. For example, the sound from a clarinet is produced by a vibrating reed, the sound from a drum arises from the vibration of the taut drum head, the sound from a piano is produced by vibrating strings, and the sound from a singer originates from vibrating vocal cords.

Sound waves are longitudinal waves traveling through a medium, such as air. In order to investigate how sound waves are produced, we shall focus our attention on the tuning fork, a common device for producing pure musical notes.

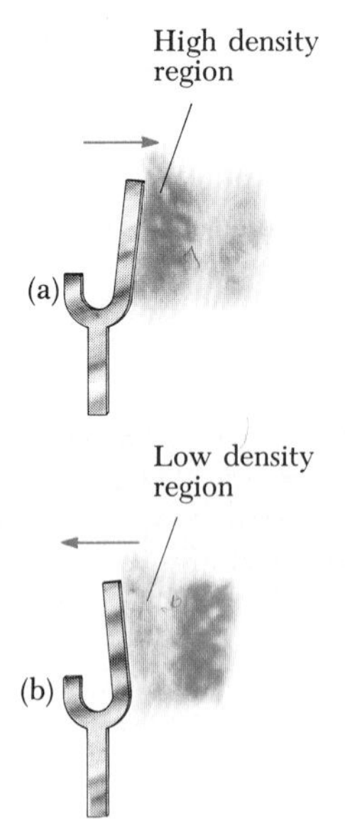

A tuning fork consists of two metal prongs, or tines, that vibrate when struck. The vibration of these tines disturbs the air near them, as shown in Figure 15.1. (Note that the amplitude of vibration of the tine in Figure 15.1 has been greatly exaggerated for clarity.) When a tine swings to the right, as in Figure 15.1a, the air molecules in front of the tine are forced closer together than normal. Such a region of high molecular density and high air pressure is called a **compression** or **condensation.** This compression moves to the right, moving away from the fork like a ripple on a pond. When the tine swings to the left, as in Figure 15.1b, the molecules to the right of the tine spread apart and the density and air pressure in this region are now lower than normal. Such a region of lower than normal density is called a **rarefaction.** Molecules to the right of the rarefaction move to the left, Hence, the rarefaction itself moves to the right, following the previously produced compression.

As the tuning fork continues to vibrate, a succession of condensations and rarefactions form and spread out from the fork. The resultant pattern in the air is somewhat like that pictured in Figure 15.2a. We can use a sinusoidal curve to represent a sound wave, as shown in Figure 15.2b. Notice that there are crests in the sinusoidal wave at the points where the sound wave has condensations and troughs where the sound wave has rarefactions. The molecular motion of the sound waves is superposed on the random thermal motion of the atoms and molecules discussed in Chapter 11.

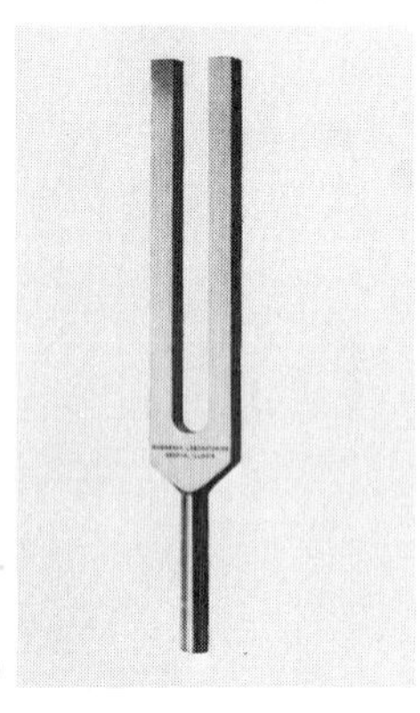

FIGURE 15.1 A vibrating tuning fork. (a) As the right tine of the fork moves to the right, a high-density region (condensation) of air is formed in front of it. (b) As the right tine moves to the left, a low-density region (rarefaction) of air is formed to the right of it. (c) Photograph of a tuning fork. (Courtesy of Riverbank Laboratories.)

15.2 CHARACTERISTICS OF SOUND WAVES

The motion of the medium particles in a **longitudinal** wave is *back and forth along the direction in which the wave travels*. This is in contrast to a **transverse** wave, in which the vibrations of the medium are *at right angles to the direction of travel of the wave.*

As noted earlier, the general motion of air molecules near a vibrating object is such that they move back and forth between regions of compression and rarefaction. This back-and-forth motion of the air molecules in the direction of the disturbance is characteristic of a longitudinal wave.

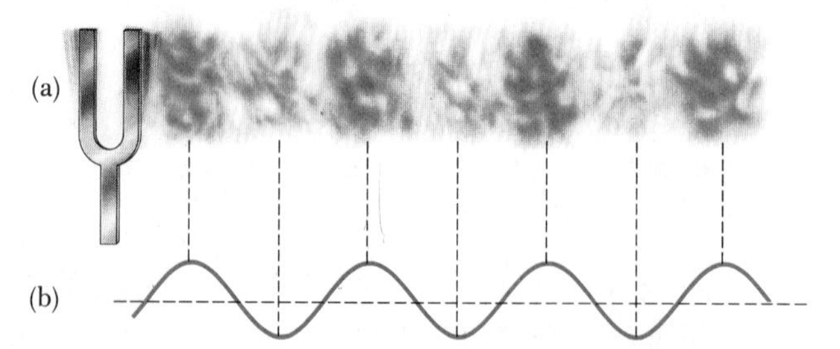

FIGURE 15.2 (a) As the tuning fork vibrates, a series of condensations and rarefactions move outward, away from the fork. (b) The crests of the wave correspond to condensations, and the troughs correspond to rarefactions.

Categories of Sound Waves

There are three categories of sound waves covering different ranges of frequencies. **Audible waves** are longitudinal waves that lie within the range of sensitivity of the human ear, approximately 20 Hz to 20,000 Hz. **Infrasonic waves** are low-frequency longitudinal waves with frequencies below the audible range. Earthquake waves are an example. **Ultrasonic waves** are longitudinal waves with frequencies above the audible range. For example, certain types of whistles produce ultrasonic waves. Even though the human ear is insensitive to the waves emitted by these whistles, certain animals, such as dogs, are able to hear frequencies in this region.

15.3 THE SPEED OF SOUND

The speed of a sound wave depends on the compressibility and inertia of the medium through which the wave is traveling. If the medium has a bulk modulus B and an equilibrium density ρ, the **speed of sound** in that medium is

$$v = \sqrt{\frac{B}{\rho}} \tag{15.1}$$

Recall from Chapter 9 that bulk modulus is defined as the ratio of the change in pressure, ΔP, to the resulting fractional change in volume, $\Delta V/V$:

$$B = -\frac{\Delta P}{\Delta V/V} \tag{15.2}$$

Note that B is always positive because an increase in pressure (positive ΔP) results in a decrease in volume. Hence the ratio $\Delta P/\Delta V$ is always negative.

It is interesting to compare Equation 15.1 with the expression for the speed of transverse waves on a string, $v = \sqrt{F/\mu}$, discussed in Chapter 14. In both cases, the wave speed depends on an elastic property of the medium (B or F) and on an inertial property of the medium (ρ or μ). In fact, the speed of all mechanical waves follows an expression of the general form

$$v = \sqrt{\frac{\text{elastic property}}{\text{inertial property}}}$$

Another example of this general form is the **speed of a longitudinal wave in a solid,** which is given by the expression

$$v = \sqrt{\frac{Y}{\rho}} \tag{15.3}$$

where Y is the Young's modulus of the solid, defined as the longitudinal stress divided by the longitudinal strain (Chapter 9), and ρ is its density.

Table 15.1 lists the speed of sound in various media. As you can see, the speed of sound is much higher in solids than in gases. This makes sense because the molecules in a solid are closer together than in a gas and hence respond more rapidly to a disturbance. In general, sound travels more slowly

TABLE 15.1 Speed of Sound in Various Media

Medium	v(m/s)
Gases	
Air (0°C)	331
Air (100°C)	366
Hydrogen (0°C)	1290
Oxygen (0°C)	317
Helium (0°C)	972
Liquids at 25°C	
Water	1490
Methyl alcohol	1140
Seawater	1530
Solids	
Aluminum	5100
Copper	3560
Iron	5130
Lead	1320
Vulcanized rubber	54

in liquids than in solids because liquids are more compressible and hence have a smaller bulk modulus.

The velocity of sound also depends on the temperature of the medium. For sound traveling through air, the relationship between the speed of sound and temperature is

$$v = (331 \text{ m/s})\sqrt{1 + \frac{T}{273}} \tag{15.4}$$

where 331 m/s is the speed of sound in air at 0°C and T is the temperature in degrees Celsius. Using the equation above, you should show that at 20°C, the speed of sound in air is approximately 343 m/s.

EXAMPLE 15.1 Sound Waves in a Solid Bar

If a solid bar is struck at one end with a hammer, a longitudinal pulse will propagate down the bar. Find the speed of sound in a bar of aluminum, which has a Young's modulus of 7.0×10^{10} Pa and a density of 2.7×10^3 kg/m^3.

Solution From Equation 15.3 we find that

$$v_{\text{Al}} = \sqrt{\frac{Y}{\rho}} = \sqrt{\frac{7.0 \times 10^{10} \text{ Pa}}{2.7 \times 10^3 \text{ kg/m}^3}} \approx 5100 \text{ m/s}$$

This is a typical value for the speed of sound in solids (see Table 15.1).

EXAMPLE 15.2 Speed of Sound in a Liquid

Find the speed of sound in water, which has a bulk modulus of about 2.1×10^9 Pa and a density of about 10^3 kg/m^3.

Solution From Equation 15.1, we find

$$v_{\text{water}} = \sqrt{\frac{B}{\rho}} = \sqrt{\frac{2.1 \times 10^9 \text{ Pa}}{1 \times 10^3 \text{ kg/m}^3}} \approx 1500 \text{ m/s}$$

15.4 ENERGY AND INTENSITY OF SOUND WAVES

As the tines of a tuning fork move back and forth through the air, they exert a force on the air and cause it to move. In other words, tines do work on the air. The fact that the fork pours energy into the air as sound energy is one of the reasons that the vibration of the fork slowly dies out. (Other factors, such as the energy lost to friction as the tines bend, also are responsible for the diminution of movement.)

We define the intensity I of a wave to be the rate at which sound energy flows through a unit area A perpendicular to the direction of travel of the wave.

In equation form this is

$$I = \frac{1}{A}\frac{\Delta E}{\Delta t} \tag{15.5}$$

Equation 15.5 can be written in an alternative form if you recall that the rate of transfer of energy is defined as power. Thus,

$$I \equiv \frac{\text{power}}{\text{area}} = \frac{P}{A} \tag{15.6}$$

Intensity of a wave

P is the sound power passing through A measured in watts, and the intensity has units of W/m^2.

The faintest sounds the human ear can detect at a frequency of 1000 Hz have an intensity of about 10^{-12} W/m^2. This intensity is called the **threshold of hearing.** The loudest sounds the ear can tolerate have an intensity of about 1 W/m^2 (the **threshold of pain**). At the threshold of hearing, the increase in pressure in the ear is approximately 3×10^{-5} Pa over normal atmospheric pressure. Since atmospheric pressure is about 10^5 Pa, this means the ear can detect pressure fluctuations as small as about 3 parts in 10^{10}! Also, at the threshold of hearing, the maximum displacement of an air molecule is about 1×10^{-11} m. This is a remarkably small number! If we compare this result with the diameter of a molecule (about 10^{-10} m), we see that the ear is an extremely sensitive detector of sound waves.

In a similar manner, one finds that the loudest sounds the human ear can tolerate correspond to a pressure increase of about 29 Pa over normal atmospheric pressure and a maximum displacement of air molecules of 1×10^{-5} m.

Intensity Levels in Decibels

In the previous section it was pointed out that the human ear can detect a wide range of intensities, with the loudest tolerable sounds having intensities about 10^{12} times greater than those of the faintest detectable sounds. However, the loudness of the most intense sound is not perceived as being 10^{12} times greater than that of the faintest sound. This is because the sensation of loudness is approximately logarithmic in the human ear. The relative loudness of a sound is called the **intensity level** or **decibel level,** β. This unit is named after the inventor of the telephone, Alexander Graham Bell (1847–1922) and is defined as

$$\beta \equiv 10 \log\left(\frac{I}{I_0}\right) \tag{15.7}$$

Intensity levels

The constant I_0 is the reference intensity level, taken to be the sound intensity at the threshold of hearing ($I_0 = 10^{-12}$ W/m^2), and I is the intensity at the level β, where β is measured in decibels (dB). On this scale, the threshold of pain ($I = 1$ W/m^2) corresponds to an intensity level of $\beta = 10 \log(1/10^{-12}) = 10 \log(10^{12}) = 120$ dB. Likewise, the threshold of hearing corresponds to $\beta = 10 \log(1/1) = 0$ dB. Nearby jet airplanes can create intensity levels of 150 dB, and subways and riveting machines have levels of 90 to 100 dB. The electronically amplified sound heard at rock concerts can be at levels of up to 120 dB, the threshold of pain. Prolonged exposure to such high intensity levels can produce serious damage to the ear. Ear plugs are recommended whenever intensity levels exceed 90 dB. Recent evidence suggests that noise pollution, which is common in most large cities and in some industrial envi-

TABLE 15.2 Intensity Levels in Decibels for Some Sources

Source of Sound	β (dB)
Nearby jet airplane	150
Jackhammer, machine gun	130
Siren, rock concert	120
Subway, power mower	100
Busy traffic	80
Vacuum cleaner	70
Normal conversation	50
Mosquito buzzing	40
Whisper	30
Rustling leaves	10
Threshold of hearing	0

ronments, may be a contributing factor to high blood pressure, anxiety, and nervousness. Table 15.2 gives some idea of the intensity levels of various sounds.

EXAMPLE 15.3 Intensity Levels of Sound

Calculate the intensity level of a sound wave having an intensity of (a) 10^{-12} W/m^2, (b) 10^{-11} W/m^2, and (c) 10^{-10} W/m^2.

Solution (a) For an intensity of 10^{-12} W/m^2, the intensity level in decibels is

$$\beta = 10 \log\left(\frac{10^{-12}\ \text{W/m}^2}{10^{-12}\ \text{W/m}^2}\right) = 10 \log(1) = 0 \text{ dB}$$

This answer should have been obvious without calculation because an intensity of 10^{-12} W/m^2 corresponds to the threshold of hearing.

(b) In this case, the intensity is exactly 10 times greater than in part (a). The intensity level is

$$\beta = 10 \log\left(\frac{10^{-11}\ \text{W/m}^2}{10^{-12}\ \text{W/m}^2}\right) = 10 \log(10) = 10 \text{ dB}$$

(c) Here the intensity is 100 times greater than the intensity at the threshold of hearing and the intensity level is

$$\beta = 10 \log\left(\frac{10^{-10}\ \text{W/m}^2}{10^{-12}\ \text{W/m}^2}\right) = 10 \log(100) = 20 \text{ dB}$$

Note the pattern in these answers. A sound with an intensity level of 10 dB is 10 times more intense than the 0-dB sound, and a sound with an intensity level of 20 dB is 100 times more intense than a 0-db sound. This pattern is continued throughout the decibel scale. In short, on the decibel scale *an increase of 10 dB means that the intensity of the sound increases by a factor of 10.* For example, a 50-dB sound is 10 times more intense than a 40-dB sound and a 60-dB sound is 100 times more intense than the 40-dB level.

EXERCISE Determine the intensity level of a sound wave whose intensity is equal to 5×10^{-7} W/m^2.

Answer 57 dB.

EXAMPLE 15.4 The Noisy Typewriter

A rather noisy typewriter produces a sound intensity of 10^{-5} W/m^2. Find the decibel level of this machine and calculate the new decibel level when a second identical typewriter is added to the office.

Solution The decibel level of the single typewriter is

$$\beta = 10 \log\left(\frac{10^{-5}\ \text{W/m}^2}{10^{-12}\ \text{W/m}^2}\right) = 10 \log(10^7) = 70 \text{ dB}$$

Adding the second typewriter doubles the energy input into sound and hence doubles the intensity. The new decibel level is

$$\beta = 10 \log\left(\frac{2 \times 10^{-5}\ \text{W/m}^2}{10^{-12}\ \text{W/m}^2}\right) = 73 \text{ dB}$$

Federal regulations now demand that no office or factory worker can be exposed to noise levels that average more than 90 dB over an 8-h day. The results in this example read like one of the old jokes that start "There is some good news and some bad news." First the good news. Imagine that you are a manager analyzing the noise conditions in your office. One typewriter in the office produces a noise level of 70 db. When you add a second typewriter, the noise level increases by only 3 dB. Because of the logarithmic nature of decibel levels, doubling the intensity does not double the decibel level; in fact, it alters it by a surprisingly small amount. This means that additional equipment can be added to an office or factory without appreciably altering the decibel level of the environment.

Now the bad news. The results also work in reverse. As you remove noisy machinery, the decibel level is not lowered appreciably. For example, consider an office with 60 typewriters producing a noise level of 93 dB, which is 3 dB above the maximum allowed. In order to reduce the noise level by 3 db, half the machines would have to be removed! That is, you would have to remove 30 typewriters to reduce the noise level to 90 dB. To reduce the level another 3 db, you would have to remove half of the remaining machines, and so on.

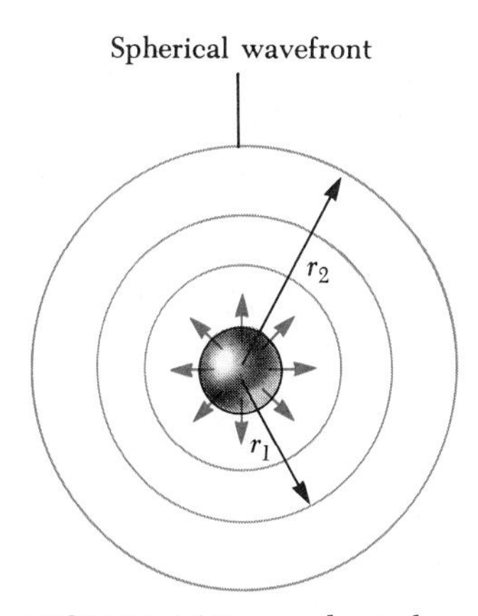

FIGURE 15.3 A spherical wave propagating radially outward from an oscillating sphere. The intensity of the spherical wave varies as $1/r^2$.

15.5 SPHERICAL AND PLANE WAVES

If a small spherical object oscillates such that its radius changes periodically with time, a spherical sound wave will be produced (Fig. 15.3). The wave moves outward from the source at a constant speed.

Because all points on the vibrating sphere behave in the same way, we conclude that the energy in a spherical wave propagates equally in all directions. That is, no one direction is preferred over any other. If P_{av} is the average power emitted by the source, then at any distance r from the source, this power must be distributed over a spherical surface of area $4\pi r^2$ assuming no absorption in the medium. (Recall that $4\pi r^2$ is the surface area of a sphere.) Hence, the **intensity** of the sound at a distance r from the source is

$$I = \frac{\text{average power}}{\text{area}} = \frac{P_{av}}{A} = \frac{P_{av}}{4\pi r^2} \tag{15.8}$$

This shows that the intensity of a wave decreases with increasing distance from its source, as you might expect. The fact that I varies as $1/r^2$ is a result of the assumption that the small source (sometimes called a **point source**) emits a spherical wave. Since the average power is the same through any spherical surface centered at the source, we see that the intensities at distance r_1 and r_2 (Fig. 15.3) from the center of the source are

$$I_1 = \frac{P_{av}}{4\pi r_1^2} \qquad I_2 = \frac{P_{av}}{4\pi r_2^2}$$

Therefore, the ratio of intensities at these two spherical surfaces is

$$\frac{I_1}{I_2} = \frac{r_2^2}{r_1^2}$$

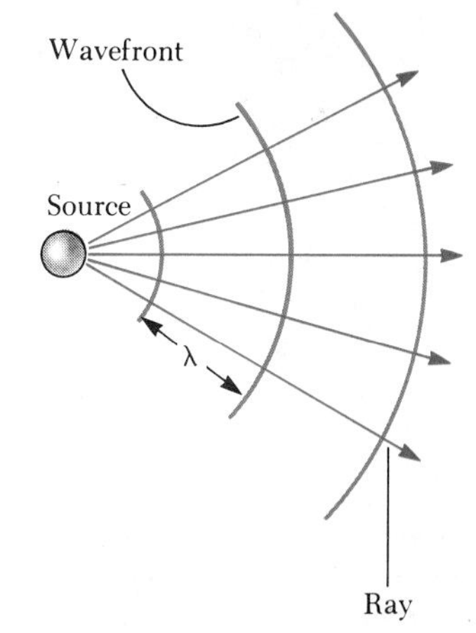

FIGURE 15.4 Spherical waves emitted by a point source. The circular arcs represent the spherical wavefronts concentric with the source. The rays are radial lines pointing outward from the source, perpendicular to the wavefronts.

It is useful to represent spherical waves graphically by a series of circular arcs concentric with the source representing part of a spherical surface, as in Figure 15.4. We call such a surface a **wavefront.** The distance between adjacent wavefronts equals the wavelength, λ. The radial lines pointing outward from the source and cutting the arcs perpendicularly are called **rays.**

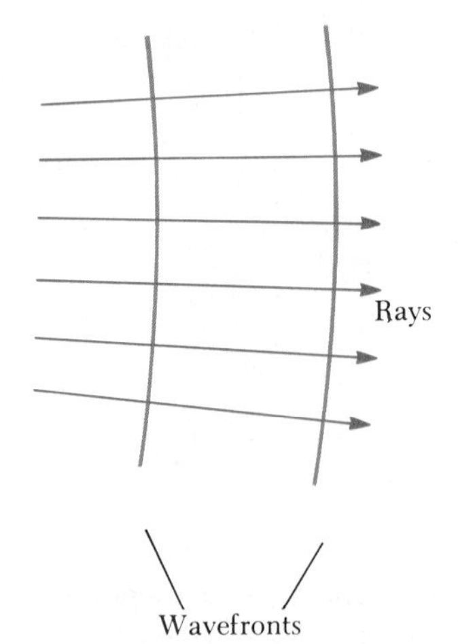

FIGURE 15.5 Far away from a point source, the wavefronts are nearly parallel planes and the rays are nearly parallel lines perpendicular to the planes. Hence, a small segment of a spherical wavefront is approximately a plane wave.

Now consider a small portion of the wavefront that is at a *large* distance (large relative to λ) from the source, as in Figure 15.5. In this case, the rays are nearly parallel to each other and the wavefronts are very close to being planes. Therefore, at distances from the source that are large relative to the wavelength, we can approximate the wavefronts by parallel planes. We call such waves plane waves. Any small portion of a spherical wave that is far from the source can be considered a **plane wave.** Figure 15.6 illustrates a plane wave propagating along the x axis. If x is taken to be the direction of the wave motion (or rays) in Figure 15.6, then the wavefronts are parallel to the plane containing the y and z axes.

EXAMPLE 15.5 Intensity Variations of a Point Source

A small source emits sound waves with a power output of 80 W. (a) Find the intensity 3 m from the source.

Solution A small source emits energy in the form of spherical waves (Fig. 15.3). Let P_{av} be the average power output of the source. At a distance r from the source, the power is distributed over the surface area of a sphere, $4\pi r^2$. Therefore, the intensity at a distance r from the source is given by Equation 15.8. Since $P_{av} = 80$ W and $r = 3$ m, we find that

$$I = \frac{P_{av}}{4\pi r^2} = \frac{80\text{ W}}{4\pi(3\text{ m})^2} = 0.71\text{ W/m}^2$$

which is close to the threshold of pain.

(b) Find the distance at which the sound level is 40 dB.

Solution We can find the intensity at the 40-dB intensity level by using Equation 15.7 with $I_0 = 10^{-12}$ W/m^2:

$$40 = 10\log(I/I_0)$$
$$4 = \log(I/I_0)$$
$$I = 10^4 I_0 = 10^{-8}\text{ W/m}^2$$

When this value for I is used in Equation 15.8, solving for r gives

$$r = \left(\frac{P_{av}}{4\pi I}\right)^{1/2} = \left(\frac{80\text{ W}}{4\pi \times 10^{-8}\text{ W/m}^2}\right)^{1/2} = 2.5 \times 10^4\text{ m}$$

which is approximately 15 mi!

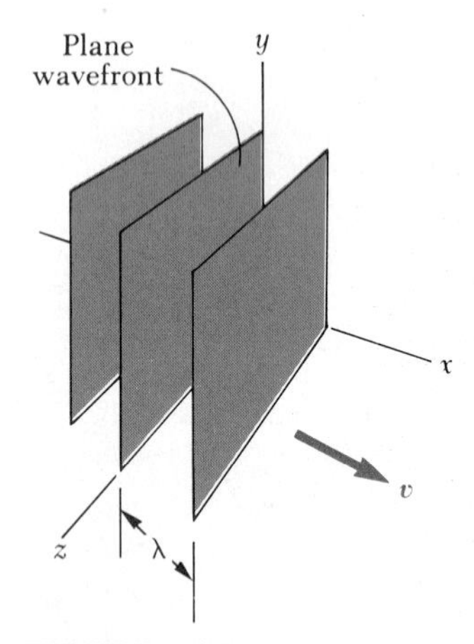

FIGURE 15.6 Representation of a plane wave moving in the positive x direction with a speed v. The wavefronts are planes parallel to the yz plane.

15.6 THE DOPPLER EFFECT

When a car or truck is moving while its horn is blowing, the frequency of the sound you hear is higher as the vehicle approaches you and lower as it moves away from you. This is one example of the **Doppler effect,** named for the Austrian physicist Christian Doppler (1803–1853), who discovered the effect.

In general, a Doppler effect is experienced whenever there is relative motion between source and observer. When the source and observer are moving toward each other, the frequency heard by the observer is higher than the frequency of the source. When the source and observer are moving away from each other, the observer hears a frequency lower than the source frequency.

Although the Doppler effect is most commonly experienced with sound waves, it is a phenomenon common to all waves. For example, the frequencies of light waves are also shifted by relative motion.

First let us consider the case where the observer is moving and the sound source is stationary. For simplicity, we shall assume that the air is also stationary and all velocity measurements are made relative to this stationary medium. Figure 15.7 describes the situation when the observer is moving with a speed v_0 toward the source (considered a point source), which is at rest ($v_s = 0$).

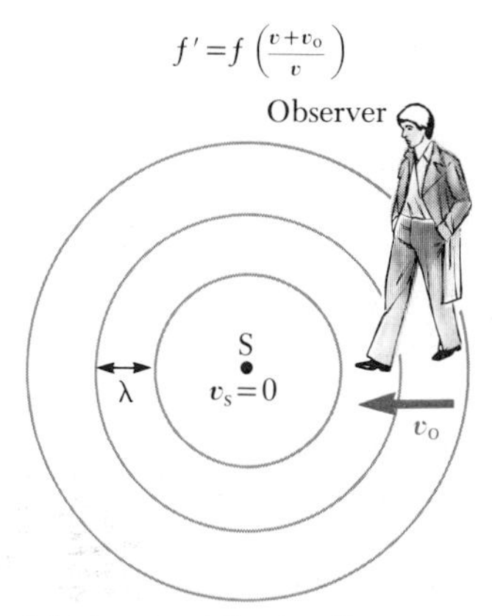

FIGURE 15.7 An observer moving with a speed v_0 *toward* a stationary point source hears a frequency f' that is *greater* than the source frequency f.

We shall take the frequency of the source to be f, the wavelength to be λ, and the speed of sound in air to be v. Clearly, if both observer and source were stationary, the observer would detect f wavefronts per second. (That is, when $v_0 = 0$ and $v_s = 0$, the observed frequency equals the source frequency.) When the observer is moving toward the source, he or she moves a distance $v_0 t$ in a time of t seconds. In this time, *the observer detects an additional number of wavefronts.* The number of extra wavefronts detected will be equal to the distance traveled, $v_0 t$, divided by the number of wavelengths in this distance, λ. Thus, we have

$$\text{Additional wavefronts detected} = \frac{v_0 t}{\lambda}$$

The number of additional wavefronts detected *per second* is v_0/λ. Hence, the frequency f' heard by the observer is *increased* and given by

$$f' = f + \frac{v_0}{\lambda}$$

Using the fact that $\lambda = v/f$, we see that $v_0/\lambda = (v_0/v)f$. Hence f' can be expressed as

$$f' = f\left(\frac{v + v_0}{v}\right) \qquad (15.9)$$

An observer traveling *away* from the source, as in Figure 15.8, *detects fewer wavefronts per second.* Thus, from Equation 15.9, it follows that the frequency heard by the observer in this case is *lowered* and given by

$$f' = f\left(\frac{v - v_0}{v}\right) \qquad (15.10)$$

We can incorporate these two equations into one:

$$\boxed{f' = f\left(\frac{v \pm v_0}{v}\right)} \qquad (15.11)$$

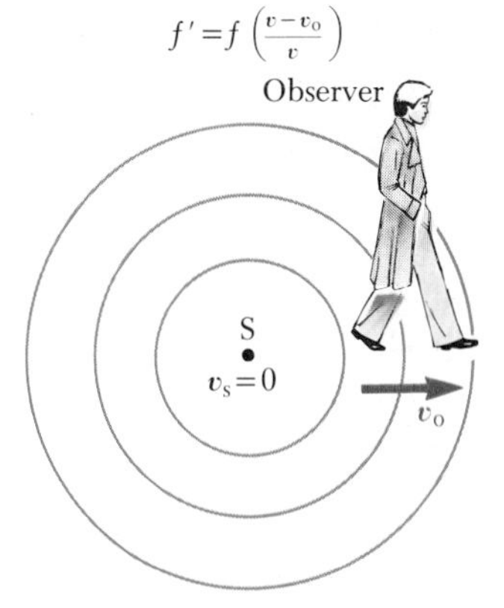

FIGURE 15.8 An observer moving with a speed v_0 *away* from a stationary point source hears a frequency f' that is *lower* than the source frequency.

This general equation applies when an observer is moving with a speed v_0 relative to a stationary source. *The positive sign is used when the observer is moving toward the source and the negative sign is used when the observer is moving away from the source.*

Now consider the situation in which the source is in motion and the observer is at rest. If the source is moving directly toward observer A in

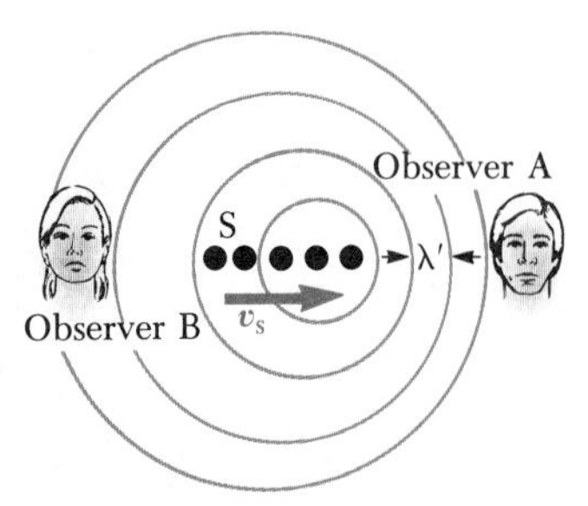

FIGURE 15.9 A source moving with a speed v_s toward a stationary observer A and away from stationary observer B. Observer A hears an *increased* frequency, while observer B hears a *decreased* frequency.

Figure 15.9, the wavefronts heard by A are closer together because the source is moving in the direction of the outgoing wave. As a result, the wavelength λ' measured by observer A is shorter than the true wavelength, λ, of the source. During each vibration, which lasts for a time T (the period), the source moves a distance $v_sT = v_s/f$ and *the wavelength is shortened by this amount.* Therefore, the observed wavelength, λ', is given by

$$\lambda' = \lambda - \frac{v_s}{f}$$

Since $\lambda = v/f$, the frequency heard by observer A is

$$f' = \frac{v}{\lambda'} = \frac{v}{\lambda - \dfrac{v_s}{f}} = \frac{v}{\dfrac{v}{f} - \dfrac{v_s}{f}} = f\left(\frac{v}{v - v_s}\right) \tag{15.12}$$

That is, *the observed frequency is increased when the source is moving toward the observer.*

In Figure 15.9 the source is moving away from observer B, who is at rest to the left of the source. Thus observer B measures a wavelength that is *greater* than λ and hears a *decreased* frequency given by

$$f' = f\left(\frac{v}{v + v_s}\right) \tag{15.13}$$

Combining Equations 15.12 and 15.13, we can express the general relationship for the observed frequency when the source is moving and the observer is at rest:

$$\boxed{f' = f\left(\frac{v}{v \mp v_s}\right)} \tag{15.14}$$

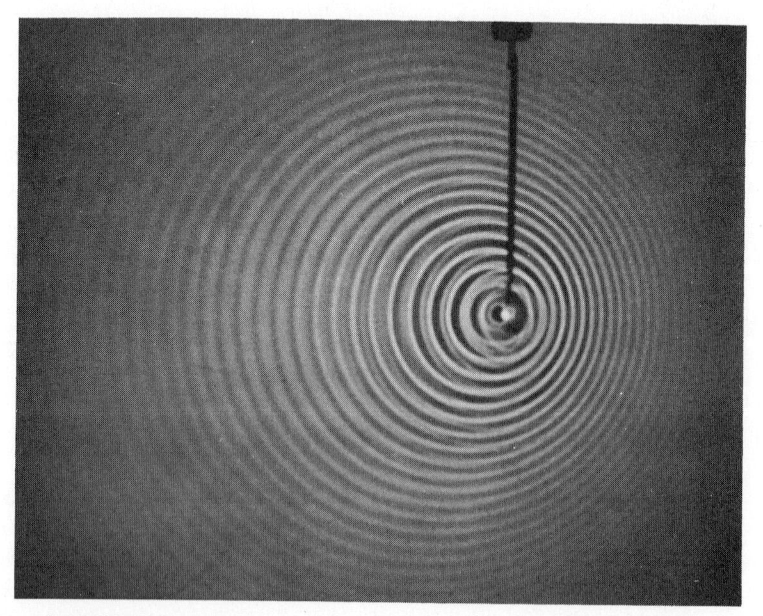

The Doppler effect in water observed in a ripple tank. (Courtesy Educational Development Center, Newton, Mass.)

Finally, if both the source and the observer are in motion, one finds the following general relationship for the observed frequency:

$$f' = f\left(\frac{v \pm v_o}{v \mp v_s}\right) \qquad (15.15)$$

In this expression, the upper signs ($+v_o$ and $-v_s$) refer to motion of either source or observer in a direction toward the other and the lower signs ($-v_o$ and $+v_s$) refer to motion of one in a direction away from the other.

EXAMPLE 15.6 Listen, but Don't Stand on the Track

A train moving at a speed of 40 m/s sounds its whistle, which has a frequency of 500 Hz. Determine the frequency heard by a stationary observer as the train approaches.

Solution We can use Equation 15.12 to get the apparent frequency as the train approaches the observer. With v = 345 m/s as the speed of sound in air, we have

$$f' = f\left(\frac{v}{v - v_s}\right) = (500\text{ Hz})\left(\frac{345\text{ m/s}}{345\text{ m/s} - 40\text{ m/s}}\right) = 566\text{ Hz}$$

EXERCISE Determine the frequency heard by the stationary observer as the train recedes.

Answer: 448 Hz.

EXAMPLE 15.7 The Noisy Siren

An ambulance travels down a highway at a speed of 75 mi/h, its siren emitting sound at a frequency of 400 Hz. What is the frequency heard by a passenger in a car traveling at 55 mi/h in the opposite direction as the car (a) approaches and (b) moves away from the ambulance?

Solution Let us take the velocity of sound in air to be v = 345 m/s and note that 1 mi/h = 0.447 m/s. Therefore, v_s = 75 mi/h = 33.5 m/s and v_o = 55 mi/h = 24.6 m/s. We can use Equation 15.15 in both cases.

(a) As the ambulance and car approach each other, the observed frequency is

$$f' = f\left(\frac{v + v_o}{v - v_s}\right) = (400\text{ Hz})\left(\frac{345\text{ m/s} + 24.6\text{ m/s}}{345\text{ m/s} - 33.5\text{ m/s}}\right) = 475\text{ Hz}$$

(b) As the two vehicles recede from each other, the passenger in the car hears a frequency

$$f' = f\left(\frac{v - v_o}{v + v_s}\right) = (400\text{ Hz})\left(\frac{345\text{ m/s} - 24.6\text{ m/s}}{345\text{ m/s} + 33.5\text{ m/s}}\right) = 339\text{ Hz}$$

Shock Waves

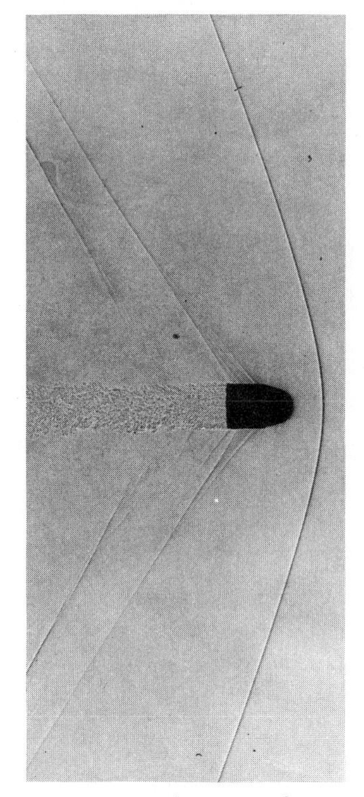

A bullet traveling in air faster than the speed of sound. Note the shape of the shock waves accompanying the bullet. See also Color Plate 10.

Now let us consider what happens when the source velocity, v_s, *exceeds* the wave velocity, v. This situation is described graphically in Figure 15.10. The circles represent spherical wavefronts emitted by the source at various times during its motion. At t = 0, the source is at point S_0, and at some later time t, the source is at point S_n. In the time t, the wavefront centered at S_0 reaches a radius of vt. In this same interval, the source travels a distance v_st to S_n. At

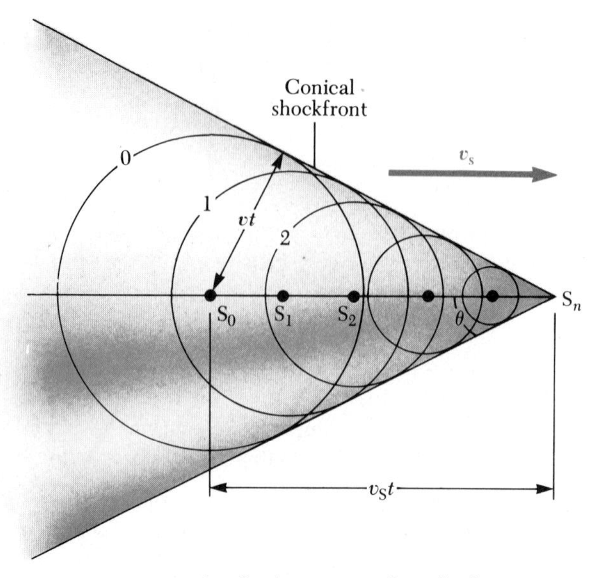

FIGURE 15.10 Representation of a shock wave produced when a source moves from S_0 to S_n with a speed v_s that is *greater* than the wave speed, v, in that medium. The envelope of the wavefronts form a cone whose half-angle is given by $\sin\theta = v/v_s$.

the instant the source is at S_n, the waves just beginning to be generated at this point have wavefronts of zero radius. The line drawn from S_n to the wavefront centered on S_0 is tangent to all other wavefronts generated at intermediate times. All such tangent lines lie on the surface of a cone. The angle θ between one of these tangent lines and the direction of travel is given by

$$\sin\theta = v/v_s$$

The ratio v_s/v is referred to as the **Mach number.** The conical wavefront produced when $v_s > v$ (supersonic speeds) is known as a **shock wave.** An interesting example of shock waves is the V-shaped wavefronts produced by a boat (the bow wave) when the boat's speed exceeds the speed of the water waves.

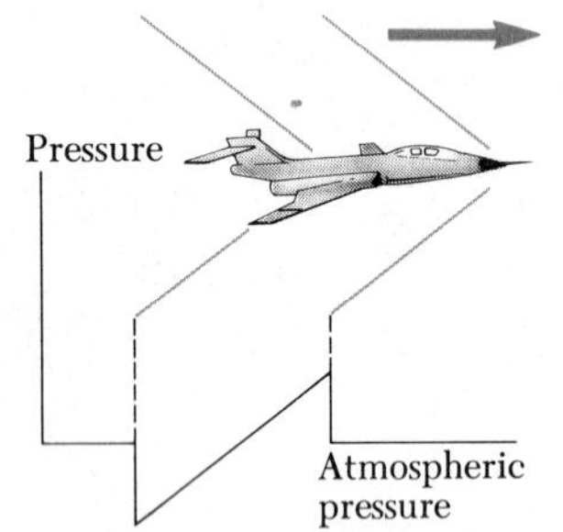

FIGURE 15.11 The two shock waves produced by the nose and tail of a jet airplane traveling at supersonic speed.

Jet airplanes traveling at supersonic speeds produce shock waves, which are responsible for the loud explosion, or sonic boom, one hears. The shock wave carries a great deal of energy concentrated on the surface of the cone, with correspondingly large pressure variations. Such shock waves are unpleasant to hear and can damage buildings when aircraft fly supersonically at low altitudes. In fact, an airplane flying at supersonic speeds produces a double boom because two shock waves are formed, one from the nose of the plane and one from the tail (Fig. 15.11).

15.7 INTERFERENCE IN SOUND WAVES

The fact that sound waves can be made to interfere with each other can be demonstrated with the device shown in Figure 15.12. Sound from a loudspeaker at S is sent into a tube at P, where there is a T-shaped junction. Half the sound intensity travels in one direction and half in the opposite direction. Thus, the sound waves that reach the receiver at R travel along two different paths. If the two paths are of the same length, a crest of the wave that enters the junction will separate into two halves, travel the two paths, and then combine again at the receiver. The upper and lower waves reunite at the receiver such that constructive interference takes place, and thus a loud sound is heard at the detector.

However, suppose one of the path lengths is adjusted by sliding the upper U-shaped tube upward so that the upper path is half a wavelength *longer* than the lower path. In this case, an entering sound wave splits and travels the two paths as before but now the wave along the upper path must travel a distance equivalent to half a wavelength farther than the wave traveling along the lower path. As a result, the crest of one wave meets the trough of the other when they merge at the receiver. Since this is the condition for destructive interference, no sound is detected at the receiver.

You should be able to predict what will be heard if the upper path is adjusted to one full wavelength longer than the lower path. In this case, constructive interference of the two waves occurs and a loud sound is detected at the receiver.

There are many other examples of interference phenomena in nature. In a later chapter, we shall describe several interesting interference effects involving light waves.

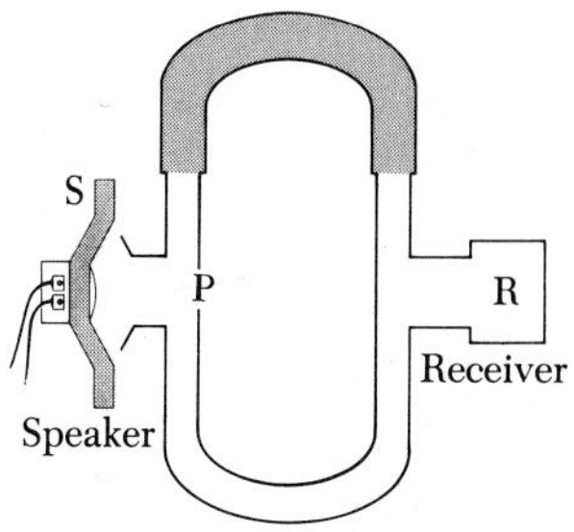

S
P
R
Receiver
Speaker

FIGURE 15.12 An acoustical system for demonstrating interference of sound waves. Sound from the speaker enters the tube and splits into two parts at P. The two waves combine at the opposite side and are detected at R. The upper path length is varied by the sliding section.

EXAMPLE 15.8 Interference from Two Loudspeakers

Two loudspeakers are placed as in Figure 15.13 and driven by the same source at a frequency of 2000 Hz. The top speaker is then moved to the left to position *A*. At this location, an observer at a large distance from the speakers and directly in front of them notices that the intensity of the sound from the two sources has decreased to a minimum. How far back has the speaker been moved? Assume that the speed of sound in air is 345 m/s.

Solution Initially, both speakers are at the same distance from the observer. Hence the sound from each speaker must travel the same distance and the observer hears a loud sound corresponding to constructive interference. When the top speaker is at position *A*, its sound must travel farther to reach the observer. Since the observer notices a minimum in the sound level when the top speaker is at *A*, destructive interference is taking place between the two separate sound signals. This means that the top speaker has been moved half a wavelength. With the speed of sound in air equal to 345 m/s, we can calculate the wavelength:

$$\lambda = \frac{v}{f} = \frac{345 \text{ m/s}}{2000 \text{ s}^{-1}} = 0.173 \text{ m}$$

Therefore, the distance moved is half this value, or 0.0865 m.

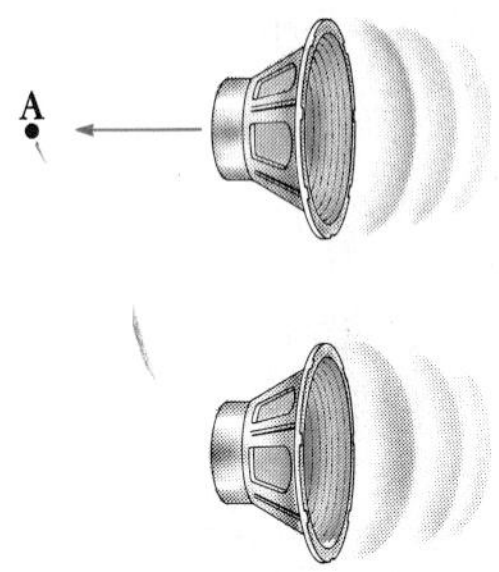

A

FIGURE 15.13 (Example 15.8) Two loudspeakers driven by the same source produce an interference pattern.

15.8 STANDING WAVES

If a stretched string is clamped at both ends, traveling waves will reflect from the fixed ends, creating waves traveling in both directions. The incident and reflected waves will combine according to the superposition principle. For example, if the string is vibrated at exactly the right frequency, a crest moving toward one end and a reflected trough will meet at some point along the

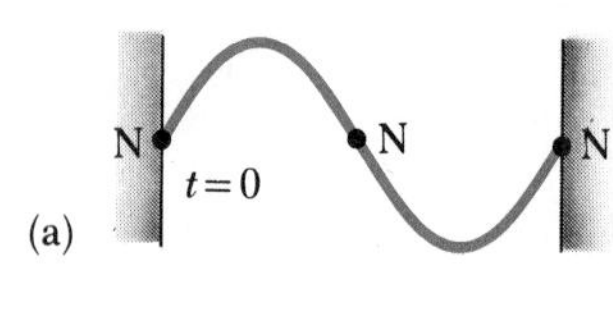

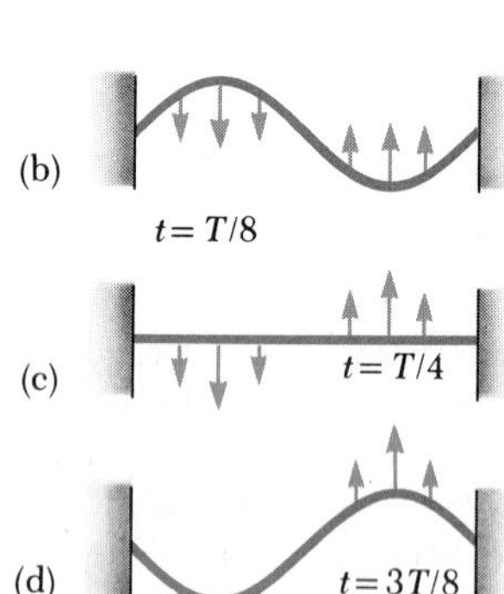

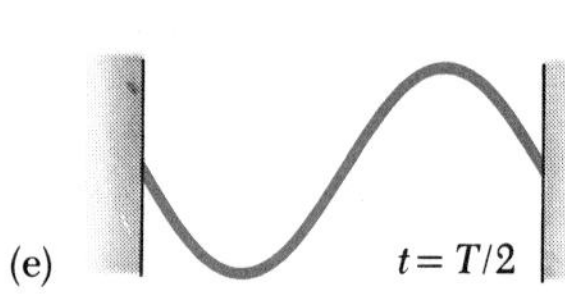

FIGURE 15.14 A standing wave pattern in a stretched string showing snapshots of the string during one half of a cycle.

string. The two waves cancel at this point, which is called a **node.** The resulting pattern on the string is one in which the wave appears to stand still, and we have what is called a **standing wave** on the string. There is no motion in the string at the nodes, but at a distance equal to 0.25λ away from each node, the string vibrates with a large amplitude. These points are called **antinodes.** The distance between successive nodes, or between successive antinodes, is equal to 0.5λ.

The oscillation of a standing wave during one half of a cycle is shown in Figure 15.14. *Notice that all points on the string oscillate vertically with the same frequency except for the node, which is stationary.* (The points at which the string is attached to the wall are also nodes, which are labeled N in Figure 15.14a.) Furthermore, the various points have different amplitudes of motion.

Consider a string of length L that is fixed at both ends, as in Figure 15.15. The string has a number of natural patterns of vibration, called normal modes. Three of these are pictured in Figures 15.15b, 15.15c, and 15.15d. Each of these has a characteristic frequency, which we shall now calculate.

First, note that *the ends of the string must be nodes because these points are fixed.* If string is displaced at its midpoint and released, the vibration shown in Figure 15.15b can be produced, in which case the center of the string is an antinode. For this normal mode, the length of the string equals $\lambda/2$ (the distance between nodes). Thus,

$$L = \frac{\lambda_1}{2} \quad \text{or} \quad \lambda_1 = 2L$$

and the frequency of this vibration is

$$f_1 = \frac{v}{\lambda_1} = \frac{v}{2L} \tag{15.16}$$

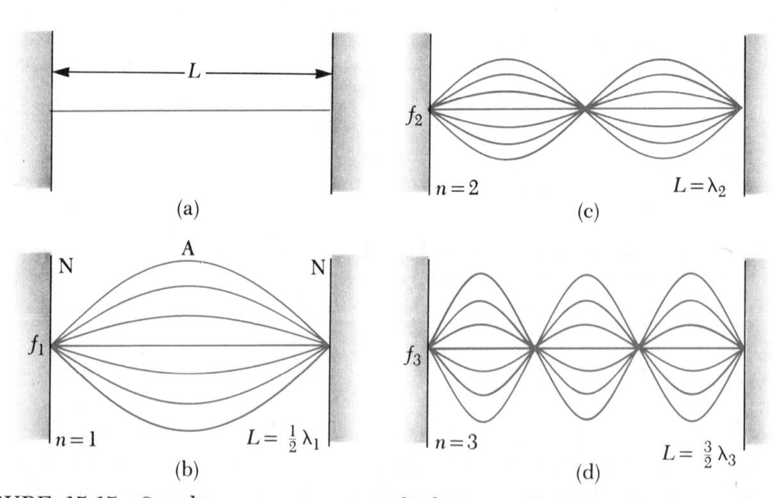

FIGURE 15.15 Standing waves in a stretched string of length L fixed at both ends. The normal frequencies of vibration form a harmonic series: (b) the fundamental frequency, or first harmonic, (c) the second harmonic, and (d) the third harmonic.

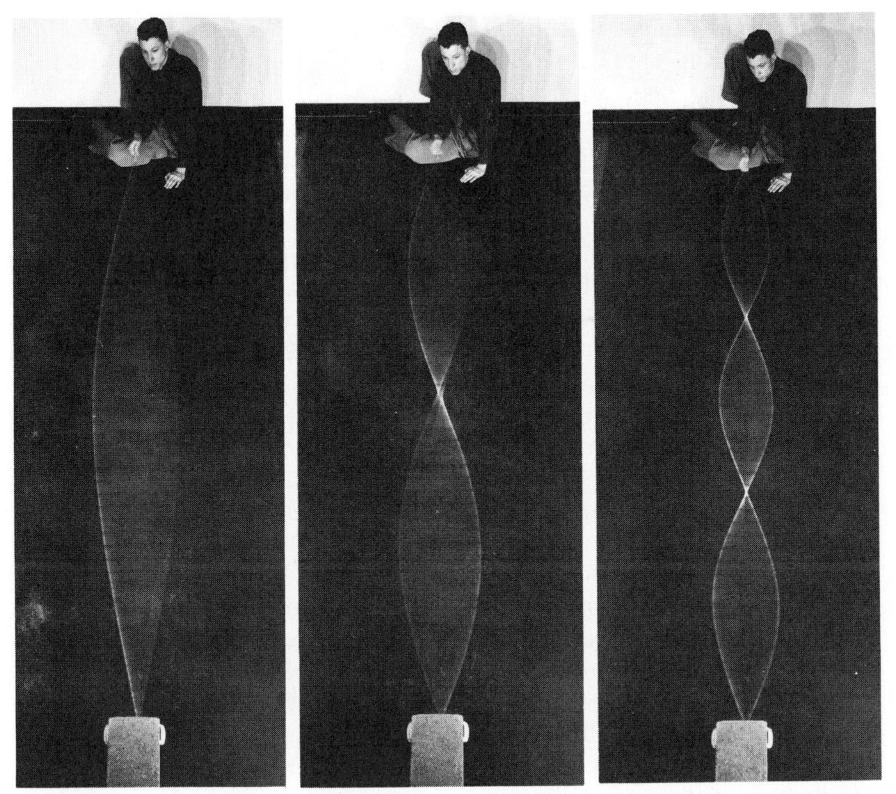

Standing waves. As one end of the stretched string is moved from side to side with increasing frequency, patterns with more and more loops are formed. Only certain definite frequencies produce fixed patterns. (Courtesy Educational Development Center, Newton, Mass.)

In Chapter 14, the speed of a wave on a string was given as $v = \sqrt{F/\mu}$, where F is the tension in the string and μ is its mass per unit length. Thus, we can express Equation 15.16 as

$$f_1 = \frac{1}{2L}\sqrt{\frac{F}{\mu}} \tag{15.17}$$

Fundamental frequency

This lowest frequency of vibration is called the **fundamental frequency.**

The next normal mode, of wavelength λ_2 (Fig. 15.15c), occurs when the length of the string equals one wavelength, that is when $\lambda_2 = L$. Hence

$$f_2 = \frac{v}{L} = \frac{2v}{2L} = 2f_1 \tag{15.18}$$

Note that this frequency is equal to *twice* the fundamental frequency. You should convince yourself that the next highest frequency of vibration, shown in Figure 15.15d, is given by

$$f_3 = \frac{3v}{2L} = 3f_1 \tag{15.19}$$

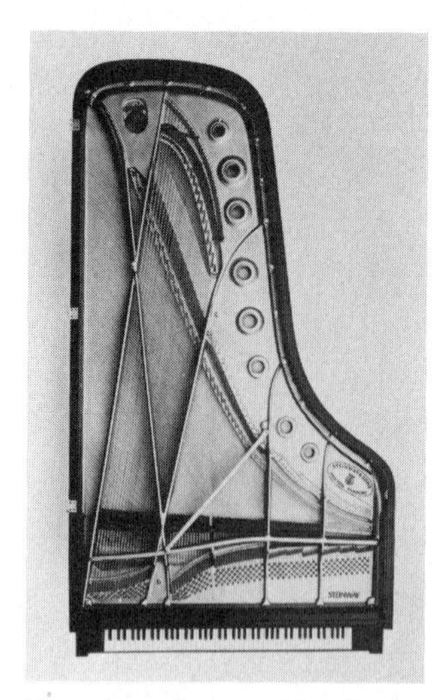

Top view of a concert grand piano. Why do you suppose the sounding board has this unusual shape?

In general, the characteristic frequencies are given by

$$f_n = \frac{nv}{2L} = \frac{n}{2L}\sqrt{\frac{F}{\mu}} \tag{15.20}$$

where $n = 1, 2, 3, \ldots$. In other words, the frequencies are integral multiples of the fundamental frequency. The set of frequencies f_1, $2f_1$, $3f_1$, and so on form a **harmonic series.** The fundamental, f_1, corresponds to the **first harmonic;** the frequency $f_2 = 2f_1$ corresponds to the **second harmonic,** and so on.

When a stretched string is distorted to a shape that corresponds to any one of its harmonics, after being released it will only vibrate at the frequency of that harmonic. If the string is struck or bowed, however, the resulting vibration will include frequencies of various harmonics, including the fundamental. Waves not in the harmonic series are quickly damped out on a string fixed at both ends. In effect, the string "selects" the normal-mode frequencies when disturbed. As we shall see later, the presence of several harmonics on a string gives the stringed instruments their characteristic sound, which enables us to distinguish one from another even when they are sounding identical fundamental frequencies.

The frequency of a string on a musical instrument can be changed either by varying the tension or by changing the length. For example, the tension in guitar and violin strings is varied by turning pegs on the neck of the instrument. As the tension is increased, the frequency of the normal modes increases according to Equation 15.20. Once the instrument is tuned, the musician varies the frequency by moving his or her fingers along the neck, thereby changing the length of the vibrating portion of the string. As the length is shortened, the frequency increases, as Equation 15.20 indicates.

Finally, note from Equation 15.20 that if one has a string of fixed length, one can cause it to vibrate at a lower fundamental frequency by increasing its mass per unit length. This is achieved in the bass strings of guitars and pianos by wrapping the strings with a metal winding.

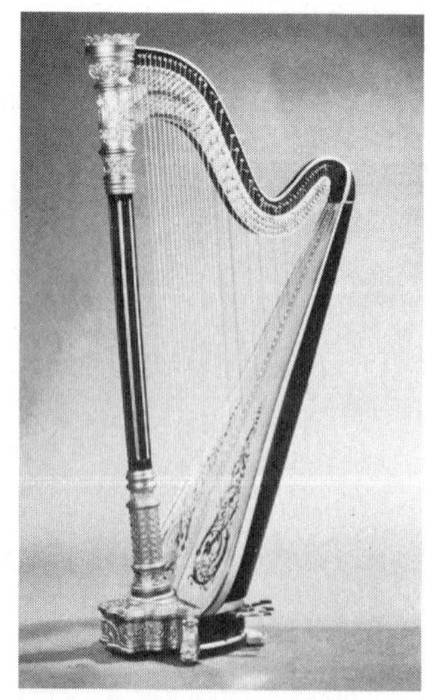

Concert style harp. (Photo Lyon & Healy Harps, Chicago)

EXAMPLE 15.9 Harmonics of a Stretched String

Find the first four harmonics for a string 1 m long if the string has a mass per unit length of 2×10^{-3} kg/m and is under a tension of 80 N.

Solution The speed of wave on the string is

$$v = \sqrt{\frac{F}{\mu}} = \sqrt{\frac{80 \text{ N}}{2 \times 10^{-3} \text{ kg/m}}} = 200 \text{ m/s}$$

The fundamental frequency can be found by using Equation 15.16:

$$f_1 = \frac{v}{2L} = \frac{200 \text{ m/s}}{2(1 \text{ m})} = 100 \text{ Hz}$$

The frequencies of the next three modes are $f_2 = 2f_1$, $f_3 = 3f_1$, and $f_4 = 4f_1$. Thus, $f_2 = 200$ Hz, $f_3 = 300$ Hz, and $f_4 = 400$ Hz.

EXERCISE Find the tension in the string if the fundamental frequency is increased to 120 Hz.

Answer: 115 N.

15.9 FORCED VIBRATIONS AND RESONANCE

In Chapter 14 we learned that the energy of a damped oscillator decreases in time because of friction. It is possible to compensate for this energy loss by applying an external force that does positive work on the system.

For example, suppose a mass-spring system having some natural frequency of vibration f_0 is pushed back and forth with a periodic force whose frequency is f. The system will vibrate at the frequency f of the driving force. This type of motion is referred to as a **forced vibration.** The amplitude of the motion reaches a maximum when the frequency of the driving force equals the natural frequency of the system, f_0, called the **resonant frequency** of the system. Under this condition, the system is said to be in **resonance.**

In Section 15.8 we learned that a stretched string is able to vibrate in one or more of its natural modes. Here again, if a periodic force is applied to the string, the amplitude of vibration will increase as the frequency of the applied force approaches one of the natural frequencies of vibration.

Resonance vibrations occur in a wide variety of circumstances. One experiment that demonstrates a resonance condition is illustrated in Figure 15.16. Several pendulums of different lengths are suspended from a flexible beam. If one of them, such as A, is set into motion, the others will begin to oscillate because they are coupled by vibrations in the flexible beam. Pendulum C, whose length is the same as that of A, will oscillate with the greatest amplitude since its natural frequency matches that of pendulum A (the driving force).

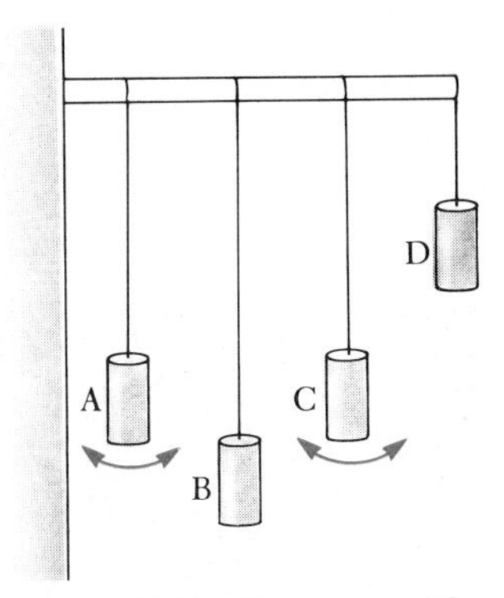

Figure 15.16 Resonance: If pendulum A is set into oscillation, only pendulum C, whose length is the same as that of A, will eventually oscillate with large amplitude, or resonate.

Another apparatus that can easily be set up in the laboratory is a stretched string fixed at one end and connected at the opposite end to a vibrating blade, as in Figure 15.17. As the blade oscillates, transverse waves sent down the string are reflected at the fixed end. As we found in Section 15.8, the string has natural frequencies that are determined by its mass per unit length and the tension (Eq. 15.20). When the frequency of the vibrating blade equals one of the natural frequencies of the string, standing waves will be produced and the string will vibrate with a large amplitude.

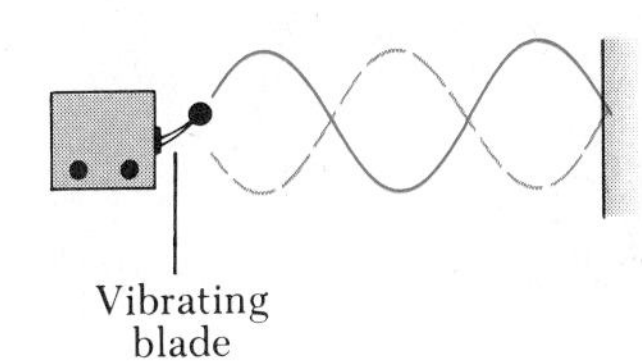

FIGURE 15.17 Standing waves can be set up in a stretched string by connecting one end of the string to a vibrating blade. When the blade vibrates at one of the natural frequencies of the string, large amplitude standing waves are created.

Another simple example of resonance is a child being pushed on a swing, which is essentially a pendulum with a natural frequency that depends on the length. The swing is kept in motion by a series of approximately timed pushes. In order to increase the height of the swing, you must push it each time it returns to your hands. This corresponds to a frequency equal to the natural frequency of the swing. If the energy you put into the system per cycle of motion exactly equals the energy lost due to friction, the amplitude remains constant.

A wine glass shattered by the amplified sound of a human voice. (Courtesy Memorex Corporation)

Opera singers have been known to set crystal goblets into audible vibration with their powerful voices. This is yet another example of resonance: the sound waves emitted by the singer are able to set up large-amplitude vibrations in the glass. If a highly amplified sound wave has the right frequency,

FIGURE 15.18 The collapse of the Tacoma Narrows suspension bridge in 1940 is a vivid demonstration of mechanical resonance. High winds set up standing waves in the bridge, causing it to oscillate at a frequency near one of its natural frequencies. Once established, this resonance condition led to the bridge's collapse. (United Press International Photo)

the amplitude of forced vibrations in the glass will increase to the point where the glass becomes heavily strained and shatters. In this case, resonance occurs when the wavelength of the emitted sound wave equals the circumference of the glass.

A striking example of structural resonance occurred in 1940, when the Tacoma Narrows bridge in the state of Washington was set into oscillation by the wind (Fig. 15.18). The amplitude of these oscillations increased steadily until the bridge was ultimately destroyed.

15.10 STANDING WAVES IN AIR COLUMNS

Standing longitudinal waves can be set up in a tube of air, such as an organ pipe, as the result of interference between sound waves traveling in opposite directions. The relationship between the incident wave and the reflected wave depends on whether the reflecting end of the tube is open or closed. A portion of the sound wave is reflected back into the tube at the open end. *If the reflecting end is closed, a node must exist at this end because the movement of air molecules is restricted. If the end is open, the air molecules have complete freedom of motion and an antinode exists.*

The first three modes of vibration of a pipe open at both ends are shown in Figure 15.19a. When air is directed against an edge at the left, longitudinal standing waves are formed and the pipe vibrates at its natural frequencies. Note that for the fundamental frequency the wavelength is twice the length of the pipe and hence $f_1 = v/2L$. Similarly, one finds that the frequencies of the second and third harmonics are $2f_1, 3f_1, \ldots$. Thus,

in a pipe open at both ends, the natural frequencies of vibration form a series in which all **harmonics are present** and are equal to integral multiples of the fundamental.

We can express this harmonic series as

$$f_n = n\frac{v}{2L} \qquad n = 1, 2, 3, \ldots \tag{15.21}$$

where v is the speed of sound in air.

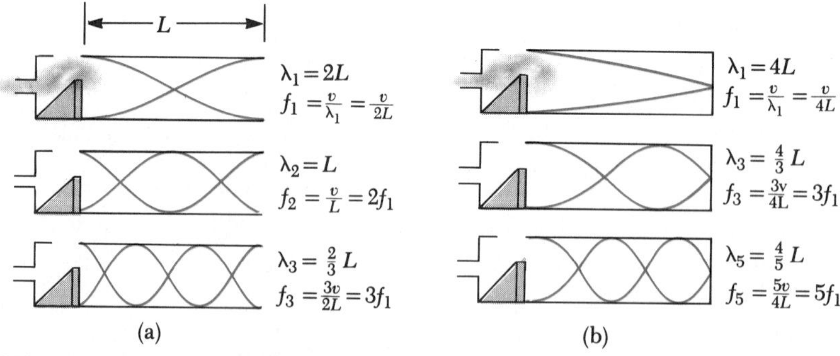

FIGURE 15.19 (a) Standing longitudinal waves in an organ pipe open at both ends. The natural frequencies $f_1, 2f_1, 3f_1 \ldots$ form a harmonic series. (b) Standing longitudinal waves in an organ pipe closed at one end. Only *odd* harmonics are present, and the natural frequencies are $f_1, 3f_1, 5f_1$, etc.

If a pipe is closed at one end and open at the other, the closed end is a node (Fig. 15.19b). In this case, the wavelength for the fundamental mode is four times the length of the tube. Hence, $f_1 = v/4L$ and the frequencies of the third and fifth harmonics are $3f_1, 5f_1, \ldots$. That is,

in a pipe closed at one end and open at the other, **only odd harmonics are present,**

and these are given by

$$f_n = n\frac{v}{4L} \qquad n = 1, 3, 5, \ldots \tag{15.22}$$

EXAMPLE 15.10 Harmonics of a Pipe

A pipe has a length of 2.46 m. (a) Determine the frequencies of the first three harmonics if the pipe is open at each end. Take 345 m/s as the speed of sound in air.

Solution The fundamental frequency of a pipe open at both ends can be found from Equation 15.21, with $n = 1$.

$$f_1 = \frac{v}{2L} = \frac{345 \text{ m/s}}{2(2.46 \text{ m})} = 70 \text{ Hz}$$

Since all harmonics are present in a pipe open at both ends, the second and third harmonics have frequencies given by $f_2 = 2f_1 = 140$ Hz and $f_3 = 3f_1 = 210$ Hz.

(b) What are the lowest possible frequencies if the pipe is closed at one end and open at the other?

Solution The fundamental frequency of a pipe closed at one end can be found from Equation 15.22, with $n = 1$.

$$f_1 = \frac{v}{4L} = \frac{345 \text{ m/s}}{4(2.46 \text{ m})} = 35 \text{ Hz}$$

In this case, only odd harmonics are present, and so the third and fifth harmonics have frequencies given by $f_3 = 3f_1 = 105$ Hz and $f_5 = 5f_1 = 175$ Hz.

EXERCISE If the pipe is open at one end, how many harmonics are possible in the normal hearing range, 20 Hz to 20,000 Hz?

Answer: 286.

EXAMPLE 15.11 Resonance in a Tube of Variable Length

A simple apparatus for demonstrating resonance in a tube is described in Figure 15.20a. A long tube open at both ends is partially submerged in a beaker of water, and a vibrating tuning fork of unknown frequency is placed near the top. The length of the air column, L, is adjusted by moving the tube vertically. The sound waves generated by the fork are reinforced when the length of the air column corresponds to one of the resonant frequencies of the tube. The smallest value of L for which a peak occurs in the sound intensity is 9 cm. For this measurement, determine the frequency of the tuning fork and the value of L for the next two resonant vibrations.

Solution Once the tube is in the water, this setup represents a pipe closed at one end, and the fundamental has a frequency of $v/4L$ (Fig. 15.20b). If we take $v = 345$ m/s for the speed of sound in air and $L = 0.09$ m, we get

$$f_1 = \frac{v}{4L} = \frac{345 \text{ m/s}}{4(0.09 \text{ m})} = 958 \text{ Hz}$$

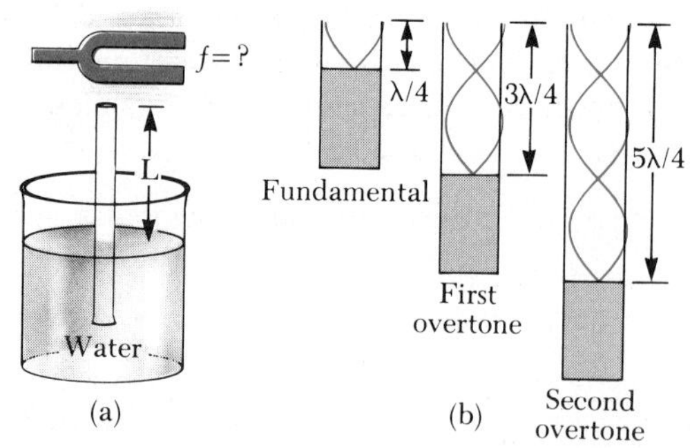

FIGURE 15.20 (Example 15.11) (a) Apparatus for demonstrating the resonance of sound waves in a tube closed at one end. The length, L, of the air column is varied by moving the tube vertically while it is partially submerged in water. (b) The first three normal frequencies of vibration for the system.

The fundamental wavelength of the pipe is given by $\lambda = 4L = 0.36$ m. Since the frequency of the source is constant, we see that the next resonance positions (Fig. 15.20b) correspond to lengths of $3\lambda/4 = 0.27$ m and $5\lambda/4 = 0.45$ m.

This arrangement is often used to measure the speed of sound, in which case the frequency of the tuning fork and the lengths at which resonance occurs must be known.

*15.11 Beats

The interference phenomena we have been dealing with up until now involve the superposition of two or more waves with the same frequency traveling in opposite directions. Let us now consider another type of interference effect which results from the superposition of two waves with slightly different frequencies. In this situation, the waves at some fixed point are periodically in and out of phase corresponding to an alternation in time between constructive and destructive interference. In order to understand this phenomenon, consider Figure 15.21. The two waves shown in Figure 15.21a represent the waves emitted by two tuning forks having slightly different frequencies, while

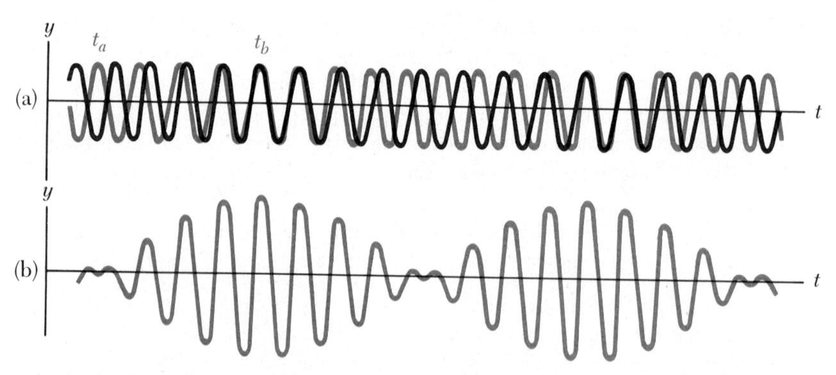

FIGURE 15.21 Beats are formed by the combination of two waves of slightly different frequencies traveling in the same direction. (a) The individual waves. (b) The combined wave has an amplitude that oscillates in time.

Figure 15.21b shows the superposition of these two waves. At some time indicated as t_a in Figure 15.21a, the two waves are out of phase, and destructive interference occurs as demonstrated by the resultant curve in Figure 15.21b. At some later time, however, the vibrations of the two forks move into step with one another. At the time indicated as t_b in Figure 15.21a, the two forks sumultaneously emit compressions, and constructive interference results as demonstrated by the curve in Figure 15.21b. As time commences further, the vibrations of the two forks will continue to move back out of phase, then into phase, and so on. Consequently, a listener at some fixed point will hear an alternation in loudness, known as **beats.** The number of beats per second, or beat frequency, equals the difference in frequency between the two sources. One can use beats to tune a stringed instrument, such as a piano, by beating a note on the instrument against a note of known frequency. The string can then be adjusted to the desired frequency by tightening or loosening it until no beats are heard.

EXAMPLE 15.12 Sour Notes

A particular piano string is supposed to vibrate at a frequency of 440 Hz. In order to check its frequency, a tuning fork known to vibrate at a frequency of 440 Hz is sounded at the same time the piano key is struck, and a beat frequency of 4 beats per second is heard. Find the possible frequencies at which the string could be vibrating.

Solution The number of beats per second is equal to the difference in frequency between the two sound sources. In this case, since one of the source frequencies is 440 Hz, four beats per second would be heard if the frequency of the string (the second source) were either 444 Hz or 436 Hz.

*15.12 Quality of Sound

The sound wave patterns produced by most musical instruments are very complex. Some characteristic waveforms produced by a tuning fork, a flute, and a clarinet, each playing the same note, are shown in Figure 15.22. Although each instrument has its own characteristic pattern, Figure 15.22 shows that each of the waveforms is periodic. Note that the tuning fork produces only one harmonic (the fundamental frequency) but the two instruments emit a mixture of harmonics. Figure 15.23 shows the harmonics of the waveforms of Figure 15.22. In Figure 15.23b, we see that, when this note is played on the flute, part of the sound consists of a vibration at the fundamental frequency, an even higher intensity is contributed by the second harmonic, the fourth harmonic produces about the same intensity as the fundamental, and so on. These sounds add together according to the principle of superposition to give the complex waveform for the flute shown in Figure 15.22b. Figure 15.23c shows that the clarinet emits a certain intensity at a frequency of the first harmonic, about half as much intensity at the frequency of the second harmonic, and so forth. The resultant superposition of these frequencies produces the pattern for the clarinet shown in Figure 15.22c. Note from Figure 15.23a that the tuning fork emits sound intensity only at the frequency of the first harmonic.

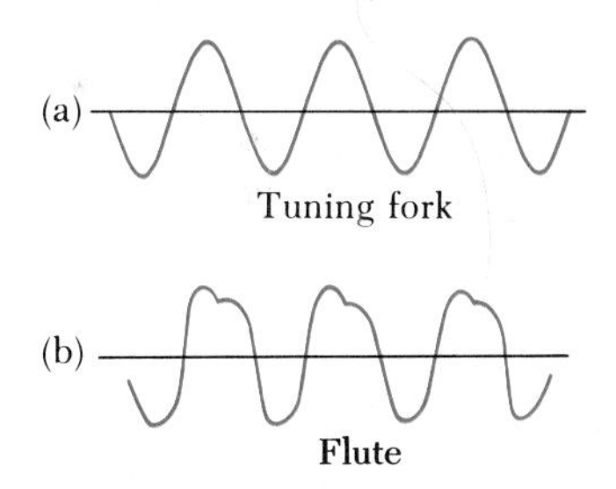

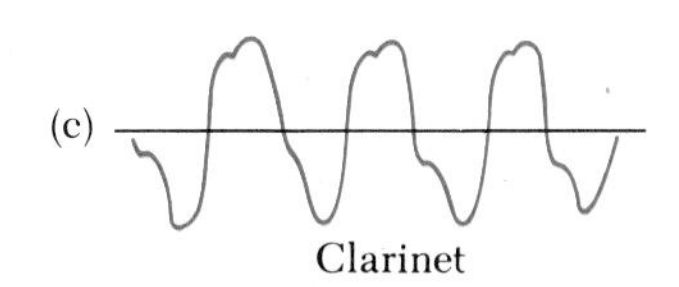

FIGURE 15.22 Waveforms produced by (a) a tuning fork, (b) a flute, and (c) a clarinet, each playing the same note. (Adpated from C.A. Culver, *Musical Acoustics*, 4th ed., New York, McGraw-Hill, 1956)

In music, the mixture of harmonics that produces the characteristic sound of any instrument is referred to as the *quality*, or *timbre*, of the sound.

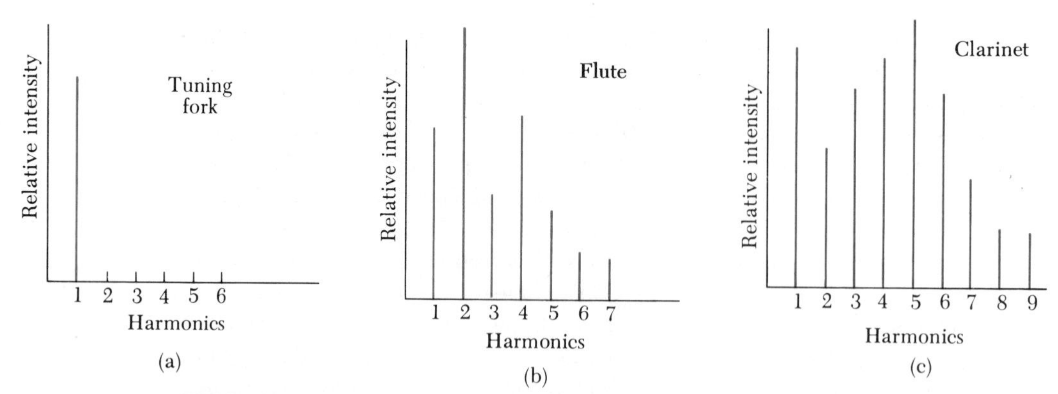

FIGURE 15.23 Harmonics of the waveforms shown in Figure 15.22. Note the variations in intensities of the various harmonics. (Adapted from C. A. Culver, *Musical Acoustics*, 4th ed., New York, McGraw-Hill, 1956)

We say that the note C on a flute differs in quality from the same C on a clarinet. Instruments such as the bugle, trumpet, violin, and tuba are rich in harmonics. The musician can cause one or another of these harmonics to be emphasized by lip control and thus play different musical notes with the same valve openings.

*15.13 THE EAR

The human ear is divided into three regions: the outer ear, the middle ear, and the inner ear (Fig. 15.24a). The outer ear consists of the ear canal (open to the atmosphere), which terminates on the eardrum (tympanum). Sound waves travel down the ear canal to the eardrum, which vibrates in and out in phase with the pushes and pulls caused by the alternating high and low pressures of the sound wave. Behind the eardrum are three small bones of the

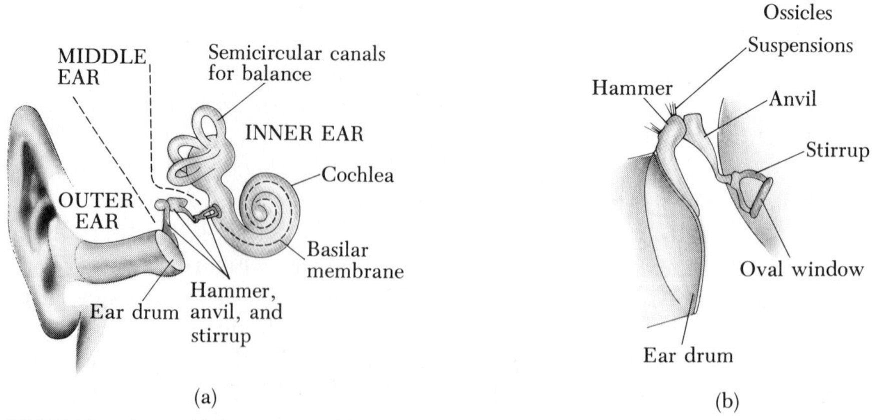

FIGURE 15.24 (a) Structure of the human ear. (b) The three tiny bones (ossicles) that connect the eardrum to the window of the cochlea act as a double lever system to decrease the amplitude of vibration and hence increase the pressure on the fluid in the cochlea.

middle ear, called the hammer, the anvil, and the stirrup because of their shapes (Fig. 15.24b). These bones transmit the vibration to the inner ear, which contains the cochlea, a snail-shaped tube about 1 in. long. The cochlea makes contact with the stirrup of the oval window and is divided along its length by the basilar membrane, which consists of small hairs and nerve fibers. This membrane varies in mass per unit length and tension along its length, and different portions of it resonate at different frequencies. (Recall that the natural frequency of a string depends on its mass per unit length and on the tension on it.) Along the basilar membrane are numerous nerve endings, which sense the vibration of the membrane and in turn transmit impulses to the brain. The brain interprets the impulses as sounds of varying frequency, depending on the location along the basilar membrane of the impulse-transmitting nerves and on the rate at which the impulses are transmitted.

Figure 15.25 is a frequency response curve for an average human ear. It shows that the decibel level at which pain is experienced does not vary greatly from 120 dB regardless of the frequency of the tone. However, the threshold of hearing is very strongly dependent on frequency. The easiest frequencies to hear are around 3300 Hz, whereas frequencies about 12,000 Hz or below about 50 Hz must be relatively intense to be heard.

The exact mechanism by which sound waves are amplified and detected by the ear are rather complex and not fully understood. However, we can give a qualitative description of the amplification mechanisms. The small bones in the middle ear represent an intricate lever system that increases the force on the oval window over a given force on the eardrum. The pressure is greatly magnified because the surface area of the eardrum is about 20 times that of the oval window (in analogy with the hydraulic press). The middle ear, together with the eardrum and oval window, in effect acts as a matching net-

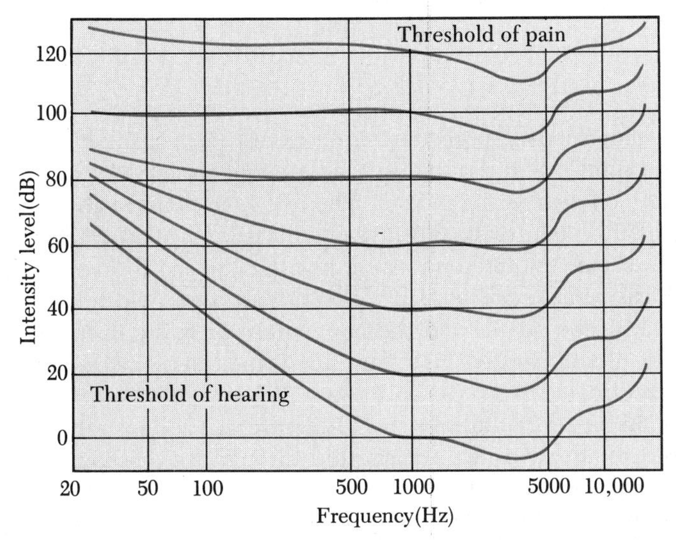

FIGURE 15.25 Curves of intensity level versus frequency for sounds that are perceived to be equally intense by a person with average hearing. Note that the ear is most sensitive at a frequency of about 3300 Hz.

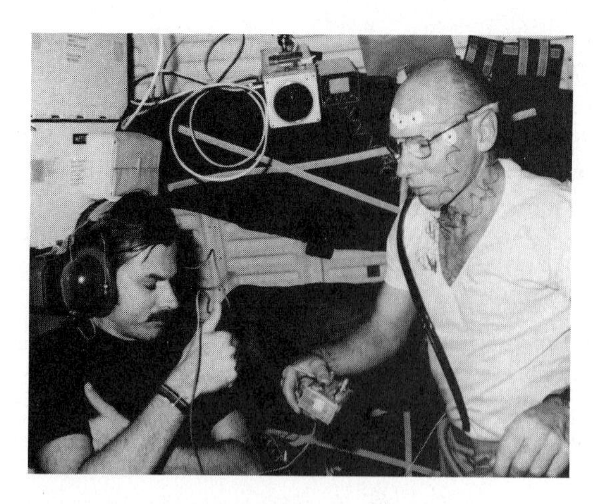

An audiometry test in progress on the space shuttle Challenger during its second mission. (Courtesy NASA)

work between the air in the outer ear and the liquid in the inner ear. The overall energy transfer between the outer ear and inner ear is highly efficient, with pressure amplification factors of several thousand. In other words, pressure variations in the inner ear are much greater than those in the outer ear.

The ear has its own built-in protection against loud sounds. The muscles connecting the three bones to the walls of the middle ear control the volume of the sound by changing the tension on the bones as sound builds up, thus hindering their ability to transmit vibrations. In addition, the eardrum becomes stiffer. These two occurrences cause the ear to be less sensitive to loud sounds. There is a time delay between the onset of loud sound and the ear's protective reaction, however, so that a very sudden loud sound can still damage the ear.

*Audiometry

Audiometry is a common medical procedure used to test for hearing loss. In this technique, sound signals of known intensity and frequency are introduced through a headset into one ear of the patient. The frequencies normally used are 125, 250, 500, 750, 1000, 2000, 3000, 4000, 6000, and 8000 Hz. The results of the test are usually plotted on a graph like the one shown in Figure 15.26. The vertical axis represents the decibel level that the sound must be raised above threshold to be heard, and the horizontal axis is the frequency of the sound. The results illustrated in Figure 15.26 indicate some hearing loss from 3000 Hz upward.

When hearing loss is severe in one ear, there are difficulties with the hearing test because sound waves are transmitted through the bones of the head to the normal ear. In order to overcome this difficulty, a technique called masking is often used. The principle behind the masking procedure is to apply a masking sound signal to the normal ear so that it will be occupied by this stimulus while the test is being carried out on the impaired ear. This inhibits the crossover effect and produces more reliable results.

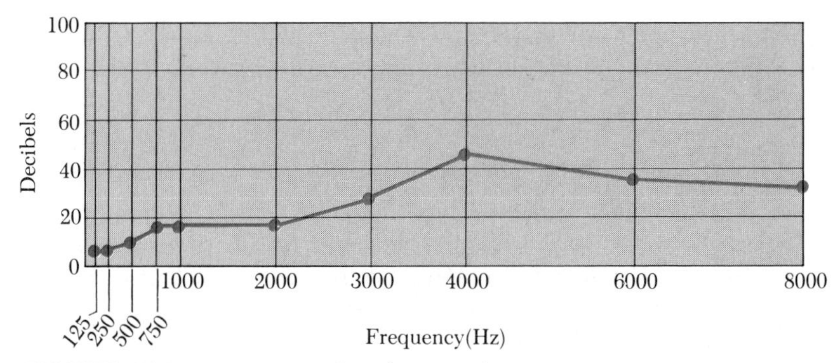

FIGURE 15.26 Hearing analysis by an audiometer.

*15.14 ULTRASOUND AND ITS APPLICATION

Ultrasonic waves are sound waves whose frequencies are in the range of 20 kHz to 100 kHz, which is beyond the audible range. Because of their high frequency, and corresponding short wavelengths, ultrasonic waves can be used to produce images of small objects and are currently in wide use in medical applications, both as a diagnostic tool and in certain treatments. Various internal organs in the body can be examined through the images produced by the reflection and absorption of ultrasonic waves. Although ultrasonic waves are far safer than x-rays, their images do not always provide as much detail. On the other hand, certain organs, such as the liver and the spleen, are invisible to x-rays but can be diagnosed with ultrasonic waves.

It is possible to measure the speed of blood flow in the body using a device called an ultrasonic flow meter. The technique makes use of the Doppler effect. By comparing the frequency of the waves scattered by the blood vessels with the incident frequency, one can obtain the speed of blood flow.

The technique used to produce ultrasonic waves for clinical use is illustrated in Figure 15.27. Electrical contacts are made to the opposite faces of a crystal, such as quartz or strontium titanate. If an alternating voltage of very high frequency is applied to these contacts, the crystal will vibrate at the same frequency as the applied voltage. As the crystal vibrates, it emits a beam of ultrasonic waves. This technique for producing sound waves is not new. At one time, almost all of the headphones used in radio reception produced their sound in this manner. This method of transforming electrical energy into mechanical energy is called the **piezoelectric effect.** This effect is also reversible. That is, if some external source causes the crystal to vibrate, an alternating voltage is produced across the crystal. Hence, a single crystal can be used to both transmit and receive ultrasonic waves.

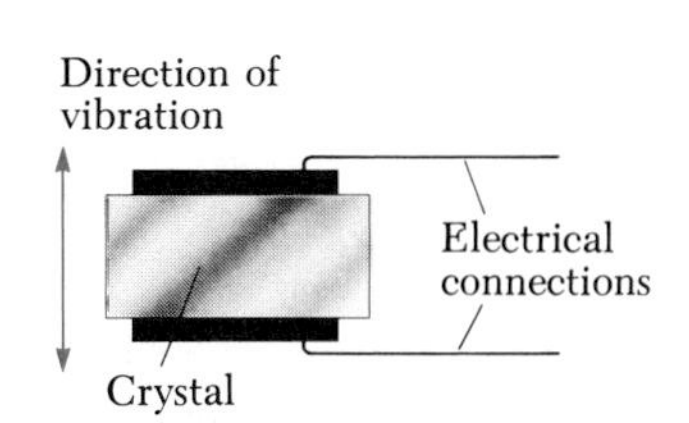

FIGURE 15.27 An alternating voltage applied to the faces of a piezoelectric crystal causes the crystal to vibrate.

The production of electric voltages by a vibrating crystal is a technique that has been used for years in stereo and hi-fi equipment. In this application, a phonograph needle is attached to the crystal and the vibrations of the needle as it rides in the groove of the record are translated by the crystal into an alternating voltage. This voltage is then amplified and used to drive the system's speakers.

The primary physical principal that makes ultrasound imaging possible is the fact that a sound wave is partially reflected whenever it is incident on a boundary between two materials having different densities. It is found that, if

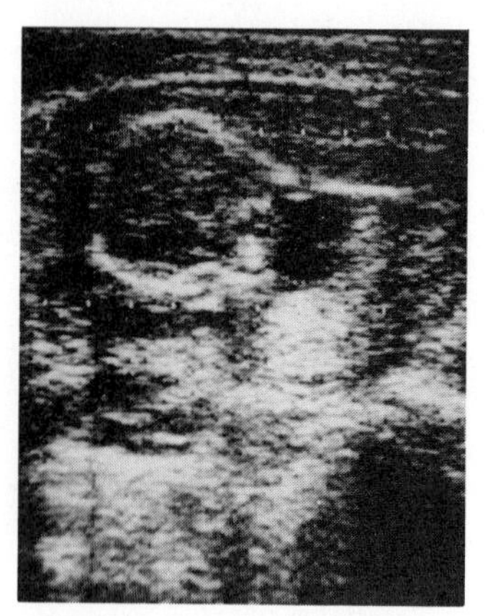

Sonogram of fetus at 12 weeks. (Courtesy of Samuel Breckinridge Field)

a sound wave is traveling in a material of density ρ_i and strikes a material of density ρ_t, the percentage of the incident sound wave reflected, *RP*, is given by

$$RP = \left(\frac{\rho_i - \rho_t}{\rho_i + \rho_t}\right)^2 \times 100 \qquad (15.23)$$

This equation assumes that the incident sound wave travels perpendicular to the boundary and that the speed of sound is approximately the same in both materials. This latter assumption holds very well for the human body since the speed of sound does not vary much in the various organs of the body.

Physicians commonly use ultrasonic waves to observe a fetus. This technique offers far less risk than x-rays, which can be genetically dangerous to the fetus and can produce birth defects. First the abdomen of the mother is coated with a liquid, such as mineral oil. If this is not done, most of the incident ultrasonic waves from the piezoelectric source will be reflected at the boundary between the air and the skin of the mother. Mineral oil has a density similar to that of skin, and as Equation 15.23 indicates, a very small fraction of the incident ultrasonic wave is reflected when $\rho_i \approx \rho_t$. The ultrasound energy is emitted as pulses rather than as a continuous wave so that the same crystal can be used as a detector as well as a transmitter. The source-receiver is then passed over a particular line along the mother's abdomen. The reflected sound waves picked up by the receiver are converted to an electric signal, which forms an image along a line on a fluorescent screen. The sound source is then moved a few centimeters on the mother's body,and the process is repeated. The reflected signal produces a second line on the fluorescent screen. In this fashion a complete scan of the fetus can be made. Difficulties with the pregnancy, such as the likelihood of abortion or of breech birth, are easily detected with this technique. Also, such fetal abnormalities as spina bifida and water on the brain are readily observable.

Another interesting application of ultrasound is the ultrasonic ranging unit designed by the Polaroid Corporation. This device is used in some of their cameras to provide an almost instantaneous measurement of the distance between the camera and object to be photographed. The principal component of this device is a crystal that acts as both a loudspeaker and a microphone. A pulse of ultrasonic waves is transmitted from the transducer to the object to be photographed. The object reflects part of the signal, producing an echo that is detected by the device. The time interval between the outgoing pulse and the detected echo is then electronically converted to a distance value, since the speed of sound is a known quantity.

SUMMARY

Sound waves are longitudinal waves. *Audible waves* are sound waves with a frequency between 20 Hz and 20,000 Hz. *Infrasonic waves* have frequencies below the audible range, and ultrasonic waves have frequencies above this range.

The speed of sound in a medium of bulk modulus B and density ρ is

$$v = \sqrt{\frac{B}{\rho}} \tag{15.1}$$

The velocity of sound depends on the temperature of the medium. The relationship between temperature and speed for sound in air is

$$v = (331 \text{ m/s})\sqrt{1 + \frac{T}{273}} \tag{15.4}$$

where T is the temperature in degrees Celsius and 331 m/s is the speed of sound in air at 0°C.

The **intensity level** of a sound wave in decibels is given by

$$\beta \equiv 10 \log\left(\frac{I}{I_0}\right) \tag{15.7}$$

The constant I_0 is a reference intensity level, usually taken to be at the threshold of hearing ($I_0 = 10^{-12}$ W/m^2), and I is the intensity at the level β where β is measured in **decibels** (dB).

The **intensity** of a *spherical wave* produced by a point source is proportional to the average power emitted and inversely proportional to the square of the distance from the source:

$$I = \frac{P_{av}}{4\pi r^2} \tag{15.8}$$

The change in frequency heard by an observer whenever there is relative motion between the frequency sound and observer is called the **Doppler effect.** If the observer is moving with a speed v_o and the source is at rest, the observed frequency, f', is

$$f' = f\left(\frac{v \pm v_o}{v}\right) \tag{15.11}$$

where the positive sign is used when the observer is moving toward the source and the negative sign refers to motion of the observer away from the source.

If the source is moving with a speed v_s and the observer is at rest, the observed frequency is

$$f' = f\left(\frac{v}{v \mp v_s}\right) \tag{15.14}$$

where v_s refers to motion toward the observer and $-v_s$ refers to motion away from the observer.

When the observer and source are both moving, the observed frequency is

$$f' = f\left(\frac{v \pm v_o}{v \mp v_s}\right) \tag{15.15}$$

When waves interfere, the resultant wave is found by adding the individual waves together point by point. When crest meets crest and trough meets trough, the waves undergo **constructive interference.** When crest meets trough, **destructive interference** occurs.

Standing waves are formed when two waves having the same frequency, amplitude, and wavelength travel in opposite directions through a medium. One can set up standing waves of specific frequencies in a stretched string. The natural frequencies of vibration of a stretched string of length L and fixed at both ends are

$$f_n = \frac{n}{2L}\sqrt{\frac{F}{\mu}} \tag{15.20}$$

where F is the tension in the string and μ is its mass per unit length. The natural frequencies of vibration form a **harmonic series,** that is, the frequencies are integral multiples of the fundamental (lowest) frequency.

A system capable of oscillating is said to be in **resonance** with some driving force whenever the frequency of the driving force matches one of the natural frequencies of the system. When the system is resonating, it oscillates with maximum amplitude.

Standing waves can be produced in a tube of air. If the reflecting end of the tube is *open,* the natural frequencies of vibration are

$$f_n = n\frac{v}{2L} \qquad n = 1, 2, 3, \ldots \tag{15.21}$$

If the tube is *closed* at the reflecting end, the natural frequencies of vibration are

$$f_n = n\frac{v}{4L} \qquad n = 1, 3, 5, \ldots \tag{15.22}$$

The phenomenon of **beats** is an interference effect occurring when two waves of slightly different frequencies travel in the same direction. For sound waves, the loudness of the resultant sound changes with time.

ADDITIONAL READING

L. N. Baranek, "Noise," *Sci. American,* December 1966, p. 66.

C. A. Culver, *Musical Acoustics,* New York, McGraw-Hill, 1957.

N. H. Fletcher and S. Thwaites, "The Physics of Organ Pipes," *Sci. American,* January 1983, p. 94.

A. J. Hudspeth, "The Hair Cells of the Inner Ear," *Sci. American,* January 1983, p. 54.

C. M. Hutchins, "The Acoustics of Violin Plates," *Sci. American,* October 1981, p. 170.

C. M. Hutchins et al., *The Physics of Music,* New York, Freeman, 1978 (a collection of readings from *Scientific American*).

G. Loeb, "The Functional Replacement of the Ear," *Sci. American,* February 1985, p. 104.

J. Monforte, "The Digital Reproduction of Sound," *Sci. American,* December 1984, p. 78.

B. Patterson, "Musical Dynamics," *Sci. American,* November 1974, p. 78.

C. Shadle, "Experiments on the Acoustics of Whistling," *The Physics Teacher,* March 1983, p. 148.
J. Sundberg, "The Acoustics of the Singing Voice," *Sci. American,* March 1977, p. 82.
G. Von Bekesy, "The Ear," *Sci. American,* August 1957, p. 66.
H. E. White and D. H. White, *Physics and Music,* Philadelphia, Saunders, 1980.

QUESTIONS

1. As a result of a distant explosion, an observer senses a ground tremor and then hears the explosion. Explain the time difference.
2. If a bell is ringing inside a glass container, we cease to hear the sound when the air is pumped out but we can still see the bell vibrating. What difference does this indicate in the properties of sound waves and light waves?
3. Explain why sound travels faster in warm air than in cool air.
4. How could you determine by listening to a band or orchestra that the speed of sound is the same for all frequencies?
5. Why does a vibrating guitar string sound louder when placed on the instrument than it would if allowed to vibrate in the air while off the instrument?
6. Of the following sounds, which is most likely to have an intensity level of 60 dB: a rock concert, the turning of a page in this text, normal conversation, a cheering crowd at a football game, or background noise at a church?
7. Estimate the decibel level of each of the sounds in Question 6.
8. If the distance from a point source of sound is tripled, by what factor does the sound intensity decrease? Assume there are no reflections from nearby objects to affect your results.
9. An airplane mechanic notices that the sound from a twin-engine aircraft rapidly varies in loudness when both engines are running. What could be causing this variation from loud to soft?
10. At certain speeds, an automobile driven on a washboard road will vibrate disastrously and lose traction and braking effectiveness. At other speeds, either lesser or greater, the vibration is more manageable. Explain. Why are "rumble strips," which work on this same principle, often used just before stop signs?
11. A binary star system consists of two stars revolving about each other. If we observe the light reaching us from one of these stars as it makes one complete revolution about the other, what does the Doppler effect predict will happen to this light?
12. How could an object move with respect to an observer such that the sound from it is not shifted in frequency?
13. Why is the intensity of an echo less than that of the original sound?
14. Does the wind alter the frequency of sound heard by an observer who is at rest relative to the source of sound? Explain.
15. How is the natural frequency of vibration of an organ pipe altered as room temperature increases?
16. A person who has just inhaled helium speaks with a high-pitched voice and sounds like Donald Duck. Why does this occur?
17. A soft drink bottle will resonate as air is blown across the top. What happens to the resonant frequency as the level of fluid in the bottle decreases?
18. If we observe light reaching us from distant galaxies, we find that the light has been shifted toward longer wavelengths. What does this tell us about the motion of these galaxies?
19. After a snowfall sounds seem muffled or quiet. Why does this occur?
20. Why can a boat easily produce a bow wave in water, while an airplane must fly very fast to produce a shock wave in air?

PROBLEMS

Section 15.2 Characteristics of Sound Waves

Section 15.3 The Speed of Sound

Unless otherwise stated, use 345 m/s as the speed of sound in air for all problems below.

1. What is the speed of sound in air at (a) 27°C, (b) 100°C, (c) 200°C?

2. The limits of human hearing extend from approximately 20 Hz to 20,000 Hz. Find the wavelengths of these extremes at a temperature of 27°C.

3. A group of hikers hear an echo 3 s after they shout. If the temperature is 22°C, how far away is the mountain that reflected the sound wave?

4. A dolphin swimming in seawater at a temperature of 25°C, emits a sound directed toward the bottom of the ocean 150 m below. What period of time passes before it hears an echo?

5. A sound wave propagating in air has a frequency of 4000 Hz. Calculate the percent change in wavelength when the wavefront, initially in a region where T = 27°C, enters a region where T = 10°C.

6. The largest value ever achieved for the velocity of sound in air is about 10^4 m/s, and the highest frequency ever produced is about 2×10^{10} Hz. Find the wavelength of this wave.

7. Use the values for Young's modulus and the density of copper found in Chapter 9 to find the velocity of sound in this material.

8. If the density of methyl alcohol is about 0.8×10^3 kg/m^3, find an approximate value for the bulk modulus of this liquid at 25°C.

Section 15.4 Energy and Intensity of Sound Waves

9. What is the intensity level in decibels of a sound wave of intensity (a) 10^{-6} W/m^2 and (b) 10^{-5} W/m^2?

10. What is the intensity of a sound whose intensity level is (a) 40 dB, (b) 100 dB?

11. The area of a typical eardrum is about 5.0×10^{-5} m^2. Calculate the sound power (the energy per second) incident on an eardrum at (a) the threshold of hearing, (b) the threshold of pain.

• **12.** A noisy machine in a factory produces a dB rating of 80 dB. How many identical machines could you add to the factory without exceeding the 90 dB limit?

• **13.** Intensity is defined as power per unit area. As a result, show that the dB level of two sounds can be related to the ratio of two powers as

$$\text{dB} = 10 \log(P_1/P_0).$$

• **14.** On a work day the average decibel level of a busy street is 70 dB with 100 cars passing a point every minute. If the number of cars is reduced to 25 every minute on a weekend, what is the decibel level of the street?

Section 15.5 Spherical and Plane Waves

15. A stereo speaker which we shall consider as a small source emits sound waves with a power output of 100 W. (a) Find the intensity at a distance of 10 m from the source. (b) Find the intensity level in dB at this distance. (c) At what distance would you have to be such that the sound was at the threshold of pain, 120 dB?

16. A man shouting loudly produces a dB level of about 70 dB at a distance of 5 m. How many watts of power does the man emit? (Treat the man as a point source.)

17. The sound level 3 m from a point source is 120 dB. At what distance will the sound level be 100 dB?

18. The dB level at a distance of 2 m from a sound source is 100 dB. What is the dB level at distances of (a) 4 m, (b) 6 m, (c) 8 m?

Section 15.6 Doppler Effect

19. A train at rest emits a sound at a frequency of 1000 Hz. An observer in a car travels away from the sound at a speed of 30 m/s. What is the frequency heard by the observer?

20. The observer of Problem 19 stops the car. Determine the frequency he hears when (a) the train approaches him at a speed of 30 m/s, (b) when the train recedes from him at a speed of 30 m/s.

21. An ambulance siren emits a note of 500 Hz when the ambulance is at rest. When the ambulance is in motion at a speed of 10 m/s, what is the frequency heard by an observer traveling at (a) 15 m/s toward the source, and (b) 15 m/s after the observer has passed the ambulance and recedes from it?

22. An alert physics student stands beside the tracks as a train rolls slowly past. He notes that the frequency of the train whistle is 442 Hz when the train is approaching him and is 441 Hz when the train is receding from him. From this he is able to find the speed of the train. What value does he find?

23. At what angle should a supersonic aircraft fly so that the angle θ shown in Figure 15.10 will be (a) 40°, (b) 30°?

24. The Concorde flies at mach 1.5, which means the speed of the plane is 1.5 times the speed of sound in air. What is the angle between the direction of propagation of the shock wave and the direction of the plane's velocity?

Section 15.7 Interference in Sound Waves

25. Two loudspeakers are placed above and below one another as in Figure 15.13 and driven by the same source at a frequency of 500 Hz. (a) What minimum distance should the top speaker be moved back in order to have destructive interference between the two speakers? (b) If the top speaker is moved back twice the distance calculated in part (a), will constructive or destructive interference occur?

26. A source of unknown frequency is attached to the two loudspeakers of Problem 25. It is found that when the upper loudspeaker is moved back 0.5 m, **constructive** interference occurs between the two. What is the frequency of the sound generator?

27. The sound interferometer shown in Figure 15.12 is driven by a speaker emitting a 400 Hz note. If **destructive** interference occurs at a particular instant, how much must the path length in the U-shaped tube be increased in order to hear (a) constructive interference, (b) destructive interference once again?

• **28.** The two speakers of Figure 15.13 are 0.6 m apart, and both emit a frequency of 1000 Hz. An observer on the center line between the speakers begins to walk along a line parallel to a line joining the speakers. When he is 5 m from the closest speaker, he hears destructive interference. How far is he from the most distant speaker?

Section 15.8 Standing Waves

29. A string 50 cm long has a mass per unit length equal to 20×10^{-5} kg/m. To what tension should this string be stretched if its fundamental frequency is to be (a) 20 Hz and (b) 4500 Hz?

30. A wire of mass 0.3 g is stretched between two points 70 cm apart. If the tension in the wire is 600 N, find the first, second, and third harmonics of this string.

31. (a) To what tension would the wire of Problem 30 have to be stretched in order for the frequency of the fundamental to be 20,000 Hz? (b) Does it seem likely that this could be accomplished?

32. A stretched string of length L is observed to vibrate in five equal segments when driven by a 630 Hz oscillator. What oscillator frequency will set up a standing wave such that the string vibrates in three segments?

Section 15.9 Forced Vibrations and Resonance

33. A child's swing has a length of 2 m. List three frequencies at which this swing could be pushed to set up resonant vibrations in the swing.

34. If the circumference of the mouth of a crystal goblet is 10 cm, at what fundamental frequency could an opera singer have to sing in order to set the glass into a resonant vibration?

Section 15.10 Standing Waves in Air Columns

35. Middle C has a frequency of 264 Hz. (a) What is the length of the shortest pipe open at both ends which would produce a sound of this frequency? (b) What would be the required length if the pipe is closed at one end?

36. A closed organ pipe is 3 m long. At what frequencies between 20 Hz and 20,000 Hz will this pipe resonate?

37. A pipe open at both ends has a fundamental frequency of 300 Hz when the temperature is 0°C. (a) What is the length of the pipe? (b) What is the fundamental frequency at a temperature of 30°C?

38. A tuning fork is sounded above a resonating tube as in Figure 15.20. The first resonant point is 0.08 cm from the top of the tube, and the second is at 0.24 m. (a) Where would the third resonant point be located? (b) What is the frequency of the tuning fork?

Section 15.11 Beats

39. A piano tuner strikes a 440 Hz tuning fork at the instant she strikes a piano key that should emit a tone of 440 Hz and hears a beat frequency of 2 Hz. What are the possible frequencies the piano key could be emitting?

• **40.** A flute is designed such that it plays a frequency of 264 Hz, middle C, when all the holes are covered and the temperature is 20°C. (a) Consider the flute as a pipe open at both ends and find the length of the flute, assuming that the frequency above is the fundamental. (b) A second player nearby in a colder room also attempts to play middle C on an identical flute. A beat frequency of 3 Hz is heard. What is the temperature of the room?

• **41.** The G string on a violin has a fundamental frequency of 196 Hz. It is 30 cm long and has a mass of 0.5 g. A nearby violinist fingers her identical violin until a beat frequency of 2 Hz is heard between the two. What is the length of her violin string?

Section 15.13 The Ear

42. If the ear canal of a person can be considered as acting like an organ pipe closed at one end that resonates at a fundamental frequency of 3000 Hz, what is the length of the canal? Use normal body temperature for your determination of the speed of sound in the canal.

ADDITIONAL PROBLEMS

43. Some studies indicate that the upper limit of hearing is determined by the diameter of the eardrum. The wavelength of the sound wave and the diameter of the eardrum are approximately equal at this upper limit of frequency. If this is so, what is the diameter of the eardrum of a person capable of hearing 20,000 Hz? (Assume a body temperature of 37°C.)

44. What is the lowest frequency of the standing wave of sound that can be set up between two walls that are 8 m apart if the temperature is 22°C?

45. A commuter train passes a passenger platform at a constant speed of 40 m/s. The train horn is sounded at a frequency of 320 Hz when the train is at rest. (a) What is the frequency observed by a person on the platform as the train approaches and (b) as the train recedes from him? (c) What wavelength does the observer find in each case?

46. A standing wave is established in a string that is 240 cm long and fixed at both ends. The string vibrates in four segments when driven at 120 Hz. (a) Determine the wavelength. (b) What is the fundamental frequency?

47. A string of length L, mass per unit length μ, and tension F is vibrating at its fundamental frequency. What effect will the following have on the fundamental frequency? (a) The length of the string is doubled with all other factors held constant. (b) The mass per unit length is doubled with all other factors held constant. (c) The tension is doubled with all other factors held constant.

• **48.** Two point sound sources have measured intensities of $I_1 = 100\ \text{W/m}^2$ and $I_2 = 200\ \text{W/m}^2$. By how many decibels is the level of source 1 lower than the level of source 2? (Assume that the observer is the same distance from both sources.)

• **49.** A string of 8 g mass and 5 m length has one end attached to a wall while the other end is draped over a pulley and attached to a hanging mass of 4 kg. If this string is plucked, what is the fundamental frequency of vibration?

• **50.** A typical sound level for a buzzing mosquito is 40 dB, and normal conversation is approximately 50 dB. How many buzzing mosquitoes will produce a sound intensity equal to that of normal conversation?

• **51.** When high-energy charged particles move through a transparent medium with a velocity greater than the velocity of light in that medium, a shock wave, or bow wave, of light is produced. This phenomenon is called the *Cerenkov effect* and can be observed in the vicinity of the core of a swimming pool nuclear reactor due to high-speed electrons moving through the water. In a particular case, the Cerenkov radiation produces a wavefront with a cone angle of 53°. Calculate the velocity of the electrons in the water. (Use 2.25×10^8 m/s as the velocity of light in water.)

• **52.** Refer to Table 15.2, which gives dB values for representative sounds from various sources. Determine the resultant intensity in dB when a vacuum cleaner and a power mower are operated against a background of busy traffic.

• **53.** A variable-length air column is placed just below a vibrating wire fixed at both ends. The length of the air column is gradually increased from zero until the first position of resonance is observed at $L = 34$ cm. The wire is 120 cm in length and is vibrating in its third harmonic. If the speed of sound in air is 340 m/s, what is the speed of transverse waves in the wire?

• **54.** Two pipes are each open at one end and are of adjustable length. Each has a fundamental frequency of 480 Hz at 300 K. The air temperature is increased in one pipe to 305 K. If the two pipes are sounded together, what beat frequency will result?

•• **55.** Two trains when at rest have sirens which emit a frequency of 300 Hz. The two trains travel toward one another and toward an observer stationed between them. One of the trains moves at 30 m/s, and the observer hears a beat frequency of 3 beats per second. What is the velocity of the second train, which travels faster than the first?

•• **56.** By proper excitation, it is possible to produce both longitudinal and transverse waves in a long metal rod. In one particular case, the rod is 150 cm in length and 0.2 cm in radius and has a mass of 50.9 g. Young's modulus for the material is $6.8 \times 10^{11}\ \text{dynes/cm}^2$. Determine the required tension in the rod so that the ratio of the speed of longitudinal waves to the speed of transverse waves is 8.

•• **57.** A speaker at the front of a room and an identical speaker at the rear of the room are being driven by the same sound source at 456 Hz. A student walks at a uniform rate of 1.5 m/s along the length of the room. How many beats does the student hear per second?

•• **58.** A student located several meters in front of a smooth reflecting wall is holding a board on which a wire is fixed at each end. The wire, vibrating in its third harmonic, is 75 cm long, has a mass of 2.25 g, and is under a tension of 400 N. A second student is moving toward the wall and hears 8.3 beats per second. At what speed does the moving student approach the wall? Use 340 m/s as the speed of sound in air.

•• **59.** Two ships are moving along a line due east. The trailing vessel has a speed relative to a land-based observation point of 64 km/h, and the leading ship has a speed of 45 km/h relative to that station. The two ships are in a region of the ocean where the current is moving uniformly due west at 10 km/h. The trailing ship transmits a sonar signal at a frequency of 1200 Hz. What frequency is monitored by the leading ship? (Use 1520 m/s as the speed of sound in ocean water.)

•• **60.** Three metal rods are located relative to each other as shown in Figure 15.28, where $L_1 + L_2 = L_3$. Values of the density and Young's modulus for the three materials are $\rho_1 = 2.7 \times 10^3$ kg/m^3, $Y_1 = 7 \times 10^{10}$ Pa, $\rho_2 = 11.3 \times 10^3$ kg/m^3, $Y_2 = 1.6 \times 10^{10}$ Pa, $\rho_3 = 8.8 \times 10^3$ kg/m^3, $Y_3 = 11 \times 10^{10}$ Pa. (a) If $L_3 = 1.5$ m, what must the ratio L_1/L_2 be if a sound wave is to travel the length of rods 1 and 2 in the same time required to travel the length of rod 3?

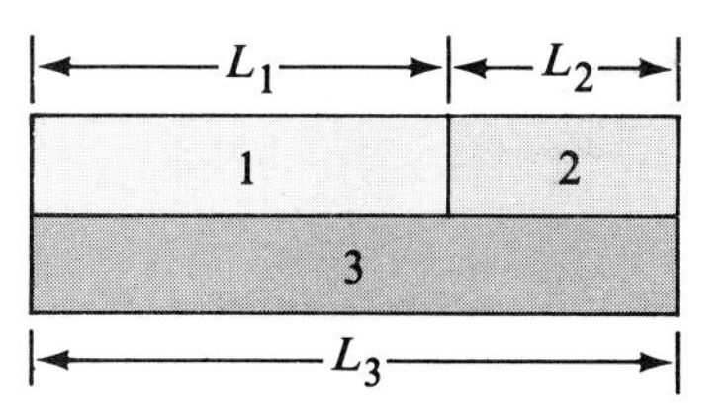

FIGURE 15.28 (Problem 60)

Essay

Physics of the Guitar

William R. Savage *Department of Physics and Astronomy, The University of Iowa.*

The guitar is a musical instrument based on the vibrational properties of a string fixed at each end. The six, or sometimes 12, strings are connected to a bridge glued on a resonator and a beam with tension-adjusting screw mechanisms. The resonator consists of a box made of thin wood with a circular opening called the *rose* under the strings. The beam supports a fingerboard fitted with frets located in fixed positions that aid the formation of definite frequencies of vibration. Pressing the strings against the frets changes the length and hence the pitch of the sound produced by plucking the string.

The sound produced by the guitar is determined in part by the mechanical properties of the string and the effect of these properties on the waves produced on the string, the coupling of the string to the vibrations of the body of the instrument, and the vibrations of the air in the cavity.

The Ideal String

The vibrations of an ideal string fastened between rigid supports serves as the model for the behavior of guitar strings. An ideal string would be flexible with no transverse rigidity; extending the string should not change the tension. The shape of the plucked string is a triangle with the original equilibrium position of the string as the base and the displacement of the pluck point from the equilibrium as the height of the apex. The string is tuned by adjusting the tension using a simple screw mechanism. At the pluck point there is a component of force directed toward the equilibrium position of the string. At the ends of the string the net force is zero, because a force exerted on the string by the support is equal and opposite to the tension, in accordance with Newton's second law. When the string is released, the pluck point is the only part of the string to be accelerated toward the equilibrium position. The ensuing string motion results in two kinks which move toward opposite ends of the string. The net force on an end support is constant until the kink arrives. When the kink strikes an end support, the force suddenly changes direction. The force of the support on the string reflects the displacement of the string. The string waves incident on the support are reflected back and forth until they die out or are stopped.

Applications of Fourier's Theorem

The French mathematician Jean Baptiste Fourier (1768–1830) discovered a powerful theorem about the description of periodic functions. They can be expressed as the sum or superposition of a set of sinusoids each with the appropriate frequency, amplitude, and phase. This theorem can be applied to the description of the triangular shape of the plucked string because it is a vibration with a definite frequency. Stopping the string at one of the frets will produce a different frequency, and that vibration will have a unique set of sinusoidal components. The motion of the string can be described in terms of the component sinusoidal waves.

The description is somewhat paradoxical. The reflections of the disturbance from the ends creates a collection

FIGURE 1 Acoustic/electric guitar. (Photo courtesy of Fender Musical Instruments)

of waves moving in both directions on the string. The sinusoidal components that travel back and forth and have the same amplitude and frequency superimpose or add to give a standing wave that does not appear to move from end to end. However, the waves must have the proper frequency to form a standing wave with null points at each end; otherwise they will not combine properly as standing waves and disappear very rapidly by cancellation. The string cannot move at the rigid end support. Such a motionless position is a *null point* or *node* in the standing wave.

The formation of the nodes at the support permits the use of symmetry principles in selecting the sinusoids that describe the string motion. The lowest frequency sinusoid or fundamental is symmetric about the center and has a half wavelength equal to the space between the end supports. The next sinusoid, with a higher frequency, is antisymmetric with respect to the center of the string and has a wavelength equal to the separation of the end supports or the vibrational length of the string. At an instant of time the position of the right half wavelength could be up and the left half wavelength could be down. As the wavelengths shorten by half wavelength steps, the waveforms are successively symmetrical and antisymmetrical. The frequencies are quantized by being related to each other as integral multiples of the lowest or fundamental. The harmonic or mode number serves as a type of quantum number. Each of the sinusoidal vibrations meeting the rules for forming a component of a standing wave is called a *mode* of the vibration. These individual components, especially those with the lowest frequency, can be heard and are called *partials* by the musician. Because the modes are related by integers, they have been called *harmonics* from the time of the early Greek philosophers such as Pythagorus (ca. 530 B.C). The standing wave can be represented by the superposition of members of a set of harmonic sinusoids which have the proper amplitude and phase.

Galileo and Mersenne's Experiments

Both Galileo Galilei (1564–1642) and Marin Mersenne (1588–1648) studied the relationship between frequency, length, diameter, density, and tension in vibrating strings. Galileo's father, Vincensio Galilei, was a lute builder and player of considerable skill. Galileo is identified with physics, and Mersenne is identified with musical acoustics, mathematics, and theology. Working independently, Galileo and Mersenne found experimentally that the fundamental frequency of sounds of a vibrating string is in an inverse ratio to the length, in an inverse ratio to the diameter when the strings are of the same material, in an inverse ratio to the square root of the density of the strings, and in a direct ratio to the square root of the tension in the string. This agrees with the observation that the vibration rate of a simple harmonic oscillator is related directly to the square root of the elastic force and inversely to the square root of inertia or mass of the object.

In 1636, Mersenne discovered that the observed frequencies and those predicted by his formulas did not agree exactly. The difference was very small but observable by a careful listener. Mersenne is credited with the violin-playing technique called harmonics. The string is lightly touched at a place where the player wishes to eliminate modes with nodes at the point. When the string is divided into two equal parts by a light touch at the center, it is apparent that the higher mode does not vibrate at a frequency exactly an octave higher than the full string. The shorter the string and the larger the diameter, the greater is the discrepancy. The model assumes an ideal string that can be best approximated by a long slim string rather than by a short heavy string.

Resonance

Resonance is an enhanced vibration of a system at one of its mode frequencies when the forcing vibration is of the same frequency. The elasticity and low damping of the string result in a set of very narrow mode frequencies. The resonances of the guitar body and air cavity are broad. A property of a broad resonance is that a wide range of frequencies near the resonance can initiate the resonance response. A typical acoustic spectrum of a string plucked at ⅛ its length is shown in Figure 2. Vibrational energy is stored in the vibration and is available for radiation. The result is a louder sound as more energy is available to leak from the string into the air.

If the ends of the guitar string were rigidly fixed, no sound would leave the string. The string on its own could not move much air and, as a result, little sound would be radiated. The end of the string that attaches to the box transmits forces and vibrational energy to the guitar through the bridge. The *bridge* is a small piece of hard dense wood with a narrow strip of hard material such as wood, bone, or plastic for the strings to press against. The bridge transmits the periodic force of the string to the top plate of the guitar. The motion is subsequently transmitted to the ribs, back plate, and air cavity.

Unlike the vibrations of the string, the vibrations of the rest of the instrument are too small to be seen. However, if the body of the guitar is lightly touched while the instrument is played, the vibrations of the body and neck can easily be felt. The strength of these vibrations at different locations can be detected by the sensitive finger tips. A microphone can be held within a few millimeters of the surface, and the signal strength will be related to the vibrational amplitude at its location.

A mode can be identified as a stationary pattern of vibration with fixed nodal lines. Mode shapes and frequencies can be adjusted by thinning the wood and using one of the numerous barring patterns. The results are guitars of differing tonal qualities. For a well-constructed guitar most of the energy of the string is transferred to the body and air cavity.

The string spectrum is modified by the resonances and radiation efficiency of the body of the instrument. The body can have several distinct low-frequency resonances. The value of the frequency of the lowest wood resonance is crucial for well-designed instruments. It should not be close enough to conflict with a string mode. The air cavity can have a number of resonant modes for the oscillations of the air inside. The lowest frequency resonant mode is known as a *Helmholtz oscillation*. This mode, where the air pressure inside the cavity is uniform, is named for Hermann von Helmholtz (1821–1894), an early researcher in musical acoustics and the process of hearing. The resonance is lower than that of the wood. If this resonance frequency were close to a string resonance, an undesirable loudness or beating would occur.

The discussion above applies to the acoustic guitar where the normal wood and air resonances of the instrument are used to amplify and radiate the vibrations of the string. A significant recent development in plucked string instruments of the guitar family is the use of electronic pickups. In the electric guitar, one or more transducers is mounted on the wood plates of the guitar to

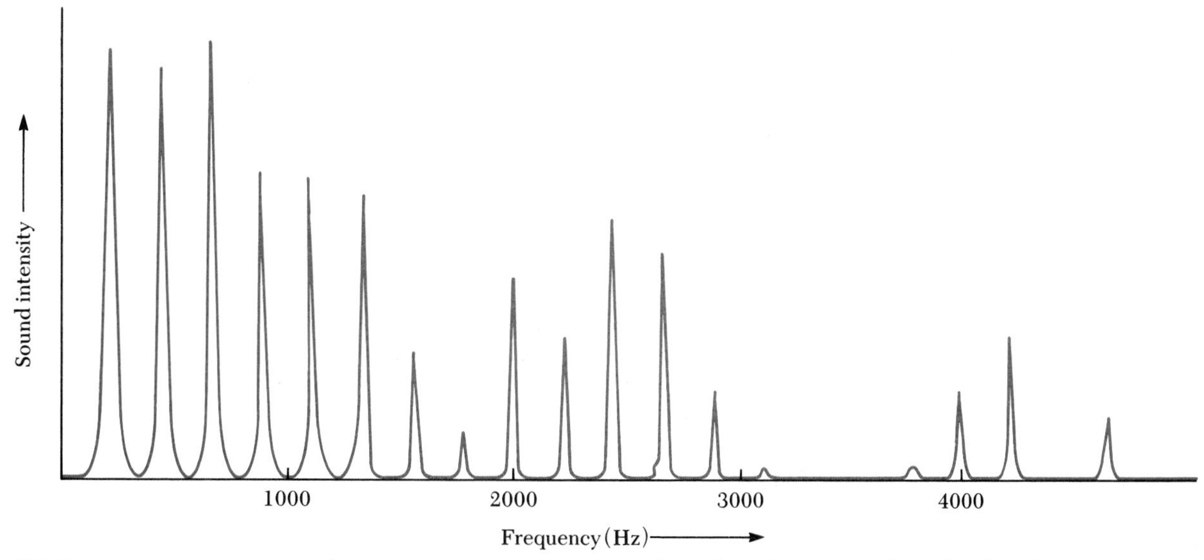

FIGURE 2 Acoustical spectral components of a guitar string plucked at about one eighth of its length.

convert the sound vibration into an electronic signal. Often the body of the guitar is solid without resonances. The electrical signal is sent to a preamplifier where the strength of the sinusoidal components can be adjusted. The signal is then delivered to a power amplifier, and finally to a speaker system. Many electronic pickups for guitars use magnetic motion detectors. Such pickups use the change in magnetic flux in an electrical circuit to detect the velocity of the string. These pickups use the principle, discovered by Michael Faraday (1791–1867), that a change in flux through an electrical coil results in an *emf* or *voltage*. This topic will be discussed in detail in Chapter 21. The velocity of the string has the same spectral distribution and components as the displacement. The main difference is that the magnetic pickup will produce no signal for any mode with a velocity node at its location. As a result, the position of the pickup point is important in achieving a guitar sound. Acoustic guitars with top plates and an air cavity can use a microphone as the pickup. Some pickups use the change in electrical resistance or voltage developed by the piezoelectric effect (see Section 15.14) to produce the electrical signal for amplification. Other pickups use physical changes that convert mechanical motion to electrical signals as the transducer operating principle. Because of the different pathway between the string vibrations and the sound in air, acoustic and electric guitars can differ from each other in the sounds they produce.

Part 4

ELECTRICITY AND MAGNETISM

For the sake of persons of . . . different types, scientific truth should be presented in different forms, and should be regarded as equally scientific, whether it appears in the robust form and the vivid coloring of a physical illustration, or in the tenuity and paleness of a symbolic expression.

JAMES CLERK MAXWELL

We now begin the study of that branch of physics concerned with electric and magnetic phenomena. The laws of electricity and magnetism play a central role in the operation of various devices such as radios, televisions, electric motors, computers, high-energy accelerators, and a host of electronic devices used in medicine. However, more fundamentally, we now know that the interatomic and intermolecular forces that are responsible for the formation of solids and liquids are electric in origin. Furthermore, such forces as the pushes and pulls between objects and the elastic force in a spring arise from electric forces at the atomic level.

The ancient Greeks observed electric and magnetic phenomena as early as 700 B.C. They found that a piece of amber, when rubbed, becomes electrified and attracts pieces of straw or feathers. The existence of magnetic forces was known from observations that a naturally occurring stone called *magnetite* (Fe_2O_3) is attracted to iron. (The word *electric* comes from the Greek word for amber, *elecktron*. The word *magnetic* comes from the name of the country where magnetite was found, *Magnesia.)*

In 1600, William Gilbert discovered that electrification was not limited to amber but is a general phenomenon. Scientists went on to electrify a variety of objects, including chickens and people! Experiments by Charles Coulomb confirmed the inverse-square force law for electricity.

It was not until the early part of the 19th century that scientists established that electricity and magnetism are, in fact, related phenomena. In 1820, Hans Oersted discovered that a compass needle is deflected when placed near a wire carrying an electric current. A few years later, Michael Faraday showed that, when a wire is moved near a magnet (or, equivalently, when a magnet is moved near a wire), an electric current is observed in the wire. James Clerk Maxwell used these observations and other experimental facts as a basis for formulating the laws of electromagnetism as we now know them. (*Electromagnetism* is a name given to the combined fields of electricity and magnetism.) Shortly thereafter, Heinrich Hertz verified Maxwell's predictions by producing electromagnetic waves in the laboratory. This was followed by such practical developments as radio and television.

Maxwell's contributions to the science of electromagnetism were especially significant because the laws he formulated are basic to *all* forms of electromagnetic phenomena. His work is comparable in importance to Newton's discovery of the laws of motion and the theory of gravitation.

CHAPTER 16

Electric Forces and Electric Fields

Sparks from the top of a Van de Graaff generator to its containment vessel. The spark lengths are about 11 ft. (Courtesy National Electrostatics Corp.)

The earliest known study of electricity was conducted by the Greeks about 700 B.C. By modern standards, their contributions to the field were modest. However, from these roots have sprung the enormous electrical distribution systems and sophisticated electronic instruments that are so much a part of our world today. It all apparently began when someone noticed that a fossil material called amber, when rubbed with wool, would attract small objects. Since then we have learned that this phenomenon is not restricted to amber and wool but occurs (to some degree) when almost any two nonmetallic substances are rubbed together.

In this chapter, we shall use this effect of charging by friction to begin an investigation of electric forces. We shall then discuss Coulomb's law, which is the fundamental law of force between any two charged particles. The concept of an electric field associated with charges will then be introduced, and its effects on other charged particles will be described. Finally, we shall end with a brief discussion of the Van de Graaff generator and the oscilloscope.

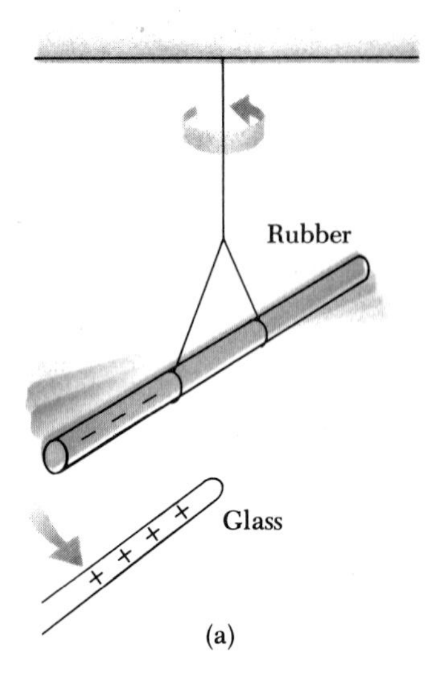

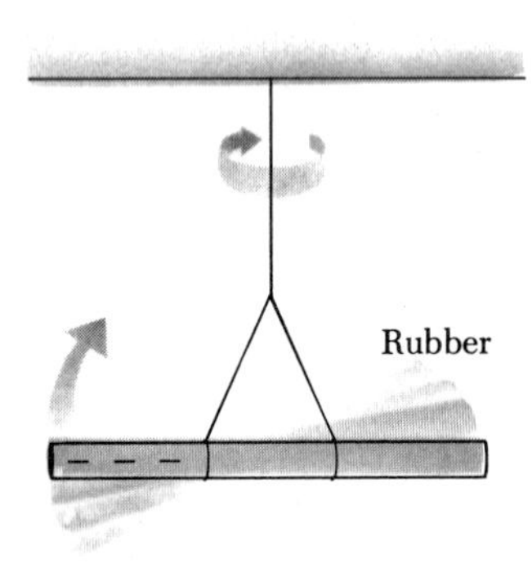

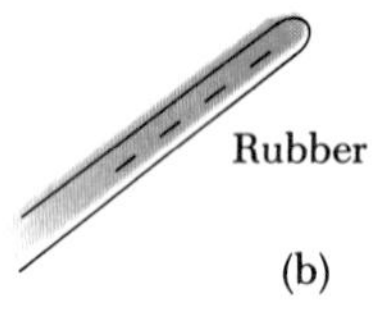

FIGURE 16.1 (a) A negatively charged rubber rod, suspended by a thread, is attracted to a positively charged glass rod. (b) A negatively charged rubber rod is repelled by another negatively charged rubber rod.

16.1 PROPERTIES OF ELECTRIC CHARGES

There are many simple experiments you can do to demonstrate the existence of electric forces and charges. For example, after running a comb through your hair, you will find that the comb will attract bits of paper. The attractive force is often strong enough to suspend the paper from the comb. The same effect occurs with other rubbed materials, such as glass, hard rubber, and plastic.

Another simple experiment is to rub an inflated balloon with wool (or across your hair). On a dry day, the rubbed balloon will then stick to the wall of a room, often for hours. When materials behave this way, they are said to have become **electrically charged.** You can give your body an electric charge by vigorously rubbing your shoes on a wool rug or by sliding across a car seat. You can then feel, and remove, the charge on your body by lightly touching another person. Under the right conditions, a visible spark is seen when you touch, and a slight tingle will be felt by both parties. (Experiments such as these work best on a dry day because excessive moisture in the air can provide a pathway for charge to leak off a charged object.)

A series of simple experiments can demonstrate that there are two kinds of electric charge, which were given the names **positive** and **negative** by Benjamin Franklin (1706–1790). Figure 16.1 illustrates the approach. A hard rubber (or plastic) rod that has been rubbed with fur (or an acrylic material) is suspended by a piece of string. When a glass rod that has been rubbed with silk is brought near the rubber rod, the rubber rod will be attracted toward the glass rod (Fig. 16.1a). If two charged rubber rods (or two charged glass rods) are brought near each other, as in Figure 16.1b, the force between them will be repulsive. This observation shows that the rubber and glass have different kinds of charge. We use the convention suggested by Franklin, wherein the electric charge on the glass rod is called positive and that on the rubber rod is called negative. On the basis of observations such as these, we conclude that *like charges repel one another and unlike charges attract one another.* It should be emphasized here that there is a natural tendency for charge to be transferred between unlike materials. Rubbing the two materials together serves to increase the area of contact, and thus, to enhance the charge transfer process.

Another important characteristic of charge is that *electric charge is always conserved.* That is, when two initially neutral objects are charged by being rubbed together, charge is not created in the process. The objects become charged because *negative charge is transferred* from one object to the other. One object gains some amount of negative charge while the other loses an equal amount of negative charge and hence is left with a positive charge. For example, when a glass rod is rubbed with silk, the silk obtains a negative charge that is equal in magnitude to the positive charge on the glass rod. We now know that what is happening in the process is that negatively charged electrons are transferred from the glass to the silk in the rubbing process. Likewise, when rubber is rubbed with fur, electrons are transferred from the fur to the rubber. An *uncharged object* contains an enormous number of electrons (of the order of 10^{23} electrons). However, for every negative electron there is a positively charged proton present; hence an uncharged object has no net charge of either sign.

In 1909, Robert Millikan (1886–1953) discovered that if an object is charged, its charge is always a multiple of a fundamental unit of charge, which we designate by the symbol e. In modern terms, the charge is said to be **quantized.** This means that charge occurs as discrete bundles in nature. Thus, an object may have a charge of $\pm e$, $\pm 2e$, $\pm 3e$, etc., but never a charge of $\pm 1.5e$. Recent theoretical developments have suggested the existence of fundamental particles called **quarks,** having charges of $\pm e/3$ or $\pm 2e/3$. Although many experiments have been conducted to detect such fractionally charged particles, their results remain inconclusive. Other experiments in Millikan's time showed that the electron has a charge $-e$ and the proton has an equal and opposite charge $+e$. Some particles, such as a neutron, have no charge. A neutral atom (one with no net charge) contains as many protons as electrons. The value of e is now known to be 1.60219×10^{-19} C. (The unit of electric charge, the **coulomb,** abbreviated C, will be defined more precisely in a later section.)

16.2 INSULATORS AND CONDUCTORS

It is convenient to classify substances in terms of their ability to conduct electric charge.

Conductors are materials through which electric charges can move under the influence of electric forces; **insulators** are materials that do not readily allow charges to move.

Materials such as glass and rubber are insulators. When insulators are charged by rubbing, only the rubbed area becomes charged and there is no tendency for the charge to move into other regions of the material. In contrast, materials such as copper, aluminum, and silver are good conductors. When such materials are charged in some small region the charge quickly spreads over the entire surface of the material. If you hold a copper rod in your hand and rub the rod with wool or fur, it will not attract a piece of paper. This might suggest that a metal cannot be charged. If you hold the copper rod with an insulator, however, so that your hand does not touch the copper and then rub, the rod will become charged and attract the paper. This is explained by noting that in the first case the electric charges transferred to the rod by rubbing readily move from the copper through your body and finally into the earth. In the second case the insulator prevents the flow of charge to earth.

Semiconductors are a third class of materials, and their electrical properties are intermediate between those of insulators and conductors. Silicon and germanium are two semiconducting materials widely used in a variety of electronic components which are used in calculators and computers. The ability of semiconductors to conduct electricity can be changed over many orders of magnitude by adding controlled amounts of foreign atoms, or impurities, to the materials by a process known as doping. This will be discussed in more detail in Chapter 31.

Charging by Induction

When a conductor is connected to earth by means of a conducting wire or copper pipe, it is said to be **grounded.** The earth can be considered as a infinite reservoir for electrons; this means that it can accept or supply an unlimited number of electrons. With this in mind, we can understand how to charge a conductor by a process known as **induction.**

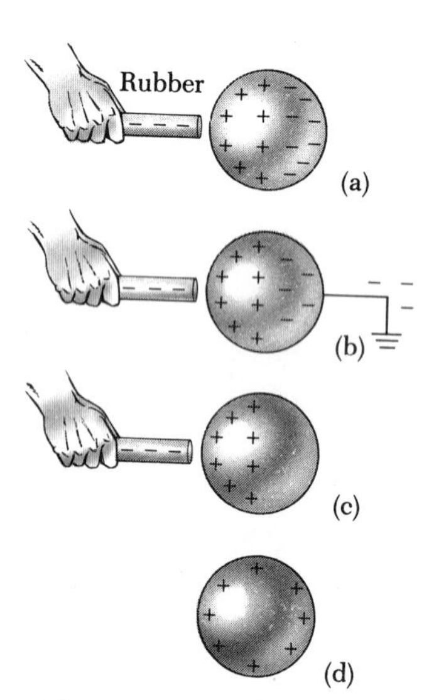

FIGURE 16.2 Charging a metal object by induction. (a) The charge on a neutral metal sphere is redistributed when a charged rubber rod is placed near the sphere. (b) The sphere is grounded, and some of the electrons leave the sphere through the ground wire. (c) The ground connection is removed, and the sphere is left with excess positive charge. (d) When the rubber rod is moved away, the sphere becomes uniformly charged.

In order to understand the process of charging by induction, consider a negatively charged rubber rod brought near a neutral (uncharged) conducting sphere that is insulated so that there is no conducting path to ground (Fig. 16.2). The repulsive force between the electrons in the rod and those in the sphere will cause a redistribution of charge on the sphere such that some electrons move to the side of the sphere farthest away from the rod (Fig. 16.2a). The region of the sphere nearest the negatively charged rod will have an excess of positive charge because of the migration of electrons away from this location. If a grounded conducting wire is then connected to the sphere, as in Figure 16.2b, some of the electrons will leave the sphere and travel to the earth. If the wire to ground is then removed (Fig. 16.2c), the conducting sphere is left with an excess of induced positive charge. Finally, when the rubber rod is removed from the vicinity of the sphere (Fig. 16.2d), the induced positive charge will remain on the ungrounded sphere.

Note that the charge remaining on the sphere is uniformly distributed over its surface because electrons from other parts of the sphere are attracted to the region containing the high density of positive charge. The effect of this migration of electrons is the same as if the positive charge moved around on the sphere.

In the process of inducing a charge on the sphere, the charged rubber rod loses none of its negative charge since it never came in contact with the sphere. Thus we see that *charging an object by induction requires no contact with the object inducing the charge.* This is in contrast to charging an object by rubbing, which does require contact between the two objects.

A process which is very similar to that of charging by induction in conductors also takes place in insulators. In most neutral atoms or molecules the center of positive charge coincides with the center of negative charge. However, in the presence of a charged object, these centers may shift slightly, resulting in more positive charge on one side of the molecule than on the other. This effect, known as **polarization,** will be discussed more completely in a later section. This realignment of charge within individual molecules produces an induced charge on the surface of the insulator as shown in Figure 16.3. With these ideas, you should be able to explain why a comb that has been rubbed through hair will attract bits of neutral paper, or why a balloon that has been rubbed against your clothing is able to stick to a neutral wall.

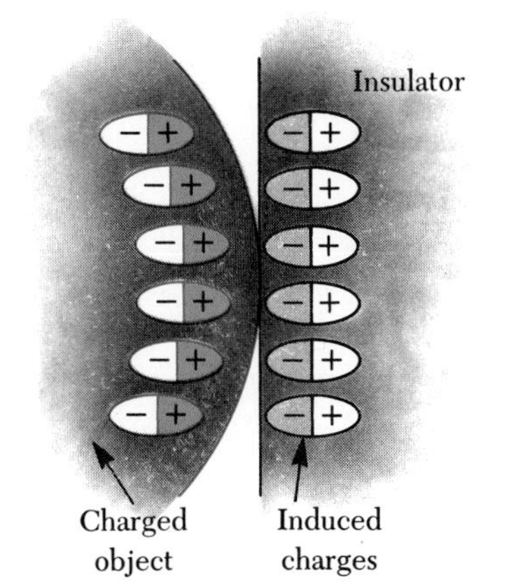

FIGURE 16.3 The charged object on the left induces charges on the surface of an insulator at the right.

16.3 COULOMB'S LAW

In 1785, Charles Augustin Coulomb (1736–1806) established the fundamental law of electric force between two stationary charged particles. Experiments show that an **electric force** has the following properties:

1. It is inversely proportional to the square of the separation, r, between the two particles and is along the line joining them.
2. It is proportional to the product of the magnitudes of the charges, $|q_1|$ and $|q_2|$, on the two particles.
3. It is attractive if the charges are of opposite sign and repulsive if the charges have the same sign.

From these observations, we can express the magnitude of the electric force that each charge exerts on the other as

$$F = k\frac{|q_1|\,|q_2|}{r^2} \qquad (16.1)$$

Coulomb's law

where k is a constant called the **Coulomb constant.**

The value of the Coulomb constant in Equation 16.1 depends on our choice of units. The unit of charge in SI units is the **coulomb** (C), which is defined in terms of a unit current called the **ampere** (A), where current is defined as the rate of flow of charge. (The ampere will be defined in Chapter 18.) When the current in a wire is 1 A, the amount of charge that flows past a given point in the wire in 1 s is 1 C. From experiment, we know that the Coulomb constant in SI units has the value

$$k = 8.9875 \times 10^9\ \text{N}\cdot\text{m}^2/\text{C}^2 \qquad (16.2)$$

To simplify our calculations, we shall use the approximate value

$$k \approx 9.0 \times 10^9\ \text{N}\cdot\text{m}^2/\text{C}^2 \qquad (16.3)$$

The charge on the proton has a magnitude of $e = 1.6 \times 10^{-19}$ C. Therefore, it would take $1/e = 6.3 \times 10^{18}$ protons to create a total charge of $+1$ C. Likewise, 6.3×10^{18} electrons would have a total charge of -1 C. This can be compared with the number of free electrons in 1 cm^3 of copper, which is of the order of 10^{23}. Thus, we see that 1 C is a substantial amount of charge. In typical electrostatic experiments, where a rubber or glass rod is charged by friction, a net charge of the order of 10^{-6} C ($= 1\ \mu$C) is obtained. In other words, only a very small fraction of the total available charge is transferred between the rod and rubbing material.

TABLE 16.1 Charge and Mass of the Electron, Proton, and Neutron

Particle	Charge (C)	Mass (kg)
Electron	-1.60×10^{-19}	9.11×10^{-31}
Proton	$+1.60 \times 10^{-19}$	1.67×10^{-27}
Neutron	0	1.67×10^{-27}

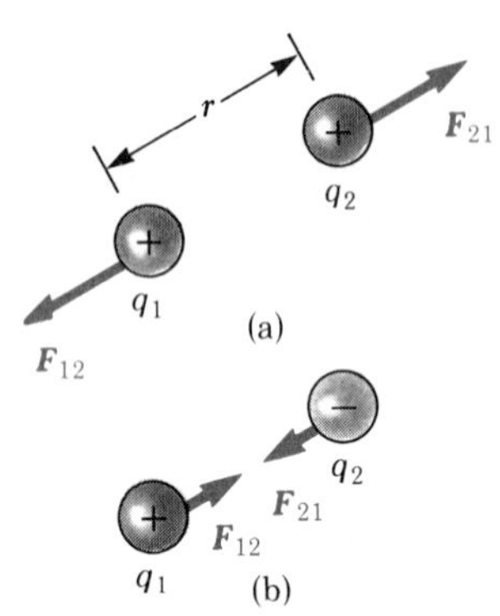

FIGURE 16.4 Two point charges separated by a distance r exert a force on each other given by Coulomb's law. The force on q_1 is equal in magnitude and opposite in direction to the force on q_2. (a) When the charges are of the same sign, the force is repulsive. (b) When the charges are of opposite sign, the force is attractive.

The charges and masses of the electron, proton, and neutron are given in Table 16.1.

When dealing with Coulomb's force law, you must remember that force is a vector quantity and must be treated accordingly. Furthermore, you should note that Coulomb's law applies exactly only to point charges or particles. The electric force of repulsion between two positively charged particles is shown in Figure 16.4a. Electric forces obey Newton's third law, and hence the forces $\mathbf{F}_{12}$ and $\mathbf{F}_{21}$ are equal in magnitude but opposite in direction. (The notation $\mathbf{F}_{12}$ denotes the force on particle 1 exerted by particle 2. Likewise, $\mathbf{F}_{21}$ is the force on particle 2 exerted by particle 1.) The attractive force between two unlike charges is shown in Figure 16.4b.

The Coulomb force is the second example we have seen of an action-at-a-distance force. One charge exerts a force on another charge even though *there is no physical contact between them.* As you recall, the other example of a force of this nature is the force of gravitational attraction. It should be noted that the mathematical form of the Coulomb force is the same as that of the gravitational force. That is, they are both inversely proportional to the square of the distance of separation. However, there are some important differences between electric and gravitational forces. Electric forces can be either attractive or repulsive, but gravitational forces are always attractive. Furthermore, gravitational forces are considerably weaker, as shown in the following example.

EXAMPLE 16.1 The Electric Force and the Gravitational Force

The electron and proton of a hydrogen atom are separated (on the average) by a distance of about 5.3×10^{-11} m. Find the magnitude of the electric force and the gravitational force that each particle exerts on the other.

Solution From Coulomb's law, we find that the attractive electric force has the magnitude

$$F_e = k\frac{|e|^2}{r^2} = \left(9.0 \times 10^9\ \frac{\text{N}\cdot\text{m}^2}{\text{C}^2}\right)\frac{(1.6 \times 10^{-19}\ \text{C})^2}{(5.3 \times 10^{-11}\ \text{m})^2} = 8.2 \times 10^{-8}\ \text{N}$$

From Newton's universal law of gravity and Table 16.1, we find that the gravitational force has a magnitude given by

$$F_g = G\frac{m_e m_p}{r^2} = \left(6.7 \times 10^{-11}\ \frac{\text{N}\cdot\text{m}^2}{\text{kg}^2}\right)\frac{(9.11 \times 10^{-31}\ \text{kg})(1.67 \times 10^{-27}\ \text{kg})}{(5.3 \times 10^{-11}\ \text{m})^2}$$

$$= 3.6 \times 10^{-47}\ \text{N}$$

The ratio $F_e/F_g \approx 3 \times 10^{39}$. Thus the gravitational force between the charged atomic particles is negligible relative to the electric force.

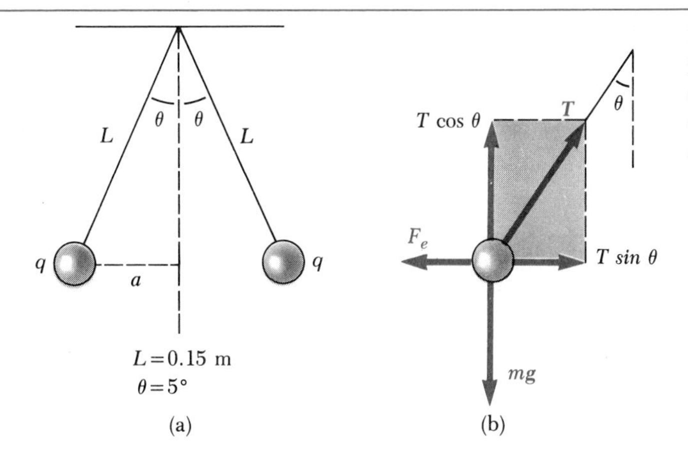

FIGURE 16.5 (Example 16.2) (a) Two identical pieces of aluminum foil, each with the same charge q, suspended in equilibrium by strings. (b) The free-body diagram for the foil on the left side.

EXAMPLE 16.2 Find the Charge on the Aluminum Foils

Two pieces of charged aluminum foil, each having a mass of 3×10^{-2} kg, hang in equilibrium as shown in Figure 16.5a. If the length of each string is 0.15 m and the angle $\theta = 5°$, find the magnitude of the charge on each foil, assuming the foils have identical charges.

Solution From the right triangle in Figure 16.5a, we see that $\sin\theta = a/L$. From the known length of the string and the angle the string makes with the vertical, the distance a is calculated to be

$$a = L \sin\theta = (0.15 \text{ m}) \sin 5° = 0.013 \text{ m}$$

Therefore, the separation of the foils is $2a = 0.026$ m.

The forces acting on one of the foils are shown in Figure 16.5b. Because the foil is in equilibrium, the resultants of the forces in the horizontal and vertical directions must separately add up to zero:

$$(1) \quad \sum F_x = T\sin\theta - F_e = 0$$

$$(2) \quad \sum F_y = T\cos\theta - mg = 0$$

From (2), we see that $T = mg/\cos\theta$, and so T can be eliminated from (1) if we make this substitution. This gives a value for the electric force, F_e:

$$(3) \quad F_e = mg\tan\theta = (3 \times 10^{-2} \text{ kg})(9.80 \text{ m/s}^2)\tan(5°) = 2.57 \times 10^{-2} \text{ N}$$

From Coulomb's law (Eq. 16.1), the electric force between the charges has magnitude given by

$$F_e = k\frac{|q|^2}{r^2}$$

where $r = 2a = 0.026$ m and $|q|$ is the magnitude of the charge on each foil. Note that the term $|q|^2$ arises here because we have assumed that the charge is the same on both foils. This equation can be solved for $|q|^2$ to give the charge as follows:

$$|q|^2 = \frac{F_e r^2}{k} = \frac{(2.57 \times 10^{-2} \text{ N})(0.026 \text{ m})^2}{9 \times 10^9 \text{ N}\cdot\text{m}^2/\text{C}^2}$$

$$|q| = 4.4 \times 10^{-8} \text{ C}$$

Exercise If the charge on the foils is negative, how many electrons have to be added to the foils to give it a net charge of -4.4×10^{-8} C?

Answer 2.7×10^{11} electrons.

The Principle of Superposition

A problem often encountered is one in which more than two charges are present and we need to find the net electric force on one of the charges. This can be accomplished by noting that the electric force between any pair of charges is given by Equation 16.1. Therefore, the resultant force on any one charge equals the vector sum of the forces exerted by the various individual charges present. This is another example of the **principle of superposition.** For example, if you have three charges and you want to find the force on charge 1 exerted by charges 2 and 3, you first find the force exerted on charge 1 by charge 2 and the force exerted on charge 1 by charge 3. You then add these two forces together vectorially to get the resultant force on charge 1. The following numerical example illustrates this procedure.

EXAMPLE 16.3 Using the Superposition Principle

Consider three point charges located at the corners of a triangle, as in Figure 16.6, where $q_1 = 6 \times 10^{-9}$ C, $q_3 = 5 \times 10^{-9}$ C, $q_2 = -2 \times 10^{-9}$ C, and the distances of separation are shown in the figure. Find the resultant force on q_3.

Solution First, note the directions of the individual forces exerted on q_3 by q_1 and q_2. The force exerted by q_2 is attractive since q_2 and q_3 have opposite signs. The force exerted by q_1 is repulsive because both q_1 and q_3 are positive.

Now let us calculate the magnitude of the forces on q_3. The magnitude of the force exerted by q_2 is

$$F_{32} = k\frac{|q_3|\,|q_2|}{r^2} = (9 \times 10^9 \text{ N} \cdot \text{m}^2/\text{C}^2)\frac{(5 \times 10^{-9}\text{ C})(2 \times 10^{-9}\text{ C})}{(4\text{ m})^2}$$

$$= 5.63 \times 10^{-9}\text{ N}$$

The magnitude of the force exerted by q_1 is

$$F_{31} = k\frac{|q_3|\,|q_1|}{r^3} = (9 \times 10^9 \text{ N} \cdot \text{m}^2/\text{C}^2)\frac{(5 \times 10^{-9}\text{ C})(6 \times 10^{-9}\text{ C})}{(5\text{ m})^2}$$

$$= 1.08 \times 10^{-8}\text{ C}$$

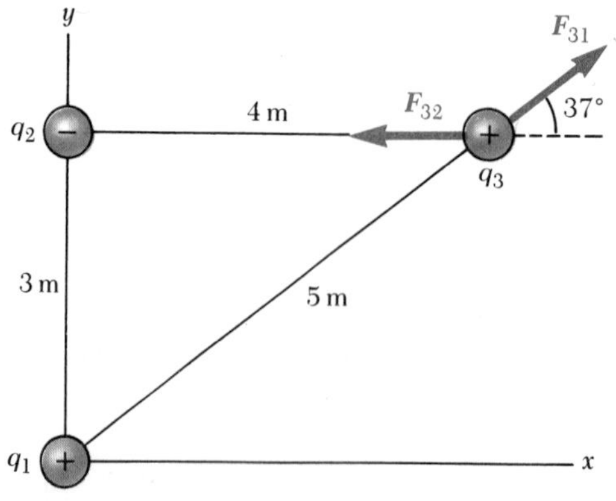

FIGURE 16.6 (Example 16.3) The force exerted on q_3 by q_1 is $\mathbf{F}_{31}$. The force exerted on q_3 by q_2 is $\mathbf{F}_{32}$. The *total* force, $\mathbf{F}_3$, on q_3 is the *vector* sum $\mathbf{F}_{31} + \mathbf{F}_{32}$.

The force $\mathbf{F}_{31}$ makes an angle of 37° with the x axis. Therefore, the x component of this force has a magnitude $F_{31} \cos 37° = 8.63 \times 10^{-9}$ N, and the y component has a magnitude $F_{31} \sin 37° = 6.50 \times 10^{-9}$ N. The force $\mathbf{F}_{32}$ is in the negative x direction. Hence, the x and y components of the resultant force on q_3 are

$$F_x = 8.63 \times 10^{-9}\text{ N} - 5.63 \times 10^{-9}\text{ N} = 3.00 \times 10^{-9}\text{ N}$$

$$F_y = 6.50 \times 10^{-9}\text{ N}$$

The magnitude of the resultant force on the charge q_3 is therefore

$$\sqrt{(3.00 \times 10^{-9}\text{ N})^2 + (6.50 \times 10^{-9}\text{ N})^2} = 7.16 \times 10^{-9}\text{ N}$$

and the force vector makes an angle of 65.2° with the x axis.

16.4 EXPERIMENTAL VERIFICATION OF COULOMB'S FORCE LAW

Electric forces between charged objects were measured by Coulomb with a torsion balance (Fig. 16.7). The apparatus consists of two small spheres fixed to the ends of a light horizontal rod made of an insulating material and suspended by a silk thread. Sphere A is given a charge, and charged object B is brought near sphere A. The attractive (or repulsive) force between the two charged objects causes the rod to rotate and to twist the suspension. The angle through which the rod rotates is measured by the deflection of a light beam reflected from a mirror attached to the suspension. The rod will rotate through some angle against the restoring force of the twisted thread before reaching equilibrium. The value of the angle of rotation increases as the charge on the objects increases. Thus, the angle of rotation provides a quantitative measure of the electric force of attraction or repulsion. Although Coulomb was unable to establish the inverse square nature of his law with very much precision using this apparatus, scientists in later years have shown that the exponent of r is 2 to within a very small uncertainty.

16.5 THE ELECTRIC FIELD

Two different action-at-a-distance forces have been introduced into our discussions so far, the gravitational force and the electrostatic force. As we pointed out earlier, these forces are capable of acting through space and produce an effect even when there is no physical contact between the objects involved. There are a variety of ways by which we can discuss action-at-a-distance forces but a method developed by Michael Faraday (1791–1867) is of such practical value that we shall devote much attention to it in the next several chapters. In this approach, an **electric field** is said to exist in the region of space around a charged object. Forces of an electrical nature arise when another charged object enters this electric field. As an example, con-

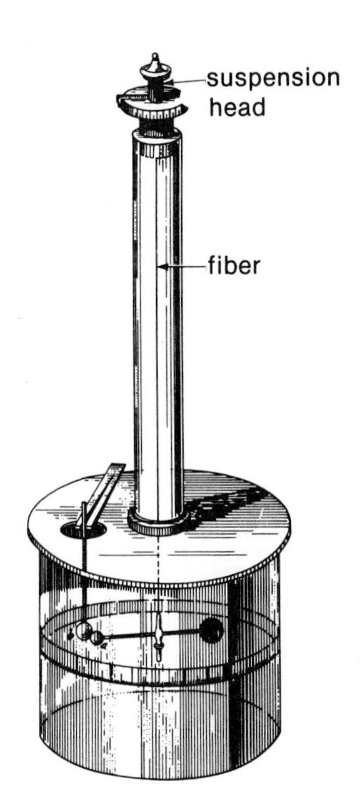

FIGURE 16.7 Coulomb's torsion balance, which was used to establish the inverse-square law for the electrostatic force between two charges. (Taken from Coulomb's 1785 memoirs to the French Academy of Sciences)

sider Figure 16.8, which shows an object with a small positive charge q_0 placed near a second object with a larger positive charge Q.

We define the strength of the electric field at the location of the smaller charge to be the magnitude of the electric force acting on it divided by the magnitude of its charge:

Electric field

$$E \equiv \frac{|\mathbf{F}|}{|q_0|} \tag{16.4}$$

Note that this is the electric field at the location of q_0 produced by the charge Q, not the field produced by q_0. The electric field is a vector quantity having the SI units of newtons per coulomb (N/C). *The direction of* **E** *at a point is defined to be the direction of the electric force that would be exerted on a small positive charge placed at that point.* Thus, in Figure 16.8, the direction of the electric field is horizontal and to the right.

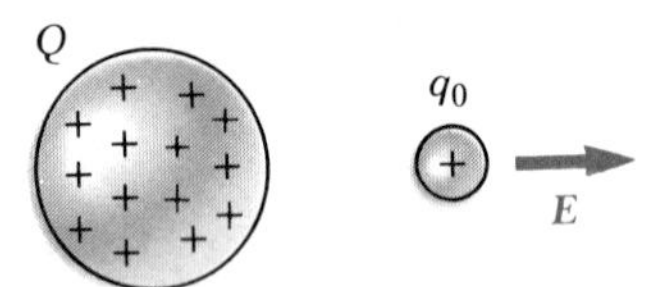

FIGURE 16.8 A small object with a positive charge q_0 placed near an object with a larger positive charge Q experiences an electric field **E** directed as shown. The magnitude of the electric field is defined as the ratio of the electric force on q_0 divided by the charge q_0.

The electric field at point A in Figure 16.9a is vertical and downward because at this point a positive charge would experience a force of attraction toward the negatively charged sphere.

As presently stated, the definition for electric field has a serious difficulty. To illustrate the problem, consider the positively charged conducting sphere shown in Figure 16.9b. The field in the region surrounding the sphere could be explored by introducing a test charge q_0 at a point such as P, finding the electric force on this charge, and then dividing this force by the magnitude of the charge on the test charge. Difficulties arise, however, when the magnitude of the test charge is large enough to influence the charge on the conducting sphere. For example, a strong test charge could cause a rearrangement of the charges on the sphere as shown in Figure 16.9c. As a result, the force exerted on the test charge is different from what it would be if the movement of charge on the sphere had not taken place. Furthermore, the strength of the measured electric field differs from what it would be in the absence of the test charge. To take care of this problem, we simply require that the test charge be small enough to have no effect at all on the charges on the sphere.

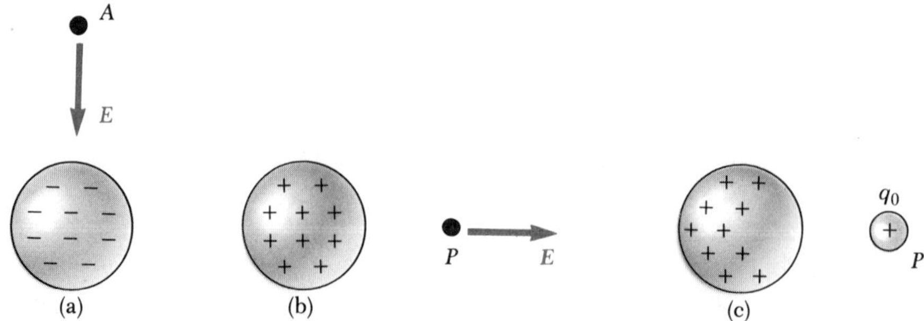

FIGURE 16.9 (a) The electric field at A due to the negatively charged sphere is downward, toward the negative charge. (b) The electric field at P due to the positively charged conducting sphere is to the right, away from the positive charge. (c) A test charge q_0 placed at P will cause a rearrangement of charge on the sphere unless q_0 is very small compared with the charge on the sphere.

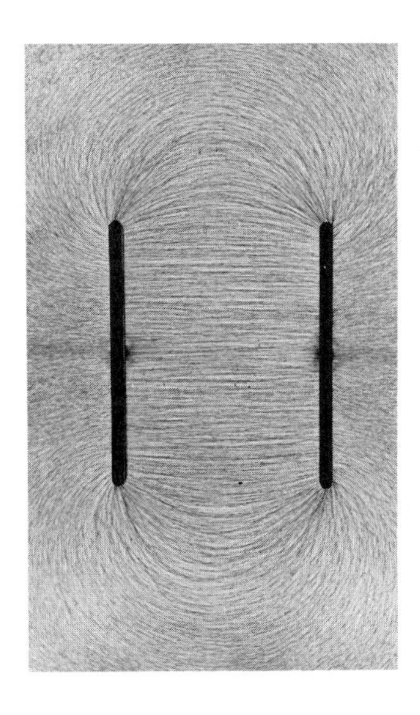

Electric field pattern of two oppositely charged conducting parallel plates. Small pieces of thread on an oil surface align with the electric field. Note the nonuniform nature of the electric field at the ends of the plates. Such end effects can be neglected if the plate separation is small compared to the length of the plates. (Courtesy of Harold M. Waage, Princeton University.)

Consider a point charge q located a distance r from a test charge q_0. According to Coulomb's law, the *magnitude* of the force on the test charge is

$$F = k\frac{|q|\,|q_0|}{r^2}$$

Because the magnitude of the electric field at the position of the test charge is defined as $E = F/q_0$, we see that the *magnitude* of the electric field due to the charge q at the position of $|q_0|$ is

$$\boxed{E = k\frac{|q|}{r^2}} \qquad (16.5)$$

Electric field due to a charge q

If q is *positive*, as in Figure 16.10a, the field at P due to this charge is *radially outward* from q. If q is *negative*, as in Figure 16.10b, the field at P is directed *toward* q.

The principle of superposition holds when calculating the electric field due to a group of point charges. We first use Equation 16.5 to calculate the electric field produced by each charge individually at a point and then add these electric fields together as vectors.

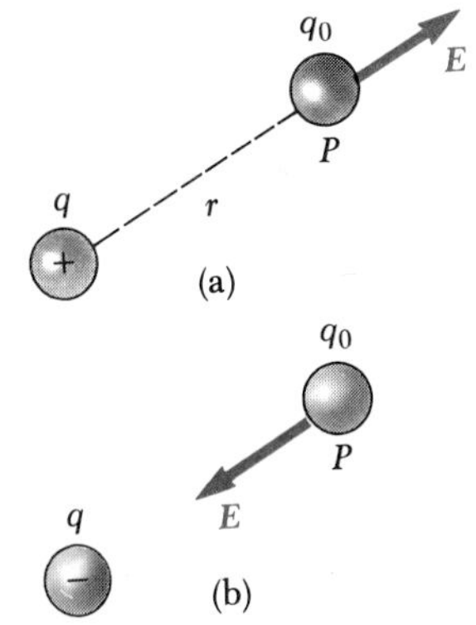

FIGURE 16.10 A test charge q_0 at P is a distance r from a point charge q. (a) If q is positive, the electric field at P points radially *outward* from q. (b) If q is negative, the electric field at P points radially *inward* toward q.

EXAMPLE 16.4 Electric Force on a Proton

Find the electric force on a proton placed in an electric field of 2×10^4 N/C that is directed along the positive x axis.

Solution Because the charge on a proton is $+e = +1.6 \times 10^{-19}$ C, the electric force acting on it is

$$F = eE = (1.6 \times 10^{-19}\ \text{C})(2 \times 10^4\ \text{N/C}) = 3.2 \times 10^{-15}\ \text{N}$$

where the force is in the positive x direction. The weight of the proton has the value $mg = (1.67 \times 10^{-27}\ \text{kg})(9.8\ \text{m/s}^2) = 1.6 \times 10^{-26}$ N. Hence, we see that the magnitude of the gravitational force is negligible compared with that of the electric force.

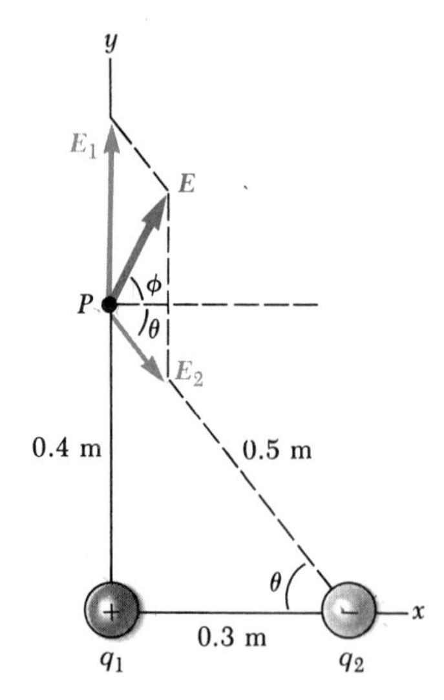

FIGURE 16.11 (Example 16.5) The total electric field **E** at *P* equals the vector sum $\mathbf{E}_1 + \mathbf{E}_2$, where $\mathbf{E}_1$ is the field due to the positive charge q_1 and $\mathbf{E}_2$ is the field due to the negative charge q_2.

EXAMPLE 16.5 Electric Field Due to Two Point Charges

A charge $q_1 = 7\ \mu\text{C}$ is located at the origin, and a charge $q_2 = -5\ \mu\text{C}$ is located on the x axis 0.3 m from the origin (Fig. 16.11). Find the electric field at the point P, which has coordinates (0, 0.4) m.

Solution First let us find the magnitudes of the electric fields due to each charge. The fields at P, $\mathbf{E}_1$ due to the 7-μC charge and $\mathbf{E}_2$ due to the −5-μC charge, are shown in Figure 16.11. Their magnitudes are

$$E_1 = k\frac{|q_1|}{r_1^2} = (9.0 \times 10^9\ \text{N}\cdot\text{m}^2/\text{C}^2)\frac{(7 \times 10^{-6}\ \text{C})}{(0.4\ \text{m})^2} = 3.9 \times 10^5\ \text{N/C}$$

$$E_2 = k\frac{|q_2|}{r_2^2} = (9.0 \times 10^9\ \text{N}\cdot\text{m}^2/\text{C}^2)\frac{(5 \times 10^{-6}\ \text{C})}{(0.5\ \text{m})^2} = 1.8 \times 10^5\ \text{N/C}$$

The vector $\mathbf{E}_1$ has an x component of zero. The vector $\mathbf{E}_2$ has an x component given by $E_2 \cos\theta = E_2(0.3/0.5) = \frac{3}{5}E_2 = 1.1 \times 10^5$ N/C and a negative y component given by $-E_2 \sin\theta = -E_2\,(0.4/0.5) = -\frac{4}{5}E_2 = -1.4 \times 10^5$ N/C. Hence, the resultant component in the x direction is

$$E_x = 1.1 \times 10^5\ \text{N/C}$$

and the resultant component in the y direction is

$$E_y = E_{y1} + E_{y2} = 3.9 \times 10^5\ \text{N/C} - 1.4 \times 10^5\ \text{N/C} = 2.5 \times 10^5\ \text{N/C}$$

From the Pythagorean theorem ($E = \sqrt{E_x^2 + E_y^2}$), we find that **E** has a magnitude of 2.7×10^5 N/C and makes an angle ϕ of 66° with the positive x axis.

EXERCISE Find the force on a positive test charge of 2×10^{-8} C placed at P.

Answer 5.4×10^{-3} N in the same direction as **E**.

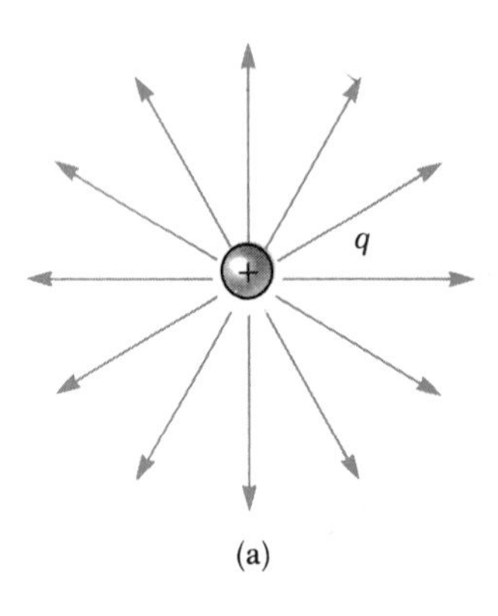

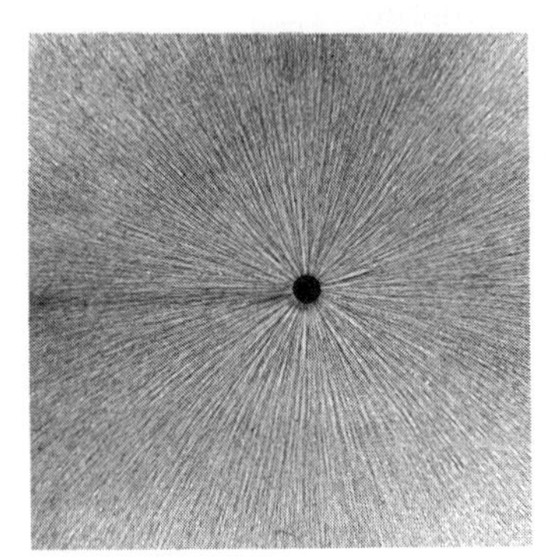

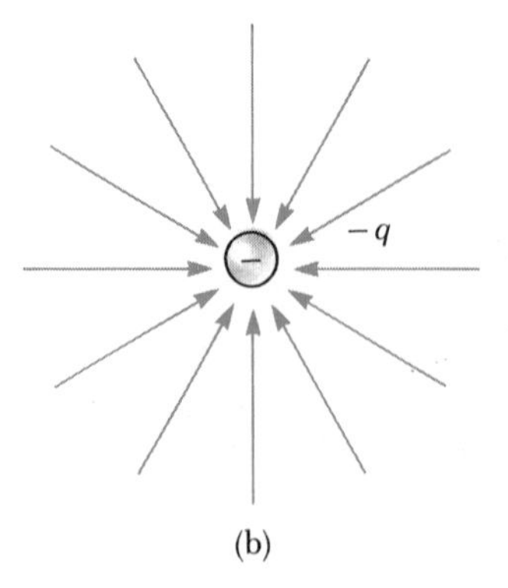

FIGURE 16.12 The electric field lines for a point charge. (a) For a positive point charge, the lines are radially outward. (b) For a negative point charge, the lines are radially inward. Note that the figures show only those field lines that lie in the plane containing the charge. (c) The dark areas are small pieces of thread suspended in oil, which align with the electric field produced by a small charged conductor at the center. (Photo courtesy of Harold M. Waage, Princeton University)

16.6 ELECTRIC FIELD LINES

A convenient aid for visualizing electric field patterns is to draw lines pointing in the direction of the electric field vector at any point. These lines, called **electric field lines,** are related to the electric field in any region of space in the following manner: (1) the electric field vector $\boldsymbol{E}$ is tangent to the electric field lines at each point and (2) the number of lines per unit area through a surface perpendicular to the lines is proportional to the strength of the electric field in a given region; this means that $\boldsymbol{E}$ is large when the field lines are close together and small when they are far apart.

Some representative electric field lines for a single positive point charge are shown in Figure 16.12a. Note that in this two-dimensional drawing we show only the field lines that lie in the plane containing the point charge. The lines are actually directed radially outward from the charge in *all* directions, somewhat like the needles of a porcupine. Since a positive test charge placed in this field would be repelled by the charge q, the lines are directed radially away from the positive charge. Similarly, the electric field lines for a single negative point charge are directed toward the charge (Fig. 16.12b). In either case, the lines are along the radial direction and extend all the way to infinity. Note that the lines are closer together as they get near the charge, indicating that the strength of the field is increasing. Equation 16.5 verifies that this should indeed be the case.

The rules for drawing electric field lines for any charge distribution are as follows:

1. The lines must begin on positive charges and terminate on negative charges, or at infinity in the case of a lone charge.
2. The number of lines drawn leaving a positive charge or approaching a negative charge is proportional to the magnitude of the charge.
3. No two field lines can cross.

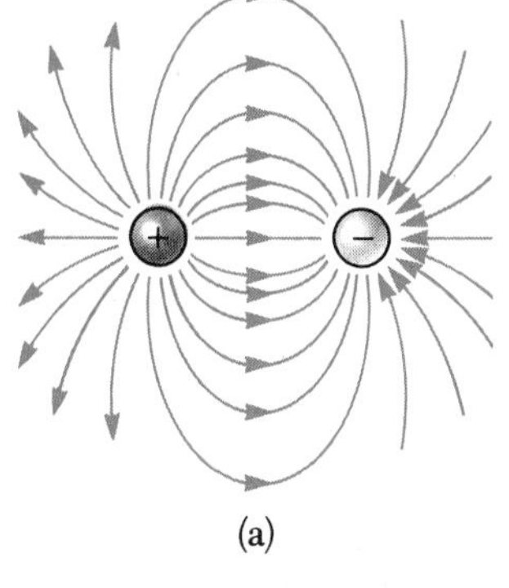

(a)

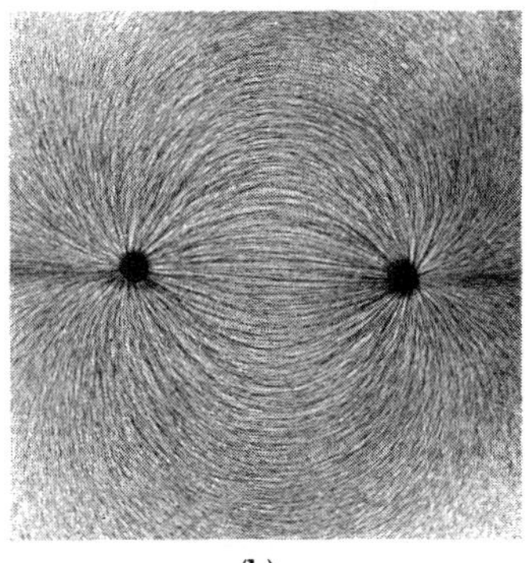

(b)

FIGURE 16.13 (a) The electric field lines for two equal and opposite point charges (an electric dipole). Note that the number of lines leaving the positive charge equals the number terminating at the negative charge. (b) The photograph was taken using small pieces of thread suspended in oil, which align with the electric field produced by two charged conductors. (Photo courtesy of Harold M. Waage, Princeton University)

The electric field lines for two point charges of equal magnitude but of opposite sign are shown in Figure 16.13. This charge configuration is called an **electric dipole.** In this case the number of lines that begin at the positive charge must equal the number that terminate at the negative charge. At points very near the charges, the lines are nearly radial. The close spacing of lines between the charges indicates a strong electric field in this region.

Figure 16.14 shows the electric field lines in the vicinity of two equal positive charges. Again, the lines are nearly radial at points close to either

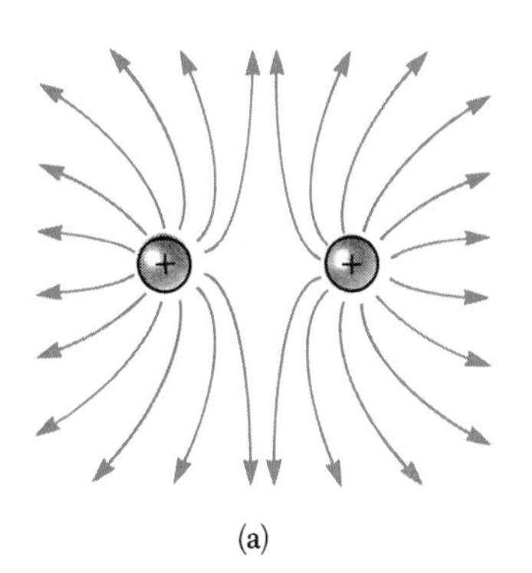

(a)

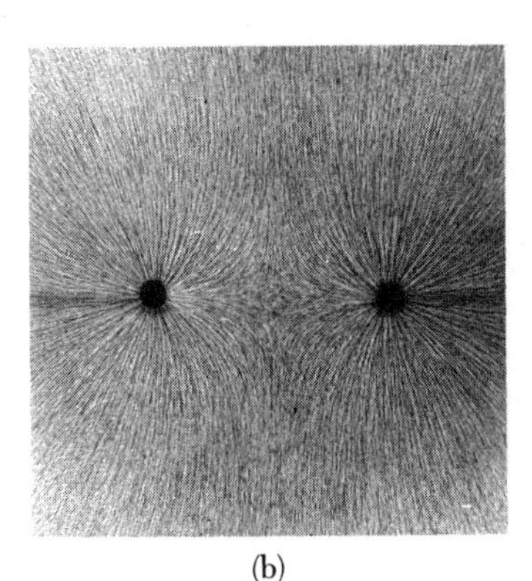

(b)

FIGURE 16.14 (a) The electric field lines for two positive point charges. (b) The photograph was taken using small pieces of thread suspended in oil, which align with the electric field produced by two charged conductors. (Photo courtesy of Harold M. Waage, Princeton University)

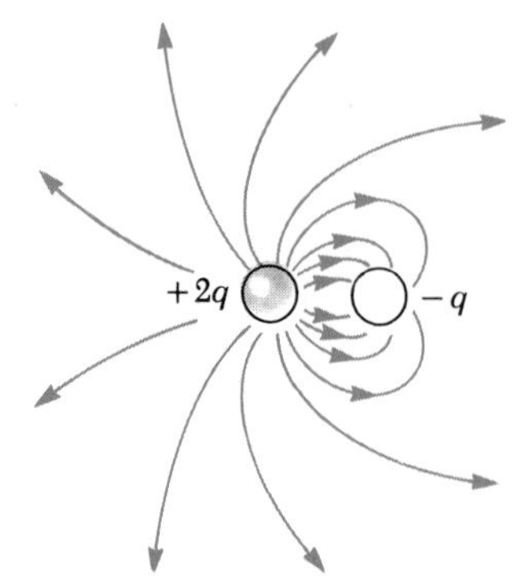

FIGURE 16.15 The electric field lines for a point charge $+2q$ and a second point charge $-q$. Note that two lines leave the charge $+2q$ for every one that terminates on the charge $-q$.

charge. The same number of lines are shown emerging from each charge because the charges are equal in magnitude. The lines are not close together between the charges, indicating that this is a region of low electric field strength.

Finally, in Figure 16.15 we sketch the electric field lines associated with a positive charge $+2q$ and a negative charge $-q$. In this case, the number of lines leaving the charge $+2q$ is twice the number entering the charge $-q$. Hence, only half the lines that leave the positive charge end on the negative charge. The remaining half terminate at infinity.

16.7 CONDUCTORS IN ELECTROSTATIC EQUILIBRIUM

A good electric conductor, such as copper, contains charges (electrons) that are not bound to any atom and are free to move about within the material but cannot easily leave. When there is no net motion of charge within a conductor or on its surface, the conductor is said to be in **electrostatic equilibrium.** The electric field in and around a conductor in electrostatic equilibrium has some important characteristics, which we shall now consider. As we shall see, a conductor in *electrostatic equilibrium* has the following properties:

1. The electric field is zero everywhere inside the conductor.
2. Any excess charge on an isolated conductor resides entirely on its surface.
3. The electric field just outside a charged conductor is perpendicular to the conductor's surface.
4. On an irregularly shaped conductor, charge tends to accumulate at locations where the curvature of the surface is greatest, that is, at sharp points.

The first property can be understood by examining what would happen if it were *not* true. If there were an electric field inside a conductor, the free charge there would move and a flow of charge, or current, would be created. However, if there were a net movement of charge, we would no longer have electrostatic equilibrium.

Property 2 arises because of property 1. If there were a net number of excess charges inside a conductor, there would be lines of the electric field originating or terminating on these charges. However, this cannot be the case since there can be no electric field inside a conductor in electrostatic equilibrium. If by some means an excess charge is placed inside a conductor, it quickly migrates to the surface.

Property 3 can be understood by again considering what would happen if it were not true. In Figure 16.16 we assume that the electric field is not perpendicular to the surface. In this case, the electric field would have a

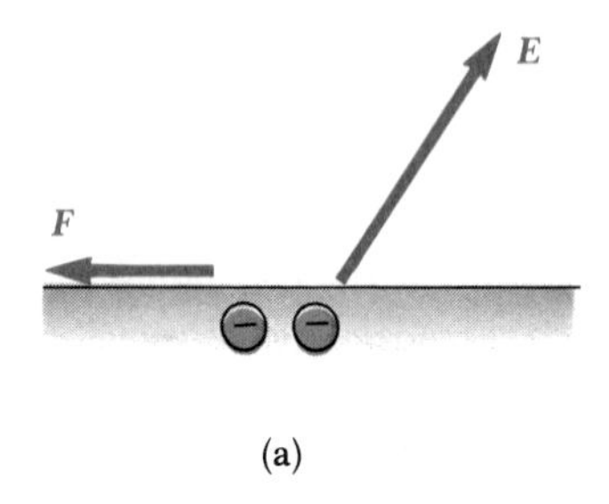

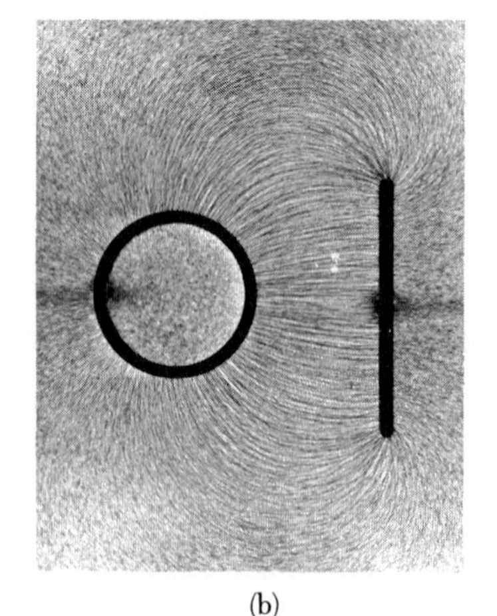

FIGURE 16.16 (a) Negative charges at the surface of a conductor. If the electric field were at an angle to the surface, as shown, an electric force would be exerted on the charges along the surface and they would move to the left. Since the conductor is assumed to be in electrostatic equilibrium, $\boldsymbol{E}$ cannot have a component along the surface and hence must be perpendicular to it. (b) Electric field pattern of a charged conducting plate near an oppositely charged conducting cylinder. Small pieces of thread suspended in oil align with the electric field lines. Note that (1) the electric field lines are perpendicular to the conductors and (2) there are no lines inside the cylinder ($\boldsymbol{E} = 0$). (Courtesy of Harold M. Waage, Princeton University)

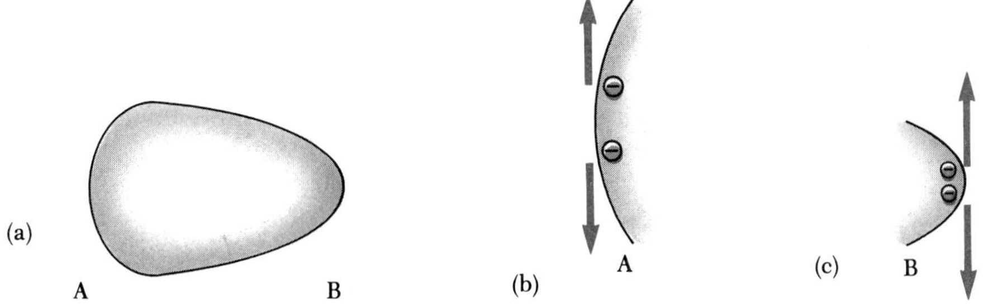

FIGURE 16.17 (a) The conductor shaped as shown has a flatter end A and a sharp surface B. Excess charge placed on a conductor resides entirely at its surface and is distributed such that (b) there is less charge per unit area on the flatter surface and (c) there is a large charge per unit area on the sharp surface.

component along the surface, and this component would cause the free charges of the conductor to move (to the left in the figure). If the charges move, however, a current is created and we no longer have electrostatic equilibrium. Hence, $\boldsymbol{E}$ must be perpendicular to the surface.

To see why property 4 must be true, consider Figure 16.17a, which shows a conductor that is relatively flat at one end and sharply pointed at the other. Any excess charge placed on the object moves to its surface. Figure 16.17b shows the forces between two of these charges at the flatter end of the object. These forces are predominantly directed parallel to the surface. Thus, the charges move apart until repulsive forces from other nearby charges create an equilibrium situation. At the sharp end, however, the forces of repulsion between two charges are directed predominantly away from the surface, as in Figure 16.17c. As a result, there is less tendency for the charges to move apart along the surface here and the amount of charge per unit area is greater than at the flat end. The cumulative effect of many such outward forces by nearby charges at the sharp end produces a large force directed away from the surface that could be great enough to cause charges to leap from the surface into the surrounding air.

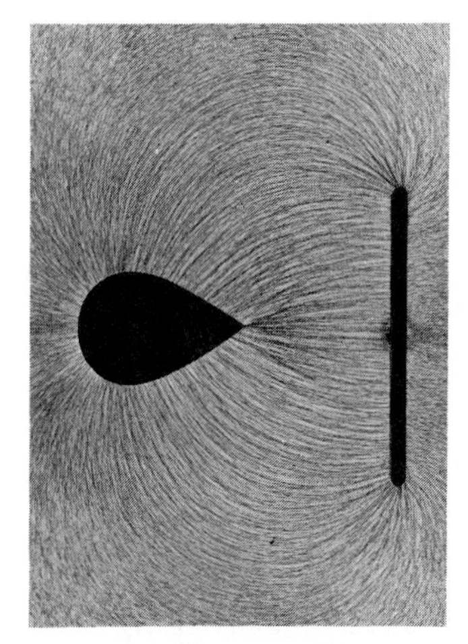

Electric field pattern of a charged conducting plate near an oppositely charged pointed conductor. Small pieces of thread suspended in oil align with the electric field lines. Note that the electric field is most intense near the pointed part of the conductor where the radius of curvature is the smallest. Also, the lines are perpendicular to the conductors. (Courtesy of Harold M. Waage, Princeton University.)

Property 4 indicates that if a metal rod having sharp points is attached to a house, most of the charge sprayed off of or onto the house will pass through these points, thus eliminating the induced charge on the house produced by storm clouds. In addition, a lightning discharge striking the house can pass through the metal rod and be safely carried to the ground through wires leading from the rod to the earth. Lightning rods using this principle were first developed by Benjamin Franklin. It is an interesting sidelight to American history to note that some foreign countries could not accept the fact that such a worthwhile idea could originate in this country. As a result, they "improved" the design by eliminating the sharp points. This modification in design drastically reduced the efficiency of the lightning rod.

16.8 FARADAY'S ICE-PAIL EXPERIMENT

Many experiments have been performed to show that the net charge on a conductor resides on its surface. The experiment we shall describe here was first performed by Michael Faraday. A metal ball having a positive charge was lowered at the end of a silk thread into an uncharged hollow conductor insulated from ground, as in Figure 16.18a. (The experiment is referred to as Faraday's ice-pail experiment because he used a metal ice pail as the hollow conductor.) As the ball was lowered into the pail, the needle on an electrometer attached to the outer surface of the pail was observed to deflect. (An electometer is a device used to measure charge.) The needle deflected because the charged ball induced a negative charge on the inner wall of the pail, which left an equal positive charge on the outer wall, as indicated in Figure 16.18b.

Faraday also noted that the needle deflection did not change when the ball touched the inner surface of the pail (Fig. 16.18c). After the ball had touched the inside of the pail and then been removed, the electrometer reading still remained unchanged (Fig. 16.18d). Furthermore, Faraday found that the ball was now uncharged. Apparently, when the ball touched the inside of the pail, the excess positive charge on the ball was neutralized by the induced negative charge on the inner surface of the pail.

Faraday concluded that since the electrometer deflection did not change when the charged ball touched the inside of the pail, the negative charge induced on the inside surface of the pail was just enough to neutralize the positive charge on the ball. As a result of these investigations, he concluded that a charged object suspended inside a metal container will cause a rearrangement of charge on the container in such a manner that the sign of the charge on the inside surface of the container is *opposite* the sign of the charge on the suspended object. This produces a charge on the outside surface of the container of the same sign as that on the suspended object.

Faraday also found that if the electrometer was connected to the inside surface of the pail after the experiment had been run, the needle showed no deflection. Thus, the *excess* charge acquired by the pail when contact was made between ball and pail appeared on the outer surface of the pail.

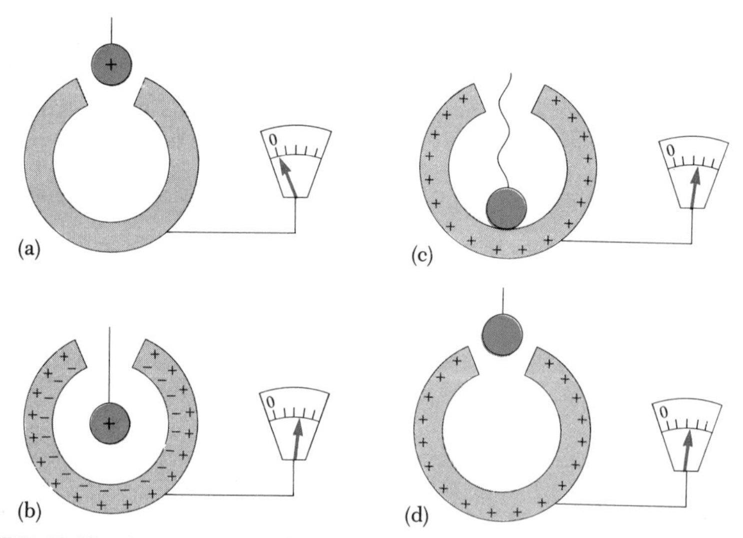

FIGURE 16.18 An experiment showing that any charge transferred to a conductor resides on its surface in electrostatic equilibrium. The hollow conductor is insulated from ground, and the small metal ball is supported by an insulating thread.

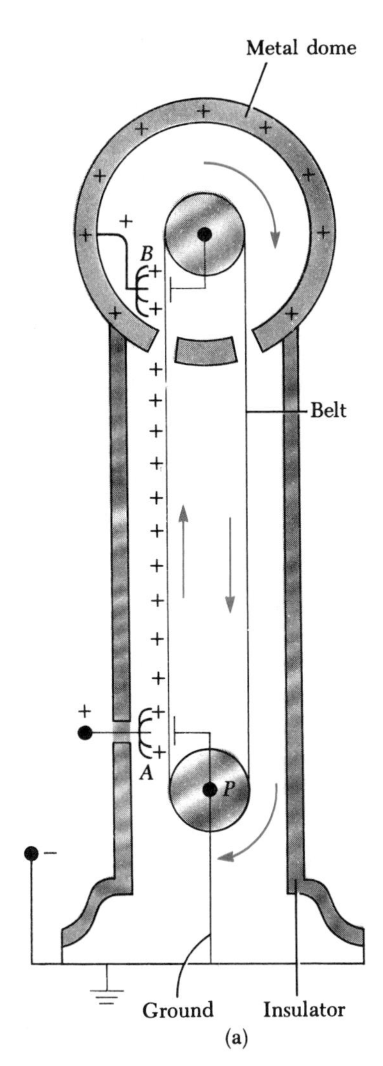

(a)

(b)

FIGURE 16.19 (a) Diagram of a Van de Graaff generator. Charge is transferred to the dome by means of a rotating belt. The charge is deposited on the belt at point A and transferred to the dome at point B.
(b) Photograph of a Van de Graaff generator.

*16.9 THE VAN DE GRAAFF GENERATOR

In 1929 Robert J. Van de Graaff designed and built an electrostatic generator that is used extensively in nuclear physics research. The principles of its operation can be understood with the knowledge of electric fields and charges we have developed in this chapter. The basic construction details of this device are shown in Figure 16.19a. A motor-driven pulley, P, moves a belt past positively charged comb-like metallic needles located at A. Negative charges are attracted to these needles from the belt, leaving the left side of the belt with a net positive charge. These positive charges attract electrons onto the belt as it moves past a second comb of needles at B, increasing the excess positive charge on the dome. Since the electric field inside the metal dome is negligible, the positive charge on it can easily be increased regardless of how much charge is already present. The result is that the dome is left with a large amount of positive charge. The dome will continue to accumulate charge in this fashion until electric discharge occurs through the air. The charge collected can be increased by increasing the radius of the dome and by placing the entire system in a container filled with high-pressure gas.

If protons (or other charged particles) are introduced into a tube attached to the dome, the large electric field of the dome exerts a repulsive force on the protons, causing them to accelerate to energies high enough to initiate nuclear reactions between the protons and various target nuclei.

*16.10 THE OSCILLOSCOPE

The oscilloscope is an electronic instrument widely used in making electrical measurements. The main component of the oscilloscope is the cathode ray tube (CRT), shown in Figure 16.20. This tube is commonly used to obtain a visual display of electronic information for other applications, including radar systems, television receivers, and computers. The CRT is essentially a vacuum tube in which electrons are accelerated and deflected under the influence of electric fields.

The electron beam is produced by an assembly called an *electron gun*, located in the neck of the tube. The assembly shown in Figure 16.20 consists of a heater (H), a cathode (C), and a positively charged anode (A). An electric current through the heater causes its temperature to rise, which in turn heats the cathode. The cathode reaches temperatures high enough to cause electrons to be "boiled off." Although they are not shown in the figure, the electron gun also includes an element that focuses the electron beam and one that controls the number of electrons reaching the anode (that is, a brightness control). The anode has a hole in its center that allows the electrons to pass through without striking the anode. These electrons, if left undisturbed, travel in a straight-line path until they strike the face of the CRT. The screen at the front of the tube is coated with a fluorescent material that emits visible light when bombarded with electrons. This results in a visible spot on the screen of the CRT.

The electrons are moved in various directions by two sets of deflection plates placed at right angles to each other in the neck of the tube. In order to understand how the deflection plates operate, first consider the horizontal deflection plates in Figure 16.20. External electric circuits can change the amount of charge present on these plates, with positive charge being placed on one plate and negative on the other. (In Chapter 17 we shall see that this

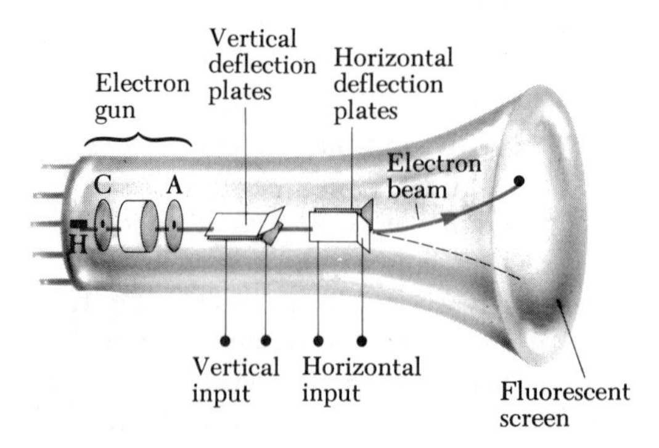

FIGURE 16.20 Diagram of a cathode ray tube. Electrons leaving the hot cathode (C) are accelerated to the anode (A). The charge plates deflect the beam.

can be accomplished by applying a voltage across the plates.) This increasing charge creates an increasing electric field between the plates, which causes the electron beam to be deflected from its straight-line path. The tube face is slightly phosphorescent and therefore glows briefly after the electron beam moves from one point to another on the screen. Slowly increasing the charge on the horizontal plates causes the electron beam to move gradually from the center toward the side of the screen. Because of the phosphorescence, however, one sees a horizontal line extending across the screen instead of the simple movement of the dot. The horizontal line can be maintained on the screen by rapid, repetitive tracing.

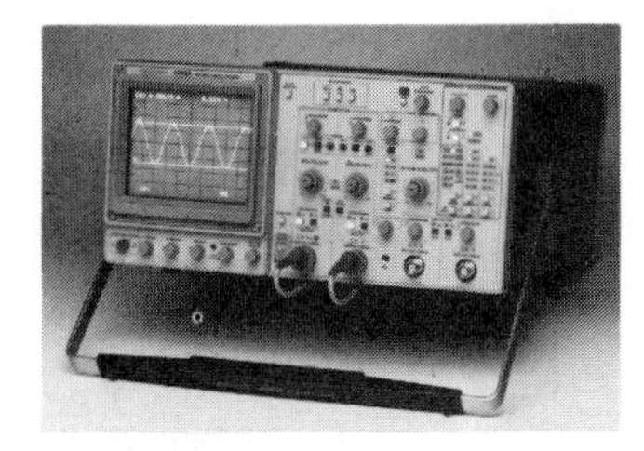

A typical laboratory oscilloscope, showing a waveform similar to that produced by a tuning fork. The horizontal lines mark the upper and lower limits of the waveform. (Courtesy of Tektronix, Inc.)

The vertical deflection plates act in exactly the same way as the horizontal plates, except that changing the charge on them causes a vertical line on the tube face. In practice, the horizontal and vertical deflection plates are used simultaneously. To see how the oscilloscope can display visual information, let us examine how we could observe the sound wave from a tuning fork on the screen. For this purpose, the charge on the horizontal plates changes in such a manner that the beam sweeps across the face of the tube at a constant rate. The tuning fork is then sounded into a microphone, which changes the sound signal to an electric signal that is applied to the vertical plates. The combined effect of the horizontal and vertical plates causes the beam to sweep the tube horizontally and up and down at the same time, with the vertical motion corresponding to the tuning fork signal. A pattern like that shown in the photograph is swept out on the screen.

SUMMARY

Electric charges have the following important properties:

1. Unlike charges attract one another, and like charges repel one another.
2. Electric charge is always conserved. That is, charge is never created or destroyed.
3. The smallest unit of negative charge is the charge on an electron, which has the value -1.60×10^{-19} C. The charge on the proton is identical in magnitude, but is positive ($+1.60 \times 10^{-19}$ C).

Conductors are materials through which charges can move freely. Some examples of good electric conductors are copper, aluminum, and silver. **Insulators** are materials that do not readily allow charge to move through them. Some examples of good electric insulators are glass, rubber, and wood.

Coulomb's law states that the electric force between two stationary charged particles separated by a distance r has a magnitude given by

$$F = k\frac{|q_1|\,|q_2|}{r^2} \tag{16.1}$$

where $|q_1|$ and $|q_2|$ are the magnitudes of the charges on the particles in coulombs and k is the **Coulomb constant,** which has the value

$$k \approx 9.0 \times 10^9 \text{ N} \cdot \text{m}^2/\text{C}^2 \tag{16.3}$$

The magnitude of the **electric field *E*** at some point in space is defined as the magnitude of the electric force that acts on a small positive charge placed at that point divided by the magnitude of its charge, $|q_0|$.

$$E \equiv \frac{|\mathbf{F}|}{|q_0|} \tag{16.4}$$

The direction of the electric field at a point in space is defined to be in the direction of the electric force that would be exerted on a small positive charge placed at that point.

The electric field due to a *point charge q*, at a distance r from the point charge, has a magnitude

$$E = k\frac{|q|}{r^2} \tag{16.5}$$

Electric field lines are useful for describing the electric field in any region of space. These lines are related to the electric field as follows: (1) the electric field is tangent to the electric field lines at every point and (2) the number of electric field lines per unit area through a surface perpendicular to the lines is proportional to the strength of the electric field in that region.

A **conductor in electrostatic equilibrium** has the following properties:

1. The electric field is zero everywhere inside the conductor.
2. Any excess charge on an isolated conductor must reside entirely on its surface.
3. The electric field just outside a charged conductor is perpendicular to the conductor's surface.
4. On an irregularly shaped conductor, charge tends to accumulate where the curvature of the surface is greatest, that is, at sharp points.

ADDITIONAL READING

W. F. Magie, *Source Book in Physics*, Cambridge, Mass., Harvard University Press, 1963. This includes extracts from the works of Coulomb and others.

H. W. Meyer, *History of Electricity and Magnetism*, Cambridge, Mass., MIT Press, 1971.

A. D. Moore, "Electrostatics," *Sci. American*, March 1972, p. 46.

A. F. Moyer, "Benjamin Franklin: Let the Experiment Be Made," *Am. J. Phys.* 44:536, 1976.

R. W. Purcell, "Bolts From the Blue," *The Sciences*, July/August 1986, p. 48.

D. Roller and D. H. D. Roller, *The Development of the Concept of Electric Charge*, Cambridge, Mass., Harvard University Press, 1954.

QUESTIONS

1. When the air is very dry, sparks are often observed (or heard) when clothes are removed in the dark. Explain.
2. A balloon is negatively charged by rubbing and then clings to a wall. Does this mean that the wall is positively charged? Why does the balloon eventually fall?
3. If a metal object receives a positive charge, does its mass increase, decrease, or stay the same? What happens to the mass if the object is given a negative charge?
4. A charged comb will often attract small bits of dry paper that fly away when they touch the comb. Explain.
5. A student stands on a piece of insulating material, places his hand on top of a Van de Graaff generator, and then turns on the generator. Will he be shocked? Why or why not? He will find that his hair

stands on end as in the photograph. Why does this occur?

Question 5.

6. Would life be different if the electron were positively charged and the proton were negatively charged? Could we identify a difference? Would any physical laws be different?
7. Assume that someone proposes a theory that says people are bound to the earth by electric forces rather than by gravity. How could you prove this theory wrong?
8. Are the occupants of a steel-frame building safer than those in a wood-frame house during an electrical storm or vice versa? Explain.
9. Why should a ground wire be connected to the metal support rod for a television antenna?
10. It has been reported that in some instances people near where a lightning bolt strikes the earth have had their clothes thrown off. Explain why this might happen.
11. A light piece of aluminum foil is draped over a wooden rod. When a rod with a positive charge is brought close to the foil, the foil leaves stand apart. Why? What kind of charge is on the foil?
12. Why is it more difficult to charge an object by friction on a humid day than on a dry day?
13. Compare and contrast Newton's law of universal gravitation with Coulomb's force law.
14. Why should you get out of a swimming pool during a lightning storm?
15. Why is it not a good idea to seek shelter under a tree during a lightning storm?
16. Suppose someone tells you that it is the rubber tires on a car that makes a car safe during a lightning storm. What would be your response? Do the relatively new steel-belted tires on a car affect the safety of an occupant in a car?

PROBLEMS

Section 16.3 Coulomb's Law

1. Calculate the net charge on a substance consisting of (a) 5×10^{14} electrons and (b) a combination 7×10^{13} protons and 4×10^{13} electrons.

2. Two point charges of magnitude 3×10^{-9} C and 6×10^{-9} C are separated by a distance of 0.3 m. Find the electric force of repulsion between them.

3. A 4.5×10^{-9} C charge is located 3.2 m from a -2.8×10^{-9} C charge. Find the electrostatic force exerted by one charge on the other.

4. Two conducting spheres are placed with a distance of 0.3 m between their centers. One is given a charge of 12×10^{-9} C and the other a charge of -18×10^{-9} C. (a) Find the electrostatic force exerted on one sphere by the other. (b) The spheres are connected by a conducting wire. After equilibrium has occurred, find the electrostatic force between the two.

5. The moon and earth are bound together by gravity. If, instead, the force of attraction was the result of each having a charge of the same magnitude, yet opposite in sign, find the quantity of charge that would have to be placed on each to produce the required force.

6. An electron is released above the surface of the earth. A second electron directly below it exerts an electrostatic force on the first just great enough to cancel the gravitational force on it. How far below the first electron is the second?

7. A 2.2×10^{-9} C charge is located on the x axis at $x = -1.5$ m, a 5.4×10^{-9} C charge is located on the x

axis at $x = 2$ m, and a 3.5×10^{-9} C charge is located at the origin. Find the net force on the 3.5×10^{-9} C charge.

8. Three charges are arranged as shown in Figure 16.21. Find the magnitude and direction of the electrostatic force on the charge at the origin.

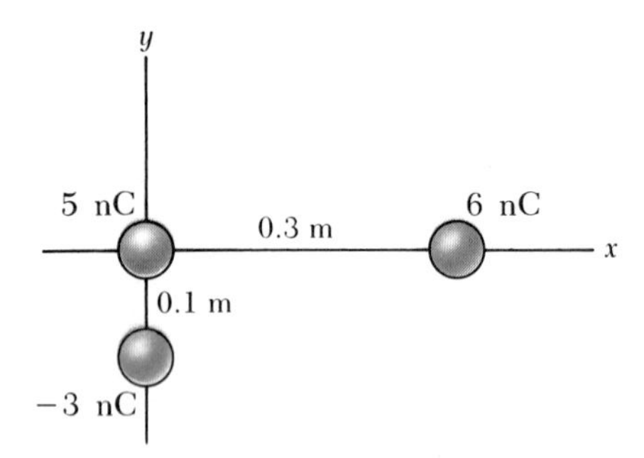

FIGURE 16.21 Problems 8 and 26

• **9.** Three charges are arranged as shown in Figure 16.22. Find the magnitude and direction of the electrostatic force on the 6 nC charge.

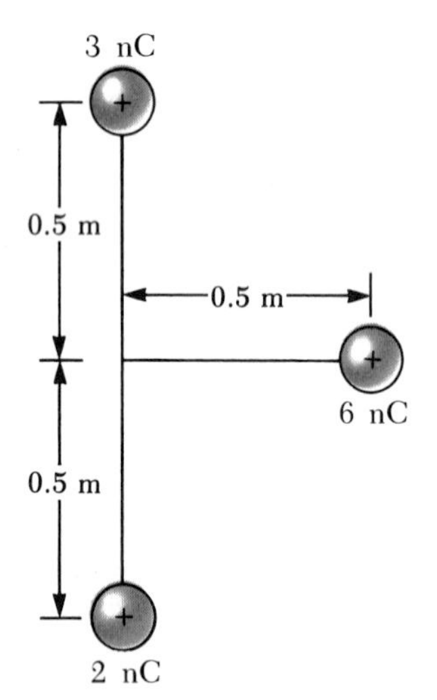

FIGURE 16.22 Problem 9

• **10.** In the Bohr theory of the hydrogen atom, an electron moves in a circular orbit about a proton, where the radius of the orbit is 0.51×10^{-10} m. (a) Find the electrostatic force between the two. (b) If this force serves as the centripetal force on the electron, what is the speed of the electron?

• **11.** Two point charges of equal magnitude repel each other with a force of 2 N when separated by 5 cm. Find the magnitude of the charge on each.

• **12.** Find (a) the total number of electrons in one gram of copper and (b) the total charge of all these electrons.

• **13.** A charge of 2×10^{-9} C is placed at the origin, and a charge of 4×10^{-9} C is placed at $x = 1.5$ m. Find the location of the point between the two charges where a charge of 3×10^{-9} C should be placed so that the net electric force on it is zero.

• **14.** Two pieces of aluminum foil each of mass 0.2 g are suspended as pendulums from a common point. Each is given the same electric charge, and it is found that the two come to equilibrium at an angle of 5° with the vertical as shown in Figure 16.5. If the length of each pendulum is 30 cm, what is the magnitude of the charge on each piece of foil?

• **15.** A charge of 6×10^{-9} C and a charge of -3×10^{-9} C are separated by a distance of 60 cm. Find the position at which a third charge of 12×10^{-9} C can be placed so that the net electrostatic force on it is zero.

•• **16.** The total charge on two spheres is 600 μC. The two charges are placed 0.9 m apart and the force of repulsion between the two is 30 N. What is the charge on each?

Section 16.5 The Electric Field

17. The electric force on a point charge of 5×10^{-9} C at some point is 3.8×10^{-3} N in the positive x direction. What is the magnitude of the electric field at this location?

18. The magnitude of the electric field at a certain location is 500 N/C and the field is directed east to west. (a) Find the magnitude and direction of the force acting on a proton placed at this point. (b) Repeat for an electron located at the point.

19. Find the magnitude and direction of the electric field at a distance of 10 cm from an electron.

20. The electric field at a distance of 0.8 m from a certain charge is found to have a magnitude of 200 N/C. What is the magnitude of the charge?

21. What is the magnitude and direction of the electric field set up by the proton at the position of the electron in the hydrogen atom? (See Problem 10.)

22. Find the electric field at a point midway between two charges of (a) $+30 \times 10^{-9}$ C and $+60 \times 10^{-9}$ C separated by 30 cm. (b) Repeat when the magnitude of the second charge is -60×10^{-9} C.

23. A piece of aluminum foil of mass 5×10^{-2} kg is suspended by a string in an electric field directed vertically upward. If the charge on the foil is 3 μC, find the strength of the field that will reduce the tension in the string to zero.

• **24.** An electron is accelerated by a constant electric field of magnitude 300 N/C. (a) Find the acceleration of the electron. (b) Use the equations of motion with constant acceleration to find its speed after 10^{-8} s, assuming it starts from rest.

• **25.** A constant electric field directed along the positive x axis has a strength of 2000 N/C. Find (a) the force exerted on the proton by the field, (b) the acceleration of the proton, and (c) the time required for the proton to reach a speed of 10^6 m/s, assuming it starts from rest.

• **26.** Find the net electric field exerted on the charge at the origin in Figure 16.21.

• **27.** Positive charges are located at three corners of a rectangle as shown in Figure 16.23. Find the electric field at the fourth corner.

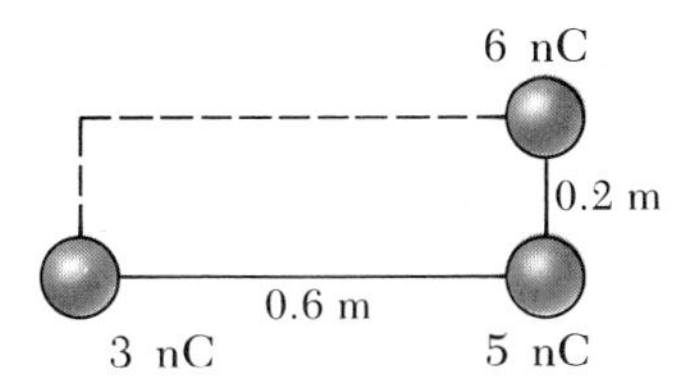

FIGURE 16.23 Problem 27

• **28.** Repeat Problem 27 if the 5 nC charge is replaced by a negative charge of the same magnitude.

• **29.** Three charges are at the corners of an equilateral triangle as shown in Figure 16.24. Calculate the electric field at a point directly below the 8 nC charge and on the x axis.

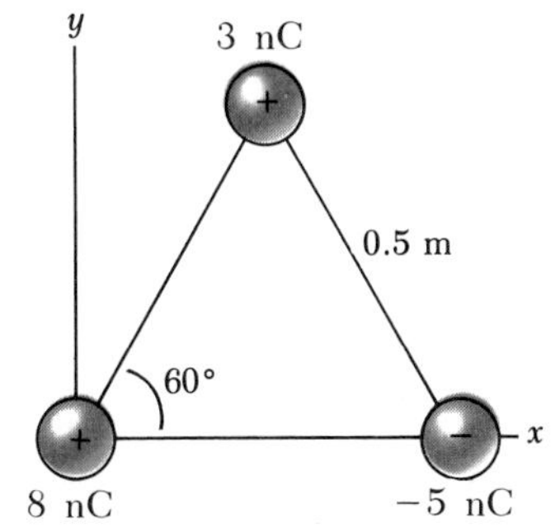

FIGURE 16.24 Problem 29

• **30.** What is the strength of the electric field that would give a proton an acceleration equal to that of gravity?

• **31.** In Figure 16.25, determine the point (other than infinity) at which the total electric field is zero.

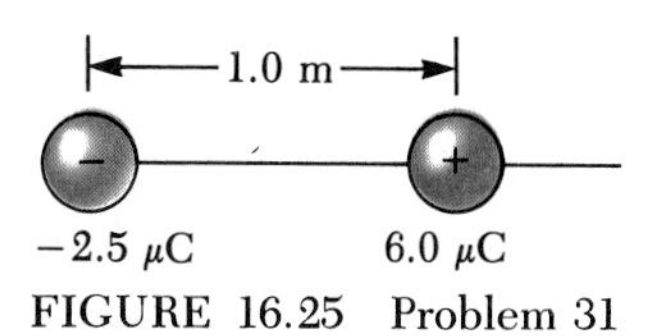

FIGURE 16.25 Problem 31

• **32.** Repeat Problem 31 for the case in which the 6 μC charge is replaced by a charge of −6 μC.

Section 16.6 Electric Field Lines

Section 16.7 Conductors in Electrostatic Equilibrium

33. (a) Sketch the electric field lines around an isolated 1 μC positive point charge. (b) Sketch the electric field pattern around an isolated negative point charge of magnitude −2 μC.

34. (a) Sketch the electric field pattern around two positive point charges of magnitude 1 μC placed close together. (b) Sketch the electric field pattern around two negative point charges of magnitude −2 μC placed close together. (c) Sketch the pattern around a 1 μC positive point charge and a −2 μC charge placed close together.

35. Figure 16.26 shows the electric field lines for two point charges separated by a small distance. (a) Determine the ratio q_1/q_2. (b) What are the signs of q_1 and q_2?

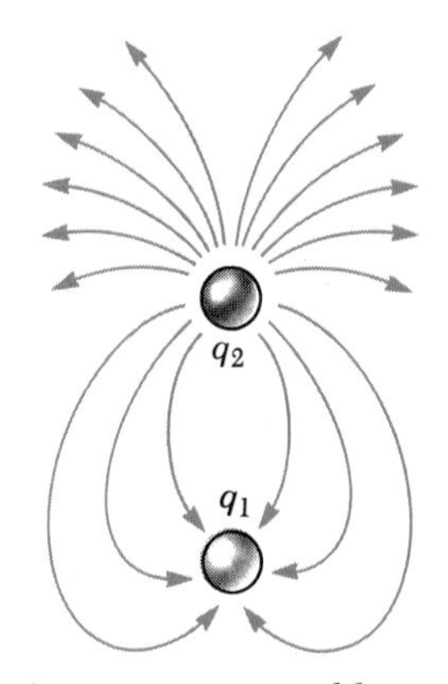

FIGURE 16.26 Problem 35

36. (a) Sketch the electric field pattern set up by a positively charged hollow sphere. Include a sketch of the lines both inside and outside the sphere. (b) A conducting cube is given a positive charge. Sketch the electric field pattern both inside and outside the cube.

37. Consider a rod of finite length having a uniform negative charge. Sketch the pattern of the electric field lines in a plane containing the rod.

38. Two point charges are a small distance apart. (a) Sketch the electric field lines for the two if one has a charge four times that of the other and both charges are positive. (b) Repeat if both charges are negative.

Section 16.8 Faraday's Ice-Pail Experiment

39. Refer to Figure 16.18. The charge lowered into the center of the hollow conductor has a magnitude of 5 μC. Find the magnitude and sign of the charge on the inside and outside of the hollow conductor when the charge is as shown in (a) Figure 16.18a, (b) Figure 16.18b, (c) Figure 16.18c, and (d) Figure 16.18d.

40. Repeat Problem 39 if the charge has a magnitude of −5 μC.

Section 16.9 The Van de Graaff Generator

41. The dome of a Van de Graaff generator receives a charge of 2×10^{-4} C. Find the strength of the electric field (a) inside the dome, (b) at the surface of the dome, assuming it has a radius of 1 m, and (c) 4 m from the center of the dome. (*Hint:* Refer to Section 16.7 to review properties of conductors in electrostatic equilibrium. Also use the fact that the points on the surface are outside a spherically symmetric charge distribution; the total charge may be considered as being located at the center of the sphere.)

42. A Van de Graaff generator is charged such that the electric field at its surface is 3×10^4 N/C. Find (a) the electric force exerted on a proton released at its surface and (b) the acceleration of the proton at this instant of time.

ADDITIONAL PROBLEMS

• **43.** A point charge $q_1 = -3.5 \times 10^{-9}$ C is located on the y axis at $y = 0.12$ m, a charge $q_2 = -1.8 \times 10^{-9}$ C is located at the origin, and a charge $q_3 = 2.6 \times 10^{-9}$ C is located on the x axis at $x = -0.12$ m. Find the resultant force on q_3.

• **44.** A proton is accelerated from rest by an electric field of 400 N/C. (a) Find its acceleration, (b) its speed after 10^{-8} s, and (c) its kinetic energy at this time.

• **45.** Three point charges are aligned along the x axis as shown in Figure 16.27. Find the electric field at (a) the position (2, 0), and (b) the position (0, 2).

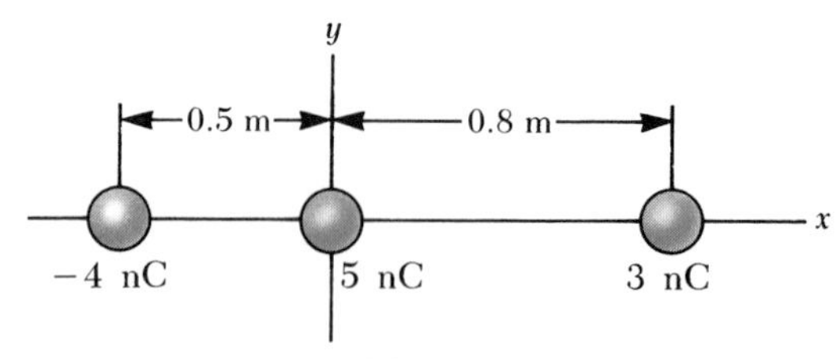

FIGURE 16.27 Problem 45

• **46.** Two charges of equal magnitude 2 nC but of opposite sign are connected together at a distance of separation of 3×10^{-8} m and placed in a uniform upward electric field of 200 N/C, as shown in Figure 16.28a. (a) Find the net torque on the charges. (b) Repeat when the charges are aligned with respect to the field as shown in Figure 16.28b.

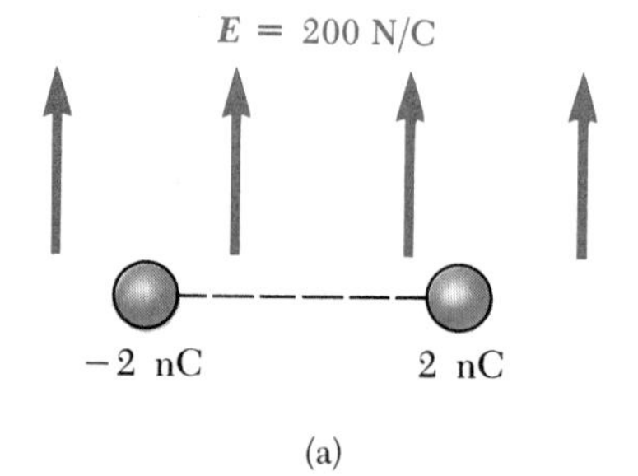

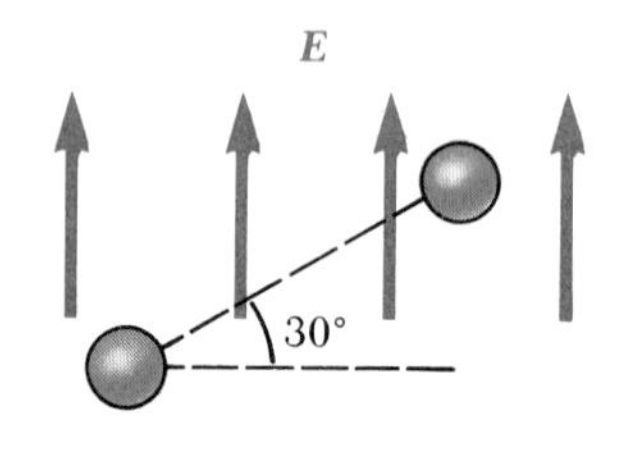

FIGURE 16.28 Problem 46

• **47.** Three point charges lie along the y axis. A charge $q_1 = -3$ μC is at $y = 5.0$ m, and a charge $q_2 = 8$ μC is at $y = 2.0$ m. Where must a positive charge q_3 be placed such that the resultant force on it is zero?

• **48.** A proton is shot vertically upward with a speed of 2×10^5 m/s in a downward-directed electric field of 500 N/C. How high will it rise?

•• **49.** Two small spheres each of mass 2 g are suspended by light strings 10 cm in length (Fig. 16.29). A uniform electric field is applied in the x direction. If the spheres have charges of -5×10^{-8} C and $+5 \times 10^{-8}$ C, determine the electric field intensity that enables the spheres to be in equilibrium at $\theta = 10°$.

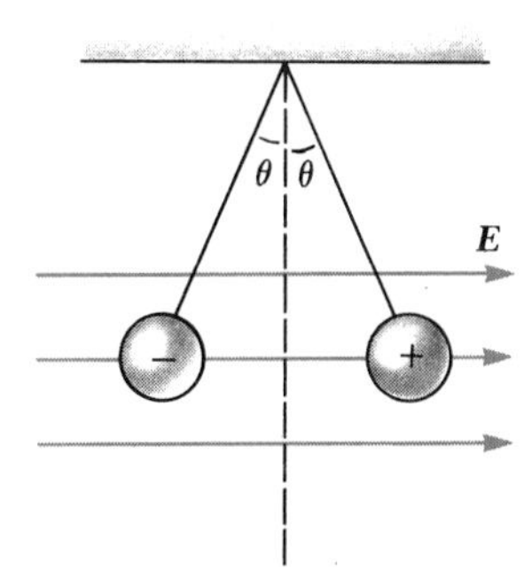

FIGURE 16.29 Problem 49

•• **50.** Four point charges are situated at the corners of a square of sides a as in Figure 16.30. Find the expression for the resultant force on the positive charge q.

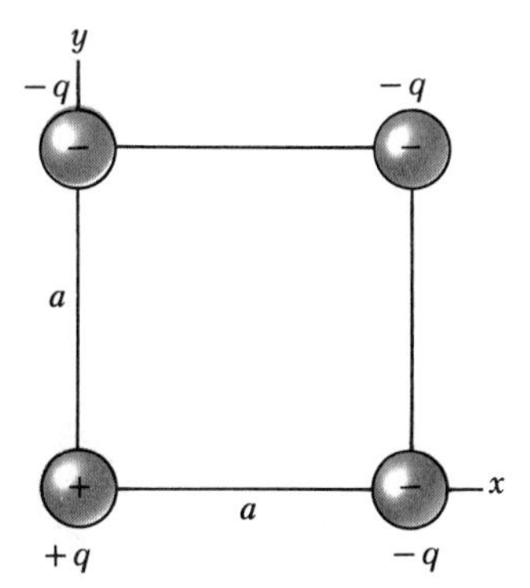

FIGURE 16.30 Problem 50

• **51.** An electron traveling with an initial speed equal to 4.0×10^6 m/s enters a region of a uniform electric field of magnitude 2.5×10^4 N/C. The direction of travel of the electron is in the same direction as the field. (a) Find the acceleration of the electron. (b) Determine the time it takes for the electron to come to rest after it enters the field. (c) How far does the electron move in the electric field before coming to rest?

• **52.** Two point charges like those shown in Figure 16.31 are called an electric dipole. Show that the electric field at a distant point along the x axis is given by the expression $E_x = 4kqa/x^3$.

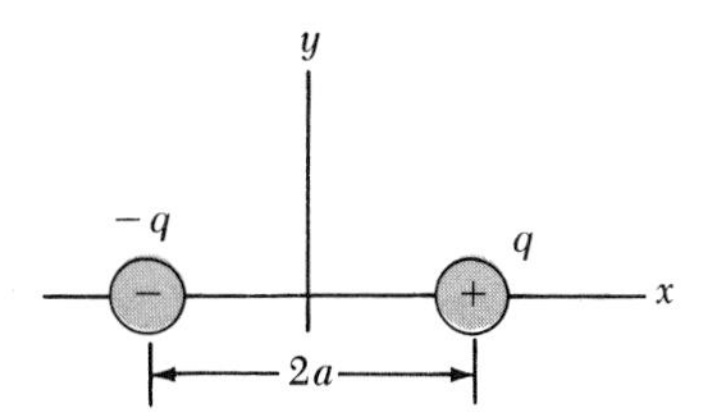

FIGURE 16.31 Problem 52

•• **53.** A charged cork ball of mass 1 g is suspended on a light string in the presence of a uniform electric field as in Figure 16.32. The electric field has components, $E_x = 3 \times 10^5$ N/C, and $E_y = 5 \times 10^5$ N/C. The ball is in equilibrium at $\theta = 37°$. Find (a) the charge on the ball, and (b) the tension in the string.

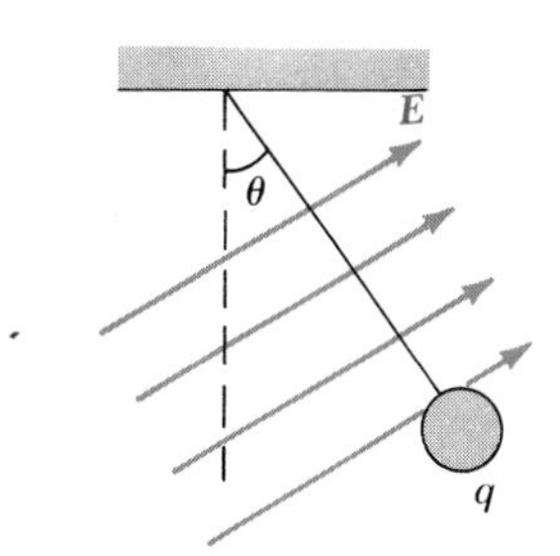

FIGURE 16.32 Problem 53

•• **54.** Two equal positive charges q are located on the x axis at $x = a$ and $x = -a$. Show that the field along the y axis is in the y direction and is given by the relation $E_y = 2kqy(y^2 + a^2)^{-3/2}$.

CHAPTER 17

Electrical Energy and Capacitance

Part of a large power distribution center.

The concept of potential energy was first introduced in Chapter 5. A potential energy function can be defined for all conservative forces, such as the force of gravity. By using the law of conservation of energy, we were often able to avoid working directly with forces when solving various problems. In this chapter we shall see that the energy concept is also useful in the study of electricity. Because the Coulomb force is conservative, we shall be able to define an electrical potential energy corresponding to the Coulomb force.

This concept of potential energy will be of value to us, but perhaps of even more value will be a quantity called electric potential, defined as the potential energy per unit charge.

Electric potential is of great practical value in dealing with electric circuits. For example, when one speaks of a voltage applied between two points, one is actually referring to an electric potential difference between those points. We shall take our first steps toward an investigation of circuits with our discussion of electric potential, and we shall carry this forward with an investigation of a common circuit element called a capacitor.

17.1 POTENTIAL DIFFERENCE AND ELECTRIC POTENTIAL

In Chapter 5 we showed that the gravitational force is a conservative force. As you may recall, this means that the work this force does on an object depends only on the initial and final positions of the object and not on the path followed between the two points. Furthermore, since the gravitational force is conservative, it is possible to define a potential energy function, which we call gravitational potential energy. Because the Coulomb force is of the same form as the universal law of gravity, it follows that *the electrostatic force is also conservative*. Therefore, it is possible to define an electrical potential energy function associated with this force.

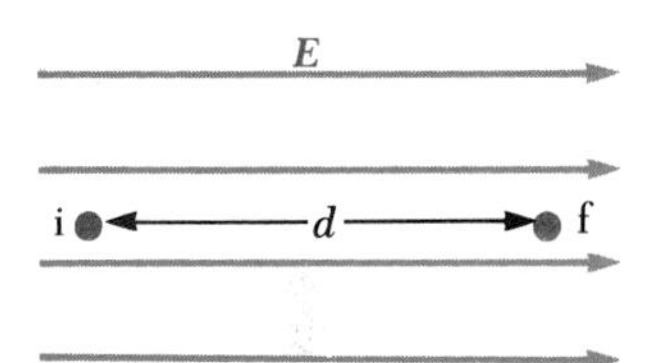

FIGURE 17.1 When a charge q moves in a uniform electric field $\boldsymbol{E}$ from point i to point f, the work done on the charge by the electric force is qEd.

We shall consider potential energy from the point of view of the particular situation shown in Figure 17.1. Imagine a small positive charge placed at point i in a uniform electric field of magnitude E. As the charge moves from point i to point f under the influence of the electric force, qE, exerted on it, the work done on the charge by the electric force is

$$W = Fd = qEd$$

where d is the distance between i and f.

By definition, *the work done by a conservative force equals the negative of the change in potential energy*, ΔPE. Therefore we find the change in electrical potential energy to be

$$\boxed{\Delta PE = -W = -qEd} \qquad (17.1)$$

Change in potential energy

Note that although potential energy can be defined for any electric field, Equation 17.1 is valid only for the case of a *uniform* electric field. We shall examine situations in which the electric field is not uniform in subsequent sections.

We shall have occasion to use electrical potential energy often in the coming pages, but of even more practical importance in the study of electricity is the concept of electric potential.

The potential difference between points i and f, $V_f - V_i$, is defined as the change in potential energy (final value minus initial value) of a charge q moved from i to f, divided by the charge.

$$\boxed{\Delta V \equiv V_f - V_i = \frac{\Delta PE}{q}} \qquad (17.2)$$

Potential difference

Potential difference should not be confused with potential energy. The change in potential between two points is proportional to the change in potential energy between the points, and we see from Equation 17.2 that the two are related as $\Delta PE = q\Delta V$. Because potential energy is a scalar quantity, *electric potential is also a scalar quantity.* From Equation 17.2 we see that potential difference is a measure of energy per unit charge. Thus, the SI units of potential are joules per coulomb, defined to be equal to a unit called the volt (V)[1]:

$$1 \text{ V} \equiv 1 \text{ J/C} \tag{17.3}$$

This says that 1 J of work must be done to move a 1-C charge between two points that are at a potential difference of 1 V. In the process, the 1-C charge gains (or loses) 1 J of energy as it moves through a potential difference of 1 V. Dividing Equation 17.1 by q gives

$$\frac{\Delta PE}{q} = V_f - V_i = -Ed \tag{17.4}$$

This equation shows that potential difference also has units of electric field times distance. From this, it follows that the SI unit of electric field (N/C) can also be expressed as volts per meter:

$$1 \text{ N/C} = 1 \text{ V/m}$$

Let us examine the changes in energy associated with movements in the electric field pictured in Figure 17.2. Because the positive charge q tends to move in the direction of the electric field, we would have to apply an upward external force on the charge to move it from i to f. Since work is done on the charge, this means that *a positive charge will gain electrical potential energy when it is moved in a direction opposite the electric field.* This is analogous to a mass gaining gravitational potential energy when it rises to higher elevations in the presence of gravity. If a positive charge is released from rest at point i, it experiences a force qE in the direction of the field (downward in Figure 17.2). Therefore, it accelerates downward, gaining kinetic energy. *As it gains kinetic energy, it loses an equal amount of potential energy.*

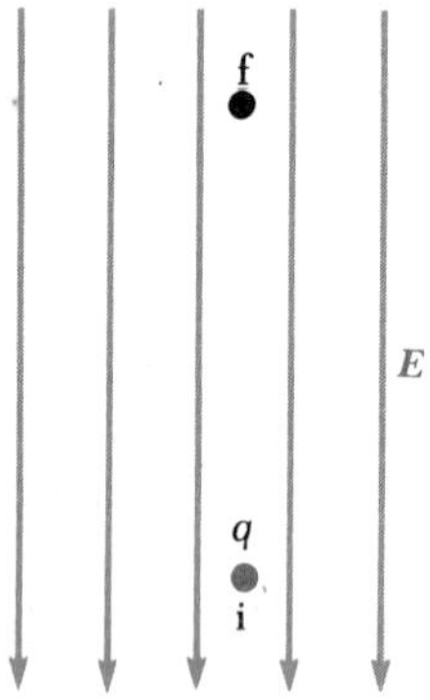

On the other hand, if the test charge q is negative, the situation is reversed. *A negative charge loses electrical potential energy when it moves in the direction opposite the electric field.* That is, a negative charge released from rest in the field $\boldsymbol{E}$ accelerates in a direction opposite the field.

f

E

q

i

FIGURE 17.2 If we wish to move a positive charge q from i to f, against the external electric field $\boldsymbol{E}$, we have to apply an upward force on the charge. The electric potential energy of the positive charge increases as it is moved in the direction opposite the electric field.

Let us pause briefly to discuss a situation that illustrates the concept of potential difference. Consider the common 12 V automobile battery. Such a battery maintains a potential difference across its terminals, where the positive terminal is 12 V higher in potential than the negative terminal. In practice, the negative terminal is usually connected to the metal body of the car, which can be considered at a potential of zero volts. The battery becomes a useful device when connected by conducting wires to such things as light bulbs, a radio, powered windows, and so forth. Now consider a charge of +1 C which is to be moved around a circuit which contains the battery connected

[1]Note that the symbol V (italic letter) represents potential, while V (Roman letter) is the symbol for the units of this quantity (volts). The two symbols should not be confused with each other.

to some external devices. As the charge is moved inside the battery from the negative terminal (at 0 V) to the positive terminal (at 12 V), the work done on the charge by the battery is 12 J. Thus, every coulomb of positive charge that leaves the positive terminal of the battery carries an energy of 12 J. As this charge moves through the external circuit toward the negative terminal, it gives up its 12 J of electrical energy to the external devices. When the charge reaches the negative terminal, its electrical energy is zero. At this point, the battery takes over and again replenishes 12 J of energy to the charge as it is moved from the negative to the positive terminal, enabling it to make another transit of the circuit. The actual amount of charge that leaves the battery depends on the properties of the external devices as we shall see in the next chapter.

EXAMPLE 17.1 The Field Between Two Parallel Plates of Opposite Charge

Figure 17.3 illustrates a situation in which a constant electric field can be set up. In this situation, a 12-V battery is connected between two parallel metal plates separated by 0.3 cm. Find the strength of the electric field.

Solution The electric field is uniform (except near the edges of the metal plates), and thus the relationship between potential difference and the magnitude of the field is given by Equation 17.4:

$$V_f - V_i = -Ed$$

As noted above, chemical forces inside a battery maintain one electrode, called the positive terminal, at a higher potential than a second electrode, the negative terminal. Thus, in Figure 17.3, plate f, which is connected to the negative terminal, must be at a lower potential than plate i, which is connected to the positive terminal. That is,

$$V_f - V_i = -12 \text{ V}$$

This gives a value of E of

$$E = -\frac{(V_f - V_i)}{d} = -\frac{(-12 \text{ V})}{0.3 \times 10^{-2} \text{ m}} = 4 \times 10^3 \text{ V/m}$$

The direction of this field is from the positive plate to the negative plate. A device consisting of two plates separated by a small distance is called a *parallel-plate capacitor*. We shall examine these devices later in this chapter.

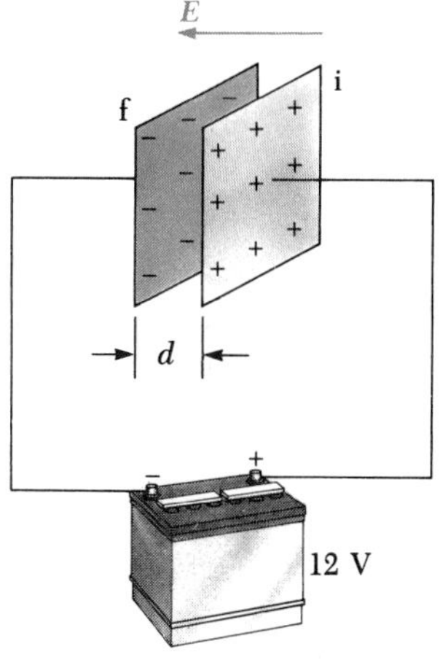

FIGURE 17.3 (Example 17.1) A 12-V battery connected to two parallel plates. The electric field $\boldsymbol{E}$ between the plates has a magnitude given by the potential difference divided by the plate separation, d.

EXAMPLE 17.2 Motion of a Proton in a Uniform Electric Field

A proton is released from *rest* in a uniform electric field of magnitude 8×10^4 V/m directed along the positive x axis (Fig. 17.4). The proton undergoes a displacement of 0.5 m in the direction of the field. (a) Find the *change* in the electric potential of the proton as a result of this displacement.

Solution From Equation 17.4, we have

$$\Delta V = V_f - V_i = -Ed = -(8 \times 10^4 \text{ V/m})(0.5 \text{ m}) = -4 \times 10^4 \text{ V}$$

Thus, the electric potential of the proton *decreases* as it moves from i to f.

(b) Find the change in potential energy of the proton for this displacement.

Solution

$$\Delta PE = q\Delta V = e\Delta V = (1.6 \times 10^{-19} \text{ C})(-4 \times 10^4 \text{ V}) = -6.4 \times 10^{-15} \text{ J}$$

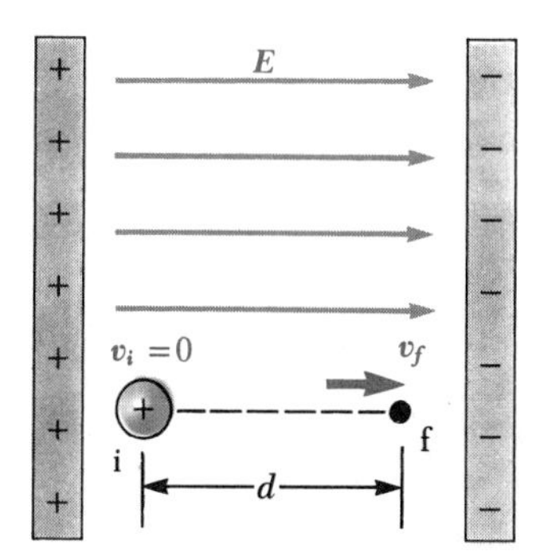

FIGURE 17.4 (Example 17.2) A proton accelerates from i to f in the direction of the uniform electric field.

The negative sign here means that the potential energy of the proton decreases as it moves in the direction of the electric field. This makes sense since, as the proton *accelerates* in the direction of the field, it gains kinetic energy and at the same time loses potential energy (mechanical energy is conserved).

(c) Find the speed of the proton after it has been moved 0.5 m, starting from rest.

Solution If there are no forces acting on the proton other than the conservative electrical force, we can apply the principle of conservation of mechanical energy in the form

$$KE_i + PE_i = KE_f + PE_f$$

In our case, $KE_i = 0$, hence the expression above gives

$$KE_f = PE_i - PE_f = -\Delta PE$$

With this equation and the results of part (b), we find

$$\frac{1}{2}mv_f^2 = 6.4 \times 10^{-15}\ \text{J}$$

and

$$v_f^2 = \frac{2(6.4 \times 10^{-15}\ \text{J})}{1.67 \times 10^{-27}\ \text{kg}} = 7.66 \times 10^{12}\ \text{m}^2/\text{s}^2$$

$$v_f = 2.77 \times 10^6\ \text{m/s}$$

17.2 ELECTRIC POTENTIAL AND POTENTIAL ENERGY DUE TO POINT CHARGES

In electric circuits, a point of zero potential is often defined by grounding (connecting to earth) some point in the circuit. For example, if the negative plate in Example 17.1 had been grounded, its potential would be considered to have a value of zero and the positive plate to have a value of 12 V. It is also possible to define the potential at a point in space due to a point charge. In this case, the point of zero potential is taken to be at an infinite distance from the charge. With this choice, the methods of calculus can be used to show that *the electric potential due to a point charge q at any distance r from the charge is given by*

Electric potential due to a point charge

$$V = k\frac{q}{r} \tag{17.5}$$

The electric potential of two or more charges is obtained by applying the **superposition principle.** That is, *the total potential at some point P due to several point charges is the algebraic sum of the potentials due to the individual charges.* This is similar to the method used in Chapter 16 to find the resultant electric field at a point in space. However, note that in the case of potentials, one must evaluate an *algebraic sum* of individual potentials to obtain the total potential since *potentials are scalar quantities.* Thus, it is much easier to evaluate the electric potential at some point due to several charges than to evaluate the electric field, which is a vector quantity.

We now consider the potential energy of interaction of a system of two charged particles. If V_1 is the electric potential due to charge q_1 at a point P, then the work required to bring a charge q_2 from infinity to the point P without acceleration is q_2V_1. By definition, this work equals the potential energy, PE, of the two-particle system when the particles are separated by a distance r_{12} (Fig. 17.5)

Therefore, we can express the potential energy of the pair of charges as

$$PE = q_2V_1 = k\frac{q_1q_2}{r_{12}} \tag{17.6}$$

Potential energy of a pair of charges

Note that if the charges are of the same sign, PE is positive. This is consistent with the fact that like charges repel, and so positive work must be done on the system to bring two charges near one another. Conversely, if the charges are of opposite sign, the force is attractive and PE is negative. This means that negative work must be done to bring unlike charges close together.

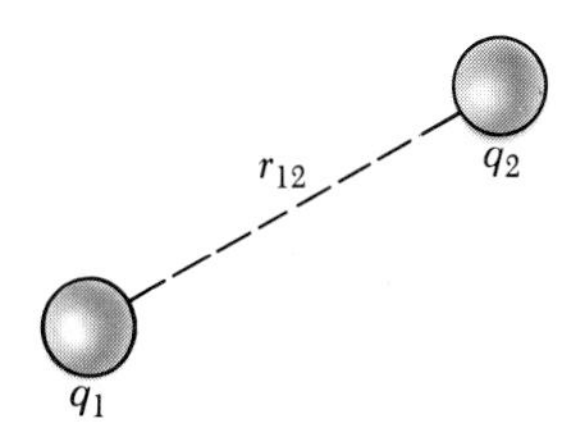

FIGURE 17.5 If two point charges q_1 and q_2 are separated by a distance r_{12}, the potential energy of the pair of charges is kq_1q_2/r_{12}.

EXAMPLE 17.3

A 5-μC point charge is located at the origin, and a point charge of −2 μC is located on the x axis at the position (3, 0) m, as in Figure 17.6. (a) If the potential is taken to be zero at infinity, find the total electric potential due to these charges at point P, whose coordinates are (0, 4) m.

Solution The total potential at P due to the two charges is

$$V_P = k\left(\frac{q_1}{r_1} + \frac{q_2}{r_2}\right)$$

From the data given, we find

$$V_P = \left(9 \times 10^9 \frac{\text{N}\cdot\text{m}^2}{\text{C}^2}\right)\left(\frac{5 \times 10^{-6}\text{ C}}{4\text{ m}} - \frac{2 \times 10^{-6}\text{ C}}{5\text{ m}}\right) = 7.65 \times 10^3\text{ V}$$

(b) How much work is required to bring a third point charge of 4 μC from infinity to P?

Solution

$$W = q_3V_P = (4 \times 10^{-6}\text{ C})(7.65 \times 10^3\text{ V})$$

Since 1 V = 1 J/C, W reduces to

$$W = 3.06 \times 10^{-2}\text{ J}$$

EXERCISE Find the magnitude and direction of the electric field at the point P.

Answer 2.28×10^3 N/C at an angle of 79.1° with the x axis.

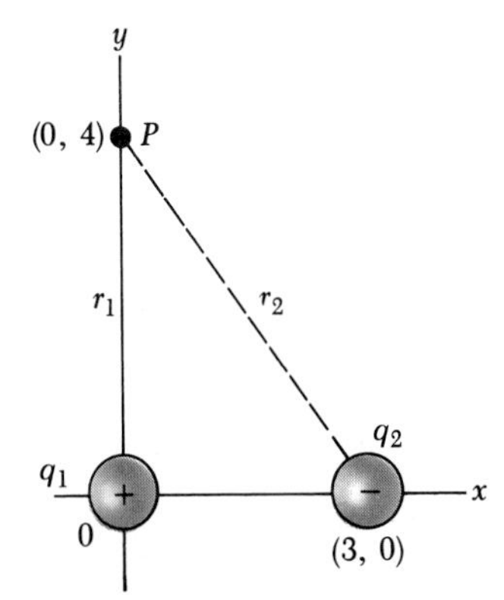

FIGURE 17.6 (Example 17.3) The electric potential at point P due to the point charges q_1 and q_2 is the algebraic sum of the potentials due to the individual charges.

17.3 POTENTIALS AND CHARGED CONDUCTORS

In order to examine the potential at all points on a charged conductor, let us combine Equations 17.1 and 17.2. From Equation 17.1, we see that the work done on a charge by electric forces is related to the change in electrical potential energy of the charge as

$$W = -\Delta PE$$

Furthermore, from Equation 17.2 we see that the change in potential energy between two points i and f is related to the potential difference between these points as

$$\Delta PE = q(V_f - V_i)$$

Combining these two equations, we find

Work

$$\boxed{W = -q(V_f - V_i)} \tag{17.7}$$

As we see from this result, no work is required to move a charge between two points that are at the same potential. That is, $W = 0$ when $V_f = V_i$.

In Chapter 16 we found that, when a conductor is in electrostatic equilibrium, a net charge placed on it resides entirely on its surface. Furthermore, we showed that the electric field just outside the surface of a charged conductor in electrostatic equilibrium is perpendicular to the surface and that the field inside the conductor is zero. We shall now show that *every point on the surface of a charged conductor in electrostatic equilibrium is at the same potential.*

Consider a surface path connecting any two points i and f on a charged conductor, as in Figure 17.7. The electric field, **E**, is always perpendicular to the displacement along this path; therefore no work is done if a charge is moved between these points. From Equation 17.7, we see that if the work done is zero, the difference in potential, $V_f - V_i$, is also zero. Therefore,

the electric potential is a constant everywhere on the surface of a charged conductor in equilibrium.

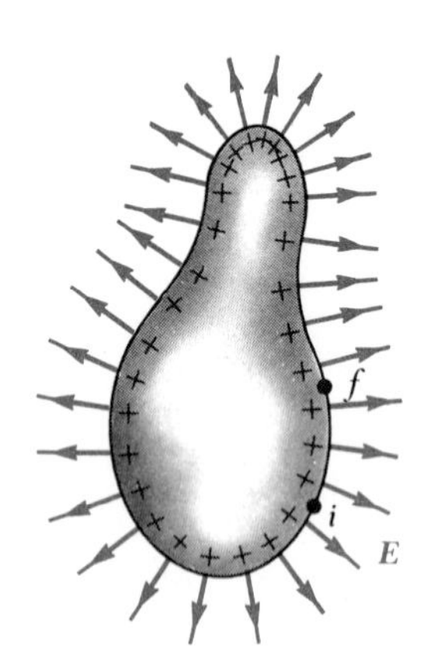

FIGURE 17.7 An arbitrarily shaped conductor with an excess positive charge. When the conductor is in electrostatic equilibrium, all of the charge resides at the surface, **E** = 0 inside the conductor, and the electric field just outside the conductor is perpendicular to the surface. The potential is constant inside the conductor and is equal to the potential at the surface.

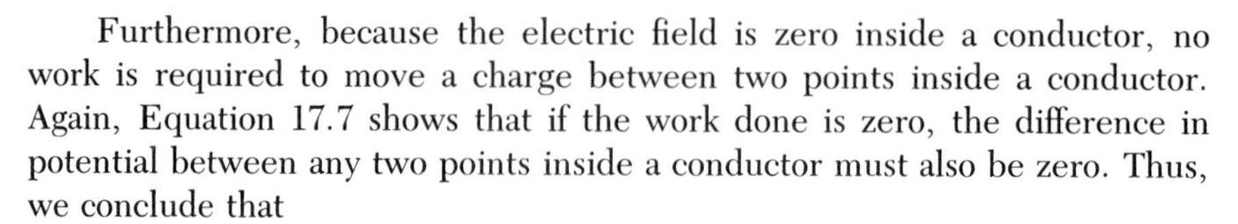

Furthermore, because the electric field is zero inside a conductor, no work is required to move a charge between two points inside a conductor. Again, Equation 17.7 shows that if the work done is zero, the difference in potential between any two points inside a conductor must also be zero. Thus, we conclude that

the electric potential is constant everywhere inside a conductor and equal to its value at the surface.

Consequently, no work is required to move a charge from the interior of a charged conductor to its surface. (Note that the potential is not zero inside a conductor even though the electric field is zero in its interior.)

The Electron Volt

A unit of energy commonly used in atomic and nuclear physics is the **electron volt** (eV).

The electron volt is defined as the energy that an electron (or proton) gains when accelerated through a potential difference of 1 V.

Since 1 V = 1 J/C and since the magnitude of charge on the electron or proton is 1.6×10^{-19} C, we see that the electron volt is related to the joule through the relation

$$1 \text{ eV} = 1.6 \times 10^{-19} \text{ C} \cdot \text{V} = 1.6 \times 10^{-19} \text{ J} \tag{17.8}$$

EXAMPLE 17.4 An Electron in a Television Tube

An electron in the beam of a typical television picture tube is accelerated through a potential difference of 20,000 V before it strikes the face of the tube. What is the energy of this electron in electron volts, and what is its speed when it strikes the screen?

Solution The initial potential energy of the electron in joules is

$$PE_i = qV = (1.60 \times 10^{-19} \text{ C})(20 \times 10^3 \text{ V}) = 3.20 \times 10^{-15} \text{ J}$$

Converting this to electron volts, we have

$$PE_i = (3.20 \times 10^{-15} \text{ J})\left(\frac{1 \text{ eV}}{1.6 \times 10^{-19} \text{ J}}\right) = 20{,}000 \text{ eV}$$

An inspection of the definition of the electron volt shows that this answer could have been obtained without any calculations.

The speed of the electron is found from the law of conservation of energy by equating the initial electrical potential energy of 3.20×10^{-15} J to the final kinetic energy, where the mass of the electron is 9.11×10^{-31} kg:

$$PE_i = KE_f$$

$$3.20 \times 10^{-15} \text{ J} = \frac{1}{2}(9.11 \times 10^{-31})v^2$$

$$v = 8.38 \times 10^7 \text{ m/s}$$

This is about 28% of the speed of light, 3×10^8 m/s.

---

*17.4 APPLICATIONS OF ELECTROSTATICS

The principles of electrostatics are used in various applications. Before we move to a new topic in electricity, let us pause to discuss a few of them briefly. One application, the Van de Graaff generator, was discussed in Chapter 16. We shall now examine electrostatic precipitators, which are used to reduce the level of atmospheric pollution, and the xerography process, which has revolutionized imaging process technology.

The Electrostatic Precipitator

One important application of electric discharge in gases is a device called an *electrostatic precipitator*. This device is used to remove particulate matter from combustion gases, thereby reducing air pollution. It is especially useful in coal-burning power plants and in industrial operations that generate large quantities of smoke. Current systems are able to eliminate approximately 90% (by mass) of the ash and dust from the smoke. Unfortunately, a very high

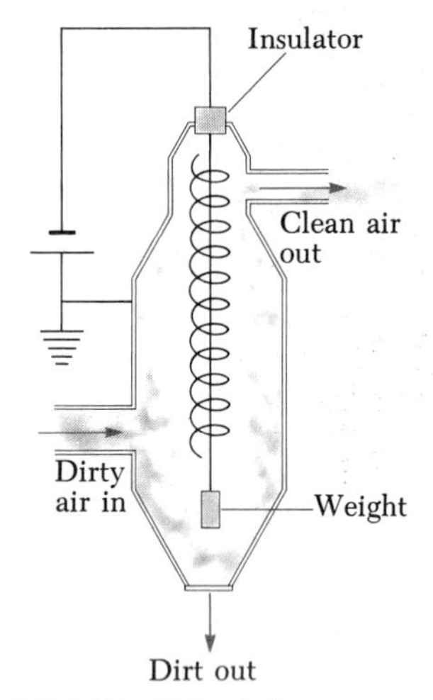

FIGURE 17.8 Schematic diagram of an electrostatic precipitator. The high voltage maintained on the central wire creates an electric discharge in the vicinity of the wire.

percentage of the lighter particles still escape, and these contribute significantly to smog, haze, and so forth.

Figure 17.8 shows the basic idea of the electrostatic precipitator. A high voltage (typically 40 kV to 100 kV) is maintained between a wire running down the center of a duct and the outer wall, which is grounded. The wire is maintained at a negative potential with respect to the wall, and so the electric field is directed towards the wire. The electric field near the wire reaches high enough values to cause a discharge around the wire and the formation of positive ions, electrons, and negative ions, such as O_2^-. As the electrons and negative ions are accelerated toward the outer wall by the nonuniform electric field, the dirt particles in the streaming gas become charged by collisions and ion capture. Since most of the charged dirt particles are negative, they are also drawn to the outer wall by the electric field. When the duct is shaken, the particles fall loose and are collected at the bottom.

In addition to reducing the amount of harmful gases and particulate matter in the atmosphere, the electrostatic precipitator also recovers valuable materials from the stack in the form of metal oxides.

Xerography

The process of xerography is widely used for making photocopies of letters, documents, and other printed materials. The basic idea for the process was developed by Chester Carlson, who was granted a patent for his invention in 1940. In 1947, the Xerox Corporation launched a full-scale program to develop automated duplicating machines using this process. The huge success of this development is quite evident; today, practically all offices and libraries have one or more duplicating machines, and the capabilities of modern machines are on the increase.

Some features of the xerographic process involve simple concepts from electrostatics and optics. However, the one idea that makes the process unique is the use of photoconductive material to form an image. (A photoconductor is a material that is a poor conductor of electricity in the dark but becomes a reasonably good electric conductor when exposed to light.)

The sequence of steps used in the xerographic process is illustrated in Figure 17.9. First, the surface of a plate or drum is coated with a thin film of

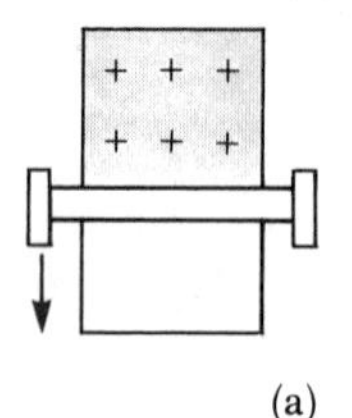
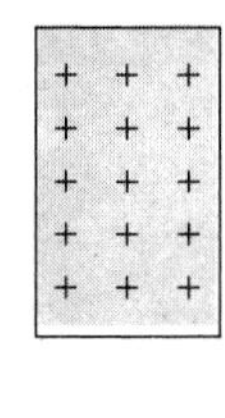
(a)

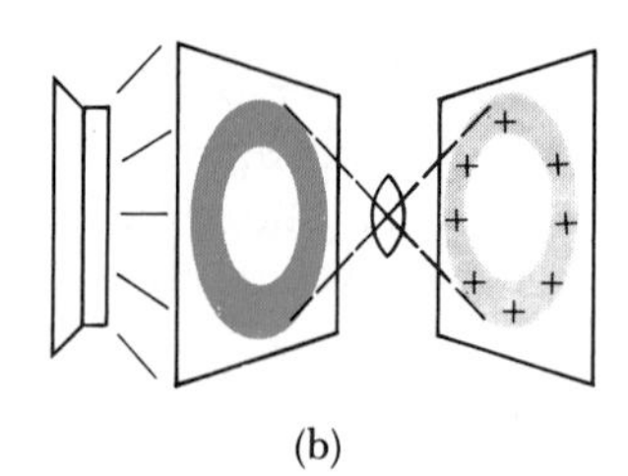
(b)

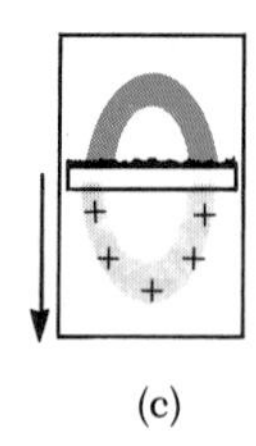
(c)

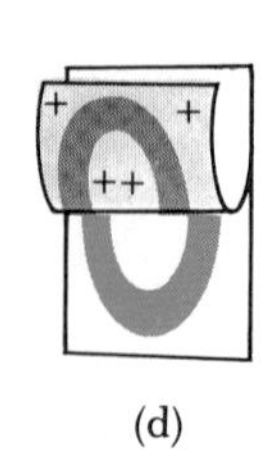
(d)

FIGURE 17.9 The xerographic process. (a) The photoconductive surface is positively charged. (b) Through the use of a light source and lens, a hidden image is formed on the charged surface in the form of positive charges. (c) The surface containing the image is covered with a negatively charged powder, which adheres only to the image area. (d) A piece of paper is placed over the surface and given a charge. This transfers the image to the paper, which is then heated to "fix" the powder to the paper.

the photoconductive material (usually selenium or some compound of selenium) and the photoconductive surface is given a positive electrostatic charge in the dark (see Fig. 17.9a). The page to be copied is then projected onto the charged surface (see Fig. 17.9b). The photoconducting surface becomes conducting only in areas where light strikes. In these areas, the light produces charge carriers in the photoconductor, which neutralize the positively charged surface. The charges remain on those areas of the photoconductor not exposed to light, however, leaving a hidden image of the object in the form of a positive surface charge distribution.

Next, a negatively charged powder called a *toner* is dusted onto the photoconducting surface (see Fig. 17.9c). The charged powder adheres only to those areas of the surface that contain the positively charged image. At this point, the image becomes visible. The image is then transferred to the surface of a sheet of positively charged paper.

Finally, the toner material is "fixed" to the surface of the paper through the application of heat (see Fig. 17.9d). This results in a permanent copy of the original.

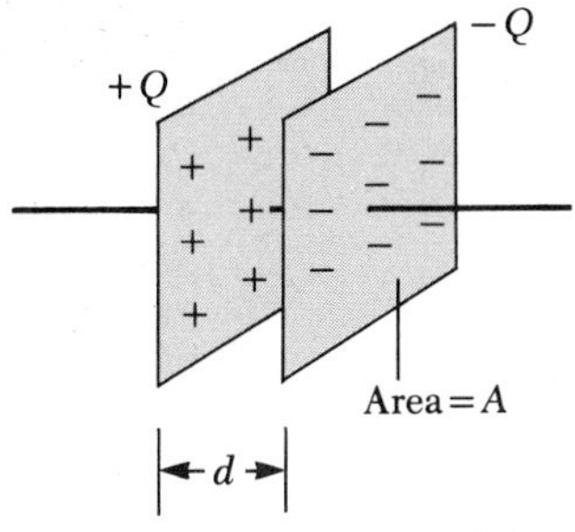

FIGURE 17.10 A parallel-plate capacitor consists of two parallel plates, each of area A, separated by a distance d. The plates carry equal and opposite charge.

17.5 THE DEFINITION OF CAPACITANCE

A **capacitor** is a device used in a variety of electric circuits. For example, capacitors are used (1) to tune the frequency of radio receivers, (2) to eliminate sparking in automobile ignition systems, and (3) as short-term energy-storing devices in electronic flash units. A typical design for a capacitor is shown in Figure 17.10. It consists of two parallel metal plates separated by a distance d. The plates are given equal and opposite charges. This is usually accomplished by connecting the two plates to the opposite terminals of a battery or some other voltage source.

The capacitance, C, of a capacitor is defined as the ratio of the magnitude of the charge on either conductor to the magnitude of the potential difference between them:

$$C \equiv \frac{Q}{V} \qquad (17.9)$$

Note, that, by definition, capacitance is always a positive quantity. From this equation, we see that capacitance has SI units of coulombs per volt. The SI unit of capacitance is called the **farad** (F), in honor of Michael Faraday. That is

$$1 \text{ F} \equiv 1 \text{ C/V}$$

The farad is a very large unit of capacitance. In practice, typical devices have capacitances ranging from microfarads ($1\ \mu\text{F} = 10^{-6}$ F) to picofarads ($1 \text{ pF} = 10^{-12}$ F).

The large capacitor on the left can store large amounts of electric charge. The variable capacitor on the right could be used in the tuning circuit of a radio and has a small capacitance. (James Madison University)

EXAMPLE 17.5 The Charge on the Plates of a Capacitor
A 3-μF capacitor is connected to a 12-V battery. How much charge is put on each plate by the battery?

Solution The definition of capacitance, Equation 17.9, gives us

$$Q = CV = (3 \times 10^{-6}\ \text{F})(12\ \text{V}) = 36\ \mu\text{C}$$

17.6 THE PARALLEL-PLATE CAPACITOR

The capacitance of a device depends on the physical characteristics and geometric arrangement of the conductors. For example, the capacitance of a parallel-plate capacitor whose plates are separated by air (Fig. 17.10) is

Capacitance of a parallel-plate capacitor

$$C = \epsilon_0 \frac{A}{d} \tag{17.10}$$

where A is the area of one of the plates, d is the distance of separation of the plates, and ϵ_0 (Greek letter epsilon) is a constant called the **permittivity of free space,** which has the value

$$\epsilon_0 = 8.85 \times 10^{-12}\ \text{C}^2/\text{N} \cdot \text{m}^2$$

The permittivity of free space is related to the Coulomb constant, k, as

$$k = \frac{1}{4\pi\epsilon_0}$$

Although we shall not derive Equation 17.10, we shall attempt to make it seem plausible. As you can see from the definition of capacitance, $C \equiv Q/V$, the amount of charge a given capacitor is able to store for a given potential difference across its plates increases as the capacitance increases. Therefore, it seems reasonable that a capacitor constructed from plates having a large area should be able to store a large charge. Furthermore, if the oppositely charged plates are close together, the attractive force between them will be large. Therefore, for a given potential difference, the charge on the plates increases with decreasing plate separation.

EXAMPLE 17.6 Calculating C for a Parallel-Plate Capacitor
A parallel-plate capacitor has an area of $A = 2\ \text{cm}^2 = 2 \times 10^{-4}\ \text{m}^2$ and a plate separation of $d = 1\ \text{mm} = 10^{-3}\ \text{m}$. Find its capacitance.

Solution From $C = \epsilon_0 A/d$ we find

$$C = \epsilon_0 \frac{A}{d} = (8.85 \times 10^{-12}\ \text{C}^2/\text{N} \cdot \text{m}^2)\left(\frac{2 \times 10^{-4}\ \text{m}^2}{1 \times 10^{-3}\ \text{m}}\right)$$

$$= 1.77 \times 10^{-10}\ \text{F} = 177\ \text{pF}$$

Exercise Show that 1 $\text{C}^2/\text{N} \cdot \text{m}$ equals 1 F.

17.7 COMBINATIONS OF CAPACITORS

Two or more capacitors are often combined in circuits in several ways. The equivalent capacitance of certain combinations can be calculated using methods described in this section. The symbol that is commonly used to represent a capacitor in a circuit is —| |—, or sometimes —) |—. The circuit symbol —▌|— is used to designate a battery (or any other direct current source). The positive terminal of the battery is at the higher potential and is represented by the longer vertical line in the battery symbol.

Parallel Combination

Two capacitors connected as shown in Figure 17.11a are known as a *parallel combination* of capacitors. The left plate of each capacitor is connected by a conducting wire to the positive terminal of the battery and the left plates are therefore at the same potential. Likewise, the right plates are connected to the negative terminal of the battery. When the capacitors are first connected in the circuit, electrons are transferred from the left plates through the battery to the right plates, leaving the left plates positively charged and the right plates negatively charged. The energy source for this charge transfer is the internal chemical energy stored in the battery, which is converted to electrical energy. The flow of charge ceases when the voltage across the capacitors is equal to that of the battery. The capacitors reach their maximum charge when the flow of charge ceases. Let us call the maximum charges on the two capacitors Q_1 and Q_2. Then the *total charge*, Q, stored by the two capacitors is

$$Q = Q_1 + Q_2 \tag{17.11}$$

Suppose we wish to replace these two capacitors by one equivalent capacitor having a capacitance C_{eq}. This equivalent capacitor must have exactly

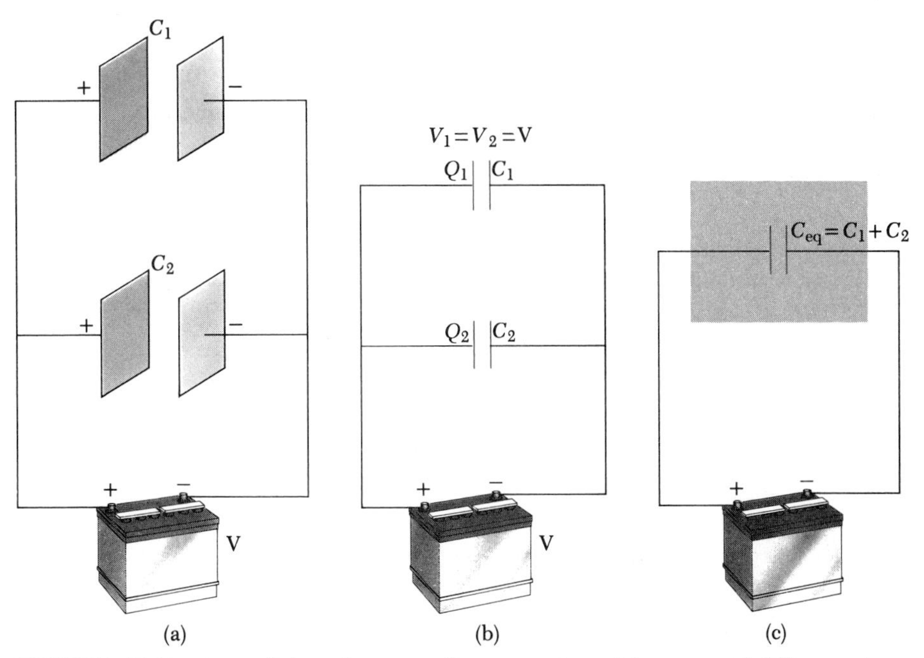

FIGURE 17.11 A parallel combination of two capacitors. The potential difference is the same across each capacitor, and the equivalent capacitance is $C_{eq} = C_1 + C_2$.

the same external effect on the circuit as the original two. That is, it must store Q units of charge. We also see from Figure 17.11b that

the potential difference across each capacitor in the parallel circuit is the same and is equal to the voltage of the battery, V.

From Figure 17.11c, we see that the voltage across the equivalent capacitor is also V. Thus, we have

$$Q_1 = C_1V \qquad Q_2 = C_2V$$

and, for the equivalent capacitor,

$$Q = C_{eq}V$$

Substituting these relations into Equation 17.11 gives

$$C_{eq}V = C_1V + C_2V$$

or

$$\boxed{C_{eq} = C_1 + C_2} \qquad \begin{pmatrix}\text{parallel}\\ \text{combination}\end{pmatrix} \qquad (17.12)$$

If we extend this treatment to three or more capacitors connected in parallel, the equivalent capacitance is found to be

$$\boxed{C_{eq} = C_1 + C_2 + C_3 + \cdots} \qquad \begin{pmatrix}\text{parallel}\\ \text{combination}\end{pmatrix} \qquad (17.13)$$

Thus we see that *the equivalent capacitance of a parallel combination of capacitors is larger than any of the individual capacitances.*

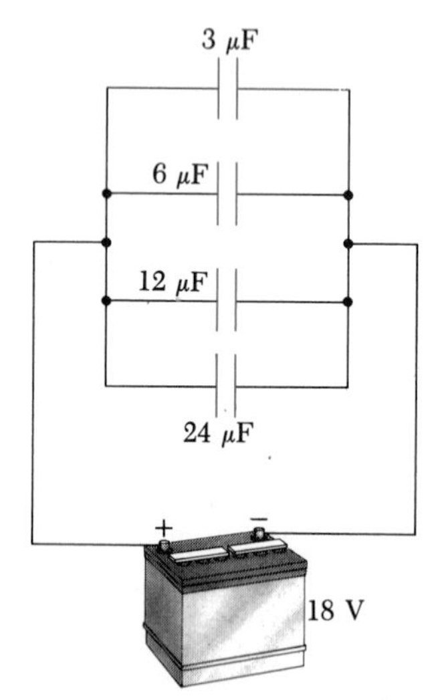

FIGURE 17.12 (Example 17.7) Four capacitors connected in parallel.

EXAMPLE 17.7 Four Capacitors Connected in Parallel

Obtain the capacitance of the single capacitor that is equivalent to the parallel combination of capacitors shown in Figure 17.12 and find the charge on the 12-μF capacitor.

Solution The equivalent capacitance is found by use of Equation 17.13:

$$C_{eq} = C_1 + C_2 + C_3 + C_4$$
$$= 3\ \mu\text{F} + 6\ \mu\text{F} + 12\ \mu\text{F} + 24\ \mu\text{F} = 45\ \mu\text{F}$$

The potential difference across the 12-μF capacitor is equal to the voltage of the battery, and so

$$Q = CV = (12 \times 10^{-6}\ \text{F})(18\ \text{V}) = 216 \times 10^{-6}\ \text{C} = 216\ \mu\text{C}$$

Series Combination

Now consider two capacitors connected in *series*, as illustrated in Figure 17.13a.

For this series combination of capacitors, the magnitude of the charge must be the same on all the plates.

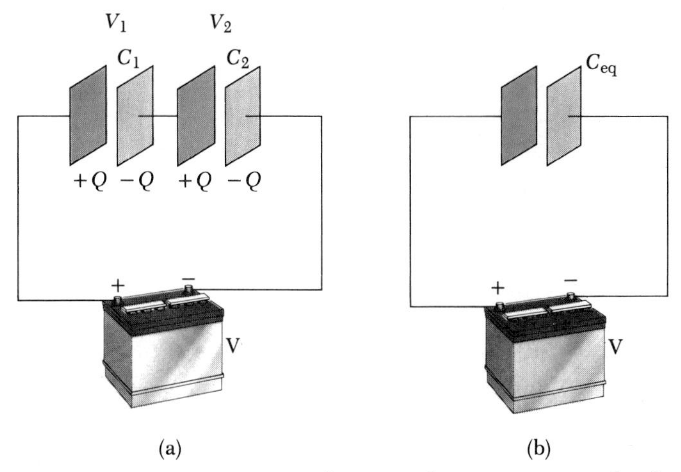

FIGURE 17.13 A series combination of two capacitors. The charge on each capacitor is the same, and the equivalent capacitance can be calculated from the relation $1/C_{eq} = (1/C_1) + (1/C_2)$.

To see why this must be true, let us consider the charge transfer process in some detail. We start with uncharged capacitors and follow what happens just after a battery is connected to the circuit. When the battery is connected, electrons are transferred from the left plate of C_1 to the right plate of C_2 through the battery. As this negative charge accumulates on the right plate of C_2, an equivalent amount of negative charge is forced off the left plate of C_2, leaving it with an excess positive charge. The negative charge leaving the left plate of C_2 accumulates on the right plate of C_1, where again an equivalent amount of negative charge is pushed off the left plate. The result of this is that *all of the right plates gain a charge of* $-Q$ *while all the left plates have a charge of* $+Q$.

Suppose an equivalent capacitor performs the same function as the series combination. After it is fully charged, *the equivalent capacitor must end up with a charge of* $-Q$ *on its right plate and* $+Q$ *on its left plate.* By applying the definition of capacitance to the circuit shown in Figure 17.13b, we have

$$V = \frac{Q}{C_{eq}}$$

where V is the potential difference between the terminals of the battery and C_{eq} is the equivalent capacitance. From Figure 17.13a, we see that

$$V = V_1 + V_2 \tag{17.14}$$

where V_1 and V_2 are the potential differences across capacitors C_1 and C_2.

In general, the potential difference across any number of capacitors (or other circuit elements) in series is equal to the sum of the potential differences across the individual capacitors.

Since $Q = CV$ can be applied to each capacitor, the potential difference across each is given by

$$V_1 = \frac{Q}{C_1} \qquad V_2 = \frac{Q}{C_2}$$

Substituting these expressions into Equation 17.14, and noting that $V = Q/C_{eq}$, we have

$$\frac{Q}{C_{eq}} = \frac{Q}{C_1} + \frac{Q}{C_2}$$

Cancelling Q, we arrive at the relationship

$$\boxed{\frac{1}{C_{eq}} = \frac{1}{C_1} + \frac{1}{C_2}} \quad \left(\begin{array}{l}\text{series}\\ \text{combination}\end{array}\right) \tag{17.15}$$

If this analysis is applied to three or more capacitors connected in series, the equivalent capacitance is found to be

$$\boxed{\frac{1}{C_{eq}} = \frac{1}{C_1} + \frac{1}{C_2} + \frac{1}{C_3} + \cdots} \quad \left(\begin{array}{l}\text{series}\\ \text{combination}\end{array}\right) \tag{17.16}$$

As we shall see in the following example, this shows that *the equivalent capacitance of a series combination is always less than any individual capacitance in the combination.*

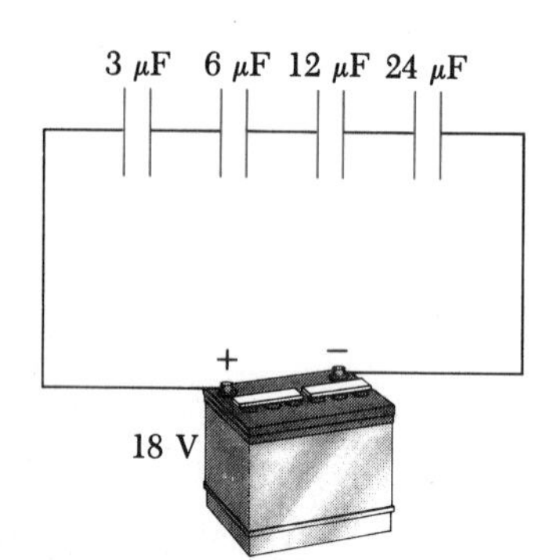

FIGURE 17.14 (Example 17.8) Four capacitors connected in series.

EXAMPLE 17.8 Four Capacitors Connected in Series

Four capacitors are connected in series with a battery, as in Figure 17.14. Find (a) the capacitance of the equivalent capacitor and (b) the charge on the 12-μF capacitor.

Solution (a) The equivalent capacitance is found from Equation 17.16:

$$\frac{1}{C_{eq}} = \frac{1}{3\ \mu\text{F}} + \frac{1}{6\ \mu\text{F}} + \frac{1}{12\ \mu\text{F}} + \frac{1}{24\ \mu\text{F}}$$

$$C_{eq} = 1.6\ \mu\text{F}$$

Note that the equivalent capacitance is less than the capacitance of any of the individual capacitors in the combination.

(b) To find the charge on the 12-μF capacitor, we can find the charge on the equivalent capacitor:

$$Q = C_{eq}V = (1.6 \times 10^{-6}\ \text{F})(18\ \text{V}) = 28.8\ \mu\text{C}$$

This is also the charge on each of the capacitors it replaced. Thus, the charge on the 12-μF capacitor in the original circuit is 28.8 μC.

EXAMPLE 17.9 Equivalent Capacitance

Find the equivalent capacitance between a and b for the combination of capacitors shown in Figure 17.15a. All capacitances are in μF.

Solution Using Equations 17.13 and 17.16, we reduce the combination step by step as indicated in the figure. The 1-μF and 3-μF capacitors are in *parallel* and combine according to $C_{eq} = C_1 + C_2$. Their equivalent capacitance is 4 μF. Likewise, the 2-μF and 6-μF capacitors are also in *parallel* and have an equivalent capacitance of 8 μF. The upper branch in Figure 17.15b now consists of two 4-μF capacitors in *series*, which combine according to

$$\frac{1}{C_{eq}} = \frac{1}{C_1} + \frac{1}{C_2} = \frac{1}{4\ \mu\text{F}} + \frac{1}{4\ \mu\text{F}} = \frac{1}{2\ \mu\text{F}}$$

$$C_{eq} = 2\ \mu\text{F}$$

Likewise, the lower branch in Figure 17.15b consists of two 8-μF capacitors in *series*, which give an equivalent capacitance of 4 μF. Finally, the 2-μF and 4-μF capacitors in Figure 17.15c are in *parallel* and have an equivalent capacitance of 6 μF. Hence, the equivalent capacitance of the circuit is 6 μF.

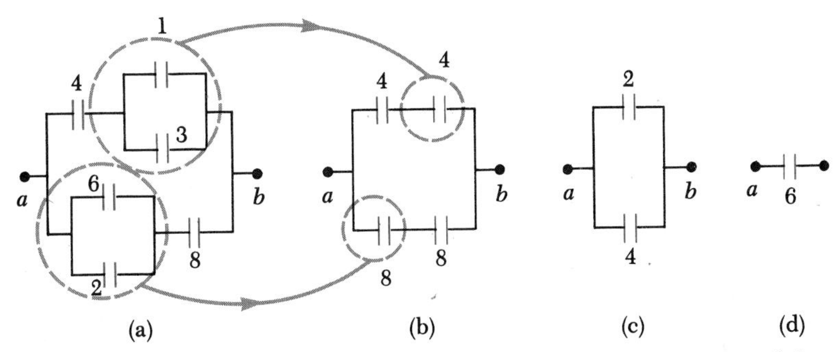

FIGURE 17.15 (Example 17.9) To find the equivalent capacitance of the circuit in (a), the circuit is reduced in steps as indicated in (b), (c), and (d), using the series and parallel rules described in the text.

17.8 ENERGY STORED IN A CHARGED CAPACITOR

Almost everyone that works with electronic equipment has at some time verified that a capacitor is able to store energy. If the plates of a charged capacitor are connected together by a conductor, such as a wire, charge will transfer from one plate to the other until the two are uncharged. The discharge can often be observed as a visible spark. If you should accidentally touch the opposite plates of a charged capacitor, your fingers would act as a pathway by which the capacitor can discharge, which results in an electric shock. The degree of shock you would receive depends on the capacitance and voltage applied to the capacitor. Such a shock could be fatal, where high voltages and large quantities of charge are present, such as in the power supply of a television set.

If a capacitor is initially uncharged (both plates neutral), so that the plates are at the same potential, almost no work is required to transfer a small amount of charge ΔQ from one plate to the other. However, once this charge has been transferred, a small potential difference $\Delta V = \Delta Q/C$ will appear between the plates. Therefore, work must be done to transfer additional charge through this potential difference. As more charge is transferred from one plate to the other, the potential difference increases in proportion. If the potential difference at any instant during the charging process is V, the work required to move more charge ΔQ through this potential difference is $V\Delta Q$, that is,

$$\Delta W = V\Delta Q$$

We know that $V = Q/C$ for a capacitor that has a total charge Q. Therefore, a plot of voltage versus charge gives a straight line whose slope is $1/C$, as shown in Figure 17.16. Since the work ΔW is the area of the shaded rectangle, the total work done in charging the capacitor to a final voltage V is the area under the voltage-charge curve, which in this case equals the area under

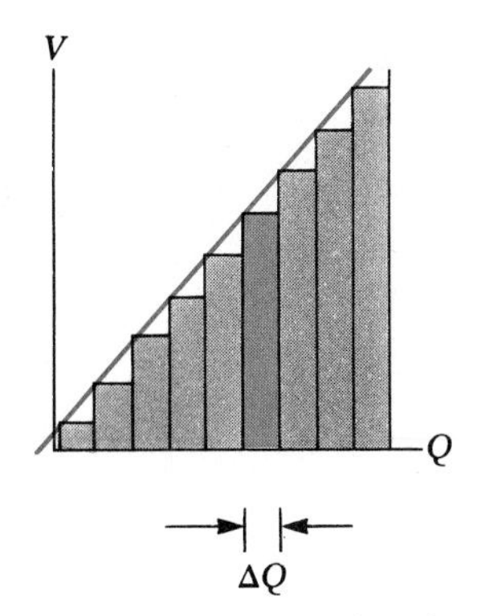

FIGURE 17.16 A plot of voltage versus charge for a capacitor is a straight line whose slope is $1/C$. The work required to move a charge ΔQ through a potential difference V across the capacitor plates is $\Delta W = V\Delta Q$, which equals the area of the colored rectangle. The *total work* required to charge the capacitor to a final charge Q is the area under the straight line, which equals $QV/2$.

the straight line. Since the area under this line is the area of a triangle (which is one half the product of the base and height), the total work done is

$$W = \frac{1}{2}QV \tag{17.17}$$

Note that this is also the energy stored in the capacitor since the work required to charge the capacitor will equal the energy stored in the capacitor after it is charged (neglecting frictional losses). From the definition of capacitance, we find $Q = CV$, hence we can express the energy stored as

$$\text{Energy stored} = \frac{1}{2}QV = \frac{1}{2}CV^2 = \frac{Q^2}{2C} \tag{17.18}$$

This result applies to any capacitor. In practice, there is a limit to the maximum energy (or charge) that can be stored because electrical breakdown will ultimately occur between the plates of the capacitor at a sufficiently large value of V. For this reason, capacitors are usually labeled with a maximum operating voltage.

EXAMPLE 17.10 Energy Stored in a Charged Capacitor

Find the amount of energy stored in a 5-μF capacitor when it is connected across a 120-V line.

Solution Using Equation 17.18, we have

$$\text{Energy stored} = \frac{1}{2}CV^2 = \frac{1}{2}(5 \times 10^{-6}\ \text{F})(120\ \text{V})^2 = 3.6 \times 10^{-2}\ \text{J}$$

---

*17.9 CAPACITORS WITH DIELECTRICS

A **dielectric** is an insulating material, such as rubber, glass, or waxed paper. When a dielectric is inserted between the plates of a capacitor, the capacitance increases. If the dielectric completely fills the space between the plates, the capacitance increases by a factor κ, called the **dielectric constant.**

The following experiment can be performed to illustrate the effect of a dielectric in a capacitor. Consider a parallel-plate capacitor of charge Q_0 and capacitance C_0 in the absence of a dielectric. The potential difference across the capacitor plates can be measured, and is found to be given by $V_0 = Q_0/C_0$ (Fig. 17.17a). Because the capacitor is not connected to an external circuit,

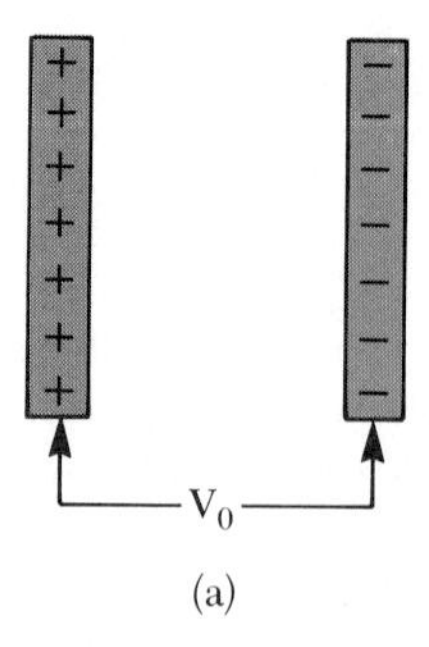

(a)

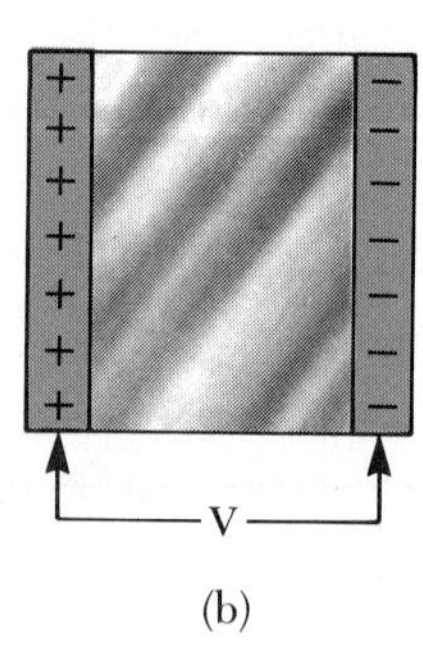

(b)

FIGURE 17.17 (a) With air between the plates, the voltage across the capacitor is V_0, the capacitance is C_0, and the charge is Q_0. (b) With a dielectric between the plates, the charge remains at Q_0, but the voltage and capacitance both change.

there is no pathway for charge to leave or to be added to the plates. If a dielectric is now inserted between the plates as in Figure 17.17b, it is found that the voltage across the plates *decreases* by a factor κ to a value V, where

$$V = \frac{V_0}{\kappa}$$

Since V is less than V_0, we see that κ must be greater than 1. Because the charge Q_0 on the capacitor does not change, we conclude that the capacitance in the presence of the dielectric, C, must change to the value

$$C = \frac{Q_0}{V} = \frac{Q_0}{V_0/\kappa} = \frac{\kappa Q_0}{V_0}$$

or

$$\boxed{C = \kappa C_0} \qquad (17.19)$$

According to this result, the capacitance *increases* by a factor κ when the dielectric completely fills the region between the plates. For a parallel-plate capacitor, where the capacitance in the absence of a dielectric is $C_0 = \epsilon_0 A/d$, we can express the capacitance in the presence of a dielectric as

$$\boxed{C = \kappa \epsilon_0 \frac{A}{d}} \qquad (17.20)$$

From this result, it would appear that the capacitance could be made very large by decreasing d, the distance between the plates. In practice, the lowest value of d is limited by the electric discharge that could occur through the dielectric material separating the plates. For any given plate separation, there is a maximum electric field that could be produced in the dielectric before it breaks down and begins to conduct. This maximum electric field is called the **dielectric strength,** and for air its value is about 3×10^6 V/m. Most insulating materials have dielectric strengths greater than that of air, as indicated by the values in Table 17.1.

TABLE 17.1 Dielectric Constants and Dielectric Strengths of Various Materials at Room Temperature

Material	Dielectric Constant κ	Dielectric Strength (V/m)
Vacuum	1.00000	—
Air	1.00059	3×10^6
Bakelite	4.9	24×10^6
Fused quartz	3.78	8×10^6
Pyrex glass	5.6	14×10^6
Polystyrene	2.56	24×10^6
Teflon	2.1	60×10^6
Neoprene rubber	6.7	12×10^6
Nylon	3.4	14×10^6
Paper	3.7	16×10^6
Strontium titanate	233	8×10^6
Water	80	—
Silicone oil	2.5	15×10^6

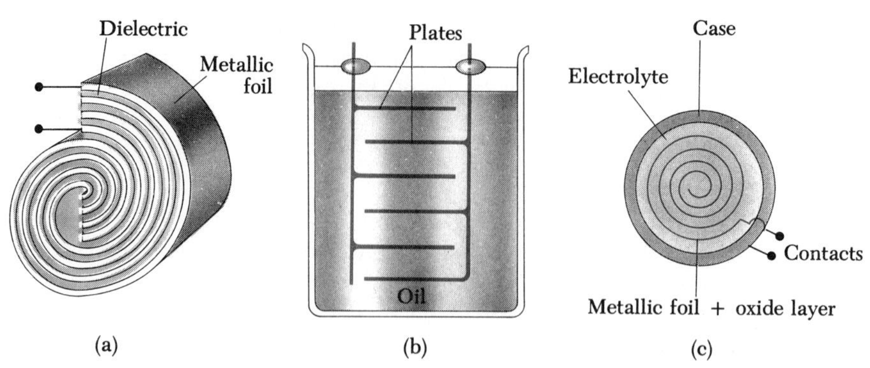

FIGURE 17.18 Three commercial capacitor designs: (a) a tubular capacitor whose plates are separated by paper and then rolled into a cylinder, (b) a high-voltage capacitor consisting of many parallel plates separated by oil, and (c) an electrolytic capacitor.

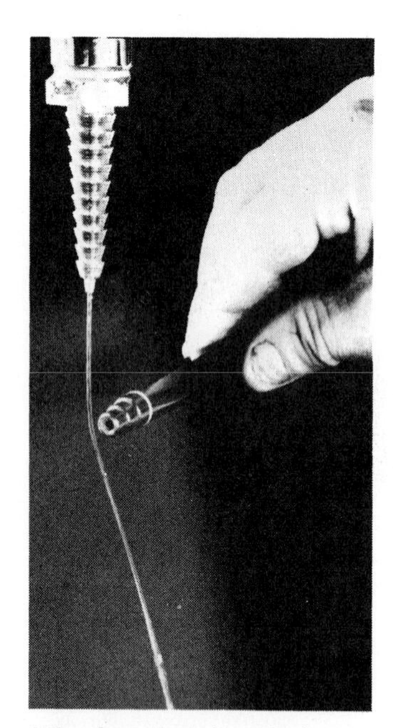

Why does this electrified ballpoint pen bend the stream of water? (From M. Merken, *Physical Science with Modern Applications*, 3rd ed., Saunders College Publishing, 1985, p. 162.)

Commercial capacitors are often made using metal foil interlaced with thin sheets of paraffin-impregnated paper or mylar, which serves as the dielectric material. These alternate layers of metal foil and dielectric are then rolled into the shape of a cylinder to form a small package (Fig. 17.18a). High-voltage capacitors commonly consist of a number of interwoven metal plates immersed in silicone oil (Fig. 17.18b). Small capacitors are often constructed from ceramic materials. Variable capacitors (typically 10 to 500 pF) usually consist of two interwoven sets of metal plates, one fixed and the other movable, with air as the dielectric.

An electrolytic capacitor is often used to store large amounts of charge at relatively low voltages. This device, shown in Figure 17.18c, consists of a metal foil in contact with an electrolyte—a solution that conducts electricity by virtue of the motion of ions contained in the solution. When a voltage is applied between the foil and the electrolyte, a thin layer of metal oxide (an insulator) is formed on the foil, and this layer serves as the dielectric. Very large values of capacitance can be obtained because the dielectric layer is very thin.

When electrolytic capacitors are used in circuits, the polarity (the plus and minus signs on the device) must be installed properly. If the polarity of the applied voltage is opposite what is intended, the oxide layer will be removed and the capacitor will conduct electricity rather than store charge. Furthermore, reversing the polarity could result in such a large current that the capacitor could either burn or produce steam and explode.

EXAMPLE 17.11 A Paper-Filled Capacitor

A parallel-plate capacitor has plates of dimensions 2 cm by 3 cm. The plates are separated by a 1 mm thickness of paper. (a) Find the capacitance of this device, and (b) find the maximum charge that can be placed on the capacitor.

Solution (a) Since $\kappa = 3.7$ for paper (Table 17.1), we get

$$C = \kappa\epsilon_0\frac{A}{d} = 3.7\left(8.85 \times 10^{-12}\ \frac{\mathrm{C}^2}{\mathrm{N\cdot m^2}}\right)\left(\frac{6 \times 10^{-4}\ \mathrm{m}^2}{1 \times 10^{-3}\ \mathrm{m}}\right)$$
$$= 19.6 \times 10^{-12}\ \mathrm{F} = 19.6\ \mathrm{pF}$$

(b) From Table 17.1, we see that the dielectric strength of paper is 16×10^6 V/m. Because the paper thickness is 1 mm, the maximum voltage that can be applied before electrical breakdown occurs is

$$V_{max} = E_{max}d = (16 \times 10^6 \text{ V/m})(1 \times 10^{-3} \text{ m}) = 16 \times 10^3 \text{ V}$$

Hence, the maximum charge that can be placed on the capacitor is

$$Q_{max} = CV_{max} = (19.6 \times 10^{-12} \text{ F})(16 \times 10^3 \text{ V}) = 0.31 \ \mu\text{C}$$

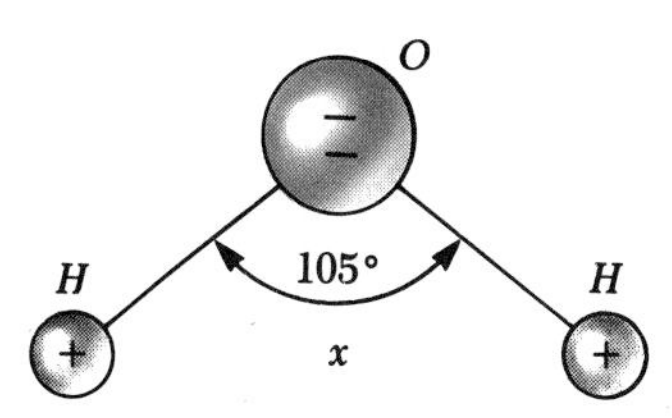

FIGURE 17.19 The water molecule, H_2O, has a permanent polarization resulting from its bent geometry.

An Atomic Description of Dielectrics

Our explanation of why a dielectric increases the capacitance of a capacitor is based on an atomic description of the material. In order to do this, we shall first have to explain a property of some molecules called **polarization.** Molecules are said to be polarized when there is a separation between the "center of gravity" of the negative charges and that of the positive charges that make up the molecule. In some molecules, such as water, this condition is always present. To see why this is the case, consider the geometry of a water molecule (Fig. 17.19). The molecule is arranged so that the negative oxygen atom is bonded to the positively charged hydrogen atoms with a 105° angle between the two bonds. The center of negative charge is at the oxygen atom, and the center of positive charge lies at a point midway along the line joining the hydrogen atoms (point x in the diagram). Materials composed of molecules that are permanently polarized in this fashion have large dielectric constants, and, indeed, Table 17.1 shows that the dielectric constant of water is quite large ($\kappa = 80$).

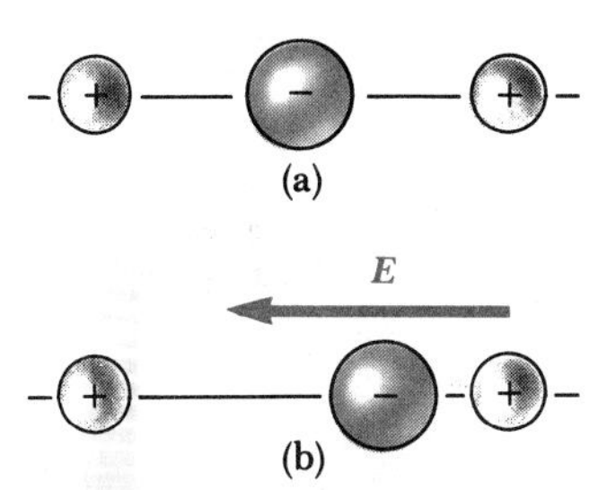

FIGURE 17.20 (a) A symmetrical molecule has no permanent polarization. (b) An external electric field induces a polarization in the molecule.

A symmetrical molecule (Fig. 17.20a) must have no permanent polarization, but a polarization can be induced by an external electric field. A field directed to the left, as in Figure 17.20b, would cause the center of positive charge to shift to the left from its initial position and the center of negative charge to shift to the right. This *induced polarization* is the effect that predominates in most materials used as dielectrics in capacitors.

To understand why the polarization of a dielectric can affect the capacitance, consider Figure 17.21, which shows a slab of dielectric placed between the plates of a parallel-plate capacitor. The dielectric becomes polarized as shown because it is in the electric field that exists between the metal plates. Notice that a net positive charge appears on the surface of the dielectric that is adjacent to the negatively charged metal plate. The presence of the positive charge on the dielectric effectively cancels some of the negative charge on the metal. Therefore, more negative charge can be stored on the capacitor plates for a given applied voltage. From the definition of capacitance, $C = Q/V$, we see that the capacitance must increase since the plates are able to store more charge for a given voltage.

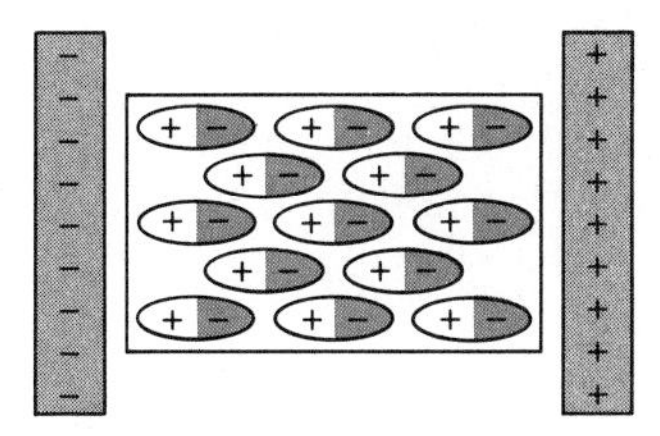

FIGURE 17.21 When a dielectric is placed between the plates of a charged parallel-plate capacitor, the dielectric becomes polarized. This creates a net positive induced charge on the left side of the dielectric and a net negative induced charge on the right side. As a result, the capacitance of the device increases by the factor κ.

*17.10 LIVING CELLS AS CAPACITORS

Although we shall examine the subject of electrical phenomena in the human body in detail in Chapter 19 for now let us consider a feature of living cells that gives them capacitor-like characteristics.

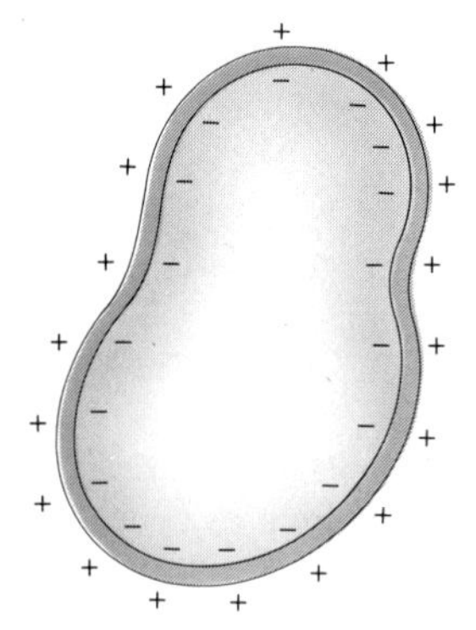

FIGURE 17.22 A living cell is equivalent to a small capacitor separated by a dielectric, which is the membrane wall of the cell.

As shown in Figure 17.22, the presence of charged ions in a cell and in the fluid surrounding the cell sets up a charge distribution across the membrane wall. From this charge distribution, we see that the cell is equivalent to a small capacitor separated by a dielectric, with the membrane wall acting as the dielectric. The potential difference across this "capacitor" can be measured using the technique described in Figure 17.23. A tiny probe is forced through the cell wall, and a second probe is placed in the extracellular fluid. With this technique, typical potential differences across the cell wall are found to be of the order of 100 mV.

The charge distribution of a cell can be understood on the basis of the theory of transport through selectively permeable membranes, discussed in Chapter 10. The primary ionic constituents of a cell are potassium ions (K^+) and chloride ions (Cl^-). The cell wall is highly permeable to K^+ ions but only moderately permeable to Cl^- ions. Since the K^+ ions can cross the cell wall with ease, the charge distribution on a cell wall is determined primarily by the movement of these ions.

In order to explain the process, let us assume that the cellular fluid is electrically neutral at some point. Under normal circumstances, the concentration of K^+ ions outside the cell is much smaller than inside the cell. Therefore, K^+ ions will diffuse out of the cell, leaving behind a net negative charge. As diffusion commences, electrostatic forces of attraction across the membrane wall produce a layer of positive charge on the exterior surface of the wall and a layer of negative charge on the interior of the wall. At the same time, the electric field set up by these charges *impedes* the flow of additional K^+ ions diffusing out of the cell. Because of this impeding effect to diffusion, an equilibrium is finally established such that there is no net movement of K^+ ions through the cell wall. When this equilibrium situation is reached, the potential difference across the cell wall is given by the **Nernst potential,**

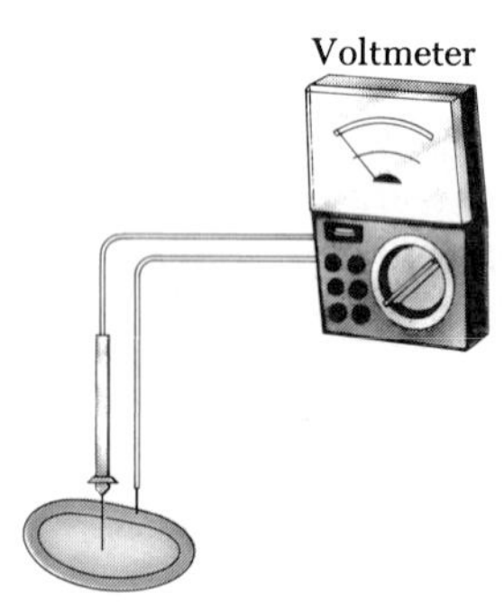

FIGURE 17.23 An experimental technique for measuring the potential difference across the walls of a living cell.

$$V_N = \frac{kT}{q}\ln\left(\frac{c_i}{c_o}\right) \qquad (17.21)$$

where k is Boltzmann's constant (1.38×10^{-23} J/K), T is the kelvin temperature, q is 1.6×10^{-19} C, and c_i and c_o are the concentrations of K^+ ions inside and outside the cell. At normal body temperatures ($T = 310$ K), this equation becomes

$$V_N = (26.7\text{ mV})\ln\left(\frac{c_i}{c_o}\right) \qquad (17.22)$$

The electrical balance achieved in the cell is necessary in order for it to function properly. A threat to this balance occurs when the extracellular fluid contains a higher concentration of sodium ions (Na^+) than is normally found inside a cell. Although the cell wall is not highly permeable to these ions, a fraction are able to penetrate the wall. This encroachment by Na^+ ions inside the cell will gradually deplete the negative charge on the interior of the cell wall. If this occurs, more K^+ ions will escape, thus endangering the cell. Fortunately, the cell is able to prevent this by effectively "pumping" Na^+ ions out of the cell while "pumping" K^+ ions into the cell. The mechanism for these processes is not fully understood.

EXAMPLE 17.12 Concentration Ratio

Find the ratio of the concentration of K^+ ions inside a cell to the concentration outside the cell if the Nernst potential is measured to be 90 mV.

Solution Solving Equation 17.22 for $\ln(c_i/c_o)$, we obtain

$$\ln\left(\frac{c_i}{c_o}\right) = \frac{V_N}{26.7 \times 10^{-3}} = \frac{90 \times 10^{-3}\text{ V}}{26.7 \times 10^{-3}\text{ V}} = 3.37$$

The number whose natural logarithm is 3.37 is 29.1. (Use your calculator to verify this.) Thus, the required ratio is

$$\frac{c_i}{c_o} = 29.1$$

SUMMARY

The **difference in potential** between two points A and B is defined as

$$V_f - V_i \equiv \frac{\Delta PE}{q} \quad (17.2)$$

where ΔPE is the *change* in electrical potential energy experienced by a charge q as it moves between i and f. The units of potential difference are joules per coulomb, defined to be equal to a unit called the **volt**; that is, 1 J/C = 1 V.

The **potential difference** between two points A and B in a *uniform electric field* $\mathbf{E}$ is

$$V_f - V_i = -Ed \quad (17.4)$$

where d is the distance between i and f and E is the strength of the electric field in that region.

The **electric potential** due to a point charge, at a distance r from the point charge, is

$$V = k\frac{q}{r} \quad (17.5)$$

The **electrical potential energy** of a pair of point charges separated by a distance r_{12} is

$$PE = k\frac{q_1q_2}{r_{12}} \quad (17.6)$$

Every point on the surface of a charged conductor in electrostatic equilibrium is at the same potential. Furthermore, the potential is constant everywhere inside the conductor and is equal to its value on the surface.

The **electron volt** is defined as the energy that an electron (or proton) gains when accelerated through a potential difference of 1 V. The conversion between electron volts and joules is

$$1\text{ eV} = 1.6 \times 10^{-19}\text{ J} \quad (17.8)$$

A **capacitor** consists of two metal plates whose charges are equal in magnitude but opposite in sign. The **capacitance** (C) of any capacitor is defined to be the ratio of the magnitude of the charge Q on either plate to the potential difference V between them:

$$C \equiv \frac{Q}{V} \tag{17.9}$$

Capacitance has units of coulombs per volt, which is defined to be equal to the **farad:** 1 C/V $\equiv$ 1 F.

The capacitance of two *parallel metal plates* of area A separated by a distance d is

$$C = \epsilon_0 \frac{A}{d} \tag{17.10}$$

where ϵ_0 is a constant called the **permittivity of free space** and has the value

$$\epsilon_0 = 8.85 \times 10^{-12} \text{ C}^2/\text{N} \cdot \text{m}^2$$

The **equivalent capacitance** of a **parallel** combination of capacitors is

$$C_{eq} = C_1 + C_2 + C_3 + \cdots \tag{17.13}$$

If two or more capacitors are connected in **series**, the **equivalent capacitance** of the series combination is

$$\frac{1}{C_{eq}} = \frac{1}{C_1} + \frac{1}{C_2} + \frac{1}{C_3} + \cdots \tag{17.16}$$

Three equivalent expressions for calculating the **energy stored** in a charged capacitor are

$$\text{Energy stored} = \frac{1}{2} QV = \frac{1}{2} CV^2 = \frac{Q^2}{2C} \tag{17.18}$$

When a nonconducting material, called a **dielectric,** is placed between the plates of a capacitor, the capacitance increases by a factor κ, which is called the **dielectric constant** and is a property of the dielectric material. The capacitance of a parallel-plate capacitor filled with a dielectric is

$$C = \kappa \epsilon_0 \frac{A}{d} \tag{17.20}$$

ADDITIONAL READING

Alan H. Cromer, *Physics for the Life Sciences,* New York, McGraw-Hill, 1974, Chap. 17.

A. Einstein and L. Infeld, *The Evolution of Physics,* New York, Simon and Schuster, 1938.

A. L. Stanford Jr., *Foundations of Biophysics,* New York, Academic Press, 1975. Electric dipoles.

QUESTIONS

1. A constant electric field is parallel to the x axis. In what direction can a charge be displaced in this field without any external work being done on the charge?
2. Would a free proton move from a point of high potential to a point of low potential or vice versa? What about a free electron?
3. If the electric potential at some point is zero, can you conclude that there are no charges in the vicinity of that point? Explain.
4. We said in Chapter 16 that electric field lines never intersect. Do equipotential surfaces ever intersect?
5. Why is it important to avoid sharp edges, or points, on conductors used in high-voltage equipment?
6. Two capacitors have the same potential difference across them, but one has a large capacitance and one a small capacitance. When you touch the leads to the capacitors, you get a greater shock when you touch the one with high capacitance. From this information, what causes the sensation of shock?
7. What happens to the charge on a capacitor if the potential difference between the conductors is doubled?
8. A pair of capacitors is connected in parallel while an identical pair is connected in series. Which pair would be more dangerous to handle after being connected to the same voltage source?
9. If the potential difference across a capacitor is doubled, by what factor does the energy stored change?
10. Why is it dangerous to touch the terminals of a high-voltage capacitor even after the applied voltage has been turned off? What could be done to make the capacitor safe to handle after the voltage source has been removed?
11. If you were asked to design a capacitor where small size and large capacitance were required, what factors would be important in your design?
12. A capacitor is designed such that one plate is large while the other is small. Do each of the plates have the same charge when connected to a battery?
13. Two charges have the same electric potential energy. Do they also have the same electric potential?
14. Two parallel plates are uncharged. Does the set of plates have a capacitance? Explain.

PROBLEMS

Section 17.1 Potential Difference and Electric Potential

1. (a) How much work is done on a proton by a uniform electric field of 200 N/C as the charge moves a distance of 2 cm in the field? (b) What is the difference in potential energy between these two points?

2. The difference in potential between the accelerating plates of a TV set is about 25,000 V. If the distance between these plates is 1.5 cm, find the magnitude of the uniform electric field in this region.

3. How much work is done (by a battery, generator, or some other source of electrical energy) in moving Avogadro's number of electrons from a point where the electric potential is 6 V to a point where the electric potential is -10 V?

4. Show that the units of N/C and V/m are equivalent.

5. An ion, after being accelerated through a potential difference of 60 V, experiences an increase of potential energy of 1.92×10^{-17} J. Calculate the charge on the ion.

6. A proton is between two plates, separated by a distance of 2 cm, across which there is a potential difference of 5000 V. (a) Find the electrical force exerted on the proton, and (b) its acceleration.

7. (a) Through what potential difference would one need to accelerate an electron in order for it to achieve a speed of 60% of the speed of light, starting from rest? The speed of light is 3×10^8 m/s. (b) Repeat your calculation for a proton. (Do not consider relativistic effects.)

8. An electron moves from one plate to another across which there is a potential difference of 2000 V. (a) Find the velocity with which it strikes the positive plate. (b) Repeat part (a) for a proton moving from the positive toward the negative plate.

Section 17.2 Electric Potential and Potential Energy Due to Point Charges

Section 17.3 Potentials and Charged Conductors

9. At what distance from a point charge of 6 μC would the potential equal 2.7×10^4 V?

10. (a) Find the potential at a distance of 1 cm from a proton. (b) What is the potential difference between two points 1 cm and 2 cm from a proton? (c) Repeat parts (a) and (b) for an electron.

11. In the Bohr model of the hydrogen atom an electron circles a proton in an orbit of radius 0.51×10^{-10} m. Find the potential at the position of the electron.

12. Two point charges are located along the y axis. One charge of magnitude 3×10^{-9} C is at the origin and a second of magnitude 6×10^{-9} C is at the point $y = 30$ cm. (a) Calculate the potential at $y = 60$ cm, and at (b) $y = -60$ cm. (c) Repeat this problem assuming the 6×10^{-9} C charge is replaced by a -6×10^{-9} C charge.

13. A 6×10^{-9} C charge is located at the point (0, 0.15) where the coordinates are in meters, and a second charge, -4×10^{-9} C, is at $(-0.45, 0)$. Find the potential at (a) $(0, -0.45)$ and (b) $(0.15, 0)$.

14. A point charge of 9×10^{-9} C is located at the origin. How much work is required to bring a positive charge of 3×10^{-9} C from infinity to the location $x = 30$ cm?

15. A charge of -3×10^{-9} C is located at the origin of a coordinate system while a charge of 8×10^{-9} C is on the x axis at $x = 2$ m. At what two location(s) on the x axis is the electric potential equal to zero?

16. How much work is required to move an electron from a point 50 cm distant from a proton to a point 100 cm away from the proton?

17. Calculate the speed of (a) an electron that has a kinetic energy of 1 eV and (b) a proton that has a kinetic energy of 1 eV.

18. A proton is accelerated from rest through a potential difference of 25,000 V. (a) What is the kinetic energy of the proton in eV? (b) What is the speed of the proton after this acceleration?

Section 17.5 The Definition of Capacitance

Section 17.6 The Parallel-Plate Capacitor

19. (a) How much charge is on each plate of a 4 μF capacitor when it is connected to a 12 V battery? (b) If this same capacitor is connected to a 1.5 V battery, what charge is stored?

20. To what difference in potential would a 2 μF capacitor have to be connected in order to store 98 μC of charge?

21. The plates of a parallel-plate capacitor are separated by 0.1 mm. If the material between the plates is air, what plate area is required to provide a capacitance of 2 pF?

22. The plates of a parallel-plate capacitor are separated by 0.2 mm. What plate area is required to produce a capacitance of 1 F?

23. A parallel-plate capacitor has an area of 2 cm^2 and a plate separation of 2 mm. How much charge does this capacitor store when connected to a 6 V battery? (There is air between the plates.)

24. A parallel-plate capacitor has an area of 5 cm^2 and the plates are separated by 1 mm with air between the plates. It stores a charge of 400 pC. (a) What is the potential difference across the plates of the capacitor? (b) What is the magnitude of the uniform electric field in the region between the plates?

Section 17.7 Combinations of Capacitors

25. Two capacitors, $C_1 = 5$ μF and $C_2 = 12$ μF are connected in parallel, and the resulting combination is connected to a 9-V battery. What is the value of the equivalent capacitance of the combination?

26. (a) What is the charge on and the potential difference across each capacitor in Problem 25? (b) What is the charge stored on each of these capacitors?

27. The two capacitors of Problem 25 are now connected in series and to a 9-V battery. Find (a) the value of the equivalent capacitance of the combination and (b) the charge on and the voltage across each.

28. Three capacitors, $C_1 = 5$ μF, $C_2 = 4$ μF, and $C_3 = 9$ μF, are connected together. (a) Find the effective capacitance of the group if they are all in parallel. (b) Find the effective capacitance of the group if they are all in series.

29. The combination of capacitors in Problem 28 is connected to a 12-V battery. Find the charge on and the potential difference across each of the capacitors (a) when they are connected in parallel, and (b) when they are connected in series.

• **30.** (a) Find the equivalent capacitance of the group of capacitors shown in Figure 17.24. (b) Find the charge on and the potential difference across each.

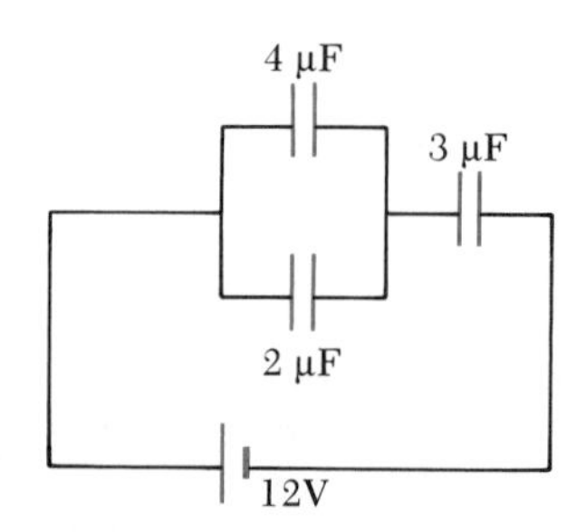

FIGURE 17.24 Problems 30 and 38

• **31.** Consider the group of capacitors shown in Figure 17.25. Find (a) the equivalent capacitance of the group, and (b) the charge and potential difference across each capacitor.

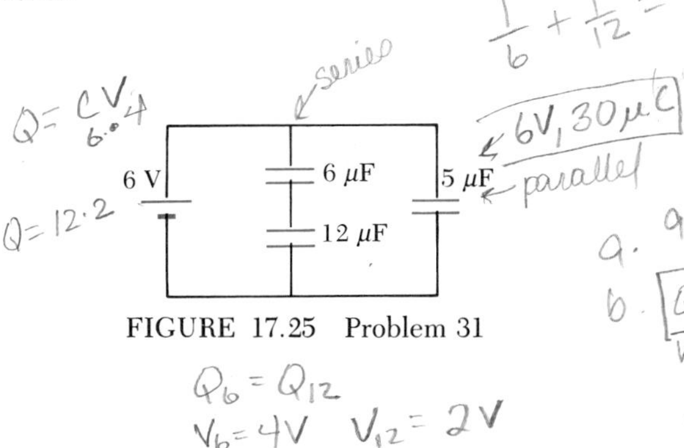

FIGURE 17.25 Problem 31

• **32.** Consider the combination of capacitors shown in Figure 17.26. (a) What is the equivalent capacitance of the group? (b) Determine the charge on each capacitor.

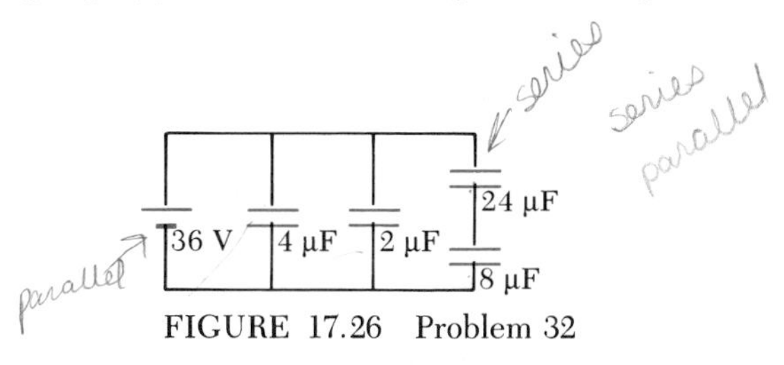

FIGURE 17.26 Problem 32

• **33.** Find the equivalent capacitance of the group of capacitors shown in Figure 17.27.

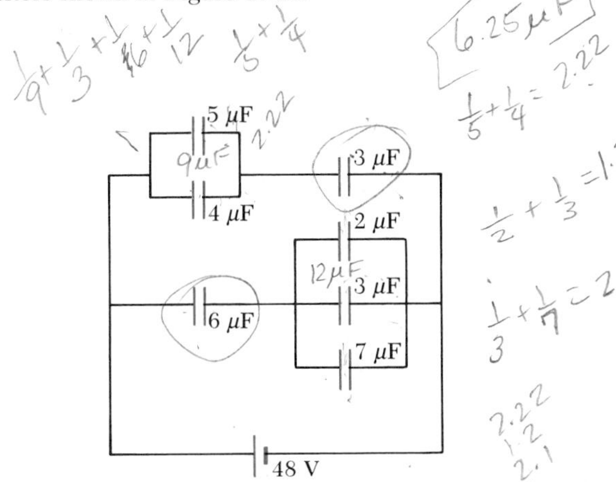

FIGURE 17.27 Problem 33

• **34.** Find the equivalent capacitance of the group of capacitors shown in Figure 17.28.

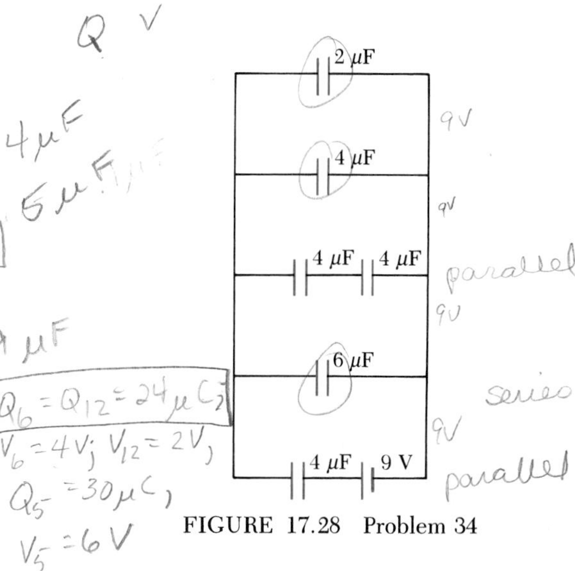

FIGURE 17.28 Problem 34

Section 17.8 Energy Stored in a Charged Capacitor

35. A capacitor is connected to a 120-V source and holds a charge of 36 μC. (a) What is the capacitance of the capacitor? (b) Find the energy stored by the capacitor by the use of at least two different equations.

36. (a) A 3-μF capacitor is connected to a 12-V battery. How much energy is stored in the capacitor? (b) If the capacitor had been connected to a 6-V battery how much energy would have been stored?

37. A parallel-plate capacitor has plates of area 2 cm^2, a distance of separation of 5 mm, and air between the plates. If a 12-V battery is connected to this capacitor, how much energy does it store?

38. Find the energy stored by each capacitor in Figure 17.24.

Section 17.9 Capacitors with Dielectrics

39. Find the capacitance of a parallel-plate capacitor that uses Bakelite as a dielectric if the plates each have an area of 5 cm^2 and the plate separation is 2 mm.

40. A capacitor that has air between its plates is connected across a potential difference of 12 V and stores 48 μC of charge. It is then disconnected from the source while still charged. (a) Find the capacitance of the capacitor. (b) If a piece of teflon is now inserted between the plates, find the voltage and charge on the capacitor, and (c) its new capacitance.

41. A 3-μF capacitor is connected to a 12-V battery. If the material between the plates is initially air, by what factor will the energy stored change if a sheet of nylon is placed between the plates?

42. (a) How much charge can be placed on a capacitor with air between the plates before it breaks down if the area of the plates is 5 cm^2? (b) If polystyrene is used between the plates instead of air, find the maximum charge.

ADDITIONAL PROBLEMS

43. A proton moving in a uniform electric field experiences an increase in kinetic energy of 9×10^{-18} J after being displaced 1 cm in a direction parallel to the field. What is the magnitude of the electric field?

44. At a distance r away from a point charge q, the electric potential is $V = 600$ V and the magnitude of the electric field is $E = 200$ N/C. Determine the values of q and r.

45. Find the potential at point P for the rectangular grouping of charges shown in Figure 17.29.

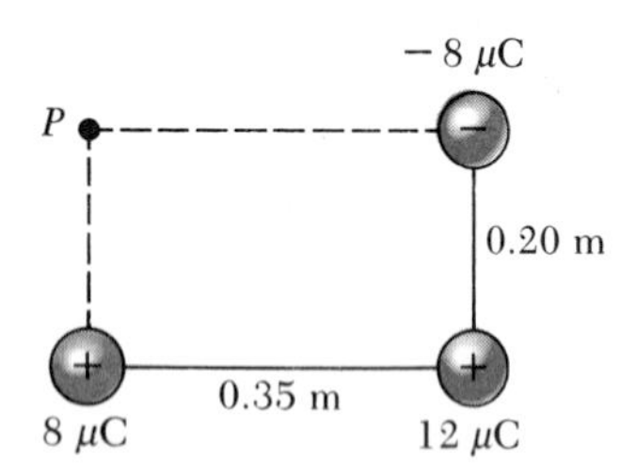

FIGURE 17.29 Problem 45

46. A capacitor with air between the plates is charged to 100 V and then disconnected from the battery. When a piece of glass is placed between the plates, the voltage across the capacitor drops to 25 V. What is the dielectric constant of the glass? Assume the glass completely fills the space between the plates.

• **47.** Find the equivalent capacitance of the group of capacitors shown in Figure 17.30.

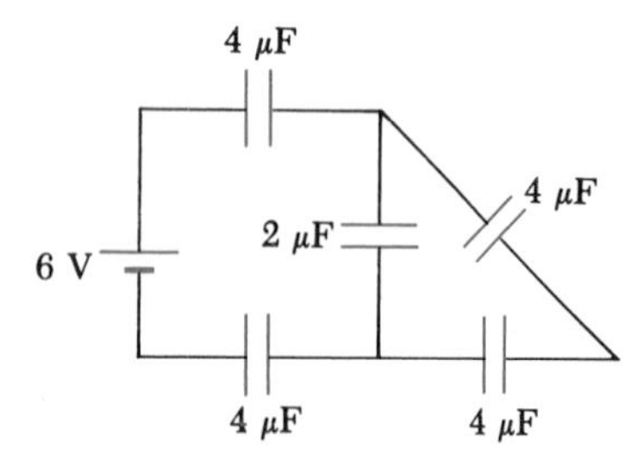

FIGURE 17.30 Problems 47 and 48

48. Find the total energy stored by the grouping of capacitors in Problem 47.

49. The three charges shown in Figure 17.31 are at the vertices of an isoceles triangle. Calculate the electric potential at the midpoint of the base, taking the charge $q = 5 \times 10^{-9}$ C.

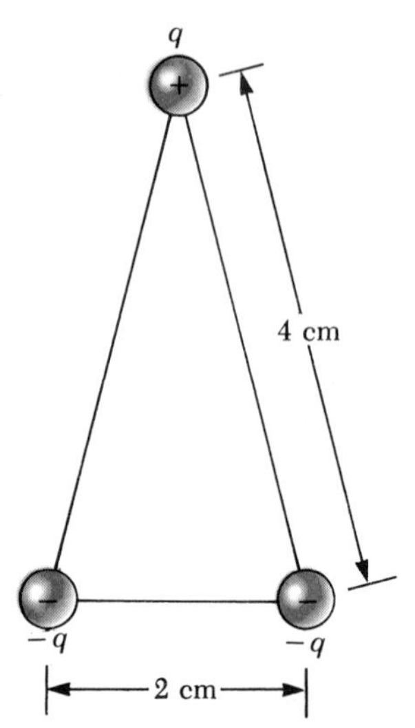

FIGURE 17.31 Problem 49

50. Suppose that the extracellular fluid of a cell has a potassium ion concentration of 0.0045 mole/liter and the intracellular fluid has a potassium ion concentration of 0.138 mole/liter. Calculate the potential difference across the cell membrane. Neglect the diffusion of any other ions across the membrane and assume a temperature of 310 K.

51. A parallel-plate capacitor is to be constructed using Pyrex glass as a dielectric. If the capacitance of the device is to be 0.2 μF and it is to be operated at 6000 V, (a) calculate the minimum plate area required. (b) What is the energy stored in the capacitor at the operating voltage?

• **52.** A commercial capacitor is made by rolling two strips of aluminum separated by two strips of paraffin-coated paper into a cylinder. Each strip of foil and paper is 5 cm wide. The foil is 0.0005 cm thick, and the paper is 0.002 cm thick and has a dielectric constant of 3. What length should the strips be if a capacitance of 2×10^{-8} F is desired? (Use the parallel plate equation.)

• **53.** When a certain air-filled parallel-plate capacitor is connected across a battery, it acquires a charge (on each plate) of 150 μC. While the battery connection is maintained, a dielectric slab is inserted into and fills the region between the plates. This results in the accumulation of an additional charge of 200 μC on each plate. What is the dielectric constant of the dielectric slab?

• **54.** It is possible to obtain large potential differences by first charging a group of capacitors connected in parallel and then activating a switching arrangement that in effect disconnects the capacitors from the charging source and reconnects them in a series arrangement. The group of charged capacitors is then discharged in series. What is the maximum potential difference that can be obtained in this manner by using ten capacitors each of 500 μF and a charging source of 800 V?

• **55.** A 2-μF capacitor charged to 200 V and a 4-μF capacitor charged to 400 V are connected to each other, with the positive plate of each connected to the negative plate of the other. (a) What is the final value of the charge that resides on each capacitor? (b) What is the potential difference across each capacitor after they have been connected?

• **56.** Figure 17.32 shows two capacitors in series. The rigid center section of length b is movable vertically, and the area of each plate is A. Show that the capacitance of the series combination is independent of the position of the center section and is given by

$$C = \frac{\epsilon_0 A}{a - b}$$

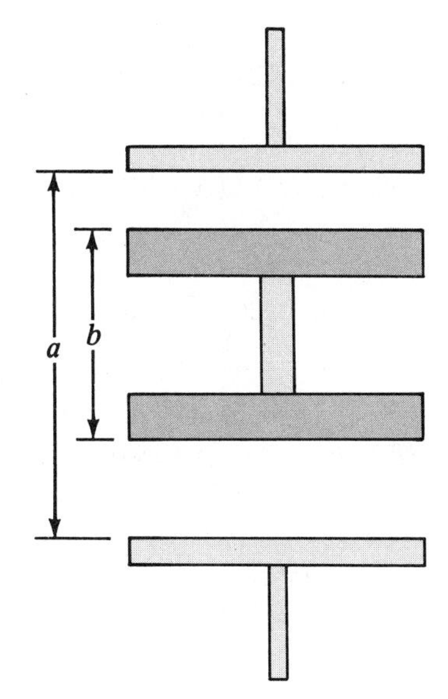

FIGURE 17.32 Problem 56

•• **57.** Capacitors $C_1 = 4$ μF and $C_2 = 2$ μF are charged as a series combination across a 100-V battery. The two capacitors are disconnected from the battery and from each other. They are then connected positive plate to positive plate and negative plate to negative plate. Calculate the resulting charge on each capacitor.

•• **58.** A capacitor $C_1 = 4$ μF is charged to a potential difference of 800 V. The capacitor is then removed from the charging source, and each plate of the charged capacitor is connected to one of the plates of an uncharged capacitor $C_2 = 6$ μF. What is the resulting charge on each capacitor?

•• **59.** Capacitors $C_1 = 6$ μF and $C_2 = 2$ μF are charged as a parallel combination across a 250-V battery. The capacitors are disconnected from the battery and from each other. They are then connected positive plate to negative plate and negative plate to positive plate. Calculate the resulting charge on each capacitor.

•• **60.** Consider an array of eight equal negative charges located so as to define the corners of a cube of edge length $d = 0.15$ m. If each of the eight charges has a charge $q = -6$ μC, determine the potential at the center of the cube.

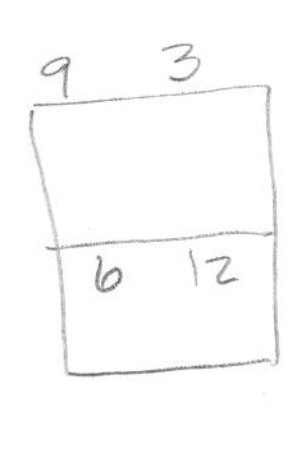

CHAPTER 18

Current and Resistance

Photograph of a large-scale integrated circuit that contains more than 65,000 devices on a single chip. (Dimensions are about 1 cm by 1 cm.) This particular chip is part of a 32-bit microprocessor (the APX 432) that integrates 200,000 transistors and can execute two million instructions per second. (Courtesy Intel Corporation)

There are many practical applications and devices based on the principles of static electricity, but electricity truly became an inseparable part of our daily lives when scientists learned how to control the flow of electric charges. Electric currents power our lights, radios, television sets, air conditioners, and refrigerators; they ignite the gasoline in automobile engines, travel through miniature components making up the chips of microcomputers, and perform countless other invaluable tasks.

In this chapter we shall define current and discuss some of the factors that contribute to the resistance to the flow of charge in conductors. We shall discuss energy transformations in electric circuits. These topics will form a foundation for additional work with circuits in later chapters. Finally, a most interesting essay on the exciting topic of superconductivity follows this chapter.

18.1 ELECTRIC CURRENT

Whenever electric charges move, a *current* is said to exist. To define current more precisely, suppose the charges are moving perpendicular to a surface of area A as in Figure 18.1. This area could be the cross-sectional area of a wire, for example. The **current** is *the rate at which charge flows through this surface.* If ΔQ is the amount of charge that passes through this area in a time interval Δt, the current, I, is equal to the ratio of the charge to the time interval:

$$I \equiv \frac{\Delta Q}{\Delta t} \quad (18.1)$$

Current

The SI unit of current is the **ampere** (A), where

$$1 \text{ A} = 1 \text{ C/s} \quad (18.2)$$

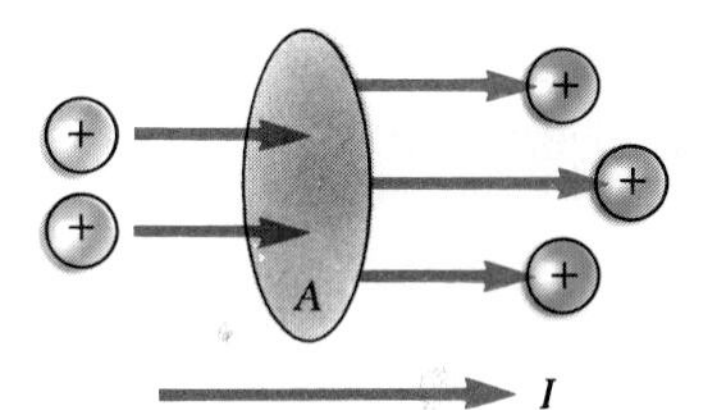

FIGURE 18.1 Charges in motion through an area A. The time rate of flow of charge through the area is defined as the current I. The direction of the current is in the direction of flow of positive charges.

Thus, 1 A of current is equivalent to 1 C of charge passing through the cross-sectional area in a time interval of 1 s. In practice, smaller units of current are often used, such as the milliampere (1 mA $= 10^{-3}$ A) and the microampere (1 μA $= 10^{-6}$ A).

The charges flowing through a surface as in Figure 18.1 can be positive, negative, or both. It is *conventional to choose the direction of the current to be in the direction of flow of positive charge.* In a metallic conductor such as copper, the current is due to the motion of the negatively charged electrons. Therefore, when we speak of current in such a conductor, *the direction of the current will be opposite the flow of electrons.*

Students studying physics for the first time often wonder why the direction of current is defined in this manner. The primary reason for this choice of direction lies in a lack of understanding of what actually moves when there is a current in a wire. In the early study of electric circuits, the direction of current was taken to be that of the motion of positive charges. Even though we now know that it is negative charges (electrons) that move in a conductor, the historical legacy of a current being in the direction of motion of positive charges remains with us.

If one considers a beam of positively charged protons in an accelerator, like the Van de Graaff accelerator discussed in Chapter 16, the current is in the direction of motion of the protons. In some cases, the current is the result of the flow of both positive and negative charges. This occurs, for example, in semiconductors, which we shall discuss in Chapter 31. It is common to refer to a moving charge (whether it is positive or negative) as a *mobile charge carrier.* For example, the mobile charge carriers in a metal are electrons.

EXAMPLE 18.1 The Current in a Wire

A charge of 6×10^{-3} C passes through a cross-sectional area of a wire in 2 s. What is the current in the wire?

Solution From Equation 18.1, we have

$$I = \frac{\Delta Q}{\Delta t} = \frac{6 \times 10^{-3}\ \text{C}}{2\ \text{s}} = 3 \times 10^{-3}\ \text{A} = 3\ \text{mA}$$

EXERCISE If a wire carries a current of 5 A, how much charge moves through a cross section in 4 s?

Answer 20 C.

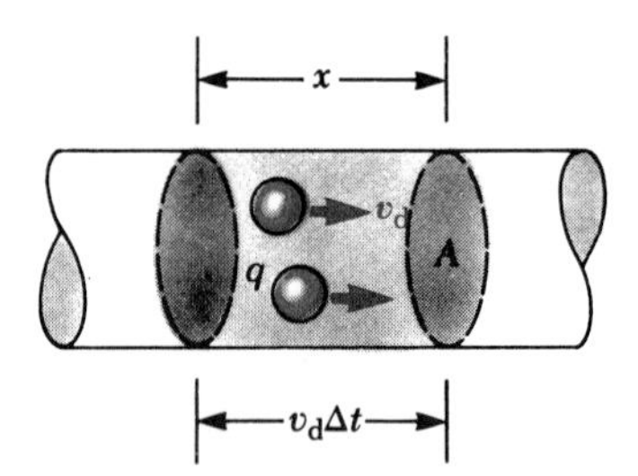

FIGURE 18.2 A section of a uniform conductor of cross-sectional area A. The charge carriers move with a speed v_d, and the distance they travel in a time Δt is given by $x = v_d\Delta t$. The number of mobile charge carriers in the section of length x is given by $nAv_d\Delta t$, where n is the number of mobile charge carriers per unit volume.

18.2 CURRENT AND DRIFT VELOCITY

It is instructive to relate current to the motion of the charged particles. To illustrate this point, consider the current in a conductor of cross-sectional area A (Fig. 18.2). We shall focus our attention on a small segment of the conductor of length x; the volume of this segment is Ax. If n represents the number of mobile charge carriers per unit volume, then the total number of mobile charge carriers in this volume element is nAx. Therefore, the total charge ΔQ that is free to move in this volume is

$$\Delta Q = \text{number of charges} \times \text{charge per particle} = (nAx)(q)$$

where q is the charge on each particle. Let us choose a time interval Δt such that all the charges in the segment of length x will move through the cross-sectional area A. If the charge carriers move with a speed v_d, the distance they move in a time Δt is $x = v_d\Delta t$. Therefore, we can write ΔQ in the form

$$\Delta Q = (nAv_d\Delta t)(q)$$

If we divide both sides of this equation by Δt, we see that the current in the conductor is given by

$$I = \frac{\Delta Q}{\Delta t} = nqv_dA \qquad (18.3)$$

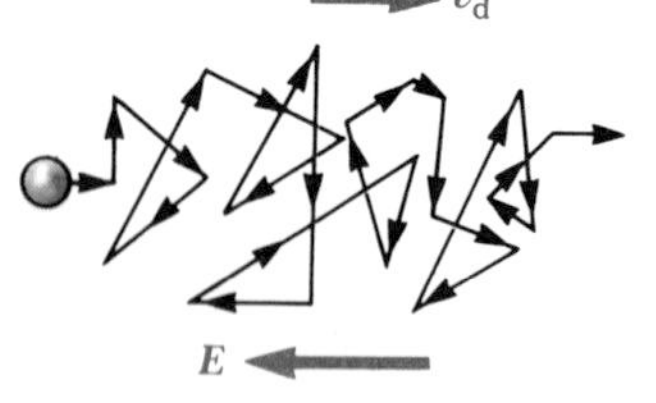

FIGURE 18.3 A schematic representation of the zigzag motion of a charge carrier in a conductor. The changes in direction are due to collisions with atoms in the conductor. Note that the net motion of electrons is opposite the direction of the electric field.

The velocity of the charge carriers, v_d, is actually an average velocity and is called the **drift velocity.** To understand the meaning of drift velocity, consider a conductor in which the charge carriers are free electrons. In such a conductor, the electrons undergo random motion similar to that of gas molecules. When a battery is connected across the ends of the conductor, an electric field is set up in the conductor, which exerts an electric force on the electrons. This electric force causes the electrons to drift in a direction opposite that of the electric field (the drift is opposite the field because the electrons are negatively charged). This motion of charge constitutes a current. In reality, the electrons do not simply move in straight lines along the conductor. Instead, they undergo repeated collisions with the metal atoms, which results in a complicated zigzag motion (Fig. 18.3). Despite the collisions, the electrons do move slowly (drift) along the conductor, and their average velocity is what is called the drift velocity. However, if the voltage across the conductor is zero, the drift velocity is zero.

Before we end our discussion of current, it is instructive to clarify what might appear to be a discrepancy regarding a property of electric fields in conductors. We stated above that an electric field is set up in a conductor when a voltage is applied across its ends, but in Chapter 16 we stated that the electric field is zero inside a conductor. There is no conflict between these statements because an electric field *can* exist in a conductor when we are dealing with charges in motion. This is in contrast to the discussion of Chapter 16, where electrostatic conditions were assumed (the charges are at rest). In the latter case, the electric field inside the conductor must indeed be zero.

EXAMPLE 18.2 The Drift Velocity in a Copper Wire

A copper wire of cross-sectional area 3×10^{-6} m^2 carries a current of 10 A. Assuming that each copper atom contributes one free electron to the metal, find the drift velocity of the electrons in this wire. The density of copper is 8.95 g/cm^3.

Solution From the periodic table of the elements, we find that the atomic weight of copper is 63.5 g/mole. Recall that one atomic mass of any substance contains Avogadro's number of atoms, 6.02×10^{23} atoms. Knowing the density of copper enables us to calculate the volume occupied by 63.5 g of copper:

$$V = \frac{m}{\rho} = \frac{63.5 \text{ g}}{8.95 \text{ g/cm}^3} = 7.09 \text{ cm}^3$$

If we now assume that each copper atom contributes one free electron to the body of the material, we have

$$n = \frac{6.02 \times 10^{23} \text{ electrons}}{7.09 \text{ cm}^3} = 8.48 \times 10^{22} \text{ electrons/cm}^3$$

$$= \left(8.48 \times 10^{22} \frac{\text{electrons}}{\text{cm}^3}\right)\left(10^6 \frac{\text{cm}^3}{\text{m}^3}\right)$$

$$= 8.48 \times 10^{28} \text{ electrons/m}^3$$

From Equation 18.3, we find that the drift velocity is

$$v_d = \frac{I}{nqA} = \frac{10 \text{ C/s}}{(8.48 \times 10^{28} \text{ electrons/m}^3)(1.6 \times 10^{-19} \text{ C})(3 \times 10^{-6} \text{ m}^2)}$$

$$= 2.46 \times 10^{-4} \text{ m/s}$$

Example 18.2 shows that typical drift velocities are very small. In fact, the drift velocity is much smaller than the average velocity between collisions. For instance, electrons traveling with this velocity would take about 68 min to travel 1 m! In view of this low speed, you might wonder why a light turns on almost instantaneously when a switch is thrown. This can be explained by considering the flow of water through a pipe. If a drop of water is forced in one end of a pipe that is already filled with water, a drop must be pushed out the other end of the pipe. While it may take individual drops of water a long time to make it through the pipe, a flow initiated at one end produces a similar flow at the other end very quickly. In a conductor, the electric field that drives the free electrons travels through the conductor with a speed close to that of light. Thus, when you flip a light switch, the message for the electrons to start moving through the wire (the electric field) reaches them at a speed of the order of 10^8 m/s.

18.3 OHM'S LAW

For many conductors, the current is proportional to the potential difference, V, across the conductor, that is, $I \propto V$. This says that if one connects one wire to a 3-V battery and an identical wire to a 1.5-V battery, the current in the first wire will be twice the current in the second wire.

An additional factor that affects the flow of charges through a wire is the **resistance,** R. The current in a wire is inversely proportional to the resistance of the wire, that is, $I \propto 1/R$. According to this relationship, if we double the resistance of a circuit without changing the applied voltage, the current is reduced to one half its original value.

George Simon Ohm (1787–1854) was the first to note that the current in a conductor is directly proportional to the applied voltage as

$$I \propto V$$

Later, a proportionality constant R was introduced to provide the following useful equation:

Ohm's law

$$\boxed{V = IR} \tag{18.4}$$

This equation is called **Ohm's law.** R is called the **resistance** of the conductor and has the units of **ohms.** The symbol Ω (Greek letter omega) is usually used to represent ohms. For example, the filament of a lightbulb might have a resistance of 100 Ω when hot.

Ohm's law is not a fundamental law of nature. Rather, *it is based on experimental observation and is valid only for certain materials.* Materials that obey Ohm's law are said to be *ohmic,* and those that do not are called *non-ohmic.*

From Equation 18.4, we see that resistance has the units of volts per ampere. One volt per ampere is defined to be one ohm:

$$\boxed{1\ \Omega \equiv 1\ \text{V/A}}$$

That is, if a potential difference of 1 V across a conductor produces a current of 1 A, the resistance of the conductor is equal to 1 Ω.

EXAMPLE 18.3 The Resistance of a Steam Iron

All electric devices are required to have an identifying plate that specifies their electrical characteristics. For example, the plate on a certain steam iron states that the iron carries a current of 6 A when connected to a 120-V source. What is the resistance of the steam iron?

Solution From Ohm's law, we find the resistance to be

$$R = \frac{V}{I} = \frac{120\ \text{V}}{6\ \text{A}} = 20\ \Omega$$

EXERCISE The resistance of a hot plate is equal to 40 Ω. How much current does the plate carry when connected to a 120-V source?

Answer 3 A.

18.4 RESISTIVITY

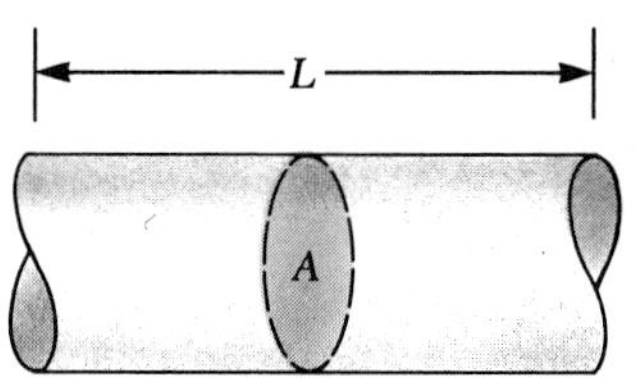

FIGURE 18.4 A conductor of length L and cross-sectional area A.

In an earlier section we pointed out that electrons do not move in straight-line paths through a conductor. Instead, they undergo repeated collisions with the metal atoms. The state of affairs can be pictured as follows. Consider a conductor with a voltage applied between its ends. An electron gains speed as the electric force associated with the internal electric field accelerates the electron, giving it a velocity in the direction opposite the direction of the electric field. A collision with an atom randomizes the electron's velocity, thus reducing its velocity in the direction opposite the field. The process then repeats itself. The net effect these collisions have on the electron is somewhat like a force of internal friction. This is the origin of the resistance of a material. If a conductor has a length L and cross-sectional area A, as shown in Figure 18.4, its resistance is given by

$$R = \rho\frac{L}{A} \tag{18.5}$$

where ρ is an intrinsic property of the conductor called the **resistivity.**[1] Every material has a characteristic resistivity, which depends on the electronic structure of the material and on temperature. Good electric conductors have very low resistivities, and good insulators have very high resistivities. Table 18.1 gives the resistivities of a variety of materials at 20°C. Because resistance values are in ohms, resistivity values must be in ohm-meters.

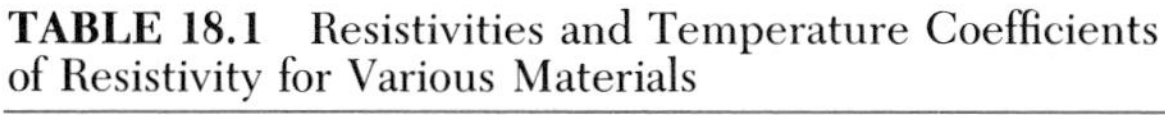

TABLE 18.1 Resistivities and Temperature Coefficients of Resistivity for Various Materials

Material	**Resistivity**[a] **($\Omega \cdot$ m)**	**Temperature coefficient**[a] **of resistivity**[a] **[(C°)$^{-1}$]**
Silver	1.59×10^{-8}	3.8×10^{-3}
Copper	1.7×10^{-8}	3.9×10^{-3}
Gold	2.44×10^{-8}	3.4×10^{-3}
Aluminum	2.82×10^{-8}	3.9×10^{-3}
Tungsten	5.6×10^{-8}	4.5×10^{-3}
Iron	10×10^{-8}	5.0×10^{-3}
Platinum	11×10^{-8}	3.92×10^{-3}
Lead	22×10^{-8}	3.9×10^{-3}
Nichrome[b]	150×10^{-8}	0.4×10^{-3}
Carbon	3.5×10^{5}	-0.5×10^{-3}
Germanium	0.46	-48×10^{-3}
Silicon	640	-75×10^{-3}
Glass	10^{10}–10^{14}	
Hard rubber	10^{13}–10^{6}	
Sulfur	10^{15}	
Quartz (fused)	75×10^{16}	

[a]All values at 20°C.
[b]A nickel-chromium alloy commonly used in heating elements.

[1]The symbol ρ used for resistivity should not be confused with the same symbol used earlier in the text for density. Very often, the same symbol is used to represent different quantities.

Equation 18.5 shows that the resistance of a cylindrical conductor is proportional to its length and inversely proportional to its cross-sectional area. Therefore, if the length of a wire is doubled, its resistance doubles. As its cross-sectional area increases, its resistance decreases.

EXAMPLE 18.4 The Resistance of a Conductor

Calculate the resistance of a piece of aluminum that is 10 cm long and has a cross-sectional area of 10^{-4} m^2. Repeat the calculation for a piece of hard rubber of resistivity 10^{13} $\Omega \cdot$ m.

Solution From Equation 18.5 and Table 18.1, we can calculate the resistance of the aluminum bar:

$$R = \rho\frac{L}{A} = (2.82 \times 10^{-8}\ \Omega \cdot \text{m})\left(\frac{0.1\ \text{m}}{10^{-4}\ \text{m}^2}\right) = 2.82 \times 10^{-5}\ \Omega$$

Similarly, for rubber we find

$$R = \rho\frac{L}{A} = (10^{13}\ \Omega \cdot \text{m})\left(\frac{0.1\ \text{m}}{10^{-4}\ \text{m}^2}\right) = 10^{16}\ \Omega$$

As you might expect, aluminum has a much lower resistance than rubber. This is the reason that aluminum is a good conductor and rubber is a poor conductor.

EXAMPLE 18.5 The Resistance of Nichrome Wire

(a) Calculate the resistance per unit length of a 22-gauge nichrome wire of radius 0.321 mm.

Solution The cross-sectional area of this wire is

$$A = \pi r^2 = \pi(0.321 \times 10^{-3}\ \text{m})^2 = 3.24 \times 10^{-7}\ \text{m}^2$$

The resistivity of nichrome is 1.5×10^{-6} $\Omega \cdot$ m (Table 18.1). Thus, we can use Equation 18.5 to find the resistance per unit length:

$$\frac{R}{L} = \frac{\rho}{A} = \frac{1.5 \times 10^{-6}\ \Omega \cdot \text{m}}{3.24 \times 10^{-7}\ \text{m}^2} = 4.6\ \Omega/\text{m}$$

(b) If a potential difference of 10 V is maintained across a 1-m length of the nichrome wire, what is the current in the wire?

Solution Since a 1-m length of this wire has a resistance of 4.6 Ω, Ohm's law gives

$$I = \frac{V}{R} = \frac{10\ \text{V}}{4.6\ \Omega} = 2.2\ \text{A}$$

The resistance of nichrome wire is about 100 times larger than that of copper wire. Specifically, a copper wire of the same gauge would have a resistance per unit length of only 0.052 Ω/m. A 1-m length of 22-gauge copper wire would carry the same current (2.2 A) with an applied voltage of only 0.11 V.

Because of its high resistivity, nichrome is often used for heating elements in toasters, irons, and electric heaters.

EXERCISE What is the resistance of a 6-m length of 22-gauge nichrome wire? How much current does it carry when connected to a 120-V source?

Answer: 28 Ω, 4.3 A.

18.5 TEMPERATURE VARIATION OF RESISTANCE

The resistivity, and hence the resistance, of a conductor depends on a number of factors. One of the most important of these factors is the temperature of the metal. For most metals, resistivity increases with increasing temperature. This increase can be understood as follows. As the temperature of the material increases, its constituent atoms vibrate with larger and larger amplitudes. Just as it is more difficult to weave one's way through a crowded room when the people are in motion than it is when they are standing still, so do the electrons find it more difficult to pass the atoms moving with large amplitudes.

The resistivity of a conductor varies with temperature over a limited range according to the expression

$$\boxed{\rho = \rho_0[1 + \alpha(T - T_0)]} \tag{18.6}$$

where ρ is the resistivity at some temperature T (in °C), ρ_0 is the resistivity at some reference temperature T_0 (usually taken to be 20°C), and α is a parameter called the **temperature coefficient of resistivity.** The temperature coefficients for various materials are given in Table 18.1.

Since the resistance of a conductor with uniform cross section is proportional to the resistivity according to Equation 18.5 ($R = \rho L/A$), the temperature variation of resistance can be written

$$\boxed{R = R_0[1 + \alpha(T - T_0)]} \tag{18.7}$$

Precise temperature measurements are often made using this property, as shown in the following example.

EXAMPLE 18.6 A Platinum Resistance Thermometer

A resistance thermometer made from platinum has a resistance of 50.0 Ω at 20°C. When the device is immersed in a vessel containing melting indium, its resistance increases to 76.8 Ω. From this information, find the melting point of indium. Note that $\alpha = 3.92 \times 10^{-3}\ (\text{C}°)^{-1}$ for platinum.

Solution If we solve Equation 18.7 for $T - T_0$, we obtain

$$T - T_0 = \frac{R - R_0}{\alpha R_0} = \frac{76.8\ \Omega - 50.0\ \Omega}{[3.92 \times 10^{-3}\ (\text{C}°)^{-1}][50.0\ \Omega]} = 137\ \text{C}°$$

Since $T_0 = 20°\text{C}$, we find that the melting point is $T = 157°\text{C}$.

---

*The Carbon Microphone

The carbon microphone is commonly used in the mouthpiece of a telephone. The construction of this device is illustrated in Figure 18.5. A flexible steel diaphragm is placed in contact with carbon granules inside a container. The carbon granules serve as the primary resistance medium in a circuit containing a source of current (shown here as a battery) and a transformer, whose operation will be described in a later chapter.

The magnitude of the current in the circuit changes when a sound wave strikes the diaphragm. When a compression arrives at the microphone, the diaphragm flexes inward, causing the carbon granules to press together into a

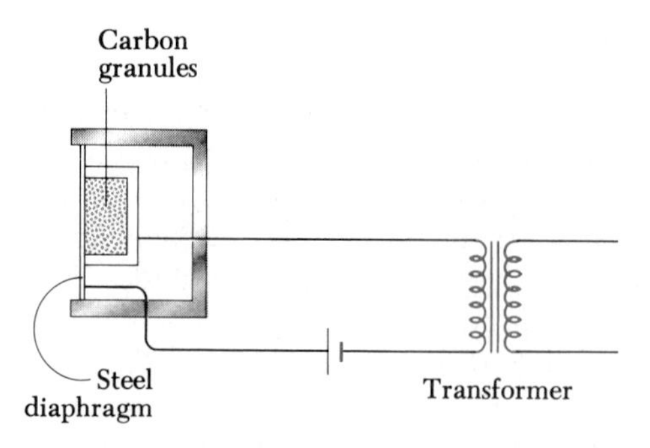

FIGURE 18.5 Diagram of a carbon microphone.

smaller-than-normal volume, corresponding to a decrease in the length of the resistance medium. This results in a lower circuit resistance and hence an increase in current in the circuit. When a rarefaction arrives at the microphone, the diaphragm relaxes, and the carbon granules become more loosely packed. This results in an increase in the circuit resistance and a corresponding decrease in current. These variations in current, which follow the changes of the sound wave, are sent through the transformer into the transmission line of the telephone company. A speaker in the listener's earpiece then converts these electric signals back to a sound wave.

The carbon microphone has a very poor frequency response. It reproduces frequencies below 4000 Hz adequately and is therefore suitable for speech transmission, since the critical frequencies in normal conversation are usually below this value. However, its capabilities fall off dramatically at higher frequencies, rendering it useless for high-fidelity purposes, which require reliable sound reproduction at all frequencies between 20 Hz and 20,000 Hz.

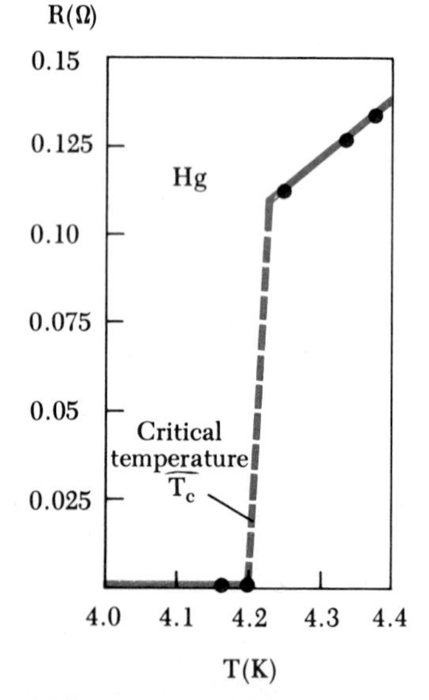

FIGURE 18.6 Resistance versus temperature for mercury. The graph follows that of a normal metal above the critical temperature T_c. The resistance drops to zero at the critical temperature, which is 4.2 K for mercury, and remains at zero for lower temperatures.

18.6 SUPERCONDUCTIVITY

There is a class of metals and compounds whose resistance goes virtually to *zero* below a certain temperature, T_c, called the *critical temperature*. These materials are known as **superconductors.** The resistance-temperature graph for a superconductor follows that of a normal metal at temperatures above T_c (Fig. 18.6). When the temperature is at or below T_c, the resistance drops suddenly to zero. This phenomenon was discovered in 1911 by the Dutch physicist H. Kamerlingh-Onnes when he was working with mercury, which is a superconductor below 4.2 K. Recent measurements have shown that the resistivities of superconductors below T_c are less than $4 \times 10^{-25}\ \Omega \cdot \text{m}$, which is around 10^{17} times smaller than the resistivity of copper and considered to be *zero* in practice.

Today there are thousands of known superconductors. Such common metals as aluminum, tin, lead, zinc, and indium are superconductors. Table 18.2 lists the critical temperatures of several superconductors. The value of T_c is sensitive to chemical composition, pressure, and crystalline structure. It is interesting to note that copper, silver, and gold, which are excellent conductors, do not exhibit superconductivity.

One of the most important recent developments in physics that has created much excitement in the scientific community has been the discovery of high temperature copper-oxide-based superconductors. The excitement began with a 1986 publication by Georg Bednorz and K. Alex Müller, two scientists working at the IBM Zurich Research Laboratory in Switzerland, who reported evidence for superconductivity in the 30 K range in an oxide of barium, lanthanum, and copper. Bednorz and Müller were awarded the Nobel Prize in 1987 for their incredible and important discovery. Shortly thereafter, a new family of compounds was open for investigation, and research activity in the field of superconductivity proceeded vigorously. In early 1987, groups at the University of Alabama at Huntsville and the University of Houston announced the discovery of superconductivity at about 92 K in an oxide of yttrium, barium, and copper ($YBa_2Cu_3O_7$). Late in 1987, teams of scientists from Japan and the United States reported superconductivity at 105 K in an oxide of bismuth, strontium, calcium, and copper. Most recently, scientists have reported superconductivity as high as 125 K in an oxide containing thallium. At this point, one cannot rule out the possibility of room temperature superconductivity, and the search for novel superconducting materials continues. These developments are very exciting and important both for scientific reasons and because practical applications become more probable and widespread as the critical temperature is raised. The interesting essay at the end of this chapter discusses superconductivity and the new generation of superconductors in more detail.

TABLE 18.2 Critical Temperatures for Various Superconductors

Material	T_c (K)
Nb_3Ge	23.2
Nb_3Sn	18.05
Nb	9.46
Pb	7.18
Hg	4.15
Sn	3.72
Al	1.19
Zn	0.88
$YBa_2Cu_3O_7$	90
Bi-Sr-Ca-Cu-O	105
Tl-Ba-Ca-Cu-O	125

One of the truly remarkable features of superconductors is the fact that, once a current is set up in them, it will persist *without any applied voltage* (since $R = 0$). In fact, steady currents have been observed to persist in superconducting loops for many years with no apparent decay!

One important and useful application of superconductivity has been the construction of superconducting magnets in which the magnetic field intensities are about ten times greater than those of the best normal electromagnets. Such superconducting magnets are being considered as a means of storing energy. The idea of using superconducting power lines for transmitting power efficiently is also receiving some consideration. Modern superconducting electronic devices consisting of two thin-film superconductors separated by a thin insulator have been constructed. These devices include magnetometers (a magnetic-field measuring device) and various microwave devices.

18.7 ELECTRICAL ENERGY AND POWER

If a battery is used to establish an electric current in a conductor, there is a continuous transformation of the chemical energy stored in the battery into potential energy in the charge carriers. This extra energy added to the charge carriers is lost when they collide with the atoms of the conductor. This results in a temperature increase in the conductor. In other words, the chemical energy stored in the battery is continuously transformed into thermal energy.

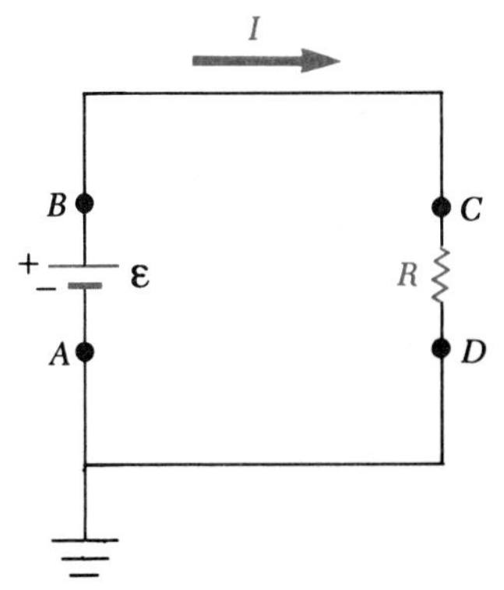

FIGURE 18.7 A circuit consisting of a battery and resistance R. Positive charge flows in the clockwise direction, from the positive to the negative terminal of the battery. Point A is grounded.

Consider a simple circuit consisting of a battery whose terminals are connected to a resistor, as shown in Figure 18.7. Let us now follow a positive quantity of charge ΔQ as it moves around the circuit from point A through the battery and resistor and back to A. Point A is a reference point that is grounded (ground symbol is ⏚), and its potential is zero. As the charge moves from A to B through the battery, its electrical potential energy increases by an amount $V\Delta Q$ (where V is the potential at B) while the chemical potential energy in the battery decreases by the same amount. However, as the charge moves from C to D through the resistor, it loses this electrical potential energy as it undergoes collisions with the atoms of the resistor. The electrical potential energy is converted to thermal energy. Note that there is no loss in energy for the paths BC and DA because the connecting wires are assumed to have zero resistance. (In all future circuit diagrams, straight lines will indicate connecting wires with zero resistance.) When the charge returns to the negative terminal of the battery, *it must have the same potential energy (zero) as it had at the start.* If this were not the case, the charge would gain energy during each trip around the circuit, which would be a violation of the law of conservation of energy.

The electrical potential energy that is transformed into heat as the charge ΔQ flows through the resistor is $V\Delta Q$. Thus, if a time Δt is required for the charge to move through the resistor, the rate of conversions of electrical energy to heat is

$$\frac{\Delta Q}{\Delta t}V = IV$$

where I is the current in the circuit. Of course, the charge regains this energy when it passes through the battery. Since the rate at which the charge loses energy equals the **power,** P, lost in the resistor, we have

Power

$$\boxed{P = IV} \tag{18.8}$$

In this case, the power is supplied to a resistor by a battery. However, Equation 18.8, often referred to as **Joule's law,** can be used to calculate the power transferred to any device carrying a current I and having a potential difference V between its terminals.

By combining Equation 18.8 and the voltage-current relationship for a resistor, $V = IR$, we can express the power lost in the resistor in either of two alternative forms:

Power

$$\boxed{P = I^2R = \frac{V^2}{R}} \tag{18.9}$$

When I is in amperes, V in volts, and R in ohms, the SI unit of power is the **watt** (W). (Recall that 1 W = 1 J/s.)

A battery or any other device that provides electrical energy is called a seat of electromotive force, usually referred to as an **emf,** but the phrase *electromotive force* is an unfortunate one because it does not really describe a force but actually refers to a potential difference in volts.

Neglecting the internal resistance of the battery, the potential difference between points A and B in Figure 18.7 is equal to the emf, $\mathcal{E}$, of the battery. That is $V = V_B - V_A = \mathcal{E}$, and the current in the circuit (from Ohm's law) is given by $I = V/R = \mathcal{E}/R$. Since $V = \mathcal{E}$, the power supplied by the emf can be expressed as $P = I\mathcal{E}$, which, of course, equals the power lost in the resistor, I^2R.

EXAMPLE 18.7 The Power Consumed by an Electric Heater

An electric heater is constructed by applying a potential difference of 50 V to a nichrome wire of total resistance 8 Ω. Find the current carried by the wire and the power rating of the heater.

Solution Since $V = IR$, we have

$$I = \frac{V}{R} = \frac{50\ \text{V}}{8\ \Omega} = 6.25\ \text{A}$$

We can find the power rating using $P = I^2R$:

$$P = I^2R = (6.25\ \text{A})^2(8\ \Omega) = 313\ \text{W}$$

EXERCISE If we were to double the applied voltage to the heater, what would happen to the current and power?

Answer The current would double, and the power would quadruple.

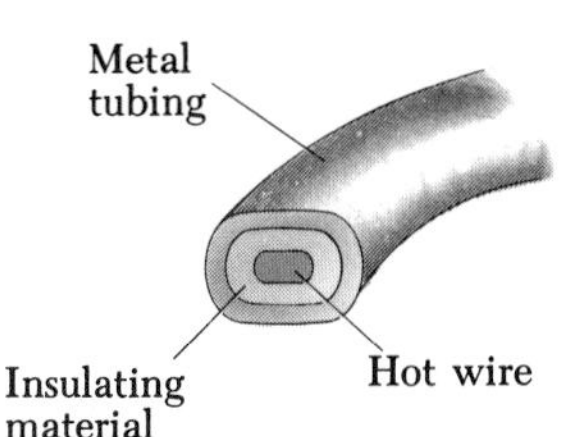

FIGURE 18.8 The cross-section of a heating element used in an electric range.

EXAMPLE 18.8 Electrical Rating of a Lightbulb

A lightbulb is rated at 120 V and 75 W. That is, its operating voltage is 120 V and it has a power rating of 75 W. The bulb is powered by a 120-V direct current power supply. Find the current in the bulb and its resistance.

Solution Since we know that the power rating of the bulb is 75 W and the operating voltage is 120 V, we can use $P = IV$ to find the current:

$$I = \frac{P}{V} = \frac{75\ \text{W}}{120\ \text{V}} = 0.625\ \text{A}$$

Using Ohm's law, $V = IR$, the resistance is calculated to be

$$R = \frac{V}{I} = \frac{120\ \text{V}}{0.625\ \text{A}} = 192\ \Omega$$

EXERCISE What would the resistance be in a lamp rated at 120 V and 100 W?

Answer 144 Ω.

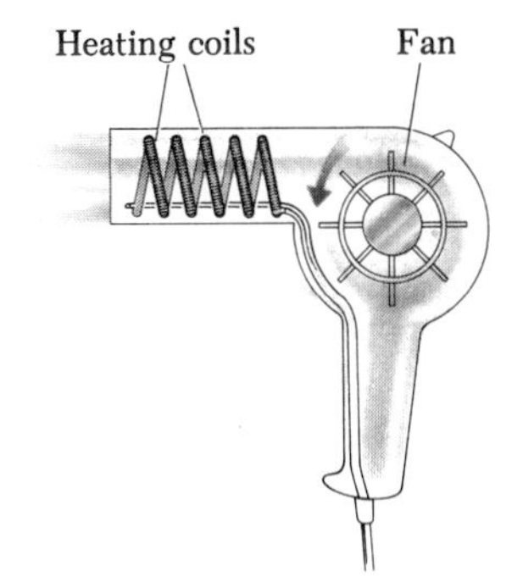

FIGURE 18.9 Diagram of a hair dryer. Warm air is produced by blowing air from a fan past the heating coils.

*18.8 ENERGY CONVERSION IN HOUSEHOLD CIRCUITS

The heat generated when current flows through a resistive material is used in many common devices. A cross-sectional view of the spiral heating element of an electric range is shown in Figure 18.8. The material through which the current passes is surrounded by an insulating substance in order to prevent the current from flowing through the cook to the earth when he or she touches the pan. A material that is a good conductor of heat surrounds the insulator.

Figure 18.9 shows a common hair dryer, in which a fan blows air past heating coils. In this case the warm air can be used to dry hair, but on a broader scale this same principle is used to dry clothes and to heat buildings.

A final example of a household appliance that uses the heating effect of electric currents is the steam iron shown in Figure 18.10. A heating coil warms the bottom of the iron and simultaneously turns water to steam, which is sprayed from jets located in the bottom of the iron.

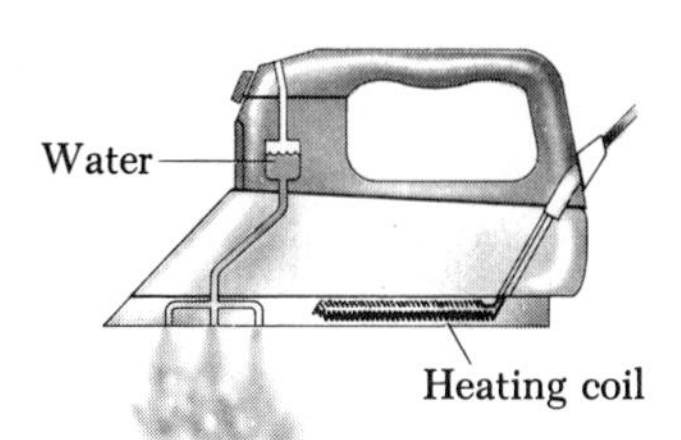

FIGURE 18.10 Diagram of a steam iron. Water is turned into steam by heat from the heating coil.

Regardless of the way in which you use electrical energy in your home, you ultimately must pay for it or risk having your power turned off. The unit of energy the electric company uses to calculate consumption, the **kilowatt-hour,** is defined in terms of the unit of power. One kilowatt-hour (kWh) is the energy converted or consumed in 1 h at the constant rate of 1 kW. The numerical value of 1 kWh is

$$1 \text{ kWh} = (10^3 \text{ W})(3600 \text{ s}) = 3.6 \times 10^6 \text{ J} \tag{18.10}$$

On your electric bill, the amount of electricity used is usually stated in multiples of kWh.

EXAMPLE 18.9 The Cost of Operating a Lightbulb

How much does it cost to burn a 100-W lightbulb for 24 h if electricity costs eight cents per kilowatt-hour?

Solution A 100-W lightbulb is equivalent to a 0.1-kW bulb. Since the energy consumed equals power × time, the amount of energy you must pay for, expressed in kWh, is

$$\text{Energy} = (0.10 \text{ kW})(24 \text{ h}) = 2.4 \text{ kWh}$$

If energy is purchased at eight cents per kWh, the cost is

$$\text{Cost} = (2.4 \text{ kWh})(\$0.08/\text{kWh}) = \$0.19$$

That is, it will cost 19 cents to operate the lightbulb for one day. This is a small amount, but when larger and more complex devices are being used, the costs go up rapidly.

Demands on our energy supplies have made it necessary to be aware of the energy requirements of our electric devices. This is true not only because they are becoming more expensive to operate but also because, with the dwindling of the coal and oil resources that ultimately supply us with electrical energy, increased awareness of conservation becomes necessary. On every electric appliance is a label that contains the information you need to calculate the power requirements of the appliance. The power consumption in watts is often stated directly, as on a lightbulb. In other cases, the amount of current used by the device and the voltage at which it operates are given. This information and Equation 18.8 are sufficient to calculate the operating cost of any electric device.

EXERCISE If electricity costs eight cents per kilowatt-hour, what does it cost to operate an electric oven, which operates at 20 A and 220 V, for 5 h?

Answer \$1.76.

SUMMARY

The **electric current,** I, in a conductor is defined as

$$I \equiv \frac{\Delta Q}{\Delta t} \tag{18.1}$$

where ΔQ is the charge that passes through a cross section of the conductor in a time Δt. The SI unit of current is the **ampere** (A), where I A = 1 C/s.

By convention, the direction of current is in the direction of flow of positive charge.

The current in a conductor is related to the motion of the charge carriers through the relationship

$$I = nqv_dA \tag{18.3}$$

where n is the number of mobile charge carriers per unit volume, q is their charge, v_d is the drift velocity of the charges, and A is the cross-sectional area of the conductor.

Ohm's law provides a relationship between the voltage drop across a conductor (V), the current through it (I), and the resistance of the conductor (R):

$$V = IR \tag{18.4}$$

If a conductor has a length L and cross-sectional area A, its **resistance** is

$$R = \rho\frac{L}{A} \tag{18.5}$$

where ρ is an intrinsic property of the conductor called the **electrical resistivity.** The SI unit of resistance is the **ohm** (Ω), and the SI unit of resistivity is the **ohm-meter** ($\Omega \cdot$ m).

The resistivity of a conductor varies with temperature over a limited temperature range according to the expression

$$\rho = \rho_0[1 + \alpha(T - T_0)] \tag{18.6}$$

where α is the **temperature coefficient of resistivity** and ρ_0 is the resistivity at some reference temperature T_0 (usually taken to be 20° C).

The resistance of a conductor varies with temperature according to the expression

$$R = R_0[1 + \alpha(T - T_0)] \tag{18.7}$$

If a potential difference V is maintained across a conductor, the **power,** or rate at which energy is supplied to the conductor, is given by

$$P = IV$$

Using the fact that $V = IR$ for a resistor, the power lost in a resistor can be expressed as

$$P = I^2R = \frac{V^2}{R} \tag{18.9}$$

A **kilowatt-hour** is the amount of energy converted or consumed in one hour by a device being supplied with power at the rate of 1 kW. This is equivalent to

$$1 \text{ kWh} = 3.6 \times 10^6 \text{ J} \tag{18.10}$$

ADDITIONAL READING

M. Azbel et al., "Conduction Electrons in Metals," *Sci. American*, January 1973, p. 88.

H. Ehrenreich, "The Electrical Properties of Materials," *Sci. American*, September 1967, p. 194.

T. Geballe, "Superconductors in Electric Power Technology," *Sci. American*, November 1980, p. 138.

R. T. Matthias, "Superconductivity," *Sci. American*, November 1957, p. 92.

D. Snowden, "Superconductors in Power Transmission," *Sci. American*, April 1972, p. 84.

QUESTIONS

1. What would happen to the drift velocity of the electrons in a wire and to the current in the wire if the electrons could move freely without resistance through the wire?
2. What is the difference between resistance and resistivity?
3. If charges flow very slowly through a metal, why does it not require several hours for a light to come on when you throw a switch?
4. What factors affect the resistance of a conductor?
5. Two wires, A and B, are made of the same metal and have equal lengths, but the resistance of wire A is three times greater than that of wire B. What is the ratio of their cross-sectional areas? How do their radii compare?
6. We have seen that an electric field must exist inside a conductor that carries a current. How is this possible in view of the fact that, in our study of electrostatics, we concluded that the electric field must be zero inside a conductor?
7. When the voltage across a certain conductor is doubled, the current is observed to increase by a factor of 3. What can you conclude about the conductor?
8. When incandescent lamps burn out, they usually do so just after they are switched on. Why?
9. Two lightbulbs operate from 120 V, but one has a power rating of 25 W and the other a power rating of 100 W. Which bulb has the higher resistance? Which carries the greater current?
10. Two conductors of the same length and radius are connected across the same potential difference. One conductor has twice the resistance of the other. Which conductor will dissipate more power?
11. Use the atomic theory of matter to explain why the resistance of a material should increase as its temperature increases.
12. A typical monthly utility rate structure might be: $1.60 for the first 16 kWh, 7.05 cents/kWh for the next 34 hWh used, 5.02 cents/kWh for the next 50 kWh, 3.25 cents/kWh for the next 200 kWh, 2.95 cents/kWh for the next 200 kWh, 2.35 cents/kWh for all in excess of 400 kWh. Based on these rates, what would be the charge for 227 kWh? From the standpoint of encouraging conservation of energy, what is wrong with this pricing method?
13. Some homes have light dimmers which you operate by rotating a knob. What are you changing in the electric circuit when you rotate the knob?
14. If materials could be produced that would be superconducting at room temperature, list some ways that such materials could benefit mankind.
15. What is the difference between the drift velocity of an electron and its velocity?
16. Suppose you charge a pocket comb by friction and then walk around the room carrying the comb. Are you producing an electric current?

PROBLEMS

Section 18.1 Electric Current

Section 18.2 Current and Drift Velocity

1. The compressor on an air conditioner draws 90 A when it starts up. If the start-up time is about 1/2 s, how much charge passes a cross-sectional area of the circuit in this time?

2. How many electrons pass through the cross-sectional area in Problem 1 during the 1/2 s interval?

3. If Avogadro's number of electrons pass by a given cross-sectional area in one hour, find the current in the conductor.

4. An aluminum wire having a cross-sectional area of 4×10^{-6} m^2 carries a current of 5 A. Find the drift velocity of the electrons in the wire. The density of aluminum is 2.7 g/cm^3.

5. In a particular television picture tube, the measured beam current is 60 μA. How many electrons strike the screen every second?

• **6.** A typical aluminum wire has about 1.5×10^{20} free electrons in every centimeter of its length, and the drift speed of these electrons is about 0.03 cm/s. (a) How many electrons pass through a cross-sectional area of the wire every second? (b) What is the current in the wire?

Section 18.3 Ohm's Law

Section 18.4 Resistivity

7. Find the resistance of a piece of silver 1 m long if its cross-sectional area is 10^{-5} m^2.

8. An electrician finds that the resistance of a 1 m length of wire is 0.20 Ω. If he plans to use 120 m of this wire in wiring a house, what is the total resistance of all the wiring used?

9. Calculate the diameter of a 2-cm length of tungsten filament in a small lightbulb if its resistance is 0.05 Ω.

10. A lightbulb has a resistance of 240 Ω when operating at a voltage of 120 V. What is the current through the lightbulb?

11. A typical color television draws about 2.5 A when connected to a 120 V source. What is the effective resistance of the TV set?

12. When operating at 120 V, a certain device draws a current of 0.5 A. What current would be drawn if (a) the operating voltage is lowered to 90 V, (b) the voltage is raised to 130 V?

13. A sales representative promoting energy-efficient products advertises that a certain product has a resistance of 20 Ω and draws a current of only 0.6 A when connected to a 120 V source. Would you buy this product? Why or why not?

14. (a) Calculate the resistance of a piece of nichrome wire 200 cm long with a cross-sectional area of 0.02 cm^2. (b) What current would the wire carry if connected to a 3-V flashlight battery?

15. A potential difference of 12 V is found to produce a current of 0.4 A in a 3.2-m length of wire having a uniform radius of 0.4 cm. (a) Calculate the resistivity of the wire. (b) What is the resistance of the wire?

• **16.** A 2-m piece of iron wire carries a current of 0.25 A when connected to a 6-V battery. What length of gold wire would carry the same current when connected to the battery if the gold wire has the same radius as the iron?

• **17.** A rectangular block of copper has sides of length 10 cm, 20 cm, and 40 cm. If the block is connected to a 6-V source across opposite faces of the rectangular block, what is (a) the maximum current that can be drawn, and (b) the minimum current?

• **18.** A battery establishes a uniform electric field of 100 V/m in a wire of radius 0.5 cm. If the current in the wire is 5 A, what is the resistivity of the material?

Section 18.5 Temperature Variation of Resistance

19. If a silver wire has a resistance of 10 Ω at 20°C, what resistance will it have at 40°C? Neglect any change in length or cross-sectional area due to the change in temperature.

20. The copper wire used in a house has a cross-sectional area of 3 mm^2. If 10 m of this wire is used to wire a circuit in the house at 20°C, find the resistance of the wire at a temperature of (a) 30°C, and (b) 10°C.

21. At 40° C, the resistance of a segment of gold wire is 100 Ω. When the wire is placed in a liquid bath, the resistance decreases to 97 Ω. What is the temperature of the bath?

22. At 20°C the resistance of the carbon resistor in an electric circuit is 200 Ω and it is connected to a 5-V battery. What is the current in the circuit when the temperature of the carbon rises to 80°C?

• **23.** How much would the temperature of a copper wire have to be increased to raise its resistance by 20% over the value it had at 20°C?

• **24.** A certain lightbulb has a tungsten filament with a resistance of 19 Ω when cold and 140 Ω when hot. Assume that Equation 18.7 can be used over the large temperature range involved here and find the temperature of the filament when hot.

• **25.** A copper wire of length 100 cm and radius 0.5 cm has a potential difference across it sufficient to produce a current of 3 A at 20°C. (a) What is the potential difference? (b) If the temperature of the wire is increased to 200°C, what potential difference is now required to produce a current of 3 A?

• **26.** At what temperature will tungsten have a resistivity four times the resistivity of copper at room temperature?

Section 18.7 Electrical Energy and Power

27. A toaster is rated at 600 W when connected to a 120-V source. What current does the toaster carry and what is its resistance?

28. The power supplied to a typical black-and-white television set is 90 W when the set is connected to 120 V. (a) How much energy does this set consume in one hour? (b) A color television set draws about 2.5 A when connected to 120 V. How long is required for it to consume the same energy as does the black-and-white model in one hour?

29. How much current is being supplied by a 240-V generator delivering 120 kW of power?

30. The sticker on a compact disk player says that it draws 300 mA of current at 9 V. What power does it dissipate?

31. The tungsten heating element in a 1500-W heater is 3 m long, and the resistor is to be connected to a 120-V source. What is the cross-sectional area of the wire?

32. How many 100-W lightbulbs can you use in a 120-V circuit without tripping a 15-A circuit breaker? (Note that the bulbs are connected in parallel.)

• 33. A 0.25-hp motor is connected to a 120-V line. What current is drawn?

• 34. What is the required resistance of an immersion heater that will increase the temperature of 1.5 kg of water from 10°C to 50°C in 10 min while operating at 120 V?

• 35. In a hydroelectric installation, a turbine delivers 2000 hp to a generator, which in turn converts 90% of the mechanical energy to electrical energy. Under these conditions, what current will the generator deliver at a potential difference of 3000 V?

•• 36. Two conductors made of the same material are connected across a common potential difference. Conductor A has twice the diameter and twice the length of conductor B. What is the ratio of the power delivered to the two conductors?

Section 18.8 Energy Conversion in Household Circuits

37. How much does it cost to watch a complete 21-hour-long world series on a 90-W black-and-white television set? Assume electricity costs 7 cents per kWh.

38. A certain household uses an average of 2500 kWh of energy per month. If coal supplies 7×10^6 cal/kg, how much coal must be burned in a 40% efficient generator to supply this much energy?

39. If electricity costs 8 cents per kWh, estimate how much it costs a person to dry his hair with a 1500-W blow dryer during a year's time.

• 40. A small motor draws a current of 1.75 A from a 120-V line. The output power of the motor is 0.2 hp. (a) At a rate of \$0.06/kWh, what is the cost of operating the motor for 4 h? (b) What is the efficiency of the motor?

ADDITIONAL PROBLEMS

41. Storage batteries are often rated in terms of the amount of charge they can deliver. How much charge can a 90 amp-hour battery deliver?

42. A steam iron draws 6 A from a 120-V line. How many joules of heat are produced in 20 min?

43. How much does it cost at \$0.08/kWh to run the steam iron of Problem 42 for twenty minutes?

44. The headlights on a car are rated at 80 W. If these are connected to a fully charged 90 amp · h, 12-V battery, how long will it take the battery to completely discharge?

• 45. A particular wire has a resistivity of $3 \times 10^{-8}\ \Omega \cdot$ m and a cross-sectional area of 4×10^{-6} m^2. A length of this wire is to be used as a resistor that will develop 48 W of power when connected across a 20-V battery. What length of wire is required?

• 46. Birds resting comfortably on a high-voltage power line are a common sight. The copper wire on which a bird stands is 2.2 cm in radius and carries a current of 50 A. If the bird's feet are 4 cm apart, calculate the potential difference across his body.

• 47. A length of metal wire has a radius of 5×10^{-3} m and a resistance of 0.1 Ω. When the potential difference across the wire is 15 V, the electron drift velocity is found to be 3.17×10^{-4} m/s. Based on these data, calculate the density of free electrons in the wire.

• 48. Two conductors of the same length and same material are connected in series (one after the other) across a difference in potential. The wires have different cross-sectional areas. (a) When in series, the current is the same in both wires. Does the wire of larger cross-sectional area dissipate more or less power than the smaller wire? (b) Repeat the problem if the wires are connected in parallel (so that the voltage is the same across each).

• 49. What is the required resistance of an immersion heater that will increase the temperature of 1.5 kg of water from 10°C to 50°C in 10 min while operating at a voltage of 110 V?

• 50. The current through a resistor decreases by 3 A when the voltage applied across the resistor decreases from 12 V to 6 V. Find the resistance of the resistor.

• 51. (a) A 115-g mass of aluminum is formed into a right circular cylinder shaped so that the diameter of the cylinder equals its height. Calculate the resistance between the top and bottom faces of the cylinder at 20°C. (b) Calculate the resistance between opposite faces if the same mass of aluminum is formed into the cube.

• **52.** A resistor is constructed by forming a material of resistance $3.5 \times 10^5\ \Omega \cdot m$ into the shape of a hollow cylinder of length 4 cm and inner and outer radii of 0.5 cm and 1.2 cm, respectively. In use, a potential difference is applied between the ends of the cylinder, producing a current parallel to the length of the cylinder. Find the resistance of the cylinder.

•• **53.** The current in a conductor varies in time as shown in Figure 18.11. (a) How many coulombs of charge pass through a cross section of the conductor in the interval $t = 0$ to $t = 5$ s? (b) What constant current would transport the same total charge during the 5-s interval as does the actual current?

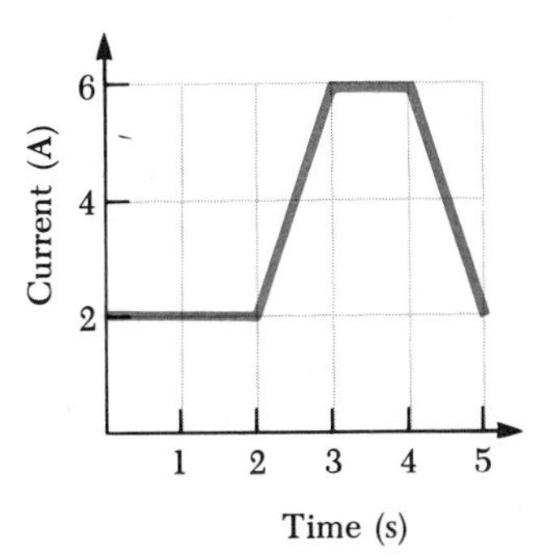

Figure 18.11 Problem 53

•• **54.** A wire of initial length L_0 and radius r_0 has a measured resistance of 1 Ω. The wire is drawn under tensile stress to a new uniform radius of $r = 0.25r_0$. What is the new resistance of the wire?

•• **55.** A 50-g sample of a conducting material is all that is available. The resistivity of the material is measured to be $11 \times 10^{-8}\ \Omega \cdot m$, and the density is 7.86 g/cm^3. It is desired to shape the material into a wire that has a total resistance of 1.5 Ω. (a) What length is required, and (b) what must be the diameter of the wire?

•• **56.** (a) A sheet of copper ($\rho = 1.7 \times 10^{-8}\ \Omega \cdot m$) is 2 mm thick and has surface dimensions of 8 cm × 24 cm. If the long edges are joined to form a hollow tube 24 cm in length, what is the resistance between the ends? (b) What mass of copper would be required to manufacture a spool of copper cable 1500 m in length and having a total resistance of 4.5 Ω?

•• **57.** An engineer is in need of a resistor that is to have zero overall temperature coefficient of resistance at 20°C. The design is a composite of right circular cylinders of two materials, as in Figure 18.12. The ratio of the resistivities of the two materials is $\rho_1/\rho_2 = 3.2$, and the ratio of the lengths of the sections is $L_1/L_2 = 2.6$. The radius is uniform throughout. Assuming that the temperature of the two sections remains equal, calculate α_1/α_2, the required ratio of temperature coefficients of resistivity of the two materials.

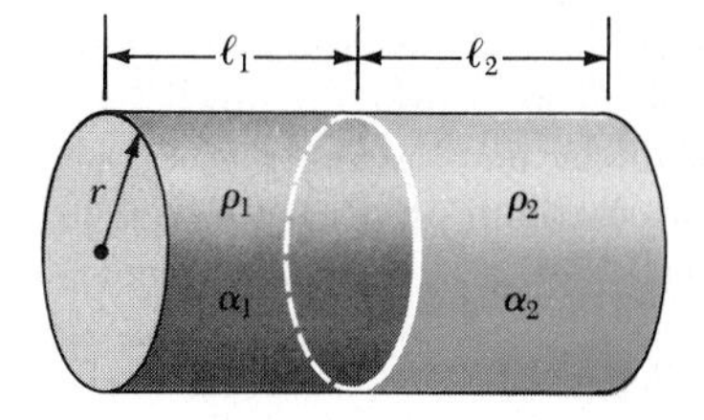

Figure 18.12 Problem 57

Essay

Superconductivity, Then and Now

David Markowitz *University of Connecticut*

Understanding Superconductivity

As temperature is lowered to a few degrees above absolute zero, many metals undergo a change of phase into what is called a superconducting state. A superconductor has many unusual thermal, electrical, magnetic, and optical properties. The most striking property of a superconductor is its electrical resistance. (The electrical resistance of materials is discussed in more detail in Section 18.6.) Above a temperature T_c, called the critical temperature, the metal, perhaps in the form of a wire, is an ordinary good conductor and has some electrical resistance. At or below the critical temperature, however, the resistance of the wire falls abruptly to zero, as shown in Figure 1.

The critical temperature is a property of a particular metallic substance, just as the freezing point of fresh water is 0°C, whether the water is in a tray in your freezer or in a pond in winter. For example, the critical temperature for mercury is about 4.2 K, which is the boiling point of liquid helium. Above T_c the metal is said to be in the normal (N) state (or phase), and below T_c it is said to be in the superconducting (S) state. This is analogous

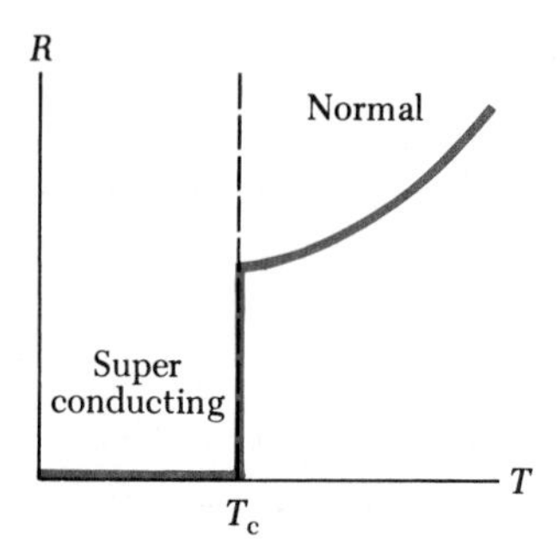

FIGURE 1 Resistance versus temperature for a pure superconductor, such as mercury, lead, or aluminum. The metal behaves normally down to the critical temperature, T_c, below which the resistance drops to zero.

to the fact that a substance is in the liquid state above its freezing temperature and in the solid state below the temperature.

There are many aspects to superconducting materials and many views that could be taken in discussing their properties. In the first section of this essay, we shall not emphasize the electrical and magnetic properties of superconductors. It is sufficient to note that in electrical conduction, the conduction electrons inside a metal are not bound to individual atoms, but rather are free to move through the metal over long distances as they contribute to the electric current.

We begin by emphasizing that aspect of superconductors that contributes to their thermodynamic properties. We shall show that the change from the N state to the S state is a change of phase, just as the freezing of water is a change of phase from the liquid state to the solid state. However, as you will see, the N to S phase change in a superconductor is a very unusual kind of "freezing."

When the temperature of a material is lowered while the material remains in one phase, its properties will change gradually with temperature. For example, when the temperature of water is steadily lowered from 100°C to 0°C, its density will gradually change. On the other hand, there is a dramatic decrease in the density of water as it freezes at 0°C even though there is no change in its temperature during that process. This is the reason that water expands upon freezing. In analogy, the behavior of the resistance of a superconductor at the critical temperature, as shown in Figure 1, is indicative of a phase transition.

Many metals exhibit superconducting behavior, particularly those that have more than one or two valence electrons per atom. Moreover, alloys made from these metals also exhibit superconducting behavior. It is the valence electrons of the metal atoms that are the conduction electrons. Apparently the phenomenon of superconductivity does not depend on the detailed nature of any single metal, but rather requires some very general and widespread properties, such as the metal's having a sufficient number of conduction electrons. The explanation of what causes the N to S transition should likewise be a very general one.

Heike Kamerlingh-Onnes (1853–1926), a Dutch physicist, was the first to produce liquified helium in the early part of this century. Because helium is a liquid below 4.2 K, it may be used as a coolant for any material below this temperature. Kamerlingh-Onnes developed the helium liquifier because of his interest in the properties of other materials at those temperatures. As it turned out, he first observed the phenomenon of superconductivity in mercury, which has (as fate would have it) a critical temperature of about 4.2 K. At that time, Kammerlingh-Onnes wished to test a proposed theory that predicted the behavior of metals near absolute zero. We shall first briefly discuss the shortcomings of this incorrect theory before describing some features of the correct one.

The conduction electrons in a metal can be viewed as a fluid of charged particles. Each electron moves inside its container (the block of metal) and bumps into neighboring electrons (via mutual electrical repulsion). The positions and motions of these particles are both disordered, as shown in Figure 2a. An electric current is established when the fluid flows in a particular direction. (The motion of charges in a preferred direction constitutes a current.) Every fluid that was known early in this century formed a solid (a crystal lattice) at a sufficiently low temperature. Lattice ions in a solid do not flow but simply vibrate about some mean position, as if they are held in the lattice by attachment to springs.

So the incorrect theory is as follows. At very low temperatures, conduction electrons might freeze to form an electron lattice. Since they are not allowed to move through the metal in that state, the conduction current would be zero and the electrical resistance would be infinite. This is in complete contradiction with the experimental result. The experiment says that the resistance of a superconductor is *zero* at or below T_c. Thus, we must conclude that the electrons do not freeze into a lattice.

The correct explanation begins again with the electron fluid. Clearly, some sort of "freezing" must be taking place in a superconductor at or below T_c, but what is meant by *"freezing"* as used in this context? Ordinary freezing means that an orderly arrangement of the positions of particles is established—a place for each particle and each particle in its place. Note that each particle vibrates in place, but the vibrations are not orderly, as indicated in Figure 2b. This is what happens when water becomes ice. On the other hand, this ordinary type of freezing *does not* occur for conduction electrons in a metal.

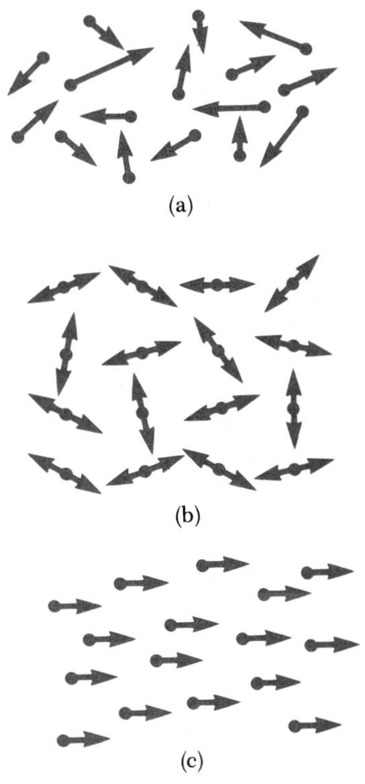

FIGURE 2 (a) A fluid of particles. The particles are arranged neither in position nor in velocity. This represents a normal fluid. (B) An orderly arrangement of positions of particles. The particles vibrate around these positions, and these vibrations are not orderly. (c) An orderly arrangement of velocities of particles. The particles do not care where their positions are, as long as they maintain the exact same velocity.

In order to gain some understanding of the type of "freezing" that occurs for conduction electrons in a superconductor, we must introduce a totally different type of order. The order we are referring to is an orderly arrangement of the velocities of the particles. Consider a system of particles whose positions are random, but all having the same velocity. Such a system is ordered in the sense that the relative positions of all the particles will remain the same as long as they all maintain the same velocity, as shown in Figure 2c.

The following familiar example may be useful for understanding the concept and consequences of an orderly arrangement of velocities. On a moderately crowded highway, if a few cars travel much faster or much slower than the other cars, then traffic is greatly hindered, resulting in major traffic jams and perhaps an occasional collision. However, suppose all cars travel at nearly the same velocity. In this case, the traffic will flow very smoothly, regardless of the spacing between cars and regardless of their common velocity. Ideally, all cars should maintain exactly the same velocity in order to optimize the flow of traffic and eliminate collisions.

Although it is very difficult (if not impossible) for cars to maintain a common velocity on a highway, the situation is quite different for electrons in a superconductor. For a very subtle reason that requires an understanding of modern atomic physics, *individual* electrons in a superconductor are unable to all travel at the same velocity, but *pairs* of electrons are able to do so. In the superconducting state, the particles which have a common velocity are actually electron pairs. Above T_c, the center of mass of any pair of electrons moves randomly through the material. In fact, it is meaningless to refer to an electron pair in the normal state; they simply do not exist as pairs above T_c. Below T_c, however, electron pairs do exist, and *the center of mass of each pair travels at exactly the same velocity.*

We have attempted to explain, in very general terms, the meaning of the phase change upon going from the N state to the S state. It is a "freezing" of the substance (metal or fluid), not in position space, as with ice, but in velocity space, as with cars all traveling at exactly 55 mi/h down a highway.

Superconductivity: The Next Generation

There have been roughly 80 years of research into the phenomenon of superconductivity and into the development of more interesting and useful superconducting materials. The significant properties and the materials that exhibit them are extensive and varied. Understanding many of them involves advanced and detailed physical reasoning and mathematical manipulation. Recent exciting developments make it more necessary than formerly that we attend to some of these topics.

One message this entire book conveys is that scientists have a passion for studying systems hotter, colder, larger, smaller, and so on, than previously achieved. It should not surprise you that during the era of superconductivity investigators have attempted to make materials with ever higher critical temperatures. There are many motivations for this, besides intrinsic scientific interest. A practical consideration is that the colder you need to make something, the more expensive it is to do it.

The first materials found to be superconducting had their transitions in the vicinity of approximately 4 K. These included common metals, such as tin, lead, and mercury. For 75 years (out of the 80) all the attempts to discover higher T_c superconductors pushed the record to a few degrees above 20 K. The higher critical temperatures are obtained by making alloys rather than by using chemically pure elements.

Although an increase factor of only 5 in T_c (20 K compared to 4 K) was shown by those 75 years of activity, something very important was achieved. At 20 K hydrogen liquefies. If your scientific or technological work can be performed at "liquid hydrogen temperatures" rather than at "liquid helium temperatures," great advantages ensue, ranging from ease of handling to far lower expense. Remember, hydrogen is abundant; every water molecule contains two hydrogen atoms. Helium is relatively rare. Every helium molecule in the atmosphere has to be separated from thousands of times as many other molecules (chiefly nitrogen and oxygen), some of which are chemically like helium (chiefly neon and argon).

For excellent reasons, during those 75 years the materials scientists concentrated almost exclusively upon were metals: metallic elements and metallic alloys. The best reasons for doing so are as follows: Metals are already conductors in their normal state—they possess "free electrons" that conduct current and hence are called "conduction electrons." Nonmetals, as a rule, do not possess free electrons and are good electrical insulators. Furthermore, metals do not appear to be structure-dependent in many of their properties, including conduction. You may remember from a chemistry course that certain types of chemical bond place fairly strict requirements on material structure. Most notably, the covalent bond requires specific angles between bond directions; reasonably, these are called "directional bonds." Metallic bonds are formed between atoms that are, quite simply, bathed in free electrons—the more free electrons they have, the better they like it. Metals can be pounded into various shapes and can be drawn into wires of the desirable fineness. Covalent crystals cleave when they are pounded and cannot be drawn into wires. Ionic crystals resemble covalent crystals much more than they do metals.

Metallic superconductivity at low temperatures seems to be largely structure-neutral. True, certain types of structure are represented by superconductors in the 20-K range, but all attempts to vary the composition while keeping the structure, and thereby to push the transition close to 30 K, have failed with metals.

It would seem to be quite unpromising to forsake the realm of metals and enter that of insulators for the purpose of reaching high T_c. Yet that is the program that has met with success over just the past few years and has pushed T_c above 120 K (as of 1988). We cannot guess how high the record T_c will be established by the time students read this text. Since nitrogen liquefies at 77 K, it is clear that superconductivity has entered the liquid nitrogen temperature range. Thus, instead of being confined to research labs with large budgets, the phenomenon has entered, with a sudden bound, the curriculum of enterprising high schools. Figure 3 shows a magnetic disk levitating above a superconducting block immersed in liquid nitrogen, a common sight in many high school laboratories. (Why the disk levitates will be discussed.)

The "new superconductivity"—the high T_c variety—is as surprising as it can be. It contradicts virtually all the expectations of the preceding decades of research. First of all, the new superconductors are ceramics, a class of materials already valuable for many reasons excluding conduction. Ceramics are used to make pottery, not wires. Second, the properties of the materials are strongly structure-dependent, and superconductivity is favored by a specific structure type: the layered compound. To give you an idea, instead of resembling pound cake, the layered compound looks more like seven-layer cake (or, rather, million-layer cake). Whereas metals look about the same and, therefore, behave in about the same way in all directions inside the material, these ceramics have strongly directional properties. The primary directions are along the parallel layers and perpendicular to the layers. Third, the properties of the materials are crucially dependent on composition, which means on relative proportion of each type of atom. Most notably, the composition is decisive for the occurrence of superconductivity and in an unsuspected way. The proportion of a chemical element known for its prevalence in insulators, namely oxygen, is an important factor.

Almost all oxides form via the binding of electrons: this is, in fact, what oxidation is all about. When an oxygen atom seizes the electrons of two hydrogen atoms, we get a strongly bound water molecule. To the contrary, the ceramic oxides, or at least a subset of them, are conductors. Furthermore, the highest temperature superconductors synthesized as of spring 1988 (at this writing) are in this class of material.

The startling new discovery was made by Alex Müller and Georg Bednorz at the IBM Zurich Research Laboratory in Switzerland. Very quickly, it was con-

FIGURE 3 Photograph of a small permanent magnet levitated above a disc of the superconductor $YBa_2Cu_3O_7$ which is at a temperature of 77 K. (Photo courtesy of Som Tyagi and Michel Barsoum, Drexel University)

firmed by research groups in the United States and Japan. Almost as quickly, it won the Nobel Prize in Physics (in 1987) for the two Swiss originators. The Swedish Academy acknowledged Bednorz and Müller as the discoverers of this specific superconductivity and as the inspirers of the growing population of physicists worldwide working to push T_c ever higher.

The most important result of Bednorz and Müller's research was the onset of superconductivity above 30 K in samples of the material represented by the chemical formula $La_{5-x}Ba_xCu_5O_{5(3-y)}$ for $x = 1$ and 0.75 and y positive. The x and y values represent a restricted range, rather than a single value. By replacing barium with strontium, other experimentalists raised T_c to about 40 K. Research groups at the University of Alabama at Huntsville and the University of Houston stunned the scientific community by more than doubling the previous high to T_c above 90 K in the oxide material $RBa_2Cu_3O_{7-y}$, where R is any of several rare earth elements. (Look this up in the periodic table provided in the endpapers.) Based on the chemical composition of one R atom connected to two Ba and three Cu, the material is nicknamed 1-2-3. The oxygen atoms are crucial, in a way still to be understood. Evidently, they somehow allow just the right number of electrons to become conducting and, at the same time, create the environment that is hospitable to the superconducting transition. There is no agreement in the scientific community on how they do that.

The amazing nature of these discoveries has been felt by the world of science and technology but is not so evident to students. In groping for an analogy, we should consider something like this. Suppose, after achieving a remarkable sports record, such as the 4-minute mile, runners in the peak of condition find it ever harder to push the record down to 3 minutes, 50 seconds. Then a completely different type of athlete, say a shot-putter with huge upper-body strength, rather than speed and stamina of legs, breaks the record handily, and suddenly shot-putters around the world, with their newly found ability, are doing 3-minute and even 2-minute miles.

To return to physics, rather than bold analogy, the structures of the old and the new superconductors are compared in Figure 4. Figure 4a shows the atomic arrangement of a simple crystal such as tin; the general scheme is a variation on the cubic array, the most common type of elemental metallic crystal. Figure 4b shows several layers of the ceramic oxide called 1-2-3 above.

Although the zero electrical resistance is the most

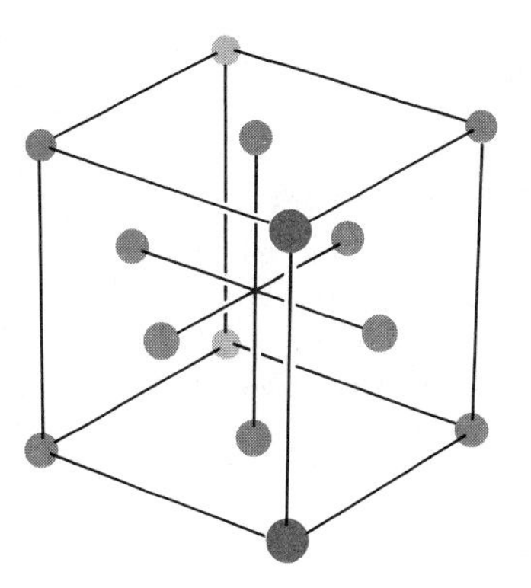

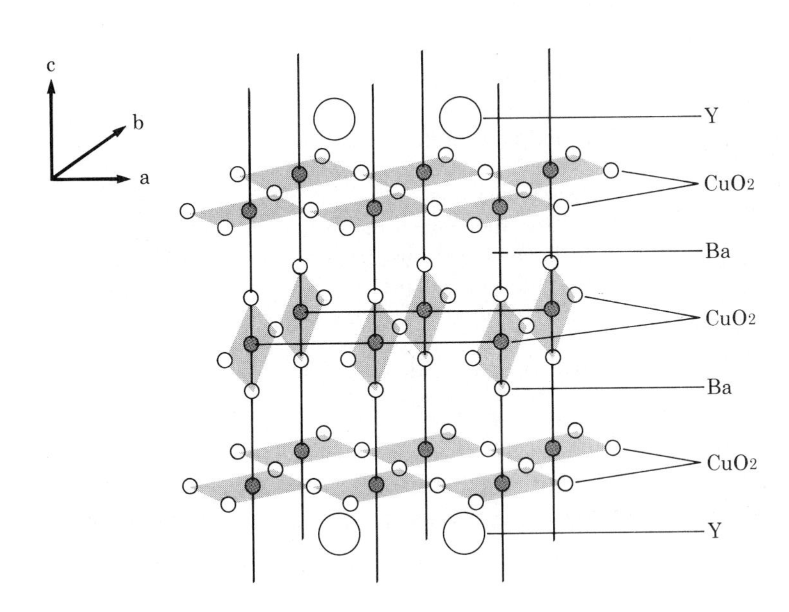

FIGURE 4 (a) Some points from a face-centered Bravais lattice. (b) The double-layered perovskite with its orthorhombic structure ($a \neq b \neq c$). The compound $YBa_2Cu_3O_7$ has this structure.

striking and well-known superconducting property, numerous other properties (magnetic, optical, and so on) are intimately linked to zero resistance and, to an extent, to the structure and composition of the material. Some of these properties are as interesting scientifically and useful technologically as the resistance. Most notably, a superconductor expels magnetic field from its interior: if a magnet is brought near a superconductor, then the magnet is repelled by the superconductor. This is the reason for the magnetic levitation effect shown previously in Fig. 3.

This magnetic effect may be used on a far larger scale. Figure 5 shows a train supported on large magnets, hovering just above superconducting tracks. Japan and West Germany are leaders in this endeavor, which is in the developmental stage. The tracks must be kept colder than T_c at all times. The cost of doing this needs to be more than offset by the resulting economies. These include time-saving higher speed, energy-saving frictionless travel along the tracks, including braking, and the great reduction of wear-and-tear on equipment that friction and other heat- and stress-producing influences would cause.

What of the future? Technologically, the benefits have not yet arrived because of several factors. The most important electrical use of superconducting materials is the wiring of superconducting magnets to provide high magnetic fields without suffering energy loss to heat. But (1) high T_c ceramics are brittle and break when you attempt to draw them into wires, and (2) high magnetic fields overwhelm superconductivity, so that, instead of field expulsion, the material gives up and turns into a normal conductor (even at temperature far below T_c) with ordinary unfavorable resistance properties. We need to find superconductors that can be formed into special shapes and that are more compatible with magnetism. There is clearly a long way to go in this program.

The large-scale magnetic uses are further along but not yet economical. They require large initial investments of materials, some of them not abundant, and energy, in order to obtain great eventual savings. As yet, the savings do not look great enough.

The most significant looked for scientific advance would be the discovery of a superconductor with T_c around room temperature. In that case, major uses for such a superconductor would not require temperatures much colder than a household refrigerator. Perhaps, in the next edition of this text, you will be reading about prospects for train travel at airplane speeds, purchases of rolls of wire made of "new improved 1-2-3" (as advertisers like to put it), and assembling superconducting components of devices to be operated inside a home freezer.

FIGURE 5 Using electromagnetic induction, this experimental train without wheels levitates off the track.

SUGGESTED READING

Physics Today, December 1987, p. 17.*
Science, September 4, 1987, p. 1133.*
Time, May 11, 1987, p. 64.
Scientific American, December 1987, p. 46.
Discover, January 1988, p. 24.

ESSAY QUESTIONS

1. Why do electrons inside a normally conducting material (like tin) prefer to turn superconducting below T_c? Think of why water turns to ice below 0°C.
2. What behavior of electric current might link the property of zero resistance with the property of magnetic expulsion?
3. A magnetic disk is repelled if placed near a fixed chunk of superconductor. Will a superconducting disk be repelled if placed near a fixed chunk of magnet?
4. What is there about the structures on the atomic level that allows tin to be drawn and pounded into any desired shape, but prevents a ceramic like 1-2-3 from undergoing the same type of behavior?
5. If they could be made economical, what other uses for high T_c superconductors do you envision?

*These references are more technical than the others. Needless to say articles are appearing at a rapid rate.

CHAPTER 19

Direct Current Circuits

A 15-ft-long vacuum vessel in which a deuterium plasma has been heated to 20,000,000°C. (Photo, U.S. Atomic Energy Commission)

This chapter is concerned with the analysis of some simple circuits whose elements include batteries, resistors, and capacitors in various combinations. In some cases, Ohm's law is sufficient to enable us to find out what we need to know about a circuit. However, as the circuits get more complex, we shall make use of two rules known as Kirchhoff's rules. These rules arise from the laws of conservation of energy and conservation of charge. Also following this chapter is an essay on electric current in the nervous system.

19.1 SOURCES OF ELECTROMOTIVE FORCE

A current can be maintained in a closed circuit through the use of a source of energy, called an *electromotive force* (emf). **A source of emf is any device that will transform nonelectrical energy into electrical energy.** Probably the most important sources of emf are the battery, which converts chemical energy into electrical energy, and the electric generator, which converts mechanical energy into electrical energy. In addition, there are many other sources of emf that are important to certain applications. As examples, quartz crystals discussed in Chapter 15 transform mechanical energy into an emf, and as we shall discuss in Chapter 23, photovoltaic cells convert light energy into electrical energy. One can think of a source of emf as a "charge pump" that forces charges to move around a circuit. The emf, $\mathcal{E}$, of a source describes the work done per unit charge, and hence the SI unit of emf is the volt. Let us now examine two sources of emf, the common flashlight battery, and the thermocouple. We shall consider the generator after we have discussed alternating currents.

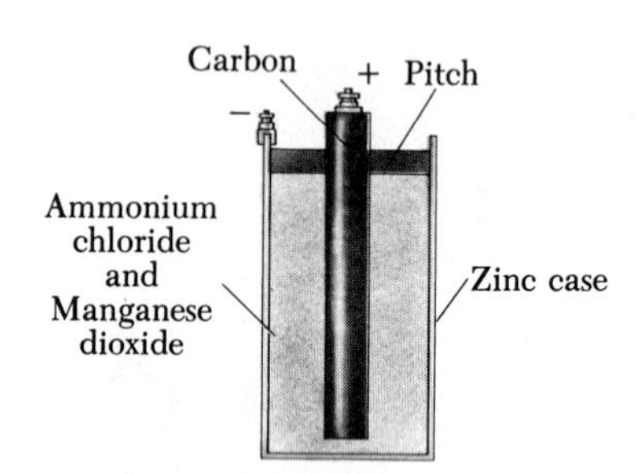

FIGURE 19.1 Cross-sectional view of a dry cell battery.

Flashlight batteries are produced in a variety of shapes and sizes, but they all work in basically the same way. Figure 19.1 is a diagram of the interior of such a battery. In this particular battery, often referred to as a dry cell, the zinc case serves as the negative terminal, while the carbon rod down its center serves as the positive terminal. The space between the electrodes contains a paste-like mixture of manganese dioxide, ammonium chloride, and carbon.

When these materials are assembled in this fashion, two chemical reactions take place; one occurs at the zinc case, the other at the manganese dioxide layer surrounding the carbon rod. Positively charged zinc ions (Zn^{2+}) leave the case and enter the ammonium chloride paste, where they combine with chloride ions (Cl^{-}). (The chloride ions are present because a small percentage of the ammonium chloride dissociates, leaving some free chloride ions in the solution.) As each zinc ion is removed from the case, it leaves behind two electrons. As additional zinc ions leave the case, more electrons accumulate, leaving the zinc case with a net negative charge.

When a chloride ion breaks free from the ammonium chloride molecule, the remnant portion of the molecule becomes singly ionized. This positively charged ion is neutralized by the manganese dioxide, which supplies the needed electrons. As a result, the carbon rod surrounded by its manganese dioxide layer ends up with a net positive charge.

These chemical reactions and thus the charge separation do not continue without limit. The zinc case ultimately achieves such a strong negative charge that the zinc ions can no longer escape. A similar charge saturation occurs at the carbon rod.

A **thermocouple** is a source of emf that is frequently used as a thermometer. The thermocouple is essentially a junction formed by bonding two different metals or alloys together. Figure 19.2 shows a thermocouple consisting of two copper-iron junctions. The junction on the right, called the test junc-

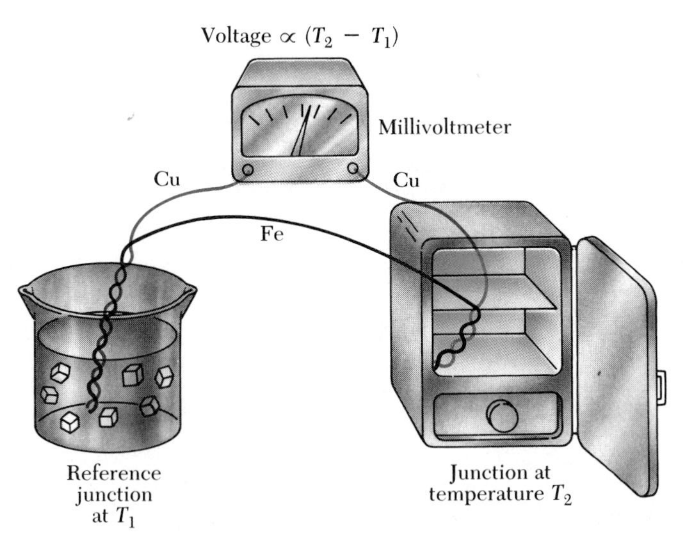

FIGURE 19.2 A thermocouple used to measure temperature. The voltage developed between the two junctions is proportional to the difference in temperature between the reference junction at T_1 and the junction at temperature T_2.

tion, is placed in thermal contact with a material whose temperature is to be measured, while the junction on the left is maintained at some constant reference temperature (usually a water-ice mixture). When the reference temperature is different than that of the test junction, an emf is set up in the circuit. The emf is proportional to the temperature difference between the two junctions, and therefore its value can be used to measure an unknown temperature. In practice, one usually uses thermocouples for which calibration curves are available.

19.2 RESISTORS IN SERIES

Figure 19.3 shows two resistors, R_1 and R_2, connected to a battery in a circuit called a series circuit. In such a circuit, there is only one pathway for the current. Charges must pass through both resistors and the battery as they traverse the circuit. Hence, all charges in a series circuit must follow the same conducting path.

Note that the current is the same through each resistor in a series circuit because any charge that flows through R_1 must also flow through R_2.

This is analogous to water flowing through a pipe with two constrictions, corresponding to R_1 and R_2. Whatever volume of water flows in one end in a given time interval must exit the opposite end.

Since the potential drop from a to b in Figure 19.3b equals IR_1 and the potential drop from b to c equals IR_2, we see that the potential drop from a to c is

$$V = IR_1 + IR_2$$

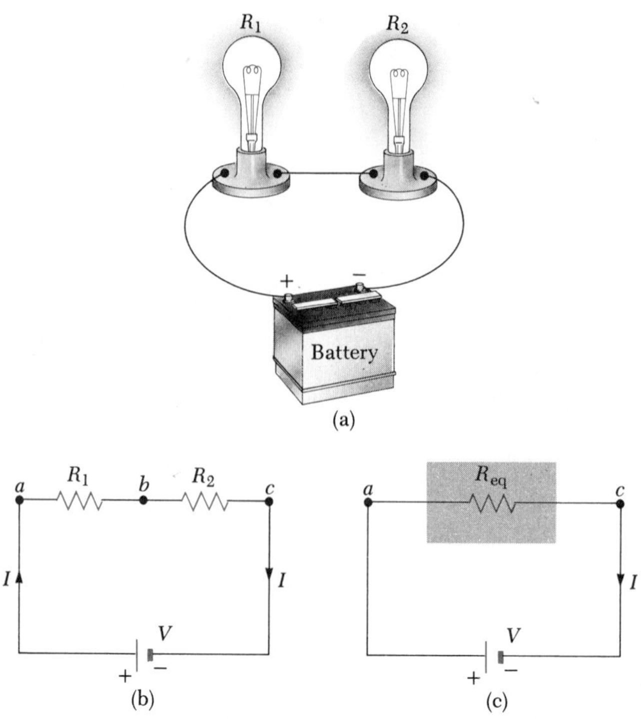

FIGURE 19.3 Series connection of two resistors, R_1 and R_2. The current in each resistor is the same, and the equivalent resistance of the combination is given by $R_{eq} = R_1 + R_2$.

Figure 19.3c shows an equivalent resistor, R_{eq}, that can replace the two resistors of the original circuit. Applying Ohm's law to this resistor, we have

$$V = IR_{eq}$$

Equating these two expressions, we have

$$IR_{eq} = IR_1 + IR_2$$

or

$$\boxed{R_{eq} = R_1 + R_2} \qquad \text{(series combination)} \qquad (19.1)$$

The resistance R_{eq} is equivalent to the series combination $R_1 + R_2$ in the sense that the circuit current is unchanged when R_{eq} replaces R_1 and R_2.

An extension of this analysis shows that the equivalent resistance of three or more resistors connected in series is

$$\boxed{R_{eq} = R_1 + R_2 + R_3 + \cdots} \qquad \text{(series combination)} \qquad (19.2)$$

From this, we see that *the equivalent resistance of a series combination of resistors is always greater than any individual resistance.*

Note that if the filament of one lightbulb in the series circuit shown in Figure 19.3a were to break, or burn out, the circuit would no longer be complete and the second bulb would also go out. Some Christmas tree light sets

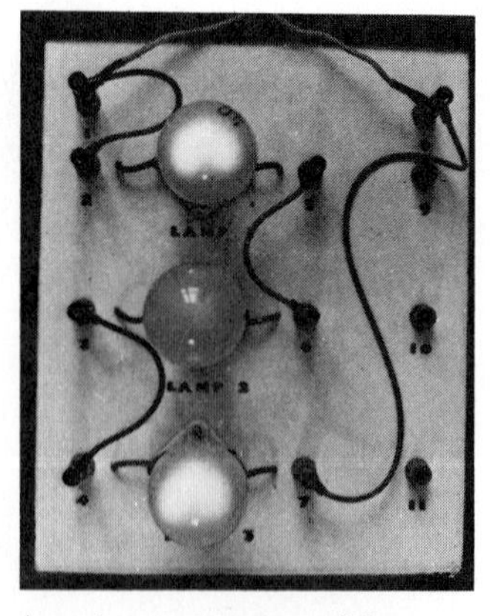

A series connection of three lamps with different power ratings. Why is the illumination of the middle lamp barely visible?

are connected in this way, and the agonizing experience of determining which bulb in a set is the culprit that has burned out is a familiar one. Frustrating experiences such as this illustrate how inconvenient it would be to have all appliances in a house connected in series.

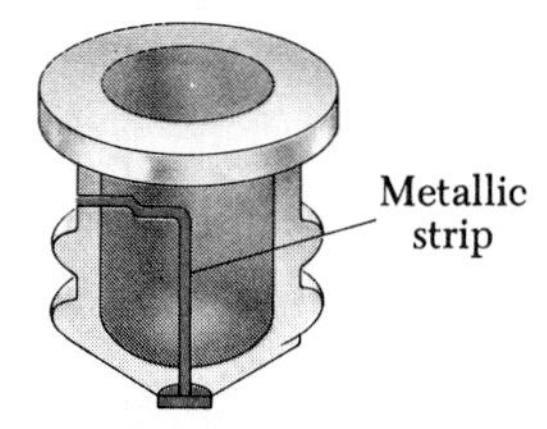

FIGURE 19.4 A cutaway view of a fuse.

In many circuits, fuses are used in series with other circuit elements for safety purposes. A cutaway view of a fuse is shown in Figure 19.4. The current passing through the fuse flows through a small metallic strip. If the magnitude of the current is large enough, it causes the strip to melt. If a circuit does not include a fuse, excessive currents could damage circuit elements, overheat wires, and perhaps cause a fire.

In modern home construction, circuit breakers are used in place of fuses. When the current in a circuit exceeds some value (typically 15 A), the circuit breaker acts as a switch and opens the circuit. One design for a circuit breaker is shown in Figure 19.5. In this design, current passes through a bimetallic strip, the top of which bends to the left when excessive current heats it. If the strip bends far enough to the left, it settles into a groove in the spring-loaded metal bar. When this occurs, the bar drops sufficiently downward to open the circuit at the contact point. The bar also flips the switch to indicate that the circuit breaker is not operational. After the overload is removed, the switch can be flipped back on. Circuit breakers using this design have the disadvantage that some time is required for the heating of the strip. Thus the circuit is not opened immediately when it is overloaded. For this reason, many circuit breakers use electromagnets, which we shall discuss later.

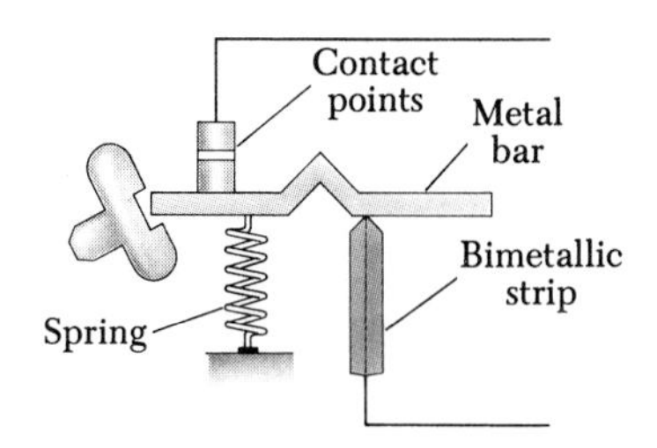

FIGURE 19.5 A circuit breaker uses a bimetallic strip for its operation.

EXAMPLE 19.1 Four Resistors in Series

Four resistors are arranged as shown in Figure 19.6a. Find (a) the equivalent resistance and (b) the current in the circuit if the emf of the battery is 6 V.

Solution (a) The equivalent resistance is found from Equation 19.2 to be

$$R_{eq} = R_1 + R_2 + R_3 + R_4 = 2\ \Omega + 4\ \Omega + 5\ \Omega + 7\ \Omega = 18\ \Omega$$

(b) If we apply Ohm's law to the equivalent resistor in Figure 19.6b, we find the current in the circuit to be

$$I = \frac{V}{R_{eq}} = \frac{6\ \text{V}}{18\ \Omega} = \frac{1}{3}\ \text{A}$$

EXERCISE If $\frac{1}{3}$ A flows through the equivalent resistor, this must also be the current in each resistor of the original circuit. Find the voltage drop across each resistor.

Answer $V_{2\Omega} = \frac{2}{3}$ V, $V_{4\Omega} = \frac{4}{3}$ V, $V_{5\Omega} = \frac{5}{3}$ V, $V_{7\Omega} = \frac{7}{3}$ V.

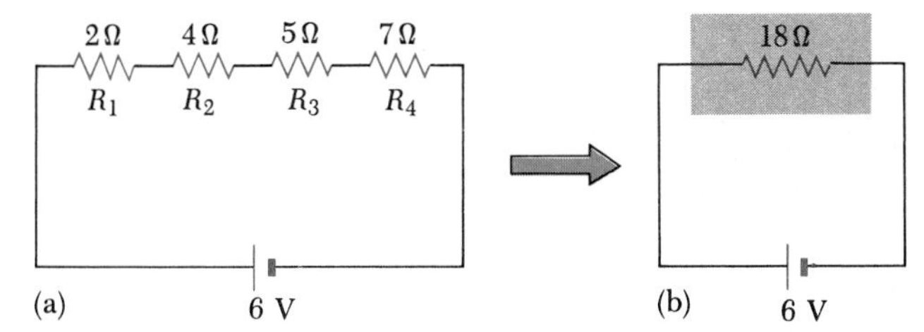

FIGURE 19.6 (Example 19.1) (a) Four resistors connected in series. (b) The equivalent resistance of the circuit in (a).

19.3 RESISTORS IN PARALLEL

Now consider two resistors connected in parallel, as shown in Figure 19.7.

When resistors are connected in parallel, the potential difference is the same across each resistor.

This must be true because the left side of each resistor is connected to a common point, the positive side of the battery, point a in Figure 19.7b, and the right side of each resistor is connected to a common point, the negative terminal of the battery in Figure 19.7b.

However, the current in each resistor is in general not the same. The currents will be the same only if the resistors have the same resistance. For example, three identical bulbs connected in parallel with a battery would each draw the same current. When the current, I, reaches point a (called a junction) in Figure 19.7b, it splits into two parts, I_1 going through R_1 and I_2 going through R_2. If R_1 is greater than R_2, then I_1 will be less than I_2. That is, *the charge will tend to follow the path of least resistance. Since charge is conserved, the current I that enters point a must equal the total current leaving this point, $I_1 + I_2$.* That is,

$$I = I_1 + I_2$$

This is a special case of one of two Kirchhoff rules discussed in the next section. The potential drop must be the same across each resistor and must also equal the potential difference across the battery. Ohm's law applied to each resistor gives

$$I_1 = \frac{V}{R_1} \qquad I_2 = \frac{V}{R_2}$$

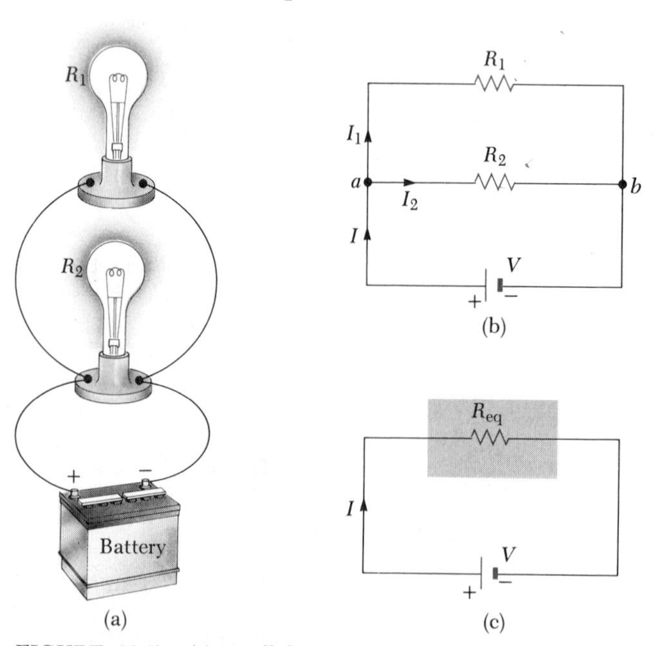

FIGURE 19.7. (a) Parallel connection of two resistors. (b) Circuit diagram for the parallel combination. (c) The voltage across each resistor is the same, and the equivalent resistance of the combination is given by $1/R_{eq} = 1/R_1 + 1/R_2$.

If we apply Ohm's law to the equivalent resistor in Figure 19.7c, we have

$$I = \frac{V}{R_{eq}}$$

When these expressions for the current are substituted into the equation $I = I_1 + I_2$ and V is cancelled, we obtain

$$\frac{1}{R_{eq}} = \frac{1}{R_1} + \frac{1}{R_2} \quad \text{(parallel combination)} \quad (19.3)$$

An extension of this analysis to three or more resistors in parallel gives the following general expression for the equivalent resistance:

$$\frac{1}{R_{eq}} = \frac{1}{R_1} + \frac{1}{R_2} + \frac{1}{R_3} + \cdots \quad \text{(parallel combination)} \quad (19.4)$$

From this result, you should be able to show that *the equivalent resistance of two or more resistors connected in parallel is always less than the smallest resistance in the group.*

Household circuits are always wired such that the lights and appliances are connected in parallel. In this manner, each device operates independently of the others, so that if one is switched off, the others remain on. Equally important, each device operates on the same voltage.

EXAMPLE 19.2 Three Resistors in Parallel

Three resistors are connected in parallel as in Figure 19.8. A potential difference of 18 V is maintained between points a and b. (a) Find the current in each resistor.

Solution The resistors are in parallel, and so the potential difference across each is 18 V. Let us apply $V = IR$ to find the current in each resistor:

$$I_1 = \frac{V}{R_1} = \frac{18\text{ V}}{3\ \Omega} = 6\text{ A}$$

1.64

$$I_2 = \frac{V}{R_2} = \frac{18\text{ V}}{6\ \Omega} = 3\text{ A}$$

$$I_3 = \frac{V}{R_3} = \frac{18\text{ V}}{9\ \Omega} = 2\text{ A}$$

(b) Calculate the power dissipated by each resistor and the total power dissipated by the three resistors.

Solution Applying $P = I^2R$ to each resistor gives

$$3\ \Omega: \quad P_1 = I_1^2R_1 = (6\text{ A})^2(3\ \Omega) = 108\text{ W}$$
$$6\ \Omega: \quad P_2 = I_2^2R_2 = (3\text{ A})^2(6\ \Omega) = 54\text{ W}$$
$$9\ \Omega: \quad P_3 = I_3^2R_3 = (2\text{ A})^2(9\ \Omega) = 36\text{ W}$$

This shows that the smallest resistor dissipates the most power since it carries the most current. (Note that you can also use $P = V^2/R$ to find the power dissipated by each resistor.) Summing the three quantities gives a total power of 198 W.

(c) Calculate the equivalent resistance of the three resistors.

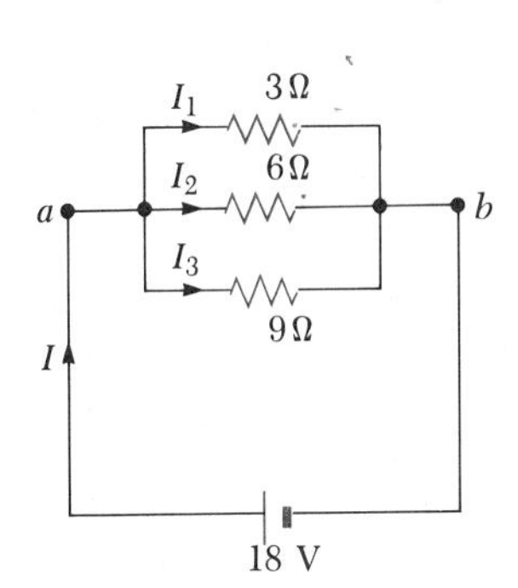

FIGURE 19.8 (Example 19.2) Three resistors connected in parallel. The voltage across each resistor is 18 V.

Solution We can use Equation 19.4 to find R_{eq}:

$$\frac{1}{R_{eq}} = \left(\frac{1}{3} + \frac{1}{6} + \frac{1}{9}\right) \Omega^{-1}$$

$$R_{eq} = \frac{18}{11}\ \Omega$$

EXERCISE Use the result for R_{eq} to calculate the total power dissipated in the circuit.

Answer 198 W.

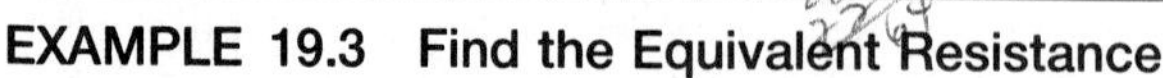

EXAMPLE 19.3 Find the Equivalent Resistance

Four resistors are connected as shown in Figure 19.9a. (a) Find the equivalent resistance between points *a* and *c*.

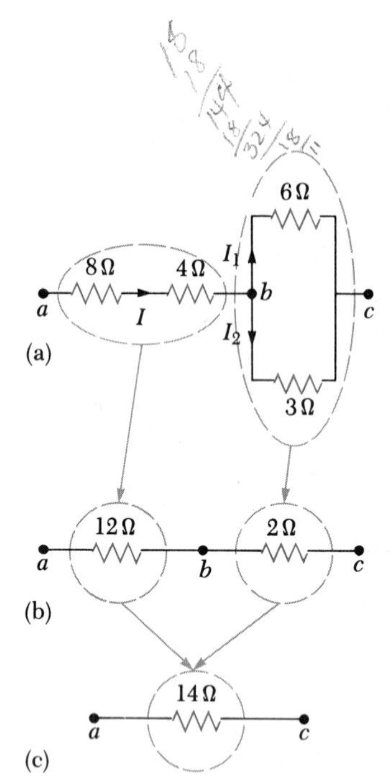

FIGURE 19.9 (Example 19.3) The four resistors shown in (a) can be reduced in steps to an equivalent 14-Ω resistor.

Solution The circuit can be reduced in steps, as shown in Figure 19.9b. The 8-Ω and 4-Ω resistors are in series, and so the equivalent resistance between *a* and *b* is 12 Ω (Eq. 19.2). The 6-Ω and 3-Ω resistors are in parallel, and so from Equation 19.4 we find that the equivalent resistance from *b* to *c* is 2 Ω. Hence, the equivalent resistance from *a* to *c* is 14 Ω.

(b) What is the current in each resistor if a 42-V battery is placed between *a* and *c*?

Solution The current I in the 8-Ω and 4-Ω resistors is the same since they are in series. Let us use Ohm's law and the results from (a) to get

$$I = \frac{V_{ac}}{R_{eq}} = \frac{42\ \text{V}}{14\ \Omega} = 3\ \text{A}$$

When this current enters the junction at b, it splits; part of the current goes through the 6-Ω resistor (I_1), and part goes through the 3-Ω resistor (I_2). Since the potential difference across these resistors, V_{bc}, is the *same* (they are in parallel), we see that $(6\ \Omega)(I_1) = (3\ \Omega)(I_2)$, or $I_2 = 2I_1$. Using this result and the fact that $I_1 + I_2 = 3$ A, we find that $I_1 = 1$ A and $I_2 = 2$ A. We could have guessed this from the start by noting that the current through the 3-Ω resistor has to be twice the current through the 6-Ω resistor in view of their relative resistances and the fact that the same voltage is applied in both.

As a final check, note that $V_{bc} = 6I_1 = 3I_2 = 6$ V and $V_{ab} = 12I = 36$ V; therefore, $V_{ac} = V_{ab} + V_{bc} = 42$ V, as it must.

19.4 KIRCHHOFF'S RULES

As Section 19.3 indicates, we can analyze simple circuits using Ohm's law and the rules for series and parallel combinations of resistors. However, there are many ways in which resistors can be connected such that the circuits formed cannot be reduced to a single equivalent resistor. The procedure for analyzing more complex circuits is greatly simplified by the use of two simple rules called **Kirchhoff's rules:**

1. The sum of the currents entering any junction must equal the sum of the currents leaving that junction. (A junction is any point in a circuit where the current can split.) This rule is often referred to as the **junction rule.**
2. The sum of the potential differences across each element around any closed circuit loop must be zero. This rule is usually called the **loop rule.**

The first rule is a statement of *conservation of charge*. Whatever current enters a given point in a circuit must leave that point because charge cannot build up or disappear at a point. If we apply this rule to the junction shown in Figure 19.10a, we get

$$I_1 = I_2 + I_3$$

Figure 19.10b represents a mechanical analog to this situation in which water flows through a branched pipe with no leaks. The flow rate into the pipe equals the total flow rate out of the two branches.

The second rule is equivalent to the law of *conservation of energy*. Any charge that moves around any closed loop in a circuit (it starts and ends at the same point) must gain as much energy as it loses. Its energy may decrease in the form of a potential drop, $-IR$, across a resistor or as a result of flowing backward through a source of emf, that is, from the positive to the negative terminal inside the battery. In the latter case, electrical energy is converted to chemical energy as the battery is charged.

When applying Kirchhoff's rules, there are two decisions that you must make at the beginning of the problem.

1. You must assign symbols and directions to the currents in the various branches of the circuit. If you should happen to guess the wrong direction for a current, the end result will be a negative answer for that current, but its magnitude will be correct.
2. When applying the loop rule, you must choose a direction (clockwise or counterclockwise) for moving around the loop. As you traverse the loop, record voltage drops and increases according to the rules given below.

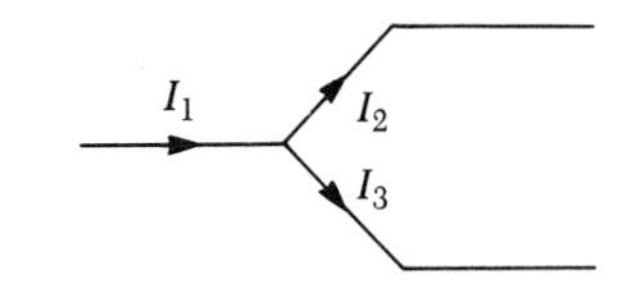

(a)

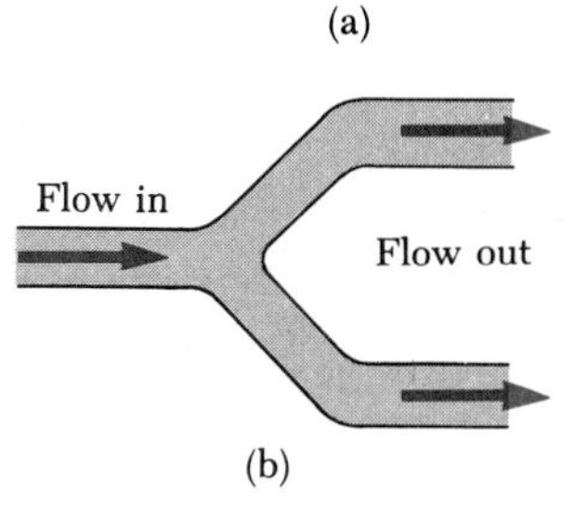

(b)

FIGURE 19.10 (a) A schematic diagram illustrating Kirchhoff's junction rule. Conservation of charge requires that whatever current enters a junction must leave that junction. Therefore, in this case, $I_1 = I_2 + I_3$. (b) A mechanical analog of the junction rule: the flow out must equal the flow in.

As an aid in applying the loop rule, the following points should be noted. These points are summarized in Figure 19.11, where it is assumed we are moving from point *a* toward point *b*:

1. If a resistor is traversed in the direction of the current, the change in potential across the resistor is $-IR$ (Fig. 19.11a).
2. If a resistor is traversed in the direction opposite the current, the change in potential across the resistor is $+IR$ (Fig. 19.11b).
3. If a source of emf is traversed in the direction of the emf (from $-$ to $+$ on the terminals), the change in potential is $+\mathcal{E}$ (Fig. 19.11c).
4. If a source of emf is traversed in the direction opposite the emf (from $+$ to $-$ on the terminals), the change in potential is $-\mathcal{E}$ (Fig. 19.11d).

(a) $\Delta V = V_b - V_a = -IR$

(b) $\Delta V = V_b - V_a = +IR$

(c) $\Delta V = V_b - V_a = +\mathcal{E}$

(d) $\Delta V = V_b - V_a = -\mathcal{E}$

FIGURE 19.11 Rules for determining the potential changes across a resistor and a battery.

There are limitations on the number of times you can use the junction rule and the loop rule. The junction rule can be used as often as needed so long as each time you write an equation, you include in it a current that has not been used in a previous junction rule equation. In general, the number of times the junction rule can be used is one fewer than the number of junction points in the circuit. The loop rule can be used as often as needed so long as a new circuit element (resistor or battery) or a new current appears in each new equation. In general, the number of independent equations you need must at least equal the number of unknowns in order to solve a particular circuit problem.

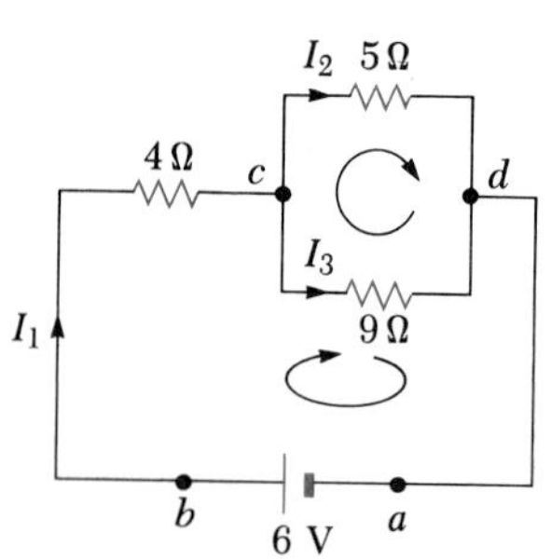
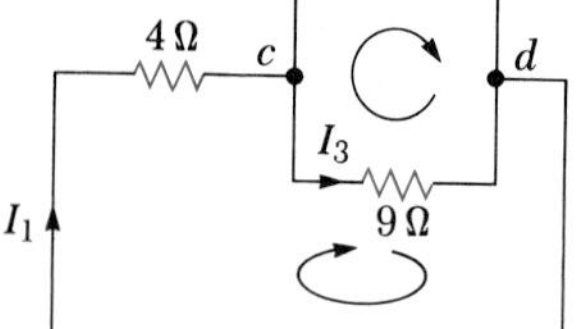

FIGURE 19.12 (Example 19.4) A multiloop circuit.

EXAMPLE 19.4 Applying Kirchhoff's Rules

Find the currents in the circuit shown in Figure 19.12.

Solution Although this circuit could be handled much more easily by Ohm's law and the rules for series and parallel circuits, we shall use Kirchhoff's rules to demonstrate the techniques. Also, because students often find the application of Kirchhoff's rules confusing, we shall go into considerable detail in this example.

The first step is to assign currents to each branch of the circuit; these are our unknowns and are labeled I_1, I_2, and I_3 in Figure 19.12. Furthermore, it is necessary to guess a direction for these currents. The experience that you have with circuits like this should tell you that the directions of all three have been chosen correctly. However, recall that if a current direction were chosen incorrectly, the numerical answer will turn out negative, but the magnitude will be correct. This point will be demonstrated in the next example.

We now apply Kirchhoff's rules. First we can apply the junction rule using either point c or point d, the only two junctions in the circuit. Let us choose point c. The net current into this junction is I_1, and the net current leaving the junction is $I_2 + I_3$. Thus, the junction rule applied to point c gives

$$I_1 = I_2 + I_3$$

Recall that you can apply the junction rule over and over until you reach a situation in which no new currents appear in an equation. We have reached that point with one application in this example. If we apply the junction rule at point d, we find that $I_1 = I_2 + I_3$, exactly the same equation.

We have three unknowns in our problem, I_1, I_2, and I_3; thus we need two more independent equations before we can find a solution. These are obtained by applying the loop rule to the two loops indicated in the figure. Note that there are actually three loops in the circuit, but these two are sufficient for us to complete the problem. (Where is the loop that we do not use?)

When applying the loop rule, we must first choose the loops to be traversed, and then the direction in which to traverse them. We have selected the two loops indicated in the figure, and have decided to traverse both of them in the clockwise direction. Although other choices could have been selected, the final result would be the same. Also, note that there are actually three loops in the circuit. (The third is the outside loop that includes the 5 Ω resistor.) Any two of these three loops could have been selected, but only two are needed to solve the problem. In fact, the third would give redundant information.

Starting at point a and moving clockwise around the large loop, we encounter the following voltage changes (see Figure 19.11 for the basic rules):

1. Going from a to b, we encounter a voltage rise of 6 V.
2. Going from b to c through the 4-Ω resistor results in a voltage drop of $-4I_1$.
3. Going from c to d through the 9-Ω resistor results in a voltage drop of $-9I_3$.

No voltage change occurs going from d back to a. Now that we have made a complete traversal of the loop, we can equate the sum of the voltage changes to zero:

$$6 - 4I_1 - 9I_3 = 0$$

Moving clockwise around the small loop from point c, we have the following:

1. Going from c to d through the 5-Ω resistor results in a voltage drop of $-5I_2$.
2. Going from d to c through the 9-Ω resistor results in a voltage rise of $+9I_3$.

We find

$$-5I_2 + 9I_3 = 0$$

Thus we have the following three equations to be solved for the three unknowns:

$$I_1 = I_2 + I_3$$
$$6 - 4I_1 - 9I_3 = 0$$
$$-5I_2 + 9I_3 = 0$$

If you need help in solving three equations with three unknowns, see Example 19.5. You should be able to show that the following answers are obtained:

$$I_1 = 0.83 \text{ A} \qquad I_2 = 0.53 \text{ A} \qquad I_3 = 0.30 \text{ A}$$

You should now solve this problem using the methods for series and parallel circuits to show that the answers given can be found by this alternate method. However, the next example can be solved only by applying Kirchhoff's rules.

EXAMPLE 19.5

Find the currents I_1, I_2, and I_3 in the circuit shown in Figure 19.13.

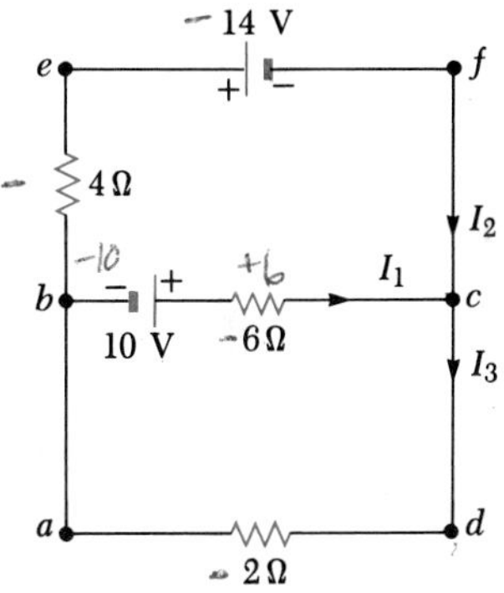

FIGURE 19.13 (Example 19.5) A circuit containing three loops.

Solution We shall choose the directions of the currents as shown in Figure 19.13. Kirchhoff's junction rule applied at c gives

$$(1) \qquad I_1 + I_2 = I_3$$

There are *three* loops in the circuit, *abcda*, *befcb*, and *aefda* (the outer loop), but we need only two loop equations to determine the unknown currents. The third loop equation would give no new information. When Kirchhoff's second rule is applied to loops *abcda* and *befcb* traversed in the clockwise direction, we obtain the following expressions:

(2) Loop *abcda*: $10 - 6I_1 - 2I_3 = 0$

(3) Loop *befcb*: $-14 - 10 + 6I_1 - 4I_2 = 0$

Note that, in loop *befcb*, a positive sign is obtained when traversing the 6-Ω resistor since the current, I_1, is in the opposite direction. A loop equation for *aefda* gives $-14 = 2I_3 + 4I_2$, which is just the sum of (2) and (3).

In (1), (2), and (3) we now have three independent expressions with three unknowns. We can solve the problem as follows. If (1) is substituted into (2), we have

$$10 - 6I_1 - 2(I_1 + I_2) = 0$$
$$(4) \qquad 10 = 8I_1 + 2I_2$$

Let us divide each term in (3) by 2 and rearrange the equation to obtain

$$(5) \qquad -12 = -3I_1 + 2I_2$$

Subtracting (5) from (4) eliminates I_2, giving

$$22 = 11I_1$$
$$I_1 = 2 \text{ A}$$

This value of I_1 can be inserted into (5) to give a value for I_2:

$$2I_2 = 3I_1 - 12 = 3(2) - 12 = -6$$
$$I_2 = -3 \text{ A}$$

The fact that I_2 is negative indicates only that we chose the *wrong* direction for this current.

EXERCISE Calculate the value of I_3.

Answer -1 A.

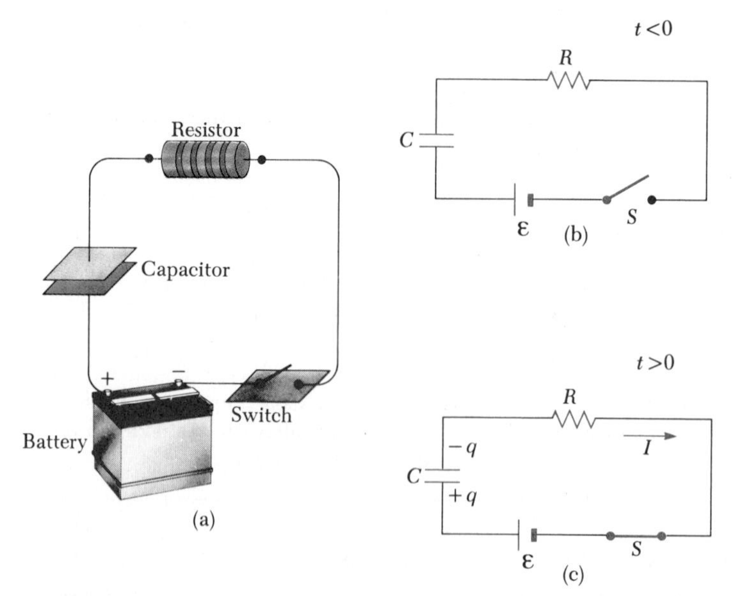

FIGURE 19.14 (a) A capacitor in series with a resistor, battery, and switch. (b) Circuit diagram representing this system before the switch is closed ($t < 0$). (c) Circuit diagram after the switch is closed ($t > 0$).

19.5 *RC* CIRCUITS

So far we have been concerned with circuits with constant currents. We shall now consider direct current circuits containing capacitors, in which the current varies with time. Consider the series circuit shown in Figure 19.14. Let us assume that the capacitor is initially uncharged. There is no current when the switch is open (Fig. 19.14b). If the switch is closed at $t = 0$, a current will pass through the resistor and the capacitor will begin to charge (Fig. 19.14c). Note that, during charging, charges do not jump across the plates of the capacitor because the gap between the plates represents an open circuit. This charge transfer ceases when the voltage drop across the capacitor is equal to the emf of the battery, $\mathcal{E}$. Thus, from the definition of capacitance, the

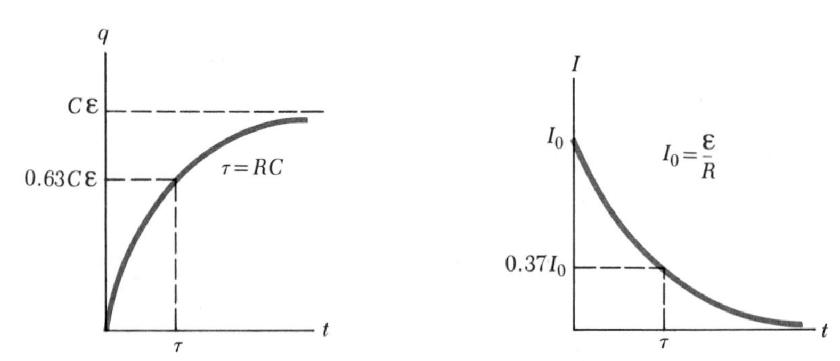

FIGURE 19.15 (a) Plot of the charge on the capacitor versus time for the circuit shown in Figure 19.14. After one time constant, τ, the charge is 0.63 of the maximum value, $C\mathcal{E}$. The charge approaches its maximum value as t approaches infinity. (b) Plot of the current versus time for the *RC* circuit shown in (a). At $t = 0$, the current has its maximum value, $I_0 = \mathcal{E}/R$ and decays to zero exponentially as t approaches infinity. After one time constant, τ, the current decreases to 0.37 of its initial value.

maximum charge on the capacitor is given by $Q = C\mathcal{E}$. Once the capacitor reaches its maximum charge, the current in the circuit is zero.

It is found that the charge on the capacitor increases with time in a manner pictured by Figure 19.15a. Note that the charge is zero at time $t = 0$ and approaches the maximum value of $C\mathcal{E}$ as t approaches infinity. Figure 19.15b shows the current in the circuit as a function of time. Because the capacitor is uncharged at $t = 0$, the only element in the circuit which inhibits the current is the resistor R. Thus, from Ohm's law, the current in the circuit at $t = 0$ is given by $I_0 = \mathcal{E}/R$. The current decreases exponentially to zero as t approaches infinity, as pictured in Figure 19.15b.[1]

It takes an infinite amount of time for a capacitor to become fully charged, but the capacitor has stored a substantial fraction of its final charge in a very short period of time. A quantity called the *time constant*, τ (Greek letter tau), is used to describe the rate at which the charge on the capacitor varies with time. For a circuit containing a capacitor and resistor in series, $\tau = RC$. The time constant represents *the time for the charge to increase from zero to 63% of its maximum value*. That is, after one time constant, the charge on the capacitor has risen from zero to $0.63C\mathcal{E}$. Also in one time constant, the current in the circuit has fallen from its initial value of I_0 to $0.37I_0$.

The following dimensional analysis shows that τ has the unit of time:

$$[\tau] = [RC] = \left[\frac{V}{I} \times \frac{Q}{V}\right] = \left[\frac{Q}{Q/T}\right] = [T]$$

Now consider the circuit in Figure 19.16, consisting of a capacitor with an initial charge Q, a resistor, and a switch. When the switch is open (Fig. 19.16a), there is a potential difference Q/C across the capacitor and zero potential difference across the resistor because $I = 0$. If the switch is closed at $t = 0$, the capacitor begins to discharge through the resistor. The charge does

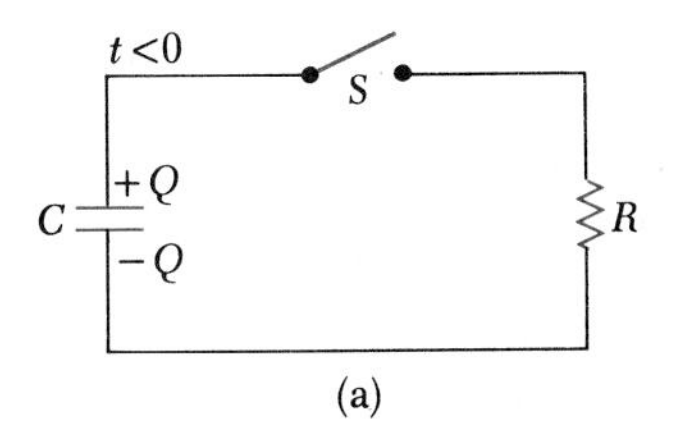

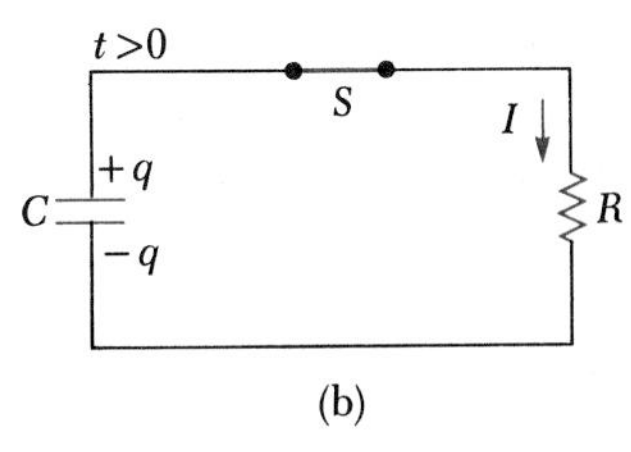

FIGURE 19.16 (a) A charged capacitor connected to a resistor and a switch, which is open at $t < 0$. (b) After the switch is closed, a nonsteady current is set up in the direction shown and the charge on the capacitor decreases exponentially with time.

[1]The methods of calculus can be used to show that the current in an RC circuit as a function of time is given by

$$I = I_0 e^{-t/RC} = \frac{\mathcal{E}}{R} e^{-t/RC}$$

where e is the base of the natural logarithm, $e = 2.718$, and $I_0 = \mathcal{E}/R$. The charge on the capacitor as a function of time is given by

$$q = Q(1 - e^{-t/RC}) = C\mathcal{E}\,(1 - e^{-t/RC})$$

where $Q = C\mathcal{E}$ is the maximum charge on the capacitor.

As an example of how to use the two equations above, assume that a circuit like that in Figure 19.14 has $\mathcal{E} = 12$ V, $C = 5\ \mu$F, and $R = 8 \times 10^5\ \Omega$. Find the current in the circuit and the charge on the capacitor at $t = 2$ s.

Since the time constant is $\tau = RC = (8 \times 10^5\ \Omega)(5 \times 10^{-6}\ \text{F}) = 4$ s, the current is found as follows:

$$I = \frac{\mathcal{E}}{R} e^{-t/RC} = \left(\frac{12\ \text{V}}{8 \times 10^5\ \Omega}\right) e^{-(2\ \text{s}/4\ \text{s})}$$
$$= (1.5 \times 10^{-5})e^{-0.5}\ \text{A} = (1.5 \times 10^{-5})(0.607)$$
$$= 9.1 \times 10^{-6}\ \text{A} = 9.1\ \mu\text{A}$$

The charge on the capacitor is given by

$$q = C\mathcal{E}\,(1 - e^{-t/RC}) = (5 \times 10^{-6}\ \text{F})(12\ \text{V})(1 - e^{-0.5})$$
$$= (6 \times 10^{-5})(1 - 0.607)\ \text{C} = 2.36 \times 10^{-5}\ \text{C} = 0.236\ \mu\text{C}$$

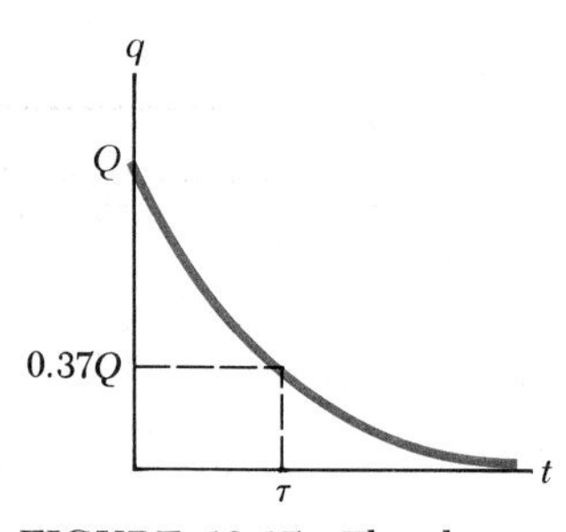

FIGURE 19.17 The charge on the capacitor of Figure 19.16 as a function of time.

not disappear instantaneously but decays exponentially to zero, as shown in Figure 19.17. Likewise, the current decays to zero exponentially. The time constant, $\tau = RC$, gives the time it takes both current and charge to fall to 37% of their initial values.

EXAMPLE 19.6 Charging a Capacitor in an *RC* Circuit

An uncharged capacitor and a resistor are connected in series to a battery, as in Figure 19.18. If $\mathcal{E} = 12$ V, $C = 5$ μF, and $R = 8 \times 10^5\ \Omega$, find the time constant of the circuit, the maximum charge on the capacitor, and the charge on the capacitor after one time constant.

Solution The time constant of the circuit is

$$\tau = RC = (8 \times 10^5\ \Omega)(5 \times 10^{-6}\ \text{F}) = 4\ \text{s}$$

The maximum charge on the capacitor is

$$Q = C\mathcal{E} = (5 \times 10^{-6}\ \text{F})(12\ \text{V}) = 60\ \mu\text{C}$$

After a time equal to one time constant, the charge on the capacitor is 63% of its maximum value:

$$q = 0.63Q = 0.63(60 \times 10^{-6}\ \text{C}) = 37.8\ \mu\text{C}$$

EXERCISE Calculate the maximum current in the circuit and the current in the circuit after one time constant has elapsed.

Answer 15 μA; 5.6 μA.

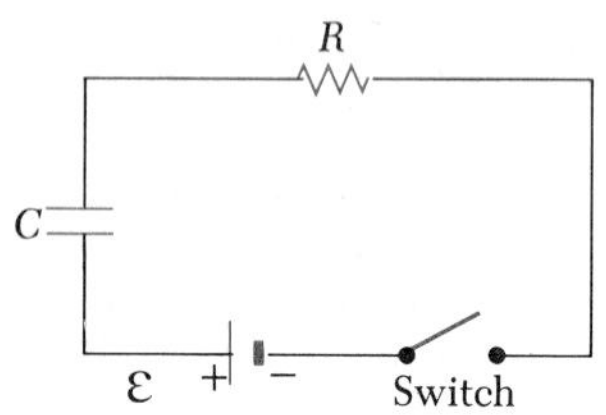

FIGURE 19.18 (Example 19.6) The switch of this series *RC* circuit is closed at $t = 0$.

19.6 MEASUREMENT OF RESISTANCE

The Voltmeter-Ammeter Method

From Ohm's law, $V = IR$, it is easy to see that the resistance of a conductor can be obtained if the current through the conductor and the potential difference across it are measured simultaneously. A device that measures current is called an **ammeter,** and one that measures potential difference across two terminals is called a **voltmeter.** A multimeter is an instrument that is designed (with proper switch settings) to be used as either an ammeter, a voltmeter, or an ohmeter, a device to measure resistance. Figure 19.19 is a photo of a typical multimeter. We shall describe the construction of these instruments

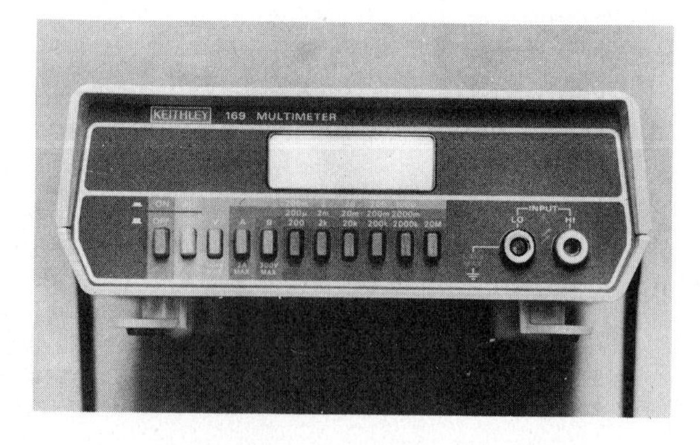

FIGURE 19.19 Voltages, currents, and resistances are frequently measured by digital multimeters like the one shown in this photograph.

in Chapter 20. For now, we shall describe a few of their characteristics and the procedure for using them.

An ammeter is constructed so that it must be placed in series with the current that it is to measure (Fig. 19.20). When using an ammeter to measure direct currents, one must be sure to connect it such that current enters at the positive terminal of the instrument and exits at the negative terminal. *Ideally, an ammeter should have zero resistance so as not to alter the current being measured.* Since any ammeter always has some resistance, its presence in the circuit will slightly reduce the current from its value when the ammeter is not present. If its resistance is known, one can calibrate the ammeter to give the value of the current that would be present in the absence of the ammeter.

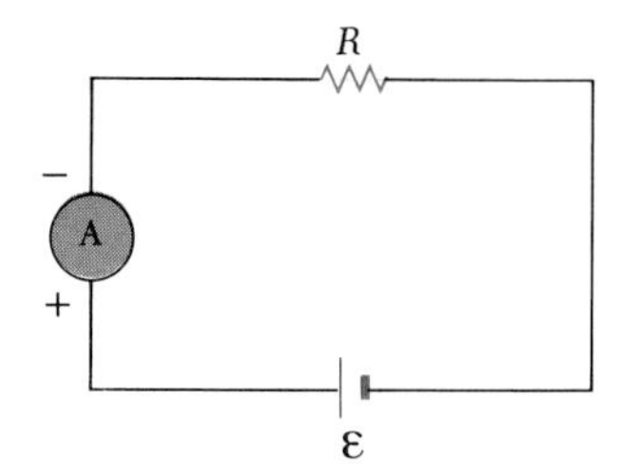

FIGURE 19.20 The current in a circuit can be measured with an ammeter Ⓐ connected in series with the resistor and battery. An ideal ammeter has zero resistance.

The potential difference across a resistor can be measured with a voltmeter connected in parallel with the resistor, as in Figure 19.21. Again it is necessary to observe the polarity of the instrument. The positive terminal of the voltmeter must be connected to the end of the resistor at the higher potential, and the negative terminal to the low-potential end of the resistor. *An ideal voltmeter has infinite resistance so that no current will pass through it.* In practice, the resistance of a voltmeter should be large compared with the resistance to be measured.

The ammeter-voltmeter method is not recommended for precise measurements because most meters cannot be read to more than three-digit accuracy.

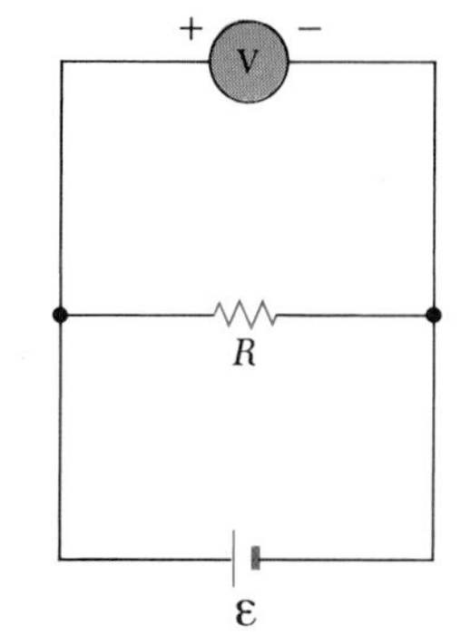

FIGURE 19.21 The potential difference across a resistor can be measured with a voltmeter Ⓥ connected in parallel with the resistor. An ideal voltmeter has infinite resistance and does not affect the circuit.

The Wheatstone Bridge

Unknown resistances can be accurately measured using a circuit known as a **Wheatstone bridge.** This circuit consists of the unknown resistance, R_x; a sensitive current detector called a galvanometer; three known resistances, R_1, R_2, and R_3, and a source of emf (Fig. 19.22). When the bridge is being used, R_1 is varied until the galvanometer reading is zero, that is, until there is no current from a to b in Figure 19.22. (Note that the symbol used for R_1 is ⩘, which indicates that it is a variable resistor.) When there is no current through the galvanometer, the bridge is said to be *balanced* and the potential at point a must equal the potential at point b (otherwise there would be a current between these points). This means that the potential difference across R_1 must equal the potential difference across R_x and the potential difference across R_2 must equal the potential difference across R_3. From these considerations, we see that

(1) $I_1R_1 = I_2R_x$

(2) $I_1R_2 = I_2R_3$

Dividing (1) by (2) and solving for R_x, we find

$$R_x = \frac{R_1R_3}{R_2} \tag{19.5}$$

Since R_1, R_2, and R_3 are known quantities, R_x can be calculated.

When high resistances are to be measured (above $10^5\ \Omega$), it becomes difficult to use the Wheatstone bridge. As a result of recent advances in solid-

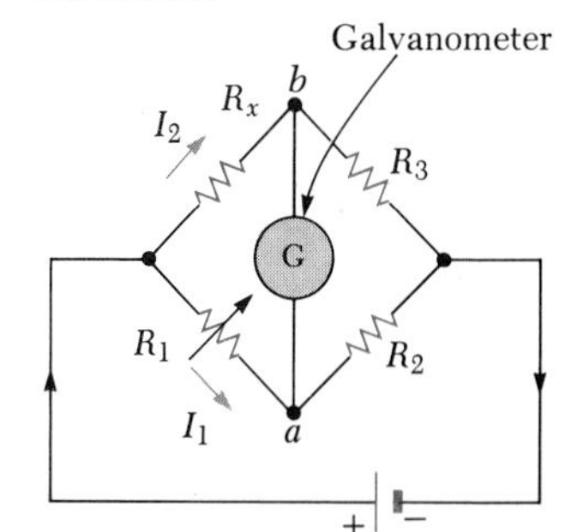

FIGURE 19.22 Circuit diagram for a Wheatstone bridge. This circuit can be used to measure an unknown resistance R_x in terms of known resistances R_1, R_2, and R_3. When the bridge is balanced, there is no current in the galvanometer.

state technology, modern electronic instruments are capable of measuring resistances as high as 10^{12} Ω. Such instruments are designed to have an extremely high effective resistance (as high as 10^{10} Ω) between the input terminals.

EXAMPLE 19.7 Poorly Designed Meters

(a) A poorly designed ammeter has a resistance of 6 Ω and is to be used to measure the current in the circuit shown in Figure 19.23a. Find the current in the absence of the meter and the current that the meter will read when inserted in the circuit.

Solution In the absence of the ammeter, the current in the circuit is found from Ohm's law as

$$I = \frac{V}{R} = \frac{6\text{ V}}{12\ \Omega} = \frac{1}{2}\text{ A}$$

When the ammeter is inserted in series as in Figure 19.23b, the total resistance in the circuit is 18 Ω, so the current in the circuit is

$$I = \frac{V}{R} = \frac{6\text{ V}}{18\ \Omega} = \frac{1}{3}\text{ A}$$

No measuring instrument should affect the value of the quantity it is designed to measure. In this case, the ammeter has reduced the current from 1/2 A to 1/3 A because its resistance is too high. Typical ammeters have resistances of a fraction of an ohm.

(b) A poorly designed voltmeter whose resistance is 6 Ω is to be used to measure the voltage across one of the 6 Ω resistors in the circuit of Figure 19.23a. Find the voltage across the resistor before and after insertion of the meter.

Solution In the absence of the voltmeter, the current in the circuit is 1/2 A, and the voltage drop across one of the 6 Ω resistors is found from Ohm's law to be (1/2 A)(6 Ω) = 3 V. When the 6 Ω voltmeter is connected across one of the resistors as in Figure 19.23c, the equivalent resistance of the circuit becomes 9 Ω. (You should use the rules for combining resistors in parallel and series to verify this value.) Thus, the current in the circuit now becomes

$$I = \frac{V}{R} = \frac{6\text{ V}}{9\ \Omega} = \frac{2}{3}\text{ A}$$

and the voltage across the voltmeter is

$$V = IR = \left(\frac{2}{3}\text{ A}\right)(3\ \Omega) = 2\text{ V}$$

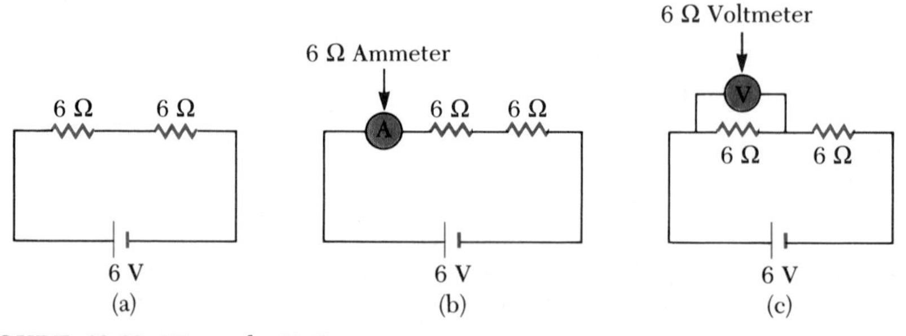

FIGURE 19.23 Example 19.7

Again, the meter has affected the quantity it is to measure, which in this case is voltage, because its resistance is too low. Properly designed voltmeters have very high resistances, typically several thousand ohms or higher.

*19.7 THE ELECTRONIC BALANCE

A common device used in many laboratories to measure the mass of an object with digital readout is an electronic balance. The basic circuit used in many of these instruments is a Wheatstone bridge circuit designed such that two of the resistors in the bridge are strain gages. (The normal spelling is gauge, but in this application the spelling is always gage.) The strain gage shown in Figure 19.24a consists of a piece of wire bonded to a flexible plastic backing. The typical resistance of such a gage for use in a balance is around 100 Ω. Figure 19.24b shows how two of these gages are used to measure mass. One gage is glued to the top of a flexible piece of metal and the other is glued to the lower side. When the metal strip is bent, the length of the wire on the top gage increases and its cross-sectional area decreases. Thus, from the expression $R = \rho\ell/A$ we see that the resistance of this gage increases. Likewise, when the metal strip is bent, the length of the wire on the bottom gage decreases, its cross-sectional area increases, causing its resistance to decrease. These changes in resistance are detected by a Wheatstone bridge circuit like that shown in Figure 19.24c. The resistors R_1 and R_2 are selected such that the bridge is balanced when there is no mass on the metal strip. When the bridge is balanced, there is no potential difference between the points a and b. However, when the strip is bent by the mass to be measured, the resistance of

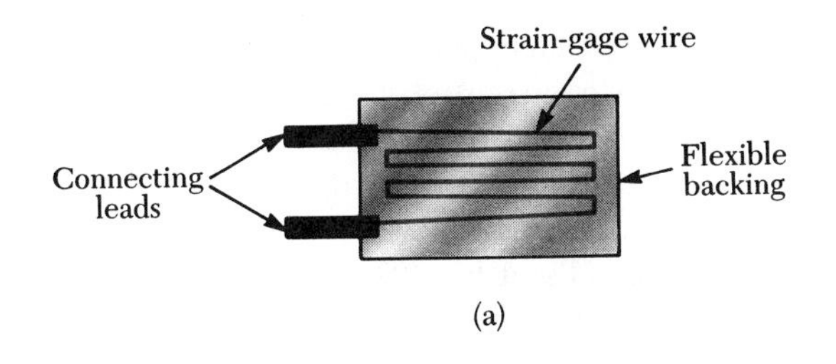

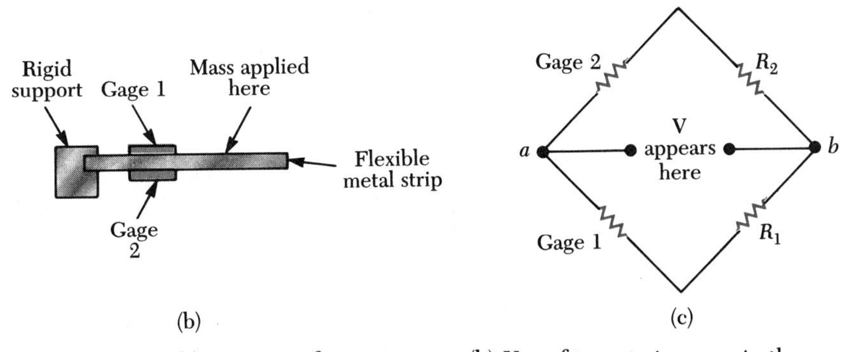

FIGURE 19.24 (a) Diagram of a strain gage. (b) Use of two strain gages in the design of an electronic balance. (c) Wheatstone bridge circuit used in the electronic balance.

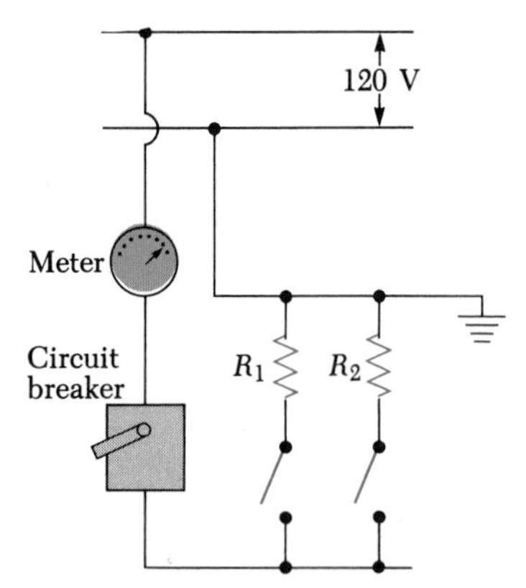

FIGURE 19.25 Wiring diagram for a household circuit. The resistances R_1 and R_2 represent appliances or other electrical devices which operate with an applied voltage of 120 V.

each gage changes and the bridge is no longer balanced. Consequently, a potential difference appears between points a and b which is proportional to the magnitude of the applied mass. Suitable electronic circuits are used to measure this potential difference and convert it to a digital readout of the mass.

*19.8 HOUSEHOLD CIRCUITS

Household circuits represent a very practical application of some of the ideas we have presented in this chapter. In a typical installation, the utility company distributes electric power to individual houses with a pair of wires, or power lines. A circuit in a house is then connected in parallel to these lines, as shown in Figure 19.25. The potential difference between the two wires is about 120 V. (These are alternating currents and voltages, but for the present discussion we shall assume that they are direct currents and voltages.) One of the wires is connected to ground, and the other wire, sometimes called the "live" wire, has an average potential of 120 V. A meter and circuit breaker (or in older installations, a fuse) are connected in series with the wire entering the house, as indicated in Figure 19.25.

The wire and circuit breaker are carefully selected to meet the current demands of a circuit. If the circuit is to carry currents as large as 30 A, a heavy-duty wire and appropriate circuit breaker must be used. Household circuits normally used to power lamps and small appliances often require only 15 A. Therefore, each circuit has its own circuit breaker to accommodate various load expectations.

As an example, consider a circuit containing a toaster, a microwave oven, and a heater (represented by R_1, R_2, . . . in Fig. 19.25). We can calculate the current through each appliance using the equation $P = IV$. The toaster, rated at 1000 W, would draw a current of 1000/120 = 8.33 A. The microwave oven, rated at 800 W, would draw a current of 6.67 A, and the heater, rated at 1300 W, would draw a current of 10.8 A. If the three appliances are operated simultaneously, they will draw a total current of 25.8 A. Therefore the breaker should be able to handle at least this much current or else it will be tripped. Alternatively, one could operate the toaster and microwave oven on one 15-A circuit and the heater on a separate 15-A circuit.

Many heavy-duty appliances, such as electric ranges and clothes dryers, require 240 V for their operation. The power company supplies this voltage by providing a third wire, also considered live, which is 120 V below ground potential (Fig. 19.26). Therefore, the potential difference between this wire and the other live wire (which is 120 V above ground potential) is 240 V. An appliance that operates from a 240-V line requires half the current of one operating from a 120-V line; therefore smaller wires can be used in the higher-voltage circuit without overheating of the wires becoming a problem.

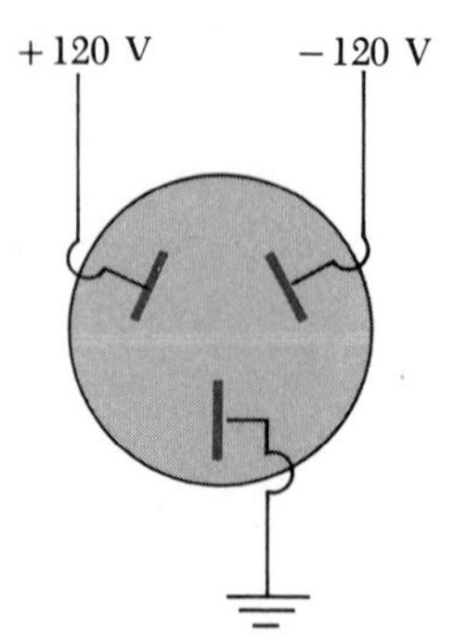

FIGURE 19.26 Power connections for a 240-V appliance.

*19.9 ELECTRICAL SAFETY

A person can be electrocuted by touching a live wire (which commonly happens because of a frayed cord or other exposed conductors) while in contact with ground. The ground contact might be made either by the person touching a water pipe (which is normally at ground potential) or by standing on ground with wet feet, since water is a good electric conductor. Such situations should be avoided at all costs.

Electric shock can result in fatal burns, or it can cause the muscles of vital organs, such as the heart, to malfunction. The degree of damage to the body depends on the magnitude of the current, the length of time it acts, and the part of the body through which it passes. Currents of 5 mA or less can cause a sensation of shock but ordinarily do little or no damage. If the current is larger than about 10 mA, the hand muscles contract and the person may be unable to release the live wire. If a current of about 100 mA passes through the body for only a few seconds, the result could be fatal. Such large currents will paralyze the respiratory muscles and prevent breathing. In some cases, currents of about 1 A through the body may produce serious (and sometimes fatal) burns.

Many 120-V outlets are designed to take a three-pronged power cord. (This feature is required in all new electrical installations.) One of these prongs is the live wire and two are common with ground. The additional ground connection is provided as a safety feature. Many appliances contain a three-pronged 120-V power cord with one of the ground wires connected directly to the casing of the appliance. If the live wire is accidentally shorted to ground (which often occurs when the wire insulation wears off), the current will take the low-resistance path through the appliance to ground. In contrast, if the casing of the appliance is not properly grounded and a short occurs, a person in contact with the appliance will experience an electric shock since the body will provide the low-resistance path to ground.

*19.10 VOLTAGE MEASUREMENTS IN MEDICINE

Electrocardiograms

Every action involving the body's muscles is initiated by electrical activity. The voltages produced by muscular action in the heart are particularly important to physicians. Voltage pulses cause the heart to beat, and the waves of electrical excitation associated with the heart beat are conducted through the body via the body fluids. These voltage pulses are large enough to be detected by suitable monitoring equipment attached to the skin. Standard electric devices can be used to record these voltage pulses because the amplitude of a typical pulse associated with heart activity is of the order of 1 mV. These voltage pulses are recorded on an instrument called an **electrocardiograph,** and the pattern recorded by this instrument is called an **electrocardiogram** (EKG). In order to understand the information contained in an EKG pattern, it is useful to first describe the underlying principles concerning electrical activity in the heart.

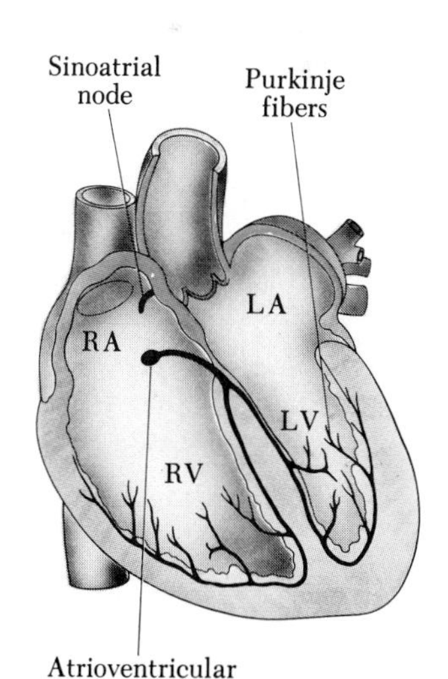

FIGURE 19.27 The electrical conduction system of the human heart. (RA: right atrium; LA: left atrium; RV: right ventricle; LV: left ventricle.)

The right atrium of the heart contains a specialized set of muscle fibers called the SA (sinoatrial) node, which initiate the heartbeat (Fig. 19.27). Electric impulses that originate in these fibers gradually spread from cell to cell throughout the right and left atrial muscles, causing them to contract. The pulse that passes through the muscle cells is often called a *depolarization wave* because of its effect on individual cells. If one were to examine an in-

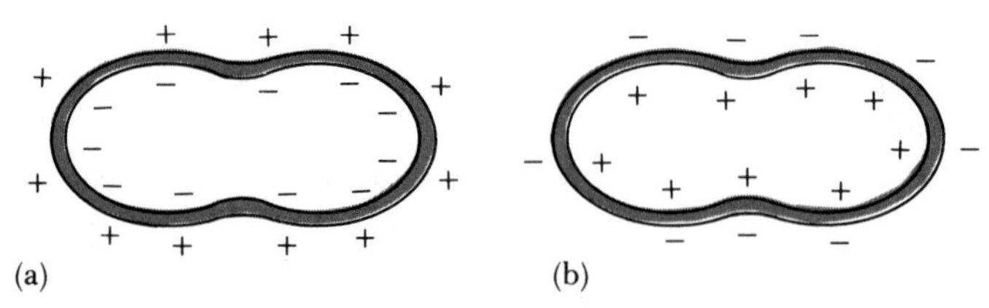

FIGURE 19.28 (a) Charge distribution of a muscle cell in the atrium before the depolarization wave has passed through the cell. (b) Charge distribution after the wave has passed.

dividual muscle cell, one would find an electric charge distribution on its surface, as shown in Figure 19.28a. (See Section 17.10 for an explanation of how this charge distribution arises.) The impulse generated by the SA node momentarily changes the cell's charge distribution to that shown in Figure 19.28b. The positively charged ions on the surface of the cell are temporarily able to diffuse through the membrane wall such that the cell attains an excess positive charge on its inside surface. As the depolarization wave travels from cell to cell throughout the atria, the cells recover to the charge distribution shown in Figure 19.28a. When the impulse reaches the AV (atrioventricular) node (Fig. 19.27), the muscles of the atria begin to relax, and the pulse is directed by the AV node to the ventricular muscles. The muscles of the ventricles then contract as the depolarization wave spreads through the ventricles along a group of fibers called the *Purkinje fibers*. The ventricles then relax after the pulse has passed through. At this point, the SA node is again triggered and the cycle is repeated.

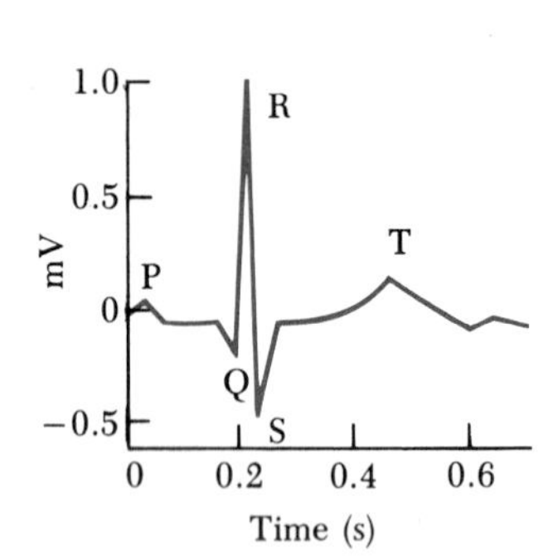

FIGURE 19.29 An EKG response for a normal heart.

A sketch of the electrical activity registered on an EKG for one beat of a normal heart is shown in Figure 19.29. The pulse indicated by P occurs just before the atria begin to contract. The QRS pulse occurs in the ventricles just before they contract, and the T pulse occurs when the cells in the ventricles begin to recover. EKGs for an abnormal heart are shown in Figure 19.30. The QRS portion of the pattern shown in Figure 19.30a is wider than normal. This indicates that the patient may have an enlarged heart. Figure 19.30b indicates that there is no relationship between the P pulse and the QRS pulse. This suggests a blockage in the electrical conduction path between the SA and AV nodes. This can occur when the atria and ventricles beat independently. Finally, Figure 19.30c shows a situation in which there is no P pulse and an irregular spacing between the QRS pulses. This is symptomatic of irregular atrial contraction, which is called fibrillation. In this situation, the atrial and ventricular contractions are irregular.

Electroencephalography

The electrical activity of the brain can be measured with an instrument called an **electroencephalograph** in much the same way that an electrocardiograph measures the electrical activity of the heart. The voltage pattern measured by an electroencephalograph is referred to as an **electroencephalogram** (EEG). An EEG pattern is recorded by placing electrodes on the patient's scalp. While an EKG voltage pulse is typically 1 mV, a typical voltage amplitude

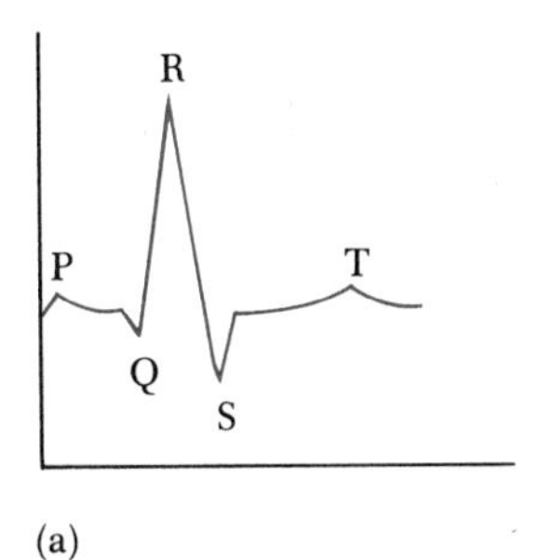

(a)

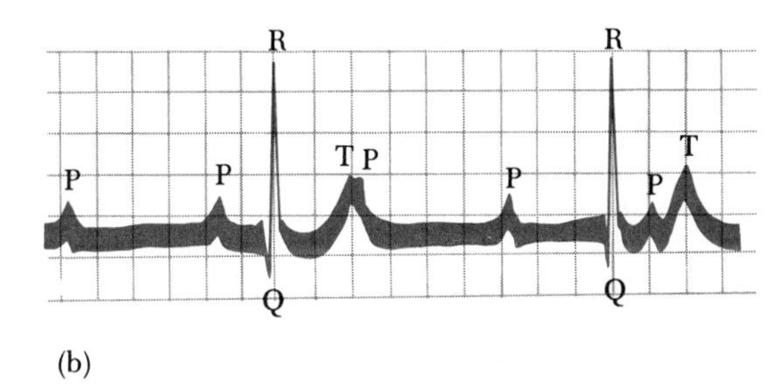

(b)

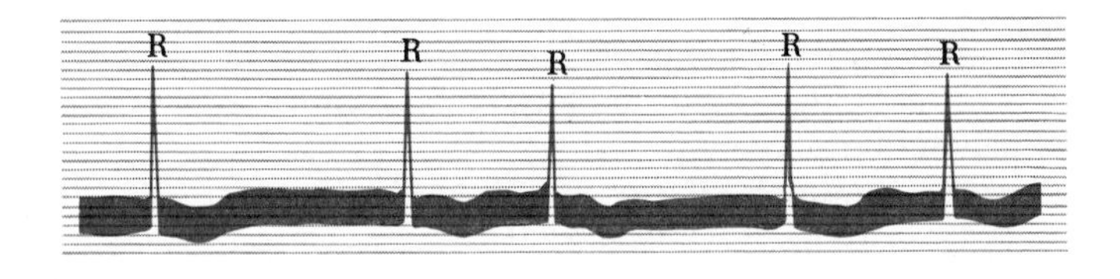

(c)

FIGURE 19.30 Abnormal EKGs.

associated with brain activity is only a few *micro*volts and is therefore more difficult to measure.

The EEGs in Figure 19.31 represent the brain wave pattern of a patient awake and then in various stages of sleep. In Figure 19.31b the patient begins to fall into a light sleep, and in Figure 19.31d the patient is in a deep sleep. Figure 19.31c is the EEG during a type of sleep called REM (rapid eye movement) sleep, which occurs approximately every 2 h. In this stage, the brain activity is quite similar to that of the patient while awake. It is interesting to note that a person in this stage of sleep is extremely difficult to awaken. Apparently, REM sleep is necessary for psychological well-being. Anyone who is deprived of this stage of sleep for an extended period of time becomes extremely fatigued and irritable.

The EEG is an extremely important diagnostic tool for detecting epilepsy, brain tumors, brain hemorrhage, meningitis, and so forth. For example, the EEG pattern of a patient suffering an epileptic seizure would show very little structure, indicating that the brain activity is greatly reduced.

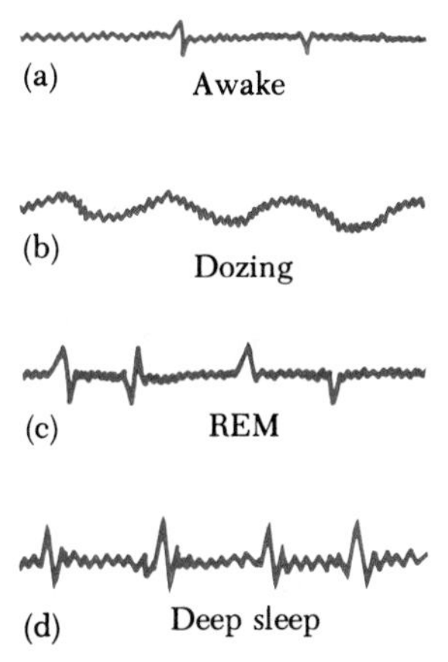

FIGURE 19.31 Brain waves from an individual in various stages of sleep.

Brain waves are often discussed in terms of their frequencies. Frequencies of about 10 Hz are referred to as *alpha waves*, frequencies between 10 Hz and 60 Hz are called *beta waves*, and those below 10 Hz are called *delta waves*. As a person falls asleep, the frequency of brain wave activity generally decreases. For example, the brain wave frequency for a person in a very deep sleep may drop as low as 1 or 2 Hz.

SUMMARY

A source of electromotive force (emf) is any device that will transform nonelectrical energy into electrical energy.

The **equivalent resistance** of a set of resistors connected in **series** is

$$R_{eq} = R_1 + R_2 + R_3 + \ldots \tag{19.2}$$

The **equivalent resistance** of a set of resistors connected in **parallel** is

$$\frac{1}{R_{eq}} = \frac{1}{R_1} + \frac{1}{R_2} + \frac{1}{R_3} + \ldots \tag{19.4}$$

Complex circuits are conveniently solved by using two rules called **Kirchhoff's rules:**

1. The sum of the currents entering any junction must equal the sum of the currents leaving that junction.
2. The sum of the potential differences across each element around any closed circuit loop must be zero.

The first rule is a statement of **conservation of charge.** The second is a statement of **conservation of energy.**

If a capacitor is charged by a battery through a resistor, it is found that the current approaches zero as the capacitor is being charged. The **time constant,** $\tau = RC$, represents the time it takes the charge to increase from zero to 63% of its maximum value.

A **Wheatstone bridge** is a circuit that can be used to measure an unknown resistance.

ADDITIONAL READING

P. F. Baker, "The Nerve Axon," *Sci. American*, March 1966, p. 74.

P. Davidovits, *Physics in Biology and Medicine*, Englewood Cliffs, N.J., Prentice-Hall, 1977.

B. Katz, "The Nerve Impulse," *Sci. American*, November 1952, p. 55.

K. Kordesch and K. Tomantschger, "Primary Batteries," *The Physics Teacher*, January 1981, p. 12.

T. F. Robinson, S. M. Factor, and E. H. Sonnenblick, "The Heart as a Suction Pump," *Sci. American*, June 1986, p. 84.

A. M. Scher, "The Electrocardiogram," *Sci. American*, November 1961, p. 132.

G. M. Shepherd, "Microcircuits in the Nervous System," *Sci. American*, February 1978, p. 92.

A. K. Solomon, "Pumps in the Living Cell," *Sci. American*, August 1962, p. 100.

QUESTIONS

1. Two sets of Christmas tree lights are available. In set A, when one bulb is removed or burns out, all the others remain illuminated. In set B, when one bulb is removed, the remaining bulbs will not operate. Explain the difference in the wiring for the two sets.
2. Are the two headlights on a car wired in series or in parallel? How can you tell? Suppose you rewired your headlights so that they were in series. What would be the effect (even before one burns out)?
3. An incandescent lamp connected to a 120-V source with a short extension cord will provide more illumination than the same lamp connected to the same source with a long extension cord. Explain.
4. Is the direction of current through a battery always from negative to positive on the terminals?
5. Embodied in Kirchhoff's rules are two conservation laws. What are they?
6. A capacitor, a battery, a switch, and a lightbulb are connected in series. Describe what happens to the bulb after the switch is closed. Assume the capacitor is initially uncharged and assume the light will illuminate when connected directly across the battery terminals.
7. When electricians work with potentially live wires, they often use the backs of their hands or fingers to move the wires. Why do you suppose they use this technique?
8. Would a fuse work successfully if it were placed in parallel with the device it is supposed to protect?
9. Given three lightbulbs and a battery, sketch as many different electric circuits as you can.
10. Suppose you fall from a building and on the way down grab a high-voltage wire. Assuming that the wire holds you, will you be electrocuted? If the wire then breaks, should you continue to hold onto an end of the wire as you fall?
11. Why is it dangerous to turn on a light when you are in the bathtub?
12. Why is it possible for a bird to perch on a high-voltage wire without being electrocuted?
13. What procedure would you use to try to save a person who is "frozen" to a live high-voltage wire without endangering your own life?
14. If it is the current flowing through the body that determines how serious a shock will be, why do we see warnings of high voltage rather than high current near electric equipment?
15. Suppose you are flying a kite when it strikes a high-voltage wire. What factors determine how great a shock you receive?
16. Does it matter whether a fuse is placed in the high-voltage line of a household circuit or in the ground line?
17. Car batteries are often rated in ampere-hours. Does this designate the amount of current, power, energy, or charge that can be drawn from the battery?

PROBLEMS

Section 19.2 Resistors in Series

Section 19.3 Resistors in Parallel

1. A 4 Ω, an 8 Ω, and a 12 Ω resistor are connected in series with a 24-V battery. What is (a) the equivalent resistance, and (b) the current in each resistor?

2. A 9 Ω and a 6 Ω resistor are connected in series to a power supply. The current in the circuit is found to be 1/3 A. What is the voltage setting of the power supply?

3. The resistors of Problem 1 are connected in parallel across a 24-V battery. Find (a) the equivalent resistance, and (b) the current in each resistor.

4. A 9 Ω and a 6 Ω resistor are connected in parallel across a power supply. If the current through the equivalent resistor is 1/3 A, find the setting of the power supply.

5. Find the equivalent resistance of the circuit shown in Figure 19.32.

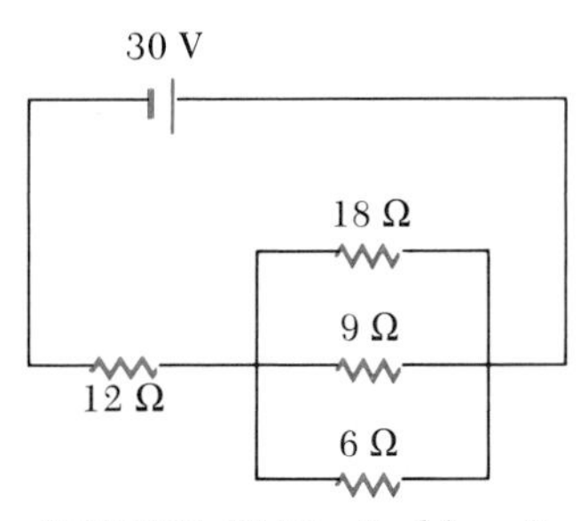

FIGURE 19.32 Problem 5

6. An 18 Ω and a 6 Ω resistor are connected in series across an 18-V battery. (a) Find the current through each resistor and the voltage drop across each resistor. (b) Repeat part (a) for the situation in which the resistors are connected in parallel.

7. Find the equivalent resistance of the circuit shown in Figure 19.33.

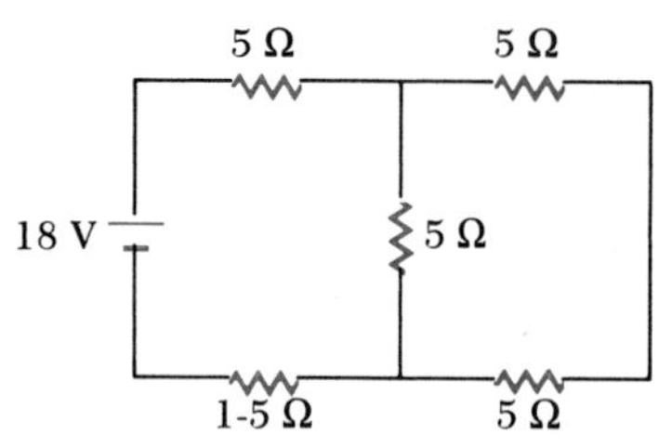

FIGURE 19.33 Problem 7

8. A 9 Ω and a 6 Ω resistor are connected in series with a power supply. (a) The voltage drop across the 6 Ω resistor is measured and found to be 12 V. Find the voltage output of the power supply. (b) The two resistors are connected in parallel across a power supply and the current through the 9 Ω resistor is found to be 1/4 A. Find the voltage setting of the power supply.

9. Find the equivalent resistance of the circuit shown in Figure 19.34.

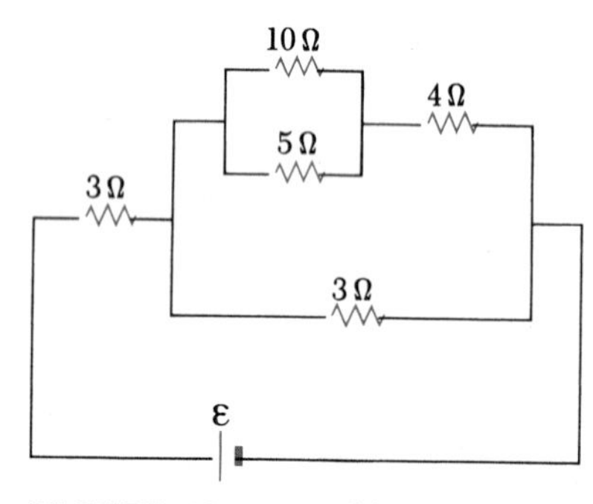

FIGURE 19.34 Problems 9 and 10

• 10. If the total power supplied to the circuit shown in Figure 19.34 is 4 W, find the emf of the battery.

• 11. Find the current in the 12 Ω resistor in Figure 19.35.

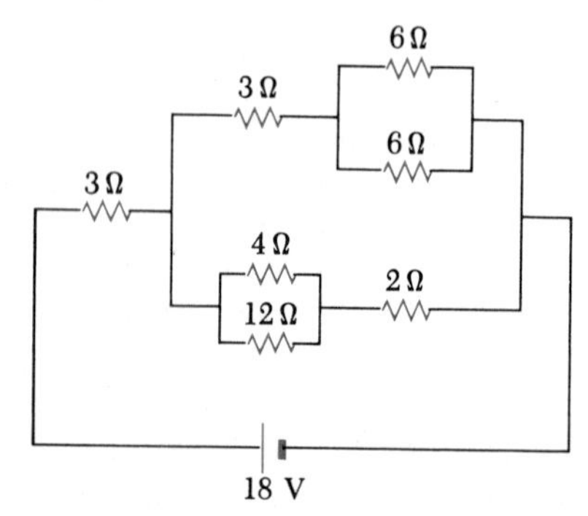

FIGURE 19.35 Problem 11

• 12. Two resistors A and B are connected in series to a 6-V battery. A voltmeter connected across resistor A measures a voltage of 4 V. When the two resistors are connected in parallel across the 6-V battery, the current through B is found to be 2 A. Find the resistance of A and B.

Section 19.4 Kirchhoff's Rules

13. Find the current through each resistor of Figure 19.36 by (a) the rules for resistors in series and parallel, and (b) by the use of Kirchhoff's rules.

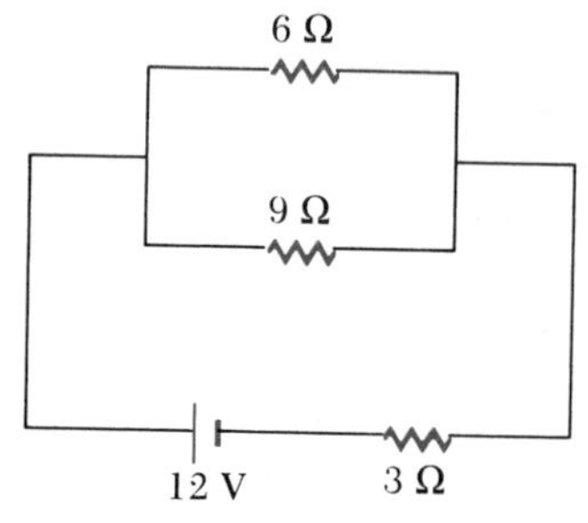

FIGURE 19.36 Problem 13

• 14. Find the current in each resistor in Figure 19.37.

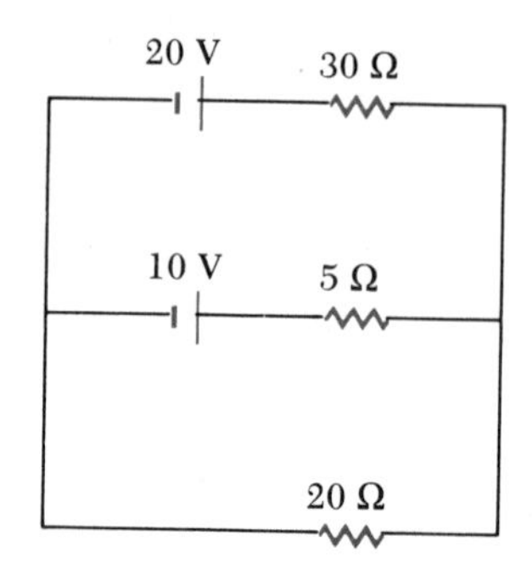

FIGURE 19.37 Problem 14

• 15. Find the current in each branch of the circuit in Figure 19.38 and the value of ℰ.

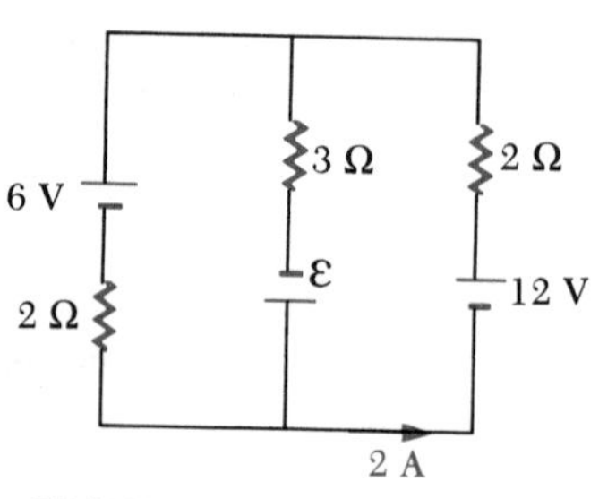

FIGURE 19.38 Problem 15

• **16.** Find the potential difference across each resistor shown in Figure 19.39.

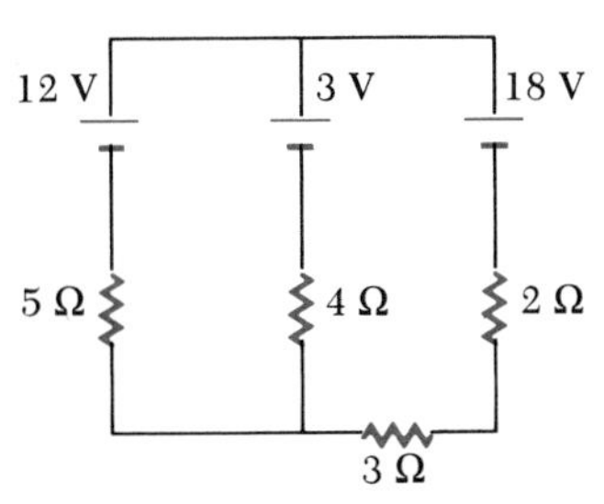

FIGURE 19.39 Problem 16

Section 19.5 *RC* Circuits

17. Consider a series *RC* circuit for which $C = 6\ \mu F$, $R = 2 \times 10^6\ \Omega$, and $\mathcal{E} = 20$ V. Find (a) the time constant of the circuit, (b) the maximum charge on the capacitor after a switch in the circuit is closed, and (c) the current in the circuit at the instant just after the switch in the circuit is closed.

• **18.** An uncharged capacitor and a resistor are connected in series to a source of emf. If $\mathcal{E} = 9$ V, $C = 20\ \mu F$, and $R = 100\ \Omega$, find (a) the time constant of the circuit, (b) the maximum charge on the capacitor, (c) the maximum current in the circuit, (d) the charge on the capacitor after one time constant, and (e) the current in the circuit after one time constant.

• **19.** An uncharged capacitor, $C = 10\ \mu F$, and a resistor are connected in series to an emf of 9 V. (a) What should the resistance equal to give a time constant of 5 s? (b) What is the current in this circuit, after two time constants have elapsed, if the resistance has the same value found in (a) above?

• **20.** (This problem assumes coverage of the material in Footnote 1 in Section 19.5.) A series circuit contains an emf of 9 V, a resistor of $3 \times 10^6\ \Omega$, and a capacitor of 2 μF. Find (a) the charge on the capacitor after 10 s, and (b) the current in the circuit at this time.

Section 19.6 Measurement of Resistance

Section 19.7 The Electronic Balance

21. (a) Calculate the potential difference across the 6 Ω resistor shown in Figure 19.40. (b) If you attempt to measure this potential difference by connecting across it a voltmeter that has a resistance of 1 Ω, what voltage will your meter read? (c) Repeat your calculation in part (b) assuming a voltmeter resistance of $1 \times 10^6\ \Omega$.

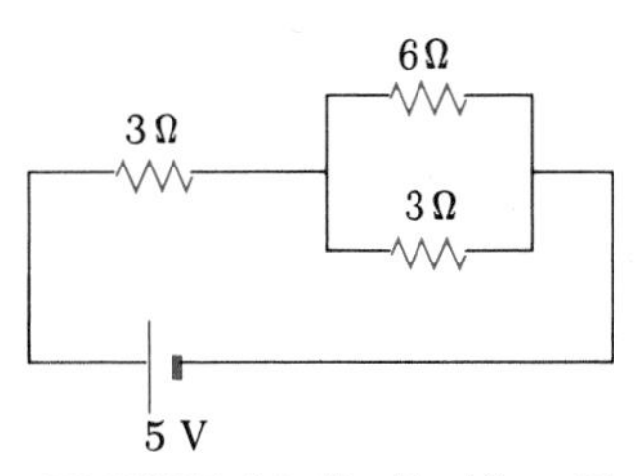

FIGURE 19.40 Problem 21

22. A circuit consists of an emf of 12 V connected to a 6-Ω resistor. (a) Calculate the current in the circuit. (b) Assume you attempt to measure this current by using an ammeter having a resistance of 100 Ω. What current will your meter read? (c) Repeat your calculations in part (b) assuming an ammeter resistance of 0.1 Ω.

23. The Wheatstone bridge in Figure 19.22 is balanced when $R_1 = 15\ \Omega$, $R_2 = 25\ \Omega$, and $R_3 = 40\ \Omega$. Calculate the value of R_x.

• **24.** The resistors R_1 and R_2 in Figure 19.22 are replaced by a 1-m length of nichrome wire between points *a* and *c* in Figure 19.41. Find an expression for R_x in terms of the resistor R_3, the length of wire between points a and b in Figure 19.41, and the length between points *b* and *c* when the bridge is balanced.

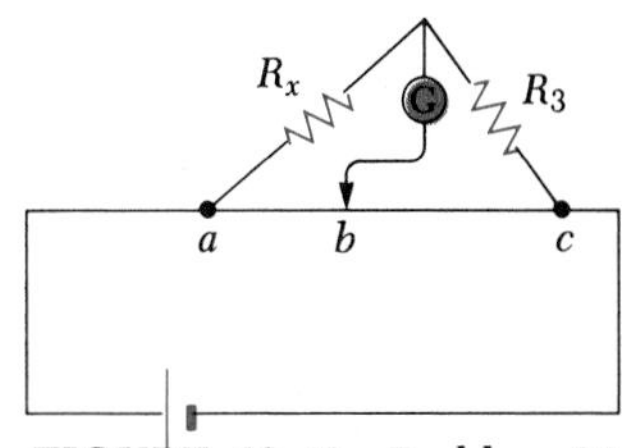

FIGURE 19.41 Problem 24

• **25.** A Wheatstone bridge of the type shown in Figure 19.22 is used to measure an unknown resistance. The resistor shown in the circuit as $R_3 = 1$ kΩ. If the bridge is balanced by adjusting R_1 such that $R_1 = 2.5R_2$, what is the resistance of the unknown resistor?

• **26.** In Figure 19.24 the positive terminal of a 5-V power supply is connected to the point where gage 1 and R_1 join, and the negative terminal is connected to the point where gage 2 and R_2 join. Initially, both gages have a resistance of 120 Ω and R_1 and R_2 are precision 1000 Ω resistors. What is the potential difference between points *a* and *b* when gage 1 is strained such that its resistance becomes 121 Ω and gage 2 is strained such that its resistance becomes 119 Ω?

Section 19.8 Household Circuits

27. An electric heater is rated at 1300 W, a toaster is rated at 1000 W, and an electric grill is rated at 1500 W.

The three appliances are connected in parallel to a common 120-V circuit. (a) How much current does each appliance draw? (b) Is a 30-A circuit sufficient in this situation? Explain.

28. How many 100-W lightbulbs can you connect in a 120-V house circuit without tripping a 20-A circuit breaker?

29. A heating element in a stove is designed to dissipate 3000 W when connected to 240 V. (a) Calculate the current in this element if it is connected to 120 V, and (b) calculate the power it dissipates at this voltage.

30. A lamp ($R = 150\ \Omega$), an electric heater ($R = 25\ \Omega$), and a fan ($R = 50\ \Omega$), are connected in parallel across a 120-V line. (a) What total current is supplied to the circuit? (b) What is the voltage across the fan? (c) What is the current in the lamp? (d) What power is expended in the heater?

ADDITIONAL PROBLEMS

31. If you have a 2 Ω, a 4 Ω, and a 6 Ω resistor, how many different resistance values can you obtain from the set? Show how you would connect them to get each resistance value.

32. The meter in Figure 19.42 reads 1 mA when a 2-V power supply is connected between points A and B. (a) Find the resistance R_1. (b) The meter continues to read 1 mA if a 20-V power supply is now connected between points A and C. Find R_2. (c) The meter continues at its reading of 1 mA when a 200-V power supply is connected between points A and D. Find R_3.

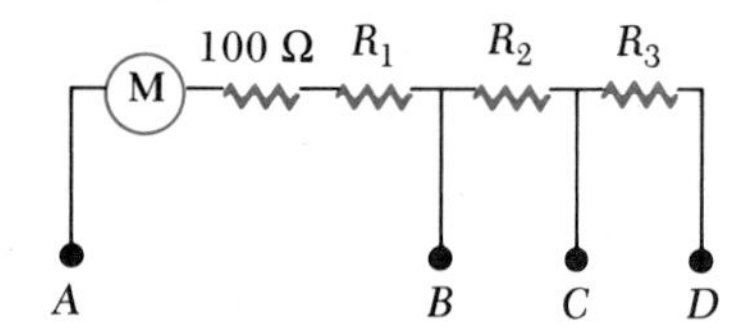

FIGURE 19.42 Problem 32

33. In Figure 19.43, $R_1 = 0.1\ \Omega$, $R_2 = 1.0\ \Omega$, and $R_3 = 10.0\ \Omega$. Find the equivalent resistance of the circuit and the current in each resistor when a 5-V power supply is connected between (a) points A and B, (b) points A and C, and (c) points A and D.

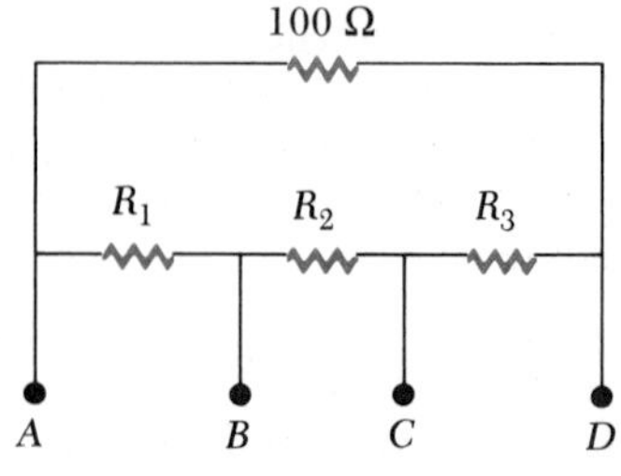

FIGURE 19.43 Problem 33

• **34.** Find the equivalent resistance of the circuit shown in Figure 19.44.

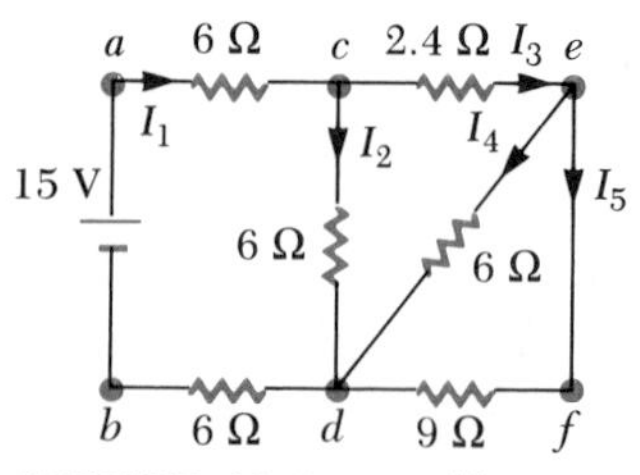

FIGURE 19.44 Problem 34

• **35.** For the circuit of Problem 34, find (a) each current in the circuit, (b) the potential difference across each resistor, and (c) the power dissipated by each resistor.

• **36.** If the current in the 5-Ω resistor of Figure 19.45 is 3 A, find the current in all the other resistors and the emf of the battery.

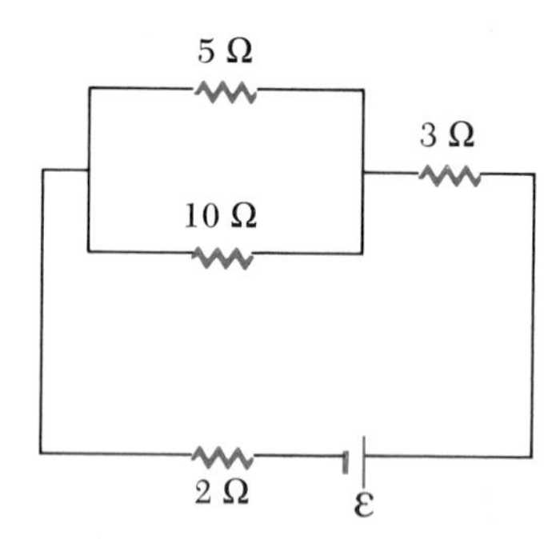

FIGURE 19.45 Problem 36

• **37.** A series *RC* circuit has a time constant of 0.96 s. The battery has an emf of 48 V, and the maximum current in the circuit is 500 mA. What is (a) the value of the capacitance and (b) the charge stored in the capacitor 1.92 s after the switch is closed?

• **38.** In the circuit of Figure 19.46, find the power delivered or absorbed by each battery and the power dissipated in each resistor.

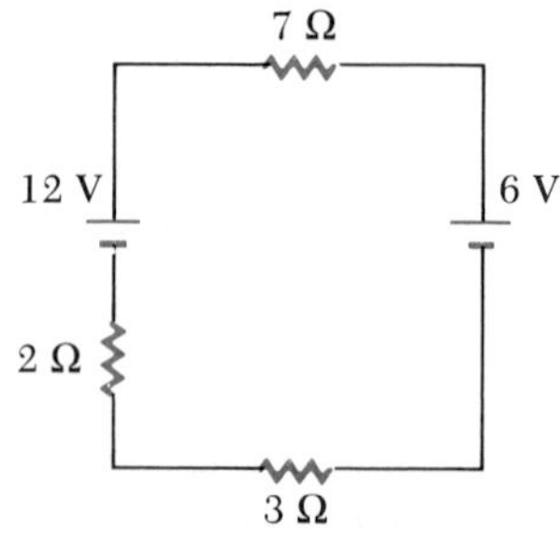

FIGURE 19.46 Problem 38

• **39.** Consider the circuit shown in Figure 19.47. Find the values of I_1, I_2, and I_3.

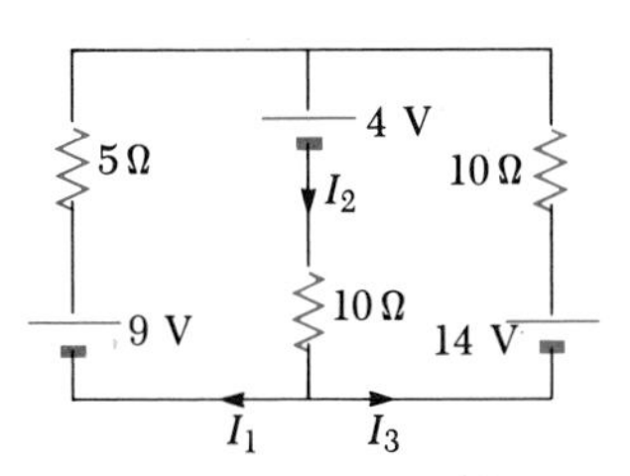

FIGURE 19.47 Problem 39

• **40.** In a certain circuit, a current of 1.00 A is drawn from a battery. The current then divides and passes through two resistors in parallel. One of the resistors has a value of 41 Ω, and the current through it is 0.10 A. What is the value of the other resistor?

• **41.** (a) Calculate the value of R for the circuit shown in Figure 19.48. (b) Determine the currents in the 6-Ω and 8-Ω resistors.

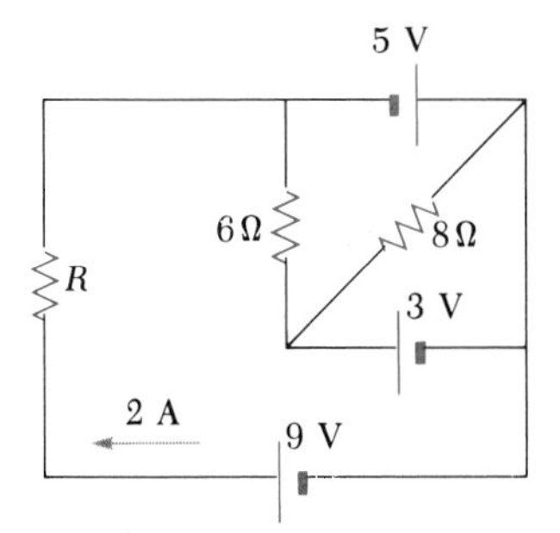

FIGURE 19.48 Problem 41

• **42.** Find the equivalent resistance between points a and b in Figure 19.49.

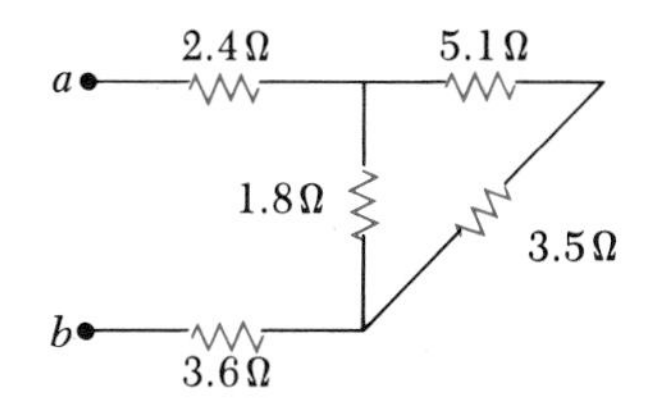

FIGURE 19.49 Problem 42

•• **43.** Consider the circuit shown in Figure 19.50. Calculate (a) the equivalent resistance of the circuit, and (b) the power dissipated by the entire circuit. Find (c) the current in the 5-Ω resistor.

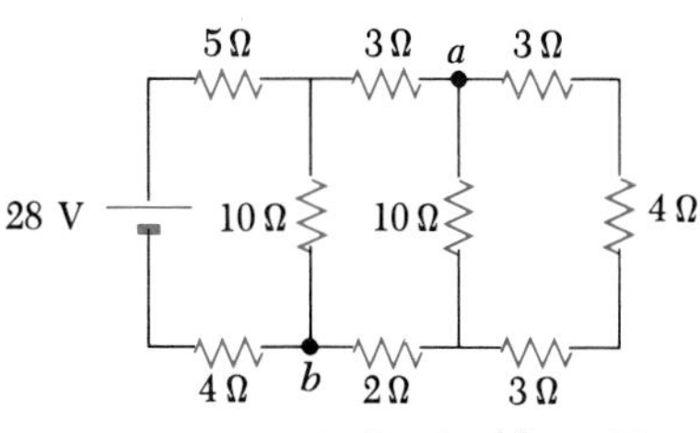

FIGURE 19.50 Problem 43

•• **44.** (This problem assumes coverage of Footnote 1.) An emf of 10 V is connected to a series RC circuit of a resistor of 2×10^6 Ω and a capacitor of 3 μF. Find the time taken for the charge on the capacitor to reach 90% of its final value.

•• **45.** Consider the circuit shown in Figure 19.51. In this circuit, the 4-V battery and the 2-V battery have internal resistances of 0.5 Ω and 0.25 Ω, respectively. (a) Find the current in the circuit. (b) Find the power dissipated as heat in the circuit. (c) Find the power absorbed by the 2-V battery as it charges, including the power dissipated as heat within the battery.

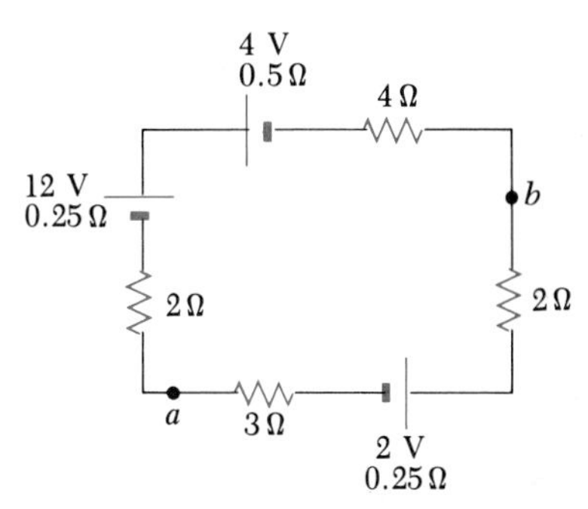

FIGURE 19.51 Problem 45

•• **46.** When a battery of unknown emf is connected to a 5-Ω resistor, the current in the circuit is 0.3 A. If the battery is now connected to an 8-Ω resistor, the current is 0.2 A. What is the emf of the battery and its internal resistance? (*Note:* The internal resistance of a battery may be represented in a circuit diagram as a resistor in series with the battery. The internal resistance of a dry cell depends on the condition of the electrolyte and increases as the cell ages.)

•• **47.** The current in a circuit that has a resistance of R_1 is 2 A. The current is reduced to 1.6 A when an additional resistor R_2 = 3 Ω is added in series with R_1. What is the value of R_1?

•• **48.** Calculate the current in the 6-V battery in Figure 19.52.

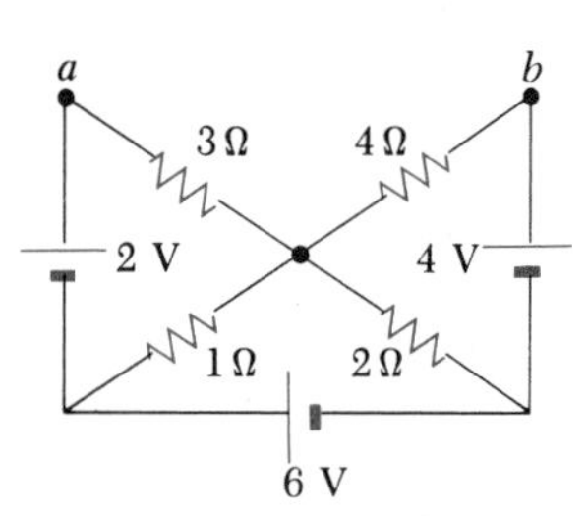

FIGURE 19.52 Problem 48

•• **49.** Consider the circuit shown in Figure 19.53. What are the expected readings of the ammeter and voltmeter?

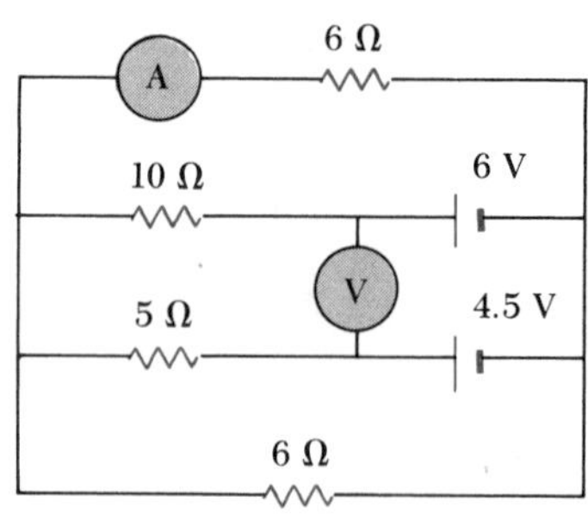

FIGURE 19.53 Problem 49

•• **50.** A voltage V is applied to a series configuration of n resistors each of value R. The circuit components are reconnected into a parallel configuration and the voltage V is again applied. Show that the power consumed by the series configuration is $1/n^2$ times the power consumed by the parallel combination.

Essay

Current in the Nervous System

Paul Davidovits *Boston College*

The most remarkable use of electrical phenomena in living organisms is found in the nervous system of animals. Specialized cells in the body called **neurons** form a complex network that receives, processes, and transmits information from one part of the body to another. The center of this network is located in the brain, which has the ability to store and analyze information. Based on this information, the nervous system controls parts of the body.

The nervous system is very complex: the human nervous system, for example, consists of about 10^{10} interconnected neurons. Some aspects of the nervous system are well known. During the past 35 years, the method of signal propagation through the nervous system has been firmly established. The messages are electric pulses transmitted by neurons. When a neuron receives an appropriate stimulus, it produces electric pulses that are propagated along its cable-like structure. The strength of the stimulus is conveyed by the number of pulses produced. When the pulses reach the end of the "cable," they activate other neurons or muscle cells.

The neurons, which are the basic units of the nervous system, can be divided into three classes: sensory neurons, motoneurons, and interneurons. The sensory neurons receive stimuli from sensory organs that monitor the external and internal environment of the body. Depending on their specialized functions, the sensory neurons convey messages about factors such as heat, light, pressure, muscle tension, and odor to higher centers in the nervous system. The motoneurons carry messages that control the muscle cells. The messages are based on the information provided by the sensory neurons and by the brain. The interneurons transmit information from one neuron to another.

Each neuron consists of a cell body to which are attached input ends called **dendrites** and a long tail called the **axon,** which propagates the signal away from the cell (Fig. 1). The far end of the axon branches into nerve end-

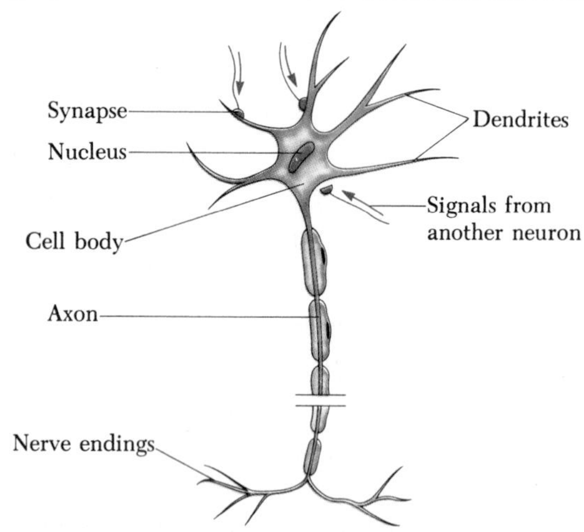

FIGURE 1 Diagram of a neuron

ings that transmit the signal across small gaps to other neurons or to muscle cells. A simple sensory-motoneuron circuit is shown in Figure 2. A stimulus from a muscle produces nerve impulses that travel to the spine. Here the signal is transmitted to a motoneuron, which in turn sends impulses to control the muscle.

The axon, which is an extension of the neuron cell, conducts the electric impulses away from the cell body. Some axons are extremely long. In humans, for example, the axons connecting the spine with the fingers and toes are more than 1 m long. The neuron can transmit messages because of the special electrical characteristics of the axon. Most of the information about the electrical and chemical properties of the axon is obtained by inserting small needle-like probes into the axon. With such probes it is possible to measure currents flowing in the axon and to sample its chemical composition. Such experiments are usually difficult to run because the diameter of most axons is very small. Even the largest axons in the human nervous system have a diameter of only about 20×10^{-4} cm. The giant squid, however, has an axon with a diameter of about 0.5 mm, which is large enough for the convenient insertion of probes. Much of the information about signal transmission in the nervous system has come from experiments with the squid axon.

In the aqueous environment of the body, salts and other molecules dissociate into positive and negative ions. As a result, body fluids are relatively good conductors of electricity. The inside of the axon is filled with an ionic fluid that is separated from the surrounding body fluid by a thin membrane that is only about 5 nm to 10 nm thick.

The resistivities of the internal and external fluids are about the same, but their chemical compositions are substantially different. The external fluid is similar to seawater. Its ionic solutes are mostly positive sodium ions and negative chloride ions. Inside the axon, the positive ions are mostly potassium ions and the negative ions are mostly large negatively charged organic ions.

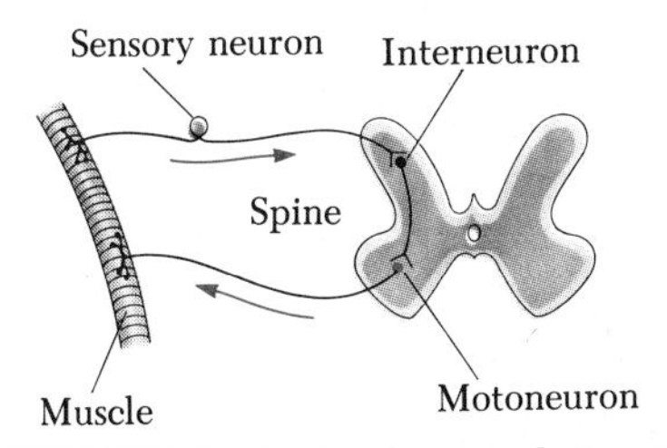

FIGURE 2 A simple neural circuit

Since there is a large concentration of sodium ions outside the axon and a large concentration of potassium ions inside, we may ask why the concentrations are not equalized by diffusion. In other words, why don't the sodium ions leak into the axon and the potassium ions leak out of it? The answer lies in the properties of the axon membrane.

In the resting condition, when the axon is not conducting an electric pulse, the axon membrane is highly permeable to potassium ions, slightly permeable to sodium ions, and impermeable to the large organic ions. Thus, while sodium ions cannot easily leak into the axon, potassium ions can certainly leak out of it. As the potassium ions leak out of the axon, however, they leave behind large negative ions, which cannot follow them through the membrane. As a result, a negative potential is produced inside the axon with respect to the outside. The negative potential, which has been measured to about 70 mV, holds back the outflow of potassium ions so that, at equilibrium, the concentration of ions is as we have stated.

The mechanism for the production of an electric signal by the neuron is conceptually remarkably simple. When a neuron receives an appropriate stimulus, which may be heat, pressure, or a signal from another neuron, the properties of its membrane change. As a result, sodium ions first rush into the cell while potassium ions flow out of it. This flow of charged particles constitutes an electric current signal which propagates along the axon to its destination.

Although the axon is a highly complex structure, its main electrical properties can be represented by the standard electric circuit concepts of resistance and capacitance. The propagation of the signal along the axon is then well described by the techniques of electric circuit analysis discussed in the text.

SUGGESTED READING

1. Katz, B. "How Cells Communicate," *Scientific American*, September 1981, p. 208.
2. Katz, B. *Nerve Muscle and Synapse,* New York, McGraw-Hill Inc., 1966.

ESSAY QUESTIONS

1. Why has the nervous system developed to utilize sodium and potassium ions to conduct electrical signals?
2. In the nervous system the strength of a stimulus is conveyed by the number of pulses produced rather than by the amplitude of the signal. What is the advantage of this arrangement?

CHAPTER 20

Magnetism

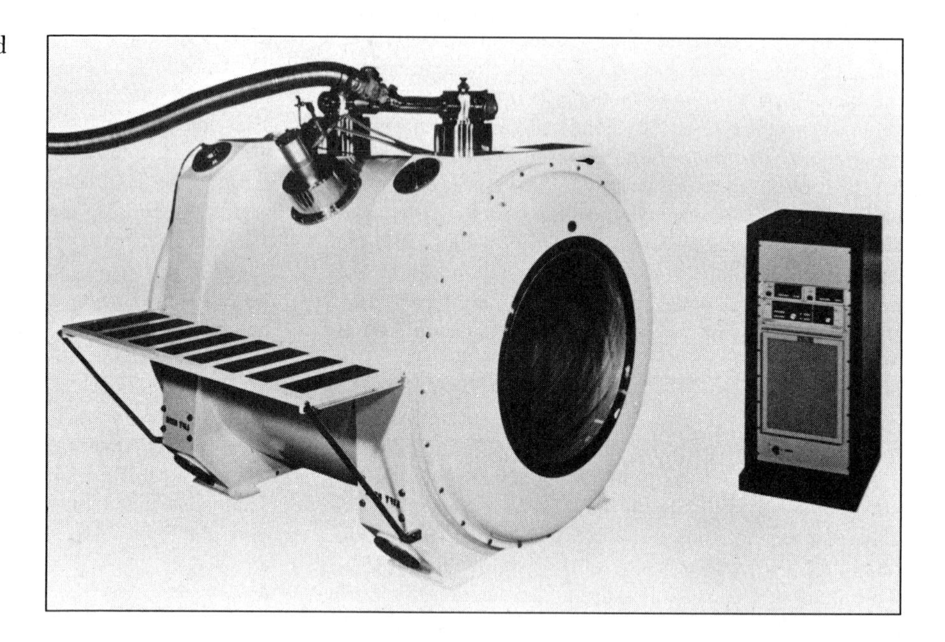

Superconducting magnet used in whole body magnetic resonance imaging.

The list of important technological applications of magnetism is very long. For instance, large electromagnets are used to pick up heavy loads. Magnets are also used in such devices as meters, motors, and loudspeakers. Magnetic tapes are routinely used in sound and video recording equipment, and for computer memory, and magnetic recording material is used on computer disks.

The phenomenon of magnetism was known to the Greeks as early as 800 B.C. and perhaps even earlier to the Chinese. The Greeks discovered that certain stones, now called magnetite, attract pieces of iron. Legend ascribes the name *magnetite* to the shepherd Magnes, "the nails of whose shoes and the tip of whose staff stuck fast in a magnetic field while he pastured his flocks."

As we investigate magnetism in this chapter, you will find that the subject cannot be divorced from electricity. For example, magnetic fields affect moving charges and moving charges produce magnetic fields. Ultimately we shall find that the source of all magnetic fields is electric current, whether it be the current in a wire or the current produced by the motion of charges within atoms or molecules.

20.1 MAGNETS

Most people have had some experiences with magnets in one form or another. You are most likely familiar with the common iron horseshoe magnet which is able to pick up iron objects such as paper clips and nails. In the discussion that follows, we shall assume that the magnet has the shape of a bar. Iron objects are most strongly attracted to the ends of such a bar magnet, called the **poles** of the magnet. One end is called the **north pole,** and the other the **south pole.** These poles received their names because of the behavior of a magnet in the presence of the earth's magnetic field. If a bar magnet is suspended from its midpoint by a piece of string so that it can swing freely in a horizontal plane, it will rotate until its north pole points to the north of the earth, and the south pole of the magnet points to the south of the earth. The same idea is used to construct a simple compass. Magnetic poles also exert attractive or repulsive forces on each other similar to electrical forces between charged objects. In fact, simple experiments with two bar magnets show that

like poles repel each other and unlike poles attract each other.

Although the force between two magnetic poles is similar to the force between two electric charges, there is an important difference. Electric charges can be isolated (witness the proton and the electron), but magnetic poles cannot be isolated. In fact, no matter how many times a permanent magnet is cut, each piece will always have a north and a south pole. Thus, magnetic poles always occur in pairs. There is some theoretical basis for speculating that magnetic monopoles (isolated north or south poles) may exist in nature, and attempts to detect them is currently an active experimental field of investigation. However, none of these attempts has proven successful.

There is yet another similarity between electric and magnetic effects, which is concerned with methods for making permanent magnets. In Chapter 16 we learned that when two materials such as rubber and wool are rubbed together, each becomes charged, one positive and the other negative. In a somewhat analogous fashion, an unmagnetized piece of iron can be magnetized by stroking it with a magnet. Magnetism can be induced in iron (and other materials) by other means. For example, if a piece of unmagnetized iron is placed near a strong permanent magnet, the piece of iron will eventually become magnetized. The process of magnetizing the piece of iron in the presence of a strong external field can be accelerated by either heating and cooling the iron or by hammering the iron. Naturally occuring magnetic materials, such as magnetite, achieve their magnetism in this manner since they have been subjected to the earth's magnetic field over very long periods of time. The extent to which a piece of material will retain its magnetism depends on whether it is classified as being magnetically **hard** or **soft.** Soft magnetic materials, such as iron, are easily magnetized but also tend to easily lose their magnetism. In contrast, hard magnetic materials such as cobalt and nickel are difficult to magnetize, but tend to retain their magnetism.

In earlier chapters we found it convenient to describe the interaction between charged objects in terms of electric fields. Recall that an electric field surrounds any electric charge. The region of space surrounding a moving charge also includes a magnetic field. Thus, a moving charge is surrounded by both electric and magnetic fields. There is also a magnetic field surround-

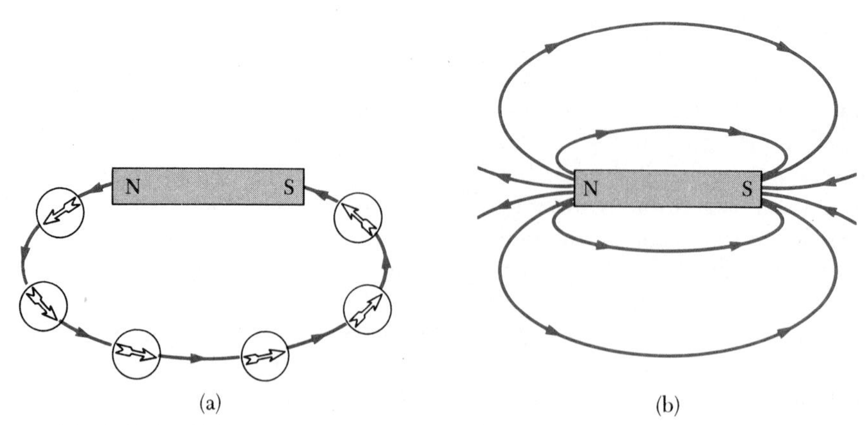

FIGURE 20.1 (a) Tracing the magnetic field of a bar magnet. (b) Several magnetic field lines of a bar magnet.

ing any magnetized material. The connection between this magnetic field and the motion of charged particles will be discussed in a later section. We shall use the symbol $\boldsymbol{B}$ to represent a magnetic field. In order to describe any type of field, we must define (1) the magnitude or strength of the field, and (2) the direction of the field. The direction of the magnetic field at any location can be defined as follows:

The direction of the magnetic field, $\boldsymbol{B}$, at any point is taken to be the direction that the north pole of a compass needle points at this location.

Figure 20.1a shows how the magnetic field of a bar magnet can be traced with the aid of a compass. Several magnetic field lines of a bar magnet traced out in this manner are shown in Figure 20.1b. The magnetic field patterns of a bar magnet and two bar magnets can be displayed by small iron filings as shown in the photographs on the opposite page.

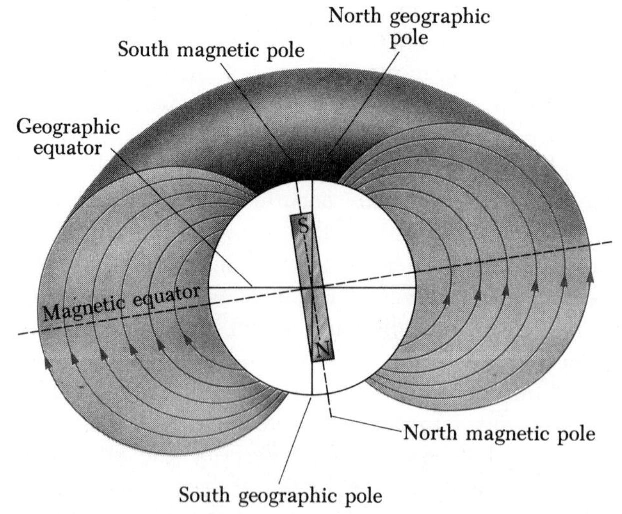

FIGURE 20.2 The earth's magnetic field lines. Note that the magnetic south is at the north geographic pole, while the magnetic north pole is at the south geographic pole.

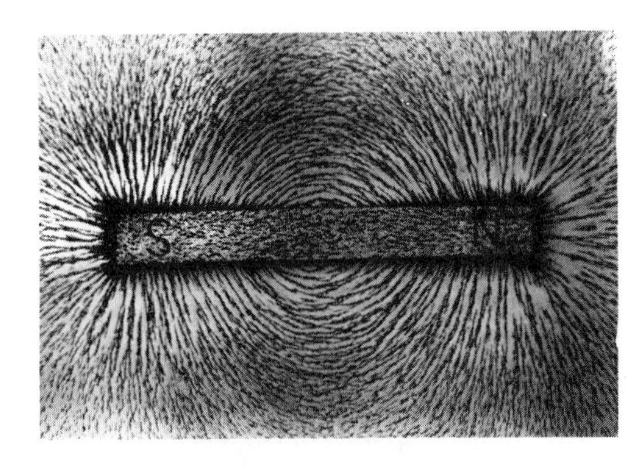

Magnetic field patterns of a bar magnet as displayed by small iron filings on a sheet of paper.

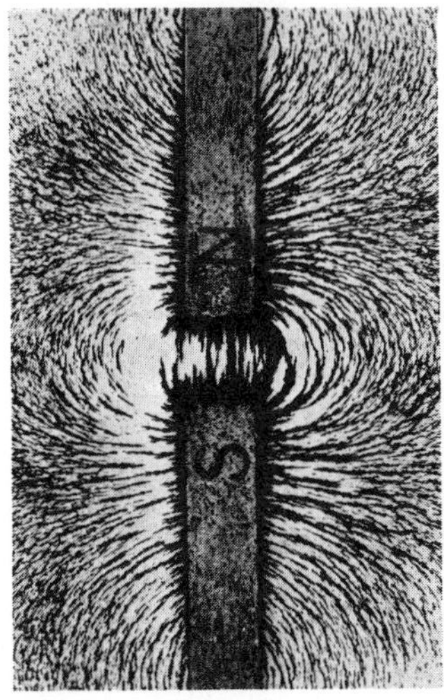

Magnetic field pattern surrounding two bar magnets, as displayed with iron filings on a sheet of paper.

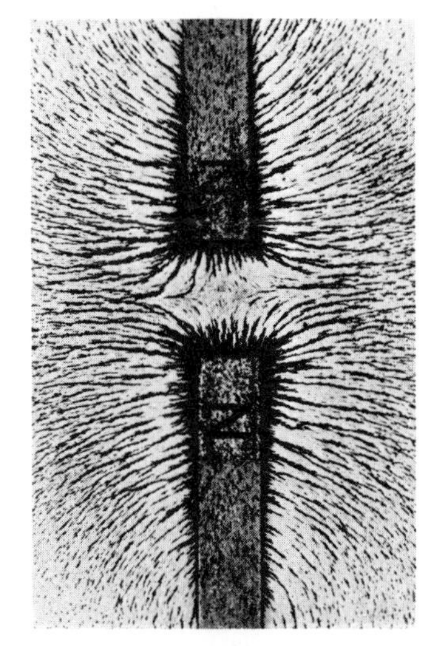

Magnetic field pattern demonstrates the repulsion between two like poles. (Courtesy of Hugh Strickland and Jim Lehman, James Madison University)

20.2 MAGNETIC FIELD OF THE EARTH

When we speak of a small bar magnet as having a north and a south pole, we should more properly say that it has a "north-seeking" and a "south-seeking" pole. By this we mean that if such a magnet is used as a compass, one end will seek, or point to, the north geographic pole of the earth. Thus, we conclude that *the magnetic north pole corresponds to the south geographic pole and the magnetic south pole corresponds to the north geographic pole.* In fact, the configuration of the earth's magnetic field, pictured in Figure 20.2, is very much like that which would be achieved by burying a bar magnet deep in the interior of the earth.

If a compass needle is suspended in bearings that allow it to rotate in the vertical plane as well as in the horizontal plane, the needle is horizontal with respect to the earth's surface only near the equator. As the device is moved northward, the needle rotates such that it points more and more toward the surface of the earth. Finally, at a point just north of Hudson Bay in Canada, the north pole of the needle would point directly downward. This location, first found in 1832, is considered to be the location of the south magnetic pole of the earth. This site is approximately 1300 mi from the earth's geographic north pole and varies with time. Similarly, the magnetic north pole of the earth is about 1200 miles away from the earth's geographic south pole. Thus, it is only approximately correct to say that a compass needle points north. The difference between true north, defined as the geographic north pole, and north indicated by a compass varies from point to point on the earth, and the difference is referred to as *magnetic declination.* For example, along a line through Florida and the Great Lakes, a compass indicates true north, whereas in Washington state, it aligns 25° east of true north.

Although the magnetic field pattern of the earth is similar to that which would be set up by a bar magnet deep within the earth, it is easy to understand why the source of the earth's field cannot be large masses of permanently magnetized material. The earth does have large deposits of iron ore deep beneath its surface, but the high temperatures in the earth's core prevent the iron from retaining any permanent magnetization. It is considered more likely that the true source is charge-carrying convection currents in the earth's core. Charged ions or electrons circling in the liquid interior could produce a magnetic field. There is also some evidence to indicate that the strength of a planet's field is related to the planet's rate of rotation. For example, Jupiter rotates faster than the earth, and recent space probes indicate that Jupiter's magnetic field is stronger than ours. Venus, on the other hand, rotates more slowly than the earth, and its magnetic field is found to be weaker. Investigation into the cause of the earth's magnetism remains open.

There is an interesting sidelight concerning the earth's magnetic field. It has been found that the direction of the field has been reversed several times during the last million years. Evidence for this is provided by basalt (a type of rock that contains iron) that is spewed forth by volcanic activity on the ocean floor. As the lava cools, it solidifies and retains a picture of the earth's magnetic field direction. The rocks can be dated by other means to provide the evidence for these periodic reversals of the magnetic field.

20.3 MAGNETIC FIELDS

Experiments show that a stationary charged particle does not interact with a static magnetic field. However, *a charged particle experiences a force when moving through a magnetic field.* This force has its maximum value when the charge moves perpendicular to the magnetic field lines, decreases in value at other angles, and becomes zero when the particle moves along the field lines. We shall make use of these observations in describing the characteristics of the magnetic field.

In our discussion of electricity, the electric field at some point was defined as the electric force per unit charge acting on some test charge placed at that point. In a similar manner, we shall describe the properties of the magnetic field, $\boldsymbol{B}$, at some point, in terms of the magnetic force exerted on a test charge at that point. Our test object is assumed to be a charge q moving with velocity $\boldsymbol{v}$. It is found that the strength of the magnetic force on the particle is proportional to the magnitude of the charge q, the magnitude of the velocity $\boldsymbol{v}$, the strength of the external magnetic field, $\boldsymbol{B}$, and the sine of the angle θ between the direction of $\boldsymbol{v}$ and the direction of $\boldsymbol{B}$. That is, $F \propto qvB \sin \theta$. Taking the constant of proportionality to be one, we obtain the equation

Magnetic force

$$F = qvB \sin \theta \tag{20.1}$$

This expression is used to define the magnitude of the magnetic field as

Magnetic field defined

$$B \equiv \frac{F}{qv \sin \theta} \tag{20.2}$$

If F is in newtons, q in coulombs, and v in meters per second, the SI unit of magnetic field is the **tesla** (T), also called the **weber per square meter** (that is, $1 \text{ T} = 1 \text{ Wb/m}^2$). Thus, if a 1 C charge moves through a magnetic field of magnitude 1 T with a velocity of 1 m/s in a direction perpendicular to the field ($\sin \theta = 1$), the magnetic force exerted on the charge is 1 N. We can express the units of $\boldsymbol{B}$ as

$$[\text{B}] = \text{T} = \frac{\text{Wb}}{\text{m}^2} = \frac{\text{N}}{\text{C} \cdot \text{m/s}} = \frac{\text{N}}{\text{A} \cdot \text{m}} \tag{20.3}$$

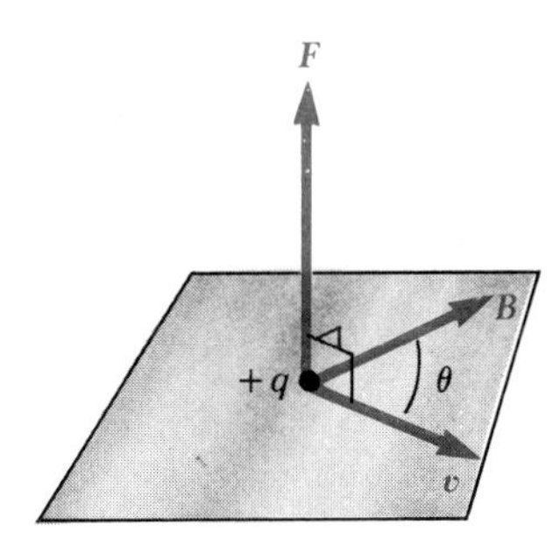

FIGURE 20.3 The direction of the magnetic force $\boldsymbol{F}$ on a charged particle moving with a velocity $\boldsymbol{v}$ in the presence of a magnetic field $\boldsymbol{B}$ is perpendicular to both $\boldsymbol{v}$ and $\boldsymbol{B}$.

In practice, the cgs unit for magnetic field, the **gauss** (G), is often used. The gauss is related to the tesla through the conversion

$$1 \text{ T} = 10^4 \text{ G}$$

Conventional laboratory magnets can produce magnetic fields as large as about 25,000 G, or 2.5 T. Superconducting magnets that can generate magnetic fields as high as 3×10^5 G, or 30 T, have been constructed. These values can be compared with the value of the earth's magnetic field near its surface, which is about 0.5 G, or 0.5×10^{-4} T.

From Equation 20.1, we see that the force on a charged particle moving in a magnetic field has its maximum value when the particle moves *perpendicular* to the magnetic field, corresponding to $\theta = 90°$, so that $\sin \theta = 1$. The magnitude of this maximum force has the value

$$F_{\text{max}} = qvB \tag{20.4}$$

Also, note from Equation 20.1 that F is zero when $\boldsymbol{v}$ is parallel to $\boldsymbol{B}$ (corresponding to $\theta = 0°$ or $180°$). Thus, there is no magnetic force exerted on a charged particle when it moves in the direction of the magnetic field or opposite the field.

Experiments show that the direction of the magnetic force is always perpendicular to both $\boldsymbol{v}$ and $\boldsymbol{B}$, as shown in Figure 20.3. In order to determine the direction of the force, we shall employ the following right-hand rule:

Hold your right hand open as illustrated in Figure 20.4, and then place your fingers in the direction of $\boldsymbol{B}$, with your thumb pointing in the direction of $\boldsymbol{v}$. The force $\boldsymbol{F}$ on the opposite charge is directed *out* of the palm of your hand.

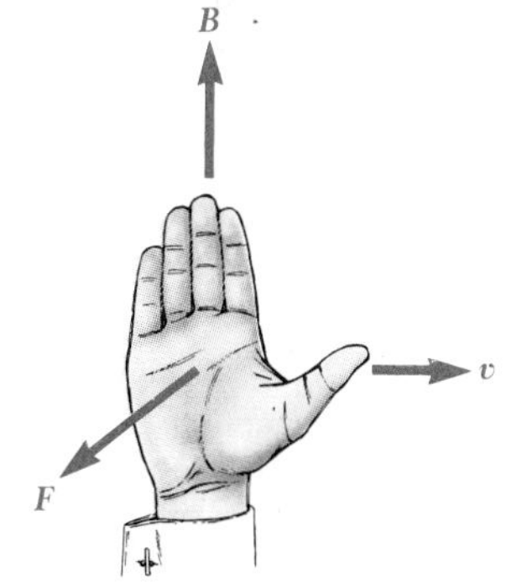

FIGURE 20.4 The right-hand rule for determining the direction of the magnetic force on a positive charge moving with a velocity $\boldsymbol{v}$ in a magnetic field $\boldsymbol{B}$. With your thumb in the direction of $\boldsymbol{v}$ and your four fingers in the direction of $\boldsymbol{B}$, the force is directed out of the palm of your hand.

If the charge is negative rather than positive, the force will be directed *opposite* that shown in Figure 20.4. That is, if q is negative, simply use the right-hand rule to find the direction of F for positive q and then reverse this direction to get the correct direction for the negative charge.

EXAMPLE 20.1 A Proton Traveling in the Earth's Magnetic Field
A proton moves with a speed of 10^5 m/s through the earth's magnetic field, which has a value of 50 μT at a particular location. When the proton moves eastward, the magnetic force acting on it is a maximum, and when it moves northward, no magnetic force acts on it. What is the strength of the magnetic force, and what is the direction of the magnetic field?

Solution The magnitude of the force can be found from Equation 20.4:

$$F_{\max} = qvB = (1.6 \times 10^{-19}\ \text{C})(10^5\ \text{m/s})(50 \times 10^{-6}\ \text{T}) = 8.0 \times 10^{-19}\ \text{N}$$

The direction of the magnetic field cannot be determined precisely from what is given in this problem. Since no magnetic force acts on a charged particle when it is moving parallel to the field, all that we are able to say for sure is that the magnetic field is directed either northward or southward.

EXERCISE Calculate the gravitational force on the proton, and compare it with the magnetic force. Note that the mass of the proton is equal to 1.67×10^{-27} kg.

Answer 1.6×10^{-26} N.

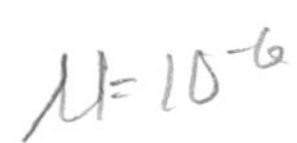

EXAMPLE 20.2 A Proton Moving in a Strong Magnetic Field
A proton moves with a velocity of 8×10^6 m/s along the x axis. It enters a region where there is a magnetic field of magnitude 2.5 T, directed at an angle of 60° to the x axis and lying in the xy plane (Fig. 20.5). Calculate the initial force and acceleration of the proton.

Solution From Equation 20.1 we get

$$F = qvB \sin\theta = (1.6 \times 10^{-19}\ \text{C})(8 \times 10^6\ \text{m/s})(2.5\ \text{T})(\sin 60°)$$
$$= 2.77 \times 10^{-12}\ \text{N}$$

Use the right-hand rule and note that the charge is positive to see that the force is in the positive z direction. You should verify that the units of F in the calculation reduce to newtons.

Since the mass of the proton is 1.67×10^{-27} kg, its initial acceleration is

$$a = \frac{F}{m} = \frac{2.77 \times 10^{-12}\ \text{N}}{1.67 \times 10^{-27}\ \text{kg}} = 1.66 \times 10^{15}\ \text{m/s}^2$$

in the positive z direction.

EXERCISE Calculate the acceleration of an electron that moves through the same magnetic field at the same speed as the proton. The mass of an electron is equal to 9.11×10^{-31} kg.

Answer 3.04×10^{18} m/s^2.

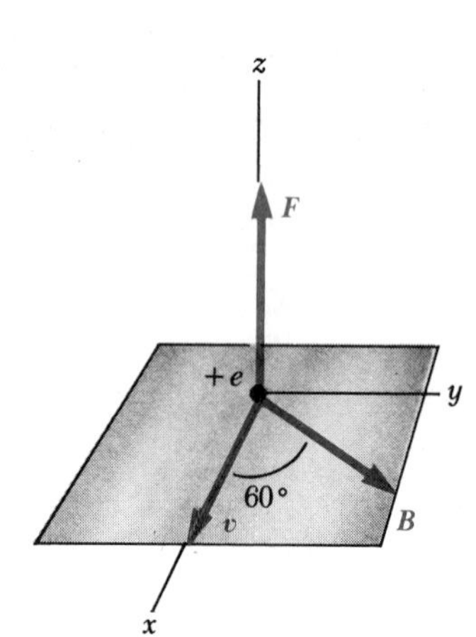

FIGURE 20.5 (Example 20.2) The magnetic force $\boldsymbol{F}$ on a proton is in the positive z direction when $\boldsymbol{v}$ and $\boldsymbol{B}$ lie in the xy plane.

20.4 MAGNETIC FORCE ON A CURRENT-CARRYING CONDUCTOR

If a force is exerted on a single charged particle when it moves through a magnetic field, it should not surprise you to find that a current-carrying wire also experiences a force when placed in a magnetic field. This follows from the fact that the current represents a collection of many charged particles in motion; hence, the resultant force on the wire is due to the sum of the individual forces on the charged particles. The force on the particles is transmitted to the "bulk" of the wire through collisions with the atoms making up the wire.

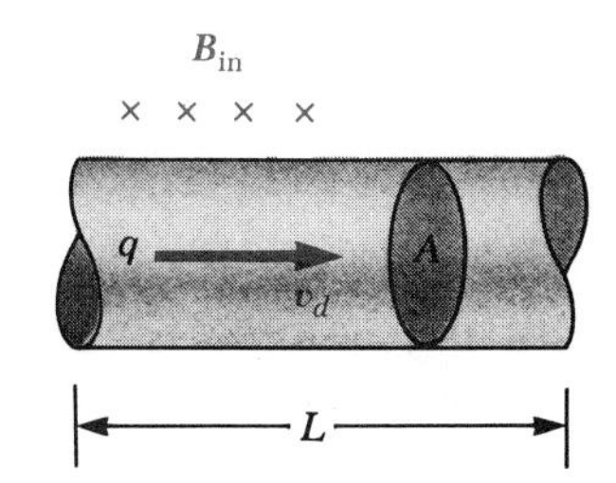

FIGURE 20.6 A section of a wire containing moving charges in an external magnetic field **B**.

Consider a straight segment of wire of length L and cross-sectional area A, carrying a current I in a uniform external magnetic field $\mathbf{B}$, as in Figure 20.6. Let us assume that the magnetic field is perpendicular to the wire and is directed into the page. (Note that the direction of magnetic fields directed into the page are represented by crosses and fields coming out of the page are represented by dots. The crosses can be thought of as the tail feathers of arrows moving away from you, while the dots represent the points of arrows coming toward you.) Because of the external magnetic field, each charge carrier in the wire experiences a force $F_{max} = qv_dB$, where v_d is the drift velocity of the charge. This force causes the charge carrier to be deflected upward if it is positive and downward if it is negative. This single force would be imperceptible by itself, but every charge carrier in the wire has a similar force on it. Hence, the combined effect of all these forces on the charge carriers (that is, the net magnetic force) can be great enough to move the wire. To find the total force on the wire, we multiply the force on one charge carrier by the number of carriers in the segment. Since the volume of the segment is AL, the number of carriers is nAL, where n is the number of carriers per unit volume. Hence, the total magnetic force on the wire of length L is

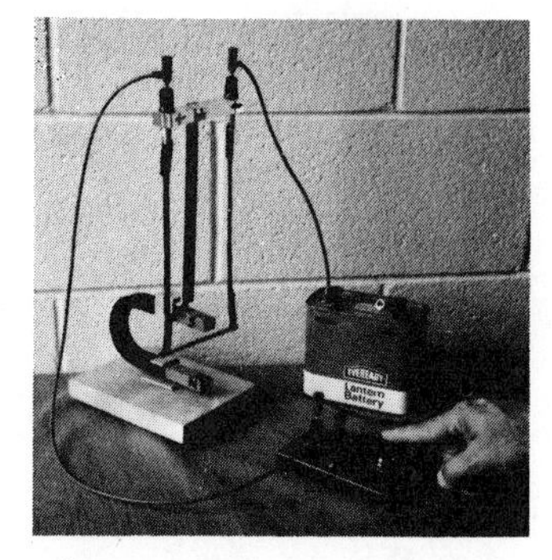

This apparatus demonstrates the force on a current-carrying conductor in an external magnetic field. Why does the bar swing *away* from the magnet after the switch is closed? (Courtesy of Jim Lehman, James Madison University)

$$\text{Total force} = (\text{force on each charge carrier})(\text{total number of carriers})$$
$$F_{max} = (qv_dB)(nAL)$$

From Chapter 18, however, we know that the current in the wire is given by $I = nqv_dA$. Therefore, F_{max} can be expressed as

$$F_{max} = BIL \quad (20.5)$$

This equation can be used only when *the current and magnetic field are at right angles to each other.*

If the wire is not perpendicular to the field, but is at some arbitrary angle, as in Figure 20.7, the magnitude of the magnetic force on the wire is

$$F = BIL \sin\theta \quad (20.6)$$

where θ is the angle between $\mathbf{B}$ and the direction of the current. The direction of this force can be obtained using the right-hand rule. However, in this case, you must place your thumb in the direction of the current rather than in the direction of $\mathbf{v}$. In Figure 20.7, the direction of the magnetic force on the wire is out of the page.

Finally, you should note that when the current is either in the direction of the field or opposite the direction of the field, the magnetic force on the wire is zero.

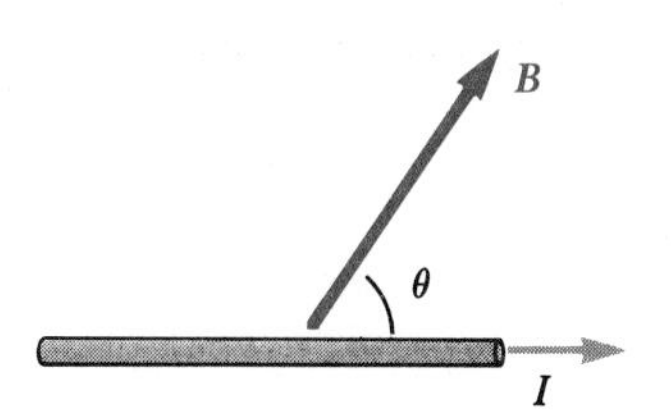

FIGURE 20.7 A wire carrying a current I in the presence of an external magnetic field **B** which makes an angle θ with the wire.

EXAMPLE 20.3 A Current-Carrying Wire in the Earth's Magnetic Field

A wire carries a current of 20 A from east to west. Assume that the magnetic field of the earth is horizontal at this location, directed from south to north, and that it has a magnitude of 0.5×10^{-4} T. Find the force on a 30-m length of wire. How is the force changed if the current runs west to east?

Solution Because the direction of the current and the direction of the magnetic field are at right angles, we can use Equation 20.5. The magnitude of the force is

$$F_{max} = BIL = (0.5 \times 10^{-4}\ \text{T})(20\ \text{A})(30\ \text{m}) = 3 \times 10^{-2}\ \text{N}$$

The right-hand rule shows that the force on the wire is directed toward the earth.

If the direction of the current is from west to east, the magnitude of the force remains the same but its direction will be upward, away from the earth.

EXERCISE If the current flows north to south, what is the magnetic force on the wire?

Answer Zero.

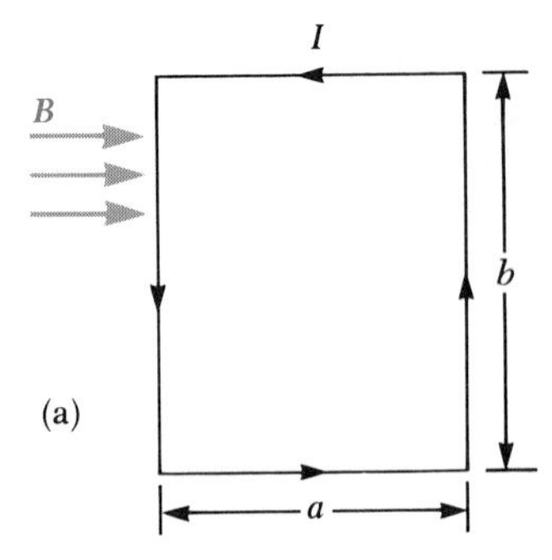

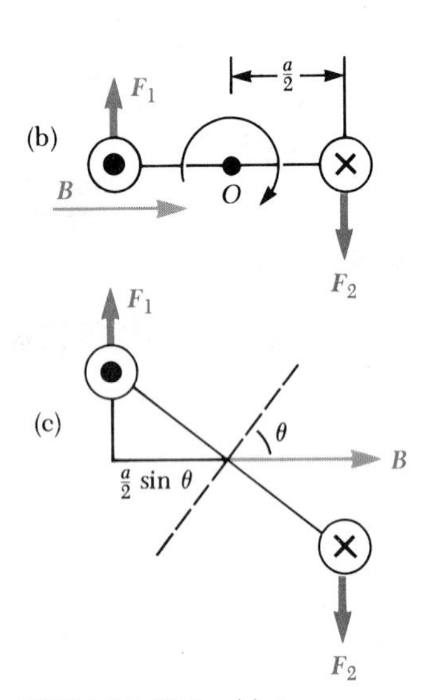

FIGURE 20.8 (a) A rectangular loop in a uniform magnetic field $\boldsymbol{B}$. There are no magnetic forces on the sides of width a parallel to $\boldsymbol{B}$, but there are forces acting on the sides of length b. (b) An end view of the rectangular loop shows that the forces $\boldsymbol{F}_1$ and $\boldsymbol{F}_2$ on the sides of length b create a torque that tends to twist the loop clockwise as shown. (c) If $\boldsymbol{B}$ is at an angle θ with respect to a line perpendicular to the plane of the loop, the torque is given by $IAB \sin \theta$.

20.5 TORQUE ON A CURRENT LOOP

In the previous section we showed how a force is exerted on a current-carrying conductor when the conductor is placed in an external magnetic field. With this as a starting point, we shall show that a torque is exerted on a current loop placed in a magnetic field. The results of this analysis will be of great practical value when we discuss the galvanometer in this chapter and in a discussion of generators in a future chapter.

Consider a rectangular loop carrying a current I in the presence of an external uniform magnetic field directed as shown in Figure 20.8a. The forces on the sides of length a are zero because these wires are parallel to the field. The magnitude of the forces on the sides of length b, however, is

$$F_1 = F_2 = BIb$$

The direction of $\boldsymbol{F}_1$, the force on the left side of the loop, is out of the paper and that of $\boldsymbol{F}_2$, the force on the right side of the loop, is into the paper. If we were to view the loop from below the loop, as in Figure 20.8b, we would see the forces directed as shown. If we assume that the loop is pivoted so that it can rotate about point O, we see that these two forces produce a torque about O that rotates the loop clockwise. The magnitude of this torque, τ_{max}, is

$$\tau_{max} = F_1\frac{a}{2} + F_2\frac{a}{2} = (BIb)\frac{a}{2} + (BIb)\frac{a}{2} = BIab$$

where the moment arm about O is $a/2$ for both forces. Since the area of the loop is $A = ab$, the torque can be expressed as

$$\boxed{\tau_{max} = BIA} \qquad (20.7)$$

Note that this result is valid only when the magnetic field is parallel to the plane of the loop, as shown in the view of the loop pictured in Figure 20.8b. If the field makes an angle θ with respect to a line perpendicular to the plane of the loop, as shown in Figure 20.8c, the moment arm for each force is given by $(a/2) \sin \theta$. An analysis like that just used shows that the magnitude of the torque is given by

$$\boxed{\tau = BIA \sin \theta} \qquad (20.8)$$

This result shows that the torque has the *maximum* value BIA when the field is parallel to the plane of the loop ($\theta = 90°$) and is *zero* when the field is perpendicular to the plane of the loop ($\theta = 0$). As we see in Figure 20.8c, the loop tends to rotate to smaller values of θ (that is, such that the normal to the plane of the loop rotates toward the direction of the magnetic field).

Although this analysis is for a rectangular loop, a more general derivation would indicate that Equation 20.8 applies regardless of the shape of the loop. Furthermore, the torque on a loop with N turns is equal to

$$\tau = NBIA \sin\theta \qquad (20.9)$$

EXAMPLE 20.4 The Torque on a Circular Loop in a Magnetic Field

A circular loop of wire of radius 50 cm is oriented at an angle of 30° to a magnetic field of 0.5 T, as shown in an end view in Figure 20.9. The current in the loop is 2 A in the direction shown. Find the magnitude of the torque at this instant.

Solution Regardless of the shape of the loop, Equation 20.8 is valid:

$$\tau = BIA \sin\theta = (0.5\ \text{T})(2\ \text{A})[\pi(0.5\ \text{m})^2](\sin 30°) = 0.393\ \text{N}\cdot\text{m}$$

EXERCISE Find the torque on the loop if it has three turns rather than one.

Answer The torque is three times greater than on the one-turn loop, or 1.18 N · m.

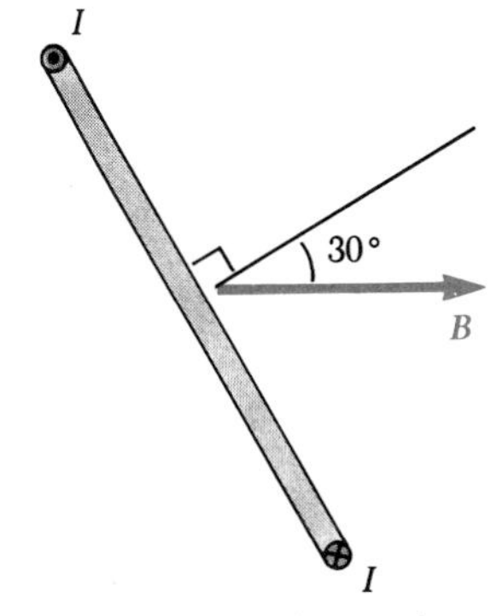

FIGURE 20.9 (Example 20.4) The edge view of a circular current loop in an external magnetic field ***B***.

20.6 THE GALVANOMETER AND ITS APPLICATIONS

The Galvanometer

The *galvanometer* is a device used in the construction of both ammeters and voltmeters. The basic operation of this instrument makes use of the fact that a torque acts on a current loop in the presence of a magnetic field. The main components of a galvanometer are shown in Figure 20.10. It consists of a coil of wire mounted such that it is free to rotate on a pivot in a magnetic field provided by a permanent magnet. The torque experienced by the coil is proportional to the current through it. This means that the larger the current, the larger the torque and the more the coil will rotate before the spring tightens enough to stop the movement. Hence, the amount of deflection is proportional to the current. Once the instrument is properly calibrated, it can be used in conjunction with other circuit elements to measure either currents or potential differences.

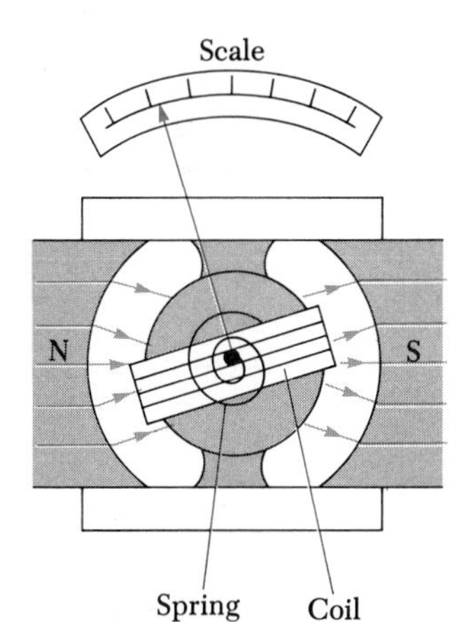

FIGURE 20.10 The principal components of a galvanometer. When current passes through the coil, situated in a magnetic field, the magnetic torque causes the coil to twist. The angle through which the coil rotates is proportional to the current through it.

A Galvanometer Is the Basis of an Ammeter

A typical off-the-shelf galvanometer is usually not suitable for use as an ammeter (a current-measuring device). One of the main reasons for this is that a typical galvanometer has a resistance of about 60 Ω. An ammeter resistance this large could considerably alter the current in the circuit in which it is placed. This can be easily understood by considering the following example. Suppose you were to construct a simple series circuit containing a 3-V battery and a 3-Ω resistor. The current in such a circuit is 1 A. However, if you include a 60-Ω galvanometer in the circuit in an attempt to measure the current, the total resistance of the circuit is now 63 Ω and the current is reduced to 0.048 A.

A second factor that limits the use of a galvanometer as an ammeter is the fact that a typical galvanometer will give a full-scale deflection for very low currents, of the order of 1 mA or less. Consequently, such a galvanometer cannot be used directly to measure currents greater than this.

Now let us assume we wish to convert a 60-Ω, 1-mA galvanometer to an ammeter that deflects full scale when 2 A passes through it. In spite of the factors just described, this can be accomplished by simply placing a resistor, R_p, in *parallel* with the galvanometer, as in Figure 20.11. (The galvanometer and parallel resistor combination is called an ammeter.) The size of the resistor must be selected such that when 2 A passes through the ammeter, only 0.001 A passes through the galvanometer and the remaining 1.999 A passes through the resistor R_p, sometimes called the *shunt resistor*. Because the galvanometer and shunt resistor are in parallel, the potential difference across each is the same. Thus, using Ohm's law we get

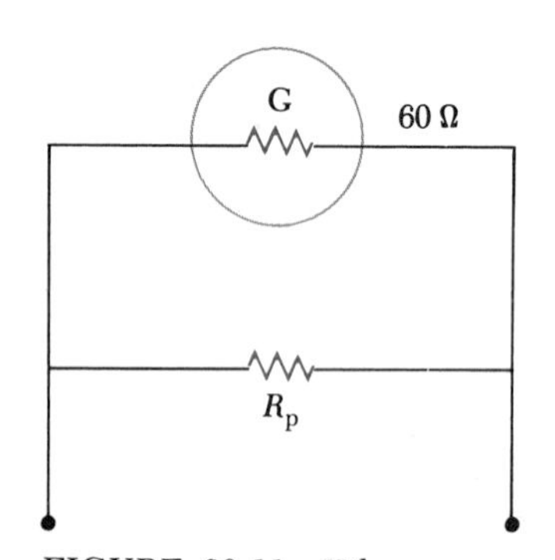

FIGURE 20.11 When a galvanometer is to be used as an ammeter, a resistor R_p is connected in parallel with the galvanometer, Ⓖ.

$$(0.001 \text{ A})(60\ \Omega) = (1.999 \text{ A})R_p$$
$$R_p = 0.03\ \Omega$$

Notice that the shunt resistance, R_p, is extremely small. Thus, we see that the configuration shown in Figure 20.11 solves both problems associated with converting a galvanometer to an ammeter. The ammeter just described is able to measure a large current (2 A), and it has a low resistance, on the order of 0.03 Ω. (Recall that the equivalent resistance of two resistors in parallel is always *less* than the value of the individual resistors.)

A Galvanometer Is the Basis of a Voltmeter

With the proper modification, the basic galvanometer can also be used to measure potential differences in a circuit. In order to understand how this can be accomplished, let us first calculate the largest voltage that can be measured with a galvanometer. If the galvanometer has a resistance of 60 Ω and gives a maximum deflection for a current of 1 mA, the largest voltage it can measure is

$$V_{max} = (0.001 \text{ A})(60\ \Omega) = 0.06 \text{ V}$$

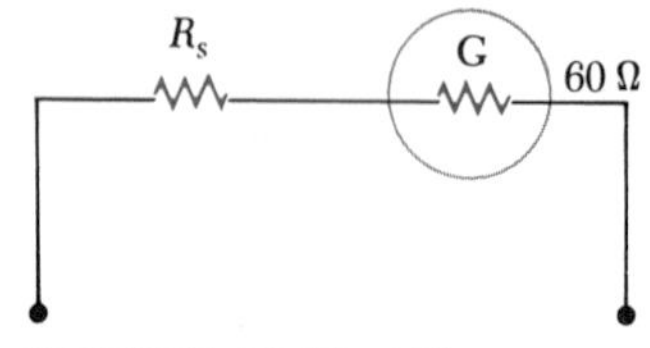

FIGURE 20.12 When a galvanometer is to be used as a voltmeter, a resistor R_s is connected in series with the galvanometer, Ⓖ.

From this result, we see that some modification is required to enable this device to measure larger voltages. Furthermore, a voltmeter must have a very high resistance in order to insure that it will not disturb the circuit in which it is placed. The basic galvanometer, with a resistance of only 60 Ω, is not acceptable for direct voltage measurements.

The circuit in Figure 20.12 shows the basic modification that must be made to convert a galvanometer to a voltmeter. Let us assume we want to construct a voltmeter capable of measuring a maximum voltage of 100 V. In this situation, a resistor, R_s, is placed in *series* with the galvanometer. The value of R_s is found by noting that a current of 1 mA must pass through the galvanometer when the voltmeter is connected across a potential difference of 100 V. Applying Ohm's law to this circuit gives

$$100\ \text{V} = (0.001\ \text{A})(R_s + 60\ \Omega)$$
$$R_s = 99{,}940\ \Omega$$

This result shows that this voltmeter has a very high resistance.

When a voltmeter is constructed with several available ranges, one selects various values of R_s by using a switch that can be connected to a preselected set of resistors. The required value of R_s increases as the maximum voltage to be measured increases.

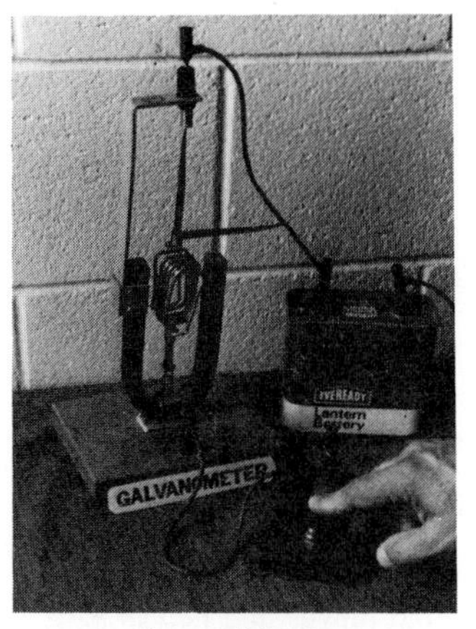

Large-scale model of a galvanometer movement. Why does the coil rotate about the vertical axis after the switch is closed? (Courtesy of Jim Lehman, James Madison University)

20.7 MOTION OF A CHARGED PARTICLE IN A MAGNETIC FIELD

Consider the case of a positively charged particle moving in a uniform magnetic field such that the direction of the particle's velocity is *perpendicular to the field*, as in Figure 20.13. The label $\boldsymbol{B}_{in}$ indicates that $\boldsymbol{B}$ is directed into the page. Application of the right-hand rule at point P shows that the direction of the magnetic force, $\boldsymbol{F}$, at this location is upward. This causes the particle to alter its direction of travel and to follow a curved path. Application of the right-hand rule at any point shows that *the magnetic force is always toward the center of the circular path;* therefore the magnetic force is effectively a centripetal force that changes only the direction of $\boldsymbol{v}$ and not its magnitude. Since $\boldsymbol{F}$ is a centripetal force, we can equate its magnitude, qvB in this case, to the mass of the particle multiplied by the centripetal acceleration, v^2/r. From Newton's second law, we find that

$$F = qvB = \frac{mv^2}{r}$$

which gives

$$\boxed{r = \frac{mv}{qB}} \qquad (20.10)$$

This says that the radius of the path is proportional to the momentum, mv, of the particle and is inversely proportional to the magnetic field.

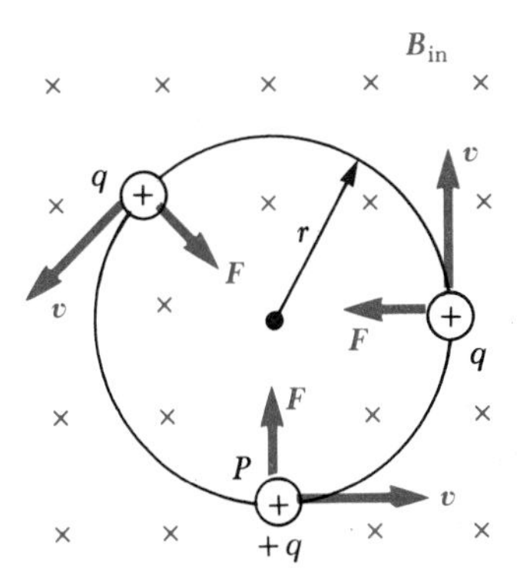

FIGURE 20.13 When the velocity of a charged particle is perpendicular to a uniform magnetic field, the particle moves in a circle whose plane is perpendicular to $\boldsymbol{B}$, which is directed into the page. (The crosses represent the tail of the magnetic field vector). The magnetic force $\boldsymbol{F}$ on the charge is always directed toward the center of the circle.

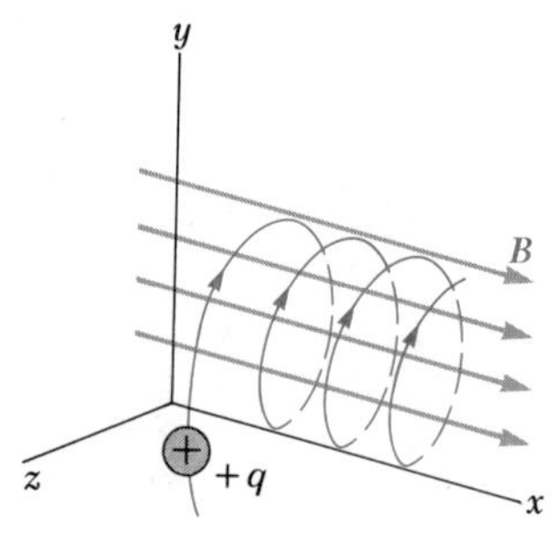

FIGURE 20.14 A charged particle having a velocity vector that has a component parallel to a uniform magnetic field moves in a helical path.

If the initial direction of the velocity of the charged particle is not perpendicular to the magnetic field but instead is directed at an angle to the field, as shown in Figure 20.14, the path followed by the particle is a spiral (called a helix) along the magnetic field lines.

When charged particles move in a nonuniform magnetic field, the motion is rather complex. For example, in a magnetic field that is strong at the ends and weak in the middle, as in Figure 20.15, the particles can oscillate back and forth between the end points. Such a field can be produced by two current loops as in Figure 20.15. In this case, a charged particle starting at one end will spiral along the field lines until it reaches the other end, where it reverses its path and spirals back. This configuration is known as a *magnetic bottle* because charged particles can be trapped in it. This concept has been used to confine very hot gases (T about 10^6 K) consisting of electrons and positive ions, known as a **plasma.** Such a plasma-confinement scheme could play a crucial role in achieving a controlled nuclear fusion process, which could supply us with an almost endless source of energy. Unfortunately, the magnetic bottle has its problems. If a large number of particles are trapped, collisions between the particles cause them to eventually "leak" from the system. We shall discuss nuclear fusion and its prospects as an energy resource in Chapter 33.

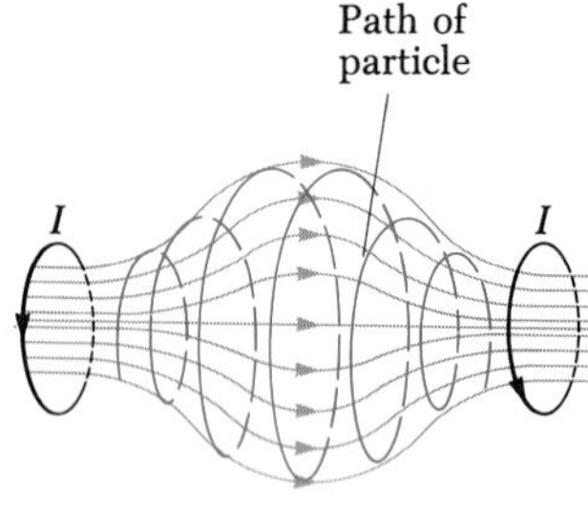

FIGURE 20.15 A charged particle moving in a nonuniform magnetic field (a magnetic bottle) spirals about the field and oscillates between the end points.

The Van Allen radiation belts consist of charged particles (mostly electrons and protons) surrounding the earth in doughnut-shaped regions (Fig. 20.16). These radiation belts were discovered in 1958 by a team of researchers under the direction of James Van Allen, using data gathered by instrumentation aboard the Explorer I satellite. The charged particles, trapped by the earth's nonuniform magnetic field, spiral around the earth's field lines from pole to pole. These particles originate mainly from the sun, but some come from stars and other heavenly objects, and are given the name *cosmic rays.* Most cosmic rays are deflected by the earth's magnetic field and never reach the earth. However, some become trapped, and these make up the Van Allen belts. When these charged particles are in the earth's atmosphere over the poles, they often collide with other atoms, causing them to emit visible light. This is the origin of the beautiful Aurora Borealis, or Northern Lights. A similar phenomenon seen in the southern hemisphere is called the Aurora Australis.

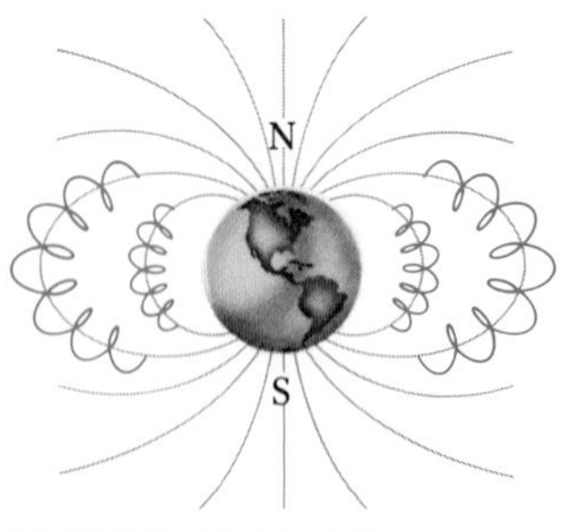

FIGURE 20.16 The Van Allen belts are made up of charged particles (electrons and protons) trapped by the earth's nonuniform magnetic field.

EXAMPLE 20.5 A Proton Moving Perpendicular to a Uniform Magnetic Field

A proton is moving in a circular orbit of radius 14 cm in a uniform magnetic field of magnitude 0.35 T directed perpendicular to the velocity of the proton. Find the orbital speed of the proton.

Solution From Equation 20.10, we get

$$v = \frac{qBr}{m} = \frac{(1.6 \times 10^{-19}\ \text{C})(0.35\ \text{T})(14 \times 10^{-2}\ \text{m})}{1.67 \times 10^{-27}\ \text{kg}} = 4.7 \times 10^6\ \text{m/s}$$

EXERCISE If an electron moves perpendicular to the same magnetic field with this speed, what is the radius of its circular orbit?

Answer 7.6×10^{-5} m.

EXAMPLE 20.6 The Mass Spectrometer
Two singly ionized atoms move out of a slit at point S in Figure 20.17 and into a magnetic field of 0.1 T. They each have a velocity of 10^6 m/s. The nucleus of the first atom contains one proton and has a mass of 1.68×10^{-27} kg, and the nucleus of the second atom contains a proton and a neutron and has a mass of 3.36×10^{-27} kg. Atoms with the same chemical properties but different masses are called isotopes. The two isotopes here are hydrogen and deuterium. Find their distance of separation when they strike a photographic plate at P.

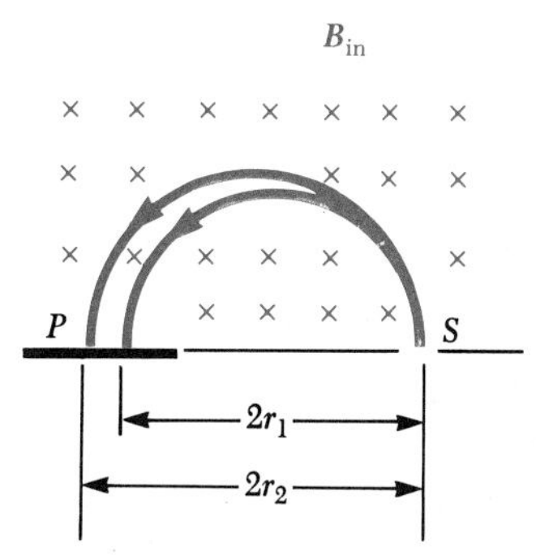

FIGURE 20.17 (Example 20.6) Two isotopes leave the slit at point S and travel in different circular paths before striking a photographic plate at P.

Solution The radius of the circular path followed by the lighter isotope, hydrogen, is

$$r_1 = \frac{m_1 v}{qB} = \frac{(1.68 \times 10^{-27}\text{ kg})(10^6\text{ m/s})}{(1.6 \times 10^{-19}\text{ C})(0.1\text{ T})} = 0.11\text{ m}$$

The radius of the path of the heavier isotope, deuterium, is

$$r_2 = \frac{m_2 v}{qB} = \frac{(3.36 \times 10^{-27}\text{ kg})(10^6\text{ m/s})}{(1.6 \times 10^{-19}\text{ C})(0.1\text{ T})} = 0.21\text{ m}$$

The distance of separation is

$$x = 2r_2 - 2r_1 = 0.20\text{ m}$$

The concepts used in this example underlie the operation of a device called a **mass spectrometer.** This instrument is sometimes used to separate isotopes according to their mass-to-charge ratio but it is used more often to measure masses.

20.8 MAGNETIC FIELD OF A LONG, STRAIGHT WIRE AND AMPÈRE'S LAW

The relationship between electricity and magnetism was discovered in 1819 when, during a lecture demonstration, the Danish scientist Hans Oersted (1777–1851) found that an electric current in a wire deflected a nearby compass needle. This discovery, linking a magnetic field with an electric current, was the beginning of our understanding of the origin of magnetism.

Figure 20.18 illustrates an experiment one can perform to show that a

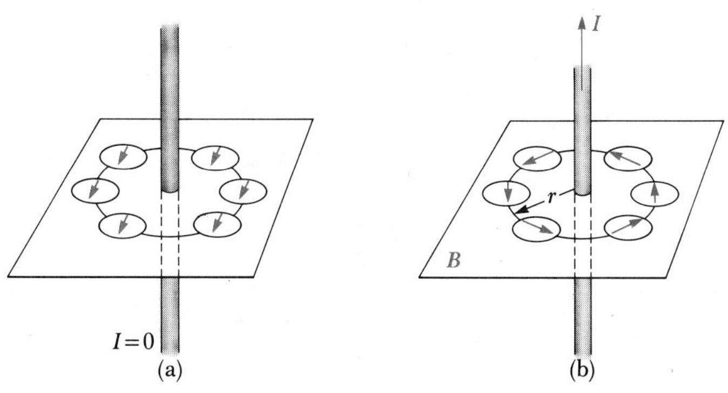

FIGURE 20.18 (a) When there is no current in the vertical wire, all compass needles point in the same direction. (b) When the wire carries a large current I, the compass needles deflect in a direction tangent to the circle, which is the direction of the magnetic field due to the current.

current produces a magnetic field. Several compasses are placed in a horizontal plane near a long vertical wire. The compasses are positioned on a circular path concentric with the wire. When there is no current in the wire, all compasses point in the same direction, as in Figure 20.18a (that of the earth's field), as one would expect. However, when there is a steady current flowing upward, as in Figure 20.18b, the compasses point as shown. The fact that the compasses point in a direction tangent to the circle indicates that the magnetic field lines of the wire are circles concentric with the wire. If the current in the wire is reversed, the compasses point in the opposite direction, indicating that the direction of the magnetic field is also reversed.

A convenient rule for determining the direction of a magnetic field produced by a current in a wire is the following:

Grasp the wire with the right hand, with the thumb along the direction of the current, as shown in Figure 20.19a. The four fingers wrap around the wire in the direction of the magnetic field lines.

Shortly after Oersted's discovery, scientists were able to arrive at an expression that gives the strength of the magnetic field due to a long, straight wire. The magnetic field strength at a distance r from a wire carrying a current I is

$$B = \frac{\mu_0 I}{2\pi r} \tag{20.11}$$

This result shows that the magnitude of the magnetic field is proportional to the current and decreases as the distance from the wire increases, as one might intuitively expect. The proportionality constant μ_0, called the **permeability of free space,** is defined to have the value

$$\mu_0 \equiv 4\pi \times 10^{-7}\ \mathrm{T \cdot m/A} \tag{20.12}$$

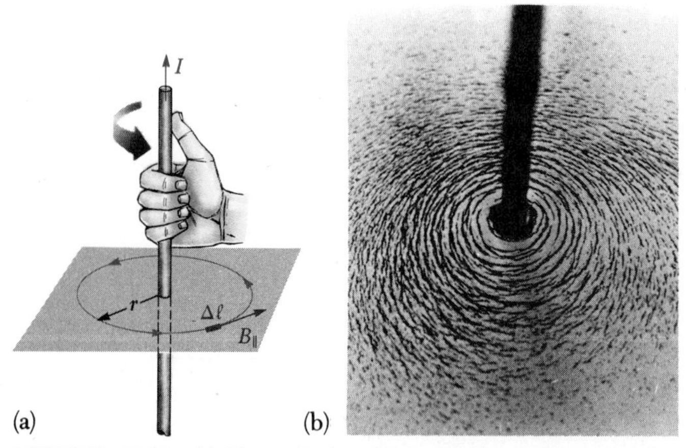

FIGURE 20.19 (a) The right-hand rule for determining the direction of the magnetic field due to a long, straight wire carrying a current. Note that the magnetic field lines form circles around the wire. (b) Circular magnetic field lines surrounding a current-carrying wire, as displayed with iron filings. (Courtesy of J. Lehman, James Madison University)

*Ampère's Law

Equation 20.11 enables us to calculate the magnetic field due to a long, straight wire carrying a current. A generalization of this result was proposed by the French scientist André Marie Ampère (1775–1836). This scheme provides a relation between the current in an arbitrarily shaped wire and the magnetic field produced by the wire.

Consider a circular path surrounding a current, as in Figure 20.19a. The path can be divided into many short segments, each of length $\Delta\ell$. Let us now multiply one of these lengths by the component of the magnetic field parallel to that segment, where the product is labeled $B_{\|}\Delta\ell$. According to Ampère, the sum of all such products over the closed path is equal to μ_0 times the net current I that passes through the surface bounded by the closed path. This satement, known as **Ampère's circuital law**, can be written

$$\boxed{\sum B_{\|}\Delta\ell = \mu_0 I} \qquad (20.13)$$

Ampère's circuital law

where $\Sigma B_{\|}\Delta\ell$ means that we take the sum over all the products $B_{\|}\Delta\ell$ around the closed path.

We can use Ampère's circuital law to derive the magnetic field due to a long, straight wire carrying a current I. As we discussed earlier, the magnetic field lines of this configuration form circles with the wire at their centers, as shown in Figure 20.19a. The magnetic field is tangent to this circle at every point and has the same value, $B_{\|}$, over the entire circumference of a circle of radius r. We now calculate the sum $\Sigma B_{\|}\Delta\ell$ over a circular path and note that $B_{\|}$ can be removed from the sum (since it has the same value for each element on the circle). Equation 20.13 then gives

$$\sum B_{\|}\Delta\ell = B_{\|}\sum\Delta\ell = B_{\|}(2\pi r) = \mu_0 I$$

If we divide both sides by $2\pi r$, we obtain

$$B = \frac{\mu_0 I}{2\pi r}$$

This is identical to Equation 20.11, which is the magnetic field of a long, straight current.

Ampère's circuital law is extremely important because it provides an elegant and simple method for calculating the magnetic fields of current configurations with high symmetry. However, it cannot be used to calculate magnetic fields for complex current configurations which lack symmetry.

EXAMPLE 20.7 The Magnetic Field of a Long Wire

A long, straight wire carries a current of 5 A. Find the magnitude of the magnetic field 4 cm from the wire.

Solution From Equation 20.11, we find

$$B = \frac{\mu_0 I}{2\pi r} = \frac{(4\pi \times 10^{-7}\ \text{T·m/A})(5\ \text{A})}{2\pi(4 \times 10^{-2}\ \text{m})}$$

$$= 2.5 \times 10^{-5}\ \text{T}$$

EXERCISE At what distance from the wire does the field have a magnitude equal to 0.5×10^{-5} T?

Answer 20 cm.

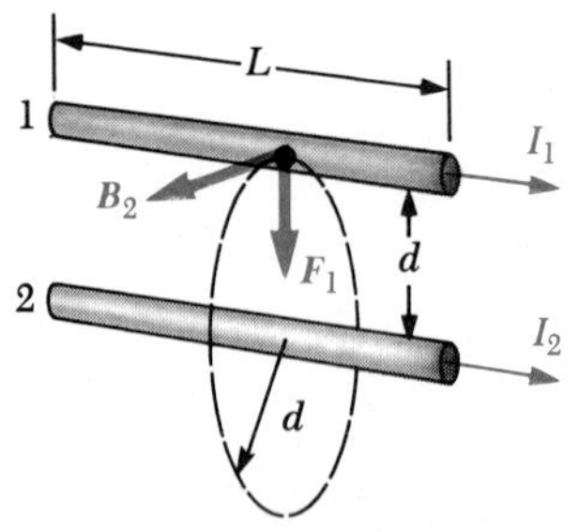

FIGURE 20.20 Two parallel wires, each carrying a steady current, exert a force on each other. The field $\boldsymbol{B}_2$ at wire 1 due to wire 2 produces a force on wire 1 given by $F_1 = I_1LB_2$. The force is attractive if the currents flow in the same direction, as shown, and repulsive if the two currents flow in opposite directions.

20.9 MAGNETIC FORCE BETWEEN TWO PARALLEL CONDUCTORS

As we have seen, a magnetic force acts on a current-carrying conductor when the conductor is placed in an external magnetic field. Since a current in a conductor creates its own magnetic field, it is easy to understand that two current-carrying wires placed close together will exert magnetic forces on each other. Consider two long, straight, parallel wires separated by a distance d and carrying currents I_1 and I_2 in the same direction, as shown in Figure 20.20. Let us determine the force on one wire due to a magnetic field set up by the other wire.

Wire 2, which carries a current I_2, sets up a magnetic field $\boldsymbol{B}_2$ at wire 1. The direction of $\boldsymbol{B}_2$ is perpendicular to the wire, as shown in the figure. Using Equation 20.11, we see that the magnitude of this magnetic field is

$$B_2 = \frac{\mu_0 I_2}{2\pi d}$$

According to Equation 20.5, the magnetic force on wire 1 in the presence of the field $\boldsymbol{B}_2$ due to I_2 is

$$F_1 = B_2 I_1 L = \left(\frac{\mu_0 I_2}{2\pi d}\right) I_1 L = \frac{\mu_0 I_1 I_2 L}{2\pi d}$$

We can rewrite this in terms of the force per unit length as

$$\boxed{\frac{F_1}{L} = \frac{\mu_0 I_1 I_2}{2\pi d}} \qquad (20.14)$$

The direction of $\boldsymbol{F}_1$ is downward, toward wire 2, as indicated by the right-hand rule. If one considers the field set up at wire 2 due to wire 1, the force $\boldsymbol{F}_2$ on wire 2 is found to be equal to and opposite $\boldsymbol{F}_1$. This is what one would expect from Newton's third law of action-reaction.

When the currents are in opposite directions in the two wires, the forces are reversed and the wires repel one another. Hence, we find that *parallel conductors carrying currents in the same direction attract each other and parallel conductors carrying currents in opposite directions repel each other.*

The force between two parallel wires carrying a current is used to define the SI unit of current, the **ampere,** as follows:

If two long, parallel wires 1 m apart carry the same current and the force per unit length on each wire is 2×10^{-7} N/m, then the current is defined to be 1 A.

The SI unit of charge, the **coulomb,** can now be defined in terms of the ampere as follows:

If a conductor carries a steady current of 1 A, then the quantity of charge that flows through any cross section in 1 s is 1 C.

EXAMPLE 20.8 Levitating a Wire
Two wires, each having a weight per unit length of 10^{-4} N/m, are strung parallel to one another above the surface of the earth, with one being placed directly above the other. The wires are aligned in a north-south direction so that the earth's magnetic field will not affect them. When their distance of separation is 0.1 m, what must the current be in each in order for the lower wire to levitate the upper wire? Assume that each wire carries the same current traveling in opposite directions.

Solution If the upper wire is to float, it must be in equilibrium under the action of two forces: its weight and magnetic repulsion. In our example, the weight per unit length (10^{-4} N/m) must be equal and opposite the magnetic force per unit length given in Equation 20.14. Since the currents are the same, we have

$$\frac{F_1}{L} = \frac{mg}{L} = \frac{\mu_0 I^2}{2\pi d}$$

$$10^{-4}\ \text{N/m} = \frac{(4\pi \times 10^{-7}\ \text{T·m/A})(I^2)}{(2\pi)(0.1\ \text{m})}$$

We solve for the current to find

$$I = 7.1\ \text{A}$$

EXERCISE If the current in each wire is increased by a factor of 2, what is the equilibrium separation of the two wires?

Answer 0.4 m.

20.10 MAGNETIC FIELD OF A CURRENT LOOP

The strength of the magnetic field set up by a piece of wire carrying a current can be enhanced at a specific location if the wire is formed into a loop. This can be understood by considering the effect of several small segments of the current loop, as in Figure 20.21. The small segment labeled Δx_1 at the top of the loop produces at the loop's center a magnetic field of magnitude B_1 and directed out of the page. The direction of $\boldsymbol{B}$ can be verified using the right-hand rule for a long straight wire. Grab the wire with your right hand with your thumb pointing in the direction of the current. Your fingers curl around in the direction of $\boldsymbol{B}$.

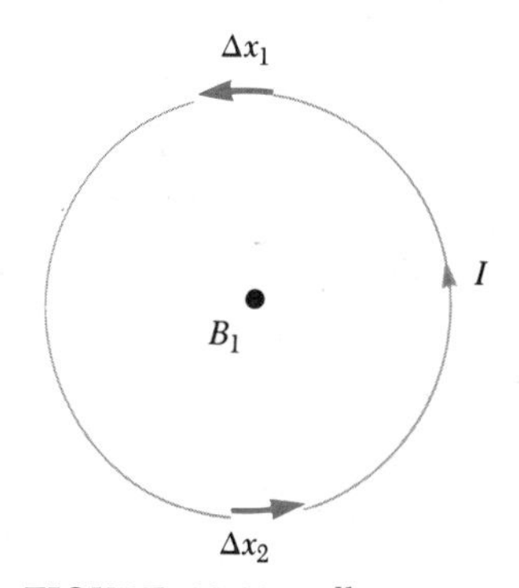

FIGURE 20.21 All segments of the current loop produce a magnetic field at the center of the loop directed *out of the page.*

A segment at the bottom of the loop, Δx_2, also contributes to the field at the center, thus increasing its strength. The magnitude of the field produced at the center of the current loop by the segment Δx_2 has the same magnitude as B_1 and is also directed out of the page. Similarly, all other such segments of the current loop contribute to the field. The net effect is to produce a magnetic field for the current loop as pictured in Figure 20.22a. The magni-

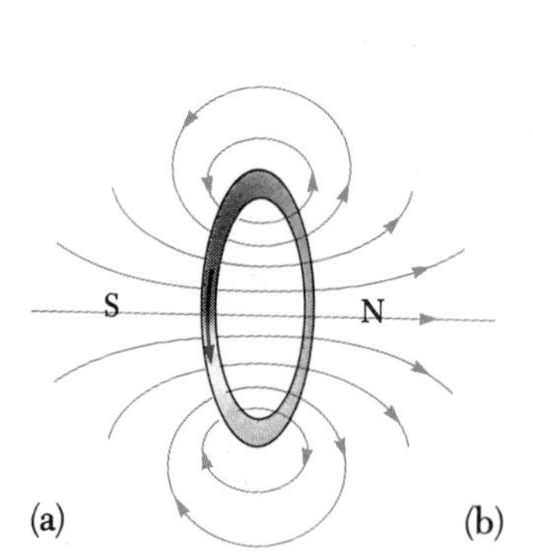

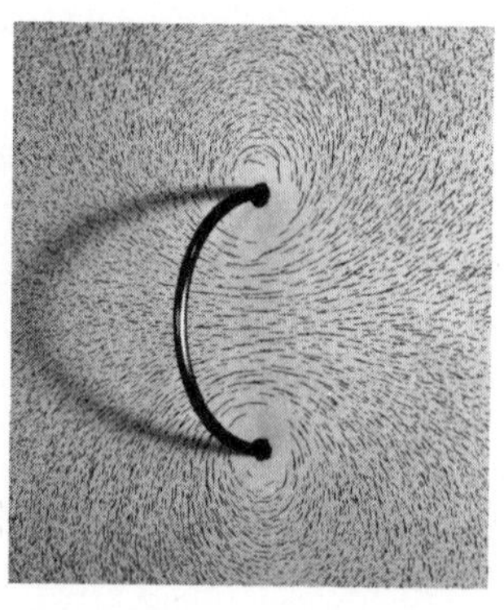

FIGURE 20.22 (a) Magnetic field lines for a current loop. Note that the magnetic field lines of the current loop resemble those of a bar magnet. (b) Field lines of a current loop as displayed by iron filings. (Photo Education Development Center, Newton, Mass.)

tude of the magnetic field at the center of a current loop of radius R is given by

$$B = \frac{\mu_0 I}{2R} \tag{20.15}$$

If the loop contains N turns, the field at the center has a magnitude $N\mu_0 I/2R$.

Notice in Figure 20.22a that the magnetic field lines enter at the left side of the current loop and exit at the right. Thus, one side of the loop acts as though it were the north pole of a magnet and the other acts as a south pole. The fact that the field set up by such a current loop bears a striking resemblance to the field of a bar magnet will be of more than casual interest to us in a future section.

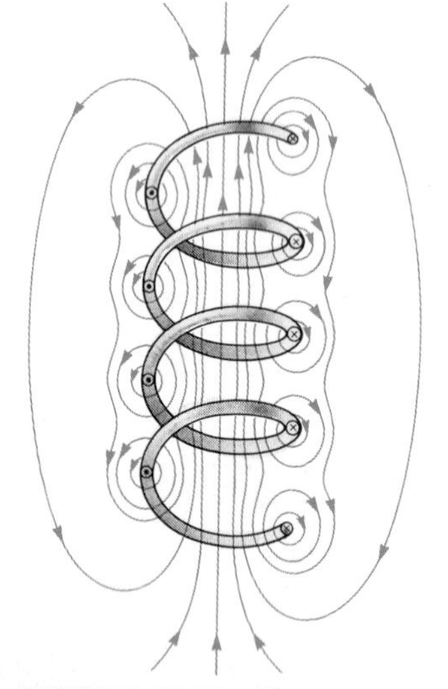

FIGURE 20.23 The magnetic field lines for a loosely wound solenoid. (Adapted from D. Halliday and R. Resnick, *Physics*, New York, Wiley, 1978)

20.11 MAGNETIC FIELD OF A SOLENOID

If a long, straight wire is wrapped into a coil of several closely spaced loops, the resulting device is a **solenoid** or, as it is often called, an **electromagnet.** This device is important in many applications since it acts like a magnet only when it carries a current. As we shall see, the magnetic field inside a solenoid increases with the current and is proportional to the number of coils per unit length.

Figure 20.23 shows the magnetic field lines of a loosely wound solenoid of length L, and total number of turns, N. Note that the field lines inside the solenoid are nearly parallel, uniformly spaced, and close together. This indicates that the field inside the solenoid is nearly uniform and strong. The exterior field at the sides of the solenoid is nonuniform and is much weaker than the field inside.

If the turns are closely spaced, the field lines are as shown in Figure 20.24a. In this case, the field lines enter at one end of the solenoid and emerge at the other. This means that one end of the solenoid acts as though it were the north pole of a magnet while the other end acts as though it were the south pole. If the length of the solenoid is much greater than its radius,

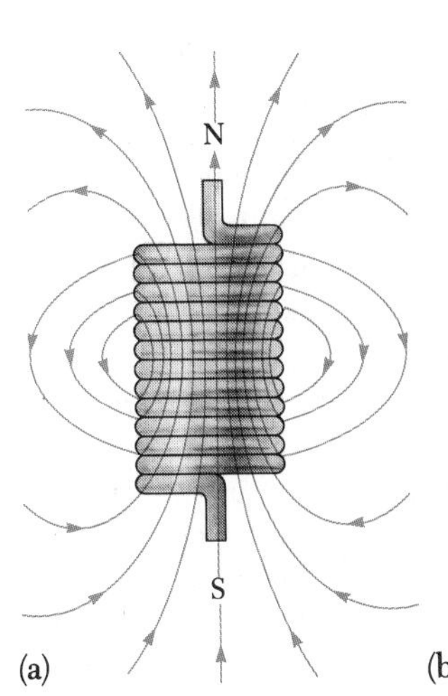

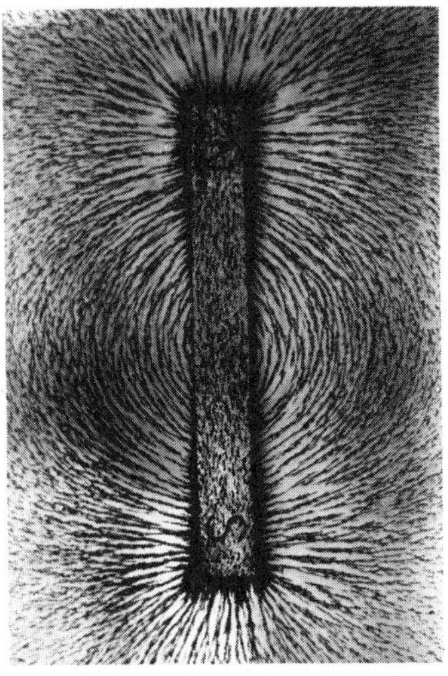

FIGURE 20.24 (a) Magnetic field lines for a tightly wound solenoid of finite length carrying a steady current I. The field inside the solenoid is nearly uniform and strong. Note that the field lines resemble those of a bar magnet, so that the solenoid effectively has north and south poles. (b) Magnetic field pattern of a bar magnet, as displayed by small iron filings on a sheet of paper. (Courtesy of Jim Lehman, James Madison University)

the field outside the solenoid becomes very weak relative to the field inside. This occurs because the lines that leave the north end of the solenoid spread out over a wide region before returning to enter the south end. Hence, we see that the magnetic field lines outside are widely separated, indicative of a weak field. This is in contrast to a much stronger field inside the solenoid, where the lines are close together. Also, the field inside the solenoid has a constant magnitude at all points far from its ends. The expression for the field inside the solenoid is

$$\boxed{B = \mu_0 n I} \qquad (20.16)$$

The field inside a solenoid

where $n = N/L$ is the number of turns per unit length.

EXAMPLE 20.9 The Magnetic Field Inside a Solenoid

A solenoid consists of 100 turns of wire and has a length of 10 cm. Find the magnetic field inside the solenoid when it carries a current of 0.5 A.

Solution The number of turns per unit is

$$n = \frac{N}{L} = \frac{100 \text{ turns}}{0.1 \text{ m}} = 1000 \text{ turns/m}$$

so

$$B = \mu_0 n I = (4\pi \times 10^{-7} \text{ T}\cdot\text{m/A})(1000 \text{ turns/m})(0.5 \text{ A}) = 6.28 \times 10^{-4} \text{ T}$$

EXERCISE How many turns should the solenoid have (assuming it carries the same current) if the field inside is to be increased by a factor of 5?

Answer 500 turns.

This strong electromagnet lifts pellets of iron oxide used in a steel-making process.

*20.12 MAGNETIC DOMAINS

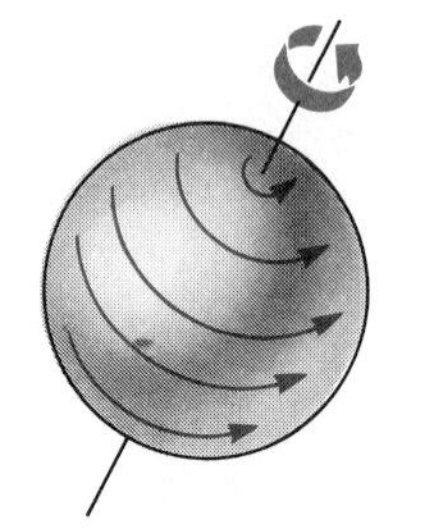

FIGURE 20.25 Model of a spinning electron.

The magnetic field produced by a current in a coil of wire gives us a hint as to what might cause certain materials to exhibit strong magnetic properties. A single coil like that of Figure 20.22 has a north and a south pole, but if this is true for a coil of wire, it should also be true for any current confined to a circular path. In particular, an individual atom should act like a magnet because of the motion of the electrons about the nucleus. Each electron, with its charge of 1.6×10^{-19} C, circles the atom once in about 10^{-16} s. If we divide the electronic charge by this time interval, we see that the orbiting electron is equivalent to a current of 1.6×10^{-3} A. Such a current would produce a magnetic field of the order of 20 T at the center of the circular path. From this result, we see that a very strong magnetic field would be produced if several of these atomic magnets could be aligned inside a material. However, this does not occur because the simple model we have described is not the complete story. A thorough analysis of atomic structure shows that the magnetic field produced by one electron in an atom is often canceled by an oppositely revolving electron in the same atom. The net result is that *the magnetic effect produced by the electrons orbiting the nucleus is either zero or very small for most materials.*

The magnetic properties of many materials is explained by the fact that an electron also spins on its axis like a top as it circles in its orbit, as indicated in Figure 20.25. (This classical description should not be taken literally. The property of spin can be understood only by relying on the methods of quantum mechanics, which we shall not discuss here.) The spinning electron represents a charge in motion which produces a magnetic field. The field due to the spinning motion is generally stronger than the field due to the orbital motion. In atoms containing many electrons, the electrons usually pair up with their spins opposite each other, so that their fields cancel each other. Hence most substances are not magnets. However, for certain strongly magnetic materials such as iron, cobalt, and nickel, the magnetic fields produced by the electron spins do not cancel completely. Such strongly magnetic materials whose spins do not completely cancel are said to be **ferromagnetic.** In these materials there is a strong coupling between neighboring atoms to form large groups of atoms whose spins are aligned. Typically, the size of one of these groups, called a **domain,** ranges from about 10^{-4} cm to 0.1 cm. In an unmagnetized substance, the domains are randomly oriented, as shown in Figure 20.26a. When an external field is applied, as in Figure 20.26b, the magnetic field of each domain tends to align with the external field, resulting in a magnetized substance.

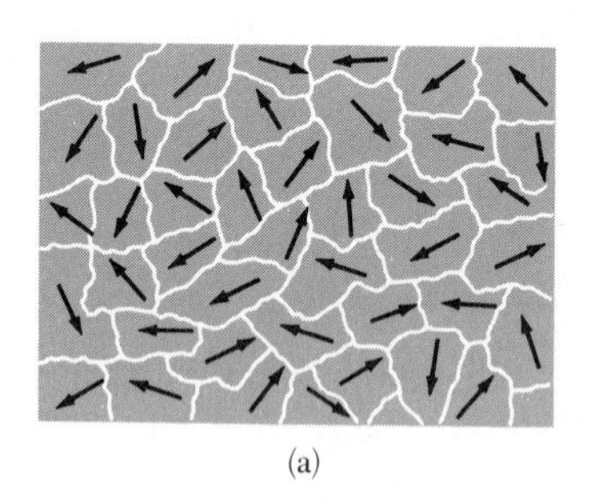

(a)

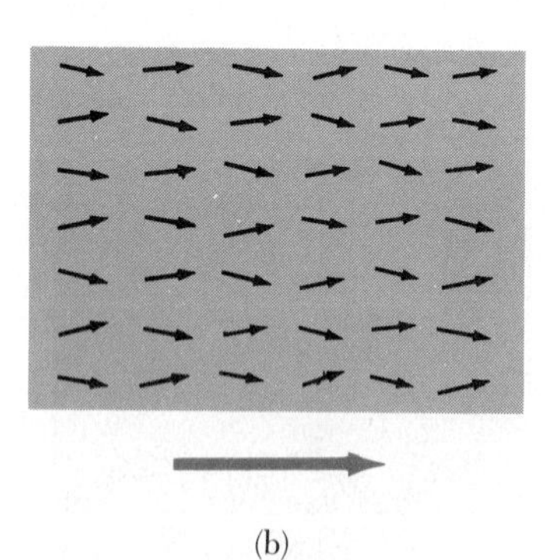

(b)

FIGURE 20.26 (a) Random orientation of domains in an unmagnetized substance. (b) When an external magnetic field is applied, the domains tend to align with the magnetic field.

Two effects, both dependent on the domains, are responsible for the material's becoming magnetized. In some cases, the domains rotate slightly to become more closely aligned with the external field. This is pictured in Figure 20.27b. A second effect, which occurs in some substances, is one in which domains that are already aligned with the field tend to grow at the expense of the others, as shown in Figure 20.27c.

In what are called hard magnetic materials, domain alignment persists after the external field is removed, and such materials are referred to as **permanent magnets.** In soft magnetic materials, such as iron, once the external field is removed, thermal agitation produces domain motion and the material quickly returns to an unmagnetized state.

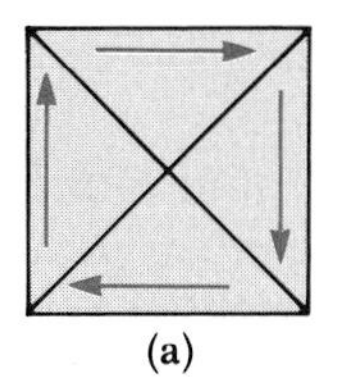

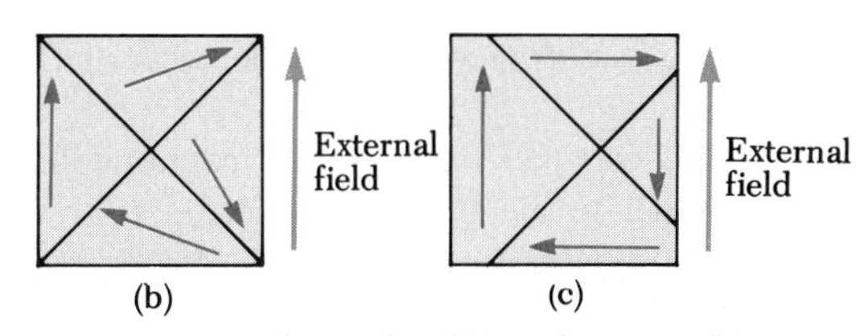

FIGURE 20.27 (a) An unmagnetized sample of four domains. (b) Magnetization due to rotation of the domains. (c) Magnetization due to a shift in domain boundaries.

The alignment of domains explains why the strength of an electromagnet is increased dramatically by the insertion of an iron core into the magnet's center. The magnetic field produced by the current in the loops causes alignment of the domains, thus producing a large net external field. The use of iron, a soft magnetic material, as a core is also advantageous since the iron loses its magnetism almost instantaneously after the current in the coils is turned off.

SUMMARY

The **magnetic force** that acts on a charge q moving with a velocity $\boldsymbol{v}$ in a magnetic field $\boldsymbol{B}$ has a magnitude given by

$$F = qvB \sin \theta \tag{20.1}$$

where θ is the angle between $\boldsymbol{v}$ and $\boldsymbol{B}$.

In order to find the direction of this force, a rule called the **right-hand rule** applies: Using your open right hand, place your fingers in the direction of $\boldsymbol{B}$ and point your thumb along the direction of the velocity $\boldsymbol{v}$. The force $\boldsymbol{F}$ on a positive charge is directed out of the palm of your hand.

If the charge is *negative* rather than positive, the force is directed opposite the force given by the right-hand rule.

The SI unit of magnetic field is the **tesla** (T), also called the weber per square meter (Wb/m^2). An additional commonly used unit for magnetic field is the **gauss** (G), where $1 \text{ T} = 10^4 \text{ G}$.

If a straight conductor of length L carries a current I, the magnetic force on that conductor when it is placed in a uniform external magnetic field, $\boldsymbol{B}$, is

$$F = BIL \sin \theta \tag{20.6}$$

The right-hand rule also gives the direction of the magnetic force on the conductor. In this case, however, you must place your thumb in the direction of the current rather than in the direction of $\boldsymbol{v}$.

The torque, τ, on a current-carrying loop of wire in a magnetic field $\boldsymbol{B}$ has a magnitude given by

$$\tau = BIA \sin \theta \tag{20.8}$$

where I is the current in the loop and A is its cross-sectional area. The angle θ is the angle between $\boldsymbol{B}$ and a line drawn perpendicular to the plane of the loop.

The galvanometer is a device used in the construction of both ammeters and voltmeters.

If a charged particle moves in a uniform magnetic field such that its initial velocity is perpendicular to the field, the particle will move in a circular path whose plane is perpendicular to the magnetic field. The radius r of the circular path is

$$r = \frac{mv}{qB} \tag{20.10}$$

where m is the mass of the particle and q is its charge.

The magnetic field at a distance r from a **long, straight wire** carrying a current I has a magnitude

$$B = \frac{\mu_0 I}{2\pi r} \tag{20.11}$$

where $\mu_0 = 4\pi \times 10^{-7}$ T·m/s is the **permeability of free space.** The magnetic field lines around a long, straight wire are circles concentric with the wire.

Ampère's law can be used to find the magnetic field around certain simple current-carrying conductors. It is stated as

$$\sum B_{\parallel} \Delta \ell = \mu_0 I \tag{20.13}$$

where $B_{\parallel}$ is the component of $\mathbf{B}$ tangent to a small current element of length $\Delta \ell$ that is part of a closed path and I is the total current that penetrates the closed path.

The force per unit length between two parallel wires separated by a distance d and carrying currents I_1 and I_2 has a magnitude given by

$$\frac{F}{L} = \frac{\mu_0 I_1 I_2}{2\pi d} \tag{20.14}$$

The force is attractive if the currents are in the same direction and repulsive if they are in opposite directions.

The magnetic field inside a solenoid has a magnitude

$$B = \mu_0 n I \tag{20.16}$$

where n is the number of turns of wire per unit length, $n = N/L$.

ADDITIONAL READING

S. Banerjee, "Polar Flip-Flops," *The Sciences*, November/December 1984, p. 24.

J. J. Becker, "Permanent Magnets," *Sci. American*, December 1970, p. 92.

F. Bitter, *Magnets: The Education of a Physicist*, Science Study Series, Garden City, N.Y., Doubleday, 1959.

C. Carrigan and D. Gubbins, "The Source of the Earth's Magnetic Field," *Sci. American*, February 1979, p. 118.

R. A. Carrigan, Jr. and W. P. Trower, "Superheavy Magnetic Monopoles," *Sci. American*, April 1982, p. 106.

B. Dibner, "*Oersted and the Discovery of Electromagnetism,*" Blaisdell, 1962.
S. Felch, "Searches for Magnetic Monopoles and Fractional Electric Charge," *The Physics Teacher*, March 1984, p. 142.
H. H. Kolm and A. J. Freeman, "Intense Magnetic Fields," *Sci. American*, April 1965, p. 66.
A. Nier, "The Mass Spectrometer," *Sci. American*, March 1953, p. 68.

QUESTIONS

1. Two charged particles are projected into a region where there is a magnetic field perpendicular to their velocities. If the particles are deflected in opposite directions, what can you say about them?
2. List several similarities and differences in electric and magnetic forces.
3. If a charged particle moves in a straight line through some region of space, can you say that the magnetic field in that region is zero?
4. A magnet attracts a piece of iron. The iron can then attract another piece of iron. Explain, on the basis of alignment of the domains, what happens in each piece of iron.
5. How can the motion of a charged particle be used to distinguish between a magnetic field and an electric field? Give a specific example to justify your argument.
6. You are an astronaut stranded on a planet with no test equipment or minerals around. The planet does not even have a magnetic field. You have two iron bars in your possession; one is magnetized, one is not. How could you determine which is magnetized?
7. Why does the picture on a television screen become distorted when a magnet is brought near the screen?
8. Can a magnetic field set a resting electron into motion? If so, how?
9. Is it possible to orient a current loop in a uniform magnetic field such that the loop will not tend to rotate?
10. How can a current loop be used to determine the presence of a magnetic field in a given region of space?
11. Why will hitting a magnet with a hammer cause its magnetism to be reduced?
12. Will a nail be attracted to either pole of a magnet? Explain what is happening inside the nail.
13. The north-seeking pole of a magnet is attracted toward the geographic north pole of the earth. Yet, like poles repel. What is the way out of this dilemma?
14. Suppose an electron is chasing a proton up this page when suddenly a magnetic field is applied perpendicular to the page. What will happen to the particles?
15. A Hindu ruler once suggested that he be entombed in a magnetic coffin with the polarity arranged such that he would be forever suspended between heaven and earth. Is such magnetic levitation possible? Discuss.
16. Two wires carrying equal but opposite currents are twisted together in the construction of a circuit. Why does this technique reduce stray magnetic fields?
17. Describe the change in the magnetic field inside a solenoid carrying a steady current if (a) the length of the solenoid is doubled but the number of turns remains the same and (b) the number of turns is doubled but the length remains the same.
18. Which way would a compass point if you were at the north magnetic pole?
19. If a solenoid were suspended by a string so that it could rotate freely, could it be used as a magnet when it carries a direct current? Could it also be used if the current were alternating?
20. It is found that charged particles from outer space, called cosmic rays, strike the earth more frequently at the poles than at the equator. Why does this occur?
21. A charged particle moves in a circular path because of an applied magnetic field. Does the particle gain energy from the magnetic field?

PROBLEMS

Section 20.3 Magnetic Fields

1. A proton is moving at right angles to a magnetic field of 2 T. What speed does the proton have if the magnetic force on it has a magnitude of 6×10^{-11} N?

2. A proton moves at right angles to a magnetic field of 0.1 T with a speed of 2×10^{7} m/s. Find the magnitude of the acceleration of the proton.

3. A proton travels with a speed of 3×10^6 m/s at an angle of 37° with respect to the direction of a magnetic field of 0.3 T in the $+y$ direction. What is (a) the magnitude of the magnetic force on the proton, and (b) its acceleration?

4. A proton moves with a speed of 2.5×10^6 m/s horizontally at right angles to a magnetic field. (a) What strength field is required to just balance the weight of the proton and keep it moving horizontally? (b) Should the direction of the magnetic field be in a horizontal or a vertical plane?

5. (a) Find the direction of the force on a proton moving as shown through the magnetic fields pictured in Figure 20.28. (b) Repeat part (a) assuming the moving particle is an electron.

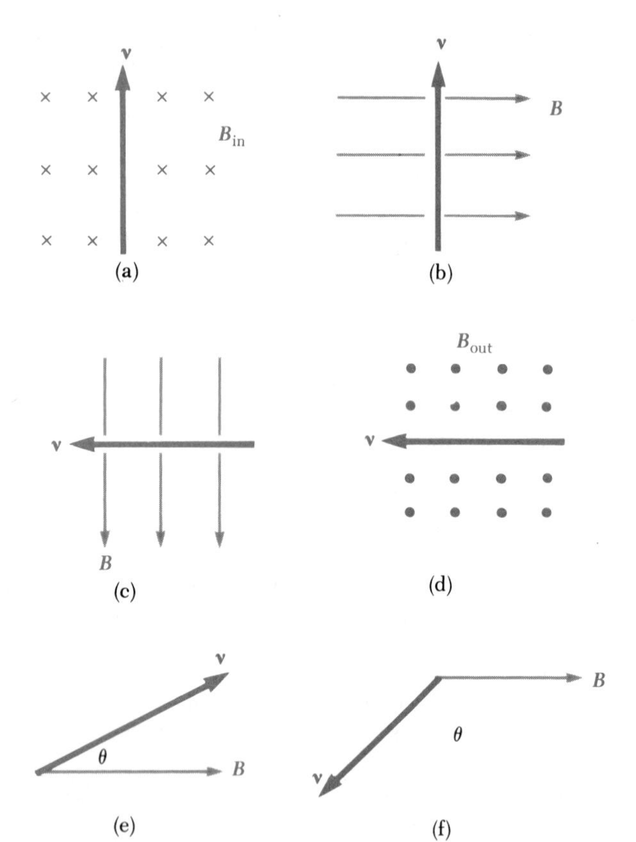

FIGURE 20.28 Problems 5 and 13. For problem 13, replace the velocity vector by a current in that direction.

6. Find the direction of the magnetic field on the positively charged particle moving in the various situations shown in Figure 20.29 if the direction of the magnetic force acting on it is as indicated.

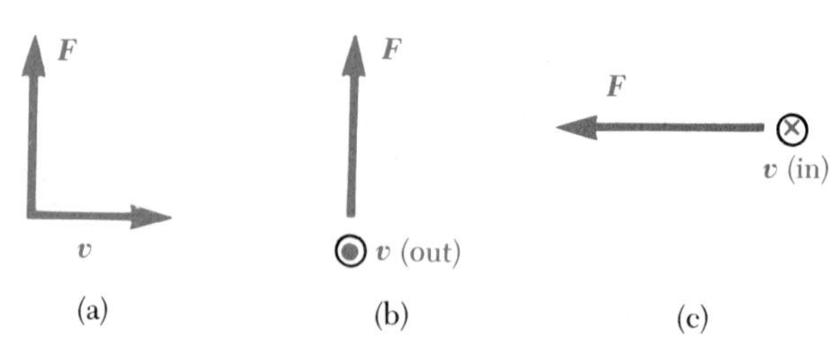

FIGURE 20.29 Problems 6 and 14. For Problem 14, replace the velocity vector by a current in that direction.

• 7. A proton moving with a speed of 5×10^7 m/s through a magnetic field of 2 T experiences a magnetic force of 3×10^{-12} N. What is the angle between the proton's velocity and the field?

•• 8. Show that the work done by the magnetic force on a charged particle moving in a uniform magnetic field is zero for any displacement of the particle.

Section 20.4 Magnetic Force on a Current-Carrying Conductor

9. Calculate the force on a 2-m length of conductor carrying a current of 10 A in a region where a uniform magnetic field has a magnitude of 1.2 T and is directed perpendicular to the conductor.

10. A wire carries a current of 10 A in a direction of 30° with respect to the direction of a magnetic field of strength 0.3 T. Find the magnetic force on a 5-m length of the wire.

11. A vertical wire carries a current of 10 A directed upward at a location where the magnetic field of the earth is horizontal and has a value of 5×10^{-5} T. If the wire is 20 m long, find the magnitude and direction of the net magnetic force on it.

12. A current I = 15 A is directed along the positive x axis and perpendicular to a magnetic field. The conductor experiences a magnetic force per unit length of 0.12 N/m in the negative y direction. Calculate the magnitude and direction of the magnetic field in the region through which the current passes.

13. In Figure 20.28 assume the velocity vector shown in each case is replaced with a wire carrying a current in the direction of the velocity vector. Find the direction of the magnetic force acting on the wire for each case.

14. In Figure 20.29 assume the velocity vector shown in each case is replaced with a wire carrying a current in the direction of the velocity vector. Find the direction of the magnetic field that will produce the magnetic force shown in each case.

• **15.** A thin copper rod 1 m long has a mass of 50 g. What is the minimum current in the rod that will cause it to float in a magnetic field of 2 T?

•• **16.** An unusual message delivery system is pictured in Figure 20.30. A 15-cm length of conductor which is free to move is held in place between two thin conductors. When a 5-A current is directed as shown in the figure, the wire segment moves upward at a constant velocity. If the mass of the wire is 0.15 kg, find the magnitude and direction of the minimum magnetic field required to move the wire. (The wire slides without friction on the two vertical conductors.)

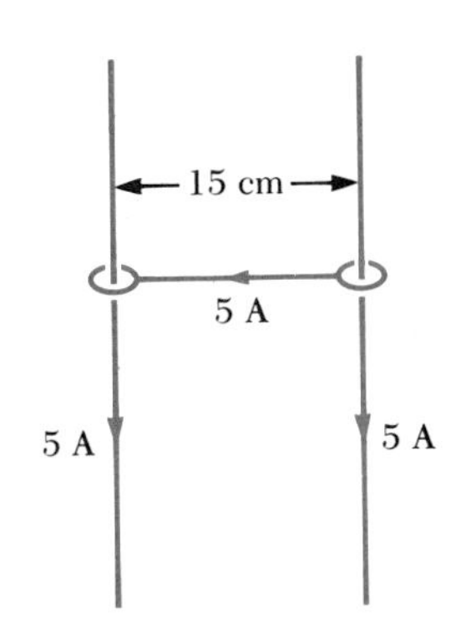

FIGURE 20.30 Problem 16

Section 20.5 Torque on a Current Loop

17. A rectangular coil of 500 turns has dimensions of 4 cm by 5 cm and is suspended in a magnetic field of 0.65 T. What is the current in the coil if the maximum torque exerted on it by the magnetic field is 0.18 N · m?

18. A single circular loop of wire of radius 50 cm and carrying a current of 2 A is in a magnetic field of 0.4 T. (a) Find the maximum torque which acts on this loop. (b) Find the angle that the plane of the loop makes with respect to the field when the torque is one-half the value found in part (a).

Section 20.6 The Galvanometer and its Applications

19. A 50 Ω, 10 mA galvanometer is to be converted to an ammeter that reads 3 A at full-scale deflection. What value of R_p should be placed in parallel with the coil?

20. An ammeter reads 10 A at full-scale deflection. The meter was constructed by inserting a resistor in parallel with a 50-Ω galvanometer coil that will deflect full scale when the voltage across it is 50 mV. What is the value of the parallel resistance that should be used?

21. A 40 Ω, 2 mA galvanometer is to be converted to a voltmeter that reads 150 V at full-scale deflection. What value of series resistance should be used with the galvanometer coil?

22. A galvanometer has an internal resistance of 100 Ω and deflects full scale for a current of 100 μA. This galvanometer is to be used to construct a multirange voltmeter as shown in Figure 20.31. Find the value of R_1, R_2, and R_3 which will enable the meter to give the full-scale readings indicated on the figure.

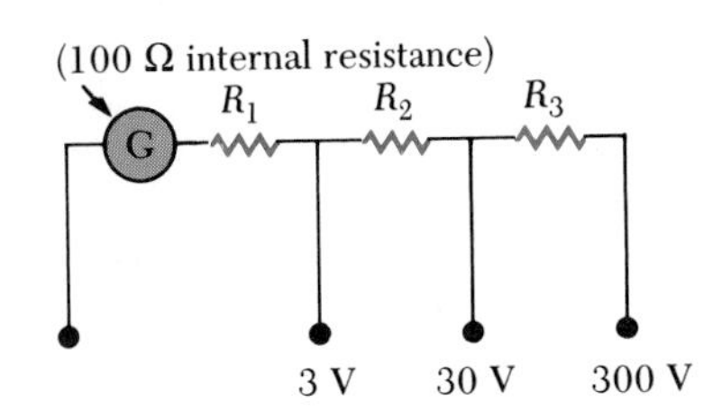

FIGURE 20.31 Problem 22

23. The galvanometer of Problem 22 is to be converted to a multirange ammeter using the circuit shown in Figure 20.32. Find the values of R_1, R_2, and R_3 which will give the full-scale readings indicated on the figure.

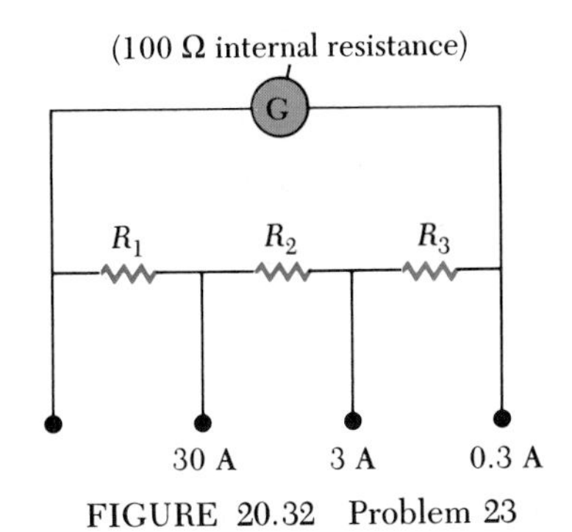

FIGURE 20.32 Problem 23

Section 20.7 Motion of a Charged Particle in a Magnetic Field

24. A proton moves with a velocity of 2×10^5 m/s perpendicular to a uniform magnetic field of strength 0.2 T. What is the radius of the path?

25. A singly charged positive ion of mass 6.68×10^{-27} kg moves clockwise in a circular path of radius 3 cm with a speed of 10^4 m/s. Find the direction and strength of the magnetic field.

26. What is the radius of the circular path of an electron with a kinetic energy of 1.6×10^{-19} J moving perpendicular to a 0.2 T magnetic field?

• **27.** Two ions of mass 6.64×10^{-27} kg move out of the slit of a mass spectrometer and into a region where the magnetic field is 0.2 T. They each have a speed of 10^6 m/s, but one ion is singly charged while the other is doubly charged. Find (a) the radius of the circular path followed by each in the field, and (b) the distance of separation when they have moved through one-half their circular path and strike a piece of photographic paper.

• **28.** Figure 20.33 shows a diagram of a device called a velocity selector. In this device, particles of a specific velocity pass through undeflected while those with greater or smaller velocities are deflected either upward or downward. In this device, an electric field is directed perpendicular to a magnetic field. This produces on the charged particle an electric force and a magnetic force which are in opposite directions, and hence cancel. Show that particles with a speed of $v = E/B$ will pass through undeflected.

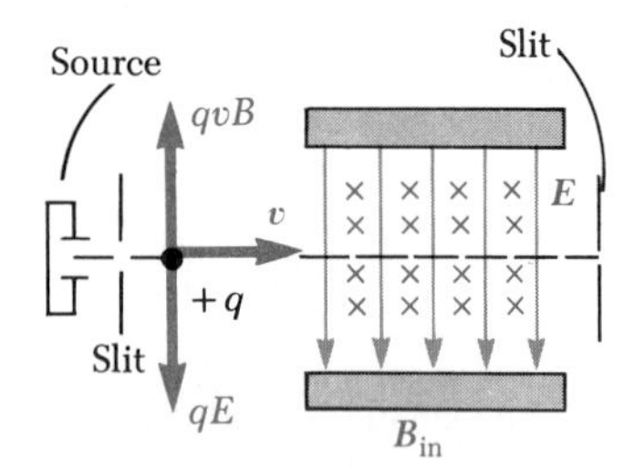

FIGURE 20.33 Problem 28

• **29.** A beam of protons is fired through crossed electric and magnetic fields as described in Problem 28. If protons of speed 10^5 m/s are to be selected, find the strength of the required magnetic field when the electric field is 5×10^5 N/C.

• **30.** A singly charged positive ion has a mass equal to 2.5×10^{-26} kg. After being accelerated through a potential difference of 250 V, the ion enters a magnetic field of 0.5 T along a direction perpendicular to the field. Calculate the radius of the path of the ion in the field.

Section 20.9 Magnetic Field of a Long, Straight Wire and Ampère's Law

31. Find the magnetic field strength at a distance of (a) 10 cm from a long, straight wire carrying a current of 5 A. Repeat the calculation for distances of (b) 50 cm, and (c) 2 m.

32. At what distance from a long, straight wire carrying a current of 5 A is the magnetic field due to the wire equal to the strength of the earth's field, approximately 5×10^{-5} T?

33. Find the direction of the current in the wire shown in Figure 20.34 that would produce a magnetic field directed as shown in each case.

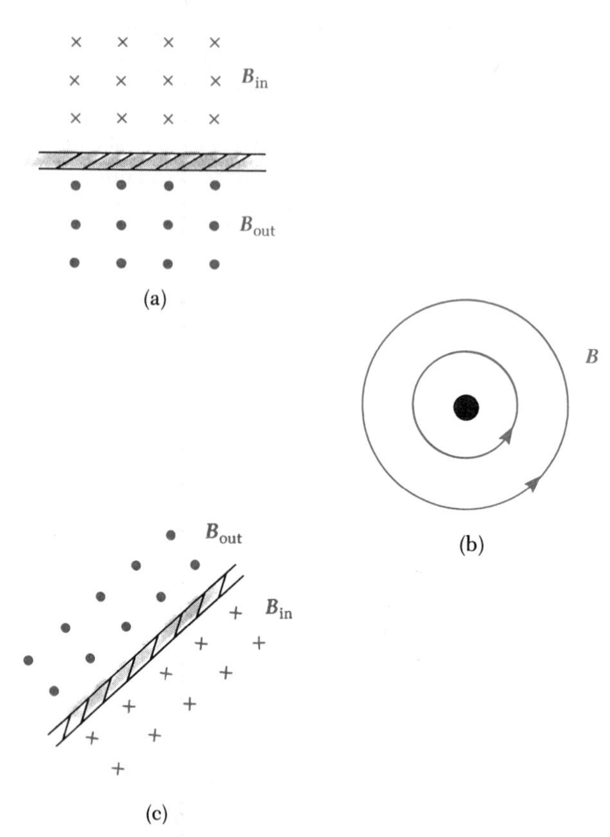

FIGURE 20.34 Problem 33

• **34.** The two wires shown in Figure 20.35 each carry currents of 5 A in opposite directions and are separated by 10 cm. Find (a) the direction and magnitude of the net magnetic field at a point midway between the wires, (b) at a point P_1, that is, 10 cm to the right of the wire on the right, and (c) at a point P_2, that is, 20 cm to the left of the wire on the left.

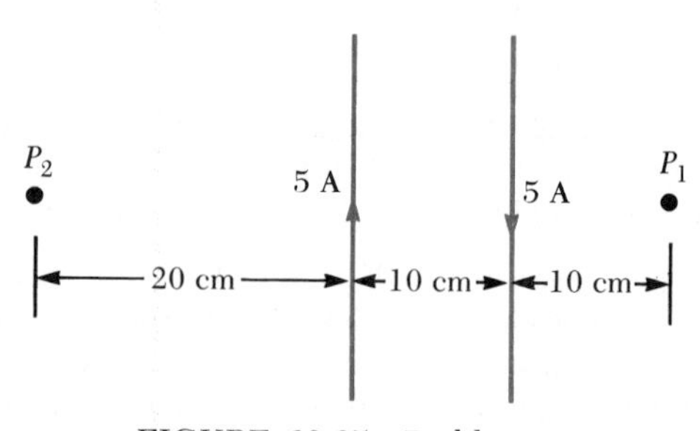

FIGURE 20.35 Problem 34

• **35.** The two wires shown in Figure 20.36 carry currents of 3 A and 5 A in the direction indicated. (a) Find the direction and magnitude of the magnetic field at a point midway between the wires. (b) Find the magnitude and direction of the magnetic field at a point 20 cm above the wire carrying the 5 A current.

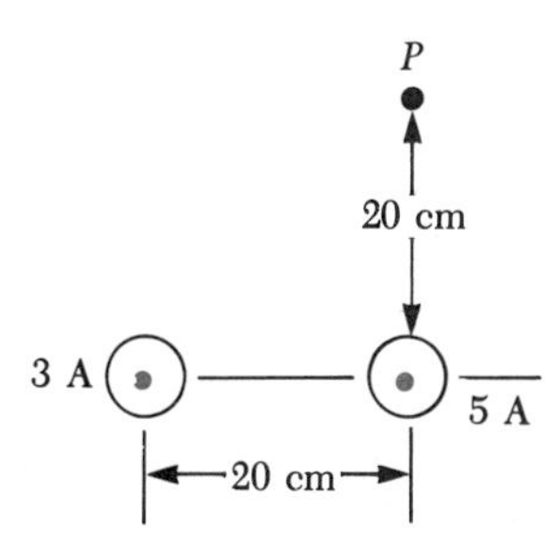

FIGURE 20.36 Problem 35

Section 20.10 Magnetic Force Between Two Parallel Conductors

36. Two long, straight wires separated by a distance of 0.3 m carry currents in the same direction. If the current in one wire is 5 A and the current in the other is 8 A, find the magnitude and direction of the force that one exerts on a 2-m length of the other.

37. Two parallel wires are 10 cm apart and each carries a current of 10 A. (a) If the currents are in the same direction, find the force per unit length exerted on one of the wires by the other. Are the wires attracted or repelled? (b) Repeat the problem if the currents are in opposite directions.

• **38.** A wire two meters long and weighing 0.08 N per meter is suspended directly above a second wire. The top wire carries a current of 30 A and the bottom wire carries a current of 60 A. Find the distance of separation between the wires such that the top wire will be held in place by magnetic repulsion.

Section 20.11 Magnetic Field of a Current Loop

Section 20.12 Magnetic Field of a Solenoid

39. A single-turn loop of wire of radius 20 cm carries a current of 10 A. (a) Find the magnetic field at the center of the loop. (b) What is the magnetic field if the loop consists of 7 closely spaced turns?

40. A hydrogen atom consists of an electron circling a proton at a distance of 0.51×10^{-10} m. If the electron makes approximately 10^{13} complete revolutions about the proton every second, find the magnetic field at the proton as set up by the electron. (*Hint:* Treat the orbiting electron as an equivalent current loop.)

41. What current in a solenoid 15 cm long wound with 100 turns would produce a magnetic field at the center equal to that of the earth, which is about 5×10^{-5} T?

42. Consider the solenoid described in Problem 41. If it carries the same current, how many turns must it have to increase the strength of the magnetic field by a factor of 3?

43. A single-turn square loop of wire, 2 cm on a side, carries a current of 0.2 A. The loop is inside a solenoid, with the plane of the loop perpendicular to the magnetic field of the solenoid. The solenoid has 30 turns per centimeter and carries a current of 15 A. Find the force on each side of the loop and the torque acting on it.

•• **44.** Use the path shown in Figure 20.37 and Ampère's Law to derive the expression for the magnetic field inside a solenoid of n turns per unit length and current I. (*Hint:* assume the strength of the magnetic field is zero outside the solenoid, a good assumption.)

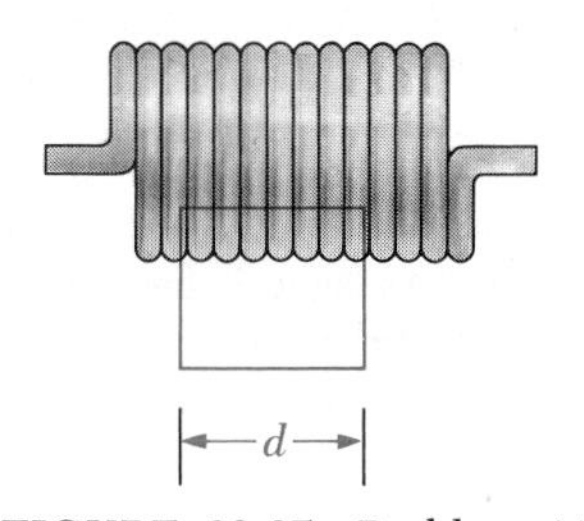

FIGURE 20.37 Problem 44

ADDITIONAL PROBLEMS

45. What current is required in the windings of a solenoid that has 500 turns uniformly distributed over a length of 0.2 m in order to produce a magnetic field of 1.2×10^{-4} T at the center of the solenoid?

46. An electron moves at a speed of 0.6 the speed of light along the $+x$ direction. A magnetic field of strength 0.6 T is in the $-y$ direction. Find the magnitude and direction of the force on the electron.

47. What magnetic field would be required to constrain an electron whose kinetic energy is 400 eV to a circular path of radius 0.8 m?

48. A circular coil consisting of a single loop of wire has a radius of 30 cm and carries a current of 25 A. It is placed in an external magnetic field of 0.3 T. Find the torque on the wire when the plane of the coil makes an angle of 35° with the direction of the field.

• **49.** Two species of singly charged positive ions of mass 20×10^{-27} kg and 23.4×10^{-27} kg enter a magnetic field at the same location with a speed of 10^5 m/s. If the strength of the field is 0.2 T, and they move perpendicular to the field, find their distance of separation after completing one half of their circular path.

• **50.** A conductor suspended by two cords as in Figure 20.38 has a mass per unit length of 0.04 kg/m. What current must exist in the conductor in order for the tension in the supporting wires to be zero, if the magnetic field over the region is 3.6 T and directed into the page? What is the required direction of the current?

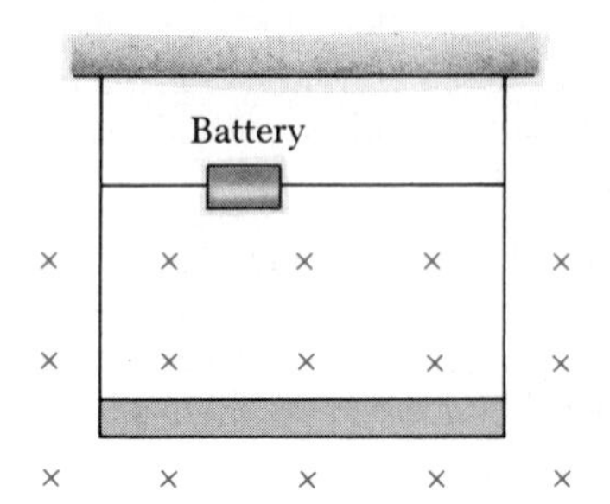

FIGURE 20.38 Problem 50

• **51.** Two long, straight wires cross one another at right angles, as shown in Figure 20.39. (a) Find the direction and magnitude of the magnetic field at point *P*, which is in the same plane as the two wires. (b) Find the magnetic field at a point 30 cm above the point of intersection. This will be a point 30 cm out of the page toward you.

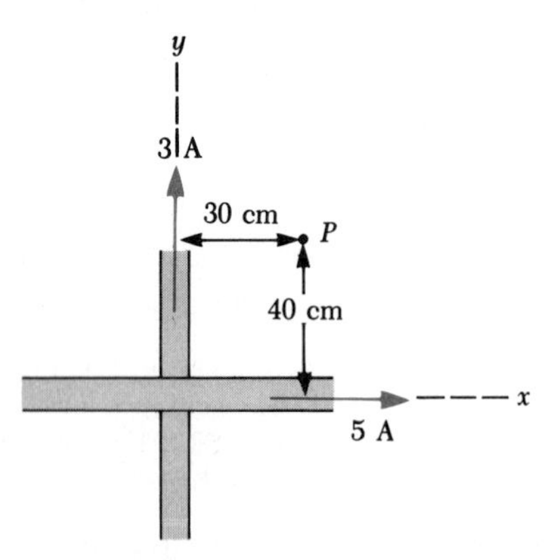

FIGURE 20.39 Problem 51

• **52.** A rectangular coil of 150 turns and area 0.12 m^2 is in a uniform magnetic field of 0.15 T. Measurements indicate that the maximum torque exerted on the loop by the field is 6×10^{-4} N · m. (a) Calculate the current in the coil. (b) Would the value found for the required current be different if the 150 turns of wire were used to form a single-turn coil of larger area? Explain.

• **53.** A rectangular loop of wire carrying a current of 2 A is suspended vertically and attached to the right arm of a balance. After the system is balanced, an external magnetic field, ***B***, is introduced. The field threads the lower end of the loop in a direction perpendicular to the wire. If the width of the loop is 20 cm and it takes 13.5 g of added mass on the left arm to rebalance the system, determine ***B***.

• **54.** A conducting wire of circular cross section formed of a material that has a mass density of 2.7 g/cm^3 is placed in a uniform magnetic field with the axis of the wire perpendicular to the direction of the field. A current density of 2.4×10^6 A/m^2 is established in the wire and the magnetic field increased until the magnetic force on the wire just balances the gravitational force. Calculate the value of ***B*** when this condition is met.

• **55.** Two parallel conductors carry currents in opposite directions as shown in Figure 20.40. One conductor carries a current of 10 A. Point *A* is at the midpoint between the wires and point *C* is a distance of 5 cm to the right of the 10-A current. *I* is adjusted so that the magnetic field at *C* is zero. Find (a) the value of the current *I* and (b) the value of the magnetic field at *A*.

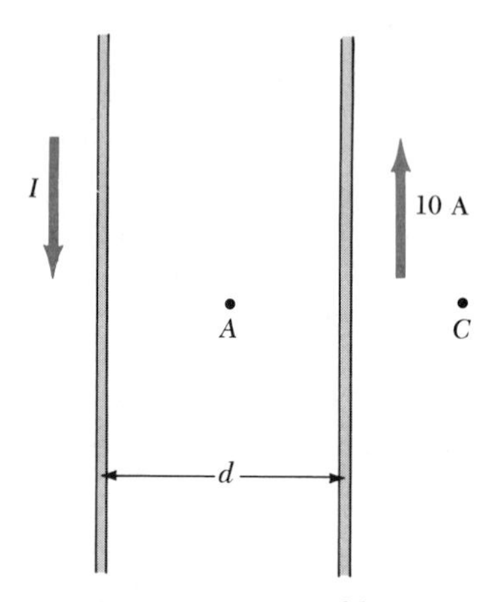

FIGURE 20.40 Problem 55

•• **56.** Four long, parallel conductors all carry a current of 4 A. An end view of the conductors is shown in Figure 20.41. The current direction is out of the page at points *A* and *B* (indicated by the dots) and into the page at *C* and *D* (indicated by the crosses). Calculate the magnitude and direction of the magnetic field at point *P*, located at the center of the square of edge length 0.2 m.

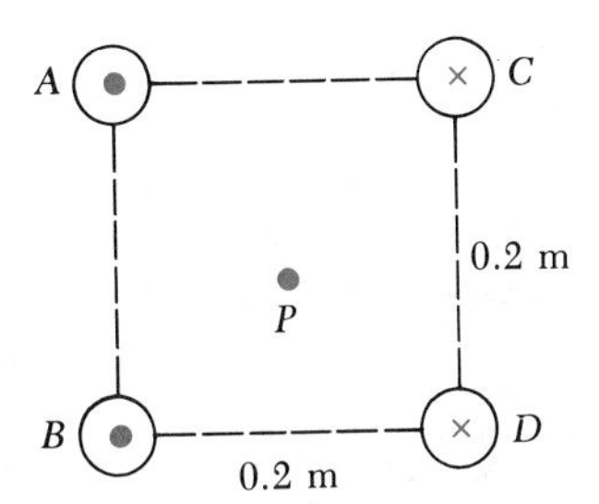

FIGURE 20.41 Problem 56

•• **57.** An electron circles at a speed of 10^4 m/s in a radius of 2 cm in a solenoid. The magnetic field of the solenoid is perpendicular to the plane of the electron's path. Find (a) the strength of the magnetic field inside the solenoid and (b) the current in the solenoid if it has 25 turns per centimeter.

•• **58.** Consider a particle of mass m and charge q moving with a velocity $\boldsymbol{v}$. The particle enters a region perpendicular to a magnetic field $\boldsymbol{B}$. Show that, while in the region of the magnetic field, the kinetic energy on the particle is proportional to the square of the radius of the orbit.

•• **59.** For the arrangement shown in Figure 20.42, the current in the straight conductor has the value $I_1 = 5$ A and lies in the plane of the rectangular loop, which carries a current $I_2 = 10$ A. The dimensions are $c = 0.1$ m, $a = 0.15$ m, and $\ell = 0.45$ m. Find the magnitude and direction of the *net force* exerted on the rectangle by the magnetic field of the straight current-carrying conductor.

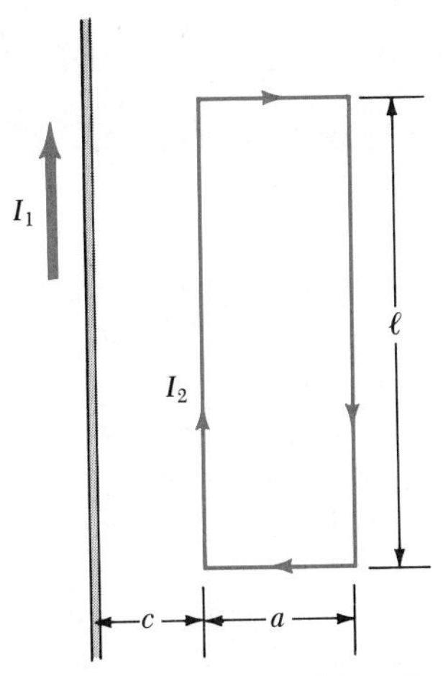

FIGURE 20.42 Problem 59

•• **60.** A straight wire of mass 10 g and length 5 cm is suspended from two identical springs which, in turn, form a closed circuit (Fig. 20.43). The springs stretch a distance of 0.5 cm under the weight of the wire. The circuit has a total resistance of 12 Ω. When a magnetic field is turned on, directed out of the page (indicated by the dots in Figure 20.43), the springs are observed to stretch an additional 0.3 cm. What is the strength of the magnetic field? (The upper portion of the circuit is fixed.)

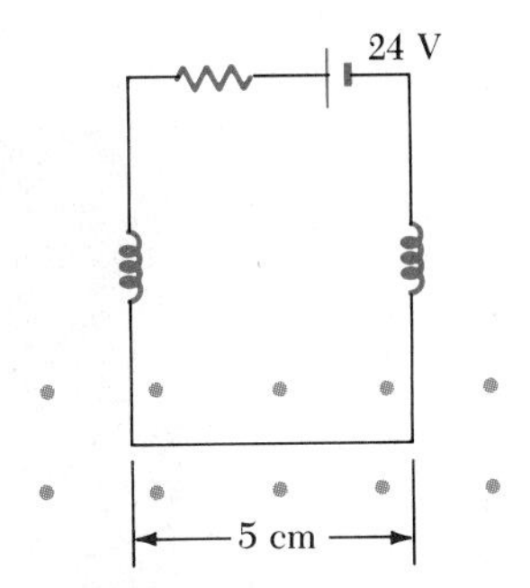

FIGURE 20.43 Problem 60

•• **61.** When the spring on a galvanometer is put under tension, a restoring torque given by $\tau = K\theta$ is set up, where K is called the torque constant and θ is the angle through which the end of the spring is turned. A coil is wrapped so that the loop is equivalent to a straight wire whose length perpendicular to the field is 0.7 m. The radius of the coil is 1.5 cm, and the radial magnetic field has a magnitude of 0.5 T. The torque constant for the spring is 2×10^{-7} N · m/rad. What value of current will produce a full-scale deflection of $\pi/4$ rad?

CHAPTER 21

Induced Voltages and Inductance

This powerhouse at Watts Bar dam in Tennessee has a capacity of 150,000 kW. (Courtesy Tennessee Valley Authority)

In 1819, Hans Christian Oersted discovered that a magnetic compass experiences a force in the vicinity of an electric current. Although there had long been speculation that such a relationship existed, this was the first evidence of a link between electricity and magnetism. Because nature is often symmetric, the discovery that electric currents produce magnetic fields led scientists to suspect that magnetic fields could produce electric currents. Indeed, experiments conducted by Michael Faraday in England and independently by Joseph Henry in the United States in 1831 showed that an electric current could be induced in a circuit by a changing magnetic field. The results of these experiments led to a very basic and important law known as Faraday's law. We shall discuss several practical applications for Faraday's law, one of which is the production of electrical energy in power generation plants throughout the world.

21.1 INDUCED EMF AND MAGNETIC FLUX

Induced emf

We begin this chapter by describing an experiment, first conducted by Faraday, that demonstrates that a current can be produced by a changing magnetic field. The apparatus shown in Figure 21.1 consists of a coil connected to a switch and a battery. We shall refer to this coil as the *primary coil* and to the corresponding circuit as the primary circuit. The coil is wrapped around an iron ring to intensify the magnetic field produced by the current through the coil. A second coil, at the right, is wrapped around the iron ring and is connected to a galvanometer. We shall refer to this as the *secondary coil* and to the corresponding circuit as the secondary circuit. Note that there is no battery in the secondary circuit. The only purpose of this circuit is to detect any current that might be produced by a magnetic field.

Michael Faraday (1791–1867) "It appeared very extraordinary, that as every electric current was accompanied by a corresponding intensity of magnetic action at right angles to the current, good conductors of electricity, when placed within the sphere of this action, should have any current induced through them, or some sensible effect produced equivalent in force to such a current."

At first sight, you might guess that no current would ever be detected in the secondary circuit. However, something quite amazing happens when the switch in the primary circuit is suddenly closed or opened. At the instant the switch in the primary circuit is closed, the galvanometer in the secondary circuit deflects in one direction and then returns to zero. When the switch is opened, the galvanometer deflects in the opposite direction and again returns to zero. Finally, the galvanometer reads zero when there is a steady current in the primary circuit.

From observations such as these, Faraday concluded that *an electric current can be produced by a changing magnetic field.* A current cannot be produced by a steady magnetic field. The current that is produced in the secondary circuit occurs for only an instant while the magnetic field through the secondary coil is changing. In effect, the secondary circuit behaves as though there were a source of emf connected to it for a short instant. It is customary to say that *an induced emf is produced in the secondary circuit by the changing magnetic field.*

Magnetic Flux

In order to provide a quantitative description of induced emfs, it is first necessary to fully understand what factors affect the phenomenon. As we shall

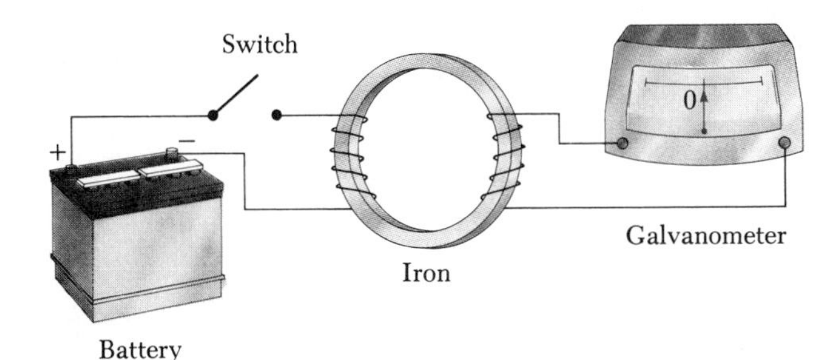

FIGURE 21.1 Faraday's experiment: When the switch in the primary circuit at the left is closed, the galvanometer in the secondary circuit at the right deflects momentarily. The emf induced in the secondary circuit is caused by the changing magnetic flux through the coil in this circuit.

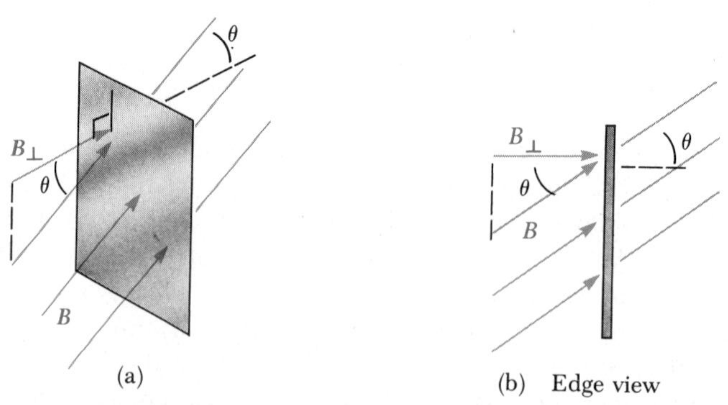

FIGURE 21.2 (a) A uniform magnetic field $\boldsymbol{B}$ making an angle θ with the normal to the plane of a loop of wire of area **A.** (b) Edge view of the loop.

see later, the induced emf is in general produced by a change in a quantity called the magnetic flux, rather than simply by a change in the magnetic field.

Consider a loop of wire in the presence of a uniform magnetic field, $\boldsymbol{B}$. If the the loop of wire has an area A, the **magnetic flux, Φ,** through the loop is defined as

Magnetic flux

$$\Phi \equiv B_{\perp}A = BA\cos\theta \tag{21.1}$$

where $B_\perp$ is the component of $\boldsymbol{B}$ perpendicular to the plane of the loop, as in Figure 21.2a, and θ is the angle between $\boldsymbol{B}$ and the direction perpendicular to the plane of the loop. An end view of the loop and the penetrating magnetic field lines is shown in Figure 21.2b. When the field is perpendicular to the plane of the loop, as in Figure 21.3a, $\theta = 0$ and Φ has a maximum value given by

$$\Phi_{\max} = BA \qquad (\boldsymbol{B} \text{ perpendicular to loop}) \tag{21.2}$$

When the plane of the loop is parallel to $\boldsymbol{B}$, as in Figure 21.3b, $\theta = 90°$ and $\Phi = 0$. Since B has the units of T, or Wb/m^2, the units of flux are $T \cdot m^2$, or Wb.

We can emphasize the significance of Equation 21.1 by first drawing magnetic field lines as in Figure 21.3. The number of lines per unit area increases as the field strength increases. The value of the magnetic flux *is proportional to the total number of lines passing through the loop.* Thus, we see that a maximum number of lines pass through the loop when its plane is perpendicular to the field, as in Figure 21.3a, and so the flux has its maximum

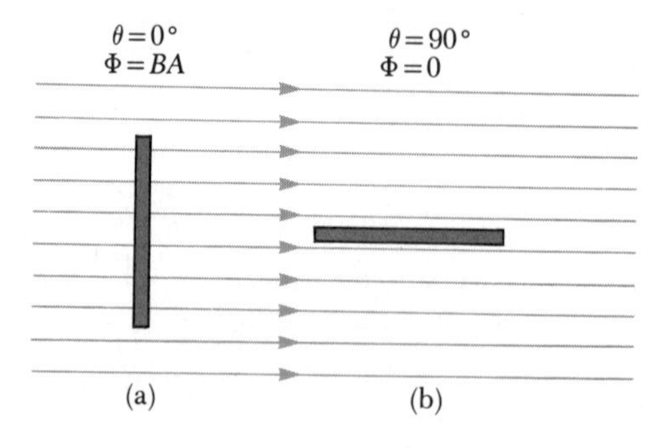

FIGURE 21.3 The edge view of a loop in a uniform magnetic field. (a) When the field lines are perpendicular to the plane of the loop, the magnetic flux through the loop is a maximum, and equal to $\phi = BA$. (a) When the field lines are parallel to the plane of the loop, the magnetic flux through the loop is zero.

value. As Figure 21.3b shows, no lines pass through the loop when its plane is parallel to the field, and so here $\Phi = 0$.

21.2 FARADAY'S LAW OF INDUCTION

The usefulness of the concept of magnetic flux will become obvious as we describe another simple experiment that demonstrates the basic idea of electromagnetic induction. Consider a loop of wire connected to a galvanometer, as in Figure 21.4. If a magnet is moved toward the loop, the galvanometer needle will deflect in one direction, as in Figure 21.4a. If the magnet is moved away from the loop, the galvanometer needle will deflect in the opposite direction, as in Figure 21.4b. If the magnet is held stationary and the loop is moved either toward or away from the magnet, the needle will also deflect. From these observations, one concludes that *a current is set up in the circuit as long as there is relative motion between the magnet and the loop.*

These results are quite remarkable in view of the fact that *a current is set up in the circuit even though the circuit contains no batteries!* We call such a current an *induced current* because it is produced by an *induced emf.*

This experiment has something in common with the Faraday experiment discussed in Section 21.1. In both cases, an emf is induced in a current when the magnetic flux through the circuit changes with time. In fact, one can make the following general statement that summarizes such experiments involving induced currents and emfs:

The instantaneous emf induced in a circuit is equal to the rate of change of magnetic flux through the circuit.

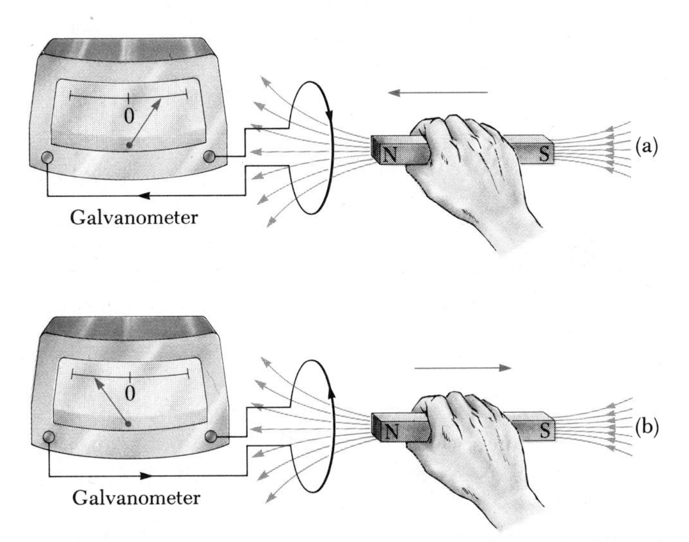

FIGURE 21.4 (a) When a magnet is moved toward a loop of wire connected to a galvanometer, the galvanometer deflects as shown. This shows that a current is induced in the loop. (b) When the magnet is moved away from the loop of wire, the galvanometer deflects in the opposite direction, indicating that the induced current is opposite that shown in (a).

If a circuit contains N loops and the flux changes by an amount $\Delta\Phi$ during a time Δt, the average induced emf during the time Δt is

Faraday's law

$$\mathcal{E} = -N\frac{\Delta\Phi}{\Delta t} \tag{21.3}$$

This is a statement of **Faraday's law of induction.** The minus sign is included to indicate the polarity of the induced emf, which can be found by use of **Lenz's law:**

Lenz's law

The polarity of the induced emf is such that it produces a current whose magnetic field opposes the change in magnetic flux through the loop. That is, the induced current tends to maintain the original flux through the circuit.

We shall consider several applications of Lenz's law in Section 21.4.

EXAMPLE 21.1 Application of Faraday's Law

A coil is wrapped with 200 turns of wire on a square frame of sides 18 cm. Each turn has the same area, equal to that of the frame, and the total resistance of the coil is 2 Ω. A uniform magnetic field is applied perpendicular to the plane of the coil. If the field changes uniformly from 0 to 0.5 T in 0.8 s, find the magnitude of the induced emf in the coil while the field is changed.

Solution The area of the coil is $(0.18\ \text{m})^2 = 0.0324\ \text{m}^2$. The magnetic flux through the coil at $t = 0$ is zero because $B = 0$. At $t = 0.8$ s, the magnetic flux through the coil is

$$\Phi_f = BA = (0.5\ \text{T})(0.0324\ \text{m}^2) = 0.0162\ \text{T}\cdot\text{m}^2$$

Therefore, the *change* in flux through the coil during the 0.8-s interval is

$$\Delta\Phi = \Phi_f - \Phi_i = 0.0162\ \text{T}\cdot\text{m}^2$$

Faraday's law of induction enables us to find the magnitude of the induced emf:

$$|\mathcal{E}| = N\frac{\Delta\Phi}{\Delta t} = (200\ \text{turns})\left(\frac{0.0162\ \text{T}\cdot\text{m}^2}{0.8\ \text{s}}\right) = 4.05\ \text{V}$$

(Note that $1\ \text{T}\cdot\text{m}^2 = 1\ \text{V}\cdot\text{s}$.)

EXERCISE Find the magnitude of the induced current in the coil while the field is changing.

Answer 2.03 A.

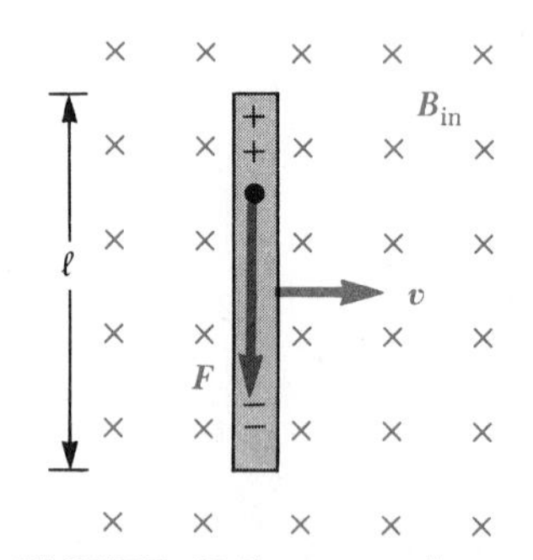

FIGURE 21.5 A straight conductor of length ℓ moving with a velocity v through a uniform magnetic field B directed perpendicular to v. The vector F is the force on an electron in the conductor. An emf equal to $B\ell v$ is induced between the ends of the bar.

21.3 MOTIONAL EMF

In Section 21.2, we considered a situation in which an emf is induced in a circuit when the magnetic field changes with time. In this section we describe what is called **motional emf,** which is the emf induced in a conductor moving through a magnetic field.

First consider a straight conductor of length ℓ moving with constant velocity through a uniform magnetic field directed into the paper, as in Figure 21.5. For simplicity, we shall assume that the conductor is moving perpendicular to the field. The electrons in the conductor will experience a force directed downward along the conductor and having a magnitude $F = qvB$.

Because of this magnetic force, the electrons will move to the lower end and accumulate there, leaving a net positive charge at the upper end. An electric field is therefore produced in the conductor as a result of this charge separation. The charge at the ends build up until the downward magnetic force, qvB, is balanced by the upward electric force, qE. At this point, charge stops flowing and the condition for equilibrium requires that

$$qE = qvB \qquad \text{or} \qquad E = vB$$

Since the electric field is constant, the electric field produced in the conductor is related to the potential difference across the ends according to the relation $V = E\ell$. Thus,

$$V = E\ell = B\ell v$$

Since there is an excess of positive charge at the upper end of the conductor and an excess of negative charge at the lower end, the upper end is at a higher potential than the lower end. Thus,

a potential difference is maintained across the conductor as long as there is motion through the field. If the motion is reversed, the polarity of the potential difference is also reversed.

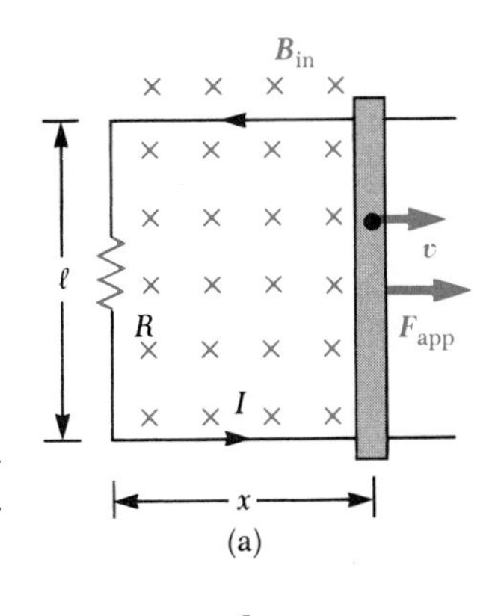

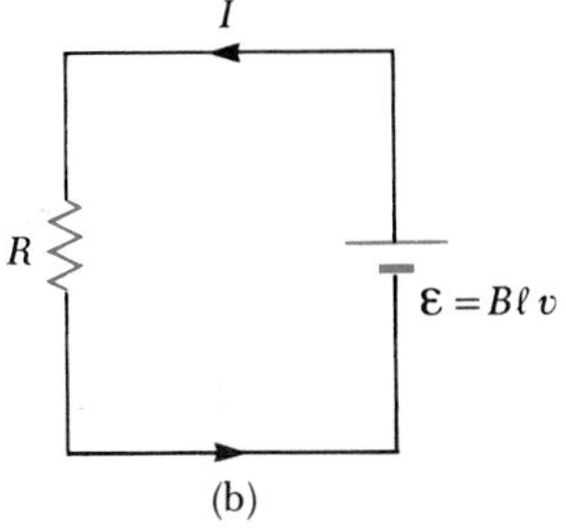

FIGURE 21.6 (a) A conducting bar sliding with a velocity $\boldsymbol{v}$ along two conducting rails under the action of an applied force $\boldsymbol{F}_{app}$. The magnetic force opposes the motion, and a counterclockwise current is induced in the loop. (b) The equivalent circuit of (a).

A more interesting situation occurs if we now consider what happens when the moving conductor is part of a closed conducting path. This situation is particularly useful for illustrating how a changing magnetic flux can cause an induced current in a closed circuit. Consider a circuit consisting of a conducting bar of length ℓ sliding along two fixed parallel conducting rails, as in Figure 21.6a. For simplicity, we shall assume that the moving bar has zero resistance and that the stationary part of the circuit has a resistance R. A uniform and constant magnetic field $\boldsymbol{B}$ is applied perpendicular to the plane of the circuit. As the bar is pulled to the right with a velocity $\boldsymbol{v}$, under the influence of an applied force $\boldsymbol{F}_{app}$, the free charges in the bar experience a magnetic force along the length of the bar. This force in turn sets up an induced current because the charges are free to move in a closed conducting path. In this case, the changing magnetic flux through the loop and the corresponding induced emf across the moving bar arise from the change in area of the loop as the bar moves through the magnetic field.

Let us assume that the bar moves a distance Δx, as shown in Figure 21.7. The increase in flux, $\Delta\Phi$, through the loop is the amount that now passes through the shaded portion of the circuit:

$$\Delta\Phi = BA = B\ell\Delta x$$

Let us use Faraday's law, noting that there is one loop ($N = 1$); we find that the induced emf has a magnitude

$$\mathcal{E} = \frac{\Delta\Phi}{\Delta t} = B\ell\frac{\Delta x}{\Delta t} = B\ell v \tag{21.4}$$

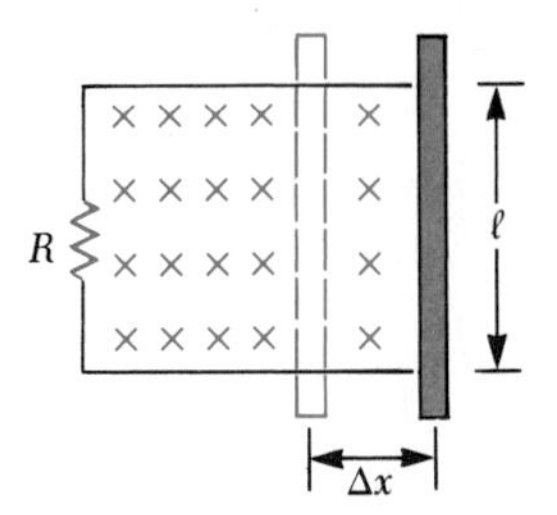

FIGURE 21.7 The magnetic flux through the loop increases as the bar moves to the right.

This induced emf is often called a motional emf because it arises from the motion of a conductor through a magnetic field.

Furthermore, if the resistance of the circuit is R, the magnitude of the induced current is

Induced current

$$I = \frac{\mathcal{E}}{R} = \frac{B\ell v}{R} \tag{21.5}$$

The equivalent circuit diagram for this example is shown in Figure 21.6b.

EXAMPLE 21.2 The Electrified Airplane Wing

An airplane with a wing span of 50 m flies horizontal to the earth at a location where the downward component of the earth's magnetic field is 0.6×10^{-4} T. Find the difference in potential between the tips of the wing when the velocity of the plane is 250 m/s.

Solution Because the plane is flying horizontally, we do not have to concern ourselves with the horizontal component of the earth's field. Thus we find

$$\mathcal{E} = B\ell v = (0.6 \times 10^{-4}\ \text{T})(50\ \text{m})(250\ \text{m/s}) = 0.75\ \text{V}$$

21.4 LENZ'S LAW REVISITED

In order to obtain a better understanding of Lenz's law, let us return to the example of a bar moving to the right on two parallel rails in the presence of a uniform magnetic field directed into the paper (Fig. 21.8a). As the bar moves to the right, the magnetic flux through the circuit increases with time because the area of the loop increases. Lenz's law says that the induced current must be in a direction such that the flux *it* produces opposes the change in the external magnetic flux. Since the flux due to the external field is increasing *into* the paper, the induced current, if it is to oppose the change, must produce a flux *out* of the paper. Hence, the induced current must be counterclockwise when the bar moves to the right to give a counteracting flux out of the paper. (Use the right-hand rule to verify this direction.) On the other hand, if the bar is moving to the left, as in Figure 21.8b, the magntic flux through the loop decreases with time. Since the flux is into the paper, the

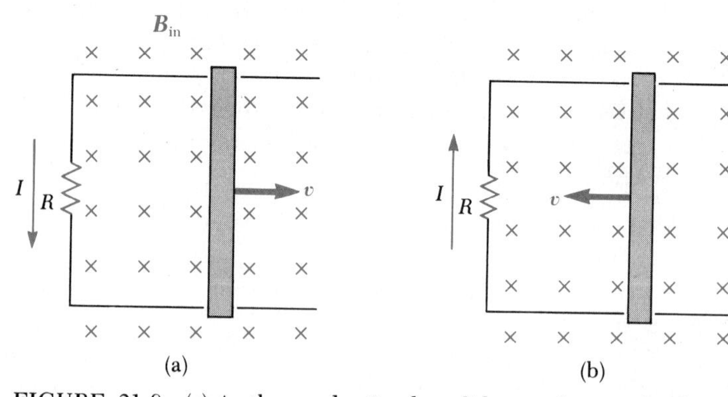

FIGURE 21.8 (a) As the conducting bar slides on the two fixed conducting rails, the magnetic flux through the loop increases in time. By Lenz's law, the induced current must be *counterclockwise* so as to produce a counteracting flux *out of the paper*. (b) When the bar moves to the left, the induced current must be *clockwise*. Why?

induced current has to be clockwise to produce a flux into the paper. In either case, the induced current tends to maintain the original flux through the circuit.

Let us look at this situation from the viewpoint of energy considerations. Suppose that the bar is given a slight push to the right. In the above analysis, we found that this motion leads to a counterclockwise current in the loop. Let us see what happens if we assume that the current is clockwise. For a clockwise current I, the direction of the magnetic force, $BI\ell$, on the sliding bar would be to the right. This force would accelerate the rod and increase its velocity. This, in turn, would cause the area of the loop to increase more rapidly, thus increasing the induced current, which would increase the force, which would increase the current, which would. . . . In effect, the system would acquire energy with zero input energy. This is clearly inconsistent with all experience and with the law of conservation of energy. Thus, we are forced to conclude that the current must be counterclockwise.

Consider another situation, one in which a bar magnet is moved to the right toward a stationary loop of wire, as in Figure 21.9a. As the magnet moves to the right toward the loop, the magnetic flux through the loop increases with time. To counteract this increase in flux to the right, the induced current produces a flux to the left, as in Figure 21.9b; hence the induced current is in the direction shown. Note that the magnetic field lines associated with the induced current oppose the motion of the magnet. Therefore, the left face of the current loop is a north pole and the right face is a south pole.

On the other hand, if the magnet were moving to the left, its flux through the loop, which is toward the right, would decrease in time. Under these circumstances, the induced current in the loop would be in a direction such as to set up a field through the loop directed from left to right in an effort to maintain a constant number of flux lines. Hence, the induced current in the loop would be opposite that shown in Figure 21.9b. In this case, the left face of the loop would be a south pole and the right face would be a north pole.

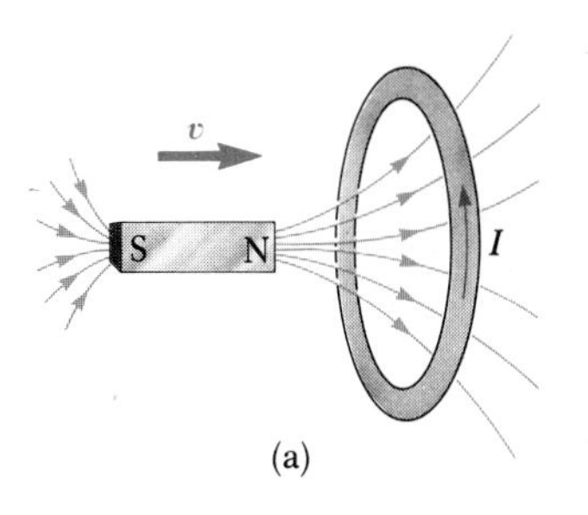

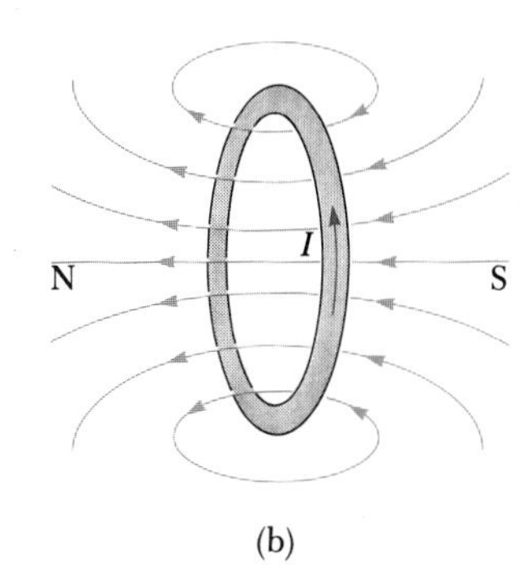

FIGURE 21.9 (a) When the magnet is moved toward the stationary conducting loop, a current is induced in the direction shown. (b) This induced current produces its own flux to the left to counteract the increasing external flux to the right.

EXAMPLE 21.3 Application of Lenz's Law

A coil of wire is placed near an electromagnet as shown in Figure 21.10a. Find the direction of the induced current in the coil (a) at the instant the switch is closed, (b) after the switch has been closed for several seconds, and (c) when the switch is opened.

Solution (a) When the switch is closed, the situation changes from a condition in which no lines of flux pass through the coil to one in which lines of flux pass through in the direction shown in Figure 21.10b. To counteract this change in the number of

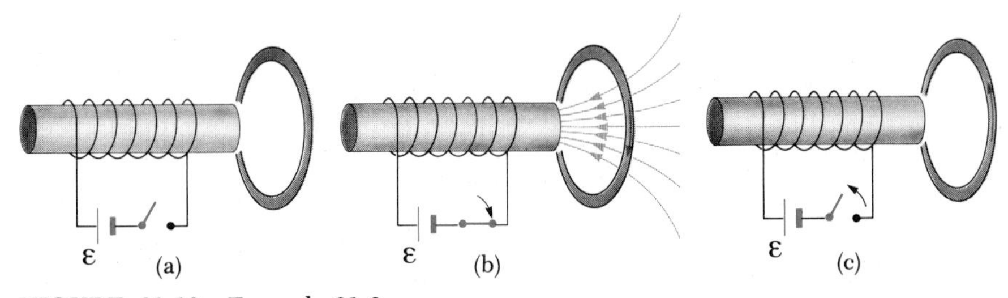

FIGURE 21.10 Example 21.3.

lines, the coil must set up a field from left to right in the figure. This requires a current directed as shown in Figure 21.10b.

(b) After the switch has been closed for several seconds, there is no change in the number of lines through the loop; hence the induced current is zero.

(c) Opening the switch causes the magnetic field to change from a condition in which flux lines thread through the coil from right to left to a condition of zero flux. The induced current must then be as shown in Figure 21.10c, so as to set up its own field from right to left.

*Tape Recorders

One interesting practical use of induced currents and emfs is the tape recorder. There are many different types of tape recorders, but the basic principles are the same for all. In this device, a magnetic tape moves past a recording head and a playback head, as in Figure 21.11a. The tape is a plastic ribbon coated with iron oxide or chromium oxide. We shall examine the function of both heads, starting with the recording head.

The steps of recording are illustrated in Figure 21.11b. The recording process uses the fact that a current passing through an electromagnet produces a magnetic field. A sound wave sent into a microphone is transformed into an electric current. This signal is amplified and allowed to pass through a wire coiled around a doughnut-shaped piece of iron, which functions as the recording head. The iron ring and the wire constitute an electromagnet, which contains the lines of the magnetic field completely inside the iron except at the point where a slot is cut in the ring. At this location the magnetic field fringes out of the iron. This field magnetizes the small pieces of iron oxide embedded in the tape. Thus, as the tape moves past the slot, it becomes magnetized in a pattern that reproduces both the frequency and the intensity of the sound signal entering the microphone.

To reconstruct the sound signal, the tape is allowed to pass through a recorder with the playback head in operation. This head is very similar to the recording head in that it consists of a wire-wound doughnut-shaped piece of iron with a slot in it. When the tape moves past this head, the varying magnetic fields on the tape produce changing field lines through the wire coil. These changing lines induce a current in the coil that corresponds to the current in the recording head that produced the tape originally. This changing electric current can be amplified and used to drive a speaker. Playback is thus an example of induction of a current by a moving magnet.

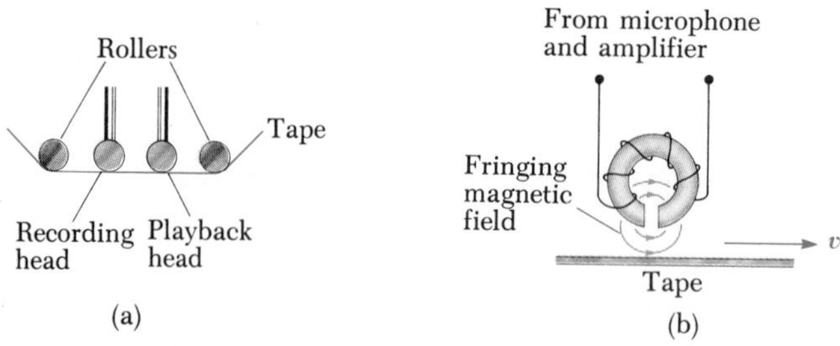

FIGURE 21.11 (a) The head of a magnetic tape recorder. (b) The fringing magnetic field magnetizes the tape during recording.

21.5 GENERATORS AND MOTORS

Generators and motors are important practical devices that operate on the principle of electromagnetic induction. First, let us consider the **alternating current generator** (or ac generator), a device that converts mechanical energy to electrical energy. In its simplest form, the ac generator consists of a loop of wire rotated by some external means in a magnetic field (Fig. 21.12a). In commercial power plants, the energy required to rotate the loop can be derived from a variety of sources. For example, in a hydroelectric plant, falling water directed against the blades of a turbine produces the rotary motion; in a coal-fired plant, the heat produced by burning coal is used to convert water to steam and this steam is directed against the turbine blades. As the loop rotates, the magnetic flux through it changes with time, inducing an emf and a current in an external circuit. The ends of the loop are connected to slip rings that rotate with the loop. Connections to the external circuit are made by stationary brushes in contact with the slip rings.

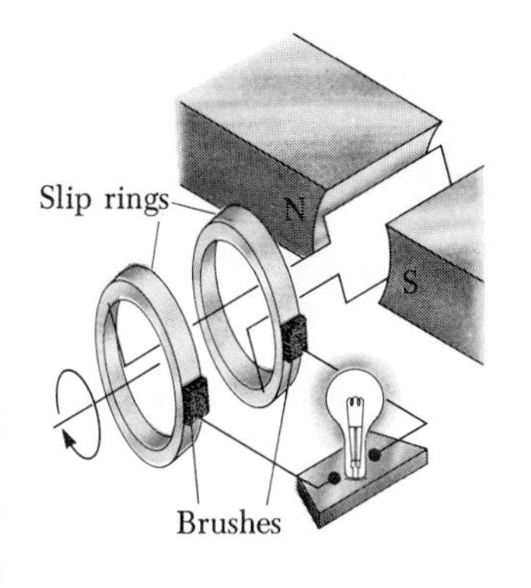

(a)

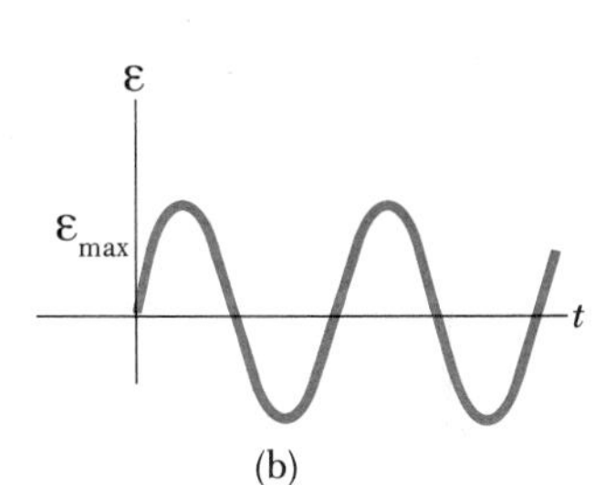

(b)

FIGURE 21.12 (a) Schematic diagram of an ac generator. An emf is induced in a coil which rotates by some external means in a magnetic field. (b) The alternating emf induced in the loop plotted versus time.

We can derive an expression for the emf generated in the rotating loop by making use of the expression for motional emf, $\mathcal{E} = B\ell v$. Figure 21.13a shows a loop of wire rotating clockwise in a uniform magnetic field directed to the right. The magnetic force (qvB) on the charges in wires AB and CD is not along the length of the wires. (The force on the electrons in wires AB and CD is perpendicular to the wires, rather than along their length.) Hence, an emf is generated only in wires BC and AD. At any instant, wire BC has a velocity $\mathbf{v}$ at an angle θ with the magnetic field as shown in Figure 21.13b. (Note that the component of velocity parallel to the field has no effect on the charges in the wire whereas the component of velocity perpendicular to the field produces a magnetic force on the charges that moves electrons from C to B.) The emf generated in wire BC is equal to $B\ell v_\perp$, where ℓ is the length of the wire and $v_\perp$ is the component of velocity perpendicular to the field. An emf of $B\ell v_\perp$ is also generated in wire DA, and the sense of this emf is the same as in wire BC. Since $v_\perp = v \sin\theta$, the total emf is

$$\mathcal{E} = 2B\ell v_\perp = 2B\ell v \sin\theta \qquad (21.6)$$

If the loop rotates with a constant angular velocity ω, we can use the relation $\theta = \omega t$ in Equation 21.6. Furthermore, since every point on wires BC and DA rotates in a circle about the axis of rotation with the same angular velocity

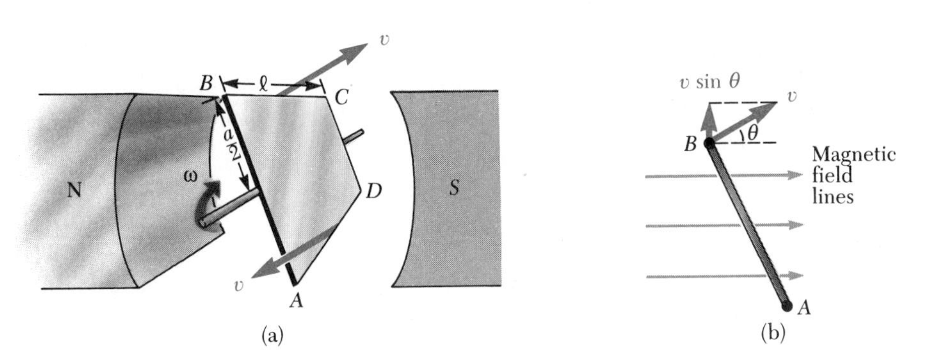

FIGURE 21.13 (a) A loop rotating at constant angular velocity in an external magnetic field. The emf induced in the loop varies sinusoidally with time. (b) An edge view of the rotating loop.

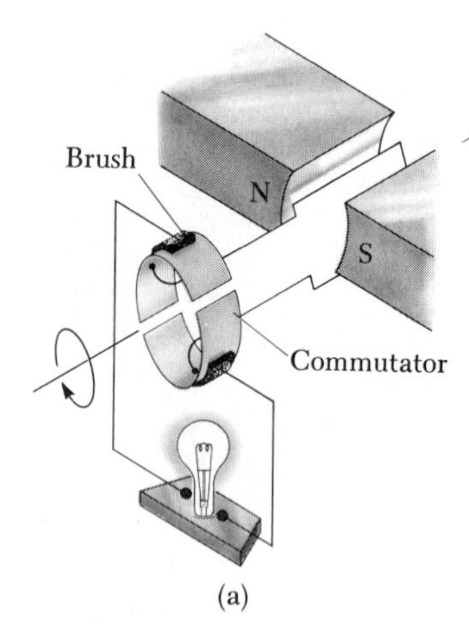

(a)

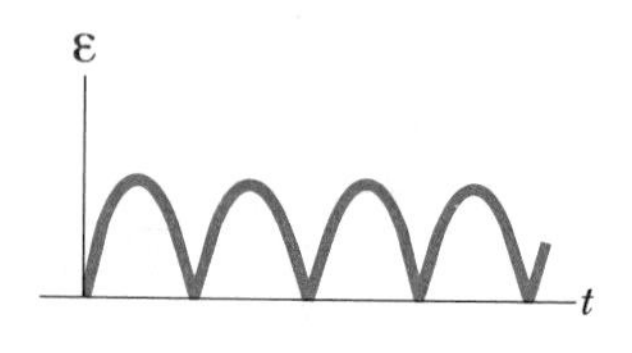

(b)

FIGURE 21.14
(a) Schematic diagram of a dc generator. (b) The emf fluctuates in magnitude but always has the same polarity.

ω, we have $v = r\omega = (a/2)\omega$, where a is the length of sides AB and CD. Therefore, Equation 21.6 reduces to

$$\mathcal{E} = 2B\ell\left(\frac{a}{2}\right)\omega \sin \omega t = B\ell a\omega \sin \omega t$$

If the loop has N turns, the emf is N times larger because each loop has the same emf induced in it. Furthermore, since the area of the loop is $A = \ell a$, the total emf is

$$\boxed{\mathcal{E} = NBA\omega \sin \omega t} \tag{21.7}$$

This result shows that the emf varies sinusoidally with time, as plotted in Figure 21.12b. Note that the maximum emf has the value

$$\boxed{\mathcal{E}_{max} = NBA\omega} \tag{21.8}$$

which occurs when $\omega t = 90°$ or $270°$. In other words, $\mathcal{E} = \mathcal{E}_{max}$ when the plane of the loop is parallel to the magnetic field. Furthermore, the emf is zero when $\omega t = 0$ or $180°$, that is, when the magnetic field is perpendicular to the plane of the loop. The frequency of rotation for commercial generators in this country and Canada is 60 Hz, whereas in some European countries, 50 Hz is used. (Recall that $\omega = 2\pi f$, where f is the frequency in hertz.)

The **direct current** (dc) **generator** is illustrated in Figure 21.14a. The components are essentially the same as those of the ac generator, except that the contacts to the rotating loop are made by a split ring, or commutator. In this design, the output voltage always has the same polarity and the current is a pulsating direct current, as in Figure 21.14b. The reason for this can be understood by noting that the contacts to the split ring reverse their roles every half cycle. At the same time, the polarity of the induced emf reverses; hence the polarity of the split ring remains the same.

A pulsating dc current is not suitable for most applications. To obtain a more steady dc current, commercial dc generators use many loops and commutators distributed around the axis of rotation so that the sinusoidal pulses from the various loops are out of phase. When these pulses are superimposed, the dc output is almost free of fluctuations.

EXAMPLE 21.4 Emf Induced in an ac Generator

An ac generator consists of eight turns of wire of area $A = 0.09\ \text{m}^2$ and total resistance $12\ \Omega$. The loop rotates in a magnetic field of 0.5 T at a constant frequency of 60 Hz. (a) Find the maximum induced emf.

Solution First note that $\omega = 2\pi f = 2\pi(60\ \text{Hz}) = 377\ \text{s}^{-1}$. When we substitute the appropriate numerical values into Equation 21.8, we obtain

$$\mathcal{E}_{max} = NAB\omega = 8(0.09\ \text{m}^2)(0.5\ \text{T})(377\ \text{s}^{-1}) = 136\ \text{V}$$

(b) What is the maximum induced current?

Solution From Ohm's law and the results to (a), we find that the maximum induced current is

$$I_{max} = \frac{\mathcal{E}_{max}}{R} = \frac{136\ \text{V}}{12\ \Omega} = 11.3\ \text{A}$$

(c) Determine the time variation of the induced emf.

Solution We can use Equation 21.7 to obtain the time variation of $\mathcal{E}$:

$$\mathcal{E} = \mathcal{E}_{max} \sin \omega t = 136 \sin 377t \text{ V}$$

EXERCISE Determine the time variation of the induced current.

Answer: $I = 11.3 \sin 377t$ A.

A turbine being lowered into place at TVA's Wheeler Dam on the Tennessee River. (Courtesy of the Tennessee Valley Authority)

*Motors and Back emf

Motors are devices that convert electrical energy to mechanical energy. Essentially, *a motor is a generator run in reverse.* Instead of a current being generated by a rotating loop, a current is supplied to the loop by a source of emf and the magnetic force on the current-carrying loop causes it to rotate.

A motor can perform useful mechanical work when a shaft connected to its rotating coil is attached to some external device. As the coil in the motor rotates, however, the changing magnetic flux through it induces an emf that always acts to reduce the current in the coil. If this were not the case, Lenz's law would be violated. The phrase **back emf** is used to indicate an emf that tends to reduce the supplied current. The back emf increases in magnitude as the rotational speed of the coil increases. We can picture the state of affairs in such situations as the equivalent circuit shown in Figure 21.15. For illustrative purposes, we assume in Figure 21.15 that the external power source attempting to drive current through the coil of the motor has a voltage of 120 V, that the coil has a resistance of 10 Ω, and that the back emf induced in the coil at this instant is 70 V. Thus, the voltage available to supply current equals the difference between the supply voltage and the back emf, 50 V in this case. So we see that the current is limited by the back emf.

When a motor is first turned on, there is initially no back emf and the current is very large because it is limited only by the resistance of the coil. As the coil begins to rotate, the induced back emf opposes the applied voltage and the current in the coil is reduced. If the mechanical load increases, the motor will slow down, which causes the back emf to decrease. This reduction in the back emf increases the current in the coil and therefore also increases the power needed from the external voltage source. For this reason, the power requirements are greater for starting a motor and for running it under heavy loads. If the motor is allowed to run under no mechanical load, the back emf reduces the current to a value just large enough to overcome energy losses by heat and friction.

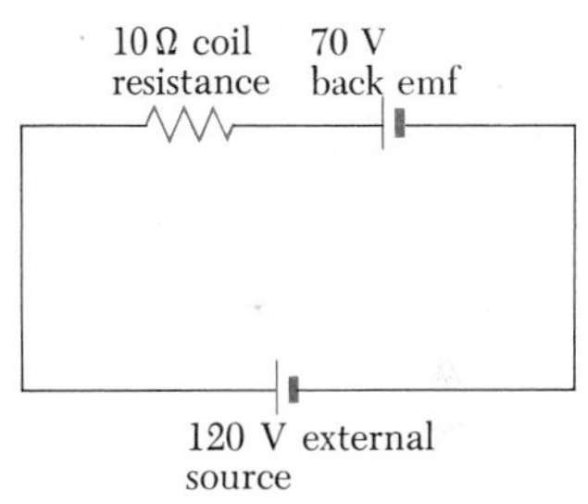

FIGURE 21.15 A motor can be represented as a resistance plus a back emf.

EXAMPLE 21.5 The Induced Current in a Motor

Assume that a motor having coils with a resistance of 10 Ω is supplied by a voltage of 120 V. When the motor is running at its maximum speed, the back emf is 70 V. Find the current in the coils (a) when the motor is first turned on and (b) when the motor has reached maximum speed.

Solution (a) When the motor is first turned on, the back emf is zero. (The coils are motionless.) Thus the current in the coils is a maximum and equal to

$$I = \frac{\mathcal{E}}{R} = \frac{120 \text{ V}}{10\ \Omega} = 12 \text{ A}$$

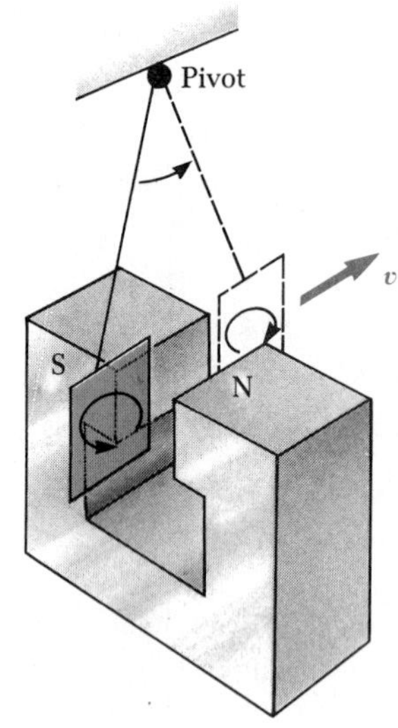

FIGURE 21.16 An apparatus that demonstrates the formation of eddy currents in a conductor moving through a magnetic field. As the plate enters or leaves the field, the changing magnetic flux sets up an induced emf, which causes the eddy currents in the plate.

(b) At the maximum speed, the back emf has its maximum value. Thus, the effective supply voltage is now that of the external source minus the back emf. Hence, the current is reduced to

$$I = \frac{\mathcal{E} - \mathcal{E}_{\text{back}}}{R} = \frac{120\text{ V} - 70\text{ V}}{10\ \Omega} = \frac{50\text{ V}}{10\ \Omega} = 5\text{ A}$$

EXERCISE If the current in the motor is 8 A at some instant, what is the back emf at this time?

Answer 40 V.

*21.6 EDDY CURRENTS

As we have seen, an emf and a current are induced in a circuit by a changing magnetic flux. In the same manner, circulating currents called **eddy currents** are set up in pieces of metal moving through a magnetic field. This can easily be demonstrated by allowing a flat metal plate at the end of a bar to swing through a magnetic field (Fig. 21.16). The metal should be a nonmagnetic material, such as aluminum or copper. As the plate enters the field, the changing flux creates an induced emf in the plate, which in turn causes the free electrons in the metal to move, producing the swirling eddy currents. According to Lenz's law, the direction of the eddy currents must oppose the change that causes them. For this reason, the eddy currents must produce effective magnetic poles on the plate, which are repelled by the poles of the magnet, thus giving rise to a repulsive force that opposes the swinging motion of the plate.

As indicated in Figure 21.17, when the magnetic field is into the paper, the eddy current is counterclockwise as the swinging plate enters the field at position 1. This is because the external flux into the paper is increasing, and hence by Lenz's law the induced current must provide a flux out of the paper. The opposite is true as the plate leaves the field at position 2, where the current is clockwise. Since the induced eddy current always produces a retarding force when the plate enters or leaves the field, the swinging plate quickly comes to rest.

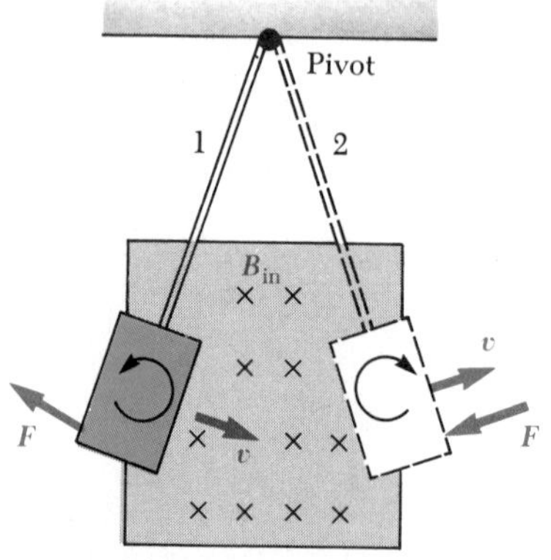

FIGURE 21.17 As the conducting plate enters the magnetic field in position 1, the eddy currents are counterclockwise. At position 2, however, the currents are clockwise. In either case, the magnetic force retards the motion of the plate.

If slots are cut in the metal plate, as in Figure 21.18, the eddy currents and the corresponding retarding force are greatly reduced. This can be understood by noting that the cuts in the plate are open circuits for any large current loops that might otherwise be formed.

The braking system on many rapid transit cars make use of electromagnetic induction and eddy currents. An electromagnet, which can be energized with a current, is positioned near the steel rails. The braking action occurs when a large current is passed through the electromagnet. The relative motion of the magnet and rails induces eddy currents in the rails, and the direction of these currents produces a drag force on the moving vehicle. The loss is mechanical energy of the vehicle is transformed into heat. Because the eddy currents decrease steadily in magnitude as the vehicle slows, the braking effect is quite smooth.

Eddy currents are often undesirable because they dissipate energy in the form of heat. To reduce this energy loss, moving conducting parts are often laminated, that is, built up in thin layers separated by a nonconducting material, such as lacquer or metal oxide. This layered structure increases the resistance of the possible paths of the eddy currents and effectively confines the currents to individual layers. Such a laminated structure is used in the cores of transformers and motors to minimize eddy currents and thereby increase the efficiency of those devices.

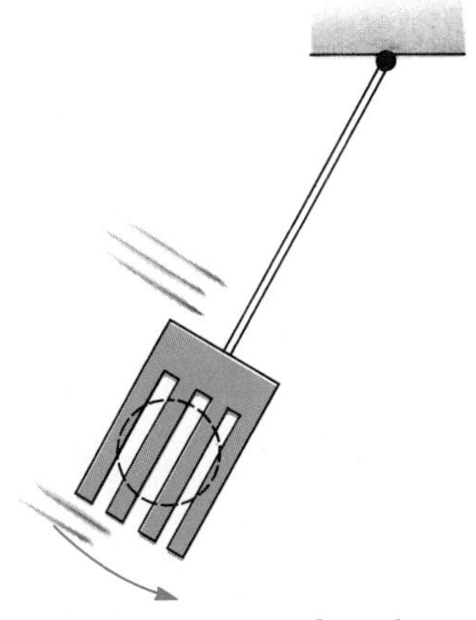

FIGURE 21.18 When slots are cut in the conducting plate, the eddy currents are reduced and the plate swings more freely through the magnetic field.

21.7 SELF-INDUCTANCE

Consider a circuit consisting of a switch, a resistor, and a source of emf, as in Figure 21.19. When the switch is closed, the current does not immediately change from zero to its maximum value, $\mathcal{E}/R$. The law of electromagnetic induction, Faraday's law, prevents this from occurring. What happens is the following. As the current increases with time, the magnetic flux through the loop due to this current also increases. This increasing flux induces an emf in the circuit that opposes the change in magnetic flux. By Lenz's law, the induced electric field in the loop must therefore be opposite the direction of flow of current. That is, the induced emf is in the direction indicated by the dashed battery in Figure 21.19. The net potential difference across the resistor is the emf of the battery minus the opposing induced emf. As the magnitude of the current increases, the *rate* of increase becomes smaller and hence the induced emf decreases. This opposing emf results in a gradual increase in the current. For the same reason, when the switch is opened, the current gradually decreases to zero. This effect is called **self-induction** because the changing flux through the circuit arises from the circuit itself. The emf that is set up in this case is called a **self-induced emf.**

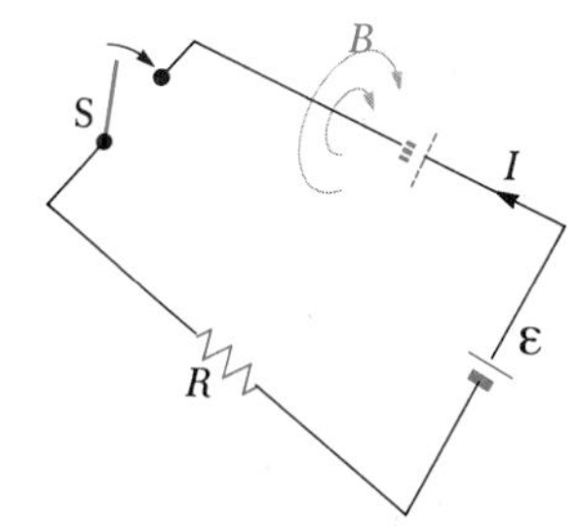

FIGURE 21.19 After the switch in the circuit is closed, the current produces its own magnetic flux through the loop. As the current increases toward its equilibrium value, the flux changes in time and induces an emf in the loop.

As a second example of self-inductance, consider Figure 21.20, which shows a coil wound on a cylindrical form. (A practical device would have several hundred turns.) Assume that the current changes with time. When the current is in the direction shown, a magnetic field is set up inside the coil directed from right to left. As a result, some lines of magnetic flux pass through the cross-sectional area of the coil. The flux through the coil changes as the current changes with time, which induces an emf in the coil. Application of Lenz's law shows that the direction of this induced emf is such as to oppose the change in the current. That is, if the current is increasing, the induced emf will be as pictured in Figure 21.20b and if the current is decreasing, the induced emf will be as shown in Figure 21.20c.

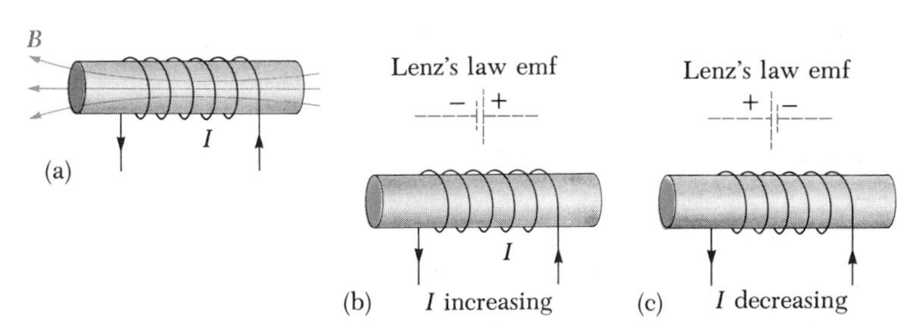

FIGURE 21.20 (a) A current in the coil produces a magnetic field directed to the left. (b) If the current increases, the coil acts as a source of emf directed as shown by the dashed battery. (c) The emf of the coil changes its polarity if the current decreases.

To obtain a quantitative description of self-inductance, first note that Faraday's law says that the induced emf is as given by Equation 21.3:

$$\mathcal{E} = -N\frac{\Delta\Phi}{\Delta t}$$

The magnetic flux is proportional to the magnetic field, which is proportional to the current in the circuit. Thus, *the self-induced emf is always proportional to the time rate of change of the current:*

Self-induced emf

$$\boxed{\mathcal{E} \equiv -L\frac{\Delta I}{\Delta t}} \qquad (21.9)$$

where L is a proportionality constant called the **inductance** of the device. The inductance depends on the cross-sectional area of the coils and other quantities, all of which can be grouped under the general heading of geometric factors. The SI unit of inductance is the **henry (H)**, which from Equation 21.9 is seen to be equal to 1 volt-second per ampere:

$$\boxed{1\ \text{H} = 1\ \text{V·s/A}}$$

Examples 21.6 and 21.7 discuss some simple situations for which self-inductances are easily evaluated. When evaluating self-inductances, it is often convenient to equate Equations 21.3 and 21.9 to find an expression for L:

$$N\frac{\Delta\Phi}{\Delta t} = L\frac{\Delta I}{\Delta t}$$

Inductance

$$\boxed{L = N\frac{\Delta\Phi}{\Delta I} = \frac{N\Phi}{I}} \qquad (21.10)$$

where we assume $\Phi = 0$ and $I = 0$ at $t = 0$.

EXAMPLE 21.6 Inductance of a Solenoid

Find the inductance of a uniformly wound solenoid with N turns and length ℓ. Assume that ℓ is large compared with the radius and that the core of the solenoid is air.

Solution In this case, we can take the interior field to be uniform and given by Equation 20.16:

$$B = \mu_0 nI = \mu_0\frac{N}{\ell}I$$

where n is the number of turns per unit length, N/ℓ. The flux through each turn is

$$\Phi = BA = \mu_0\frac{N}{\ell}AI$$

where A is the cross-sectional area of the solenoid. From this expression and Equation 21.10, we find that

$$L = \frac{N\Phi}{I} = \frac{\mu_0 N^2 A}{\ell} \qquad (21.11)$$

This shows that L depends on geometric factors and μ_0 and is proportional to the square of the number of turns. Since $N = n\ell$, we can also express the result in the form

$$L = \mu_0 \frac{(n\ell)^2}{\ell} A = \mu_0 n^2 A\ell = \mu_0 n^2(\text{volume}) \qquad (21.12)$$

where $A\ell$ is the volume of the solenoid.

EXAMPLE 21.7 Calculating Inductance and Self-induced emf

(a) Calculate the inductance of a solenoid containing 300 turns if the length of the solenoid is 25 cm and its cross-sectional area is 4 cm^2 = 4×10^{-4} m^2.

Solution Using Equation 21.11, we get

$$L = \frac{\mu_0 N^2 A}{\ell} = (4\pi \times 10^{-7}\ \text{T·m/A}) \frac{(300)^2(4 \times 10^{-4}\ \text{m}^2)}{25 \times 10^{-2}\ \text{m}}$$

$$= 1.81 \times 10^{-4}\ \text{T·m}^2\text{/A} = 0.181\ \text{mH}$$

(b) Calculate the self-induced emf in the solenoid described in (a) if the current through it is decreasing at the rate of 50 A/s.

Solution Equation 21.9 ($\mathcal{E} = -L$hair $\Delta I/\Delta t$) can be combined with $\Delta I/\Delta t = -50$ A/s to obtain

$$\mathcal{E} = -L\frac{\Delta I}{\Delta t} = -(1.81 \times 10^{-4}\ \text{H})(-50\ \text{A/s}) = 9.05\ \text{mV}$$

---

*21.8 *RL* CIRCUITS

A circuit element that has a large inductance, such as a closely wrapped coil of many turns, is called an **inductor.** The circuit symbol for an inductor is ꝏ. We shall always assume that the self-inductance of the remainder of the circuit is negligible compared with that of the inductor.

In order to gain some insight into the effect of an inductor in a circuit, consider the two circuits shown in Figure 21.21. Figure 21.21a consists of a resistor connected to the terminals of a battery. Applying Ohm's law to this circuit gives

$$\boxed{\mathcal{E} = RI} \qquad (21.13)$$

In the past, we have interpreted resistance as a measure of the *opposition to the current.*

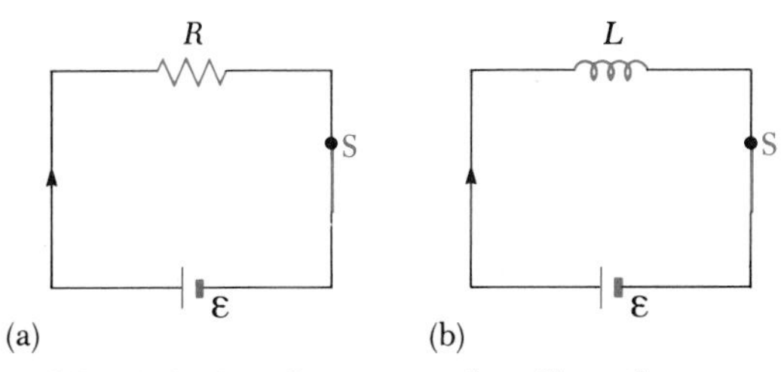

FIGURE 21.21 Comparing the effect of a resistor to that of an inductor in a simple circuit.

Now consider the circuit shown in Figure 21.21b, consisting of an inductor connected to the terminals of a battery. At the instant the switch in this circuit is closed, the emf of the battery is equal to the back emf generated in the coil. Thus we have

Back emf

$$\mathcal{E} = -L\frac{\Delta I}{\Delta t} \tag{21.14}$$

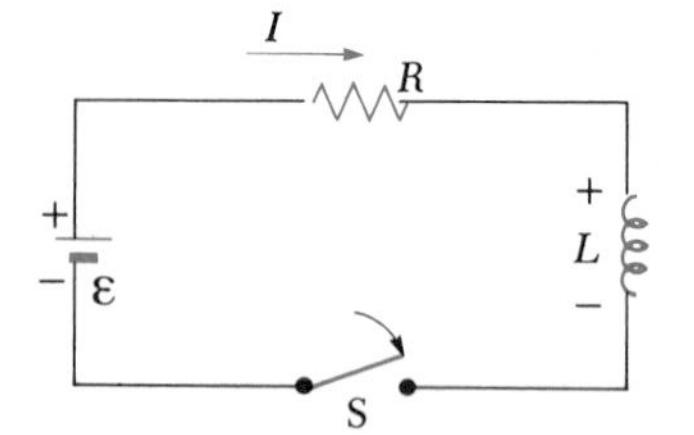

FIGURE 21.22 A series *RL* circuit. As the current increases toward its maximum value, the inductor produces an emf that opposes the increasing current.

By comparing this equation with Equation 21.13, we see that the position of R is now occupied by L and the position of I is now occupied by $\Delta I/\Delta t$. As a result, we can interpret L as a measure of the *opposition to the rate of change in current.*

Consider the circuit consisting of a resistor, inductor, and battery shown in Figure 21.22. Suppose the switch is closed at $t = 0$. The current will begin to increase, but the inductor will produce an emf that opposes the increasing current. Thus, the current is unable to change from zero to its maximum value of $\mathcal{E}/R$ instantaneously. This induced emf has its maximum value when the switch is first closed. That is, Equation 21.14 shows that the induced emf is a maximum when the current is changing at its most rapid rate, which occurs when the switch is first closed. As the current approaches its steady-state value, the back emf of the coil falls off because the current is changing more slowly with time. Finally, when the current reaches its steady-state value, the rate of change is zero and the back emf is also zero. Figure 21.23 represents a plot of current in the circuit as a function of time.[1] This plot is very similar to that of the charge on the plate of a capacitor as a function of time discussed in Section 19.5. In that case, we found it convenient to introduce a quantity called the time constant of the circuit, which told us something about the time required for the capacitor to approach its steady-state charge. In the same fashion, time constants are defined for circuits containing resistors and inductors. The **time constant** in an *RL* circuit is defined as

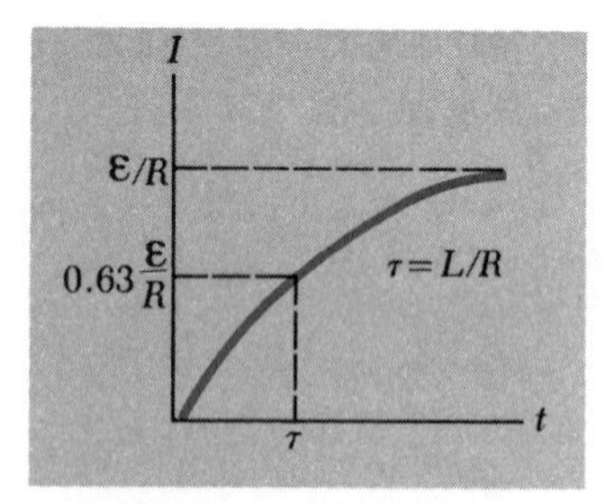

FIGURE 21.23 Plot of the current versus time for the *RL* circuit shown in Figure 21.22. The switch is closed at $t = 0$, and the current increases toward its maximum value, $\mathcal{E}/R$. The time constant, τ, is the time it takes the current to reach 63% of its maximum value.

$$\tau \equiv \frac{L}{R} \tag{21.15}$$

The time required for the current in the circuit to reach 63% of its final value, $\mathcal{E}/R$, is the time constant, τ, for the circuit.

EXAMPLE 21.8 The Time Constant for an *RL* Circuit

The circuit shown in Figure 21.24 consists of a 30-mH inductor, a 6-Ω resistor, and a 12-V battery. The switch is closed at $t = 0$. Find (a) the time constant of the circuit and (b) the current after one time constant has elapsed.

Solution (a) The time constant is given by Equation 21.15:

$$\tau = \frac{L}{R} = \frac{30 \times 10^{-3}\text{ H}}{6\ \Omega} = 5\text{ ms}$$

[1]The equation for the current in the circuit as a function of time is

$$I = \frac{\mathcal{E}}{R}(1 - e^{-Rt/L})$$

(b) After a time equal to one time constant, the current in the circuit has risen to 63% of its final value. Thus, the current is

$$I = 0.63\frac{\mathcal{E}}{R} = (0.63)\left(\frac{12\text{ V}}{6\ \Omega}\right) = 1.26\text{ A}$$

EXERCISE What is the voltage drop across the resistor (a) at $t = 0$ and (b) after one time constant?

Answer (a) 0 (b) 7.56 V.

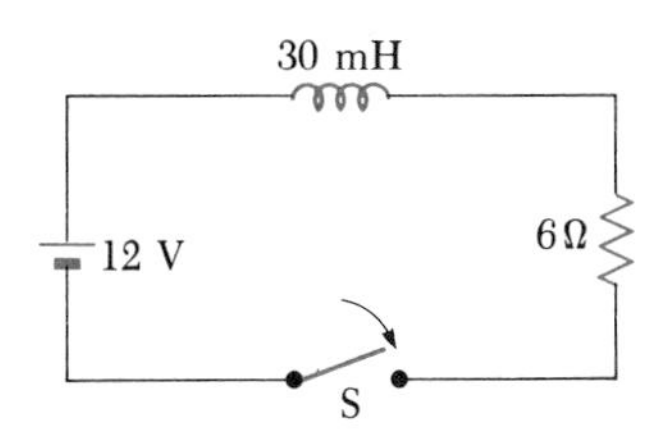

FIGURE 21.24 Example 21.8.

21.9 ENERGY STORED IN A MAGNETIC FIELD

The induced emf set up by an inductor prevents a battery from establishing a current in a circuit instantaneously. Hence, a battery has to do work to produce a current. We can think of this work needed to produce the current as being energy stored by the inductor in its magnetic field. In a manner quite similar to that we used in Section 17.8 to find the energy stored by a capacitor, we find that the energy stored by an inductor is

$$PE_L = \frac{1}{2}LI^2 \quad (21.16)$$

Energy stored in an inductor

Note that the result is similar in form to the expression for the energy stored in a capacitor, which is

$$PE_C = \frac{1}{2}CV^2$$

Energy stored in a capacitor

SUMMARY

The **magnetic flux, Φ**, through a closed loop is defined as

$$\Phi \equiv BA\cos\theta \quad (21.1)$$

magnetic field
angle btwn B
cross sectional area

where B is the strength of the uniform magnetic field, A is the cross-sectional area of the loop, and θ is the angle between $\mathbf{B}$ and the direction perpendicular to the plane of the loop.

Faraday's law of induction states that the instantaneous emf induced in a circuit is equal to the rate of change of magnetic flux through the circuit:

$$\mathcal{E} = -N\frac{\Delta\Phi}{\Delta t} \quad (21.3)$$

N= # of loops

where N is the number of loops in the circuit.

Lenz's law states that the polarity of the induced emf is such that it produces a current whose magnetic field opposes the *change* in magnetic flux through a circuit.

If a conducting bar of length ℓ moves through a magnetic field with a speed v such that $\boldsymbol{B}$ is perpendicular to the bar, the emf induced in the bar, often called a **motional emf,** is

$$\mathcal{E} = B\ell v \tag{21.4}$$

When a coil of wire having N turns, each of area A, rotates with constant angular velocity ω in a uniform magnetic field $\boldsymbol{B}$, the emf induced in the coil is

$$\mathcal{E} = NAB\omega \sin \omega t \tag{21.7}$$

When the current in a coil changes with time, an emf is induced in the coil according to Faraday's law. This **self-induced emf** is defined by the expression

$$\mathcal{E} \equiv -L\frac{\Delta I}{\Delta t} \tag{21.9}$$

where L is the **inductance** of the coil. Inductance has the SI unit of henry (H), where 1 H = 1 V·s/A.

The **inductance,** L, of a coil can be found from the expression

$$L = \frac{N\Phi}{I} \tag{21.10}$$

where N is the number of turns on the coil, I is the current in the coil, and Φ is the magnetic flux through the coil.

If a resistor and inductor are connected in series to a battery and a switch is closed at $t = 0$, the current in the circuit does not rise instantly to its maximum value. After one **time constant,** $\tau = L/R$, the current in the circuit is 63% of its final value, $\mathcal{E}/R$.

The **energy stored** in the magnetic field of an inductor carrying a current I is

$$PE_L = \frac{1}{2}LI^2 \tag{21.16}$$

ADDITIONAL READING

H. Kondo, "Michael Faraday," *Sci. American*, October 1953, p. 90.

D. K. C. McDonald, *Faraday, Maxwell and Kelvin*, New York, Doubleday Anchor, 1964.

H. L. Sharlin, "From Faraday to the Dynamo," *Sci. American*, May 1961, p. 107.

G. Shiers, "The Induction Coil," *Sci. American*, May 1971, p. 80.

QUESTIONS

1. A circular loop is located in a uniform and constant magnetic field. Describe how an emf can be induced in the loop in this situation.
2. Will dropping a magnet down a long copper tube produce a current in the tube? Explain.

3. A spacecraft circling the earth has a coil of wire in it. An astronaut notes that there is a current in the coil although no battery is connected to it and there are no magnets on the spacecraft. What is causing the current?
4. What happens when the coil of a generator is rotated at a faster rate?
5. Could a current be induced in a coil by rotating a magnet inside the coil? If so, how?
6. As the conducting bar in Figure 21.25 moves to the right, an electric field directed downward is set up in the conductor. If the bar were moving to the left, explain why the electric field would be upward.

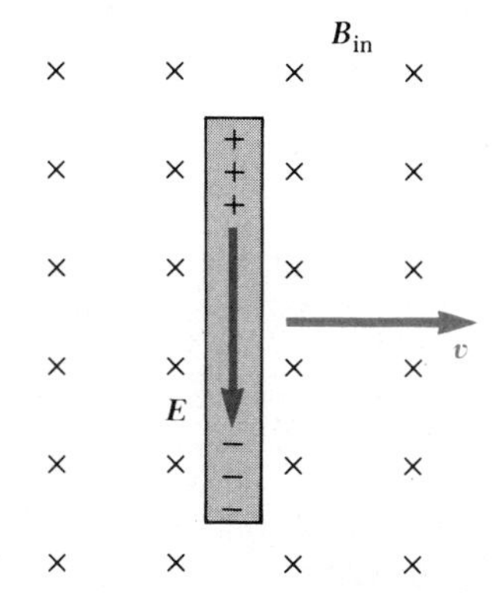

FIGURE 21.25 Questions 6 and 7

7. As the bar in Figure 21.25 moves perpendicular to the magnetic field, is an external force required to keep it moving with constant velocity? Explain.
8. We saw in the last chapter that a steady magnetic field does not change the energy of a charge moving in it. Can a changing magnetic field affect the energy of a charge?
9. Magnetic storms on the sun can cause difficulties with communications here on the earth. Why do these sunspots affect us in this way?
10. Would you expect the tape from a tape recorder to be attracted to a magnet? (Try it, but not with a recording you wish to save.)
11. A circuit containing a coil, resistor, and battery is in a steady state, that is, the current has reached a constant value. Does the coil have an inductance? Does the coil affect the value of the current in the circuit?
12. Suppose the switch in a circuit containing an inductor, a battery, and a resistor has been closed for a long time and is suddenly opened. Does the current instantaneously drop to zero? Why does a spark usually appear at the switch contacts when the switch is opened?
13. If the current in an inductor is doubled, by what factor does the stored energy change?
14. The bar in Figure 21.26 moves to the right with a velocity v, and a uniform, constant magnetic field is directed *outward* (represented by the dots). Why is the induced current clockwise? If the bar were moving to the left, what would be the direction of the induced current?

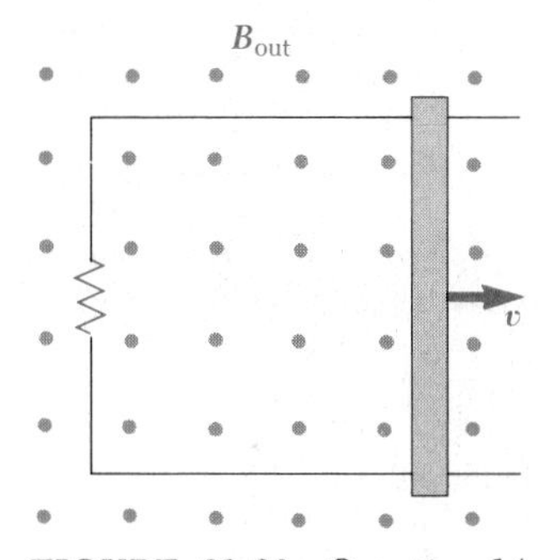

FIGURE 21.26 Question 14

15. A piece of aluminum is dropped vertically downward between the poles of an electromagnet. Does the magnetic field affect the velocity of the aluminum?
16. A bar magnet is dropped toward a conducting ring laying on the floor. As the magnet falls toward the ring does it move as a freely falling body?

PROBLEMS

Section 21.1 Induced Emf and Magnetic Flux

1. A magnetic field of strength 0.25 T is directed perpendicular to a circular loop of wire of radius 30 cm. Find the magnetic flux through the area.

2. Find the flux of the earth's magnetic field of magnitude 5×10^{-5} T through a square loop of area 20 cm^2 when the field is (a) perpendicular to the plane of the loop, (b) when the field makes a 30° angle with the normal to the plane of the loop, and (c) when the field makes a 90° angle to the normal to the plane.

3. A square loop 2 m on a side is placed in a magnetic field of strength 0.3 T. If the field makes an angle of 50° with the normal to the plane of the loop, as in Figure 21.2, determine the magnetic flux through the loop.

4. A long straight wire lies in the plane of a circular coil of radius 0.010 m. The wire carries a current of 2 A and is placed along a diameter of the coil. (a) What is the net flux through the coil? (b) If the wire passes through the center of the coil and is perpendicular to the plane of the coil, find the net flux through the coil.

• **5.** A solenoid 4 cm in diameter and 20 cm in length has 250 turns and carries a current of 15 A. Calculate the magnetic flux through the circular cross-sectional area of the solenoid.

• **6.** The solenoid of Problem 5 is surrounded by a single loop of wire 10 cm in diameter. The loop is positioned perpendicular to and centered on the axis of the solenoid as shown in Figure 21.27. Find the magnetic flux through the loop of wire. (The field outside the solenoid is small enough to be negligible.)

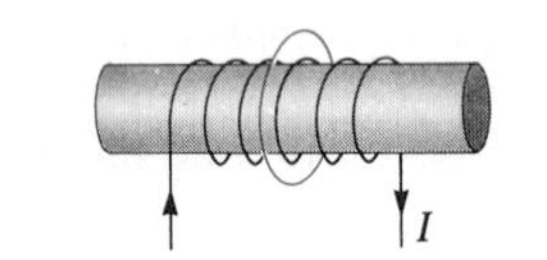

FIGURE 21.27 Problem 6

Section 21.2 Faraday's Law of Induction

7. The magnetic flux through a loop consisting of two turns of wire is changing at the rate of 3 Wb/s. Find the magnitude of the induced emf in the loop.

8. A coil of radius 20 cm is placed in an external magnetic field of strength 0.20 T such that the plane of the coil is perpendicular to the field. The coil is pulled out of the field in a time of 0.3 s. Find the average induced emf during this time.

9. A coil has a radius of 15 cm and is placed in an external magnetic field of strength 0.25 T with the plane of the coil perpendicular to the field direction. The field increases to 0.50 T in a time of 0.7 s. (a) Find the magnitude of the average induced emf during this time. (b) If the average current during this time was 0.8 A, find the resistance of the coil.

10. A loop of wire of radius 0.30 m lies such that an external magnetic field of strength +0.30 T is perpendicular to the loop. The field changes to −0.20 T in a time of 1.5 s. (The plus and minus signs here refer to opposite directions through the coil.) Find the magnitude of the average induced emf in the coil during this time.

11. The flexible loop shown in Figure 21.28 has a radius of 12 cm and is in a magnetic field of strength 0.15 T. The loop is grasped at points A and B and stretched until it closes. If it takes 0.2 s to close the loop, find the magnitude of the average induced emf in it during this time.

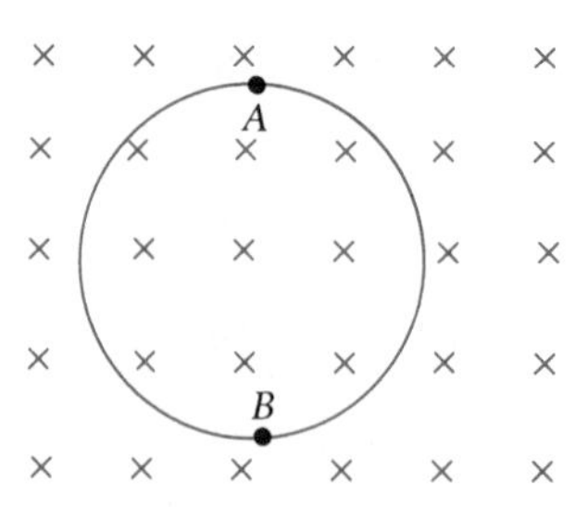

FIGURE 21.28 Problem 11

• **12.** A 300-turn solenoid with a length of 20 cm and a radius of 1.5 cm carries a current of 2 A. A second coil of 4 turns is wrapped tightly about this solenoid such that it can be considered to have the same radius as the solenoid. (a) Find the change in the magnetic flux through the coil, and (b) the magnitude of the average induced emf in the coil when the current in the solenoid increases to 5 A in a period of 0.9 s.

• **13.** A square coil of wire having a single turn 1 cm on a side is placed inside a solenoid that has a circular cross section of radius 3 cm, as shown in Figure 21.29. The solenoid is 20 cm long and wound with 100 turns of wire. (a) When the current in the solenoid is 3 A, find the flux through the coil. (b) If the current in the solenoid is reduced to zero in 3 s, find the magnitude of the average induced emf in the coil.

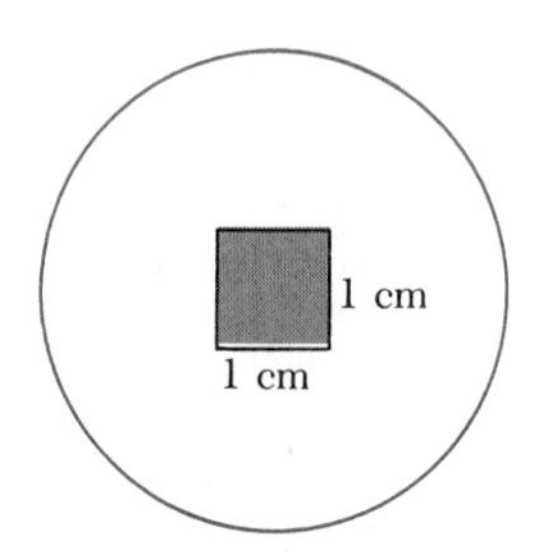

FIGURE 21.29 Problem 13

•• **14.** A coil formed by wrapping 50 turns of wire in the shape of a square is positioned in a magnetic field so that the normal to the plane of the coil makes an angle of 30° with the direction of the field. It is observed that, if the magnitude of the magnetic field is increased uniformly from 200 μT to 600 μT in 0.4 s, an emf of 80 mV is induced in the coil. What is the total length of the wire forming the coil?

Section 21.3 Motional Emf

15. A sliding bar moves with a velocity of 20 cm/s toward the right along a U-shaped piece of metal, as shown in Figure 21.7. The magnetic field is out of the page and has a strength of 0.2 T. Find the induced emf in the circuit if the length of the bar is 30 cm.

16. An airplane with a wingspan of 40 m flies horizontal to the earth at a location where the downward component of the earth's magnetic field is 0.6×10^{-4} T. At what speed would the plane have to fly to produce a difference in potential of 1.5 V between the tips of its wings? Is this likely to occur in a typical airplane flight?

17. Over a region where the vertical component of the earth's magnetic field is 40 μT, a 50-cm length of wire is moved with a speed of 5 m/s along U-shaped rails as shown in Figure 21.7. Find (a) the induced emf in the circuit, and (b) the induced current in the circuit if there is a resistor of 5 Ω in the U-shaped track. The resistance of the wire and track can be considered negligible.

• **18.** For the circuit described in Problem 17, find (a) the average power supplied to the circuit by the induced emf, (b) the power dissipated in the 5-Ω resistor, (c) the force required to keep the length of wire moving at a constant speed, and (d) the power supplied to the circuit by the agent exerting the force calculated in part (c)?

• **19.** A steel beam 12 m in length is accidentally dropped by a construction crane from a height of 9 m. The horizontal component of the earth's magnetic field over the region is 18 μT. What is the induced emf in the beam just before impact with the earth assuming its long dimension remains in a horizontal plane and oriented perpendicular to the horizontal component of the earth's magnetic field?

Section 21.4 Lenz's Law Revisited

20. A bar magnet is held above a loop of wire in a horizontal plane as shown in Figure 21.30. The south end of the magnet is toward the loop of wire. The magnet is dropped toward the loop. Find the direction of the current through the resistor (a) while the magnet is falling toward the loop, and (b) after the magnet has passed through the loop and moves away from it.

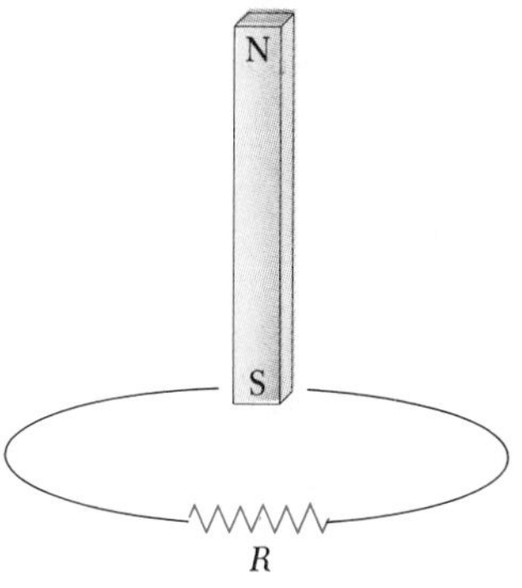

FIGURE 21.30 Problem 20

21. A bar magnet is positioned near a coil of wire as shown in Figure 21.31. What is the direction of the current through the resistor when the magnet is moved (a) to the left and (b) to the right?

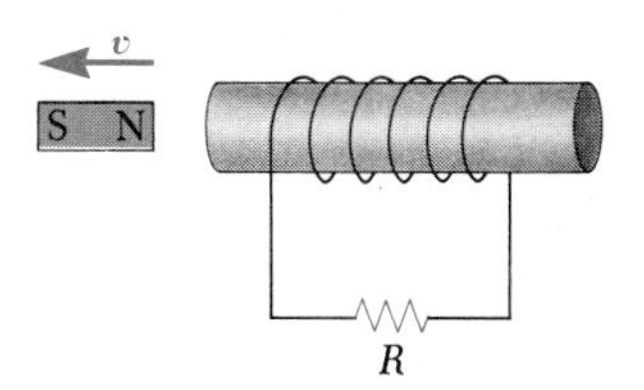

FIGURE 21.31 Problem 21

22. Find the direction of the current through the resistor in Figure 21.32 (a) at the instant the switch is closed, (b) after the switch has been closed for several minutes, and (c) at the instant the switch is opened.

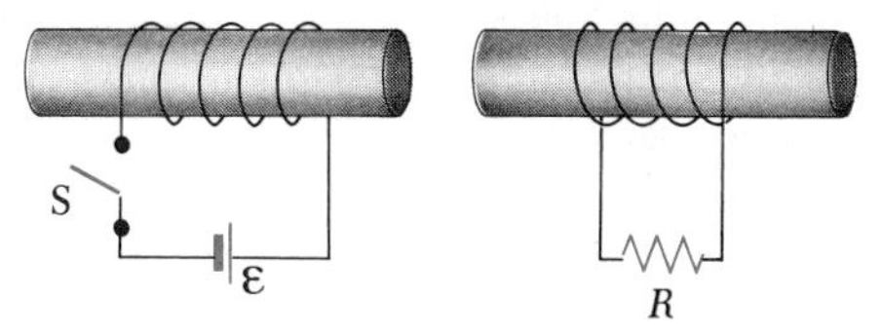

FIGURE 21.32 Problem 22

23. Find the direction of the current in the resistor R in Figure 21.33 when the following steps are taken in the order given. (a) The switch is closed. (b) The variable resistance in series with the battery is decreased. (c) The circuit containing resistor R is moved to the left. (d) The switch is opened.

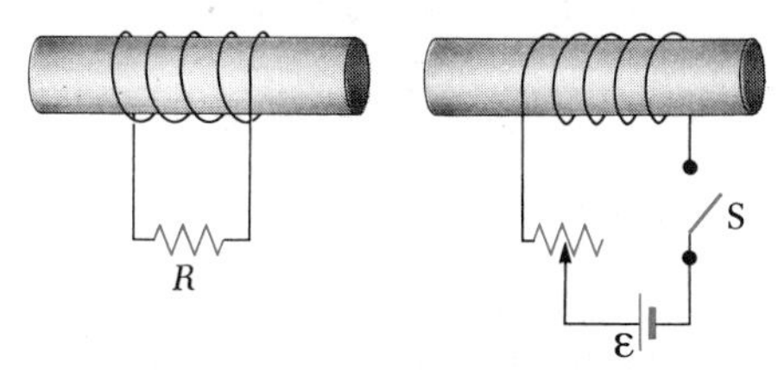

FIGURE 21.33 Problem 23

24. What is the direction of the current induced in the resistor at the instant the switch in the circuit of Figure 21.34 is closed?

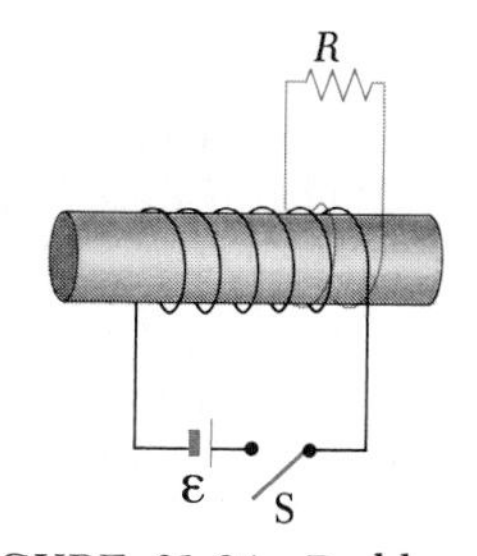

FIGURE 21.34 Problem 24

25. What is the direction of the current induced in the resistor when the current in the long, straight wire shown in Figure 21.35 decreases rapidly to zero?

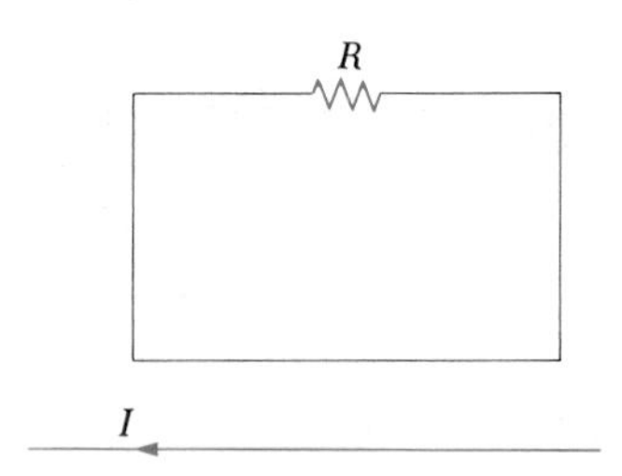

FIGURE 21.35 Problem 25

26. A square loop of wire is placed perpendicular to a magnetic field directed as shown in Figure 21.36a. Find the *direction* of the induced current through R while the loop is being rotated from the position shown in (a) to that shown in Figure 21.36b.

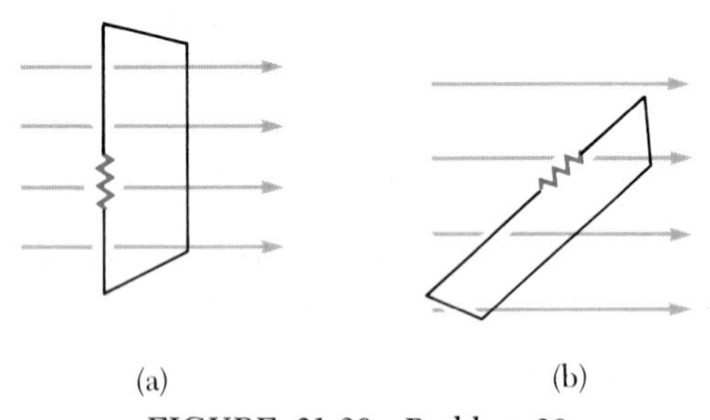

FIGURE 21.36 Problem 26

Section 21.5 Generators and Motors

27. The alternating voltage of a generator is represented by the equation $\mathcal{E} = 240 \sin 500t$, where $\mathcal{E}$ is in volts and t is in seconds. Find the frequency of the voltage and the maximum voltage output of the source.

28. Coils rotating in a magnetic field are often used to measure unknown magnetic fields. As an example, consider a coil of radius 1 cm with 50 turns that is rotated about an axis perpendicular to the field at a rate of 20 Hz. If the maximum induced emf in the coil is 3 V, find the strength of the magnetic field.

29. A 500-turn circular coil of radius 20 cm is rotating about an axis perpendicular to a magnetic field of 0.1 T. What angular velocity will produce a maximum induced emf of 2 mV?

30. An ac generator with its terminals shorted together consists of 40 turns of wire of area 0.12 m^2 and with total resistance 30 Ω. The loop rotates in a magnetic field of 0.1 T at a constant frequency of 60 Hz. (a) Find the maximum induced emf, (b) the maximum induced current, (c) an expression for the time variation of $\mathcal{E}$, and (d) an expression for the time variation of the induced current.

31. When the coil of a motor is rotating at maximum speed, the current in the windings is 4 A. When the motor is first turned on, the current in the windings is 11 A. If the motor is operated at 120 V, find the back emf in the coil and the resistance of the windings.

32. A motor has coils with a resistance of 30 Ω and operates from a voltage of 240 V. When the motor is operating at its maximum speed, the back emf is 145 V. Find the current in the coils (a) when the motor is first turned on and (b) when the motor has reached maximum speed. (c) If the current in the motor is 6 A at some instant, what is the back emf at this time?

•• 33. In a model ac generator, a 500-turn rectangular coil of dimensions 8 cm by 20 cm rotates at 120 rev/min in a uniform magnetic field of 0.6 T. (a) What is the maximum emf induced in the coil? (b) What is the instantaneous value of the emf in the coil at $t = (\pi/32)$ s? Assume that the emf is zero at $t = 0$. (c) What is the smallest value of t for which the emf will have its maximum value?

Section 21.7 Self-Inductance

34. A coil has an inductance of 3 mH and a current through it changes from 0.2 A to 1.5 A in a time of 0.2 s. Find the magnitude of the average induced emf in the coil during this time.

35. A Slinky toy spring has a radius of 4 cm and an inductance of 275 μH when extended to a length of 1.5 m. What is the total number of turns in the spring?

36. A solenoid of radius 2.5 cm has 400 turns and a length of 20 cm. (a) Find the inductance of the solenoid, and (b) the rate at which current must change through it to produce an emf of 75 mV.

37. At the instant the emf is 75 mV for the solenoid of Problem 36, find the rate of change of magnetic flux through the cross-sectional area of the solenoid.

• 38. An emf of 24 mV is induced in a 500-turn coil at an instant when the current is 4 A and is changing at a rate of 10 A/s. What is the total magnetic flux through the coil?

Section 21.8 *RL* Circuits

39. A 6-V battery is connected in series with a resistor and an inductor. The series circuit has a time constant of 600 μs, and the maximum current is 300 mA. What is the value of the inductance?

• **40.** A 25-mH inductor, an 8-Ω resistor and a 6-V battery are connected in series. The switch is closed at $t = 0$. Find the voltage drop across the resistor (a) at $t = 0$ and (b) after one time constant has passed. Also, find the voltage drop across the inductor (c) at $t = 0$ and (d) after one time constant has elapsed.

• **41.** The switch in a series *RL* circuit in which $R = 6\ \Omega$, $L = 3$ H, and $\mathcal{E} = 24$ V is closed at $t = 0$. (a) What is the maximum current in the circuit? (b) What is the current when $t = 0.5$ s?

• **42.** This problem assumes coverage of footnote 1. Find the current in the circuit of Problem 41 at $t = 0.7$ s.

Section 21.9 Energy Stored in a Magnetic Field

43. How much energy is stored in a 70-mH inductor at the instant when the current is 2 A?

• **44.** A 24-V battery is connected in series with a resistor and an inductor, where $R = 8\ \Omega$ and $L = 4$ H. What energy is stored in the inductor (a) when the current reaches its maximum value, and (b) one time constant after the switch is closed?

ADDITIONAL PROBLEMS

45. A 50 turn-rectangular coil of dimensions 0.2 m by 0.3 m is rotated at 90 rad/s in a magnetic field so that the axis of the coil is perpendicular to the direction of the field. The maximum emf induced in the coil is 0.5 V. What is the magnitude of the field?

46. Show that the units of the inductive time constant are seconds.

47. A tightly wound circular coil has 50 turns, each of radius 0.2 m. A uniform magnetic field is introduced perpendicular to the plane of the coil. If the field increases in strength from 0 to 0.3 T in 0.4 s, what average emf is induced in the windings of the coil?

48. Find the self-inductance of a coil if an emf of 100 mV is produced when the current changes from 0 to 1.5 A in a time of 0.3 s.

49. An automobile starter motor draws a current of 3.5 A from a 12-V battery when operating at normal speed. A broken pulley locks the armature in position and the current increases to 18 A. What is the back emf of the motor?

• **50.** A circular coil of radius 15 cm and 5 turns is oriented with the plane of the coil perpendicular to a uniform magnetic field of 0.15 T. This field increases at a uniform rate of 0.20 T in 3 s. If the resistance of the coil is 8 Ω, find the amount of charge that passes through the coil during this 3-s interval.

• **51.** Consider a 100-turn rectangular coil of cross-sectional area 0.06 m^2 rotating with an angular velocity of 20 rad/s about an axis perpendicular to a magnetic field of 2.5 T. Plot the induced voltage as a function of time over one complete period of rotation.

• **52.** A 300-turn inductor has a radius of 5 cm and a length of 20 cm. Find the energy stored in it when the current is 0.5 A.

• **53.** An open hemispherical surface of radius 0.1 m is located in a magnetic field of 0.15 T. The circular cross section of the surface is perpendicular to the direction of the field. Calculate the magnetic flux through the surface.

• **54.** The coil of Problem 52 is connected in a series circuit with a 1000 Ω resistor and a 12-V battery. Find the energy stored in the inductor (a) after one time constant, and (b) after two time constants.

• **55.** Shown in Figure 21.37 is a graph of the induced emf versus time for a coil of *N* turns rotating with angular velocity ω in a uniform magnetic field directed perpendicular to the axis of rotation of the coil. Copy this sketch (to a larger scale), and on the same set of axes show the graph of emf versus *t* when (a) the number of turns in the coil is doubled, (b) the angular velocity is doubled, and (c) the angular velocity is doubled while the number of turns in the coil is halved.

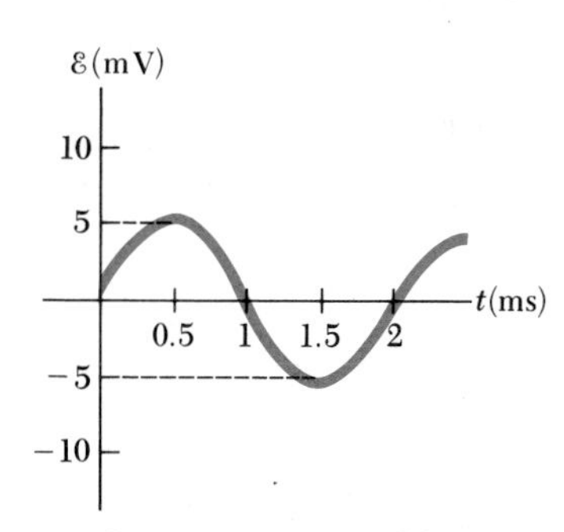

FIGURE 21.37 Problem 55

• **56.** Two coils, A and B, are wound using equal lengths of wire. Each coil has the same number of turns per unit length, but coil A has twice as many turns as coil B. What is the ratio of the self-inductance of A to the self-inductance of B? (Note: The radii of the two coils are not equal.)

•• **57.** A single-turn, circular loop of radius 0.2 m is coaxial with a long solenoid of radius 0.05 m, length 0.8 m, and having 1600 turns, as in Figure 21.38. The variable resistor is changed so that the solenoid current decreases linearly from 6 A to 1.5 A in 0.2 s. Calculate the induced emf in the circular loop. (The field just outside the solenoid is small enough to be negligible.)

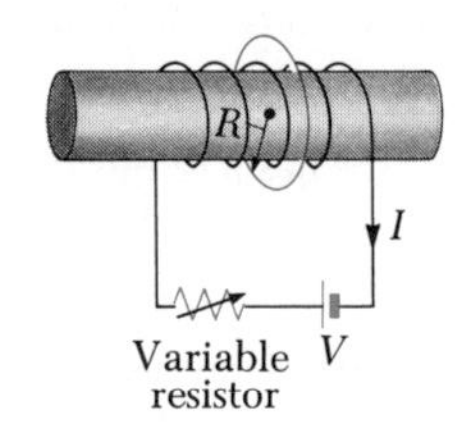

FIGURE 21.38 Problem 57

•• **58.** A series circuit contains a 12-V battery, a 2000-Ω resistor, and a 3-mH inductor. If the switch to the battery is closed at $t = 0$, find the time required for the current in the circuit to reach 90% of its final value.

•• **59.** A rectangular coil of 50 turns, length 0.15 m, and width 0.25 m, and of resistance 12 Ω rotates with an angular velocity 20 rad/s about the y axis in a region where a uniform magnetic field of 1.4 T is directed along the x axis. The rotation is initiated when the plane of the coil is perpendicular to the direction of the magnetic field. Calculate (a) the maximum induced emf in the coil, (b) the value of the induced emf at $t = 0.05$ s, and (c) the torque exerted on the coil by the magnetic field at the instant when the emf is a maximum.

•• **60.** A horizontal wire is free to slide on the vertical rails of a conducting frame as in Fig. 21.39. The wire has a mass m and length ℓ, and the resistance of the circuit is R. If a uniform magnetic field is directed perpendicular to the frame, what is the terminal velocity of the wire as it falls under the force of gravity? (Neglect friction.)

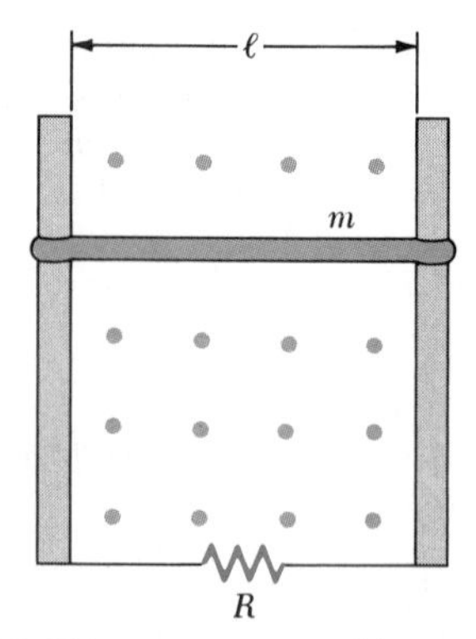

FIGURE 21.39 Problem 60

•• **61.** A thin metal strip is allowed to slide down parallel frictionless rails of negligible resistance connected at the bottom end and elevated at an angle of 30° above the horizontal as in Fig. 21.40. A uniform magnetic field of 2.60 T is directed vertically upward throughout the region. The strip has a mass $m = 40$ g, resistance $R = 30$ Ω, and length between the rails $\ell = 0.2$ m. Calculate the terminal speed achieved by the strip sliding along the incline.

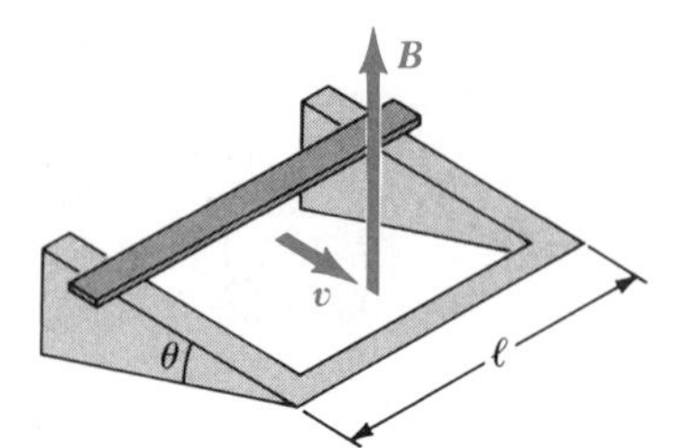

FIGURE 21.40 Problem 61

Essay

Exponential Growth

Albert A. Bartlett *University of Colorado*

If something is growing in size at a constant rate, such as 5 percent per year, we refer to the growth as steady growth. It is also referred to as exponential growth because the size, N, of the growing quantity at some time t in the future is related to its present size, N_0 (at time $t = 0$), by the exponential function

$$N = N_0 e^{kt} \tag{E.1}$$

where $e = 2.718\ldots$ is the base of the natural logarithms and k is the annual percent growth rate, P, divided by 100:

$$k = \frac{P}{100} \tag{E.2}$$

Note that if k is positive, N increases exponentially with time (exponential growth). If k is negative, N decreases exponentially with time (exponential decay). Some examples of exponential decay are radioactive decay (Section 32.4) and the decay of the quantity of charge on the plates of a capacitor as the capacitor is discharging through a resistor (Section 19.5). This essay describes several examples of exponential growth.

We have steady growth whenever the rate of change of a quantity is proportional to the quantity,

$$\boxed{\frac{\Delta N}{\Delta t} = kN} \tag{E.3}$$

Here ΔN is the small change in N during the short time interval Δt. If we can show that a quantity N obeys Equation E.3, then it follows automatically that N will increase in accord with Equation E.1.

If N is the number of dollars in a savings account, then ΔN is the interest added to the account in a time Δt, and k is the interest rate. If I have \$250 in the bank at an annual interest rate of 3% ($k = 0.03$), then the amount added to the account in $\Delta t = 1$ year is

$$\frac{\Delta N}{\Delta t} = 0.03 \times 250 = \$7.50.$$

This is in accord with Equation E.3, so the sum in the account will grow according to the relation

$$N = 250\, e^{0.03t}$$

and when $t = 1$ year this gives

$$N = 250 \times 1.03045 = 257.61.$$

Note that if the interest is compounded once a year, at the end of the year the account will contain \$250 + \$7.50 = \$257.50. However, if the interest is "compounded continuously," the amount follows Equation E.1, and the amount at the end of the year is \$257.61. With the advent of computers, banks are able to compound their interest continuously or "daily."

It has been shown that the number of miles of highways in the United States obeys Equation E.3 so that the number of miles of highway will grow exponentially in accord with Equation E.1.[1]

In steady growth, it takes a fixed length of time for a quantity to grow by a fixed fraction such as 5 percent. From this it follows that it takes a fixed longer length of time for that quantity to grow by 100 percent. Let us calculate the time required for the quantity N to double in value, which is called the **doubling time,** T_2. We can obtain an expression for T_2 by writing Equation E.1 as $N/N_0 = e^{kt}$ and taking the natural logarithm of each side:

$$\ln\left(\frac{N}{N_0}\right) = kt.$$

[1] A. A. Bartlett, *Civil Engineering*. December 1969, pgs 71-72.

If we set $N = 2N_0$ (that is, we double N_0), then T_2 (which is the time t when $N = 2N_0$) is

$$T_2 = \frac{\ln(2N_0/N_0)}{k} = \frac{\ln 2}{k} = \frac{0.693}{k}.$$

Since $k = P/100$, this becomes[2]

$$\boxed{T_2 \approx \frac{70}{P}.} \tag{E.4}$$

EXAMPLE E.1 Compound Interest

Suppose you put \$1 in a savings account at 10 percent annual interest to be compounded continuously for 200 years. How much money would be in the account at the end of the period?

Solution: In this case, $N_0 = \$1$, $k = 10\%/100 = 0.10$ per year, and $t = 200$ years. Therefore, from Equation E.1 we have

$$N = \$1 \times e^{(0.10)(200)} = (\$1) \times e^{20} = \$4.85 \times 10^8$$
$$= \$485 \text{ million!}$$

Now you can see why a famous financier once said that he could not name the seven wonders of the world but surely the eighth wonder would have to be compound interest!

EXAMPLE E.2 The Consequences of Inflation

Let us use the doubling time instead of the quantity e to estimate the consequences of an annual inflation rate of 14 percent that continued for 50 years.

Solution: First, we calculate T_2 from Equation E.3:

$$T_2 = \frac{70}{14} = 5 \text{ years.}$$

This inflation rate will cause prices to double every five years!

In the next step, we calculate the number of doubling times in 50 years:

$$\text{Number of doublings} = \frac{50 \text{ years}}{5 \text{ years/doubling}} = 10.$$

[2] If you wanted to calculate T_3, the time it takes N to triple, it is $T_3 = (100 \ln 3)/P$.

Finally, we count up the consequence of each doubling by making use of the following table:

Number of Doublings	Price Increase Factor
1	$2 = 2^1$
2	$4 = 2^2$
3	$8 = 2^3$
4	$16 = 2^4$
5	$32 = 2^5$
6	$64 = 2^6$
7	$128 = 2^7$
8	$256 = 2^8$
9	$512 = 2^9$
10	$1024 = 2^{10}$

The table shows us that, in ten doubling times, prices will increase by a factor of 1024, which is approximately 1000. Thus, in 50 years of 14 percent annual inflation, the cost of a $4 ticket to the movies would increase to roughly $4000!

It is very convenient to remember that 10 doublings give an increase by a factor of 10^3, that 20 doublings give an increase of a factor of 10^6, that 30 doublings give an increase by a factor of 10^9, and so on.

EXAMPLE E.3 The Increasing Rate of Energy Consumption

For many years before 1975, consumption of electrical energy in the United States grew steadily at a rate of about 7 percent per year. By what factor would consumption increase if this growth rate continued for 40 years?

Solution: In this case, $P = 7$, and so the doubling time from Equation E.3 is

$$T_2 = \frac{70}{7} = 10 \text{ years}$$

and the number of doublings in 40 years is

$$\text{Number of doublings} = \frac{40 \text{ years}}{10 \text{ years/doubling}} = 4.$$

Therefore, in 40 years the amount of power consumed would be $2^4 = 16$ times the amount used today. That is, 40 years from now we would need 16 times as many electric generating plants as we have at the present. Furthermore, if those additional plants are similar to today's, then each day they would consume 16 times as much fuel as is used by our present plants and there would be 16 times as much pollution and waste heat to contend with!

Populations tend to grow steadily. In July of 1987 we saw reports that the population of the earth had reached 5×10^9 people. The world birth rate was estimated to be 28 per 1000 each year while the annual death rate was estimated to be 11 per 1000. Thus for every 1000 people, the population increase each year is $28 - 11 = 17$. For this growth rate we find

$$k = \frac{17}{1000} = 0.017 \text{ per year}$$

$$P = 100k = 1.7\% \text{ per year.}$$

This growth rate seems so small that many people regard it as trivial and inconsequential. A proper perspective of this rate appears only when we calculate the doubling time:

$$T_2 = \frac{70}{1.7} = 41 \text{ years.}$$

This simple calculation indicates that it is most likely that the world population will double within the life expectancy of today's students! At the most elemental level, this means that we have approximately 41 years to double world food production.

EXAMPLE E.4 Annual Increase in World Population

What is the annual increase in the earth's population? Since $\Delta N \ll N$ we can get a good answer from Equation E.3

$$\frac{\Delta N}{\Delta t} = 0.017 \times 5 \times 10^9 = 85 \text{ million per year.}$$

This annual increase in the world population is roughly one third of the population of the United States.

Some illuminating calculations can be made based on the assumption that this rate of growth has been constant and will remain constant. These calculations will demonstrate that the growth rate has not been constant at this value in the past and cannot remain this high for very long.

EXAMPLE E.5 When did Adam and Eve Live?

If we use Equation E.1, setting $N = 5 \times 10^9$, $N_0 = 2$ (Adam and Eve) and $k = 0.017$, we have

$$5 \times 10^9 = 2e^{0.017t}$$

This gives $t = 1273$ years ago, or about 714 A.D.! This result proves that through essentially all of human history the population growth rate was very much smaller than it is today. It must have been near zero through most of human history.

EXAMPLE E.6

The land area of the continents (excluding Antarctica) is 1.24×10^{14} m^2. How long would it take for the population to reach a density of one person per square meter on the continents?

$$1.24 \times 10^{14} = 5 \times 10^9 \, e^{0.017t}.$$

Solving, we find t is slightly less than 600 years!

EXAMPLE E.7

How long would it take for the mass of people to equal the mass of the earth (5.98×10^{24} kg)? (Assume that the mass of an average person is 65 kg.)

$$5.98 \times 10^{24} = (5 \times 10^9 \times 65) \, e^{0.017t}.$$

This gives a value of t of about 1800 years!

These last two examples prove that the growth rate of world population can not stay as high as it presently is for any extended period of time.

Although world agricultural production has been just barely keeping pace with world population growth, millions are malnourished and many people are starving.

However, we will not have to double food production in 41 years if we can lower the worldwide birth rate. If we fail to lower the birth rate and if we also fail to double world food production in 41 years, then the death rate will rise. Dramatic increases in world food production in recent decades are due almost exclusively to the rapid growth of the use of petroleum for powering machinery and for manufacturing fertilizers and insecticides.

Indeed, it has been noted that "modern agriculture is the use of land to convert petroleum into food." The student must wonder how much longer we can continue the long history of approximately steady population growth when our food supplies are tied so closely to dwindling supplies of petroleum.

This brief introduction to the arithmetic of steady growth enables us to understand that, in all biological systems, the normal condition is steady-state condition, where the birth rate is equal to the death rate. Growth is a short-term transient phenomenon that can never continue for more than a short period of time. Yet in the United States, business and government leaders at all levels, from local communities to Washington, D.C., would have us believe that steady growth forever is a goal we can achieve. They would have us believe that we should continue our population growth (the U.S. population increases by 2 million people per year) and the growth in our rates of consumption of natural resources.

In contrast to all this optimism, please remember that someone once noted that "The greatest shortcoming of the human race is our inability to understand the exponential function."[3]

[3]See A. A. Bartlett, "The Exponential Function," *The Physics Teacher*, October 1976 to January 1979, and A. A. Bartlett, "The Forgotten Fundamentals of the Energy Crisis," *Am. J. Physics 46*, 876 (1978).

CHAPTER 22

Alternating Current Circuits

Power lines. (Photo by George E. Jones III)

INTRODUCTION

It is important to understand the basic principles of alternating current (ac) circuits because they are so much a part of our everyday life. Every time we turn on a television set or a stereo, or any of a multitude of other electric appliances, we are calling on alternating currents to provide the power to operate them. We shall begin our study of ac circuits by examining the characteristics of circuits containing a source of emf and a single circuit element: a resistor, or a capacitor, or an inductor. Then we shall examine what happens when these elements are connected in combination with each other. Our discussion will be limited to situations in which the elements are arranged in simple series configurations. We shall conclude the chapter with a discussion of transformers and power transmission.

22.1 RESISTORS IN AN AC CIRCUIT

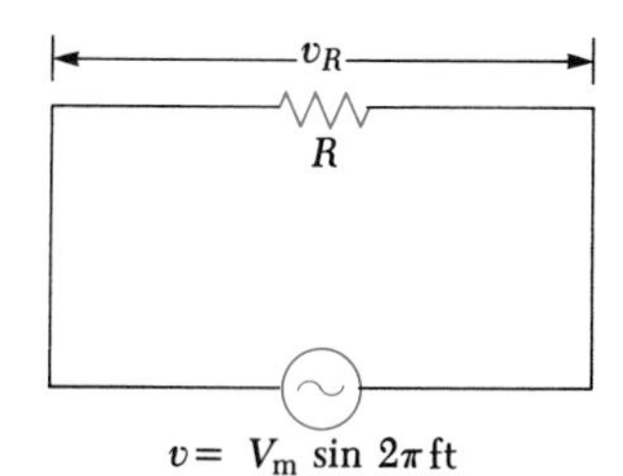

FIGURE 22.1 A series circuit consisting of a resistor R connected to an ac generator, designated by the symbol —⊝—.

An ac circuit consists of combinations of circuit elements and an ac generator, which provides the alternating current. We have seen that the output of an ac generator is sinusoidal and varies with time according to

$$v = V_m \sin 2\pi ft \tag{22.1}$$

where v is the instantaneous voltage, V_m is the maximum voltage of the ac generator, and f is the frequency at which the voltage changes, measured in hertz. We shall first consider a simple circuit consisting of a resistor and an ac generator (designated by the symbol —⊝—), as in Figure 22.1. The voltage across the resistor and the current are shown in Figure 22.2.

Let us briefly discuss the current versus time curve in Figure 22.2. At point a on the curve, the current has a maximum value in one direction, arbitrarily called the positive direction. Between points a and b, the current is decreasing in magnitude but is still in the positive direction. At point b, the current is momentarily zero, and then begins to increase in the opposite (negative) direction between points b and c. At point c, the current has reached its maximum value in the negative direction.

Note that the current and voltage are in step with each other since they vary identically with time.

Since the current and the voltage reach their maximum values at the same time, they are said to be in phase.

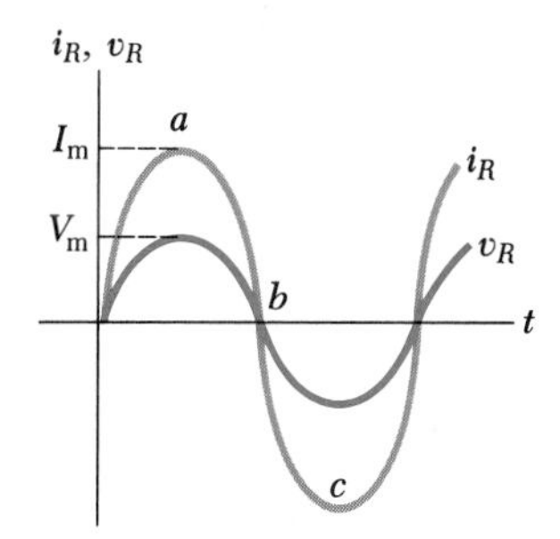

FIGURE 22.2 Plot of current and voltage across a resistor versus time.

Note that *the average value of the current over one cycle is zero.* That is, the current is maintained in one direction (the positive direction) for the same amount of time and at the same magnitude as it is in the opposite direction (the negative direction). However, the direction of the current has no effect on the behavior of the resistor in the circuit. This can be understood by realizing that collisions between electrons and the fixed atoms of the resistor result in an increase in the temperature of the resistor. Although this temperature increase depends on the magnitude of the current, it is independent of its direction.

This discussion can be made quantitative by recalling that the rate at which electrical energy is converted to heat in a resistor, which is the power P, is given by

Power

$$P = i^2R$$

where i is the instantaneous current in the resistor. Since the heating effect of a current is proportional to the *square* of the current, it makes no differ-

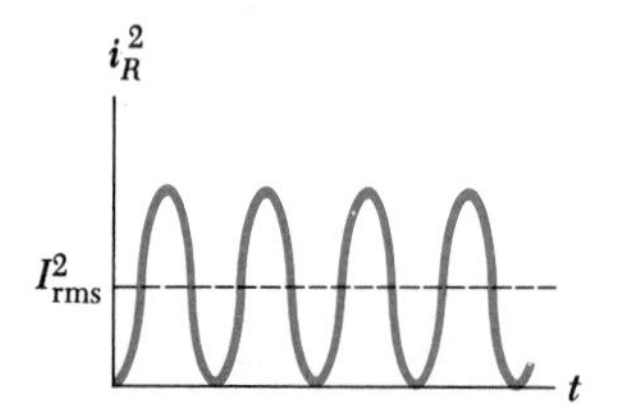

FIGURE 22.3 A plot of the square of the current in a resistor versus time. The rms current is the square root of the average of the square of the current.

ence whether the current is direct or alternating, that is, whether the sign associated with the current is positive or negative. However, the heating effect produced by an alternating current having a maximum value of I_m *is not the same* as that produced by a direct current of the same value. This is explained by the fact that the alternating current is at this maximum value for only a very brief instant of time during a cycle. What is of importance in an ac circuit is an average value of current referred to as the rms current. The term *rms* refers to *root mean square*, which simply means that one takes the square root of the average value of the square of the current. Since i^2 varies as $\sin^2 2\pi ft$, one can show[1] that the average value of i^2 is $\frac{1}{2}I_m^{\ 2}$ (Fig. 22.3). Therefore, the rms current, I, is related to the maximum value of the alternating current, I_m, as

rms current

$$I = \frac{I_m}{\sqrt{2}} = 0.707I_m \tag{22.2}$$

This equation says that an alternating current whose maximum value is 2 A will produce the same heating effect in a resistor as a direct current of $\frac{2}{\sqrt{2}}$ A. Thus, we can say that the average power dissipated in a resistor that carries an alternating current I is $P_{av} = I^2R$, where I is the rms current.

Alternating voltages are also best discussed in terms of rms voltages, and the relationship here is identical to the above, that is, the rms voltage, V, is related to the maximum value of the alternating voltage, V_m, as

rms voltage

$$V = \frac{V_m}{\sqrt{2}} = 0.707V_m \tag{22.3}$$

When one speaks of measuring an ac voltage of 120 V from an electric outlet, one is really referring to an rms voltage of 120 V. A quick calculation using Equation 22.3 shows that such an ac voltage actually has a peak value of about 170 V. In this chapter we shall use rms values when discussing alternating currents and voltages. One reason for this is that ac ammeters and voltmeters are designed to read rms values. Furthermore, we shall find that if we use rms values, many of the equations we use will have the same form

[1]The fact that the square root of the average value of the square of the current is equal to $I_m/\sqrt{2}$ can be shown as follows. The current in the circuit varies with time according to the expression $i = I_m \sin 2\pi ft$, so that $i^2 = I_m^{\ 2} \sin^2 2\pi ft$. Therefore we can find the average value of i^2 by calculating the average value of $\sin^2 2\pi ft$. Note that a graph of $\cos^2 2\pi ft$ versus time is identical to a graph of $\sin^2 2\pi ft$ versus time, except that the points are shifted on the time axis. Thus, the time average of $\sin^2 2\pi ft$ is equal to the time average of $\cos^2 2\pi ft$ when taken over one or more cycles. That is,

$$(\sin^2 2\pi ft)_{av} = (\cos^2 2\pi ft)_{av}$$

With this fact and the trigonometric identity $\sin^2 \theta + \cos^2 \theta = 1$, we get

$$(\sin^2 2\pi ft)_{av} + (\cos^2 2\pi ft)_{av} = 2(\sin^2 2\pi ft)_{av} = 1$$

$$(\sin^2 2\pi ft)_{av} = \frac{1}{2}$$

When this result is substituted in the expression $i^2 = I_m^{\ 2} \sin^2 2\pi ft$, we get $(i^2)_{av} = I^2 = I_m^{\ 2}/2$, or $I = I_m/\sqrt{2}$, where I is the rms current.

as those used in the study of direct current (dc) circuits. Table 22.1 summarizes the notation that will be used in this chapter.

TABLE 22.1 Notation Used in This Chapter

	Voltage	Current
Instantaneous value	v	i
Maximum value	V_m	I_m
rms value	V	I

Consider the series circuit shown in Figure 22.1, consisting of a resistor connected to an ac generator. A resistor limits the current in an ac circuit in the same way it does in a dc circuit. Therefore, Ohm's law is valid for an ac circuit, and we have

$$\boxed{V_R = IR} \tag{22.4}$$

That is, the rms voltage across a resistor is equal to the rms current in the circuit times the resistance. This equation also applies if maximum values of current and voltage are used. That is, the maximum voltage drop across a resistor is equal to the maximum current in the resistor times the resistance.

EXAMPLE 22.1 What is the rms Current?

An ac voltage source has an output given by $v = 200 \sin 2\pi ft$. This source is connected to a 100-Ω resistor as in Figure 22.1. Find the rms current through the resistor.

Solution Compare the expression for the voltage output given above with the general form, $v = V_m \sin 2\pi ft$. We see that the maximum output voltage of the device is 200 V. Thus, the rms voltage output of the source is

$$V = \frac{V_m}{\sqrt{2}} = \frac{200\text{ V}}{\sqrt{2}} = 141\text{ V}$$

Ohm's law can be used in resistive ac circuits as well as in dc circuits. The calculated rms voltage can be used with Ohm's law to find the rms current in the circuit:

$$I = \frac{V}{R} = \frac{141\text{ V}}{100\ \Omega} = 1.41\text{ A}$$

EXERCISE Find the maximum current in the circuit.

Answer 2 A.

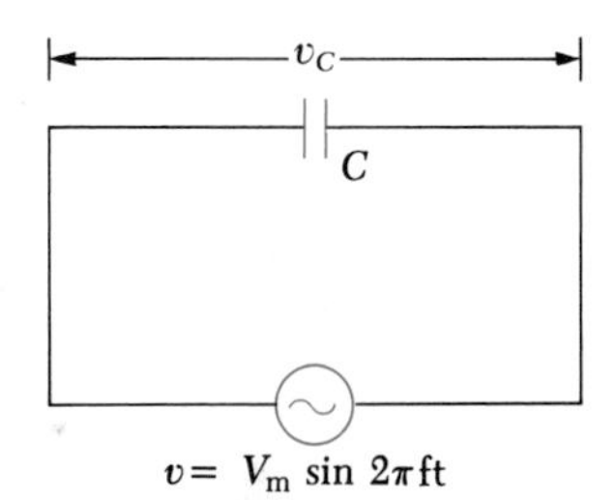

FIGURE 22.4 A series circuit consisting of a capacitor C connected to an ac generator.

22.2 CAPACITORS IN AN AC CIRCUIT

In order to consider the effect of a capacitor on the behavior of a circuit containing an ac voltage source, let us first recall what happens when a capacitor is placed in a circuit containing a dc source, such as a battery. At the instant a switch is closed in a series circuit containing a battery, a resistor, and a capacitor, there is no charge on the plates of the capacitor. Therefore, the motion of charge through the circuit is relatively free, and initially there is a large current in the circuit. As more charge accumulates on the capacitor, the voltage across it increases, opposing the current. After some time interval has elapsed, which depends on the time constant, RC, the current approaches zero. From this, we see that a capacitor in a dc circuit will limit, or impede, the current such that it approaches zero after a brief time.

Now consider the simple series circuit shown in Figure 22.4, consisting of a capacitor connected to an ac generator. Let us sketch a curve of current versus time and one of voltage versus time and then attempt to make the graphs seem reasonable. These graphs are given in Figure 22.5. First, note that the segment of the current curve from a to b indicates that the current

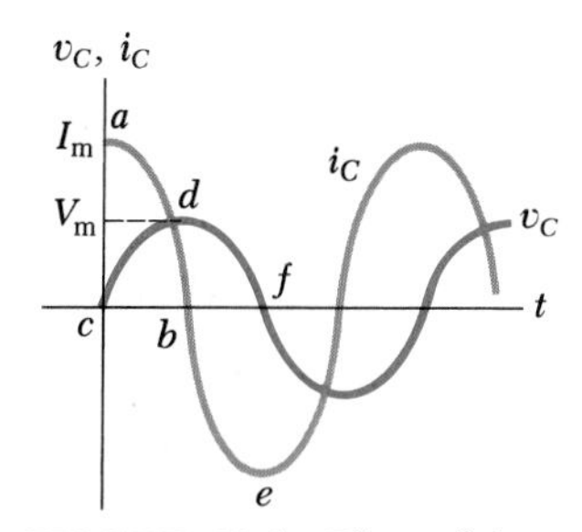

FIGURE 22.5 Plots of the current and voltage across a capacitor versus time in an ac circuit. The voltage lags the current by 90°.

starts out at a rather large value. This can be understood by recognizing that there is no charge on the capacitor at $t = 0$; consequently, there is nothing in the circuit except the resistance of the wires to hinder the flow of charge at this instant. However, the current decreases as the voltage across the capacitor increases, from c to d on the voltage curve. When the voltage is at point d, the current reverses and begins to increase in the opposite direction (from b to e on the current curve). During this time, the voltage across the capacitor decreases in going from d to f because the plates are now losing the charge that they had accumulated earlier. The remainder of the cycle for both voltage and current is a repeat of what happened during the first half cycle. The current reaches a maximum value in the opposite direction at point e on the current curve and then decreases as the voltage across the capacitor builds up.

Note that the current and voltage are not in step with each other, as they are in a purely resistive circuit. The curves of Figure 22.5 indicate that

when an alternating voltage is applied across a capacitor, the voltage reaches its maximum value one quarter of a cycle after the current reaches its maximum value. In this situation, it is common to say that the voltage always lags the current through a capacitor by 90°.

The impeding effect of a capacitor to the flow of current in an ac circuit is expressed in terms of a factor called the **capacitive reactance,** X_C, defined as

Capacitive reactance

$$X_C \equiv \frac{1}{2\pi f C} \tag{22.5}$$

It is left as a problem to show that when C is in farads and f is in hertz, the unit of X_C is the ohm.

Let us examine whether or not Equation 22.5 is reasonable. If we have a dc source (a dc source can be considered as an ac source with zero frequency), X_C is infinitely large. This means that a capacitor will impede the flow of direct current the same way a resistor of infinitely large resistance would. The current in such a circuit is zero. Indeed, we found that to be the case in Chapter 17. On the other hand, Equation 22.5 predicts that as the frequency increases, the capacitive reactance decreases. This means that before the charge on a capacitor has time to build up to the point where the current is zero, the direction of current flow has reversed.

The analogy between capacitive reactance and resistance allows us to write an equation that has the same form as Ohm's law to describe ac circuits containing capacitors. This equation relates the rms voltage and rms current in the circuit to the reactance as

$$V_C = IX_C \tag{22.6}$$

EXAMPLE 22.2 A Purely Capacitive ac Circuit

An 8-μF capacitor is connected to the terminals of an ac generator whose rms voltage is 150 V and whose frequency is 60 Hz. Find the capacitive reactance and the rms current in the circuit.

Solution From Equation 22.5 and the fact that $2\pi f = 377\ \text{s}^{-1}$, we have

$$X_C = \frac{1}{2\pi f C} = \frac{1}{(377\ \text{s}^{-1})(8 \times 10^{-6}\text{F})} = 332\ \Omega$$

If we substitute this result into Equation 22.6, we find that

$$I = \frac{V_C}{X_C} = \frac{150\ \text{V}}{332\ \Omega} = 0.452\ \text{A}$$

EXERCISE If the frequency is doubled, what happens to the capacitive reactance and the current?

Answer X_C is halved, and I is doubled.

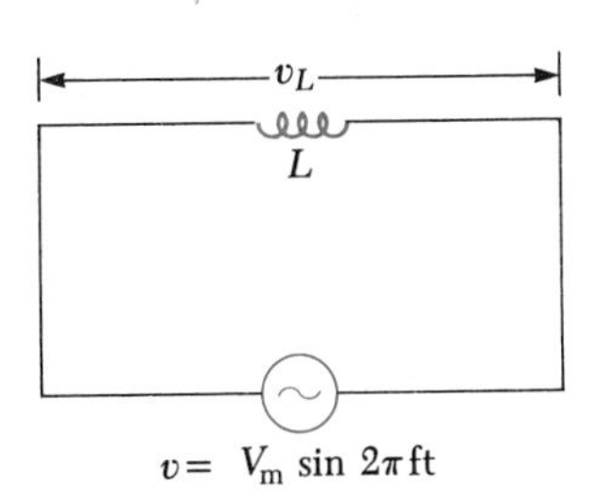

FIGURE 22.6 A series circuit consisting of an inductor L connected to an ac generator.

22.3 INDUCTORS IN AN AC CIRCUIT

Now consider an ac circuit consisting only of an inductor connected to the terminals of an ac generator, as in Figure 22.6. (In any real circuit, there is some resistance in the wire forming the inductive coil.) The changing current output of the generator produces a back emf in the coil of magnitude

$$v_L = L\frac{\Delta I}{\Delta t} \tag{22.7}$$

Thus, the current in the circuit is impeded by the back emf of the inductor. The effective resistance of the coil in an ac circuit is measured by a quantity called the **inductive reactance,** X_L, defined by

Inductive reactance

$$\boxed{X_L \equiv 2\pi f L} \tag{22.8}$$

It is left as a problem to show that when f is in hertz and L is in henries, the unit of X_L is the ohm. Note that the inductive reactance increases with increasing frequency and increasing inductance.

In order to understand the meaning of inductive reactance, let us compare this equation of X_L with Equation 22.7. First, note from Equation 22.8 that the inductive reactance depends on the inductance, L. This seems reasonable because the back emf (Eq. 22.7) is large for large values of L. Second, note that the inductive reactance depends on the frequency, f. This, too, seems reasonable because the back emf depends on $\Delta I/\Delta t$, a quantity that is large when the current is changing at a rapid rate, as it would be for large frequencies.

With inductive reactance defined in this manner, we can write an equation for the voltage across the coil or inductor that has the same form as Ohm's law:

$$\boxed{V_L = IX_L} \tag{22.9}$$

where V_L is the rms voltage drop across the coil and I is the rms current in the coil.

The instantaneous voltage and instantaneous current across the coil as functions of time are shown in Figure 22.7.

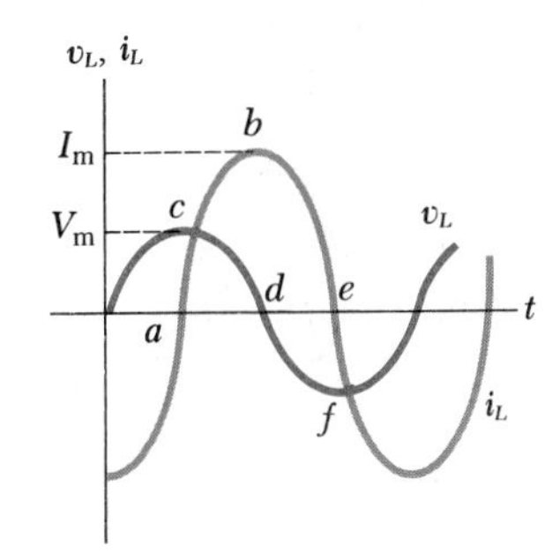

FIGURE 22.7 Plots of the current and voltage across an inductor versus time in an ac circuit. The voltage leads the current by 90°.

When a sinusoidal voltage is applied across an inductor, the voltage reaches its maximum value one quarter of an oscillation period before the current reaches its maximum value. In this situation, we say that the voltage always leads the current by 90°.

To understand why this phase relationship between voltage and current should exist, let us examine a few points on the curves of Figure 22.7. Note that at point *a* on the current curve, the current is beginning to increase in the positive direction. At this instant, the rate of change of current is at a maximum, and consequently we see from Equation 22.7 that the voltage across the inductor is also at a maximum at this time. As the current rises, between points *a* and *b* on the curve, $\Delta I/\Delta t$ (the slope of the current curve) gradually decreases until it reaches zero at point *b*. As a result, the voltage across the inductor is decreasing during this same time interval, as the segment between *c* and *d* on the voltage curve indicates. Immediately after point *b*, the current begins to decrease, although it is still in the same direction as during the previous quarter cycle. As the current decreases to zero (from *b* to *e* on the curve), a voltage is again induced in the coil (*d* to *f*), but the sense of this voltage is opposite the sense of the voltage induced between *c* and *d*. This occurs because back emfs are always directed to oppose the change in the current. We could continue to examine other segments of the curves, but no new information would be gained since the current and voltage variations are repetitive.

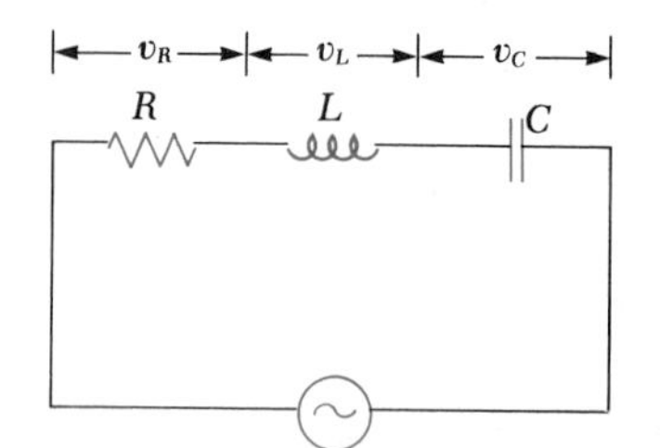

FIGURE 22.8 A series circuit consisting of a resistor, an inductor, and a capacitor connected to an ac generator.

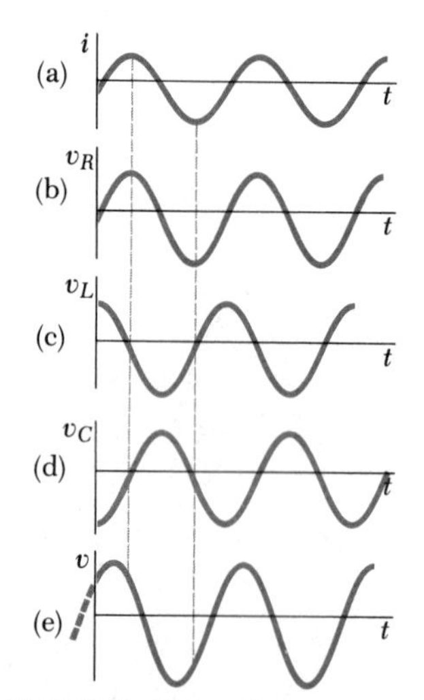

FIGURE 22.9 Phase relations in the series *RLC* circuit shown in Figure 22.8.

EXAMPLE 22.3 A Purely Inductive ac Circuit

In a purely inductive ac circuit (Fig. 22.6), $L = 25$ mH and the rms voltage is 150 V. Find the inductive reactance and rms current in the circuit if the frequency is 60 Hz.

Solution First, note that $2\pi f = 2\pi(60) = 377\ \text{s}^{-1}$. Equation 22.8 then gives

$$X_L = 2\pi f L = (377\ \text{s}^{-1})(25 \times 10^{-3}\ \text{H}) = 9.43\ \Omega$$

Substituting this result into Equation 22.9 gives

$$I = \frac{V_L}{X_L} = \frac{150\ \text{V}}{9.43\ \Omega} = 15.9\ \text{A}$$

EXERCISE Calculate the inductive reactance and rms current in the circuit if the frequency is 6 Hz.

Answer $X_L = 943\ \Omega$, $I_m = 0.159$ A.

22.4 THE *RLC* SERIES CIRCUIT

In the previous sections, we examined the effects that an inductor, a capacitor, and a resistor have when connected separately across an ac voltage source. We shall now consider what happens when combinations of these devices are used.

Figure 22.8 shows a circuit containing a resistor, an inductor, and a capacitor connected in series across an ac generator. We shall assume that the current varies sinusoidally with time, as indicated in Figure 22.9a. Thus,

$$i = I_m \sin 2\pi f t$$

Earlier we learned that the voltage across each element may or may not be in phase with the current. The instantaneous voltages across the three elements, shown in Figure 22.9, have the following phase relations to the instantaneous current:

1. The instantaneous voltage across the resistor, v_R, is *in phase* with the instantaneous current. (See Fig. 22.9b.)
2. The instantaneous voltage across the inductor, v_L, *leads* the current by 90°. (See Fig. 22.9c.)
3. The instantaneous voltage across the capacitor, v_C, lags the current by 90°. (See Fig. 22.9d.)

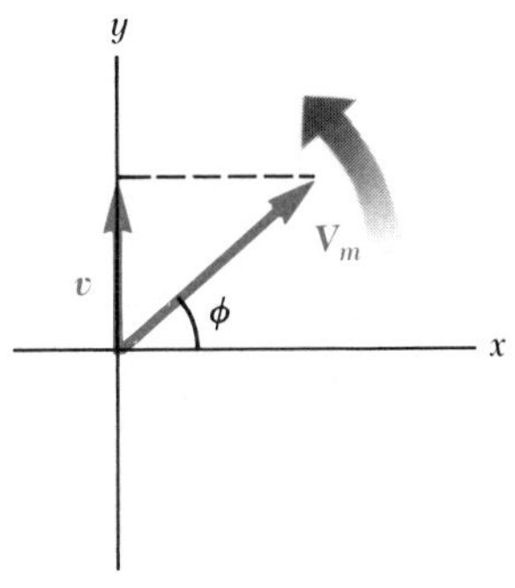

FIGURE 22.10 Phasor diagram for the voltage in an ac circuit, where φ is the phase angle between the voltage and current, and v is the instantaneous voltage.

The net instantaneous voltage, v, across all three elements is the sum of the instantaneous voltages across the separate elements:

$$v = v_R + v_C + v_L \tag{22.10}$$

This net voltage can be obtained graphically as shown in Figure 22.9e. Note that the instantaneous voltage across the resistor (Fig. 22.9b) is in phase with the current, the instantaneous voltage across the inductor (Fig. 22.9c) leads the current by a quarter cycle, and the instantaneous voltage across the capacitor (Fig. 22.9d) lags the current by a quarter cycle. Because both v_L and v_C are out of phase with v_R, the net voltage is a complicated function of the voltage amplitudes across the individual elements.

Rather than following the analytical procedure of adding the individual instantaneous voltages to find the net voltage, it is helpful to use another technique involving vectors. In this technique, we represent the voltage across each element by a rotating vector as in Figure 22.10. Such a diagram is called a **phasor diagram,** and the rotating vectors are referred to as phasors. This particular diagram represents the circuit voltage given by the expression $v = V_m \sin(2\pi ft + \phi)$, where V_m is the maximum voltage (the amplitude of the phasor) and φ is the angle between the phasor and the $+x$ axis. The phasor can be viewed as a vector of magnitude V_m rotating at a constant frequency f, such that its projection along the y axis is the instantaneous voltage in the circuit. Since φ is the phase angle between the voltage and current in the circuit, the phasor for the current (not shown in Figure 22.10) would lie along the $+x$ axis, and would be expressed by the relation $i = I_m \sin(2\pi ft)$.

The phasor diagrams given in Figure 22.11 are useful for analyzing the series *RLC* circuit. These diagrams are plotted such that voltages in phase with the current are represented by vectors along the $+x$ axis, while voltages out of phase with the current lie along other axes. Thus, V_R is horizontal and to the right because the voltage across the resistor is in phase with the current. Likewise, V_L is represented by a phasor along the $+y$ axis since V_L leads the current by 90°. Finally, V_C is along the $-y$ axis since V_C lags the current by 90°. If the phasors are added as vector quantities, Figure 22.11a shows that the only x component for the voltages is V_R, while the net y component is $V_L - V_C$. It is convenient to now add the phasors vectorially as shown in Figure 22.11b, where V is the total rms voltage in the circuit. The right tri-

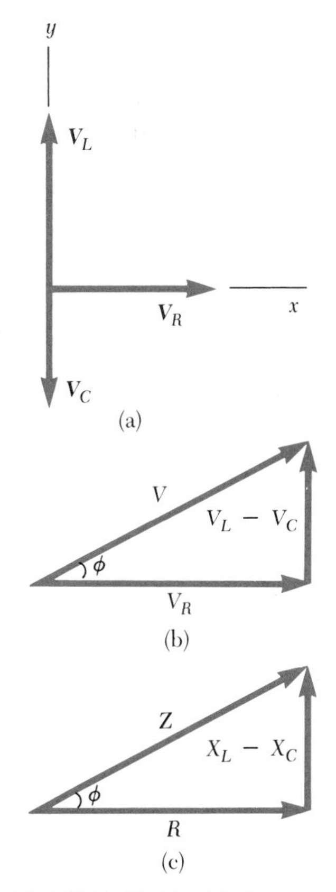

FIGURE 22.11 (a) Phasor diagram for the *RLC* circuit. (b) Adding the phasors as vectors gives $V = \sqrt{V_R^2 + (V_L - V_C)^2}$. (c) The reactance triangle which gives the impedance relation, $Z = \sqrt{R^2 + (X_L - X_C)^2}$.

angle in Figure 22.11b gives the following equations for the total voltage and phase angle:

$$V = \sqrt{V_R^2 + (V_L - V_C)^2} \tag{22.11}$$

$$\tan \phi = \frac{V_L - V_C}{V_R} \tag{22.12}$$

where all voltages have rms values. Note that although we choose to use rms voltages in our analysis, the equations above apply equally well to peak voltages because the two quantities are related to each other by the same factor for all circuit elements. The result for the total voltage V, as given by Equation 22.11, reinforces the fact that *the rms voltages across the resistor, capacitor, and inductor are not in phase, so one cannot simply add the individual voltages across the separate elements to get the voltage across the combination of elements.*

We can write Equation 22.11 in the form of Ohm's law using the relations $V_R = IR$, $V_L = IX_L$, and $V_C = IX_C$, where I is the rms current in the circuit:

$$V = I\sqrt{R^2 + (X_L - X_C)^2} \tag{22.13}$$

It is convenient to define a parameter called the **impedance,** Z, of the circuit as

Impedance

$$Z \equiv \sqrt{R^2 + (X_L - X_C)^2} \tag{22.14}$$

so that Equation 22.13 becomes

$$V = IZ \tag{22.15}$$

Note that Equation 22.15 is in the form of Ohm's law, $V = IR$, where R is replaced by the impedance, which has the unit of ohm. Equation 22.15 can be regarded as a generalized form of Ohm's law applied to a series ac circuit. Note that the current in the circuit depends upon the resistance, the inductance, the capacitance, *and* the frequency since the reactances are frequency-dependent.

It is useful to represent the impedance Z by a vector diagram, as shown in Figure 22.11c. A right triangle is constructed whose right side is the quantity $X_L - X_C$, whose base is R, and whose hypotenuse is Z. Applying the pythagorean theorem to this triangle, we see that

$$Z = \sqrt{R^2 + (X_L - X_C)^2}$$

Furthermore, we see from the vector diagram that the phase angle, ϕ, between the current and the voltage is given by

Phase angle ϕ

$$\tan \phi = \frac{X_L - X_C}{R} \tag{22.16}$$

The physical significance of the phase angle will become apparent in Sections 22.5 and 22.6.

Figure 22.12 gives impedance values and phase angles for various series circuits containing different combinations of circuit elements.

Circuit Elements	Impedance, Z	Phase angle, ϕ
R	R	0°
C	X_C	−90°
L	X_L	+90°
R C	$\sqrt{R^2+X_C^2}$	Negative, between −90° and 0°
R L	$\sqrt{R^2+X_L^2}$	Positive, between 0° and 90°
R L C	$\sqrt{R^2+(X_L-X_C)^2}$	Negative if $X_C > X_L$ Positive if $X_C < X_L$

FIGURE 22.12 The impedance values and phase angles for various circuit element combinations. In each case, an ac voltage (not shown) is applied across the combination of elements (that is, between the two dots).

There are also many parallel alternating current circuits which are useful in everyday applications. We shall not discuss them here, however, because the analysis of such circuits is beyond the scope of this text.

EXAMPLE 22.4 Analyzing a Series *RLC* ac Circuit

Analyze a series *RLC* ac circuit for which $R = 250\ \Omega$, $L = 0.6$ H, $C = 3.5\ \mu$F, $f = 60$ Hz, and $V = 150$ V.

Solution The reactances are given by $X_L = 2\pi fL = 226\ \Omega$ and $X_C = 1/2\pi fC = 758\ \Omega$. Therefore, the impedance is

$$Z = \sqrt{R^2 + (X_L - X_C)^2} = \sqrt{(250\ \Omega)^2 + (226\ \Omega - 758\ \Omega)^2} = 588\ \Omega$$

The rms current is

$$I = \frac{V}{Z} = \frac{150\ \text{V}}{588\ \Omega} = 0.255\ \text{A}$$

The phase angle between the current and voltage is

$$\phi = \tan^{-1}\left(\frac{X_L - X_C}{R}\right) = \tan^{-1}\left(\frac{226\ \Omega - 758\ \Omega}{250\ \Omega}\right) = -64.8°$$

Since the circuit is more capacitive than inductive (that is, $X_C > X_L$), ϕ is negative. A negative phase angle means that the current leads the applied voltage.

The rms voltages across the elements are

$$V_R = IR = (0.255\ \text{A})(250\ \Omega) = 63.8\ \text{V}$$
$$V_L = IX_L = (0.255\ \text{A})(226\ \Omega) = 57.6\ \text{V}$$
$$V_C = IX_C = (0.255\ \text{A})(758\ \Omega) = 193\ \text{V}$$

Note that the sum of the three rms voltages, $V_R + V_L + V_C$, is 314 V, which is much larger than the rms voltage of the generator, 150 V. The sum of 314 V is a meaningless quantity because, when alternating voltages are added, *both their amplitude and their phase* must be taken into account. That is, the voltages must be added in a way that takes account of the different phases. The relationship between V, V_R, and V_C is given by Equation 22.11. You should use the values found above to verify this equation.

22.5 POWER IN AN AC CIRCUIT

As we shall see in this section, *there are no power losses associated with capacitors and pure inductors in an ac circuit.* (A pure inductor is defined as one with no resistance or capacitance.) First, let us analyze the power dissipated in an ac circuit containing only a generator and a capacitor.

When the current begins to increase in one direction in an ac circuit, charge begins to accumulate on the capacitor and a voltage drop appears across it. When the voltage drop across the capacitor reaches its maximum value, the energy stored in the capacitor is

$$PE_C = \tfrac{1}{2}CV_m^2$$

However, this energy storage is only momentary. When the current reverses direction, this charge leaves the capacitor plates. In this discharging process, the charge returns to the voltage source. Thus in one cycle, the capacitor is being charged half of the time and the charge is being returned to the voltage source during the other half of the time. Therefore, *the average power supplied by the source is zero.* In other words, *a capacitor in an ac circuit does not dissipate energy.*

Similarly, the source must do work against the back emf of the inductor, which carries a current. When the current reaches its maximum value, the energy stored in the inductor is a maximum and is given by

$$PE_L = \tfrac{1}{2}LI_m^2$$

When the current begins to decrease in the circuit, this stored energy is returned to the source as the inductor attempts to maintain the current in the circuit.

The only element in an *RLC* circuit that dissipates energy is the resistor.

The average power lost in a resistor is

$$\boxed{P_{av} = I^2R} \tag{22.17}$$

where I is the rms current in the circuit. An alternative equation for the average power dissipated in an ac circuit can be found by substituting (from Ohm's law) $R = V_R/I$ into Equation 22.17:

$$P_{av} = IV_R$$

It is convenient to refer to the voltage triangle which shows the relationship between V, V_R, and $V_L - V_C$ as in Figure 22.10b. From this figure, we see that the voltage drop across a resistor can be written in terms of the voltage of the source:

$$V_R = V \cos \phi$$

Thus, the average power dissipated in an ac circuit is

Average power

$$\boxed{P_{av} = IV \cos \phi} \tag{22.18}$$

where the quality $\cos \phi$ is called the **power factor.**

EXAMPLE 22.5 Calculate the Average Power

Calculate the average power delivered to the series *RLC* circuit described in Example 22.4.

Solution We are given that the rms voltage supplied to the circuit is 150 V, and we have calculated that the rms current in the circuit is 0.255 A and the phase angle ϕ is $-64.8°$. Thus the power factor, $\cos \phi$, is 0.426. From these values, we calculate the average power using Equation 22.18:

$$P_{av} = IV \cos \phi = (0.255 \text{ A})(150 \text{ V})(0.426) = 16.3 \text{ W}$$

The same result can be obtained using Equation 22.17.

22.6 RESONANCE IN A SERIES *RLC* CIRCUIT

A series *RLC* circuit is said to be a **resonance** when the current has its maximum value. In general, the current in this series circuit can be written as

$$I = \frac{V}{Z} = \frac{V}{\sqrt{R^2 + (X_L - X_C)^2}} \tag{22.19}$$

From this, we see that the current has its *maximum* value when the impedance has its *minimum* value. This occurs when $X_L = X_C$. In such a circumstance, the impedance of the circuit reduces to $Z = R$. The frequency, f_0, at which this occurs is called the **resonance frequency** of the circuit. To find f_0, we set $X_L = X_C$, which gives, from Equations 22.5 and 22.8,

$$2\pi f_0 L = \frac{1}{2\pi f_0 C}$$

$$\boxed{f_0 = \frac{1}{2\pi\sqrt{LC}}} \tag{22.20}$$

Resonance frequency

A plot of current as a function of frequency is shown in Figure 22.13 for a circuit containing a fixed value of capacitance and inductance. From Equation 22.19, one must conclude that the current would become infinite at resonance when $R = 0$. Although Equation 22.19 predicts this result, real circuits always have some resistance, which limits the value of the current.

The receiving circuit of a radio is an important application of a series resonance circuit. The radio is tuned to a particular station (which transmits a specific radio-frequency signal) by varying a capacitor, which changes the resonance frequency of the receiving circuit. When the resonance frequency of the circuit matches that of the incoming radio wave, the current in the receiving circuit increases. The techniques for removing the information from the radio wave will be discussed in Chapter 23.

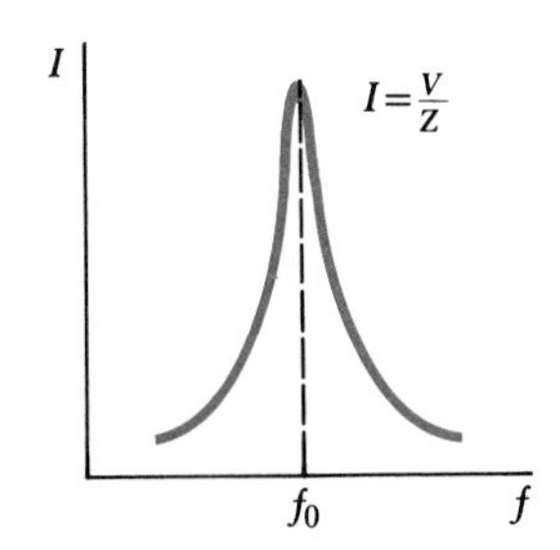

FIGURE 22.13 Current amplitude in a series *RLC* circuit plotted versus frequency of the generator voltage. Note that the current reaches its maximum value at the resonance frequency, f_0.

EXAMPLE 22.6 The Capacitance of a Circuit in Resonance

Consider a series *RLC* circuit for which $R = 150\ \Omega$, $L = 20$ mH, $V = 20$ V, and $2\pi f = 5000\ \text{s}^{-1}$. Determine the value of the capacitance for which the rms current is a maximum.

Solution The current is a maximum at the resonance frequency, f_0, which should be made to match the driving frequency of 5000 s^{-1} in this problem:

$$2\pi f_0 = 5 \times 10^3\ s^{-1} = \frac{1}{\sqrt{LC}}$$

$$C = \frac{1}{(25 \times 10^6\ s^{-2})\ L} = \frac{1}{(25 \times 10^6\ s^{-2})(20 \times 10^{-3}\ H)} = 2.0\ \mu F$$

EXERCISE Calculate the maximum rms current in the circuit.

Answer 0.133 A.

---

*22.7 THE TRANSFORMER

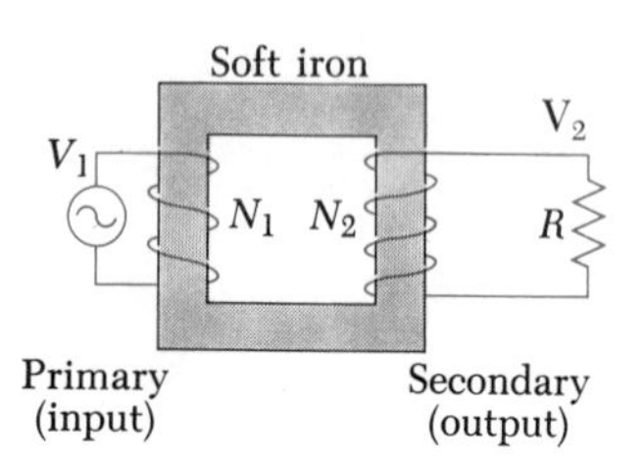

FIGURE 22.14 An ideal transformer consists of two coils wound on the same soft iron core. An ac voltage V_1 is applied to the primary coil, and the output voltage V_2 is observed across the load resistance R.

There are many situations in which it is desirable or necessary to change a small ac voltage to a larger one or vice versa. Before we examine a few cases in which such changes are made, let us consider the device that makes such conversions possible, the **ac transformer.**

In its simplest form, the ac transformer consists of two coils of wire wound around a core of soft iron, as in Figure 22.14. The coil on the left, which is connected to the input ac voltage source and has N_1 turns, is called the *primary* winding (or primary). The coil on the right, which is connected to a resistor R and consists of N_2 turns, is called the *secondary*. The purpose of the common iron core is to increase the magnetic flux and to provide a medium in which nearly all the flux through one coil passes through the other.

When an input ac voltage V_1 is applied to the primary, the induced voltage across it is given by

$$V_1 = -N_1 \frac{\Delta\Phi}{\Delta t} \tag{22.21}$$

where Φ is the magnetic flux through each turn. If we assume that no flux leaks from the iron core, then the flux through each turn of the primary equals the flux through each turn of the secondary. Hence, the voltage across the secondary coil is

$$V_2 = -N_2 \frac{\Delta\Phi}{\Delta t} \tag{22.22}$$

The term $\Delta\Phi/\Delta t$ is common to Equations 22.21 and 22.22. Thus, we see that

$$\boxed{V_2 = \frac{N_2}{N_1} V_1} \tag{22.23}$$

When N_2 is greater than N_1, and thus V_2 exceeds V_1, the transformer is referred to as a *step-up transformer*. When N_2 is less than N_1, making V_2 less than V_1, we speak of a *step-down transformer*.

It should be clear that a voltage is generated across the secondary only when there is a *change* in the number of flux lines passing through it. Thus,

This transformer steps the ac voltage down to 120 V at the customer's utility pole.

the input current in the primary must change with time, which is what happens when an alternating current is used. However, when the input at the primary is a direct current, there is a voltage output at the secondary only at the instant when a switch in the primary circuit is opened or closed. Once the current in the primary reaches a steady value, the output voltage at the secondary is zero.

At first thought, it seems that a transformer is a device in which it is possible to get something for nothing. For example, a step-up transformer can change an input voltage from, say, 10 V to 100 V. This says that each 1 C of charge leaving the secondary has 100 J of energy while each 1 C of charge entering the primary has only 10 J of energy. However, this is not an example of a breakdown in the law of conservation of energy since *the power input to the primary is equal to the power output at the secondary*, that is,

$$\boxed{I_1 V_1 = I_2 V_2} \tag{22.24}$$

Thus, we see that if the voltage at the secondary is ten times that at the primary, the current at the secondary is reduced by a factor of 10. Equation 22.24 assumes an **ideal transformer,** in which there are no power losses between the primary and the secondary. Typical transformers have power efficiencies ranging from 90% to 99%. Power losses are due to such factors as eddy currents induced in the iron core of the transformer. These currents dissipate energy in the form of an I^2R loss.

When electric power is transmitted over large distances, it is economical to use a high voltage and low current because the power lost via resistive heating in the transmission lines varies as I^2R. This means that if a utility company can reduce the current by a factor of 10, for example, the power loss will be reduced by a factor of 100. In practice, the voltage is stepped up to around 230,000 V at the generating station, then stepped down to around 20,000 V at a distribution station, and finally stepped down to 120 V at the customer's utility pole.

A situation in which a transformer is used with a dc source is the electrical system of automobiles. In an automobile engine, spark plugs produce a spark to ignite a gasoline-air mixture in the cylinders. In order for a plug to work, there must be a voltage high enough to cause the spark to jump across

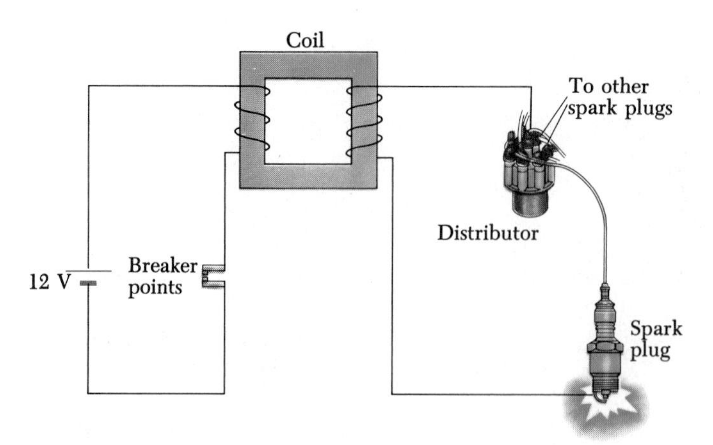

FIGURE 22.15 Circuit diagram of an automobile ignition system.

the gap in the plug. The technique for creating this high voltage is indicated in Figure 22.15. The primary coil of the transformer is connected to the 12-V battery of the car. The secondary is connected to the spark plugs through the distributor. The distributor is a mechanical device that connects, in turn, each spark plug to the secondary coil at the instant the plug is supposed to fire. At this instant, a switch (called the "breaker points" or simply the "points") breaks the circuit between the battery and the primary. This interrupts the current in the primary and causes a voltage to be induced in the secondary. The transformer used is a step-up transformer, and the resulting voltage in the secondary is quite high (about 20,000 V), high enough to cause a spark to jump when the voltage is applied across the gap of the spark plug.

EXAMPLE 22.7 A Step-Up Transformer

A generator produces 10 A of current at 400 V. The voltage is stepped up to 4500 V by an ideal transformer and transmitted a long distance through a power line of total resistance 30 Ω. (a) Determine the percentage of power lost.

Solution From Equation 22.24, we find that the current in the transmission line is

$$I_2 = \frac{I_1 V_1}{V_2} = \frac{(10\ \text{A})(400\ \text{V})}{4500\ \text{V}} = 0.89\ \text{A}$$

Hence, the power lost in the transmission line is

$$P_{\text{lost}} = I_2^{\,2}R = (0.89\ \text{A})^2(30\ \Omega) = 24\ \text{W}$$

Since the output power of the generator is $P = IV = (10\ \text{A})(400\ \text{V}) = 4000\ \text{W}$, the percentage of power lost is

$$\%\ \text{power lost} = \frac{24}{4000} \times 100 = 0.6\%$$

(b) What percentage of the original power would be lost in the transmission line if the voltage were not stepped up?

Solution If the voltage were not stepped up, the current in the transmission line would be 10 A and the power lost in the line would be $I^2R = (10\ \text{A})^2(30\ \Omega) = 3000\ \text{W}$. Hence, the percentage of power lost would be

$$\%\ \text{power lost} = \frac{3000}{4000} \times 100 = 75\%$$

This example illustrates the advantage of high-voltage transmission lines.

EXERCISE If the transmission line is cooled so that the resistance is reduced to 5 Ω, how much power will be lost in the line if it carries a current of 0.89 A?

Answer 4 W.

SUMMARY

If an ac circuit consists of a generator and a resistor, *the current in the circuit is in phase with the voltage.* That is, the current and voltage reach their maximum values at the same time.

In discussing voltages and currents in ac circuits, **rms values** of voltages are usually used. One reason for this is that ac ammeters and voltmeters are designed to read rms values. The rms values of voltage and currents (V and I) are related to the maximum value of these quantities (V_m and I_m) as follows:

$$V = \frac{V_m}{\sqrt{2}} \qquad I = \frac{I_m}{\sqrt{2}} \qquad (22.2) \quad (22.3)$$

The rms voltage across a resistor is related to the rms current through the resistor by **Ohm's law:**

$$V_R = IR \qquad (22.4)$$

If an ac circuit consists of a generator and a capacitor, *the voltage lags the current by 90°.* That is, the voltage reaches its maximum value one quarter of a period after the current reaches its maximum value.

The impeding effect of a capacitor on the flow of current in an ac circuit is given by the **capacitive reactance,** X_C, defined as

$$X_C \equiv \frac{1}{2\pi fC} \qquad (22.5)$$

where f is the frequency of the ac generator.

The rms voltage across and the rms current through a capacitor are related as

$$V_C = IX_C \qquad (22.6)$$

If an ac circuit consists of a generator and an inductor, *the voltage leads the current by 90°.* That is, the voltage reaches its maximum value one quarter of a period before the current reaches its maximum value.

The effective impedance of a coil in an ac circuit is measured by a quantity called the **inductive reactance,** X_L, given by

$$X_L \equiv 2\pi fL \qquad (22.8)$$

The rms voltage drop across a coil is related to the rms current through the coil by

$$V_L = IX_L \qquad (22.9)$$

In an *RLC* series ac circuit, the applied rms voltage, V, is related to the rms voltages across the resistor V_R, capacitor V_C, and inductor V_L, as

$$V = \sqrt{V_R^{\,2} + (V_L - V_C)^2} \tag{22.11}$$

If an ac circuit contains a resistor, an inductor, and a capacitor, the effective resistance of the circuit is given by the **impedance,** Z, of the circuit, defined as

$$Z \equiv \sqrt{R^2 + (X_L - X_C)^2} \tag{22.14}$$

The relationship between the rms voltage supplied to an *RLC* circuit and the rms current in the circuit is

$$V = IZ \tag{22.15}$$

In an *RLC* series ac circuit, the applied rms voltage and current are out of phase. The **phase angle,** ϕ, between the current and voltage is given by

$$\tan \phi = \frac{X_L - X_C}{R} \tag{22.16}$$

The **average power** delivered by the generator in an *RLC* ac circuit is

$$P_{\text{av}} = IV \cos \phi \tag{22.18}$$

where the constant $\cos \phi$ is called the **power factor.**

ADDITIONAL READING

L. Barthold and H. G. Pfeiffer, "High-Voltage Transmission," *Sci. American,* May 1964, p. 38.

T. Geballe and J. K. Hulm, "Superconductors in Electric-Power Technology," *Sci. American,* November 1980, p. 138.

A. L. Hammond, W. D. Metz, and T. H. Maugh, *Energy and the Future,* American Association for the Advancement of Science, 1973.

J. J. O'Neill, *Prodigal Genius, The Life of Nikola Tesla,* Ives Washburn, 1944.

QUESTIONS

1. How can the average value of a current be zero and yet the square root of the average value squared not be zero?
2. Explain why the reactance of a capacitor decreases with increasing frequency whereas the reactance of an inductor increases with increasing frequency.
3. What is meant by the statement, "The voltage across an inductor leads the current by 90°"?
4. Is the voltage applied to a circuit always in phase with the current through a resistor in the circuit?
5. Does the phase angle depend on frequency? What is the phase angle when the inductive reactance equals the capacitive reactance?
6. If the frequency is doubled in a series *RLC* circuit, what happens to the resistance, the inductive reactance, and the capacitive reactance?
7. Would an inductor and a capacitor used together in an ac circuit dissipate any power?
8. Show that the maximum current in an *RLC* circuit occurs when the circuit is in resonance.

9. What is the impedance of an *RLC* circuit at the resonance frequency?
10. Will a transformer operate if a battery is used for the input voltage across the primary?
11. Why are the primary and secondary coils of a transformer wrapped on an iron core that passes through both coils?
12. At one time there were debates in this country concerning whether we should use ac or dc for transmitting electrical power. What disadvantage would dc have in the transmission of power over a long distance?
13. If it were not for the eye's persistence of vision, one could see a lightbulb operated with an ac voltage flicker on and off. Would you see 60 or 120 flickers per second?
14. A memory device often used with alternating current circuits is ELI the ICE man. ICE means that in a capacitive circuit, the current leads the voltage. What does ELI mean?

PROBLEMS

Section 22.1 Resistors in an ac Circuit

1. An rms voltage of 100 V is applied to a device that has a resistance of 5 Ω. Find (a) the maximum voltage applied, (b) the maximum current, and (c) the rms current supplied.

2. What are the maximum values of a 6 A rms current and a 240 V rms voltage?

• **3.** An ac voltage source has an output given by $v = 150 \sin 377t$. Find (a) the rms voltage output, (b) the frequency of the source, and (c) the voltage at $t = 1/120$ s. (d) Find the maximum current in the circuit when the generator is connected to a 50 Ω resistor.

• **4.** For a particular ac generator, the output voltage is $v = 0.25V_m$ when $t = 0.002$ s. What is the operating frequency of the generator?

Section 22.2 Capacitors in an ac Circuit

5. Show that the SI unit of capacitive reactance is the ohm.

6. A certain capacitor in a circuit has a capacitive reactance of 30 Ω when the frequency is 120 Hz. What capacitive reactance does the capacitor have at a frequency of 10,000 Hz?

7. What rms current will be delivered by an ac generator whose rms output voltage is 48 V at a frequency of 90 Hz when connected across a 3.7 μF capacitor?

8. The current in a circuit containing a 4 μF capacitor is 0.3 A when connected to a generator whose rms output is 30 V. What is the frequency of the source?

9. The generator in a purely capacitive ac circuit has an angular frequency of 120 π rad/s. If $V_m = 140$ V and $C = 6$ μF, what is the rms current in the circuit?

10. What value of capacitor must be inserted in a 60-Hz circuit in series with a generator of 170 V maximum output voltage to produce an rms current output of 0.75 A?

Section 22.3 Inductors in an ac Circuit

11. Find the inductive reactance of a 3-mH inductor at a frequency of (a) 10 Hz and (b) 10^6 Hz.

12. Show that the inductive reactance, X_L, has the SI unit of ohm.

13. In a purely inductive circuit, the rms voltage is 120 V. (a) If the rms current is 10 A at a frequency of 60 Hz, calculate the inductance. (b) At what frequency will the rms current be reduced to 5 A?

14. In a purely inductive circuit, $L = 30$ mH, the frequency is 60 Hz, and the maximum voltage is 200 V. Find (a) the rms current in the circuit. (b) Repeat for a frequency of 6000 Hz.

15. Calculate the inductance of a coil that has an inductive reactance of 40 Ω at an angular frequency of 754 rad/s.

Section 22.4 The *RLC* Series Circuit

16. A 40 μF capacitor is connected to a 50 Ω resistor and a generator whose rms output is 30 V at 60 Hz. What is (a) the rms current in the circuit, (b) the voltage drop across the resistor, (c) the voltage drop across the capacitor, and (d) the phase angle for the circuit? (e) Sketch the phasor diagram for this circuit.

17. A 10 μF capacitor and a 2-H inductor are connected in series with a 60-Hz source whose rms output is 100 V. Find (a) the rms current in the circuit, (b) the voltage drop across the inductor, (c) the voltage drop across the capacitor, and (d) the phase angle for the circuit. (e) Sketch the phasor diagram for this circuit.

18. A pure 20-mH inductor is connected in series with a 20-Ω resistor and a 60-Hz, 100-V rms source. Find (a) the rms current in the circuit, (b) the voltage drop across the inductor, (c) the voltage drop across the resistor, and (d) the phase angle for this circuit. (e) Sketch the phasor diagram for this circuit.

19. A 60-Hz source with an rms output of 100 V is connected to a series combination of a 30-Ω resistor, a 3-μF capacitor, and a 2-H inductor. Find (a) the impedance of the circuit, (b) the rms current in the circuit, (c) the voltage drop across each element, and (d) the phase angle. (e) Sketch the phasor diagram for this circuit.

20. A 50-Ω resistor, a 0.1-H inductor, and a 10-μF capacitor are connected in series to a 60-Hz source. The rms current in the circuit is 2.75 A. Find the rms voltage across (a) the resistor, (b) the inductor, (c) the capacitor, and (d) the *RLC* combination. (e) Sketch the phasor diagram for this circuit.

• **21.** A 60-Ω resistor, a 3-μF capacitor, and a 0.4-H inductor are connected in series to a 90-V, 60-Hz source. Find (a) the voltage drop across the *LC* combination and (b) the voltage drop across the *RC* combination.

• **22.** A 50-Ω resistor is connected in series with a 1.5-μF capacitor and a 60-Hz, 120-V source. (a) Find the current in the circuit. (b) What value inductor must be inserted in the circuit to reduce the current to one-half that found in (a)?

• **23.** A 100-Ω resistor and a 3-μF capacitor are connected in series. What is the current in the circuit (a) when a dc source of 10 V is connected to the circuit, and (b) when a 10-V, 60-Hz source is connected?

• **24.** A 20-Ω resistor, a 40-μF capacitor and a 0.2-H inductor are connected in series with a 60-Hz, 70-V source. The inductor is found to have a resistance of 15 Ω. Find (a) the current in the circuit, and (b) the voltage drop across the inductor.

Section 22.5 Power in an ac Circuit

25. A series ac circuit contains a 50-Ω resistor, a 1.5-μF capacitor, a 0.2-H inductor, and a 60-Hz generator which has an rms output of 90 V. Find the average power delivered to the circuit.

26. A 50-Ω resistor is connected to a 30-μF capacitor and to a 60-Hz, 100-V rms source. (a) Find the power factor and the average power delivered to the circuit. (b) Repeat part (a) when the capacitor is replaced with a 0.3-H inductor.

27. A series *RLC* circuit has a resistance of 80 Ω and an impedance of 180 Ω. What average power will be delivered to the circuit when the rms voltage is 120 V?

• **28.** In a certain *RLC* circuit, the rms current is 6 A, the rms voltage is 240 V, and the current leads the voltage by 53°. (a) What is the total resistance of the circuit? (b) Calculate the total reactance $X_L - X_C$. (c) Find the average power dissipated in the circuit.

• **29.** A series *RLC* circuit takes 108 W from a 110-V line. The applied voltage leads the voltage across the resistor by 37°. (a) What is the value of the current in the circuit? (b) What is the impedance of the circuit?

• **30.** An inductor and a resistor are connected in series. When connected to a 60-Hz, 90-V source the voltage drop across the resistor is found to be 50 V and the power dissipated in the circuit is 14 W. Find (a) the value of the resistance and (b) the value of the inductance.

Section 22.6 Resonance in a Series *RLC* Circuit

31. A 40-Ω resistor, a 2-μF capacitor, and a 2-mH inductor are connected in series to a source with an rms output of 60 V. Find (a) the frequency that would produce resonance, (b) the current at resonance, and (c) the current in the circuit when the frequency is one-half the resonance frequency.

32. A resonant circuit in a radio receiver is tuned to a certain station when the inductor has a value of 0.2 mH and the capacitor has a value of 30 pF. Find the frequency of the radio station and the wavelength sent out by the station.

33. The resonance frequency of a series *RLC* circuit is found to be 4.2×10^4 Hz. If the capacitance of the circuit is 300 pF, what is the value of the inductance?

• **34.** A series circuit contains a 3-H inductor, a 3-μF capacitor, and a 30-Ω resistor connected to a 120-V rms source of variable frequency. Find the power delivered to the circuit when the frequency of the source is (a) the resonance frequency, (b) one-half the resonance frequency, (c) one-fourth the resonance frequency, (d) two times the resonance frequency, and (e) four times the resonance frequency. From your calculations can you draw a conclusion about the frequency at which the maximum power is delivered to the circuit?

• **35.** The *Q* value of an *RLC* circuit is defined as the voltage drop across the inductor (or capacitor) at resonance divided by the voltage drop across the resistor. The larger the *Q* value, the sharper, or narrower, is the curve of power versus frequency. Figure 22.16 shows such curves for small *R* and for large *R*. (a) Show that the *Q* value is given by the expression $Q = 2\pi f_0 L/R$, where f_0 is the resonance frequency. (b) Calculate the Q value for the circuit of Problem 34.

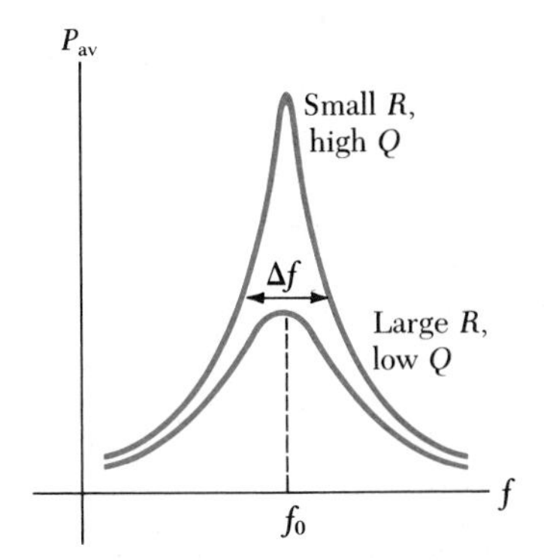

FIGURE 22.16 Problems 35 and 36

• **36.** The curves of Figure 22.16 are characterized by the width at half-maximum, Δf. This width is the difference in frequency between the two points on the curve where the power is one-half the maximum value. It can be shown that this width is related to the Q value (when the Q value is large) as $\Delta f = f_0/Q$. (a) Find Δf for the circuit in the Problem 34. (b) Find the Q value and Δf for the circuit in Problem 34 if R is replaced by a 300-Ω resistor.

Section 22.7 The Transformer

37. An ideal transformer has 150 turns on the primary winding and 600 turns on the secondary. If the primary is connected across a 110-V rms generator, what is the rms output voltage at the secondary?

38. A transformer at a public utility reduces the voltage on a line from 360,000 V to 3,600 V. The primary has 10,000 turns. (a) How many turns are there on the secondary? (b) If the current in the secondary is 600 A, find the current in the primary. Assume an ideal transformer.

39. The primary current of an ideal transformer is 6.5 A when the primary voltage is 96 V. Calculate the voltage across the secondary when a current of 0.8 A is delivered to a resistor.

40. A transformer on a pole near a factory steps the voltage down from 3600 V to 120 V. The transformer is to deliver 1000 kW to the factory at 90% efficiency. Find (a) the power delivered to the primary, (b) the current in the primary, and (c) the current in the secondary.

• **41.** Each inhabitant of a city of 20,000 people turns on a 100-W lightbulb at the same time. Assume no other power in the city is being used. (a) If the utility company were to furnish this total power at 120 V, calculate the current in the power lines from the utility to the city. (b) Calculate this current if the power company first steps up the voltage to 200,000 V. (c) How much heat is lost in each 1-m length of the power lines if the resistance of the lines is 5×10^{-4} Ω? Repeat this calculation for situation (a) and for situation (b). (d) If an individual line from the utility can handle only 100 A, how many lines are required to handle the current for each situation described?

ADDITIONAL PROBLEMS

42. A 300-Ω resistor is in series with a 0.3-H inductor, a 15-μF capacitor, and a 50-Hz, 50-V generator. Find (a) the current in the circuit, (b) the voltage drop across the resistor-capacitor combination, and (c) the power delivered to the circuit.

43. A 250-mH inductor is connected across an ac source of $V_m = 90$ V. (a) At what frequency will the inductive reactance equal 20 Ω? (b) What is the rms value of current in the inductor at the frequency found in part (a)?

44. A 0.25-H inductor is connected to a capacitor and a 30-Ω resistor along with a 60-Hz, 30-V generator. (a) To what value would the capacitor have to be adjusted to produce resonance? (b) At resonance find the voltage drop across the capacitor and inductor combination.

45. A coil with an inductance of 15.3 mH and a resistance of 5 Ω is connected to a variable frequency ac generator. At what frequency would the voltage across the coil lead the current by 60°?

46. The AM band extends from approximately 500 kHz to 1600 kHz. If a 2-μH inductor is used in a tuning circuit for a radio, what are the extremes that a capacitor must reach in order to cover the complete band of frequencies?

47. A resistor, inductor, and capacitor are connected in series. The rms voltages are 75 V across the resistor, 75 V across the capacitor, and 150 V across the inductor. Find (a) the voltage of the source, and (b) the phase angle.

48. A transformer to be used with a small portable cassette player has a flexible design that enables it to deliver either 3 V, 6 V, or 9 V when connected to a 60-Hz, 120-V rms source. If the primary has 200 turns, (a) find the total number of turns in the secondary, and (b) the number of turns that must be tapped to produce each of the voltages above.

• **49.** A particular inductor has appreciable resistance. When the inductor is connected to a 12-V battery, the current through the inductor is 3 A. When it is connected to an ac source whose rms output is 12 V and whose frequency is 60 Hz, the current drops to 2 A.

What is (a) the impedance at 60 Hz and (b) the inductance of the inductor?

• **50.** A 0.7-H inductor is connected in series with a fluorescent lamp to limit the current drawn by the lamp. If the combination is connected to a 60-Hz, 120-V line, and if the voltage across the lamp is to be 40 V, what is the current in the circuit? (The lamp is a pure resistive load.)

• **51.** A small transformer is used to supply an ac voltage of 6 V to a model railroad lighting circuit. The primary has 220 turns and is connected to a standard 110-V, 60-Hz line. Although the resistance of the primary may be neglected, it has an inductance of 150 mH. (a) How many turns are required on the secondary winding? (b) If the transformer is left plugged in, what current will be drawn by the primary when the secondary is open? (c) What power will be drawn by the primary when the secondary is open?

• **52.** A 200-Ω resistor is connected in series with a 5-μF capacitor and a 60-Hz, 120-V rms line. If electrical energy costs $0.08 per kWh, how much does it cost to leave this circuit connected for 24 h?

• **53.** A 100-Ω resistor is connected in a series circuit to a 0.6-H inductor and a 25-μF capacitor. This *RLC* combination is then connected across a voltage source that varies as $v = 80 \sin(1000t/\pi)$ V. Calculate (a) the maximum current, (b) the phase angle, (c) the power factor, (d) the rms voltage across the inductor, and (e) the power delivered to the circuit.

• **54.** The current in a purely resistive circuit equals 70% of the maximum current at $t = 0.003$ s. What is the frequency of the generator?

• **55.** A series *RLC* circuit has a resonance frequency of $2000/\pi$ Hz. When operating at a frequency $\omega > \omega_0$, $X_L = 12\ \Omega$ and $X_C = 8\ \Omega$. Calculate the values of L and C for the circuit.

•• **56.** An inductor is to be made of a 5-m length of copper wire of diameter 1 mm wound on a coil of radius 3 cm. (The coil has an air core.) Estimate the length of the completed solenoid and number of turns on it. Find (b) the inductance of the completed coil, and (c) the resistance of the completed coil. (d) Find the current that would be drawn when this device is connected to a 60-Hz, 20-V rms source.

•• **57.** Two connections allow contact with two circuit elements in series inside a box, but it is not known whether the circuit elements are R, L, or C. In an attempt to find what is inside the box, you perform the following set of measurements. When a 3-V dc power supply is connected across the terminals, there is a direct current of 300 mA in the circuit. When a 3-V, 60-Hz source is connected, the current becomes 200 mA. (a) What are the two elements in the box? (b) What are their values of R, L, or C?

•• **58.** A transmission line with a resistance per unit length of 4.5×10^{-4} Ω/m is to be used to transmit 5000 kW of power over a distance of 400 miles (6.44×10^5 m). The output voltage of the generator is 4500 V. (a) What is the line loss if a transformer is used to step up the voltage to 500 kV? (b) What fraction of the input power is lost to the line under these circumstances? (c) What difficulties would be encountered by attempting to transmit the 5000 kW of power at the generator voltage of 4500 V?

•• **59.** Suppose you wish to use a transformer as an impedance-matching device between an audio amplifier that has an output impedance of 8000 Ω and a speaker that has an input impedance of 8 Ω. What should be the ratio of primary to secondary turns on the transformer?

•• **60.** A power transmission line consists of two parallel wires each having a cross-sectional area A. (One serves as a return line, so the effective length is 2ℓ.) Show that the power lost as heat in the conductors is given by

$$P_{\text{loss}} = \frac{2\rho\ell P_L^{\,2}}{AV^2 \cos^2 \phi}$$

where $\cos\phi$ is the power factor, P_L is the power supplied to the load, and ρ is the resistivity of the conductor. (Note that V represents an rms value in this problem.)

CHAPTER 23

Electromagnetic Waves

Motorized polar mount for 5- and 7-m antennas allows broadcasters and cable TV system operators to switch from one satellite to another in a matter of seconds. (Courtesy of Microdyne Corporation)

As we go about our daily activities, we observe thousands of objects, such as books, trees, furniture, and flowers, to name just a few. However, what we are not so aware of are the invisible atoms and molecules that constitute the air we breathe, and the waves called **electromagnetic waves** that permeate our environment. Electromagnetic waves in the form of visible light enable us to view the world around us; infrared waves warm our environment; radio-frequency waves carry our favorite television and radio programs; the list goes on and on.

The purpose of this chapter is to study the properties of these various types of electromagnetic waves. In a sense this is a transitional chapter, in that it takes us from the study of electricity and magnetism to the study of light. However, as we shall see, these subjects are related. Light is one part of the spectrum of electromagnetic radiation, a type of wave motion consisting of oscillating electric and magnetic fields.

Thus far, the waves we have described have been mechanical waves, which propagate as the result of the disturbance of a medium. By definition, mechanical disturbances, such as sound waves, water waves, and waves on a string, can exist only if a medium is present. Electromagnetic waves, on the other hand, can propagate through a vacuum.

23.1 MAXWELL'S PREDICTIONS

During the early stages of their study and development, electric and magnetic phenomena were thought to be unrelated. In 1865, however, James Clerk Maxwell (1831–1879) provided a mathematical theory that showed a close relationship between all electric and magnetic phenomena. Additionally, his theory predicted that electric and magnetic fields can move through space as waves. The theory developed by Maxwell is based upon the following four pieces of information:

James Clerk Maxwell (1831–1879) "I am not attempting to establish any physical theory of a science in which I have not made a single experiment worthy of the name. . . . The theory of electromagnetism including the induction of electric currents, which I have deduced mathematically from certain ideas due to Faraday, I reserve for future communication." (Photo courtesy of AIP Niels Bohr Library)

1. Electric fields originate on positive charges and terminate on negative charges. The electric field due to a point charge can be determined at a location by applying Coulomb's force law to a test charge placed at that location.
2. Magnetic field lines always form closed loops, that is, they do not begin or end anywhere.
3. A varying magnetic field induces an emf and hence an electric field. You should recognize this as a statement of *Faraday's law* (Chap. 21).
4. Magnetic fields are generated by moving charges (or currents), as summarized in *Ampère's law* (Chap. 20).

Let us examine these statements further in order to understand their significance and Maxwell's contributions to the theory of electromagnetism. The first statement is a consequence of the nature of the electrostatic force between charged particles, given by Coulomb's law. Embodied in this statement is a recognition of the fact that *free charges (electric monopoles) exist in nature.*

The second statement says that magnetic fields do not originate or terminate at any point; instead, they form continuous loops. For example, the magnetic field lines around a long, straight wire are closed circles and the magnetic field lines of a bar magnet form closed loops. In the latter example, an individual line leaves the magnet at the north pole, loops around to enter the magnet at the south pole, and then continues through the interior of the magnet back to the north pole. (It does not start at the north pole and end on the south pole.) This second statement implies that *free magnetic monopoles do not exist.* Although the idea of magnetic monopoles is embodied in modern field theory, no monopole has yet been isolated conclusively.

The third statement is equivalent to Faraday's law of induction, and the fourth statement is equivalent to Ampere's law.

In one of the greatest theoretical developments of the 19th century, Maxwell used these four statements within a corresponding mathematical framework to prove that electric and magnetic fields play a symmetrical role in nature. It was known from experiments that a changing magnetic field produces an electric field according to Faraday's law. Maxwell suspected that nature should be symmetrical, and therefore hypothesized that *a changing electric field should produce a changing magnetic field.* This hypothesis could not be proven experimentally at the time it was developed because the magnetic fields generated by changing electric fields are generally very weak and therefore difficult to detect.

Heinrich Hertz (1857–1894) "The described experiments appear, at least to me, in a high degree suited to remove doubt in the identity of light, heat radiation, and electrodynamic wave motion."

In order to justify his hypothesis, Maxwell searched for other phenomena that might be explained with his theory. Thus, he turned his attention to the motion of rapidly oscillating charges, such as those in a conducting rod connected to an alternating voltage. Such oscillating charges experience accelerations and according to Maxwell's predictions, generate changing electric and magnetic fields. According to this theory, the changing fields produced by the oscillating charges result in electromagnetic disturbances that travel through space as waves, similar to the spreading water waves created by a pebble thrown into a pool. One views the waves sent out by the oscillating charges as fluctuating electric and magnetic fields; hence, they are called *electromagnetic waves.* Maxwell calculated the speed of these waves to be equal to the speed of light, $c = 3 \times 10^8$ m/s. Thus, he concluded that light waves are electromagnetic in nature. That is, light waves and other electromagnetic waves consist of fluctuating electric and magnetic fields traveling through space with a velocity of 3×10^8 m/s! This was truly one of the greatest discoveries of science and one that had a profound influence on later developments.

23.2 HERTZ'S DISCOVERIES

In 1887, Heinrich Hertz (1857–1894) first generated and detected electromagnetic waves in a laboratory setting. In order to appreciate the details of his experiment, let us first re-examine the properties of an *LC* circuit, first discussed in Chapter 22. In such a circuit, a charged capacitor is connected to an inductor, as in Figure 23.1. When the switch is closed, oscillations occur in the current in the circuit and in the charge on the capacitor. If the resistance of the circuit is neglected, no energy is lost to heat and the oscillations continue.

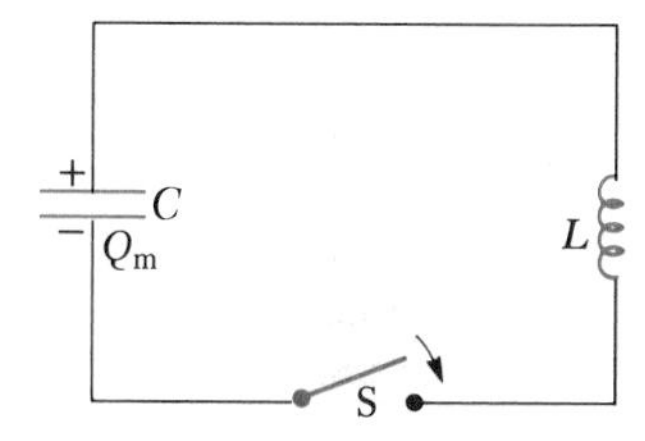

FIGURE 23.1 A simple *LC* circuit. The capacitor has an initial charge Q_m, and the switch is closed at $t = 0$.

In the following analysis, we shall neglect the resistance in the circuit. Let us assume that the capacitor has an initial charge Q_m and that the switch is closed at $t = 0$. It is convenient to describe what happens from an energy viewpoint. When the capacitor is fully charged, the total energy in the circuit is stored in the electric field of the capacitor and is equal to $Q_m^2/2C$. At this time, the current is zero and so there is no energy stored in the inductor. As the capacitor begins to discharge, the energy stored in its electric field de-

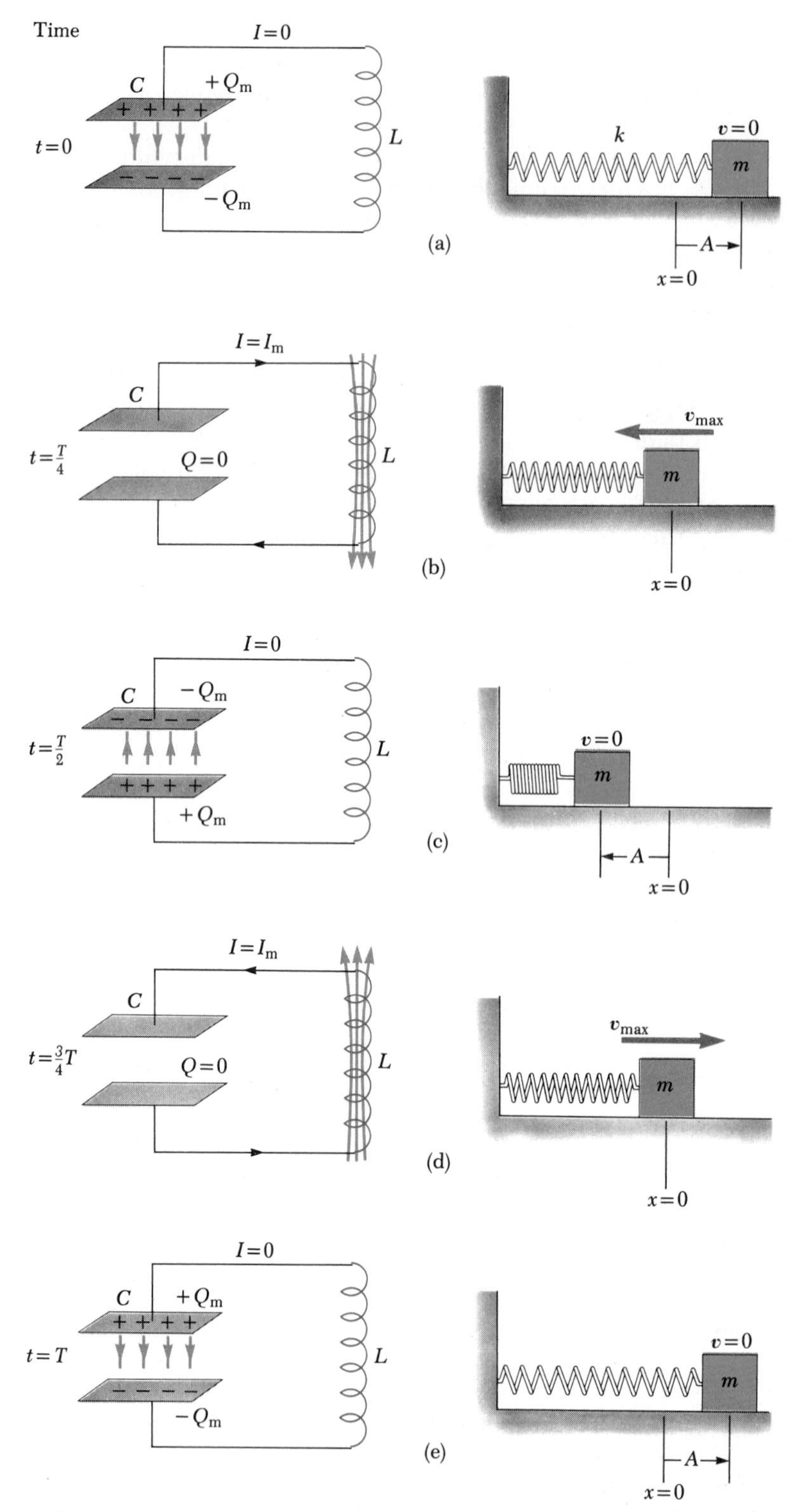

FIGURE 23.2 Various stages of energy transfer in an LC circuit with zero resistance. The capacitor has a charge Q_m at $t = 0$, when the switch is closed. The mechanical analog of this circuit, the mass-spring system, is shown at the right.

creases. At the same time, the current increases and an amount of energy equal to $LI^2/2$ is now stored in the magnetic field of the inductor. Thus, we see that energy is transferred from the electric field of the capacitor to the magnetic field of the inductor. When the capacitor is fully discharged, it stores no energy. At this time, the current reaches its maximum value and all of the energy is stored in the inductor. This process then repeats in the reverse direction. The energy continues to transfer between the inductor and the capacitor, corresponding to oscillations in the current and charge.

A representation of this energy transfer is shown in Figure 23.2. The circuit behavior is analogous to the oscillating mass-spring system studied in Chapter 14. The potential energy stored in a stretched spring, $kx^2/2$, is analogous to the potential energy stored in the capacitor, $Q_m{}^2/2C$. The kinetic energy of the moving mass, $mv^2/2$, is analogous to the energy stored in the inductor, $LI^2/2$, which requires the presence of moving charges. In Figure 23.2a, all of the energy is stored as potential energy in the capacitor at $t = 0$ (because $I = 0$). In Figure 23.2b, all of the energy is stored as "kinetic" energy in the inductor, $LI_m{}^2/2$, where I_m is the maximum current. At intermediate points, part of the energy is potential energy and part is kinetic energy.

As we saw in Chapter 22, the frequency of oscillation of an LC circuit is called the *resonance frequency* of the circuit and is given by

$$f_0 = \frac{1}{2\pi\sqrt{LC}} \tag{23.1}$$

The circuit used by Hertz in his investigations of electromagnetic waves is similar to that discussed above and is shown schematically in Figure 23.3. An induction coil (a large coil of wire) is connected to two metal spheres with a narrow gap between them to form a capacitor. Oscillations are initiated in the circuit by sending short voltage pulses via the coil to the spheres, charging one positive, the other negative. Because L and C are quite small in this circuit, the frequency of oscillation is quite high, $f \approx 100$ MHz. This circuit is called a transmitter because it produces electromagnetic waves.

Hertz placed a second circuit, the receiver, several meters from the transmitter circuit. This receiver circuit, which consisted of a single loop of wire connected to two spheres, has its own effective inductance, capacitance, and natural frequency of oscillation. Hertz found that energy was being sent from the transmitter to the receiver when the resonance frequency of the receiver was adjusted to match that of the transmitter. The energy transfer was detected when the voltage across the spheres in the receiver circuit became high enough to produce ionization in the air, which caused sparks to appear in the air gap separating the spheres. Hertz's experiment is analogous to the mechanical phenomenon in which a tuning fork picks up the vibrations from another, identical vibrating tuning fork.

Hertz assumed that the energy transferred from the transmitter to the receiver was carried in the form of waves, which are now known to be electromagnetic waves. In a series of experiments, he also showed that the radiation generated by the transmitter exhibited the wave properties of interference, diffraction, reflection, refraction, and polarization. As we shall see shortly, all of these properties are exhibited by light. Thus, it became evident

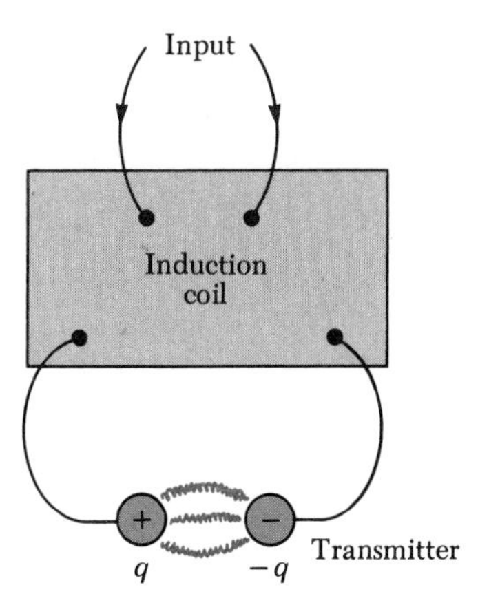

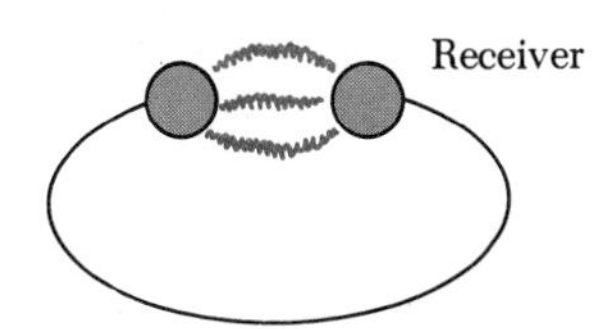

FIGURE 23.3 Schematic diagram of Hertz's apparatus for generating and detecting electromagnetic waves. The transmitter consists of two spherical electrodes connected to an induction coil, which provides short voltage surges to the spheres, setting up oscillations in the discharge. The receiver is a nearby loop containing a spark gap.

that these waves had properties similar to those of light waves and differed only in frequency and wavelength.

Perhaps the most convincing experiment performed by Hertz was his measurement of the velocity of the waves from the transmitter, accomplished as follows. Waves of known frequency from the transmitter were reflected from a metal sheet such that an interference pattern was set up, much like the standing wave pattern on a stretched string. As we saw in our discussion of standing waves, the distance between nodes is $\lambda/2$. Thus, Hertz was able to determine the wavelength, λ. Using the relationship $v = \lambda f$, Hertz found that v was close to 3×10^8 m/s, the known speed of visible light. Thus, Hertz's experiments provided the first evidence in support of Maxwell's theory.

23.3 THE PRODUCTION OF ELECTROMAGNETIC WAVES BY AN ANTENNA

In the previous section, we found that the energy stored in an *LC* circuit is continuously being transferred between the electric field of the capacitor and the magnetic field of the inductor. However, this energy transfer continues for prolonged periods of time only when the changes occur slowly. If the current alternates rapidly, the circuit loses some of its energy in the form of electromagnetic waves. In fact, electromagnetic waves are radiated by *any* circuit carrying an alternating current. The fundamental mechanism responsible for this radiation is the acceleration of a charged particle.

Whenever a charged particle undergoes an acceleration, it must radiate energy.

An alternating voltage applied to the wires of an antenna forces an electric charge in the antenna to oscillate. This is a common technique for accelerating charged particles and is the source of the radio waves emitted by the antenna of a radio station.

Figure 23.4 illustrates the production of an electromagnetic wave by oscillating electric charges in an antenna. Two metal rods are connected to an ac generator, which causes charges to oscillate between the two rods. The output voltage of the generator is sinusoidal. At $t = 0$, the upper rod is given a maximum positive charge and the bottom rod an equal negative charge, as in Figure 23.4a. The electric field near the antenna at this instant is also shown in Figure 23.4a. As the charges oscillate, the rods become less charged, the field near the rods decreases in strength, and the downward-directed maximum electric field produced at $t = 0$ moves away from the rod. When the charges are neutralized, as in Figure 23.4b, the electric field has dropped to zero. This occurs at a time equal to one quarter of the period of oscillation. Continuing in this fashion, the upper rod soon obtains a maximum negative charge and the lower rod becomes positive, as in Figure 23.4c, resulting in an electric field directed upward. This occurs after a time equal to one half the period of oscillation. The oscillations continue as indicated in Figure 23.4d. Note that the electric field near the antenna oscillates in phase with the charge distribution. That is, the field points down when the upper rod is positive and up when the upper rod is negative. Furthermore, the magnitude of the field at any instant depends on the amount of charge on the rods at that instant.

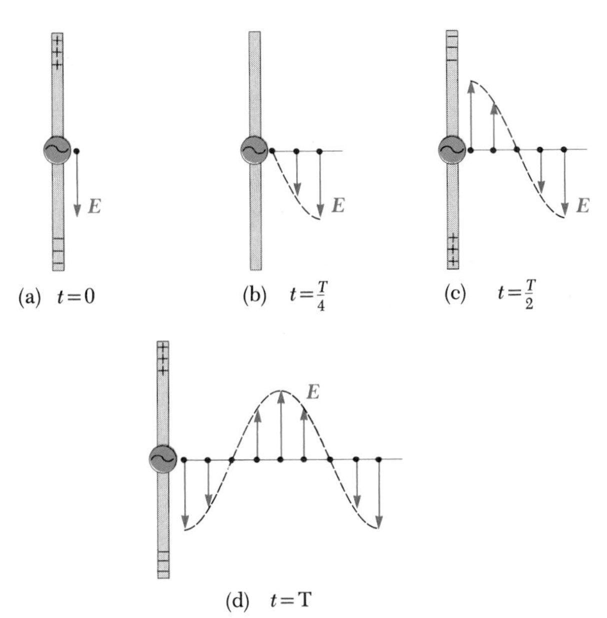

FIGURE 23.4 The electric field set up by oscillating charges in an antenna. The field moves away from the antenna at the speed of light.

As the charges continue to oscillate (and accelerate) between the rods, the electric field set up by the charges moves away from the antenna at the speed of light. Figure 23.4 shows the electric field pattern at various times during the oscillation cycle. As you can see, one cycle of charge oscillation produces one full wavelength in the electric field pattern.

Since the oscillating charges create a current in the rods, a magnetic field is also generated when the current in the rods is upwards, as shown in Figure 23.5. The magnetic field lines circle the antenna and *are perpendicular to the electric field at all points.* As the current changes with time, the magnetic field lines spread out from the antenna. At large distances from the antenna, the strengths of the electric and magnetic fields become very weak. However, at these large distances, it is necessary to take into account the facts that (1) a changing magnetic field produces a changing electric field, and (2) a changing electric field produces a changing magnetic field, as predicted by Maxwell. These induced electric and magnetic fields are in phase; at any point, the two fields reach their maximum values at the same instant. This is illustrated at one instant of time in Figure 23.6. You should make note of the facts that

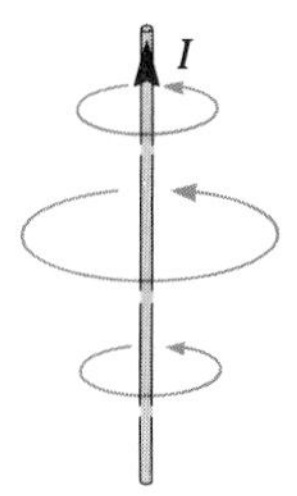

FIGURE 23.5 Magnetic field lines around an antenna carrying a changing current. Why do the circles have different radii?

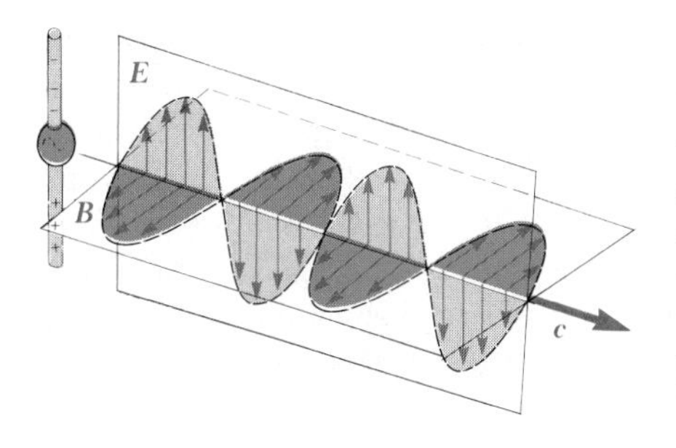

FIGURE 23.6 An electromagnetic wave sent out by oscillating charges in an antenna. This represents the wave at one instant of time. Note that the electric field is perpendicular to the magnetic field, and both are perpendicular to the direction of wave propagation.

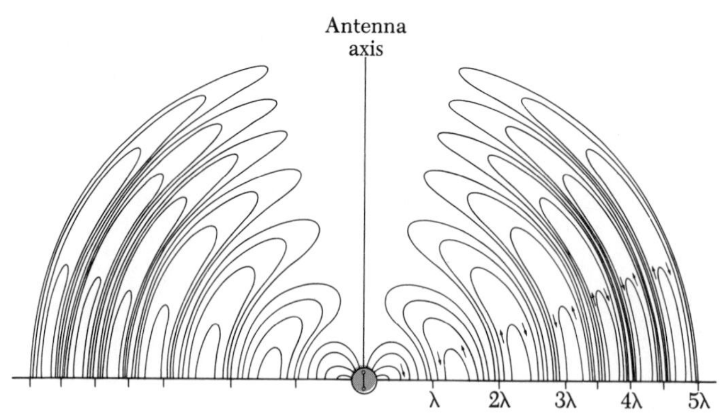

FIGURE 23.7 Electric field lines surrounding an antenna at a given instant. The radiation fields propagate outward with the speed of light.

(1) these fields are perpendicular to each other and (2) both fields are perpendicular to the direction of motion of the wave. This second fact is a property characteristic of transverse waves. Hence, we see that *an electromagnetic wave is a transverse wave.*

A detailed analysis of the radiation from an antenna shows that *energy flows outward from the antenna at all times.* The electric field lines produced by the oscillating charges in an antenna at some instant are shown in Figure 23.7. The intensity of the radiation pattern, and the power radiated, are a maximum in a plane that is perpendicular to the antenna and passing through its midpoint. Furthermore, the power radiated is zero along the axis of the antenna.

23.4 PROPERTIES OF ELECTROMAGNETIC WAVES

As we have seen, Maxwell performed a detailed analysis that predicted the existence and properties of electromagnetic waves. We have already examined some of these properties, and in this section we shall summarize what we have found out about these waves thus far. Additionally, we shall consider some properties we have not yet discussed. In our discussion here and in future sections, we shall often make reference to a type of wave called a **plane wave.** A plane electromagnetic wave is a wave in which the oscillating fields associated with the wave are uniform over a plane at any given time. Such a wave is pictured in Figure 23.8 at a given instant of time. In this case, the oscillations of the electric and magnetic fields take place in planes perpendicular to the x axis, which corresponds to the direction of travel for the wave.

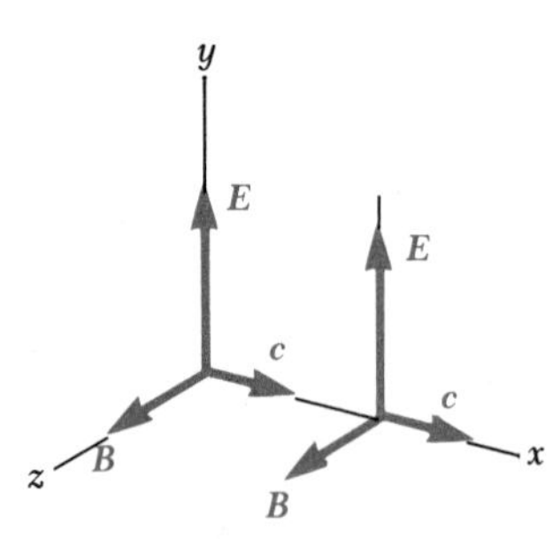

FIGURE 23.8 A plane electromagnetic wave traveling in the positive x direction. The electric field is along the y direction, and the magnetic field is along the z direction.

Electromagnetic waves are transverse waves since the electric and magnetic fields are perpendicular to the direction of travel of the wave. In Figure 23.8, the vibration of the electric field portion of the wave is taken to be in the y direction and the vibration of the magnetic field is in the z direction. Both vibrations are perpendicular to the direction of travel of the wave, which is along the x axis. Figure 23.9 is a pictorial representation of an electromagnetic wave moving in the x direction. This sketch is a "snapshot" of the wave, that is, a representation of its appearance at some instant of time. In this representation, the strength of the electric field is a maximum at the origin,

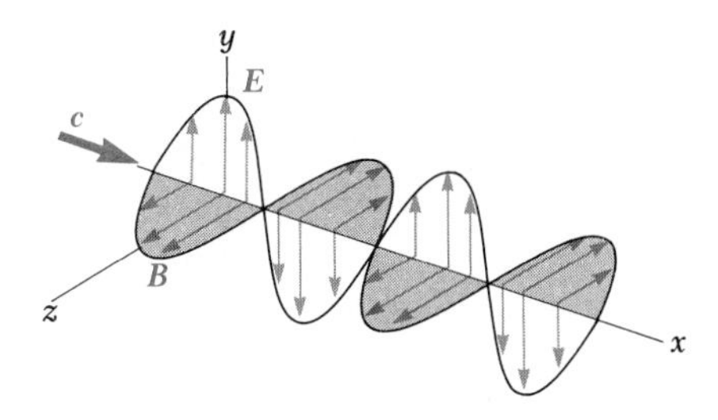

FIGURE 23.9 Representation of a sinusoidal, plane electromagnetic wave moving in the positive x direction with a speed c. The drawing represents a snapshot, that is, the wave at some instant. Note the sinusoidal variations of $\mathbf{E}$ and $\mathbf{B}$ with x.

and the direction of the electric field is in the positive y direction. As one moves out along the x axis, the magnitude of $\mathbf{E}$ decreases sinusoidally to zero. As one proceeds farther along the x axis, the direction of $\mathbf{E}$ reverses and points in the negative y direction and the magnitude begins to increase. The magnetic field follows a similar pattern, except its vibrations take place along the z axis in Figure 23.9.

Electromagnetic waves travel with the speed of light. In fact, it can be shown that the speed of an electromagnetic wave is related to the permeability and permittivity of the medium through which it travels. Maxwell found this relationship for free space to be

$$c = \frac{1}{\sqrt{\mu_0 \epsilon_0}} \tag{23.2}$$

Speed of light

where c is the speed of light, $\mu_0 = 4\pi \times 10^{-7}\ \mathrm{N \cdot s^2/C^2}$ is the permeability constant of free space, and $\epsilon_0 = 8.85 \times 10^{-12}\ \mathrm{C^2/N \cdot m^2}$ is the permittivity of free space. Substituting these values into Equation 23.2, we find that

$$c = 2.99792 \times 10^8\ \mathrm{m/s} \tag{23.3}$$

Since electromagnetic waves travel with a speed that is precisely the same as the speed of light in vacuum, one is led to believe (correctly) that *light is an electromagnetic wave.*

The ratio of the electric to the magnetic field in an electromagnetic field equals the speed of light. That is,

$$\frac{E}{B} = c \tag{23.4}$$

Electromagnetic waves carry energy as they travel through space, and this energy can be transferred to objects placed in the path of the waves. The average rate at which energy passes through an area perpendicular to the direction of travel of a wave, or the average power per unit area, is given by

$$\text{Average power per unit area} = \frac{E_\mathrm{m} B_\mathrm{m}}{2\mu_0} \tag{23.5}$$

Since $E = cB = B/\sqrt{\mu_0 \epsilon_0}$, this can also be expressed as

$$\text{Average power per unit area} = \frac{E_\mathrm{m}^{\,2}}{2\mu_0 c} = \frac{c}{2\mu_0} B_\mathrm{m}^{\,2} \tag{23.6}$$

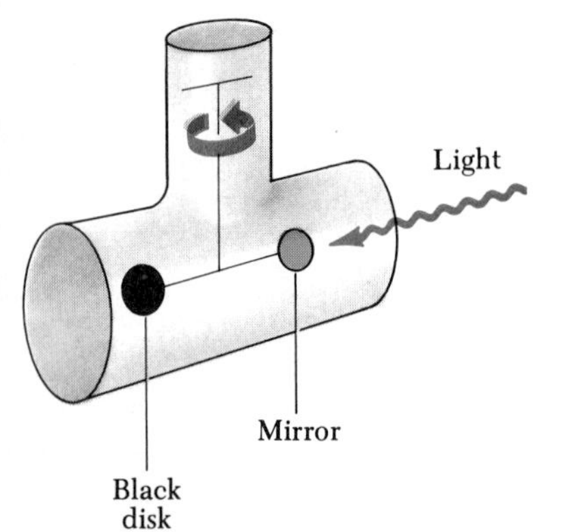

FIGURE 23.10 An apparatus for measuring the pressure of light. In practice, the system is contained in a high vacuum.

Note that the power per unit area in these expressions represents the *average* power per unit area. Also note that the values to be used for E and B in these equations are the *maximum* values. It is of some interest to note that a detailed analysis would show that the energy carried by an electromagnetic wave is shared equally by the electric and magnetic fields.

Electromagnetic waves transport momentum as well as energy. This is demonstrated by the fact that radiation pressure is exerted on a surface when an electromagnetic wave impinges upon it. Although radiation pressures are very small (about 5×10^{-6} N/m^2 for direct sunlight), they have been measured using a device like the one pictured in Figure 23.10. Light is allowed to strike a mirror and a black disk connected to each other by a horizontal bar suspended from a fine fiber. Light striking the black disk is completely absorbed, and so *all* of the momentum of the light is transferred to the disk. Light striking the mirror head on is totally reflected; hence the momentum transfer to the mirror is *twice* as great as that transmitted to the disk. As a result, the horizontal bar supporting the disks twists counterclockwise as seen from above. The bar comes to equilibrium at some angle under the action of the torques due to the radiation pressure and the torque produced as the fiber twists. The radiation pressure can be determined by measuring the angle at which equilibrium occurs. The apparatus must be placed in a high vacuum to eliminate the effects of air currents.

In summary, electromagnetic waves traveling through free space have the following properties:

1. Electromagnetic waves travel with the speed of light.
2. Electromagnetic waves are transverse waves since the electric and magnetic fields are perpendicular to the direction of propagation of the wave and to each other.
3. The ratio of the electric to the magnetic field in an electromagnetic wave equals the speed of light.
4. Electromagnetic waves carry both energy and momentum, which can be delivered to a surface.

EXAMPLE 23.1 Solar Energy

The sun delivers an average power per unit area of about 700 W/m^2 to the earth's surface. Calculate the total power incident on a roof of dimensions 8 m by 20 m. Assume the radiation is incident *normal* to the roof (sun directly overhead).

Solution The power per unit area, or light intensity, is 700 W/m^2. For normal incidence, we get

$$\text{Power} = (700\ \text{W/m}^2)(8 \times 20\ \text{m}^2) = 1.12 \times 10^5\ \text{W}$$

Note that if this power could *all* be converted to electric power, it would provide more than enough power for the average home. Unfortunately, solar energy is not easily harnessed, and the prospects for large-scale conversion are not as bright as they may appear from this simple calculation. For example, the conversion efficiency from solar to electrical energy is far less than 100% (typically, it is 10% for photovoltaic cells). Roof systems for converting solar energy to thermal energy have been built with efficiencies of around 50%; however, there are other practical problems with solar energy that must be considered, such as overcast days, geographic location, and energy storage.

EXERCISE How much solar energy (in joules) is incident on the roof in 1 h?

Answer 4.03×10^8 J.

23.5 THE SPECTRUM OF ELECTROMAGNETIC WAVES

In 1887, when Hertz successfully generated and detected the electromagnetic waves that had been predicted by Maxwell, the only electromagnetic waves recognized were those that have since come to be known as radio waves and those in the frequency range of visible light. It is now known that there are electromagnetic waves of many other frequencies and wavelengths.

Because all electromagnetic waves travel through a vacuum with a speed c, their frequency and wavelength are related by the important expression

$$c = f\lambda \tag{23.7}$$

The various types of electromagnetic waves are listed in Figure 23.11. Note the wide range of frequencies and wavelengths. For instance, a radio wave of frequency 5 MHz (a typical value) has a wavelength in vacuum of

$$\lambda = \frac{c}{f} = \frac{3 \times 10^8 \text{ m/s}}{5 \times 10^6 \text{ s}^{-1}} = 60 \text{ m}$$

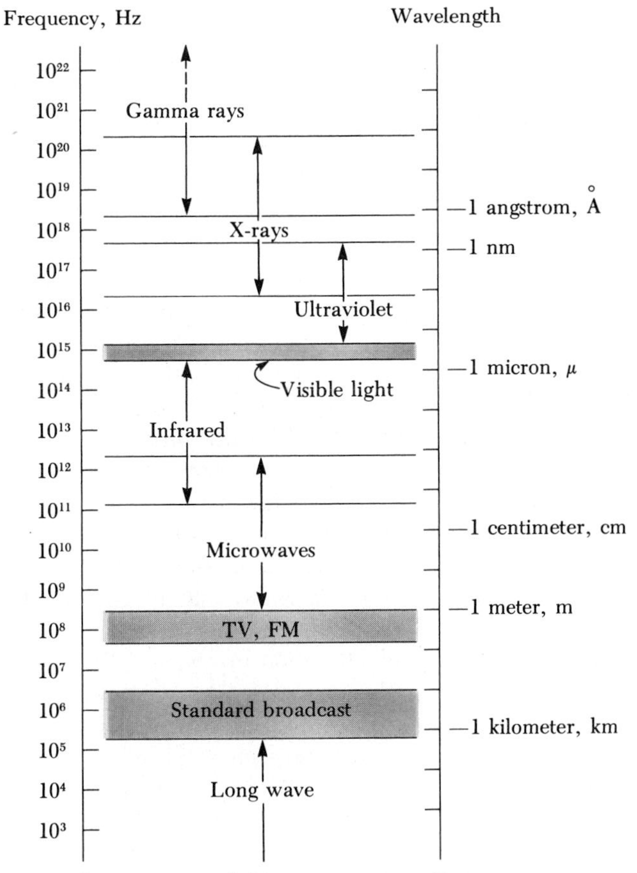

FIGURE 23.11 The spectrum of electromagnetic radiation.

In the sections that follow, we shall give a brief description of these various waves in order of decreasing wavelength. There is no sharp dividing point between one kind of wave and the next.

EXAMPLE 23.2 An Electromagnetic Wave

A plane electromagnetic sinusoidal wave of frequency 40 MHz travels in free space in the x direction, as in Figure 23.8. At some point and at some instant, the electric field has its *maximum* value of 750 N/C and is along the y axis. Determine the wavelength and period of the wave.

Solution Since $c = \lambda f$ and $f = 40 \text{ MHz} = 4 \times 10^7 \text{ s}^{-1}$, we get

$$\lambda = \frac{c}{f} = \frac{3 \times 10^8 \text{ m/s}}{4 \times 10^7 \text{ s}^{-1}} = 7.5 \text{ m}$$

The period of the wave, T, equals the inverse of the frequency:

$$T = \frac{1}{f} = \frac{1}{4 \times 10^7 \text{ s}^{-1}} = 2.5 \times 10^{-8} \text{ s}$$

(b) Calculate the magnitude and direction of the magnetic field when $E = 750$ N/C and is in the y direction.

Solution From Equation 23.4 we see that

$$B = \frac{E}{c} = \frac{750 \text{ N/C}}{3 \times 10^8 \text{ m/s}} = 2.5 \times 10^{-6} \text{ T}$$

Because the magnetic and electric fields must be perpendicular to each other and both must be perpendicular to the direction of wave propagation (x in this case), we conclude that $\boldsymbol{B}$ is along the z axis.

Radio Waves

This radiotelescope is used to detect radio-frequency radiation from stars and other objects in the universe. See also Color Plate 18. (Photo California Institute of Technology)

Radio waves are produced by charges accelerating through conducting wires, as discussed in Section 23.3. They get their name because their principal use is as a carrier of information from a radio transmitting station to your home receiver. Because communication via radio signals is so important in today's world, we shall examine the fundamentals of this process.

All domestic radio stations are assigned a *carrier frequency* by the federal government. This assigned frequency is the basic frequency of the electromagnetic wave the station is allowed to broadcast. If the station is an AM *(amplitude modulation)* station, its carrier frequency will be between 530 and 1605 kHz, where FM *(frequency modulation)* stations operate at frequencies from 88 to 108 MHz.

The term *modulation* means alteration or change. To visualize the process for an AM station, let the carrier wave be represented as in Figure 23.12a. The sound signal to be transmitted by this carrier wave is converted by a microphone to an electric signal and superimposed on the carrier wave. For example, if the audio signal is a 100-Hz tone, the amplitude of the carrier wave would fluctuate at a rate of 100 times per second. Figure 23.12b shows a waveform representing an arbitrary audio signal, and Figure 23.12c represents the carrier wave with the audio signal impressed on it.

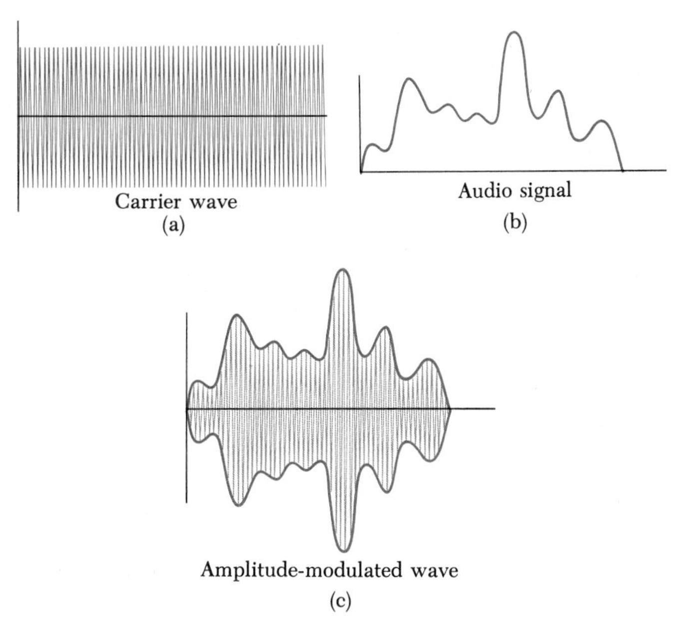

FIGURE 23.12 Amplitude modulation.

The modulated carrier wave travels through space as an unnoticed presence in our environment until it is intercepted by an antenna. There are two common types of antennas used to detect these waves. The one most often used in cars consists of a metal rod extended into the air. In this type of antenna, the changing electric field in the carrier wave causes electrons in the rod to oscillate. The amplitude of these oscillations varies with the strength of the electric field in the modulated carrier wave. That is, the strength of the receiving antenna current varies with a frequency matching that of the audio wave impressed on the carrier wave.

The second type of antenna is a coil of wire built into the back of the radio receiver. The changing magnetic field portion of the carrier wave induces a current in the coil that emulates the variations in amplitude of the carrier wave.

Impinging on the receiving antenna are not one but several such electromagnetic waves, one from every radio station within range. Figure 23.13 illustrates one method in which the tuner of the receiver selects the desired

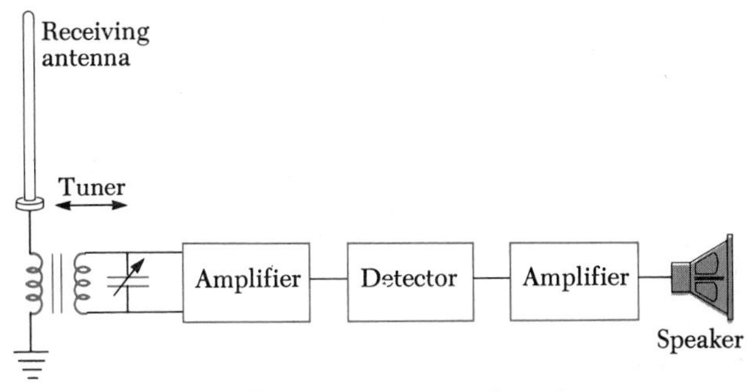

FIGURE 23.13 The various stages of a radio receiver.

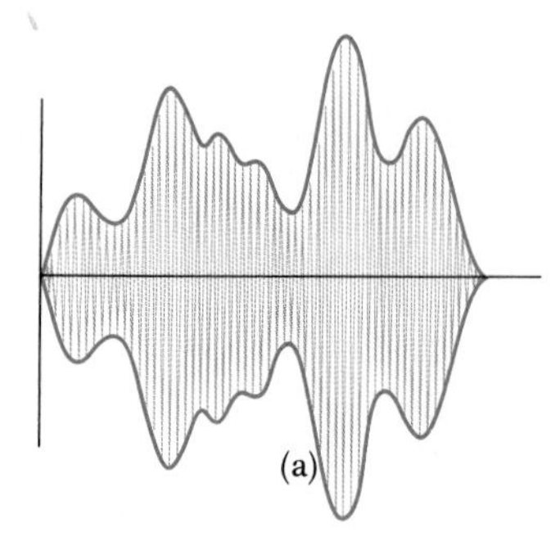

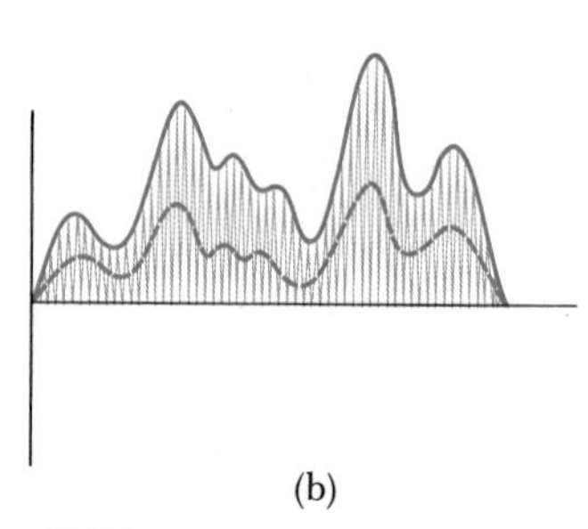

FIGURE 23.14 (a) The average value of this electrical signal is zero. (b) The dashed colored line represents the average of the signal shown by the solid line.

signal. The oscillating antenna current induces a current in the tuner coil via the coil shown on the left. As we have seen, a circuit consisting of an inductor and a capacitor can serve as a resonant circuit and enhance a certain frequency while suppressing others. As you tune your radio, you vary the capacitance in the tuner until the circuit is in resonance at a frequency equal to the carrier wave of the desired radio station signal.

The current delivered to the receiver by the antenna is very weak and must be strengthened by an amplifier circuit (Fig. 23.13) before it is delivered to the next stage of the radio receiver, the *detector*, which removes the audio information encoded on the carrier wave. The detector in most radios is either a vacuum tube or a transistor, but we can understand the principle of what occurs by considering our detector to be a diode. A diode is a solid-state device that allows a current in only one direction. If one attempts to feed the high-frequency current from the antenna to a speaker, the speaker will not respond because it is a mechanical device and thus has too much inertia to vibrate at the high frequencies of radio waves. Instead, the speaker will respond to the average value of the current; however, for an input signal like that shown in Figure 23.14a, the average value of the current in the circuit is zero. Allowing the current to pass through a diode clips off half the current, as shown in Figure 23.14b. The average value for this current is represented by the heavy dashed line in the figure. This average value follows the same pattern as the audio signal on the carrier wave. The speaker is able to follow these lower-frequency variations and thus reproduce the original sound.

In frequency modulation (FM), the frequency is altered, as pictured in Figure 23.15. Figure 23.15a shows the unmodulated carrier wave, Figure 23.15b depicts an audio signal, and Figure 23.15c shows the carrier wave after it has been modulated. The wider-than-normal portions in Figure 23.15c indicate that the carrier has a lower-than-normal frequency at these locations, and the closely spaced portions indicate points where the frequency is higher than normal. For example, if an FM station is assigned a carrier frequency of 98.6 MHz, this represents the basic frequency the station is allowed to transmit, but this frequency can vary between 98.4 MHz and 98.8 MHz. For example, if the station is transmitting a frequency of 100 Hz, the frequency of the carrier wave varies 100 times per second within its allowed range. The function of the radio receiver is, as before, to select the desired station, separate the audio signal from the carrier, and reproduce the audio signal with a speaker.

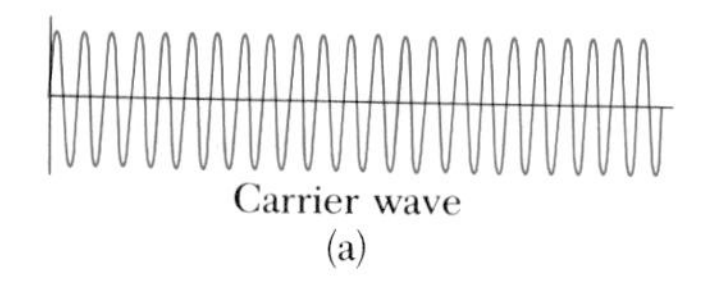

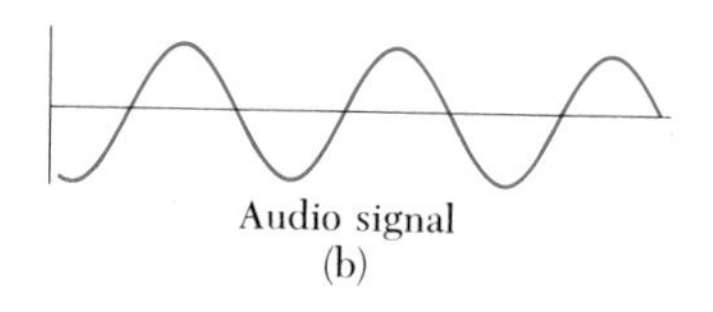

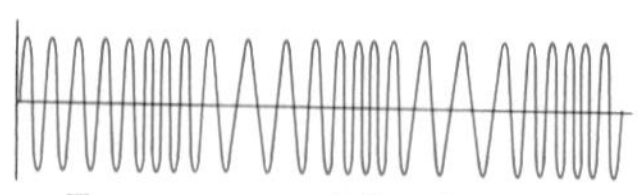

FIGURE 23.15 Frequency modulation.

EXAMPLE 23.3 Tuning a Radio

The tuner coil in Figure 23.13 has an inductance of 5 μH. If you desire to listen to a radio station that broadcasts at a frequency of 1400 kHz, to what value will you have to adjust the capacitor?

Solution The frequency at which the LC circuit will resonate is given by Equation 23.1. Solving this equation for C and substituting, we find

$$C = \frac{1}{4\pi^2 f_0^2 L} = \frac{1}{4\pi^2(1400 \times 10^3\ \text{s}^{-1})^2(5 \times 10^{-6}\ \text{H})}$$
$$= 2.6 \times 10^{-9}\ \text{F} = 2.6\ \text{pF}$$

EXERCISE What is the wavelength of the carrier wave transmitted by this radio station?

Answer

Radar detectors on the tower of a ship. (Courtesy of Tom Penegor)

Microwaves

Microwaves are short-wavelength radio waves that have wavelengths between about 1 mm and 30 cm and are generated by electronic devices. Microwave ovens represent an interesting domestic application of these waves. Food can be cooked by this radiation because one of the natural rotational frequencies of water molecules is in the range of microwave frequencies. When microwave radiation of this frequency is directed at something to be cooked, resonance is set up in the water molecules in the food. This energy added to the water heats the food internally.

Microwaves are also used in communication, particularly in the transmission of long-distance telephone messages. In this process, the information to be transmitted is encoded on microwaves by either amplitude or frequency modulation and carried from point to point via microwave relay towers.

In a recent proposal, it was suggested that solar energy could be harnessed by placing a solar collector in space and then beaming down the collected energy by microwave to a receiving station on earth. The obvious advantage to this scheme is that the sun shines on the collector continuously, and there are no problems with cloud cover.

One of the most common applications of microwaves is in the radar (radio detection and ranging) systems used in aircraft navigation and police cars. The principles of radar are quite similar to those used in radio broadcasting. As in broadcasting, an antenna is used to transmit an electromagnetic wave. The wave produced is not a continuous wave, but consists of short bursts, or pulses. After the emission of a pulse, the transmitting antenna is deactivated as a transmitter momentarily and used as a receiver to detect any reflected pulses. A signal, or "blip," proportional to the distance of the reflecting object can easily be displayed on a fluorescent screen or calculated as outlined in the following example.

EXAMPLE 23.4 Measuring Distances with Radar

A radar receiver indicates that a transmitted pulse is returned as an echo 20 μs after transmission. How far away is the reflecting object?

Solution Traveling with a speed of 3×10^8 m/s for 20 μs, the pulse travels a total distance

$$x = ct = (3 \times 10^8 \text{ m/s})(20 \times 10^{-6} \text{ s}) = 6 \times 10^3 \text{ m}$$

Because the pulse must travel to the reflecting object and back, the distance of the object from the transmitter is half this value, or 3×10^3 m.

Infrared Waves

Infrared waves (sometimes called heat waves) have wavelengths ranging from about 1 mm to the longest wavelength of visible light, 7×10^{-7} m. These waves, produced by hot objects, are readily absorbed by most materials. The infrared energy absorbed by a substance appears as heat because the energy agitates the atoms of the substance, increasing their vibrational or translational motion, which is detected as a temperature rise.

Photographic film made with an emulsion sensitive to infrared waves has found increased usage. One advantage is that infrared rays can penetrate fog and haze better than visible light. Additionally, when these films are used in cameras on airplanes or satellites, the developed film provides more information about land areas than photographs taken with visible light. This occurs because there are distinct differences between the amount of radiation emitted by the ground, by water, and by vegetation. It is possible, for example, to distinguish between a field of corn and a field of wheat because the two emit different wavelengths of infrared radiation. Remote disease detection is also feasible since a diseased crop differs from a healthy crop in the amount of infrared radiation emitted.

There are special telescopes that can immediately change an image produced by infrared radiation into a visible image. The visible image produced in this way, however, is considerably different from a normal image. For example, the exposed portions of a person's body would emit more infrared radiation than would the cooler clothes that cover most of the body. This means that the uncovered areas would appear much brighter than the rest of the body. Slightly revised versions of this process enable doctors to diagnose cancerous growths. An infrared scan may reveal hot spots in a patient's body. These hotter-than-normal regions are characteristic of malignant growths.

Visible Light

Visible light, the most familiar form of electromagnetic waves, may be defined as that part of the electromagnetic spectrum that the human eye can detect. Light is produced by the rearrangement of electrons in atoms and molecules, as we shall discuss in Chapters 29 and 30. The various wavelengths of visible light are classified by color, ranging from violet ($\lambda \approx 4 \times 10^{-7}$ m) to red ($\lambda \approx 7 \times 10^{-7}$ m). The eye's sensitivity is a function of wavelength, the sensitivity being a maximum at a wavelength of about 5.6×10^{-7} m (yellow-green).

Light is the basis of the science of optics and optical instruments, which we shall deal with in Chapters 24 through 27. The units often used to designate short wavelengths and distances are abbreviated as follows:

$$1 \text{ micron } (\mu\text{m}) = 10^{-6} \text{ m}$$
$$1 \text{ nanometer (nm)} = 10^{-9} \text{ m}$$
$$1 \text{ angstrom } (\text{Å}) = 10^{-10} \text{ m}$$

In these units, the wavelengths of visible light range from 0.4 to 0.7 μm, or 400 to 700 nm, or 4000 to 7000 Å.

Ultraviolet Light

Ultraviolet light is the name given to wavelengths ranging from approximately 3.8×10^{-7} m (380 nm) down to 6×10^{-8} m (60 nm). The sun is an important source of ultraviolet light, which is the main cause of suntans. Most of the ultraviolet light from the sun is absorbed by atoms in the upper atmosphere, or stratosphere. This is fortunate because ultraviolet light in larger quantities is harmful to humans. One important constituent of the stratosphere is ozone (O_3), which results from reactions between oxygen and ultraviolet radiation. This ozone shield converts lethal high-energy ultraviolet radiation to infrared radiation, which in turn warms the stratosphere. Recently there has been a great deal of controversy concerning the possible depletion of the protective ozone layer as a result of the release into the air of Freons used in aerosol spray cans and as refrigerants.

The tanning process of the human body occurs as a result of a protective mechanism of the body against the harmful effects of ultraviolet rays. Normal skin contains melanin pigment, which gives the skin a brownish coloration. Excessive ultraviolet light stimulates the body to produce more melanin pigment, resulting in a rosy tan. Ultraviolet light in moderation may give the body a healthy glow, but if carried to extremes, tanning may, over a period of years, lead to unsightly thickening of the skin and even to skin cancer.

Ultraviolet radiation is often used to kill germs. Ultraviolet lamps are used to sterilize hospital operating rooms and surgical instruments. Low-intensity ultraviolet lamps are sometimes placed above grocery meat counters to reduce spoilage.

Ordinary glass provides an effective shield against ultraviolet light. As a result, one cannot get a tan when separated from the sun by a glass window. Some light sources inherently produce large amounts of ultraviolet radiation; mercury lamps are a case in point. These lamps are enclosed in a glass shield if they are to be used for the visible light they produce. If one is interested in obtaining the ultraviolet light, quartz housings are used because ultraviolet light can readily pass through quartz.

X-rays

X-rays are electromagnetic waves with wavelengths in the range from about 10^{-8} m (10 nm) down to 10^{-13} m (10^{-4} nm). X-rays are produced when high-energy electrons decelerate after striking a metal target. These rays are used as a treatment for certain forms of cancer, but their most widely known use is as a diagnostic tool in medicine and dentistry. If a beam of x-rays is focused on a hand, for example, the penetrating power of the rays is sufficient to allow the rays to pass right through the hand and expose a photographic film on which the hand rests. Because it is easier for the waves to penetrate fatty tissue than bone, the exposed film shows the bones standing out in vivid detail. Furthermore, since x-rays damage or destroy living tissues and organisms, care must be taken to avoid unnecessary exposure or overexposure. We shall discuss x-rays in more detail in Chapter 30.

Gamma Rays

Gamma rays are electromagnetic waves emitted by radioactive nuclei and during certain nuclear reactions. They have wavelengths ranging from about 10^{-10} m to less than 10^{-14} m. They are highly penetrating and produce serious damage when absorbed by living tissues. Consequently, those working near such dangerous radiation must be protected with heavily absorbing materials, such as thick layers of lead. Under controlled conditions, gamma rays can be used to kill cancerous cells.

Note in Figure 23.11 that the wavelength ranges of x-rays and gamma rays overlap. This means that the two are essentially the same in their physical nature. They are divided according to their origin rather than to some physical quantity, such as wavelength. We shall discuss the origin of gamma rays in greater detail in Chapter 30. Finally, note that the boundary between ultraviolet rays and x-rays is as blurred as that between x-rays and gamma rays. Again, the distinction between the two is based on how they are produced.

*23.6 TELEVISION

In this section we shall describe the underlying physical principles of television transmission and reception. As we shall see, the techniques used in this form of communication include a multitude of phenomena from the fields of electricity and optics.

Television signals, like radio signals, are carried from the transmitting station to our homes via electromagnetic carrier waves. However, television signals are sent out on two carrier waves; one carries the audio (sound) signal, and the other carries the video (picture) signal. The basic idea of sound transmission by television carrier waves is identical to that for FM radio wave transmission, discussed in Section 23.5. Therefore, let us confine our discussion to the video portion of the television signal.

Video signals displayed on your television screen use two forms of optical deception, one of which is used in motion pictures, the other in photography. The optical deception used in motion pictures involves the fact that a series of still pictures scanned rapidly by the eye are perceived as continuous motion in the scene that was photographed. A theater movie projector flashes 24 different still pictures before your eyes each second. Each picture is only slightly different from the preceding one, and the persistence of eye vision blends them together as continuous action. In the case of television signals, 30 different pictures appear on the television screen each second.

The second optical deception used in video display can be understood by careful examination of a photograph in any newspaper. Close scrutiny of such a photograph with a good magnifying glass will reveal that the photograph is not continuous. Rather, it consists of a large number of closely spaced dots. At large distances from the photograph, the eye is not able to distinguish the individual dots and therefore is tricked into thinking that the photograph is a continuous scene. A television camera reduces an individual picture into a series of tiny dots similar to the scheme used in photographs. The camera produces 30 complete pictures each second, one dot at a time, which are sent out on the carrier wave. Your television set receives these strings of dots and reassembles them into a picture on the screen of the picture tube.

The basic element in a television camera is an image orthicon tube (Fig. 23.16), which generates the video signal. Light from the object being tele-

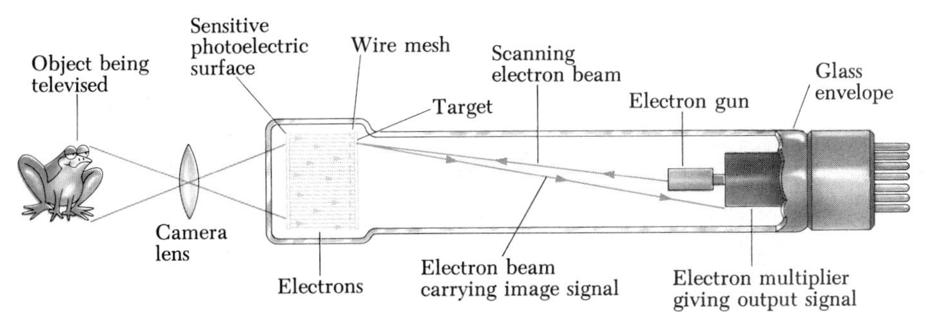

FIGURE 23.16 An image orthicon tube, which is the basic element of a television camera. (Adapted from J. Wilson, *Technical College Physics*, Saunders College Publishing Co., 1982).

vised is focused by lenses onto a light-sensitive surface that releases electrons. Few electrons are released in regions where the surface is dark, and many are released in highly illuminated regions. Thus, an image of the object appers on the surface, which is in fact an electron image. The surface is then scanned with an electron beam, which "reads" the information by converting the image to electric signals.

The signals produced in this manner are transmitted to the receiver, along with the FM audio signal, via the electromagnetic carrier wave (Fig. 23.17). The picture tube (cathode ray tube) in the home receiver is synchronized with the message sent to it by the carrier wave. The electron beam in the picture tube scans the fluorescent screen, producing hundreds of horizontal lines in a fraction of a second. Signals from the television camera are used to control the variations in the electron beam striking the screen. In this manner, an image is reproduced on the screen of the picture tube as a mosaic of light and dark dots. The dots are perceived by the eye as a complete black-and-white picture.

Color television cameras and sets operate in basically the same way, except that a color camera contains three image tubes instead of one. A series

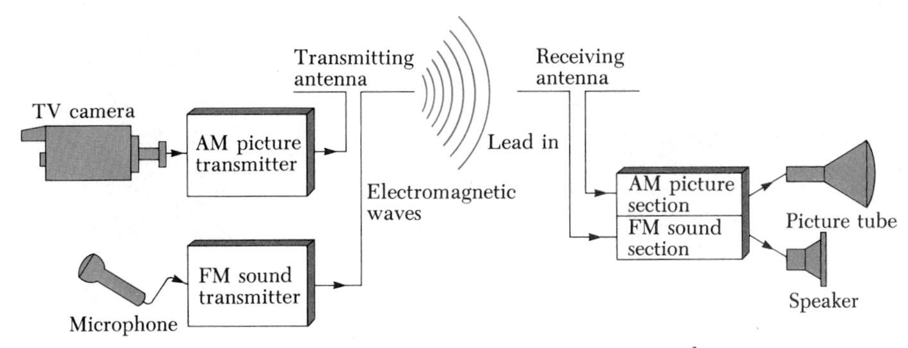

FIGURE 23.17 Simplified diagram of television transmission and reception. Television combines both a picture AM signal and an audio FM signal. (Adapted from J. Wilson, *Technical College Physics*, Saunders College Publishing Co., 1982).

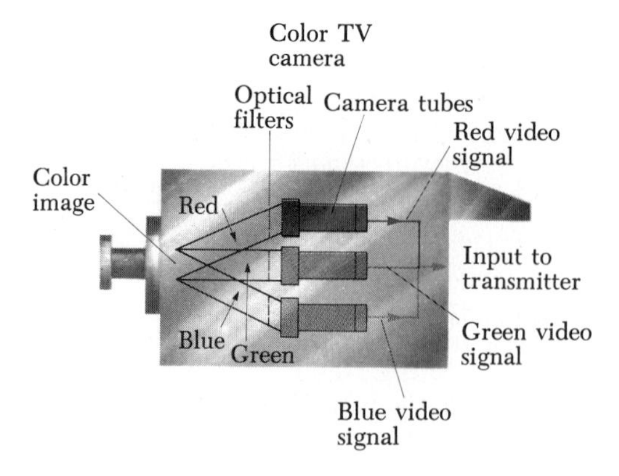

FIGURE 23.18 In the color camera, the incoming light beam is separated into three light beams of primary colors. In the receiving set, the beams are recombined to produce a colored image. (Adapted from J. Wilson, *Technical College Physics*, Saunders College Publishing Co., 1982).

of filters and mirrors (Fig. 23.18) separate the light from the scene being televised into the three primary colors, red, green, and blue. Information associated with the strength of each color signal is then transmitted to the home receiver. The color picture tube in the receiver has three electron guns (Fig. 23.19a), each of which produces an electron beam whose intensity depends on the intensity of the red, green, or blue signal reaching it. These three electron beams are directed to the fluorescent screen on the tube face (Fig. 23.19b). The signal from the blue electron gun strikes a dot on the screen, which causes it to give off blue light. Likewise, the red and green guns produce their respective colors. Because these dots are very close together, the eye cannot distinguish them individually from a normal viewing distance. Instead, the eye blends the three colors together to produce a color that depends on the intensity of the light being emitted by each dot. If each dot emits the same light intensity, the colors blend together to produce a white spot at that location. Any color can be reproduced at a given screen location by a proper combination of the three primary colors.

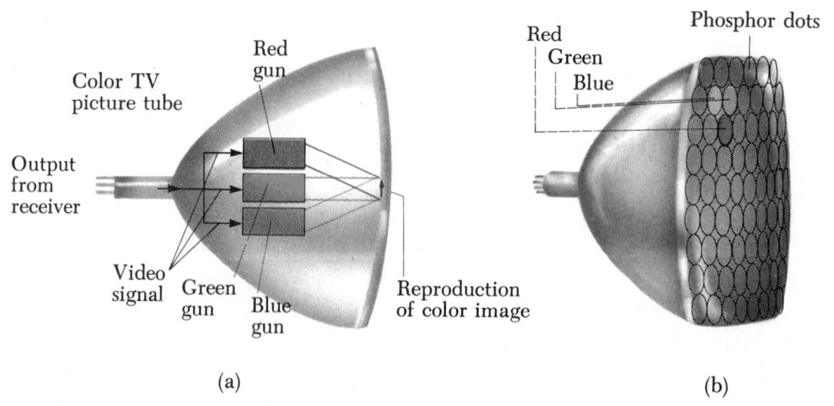

FIGURE 23.19 (a) The color television picture tube has three electron guns. (b) The screen of a color television picture tube contains three sets of colored dots. The size of the dots has been exaggerated for clarity.

SUMMARY

Electromagnetic waves were predicted by James Clerk Maxwell and later generated and detected by Heinrich Hertz. These waves have the following properties:

1. Electromagnetic waves are transverse waves since the electric and magnetic fields are perpendicular to the direction of travel.
2. Electromagnetic waves travel with the speed of light.
3. The ratio of the electric to the magnetic field in an electromagnetic field equals the speed of light, that is

$$\boxed{\frac{E}{B} = c} \qquad (23.4)$$

4. Electromagnetic waves carry energy as they travel through space. The average power per unit area is

$$\boxed{\frac{E_m B_m}{2\mu_0} = \frac{E_m{}^2}{2\mu_0 c} = \frac{c}{2\mu_0} B_m{}^2} \qquad (23.5,\ 23.6)$$

   where E_m and B_m are the maximum values of the electric and magnetic fields.
5. Electromagnetic waves transport momentum as well as energy. The speed, c, frequency, f, and wavelength, λ, of an electromagnetic wave are related as

$$\boxed{c = f\lambda} \qquad (23.7)$$

The electromagnetic spectrum includes waves covering a broad range of frequencies and wavelengths. These waves have a variety of applications and characteristics, depending on their frequency or wavelength.

ADDITIONAL READING

A. H. W. Beck, *Words and Waves*, World University Library, New York, McGraw-Hill, 1967.

J. Hertz, *Heinrich Hertz: Memoirs, Letters, Diaries*, San Francisco Press, 1977.

O. Kedem and U. Ganiel, "Solar Energy, How Much Do We Receive?" *The Physics Teacher*, December 1983, p. 573.

D. K. C. McDonald, *Faraday, Maxwell and Kelvin*, Science Study Series, Garden City, N.Y., Doubleday, 1964.

J. R. Pierce, *Electrons and Waves*, Science Study Series, Garden City, N.Y., Doubleday, 1964.

QUESTIONS

1. For a given incident energy of an electromagnetic wave, why is the radiation pressure on a perfectly reflecting surface twice as large as the pressure on a perfectly absorbing surface?
2. Does the wire connected to a battery emit an electromagnetic wave? Explain.
3. If you charge a comb by running it through your hair and then hold the comb next to a bar magnet, do the electric and magnetic fields produced constitute an electromagnetic wave?
4. If the fundamental source of a sound wave is a vibrating object, what is the fundamental source of an electromagnetic wave?

5. How does a microwave oven cook food?
6. An empty plastic or glass dish removed from a microwave oven is cool to the touch. How can this be possible?
7. Certain orientations of the receiving antenna of a television set give better reception than others. Furthermore, the best orientation varies from station to station. Explain these observations.
8. Often when you touch the indoor antenna on a television receiver, the reception instantly improves. Why?
9. List as many similarities and differences as you can between sound waves and light waves.
10. What does a radio wave do to the charges in the receiving antenna to provide a signal for your car radio?
11. When light (or other electromagnetic radiation) travels across a given region, what is it that moves?
12. Why should an infrared photograph of a person look different from a photograph taken with visible light?
13. Radio stations often advertise "instant news." If what they mean is that you hear the news at the instant they speak it, is their claim true? About how long would it take for a message to travel across this country by radio waves, assuming that these waves could travel this great distance and still be detected?
14. Suppose a creature from another planet had eyes that were sensitive to infrared radiation. Describe what he would see if he looked around the room you are now in. That is, what would be bright and what would be dim?
15. Why would you expect to find less static on the FM radio band than on the AM band?

PROBLEMS

Section 23.2 Hertz's Discoveries

1. If the coil in the resonant circuit of a radio has an inductance of 2 μH, what range of values must the tuning capacitor have in order to cover the complete range of FM frequencies? (See the section on Radio Waves to find the FM range of frequencies.)

2. A standing wave interference pattern is set up by radio waves between two metal sheets located 2 m apart. This is the shortest distance between the plates which will produce a standing wave pattern. What is the fundamental frequency?

Section 23.4 Properties of Electromagnetic Waves

3. Experimenters at the National Bureau of Standards have made precise measurements of the speed of light using the fact that, in vacuum, the velocity of electromagnetic waves is $c = 1/\sqrt{\mu_0\epsilon_0}$, where $\mu_0 = 4\pi \times 10^{-7}$ N · s^2/C^2 and $\epsilon_0 = 8.854 \times 10^{-12}$ C^2/N·m^2. What value (to four significant figures) does this give for the speed of light in vacuum?

4. Verify that Equation 23.2 has the proper dimensions.

5. An electromagnetic wave in vacuum has an electric field amplitude of 150 V/m. Calculate the amplitude of the accompanying magnetic field.

6. Find the average power per unit area associated with the wave described in Problem 5.

7. The sun delivers an average power of 1340 W/m^2 to the top of the earth's atmosphere. Find the magnitudes of $\boldsymbol{E}_m$ and $\boldsymbol{B}_m$ for the electromagnetic waves at the top of the atmosphere.

8. How much solar energy is incident on a collector with dimensions 20 m by 30 m at the top of the earth's atmosphere in 8 h? (See Problem 7.)

9. A particular electromagnetic wave traveling in vacuum has a magnetic field intensity of 1.5×10^{-7} T. Find (a) the electric field intensity, and (b) the average power per unit area associated with the wave.

10. An electromagnetic wave is traveling west with the electric field vibrating at a frequency of 10^{14} Hz perpendicular to the surface of the earth. What is (a) the frequency of vibration of the magnetic field, and (b) the direction of vibration?

11. The light from a 5 mW laser is spread out in a cylindrical beam whose diameter is about 0.5 cm. What are the peak values of $\boldsymbol{E}$ and $\boldsymbol{B}$ in this beam?

12. A plane electromagnetic wave has an average power per unit area of 300 W/m^2. A flat, rectangular surface of dimensions 20 cm by 40 cm is placed perpendicular to the direction of the plane wave. If the surface absorbs half the energy and reflects half, calculate the net energy delivered to the surface in 1 min.

Section 23.5 The Spectrum of Electromagnetic Waves

13. What is the frequency of an electromagnetic wave that has a wavelength of (a) 2 m, (b) 20 m, and (c) 200 m?

14. Locate the position in the electromagnetic spectrum of the waves that have the following frequencies, and find the wavelength associated with each of these frequencies: (a) 10^6 Hz, (b) 10^8 Hz, (c) 10^{10} Hz, (d) 10^{13} Hz, (e) 10^{15} Hz, (f) 10^{17} Hz, and (g) 10^{21} Hz.

15. A wave has a wavelength of 4×10^{-7} m and a speed of 2.5×10^8 m/s in a particular material. (a) What is the frequency of this wave in the material? (b) The frequency of a wave does not change as the wave moves from one material to another. Find the wavelength of this wave in vacuum.

16. A singer's voice is transmitted by a radio wave to a person 100 km away. (a) How much time passes before the distant listener hears the sound? (b) By the time the radio message reaches a listener, how far from the singer has the sound wave moved in the auditorium? Assume that the speed of sound is 345 m/s.

17. The moon is approximately 250,000 mi from us and the sun is approximately 93,000,000 mi away. How much time is required for an electromagnetic wave to reach us from (a) the sun and (b) the moon?

18. A tuning circuit in a radio receiver is set such that $L = 5.4\ \mu$H and $C = 3.1$ pF. What is the wavelength of the wave being detected?

19. A radar pulse returns to the receiver after a total travel time of 4×10^{-4} s. How far away is the object that reflected the wave?

20. An airplane is traveling at 50 m/s and is 6000 m from a ship. (a) What is the total time of flight of the radar wave from the ship to the plane and back? (b) What distance does the plane move in this time?

ADDITIONAL PROBLEMS

21. At one location on the earth, the amplitude of the magnetic field due to solar radiation is 2.4 μT. From this value calculate the magnitude of the electric field due to solar radiation.

22. Find the frequency of waves having the following wavelengths in vacuum: (a) 1 micron, (b) 1 nanometer, and (c) 1 angstrom.

23. What value of inductance should be used in series with a capacitor of 1.50 pF to form an oscillating circuit that will radiate a wavelength of 5.25 m?

24. Show that the ratio E/B has units of m/s.

25. The wave number, k, of a wave of wavelength λ is defined by the equation $k = 2\pi/\lambda$. What is the wave number corresponding to light of frequency 4.8×10^{14} Hz?

26. The nearest star to us is approximately 4 lightyears away, where 1 lightyear is the distance light travels in one year. Find the distance to this star in meters.

27. As the spaceship Voyager neared Jupiter a radio command from earth was sent to it to turn on its TV cameras. If Jupiter is 630×10^6 km away, what is the minimum time required for a TV signal to be received on earth?

• **28.** An incandescent lamp is radiating uniformly in all directions at 15 W. Calculate the maximum values of the electric and magnetic fields (a) 1 m and (b) 5 m from the source. (*Hint:* Use the formula for the surface area of a sphere, $A = 4\pi r^2$, to find the area on which the total energy is incident in each case, and then use Equation 23.6.)

• **29.** What power must be radiated uniformly in all directions by a source if the amplitude of the magnetic field is 7×10^{-8} T at a distance of 2 m? See the hint in Problem 28.

• **30.** A microwave transmitter emits electromagnetic waves of a single wavelength. The maximum electric field 1 km from the transmitter is 6.0 V/m. Assuming the transmitter is a point source and neglecting waves reflected from the earth, calculate (a) the maximum magnetic field at this distance and (b) the total power emitted by the transmitter.

Part 5

LIGHT AND OPTICS

"I procured me a Triangular glass-Prisme to try therewith the celebrated Pheaenomena of Colours. . . . *I placed my Prisme at his entrance (the sunlight), that it might thereby be refracted to the opposite wall. It was a very pleasing divertisement to view the vivid and intense colours produced thereby; . . . I have often with admiration beheld that all the colours of the Prisme being made to converge, and thereby to be again mixed, as they were in the light before it was incident upon the Prisme, reproduced light, entirely and perfectly white, and not at all sensibly differing from the direct light of the sun. . . ."*

ISAAC NEWTON

Scientists have long been intrigued by the nature of light, and philosophers have had endless arguments concerning the proper definition and perception of light. It is important to understand the nature of light because it is one of the basic ingredients of life on earth. Plants convert light energy from the sun to chemical energy through photosynthesis. Light is the means by which we are able to transmit and receive information from objects around us and throughout the universe.

The nature and properties of light have been a subject of great interest and speculation since ancient times. The Greeks believed that light consisted of tiny particles (corpuscles) that were emitted by a light source and then stimulated the perception of vision upon striking the observer's eye. Newton used this corpuscular theory to explain the reflection and refraction of light. In 1670, one of Newton's contemporaries, the Dutch scientist Christian Huygens, was able to explain many properties of light by proposing that light was wave-like in character. In 1801, Thomas Young showed that light beams can interfere with one another, giving strong support to the wave theory. In 1865, Maxwell developed a brilliant theory that electromagnetic waves travel with the speed of light (Chap. 23). By this time, the wave theory of light seemed to be on firm ground.

However, at the beginning of the 20th century, Max Planck and Albert Einstein returned to the corpuscular theory of light in order to explain the radiation emitted by hot objects and the electrons emitted by a metal exposed to light (the photoelectric effect). We shall discuss these and other topics in modern physics in the last part of this text.

Today, scientists view light as having a dual nature. Experiments can be devised which will display either its particle-like or its wave-like nature.

In this part of the book, we shall concentrate on those aspects of light that are best understood through the wave model. First, we shall discuss the reflection of light at the boundary between two media and the refraction (bending) of light as it travels from one medium into another. We shall use these ideas to study the refraction of light as it passes through lenses and the reflection of light from mirrored surfaces. Finally, we shall describe how lenses and mirrors can be used to view objects with such instruments as cameras, telescopes, and microscopes.

CHAPTER 24

Reflection and Refraction of Light

Reflections in an Alpine lake. (Courtesy of Swiss National Tourist Office)

24.1 THE NATURE OF LIGHT

Until the beginning of the 19th century, light was considered to be a stream of particles that were emitted by a light source and stimulated the sense of sight upon entering the eye. The chief architect of this particle theory of light was Newton. With this theory, Newton was able to provide a simple explanation of some known experimental facts concerning the nature of light, namely, the laws of reflection and refraction.

Christian Huygens (1629–1695)

Most scientists accepted Newton's particle theory of light. However, during Newton's lifetime another theory was proposed, one that argued that light might be some sort of wave motion. In 1678, a Dutch physicist and astronomer, Christian Huygens (1629–1695), showed that a wave theory of light could explain the laws of reflection and refraction. The wave theory did not receive immediate acceptance for several reasons. All the waves known at the time (sound, water, and so on) traveled through some sort of medium, but light from the sun could travel to us through space. Furthermore, it was argued that if light were some form of wave motion, the waves could bend around obstacles; hence, we should be able to see around corners. It is now known that light does indeed bend around the edges of objects. This phenomenon, known as *diffraction*, is not easy to observe because light waves have such short wavelengths. Thus, although experimental evidence for the diffraction of light was discovered by Francesco Grimaldi (1618–1663) around 1660, most scientists rejected the wave theory and adhered to Newton's particle theory for more than a century. This was, for the most part, due to Newton's great reputation as a scientist.

The first clear demonstration of the wave nature of light was provided in 1801 by Thomas Young (1773–1829), who shows that, under appropriate conditions, light exhibits interference behavior. That is, at certain points in the vicinity of two sources, light waves can combine and cancel each other by destructive interference. Such behavior could not be explained at that time by a particle theory because there was no conceivable way by which two or more particles could come together and cancel one another. Several years later, a French physicist, Augustin Fresnel (1788–1829), performed a number of detailed experiments dealing with interference and diffraction phenomena. In 1850, Jean Foucault (1791–1868) provided further evidence of the inadequacy of the particle theory by showing that the speed of light in liquids is less than in air. According to the particle model of light, the speed of light would be higher in glasses and liquids than in air. Additional developments during the 19th century led to the general acceptance of the wave theory of light.

The most important development concerning the theory of light was the work of Maxwell, who in 1865 predicted that light was a form of high-frequency electromagnetic wave (Chap. 23). His theory predicted that these waves should have a speed of about 3×10^8 m/s. Within experimental error, this value is equal to the speed of light. As discussed in Chapter 23, Hertz provided experimental confirmation of Maxwell's theory in 1887 by producing and detecting electromagnetic waves. Furthermore, Hertz and other investigators showed that these *waves exhibited reflection, refraction, and all the other characteristic properties of waves.*

Although the classical theory of electricity and magnetism was able to explain most known properties of light, some subsequent experiments could not be explained by assuming that light was a wave. The most striking of these is the *photoelectric effect*, also discovered by Hertz. The photoelectric effect is the ejection of electrons from a metal whose surface is exposed to light. As one example of the difficulties that arose, experiments showed that the kinetic energy of an ejected electron is *independent* of the light intensity. This was in contradiction of the wave theory, which held that a more intense beam of light should add more energy to the electron. An explanation of this phenomenon was proposed by Einstein in 1905. Einstein's theory used the concept of quantization developed by Max Planck (1858–1947) in 1900. The quantization model assumes that the energy of a light wave is present in bundles of energy called *photons;* hence the energy is said to be *quantized.* (Any quantity that appears in discrete units is said to be quantized. For example, electric charge is quantized because it always appears in multiples of 1.6×10^{-19} C.) Einstein modified and reinterpreted the work of Planck to show that the energy of a photon is proportional to the frequency of the electromagnetic wave:

Energy of a photon

$$\boxed{E = hf} \tag{24.1}$$

where $h = 6.63 \times 10^{-34}$ J·s is *Planck's constant.* It is important to note that this theory retains some features of both the wave theory and the particle theory of light. As we shall discuss later, the photoelectric effect is the result of energy transfer from a single photon to an electron in the metal. That is, the electron interacts with one photon of light as if it, the electron, had been struck by a particle. Yet this photon has wave-like characteristics because its energy is determined by the frequency (wave-like quantity).

In view of these developments, one must regard light as having a *dual nature.* That is, *in some cases light acts like a wave and in others it acts like a particle.* Classical electromagnetic wave theory provides an adequate explanation of light propagation and of the effects of interference, whereas the photoelectric effect and other experiments involving the interaction of light with matter are best explained by assuming that light is a particle. Light is light, to be sure. However, the question, "Is light a wave or a particle?" is an inappropriate one. Sometimes it acts like a wave, sometimes like a particle. Fortunately, it never acts like both in the same experiment. In the next few chapters, we shall investigate the wave nature of light.

24.2 MEASUREMENTS OF THE SPEED OF LIGHT

Light travels at such a high speed ($c \approx 3 \times 10^8$ m/s) that early attempts to measure its speed were unsuccessful. Galileo attempted to measure the speed of light by positioning two observers in towers separated by about 5 miles. Each observer carried a shuttered lantern. One observer would open his lantern first, and then the other would open his lantern at the moment he saw the light from the first lantern. The velocity could then be obtained, in principle, knowing the transit time of the light beams between lanterns. However, at a speed of 186,000 mi/s, light would travel the ten-mile roundtrip in a time of approximately 54 μs. Thus, it is impossible to measure the speed of light in this manner because the transit time is very small compared with the reaction time of the observers.

Roemer's Method

The first successful estimate of the speed of light was made in 1675 by the Danish astronomer Ole Roemer (1644–1710). His technique involved astronomical observations of one of the moons of Jupiter, called Io. At that time only four of Jupiter's moons had been discovered, and the periods of their orbits were known. Io, the innermost moon, has a period of 42.5 h, and its orbit, the orbit of Jupiter, and the orbit of the earth all lie in approximately the same plane. As a result, Io goes into eclipse behind Jupiter every 42.5 h as seen from earth. Using the orbital motion of Io as a clock, Roemer expected to find a constant period over long time intervals. Instead, Roemer observed a systematic variation in Io's period. He found that the periods were larger than average when earth receded from Jupiter and smaller than average when earth approached Jupiter. If Io had a constant period, Roemer should have been able to observe a particular eclipse and be able to predict when the next eclipse would occur. However, when Roemer checked to see if the second eclipse did occur at the predicted time, he often found that his predictions did not agree with the actual time of the eclipse. For example, consider the situation shown in Figure 24.1. Suppose an eclipse had occurred when the earth was at position E1. Knowing the period of Io, Roemer was able to predict when an eclipse should occur three months later when the earth was at position E2. However, the eclipse at E2 actually occurred at a time approximately 600 s later than the predicted time. Roemer attributed this discrepancy to the fact that the distance between the earth and Jupiter was changing between these observations. In three months (one quarter of the earth's period), the earth has moved through one-fourth of its orbit from E1 to E2, as shown in Figure 24.1. In the same time interval, Jupiter, whose period is about 12 years, has moved a much shorter distance from J1 to J2. Therefore as the earth moves from E1 to E2, light from Jupiter has to travel an additional distance equal to the radius of the earth's orbit.

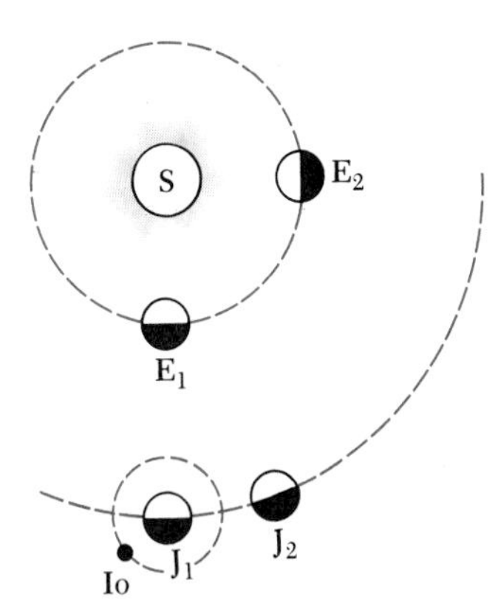

FIGURE 24.1 As the earth moves from E1 to E2, Jupiter moves only from J1 to J2.

Using data available at that time, Roemer would have estimated the speed of light to be about 2.1×10^8 m/s. The large discrepancy between this value and the currently accepted value of 3×10^8 m/s is due to a large error in the assumed radius of the earth's orbit. This experiment is important historically because it demonstrated that light does have a finite speed and established a rough estimate of the magnitude of this speed.

Fizeau's Technique

The first successful method of measuring the speed of light using purely terrestrial techniques was developed in 1849 by Armand H. L. Fizeau (1819–1896). Figure 24.2 represents a simplified diagram of his apparatus. The basic idea is to measure the total time it takes light to travel from some point to a distant mirror and back. If d is the distance between the light source and the mirror and if the transit time for one round trip is t, then the speed of light is $c = 2d/t$. To measure the transit time, Fizeau used a rotating toothed wheel, which converts an otherwise continuous beam of light into a series of light pulses. Additionally, the rotation of the wheel controls what an observer at the light source sees. For example, if the light passing the opening at point A in Figure 24.2 should return at the instant that tooth B had rotated into position to cover the return path, the light would not reach the observer. At a faster rate of rotation, the opening at point C could move into position to allow the reflected beam to pass and reach the observer. Knowing the distance d, the number of teeth in the wheel, and the angular velocity of the

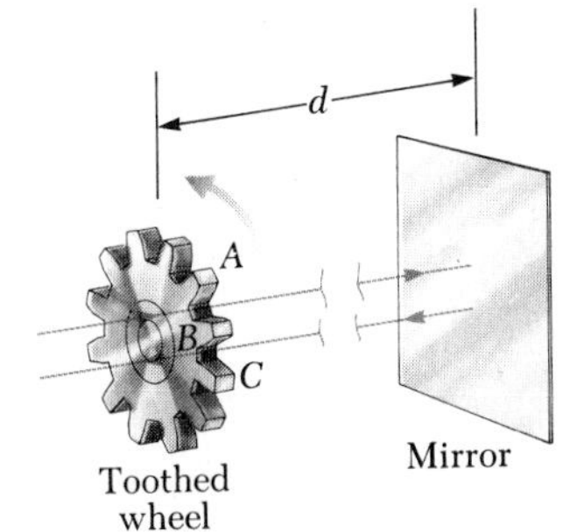

FIGURE 24.2 Fizeau's method for measuring the speed of light using a rotating, toothed wheel.

wheel, Fizeau arrived at a value of $c = 3.1 \times 10^8$ m/s. Similar measurements made by subsequent investigators yielded more accurate values for c, approximately 2.9977×10^8 m/s.

A variety of other, more accurate measurements have been reported for the value of c. A recent value obtained using laser techniques is

$$c = 2.997924574(12) \times 10^8 \text{ m/s}$$

where the (12) indicates the uncertainty in the last two digits. The number of significant figures here is certainly impressive. In fact, the speed of light has been determined with such high accuracy that it is now used to define the SI unit of length, the meter. As noted in Chapter 1, the meter is defined as the distance traveled by light in a vacuum during a time of 1/299,792,458 s.

EXAMPLE 24.1 Meauring the Speed of Light with Fizeau's Toothed Wheel

Assume the toothed wheel of the Fizeau experiment has 360 teeth and is rotating with a speed of 27.5 rev/s when the light from the source is extinguished, that is, when a burst of light passing through opening A in Figure 24.2 is blocked by tooth B on return. If the distance to the mirror is 7500 m, find the speed of light.

Solution If the wheel has 360 teeth, it will turn through an angle of 1/720 rev in the time that passes while the light makes its round trip. From the definition of angular velocity, we see that the time is

$$t = \frac{\theta}{\omega} = \frac{(1/720) \text{ rev}}{27.5 \text{ rev/s}} = 5.05 \times 10^{-5} \text{ s}$$

Hence, the speed of light is

$$c = \frac{2d}{t} = \frac{2\,(7500 \text{ m})}{5.05 \times 10^{-5} \text{ s}} = 2.97 \times 10^8 \text{ m/s}$$

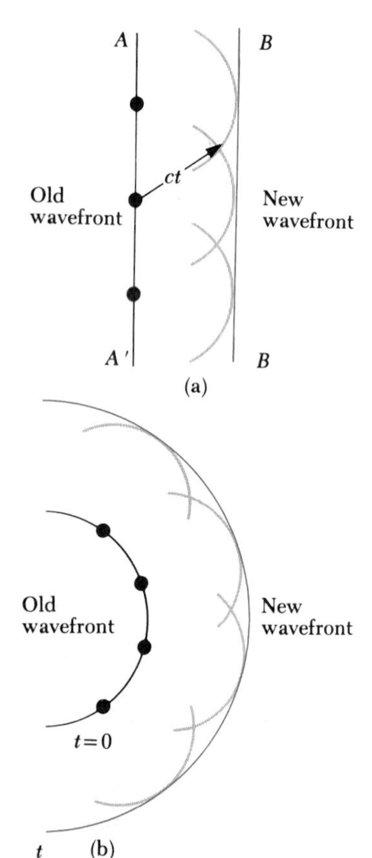

FIGURE 24.3 Huygens' construction for (a) a plane wave propagating to the right, and (b) a spherical wave.

24.3 HUYGENS' PRINCIPLE

We shall develop the laws of reflection and refraction by using a geometric method proposed by Huygens in 1678. Huygens assumed that light is some form of wave motion rather than a stream of particles. He had no knowledge of the nature of light or of its electromagnetic character. Nevertheless, his simplified wave model is adequate for understanding many practical aspects of the propagation of light.

Huygens' principle is a geometric construction for determining the position of a new wavefront at some instant from the knowledge of an earlier wavefront. (A wavefront is a surface passing through those points of a wave that are behaving identically. For instance, a wavefront could be a surface over which the wave is at a crest.) In Huygens' construction,

all points on a given wavefront are taken as point sources for the production of spherical secondary waves, called wavelets, which propagate outward with speeds characteristic of waves in that medium. After some time has elapsed, the new position of the wavefront is the surface tangent to the wavelets.

Figure 24.3 illustrates two simple examples of Huygens' construction. First, consider a plane wave moving through free space, as in Figure 24.3a.

FIGURE 24.4 Water waves in a ripple tank, which demonstrates Huygens' wavelets. A plane wave is incident on a barrier with a small opening. The opening acts as a source of circular wavelets. (Photograph courtesy of Education Development Center, Newton, MA)

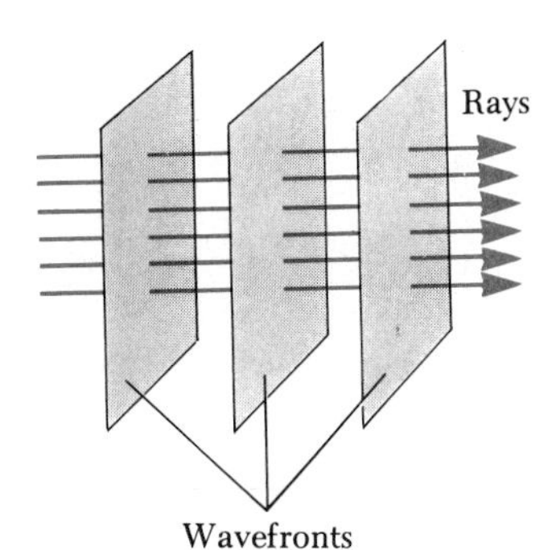

FIGURE 24.5 A plane wave traveling to the right. Note that the rays, corresponding to the direction of wave motion, are straight lines perpendicular to the wavefronts.

At $t = 0$, the wavefront is indicated by the plane labeled AA'. In Huygens' construction, each point on this wavefront is considered a point source. For clarity, only a few points on AA' are shown. With these points as sources for the wavelets, we draw circles each of radius ct, where c is the speed of light in vacuum and t is the time of propagation from one wavefront to the next. The surface drawn tangent to these wavelets is the plane BB', which is parallel to AA'. In a similar manner, Figure 24.3b shows Huygens' construction for an outgoing spherical wave.

A convincing demonstration of Huygens' principle is obtained with water waves in a shallow tank (called a ripple tank), as in Figure 24.4. Plane waves produced below the slit emerge above the slit as two-dimensional circular waves propagating outward.

The Ray Approximation in Geometric Optics

In studying geometric optics here and in Chapter 25, we shall use what is called the *ray approximation*. To understand this approximation, first recall that the direction of energy flow of a wave, corresponding to the direction of wave propagation, is called a ray. The rays of a given wave are straight lines that are perpendicular to the wavefronts, as illustrated in Figure 24.5 for a plane wave. In the ray approximation, we assume that a wave moving through a medium travels in a straight line in the direction of its rays. That is, a ray is a line drawn in the direction in which the light is traveling. For example, a beam of sunlight passing through a darkened room traces out the path of a ray.

24.4 REFLECTION AND REFRACTION

Reflection of Light

When a light traveling in a medium encounters a boundary leading into a second medium, part of the incident ray is reflected back into the first me-

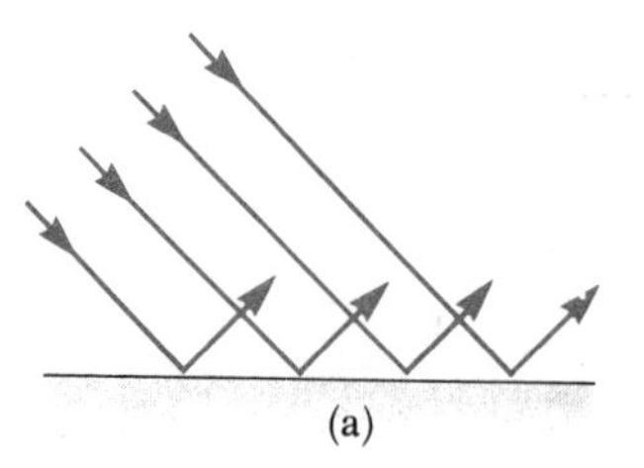

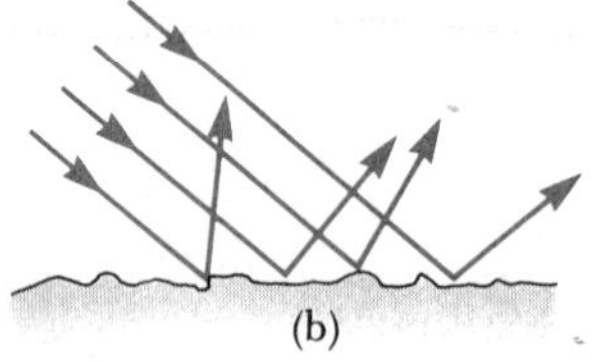

FIGURE 24.6 Schematic representation of (a) specular reflection, where the reflected rays are parallel, and (b) diffuse reflection, where the reflected rays travel in random directions.

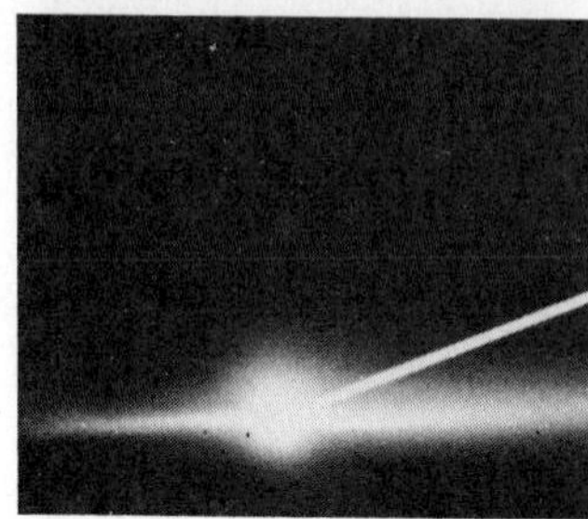

(Top) The light from the laser undergoes specular reflection at the smooth surface. (Bottom) At the rough surface, the light from the laser is diffusely reflected.

dium. Figure 24.6a shows several rays of a beam of light incident on a smooth, mirror-like, reflecting surface. The reflected rays are parallel to each other, as indicated in the figure. Reflection of light from such a smooth surface is called *specular reflection.* On the other hand, if the reflecting surface is rough, as in Figure 24.6b, the surface will reflect the rays in various directions. Reflection from any rough surface is known as *diffuse reflection.* A surface will behave as a smooth surface as long as the surface variations are small compared with the wavelength of the incident light.

For instance, consider the two types of reflection one can observe from a road's surface while driving a car at night. When the road is dry, light from oncoming vehicles is scattered off the road in different directions (diffuse reflection) and the road is quite visible. On a rainy night, when the road is wet, the road irregularities are filled with water. Because the water surface is quite smooth, the light undergoes specular reflection. In this book, we shall concern ourselves only with specular reflection, and we shall use the term *reflection* to mean specular reflection.

Consider a light ray traveling in air and incident at an angle on a flat, smooth surface, as in Figure 24.7. The incident and reflected rays make angles θ_1 and θ_1', respectively, with a line drawn perpendicular to the surface at the point where the incident ray strikes the surface. We shall call this line the *normal* to the surface. Experiments show that *the angle of reflection equals the angle of incidence*, that is,

Normal

θ_1 θ'_1

FIGURE 24.7 According to the law of reflection, $\theta_1 = \theta_1'$.

$$\boxed{\theta_1' = \theta_1} \tag{24.2}$$

EXAMPLE 24.2 The Double-Reflecting Light Ray

Two mirrors make an angle of 120° with each other, as in Figure 24.8. A ray is incident on mirror M_1 at an angle of 65° to the normal. Find the direction of the ray after it is reflected from mirror M_2.

Solution From the law of reflection, we see that the first reflected ray also makes an angle of 65° with the normal. Thus, it follows that this same ray makes an angle of 90° − 65°, or 25°, with the horizontal. From the triangle made by the first reflected ray and the two mirrors, we see that the first reflected ray makes an angle of 35° with M_2 (since the sum of the interior angles of any triangle is 180°). This means that this ray makes an angle of 55° with the normal to M_2. Hence, from the law of reflection, it follows that the second reflected ray makes an angle of 55° with the normal to M_2.

FIGURE 24.8 (Example 24.2) Mirrors M_1 and M_2 make an angle of 120° with each other.

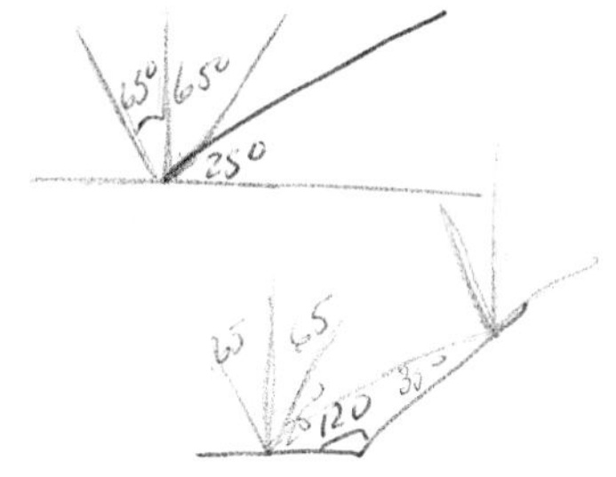

Refraction of Light

When a ray of light traveling through a transparent medium encounters a boundary leading into another transparent medium, as in Figure 24.9a, part of the ray is reflected and part enters the second medium. The ray that enters the second medium is bent at the boundary and is said to be *refracted*. The incident ray, the reflected ray, and the refracted ray all lie in the same plane. The **angle of refraction,** θ_2 in Figure 24.9a, depends on the properties of the two media and on the angle of incidence through the relationship

$$\frac{\sin\theta_2}{\sin\theta_1} = \frac{v_2}{v_1} = \text{constant} \tag{24.3}$$

where v_1 is the speed of light in medium 1 and v_2 is the speed of light in medium 2. The experimental discovery of this relationship is usually credited to Willebrord Snell (1591–1626) and is therefore known as **Snell's law.** In Section 24.9, we shall verify the laws of reflection and refraction using Huygens' principle.

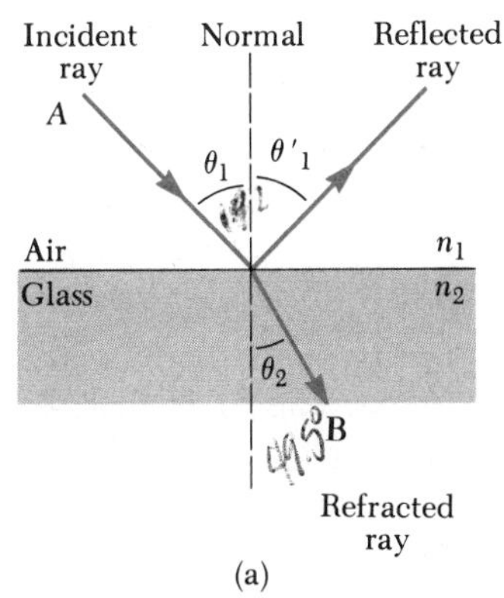

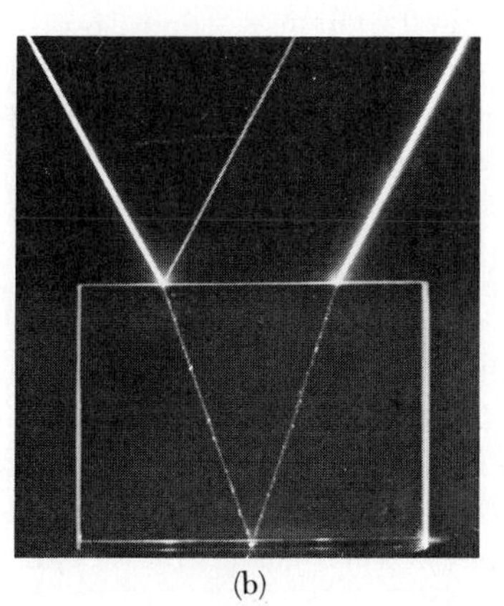

FIGURE 24.9 (a) A ray obliquely incident on an air-glass interface. The refracted ray is bent toward the normal since $n_2 > n_1$ and $v_2 < v_1$. (b) Light incident on the lucite block bends when it enters the block and also when it leaves the block.

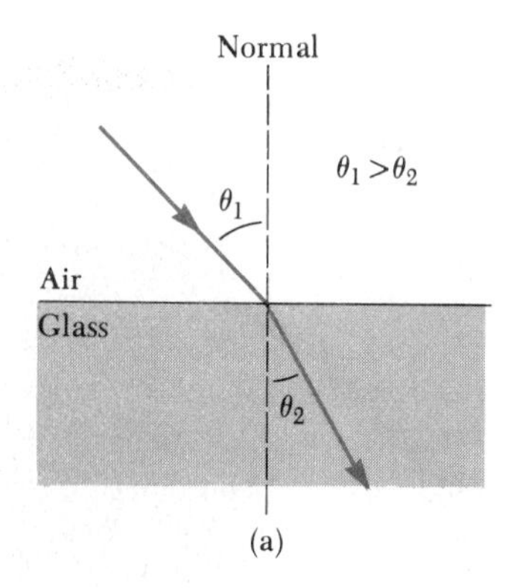

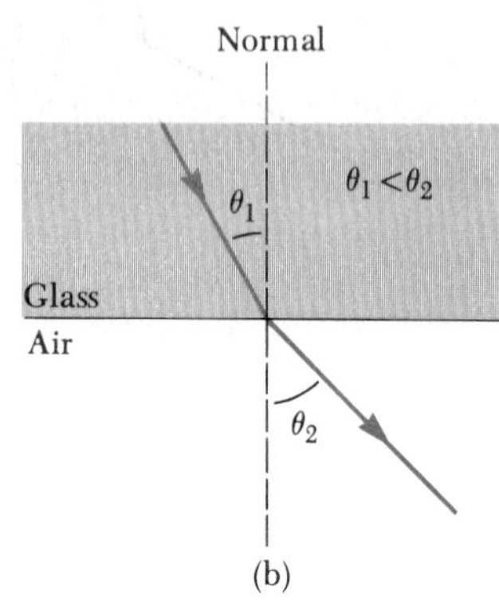

FIGURE 24.10 (a) When the light beam moves from air into glass, its path is bent toward the normal. (b) When the beam moves from glass into air, its path is bent away from the normal.

It is found that *the path of a light ray through a refracting surface is reversible.* For example, the ray in Figure 24.9a, travels from point A to point B. If the ray originated at B, it would follow the same path to reach point A. In the latter case, however, the reflected ray would be in the glass.

When light moves from a material in which its speed is high to a material in which its speed is lower, the angle of refraction, θ_2, is less than the angle of incidence, as shown in Figure 24.10a. If the ray moves from a material in which it moves slowly to a material in which it moves more rapidly, it is bent away from the normal, as shown in Figure 24.10b.

The behavior of light as it passes from air into another substance and then re-emerges into air is often a source of confusion to students. Let us take a look at what happens and see why this behavior is so different from other occurrences in our daily lives. When light travels in air, its speed is about 3×10^8 m/s, and its speed is reduced to about 2×10^8 m/s upon entering a block of glass. When the light re-emerges into air, its speed instantaneously increases to its original value of 3×10^8 m/s. This is far different from what happens, for example, when a bullet is fired through a block of wood. In this case, the speed of the bullet is reduced as it moves through the wood because some of its original energy is used to tear apart the fibers of the wood. When the bullet enters the air once again, it emerges at the speed it had just before leaving the block of wood.

In order to see why light behaves as it does, consider Figure 24.11, which represents a beam of light entering a piece of glass from the left. Once inside the glass, the light may encounter an electron bound to an atom, indicated as point A in the figure. Let us assume that light is absorbed by the atom, which causes the electron to oscillate. The oscillating electron then acts as an antenna and radiates the beam of light toward an atom at point B, where the light is again absorbed by an atom at that point. The details of these absorptions and emissions are best explained in terms of quantum mechanics, a subject we shall study in Chapter 29. For now, it is sufficient to think of the process as one in which the light passes from one atom to another through the glass. (The situation is somewhat analogous to a relay race in which a baton is passed between runners on the same team.) Although light travels from one atom to another with a speed of 3×10^8 m/s, the processes of absorption and emission of light by the atoms take time. Enough time is required, in fact, to lower the speed of light in glass to 2×10^8 m/s. Once the light emerges into the air, the absorptions and emissions cease and its speed returns to the original value.

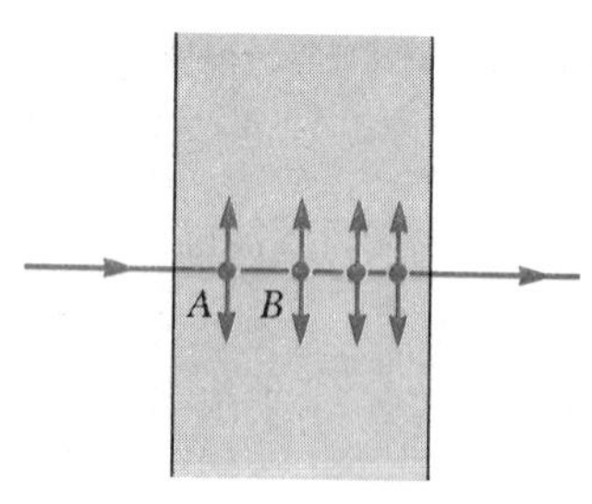

FIGURE 24.11 Light passing from one atom to another in a material medium.

24.5 THE LAW OF REFRACTION

When light passes from one medium to another, it is refracted because the speed of light is different in the two media. In general, one finds that the speed of light in any material is less than the speed of light in vacuum. In fact, *light travels at its maximum speed in vacuum.* It is convenient to define the **index of refraction,** n, of a medium to be the ratio

Index of refraction

$$n \equiv \frac{\text{speed of light in vacuum}}{\text{speed of light in a medium}} = \frac{c}{v} \qquad (24.4)$$

From this definition, we see that the index of refraction is a dimensionless number greater than unity since v is always less than c. Furthermore, n is equal to unity for a vacuum. The indices of refraction for various substances measured with respect to vacuum are listed in Table 24.1.

As light travels from one medium to another, *the frequency of the light does not change.* To see why this is so, consider Figure 24.12. Wavefronts pass an observer at point A in medium 1 with a certain frequency and are incident on the boundary between medium 1 and medium 2. The frequency with which the wavefronts pass an observer at point B in medium 2 must equal the frequency at which they arrive at point A in medium 1. If this were not the case, wavefronts would either be piling up at the boundary or they would be destroyed or created at the boundary. Since there is no mechanism for this to occur, the frequency must be a constant as a light ray passes from one medium into another.

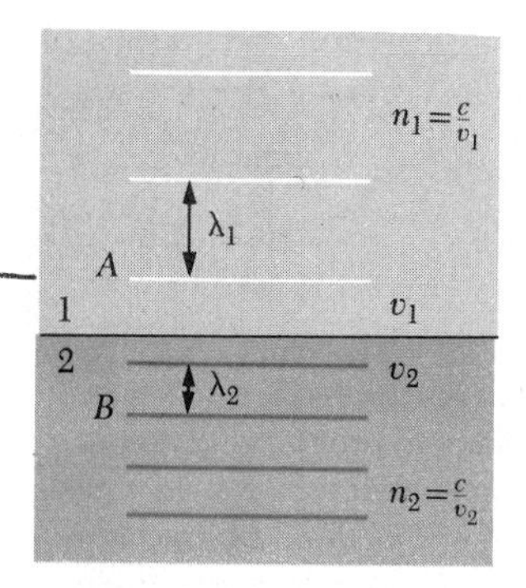

FIGURE 24.12 As the wave moves from medium 1 to medium 2, its wavelength changes but its frequency remains constant.

Therefore, because the relation $v = f\lambda$ must be valid in both media and because $f_1 = f_2 = f$, we see that

$$v_1 = f\lambda_1 \quad \text{and} \quad v_2 = f\lambda_2$$

where the subscripts refer to the two media. A relationship between index of refraction and wavelength can be obtained by dividing these two equations and making use of the definition of the index of refraction given by Equation 24.4:

$$\frac{\lambda_1}{\lambda_2} = \frac{v_1}{v_2} = \frac{c/n_1}{c/n_2} = \frac{n_2}{n_1} \tag{24.5}$$

which gives

$$\lambda_1 n_1 = \lambda_2 n_2 \tag{24.6}$$

TABLE 24.1 Index of Refraction for Various Substances Measured with Light of Vacuum Wavelength $\lambda_0 = 589$ nm

Substance	Index of Refraction	Substance	Index of Refraction
Solids at 20°C		Liquids at 20°C	
Diamond (C)	2.419	Benzene	1.501
Fluorite (CaF_2)	1.434	Carbon disulfide	1.628
Fused quartz (SiO_2)	1.458	Carbon tetrachloride	1.461
Glass, crown	1.52	Ethyl alcohol	1.361
Glass, flint	1.66	Glycerine	1.473
Ice (H_2O) (at 0°C)	1.309	Water	1.333
Polystyrene	1.49	Gases at 0° C, 1 atm	
Sodium chloride (NaCl)	1.544	Air	1.000293
Zircon	1.923	Carbon dioxide	1.00045

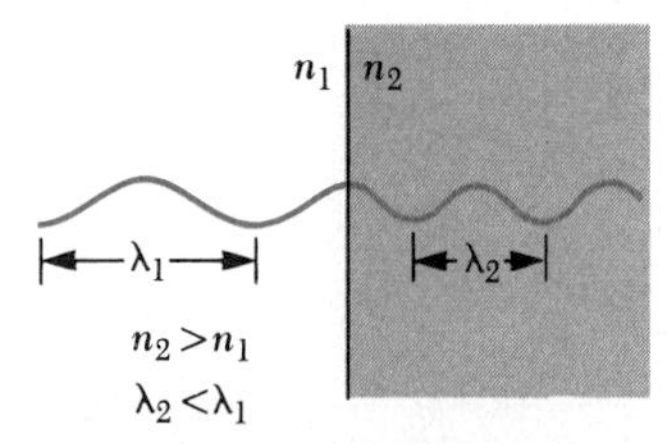

FIGURE 24.13 Schematic diagram of the *reduction* in wavelength when light travels from a medium of low index of refraction to one of higher index of refraction.

If medium 1 is a vacuum, or for all purposes air, then $n_1 = 1$. Hence, it follows from Equation 24.6 that the index of refraction of any medium can be expressed as the ratio

$$n = \frac{\lambda_0}{\lambda_n} \tag{24.7}$$

where λ_0 is the wavelength of light in vacuum and λ_n is the wavelength in the medium whose index of refraction is n. A schematic representation of this reduction in wavelength is shown in Figure 24.13.

We are now in a position to express Snell's law (Eq. 24.3) in an alternative form. If we substitute Equation 24.5 into Equation 24.3, we get

Snell's law

$$n_1 \sin \theta_1 = n_2 \sin \theta_2 \tag{24.8}$$

This is the most widely used and practical form of **Snell's law.**

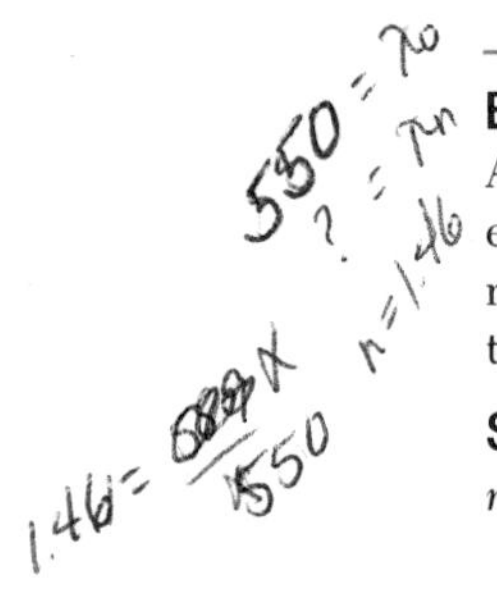

EXAMPLE 24.3 An Index of Refraction Measurement

A beam of light of wavelength 550 nm traveling in air is incident on a slab of transparent material. The incident beam makes an angle of 40° with the normal, and the refracted beam makes an angle of 26° with the normal. Find the index of refraction of the material.

Solution Snell's law of refraction (Eq. 24.8) together with the given data, $\theta_1 = 40°$, $n_1 = 1.00$ for air, and $\theta_2 = 26°$, gives

$$n_1 \sin \theta_1 = n_2 \sin \theta_2$$

$$n_2 = \frac{n_1 \sin \theta_1}{\sin \theta_2} = (1.00)\frac{\sin 40°}{\sin 26°} = \frac{0.643}{0.440} = 1.46$$

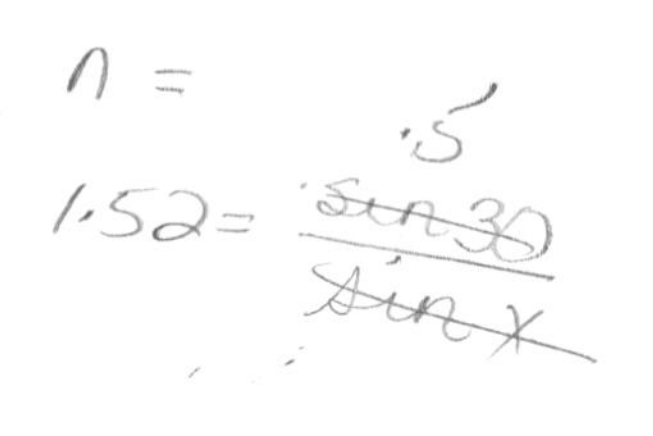

If we compare this value with the data in Table 24.1, we see that the material is probably fused quartz.

EXERCISE What is the wavelength of light in the material?

Answer 377 nm

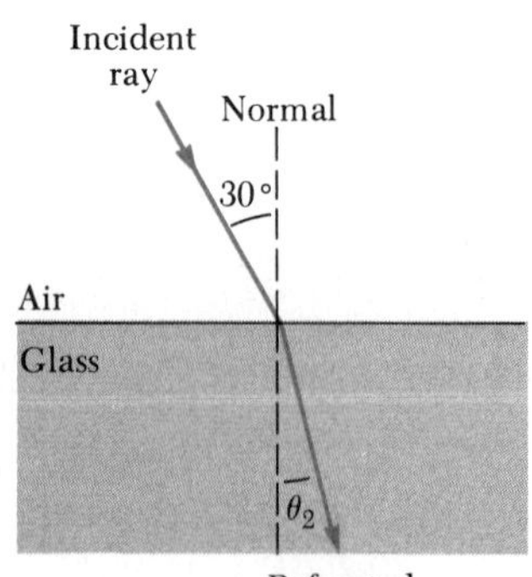

FIGURE 24.14 (Example 24.4) Refraction of light by glass.

EXAMPLE 24.4 Angle of Refraction for Glass

A light ray of wavelength 589 nm (produced by a sodium lamp) traveling through air is incident on a smooth, flat slab of crown glass at an angle of 30° to the normal, as sketched in Figure 24.14. Find the angle of refraction, θ_2.

Solution Snell's law given by Equation 24.8 can be rearranged as

$$\sin \theta_2 = \frac{n_1}{n_2} \sin \theta_1$$

From Table 24.1, we find that $n_1 = 1.00$ for air and $n_2 = 1.52$ for crown glass. Therefore, the unknown refraction angle is

$$\sin \theta_2 = \left(\frac{1.00}{1.52}\right)(\sin 30°) = 0.329$$

$$\theta_2 = \sin^{-1}(0.329) = 19.2°$$

Thus we see that the ray is bent *toward* the normal, as expected.

EXERCISE If the light ray moves from inside the glass toward the glass-air interface at an angle of 30° to the normal, determine the angle of refraction.

Answer 49.5° *away* from the normal.

EXAMPLE 24.5 The Speed of Light in Quartz

Light of wavelength 589 nm in vacuum passes through a piece of fused quartz whose index of refraction is (n = 1.458). (a) Find the speed of light in quartz.

Solution The speed of light in quartz can be easily obtained from Equation 24.4:

$$v = \frac{c}{n} = \frac{3 \times 10^8 \text{ m/s}}{1.458} = 2.058 \times 10^8 \text{ m/s}$$

It is interesting to note that the speed of light in vacuum, 3×10^8 m/s, is an upper limit on the speed of material objects. In our treatment of relativity in Chapter 28, we shall find that this upper limit is consistent with experimental observations. However, it is possible for a particle moving in a medium to have a speed which exceeds the speed of light in that medium. For example, it is theoretically possible for a particle to travel through fused quartz at a speed greater than 2.06×10^8 m/s, but it must have a speed less than 3×10^8 m/s when moving in a vacuum.

(b) What is the wavelength of this light in quartz?

Solution We can use $\lambda_n = \lambda_0/n$ (Eq. 24.7) to calculate the wavelength in quartz, noting that we are given λ_0 = 589 nm = 589×10^{-9} m:

$$\lambda_n = \frac{\lambda_0}{n} = \frac{589 \text{ nm}}{1.458} = 404 \text{ nm}$$

EXERCISE Find the frequency of the light passing through the quartz.

Answer 5.09×10^{14} Hz.

EXAMPLE 24.6 Light Passing Through a Slab

A light beam passes from medium 1 to medium 2 through a thick slab of material whose index of refraction is n_2 (Fig. 24.15). Show that the emerging beam is parallel to the incident beam.

Solution First, let us apply Snell's law to the upper surface:

$$(1) \qquad \sin\theta_2 = \frac{n_1}{n_2}\sin\theta_1$$

Applying Snell's law to the lower surface gives

$$(2) \qquad \sin\theta_3 = \frac{n_2}{n_1}\sin\theta_2$$

Substituting (1) into (2) gives

$$\sin\theta_3 = \frac{n_2}{n_1}\left(\frac{n_1}{n_2}\sin\theta_1\right) = \sin\theta_1$$

That is, $\theta_3 = \theta_1$, and so the layer does not alter the direction of the beam. It does, however, produce a displacement of the beam. The same result is obtained when light passes through multiple layers of materials.

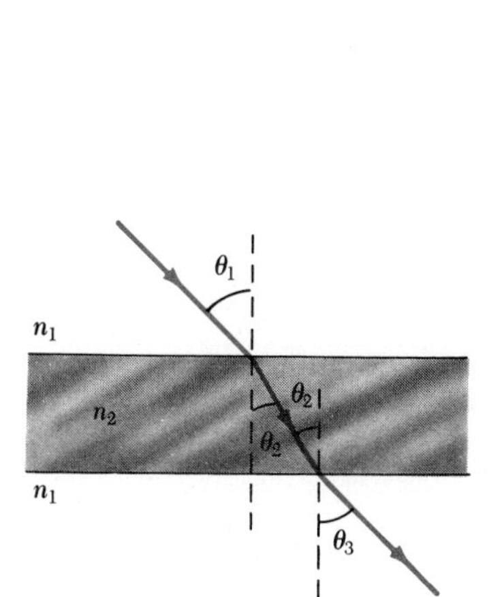

FIGURE 24.15 (Example 24.6) When light passes through a flat slab of material, the emerging beam is parallel to the incident beam, and therefore $\theta_1 = \theta_3$.

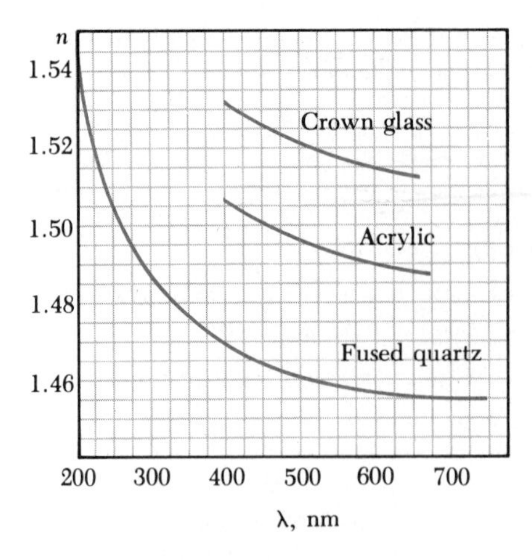

FIGURE 24.16 Variations of index of refraction with wavelength for three materials.

24.6 DISPERSION AND PRISMS

An important property of the index of refraction is that it is different for different wavelengths of light. A graph of the index of refraction for three materials is shown in Figure 24.16. Since n is a function of wavelength, Snell's law indicates that light of *different wavelengths* will be bent at *different angles* when incident on a refracting material. As we see from Figure 24.16, the index of refraction decreases with increasing wavelength. This means that blue light will bend more than red light when passing into a refracting material.

Any substance in which n varies with wavelength is said to exhibit **dispersion**. To understand the effects that dispersion can have on light, let us consider what happens when light strikes a prism, as in Figure 24.17a. A single ray of light incident on the prism from the left emerges bent away from its original direction of travel by an angle δ, called the **angle of deviation.** Now suppose a beam of white light (a combination of all visible wavelengths)

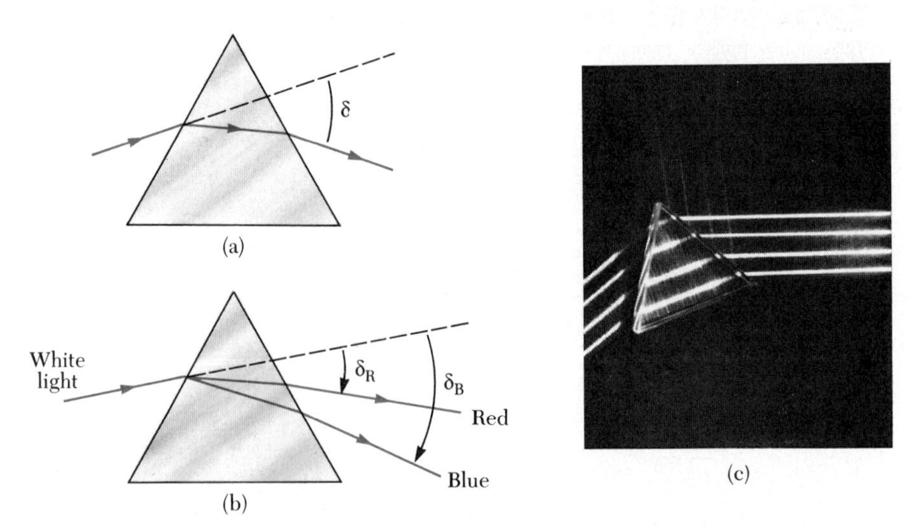

FIGURE 24.17 (a) A prism refracts a light ray and deviates the light through an angle δ. (b) When light is incident on a prism, the blue light is bent more than the red. (c) Photograph of light being refracted by a prism.

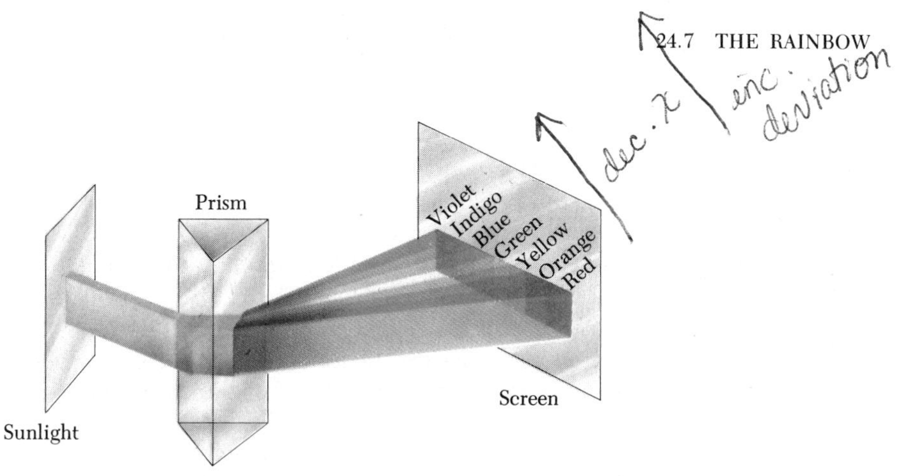

FIGURE 24.18 Dispersion of white light by a prism. Since n varies with wavelength, the prism disperses the white light into its various spectral components, or colors.

is incident on a prism, as in Figure 24.17b. The blue component of the incident beam is bent more than the red component, and the rays that emerge from the second face of the prism spread out in a series of colors known as a **spectrum,** as shown in Figure 24.18 and Plate 19. These colors, in order of decreasing wavelength, are red, orange, yellow, green, blue, indigo, and violet. Clearly, the angle of deviation, δ, depends on the wavelength of a given color. Violet light deviates the most, red light deviates the least, and the remaining colors in the visible spectrum fall between these extremes. When light is spread out by a substance such as the prism, the light is said to be dispersed into a spectrum.

A prism is often used in an instrument known as a **prism spectrometer,** the essential elements of which are shown in Figure 24.19. The instrument is commonly used to study the wavelengths emitted by a light source, such as a sodium vapor lamp. Light from the source is sent through a narrow, adjustable slit to produce a parallel, or collimated, beam. The light then passes through the prism and is dispersed into a spectrum. The refracted light is observed through a telescope. The experimenter sees an image of the slit through the eyepiece of the telescope. The telescope can be moved or the prism can be rotated in order to view the various wavelengths, which have different angles of deviation.

All hot low-pressure gases emit their own characteristic spectrum. Thus, one use of a prism spectrometer is to identify gases. For example, sodium emits only two wavelengths in the visible spectrum; these are two closely spaced yellow lines. Thus, a gas emitting these and only these colors can be identified as sodium. Likewise, mercury vapor has its own characteristic spectrum, consisting of four prominent wavelengths—orange, green, blue, and violet lines—along with some wavelengths of lower intensity. The particular wavelengths emitted by a gas serve as "fingerprints" of that gas.

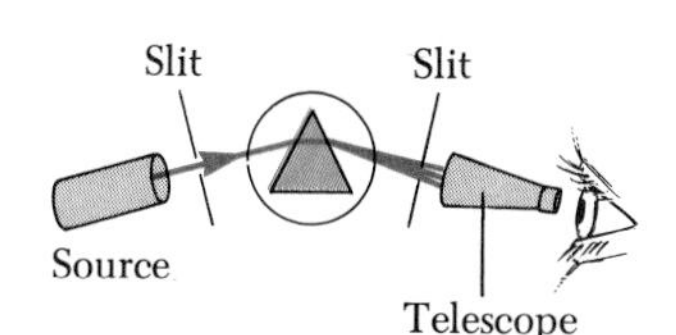

FIGURE 24.19 Diagram of a prism spectrometer. The various colors in the spectrum are viewed through a telescope.

*24.7 THE RAINBOW

The dispersion of light into a spectrum is demonstrated most vividly in nature through the formation of a rainbow, often seen by an observer located be-

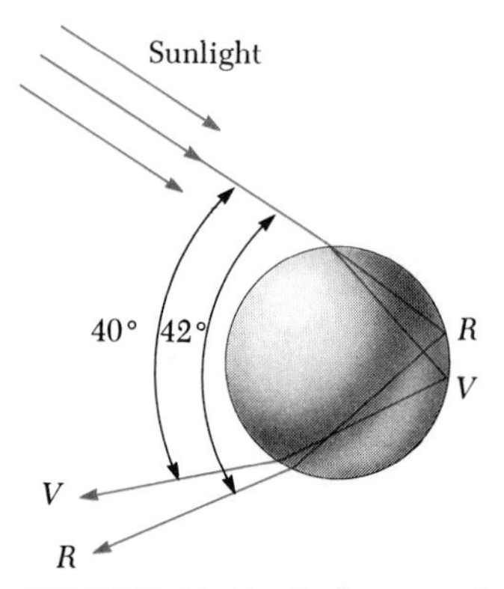

FIGURE 24.20 Refraction of sunlight by a spherical raindrop.

tween the sun and a rain shower (see Color Plate 21 for a good photograph of a rainbow). To understand how a rainbow is formed, consider Figure 24.20. A ray of light passing overhead strikes a drop of water in the atmosphere. The light is refracted and reflected as follows. It is first refracted at the front surface of the drop, with the violet light deviating the most and the red light the least. At the back surface of the drop, the light is reflected and returns to the front surface, where it again undergoes refraction as it moves from water into air. The rays leave the drop such that the angle between the incident white light and the returning violet ray is 40°, while the angle between the white light and the returning red ray is 42° (Fig. 24.20). This small angular difference between the returning rays causes us to see the bow.

Now consider an observer viewing a rainbow, as in Figure 24.21. If a raindrop high in the sky is being observed, the red light returning from the drop is able to reach the observer because it is deviated the most, but the violet light passes over the observer because it is deviated the least. Hence, the observer sees this drop as being red. Similarly, a drop lower in the sky would direct violet light toward the observer and appear to be violet. (The red light from this drop would strike the ground and not be seen.) The remaining colors of the spectrum would reach the observer from raindrops lying between these two extreme positions.

Passengers in an airplane at high altitude can see light directed toward them returning from drops above and below the plane. The rainbow the passengers see forms a great circle (with no end for the pot of gold). At ground level, the lower part of the circle is not formed, hence we see a half-circle rainbow.

*24.8 COLOR

In Chapter 23 we found that the wavelengths in the visible portion of the electromagnetic spectrum range from 400 nm to 700 nm, with the shortest wavelength corresponding to violet and the longest corresponding to red. Each wavelength between these two extremes has its own characteristic hue, or shade. The spectrum of colors can be remembered by the boy's name

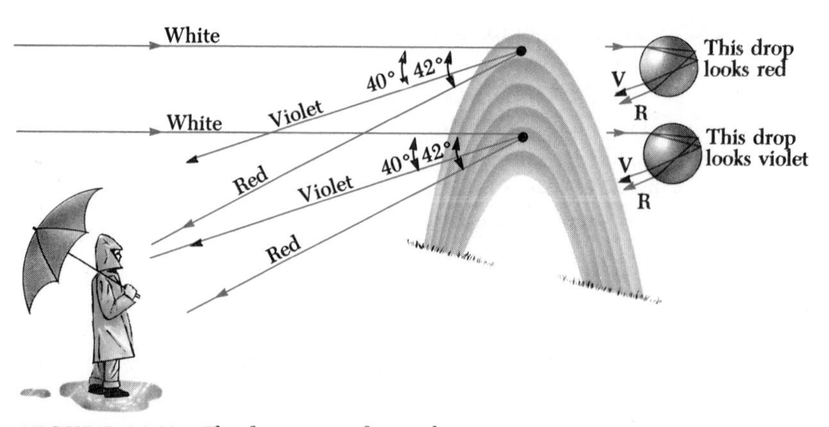

FIGURE 24.21 The formation of a rainbow.

"ROY G BIV," where each letter represents one of the prominent spectral colors: red, orange, yellow, green, blue, indigo, and violet. (Indigo is not really distinct from violet, but it does allow the boy's last name to have a vowel.) If all the colors of the visible spectrum are incident on the eye simultaneously, the mixture is interpreted as white by the eye. On the other hand, black denotes the complete absence of color.

A thorough study of color and of why we see colors as we do would constitute a rather long chapter or perhaps even a textbook to do it justice. In this section, we shall present some concepts that will help you have some understanding of the colors in our world.

The color of an object depends on one or more of the following three processes:

1. The emission of light by the object
2. The reflection of light by the object
3. The transmission of light by the object

Emission of Light

Any object whose atoms have been sufficiently excited by the absorption of energy will emit light. For example, if you heat an iron poker to the point where it glows, the added heat energy causes a rearrangement of the electrons in the atoms of the poker. Light is emitted by the atoms as they return to the state they were in before the rearrangement took place. Hence, the light we see when the poker glows arises from this emission.

Likewise, the light from a neon sign in a store window is emitted by the excited gas atoms contained in the tubes. The colors emitted by a heated gas depend on the characteristics of atoms contained in the tubes.

A more complete discussion of emission requires a knowledge of the concepts of quantum physics, which we shall present in Chapter 29.

Transmission of Light

The color of a piece of transparent material depends, to a large extent, on the wavelengths of light that the material transmits. A piece of colored glass receives its color by absorbing most of the wavelengths of light incident upon it while allowing others to pass through. Those wavelengths that pass through (or are transmitted) give the glass a distinctive color. For example, a piece of red glass in the stained glass window of a church appears red because it absorbs all colors in the visible spectrum except those the eye interprets as red. Because the various wavelengths of light carry energy, a piece of glass is slightly warmed when it absorbs light. A clear piece of glass, such as a window pane, allows all colors in the visible spectrum to pass through and so has no characteristic color.

Pieces of colored glass are often used as optical filters to select a particular wavelength for certain experiments. A color television camera uses three filters to separate the light from a scene into the three primary colors—red, blue, and green. (See Section 23.6 for a discussion of the television camera.)

Reflection of Light

Most of the objects around us achieve their color by reflection. For example, a red rose neither emits nor transmits visible light. Rather, its color arises because it reflects only those wavelengths in the visible spectrum that are interpreted by the eye as red. Thus, when a red rose is placed in sunlight, it absorbs all the colors in the visible spectrum except red. Now imagine that we take our red rose into a room that is illuminated with only a green light source. There is no red light available from the source, and hence the rose will not reflect any light at all. For this reason, the rose will appear to be black when viewed with green light. On the other hand, the stem and leaves of the rose will appear to have their natural green color when viewed in green light.

Before you read the next few sentences, see if you can guess what the American flag would look like if you observed it in a room illuminated with only a blue light source. The red stripes would appear black because there is no red light for them to reflect. The white stripes and stars would appear blue because all the colors in the spectrum are reflected by a white object. Finally, the blue square in the corner would appear blue because there is blue light in the room for it to reflect.

The Additive Primaries and Color Vision

As we have seen, when all the colors of the visible spectrum are mixed together, the light appears to be white. However, if equal intensities of red, green, and blue light are mixed together, the result is also interpreted by the eye as white. In fact, *all of the colors in the visible spectrum can be generated by mixing together proper proportions of these three colors.* Because of their ability to reproduce the sensation of any color when added to each other in varying amounts, red, green, and blue are called the **additive primaries.**

You can demonstrate some of the properties of the additive primaries by placing a piece of red cellophane over the face of one flashlight, a piece of green cellophane over the face of a second flashlight, and a piece of blue cellophane over the face of a third flashlight. When the three beams are superimposed on a white wall, the illuminated area will appear as in Figure 24.22 and Color Plate 24. The area where the green light overlaps the red light appears yellow, a magenta area appears where the red light overlaps the blue, and a cyan (bluish green) area appears where the green light overlaps the blue. However, the central region, where all the three colors overlap, is white.

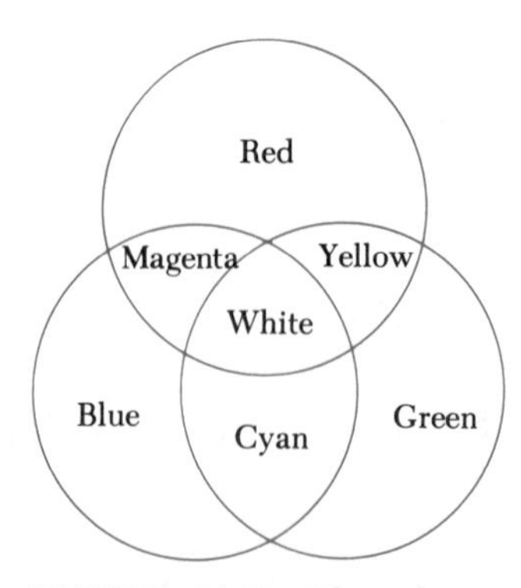

FIGURE 24.22 The color wheel. (See Color Plate 24 for a four-color photo of the additive primary colors.)

It is perhaps in order to remark here that any textbook on art seems to disagree with this last paragraph. Such books claim that if you mix red, green, and blue paint, the resultant color will be black or muddy brown, rather than white as we have stated. Actually, both claims are correct. The art textbook is referring to the colors one obtains when mixing various pigments of paint, whereas we are referring to the mixing of light of different wavelengths. In mixing colored pigments rather than colored lights, the three colors that combine in various proportions to give all the other colors of the spectrum are red, yellow, and blue. Because this phenomenon depends on the ability of the pigments to absorb the various wavelengths of white light, the red, yellow, and blue are called the **subtractive primaries.**

The color television set, discussed in Section 23.6, is a good example of how one can make practical use of color mixing. The colors observed on the screen of a color television set are produced when the electron beams strike the screen causing the dots to fluoresce in various colors. If three dots at a particular location on the screen are emitting equal intensities of light, the blue, red, and green light coming from this location appear to be white light. If the blue dot at another location is emitting no light while the red and green dots are emitting light of equal intensity, that spot appears to be yellow. (See the color wheel, Color Plate 24.

The light sensors on the retina of the eye are of two types, rods and cones. The rods are sensitive enough to respond to dim light and to small variations in light intensity. However, they cannot distinguish between the various wavelengths, or colors, of light. The cones are the sensors that enable us to distinguish colors. There are three types of cones, each sensitive to one of the three additive primary colors. The cones sensitive to red light respond when illuminated by red, orange, yellow, and, to a lesser extent, green light. The signal sent to the brain from the red-sensitive cones is the same regardless of which of these colors strikes the cones. Likewise, the green-sensitive cones respond primarily to green and yellow light and, to a lesser extent, blue and orange light. Finally, the blue-sensitive cones are sensitive to blue and violet light.

Now let us assume that light entering the eye stimulates only the red-sensitive cones, which in turn causes a message to be sent to the brain. Based on this information alone, the brain would have no way of determining the color of the light. In addition, however, the brain recognizes that no signals are being sent by the green-sensitive cones. Hence, the brain concludes that the light must be red since it receives signals from the red-sensitive cones only. If pure yellow light enters the eye, it stimulates both the red-sensitive and the green-sensitive cones. The brain interprets such stimulation as yellow. It is interesting that light of this single wavelength appears to be yellow, even though we see the same yellow color when a combination of green and red light strikes the retina. Such a combination appears to be yellow even though there is no yellow light present. This is explained by the fact that red and green wavelengths stimulate both the red-sensitive and green-sensitive cones in the same manner as light of a single yellow wavelength.

Most animals are completely color blind. Except for the other primates and a few other species, notably bees, the human being is the only animal capable of seeing the world in full color. Complete lack of color vision is very rare in humans, but partial color blindness is common. About one man in 12 and one woman in 120 have what is known as red-green blindness. The eyes of such individuals respond in the same way to red and green light, and therefore the brain cannot distinguish these colors.

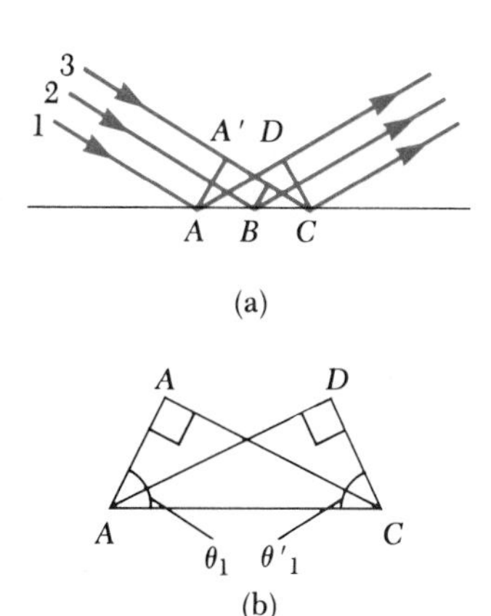

FIGURE 24.23 (a) Huygens' construction for proving the law of reflection. (b) Triangle ADC is identical to triangle $AA'C$.

24.9 HUYGENS' PRINCIPLE APPLIED TO REFLECTION AND REFRACTION

The laws of reflection and refraction were stated earlier in this chapter without proof. We shall now derive these laws using Huygens' principle. Figure 24.23a will be used in our consideration of the law of reflection. The line AA'

represents a wavefront of the incident light. As ray 3 travels from A' to C, ray 1 reflects from A and produces a spherical wavelet of radius AD. (Recall that the radius of a Huygens' wavelet is equal to vt.) Since the two wavelets having radii $A'C$ and AD are in the same medium, they have the same velocity, v, and thus $AD = A'C$. Meanwhile, the spherical wavelet centered at B has spread only half as far as the one centered at A since ray 2 strikes the surface later than ray 1.

From Huygens' principle, we find that the reflected wavefront is CD, a line tangent to all the outgoing spherical wavelets. The remainder of the analysis depends upon geometry, as summarized in Figure 24.23b. Note that the right triangles ADC and $AA'C$ are congruent because they have the same hypotenuse, AC, and because $AD = A'C$. From Figure 24.23b we have

$$\sin \theta_1 = \frac{A'C}{AC} \quad \text{and} \quad \sin \theta_1' = \frac{AD}{AC}$$

Thus,

$$\sin \theta_1 = \sin \theta_1'$$
$$\theta_1 = \theta_1'$$

which is the law of reflection.

Now let us use Huygens' principle and Figure 24.24a to derive Snell's law of refraction. Note that in the time interval Δt, ray 1 moves from A to B and ray 2 moves from A' to C. The radius of the outgoing spherical wavelet centered at A is equal to $v_2\Delta t$. The distance $A'C$ is equal to $v_1\Delta t$. Geometric considerations show that angle $A'AC$ equals θ_1 and angle ACB equals θ_2. From triangles $AA'C$ and ACB, we find that

$$\sin \theta_1 = \frac{v_1 \Delta t}{AC} \quad \text{and} \quad \sin \theta_2 = \frac{v_2 \Delta t}{AC}$$

If we divide these two equations, we get

$$\frac{\sin \theta_1}{\sin \theta_2} = \frac{v_1}{v_2}$$

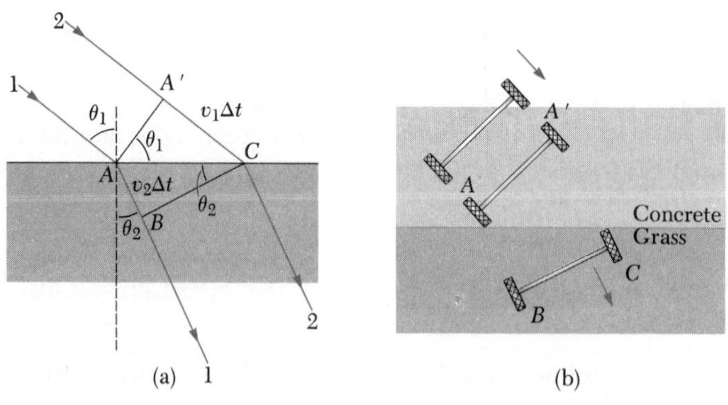

FIGURE 24.24 (a) Huygens' construction for proving the law of refraction. (b) A mechanical analog of refraction.

But from Equation 24.4 we know that $v_1 = c/n_1$ and $v_2 = c/n_2$. Therefore,

$$\frac{\sin \theta_1}{\sin \theta_2} = \frac{c/n_1}{c/n_2} = \frac{n_2}{n_1}$$

$$n_1 \sin \theta_1 = n_2 \sin \theta_2$$

which is the law of refraction.

A mechanical analog of refraction is shown in Figure 24.24b. The wheels on a device such as a lawnmower change their direction as they move from a concrete surface to a grass surface.

24.10 TOTAL INTERNAL REFLECTION

An interesting effect called *total internal reflection* can occur when light attempts to move from a medium having a *high* index of refraction to one having a *lower* index of refraction. Consider a light beam traveling in medium 1 and meeting the boundary between medium 1 and medium 2, where n_1 is greater than n_2 (Fig. 24.25). Various possible directions of the beam are indicated by rays 1 through 5. Note that the refracted rays are bent away from the normal because n_1 is greater than n_2. At some particular angle of incidence, θ_c, called the **critical angle,** the refracted light ray will move parallel to the boundary so that $\theta_2 = 90°$ (Fig. 24.25b). *For angles of incidence greater than* θ_c, the beam is entirely reflected at the boundary. Ray 5 in Figure 24.25a shows this occurrence. This ray is reflected at the boundary as though it had struck a perfectly reflecting surface. This ray, and all those like it, obey the law of reflection; that is, the angle of incidence equals the angle of reflection.

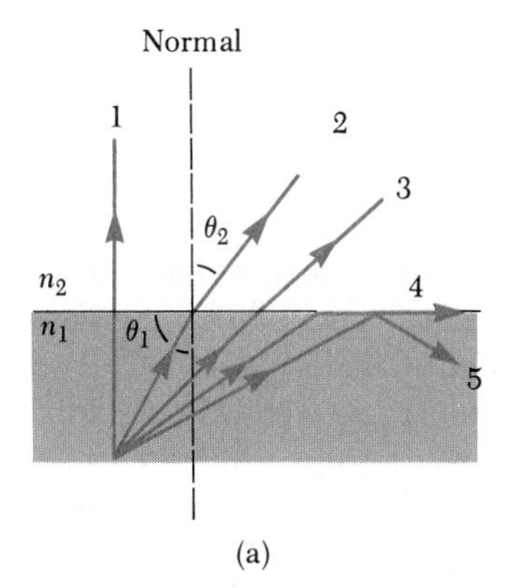

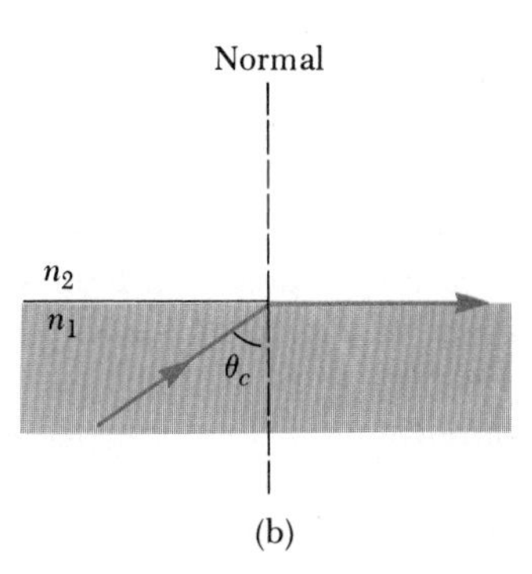

FIGURE 24.25 (a) A ray traveling from a medium of index n_1 to a medium of index n_2, where $n_1 > n_2$. As the angle of incidence increases, the angle of refraction increases until θ_2 is 90° (ray 4). For even larger angles of incidence, total internal reflection occurs (ray 5). (b) The angle of incidence producing an angle of refraction equal to 90° is often called the *critical angle,* θ_c.

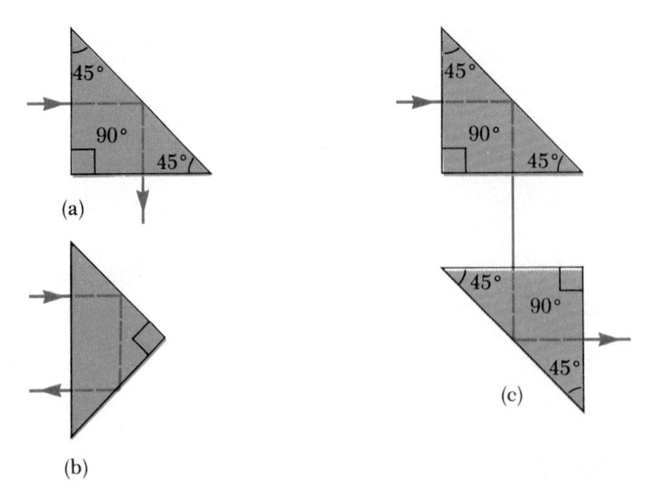

FIGURE 24.26 Internal reflection in a prism. (a) The ray is deviated by 90°. (b) The direction of the ray is reversed. (c) Two prisms used as a periscope.

We can use Snell's law to find the critical angle. When $\theta_1 = \theta_c$, $\theta_2 = 90°$ and Snell's law (Eq. 24.8) gives

$$n_1 \sin \theta_c = n_2 \sin 90° = n_2$$

Critical angle

$$\boxed{\sin \theta_c = \frac{n_2}{n_1}} \qquad \text{for } n_1 > n_2 \qquad (24.9)$$

Note that this equation can be used only when n_1 is greater than n_2. That is,

total internal reflection occurs only when light attempts to move from a medium of high index of refraction to a medium of lower index of refraction.

If n_1 were less than n_2, Equation 24.9 would give $\sin \theta_c > 1$, which is an absurd result because the sine of an angle can never be greater than unity.

The critical angle is small for substances with a large index of refraction, such as diamond, where $n = 2.42$ and $\theta_c = 24°$. For crown glass, $n = 1.52$ and $\theta_c = 41°$. In fact, this property combined with proper faceting causes diamonds and crystal glass to sparkle.

One can use a prism and the phenomenon of total internal reflection to alter the direction of travel of a light beam. Two such possibilities are illustrated in Figure 24.26. In one case the light beam is deflected by 90° (Fig. 24.26a), and in the second case the path of the beam is reversed (Fig. 24.26b). A common application of total internal reflection is in a submarine periscope. In this device, two prisms are arranged as in Figure 24.26c so that an incident beam of light follows the path shown and one is able to "see around corners."

EXAMPLE 24.7 A View from the Fish's Eye

(a) Find the critical angle for a water-air boundary if the index of refraction of water is 1.33.

Solution Applying Equation 24.9, we find the critical angle to be

$$\sin \theta_c = \frac{n_2}{n_1} = \frac{1}{1.33} = 0.752$$

$$\theta_c = 48.8°$$

(b) Use the results of (a) to predict what a fish will see if it looks upward toward the water surface at an angle of 40°, 48.8°, and 60°.

Solution Because the path of a light ray is reversible, the fish can see out of the water if it looks toward the surface at an angle less than the critical angle. Thus, at 40°, the fish can see into the air above the water. At an angle of 48.8°, the critical angle for water, the light that reaches the fish has to skim along the water surface before being refracted to the fish's eye. At angles greater than the critical angle, the light reaching the fish comes via internal reflection at the surface. Thus, at 60°, the fish sees a reflection of some object on the bottom of the pool.

Light fibers used in the construction of a decorative lamp. (Courtesy of Poly-Optics)

Fiber Optics

Another interesting application of total internal reflection is the use of glass or transparent plastic rods to "pipe" light from one place to another. As indicated in Figure 24.27, light is confined to traveling within the rods, even around gentle curves, as the result of successive internal reflections. Such a "light pipe" will be flexible if thin fibers are used rather than thick rods. If a bundle of parallel fibers is used to construct an optical transmission line, images can be transferred from one point to another.

This technique is used in a sizable industry known as *fiber optics*. There is very little light intensity lost in these fibers as a result of reflections on the sides. Any loss in intensity is due essentially to reflections from the two ends and absorption by the fiber material. These devices are particularly useful when one wishes to view an image produced at inaccessible locations. For example, physicians often use this technique to examine internal organs of the body. The field of fiber optics is finding increasing use in telecommunications, since the fibers can carry a much higher volume of telephone calls or other forms of communication than electrical wires. The essay in this chapter discusses the use of fiber optics in the expanding field of telecommunications.

SUMMARY

Huygens' principle states that all points on a wavefront are point sources for the production of spherical secondary waves called wavelets. These wavelets propagate outward at a speed characteristic of waves in a particular medium. After some time has elapsed, the new position of the wavefront is the surface tangent to the wavelets.

The **index of refraction** of a material, n, is defined as

$$n \equiv \frac{c}{v} \quad (24.4)$$

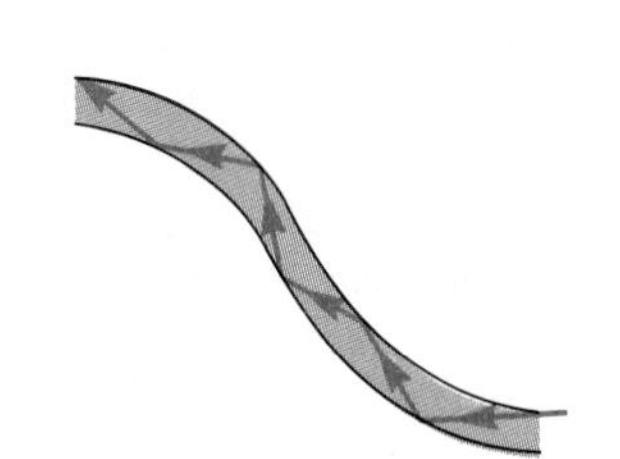

FIGURE 24.27 Light travels in a curved, transparent rod by multiple internal reflections.

where c is the speed of light in vacuum and v is the speed of light in the material. In general, n varies with wavelength as

$$n = \frac{\lambda_0}{\lambda_n} \tag{24.7}$$

where λ_0 is the wavelength of the light in vacuum and λ_n is its wavelength in the material.

The **law of reflection** states that a wave reflects from a surface such that the *angle of reflection*, θ_1', equals the *angle of incidence*, θ_1.

The **law of refraction**, or **Snell's law**, states that

$$n_1 \sin \theta_1 = n_2 \sin \theta_2 \tag{24.8}$$

Total internal reflection can occur when light attempts to move from a material having a high index of refraction to one having a lower index of refraction. The *maximum angle of incidence*, θ_c, for which light can move from a medium of index n_1 into a medium of index of refraction n_2, where n_1 is greater than n_2, is called the **critical angle** and is given by

$$\sin \theta_c = \frac{n_2}{n_1} \quad \text{for } n_1 > n_2 \tag{24.9}$$

ADDITIONAL READING

W. S. Boyle, "Light-Wave Communications," *Sci. American*, August 1977, p. 40.

W. Bragg, *The Universe of Light*, Dover, 1959.

A. B. Fraser and W. H. Mach, "Mirages," *Sci. American*, January 1976, p. 102.

N. S. Kapany, "Fiber Optics," *Sci. American*, November 1960, p. 72.

E. A. Lacy, *Fiber Optics*, Englewood Cliffs, N.J., Prentice-Hall, 1982.

E. H. Land, "The Retinex Theory of Color Vision," *Sci. American*, December 1977, p. 108.

"Light," *Sci. American*, September 1968 (entire issue).

D. F. Mandoli and W. R. Briggs, "Fiber Optics in Plants," *Sci. American*, August 1984, p. 90.

E. W. Stark, "Diffuse Reflection: Uses That Affect Our Lives," *The Physics Teacher*, March 1986, p. 144.

A. C. S. van Heel and C. H. F. Velzel, *What Is Light*, World University Library, New York, McGraw-Hill, 1968, Chap. 1.

S. Williamson and H. Cummins, *Light and Color in Nature and Art*, John Wiley and Son, 1983.

QUESTIONS

1. Under certain circumstances, sound can be heard over extremely long distances. This frequently happens over a body of water, where the air near the water surface is cooler than the air higher up. Explain how the refraction of sound waves in such a situation could increase the distance over which the sound can be heard.
2. Why do astronomers looking at distant galaxies talk about looking backward in time?
3. A solar eclipse occurs when the moon gets between the earth and the sun. Use a diagram to show why some areas of the earth see a total eclipse, other areas see a partial eclipse, and most areas see no eclipse.
4. As light travels from one medium to another, does its wavelength change? Does its frequency change? Does its velocity change? Explain.
5. Some department stores have their windows slanted slightly inward at the bottom. This is to decrease the glare from streetlights or the sun, which would make it difficult for shoppers to see the display inside. Draw a sketch of a light ray reflecting off such a window to show how this technique works.

6. Suppose you are told only that two colors of light (X and Y) are sent through a prism and that X is bent more than Y. Which color travels more slowly in the glass of the prism?
7. The level of water in a clear, colorless glass is easily observed with the naked eye. The level of liquid helium in a clear glass is extremely difficult to see with the naked eye. Explain.
8. Why does a diamond show flashes of color when observed under ordinary white light?
9. Explain why a diamond sparkles more than a glass crystal of the same shape and size.
10. Redesign the periscope of Figure 24.26c so that it can show you where you have been rather than where you are going.
11. Explain why an oar in the water appears to be bent.
12. Light from a helium-neon laser beam (λ = 632.8 nm) is incident on a block of Lucite, as shown in Figure 24.28. From this photograph, how would you estimate the index of refraction of the Lucite at this wavelength?

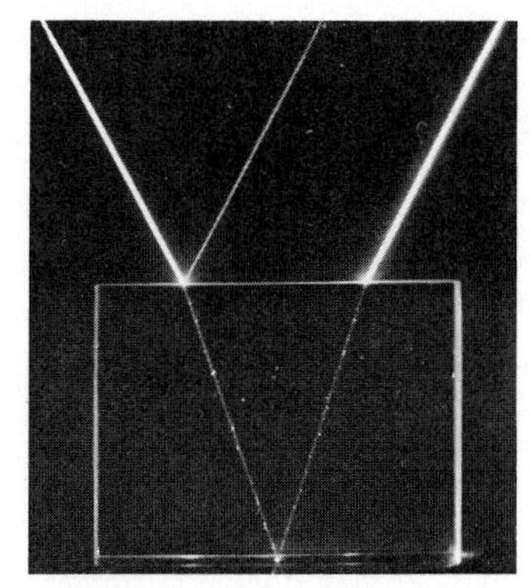

FIGURE 24.28 Question 12

13. Is it possible to have total internal reflection for light incident from air on water? Explain.

PROBLEMS

Section 24.1 The Nature of Light

Section 24.2 Measurements of the Speed of Light

1. As a result of his observations, Roemer concluded that the time interval between successive eclipses of the moon Io by the planet Jupiter increased by 22 minutes during a six-month period as the earth moved from a point in its orbit on the side of the sun nearer Jupiter to a point on the side opposite Jupiter. Using 1.5×10^8 km as the average radius of the earth's orbit about the sun, calculate the speed of light from these data.

2. Use Roemer's value of 22 min discussed in Problem 1 and the presently accepted value of c (2.998×10^8 m/s) to find an average value for the distance between the earth and the sun.

3. If the Fizeau experiment is performed such that the round-trip distance for the light is 40 m, find the two lowest speeds of rotation that allow the light to pass through the notches. Assume that the wheel has 360 teeth and that the speed of light is 3×10^8 m/s. Repeat for a round-trip distance of 4000 m.

4. Albert A. Michelson very carefully measured the speed of light using an alternative version of the technique developed by Fizeau. Figure 24.29 shows the approach used. Light was reflected from one face of a rotating eight-sided mirror toward a stationary mirror 35 km away. At certain rates of rotation, the returning beam of light would be directed toward the eye of an observer as shown. (a) What minimum angular velocity must the rotating mirror have in order that side A will have rotated to position B so that the light will be reflected to the eye? (b) What is the next highest angular velocity that will enable the source of light to be seen?

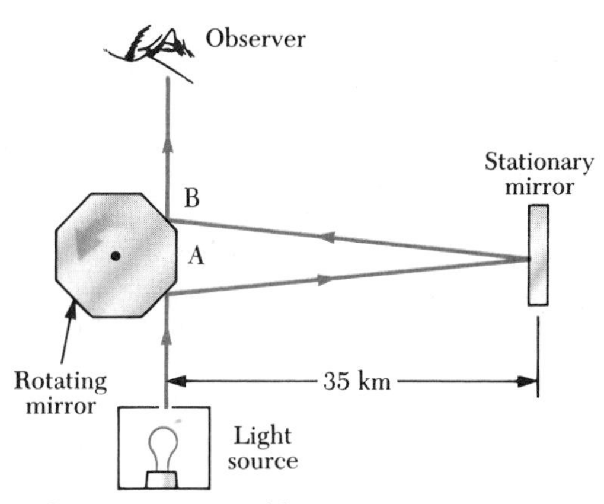

FIGURE 24.29 Problem 4

Section 24.4 Reflection and Refraction

Section 24.5 The Law of Refraction

5. Find the speed of light in benzene.

6. Find the speed of light in (a) flint glass, (b) water, and (c) zircon.

7. The wavelength of sodium light in air is 589 nm. Find its wavelength in ethyl alcohol.

8. Light of wavelength 436 nm in air enters a fishbowl filled with water then exits through the crown glass wall of the container. What is the wavelength of the light (a) while in the water, and (b) while in the glass?

9. A light ray in air is incident on a water surface at an angle of 30° with respect to the normal to the surface. What is the angle of the refracted ray relative to the normal to the surface?

10. A ray of light is incident on the surface of a block of clear ice at an angle of 40° with the normal. Part of the light is reflected and part is refracted. Find the angle between the reflected and refracted light.

11. A ray of light strikes a flat block of glass ($n = 1.50$) of thickness 2 cm at an angle of 30° with the normal (Fig. 24.30). Trace the light beam through the glass and find the angles of incidence and refraction at each surface.

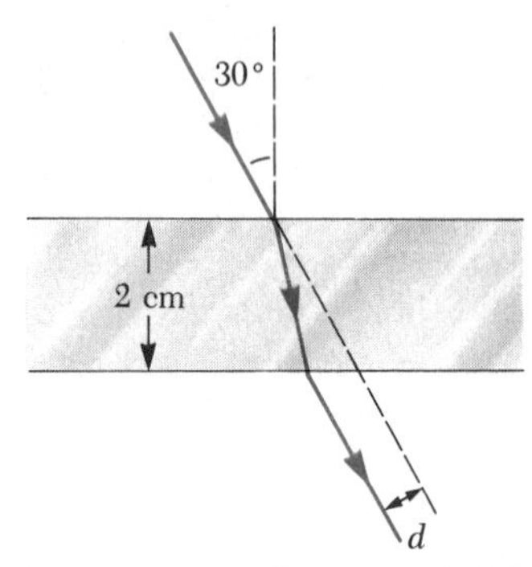

FIGURE 24.30 Problems 11, 12, and 16

12. Find the time for the light to pass through the glass block described in Problem 11.

13. A light ray initially in water enters a transparent substance at an angle of 37° with respect to the normal, and the transmitted ray is refracted at an angle of 25°. Calculate the speed of light in the transparent substance.

14. You are standing with a mirror at the center of a giant floral clock. Someone at 12 o'clock shines a beam of light toward you, and you want to use the mirror to reflect the beam toward an observer at 5 o'clock. What should be the angle of incidence to achieve this?

15. A ray of light strikes the midpoint of one face of an equiangular glass prism ($n = 1.5$) at an angle of incidence of 30°. (a) Trace the path of the light ray through the glass and find the angles of incidence and refraction at each surface. (b) If a small fraction of light is also reflected at each surface, find the angle of incidence and angle of reflection at these surfaces.

• **16.** When the light of Problem 11 passes through the glass block, it is shifted laterally by a distance d (Fig. 24.30). Find the value of d.

• **17.** (a) What is the speed of light in crown glass if the wavelength in vacuum is 589 nm? (b) What thickness of crown glass will equal 100 wavelengths of this light (measured in the glass)? (Make use of the curve for crown glass in Figure 24.16.)

• **18.** A submarine is 300 m horizontally out from the shore and 100 m beneath the surface of the water. A laser beam is sent from the sub such that it strikes the surface of the water at a point 210 m from the shore. If the beam just strikes the top of a building standing directly at the water's edge, find the height of the building.

• **19.** How far does a beam of light travel in water in the same time it takes it to travel 10 m in glass of index of refraction 1.5?

•• **20.** A drinking glass is 4 cm wide at the bottom, as shown in Figure 24.31. When an observer's eye is placed as shown, the observer sees the edge of the bottom of the glass. When this glass is filled with water, the observer sees the center of the bottom of the glass. Find the height of the glass.

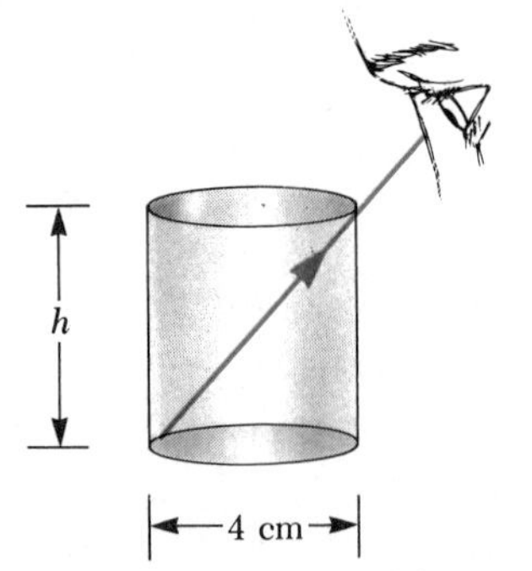

FIGURE 24.31 Problem 20

Section 24.6 Dispersion and Prisms

21. A certain kind of glass has an index of refraction of 1.6500 for blue light at wavelength 430 nm and an index of 1.615 for red light of wavelength 680 nm. If a beam containing these two colors is incident at an angle of 30° on a piece of this glass, what is the angle between the two beams inside the glass?

22. Light of wavelength 400 nm is incident at an angle of 45° on acrylic and is refracted as it passes into the material. What wavelength of light incident on fused quartz at an angle of 45° would be refracted at exactly this same angle? (See Fig. 24.16.)

•• **23.** Light of wavelength 700 nm is incident on the face of a fused quartz prism at an angle of 75° with respect to the normal to the surface. The apex angle of the prism is 60°. Use the value of n from Figure 24.16 and calculate the angle (a) of refraction at the first surface, (b) of incidence at a second surface, (c) of refraction at the second surface, and (d) between the incident and emerging rays.

•• **24.** The glass described in Problem 21 is shaped into an equiangular prism and the red and blue wavelengths of Problem 21 are directed toward one face at an angle of 75° with the normal. (a) Find the angles δ_R and δ_B shown in Figure 24.17. (b) Calculate the dispersion of $\delta_B - \delta_R$ produced.

Section 24.10. Total Internal Reflection

25. Calculate the critical angle for the following materials when surrounded by air: (a) diamond, (b) fluorite, (c) ice. Assume that $\lambda = 589$ nm.

26. Repeat Problem 25 when the materials are surrounded by water.

27. A beam of light is incident from air on the surface of a liquid. If the angle of incidence is 30° and the angle of refraction is 22°, find the critical angle for the liquid when surrounded by air.

28. A jewel thief hides a diamond by placing it on the bottom of a public swimming pool. He places a circular piece of metal on the surface of the water directly above and centered on the diamond. If the pool is 2 m deep, find the diameter of the piece of metal such that the diamond cannot be seen.

• **29.** A light ray is incident perpendicular to the long face (the hypotenuse) of a 45°–45°–90° prism surrounded by air, as shown in Figure 24.26b. Calculate the minimum index of refraction of the prism for which the ray will follow the path shown.

• **30.** Repeat Problem 29 if the prism is surrounded by water.

•• **31.** A large Lucite cube ($n = 1.59$) has a small air bubble (a defect in the casting process) below one surface. When a penny (diameter 1.9 cm) is placed directly over the bubble, it cannot be seen at any angle from the opposite site of the cube. However, when a dime (diameter 1.75 cm) is placed directly over the bubble, it can be seen from the opposite surface of the cube. What is the range of the possible depths of the air bubble beneath the surface?

ADDITIONAL PROBLEMS

32. A layer of ice floats on water. If light is incident on the upper surface of the ice at an angle of 30°, what is the angle of refraction in the water?

33. A beam of light is incident at an angle of 37° on the surface of a block of transparent material. The angle of refraction is found to be 25°. What is the speed of light in the material?

34. Light is incident on the surface of a prism, $n = 1.8$, as shown in Figure 24.26a. If the prism is surrounded by a fluid, what is the maximum index of refraction of the fluid that would still cause total internal reflection?

35. The sun is 10° above the horizon. If you are swimming beneath the surface of a pool of water, at what angle above the horizon does the sun appear to be?

• **36.** A narrow beam of light is incident from air onto a glass surface of index of refraction 1.56. Find the angle of incidence for which the corresponding angle of refraction will be one-half the angle of incidence. (*Hint:* you might want to use the trigonometric identity $\sin 2\theta = 2 \sin \theta \cos \theta$.)

• **37.** A thick plate of flint glass ($n = 1.66$) rests on top of a thick plate of transparent acrylic ($n = 1.50$). A beam of light is incident on the top surface of the flint glass at an angle θ_i. The beam passes through the glass and the acrylic and emerges from the acrylic at an angle of 40° with respect to the normal. Calculate the value of θ_i. A sketch of the light path through the two plates of refracting material would be helpful.

• **38.** One technique to measure the angle of a prism is shown in Figure 24.32. A parallel beam of light is directed on the angle such that the beam reflects from opposite sides. Show that the angular separation of the two beams is given by $B = 2A$.

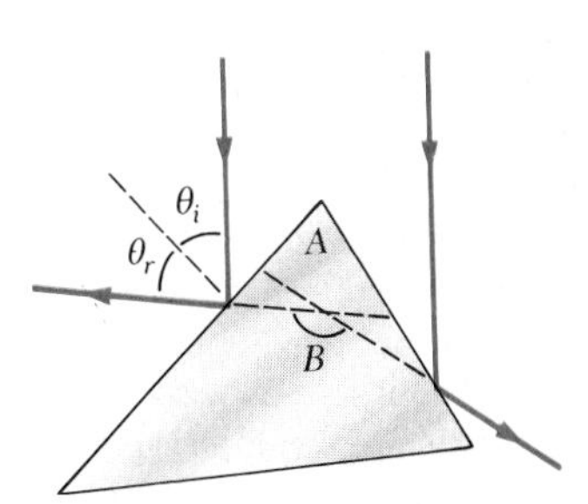

FIGURE 24.32 Problem 38

• **39.** Repeat Example 24.2 for the case in which the angle between the two mirrors is 60°.

• **40.** Repeat Example 24.2 for the case in which the angle is 90°. Show that the ray of light is always reflected from the second mirror so that it travels opposite to the original direction.

• **41.** Light of wavelength λ_0 in vacuum has a wavelength of 438 nm in water and a wavelength of 390 nm in benzene. What is the index of refraction of water relative to benzene at the wavelength λ_0?

• **42.** A light ray of wavelength 589 nm is incident at an angle θ on the top surface of a block of polystyrene, as shown in Fig. 24.33. (a) Find the maximum value of θ for which the refracted ray will undergo total internal reflection at the left vertical face of the block. (b) Repeat the calculation for the case in which the polystyrene block is immersed in water. (c) What happens if the block is immersed in carbon disulfide?

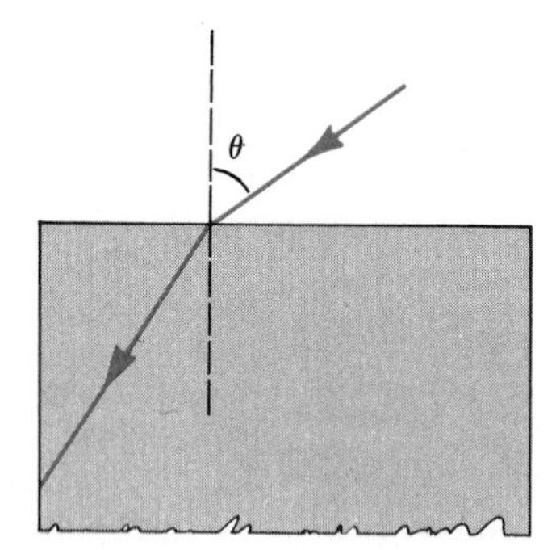

FIGURE 24.33 Problem 42

•• **43.** For this problem refer to Figure 24.17. For various angles of incidence, it can be shown that the angle δ is a minimum when the ray passes through the glass such that the ray is parallel to the base. A measurement of this minimum angle of deviation enables one to find the index of refraction of the prism material. Show that n is given by the expression

$$n = \frac{\sin\left[\frac{1}{2}(A + \delta_{\text{min}})\right]}{\sin\left(\frac{A}{2}\right)}$$

where A is the apex angle of the prism.

•• **44.** A light ray is incident on a prism and refracted at the first surface as shown in Figure 24.34. Let ϕ represent the apex angle of the prism and n its index of refraction. Find in terms of n and ϕ the smallest allowed value of the angle of incidence at the first surface for which the refracted ray will not undergo internal reflection at the second surface.

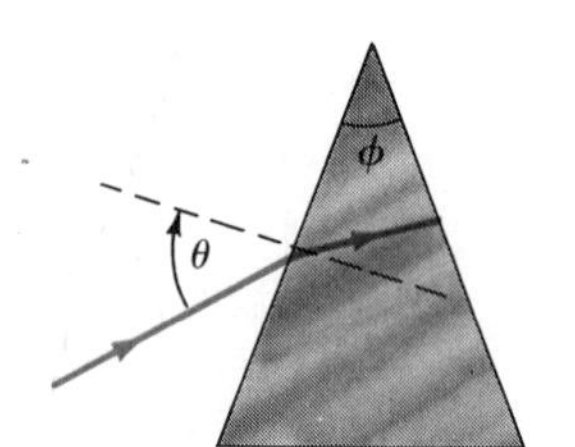

FIGURE 24.34 Problem 44

Essay

Fiber Optics: A Revolution in Communications

Edward A. Lacy

From the invention of the telephone in the late 19th century until the 1970s, telephone communication depended on the movement of electrons back and forth over metallic wires. Now, however, in this age of optical communication, light waves (photons) guided by tiny fibers of glass are being used to accomplish the same purpose.

Fiber-optic lightwave technology has evolved into a communication system of enormous capabilities. The full potential of fiber-optic communication systems is still being determined, but its effect on our lives may be as great as that of computers and integrated semiconductor circuits. It has become a billion-dollar industry in just a few years.

As the United States and the world are being rewired with fiber-optic cables, the 1980s have been called the "decade of glass" as technology pushes toward "glass-wired" communities.

The exciting feature of fiber-optic systems is their ability to transmit thousands of telephone conversations, several television programs, and numerous computer data signals via one or two flexible, hair-thin threads of optical fiber. Neither conventional copper wires nor special coaxial cables can transmit so much information. Even microwave relays and satellites are being replaced in some cases by fiber-optic systems.

With their tremendous information-carrying capacity, which is called *bandwidth*, fiber-optic systems will

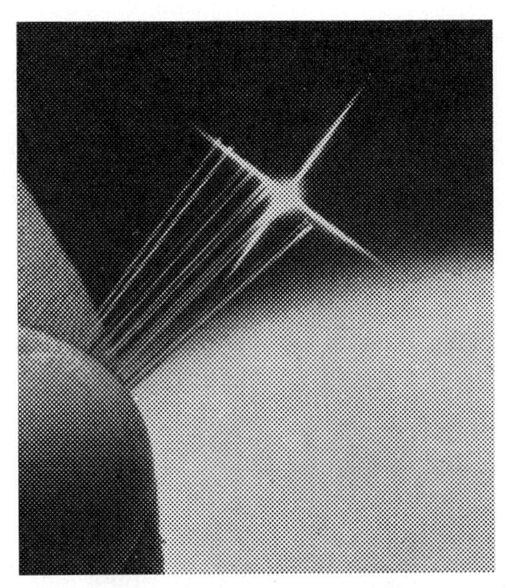

FIGURE 1 These light guides are used by the Bell System in transmitting telephone calls. (Courtesy of Bell Systems)

undoubtedly make practical such broadband services as high-definition television and two-way television which were too costly before the development of fiber optics. These systems will also allow word processors and image transmitting-and-receiving equipment to operate more efficiently in offices of the future.

The wideband world of fiber will make possible the Broadband Integrated Services Digital Network (BISDN) now being developed. With the BISDN, there will be an era of Universal Information Services which will permit high-quality television, voice, and data signals to be transmitted between businesses and homes.

Present local area networks (LANs) interconnect computers at a common geographic site. Because of their low bandwidth requirements, the LANs typically do not use fiber optics. But as LANs are interconnected, the higher bandwidth of fiber-optic systems becomes essential.

Because fiber-optic systems are immune to electromagnetic interference, produced for example by nearby lightning and large electric motors cycling on and off, they give accurate transmission of data—about 100 times better than transmission over copper cables. For information transmitted in bits, the accuracy is typically one error in 100 million bits!

Another significant advantage of fiber-optic systems is their much smaller, lighter-weight cables. An optical fiber with its protective jacket may be typically 0.635 cm in diameter, and yet it can replace a 7.62-cm diameter bundle of copper wires now used to carry the same number of telephone conversations and other signals. The importance of this dramatic decrease in size is not obvious in uncongested rural areas; in major cities, however, where telephone cables must be placed underground, conduits are so crammed that they can scarcely accommodate a single additional copper cable.

Because of the size reduction, there is a corresponding weight reduction; for example, 94.5 kg of copper wire can be replaced with 3.6 kg of optical fiber. Weight reduction is important for the military services as it allows faster deployment of communication cables on battlefields. On the civilian side, it is important in huge jet aircraft which use a surprising amount of copper cables between the various equipment and instrumentation on board. By replacing these cables with optical cables, up to 1000 pounds of weight may be saved, thereby giving better fuel consumption.

Fiber-optic systems provide more private communication than do satellites, as anyone with a satellite receiver can intercept satellite transmissions. Fiber-optic cables are so difficult to "tap" that they are generally considered to be "secure," an important feature for the military. To protect satellite communication from eavesdropping, the transmissions must be encrypted or encoded, which results in additional costs.

Because of these numerous advantages, dozens of major fiber-optic systems are now in operation around the world, and numerous others are being constructed. By the end of 1987, more than 3 million kilometers of optical fibers had been installed in the United States alone. AT&T is now completing a transatlantic underwater fiber-optic system that will carry 40,000 telephone conversations simultaneously. A transpacific underwater cable connecting California, Hawaii, and Japan by fiber optics is scheduled for completion by 1989.

A fiber-optic communication system has three major components: a *transmitter* that converts electrical signals to light signals, an *optical fiber* for guiding the signals, and a *receiver* that captures the light signals at the other end of the fiber and reconverts them to electric signals.

The light source in the transmitter can be either a semiconductor laser or a light-emitting diode (LED). With either device, the light emitted is an invisible *infrared* signal which has a wavelength of 1300 nm. This wavelength is used because optical cables pass such signals with the least resistance. Wavelengths that are either less or greater than this number are significantly attenuated as they cannot pass through the "window" at 1300 nanometers.

The lasers and LEDs used in fiber optics are tiny units, less than half the size of a thumbnail. The light waves they produce are *modulated* in order to transmit the audio, television, or computer data signal.

By continuously varying the intensity of the light beam from the laser or the LED in step with the signal to be transmitted, *analog* modulation is achieved. In *digital* modulation the laser or LED is flashed on and off at an extremely fast rate, up to billions of times per second.

In digital modulation, a pulse of light represents the number 1, and the absence of light represents 0. In a sense, instead of flashes of light traveling down the fiber, 1s and 0s are moving down the path. With computer-type equipment, any communication can be represented by a particular pattern or code of these 1s and 0s. If the receiver is programmed to recognize such digital patterns, it can reconstruct the original signal from the 1s and 0s it receives.

Digital modulation is expressed in bits (short for binary digit) per second, megabits (1,000,000 bits) per second, or gigabits (1,000,000,000 bits) per second, where a bit is a 1 or a 0. Engineers have demonstrated a fiber-optic system that can transmit 27 gigabits per second. At this rate, the information in 700 novels, for example, could be transmitted in 1 s!

As remarkable as these bandwidths are, engineers are pursuing other techniques to give even greater bandwidth. In *wavelength division multiplexing* (also called color multiplexing) the outputs of several lasers with different wavelengths are combined and sent through one optical fiber. In *coherent detection*, a very stable frequency laser sends a continuous beam whose frequency is varied by the messages that are being transmitted. At the receiver, this lightbeam is combined with a lightbeam that has been generated at the receiver. The frequencies are designed to be slightly different so that when they are combined there will be a new signal which is the difference of the two. The new signal can be processed, as in superheterodyne radio and television sets, much easier than the original incoming lightbeam.

As you might suspect, the equipment used in digital modulation, such as encoders, is much more complicated than that used in analog modulation. Digital modulation also requires more bandwidth than analog modulation to send the same message. The former is, however, far more popular because it allows greater transmission distance with the same power and less expensive switching equipment. Thus, even though digital telecommunication is only a minority of present telecommunication now, it is rapidly replacing analog transmission, which is used mainly for television signals.

Optical fibers may be glass or plastic. *Plastic* fibers are used for some short range data links, sensors, and light pipes. But their high attenuation, which is expressed in decibels per kilometer (dB/km), prevents their use in telecommunication circuits. Whereas a plastic fiber may have a loss of 100 dB/km, present *glass* optical fibers have losses less than 1 dB/km because of their ultrapurity (99.99%). (Fluoride glass fibers being investigated may have losses as low as 0.001 dB/km.) If a window pane of 1 km thickness was made of such glass, it would be as transparent as an ordinary pane of glass.

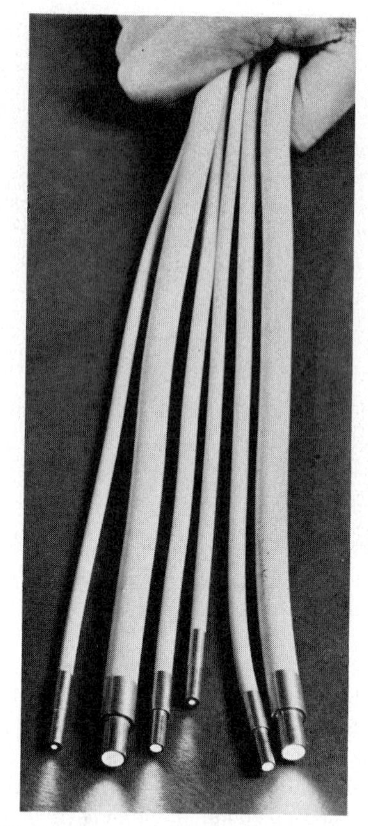

FIGURE 2 Light pipe. (Edmund Scientific Co.)

Despite this purity, the light signals eventually become dim or distorted and must be regenerated by devices called *repeaters* (or regenerators).

A *repeater* is a combination of a fiber-optic receiver and a fiber-optic transmitter. The receiver decodes the signal and triggers the transmitter to produce an identical version, only now the signal has greater strength and purity (and is in better time synchronism in the case of digital signals). Repeaters are typically placed about 30 km apart, but in the newer systems they may be separated as much as 200 km or more. Whereas present repeaters convert the light signals to electrical signals and then back to light signals, *all-optical* repeaters being investigated would convert weak light signals directly to strong light signals, skipping the conversion from light signals to electric signals to light signals. By doing this, these repeaters could theoretically be made much more sensitive and thus placed farther apart.

The center of each fiber is the *core*, which carries the light signal; a layer of glass or plastic, called the *cladding*, surrounds the core and keeps the light in the core.

Surrounding the cladding is a polyurethane jacket that protects the fiber from abrasion, crushing, and chemicals. From 1 to 144 fibers may be grouped to form a cable. (Because so many thousands of telephone conversations and other signals are carried by a single fiber-optic *cable*, it is a small catastrophe when a gopher gnaws through an underground cable or when road construction crews accidentally dig up such a cable. Telephone engineers compare such a loss to the failure of a microwave tower in a major microwave route. An armor may be added to the cable to prevent gophers from biting through but only diligence can prevent accidental dig-ups.)

In large-core fibers, light pulses can take numerous paths (called *modes*) as they bounce back and forth down the fiber. Because the different paths are not equal in length, some pulses will take longer to travel down the fiber, causing some of the pulses to overlap and thereby cause distortion. These *multimode* fibers were widely used in the early days of fiber optics but now are used mainly for certain short-distance links such as local area networks.

In the newer fibers, the core is much smaller: only 8 micrometers in diameter, about one sixth the thickness of a sheet of paper! Because of the small diameter, only one light path is possible—straight down the core with no zigzagging. Because there is only one path, there is much less distortion, giving these *single-mode* fibers a much higher bandwidth than multimode fibers.

At the end of the fiber, a photodiode converts the light signals to electric signals, which are then amplified and decoded, if necessary, to reform the signals originally transmitted.

While long-distance fiber-optic systems have received most of the attention, fiber-optic links are also quite useful for very short distances, such as *between* large computer mainframes and their peripheral terminals and printers. *Within* computers, fiber optics is being used to carry signals between circuit boards. As computers are being built to perform even faster calculations, it is being found that copper wires cannot carry the computer signals fast enough. For this reason, engineers are investigating the use of fiber-optic connections at even smaller levels such as between microchips.

In a different use of fiber optics, optical fibers are being used as sensors to detect pressure, liquid level, temperature, magnetic fields, acidity, and other stimuli. In these applications, the fibers offer the advantages of compactness, sensitivity, and immunity to hostile environments as compared with other sensors. The sensors are short lengths of optical fibers with intentionally small lateral deformations called microbends. At these microbends some of the light radiates from the fiber. The behavior of this light can be influenced by temperature, acceleration, and other parameters to be sensed.

While most optical fibers are made of glass, *plastic* fibers serve a useful role in automobile displays, electronic billboards, and short-range data links. In the first applications, the fiber is used simply as a light pipe. In automobiles, a fiber is run from a lamp to the dashboard so that the driver can monitor the condition of out-of-sight lamps and thus detect when a lamp fails. Plastic fibers have excessive losses when used for distances greater than 1 km, but the losses can be tolerated in the short distance within an auto or billboard. Plastic fibers are more flexible than glass fibers and less expensive, features that are particularly useful for automobiles.

Fiber-optic or lightwave technology's future seems insured but if *superconductors* can be built to perform at room temperature and can be produced and handled as wires, they may replace fiber-optic cables as they would offer even lower line losses than fiber optics.

SUGGESTED READING

Edward A. Lacy, *Fiber Optics*, Englewood Cliffs, New Jersey, Prentice-Hall, Inc., 1982. (An introductory text)

E. E. Basch, *Optical-Fiber Transmission*, Howard W. Sams & Co., 1987. (A more advanced reference)

C. David Chaffee, *The Rewiring of America: The Fiber Optics Revolution*, San Diego, Academic Press, 1987.

ESSAY QUESTIONS

1. What is the theoretical limit for bandwidth in fiber optics?
2. How will business and society be affected if the cost of transferring information becomes insignificant?
3. What will be the role of fiber sensors as invasive medical probes?
4. Why are fiber-optic cables difficult to "tap" (as in eavesdropping)?
5. What areas of technology had to be developed for fiber optics to be possible?

ESSAY PROBLEMS

1. If fluoride glass fibers are perfected, what would be the attenuation of such a cable between Hawaii and Japan?
2. With plastic fibers, what would be the approximate attenuation between the taillights and the dashboard of an automobile?

CHAPTER 25

Mirrors and Lenses

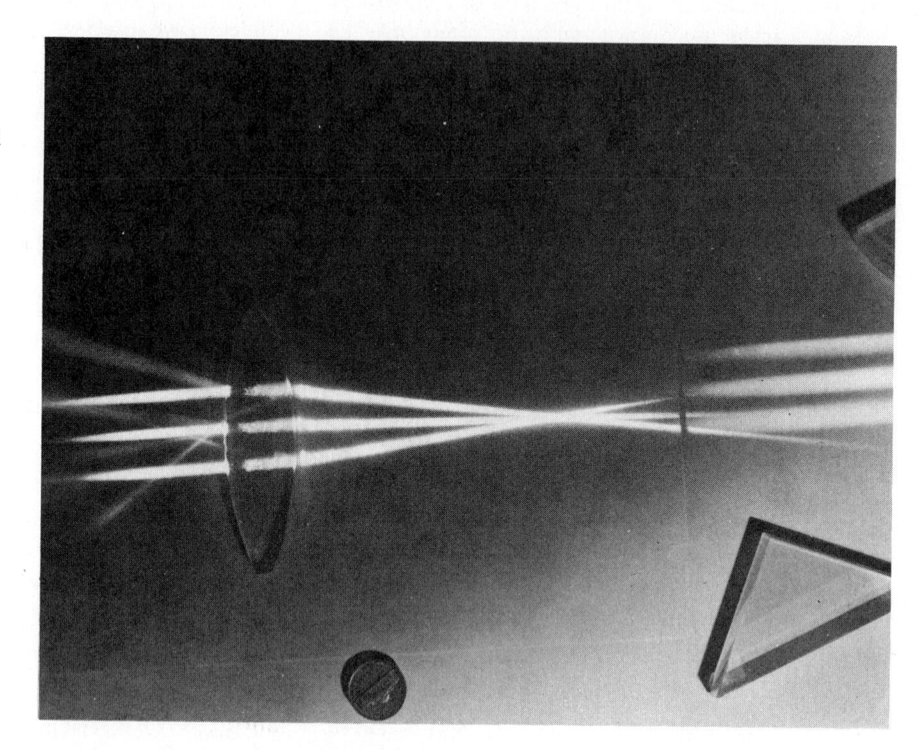

Light rays from a source at the focal point are made parallel by a convex lens. (Photo courtesy of Bausch and Lomb Optical Co.)

Mirrors and lenses are basic components of such ordinary devices as cameras, telescopes, and microscopes. In Chapter 27, we shall examine the construction and properties of such optical instruments. This chapter is concerned solely with the formation of images when light rays fall on plane and spherical surfaces. These surfaces are either reflecting, as in the case of mirrors, or refracting, as in the case of lenses. Our studies will show that the images formed are the result of reflection or refraction.

25.1 PLANE MIRRORS

One of the objectives of this chapter will be to discuss the manner in which optical elements such as lenses and mirrors form images. We shall begin this investigation by considering the simplest possible mirror, the plane mirror. Throughout our discussion of both mirrors and lenses, we shall use the ray model of light.

Consider a point source of light placed at O in Figure 25.1, a distance s in front of a plane mirror. The distance s is often referred to as the **object distance.** Light rays leave the source and are reflected from the mirror. After reflection, the rays diverge (spread apart), but they appear to the viewer to come from a point I located behind the mirror. The eye and the brain treat all light rays which enter the eye as though the rays follow straight-line paths. As a result, the eye follows the rays backward along a straight-line path to the point from which they appear to originate, which is point I in Figure 25.1. Point I is called the **image** of the object at O. Regardless of the system under study, images are always formed in the same way. *Images are formed at the point where rays of light actually intersect or at the point from which they appear to originate.* Since the rays in Figure 25.1 appear to originate at I, which is a distance s' behind the mirror, this is the location of the image. The distance s' is often referred to as the **image distance.**

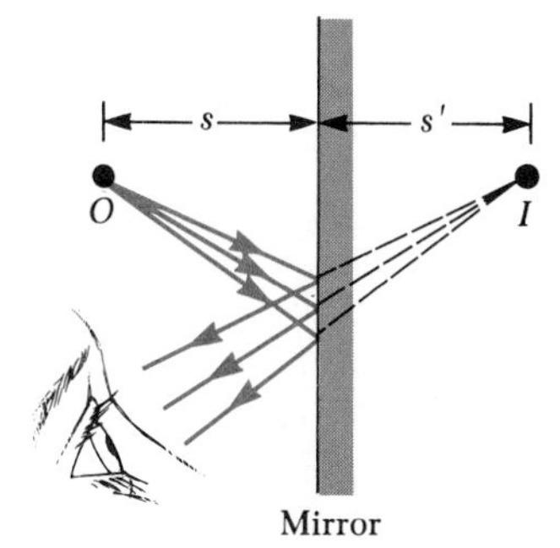

FIGURE 25.1 An image formed by reflection from a plane mirror. The image point, I, is located behind the mirror at a distance s', which is equal to the object distance, s.

Images are classified as real or virtual. *A real image is one in which light actually intersects, or passes through, the image point; a virtual image is one in which the light does not really pass through the image point but appears to diverge from that point.* The image formed by the plane mirror in Figure 25.1 is a virtual image. The images seen in plane mirrors *are always virtual* for real objects. Real images can always be displayed on a screen (as at a movie), but virtual images cannot be displayed on a screen.

We shall examine some of the properties of the images formed by plane mirrors by using the simple geometric techniques shown in Figure 25.2. In order to find out where an image is formed, it is always necessary to follow at least two rays of light as they reflect from the mirror. One of those rays starts at P, follows a horizontal path to the mirror, and reflects back on itself. The second ray follows the oblique path PR and reflects as shown. An observer to the left of the mirror would trace the two reflected rays back to the point from which they appear to have originated, that is, point P'. A continuation of this process for points on the object other than P would result in a virtual image (drawn as an uncolored arrow) to the right of the mirror. Since triangles PQR and $P'QR$ are congruent, $PQ = P'Q$. Hence, we conclude that *the image formed by an object placed in front of a plane mirror is as far behind the mirror as the object is in front of the mirror.* Geometry also shows that the object height, h, equals the image height, h'. Let us define **lateral magnification, M,** as follows:

$$M \equiv \frac{\text{image height}}{\text{object height}} = \frac{h'}{h} \tag{25.1}$$

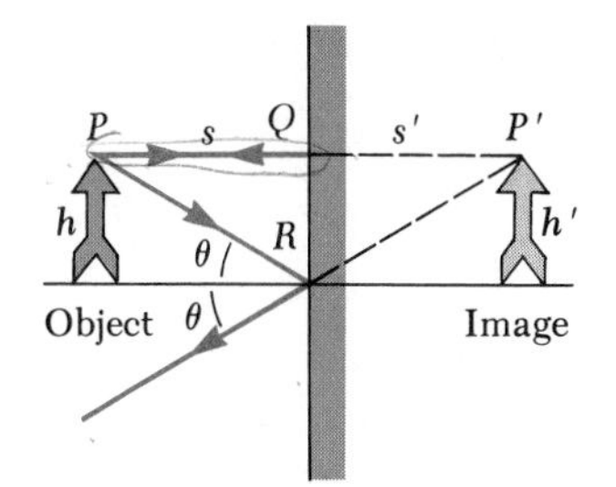

FIGURE 25.2 Geometric construction used to locate the image of an object placed in front of a plane mirror. Because the triangles PQR and $P'QR$ are congruent, $s = s'$, and $h = h'$.

Note that $M = 1$ for a plane mirror becuse $h' = h$ in this case.

The image formed by a plane mirror has one more important property, that of right-left reversal between image and object. This reversal can be seen by standing in front of a mirror and raising your right hand. The image you see raises its left hand. Likewise, your hair appears to be parted on the opposite side and a mole on your right cheek appears to be on your left cheek.

Thus, we conlcude that the image formed by a plane mirror has the following properties:

1. The image is as far behind the mirror as the object is in front.

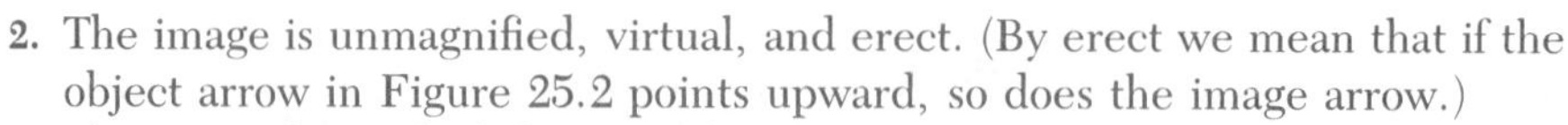
2. The image is unmagnified, virtual, and erect. (By erect we mean that if the object arrow in Figure 25.2 points upward, so does the image arrow.)
3. The image has right-left reversal.

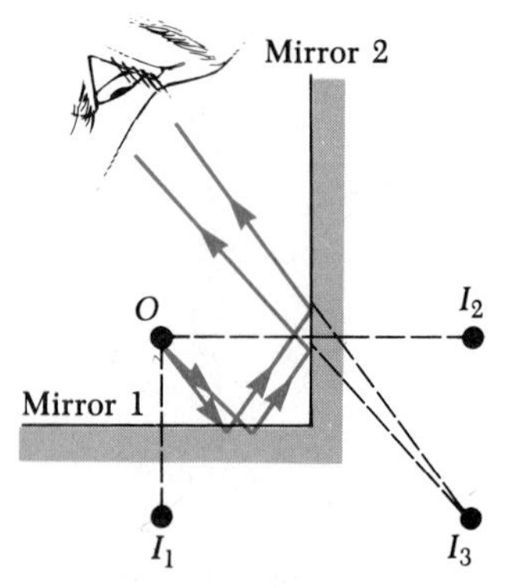

FIGURE 25.3 (Example 25.1) When an object is placed in front of two perpendicular mirrors as shown, three images are formed.

EXAMPLE 25.1 Multiple Images Formed by Two Mirrors

Two plane mirrors are at right angles to each other, as in Figure 25.3, and an object is placed at point O. In this situation, multiple images are formed. Locate the positions of these images.

Solution The image of the object is at I_1, in mirror 1 and at I_2 in mirror 2. In addition, a third image is formed at I_3, which will be considered to be the image of I_1 in mirror 2 or, equivalently, the image of I_2 in mirror 1. That is, the image at I_1 (or I_2) serves as the object for I_3. When viewing I_3, note that the rays reflect twice after leaving the object at O.

EXERCISE Sketch the rays corresponding to viewing the images at I_1 and I_2 and show that the light is reflected only once in these cases.

25.2 IMAGES FORMED BY SPHERICAL MIRRORS

Concave Mirrors

A **spherical mirror,** as its name implies, has the shape of a segment of a sphere. Figure 25.4 shows a spherical mirror with light reflecting from its silvered inner surface. Such a mirror, in which light is reflected from the inner, concave surface, is called a **concave mirror.** The mirror has a radius of curvature R, and its center of curvature is located at point C. Point V is the center of the spherical segment, and a line drawn from C to V is called the **principal axis** of the mirror.

Now consider a point source of light placed at point O in Figure 25.4, located on the principal axis and outside point C. Several diverging rays originating at O are shown. After reflecting from the mirror, these rays converge and meet at I, called the **image point.** The rays then continue to diverge from I as if there were an object there. As a result, we have a real image formed. *Real images are always formed at a point when reflected light actually passes through that point.*

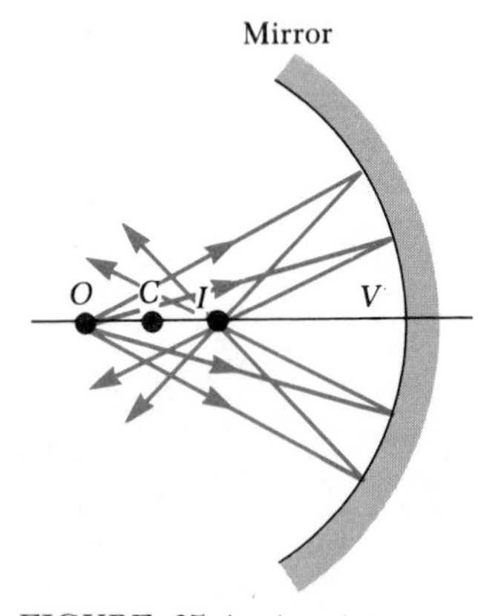

FIGURE 25.4 A point object placed at O outside the center of curvature of a concave spherical mirror forms a real image at I as shown. If the rays diverge from O at small angles, they all reflect through the same image point.

In what follows, we shall assume that all rays that diverge from the object make a small angle with the principal axis. All such rays reflect through the image point, as in Figure 25.4. Rays that are far from the principal axis, as in Figure 25.5, converge to other points on the principal axis, producing a blurred image. This effect, called **spherical aberration,** is present to some extent for any spherical mirror and will be discussed in Section 25.8.

We can use the geometry shown in Figure 25.6 to calculate the image distance, s', from a knowledge of the object distance, s, and radius of curvature, R. By convention, these distances are measured from point V. Figure 25.6 shows two rays of light leaving the tip of the object. One of these rays passes through the center of curvature, C, of the mirror, hitting the mirror head on (perpendicular to the mirror surface) and reflecting back on itself. The second ray strikes the mirror at the center, point V, and reflects as shown, obeying the law of reflection. The image of the tip of the arrow is located at the point where these two rays intersect. From the largest triangle in Figure 25.6 we see that $\tan\theta = h/s$, while the smallest triangle gives $\tan\theta = -h'/s'$. The negative sign signifies that the image is inverted, and so h' is negative. Thus, from Equation 25.1 and these results, we find that the magnification of the mirror is

$$M = \frac{h'}{h} = -\frac{s'}{s} \qquad (25.2)$$

Magnification

We also note from two other triangles in the figure that

$$\tan\alpha = \frac{h}{s - R} \quad \text{and} \quad \tan\alpha = -\frac{h'}{R - s'}$$

from which we find that

$$\frac{h'}{h} = -\frac{R - s'}{s - R} \qquad (25.3)$$

We equate Equation 25.2 to Equation 25.3 to give

$$\frac{R - s'}{s - R} = \frac{s'}{s}$$

Simple algebra reduces this to

$$\frac{1}{s} + \frac{1}{s'} = \frac{2}{R} \qquad (25.4)$$

This expression is called the **mirror equation.**

If the object is very far from the mirror, that is, if the object distance, s, is large enough compared with R that s can be said to approach infinity, then $1/s \approx 0$, and we see from Equation 25.4 that $s' \approx R/2$. That is, when the object is very far from the mirror, *the image point is halfway between the*

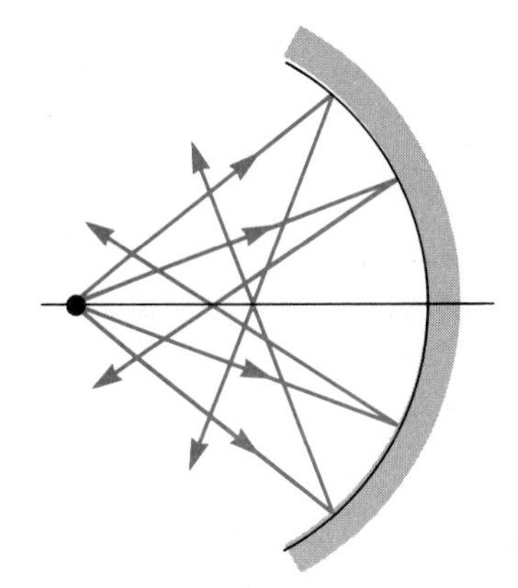

FIGURE 25.5 Rays at large angles from the horizontal axis reflect from a spherical, concave mirror to intersect the optic axis at different points, resulting in a blurred image. This is called *spherical aberration.*

blurred image

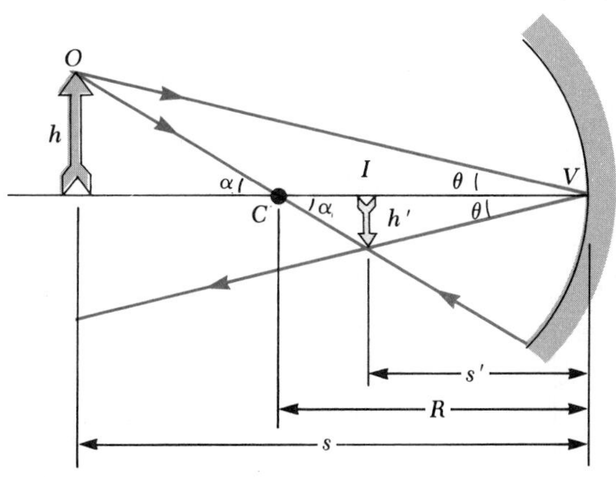

FIGURE 25.6 Ray diagram for a spherical concave mirror, where the object, at O, lies outside the center of curvature, C.

$\tan\theta = \frac{h}{s}$ $\tan\theta = -\frac{h'}{s'}$

center of curvature and the center of the mirror, as in Figure 25.7a. Note that the rays are essentially parallel in this figure because the source is assumed to be very far from the mirror. We call the image point in this special case the **focal point,** F, and the image distance the **focal length,** f, where

Focal length

$$f = \frac{R}{2} \tag{25.5}$$

The mirror equation can therefore be expressed in terms of the focal length:

Mirror equation

$$\frac{1}{s} + \frac{1}{s'} = \frac{1}{f} \tag{25.6}$$

Note that objects at infinity are always focused at the focal point.

You should be alerted to the fact that in order to apply the equations of this section to specific problems, you will have to follow a sign convention for the quantities s, s', f, and R. We shall summarize this sign convention, which is applicable to all spherical mirrors, in the next section.

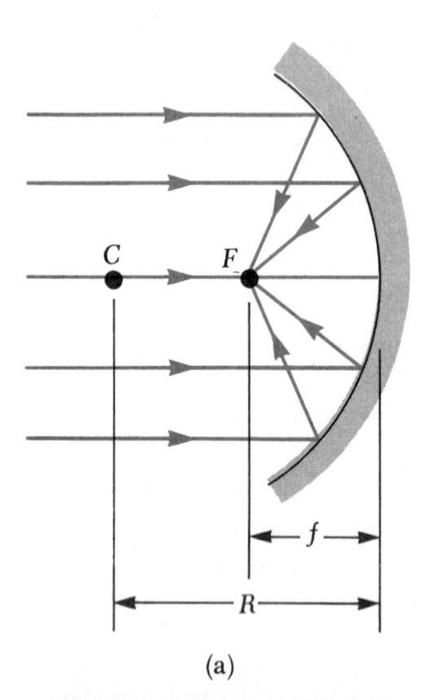

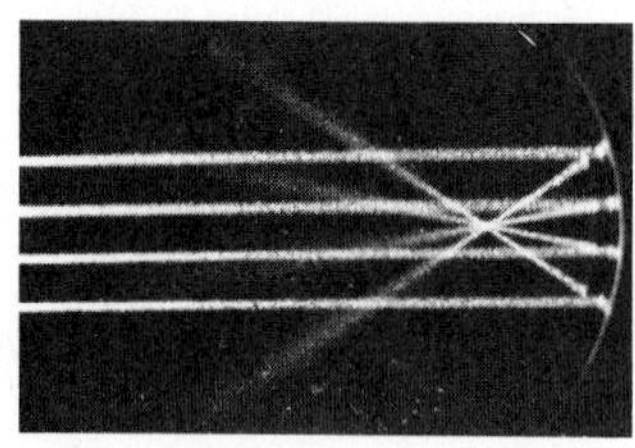

(b)

FIGURE 25.7 (a) Light rays from a distant object ($s = \infty$) reflect from a concave mirror through the focal point, F. In this case, the image distance $s' = R/2 = f$, where f is the focal length of the mirror. (b) Photograph of the reflection of parallel rays from a concave mirror. (Courtesy of Jim Lehman, James Madison University)

25.3 CONVEX MIRRORS AND SIGN CONVENTIONS

Figure 25.8 shows the formation of an image by a **convex mirror,** that is, one silvered such that light is reflected from the outer, convex surface. This is sometimes called a **diverging mirror** because the rays from any point on the object diverge after reflection as though they were coming from some point behind the mirror. Note that the image in Figure 25.8 is virtual rather than real because it lies behind the mirror at the location from which the reflected rays appear to originate. Furthermore, whenever the object is located on the left side of the mirror, the image will always be erect, virtual, and smaller than the object, as shown in the figure. Convex mirrors are commonly used at various vantage points around a place of business in order to spot shoplifters in action. They also are used as rearview mirrors in some vans and trucks.

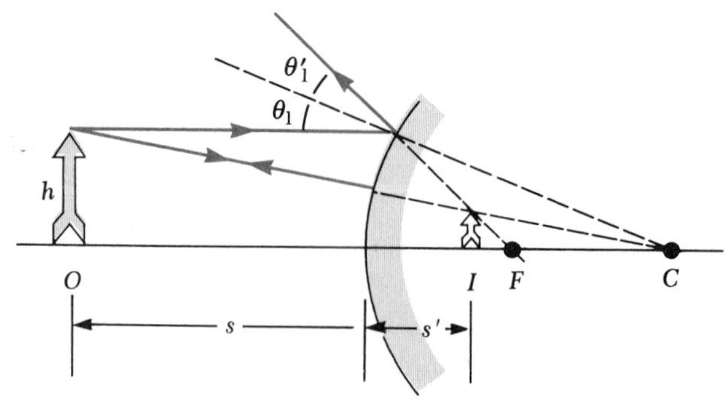

FIGURE 25.8 Ray diagram for a spherical convex mirror. Note that the image is virtual and erect.

TABLE 25.1 Sign Convention for Mirrors

s is + if the object is in front of the mirror (real object).
s is − if the object is in back of the mirror (virtual object).

s' is + if the image is in front of the mirror (real image).
s' is − if the image is in back of the mirror (virtual image).

Both f and R are + if the center of curvature is in front of the mirror (concave mirror).
Both f and R are − if the center of curvature is in back of the mirror (convex mirror).

If M is positive, the image is erect.
If M is negative, the image is inverted.

We shall not attempt to derive any equations for convex spherical mirrors. The results of such derivations would show that the equations developed for concave mirrors can be used if we adhere to a sign convention that we shall now summarize.

We can use Equations 25.2, 25.4, and 25.6 for either concave or convex mirrors if we adhere to the following procedure. Let us refer to the region in which light rays move as the *front side* of the mirror and the other side, where virtual images are formed, as the *back side*. For example, in Figures 25.6 and 25.8, the side to the left of the mirrors is the front side and the side to the right of the mirrors is the back side. Figure 25.9 is useful in understanding the rules for object and image distances, and Table 25.1 summarizes the sign conventions for all the necessary quantities.

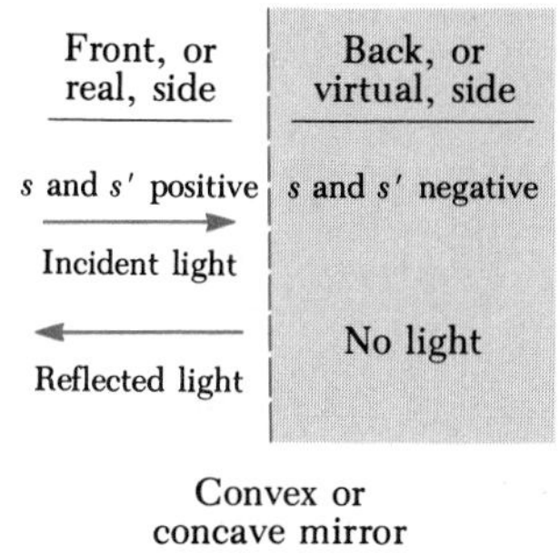

FIGURE 25.9 Diagram for describing the signs of s and s' for convex or concave mirrors.

The following examples should help you become familiar with these sign conventions.

EXAMPLE 25.2 The Image for a Concave Mirror

Assume that a certain concave spherical mirror has a focal length of 10 cm. Find the location of the image for object distances of (a) 25 cm, (b) 10 cm, and (c) 5 cm. Describe the image in each case.

Solution (a) For an object distance of 25 cm, we find the image distance using the mirror equation:

$$\frac{1}{s} + \frac{1}{s'} = \frac{1}{f}$$

$$\frac{1}{25\text{ cm}} + \frac{1}{s'} = \frac{1}{10\text{ cm}}$$

$$s' = 16.7\text{ cm}$$

The magnification is given by Equation 25.2:

$$M = -\frac{s'}{s} = -\frac{16.7\text{ cm}}{25\text{ cm}} = -0.67$$

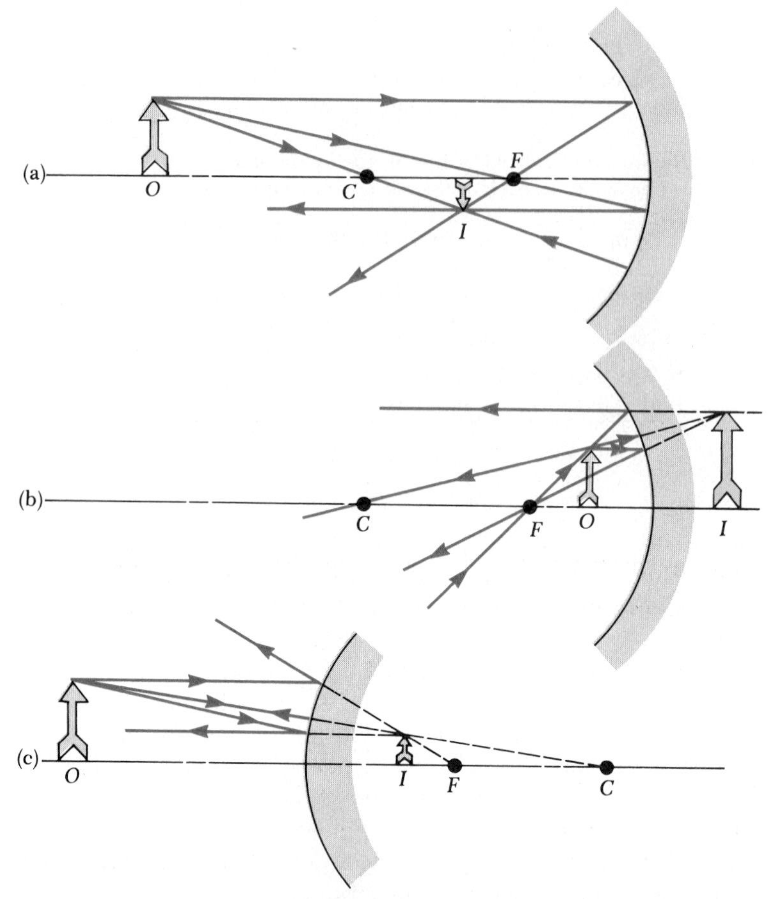

FIGURE 25.10 Ray diagrams for spherical mirrors. (a) The object is located outside the center of curvature of a spherical concave mirror. (b) The object is located between the spherical concave mirror and the focal point F. (c) The object is located in front of a spherical convex mirror.

Thus, the image is smaller than the object. Furthermore, the image is inverted because M is negative. Finally, because s' is positive, the image is located on the front side of the mirror and is real. This situation is pictured in Figure 25.10a.

(b) When the object distance is 10 cm, the object is located at the focal point. Substituting the values $s = 10$ cm and $f = 10$ cm into the mirror equation, we find

$$\frac{1}{10 \text{ cm}} + \frac{1}{s'} = \frac{1}{10 \text{ cm}}$$

$$s' = \infty$$

Thus, we see that rays of light originating from an object located at the focal point of a concave mirror are reflected such that the image is formed at an infinite distance from the mirror; that is, the rays travel parallel to one another after reflection.

(c) When the object is at the position $s = 5$ cm, it is inside the focal point of the mirror. In this case, the mirror equation gives

$$\frac{1}{5 \text{ cm}} + \frac{1}{s'} = \frac{1}{10 \text{ cm}}$$

$$s' = -10 \text{ cm}$$

That is, the image is virtual since it is located behind the mirror. The magnification is

$$M = -\frac{s'}{s} = -\left(\frac{-10 \text{ cm}}{5 \text{ cm}}\right) = 2$$

From this, we see that the image is magnified by a factor of 2, and the positive sign indicates that the image is erect (Fig. 25.10b).

Note the characteristics of the images formed by a concave spherical mirror. When the object is outside the focal point, the image is inverted and real; at the focal point, the image is formed at infinity; inside the focal point, the image is erect and virtual.

EXERCISE If the object distance is 20 cm, find the image distance and the magnification of the mirror.

Answer $s' = 20$ cm, $M = -1$.

EXAMPLE 25.3 The Image for a Convex Mirror

An object 3 cm high is placed 20 cm from a convex mirror having a focal length of 8 cm. Find (a) the position of the final image and (b) the magnification of the mirror.

Solution (a) Since the mirror is convex, its focal length is negative. To find the image position, we use the mirror equation:

$$\frac{1}{s} + \frac{1}{s'} = \frac{1}{f}$$

$$\frac{1}{20 \text{ cm}} + \frac{1}{s'} = \frac{1}{-8 \text{ cm}}$$

$$s' = -5.71 \text{ cm}$$

The negative value of s' indicates that the image is virtual, or behind the mirror, as in Figure 25.10c.

(b) The magnification of the mirror is

$$M = -\frac{s'}{s} = -\left(\frac{-5.71 \text{ cm}}{20 \text{ cm}}\right) = 0.286$$

The image is erect because M is positive.

EXERCISE Find the height of the image.

Answer 0.857 cm.

EXAMPLE 25.4

A convex spherical mirror has a focal length of -10 cm. Find the image distance and describe the image when the object distance is 25 cm.

Solution We can use the mirror equation to find the image distance:

$$\frac{1}{25 \text{ cm}} + \frac{1}{s'} = \frac{1}{-10 \text{ cm}}$$

$$s' = -7.14 \text{ cm}$$

The negative sign tells us that the image is 7.14 cm behind the mirror and hence is virtual.

The magnification is

$$M = -\frac{s'}{s'} = -\left(\frac{-7.14 \text{ cm}}{25 \text{ cm}}\right) = 0.286$$

This result indicates that the image is smaller than the object and erect. In fact, when an object is placed in front of a convex mirror, the image will always be erect, virtual, and smaller than the object, as in Figure 25.10c.

EXERCISE Find the image distance and magnification when the object distance is 5 cm.

Answer $s' = -3.33$ cm, $M = 0.666$.

EXAMPLE 25.5 Who is Fairest of Them All?

When a woman stands 40 cm in front of a cosmetic mirror, the erect image is twice the size of her face. What is the focal length of this mirror?

Solution The magnification equation gives us a relationship between the object and image distances:

$$M = -\frac{s'}{s} = 2$$

$$s' = -2s = -2(40 \text{ cm}) = -80 \text{ cm}$$

First, note that a virtual image is formed because the woman is able to see her erect image in the mirror. This explains why the image distance is negative. Substitute $s' = -80$ cm into the mirror equation to give

$$\frac{1}{40 \text{ cm}} - \frac{1}{80 \text{ cm}} = \frac{1}{f}$$

$$f = 80 \text{ cm}$$

The positive sign for the focal length indicates that the mirror is concave, a fact that we already knew because the mirror magnified the object (a convex mirror would have produced an image of diminished size).

25.4 IMAGES FORMED BY REFRACTION

We shall now consider the process by which images are formed by refraction at a spherical surface. Consider two transparent media with indices of refraction n_1 and n_2, separated by a spherical surface of radius R (Fig. 25.11). Let us assume that the medium to the right has a higher index of refraction than the one to the left, that is, $n_2 > n_1$. This would be the situation for light entering a curved piece of glass from air or for light entering the water in a fish bowl from air. Note that the rays originating at the object location, O, are refracted at the spherical surface and then converge to the image point, I. You can begin with Snell's law of refraction and use simple geometric techniques to show that the object distance, image distance, and radius of curvature are related by the equation

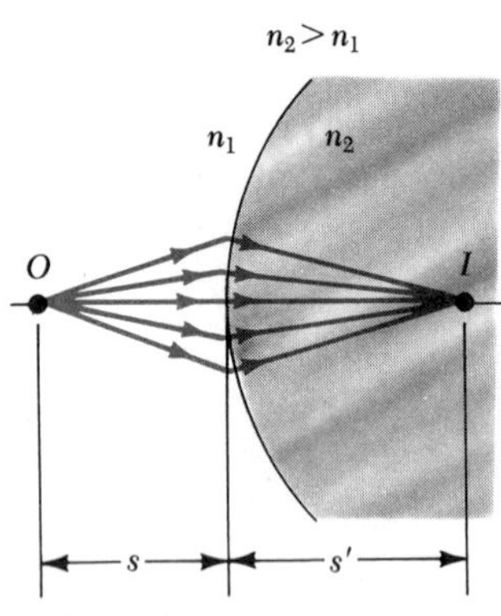

FIGURE 25.11 Image formed by refraction at a spherical surface. Rays making small angles with the optic axis diverge from a point object at O, and pass through the image point I.

$$\frac{n_1}{s} + \frac{n_2}{s'} = \frac{n_2 - n_1}{R} \tag{25.7}$$

Furthermore, the magnification of a refracting surface is given by

$$M = \frac{h'}{h} = -\frac{n_1 s'}{n_2 s} \tag{25.8}$$

As was the case for mirrors, we must use a sign convention if we are to apply these equations to a variety of circumstances. First note that real images are formed on the side of the surface that is *opposite* the side from which the light comes, in contrast to mirrors, where real images are formed on the *same* side of the reflecting surface. Therefore, *the sign convention for spherical refracting surfaces is the same as for mirrors, recognizing the change in sides of the surface for real and virtual images.* For example, in Figure 25.11, s, s', and R are all positive.

The sign convention for spherical refracting surfaces is summarized in Table 25.2. The same sign convention will be used for thin lenses, which will be discussed in the next section. As with mirrors, we assume that the front of the refracting surface is the side from which the light approaches the surface.

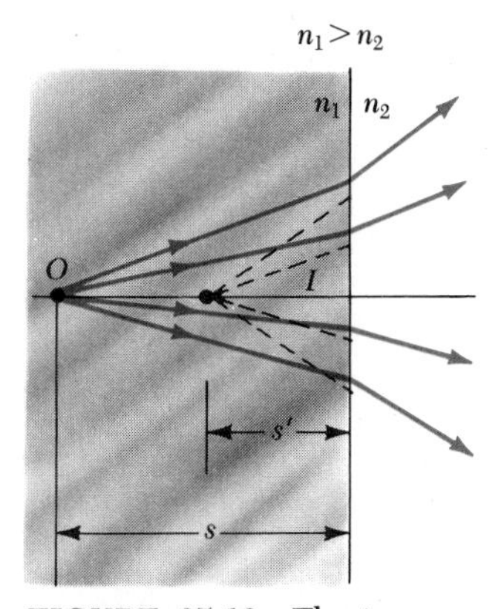

FIGURE 25.12 The image formed by a plane refracting surface is virtual; that is, it forms to the left of the refracting surface.

Plane Refracting Surfaces

If the refracting surface is a plane, then R approaches infinity and Equation 25.7 reduces to

$$\frac{n_1}{s} = -\frac{n_2}{s'}$$

or

$$s' = -\frac{n_2}{n_1}s \qquad (25.9)$$

From this equation we see that the sign of s' is opposite that of s. Thus, *the image formed by a plane refracting surface is on the same side of the surface as the object.* This is illustrated in Figure 25.12 for the situation in which n_1 is greater than n_2, where a virtual image is formed between the object and the surface. If n_1 is less than n_2, the image will still be virtual but will be formed to the left of the object.

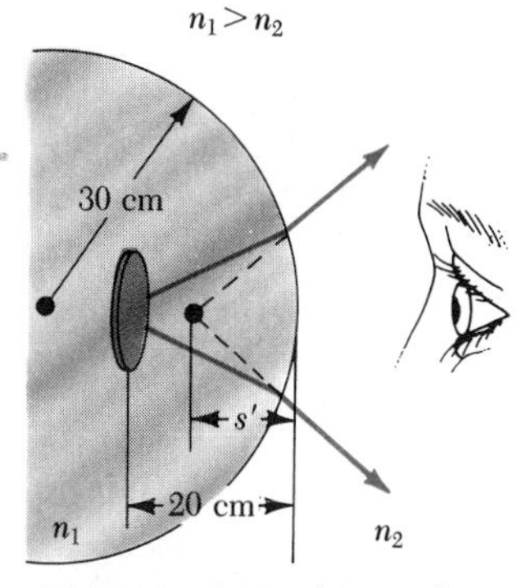

FIGURE 25.13 (Example 25.6) A coin embedded in a glass ball forms a virtual image between the coin and the glass surface.

EXAMPLE 25.6 Gaze into the Crystal Ball

A coin 2 cm in diameter is embedded in a solid glass ball of radius 30 cm (Fig. 25.13). The index of refraction of the ball is 1.5, and the coin is 20 cm from the surface. Find the position and height of the image.

Solution First, note that the rays originating from the object are refracted away from the normal at the surface and diverge outward. Hence, the image is formed in

TABLE 25.2 Sign Convention for Refracting Surfaces

s is + if the object is in front of the surface (real object).
s is − if the object is in back of the surface (virtual object).
s' is + if the image is in back of the surface (real image).
s' is − if the image is in front of the surface (virtual image).
R is + if the center of curvature is in back of the surface.
R is − if the center of curvature is in front of the surface.

the glass and is virtual. Applying Equation 25.7 and taking $n_1 = 1.5$, $n_2 = 1$, $s = 20$ cm, and $R = -30$ cm, we get

$$\frac{n_1}{s} + \frac{n_2}{s'} = \frac{n_2 - n_1}{R}$$

$$\frac{1.5}{20 \text{ cm}} + \frac{1}{s'} = \frac{1 - 1.5}{-30 \text{ cm}}$$

$$s' = -17 \text{ cm}$$

The negative sign indicates that the image is in the same medium as the object (the side of incident light), in agreement with our ray diagram. Since the image is in the same medium as the object, it must be virtual.

To find the image height, we first use Equation 25.8 for the magnification:

$$M = -\frac{n_1 s'}{n_2 s} = -\frac{1.5(-17 \text{ cm})}{1(20 \text{ cm})} = \frac{h'}{h} = 1.28$$

Therefore,

$$h' = 1.28h = (1.28)(2 \text{ cm}) = 2.56 \text{ cm}$$

The positive value for M indicates an erect image.

EXAMPLE 25.7 The One That Got Away

A small fish is swimming at a depth d below the surface of a pond (Fig. 25.14). What is the *apparent depth* of the fish as viewed from directly overhead?

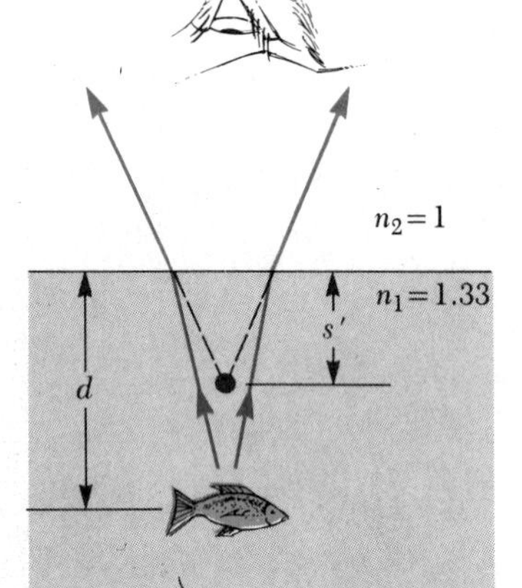

FIGURE 25.14 (Example 25.7) The apparent depth s' of the fish is less than the true depth d.

Solution In this example, the refracting surface is a plane, and so R is infinite. Hence, we can use Equation 25.9 to determine the location of the image. Using the facts that $n_1 = 1.33$ for water and $s = d$ gives

$$s' = -\frac{n_2}{n_1}s = -\frac{1}{1.33}d = -0.75d$$

Again, since s' is negative, the image is virtual, as indicated in Figure 25.14. The apparent depth is three fourths the actual depth. For instance, if $d = 4$ m, $s' = -3$ m.

EXERCISE If the fish is 10 cm long, how long is its image?

Answer 10 cm.

---

*25.5 ATMOSPHERIC REFRACTION

Images formed by refraction in our atmosphere lead to some interesting results. In this section, we shall look at two examples. One situation that occurs daily is that we are able to see the sun at dusk even though it has passed below the horizon. Figure 25.15 shows why this occurs. Rays of light from the sun strike the earth's atmosphere, represented by the shaded area around the earth. These rays are bent as they pass into a medium having an index of refraction different from that of the almost empty space in which they were traveling. The bending in this situation is somewhat different from those cases previously considered because here there is a gradual, continuous bending as the light moves through the atmosphere toward an observer at point O. This occurs because the light moves through layers of air having a continuously

Color Plate 19 Dispersion of white light. White light is dispersed into a spectrum of visible colors by a prism. The various colors are refracted at different angles because the index of refraction of the glass depends on wavelength. Refraction at the first surface produces a different path for each color passing into the glass. The path separation between the various colors is enhanced following refraction from the second surface as the light exits the prism. Note that the blue light deviates the most, while red light deviates the least, as discussed in Section 24.6. (Courtesy of Bausch and Lomb)

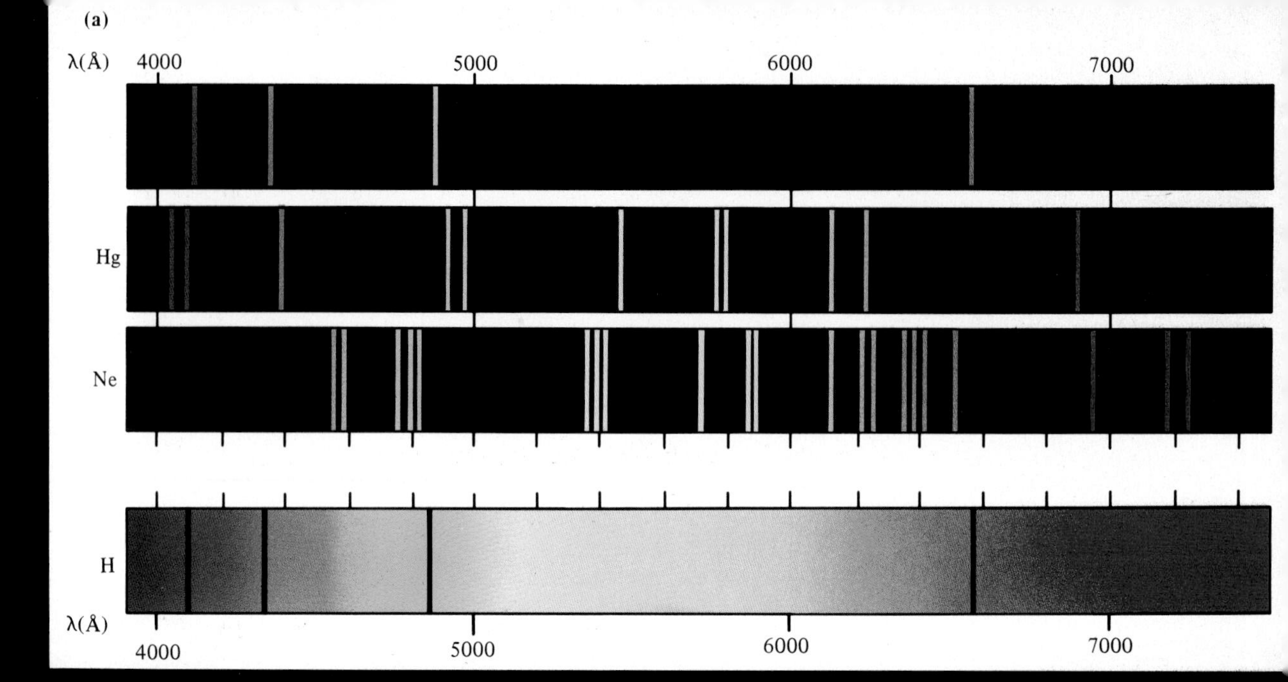

Color Plate 20 Visible spectra. (a) Line spectra produced by emission in the visible range for the elements hydrogen, helium, and neon. (b) The absorption spectrum for hydrogen. Note that the dark absorption lines occur at the same wavelengths as the emission lines for hydrogen shown in part (a). (K. W. Whitten, K. D. Gailey, and R. E. Davis, *General Chemistry,* 3rd ed., Saunders College Publishing, 1987)

Color Plate 21 Rainbow. A primary rainbow is formed as sunlight undergoes refraction and internal reflection in water droplets in the atmosphere (see Section 24.7 for more details). Under certain conditions, a more faint secondary rainbow can also be seen whose order of colors is reversed from that of the primary rainbow. (Courtesy of Hawaii Visitors Bureau)

Color Plate 22 Thin film interference. A thin film of oil on water displays interference, as shown by the pattern of colors when white light is incident on the film (see Section 26.4). The film thickness varies in the vicinity of the blade, thereby producing the interesting color pattern. (Peter Aprahamian, Science Photo Library)

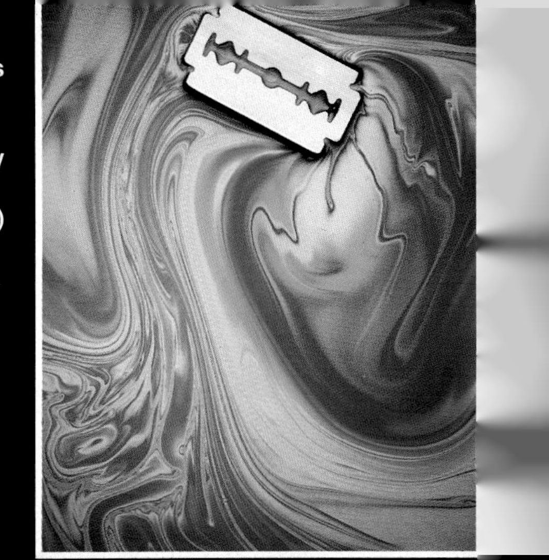

Color Plate 23 Interference in soap bubbles. White light incident on soap bubbles (thin films of soap) forms a beautiful pattern of colors as a result of interference in the films, as described in Section 26.4. (Peter Aprahamian, Science Photo Library)

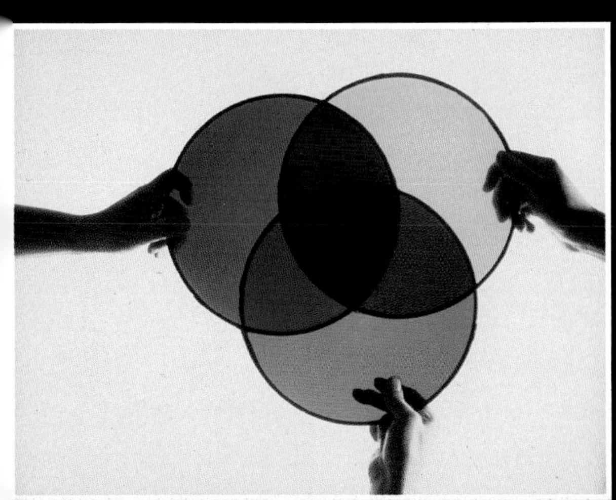

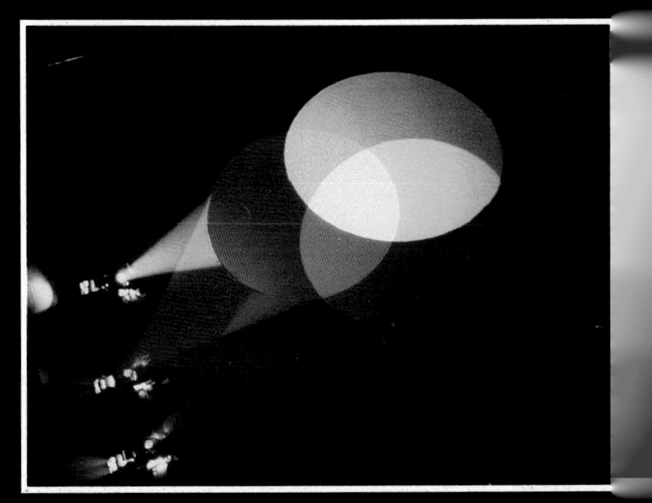

Color Plate 24 Color mixing. Different colors in the visible spectrum can be produced by an additive or subtractive process. (a) Additive mixing is demonstrated using the three primary colors, red, blue, and green. The overlap areas involving a combination of any two primary colors are yellow (red + green), cyan (green + blue), and magenta (red + blue). The central region where all three primary colors overlap is white (see Section 24.8). (Fritz Goro, Life Magazine) (b) Subtractive mixing occurs when white light passes through colored filters (yellow, magenta, and cyan) that allow certain wavelengths to pass. For example, when white light passes through a combination of the yellow and magenta filters, every color is removed except red, so the color we see is

Color Plate 25 Calcite splits a laser beam. As light from a laser is passed through a crystal of calcite ($CaCO_3$), the light undergoes double refraction and the beam splits into two parts: an ordinary ray and an extraordinary ray. This occurs because the crystal is optically anisotropic. The two rays correspond to two different indices of refraction, and therefore travel through the crystal with different speeds, as discussed further in Section 26.7. Because of this property, a double image of an object is seen when viewed through the crystal, as shown in Figure 26.26. (Courtesy U.S. Laser Corp., Wyckoff, N.J., and Rudolph Instruments, Fairfield, N.J.)

Color Plate 26 Lasers in industry. Lasers are used in a multitude of applications, as discussed in the essay in Chapter 30. In this photograph, scientists use a laser interferometer to test the surface finish of a metal mirror.

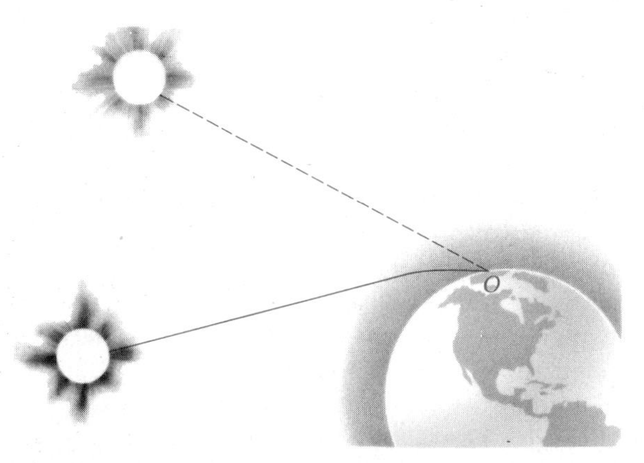

FIGURE 25.15 Because of refraction, an observer at O sees the sun even though it has fallen below the horizon.

changing index of refraction. When the rays reach the observer, the eye will follow them back along the direction from which they appear to have come; this is indicated by the dashed path in the figure. The end result is that the sun is seen to be above the horizon even after it has fallen below it.

The **mirage** is another phenomenon of nature produced by refraction in the atmosphere. A mirage can be observed when the ground is so hot that the air directly above it is warmer than the air at higher elevations. The desert is, of course, a region in which such circumstances prevail, but mirages are also seen on heated roadways during the summer. The layers of air at different heights above the earth have different densities and different refractive indices. The effect that this can cause is pictured in Figure 25.16. In this situation the observer is able to see a tree, for example, in two different ways. One group of light rays reaches the observer by the straight-line path A, and the eye traces these back to see the tree in the normal fashion. In addition, a second group of rays travels along the curved path B. These rays are directed toward the ground and are bent as a result of refraction. Consequently, the observer also sees an inverted image of the tree as these rays are traced back to the point from which they appear to originate. An erect image and an inverted image are seen when one observes the image of a tree formed in a reflecting pool of water; therefore, the observer calls upon this past experience and concludes that a pool of water must be in front of the tree.

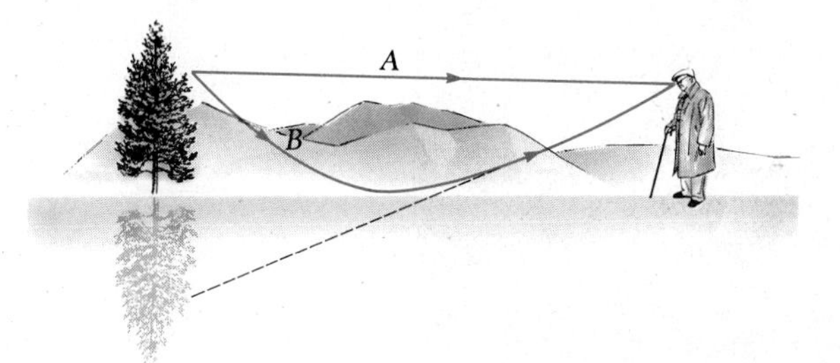

FIGURE 25.16 A mirage is produced by the bending of light rays in the atmosphere when there are large temperature differences between the ground and the air.

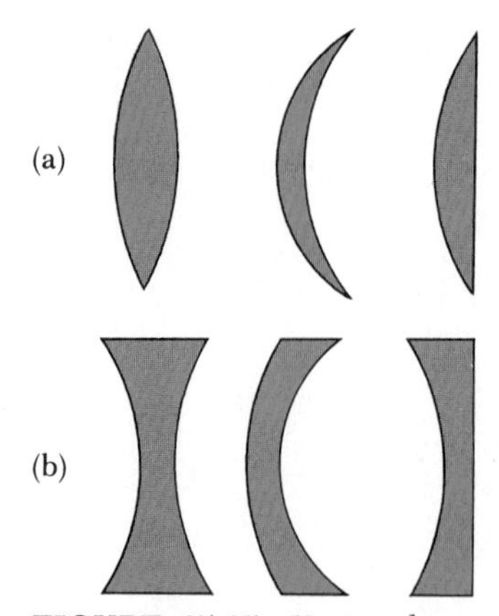

FIGURE 25.17 Various lens shapes: (a) Converging lenses have a positive focal length and are thickest at the middle. (b) Diverging lenses have a negative focal length and are thickest at the edges.

25.6 THIN LENSES

A typical **thin lens** consists of a piece of glass or plastic ground so that its two refracting surfaces are segments of spheres. Lenses are commonly used to form images by refraction in optical instruments, such as cameras, telescopes, and microscopes. The equation that relates object distances and image distances for a lens is virtually identical to the mirror equation derived earlier, and the method used to derive the equation is similar.

Figure 25.17 shows some representative shapes for lenses. Note that we have placed these lenses in two groups. Those in Figure 25.17a are thicker at the center than at the rim, and those in Figure 25.17b are thinner at the center than at the rim. The first group are examples of **converging lenses,** and the second group are called **diverging lenses.** The reason for these names will become apparent shortly.

As was the case for mirrors, it is convenient to define a point called the **focal point** for a lens. For example, in Figure 25.18a a group of rays parallel to the axis pass through the focal point, F, after being converged by the lens. The distance from the focal point to the lens is called the **focal length,** f. *The focal length is the image distance that corresponds to an infinite object dis-*

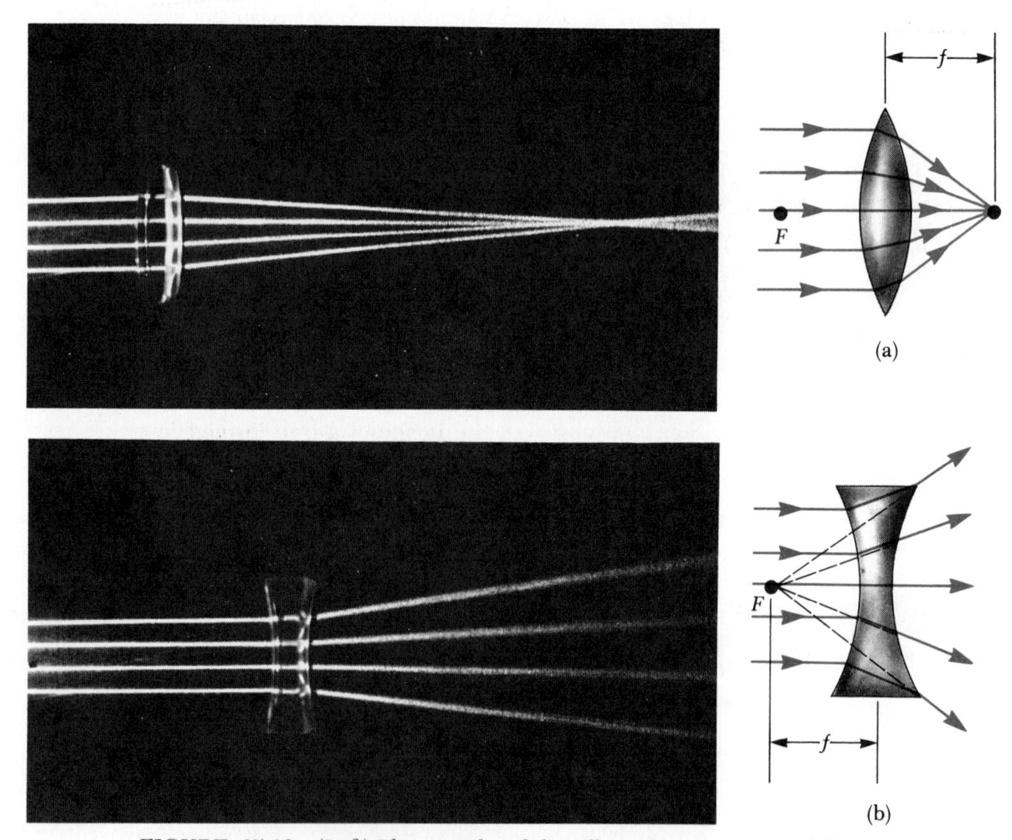

FIGURE 25.18 (Left) Photographs of the effect of converging and diverging lenses on parallel rays. (Courtesy of Jim Lehman, James Madison University) (Right) The principal focal points of (a) the biconvex lens and (b) the biconcave lens.

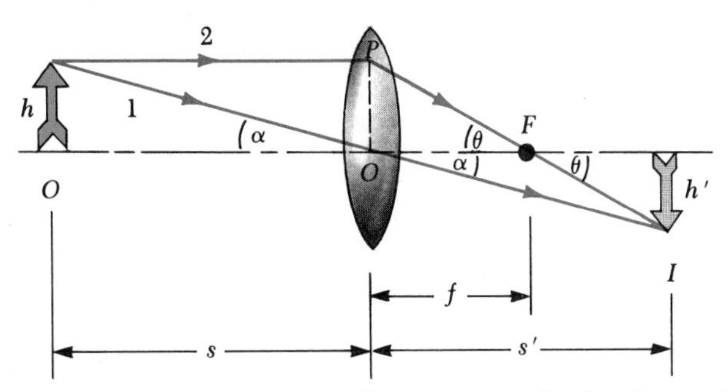

FIGURE 25.19 Geometrical construction for developing the thin lens equation.

tance. Recall that we are considering the lens to be very thin. As a result, it makes no difference whether we consider the focal length to be the distance from the focal point to the surface of the lens or the distance from the focal point to the center of the lens because the difference in lengths is negligible.

Rays parallel to the axis diverge after passing through a lens of the shape shown in Figure 25.18b. In this case, the focal point is defined to be the point from which the diverged rays appear to originate, labeled F in the figure. Figures 25.18a and 25.18b indicate why the names *converging* and *diverging* are applied to these lens. A converging lens brings together, or converges, parallel rays of light; a diverging lens spreads apart, or diverges, parallel rays.

Consider a ray of light that passes through the center of a lens, shown as ray 1 in Figure 25.19. If we followed this ray through the lens by applying Snell's law at both surfaces, we would find that this ray is deflected from its original direction of travel by a distance δ, shown in Figure 25.20. We shall make here what is called the *thin-lens approximation*, that is, *the thickness of the lens is assumed to be negligible*, and as a result the distance δ becomes vanishingly small. Thus, this ray will pass through the lens undeflected. Ray 2 in Figure 25.19 is parallel to the principal axis of the lens, and as a result it passes through the focal point, F, after refraction. The point at which these rays intersect is the point at which the image is formed.

We first note that the tangent of the angle α can be found by using the triangle in Figure 25.19:

$$\tan \alpha = \frac{h}{s} \qquad \text{or} \qquad \tan \alpha = -\frac{h'}{s'}$$

from which

$$\boxed{M = \frac{h'}{h} = -\frac{s'}{s}} \tag{25.10}$$

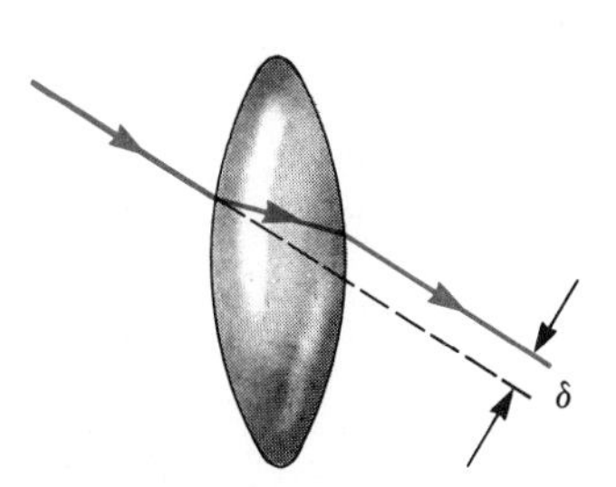

FIGURE 25.20 A ray passing through the center of the lens is deviated from its path by a distance δ.

Thus, the equation for magnification by a lens is the same as the equation for magnification by a mirror. We also note from Figure 25.19 that the tangent of θ is

$$\tan \theta = \frac{PO}{f} \qquad \text{or} \qquad \tan \theta = -\frac{h'}{s' - f}$$

TABLE 25.3 Sign Convention for Thin Lenses

s is + if the object is in front of the lens.
s is − if the object is in back of the lens.
s' is + if the image is in back of the lens.
s' is − if the image is in front of the lens.
R_1 and R_2 are + if the center of curvature is in back of the lens.
R_1 and R_2 are − if the center of curvature is in front of the lens.
f is + for a converging lens.
f is − for a diverging lens.

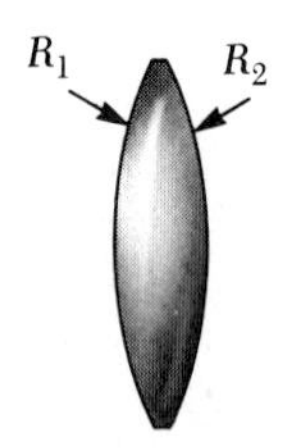

FIGURE 25.21 Diagram for obtaining the signs of s and s' for a thin lens or a refracting surface.

However, the height PO used in the first of these equations is the same as h, the height of the object. Therefore we get

$$\frac{h}{f} = -\frac{h'}{s' - f}$$

$$\frac{h'}{h} = -\frac{s' - f}{f}$$

Using this in combination with Equation 25.10 gives

$$\frac{s'}{s} = \frac{s' - f}{f}$$

which reduces to

Thin lens equation

$$\boxed{\frac{1}{s} + \frac{1}{s'} = \frac{1}{f}} \qquad (25.11)$$

This equation, called the **thin lens equation,** can be used with either converging or diverging lenses if we adhere to a set of sign conventions. Figure 25.21 is useful for obtaining the signs of s and s', and the complete sign convention for lenses is given in Table 25.3. Note that a *converging lens has a positive focal length* under this convention, and a *diverging lens has a negative focal length.* Hence the names *positive* and *negative* are often given to these lenses.

The focal length for a lens in air is related to the curvatures of its front and back surfaces and to the index of refraction n of the lens material as

Lens makers' equation

$$\boxed{\frac{1}{f} = (n - 1)\left(\frac{1}{R_1} - \frac{1}{R_2}\right)} \qquad (25.12)$$

where R_1 is the radius of curvature of the front surface of the lens and R_2 is the radius of curvature of the back surface. Equation 25.12 enables one to calculate the focal length from the known properties of the lens, and is called the **lens makers' equation.**

Virtual side	Real side
s positive s' negative	s negative s' positive
Incident light →	Refracted light →

Concave or convex thin lens or refracting surfaces

FIGURE 25.22 (Example 25.8)

EXAMPLE 25.8 The Lens Makers' Equation

A lens shaped as shown in Figure 25.22 has an index of refraction of 1.50. The magnitude of the radius of curvature of the front surface is 10 cm, and the radius of curvature of the back surface has a magnitude of 15 cm. Find the focal length of the lens.

Solution From the sign conventions in Table 25.3 we find that $R_1 = +10$ cm and $R_2 = -15$ cm. Thus, using the lens makers' equation, we have

$$\frac{1}{f} = (n-1)\left(\frac{1}{R_1} - \frac{1}{R_2}\right) = (1.5 - 1)\left(\frac{1}{10\text{ cm}} - \frac{1}{-15\text{ cm}}\right)$$

$$f = 12\text{ cm}$$

EXAMPLE 25.9 An Image Formed by a Converging Lens

A converging lens of focal length 10 cm forms an image of an object placed (a) 30 cm, (b) 10 cm, and (c) 5 cm from the lens. Find the image distance and describe the image in each case.

Solution (a) The thin lens equation, Equation 25.11, can be used to find the image distance:

$$\frac{1}{s} + \frac{1}{s'} = \frac{1}{f}$$

$$\frac{1}{30\text{ cm}} + \frac{1}{s'} = \frac{1}{10\text{ cm}}$$

$$s' = 15\text{ cm}$$

f = 10cm

The positive sign for the image distance tells us that the image is on the real side of the lens (Fig. 25.21). The magnification of the lens is

$$M = -\frac{s'}{s} = -\frac{15\text{ cm}}{30\text{ cm}} = -0.50$$

Thus, the image is reduced in size by one half, and the negative sign for M tells us that the image is inverted. The situation is like that pictured in Figure 25.23a.

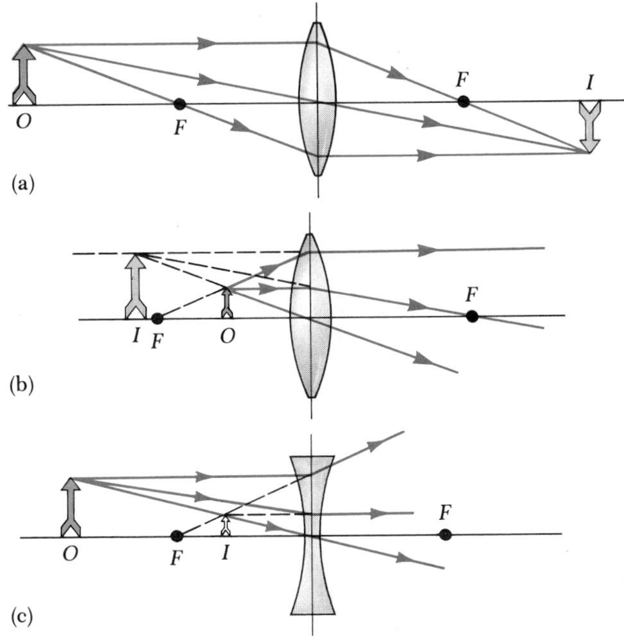

FIGURE 25.23 (Examples 25.9 and 25.10) Ray diagrams for locating the image of an object. (a) The object is located outside the focal point of a converging lens. (b) The object is located inside the focal point of a converging lens. (c) The object is located outside the focal point of a diverging lens.

(b) No calculation should be necessary for this case because we know that, when the object is placed at the focal point, the image will be formed at infinity. This is readily verified by substituting $s = 10$ cm into the lens equation.

(c) We now move inside the focal point, to an object distance of 5 cm. In this case, the lens equation gives

$$\frac{1}{5 \text{ cm}} + \frac{1}{s'} = \frac{1}{10 \text{ cm}}$$

$$s' = -10 \text{ cm}$$

and

$$M = -\frac{s'}{s} = -\left(\frac{-10 \text{ cm}}{5 \text{ cm}}\right) = 2$$

The negative image distance tells us that the image is formed on the side of the lens from which the light is incident, the virtual side (Fig. 25.22). The image is enlarged, and the positive sign for M tells us that the image is erect, as shown in Figure 25.23b.

You should note that there are two general cases for a converging lens. When the object is outside the focal point ($s > f$), the image is real, inverted, and smaller than the object. When the object is inside the focal point ($s < f$), the image is virtual, erect, and enlarged.

EXAMPLE 25.10 The Case of a Diverging Lens

Repeat the problem of Example 25.9 for a *diverging* lens of focal length 10 cm.

Solution (a) Let us apply the lens equation with an object distance of 30 cm:

$$\frac{1}{s} + \frac{1}{s'} = \frac{1}{f}$$

$$\frac{1}{30 \text{ cm}} + \frac{1}{s'} = -\frac{1}{10 \text{ cm}}$$

$$s' = -7.5 \text{ cm}$$

The magnification is

$$M = -\frac{s'}{s'} = -\left(\frac{-7.5 \text{ cm}}{30 \text{ cm}}\right) = 0.25$$

Thus, the image is virtual, smaller than the object, and erect.

(b) When the object is at the focal point, $s = 10$ cm, we have

$$\frac{1}{10 \text{ cm}} + \frac{1}{s'} = -\frac{1}{10 \text{ cm}}$$

$$s' = -5 \text{ cm}$$

and

$$M = -\frac{s'}{s'} = -\left(\frac{-5 \text{ cm}}{10 \text{ cm}}\right) = 0.50$$

(c) When the object is inside the focal point, at 5 cm, we have

$$\frac{1}{5 \text{ cm}} + \frac{1}{s'} = -\frac{1}{10 \text{ cm}}$$

$$s' = -3.33 \text{ cm}$$

and

$$M = -\left(\frac{-3.33 \text{ cm}}{5 \text{ cm}}\right) = 0.66$$

Again, we have a virtual image that is smaller than the object and erect.

You should make note of the fact that, for a diverging lens, the image is always virtual and erect. Figure 25.23c shows a representative situation.

25.7 MULTIPLE LENS SYSTEMS

Many practical optical systems consist of two or more lenses in a row. In order to solve problems involving several lenses, one can apply Equation 25.11 to each lens in turn. For example, if two lenses are used to form an image, the system can be treated by finding the image of the first lens imagining that the second lens were not present. Next, it is important to note that

the image of the first lens is treated as the object of the second lens.

The image of the second lens is the final image of the system. The same procedure can be extended to three or more lenses in a row. *The overall magnification of the system is the product of the magnifications for each individual lens.* The techniques for solving such problems are best shown through the following example problems.

EXAMPLE 25.11 Two Lenses in a Row

Two converging lenses are placed 20 cm apart as shown in Figure 25.24. If the first lens has a focal length of 10 cm and the second 20 cm, find the location of the final image formed of an object 30 cm in front of the first lens. Also, find the magnification of the system.

Solution The location of the image formed by the first lens is found by the thin lens equation as

$$\frac{1}{30 \text{ cm}} + \frac{1}{s'} = \frac{1}{10 \text{ cm}}$$

$$s' = 15 \text{ cm}$$

and the magnification of this lens is

$$M_1 = -\frac{s'}{s} = -\frac{15 \text{ cm}}{30 \text{ cm}} = -\frac{1}{2}$$

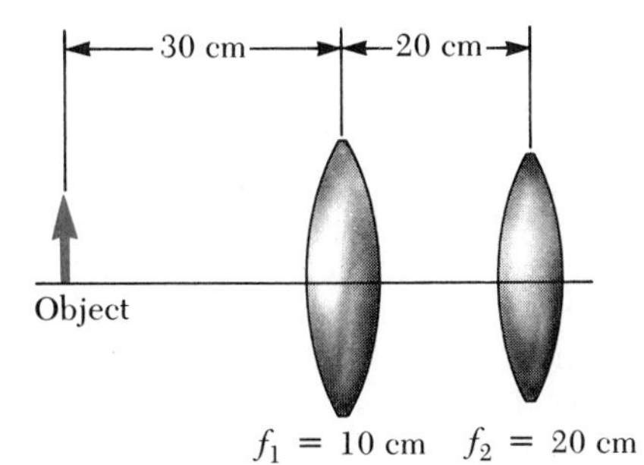

FIGURE 25.24 (Examples 25.11 and 25.12)

The image formed by this lens becomes the object for the second lens. Thus, the object distance for the second lens is 5 cm. We now apply the thin lens equation again to find the location of the final image.

$$\frac{1}{5\text{ cm}} + \frac{1}{s'} = \frac{1}{20\text{ cm}}$$

$$s' = -6.67\text{ cm}$$

The magnification of the second lens is

$$M_2 = -\frac{s'}{s} = -\frac{(-6.67\text{ cm})}{5\text{ cm}} = 1.33$$

Thus, the final image is located 6.67 cm to the left of the second lens, and the overall magnification of the system is

$$M = M_1 M_2 = \left(-\frac{1}{2}\right)(1.33) = -0.667$$

The negative sign indicates that the final image is inverted with respect to the initial object.

EXAMPLE 25.12 Move the Lenses Closer Together

Find the location of the final image in Example 25.11 if the two lenses are separated by 10 cm.

Solution As above, the first lens attempts to form an image 15 cm to the right of the lens. However, before this image is formed, the light rays are intercepted by the second lens. Regardless of this, the second lens treats the image formed (or trying to be formed) by the first lens as its object. Thus, the object position for the second lens is 5 cm to its right. This is a virtual object, and the thin lens equation becomes

$$\frac{1}{-5\text{ cm}} + \frac{1}{s'} = \frac{1}{20}$$

$$s' = 4\text{ cm}$$

Thus, the final image is formed 4 cm to the right of the second lens.

EXERCISE Find the overall magnification of the system.

Answer $M = 0.40$

---

*25.8 LENS ABERRATIONS

One of the basic problems of lenses and lens systems is the imperfect quality of the images. The simple theory of mirrors and lenses assumes that rays make small angles with the optic axis. In this simple model, all rays leaving a point source focus at a single point, producing a sharp image. Clearly, this is not always true. For those cases where the approximations used in this theory do not hold, imperfect images are formed.

If one wishes to perform a precise analysis of image formation, it is necessary to trace each ray using Snell's law at each refracting surface. This procedure shows that the rays from a point object do *not* focus at a single point. This means there is no single point image; instead, the image is *blurred*. The departures of real (imperfect) images from the ideal image predicted by the simple theory are called **aberrations.** Two types of aberrations will now be described.

Spherical Aberrations

Spherical aberrations result from the fact that the focal points of light rays far from the optic axis of a spherical lens (or mirror) are different from the focal points of rays passing through the center. Figure 25.25 illustrates spherical aberration for parallel rays passing through a converging lens. Rays near the middle of the lens have longer focal lengths than rays at the edges. Hence, there is no single focal length for a lens. Many optical devices are equipped with an adjustable aperture to reduce spherical aberration when possible. (An aperture is an opening that only allows light rays to pass that are close to the lens axis.) Sharper images are produced as the aperture size is reduced, since only the central portion of the lens is exposed to the incident light. At the same time, however, less light is imaged.

In the case of mirrors, one can eliminate, or at least minimize, spherical aberration by using a parabolic surface rather than a spherical surface. Parabolic surfaces are not often used, however, because they are very expensive to make. Parallel light rays incident on such a surface focus at a common point. Parabolic reflecting surfaces are used in astronomical telescopes to enhance the image intensity. They are also used in flashlights, where a parallel light beam is produced from a small lamp placed at the focal point of the surface.

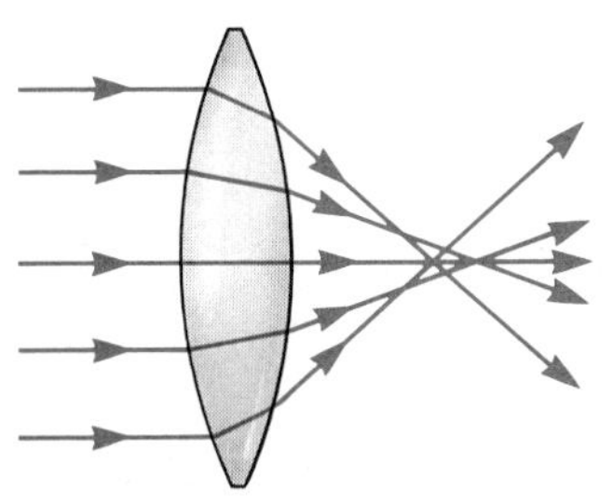

FIGURE 25.25 Spherical aberration produced by a converging lens. Does a diverging lens produce spherical aberration? (Angles are greatly exaggerated for clarity.)

Chromatic Aberrations

The fact that different wavelengths of light refracted by a lens focus at different points gives rise to chromatic aberrations. In Chapter 24, we described how the index of refraction of a material varies with wavelength. When white light passes through a lens, one finds, for example, that violet light rays are refracted more than red light rays (Fig. 25.26). From this we see that the focal length is larger for red light than for violet light. Other wavelengths (not shown in Figure 24.26) would have intermediate focal points. The chromatic aberration for a diverging lens is opposite that for a converging lens. Chromatic aberration can be greatly reduced by using a combination of a converging and diverging lens made from two different types of glass.

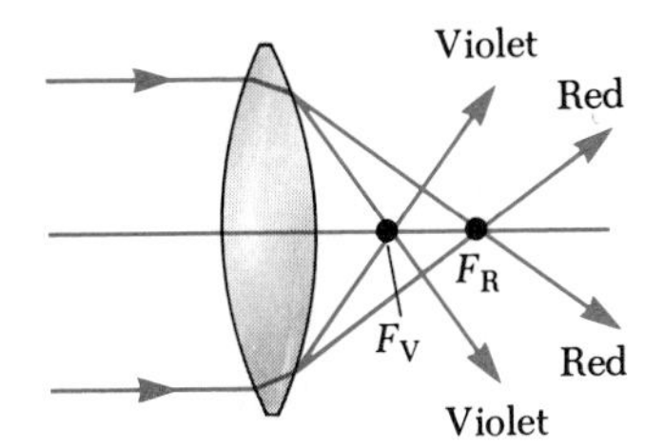

FIGURE 25.26 Chromatic aberration produced by a converging lens. Rays of different wavelengths focus at different points. (Angles are greatly exaggerated for clarity.)

SUMMARY

Images are formed where rays of light intersect or at the point from which they appear to originate. A **real image** is one in which light intersects, or passes through, an image point. A **virtual image** is one in which the light does not pass through the image point but appears to diverge from that point.

The image formed by a plane mirror has the following properties:

1. The image is as far behind the mirror as the object is in front.
2. The image is unmagnified, virtual, and erect.
3. The image has right-left reversal.

The **magnification,** M, of a mirror is defined as the ratio of **image height,** h', to **object height,** h, which is equal to the negative of the ratio of image distance, s', to object distance, s:

$$\boxed{M = \frac{h'}{h'} = -\frac{s'}{s}} \tag{25.2}$$

The **object distance** and **image distance** for a spherical mirror of radius R are related by the **mirror equation:**

$$\boxed{\frac{1}{s} + \frac{1}{s'} = \frac{1}{f}} \tag{25.6}$$

where $f = R/2$ is the **focal length** of the mirror.

An image can be formed by refraction at a spherical surface of radius R. The object and image distances for refraction from such a surface are related as

$$\boxed{\frac{n_1}{s} + \frac{n_2}{s'} = \frac{n_2 - n_1}{R}} \tag{25.7}$$

The **magnification** of a refracting surface is

$$\boxed{M = \frac{h'}{h'} = -\frac{n_1 s'}{n_2 s}} \tag{25.8}$$

where the object is located in the medium of index of refraction n_1 and the image is formed in the medium whose index of refraction is n_2.

For a **thin lens** the **magnification** is

$$\boxed{M = \frac{h'}{h'} = -\frac{s'}{s'}} \tag{25.10}$$

and the object and image distances are related by the **thin lens equation:**

$$\boxed{\frac{1}{s} + \frac{1}{s'} = \frac{1}{f}} \tag{25.11}$$

Aberrations are responsible for the formation of imperfect images by lenses and mirrors. **Spherical aberration** results from the fact that the focal points of light rays far from the optic axis of a spherical lens or mirror are different from those of rays passing through the center. **Chromatic aberration** arises from the fact that light rays of different wavelengths focus at different points when refracted by a lens.

ADDITIONAL READING

A. B. Fraser and M. W. Hirsch, *Mirages*, Springer-Verlag, 1975.

T. B. Greenslade, "Multiple Images in Plane Mirrors," *The Physics Teacher*, January 1982, p. 29.

R. C. Jones, "How Images Are Detected," *Sci. American*, September 1968, p. 111.

F. D. Smith, "How Images Are Formed," *Sci. American*, September 1968, p. 97.

W. Tape, "The Topology of Mirages," *Sci. American*, June 1985, p. 120.

D. E. Thomas, "Mirror Images," *Sci. American*, December 1980, p. 206.

P. K. Tien, "Integrated Optics," *Sci. American*, April 1974, p. 28.

A. C. S. van Heel and C. H. F. Velzel, *What is Light?* World University Library, New York, McGraw-Hill, 1968, Chaps. 1 and 2.

QUESTIONS

1. It is well known that underwater objects appear blurred and out of focus to a swimmer not wearing goggles. On the other hand, the use of goggles provides the swimmer with a clear view of underwater objects. Explain, basing your explanation on the fact that the indices of refraction of the cornea, water, and air are 1.376, 1.333, and 1.032, respectively.
2. The rear-view mirror on late model cars warns the user that objects may be closer than they appear. What kind of mirror is being used and why was that type selected?
3. Explain why a fish in a spherical goldfish bowl appears larger than it really is.
4. Why does a clear stream always appear to be shallower than it actually is?
5. A person spearfishing from a boat sees a fish located 3 m from the boat at a depth of 1 m. In order to spear the fish, should the person aim at, above, or below the fish?
6. Lenses used in eyeglasses, whether converging or diverging, are always designed such the the middle of the lens curves away from the eye, like the center lenses on Figure 25.17a and 25.17b. Why?
7. A mirage is formed when the air gets gradually cooler as the height above the ground increases. What might happen if the air becomes gradually warmer as the height is increased? This often happens over bodies of water or snow-covered ground: the effect is called looming.
8. Explain why a mirror cannot give rise to a chromatic aberration.
9. If a cylinder of solid glass or clear plastic is placed above the words LEAD OXIDE and the words viewed from above, as in Figure 25.27, LEAD appear inverted but OXIDE does not. Explain.

FIGURE 25.27 Question 9 (Photograph courtesy of Jim Lehman, James Madison University)

10. What is the magnification of a plane mirror? What is its focal length?
11. When you look in a mirror, the image of your left and right sides is reversed, yet the image of your head and legs is not reversed. Explain.

PROBLEMS

Section 25.1 Plane Mirrors

1. Use Figure 25.2 to give a geometric proof that the virtual image formed by a plane mirror is the same distance behind the mirror as the object is in front of it.

2. (a) What is the minimum length for a plane mirror that allows a person 6 ft tall to see his full height? (b) Does the distance from the mirror affect your answer?

(*Hint:* A diagram showing the path followed by the rays which leave the person would help. Follow those that leave the top of the head and the feet, strike the mirror, and then enter the eye.)

3. A person walks into a room that has two plane mirrors on opposite walls which produce multiple images. When the person is 5 ft from the mirror on the left wall and 10 ft from the mirror on the right wall, find the distance from the person to the first three images seen in the mirror on the left.

• **4.** Consider the case in which a light ray A is incident on mirror 1 in Figure 25.3. The reflected ray is incident on mirror 2 and subsequently reflected as ray B. Let the angle of incidence (with respect to the normal) on mirror 1 be 53°. Determine the angle between the two rays.

Section 25.2 Images Formed by Spherical Mirrors

Section 25.3 Convex Mirrors and Sign Conventions

Note that in the problems below, the algebraic signs are not given. We leave it to you to determine the correct algebraic sign to use with each quantity from an analysis of the problem and the sign conventions in Table 25.1.

5. A concave spherical mirror has a focal length of 20 cm. Find the location of the image for object distances of (a) 40 cm, (b) 20 cm, and (c) 10 cm. For each case, state whether the image is real or virtual and erect or inverted. Find the magnification for each case.

6. A concave spherical mirror has a radius of curvature of 20 cm. Find the location of the image for object distances of (a) 40 cm, (b) 20 cm, and (c) 10 cm. For each case, state whether the image is real or virtual and erect or inverted. Find the magnification for each case.

7. Repeat Problem 5 for a convex spherical mirror.

8. Repeat Problem 6 for a convex spherical mirror.

9. A virtual image is formed 20 cm from a concave mirror, having a radius of curvature of 40 cm. Find the position of the object.

10. A convex spherical mirror of focal length 15 cm is to form an image 10 cm from the mirror. Where should the object be placed?

11. A spherical Christmas tree ornament is 6 cm in diameter. What is the magnification of an object placed 10 cm away from the ornament?

12. A convex mirror has a focal length of 20 cm. Determine the object location for which the image will be one half the size of the object.

• **13.** The real image height of a concave mirror is observed to be four times larger than the object height when the object is 30 cm in front of a mirror. What is the radius of curvature of the mirror?

• **14.** A concave make-up mirror is designed so that a person 25 cm in front of the mirror sees an erect image magnified by a factor of two. What is the radius of curvature of the mirror?

• **15.** A convex spherical mirror with radius of curvature of 10 cm produces a virtual image one third the size of the object. Where is the object located?

• **16.** A child holds a candy bar 10 cm in front of a convex mirror and notices that the image is reduced by one half. What is the radius of curvature of the mirror?

• **17.** A concave mirror has a focal length of 40 cm. Determine the object position for which the resulting image will be erect and four times the size of the object.

• **18.** An object 2 cm in height is placed 3 cm in front of a concave mirror. If the image is 5 cm in height and virtual, what is the focal length of the mirror?

• **19.** What type of mirror is required to form an image on a wall 2 m from the mirror of an object placed 10 cm in front of the mirror? What is the magnification of the image?

• **20.** Under certain limiting conditions Equation 25.4 can be used for a plane mirror. (a) What value must be assumed for R if the mirror is plane? (b) What is the relationship between the object and image distance in this limiting case? (c) Does the result of (b) agree with the discussion given previously for the plane mirror?

Section 25.4 Images Formed by Refraction

21. A swimming pool is 2 m deep. How deep does it appear to be (a) when completely filled with water, (b) when filled half-way with water?

22. A cubical block of ice 50 cm on a side is placed on a level floor over a speck of dust. Find the location of the image of the speck if the index of refraction of ice is 1.309.

23. A colored marble is dropped into a large tank filled with benzene ($n = 1.50$). What is the depth of the tank if the apparent depth of the marble when viewed from directly above the tank is 35 cm?

24. A paperweight is made of a solid glass hemisphere of index of refraction 1.50. The radius of the circular cross-section is 4 cm. The center of the hemisphere is placed directly over a line drawn to a length of 2.5 mm on a sheet of paper. What is the length of the line as seen looking vertically down on the hemisphere?

25. One end of a long glass rod ($n = 1.50$) is formed into the shape of a convex surface of radius 8 cm. An object is located in air along the axis of the rod. Find the image position corresponding to the following object positions, (a) 20 cm, (b) 8 cm, (c) 4 cm, and (d) 2 cm.

26. Repeat Problem 25 assuming the rod is placed in water instead of air.

27. Calculate the image positions for the object positions of Problem 25 assuming the end of the rod has a concave shape of radius 8 cm.

28. Repeat Problem 27 if the object is in water rather than air.

Section 25.6 Thin Lenses

29. A converging lens has a focal length of 20 cm. Find the location of the image for object distances of (a) 40 cm, (b) 20 cm, and (c) 10 cm. For each case state whether the image is real or virtual and erect or inverted. Find the magnification for each case.

30. A glass converging lens ($n = 1.50$) is designed to look like the lens shown in Figure 25.23a. The radius of the first surface is 15 cm and the radius of the second surface is 10 cm. (a) Find the focal length of the lens. Determine the position of the image for object distances of (b) infinity, (c) $3f$, (d) f, and (e) $f/2$.

31. A diverging lens has a focal length of 20 cm. Find the location of the image for object distances of (a) 40 cm, (b) 20 cm, and (c) 10 cm. For each case state whether the image is real or virtual and erect or inverted. Find the magnification for each case.

32. A diverging lens ($n = 1.50$) is shaped like that shown in Figure 25.23c. The radius of the first surface is 15 cm and the radius of the second surface is 10 cm. (a) Find the focal length of the lens. Determine position of the image for object distances of (b) infinity, (c) $3|f|$, (d) $|f|$, (e) $|f|/2$.

33. Repeat Problem 32 for the case in which the radius of the first surface is 10 cm and the radius of the second surface is 15 cm.

34. An object is placed 30 cm from a lens and a virtual image is formed 15 cm from the lens. (a) What is the focal length of the lens? (b) Is the lens converging or diverging?

35. A diverging lens is used to form a virtual image of an object. The object is 80 cm to the left of the lens, and the image is 40 cm to the left of the lens. Determine the focal length of the lens.

• **36.** A diverging lens is to be used to produce a virtual image one third the size of the object. Where should the object be placed?

• **37.** An object is placed 50 cm from a screen. Where should a converging lens with a 10 cm focal length be placed in order to form an image on the screen? Find the magnification(s).

Section 25.7 Multiple Lens Systems

38. Two converging lenses, each of focal length 15 cm, are placed 40 cm apart, and an object is placed 30 cm in front of the first. Where is the final image formed, and what is the magnification of the system?

39. Repeat Problem 38 for the case in which the two lenses are placed (a) 20 cm apart, and (b) 30 cm apart.

40. An object is placed 20 cm to the left of a converging lens of focal length 25 cm. A diverging lens of focal length 10 cm is located 25 cm to the right of the converging lens. Find the position of the final image and its magnification.

41. Repeat Problem 40 for the case in which the light first encounters the diverging lens.

ADDITIONAL PROBLEMS

42. A concave mirror has a focal length of 30 cm. (a) What is its radius of curvature? Locate and describe the image when the object is at (b) 100 cm, and (c) 10 cm.

43. A planoconvex lens ($n = 1.50$) is flat on one side. What must be the radius of curvature of the curved side to produce a converging lens of focal length 20 cm?

44. Repeat Problem 43 for a planoconcave lens having a focal length of -20 cm.

45. A converging lens has a focal length of 20 cm. Find the position of the image for object distances of (a) 50 cm, (b) 30 cm, and (c) 10 cm. (d) For each object distance, determine the magnification of the lens and whether the image is erect or inverted.

• **46.** An object placed 10 cm from a concave spherical mirror produces a real image 8 cm from the mirror. If the object is moved to a new position 20 cm from the mirror, what is the position of the image? Is the final image real or virtual?

• **47.** A converging lens of focal length 20 cm is separated by 50 cm from a converging lens of focal length 5 cm. (a) Find the final position of the image of an object placed 40 cm in front of the first lens. (b) If the height of the object is 2 cm, what is the height of the final image? Is it real or virtual? (c) Determine the image position of an object placed 5 cm in front of the two lenses in contract.

• **48.** A real object is located at the zero end of a meter stick. A concave mirror is located at the 100-cm end of the meter stick forms an image of the object at the 70-cm position. A convex mirror placed at the 60-cm position forms a final image at the 10-cm point. What is the radius of curvature of the convex mirror?

• **49.** An object is located 36 cm to the left of a biconvex lens of index of refraction 1.5. The left surface of the lens has a radius of curvature of 20 cm. The right surface of the lens is to be shaped so that a real image will be formed 72 cm to the right of the lens. What is the requred radius of curvature of the second surface?

• **50.** An object is placed 12 cm to the left of a diverging lens of focal length −6 cm. A converging lens of focal length 12 cm is placed a distance *d* to the right of the diverging lens. Find the distance *d* such that the final image is at infinity.

• **51.** A colored marble is dropped in a large tank filled with benzene ($n = 1.50$). If the marble has a diameter of 1.5 cm, what is its apparent diameter when viewed from directly above, outside the tank? The apparent depth of the marble is 35 cm when viewed from directly above.

•• **52.** An object is placed 12 cm to the left of a diverging lens of focal length −6 cm. A converging lens of focal length 12 cm is placed a distance *d* to the right of the diverging lens. For what value of *d* is the final image at infinity?

•• **53.** Use the fact that the image formed by the first surface becomes the object for the second surface to work this problem. Figure 25.28 shows a piece of glass whose index of refraction is 1.50. The ends are hemispheres with radii 2 cm and 4 cm, and the centers of the hemispherical ends are separated by a distance of 8 cm. A point object is located in air 1 cm from the left end of the glass. Find the location of the image of the object due to refraction at the two spherical surfaces.

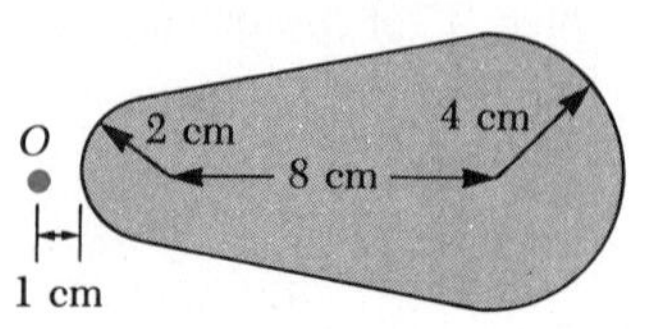

FIGURE 25.28 Problem 53

•• **54.** Repeat Problem 53 for the case in which the distance between the centers of the hemispherical ends is 4 cm and the object is placed 28 cm to the left of the first surface.

•• **55.** Find the object distances (in terms of *f*) of a thin converging lens of focal length *f* if (a) the image is real and the image distance is four times the focal length, and (b) the image is virtual and the image distance is three times the focal length. (c) Calculate the magnification of the lens for cases (a) and (b).

CHAPTER 26

Wave Optics

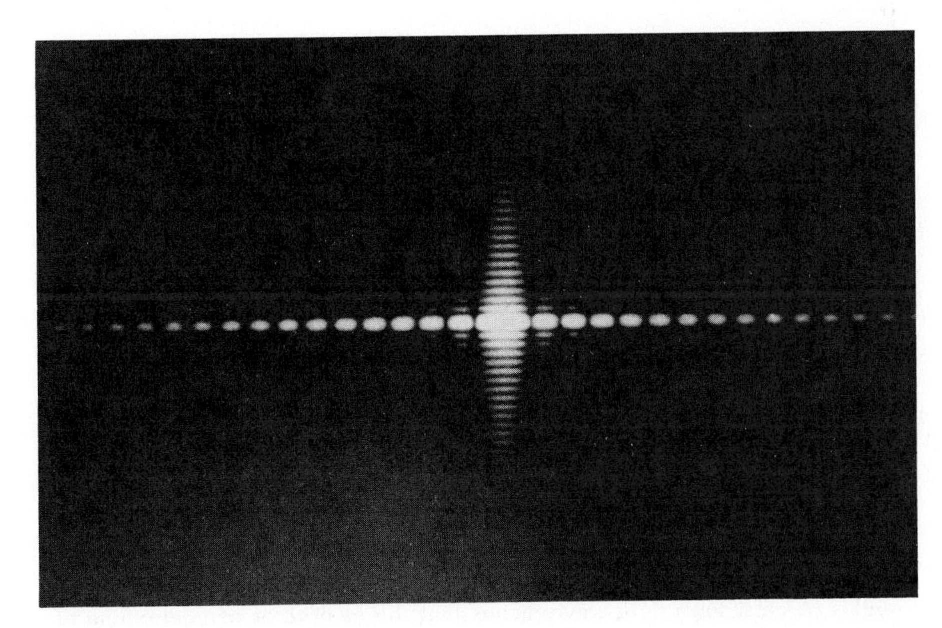

Diffraction pattern of a rectangular aperture. (From M. Cagnet, M. Francon, and J.C. Thierr, *Atlas of Optical Phenomena*, Berlin, Springer-Verlag, 1962)

Our discussion of light thus far has been concerned with examining what happens when light passes through a lens or reflects from a mirror. This part of optics is often called geometric optics because an explanation of such phenomena relies on a geometric analysis of the light rays. We shall now expand our study of light into an area called wave optics. The three primary topics we shall examine in this chapter are interference, diffraction, and polarization. These phenomena cannot be adequately explained with ray optics, but wave theory leads to a satisfying description.

26.1 CONDITIONS FOR INTERFERENCE

In our discussion of wave interference in Section 14.12, we found that two waves could add together constructively or destructively. In constructive interference, the amplitude of the resultant wave is greater than that of either of the individual waves, while in destructive interference, the resultant amplitude is less than that of either of the individual waves. Light waves also undergo interference. Fundamentally, all interference associated with light waves arises as a result of the combining of the electric and magnetic field vectors that constitute the individual waves.

Interference effects in light waves are not easy to observe because of the short wavelengths involved (about 4×10^{-7} m to about 7×10^{-7} m). In order to observe sustained interference in light waves, the following conditions must be met:

1. The sources must be coherent. (We shall discuss this new term below.)
2. The sources must be monochromatic, that is, of a single wavelength.
3. The superposition principle must apply.

We shall now describe the characteristics of **coherent sources.** As we have said, two sources (producing two traveling waves) are needed to create interference. However, in order to produce a stable interference pattern, *the individual waves must maintain a constant phase relationship with one another.* A constant phase relationship implies that the two waves move together throughout their oscillations. For example, both waves may always be at a crest, or one wave might always be at a crest when the other is at a trough, or one wave might always be at a crest with the other is at some other point in its oscillation. As an example, the sound waves emitted by two side-by-side loudspeakers driven by a single amplifier can produce interference because the two speakers respond to the amplifier in the same way at the same time. Now, if two separate light sources are placed side by side, no interference effects are observed because in this case the light waves emitted by the two sources undergo independent random frequency and amplitude fluctuations and hence do not maintain a constant phase relationship with each other over the time of observation. For constructive interference to occur, two waves must overlap such that crests meet crests and troughs meet troughs. (A crest for a light wave means that the electric or magnetic field of the wave is a maximum in some direction. A trough occurs when the fields are at a maximum in the opposite direction.) If for some reason one of the waves undergoes some random change, perfect overlap is not maintained. Light from an ordinary light source undergoes such random changes about once every 10^{-8} s. Therefore the conditions for constructive interference, destructive interference, or some intermediate state last for times of the order of 10^{-8} s. The result is that no interference effects are observed since the eye cannot follow such short-term changes. Such light sources are said to be **non-coherent.**

A common method for producing two coherent light sources is to use one monochromatic source to illuminate a screen containing two small openings (usually in the shape of slits). The light emerging from the two slits is coherent because a single source produces the original light beam and the two slits serve only to separate the original beam into two parts (which, after all, is what was done to the sound signal discussed above). A random change in the light emitted by the source will occur in the two separate beams at the same time, and interference effects can be observed. An alternative to the use of conventional light sources is the laser, which will be discussed in more detail in Chapter 30. One of the most interesting and useful properties of this device is the highly coherent nature of the light it produces. In fact, one of the most unique applications of interference phenomena which uses laser light is the production of three-dimensional images called holograms.

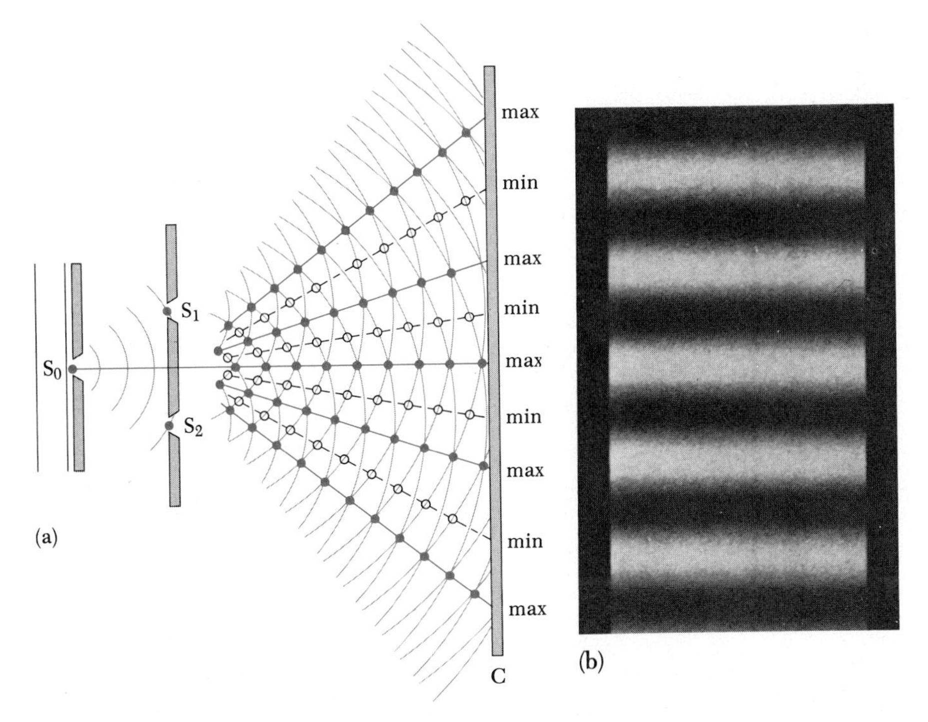

FIGURE 26.1 (a) Schematic diagram of Young's double-slit experiment. The narrow slits act as sources of waves. Slits S_1 and S_2 behave as coherent sources which produce an interference pattern on screen C. (Note that this is not to scale.) (b) The fringe pattern formed on screen C could look like this.

26.2 YOUNG'S DOUBLE-SLIT INTERFERENCE

The phenomenon of interference in light waves from two sources was first demonstrated by Thomas Young in 1801. A schematic diagram of the apparatus used in this experiment is shown in Figure 26.1a. Light is incident on a screen, which is provided with a narrow slit S_0. The waves emerging from this slit arrive at a second screen, which contains two narrow, parallel slits, S_1 and S_2. These two slits serve as a pair of coherent light sources because waves emerging from them originate from the same wavefront and therefore are always in phase. The light from the two slits produces a visible pattern on screen C; the pattern consists of a series of bright and dark parallel bands called **fringes** (Fig. 26.1b). When the light from slits S_1 and S_2 arrives at a point on the screen such that constructive interference occurs at that location, a bright line appears. When the light from the two slits combines destructively at any location on the screen, a dark line results. Figure 26.2 is a photograph of an interference pattern produced by two coherent vibrating sources in a water tank.

FIGURE 26.2 Water waves set up by two vibrating sources produce an effect similar to Young's double-slit interference. The waves interfere constructively at X and destructively at Y.

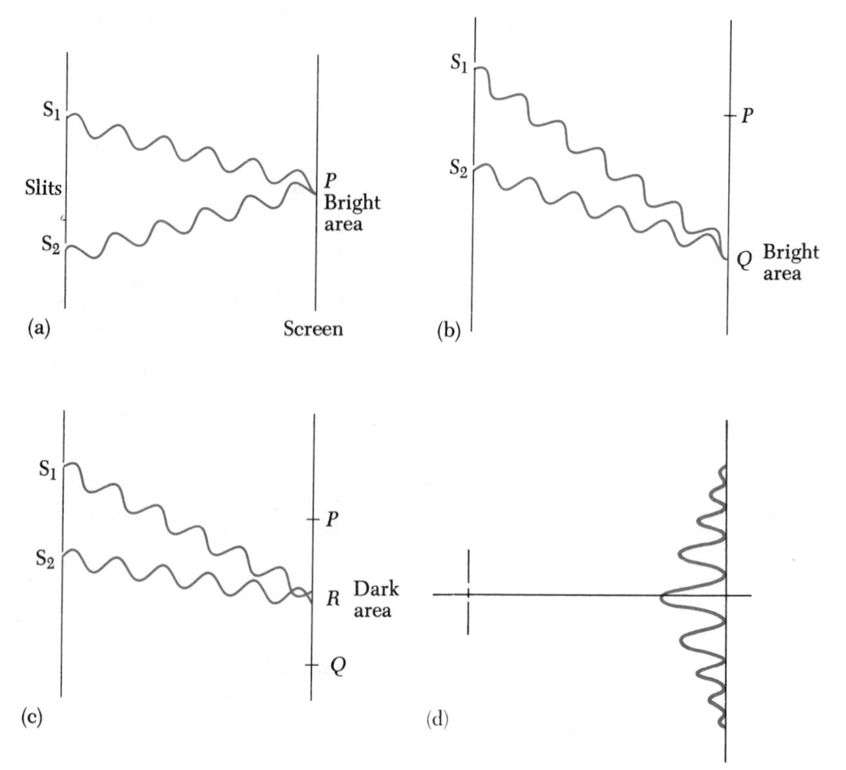

FIGURE 26.3 (a) Constructive interference occurs at P when the waves combine. (b) Constructive interference also occurs at Q. (c) Destructive interference occurs at R when the wave from the upper slit falls one-half wavelength behind the wave from the lower slit. (d) The intensity of the fringes decreases as one moves to higher orders. (Note that these figures are not drawn to scale.)

Figure 26.3 is a schematic diagram of some of the ways the two waves can combine at the screen. In Figure 26.3a, the two waves, which leave the two slits in phase, strike the screen at the central point P. Since these waves travel an equal distance, they arrive in phase at P, and as a result constructive interference occurs at this location and a bright area is observed. In Figure 26.3b, the two light waves again start in phase, but the upper wave has to travel one wavelength farther to reach point Q on the screen. Since the upper wave falls behind the lower one by exactly one wavelength, they still arrive in phase at Q, and so a second bright light appears at this location. Now consider point R, midway between P and Q in Figure 26.3c. At this location, the upper wave has fallen half a wavelength behind the lower wave. This means that the trough from the bottom wave overlaps the crest from the upper wave, giving rise to destructive interference at R. For this reason, one observes a dark region at this location. Figure 26.3d shows the intensity distribution on the screen. Notice that the central fringe is most intense and that the intensity decreases for higher order fringes.

We can obtain a quantitative description of Young's experiment with the help of Figure 26.4. Consider a point P on the viewing screen; the screen is located a perpendicular distance L from the screen containing slits S_1 and S_2, which are separated by a distance d. Let us assume that the source is monochromatic. Under these conditions, the waves emerging from S_1 and S_2 have the same frequency and amplitude and are in phase. The light intensity on the screen at P is the resultant of the light coming from both slits. Note that

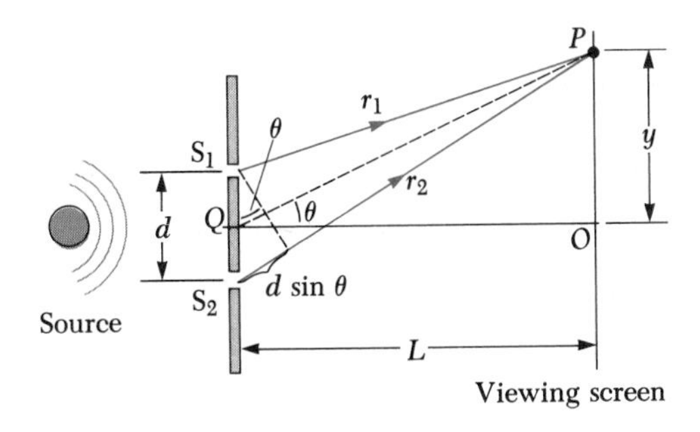

FIGURE 26.4 Geometric construction for describing Young's double-slit experiment. The path difference between the two rays is $r_2 - r_1 = d \sin \theta$. (Note that this figure is not to scale.)

a wave from the lower slit travels farther than a wave from the upper slit by an amount equal to $d \sin \theta$. This distance is called the **path difference,** δ, where

$$\boxed{\delta = r_2 - r_1 = d \sin \theta} \tag{26.1}$$

Path difference

This equation assumes that r_1 and r_2 are parallel, which is approximately true because L is much greater than d. As noted earlier, the value of this path difference will determine whether or not the two waves are in phase when they arrive at P. If the path difference is either zero or some integral multiple of the wavelength, the two waves are in phase at P and constructive interference results. Therefore, the condition for bright fringes, or **constructive interference,** at P is given by

$$\boxed{\delta = d \sin \theta = m\lambda} \qquad m = 0, \pm 1, \pm 2, \ldots \tag{26.2}$$

Constructive interference

The number m is called the **order number** of the fringe. The central bright fringe at $\theta = 0$ ($m = 0$) is called the *zeroth-order maximum*. The first maximum on either side, when $m = \pm 1$, is called the *first-order maximum*, and so forth.

Similarly, when the path difference is an odd multiple of $\lambda/2$, the two waves arriving at P will be 180° out of phase and will give rise to destructive interference. Therefore, the condition for dark fringes, or **destructive interference,** at P is given by

$$\boxed{\delta = d \sin \theta = \left(m + \frac{1}{2}\right)\lambda} \qquad m = 0, \pm 1, \pm 2, \ldots \tag{26.3}$$

Destructive interference

It is useful to obtain expressions for the positions of the bright and dark fringes measured vertically from O to P. We shall assume the $L \gg d$ (Fig. 26.4), that is, the distance from the slits to the screen is much larger than the distance between the two slits. This situation prevails in practice because L is often of the order of 1 m while d is a fraction of a millimeter. Under these conditions, θ is small, and so we can use the approximation $\sin \theta \approx \tan \theta$. From the triangle OPQ in Figure 26.4, we see that

$$\sin \theta \approx \tan \theta = \frac{y}{L} \tag{26.4}$$

Using this result together with Equation 26.2, we see that the positions of the *bright fringes* measured from O are given by

$$y_{\text{bright}} = \frac{\lambda L}{d} m \tag{26.5}$$

Similarly, using Equations 26.3 and 26.4, we find that the *dark fringes* are located at

$$y_{\text{dark}} = \frac{\lambda L}{d}\left(m + \frac{1}{2}\right) \tag{26.6}$$

As we shall demonstrate in Example 26.1, Young's double-slit experiment provides a method for measuring the wavelength of light. In fact, Young used this technique to make the first measurement of the wavelength of light. Additionally, the experiment gave the wave model of light a great deal of credibility. It was inconceivable that particles of light coming through the slits could cancel each other in a way that would explain the regions of darkness. Today we still use the phenomenon of interference to explain wave-like behavior in many observations.

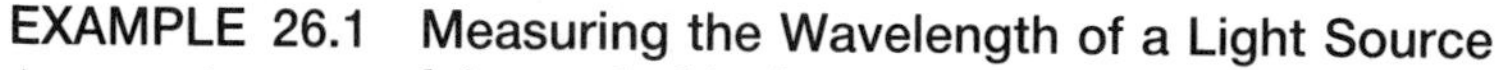

EXAMPLE 26.1 Measuring the Wavelength of a Light Source

A screen is separated from a double-slit source by 1.2 m. The distance between the two slits is 0.03 mm. The second-order bright fringe ($m = 2$) is measured to be 4.5 cm from the center line. Determine (a) the wavelength of the light and (b) the distance between adjacent bright fringes.

Solution (a) We can use Equation 26.5 with $m = 2$, $y_2 = 4.5 \times 10^{-2}$ m, $L = 1.2$ m, and $d = 3 \times 10^{-5}$ m:

$$\lambda = \frac{y_2 d}{mL} = \frac{(4.5 \times 10^{-2} \text{ m})(3 \times 10^{-5} \text{ m})}{2(1.2 \text{ m})} = 5.6 \times 10^{-7} \text{ m} = 560 \text{ nm}$$

(b) The location of the third-order bright fringe is found by using the results of (a) as follows:

$$y_3 = \frac{\lambda L}{d} m = \frac{(5.6 \times 10^{-7} \text{ m})(1.2 \text{ m})}{3 \times 10^{-5} \text{ m}}(3) = 6.7 \times 10^{-2} \text{ m}$$

Thus, the distance between the second and third bright fringes is

$$y_3 - y_2 = 6.7 \times 10^{-2} \text{ m} - 4.5 \times 10^{-2} \text{ m} = 2.2 \times 10^{-2} \text{ m} = 2.2 \text{ cm}$$

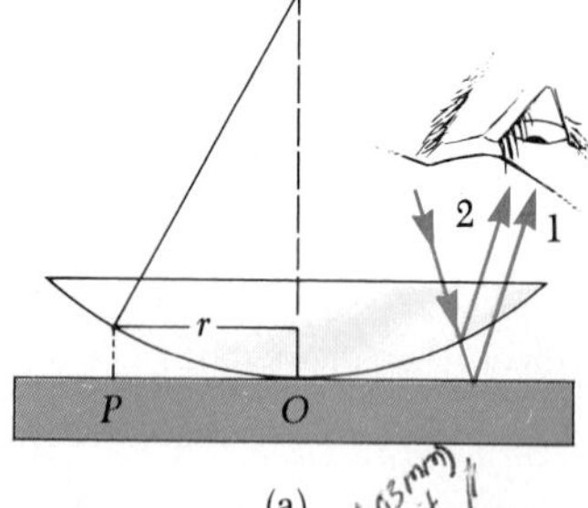

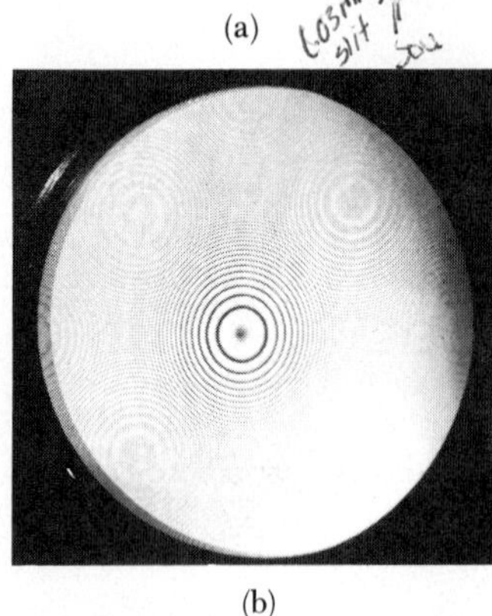

FIGURE 26.5 (a) The combination of rays reflected from the glass plate and the curved surface of the lens give rise to an interference pattern known as Newton's rings. (b) Photograph of Newton's rings. (Courtesy of Bausch & Lomb Optical Co.)

26.3 NEWTON'S RINGS

An interesting interference effect can be observed by placing a converging lens that is flat on one side on a glass surface, as in Figure 26.5a. When monochromatic light falls on the surface from above, a pattern of light and dark rings is seen, as shown in Figure 26.5b. These circular fringes, discovered by Newton, are referred to as **Newton's rings.** Newton's particle model of light could not explain the origin of these rings.

The interference effect is due to the combination of ray 1, reflected from the flat glass plate, with ray 2, reflected from the lower part of the lens (Fig.

26.5a). At first thought, it would seem that the central spot in the pattern should be bright. At this location indicated as point O in Figure 26.5a, the extra distance that ray 1 has to travel is extremely small. It is not exactly zero, because there is always a thin layer of air trapped between the two surfaces. Therefore, when rays 1 and 2 recombine, one would think that they would interfere constructively, producing a bright spot at the center. The photograph of Newton's rings shown in Figure 26.5b shows that this is not the case. Instead, a dark spot is observed at the center. Early observers of this phenomenon used great care to polish and clean the surfaces in order to bring them closer together, but the more they polished and cleaned, the more evident it became that the central spot was indeed dark. This difficulty was resolved when it was realized that

a light wave undergoes a phase change of 180° upon reflection from a medium having an index of refraction greater than the medium in which the wave is traveling.

In the case of a light wave, a 180° phase shift means that a crest (a maximum in the electric field) incident on a surface is reflected as a trough (a minimum in the electric field).

It is useful to draw an analogy between reflected light waves and the reflection of a transverse wave on a stretched string (Sec. 14.13). The reflected pulse on a string undergoes a phase change of 180° when it is reflected from a rigid support. Figure 26.6a compares this situation with that in which a light wave traveling in a medium of index of refraction n_1 is reflected from

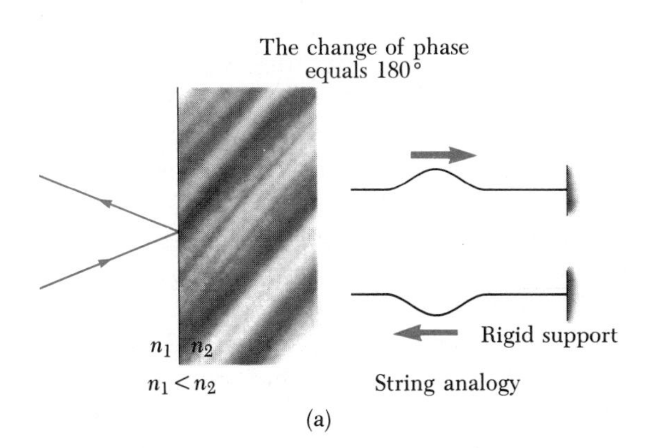

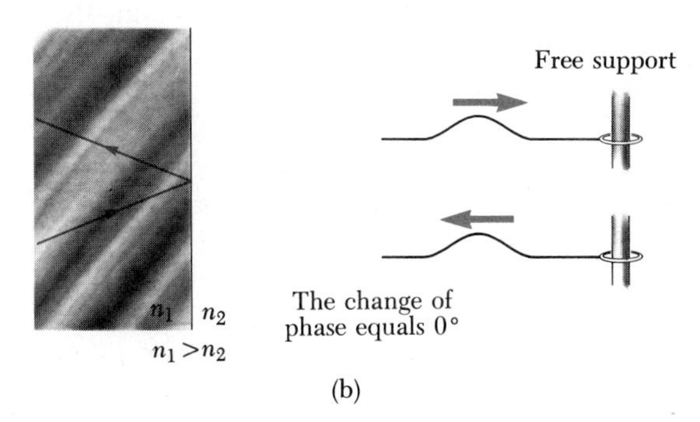

FIGURE 26.6 (a) A ray reflecting from a medium of higher refractive index undergoes a 180° phase change. The right side shows the analogy with a reflected pulse on a string. (b) A ray reflecting from a medium of lower refractive index undergoes *no* phase change.

FIGURE 26.7 This asymmetrical interference pattern indicates imperfections in the lens. (From Physical Science Study Committee, *College Physics*, Lexington, MA, D.C. Heath and Co., 1968)

a medium of index n_2, where n_1 is less than n_2. (An example of this would be light traveling through air and reflecting off a glass plate.) In this case the light undergoes a 180° phase change upon reflection. On the other hand, there is no phase change if a pulse on a string reflects from a support that is free to move, as in Figure 26.6b. Likewise, when light traveling in a medium of index of refraction n_1 reflects off a boundary with a lower index of refraction n_2, as in Figure 26.6b, the reflected wave undergoes no phase change.

In view of this information, we are now in a position to explain the pattern observed in Newton's ring experiment. Ray 2 in Figure 26.5a is traveling in glass and reflects off the glass-air interface between the lens and plate. This ray does not undergo a phase shift upon reflection. However, ray 1 travels through the air between the plate and the lens before it reflects off the plate, which has a higher index of refraction than the air. Therefore, ray 1 undergoes a 180° phase shift upon reflection. At point O, ray 1 does not travel appreciably farther than ray 2, but it does undergo a 180° phase shift because of reflection. Consequently when rays 1 and 2 recombine after reflection, they undergo destructive interference.

Let us now consider what happens to other rays of light striking at points away from the center of the convex lens. In this case, the extra distance that ray 1 travels relative to ray 2 becomes important. For example, suppose the extra distance of travel is $\lambda/2$. In this case, ray 1 loses 180° upon reflection, corresponding to half a wavelength, and then loses another half wavelength because of the extra distance it must travel. Rays 1 and 2 therefore combine in phase upon reflection, and a bright spot appears. Actually, constructive interference is occurring at all points P located the same distance r from the center of the lens, as in Figure 26.5. Therefore, what one really observes is a bright ring of light. Under what conditions would one see the first dark ring? The second bright ring?

One of the important uses of Newton's rings is in the testing of optical lenses. A circular pattern like that pictured in Figure 26.5b is obtained only when the lens is ground to a perfectly symmetric curvature. Variations from such symmetry might produce a pattern like that in Figure 26.7. These variations give an indication of how the lens must be ground and polished in order to remove the imperfections.

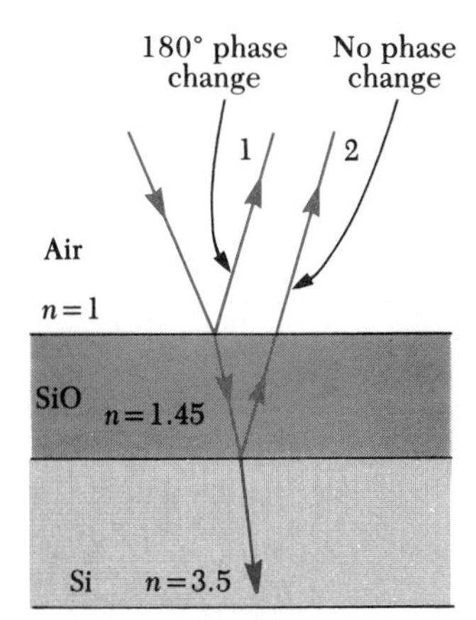

FIGURE 26.8 Interference observed in light reflected from a thin film is due to a combination of rays reflected from the upper and lower surfaces.

26.4 INTERFERENCE IN THIN FILMS

Everyone has seen the vivid colors produced when light shines on a thin film of oil floating on water. Likewise, the film on a soap bubble can produce striking colors, as shown on the cover of this textbook and Color Plate 23. The explanation of how these colors are produced follows much the same argument used to explain Newton's rings. The interference in this case is caused by the interference of waves reflected from the opposite surfaces of the film.

Consider a thin film of uniform thickness t and index of refraction n, as in Figure 26.8. Let us assume that the light rays traveling in air are nearly normal to the surface of the film. In order to determine whether the reflected rays interfere constructively or destructively, we must first note the following facts:

1. There is a phase change of 180° upon reflection if the reflecting medium has a higher index of refraction than the medium in which the wave is traveling.
2. The wavelength of light λ_n in a medium whose index of refraction is n (Chap. 24) is

$$\boxed{\lambda_n = \frac{\lambda}{n}} \qquad (26.7)$$

where λ is the wavelength of light in free space.

Let us apply these facts to the film shown in Figure 26.8. We find that ray 1, which is reflected from the upper surface (A), undergoes a phase change of 180° with respect to the incident wave. Ray 2, which is reflected from the lower surface (B), undergoes no phase change with respect to the incident wave. Therefore rays 1 and 2 are 180° out of phase following reflection. However, we must also consider that ray 2 travels an extra distance equal to $2t$ before the waves recombine. For example, if $2t = \lambda_n/2$, rays 1 and 2 will recombine in phase and constructive interference will result.

There are many permutations and combinations of surfaces which may be encountered in problems dealing with thin film interference. For example, suppose the film shown in Figure 26.8 is water, with air above its upper surface and glass below its lower surface. In this situation, rays 1 and 2 would both undergo a 180° phase change. Because of the wide variety of situations which could occur, it is impossible to derive a single equation that will enable you to solve such problems. Instead, each situation must be examined to determine whether or not phase changes occur upon reflection, and the conditions for constructive or destructive interference.

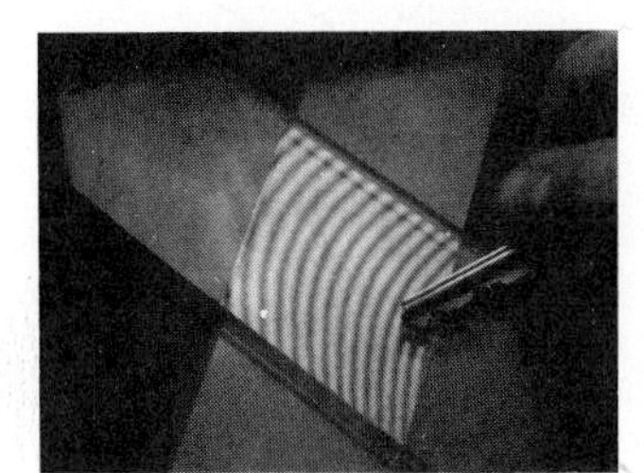

The thin film of air between two glass plates is responsible for the interference pattern. The lines are curved because pressure from the key bends the glass slightly, thus changing the thickness of the air film.

EXAMPLE 26.2 Interference in a Soap Film

Calculate the minimum thickness of a soap bubble film (n = 1.46) that will result in constructive interference in the reflected light if the film is illuminated with light whose wavelength in air is 600 nm.

Solution This situation is the same as that shown in Figure 26.8. Because there is air on each side of the soap film, ray 1 undergoes a phase change of 180°, while ray 2 does not undergo a phase change. In order for the two waves to be in phase (the condition for constructive interference), ray 2 must travel a minimum extra distance equal to $\lambda_n/2$. Setting $2t = \lambda_n/2$, and using Equation 26.7, $\lambda_n = \lambda/n$, we get

$$t = \frac{\lambda}{4n} = \frac{600 \text{ nm}}{4(1.46)} = 103 \text{ nm}$$

EXERCISE What other film thicknesses will produce constructive interference?

Answer $3t$, $5t$, $7t$, and so on.

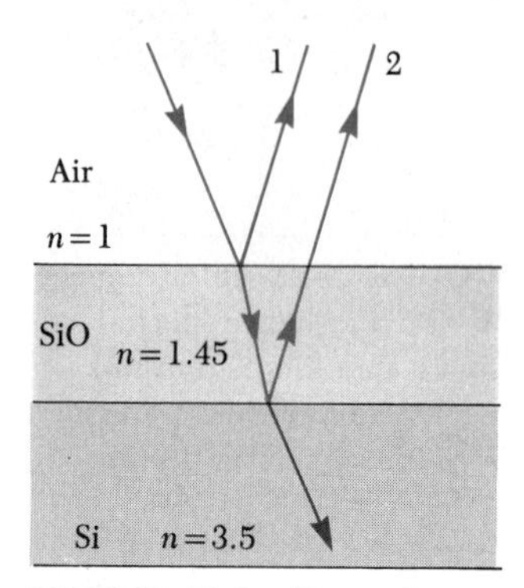

FIGURE 26.9 (Example 26.3) Reflective losses from a silicon solar cell are minimized by coating it with a thin film of silicon monoxide, SiO.

EXAMPLE 26.3 Nonreflecting Coatings for Solar Cells

Solar cells are often coated with a transparent thin film, such as silicon monoxide (SiO, n = 1.45), in order to minimize reflective losses from the surface (Fig. 26.9). A silicon solar cell (n = 3.5) is coated with a thin film of silicon monoxide for this purpose. Determine the minimum thickness of the film that will produce the least reflection of a wavelength of 550 nm, which is the center of the visible spectrum.

Solution The reflected light is a minimum when rays 1 and 2 meet the condition of destructive interference. Note that *both* rays undergo a 180° phase change upon reflection in this case. Hence, the net change in phase because of reflection is zero and the condition for a reflection *minimum* is $2t = \lambda_n/2$ or, using Equation 26.7

$$t = \frac{\lambda}{4n} = \frac{550 \text{ nm}}{4(1.45)} = 94.8 \text{ nm}$$

Typically, such antireflecting coatings reduce the reflective loss from 30% (with no coatings) to 10% (with coating). Such coatings will therefore increase the cell's efficiency since more light will be available to create charge carriers in the cell. Note that, in reality, the coating will never be perfectly nonreflecting because the required thickness is wavelength-dependent and the incident light covers a wide range of wavelengths.

Glass lenses used in cameras and other optical instruments are usually coated with a transparent thin film, such as magnesium fluoride (MgF_2), to reduce or eliminate unwanted reflection.

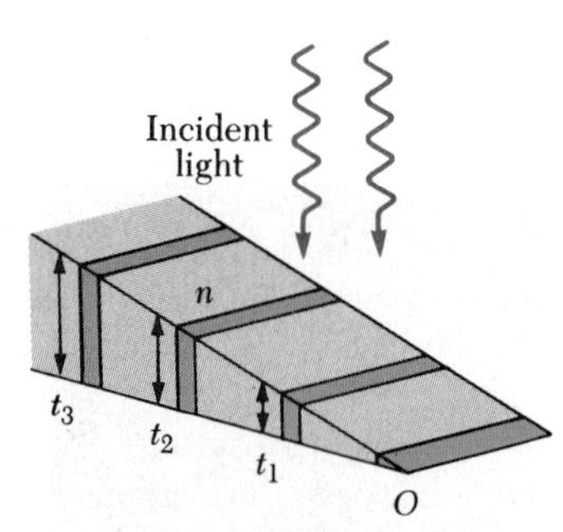

FIGURE 26.10 (Example 26.4) Interference bands in reflected light can be observed by illuminating a wedge-shaped film with monochromatic light. The dark areas correspond to positions of destructive interference.

EXAMPLE 26.4 Interference in a Wedge-Shaped Film

A thin, wedge-shaped film of refractive index n is illustrated with monochromatic light of wavelength λ, as illustrated in Figure 26.10. Describe the interference pattern observed for this case.

Solution The interference pattern is that of a thin film of variable thickness surrounded by air. Hence, the pattern will be a series of alternating bright and dark parallel bands. A dark band corresponding to destructive interference appears at point O, the apex, since the upper reflected ray undergoes a 180° phase change while the lower one does not. Other dark bands appear when $t_1 = \lambda/2n$, $t_2 = \lambda/n$, $t_3 = 3\lambda/2n$, and so on. Similarly, bright bands will be observed for thicknesses of $\lambda/4n$, $3\lambda/4n$, $5\lambda/4n$, and so on. If white light is used, bands of different colors will be observed at different points, corresponding to the different wavelengths of light.

26.5 DIFFRACTION

Suppose a light beam is incident on two slits, as in Young's double-slit experiment. If a light truly traveled in straight-line paths after passing through the slits, as in Figure 26.11a, the waves would not overlap and no interference pattern would be seen. Instead, Huygens' principle requires that the waves spread out from the slits as shown in Figure 26.11b. In other words, the light deviates from a straight-line path and enters the region that would otherwise be shadowed. **Diffraction** is defined as the bending of light when it passes an obstacle or barrier. Note that the phenomenon of diffraction should not be confused with the refraction of light discussed earlier. Refraction is concerned with the bending of light as it passes from one material to another having a different index of refraction. In general, diffraction occurs when waves pass through small openings, around obstacles, or by relatively sharp edges. The diffraction of sound waves is easy to experience. For example, if a radio located in another room is turned on, you can hear the sound waves it emits even if you are not standing in front of your doorway. On the other hand, the diffraction of light is not as easy to detect because the bending is usually very small. The most common way of detecting the diffraction of light is by observing interference patterns in the deviated light.

As an example of diffraction, consider the following. When an opaque object is placed between a point source of light and a screen, as in Figure 26.12a, the boundary between the shadowed and illuminated regions on the screen is not sharp. A careful inspection of the boundary shows that a small amount of light bends into the shadowed region. The region outside the shadow contains alternating light and dark bands, as in Figure 26.12b. A plot of the intensity of the light versus distance from the edge of the shadow is shown in Figure 26.12c. The dashed line in the figure shows how the light intensity would vary if no diffraction occurred. In this case, the region behind the obstacle would be completely shadowed, and the region outside the barrier would be uniformly illuminated. When the effects of diffraction are con-

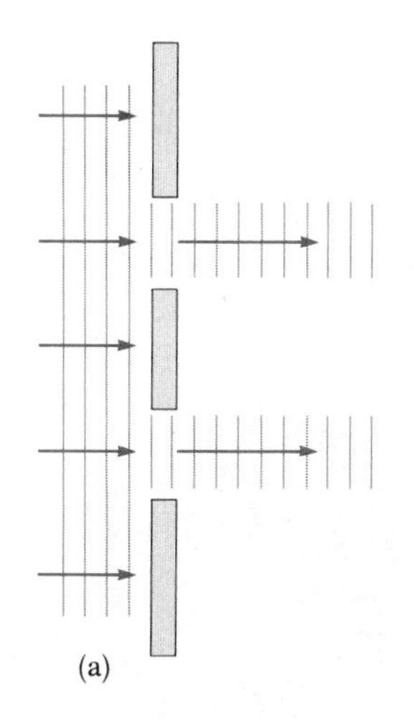

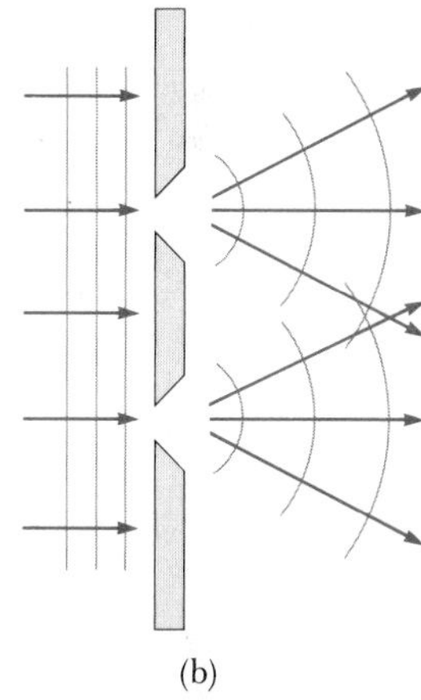

FIGURE 26.11 (a) If light did not spread out after passing through the slits, no interference would occur. (b) The light from the two slits overlaps as it spreads out, producing interference fringes.

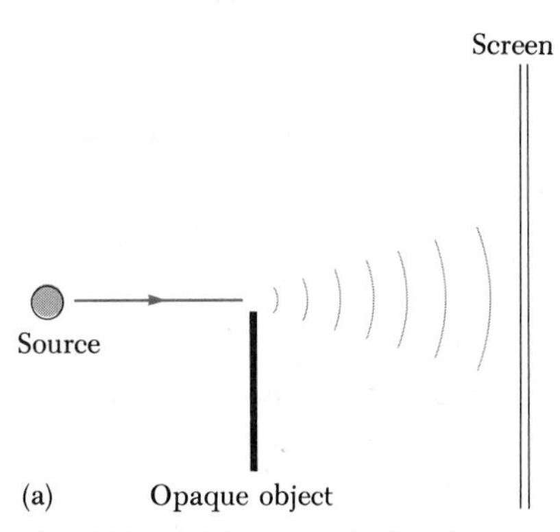

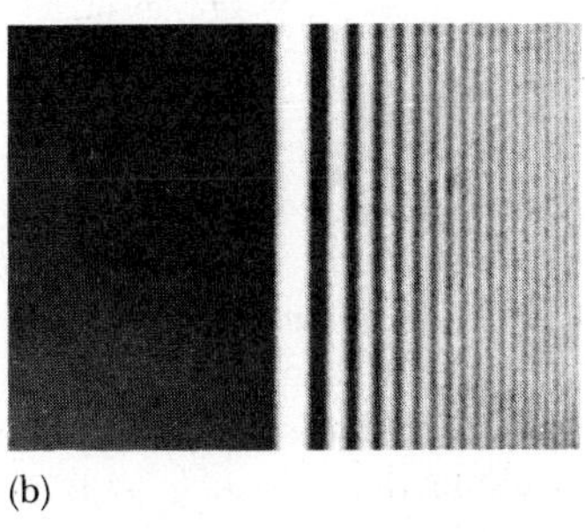

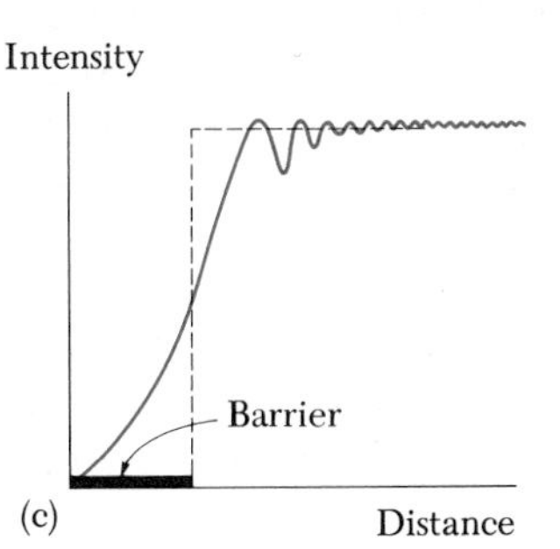

FIGURE 26.12 (a) Light bends around the opaque object. (b) The result is a series of dark and light bands in the region outside the shadow. (c) Intensity variation with distance from the edge of the shadow. The dashed line indicates the intensity pattern if no diffraction occurred.

FIGURE 26.13 Diffraction pattern of a penny, taken with the penny midway between the screen and source. (Courtesy of P.M. Rinard, from *Am. J. Phys.*, **44**:70, 1976)

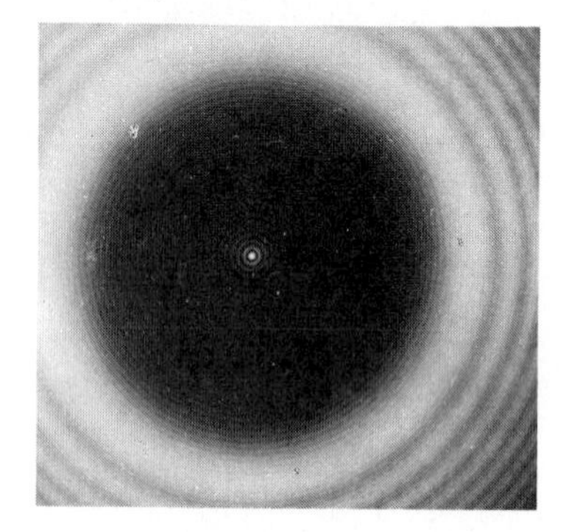

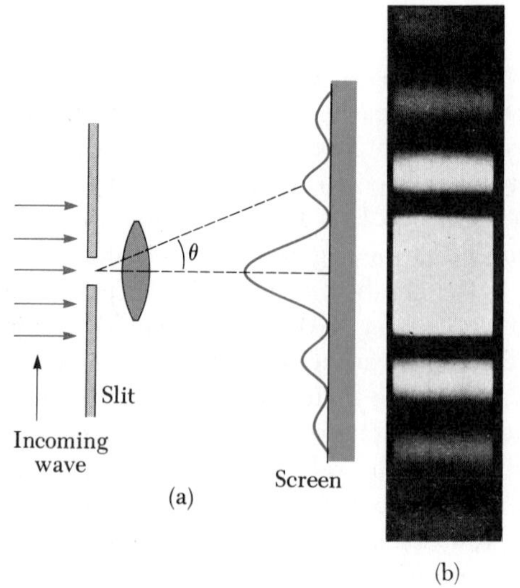

FIGURE 26.14 (a) Fraunhofer diffraction pattern of a single slit. The parallel rays are brought into focus on the screen with a converging lens. The pattern consists of a central bright region flanked by much weaker maxima. (Note that this is not to scale.) (b) Photograph of a single-slit Fraunhofer diffraction pattern. (From M. Cagnet, M. Francon, and J.C. Thierr, *Atlas of Optical Phenomena*, Berlin, Springer-Verlag, 1962, plate 18)

sidered, the first bright band has an intensity greater than the intensity of the region under uniform illumination. Note that the intensity of the light in these bands gradually diminishes, approaching that of uniform illumination as one moves further away from the barrier. Effects of this type were first reported in the 17th century by Francesco Grimaldi.

Figure 26.13 shows the shadow of the diffraction pattern of a penny. There is a bright spot at the center, circular fringes near the shadow's edge, and another set of fringes outside the shadow. This particular type of diffraction pattern was first observed in 1818 by Dominique Arago (1756–1853). The bright spot at the center of the shadow can be explained only through the use of the wave theory of light, which predicts constructive interference at this point. This was certainly a most dramatic experimental proof of the wave nature of light.

Diffraction phenomena are usually classified as being of two types, which are named after the men who first explained them. The first type, called **Fraunhofer diffraction,** occurs when the rays reaching a point are approximately parallel. This can be achieved experimentally either by placing the observing screen far from the opening or by using a converging lens to focus the parallel rays on the screen, as in Figure 26.14a. Note that a bright fringe is observed along the axis at $\theta = 0$, with alternating dark and bright fringes on either side of the central bright fringe. Figure 26.14b is a photograph of a single-slit Fraunhofer diffraction pattern.

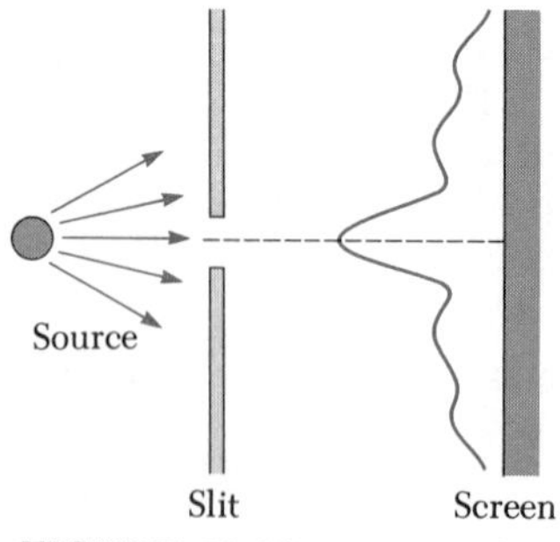

FIGURE 26.15 A Fresnel diffraction pattern of a single slit is observed when the incident rays are not parallel and the observing screen is at a finite distance from the slit. (Note that this is not to scale.)

When the observing screen is placed a finite distance from the slit and no lens is used to focus parallel rays, the observed pattern is called a **Fresnel diffraction** pattern (Fig. 26.15). The diffraction patterns shown in Figures 26.12b and 26.13 are examples of Fresnel diffraction. Another example, shown in Figure 26.16, is the diffraction pattern of a rectangular slit. Fresnel diffraction is rather complex to treat quantitatively. Therefore, the following discussion will be restricted to Fraunhofer diffraction.

26.6 SINGLE-SLIT DIFFRACTION

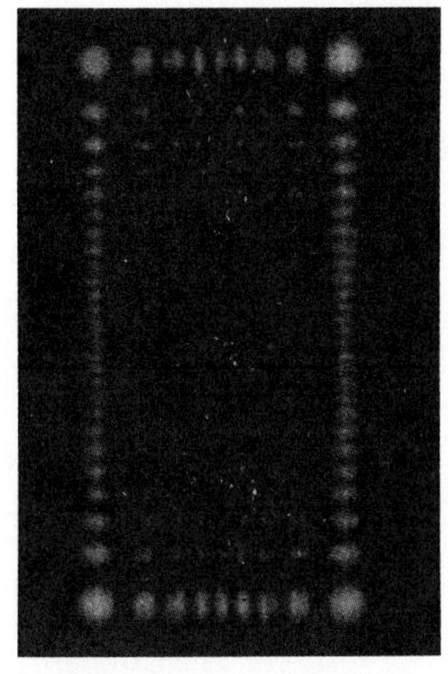

FIGURE 26.16 Fresnel diffraction pattern for a rectangular aperture. (From M. Cagnet, M. Francon, and J.C. Thierr, *Atlas of Optical Phenomena*, Berlin, Springer-Verlag, 1962, plate 34)

Let us discuss the nature of the Fraunhofer diffraction pattern produced by a single slit. We can deduce some important features of this problem by examining waves coming from various portions of the slit, as shown in Figure 26.17. According to Huygens' principle, *each portion of the slit acts as a source of waves.* Hence, *light from one portion of the slit can interfere with light from another portion,* and the resultant intensity on the screen will depend on the direction θ.

To analyze the diffraction pattern, it is convenient to divide the slit in two halves, as in Figure 26.17. All the waves that originate from the slit are in phase. Consider waves 1 and 3, which originate from the bottom and center of the slit, respectively. Wave 1 travels farther than wave 3 by an amount equal to the path difference $(a/2)\sin\theta$, where a is the width of the slit. Similarly, the path difference between waves 2 and 4 is also $(a/2)\sin\theta$. If this path difference is exactly one half of a wavelength (corresponding to a phase difference of 180°), the two waves cancel each other and destructive interference results. This is true, in fact, for any two waves that originate at points separated by half the slit width because the phase difference between two such points is 180°. Therefore, waves from the upper half of the slit interfere *destructively* with waves from the lower half of the slit when

$$\frac{a}{2}\sin\theta = \frac{\lambda}{2}$$

or when

$$\sin\theta = \frac{\lambda}{a}$$

If we divide the slit into four parts rather than two and use similar reasoning, we find that the screen is also dark when

$$\sin\theta = \frac{2\lambda}{a}$$

Likewise, we can divide the slit into six parts and show that darkness occurs on the screen when

$$\sin\theta = \frac{3\lambda}{a}$$

Therefore, the general condition for **destructive interference** is

$$\boxed{\sin\theta = m\frac{\lambda}{a}} \qquad m = \pm 1, \pm 2, \pm 3, \cdots \tag{26.8}$$

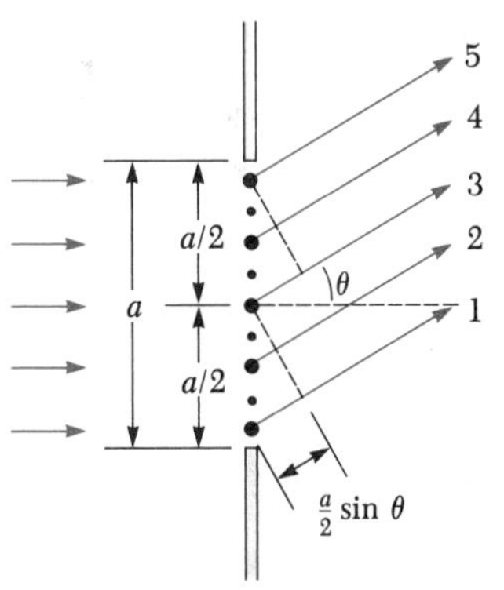

FIGURE 26.17 Diffraction of light by a narrow slit of width a. Each portion of the slit acts as a point source of waves. The path difference between rays 1 and 3 or between rays 2 and 4 is equal to $(a/2)\sin\theta$. (Note that this is not to scale, and that the rays are assumed to converge at a distant point.)

Equation 26.8 gives the values of θ for which the diffraction pattern has zero intensity, that is, where a dark fringe is formed. However, it tells us nothing about the variation in intensity along the screen. The general features

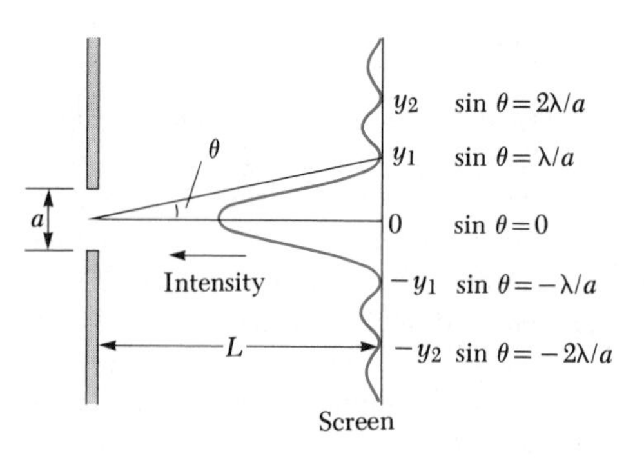

FIGURE 26.18 Positions of the various minima for the Fraunhofer diffraction pattern of a single slit of width a. (Note that this is not to scale.)

of the intensity distribution along the screen are shown in Figure 26.18. A broad central bright fringe is observed, flanked by much weaker, alternating bright fringes. The various dark fringes (points of zero intensity) occur at the values of θ that satisfy Equation 26.8. The position of the points of constructive interference lie approximately halfway between the dark fringes. Note that the central bright fringe is twice as wide as the weaker bright fringes.

In this section we referred to the alternating dark and bright bands on the screen as a diffraction pattern, while in our discussion of Young's double-slit experiment we referred to a similar pattern as an interference pattern. Also, you should note that the derivation of the equations associated with interference and diffraction effects are similar in that they consider the addition of waves that are either in or out of phase. One can distinguish the effects of interference and diffraction as follows. When the waves to be added come from two or more openings, as in Young's experiment, the resulting pattern is called an interference pattern. When various portions of a single wave interfere, as in this section, the result is a diffraction pattern.

EXAMPLE 26.5 Where are the Dark Fringes?

Light of wavelength 580 nm is incident on a slit of width 0.30 mm. The observing screen is placed 2 m from the slit. Find the positions of the first dark fringes and the width of the central bright fringe.

Solution The first dark fringes that flank the central bright fringe correspond to $m = \pm 1$ in Equation 26.8. Hence, we find that

$$\sin\theta = \pm\frac{\lambda}{a} = \pm\frac{5.8 \times 10^{-7}\ \text{m}}{0.3 \times 10^{-3}\ \text{m}} = \pm 1.93 \times 10^{-3}$$

From the triangle in Figure 26.18, we see that $\tan\theta = y_1/L$. Since θ is very small, we can use the approximation $\sin\theta \approx \tan\theta$, so that $\sin\theta \approx y_1/L$. Therefore, the positions of the first minima measured from central axis are

$$y_1 \approx L\sin\theta = \pm L\frac{\lambda}{a} = \pm 3.86 \times 10^{-3}\ \text{m}$$

The positive and negative signs correspond to the first dark fringes on either side of the central bright fringe. Hence, the width of the central bright fringe is given by $2|y_1| = 7.72 \times 10^{-3}\ \text{m} = 7.72\ \text{mm}$. Note that this value is *much larger* than the width of the slit. However, as the width of the slit is *increased*, the diffraction pattern will *narrow*, corresponding to smaller values of θ. In fact, for *large* values of a, the various maxima and minima will be so closely spaced that the only pattern observed will be a large central bright area that resembles the geometric image of the slit.

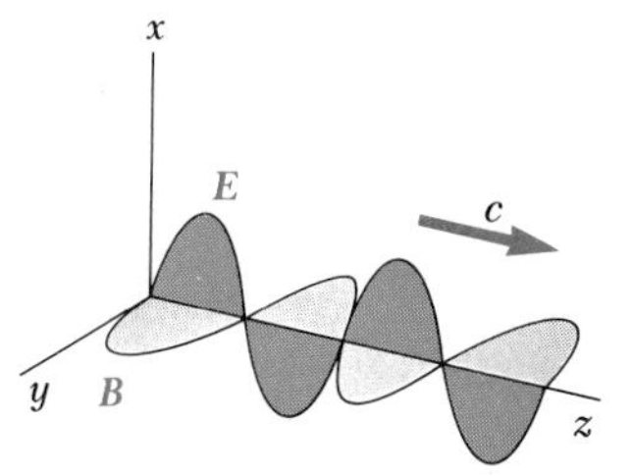

FIGURE 26.19 Schematic diagram of a polarized electromagnetic wave propagating in the z direction. The electric field vector $\boldsymbol{E}$ vibrates in the xz plane, and the magnetic field vector $\boldsymbol{B}$ vibrates in the yz plane.

26.7 POLARIZATION OF LIGHT WAVES

In Chapter 23 we characterized electromagnetic waves as transverse waves because *the oscillating electric and magnetic fields vibrate perpendicular to the direction in which the wave travels.* Figure 26.19 shows a polarized electromagnetic wave moving in the positive z direction. The phenomenon of **polarization,** which will be described in this section, is firm evidence of the transverse nature of electromagnetic waves.

An ordinary beam of light consists of a large number of waves emitted by the atoms or molecules contained in the light source. Each atom produces a wave having a particular orientation of the electric field vector. The result is a huge collection of waves with electric fields oscillating in all conceivable transverse directions. We can picture this state of affairs using Figure 26.20a, which represents an **unpolarized** light wave. In contrast to this situation, a wave is said to be **linearly polarized** if only one of these directions of vibration of the electric field vector exists at a particular point, as in Figure 26.20b. (Sometimes such a wave is described as being plane-polarized or simply polarized.)

It is possible to obtain a linearly polarized beam from an unpolarized beam by removing all waves from the beam except those whose electric field vectors oscillate along a single direction. We shall now discuss four processes for producing polarized light from unpolarized light: (1) selective absorption, (2) reflection, (3) double refraction, and (4) scattering.

Polarization by Selective Absorption

The most common technique for obtaining polarized light is to use a material that will transmit only those waves whose electric field vectors vibrate in a plane parallel to a certain direction and will absorb those waves whose electric field vectors vibrate in other directions. Any material that has the property of transmitting light with the electric field vector vibrating in only one direction is called a **dichroic** substance.

In 1938, E. H. Land discovered a material, which he called **Polaroid,** that polarizes light through selective absorption by oriented molecules. This material is fabricated in thin sheets of long-chain hydrocarbons, such as polyvinyl alcohol. The sheets are stretched during manufacture so that the long-chain molecules align. After a sheet is dipped into a solution containing iodine, the molecules can conduct electric charges along their hydrocarbon chains. As a result, the molecules readily *absorb* light whose electric field vector is *parallel* to their length and *transmit* light whose electric field vector is *perpendicular* to their length. It is common to refer to the direction perpendicular to the molecular chains as the *transmission axis.* In an ideal polarizer, all light with the electric field vector parallel to the transmission axis is transmitted and all light with the electric field vector perpendicular to the transmission axis is absorbed.

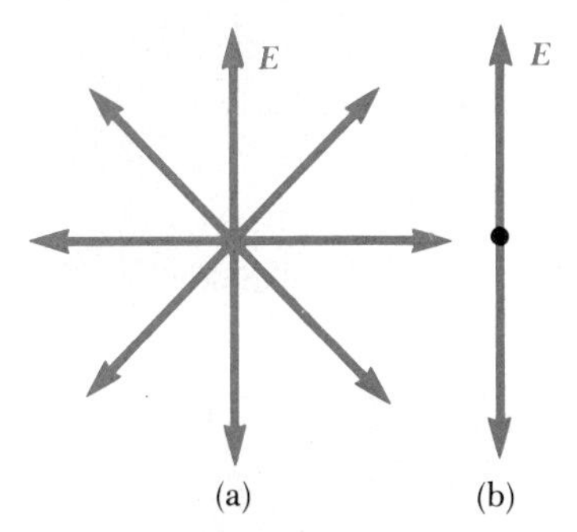

FIGURE 26.20 (a) An unpolarized light beam viewed along the direction of propagation (perpendicular to the page). The transverse electric field vector can vibrate in any direction with equal probability. (b) A linearly polarized light beam with the electric field vector vibrating in the vertical direction.

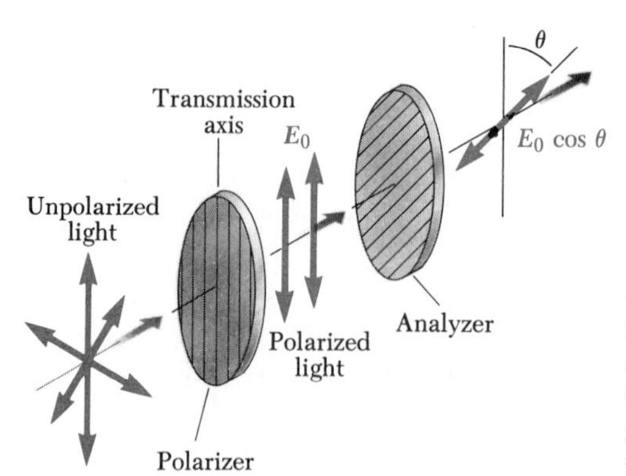

FIGURE 26.21 Two polarizing sheets whose transmission axes make an angle θ with each other. Only a fraction of the polarized light incident on the analyzer is transmitted.

Figure 26.21 represents an unpolarized beam of light incident on a polarizing sheet, called a **polarizer,** where the transmission axis is indicated by the straight lines on the polarizer. The light that passes through this sheet is plane-polarized vertically as shown, where the transmitted electric field vector is labeled $\mathbf{E}_0$. A second polarizing sheet, called an **analyzer,** intercepts this beam with its transmission axis at an angle θ to the axis of the polarizer. The component of $\mathbf{E}_0$ *perpendicular* to the axis of the analyzer is completely *absorbed*, and the component of $\mathbf{E}_0$ *parallel* to the axis of the analyzer is *transmitted.*

The transmitted intensity is a maximum when the transmission axes of the polarizer and analyzer are parallel to each other ($\theta = 0°$ or $180°$) and is zero when the transmission axes are perpendicular to each other (corresponding to complete absorption by the analyzer). This variation in transmitted intensity through a pair of polarizing sheets is illustrated in Figure 26.22.

FIGURE 26.22 Two crossed polarizing sheets. (Courtesy of Henry Leap, James Madison University)

Polarization by Reflection

Another method for obtaining polarized light is by reflection. When an unpolarized light beam is reflected from a surface, the reflected light is completely polarized, partially polarized, or unpolarized, depending on the angle of incidence. If the angle of incidence is either 0° or 90° (normal or grazing angles), the reflected beam is unpolarized. However, for intermediate angles of incidence, the reflected light is polarized to some extent. In fact, for one particular angle of incidence, the reflected beam is completely polarized.

Suppose an unpolarized light beam is incident on a surface, as in Figure 26.23a. The beam can be described by two electric field components, one parallel to the surface (the dots) and the other perpendicular to the first component and to the direction of propagation (the arrows). It is found that the parallel component reflects more strongly than the other components, and this results in a partially polarized reflected beam. Furthermore, the refracted beam is also partially polarized.

Now suppose that the angle of incidence, θ_1, is varied until the angle between the reflected and refracted beams is 90° (Fig. 26.23b). At this particular angle of incidence, the reflected beam is completely polarized, with its electric field vector parallel to the surface, and the refracted beam is partially

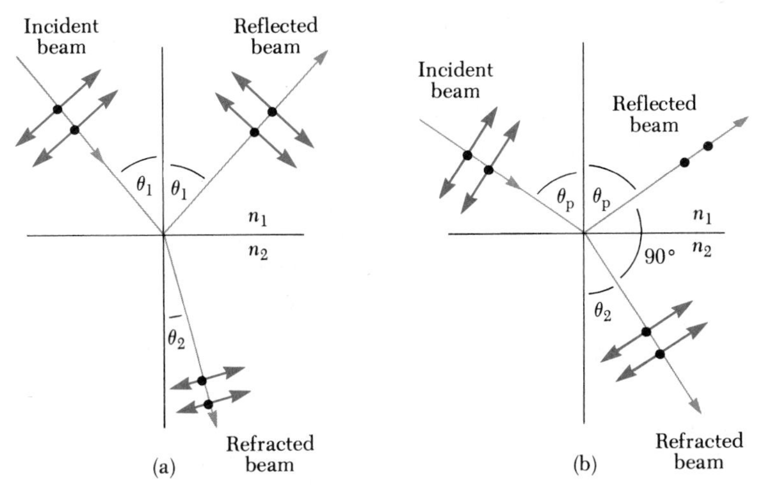

FIGURE 26.23 (a) When unpolarized light is incident on a reflecting surface, the reflected and refracted beams are partially polarized. (b) The reflected beam is completely polarized when the angle of incidence equals the polarizing angle, θ_p, which satisfies the equation $n = \tan \theta_p$.

polarized. The angle of incidence at which this occurs is called the **polarizing angle,** θ_p.

An expression can be obtained relating the polarizing angle to the index of refraction of the reflecting surface. From Figure 26.23b, we see that, at the polarizing angle, $\theta_p + 90° + \theta_2 = 180°$, so that $\theta_2 = 90° - \theta_p$. Using Snell's law, we have

$$n = \frac{\sin \theta_1}{\sin \theta_2} = \frac{\sin \theta_p}{\sin \theta_2}$$

Because $\sin \theta_2 = \sin(90° - \theta_p) = \cos \theta_p$, the expression for n can be written

$$n = \frac{\sin \theta_p}{\cos \theta_p} = \tan \theta_p \qquad (26.9)$$

Brewster's law

This expression is called **Brewster's law,** and the polarizing angle, θ_p, is sometimes called **Brewster's angle** after its discoverer, David Brewster (1781–1868). Because n varies with wavelength for a given substance, Brewster's angle is also a function of wavelength.

Polarization by reflection is a common phenomenon. Sunlight reflected from water, glass, or snow is partially polarized. If the surface is horizontal, the electric field vector of the reflected light will have a strong horizontal component. Sunglasses made of polarizing material reduce the glare of reflected light. The transmission axes of the lenses are oriented vertically to absorb the strong horizontal component of the reflected light.

EXAMPLE 26.6 Brewster's Angle for Water

At what angle of incidence will unpolarized light from the sun become completely polarized upon reflection from a pool of water?

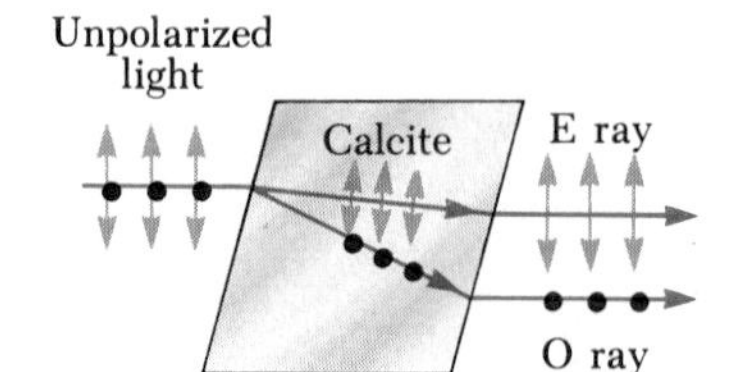

FIGURE 26.24 When an unpolarized light beam is incident on a calcite crystal, it splits into an ordinary (O) ray and an extraordinary (E) ray. The two rays are polarized in mutually perpendicular directions. (Note that this is not to scale.)

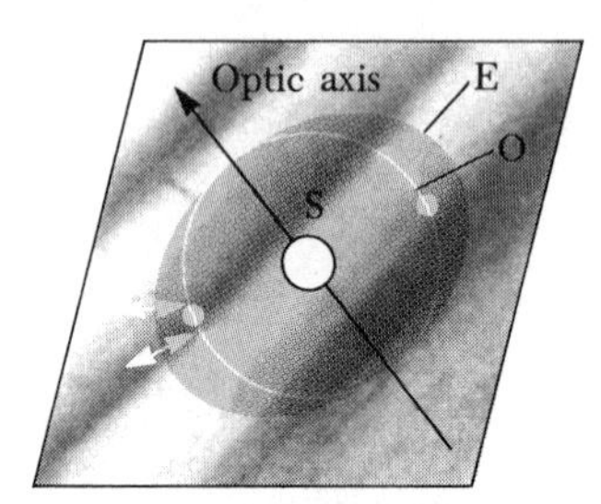

FIGURE 26.25 A point source S inside a doubly refracting crystal produces a spherical wavefront corresponding to the O ray and an elliptical wavefront corresponding to the E ray. The two waves propagate with the same velocity along the optic axis.

FIGURE 26.26 A calcite crystal produces a double image because it is a birefringent (double-refracting) material. (Courtesy of Henry Leap, James Madison University)

Solution In this case, $n = 1.33$, and Equation 26.9 gives Brewster's angle as

$$\theta_p = \tan^{-1} 1.33 = 53°$$

Polarization by Double Refraction

When light travels through an amorphous material, such as glass, it travels with a speed that is the same in all directions. That is, glass has a single index of refraction. However, in certain crystalline materials, such as calcite and quartz, the speed of light is *not* the same in all directions. Such materials are characterized by two indices of refraction. Hence, they are often referred to as **double-refracting** or **birefringent** materials.

When an unpolarized beam of light enters a calcite crystal, the beam splits into two plane-polarized rays that travel with different velocities, corresponding to two different angles of refraction, as in Figure 26.24 and Color Plate 25. The two rays are polarized in two mutually perpendicular directions, as indicated by the dots and arrows. One ray, called the **ordinary** (O) **ray,** is characterized by an index of refraction, n_O, that is the *same* in all directions. This means that if one could place a point source of light inside the crystal, as in Figure 26.25, the ordinary waves would spread out from the source as spheres.

The second plane-polarized ray, called the **extraordinary** (E) **ray,** travels with *different* speeds in different directions and hence is characterized by an index of refraction, n_E, that *varies* with the direction of propagation. A point source of light inside such a crystal would send out an extraordinary wave having wavefronts that are elliptical in cross section (Fig. 26.25). Note from Figure 26.25 that there is one direction, called the **optic axis,** along which the ordinary and extraordinary rays have the *same* velocity, corresponding to the direction for which $n_O = n_E$. The difference in velocity for the two rays is a maximum in the direction perpendicular to the optic axis. For example, in calcite $n_O = 1.658$ at a wavelength of 589.3 nm and n_E varies from 1.658 along the optic axis to 1.486 perpendicular to the optic axis.

If one places a piece of calcite on a sheet of paper and then looks through the crystal at any writing on the paper, two images of the writing are seen, as shown in Figure 26.26. As can be seen from Figure 26.24, these two images correspond to one formed by the ordinary ray and the second formed by the extraordinary ray. If the two images are viewed through a sheet of rotating polarizing glass, they will alternately appear and disappear because the ordinary and extraordinary rays are plane-polarized along mutually perpendicular directions.

Polarization by Scattering

When light is incident on a system of particles, such as a gas, the electrons in the medium can absorb and reradiate part of the light. The absorption and reradiation of light by the medium, called **scattering,** is what causes sunlight reaching an observer on the earth from straight overhead to be partially polarized. You can observe this effect by looking directly up through a pair of sunglasses made of polarizing glass. At certain orientations of the lenses, less light passes through than at others.

Figure 26.27 illustrates how the sunlight becomes partially polarized. The left side of the figure shows an incident unpolarized beam of sunlight on the verge of striking an air molecule. When this beam strikes the air molecule, it sets the electrons of the molecule into vibration. These vibrating charges act like the vibrating charges in an antenna, except that these charges are vibrating in a complicated pattern. The horizontal part of the electric field vector in the incident wave causes the charges to vibrate horizontally, and the vertical part of the vector simultaneously causes them to vibrate vertically. A polarized wave is directed to an observer as shown in Figure 26.27.

Some other atmospheric phenomena that are explained by scattering are discussed in the essay that appears below.

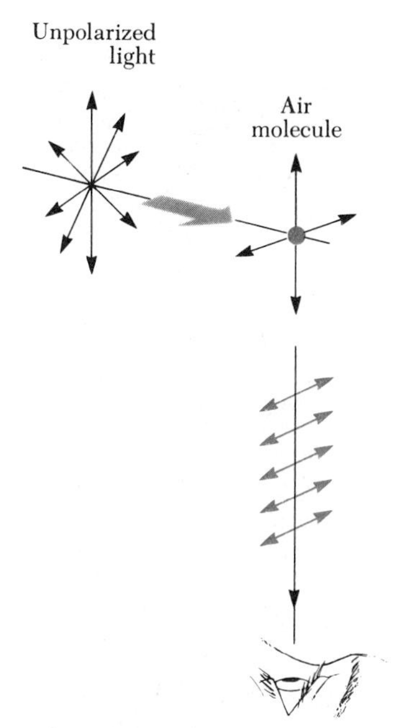

FIGURE 26.27 The scattering of unpolarized sunlight by air molecules. The light observed at right angles is plane polarized because the vibrating molecule has a horizontal component of vibration.

Uses of Polarized Light

Many important practical applications of polarized light involve the use of certain materials which display the property of **optical activity.** A substance is said to be optically active if it rotates the plane of polarization of transmitted light. In order to understand how this process occurs, suppose unpolarized light is incident on a polarizer from the left, as in Figure 26.28a. The transmitted light is polarized vertically as shown. If this light is then incident on an analyzer whose axis is perpendicular to that of the polarizer, no light emerges from it. If an optically active material is placed between the polarizer and analyzer as in Figure 26.28b, the material causes the direction of the polarized beam to rotate through an angle θ. As a result, some light is able to pass through the analyzer. The angle through which the light is rotated by the material can be found by rotating the polarizer until the light is again extinguished. It is found that the angle of rotation depends on the length of the sample and on the concentration if the substance is in solution. One op-

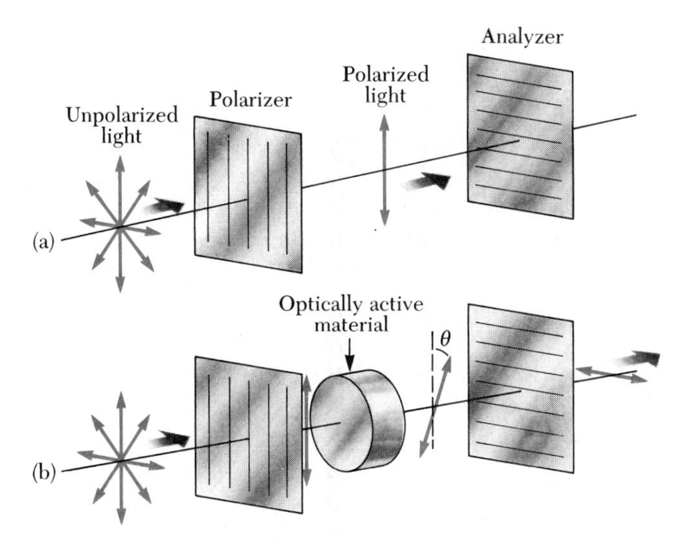

FIGURE 26.28 (a) When crossed polarizers are used, none of the polarized light can pass through the analyzer. (b) An optically active material rotates the direction of polarization through an angle θ, enabling some of this light to pass through the analyzer.

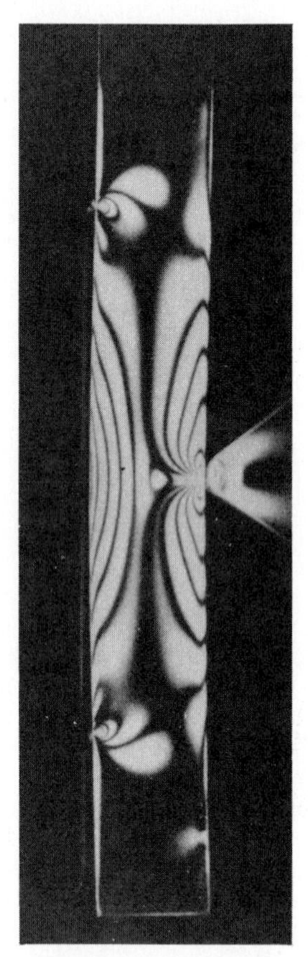

The distribution in strain in a beam shown in a plastic model under stress. The test piece is viewed between two crossed polarizers. The strain pattern disappears when the stress is released.

tically active material is a solution of common sugar dextrose. A standard method for determining the concentration of sugar solutions is to measure the rotation produced by a fixed length of the solution.

Optical activity occurs in a material because of an asymmetry in the shape of its constituent molecules. For example, some proteins are optically active because of their spiral shape. Environmental influences such as temperature changes can affect the shape and properties of these molecules. Studies of the changes in optical activity provides a great deal of information on such properties and their relation to the environment.

Some materials, such as glass and plastic, become optically active when placed under stress. If polarized light is passed through an unstressed piece of plastic and then through an analyzer whose axis is parallel to that of the polarizer, all of the polarized light is transmitted. However, if the plastic is placed under stress, the regions of greatest stress produce the largest angles of rotation of polarized light. As a result, one observes a series of light and dark bands in the transmitted light, with the darkest bands corresponding to points of greatest stress. Engineers often use this technique to assist in the design of various structures ranging from bridges to small tools. A plastic model is built and analyzed under different load conditions to determine positions of potential weakness and failure under stress. If the design is poor, the pattern of light and dark bands will indicate the points of greatest weakness and the design can be corrected at an early stage of development. An example of the strain distribution in a plastic rectangular beam under stress is shown in the photograph on the left.

SUMMARY

Interference occurs when two or more light waves overlap at a given point. A sustained interference pattern is observed if (1) the sources are coherent (that is, they maintain a constant phase relationship with one another), (2) the sources are monochromatic (of a single wavelength), and (3) the superposition principle is applicable.

In **Young's double-slit experiment,** two slits separated by a distance d are illuminated by a monochromatic light source. An interference pattern consisting of bright and dark fringes is observed on a screen a distance L from the slits. The condition for **bright fringes** (constructive interference) is

$$\boxed{d \sin \theta = m\lambda} \qquad m = 0, \pm 1, \pm 2, \ldots \tag{26.2}$$

The condition for **dark fringes** (destructive interference) is

$$\boxed{d \sin \theta = \left(m + \frac{1}{2}\right)\lambda} \qquad m = 0, \pm 1, \pm 2, \ldots \tag{26.3}$$

The number m is called the **order number** of the fringe.

An electromagnetic wave undergoes a phase change of 180° upon reflection from a medium whose index of refraction is higher than that of the medium in which the wave is traveling.

The wavelength of light, λ_n, in a medium whose refractive index is n is

$$\lambda_n = \frac{\lambda}{n} \qquad (26.7)$$

where λ is the wavelength of the light in free space.

The **diffraction pattern** produced by a single slit on a distant screen consists of a central bright maximum and alternating bright and dark regions of lower intensity. The angles θ at which the diffraction pattern has zero intensity (regions of constructive interference) are given by

$$\sin\theta = \frac{m\lambda}{a} \qquad m = \pm 1,\ \pm 2,\ \pm 3,\ \ldots \qquad (26.8)$$

where a is the width of the slit and λ is the wavelength of the light incident on the slit.

Unpolarized light can be polarized by selective absorption, reflection, double refraction, and scattering.

In general, light reflected from an amorphous material, such as glass, is partially polarized. However, the reflected light is completely polarized when the angle of incidence is such that the angle between the reflected and refracted beams is 90°. This angle of incidence, called the **polarizing angle, θ_p,** satisfies **Brewster's law,** given by

$$n = \tan\theta_p \qquad (26.9)$$

where n is the index of refraction of the reflecting medium.

ADDITIONAL READING

P. Baumeister and G. Pincus, "Optical Interference Coatings," *Sci. American*, December 1970, p. 59.

C. Bohrens and A. Fraser, "Colors of the Sky," *The Physics Teacher*, May 1985, p. 267.

H. M. Nussensvieg, "The Theory of the Rainbow," *Sci. American*, April 1977, p. 116.

P. M. Rinard, "Large Scale Diffraction Patterns from Circular Objects," *American Journal of Physics*, January 1976, p. 70.

Scientific American, September 1968. The entire issue is devoted to light.

W. A. Shurcliffe and S. S. Ballard, *Polarized Light*, Van Nostrand, 1964.

A. C. S. van Heel and C. H. F. Velzel, *What Is Light?* World University Library, New York, McGraw-Hill, 1968, Chaps. 3, 4, and 5.

H. C. Von Baeyer, "Rainbows, Whirlpools, and Clouds," *The Scientist*, July/August 1984, p. 24.

R. Wehner, "Polarized Light Navigation by Insects," *Sci. American*, July 1976, p. 106.

A. Wood and F. Oldham, *Thomas Young, Natural Philosopher, 1773–1829*, Cambridge University Press, 1954.

QUESTIONS

1. What is the necessary condition on path length difference between two waves that interfere (a) constructively and (b) destructively?
2. Explain why two flashlights held close together will not produce an interference pattern on a distant screen.
3. A simple way of observing an interference pattern is to look at a distant light source through a stretched handkerchief or an opened umbrella. Explain how this works.
4. If Young's double-slit experiment were performed under water, how would the observed interference pattern be affected?
5. In order to observe interference in a thin film, why must the film thickness be thin compared with the wavelengths of visible light? (*Hint:* How far apart are

the two reflected waves when they attempt to interfere if the film is thick?)

6. A soap bubble appears black just before it breaks. Explain this phenomenon in terms of the phase changes that occur upon reflection from the two surfaces.
7. Observe the shadow of your book when it is held a few inches above a table with a lamp several feet above the book. Why is the shadow of the book fuzzy at the edges?
8. Describe the change in width of the central maximum of a single-slit diffraction pattern as the slit is made narrower.
9. Certain sunglasses use a polarizing material to reduce the intensity of light reflected from shiny surfaces, such as water or the hood of a car. What orientation of polarization should the material have in order to be most effective?
10. Show, by a sketch, that the waves emitted by a radio antenna are polarized.
11. Can a sound wave be polarized? Explain.
12. Why is the sky black when viewed from the moon?
13. Washed dishes that are not rinsed well often have colored bands or rings across them. Discuss the interference effect which causes this.
14. If you observe a distant streetlight between two fingers pinched together, you will observe a pattern of alternating light and dark bands. Is this an interference or a diffraction pattern?
15. Light falls on a slit that is twice as wide as it is high. What kind of pattern will be formed on a screen? In which direction will the beam be the most spread out?

PROBLEMS

Section 26.2 Young's Double-Slit Interference

1. Monochromatic light falls on a screen 1.5 m from two slits spaced 1.3 mm apart. The first- and second-order bright fringes are found to be 0.5 mm apart. What is the wavelength of the light?

2. If the distance between two slits is 0.05 mm and the distance to a screen is 2.5 m, find the spacing between the first- and second-order bright fringes for yellow light of 600 nm wavelength.

3. A pair of narrow, parallel slits separated by 0.25 mm are illuminated by the green component from a mercury vapor lamp ($\lambda = 546.1$ nm). The interference pattern is observed on a screen 1.2 m from the plane of the parallel slits. Calculate the distance (a) from the central maximum to the first bright region on either side of the central maximum and (b) between the first and second dark bands in the interference pattern.

4. Light of wavelength 575 nm falls on a double slit and the first bright fringe is seen at an angle of 16.5°. Find the distance between the slits.

5. Light of wavelength 460 nm falls on two slits spaced 0.3 mm apart. What is the required distance from the slits to a screen if the spacing between the first and second dark fringes is to be 4 mm?

6. White light spans the wavelength range between about 400 nm and 700 nm. If white light passes through two slits 0.30 mm apart and falls on a screen 1.5 m from the slits, find the distance between the first-order violet and red fringes.

• **7.** The yellow component of light from a helium discharge tube ($\lambda = 587.5$ nm) is allowed to fall on a plate containing parallel slits that are 0.2 mm apart. A screen is located such that the second bright fringe in the interference pattern is at a distance equal to 10 slit spacings from the central maximum. What is the distance between the plate and the screen?

• **8.** Waves from a radio station have a wavelength of 300 m. They arrive by two paths at a home receiver 20 km from the transmitter. One path is a direct path, and the second is by reflection from a mountain directly behind the home receiver. What is the minimum distance from the mountain to the receiver such that destructive interference occurs at the receiver? (Assume no phase change occurs on reflection from the mountain.)

• **9.** Waves broadcast by a 1500-kHz radio station arrive at a home receiver by two paths. One is a direct path, and the second is from reflection off an airplane directly above the home receiver. The airplane is approximately 100 m above the home receiver, and the direct distance from the station to the home is 20 km. What is the exact height of the airplane if destructive interference is occurring? (Assume no phase change occurs on reflection from the plane.)

• **10.** The waves from a radio station can reach a home receiver by two different paths. One is a straight line path from the transmitter to the home, a distance of 30 km. The second path is by reflection from the ionosphere (a layer of ionized air molecules near the top of the atmosphere). Assume this reflection takes place at a point midway between receiver and transmitter. If the wavelength broadcast by the radio station is 350 m, find the minimum height of the ionospheric layer that would produce destructive interference between the direct and reflected beams. (Assume no phase changes on reflection.)

• **11.** Radio waves of wavelength 250 m from a star reach a radio telescope by two separate paths. One is a direct path to the receiver, which is situated on the edge of a cliff by the ocean. The second is by reflection off the water. The first minimum of destructive interference occurs when the star is 25° above the horizon. Find the height of the cliff. (Assume no phase change on reflection.)

• **12.** One of the bright bands in Young's interference pattern is located 12 mm from the central maximum. The screen is 119 cm from the pair of slits that serve as sources. The slits are 0.241 mm apart and are illuminated by the blue light from a hydrogen discharge tube (λ = 486 nm). How many bright lines are observed between the central maximum and the 12 mm position?

Section 26.3 Newton's Rings

Section 26.4 Interference in Thin Films

13. Suppose the film shown in Figure 26.8 has an index of refraction of 1.36 and is surrounded by air on both sides. Find the minimum thickness other than zero which will produce constructive interference in the reflected light when illuminated by light of wavelength 500 nm.

14. Repeat Problem 13 for the case in which the thin film is resting on a glass plate of index 1.50.

15. Two parallel glass plates are placed in contact and illuminated from above with light of wavelength 580 nm. As the plates are slowly moved apart, darkness occurs at certain separations when the reflected light is viewed. What are the distances (other than, perhaps, zero) of the first three of these separations?

16. Repeat Problem 8 if a phase change of 180° occurs on reflection.

17. A beam of light of wavelength 580 nm passes through two closely spaced glass plates, as shown in Figure 26.29. For what minimum nonzero value of the plate separation, d, will the transmitted light be bright?

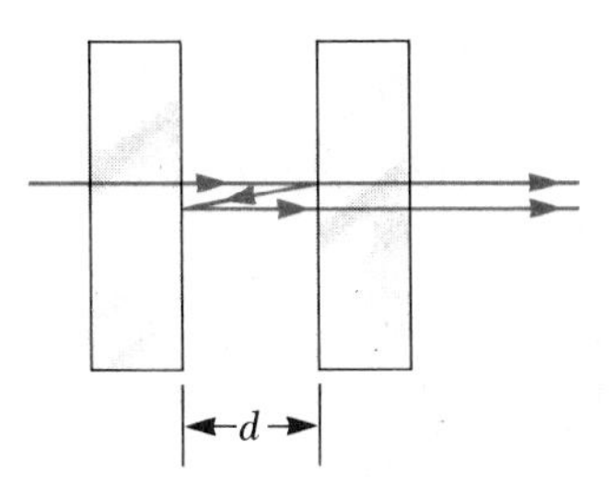

FIGURE 26.29 Problem 17

18. A planoconvex lens rests with its curved side on a flat glass surface and is illuminated from above by light of wavelength 500 nm. It is observed that there is a dark spot at the center surrounded by 19 concentric dark rings (with bright rings in between). How much thicker is the air wedge at the position of the 19th dark ring than at the center?

19. An oil film 500 nm thick floats on water. It is illuminated with white light in the direction perpendicular to the film. What wavelengths will be strongly reflected in the range from 300 nm to 700 nm? Take n = 1.46 for oil.

20. A thin film of glass (n = 1.50) floats on a liquid of n = 1.35. When illuminated by light of λ = 580 nm incident from air above the glass, find the minimum thickness other than zero that will produce destructive interference in the reflected light.

• **21.** Two rectangular optically flat plates (n = 1.52) are in contact along one end and are separated along the other end by a sheet of paper that is 4×10^{-3} cm thick (Fig. 26.30). The top plate is illuminated by monochromatic light of wavelength 546.1 nm. Calculate the number of dark parallel bands crossing the top plate (including the dark band at zero thickness along the edge of contact between the two plates).

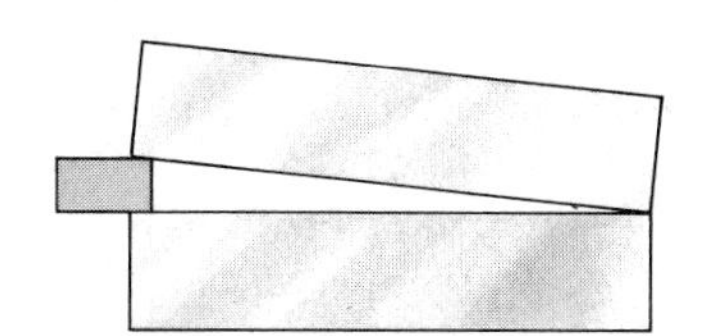

FIGURE 26.30 Problems 21, 22, and 23

• **22.** A hair replaces the sheet of paper in Problem 21, and when the reflected light is viewed, it is found that seven dark bands (including the one along the edge of contact) cross the top plate, with the seventh being directly over the hair. What is the thickness of the hair? The wavelength of light remains the same.

• **23.** An air wedge is formed between two thick glass plates in a manner like that described in Problem 21. Light of wavelength 434 nm is incident vertically on the top plate. In this case, there are 20 bright parallel interference fringes across the top plate. Calculate the thickness of the paper separating the plates.

•• **24.** A flat piece of glass is supported horizontally above the flat end of a metal rod of length 10 cm which has its lower end rigidly fixed. The thin film of air between the rod and glass is observed to be bright when illuminated by light of wavelength 500 nm. As the temperature is slowly increased by 25° C, the film changes from bright to dark and back to bright 200 times. What is the coefficient of linear expansion of the metal?

•• **25.** If the Newton's ring arrangement is such that the radius of curvature of the lens R is much greater than the thickness of the air gap, the radius of the dark rings is given by $r = \sqrt{m\lambda R}$, where m is an integer counting the rings, starting with 0 at the central spot and increasing outward, and λ is the wavelength of the light used. Show that this relationship for r is valid.

Section 26.6 Single-Slit Diffraction

26. Light of wavelength 600 nm falls on a slit 0.4 mm wide and forms a diffraction pattern on a screen 1.5 m away. (a) Find the position of the first dark band on each side of the central maximum. (b) Find the width of the central maximum.

27. A screen is placed 50 cm from a single slit, which is illuminated with light of wavelength 680 nm. If the distance between the first and third minima in the diffraction pattern is 3.0 mm, what is the width of the slit?

28. Light of wavelength 550 nm falls on a single slit of width 0.2 mm. Find the angle between the first dark bands on each side of the central maximum.

29. Light of wavelength 587.5 nm illuminates a single slit 0.75 mm in width. (a) At what distance from the slit should a screen be located if the first minimum in the diffraction pattern is to be 0.85 mm from the central maximum? (b) Calculate the width of the central maximum.

30. A slit of width 0.5 mm is illuminated with light of wavelength 500 nm, and a screen is located 120 cm in front of the slit. Find the widths of the first and second maxima on each side of the central maximum.

Section 26.7 Polarization of Light Waves

31. Determine Brewster's angle for light incident on benzene.

32. Equation 26.9 assumes the incident light is in air. If the light is incident from a medium of index n_1 on a medium of index n_2, use the procedure used to derive Equation 26.9 to show that $\tan \theta_p = n_2/n_1$.

33. Use the results of Problem 32 to find Brewster's angle when light is reflected off a piece of glass ($n = 1.50$) submerged in water.

34. At what angle above the horizon is the sun if light from it is completely polarized upon reflection from water?

• **35.** The angle of incidence of a light beam in air onto a reflecting surface is continuously variable. The reflected ray is found to be completely polarized when the angle of incidence is 48°. (a) What is the index of refraction of the reflecting material? (b) If some of the light passes into the material, what is the angle of refraction?

• **36.** The index of refraction of the ordinary ray in a calcite crystal is 1.66, and the index of the extraordinary ray is 1.49. Find the angle between these two rays inside the crystal if the incident beam strikes the refracting surface at an angle of incidence of 20°.

37. The critical angle for total internal reflection for sapphire surrounded by air is 34.4°. Calculate the polarizing angle for sapphire if the light is incident from the air.

• **38.** For a particular transparent medium surrounded by air, show that the critical angle for internal reflection and the polarizing angle are related by $\cot \theta_p = \sin \theta_c$.

ADDITIONAL PROBLEMS

39. When a monochromatic beam of light is incident from air at an angle of 37° with the normal on the surface of a glass block, one observes that the refracted ray is directed at 22° with the normal. What angle of incidence from air would result in total polarization of the reflected beam?

40. The transmitting antenna on a submarine is 5 m above the ocean surface when the ship is surfaced. The captain wishes to transmit a message to a receiver on a 90 m tall cliff at the ocean shore. If his signal is to be completely polarized by reflection off the ocean surface, how far must the ship be from the shore?

41. Monochromatic light incident on a slit of width 0.05 mm forms a diffraction pattern on a screen 2 m away. The second-order dark fringe is observed at an angle of 1.56° (see Fig. 26.18). Calculate the wavelength of the light.

• **42.** Light of wavelength 546 nm (the intense green line from a mercury source) produces a Young's interference pattern in which the second minimum from the central maximum is along a direction that makes an angle of 18 min of arc relative to the direction to the central maximum. What is the distance between the parallel slits?

• **43.** A light beam containing light of wavelengths λ_1 and λ_2 is incident on a set of parallel slits. In the interference pattern, the fourth bright line of the λ_1 light occurs at the same position as the fifth bright line of the λ_2 light. If λ_1 is known to be 540 nm, what is the value of λ_2?

• **44.** When a beam of light passes through a pair of polarizing disks (polarizer and analyzer), the intensity of the transmitted beam is a maximum when the transmission axes of the polarizer and the analyzer are parallel to each other ($\theta = 0°$ or $180°$) and zero when the axes are perpendicular ($\theta = 90°$). For any value of θ between 0° and 90°, the transmitted intensity is given by **Malus' law:**

$$I = I_i \cos^2 \theta$$

where I_i is the intensity of the light incident on the analyzer after the light has passed through the polarizer. (a) Polarized light is incident on the analyzer such that the direction of polarization makes an angle of 45° with respect to the transmission axes. What is the ratio I/I_i? (b) What should the angle between the transmission axes be to make $I/I_i = 1/3$?

• **45.** Three polarizing disks whose planes are parallel are centered on a common axis. The directions of the transmission axes relative to the common vertical direction are shown in Figure 26.31. A plane-polarized beam of light with the plane of polarization parallel to the vertical reference direction is incident from the left on the first disk with intensity $I_i = 10$ units (arbitrary). Calculate the transmitted intensity, I_f when $\theta_1 = 20°$, $\theta_2 = 40°$, and $\theta_3 = 60°$. (*Hint:* Refer to Problem 44.)

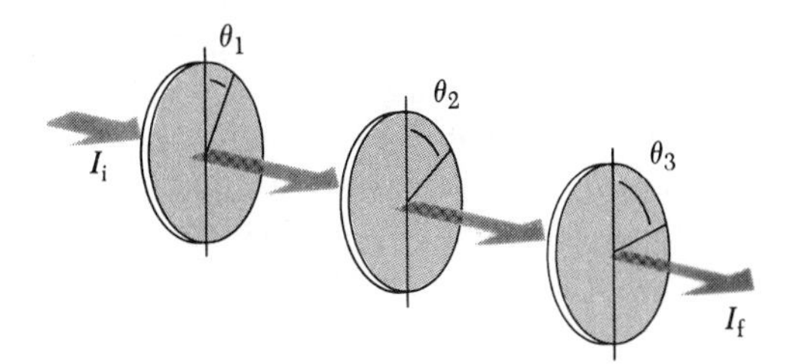

FIGURE 26.31 Problem 45

• **46.** Suppose that a slit 6 cm wide is placed in front of a microwave source operating at a frequency of 7.5 GHz. Calculate the angle (measured from the central maximum) where the first minimum in the diffraction pattern occurs.

• **47.** A diffraction pattern is produced on a screen 140 cm from a single slit using monochromatic light of wavelength 500 nm. The distance from the center of the central maximum to the first-order maximum is 3 mm. Calculate the slit width. (*Hint:* Assume the first-order maximum is halfway between the first- and second-order minima.)

• **48.** In a double-slit arrangement, $d = 0.15$ mm, $L = 140$ cm, $\lambda = 643$ nm, and $y = 1.8$ cm. (a) What is the path difference δ for the two slits at this y location? (b) Express this path difference in terms of the wavelength. (c) Will the interference occurring correspond to a maximum, a minimum, or an intermediate condition?

• **49.** A glass plate ($n = 1.61$) is covered with a thin uniform layer of oil ($n = 1.2$). A light beam in air of variable wavelength is normally incident on the oil surface. Observation of the reflected beam shows destructive interference at 500 nm and constructive interference at 750 nm. Calculate the thickness of the oil film from this information.

•• **50.** A piece of transparent material having an index of refraction n is cut into the shape of a wedge as shown in Figure 26.32. The angle of the wedge is small, and monochromatic light of wavelength λ is normally incident from above. If the height of the wedge is h and the length is ℓ, show that bright fringes occur at the positions

$$x = \frac{\lambda \ell \left(m + \frac{1}{2}\right)}{2hn}$$

and dark fringes occur at the positions $x = \dfrac{\lambda \ell m}{2hn}$ where $m = 0, 1, 2, \ldots$ and x is measured as shown.

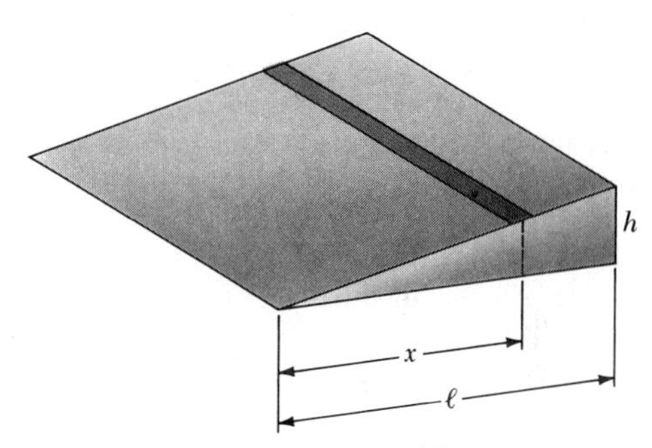

FIGURE 26.32 Problem 50

•• **51.** The discussion of the techniques for determining constructive and destructive interference by reflection from a thin film in air has been confined to rays striking the film at nearly normal incidence. Assume that a ray is incident at an angle of 30° (relative to the normal) on a film with index of refraction 1.38. Calculate the minimum thickness for constructive interference if the light is sodium light having a wavelength of 590 nm.

•• **52.** Figure 26.33 illustrates the formation of an interference pattern by the Lloyd's mirror method. Light from a source S strikes the screen by two different pathways. One is a direct path, and the second is the path followed as light reflects from a horizontal mirror before arriving at the screen. The interference is as though light from two different sources, S and S', had interfered as in the Young's double-slit arrangement. Assume the actual source S and the virtual source S' are in a plane 25 cm to the left of the mirror and the screen is a distance L = 120 cm to the right of this plane. The source S is a distance d = 2.5 mm above the top surface of the mirror and the light is monochromatic with λ = 620 nm. Determine the distance of the first bright fringe above the surface of the mirror.

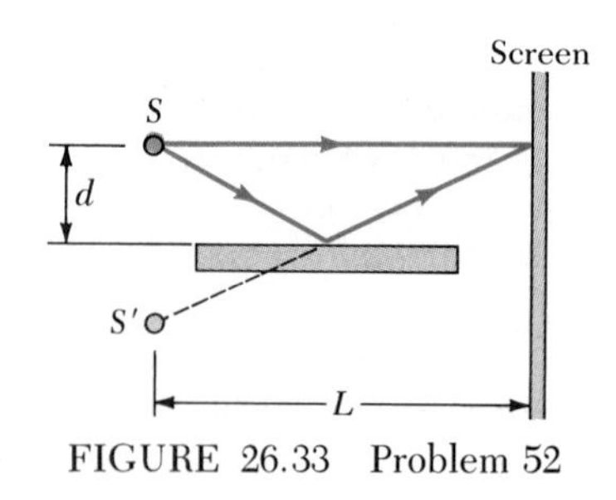

FIGURE 26.33 Problem 52

Essay

Atmospheric Physics

Milton Kerker *Clarkson University*

The blue of the sky and the brilliance of the sunsets have long fascinated scientists as well as poets. Why, after all, is the sky blue and why are sunsets so spectacularly red?

As scientists learned more about the nature of light, it became apparent that the color of the sky is the result of light scattering. Everyone has seen how a searchlight beam makes fog visible on a dark night. This happens because the light is scattered by the water droplets of which the fog is composed. In a similar manner, the sky looks blue because rays of sunlight are deflected in all directions by the oxygen and nitrogen molecules of the atmosphere. Such deflections of light, by molecules or any other objects, is called light scattering.

To an observer in space above the atmosphere, the heavens are black and our "sky" consists of a blue rim enveloping the globe as shown in Color Plate 3. Yuri Gagarin, the first astronaut, described this spectacle as follows: "The sky is very dark. The earth is bluish." The American astronauts who traveled to the moon found that, on that atmosphereless rock, the blinding white sun and the stars burn in a jet black sky.

As you have learned in your study of light, ordinary light is a mixture of colors, each pure color being characterized by its wavelength. In a rainbow, the white light of the sun is dispersed according to wavelength, that of blue light being the shortest and that of red light being the longest. More than a century ago, Lord Rayleigh, an English physicist, was able to explain the blueness of the sky by showing theoretically that very small particles, such as the molecules of air, preferentially scatter short-wavelength light. The blue rays of sunlight, having the shorter wavelengths, are scattered more than the other colors, and thus the sky appears blue.

At sunset, the sun is actually below the horizon and is visible only because the rays from it follow a curved and very extended path through the atmosphere. Because this path is so much longer than when the sun is overhead, there is much more scattering of the blue part of the light. The result is that what comes through to us is the familiar reddish tint of sunset.

But nature is even more colorful than this simple picture. In addition to the gaseous components of the air, there are particulate impurities present—sea salt from the ocean, smoke from forest fires, volcanic ash, and the debris belching forth from industrial smokestacks, automobiles, and bombs. These add color. The scattering law of Rayleigh applies only to molecular scattering. For particles as large as one hundred thousandth of an inch, both the scattered and the transmitted light may take on a variety of colors which, in turn, depend upon the particular size, structure, and composition of the scattering particles.

A most dramatic illustration of the influence of dust upon atmospheric optics occurred after the famous 1883 explosion of the volcanic island Krakatoa in Indonesia. The tidal waves caused by this eruption washed over many of the islands of the Dutch East Indies, causing the death of about 36,000 people, and even battered the coast of California many thousands of miles away. The volcanic dust traveling in the upper atmosphere spread all over the globe, settling over a period of several years and causing the appearance of weirdly brilliant sunsets as far away as northern Europe. In 1950 there was a similar effect when tremendous smoke clouds from forest fires in Canada floated across the Atlantic to Scandinavia, causing the sun, moon, and stars to appear blue.

If particulate matter scatters light according to the size, structure, and composition of the matter, why not learn about particles by analysis of scattered light? There are dust particles out in space as well as in the terrestrial atmosphere. Interstellar dust is bathed in starlight; it is the end product of the material manufactured and spewed out by the stars and may well be the stuff out of which new stars are formed. In any case, there is a fantastic amount of this dust, and our knowledge of it comes from observations of scattered starlight.

The trick is to recognize these observations with what is known about the optical properties of small particles containing various chemical species. The particles may have odd shapes, and they may be aligned by magnetic fields in space. Moreover, they may be pushed about and caused to rotate by the pressure of the starlight. The story is far from complete at the moment, but with recent developments in scattering theory, the picture may soon become much clearer.

Here on earth, the mundane scientific problems studied by light scattering are hardly less important. There is a class of molecules known as high polymers. Rubber, starch, nylon, and myriads of compounds found in living organisms are high polymers. Among the latter are proteins and nucleic acids, which constitute the chemical essence of terrestrial life.

When an intense beam of light is passed through a solution of a high polymer, the path of the beam is illuminated because of light scattering by the polymer molecules. Analysis of this scattered light can lead to information such as the size and shape of the molecules. Such molecules are complex, dynamic systems—they move to and fro, they tumble, they twist, and they bend. These particle dynamics can also be measured by light scattering, leading to greater insights into the chemical and biological functions of particular molecules.

Between the heavens and the earth there is the atmosphere—the seat of our sky and of the hydrometeors. Hydrometeors are the cloud and rain droplets, the snowflakes, the hailstones—those things that make for stormy weather. They, too, can be studied by scattering experiments.

Commonplace though it is and despite its crucial economic importance, the phenomenon of rainfall is still incompletely understood. It does appear that many rainstorms, even in midsummer, are triggered by the formation of snow from supercooled clouds and that this snow melts on the way down. However, because the upper reaches of the atmosphere, where these processes take place, are not directly accessible to observation, the picture of just what happens and where it happens is quite spotty.

One of the most useful tools for studying this problem is radar scattering, which is very much akin to light scattering. We have already pointed out how light is characterized by wavelength. Actually, visible light is only one form of electromagnetic radiation. There is also electromagnetic radiation of shorter wavelengths, such as gamma radiation, x-radiation, ultraviolet radiation, microwaves, and radiowaves. Radar is an electronic system for sending out a pulse of microwaves and then "listening" for the echo resulting from the backscattering by objects in the path of the microwaves. Nature has devised things so that the scattering of microwaves by hydrometeors is completely analogous to the scattering of light by molecules and small particles.

With the advent of the laser, another technique, analogous to radar and called *lidar*, has been developed for studying our atmosphere. A pulse of light rather than microwaves is radiated, and its echo is detected. Lidar signals depend upon the scattering of the laser beam by dust particles in the atmosphere, and such signals have now become an important tool for determining the presence of dust particles both in the upper atmosphere and in polluted cities.

Scientists have recently initiated a new area of research—the use of light scattering as a tool for studying biological cells. Light scattering has several advantages in this application. Because it is nondestructive, light scattering can be used to study living cells without perturbing their normal functions. Light-scattering signals are sensitive to the internal structure of cells, thereby providing information about those structures.

Because the signals are obtained almost at the speed of light, it is possible to "view" many cells very rapidly. Thus, a stream of cells can be forced single file through a laser beam so rapidly that individual light-scattering signals from as many as 10,000 cells can be obtained and processed every second. This has provided a powerful tool for scanning samples of cells to determine, for instance, the fraction of cancerous cells present.

ESSAY QUESTIONS

1. Account for the physics underlying the expression, "once in a blue moon."
2. In what way are there similarities between the study of interstellar dust and the laboratory study of biological molecules such as DNA?
3. The meteorologist uses both radar and lidar to study the physics of the atmosphere. In what way do the particles studied by each of these techniques differ?

CHAPTER 27

Optical Instruments

(Left) A telephoto lens designed to mount on a 35-mm camera body. (Middle) A small Maksutov telescope. (Right) Binoculars. (Courtesy of Lloyd Black)

We make use of devices made from lenses, mirrors, or other optical components every time we put on a pair of eyeglasses to read the daily newspaper, take a photograph, look at the heavens through a telescope, and so on. In this chapter we shall examine how these and other optical instruments work. For the most part, our analyses will involve the laws of reflection and refraction and the procedures of geometric optics. However, for some instruments and in some instances we must use the wave nature of light to explain certain phenomena.

27.1 THE CAMERA

The single-lens photographic **camera** is a simple optical instrument whose essential features are shown in Figure 27.1a. It consists of a lighttight box, a converging lens that produces a real image, and a film behind the lens to receive the image. Focusing is accomplished by varying the distance between lens and film with an adjustable bellows in older style cameras or some other mechanical arrangement in newer models. For proper focusing, or sharp images, the lens-to-film distance will depend on the object distance as well as on the focal length of the lens. The shutter, located behind the lens, is a mechanical device that is opened for selected time intervals. With this arrangement, one can photograph moving objects by using short exposure times or dark scenes (low light levels) by using long exposure times. If this arrangement were not available, it would be impossible to take stop-action photographs. For example, a rapidly moving racecar could move enough in the interval that the shutter was open to produce a blurred image. Typical shutter speeds are 1/30, 1/60, 1/125, and 1/250 s. A stationary object is normally shot with a shutter speed of 1/60 s.

More expensive cameras like that shown in Figure 27.1b also have an aperture of adjustable diameter either behind or in front of the lens to provide further control of the intensity of the light reaching the film. When small aperture diameters are used, only light from the central portion of the lens reaches the film and so the aberration is reduced somewhat. (See Chapter 25 for a discussion of aberration.)

The brightness of the image focused on the film depends on the focal length and diameter of the lens. Clearly, the amount of light that reaches the film increases as the size of the lens increases. That is, as the lens diameter increases, the brightness of the image formed on the film increases. We can see that the focal length of the lens affects the brightness of the image formed by considering the magnification equation for a thin lens:

$$M = \frac{h'}{h} = \frac{s'}{s}$$

$$h' = h\frac{s'}{s}$$

where h and h' are the object and image heights, respectively, and s and s' are the object and image distances. When s is large, s' is approximately equal to the focal length, f. Thus we have

$$h' = \frac{h}{s}f$$

From this result, we see that a lens having a short focal length produces a small image, corresponding to a small value of h'.

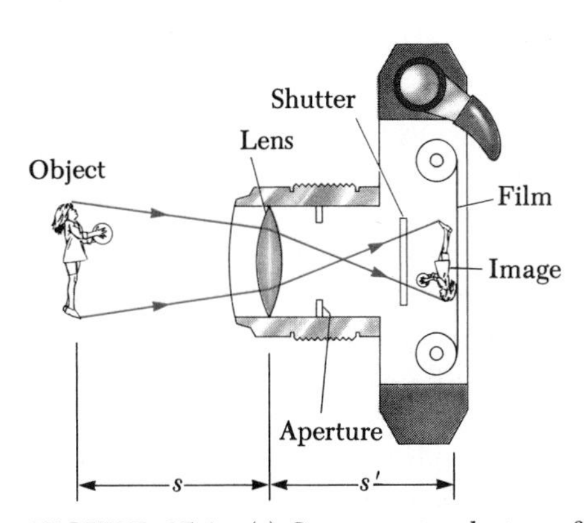

FIGURE 27.1 (a) Cross-sectional view of a simple camera. (b) Cutaway view of a modern 35-mm camera.

A small image will be brighter than a larger one because all of the incoming light is spread over a much smaller area. Because the brightness of the image depends on f and D, the diameter of the lens, a quantity called the ***f*-number** is defined as

$$f\text{-number} \equiv \frac{f}{D} \tag{27.1}$$

The f-number is a measure of the "light-concentrating" power of a lens and determines what is called the speed of the lens. A fast lens has a small f-number and is usually one with a small focal length and large diameter. Camera lenses are often marked with various f-numbers, such as $f/2.8$, $f/4$, $f/5.6$, $f/8$, $f/11$, $f/16$. These f-numbers are obtained by adjusting the aperture, which effectively changes D. When the f-number is increased by one position (or one "stop"), the light admitted decreases by a factor of 2. Likewise, the shutter speed is changed in steps whose factor is 2. The smallest f-number corresponds to the case where the aperture is wide open and as much as possible of the lens area is in use. Fast lenses, with f-numbers as low as 1.2, are more expensive because it is more difficult to keep aberrations acceptably small. Simple cameras for routine snapshots usually have a fixed focal length and fixed aperture size, with f-number of about $f/11$.

EXAMPLE 27.1 Choosing the *f*-Number

Suppose you are using a single-lens 35-mm cameras (where 35 mm is the width of the film strip), and the camera has only two f-numbers, $f/2.8$ and $f/22$. Which f-number would you use on a cloudy day and why?

Solution Substituting the given f-numbers in Equation 27.1, we have

$$2.8 = \frac{f_1}{D_1} \qquad \text{and} \qquad 22 = \frac{f_2}{D_2}$$

The focal length of the camera is fixed ($f_1 = f_2$ in the above), but the diameter of the aperture is not. On a cloudy day, you should make the shutter opening as large as possible. As these equations indicate, the largest value of D produces the smallest f-number. Thus, you should use the 2.8 setting.

27.2 THE EYE

The eye is an extremely complex part of the body, and because of its complexity, certain defects often arise which can cause the impairment of vision. In these cases, external aids, such as eyeglasses, are often used. In this section we shall describe the parts of the eye, their purpose, and some of the corrections that can be made when the eye does not function properly. You will find that the eye has much in common with the camera. Like the camera, the eye gathers light and produces a sharp image. However, the mechanisms by which the eye controls the amount of light admitted and adjusts itself to produce correctly focused images are far more complex, intricate, and effective than those in the most sophisticated camera. In all respects, the eye is an architectural wonder.

Figure 27.2 shows the essential parts of the eye. The front is covered by a transparent membrane called the *cornea.* This is followed by a clear liquid region (the *aqueous humor*), a variable aperture (the *iris* and *pupil*), and the *crystalline lens.* Most of the refraction occurs in the cornea because the liquid medium surrounding the lens has an average index of refraction close to that of the lens. The iris, which is the colored portion of the eye, is a muscular diaphragm that controls the size of the pupil. The iris regulates the amount of light entering the eye by dilating the pupil in light of low intensity and contracting the pupil in high-intensity light. The *f*-number range of the eye is about *f*/2.8 to *f*/16.

Light entering the eye is focused by the cornea-lens system onto the back surface of the eye, called the *retina.* The surface of the retina consists of millions of sensitive structures called *rods* and *cones.* When stimulated by light, these receptors send impulses via the optic nerve to the brain, where an image is perceived. By this process, a distinct image of an object is observed when the image falls on the retina.

The eye focuses on a given object by varying the shape of the pliable crystalline lens through an amazing process called **accommodation.** An important component in accommodation is the *ciliary muscle,* which is attached to the lens. When the eye is focused on distant objects, the ciliary muscle is relaxed. For an object distance of infinity, the focal length of the eye (the distance between the lens and the retina) is about 1.7 cm. The eye focuses on nearby objects by tensing the ciliary muscle. This action effectively decreases the focal length by slightly decreasing the radius of curvature of the lens, which allows the image to be focused on the retina. This lens adjustment takes place so swiftly that we are not even aware of the change. Again in this respect, even the finest electronic camera is a toy compared with the eye. It is evident that there is a limit to accommodation because objects that are very close to the eye produce blurred images.

The **near point** represents the closest distance for which the lens will produce a sharp image on the retina. This distance usually increases with age and has an average value of around 25 cm.

Typically, at age ten the near point of the eye is about 18 cm. This increases to about 25 cm at age 20, to 50 cm at age 40, and to 500 cm or greater at age 60.

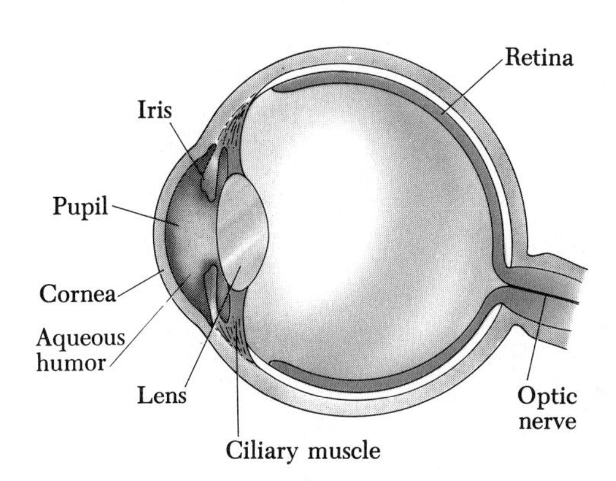

FIGURE 27.2 Essential parts of the eye. Note the similarity between the eye and the simple camera. Can you correlate the essential parts of the eye with those of the camera?

Defects of the Eye

Although the eye is one of the most remarkable organs in the body, it often does not function properly. The eye may have several abnormalities, which can often be corrected with eyeglasses, contact lenses, or surgery.

When the relaxed eye produces an image of a distant object *behind* the retina, as in Figure 27.3a, the abnormality is known as **hyperopia,** and the person is said to be farsighted. With this defect, distant objects are seen clearly but near objects are blurred. Either the hyperopic eye is too short or the ciliary muscle is unable to change the shape of the lens enough to properly focus the image. The condition can be corrected with a converging lens, as shown in Figure 27.3b.

Another condition, known as **myopia,** or nearsightedness, occurs either when the eye is longer than normal or when the maximum focal length of the lens is insufficient to produce a clearly formed image on the retina. In this case, light from a distant object is focused in front of the retina (Fig. 27.4a). The distinguishing feature of this imperfection is that distant objects are not seen clearly. Nearsightedness can be corrected with a diverging lens, as in Figure 27.4b.

Beginning with middle age, most people lose some of their accommodation power as a result of a weakening of the ciliary muscle and a hardening of the lens. This causes an individual to become farsighted, and the condition can be corrected with converging lenses.

A person may also have an eye defect known as **astigmatism,** in which light from a point source produces a line image on the retina. This defect arises either when the cornea or the crystalline lens or both are not perfectly spherical. Astigmatism can be corrected with lenses having different curvatures in two mutually perpendicular directions. A cylindrical lens (a segment of a cylinder) is typically used for this purpose.

Bifocals contain lenses that can correct two different defects with one pair of eyeglasses. For example, the lower half of the lens could be ground to

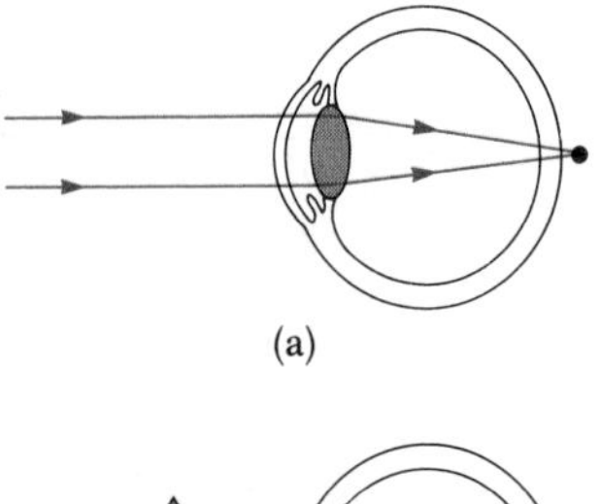

(a)

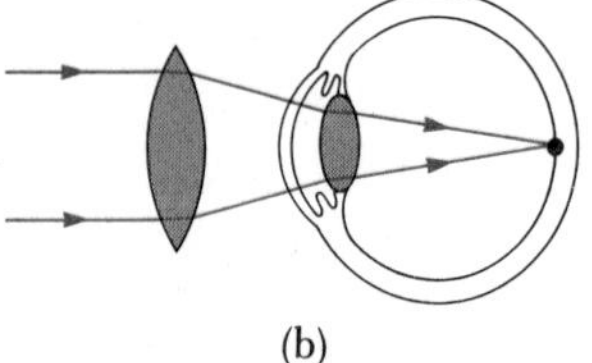

(b)

FIGURE 27.3 (a) A hyperopic eye (farsightedness) is slightly shorter than normal, hence the image of a distant object focuses behind the retina. (b) The condition can be corrected with a converging lens.

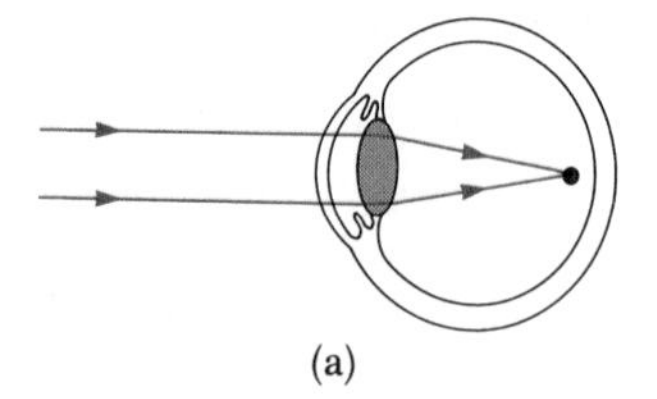

(a)

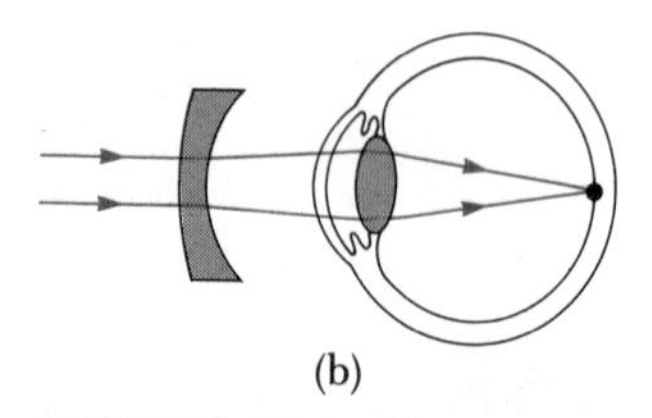

(b)

FIGURE 27.4 (a) A myopic eye (nearsightedness) is slightly longer than normal, hence the image of a distant object focuses in front of the retina. (b) The condition can be corrected with a diverging lens.

correct for farsightedness as an aid to reading and the upper half ground to correct a problem such as astigmatism.

The eye is also subject to several diseases. One disease, which usually occurs in old age, is the formation of **cataracts,** where the lens becomes partially or totally opaque. One remedy for cataracts is surgical removal of the lens. Another disease, called **glaucoma,** arises from an abnormal increase in fluid pressure inside the eyeball. The pressure increase can lead to a swelling of the lens and to strong myopia. There is a chronic form of glaucoma in which the pressure increase causes a reduction in blood supply to the retina. This can eventually lead to blindness because the nerve fibers of the retina eventually die. If the disease is discovered early enough, it can be treated with medicine or surgery.

Optometrists and ophthalmologists usually prescribe lenses measured in **diopters.**

The **power,** P, of a lens in diopters equals the inverse of the focal length in meters, that is $P = 1/f$.

For example, a converging lens whose focal length is $+20$ cm has a power of $+5$ diopters, and a diverging lens whose focal length is -40 cm has a power of -2.5 diopters.

EXAMPLE 27.2 Prescribing a Lens

The near point of an eye is 50 cm. (a) What focal length lens should be used so that the eye can clearly see an object 25 cm away?

Solution The thin lens equation (Eq. 25.11) enables us to solve this problem. We have placed our object at 25 cm, and we want our lens to form an image at the closest point that the eye can see clearly. This corresponds to the near point, 50 cm. Applying the thin lens equation, we have

$$\frac{1}{25\text{ cm}} + \frac{1}{(-50\text{ cm})} = \frac{1}{f}$$

$$f = 50\text{ cm}$$

Why did we use a negative sign for the image distance? Note that the focal length is positive, indicating the need for a converging lens to correct farsightedness problems such as this.

(b) What is the power of this lens?

Solution The power of the lens is the reciprocal of the focal length in meters:

$$P = \frac{1}{f} = \frac{1}{0.5\text{ m}} = 2\text{ diopters}$$

EXAMPLE 27.3 A Case of Nearsightedness

A particular nearsighted person is unable to see objects clearly when they are beyond 50 cm (the far point of the eye). What should the focal length of the lens prescribed to correct this problem be?

Solution The purpose of the lens in this instance is to "move" an object from infinity to a distance where it can be seen clearly. From the thin lens equation, we have

$$\frac{1}{s} + \frac{1}{s'} = \frac{1}{\infty} + \frac{1}{(-50\text{ cm})} = \frac{1}{f}$$

$$f = -50\text{ cm}$$

Why did we use a negative sign for the image distance? As you should have suspected, the lens must be a diverging lens (negative focal length) to correct nearsightedness.

EXERCISE What is the power of this lens?

Answer -2 diopters.

---

*The Ophthalmoscope

One of the earliest optical instruments to be used for medical purposes is the **ophthalmoscope,** invented almost a century ago. This device, which is simply a mirror with a hole in the middle, is used to examine the retina of the eye, and Figure 27.5 indicates how it works. Light rays from the source are reflected by a small mirror into the eye of the patient. If the eye is relaxed, the light is reflected from the retina and the rays emerge from the eye traveling in parallel lines. This parallel beam of light then passes through a small hole in the center of the mirror and into the eye of the examiner. This instrument can be used to examine the retina for abnormal conditions, such as detachment. It can also be used to give limited information about the refractive properties of the eye lens. If the lens has a problem associated with refraction, a set of lenses on a wheel attached to the mirror of the ophthalmoscope can be used to bring the retina into focus.

27.3 THE SIMPLE MAGNIFIER

The **simple magnifier** is one of the simplest and most basic of all optical instruments because it consists of only a single converging lens. As the name implies, this device is used to increase the apparent size of an object. Suppose an object is viewed at some distance s from the eye, as in Figure 27.6. Clearly, the size of the image formed at the retina depends on the angle θ

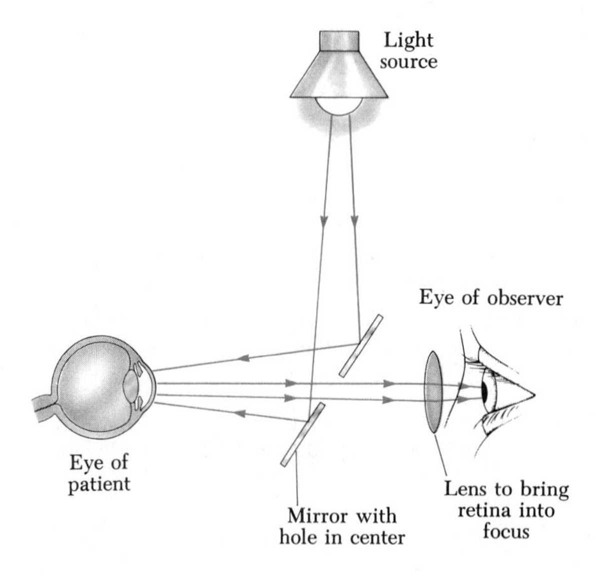

FIGURE 27.5 The ophthalmoscope.

subtended by the object at the eye. As the object moves closer to the eye, θ increases and a larger image is observed. However, a normal eye is unable to focus on an object closer than about 25 cm, the near point (Fig. 27.7a). Try it! Therefore, θ is maximum at the near point.

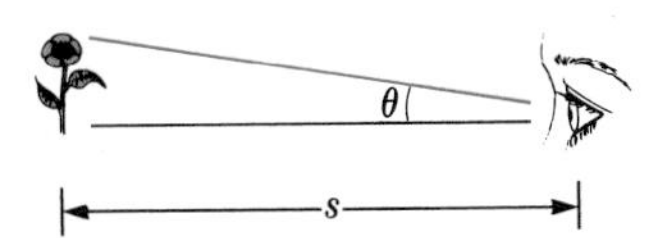

FIGURE 27.6 The size of the image formed on the retina depends on the angle θ subtended at the eye.

To further increase the apparent angular size of an object, a converging lens can be placed in front of the eye with the object located at point O, just inside the focal point of the lens, as in Figure 27.7b. At this location, the lens forms a virtual, erect, and enlarged image, as shown. Clearly, the lens increases the angular size of the object. We define the **angular magnification,** m, as the ratio of the angle subtended by the object with a lens in use (angle θ in Figure 27.7b) to that subtended by the object when it is placed at the near point with no lens (angle θ_0 in Figure 27.7a):

Angular magnification

$$m \equiv \frac{\theta}{\theta_0} \tag{27.2}$$

The angular magnification is a maximum when the image is at the near point of the eye, that is, when $s' = -25$ cm (see Fig. 27.7b). The object distance corresponding to this image distance can be calculated from the thin lens formula:

$$\frac{1}{s} + \frac{1}{-25\text{ cm}} = \frac{1}{f}$$

$$s = \frac{25f}{25 + f} \tag{27.3}$$

where f is the focal length in centimeters. Let us now make the small angle approximation as follows:

$$\tan\theta_0 \approx \theta_0 \approx \frac{h}{25} \quad \text{and} \quad \tan\theta \approx \theta \approx \frac{h}{s} \tag{27.4}$$

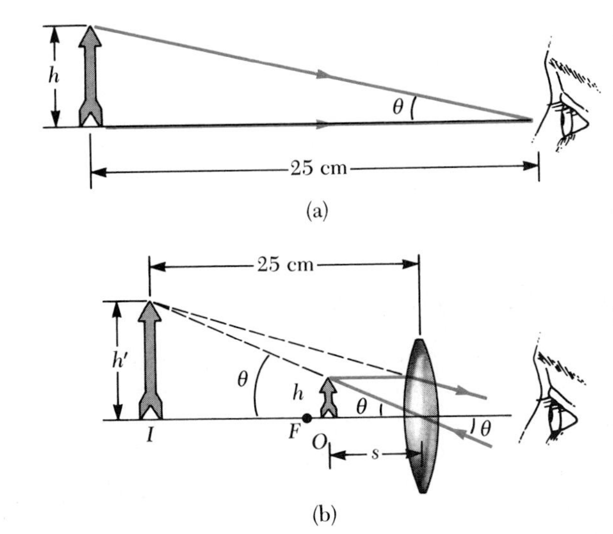

FIGURE 27.7 (a) An object placed at the near point (s = 25 cm) subtends an angle θ_0 at the eye, where $\theta_0 \approx h/25$. (b) An object placed near the focal point of a converging lens produces a magnified image, which subtends an angle $\theta \approx h'/25$ at the eye.

Thus, Equation 27.2 becomes

$$m = \frac{\theta}{\theta_0} = \frac{h/s}{h/25} = \frac{25}{s} = \frac{25}{25f/(25 + f)}$$

$$\boxed{m = 1 + \frac{25 \text{ cm}}{f}} \quad (27.5)$$

The magnification given by Equation 27.5 is the ratio of the angular size seen with the lens to the angular size seen when the object is viewed at the near point of the eye with no lens. Actually, the eye can focus on an image formed anywhere between the near point and infinity. However, the eye is more relaxed when the image is at infinity (Sec. 27.2). In order for the image formed by the magnifying lens to appear at infinity, the object has to be placed at the focal point of the lens, that is, $s=f$. In this case, Equation 27.4 becomes

$$\theta_0 \approx \frac{h}{25} \quad \text{and} \quad \theta \approx \frac{h}{f}$$

and the magnification is

$$\boxed{m = \frac{\theta}{\theta_0} = \frac{25 \text{ cm}}{f}} \quad (27.6)$$

With a single lens, it is possible to obtain angular magnifications up to about 4 without serious aberrations. Magnifications up to about 20 can be achieved by using a second lens to correct for aberrations.

EXAMPLE 27.4 Maximum Magnification of a Lens
What is the maximum magnification of a lens having a focal length of 10 cm, and what is the magnification of this lens when the eye is relaxed?

Solution The maximum magnification occurs when the image formed by the lens is located at the near point of the eye. Under these circumstances, Equation 27.5 gives us the magnification as

$$m = 1 + \frac{25 \text{ cm}}{f} = 1 + \frac{25 \text{ cm}}{10 \text{ cm}} = 3.5$$

When the eye is relaxed, the image is at infinity. In this case, we use Equation 27.6:

$$m = \frac{25}{f} = \frac{25 \text{ cm}}{10 \text{ cm}} = 2.5$$

27.4 THE COMPOUND MICROSCOPE

A simple magnifier provides only limited assistance in inspecting the minute details of an object. Greater magnification can be achieved by combining two lenses in a device called a compound microscope, a schematic diagram of which is shown in Figure 27.8a. It consists of an objective lens with a very short focal length, f_o (where $f_o < 1$ cm), and an ocular, or eyepiece lens, having a focal length, f_e, of a few centimeters. The two lenses are separated

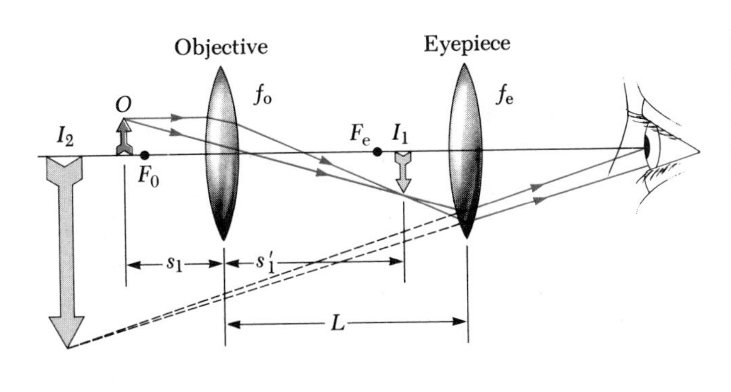

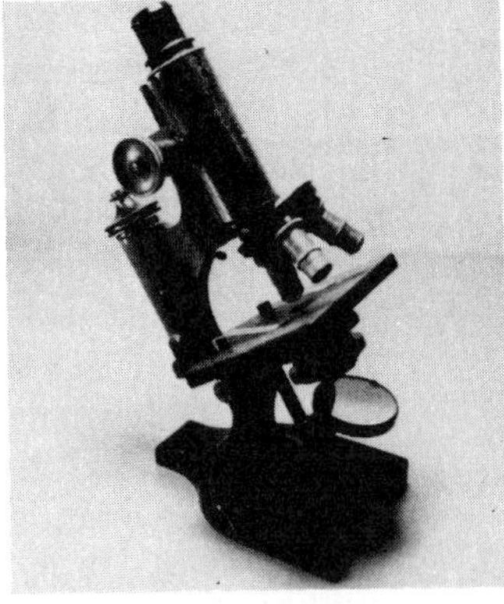

FIGURE 27.8 (a) Diagram of a compound microscope, which consists of an objective lens and an eyepiece, or ocular, lens. (b) An old-fashioned compound microscope. The three-objective turret allows the user to switch to several different powers of magnification. Combinations of oculars with different focal lengths and different objectives can produce a wide range of magnifications. (Photo by Lloyd Black)

by a distance L, where L is much greater than either f_o or f_e. The object, O, which is placed just outside the focal length of the objective, forms a real, inverted image at I_1, which is at or just inside the focal point of the eyepiece. The image at I_1 is real and much enlarged. (For clarity, the enlargement of I_1 is not shown in Figure 27.8.) The eyepiece, which serves as a simple magnifier, uses the image at I_1 as its object, and produces an image at I_2. The image which is seen by the eye at I_2 is virtual, inverted, and very much enlarged.

The lateral magnification, M_1, of the first image is $-s_1'/s_1$. Note in Figure 27.8 that s_1' is approximately equal to L. This occurs because the object is placed close to the focal point of the objective lens which ensures that the image formed will be far from the objective lens. Furthermore, since the object is very close to the focal point of the objective lens, $s_1 \approx f_0$. This gives a magnification for the objective of

$$M_1 = -\frac{s_1'}{s_1} \approx -\frac{L}{f_o}$$

The angular magnification of the eyepiece for an object (corresponding to the image at I_1) placed at the focal point is found from Equation 27.6 to be

$$m_e = \frac{25 \text{ cm}}{f_e}$$

The overall magnification of the compound microscope is defined as the product of the lateral and angular magnification:

$$\boxed{M = M_1 m_e = -\frac{L}{f_o}\left(\frac{25 \text{ cm}}{f_e}\right)} \qquad (27.7)$$

The negative sign indicates that the image is inverted with respect to the object.

The microscope has extended our vision into the previously unknown depths of incredibly small objects. The capabilities of this instrument have steadily increased with improved techniques in the precision grinding of lenses. A question that is often asked about microscopes is, "If you were extremely patient and careful, would it be possible to construct a microscope that would enable you to see an atom?" The answer to this equestion is no, as long as visible light is used to illuminate the object. The reason is that, in order to be seen, the object under a microscope must be at least as large as a wavelength of light. An atom is many times smaller then the wavelengths of visible light, and so its mysteries have to be probed using other techniques.

The wavelength dependence of the "seeing" ability of a wave can be illustrated by water waves set up in a bathtub in the following manner. Suppose you vibrate your hand in the water until waves having a wavelength of about 6 in. are moving along the surface. If you fix a small object, such as a toothpick, in the path of the waves, you will find that the waves are not disturbed appreciably by the toothpick but instead continue along their path, oblivious of the small object. Now suppose you fix a larger object, such as a toy sailboat, in the path of the waves. In this case, the waves are considerably "disturbed" by the object. In the first case, the toothpick is much smaller than the wavelength of the waves, and as a result the waves do not "see" the toothpick. In the second case, the toy sailboat is about the same size as the wavelength of the waves and hence the sailboat creates a disturbance. Light waves behave in this same general way. The ability of an optical microscope to view an object depends on the size of the object relative to the wavelength of the light used to observe it. Hence, one will never be able to observe atoms or molecules with such a microscope, since their dimensions are small ($\approx$ 0.1 nm) relative to the wavelength of the light ($\approx$ 500 nm).

EXAMPLE 27.5 Magnifications of a Microscope

A certain microscope has two possible objectives that can be used. One has a focal length of 20 mm, and the second has a focal length of 2 mm. Also available are two eyepieces of focal lengths 2.5 cm and 5 cm. If the length of the microscope is 18 cm, what magnifications are possible?

Solution The solution consists of applying Equation 27.7 to four different combinations of lenses. For the combination of the two long focal lengths, we have

$$M = -\frac{L}{f_o}\left(\frac{25\text{ cm}}{f_e}\right) = -\frac{18}{2}\left(\frac{25}{5}\right) = -45$$

The combination of the 20-mm objective and the 2.5-cm eyepiece gives

$$M = -\frac{18}{2}\left(\frac{25}{2.5}\right) = -90$$

The 2-mm and 5-cm combination produces

$$M = -\frac{18}{0.2}\left(\frac{25}{5}\right) = -450$$

and finally the two short focal lengths give

$$M = -\frac{18}{0.2}\left(\frac{25}{2.5}\right) = -900$$

27.5 THE TELESCOPE

There are two fundamentally different types of **telescopes,** both designed to aid in viewing distant objects, such as the planets in our solar system. The two classifications are (1) the **refracting telescope,** which uses a combination of lenses to form an image, and (2) the **reflecting telescope,** which uses a curved mirror and a lens to form an image.

First, let us consider the refracting telescope. In this device, two lenses are arranged such that the objective forms a real, inverted image of the distant object very near the focal point of the eyepiece (Fig. 27.9). Furthermore, the image at I_1 is formed at the focal point of the objective because the object is essentially at infinity. Hence, the two lenses are separated by a distance $f_o + f_e$, which corresponds to the length of the telescope's tube. The eyepiece finally forms, at I_2, an enlarged, inverted image of the image at I_1.

The angular magnification of the telescope is given by θ/θ_o, where θ_o is the angle subtended by the object at the objective and θ is the angle subtended by the final image. From the triangles in Figure 27.9, and for small angles, we have

$$\theta \approx \frac{h'}{f_e} \quad \text{and} \quad \theta_o \approx \frac{h'}{f_o}$$

Hence, the angular magnification of the telescope can be expressed as

$$m = \frac{\theta}{\theta_o} = \frac{h'/f_e}{h'/f_o} = \frac{f_o}{f_e} \tag{27.8}$$

This says that the angular magnification of a telescope equals the ratio of the objective focal length to the eyepiece focal length. Here again, the magnification is the ratio of the angular size seen with the telescope to the angular size seen with the unaided eye.

In some applications, such as observing nearby objects like the sun, moon, or planets, magnification is important. However, stars are so far away that they always appear as small points of light regardless of how much magnification is used. Large research telescopes used to study very distant objects

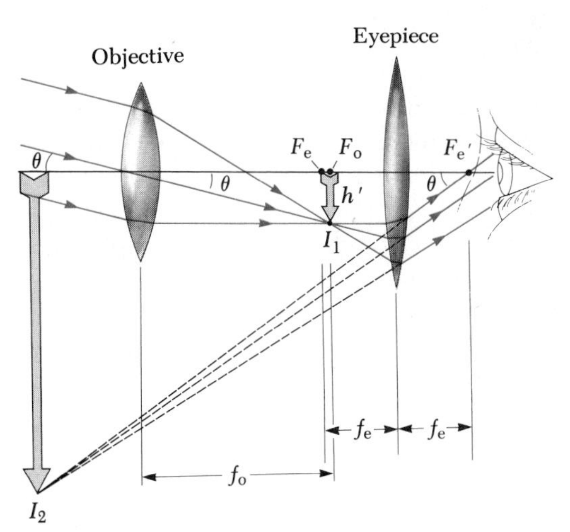

FIGURE 27.9 Diagram of a refracting telescope, with the object at infinity.

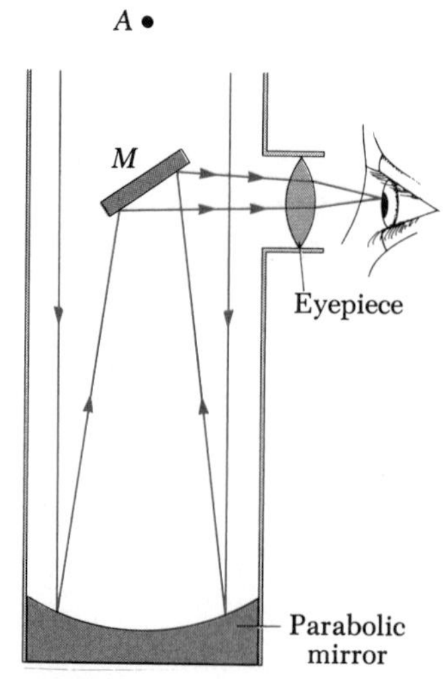

FIGURE 27.10 A reflecting telescope with a Newtonian focus.

must have a large diameter in order to gather as much light as possible. It is difficult and expensive to manufacture large lenses for refracting telescopes. Another difficulty with large lenses is that their large weight leads to sagging, which is an additional source of aberration. These problems can be partially overcome by replacing the objective lens with a reflecting, concave mirror. Figure 27.10 shows the design for a typical reflecting telescope. Incoming light rays pass down the barrel of the telescope and are reflected by a parabolic mirror at the base. These rays converge toward point A in the figure, where an image would be formed. However, before this image is formed, a small flat mirror at point M reflects the light toward an opening in the side of the tube that passes into an eyepiece. This particular design is said to have a Newtonian focus because it was Newton who developed it. Note that the light never passes through glass in the reflecting telescope (except through the small eyepiece). As a result, problems associated with chromatic aberration are virtually eliminated. Also, difficulties arising from spherical aberration are reduced because of the parabolic shape of the mirror.

The largest telescope in the world is the 6-m-diameter reflecting telescope on Mount Pastukhov in the Caucasus, Soviet Union. The largest reflecting telescope in this country is the 5-m-diameter instrument on Mount Palomar in California. In contrast, the largest refracting telescope in the world, which is located at the Yerkes Observatory in Williams Bay, Wisconsin, has a diameter of only 1 m.

EXAMPLE 27.6 Magnification of a Reflecting Telescope

A reflecting telescope has an 8-in.-diameter objective mirror with a focal length of 1500 mm. What is the magnification of this telescope when an eyepiece having an 18-mm focal length is used?

Solution The equation for finding the magnification of a reflector is the same as that for a refractor. Thus, Equation 27.8 gives

$$m = \frac{f_o}{f_e} = \frac{1500 \text{ mm}}{18 \text{ mm}} = 83$$

27.6 RESOLUTION OF SINGLE-SLIT AND CIRCULAR APERTURES

The ability of optical systems such as microscopes and telescopes to distinguish between closely spaced objects is limited because of the wave nature of light. To understand this difficulty, consider Figure 27.11, which shows two light sources far from a narrow slit of width a. The sources can be considered as two point sources, S_1 and S_2, that are *not* coherent. For example, they could be two distant stars. If no diffraction occurred, one would observe two distinct bright spots (or images) on the screen at the right in the figure. However, because of diffraction, each source is imaged as a bright central region flanked by weaker bright and dark rings. What is observed on the screen is the sum of two diffraction patterns, one from S_1, and the other from S_2.

If the two sources are separated such that their central maxima do not overlap, as in Figure 27.11a, their images can be distinguished and are said to be *resolved*. If the sources are close together, however, as in Figure

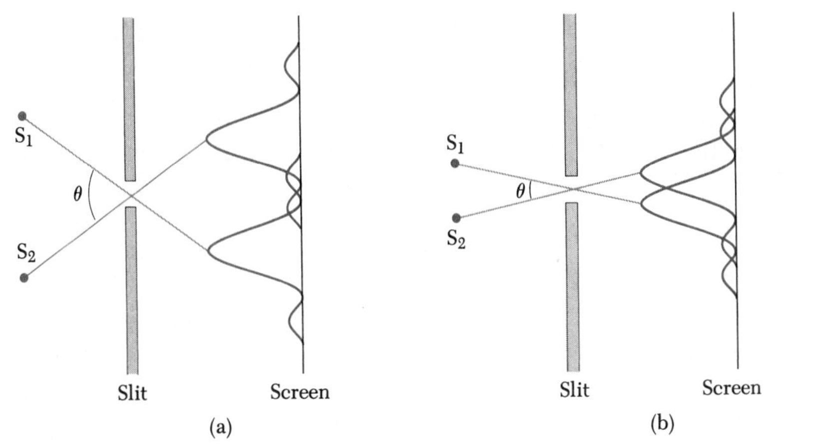

FIGURE 27.11 Two point sources at some distance from a small aperture each produce a diffraction pattern. (a) The angle subtended by the sources at the aperture is large enough so that the diffraction patterns are distinguishable. (b) The angle subtended by the sources is so small that the diffraction patterns are not distinguishable. (Note that the angles are greatly exaggerated.)

27.11b, the two central maxima may overlap and the images are *not resolved*. To decide when two images are resolved, the following condition applied to their diffraction patterns is often used:

When the central maximum of one image falls on the first minimum of another image, the images are said to be just resolved. This limiting condition of resolution is known as **Rayleigh's criterion.**

Figure 27.12 shows the diffraction patterns for three situations. When the objects are far apart, their images are well resolved (Fig. 27.12a). The

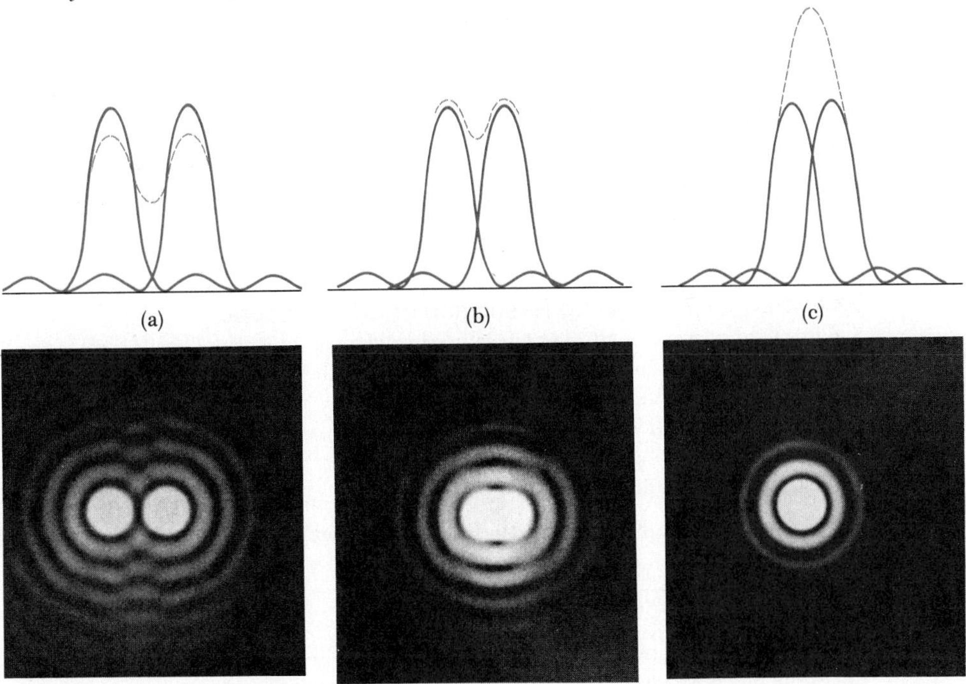

FIGURE 27.12 The diffraction patterns of two point sources (solid curves) and the resultant pattern (dashed curve) for various angular separations of the sources. (a) The sources are far apart, and the patterns are well resolved. (b) The sources are closer together, and their patterns are just resolved. (c) The sources are so close together that their patterns are not resolved. (From M. Cagnet, M. Francon, and J.C. Thierr, *Atlas of Optical Phenomena*, Berlin, Springer-Verlag, 1962, plate 16).

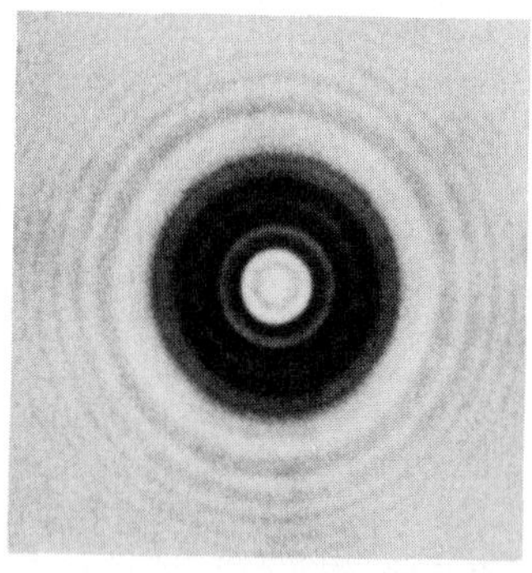

FIGURE 27.13 The Fresnel diffraction pattern of a circular aperture consists of a central bright disk surrounded by concentric bright and dark rings. (From M. Cagnet, M. Francon, and J.C. Thierr, *Atlas of Optical Phenomena,* Berlin, Springer-Verlag, 1962, plate 34)

images are just resolved when their angular separation satisfies Rayleigh's criterion (Fig. 27.12b). Finally, the images are not resolved in Figure 27.12c.

From Rayleigh's criterion, we can determine the minimum angular separation, θ_m, subtended by the source at the slit such that their images will be just resolved. In Chapter 26, we found that the first minimum in a single-slit diffraction pattern occurs at the angle that satisfies the relationship

$$\sin \theta = \frac{\lambda}{a}$$

where a is the width of the slit. According to Rayleigh's criterion, this expression gives the smallest angular separation for which the two images will be resolved. Because $\lambda \ll a$ in most situations, $\sin \theta$ is small and we can use the approximation $\sin \theta \approx \theta$. Therefore, the limiting angle of resolution for a slit of width a is

$$\boxed{\theta_m \approx \frac{\lambda}{a}} \tag{27.9}$$

where θ_m is expressed in radians. Hence, the angle subtended by the two sources at the slit must be *greater* than λ/a if the images are to be resolved.

Many optical systems use circular apertures rather than slits. The diffraction pattern of a circular aperture, illustrated in Figure 27.13, consists of a central circular bright disk surrounded by progressively fainter rings. Analysis shows that the limiting angle of resolution of the circular aperture is

Resolution for a circular aperture

$$\boxed{\theta_m = 1.22 \frac{\lambda}{D}} \tag{27.10}$$

where D is the diameter of the aperture. Note that Equation 27.10 is similar to Equation 27.9 except for the factor of 1.22, which arises from a complex mathematical analysis of diffraction from the circular aperture.

EXAMPLE 27.7 Limiting Resolution of a Microscope

Sodium light of wavelength 589 nm is used to view an object under a microscope. If the aperture of the objective is 0.9 cm, (a) find the limiting angle of resolution. (b) Using visible light of any wavelength you desire, what is the maximum limit of resolution for this microscope? (c) Suppose water of index of refraction 1.33 fills the space between the object and the objective. What effect would this have on the resolving power of the microscope?

Solution (a) From Equation 27.10, we find the limiting angle of resolution to be

$$\theta_m = 1.22 \left(\frac{589 \times 10^{-9} \text{ m}}{0.9 \times 10^{-2} \text{ m}} \right) = 8 \times 10^{-5} \text{ rad}$$

This means that any two points on the object subtending an angle less than 8×10^{-5} rad at the objective cannot be distinguished in the image.

(b) To obtain the maximum angle of resolution, we have to use the shortest wavelength available in the visible spectrum. Violet light of wavelength 400 nm gives us a limiting angle of resolution of

$$\theta_m = 1.22 \left(\frac{400 \times 10^{-9} \text{ m}}{0.9 \times 10^{-2} \text{ m}} \right) = 5.4 \times 10^{-5} \text{ rad}$$

(c) In this case, the wavelength of the sodium light in the water is found by $\lambda_w = \lambda_a/n$ (Chap. 23). Thus, we have

$$\lambda_w = \frac{\lambda_a}{n} = \frac{589 \text{ nm}}{1.33} = 443 \text{ nm}$$

The limiting angle of resolution at this wavelength is

$$\theta_m = 1.22 \left(\frac{443 \times 10^{-9} \text{ m}}{0.9 \times 10^{-2} \text{ m}} \right) = 6 \times 10^{-5} \text{ rad}$$

EXAMPLE 27.8 Resolution of a Telescope

The Hale telescope at Mount Palomar has a diameter of 200 in. What is its limiting angle of resolution at a wavelength of 600 nm?

Solution Because $D = 200$ in. $= 5.08$ m and $\lambda = 6 \times 10^{-7}$ m, Equation 27.10 gives

$$\theta_m = 1.22 \frac{\lambda}{D} = 1.22 \left(\frac{6 \times 10^{-7} \text{ m}}{5.08 \text{ m}} \right) = 1.44 \times 10^{-7} \text{ rad} = 0.03 \text{ s of arc}$$

Therefore, any two stars that subtend an angle greater than or equal to this value will be resolved (assuming ideal atmospheric conditions).

It is interesting to compare this value with the resolution of a large radiotelescope, such as the system at Arecibo, Puerto Rico, which has a diameter of 1000 ft (305 m). This telescope detects radio waves at a wavelength of 0.75 m. The corresponding minimum angle of resolution is calculated to be 3×10^{-3} rad (10 min 19 s of arc), which is more than 10,000 times larger than the calculated minimum angle for the Hale telescope.

The Hale telescope can never reach its diffraction limit. Instead, its limiting angle of resolution is always set by atmospheric blurring. This seeing limit is usually about 1 s of arc and is never smaller than about 0.1 s of arc. (This is one of the reasons for the current interest in a large space telescope.)

EXAMPLE 27.9 Compare the Two Telescopes

Two telescopes have the following properties:

Telescope	Diameter of Objective (in.)	Focal Length of Objective (mm)	Focal Length of Eyepiece (mm)
A	6	1000	6
B	8	1250	25

(a) Which has the better resolving power? (b) Which has the greater light-gathering ability? (c) Which will produce the larger magnification?

Solution (a) The telescope with the larger objective has the greatest ability to discriminate between nearby objects. Thus, we should choose telescope B.

(b) The telescope with the larger objective is able to collect more light. Hence, B is again the choice.

(c) The magnification of telescope A is

$$m = \frac{f_o}{f_e} = \frac{1000 \text{ mm}}{6 \text{ mm}} = 167$$

and for telescope B, we have

$$m = \frac{1250 \text{ mm}}{25 \text{ mm}} = 50$$

Thus, select telescope A if magnification is the most important consideration.

27.7 THE MICHELSON INTERFEROMETER

The camera and telescope are examples of optical instruments that are common in our daily activities. In contrast, the Michelson interferometer is not familiar to most people. It is, however, an optical instrument that has great scientific importance. The interferometer, invented by the American physicist A. A. Michelson (1852–1931), is an ingenious device that splits a light beam into two parts and then recombines them to form an interference pattern. The device is used for obtaining accurate length measurements.

A schematic diagram of the interferometer is shown in Figure 27.14. A beam of light provided by a monochromatic source is split into two rays by a partially silvered mirror M inclined at an angle of 45° relative to the incident light beam. One ray is reflected vertically upward to mirror M_1, and the second ray is transmitted horizontally through mirror M to mirror M_2. Hence, the two rays travel separate paths, l_1 and l_2. After reflecting from mirrors M_1 and M_2, the two rays eventually recombine to produce an interference pattern, which can be viewed through a telescope. The glass plate P, equal in thickness to mirror M, is placed in the path of the horizontal ray in order to insure that the two rays travel the same distance through glass.

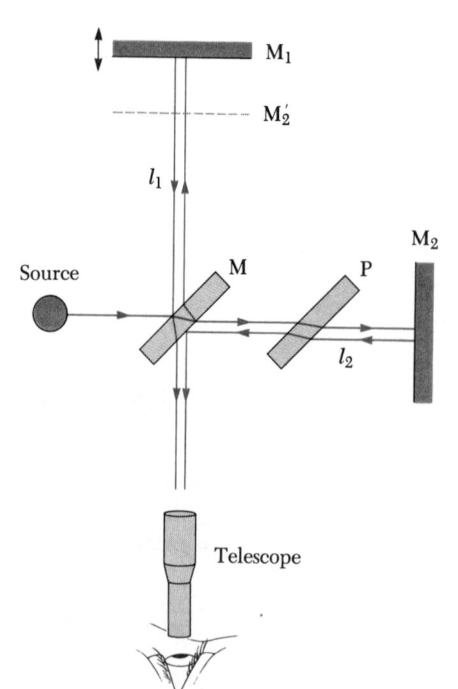

FIGURE 27.14 Diagram of the Michelson interferometer. A single beam is split into two rays by the half-silvered mirror M. The path difference between the two rays is varied with the adjustable mirror M_1.

The interference pattern for the two rays is determined by the difference in their path lengths. When the two rays are viewed as shown, the image of M_2 is at M_2' parallel to M_1. Hence, the space between M_2' and M_1 forms the equivalent of a parallel air film. The effective thickness of the air film is varied by moving mirror M_1 in the direction indicated by the arrows in Figure 27.14 with a finely threaded screw. If one of the mirrors is tipped slightly with respect to the other, the thin film between the two is wedge shaped, and an interference pattern consisting of parallel fringes is set up, as described in Chapter 26, Example 26.4. Now suppose we focus on one of the dark lines with the crosshairs of a telescope. As the mirror M_1 is moved to make the path length l_1 longer, the thickness of the wedge increases. When the thickness increases by $\lambda/4$, the destructive interference which initially produced the dark fringe has changed to constructive interference, and we now observe a bright fringe at the location of the crosshairs. The term *fringe shift* is used to describe the change in a fringe from dark to light or light to dark. Thus, successive light and dark fringes are formed each time M_1 is moved a distance of $\lambda/4$. The wavelength of light can be measured by counting the number of fringe shifts for a measured displacement of M_1. Conversely, if the wavelength is accurately known (as with a laser beam), mirror displacements can be measured to within a fraction of the wavelength. Because the interferometer can measure displacements precisely, it is often used to make highly accurate measurements of the dimensions of mechanical components.

If the mirrors are perfectly aligned, rather than tipped with respect to one another, the path difference is slightly different for different angles of view. This results in an interference pattern that resembles Newton's rings. The pattern can be used in a fashion similar to that described for tipped mirrors. To do so, one concentrates on the center spot in the interference pattern. For example, suppose the spot is initially dark, indicating that destructive interference is occurring. If M_1 is now moved by a distance of $\lambda/4$, this central spot changes to a light region, corresponding to a fringe shift.

27.8 THE DIFFRACTION GRATING

The diffraction grating, a very useful device for analyzing light sources, consists of a large number of equally spaced parallel slits. A grating can be made by scratching parallel lines on a glass plate with a precision machining technique. The spaces between the scratches are transparent to the light and hence act as separate slits. A typical grating contains several thousand lines per centimeter. For example, a grating ruled with 5000 lines/cm has a slit spacing d equal to the inverse of this number, hence $d = (1/5000)$ cm $= 2 \times 10^{-4}$ cm.

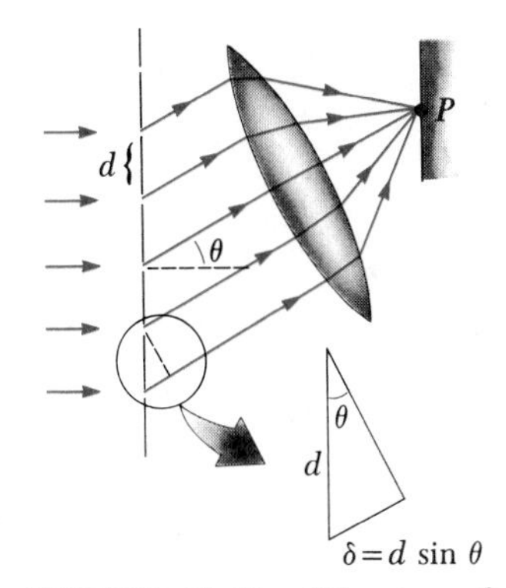

FIGURE 27.15 Side view of a diffraction grating. The slit separation is d, and the path difference between adjacent slits is $d \sin \theta$.

A schematic diagram of a section of a plane diffraction grating is illustrated in Figure 27.15. A plane wave is incident from the left, normal to the plane of the grating. A converging lens can be used to bring the rays together at point P. The intensity of the pattern on the screen is the result of the combined effects of interference and diffraction. Each slit produces diffraction, and the diffracted beams in turn interfere with each other to produce the pattern. Moreover, each slit acts as a source of waves, where all waves start at the slits in phase. However, for some arbitrary direction θ measured

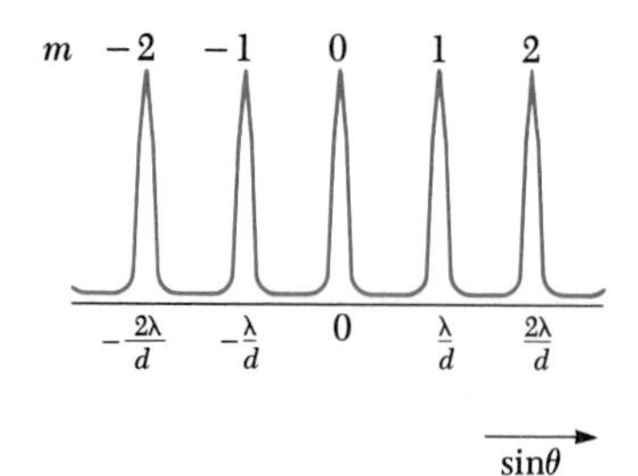

FIGURE 27.16 Intensity versus sin θ for the diffraction grating. The zeroth-, first-, and second-order principal maxima are shown.

from the horizontal, the waves must travel *different* path lengths before reaching a particular point P on the screen. From Figure 27.15, note that the path difference between waves from any two adjacent slits is equal to $d \sin \theta$. If this path difference equals one wavelength or some integral multiple of a wavelength, waves from all slits will be in phase at P and a bright line will be observed. Therefore, the condition for **maxima** in the interference pattern at the angle θ is

$$\boxed{d \sin \theta = m\lambda} \qquad m = 0, 1, 2, \ldots \tag{27.11}$$

This expression can be used to calculate the wavelength from a knowledge of the grating space and the angle of deviation, θ. The integer m represents the **order number** of the diffraction pattern. If the incident radiation contains several wavelengths, each wavelength will deviate through a specific angle which can be found from Equation 27.11. All wavelengths are focused at $\theta = 0$, corresponding to $m = 0$. This is called the *zeroth-order maximum.* The *first-order maximum,* corresponding to $m = 1$, is observed at an angle that satisfies the relationship $\sin \theta = \lambda/d$; the *second-order maximum,* corresponding to $m = 2$, is observed at a larger angle θ, and so on. A sketch of the intensity distribution for the various orders produced by a diffraction grating is shown in Figure 27.16. Let us compare the diffraction pattern of the grating with the pattern produced by two slits (see Young's double slit experiment discussed in Chapter 26). The two patterns differ as follows: (1) The bright lines of the diffraction grating are sharper (narrower), and the dark areas are broader. (2) The lines of the two-slit pattern are more intense because the light has to travel through two slits rather than many slits.

A simple arrangement that can be used to measure the various orders of a diffraction pattern is shown in Figure 27.17. This is a form of diffraction grating spectrometer. The light to be analyzed passes through a slit and is formed into a parallel beam by a lens. The light then strikes the grating at a 90° angle. The diffracted light leaves the grating at angles that satisfy Equation 27.11. A telescope is used to view the image of the slit. The wavelength can be determined by measuring the angles at which the images of the slit appear for the various orders.

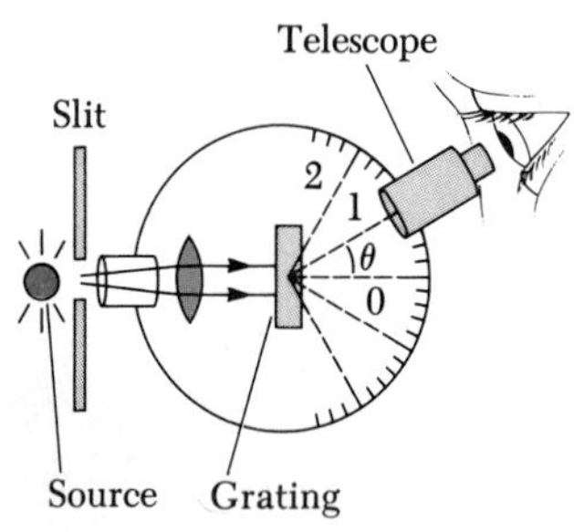

FIGURE 27.17 Diagram of a diffraction grating spectrometer. The collimated beam incident on the grating is diffracted into the various orders at the angles θ which satisfy the equation $d \sin \theta = m\lambda$, where $m = 0, 1, 2, 3, \ldots$

EXAMPLE 27.10 The Orders of a Diffraction Grating

Monochromatic light from a helium-neon laser ($\lambda = 632.8$ nm) is incident normally on a diffraction grating containing 6000 lines/cm. Find the angles at which one would observe the first-order maximum, the second-order maximum, and so forth.

Solution First we must calculate the slit separation, which is equal to the inverse of the number of lines per centimeter:

$$d = \frac{1}{6000} \text{ cm} = 1.667 \times 10^{-4} \text{ cm} = 1667 \text{ nm}$$

For the first-order maximum ($m = 1$), we get

$$\sin \theta_1 = \frac{\lambda}{d} = \frac{632.8 \text{ nm}}{1667 \text{ nm}} = 0.3796$$

$$\theta_1 = 22.31°$$

For $m = 2$ we find that

$$\sin \theta_2 = \frac{2\lambda}{D} = \frac{2(632.8 \text{ nm})}{1667 \text{ nm}} = 0.7592$$

$$\theta_2 = 49.39°$$

However, for $m = 3$ we find that $\sin \theta_3 = 1.139$. Since $\sin \theta$ cannot exceed unity, this does not represent a realistic solution. Hence, only zeroth-, first-, and second-order maxima will be observed for this situation.

Resolving Power of the Diffraction Grating

The diffraction grating is most useful for making accurate wavelength measurements. Like the prism, the diffraction grating can be used to disperse a spectrum into its components. Of the two devices, the grating is more precise if one wants to distinguish between two closely spaced wavelengths. We say that the grating spectrometer has a higher resolution than the prism spectrometer. If λ_1 and λ_2 are two nearly equal wavelengths between which the spectrometer can just barely distinguish, the **resolving power,** R, of the grating is defined as

$$R \equiv \frac{\lambda}{\lambda_2 - \lambda_1} = \frac{\lambda}{\Delta\lambda} \tag{27.12}$$

where $\lambda \approx \lambda_1 \approx \lambda_2$ and $\Delta\lambda = \lambda_2 - \lambda_1$. Thus, we see that a grating with a high resolving power can distinguish small differences in wavelength. Furthermore, if N lines of the grating are illuminated, it can be shown that the resolving power in the mth-order diffraction equals the product Nm:

$$R = Nm \tag{27.13}$$

Resolving power of a grating

Thus, the resolving power increases with increasing order number. Furthermore, R is large for a grating with a large number of illuminated slits. Note that for $m = 0$, $R = 0$, which signifies that *all wavelengths are indistinguishable* for the zeroth-order maximum (all wavelengths fall at the same point on the screen). However, consider the second-order diffraction pattern of a grating that has 5000 rulings illuminated by the light source. The resolving power of such a grating in second order is $R = 5000 \times 2 = 10{,}000$. Therefore, the *minimum* wavelength separation between two spectral lines that can be just resolved, assuming a mean wavelength of 600 nm, is $\Delta\lambda = \lambda/R = 6 \times 10^{-2}$ nm. For the third-order principal maximum $R = 15{,}000$ and $\Delta\lambda = 4 \times 10^{-2}$ nm, and so on.

EXAMPLE 27.11 Resolving the Sodium Spectral Lines

Two strong lines in the spectrum of sodium have wavelengths of 589.00 nm and 589.59 nm. (a) What must the resolving power of a grating be in order to distinguish these wavelengths?

Solution From Equation 27.12, we find

$$R = \frac{\lambda}{\Delta\lambda} = \frac{589 \text{ nm}}{589.59 \text{ nm} - 589.00 \text{ nm}} = \frac{589}{0.59} = 998$$

(b) In order to resolve these lines in the second-order spectrum, how many lines of the grating must be illuminated?

Solution From Equation 27.13 and the results to (a), we find that

$$N = \frac{R}{m} = \frac{998}{2} = 499 \text{ lines}$$

SUMMARY

The light-concentrating power of a lens of focal length f and diameter D is determined by the ***f*-number**, defined as

$$f\text{-number} \equiv \frac{f}{D} \tag{27.1}$$

The smaller the f-number of a lens, the greater the brightness of the image formed.

Hyperopia (farsightedness) is a defect of the eye that results either when the eyeball is too short or when the ciliary muscle is unable to change the shape of the lens enough to form a properly focused image. **Myopia** (nearsightedness) is caused either when the eye is longer than normal or when the maximum focal length of the lens is insufficient to produce a clearly focused image on the retina.

The **power** of a lens in **diopters** equals the inverse of the focal length in meters.

The **angular magnification of a lens** is defined as

$$m \equiv \frac{\theta}{\theta_o} \tag{27.2}$$

where θ is the angle subtended by an object at the eye with a lens in use and θ_0 is the angle subtended by the object when it is placed at the near point of the eye and no lens is used. The **maximum magnification of a lens** is

$$m = 1 + \frac{25 \text{ cm}}{f} \tag{27.5}$$

When the eye is relaxed, the magnification is

$$m = \frac{25 \text{ cm}}{f} \tag{27.6}$$

The overall **magnification of a compound microscope** of length L is equal to the product of the magnification produced by the objective of focal length f_o and the magnification produced by the eyepiece of focal length f_e:

$$M = -\frac{L}{f_o}\left(\frac{25 \text{ cm}}{f_e}\right) \tag{27.7}$$

The **magnification of a telescope** is given by

$$m = \frac{f_o}{f_e} \tag{27.8}$$

where f_o is the focal length of the objective and f_e is the focal length of the eyepiece.

Two images are said to be **resolved** when the central maximum of the diffraction pattern for one image falls on the first minimum of the other image. This limiting condition of resolution is known as **Rayleigh's criterion.** The limiting angle of resolution for a **slit** of width a is

$$\theta_m \approx \frac{\lambda}{a} \tag{27.9}$$

The limiting angle of resolution of a **circular aperture** is

$$\theta_m = 1.22\frac{\lambda}{D} \tag{27.10}$$

where D is the diameter of the aperture.

A **diffraction grating** consists of a large number of equally spaced, identical slits. The condition for **maximum intensity** in the interference pattern of a diffraction grating is

$$d \sin\theta = m\lambda \qquad m = 0, 1, 2, \ldots \tag{27.11}$$

where d is the spacing between adjacent slits and m is the order number of the diffraction pattern. The **resolving power** of a diffraction grating in the mth order is

$$R = Nm \tag{27.13}$$

where N is the number of rulings on the grating.

ADDITIONAL READING

J. Bahcall and L. Spitzer, Jr., "The Space Telescope," *Sci. American*, July 1982, p. 40.

E. Bandana, "The Mystery of Myopia," *The Sciences*, Nov./Dec. 1985, p. 46.

D. Marr, *Vision*, W. H. Freeman, 1982.

C. R. Michael, "Retinal Processing of Visual Images," *Sci. American*, May 1969, p. 105.

U. Neisser, "The Processes of Vision," *Sci. American*, September 1968, p. 204.

W. H. Price, "Photographic Lens," *Sci. American*, August 1976, p. 72.

M. Ruiz, "Camera Optics," *The Physics Teacher*, September 1982, p. 372.

R. S. Shankland, "Michelson and His Interferometer," *Physics Today*, April 1976, p. 72.

G. Wald, "Eye and Camera," *Sci. American*, August 1950, p. 32.

T. D. Walker, *Light and Its Uses*, W. H. Freeman, 1980.

QUESTIONS

1. Compare and contrast the eye and a camera. What parts of the camera correspond to the iris, the retina, and the cornea of the eye?
2. Estimate the shutter speed and f-number necessary to photograph a center driving for the goal in a well-lighted basketball gymnasium. Repeat for a racecar

during the final lap on a cloudy day. Why will the latter probably be blurred anyway?

3. If you want to use a converging lens to set fire to a piece of paper, why should the light source be farther away than the focal point of the lens?
4. If you want to examine the fine detail in an object, could you hold the object farther away than the focal length of a lens used as a simple magnifier? Why or why not?
5. Explain why it is theoretically impossible to see an object as small as an atom regardless of how good a light microscope is used.
6. A pinhole camera can be constructed by punching a small hole in one side of a cardboard box. If the opposite side is cut out and replaced with tissue paper, the image formed by the box can be examined. When the box is pointed toward an open window, you should be able to see an image on the tissue paper. No lens is involved here. Can you explain why the image is formed?
7. What difficulty would astronauts encounter in finding their landing site on the moon if astronomers were not aware of the properties of the telescope?
8. Large telescopes are usually reflecting rather than refracting. List several reasons why this is true.
9. The optic nerve and the brain invert the image formed on the retina. Why do we not see everything upside down?
10. Assuming that the headlights of a car are point sources, estimate the maximum distance from an observer to the car at which the headlights are distinguishable from each other.
11. The diffraction grating effect is easily observed with everyday equipment. For example, a long-playing record can be held so that light is reflected from it at a small glancing angle. When the record is held this way, various colors in the reflected light can be seen. Explain how this works.
12. In an experiment similar to that discussed in Question 11, a laser beam incident at a shallow angle on a machinist's ruler that has a finely calibrated scale will give rise to a diffraction pattern on a screen. Discuss how you can use this technique to obtain a measure of the wavelength of the laser light.
13. A classic science-fiction story, "The Invisible Man," tells of a person who becomes invisible by changing the index of refraction of his body to that of air. This story has been criticized by students who know how the eye works by claiming that the invisible man would be unable to see. On the basis of your knowledge of the eye, could he see or not?
14. Explain why you could not use a diverging lens as a simple magnifier.
15. Can a diffraction grating be used for infrared, ultraviolet, or *x*-rays? Could one use a prism to examine the spectrum of these parts of the electromagnetic spectrum?
16. In what respect are the optical system of a microscope and a telescope (a) similar and (b) different?

PROBLEMS

Section 27.1 The Camera

1. The focal length of the lens in a simple camera is 10 cm, and the size of the image formed on the film is to be 35 mm high. How far would a person 2 m tall have to stand from the camera so that the image will fit on the film?

2. A camera used by a professional photographer to shoot portraits has a focal length of 25 cm. The photographer takes a portrait of a person 1.5 m in front of the camera. Where will the image be formed, and what is the magnification?

3. A certain lens forms a real image 15 cm to its right. The object is 20 cm to the left of the lens. If the lens has a diameter of 2 cm, find its *f*-number.

4. The *f*-numbers of a simple camera are *f*/4, *f*/8, and *f*/16. What are the possible aperture openings for this camera if the focal length is 12 cm?

5. A camera is found to give proper film exposure when it is set at *f*/16 and the shutter is open for 1/32 s. Determine the correct exposure time if a setting of *f*/8 is used. Assume the lighting conditions are the same.

6. A camera is being used with the correct exposure at *f*/4 and a shutter speed of 1/32 s. In order to "stop" a fast-moving subject, the shutter speed is changed to 1/256 s. Find the new *f*-number setting that should be used to maintain satisfactory exposure, assuming no change in lighting conditions.

Section 27.2 The Eye

7. The near point of an eye is 100 cm. (a) What focal length lens should be used so that the eye can see clearly an object 25 cm in front of it? (b) What is the power of the lens?

8. A certain farsighted person uses lenses with a power of 2.9 diopters. What is her near point?

9. A particular nearsighted person is unable to see objects clearly when they are beyond 100 cm (the far point of the eye). What power lens should be used to correct the eye?

10. An individual is nearsighted such that his near point is 13 cm and his far point is 20 cm. (a) What lens power is needed to correct his nearsightedness? (b) When the lenses are in use, what is the person's near point?

11. A person required a lens with a power of −3 diopters in order to see distant objects clearly. What is the far point of this person's eye?

12. A person is to be fitted with bifocals. She can see clearly when the object is between 30 cm and 1.5 m from the eye. (a) The upper portion of the bifocals should be designed to enable her to see distant objects clearly. What power should the upper portion have? (b) The lower portion of the bifocals should enable her to see objects comfortably at 25 cm. What power should the lower portion have?

Section 27.3 The Simple Magnifier

13. A biology student uses a simple magnifier to examine the structural features of the wing of an insect. The lens is mounted in a frame 3.5 cm above the work surface, and the image is formed 25 cm from the eye. What is the focal length of the lens?

14. A stamp collector uses a 7.5-cm focal length lens as a simple magnifier. The virtual image produced is formed at the normal near point (25 cm). (a) How far from the lens should the stamp be placed, and (b) what is the expected magnification?

15. (a) What is the angular magnification of a lens having a focal length of 25 cm? (b) What is the magnification of this lens when the eye is relaxed?

• **16.** A boy scout uses a certain lens to set a fire by using the sun as an object and placing some kindling 5 cm from the lens. The boy scout has a near point of 15 cm. When used as a simple magnifier, (a) what is the maximum magnification that can be achieved, and (b) what is the magnification when the eye is relaxed? (*Caution:* The equations derived in the text for a simple magnifier assume a "normal" eye.)

Section 27.4 The Compound Microscope

Section 27.5 The Telescope

17. The desired overall magnification of a compound microscope is 140. The objective alone produces a lateral magnification of 12. Determine the required focal length of the eyepiece. Assume the final image will be 25 cm from the eye.

18. The length of microscope tube is 15 cm. The focal length of the objective lens is 1 cm, and the focal length of the eyepiece is 2.5 cm. What is the magnification of the microscope, assuming that it is adjusted so the eye is relaxed?

19. An astronomical telescope has an objective with a focal length of 75 cm and an eyepiece with a focal length of 4 cm. What is the magnifying power of this instrument?

20. A telescope adjusted for very distant objects has an objective with a focal length of 100 cm and an eyepiece with a focal length of 10 cm. A bee buzzes in front of the telescope at a distance of 150 cm from the objective. (a) Use the thin lens equation twice (once for each lens) to find the location of the final image of the bee. (b) On the basis of your answer to (a), could you see the bee through the telescope? Why or why not?

21. A microscope is to have a magnification of 800. If the tube length is 15 cm, the focal length of the objective is 0.5 cm, what focal length eyepiece should be selected?

22. The objective lens in a microscope with a 20-cm long tube has a magnification of 50 and the eyepiece has a magnification of 20. What is the focal length (a) of the objective, and (b) of the eyepiece? (c) What is the overall magnification of the microscope?

23. What is the magnification of a telescope that uses a 2 diopter objective and a 30 diopter eyepiece?

24. A certain telescope has a 5-inch diameter and an objective lens of focal length 1250 mm. (a) What is the *f*-number of this lens? (b) What is the magnification of this telescope when used with a 25-mm eyepiece?

• **25.** A certain telescope has an objective of focal length 1500 cm. If one uses the moon as an object, a 1-cm length of the image formed by the objective corresponds to what distance in miles on the moon? Assume the earth-moon distance is 3.8×10^8 m.

• **26.** A person decides to use an old pair of eyeglasses to make some optical instruments. He knows that the near point in his left eye is 50 cm and that the near point in his right eye is 100 cm. (a) What is the maximum magnification of the telescope he produces? (b) If he places the lens 10 cm apart, what is the maximum magnification of the microscope he can produce? (Go back to basics and use the thin-lens equation to solve part (b).)

Section 27.6 Resolution of Single-Slit and Circular Apertures

27. If the distance from the earth to the moon is 250,000 mi, what diameter telescope objective would be required to resolve a moon crater 300 m in diameter? Assume a wavelength of 500 nm.

28. Two telescopes have the following characteristics. Telescope A has a diameter of 5 in, a focal length for the objective of 1200 mm and is used with an 18 mm eyepiece. Telescope B has a diameter of 3 in, a focal length for the objective of 1000 mm, and is used with a 25 mm eyepiece. (a) What is the limiting angle of resolution for each telescope when used with 600 nm light? (b) What is the magnification of each telescope?

29. What is the minimum distance between two points that will permit them to be resolved at 1 km (a) using a terrestrial telescope with a 6.5-cm diameter objective (assume $\lambda = 550$ nm) and (b) using the unaided eye (assume a pupil diameter of 2.5 mm)?

30. Two stars are 8 lightyears away and can just be resolved by a 20 in. telescope which is equipped with a filter that only allows light of wavelength 500 nm to pass. What is the distance between the two stars?

31. A light source that emits light of wavelength 580 nm is lowered into water ($n = 1.33$). If this source is viewed through a telescope that is in the water and has a 5-cm diameter objective, find the minimum angle of resolution. Assume that the barrel of the telescope is in water.

32. Two radio telescopes are used in a special technique called long-baseline interferometry, which causes the two to act as a single telescope having an effective diameter of 1000 mi. At what distance from this telescope would two objects 250,000 mi apart be at the limit of resolution? Assume that the wavelength of the radio waves from these objects is about 1 m.

• **33.** (a) Calculate the limiting angle of resolution for the eye, assuming a pupil diameter of 2 mm, a wavelength of 500 nm in air, and an index of refraction for the eye equal to 1.33. (b) What is the maximum distance from the eye at which two points separated by 1 cm could be resolved?

• **34.** Two motorcycles separated laterally by 2 m are approaching an observer who is holding an infrared detector sensitive to light of wavelength 885 nm. What aperture diameter is required in the detector if the two headlights are to be resolved at a distance of 10 km?

Section 27.7 The Michelson Interferometer

35. A Michelson interferometer is used to measure an unknown wavelength. The mirror in one arm of the instrument is moved 0.12 mm as 481 dark fringes are counted. Determine the wavelength of the light used.

36. An interferometer is used to measure the length of an insect. The wavelength of the light used is 650 nm. As one arm of the interferometer is moved from one end of the insect to the other, 155 dark fringes are counted. How long is the insect?

37. Light of wavelength 550 nm is used to calibrate a Michelson interferometer. By use of a micrometer screw, the platform on which one mirror is mounted is moved 0.18 mm. How many dark fringe shifts are counted?

•• **38.** The Michelson interferometer can be used to measure the index of refraction of a gas by placing an evacuated transparent tube along one arm of the device. Fringe shifts occur as the gas is slowly added to the tube. Assume that the tube is 5 cm long and that 60 fringe shifts occur as the pressure of the gas in the tube increases to atmospheric pressure. What is the index of refraction of the gas? (*Hint:* The fringe shifts occur because the wavelength of the light changes inside the gas-filled tube.)

•• **39.** A thin sheet of transparent material has an index of refraction of 1.40 and is 15 μm thick. When it is inserted along one arm of an interferometer in the path of 600 nm light, how many fringe shifts occur in the pattern?

Section 27.8 The Diffraction Grating

40. The second-order image formed by a diffraction grating having 6000 lines/cm is observed at an angle of 29°. What is the wavelength of the light used?

41. A diffraction grating is calibrated by using the 546.1 nm line of mercury vapor. It is found that the first-order line is located at an angle of 21°. Calculate the number of lines per centimeter on this grating.

42. A grating having 1500 slits per centimeter is illuminated with light of wavelength 500 nm. (a) What is the highest order number that can be observed? (b) Repeat for a grating of 15,000 slits per centimeter.

43. Light from a hydrogen source is incident on a diffraction grating. The incident light contains four wavelengths: $\lambda_1 = 410.1$ nm, $\lambda_2 = 434.0$ nm, $\lambda_3 = 486.1$ nm, and $\lambda_4 = 656.3$ nm. There are 410 lines/mm in the grating. Calculate the angles between (a) λ_1 and λ_4 in the first-order spectrum and (b) λ_1 and λ_3 in the third-order spectrum.

44. Visible light ranging from 400 nm to 700 nm is focused on a diffraction grating. The entire first order is observed, but no second-order spectrum is seen. What is the maximum spacing between lines on this grating?

45. A diffraction grating is 3 cm wide and contains lines uniformly spaced at 775 nm. Determine the minimum wavelength difference that can be resolved in first order at 600 nm by this grating. Assume that the full grating width is illuminated.

• **46.** The H_α line in hydrogen has a wavelength of 656.2 nm. This line differs in wavelength from the corresponding spectral line in deuterium (the heavy stable isotope of hydrogen) by 0.18 nm. (a) Determine the minimum number of lines a grating must have in order to resolve these two wavelengths in the first order. (b) Repeat part (a) for the second order.

ADDITIONAL PROBLEMS

47. A telescope has an objective of focal length 100 cm and an eyepiece of focal length 1.5 cm. (a) What is the distance between the two lenses when the telescope is used to view a distant object? (b) What is the angular magnification of the telescope?

48. The near point of an eye is 75 cm. (a) What power lens should be prescribed so that the eye can see an object clearly at 25 cm? (b) If the user can actually see an object clearly at 26 cm rather than 25 cm, by how many diopters did the lens grinder miss the prescription?

49. The 546.1-nm line in mercury is measured at an angle of 81° in the third-order spectrum of a diffraction grating. Calculate the number of lines per millimeter for the grating.

50. The far point of an eye is 75 cm. What power lens should be prescribed?

51. A certain lens forms a real image 20 cm from a lens when an object is placed 40 cm from the lens. What is the maximum magnification produced by this lens when used as a simple magnifier?

• **52.** The wavelengths of the sodium doublet are $\lambda_1 =$ 589.0 nm and $\lambda_2 = 589.6$ nm. Determine the minimum number of lines in a grating that will allow resolution of the sodium doublet in (a) the first order and (b) the third order.

• **53.** Light consisting of two wavelength components is incident on a grating. The shorter wavelength component has a wavelength of 440 nm. The third-order image of this component is coincident with the second-order image of the longer wavelength component. Determine the wavelength of the longer wavelength component.

• **54.** A grating is 4 cm wide and is entirely illuminated with monochromatic light of wavelength 577 nm. The second-order maximum is formed at an angle of 41.25°. What is the total number of lines in this grating?

• **55.** If a typical eyeball is 2 cm long and has a pupil opening which can range from about 2 mm to 6 mm, what is (a) the focal length of the eye when focused on objects 1 m away, (b) the smallest *f*-number of the eye when focused at objects 1 m away, and (c) the largest *f*-number of the eye when focused at objects 1 m away.

• **56.** The 501.5-nm line in helium is observed at an angle of 30° in the second-order spectrum of a diffraction grating. Calculate the angular deviation of the 667.8-nm line in helium in the first-order spectrum for the same grating.

• **57.** A fringe pattern is established in the field of view of a Michelson interferometer using light of wavelength 580 nm. A parallel-faced sheet of transparent material 2.5 μm thick is placed in front of one of the mirrors perpendicular to the incident and reflected light beams. An observer counts a fringe shift of six dark fringes. What is the index of refraction of the sheet?

• **58.** A person uses a +2 diopter lens in order to comfortably see an object placed 25 cm away. By the time he decides to have his prescription changed, he must hold an object 35 cm away to see it. What power lens does he now need?

• **59.** If the aqueous humor of the eye has an index of refraction of 1.34 and the distance from the vertex of the cornea to the retina is 2 cm, what is the radius of curvature of the cornea for which distant objects will be focused on the retina? (Assume, incorrectly, that all refraction occurs in the aqueous humor.)

Part 6

MODERN PHYSICS

"The scientist does not study nature because it is useful; he studies it because he delights in it, and he delights in it because it is beautiful. If nature were not beautiful, it would not be worth knowing, and if nature were not worth knowing, life would not be worth living."

HENRI POINCARÉ

At the end of the 19th century, scientists believed that they had learned most of what there was to know about physics. Newton's laws of motion and his universal theory of gravitation, Maxwell's theoretical work in unifying electricity and magnetism, and the laws of thermodynamics and kinetic theory were highly successful in explaining a wide variety of phenomena.

However, at the turn of the 20th century, a major revolution shook the world of physics. In 1900 Planck provided the basic ideas that led to the formulation of the quantum theory, and in 1905 Einstein formulated his brilliant special theory of relativity. The excitement of the times is captured in Einstein's own words: "It was a marvelous time to be alive." Both ideas were to have a profound effect on our understanding of nature. Within a few decades, these theories inspired new developments and theories in the fields of atomic physics, nuclear physics, and condensed matter physics.

The discussion of modern physics in this last part of the text will begin with a treatment of the special theory of relativity in Chapter 28. Although the concepts underlying this theory often violate our common sense, the theory provides us with a new and deeper view of physical laws. Next we shall discuss various developments in quantum theory (Chapter 29), which provides us with a successful model for understanding electrons, atoms, and molecules. The last five chapters of the text are concerned with applications of quantum theory. Chapters 30 and 31 discuss the structure and properties of atoms and molecules in terms of concepts from quantum mechanics. Chapter 32 is concerned with the structure and properties of the atomic nucleus, and Chapter 33 discusses many practical applications of nuclear physics, including nuclear reactors, and the interaction of radiation with matter. We conclude the text with a discussion of elementary particles in Chapter 34.

You should keep in mind that, although modern physics has been developed during this century and has led to a multitude of important technological achievements, the story is still incomplete. Discoveries will continue to evolve during our lifetime, many of which will deepen or refine our understanding of nature and the world around us. It is still "a marvelous time to be alive."

CHAPTER 28

Relativity

Albert Einstein (1879–1955)

28.1 INTRODUCTION

Light waves and other forms of electromagnetic radiation travel through a vacuum with a speed of $c = 3.00 \times 10^8$ m/s. As we shall see in this chapter, the speed of light is the upper limit for the speed of particles and mechanical waves.

Most of our everyday experiences and observations deal with objects that move with speeds much less than the speed of light. Newtonian mechanics and the early ideas on space and time were formulated to describe the motion of such objects. As we saw in the chapters on mechanics, this formalism is very successful in describing a wide range of phenomena. Although Newtonian mechanics works very well at low speeds, it fails when applied to particles whose speeds approach that of light. Experimentally, one can test the predictions of Newtonian theory at high speeds by accelerating an electron through a large electric potential difference. For example, it is possible to accelerate an electron to a speed of $0.99c$ by using a potential difference of several million volts. According to Newtonian mechanics, if the potential difference (as well as the corresponding energy) is increased by a factor of 4, then the speed of the electron should be doubled to $1.98c$. However, experiments show that the speed of the electron always remains *less* than the speed of light, regardless of the size of the accelerating voltage. Since Newtonian mechanics places no upper limit on the speed that a particle can attain, it is contrary to modern experimental results and is clearly a limited theory.

In 1905, at the age of only 26, Einstein published his *special theory of relativity:*

> The relativity theory arose from necessity, from serious and deep contradictions in the old theory from which there seemed no escape. The strength of the new theory lies in the consistency and simplicity with which it solves all these difficulties, using only a few very convincing assumptions.[1]

Although Einstein made many other important contributions to science, his theory of relativity alone represents one of the greatest intellectual achievements of the 20th century. With this theory, one can correctly predict experimental observations over the range from $v = 0$ to velocities approaching the speed of light. Newtonian mechanics, which was accepted for more than 200 years, is in fact a specialized case of Einstein's theory. This chapter introduces the special theory of relativity, with emphasis on some of the consequences of the theory. A discussion of general relativity and some of its consequences is presented in the essay which follows this chapter.

As we shall see, the special theory of relativity is based on two basic postulates:

1. The law of physics are the same in all inertial reference systems.
2. The speed of light in vacuum is always measured to be 3×10^8 m/s, and the measured value is independent of the motion of the observer or of the motion of the source of light.

Special relativity covers such phenomena as the slowing down of clocks and the contraction of lengths in moving reference frames as measured by a stationary observer. In addition to these topics, we shall also discuss the relativistic forms of momentum and energy, terminating the chapter with the famous mass-energy equivalence formula, $E = mc^2$.

[1] A. Einstein and L. Infeld, *The Evolution of Physics,* New York, Simon and Schuster, 1961.

28.2 THE PRINCIPLE OF RELATIVITY

In order to describe a physical event, it is necessary to establish a *frame of reference.* For example, when you perform an experiment in a laboratory, you must select a coordinate system, or frame of reference, that is at rest with respect to the laboratory. However, suppose an observer in a passing car moving at a constant velocity with respect to the lab were to observe your experiment. Would the observations made by the moving observer differ dramatically from yours? That is, if you found Newton's first law to be valid in your frame of reference, would the moving observer also agree to its validity? This type of question was addressed briefly in Chapter 3, and there we stated that

according to the principle of Newtonian relativity, the laws of mechanics are the same in all inertial frames of reference. Inertial frames of reference are those coordinate systems which are at rest with respect to one another or which move at constant velocity with respect to one another.

Thus, for the situation described above, the laboratory coordinate system and the coordinate system of the moving car are both inertial reference frames. Consequently, if the laws of mechanics are found to be true in the lab, the person in the car must agree with this observation.

Let us describe a common observation to illustrate the equivalence of the laws of mechanics in different inertial frames. Consider an airplane in flight moving with a constant velocity as in Figure 28.1a. If a passenger in the airplane throws a ball straight up in the air, the passenger observes that the ball moves in a vertical path. The motion of the ball is precisely the same as if the ball were thrown while at rest on earth. The law of gravity and the equations of motion under constant acceleration are obeyed whether the airplane is at rest or in uniform motion. Now consider the same experiment when viewed from another observer at rest on the earth. This stationary ob-

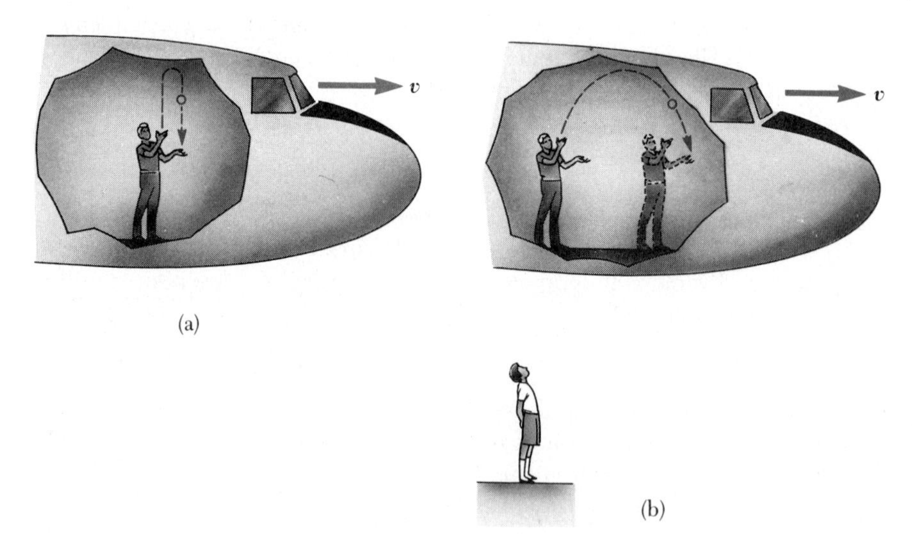

FIGURE 28.1 (a) The observer on the airplane sees the ball move in a vertical path when thrown upwards. (b) The earth observer views the path of the ball to be a parabola.

server views the path of the ball to be a parabola as in Figure 28.1b. Furthermore, according to this observer, the ball has a velocity to the right equal to the velocity of the plane. Although the two observers disagree on certain aspects of the experiment, both will agree that the motion of the ball obeys the law of gravity and the laws of motion. Thus, we draw the following important conclusion:

There is no preferred frame of reference for describing the laws of mechanics.

28.3 THE MICHELSON-MORLEY EXPERIMENT

Many experiments similar to the one described in the previous section show us that the laws of mechanics are the same in all inertial frames of reference. When similar inquiries are made into the laws of other branches of physics, the results are contradictory. In particular, one finds that the laws of electricity and magnetism depend on the frame of reference used. One might argue that it is these laws that are wrong, but this is difficult to accept because the laws are in total agreement with all known experimental results. The Michelson-Morley experiment was an attempt to find an answer to this dilemma.

The Michelson-Morley experiment was designed because of a misconception early physicists had concerning the manner in which light propagates. The properties of mechanical waves, such as water or sound waves, were well known, and all of these require a medium to support the disturbances. In the 19th century, physicists thought that electromagnetic waves also required a medium through which to propagate. They proposed that such a medium existed, and they named it the **luminiferous ether.** This ether was assumed to be present everywhere, even in a vacuum, and to have the unusual property of being a massless but rigid medium. Indeed, this is a strange concept. Additionally, it was found that the troublesome laws of electricity and magnetism would take on their simplest form in a frame of reference *at rest* with respect to this ether. This frame was called the **absolute frame.** The laws of electricity and magnetism would be valid but would have to be modified in any reference frame moving with respect to it.

As a result of the importance attached to this absolute frame, it became of considerable interest in physics to prove its existence by experiment. However, all attempts to detect the presence of the ether (and hence the absolute frame) proved futile!

The most famous experiment designed to show the presence of the ether was performed in 1887 by A. A. Michelson (1852–1931) and E. W. Morley (1838–1923). Later in this section we shall discuss the Michelson-Morley experiment in more detail. We should state at the outset that the outcome of the experiment was *negative*, thus contradicting the ether hypothesis. The experiment was designed to determine the velocity of the earth with respect to the ether, and the experimental tool used was the interferometer discussed in Section 27.7. Suppose one of the arms of the interferometer is aligned along the direction of the motion of the earth through space. The earth moving through the ether would be equivalent to the ether flowing past the earth in the opposite direction. This "ether wind" blowing in the direction opposite the earth's motion should cause the speed of light as measured in the earth's

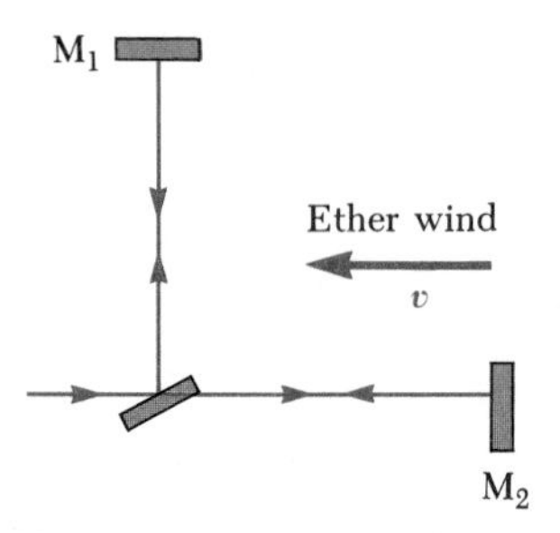

FIGURE 28.2 According to the ether wind theory, the speed of light should be $c - v$ as the beam approaches mirror M_2 and $c + v$ after reflection.

frame of reference to be $c - v$ as the light approaches mirror M_2 in Figure 28.2 and $c + v$ after reflection, where c is the speed of light in the ether frame and v is the speed of the earth through space and hence the speed of the ether wind. The incident and reflected beams of light would recombine, and an interference pattern consisting of alternating dark and bright bands would be formed.

During the experiment, the interference pattern was observed while the interferometer was rotated through an angle of 90°. The idea was that this rotation would change the speed of the ether wind along the direction of the arms of the interferometer, and consequently the fringe pattern would shift slightly but measurably. Measurements failed to show any change in the interference pattern! The Michelson-Morley experiment was repeated by other researchers under various conditions and at different locations, but the results were always the same: *no fringe shift of the magnitude required was ever observed.*

The problem that arises can be understood by considering the following simple situation. Suppose an observer riding in the back of a truck moving at a speed of 50 mi/h throws a baseball in the direction of travel of the truck such that the ball leaves his hand with a speed of 30 mi/h with respect to the truck. What speed will a stationary observer by the side of the road find for the baseball? The answer is quite simple. The stationary observer would add the speed of the truck to the speed of the baseball and conclude that its speed with respect to his frame of reference is 80 mi/h. (If the ball were thrown in the direction opposite that of the truck's motion, the stationary observer would find a speed of 20 mi/h.) This simple calculation is called a *transformation* because we have transformed a measurement made in one reference frame to a measurement made in another. The transformation we have described is in accord with our common sense that a velocity measurement made in one reference frame will be different in another.

Now let us consider a transformation involving the measurement of the speed of light. Suppose an astronaut traveling in a rocket ship at a speed of 1×10^8 m/s turns on a flashlight pointed in the direction of motion of the ship. The astronaut measures the speed of the light to be 3×10^8 m/s. If we assume that velocities add vectorially as in the case of the moving baseball, a stationary observer should measure the speed of the light beam to be the sum of the rocket's speed and the speed of light, or 4×10^8 m/s. This result would mean that the two observers in the different reference frames would get different results when attempting to measure the same electromagnetic effect. As we shall see later, the assumption that velocities add vectorially is not valid when the relative velocity between reference frames is very large. Furthermore, Maxwell's equations, which form the basis of all electromagnetic phenomena, take on a different form when transformed from one inertial frame to another. In fact, they take on their simplest form in a reference frame where the speed of light is 3×10^8 m/s. This strange result is at odds with the fact that the laws of physics must be the same in all inertial frames of reference.

The negative result of the Michelson-Morley experiment was in contradiction with the ether hypothesis. This result also meant that it was impossible to measure the absolute (orbital) velocity of the earth with respect to the ether frame. From a theoretical viewpoint, this meant that it was impossible to find the absolute frame. However, as we shall see in the next section, Einstein offered a postulate that places a different interpretation on these negative results. In later years, when more was known about the nature of light, the idea of an ether that permeated all space was relegated to the ash heap of worn-out concepts. Light is now understood to be *an electromagnetic wave that requires no medium for its propagation.* As a result, the idea of having an ether in which these waves could travel became an unnecessary construct.

*Details of the Michelson-Morley Experiment

As we mentioned earlier, the Michelson-Morley experiment was designed to detect the motion of the earth with respect to the ether. Before we examine the details of this important, historical experiment, it is instructive to first consider a race between two airplanes as shown in Figure 28.3a. One airplane flies from point O to point A perpendicular to the wind, while the second airplane flies from point O to point B parallel to the direction of the wind. We shall assume that they start at O at the same time, travel the same distance L, with the same cruising speed c with respect to the wind, and return to O. Which airplane will win the race? In order to answer this question, we shall first calculate the time of flight for both airplanes.

First, consider the airplane that moves along path I parallel to the wind. As it moves to the right, its speed is enhanced by the wind and its velocity with respect to the earth is $c + v$. On its return journey as it moves to the left, it must fly opposite the wind, hence its speed with respect to the earth is $c - v$. Thus, the times of flight to the right and to the left are, respectively,

$$t_R = \frac{L}{c + v} \quad \text{and} \quad t_L = \frac{L}{c - v}$$

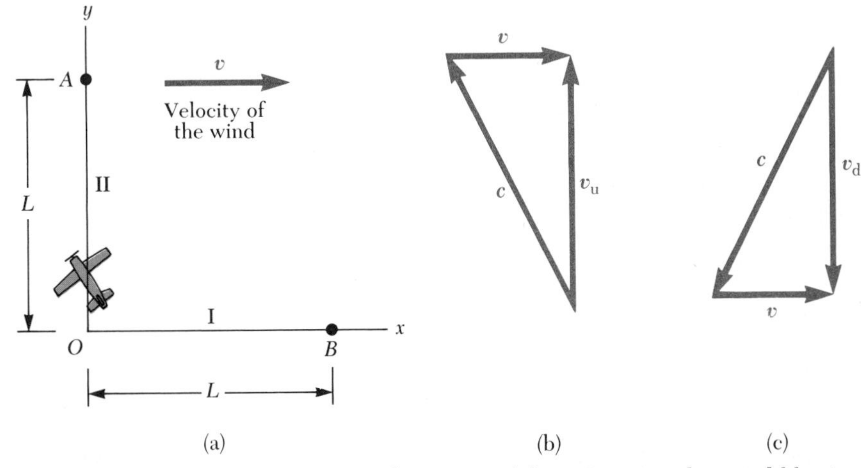

FIGURE 28.3 (a) If an airplane wishes to travel from O to A with a wind blowing to the right, it must head into the wind at some angle. (b) Vector diagram for determining the heading of the airplane for the trip from O to A. (c) Vector diagram for determining the heading for the trip from A to O.

and the total time of flight for the airplane moving along path I is

$$t_1 = t_R + t_L = \frac{L}{c+v} + \frac{L}{c-v} = \frac{2Lc}{c^2 - v^2}$$

$$= \frac{2L}{c\left(1 - \frac{v^2}{c^2}\right)} \tag{28.1}$$

Now consider the airplane flying along path II. If the pilot points the airplane directly towards point A, the airplane will be blown off course by the wind and will not reach its destination. To compensate for the wind, the pilot must point the airplane into the wind at some angle as shown in Figure 28.3a. This angle must be selected such that the vector sum of $\boldsymbol{c}$ and $\boldsymbol{v}$ leads to a velocity vector pointed directly toward A. The resultant vector diagram is shown in Figure 28.3b, where $\boldsymbol{v}_{\text{u}}$ is the velocity of the airplane with respect to the ground as it moves from O to A. From the Pythagorean theorem, the magnitude of the vector $\boldsymbol{v}_{\text{u}}$ is

$$v_u = \sqrt{c^2 - v^2} = c\sqrt{1 - \frac{v^2}{c^2}}$$

Likewise, on the return trip from A to O, the pilot must again head into the wind such that the airplane's velocity with respect to the earth, $\boldsymbol{v}_{\text{d}}$, will be directed toward O, as shown in Figure 28.3c. From this figure, we see that

$$v_{\text{d}} = \sqrt{c^2 - v^2} = c\sqrt{1 - \frac{v^2}{c^2}}$$

Thus, the total time of flight for the trip along path II is

$$t_2 = \frac{L}{v_{\text{u}}} + \frac{L}{v_{\text{d}}} = \frac{L}{c\sqrt{1 - \frac{v^2}{c^2}}} + \frac{L}{c\sqrt{1 - \frac{v^2}{c^2}}}$$

$$= \frac{2L}{c\sqrt{1 - \frac{v^2}{c^2}}} \tag{28.2}$$

Comparing Equations 28.1 and 28.2, we see that the airplane flying along path II wins the race. The difference in flight times is given by

$$\Delta t = t_1 - t_2 = \frac{2L}{c}\left[\frac{1}{\left(1 - \frac{v^2}{c^2}\right)} - \frac{1}{\sqrt{1 - \frac{v^2}{c^2}}}\right]$$

This expression can be simplified using the following binomial expansions after dropping all terms higher than second-order:

$$\left(1 - \frac{v^2}{c^2}\right)^{-1} \approx 1 + \frac{v^2}{c^2}$$

and

$$\left(1 - \frac{v^2}{c^2}\right)^{-1/2} \approx 1 + \frac{1}{2}\frac{v^2}{c^2}$$

This gives

$$\Delta t = \frac{Lv^2}{c^3} \tag{28.3}$$

The analogy between this airplane race and the Michelson-Morley experiment is shown in Figure 28.4a. Two beams of light travel along two arms of an interferometer. In this case, the "wind" is the ether blowing across the earth from left to right as the earth moves through the ether from right to left. Because the speed of the earth in its orbital path is approximately equal to 3×10^4 m/s, the speed of the wind should be at least this great. The two light beams start out in phase and return to form an interference pattern. Let us assume that the interferometer is adjusted for parallel fringes and that a telescope is focused on one of these fringes. The time difference between the two light beams gives rise to a phase difference between the beams, producing an interference pattern when they combine at the position of the telescope. The difference in the pattern is detected by rotating the interferometer through 90° in a horizontal plane, such that the two beams exchange roles (see Fig. 28.4b). This results in a net time shift of twice the time difference given by Equation 28.3. Thus, the net time difference is

$$\Delta t_{net} = 2\,\Delta t = \frac{2Lv^2}{c^3} \tag{28.4}$$

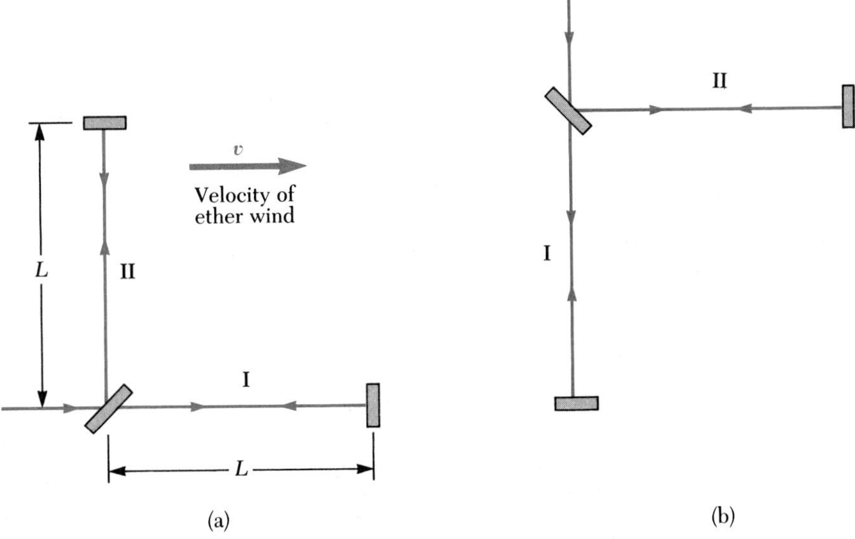

FIGURE 28.4 (a) Top view of the Michelson-Morley interferometer, where v is the velocity of the ether and L is the length of each arm. (b) When the interferometer is rotated by 90°, the role of each arm is reversed.

The path difference which corresponds to this time difference is

$$\Delta d = c\,\Delta t_{\text{net}} = \frac{2Lv^2}{c^2} \tag{28.5}$$

In the first experiments by Michelson and Morley, each light beam was reflected by mirrors many times to give an increased effective path length L of about 11 meters. Using this value, and taking v to be equal to 3×10^4 m/s gives a path difference of

$$\Delta d = \frac{2(11 \text{ m})(3 \times 10^4 \text{ m/s})^2}{(3 \times 10^8 \text{ m/s})^2} = 2.2 \times 10^{-7} \text{ m}$$

This extra distance of travel should produce a noticeable shift in the fringe pattern. Specifically, calculations show that if one views the pattern while the interferometer is rotated through 90°, a shift of about 0.4 fringes should be observed. The instrument used by Michelson and Morley had the capability of detecting a shift in the fringe pattern as small as 0.01 fringes. However, *they detected no shift in the fringe pattern.* Since then, the experiment has been repeated many times by various scientists under various conditions and no fringe shift has ever been detected. Thus, it was concluded that one cannot detect the motion of the earth with respect to the ether.

Many efforts were made to explain the null results of the Michelson-Morley experiment. For example, perhaps the earth drags the ether with it in its motion through space. To test this assumption, interferometer measurements were made at various altitudes, but again no fringe shift was detected. In the 1890's, G. F. Fitzgerald and H. A. Lorentz tried to explain the null results by making the following ad hoc assumption. They proposed that the length of an object moving along the direction of the ether wind would contract by a factor of $\sqrt{1 - v^2/c^2}$. The net result of this contraction would be a change in length of one of the arms of the interferometer such that no path difference would occur as it was rotated.

No experiment in the history of physics has received such valiant efforts to try to explain the absence of an expected result as did the Michelson-Morley experiment. The stage was set for the brilliant Albert Einstein, who solved the problem in 1905 with his special theory of relativity.

28.4 EINSTEIN'S POSTULATES

The special theory of relativity deals with situations involving inertial reference frames, that is, *frames which are moving at constant velocities relative to one another*, and is based on two postulates. First, Einstein postulated that

the laws of physics are the same in every inertial frame of reference.

In Einstein's own words, "The same laws of electrodynamics and optics will be valid for all frames of reference for which the equations of mechanics hold good." This is, in effect, a generalization of Newton's principle of relativity, which applies only to the laws of mechanics.

Now consider Einstein's second postulate, which states that

the speed of light has the same value for all observers, independent of their motion or of the motion of the light source.

Albert Einstein (1879–1955), one of the greatest physicists of all times, was born in Ulm, Germany. Because he was unable to obtain an academic position following graduation from the Swiss Federal Polytechnic School in 1901 at the age of 22, he accepted a job at the Swiss Patent Office in Berne. During his spare time, he continued his studies in theoretical physics. In 1905, at the age of 26, he published four scientific papers that revolutionized physics. One of these papers, which won him the Nobel prize in 1921, dealt with the photoelectric effect, and another was concerned with Brownian motion. The remaining two papers were concerned with what is now considered his most important contribution of all, the special theory of relativity. In 1916, Einstein published his work on the general theory of relativity.

Here we are faced with a fundamental problem. We can demonstrate the nature of the problem by considering a light pulse sent out by an observer in a boxcar moving with a velocity v (Fig. 28.5). The light pulse has a velocity c relative to observer S′ in the boxcar. According to the ideas of Newtonian relativity, the speed of the pulse relative to stationary observer S outside the boxcar should be $c + v$. This is in obvious contradiction to Einstein's second postulate, which states that the velocity of the light pulse is the same for all observers. According to Einstein's theory, the stationary and the moving observer should both measure the same velocity for the light pulse. This conclusion seems strange because it contradicts our intuition, or what we often call common sense. However, common sense ideas are based on everyday experiences, which do not involve speed-of-light measurements.

FIGURE 28.5 A pulse of light is sent out by a person in a moving boxcar. According to Newtonian relativity, the speed of the pulse should be $c + v$ relative to a stationary observer.

Although the Michelson-Morley experiment was performed before Einstein published his work on relativity, it is not clear whether or not Einstein was aware of the details of the experiment. Nonetheless, the second postulate explains the null result of the experiment for, in effect, the second postulate means that the premises of the experiment were incorrect. For example, in explaining the expected results, we stated that when light traveled against the ether wind its speed was $c - v$. However, if the state of motion of the observer or of the source has no influence on the value found for the speed of light, one will always measure the value to be c. Likewise, after reflection from the mirror, the light makes the return trip with a speed of c, not $c + v$. Thus, if we accept Einstein's second postulate, the motion of the earth should not influence the fringe pattern observed in the Michelson-Morley experiment and a null result should be expected.

If we accept Einstein's postulates, we must conclude that *relative motion is unimportant when measuring the speed of light.* In effect, Einstein altered our concepts of space and time in such a way as to give the same result for the speed of light measured by any observer located in any inertial frame.

28.5 CONSEQUENCES OF SPECIAL RELATIVITY

Almost everyone who has dabbled even superficially in science is aware of some of the startling predictions that arise because of Einstein's approach to relative motion. As we examine some of the consequences of relativity in this section, we shall find that they conflict with some of our basic notions of space and time. We shall restrict our discussion to the concepts of length, time, and simultaneity, which are quite different in relativistic mechanics from what they are in Newtonian mechanics. For example, we shall see that *the distance between two points and the time interval between two events depend on the frame of reference in which they are measured.* That is, *in relativity, there is no such thing as absolute length or absolute time. Furthermore, events at different locations that occur simultaneously in one frame are not simultaneous in another frame.*

Simultaneity and the Relativity of Time

A basic premise of Newtonian mechanics is that there is a universal time scale that is the same for all observers. In fact, Newton wrote, "Absolute, true, and mathematical time, of itself, and from its own nature, flows equably without relation to anything external." In his special theory of relativity, Einstein abandoned this assumption. According to Einstein, *time interval measurements depend on the reference frame in which they are made.*

Einstein devised the following thought experiment to illustrate this point. A boxcar moves with uniform velocity, and two lightning bolts strike its ends, as in Figure 28.6a, leaving marks on the boxcar and the ground. The marks left on the boxcar are labeled A' and B', and those on the ground are labeled A and B. An observer at O' moving with the boxcar is midway between A' and B', and an observer on the ground at O is midway between A and B. The events seen by the observers are the light signals from the lightning bolts.

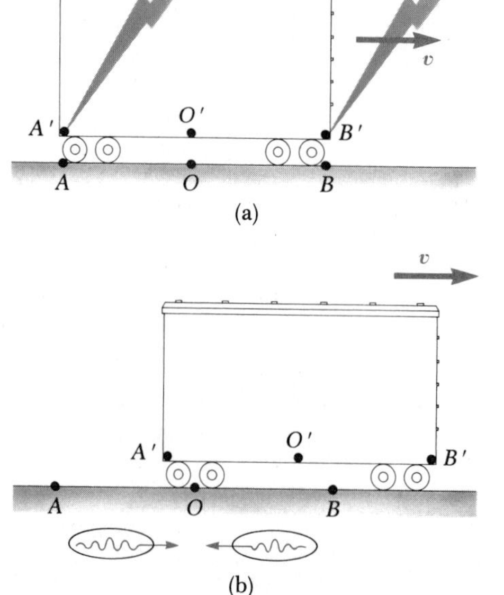

FIGURE 28.6 Two lightning bolts strike the ends of a moving boxcar. (a) The events appear to be simultaneous relative to the stationary ground observer at O who is midway between the events. (b) The events do not appear to be simultaneous to an observer at O' on the train, who claims that the front of the train is struck *before* the rear.

Let us assume that the two light signals reach the observer at O at the same time, as indicated in Figure 28.6b. This observer realizes that the light signals have traveled at the same speed over distances of equal length. Thus, the observer at O rightly concludes that the events at A and B occurred simultaneously. Now consider the same events as viewed by the observer on the boxcar at O'. By the time the light has reached the observer at O, the observer at O' has moved as indicated in Figure 28.6b. Thus, the light signal from B' has already swept past O', while the light from A' has not yet reached O'. According to Einstein's second postulate, the observer at O' must find that light travels at the same speed as that measured by the observer at O. Therefore, the observer at O' concludes that the light reaches the front of the boxcar before it reaches the back. This thought experiment clearly demonstrates that the two events which appear to be simultaneous to the observer at O do not appear to be simultaneous to the observer at O'. In other words,

two events that are simultaneous in one reference frame are in general not simultaneous in a second frame moving with respect to the first. That is, simultaneity is not an absolute concept.

At this point, you might wonder which observer is right concerning the two events. The answer is that *both are correct* because the principle of relativity states that *there is no preferred inertial frame of reference.* Although the two observers reach different conclusions, both are correct in their own reference frame because the concept of simultaneity is not absolute.

Time Dilation

Consider a vehicle moving to the right with a speed v, as in figure 28.7a. A perfectly reflecting mirror is fixed to the ceiling of the vehicle, and an ob-

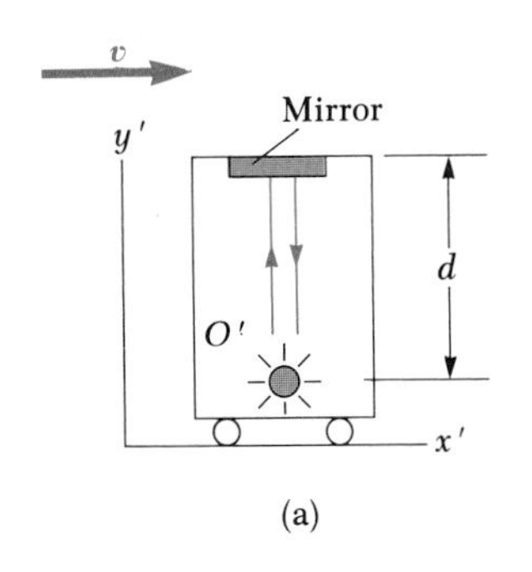

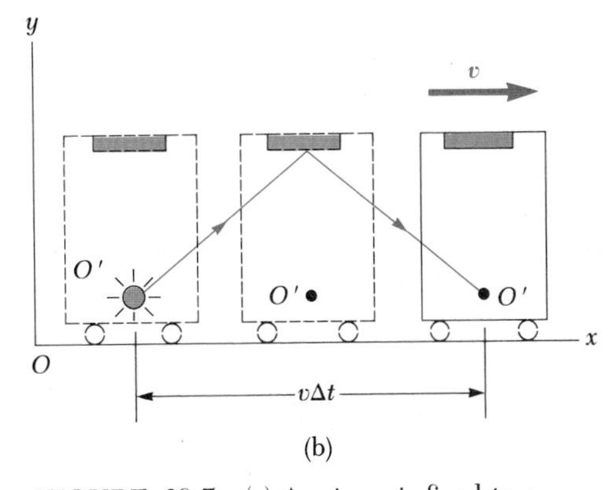

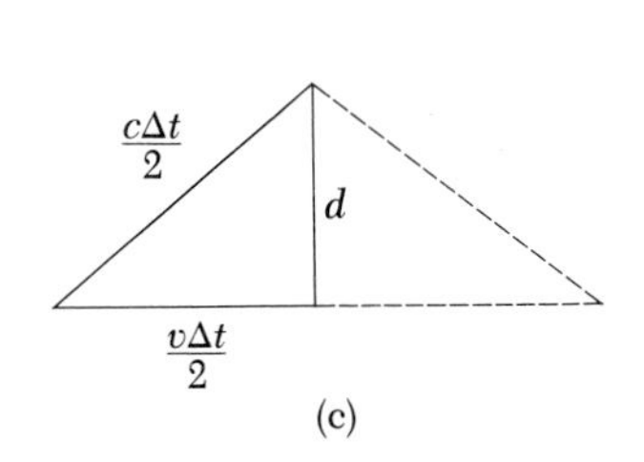

FIGURE 28.7 (a) A mirror is fixed to a moving boxcar and a light pulse leaves O' at rest in the boxcar. (b) Relative to an observer on earth, the mirror and O' move with a speed v. Note that the distance the pulse travels is greater than $2d$ as measured by the observer. (c) The right triangle is useful for calculating the relationship between Δt and $\Delta t'$.

server at O' at rest in this system holds a flash gun a distance d below the mirror. At some instant, the flash gun goes off and a pulse of light is released. Because the light pulse has a speed c, the time it takes it to travel from the observer to the mirror and back again can be found from the definition of velocity:

$$\Delta t' = \frac{\text{distance traveled}}{\text{velocity}} = \frac{2d}{c} \tag{28.6}$$

where the prime notation indicates that this is the time measured by the observer in the reference frame of the moving vehicle.

Now consider the same set of events as viewed by an observer at O in a stationary frame (Fig. 28.7b). According to this observer, the mirror and flash gun are moving to the right with a speed v. The sequence of events just described would appear entirely different to this stationary observer. By the time the light from the flash gun reaches the mirror, the mirror will have moved a distance $v\Delta t/2$, where Δt is the time it takes the light pulse to travel from O' to the mirror and back, as measured by the stationary observer. In other words, the stationary observer concludes that, because of the motion of the system, the light, if it is to hit the mirror, will leave the flash gun at an angle with respect to the vertical. Comparing Figures 28.7a and 28.7b, we see that the light must travel farther in the stationary frame than in the moving frame.

Now, according to Einstein's second postulate, the speed of light must be c as measured by both observers. Therefore, it follows that the time interval Δt, measured by the observer in the stationary frame, is *longer* than the time interval $\Delta t'$, measured by the observer in the moving frame. To obtain a relationship between Δt and $\Delta t'$, it is convenient to use the right triangle shown in Figure 28.7c. The Pythagorean theorem applied to this triangle gives

$$\left(\frac{c\Delta t}{2}\right)^2 = \left(\frac{v\Delta t}{2}\right)^2 + d^2$$

Solving for Δt gives

$$\Delta t = \frac{2d}{\sqrt{c^2 - v^2}} = \frac{2d}{c\sqrt{1 - v^2/c^2}} \tag{28.7}$$

Because $\Delta t' = 2d/c$, we can express Equation 28.7 as

Time dilation

$$\Delta t = \frac{\Delta t'}{\sqrt{1 - v^2/c^2}} = \gamma \Delta t' \tag{28.8}$$

where $\gamma = 1/\sqrt{1 - v^2/c^2}$. This result says that the time interval measured by the observer in the stationary frame is *longer* than that measured by the observer in the moving frame (because γ is always greater than unity).

For example, suppose an observer in a moving vehicle has a clock which he uses to measure the time required for the light flash to leave the gun and return. Let us assume that the measured time interval in this frame of reference, $\Delta t'$, is one second. (This would require a very tall vehicle.) Now let us find the time interval as measured by a stationary observer using an identical clock. If the vehicle is traveling at half the speed of light ($v = 0.5c$), then $\gamma = 1.15$, and according to Equation 28.6 $\Delta t = \gamma\Delta t' = 1.15\ (1\ \text{s}) = 1.15\ \text{s}$. Thus, when the observer on the moving vehicle claims that 1 s has passed, a stationary observer claims that 1.15 s has passed. From this we may conclude that

according to a stationary observer, a moving clock runs slower than an identical stationary clock by a factor of γ^{-1}. This effect is known as **time dilation.**

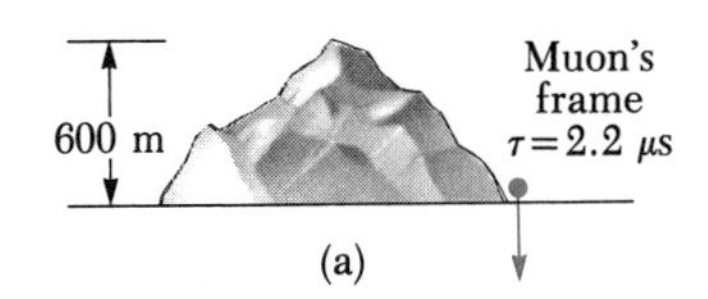

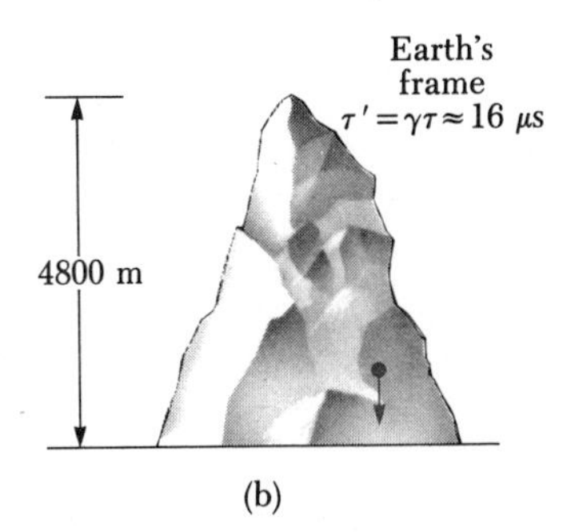

FIGURE 28.8 (a) Muons traveling with a speed $0.99c$ travel a distance of about 4800 m as measured by an observer on earth. (b) The muons travel only about 600 m as measured in the muons' reference frame, where their lifetime is about 2.2 μs. Because of time dilation, the muons' lifetime is longer as measured by the observer on earth.

The time interval $\Delta t'$ in Equation 28.8 is called the **proper time.** In general, **proper time** is defined as *the time interval between two events as measured by an observer who sees the events occur at the same place.* In our case, the observer at O' measures the proper time. That is, *proper time is always the time measured by an observer moving along with the clock.*

We have seen that moving clocks run slow by a factor of γ^{-1}. This is true for ordinary mechanical clocks as well as for the light clock just described. In fact, we can generalize these results by stating that *all physical processes, including chemical and biological reactions, slow down relative to a stationary clock when they occur in a moving frame.* For example, the heartbeat of an astronaut moving through space has to keep time with a clock inside the spaceship. Both the spaceship clock and the heartbeat are slowed down relative to a stationary clock. The astronaut would not have any sensation of life slowing down in the spaceship.

Time dilation is a very real phenomenon that has been verified by various experiments. For example, muons are unstable elementary particles that have a charge equal to that of the electron and a mass 207 times that of the electron. Muons can be produced by the absorption of cosmic radiation high in the atmosphere. These unstable particles have a lifetime of only 2.2 μs when measured in a reference frame at rest with respect to them. If we take 2.2 μs as the average lifetime of a muon and assume that their speed is close to the speed of light, we find that these particles can travel only about 600 m before they decay into something else (Fig. 28.8a). Hence, they could never reach the earth from the upper atmosphere where they are produced. However, experiments show that a large number of muons *do* reach the earth, and the phenomenon of time dilation explains how. Relative to an observer on earth, the muons have a lifetime equal to $\gamma\tau$, where $\tau = 2.2\ \mu\text{s}$ is the lifetime in a frame of reference traveling with the muons. For example, for $v = 0.99c$, $\gamma \approx 7.1$ and $\gamma\tau \approx 16\ \mu\text{s}$. Hence, the average distance traveled as measured by an observer on earth is $\gamma v\tau \approx 4800$ m, as indicated in Figure 28.8b.

The results of an experiment reported by Hafele and Keating provided direct evidence for the phenomenon of time dilation.[2] The experiment involved the use of very stable cesium-beam atomic clocks. Time intervals measured with four such clocks in jet flight were compared with time intervals measured by reference atomic clocks at the U.S. Naval Observatory. (Because of the earth's rotation about its axis, a ground-based clock is not in a true inertial frame.) Time intervals measured with the flying clocks were compared with time intervals measured with the earth-based reference clocks. In order

[2] J. C. Hafele and R. E. Keating, "Around the World Atomic Clocks: Relativistic Time Gains Observed," *Science*, July 14, 1972, p. 168.

to compare the results with the theory, many factors had to be considered, including periods of acceleration and deceleration relative to the earth, variations in direction of travel, and the weaker gravitational field experienced by the flying clocks. Their results were in good agreement with the predictions of the special theory of relativity. In their paper, Hafele and Keating report the following: "Relative to the atomic time scale of the U.S. Naval Observatory, the flying clocks lost 59 ± 10 ns during the eastward trip and gained 273 ± 7 ns during the westward trip. . . . These results provide an unambiguous empirical resolution of the famous clock paradox with macroscopic clocks."

EXAMPLE 28.1 What Is the Period of the Pendulum?

The period of a pendulum is measured to be 3 s in the inertial frame of the pendulum. What is the period when measured by an observer moving with a speed of $0.95c$ with respect to the pendulum?

Solution In this case, the proper time is equal to 3 s. We can use Equation 28.8 to calculate the period measured by the moving observer:

$$T = \gamma T' = \frac{1}{\sqrt{1 - \frac{(0.95c)^2}{c^2}}} T' = (3.2)(3\text{ s}) = 9.6\text{ s}$$

That is, the observer moving with a speed of $0.95c$ observes that the pendulum slows down.

Length Contraction

We have seen that measured time intervals are not absolute, that is, the time interval between two events depends on the frame of reference in which it is measured. Likewise, the measured distance between two points depends on the frame of reference. The **proper length** of an object is defined as *the length of the object measured in the reference frame in which the object is at rest.* The length of an object measured in a reference frame in which the object is moving is always less than the proper length. This effect is known as **length contraction.**

To understand length contraction quantitatively, let us consider a spaceship traveling with a speed v from one star to another, as seen by two observers. An observer at rest on earth (and also assumed to be at rest with respect to the two stars) measures the distance between the stars to be L' (where L' is the proper length), and according to this observer, it takes a time $\Delta t = L'/v$ for the spaceship to complete the voyage. What does an observer in the spaceship measure? Because of time dilation, the space traveler measures a smaller time of travel: $\Delta t' = \Delta t/\gamma$. The observer in the spaceship claims to be at rest and sees the destination star as moving toward the ship with speed v. Since the space traveler reaches the star in the time $\Delta t'$, he or she concludes that the distance, L, between the stars is shorter than L'. This distance is given by

$$L = v\,\Delta t' = v\frac{\Delta t}{\gamma}$$

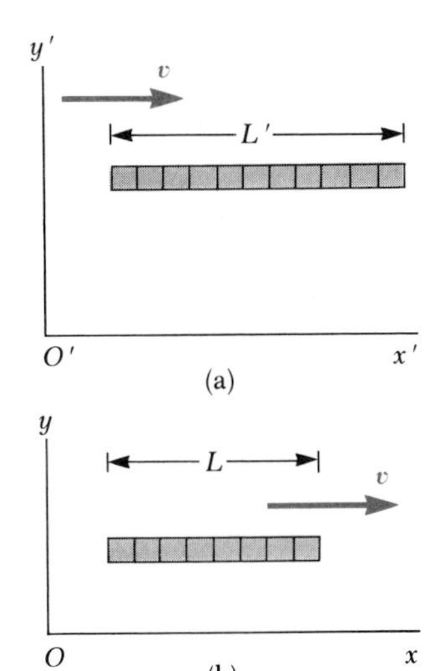

FIGURE 28.9 A meter stick moves to the right with a speed v. (a) The meter stick as viewed by a frame attached to it. (b) The stick as seen by an observer in a rest frame. The length measured in the rest frame is *shorter* than the proper length L', by a factor $\sqrt{1 - v^2/c^2}$.

Since $L' = v\,\Delta t$, we see that

$$L = \frac{L'}{\gamma}$$

or,

$$L = L'\sqrt{1 - v^2/c^2} \tag{28.9}$$

Length contraction

According to this result, which is illustrated in Figure 28.9, if an observer at rest with respect to an object measures its length to be L', an observer moving with a relative speed v with respect to the object will find it to be shorter than its rest length by the factor $\sqrt{1 - v^2/c^2}$.

You should note that *the length contraction takes place only along the direction of motion.* For example, suppose a spaceship as seen in a rest frame is in the shape of a cube 1 km on a side, as in Figure 28.10a. When this ship is in motion with a speed v relative to an observer at rest, measurements made by the observer will indicate that the ship has the shape shown in Figure 28.10b.

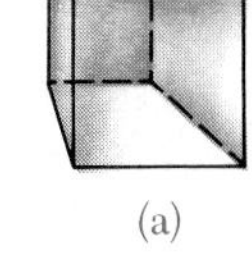

(a)

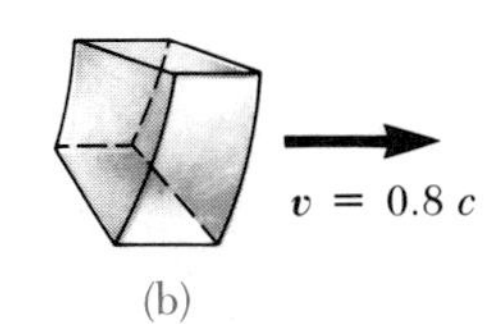

(b)

FIGURE 28.10 A spaceship in the shape of a cube when viewed in a rest frame as in (a) takes on the shape shown in (b) when in motion to the right.

EXAMPLE 28.2 The Contraction of a Spaceship

A spaceship is measured to be 100 m long while it is at rest with respect to an observer. If this spaceship now flies by the observer with a speed of 0.99c, what length will the observer find for the spaceship?

Solution From Equation 28.9, the length measured by an observer in the spaceship is

$$L = L'\sqrt{1 - v^2/c^2} = (100\text{ m})\sqrt{1 - \frac{(0.99c)^2}{c^2}} = 14\text{ m}$$

EXERCISE If the ship moves past the observer with a speed of $0.01c$, what length will the observer measure?

Answer 99.99 m.

EXAMPLE 28.3 How High is the Spaceship?

An observer on earth sees a spaceship at an altitude of 435 m moving downward toward the earth with a speed of $0.97c$. What is the altitude of the spaceship as measured by an observer in the spaceship?

Solution The moving observer in the ship finds the altitude to be

$$L = L'\sqrt{1 - v^2/c^2} = (435\text{ m})\sqrt{1 - \frac{(0.97c)^2}{c^2}} = 106\text{ m}$$

EXAMPLE 28.4 The Triangular Spaceship

A spaceship in the form of a triangle flies by an observer with a speed of $0.95c$. When the ship is at rest (Fig. 28.11a), the distances x and y are found to be 50 m and 25 m, respectively. What is the shape of the ship as seen by an observer at rest when the ship is in motion along the direction shown in Figure 28.11b?

Solution The observer sees the horizontal length of the ship to be contracted to a length of

$$L = L'\sqrt{1 - v^2/c^2} = (50\text{ m})\sqrt{1 - \frac{(0.95c)^2}{c^2}} = 15.6\text{ m}$$

The 25-m vertical height is unchanged because it is perpendicular to the direction of relative motion between the observer and the spaceship. Figure 28.11b represents the shape of the spaceship as seen by the observer at rest.

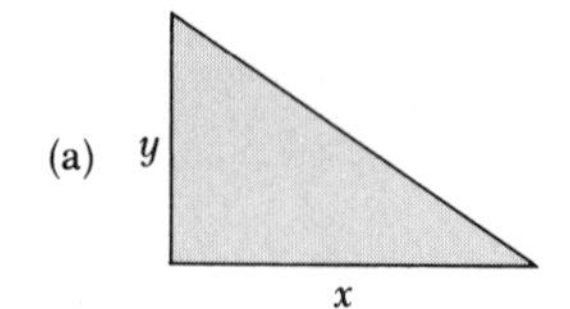

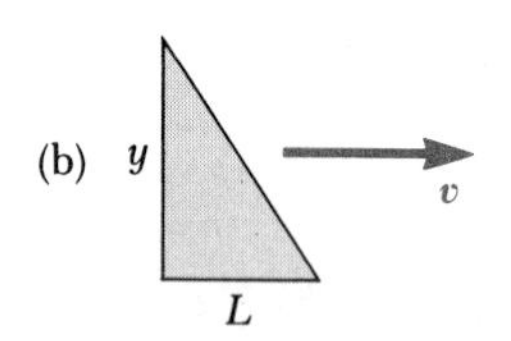

FIGURE 28.11 (Example 28.4) (a) When the spaceship is at rest, its shape is as shown. (b) The spaceship appears to look like this when it moves to the right with a speed v. Note that only its x dimension is contracted in this case.

28.6 RELATIVISTIC MOMENTUM

In order to properly describe the motion of particles within the framework of special relativity, we must generalize Newton's laws and the definitions of momentum and energy. These generalized definitions reduce to the classical (nonrelativistic) definitions when v is much less than c.

First, recall that conservation of momentum states that when two objects collide, the total momentum of the system remains constant, assuming that the objects are isolated (that is, they interact only with each other). If one analyzes such a collision within the framework of Einstein's postulates of relativity, one finds that momentum is not conserved if one uses the classical definition of momentum, $p = m_0v$, where the subscript on m_0 is used in discussions of relativity to indicate the mass of an object at rest or moving at low speed. However, according to the principle of relativity, momentum must be conserved in all systems. In view of this condition, it is necessary to modify the definition of momentum to satisfy the following conditions:

1. The relativistic momentum must be conserved in all collisions.
2. The relativistic momentum must approach the classical value, m_0v, as the quantity v/c approaches zero.

The correct relativistic equation for momentum that satisfies these conditions is

$$p \equiv \frac{m_0 v}{\sqrt{1 - v^2/c^2}} = \gamma m_0 v \qquad (28.10)$$

Momentum

where m_0 is the mass of the particle and v is its velocity. The theoretical proof of this generalized expression for momentum is beyond the scope of this text. Note that when v is much less than c, the denominator of Equation 28.10 approaches unity and so p approaches $m_0 v$. Therefore, the relativistic equation for momentum reduces to the classical expression when v is small compared with c.

EXAMPLE 28.5 The Relativistic Momentum of an Electron

An electron, which has a mass of 9.11×10^{-31} kg, moves with a speed of $0.75c$. Find its relativistic momentum and compare this value with the momentum calculated from the classical expression.

Solution From Equation 28.10, with $v = 0.75c$, we have

$$p = \frac{m_0 v}{\sqrt{1 - v^2/c^2}}$$

$$= \frac{(9.11 \times 10^{-31}\ \text{kg})(0.75 \times 3 \times 10^8\ \text{m/s})}{\sqrt{1 - (0.75c)^2/c^2}}$$

$$= 3.10 \times 10^{-22}\ \text{kg} \cdot \text{m/s}$$

The classical expression gives

$$\text{Momentum} = m_0 v = 2.05 \times 10^{-22}\ \text{kg·m/s}$$

The (correct) relativistic result is 50% greater than the classical result!

28.7 MASS AND THE ULTIMATE SPEED

We have seen that the values of length and time intervals depend on the reference frame in which they are measured. On this basis, one must expect mass to be a relative quantity also. In fact, Einstein showed that *the observed mass of an object increases with speed according to the relation*

$$m \equiv \frac{m_0}{\sqrt{1 - v^2/c^2}} \qquad (28.11)$$

where m is the mass of the object as measured by an observer moving with a speed v and m_0 is the mass of the object as measured by an observer at rest with respect to the object. We shall refer to m_0 as the **rest mass** of the object. From this definition, the relativistic momentum given by Equation 28.10 can be interpreted as being the product of the relativistic mass, m, and the velocity of the object; that is, $p = mv$. Equation 28.11 indicates that *the mass of an object increases as its speed increases,* as shown in Figure 28.12. This prediction has been borne out through observations on elementary particles, such as energetic electrons.

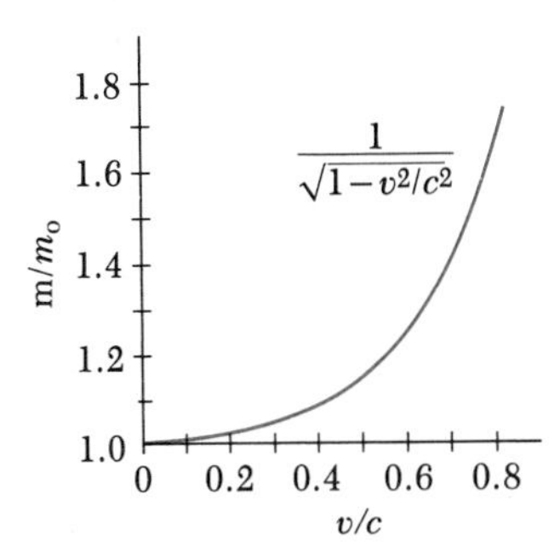

FIGURE 28.12 Variation of mass with speed. This theoretical curve is in excellent agreement with experimental data on particles such as electrons.

It is also possible to see from Equation 28.11 that the ultimate speed of an object is the speed of light. The equation indicates that, as an object is accelerated, its mass increases, and hence the unbalanced force on the object has to increase in proportion in order to maintain a constant acceleration. Finally, when $v = c$, the mass becomes infinite. This means that an infinite amount of energy would be required to accelerate the object to the speed of light. Thus, *the speed of light is the ultimate speed for any material object.*

EXAMPLE 28.6 The Mass of a Speedy Ball

Superman, who has an exceptionally strong arm, throws a fast ball with a speed of $0.9c$. If the rest mass of the ball is 0.5 kg, what is its mass in flight?

Solution This is a straightforward application of Equation 28.11. The relativistic mass is

$$m = \frac{m_0}{\sqrt{1 - v^2/c^2}} = \frac{0.5 \text{ kg}}{\sqrt{1 - \frac{(0.9c)^2}{c^2}}} = 1.15 \text{ kg}$$

28.8 RELATIVISTIC ADDITION OF VELOCITIES

Imagine a motorcycle rider moving with a speed of $0.8c$ past a stationary observer, as shown in Figure 28.13. If the rider tosses a ball in the forward direction with a speed of $0.7c$ relative to himself, what is the speed of the ball as seen by the stationary observer at the side of the road? Common sense, and the ideas of Newtonian relativity, says that the speed should be the sum of the two speeds, or $1.5c$. This answer must be incorrect since it contradicts the assertion in the last section that no material object can travel faster than the speed of light.

Einstein resolved this dilemma by deriving an equation for the relativistic addition of velocities. For one-dimensional motion, this equation is

Velocity addition

$$u = \frac{v + u'}{1 + \frac{vu'}{c^2}} \tag{28.12}$$

where u is the velocity of an object as seen in one frame of reference, u' is the velocity of the same object in a different frame of reference, and v is the relative velocity of the two frames. Let us apply this equation to the case of the speedy motorcycle rider and the stationary observer.

The velocity of the motorcycle with respect to the stationary observer is $0.8c$. Thus, $v = 0.8c$. The velocity of the ball in the frame of the motorcyclist is $0.7c$. That is, $u' = 0.7c$. Therefore, the velocity, u, of the ball relative to the stationary observer is

$$u = \frac{0.8c + 0.7c}{1 + \frac{(0.8c)(0.7c)}{c^2}} = 0.96c$$

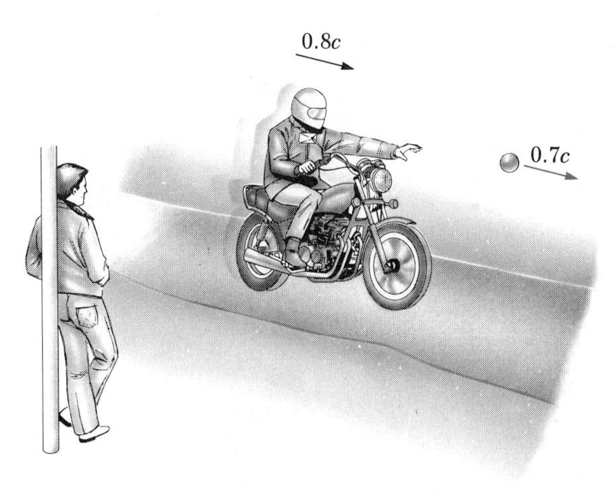

FIGURE 28.13 A motorcycle moves past a stationary observer with a speed $0.8c$ and throws a ball in the direction of motion with a speed $0.7c$ relative to himself.

EXAMPLE 28.7 Measuring the Speed of a Light Beam

Suppose that the motorcyclist moving with a speed of $0.8c$ turns on a beam of light which moves away from the motorcycle with a speed of c in the same direction as the moving motorcycle. What would the stationary observer measure for the speed of the beam of light?

Solution In this case, $v = 0.8c$ and $u' = c$. We can use these values in Equation 28.12 to give

$$u = \frac{0.8c + c}{1 + \frac{(0.8c)(c)}{c^2}} = c$$

This is consistent with the statement made earlier that *all observers measure the speed of light to be c regardless of the motion of the source of light.*

28.9 RELATIVISTIC ENERGY

We have seen that the definition of momentum required generalization to make it compatible with the principle of relativity. Likewise, the definition of kinetic energy requires modification in relativistic mechanics. In view of the definition of relativistic momentum, one might guess that the classical expression for kinetic energy, $mv^2/2$, could be translated to a relativistic expression by substituting the value for m defined in Equation 28.11. This, however, is not correct. Einstein found that the correct expression for the **kinetic energy** of an object is

$$\boxed{KE = mc^2 - m_0c^2} \quad (28.13)$$

Kinetic energy

The term m_0c^2, which is independent of the speed of the object, is called the **rest energy** of the object. The term mc^2, which depends on the object speed, is therefore the sum of the kinetic and rest energies. We define mc^2 to be the **total energy, E,** that is,

$$E = mc^2 = KE + m_0c^2 \tag{28.14}$$

This is Einstein's famous *mass-energy equivalence equation.* If we make use of Equation 28.11, we can express the total energy as[3]

Total energy

$$E = \frac{m_0c^2}{\sqrt{1 - v^2/c^2}} \tag{28.15}$$

The relation $E = mc^2$ shows that *mass is a form of energy.* Furthermore, this result shows that *even a small mass corresponds to an enormous amount of energy.* The concept stated in Equation 28.15 has revolutionized the field of nuclear physics. The validity of the relationship between mass and energy has been proved beyond question.

In many situations, the momentum or energy of a particle is known rather than its speed. It is therefore useful to have an expression relating the total energy, E, to the relativistic momentum, p. This is accomplished by using the expressions $E = mc^2$ and $p = mv$. By squaring these equations and subtracting, we can eliminate v. The result, after some algebra, is

$$E^2 = p^2c^2 + (m_0c^2)^2 \tag{28.16}$$

When the particle is at rest, $p = 0$, and so $E = m_0c^2$. That is, by definition, *the total energy of a particle at rest equals its rest energy.* As we shall discuss in Chapter 29, it is well established that there are particles that have zero rest mass, such as photons. If we set m_0 equal to zero in Equation 28.16, we see that the *total energy of a photon* can be expressed as

$$E = pc \tag{28.17}$$

This equation is an exact expression relating energy and momentum for photons, which always travel at the speed of light.

[3]At first glance, Equation 28.13 does not look even remotely similar to the classical expression for the kinetic energy. However, we shall show that the two equations are equivalent in the limit of low velocities. First, let us substitute the expression for the total energy from Equation 28.15 into Equation 28.13 to find

$$KE = mc^2 - m_0c^2 = \frac{m_0c^2}{\sqrt{1 - \frac{v^2}{c^2}}} - m_0c^2$$

$$= m_0c^2\left[\left(1 - \frac{v^2}{c^2}\right)^{-1/2} - 1\right]$$

In the limit of low velocities, v becomes very small compared to c, hence the first term in the brackets can be approximated by the first two terms in the binomial series expansion for $\left(1 - \frac{v^2}{c^2}\right)^{-1/2}$ to give

$$KE \approx m_0c^2\left[1 + \frac{1}{2}\frac{v^2}{c^2} - 1\right] = \frac{1}{2}m_0v^2$$

When dealing with electrons or other subatomic particles, it is convenient to express the energy in electron volts (eV) because the particles are usually given energy by acceleration through a potential difference. Recall that

$$1 \text{ eV} = 1.60 \times 10^{-19} \text{ J}$$

For example, the mass of an electron is 9.11×10^{-31} kg. Hence, its rest energy is

$$m_0c^2 = (9.11 \times 10^{-31} \text{ kg})(3 \times 10^8 \text{ m/s})^2 = 8.2 \times 10^{-14} \text{ J}$$

Converting this to electron volts, we have

$$m_0c^2 = (8.2 \times 10^{-14} \text{ J})\left(\frac{1 \text{ eV}}{1.60 \times 10^{-19} \text{ J}}\right) = 0.511 \text{ MeV}$$

where 1 MeV $= 10^6$ eV.

EXAMPLE 28.8 The Energy Contained in a Baseball

If a 0.5-kg baseball cound be converted completely to energy, how much energy would be released?

Solution The energy equivalent of the baseball is found from Equation 28.14 (with $KE = 0$):

$$E = m_0c^2 = (0.5 \text{ kg})(3 \times 10^8 \text{ m/s})^2 = 4.5 \times 10^{16} \text{ J}$$

This is enough energy to keep a 100-W lightbulb burning for approximately ten million years. However, you should be aware that it is impossible to achieve complete conversion from mass to energy in practical situations. For example, mass is converted to energy in nuclear power plants, but only a small fraction of the mass actually undergoes conversion.

EXAMPLE 28.9 The Energy of a Speedy Electron

An electron moves with a speed of $v = 0.85c$. Find its total energy and kinetic energy in electron volts.

Solution The fact that the rest energy of an electron is 0.511 MeV, along with Equation 28.15, gives

$$E = \frac{m_0c^2}{\sqrt{1 - v^2/c^2}} = \frac{0.511 \text{ MeV}}{\sqrt{1 - \frac{(0.85c)^2}{c^2}}}$$

$$= 1.90(0.511 \text{ MeV}) = 0.970 \text{ MeV}$$

The kinetic energy is obtained by subtracting the rest energy from the total energy:

$$KE = E - m_0c^2 = 0.970 \text{ MeV} - 0.511 \text{ MeV} = 0.459 \text{ MeV}$$

EXAMPLE 28.10 The Energy of a Speedy Proton

The total energy of a proton is three times its rest energy. (a) Find the proton's rest energy in electron volts.

Solution

$$\text{Rest energy} = m_0c^2 = (1.67 \times 10^{-27}\ \text{kg})(3 \times 10^8\ \text{m/s})^2$$

$$= (1.50 \times 10^{-10}\ \text{J})\left(\frac{1\ \text{eV}}{1.60 \times 10^{-19}\ \text{J}}\right) = 938\ \text{MeV}$$

(b) With what speed is the proton moving?

Solution Since the total energy, E, is three times the rest energy, Equation 28.15 gives

$$E = mc^2 = 3m_0c^2 = \frac{m_0c^2}{\sqrt{1 - v^2/c^2}}$$

$$3 = \frac{1}{\sqrt{1 - v^2/c^2}}$$

Solving for v gives

$$1 - \frac{v^2}{c^2} = \frac{1}{9}$$

$$\frac{v^2}{c^2} = \frac{8}{9}$$

$$v = \frac{\sqrt{8}}{3}c = 2.83 \times 10^8\ \text{m/s}$$

(c) Determine the kinetic energy of the proton in electron volts.

Solution

$$KE = E - m_0c^2 = 3m_0c^2 - m_0c^2 = 2m_0c^2$$

Since $m_0c^2 = 938$ MeV, $KE = 1876$ MeV.

SUMMARY

The two basic postulates of the **special theory of relativity** are

1. The laws of physics are the same for all observers moving at constant velocity with respect to each other.
2. The speed of light is the same for all inertial observers, independent of their motion or of the motion of the source of light.

Some of the consequences of the special theory of relativity are as follows:

1. Clocks in motion relative to an observer slow down. This is known as **time dilation.** The relationship between time intervals in the moving and at rest systems is

$$\boxed{\Delta t = \gamma \Delta t'} \qquad (28.8)$$

where Δt is the time interval in the system in relative motion with respect to the clock, $\gamma = 1/\sqrt{1 - v^2/c^2}$, and $\Delta t'$ is the time interval in the system moving with the clock.

2. The length of an object in motion is *contracted* in the direction of motion. The equation for **length contraction** is

$$L = L'\sqrt{1 - v^2/c^2} \tag{28.9}$$

where L is the length in the system in motion relative to the object, and L' is the length in the system in which the object is at rest.
3. Events that are simultaneous for one observer are not simultaneous for another observer in motion relative to the first.

The relativistic expression for the **momentum** of a particle moving with a velocity v is

$$p \equiv \frac{m_0 v}{\sqrt{1 - v^2/c^2}} = \gamma m_0 v \tag{28.10}$$

The **mass** of an object increases as its speed increases as

$$m \equiv \frac{m_0}{\sqrt{1 - v^2/c^2}} \tag{28.11}$$

where m_0 is the rest mass of the object, that is, its mass as measured by an observer at rest with respect to it.

The relativistic expression for the addition of velocities is

$$u = \frac{v + u'}{1 + \dfrac{vu'}{c^2}} \tag{28.12}$$

where u is the velocity of an object as seen in one frame of reference, u' is the velocity of the same object in a different frame of reference, and v is the relative velocity of the two frames.

The relativistic expression for the **kinetic energy** of an object is

$$KE = mc^2 - m_0c^2 \tag{28.13}$$

where m_0c^2 is the **rest energy** of the object.

The **total energy** of a particle is

$$E = \frac{m_0c^2}{\sqrt{1 - v^2/c^2}} \tag{28.15}$$

This is Einstein's famous mass-energy equivalence equation.

The relativistic momentum is related to the total energy through the equation

$$E^2 = p^2c^2 + (m_0c^2)^2 \tag{28.16}$$

ADDITIONAL READING

H. Bondi, *Relativity and Common Sense*, Science Study Series, Garden City, N.Y., Doubleday, 1964.

J. Bronowski, "The Clock Paradox," *Sci. American*, February 1963, p. 134.

R. W. Clark, *Einstein: The Life and Times*, New York, World Publishing, 1971.

J. Crelinsten, "Relativity, Einstein, Physicists, and the Public," *The Physics Teacher*, February 1980, p. 115.

J. Crelinsten, "Physicists Receive Relativity: Revolution and Reaction," *The Physics Teacher*, February 1980, p. 187.

A. Einstein, *Ideas and Opinions*, New York, Crown, 1954.

A. Einstein, *Out of My Later Years*, Secaucus, N.J., Citadel Press, 1973.

G. Gamow, *Mr. Tomkins in Wonderland*, New York, Cambridge University Press, 1939.

R. D. Henry, "Special Relativity Made Transparent," *The Physics Teacher*, December 1985, p. 536.

L. Infeld, *Albert Einstein*, New York, Scribner's, 1950.

P. K. MacKeown, "Gravity is Geometry," *The Physics Teacher*, December 1984, p. 557.

J. Schwinger, *Einstein's Legacy*, Scientific American Library, New York, N.Y., W. H. Freeman and Co., 1985.

R. S. Shankland, "The Michelson-Morley Experiment," *Sci. American*, November 1964, p. 107.

G. J. Whitrow, ed., *Einstein, The Man and His Achievement*, New York, Dover, 1967.

QUESTIONS

1. What one measurement will two observers in relative motion always agree upon?
2. It is said that Einstein, in his teenage years, asked the question, "What would I see in a mirror if I carried it in my hands and ran at the speed of light?" How would you answer this question?
3. A spaceship in the shape of a sphere moves past an observer on earth with a speed of $0.5c$. What shape will the observer see as the spaceship moves past?
4. List some ways in which our day-to-day lives would change if the speed of light were 100 mi/h.
5. An astronaut moves away from the earth at a speed close to the speed of light. If an observer on earth could measure the astronaut's size and pulse rate, what changes (if any) would he or she measure? Would the astronaut measure any changes?
6. Since mass is a form of energy, can we conclude that a compressed spring has more mass than the same spring when it is not compressed? On the basis of your answer, which has more mass, a spinning planet or an otherwise identical but nonspinning planet?
7. Suppose astronauts were paid according to the time spent traveling in space. After a long voyage at a speed near that of light, a crew of astronauts return and open their pay envelopes. What will their reaction be?
8. Give a physical argument showing that it is impossible to accelerate an object of mass m to the speed of light, even with a continuous force acting on it.
9. What happens to the density of an object as its speed increases?
10. Consider the incorrect statement, "Matter can neither be created nor destroyed." How would you correct this statement in view of the special theory of relativity?

PROBLEMS

Section 28.3 The Michelson-Morley Experiment

1. Two airplanes fly paths I and II specified in Figure 28.3a. Both planes have air speeds of 100 m/s and fly for a distance of 200 km. The wind blows at 20 m/s in the direction shown in the figure. Find (a) the time of flight to each city, (b) the time to return, and (c) the difference in total flight times.

2. In one version of the Michelson-Morley experiment, the lengths L in Figure 28.4 were 28 m. Take v to be 3×10^4 m/s and find (a) the time difference caused by rotation of the interferometer and (b) the expected fringe shift, assuming that the light used has a wavelength of 550 nm.

Section 28.5 Consequences of Special Relativity

3. An astronaut at rest on earth has a heartbeat rate of 70 beats/min. When the astronaut is traveling in a spaceship at $0.9c$ what will be this rate as measured (a) by an observer also in the ship and (b) by an observer at rest on the earth?

4. The period of a pendulum is found to be 2 s when measured in the inertial frame of the pendulum. An observer moves by at a high rate of speed and concludes that the period is 7 s. How fast is he moving?

5. With what speed will a clock have to be moving in order to run at a rate that is one-half the rate of a clock at rest?

6. The average lifetime of a pi meson in its own frame of reference is 2.6×10^{-8} s. (This is the proper lifetime.) If the meson moves with a speed of $0.98c$, what is (a) its mean lifetime as measured by an observer on earth and (b) the average distance it travels before decaying as measured by an observer on earth? (c) What distance would it travel if time dilation did not occur?

7. How fast must a meter stick be moving if its length is observed to shrink to 0.5 m?

8. A cube is 2 cm on a side when at rest. (a) What shape does it take on when moving past an observer at a speed of 2.5×10^8 m/s, and (b) what is the length of each side?

9. The nearest stars to earth are approximately 4 light-years away. How fast would you have to travel in a spaceship to cause that distance to shrink to 1.5 light-years?

10. In a high-energy physics experiment, a certain elementary particle is created which travels a distance of 0.15 m at a speed of 2.2×10^8 m/s before it decays into two other particles. Find the lifetime of this particle in its frame of reference.

11. A friend travels by you in a spaceship at a high speed. He tells you that his ship is 20 m long and that the identical ship you are sitting in is 19 m long. According to your observations, (a) how long is your ship, (b) how long is his ship, and (c) what is the speed of your friend's ship?

12. The nearest star to earth is approximately 4 light-years away. If you travel at a speed of 2.5×10^8 m/s in a spaceship, how long does it take to get there (a) according to an earthbound observer and (b) according to an observer on the spaceship?

• **13.** An atomic clock is placed on a jet airplane. The clock measures a time interval of 3600 s when the jet moves with a speed of 400 m/s. What corresponding time interval does an identical clock held by an observer on the ground measure? (*Hint:* For $v/c \ll 1$, $\gamma \approx 1 + v^2/2c^2$.)

• **14.** A 2-m long car is traveling at a speed of 65 mi/h. What is the decrease in its length as seen by a stationary observer? See hint in Problem 13.

Section 28.6 Relativistic Momentum

15. Calculate the momentum of an electron moving with a speed of (a) $0.01c$, (b) $0.5c$, (c) $0.9c$.

16. Repeat Problem 15 for a proton.

17. An electron has a momentum that is 90% larger than its classical momentum. (a) Find the speed of the electron. (b) How would your result change if the particle were a proton?

• **18.** An electron has a speed $v = 0.9c$. At what speed will a proton have a momentum equal to that of the electron?

Section 28.7 Mass and the Ultimate Speed

19. At what speed must a 0.15-kg baseball be thrown to have a mass of 0.4 kg?

20. At what speed is the mass of a particle equal to twice its rest mass? Does your answer depend on the initial mass of the particle?

21. At low speeds, a charged object in a magnetic field moves in a circular path of radius $r = mv/qB$. If an electron moves in an orbit of radius 10 cm with a speed of 10^5 m/s, what will be the radius when its speed is $0.96c$?

22. A synchrocyclotron is a particle accelerator capable of accelerating protons to a speed of $0.66c$. At this speed what is the mass of a proton?

• **23.** A cube of mass 8 kg is 0.5 m on a side. (a) What is its density (mass per unit volume) as seen by an observer when the cube is moving away from the observer with a speed of $0.9c$? (b) What is the density as measured by an observer moving with the cube?

• **24.** An electron is accelerated along a straight path in a linear accelerator. The mass of the electron becomes 10,000 times its rest mass when it reaches its final speed. (a) What is its final speed? (b) If the length of the accelerator is 3500 m, what is its length in the frame of reference of the electron?

Section 28.8 Relativistic Addition of Velocities

25. A space vehicle is moving with a speed of $0.75c$ with respect to an external observer. An atomic particle is projected with a speed of $0.9c$ in the same direction as the ship's velocity with respect to an observer inside the vehicle. What is the speed of the projectile as seen by the external observer?

26. An astronaut in a spaceship travels at a speed of $0.8c$ away from an observer at rest on the earth. He turns around and shines a beam of light toward the stationary observer. What does the stationary observer measure for the speed of the light?

• **27.** An electron moves to the right with a speed of $0.9c$ relative to the laboratory frame. A proton moves to the right with a speed of $0.7c$ relative to the electron. Find the speed of the proton relative to the laboratory frame.

• **28.** Spaceship R is moving toward the right with a speed of 0.7c with respect to the earth. A second spaceship, L, moves to the left with the same speed with respect to the earth. What is the speed of L with respect to R?

Section 28.9 Relativistic Energy

29. A proton moves with a speed of $0.95c$. Calculate its (a) rest energy, (b) total energy, and (c) kinetic energy.

30. Find the rest energy of a proton in eV.

31. Find the speed of a particle whose total energy is 50% greater than its rest energy.

32. A mass of 0.5 kg is converted completely into energy. (a) How much energy is produced and (b) how long will this much energy keep a 100-W lightbulb burning?

33. The sun radiates approximately 4×10^{26} J of energy into space each second. (a) How much mass is converted into energy each second? (b) If the mass of the sun is 2×10^{30} kg, how long can the sun survive if the energy creation continues at the present rate?

34. What are the momentum and kinetic energy of a proton moving with a speed of $0.95c$?

35. A proton in a high-energy accelerator is given a kinetic energy of 50 GeV. Determine the (a) momentum and (b) speed of the proton.

• **36.** Assuming the mass of an electron could be converted completely into energy, how many electrons would have to be destroyed in order to run a 100-W lightbulb for one hour?

ADDITIONAL PROBLEMS

37. An electron has a kinetic energy five times greater than its rest energy. Find (a) its total energy and (b) its speed.

38. What is the mass and speed of a proton that has been accelerated through a difference of potential of (a) 500 V and (b) 5×10^8 V?

39. A radioactive nucleus moves with a speed v relative to a laboratory observer. The nucleus emits an electron in the positive x direction with a speed of $0.7c$ relative to the decaying nucleus and with a speed of $0.85c$ in the $+x$ direction relative to the laboratory observer. What is the value of v?

40. Energy reaches the upper atmosphere of the earth from the sun at the rate of 1.79×10^{17} W. If all of this energy were absorbed by the earth and converted into mass, how much would the mass of the earth increase in one year?

• **41.** An electron has a total energy which is equal to five times its rest energy. (a) What is its momentum? (b) Repeat for a proton.

• **42.** Find the total energy of a proton moving with a speed of (a) $0.5c$ and (b) $0.95c$.

• **43.** An alarm clock is set to sound in 10 h. At $t = 0$ the clock is placed in a spaceship moving with a speed of $0.75c$ (relative to the earth). What distance, as determined by an earth observer, does the spaceship travel before the alarm clock sounds?

• **44.** Determine the energy required to accelerate an electron from (a) $0.50c$ to $0.75c$ and (b) $0.90c$ to $0.99c$.

• **45.** The muon is an unstable particle that spontaneously decays into an electron and two neutrinos. If the number of nuons at $t = 0$ is N_0, the number at time t is given by $N = N_0 e^{-t/\tau}$ where τ is the mean lifetime, equal to 2.2 μs. Suppose the muons move at a speed of $0.95c$ and there are 5×10^4 muons at $t = 0$. (a) What is the observed lifetime of the muons? (b) How many muons remain after traveling a distance of 3 km?

•• **46.** A rod of length L_0 moves with a speed v along the horizontal direction. The rod makes an angle of θ_0 with respect to the axis of a coordinate system moving with the rod. (a) Show that the length of the rod as measured by a stationary observer is given by

$$L = L_0\left[1 - \left(\frac{v^2}{c^2}\right)\cos^2\theta_0\right]^{1/2}$$

(b) Show that the angle that the rod makes with the axis as seen by the stationary observer is given by the expression $\tan\theta = \gamma \tan\theta_0$. These results show that the rod is both contracted and rotated. (Take the lower end of the rod to be at the origin of the moving coordinate system.)

•• **47.** If a light source moves with a speed v relative to an observer, there is a shift in the observed frequency analogous to the Doppler effect for sound waves. Show that the observed frequency f_0 is related to the true frequency through the expression

$$f_0 = \sqrt{\frac{c \pm v_s}{c \mp v_s}}\,f$$

where the upper signs correspond to the source approaching the observer and the lower signs correspond to the source receding from the observer. (*Hint:* In the moving frame the period is the proper time interval and is given by $T = 1/f$. Furthermore, the wavelength measured by the observer is $\lambda_0 = (c - v_s)T_0$, where T_0 is the period measured in the stationary frame.)

Essay

The Renaissance of General Relativity

Clifford M. Will *McDonnell Center for the Space Sciences, Washington University*

During the two decades from 1960 to 1980, the subject of general relativity experienced a rebirth. Despite its enormous influence on scientific thought in its early years, by the late 1950s, general relativity had become a sterile, formalistic subject, cut off from the mainstream of physics. It was thought to have very little observational contact, and was believed to be an extremely difficult subject to learn and comprehend.

Yet by 1970, general relativity had become one of the most active and exciting branches of physics. It took on new roles both as a theoretical tool of the astrophysicist, and as a playground for the elementary-particle physicist. New experimental tests verified its predictions in unheard-of ways, and to remarkable levels of precision. Fields of study were created, such as "black-hole physics" and "gravitational-wave astronomy," that brought together the efforts of theorists and experimentalists. One of the most remarkable and important aspects of this renaissance of relativity was the degree to which experiment and observation motivated and complemented theoretical advances.

This was not always the case. In deriving general relativity during the final months of 1915, Einstein himself was not particularly motivated by a desire to account for observational results. Instead, he was driven by purely theoretical ideas of elegance and simplicity. His goal was to produce a theory of gravitation that incorporated in a natural way both the special theory of relativity that dealt with physics in inertial frames, and the principle of equivalence, the proposal that physics in a frame falling freely in a gravitational field was in some sense equivalent to physics in an inertial frame.

Once the theory was formulated, however, he did try to confront it with experiment by proposing three tests. One of these tests was an immediate success: the explanation of the anomalous advance in the perihelion of Mercury of 43 arcseconds per century, a problem that had bedeviled celestial mechanicians of the latter part of the 19th century. The next test, the deflection of light by the Sun, was such a success that it produced what today would be called a "media event." The measurements of the deflection, amounting to 1.75 arcseconds for a ray that grazes the Sun, by two teams of British astronomers in 1919, made Einstein an instant international celebrity. However, these measurements were not all that accurate, and subsequent measurements weren't much better. The third test, actually proposed by Einstein in 1907, was the gravitational redshift of light, but it was a test that remained unfulfilled until 1960, by which time it was no longer viewed as a true test of general relativity.

The turning point for general relativity came in the early 1960s, when discoveries of unusual astronomical objects such as quasars demonstrated that the theory would have important applications in astrophysical situations. Theorists found new ways to understand the theory and its observable consequences. Finally, the technological revolution of the last quarter century, together with the development of the interplanetary space program, provided new high-precision tools to carry out experimental tests of general relativity.

After 1960, the pace of research in general relativity and in an emerging field called "relativistic astrophysics" began to accelerate. New advances, both theoretical and observational, came at an ever increasing rate. They included the discovery of the cosmic background radiation; the analysis of the synthesis of helium from hydrogen in the big bang; observations of pulsars and of black-hole candidates; the development of the theory of relativistic stars and black holes; the theoretical study of gravitational radiation and the beginning of an experimental program to detect it; improved versions of old tests of general relativity, and brand new tests, discovered after 1960; the discovery of the binary pulsar, which provided evidence for gravity waves; the analysis of quantum effects outside black holes and of black hole evaporation; the discovery of a gravitational lens; and the beginnings of a unification of gravitation theory with the other interactions and with quantum mechanics.

The Principle of Equivalence

The foundations of general relativity are actually quite old, dating back to the equivalence principle of Galileo and Newton. In Newton's view, the principle of equivalence stated that all objects accelerate at the same rate in a gravitational field regardless of their mass or composition. This equality has been verified abundantly over the years, including classic experiments by the Hungarian physicist Baron Lorand von Eötvös around 100 years ago, and recent experiments at Princeton and Moscow State Universities. The accuracy of these tests is better than one part in 10^{11}. Einstein's insight was the recognition that, to an observer inside a freely falling laboratory, not only should objects float as if gravity were absent as a consequence of this equality, but also *all* laws of nongravitational physics, such as electromagnetism and quantum mechanics, should behave as if gravity were truly absent. It turns out that a consequence of this idea is that space-time is curved; in other words, that the effects of gravity are *indistinguishable* from the effects of being in curved space-time.

Gravitational Redshift

Another consequence of Einstein's insight is the gravitational redshift effect, which is a frequency shift between two identical clocks (meaning any device that produces a signal at a well-defined, steady frequency), at different heights in a gravitational field. For small differences in height h between clocks, the shift in the frequency Δf is given by

$$\frac{\Delta f}{f} = \frac{gh}{c^2}$$

where g is the local gravitational acceleration, and c is the speed of light. If the receiver is at a lower height than the emitter, the received signal is shifted to higher frequencies ("blueshift"), while if the receiver is higher, the signal is shifted to lower frequencies ("redshift"). The generic name for the effect is "gravitational redshift."

The first and most famous high-precision redshift measurement was the Pound-Rebka experiment of 1960, which measured the frequency shift of gamma-ray photons from the decay of iron-57 (^{57}Fe) as they ascended or descended the Jefferson Physical Laboratory tower at Harvard University.

The most precise gravitational redshift experiment performed to date was a rocket experiment carried out in June 1976. A "hydrogen maser" atomic clock was flown on a Scout D rocket to an altitude of 10,000 km, and its frequency compared to a similar clock on the ground using radio signals. After the effects of the rocket's motion were taken into account, the observations confirmed the gravitational redshift to 0.02%.

Because of these kinds of experiments, physicists are now convinced that spacetime *is* curved, and that the correct theory of gravity must be based on curved spacetime. That does not automatically imply general relativity, however, since it is not the only such theory. To test the specific predictions of general relativity for the amount and nature of space-time curvature, other experiments are needed.

Deflection of Light and Time Delay

The first of these is a test that made Einstein's name a household word: the deflection of light. According to general relativity, a light ray which passes the Sun at a distance d is deflected by an angle $\Delta\theta = 1.75''/d$, where d is measured in units of the solar radius, and the notation $''$ denotes seconds of arc.

The prediction of the bending of light by the Sun was one of the great successes of general relativity. Confirmation by the British astronomers Eddington and Crommelin of the bending of optical starlight observed during a total solar eclipse in the first months following World War I helped make Einstein a celebrity. However, those measurements had only 30% accuracy, and succeeding eclipse experiments weren't much better.

However, the development of long-baseline radio interferometry produced a method for greatly improved determinations of the deflection of light. Radio interferometry is a technique of combining widely separated radio telescopes in such a way that they can measure angular separations and changes in angles between radio sources as small as 10^{-4} arcseconds. Coupled with this technological advance is a series of heavenly coincidences: each year groups of strong quasars pass near the Sun (as seen from the Earth). The idea is to measure the differential deflection of radio waves from one quasar relative to those from another as they pass near the Sun. A number of measurements of this kind occurred almost annually over the period from 1969 to 1975, yielding a confirmation of the predicted deflection to 1.5%.

Closely related to light deflection is the "Shapiro time delay," a retardation of light signals that pass near the Sun. For instance, for a signal that grazes the Sun on a round trip from Earth to Mars at superior conjunction (when Mars is on the far side of the Sun), the round trip travel time is increased over what Newtonian theory would give by about 250 μs. The effect decreases with increasing distance of the signal from the Sun.

In the two decades following radio-astronomer Irwin Shapiro's 1964 discovery of this effect, several high-precision measurements have been made using the technique of radar-ranging to planets and spacecraft. Three types of targets were employed: planets such as Mercury and Venus, free flying spacecraft such as Mariners 6 and 7, and combinations of planets and spacecraft, known as "anchored spacecraft," such as the Mariner 9 Mars orbiter and the 1976 Viking Mars landers and orbiters. The Viking experiments produced dramatic results, agreeing with the general relativistic prediction to one part in a thousand. This corresponded to a measurement accuracy in the Earth-Mars distance of 30 meters.

Perihelion Shift

The explanation of the anomalous perihelion shift of Mercury's orbit was another of the triumphs of general relativity. This had been an unsolved problem in celestial mechanics for over half a century, since the announcement by Le Verrier in 1859 that, after the perturbing effects of the planets on Mercury's orbit had been accounted for, there remained in the data an unexplained advance in the perihelion of Mercury. The modern value for this discrepancy is 42.98 arcseconds per century and general relativity accounted for it in a natural way. Radar measurements of the orbit of Mercury since 1966 have led to improved accuracy, so that the relativistic advance is known to be about 0.5%.

Other measurements carried out since 1960 have given further support to general relativity. Observations of the motion of the Moon using laser ranging to a collection of specially designed reflectors, deposited on the lunar surface during the Apollo and Luna missions, have shown that the Moon and the Earth accelerate toward the Sun equally to one part in 10^{11}, an important confirmation of the equivalence principle applied to planetary sized bodies. Other measurements of planetary and lunar orbits have shown that the gravitational constant is a true constant of nature; it does not vary with time as the universe ages. Ambitious experimenters are currently designing an experiment using supercooled quartz gyroscopes to be placed in Earth orbit, to try to detect an important general relativistic effect called the "dragging of inertial frames," caused by the rotation of the Earth.

General relativity has passed every experimental test to which it has been put, and many alternative theories have fallen by the wayside. Most physicists now take the theory for granted, and look to see how it can be used as a practical tool in physics and astronomy.

Gravitational Radiation

One of these new tools is gravitational radiation, a subject that is almost as old as general relativity itself. By 1916, Einstein had succeeded in showing that the field equations of general relativity admitted wavelike solutions analogous to those of electromagnetic theory. For example, a dumbell rotating about an axis passing at right angles through its handle will emit gravitational waves that travel at the speed of light. They also carry energy away from the rotating dumbell, just as electromagnetic waves carry energy away from a light source.

In 1968, Joseph Weber made the stunning announcement that he had detected gravitational radiation of extraterrestrial origin using massive aluminum bars as detectors. A passing gravitational wave acts as an oscillating gravitational force field that alternately compresses and extends the bar in the lengthwise direction. However, subsequent observations by other researchers using bars with sensitivity that was claimed to be better than Weber's failed to confirm Weber's results. His results are now generally regarded as a false alarm, although there is still no good explanation for the "events" that he recorded in his bars.

Nevertheless, Weber's experiments did initiate the program of gravitational-wave detection, and inspired other groups to build better detectors. Currently a dozen laboratories around the world are engaged in building and improving upon the basic "Weber bar" detector, striving to reduce noise from thermal, electrical and environmental sources as much as possible to have a hope of detecting the very weak oscillations produced by a gravitational wave. For a bar of one meter length, the challenge is to detect a variation in length smaller than 10^{-20} meters, or 10^{-5} of the radius of a proton.

Another important type of detector is the laser interferometer, currently under development at several laboratories. This device operates on the same principle as the interferometer used by Michelson and Morley, but uses a laser as the light source. A passing gravitational wave will change the length of one arm of the apparatus relative to the other, and cause the interference pattern to vary. One ambitious proposal is to build two independent interferometers, each with arm lengths of 4 km, and separated by 1000 km.

It is hoped that these detectors will eventually be sensitive enough to detect gravitational waves from many sources, both in our galaxy and in distant galaxies, such as collapsing stars, double star systems, colliding black holes, and possibly even gravitational waves left over from the big bang. When this happens, a new field of "gravitational-wave astronomy" will be born.

Although gravitational radiation has not been detected directly, we know that it exists, through a remarkable system known as the binary pulsar. Discovered in 1974 by radio astronomers Russell Hulse and Joseph Taylor, it consists of a pulsar (which is a rapidly spinning neutron star) and a companion star in orbit around each other. Although the companion has not been seen directly, it is also believed to be a neutron star.

The pulsar acts as an extremely stable clock, its pulse period of approximately 59 milliseconds drifting by only a quarter of a nanosecond per year. By measuring the arrival times of radio pulses at Earth, observers were able to determine the motion of the pulsar about its companion with amazing accuracy. For example, the accurate value for the orbital period is 27906.98163 seconds, and the orbital eccentricity is 0.617127.

Like a rotating dumbell, an orbiting binary system should emit gravitational radiation, and in the process lose some of its orbital energy. This energy loss will cause the pulsar and its companion to spiral in toward each other, and the orbital period to shorten. According to general relativity, the predicted decrease in the orbital period is 75 microseconds per year. The observed decrease rate is in agreement with the prediction to about

4%. This confirms the existence of gravitational radiation and the general relativistic equations that describe it.

Black Holes

One of the most important and exciting aspects of the relativity renaissance is the study of and search for black holes. However, the first glimmerings of the black hole idea date back to the 18th century, in the writings of a British amateur astronomer, the Reverend John Michell. Reasoning on the basis of the corpuscular theory that light would be attracted by gravity in the same way as ordinary matter is attracted, he noted that light emitted from the surface of a body such as the Earth or the Sun would be reduced in velocity by the time it reached great distances (Michell of course did not know special relativity). It would therefore be possible for a body to be sufficiently massive and compact to prevent light from escaping from its surface.

Although the general relativistic solution for a non-rotating black hole was discovered by Karl Schwarzschild in 1916, and a calculation of gravitational collapse to a black hole state was performed by J. Robert Oppenheimer and Hartland Snyder in 1939, black hole physics didn't really begin until the middle 1960s, when astronomers confronted the problem of the energy output of the quasars, and started to take black holes seriously.

A black hole is formed when a star has exhausted the thermonuclear fuel necessary to produce the heat and pressure that support it against gravity. The star begins to collapse, and if it is massive enough, it continues to collapse until the radius of the star approaches a value called the gravitational radius or Schwarzschild radius. In the non-rotating spherical case, this radius has a value given by $2GM/c^2$, where M is the mass of the star. For a body of one solar mass, the gravitational radius is about 3 km, for a body of the mass of the Earth, it is about 9 mm. An observer sitting on the surface of the star sees the collapse continue to smaller and smaller radii, until both star and observer reach the origin $r = 0$, with consequences too horrible to describe in detail. On the other hand, an observer at great distances observes the collapse to slow down as the radius approaches the gravitational radius, a result of the gravitational redshift of the light signals sent outward. The distant observer never sees any signals emitted by the falling observer once the latter is inside the gravitational radius. This is because any signal emitted inside can never escape the sphere bounded by the gravitational radius, called the "event horizon."

In 1974, Stephen Hawking discovered that the laws of quantum mechanics applied to the physics outside a black hole required it to evaporate by the creation of particles with a thermal energy spectrum, and to have an associated temperature and entropy. The temperature of a Schwarzschild black hole is $T = hc^3/8\pi kGM$, where h is Planck's constant, and k is Boltzmann's constant. This discovery demonstrated a remarkable connection between gravity, thermodynamics, and quantum mechanics, that helped renew the theoretical quest for a grand synthesis of all the fundamental interactions. On the other hand, for black holes of astronomical masses, the evaporation is completely negligible, since for a solar-mass black hole, $T \approx 10^{-6}$ K.

Although a great deal is known about black holes in theory, rather less is known about them observationally. There are several instances in which the evidence for the existence of black holes is impressive, but in all cases it is indirect. For instance, in the x-ray source Cygnus X1, the source of the x-rays is believed to be a black hole with a mass larger than about 6 solar masses in orbit around a giant star. The x-rays are emitted by matter pulled from the surface of the companion star and sent into a spiralling orbit around the black hole. Similarly, there is evidence in the centers of certain galaxies, such as M87, and possibly even our own, of black holes of between 10^2 and 10^8 solar masses. Accretion of matter onto supermassive rotating black holes may produce the jets of outflowing matter that are observed in many quasars and active galactic nuclei. These and other astrophysical processes that might aid in the detection of black holes are being studied by relativists and astrophysicists.

Cosmology

The other area in which there has been a renaissance for general relativity is cosmology. Although Einstein in 1917 first used general relativity to calculate a model for the universe as a whole, the subject was not considered a serious branch of physics until the 1960s, when astronomical observations lent firm credence to the idea that the universe was expanding from a "big bang." In 1965 came the discovery of the cosmic background radiation by Arno Penzias and Robert Wilson. This radiation is the remains of the hot electromagnetic black-body radiation that once dominated the universe in its earlier phase, now cooled to 3 kelvins by the subsequent expansion of the universe. Next came calculations of the amount of helium that would be synthesized from hydrogen by thermonuclear fusion in the very early universe, around 1000 seconds after the big bang. The amount, approximately 25% by weight, was in agreement with the abundances of helium observed in stars and in interstellar space. This was an important confirmation of the hot big-bang picture, because the amount of helium believed to be produced by fusion in the interiors of stars is woefully inadequate to explain the observed abundances.

Today, the general relativistic hot big-bang model of the universe has broad acceptance, and cosmologists now focus their attention on more detailed issues, such as how galaxies and other large-scale structures formed out of the hot primordial soup, and on what the universe might have been like earlier than 1000 seconds, all the way back to 10^{-36} s (and some brave cosmologists are going back even further) when the laws of elementary-particle physics may have played a major role in the evolution of the universe.

Acceptance of General Relativity

One of the outgrowths of the renaissance of general relativity that has occurred since 1960 has been a change in attitude about the importance and use of the theory. Its importance as a fundamental theory of the nature of spacetime and gravitation has not been diminished in the least; if anything it has been enhanced by the flowering of research in the subject that has taken place. Its importance as a foundation for other theories of physics has been strengthened by current searches for unified quantum theories of nature that incorporate gravity along with the other interactions.

But the real change in attitude about general relativity has been in its use as a tool in the real world. In astrophysical situations, general relativity plays a central role in the study of neutron stars, black holes, relativistic binary star systems, and the universe as a whole. Gravitational radiation may one day provide a completely new observational tool for exploring and examining the cosmos.

Relativity even plays a role in everyday life. For example, the gravitational redshift effect on clocks *must* be taken into account in satellite-based navigation systems, such as the U.S. Air Force's Global Positioning System, in order to achieve the required positional accuracy of a few meters or time transfer accuracy of a few nanoseconds.

To general relativists, always eager to find practical consequences of their subject, these have been very welcome developments!

SUGGESTED READING

P. C. W. Davies, *The Search for Gravity Waves*, Cambridge, England, Cambridge University Press, 1980.

G. Greenstein, *Frozen Star: Of Pulsars, Black Holes and the Fate of Stars*, New York, Freundlich Books, 1984.

S. Weinberg, *The First Three Minutes: A Modern View of the Origin of the Universe*, New York, Basic Books, 1977.

C. M. Will, *Was Einstein Right? Putting General Relativity to the Test*, New York, Basic Books, 1986.

ESSAY QUESTIONS

1. Two stars are a certain distance from each other as seen in the night sky. Half a year later, the stars are now overhead during the day, and the Sun is now located midway between the two stars. Because of the deflection of light by the Sun, do the stars now appear closer together or farther apart?
2. When gravitational waves are emitted by the binary pulsar, the double-star system loses energy. As a consequence, its orbital velocity increases, and thus its kinetic energy increases. How is this possible?
3. The temperature of a black hole is inversely proportional to its mass, so that when it gains energy (which is equivalent to mass), its temperature goes down, and when it loses energy, its temperature goes up. What do you conclude about the sign of its heat capacity? Can a black hole ever be in thermal equilibrium with an infinite reservoir at a fixed temperature?

ESSAY PROBLEMS

1. What is the gravitational frequency shift between two atomic clocks separated by 1000 km, both at sea level?
2. A pair of identical twin scientists set out to test the gravitational redshift effect in the following way. One climbs to the top of Mount Baldy and lives there for a year, while the other remains behind in the laboratory at the base of the mountain. When they are reunited after one year, one of the twins is slightly older than the other. Which one is older and why? If Mount Baldy is 5000 m high, what is the difference in age between the twins after one year?
3. Calculate the deflection of light for light rays that pass the Sun at distances of 2 solar radii, 10 solar radii, and 100 solar radii.
4. Gravitational waves travel with the same speed as light, 3×10^{10} cm/s. Calculate the wavelength of gravitational waves emitted by the following sources: (a) a star collapsing to form a black hole, emitting waves with a frequency around 1000 Hz, and (b) the primary pulsar, emitting waves with a period of four hours (one half the orbital period).
5. What is the fractional decrease in the orbital period of the binary pulsar per year? Estimate how long the binary system will live before gravitational radiation will reduce the orbital period to zero (in other words will bring the stars so close together that they will merge into one object).

6. Calculate the mass of a black hole (in solar mass units) whose Schwarzschild radius is equal to the radius of the Earth's orbit around the Sun (1.5×10^8 km). One solar mass is about 2×10^{33} g.
7. Show that a billion-ton black hole has a Schwarzschild radius comparable to the radius of a proton (1 ton $\approx 10^6$ g, proton radius $\approx 10^{-13}$ cm). What kind of object on Earth weighs a billion tons?
8. Determine the mass and radius of a black hole whose apparent density is equal to that of water, 1 gm/cm^3. Assume that the volume of a black hole is given by $4\pi R^3/3$, and express the mass in solar mass units.
9. Calculate the mass of a black hole whose temperature is room temperature.

1, 2, 4, 8, 9, 11, 12

CHAPTER 29

Quantum Physics

First Solvay Congress, Brussels, October 1911, attended by many famous scientists, including Albert Einstein and Marie Curie. (International Institute of Physics and Chemistry, Courtesy AIP Niels Bohr Library)

In the previous chapter, we discussed why Newtonian mechanics must be replaced by Einstein's special theory of relativity when we are dealing with particles whose speeds are comparable to the speed of light. Although many problems were indeed resolved by the theory of relativity in the early part of the 20th century, many experimental and theoretical problems remained unsolved. Attempts to apply the laws of classical physics to explain the behavior of matter on the atomic level were totally unsuccessful. Various phenomena, such as blackbody radiation, the photoelectric effect, and the emission of sharp spectral lines by atoms in a gas discharge tube, could not be understood within the framework of classical physics. We shall describe these phenomena because of their importance in subsequent developments.

Another revolution took place in physics between 1900 and 1930. This was the era of a new and more general formulation called *quantum mechanics*. This new approach was highly successful in explaining the behavior of atoms, molecules, and nuclei. As with relativity, the quantum theory requires a modification of our ideas concerning the physical world.

The earliest and most basic ideas of quantum theory were introduced by Planck, and most of the subsequent mathematical developments, interpretations, and improvements were made by a number of distinguished physicists,

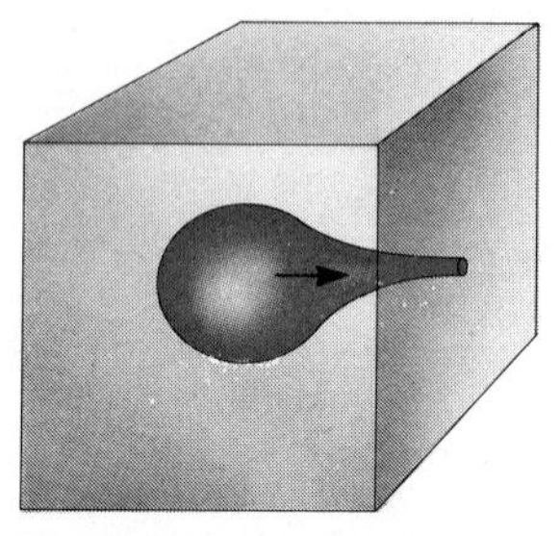

FIGURE 29.1 A good approximation to a black body is a cavity radiating energy through a small hole drilled in one side of a solid block of metal.

including Einstein, Bohr, Schrödinger, de Broglie, Heisenberg, Born, and Dirac. An extensive study of quantum theory is certainly beyond the scope of this book. This chapter is simply an introduction to the underlying ideas of quantum theory and the wave-particle nature of matter. We shall also discuss some simple applications of quantum theory, including the photoelectric effect, the Compton effect, and x-rays.

29.1 BLACKBODY RADIATION AND PLANCK'S HYPOTHESIS

An object at any temperature is known to emit radiation sometimes referred to as **thermal radiation.** The characteristics of this radiation depend on the temperature and properties of the object. At low temperatures, the wavelengths of the thermal radiation are mainly in the infrared region and hence are not observed by the eye. As the temperature of the object is increased, it eventually begins to glow red. At sufficiently high temperatures, it appears to be white, as in the glow of the hot tungsten filament of a lightbulb. A careful study of thermal radiation shows that it consists of a continuous distribution of wavelengths from the infrared, visible, and ultraviolet portions of the spectrum.

From a classical viewpoint, thermal radiation (electromagnetic waves) originates from accelerated charged particles near the surface of the object, which emit radiation much like small antennas. The thermally agitated charges can have a distribution of accelerations, which accounts for the continuous spectrum of radiation emitted by the object. By the end of the 19th century, it had become apparent that the classical theory of thermal radiation was inadequate. The basic problem was in understanding the observed distribution of wavelengths in the radiation emitted by a black body. By definition, a black body is an ideal system that absorbs *all* radiation incident on it. A good approximation to a black body is the inside of a hollow object, as shown in Figure 29.1. The nature of the radiation emitted through a small hole leading to the cavity depends only on the temperature of the cavity walls.

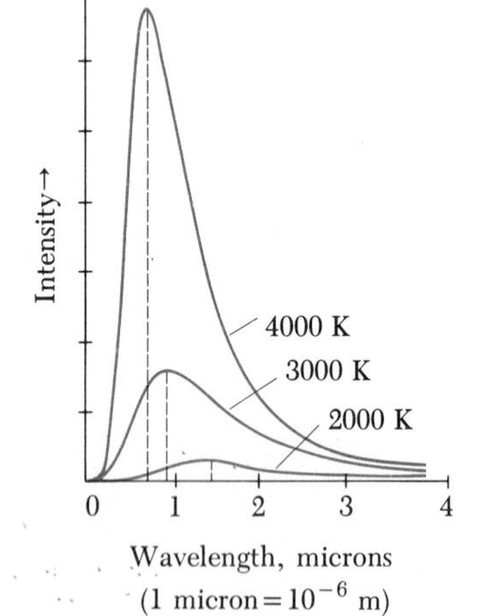

FIGURE 29.2 Intensity of blackbody radiation versus wavelength at three different temperatures. Note that the total radiation emitted (the area under a curve) increases with increasing temperature.

Experimental data for the distribution of energy for blackbody radiation at three different temperatures are shown in Figure 29.2. Note that the radiated energy varies with wavelength and temperature. As the temperature of the black body increases, the total amount of energy it emits increases. Also, with increasing temperatures, the peak of the distribution shifts to shorter wavelengths. This shift was found to obey the following relationship, called **Wien's displacement law:**

$$\lambda_{max} T = 0.2898 \times 10^{-2} \text{ m·K} \tag{29.1}$$

where λ_{max} is the wavelength at which the curve peaks and T is the absolute temperature of the object emitting the radiation.

Early attempts to use classical ideas to explain the shape of the curves shown in Figure 29.2 failed. Figure 29.3 shows an experimental plot of the blackbody radiation spectrum (solid curve) together with the theoretical picture of what this curve should look like based on classical theories (dashed curve). At long wavelengths, classical theory is in good agreement with the experimental data. However, at short wavelengths there is major disagreement between theory and experiment. This can be seen by noting that as λ approaches zero, the classical theory predicts that the amount of energy being radiated should increase, that is, short-wavelength radiation should be most intense and all objects should have a blue glow. This is contrary to the experimental data, which show that as λ approaches zero, the amount of energy carried by short-wavelength radiation also approaches zero. This contradiction is often called the **ultraviolet catastrophe.**

In 1900, Planck discovered a formula for blackbody radiation that was in complete agreement with experiments at all wavelengths. Figure 29.3 shows that Planck's law fits the experimental results quite well. In his theory, Planck made two assumptions, which at the time were bold and controversial, concerning the nature of the oscillating charges of the cavity walls:

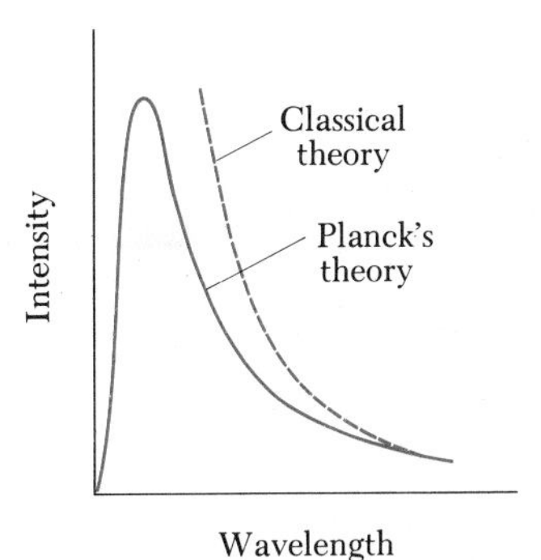

FIGURE 29.3 Comparison of the Planck theory with the classical theory for the distribution of blackbody radiation.

1. The vibrating molecules which emitted the radiation could have only certain *discrete* amounts of energy, E_n, given by

$$E_n = nhf \tag{29.2}$$

where n is a positive integer called a **quantum number** and f is the frequency of vibration of the molecules. The energies of the molecule are said to be *quantized*, and the allowed energy states are called *quantum states*. The factor h is a constant, known as **Planck's constant,** given by

$$h = 6.626 \times 10^{-34}\ \text{J}\cdot\text{s} \tag{29.3}$$

2. The molecules emit energy in discrete units of light energy called **quanta** (or **photons,** as they are now called). They do so by "jumping" from one quantum state to another. If the quantum number changes by one unit, Equation 29.2 shows that the amount of energy radiated by the molecule

Max Planck (1858–1947) was a German physicist who introduced the concept of a "quantum of action" (Planck's constant, h) in an attempt to explain the spectral distribution of blackbody radiation, which laid the foundations for quantum theory. He was awarded the Nobel prize in 1918 for his discovery of the quantized nature of energy. "My futile attempts to fit the elementary quantum of action into the classical theory continued for a number of years, and they cost me a great deal of effort. . . . I now knew for a fact that the elementary quantum of action played a far more significant part in physics than I had originally been inclined to suspect." (AIP Niels Bohr Library, W.F. Meggers Collection)

equals hf. Hence, the energy of a light quantum, corresponding to the energy difference between two adjacent levels, is given by

$$\boxed{E = hf} \tag{29.4}$$

The molecule will radiate or absorb energy only when it changes quantum states. If it remains in one quantum state, no energy is absorbed or emitted.

The key point in Planck's theory is the radical assumption of quantized energy states. This development marked the birth of the quantum theory. At that time, most scientists, including Planck, did not consider the quantum concept to be realistic. Hence, Planck and others continued to search for a more rational explanation of blackbody radiation. However, subsequent developments showed that a theory based on the quantum concept (rather than on classical concepts) had to be used to explain a number of phenomena at the atomic level.

EXAMPLE 29.1 Thermal Radiation from the Human Body

The temperature of the skin is approximately 35° C. What is the wavelength at which the peak occurs in the radiation emitted from the skin?

Solution From Wien's displacement law, we have

$$\lambda_{max} T = 0.2898 \times 10^{-2}\ \text{m}\cdot\text{K}$$

Solving for λ_{max}, noting that 35° C corresponds to an absolute temperature of 308 K, we have

$$\lambda_{max} = \frac{0.2898 \times 10^{-2}\ \text{m}\cdot\text{K}}{308\ \text{K}} = 940\ \mu\text{m}$$

This radiation is in the infrared region of the spectrum.

Exercise (a) Find the wavelength corresponding to the peak of the radiation curve for the heating element of an electric oven at a temperature of 1200 K. Note that although this radiation peak lies in the infrared, there is enough visible radiation at this temperature to give the element a red glow. (b) Calculate the wavelength corresponding to the peak of the radiation curve for an object whose temperature is 5000 K, an approximate temperature for the surface of the sun.

Answer (a) 2.415 μm (b) 580 nm, in the visible region.

EXAMPLE 29.2 The Quantized Oscillator

A 2-kg mass is attached to a massless spring of force constant $k = 25$ N/m. The spring is stretched 0.4 m from its equilibrium position and released. (a) Find the total energy and frequency of oscillation according to classical calculations. (b) Assume that the energy is quantized and find the quantum number, n, for the system. (c) How much energy would be carried away in a one-quantum change?

Solution (a) The total energy of a simple harmonic oscillator having an amplitude A is $\frac{1}{2} KA^2$. Therefore,

$$E = \frac{1}{2} KA^2 = \frac{1}{2}(25\ \text{N/m})(0.4\ \text{m})^2 = 2.0\ \text{J}$$

The frequency of oscillation is

$$f = \frac{1}{2\pi}\sqrt{\frac{K}{m}} = \frac{1}{2\pi}\sqrt{\frac{25\ \text{N/m}}{2\ \text{kg}}} = 0.56\ \text{Hz}$$

(b) If the energy is quantized, we have $E_n = nhf$, and from the result of (a), we have

$$E_n = nhf = n(6.63 \times 10^{-34}\ \text{J·s})(0.56\ \text{Hz}) = 2.0\ \text{J}$$

Therefore,

$$n = 5.4 \times 10^{33}$$

(c) The energy carried away in a one-quantum change of energy is

$$E = hf = (6.63 \times 10^{-34}\ \text{J·s})\ (0.56\ \text{Hz}) = 3.7 \times 10^{-34}\ \text{J}$$

The energy carried away by a one-quantum change in energy is such a small fraction of the total energy of the oscillator that we could not expect to see such a small change in the system. Thus, even though the decrease in energy of a spring-mass system is quantized and does decrease by small quantum jumps, our senses perceive the decrease as continuous. Quantum effects become important and measurable only on the submicroscopic level of atoms and molecules.

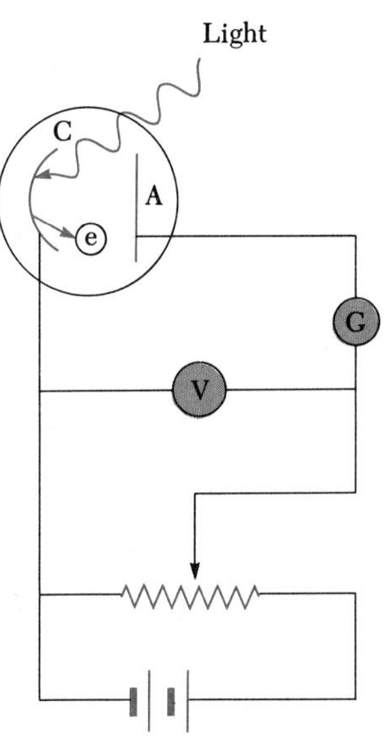

FIGURE 29.4 Circuit diagram for observing the photoelectric effect. When light strikes the plate C, photoelectrons are ejected from the plate. Electrons collected at A constitute a current in the circuit.

EXAMPLE 29.3 The Energy of a "Yellow" Photon

Yellow light having a frequency of approximately 6×10^{14} Hz is the predominant frequency in sunlight. What is the energy carried by a quantum of this light?

Solution The energy carried by one quantum of light is given by Equation 29.4:

$$E = hf = (6.63 \times 10^{-34}\ \text{J·s})\ (6 \times 10^{14}\ \text{Hz}) = 4 \times 10^{-19}\ \text{J} = 2.5\ \text{eV}$$

29.2 THE PHOTOELECTRIC EFFECT

In the latter part of the 19th century, experiments showed that, when light is incident on certain metallic surfaces, electrons are emitted from the surfaces. This phenomenon is known as the **photoelectric effect,** and the emitted electrons are called **photoelectrons.** The first discovery of this phenomenon was made by Hertz, who was also the first to produce the electromagnetic waves predicted by Maxwell.

Figure 29.4 is a schematic diagram of an apparatus in which the photoelectric effect can occur. An evacuated glass or quartz tube contains a metal plate, C, connected to the negative terminal of a battery. Another metal plate, A, is maintained at a positive potential by the battery. When the tube is kept in the dark, the galvanometer, G, reads zero, indicating that there is no current in the circuit. However, when monochromatic light of the appropriate wavelength shines on plate C, a current is detected by the galvanometer, indicating a flow of charges across the gap between C and A. The current associated with this process arises from electrons emitted from the negative plate and collected at the positive plate.

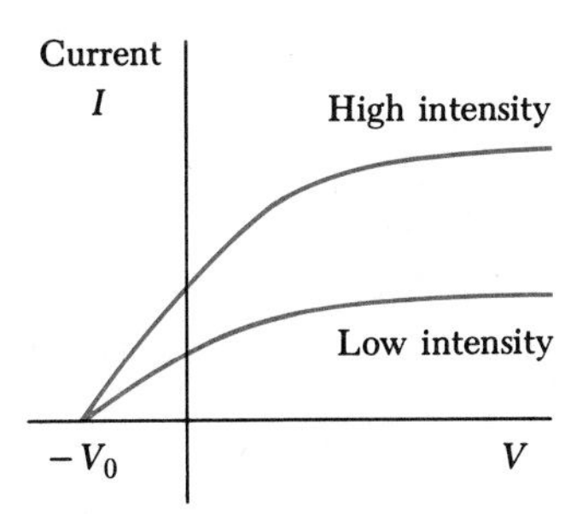

FIGURE 29.5 Photoelectric current versus voltage for two light intensities. The current increases with intensity but reaches a saturation level for large values of V. At voltages equal to or less than $-V_0$, the current is zero.

A plot of the photoelectric current versus the potential difference, V, between A and C is shown in Figure 29.5 for two light intensities. Note that for large values of V, the current reaches a maximum value, corresponding to the case where all photoelectrons are collected at A. In addition, the current increases as the incident light intensity increases, as you might expect. Finally, when V is negative, that is, when the battery in the circuit is reversed to make C positive and A negative, the photoelectrons are repelled by the negative plate A. Only those electrons having a kinetic energy greater than eV will reach A, where e is the charge on the electron. Furthermore, if V is less than or equal to V_0, called the **stopping potential,** no electrons will reach A and the current will be zero. The stopping potential is *independent* of the radiation intensity. The maximum kinetic energy of the photoelectrons is related to the stopping potential through the relation

$$KE_{max} = eV_0 \quad (29.5)$$

Several features of the photoelectric effect could not be explained with classical physics or with the wave theory of light. The major observations that were not understood are as follows:

1. No electrons are emitted if the incident light frequency falls below some **cutoff frequency,** f_c, which depends on the material being illuminated. For example, in the case of sodium, $f_c = 5.50 \times 10^{14}$ Hz. This is inconsistent with the wave theory, which predicts that the photoelectric effect should occur at any frequency, provided the light intensity is high enough.
2. If the light frequency exceeds the cutoff frequency, a photoelectric effect is observed and the number of photoelectrons emitted is proportional to the light intensity. However, the maximum kinetic energy of the photoelectrons is independent of light intensity, a fact that cannot be explained by the concepts of classical physics.
3. The maximum kinetic energy of the photoelectrons increases with increasing light frequency.
4. Electrons are emitted from the surface almost instantaneously (less than 10^{-9} s after the surface is illuminated), even at low light intensities. Classically, one would expect that the electrons would require some time to absorb the incident radiation before they acquire enough kinetic energy to escape from the metal.

A successful explanation of the photoelectric effect was given by Einstein in 1905, the same year he published his special theory of relativity. In his photoelectric paper, for which he received the Nobel price in 1921, Einstein extended Planck's concept of quantization to electromagnetic waves. He assumed that light (or any electromagnetic wave) of frequency f can be considered to be a stream of photons. Each photon has an energy E, given by

Energy of a photon

$$E = hf \quad (29.6)$$

where h is Planck's constant. Thus, Einstein considered light to be much like a stream of particles traveling through space (rather than a wave), where each "particle" could be absorbed as a unit by an electron. Furthermore, Einstein argued that when the photon's energy is transferred to an electron in a metal, the energy acquired by the electron must be hf. However, the electron must also pass through the metal surface in order to be emitted, and some energy is required to overcome this barrier. The amount of energy, ϕ, required to escape the metal is known as the **work function** of the substance and is of the order of a few electron volts for metals. For example, the work function for zinc is about 3.0 eV. Hence, in order to conserve energy, the maximum kinetic energy of the ejected photoelectrons is the difference between the photon energy and the work function of the metal, or

$$KE_{max} = hf - \phi \tag{29.7}$$

Photoelectric effect equation

That is, the excess energy $hf - \phi$ equals the maximum kinetic energy the liberated electron can have outside the surface.

With the photon theory of light, one can explain the features of the photoelectric effect that cannot be understood using classical concepts. These are briefly described in the order they were introduced earlier:

1. The fact that the photoelectric effect is not observed below a certain cutoff frequency follows from the fact that the energy of the photon must be greater than or equal to ϕ. If the energy of the incoming photon is not equal to or greater than ϕ, the electrons will never be ejected from the surface, regardless of the intensity of the light.
2. The fact that KE_{max} is independent of the light intensity can be understood with the following argument. If the light intensity is doubled, the number of photons is doubled, which doubles the number of photoelectrons emitted. However, their kinetic energy, which equals $hf - \phi$, depends only on the light frequency and the work function, not on the light intensity.
3. The fact that KE_{max} increases with increasing frequency is easily understood with Equation 29.7.
4. Finally, the fact that the electrons are emitted almost instantaneously is consistent with the particle theory of light, in which the incident energy appears in small packets and there is a one-to-one interaction between photons and electrons. This is in contrast to having the energy of the photons distributed uniformly over a large area.

A final confirmation of Einstein's theory is a test of the prediction of a linear relationship between f and KE_{max}. Indeed, such a linear relationship is observed, as sketched in Figure 29.6. The slope of such a curve gives a value for h. The intercept on the horizontal axis gives the cutoff frequency, which is related to the work function through the relation $f_c = \phi/h$. This corresponds to a **cutoff wavelength** of

$$\lambda_c = \frac{c}{f_c} = \frac{c}{\phi/h} = \frac{hc}{\phi} \tag{29.8}$$

where c is the speed of light (3.00×10^8 m/s). Wavelengths *greater* than λ_c incident on a material with a work function ϕ do not result in the emission of photoelectrons.

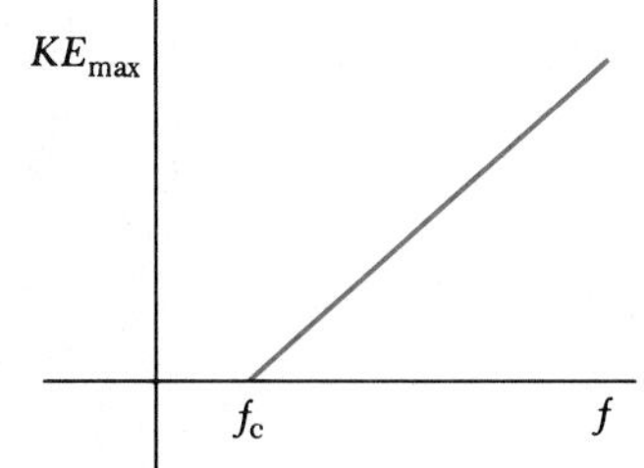

FIGURE 29.6 A sketch of KE_{max} versus frequency of incident light for photoelectrons in a typical photoelectric effect experiment. Photons with frequency less than f_c do not have sufficient energy to eject an electron from the metal.

EXAMPLE 29.4 The Photoelectric Effect for Sodium

A sodium surface is illuminated with light of wavelength 300 nm. The work function for sodium is 2.46 eV. Find (a) the kinetic energy of the ejected photoelectrons and (b) the cutoff wavelength for sodium.

Solution (a) The energy of the illuminating light beam is

$$E = hf = \frac{hc}{\lambda} = \frac{(6.63 \times 10^{-34}\ \text{J·s})(3.00 \times 10^{8}\ \text{m/s})}{300 \times 10^{-9}\ \text{m}}$$

$$= 6.63 \times 10^{-19}\ \text{J} = \frac{6.63 \times 10^{-19}\ \text{J}}{1.60 \times 10^{-19}\ \text{J/eV}} = 4.14\ \text{eV}$$

where we have used the conversion 1 eV = 1.6×10^{-19} J. Using Equation 29.7 gives

$$KE_{\text{max}} = hf - \phi = 4.14\ \text{eV} - 2.46\ \text{eV} = 1.68\ \text{eV}$$

(b) The cutoff wavelength can be calculated from Equation 29.8 after we convert ϕ from electron volts to joules:

$$\phi = 2.46\ \text{eV} = (2.46\ \text{eV})(1.6 \times 10^{-19}\ \text{J/eV}) = 3.94 \times 10^{-19}\ \text{J}$$

Hence

$$\lambda_c = \frac{hc}{\phi} = \frac{(6.63 \times 10^{-34}\ \text{J·s})(3.00 \times 10^{8}\ \text{m/s})}{3.94 \times 10^{-19}\ \text{J}}$$

$$= 5.05 \times 10^{-7}\ \text{m} = 505\ \text{nm}$$

This wavelength is in the green region of the visible spectrum.

29.3 APPLICATIONS OF THE PHOTOELECTRIC EFFECT

The photoelectric cell shown in Figure 29.4 acts much like a switch in an electric circuit in that it produces a current in an external circuit when light of sufficiently high frequency falls on the cell but does not allow a current in the dark. Many practical devices in our everyday lives depend on the photoelectric effect. One of the first practical uses was as the detector in the light meter of a camera. Light reflected from the object to be photographed strikes a photoelectric surface, causing it to emit electrons, which then pass through a very sensitive ammeter. The magnitude of the current depends on the intensity of the light. Modern solid-state devices have now replaced light meters that used the photoelectric effect.

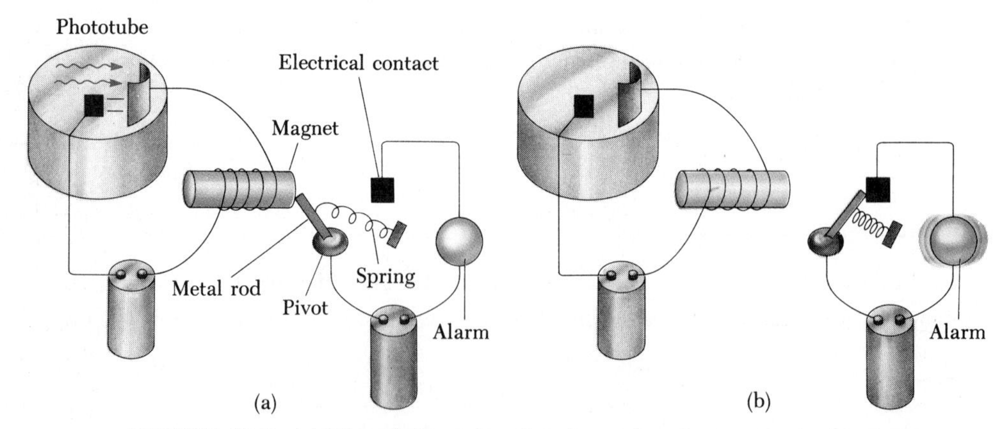

FIGURE 29.7 (a) When light strikes the phototube, the current in the circuit on the left energizes the magnet, breaking the burglar alarm circuit on the right. (b) When the light source is removed, the alarm circuit is closed, and the alarm sounds.

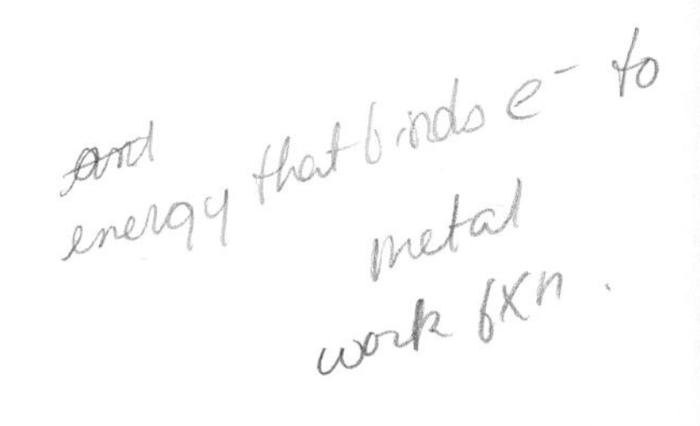

A second example of the use of the photoelectric effect is the burglar alarm. This device often uses ultraviolet rather than visible light in order to make the presence of the beam less obvious. A beam of light passes from the source to a photosensitive surface; the current produced is then amplified and used to energize an electromagnet which attracts a metal rod, as in Figure 29.7a. If an intruder breaks the light beam, the electromagnet switches off and the spring pulls the iron rod to the right (Fig. 29.7b). In this position, a completed circuit allows current to pass and the alarm system is activated.

Figure 29.8 shows how the photoelectric effect is used to produce the sound on a movie film. The sound track is located along the side of the film in the form of an optical pattern of light and dark lines. A beam of light in the projector is directed through the soundtrack toward a phototube. The variation in shading on the sound track varies the light intensity falling on the plate of the phototube, thus changing the current in the circuit. This changing current electrically simulates the original sound wave and reproduces it in the speaker.

29.4 X-RAYS

In 1895 at the University of Wurzburg, Wilhelm Roentgen (1845–1923) was studying electrical discharges in low-pressure gases when he noted that a fluorescent screen glowed even when placed several meters from the gas discharge tube and even when black cardboard was placed between the tube and the screen. He concluded that the effect was caused by a mysterious type of radiation, which he called **x-rays** because of their unknown nature. Subsequent study showed that these rays traveled at or near the speed of light and that they could not be deflected by either electric or magnetic fields. This last fact indicated that x-rays did not consist of beams of charged particles, although the possibility that they were beams of uncharged particles remained.

In 1912 Max von Laue (1879–1960) suggested that one should be able to diffract x-rays by using the regular atomic spacings of a crystal lattice as a diffraction grating, just as visible light is diffracted by a ruled grating. Shortly thereafter, researchers demonstrated that such a diffraction pattern could be observed, similar to that shown in Figure 29.9 for NaCl. The wavelengths of the x-rays were then determined from the diffraction data and the known

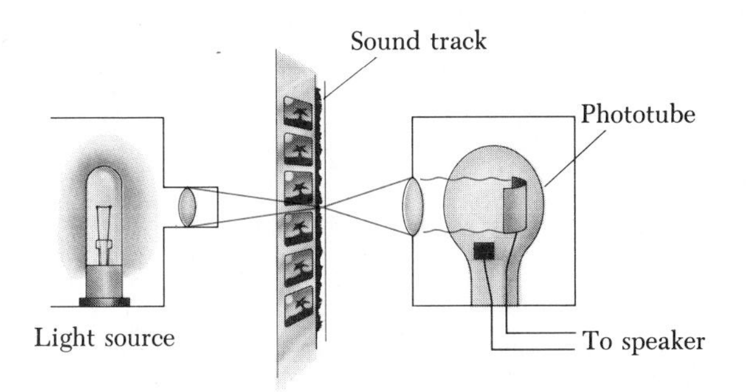

FIGURE 29.8 The shading on the sound track varies the light intensity reaching the phototube and hence the current to the speaker.

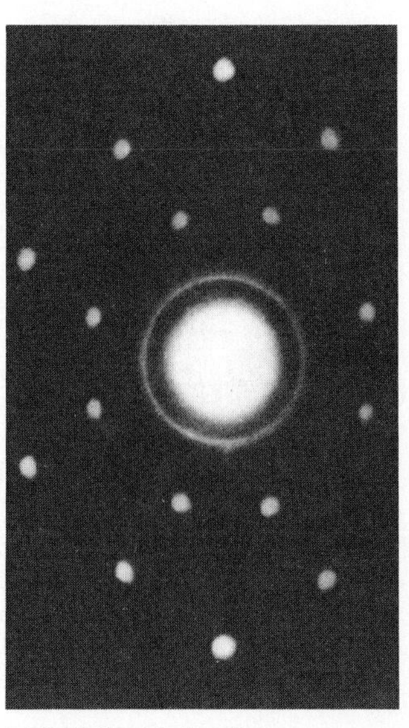

FIGURE 29.9 X-ray diffraction pattern of NaCl.

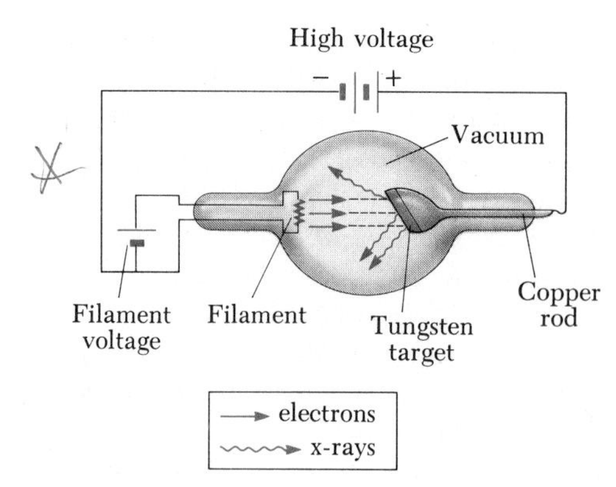

FIGURE 29.10 Diagram of an x-ray tube.

values of the spacing between atoms in the crystal. X-ray diffraction has proved to be an invaluable technique for understanding the structure of matter. We shall discuss this subject in more detail in the next section.

Typical x-ray wavelengths are about 0.1 nm, which is of the order of the atomic spacing in a solid. We now know that x-rays are a part of the electromagnetic spectrum and are characterized by frequencies higher than those of ultraviolet radiation and having the ability to penetrate most materials with relative ease.

X-rays are produced when high-speed electrons are suddenly decelerated, for example, when a metal target is struck by electrons that have been accelerated through a potential difference of several thousand volts. Figure 29.10 shows a schematic diagram of an x-ray tube. A current in the filament causes electrons to be boiled off, and these freed electrons are accelerated toward a dense metal target, such as tungsten, which is held at a higher potential than the filament.

Figure 29.11 represents a plot of x-ray intensity versus wavelength for the spectrum of radiation emitted by an x-ray tube. Note that there are two distinct patterns. One pattern is a continuous broad spectrum that depends on the voltage applied to the tube. Superimposed on this pattern is a series of sharp, intense lines that depend on the nature of the target material. The accelerating voltage must exceed a certain value, called the **threshold voltage,** in order to observe these sharp lines, which represent radiation emitted by the target atoms as their electrons undergo rearrangements. We shall discuss this further in Chapter 30. The continuous radiation is sometimes called **bremsstrahlung,** a German word meaning braking radiation. The term arises

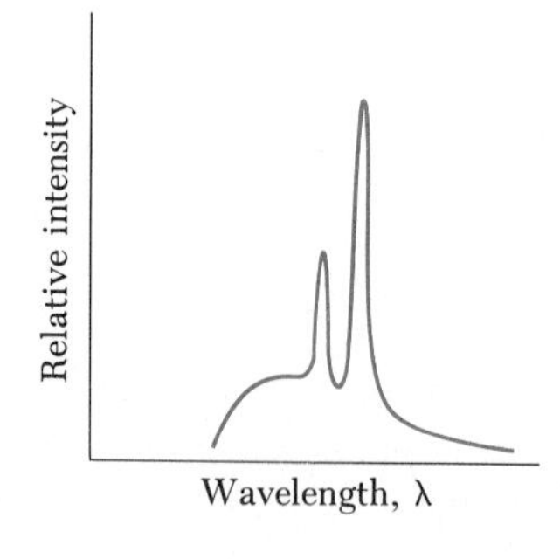

FIGURE 29.11 A typical plot of radiation intensity versus wavelength for the output of an x-ray tube.

from the nature of the mechanism responsible for the radiation. That is, electrons emit radiation when they undergo a deceleration inside the target.

The deceleration of the electrons by the target produces an effect similar to an inverse photoelectric effect. In the photoelectric process, a quantum of radiant energy is absorbed by an electron in a metal and the electron gains enough energy to escape the metal. In the case of x-ray production, the inverse of this process occurs as shown in Figure 29.12. As an electron passes close to a positively charged nucleus contained in a target material, it is deflected from its path because of its electrical attraction to the nucleus, and hence experiences an acceleration. An analysis from classical physics shows that any charged particle will radiate energy in the form of electromagnetic radiation when it is accelerated. (An example of this is the production of electromagnetic waves by accelerated charges in a radio antenna as described in Chapter 23.) According to quantum theory, this radiation must appear in the form of photons. Since the radiated photon shown in Figure 29.12 carries energy, the electron must lose kinetic energy because of its encounter with the target nucleus. Let us consider an extreme example in which the electron loses all of its energy in a single collision. In this case, the initial energy of the electron (eV) is transformed completely into the energy of the photon (hf_{max}). In equation form we have

$$eV = hf_{max} = \frac{hc}{\lambda_{min}} \tag{29.9}$$

where eV is the energy of the electron after it has been accelerated through a potential difference of V volts and e is the charge on the electron. This says that the shortest-wavelength radiation that can be produced is

$$\lambda_{min} = \frac{hc}{eV} \tag{29.10}$$

The reason that all the radiation produced does not have this particular wavelength is because many of the electrons are not stopped in a single collision. This results in the production of the continuous spectrum of wavelengths.

As noted earlier, x-rays are extremely penetrating and can produce burns or other complications if proper precautions are not taken by anyone exposed to them. Between 1930 and 1950, an x-ray device called a fluoroscope was widely used in shoe stores to examine the bone structure of the foot. Physicians used similar devices to examine the skeletal structure of their patients.

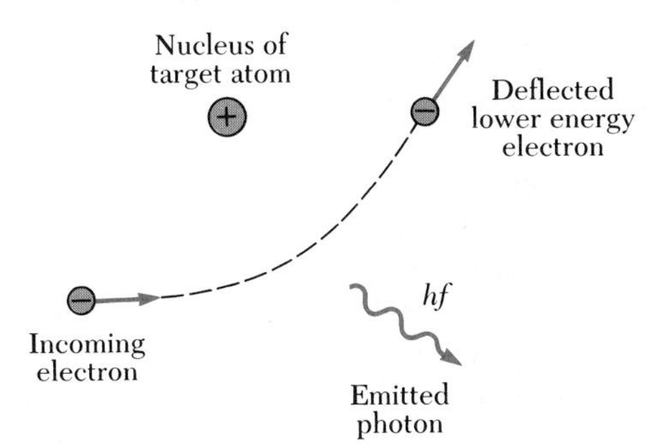

FIGURE 29.12 An electron passing near a charged target atom experiences an acceleration and a photon is emitted in the process.

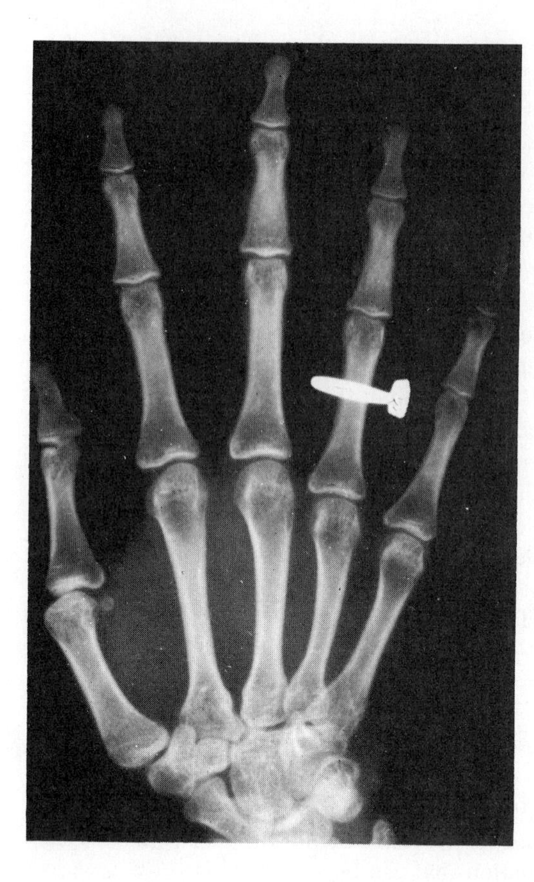

FIGURE 29.13 X-ray photograph of a human hand.

Such devices are no longer in use since they are now known to be a health hazard. An x-ray photograph of a human hand is shown in Figure 29.13.

EXAMPLE 29.5 The Minimum X-ray Wavelength

Calculate the minimum wavelength produced when electrons are accelerated through a potential difference of 100,000 V, a not-uncommon voltage for an x-ray tube.

Solution From Equation 29.10, we have

$$\lambda_{\min} = \frac{(6.63 \times 10^{-34}\ \mathrm{J\cdot s})\,(3.00 \times 10^{8}\ \mathrm{m/s})}{(1.60 \times 10^{-19}\ \mathrm{C})(10^{5}\ \mathrm{V})} = 1.24 \times 10^{-11}\ \mathrm{m}$$

*29.5 DIFFRACTION OF X-RAYS BY CRYSTALS

In Chapter 27 we described how a diffraction grating can be used to measure the wavelength of light. In principle, the wavelength of *any* electromagnetic wave can be measured if a grating having the proper line spacing can be found. The spacing between lines must be approximately equal to the wavelength of the radiation to be measured. X-rays are electromagnetic waves with wavelengths of the order of 0.1 nm. It would be impossible to construct a grating with such a small spacing. As noted in the previous section, Max von

Laue suggested that the regular array of atoms in a crystal could act as a three-dimensional grating for observing the diffraction of X-rays.

One experimental arrangement for observing x-ray diffraction is shown in Figure 29.14. A narrow beam of x-rays with a continuous range of wavelengths is incident on a crystal such as sodium chloride. The diffracted radiation is very intense in certain directions, corresponding to constructive interference from waves reflected from layers of atoms in the crystal. The diffracted radiation is detected by a photographic film and forms an array of spots known as a "Laue pattern." The crystal structure is determined by analyzing the positions and intensities of the various spots in the pattern.

The arrangement of atoms in a crystal of NaCl is shown in Figure 29.15. The smaller colored spheres represent Na^+ ions and the larger spheres represent Cl^- ions. The spacing between successive Na^+ (or Cl^-) ions in this cubic structure, denoted by the symbol a in Figure 29.15, is approximately 0.563 nm.

A careful examination of the NaCl structure shows that the ions lie in various planes. The shaded areas in Figure 29.15 represent one example in which the atoms lie in equally spaced planes. Now suppose an x-ray beam is incident at an angle θ on one of the planes, as in Figure 29.16. The beam can be reflected from both the upper and lower plane of atoms. However, the geometric construction in Figure 29.16 shows that the beam reflected from the lower surface travels farther than the beam reflected from the upper surface by a distance of $2d \sin \theta$. The two portions of the reflected beam will combine to produce constructive interference when this path difference equals some integral multiple of the wavelength λ. The same is true for reflection from the entire family of parallel planes. (Note the similarity between this analysis and that used to describe thin film interference.) The condition for constructive interference is given by

$$2d \sin \theta = m\lambda \qquad (m = 1, 2, 3, \ldots) \tag{29.11}$$

Bragg's law

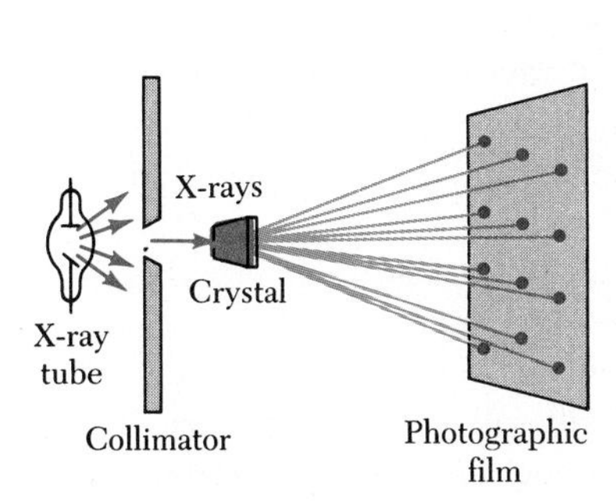

FIGURE 29.14 Schematic diagram of the technique used to observe the diffraction of x-rays by a single crystal. The array of spots formed on the film by the diffracted beams is called a Laue pattern.

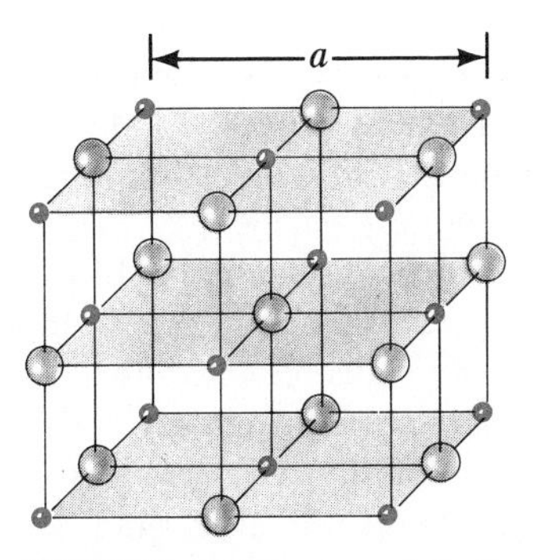

FIGURE 29.15 A model of the cubic crystalline structure of sodium chloride. The larger spheres represent the Cl^- ions, and the smaller colored spheres represent the Na^+ ions. The length of the cube edge is $a = 0.563$ nm.

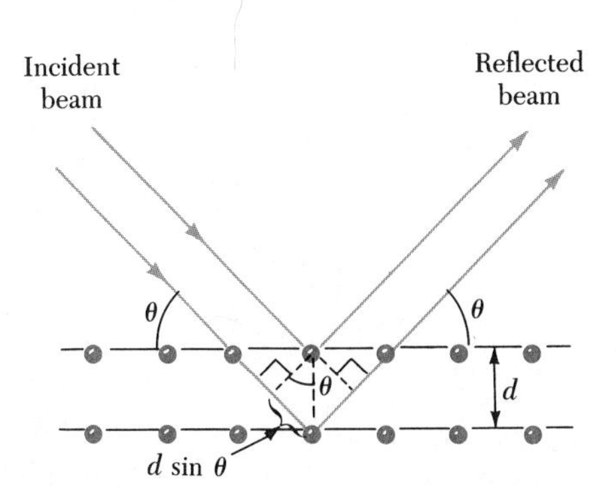

FIGURE 29.16 A two-dimensional description of the reflection of an x-ray beam from two parallel crystalline planes separated by a distance d. The beam reflected from the lower plane travels farther than the one reflected from the upper plane by an amount equal to $2d \sin \theta$.

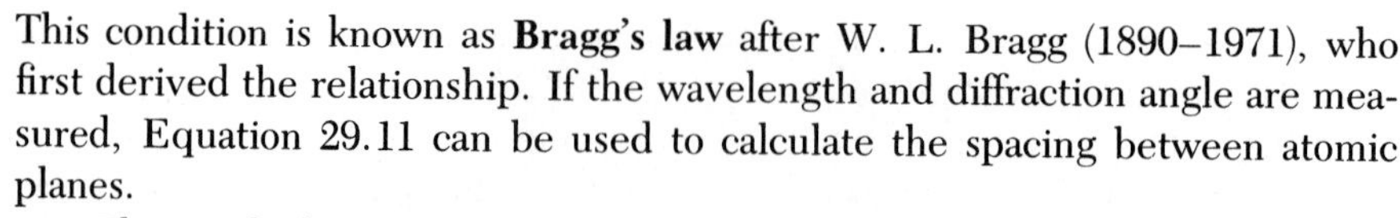

This condition is known as **Bragg's law** after W. L. Bragg (1890–1971), who first derived the relationship. If the wavelength and diffraction angle are measured, Equation 29.11 can be used to calculate the spacing between atomic planes.

The method of x-ray diffraction to determine crystalline structures was thoroughly developed in England by W. H. Bragg, and his son W. L. Bragg, who shared a Nobel prize in 1915 for their work. Since then, thousands of crystalline structures have been investigated. Recently, the technique of x-ray structural analysis has been used to unravel the mysteries of such complex organic systems as the important DNA molecule.

EXAMPLE 29.6 Reflection from Calcite

If the spacing between certain planes in a crystal of calcite is 0.314 nm, find the angles of incidence at which first and third order interference will occur for x-rays of wavelength 0.07 nm.

Solution For first order interference, the value of m in Equation 29.11 is 1. Thus, the angle of incidence corresponding to this order of interference is found as follows:

$$\sin\theta = \frac{m\lambda}{2d} = \frac{(0.07\text{ nm})}{2(0.314\text{ nm})} = 0.111$$

$$\theta = 6.37°$$

In third order interference, $m = 3$, and we find

$$\sin\theta = \frac{m\lambda}{2d} = \frac{3(0.07\text{ nm})}{2(0.314\text{ nm})} = 0.334$$

$$\theta = 19.5°$$

29.6 COMPTON SCATTERING

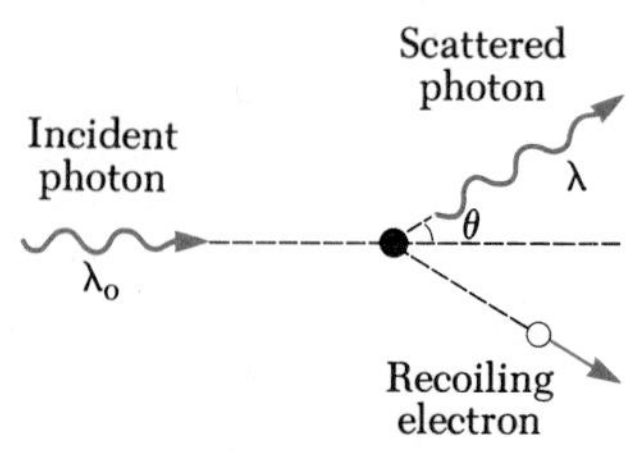

FIGURE 29.17 Diagram representing Compton scattering of a photon by an electron. The scattered photon has less energy (or longer wavelength) than the incident photon.

Further justification for the photon theory of light came from an experiment conducted by Arthur H. Compton in 1923. In his experiment, Compton directed a beam of x-rays of wavelength λ_0 toward a block of graphite. He found that the scattered x-rays had a slightly longer wavelength, λ, than the incident x-rays, and hence the energies of the scattered rays were lower. The amount of energy reduction depended on the angle at which the x-rays were scattered. The change in wavelength, $\Delta\lambda$, between a scattered x-ray and an incident x-ray is called the **Compton shift.**

In order to explain this effect, Compton assumed that if a photon behaves like a particle, its collision with other particles is similar to that between two billiard balls. Hence, both energy and momentum must be conserved. If the incident photon collides with an electron initially at rest, as in Figure 29.17, the photon transfers some of its energy and momentum to the electron. Consequently, the energy and frequency of the scattered photon are lowered and its wavelength increases. Applying relativistic energy and momentum conservation to the collision described in Figure 29.17, one finds that the shift in wavelength of the scattered photon is given by

The Compton shift formula

$$\Delta\lambda = \lambda - \lambda_0 = \frac{h}{m_0 c}(1 - \cos\theta) \tag{29.12}$$

where m_0 is the rest mass of the electron and θ is the angle between the directions of the scattered and incident photons. The quantity h/m_0c is called the **Compton wavelength** and has a value $h/m_0c = 0.00243$ nm. Note that the Compton wavelength is very small relative to the wavelengths of visible light and hence would be very difficult to detect if one were to use visible light. Furthermore, note that the Compton shift depends on the scattering angle, θ, and not on the wavelength. Experimental results for x-rays scattered from various targets strongly support the photon concept.

EXAMPLE 29.7 The Compton Shift for Carbon

X-rays of wavelength $\lambda_0 = 0.20$ nm are scattered from a block of carbon. The scattered x-rays are observed at an angle of 45° to the incident beam. Calculate the wavelength of the scattered x-rays at this angle.

Solution The shift in wavelength of the scattered x-rays is given by Equation 29.12. Taking $\theta = 45°$, we find that

$$\Delta\lambda = \frac{h}{m_0c}(1 - \cos\theta)$$

$$= \frac{6.63 \times 10^{-34}\ \text{J}\cdot\text{s}}{(9.11 \times 10^{-31}\ \text{kg})(3.00 \times 10^{8}\ \text{m/s})}(1 - \cos 45°)$$

$$= 7.11 \times 10^{-13}\ \text{m} = 0.000711\ \text{nm}$$

Hence, the wavelength of the scattered x-ray at this angle is

$$\lambda = \Delta\lambda + \lambda_0 = 0.200711\ \text{nm}$$

29.7 PAIR PRODUCTION AND ANNIHILATION

In the photoelectric and Compton effects, the energy of a photon is transformed into kinetic and potential energy of an electron. When a photon interacts with matter through the photoelectric effect, an electron is removed from an atom and the photon disappears. In the Compton effect, a photon is scattered off an electron (or a nucleus) and loses some energy in the process. We shall now describe a process in which the energy of a photon is converted completely into rest mass. This is a striking verification of the equivalence of mass and energy as predicted by Einstein's theory of relativity.

A common process in which a photon creates matter is called **pair production**, illustrated in Figure 29.18. In this process, an electron and a positron are simultaneously produced, while the photon disappears. (Note that the positron is a positively charged particle having the same mass as an electron. The positron is often called the antiparticle of the electron.) In order for pair production to occur, energy, momentum, and charge must all be conserved during the process. Note that it is impossible for a photon to produce a single electron because the photon has zero charge, and charge would not be conserved in the process. The *minimum* energy that a photon must have to produce an electron-positron pair can be found using conservation of energy by equating the photon energy to the total rest energy of the pair. That is,

$$hf_{\text{min}} = 2m_0c^2 \qquad (29.13)$$

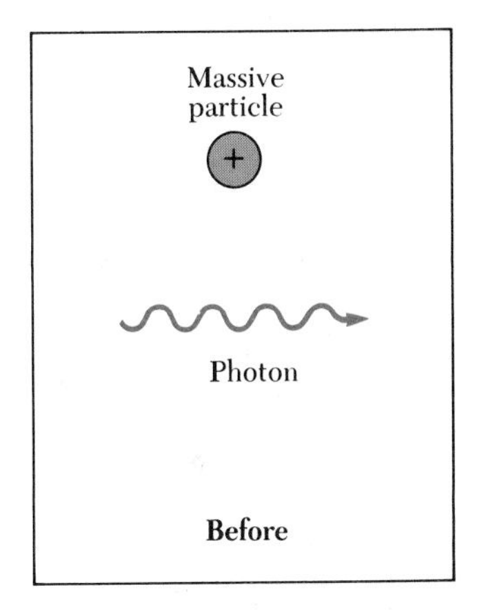

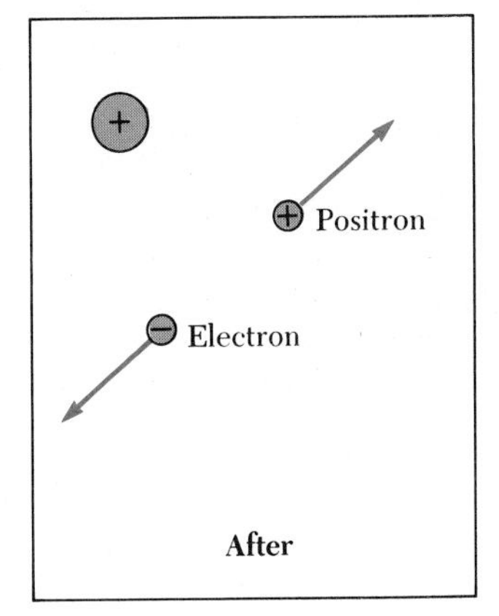

FIGURE 29.18 Representation of the process of pair production.

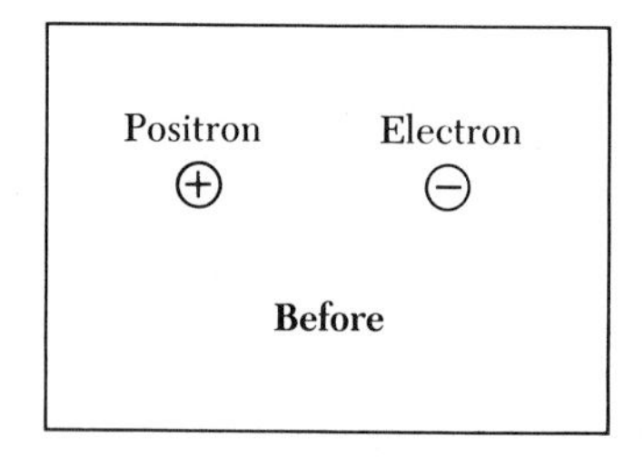

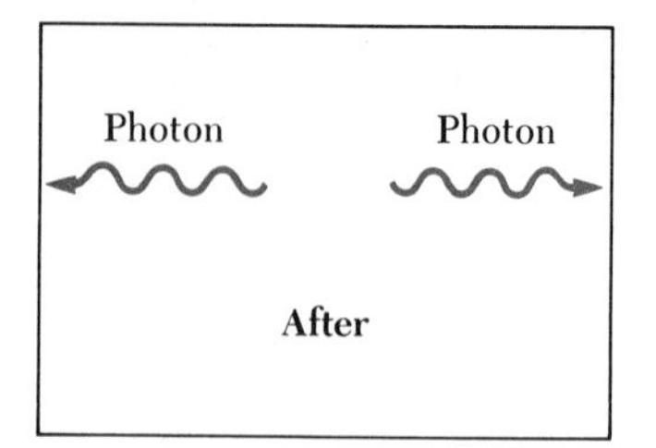

FIGURE 29.19
Representation of the process of pair annihilation.

Since the rest energy of an electron is $m_0c^2 = 0.51$ MeV (see Chapter 28), the minimum energy required for pair production is 1.02 MeV. The wavelength of a photon carrying this much energy is 0.0012 nm. Photons with such short wavelengths are in the gamma-ray (or very short x-ray) region of the spectrum.

Pair production cannot occur in a vacuum but can only take place in the presence of a massive particle such as an atomic nucleus. The massive particle must participate in the interaction in order that energy and momentum be conserved simultaneously.

Pair annihilation is a process in which an electron-positron pair produces two photons, which is the inverse of pair production. Figure 29.19 is one example of pair annihilation in which an electron and positron initially at rest combine with each other, disappear, and create two photons. Since the initial momentum of the pair is zero, it is impossible to produce a single photon. Momentum can only be conserved if two photons are produced moving in opposite directions, both with the same momentum and energy. We shall discuss particles and their antiparticles further in Chapter 34.

29.8 PHOTONS AND ELECTROMAGNETIC WAVES

An explanation of a phenomenon such as the photoelectric effect presents very convincing evidence in support of the photon (or particle) concept of light. An obvious question that arises at this point is, "How can light be considered a photon when it exhibits wave-like properties?" On the one hand, we describe light in terms of photons having energy and momentum. On the other hand, we must also recognize that light and other electromagnetic waves exhibit interference and diffraction effects, which are consistent only with a wave interpretation. Which model is correct? Is light a wave or a particle? The answer depends on the specific phenomenon being observed. Some experiments can be better, or solely, explained on the basis of the photon concept, whereas others are best described, or can be described only, with a wave model. The end result is that *we must accept both models and admit that the true nature of light is not describable in terms of a single classical picture—light has a dual nature.* However, you should recognize that the same light beam that can eject photoelectrons from a metal can also be diffracted by a grating. In other words, *the photon theory and the wave theory of light complement each other.*

Light has a dual nature

We can perhaps understand why photons are compatible with electromagnetic waves in the following manner. We may suspect that long-wavelength radio waves do not exhibit particle characteristics. Consider, for instance, radio waves at a frequency of 2.5 MHz. The energy of a photon having this frequency is only about 10^{-8} eV. From a practical viewpoint, this energy is too small to be detected as a single photon. A sensitive radio receiver might require as many as 10^{10} of these photons to produce a detectable signal. Such a large number of photons would appear, on the average, as a continuous wave. With such a large number of photons reaching the detector every second, it would be unlikely that any graininess would appear in the detected signal. That is, we would not be able to detect the individual photons striking the antenna.

Now consider what happens as we go to higher frequencies, or shorter wavelengths. In the visible region, it is possible to observe both the photon and the wave characteristics of light. As we mentioned earlier, a light beam shows interference phenomena and at the same time can produce photoelectrons, which can be understood best by using Einstein's photon concept. At even higher frequencies and correspondingly shorter wavelengths, the momentum and energy of the photon increase. Consequently, the photon nature of light becomes more evident than its wave nature. For example, an x-ray photon is easily detected as a single event. However, as the wavelength decreases, wave effects, such as interference and diffraction, become more difficult to observe. Very indirect methods are required to detect the wave nature of very-high-frequency radiation, such as gamma rays.

All forms of electromagnetic radiation can be described from two points of view. At one extreme, electromagnetic waves describe the overall interference pattern formed by a large number of photons. At the other extreme, the photon description is natural when we are dealing with a highly energetic photon of very short wavelength. Hence,

light has a dual nature; it exhibits both wave and photon characteristics.

29.9 THE WAVE PROPERTIES OF PARTICLES

Students first introduced to the dual nature of light often find the concept very difficult to accept. In the world around us, we are accustomed to regarding such things as a thrown baseball solely as particles and such things as sound waves solely as forms of wave motion. Every large-scale observation can be interpreted by considering either a wave explanation or a particle explanation, but in the world of photons and electrons, such distinctions are not as sharply drawn. Even more disconcerting is the fact that, under certain conditions, *particles such as electrons also exhibit wave characteristics.*

In 1924, the French physicist Louis de Broglie wrote a doctoral dissertation in which his thesis was that *because photons have wave and particle characteristics, perhaps all forms of matter have wave as well as particle properties.* This was a highly revolutionary idea with no experimental confirmation. However, the suggestion received immediate attention by the scientific community and played an important role in the subsequent development of quantum mechanics.

In Chapter 28, we found that the relationship between energy and momentum for a photon, which has a rest mass of zero, is $p = E/c$. We also know that the energy of a photon is

$$E = hf = \frac{hc}{\lambda} \qquad (29.14)$$

Energy of a photon

Thus, the momentum of a photon can be expressed as

$$p = \frac{E}{c} = \frac{hc}{c\lambda} = \frac{h}{\lambda} \qquad (29.15)$$

Momentum of a photon

From this equation we see that the photon wavelength can be specified by its momentum, or $\lambda = h/p$. De Broglie suggested that

material particles of momentum p should also have wave properties and a corresponding wavelength.

Because the momentum of a particle of mass m and velocity v is $p = mv$, the **de Broglie wavelength** of the particle is

De Broglie wavelength

$$\lambda = \frac{h}{p} = \frac{h}{mv} \tag{29.16}$$

Furthermore, in analogy with photons, de Broglie postulated that the frequencies of matter waves (that is, waves associated with real particles) obey the Einstein relation $E = hf$, so that

Frequency of a photon

$$f = \frac{E}{h} \tag{29.17}$$

The dual nature of matter is quite apparent in these two equations. That is, each equation contains both particle concepts (mv and E) and wave concepts (λ and f). The fact that these relationships are established experimentally for photons makes the de Broglie hypothesis that much easier to accept.

De Broglie's proposal that all particles exhibit both wave and particle properties was first regarded as pure speculation. If particles such as electrons had wave properties, then under the correct conditions they should exhibit interference phenomena. Three years later, in 1927, C. J. Davisson and L. Germer of the United States succeeded in measuring the wavelength of electrons. Their important discovery provided the first experimental confirmation of the matter waves proposed by de Broglie. In their original experiment, low-energy electrons (about 54 eV) were scattered from a single crystal of nickel. The electrons were scattered in preferred directions, as evidenced by intensity peaks at certain scattering angles. The results were explained by recognizing that the regularly spaced planes of atoms in the crystal acted as a diffraction grating for the electron waves, as discussed in Section 29.5.

The problem of understanding the dual nature of both matter and radiation is conceptually difficult because the two models seem to contradict each other. This problem as it applies to light was discussed earlier. Niels Bohr helped to resolve this problem in his principle of complementarity, which states that *the wave and particle models of either matter or radiation complement each other.* Neither model can be used exclusively to adequately describe matter or radiation. A complete understanding is obtained only if the two models are combined in a complementary manner.

Louis de Broglie (1892–1987), a French physicist, was awarded the Nobel prize in 1929 for his discovery of the wave nature of electrons. "It would seem that the basic idea of quantum theory is the impossibility of imagining an isolated quantity of energy without associating with it a certain frequency." (AIP Niels Bohr Library)

EXAMPLE 29.8 The Wavelength of an Electron

Calculate the de Broglie wavelength for an electron ($m = 9.11 \times 10^{-31}$ kg) moving with a speed of 10^7 m/s.

Solution Equation 29.16 gives

$$\lambda = \frac{h}{mv} = \frac{6.63 \times 10^{-34}\ \text{J}\cdot\text{s}}{(9.11 \times 10^{-31}\ \text{kg})(10^7\ \text{m/s})} = 7.28 \times 10^{-11}\ \text{m}$$

This wavelength corresponds to that of x-rays in the electromagnetic spectrum.

EXERCISE Find the de Broglie wavelength of a proton ($m = 1.67 \times 10^{-27}$ kg) moving with a speed of 10^7 m/s.

Answer 3.97×10^{-14} m.

EXAMPLE 29.9 The Wavelength of a Rock

A rock of mass 50 g is thrown with a speed of 40 m/s. What is the de Broglie wavelength of the rock?

Solution From Equation 29.16, we have

$$\lambda = \frac{h}{mv} = \frac{6.63 \times 10^{-34}\ \text{J}\cdot\text{s}}{(50 \times 10^{-3}\ \text{kg})\ (40\ \text{m/s})} = 3.32 \times 10^{-34}\ \text{m}$$

Notice that this wavelength is much smaller than the size of any possible aperture through which the rock could pass. This means that we could not observe diffraction effects, and as a result the wave properties of large-scale objects cannot be observed.

---

*29.10 THE ELECTRON MICROSCOPE

A practical device that relies on the wave characteristics of electrons is the **electron microscope** (Fig. 29.20), which is in many respects similar to an ordinary compound microscope. One important difference between the two is that the electron microscope has a much greater resolving power because electrons can be accelerated to very high kinetic energies, giving them a very short wavelength. Any microscope is capable of detecting details that are comparable in size to the wavelength of the radiation used to illuminate the object. Typically, the wavelengths of electrons are about 100 times shorter than those of the visible light used in optical microscopes. As a result, electron microscopes are able to distinguish details about 100 times smaller.

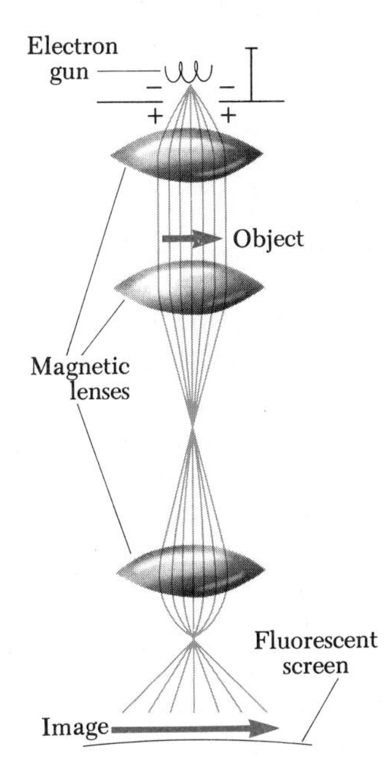

FIGURE 29.20 Diagram of an electron microscope. The "lenses" which control the electron beam are magnetic deflection coils.

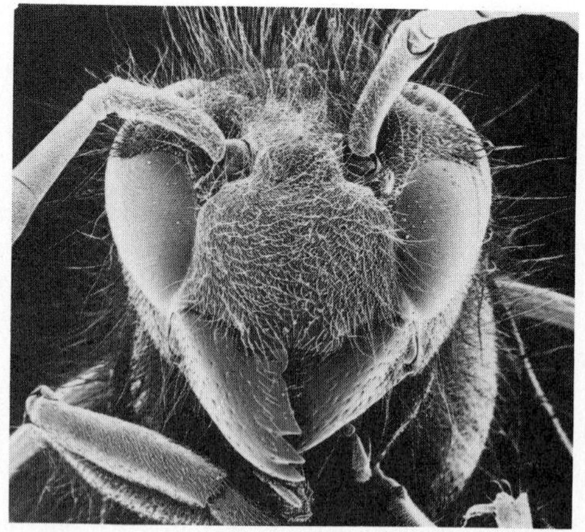

FIGURE 29.21 (a) A scanning electronmicrograph (SEM) of a grain weevil emerging from barley at magnification X70. (Prof. G.F. Leedale, Biophoto Assoc.) (b) SEM of the head of a wasp *(Vespula vulgaris)* at a magnification of X46. (Biophoto Assoc.). (c) SEM of insect eggs on the undersurface of a leaf (X40) with one of the eggs dissected open to show an insect embryo. (Courtesy of Dr. Karen A. Holbrook.)

In operation, a beam of electrons falls on a thin slice of the material to be examined. The section to be examined must be very thin, typically a few hundred angstroms, in order to minimize undesirable effects, such as absorption or scattering of the electrons. The electron beam is controlled by electrostatic or magnetic deflection, which acts on the charges to focus the beam to an image. Rather than examining the image through an eyepiece as in an ordinary microscope, a magnetic lens forms an image on a fluorescent screen. The fluorescent screen is necessary because the image produced would not otherwise be visible. Some examples of photographs taken by electron microscopes are shown in Figure 29.21.

29.11 THE WAVE FUNCTION

De Broglie's revolutionary idea that particles should have a wave nature soon moved out of the realm of skepticism to the point where it was viewed as a necessary concept in understanding the subatomic world. In 1926, the Austrian-German physicist Erwin Schrödinger proposed a wave equation that described the manner in which matter waves change in space and time. The Schrödinger wave equation represents a key element in the theory of quantum mechanics. Its role is as important in quantum mechanics as that played by Newton's laws of motion in classical mechanics. Schrödinger's equation has been successfully applied to the hydrogen atom and to many other microscopic systems. Its importance in most aspects of modern physics cannot be overemphasized.

We shall not go through a mathematical derivation of Schrödinger's wave equation, nor shall we even state the equation here since it involves mathematical operations beyond the scope of this text. Suffice it to say that, when one attempts to solve the Schrödinger equation, the basic entity one is seeking to determine is a quantity Ψ, called the **wave function.** Each particle is represented by a wave function Ψ that depends on both the position of the object and time. Once Ψ is found, what information does it give concerning the particle? To answer this question, let us consider an analogy with light.

In Chapter 26 we discussed Young's double-slit experiment and explained experimental observations of the interference pattern solely in terms of the wave nature of light. Let us now discuss this same experiment in terms of both the wave and particle nature of light.

Erwin Schrödinger (1887–1961), a German scientist, shared the Nobel prize in 1933 with P.A.M. Dirac for the discovery of new productive forms of atomic theory. (AIP Niels Bohr Library, Photograph by Francis Simon)

First, recall from Chapter 23 that the intensity of a light beam is proportional to the square of the electric field strength E associated with the beam. That is, $I \propto E^2$. According to the wave model of light, there are certain points on the viewing screen where the net electric field is zero due to destructive interference of waves from the two slits. Because E is zero at these points, the intensity is also zero, and the screen is dark at these locations. Likewise, at points on the screen where constructive interference occurs, E is large, as is the intensity, hence these locations are bright.

Now consider the same experiment when light is viewed as having a particle nature. It seems reasonable to assume that the number of photons reaching a point on the screen per second increases as the intensity (brightness) increases. Thus, the number of photons which strikes a unit area on the screen each second is proportional to the square of the electric field, or $N \propto E^2$. Now let us consider the behavior of a single photon from a probabilistic viewpoint. What will be the fate of the photon as it moves through the slits in Young's experiment? From a probabilistic point of view, a photon has a high probability of striking the screen at a point where the intensity (and E^2) is high, and a low probability of striking the screen where the intensity is low.

When describing particles rather than photons, Ψ plays the role of the amplitude, rather than E. Using an analogy with the description of light, we make the following interpretations of Ψ for particles:

1. If a wave function Ψ is used to represent a large group of particles, the value of Ψ^2 at some location at a given time is proportional to the probability of finding a large number of particles at that location at that time.
2. If Ψ is a wave function used to describe a single particle, the value of Ψ^2 at some location at a given time is proportional to the probability of finding the particle at that location at that time.

29.12 THE UNCERTAINTY PRINCIPLE

If you were to measure the position and velocity of a particle at any instant, you would always be faced with reducing the experimental uncertainties in your measurements as much as possible. According to classical mechanics, there is no fundamental barrier to an ultimate refinement of the apparatus and/or experimental procedures. That is, it would be possible, in principle, to make such measurements with arbitrarily small uncertainty or with infinite accuracy. Quantum theory predicts, however, that

it is fundamentally impossible to make simultaneous measurements of a particle's position and velocity with infinite accuracy.

This statement, known as the **uncertainty principle,** was first derived by Werner Heisenberg in 1927.

Consider a particle moving along the x axis and suppose that Δx and Δp represent the uncertainty in the measured values of the particle's position and momentum, respectively, at some instant. The uncertainty principle says that the product $\Delta x \Delta p$ is never less than a number of the order of Planck's constant. More specifically,

Uncertainty principle

$$\boxed{\Delta x \Delta p \geq \frac{h}{4\pi}} \qquad (29.18)$$

That is, *it is physically impossible to predict simultaneously the exact position and exact momentum of a particle.* If Δx is made very small, Δp will be large, and vice versa.

In order to understand the uncertainty principle, consider the following thought experiment. Suppose you wish to measure the position and momentum of an electron as accurately as possible. You might be able to do this by viewing the electron with a powerful microscope. In order for you to see the electron, and thus determine its location, at least one photon of light must bounce off the electron and pass through the microscope into your eye. This incident photon is shown moving toward the electron in Figure 29.22a. When the photon strikes the electron, as in Figure 29.22b, the photon transfers some of its energy and momentum to the electron. Thus, in the process of attempting to locate the electron very accurately (that is, by making Δx very small), we have caused a rather large uncertainty in its momentum. In other words, *the measurement procedure itself limits the accuracy to which we can determine position and momentum simultaneously.*

Werner Heisenberg (1901–1976), a German physicist, formulated the uncertainty principle, which provides a basis for quantum mechanics. He was awarded the Nobel prize in 1932 for the creation of quantum mechanics. "Since the measuring device has been constructed by the observer. . . we have to remember that what we observe is not nature in itself but nature exposed to our method of questioning." (AIP Niels Bohr Library, Bainbridge Collection)

Let us analyze the collision between the photon and the electron by first noting that the incoming photon has momentum h/λ. As a result of the collision, the photon transfers part or all of its momentum to the electron. Thus, the uncertainty in the electron's momentum after the collision is at least as great as the momentum of the incoming photon. That is, $\Delta p = h/\lambda$. Furthermore, since light also has wave properties, we would expect the uncertainty in the position of the electron to be of the order of one wavelength of the light being used to view it, because of diffraction effects. Thus, $\Delta x = \lambda$. Multiplying these two uncertainties gives

$$\Delta x \Delta p = \lambda \left(\frac{h}{\lambda}\right) = h$$

This represents the minimum in the products of the uncertainties. Since the uncertainty can always be greater than this minimum, we have

$$\Delta x \Delta p \geqslant h$$

This agrees with Equation 29.18 (apart from a small numerical factor introduced by Heisenberg's more precise analysis).

Another form of the uncertainty principle which applies to the simultaneous measurement of energy and time is

$$\boxed{\Delta E \Delta t \geqslant h/4\pi} \qquad (29.19)$$

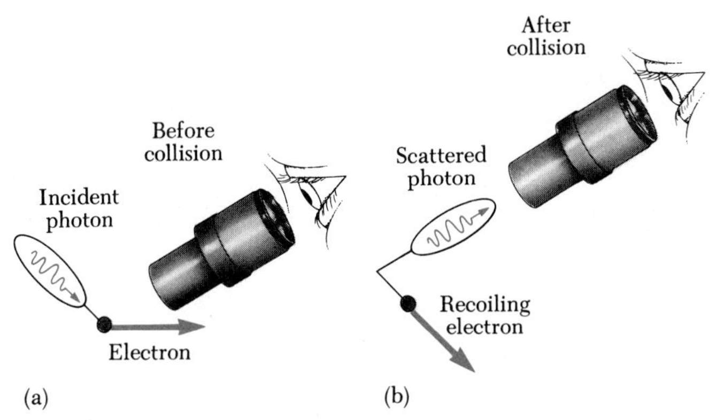

FIGURE 29.22 A thought experiment for viewing an electron with a powerful microscope. (a) The electron is viewed before colliding with the photon. (b) The electron recoils (is disturbed) as the result of the collision with the photon.

where ΔE is the uncertainty in a measurement of the energy and Δt is the time it takes to make the measurement. One can infer from this relation that the energy of a particle cannot be measured with complete precision in an infinitely short interval of time. Thus, when an electron is viewed as a particle, the uncertainty principle tells us that (a) its position and velocity cannot both be known precisely at the same time and (b) its energy can be uncertain (or may not be conserved) for a period of time given by $\Delta t = h/(4\pi\Delta E)$.

We can arrive at the uncertainty principle in the form of Equation 29.19 by returning to Figure 29.22. Recall that the position of the electron being viewed by the microscope is of the order of the wavelength of the light being used to detect the electron. The photon used to detect the electron travels with a speed c, therefore the time it takes the photon to travel a distance Δx is given by

$$\Delta t \approx \Delta x/c \approx \lambda/c$$

As the photon collides with the electron, it can transfer all or part of its energy to the electron. Thus, the uncertainty in the energy transferred is

$$\Delta E \approx hf \approx hc/\lambda$$

The product of these two uncertainties is

$$\Delta E \Delta t \approx h$$

Again, the result of our approximate analysis agrees with Heisenberg's more accurate and detailed analysis within a small numerical factor.

Heisenberg's uncertainty principle enables us to better understand the dual wave-particle nature of light and matter. We have seen that the wave description is quite different from the particle description. Therefore, if an experiment is designed to reveal the particle character of an electron (such as the photoelectric effect), its wave character will become fuzzy. Likewise, if the experiment is designed to accurately measure the electron's wave properties (such as diffraction from a crystal), its particle character will become fuzzy.

EXAMPLE 29.10 Locating an Electron

The speed of an electron is measured to be 5.00×10^3 m/s to an accuracy of 0.003%. Find the uncertainty in determining the position of this electron.

Solution The momentum of the electron is

$$p = mv = (9.11 \times 10^{-31}\text{ kg})(5.00 \times 10^3\text{ m/s}) = 4.56 \times 10^{-27}\text{ kg·m/s}$$

Because the uncertainty in p is 0.003% of this value, we get

$$\Delta p = 0.00003p = (0.00003)(4.56 \times 10^{-27}\text{ kg·m/s}) = 1.37 \times 10^{-31}\text{ kg·m/s}$$

The uncertainty in position can now be calculated by using this value of Δp and Equation 29.18:

$$\Delta x \Delta p \geqslant \frac{h}{4\pi}$$

$$\Delta x \geqslant \frac{h}{4\pi \Delta p} = \frac{6.63 \times 10^{-34}\text{ J·s}}{4\pi\,(1.37 \times 10^{-31}\text{ kg·m/s})}$$

$$= 0.385 \times 10^{-3}\text{ m} = 0.385\text{ mm}$$

EXAMPLE 29.11 Excited States of Atoms

As we shall see in the next chapter, electrons in atoms can be found in higher states of energy called excited states for short periods of time. If the average time that an electron exists in one of these states is 10^{-8} s, what is the minimum uncertainty in energy of the excited state?

Solution From the uncertainty principle in the form of Equation 29.19, we find that the minimum uncertainty in energy is

$$\Delta E = \frac{h}{4\pi \Delta t} = \frac{(6.63 \times 10^{-34}\text{ J·s})}{4\pi\,(10^{-8}\text{ s})} = 5.28 \times 10^{-27}\text{ J}$$

$$= 3.30 \times 10^{-8}\text{ eV}$$

SUMMARY

The characteristics of **blackbody radiation** cannot be explained using classical concepts. The peak of a blackbody radiation curve is given by **Wien's displacement law:**

$$\lambda_{\text{max}} T = 0.2898 \times 10^{-2}\text{ m·K} \tag{29.1}$$

where λ_{max} is the wavelength at which the curve peaks and T is the absolute temperature of the object emitting the radiation.

Planck first introduced the quantum concept when he assumed that the vibrating molecules responsible for blackbody radiation could have only discrete amounts of energy given by

$$E_n = nhf \tag{29.2}$$

where n is a positive integer called a **quantum number** and f is the frequency of vibration of the molecule.

The **photoelectric effect** is a process whereby electrons are ejected from a metal surface when light is incident on that surface. Einstein provided a successful explanation of this effect by extending Planck's quantum hypothesis to electromagnetic waves. In this model, light is viewed as a stream of particles called photons, each with energy $E = hf$, where f is the frequency and h is **Planck's constant.** The maximum kinetic energy of the ejected photoelectrons is

$$KE_{max} = hf - \phi \quad (29.7)$$

where ϕ is the **work function** of the metal.

X-rays are produced when high-speed electrons are suddenly decelerated. When electrons are accelerated through a voltage V, the shortest-wavelength radiation that can be produced is

$$\lambda_{min} = \frac{hc}{eV} \quad (29.10)$$

The regular array of atoms in a crystal can act as a diffraction grating for x-rays. The condition for constructive interference of the diffracted rays is given by **Bragg's law**

$$2d \sin \theta = m\lambda \qquad (m = 1, 2, 3, \ldots) \quad (29.11)$$

X-rays from an incident beam are scattered at various angles by electrons in a target such as carbon. In such a scattering event, a shift in wavelength is observed for the scattered x-rays. This phenomenon is known as the **Compton shift.** Conservation of momentum applied to a photon-electron collision yields the following expression for the shift in wavelength of the scattered x-rays:

$$\Delta\lambda = \lambda - \lambda_0 = \frac{h}{m_0 c}(1 - \cos \theta) \quad (29.12)$$

where m_0 is the rest mass of the electron, c is the speed of light, and θ is the scattering angle.

Pair production is a process in which the energy of a photon is converted into rest mass. In this process, the photon disappears as an electron-positron pair is created. Likewise, the energy of an electron-positron pair can be converted into electromagnetic radiation by the process of **pair annihilation.**

De Broglie proposed that all matter has both a particle and a wave nature. The **de Broglie wavelength** of any particle of mass m and speed v is

$$\lambda = \frac{h}{p} = \frac{h}{mv} \quad (29.16)$$

De Broglie also proposed that the frequencies of the waves associated with particles obey the Einstein relationship, $E = hf$.

In the theory of **quantum mechanics,** each particle is described by a quantity Ψ called the **wave function.** The probability of finding the particle at a particular point at some instant is proportional to $|\Psi^2|$. Quantum mechanics is very successful in describing the behavior of atomic and molecular processes.

According to Heisenberg's **uncertainty principle,** it is impossible to measure simultaneously the exact position and exact momentum of a particle. If Δx is the uncertainty in the measured position and Δp the uncertainty in the momentum, the product $\Delta x \Delta p$ is given by

$$\boxed{\Delta x \Delta p \geqslant \frac{h}{4\pi}} \tag{29.18}$$

or by

$$\boxed{\Delta E \Delta t \geqslant \frac{h}{4\pi}} \tag{29.19}$$

where ΔE is the uncertainty in the energy of the particle and Δt is the uncertainty in the time it takes to measure the energy.

ADDITIONAL READING

A. Baker, *Modern Physics and Antiphysics*, Reading, Mass., Addison-Wesley, 1970.

G. Binnig and H. Rohrer, "The Scanning Tunneling Microscope," *Sci. American*, August 1985, p. 50.

L. de Broglie, "The Revolution in Physics: A Non-Mathematical Survey of Quanta," Noonday Press, 1953.

W. C. Dampier, *A History of Science*, London, Cambridge University Press, 1977, Chap. 10.

G. Gamow, "The Principle of Uncertainty," *Sci. American*, January 1958, p. 51.

G. Gamow, *Thirty Years That Shook Physics*, New York, 1966. Doubleday, Anchor Books, 1966.

W. Heisenberg, *Physics and Beyond*, New York, Harper and Row, 1971 (a "biography" of quantum mechanics).

B. Hoffman, *Strange Story of the Quantum*, New York, Dover, 1959.

B. Wheaton, "Louis de Broglie and the Origin of Wave Mechanics," *The Physics Teacher*, May 1984, p. 297.

QUESTIONS

1. If the photoelectric effect is observed for one metal, can you conclude that the effect will also be observed for another metal under the same conditions?
2. Which has more energy, a photon of ultraviolet radiation or a photon of yellow light?
3. What effect, if any, would you expect the temperature of a material to have on the ease with which electrons can be ejected from it in the photoelectric effect?
4. Some stars are observed to be reddish, and some are blue. Which stars have the higher surface temperature? Explain.
5. Is light a wave or a particle? Support your answer by citing specific experimental evidence.
6. Is an electron a particle or a wave? Support your answer by citing some experimental results.
7. An electron and a proton are accelerated from rest through the same potential difference. Which particle has the longer wavelength?
8. An x-ray photon is scattered by an electron. What happens to the frequency of the scattered photon relative to that of the incident photon?
9. Why does the existence of cutoff frequency in the photoelectric effect favor a particle theory for light rather than a wave theory?
10. All objects radiate energy. Why, then, are we not able to see all objects in a dark room?
11. Why is it impossible to simultaneously measure the position and velocity of a particle with infinite accuracy?
12. When wood is stacked on a special elevated grate in a fireplace (which is commercially available), there is formed beneath the grate a pocket of burning wood whose temperature is higher than that of the burning wood at the top of the stack. Explain how this device provides more heat to the room than a conventional fire does and thus increases the efficiency of the fireplace.

PROBLEMS

Section 29.1 Blackbody Radiation and Planck's Hypothesis

1. Calculate the energy in electron volts of a photon (a) in the radio frequency range, 90 MHz, (b) in the infrared range, 10^{13} Hz, and (c) in the ultraviolet, 10^{16} Hz.

2. Calculate the energy in electron volts of a photon having a wavelength in (a) the microwave range, 5 cm, (b) the visible light range, 500 nm, and (c) the x-ray range, 10 nm.

3. If the surface temperature of the sun is 5800 K, find the wavelength which corresponds to the maximum rate of energy emission from the sun.

4. A certain light source is found to emit radiation with the peak value having a frequency of 10^{15} Hz. Find the temperature of the source.

5. A quantum of electromagnetic radiation has an energy of 2 keV. What is its wavelength?

• **6.** A 0.5-kg mass falls from a height of 3 m. If all of the energy of this mass could be converted to visible light of wavelength 500 nm, how many photons would be produced?

• **7.** Through what height would the object of Problem 6 have to fall in order to have the energy equivalent of one photon of 500 nm light?

• **8.** A 1.5-kg mass vibrates at an amplitude of 3 cm on the end of a spring of spring constant 20 N/m. If the energy of the spring is quantized, (a) find its quantum number. (b) If n changes by 1, find the fractional change in energy of the spring.

Section 29.2 The Photoelectric Effect

9. A metal has a work function of 2×10^{-19} J. If yellow light of wavelength 600 nm falls on the surface of the metal, find (a) the maximum kinetic energy of the ejected electrons, (b) the cutoff wavelength for the metal.

10. When light of wavelength 350 nm falls on a potassium surface, electrons are emitted which have a maximum kinetic energy of 1.31 eV. Find (a) the work function of potassium, (b) the cutoff wavelength, and (c) the frequency corresponding to the cutoff wavelength.

11. When cesium metal is illuminated with light of wavelength 300 nm, the photoelectrons emitted have a maximum kinetic energy of 2.23 eV. Find (a) the work function of cesium and (b) the stopping potential if the incident light has a wavelength of 400 nm.

12. When a certain metal is illuminated with light of frequency 3×10^{15} Hz, a stopping potential of 7 V is required to stop the ejected electrons. What is the work function of this metal?

13. Consider the metals lithium, iron, and mercury, which have work functions of 2.3 eV, 3.9 eV, and 4.5 eV, respectively. If light of wavelength 300 nm is incident on each of these metals, determine (a) which metals exhibit the photoelectric effect and (b) the maximum kinetic energy for the photoelectrons for those which exhibit the effect.

• **14.** When light of wavelength 253.7 nm falls on cesium, it is found that the stopping potential required is 3 V. If light of wavelength 435.8 nm is used, the stopping potential is 0.9 V. Use this information to plot a graph like that shown in Figure 29.6, and from the graph determine the cutoff frequency for cesium and its work function.

• **15.** What wavelength light would have to fall on sodium (work function 2.3 eV) if it is to emit electrons with a maximum speed of 10^6 m/s?

Section 29.4 X-rays

16. What minimum accelerating voltage would be required to produce an x-ray with a wavelength of 0.03 nm?

17. Calculate the minimum wavelength x-ray that can be produced when a target is struck by an electron that has been accelerated through a potential difference of (a) 15 kV and (b) 100 kV.

18. The extremes of the x-ray portion of the electromagnetic spectrum range from approximately 10 nm to 10^{-4} nm. Find the minimum accelerating voltages required to produce wavelengths at these two extremes.

Section 29.5 Diffraction of X-rays by Crystals

19. The spacing between certain planes in a crystal is known to be 0.3 nm. Find the smallest angle at which interference will occur for x-rays of wavelength 0.07 nm.

20. Potassium iodide has an interplane spacing of d = 0.296 nm. A monochromatic x-ray beam shows a first-order diffraction maximum when the angle of incidence is 7.6°. Calculate the x-ray wavelength.

21. A monochromatic x-ray beam is incident on a NaCl crystal surface where d = 0.353 nm. The second-order maximum in the reflected beam is found when the angle between the incident beam and the surface is 20.5°. Determine the wavelength of the x-rays.

22. X-rays of wavelength 0.14 nm are reflected from a certain crystal, and the first-order maximum occurs at an angle of 14.4°. What value does this give for the interplane spacing of this crystal?

Section 29.6 Compton Scattering

23. X-rays of wavelength 0.20 nm are scattered from a block of carbon. If the scattered radiation is detected at 90° to the incident beam, find the Compton shift.

24. Show that the Compton wavelength has the numerical value 0.00243 nm.

25. X-rays of wavelength 0.071 nm undergo Compton scattering from free electrons in carbon. What is the wavelength of the photons, which are scattered at 90° relative to the incident direction?

26. X-rays are scattered from electrons in a carbon target. The measured wavelength shift is 0.0012 nm. Calculate the scattering angle.

Section 29.7 Pair Production and Annihilation

27. Find (a) the minimum energy of the photon required to produce a proton-antiproton pair and (b) the wavelength of this radiation.

28. Repeat Problem 27 for a neutron-antineutron pair.

• **29.** How much total kinetic energy will an electron-positron pair have if produced by a photon of energy 3 MeV?

• **30.** If an electron-positron pair are produced having a total kinetic energy of 2.5 MeV, find (a) the energy of the photon which produced the pair, and (b) its frequency.

Section 29.9 The Wave Properties of Particles

31. Calculate the de Broglie wavelength of a 2000-kg car moving at 65 mi/h.

32. Calculate the de Broglie wavelength of a proton moving with a speed of (a) 10^4 m/s and (b) 10^7 m/s.

33. Calculate the de Broglie wavelength for an electron with kinetic energy (a) 50 eV and (b) 50 keV.

34. The "seeing" ability, or resolution, of radiation is determined by its wavelength. If the size of an atom is of the order of 0.1 nm, how fast must an electron travel to have a wavelength small enough to "see" an atom?

35. Find the de Broglie wavelength of a ball having a mass of 0.2 kg just before it strikes the earth after being dropped from a building 50 m tall.

36. Calculate the de Broglie wavelength of a proton that is accelerated through a potential difference of 10 MV.

• **37.** Through what potential difference would an electron have to be accelerated to give it a de Broglie wavelength of 10^{-10} m?

• **38.** Show that the de Broglie wavelength of a nonrelativistic electron accelerated from rest through a potential difference V is given by $\lambda = 1.228/\sqrt{V}$ nm, where V is in volts.

Section 29.12 The Uncertainty Principle

39. A ball of mass 50 g moves with a speed of 30 m/s. If its speed is measured to an accuracy of 0.1%, what is the minimum uncertainty in its position?

40. A 0.5-kg block rests on the icy surface of a frozen pond, which we can assume to be frictionless. If the location of the block is measured to a precision of 0.5 cm, what speed must the block acquire because of the measurement process?

• **41.** The position of an electron is known to a precision of 10^{-8} cm. What is the minimum uncertainty in the measurement of the electron's velocity?

42. The energy of an electron in a particular atom is approximately 2 eV. How long would it take to measure this energy to a precision of 1%?

•• **43.** (a) Show that the kinetic energy of a nonrelativistic particle can be written in terms of its momentum as $KE = p^2/2m$. (b) Use the results of (a) to find the smallest kinetic energy of a proton confined to a nucleus of size 10^{-15} m.

ADDITIONAL PROBLEMS

44. The current in a photocell is cut off by a retarding potential of 0.92 V for radiation of wavelength 250 nm. Find the work function for the material.

45. The material in a photocell has a work function of 2 eV. When a retarding potential is applied, the cutoff wavelength is found to be 350 nm. What is the value of the retarding potential?

46. A jungle hero of mass 70 kg swings at the end of a vine at a frequency of 0.5 Hz and has a speed of 2 m/s as he moves through the lowest point on his arc. (a) Assume the energy is quantized and find the quantum number n for the system. (b) Find the energy carried away in a one-quantum change in his energy.

47. How many photons are emitted per second by a 100-W sodium lamp if the wavelength of sodium light is 589.3 nm?

48. An x-ray tube is operated at 50,000 V. (a) Find the minimum wavelength radiation emitted by this tube.

(b) If this radiation is directed at a crystal, the first-order maximum in the reflected radiation occurs when the angle of incidence is 2.5°. What is the spacing between reflecting planes in the crystal?

• **49.** A light source of wavelength λ illumintes a metal and ejects photoelectrons with a maximum kinetic energy of 1 eV. A second light source of wavelength $\lambda/2$ ejects photoelectrons with a maximum kinetic energy of 4 eV. What is the work function of the metal?

• **50.** An electron and a proton each have a thermal kinetic energy of $3kT/2$. Calculate the de Broglie wavelength of each particle when they are at a temperature of 2000 *K*. (Recall that k is Boltzmann's constant: $k = 1.38 \times 10^{-23}$ J/K.)

• **51.** Photons of wavelength 450 nm are incident on a metal. The most energetic electrons ejected from the metal are bent into a circular arc of radius 20 cm by a magnetic field whose strength is 2×10^{-5} T. What is the work function of the metal?

Essay

The Scanning Tunneling Microscope

Roger A. Freedman and Paul K. Hansma *University of California*

The basic idea of quantum mechanics, that particles have properties of waves and vice versa, is among the strangest found anywhere in science. Because of this strangeness, and because quantum mechanics mostly deals with the very small, it might seem to you to have little practical application. As you will find in this essay, however, one of the basic phenomena of quantum mechanics—the tunneling of a particle—is at the heart of a very practical device that is one of the most powerful microscopes ever built. This device, the *scanning tunneling microscope*, or *STM*, enables us to get highly detailed images of surfaces with resolution comparable to the size of a *single atom*. Such images promise to revolutionize our understanding of structures and processes on the atomic scale.

Before discussing how the STM works, let's first look at a sample of what the STM can do. An image made by a scanning tunneling microscope of the surface of a piece of gold is shown in Figure 1. You can easily see that the surface is not uniformly flat, but is a series of terraces separated by steps that are only one atom high. Gentle corregations can be seen in the terraces, caused by subtle rearrangements of the gold atoms.

FIGURE 1 Scanning tunneling microscope image of the surface of crystalline gold. The divisions on the scale are 5 Å. Successive scans are approximately 1.5 Å apart. The figure is from G. Binnig, H. Rohrer, Ch. Gerber, and E. Stoll, *Surface Science 144*, 321 (1984).

What makes the STM so remarkable is the fineness of the detail that can be seen in images such as Figure 1. The *resolution* in this image—that is, the size of the smallest detail that can be discerned—is about 2 Å (2×10^{-10} m). For an ordinary microscope, the resolution is limited by the wavelength of the waves used to make the image. Thus an optical microscope has a resolution of no better than 2000 Å, about half the wavelength of visible light, and so could never show the detail displayed in Figure 1. Electron microscopes can have a resolution of 2 Å by using electron waves of this wavelength given by the de Broglie formula $\lambda = h/p$. The electron momentum p required to give this wavelength is 10,000 eV/c, corresponding to an electron speed of 2% of the speed of light. Electrons traveling at this speed would penetrate into the interior of the piece of gold in Figure 1 and so couldn't give us information about individual surface atoms.

The image in Figure 1 was made by Gerd Binnig, Heinrich Rohrer, and collaborators at the IBM Research Laboratory in Zurich, Switzerland. Binnig and Rohrer invented the STM and shared the 1986 Nobel Prize in Physics for their work. Such is the importance of this device that unlike most Nobel Prizes, which come dec-

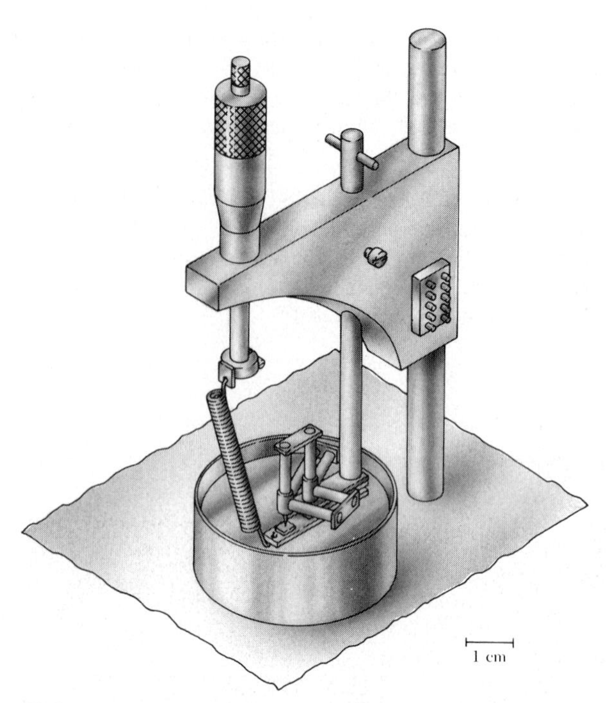

FIGURE 2. One design for a scanning tunneling microscope (STM). The sample to be studied is mounted on a plate in the cylindrical dish. The probe extends beneath the left tripod. The micrometer attached to the spring is used to position the sample.

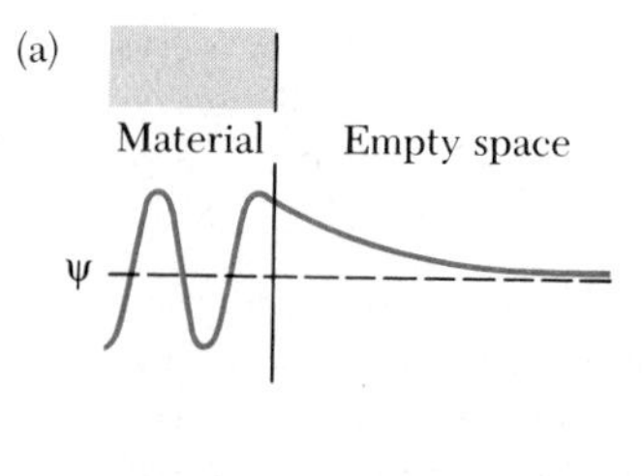

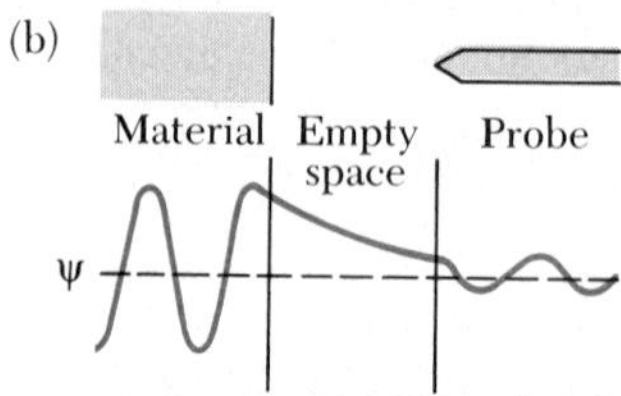

FIGURE 3 (a) The wave function of an electron in the surface of the material to be studied. The wave function extends beyond the surface into the empty region.
(b) The sharp tip of a conducting probe is brought close to the surface. The wave function of a surface electron penetrates into the tip, so that the electron can "tunnel" from surface to tip.

ades after the original work, Binnig and Rohrer received their Prize just six years after their first experiments with an STM.

One design for an STM is shown in Figure 2. The basic idea behind its operation is very simple, as shown in Figure 3. A conducting probe with a very sharp tip is brought near the surface to be studied. Because it is attracted to the positive ions in the surface, an electron in the surface has a lower total energy than would an electron in the empty space between surface and tip. The same thing is true for an electron in the tip. In classical Newtonian mechanics this would mean that electrons could not move between the surface and tip, because they would lack the energy to escape either material. But because the electrons obey quantum mechanics, they can "tunnel" across the barrier of empty space between the tip and the surface. By applying a voltage between the surface and tip, the electrons can be made to tunnel preferentially from the surface to the tip. In this way the tip samples the distribution of electrons just above the surface.

Because of the nature of tunneling, the STM is very sensitive to the distance z from tip to surface. The reason is that in the empty space between the tip and surface, the electron wave function falls off exponentially with a decay length of order 1 Å, that is, the wave function decreases by $1/e$ over that distance. For distances z greater than 10 Å (that is, beyond a few atomic diameters), essentially no tunneling takes place. This exponential behavior causes the current of electrons tunneling from surface to tip to depend very strongly on z. This sensitivity is the basis of the operation of the STM: by monitoring the tunneling current as the tip is scanned over the surface, you obtain a sensitive measure of the topography of the surface. In this way the STM can measure the height of surface features to within 0.01 Å, or approximately one one-hundredth of an atomic diameter!

The STM also has excellent lateral resolution, that is, resolution of features in the plane of the surface. This is because the tips used are *very* sharp indeed, typically only an atom or two wide at their extreme end. Thus the tip only samples the surface electrons in a very tiny region approximately 2 Å wide, and so can "see" very fine detail. You might think that making such tips would be extremely difficult, but in fact it's relatively easy: sometimes just sharpening the tip on a fine grinding stone (or even with fine sandpaper!) is enough to cause the tip atoms to rearrange by themselves into an atomically sharp configuration. (If you find this surprising, you're not alone. Binnig and Rohrer were no less surprised when they discovered this.)

There are two modes of operation for the STM, shown in Figure 4. In the *constant current mode* (Figure 4a), a convenient operating voltage (typically between 2 millivolts and 2 volts) is first established between surface

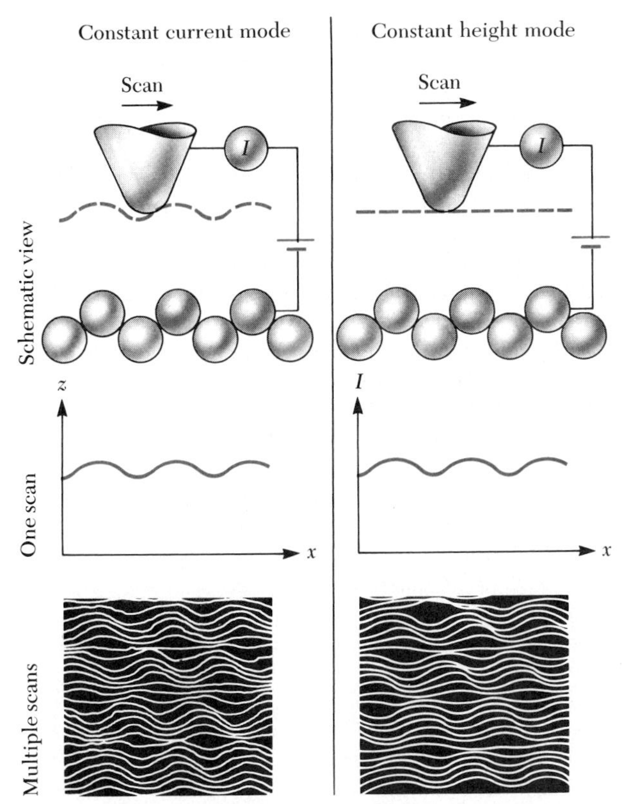

FIGURE 4 Scanning tunneling microscopes can be operated in either (a) the constant current mode or (b) the constant height mode. The images of the surface of graphite were made by Richard Sonnenfeld at the University of California at Santa Barbara. The constant height mode was first used by A. Bryant, D.P.E. Smith, and C.F. Quate, *Applied Physics Letters 48*, 832, 1986.

and tip. The tip is then brought close enough to the surface to obtain a measurable tunneling current. The tip is then scanned over the surface while the tunneling current I is measured. A feedback network changes the separation z between surface and tip to keep the current constant. An image of the surface is made by plotting z versus lateral position (x, y). The simplest scheme for plotting the image is shown in the graph below the schematic view. The height z is plotted versus the scan position x. An image consists of multiple scans displaced laterally from each other in the y direction.

The constant current mode was historically the first to be used, and has the advantage that it can be used to track surfaces that are not atomically flat (as in Figure 1). However, the feedback network requires that the scanning be done relatively slowly. As a result the sample being scanned must be held fixed in place for relatively long times to prevent image distortion.

Alternatively, in the *constant height mode* (Figure 4b) the tip is scanned across the surface at constant voltage and nearly constant height while the current is monitored. In this case the feedback network responds only rapidly enough to keep the average current constant, which means that the tip maintains the same average separation from the surface. The image is then a plot of current I versus lateral position (x, y), as shown in the graph below the schematic. Again, multiple scans along x are displayed as being laterally displaced in the y direction. The image shows the substantial variation of tunneling current as the tip passes over surface features such as individual atoms.

The constant height mode allows much faster scanning of atomically flat surfaces (100 times faster than the constant current mode) since the tip does not have to be moved up and down over the surface "terrain." This fast scanning means that making an image of a surface requires only a short "exposure time." By making a sequence of such images, researchers may be able to study in real time processes in which the surfaces rearrange themselves—in effect making an STM "movie."

Individual atoms have been imaged on a variety of surfaces, including those of so-called *layered materials* in which atoms are naturally arranged into two-dimensional layers. Figure 5 shows an example of atoms on one of these layered materials. In this image it is not only fascinating to see individual atoms, but also to note that some atoms are missing. Specifically, there are three missing atoms in Figure 5. Can you find them?

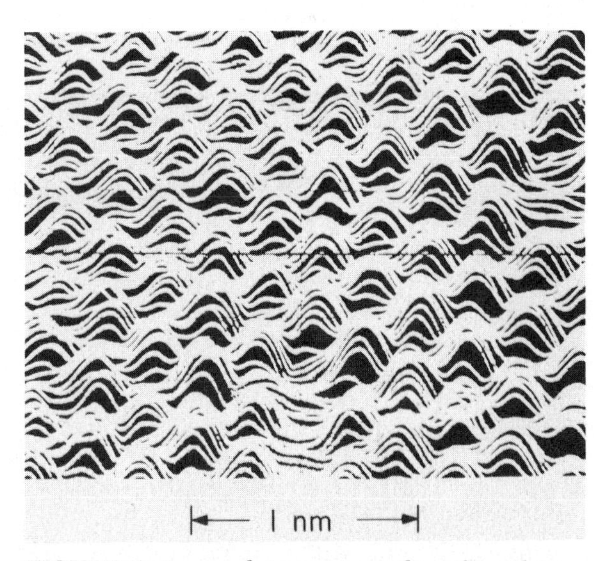

FIGURE 5 Image of atoms on a surface of tantalum disulfide (TaS_2) immersed in liquid nitrogen. 1 nm = 10^{-9} meter = 10 Å. The figure is from C.G. Slough, W.W. McNairy, R.V. Coleman, B. Drake, and P.K. Hansma, *Physical Review B 34*, 994, 1986.

Another remarkable aspect of the STM image in Figure 5 is that it was obtained with the surface and tip immersed in liquid nitrogen. While we assumed earlier in this essay that the space between the surface and tip must be empty, in fact electron tunneling can take place not just through a vacuum but also through gases and liquids—even water. This seems very surprising since we think of water, expecially water with salts dissolved in it, as a conductor. But water is only an *ionic* conductor. For electrons, water behaves as an insulator just as a vacuum behaves as an insulator. Thus electrons can flow through water only by tunneling, which makes scanning tunneling microscopy possible "under water."

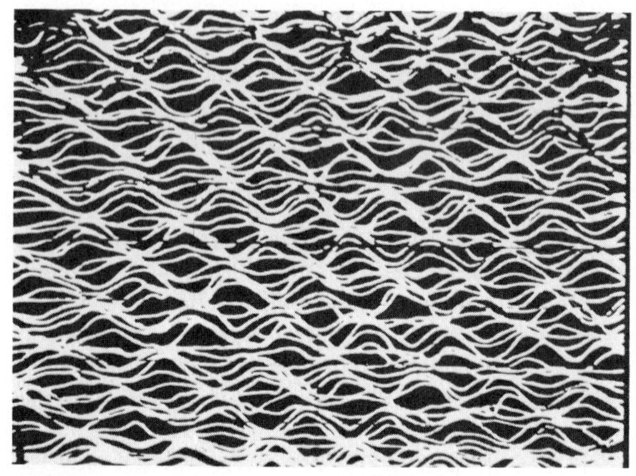

FIGURE 6 Image of a graphite electrode in an electrolyte used for silver plating. The figure is from R. Sonnenfeld and B. Schardt, *Applied Physics Letters 49*, 1172 (1986).

As an example, Figure 6 shows individual carbon atoms on a graphite surface. It was obtained for a surface immersed in a silver plating solution, which is highly conductive for ions but behaves as an insulator for electrons. (The sides of the conducting probe were sheathed with a nonconductor, so that the predominant current into the probe comes from electrons tunneling into the exposed tip. The design of STM used to make this particular image is the one shown in Figure 2). Sonnenfeld and Schardt observed atoms on this graphite surface before plating it with silver, after "islands" of silver atoms were plated onto the surface, and after the silver was electrochemically stripped from the surface. Their work illustrates the promise of the scanning tunneling microscope for seeing processes that take place on an atomic scale.

While the original STMs were one-of-a-kind laboratory devices, commercial STMs have recently become available. Figure 7 is an image of a graphite surface in air made with such a commercial STM. Note the high quality of this image and the recognizable rings of carbon atoms. You may be able to see that three of the six carbon atoms in each ring *appear* lower than the other three. All six atoms are in fact at the same level, but the three that appear lower are bonded to carbon atoms lying directly beneath them in the underlying atomic layer. The atoms in the surface layer that appear higher do not lie directly over subsurface atoms, and hence are not bonded to carbon atoms beneath them. For the higher-appearing at-

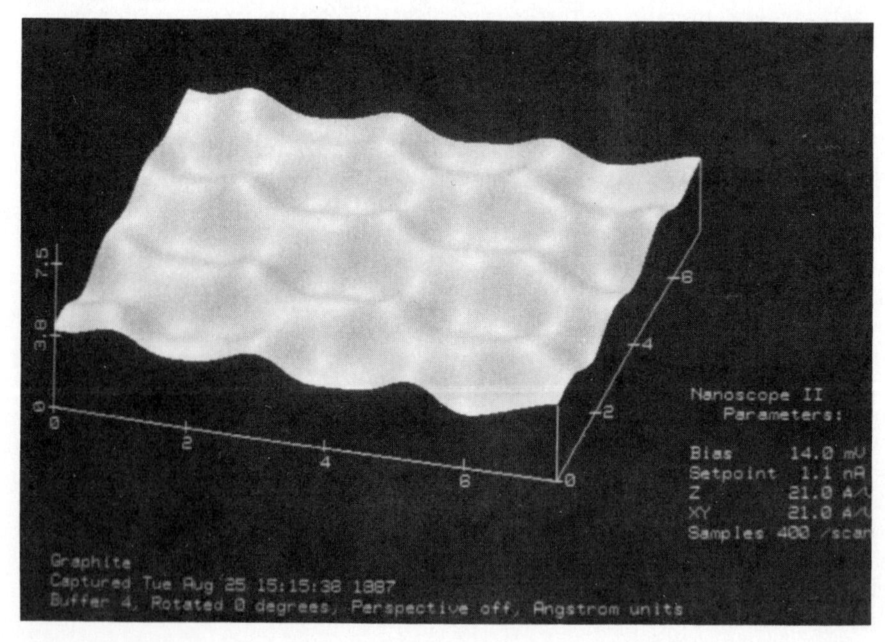

FIGURE 7 Image of a graphite surface in air, obtained with a commercial STM: the Nanoscope II from Digital Instruments in Goleta, California.

oms, some of the electron density that would have been involved in bonding to atoms beneath the surface instead extends into the space above the surface. This extra electron density makes these atoms appear higher in Figure 7, since what the STM maps is the topography of the electron distribution. (See Color Plates 28 and 29.)

The availability of commercial instruments should speed the use of scanning tunneling microscopy in a variety of applications. These include characterizing electrodes for electrochemistry (while the electrode is still in the electrolyte), characterizing the roughness of surfaces, measuring the quality of topical gratings, and even imaging replicas of biological structures.

Perhaps the most remarkable thing about the scanning tunneling microscope is that its operation is based on a quantum mechanical phenomenon—tunneling—that was well understood in the 1920s, yet the STM itself wasn't built until the 1980s. What other applications of quantum mechanics may yet be waiting to be discovered?

SUGGESTED READING

G. Binnig, H. Rohrer, Ch. Gerber, and E. Weibel, *Physical Review Letters 49*, 57 (1982). The first description of the operation of a scanning tunneling microscope.

G. Binnig and H. Rohrer, *Scientific American,* August 1985, p. 50. A popular description of the STM and its applications.

C. F. Quate, *Physics Today,* August 1986, p. 26. An overview of the field of scanning tunneling microscopy, including insights into how it came to be developed.

P. K. Hansma and J. Tersoff, *Journal of Applied Physics 61,* R1 (1987). A comprehensive review of the state of the art in scanning tunneling microscopy.

G. Binnig and H. Rohrer, *Reviews of Modern Physics 59,* 615 (1987). The text of the lecture given on the occasion of the presentation of the 1986 Nobel Prize in Physics.

ESSAY QUESTIONS

1. The density of the earth's atmosphere of air depends on altitude z according to $\rho = \rho_0 e^{-z/z_0}$, where h_0, the "scale height" of the atmosphere, is 8430 meters. What is the scale height for the electron "atmosphere" above a conductor? Give both a formula and a numerical estimate.

2. Our discussion of STM operation was based on the assumption that the conducting probe has only a single tip. In fact a probe may have a number of protrusions on its end, each of which acts as a "tip." Hence it might be expected that such a probe would give multiple STM images of the surface, greatly complicating the analysis. Explain why there is in fact no problem with multiple images, provided that the multiple tips differ in proximity to the surface by 20 Å or more.

3. The STM, while using physical concepts which date from the 1920s, was not developed until the 1980s. Suggest some reasons why the STM was not invented half a century earlier.

4. It was stated in the essay that for conventional microscopes, the resolution is limited by the wavelength of the waves used to make the image. To see whether this guideline applies to the STM, estimate the wavelength of an electron in the surface of a conductor and compare your estimate to the vertical resolution (about 0.01 Å) and lateral resolution (about 2 Å). Does the guideline apply to the two resolutions? Why or why not?

CHAPTER 30

Atomic Physics

Niels Bohr (1885–1962). (AIP Niels Bohr Library, W.F. Meggers Collection)

A large portion of this chapter is concerned with the study of the hydrogen atom. Although the hydrogen atom is the simplest atomic system, it is an especially important system to understand for several reasons:

1. Much of what is learned about the hydrogen atom with its single electron can be extended to such single-electron ions as He^+ and Li^{2+}, which are hydrogen-like in their atomic structure.
2. The hydrogen atom is an ideal system for performing precise tests of theory against experiment and for improving our overall understanding of atomic structure.
3. The quantum numbers used to characterize the allowed states of hydrogen can be used to describe the allowed states of more complex atoms. This

enables us to understand the periodic table of the elements, the explanation of which is one of the greatest triumphs of quantum mechanics.

In this chapter, we shall first discuss the Bohr model of hydrogen, which helps us understand many features of hydrogen but fails to explain many finer details of atomic structure. Following this discussion, we shall examine the hydrogen atom from the viewpoint of quantum mechanics and the quantum numbers used to characterize various atomic states. Additionally, we shall examine the physical significance of the quantum numbers and the effect of a magnetic field on certain quantum states. The Pauli exclusion principle is also presented in this chapter. This physical principle is extremely important in understanding the properties of complex atoms and the arrangement of elements in the periodic table. Finally, we shall apply our knowledge of atomic structure to describe the mechanisms involved in the operation of a laser.

30.1 EARLY MODELS OF THE ATOM

The model of the atom in the days of Newton was that of a tiny, hard, indestructible sphere. This model provided a good basis for the kinetic theory of gases. However, new models had to be devised when later experiments revealed the electrical nature of atoms. J. J. Thompson (1856–1940) suggested a model that described the atom as a volume of positive charge with electrons embedded throughout the volume, much like the seeds in a watermelon (Fig. 30.1). Thompson referred to this model as the "plum-pudding atom."

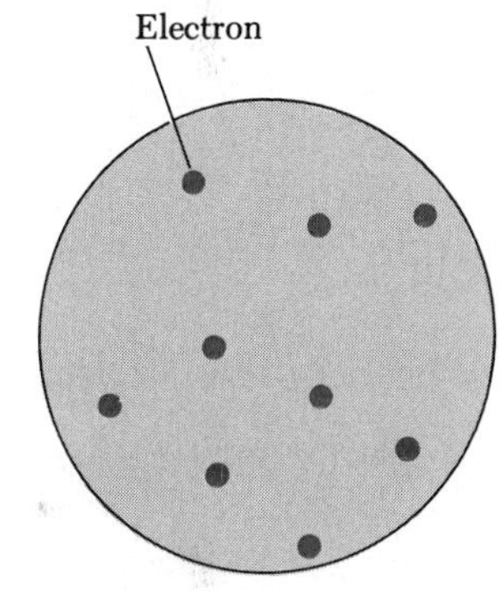

FIGURE 30.1 Thompson's "plum-pudding" model of the atom.

In 1911, Geiger and Marsden under the supervision of Ernest Rutherford (1871–1937) performed a critical experiment that showed that Thompson's plum-pudding model could not be correct. In this experiment, a beam of positively charged **alpha particles**, the nuclei of helium atoms, was projected into a thin metal foil, as in Figure 30.2a. The results of the experiment were

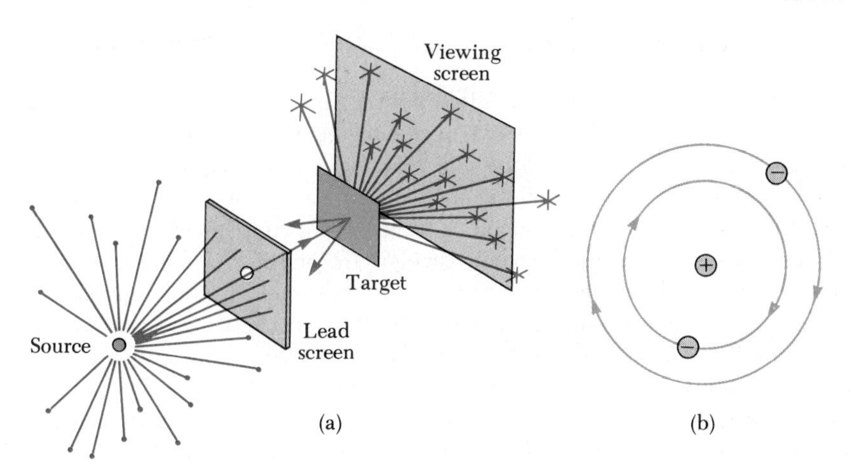

FIGURE 30.2 (a) Geiger and Marsden's technique for observing the scattering of alpha particles from a thin foil target. The source is a naturally occurring radioactive substance, such as radium. (b) Rutherford's planetary model of the atom.

astounding. It was found that most of the alpha particles passed through the foil as if it were empty space! Furthermore, many of the alpha particles that were deflected from their original direction of travel were scattered through very large angles. Some particles were even deflected backwards so as to reverse their direction of travel. When Rutherford was informed that some alpha particles were scattered backward, he wrote, "It was quite the most incredible event that has ever happened to me in my life. It was almost as incredible as if you fired a 15-inch shell at a piece of tissue paper and it came back and hit you."

Such large deflections were not expected on the basis of the plum-pudding model. According to this model, a positively charged alpha particle would never come close enough to a large enough volume of charge to cause any large-angle deflections. Rutherford explained these observations by assuming that the positive charge was concentrated in a region that was small relative to the size of the atom. He called this concentration of positive charge the **nucleus** of the atom. Any electrons belonging to the atom were assumed to be in the relatively large volume outside the nucleus. In order to explain why electrons in this outer region of the atom were not pulled into the nucleus, Rutherford developed a model similar to that of our solar system. He viewed the electrons as particles moving in orbits about the positively charged nucleus in the same manner as the planets orbit the sun, as in Figure 30.2b.

There are two basic difficulties with Rutherford's planetary model of the atom. As we shall see in the next section, an atom emits certain characteristic frequencies of electromagnetic radiation and no others. The Rutherford model is unable to explain this phenomenon. A second difficulty is that, according to classical theory, an accelerated charge must radiate energy in the form of electromagnetic waves. In the Rutherford model, the electrons undergo a centripetal acceleration because they move in circular paths about the nucleus. Therefore, an accelerating electron would rapidly lose its energy as it radiates, which would cause it to spiral quickly into the nucleus.

30.2 ATOMIC SPECTRA

Suppose an evacuated glass tube is filled with a gas such as neon, helium, or argon. If a potential difference between electrodes in the tube produces an electric current in the gas, the tube will emit light whose color is characteristic of the gas. If the emitted light is analyzed by passing it through a narrow slit and then through a spectroscope, a series of discrete lines is observed, each line corresponding to a different wavelength, or color. We refer to such a series of lines as a **line spectrum.** The wavelengths contained in a given line spectrum are characteristic of the element emitting the light. Because no two elements emit the same line spectrum, this phenomenon represents a practical technique for identifying elements in a chemical substance.

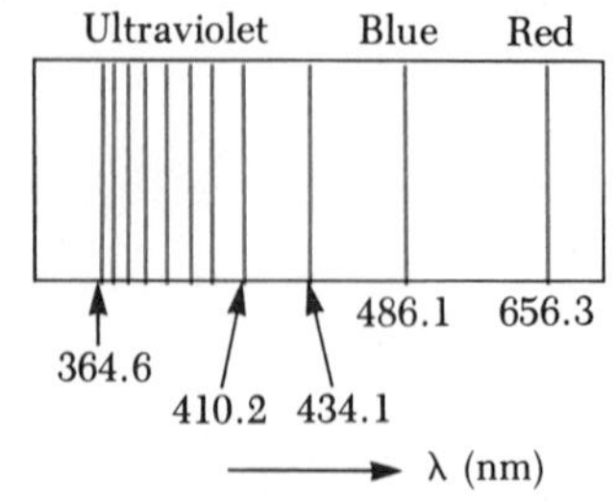

FIGURE 30.3 A series of spectral lines for atomic hydrogen. The prominent lines labeled are part of the Balmer series.

The line spectrum of hydrogen includes a series of lines in the visible region of the spectrum, shown in Figure 30.3. Four of the most prominent lines in this region occur at the wavelengths 656.3 nm, 486.1 nm, 434.1 nm, and 410.2 nm. Later it was discovered that this series of lines continued into the ultraviolet region, ending at 364.6 nm, as shown in Figure 30.3.

In 1885, Johann Balmer (1825–1898) found that the wavelengths of these lines can be described by this simple empirical equation:

$$\frac{1}{\lambda} = R\left(\frac{1}{2^2} - \frac{1}{n^2}\right) \tag{30.1}$$

Balmer series

where n may have integral values of 3, 4, 5, . . . and R is a constant, now called the **Rydberg constant.** If the wavelength is in meters, R has the value

$$R = 1.0973732 \times 10^7 \text{ m}^{-1} \tag{30.2}$$

Rydberg constant

The first line in the Balmer series, at 656.3 nm, corresponds to $n = 3$ in Equation 30.1; the line of 486.1 nm corresponds to $n = 4$, and so on.

Balmer predicted that other series of lines might exist outside the visible region which would obey the equations given below. Several years later, these lines were indeed discovered. These spectra are called the Lyman, Paschen, and Brackett series after their discoverers. The wavelengths of the lines in these series can be calculated by the following empirical formulas:

$$\frac{1}{\lambda} = R\left(1 - \frac{1}{n^2}\right) \qquad n = 2, 3, 4, \ldots \tag{30.3}$$

Lyman series

$$\frac{1}{\lambda} = R\left(\frac{1}{3^2} - \frac{1}{n^2}\right) \qquad n = 4, 5, 6, \ldots \tag{30.4}$$

Paschen series

$$\frac{1}{\lambda} = R\left(\frac{1}{4^2} - \frac{1}{n^2}\right) \qquad n = 5, 6, 7, \ldots \tag{30.5}$$

Brackett series

In addition to emitting light at specific wavelengths, an element can also absorb light at specific wavelengths. The spectral lines corresponding to this process form what is known as an **absorption spectrum.** One can obtain an absorption spectrum by passing a continuous spectrum (one containing all wavelengths) through a vapor of the element being analyzed. The absorption spectrum consists of a series of dark lines superimposed on the otherwise continuous spectrum. It is found through experiments that each line in the absorption spectrum of a given element coincides with a line in the emission spectrum of the element. That is, if hydrogen is the absorbing vapor, dark lines will appear at the visible wavelengths 656.3 nm, 486.1 nm, 434.1 nm, and 410.2 nm. (In practice, the hydrogen vapor must be very hot before it will absorb in the visible portion of the spectrum.)

The absorption spectrum of an element has many practical applications. For example, the continuous spectrum of radiation emitted by the sun must pass through the cooler gases of the solar atmosphere and through the earth's atmosphere. The various absorption lines observed in the solar spectrum have been used to identify elements in the solar atmosphere. It is interesting to note that, when the solar spectrum was first being studied, some lines were found that did not correspond to any known element. A new element had been discovered! Since the Greek word for sun is *helios*, the new element was named helium. Scientists are able to examine the light from stars other than our sun in this fashion, but elements other than those present on earth have never been detected.

30.3 THE BOHR THEORY OF HYDROGEN

At the beginning of the 20th century, scientists were perplexed by the failure of classical physics in explaining the characteristics of atomic spectra. Why did hydrogen emit only certain lines in the visible part of the spectrum? Furthermore, why did hydrogen absorb only those wavelengths which it emitted? In 1913, the Danish scientist Niels Bohr (1885–1963) provided an explanation of atomic spectra that included some features contained in the currently accepted theory. Bohr's theory contained a combination of ideas from classical physics, Planck's original quantum theory, Einstein's photon theory of light, and Rutherford's model of the atom. Bohr's model of the hydrogen atom contains some classical features as well as some revolutionary postulates that could not be justified within the framework of classical physics. The Bohr model can be applied quite successfully to such hydrogen-like ions as singly ionized helium and doubly ionized lithium. However, the theory does not properly describe the spectra of more complex atoms and ions.

The basic postulates of the Bohr model of the hydrogen atom are as follows:

Postulates of the Bohr model

1. The electron moves in circular orbits about the nucleus (the planetary model of the atom) under the influence of the Coulomb force of attraction between the electron and the positively charged nucleus (Fig. 30.4).
2. The electron can exist only in very specific orbits; hence the states are quantized (Planck's quantum hypothesis). The allowed orbits are those for which the angular momentum of the electron about the nucleus is an integral multiple of $\hbar = h/2\pi$, where h is Planck's constant. The angular momentum of the electron is $I\omega$, where $I = mr^2$ and $\omega = v/r$. Thus, the angular momentum is $I\omega = (mr^2)(v/r) = mvr$. Applying the condition that the **angular momentum is quantized,** we have

$$mvr = \frac{nh}{2\pi} = n\hbar \qquad n = 1, 2, 3, \ldots \tag{30.6}$$

3. When the electron is in one of its allowed orbits, it does not radiate energy; hence the atom is stable. Such stable orbits are called **stationary states.** (As discussed earlier, according to classical electricity and magnetism models, the accelerating electron must radiate electromagnetic waves and in doing so would quickly spiral into the nucleus and annihilate the atom.)

Niels Bohr (1885–1962), a Danish physicist, proposed the first quantum model of the atom and was an active participant in the early development of quantum mechanics. He also made important contributions to the theory of nuclear structure. He was awarded the Nobel prize in 1922 for his investigation of the structure of atoms and of the radiation emanating from them. "I remember, as if it were yesterday, the enthusiasm with which the new prospects for the whole of physical and chemical science, opened by the discovery of the atomic nucleus, were discussed in the spring of 1912 among the pupils of Rutherford." (Photo, courtesy of AIP Niels Bohr Library, Lande Collection)

4. The atom radiates energy only when the electron "jumps" from one allowed stationary orbit to another. The frequency of the radiation obeys the condition

$$hf = E_i - E_f \quad (30.7)$$

where E_i and E_f are the energies of the initial and final stationary states. This postulate states that the energy given off by an atom, $E_i - E_f$, is carried away by a photon of energy hf. Note that this expression for the photon energy agrees with that proposed by Einstein in arriving at his photoelectric effect equation (Section 29.2).

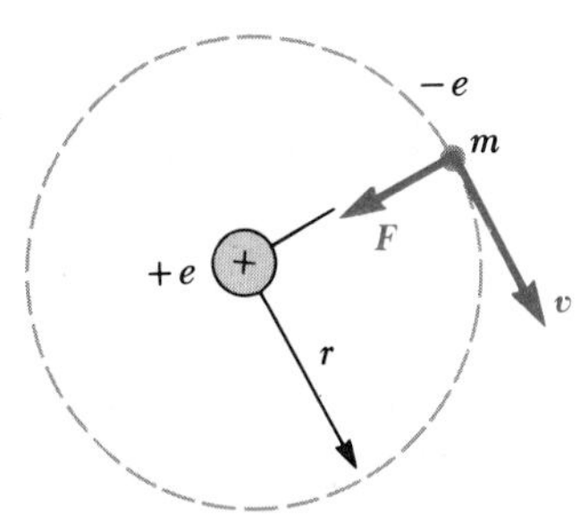

FIGURE 30.4 Diagram representing Bohr's model of the hydrogen atom. In this model, the orbiting electron is allowed to be only in specific orbits of discrete radii.

With these assumptions, we shall now calculate the allowed energies of the hydrogen atom, which we can then use to calculate the wavelengths of the spectral lines emitted by the atom. We shall use the model pictured in Figure 30.4, in which the electron travels in a circular orbit of radius r with an orbit speed v.

The electrical energy of the atom is

$$PE = k\frac{q_1 q_2}{r} = k\frac{(-e)(e)}{r} = -k\frac{e^2}{r}$$

where k is the Coulomb constant. The total energy, E, of the atom is the sum of the kinetic energy of the electron (when the nucleus is at rest) and the potential energy:

$$E = KE + PE = \frac{1}{2}mv^2 - k\frac{e^2}{r} \quad (30.8)$$

Let us apply Newton's second law to the electron. We see that the electric force of attraction on the electron, ke^2/r^2, must equal ma_r, where $a_r = v^2/r$ is the centripetal acceleration of the electron:

$$k\frac{e^2}{r^2} = m\frac{v^2}{r} \quad (30.9)$$

From this equation, we see that the kinetic energy of the electron is

$$\frac{1}{2}mv^2 = \frac{ke^2}{2r} \quad (30.10)$$

We can combine this result with Equation 30.8 and express the **total energy** of the atom as

$$E = -\frac{ke^2}{2r} \quad (30.11)$$

Total energy of the hydrogen atom

An expression for r is obtained by solving Equations 30.6 and 30.9 for v and equating the results:

$$v^2 = \frac{n^2\hbar^2}{m^2r^2} = \frac{ke^2}{mr}$$

$$r_n = \frac{n^2\hbar^2}{mke^2} \qquad n = 1, 2, 3, \ldots \quad (30.12)$$

Radii of the allowed orbits

This equation is based on the assumption that *the electron can exist only in certain allowed orbits.*

The orbit with the smallest radius, called the **Bohr radius,** a_0, corresponds to $n = 1$ and has the value

Bohr radius

$$a_0 = \frac{\hbar^2}{mke^2} = 0.529 \text{ Å} \tag{30.13}$$

A general expression for the radius of any orbit in the hydrogen atom is obtained by substituting Equation 30.13 into Equation 30.12:

$$r_n = n^2 a_0 = n^2(0.529 \text{ Å}) \tag{30.14}$$

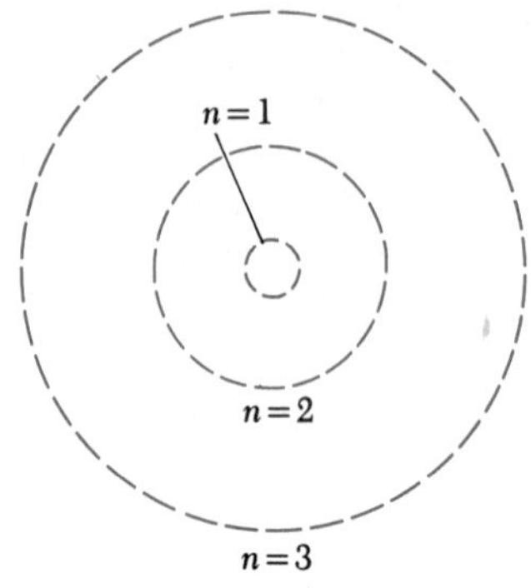

FIGURE 30.5 The first three circular orbits predicted by the Bohr model of the hydrogen atom.

A representation of the various circular orbits is shown in Figure 30.5.

Equation 30.12 may be substituted into Equation 30.11 to give the following expression for the energies of the quantum states:

$$E_n = -\frac{mk^2e^4}{2\hbar^2}\left(\frac{1}{n^2}\right) \qquad n = 1, 2, 3, \ldots \tag{30.15}$$

If we insert numerical values into Equation 30.15, we have

$$E_n = -\frac{13.6}{n^2} \text{ eV} \tag{30.16}$$

The lowest stationary energy state, or **ground state,** corresponds to $n = 1$ and has an energy $E_1 = -mk^2e^4/2\hbar^2 = -13.6$ eV. The next state, corresponding to $n = 2$, has an energy $E_2 = E_1/4 = -3.4$ eV, and so on. An energy level diagram showing the energies of these stationary states and the corresponding quantum numbers is shown in Figure 30.6. The uppermost level shown, corresponding to $n \to \infty$, represents the state for which the electron is completely removed from the atom. In this case, $E = 0$ for $r = \infty$. The minimum energy required to ionize the atom, that is, to completely remove the electron, is called the **ionization energy.** The ionization energy for hydrogen is 13.6 eV.

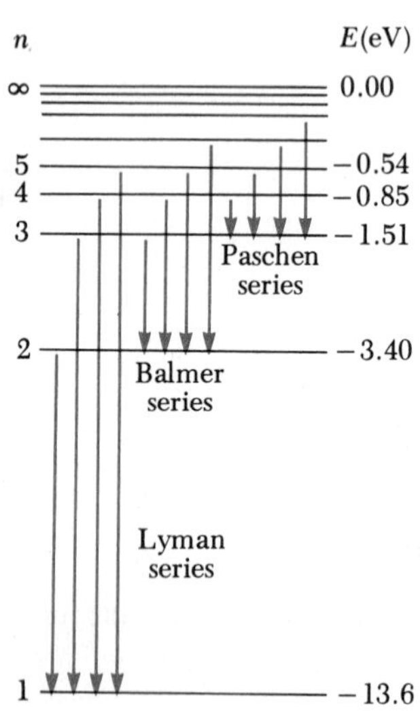

FIGURE 30.6 Energy level diagram for the hydrogen atom. Some transitions for the Lyman, Balmer, and Paschen series are shown. The quantum numbers are at the left, and the energies of the levels (in eV) are at the right.

Equations 30.7 and 30.15 and the fourth Bohr postulate show that if the electron jumps from one orbit, whose quantum number is n_i, to a second orbit, whose quantum number is n_f, it emits a photon of frequency f, given by

$$f = \frac{E_i - E_f}{h} = \frac{mk^2e^4}{4\pi\hbar^3}\left(\frac{1}{n_f^2} - \frac{1}{n_i^2}\right) \tag{30.17}$$

Finally, to compare this result with the empirical formulas for the various spectral series, we use the fact that $\lambda f = c$ and Equation 30.17 to get

$$\frac{1}{\lambda} = \frac{f}{c} = \frac{mk^2e^4}{4\pi c\hbar^3}\left(\frac{1}{n_f^2} - \frac{1}{n_i^2}\right) \tag{30.18}$$

A comparison of this result with Equation 30.1 gives the following expression for the Rydberg constant:

$$R = \frac{mk^2e^4}{4\pi c\hbar^3} \tag{30.19}$$

If we insert the known values into this expression, the theoretical value for R is found to be in excellent agreement with the value determined experimentally for the Rydberg constant. When Bohr demonstrated this agreement, it was recognized as a major accomplishment of his theory.

In order to compare Equation 30.18 with spectroscopic data, it is convenient to express it in the form

$$\frac{1}{\lambda} = R\left(\frac{1}{n_f^2} - \frac{1}{n_i^2}\right) \tag{30.20}$$

We can use this expression to evaluate the wavelengths for the various series in the hydrogen spectrum. For example, in the Balmer series, $n_f = 2$ and $n_i = 3, 4, 5, \ldots$ (Eq. 30.1). For the Lyman series, we take $n_f = 1$ and $n_i = 2, 3, 4, \ldots$. The energy level diagram for hydrogen, shown in Figure 30.6, indicates the origin of the spectral lines described earlier. The transitions between levels are represented by vertical arrows. Note that whenever a transition occurs between a state designated by n_i to one designated by n_f (where $n_i > n_f$), a photon is emitted whose frequency is $(E_i - E_f)/h$. This can be interpreted as follows. The lines in the visible part of the hydrogen spectrum arise when the electron jumps from the third, fourth, or even higher orbit to the second orbit. Likewise, the lines of the Lyman series arise when the electron jumps from the second, third, or even higher orbit to the innermost ($n_f = 1$) orbit. Hence, the Bohr theory successfully predicts the wavelengths of all observed spectral lines in hydrogen.

EXAMPLE 30.1 An Electronic Transition in Hydrogen

The electron in the hydrogen atom makes a transition from the $n = 2$ energy state to the ground state (corresponding to $n = 1$). Find the wavelength and frequency of the emitted photon.

Solution We can use Equation 30.20 directly to obtain λ, with $n_i = 2$ and $n_f = 1$:

$$\frac{1}{\lambda} = R\left(\frac{1}{n_f^2} - \frac{1}{n_i^2}\right)$$

$$\frac{1}{\lambda} = R\left(\frac{1}{1^2} - \frac{1}{2^2}\right) = \frac{3R}{4}$$

$$\lambda = \frac{4}{3R} = \frac{4}{3(1.097 \times 10^7 \text{ m}^{-1})}$$

$$= 1.215 \times 10^{-7} \text{ m} = 121.5 \text{ nm}$$

This wavelength lies in the ultraviolet region.

Since $c = f\lambda$, the frequency of the photon is

$$f = \frac{c}{\lambda} = \frac{3.00 \times 10^8 \text{ m/s}}{1.215 \times 10^{-7} \text{ m}} = 2.47 \times 10^{15} \text{ Hz}$$

EXERCISE What is the wavelength of the photon emitted by hydrogen when the electron makes a transition from the $n = 3$ state to the $n = 1$ state?

Answer $9/8R = 102.6$ nm.

EXAMPLE 30.2 The Balmer Series for Hydrogen

The Balmer series for the hydrogen atom corresponds to electronic transitions that terminate in the state of quantum number $n = 2$, as shown in Figure 30.7. (a) Find the longest-wavelength photon emitted and determine its energy.

Solution The longest-wavelength photon in the Balmer series results from the transition from $n = 3$ to $n = 2$. Using Equation 30.20 gives

$$\frac{1}{\lambda} = R\left(\frac{1}{n_f^2} - \frac{1}{n_i^2}\right)$$

$$\frac{1}{\lambda_{\text{max}}} = R\left(\frac{1}{2^2} - \frac{1}{3^2}\right) = \frac{5}{36} R$$

$$\lambda_{\text{max}} = \frac{36}{5R} = \frac{36}{5(1.097 \times 10^7 \text{ m}^{-1})} = 656.3 \text{ nm}$$

This wavelength is in the red region of the visible spectrum.

The energy of this photon is

$$E_{\text{photon}} = hf = \frac{hc}{\lambda_{\text{max}}}$$

$$= \frac{(6.626 \times 10^{-34} \text{ J·s})(3.00 \times 10^8 \text{ m/s})}{656.3 \times 10^{-9} \text{ m}}$$

$$= 3.03 \times 10^{-19} \text{ J} = 1.89 \text{ eV}$$

We could also obtain the energy of the photon by using Equation 30.7 in the form $hf = E_3 - E_2$, where E_2 and E_3 are the energy levels of the hydrogen atom, which can be calculated from Equation 30.16. Note that this is the lowest-energy photon in this series since it involves the smallest energy change.

(b) Find the shortest-wavelength photon emitted in the Balmer series.

Solution The shortest-wavelength photon in the Balmer series is emitted when the electron makes a transition from $n = \infty$ to $n = 2$. Therefore

$$\frac{1}{\lambda_{\text{min}}} = R\left(\frac{1}{2^2} - \frac{1}{\infty}\right) = \frac{R}{4}$$

$$\lambda_{\text{min}} = \frac{4}{R} = \frac{4}{1.097 \times 10^7 \text{ m}^{-1}} = 364.6 \text{ nm}$$

This wavelength is in the ultraviolet region and corresponds to the series limit.

EXERCISE Find the energy of the shortest-wavelength photon emitted in the Balmer series for hydrogen.

Answer 3.40 eV.

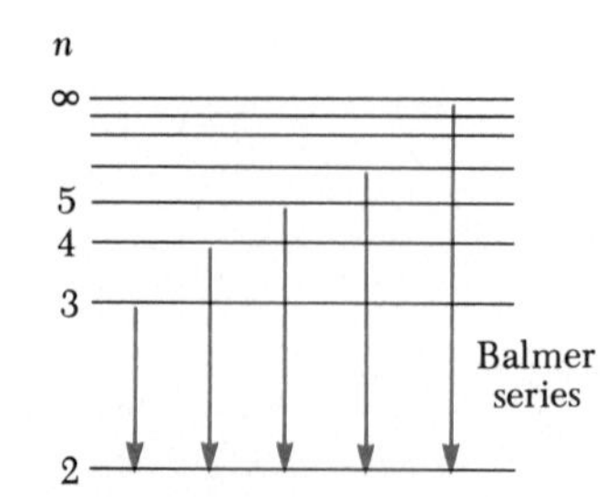

FIGURE 30.7 (Example 30.2) Transitions responsible for the Balmer series for the hydrogen atom. All transitions terminate at the $n = 2$ level.

Bohr's Correspondence Principle

In our study of relativity, we found that Newtonian mechanics cannot be used to describe phenomena that occur at speeds approaching the speed of light. Newtonian mechanics is a special case of relativistic mechanics and is usable only when v is much less than c. Similarly, *quantum mechanics is in agreement with classical physics when the quantum numbers are very large.* This principle, first set forth by Bohr, is called the **correspondence principle.**

For example, consider the hydrogen atom in an orbit for large values of n, say $n > 10{,}000$. For such large values of n, the energy differences between adjacent levels approach zero and the levels are nearly continuous. Consequently, the classical model is reasonably accurate in describing the system for large values of n. According to the classical picture, the frequency of the light emitted by the atom is equal to the frequency of revolution of the electron in its orbit about the nucleus. Calculations show that for $n > 10{,}000$, this frequency is different from that predicted by quantum mechanics by less than 0.015%.

30.4 MODIFICATION OF THE BOHR THEORY

The Bohr theory of the hydrogen atom was a tremendous success in certain areas in that it explained several features of the spectra of hydrogen that had previously defied explanation. It accounted for the Balmer series and other series, it predicted a value for the Rydberg constant which is in excellent agreement with the experimental value, it derived an expression for the radius of the atom, and it predicted the energy levels of hydrogen. Although these successes were important to scientists, it is perhaps even more important that the Bohr theory gave us a model of what the atom looks like and how it behaves. Once a basic model is constructed, refinements and modifications can be made to enlarge upon the concept and to explain finer details.

The analysis used in the Bohr theory is also successful when applied to *hydrogen-like* atoms. An atom is said to be hydrogen-like when it contains only one electron. Examples are singly ionized helium, doubly ionized lithium, triply ionized beryllium, etc. The results of the Bohr theory for hydrogen can be extended to hydrogen-like atoms by substituting Ze^2 for e^2 in the hydrogen equations, where Z is the atomic number of the element. For example, Equations 30.15 and 30.18 become

$$E_n = -\frac{mk^2 Z^2 e^4}{2\hbar^2}\left(\frac{1}{n^2}\right) \qquad n = 1, 2, 3, \ldots \tag{30.21}$$

and

$$\frac{1}{\lambda} = \frac{mk^2 Z^2 e^4}{4\pi c\hbar^3}\left(\frac{1}{n_f^2} - \frac{1}{n_i^2}\right) \tag{30.22}$$

Although many attempts were made to extend the Bohr theory to more complex (multi-electron) atoms, the results were unsuccessful. Even today, there are only approximate methods available for treating multi-electron atoms.

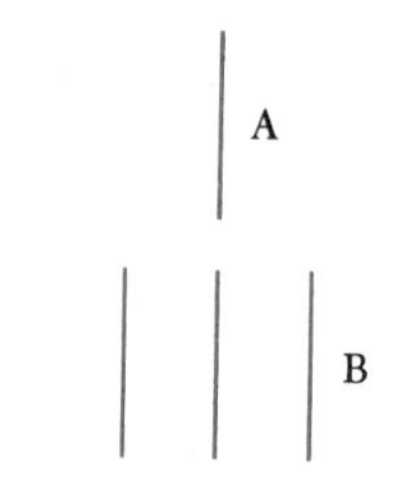

FIGURE 30.8 A single line *(A)* can split into three separate lines *(B)* in a magnetic field.

Within a few months following the publication of Bohr's theory, Arnold Sommerfeld (1868–1951) extended the results to include elliptical orbits. We shall examine his model briefly because much of the nomenclature used in this treatment is still in use today. When Bohr introduced the concept of quantization of angular momentum, the idea led to the introduction of the **principal quantum number** n, which determines the energy of the allowed states of hydrogen. Sommerfeld's theory retained n, but also introduced a new quantum number ℓ, called the **orbital quantum number,** where the values that ℓ can have range from 0 to $n - 1$, in integer steps. According to this model, an electron in any one of the allowed energy states of a hydrogen atom may move in any one of a number of orbits. For each value of n there are n possible orbits. Since $n = 1$ and $\ell = 0$ for the first energy level (ground state), there is only one possible orbit for this state. The second energy level, with $n = 2$, has two possible orbits corresponding to $\ell = 0$ and $\ell = 1$. The third energy level, with $n = 3$, has three possible orbits corresponding to $\ell = 0$, $\ell = 1$, and $\ell = 2$.

It became customary to use a specific nomenclature when referring to these states and orbits. **All states with the same principal quantum number are said to form a shell.** These shells are identified by the letters K, L, M, . . . , which designate the states for which $n = 1, 2, 3, \ldots$. Likewise, **the states having the same values of n and ℓ are said to form a subshell.** The leters $s, p, d, f, g, \ldots$ are used to designate the states for which $\ell = 0, 1, 2, 3, 4, \ldots$. These notations are summarized in Table 30.1.

You should note that the maximum number of electrons that are allowed in any given subshell is given by $2(2\ell + 1)$. For example, the p subshell ($\ell = 1$) is filled when it contains 6 electrons. This fact will be important to us later when we discuss the *Pauli exclusion principle.*

Another modification of the Bohr theory arose when it was discovered that the spectral lines of a gas are split into several closely spaced lines when the gas is placed in a strong magnetic field. (This is called the Zeeman effect after its discoverer.) Figure 30.8 shows a single spectral line being split into three closely spaced lines. This observation indicates that the energy of an electron is slightly modified when the atom is immersed in a magnetic field. In order to explain this observation, a new quantum number, m_ℓ, called the **orbital magnetic quantum number**, was introduced. The theory is in accord with experimental results when m_ℓ is restricted to values ranging from $-\ell$ to $+\ell$, in integer steps.

Finally, very high resolution spectrometers revealed that spectral lines of gases are in fact two very closely spaced lines even in the absence of an ex-

TABLE 30.1 Shell and Subshell Notations

n	Shell Symbol	ℓ	Subshell Symbol
1	K	0	s
2	L	1	p
3	M	2	d
4	N	3	f
5	O	4	g
6	P	5	h
. . .		. . .	

ternal magnetic field. This splitting was referred to as **fine structure.** In 1925, Samuel Goudsmit and George Uhlenbeck introduced the idea of an electron spinning about its own axis to explain the origin of fine structure. The results of their work introduced yet another quantum number, m_s, called the **spin quantum number.** We shall withhold further discussion of this quantum number for a later section. It is interesting to note that each of the new concepts introduced into the original Bohr theory added new quantum numbers and improved the original model. However, the most profound step forward in our current understanding of atomic structure came with the development of quantum mechanics.

EXAMPLE 30.3 Singly Ionized Helium

Singly ionized helium, He^+, a hydrogen-like system, has one electron in the 1*s* orbit when the electron is in its ground state. Find (a) the energy of the electron in the ground state, and (b) the radius of the ground state orbit.

Solution (a) From Equation 30.21, the energy of a level whose principal quantum number is n is given by

$$E_n = -\frac{mk^2Z^2e^4}{2\hbar^2}\left(\frac{1}{n^2}\right)$$

This can be expressed in eV units as

$$E_n = -\frac{Z^2(13.6)}{n^2}\text{ eV}$$

Since $Z = 2$ for helium, and $n = 1$ in the ground state, we have

$$E_1 = -4(13.6)\text{ eV} = -54.4\text{ eV}$$

(b) The radius of the ground state orbit can be found with the help of Equation 30.12. This equation must be modified in the case of a hydrogen-like atom by substituting Ze^2 for e^2, to obtain

$$r_n = \frac{n^2\hbar^2}{mkZe^2} = \frac{n^2}{Z}(0.529\text{ Å})$$

For our case, $n = 1$ and $Z = 2$, and the result is

$$r_1 = 0.265\text{ Å}$$

30.5 DE BROGLIE WAVES AND THE HYDROGEN ATOM

One of the postulates made by Bohr in this theory of the hydrogen atom was that angular momentum of the electron is quantized in units of $\hbar = h/2\pi$, or

$$mvr = n\hbar$$

For more than a decade following Bohr's publication, no one was able to explain why the angular momentum of the electron was restricted to these discrete values. Finally, de Broglie recognized a connection between his theory of the wave character of material properties and the quantization condition given above. De Broglie assumed that an electron orbit would be stable (allowed) only if it contained an integral number of electron wavelengths. Figure 30.9 demonstrates this point when five complete wavelengths are contained in one circumference of the orbit. Similar patterns can be drawn for orbits containing four wavelengths, three wavelengths, two wavelengths, etc. This situation is analogous to that of standing waves on a string discussed in Chapter 15. There we found that strings have preferred (resonant) frequencies of vibration. Figure 30.10 shows a standing wave pattern for a string fixed at each end (in this case, the pattern contains three complete wavelengths.) Now imagine that the vibrating string is removed from its supports at A and B and bent into a circular shape such that points A and B are brought together. The end result is a pattern similar to that shown in Figure 30.9a.

In general, the condition for a de Broglie standing wave in an electron orbit is that the circumference must contain an integral multiple of electron wavelengths. We can express this condition as

$$2\pi r = n\lambda \qquad (n = 1, 2, 3, \ldots) \tag{30.23}$$

De Broglie's equation for the wavelength of an electron in terms of its momentum is

$$\lambda = h/mv$$

When this expression for λ is substituted into Equation 30.23, we find

$$2\pi r = \frac{nh}{mv}$$

or

$$mvr = n\hbar$$

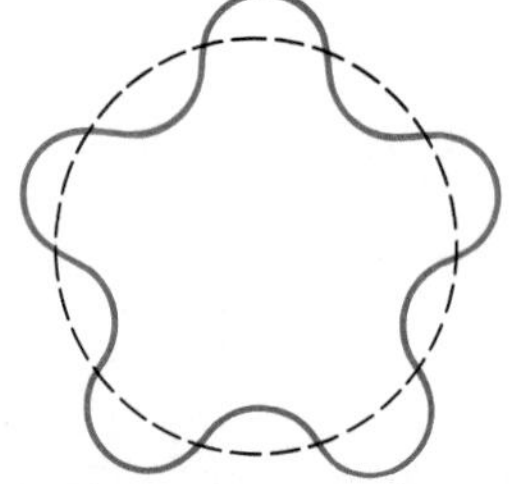

FIGURE 30.9 Standing wave pattern for an electron wave in a stable orbit of hydrogen. There are five full wavelengths in this orbit.

This is precisely the quantization of angular momentum condition imposed by Bohr in his original theory of hydrogen.

The electron orbit shown in Figure 30.9 contains five complete wavelengths and corresponds to the case where the principal quantum number n equals five. The orbit with one complete wavelength in its circumference would correspond to the first Bohr orbit, $n = 1$; the orbit with two complete wavelengths would correspond to the second Bohr orbit, $n = 2$, and so forth.

By applying the wave theory of matter to electrons in atoms, de Broglie was able to explain the appearance of integers in the Bohr theory as a natural consequence of interference. This was the first convincing argument which

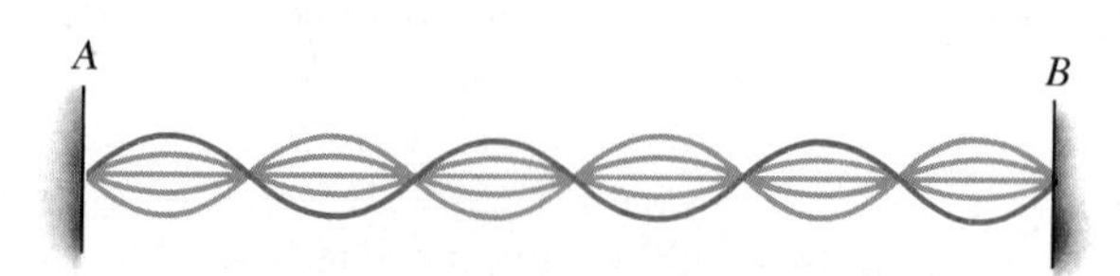

FIGURE 30.10 Standing wave pattern for a vibrating stretched string fixed at its ends. This pattern has three full wavelengths.

suggested that the wave nature of matter was at the heart of the behavior of atomic systems. Although the analysis provided by de Broglie was a promising first step, gigantic strides were subsequently made with the development of Schrödinger's wave equation and its application to atomic systems.

30.6 QUANTUM MECHANICS AND THE HYDROGEN ATOM

One of the first great achievements of quantum mechanics was the solution of the wave equation for the hydrogen atom. We shall not attempt to carry out this solution. Rather, we shall simply describe its properties and some of the implications with regard to atomic structure.

According to quantum mechanics, the energies of the allowed states are in exact agreement with the values obtained by the Bohr theory (Eq. 30.15), where the allowed energies depend only on the principal quantum number, n.

In addition to the principal quantum number, two other quantum numbers emerged from the solution, ℓ and m_ℓ. The quantum number ℓ is called the **orbital quantum number,** and m_ℓ is called the **orbital magnetic quantum number.** As pointed out in Section 30.4, these quantum numbers had already appeared in modifications made to the Bohr theory. The significance of quantum mechanics is that these quantum numbers and the restrictions placed on their values arose directly from the mathematics and not by some ad hoc assumptions to make the theory consistent with experimental observation. Because we shall need to make use of the various quantum numbers in the next several sections, the ranges of their values are repeated below.

The values of n can range from 1 to ∞.
The values of ℓ can range from 0 to $n - 1$.
The values of m_ℓ can range from $-\ell$ to ℓ in integer steps.

For example, if $n = 1$, only $\ell = 0$ and $m_\ell = 0$ are permitted. If $n = 2$, the value of ℓ may be 0 or 1; if $\ell = 0$, then $m_\ell = 0$, but if $\ell = 1$, then m_ℓ may be 1, 0, or -1. Table 30.2 summarizes the rules for determining the allowed values of ℓ and m_ℓ for a given value of n.

States that violate the rules given in Table 30.2 cannot exist. For instance, one state that cannot exist is the $2d$ state, which would have $n = 2$ and $\ell = 2$. This state is not allowed because the highest allowed value of ℓ is $n - 1$, or 1 in this case. Thus, for $n = 2$, $2s$ and $2p$ are allowed states but $2d$, $2f$, . . . are not. For $n = 3$, the allowed states are $3s$, $3p$, and $3d$.

TABLE 30.2 Three Quantum Numbers for the Hydrogen Atom

Quantum Number	Name	Allowed Values	Number of Allowed States
n	Principal quantum number	1, 2, 3, . . .	Any number
ℓ	Orbital quantum number	0, 1, 2, . . ., $n - 1$	n
m_ℓ	Orbital magnetic quantum number	$-\ell, -\ell + 1, \ldots, 0, \ldots, \ell - 1, \ell$	$2\ell + 1$

EXAMPLE 30.4 The $n = 2$ Level of Hydrogen

Determine the number of states in the hydrogen atom corresponding to the principal quantum number $n = 2$ and calculate the energies of these states.

Solution For $n = 2$, ℓ can have the values 0 and 1. For $\ell = 0$, m_ℓ can only be 0; for $\ell = 1$, m_ℓ can be -1, 0, or 1. Hence, we have a state designated as the 2*s* state associated with the quantum numbers $n = 2$, $\ell = 0$, and $m_\ell = 0$, and three states designated as 2*p* states for which the quantum numbers are $n = 2$, $\ell = 1$, $m_\ell = -1$; $n = 2$, $\ell = 1$, $m_\ell = 0$; and $n = 2$, $\ell = 1$, $m_\ell = 1$.

Because all of these states have the same principal quantum number, $n = 2$, they also have the same energy, which can be calculated using Equation 30.16, that is, $E_n = -13.6/n^2$. For $n = 2$, this gives

$$E_2 = -\frac{13.6}{2^2}\text{ eV} = -3.4\text{ eV}$$

EXERCISE How many possible states are there for the $n = 3$ level of hydrogen? For the $n = 4$ level?

Answers 9 states for $n = 3$, and 16 states for $n = 4$.

30.7 THE SPIN MAGNETIC QUANTUM NUMBER

Example 30.4 was presented to give you some practice in manipulating quantum numbers. For example, we found that there are four states corresponding to the principal quantum number $n = 2$. As we shall see in this section, there actually are *eight* such states rather than four in this case. This can be explained by requiring a fourth quantum number for each state, called the **spin magnetic quantum number,** m_s.

As pointed out in Section 30.4, the need for this new quantum number first came about because of an unusual feature in the spectra of certain gases, such as sodium vapor. Close examination of one of the prominent lines of sodium shows that this line is, in fact, two very closely spaced lines. The wavelengths of these lines occur in the yellow region at 589.0 nm and 589.6 nm. In 1925, when this was first noticed, the theory of atomic structure had not been adequately developed to explain why there were two lines rather than one. To resolve this dilemma, Samual Goudsmidt and George Uhlenbeck, following a suggestion by the Austrian physicist Wolfgang Pauli, proposed that a new quantum number, called the spin quantum number, must be added to the set of required quantum numbers in describing a quantum state.

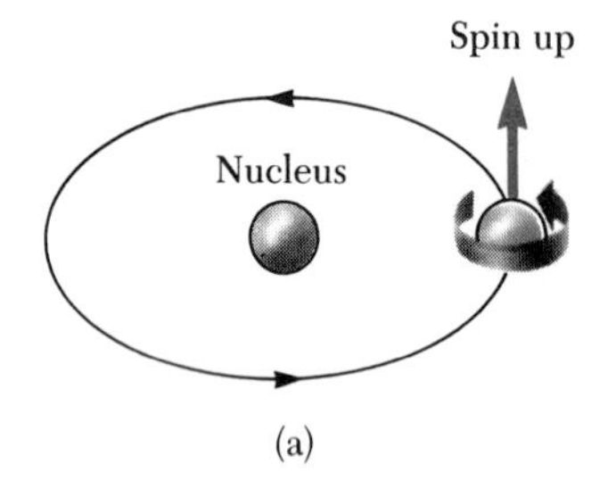

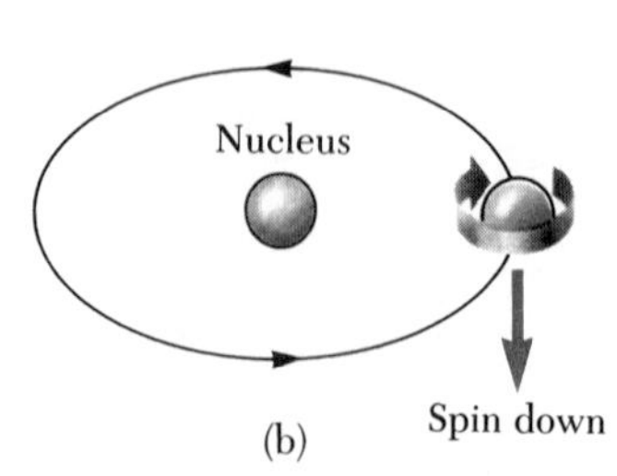

FIGURE 30.11 As an electron moves in its orbit about the nucleus, its spin can be either *(a)* up or *(b)* down.

In order to describe the spin quantum number, it is convenient (but incorrect) to think of the electron as spinning on its axis as it orbits the nucleus, just as the earth spins about its axis as it orbits the sun. According to the theory, there are only two ways that the electron can spin as it orbits the nucleus, as shown in Figure 30.11. If the direction of spin is relative to its orbital motion, as shown in Figure 30.11a, the electron is said to have "spin up." If the direction of spin is relative to its orbital motion, as in Figure 30.11b, the electron is said to have "spin down." The energy of the electron is slightly different for the two spin directions. As it turns out, this energy difference properly accounts for the observed splitting of the yellow sodium lines. The quantum numbers associated with the spin of the electron are $m_s = \frac{1}{2}$ for the spin-up state and $m_s = -\frac{1}{2}$ for the spin-down state. As we shall see in the following example, this added quantum number doubles the number of allowed states specified by the quantum numbers n, ℓ, and m_ℓ.

The classical description of electron spin described above is incorrect because quantum mechanics tells us that, since the electron cannot be precisely located in space, it cannot be considered to be spinning as pictured in Figure 30.11.

Although the classical description of electron spin described above is helpful as a mental construct, it should not be taken too seriously. It is incorrect to view the electron as a charged, spinning object with a definite size. In spite of this conceptual difficulty, all experimental evidence supports the fact that an electron has an intrinsic property that can be described by the spin magnetic quantum number.

EXAMPLE 30.5 Adding Electron Spin to Hydrogen

Determine the quantum numbers associated with the possible states in the hydrogen atom corresponding to the principal quantum number $n = 2$.

Solution With the addition of the spin quantum number, we have the following possibilities:

n	ℓ	m_ℓ	m_s	Subshell	Shell	Number of Electrons in Subshell
2	0	0	½	2*s*	L	2
2	0	0	−½			
2	1	1	½	2*p*	L	6
2	1	1	−½			
2	1	0	½			
2	1	0	−½			
2	1	−1	½			
2	1	−1	−½			

EXERCISE Show that for $n = 3$, there are 18 possible states. (This follows from the restrictions that the maximum number of electrons in the 3*s* state is 2, the maximum number in the 3*p* state is 6, and the maximum number in the 3*d* state is 10.)

30.8 ELECTRON CLOUDS

The solution of the wave equation for the wave function Ψ, discussed in Section 29.11, yields a quantity dependent on the quantum numbers n, ℓ, and m_ℓ. Let us assume that we have found a value for Ψ and see what it may tell us about the hydrogen atom. Let us choose a value of $n = 1$ for the principal quantum number, which corresponds to the lowest energy state for hydrogen. For $n = 1$, the restrictions placed on the remaining quantum numbers are that $\ell = 0$ and $m_\ell = 0$. We can now substitute these values into our expression for Ψ.

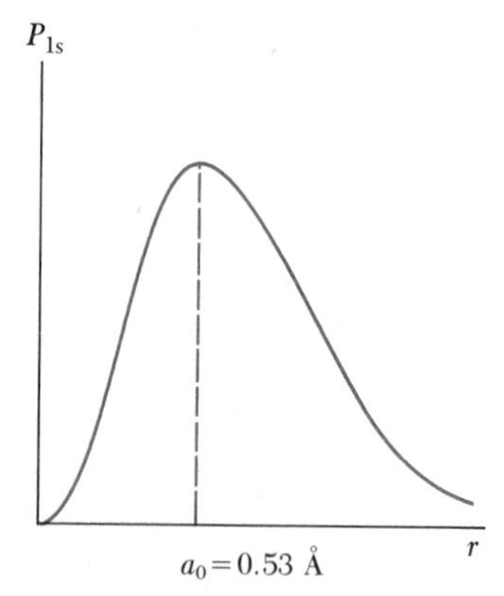

FIGURE 30.12 The probability of finding the electron versus distance from the nucleus for the hydrogen atom in the $1s$ (ground) state. Note that the probability has its maximum value when r equals the first Bohr radius, a_0.

The quantity $|\Psi|^2$ has great physical significance because it is proportional to *the probability of finding the electron at a given position.* Figure 30.12 gives the probability of finding the electron at various distances from the nucleus in the $1s$ state of hydrogen. Some useful and surprising information can be extracted from this curve. First, the curve peaks at a value of r equal to 0.53 Å, the Bohr value of the radius of the first electron orbit in hydrogen. This means that the probability of finding the electron at this distance from the nucleus is a maximum. However, as the curve indicates, there is also a probability of finding the electron at other distances from the nucleus. In other words, the electron is not confined to a particular orbital distance from the nucleus, as assumed in the Bohr model. The electron may be found at various distances from the nucleus, but *the probability of finding it at a distance corresponding to the first Bohr orbit is a maximum.* Quantum mechanics also predicts that the wave function for the hydrogen atom in the ground state is spherically symmetric, hence the electron can be found in a spherical region surrounding the nucleus. This is in contrast to the Bohr theory which confines the position of the electron to points in a plane. This result is often interpreted by viewing the electron as a cloud surrounding the nucleus. An attempt at picturing this cloud-like behavior is shown in Figure 30.13. The densest regions of the cloud represent those locations where the electron is most likely to be found.

If a similar analysis is carried out for the $n = 2$, $\ell = 1$ state of hydrogen, a peak of the probability curve is found at $4a_0$. Likewise, for the $n = 3$, $\ell = 2$ state, the curve peaks at $9a_0$. Thus, quantum mechanics predicts a most probable electron location that is in agreement with the locations predicted by the Bohr theory.

30.9 THE EXCLUSION PRINCIPLE AND THE PERIODIC TABLE

Earlier we found that the state of an electron in an atom is specified by four quantum numbers: n, ℓ, m_ℓ and m_s. For example, an electron in the ground state of hydrogen could have quantum numbers of $n = 1$, $\ell = 0$, $m_\ell = 0$, $m_s = \frac{1}{2}$. As it turns out, the state of an electron in any other atom may also be specified by this same set of quantum numbers. In fact, these four quantum numbers can be used to describe all the electronic states of an atom regardless of the number of electrons in its structure.

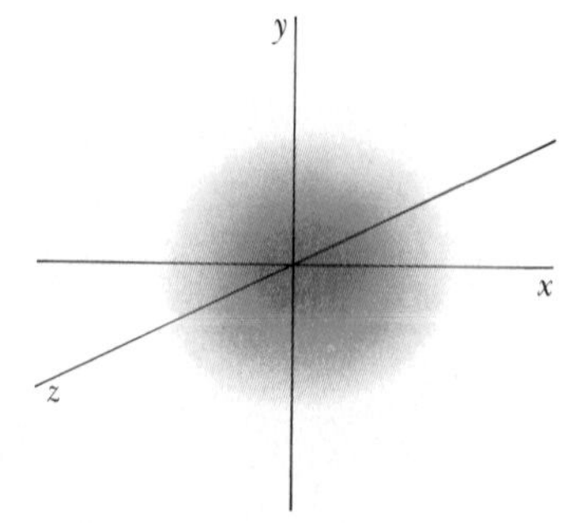

FIGURE 30.13 The spherical electron cloud for the hydrogen atom in its $1s$ state.

An obvious question that arises here is "How many electrons can have a particular set of quantum numbers?" This important question was answered by Pauli in 1925 in a powerful statement known as the **exclusion principle**:

No two electrons in an atom can ever be in the same quantum state; that is, no two electrons in the same atom can have the same set of quantum numbers n, ℓ, m_ℓ, and m_s.

It is interesting to note that if this principle were not valid, every electron would end up in the lowest energy state of the atom and the chemical behavior of the elements would be grossly modified. Nature as we know it would not exist! In reality, we can view the electronic structure of complex atoms as

Wolfgang Pauli (1900–1958), an Austrian physicist, was awarded the Nobel prize in 1945 for his discovery of the exclusion principle, also called the Pauli principle. (Photo taken by S.A. Goudsmit, AIP Niels Bohr Library)

a succession of filled levels increasing in energy, where the outermost electrons are primarily responsible for the chemical properties of the element.

As a general rule, the order of filling of an atom's subshell with electrons is as follows. Once one subshell is filled, the next electron goes into the vacant subshell that is lowest in energy. One can understand this principle by recognizing that if the atom was not in the lowest energy state available to it, it would radiate energy until it reached this state. Recall that a subshell is filled when it contains $2(2\ell + 1)$ electrons. This rule is based on the analysis of quantum numbers to be described below. Following this rule, shells and subshells are filled according to the pattern given in Table 30.3.

The exclusion principle can be illustrated by an examination of the electronic arrangement in a few of the lighter atoms.

Hydrogen has only one electron, which, in its ground state, can be described by either of two sets of quantum numbers: 1, 0, 0, $\frac{1}{2}$, or 1, 0, 0, $-\frac{1}{2}$. The electronic configuration of this atom is often designated as $1s^1$. The notation $1s$ means that we are referring to a state for which $n = 1$ and $\ell = 0$, and the superscript indicates that one electron is present in this level.

Neutral *helium* has two electrons. In the ground state, the quantum numbers for these two electrons are 1, 0, 0, $\frac{1}{2}$ and 1, 0, 0, $-\frac{1}{2}$. There are no other possible combinations of quantum numbers for this level, and we say that the K shell is filled. Helium is designated by the notation $1s^2$.

TABLE 30.3 The Number of Electrons in Filled Subshells and Shells

Shell	Subshell	Number of Electrons in Filled Subshell	Number of Electrons in Filled Shell
K ($n = 1$)	$s(\ell = 0)$	2	2
L ($n = 2$)	$s(\ell = 0)$	2	8
	$p(\ell = 1)$	6	
M ($n = 3$)	$s(\ell = 0)$	2	18
	$p(\ell = 1)$	6	
	$d(\ell = 2)$	10	
N ($n = 4$)	$s(\ell = 0)$	2	32
	$p(\ell = 1)$	6	
	$d(\ell = 2)$	10	
	$f(\ell = 3)$	14	

Neutral *lithium* has three electrons. In the ground state, two of these are in the $1s$ subshell and the third is in the $2s$ subshell because this subshell is slightly lower in energy than the $2p$ subshell. Hence, the electronic configuration for lithium is $1s^2\ 2s^1$.

A list of electronic ground-state configurations for a number of atoms is provided in Table 30.4. This listing will help us to gain an understanding of the periodic table of the elements given in the endpapers. In 1871 Dmitri Mendeleev (1834–1907), a Russian chemist, arranged the elements known at that time in a table similar to that shown in Appendix C according to their atomic weights and chemical similarities. The first table Mendeleev proposed contained many blank spaces, and he boldly stated that the gaps were there only because those elements had not yet been discovered. By noting the column in which these missing elements should be located, he was able to make rough predictions about their chemical properties. Within 20 years of this announcement, these elements were indeed discovered.

The elements in our current version of the periodic table are still arranged such that all those in a vertical column have similar chemical properties. For example, consider the elements in the last column: He (helium), Ne (neon), Ar (argon), Kr (krypton), Xe (xenon), and Rn (radon). The outstanding characteristic of these elements is that they do not normally take part in chemical reactions, that is, they do not join with other atoms to form molecules, and are therefore classified as being inert. Because of this aloofness, they are referred to as the noble gases. We can partially understand this behavior by looking at the electronic configurations shown in Table 30.4. The

TABLE 30.4 Electronic Configuration of Some Elements

Z	Symbol	Ground-State Configuration	Ionization Energy (eV)	Z	Symbol	Ground-State Configuration	Ionization Energy (eV)
1	H	$1s^1$	13.595	19	K	[Ar] $4s^1$	4.339
2	He	$1s^2$	24.581	20	Ca	$4s^2$	6.111
				21	Sc	$3d4s^2$	6.54
3	Li	[He] $2s^1$	5.390	22	Ti	$3d^24s^2$	6.83
4	Be	$2s^2$	9.320	23	V	$3d^34s^2$	6.74
5	B	$2s^22p^1$	8.296	24	Cr	$3d^54s$	6.76
6	C	$2s^22p^2$	11.256	25	Mn	$3d^54s^2$	7.432
7	N	$2s^22p^3$	14.545	26	Fe	$3d^64s^2$	7.87
8	O	$2s^22p^4$	13.614	27	Co	$3d^74s^2$	7.86
9	F	$2s^22p^5$	17.418	28	Ni	$3d^84s^2$	7.633
10	Ne	$2s^22p^6$	21.559	29	Cu	$3d^{10}4s^1$	7.724
				30	Zn	$3d^{10}4s^2$	9.391
11	Na	[Ne] $3s^1$	5.138	31	Ga	$3d^{10}4s^24p^1$	6.00
12	Mg	$3s^2$	7.644	32	Ge	$3d^{10}4s^24p^2$	7.88
13	Al	$3s^23p^1$	5.984	33	As	$3d^{10}4s^24p^3$	9.81
14	Si	$3s^23p^2$	8.149	34	Se	$3d^{10}4s^24p^4$	9.75
15	P	$3s^23p^3$	10.484	35	Br	$3d^{10}4s^24p^5$	11.84
16	S	$3s^23p^4$	10.357	36	Kr	$3d^{10}4s^24p^6$	13.996
17	Cl	$3s^23p^5$	13.01				
18	Ar	$3s^23p^6$	15.755				

Note: The bracket notation is used as a shorthand method to avoid repetition in indicating inner-shell electrons. Thus [He] represents $1s^2$, [Ne] represents $1s^22s^22p^6$, [Ar] represents $1s^22s^22p^63s^23p^6$, and so on.

element helium is one in which the electronic configuration is $1s^2$. In other words, one shell is filled. Additionally, it is found that the electrons in this filled shell are considerably separated in energy from the next available level, the 2*s* level.

The electronic configuration for neon is $1s^2 2s^2 2p^6$. Again, the outer shell is filled and there is a large difference in energy between the 2*p* level and the 3*s* level. Argon has the configuration $1s^2 2s^2 2p^6 3s^2 3p^6$. Here, the 3*p* subshell is filled and there is a wide gap in energy between the 3*p* subshell and the 3*d* subshell. We could continue this procedure through all the noble gases, but the pattern remains the same. A noble gas is formed when either a shell or a subshell is filled and there is a large gap in energy before the next possible level is encountered.

The elements in the first column of the periodic table are called the alkali metals and are characterized by the fact that they are very chemically active. Referring to Table 30.4, we can understand why these elements interact so strongly with other elements. All of these alkali metals have a single outer electron in an *s* subshell. This electron is shielded from the nucleus by all the electrons in the inner shells. Thus, it is only loosely bound to the atom and can readily be accepted by other atoms to form molecules.

All the elements in the seventh column of the periodic table (called the halogens) are also very active chemically. Note that all these elements are lacking one electron in a subshell. Consequently, these elements readily accept electrons from other atoms and form molecules.

EXAMPLE 30.6 The Quantum Number for the 2*p* Subshell
List the quantum numbers for electrons in the 2*p* subshell.

Solution For this subshell, $n = 2$ and $\ell = 1$. The magnetic quantum number can have the values -1, 0, 1, and the spin quantum number is always $+\frac{1}{2}$ or $-\frac{1}{2}$. Thus, the six possibilities are

n	ℓ	m_ℓ	m_s
2	1	-1	$-\frac{1}{2}$
2	1	-1	$\frac{1}{2}$
2	1	0	$-\frac{1}{2}$
2	1	0	$\frac{1}{2}$
2	1	1	$-\frac{1}{2}$
2	1	1	$\frac{1}{2}$

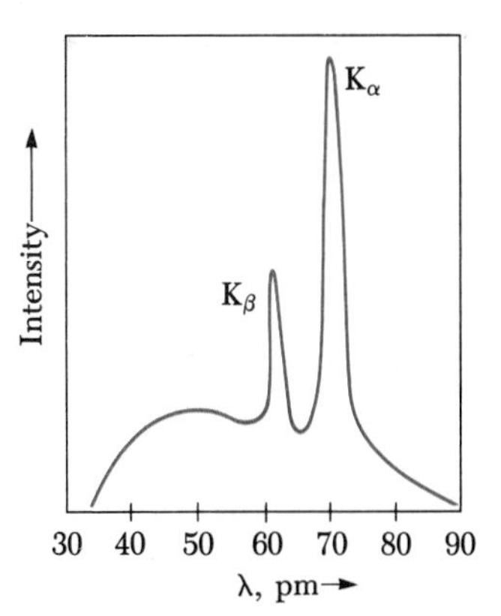

FIGURE 30.14 The x-ray spectrum of a metal target consists of a broad continuous spectrum plus a number of sharp lines which are due to *characteristic x-rays*. The data shown were obtained when 35-keV electrons bombarded a molybdenum target. Note that 1 pm = 10^{-12} m = 0.01 Å.

30.10 CHARACTERISTIC X-RAYS

In Section 29.4, we discussed the manner in which x-rays are emitted when a metal target is bombarded by high-energy electrons. The x-ray spectrum typically consists of a broad continuous band and a series of sharp lines that are dependent on the type of metal used for the target, as shown in Figure 30.14. The presence of these lines, called **characteristic x-rays,** was discovered in 1908, but their origin remained unexplained until the details of atomic structure, particularly the shell structure of the atom, were developed.

The first step in the production of characteristic x-rays occurs when a bombarding electron collides with an electron in an inner shell of a target atom with sufficient energy to remove the electron from the atom. The vacancy created in the shell can now be filled when an electron in a higher level drops down into the lower energy level containing the vacancy. The time it takes for this to happen is very short, less than 10^{-9} s. This transition is accompanied by the emission of a photon whose energy will equal the difference in energy between the two levels. Typically, the energy of such transitions is greater than 1000 eV, and the emitted x-ray photons have wavelengths in the range of 0.01 nm to 1 nm.

Let us assume that the incoming electron has dislodged an atomic electron from the innermost shell, the K shell. If the vacancy is filled by an electron dropping from the next higher shell, the L shell, the photon emitted in the process is referred to as the K_α line on the curve of Figure 30.14. If the vacancy is filled by an electron dropping from the M shell, the line produced is called the K_β line.

Other characteristic x-ray lines are formed when electrons drop from upper levels to vacancies other than those in the K shell. For example, L lines are produced when vacancies in the L shell are filled by electrons dropping from higher shells. An L_α line is produced as an electron drops from the M shell to the L shell, and an L_β line is produced by a transition from the N shell to the L shell.

We can estimate the energy of the x-rays emitted by an atom as follows. Consider two electrons in the K shell of an atom whose atomic number is Z, where Z is the number of protons in the nucleus. Each electron partially shields the other from the charge of the nucleus, Ze, and so each electron is subject to an effective nuclear charge $Z_{\text{eff}} = (Z - 1)e$. We can now use a modified form of Equation 30.15 to estimate the energy of either electron in the K shell (with $n = 1$):

$$E_K = -mZ_{\text{eff}}\frac{k^2e^4}{2\hbar^2} = -Z_{\text{eff}}^2 E_0$$

where E_0 is the ground state energy. Substituting $Z_{\text{eff}} = Z - 1$ gives

$$\boxed{E_K = -(Z - 1)^2(13.6 \text{ eV})} \tag{30.24}$$

As we shall show in the following example, one can estimate the energy of an electron in an L or M shell in a similar fashion. Taking the energy difference between these two levels, one can then calculate the energy and wavelength of the emitted photon.

EXAMPLE 30.7 Estimating the Energy of an X-Ray

Estimate the energy of the characteristic x-ray emitted from a tungsten target when an electron drops from an M shell ($n = 3$ state) to a vacancy in the K shell ($n = 1$ state).

Solution The atomic number for tungsten is $Z = 74$. Using Equation 30.24, we see that the energy of the electron in the K shell state is approximately

$$E_K = -(74 - 1)^2(13.6 \text{ eV}) = -72{,}500 \text{ eV}$$

The electron in the M shell ($n = 3$) is subject to an effective nuclear charge that depends on the number of electrons in the $n = 1$ and $n = 2$ states, which shield the nucleus. Because there are eight electrons in the $n = 2$ state and one electron in the $n = 1$ state, roughly nine electrons shield the nucleus, and so $Z_{eff} = Z - 9$. Hence, the energy of an electron in the M shell ($n = 3$), following Equation 30.15, is equal to

$$E_M = -Z^2_{eff}E_3 = -(Z - 9)^2 \frac{E_0}{3^2} = -(74 - 9)^2 \frac{(13.6 \text{ eV})}{9} = -6380 \text{ eV}$$

where E_3 is the energy of an electron in the $n = 3$ level of the hydrogen atom. Therefore, the emitted x-ray has an energy equal to $E_M - E_K = -6380 \text{ eV} - (-72{,}500 \text{ eV}) = 66{,}100$ eV. Note that this energy difference is also equal to hf, where $hf = hc/\lambda$, and where λ is the wavelength of the emitted x-ray.

EXERCISE Calculate the wavelength of the emitted x-ray for this transition.

Answer 0.0188 nm.

In 1914, Henry G. J. Moseley plotted the Z values for a number of elements versus $\sqrt{1/\lambda}$, where λ is the wavelength of the K_α line for each element. He found that such a plot produced a straight line, as in Figure 30.15. This is consistent with our rough calculations of the energy levels based on Equation 30.24. From this plot, Moseley was able to determine the Z values of other elements, which provided a periodic chart in excellent agreement with the known chemical properties of the elements.

30.11 ATOMIC TRANSITIONS

In this section we shall look at some of the basic processes involved in an atomic system. It is necessary to understand these mechanisms before we can understand the operation of a laser, which we shall examine in the next section.

We have seen that an atom will emit radiation only at certain frequencies, which correspond to the energy separation between the various allowed states. Consider an atom with many allowed energy states, labeled E_1, E_2, E_3, . . ., as in Figure 30.16. When light is incident on the atom, only those

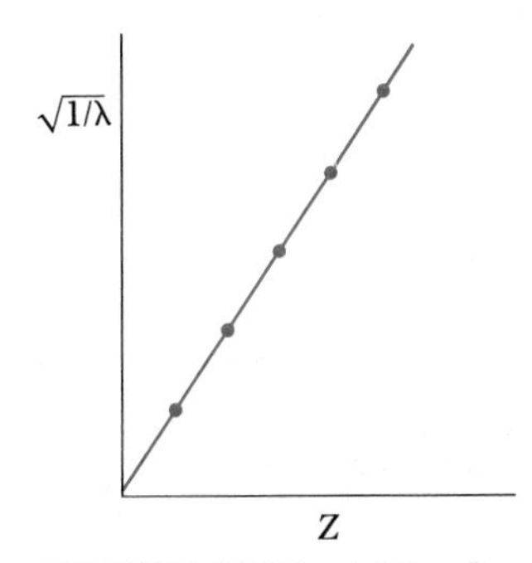

FIGURE 30.15 A Moseley plot. A straight line is obtained when $\sqrt{1/\lambda}$ is plotted versus Z for the K_α x-ray lines of a number of elements.

FIGURE 30.16 Energy level diagram of an atom with various allowed states. The lowest energy state E_1 is the ground state. All others are excited states.

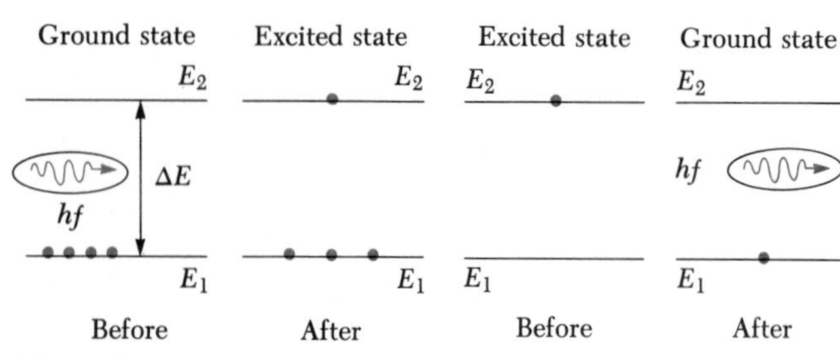

FIGURE 30.17 Diagram representing the process of *stimulated absorption* of a photon by an atom. The dots represent electrons in the various states. One electron is transferred from the ground state to the excited state when the atom absorbs one photon whose energy $hf = E_2 - E_1$.

FIGURE 30.18 Diagram representing the process of *spontaneous emission* of a photon by an atom that is initially in the excited state E_2. When the electron falls to the ground state, the atom emits a photon whose energy is $hf = E_2 - E_1$.

photons whose energy, hf, matches the energy separation ΔE between two levels can be absorbed by the atom. A schematic diagram representing this **stimulated absorption process** is shown in Figure 30.17. At ordinary temperatures, most of the atoms are in the ground state. If a vessel containing many atoms of a gaseous element is illuminated with a light beam containing all possible photon frequencies (that is, a continuous spectrum), only those photons of energies $E_2 - E_1$, $E_3 - E_1$, $E_4 - E_1$, and so on, can be absorbed. As a result of this absorption, some atoms are raised to various allowed higher energy levels called **excited states.**

Once an atom is in an excited state, there is a certain probability that it will jump back to a lower level by emitting a photon, as shown in Figure 30.18. This process is known as **spontaneous emission.** In typical cases, an atom will remain in an excited state for only about 10^{-8}s.

Finally, there is a third process, which is of importance in lasers, known as **stimulated emission.** Suppose an atom is in an excited state E_2, as in Figure 30.19, and a photon of energy $hf = E_2 - E_1$ is incident on it. The incoming photon will increase the probability that the electron will return to the ground state and thereby emit a second photon having the same energy, hf. This process of speeding up atomic transitions to lower levels is called stimulated emission. Note that there are two identical photons that result from this process, the incident photon and the emitted photon. The emitted photon will be exactly in phase with the incident photon. These photons can, in turn, stimulate other atoms to emit photons in a chain of similar processes. The many photons produced in this fashion are the source of the intense, coherent light in a laser.

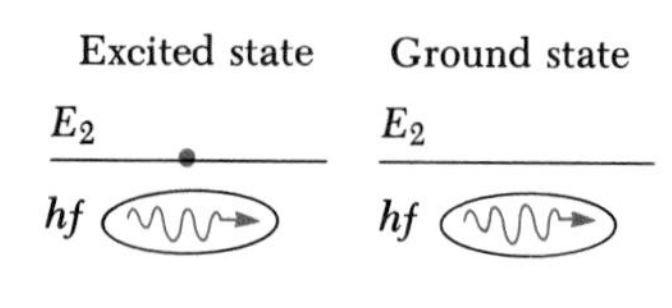

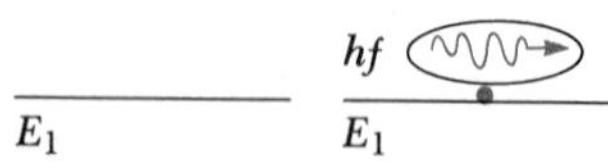

FIGURE 30.19 Diagram representing the process of *stimulated emission* of a photon by an incoming photon of energy hf. Initially, the atom is in the excited state. The incoming photon stimulates the atom to emit a second photon of energy $hf = E_2 - E_1$.

*30.12 LASERS AND HOLOGRAPHY

We have described how an incident photon can cause atomic transitions either upward (stimulated absorption) or downward (stimulated emission). Both processes are equally probable. When light is incident on a system of atoms, there is usually a net absorption of energy because there are many more atoms in the ground state than in excited states. That is, in a normal situation, there are more atoms in state E_1 ready to absorb photons than there are atoms in states E_2, E_3, . . . ready to emit photons. However, if one can invert the situation so that there are more atoms in an excited state than in the ground state, a net emission of photons can result. Such a condition is called **population inversion.** This, in fact, is the fundamental principle involved in the operation of a **laser,** an acronym for _l_ight _a_mplification by _s_timulated _e_mission of _r_adiation. The amplification corresponds to a buildup of photons in the system as the result of a chain reaction of events.

The following three conditions must be satisfied in order to achieve laser action:

1. The system must be in a state of population inversion (that is, more atoms in an excited state than in the ground state).
2. The excited state of the system must be a *metastable state*, which means its lifetime must be long compared with the usually short lifetimes of excited states. When such is the case, stimulated emission will occur before spontaneous emission.
3. The emitted photons must be confined in the system long enough to allow them to stimulate further emission from other excited atoms. This is achieved by the use of reflecting mirrors at the ends of the system. One end is made totally reflecting, and the other is slightly transparent to allow the laser beam to escape.

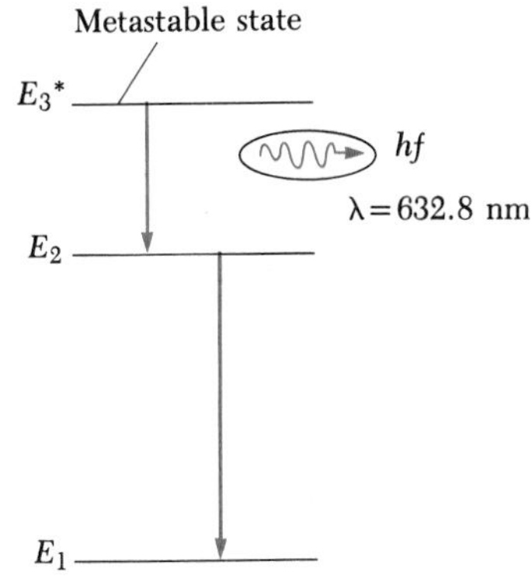

FIGURE 30.20 Energy level diagram for the neon atom, which emits photons at a wavelength of 632.8 nm through stimulated emission. The photon at this wavelength arises from the transition $E_3^* \rightarrow E_2$. This is the source of coherent light in the helium-neon gas laser.

One device that exhibits stimulated emission of radiation is the helium-neon gas laser. The energy level diagram for the neon atom in this system is shown in Figure 30.20. The mixture of helium and neon is confined to a glass tube sealed at the ends by mirrors. An oscillator connected to the tube causes electrons to sweep through the tube, colliding with the atoms of the gas and raising them into excited states. Some neon atoms are excited to state E_3 through this process and also as a result of collisions with excited helium atoms. Stimulated emission occurs as the neon atoms make a transition to state E_2 and neighboring excited atoms are stimulated to emit at the same frequency and phase. This results in the production of coherent light at a wavelength of 632.8 nm. Figure 30.21 summarizes the steps in the production of a laser beam.

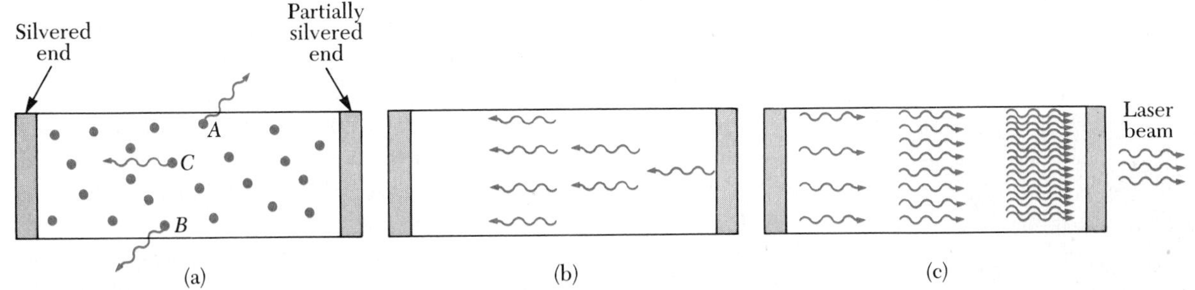

FIGURE 30.21 (a) Neon atoms in the gas are raised to a metastable excited state by collisions with electrons sweeping through the tube and by collisions with excited helium atoms. The excited neon atoms, shown as colored dots, eventually return to their ground state and emit photons. Some of the emitted photons emerge from the sides of the glass tube and are lost (atoms *A* and *B*). However, other photons are emitted along the axis of the tube (atom *C*). (b) A photon moving along the axis of the tube encounters other excited neon atoms and causes them to emit a photon. The two photons then stimulate other excited atoms causing further emission. Then the number of photons multiplies as the process continues. (c) The photon beam is reflected from the left silvered end and moves back toward the partially silvered end at the right. The process of stimulated emission continues and the number of photons in the beam increases further. After many reflections between the two ends of the tube, a small fraction of the photons emerges from the partially silvered end. These are the photons that make up the coherent laser beam.

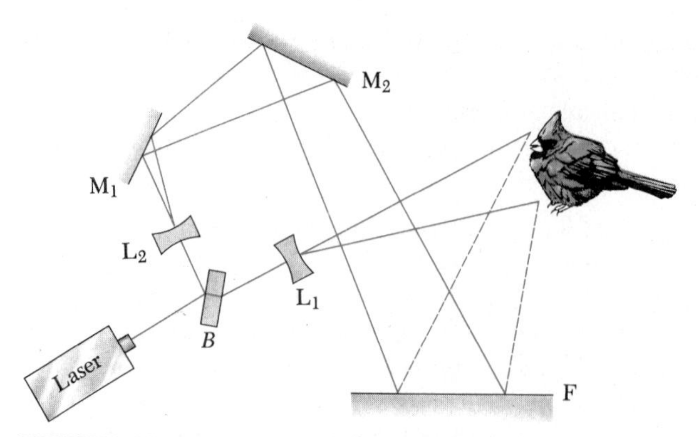

FIGURE 30.22 Experimental arrangement for producing a hologram.

A ⅜-in. titanium plate being cut with coherent radiation from a carbon dioxide laser at 90 in/min. (Courtesy of Coherent Radiation)

Since the development of the first laser in 1960, there has been a tremendous growth in laser technology. Lasers are now available that cover wavelengths in the infrared, visible, and ultraviolet regions. Applications include surgical "welding" of detached retinas, precision surveying and length measurement, a potential source for inducing nuclear fusion reactions, precision cutting of metals and other materials, and telephone communication along optical fibers. These and other applications are possible because of the unique characteristics of laser light. In addition to its being highly monochromatic and coherent, laser light is also highly directional and can be sharply focused to produce regions of extremely intense light energy.

Holography

One of the most unusual and interesting applications of the laser is in the production of three-dimensional images of an object in a process called **holography.** Figure 30.22 shows how a hologram is made. Light from the laser is split into two parts by a half-silvered mirror at B. One part of the beam reflects off the object to be photographed and strikes an ordinary photographic film. The other half of the beam is diverged by lens L_2, reflects from mirrors M_1 and M_2, and finally strikes the film. The two beams overlap to form an extremely complicated interference pattern on the film. Such an interference pattern can be produced only if the phase relationship of the two waves is maintained constant throughout the exposure of the film. This condition is met if one uses light from a laser because such light is coherent. The hologram records not only the intensity of the light but also the phase difference between the two halves of the laser beam. It is this last piece of information, the phase relationship, that causes the interference pattern that results in the three-dimensional image.

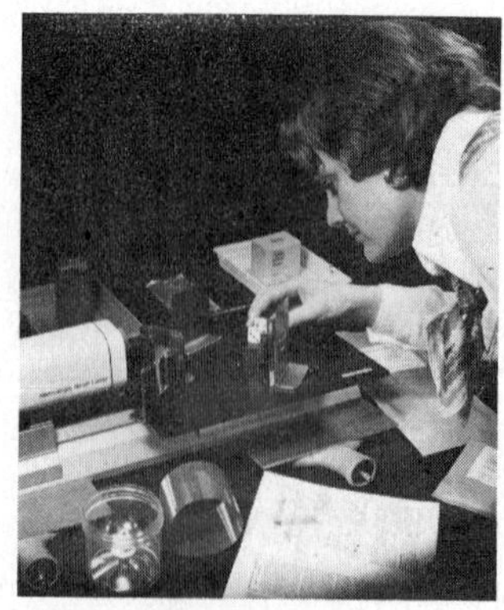

Setting up a holography experiment. (Courtesy of Metrologic)

A hologram is best viewed by allowing coherent light to pass through the developed film as one looks back along the direction from which the beam comes. A three-dimensional image of the object is observed, suspended in midair. The applications of holography promise to be many and varied. For example, someday your television set may be replaced by one using holography. Many other interesting applications of lasers are discussed in the essay in this chapter.

Color Plate 27 Magnetic levitation. A small permanent magnet floating above a disk of superconducting yttrium barium copper oxide ($YBa_2Cu_3O_7$), which has a critical temperature of about 92 K (see Section 18.5). The superconductor is maintained at a temperature of 77 K by using liquid nitrogen. This shows that a superconductor is diamagnetic; that is, it expels magnetic fields. (Courtesy of Argonne National Laboratory)

Color Plate 28 The surface of graphite. The surface of graphite as "viewed" with a scanning tunneling microscope, discussed in the essay in Chapter 29. This technique enables scientists to see small details on surfaces, with a resolution of about 2 Å. The contours seen here represent the arrangement of individual atoms on the crystal surface. (Roger A. Freedman and Paul K. Hansma, University of California at Santa Barbara)

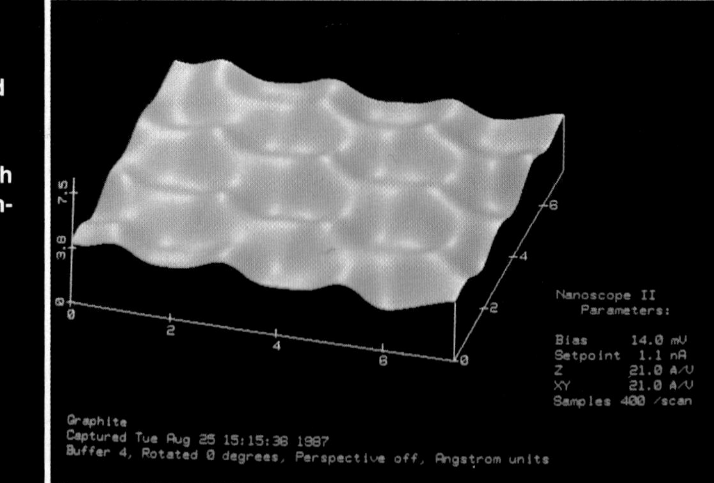

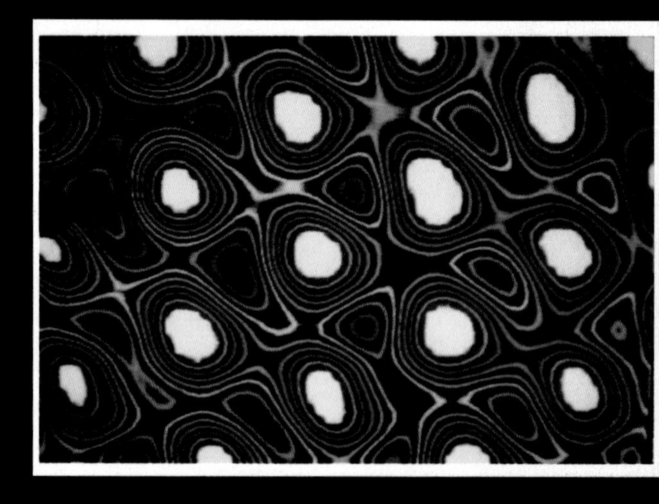

Color Plate 29 Viewing crystal surfaces. The surface of $TaSe_2$ "viewed" with a scanning tunneling microscope, described in the essay in Chapter 29. The photograph is actually a charge density wave contour of the surface, where the various colors indicate regions of different charge densities. (Courtesy of Prof. R.V. Coleman, University of Virginia)

Color Plate 30 Light-emitting semiconductor. Light emitted by the semiconductor selenium is separated into its various colors. Each color corresponds to a specific wavelength and atomic transition within the material. (Fritz Goro)

Color Plate 31 Crystalline solids. (a) A man-made cylinder of nearly pure crystalline silicon (Si), approximately 10 inches long. Such crystals are cut into wafer form and processed to make various semiconductor devices such as integrated circuits and solar cells. (b) Natural quartz (SiO_2) crystals, one of the most common minerals on earth. Quartz crystals are used to make special lenses and prisms and in certain electronic applications. (c) Amethyst crystals are an impure form of quartz crystals. Their color, which is due to iron impurities (Fe^{3+}), can range from pale lilac to deep purple, depending on the iron concentration.

olor Plate 32 owerful x-rays. The low from underwater witches produced by ne nation's most owerful x-ray source t Sandia National aboratories. High-nergy electrons are onverted to x-rays at ne center of the nachine, which in turn re used to understand neir effect on weapon ystems and other omponents. (Courtesy f Sandia National aboratories, photo by Valter Dickenman)

Color Plate 33 Solar cell research. A "tandem" solar cell using two photovoltaic materials (silicon and gallium arsenide) is being tested in the laboratory. This design improves the efficiency of the cell, since the two semiconductors use different wavelengths of the sunlight in the process of converting electromagnetic energy into electricity (see the essay in Chapter 31).

Color Plate 34 Scanning electron microscope picture of a caterpillar. (Dr. Jeremy Burgess/ Science Photo Library)

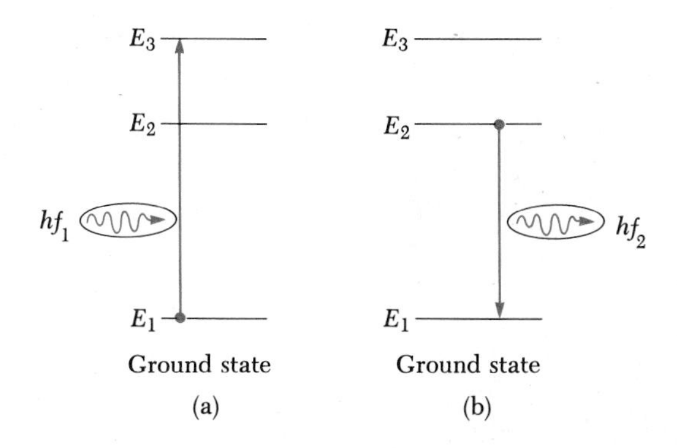

FIGURE 30.23 In the process of fluorescence, (a) an atom absorbs a photon of energy hf_1 and ends up in an excited state, E_3. (b) The atom emits a photon of energy hf_2 when the electron moves from an intermediate state, E_2, back to the ground state.

*30.13 FLUORESCENCE AND PHOSPHORESCENCE

When an atom absorbs a photon of energy and ends up in an excited state, it can return to the ground state via some intermediate states, as shown in Figure 30.23. The photons emitted by the atom will have lower energy, and therefore lower frequency, than the absorbed photon. The process of converting ultraviolet light to visible light by this means is referred to as **fluorescence.**

The common fluorescent light, which makes use of this principle, works as follows. Electrons are produced in the tube as a filament at the end is heated to sufficiently high temperatures. The electrons are accelerated by an applied voltage, and this causes them to collide with atoms of mercury vapor present in the tube. As a result of the collisons, many mercury atoms are raised to excited states. As the excited atoms drop to their normal levels, some ultraviolet photons are emitted, and these strike a phosphor-coated screen. The screen absorbs these photons and emits visible light by means of fluorescence. Different phosphors on the screen emit light of different colors. "Cool white" fluorescent lights emit nearly all the visible colors and hence the light is very white. "Warm white" fluorescent lights have a phosphor that emits more red light and thereby produces a "warm" glow. It is interesting to note that the fluorescent lights above the meat counter in a grocery store are usually "warm white" to give the meat a redder color.

Two common fluorescent materials that you may have in your medicine cabinet are Murine eye drops and Pearl Drops toothpaste. If you were to use these products and then stand under a "black light," your eyes and teeth would glow with a beautiful yellow color. (A black light is simply a lamp that emits ultraviolet light along with some visible violet-blue light.)

Fluorescence analysis is often used to identify compounds. This is made possible by the fact that every compound has a "fingerprint" associated with the specific wavelength at which it fluoresces.

As we have seen, a fluorescent material emits visible light only when ultraviolet radiation bombards it. Another class of materials, called **phosphorescent** materials, continue to glow long after the illumination has been removed. An excited atom in a fluorescent material drops to its normal level in about 10^{-8} s, but an excited atom of a phosphorescent material may remain in an excited metastable state for periods ranging from a few seconds to several hours. Eventually, the atom will drop to its normal state and emit a visible photon. For this reason, phosphorescent materials emit light long after being placed in the dark. Paints made from these substances are often used to decorate the hands of watches and clocks and to outline doors and stairways in large buildings so that these exits will be visible if there is a power failure.

SUMMARY

The **Bohr model** of the atom is successful in describing the spectra of atomic hydrogen and hydrogen-like ions. One of the basic assumptions of the model is that the electron can exist only in certain orbits such that its angular momentum, mvr, is an integral multiple of $\hbar/2\pi = h$, where h is Planck's constant. Assuming circular orbits and a Coulomb force of attraction between electron and proton, the energies of the quantum states for hydrogen are

$$E_n = -\frac{mk^2e^4}{2\hbar^2}\left(\frac{1}{n^2}\right) \qquad n = 1, 2, 3, \ldots \tag{30.15}$$

where k is the Coulomb constant, e is the charge on the electron, and n is an integer called a **quantum number.**

If the electron in the hydrogen atom jumps from an orbit whose quantum number is n_i to an orbit whose quantum number is n_f, it emits a photon of frequency f, given by

$$f = \frac{mk^2e^4}{4\pi\hbar^3}\left(\frac{1}{n_f^2} - \frac{1}{n_i^2}\right) \tag{30.17}$$

Bohr's **correspondence principle** states that quantum mechanics is in agreement with classical physics when the quantum numbers for a system are very large.

One of the many great successes of quantum mechanics is that the quantum numbers n, ℓ, and m_ℓ associated with atomic structure arise directly from the mathematics of the theory. The quantum number n is called the **principal quantum number,** ℓ is the **orbital quantum number,** and m_ℓ is the **orbital magnetic quantum number.** In addition, a fourth quantum number, called the **spin magnetic quantum number,** m_s, is needed to explain certain features of atomic structure.

An understanding of the periodic table of the elements became possible when Pauli formulated the **exclusion principle,** which states that no two electrons in an atom can ever be in the same quantum state, that is, no two electrons in the same atom can have the same set of quantum numbers, n, ℓ, m_ℓ, and m_s.

Characteristic x-rays are produced when a bombarding electron collides with an electron in an inner shell of an atom with sufficient energy to remove the electron from the atom. The vacancy thus created is filled when an electron in a higher level drops down into the level containing the vacancy.

Lasers are monochromatic, coherent light sources that work on the principle of **stimulated emission** of radiation from a system of atoms.

ADDITIONAL READING

E. N. Andrade, *Rutherford and the Nature of the Atom*, N.Y., Doubleday Anchor, 1964.

G. Gamow, *Mr. Tomkins Explores the Atom*, New York, Cambridge University Press, 1945.

W. E. Kock, *Lasers and Holography*, Garden City, N.Y., Doubleday, 1969.

A. V. La Rocca, "Laser Applications in Manufacturing," *Sci. American*, March 1982, p. 94.

Lasers and Light, San Francisco, Freeman, 1969 (readings from *Scientific American*).

R. E. Latham, "Holography in the Science Classroom," *The Physics Teacher*, October 1986, p. 395.

E. N. Leith, "White-Light Holograms," *Sci. American*, October 1976, p. 80.

R. Moore, *Niels Bohr, The Man, His Science, and the World They Changed*, N.Y., Knopf, 1966.

P. Schewe, "Lasers," *The Physics Teacher*, November 1981, p. 534.

H. Spetzler, "Interesting Example of Holographic Interferometry," *The Physics Teacher*, February 1986, p. 80.

C. L. Strong, "How to Make Holograms," *Sci. American*, February 1967, p. 122.

J. A. Wheeler, "Niels Bohr, the Man," *Physics Today*, October 1985, p. 66.

QUESTIONS

1. Discuss why the term *electron clouds* is used to describe the electronic arrangement in the quantum mechanical view of the atom.
2. What is excluded by the exclusion principle?
3. Which of the following elements would you predict to have properties most like those of silver: copper, cadmium, or palladium?
4. It was stated in the text that, if the exclusion principle were not valid, every electron would end up in the lowest energy state of the atom and the chemical behavior of the elements would be grossly modified. Explain this statement.
5. Why are the frequencies of characteristic x-rays dependent on the type of material used for the target?
6. In what ways is the introduction of the quantum numbers via quantum mechanics more satisfying than their introduction through improvements on the Bohr theory?
7. List some ways in which quantum mechanics altered our view of the atom as pictured by the Bohr theory.
8. Explain how the fact that the angular momentum of an electron can be zero does not introduce any philosophical difficulties in the quantum mechanical view of the atom.
9. Suppose that the electron in the hydrogen atom obeyed classical mechanics rather than quantum mechanics. Why would such a hypothetical atom emit a continuous spectrum rather than the observed line spectrum?
10. Does the light emitted by a neon sign constitute a continuous spectrum or only a few colors? Defend your answer.
11. Can the electron in the ground state of hydrogen absorb a photon of energy (a) less than 13.6 eV and (b) greater than 13.6 eV? Explain.
12. Suppose an electron in the ground state of an atom has its first excited level at an energy of 3 eV above its ground-state value. What could happen to this electron if it were struck by an incoming electron having an energy of (a) 3 eV and (b) 4 eV?
13. If matter has a wave nature, why is this not observable in our daily experiences?
14. Explain the difference between laser light and light from an incandescent lamp.
15. Why does a laser usually emit only one particular color of light rather than several colors or perhaps a continuous spectrum?
16. Why can a small vibration of the apparatus used to produce a hologram ruin the image?

PROBLEMS

Section 30.2 Atomic Spectra

1. Use Equation 30.3 to calculate the wavelength of the first three lines in the Lyman series. On the basis of your calculation identify the region of the electromagnetic spectrum in which these lines appear.

2. Use Equation 30.1 to calculate the wavelength of the first three lines in the Balmer series for hydrogen.

3. Use Equation 30.4 to calculate the frequency of the emissions from hydrogen for the first two lines of the Paschen series. On the basis of your calculations identify the region of the electromagnetic spectrum in which these lines appear.

4. On the basis of your calculations for Problem 2, find three frequencies that will be absorbed if a continuous spectrum is passed through heated hydrogen gas.

Section 30.3 The Bohr Theory of Hydrogen

5. A photon is emitted from a hydrogen atom, which undergoes a transition from the $n = 3$ state to the $n = 2$ state. Calculate (a) the energy, (b) the wavelength, and (c) the frequency of the emitted photon.

6. How much energy is required to ionize hydrogen when it is in (a) the ground state, (b) the state for which $n = 3$?

7. A hydrogen atom initially in its ground state ($n = 1$) absorbs a photon and ends up in the state for which $n = 3$. (a) What is the energy of the absorbed photon? (b) If the atom returns to the ground state, what photon energies could the atom emit?

8. Use Equation 30.13 to show that the radius of the ground state orbit of hydrogen is 0.529×10^{-10} m.

9. Substitute numerical values into Equation 30.19 to determine a value for R.

10. Calculate the radius of the first, second, and third Bohr orbits of hydrogen.

11. What wavelength radiation would ionize hydrogen when the atom is in its ground state?

12. Calculate the Coulomb force of attraction on the electron when it is in the ground state.

13. Calculate the angular momentum of the electron in the $n = 2$ state of hydrogen.

14. Find the shortest wavelength photon emitted in the Lyman series of hydrogen.

• **15.** An electron is in the first Bohr orbit of hydrogen. Find (a) the speed of the electron, (b) the time required for the electron to circle the nucleus, and (c) the current in amps corresponding to the motion of the electron.

• **16.** (a) Calculate the angular momentum of the moon due to its orbital motion about the earth. In your calculation, use 3.84×10^8 m as the average earth-moon distance and 2.36×10^6 s as the period of the moon in its orbit. (b) Assume that the moon is in its lowest energy state while orbiting the earth and determine the corresponding quantum number. (c) By what fraction would the earth-moon radius have to be increased to increase the quantum number by 1?

Section 30.4 Modification of the Bohr Theory

17. (a) Find the energy of the electron in the ground state of doubly ionized lithium which has an atomic number $Z = 3$. (b) Find the radius of this ground state orbit.

18. Plot an energy level diagram like that of Figure 30.6 for singly ionized helium.

19. (a) Substitute numerical values into Equation 30.22 to find a value for the Rydberg constant for singly ionized helium, He^+. (b) Use the result of part (a) to find the wavelength associated with a transition from the $n = 2$ state to the $n = 1$ state of He^+. (c) Identify the region of the electromagnetic spectrum associated with this transition.

20. Using the Sommerfeld model, (a) identify all the possible subshells that can exist for the P shell. (b) Find the maximum number of electrons that can occupy each subshell; and (c) find the maximum number of electrons that can occupy the P shell.

Section 30.6 Quantum Mechanics and the Hydrogen Atom

Section 30.7 The Spin Magnetic Quantum Number

21. List the possible sets of quantum numbers for electrons in the $3d$ subshell.

22. List the possible sets of quantum numbers for electrons in the $3p$ subshell.

23. When the principal quantum number is $n = 4$, how many different values of (a) ℓ and (b) m_ℓ are possible?

Section 30.9 The Exclusion Principle and the Periodic Table

24. Verify the electronic configurations for the ground-state given in Table 30.4 for (a) helium, (b) neon, and (c) argon.

• **25.** (a) Write out the electronic configuration for the ground state for oxygen ($Z = 8$). (b) Write out the values for the set of quantum numbers n, ℓ, m_ℓ, and m_s for each of the electrons in oxygen.

• **26.** How many different sets of quantum numbers are possible for an electron for which (a) $n = 1$, (b) $n = 2$, (c) $n = 3$, (d) $n = 4$, and (e) $n = 5$? Check your results to show that they agree with the general rule that the number of different sets of quantum numbers is equal to $2n^2$.

Section 30.10 Characteristic X-Rays

27. The K_α x-ray is the one emitted when an electron undergoes a transition from the L shell ($n = 2$) to the K shell ($n = 1$). Calculate the frequency of the K_α x-ray from a nickel target ($Z = 28$).

• **28.** Use the method illustrated in Example 30.7 to calculate the wavelength of the x-ray emitted from a molybdenum target ($Z = 42$) when an electron undergoes a transition from the L shell ($n = 2$) to the K shell ($n = 1$).

ADDITIONAL PROBLEMS

29. A laser emits a 0.5 W beam of light at a wavelength of 640 nm. How many photons does the laser emit per second?

30. (a) How much energy is required to cause an electron in hydrogen to move from the $n = 1$ state to the $n = 2$ state? (b) If the electrons gain this energy by collision with hydrogen atoms at a high temperature, find the minimum temperature of the heated hydrogen gas. The thermal energy of the heated atoms is given by $3kT/2$.

• **31.** (a) Calculate the radius of the third Bohr orbit in hydrogen. (b) Calculate the speed of an electron in this orbit. (c) From the de Broglie theory what is the wavelength of an electron that has this speed? (d) How does the wavelength of the electron compare with the circumference of the orbit for $n = 3$?

• **32.** A laser used in a holography experiment has an average output power of 5 mW. The laser beam is actually a series of pulses of electromagnetic radiation at a wavelength of 632.8 nm, each having a duration of 25 ms. Calculate (a) the energy (in joules) radiated with each pulse and (b) the number of photons in each pulse.

• **33.** Find the potential energy and kinetic energy of an electron in the ground state of the hydrogen atom.

• **34.** A hydrogen atom is in its ground state. Using the Bohr theory of the atom, calculate (a) the radius of the orbit, (b) the linear momentum of the electron, (c) the angular momentum of the electron, (d) the kinetic energy, (e) the potential energy, and (f) the total energy.

• **35.** An electron has a de Broglie wavelength equal to the diameter of the hydrogen atom. (a) What is the kinetic energy of the electron? (b) How does this energy compare with the ground-state energy of the hydrogen atom?

• **36.** In order for an electron to be confined to a nucleus, its de Broglie wavelength would have to be less than 10^{-14} m. (a) What would be the kinetic energy of an electron confined to this region? (b) On the basis of this result, would you expect to find an electron in a nucleus? Explain.

•• **37.** Use Bohr's model of the hydrogen atom to show that when the atom makes a transition from the state n to the state $n - 1$, the frequency of the emitted light is given by

$$f = \frac{2\pi^2 mk^2e^4}{h^3}\left(\frac{2n - 1}{(n - 1)^2n^2}\right)$$

•• **38.** Calculate the classical frequency for the light emitted by an atom. To do so, note that the frequency of revolution is $v/2\pi r$, where r is the Bohr radius. Show that as n approaches infinity in the equation of Problem 37, the expression given there varies as $1/n^3$ and reduces to the classical frequency. (This is an example of the correspondence principle, which requires that the classical and quantum models agree for large values of n.)

•• **39.** An electron of momentum p is at a distance r from a stationary proton. The electron has a kinetic energy $KE = p^2/2m$ and potential energy $PE = -ke^2/r$. Its total energy is $E = KE + PE$. If the electron is bound to the proton to form a hydrogen atom, its average position is at the proton, but the uncertainty in its position is approximately equal to the radius r of its orbit. The electron's average momentum will be zero, but the uncertainty in its momentum will be given by the uncertainty principle. Treat the atom as a one-dimensional system in the following: (a) Estimate the uncertainty in the electron's momentum in terms of r. (b) Estimate the electron's kinetic, potential, and total energies in terms of r.

•• **40.** A dimensionless number which often appears in atomic physics is the *fine-structure constant* α, given by

$$\alpha = \frac{ke^2}{\hbar c}$$

where k is the Coulomb constant. (a) Obtain a numerical value for $1/\alpha$. (b) In terms of α, what is the ratio of the Bohr radius, a_0, to the Compton wavelength, $\lambda = h/m_0c$? (c) In terms of α, what is the ratio of the reciprocal of the Rydberg constant, $1/R$, to the Bohr radius?

Essay

Lasers and Their Applications

Isaac D. Abella *The University of Chicago*

Lasers

The enormous growth of laser technology mentioned in Section 30.12 has stimulated a broad range of scientific and engineering applications that exploit some of the unique properties of laser light. These properties derive from the distinctive way laser light is produced in contrast to the generation of ordinary light. Laser light originates from atoms, ions or molecules through a process of *stimulated emission* of radiation. The active laser medium is contained in an enclosure or *cavity*, which organizes the normally random emission process into an intense directional, monochromatic and coherent wave. The end mirrors provide the essential optical feedback, which selectively builds up the stimulating wave along the tube axis. However, in an ordinary sodium vapor street lamp for example, the atoms spontaneously emit in random directions and at irregular times, over a broad spectrum, resulting in isotropic illumination of incoherent light.

Currently operating laser systems use a variety of gases, solids and liquids as their working laser substances. These devices are designed to emit either continuous or pulsed monochromatic beams, and operate over a broad range of the optical spectrum (ultraviolet, visible, infrared) with output powers from milliwatts (10^{-3} W) to megawatts (10^{6} W). The particular application determines the choice of laser system, wavelength, power level, or other relevant variables, since no one laser has all the desirable properties.

As noted in Section 30.12, several conditions must be satisfied for a laser to operate successfully. The requirement for population inversion, that is, more atoms in a particular excited state than in a lower state, essentially means that energy must be supplied from outside the system. Otherwise, atoms would eventually radiate, and develop an increasing probability for absorbing light. Finally, they fall to the lowest energy state and stop emitting altogether. Therefore, all laser systems must be connected to external energy sources, usually electrical, as required by conservation of energy (see Figure 1). For example, we can energize the atoms in a gas laser medium by bombarding them with electrons (so called "electrical pumping"). We can also supply energy to lamps whose light populates excited states by photon absorption ("optical pumping") for those solids or liquids which do not conduct electric charge. These pumping mechanisms tend to have low efficiency (ratio of laser energy output to electric energy supplied), typically a few percent, with the balance discharged as heat into cooling water or circulating air.

Controlling the electrical input into the laser system allows a variable laser energy output, which may be important in many applications. Thus, the argon ion laser system can emit up to about 10 watts in the green optical beam by adjustment of the electric current in the argon gas, which in turn controls the degree of population inversion.

Chemical lasers operate without direct electrical input. Several highly reactive gases are mixed in the laser chamber, and the energy released in the ensuing reaction populates the excited levels in the molecules. In this case, the reactants need to be resupplied for the laser to operate for any length of time.

Some laser systems have fluid media, containing dissolved dye molecules. The dye lasers are usually pumped to excited levels by an external laser. The advantage of this arrangement is that dye lasers can be continuously "tuned" over a wide range of wavelengths, using prisms or gratings, whereas the pump source has a fixed wavelength. Color variability is important for those cases where the laser is directed at materials whose absorption depends on wavelength. Thus, the laser can be tuned into exact coincidence with selected energy states. For example, blood does not absorb red light to any great extent, which excludes red laser light from most surgical applications on blood-rich tissue.

A variety of laser systems are in general use today. They include the 1 mW helium-neon laser, usually operating in the red at 632.8 nm (although yellow and green beams are available); the argon ion laser which operates in the green or blue and can emit up to 10 W; the carbon dioxide gas laser, which emits in the infrared at 10 μm and can produce several hundred watts; and the

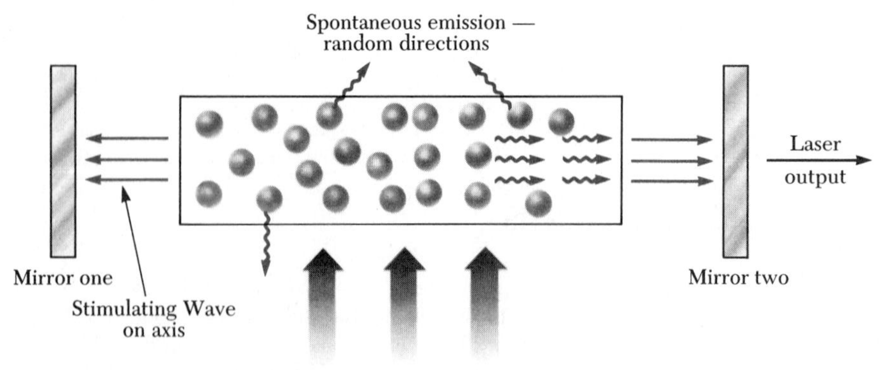

FIGURE 1 A schematic of a laser design. The tube contains atoms, which represent the active medium. An external source of energy (optical, electrical, etc.) is needed to "pump" the atoms to excited energy states. The parallel end mirrors provide the feedback of the stimulating wave.

neodymium YAG laser, a powerful solid-state optically-pumped system that emits at 1.06 μm in either continuous or pulsed mode. The recently perfected diode junction laser emits in the near infrared and operates by passage of current through a semiconductor material. The recombination radiation is essentially direct conversion of electrical energy to laser light and is a very efficient process. The diodes can emit up to 5 W and can be used to energize other laser materials.

Applications

We shall describe several applications that serve to illustrate the wide variety of laser utilization. First, lasers are being used to make precision long-range distance measurements (range-finding). It has become important, for astronomical and geophysical purposes, to measure as precisely as possible the distances from various points on the surface of the earth to a point on the moon's surface. To facilitate this, the Apollo astronauts set up a compact array, a 0.5 m square of reflector prisms, on the moon. This array allows laser pulses directed from an earth station to be retro-reflected to the same station. (See Section 24.10 on prism reflectors.) Using the known speed of light and the measured round-trip travel time of a 1 ns pulse, one can determine the earth-moon distance, 380,000 km, to a precision of better than 10 cm. Such information would be useful, for example, in making more reliable earthquake predictions and for learning more about the motions of the earth-moon system. This technique requires a high-power pulsed laser for its success, since the burst of photons must return to a collecting telescope on earth with sufficient power to be detected. Variations of this method are also used to measure the distance to inaccessible points on the earth.

FIGURE 2 (Courtesy of Laser Precision Corp.)

The low-power helium-neon laser is the basis for a technical innovation seen in many supermarkets. The laser beam can be focused with a lens to a very small bright spot, which is then reflected from an oscillating mirror producing a swiftly moving dot image. If this spot is scanned over the product identification bar-code printed on supermarket products, the variation of the reflected light can be detected and decoded. This scanner has improved speed and accuracy at the checkout counter. The spot of light must be small enough to resolve the different widths of the individual bars and bright enough to be "seen" in reflection by the optical detector below the counter. Since this laser is operated in public, the power must be low enough to be safe for any reasonable use of the system. This puts stringent limits on the type of laser used in this application.

Similarly, a laser (light emitting) diode is used to decode the digital information on the compact audio disc, the so called CD. On the compact disc, the music has been digitized as pits and grooves embedded in the plastic. The fluctuating reflection of the weak laser spot from the disc surface is detected by a photocell and decoded by digital-to-analog circuits. The high resolution of the laser beam allows the CD player to reproduce music with extremely high fidelity, without the noise or hiss associated with regular long-playing records or magnetic tape. There are also video versions of the laser disc. In all of these decoding applications, the essential laser properties that are exploited are the accurate focusing of the beam, the monochromaticity (to be able to operate in the presence of background illumination) and enough power to observe the diffusely reflected light. High power lasers could lead to damage in these cases and are not employed.

The amount of information stored on a compact audio disc as digital data is estimated to be about 1 gigabyte (10^9 characters), which is enormously high density data storage. By way of comparison, the data storage capacity on magnetic "floppy" discs in personal computers is typically 800 kilobytes (8×10^5 characters). Developments are under way to transfer this optical storage technology to computer disc drives having both read and write capability. The compact disc as a read-only device with prerecorded data requires very little modification to be used with a computer and is already in use to store encyclopedia or dictionary volumes. However, the ability to alter or erase an optical disc would require a combination of optical and magnetic methods.

Novel medical applications use the fact that different laser wavelengths can be absorbed in specific biological tissues. A widespread eye condition, glaucoma, is manifested by a high fluid pressure in the eye, which can lead to destruction of the optic nerve. A simple laser operation (iridectomy) can "burn" a tiny hole in a clogged membrane, relieving the destructive pressure. Along the same lines, a serious side effect of diabetes is the formation of weak blood vessels (neovascularization) which often leak blood into extremities. When this occurs in the eye, vision deteriorates (diabetic retinopathy), leading to blindness in diabetic patients. It is now possible to direct the green light from the argon ion laser through the clear eye lens and eye fluid, focus on the retina edges, and photo-coagulate the leaky vessels. These procedures have greatly reduced the occurrence of blindness in patients with glaucoma and diabetes.

Laser surgery is now a practical reality. Infrared light at 10 μm from a carbon dioxide laser can cut through muscle tissue, primarily by heating and evaporating the water contained in cellular material. Laser power of about 100 W is required in this technique. The advantage of the "laser knife" over conventional methods is that laser radiation cuts and coagulates at the same time, leading to substantial reduction of blood loss. In addition, the technique virtually eliminates cell migration which is very important in tumor removal. Furthermore, a laser beam can be trapped in fine glass-fiber light-guides (endoscopes) by means of total internal reflection (Sec. 24.10). The light fibers can be introduced through natural orifices, conducted around internal organs, and directed to specific interior body locations, eliminating the need for massive surgery. For example, bleeding tissues in the gastrointestinal tract can be optically cauterized by a fiberoptic endoscope inserted through the mouth. (For communications applications, see the essay on fiberoptics in Chapter 24).

Finally, we describe an application to biological and medical research. It is often important to isolate and collect unusual cells for study and growth. A laser cell-separator exploits the fact that specific cells can be tagged with fluorescent dyes. All of the cells in a sample are dropped from a tiny charged nozzle, and laser-scanned for the dye tag. If the detector is triggered by the correct light-emitting tag, a small voltage is applied to parallel plates and the falling electrically charged cell is deflected into a collection beaker. This is an efficient method for extracting the proverbial needles from the haystack.

ESSAY PROBLEMS

1. (a) In the case of the earth-moon laser range-finder, what is the round-trip time for a laser pulse? (b) What precision in timing is required, that is, how small a time change needs to be detectable, to be able to measure the distance to an error of 10 cm? What effect does the earth's atmosphere have?
2. A laser beam of wavelength $\lambda = 600$ nm is directed at the moon from a laser tube of 1 cm diameter. Does the beam spread at all? What is the diameter of the "spot" on the moon's surface? Does your answer demand a good strategy for successful lunar-array illumination? How would you do it?
3. Estimate how much chemical reagent is required to produce a chemical laser of 1 kW output, in units of moles/second. Pick a reasonable exothermic reaction rate in kcal/mole. Estimate how many kilograms each of hydrogen and chlorine gas would be needed to make an HCl laser operate for 1 hour at 1 kW output.

CHAPTER 31

Molecules and Solids

Natural formation of quartz (SiO_2) crystals. (Photo © George Whiteley)

The preceding chapter was concerned with the atomic and electronic structure of single atoms. However, except for the inert gases, single isolated atoms are generally not found in nature. Instead, elements combine to form chemical compounds which consist of individual atoms joined by chemical bonds.

In this chapter, we shall first describe the bonding mechanisms in molecules, and the various types of energy that a molecule can have. We shall then take the next logical step and show how molecules combine to form solids. Our study will enable us to explain the differences between insulating, metallic, and semiconducting solids. The chapter concludes with a discussion of semiconducting junctions, the operation of several semiconductor devices, and an essay on photovoltaic devices.

31.1 MOLECULAR BONDS

Fundamentally, the bonding mechanisms in a molecule are primarily due to electrostatic forces between atoms (or ions). When two atoms are separated by a very large distance, the dominant force between them is attractive. As

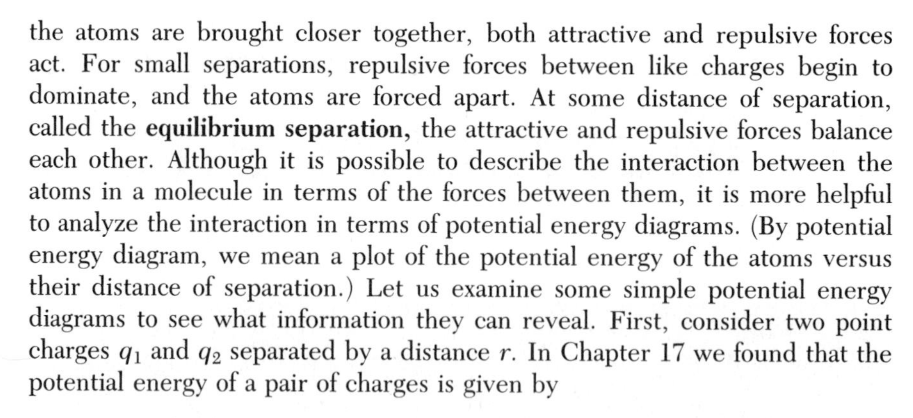

the atoms are brought closer together, both attractive and repulsive forces act. For small separations, repulsive forces between like charges begin to dominate, and the atoms are forced apart. At some distance of separation, called the **equilibrium separation,** the attractive and repulsive forces balance each other. Although it is possible to describe the interaction between the atoms in a molecule in terms of the forces between them, it is more helpful to analyze the interaction in terms of potential energy diagrams. (By potential energy diagram, we mean a plot of the potential energy of the atoms versus their distance of separation.) Let us examine some simple potential energy diagrams to see what information they can reveal. First, consider two point charges q_1 and q_2 separated by a distance r. In Chapter 17 we found that the potential energy of a pair of charges is given by

$$PE = k\frac{q_1 q_2}{r} \tag{32.1}$$

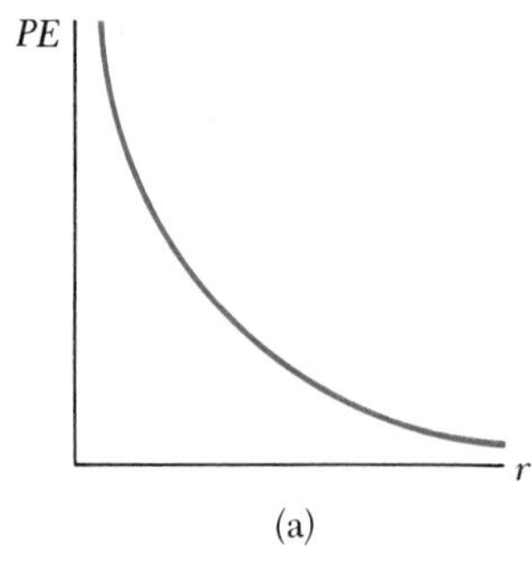

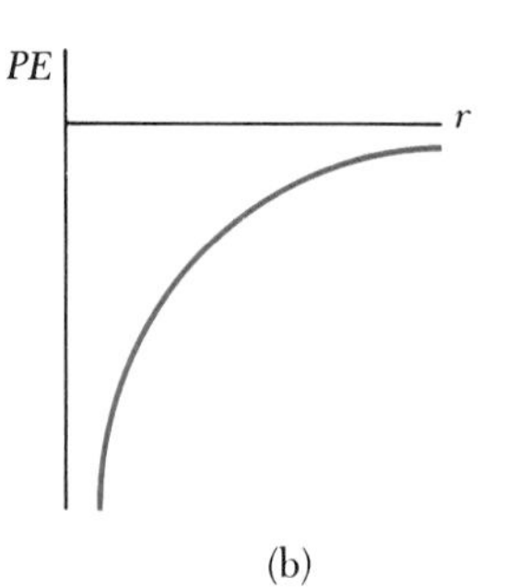

FIGURE 31.1 Plots of the potential energy versus particle separation for two charges when (a) the charges have the same sign and (b) the charges have opposite sign.

Figure 31.1a is a plot of *PE* versus r for the cases in which both charges have the same sign and thus repel one another. Note that the characteristic shape of this curve is that the potential energy increases as r decreases. Figure 31.1b is a plot of *PE* versus r for the case in which the charges have opposite signs and thus attract one another. The shape of this curve is such that the potential energy decreases as r decreases.

With this background, let us now examine the general shape of a potential energy curve for a molecule containing two atoms whose nuclei are separated by a distance r, as in Figure 31.2. For large separations, the potential energy curve decreases as r decreases, corresponding to a net attractive force between the atoms. For small separations, the potential energy curve increases as r decreases, corresponding to a net repulsive force between the atoms. At the equilibrium separation, the potential energy has its minimum value, which corresponds to the binding energy of the molecule. The binding energy is a measure of how much energy would have to be added to the molecule to separate its atoms by an infinite distance.

The following example will involve a more realistic potential energy curve that is typically encountered in practice. This example also discusses the concept of activation energy.

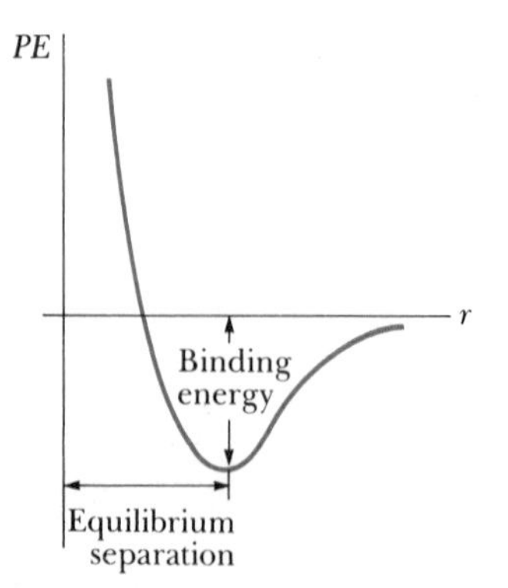

FIGURE 31.2 The potential energy versus internuclear separation for two atoms.

EXAMPLE 31.1 Activation Energy

In certain situations, the potential energy diagram for a molecule looks like that pictured in Figure 31.3. (a) Determine whether the force is attractive or repulsive at points A, B, and C on the curve.

Solution At point A the potential energy curve is increasing as r decreases, so the force is repulsive. At B the curve decreases as r decreases, so the force is attractive, and at C the curve again increases as r decreases, indicating a repulsive force. The fact that the force between the atoms is repulsive at point A means that the atoms will not come together without external assistance. Some additional energy must be added to the system to force the atoms over this potential energy barrier. The amount of energy that must be added to perform this task is called the *activation energy*.

The activation energy peak in the curve often arises because other bonds must be broken before the bond between the atoms can form. A common example of this is the formation of water from H_2 and O_2 molecules. Before the water bonds can be formed to make H_2O, the bonds between the H and O atoms of H_2 and O_2 molecules

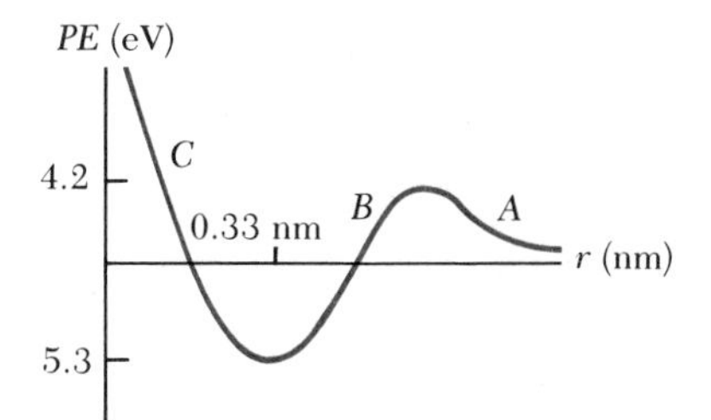

FIGURE 31.3 Example 31.1

must first be broken. Passing a current (or spark) through gaseous H_2 and O_2 will break some of the bonds in the gas, thus providing the activation energy. When these freed H and O atoms combine to form water, there is a release of energy which can provide the activation energy to cause additional bonds to be broken and the subsequent formation of more water.

(b) Find the binding energy, the activation energy, and the equilibrium separation of the molecule.

Solution From the graph, we find that the depth of the potential well, and thus the binding energy, is equal to 5.3 eV. Also, the graph shows that the activation energy is 4.2 eV, and the equilibrium separation is 0.33 nm.

31.2 TYPES OF MOLECULAR BONDS

A complete description of the binding mechanisms in molecules is a highly complex problem because it involves the mutual interactions of many particles. In this section, we shall discuss some simplified models in the following order of decreasing bond strength: the ionic bond, the covalent bond, the hydrogen bond, and the van der Waals bond. We shall also briefly describe the bonding mechanism in metals.

Ionic Bonds

Ionic bonds are due to the Coulomb attraction between oppositely charged ions. A familiar example of an ionically bonded molecule is sodium chloride, NaCl, which is common table salt. Sodium has an electronic configuration $1s^2 2s^2 2p^6 3s$. Thus, the arrangement is such that the atom has 11 electrons with 10 of them in closed shells or subshells. The electron in the outermost shell "sees" a charge of $+11e$ on the nucleus, but it is partially shielded from the nucleus by the inner core electrons which have a charge of $-10e$. As a result, the electron in the outermost shell is attracted by a net charge of only $+1e$, so it is relatively easy to remove this electron from the atom. In fact, the energy required to ionize sodium to form Na^+ is 5.1 eV. Chlorine, which has an electronic configuration $1s^2 2s^2 2p^5$, is one electron short of a closed-shell structure. Because closed-shell configurations are energetically more favorable, the Cl^- ion is more stable than the neutral Cl atom. In fact, if neutral chlorine takes on an electron to become Cl^-, there will be an energy release of 3.7 eV. The energy released when an atom takes on an electron is called the **electron affinity.** Therefore, the energy required to combine Na^+

and Cl^- to form NaCl is 5.1 − 3.7 = 1.4 eV. When the ions are brought close together, the electrostatic potential energy is greater than the 1.4 eV needed, and the ions coalesce to form NaCl.

The total energy versus the internuclear separation for NaCl is shown in Figure 31.4. Note that the total energy of the molecule has a minimum value of −4.2 eV at the equilibrium separation of about 0.24 nm. Thus, the total energy required to separate the NaCl molecule into neutral sodium and chlorine atoms is equal to 4.2 eV. Also, note that the energy at an infinite separation is 1.3 eV, which corresponds to the energy required to form the ions from neutral atoms. The equilibrium separation for an isolated NaCl molecule is 0.24 nm, compared to a value of 0.28 nm for solid NaCl. The equilibrium separation is greater in the solid state because the cubic arrangement of the ions in a NaCl crystal is such that each ion is surrounded by six nearest-neighbor ions of opposite charge.

When the ions in an isolated NaCl molecule are brought closer than their equilibrium separation, 0.24 nm, the electrons in the closed shells begin to overlap, which results in a repulsive force between the ions.

Covalent Bonds

A covalent bond between two atoms can be visualized as the sharing of electrons supplied by one or both atoms that form the molecule. Many diatomic molecules such as H_2, F_2, and CO owe their stability to covalent bonds. In the case of the H_2 molecule, the two electrons are equally shared between the nuclei, and form a so-called *molecular orbital*. To understand how this bond is formed, consider two H atoms approaching one another. Each has an electron in the ground state orbiting its nucleus, and each electron can have spin quantum numbers of either $+\frac{1}{2}$ or $-\frac{1}{2}$. If the two electrons have the same spin quantum number, a bond cannot form because this would violate the exclusion principle. However, if the spins are opposite, the two atoms are

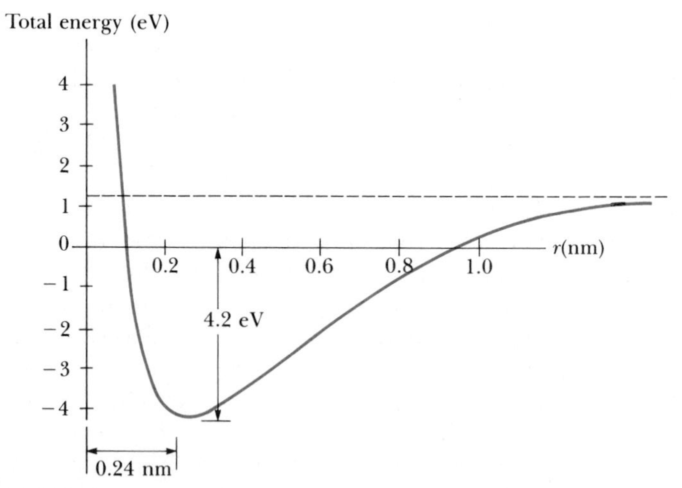

FIGURE 31.4 Total energy versus the internuclear separation for the NaCl molecule.

able to share their electrons. The two electrons are more likely to be found between the two nuclei, hence the electron density is large in this region. The formation of the molecular orbital from the *s* orbitals of the two hydrogen atoms is shown in Figure 31.5.

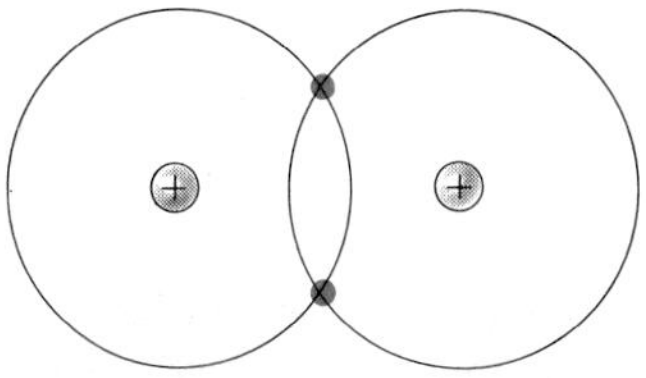

More complex stable molecules such as H_2O, CO_2, and CH_4 are also formed by covalent bonds. Consider methane, CH_4, a typical organic molecule shown schematically in Figure 31.6a. Note that four covalent bonds are formed between the carbon atom and each of the hydrogen atoms. The spatial electron distribution of the four covalent bonds is shown in Figure 31.6b. The four hydrogen nuclei are at the corners of a regular tetrahedron, with the carbon nucleus at the center. The energy required to break a bond between carbon and hydrogen is about 4.28 eV.

FIGURE 31.5 The covalent bond formed by the two 1*s* electrons of the H_2 molecule.

The Hydrogen Bond

Because hydrogen has only one electron, it is expected to form a covalent bond with only one other atom. However, in some molecules hydrogen forms a different type of bond between two atoms or ions, called a hydrogen bond. One example of a hydrogen bond, shown in Figure 31.7, is the hydrogen difluoride ion, $(HF_2)^-$. The two negative fluorine ions are bound by the positively charged proton between them. The bond is similar to the covalent bond except that it is the proton associated with hydrogen that is being shared rather than electrons. The hydrogen bond is relatively weak, with a binding energy of about 0.1 eV. Although the bond is weak, it is the mechanism responsible for linking giant biological molecules and polymers. For example, in the case of the famous DNA molecule which has a double helix structure as shown in Figure 31.8a, hydrogen bonds link the turns of the helix. DNA, or deoxyribonucleic acid, is the primary molecule in chromosomes, which are

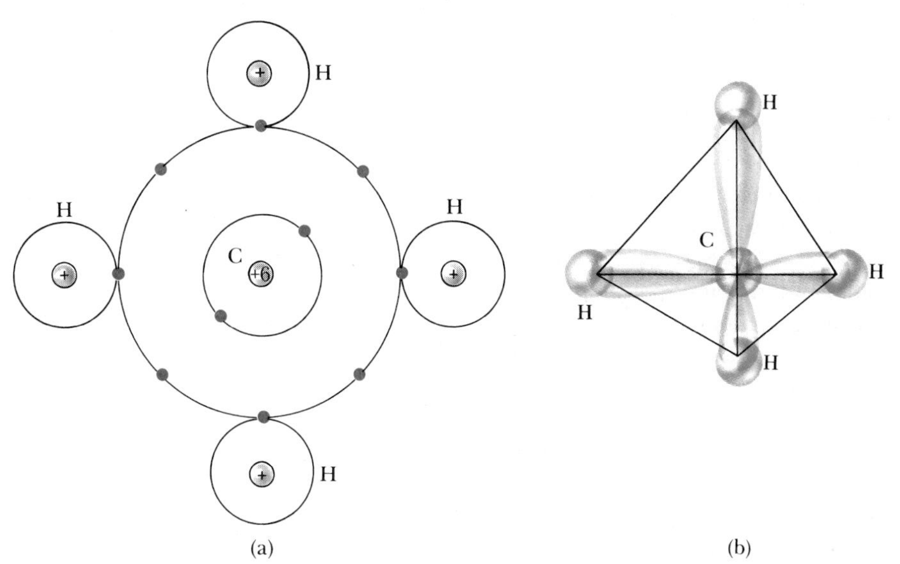

FIGURE 31.6 (a) Diagram of the four covalent bonds in the CH_4 molecule. (b) The electron distribution of the four covalent bonds in the CH_4 molecule. Note that the carbon atom is at the center of a tetrahedron with hydrogen atoms at its corners.

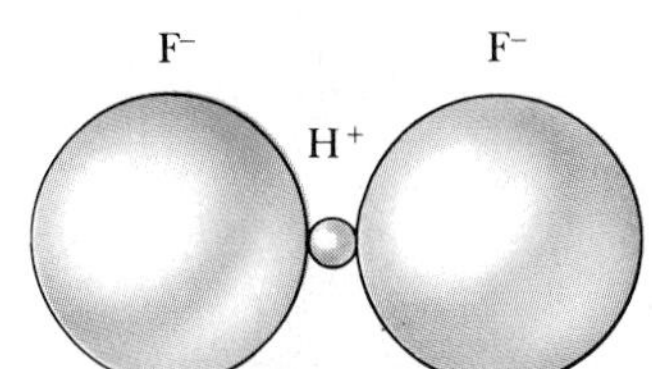

FIGURE 31.7 Hydrogen bonding in the $(HF_2)^-$ molecule ion. Note that the two negative fluorine ions are bound by the positively charged proton between them.

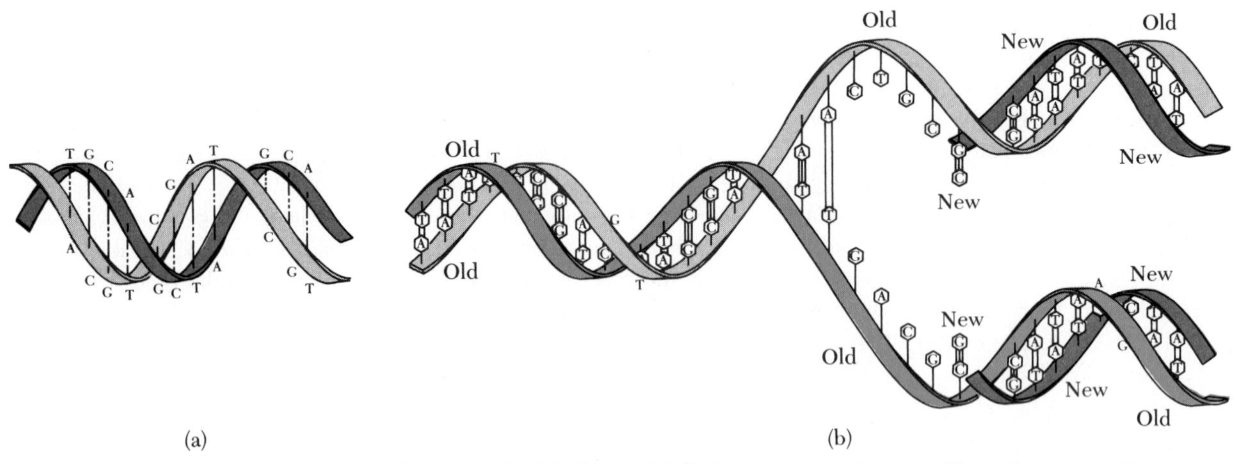

FIGURE 31.8 (a) The double helix structure of DNA. (b) As the DNA cell reproduces, the double helix separates at one end.

the carriers of genetic information. In Figure 31.8a, the hydrogen bonds connecting the double helix are labeled A–T and C–G. These subunits of the helix, called nucleotide base pairs, are composed of adenine (A), thymine (T), cytosine (C), and guanine (G). Because of the structure of these units, hydrogen bonds can only form between adenine and thymine and between cytosine and guanine. When a cell reproduces, the double helix comes apart at one end as shown in Figure 31.8b and separates. Each portion of the split double helix can now form a new double helix. The new helix must match the original exactly because all of the A–T and C–G bonds must match perfectly throughout the strand. Genetic information is passed from one generation of cells to the next through this matching of hydrogen bonds.

Van der Waals Bonds

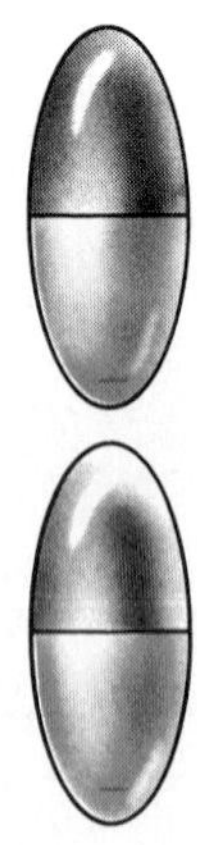

FIGURE 31.9 A representation of the dipole-dipole force between two polar molecules.

The two types of bonds discussed so far are produced by forces between atoms that cause them to combine and form molecules. **Van der Waals bonds,** on the other hand, usually occur *between molecules.* There are actually three types of van der Waals forces which we shall briefly describe. The first type, called the dipole-dipole force, is an interaction between two molecules each having permanent electric dipole moments. Such molecules are called *polar molecules.* Figure 31.9 shows a representation of a dipole-dipole interaction between two polar molecules. Examples of polar molecules are HCl and H_2O.

The second type of van der Waals force is a dipole-induced force in which a polar molecule induces a dipole moment in a nonpolar molecule. The third type of van der Waals force is a force of attraction between two nonpolar molecules. Even though a nonpolar molecule has no permanent dipole moment (when averaged over time), the separation of charges within a dynamic molecule at any instant may induce a dipole moment in a nearby molecule.

Van der Waals bonds are very weak and hence are easily broken by thermal motion of the molecules. However, at sufficiently low temperatures where thermal excitations are negligible, these bonds cause substances to condense into a liquid.

Metallic Bonds

Metallic bonds are generally weaker than ionic or covalent bonds. The valence electrons in a metal are relatively free to move throughout the material. There are a large number of such mobile electrons in a metal, typically one or two electrons per atom. The metal structure can be viewed as a "sea" or "gas" of nearly free electrons surrounded by a lattice of positive ions (Fig. 31.10). The binding mechanism in a metal is the attractive force between the positive ions and the electron gas. The free electrons are responsible for the high electrical and thermal conductivities of metals. For this reason, the free electrons are often called conduction electrons.

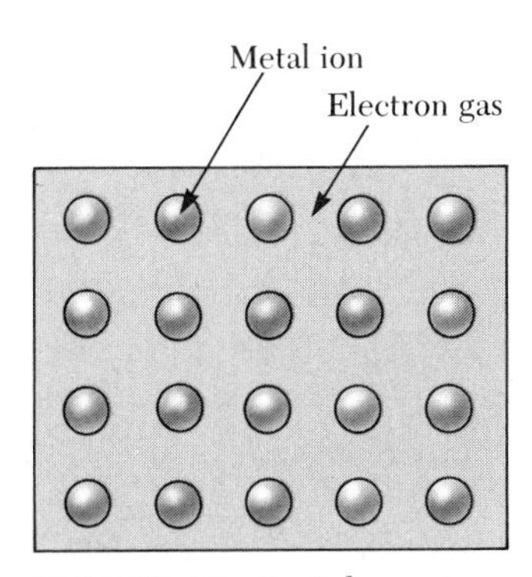

FIGURE 31.10 Schematic diagram of a metal. The colored area represents the electron gas, while the spheres represent the positive metal ion cores.

EXAMPLE 31.2 Breaking Bonds in Water

In both the solid state and the liquid state, H_2O molecules are held in place by six bonds to their nearest neighbor molecules. If the latent heat of fusion of water is 3.34×10^5 J/kg, find the average energy required to break each of the bonds in ice.

Solution The molecular weight of H_2O is 18 (see Appendix C), and thus, we know that 18 kg of ice contains a total of 6.02×10^{26} molecules. Thus, 1 kg of ice contains $6.02 \times 10^{26}/18 = 3.34 \times 10^{25}$ molecules. If each of these is bound by six bonds, we have a total number of bonds in the 1 kg of ice of

$$\left(6 \frac{\text{bonds}}{\text{molecule}}\right)(3.34 \times 10^{25} \text{ molecules}) = 2.00 \times 10^{26} \text{ bonds}$$

The latent heat of fusion of ice is the energy required to break all of these bonds in one kg of ice. Thus, the energy per bond is

$$\frac{3.34 \times 10^5 \text{ J}}{2.00 \times 10^{26} \text{ bonds}} = 1.67 \times 10^{-21} \text{ J} = 0.01 \text{ eV}$$

EXAMPLE 31.3 Energy to Break Bonds (Methane)

If the average energy associated with a bond between carbon and hydrogen is 4.28 eV in methane (CH_4), find the total amount of energy required to break all the bonds in 1 kg of methane.

Solution The molecular weight of methane (CH_4) is 16. Thus, the number of molecules in 1 kg of methane is

$$\frac{6.02 \times 10^{26}}{16} = 3.76 \times 10^{25} \text{ molecules}$$

and with four bonds per molecule, see Figure 31.6, the total number of bonds broken is

$$\left(4 \frac{\text{bonds}}{\text{molecule}}\right)(3.76 \times 10^{25} \text{ molecules}) = 1.50 \times 10^{26} \text{ bonds}$$

Thus, the total energy required is

$$\left(4.28 \frac{\text{eV}}{\text{bond}}\right)(1.50 \times 10^{26} \text{ bonds}) = 6.42 \times 10^{26} \text{ eV} = 1.03 \times 10^8 \text{ J}$$

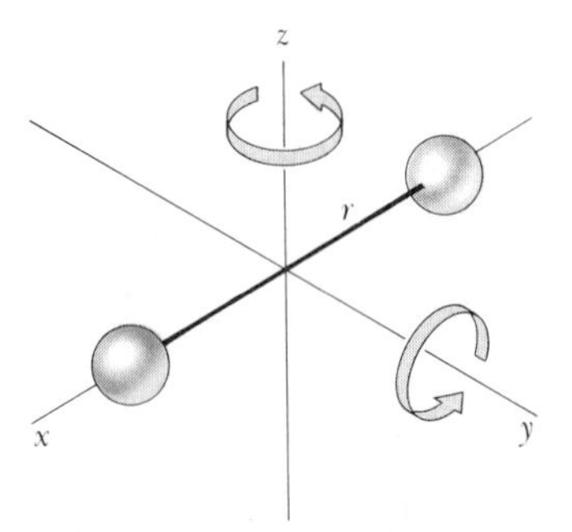

FIGURE 31.11 A diatomic molecule can rotate about the y or z axes.

31.3 THE ENERGY AND SPECTRA OF MOLECULES

As in the case of atoms, the structure and properties of molecules can be studied by examining the radiation they emit or absorb. A molecule emits radiation when it moves from an excited state to a state lower in energy, and it absorbs radiation when it moves from a low energy state to a state of higher energy. Much of the information about molecules comes through a study of their rotational and vibrational motions, and we shall study each of these separately in that which follows.

Rotational Motion of a Molecule

Let us consider the rotation of a molecule about its center of mass. We shall confine our discussion to a diatomic molecule, although the basic ideas can be extended to more complex molecules. The diatomic molecule pictured in Figure 31.11 can have rotational energy associated with its rotation about either the z or the y axis. The x axis is not considered because the atoms are assumed to be points, and thus, they have no moment of inertia about this axis. If ω is the angular velocity of rotation about one of these axes, the rotational kinetic energy of the molecule can be expressed in the form

$$E_{\text{rot}} = \frac{1}{2} I\omega^2 \tag{31.2}$$

where I is the moment of inertia of the molecule about the axis under consideration. Equation 31.2 can be written in terms of the angular momentum, $I\omega$, of the molecule as

$$E_{\text{rot}} = \frac{1}{2} I\omega^2 = \frac{(I\omega)^2}{2I} \tag{31.3}$$

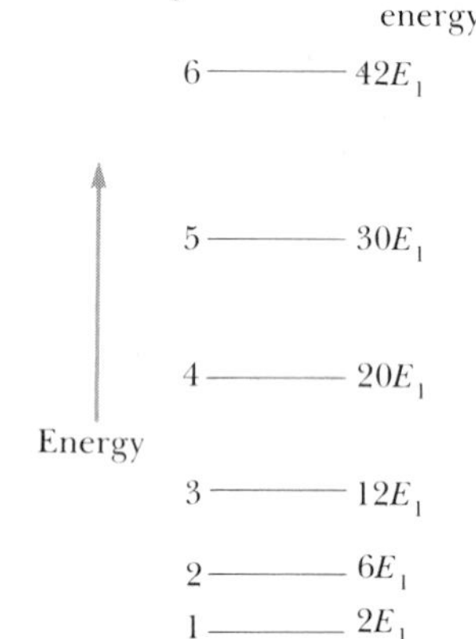

FIGURE 31.12 Allowed rotational energies of a diatomic molecule calculated using Equation 31.5, where $E_1 = \hbar^2/2I$.

Wave mechanics restricts the angular momentum values to multiples of $\hbar$ such that the allowed values of the rotational angular momentum are given by

$$I\omega = \sqrt{J'(J' + 1)}\, \hbar \qquad J' = 0, 1, 2, \ldots \tag{31.4}$$

where J' is an integer called the **rotational quantum number.** Substituting Equation 31.4 into Equation 31.3, we get an expression for the allowed values of the rotational kinetic energy:

Rotational energy of a molecule

$$\boxed{E_{\text{rot}} = \frac{\hbar^2}{2I} J'(J' + 1)} \qquad J' = 0, 1, 2, \ldots \tag{31.5}$$

Thus, we see that the rotational energy of the molecule is quantized and depends on the moment of inertia of the molecule. The allowed rotational energies of a diatomic molecule are plotted in Figure 31.12. The radiation emitted as a molecule makes a transition from one rotational level to another lies in the microwave range of frequencies which are of the order of 10^{11} Hz. The radiation emitted as a molecule undergoes such transitions can be used to find such important properties of a molecule as its moment of inertia and the spacing between atoms, as the following example shows.

EXAMPLE 31.4 Rotation of Oxygen

The O_2 molecule is found to make a transition from the lowest rotational energy state ($J' = 0$) to its first excited state ($J' = 1$) when the molecule absorbs a photon of frequency 4.30×10^{10} Hz. The molecule is shown in Figure 31.13, where the mass of each atom is 2.67×10^{-26} kg and the distance of separation is r. (a) Find the moment of inertia of the oxygen molecule.

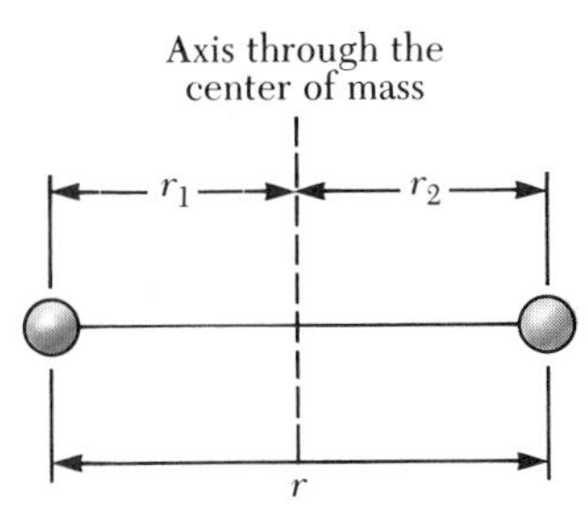

FIGURE 31.13 (Example 31.4) A diatomic molecule with an axis shown through its center of mass. For O_2, $r_1 = r_2 = r/2$.

Solution The energy of the oxygen molecule in the $J' = 1$ state is found from Eq. 31.5 to be

$$E_1 = \frac{\hbar^2}{2I}J'(J'+1) = \frac{\hbar^2}{2I}(2) = \frac{\hbar^2}{I}$$

From the same equation, we see that the rotational energy is 0 in the $J' = 0$ state. Thus, the difference in energy between these two states is

$$\Delta E = E_1 - E_0 = \frac{\hbar^2}{I}$$

The molecule acquires the energy to make this transition from the incoming photon of energy hf. Thus, we have

$$hf = \frac{\hbar^2}{I}$$

and from this we see that the moment of inertia can be found as follows.

$$I = \frac{h}{4\pi^2 f} = \frac{6.63 \times 10^{-34}\ \text{J·s}}{4\pi^2(4.30 \times 10^{10}\ \text{Hz})} = 3.91 \times 10^{-46}\ \text{kg·m}^2$$

(b) Find the distance of separation of the oxygen atoms in the molecule.

Solution Since both atoms have the same mass, the center of mass of the molecule is located midway between the two atoms, thus $r_1 = r_2 = r/2$ in Figure 31.13. We shall consider the atoms to be point masses, so the moment of inertia about the center of mass is given by

$$I = \sum m_i r_i^2 = mr_1^2 + mr_2^2 = m\left(\frac{r}{2}\right)^2 + m\left(\frac{r}{2}\right)^2 = \frac{1}{2}mr^2$$

Solving for r, the distance of separation, we have

$$r = \left(\frac{2I}{m}\right)^{1/2} = \left[\frac{2(3.91 \times 10^{-46})\ \text{kg·m}^2}{2.67 \times 10^{-26}\ \text{kg}}\right]^{1/2} = 1.71 \times 10^{-10}\ \text{m}$$

EXAMPLE 31.5 The Rotation of CO

The CO molecule has a distance of separation between the C and O atoms given by $r = 1.13 \times 10^{-10}$ m. The mass of the carbon atom is 2.00×10^{-26} kg and the mass of the oxygen atom is 2.67×10^{-26} kg. (a) Find the center of mass of this molecule.

Solution To find the center of mass of the CO molecule, we shall use a coordinate system like that shown in Figure 31.14 where the origin is selected at the position of the carbon atom. In Chapter 6, we found that the center of mass for a system such as this can be found as

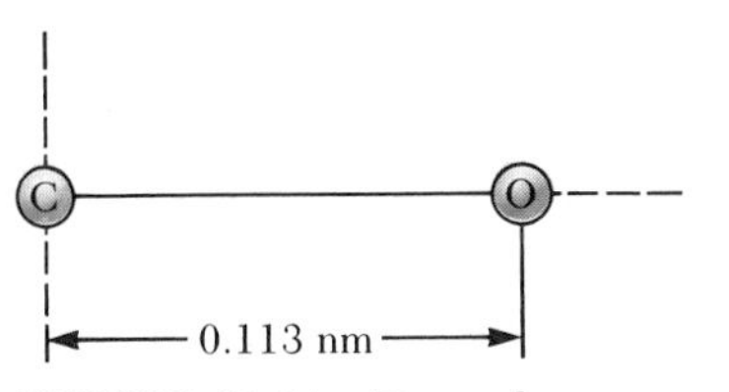

FIGURE 31.14 (Example 31.5) Representation of the CO molecule.

$$x = \frac{\sum m_i x_i}{M} = \frac{(2.00 \times 10^{-26}\ \text{kg})(0) + (2.67 \times 10^{-26}\ \text{kg})(1.13 \times 10^{-10}\ \text{m})}{4.67 \times 10^{-26}\ \text{kg}}$$

$$x = 6.46 \times 10^{-11}\ \text{m} = 0.646\ \text{nm}$$

Thus, we see that the center of mass is located 6.46×10^{-11} m from the carbon atom, which corresponds to 4.84×10^{-11} m from the oxygen atom.

(b) Find the moment of inertia of the CO molecule about its center of mass.

Solution The moment of inertia is given by

$$I = \sum m_i r_i^2 = m_1 r_1^2 + m_2 r_2^2$$

where m_1 is taken to be the mass of the carbon atom, m_2 the mass of the oxygen atom, r_1 is the distance from the center of mass to the carbon atom, and r_2 is the distance from the center of mass to the oxygen atom. Thus, we have

$$I = (2 \times 10^{-26}\ \text{kg})(6.46 \times 10^{-11}\ \text{m})^2 + (2.67 \times 10^{-26}\ \text{kg})(4.84 \times 10^{-11}\ \text{m})^2$$
$$= 1.46 \times 10^{-46}\ \text{kg}\cdot\text{m}^2$$

(c) Find the energy of the CO molecule in the $J' = 1$ state and in the $J' = 0$ state.

Solution For the $J' = 1$ state, Equation 31.5 gives

$$E_1 = \frac{\hbar^2}{2I} J'(J' + 1) = \frac{\hbar^2}{I} = \frac{(1.054 \times 10^{-34}\ \text{J}\cdot\text{s})^2}{1.46 \times 10^{-46}\ \text{kg}\cdot\text{m}^2} = 7.61 \times 10^{-23}\ \text{J}$$

The energy in the $J' = 0$ state is zero, as can be seen from Equation 31.5.

EXERCISE Find the frequency of the photon that would be emitted as the molecule makes a transition from the $J' = 1$ state to the $J' = 0$ state.

Answer 1.15×10^{11} Hz. (The photon energy hf must be equal to the difference in energies of these two states.)

Vibrational Energy of Molecules

As we mentioned earlier, molecules may also have energy associated with their vibrational motion. A molecule is a flexible structure whose atoms are bound together as though they were connected by a spring, as shown in Fig-

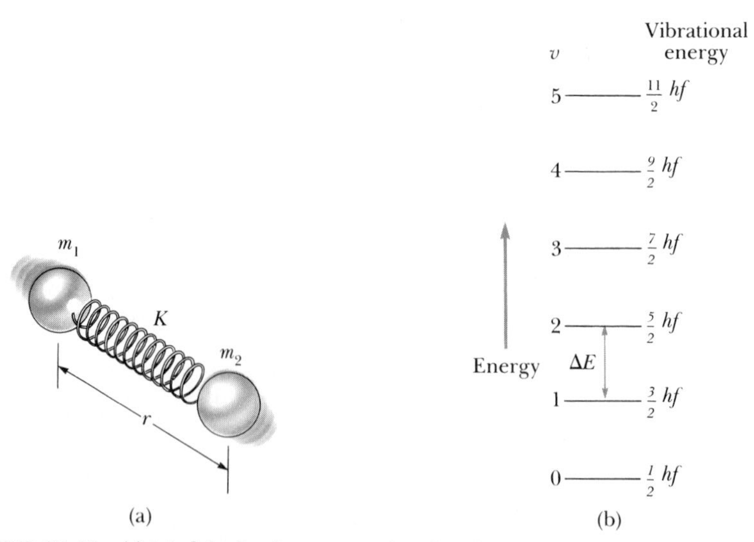

FIGURE 31.15 (a) Model of a diatomic molecule whose atoms are bonded by an imaginary spring of spring constant K. (b) Allowed vibrational energies of a diatomic molecule. Note that the spacings between adjacent levels are equal.

ure 31.15a. Hence, the molecule can be set into vibration (through collisions with other molecules or the walls of a container) and acquire vibrational energy. The problem of finding an expression for the vibrational energy for an oscillator was one of the first to be solved in quantum mechanics. The energy of the oscillator is quantized, and the allowed energies are given by

$$E_{\text{vib}} = (v + \frac{1}{2})hf \tag{31.6}$$

Vibrational energies of a molecule

where v is an integer called the **vibrational quantum number.** At ordinary temperatures, most molecules have vibrational energies corresponding to the $v = 0$ state having the lowest energy. The vibrational energy of a molecule in the $v = 0$ state is $E_{\text{vib}} = hf/2$, and is called the **zero-point energy.** In the first excited state, $v = 1$, and its energy is $3hf/2$, and so on. The allowed vibrational energies of a diatomic molecule are shown in Figure 31.15b. Note that the spacings between adjacent levels are equal, unlike the situation for the rotational levels. The frequency of the radiation emitted or absorbed during a vibrational transition is usually in the infrared region of the spectrum, with frequencies of the order of 10^{13} Hz.

31.4 BAND THEORY OF SOLIDS

An understanding of the behavior of semiconductors would not be possible without the techniques and ideas of quantum physics. Let us examine what happens to the energy levels of atoms as they are assembled to form a solid. We have already seen that the electrons of an isolated atom (one far removed from other atoms) can occupy only certain well-defined energy levels as shown in Figure 31.16a. Now suppose that two similar atoms are brought

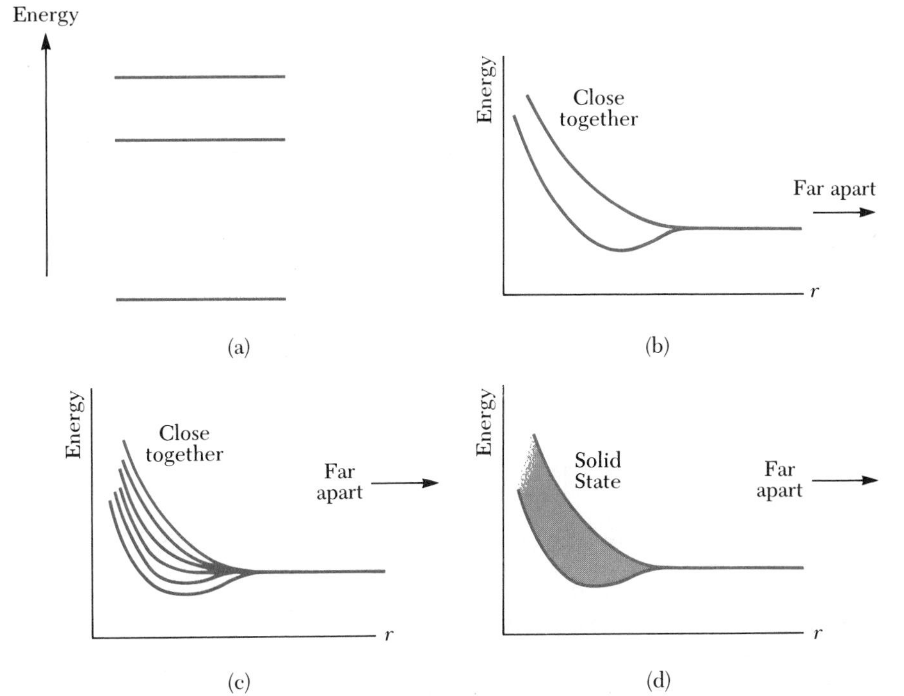

FIGURE 31.16 (a) Well-defined energy levels of an isolated atom. (b) The splitting of one of the levels when two atoms are brought together. (c) The splitting of one of the levels when six atoms are brought together. (d) The formation of a band of energies when many atoms are assembled to form a solid.

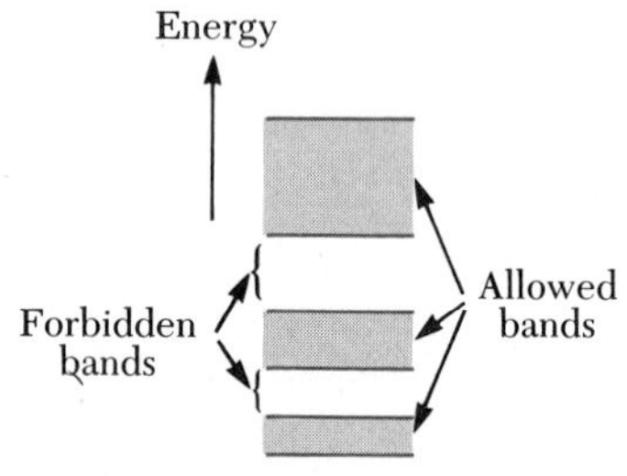

FIGURE 31.17 Energy level of a solid consists of a set of allowed bands (shaded) separated by forbidden bands.

close together as in Figure 31.16b. An original single energy level common to both is split such that two energy levels will form. If six atoms are combined, the original energy level splits into six closely spaced energy levels, as in Figure 31.16c. Finally, imagine a system of N atoms being brought together as would be the case in a solid. The atomic energy levels are split into N closely spaced levels as in Figure 31.16d. In fact, if N is very large, as it is in a solid (of the order of 10^{23} atoms/cm^3), the spacing between these levels is so small that they can be considered to form a continuous band of energies.

By extending the discussion above, it is easy to understand that each atomic energy level of the individual atoms is replaced by a continuous band of energies. As a result, the energy level diagram for a solid is a set of bands as shown in Figure 31.17. The figure shows a series of allowed bands of energies separated by *forbidden bands*. These latter bands are called forbidden because electrons in the solid are not allowed to have these particular energies.

We can use this simple band theory to explain some of the electrical properties of solids. First, consider the energy bands for an insulator shown in Figure 31.18a. At ordinary temperatures, the insulator has a band of energy levels, called the *valence band*, which is completely filled with electrons. Above this in energy lies a completely empty band, called the *conduction band*, and between the valence and conduction bands is a wide energy gap. This gap is so wide (typically greater than 10 eV) that it is virtually impossible at ordinary temperatures to move electrons from the valence band into the conduction band. Since all the levels in the valence band are filled, an individual electron in this band is essentially immobile, even when an electric field is present. Hence, there is no current and the material behaves like an insulator. From another point of view, thermal agitation is insufficient to free an electron from its covalent bonds. Electrons in the valence band remain in a bound state as opposed to becoming "free" (moving into the conduction band).

The energy band structure for a metal is shown in Figure 32.18b. In this case, the uppermost band, which is the conduction band, is only partially filled. In this band the electrons have many levels to which they can easily move; therefore an electric current is readily established when an electric

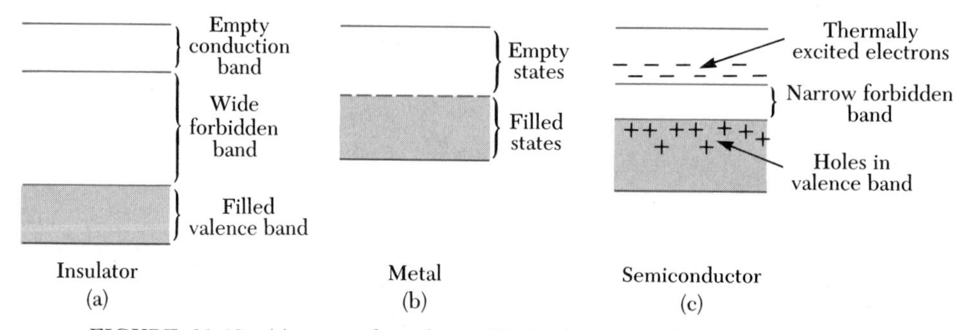

FIGURE 31.18 (a) An insulator has a filled valence band and an empty conduction band separated by a wide forbidden band. (b) The uppermost band of a metal, which is the conduction band, is partially filled with electrons. (c) A semiconductor at ordinary temperatures has a nearly filled valence band and a nearly empty conduction band. The forbidden band is very narrow, usually less than 1 eV.

field is applied. Such conduction band electrons are often referred to as *free* electrons since they are not bound to any one atom in the structure.

In a semiconductor, the last filled energy band (the valence band) is completely occupied as in the case of an insulator. However, the energy gap which separates the valence band from the conduction band is very narrow for a semiconductor. In silicon this gap is about 1 eV in width, while in germanium it is about 0.7 eV. At very low temperatures, there are very few electrons in the conduction band, while the valence band is filled. However because the energy gap is small, as the temperature increases, electrons near the top of the valence band may have enough thermal energy to be able to move into the conduction band (Fig. 31.18c). These electrons can then participate in the conduction process if a voltage is applied across the material.

When an electron moves from the valence band into the conduction band, it leaves behind a vacant crystal site, or so-called hole, in the valence band. This hole (electron-deficient site) appears as a positive ion of charge $+e$. The hole also acts as a charge carrier in the sense that a valence electron from a nearby bond can transfer into the hole without becoming a free electron. Thus we see that there are both negative and positive charge carriers in a semiconductor.

When impurities are added to semiconductors, their band structures are modified as shown in Figure 31.19. For example, when an atom with five valence electrons, such as arsenic, is added to a semiconductor, four valence electrons participate in the covalent bonds and one electron is left over. This extra electron is nearly free and has an energy level in the band diagram that lies within the energy gap, just below the conduction band (Fig. 31.19a). Such an impurity in effect donates an electron to the structure and hence is referred to as a *donor atom*, and the levels they occupy are called *donor levels*. Since the energy spacing between the donor levels and the bottom of the conduction band is very small (typically, about 0.05 eV), a small amount of thermal energy will cause an electron in these levels to move into the conduction band. If the semiconductor is doped with atoms with three valence electrons, such as indium and aluminum, the three electrons form covalent bonds with neighboring atoms, leaving an electron deficiency, or hole. Such impurities are called *acceptors* and the levels they occupy are called *acceptor levels*. The energy levels of acceptors also lie within the energy gap, just above the valence band, as indicated in Figure 31.19b. Electrons from the valence band have enough thermal energy at room temperature to fill acceptor levels, leaving behind a hole in the valence band.

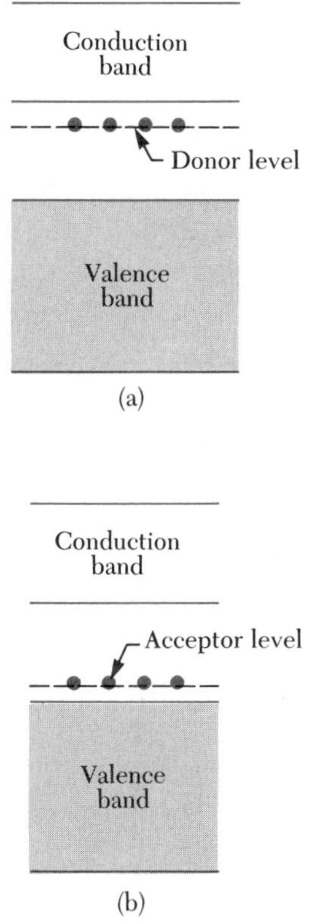

FIGURE 31.19 (a) Energy band diagram for a semiconductor containing donor impurities whose levels lie just below the bottom of the conduction band. (b) A semiconductor containing acceptor impurities has acceptor levels that are just above the top of the valence band.

*31.5 SEMICONDUCTORS

Electronic devices made from semiconducting materials are used in many facets of our everyday life. In this section, we shall describe the electrical properties of semiconductors and their role in the operation of electronic devices. In Chapter 18, we found that a good conductor contains a large number of free electrons that are able to move readily when a voltage is applied across the conductor. On the other hand, insulators have very few free electrons and hence are very poor conductors of electricity. Semiconductors are a class of

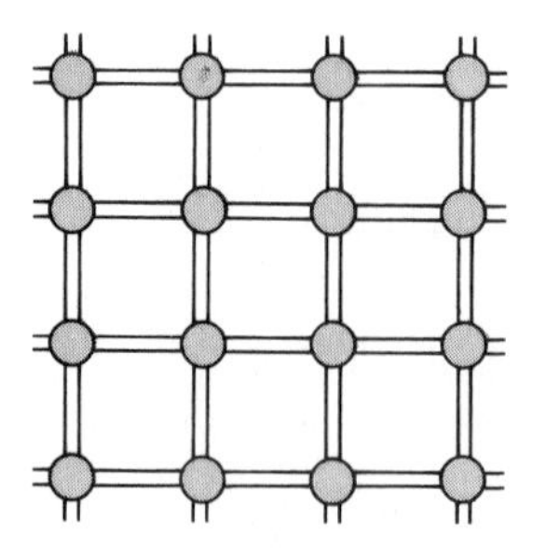

= Semiconductor atom

FIGURE 31.20 A two-dimensional model of the bonding between atoms in a semiconductor. Each horizontal or vertical line between atoms represents a shared electron.

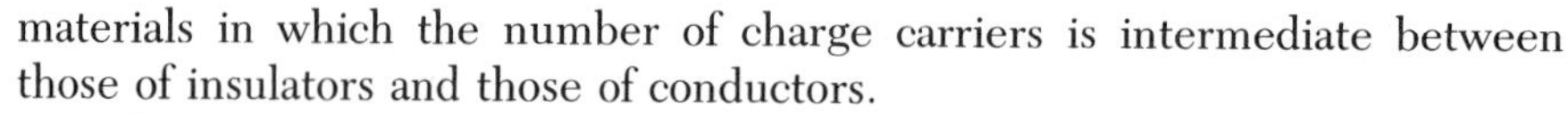

materials in which the number of charge carriers is intermediate between those of insulators and those of conductors.

The most common semiconducting materials used in electronic devices are silicon and germanium. Each of these materials is characterized by the fact that the constituent atoms have four valence electrons. In the solid state, each atom shares one of its valence electrons with each of its four nearest neighbors. A two-dimensional model for this covalent bonding is illustrated in Figure 31.20.

As the temperature of the material is increased above room temperature, the atoms vibrate more vigorously about their equilibrium positions. In this process, some electrons acquire sufficient energy to break free of their bonds and are able to move freely through the material. These electrons can then participate in the conduction process if a voltage is applied across the material. This explains why the resistance of a pure semiconductor decreases with increasing temperature.

Doped Semiconductors

Pure semiconducting materials, such as silicon, are virtually never used in electronic devices. Instead, controlled amounts of impurities, such as arsenic or indium, are added. The electrical properties of the semiconductor are substantially altered by these impurities even if the number of impurity atoms added is relatively small. The process of adding impurities is called **doping.**

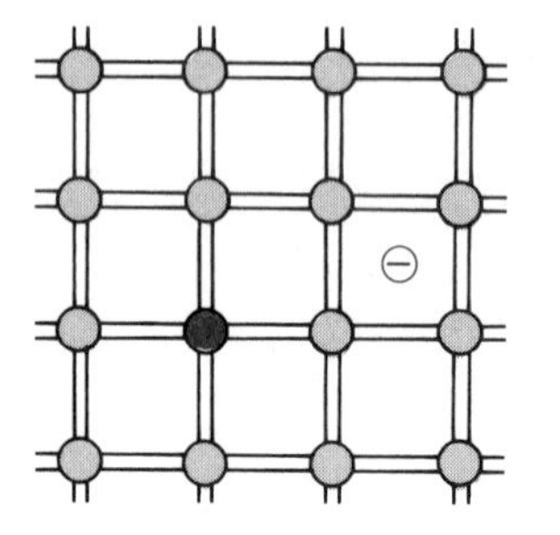

= Semiconductor atoms

= Impurity atom with five valence electrons

⊖ = Extra electron from impurity atom

FIGURE 31.21 A semiconductor material containing an impurity atom with five valence electrons.

In order to understand the effect of doping on semiconductors, consider the situation in which an impurity atom, such as arsenic, with five valence electrons, is added to a semiconductor, such as silicon. Four of the arsenic valence electrons participate in the bonding between nearest-neighbor silicon atoms, and one electron is left over (Fig. 31.21). This extra electron is loosely bound and hence easily freed from the atom. When the electron is free, it contributes to conduction in the material. Thus, the impurity atom in effect donates an electron to the material and hence is referred to as a **donor** atom. Semiconductors doped with donor atoms are called ***n*-type semiconductors** because most of the charge carriers are electrons, whose charge is negative.

If a semiconductor is doped with atoms having three valence electrons, such as indium or aluminum, the three electrons form bonds with nearest-neighbor atoms. However, there is an electron deficiency in one of the bonds with a neighbor atom (Fig. 31.22). The vacancy in the lattice structure of the solid caused by this deficiency is referred to as a **hole.** Figure 31.23 shows how holes can contribute to conduction in a semiconductor. If an electric field is applied to the semiconductor, by connecting the material to a battery, say, electrons bound to atoms near the hole can jump over to the site of the hole. As a result, the hole migrates in the direction of the electric field as if it, the hole, were a positive charge carrier (shown by the curved arrow in Fig. 31.23). Because there are more holes than electrons, materials doped in this way are known as ***p*-type semiconductors.** The impurity atoms added are referred to as **acceptors.**

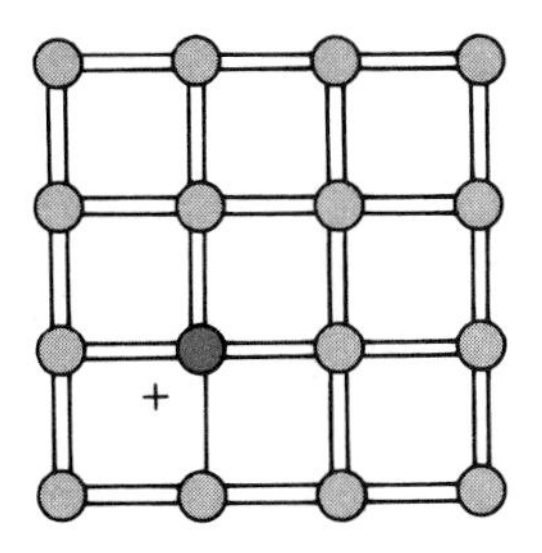

○ = Semiconductor atoms

● = Impurity atom with three valence electrons

+ = Hole, or electron deficiency in a bond

FIGURE 31.22 Two-dimensional representation of a semiconductor containing an impurity with three valence electrons.

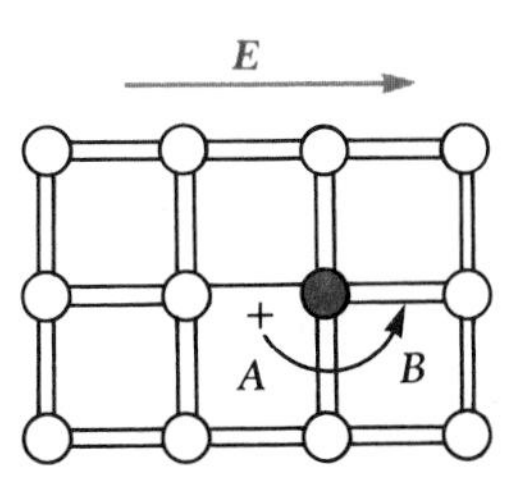

FIGURE 31.23 An electron originally at point B jumps to fill the hole originally at point A. Thus, the hole moves from A to B, in the direction of the electric field.

The *p-n* Junction

Most semiconductor devices, such as diodes and transistors, are fabricated using a semiconductor that is partly *p*-type and partly *n*-type. Figure 31.24 shows a semiconductor in which the left side has been made *p*-type and the right side has been made *n*-type. The boundary between the *p* and *n* regions is called a ***p-n* junction.** One of the outstanding features of this junction is that it is virtually swept free of charge carriers that move through it. Electrons in the *n* region diffuse through the junction into the *p* region. At the same time, holes from the *p* region diffuse through the junction into the *n* region. The electrons and holes in the junction region recombine as an electron moves into the vacant site where a hole exists. This recombination process of electrons and holes explains why this narrow region around the junction has no free charge carriers.

One might guess that this diffusion of electrons into the *p* region and of holes into the *n* region could continue until all the electrons and holes in these regions are neutralized. In order to understand why this does not occur, let us examine what happens when a hole and an electron recombine in the *n* region. The *n* region is electrically neutral because for every free electron in this region, there exists a positive charge on the atom from which it came. If an electron in the *n* region combines with a hole, effectively causing both to disappear, the positive charge present on the atom is no longer electrically balanced. Hence, the right side of the junction becomes positively charged (see Fig. 31.24). In a similar manner, the left side of the junction becomes negatively charged. An electric field is thus set up directed from right to left, as in Figure 31.24. The presence of this field prevents further diffusion of electrons and holes through the junction.

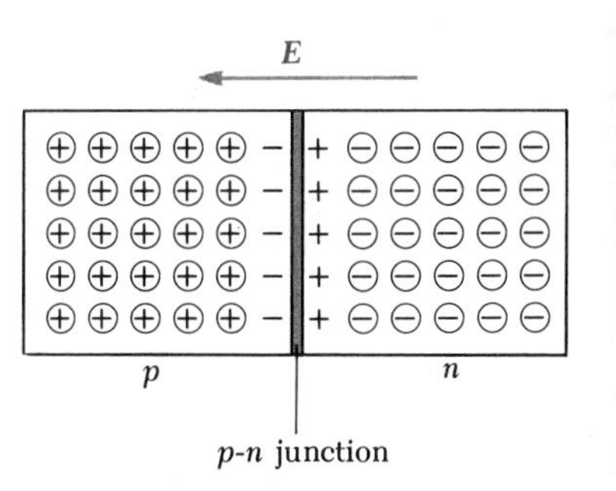

FIGURE 31.24 A semiconductor fabricated such that the left side is *p*-type and the right side is *n*-type. At the junction, an electric field is set up directed from right to left.

The *p-n* Junction Diode

The *p-n* device just described is called a ***p-n* junction diode.** It is a good conductor when a voltage applied to it attempts to drive current in one direc-

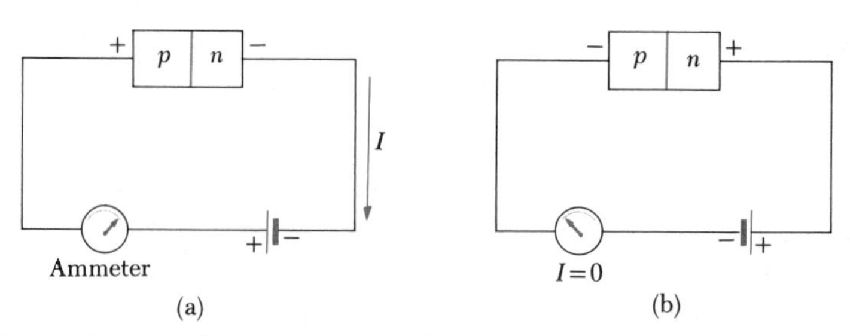

FIGURE 31.25 (a) A forward-biased diode carries a large current. (b) In a reverse-biased junction, the current is approximately zero.

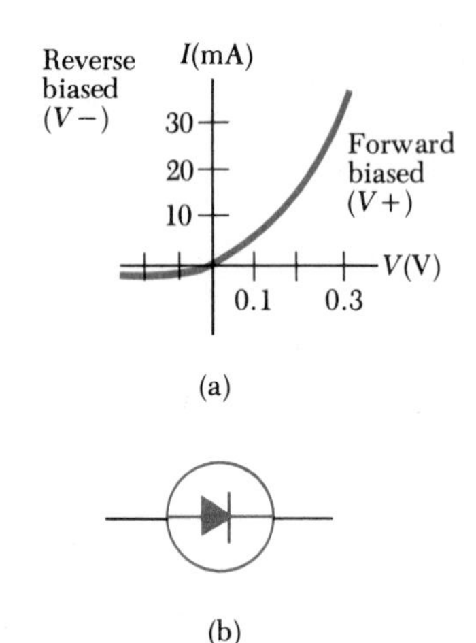

FIGURE 31.26 (a) Current versus applied voltage for a *p*-*n* junction diode. (b) Circuit symbol for a diode.

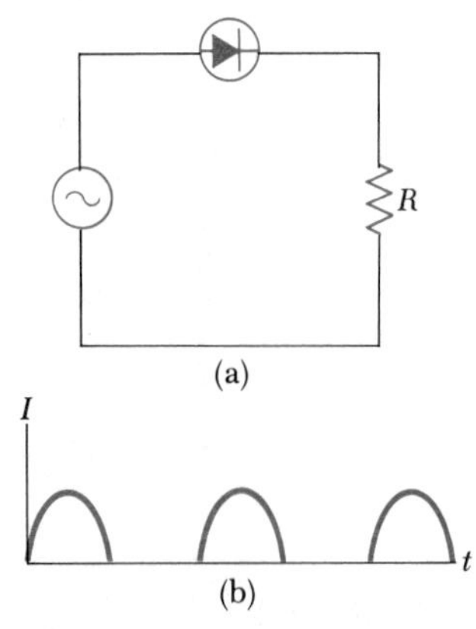

FIGURE 31.27 (a) A diode in series with a resistor allows current to flow during one half of the cycle. (b) The current versus time for the circuit in (a).

tion but a poor conductor if the voltage is reversed and attempts to drive current in the opposite direction.

Suppose a battery is connected across the junction, with the *p* side positive and the *n* side negative, as in Figure 31.25a. This situation is referred to as a **forward-biased junction.** This polarity provides an extra push for the electrons and holes in the form of an accelerating field, enabling them to overcome the opposing electric field at the junction. Thus, free electrons move readily from the *n* side to the *p* side of the junction, while holes move readily from the *p* side to the *n* side. This flow of charge across the junction constitutes a current.

Now consider what happens when the polarity of the battery is reversed, as in Figure 31.25b, where the *p* side is negative and the *n* side is positive. In this case, the junction is said to be **reverse-biased.** The electric field at the junction is now enhanced, and very few charges are able to cross the junction. This results in a very small current under reverse-bias conditions.

The current-voltage curve of a typical diode is shown in Figure 31.26a. Note that a large current passes through the diode under forward-bias conditions but there is virtually no current under reverse-bias conditions.

The circuit symbol for a diode is shown in Figure 31.26b. The arrow indicates the direction of the current through the diode.

The Diode as a Rectifier

The current from electric outlets in this country is an alternating current. However, the tubes and transistors in such electronic devices as your television set and stereo system require a direct current or voltage in order to operate properly. Therefore, it is necessary to change the alternating current from the outlet to a direct current. This process is called **rectification,** and the diode provides a simple and convenient means for achieving this goal.

We can understand how a diode rectifies a current by considering Figure 31.27a, which shows a diode placed in series with a resistor and an ac source. Because current can pass through the diode only in one direction, the alternating current in the resistor is reduced to the form shown in Figure 31.27b. The diode is said to act as a **half-wave rectifier** since there is current in the circuit during only half of each ac cycle.

When a diode is used as a half-wave rectifier, the current in the resistor is a pulsating direct current, as shown in Figure 31.27b. A closer approximation to a true direct current can be achieved by using two diodes and a transformer connected as shown in Figure 31.28a. Consider an instant at which the voltage at point A in the secondary of the transformer is higher than the voltage at B, which is higher than the voltage at C. At this instant, the current path in the circuit is from A through diode D_1, downward through the resistor and back to point B. Half a cycle later, the alternating voltage at the output of the secondary of the transformer is reversed so that point C is at a higher voltage than point B, which is at a higher voltage than point A. In this case, the current path in the circuit is from C through diode D_2, again downward through the resistor, and back to point B. Thus there is current in the circuit during the complete cycle of the ac voltage, and the direction of the current is always downward through the resistor. The current in the resistor varies with time as shown in Figure 31.28b. In this situation, the pair of diodes is said to be acting as a **full-wave rectifier.** Note that the current in the resistor is still pulsating but is closer to being a true direct current than with the half-wave rectifier.

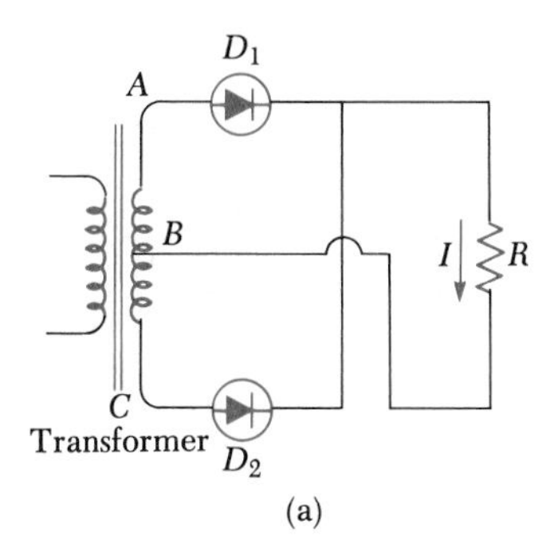

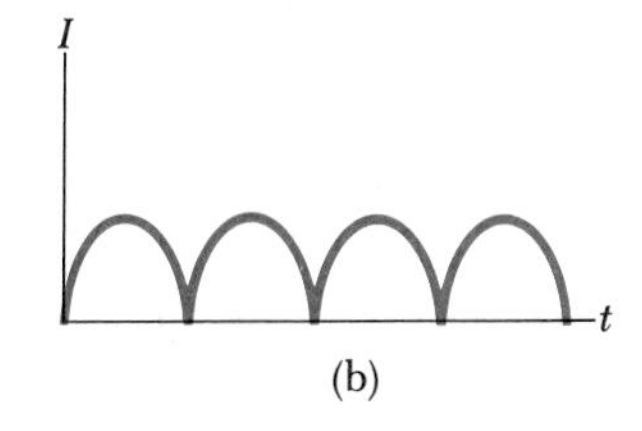

FIGURE 31.28 (a) A full-wave rectifier circuit. (b) The current versus time through the resistor in (a).

Filters

As stated earlier, the transistors and tubes in an electronic device such as a television or stereo need to be supplied with a smooth direct current or voltage. The output of a full-wave rectifier is not smooth enough. Various kinds of filters using inductors and capacitors have been developed to smooth out the pulsating output of a full-wave rectifier. In this section we shall briefly describe how a filter that uses only a single capacitor works.

Figure 31.29a shows a capacitor connected in parallel with the resistor in a full-wave rectifier circuit. In the absence of the capacitor, the current in the

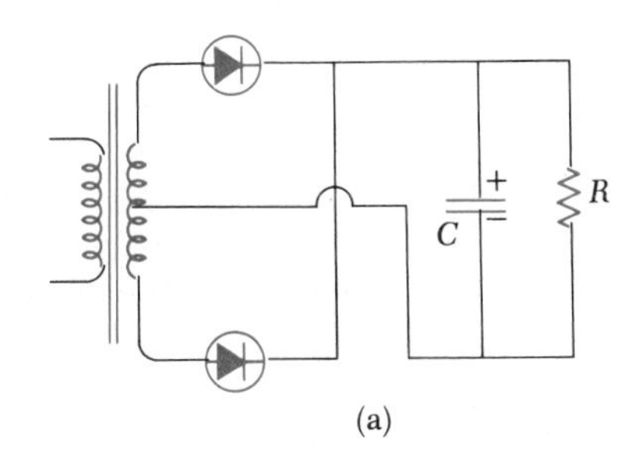

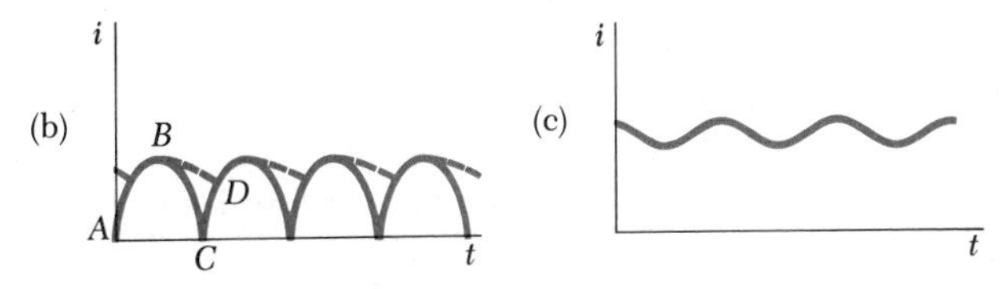

FIGURE 31.29 (a) A full-wave rectifier circuit with a filter capacitor. (b) The solid line represents the current versus time with no filter capacitor, and the dashed line represents the current when the circuit includes the capacitor. (c) A graph of current versus time for the circuit shown in (a), that is, another rendering of the dashed line in (b).

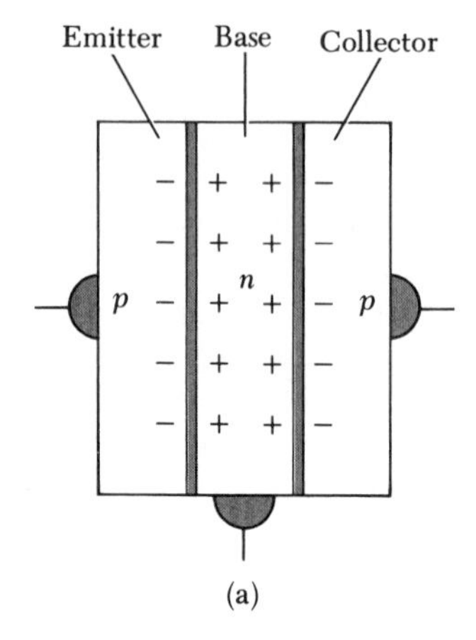

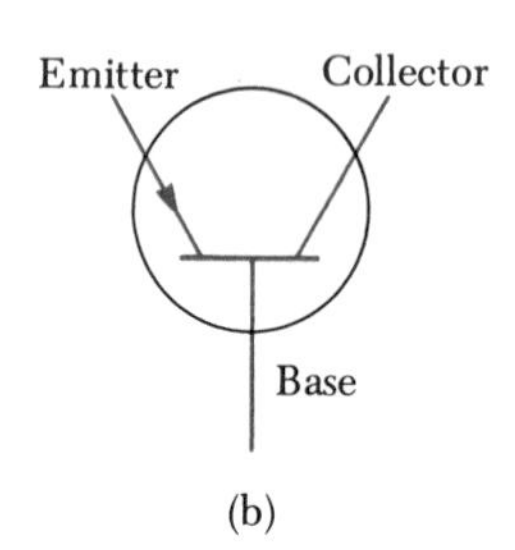

FIGURE 31.30 (a) The *pnp* transistor consists of an n region (base) sandwiched between two p regions (the emitter and collector). (b) The circuit symbol for the *pnp* transistor.

resistor would be as shown by the solid curve in Figure 31.29b. When the circuit includes the capacitor, however, we have the following. As the current in the resistor increases with time, from A to B in Figure 31.29b, the capacitor charges such that its upper plate is positive. As the current through the resistor attempts to drop off between points B and C, the capacitor begins to discharge through the resistor. As a result, the current in the resistor never reaches zero. Instead, it follows the dashed path from B to D in Figure 31.29b. This results in a current in the resistor as shown in Figure 31.29c. Although a ripple still remains in the current, it is smooth enough for most applications.

The Junction Transistor

The discovery of the transistor by John Bardeen, Walter Brattain, and William Shockley in 1948 totally revolutionized the world of electronics. For this work, these three men shared a Nobel prize in 1956. By 1960, the transistor had replaced the vacuum tube in many electronic applications. The advent of the transistor created a multibillion dollar industry that produces such popular devices as pocket radios, handheld calculators, computers, television receivers, and electronic games.

The junction transistor consists of a semiconducting material with a very narrow n region sandwiched between two p regions. This configuration is the ***pnp* transistor.** Another configuration is the ***npn* transistor,** which consists of a p region sandwiched between two n regions. Because the operation of the two transistors is essentially the same, we shall describe only the *pnp* transistor.

The structure of the *pnp* transistor, together with its circuit symbol, is shown in Figure 31.30. The outer regions of the transistor are called the **emitter** and **collector,** and the narrow central region is called the **base.** Note that the configuration contains two junctions. One junction is the interface between the emitter and the base, and the other is between the base and the collector.

Suppose a voltage is applied to the transistor such that the emitter is at a higher potential than the collector. (This is accomplished with battery V_{ec} in Figure 31.31.) If we think of the transistor as two diodes back to back, we see that the emitter-base junction is forward-biased and the base-collector junction is reverse-biased.

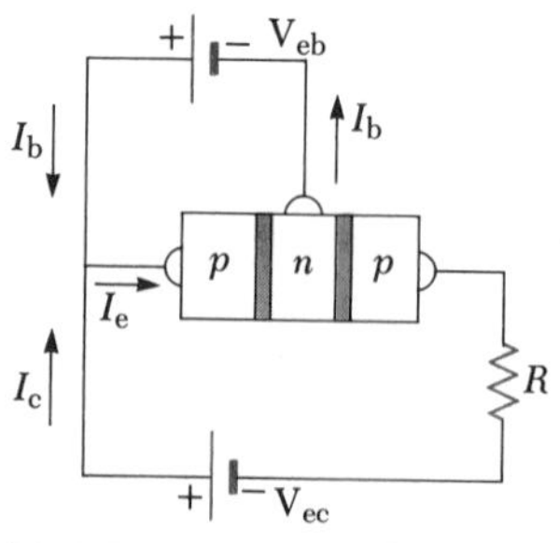

FIGURE 31.31 A bias voltage applied to the base as shown produces a small base current I_b, which is used to control the collector current I_c.

Because the p-type emitter is heavily doped relative to the base, nearly all of the current consists of holes moving across the emitter-base junction. Most of these holes do not recombine in the base because it is very narrow. The holes are finally accelerated across the reverse-biased base-collector junction, producing the current I_c in Figure 31.31.

Although only a small percentage of holes recombine in the base, those that do limit the emitter current to a small value because positive charge carriers accumulate in the base and prevent holes from flowing into this region. In order to prevent this limitation of current, some of the positive charge on the base must be drawn off; this is accomplished by connecting the base to a second battery, V_{eb} in Figure 31.31. Those positive charges that are not swept across the collector-base junction leave the base through this added pathway. This base current, I_b, is very small, but a small change in it can significantly change the collector current, I_c. For example, if the base current is increased, the holes find it easier to cross the emitter-base junction and there is a large increase in the collector current. Since the base current is very small, the collector current is approximately equal to the emitter current.

One common application of the transistor is to amplify a small, time-varying signal. The small voltage to be amplified is placed in series with the battery V_{eb} in Figure 31.31. The time-varying input signal produces a small variation in the base voltage. This results in a large change in collector current and hence a large change in voltage across the resistor.

SUMMARY

Two or more atoms combine to form molecules because of a net attractive force between the constituent atoms. The mechanisms responsible for the bonding in the molecule can be classified as follows:

1. **Ionic bond** Certain molecules have ionic bonds because of the Coulomb attraction between oppositely charged ions. Sodium chloride (NaCl) is one example of an ionically bonded molecule.
2. **Covalent bond** The covalent bond in a molecule is formed by the sharing of valence electrons of its constituent atoms. For example, the two electrons of the H_2 molecule are equally shared between the nuclei.
3. **Hydrogen bond** This type of bonding corresponds to the attraction of two negative ions by an intermediate hydrogen atom (a proton).
4. **Van der Waals Bond** This is a weak electrostatic bond between atoms which do not form ionic or covalent bonds. It is responsible for the condensation of inert gas atoms and nonpolar molecules into the liquid phase.

Bonding mechanisms in solids can be classified in a manner similar to the schemes already described for molecules. For example, the Na^+ and Cl^- ions in NaCl form ionic bonds, while the carbon atoms in diamond form covalent bonds. Another type of bonding mechanism is the metallic bond, which is generally weaker than ionic or covalent bonds. The mechanism for the metallic bond in a metal is a net attractive force between the positive ion cores and the mobile valence electrons.

In a crystalline solid, the energy levels of the system form a set of bands. Electrons are allowed to occupy only certain bands. There are energy regions between the allowed bands corresponding to states which the electrons cannot occupy.

One can best understand the properties of metals, insulators, and semiconductors in terms of the band theory of solids. The conduction band of a metal is partially filled. Therefore, there are many electrons free to move throughout the metal and contribute to the conduction current. In an insulator at low temperatures, the valence band is completely filled with electrons, while the conduction band is empty. The region between the valence band and the conduction band is called the energy gap of the material. The energy gap for an insulator is of the order of 10 eV. Because this gap is large compared to the thermal energy at ordinary temperatures, very few electrons are thermally excited into the conduction band, which explains the small electric conductivity of an insulator.

A semiconductor is a material with a small energy gap, of the order of 1 eV, and a valence band which is filled at low temperatures. Because of their small energy gap, a significant number of electrons can be thermally excited from the valence band into the conduction band as the temperature increases. The band structures and electrical properties of a semiconductor can be modified by adding donor atoms with five valence electrons (such as arsenic), or by adding acceptor atoms with three valence electrons (such as indium). A semiconductor doped with donor impurity atoms is called an n-type semiconductor, while one doped with acceptor impurity atoms is called p-type. The energy levels of these impurity atoms falls within the energy gap of the material.

ADDITIONAL READINGS

P. W. Atkins, *Molecules*, Scientific American Library, New York, N.Y., W. H. Freeman, 1987.

M. L. Cohen, V. Heine, and J. C. Phillips, "Quantum Mechanics of Materials," *Sci. American*, June 1982, p. 82.

P. C. Hinkle and R. E. McCarty, "How Cells Make ATP," *Sci American*, March 1978, p. 104.

"Microelectronics," *Sci. American*, September 1977.

QUESTIONS

1. In general, must a photon have more energy to excite a rotational energy state than it does a vibrational energy state or vice versa? Defend your answer.
2. When one examines the emission spectrum from a collection of diatomic molecules, many more lines are seen than is typical when the spectrum is from a collection of atoms. Why should this occur?
3. The H_2 molecule is stable, yet the H_3 molecule is not. Explain why, using the Pauli exclusion principle.
4. Use the band theory of solids to discuss the difference between a material that is a good insulator and one that is a good conductor of electricity.
5. Would you expect the energy gap in a solid to increase or decrease with increasing temperature? Defend your answer.
6. Would you expect the energy gap in a solid to increase or decrease with increasing pressure? Defend your answer.
7. When a photon is absorbed by a semiconductor, an electron-hole pair is said to be created. Give a physical interpretation of this statement based on the energy band model of a solid.
8. An often used, and often incorrect, rule of thumb says that a metal is opaque, a semiconductor is translucent, and an insulator is transparent. Use the energy band model of a solid to explain how this rule of thumb might have arisen.
9. Pentavalent atoms such as arsenic are donor atoms in a semiconductor such as silicon, while trivalent atoms such as indium are acceptors. Inspect the periodic table of the elements or the endpapers and determine what other elements would be considered as either donors or acceptors.
10. What is the resistance of a diode when connected in reverse bias?
11. It is often said that a transistor consists of two back-to-back diodes, with the base-emitter junction acting as one diode and the base-collector junction as the second. Which of these junctions is reverse biased, and which is forward biased?

PROBLEMS

Section 31.1 Molecular Bonds

1. Plot a potential energy diagram for an object moving at the end of a spring of spring constant $k = 150$ N/m. Recall that the potential energy for an object moving with simple harmonic motion is given by $PE = \frac{1}{2}kx^2$.

2. Plot a graph of gravitational and potential energy between any pair of particles of masses m_1 and m_2. Recall that the expression for this is given by $PE = -Gm_1m_2/r$.

3. Consider the potential energy diagram shown in Figure 31.32. Determine whether the force is positive, negative, or zero at the various points indicated.

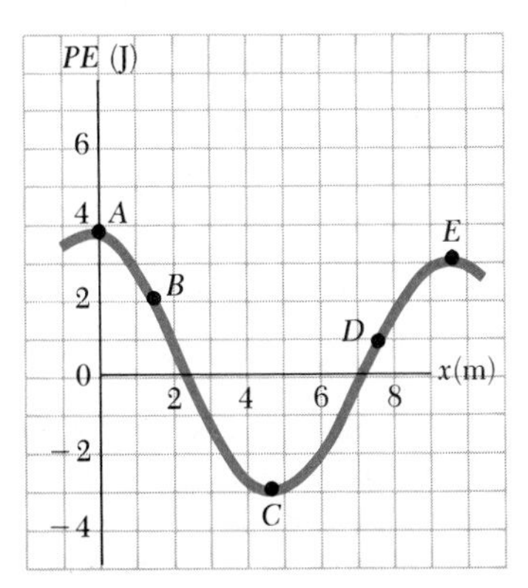

FIGURE 31.32 Problem 3

Section 31.2 Molecular Bonds: Ionic, Covalent, Hydrogen, Van der Waals, and Metallic

4. The ionization energy of K is 4.34 eV and the electron affinity of Cl is 3.7 eV. Find the energy required to form KCl.

5. The ionization energy of K is 4.34 eV and the electron affinity of F is 4.1 eV. Find the energy required to form KF.

6. Look up the heat of vaporization of water and find the energy per bond in liquid water. (We consider the water molecule to be loosely bound by six bonds to its nearest neighbor.)

7. How much methane could be broken into its components with an input energy of 3.63×10^8 J?

8. Ethane is an organic compound which is formed by two carbon atoms bound together, and with three hydrogen atoms bound to each carbon atom. The C—H bond energy is about 4.28 eV and the C—C energy bond is 3.60 eV. Find the energy input required to break existing bonds when 1 kg of ethane is burned.

Section 31.3 The Energy and Spectra of Molecules

9. Find the location of the center of mass of the HF molecule with respect to the fluorine atom if the equilibrium separation is 0.11 nm.

10. Find the location of the center of mass of the NO molecule with respect to the oxygen atom if the equilibrium separation is 0.114 nm.

11. Use the result of Problem 9 to find the moment of inertia of the HF molecule about its center of mass.

12. Use the result of Problem 10 to find the moment of inertia of the NO molecule about its center of mass.

13. Find the frequency of the photon that would cause the N_2 molecule to undergo a transition from the $J' = 0$ state to the $J' = 1$ state. The equilibrium separation is about 0.11 nm.

14. Find the energy that would have to be carried away by a photon as the HBr molecule undergoes a transition from the $J' = 1$ state to the $J' = 0$ state. The equilibrium separation is about 0.11 nm.

Section 31.5 Vibrational Energy of Molecules

15. The frequency absorbed by the HF molecule as it makes a transition from the $v = 0$ to the $v = 1$ state is 8.72×10^{13} Hz. Use this information to sketch a vibrational energy level diagram similar to that shown in Figure 31.15b.

16. Find the zero-point energy of HBr if it absorbs radiation of frequency 8.66×10^{13} Hz as it moves from the $v = 0$ state to the $v = 1$ state.

Section 31.5 Semiconductors

17. The energy gap for Si at 300 K is 1.14 eV. (a) Find the lowest frequency photon that will promote an electron from the valence band to the conduction band of silicon. (b) What is the wavelength of this photon?

18. The longest wavelength radiation that is absorbed by a certain semiconductor is 1.85 μm. What is the energy gap for this semiconductor?

19. Aluminum is added as an impurity to silicon. Is the resulting semiconductor n-type or p-type? Explain.

20. Arsenic is added as an impurity to silicon. Is the resulting semiconductor n-type or p-type? Explain.

21. Electrons from donor atoms of an n-type semiconductor can reach the conduction band by absorbing an energy of approximately 0.01 eV. (a) At what temperature does a particle have a thermal energy, kT, equivalent to 0.01 eV? (b) At what temperature would a particle have thermal energy equivalent to the energy gap in pure diamond, which is 7 eV?

ADDITIONAL PROBLEMS

22. The equilibrium separation of the atoms in a certain molecule is 0.27 nm, the depth of the potential well is 4.3 eV, and the activation energy is 3.8 eV. Sketch a graph which shows these features at the appropriate locations.

23. Propane is an organic compound which is formed as follows. Three carbon atoms are bonded together; two of the carbon atoms have three hydrogen atoms bonded to each of them and the third carbon atom has two hydrogen atoms bonded to it. The C—H bond energy is about 4.23 eV and the C—C bond energy is 3.60 eV. Find the energy required to break all the bonds in 1 kg of propane as it is burned.

24. Find the frequency of the photon that would make the HCl molecule make a transition from the $J' = 0$ to the $J' = 1$ state. The equilibrium separation is about 0.12 nm.

25. The energy gap for germanium is 0.7 eV. (a) Find the lowest frequency photon that will promote an electron from the valence band to the conduction band of germanium. (b) What is the wavelength of this photon?

26. When the following impurities are added to pure germanium, does the material become n-type or p-type: (a) phosphorus, (b) gallium, (c) indium?

Essay

Photovoltaic Conversion

John D. Meakin *University of Delaware*

The photovoltaic effect occurs when light is absorbed by a semiconductor. The energy of the photons is transferred to electrons in the valence band of the semiconductor promoting them to the conduction band, thus forming electron-hole pairs. Only photons with energies exceeding the band gap energy of the semiconductor can be effective in this process. If the semiconductor has a low band gap, a large fraction of the incident photons will be able to create electron-hole pairs but the voltage generated across the solar cell will be low. With a high band gap semiconductors, the voltage generated will be high but the current low. Analysis has shown that to convert sunlight, in which most of the photons have energies between 1 eV and 3 eV, the optimum band gap for maximum power conversion is about 1.5 eV.

In an isolated semiconductor the excited electron would recombine with a hole in the valence band emitting its excess energy as photons (photoemission) or phonons (heat) and no useful generation of electric energy would take place.

To extract the energy of the excited electrons as useable electricity requires the existence of a charge separating junction such as a p-n homojunction or diode, to prevent electron-hole recombination. In the traditional crystalline silicon solar cell the charge separating region is formed by the diffusion of specific impurities, or dopants, into a wafer of silicon, creating regions of opposite conductivity. Diffusion of carriers across the transition between the n and p regions occurs until an equilibrium is established in which the electric field created prevents further diffusion of charges. This internal electric field constitutes the charge separating barrier which is key to the operation of the solar cell. Electrons created by light in the p-type material migrate to the junction and are then swept into the n-type region. In an efficient cell this collection occurs before the electrons recombine with the stable population of holes in the p region. The energy absorbed from the light is thus converted into electrical energy which can be fed into an external circuit (the analogous situation exists for holes created on the n-type side of the junction).

Figure 1a illustrates a diode under illumination and shows the production of electron-hole pairs which are then separated by the built in electric field. Note that the I-V curve in Figure 1b is that of a diode but displaced along the current axis by the "light-generated current." If the outside circuit has zero resistance a maximum current will be produced as shown on the current-voltage plot. As the load resistance rises a voltage will be generated across the cell and the current will eventually fall to zero at the open circuit voltage. In this situation, electrons accumulate on the n-side of the junction (holes on the p-side) biasing the junction in the opposite sense to the built-in field. The maximum voltage the cell can de-

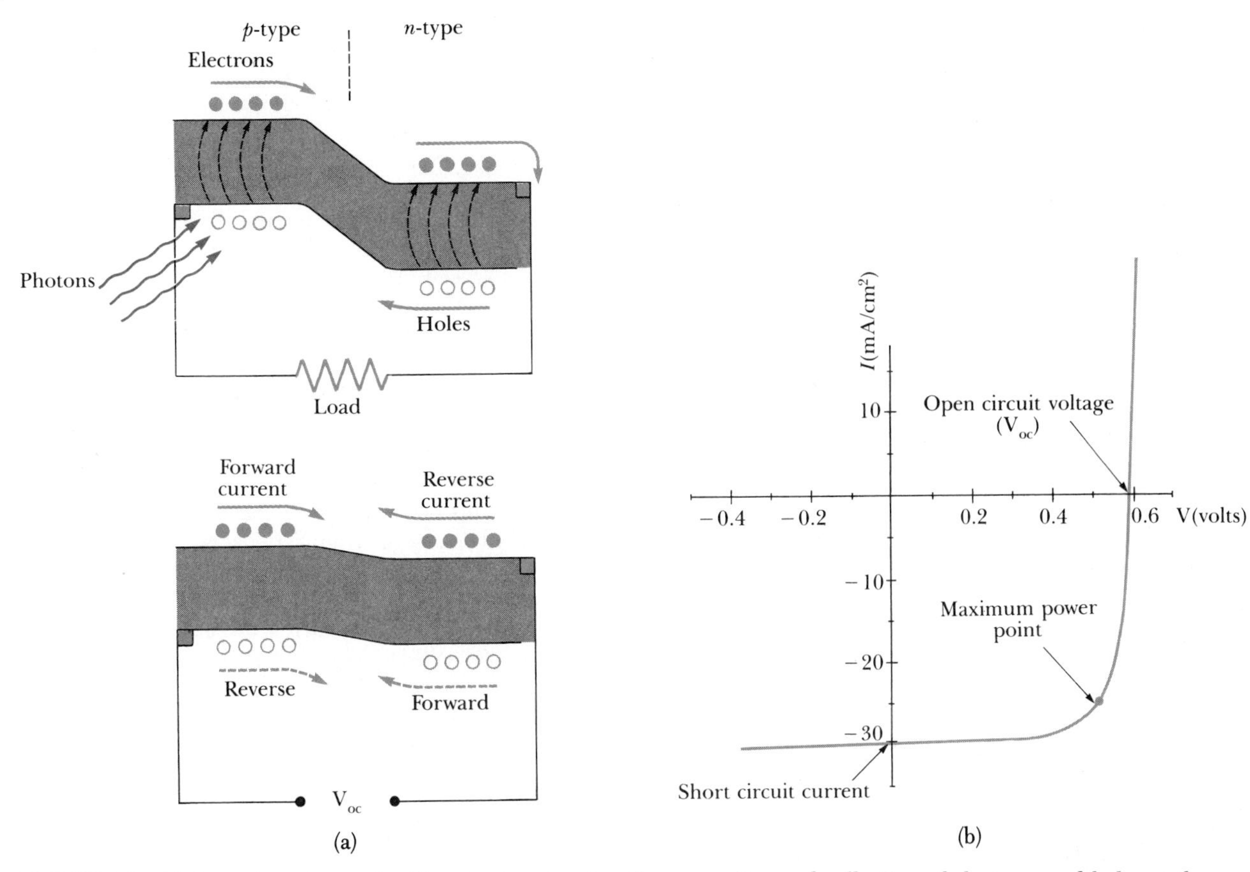

FIGURE 1 (a) A *p-n* junction under illumination showing the generation and collection of electrons and holes at the short circuit current point. As the external load increases charges accumulate in the cell reducing the height of the charge separating barrier until at open circuit there is no net current flow. (b) A typical current-voltage curve for an Si solar cell under 1 kW/m^2 insolation.

velop corresponds to the forward light generated current exactly matching the reverse current. The maximum power generation occurs where the current-voltage product is largest. Under ideal conditions a single junction solar cell should convert about 22% of the incident solar energy into electricity. This figure ignores all the losses that inevitably occur in an actual cell such as reflection from the front surface and various other electrical and optical losses.

In 1954 the first practical solar cells were made from single crystal wafers of semiconductor grade silicon with the structure shown in Figure 2. These cells converted about 6% of the total incident sunlight into electrical power; that is, they had an efficiency of 6%. Research Si cells have now reached about 20% efficiency, and commercial cells are at 15% yielding multicell modules of about 12% efficiency. Why then are solar cells not being used more extensively to generate electricity? To answer this question we must look more closely at the economics of electricity generation from solar cells.

The most significant costs are the initial capital outlays; operation and maintenance costs are small and there are no fuel costs. A simple analysis reveals the scale of investment that could be economical. A module rated at

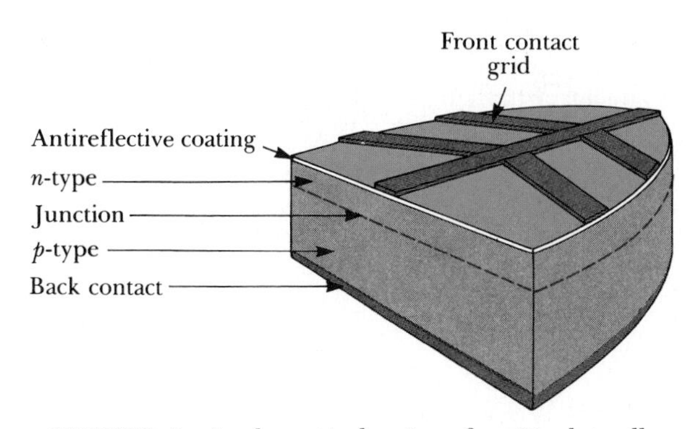

FIGURE 2 A schematic drawing of an Si solar cell.

1 kilowatt will generate on average about 5 kWh/day or about 5 × 10^4 kWh of electricity worth about $1000 over its lifetime of 20 years. Half of the initial investment will be for costs other than the modules, which must therefore be available for a few hundred dollars per kilowatt. Although the first cells produced for space use cost about $1 million/kW, extensive technical and manufacturing progress, coupled with much less stringent specifications for terrestrial use, has brought the present price down to about $5000/kW in large quantities. Is it likely that a further order of magnitude reduction can be achieved? To answer this question we must first return to the physics of solar cell operation.

The rate at which sunlight is absorbed by the solar cell is

$$F = e^{-\alpha t}$$

where F is the fraction of radiation absorbed by a thickness t of semiconductor with an absorption coefficient α. Crystalline Si has a relatively low absorption coefficient, between 10^2 and 10^4 cm^{-1}, which means that a thickness of about 100 microns is necessary to absorb most of the sunlight. The carriers generated by the photons must then be able to travel about that far before recombination occurs in order to reach the charge separating junction.

Efficient collection in crystalline silicon therefore requires very pure and perfect material. Bringing the price of Si cells down by a further factor of ten cannot be assured, although a number of ongoing developments may yield substantial reductions. These include methods to produce thin sheets of large area directly from a melt to avoid the wafering step used with signal crystals. Large grained polycrystalline material, which is less expensive than single crystals, is also in use.

However there are many semiconducting compounds in addition to the semiconducting elements, Si, Ge and grey Sn. The essential properties for a material to be a candidate for a solar cell are an appropriate band gap, say between 1.0 and 1.7 eV, and a very high absorption coefficient. The latter makes it possible to use very thin layers which in turn relaxes the limitations on purity and perfection as short collection distances (short carrier lifetimes) now apply. Any compound is a potential semiconductor if the ratio of valence electrons to atoms is 4 as is the case for the group IV elements.

Among the classes of semiconducting compounds are the II–VI (i.e., a compound AB where A is from group II and B from group VI), III–V and I–III–VI_2 families. Solar cells are being developed based on members of each of these groups, with the expectation that an improved combination of higher efficiency and lower cost will result. Representative of the III–V compounds are GaAs and InP, which in single crystal form have yielded solar cells with better than 20% efficiency. As a general rule the III–V compounds have excellent carrier properties, and indeed there is a major effort to develop them for high speed integrated circuits, but their properties are seriously degraded in thin film form, and single crystals may prove essential to maintain performance. If this should continue to be the case, a concentrating system may be essential in which an optical assembly is used to focus the energy from a large aperture onto a much smaller solar cell.

The II–VI compounds are represented by CdTe, which has an ideal band gap of about 1.5 eV and which has yielded solar cells with over 10% efficiency. A number of techniques for forming large areas of thin film are being developed including electroplating, various types of vapor deposition, and a spray pyrolysis process.

The I–III–VI_2 material receiving most attention is $CuInSe_2$. This has a rather low band gap of 1.0 eV but (as we shall discuss shortly) that may be an advantage for two junction or tandem cells. A number of research groups have demonstrated efficiencies of well over 10%, and all reports have confirmed a very high degree of stability for these cells. This is particularly important for thin film cells which have often shown unacceptably short useable lifetimes.

The reason for interest in all of these compounds is their high absorbtivity, which allows the use of very thin layers, as little as 10^{-4} cm. The carriers generated by the light need then travel equally short distances; relatively impure and imperfect materials can be used successfully.

In parallel with the development of polycrystalline materials there has been an explosive growth in the investigation and application of amorphous materials based on an alloy of Si and H, conventionally designated a-Si:H. Thin films of this material are deposited by creating a plasma in a gas containing the two elements, most frequently SiH_4. The resulting solid contains tetrahedrally bonded Si but in a disordered, noncrystalline, array. The hydrogen appears to heal any unsatisfied Si—Si bonds, eliminating energy states that would otherwise essentially eliminate the forbidden energy gap. The hydrogenated a-Si can therefore be doped, in contrast to hydrogen-free amorphous silicon.

Progress in the science and technology of amorphous thin films has been remarkable; in about a decade, a-Si:H solar cells have gone from a laboratory curiosity to a familiar component in solar calculators. Panels with areas of 1 ft^2 and larger are now entering the market for battery charging and other uses.

Scientifically and technically, the field of photovoltaic conversion remains exciting and challenging. New materials are being developed and more efficient configurations reduced to practice. Rather than having one cell to harvest the entire spectrum, stacked or tandem structures are appearing. In these systems the top cell removes the high energy photons allowing the longer wavelengths (lower energy photons) to be efficiently harvested in a second or even third cell. This is the photovoltaic equivalent of a multistage turbine. a-Si:H makes an ideal top cell, as its band gap of about 1.6 eV efficiently uses the short wavelength light while transmitting the longer wavelengths into a bottom cell. Polycrystalline $CuInSe_2$ is presently the best available material for the bottom cell because its band gap, 1.0 eV, is close to the ideal value for a two-junction device.

The present world market for solar cells, about 20 megawatts, is modest when compared to the installed electrical generating capacity, more than 600 GW in the United States alone. However production doubled each year for some time, and 10% market penetration by the year 2000 is quite possible. The economic incentive to develop alternative sources of electricity for use in the developed nations is controlled by the cost of traditional fossil fuels, typified by oil at about $20/barrel. However solar cells are ideally suited for remote locations and third world areas where grid systems do not exist, and it is already economic to use solar cells where the only alternatives are diesel generators or expendable batteries. For example, the U.S. Coast Guard is converting all remote buoys to solar cell operation.

In spite of the uncertainty in the future level of solar electricity generation, there can be no doubt that solar cells will make a significant contribution and will continue to attract both scientific and commercial interest.

SUGGESTED READING

Earl Cook, *Man, Energy, Society,* New York, W. H. Freeman, 1976.

Paul D. Maycock and Edward N. Stirewalt, *Photovoltaics,* by Brick House, 1981.

Martin A. Green, *Solar Cells,* Englewood Cliffs, N.J., Prentice Hall, 1982.

ESSAY PROBLEMS

1. Typically, 1 m^2 of land receives 2000 kWh of sunshine each year. Compare the total value of the energy generated by solar cells to the value of a typical crop such as corn. Assume that the cells have a conversion efficiency of 10% and that the electricity is worth about $0.05/kWh. A good crop yield is 150 bushels per acre, worth about $2.00/bushel.

You should find that the electricity is worth about two orders of magnitude more than the corn. However, you must also address the initial cost of the solar-cell system, and compare this with the annual cost of cultivating corn, which is about $180/acre. It is hoped that the total cost of a solar-cell system can be brought down to about $100/$m^2$; what would be the initial cost of a 1-acre system?

2. (a) Amorphous silicon cells generally contain a-Si:H films about 1 μm thick. How much Si would be needed to make enough solar cells to generate electricity at the rate of 1 kW under a noontime illumination of 1 kW/m^2? Assume that the conversion efficiency is 10% and that a-Si:H has the same density as crystalline Si, 2.3 g/cm^3. (b) A peaking generator is normally rated at about 25 MW. How much area would be needed to set up an equivalent photovoltaic plant operating at noon? You should find that over 50 acres are needed for the solar-cell system, much more than needed for a gas turbine. What factors would favor the solar array over the gas turbine? You should be able to think of environmental as well as resource-utilization considerations.

3. (a) A car normally operates at a power rating of about 20 horsepower. What area of solar cells with an efficiency of 10% would be needed to match this output under illumination of 1 kW/m^2? (b) Does simple conversion from gasoline to solar power seem feasible for a conventional car? (c) What changes would have to be made, and what technological advances would have to be achieved, in order to make a solar car possible?

• **4.** The I–V behavior of a solar cell can be represented by the expression

$$I = I_0 \left(e^{qV/kT} - 1\right) - I_{sc}$$

where I and V are the current and voltage, respectively, q is the electronic charge, k is Boltzmann's constant, and I_0 is the reverse saturation current of the diode. (a) Develop an expression for the open-circuit voltage of the cell, that is, the voltage generated when $I = 0$. (I_{sc} is always very much greater than I_0.) (b) For Si cells, I_0 is about 10^{-12} A/cm^2 and I_{sc} is about 40 mA/cm^2. Show that the open-circuit voltage at room temperature is about 0.6 V. The quantity q/kT at 300 K has the value 40 V^{-1}. (c) Draw the I–V curve for a 1-cm^2 cell, and compute the maximum power output for the preceding values of I_0 and I_{sc}. You will need to compute the product IV at various points on the I–V curve and find the maximum product value. At what efficiency do you conclude the cell is operating? It should be about 10%.

Nuclear Physics

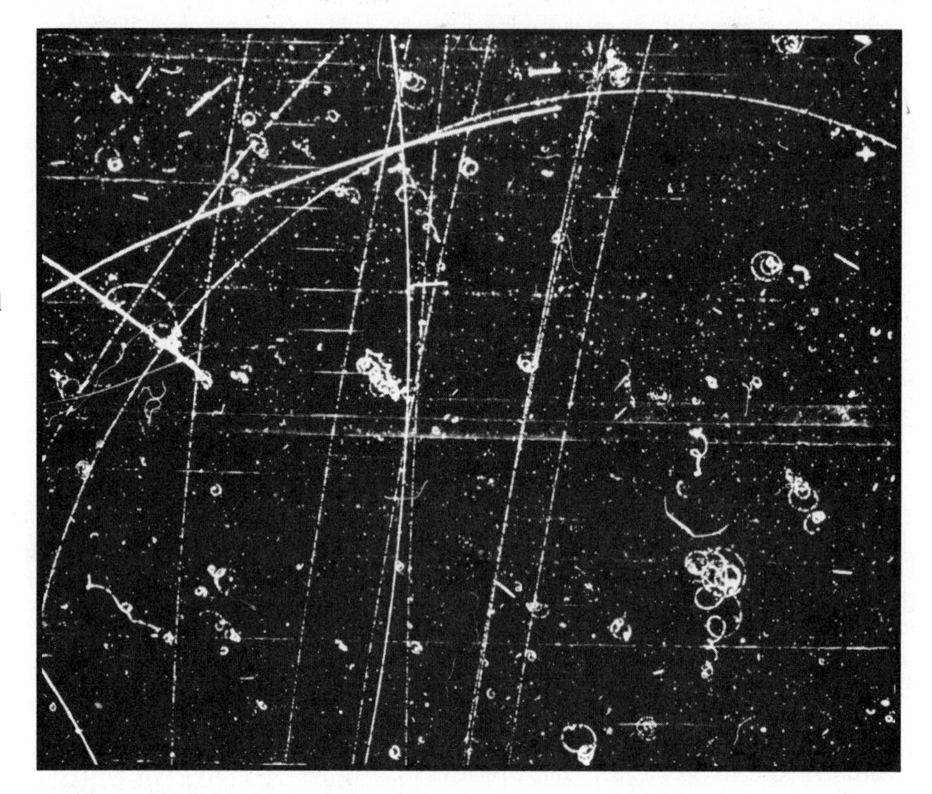

Bubble-chamber photographs such as this illustrate the interconversion of energy and matter. The curved tracks are due to charged particles moving in the presence of a magnetic field. (From G. Holton, F.J. Rutherford, and F.G. Watson, *Project Physics,* New York, Holt, Rinehart and Winston, 1981)

In 1896, the year that marks the birth of nuclear physics, Henri Becquerel (1852–1908) discovered radioactivity in uranium compounds. A great deal of activity followed this discovery as researchers attempted to understand and characterize the radiation which we now know to be emitted by radioactive nuclei. Pioneering work by Rutherford showed that the emitted radiation was of three types, which he called alpha, beta, and gamma rays. These are classified according to the nature of the electric charge they possess and according to their ability to penetrate matter. Latter experiments showed that alpha rays are helium nuclei, beta rays are electrons, and gamma rays are high-energy photons.

In 1911, Rutherford and his students Geiger and Marsden performed a number of important scattering experiments involving alpha particles. These experiments established that the nucleus of an atom can be regarded as almost a point mass and point charge and that most of the atomic mass is contained in the nucleus.

In this chapter we shall discuss the properties and structure of the atomic nucleus. We start by describing the basic properties of nuclei and follow with a discussion of the phenomenon of radioactivity. Finally, we discuss nuclear reactions and the various processes by which nuclei decay.

32.1 SOME PROPERTIES OF NUCLEI

All nuclei are composed of two types of particles: protons and neutrons. The only exception to this is the ordinary hydrogen nucleus, which is a single proton. In this section, we shall describe some of the properties of nuclei, such as their charge, mass, and radius. In doing so, we shall make use of the following quantities:

1. The **atomic number,** Z, which equals the number of protons in the nucleus
2. The **neutron number,** N, which equals the number of neutrons in the nucleus
3. The **mass number,** A, which equals the number of nucleons in the nucleus. (*Nucleon* is a generic term used to refer to either a proton or a neutron).

It will be convenient for us to have a symbolic way of representing nuclei that will show how many protons and neutrons are present. The symbol to be used is ${}^{A}_{Z}\mathrm{X}$, where X represents the chemical symbol for the element. For example, ${}^{56}_{26}\mathrm{Fe}$ has a mass number of 56 and an atomic number of 26; therefore it contains 26 protons and 30 neutrons. When no confusion is likely to arise, we shall omit the subscript Z because the chemical symbol can always be used to determine Z.

The nuclei of all atoms of a particular element contains the same number of protons but different numbers of neutrons. Nuclei which are related to this way are called **isotopes.**

The isotopes of an element have the same Z value but different N and A values.

The natural abundances of isotopes can differ substantially. For example, ${}^{11}_{6}\mathrm{C}$, ${}^{12}_{6}\mathrm{C}$, ${}^{13}_{6}\mathrm{C}$, and ${}^{14}_{6}\mathrm{C}$ are four isotopes of carbon. The natural abundance of the ${}^{12}_{6}\mathrm{C}$ isotope is about 98.9%, whereas that of the ${}^{13}_{6}\mathrm{C}$ isotope is only about 1.1%. Some isotopes do not occur naturally but can be produced in the laboratory through nuclear reactions. Even the simplest element, hydrogen, has isotopes. They are ${}^{1}_{1}\mathrm{H}$, the ordinary hydrogen nucleus; ${}^{2}_{1}\mathrm{H}$, deuterium; and ${}^{3}_{1}\mathrm{H}$, tritium.

Charge and Mass

The proton carries a single positive charge, equal in magnitude to the charge e on the electron (where $|e| = 1.6 \times 10^{-19}$ C). The neutron is electrically neutral, as its name implies. Because the neutron has no charge, it is difficult to detect.

Nuclear masses can be measured with great precision with the help of the mass spectrometer (Example 20.6) and the analysis of nuclear reactions. The proton is about 1836 times more massive than the electron, and the masses of the proton and the neutron are almost equal. It is convenient to define, for atomic masses, the **unified mass unit,** u, in such a way that the mass of the isotope ^{12}C is exactly 12 u. That is, the mass of a nucleus (or atom) is measured relative to the mass of an atom of the neutral carbon-12 isotope (the nucleus plus six electrons). Thus, the mass of ^{12}C is defined to be 12 u, where 1 u = 1.660559×10^{-27} kg. The proton and neutron each have a mass of about 1 u, and the electron has a mass that is only a small fraction of an atomic mass unit:

Mass of proton = 1.007276 u
Mass of neutron = 1.008665 u
Mass of electron = 0.000549 u

For reasons that will become apparent later, note that the mass of the neutron is greater than the combined masses of the proton and the electron.

Because the rest energy of a particle is given by $E = m_0c^2$, it is often convenient to express the particle's mass in terms of its energy equivalence. For a proton, we have

$$E = m_0c^2 = (1.67 \times 10^{-27} \text{ kg})(3 \times 10^8 \text{ m/s})^2$$
$$= 1.50 \times 10^{-10} \text{ J} = 9.39 \times 10^8 \text{ eV} = 939 \text{ MeV}$$

Following this procedure, it is found that the rest energy of an electron is 0.511 MeV.

The rest masses of the proton, neutron, and electron are given in Table 32.1. The masses and some other properties of selected isotopes are provided in Appendix B.

EXAMPLE 32.1 The Unified Mass Unit

Use Avogadro's number to show that 1 u = 1.66×10^{-27} kg.

Solution We know that 12 kg of ^{12}C contains Avogadro's number of atoms. Avogadro's number, N_A, has the value of 6.02×10^{23} atoms/g · mole = 6.02×10^{26} atoms/kg · mole.

Thus, the mass of one carbon atom is

$$\text{Mass of one } ^{12}\text{C atom} = \frac{12 \text{ kg}}{6.02 \times 10^{26} \text{ atoms}} = 1.99 \times 10^{-26} \text{ kg}$$

Since one atom of ^{12}C is defined to have a mass of 12 u, we find that

$$1 \text{ u} = \frac{1.99 \times 10^{-26} \text{ kg}}{12} = 1.66 \times 10^{-27} \text{ kg}$$

TABLE 32.1 Rest Mass of the Proton, Neutron, and Electron in Various Units

	Mass		
Particle	**kg**	**u**	**MeV/c^2**
Proton	1.6726×10^{-27}	1.007276	938.28
Neutron	1.6750×10^{-27}	1.008665	939.57
Electron	9.109×10^{-31}	5.486×10^{-4}	0.511

The Size of Nuclei

Ernest Rutherford (1871–1937). "On consideration, I realized that this scattering backward must be the result of a single collision, and when I made calculations I saw that it was impossible to get anything of that order of magnitude unless you took a system in which the greater part of the mass of the atom was concentrated in a minute nucleus. It was then that I had the idea of an atom with a minute massive center carrying a charge." (Photo courtesy of AIP Niels Bohr Library)

The size and structure of nuclei were first investigated in the scattering experiments of Rutherford, discussed in Section 30.1. In these experiments, positively charged nuclei of helium atoms (alpha particles) were directed at a thin piece of metal foil. As the alpha particles moved through the foil, they often passed near a nucleus of the metal. Because of the positive charge on both the incident particles and metal nuclei, particles were deflected from their straight-line path by the Coulomb repulsive force. In fact, some particles were even deflected back along the direction from which they had come. These particles were apparently moving directly toward a nucleus, on a head-on collision course.

Rutherford applied the principle of conservation of energy to the alpha particles and nuclei and found an expression for the nearest distance, d, that the particle can approach the nucleus when moving directly toward it before it is turned around by Coulomb repulsion. In such a head-on collision, the kinetic energy of the incoming alpha particle must be converted completely to electrical potential energy when the particle stops at the point of closest approach and turns around (Fig. 32.1). If we equate the initial kinetic energy of the alpha particle to the electrical potential energy of the system (alpha particle plus nucleus), we have

$$\frac{1}{2}mv^2 = k\frac{q_1q_2}{r} = k\frac{(2e)(Ze)}{d}$$

Solving for d, the distance of closest approach, we get

$$d = \frac{4kZe^2}{mv^2}$$

From this expression, Rutherford found that the alpha particles approached nuclei to within 3.2×10^{-14} m when the foil was made of gold. Thus, the radius of the gold nucleus must be less than this value. For silver atoms, the distance of closest approach was found to be 2×10^{-14} m. From these results, Rutherford concluded that the positive charge in an atom is concentrated in a small sphere, which he called the nucleus, whose radius is no greater than 10^{-14} m.

Since the time of Rutherford's scattering experiments, a multitude of other experiments have shown that most nuclei are approximately spherical and have an average radius given by

$$\boxed{r = r_0 A^{1/3}} \qquad (32.1)$$

where A is the mass number and r_0 is a constant equal to 1.2×10^{-15} m. Because the volume of a sphere is proportional to the cube of its radius, it follows from Equation 32.1 that the volume of a nucleus (assumed to be spherical) is directly proportional to A, the total number of nucleons. This suggests that *all nuclei have nearly the same density*, as shown in the following example.

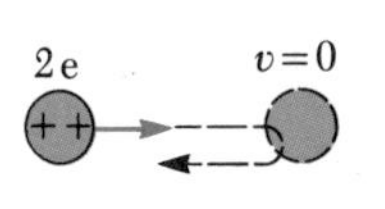

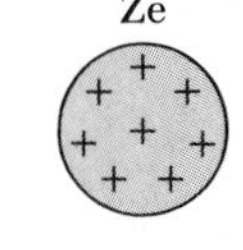

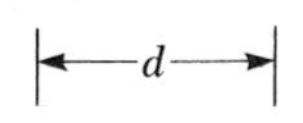

FIGURE 32.1 An alpha particle on a head-on collision course with a nucleus of charge Ze. Because of the Coulomb repulsion between the like charges, the alpha particle will stop instantaneously at a distance d from the nucleus.

EXAMPLE 32.2 The Volume and Density of a Nucleus

Find (a) an approximate expression for the mass of a nucleus of mass number A, (b) an expression for the volume of this nucleus, and (c) an expression for its density.

Solution (a) The mass of the proton is approximately equal to that of the neutron. Thus, if the mass of one of these particles is m_0, the mass of the nucleus is approximately Am_0.

(b) Assuming the nucleus is spherical and using Equation 32.1, we find that the volume is given by

$$V = \frac{4}{3}\pi r^3 = \frac{4}{3}\pi r_0^3 A$$

(c) The nuclear density can be found as follows:

$$\rho_n = \frac{\text{mass}}{\text{volume}} = \frac{Am_0}{\frac{4}{3}\pi r_0^3 A} = \frac{3m_0}{4\pi r_0^3}$$

Taking $r_0 = 1.2 \times 10^{-15}$ m and $m_0 = 1.67 \times 10^{-27}$ kg, we find that

$$\rho_n = \frac{3(1.67 \times 10^{-27}\text{ kg})}{4\pi(1.2 \times 10^{-15}\text{ m})^3} = 2.3 \times 10^{17}\text{ kg/m}^3$$

Because the density of water is only 10^3 kg/m^3, the nuclear density is about 2.3×10^{14} times greater than the density of water!

Nuclear Stability

Since the nucleus consists of a closely packed collection of protons and neutrons, you might be surprised that it can exist, given that like charges (the protons) in close proximity exert very large repulsive electric forces on each other and these forces should cause the nucleus to fly apart. However, nuclei are stable because of the presence of a force which we have not mentioned previously, called the **strong nuclear force.** This force, which is a very short-range one, is an attractive force that acts between all nuclear particles. The protons attract each other via the nuclear force, and at the same time they repel each other through the Coulomb force. However, the attractive nuclear force is much stronger than the force of Coulomb repulsion. The nuclear force also acts between neutrons and neutrons and between neutrons and protons.

There are about 400 stable nuclei; hundreds of others have been observed, but they are unstable. (A stable nucleus is one that remains as a unit essentially forever. An unstable nucleus eventually breaks apart by radioactive decay as we shall discuss in a later section.) A plot of N versus Z for a number of stable nuclei is given in Figure 32.2. Note that light nuclei are most stable if they contain an equal number of protons and neutrons, that is, if $N = Z$. For example, the helium nucleus (two protons and two neutrons) is very stable. Also note that heavy nuclei are more stable if the number of neutrons exceeds the number of protons. This can be understood by noting that, as the number of protons increases, the strength of the Coulomb force increases, which tends to break the nucleus apart. As a result, more neutrons are needed to keep the nucleus stable since neutrons experience only the attractive nuclear forces. Eventually, the repulsive forces between protons cannot be compensated by the addition of more neutrons. This occurs when $Z = 83$. Elements that contain more than 83 protons do not have stable nuclei.

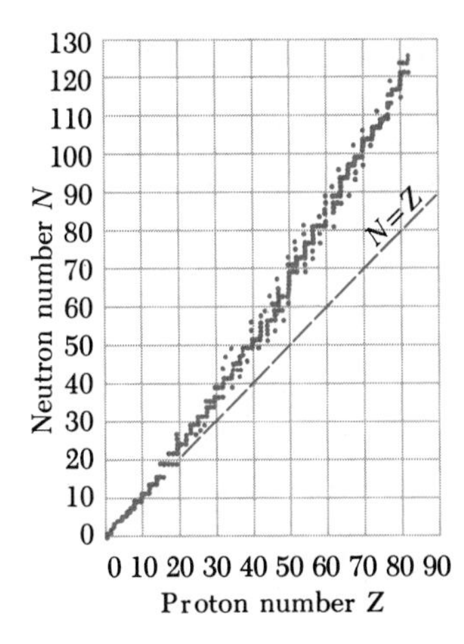

FIGURE 32.2 A plot of the neutron number N versus the proton number Z for the stable nuclei. The dashed straight line corresponding the condition $N = Z$ is called the *line of stability*.

32.2 BINDING ENERGY

The total mass of a nucleus is always less than the sum of the masses of its individual nucleons. According to the Einstein mass-energy relationship, if the mass difference, Δm, is multiplied by c^2, we obtain the binding energy of the nucleus. In other words, *the total energy of the bound system (the nucleus) is less than the total energy of the separated nucleons.* Therefore, in order to separate a nucleus into protons and neutrons, energy must be delivered to the system.

EXAMPLE 32.3 The Binding Energy of the Deuteron

Calculate the binding energy of the deuteron, which consists of a proton and a neutron, given that the mass of the deuteron is 2.014102 u.

Solution We know that the proton and neutron masses are

$$m_p = 1.007825 \text{ u}$$
$$m_n = 1.008665 \text{ u}$$

Note that the masses used for the proton and deuteron in this example are actually those of the neutral atoms, as found in Appendix B. We are able to use atomic masses for these calculations since the electron masses cancel. Therefore,

$$m_p + m_n = 2.016490 \text{ u}$$

To calculate the mass difference, we subtract the deuteron mass from this value:

$$\begin{aligned}\Delta m &= (m_p + m_n) - m_d \\ &= 2.016490 \text{ u} - 2.014102 \text{ u} \\ &= 0.002388 \text{ u}\end{aligned}$$

Since 1 u corresponds to an equivalent energy of 931.50 MeV (that is, $1 \text{ u}\cdot c^2 = 931.50$ MeV), the mass difference corresponds to the following binding energy:

$$E_b = (0.002388 \text{ u})(931.5 \text{ MeV/u}) = 2.224 \text{ MeV}$$

This result tells us that, in order to separate a deuteron into its constituent parts (a proton and a neutron), it is necessary to add 2.224 MeV of energy to the deuteron.

One way of supplying the deuteron with this energy is by bombarding it with energetic particles.

If the binding energy of a nucleus were zero, the nucleus would separate into its constituent protons and neutrons without the addition of any energy, that is, it would spontaneously break apart.

It is interesting to examine a plot of the binding energy per nucleon, E_b/A, as a function of mass number for various stable nuclei (Fig. 32.3). Note that except for the lighter nuclei, the average binding energy per nucleon is about 8 MeV. For the deuteron, the average binding energy per nucleon is $E_b/A = 2.224/2$ MeV $= 1.112$ MeV. Note that the curve in Figure 32.3 peaks in the vicinity of $A = 60$. That is, nuclei with mass numbers greater or less than 60 are not as strongly bound as those near the middle of the periodic table. As we shall see later, this fact allows energy to be released in fission and fusion reactions.

Note from Figure 32.3 that the average binding energy per nucleon is approximately constant for values of A greater than about 40. As we move to the right on the curve, we find that the number of nucleons in the nuclei increases. Thus, one might expect that an individual nucleon contained in a heavy nucleus would be tightly bound because there are many surrounding nucleons to hold it in place. Since the binding energy per nucleon can be considered as the energy needed to break these bonds, we might expect the curve in Figure 32.3 to continue to rise rather than level off. The fact that the curve does not continue to rise leads to the idea that the nuclear forces are *saturated*. This, in turn, suggests that an individual nucleon does not interact equally with all other nucleons in the nucleus but is influenced only by nearby nucleons.

32.3 NUCLEAR MODELS

Although the detailed nature of nuclear forces is still not well understood, several nuclear models have been proposed, and these are useful in understanding some experimental features concerning the nucleus. The models we

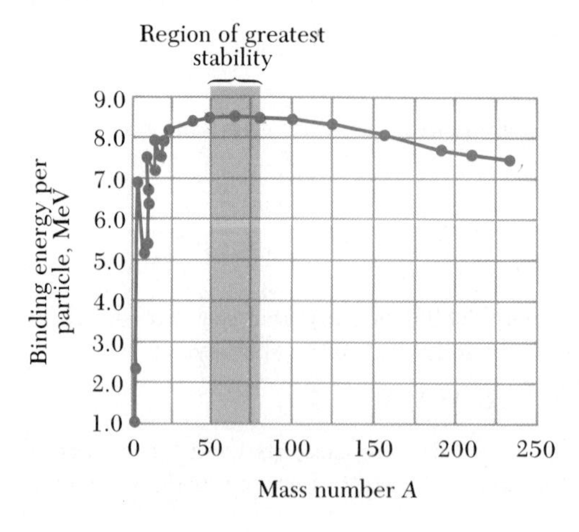

FIGURE 32.3 A plot of the binding energy per nucleon versus the mass number A for nuclei that are along the line of stability shown in Figure 32.2.

shall discuss are (1) the liquid-drop model, (2) the independent-particle model, and (3) the collective model.

Liquid-Drop Model

The **liquid-drop model** of the nucleus, proposed by Bohr in 1936, treats the nucleons as if they were molecules in a drop of liquid. The nucleons interact strongly with each other and undergo frequent collisions as they "jiggle around" within the nucleus. This is analogous to the thermally agitated motion of molecules in a liquid.

The primary reason for the success of the liquid-drop model is that it accounts very well for the nuclear binding energy. As a first approximation we may write an equation for the binding energy as

$$E_b = C_1 A - C_2 A^{2/3} - C_3 \frac{Z(Z-1)}{A^{1/3}} \tag{32.2}$$

where the C_1, C_2, and C_3 are proportionality constants. Let us discuss the significance of each term in this equation.

1. **The volume effect** The first term, $C_1 A$, is that part of the binding energy arising from the attraction between nucleons. As we showed earlier, the binding energy per nucleon is approximately constant. In order for the binding energy divided by A to be a constant, the binding energy must be proportional to A. This term is called the volume effect because the volume of the nucleus is also proportional to A.
2. **The surface effect** Since many of the nucleons will be located on the surface of the drop, they will have fewer neighbors than those in the interior of the drop. Hence, these surface nucleons will reduce the binding energy by an amount proportional to the surface area of the drop, which varies as the square of the nuclear radius or as $A^{2/3}$.
3. **The Coulomb repulsion effect** Each proton repels every other proton in the nucleus. The corresponding potential energy per pair of interacting particles is given by ke^2/r, where k is the Coulomb constant. The total Coulomb energy represents the work required to assemble Z protons from infinity to a sphere of volume V. The energy is propotional to the number of proton pairs $Z(Z-1)$ and is inversely proportional to the nuclear radius which varies as $A^{1/3}$.

In a more complete equation, another term has to be added to the binding energy formula, a term which is significant only for heavy nuclei with a large excess of neutrons. We shall not consider this term here.

The three constants in Equation 32.2 can be adjusted to fit this expression to experimental data, with reasonable results. For nuclei with $A > 15$, the constants have the values

$$C_1 = 15.7 \text{ MeV} \qquad C_2 = 17.8 \text{ MeV} \qquad C_3 = 0.71 \text{ MeV}$$

Equation 32.2 describes the binding energy of the nucleus very well. Furthermore, it provides a qualitative description of the important process of nuclear fission to be discussed in Chapter 32. However, the liquid-drop model cannot account for such features as the existence of quantized energy

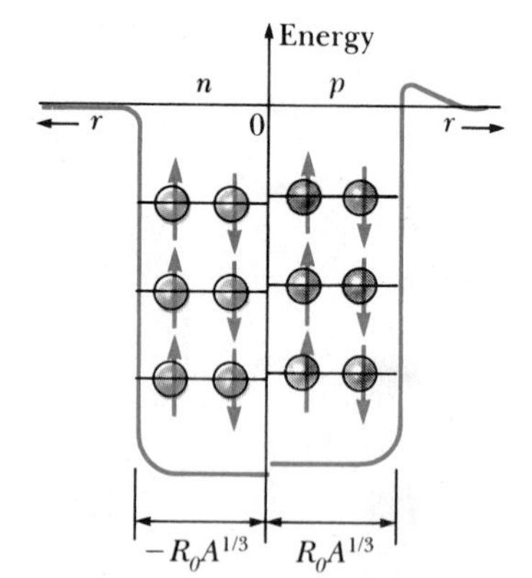

FIGURE 32.4 A nucleus containing 12 nucleons. The gray spheres represent protons, and the colored spheres represent neutrons. The energy levels for the protons are slightly higher than those for the neutrons because of the Coulomb potential in the protons. The difference in the levels increases as Z increases. Note that only two nucleons with opposite spin can occupy a given level, as required by the Pauli exclusion principle.

levels in the nucleus. Such difficulties have led to the development of other nuclear models.

The Independent-Particle Model

We have mentioned some of the limitations of the liquid-drop model and its failure to explain some features of nuclear structure. In the **independent-particle model,** often called the *shell model,* the nuclear structure is viewed in terms of orbits and shells, in analogy with the electronic structure of the atom. In this model, the nucleons exist in well-defined orbits with quantized energy states. The quantized states occupied by the nucleons can be described by a set of quantum numbers. Like the electron, both the proton and the neutron have spin ½ and obey the Pauli exclusion principle. As a result, each nuclear state can contain only two protons (or two neutrons) with opposite spin as shown in Figure 32.4. Because the protons are charged, the added Coulomb repulsion causes their energy levels to be somewhat higher than those of the neutrons.

On the basis of this model, it is easy to understand nuclear excited states in terms of nucleons moving from lower energy levels to higher levels. Also, it is possible to understand why nuclei containing an even number of protons and neutrons are more stable than others. In fact, there are 160 such stable isotopes (called even-even nuclei) and only 4 stable isotopes with an odd number of both protons and neutrons. Any particular state is filled when it contains two protons (or two neutrons) with opposite spins. *An extra proton or neutron can be added to the nucleus only at the expense of increasing the energy of the nucleus.* This increase in energy leads to a nucleus that is less stable than the original nucleus.

Certain values of Z and N correspond to nuclei with unusually high stability. These values of Z and N, called **magic numbers**, are given by

$$Z \text{ or } N = 2, 8, 20, 28, 50, 82, 126$$

For example, the alpha particle (two protons and two neutrons), with $Z = 2$ and $N = 2$ is very stable.

The Collective Model

A third model of nuclear structure, known as the **collective model,** combines some features of both the liquid-drop model and the independent-particle model. The nucleus is considered to have some "extra" nucleons moving in quantized orbits, in addition to the filled core of nucleons. Deformations can be set up in the core as a result of a strong interaction between the core and the extra nucleons, thereby setting up vibrational and rotational motions as in the liquid-drop model. This collective model has been very successful in explaining many nuclear phenomena.

32.4 RADIOACTIVITY

Marie Curie (1867–1934), a Polish scientist, shared the Nobel prize in 1903 with her husband, Pierre, and with Becquerel for their work on spontaneous radioactivity and the radiation emitted by radioactive substances. "I persist in believing that the ideas that then guided us are the only ones which can lead to true social progress. We cannot hope to build a better world without improving the individual. Toward this end, each of us must work toward his own highest development, accepting at the same time his share of responsibility in the general life of humanity."

The study of nuclear physics began when Becquerel discovered radioactivity in 1896, one year after the discovery of x-radiation. Becquerel found, quite by accident, that an ore containing uranium emits an invisible radiation that can penetrate a black paper wrapper and expose a photographic plate. Following several observations of this type under controlled conditions, he concluded that the radiation emitted by the ore was of a new type and occurred spontaneously. This process of spontaneous emission of radiation by uranium was soon called **radioactivity.** Subsequent experiments by other scientists showed that other substances were also radioactive. The most significant investigations of this type were conducted by Marie Curie and her husband, Pierre. After several years of careful and laborious chemical separation processes on tons of pitchblende, a radioactive ore, the Curies reported the discovery of two previously unknown elements, both of which were radioactive. These were named polonium and radium. Subsequent experiments, including Rutherford's famous work on alpha particle scattering, suggested that radioactivity was the result of the decay, or disintegration, of unstable nuclei.

There are three types of radiation that can be emitted by a radioactive substance: alpha (α) decay, in which the emitted particles are helium nuclei; beta (β) decay, in which the emitted "rays" are electrons; and gamma (γ) decay, in which the emitted "rays" are high-energy photons.

It is possible to distinguish these three forms of radiation by using the scheme described in Figure 32.5. The radiation from a radioactive sample is directed into a region in which there is a magnetic field, and the beam splits into three components, two bending in opposite directions and the third not changing direction. From this simple observation, one can conclude that the radiation of the undeflected beam carries no charge (the gamma ray), the component deflected upward contains positively charged particles (alpha particles), and the component deflected downward contains negatively charged particles (electrons).

The three types of radiation have quite different penetrating powers. Alpha particles barely penetrate a sheet of paper, beta particles can penetrate a few millimeters of aluminum, and gamma rays can penetrate several centimeters of lead.

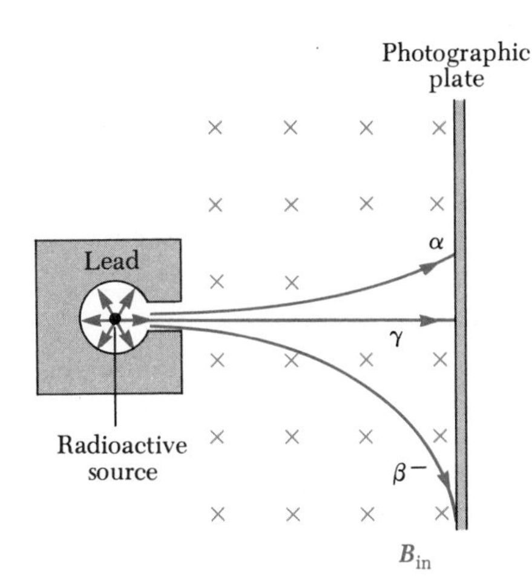

FIGURE 32.5 The radiation from a radioactive source, such as radium, can be separated into three components using a magnetic field to deflect the charged particles. The photographic plate at the right records the events.

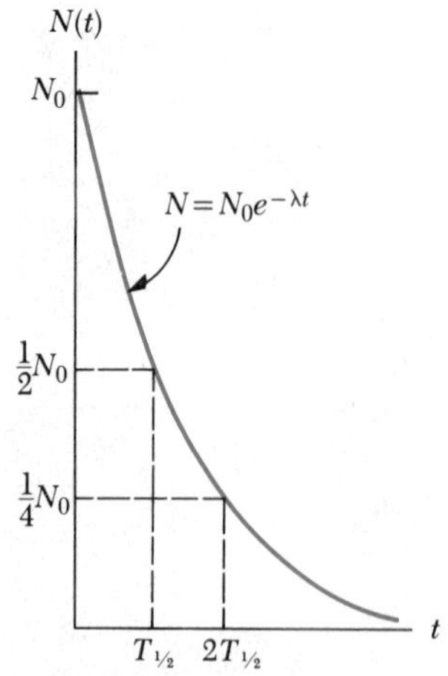

FIGURE 32.6 Plot of the exponential decay law for radioactive nuclei. The vertical axis represents the number of radioactive nuclei present at any time t, and the horizontal axis is time. The parameter $T_{1/2}$ is the half-life of the sample.

The Decay Constant and Half-Life

If a radioactive sample contains N radioactive nuclei at some instant, one finds that the number of nuclei, ΔN, that decay in a time Δt is proportional to N:

$$\Delta N = -\lambda N \Delta t \tag{32.3}$$

where λ is a constant called the **decay constant** and the negative sign signifies that N decreases with time, that is, ΔN is negative. The value of λ for any isotope determines the rate at which that isotope will decay.

The **decay rate,** or **activity,** R, of a sample is defined as the number of decays per second. From Equation 32.3, we see that the decay rate is

$$R = -\frac{\Delta N}{\Delta t} = \lambda N \tag{32.4}$$

Thus we see that isotopes with a large value of λ decay at a rapid rate and those with a small λ value decay slowly.

A general decay curve for a radioactive sample is shown in Figure 32.6. One can show from Equation 32.4 (using calculus) that the number of nuclei present varies with time according to the expression

$$N = N_0 e^{-\lambda t} \tag{32.5}$$

where N is the number of radioactive nuclei present at time t, N_0 is the number present at time $t = 0$, and $e = 2.718 \ldots$ is the base of the natural logarithm. Processes that obey Equation 32.5 are sometimes said to undergo exponential decay.[1]

The unit of activity is the **curie** (Ci), defined as

$$1 \text{ Ci} \equiv 3.7 \times 10^{10} \text{ decays/s} \tag{32.6}$$

This number of decay events per second was selected as the original activity unit because it is the approximate activity of 1 g of radium. The SI unit of activity is the **becquerel** (Bq):

$$1 \text{ Bq} = 1 \text{ decay/s} \tag{32.7}$$

Therefore, 1 Ci = 3.7×10^{10} Bq. The most commonly used units of activity are the millicurie (10^{-3} Ci) and the microcurie (10^{-6} Ci).

One of the most common terms encountered in any discussion of radioactive materials is **half-life**, $T_{1/2}$. This is because all radioactive substances follow the same general decay pattern. After a certain interval of time, half of the original number of nuclei in a sample will have decayed, then in a second time interval equal to the first, half of those nuclei remaining will have decayed, and so on.

The **half-life** is the time required for half of a given number of radioactive nuclei to decay.

[1]Other examples of exponential decays were discussed in Chapter 19 in connection with RC circuits, and in Chapter 21 in connection with RL circuits.

Let us assume that the number of radioactive nuclei present at $t = 0$ is N_0, as in Figure 32.6. The number left after one half-life, $T_{1/2}$, passes is $N_0/2$. After the second half-life, the number remaining is again reduced by one half. Hence, after a time $2T_{1/2}$, the number remaining is $N_0/4$, and so forth. Half-lives range from about 10^{-22} s to 10^{21} years.

As noted earlier, the decay constant is large for radioactive materials that decay rapidly and small for those that decay slowly. Thus, it should be evident that the half-life and the decay constant are related. The relationship between them is[2]

$$T_{1/2} = \frac{0.693}{\lambda} \qquad (32.8)$$

Half-life

EXAMPLE 32.4 How Many Nuclei Are Left?

Carbon-14, $^{14}_{6}C$, is a radioactive isotope of carbon that has a half-life of 5730 years. If you start with a sample of 1000 carbon-14 nuclei, how many will still be around in 22,920 years?

Solution In 5730 years, half the sample will have decayed, leaving 500 carbon-14 nuclei remaining. In another 5730 years (for a total elapsed time of 11,460 years), the number will be reduced to 250 nuclei. After another 5730 years (total time 17,190 years), 125 remain. Finally, after four half-lives (22,920 years), only about 62 remain.

It should be noted here that these numbers represent ideal circumstances. Radioactive decay is an averaging process over a very large number of atoms, and the actual outcome depends on statistics. Our original sample in this example contained only 1000 nuclei, certainly not a very large number. Thus, if we were actually to count the number remaining after one half-life for this small sample, it probably would not be 500. If the initial number is very large, the probability of getting extremely close to the predicted number remaining after one half-life increases greatly.

EXAMPLE 32.5 The Activity of Radium

The half-life of the radioactive nucleus $^{226}_{88}Ra$ is 1.6×10^3 years. If a sample contains 3×10^{16} such nuclei, determine the activity.

---

[2]Equation 32.8 can be derived by starting with Equation 32.5, $N = N_0e^{-\lambda t}$. After a time interval equal to one half-life, $t = T_{1/2}$, the number of radioactive nuclei remaining is $N = N_0/2$. Therefore,

$$\frac{N_0}{2} = N_0e^{-\lambda T_{1/2}}$$

Dividing both sides by N_0 gives

$$\frac{1}{2} = e^{-\lambda T_{1/2}}$$

Taking the natural logarithm of both sides of this equation eliminates the exponential factor on the right, since $\ln e = 1$:

$$\ln\frac{1}{2} = -\lambda T_{1/2}$$

Since $\ln\frac{1}{2} = -0.693$ (use your calculator to check this), we have

$$-0.693 = -\lambda T_{1/2}$$

$$T_{1/2} = \frac{0.693}{\lambda}$$

Solution First, let us calculate the decay constant, λ, using Equation 32.8 and the fact that

$$T_{1/2} = 1.6 \times 10^3 \text{ years} = (1.6 \times 10^3 \text{ years})(3.15 \times 10^7 \text{ s/year}) = 5.0 \times 10^{10} \text{ s}$$

Therefore,

$$\lambda = \frac{0.693}{T_{1/2}} = \frac{0.693}{5.0 \times 10^{10} \text{ s}} = 1.4 \times 10^{-11} \text{ s}^{-1}$$

We can calculate the activity of the sample at $t = 0$ using Equation 32.4 in the form $R_0 = \lambda N_0$, where R_0 is the decay rate at $t = 0$ and N_0 is the number of radioactive nuclei present at $t = 0$. Since $N_0 = 3 \times 10^{16}$, we have

$$R_0 = \lambda N_0 = (1.4 \times 10^{-11} \text{ s}^{-1})(3 \times 10^{16}) = 4.2 \times 10^5 \text{ decays/s}$$

Since 1 Ci $= 3.7 \times 10^{10}$ decays/s, the activity, or decay rate, is

$$R_0 = 11.3\ \mu\text{Ci}$$

32.5 THE DECAY PROCESSES

As we stated in the previous section, a radioactive nucleus spontaneously decays via three processes: alpha decay, beta decay, and gamma decay. Let us discuss these in more detail.

Alpha Decay

If a nucleus emits an alpha particle (^4_2He), it loses two protons and two neutrons. Let us adopt the historical terminology of calling the nucleus before decay the **parent nucleus** and the nucleus remaining after decay the **daughter nucleus.** The nucleus $^{238}_{92}\text{U}$ decays by alpha emission, and it is easy to visualize the process in symbolic form as follows:

$$^{238}_{92}\text{U} \longrightarrow {}^{234}_{90}\text{Th} + {}^4_2\text{He} \tag{32.9}$$

This says that a parent nucleus, $^{238}_{92}\text{U}$, emits an alpha particle, ^4_2He and thereby changes to a daughter nucleus, $^{234}_{90}\text{Th}$. Note the following facts about this reaction: (1) the atomic number (number of protons) on the left is the same as on the right ($92 = 90 + 2$) since charge must be conserved, and (2) the mass number (protons plus neutrons) on the left is the same as on the right ($238 = 234 + 4$). The half-life for ^{238}U decay is 4.47×10^9 years.

When one element changes into another, as in alpha decay, the process is called **transmutation.** In order for alpha emission to occur, the mass of the parent must be greater than the combined mass of the daughter and the alpha particle. In the decay process, this excess mass is converted into energy and appears in the form of kinetic energy in the daughter nucleus and the alpha particle. Most of the kinetic energy is carried away by the alpha particle because it is much less massive than the daughter nucleus. That is, because momentum must be conserved in the decay process, *the lighter alpha particle recoils with a much higher velocity than the daughter nucleus.* Generally, light particles carry off most of the energy in nuclear decays.

EXAMPLE 32.6 The Alpha Decay of Radium

Radium, $^{226}_{88}Ra$, decays by alpha emission. What is the daughter element formed?

Solution The decay can be written symbolically as

$$^{226}_{88}\text{Ra} \longrightarrow \text{X} + {}^{4}_{2}\text{He}$$

where X is the unknown daughter element. Requiring that the mass numbers and atomic numbers balance on the two sides of the arrow, we find that the daughter nucleus must have a mass number of 222 and an atomic number of 86:

$$^{226}_{88}\text{Ra} \longrightarrow {}^{222}_{86}\text{X} + {}^{4}_{2}\text{He}$$

The periodic table shows that the nucleus with an atomic number of 86 is radon, Rn. Thus, the process is

$$^{226}_{88}\text{Ra} \longrightarrow {}^{222}_{86}\text{Rn} + {}^{4}_{2}\text{He}$$

This decay is shown schematically in Figure 32.7.

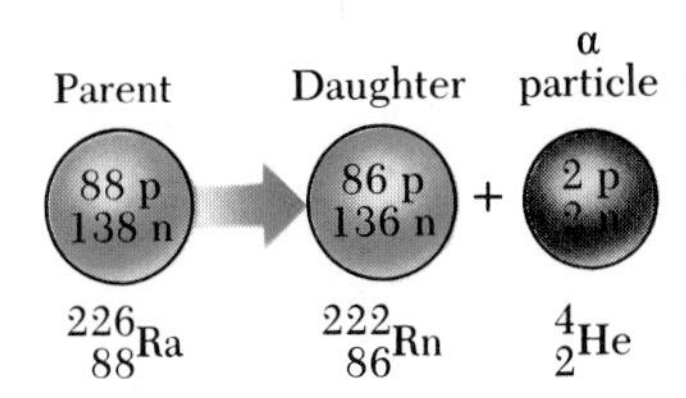

FIGURE 32.7 (Example 32.6) Alpha decay of the $^{226}_{88}Ra$ nucleus.

EXAMPLE 32.7 The Energy Liberated When Radium Decays

In Example 32.6, we showed that the $^{226}_{88}Ra$ nucleus undergoes alpha decay to $^{222}_{86}Rn$. Calculate the amount of energy liberated in this decay. Take the mass of $^{226}_{88}Ra$ to be 226.025406 u, that of $^{222}_{86}Rn$ to be 222.017574 u, and that of $^{4}_{2}He$ to be 4.002603 u, as found in Appendix B.

Solution After decay, the mass of the daughter, m_d, plus the mass of the alpha particle, m_α, is

$$m_d + m_\alpha = 222.017574 \text{ u} + 4.002603 \text{ u} = 226.020177 \text{ u}$$

Thus, calling the mass of the parent nucleus M_p, we find that the mass lost during decay is

$$\Delta m = M_p - (m_d + m_\alpha) = 226.025406 \text{ u} - 226.020177 \text{ u} = 0.005229 \text{ u}$$

Using the relationship 1 u = 931.5 MeV, we find that the energy liberated is

$$E = (0.005229 \text{ u})(931.50 \text{ MeV/u}) = 4.87 \text{ MeV}$$

Beta Decay

When a radioactive nucleus undergoes beta decay, *the daughter nucleus contains the same number of nucleons as the parent nucleus but the charge number is increased by 1.* A typical beta decay event is

$$^{14}_{6}\text{C} \longrightarrow {}^{14}_{7}\text{N} + {}^{0}_{-1}\text{e} \qquad (32.10)$$

The superscripts and subscripts on the carbon and nitrogen nuclei follow our usual conventions, but those on the electron may need some explanation. The −1 indicates that the electron has a charge whose magnitude is equal to that of the proton but negative. The 0 used for the electron's mass number indicates that the mass of the electron is almost zero relative to that of carbon and nitrogen nuclei.

The emission of electrons from a *nucleus* is surprising because, in all our previous discussions, we stated that the nucleus is composed of protons and neutrons only. This apparent discrepancy can be explained by noting that the electron that is emitted is created in the nucleus by a process in which a neutron is transformed into a proton. This can be represented by the equation

$$ {}^{1}_{0}\mathrm{n} \longrightarrow {}^{1}_{1}\mathrm{p} + {}^{0}_{-1}\mathrm{e} \qquad (32.11)$$

Let us consider the energy of the system of Equation 32.10 before and after decay. As with alpha decay, energy must be conserved in beta decay. The following example illustrates how to calculate how much energy is released in the beta decay of ${}^{14}_{6}\mathrm{C}$.

EXAMPLE 32.8 The Beta Decay of ${}^{14}_{6}\mathrm{C}$

Find the energy liberated in the beta decay of ${}^{14}_{6}\mathrm{C}$ to ${}^{14}_{7}\mathrm{N}$ as represented by Equation 32.10. Refer to Appendix B for the atomic masses.

Solution We find from Appendix B that ${}^{14}_{6}\mathrm{C}$ has a mass of 14.003242 u and ${}^{14}_{7}\mathrm{N}$ has a mass of 14.003074 u. Here, the mass difference between the initial and final states is[3]

$$\Delta m = 14.003242 \text{ u} - 14.003074 \text{ u} = 0.000168 \text{ u}$$

This corresponds to an energy release of

$$E = (0.000168 \text{ u})(931.50 \text{ MeV/u}) = 0.156 \text{ MeV}$$

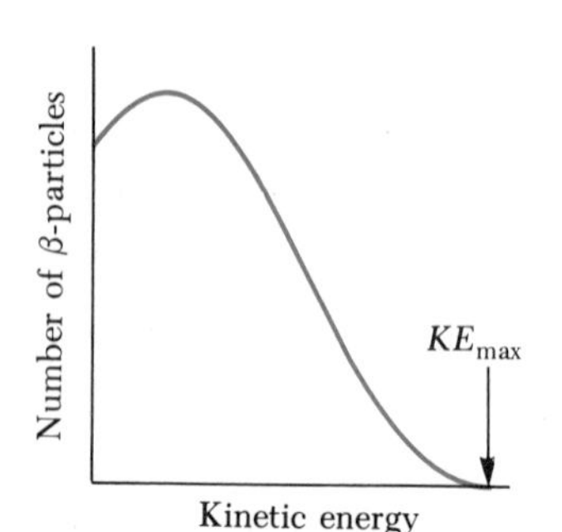

FIGURE 32.8 A typical beta-decay spectrum. The maximum kinetic energy observed for the beta particles corresponds to the value of Q for the reaction.

From Example 32.8, we see that the energy released in the beta decay of $^{14}\mathrm{C}$ is approximately 0.16 MeV. As with alpha decay, we expect the electron to carry away virtually all of this as kinetic energy because apparently it is the lightest particle produced in the decay. However, as Figure 32.8 shows, only a small number of electrons have this maximum kinetic energy, represented as KE_{max} on the graph; most of the electrons emitted have kinetic energies less than this predicted value. If the daughter nucleus and the electron are not carrying away this liberated energy, then the requirement that energy is conserved leads one to ask the question, "What accounts for the missing energy?" As an additional complication, further analysis of beta decay shows that the principles of conservation of both angular momentum and linear momentum appear to be violated!

In 1930, Pauli proposed that a third particle must be present to carry away the "missing" energy and to conserve momentum. Enrico Fermi later named this particle the **neutrino** (little neutral one) because it had to be electrically neutral and have little or no rest mass. Although it eluded detection for many years, the neutrino (symbol ν) was finally detected experimentally in 1950.

Properties of the neutrino

1. It has zero electric charge.
2. It has a rest mass smaller than that of the electron, and in fact its rest mass may be zero (although recent experiments suggest that this may not be true).
3. It interacts very weakly with matter and is therefore very difficult to detect.

[3] In beta decay, we must keep track of the electrons involved in the process. In the initial state, the neutral $^{14}\mathrm{C}$ atom has six electrons. In the final state, the $^{14}\mathrm{N}$ atom also has six electrons. Thus, we must conclude that $^{14}\mathrm{N}$ is not neutral. However, if we include the electron emitted in the decay process, we effectively have a mass equivalent to a neutral $^{14}\mathrm{N}$ atom.

Thus, with the introduction of the neutrino, we are now able to represent the beta decay process of Equation 31.10 in its correct form:

$$ {}^{14}_{6}\mathrm{C} \longrightarrow {}^{14}_{7}\mathrm{N} + {}^{0}_{-1}\mathrm{e} + \bar{\nu} \tag{32.12}$$

Where the bar in the symbol $\bar{\nu}$ indicates that this is an **antineutrino.** To explain what an antineutrino is, let us first consider the following decay.

$$ {}^{12}_{7}\mathrm{N} \longrightarrow {}^{12}_{6}\mathrm{C} + {}^{0}_{1}\mathrm{e} + \nu \tag{32.13}$$

Here we see that when ^{12}N decays into ^{12}C a particle is produced that is identical to the electron except that it has a positive charge of $+e$. This particle is called a positron. Because it is like the electron in all respects except charge, the positron is said to be **antiparticle** to the electron. We shall discuss antiparticles further in Chapter 34. For now, it suffices to say that *a neutrino is emitted in positron decay and an antineutrino is emitted in electron decay.*

Enrico Fermi (1901–1954), an Italian physicist, was awarded the Nobel prize in 1938 for his demonstrations of the existence of new radioactive elements produced by neutron radiation and for his discovery of nuclear reactions brought about by slow neutrons. "Whatever Nature has in store for mankind, unpleasant as it may be, men must accept, for ignorance is never better than knowledge."

Carbon Dating

The beta decay of ^{14}C given by Equation 32.12 is commonly used to date organic samples. Cosmic rays (high-energy particles from outer space) in the upper atmosphere cause nuclear reactions that create ^{14}C from ^{14}N. In fact, the ratio of ^{14}C to ^{12}C isotopic abundance in the carbon dioxide molecules of our atmosphere has a constant value of about 1.3×10^{-12} as determined by measuring carbon ratios in tree rings. All living organisms have the same ratio of ^{14}C to ^{12}C because they continuously exchange carbon dioxide with their surroundings. When an organism dies, however, it no longer absorbs ^{14}C from the atmosphere, and so the ratio of ^{14}C to ^{12}C decreases as the result of the beta decay of ^{14}C. It is therefore possible to measure the age of a material by measuring its activity per unit mass due to the decay of ^{14}C. Using carbon dating, samples of wood, charcoal, bone, and shell have been identified as having lived from 1000 to 25,000 years ago. This knowledge has helped us to reconstruct the history of living organisms—including humans—during this time span.

A particularly interesting example is the dating of the Dead Sea Scrolls. This group of manuscripts was first discovered by a shepherd in 1947. Translation showed them to be religious documents, including most of the books of the Old Testament. Because of their historical and religious significance, scholars wanted to know their age. Carbon dating applied to fragments of the scrolls and to the material in which they were wrapped established their age at about 1950 years.

EXAMPLE 32.9 Should We Report This to Homicide?

A 50-g sample of carbon is taken from the pelvis bone of a skeleton and is found to have a carbon-14 decay rate of 200 decays/min. It is known that carbon from a living organism has a decay rate of 15 decays/min · g and that ^{14}C has a half-life of 5730 y = 3.01×10^9 min. Find the age of the skeleton.

Solution Let us start with Equation 32.5

$$N = N_0 e^{-\lambda t}$$

and multiply both sides by λ as

$$\lambda N = \lambda N_0 e^{-\lambda t}$$

But, from Equation 32.4, we see that this is the same as

$$R = R_0 e^{-\lambda t}$$

where R is the present activity, and R_0 was the activity when the skeleton was a part of a living organism. We are given that R = 200 decays/min, and we can find R_0 as

$$R_0 = \left(15\ \frac{\text{decays}}{\text{min} \cdot \text{g}}\right)(50\ \text{g}) = 750\ \frac{\text{decays}}{\text{min}}$$

The decay constant is found from Equation 32.8 as

$$\lambda = \frac{0.693}{T_{1/2}} = \frac{0.693}{3.01 \times 10^9\ \text{min}} = 2.30 \times 10^{-10}\ \text{min}^{-1}$$

Thus, we make the following substitutions.

$$R = R_0 e^{-\lambda t}$$

$$200\ \frac{\text{decays}}{\text{min}} = \left(750\ \frac{\text{decays}}{\text{min}}\right) e^{-(2.30 \times 10^{-10}\ \text{min}^{-1})t}$$

or

$$0.266 = e^{-(2.30 \times 10^{-10}\ \text{min}^{-1})t}$$

Now, we take the natural log of both sides of the equation, to give

$$\ln(0.266) = -(2.30 \times 10^{-10}\ \text{min}^{-1})t$$

$$-1.32 = -(2.30 \times 10^{-10}\ \text{min}^{-1})t$$

$$t = 5.74 \times 10^9\ \text{min} = 10{,}930\ \text{y}$$

Gamma Decay

Very often a nucleus that undergoes radioactive decay is left in an excited energy state. The nucleus can then undergo a second decay to a lower energy state, perhaps to the ground state, by emitting one or more photons. The process is very similar to the emission of light by an atom. An atom emits radiation to release some extra energy when an electron "jumps" from a state of high energy to a state of lower energy. Likewise, the nucleus uses essentially the same method to release any extra energy it may have following a decay or some other nuclear event. In nuclear de-excitation, the "jumps" that release energy are made by protons or neutrons in the nucleus as they move from a higher energy level to a lower level (see Section 32.3). The photons emitted in such a de-excitation process are called **gamma rays,** which have very high energy relative to the energy of visible light.

A nucleus may reach an excited state as the result of a violent collision with another particle. However, it is more common for a nucleus to be in an excited state as a result of alpha or beta decay. The following sequence of events represents a typical situation in which gamma decay occurs:

$$^{12}_{5}\text{B} \longrightarrow {}^{12}_{6}\text{C}^* + {}^{0}_{-1}\text{e} \quad (32.14\text{a})$$

$$^{12}_{6}\text{C}^* \longrightarrow {}^{12}_{6}\text{C} + \gamma \quad (32.14\text{b})$$

TABLE 32.2 The Four Radioactive Series

Series	Starting Isotope	Half-life (years)	Stable End Product
Uranium	$^{238}_{92}U$	4.47×10^9	$^{206}_{82}Pb$
Actinium	$^{235}_{92}U$	7.04×10^8	$^{207}_{82}Pb$
Thorium	$^{232}_{90}Th$	1.41×10^{10}	$^{208}_{82}Pb$
Neptunium	$^{237}_{93}Np$	2.14×10^6	$^{209}_{83}Bi$

Equation 32.14a represents a beta decay in which ^{12}B decays to $^{12}C^*$, where the asterisk is used to indicate that the carbon nucleus is left in an excited state following the decay. The excited carbon nucleus then decays to the ground state by emitting a gamma ray, as indicated by Equation 32.14b. Note that gamma emission does not result in any change in either Z or A.

32.6 NATURAL RADIOACTIVITY

Radioactive nuclei are generally classified into two groups: (1) unstable nuclei found in nature, which give rise to what is called **natural radioactivity,** and (2) nuclei produced in the laboratory through nuclear reactions, which exhibit **artificial radioactivity.**

There are three series of naturally occurring radioactive nuclei (Table 32.2). Each series starts with a specific long-lived radioactive isotope whose half-life exceeds that of any of its descendants. The three natural series begin with the isotopes ^{238}U, ^{235}U, and ^{232}Th, and the corresponding stable end products are three isotopes of lead: ^{206}Pb, ^{207}Pb, and ^{208}Pb. The fourth series in Table 32.2 begins with ^{237}Np and has as its stable end product ^{209}Bi. The element ^{237}Np is a transuranic element (one having an atomic number greater than that of uranium) not found in nature and has a half-life of "only" 2.14×10^6 years.

Figure 32.9 shows the successive decays for the ^{232}Th series. Note that ^{232}Th first undergoes alpha decay to ^{228}Ra. Next, ^{228}Ra undergoes two successive β decays to ^{228}Th. The series continues and finally branches when it reaches ^{212}Bi. At this point, there are two decay possibilities. The end of the decay series is the stable isotope ^{208}Pb.

The two uranium series are somewhat more complex than the ^{232}Th series. Also, there are several naturally occurring radioactive isotopes, such as ^{14}C and ^{40}K, that are not part of either decay series.

It is interesting to note that the existence of radioactive series in nature enables our environment to be constantly replenished with radioactive elements that would otherwise have disappeared long ago. For example, because the solar system is about 5×10^9 years old, the supply of ^{226}Ra (whose half-life is only 1600 years) would have been depleted by radioactive decay long ago if it were not for the decay series that starts with ^{238}U, whose half-life is 4.47×10^9 years.

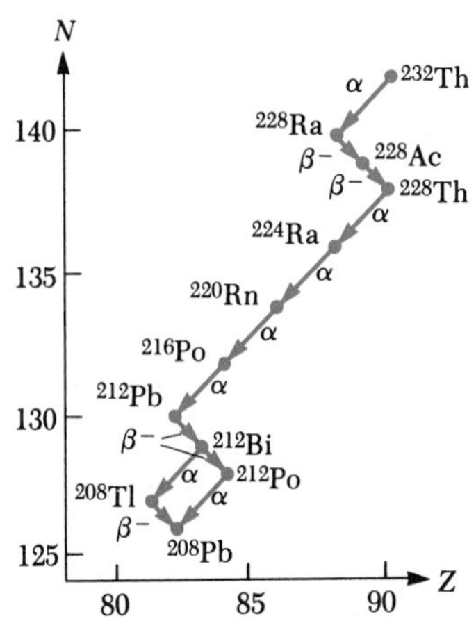

FIGURE 32.9 Decay series beginning with ^{232}Th.

32.7 NUCLEAR REACTIONS

It is possible to change the structure of nuclei by bombarding them with energetic particles. Such collisions, which change the identity or properties of the target nuclei, are called **nuclear reactions.** Rutherford was the first to observe nuclear reactions in 1919, using naturally occurring radioactive sources for the bombarding particles. He found that protons were released when alpha particles were allowed to collide with nitrogen atoms. The process can be represented symbolically as

$$^{4}_{2}He + ^{14}_{7}N \longrightarrow X + ^{1}_{1}H \tag{32.15}$$

This equation says that an alpha particle ($^{4}_{2}He$) strikes a nitrogen nucleus and produces an unknown product nucleus (X) and a proton ($^{1}_{1}H$). Balancing atomic numbers and mass numbers, as we did for radioactive decay, enables us to conclude that the unknown is characterized as $^{17}_{8}X$. Since the element whose atomic number is 8 is oxygen, we see that the reaction is

$$^{4}_{2}He + ^{14}_{7}N \longrightarrow ^{17}_{8}O + ^{1}_{1}H \tag{32.16}$$

This nuclear reaction starts with two stable isotopes, helium and nitrogen, and produces two different stable isotopes, hydrogen and oxygen.

Since the time of Rutherford, thousands of nuclear reactions have been observed, particularly following the development of charged-particle accelerators in the 1930s.

EXAMPLE 32.10 The Discovery of the Neutron

A nuclear reaction of significant historical note occurred in 1932 when Chadwick, in England, bombarded a beryllium target with alpha particles. Analysis of the experiment indicated that the following reaction occurred:

$$^{4}_{2}He + ^{9}_{4}Be \longrightarrow ^{12}_{6}C + X$$

What is X in this reaction?

Solution Balancing mass numbers and atomic numbers, we see that the unknown particle must be represented as $^{1}_{0}X$, that is, with a mass of 1 and zero charge. Hence, the particle X is the neutron, $^{1}_{0}n$. This experiment was the first to provide positive proof of the existence of neutrons.

EXAMPLE 32.11 Synthetic Elements

(a) A beam of neutrons is directed at a target of $^{238}_{92}U$. The reaction products are a gamma ray and another isotope. What is the isotope? (b) This isotope is radioactive and emits a beta particle. Write the equation symbolizing this decay and identify the resulting isotope. (c) This isotope is also radioactive and decays by beta emission. What is the end product? (d) What is the significance of these reactions?

Solution (a) Balancing input with output gives

$$^{1}_{0}n + ^{238}_{92}U \longrightarrow ^{239}_{92}U + \gamma$$

(b) The decay of ^{239}U by beta emission is

$$^{239}_{92}U \longrightarrow ^{239}_{93}Np + ^{0}_{-1}e + \bar{\nu}$$

(c) The decay of $^{239}_{93}Np$ by beta emission gives

$$^{239}_{93}Np \longrightarrow ^{239}_{94}Pu + ^{0}_{-1}e + \bar{\nu}$$

(d) The feature of these reactions that makes them of interest is the fact that uranium is the element with the greatest number of protons, 92, that exists in nature in any appreciable amount. The reactions in parts (a), (b), and (c) do occur ocasionally in nature so that minute traces of neptunium and plutonium are present. In 1940, however, researchers bombarded uranium with neutrons to produce plutonium and neptunium by the steps given above. These two elements were thus the first elements made in the laboratory, and by bombarding them with neutrons and other particles, the list of synthetic elements has been extended to include those up to atomic number 108 (and possibly 109).

32.8 *Q* VALUES

In the last section, we examined some nuclear reactions for which mass numbers and atomic numbers were balanced in the equations. We shall now consider the energy involved in these reactions, since energy is another important quantity that must be conserved.

Let us illustrate this procedure by analyzing the following nuclear reaction:

$${}^{2}_{1}\text{H} + {}^{14}_{7}\text{N} \longrightarrow {}^{12}_{6}\text{C} + {}^{4}_{2}\text{He} \tag{32.17}$$

The total amount of mass on the left side of the equation is the sum of the mass of ${}^{2}_{1}\text{H}$ (2.014102 u) and the mass of ${}^{14}_{7}\text{N}$ (14.003074 u), which equals 16.017176 u. Similarly, the mass on the right side of the equation is the sum of the mass of ${}^{12}_{6}\text{C}$ (12.000000 u) plus the mass of ${}^{4}_{2}\text{He}$ (4.002603 u), for a total of 16.002603 u. Thus, the total mass before the reaction is greater than the total mass after the reaction. The mass difference in this reaction is equal to 16.017176 u − 16.002603 u = 0.014573 u. This "lost" mass is converted to the kinetic energy of the nuclei present after the reaction. In energy units, 0.014573 u is equivalent to 13.567 MeV of kinetic energy carried away by the carbon and helium nuclei.

The amount of energy required to balance the equation is called the *Q* value of the reaction. In Equation 32.17 the *Q* value is 13.567 MeV. Nuclear reactions in which there is a release of energy, that is, positive *Q* values, are said to be **exothermic reactions.**

The energy balance sheet is not complete, however. We must also consider the kinetic energy of the incident particle before the collision. As an example, let us assume that the deuteron in Equation 32.17 has a kinetic energy of 5 MeV. Adding this to our *Q* value, we find that the carbon and helium nuclei will have a total kinetic energy of 18.567 MeV following the reaction.

Now consider the reaction

$${}^{4}_{2}\text{He} + {}^{14}_{7}\text{N} \longrightarrow {}^{17}_{8}\text{O} + {}^{1}_{1}\text{H} \tag{32.18}$$

Before the reaction, the total mass is the sum of the masses of the alpha particle and the nitrogen nucleus: 4.002603 u + 14.003074 u = 18.005704 u. After the reaction, the total mass is the sum of the masses of the oxygen nucleus and the proton: 16.999133 u + 1.007825 u = 18.006958 u. In this case, the total mass after the reaction is *greater* than the total mass before the reaction. The mass deficit is 0.001254 u, which is equivalent to an energy deficit of 1.167 MeV. This deficit is expressed by saying that the Q value of the reaction is a negative number, -1.167 MeV. Reactions with negative Q values are called **endothermic reactions.** Such reactions will not take place unless the incoming particle has at least enough kinetic energy to overcome the energy deficit.

At first, it might appear that the reaction in Equation 32.18 could take place if the incoming alpha particle had a kinetic energy of 1.167 MeV. In practice, however, the alpha particle must have more energy than this. If it had an energy of only 1.167 MeV, energy would be conserved but careful analysis would show that momentum was not. This can easily be understood by recognizing that the incoming alpha particle has some momentum before the reaction. However, if its kinetic energy was only 1.167 MeV, the products (oxygen and a proton) would be created with zero kinetic energy and, thus, zero momentum. It can be shown that, in order to conserve both energy and momentum, the incoming particle must have a minimum kinetic energy given by

$$KE_{\min} = \left(1 + \frac{m}{M}\right)|Q| \tag{32.19}$$

where m is the mass of the incident particle, M is the mass of the target, and the absolute value of the Q value is to be used. For the reaction given by Equation 32.18, we find

$$KE_{\min} = \left(1 + \frac{4.002603}{14.003074}\right)|-1.167 \text{ MeV}| = 1.501 \text{ MeV}$$

This minimum value of the kinetic energy of the incoming particle is called the **threshold energy.** The nuclear reaction shown in Equation 32.18 will not occur if the incoming alpha particle has an energy less than 1.501 MeV but will occur if the kinetic energy is equal to or greater than 1.501 MeV.

SUMMARY

Nuclei are represented symbolically as ${}^{A}_{Z}\text{X}$, where X represents the chemical symbol for the element. The quantity A is the **mass number,** which equals the total number of nucleons (neutrons plus protons) in the nucleus. The quantity Z is the **atomic number,** which equals the number of protons in the nucleus. Nuclei that contain the same number of protons but different numbers of neutrons are called **isotopes.** In other words, isotopes have the same Z value but different A values.

Most nuclei are approximately spherical, with an average radius given by

$$r = r_0 A^{1/3} \tag{32.1}$$

where A is the mass number and r_0 is a constant equal to 2×10^{-15} m.

The total mass of a nucleus is always less than the sum of the masses of its individual nucleons. This mass difference, Δm, multiplied by c^2 gives the **binding energy** of the nucleus.

There are three different models of the nucleus. These are the **liquid-drop model,** the **independent-particle model,** and the **collective model.** Each has its own sphere of usefulness. For example, the liquid-drop model is quite effective in describing the binding energy of the nucleus.

The spontaneous emission of radiation by certain nuclei is called **radioactivity.** There are three processes by which a radioactive substance can decay: alpha (α) decay, in which the emitted particles are ${}^{4}_{2}He$ nuclei; beta (β) decay, in which the emitted particles are electrons; and gamma (γ) decay, in which the emitted particles are high-energy photons.

The **decay rate,** or **activity,** R, of a sample is given by

$$R = \lambda N \tag{32.4}$$

where N is the number of radioactive nuclei at some instant and λ is a constant for a given substance called the **decay constant.**

Nuclei in a radioactive substance decay such that the number of nuclei present varies with time according to the expression

$$N = N_0 e^{-\lambda t} \tag{32.5}$$

where N is the number of radioactive nuclei present at time t, N_0 is the number at time $t = 0$, and $e = 2.718. \ . \ . \ .$

The **half-life,** $T_{1/2}$, of a radioactive substance is the time required for half of a given number of radioactive nuclei to decay. The half-life is related to the decay constant as

$$T_{1/2} = \frac{0.693}{\lambda} \tag{32.8}$$

If a nucleus decays by alpha emission, it loses two protons and two neutrons. A typical alpha decay is

$${}^{238}_{92}U \longrightarrow {}^{234}_{90}Th + {}^{4}_{2}He \tag{32.9}$$

Note that in this decay, as in all radioactive decay processes, the sum of the Z values on the left equals the sum of the Z values on the right; the same is true for the A values.

A typical beta decay is

$${}^{14}_{6}C \longrightarrow {}^{14}_{7}N + {}^{0}_{-1}e + \bar{\nu}$$

When a nucleus decays by beta emission, a particle called an **antineutrino** ($\bar{\nu}$) is also emitted. In positron decay a **neutrino** (ν) is emitted. A neutrino has zero electric charge and a small rest mass (which may be zero) and interacts weakly with matter.

Nuclei are often in an excited state following radioactive decay and release their extra energy by emitting a high-energy photon called a **gamma ray** (γ). A typical gamma ray emission is

$${}^{12}_{6}C^* \longrightarrow {}^{12}_{6}C + \gamma \tag{32.14b}$$

where the asterisk indicates that the carbon nucleus was in an excited state before gamma emission.

Nuclear reactions can occur when a bombarding particle strikes another nucleus. A typical nuclear reaction is

$$^{4}_{2}He + ^{14}_{7}N \longrightarrow ^{17}_{8}O + ^{1}_{1}H \tag{32.16}$$

In this reaction, an alpha particle strikes a nitrogen nucleus, producing an oxygen nucleus and a proton. As in radioactive decay, atomic numbers and mass numbers balance on the two sides of the arrow.

Nuclear reactions in which there is a release of energy are said to be **exothermic reactions** and are characterized by positive Q values. Reactions with negative Q values, called **endothermic reactions,** cannot occur unless the incoming particle has at least enough kinetic energy to overcome the energy deficit. In order to conserve both energy and momentum, the incoming particle must have a minimum kinetic energy, called the **threshold energy,** given by

$$\boxed{KE_{\mathrm{min}} = \left(1 + \frac{m}{M}\right)|Q|} \tag{32.19}$$

where m is the mass of the incident particle and M is the mass of the target atom.

ADDITIONAL READINGS

G. F. Bertsch, "Vibrations of the Atomic Nucleus," *Sci. American*, May 1983, p. 62.

H. Bethe, "What Holds the Nucleus Together?" *Sci. American*, September 1953, p. 201.

D. A. Bromley, "Nuclear Models," *Sci. American*, December 1978, p. 58.

B. N. Da Costa Andrade, *Rutherford and the Nature of the Atom*. Garden City, N.Y., Science Study Series, Doubleday, 1964.

S. A. Fetter and K. Tsipis, "Catastrophic Releases of Radioactivity," *Sci. American*, April 1981, p. 41.

O. Hahn, "The Discovery of Fission," *Sci. American*, February 1958, p. 76.

R. E. Hedges and J. Gowlett, "Radiocarbon Dating by Accelerator Mass Spectrometry," *Sci. American*, January 1986, p. 100.

R. E. M. Hedges, "Radioisotope Clocks in Archaeology," *Nature*, Vol. 281, No. 5725, 1979, p. 19.

M. Jacob and P. Landshoff, "The Inner Structure of the Proton," *Sci. American*, March 1980, p. 66.

J. G. Learned and D. Eichler, "A Deep-Sea Neutrino Telescope," *Sci. American*, February 1981, p. 138.

J. M. LoSecco, F. Reines, and D. Sinclair, "The Search for Proton Decay," *Sci. American*, June 1985, p. 54.

R. B. Marshak, "The Nuclear Force," *Sci. American*, March 1960, p. 98.

S. Weinberg, "The Decay of the Proton," *Sci. American*, June 1981, p. 64.

V. F. Weisskopf and E. P. Rosenbaum, "A Model of the Nucleus," *Sci. American*, December 1955, p. 261.

E. Witten, "New Ideas About Neutrino Masses," *The Physics Teacher*, February 1983, p. 78.

QUESTIONS

1. In Rutherford's experiment, assume that an alpha particle is headed directly toward the nucleus of an atom. Why does the alpha particle not make physical contact with the nucleus?
2. If the atom were indeed like the plum-pudding model, what results would Geiger and Marsden have obtained in their alpha-scattering experiment?
3. Why do heavier elements require more neutrons in order to maintain stability?
4. If film is kept in a box, alpha particles from a radioactive source outside the box cannot expose the film but beta particles can. Explain.
5. An alpha particle has twice the charge of a beta particle. Why does the former deflect less than the latter when passing between electrically charged plates, assuming they both have the same speed?
6. Pick any beta decay process and show that the neutrino must have zero charge.

7. Suppose it could be shown that the cosmic ray intensity was much greater 10,000 years ago. How would this affect present values of the age of ancient samples of once-living matter?
8. Why is carbon dating unable to provide accurate estimates of very old material?
9. Element X has several isotopes. What do these isotopes have in common? How do they differ?
10. Explain the main differences between alpha, beta, and gamma rays.
11. How many protons are there in the neucleus $^{222}_{86}Rn$? How many neutrons? How many electrons are there in the neutral atom?

PROBLEMS

Table 32.3 will be useful for many of these problems. A more complete list of atomic masses is given in Appendix B.

Section 32.1 Some Properties of Nuclei

1. Find the radius of a nucleus of (a) $^{4}_{2}He$ and (b) $^{238}_{92}U$.

TABLE 32.3 Some Atomic Masses

Element	Atomic Mass (u)
$^{4}_{2}He$	4.002603
$^{7}_{3}Li$	7.016004
$^{9}_{4}Be$	9.012182
$^{10}_{5}B$	10.012938
$^{12}_{6}C$	12.000000
$^{13}_{6}C$	13.003355
$^{14}_{7}N$	14.003074
$^{15}_{7}N$	15.000109
$^{15}_{8}O$	15.003065
$^{18}_{8}O$	17.999159
$^{18}_{9}F$	18.000937
$^{20}_{10}Ne$	19.992439
$^{23}_{11}Na$	22.989770
$^{23}_{12}Mg$	22.994127
$^{27}_{13}Al$	26.981541
$^{30}_{15}P$	29.978310
$^{40}_{20}Ca$	39.962591
$^{43}_{20}Ca$	42.958770
$^{56}_{26}Fe$	55.934939
$^{64}_{30}Zn$	63.929145
$^{64}_{29}Cu$	63.929599
$^{93}_{41}Nb$	92.906378
$^{197}_{79}Au$	196.966560
$^{202}_{80}Hg$	201.970632
$^{216}_{84}Po$	216.001790
$^{220}_{86}Rn$	220.011401
$^{238}_{92}U$	238.050786

2. Find the atomic number of a nucleus that has a radius 4.36×10^{-15} m.

3. (a) Find the weight density of nuclear matter in lb/ft^3. (See the results to Example 32.2.) (b) Find the weight in pounds that a sphere 5 cm in radius would have if it were composed completely of nuclear matter and located on the surface of the earth.

4. Find the diameter of a sphere of nuclear matter that would have a mass equal to that of the earth. Base your calculation on an earth radius of 6.37×10^6 m and an average earth density of 5.52×10^3 kg/m^3.

5. Find the nucleus that has a radius approximately equal to one-half the radius of uranium, $^{238}_{92}U$.

6. Consider the hydrogen atom to be a sphere of radius equal to the Bohr radius, 0.53×10^{-10} m, and calculate the approximate value of the ratio of the nuclear density to the atomic density.

• **7.** Use energy methods to calculate the distance of closest approach for a head-on collision between an alpha particle with an initial energy of 0.5 MeV and a gold nucleus (^{197}Au) at rest. Assume the gold nucleus remains at rest during the collision.

• **8.** (a) Find the velocity an alpha particle must have in order to come within 3.2×10^{-14} m of a gold nucleus. (b) Find the energy of the alpha particle in MeV.

• **9.** Use energy methods to find the distance of closest approach of a 0.5 MeV proton to a gold nucleus at rest.

Section 32.2 Binding Energy

10. Calculate the total binding energy for $^{20}_{10}Ne$.

11. Calculate the total binding energy for $^{40}_{20}Ca$.

12. Calculate the binding energy per nucleon for $^{93}_{41}Nb$.

13. Calculate the binding energy per nucleon for $^{197}_{79}Au$.

14. A pair of nuclei for which $Z_1 = N_2$ and $Z_2 = N_1$ are called *mirror isobars* (the atomic and neutron numbers are interchangeable). Binding energy measurements on such pairs can be used to obtain evidence of the charge independence of nuclear forces. Charge independence means that the proton-proton, proton-neutron, and neutron-neutron forces are approximately equal. Calculate the difference in binding energy for the two mirror nuclei $^{15}_{8}O$ and $^{15}_{7}N$.

15. Two isotopes having the same mass number are known as isobars. Calculate the difference in binding energy per nucleon for the isobars $^{23}_{11}Na$ and $^{23}_{12}Mg$. How do you account for this difference?

• 16. Calculate the binding energy of the last neutron in the $^{43}_{20}Ca$ nucleus. (*Hint:* You should compare the mass of $^{43}_{20}Ca$ with the mass of $^{42}_{20}Ca$ plus the mass of a neutron. The mass of $^{42}_{20}Ca$ = 41.958622 u.)

Section 32.3 Nuclear Models

17. (a) In the liquid-drop model of the nucleus, why does the surface-effect term have a minus sign? (b) The binding energy of the nucleus increases as the volume-to-surface ratio increases. Calculate this ratio for both spherical and cubical shapes and explain which shape is more plausible for nuclei.

18. Use Equation 32.2 and the numerical values given for the constants C_1, C_2, and C_3 to calculate the binding energy for $^{64}_{29}Cu$.

Section 32.4 Radioactivity

19. The half-life of a radioactive sample is 30 min. If you start with a sample containing 3×10^{16} nuclei, how many of these nuclei remain after 10 min?

20. How long will it take for a sample of polonium of half-life 140 days to decay to 1/10 its original strength?

21. Suppose that you start with 10^{-3} g of a pure radioactive substance and 2 hours later determine that only 0.25×10^{-3} g of the substance remains. What is the half-life of this substance?

22. The half-life of an isotope of phosphorus is 14 days. If a sample contains 3×10^{16} such nuclei, determine its activity.

23. How many radioactive atoms are present in a sample that has an activity of 0.2 μCi and a half-life of 8.1 days?

• 24. A freshly prepared sample of a certain radioactive isotope has an activity of 10 mCi. After 4 h, the activity is 8 mCi. (a) Find the decay constant and half-life of the isotope. (b) How many atoms of the isotope were contained in the freshly prepared sample? (c) What is the sample's activity 30 h after it is prepared?

• 25. Tritium has a half-life of 12.33 years. What percentage of the nuclei in a tritium sample will decay in 5 years?

Section 32.5 The Decay Processes

26. Complete the following radioactive decay formulas:

$$^{212}_{83}Bi \longrightarrow ? + ^{4}_{2}He$$
$$^{95}_{36}Kr \longrightarrow ? + ^{0}_{-1}e$$
$$? \longrightarrow ^{4}_{2}He + ^{140}_{58}Ce$$

27. Complete the following radioactive decay formulas:

$$^{12}_{5}B \longrightarrow ? + ^{0}_{-1}e$$
$$^{234}_{90}Th \longrightarrow ^{230}_{88}Ra + ?$$
$$? \longrightarrow ^{14}_{7}N + ^{0}_{-1}e$$

28. Find the energy released in the alpha decay of $^{232}_{92}U$.

29. Find the energy released in the alpha decay of $^{238}_{92}U$. You will find the following mass values useful:

$$M(^{238}_{92}U) = 238.050786 \text{ u}$$
$$M(^{234}_{90}Th) = 234.043583 \text{ u}$$
$$M(^{4}_{2}He) = 4.002603 \text{ u}$$

30. Is it energetically possible for a $^{8}_{4}Be$ nucleus to spontaneously decay into two alpha particles? Explain.

31. Determine which of the following suggested decays can occur spontaneously:

(a) $^{40}_{20}Ca \longrightarrow ^{0}_{1}e + ^{40}_{19}K$ (b) $^{144}_{60}Nd \longrightarrow ^{4}_{2}He + ^{140}_{58}Ce$

32. Figure 32.10 shows the steps by which $^{235}_{92}U$ decays to $^{207}_{82}Pb$. Enter the correct isotope symbol in each square.

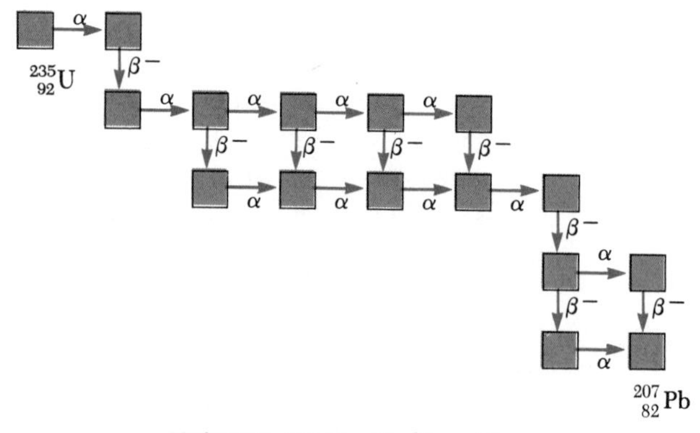

FIGURE 32.10 Problem 32.

• 33. A piece of charcoal used for cooking is found at the remains of an ancient campsite. A 1000-g sample of carbon from the wood is found to have an activity of 2000 decays per minute. Find the age of the charcoal.

Section 32.7 Nuclear Reactions

Section 32.8 Q Values

34. Find the Q value of the reaction discussed in Example 32.10.

35. The first nuclear reaction utilizing particle accelerators was performed by Cockcroft and Walton. Accelerated protons were used to bombard lithium nuclei, producing the following reaction:

$$^{1}_{1}\mathrm{H} + {}^{7}_{3}\mathrm{Li} \longrightarrow {}^{4}_{2}\mathrm{He} + {}^{4}_{2}\mathrm{He}$$

Since the masses of the particles involved in the reaction were well known, these results were used to obtain an early proof of the Einstein mass-energy relation. Calculate the Q value of the reaction.

36. Complete the following nuclear reactions:

$$? + {}^{14}_{7}\mathrm{N} \longrightarrow {}^{1}_{1}\mathrm{H} + {}^{17}_{8}\mathrm{O}$$

$$^{7}_{3}\mathrm{Li} + {}^{1}_{1}\mathrm{H} \longrightarrow {}^{4}_{2}\mathrm{He} + ?$$

37. Complete the following nuclear reactions:

$$^{27}_{13}\mathrm{Al} + {}^{4}_{2}\mathrm{He} \longrightarrow ? + {}^{30}_{15}\mathrm{P}$$

$$^{1}_{0}\mathrm{n} + ? \longrightarrow {}^{4}_{2}\mathrm{He} + {}^{7}_{3}\mathrm{Li}$$

38. The first known reaction in which the product nucleus was radioactive (achieved in 1934) was one in which $^{27}_{13}\mathrm{Al}$ was bombarded with alpha particles. Produced in the reaction were a neutron and a product nucleus. (a) What was the product nucleus? (b) Find the Q value of the reaction.

39. When a $^{6}_{3}\mathrm{Li}$ nucleus is struck by a proton, an alpha particle and a product nucleus are released. (a) What is the product nucleus? (b) Find the Q value of the reaction.

40. (a) Suppose $^{10}_{5}\mathrm{B}$ is struck by an alpha particle, releasing a proton and a product nucleus in the reaction. (a) What is the product nucleus? (b) An alpha particle and a product nucleus are produced when $^{13}_{6}\mathrm{C}$ is struck by a proton. What is the product nucleus?

41. (a) The reactions discussed in Problem 40 are often called inverse reactions. Why? (b) Show that the two reactions have the same absolute value of Q.

42. When $^{18}\mathrm{O}$ is struck by a proton, $^{18}\mathrm{F}$ and another particle are produced. (a) What is the other particle? (b) This reaction has a Q value of -2.453 MeV, and the atomic mass of $^{18}\mathrm{O}$ is 17.999160 u. What is the atomic mass of $^{18}\mathrm{F}$?

43. Find the threshold energy that the incident neutron must have to produce the reaction

$$^{1}_{0}\mathrm{n} + {}^{4}_{2}\mathrm{He} \longrightarrow {}^{2}_{1}\mathrm{H} + {}^{3}_{1}\mathrm{H}$$

ADDITIONAL PROBLEMS

44. A sample of organic material is found to contain 18 g of carbon. The investigators believe the bones to be 20,000 years old based on samples of pottery taken from the site. If so, what is the expected activity of the sample?

45. Is it energetically possible for a $^{12}_{6}\mathrm{C}$ nucleus to spontaneously decay into three alpha particles? Explain.

46. A sample of 200 mCi of a radioactive isotope is purchased by a medical supply house. If the sample has a half-life of 14 days, how long can they keep the sample before its activity is reduced to 20 mCi?

47. One method for producing neutrons for experimental use is to bombard $^{7}_{3}\mathrm{Li}$ with protons. The neutrons are emitted according to the following reaction:

$$^{1}_{1}\mathrm{H} + {}^{7}_{3}\mathrm{Li} \longrightarrow {}^{7}_{4}\mathrm{Be} + {}^{1}_{0}\mathrm{n}$$

What is the minimum kinetic energy the incident proton must have if this reaction is to occur?

• **48.** Deuterons that have been accelerated are used to bombard other deuterium nuclei, resulting in the reaction

$$^{2}_{1}\mathrm{H} + {}^{2}_{1}\mathrm{H} \longrightarrow {}^{3}_{2}\mathrm{He} + {}^{1}_{0}\mathrm{n}$$

Does this reaction require a threshold energy? If so, what is its value?

• **49.** A medical laboratory stock solution is prepared with an initial activity due to $^{24}\mathrm{Na}$ of 2.5 mCi/ml, and 10 ml of the stock solution is diluted at $t_0 = 0$ to a working solution whose total volume is 250 ml. After 48 h, a 5-ml sample of the working solution is monitored with a counter. What is the measured activity? (Note that 1 ml = 1 milliliter.)

• **50.** A by-product of some fission reactors is the isotope $^{239}_{94}\mathrm{Pu}$, which is an alpha emitter with a half-life of 24,000 years:

$$^{239}_{94}\mathrm{Pu} \longrightarrow {}^{235}_{92}\mathrm{U} + \alpha$$

Consider a sample of 1 kg of pure $^{239}_{94}\mathrm{Pu}$ at $t = 0$. Calculate (a) the number of $^{239}_{94}\mathrm{Pu}$ nuclei present at $t = 0$ and (b) the initial activity of the sample. (c) How long does the sample have to be stored if a "safe" activity level is 0.1 Bq?

•• **51.** A fission reactor is hit by a nuclear weapon and evaporates 5×10^6 Ci of $^{90}\mathrm{Sr}$ ($T_{1/2} = 28.7$ years) into the air. The $^{90}\mathrm{Sr}$ falls out over an area of 10^4 km^2. How long will it take the activity of the $^{90}\mathrm{Sr}$ to reach the agriculturally "safe" level of 2 μCi/m^2?

•• **52.** Free neutrons have a characteristic half-life of 12 min. What fraction of a group of free neutrons at thermal energy (0.04 eV) will decay before traveling a distance of 10 km?

•• **53.** The radioactive isotope ^{198}Au has a half-life of 64.8 h. A sample containing this isotope has an initial activity of 40 μCi. Calculate the number of nuclei that will decay in the time interval between $t_1 = 10$ h and $t_2 = 12$ h.

•• **54.** A piece of charcoal has a mass of 25 g and is known to be about 25,000 years old. (a) Determine the number of decays per minute expected from this sample. (b) If the radioactive background in the counter without a sample is 20 counts/min and we assume 100% efficiency in counting, explain why 25,000 years is close to the limit of dating with this technique.

•• **55.** Many radioisotopes have important industrial, medical, and research applications. One of these is ^{60}Co, which has a half-life of 5.2 years and decays by the emission of a beta particle (energy 0.31 MeV) and two gamma photons (energies 1.17 MeV and 1.33 MeV). A scientist wishes to prepare a ^{60}Co sealed source that will have an activity of at least 10 Ci after 30 months of use. What is the minimum initial mass of ^{60}Co required?

•• **56.** During the manufacture of a steel engine component, radioactive iron (^{59}Fe) is included in the total mass of 0.2 kg. The component is placed in a test engine when the activity due to the isotope is 20 μCi. After a 1000-hour test period, oil is removed from the engine and found to contain enough ^{59}Fe to produce 800 disintegrations/min per liter of oil. The total volume of oil in the engine is 6.5 liters. Calculate the total mass worn from the engine component per hour of operation. (The half-life for ^{59}Fe is 45.1 days.)

CHAPTER 33

Nuclear Physics Applications

The San Onofre Nuclear Generating Station at San Clemente, California, provides enough power–430,000 kW—to supply a city of over half a million population. A pressurized water nuclear reactor is housed in the steel containment sphere near the center of the picture. (Southern California Edison Company)

The primary thrust of this chapter is to examine some important applications of nuclear physics. We shall begin by considering the two means by which energy can be derived from nuclear reactions. These two techniques are fission, in which a nucleus of large mass number splits, or fissions, into two smaller nuclei, and fusion, in which two light nuclei fuse to form a heavier nucleus. In either case, there is a release of large amounts of energy, which can be used destructively through bombs or constructively through the production of electric power. We shall also examine several devices used to detect radiation. This will be followed by a discussion of some industrial and biological applications of radiation.

33.1 NUCLEAR FISSION

Nuclear fission occurs when a heavy nucleus, such as ^{235}U, splits, or fissions, into two smaller nuclei. In such a reaction, *the total rest mass of the products is less than the original rest mass.*

Nuclear fission was first observed in 1939 by Otto Hahn and Fritz Strassman, following some basic studies by Fermi. After bombarding uranium (Z = 92) with neutrons, Hahn and Strassman discovered among the reaction products two medium-mass elements, barium and lanthanum. Shortly thereafter, Lisa Meitner and Otto Frisch explained what had happened. The uranium nucleus had split into two nearly equal fragments after absorbing a neutron. Such an occurrence was of considerable interest to physicists attempting to understand the nucleus, but it was to have even more far-reaching consequences. Measurements showed that about 200 MeV of energy is released in each fission event, and this fact was to affect the course of human history.

The fission of ^{235}U by slow (low energy) neutrons can be represented by the equation

$$ {}^{1}_{0}n + {}^{235}_{92}U \longrightarrow {}^{236}_{92}U^{*} \longrightarrow X + Y + \text{neutrons} \tag{33.1}$$

where $^{236}U^{*}$ is an intermediate state that lasts only for about 10^{-12} s before splitting into X and Y. The resulting nuclei, X and Y, are called **fission fragments.** There are many combinations of X and Y that satisfy the requirements of conservation of mass-energy and charge. In the fission of uranium, there are about 90 different daughter nuclei that can be formed. The process also results in the production of several neutrons, typically two or three. On the average 2.47 neutrons are released per event.

A typical reaction of this type is

$$ {}^{1}_{0}n + {}^{235}_{92}U \longrightarrow {}^{141}_{56}Ba + {}^{92}_{36}Kr + 3\,{}^{1}_{0}n \tag{33.2}$$

The fission fragments, barium, and krypton, and the released neutrons have a great deal of kinetic energy following the fission event.

The breakup of the uranium nucleus can be compared to what happens to a drop of water when excess energy is added to it. All the atoms in the drop have energy, but this energy is not great enough to break up the drop. However, if enough energy is added to set the drop vibrating, it will undergo elongation and compression until the amplitude of vibration becomes large enough to cause the drop to break. In the uranium nucleus, a similar process occurs (Fig. 33.1). The sequence of events is

1. The ^{235}U nucleus captures a thermal (slow-moving) neutron.
2. This capture results in the formation of $^{236}U^{*}$, and the excess energy of this nucleus causes it to undergo violent oscillations.
3. The $^{236}U^{*}$ nucleus becomes highly distorted, and the force of repulsion between protons in the two halves of the dumbbell shape tends to increase the distortion.
4. The nucleus splits into two fragments, emitting several neutrons in the process.

Let us estimate the disintegration energy, Q, released in a typical fission process. From Figure 32.3 we see that the binding per nucleon is about 7.6

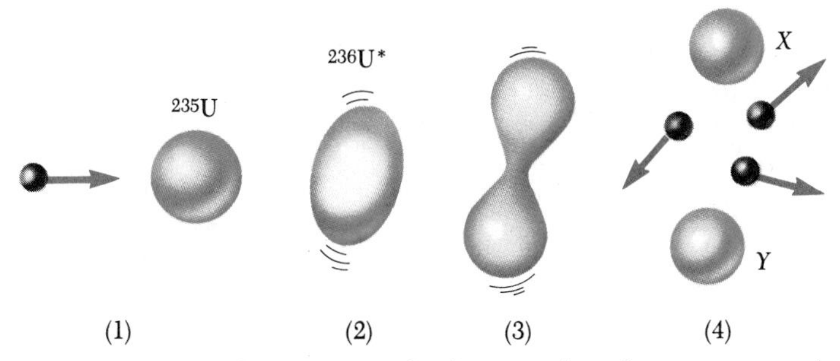

FIGURE 33.1 The stages involved in a nuclear fission event as described by the liquid-drop model of the nucleus.

MeV for heavy nuclei (those having a mass number approximately equal to 240) and about 8.5 MeV for nuclei of intermediate mass. This means that the nucleons in the fission fragments are more tightly bound and therefore have less mass than the nucleons in the original heavy nucleus. This decrease in mass per nucleon appears as released energy when fission occurs. The amount of energy released is (8.5 − 7.6) MeV per nucleon. Assuming a total of 240 nucleons, we find that the energy released per fission event is

$$Q = (240 \text{ nucleons})\left(8.5\,\frac{\text{MeV}}{\text{nucleon}} - 7.6\,\frac{\text{MeV}}{\text{nucleon}}\right) = 220 \text{ MeV}$$

This is indeed a very large amount of energy relative to the amount of energy released in chemical processes. For example, the energy released in the combustion of one molecule of the octane used in gasoline engines is about one millionth the energy released in a single fission event!

EXAMPLE 33.1 The Fission of Uranium

Two other possible ways by which ^{235}U can undergo fission when bombarded with a neutron are (1) by the release of ^{140}Xe and ^{94}Sr as fission fragments and (2) by the release of ^{132}Sn and ^{101}Mo as fission fragments. In each case, neutrons are also released. Find the number of neutrons released in each of these events.

Solution By balancing mass numbers and atomic numbers, we find that these reactions can be written

$$^{1}_{0}\text{n} + ^{235}_{92}\text{U} \rightarrow ^{140}_{54}\text{Xe} + ^{94}_{38}\text{Sr} + 2\,^{1}_{0}\text{n}$$

$$^{1}_{0}\text{n} + ^{235}_{92}\text{U} \rightarrow ^{132}_{50}\text{Sn} + ^{101}_{42}\text{Mo} + 3\,^{1}_{0}\text{n}$$

Thus, two neutrons are released in the first event and three in the second.

EXAMPLE 33.2 The Energy Released in the Fission of ^{235}U

Calculate the total energy released if 1 kg of ^{235}U undergoes fission, taking the disintegration energy per event to be Q = 208 MeV (a more accurate value than the estimate given before).

Solution We need to know the number of nuclei in 1 kg of uranium. Since A = 235, the number of nuclei is

$$N = \left(\frac{6.02 \times 10^{23} \text{ nuclei/mole}}{235 \text{ g/mole}}\right)(10^3 \text{ g}) = \boxed{2.56 \times 10^{24} \text{ nuclei}}$$

Hence the disintegration energy

$$E = NQ = (2.56 \times 10^{24} \text{ nuclei}) \left(208 \frac{\text{MeV}}{\text{nucleus}}\right)$$
$$= 5.32 \times 10^{26} \text{ MeV}$$

Since 1 MeV is equivalent to 4.45×10^{-20} kWh, $E = 2.37 \times 10^7$ kWh. This is enough energy to keep a 100-W lightbulb burning for about 30,000 years. Thus, 1 kg of ^{235}U is a relatively large amount of fissionable material.

33.2 NUCLEAR REACTORS

We have seen that, when ^{235}U undergoes fission, an average of about 2.5 neutrons are emitted per event. These neutrons can in turn trigger other nuclei to undergo fission, with the possibility of a chain reaction (Fig. 33.2). Calculations show that if the chain reaction is not controlled (that is, if it does not proceed slowly), it could result in a violent explosion, with the release of an enormous amount of energy, even from only 1 g of ^{235}U. If the energy in 1 kg of ^{235}U were released, it would be equivalent to detonating about 20,000 tons of TNT! This, of course, is the principle behind the first nuclear bomb, an uncontrolled fission reaction.

A nuclear reactor is a system designed to maintain what is called a **self-sustained chain reaction.** This important process was first achieved in 1942 by Fermi at the University of Chicago, with natural uranium as the fuel. Most reactors in operation today also use uranium as fuel. Natural uranium contains only about 0.7% of the ^{235}U isotope, with the remaining 99.3% being the ^{238}U isotope. This is important to the operation of a reactor because ^{238}U almost never undergoes fission. Instead, it tends to absorb neutrons, producing neptunium and plutonium. For this reason, reactor fuels must be artificially enriched to contain a few percent of the ^{235}U isotope.

Earlier, we mentioned that an average of about 2.5 neutrons are emitted in each fission event of ^{235}U. In order to achieve a self-sustained chain reaction, one of these neutrons, on the average, must be captured by another

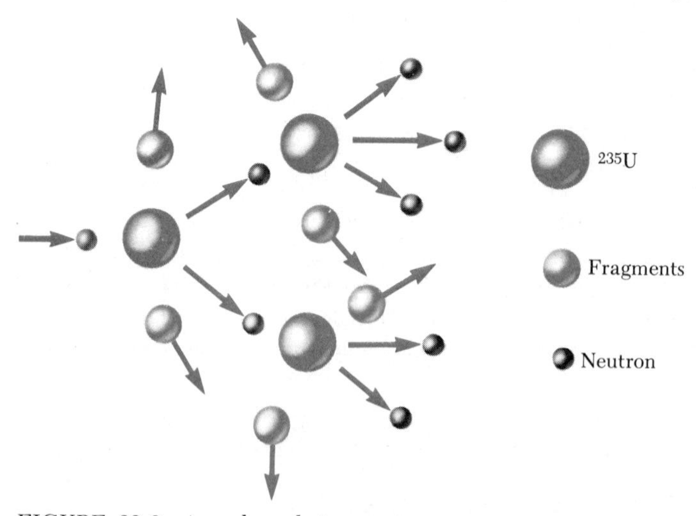

FIGURE 33.2 A nuclear chain reaction.

^{235}U nucleus and cause it to undergo fission. A useful parameter for describing the level of reactor operation is the **reproduction constant,** K, defined as *the average number of neutrons from each fission event that will cause another event.* As we have seen, K can have a maximum value of 2.5 in the fission of uranium. However, in practice K is less than this because of several factors, which we shall soon discuss.

A self-sustained chain reaction is achieved when $K = 1$. Under this condition, the reactor is said to be **critical.** When K is less than unity, the reactor is subcritical and the reaction dies out. When K is greater than unity, the reactor is said to be supercritical and a runaway reaction occurs. In a nuclear reactor used to furnish power to a utility company, it is necessary to maintain a value of K close to unity.

The basic design of a nuclear reactor is shown in Figure 33.3. The fuel elements consist of enriched uranium. The function of the remaining parts of the reactor and some aspects of its design will now be described.

Control rods
Radiation shield
Fuel elements
Moderator material

FIGURE 33.3 Cross-section of a reactor core surrounded by a radiation shield.

Neutron Leakage

In any reactor, a fraction of the neutrons produced in fission will leak out of the core before inducing other fission events. If the fraction leaking out is too large, the reactor will not operate. The percentage lost is large if the reactor is very small because leakage is a function of the ratio of surface area to volume. Therefore, a critical feature of the design of a reactor is to choose the correct surface-area-to-volume ratio so that a sustained reaction can be achieved.

Sketch of the world's first reactor. Because of wartime secrecy, there are no photographs of the completed reactor. The reactor was composed of layers of graphite interspersed with uranium. A self-sustained chain reaction was first achieved on December 2, 1942. Word of the success was telephoned immediately to Washington with this message: "The Italian navigator has landed in the New World and found the natives very friendly." The historic event took place in an improvised laboratory in the racquets court under the west stands of the University of Chicago's Stagg Field and the Italian navigator was Fermi. (Courtesy of Argonne National Laboratory)

Regulating Neutron Energies

The neutrons released in fission events are very energetic, with kinetic energies of about 2 MeV. It is found that slow neutrons are far more likely than fast neutrons to produce fission events in ^{235}U. Therefore, in order for the chain reaction to continue, the neutrons must be slowed down. This is accomplished by surrounding the fuel with a **moderator** substance.

In order to understand how neutrons are slowed, consider a collision between a light object and a very massive one. In such an event, the light object rebounds from the collision with most of its original kinetic energy. However, if the collision is between objects whose masses are nearly the same, the incoming projectile will transfer a large percentage of its kinetic energy to the target. In the first nuclear reactor ever constructed, Fermi placed bricks of graphite (carbon) between the fuel elements. Carbon nuclei are about 12 times more massive than neutrons, but after several collisions with carbon nuclei, a neutron is slowed sufficiently to increase its likelihood of fission with ^{235}U. In this design the carbon is the moderator; most modern reactors use water as the moderator.

Neutron Capture

In the process of being slowed down, the neutrons may be captured by nuclei that do not undergo fission. The most common event of this type is neutron capture by ^{238}U. The probability of neutron capture by ^{238}U is very high when the neutrons have high kinetic energies and very low when they have low kinetic energies. Thus the slowing down of the neutrons by the moderator serves the dual purpose of making them available for reaction with ^{235}U and decreasing their chances of being captured by ^{238}U.

Control of Power Level

It is possible for a reactor to reach the critical stage ($K = 1$) after all the neutron losses described above are minimized. However, a method of control is needed to maintain a K value near unity. If K rises above this value, the heat produced in the runaway reaction would melt the reactor. To control the power level, control rods are inserted into the reactor core (Fig. 33.3). These rods are made of materials, such as cadmium, that are very efficient in absorbing neutrons. By adjusting the number and position of these control rods in the reactor core, the K value can be varied and any power level within the design range of the reactor can be achieved.

A diagram of a pressurized-water reactor is shown in Figure 33.4. This type of reactor is commonly used in electric power plants in the United States. Fission events in the reactor core supply heat to the water contained in the primary (closed) system which is maintained at high pressure to keep it from boiling. This water also serves as the moderator. The hot water is pumped through a heat exchanger, and the heat is transferred to the water contained in the secondary system. The hot water in the secondary system is converted to steam, which drives a turbine-generator to create electric power. Note that the water in the secondary system is isolated from the water in the

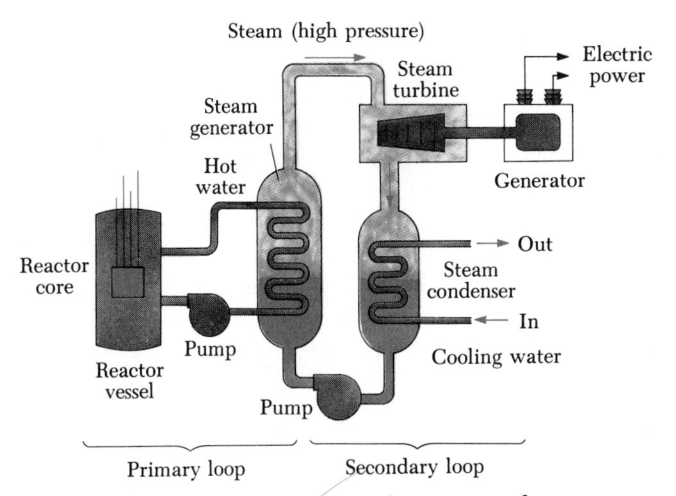

FIGURE 33.4 Main components of a pressurized-water reactor.

primary system in order to avoid contamination of the secondary water and steam by radioactive nuclei from the reactor core.

Reactor Safety

As more reactors are built around the country, there is justifiable concern about their safety and about their effect on the environment. This is particularly true in light of the 1979 near-disaster of Three Mile Island in Pennsylvania and the devastation caused by the accident at the Chernobyl reactor in Russia. The problems of reactor safety are so vast and complex that we can only touch on a few of them here.

One of the inherent dangers in a nuclear reactor is the possibility that the water flow could be interrupted. If this should occur, it is conceivable that the temperature of the reactor could increase to the point where the fuel elements would melt, which would melt the bottom of the reactor and the ground below. This possibility is referred to, appropriately enough, as the "China syndrome." Additionally, the large amounts of heat generated could lead to a high-pressure steam explosion (non-nuclear) that would spread radioactive material throughout the area surrounding the power plant. To decrease the chances of such an event as much as possible, all reactors are built with a backup cooling system that takes over if the regular cooling system fails.

Another problem of concern in nuclear fission reactors is the disposal of radioactive material when the reactor core is replaced. This waste material contains long-lived, highly radioactive isotopes and must be stored over long periods of time in such a way that there is no chance of environmental contamination. At present, sealing radioactive wastes in deep salt mines seems to be the most promising solution.

Other major concerns associated with the proliferation of nuclear power plants are the danger of sabotage at reactor sites and the danger that nuclear fuel (or waste) might be stolen during transport. In some instances, these stolen materials could be used to make an atomic bomb. The difficulties in handling and transporting such highly radioactive materials reduce the possibility of such terrorist activities.

Another consequence of nuclear power plants is thermal pollution. Water from a nearby river is often used to cool the reactor. This raises the river temperatures downstream from the reactor, which affects living organisms in and adjacent to the river. Another technique to cool the reactor water is to use evaporation towers. In this case, thermal pollution of the atmosphere is a consequence. Clearly, one must be concerned with the detrimental effects of thermal pollution when deciding on the location of nuclear power plants.

We have listed only a few of the problems associated with nuclear power reactors. Among these, the handling and disposal of radioactive wastes appears to be the chief problem. One must, of course, weigh such risks against the problems and risks associated with alternative energy sources.

33.3 NUCLEAR FUSION

Figure 32.3 shows that the binding energy for light nuclei (those having a mass number of less than 20) is much smaller than the binding energy for heavier nuclei. This suggests a possible process that is the reverse of fission. *When two light nuclei combine to form a heavier nucleus, the process is called nuclear fusion.* Because the mass of the final nucleus is less than the rest masses of the original nuclei, there is a loss of mass accompanied by a release of energy. It is important to recognize that although fusion power plants have not yet been developed, a great world-wide effort is underway to harness the energy from fusion reactions in the laboratory. Later we shall discuss the possibilities and advantages of this process for generating electric power.

Fusion in the Sun

All stars generate their energy through fusion processes. About 90% of the stars, including our own sun, fuse hydrogen, while some older stars fuse helium or other heavier elements. Stars are born in regions of space containing vast clouds of dust and gas. Recent mathematical models of these clouds indicate that star formation is triggered by shock waves passing through a cloud. These shock waves are similar to sonic booms and are produced by events such as the explosion of a nearby star, called a supernova explosion. The shock wave compresses certain regions of the cloud, causing this region to collapse under its own gravity. As the gas falls inward toward the center, the atoms gain speed, which causes the temperature of the gas to rise. Two conditions must be met before fusion reactions in the star can sustain its energy needs: (1) The temperature must be high enough (about 10^7 K for hydrogen) to allow the kinetic energy of the positively charged hydrogen nuclei to overcome their mutual Coulomb repulsion and collide, and (2) the density of nuclei must be high enough to ensure a high probability of collision.

When fusion reactions occur in a star, the energy liberated eventually becomes sufficient to prevent further collapse of the star under its own gravity. The star then continues to live out the remainder of its life under a balance between the inward force of gravity tending to cause it to collapse and the outward force due to thermal effects and radiation pressure. The proton-proton cycle is a series of three nuclear reactions which are believed to be the stages in the liberation of energy in our sun and other stars rich in hydrogen. An overall view of the proton-proton cycle is that four protons combine to form an alpha particle and two positrons, with the release of 25 MeV of energy in the process.

The three steps in the proton-proton cycle are

$$\begin{aligned} {}^{1}_{1}\mathrm{H} + {}^{1}_{1}\mathrm{H} &\rightarrow {}^{2}_{1}\mathrm{H} + {}^{0}_{1}\mathrm{e} + \nu \\ {}^{1}_{1}\mathrm{H} + {}^{2}_{1}\mathrm{H} &\rightarrow {}^{3}_{2}\mathrm{He} + \gamma \end{aligned} \tag{33.3}$$

This second reaction is followed by either

$${}^{1}_{1}\mathrm{H} + {}^{3}_{2}\mathrm{He} \rightarrow {}^{4}_{2}\mathrm{He} + {}^{0}_{1}\mathrm{e} + \nu$$

or

$${}^{3}_{2}\mathrm{He} + {}^{3}_{2}\mathrm{He} \rightarrow {}^{4}_{2}\mathrm{He} + {}^{1}_{1}\mathrm{H} + {}^{1}_{1}\mathrm{H}$$

The energy liberated is carried primarily by gamma rays, positrons, and neutrinos, as can be seen from the reactions. The gamma rays are soon absorbed by the dense gas, thus raising its temperature. The positrons combine with electrons to produce gamma rays, which in turn are also absorbed by the gas within a few centimeters. The neutrinos, however, almost never interact with matter, and hence escape from the star carrying about 2% of the energy generated with them. These energy-liberating fusion reactions are called **thermonuclear fusion reactions.** The hydrogen (fusion) bomb, first exploded in 1952, is an example of an uncontrolled thermonuclear fusion reaction.

Fusion Reactors

The enormous amount of energy released in fusion reactions suggests the possibility of harnessing this energy for useful purposes here on earth. A great deal of effort is currently under way to develop a sustained and controllable thermonuclear reactor—a fusion power reactor. Controlled fusion is often called the ultimate energy source because of the availability of its source of fuel: water. For example, if deuterium were used as the fuel, 0.12 g of it could be extracted from 1 gal of water at a cost of about four cents. Such rates would make the fuel costs of even an inefficient reactor almost insignificant. An additional advantage of fusion reactors is that comparatively few radioactive by-products are formed. As noted in Equation 33.3, the end product of the fusion of hydrogen nuclei is safe, nonradioactive helium. Unfortunately, a thermonuclear reactor that can deliver a net power output over a reasonable time interval is not yet a reality, and many difficulties must be resolved before a successful device is constructed.

We have seen that the sun's energy is based, in part, upon a set of reactions in which ordinary hydrogen is converted to helium. Unfortunately, the proton-proton interaction is not suitable for use in a fusion reactor because the event requires very high pressures and densities. The process works in the sun only because of the extremely high density of protons in the sun's interior.

The fusion reactions that appear most promising in the construction of a fusion power reactor involve deuterium and tritium, which are isotopes of hydrogen. These reactions are

$$\begin{aligned} &{}^2_1\text{H} + {}^2_1\text{H} \rightarrow {}^3_2\text{He} + {}^1_0\text{n} \qquad Q = 3.27\text{ MeV} \\ &{}^2_1\text{H} + {}^2_1\text{H} \rightarrow {}^3_1\text{H} + {}^1_1\text{H} \qquad Q = 4.03\text{ MeV} \\ &{}^2_1\text{H} + {}^3_1\text{H} \rightarrow {}^4_2\text{He} + {}^1_0\text{n} \qquad Q = 17.59\text{ MeV} \end{aligned} \tag{33.4}$$

where the Q values refer to the amount of energy released per reaction. As noted earlier, deuterium is available in almost unlimited quantities from our lakes and oceans and is very inexpensive to extract. Tritium, however, is radioactive ($T_{1/2} = 12.3$ years) and undergoes beta decay to ^{3}He. For this reason, tritium does not occur naturally to any great extent and must be artificially produced.

One of the major problems in obtaining energy from nuclear fusion is the fact that the Coulomb repulsion force between two charged nuclei must be overcome before they can fuse. The fundamental problem then is to give the two nuclei enough kinetic energy to overcome this repulsive force. This can be accomplished by heating the fuel to extremely high temperatures (about 10^8 K, far greater than the interior temperature of the sun). As you might expect, such high temperatures are not easy to obtain in the laboratory or a power plant. At these high temperatures, the atoms are ionized, and the system consists of a collection of electrons and nuclei, commonly referred to as a plasma.

In addition to the high temperature requirements, there are two other critical factors that determine whether or not a thermonuclear reactor will be successful: **plasma ion density,** n, and **plasma confinement time,** τ, the time the interacting ions are maintained at a temperature equal to or greater than the temperature required for the reaction to proceed successfully. The density and confinement time must both be large enough to ensure that more fusion energy will be released than is required to heat the plasma.

Lawson's criterion states that a net power output in a fusion reactor is possible under the following conditions:

Lawson's criterion

$$\begin{aligned} &n\tau \geqslant 10^{14}\text{ s/cm}^3 \qquad \text{Deuterium-tritium interaction} \\ &n\tau \geqslant 10^{16}\text{ s/cm}^3 \qquad \text{Deuterium-deuterium interaction} \end{aligned} \tag{33.5}$$

The problem of plasma confinement time has yet to be solved. How can one confine a plasma at a temperature of 10^8 K for times of the order of 1 s? The two basic techniques under investigation to confine plasmas are discussed following Example 33.3.

EXAMPLE 33.3 The Deuterium-Deuterium Reaction

Find the energy released in the deuterium-deuterium reaction

$${}^2_1\text{H} + {}^2_1\text{H} \rightarrow {}^3_1\text{H} + {}^1_1\text{H}$$

Solution The mass of the ${}^2_1$H atom is 2.014102 u. Thus, the total mass before the reaction is 4.028204 u. After the reaction, the sum of the masses is equal to 3.016049 u + 1.007825 u = 4.023874 u. Thus, the excess mass is 0.00433 u. In energy units, this is equivalent to 4.03 MeV.

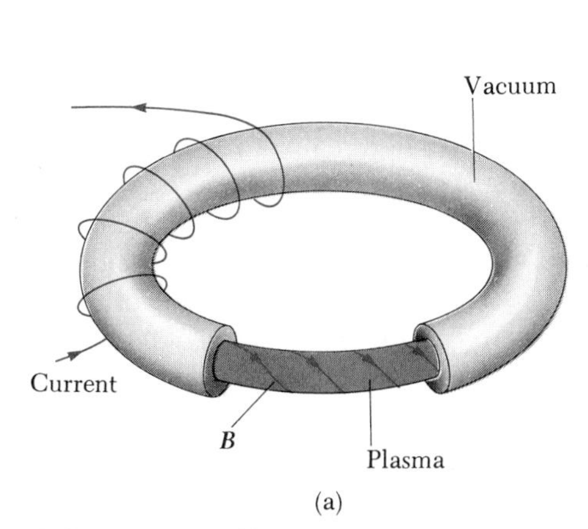

(a)

(b)

FIGURE 33.5 (a) Diagram of a tokamak used in the magnetic confinement scheme. The plasma is trapped within the spiraling magnetic field lines as shown. (b) Photograph of the Princeton tokamak fusion test reactor (TFTR), which uses magnetic field confinement. (Courtesy of the Plasma Physics Laboratory Princeton, N.J.)

Magnetic Field Confinement

Most fusion experiments use magnetic field confinement to contain a plasma. One device, called a **tokamak,** has a doughnut-shaped geometry (a toroid), as shown in Figure 33.5a. This device, first developed in the USSR, uses a combination of two magnetic fields to confine the plasma inside the doughnut. A strong magnetic field is produced by the current in the windings, and a weaker magnetic field is produced by the current in the toroid. The resulting magnetic field lines are helical, as in Figure 33.5a. In this configuration, the field lines spiral around the plasma and prevent it from touching the walls of the vacuum chamber. If the plasma comes into contact with the walls, the temperature of the plasma is reduced and impurities ejected from the walls "poison" the plasma and lead to large power losses.

In order for the plasma to reach ignition temperature, it is necessary to use some form of auxiliary heating. A recent successful and efficient auxiliary heating technique that has been used is the injection of a beam of energetic neutral particles into the plasma.

Figure 33.5b is a photograph of the largest tokamak in the U.S. fusion program.

Another scheme that has received much attention is **magnetic mirror confinement.** The idea of this technique is to trap the plasma in a cylindrical tube by producing a magnetic field like that shown in Figure 33.6. The

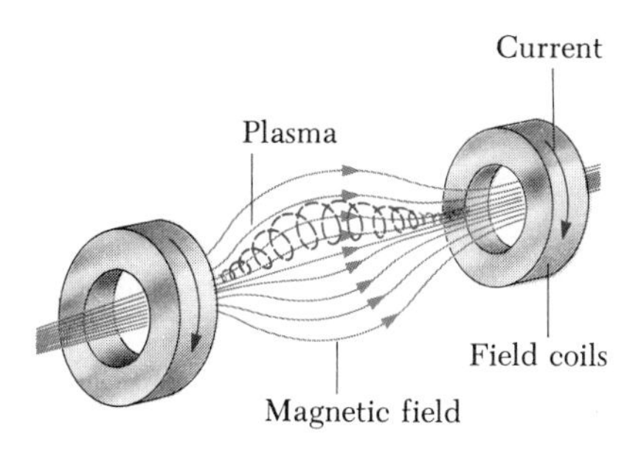

FIGURE 33.6 Magnetic mirror confinement of a plasma.

charged ions of the plasma spiral around the field lines and are thus kept out of contact with the walls of the chamber. The increased field at the end of the cylinder serve as magnetic "mirrors" for the charged particles, which are reflected back into the interior of the chamber.

If fusion power can be harnessed, it will offer several advantages over fission reactors: (1) the low cost and abundance of the deuterium fuel, (2) the absence of weapons-grade material, (3) the impossibility of runaway accidents, and (4) a lesser radiation hazard. Some of the anticipated problems and disadvantages include (1) its not yet fully established feasibility, (2) the very high projected plant costs, and (3) the anticipated high degree of thermal pollution. If these basic problems and the engineering design factors can be resolved, nuclear fusion could become a feasible source of energy within 50 years.

*33.4 RADIATION DAMAGE IN MATTER

Radiation absorbed by matter can cause severe damage. The degree and type of damage depend upon several factors, including the type and energy of the radiation and the properties of the absorbing material. For example, metals used in nuclear reactors can be severely weakened by high fluxes of energetic neutrons, a condition that often leads to metal fatigue. The damage in such situations is in the form of displacement of atoms within the metal, often resulting in major alterations in its properties.

Radiation damage in biological organisms is primarily due to ionization effects in cells. The normal function of a cell may be disrupted when highly reactive ions or radicals are formed as the result of ionizing radiation. For example, hydrogen and hydroxyl radicals produced from water molecules can induce chemical reactions that may break bonds in proteins and other vital molecules. Furthermore, the ionizing radiation may directly affect vital molecules by removing electrons from their structure. Large acute doses of radiation are especially dangerous because damage to a great number of molecules in a cell may cause death of the cell. Although the death of a single cell is usually not a problem, the death of many cells may result in irreversible damage to the organism. Also, cells that do survive the radiation may become defective. These defective cells, upon dividing, can produce more defective cells and lead to cancer.

In biological systems, it is common to separate radiation damage into two categories, somatic damage and genetic damage. **Somatic damage** is radiation damage to any cells except the reproductive cells. Such damage can lead to cancer at high radiation levels or seriously alter the characteristics of specific organisms. **Genetic damage** affects only reproductive cells. Damage to the genes in reproductive cells can lead to defective offspring. Clearly, one must be concerned about the effect of diagnostic treatments, such as x-rays and other forms of radiation exposure.

There are several units used to quantify radiation exposure and dose. The **roentgen** (R) is a unit of radiation exposure and is defined as *that amount of ionizing radiation that will produce* 2.08×10^9 *ion pairs in 1* cm^3 *of air under standard conditions.* Equivalently, the roentgen is *that amount of radiation that deposits* 8.76×10^{-3} *J of energy into 1 kg of air.*

For most applications, the roentgen has been replaced by the **rad** (which is an acronym for radiation absorbed dose), defined as follows: *one rad is that*

TABLE 33.1 RBE Factors for Several Types of Radiation

Radiation	RBE Factor
X-rays and gamma rays	1.0
Beta particles	1.0–1.7
Alpha particles	10–20
Slow neutrons	4–5
Fast neutrons and protons	10
Heavy ions	20

amount of radiation that deposits 10^{-2} J of energy into 1 kg of absorbing material.

Although the rad is a perfectly good physical unit, it is not the best unit for measuring the degree of biological damage produced by radiation. This is because the degree of biological damage depends not only on the dose but also on the type of radiation. For example, a given dose of alpha particles causes about ten times more biological damage than an equal dose of x-rays. The **RBE** (relative biological effectiveness) factor is defined as *the number of rad of x-radiation or gamma radiation that produces the same biological damage as 1 rad of the radiation being used.* The RBE factors for different types of radiation are given in Table 33.1. Note that the values are only approximate because they vary with particle energy and form of damage.

Finally, the **rem** (roentgen equivalent in man) is defined as *the product of the dose in rad and the RBE factor:*

$$\text{Dose in rem} = \text{dose in rad} \times \text{RBE} \tag{33.6}$$

According to this definition, 1 rem of any two radiations will produce the same amount of biological damage. From Table 33.1, we see that a dose of 1 rad of fast neutrons represents an effective dose of 10 rem and that 1 rad of x-radiation is equivalent to a dose of 1 rem.

Low-level radiation from natural sources, such as cosmic rays and radioactive rocks and soil, delivers to each of us a dose of about 0.13 rem/year. The upper limit of radiation dose recommended by the U.S. government (apart from background radiation and exposure related to medical procedures) is 0.5 rem/year. Many occupations involve higher levels of radiation exposure, and for individuals in these occupations an upper limit of 5 rem/year has been set for whole-body exposure. Higher upper limits are permissible for certain parts of the body, such as the hands and forearms. An acute dose of 400 to 500 rem results in a mortality rate of about 50%. The most dangerous form of exposure is ingestion or inhalation of radioactive isotopes, especially those elements the body retains and concentrates, such as ^{90}Sr. In some cases, a dose of 1000 rem can result from ingesting 1 mCi of radioactive material.

*33.5 RADIATION DETECTORS

Various devices have been developed for detecting radiation. They are used for a variety of purposes, including medical diagnoses, radioactive dating measurements, and the measurement of background radiation.

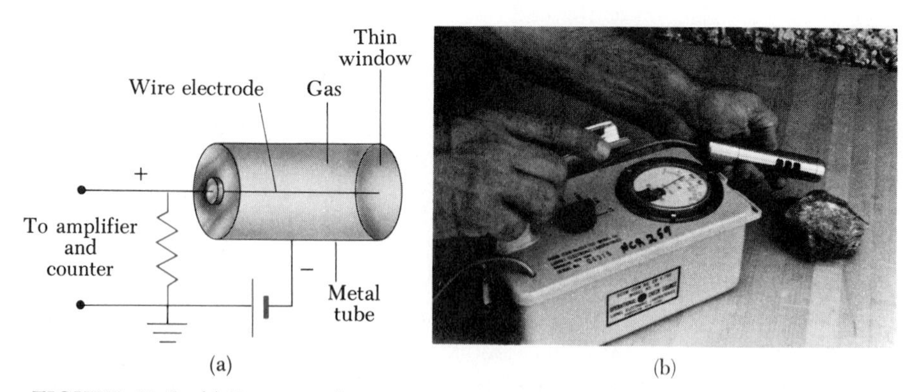

FIGURE 33.7 (a) Diagram of a Geiger counter. The voltage between the central wire and outer metal tube is usually about 1000 V. (b) Using a Geiger counter to measure the activity in a radioactive mineral. (Photographed by Jim Lehman, James Madison University)

The **Geiger counter** (Fig. 33.7) is perhaps the most common device used to detect radiation. It can be considered the prototype of all counters that make use of the ionization of a medium as the basic detection process. It consists of a cylindrical metal tube filled with gas at low pressure and a long wire along the axis of the tube. The wire is maintained at a high positive potential (about 1000 V) with respect to the tube. When a high-energy particle or photon enters the tube through a thin window at one end, some of the atoms of the gas become ionized. The electrons removed from the atoms are attracted toward the wire, and in the process they ionize other atoms in their path. This results in an avalanche of electrons, which produces a current pulse at the output of the tube. After the pulse is amplified, it can be either used to trigger an electronic counter or delivered to a loudspeaker, which clicks each time a particle enters the detector.

A **semiconductor diode detector** is essentially a reverse-biased *p-n* junction. A *p-n* junction diode is a device which passes current readily when forward-biased and impedes the flow of current when reverse-biased (see Chapter 31). As an energetic particle passes through the junction, electrons and holes are simultaneously created. The internal electric field sweeps the electrons toward the side of the junction connected to the positive side of the battery and the holes are swept toward the negative side. This creates a pulse of current that can be measured with any electronic counter. In a typical device, the duration of the pulse is about 10^{-7} s.

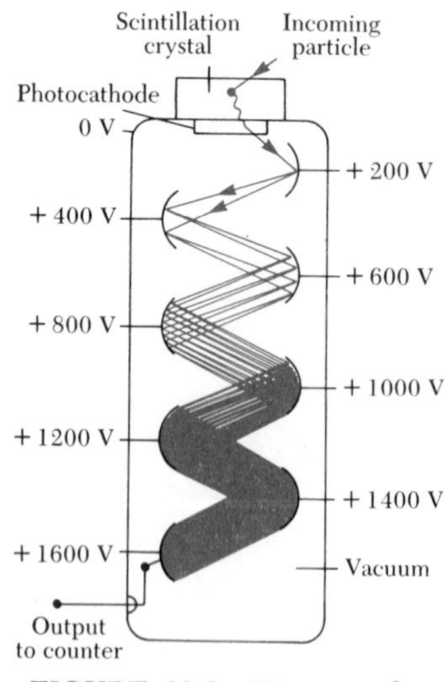

FIGURE 33.8 Diagram of a scintillation counter connected to a photomultiplier tube.

A **scintillation counter** (Fig. 33.8) usually uses a solid or liquid material whose atoms are easily excited by the incoming radiation. These excited atoms emit visible light when they return to the ground state. Common materials used as scintillators are crystals of sodium iodide and certain plastics. If such a material is attached to one end of a device called a photomultiplier tube, the photons emitted by the scintillator can be converted to an electric signal. The photomultiplier tube consists of several electrodes, called dynodes, whose potentials are increased in succession along the length of the tube, as shown in Figure 33.8. The top of the tube contains a photocathode, which emits electrons by the photoelectric effect. An emitted electron striking the first dynode has sufficient kinetic energy to eject several other electrons.

When these electrons are accelerated to the second dynode, many more electrons are ejected and an avalanche occurs. The end result is 1 million or more electrons striking the last dynode. Hence, one particle striking the scintillator produces a sizable electric pulse at the output of the photomultiplier tube, and this pulse is in turn sent to an electronic counter. A scintillator device is much more sensitive than a Geiger counter, mainly because of the higher density of the detecting medium. It is especially sensitive to gamma rays, which interact more weakly with matter than do charged particles.

The devices described so far make use of ionization processes induced by energetic particles. Various other devices can be used to view the tracks of charged particles directly. A **photographic emulsion** is the simplest example. A charged particle ionizes the atoms in a emulsion layer. The path of the particle corresponds to a family of points at which chemical changes have occurred in the emulsion. When the emulsion is developed, the particle's track becomes visible. Such devices, called **dosimeters,** are common in the film badges used in any environment where radiation levels must be monitored.

A **cloud chamber** contains a gas that has been supercooled to just below its usual condensation point. An energetic particle passing through ionizes the gas molecules along its path. These ions serve as centers for condensation of the supercooled gas. The track of the particle can be seen with the naked eye and can be photographed. A magnetic field can be applied to determine the sign and energy of the charges as they are deflected by the field.

A device called a **bubble chamber,** invented in 1952 by D. Glaser, makes use of a liquid (usually liquid hydrogen) maintained near its boiling point. Ions produced by incoming charged particles leave bubble tracks, which can be photographed. Because the density of the detecting medium of a bubble chamber is much higher than the density of a gas in a cloud chamber, the bubble chamber has a much higher sensitivity.

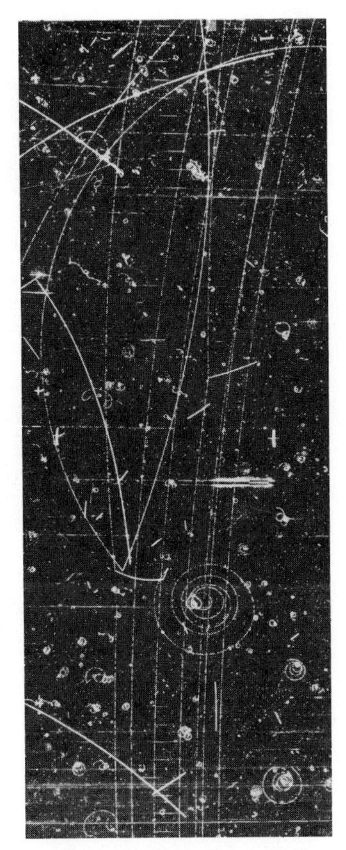

Bubble chamber photograph of the production and decay of particles. (Courtesy of Lawrence Radiation Laboratory, University of California, Berkeley, CA)

*33.6 USES OF RADIATION

Practical applications of nuclear physics are extremely widespread in manufacturing processes, medicine, and biology. Even a brief discussion of all the possibilities would fill an entire book, and to keep such a book up to date would require a number of revisions each year. In this section, we shall present a few of these applications and some of the underlying theories supporting them.

Tracing

Radioactive particles can be used to trace chemicals participating in various reactions. One of the most valuable uses of radioactive tracers is in medicine. For example, ^{131}I is an artificially produced isotope of iodine (the natural, nonradioactive isotope is ^{127}I). Iodine, which is a necessary nutrient for our bodies, is obtained largely through the intake of iodized salt and seafood. The thyroid gland plays a major role in the distribution of iodine throughout the body. In order to evaluate the performance of the thyroid, the patient drinks a very small amount of radioactive sodium iodide. Two hours later, the

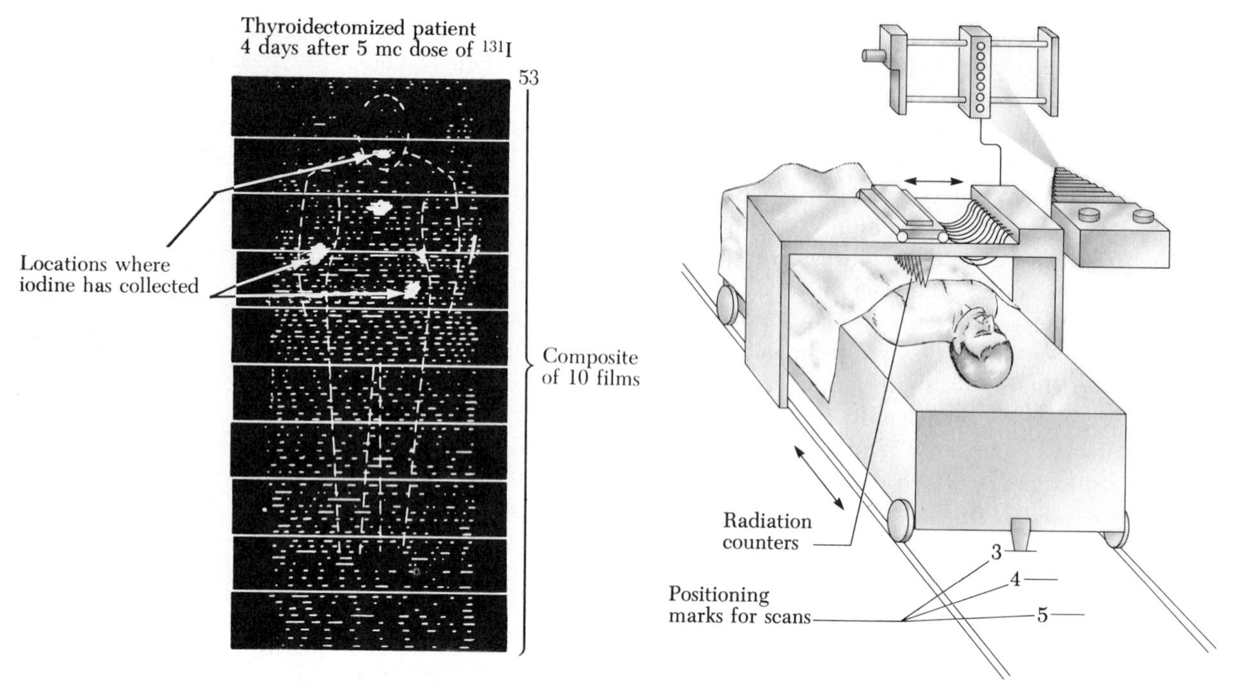

FIGURE 33.9 Scanning a patient for radioactive ^{31}I. This tracer technique is used to evaluate the condition of the thyroid gland.

amount of iodine in the thyroid gland is determined by measuring the radiation intensity at the neck area. Figure 33.9 shows a scan of the body of a patient four days after a dose of ^{131}I was administered.

A second medical application is indicated in Figure 33.10. Here a salt containing radioactive sodium is injected into a vein in the leg. The time at which the radioisotope arrives at another part of the body is detected with a radiation counter. The elapsed time is a good indication of the presence or absence of constrictions in the circulatory system.

The tracer technique is also useful in agricultural research. Suppose one wishes to determine the best method of fertilizing a plant. A certain material in the fertilizer, such as nitrogen, can be tagged with one of its radioactive isotopes. The fertilizer is then sprayed on one group of plants, sprinkled on the ground for a second group, and raked into the soil for a third. A Geiger counter is then used to track the nitrogen through the three types of plants.

Tracing techniques are as wide-ranging as human ingenuity can devise. Present applications range from checking the absorption of fluorine by teeth to checking contamination of food-processing equipment by cleansers to monitoring deterioration inside an automobile engine. In the latter case, a radioactive material is used in the manufacture of the pistons, and the oil is checked for radioactivity to determine the amount of wear on the pistons.

Activation Analysis

For centuries, a standard method of identifying the elements in a sample of material has been chemical analysis, which involves testing a portion of the

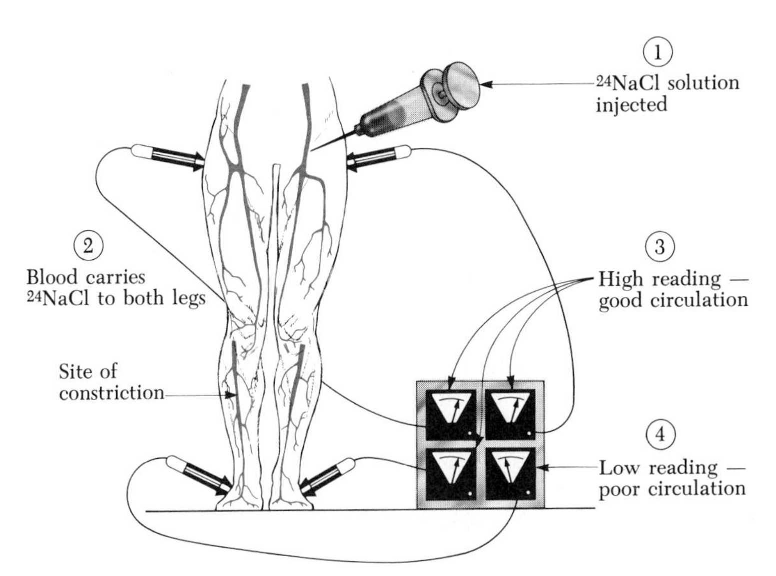

FIGURE 33.10 A tracer technique for determining the condition of the human circulatory system.

material for reactions with various chemicals. A second method is spectral analysis, which utilizes the fact that, when excited, each element emits its own characteristic set of electromagnetic wavelengths. These methods are now supplemented by a third technique, **neutron activation analysis.** Both chemical and spectral methods have the disadvantage that a fairly large sample of the material must be destroyed for the analysis. In addition, extremely small quantities of an element may go undetected by either method. Activation analysis has an advantage over the other two methods in both of these respects.

When the material under investigation is irradiated with neutrons, nuclei in the material will absorb the neutrons and be changed to different isotopes. Most of these isotopes will be radioactive. For example, ^{65}Cu absorbs a neutron to become ^{66}Cu, which undergoes beta decay:

$$ {}^{1}_{0}\text{n} + {}^{65}_{29}\text{Cu} \rightarrow {}^{66}_{29}\text{Cu} \rightarrow {}^{66}_{30}\text{Zn} + {}^{0}_{-1}\text{e} \tag{33.7} $$

The presence of the copper can be deduced because it is known that ^{66}Cu has a half-life of 5.1 min and decays with the emission of beta particles having maximum energies of 2.63 and 1.59 MeV. Also emitted in the decay of ^{66}Zn is a gamma ray having an energy of 1.04 MeV. Thus, by examining the radiation emitted by a substance after it has been exposed to neutron irradiation, one can detect extremely small traces of an element.

Neutron activation analysis is used routinely by a number of industries, but the following nonroutine example of its use is of interest. Napoleon died on the island of St. Helena in 1821, supposedly of natural causes. Over the years, suspicion has existed that his death was not all that natural. After his death, his head was shaved and locks of his hair were sold as souvenirs. In 1961, the amount of arsenic in a sample of this hair was measured by neutron activation analysis. Unusually large quantities of arsenic were found in the hair. (Activation analysis is so sensitive that very small pieces of a single hair could be analyzed.) Results showed that the arsenic was fed to him irregularly. In fact, the arsenic concentration pattern corresponded to the fluctuations in the severity of Napoleon's illness as determined from historical records.

*33.7 COMPUTED AXIAL TOMOGRAPHY (CAT SCANS)

The normal x-ray of a human body has two primary disadvantages when used as a source of clinical diagnosis. First, it is difficult to distinguish between various types of tissue in the body because they all have similar x-ray absorption properties. Second, a conventional x-ray absorption picture is indicative of the average amount of absorption along a particular direction in the body, leading to somewhat obscured pictures. To overcome these problems, a device called a CAT scanner was developed in England in 1973; it is capable of producing pictures of much greater clarity and detail than was previously obtainable.

One can understand the operation of a CAT scanner by considering the following hypothetical experiment. Suppose a box consists of four compartments, labeled *A*, *B*, *C*, and *D* as in Figure 33.11a. Each compartment has a different amount of absorbing material than any other compartment. What set of experimental procedures will enable us to determine the relative amounts of material in each compartment? The following steps outline one method that will provide this information. First, a beam of x-rays is passed through compartments *A* and *C*, as in Figure 33.11b. The intensity of the exiting radiation is reduced by absorption by some number which we assign as 8. (The number 8 could mean, for example, that the intensity of the exiting beam is reduced by eight-tenths of one percent from its initial value.) Since we do not know which of the compartments, *A* or *C*, was responsible for this reduction in intensity, half the loss is assigned to each compartment as in Figure 33.11c. Next, a beam of x-rays is passed through compartments *B* and *D*, as in Figure 33.11b. The reduction in intensity for this beam is 10, and again we assign half the loss to each compartment. We now redirect the x-ray source so that it sends one beam through compartments *A* and *B* and another through compartments *C* and *D*, as in Figure 33.11d, and again measure the absorption. Suppose the absorption through compartments *A* and *B* in this experiment is measured to be 7 units. On the basis of our first experiment, we would have guessed it would be 9 units, 4 by compartment *A* and 5 by compartment *B*. Thus, we have reduced the guessed absorption for each compartment by 1 unit so that the sum is 7 rather than 9, to give the numbers shown in Figure 33.11e. Likewise, when the beam is passed through compartments *C* and *D* as in Figure 33.11d, we may find the total absorption to be 11 as compared to our first experiment of 9. In this case, we add 1 unit of absorption to each compartment to give a sum of 11 as in Figure 33.11e. This somewhat crude procedure could be improved by measuring the absorption along other paths. However, these simple measurements are sufficient to enable us to conclude that compartment *D* contains the most absorbing material and *A* the least. A visual representation of these results can be obtained by assigning to each compartment a shade of gray corresponding to the particular number associ-

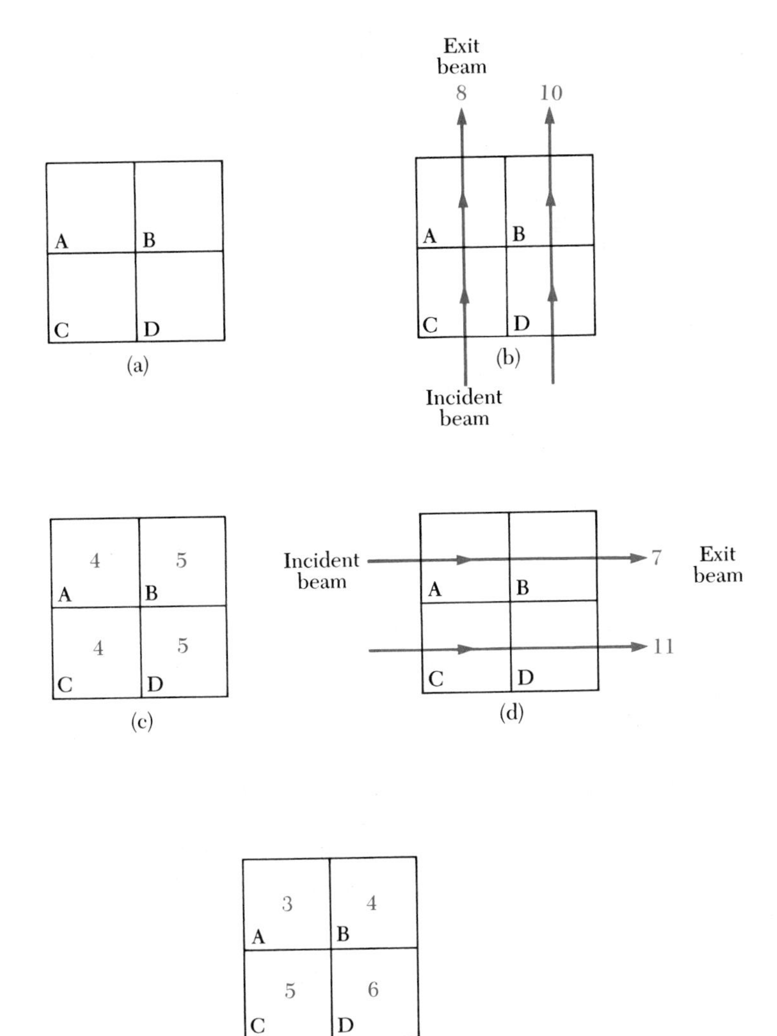

FIGURE 33.11 An experimental procedure for determining the relative amounts of x-ray absorption by four different compartments in a box.

ated with the absorption. In our example, compartment *D* would be very dark while compartment *A* would be very light.

The steps outlined above are representative of how a CAT scanner produces images of the human body. A thin slice of the body is subdivided into perhaps 10,000 compartments, rather than 4 compartments as in our simple example. The function of the CAT scanner is to determine the relative absorption in each of these 10,000 compartments and to display a picture of its calculations in various shades of gray. Note that CAT stands for **computed axial tomography.** The term axial is used because the slice of the body to be analyzed corresponds to a plane perpendicular to the head-to-toe axis. *Tomos* is the Greek word for slice and *graph* is the Greek word for picture. In a

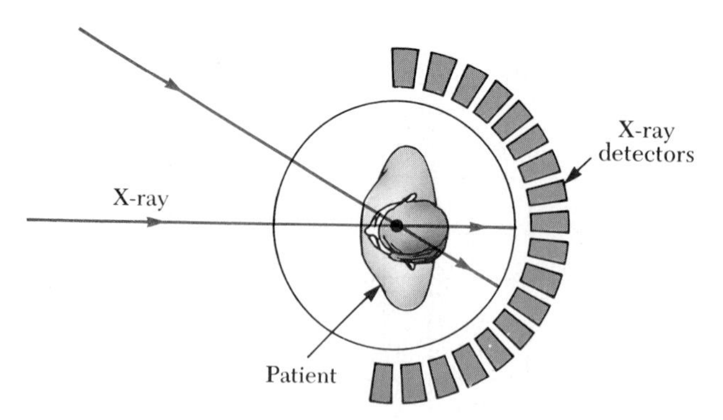

FIGURE 33.12 CAT scanner detector assembly.

typical diagnosis, the patient is placed in the position shown in Figure 33.12 and a narrow beam of x-rays is sent through the plane of interest. The emerging x-rays are detected and measured by photomultiplier tubes behind the patient. The x-ray tube is then rotated a few degrees, and the intensity is recorded again. An extensive amount of information is obtained by rotating the beam through 180 degrees at intervals of about one degree per measurement, resulting in a set of numbers assigned to each of the 10,000 "compartments" in the slice. These numbers are then converted by the computer to a photograph in various shades of gray for this segment of the body.

A brain scan of a patient can now be made in about 2 seconds, while a full-body scan requires about 6 seconds. The final result is a picture containing much greater quantitative information and clarity than a conventional x-ray photograph. Since CAT scanners use x-rays, which are an ionizing form of radiation, the technique presents a health risk to the patient being diagnosed.

SUMMARY

In **nuclear fission** and **nuclear fusion,** the total rest mass of the products is always less than the original rest mass of the reactants. Nuclear fission occurs when a heavy nucleus splits, or fissions, into two smaller nuclei. In nuclear fusion, two light nuclei combine to form a heavier nucleus.

A **nuclear reactor** is a system designed to maintain what is called a self-sustaining chain reaction. Nuclear reactors using controlled fission events are currently being used to generate electric power.

Controlled fusion events offer the hope of plentiful supplies of energy in the future. The nuclear fusion reactor is considered by many scientists to be the ultimate energy source because its fuel is water. **Lawson's criterion** states that a fusion reactor will provide a net output power if the product of the plasma ion density, n, and the plasma confinement time, τ, satisfies the following relationships:

$n\tau \geq 10^{14}$ s/cm^3	Deuterium-tritium interaction
$n\tau \geq 10^{16}$ s/cm^3	Deuterium-deuterium interaction

Devices used to detect radiation are the Geiger counter, semiconductor diode detector, scintillation counter, photographic emulsion, cloud chamber, and bubble chamber.

ADDITIONAL READING

H. Bethe, "The Necessity of Fission Power," *Sci. American*, January 1976, p. 21.

B. L. Cohen, "The Disposal of Radioactive Wastes from Fission Reactors," *Sci. American*, June 1977, p. 21.

R. W. Conn, "The Magnetic Fusion Reactors," *Sci. American*, October 1983, p. 176.

L. Fermi, *Atoms in the Family: My Life with Enrico Fermi*, Chicago, University of Chicago Press, 1954.

H. P. Furth, "Progress Toward a Tokamak Fusion Reactor," *Sci. American*, August 1979, p. 50.

R. K. Lester, "Rethinking Nuclear Power," *Sci. American*, March 1986, p. 31.

H. W. Lewis, "The Safety of Fission Reactors," *Sci. American*, March 1980, p. 3.

P. R. Moran, R. J. Nickles, and J. A. Zagzebski, "The Physics of Medical Imaging," *Physics Today*, July 1983, p. 36.

I. I. Pykett, "NMR Imaging in Medicine," *Sci. American*, May 1982, p. 78.

M. M. Ter-Pogossian, M. E. Raichle, and B. E. Sobol, "Positron-Emission Tomography," *Sci. American*, October 1980, p. 170.

A. C. Upton, "The Biological Effects of Low-Level Ionizing Radiation," *Sci. American*, February 1982, p. 41.

QUESTIONS

1. Explain the function of a moderator in a fission reactor.
2. Discuss the advantages and disadvantages of fission reactors from the point of view of safety, pollution, and resources. Make a comparison with power generated from the burning of fossil fuels.
3. In a fission reactor, nuclear reactions produce heat to drive a turbine-generator. How is this heat provided?
4. Why would a fusion reactor produce less radioactive waste than a fission reactor?
5. What factors make a fusion reaction difficult to achieve?
6. Discuss the similarities and differences between fission and fusion.
7. Discuss the advantages and disadvantages of fusion power from the point of view of safety, pollution, and resources.
8. Discuss the major problems associated with the development of a controlled fusion reactor.
9. Why is the temperature required for deuterium-tritium fusion less than that needed for deuterium-deuterium fusion? Estimate the relative importance of coulombic repulsion and nuclear attraction in each case.
10. If two radioactive samples have the same activity measured in curies, will they necessarily create the same amount of damage in a medium? Explain.
11. Radiation can be used to sterilize such things as surgical equipment and packaged foods. Why do you suppose this works?
12. One method of treating cancer of the thyroid is to insert a small radioactive source directly into the tumor. The radiation emitted by the source can destroy cancerous cells. Very often, the radioactive isotope $^{131}_{53}\mathrm{I}$ is injected into the bloodstream in this treatment. Why do you suppose iodine is used?

PROBLEMS

Section 33.1 Nuclear Fission

Section 33.2 Nuclear Reactors

1. Find the energy released in the following fission reaction:

$$^{1}_{0}\mathrm{n} + {}^{235}_{92}\mathrm{U} \rightarrow {}^{144}_{56}\mathrm{Ba} + {}^{89}_{36}\mathrm{Kr} + 3\,^{1}_{0}\mathrm{n}$$

2. Another fission reaction similar to the one in Problem 1 leads to the formation of $^{141}\mathrm{Ba}$ and $^{92}\mathrm{Kr}$ when $^{235}\mathrm{U}$ absorbs a neutron. Write down this reaction. How many neutrons are released in this reaction?

3. Find the energy released in the following fission reaction:

$${}^{1}_{0}n + {}^{235}_{92}U \rightarrow {}^{88}_{38}Sr + {}^{136}_{54}Xe + 12\,{}^{1}_{0}n$$

4. If the average energy released in a fission event is 208 MeV, find the total number of fission events required to keep a 100 W lightbulb burning for one hour.

5. How many grams of ^{235}U must undergo fission to operate a 1000 MW power plant for one day if the conversion efficiency is 30%?

6. In order to minimize neutron leakage from a reactor, the ratio of the surface area to the volume must be as small as possible. Assume that a sphere and a cube both have the same volume. Find the surface-to-volume ratio for (a) the sphere and (b) the cube. (c) Which of these shapes would have the minimum leakage?

• **7.** It has been estimated that the earth contains 10^9 tons of natural uranium that can be mined economically. Of this total, 0.7% is ^{235}U. If all the world's energy needs (7×10^{12} J/s) were to be supplied by ^{235}U fission, how long would this supply last?

• **8.** An all-electric home uses approximately 2000 kWh of electricity per month. How much ^{235}U would be required to provide this house with its energy needs for 1 year? (Assume 100% conversion efficiency and 208 MeV released per fission.)

• **9.** The first atomic bomb released an energy equivalent to 20 kilotons of TNT. If 1 ton of TNT releases about 4×10^9 J, how much uranium was lost through fission in this bomb? (Assume 208 MeV released per fission.)

Section 33.3 Nuclear Fusion

10. Find the energy released in the fusion reaction

$${}^{2}_{1}H + {}^{2}_{1}H \rightarrow {}^{3}_{1}H + {}^{1}_{1}H$$

11. Find the energy released in the fusion reaction

$${}^{2}_{1}H + {}^{3}_{1}H \rightarrow {}^{4}_{2}He + {}^{1}_{0}n$$

12. If an all-electric home uses approximately 2000 kWh of electricity per month, how many fusion events of the type described in Problem 11 would be required to keep this house running for one year?

• **13.** Another series of nuclear reactions that can produce energy in the interior of stars is the cycle described below. This process is most efficient when the central temperature in a star is above 16,000,000 K. Because the temperature at the center of the sun is only 15,000,000 K, the cycle below produces less than 10% of the sun's energy. (a) A high-energy proton is absorbed by ^{12}C. Another nucleus, A, is produced in the reaction, along with a gamma ray. Identify nucleus A. (b) Nucleus A decays through positron emission to form nucleus B. Identify nucleus B. (c) Nucleus B absorbs a proton to produce nucleus C and a gamma ray. Identify nucleus C. (d) Nucleus C absorbs a proton to produce nucleus D and a gamma ray. Identify nucleus D. (e) Nucleus D decays through positron emission to produce nucleus E. Identify nucleus E. (f) Nucleus E absorbs a proton to produce nucleus F plus an alpha particle. What is nucleus F? (*Note:* If nucleus F is not ^{12}C, the nucleus you started with, you have done something wrong and you should review the sequence of events to find your error.)

• **14.** When a star has exhausted its hydrogen fuel, it may fuse other nuclear fuels. At temperatures above 100,000,000 K, helium fusion can occur. Write the equation for the processes described below. (a) Two alpha particles fuse to produce a nucleus A and a gamma ray. What is nucleus A? (b) Nucleus A absorbs an alpha particle to produce a nucleus B and a gamma ray. What is nucleus B? (c) Find the total energy released in the reactions given in (a) and (b).

• **15.** Of all the hydrogen nuclei in the ocean, 0.0156% are deuterium. (a) How much deuterium could be obtained from 1 gal of ordinary tap water? (b) If all of this deuterium could be converted to energy through reactions such as the first reaction in Equation 33.4, how much energy could be obtained from the 1 gal of water? (c) The energy released through the burning of 1 gal of gasoline is approximately 2×10^8 J. How many gallons of gasoline would have to be burned in order to produce the same energy as the 1 gal of water?

• **16.** To understand why containment of a plasma is necessary, consider the rate at which a plasma would be lost if it were not contained. (a) Estimate the rms speed of deuterons in a plasma at 10^8 K. (b) Estimate the time such a plasma would remain in a cube 10 cm on an edge if no steps were taken to confine it.

Section 33.4 Radiation Damage in Matter

17. Two workers using an industrial x-ray machine accidentally insert their hands in the x-ray beam for the same length of time. The first worker inserts one hand in the beam, and the second worker inserts both hands. Which worker receives the larger dose in rad?

18. Assume that an x-ray technician takes an average of eight x-rays per day and receives a dose of 5 rem/year as a result. (a) Estimate the dose in rem per x-ray taken. (b) How does this result compare with the amount of low-level background radiation the technician is exposed to?

19. A person whose mass is 75 kg is exposed to a dose of 25 rad. How many joules of energy is deposited in the person's body?

20. In terms of biological damage, how many rad of heavy ions is equivalent to 100 rad of x-rays?

21. A 200-rad dose of radiation is administered to a patient in an effect to combat a cancerous growth. Assuming all of the energy deposited is absorbed by the growth, (a) calculate the amount of energy delivered. (b) Assuming the growth has a mass of 0.25 kg and a specific heat equal to that of water, calculate its temperature rise.

22. Estimate your mass and then estimate the temperature rise of your body if you were subjected to a 1000-rad whole-body radiation dose. (A dose this large is unquestionably fatal.)

Section 33.5 Radiation Detectors

23. In a photomultiplier tube, assume that there are seven dynodes with potentials of 100 V, 200 V, 300 V, . . ., 700 V. The average energy required to free an electron from the dynode surface is 10 eV. For each incident electron, how many electrons are freed (a) at the first dynode and (b) at the last dynode?

• **24.** Assume that 40 eV is required to ionize the gas in a Geiger tube and that the electric field in the counter is constant (incorrect) and of magnitude 1000 V/cm. How far would an electron released from rest by an incoming particle have to travel before it acquired enough energy to ionize another atom?

• **25.** Estimate the total number of electrons released in the chain reaction initiated by the electron in Problem 24 if the initial distance of the electron from the central wire is 0.5 cm.

ADDITIONAL PROBLEMS

26. Find the rms speed of a proton inside a star where the temperature is 3×10^7 K.

27. A particle cannot generally be localized to distances smaller than its de Broglie wavelength. This means that a slow neutron appears to be larger to a target particle than a fast neutron because the slow neutron will probably be found over a larger volume of space. For a thermal neutron at room temperature (300 K) assume that its energy is given by kT and find (a) its linear momentum and (b) the de Broglie wavelength. Compare this effective neutron size with both nuclear and atomic dimensions.

• **28.** Calculate the mass of ^{235}U required to provide the total energy requirements of a nuclear submarine during a 100-day patrol, assuming a constant power demand of 100,000 kW and a conversion efficiency of 30%. (Assume that the average energy released per fission is 208 MeV.)

• **29.** Calculate the radiation dose in rad supplied to 1 kg of water such that the energy deposited equals the water's thermal energy at 300 K. Assume that each molecule has a thermal energy kT.

• **30.** (a) Show that about 1×10^{10} J would be released by the fusion of the deuterons in 1 gal of water. Note that 1 out of every 6500 hydrogen atoms is a deuteron. (b) The average energy consumption rate of a person living in the United States is about 10^4 J/s (an average power of 10 kW). At this rate, how long would the energy needs of one person be supplied by the fusion of the deuterons in 1 gal of water? Assume that the energy released per deuteron is 1.64 MeV.

CHAPTER 34

Elementary Particles

An aerial view of the Fermi National Accelerator Laboratory, Batavia, Illinois. The largest circle is the main accelerator. Three experimental lines extend at a tangent from the accelerator. The 16-story twin-towered Wilson Hall is seen at the base of the experimental lines. (Fermilab Photo Dept.)

In this chapter, we shall examine the properties and classifications of the various known subatomic particles and the fundamental interactions that govern their behavior. We shall also discuss the current theory of elementary particles in which all matter in nature is believed to be constructed from only two families of particles, quarks and leptons. Finally, we shall discuss how clarifications of such models might help scientists understand the evolution of the universe.

34.1 INTRODUCTION

The word "atom" is from the Greek word *atomos*, which means indivisible. At one time atoms were thought to be the indivisible constituents of matter; that is, they were regarded to be elementary particles. Discoveries in the early part of the 20th century revealed that the atom is not elementary, but has as its constituents protons, neutrons, and electrons. Until 1932, physicists viewed all matter as consisting of only three constituent particles: electrons, protons, and neutrons. With the exception of the free neutron, these particles are very stable. Beginning in 1945, many new particles were discovered in experiments involving high-energy collisions between known particles. These new particles are characteristically very unstable and have very short half-lives, ranging between 10^{-6} and 10^{-23} s. So far more than 300 of these unstable, temporary particles have been catalogued.

During the last 30 years, many powerful particle accelerators have been constructed throughout the world, making it possible to observe collisions of particles with greater violence under controlled laboratory conditions, so as to reveal the subatomic world in finer detail. Up until the 1960s, physicists were bewildered by the large number and variety of subatomic particles being discovered. They wondered if the particles were like animals in a zoo with no systematic relationship connecting them, or whether a pattern was emerging that would provide a better understanding of the elaborate structure in the subnuclear world. In the last two decades, physicists have made tremendous advances in our knowledge of the structure of matter by recognizing that all particles (with the exception of electrons, photons, and a few related particles) are made of smaller particles called quarks. Thus, protons and neutrons, for example, are not truly elementary but are systems of tightly bound quarks. The quark model has reduced the bewildering array of particles to a manageable number, and has been successful in predicting new quark combinations later found in many experiments.

34.2 THE FUNDAMENTAL FORCES IN NATURE

The key to understanding the properties of elementary particles is to be able to describe the forces between them. All particles in nature are subject to four fundamental forces: strong, electromagnetic, weak, and gravitational.

The **strong force** is very short-ranged and is responsible for the binding of neutrons and protons into nuclei. This force represents the "glue" that

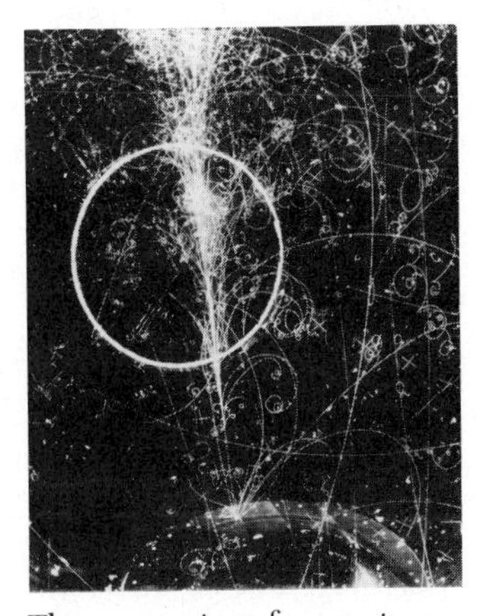

(Left) The tunnel of the main accelerator at Fermilab. The upper ring of magnets are the 400-GeV accelerator. The lower ring are superconducting magnets for the Tevatron. *(Right)* Photograph of a particle interaction in the 15-ft bubble chamber at Fermilab. The photographs showing particle "tracks" are studied by scanners and experimenters. (Fermilab Photo Dept.)

holds the nucleons together and is the strongest of all the fundamental forces. The strong force is very short-ranged and is negligible for separations greater than about 10^{-14} m (which is about the size of the nucleus). The **electromagnetic force,** which is about 10^{-2} times the strength of the strong force, is responsible for the binding of atoms and molecules. It is a long-range force that decreases in strength as the inverse square of the separation between interacting particles. The **weak force** is a short-range nuclear force that tends to produce instability in certain nuclei. It is responsible for most radioactive decay processes such as beta decay, and its strength is only about 10^{-9} times that of the strong force. (As we shall discuss later, scientists now believe that the weak and electromagnetic forces are two manifestations of a single force called the electroweak force). Finally, the **gravitational force** is a long-range force that has a strength of only about 10^{-38} times that of the strong force. Although this familiar interaction is the force that holds the planets, stars, and galaxies together, its effect on elementary particles is negligible. Thus, the gravitational force is the weakest of all the fundamental forces.

In modern physics, one often describes the interactions between particles in terms of the exchange of field particles or quanta. In the case of the familiar electromagnetic interaction, the field particles are photons. In the language of modern physics, one can say that the electromagnetic force is *mediated* by photons, which are the quanta of the electromagnetic field. Likewise, the strong force is mediated by field particles called *gluons,* the weak force is mediated by particles called the W and Z *bosons,* and the gravitational force is mediated by quanta of the gravitational field called *gravitons.* These interactions, their ranges, and their relative strengths are summarized in Table 34.1.

34.3 POSITRONS AND OTHER ANTIPARTICLES

In the 1920s, the theoretical physicist Paul Adrien Maurice Dirac (1902–1984) developed a version of quantum mechanics that incorporated special relativity. Dirac's theory was successful in explaining the origin of the electron's spin and its magnetic moment. However, Dirac was faced with one major difficulty in this theory. His relativistic wave equation required solutions corresponding to negative energy states.[1] But if negative energy states existed, one would expect an electron in a state of positive energy to make a rapid transition to one of these states, emitting a photon in the process. Dirac was able to avoid this difficulty by postulating that all negative energy states were filled. Those

TABLE 34.1 Particle Interactions

Interaction (Force)	Relative Strength	Range of Force	Mediating Field Particle
Strong	1	Short ($\approx$ 1 fm)	Gluon
Electromagnetic	10^{-2}	Long ($\propto 1/r^2$)	Photon
Weak	10^{-9}	Short ($\approx$ 1 fm)	$W^{\pm}$, Z
Gravitational	10^{-38}	Long ($\propto 1/r^2$)	Graviton

[1]P. A. M. Dirac, *The Principles of Quantum Mechanics*, 3rd ed., New York, Oxford University Press, Chapter 11, 1947.

electrons that occupy the negative energy states are called the "Dirac sea." Electrons in the Dirac sea are not directly observable because the Pauli exclusion principle does not allow them to react to external forces. However, if one of these negative energy states is vacant, leaving a hole in the sea of filled states, the hole can react to external forces and would be observable. (This is analogous to the behavior of a hole in the valence band of a semiconductor.) *The profound implication of this theory was that for every particle, there was also an antiparticle.* The antiparticle would have the same mass as the particle, but their charges would be opposite each other. For example, the electron's antiparticle (now called a *positron*) would have a mass of 0.511 MeV, and a positive charge of 1.6×10^{-19} C. Usually we shall designate an antiparticle with a bar over the symbol for the particle. Thus, the positron is denoted by $\bar{e}$ (although sometimes the notation e^+ is preferred), the antiproton is denoted by $\bar{p}$, and the antineutrino is denoted by $\bar{\nu}$.

The positron was discovered by Carl Anderson in 1932 (the same year that the neutron was discovered), and in 1936 he was awarded the Nobel Prize for his discovery. Anderson made his discovery while examining tracks created by electronlike particles of positive charge in a cloud chamber. (These early experiments used cosmic rays—mostly energetic protons passing through interstellar space—to initiate high energy reactions of the order of several GeV.) In order to discriminate between positive and negative charges, the cloud chamber was placed in a magnetic field, causing moving charges to follow curved paths. Anderson noted that some of the electronlike tracks deflected in a direction corresponding to a positively charged particle.

Since Anderson's initial discovery, the positron has been observed in a number of experiments. Perhaps the most common process for producing positrons is **pair production.** In this process, a gamma ray with sufficiently high energy collides with a nucleus and an electron-positron pair is created. Since the total rest energy of the electron-positron pair is $2m_0c^2 = 1.02$ MeV (where m_0 is the rest mass of the electron), the gamma ray must have at least this much energy to create an electron-positron pair. Thus, electromagnetic energy in the form of a gamma ray is transformed into mass in accordance with Einstein's famous relation $E = m_0c^2$. Figure 34.1 shows tracks of electron-positron pairs created by 300-MeV x-rays striking a lead sheet.

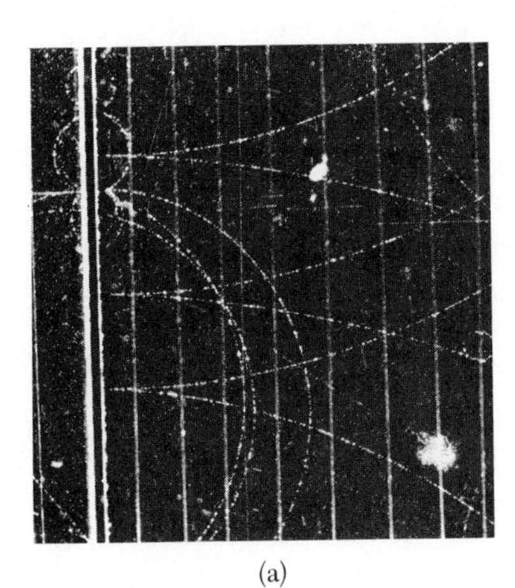

(a)

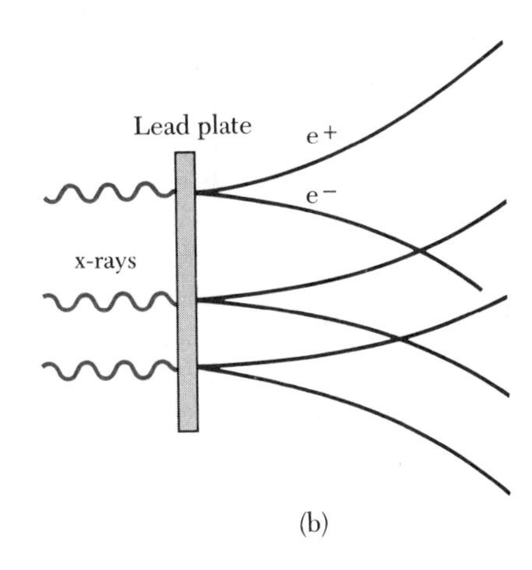

(b)

FIGURE 34.1 (a) Bubble-chamber tracks of electron-positron pairs produced by 300-MeV x-rays striking a lead sheet. (Courtesy of Lawrence Berkeley Laboratory, University of California.) (b) Sketch of the pertinent pair-production events. Note that the positrons deflect upward while the electrons deflect downward in an applied magnetic field directed into the diagram.

A process that is the reverse of pair production can also occur. Under the proper conditions, an electron and positron can annihilate and produce two photons that have a combined energy of at least 1.02 MeV. The reaction can be expressed as

$$e + \bar{e} \longrightarrow 2\gamma$$

Likewise, a proton-antiproton pair can annihilate to produce two gamma rays; however, this event is very rare.

Practically every known elementary particle has an antiparticle. Among the exceptions are the photon and the neutral pion (π^0). (Note that the π^0 and η^0 are their own antiparticles!) Following the construction of high-energy accelerators in the 1950s, many other antiparticles were discovered. These included the antiproton ($\bar{p}$) discovered by Emilio Segre and Owen Chamberlain in 1955, and the antineutron ($\bar{n}$) discovered shortly thereafter.

34.4 MESONS AND THE BEGINNING OF PARTICLE PHYSICS

The structure of matter as viewed by physicists in the mid-1930s was fairly simple. The building blocks of matter were considered to be the proton, the electron, and the neutron. Three other particles were known at the time: the gamma particle (photon), the neutrino, and the positron. These six particles were considered to be the fundamental constituents of matter. Although the accepted picture of the world was marvelously simple, no one was able to provide an answer to the following important question. If the nucleus of an atom contains many charged protons in close proximity which should strongly repel each other due to their like charges, what is the nature of the force that holds the nucleus together? Scientists recognized that this mysterious force must be much stronger than anything encountered in nature up to that time.

The first theory to explain the nature of the strong force was proposed in 1935 by the Japanese physicist Hideki Yukawa (1907–1981), an effort that later earned him the Nobel Prize. In order to understand Yukawa's theory, it is useful to first recall that *two atoms can form a covalent chemical bond by the exchange of electrons.* Similarly, in the modern views of electromagnetic interactions, *charged particles interact through the exchange of photons.* Yukawa used this same idea to explain the strong force by proposing a new particle whose exchange between nucleons in the nucleus produces the strong force with a range of about 10^{-15} m (the order of the nuclear diameter). Furthermore, he established that the range of the force is inversely proportional to the mass of this carrier particle, and predicted that the mass would be about 200 times the mass of the electron. Since the new particle would have a mass between that of the electron and proton, it was called a **meson** (from the Greek *meso,* meaning "middle").

In an effort to substantiate Yukawa's predictions, physicists began an experimental search for the meson by studying cosmic rays that enter the earth's atmosphere from interstellar space. In 1937, Carl Anderson and his collaborators discovered a particle whose mass was 106 MeV/c^2, which is about 207 times the mass of the electron. However, subsequent experiments showed that the particle interacted very weakly with matter, and hence could not be the carrier of the strong force. The puzzling situation inspired several theoreticians to propose that there are actually two mesons with slightly different masses. This idea was confirmed in 1947 with the discovery in cosmic rays of the pi meson (π), or simply *pion*, by Cecil Frank Powell (1903–1969) and Guiseppe P. S. Occhialini (1907–). The lighter meson discovered earlier by Anderson, now called a *muon* (μ), has only weak and electromagnetic interaction and plays no role in the strong interaction.

The pion comes in three varieties, corresponding to three charge states: π^+, π^-, and π^0. The π^+ and π^- particles have masses of 139.6 MeV/c^2, while the π^0 has a mass of 135.0 MeV/c^2. Pions and muons are very unstable particles. For example, the π^- first decays into a muon and an antineutrino with a mean lifetime of about 2.6×10^{-8} s. The muon then decays into an electron, a neutrino, and an antineutrino with a mean lifetime equal to about 2.2×10^{-6} s. The sequence of decays is

$$\begin{aligned} \pi^- &\longrightarrow \mu^- + \bar{\nu} \\ \mu^- &\longrightarrow e + \nu + \bar{\nu} \end{aligned} \tag{34.1}$$

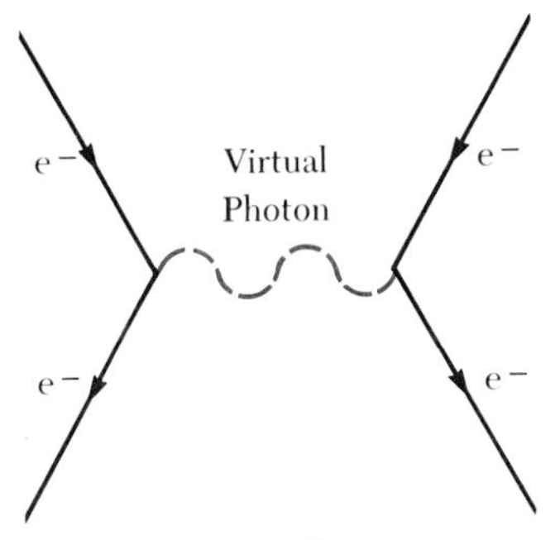

FIGURE 34.2 Feynman diagram representing how a photon mediates the electromagnetic force between two electrons.

The interaction between two particles can be represented in a simple diagram called a *Feynman diagram*, developed by Richard P. Feynman (1918–1988). Figure 34.2 is a Feynman diagram for the case of the electromagnetic interaction between two electrons. In this simple case, a single photon acts as the carrier of the electromagnetic force between the electrons. The photon transfers energy and momentum from one electron to the other in this interaction. These photons are called *virtual photons* because they can never be detected directly. This is because the photon is absorbed by the second electron very shortly after it is emitted by the first electron. The virtual photons violate the law of conservation of energy, but because of the uncertainty principle and the very short lifetime Δt, the photon's excess energy is less than the uncertainty in its energy, given by $\Delta E \approx \hbar/\Delta t$.

Now consider the pion exchange between a proton and a neutron via the strong force. The Feynman diagram for this interaction is shown in Figure 34.3. One can reason that the energy ΔE needed to create a pion of mass m_π is given by Einstein's equation $\Delta E = m_\pi c^2$. Again, the very existence of the pion violates conservation of energy by an amount, ΔE, which is permitted by the uncertainty principle only if this energy is surrendered in a time, Δt, the time it takes the pion to transfer between nucleons. From the uncertainty principle, $\Delta E\, \Delta t \approx \hbar$, we get

$$\Delta t \approx \hbar/\Delta E = \hbar/m_\pi c^2 \tag{34.2}$$

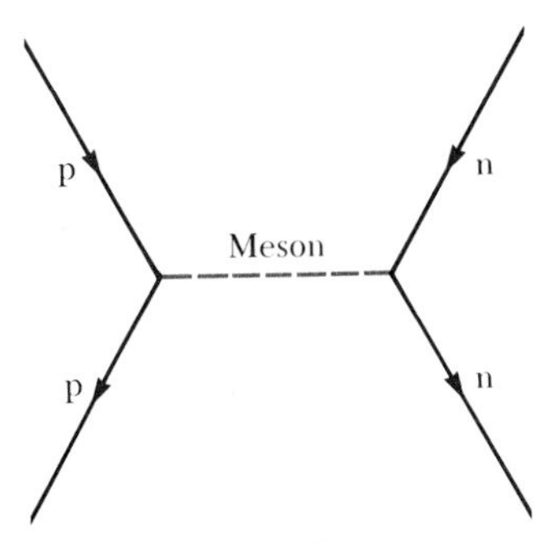

FIGURE 34.3 Feynman diagram representing a proton interacting with a neutron via the strong force. In this case, the meson mediates the strong force.

Since the pion cannot travel faster than the speed of light, the maximum distance d it can travel in a time Δt is $c\Delta t$. Using Equation 34.2, and $d = c\Delta t$, we find this maximum distance to be

$$d \approx \frac{\hbar}{m_\pi c} \tag{34.3}$$

From Chapter 32, we know that the range of the strong force has a value of about 1.5×10^{-15} m. Using this value for d in Equation 34.3, the rest energy of the pion is calculated to be

$$m_\pi c^2 \approx \frac{\hbar c}{d} = \frac{(1.05 \times 10^{-34}\ \mathrm{J \cdot s})(3 \times 10^8\ \mathrm{m/s})}{1.5 \times 10^{-15}\ \mathrm{m}}$$
$$= 2.1 \times 10^{-11}\ \mathrm{J} \cong 130\ \mathrm{MeV}$$

This corresponds to a mass of 130 MeV/c^2 (about 250 times the mass of the electron) which is in good agreement with the observed mass of the pion.

The concept we have just described is quite revolutionary, and something we have not encountered as yet. In effect, it says that a proton can change into a proton plus a pion, as long as it returns to its original state in a very short time. High energy physicists often say that a nucleon undergoes "fluctuations" as it emits and absorbs pions. As we have seen, these fluctuations are a consequence of a combination of quantum mechanics (through the uncertainty principle) and special relativity (through Einstein's energy-mass relation $E = mc^2$).

This section has dealt with the particles which mediate the strong force, namely the pions, and the mediators of the electromagnetic force, photons. The graviton, which is the mediator of the gravitational force, has yet to be observed. The particles which mediate the weak nuclear force are referred to as W^+, W^-, and Z^0. The discovery of the $W^\pm$ and Z^0 particles at CERN was announced in 1983 by Carlo Rubbia (1934–) and his associates using a proton-antiproton collider.[2] In this accelerator, protons and antiprotons that have a momentum of 270 GeV/c undergo head-on collisions with each other. In some of the collisions, $W^\pm$ and Z^0 particles are produced, which are, in turn, identified by their decay products.

34.5 CLASSIFICATION OF PARTICLES

Hadrons

All particles other than photons can be classified into two broad categories, hadrons and leptons, according to the interactions they experience. Particles that interact through the strong force are called *hadrons*. There are two classes of hadrons, known as *mesons* and *baryons*. These can be classified according to their masses and spins.

Mesons all have zero or integral spins (0 or 1), with masses that lie between the mass of the electron and the mass of the proton. All mesons are known to decay finally into electrons, positrons, neutrinos, and photons. The pion is the lightest of known mesons, with a mass of about 140 MeV/c^2 and a spin of 0. Another is the K meson, with a mass of about 500 MeV/c^2 and spin 0.

Baryons, which are the second class of hadrons, have a mass equal to or greater than the proton mass (hence the name *baryon* which means *heavy* in

[2]Carlos Rubbia, an Italian physicist, and Simon van der Meer, a Dutch physicist, both at CERN, shared the 1984 Nobel Prize in Physics for the discovery of the $W^\pm$ and Z^0 particles and the development of the proton-antiproton collider.

Greek), and their spin is always a noninteger value (1/2 or 3/2). Protons and neutrons are included in the baryon family, as are many other particles. With the exception of the proton, all baryons decay in such a way that the end products include a proton. For example, the Ξ hyperon first decays to a Λ^0 in about 10^{-10} s. The Λ^0 particle then decays to a proton and a π^- in about 3×10^{-10} s.

Today it is believed that hadrons are composed of more elemental units called *quarks*. Later we shall have more to say about the quark model. Some of the important properties of hadrons are listed in Table 34.2.

Leptons

Leptons (from the Greek *leptos* meaning *small* or *light*) are a group of particles that participate in the weak interaction. All leptons have a spin of 1/2. Included in this group are electrons, muons, and neutrinos, which are all less massive than the lightest hadron. Although hadrons have size and structure, leptons appear to be truly elementary particles with no structure (that is, pointlike).

TABLE 34.2 A Table of Some Particles and Their Properties

Category	Particle Name	Symbol	Antiparticle	Rest Mass (MeV/c^2)	B	L_e	L_μ	L_τ	S	Lifetime (s)	Principal Decay Modes[a]
Photon	Photon	γ	Self	0	0	0	0	0	0	Stable	
Leptons	Electron	e^-	e^+	0.511	0	+1	0	0	0	Stable	
	Neutrino (e)	ν_e	$\overline{\nu}_e$	0(?)	0	+1	0	0	0	Stable	
	Muon	μ^-	μ^+	105.7	0	0	+1	0	0	2.20×10^{-6}	$e^-\overline{\nu}_e\nu_\mu$
	Neutrino (μ)	ν_μ	$\overline{\nu}_\mu$	0(?)	0	0	+1	0	0	Stable	
	Tau	τ^-	τ^+	1784.	0	0	0	−1	0	$<4 \times 10^{-13}$	$\mu^-\overline{\nu}_\mu\nu_\tau$, $e^-\overline{\nu}_e\nu_t$, hadrons
	Neutrino (τ)	ν_τ	$\overline{\nu}_\tau$	0(?)	0	0	0	−1	0	Stable	
Hadrons											
Mesons	Pion	π^+	π^-	139.6	0	0	0	0	0	2.60×10^{-8}	$\mu^+\nu_\mu$
		π^0	Self	135.0	0	0	0	0	0	0.83×10^{-16}	2γ
	Kaon	K^+	K^-	493.7	0	0	0	0	+1	1.24×10^{-8}	$\mu^+\nu_\mu$, $\pi^+\pi^0$
		K_S^0	$\overline{K_S^0}$	497.7	0	0	0	0	+1	0.89×10^{-10}	$\pi^+\pi^-$, $2\pi^0$
		K_L^0	$\overline{K_L^0}$	497.7	0	0	0	0	+1	5.2×10^{-8}	$\pi^\pm e^\mp \overset{(-)}{\nu}_e$ $\pi^\pm \mu^\mp \overset{(-)}{\nu}_\mu$ $3\pi^0$
	Eta	η^0	Self	548.8	0	0	0	0	0	$<10^{-18}$	2γ, 3π
Baryons	Proton	p	$\overline{p}$	938.3	+1	0	0	0	0	Stable	
	Neutron	n	$\overline{n}$	939.6	+1	0	0	0	0	920	$pe^-\overline{\nu}_e$
	Lambda	Λ^0	$\overline{\Lambda}^0$	1115.6	+1	0	0	0	−1	2.6×10^{-10}	$p\pi^-$, $n\pi^0$
	Sigma	Σ^+	$\overline{\Sigma}^-$	1189.4	+1	0	0	0	−1	0.80×10^{-10}	$p\pi^0$, $n\pi^+$
		Σ^0	$\overline{\Sigma}^0$	1192.5	+1	0	0	0	−1	6×10^{-20}	$\Lambda^0\gamma$
		Σ^-	$\overline{\Sigma}^+$	1197.3	+1	0	0	0	−1	1.5×10^{-10}	$n\pi^-$
	Xi	Ξ^0	$\overline{\Xi}^0$	1315	+1	0	0	0	−2	2.9×10^{-10}	$\Lambda^0\pi^0$
		Ξ^-	Ξ^+	1321	+1	0	0	0	−2	1.64×10^{-10}	$\Lambda^0\pi^-$
	Omega	Ω^-	Ω^+	1672	+1	0	0	0	−3	0.82×10^{-10}	$\Xi^0\pi^0$, Λ^0K^-

[a] A notation in this column such as $p\pi^-$, $n\pi^0$ means two possible decay modes. In this case, the two possible decays are $\Lambda^0 \longrightarrow p + \pi^-$ or $\Lambda^0 \longrightarrow n + \pi^0$.

The American physicists Richard Feynman *(left)* and Murray Gell-Mann *(right)* won the Nobel prize in physics in 1965 and 1969, respectively, for their theoretical studies dealing with subatomic particles. (Photo courtesy of Michael R. Dressler)

Quite unlike the situation with hadrons, the number of known leptons is very limited. Currently, scientists believe there only are six leptons (each having an antiparticle): the electron, the muon, the tau, and a neutrino associated with each of these particles. (Note that the neutrino associated with the tau has not yet been observed in the laboratory.) The τ (tau) lepton, discovered in 1975, has a mass equal to about twice that of the proton. Associated with this heavy lepton is the τ neutrino (ν_τ). We now classify the six known leptons into three groups:

$$\begin{pmatrix} e^- \\ \nu_e \end{pmatrix} \quad \begin{pmatrix} \mu^- \\ \nu_\mu \end{pmatrix} \quad \begin{pmatrix} \tau^- \\ \nu_\tau \end{pmatrix}$$

Although neutrinos are thought to be massless, there is a possibility that they may have a small, nonzero mass. As we shall see later, a firm knowledge of the neutrino's mass could have great significance in cosmological models and the future of the universe.

34.6 CONSERVATION LAWS

In Chapter 32 we learned that conservation laws are important in understanding why certain decays or reactions occur and others do not. In general, the laws of conservation of energy, linear momentum, angular momentum, and electric charge provide us with a set of rules that all processes must follow. For example, conservation of electric charge requires that the total charge before a reaction must equal the total charge after the reaction.

A number of new conservation laws are important in the study of elementary particle decays and reactions. Two of these described in this section are the conservation of baryon number and the conservation of lepton number. Although these conservation laws have no theoretical foundation, they are supported by an abundance of empirical evidence.

Baryon Number

The conservation of baryon number implies that whenever a baryon is created in a reaction or decay, an antibaryon is also created. This can be quantified by assigning a baryon number $B = +1$ to all baryons, $B = -1$ to all antibaryons, and $B = 0$ to all other particles. Thus, the **law of conservation of baryon number** can be stated as follows:

Whenever a nuclear reaction or decay occurs, the sum of the baryon numbers before the process must equal the sum of the baryon numbers after the process.

An equivalent statement is that the net number of baryons remains constant in any process.

Note that if baryon number is absolutely conserved, the proton must be absolutely stable. If it were not for the law of conservation of baryon number, the proton could decay to a positron and a neutral pion. However, such a decay has never been observed. At the present, we can say only that the proton has a half-life of at least 10^{31} years (which could be compared with the estimated age of the universe, which is about 10^{10} years). In one recent version of the recent grand unified theory, or GUT, physicists have predicted that the proton is actually unstable. According to this theory, the baryon number (sometimes called the baryonic charge) cannot be absolutely conserved, whereas electric charge is always conserved.

EXAMPLE 34.1 Checking Baryon Numbers

Determine whether or not each of the following reactions can occur based on the law of conservation of baryon number.

$$p + n \longrightarrow p + p + n + \bar{p} \quad (1)$$

$$p + n \longrightarrow p + p + \bar{p} \quad (2)$$

Solution First, let us check reaction (1). Recall that $B = +1$ for baryons and $B = -1$ for antibaryons. Hence the left side of (1) gives a total baryon number of $1 + 1 = 2$. The right side of (1) gives a total baryon number of $1 + 1 + 1 + (-1) = 2$. Hence the reaction can occur provided the incoming proton has sufficient energy.

Now let us examine reaction (2). The left side of (2) again gives a total baryon number of $1 + 1 = 2$. However, the right side of (2) gives a total number of $1 + 1 + (-1) = 1$. Since the baryon number is not conserved, the reaction cannot occur.

Lepton Number

Recall that there are three varieties of leptons: the electron, the muon, and the recently discovered tau lepton. Each of these is accompanied by a neutrino. There are three separate conservation laws involving lepton numbers, one for each variety of lepton. The **law of conservation of electron-lepton number** states that

the sum of the electron-lepton numbers before a reaction or decay must equal the sum of the electron-lepton numbers after the reaction or decay.

The electron (e^-) and the electron neutrino (ν_e) are assigned a positive lepton number $L_e = +1$, the antileptons e^+ and $\bar{\nu}_e$ are assigned a negative lepton number $L_e = -1$, and all others have $L_e = 0$. For example, consider the decay of the neutron,

Neutron decay

$$n \longrightarrow p + e^- + \bar{\nu}_e$$

Before the decay, the electron-lepton number is $L_e = 0$, while after the decay the electron-lepton number is $0 + 1 + (-1) = 0$. Thus, the electron-lepton number is conserved. It is important to recognize that the baryon number must also be conserved. This can easily be seen by noting that before the decay $B = +1$, while after the decay the baryon number is $+1 + 0 + 0 = +1$.

Similarly, when a decay involves muons, the muon-lepton number, L_μ, is conserved. The μ^- and the ν_μ are assigned positive numbers, $L_\mu = +1$, the antimuons μ^+ and $\bar{\nu}_\mu$ are assigned negative numbers, $L_\mu = -1$, while all others have $L_\mu = 0$. Finally, the tau-lepton number, L_τ, is conserved, and similar assignments can be made for the τ lepton and its neutrino, ν_τ.

EXAMPLE 34.2 Checking Lepton Numbers

Determine which of the following decay schemes can occur on the basis of conservation of lepton-number.

$$\mu^- \longrightarrow e^- + \bar{\nu}_e + \nu_\mu \quad (1)$$

$$\pi^+ \longrightarrow \mu^+ + \nu_\mu + \nu_e \quad (2)$$

Solution First let us examine decay (1). Since this decay involves both a muon and an electron, L_μ and L_e must both be conserved. Before the decay, $L_\mu = +1$ and $L_e = 0$. After the decay, $L_\mu = 0 + 0 + 1 = +1$, and $L_e = +1 - 1 + 0 = 0$. Thus, both numbers are conserved, and on this basis the decay mode is possible.

Now consider decay (2). Before the decay, $L_\mu = 0$ and $L_e = 0$. After the decay, $L_\mu = -1 + 1 + 0 = 0$, but $L_e = +1$. Thus, the decay is not possible because the electron-lepton number is not conserved.

EXERCISE Determine whether the decay $\mu^- \longrightarrow e^- + \bar{\nu}_e$ can occur.

Answer No. The muon-lepton number is $+1$ before the decay and is 0 after the decay. Thus, the muon-lepton number is not conserved.

34.7 STRANGE PARTICLES AND STRANGENESS

Many particles discovered in the 1950s were produced by the strong interaction of pions with protons and neutrons in the atmosphere. A group of these particles, namely the K, Λ, and Σ, were found to exhibit unusual properties in their production and decay, and hence were called *strange particles*. One unusual property is that they are always produced in pairs. For example, when a pion collides with a proton, two neutral strange particles are produced with high probability (see Fig. 34.4) following the reaction

$$\pi^- + p \longrightarrow K^0 + \Lambda^0$$

On the other hand, the reaction $\pi^- + p \longrightarrow K^0 + n$ never occurred, even though no known conservation laws were violated and the energy of the pion was sufficient to initiate the reaction. The second peculiar feature of strange particles is that although they are produced by the strong interaction at a high rate, they do not decay into strongly interacting particles at a very high rate as one might expect. Instead, they decay very slowly, which is characteristic of the weak interaction. Their half-lives are in the range 10^{-10} s to 10^{-8} s; strongly interacting particles have lifetimes of the order of 10^{-23} s.

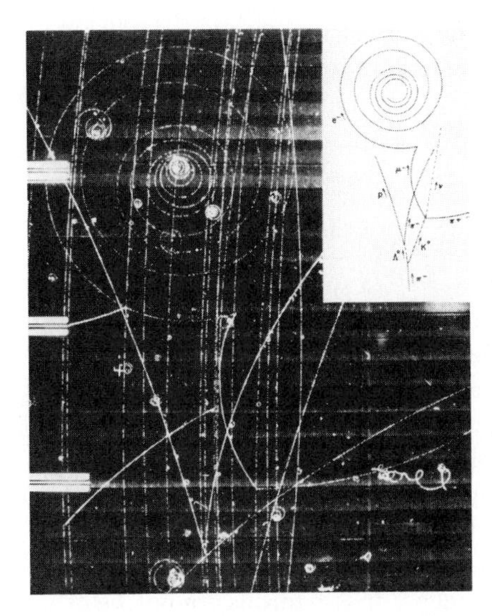

FIGURE 34.4 This bubble-chamber photograph shows many events, and the inset represents a drawing of identified tracks. The strange particles Λ^0 and K^0 are formed (at the bottom) as the π^- interacts with a proton according to $\pi^- + p \rightarrow \Lambda^0 + K^0$. (Note that the neutral particles leave no tracks, as indicated by the dashed lines.) The Λ^0 and K^0 then decay according to $\Lambda^0 \rightarrow \pi^- + p$ and $K^0 \rightarrow \pi + \mu^- + \nu_\mu$. (Courtesy of Lawrence Berkeley Laboratory, University of California, Photographic Services.)

In order to explain these unusual properties of strange particles, a new conservation law called conservation of strangeness was introduced, together with a new quantum number S called the **strangeness.** The strangeness numbers for various particles are given in Table 34.2. The production of strange particles in pairs is explained by assigning $S = +1$ to one of the particles, and $S = -1$ to the other. The nonstrange particles such as the π mesons, protons, and leptons are assigned strangeness $S = 0$. The **law of conservation of strangeness** can be stated as follows:

Whenever a nuclear reaction or decay occurs, the sum of the strangeness numbers before the process must equal the sum of the strangeness numbers after the process.

One can explain the slow decay of strange particles by assuming that the strong and electromagnetic interactions obey the law of conservation of strangeness, while the weak interaction does not. Since the decay reaction involves the loss of one strange particle, it violates strangeness conservation, and hence proceeds slowly via the weak interaction.

EXAMPLE 34.3 Is Strangeness Conserved?

(a) Determine whether the following reaction occurs on the basis of conservation of strangeness.

$$\pi^0 + n \longrightarrow K^+ + \Sigma^-$$

Solution The initial state has a total strangeness $S = 0 + 0 = 0$. Since the strangeness of the K^+ is $S = +1$, and the strangeness of the Σ^- is $S = -1$, the total strangeness of the final state is $+1 - 1 = 0$. Thus, we see that strangeness is conserved and the reaction is allowed.

(b) Show that the following reaction does not conserve strangeness, and hence cannot occur.

$$\pi^- + p \longrightarrow \pi^- + \Sigma^+$$

Solution In this case, the initial state has a total strangeness $S = 0 + 0 = 0$, while the final state has a total strangeness $S = 0 + (-1) = -1$. Thus strangeness is not conserved, so the reaction does not occur.

Exercise Show that the observed reaction $p + \pi^- \longrightarrow K^0 + \Lambda^0$ obeys the law of conservation of strangeness.

34.8 QUARKS—FINALLY

As we have noted, leptons appear to be truly elementary particles because they have no measurable size or internal structure, are limited in number, and do not seem to break down into smaller units. Hadrons, on the other hand, are complex particles having size and structure. Furthermore, we know that hadrons decay into other hadrons and are large in number. It is important to note that Table 34.2 lists only hadrons that are stable against hadronic decay. Hundreds of others have been discovered, and their properties have been determined. These facts strongly suggest that hadrons cannot be truly elementary, but have some substructure.

The Original Quark Model

In 1963 Murray Gell-Mann and George Zweig independently proposed that the known hadrons have a more elemental substructure. According to their model, all hadrons are composite systems of two or three fundamental constituents called **quarks.** Gell-Mann borrowed the word "quark" from the passage "Three quarks for Muster Mark" in James Joyce's book *Finnegan's Wake.* In the original quark model, there were three types of quarks designated by the symbols u, d, and s. These were given the arbitrary names *up*, *down*, and *sideways* (or now more commonly, *strange*). A most unusual property of quarks is that they have fractional electronic charges. The u, d, and s quarks have charges of $+2e/3$, $-e/3$, and $-e/3$, respectively. Each quark has a bar-

TABLE 34.3 Properties of Quarks and Antiquarks

Quarks								
Name	Symbol	Spin	Charge	Baryon Number	Strangeness	Charm	Bottomness	Topness
Up	u	$\frac{1}{2}$	$+\frac{2}{3}e$	$\frac{1}{3}$	0	0	0	0
Down	d	$\frac{1}{2}$	$-\frac{1}{3}e$	$\frac{1}{3}$	0	0	0	0
Strange	s	$\frac{1}{2}$	$-\frac{1}{3}e$	$\frac{1}{3}$	-1	0	0	0
Charmed	c	$\frac{1}{2}$	$+\frac{2}{3}e$	$\frac{1}{3}$	0	$+1$	0	0
Bottom	b	$\frac{1}{2}$	$-\frac{1}{3}e$	$\frac{1}{3}$	0	0	$+1$	0
Top (?)	t	$\frac{1}{2}$	$+\frac{2}{3}e$	$\frac{1}{3}$	0	0	0	$+1$

Antiquarks								
Name	Symbol	Spin	Charge	Baryon Number	Strangeness	Charm	Bottomness	Topness
Up	$\bar{u}$	$\frac{1}{2}$	$-\frac{2}{3}e$	$-\frac{1}{3}$	0	0	0	0
Down	$\bar{d}$	$\frac{1}{2}$	$+\frac{1}{3}e$	$-\frac{1}{3}$	0	0	0	0
Strange	$\bar{s}$	$\frac{1}{2}$	$+\frac{1}{3}e$	$-\frac{1}{3}$	$+1$	0	0	0
Charmed	$\bar{c}$	$\frac{1}{2}$	$-\frac{2}{3}e$	$-\frac{1}{3}$	0	-1	0	0
Bottom	$\bar{b}$	$\frac{1}{2}$	$+\frac{1}{3}e$	$-\frac{1}{3}$	0	0	-1	0
Top (?)	$\bar{t}$	$\frac{1}{2}$	$-\frac{2}{3}e$	$-\frac{1}{3}$	0	0	0	-1

yon number of 1/3 and a spin of 1/2. Furthermore, the u and d quarks have strangeness of 0, while the s quark has strangeness of -1. Other properties of quarks and antiquarks are given in Table 34.3. Note that associated with each quark is an antiquark of opposite charge, baryon number, and strangeness. The composition of all hadrons known at the time could be completely specified by three simple rules. (1) Mesons consist of one quark and one antiquark, giving them a baryon number of 0 as required. (2) Baryons consist of three quarks. (3) Antibaryons consist of three antiquarks. Table 34.4 lists the quark compositions of several mesons and baryons. For example, a π^- meson contains one $\bar{u}$ and one d quark (designated as $\bar{u}d$), with $Q = -2e/3 - e/3 = -e$, $B = -1/3 + 1/3 = 0$, and $S = 0 + 0 = 0$. A proton, on the other hand, contains two u quarks and one d quark with $Q = +1$, $B = 1$, and $S = 0$. Note that just two of the quarks, u and d, are contained in all hadrons encountered in ordinary matter (protons and neutrons). The third quark, s, is needed only to construct strange particles with a strangeness number of either $+1$ or -1. For example, the neutral particle Λ^0 has uds as its quark composition, with $Q = 0$, $B = 1$, and $S = -1$. Figure 34.5 is a pictorial representation of the quark composition of several particles.

TABLE 34.4 Quark Composition of Several Hadrons

Particle	Quark Composition
	Mesons
π^+	$u\bar{d}$
π^-	$\bar{u}d$
K^+	$u\bar{s}$
K^-	$\bar{u}s$
K^0	$d\bar{s}$
	Baryons
p	uud
n	udd
Λ^0	uds
Σ^+	uus
Σ^0	uds
Σ^-	dds
Ξ^0	uss
Ξ^-	dss
Ω^-	sss

Charm and Other Recent Developments

Although the original quark model was highly successful in classifying particles into families, there were some discrepancies between predictions of the model and certain experimental decay rates. Consequently, a fourth quark was proposed by several physicists in 1967. They argued that if there are four leptons (as was thought at the time), then there should also be four quarks because of an underlying symmetry in nature. The fourth quark, designated by c, was given the new property, or quantum number, called **charm.** The *charmed* quark would have a charge $+2e/3$, but its new property of charm would distinguish it from the other three quarks. The new quark would have a charm of $C = +1$, while its antiquark would have a charm of $C = -1$ as indicated in Table 34.3. Charm, like strangeness, would be conserved in

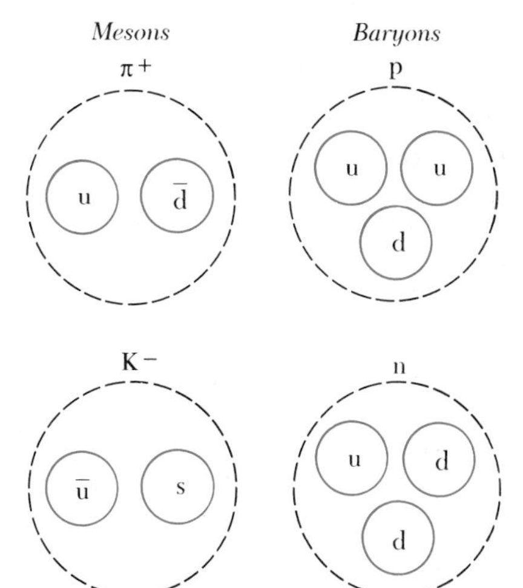

FIGURE 34.5 Quark compositions of several particles. Note that the mesons on the left contain two quarks, while the baryons on the right contain three quarks.

strong and electromagnetic interactions, but would not be conserved in weak interactions.

In 1974, a new heavy meson called the J/Ψ particle (or simply Ψ) was discovered independently by a group led by Burton Richter at the Stanford Linear Accelerator (SLAC) and another group led by Samuel Ting at the Brookhaven National Laboratory. Richter and Ting were awarded the Nobel Prize in 1976 for this work. The J/Ψ particle did not fit into the original three-quark model, but had the properties of a combination of a charmed quark and its antiquark $(c\bar{c})$. Its mass was much heavier than the other known mesons ($\sim$3100 MeV/c^2) and its lifetime was much longer than other strongly decaying particles. Soon thereafter, other related charmed mesons including D^+ and D^- were discovered corresponding to such quark combinations as $\bar{c}d$ and $c\bar{d}$, all of which have large masses and long lifetimes. In 1975, researchers at Stanford University reported strong evidence for the tau (τ) lepton, with a mass of 1784 MeV/c^2. Such discoveries led to more elaborate quark models, and the proposal of two new quarks, named *top* (t), and *bottom* (b). (Some physicists prefer the whimsical names *truth* and *beauty*.) To distinguish these quarks from the old ones, quantum numbers called *topness* and *bottomness* were assigned to these new particles, and are included in Table 34.3. In 1977, researchers at the Fermi National Laboratory under the direction of Leon Lederman reported the discovery of a very massive new meson, Υ, whose composition is considered to be $b\bar{b}$.

At this point, you are probably wondering whether or not such discoveries will ever end. How many "building blocks" of matter really exist? At the present, physicists believe that the fundamental particles in nature include six quarks and six leptons (together with their antiparticles). Some of the properties of these particles are given in Table 34.5.

It should be noted that in spite of many extensive experimental efforts, no isolated quark has ever been observed. Physicists now believe that quarks are permanently confined inside ordinary particles because of an exceptionally strong force that prevents them from escaping. This force, called the "color"

TABLE 34.5 The Fundamental Particles and Some of Their Properties

Particle	Rest Energy	Charge
Quarks		
u	360 MeV	$+\frac{2}{3}e$
d	360 MeV	$-\frac{1}{3}e$
c	1500 MeV	$+\frac{2}{3}e$
s	540 MeV	$-\frac{1}{3}e$
t (?)	~100 GeV	$+\frac{2}{3}e$
b	5 GeV	$-\frac{1}{3}e$
Leptons		
e^-	511 keV	$-e$
μ^-	107 MeV	$-e$
τ^-	1784 MeV	$-e$
ν_e	<30 eV	0
ν_μ	<0.5 MeV	0
ν_τ	<250 MeV	0

force, increases with separation distance (similar to the force of a spring); the properties of this force are discussed in the next section. The great strength of the force between quarks has been described by one author as follows:[3]

> Quarks are slaves of their own color charge, . . . bound like prisoners of a chain gang. . . . Any locksmith can break the chain between two prisoners, but no locksmith is expert enough to break the gluon chains between quarks. Quarks remain slaves forever.

34.9 THE STANDARD MODEL

Shortly after the concept of quarks was proposed, scientists recognized that certain particles had quark compositions which were in violation of the Pauli exclusion principle. Recall that quarks are fermions with spins of 1/2, and hence are expected to follow the exclusion principle. One example is the Ω^- *(sss)* baryon which contains three *s* quarks having parallel spins, giving it a total spin of 3/2. Other examples of baryons which have identical quarks with parallel spins are the Δ^{++} *(uuu)* and the Δ^- *(ddd)*. To resolve this problem, it was suggested that quarks possess a new property called **color.** This property is similar in many respects to electric charge except that it occurs in three varieties (of color) called red, green, and blue. Another term that is used to distinguish among the six quarks is **flavor.** Thus, each flavor of quark can have three colors. (The whimsical names "color" and "flavor" should not be taken literally.) *Of course, the antiquarks have the colors antired, antigreen, and antiblue.* In order to satisfy the exclusion principle, all three quarks in a baryon must have different colors. A meson consists of a quark of one color and an antiquark of the corresponding anticolor. The result is that baryons and mesons are always colorless (or white). Furthermore, the new property of color increases the number of quarks by a factor of three.

Although the concept of color in the quark model was originally conceived to satisfy the exclusion principle, it also provided a better theory for explaining certain experimental results. For example, the modified theory correctly predicts the lifetime of the π^0 meson. The theory of how quarks interact with each other is called **quantum chromodynamics,** or QCD, because it is similar in its structure to quantum electrodynamics (the theory of interaction between electric charges). In QCD, the quark is said to carry a *color charge,* in analogy to electric charge. The strong force between quarks is often called the *color force.* As mentioned earlier, the strong interaction between hadrons is mediated by massless particles called **gluons** (analogous to photons for the electromagnetic force). According to the theory, there are eight gluons, six of which have color charge. Because of their color charge, quarks can attract each other and form composite particles. When a quark emits or absorbs a gluon, its color changes. For example, a blue quark that emits a gluon may become a red quark, while the red quark that absorbs this gluon becomes a blue quark. The color force between quarks is analogous to the electric force between charges—like colors repel and opposite colors attract. Therefore, two red quarks repel each other, but a red quark will be

[3]Harald Fritzsch, *Quarks, The Stuff of Matter,* London, Allen Lane, 1983.

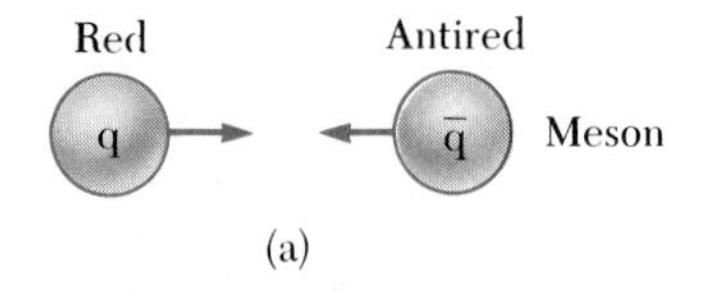

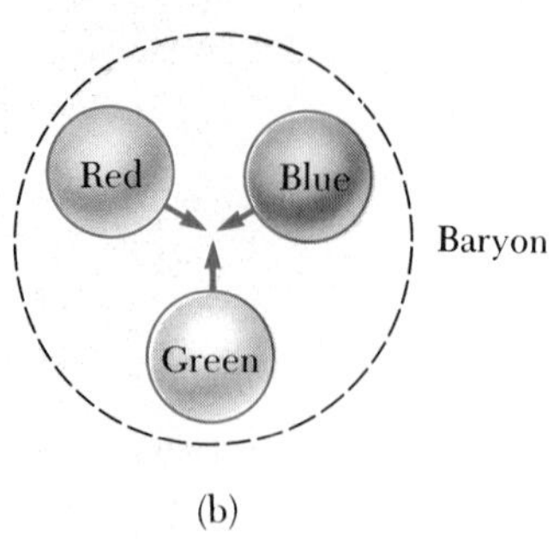

FIGURE 34.6 (a) A red quark is attracted to an antired quark. This forms a meson whose quark structure is ($q\bar{q}$). (b) Three different colored quarks attract each other to form a baryon.

attracted to an antired quark. The attraction between quarks of opposite color to form a meson ($q\bar{q}$) is indicated in Figure 34.6a. Differently colored quarks also attract each other, but with less intensity than opposite colors of quark and antiquark. For example, a cluster of red, blue, and green quarks all attract each other to form baryons as indicated in Figure 34.6b. Thus, all baryons contain three quarks each of which has a different color.

Recall that the weak force is believed to be mediated by the W^+, W^-, and Z^0 bosons (spin 1 particles). These particles are said to have *weak charge* just as a quark has color charge. Thus, each elementary particle can have mass, electric charge, color charge, and weak charge. Of course, one or more of these could be zero. Scientists now believe that the truly elementary particles are leptons and quarks, and the force mediators are the gluon, the photon, $W^\pm$, Z^0, and the graviton. (Note that quarks and leptons have spin ½ and hence are fermions, while the force mediators have spin 1 or higher, and are bosons.)

In 1979, Sheldon Glashow, Abdus Salam, and Steven Weinberg won a Nobel Prize for developing a theory that unified the electromagnetic and weak interactions. This so-called **electroweak theory** postulates that the weak and electromagnetic interactions have the same strength at very high particle energies. Thus, the two interactions are viewed as two different manifestations of a single unifying electroweak interaction. The photon and the three massive bosons ($W^\pm$ and Z^0) play a key role in the electroweak theory. The theory makes many concrete predictions, but perhaps the most spectacular is the prediction of the masses of the W and Z particles at about 82 GeV/c^2 and 93 GeV/c^2, respectively. The 1984 Nobel Prize was awarded to Carlo Rubbia and Simon van der Meer for their work leading to the discovery of these particles at just these energies at the CERN Laboratory in Geneva, Switzerland.

The combination of the electroweak theory and QCD for the strong interaction form what is referred to in high energy physics as the "Standard Model." Although the details of the Standard Model are complex, its essential ingredients can be summarized with the help of Figure 34.7. The strong force, mediated by gluons, holds quarks together to form composite particles such as protons, neutrons, and mesons. Leptons participate only in the electromagnetic and weak interactions. The electromagnetic force is mediated by photons, while the weak force is mediated by W and Z bosons. Note that all fundamental forces are mediated by spin 1 particles whose properties are given, to a large extent, by the symmetries involved in the theories. However, the Standard Model does not answer all questions. A major problem is why the photon has no mass while the W and Z bosons are massive particles. Because of their mass difference, the electromagnetic and weak forces are quite distinct at low energies, but become similar in nature at very high energies. This behavior as one goes from low to high energies, called *symmetry breaking*, leaves open the question of the origin of particle masses. In order to resolve this problem, a hypothetical particle called the *Higgs boson* has been proposed, which provides a mechanism for breaking the electroweak symmetry. The Standard Model, including the Higgs mechanism, provides a logically consistent explanation of the massive nature of the W and Z bosons. Unfortunately, the Higgs boson has not yet been found, but physicists know that its mass should be less than 1 TeV (10^{12} eV).

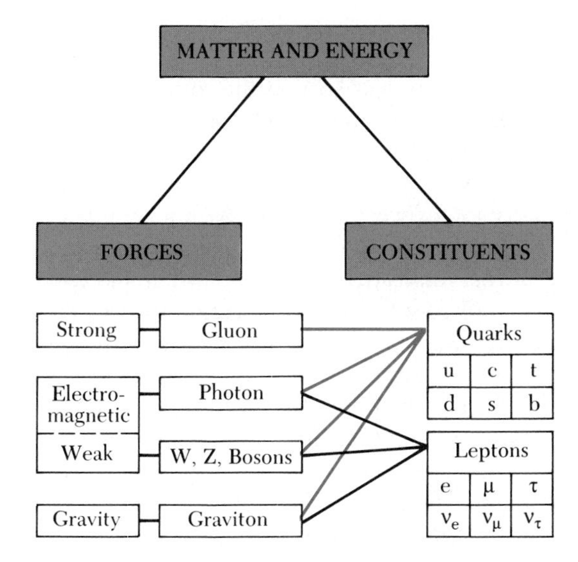

FIGURE 34.7 The Standard Model of particle physics.

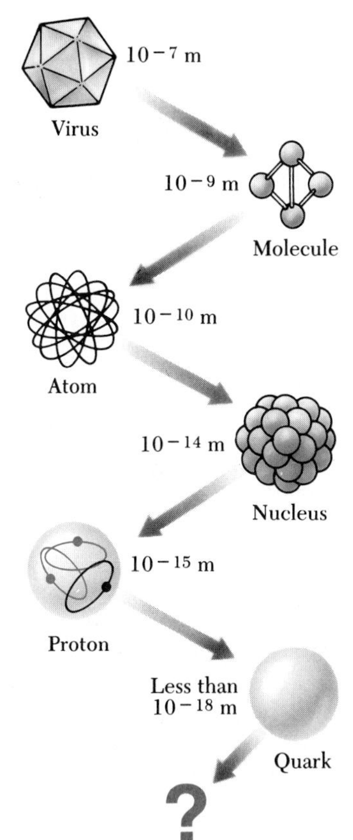

FIGURE 34.8 Looking at matter with various microscopes reveals structures ranging in size from the smallest living thing, a virus (at the top), down to a quark (at the bottom) which has not yet been observed as an isolated particle.

In order to determine whether the Higgs boson exists, two quarks of at least 1 TeV of energy must collide, but calculations show that this requires injecting 40 TeV of energy within the volume of a proton. The excess energy is needed because of the quarks and gluons contained in the proton. Although no existing accelerator can provide this energy, scientists in the United States are currently planning to construct the world's largest and most powerful particle accelerator called the Superconducting Super Collider, or SSC, which will meet this need. The SSC will accelerate protons to 20 TeV (two protons traveling in opposite directions with this energy will give the required 40 TeV), it will have a circumference of 82.944 km (about 52 miles), and will cost about 4 billion dollars! The proton energies in the SSC will be twenty times greater than is currently available at Fermilab, and physicists expect to be able to explore distances down to about 10^{-18} m (one thousandth the diameter of a proton). Thus, it will be a microscope of unparalleled power. Figure 34.8 shows the evolution of the various stages of matter that scientists have been able to investigate with various types of microscopes. The Department of Energy endorsed detailed research on the SSC in 1983, and federal funds were recently allocated to support research and development on the superconducting magnets to be used in the accelerator. The SSC is expected to be completed in the mid-1990s if federal funding and Congressional authorization are provided.

Following the success of the electroweak theory, scientists attempted to combine it with QCD in a *grand unification theory* known as GUT. In this model, the electroweak force was merged with the strong color force to form a grand unified force. One version of the theory considers leptons and quarks as members of the same family that are able to change into each other by exchanging an appropriate messenger particle. Many GUT theories predict that protons are unstable, and will decay with a half-life of about 10^{31} years. This is far greater than the age of the universe, but as yet proton decays have not been observed.

34.10 THE COSMIC CONNECTION

Today cosmologists believe that the universe was born as the result of a single violent explosion known as the Big Bang. The evolution of the four fundamental forces from the Big Bang to the present is shown in Figure 34.9. During the first 10^{-43} s (the ultra-hot epoch where $T \approx 10^{32}$ K), it is presumed that the strong, electroweak, and gravity forces were joined to form a completely unified force. In the first 10^{-32} s following the Big Bang (the hot epoch where $T \approx 10^{29}$ K), gravity broke free of this unification while the strong and electroweak forces remained as one, described by a grand unification theory. This was a period when particle energies were so great ($>10^{16}$ GeV) that very massive particles as well as light quarks, leptons, and their antiparticles existed. At the end of this fleeting instant of time, the very dense and hot universe could be squeezed into a sphere less than 10^{-4} m in diameter! Then the universe rapidly expanded and cooled during the warm epoch which covers temperatures in the range from 10^{29} to 10^{15} K (corresponding to energies ranging from 10^{16} to 10^{2} GeV). During this epoch, the strong and electroweak forces parted company, and the grand unification scheme was broken. As the universe continued to cool, about 10^{-10} s after the Big Bang, the electroweak force split in two. This resulted in the four forces in the universe as we distinguish them today. During the first microsecond of cosmic history, there were only elementary particles interacting through fundamental forces. About one microsecond after the Big Bang, the quarks condensed into protons and neu-

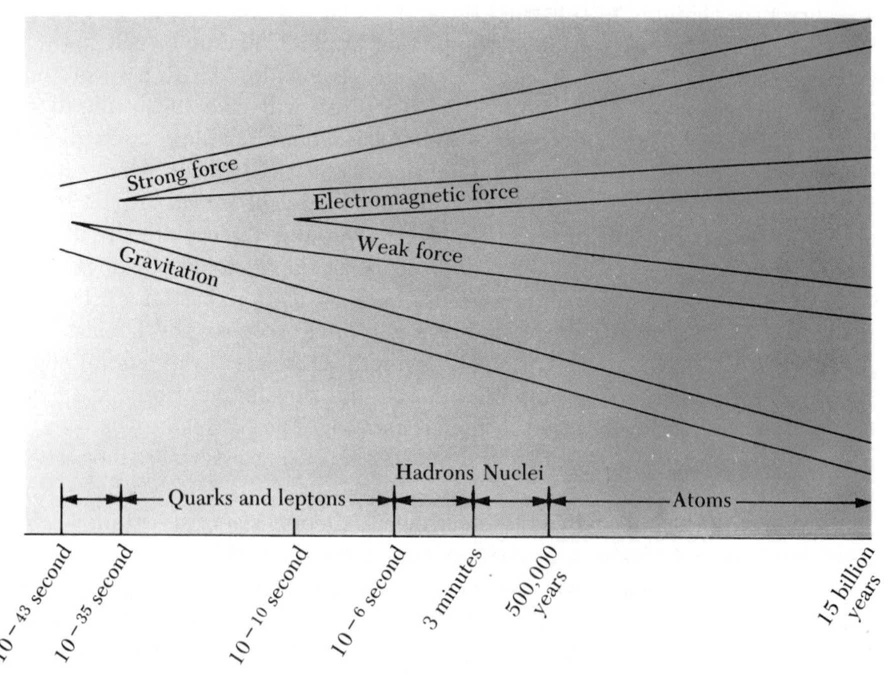

FIGURE 34.9 A brief history of the universe from the Big Bang to the present. The four forces became distinguishable during the first microsecond. Following this, all the quarks combined to form the strongly interacting particles. However, the leptons remained separate, and exist as individually observable particles to this day.

trons, and a few minutes later the sea of protons and neutrons condensed to form atomic nuclei.

While particle physicists have been exploring the realm of the very small, cosmologists have been exploring cosmic history back to the first microsecond of the Big Bang. The observation of events that occur when two particles collide in an accelerator is essential in reconstructing the early moments in cosmic history. Perhaps the key to understanding the early universe is first to understand the world of elementary particles. Cosmologists and particle physicists now find they have many common goals and are joining hands to attempt to understand the physical world at its most fundamental level.

34.11 PROBLEMS AND PERSPECTIVES

Our understanding of physics at short distances is far from complete. Particle physics is faced with many questions. Why is there so little antimatter in the universe? Do neutrinos have a small rest mass, and if so, how do they contribute to the "dark matter" of the universe? Is it possible to unify the strong and electroweak theories in a logical and consistent manner? Why do quarks and leptons form three similar but distinct families? Are muons the same as electrons (apart from their difference in mass), or do they have other subtle differences that have not been detected? Why are some particles charged and others neutral? Why do quarks carry a fractional charge? What determines the masses of the fundamental constituents? Can isolated quarks exist? The questions go on and on. Because of the rapid advances and new discoveries in the field of particle physics, by the time you read this book some of these questions will likely be resolved while others may emerge.

An important and obvious question that remains is whether leptons and quarks have a substructure. If they have a substructure, one could envision an infinite number of deeper structure levels. However, if leptons and quarks are indeed the ultimate constituents of matter, as physicists today tend to believe, we should be able to construct a final theory of the structure of matter as Einstein himself dreamed. In the view of many physicists, the end of the road is in sight, but it is anyone's guess as to how long it will take to reach that goal.

SUMMARY

There are four fundamental forces in nature: strong (hadronic), electromagnetic, weak, and gravitational. The strong force is the force between nucleons that keeps the nucleus together. The weak force is responsible for beta decay. The electromagnetic and weak forces are now considered to be manifestations of a single force called the electroweak force.

An antiparticle and a particle have the same mass, but opposite charge. Furthermore other properties may have opposite values such as lepton number and baryon number. It is possible to produce particle-antiparticle pairs in nuclear reactions if the available energy is greater than $2m_0c^2$, where m_0 is the rest mass of the particle (or antiparticle).

Particles other than photons are classified as being *hadrons* or *leptons*. Hadrons interact primarily through the strong force. We now know that hadrons have size and structure, and hence are not fundamental constituents of matter. There are two types of hadrons called *baryons* and *mesons.* Mesons have a baryon number zero and have either zero or integral spin. Baryons, which generally are the most massive particles, have a nonzero baryon number and a spin of 1/2 or 3/2. The neutron and proton are examples of baryons.

Leptons have no structure or size, and are considered to be truly elementary particles. Leptons interact only through the weak and electromagnetic forces. There are six leptons, the electron e^-, the muon μ^-, the tau τ^-, and their neutrinos ν_e, ν_μ, and ν_τ.

In all reactions and decays, quantities such as energy, linear momentum, angular momentum, electric charge, baryon number, and lepton number are strictly conserved. Certain particles have properties called *strangeness* and *charm.* These unusual properties are conserved only in those reactions and decays that occur via the strong force.

Recent theories in elementary particle physics have postulated that all hadrons are composed of smaller units known as *quarks.* Quarks have a fractional electric charge and a baryon number of 1/3. There are six flavors of quarks, up *(u),* down *(d),* strange *(s),* charmed *(c),* top *(t),* and bottom *(b).* All baryons contain three quarks, while all mesons contain one quark and one antiquark.

According to the theory of *quantum chromodynamics,* quarks have a property called *color,* and the strong force between quarks is referred to as the *color force.*

Every fundamental interaction is said to be mediated by the exchange of field particles. The electromagnetic interaction is mediated by the photon; the weak interaction is mediated by the $W^\pm$ and Z^0 bosons; the gravitational interaction is mediated by gravitons; the strong interaction is mediated by gluons.

ADDITIONAL READING

E. D. Bloom and G. J. Feldmann, "Quarkonium," *Sci. American,* May, 1982.

Frank Close, *The Cosmic Onion: Quarks and the Nature of the Universe,* The American Institute of Physics, 1986. A timely monograph on particle physics, including lively discussions of the Big Bang theory.

Harald Fritzsch, *Quarks, The Stuff of Matter,* London, Allen and Lane, 1983. An excellent introductory overview of elementary particle physics.

H. Harari, "The Structure of Quarks and Leptons," *Sci. American,* April, 1983.

J. David Jackson, Maury Tigner, and Stanley Wojcicki, "The Superconducting Supercollider," *Sci. American,* March, 1986.

Leon M. Lederman, "The Value of Fundamental Science," *Sci. American,* November, 1984.

N. B. Mistry, R. A. Poling, and E. H. Thorndike, "Particles with Naked Beauty," *Sci. American,* July, 1983.

Chris Quigg, "Elementary Particles and Forces," *Sci. American,* April, 1985.

James S. Trefil, *From Atoms to Quarks,* New York, Scribner, 1980. This is an excellent introduction to the world of particle physics.

Steven Weinberg, *The Discovery of Elementary Particles,* New York, Scientific American Library, W. H. Freeman and Company, 1983. This book emphasizes the important discoveries, experiments, and intellectual exercises which reshaped physics in the 20th century.

Steven Weinberg, "The Decay of the Proton," *Sci. American,* June, 1981.

QUESTIONS

1. Name the four fundamental interactions and the particles that mediate each interaction.
2. Discuss the quark model of hadrons, and describe the properties of quarks.
3. Discuss the differences between hadrons and leptons.
4. Describe the properties of baryons and mesons and the important differences between them.
5. Particles known as resonances have very short lifetimes, of the order of 10^{-23} s. From this information, would you guess they are hadrons or leptons? Explain.
6. The family of *K* mesons all decay into final states that contain no protons or neutrons. What is the baryon number of the *K* mesons?
7. The Ξ^0 particle decays by the weak interaction according to the decay mode $\Xi^0 \longrightarrow \Lambda^0 + \pi^0$. Would you expect this decay to be fast or slow? Explain.
8. Identify the particle decays listed in Table 34.2 that occur by the weak interaction. Justify your answers.
9. Identify the particle decays listed in Table 34.2 that occur by the electromagnetic interaction. Justify your answers.
10. Two protons in a nucleus interact via the strong interaction. Are they also subject to the weak interaction?
11. Discuss the following conservation laws: energy, linear momentum, angular momentum, electric charge, baryon number, lepton number, and strangeness. Are all of these laws based on fundamental properties of nature? Explain.
12. An antibaryon interacts with a meson. Can a baryon be produced in such an interaction? Explain.
13. Discuss the essential features of the Standard Model of particle physics.
14. How many quarks are there in (a) a baryon, (b) an antibaryon, (c) a meson, and (d) an antimeson? How do you account for the fact that baryons have half-integral spins while mesons have spins of 0 or 1? (*Hint:* Quarks have spins of ½.)
15. In the theory of quantum chromodynamics, quarks come in three colors. How would you justify the statement that "all baryons and mesons are colorless"?

PROBLEMS

34.3 Positrons and Other Antiparticles

1. Two photons are produced when a proton and antiproton annihilate each other. What is the minimum frequency and corresponding wavelength of each photon?

2. A photon produces a proton-antiproton pair according to the reaction $\gamma \longrightarrow p + \bar{p}$. What is the frequency of the photon? What is its wavelength?

34.4 Mesons and the Beginning of Particle Physics

3. One of the mediators of the weak interaction is the Z^0 boson whose mass is 96 GeV/c^2. Use this information to find an approximate value for the range of the weak interaction.

4. Occasionally, high energy muons will collide with electrons and produce two neutrinos according to the reaction $\mu^+ + e \rightarrow 2\nu$. What kind of neutrinos are these?

5. When a high-energy proton or pion traveling near the speed of light collides with a nucleus, it travels an average distance of 3×10^{-15} m before interacting. From this information, estimate the time for the strong interaction to occur.

34.6 Conservation Laws

34.7 Strange Particles and Strangeness

6. The neutral ρ meson decays by the strong interaction into two pions according to $\rho^0 \longrightarrow \pi^+ + \pi^-$ with a half-life of about 10^{-23} s. The neutral K meson also decays into two pions according to $K^0 \longrightarrow \pi^+ + \pi^-$, but with a much longer half-life of about 10^{-10} s. How do you explain these observations?

7. Determine whether or not strangeness is conserved in the following decays and reactions.

(a) $\Lambda^0 \rightarrow p + \pi^-$
(b) $\pi^- + p \rightarrow \Lambda^0 + K^0$
(c) $\bar{p} + p \rightarrow \bar{\Lambda}^0 + \Lambda^0$
(d) $\pi^- + p \rightarrow \pi^- + \Sigma^+$
(e) $\Xi^- \rightarrow \Lambda^0 + \pi^-$
(f) $\Xi^0 \rightarrow p + \pi^-$

8. Each of the following decays is forbidden. For each process, determine a conservation law that is violated.

(a) $\mu^- \longrightarrow e + \gamma$
(b) $n \longrightarrow p + e + \nu_e$
(c) $\Lambda^0 \longrightarrow p + \pi^0$
(d) $p \longrightarrow e^+ + \pi^0$
(e) $\Xi^0 \longrightarrow n + \pi^0$

9. Each of the following reactions is forbidden. Determine a conservation law that is violated for each reaction.
(a) $p + \bar{p} \longrightarrow \mu^+ + e$
(b) $\pi^- + p \longrightarrow p + \pi^+$
(c) $p + p \longrightarrow p + \pi^+$
(d) $p + p \longrightarrow p + p + n$
(e) $\gamma + p \longrightarrow n + \pi^0$

10. (a) Show that baryon number and charge are conserved in the following reactions of a pion with a proton.

$$\pi^- + p \longrightarrow K^- + \Sigma^+ \qquad (1)$$

$$\pi^- + p \longrightarrow \pi^- + \Sigma^+ \qquad (2)$$

(b) The first reaction is observed, but the second never occurs. Explain these observations.

11. The following reactions or decays involves one or more neutrinos. Supply the missing neutrinos (ν_e, ν_μ, or ν_τ).
(a) $\pi^- \longrightarrow \mu^- + ?$
(b) $K^+ \longrightarrow \mu^+ + ?$
(c) $? + p \longrightarrow n + e^+$
(d) $? + n \longrightarrow p + e$
(e) $? + n \longrightarrow p + \mu^-$
(f) $\mu^- \longrightarrow e + ? + ?$

12. Determine which of the reactions below can occur. For those that cannot occur, determine the conservation law (or laws) that each violates.
(a) $p \longrightarrow \pi^+ + \pi^0$
(b) $p + p \longrightarrow p + p + \pi^0$
(c) $p + p \longrightarrow p + \pi^+$
(d) $\pi^+ \longrightarrow \mu^+ + \nu_\mu$
(e) $n \longrightarrow p + e + \bar{\nu}_e$
(f) $\pi^+ \longrightarrow \mu^+ + n$

34.8 Quarks—Finally

13. The quark composition of the proton is *uud*, while that of the neutron in *udd*. Show that the charge, baryon number, and strangeness of these particles equal the sums of these numbers for their quark constituents.

14. The quark compositions of the K^0 and Λ^0 particles are $d\bar{s}$ and *uds*, respectively. Show that the charge, baryon number, and strangeness of these particles equal the sums of these numbers for the quark constituents.

APPENDIX A

Mathematical Review

A.1 SCIENTIFIC NOTATION

Many quantities that scientists deal with often have very large or very small values. For example, the speed of light is about 300,000,000 m/s and the ink required to make the dot over an *i* in this textbook has a mass of about 0.000000001 kg. Obviously, it is very cumbersome to read, write, and keep track of numbers such as these. We avoid this problem by using a method dealing with powers of the number 10:

$$10^0 = 1$$
$$10^1 = 10$$
$$10^2 = 10 \times 10 = 100$$
$$10^3 = 10 \times 10 \times 10 = 1000$$
$$10^4 = 10 \times 10 \times 10 \times 10 = 10{,}000$$
$$10^5 = 10 \times 10 \times 10 \times 10 \times 10 = 100{,}000$$

and so on. The number of zeros corresponds to the power to which 10 is raised, called the **exponent** of 10. For example, the speed of light, 300,000,000 m/s, can be expressed as 3×10^8 m/s.

For numbers less than one, we note the following:

$$10^{-1} = \frac{1}{10} = 0.1$$

$$10^{-2} = \frac{1}{10 \times 10} = 0.01$$

$$10^{-3} = \frac{1}{10 \times 10 \times 10} = 0.001$$

$$10^{-4} = \frac{1}{10 \times 10 \times 10 \times 10} = 0.0001$$

$$10^{-5} = \frac{1}{10 \times 10 \times 10 \times 10 \times 10} = 0.00001$$

In these cases, the number of places the decimal point is to the left of the digit 1 equals the value of the (negative) exponent. Numbers that are expressed as some power of 10 multiplied by another number between 1 and 10 are said to be in **scientific notation.** For example, the scientific notation for 5,943,000,000 is 5.943×10^9 and that for 0.0000832 is 8.32×10^{-5}.

When numbers expressed in scientific notation are being multiplied, the following general rule is very useful:

$$\boxed{10^n \times 10^m = 10^{n+m}} \qquad \text{(A.1)}$$

where n and m can be *any* numbers (not necessarily integers). For example, $10^2 \times 10^5 = 10^7$. The rule also applies if one of the exponents is negative. For example, $10^3 \times 10^{-8} = 10^{-5}$.

When dividing numbers expressed in scientific notation, note that

$$\frac{10^n}{10^m} = 10^n \times 10^{-m} = 10^{n-m} \tag{A.2}$$

EXERCISES

With help from the above rules, verify the answers to the following:

1. $86{,}400 = 8.64 \times 10^4$
2. $9{,}816{,}762.5 = 9.8167625 \times 10^6$
3. $0.0000000398 = 3.98 \times 10^{-8}$
4. $(4 \times 10^8)(9 \times 10^9) = 3.6 \times 10^{18}$
5. $(3 \times 10^7)(6 \times 10^{-12}) = 1.8 \times 10^{-4}$
6. $\dfrac{75 \times 10^{-11}}{5 \times 10^{-3}} = 1.5 \times 10^{-7}$
7. $\dfrac{(3 \times 10^6)(8 \times 10^{-2})}{(2 \times 10^{17})(6 \times 10^5)} = 2 \times 10^{-18}$

A.2 ALGEBRA

A. Some Basic Rules

When algebraic operations are performed, the laws of arithmetic apply. Symbols such as x, y, and z are usually used to represent quantities that are not specified, what are called the **unknowns.**

First, consider the equation

$$8x = 32$$

If we wish to solve for x, we can divide (or multiply) each side of the equation by the same factor without destroying the equality. In this case, if we divide both sides by 8, we have

$$\frac{8x}{8} = \frac{32}{8}$$

$$x = 4$$

Next consider the equation

$$x + 2 = 8$$

In this type of expression, we can add or subtract the same quantity from each side. If we subtract 2 from each side, we get

$$x + 2 - 2 = 8 - 2$$
$$x = 6$$

In general, if $x + a = b$, then $x = b - a$.

Now consider the equation

$$\frac{x}{5} = 9$$

If we multiply each side by 5, we are left with x on the left by itself and 45 on the right:

$$\left(\frac{x}{5}\right)(5) = 9 \times 5$$
$$x = 45$$

In all cases, *whatever operation is performed on the left side of the equality must also be performed on the right side.*

The following rules for multiplying, dividing, adding, and subtracting fractions should be recalled, where a, b, and c are three numbers:

	Rule	Example
Multiplying	$\left(\frac{a}{b}\right)\left(\frac{c}{d}\right) = \frac{ac}{bd}$	$\left(\frac{2}{3}\right)\left(\frac{4}{5}\right) = \frac{8}{15}$
Dividing	$\frac{(a/b)}{(c/d)} = \frac{ad}{bc}$	$\frac{2/3}{4/5} = \frac{(2)(5)}{(4)(3)} = \frac{10}{12}$
Adding	$\frac{a}{b} \pm \frac{c}{d} = \frac{ad \pm bc}{bd}$	$\frac{2}{3} - \frac{4}{5} = \frac{(2)(5) - (4)(3)}{(3)(5)} = -\frac{2}{15}$

EXERCISES

In the following exercises, solve for x:

	Answers
1. $a = \frac{1}{1 + x}$	$x = \frac{1 - a}{a}$
2. $3x - 5 = 13$	$x = 6$
3. $ax - 5 = bx + 2$	$x = \frac{7}{a - b}$
4. $\frac{5}{2x + 6} = \frac{3}{4x + 8}$	$x = -\frac{11}{7}$

B. Powers

When powers of a given quantity x are multiplied, the following rule applies:

$$x^n x^m = x^{n+m} \tag{A.3}$$

For example, $x^2x^4 = x^{2+4} = x^6$.

When dividing the powers of a given quantity, note that

$$\frac{x^n}{x^m} = x^{n-m} \tag{A.4}$$

For example, $x^8/x^2 = x^{8-2} = x^6$.

A power that is a fraction, such as $\frac{1}{3}$, corresponds to a root as follows:

$$x^{1/n} = \sqrt[n]{x} \tag{A.5}$$

For example, $4^{1/3} = \sqrt[3]{4} = 1.5874$. (A scientific calculator is useful for such calculations.)

Finally, any quantity x^n that is raised to the mth power is

$$(x^n)^m = x^{nm} \tag{A.6}$$

Table A.1 summarizes the rules of exponents.

TABLE A.1 Rules of Exponents

$x^0 = 1$
$x^1 = x$
$x^n x^m = x^{n+m}$
$x^n/x^m = x^{n-m}$
$x^{1/n} = \sqrt[n]{x}$
$(x^n)^m = x^{nm}$

EXERCISES

Verify the following:

1. $3^2 \times 3^3 = 243$
2. $x^5 x^{-8} = x^{-3}$
3. $x^{10}/x^{-5} = x^{15}$
4. $5^{1/3} = 1.709975$ (Use your calculator.)
5. $60^{1/4} = 2.783158$ (Use your calculator.)
6. $(x^4)^3 = x^{12}$

C. Factoring

Some useful formulas for factoring an equation are

$ax + ay + az = a(x + y + x)$	common factor
$a^2 + 2ab + b^2 = (a + b)^2$	perfect square
$a^2 - b^2 = (a + b)(a - b)$	differences of squares

D. Quadratic Equations

The general form of a quadratic equation is

$$ax^2 + bx + c = 0 \tag{A.7}$$

where x is the unknown quantity and a, b, and c are numerical factors referred to as **coefficients** of the equation. This equation has two roots, given by

$$x = \frac{-b \pm \sqrt{b^2 - 4ac}}{2a} \tag{A.8}$$

If $b^2 \geq 4ac$, the roots will be real.

EXAMPLE

The equation $x^2 + 5x + 4 = 0$ has the following roots corresponding to the two signs of the square-root term:

$$x = \frac{-5 \pm \sqrt{5^2 - (4)(1)(4)}}{2(1)} = \frac{-5 \pm \sqrt{9}}{2} = \frac{-5 \pm 3}{2}$$

that is,

$$x_+ = \frac{-5 + 3}{2} = -1 \qquad x_- = \frac{-5 - 3}{2} = -4$$

where x_+ refers to the root corresponding to the positive sign and x_- refers to the root corresponding to the negative sign.

EXERCISES

Solve the following quadratic equations:

	Answers	
1. $x^2 + 2x - 3 = 0$	$x_+ = 1$	$x_- = -3$
2. $2x^2 - 5x + 2 = 0$	$x_+ = 2$	$x_- = 1/2$
3. $2x^2 - 4x - 9 = 0$	$x_+ = 1 + \sqrt{22}/2$	$x_- = 1 - \sqrt{22}/2$

E. Linear Equations

A linear equation has the general form

$$y = ax + b \tag{A.9}$$

where a and b are constants. This equation is referred to as being linear because the graph of y versus x is a straight line, as shown in Figure A.1. The constant b, called the **intercept,** represents the value of y at which the straight line intersects the y-axis. The constant a is equal to the **slope** of the straight line and is also equal to the tangent of the angle that the line makes with the x-axis. If any two points on the straight line are specified by the coordinates (x_1, y_1) and (x_2, y_2), as in Figure A.1, then the **slope** of the straight line can be expressed

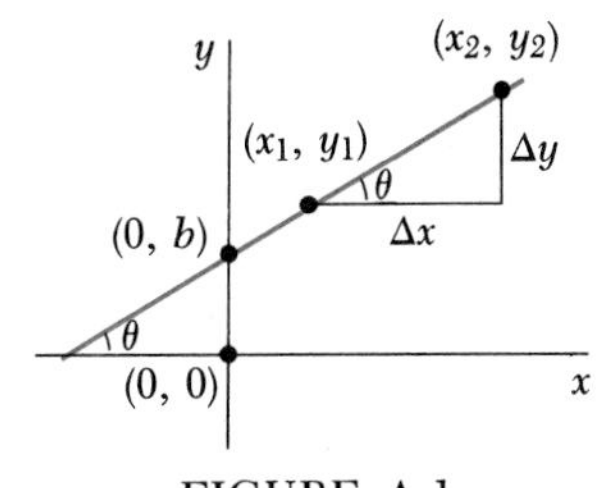

FIGURE A.1

$$\text{Slope} = \frac{y_2 - y_1}{x_2 - x_1} = \frac{\Delta y}{\Delta x} = \tan \theta \tag{A.10}$$

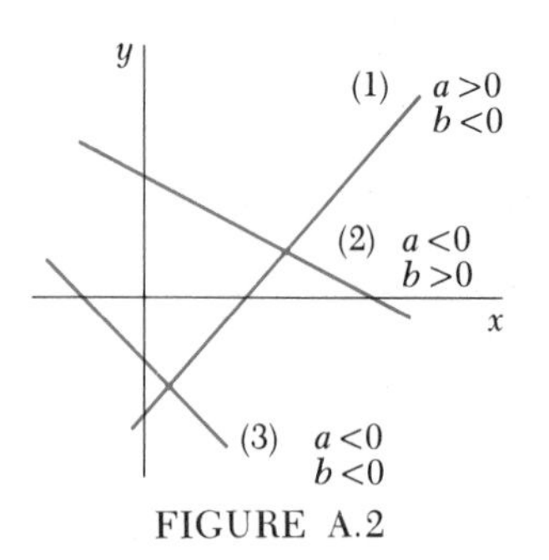

FIGURE A.2

Note that a and b can have either positive or negative values. If $a > 0$, the straight line has a *positive* slope, as in Figure A.1. If $a < 0$, the straight line has a *negative* slope. In Figure A.1, both a and b are positive. Three other possible situations are shown in Figure A.2: $a > 0$, $b < 0$; $a < 0$, $b > 0$; and $a < 0$, $b < 0$.

EXERCISES

1. Draw graphs of the following straight lines:
(a) $y = 5x + 3$ (b) $y = -2x + 4$ (c) $y = -3x - 6$

2. Find the slopes of the straight lines described in Exercise 1.
Answers: (a) 5 (b) -2 (c) -3

3. Find the slopes of the straight lines that pass through the following sets of points:
(a) $(0, -4)$ and $(4,2)$, (b) $(0,0)$ and $(2, -5)$, and (c) $(-5,2)$ and $(4, -2)$
Answers: (a) 3/2 (b) $-5/2$ (c) $-4/9$

F. Solving Simultaneous Linear Equations

Consider an equation such as $3x + 5y = 15$, which has two unknowns, x and y. Such an equation does not have a unique solution. That is, $(x = 0, y = 3)$, $(x = 5, y = 0)$, and $(x = 2, y = 9/5)$ are all solutions to this equation.

If a problem has two unknowns, a unique solution is possible only if we have *two* independent equations. In general, if a problem has n unknowns, its solution requires n independent equations. In order to solve two simultaneous equations involving two unknowns, x and y, we solve one of the equations for x in terms of y and substitute this expression into the other equation.

EXAMPLE

Solve the following two simultaneous equations:

$$(1)\quad 5x + y = -8 \qquad (2)\quad 2x - 2y = 4$$

Solution: From (2), $x = y + 2$. Substitution of this into (1) gives

$$\begin{aligned} 5(y + 2) + y &= -8 \\ 6y &= -18 \\ y &= -3 \\ x &= y + 2 = -1 \end{aligned}$$

Alternate solution: Multiply each term in (1) by the factor 2 and add the result to (2):

$$\begin{aligned} 10x + 2y &= -16 \\ 2x - 2y &= 4 \\ \hline 12x &= -12 \\ x &= -1 \\ y &= x - 2 = -3 \end{aligned}$$

Two linear equations with two unknowns can also be solved by a graphical method. If the straight lines corresponding to the two equations are plotted in a conventional coordinate system, the intersection of the two lines represents the solution. For example, consider the two equations

$$x - y = 2$$
$$x - 2y = -1$$

These are plotted in Figure A.3. The intersection of the two lines has the coordinates $x = 5$, $y = 3$. This represents the solution to the equations. You should check this solution by the analytical technique discussed above.

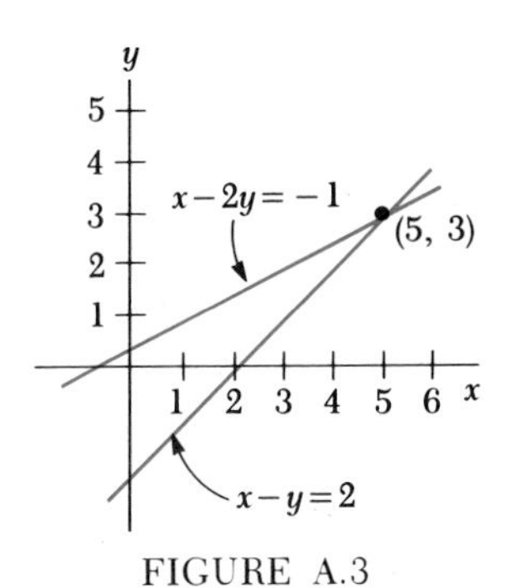

FIGURE A.3

EXERCISES

Solve the following pairs of simultaneous equations involving two unknowns:

	Answers
1. $x + y = 8$ $x - y = 2$	$x = 5$, $y = 3$
2. $98 - T = 10a$ $T - 49 = 5a$	$T = 65$, $a = 3.27$
3. $6x + 2y = 6$ $8x - 4y = 28$	$x = 2$, $y = -3$

G. Logarithms

Suppose that a quantity x is expressed as a power of some quantity a:

$$\boxed{x = a^y} \tag{A.11}$$

The number y is called the **base** number. The **logarithm** of x with respect to the base a is equal to the exponent to which the base must be raised in order to satisfy the expression $x = a^y$:

$$\boxed{y = \log_a x} \tag{A.12}$$

Conversely, the **antilogarithm** of y is the number x:

$$\boxed{x = \text{antilog}_a y} \tag{A.13}$$

In practice, the two bases most often used are base 10, called the *common* logarithm base, and base $e = 2.718\ldots$, called the *natural* logarithm base. When common logarithms are used,

$$y = \log_{10} x \qquad (\text{or } x = 10^y) \tag{A.14}$$

When natural logarithms are used,

$$y = \ln_e x \qquad (\text{or } x = e^y) \tag{A.15}$$

For example, $\log_{10} 52 = 1.716$, so that $\text{antilog}_{10} 1.716 = 10^{1.716} = 52$. Likewise, $\ln_e 52 = 3.951$, so $\text{antiln}_e 3.951 = e^{3.951} = 52$.

In general, note that you can convert between base 10 and base e with the equality

$$\ln_e x = (2.302585)\log_{10} x \tag{A.16}$$

Finally, some useful properties of logarithms are

$$\log(ab) = \log a + \log b \tag{A.17}$$

$$\log(a/b) = \log a - \log b \tag{A.18}$$

$$\log(a^n) = n \log a \tag{A.19}$$

A.3 GEOMETRY

Table A.2 gives the areas and volumes for several geometric shapes used throughout this text:

TABLE A.2 Useful Information from Geometry

Shape	Area or Volume	Shape	Area or Volume
Rectangle (w, ℓ)	Area $= \ell w$	Sphere (r)	Surface area $= 4\pi r^2$ Volume $= \frac{4\pi r^3}{3}$
Circle (r)	Area $= \pi r^2$ (Circumference $= 2\pi r$)	Cylinder (ℓ, r)	Volume $= \pi r^2 \ell$
Triangle (h, b)	Area $= \frac{1}{2}bh$	Rectangular box (h, w, ℓ)	Area $=$ $2(\ell h + \ell w + hw)$ Volume $= \ell wh$

A.4 TRIGONOMETRY

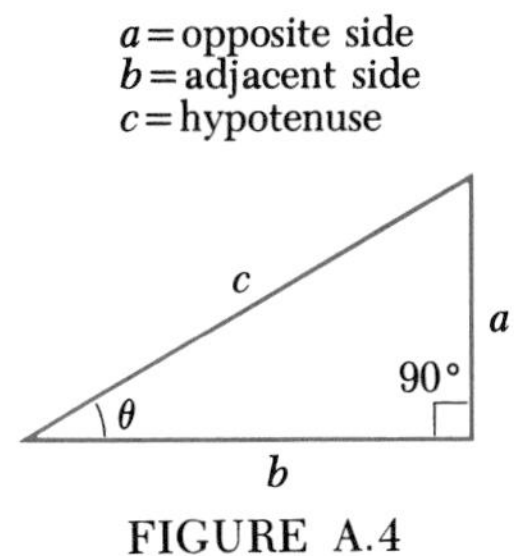

FIGURE A.4

That portion of mathematics based on the special properties of the right triangle is called trigonometry. By definition, a right triangle is one containing a 90° angle. Consider the right triangle shown in Figure A.4, where side a is opposite the angle θ, side b is adjacent to the angle θ, and side c is the hypotenuse of the triangle. The three basic trigonometric functions defined by such a triangle are the sine (sin), cosine (cos), and tangent (tan) functions. In terms of the angle θ, these functions are

$$\sin\theta = \frac{\text{side opposite } \theta}{\text{hypotenuse}} = \frac{a}{c} \tag{A.20}$$

$$\cos\theta = \frac{\text{side adjacent to } \theta}{\text{hypotenuse}} = \frac{b}{c} \tag{A.21}$$

$$\tan\theta = \frac{\text{side opposite } \theta}{\text{side adjacent to } \theta} = \frac{a}{b} \tag{A.22}$$

The Pythagorean theorem provides the following relationship between the sides of a triangle:

$$c^2 = a^2 + b^2 \tag{A.23}$$

The following represent some useful relations from trigonometry:

$$\sin^2\theta + \cos^2\theta = 1 \tag{A.24}$$

$$\sin\theta = \cos(90° - \theta) \tag{A.25}$$

$$\cos\theta = \sin(90° - \theta) \tag{A.26}$$

$$\sin 2\theta = 2\sin\theta\cos\theta \tag{A.27}$$

$$\cos 2\theta = \cos^2\theta - \sin^2\theta \tag{A.28}$$

$$\sin(\theta \pm \phi) = \sin\theta\cos\phi \pm \cos\theta\sin\phi \tag{A.29}$$

$$\cos(\theta \pm \phi) = \cos\theta\cos\phi \mp \sin\theta\sin\phi \tag{A.30}$$

EXAMPLE

Consider the right triangle in Figure A.5, in which $a = 2$, $b = 5$, and c is unknown. From the Pythagorean theorem, we have

$$c^2 = a^2 + b^2 = 2^2 + 5^2 = 4 + 25 = 29$$

$$c = \sqrt{29} = 5.39$$

To find the angle θ, note that

$$\tan\theta = \frac{a}{b} = \frac{2}{5} = 0.400$$

From a table of functions or from a calculator, we have

$$\theta = \tan^{-1}(0.400) = 21.8°$$

where $\tan^{-1}(0.400)$ is the notation for "angle whose tangent is 0.400," sometimes written as arctan(0.400).

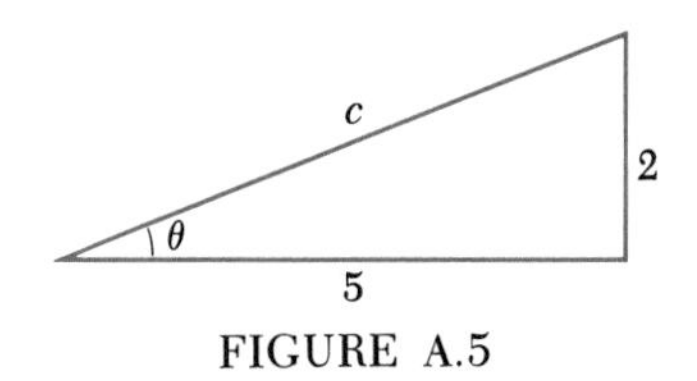

FIGURE A.5

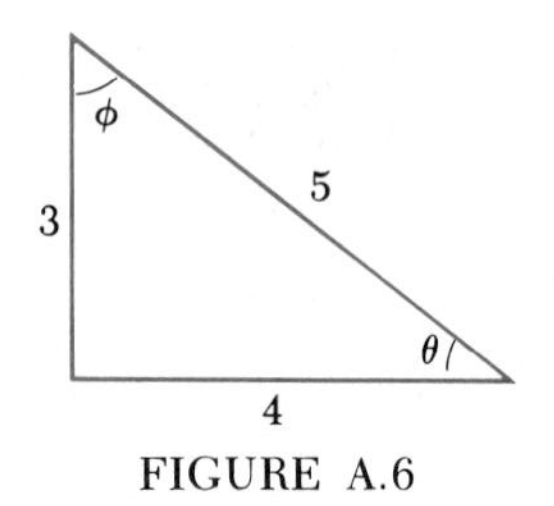

FIGURE A.6

EXERCISES

1. In Figure A.6, find (a) the side opposite θ, (b) the side adjacent to ϕ, (c) cos θ, (d) sin ϕ, and (e) tan ϕ. Answers: (a) 3, (b) 3, (c) 4/5, (d) 4/5, and (e) 4/3

2. In a certain right triangle, the two sides that are perpendicular to each other are 5 m and 7 m long. What is the length of the third side of the triangle?
Answer: 8.60 m

3. A right triangle has a hypotenuse of length 3 m, and one of its angles is 30°. What is the length of (a) the side opposite the 30° angle and (b) the side adjacent to the 30° angle? Answers: (a) 1.5 m and (b) 2.60 m

APPENDIX B

An Abbreviated Table of Isotopes

Atomic Number, Z	Element	Symbol	Mass Number, A	Atomic Mass[a]	Percent Abundance, or Decay Mode if Radioactive	Half-life if Radioactive
0	(Neutron)	n	1	1.008665	β^-	10.6 min
1	Hydrogen	H	1	1.007825	99.985	
	Deuterium	D	2	2.014102	0.015	
	Tritium	T	3	3.016049	β^-	12.33 years
2	Helium	He	3	3.016029	0.00014	
			4	4.002603	$\approx$100	
3	Lithium	Li	6	6.015123	7.5	
			7	7.016005	92.5	
4	Beryllium	Be	7	7.016930	EC, γ	53.3 days
			8	8.005305	2α	6.7×10^{-17} s
			9	9.012183	100	
5	Boron	B	10	10.012938	19.8	
			11	11.009305	80.2	
6	Carbon	C	11	11.011433	β^+, EC	20.4 min
			12	12.000000	98.89	
			13	13.003355	1.11	
			14	14.003242	β^-	5730 years
7	Nitrogen	N	13	13.005739	β^+	9.96 min
			14	14.003074	99.63	
			15	15.000109	0.37	
8	Oxygen	O	15	15.003065	β^+, EC	122 s
			16	15.994915	99.76	
			18	17.999159	0.204	
9	Fluorine	F	19	18.998403	100	
10	Neon	Ne	20	19.992439	90.51	
			22	21.991384	9.22	
11	Sodium	Na	22	21.994435	β^+, EC, γ	2.602 years
			23	22.989770	100	
			24	23.990964	β^-, γ	15.0 h
12	Magnesium	Mg	24	23.985045	78.99	
13	Aluminum	Al	27	26.981541	100	
14	Silicon	Si	28	27.976928	92.23	
			31	30.975364	β^-, γ	2.62 h
15	Phosphorus	P	31	30.973763	100	
			32	31.973908	β^-	14.28 days
16	Sulfur	S	32	31.972072	95.0	
			35	34.969033	β^-	87.4 days

Note: Data taken from *Chart of the Nuclides*, 12th ed., New York, Wiley, 1977, and from C. M. Lederer and V. S. Shirley (eds.), *Table of Isotopes*, 7th ed., New York, Wiley, 1978.

[a]Masses are those for the neutral atom, including the Z electrons.

[b]The process EC stands for "electron capture."

Atomic Number, Z	Element	Symbol	Mass Number, A	Atomic Mass[a]	Percent Abundance, or Decay Mode if Radioactive	Half-life if Radioactive
17	Chlorine	Cl	35	34.968853	75.77	
			37	36.965903	24.23	
18	Argon	Ar	40	39.962383	99.60	
19	Potassium	K	39	38.963708	93.26	
			40	39.964000	β^-, EC, γ, β^+	1.28×10^9 years
20	Calcium	Ca	40	39.962591	96.94	
21	Scandium	Sc	45	44.955914	100	
22	Titanium	Ti	48	47.947947	73.7	
23	Vanadium	V	51	50.943963	99.75	
24	Chromium	Cr	52	51.940510	83.79	
25	Manganese	Mn	55	54.938046	100	
26	Iron	Fe	56	55.934939	91.8	
27	Cobalt	Co	59	58.933198	100	
			60	59.933820	β^-, γ	5.271 years
28	Nickel	Ni	58	57.935347	68.3	
			60	59.930789	26.1	
			64	63.927968	0.91	
29	Copper	Cu	63	62.929599	69.2	
			64	63.929766	β^-, β^+	12.7 h
			65	64.927792	30.8	
30	Zinc	Zn	64	63.929145	48.6	
			66	65.926035	27.9	
31	Gallium	Ga	69	68.925581	60.1	
32	Germanium	Ge	72	71.922080	27.4	
			74	73.921179	36.5	
33	Arsenic	As	75	74.921596	100	
34	Selenium	Se	80	79.916521	49.8	
35	Bromine	Br	79	78.918336	50.69	
36	Krypton	Kr	84	83.911506	57.0	
			89	88.917563	β^-	3.2 min
37	Rubidium	Rb	85	84.911800	72.17	
38	Strontium	Sr	86	85.909273	9.8	
			88	87.905625	82.6	
			90	89.907746	β^-	28.8 years
39	Yttrium	Y	89	88.905856	100	
40	Zirconium	Zr	90	89.904708	51.5	
41	Niobium	Nb	93	92.906378	100	
42	Molybdenum	Mo	98	97.905405	24.1	
43	Technetium	Tc	98	97.907210	β^-, γ	4.2×10^6 years
44	Ruthenium	Ru	102	101.904348	31.6	
45	Rhodium	Rh	103	102.90550	100	
46	Palladium	Pd	106	105.90348	27.3	
47	Silver	Ag	107	106.905095	51.83	
			109	108.904754	48.17	
48	Cadmium	Cd	114	113.903361	28.7	
49	Indium	In	115	114.90388	95.7; β^-	5.1×10^{14} years
50	Tin	Sn	120	119.902199	32.4	
51	Antimony	Sb	121	120.903824	57.3	

Atomic Number, Z	Element	Symbol	Mass Number, A	Atomic Mass[a]	Percent Abundance, or Decay Mode if Radioactive	Half-life if Radioactive
52	Tellurium	Te	130	129.90623	34.5; β^-	2×10^{21} years
53	Iodine	I	127	126.904477	100	
			131	130.906118	β^-, γ	8.04 days
54	Xenon	Xe	132	131.90415	26.9	
			136	135.90722	8.9	
55	Cesium	Cs	133	132.90543	100	
56	Barium	Ba	137	136.90582	11.2	
			138	137.90524	71.7	
			144	143.922673	β^-	11.9 s
57	Lanthanum	La	139	138.90636	99.911	
58	Cerium	Ce	140	139.90544	88.5	
59	Praesodymium	Pr	141	140.90766	100	
60	Neodymium	Nd	142	141.90773	27.2	
			144	143.910096	α, 23.8	2.1×10^{15} years
61	Promethium	Pm	145	144.91275	EC, α, γ	17.7 years
62	Samarium	Sm	152	151.91974	26.6	
63	Europium	Eu	153	152.92124	52.1	
64	Gadolinium	Gd	158	157.92411	24.8	
65	Terbium	Tb	159	158.92535	100	
66	Dysprosium	Dy	164	163.92918	28.1	
67	Holmium	Ho	165	164.93033	100	
68	Erbium	Er	166	165.93031	33.4	
69	Thulium	Tm	169	168.93423	100	
70	Ytterbium	Yb	174	173.93887	31.6	
71	Lutecium	Lu	175	174.94079	97.39	
72	Hafnium	Hf	180	179.94656	35.2	
73	Tantalum	Ta	181	180.94801	99.988	
74	Tungsten	W	184	183.95095	30.7	
75	Rhenium	Re	187	186.95577	62.60, β^-	4×10^{10} years
76	Osmium	Os	191	190.96094	β^-, γ	15.4 days
			192	191.96149	41.0	
77	Iridium	Ir	191	190.96060	37.3	
			193	192.96294	62.7	
78	Platinum	Pt	195	194.96479	33.8	
79	Gold	Au	197	196.96656	100	
80	Mercury	Hg	202	201.97063	29.8	
81	Thallium	Tl	205	204.97441	70.5	
			210	209.990069	β^-	1.3 min
82	Lead	Pb	204	203.973044	β^-, 1.48	1.4×10^{17} years
			206	205.97446	24.1	
			207	206.97589	22.1	
			208	207.97664	52.3	
			210	209.98418	α, β^-, γ	22.3 years
			211	210.98874	β^-, γ	36.1 min
			212	211.99188	β^-, γ	10.64 h
			214	213.99980	β^-, γ	26.8 min
83	Bismuth	Bi	209	208.98039	100	
			211	210.98726	α, β^-, γ	2.15 min
			214	213.998702	β^-, α	19.7 min

Atomic Number, Z	Element	Symbol	Mass Number, A	Atomic Mass[a]	Percent Abundance, or Decay Mode if Radioactive	Half-life if Radioactive
84	Polonium	Po	210	209.98286	α, γ	138.38 days
			214	213.99519	α, γ	164 μs
85	Astatine	At	218	218.00870	α, β^-	$\approx$2 s
86	Radon	Rn	222	222.017574	α, γ	3.8235 days
87	Francium	Fr	223	223.019734	α, β^-, γ	21.8 min
88	Radium	Ra	226	226.025406	α, γ	1.60×10^3 years
			228	228.031069	β^-	5.76 years
89	Actinium	Ac	227	227.027751	α, β^-, γ	21.773 years
90	Thorium	Th	228	228.02873	α, γ	1.9131 years
			232	232.038054	100, α, γ	1.41×10^{10} years
91	Protactinium	Pa	231	231.035881	α, γ	3.28×10^4 years
92	Uranium	U	232	232.03714	α, γ	72 years
			233	233.039629	α, γ	1.592×10^5 years
			235	235.043925	0.72; α, γ	7.038×10^8 years
			236	236.045563	α, γ	2.342×10^7 years
			238	238.050786	99.275; α, γ	4.468×10^9 years
			239	239.054291	β^-, γ	23.5 min
93	Neptunium	Np	239	239.052932	β^-, γ	2.35 days
94	Plutonium	Pu	239	239.052158	α, γ	2.41×10^4 years
95	Americium	Am	243	243.061374	α, γ	7.37×10^3 years
96	Curium	Cm	245	245.065487	α, γ	8.5×10^3 years
97	Berkelium	Bk	247	247.07003	α, γ	1.4×10^3 years
98	Californium	Cf	249	249.074849	α, γ	351 years
99	Einsteinium	Es	254	254.08802	α, γ, β^-	276 days
100	Fermium	Fm	253	253.08518	EC, α, γ	3.0 days
101	Mendelevium	Md	255	255.0911	EC, α	27 min
102	Nobelium	No	255	255.0933	EC, α	3.1 min
103	Lawrencium	Lr	257	257.0998	α	$\approx$35 s
104	Unnilquadium	Rf	261	261.1087	α	1.1 min
105	Unnilpentium	Ha	262	262.1138	α	0.7 min
106	Unnilhexium		263	263.1184	α	0.9 s
107	Unnilseptium		261		α	1–2 ms

APPENDIX C

Some Useful Tables

TABLE C.1 Mathematical Symbols Used in the Text and Their Meaning

Symbol	Meaning
$=$	is equal to
$\neq$	is not equal to
$\equiv$	is defined as
$\propto$	is proportional to
$>$	is greater than
$<$	is less than
$\gg$	is much greater than
$\ll$	is much less than
$\approx$	is approximately equal to
Δx	change in x
Σx_i	sum of all quantities x_i
$\lvert x \rvert$	magnitude of x (always a positive quantity)

TABLE C.2 Standard Abbreviations of Units

Abbreviation	Unit	Abbreviation	Unit
A	ampere	J	joule
Å	angstrom	K	kelvin
atm	atmosphere	kcal	kilocalorie
Btu	British thermal unit	kg	kilogram
C	coulomb	km	kilometer
°C	degree Celsius	kmole	kilomole
cal	calorie	lb	pound
cm	centimeter	m	meter
deg	degree (angle)	min	minute
eV	electron volt	N	newton
°F	degree Fahrenheit	rev	revolution
ft	foot	s	second
G	gauss	T	tresla
g	gram	u	atomic mass unit
H	henry	V	volt
h	hour	W	watt
hp	horsepower	Wb	weber
Hz	hertz	μm	micrometer
in.	inch	Ω	ohm

TABLE C.3 The Greek Alphabet

Alpha	Α	α	Iota	Ι	ι	Rho	Ρ	ρ
Beta	Β	β	Kappa	Κ	κ	Sigma	Σ	σ
Gamma	Γ	γ	Lambda	Λ	λ	Tau	Τ	τ
Delta	Δ	δ	Mu	Μ	μ	Upsilon	Υ	υ
Epsilon	Ε	ϵ	Nu	Ν	ν	Phi	Φ	ϕ
Zeta	Ζ	ζ	Xi	Ξ	ξ	Chi	Χ	χ
Eta	Η	η	Omicron	Ο	ο	Psi	Ψ	ψ
Theta	Θ	θ	Pi	Π	π	Omega	Ω	ω

APPENDIX D

SI Units

TABLE D.1 SI Base Units

Base Quantity	SI Base Unit: Name	Symbol
Length	Meter	m
Mass	Kilogram	kg
Time	Second	s
Electric current	Ampere	A
Temperature	Kelvin	K
Amount of substance	Mole	mol

TABLE D.2 Derived SI Units

Quantity	Name	Symbol	Expression in Terms of Base Units	Expression in Terms of Other SI Units
Plane angle	Radian	rad	m/m	
Frequency	Hertz	Hz	s^{-1}	
Force	Newton	N	$kg \cdot m/s^2$	J/m
Pressure	Pascal	Pa	$kg/m \cdot s^2$	N/m^2
Energy: work	Joule	J	$kg \cdot m^2/s^2$	$N \cdot m$
Power	Watt	W	$kg \cdot m^2/s^3$	J/s
Electric charge	Coulomb	C	$A \cdot s$	
Electric potential (emf)	Volt	V	$kg \cdot m^2/A \cdot s^3$	W/A
Capacitance	Farad	F	$A^2 \cdot s^4/kg \cdot m^2$	C/A
Electric resistance	Ohm	Ω	$kg \cdot m^2/A^2 \cdot s^3$	V/A
Magnetic flux	Weber	Wb	$kg \cdot m^2/A \cdot s^2$	$V \cdot s$
Magnetic field intensity	Tesla	T	$kg/A \cdot s^2$	Wb/m^2
Inductance	Henry	H	$kg \cdot m^2/A^2 \cdot s^2$	Wb/A

Answers to Odd-numbered Problems

CHAPTER 1

3. (a) MLT^{-2}, (b) $kg \cdot m/s^2$
5. (a) 3.00×10^8 m/s, (b) 2.9979×10^8 m/s, (c) 2.997925×10^8 m/s
7. (a) 797, (b) 11, (c) 17.8
9. (a) 22 cm, (b) 67.9 cm^2
11. 3.16×10^9 y
13. 35.7 m^2
15. (a) 1 mi/h = 1.609 km/h, (b) 88.5 km/h, (c) 16.1 km/h
17. 2.95×10^2 m^3, 2.95×10^8 cm^3
19. 1.00×10^{10} lb, 4.45×10^{10} N
21. 1.3×10^{10} lb, 4.2×10^7 head (assumes 1/4 lb per burger and 300 net lb per head)
23. 4.4×10^4 pitches (assumes 30 pitches/inning and 9 inning games)
25. 2.24 m
27. 8.06 cm
29. (a) 205 m eastward, (b) 45 m westward
31. 7.92 m at 4.3° north of west
33. 15.3 m at 58° south of east
35. 25 m
37. 1320 mi at 17° north of east
39. 196 cm at 14.7° below x axis
41. (a) +50.0, +86.6, (b) −64.3, +76.6, (c) −94.0, −34.2, (d) +34.2, −94.0
43. (a) 0.677 g/cm^3, (b) 4.63×10^{17} ft^2
45. (a) 1.3×10^{10} cans/y, (b) 2×10^5 tons/y (assumes an average of one can per person each week, a population of 250 million, and 0.5 ounces of aluminum per can)
47. 8.60 m
49. $m = 1$, $n = 2$, no

CHAPTER 2

1. 6.10 m/s
3. (a) 2.50 m/s, (b) −2.27 m/s, (c) 0
5. (a) 1.17 m/s, (b) 1.40 m/s
7. (a) +4.00 m/s, (b) −4.00 m/s, (c) 0, (d) +2.00 m/s
9. +0.75 m/s, −3.00 m/s, +3.00 m/s, 0
11. +0.75 m/s^2
13. 3.73 s
15. (a) zero, +1.60 m/s^2, +0.80 m/s^2, (b) zero, +1.60 m/s^2, zero
17. (a) 1.25 m/s^2, (b) 8.00 s
19. (a) −8.00 m/s^2, (b) 100 m
21. (a) 3.00×10^{-10} s, (b) 1.24×10^{-4} m
23. Yes, the minimum acceleration required = 3.20×10^{-2} m/s^2
25. (a) 31.9 m, (b) 2.55 s, (c) 2.55 s, (d) −25 m/s
27. (a) 6.26 m/s, (b) 1.28 s
29. (a) 2.33 s, (b) −32.9 m/s
31. (a) 114 s, (b) −423 m/s
33. 35.3 m downfield
35. 34.6 m/s
37. 29.4 m/s
39. (a) 16.4 m/s, (b) 11.5 m/s
41. (a) -3.50×10^5 m/s, (b) 2.86×10^{-4} s
43. 18.0 m, 7.89 m
45. 4.16 m/s
47. 0.60 s
49. 14.0 m/s
51. (a) 2.99 s, (b) −15.4 m/s, (c) −31.3 m/s, −34.9 m/s
53. 10.8 m above the cannon
55. (a) 1524 m, (b) 36.1 m, (c) 4047 m
59. (a) $t_1 = 4.74$ s, $t_2 = 85.3$ s, (b) 200 ft/s, (c) 1.85×10^4 ft
61. Less than 265 m or more than 3476 m

CHAPTER 3

1. 19.6 N, 4.41 lb
3. (a) 534 N, (b) 54.5 kg
5. (a) 12 N, (b) 3 m/s^2
7. 25 N
9. (a) 798 N at 8.8° to right of forward direction, (b) 0.266 m/s^2 in the direction of the resultant force
11. 1.50 m
13. $F_{ball} = -375$ N, $F_{glove} = +375$ N, + direction = direction of ball's velocity

15. 23.1 m/s

17. 551 N

19. 218 N

21. (a) 36.8 N, (b) 2.45 m/s^2, (c) 1.23 m

23. $\mu_s = 0.383$, $\mu_k = 0.306$

25. $\mu_k = 0.306$

27. $\mu_k = 0.436$

29. 4.13×10^3 N

31. $\mu_s = 0.727$, $\mu_k = 0.577$

33. 3.92 m/s^2

35. $F_x = 60.6$ N, $F_y = 35.0$ N

37. (a) 84.9 N in the upward direction, (b) 84.9 N

39. 99.9 N

41. (a) 3.35 m/s^2, (b) 2.44 s

43. 50 m

45. (a) 2.73 m/s^2, (b) 10.9 N

47. 1.77 m/s^2

49. 0.685 m/s^2

51. (a) 1.67 m/s^2, 16.7 N, (b) 0.687 m/s^2, 16.7 N

53. (a) 0.682 m, (b) 3.20 m/s^2 down the incline, (c) 2.09 m/s

55. (b) $a_1 = \dfrac{2F}{(4m_1 + m_2)}$, $a_2 = \dfrac{F}{(4m_1 + m_2)}$;
(c) $T = \dfrac{2m_1F}{4m_1 + m_2}$

57. (a) 1.02 m/s^2, (b) 2.04 N, 3.06 N, 4.08 N
(c) 14 N between m_1 and m_2, 8 N between m_2 and m_3

59. (a) $T_1 = 78.0$ N, $T_2 = 35.9$ N, (b) 0.655

CHAPTER 4

1. $T_1 = T_2 = 2160$ N

3. $T_1 = T_2 = 77.8$ N

5. 50 N

7. 56.7 N

9. 133 N applied perpendicular to wrench handle

11. 168 N · m clockwise

13. $x_{cg} = 0$, $y_{cg} = +1$ m

15. $x_{cg} = +1/3$ m, $y_{cg} = +5/3$ m

17. $x_{cg} = 3.85$ cm, $y_{cg} = 6.85$ cm

19. 1.68 m

21. $T_1 = 567$ N (rope nearest worker), $T_2 = 333$ N

23. $\mu_k = 0.4$

25. $T_1 = 501$ N, $T_2 = 672$ N, $T_3 = 384$ N

27. $F = w/3$ on each front tire, $F = w/6$ on each rear tire where w = weight of car = 15,700 N

29. 6.15 m

31. 515 N

33. (a) 180 N, (b) 156 N

35. (a) $T = 1730$ N, (b) $\theta = 8.63°$

37. (a) 13.5 N · m, (b) 0

39. $T_1 = 298$ N, $T_2 = 742$ N

41. 449 N (left cable), 222 N (right cable)

43. $T = 500$ N, $H = 433$ N, $V = 250$ N

45. (a) 50 N, (b) 0.5, (c) 25 N

47. (a) $\mu = \dfrac{1}{2}\tan\theta$, (b) $T = \dfrac{1}{2}(mg)\sin\theta$

49. $T = 173$ N, $H = 163$ N, $V = 235$ N

53. $\mu_s = \sec\theta$

CHAPTER 5

1. 1.50×10^7 J

3. 1.59×10^3 J

5. (a) 900 J, (b) −900 J, (c) 0.383

7. (a) 9.07×10^5 J, (b) 9.07×10^5 J

9. 6.25×10^{-21} J

11. -2.40×10^5 J

13. 147 J

15. (a) $F = 6.3 \times 10^3$ N, $W = 2.59 \times 10^5$ J $= \Delta PE$

17. 44.3 m/s

19. 1.84 m

21. (a) 10.9 m/s, (b) 11.6 m/s

23. (a) 8.85 m/s, (b) 54.1%

25. (a) 2.29 m/s, (b) −15.6 J

27. 1.54 m

29. (a) 7.50×10^4 J, (b) 1.07×10^4 W = 14.4 hp

31. 600 W = 0.8 hp

33. 1.84×10^3 W = 2.46 hp

35. (a) 0.408 m/s, (b) 2.45×10^3 J

37. (a) 22.5 J, (b) 6.71 m/s

39. (a) 2.29 m/s, (b) 3.91 m/s, (c) 4.50 m/s

41. (a) 4π J, (b) 11.3 J, (c) 8.0 J

43. (a) 9.90 m/s, (b) 7.67 m/s

45. (a) 3.92×10^3 W, (b) 7.06×10^5 J

47. (a) 9.90 m/s, (b) −11.8 J, (c) −11.8 J

49. 9.80 m/s

51. (a) 4.43 m/s, (b) 1.47×10^5 N = 3.33×10^4 lb

53. (a) +400 J, (b) −294 J, (c) 5.94 m/s

55. (a) 1.31×10^5 J, (b) 871 W = 1.17 hp

57. (a) $W = mgy$, (b) $\Delta KE = W = mgy$, (c) $KE_f = mgy + \frac{1}{2} m{v_0}^2$

59. (a) 2.7 m/s^2, (c) 146 hp

61. (a) −5.6 J, (b) 0.152, (c) 2.29 rev

63. (a) 0.225 J, (b) 0.363 J, (c) No. The normal force varies with position, and so the frictional force also varies.

35. 14.8 kg · m/s in the direction opposite to its initial velocity

37. 0.267 m/s eastward

39. (a) 537 kg · m/s downward, (b) 380 kg · m/s downward

41. (a) −1/5 m/s, (b) −8/3 m/s

43. 56.7 m

45. 0.96 m

47. 2.78×10^3 N

49. (a) 0.566 m/s, (b) 11.1 J

51. (a) +12.0 N · s, (b) +8.00 N · s, (c) +8.00 m/s at 3 s, 5.33 m/s at 5 s

53. 0.40 N

55. 1.25×10^7 m/s at 221.8°

57. (a) $-2\,mv \sin\theta$, (b) zero, (c) $\dfrac{2\,mv \sin\theta}{t}$

59. $x = \dfrac{2\,{v_0}^2}{9\,\mu g} - \dfrac{4}{9} d$

CHAPTER 6

1. (a) 8.35×10^{-21} kg · m/s, (b) 4.50 kg · m/s, (c) 7.50×10^2 kg · m/s, (d) 1.78×10^{29} kg · m/s

3. 50.0 m/s

5. -5.40×10^5 N

7. (a) 1.73 kg · m/s in the horizontal direction, (b) 2.00 kg · m/s at 30° below the horizontal

9. (a) 9.00 m/s in the $+x$ direction, (b) 15.0 m/s in the $-x$ direction

11. 6.25×10^{-2} m/s

13. recoils westward at 2.46 m/s

15. 2.50 m/s in the same direction as the car's initial velocity

17. 3.00 m/s in the direction of the skateboard's original motion

19. 6.00 kg

21. −40 cm/s (10-g object), +10 cm/s (15-g object)

23. (a) 6×10^5 m/s (alpha particle), 1.6×10^6 m/s (proton), (b) 3.34×10^{-15} J (alpha before), 0 (proton before), 1.20×10^{-15} J (alpha after), 2.14×10^{-15} J (proton after)

25. 5.20×10^{-2} m/s horizontally

27. 48,000 N at 38.7° north of west

29. 3.98×10^{-3} or 0.398%

31. 4.67×10^6 m or 0.732 of earth's radius

33. $x_{cm} = +0.333$ m, $y_{cm} = +1.67$ m

CHAPTER 7

1. 0.524 rad, 0.785 rad, 1.05 rad, 1.57 rad, 3.14 rad, 4.71 rad, 6.28 rad

3. 1.99×10^{-7} rad/s, 9.86×10^{-1} deg/day

5. (a) 3.46 rad/s, (b) 5.18 rad

7. (a) 1.75×10^{-1} rad/s^2, (b) 5.56 rev

9. -2.26×10^2 rad/s^2

11. (a) 0.40 rad/s, (b) zero

13. (a) 3.37×10^{-2} m/s^2, (b) zero

15. (a) 3.46×10^{-1} m/s^2, (b) 1.04 m/s, (c) $a_c = 3.46 \times 10^{-1}$ m/s^2, $a_r = 0.946$ m/s^2, and $a = 1.01$ m/s^2 at an angle of 20.1° with the radius line

17. 11.7 m/s

19. (a) 146 m, (b) $7mg$ = 7 times the pilot's weight

21. (a) 68.6 N, (b) 8.57×10^{-1} m/s^2

23. $v_{min} = 3.13$ m/s

25. 3.55×10^{22} N directed toward the sun

27. 1.17×10^{-6} N

29. $g_{venus} = 0.902 g_{earth} = 8.84$ m/s^2

33. 1.49×10^{11} m from the sun

35. (a) -4.77×10^9 J, (b) 5.69×10^2 N

37. (a) 1.99×10^{12} m/s^2, (b) 2.38×10^{11} N, (c) 1.39×10^{12} J

39. (a) 2.51 m/s, (b) 7.90 m/s^2, (c) 4.00 m/s

41. (a) 7.77×10^3 m/s, (b) 5.35×10^3 s = 89.1 min

43. (a) 31.4 rad/s, (b) 2.09 m/s

45. (a) 2.34×10^{-10} N in the $-x$ direction, (b) 1.00×10^{-10} N in the $+x$ direction

47. 0.218

49. $T_1 = 108$ N, $T_2 = 55.7$ N

51. (a) 7.91×10^3 m/s, (b) 1.01×10^4 s, (c) 1.47×10^3 N

53. (a) $F_N = mg - \frac{mv^2}{r}$, (b) 17.1 m/s (38.4 mi/h)

55. $h = \frac{R}{3}$

57. (a) $v_{max} = \sqrt{Rg\left(\frac{\tan\theta + \mu}{1 - \mu\tan\theta}\right)}$

$v_{min} = \sqrt{Rg\left(\frac{\tan\theta - \mu}{1 + \mu\tan\theta}\right)}$

(b) $v_{max} = 16.6$ m/s, $v_{min} = 8.57$ m/s

CHAPTER 8

1. $28.1\ \text{kg} \cdot \text{m}^2$

3. (a) $99.0\ \text{kg} \cdot \text{m}^2$, (b) $44.0\ \text{kg} \cdot \text{m}^2$, (c) $143\ \text{kg} \cdot \text{m}^2$

5. 1.36 rad/s

7. -5.65×10^{-2} N · m

9. (a) 5.35 m/s² downward, (b) 42.8 m, (c) 8.91 rad/s²

11. (a) 1.27 N, (b) 3.18 N

13. 29.0 J

15. 276 J

17. $a_{sphere} > a_{cylinder} > a_{ring}$

19. 24.2 m

21. (a) 7.06×10^{33} J · s, (b) 2.66×10^{40} J · s

23. four times longer

25. (a) 1.91 rad/s, (b) 2.53 J (before), 6.44 J (after)

27. (a) 3.58 rad/s, (b) 539 J (work done by man to walk inward)

29. (a) $2.03 \times 10^{-46}\ \text{kg} \cdot \text{m}^2$, (b) 1.65×10^{11} Hz, (c) 121

31. $t_{sphere} = 1.44 \times 10^{-3}$ s, $t_{disk} = 1.80 \times 10^{-3}$ s, $t_{shell} = 3.60 \times 10^{-3}$ s

33. 0.167 rad/s

35. 35.6 rad/s

37. 8 rad/s

39. $\omega_{ship} = 0.94\omega_{earth}$

41. (a) 3.24 J, (b) 1.44 s, (c) Yes, a length of 2.59 m is sufficient.

43. (a) 4.5 m/s, (b) 10.1 N

45. 3220 W = 4.32 hp

47. (a) 3.12 m/s², (b) $T_1 = 26.7$ N, $T_2 = 9.37$ N

49. $2.7(R - r)$

51. 1.09 m/s², (b) $T_1 = 21.8$ N, $T_2 = 43.6$ N

CHAPTER 9

1. (a) 1.56×10^6 Pa, (b) 5.67×10^{-5} m

3. 2.38×10^{-2} mm

5. -3.57×10^{-10} m³

7. 1.45×10^{-5} m

9. $\rho = 1.05 \times 10^4$ kg/m³ made of silver

11. 1.17×10^6 Pa

13. 3.95×10^5 Pa

15. (a) 1.104×10^8 Pa, (b) 1.95×10^6 N (about 438,000 lb inward)

17. 10.5 m

19. 900 lb

21. 1.05×10^5 Pa

23. 3.33×10^3 kg/m³

25. 53.6 metric tons

27. (a) 8.57×10^3 kg/m³, (b) 714 kg/m³

29. 13.1 m

31. 4.96×10^{-4} kg

33. 0.050 N/m

35. 5.64 m

37. 0.694 mm

39. 5.5 mm

41. 8.00×10^3 N

43. 8%

45. 7.50×10^6 Pa

47. 27.9 cm (left tube), 12.1 cm (right tube) measured up from botton of u

51. 88.2 m/s²

53. (a) 2.45 mm, (b) 0.75 mm, (c) 6880 kg

57. (a) $\frac{\rho_1 h_1 + \rho_2 h_2}{h_1 + h_2}$, (b) $\frac{\rho_1 h_1 + \rho_2 h_2}{\rho_w}$, (c) $d' = d$

CHAPTER 10

1. (a) 3.18 m/s, (b) 5.66 m/s

3. 4.50×10^{-4} m³/s

5. (a) 11.0 m/s, (b) 2.64×10^4 Pa
7. 1.51×10^5 N (upward)
9. 2130 Pa
11. 2.45 m
13. 3.92×10^4 Pa
15. The water surface is 5 cm above the top hole or 17 cm above the bottom of the tank.
17. 1.51×10^5 Pa
19. 0.120 N
21. 0.83 m/s
23. RN = 2890 (unstable range)
25. 1.78×10^{-3} kg/m^3
27. 9.50×10^{-10} m^2/s
29. 1.02×10^3 kg/m^3
31. 1.36×10^{-14} N
33. 6.43×10^{-7} m = 0.643 μm
35. 32.0 m/s
39. 6.53×10^{-8} m/s
41. 21.2 cm
43. (b) $y_{max} = \frac{P_a}{\rho g}$

CHAPTER 11

1. (a) 37.0°C, 310 K, (b) −20.6°C, 253 K
3. (a) 36,000,000°F, (b) 20,000,000°C
7. 7.92 mm
9. 1.26 cm
11. (a) 263.5°C, (b) −262.2°C
13. 1.50×10^8 Pa
15. The tape will indicate that the child is 9.24×10^{-3} m shorter than he actually is.
19. 3.31 atm
21. 3/2
23. 560 K
25. (a) 1.43 kg/m^3, (b) 1.79×10^{-1} kg/m^3
27. 16.0 cm^3
29. 7.11 m
31. (a) 3.44×10^{-9} m = 3.44 nm, (b) 7.45×10^{-4} m = 0.745 mm
33. 3.65×10^4 N
35. 1.76×10^4 Pa
37. (a) 2.16×10^3 m/s, (b) 576×10^2 m/s
39. (a) 2.02×10^4 K, (b) 9.04×10^2 K
41. 1.52×10^4 Pa
43. (a) 1.35×10^{-2} cm, (b) 6.75×10^{-4} cm, (c) 3.18×10^{-2} cm^3
45. 5.05×10^{-21} J/molecule
47. (a) 343 K, (b) 14.3% of the original volume
51. 285 m/s
53. 804°C
55. 59.9°
59. 2.45 m

CHAPTER 12

1. 87.0° C
3. $(2.34 \times 10^{-2})M$ where M is your mass in kg.
5. 2.85×10^3 m
7. (a) 1.00 cal/g · C°, (b) 0.50 cal/g · C°, (c) 0.48 cal/g · C°
9. (a) 9.88×10^{-3} C°, (b) absorbed by the rough horizontal surface
11. 23.0°C
13. 0.185 kg
15. 80.8 g
17. 1.22×10^5 J
19. 4.89×10^4 J/kg
21. 11.1°C
23. (a) all ice melts, T = 40.4°C, (b) T = 0°C with 42 g of ice left over.
25. (a) 223 J/s, (b) 1.32×10^{-2} J/s, (c) 4.50×10^{-2} J/s
27. 14.1 ft^2 · F° · h/BTU
29. 16.6 J/s
31. 1.42
33. 7.23×10^{-2} J/s · m · C°
35. 9.00 cm
37. 2.44 min
39. 467 pellets
41. (a) T = 0°C with 9.62 g ice and 0.38 g of liquid water in cup, (b) T = 100°C with 8.36 g of liquid water and 1.64 g of steam in cup.
43. 44.5°C
45. (a) 2000 W, (b) 4.47° C

47. 26.9 liters

49. 20.02 cm

51. (b) 0.654 $\frac{\text{cal}}{\text{g} \cdot \text{C}^\circ}$

CHAPTER 13

1. 804 J

3. (c) More work is done in process (a) because there is a higher pressure during the expansion phase of the process than there is in (b).

5. (a) 608 J, (b) 0, (c) -405 J, (d) 0, (e) 203 J

7. (a) -152 J, (b) -248 J

9. (a) -180 J, (b) 188 J

11. (a) 5.48×10^3 J, (b) 5.48×10^3 J

13. (a) 25.9 J, (b) 8.96×10^5 J, (c) 8.96×10^5 J

15. (a) 3.08×10^{-2} m^3, (b) -3.46×10^3 J, (c) -3.46×10^3 J

17. (a) $W = 0$, $Q < 0$, $\Delta U < 0$, (b) $W = 0$, $Q > 0$, $\Delta U > 0$

19. (a) 0.067, (b) 350 J

21. (a) 10,700 J, (b) 0.533 s

23. 0.488 or 48.8%

25. 546°C

27. (a) 9.10 kW, (b) 1.19×10^4 J

29. (a) 5.12×10^{-2} or 5.12%, (b) $Q_h = 5.27 \times 10^{12}$ J/h, (c) not too practical as a power source

31. 57.2 J/K

33. -563 J/K

37. $P = 2.86 \times 10^5$ Pa

39. No. With 10^4 J of work done, he would have to expel 6×10^4 J of heat at 300 K. The claimed engine either violates the first law of thermodynamics, or does not operate in a cyclic manner ($\Delta U \neq 0$). In the latter case, the engine is incapable of continuous operation.

41. (a) 3.45×10^3 J, (b) 1.50×10^3 J, (c) 1.95×10^3 J

45. 77.8 W

47. 17.9°C

51. (a) $4P_0V_0$, (b) $4\ P_0V_0$, (c) 9.07×10^3 J

53. 2.65×10^3 metric tons per day, (b) \$7.74 million, (c) 4.06×10^4 kg/s

CHAPTER 14

1. (a) 49 N/m, (b) 11.5 cm

3. (a) 110 N

5. 2.68 m/s

7. 4

9. 0.478 m

11. 3.06 m/s

13. (a) 2.45×10^{-2} J, (b) 0.404 m/s

15. (a) zero, (b) -4.33 m/s, (c) -5 m/s, (d) zero, (e) 5 m/s

17. 3.95 N/m

19. 11.4 nanoseconds

21. (a) 20 cm, (b) 14.1 cm, (c) 0, (d) -14.1 cm, (e) -20 cm, (f) -14.1 cm, (g) 0, (h) 14.1 cm, (i) 20 cm

23. $x = (0.25\text{ m})\cos(2.47\ \pi t)$

25. 10.1 m/s^2

27. (a) $1/\sqrt{6}$, (b) 6.120 s

29. 12.3 m to 0.082 m

31. (a) 8.31 min, (b) 1.28 s

33. (a) 9 cm, (b) 20 cm, (c) 0.04 s, (d) 250 cm/s

35. (a) 30.0 N, (b) 25.8 m/s

37. 13.5 N

39. (a) zero, (b) 0.3 m

41. 127 m/s

43. (a) -500 N, -333 m/s^2; (b) -100 N, -66.7 m/s^2; (c) 0, 0; (d) 200 N, 133 m/s^2; (e) 500 N, 333 m/s^2

45. (a) 5.10×10^{-2} kg/m, (b) 19.6 m/s

47. 0.99 m

49. (a) 19.8 m/s, (b) 8.94 m

51. (a) 100 m/s, (b) 374 J

53. (a) -28 J, (b) 0.446 m

55. (a) 588 N/m, (b) 0.7 m/s

57. 0.115

CHAPTER 15

1. (a) 347 m/s, (b) 387 m/s, (c) 436 m/s

3. 516 m

5. -2.88%

7. 3.51×10^3 m/s

9. (a) 60 dB, (b) 70 dB

11. (a) 5×10^{-17} W, (b) 5×10^{-5} W

15. (a) 7.96×10^{-2} W/m^2, (b) 109 dB, (c) 2.82 m

17. 30 m

19. 913 Hz

21. (a) 537 Hz, (b) 465 Hz

23. (a) 537 m/s, (b) 690 m/s

25. (a) 34.5 cm, (b) constructive

27. 43.1 cm, (b) 86.3 cm

29. (a) 8×10^{-2} N, (b) 4.05×10^3 N

31. (a) 3.36×10^5 N, (b) No, a wire of this small mass per unit length would surely break first.

33. 0.352 Hz, 0.176 Hz, 0.117 Hz

35. (a) 0.653 m, (b) 0.327 m

37. (a) 0.552 m, (b) 317 Hz

39. 438 Hz and 442 Hz

41. 29.7 cm

43. 1.76 cm

45. (a) 362 Hz, (b) 287 Hz, (c) 0.953 m, and 1.20 m

47. (a) decreases by factor of 2, (b) decreased by a factor of $1/\sqrt{2}$, (c) increased by a factor of $\sqrt{2}$

49. 15.7 Hz

51. 2.82×10^8 m/s

53. 200 m/s

55. 32.9 m/s

57. 4 Hz

59. 1204 Hz

CHAPTER 16

1. (a) -80 μC, (b) 4.8 μC

3. 1.11×10^{-8} N (attractive)

5. 5.71×10^{13} C

7. 1.17×10^{-8} N in the $-x$ direction

9. 3.90×10^{-7} N at 11.3° below the $+x$ axis

11. 0.745 μC

13. 0.621 m

15. 1.45 m beyond the -3×10^{-9} C charge

17. 7.6×10^5 N/C, in the $+x$ direction

19. 1.44×10^{-7} N/C directed toward the electron

21. 5.54×10^{11} N/C directed away from the proton

23. 1.63×10^5 N/C

25. (a) 3.2×10^{-16} N ($+x$ direction), (b) 1.91×10^{11} m/s^2, (c) 5.23×10^{-6} s

27. 756 N/C at 70.1° above the $-x$ axis

29. 1880 N/C at 4.4° below the $+x$ axis

31. On the line connecting the 2 charges and 1.82 m to the left of the negative charge

35. (a) $|q_2| = 3|q_1|$, (b) $q_1 < 0$, $q_2 > 0$

39. (a) zero net charge on either surface of the sphere, (b) -5 μC, $+5$ μC, (c) zero, $+5$ μC, (d) zero, $+5$ μC

41. (a) zero, (b) 1.8×10^6 N/C radially outward, (c) 1.13×10^5 N/C radially outward

43. 5.33×10^{-6} N at 22.2° above the $+x$ axis.

45. (a) 24.2 N/C in the $+x$ direction, (b) 9.43 N/C at 63.5° above the $-x$ axis.

47. $x = 0$, $y = 9.79$ m

49. 4.43×10^5 N/C

51. (a) -4.4×10^{15} m/s^2, (b) 9.09×10^{-10} s, (c) 1.82×10^{-3} m

53. (a) 1.09×10^{-8} C, (b) 5.43×10^{-3} N

CHAPTER 17

1. (a) 6.40×10^{-19} J, (b) -6.40×10^{-19} J

3. 1.54×10^6 J

5. 3.2×10^{-19} C

7. (a) 9.22×10^4 V, (b) 1.69×10^8 V

9. 2.0 m

11. 28.2 V

13. (a) 33.4 V, (b) 195 V

15. $x = 0.546$ m and $x = -1.2$ m

17. (a) 5.93×10^5 m/s, (b) 1.38×10^4 m/s

19. (a) 48 μC, (b) 6 μC

21. 2.26×10^{-5} m^2

23. 5.31×10^{-12} C

25. 17 μF

27. (a) 3.53 μF, (b) $Q_1 = 31.8$ μC, $V_1 = 6.35$ V, and $Q_2 = 31.8$ μC and $V_2 = 2.65$ V

29. (a) $V_1 = V_2 = V_3 = 12$ V, $Q_1 = 60$ μC, $Q_2 = 48$ μC, and $Q_3 = 108$ μC; (b) $Q_1 = Q_2 = Q_3 = 21.4$ μC, $V_1 = 4.28$ V, $V_2 = 5.35$ V, and $V_3 = 2.38$ V

31. (a) 9 μF, (b) $Q_6 = Q_{12} = 24$ μC, $V_6 = 4$ V, $V_{12} = 2$ V, $Q_5 = 30$ μC, $V_5 = 6$ V

33. 6.25 μF

35. (a) 0.3 μF, (b) 2.16×10^{-3} J

37. 2.55×10^{-11} J

39. 10.8 pF

41. Stored energy increases by a factor of 3.4

43. 5.63×10^{3} N/C

45. 4.22×10^{5} V

47. 4/3 μF

49. -7.91×10^{3} V

51. (a) 1.73 m^2, (b) 3.6 J

53. 2.33

55. (a) 400 μC on the 2 μF, 800 μC on 4 μF, (b) 200 V

57. 177 μC on 4 μF, 88.6 μC on 2 μF

59. 750 μC on 6 μF, 250 μC on 2 μF

CHAPTER 18

1. 45 C

3. 26.8 A

5. 3.75×10^{14} electrons/s

7. 1.59×10^{-3} Ω

9. 1.69×10^{-4} m

11. 48 Ω

13. No, the claimed figures do not agree with Ohm's law. The device will draw a current of 6 A, not 0.6 A as claimed.

15. (a) 4.71×10^{-4} Ω · m, (b) 30 Ω

17. (a) 2.82×10^{8} A, (b) 1.76×10^{7} A

19. 10.8 Ω

21. 30.6°C

23. $\Delta T = 51.3$°C

25. (a) 0.649 mV, (b) 1.11 mV

27. 5A, 24 Ω

29. 500 A

31. 1.75×10^{-8} m^2

33. 1.55 A

35. 448 A

37. 13.2 cents

39. $3.65 assuming 5 min of use per day

41. 3.24×10^{5} C

43. 1.92 cents

45. 1.11×10^{3} m

47. 3.77×10^{28} electrons/m^3

49. 28.9 Ω

51. (a) 9.4×10^{-7} Ω, (b) 8.02×10^{-7} Ω

53. (a) 18 C, (b) 3.6 A

55. (a) 9.31 m, (b) 9.33×10^{-4} m

57. 0.136

CHAPTER 19

1. (a) 24 Ω, (b) 1.00 A

3. (a) 24/11 Ω, (b) $I_4 = 6$ A, $I_8 = 3$ A, $I_{12} = 2$ A

5. 15 Ω

7. 9.83 Ω

9. 5.13 Ω

11. 0.429 A

13. $I_3 = 1.82$ A, $I_6 = 1.09$ A, $I_9 = 0.727$ A

15. Left branch: $I = 1$ A down; Center branch: $I = 1$ A downward; $\mathcal{E} = 5$ V with the polarity opposite to that shown in the circuit diagram

17. (a) 12 s, (b) 120 μC, (c) 1×10^{-5} A

19. (a) 5×10^{5} Ω, (b) 2.46 μA

21. (a) 2.0 V, (b) 0.909 V, (c) 1.9999976 V

23. 24 Ω

25. 2.5 kΩ

27. (a) I through heater = 10.8 A, I in toaster = 8.33 A, I in grill = 12.5 A, (b) No, the total current needed to operate all 3 appliances at the same time is 31.6 A.

29. (a) 6.25 A, (b) 750 W

31. There are 13 distinct values. They are 2 Ω, 4 Ω, 6 Ω, 8 Ω, 10 Ω, 1.33 Ω, 1.5 Ω, 2.4 Ω, 4.4 Ω, 5.5 Ω, 7.33 Ω, 1.09 Ω, and 12 Ω.

33. (a) $R = 0.0999$ Ω, current in $R_1 = 50$ A, current in 100 Ω, R_2 and $R_3 = 0.045$ A, (b) $R = 1.09$ Ω, current in R_1 and $R_2 = 4.55$ A, current in 100 Ω and R_3 is 0.045 A, (c) $R = 9.991$ Ω, current in R_1, R_2 and $R_3 = 0.45$ A, current in 100 Ω = 0.05 A

35. (a) $I_1 = 1$ A, $I_2 = I_3 = 0.5$ A, $I_4 = 0.3$ A, $I_5 = 0.2$ A, (b) $V_{ac} = 6$ V, $V_{ce} = 1.2$ V, $V_{ed} = 1.8$ V, $V_{fd} =$ 1.8 V, $V_{cd} = 3$ V, (c) $P_{ac} = 6$ W, $P_{ce} = 0.6$ W, $P_{ed} = 0.54$ W, $P_{fd} = 0.36$ W, $P_{cd} = 1.5$ W, $P_{db} = 6$ W.

37. (a) 1×10^{-2} F, (b) 0.414 C

39. $I_1 = 0$, $I_2 = I_3 = 0.5$ A

41. (a) 7 Ω, (b) $I_6 = \frac{4}{3}$ A; $I_8 = \frac{3}{8}$ A

43. (a) 14 Ω, (b) 56 W, (c) 2 A

45. (a) 0.5 A, (b) 3 W, (c) 1.06 W

47. 12 Ω

49. 0.39 A; 1.5 V

51. (a) 5×10^{-7} T out of page, (b) 3.89×10^{-6} T in a plane parallel to the x-y plane and at 31° clockwise from the y axis

53. 0.331 T

55. (a) 30 A, (b) 160 μT

57. (a) 2.85×10^{-6} T, (b) 0.906 mA

59. 2.70×10^{-5} N toward the left

61. 60 μA

CHAPTER 20

1. 1.88×10^{8} m/s

3. (a) 8.67×10^{-14} N, (b) 5.19×10^{13} m/s^2

5. case (a) toward the left; case (b) into page; case (c) out of page; case (d) toward top of page; case (e) into page; case (f) out of page, (b) if $q < 0$, reverse each of the directions given in (a)

7. 10.8°

9. 24 N

11. 0.01 N westward

13. case (a) to the left; case (b) into page; case (c) out of page; case (d) toward top of page; case (e) into page; case (f) out of page

15. 0.245 A

17. 0.277 A

19. 0.167 Ω

21. 74,960 Ω

23. $R_1 = 3.33 \times 10^{-4}$ Ω; $R_2 = 2.998 \times 10^{-3}$ Ω; $R_3 = 2.997 \times 10^{-2}$ Ω

25. 1.39×10^{-2} T directed out of page

27. (a) $r_1 = 20.8$ cm, $r_2 = 10.4$ cm, (b) $\Delta = 20.8$ cm

29. 5 T

31. (a) 1×10^{-5} T, (b) 2×10^{-6} T, (c) 5×10^{-5} T

33. (a) right to left, (b) out of page, (c) lower left to upper right

35. (a) 4×10^{-6} T downward, (b) 6.67×10^{-6} T at 77° counterclockwise from the upward vertical

37. (a) 2×10^{-4} N/m, attracted, (b) 2×10^{-4} N/m, repelled

39. (a) 3.14×10^{-5} T, (b) 2.20×10^{-4} T

41. 59.7 mA

43. 2.26×10^{-6} N, zero

45. 38.2 mA

47. 8.44×10^{-5} T

49. 2.12 cm

CHAPTER 21

1. 7.07×10^{-2} T · m^2

3. 0.771 T · m^2

5. 2.96×10^{-5} T · m^2

7. 6 V

9. (a) 25.2 mV, (b) 3.16×10^{-2} Ω

11. 33.9 mV

13. (a) 1.88×10^{-7} T · m^2, (b) 6.28×10^{-8} V

15. 12.0 mV

17. (a) 100 μV, (b) 20 μA

19. 2.87 mV

21. (a) left to right, (b) right to left

23. (a) right to left, (b) right to left, (c) left to right, (d) left to right

25. left to right

27. 79.6 Hz, 240 V

29. 3.18×10^{-3} rad/s

31. $\mathcal{E}_{back} = 76.4$ V, $R = 10.9$ Ω

33. (a) 60.3 V, (b) 56.9 V, (c) 0.125 s

35. 256 turns

37. 1.88×10^{-4} T · m^2/s

39. 12.0 mH

41. (a) 4 A, (b) 2.52 A

43. 0.14 J

45. 1.85×10^{-3} T

47. 4.71 V

49. 9.67 V

53. 4.71×10^{-3} T · m^2

57. 444 μV

59. (a) 52.5 V, (b) 44.2 V, (c) 11.5 N · m

61. 29 m/s

CHAPTER 22

1. (a) 141 V, (b) 28.3 A, (c) 20 A
3. (a) 106 V, (b) 60 Hz, (c) zero, (d) 3 A
7. 0.100 A
9. 0.224 A
11. (a) 0.188 Ω, (b) 1.88×10^4 Ω
13. (a) 31.8 mH, (b) 120 Hz
15. 53.1 mH
17. (a) 0.205 A, (b) 154 V, (c) 54.3 V, (d) 90°
19. (a) 134 Ω, (b) 0.748 A, (c) $V_R = 22.5$ V, $V_L = 564$ V, $V_C = 662$ V, (d) −77°
21. (a) 89.7 V, (b) 108 V
23. (a) zero, (b) 11.2 mA
25. 31.6 W
27. 35.6 W
29. (a) 1.23 A, (b) 89.5 Ω
31. (a) 2520 Hz, (b) 1.5 A, (c) 0.967 A
33. 47.9 mH
35. (b) 33.3
37. 440 V
39. 780 V
41. (a) 1.67×10^4 A, (b) 10 A, (c) case a: 1.39×10^5 W/m; case b: 5×10^{-2} W/m, (d) case a: 167 lines; case b: 0.1 line
43. (a) 12.7 Hz, (b) 3.18 A
45. 90.1 Hz
47. (a) 106 V, (b) −45°
49. (a) 6 Ω, (b) 11.9 mH
51. (a) 12 turns, (b) 1.95 A, (c) zero
53. (a) 0.670 A, (b) 33.2°, (c) 0.837, (d) 90.5 V, (e) 22.5 W
55. $L = 15.4$ mH, $C = 160$ μF
57. (a) The two elements are a resistor and an inductor, (b) $R = 10$ Ω, $L = 29.7$ mH
59. 31.6

CHAPTER 23

1. 1.09 pF to 1.64 pF
3. 2.998×10^8 m/s
5. 5×10^{-7} T
7. $E_m = 1.01 \times 10^3$ N/C, $B_m = 3.35 \times 10^{-6}$ T
9. (a) 45.0 N/C, (b) 2.69 W/m^2
11. $E_m = 438$ N/C, $B_m = 1.46 \times 10^{-6}$ T
13. (a) 150 MHz, (b) 15 MHz, (c) 1.5 MHz
15. (a) 6.25×10^{14} Hz, (b) 480 nm
17. (a) 8.33 min, (b) 1.34 s
19. 60 km
21. 720 N/C
23. 5.17 μH
25. 1.0×10^7 m^{-1}
27. 70 min
29. 29.4 W

CHAPTER 24

1. 2.27×10^8 m/s
3. (a) 2.08×10^4 rev/s, 4.17×10^4 rev/s; (b) 2.08×10^2 rev/s, 4.17×10^2 rev/s
5. 2.00×10^8 m/s
7. 433 nm
9. 22.0°
11. first surface: $\theta_i = 30°$, $\theta_r = 19.5°$; second surface: $\theta_i = 19.5°$, $\theta_r = 30°$
13. 1.58×10^8 m/s
15. (a) first surface: $\theta_i = 30°$, $\theta_r = 19.5°$; second surface: $\theta_i = 40.5°$, $\theta_r = 77.1°$, (b) first surface: θ (reflection) = 30° = angle of incidence; second surface: θ (reflection) = 40.5° = angle of incidence.
17. (a) 1.98×10^8 m/s, (b) 3.89×10^{-5} m
19. 11.3 m
21. 0.39°
23. (a) 41.4°, (b) 18.4°, (c) 27.3°, (d) 42.3°
25. (a) 24.4°, (b) 44.2°, (c) 49.8°
27. 48.5°
29. $n \geq \sqrt{2}$
31. 1.08 cm $\leq h \leq$ 1.17 cm
33. 2.11×10^8 m/s
35. 42.4°
37. 40°
41. 0.890

CHAPTER 25

3. 10 ft, 30 ft, and 40 ft

5. (a) 40 cm in front of mirror, real, inverted, $M = -1$, (b) No image is formed. Reflected rays are parallel to each other. (c) 20 cm behind mirror, virtual, erect, $M = +2$

7. (a) 13.3 cm behind mirror, virtual, erect, $M = +1/3$, (b) 10 cm behind mirror, virtual, erect, $M = +1/2$, (c) 6.67 cm behind mirror, virtual, erect, $M = +2/3$.

9. 10 cm in front of the mirror

11. +0.130

13. +48 cm

15. 10 cm in front of the mirror

17. +30 cm (30 cm in front of the mirror)

19. concave mirror, $M = -20$

21. (a) 1.5 m, (b) 1.75 m

23. 52.5 cm

25. (a) 120 cm, (b) −24 cm, (c) −8 cm, (d) −3.43 cm

27. (a) −13.3 cm, (b) −8 cm, (c) −4.8 cm, (d) −2.67 cm

29. (a) Real, inverted image 40 cm past the lens, $M = -1$, (b) No image is formed. Rays leaving the lens are parallel, (c) erect, virtual image 20 cm in front of lens, $M = +2$

31. (a) Virtual, erect image 13.3 cm in front of lens, $M = +1/3$, (b) virtual, erect image 10 cm in front of lens, $M = +1/2$, (c) virtual, erect image 6.67 cm in front of lens, $M = +2/3$

33. (a) −12 cm, (b) −12 cm, (c) −9 cm, (d) −6 cm, (e) −4 cm

35. −80 cm

37. Either 13.8 cm or 36.2 cm from the object. $M = -2.62$ for $s = 13.8$ cm; $M = -0.382$ for $s = 36.2$

39. (a) Real, inverted image 6 cm past second lens, $M(\text{overall}) = -0.6$, (b) virtual, inverted image located at the second lens, $M(\text{overall}) = -1$

41. Real, inverted image 119 cm beyond second lens, $M(\text{overall}) = -1.25$

43. 10 cm

45. (a) 33.3 cm beyond the lens, (b) 60 cm beyond the lens, (c) 20 cm in front of the lens, (d) ($s = 50$ cm) $M = -0.67$; real, inverted image. ($s = 30$ cm) $M = -2$; real, inverted image. ($s = 10$ cm) $M = +2$; erect, virtual image.

47. (a) 10 cm beyond the second lens, (b) erect, real image same size as object, (c) 20 cm beyond the lens combination

49. −30 cm

51. 1.5 cm

53. A real image located 32 cm to the right of the second surface

55. $s = \frac{4f}{3}$, (b) $s = \frac{3f}{4}$, (c) $M_1 = -3$, $M_2 = 4$

CHAPTER 26

1. 433 nm

3. (a) 2.62 mm, (b) 2.62 mm

5. 2.61 m

7. 34.0 cm

9. 99.8 m

11. 148 m

13. 91.9 nm

15. 290 nm, 580 nm, 870 nm

17. 290 nm

19. 324 nm, 417 nm, and 584 nm

21. 147

23. $4.23\ \mu\text{m} \le t < 4.45\ \mu\text{m}$

27. 0.227 mm

29. (a) 1.09 m, (b) 1.70 mm

31. 56.3°

33. 48.4°

35. (a) 1.11, (b) 42°

37. 60.5°

39. 58.1°

41. 681 nm

43. 432 nm

45. 6.89 units

47. 0.35 mm

49. 313 nm

53. 9.96×10^{-8} m

CHAPTER 27

1. 5.71 m

3. 4.29

5. (1/128) s
7. (a) 33.3 cm, (b) +3.0 diopters
9. −1 diopter
11. 33.3 cm
13. 4.07 cm
15. (a) +2, (b) +1
17. 2.34 cm
19. 18.8
21. 0.938 cm
23. 15
25. 157 miles
27. 81.8 cm (32.2 in)
29. (a) 1.03 cm, (b) 26.8 cm
31. 1.06×10^{-5} rad
33. (a) 2.29×10^{-4} rad, (b) 43.7 m
35. 499 nm
37. 654
39. 20
41. 6560 lines/cm
43. (a) $\theta_4 - \theta_1 = 5.93°$, (b) $\theta_3 - \theta_1 = 6.42°$
45. 0.02 nm
47. (a) 101.5 cm, (b) 66.7
49. 603 lines/mm
51. 2.88 (1.88 if the eye is relaxed)
53. 660 nm
55. (a) 1.96 cm, (b) f/3.27, (c) f/9.8
57. 1.70
59. 5.07 mm

CHAPTER 28

1. (a) $t_I = 1667$ s, $t_{II} = 2041$ s; (b) $t'_I = 2500$ s, $t'_{II} = 2041$ s; (c) $\Delta t_{total} = 85.0$ s
3. (a) 70 beats/min, (b) 30.5 beats/min
5. $0.866c$
7. $0.866c$
9. $0.927c$
11. (a) 20 m, (b) 19 m, (c) $0.312c$
13. 3600 s + 3.2×10^{-9} s
15. (a) 2.73×10^{-24} kg · m/s, (b) 1.58×10^{-22} kg · m/s, (c) 5.64×10^{-22} kg · m/s
17. (a) $0.85c$, (b) result is the same regardless of the rest mass of the particle
19. $0.927c$
21. 1.03×10^3 m
23. (a) 337 kg/m^3, (b) 64 kg/m^3
25. $0.985c$
27. $0.982c$
29. (a) 939 Mev, (b) 3006 Mev, (c) 2067 Mev
31. $0.745c$
33. (a) 4.44×10^9 kg, (b) 1.43×10^{13} y
35. (a) 2.72×10^{-17} kg · m/s or 50.9 GeV/c, (b) $0.9998c$
37. (a) 4.92×10^{-13} J = 3.07 MeV, (b) $0.986c$
39. $0.37c$
41. (a) 2.50 MeV/c, (b) 4601 MeV/c
43. 1.22×10^{13} m
45. (a) 7.05 μs, (b) 423 muons

CHAPTER 29

1. (a) 3.73×10^{-7} eV, (b) 4.14×10^{-2} eV, (c) 41.4 eV
3. 500 nm
5. 0.622 nm
7. 8.19×10^{-20} m
9. (a) 0.822 eV, (b) 995 nm
11. (a) 1.91 eV, (b) 1.20 V
13. (a) lithium and iron, (b) 1.84 eV and 0.244 eV respectively
15. 242 nm
17. (a) 0.083 nm, (b) 0.012 nm
19. 6.7°
21. 0.124 nm
23. $\Delta\lambda = 2.43 \times 10^{-3}$ nm
25. 7.34×10^{-2} nm
27. (a) 1878 MeV, (b) 6.62×10^{-16} m
29. 1.98 MeV
31. 1.14×10^{-38} m
33. (a) 0.174 nm, (b) 5.37×10^{-12} m
35. 1.06×10^{-34} m
37. 151 V
39. 3.50×10^{-32} m
41. 5.76×10^5 m/s

43. 8.25×10^{-13} J or 5.16 MeV

45. 1.55 V

47. 2.96×10^{20} photons

49. 2.0 eV

51. 1.35 eV

CHAPTER 30

1. 121.5 nm, 102.5 nm, 97.2 nm, far ultraviolet

3. 1.60×10^{14} Hz, 2.46×10^{14} Hz, infrared

5. (a) 1.89 eV, (b) 658 nm, (c) 4.56×10^{14} Hz

7. (a) 12.09 eV, (b) 12.09 eV, 1.89 eV, 10.2 eV

9. $1.09243 \times 10^{7}\ \text{m}^{-1}$

11. 91.4 nm

13. 2.11×10^{-34} J · s

15. (a) 2.19×10^{6} m/s, (b) 1.52×10^{-16} s, (c) 1.05×10^{-3} A

17. (a) -122 eV, (b) 1.76×10^{-11} m

19. (a) $4.39 \times 10^{7}\ \text{m}^{-1}$, (b) 30.3 nm, (c) deep ultraviolet

21.

n	ℓ	m_ℓ	m_s
3	2	+2	+1/2
3	2	+2	−1/2
3	2	+1	+1/2
3	2	+1	−1/2
3	2	0	+1/2
3	2	0	−1/2
3	2	−1	+1/2
3	2	−1	−1/2
3	2	−2	+1/2
3	2	−2	−1/2

23. (a) 4 (ℓ = 3, 2, 1, and 0)
(b) 7 (−3, −2, −1, 0, 1, 2, 3)

25. (a) $1s^2 2s^2 2p^4$
(b) $n = 1, \ell = 0, m_\ell = 0, m_s = \pm 1/2$
$n = 2, \ell = 0, m_\ell = 0, m_s = \pm 1/2$
$n = 2, \ell = 1, m_\ell = -1, m_s = \pm 1/2$
$n = 2, \ell = 1, m_\ell = 0, m_s = \pm 1/2$

27. 1.88×10^{18} Hz

29. 1.61×10^{18} photons

31. (a) 0.476 nm, (b) 7.30×10^{5} m/s (c) 0.997 nm, (d) λ = (1/3)(circumference)

33. $PE = -27.2$ eV, $KE = 13.6$ eV

35. (a) 134 eV, (b) This is about 10 times greater than the ground-state energy of hydrogen (13.4 eV).

39. $\Delta p \approx \frac{\hbar}{r}$, (b) $KE = \frac{\hbar^2}{2mr^2}$, $PE = -\frac{ke^2}{r}$, $E = \frac{\hbar^2}{2mr^2} - \frac{ke^2}{r}$

CHAPTER 31

3. Point A: $F = 0$; point B: F in the positive x direction; point C: $F = 0$; point D: F in the negative x direction; point E: $F = 0$

5. 0.24 eV

7. 3.52 kg

9. 5.54×10^{-12} m from the fluorine atom

11. $1.92 \times 10^{-47}\ \text{kg} \cdot \text{m}^2$

13. 1.18×10^{11} Hz

17. (a) 2.75×10^{14} Hz, (b) 1.09 μm

19. p-type

21. (a) 116 K = −157°C, (b) 8.12×10^{4} K = 80,900°C

23. 9.07×10^{7} J

25. (a) 1.69×10^{14} Hz, (b) 1.78 μm

CHAPTER 32

1. (a) 1.9×10^{-15} m, (b) 7.4×10^{-15} m

3. (a) $1.4 \times 10^{16}\ \text{lb/ft}^3$, (b) 2.6×10^{14} lb

5. Any nucleus with $A = 30$, for example, $^{30}_{15}\text{P}$

7. 4.55×10^{-13} m

9. 2.28×10^{-13} m

11. 342 MeV

13. 7.92 MeV/nucleon

15. 0.210 MeV/nucleon (The difference is largely due to increased Coulomb repulsion caused by the extra proton in $^{23}_{12}\text{Mg}$)

17. (a) The surface-effect term must have a negative sign because the volume term over-approximates the binding energy by treating all nucleons as if they were completely surrounded by other nucleons. The nucleons on the surface are not totally surrounded and the surface term attempts to correct for this effect by subtracting off the excess which was included in the volume term.
(b) $\left(\frac{V}{A}\right)_s = 0.207\ V^{1/3}$, $\left(\frac{V}{A}\right)_c = 0.167\ V^{1/3}$

19. 2.38×10^{16} nuclei

21. 1.0 h

23. 7.47×10^9 nuclei

25. 24.5%

27. ${}^{12}_{5}\text{B} \rightarrow {}^{12}_{6}\text{C} + {}^{0}_{-1}\text{e}$
${}^{234}_{90}\text{Th} \rightarrow {}^{230}_{88}\text{Ra} + {}^{4}_{2}\text{He}$
${}^{14}_{6}\text{C} \rightarrow {}^{14}_{7}\text{N} + {}^{0}_{-1}\text{e}$

29. 4.28 MeV

31. (a) no, $Q < 0$, (b) yes, $Q = 1.91$ MeV > 0

33. 16,600 y

35. 17.35 MeV

37. ${}^{27}_{13}\text{Al} + {}^{4}_{2}\text{He} \rightarrow {}^{1}_{0}\text{n} + {}^{30}_{15}\text{P}$
${}^{1}_{0}\text{n} + {}^{10}_{5}\text{B} \rightarrow {}^{4}_{2}\text{He} + {}^{7}_{3}\text{Li}$

39. (a) ${}^{3}_{2}\text{He}$, (b) 4.02 MeV

43. $KE_{\text{min}} = 22.0$ MeV

45. No, $Q < 0$

47. 1.88 MeV

49. 0.055 mCi

51. 801 y

53. 1.60×10^8 nuclei

55. 12.2 mg

CHAPTER 33

1. 174 MeV

3. 126 MeV

5. 3.38 kg

7. 2.46×10^3 y

9. 938 g

11. 17.6 MeV

13. (a) ${}^{13}_{7}\text{N}$, (b) ${}^{13}_{6}\text{C}$, (c) ${}^{14}_{7}\text{N}$, (d) ${}^{15}_{8}\text{O}$, (e) ${}^{15}_{7}\text{N}$, (f) ${}^{12}_{6}\text{C}$

15. (a) 3.95×10^{22} deuterium nuclei, (b) 1.03×10^{10} J, (c) 51.7 gallons of gasoline

17. the second

19. 18.8 J

21. (a) 2 J/kg, (b) 4.78×10^{-4} C°

23. (a) 10, (b) 10^7

25. approximately 4100

27. (a) 3.72×10^{-24} kg · m/s, (b) 1.78×10^{-10} m, (c) the wavelength is approximately equal to the diameter of an atom, but 10^5 times the diameter of a nucleus.

29. 1.38×10^7 rad

CHAPTER 34

1. 2.26×10^{23} Hz, 1.32×10^{-15} m

3. 2×10^{-18} m

5. 1×10^{-23} s

7. (a) $-1 \neq 0$ (not conserved), (b) $0 = 0$ (conserved), (c) $0 = 0$ (conserved), (d) $0 \neq -1$ (not conserved), (e) $-2 \neq -1$ (not conserved), (f) $-2 \neq 0$ (not conserved)

9. (a) violates conservation of L_e and L_μ, (b) violates conservation of charge, (c) violates conservation of baryon number, (d) violates conservation of baryon number, (e) violates conservation of charge.

11. (a) $\pi^- \rightarrow \mu^- + \bar{\nu}_\mu$
L_μ: $0 \rightarrow 1 - 1$
(b) $\text{K}^+ \rightarrow \mu^+ + \nu_\mu$
L_μ: $0 \rightarrow -1 + 1$
(c) $\bar{\nu}_e + \text{p} \rightarrow \text{n} + \text{e}^+$
L_e: $-1 + 0 \rightarrow 0 - 1$
(d) $\nu_e + \text{p} \rightarrow \text{n} + \text{e}$
L_e: $1 + 0 \rightarrow 0 + 1$
(e) $\nu_\mu + \text{n} \rightarrow \text{p} + \mu^-$
L_μ: $1 + 0 \rightarrow 0 + 1$
(f) $\mu^- \rightarrow \text{e} + \bar{\nu}_e + \nu_\mu$
L_μ: $1 \rightarrow 0 + 0 + 1$
L_e: $0 \rightarrow 1 - 1 + 0$

13. (a)

	Proton	Quarks u	u	d	Total $(u + u + d)$
Strangeness	0	0	0	0	0
Baryon number	1	$+\frac{1}{3}$	$+\frac{1}{3}$	$+\frac{1}{3}$	1
Charge	e	$+\frac{2}{3}e$	$+\frac{2}{3}e$	$-\frac{1}{3}e$	e

(b)

	Neutron	Quarks u	d	d	Total $(u + d + d)$
Strangeness	0	0	0	0	0
Baryon number	1	$+\frac{1}{3}$	$+\frac{1}{3}$	$+\frac{1}{3}$	1
Charge	0	$+\frac{2}{3}e$	$-\frac{1}{3}e$	$-\frac{1}{3}e$	0

Index

Note: Page numbers in *italics* refer to illustrations; page numbers followed by t refer to tables; and page numbers in **bold** refer to definitions.